An Interactive Guide to Developmental Biology

Mary S. Tyler and Ronald N. Kozlowski

labs.devbio.com

Designed to complement the textbook, this unique resource helps you understand the organisms discussed in lecture and prepares you for the laboratory. *DevBio Laboratory: Vade Mecum³* is available online, which allows you the flexibility to use the software from any computer with Internet access. Access to the site is included with each new textbook.

Over 140 interactive videos and 300 labeled photographs take you through the life cycles of model organisms used in developmental biology laboratories. The easy-to-use videos provide you with the concepts, vocabulary, and motivation to enter the laboratory fully prepared. A chapter on zebrafish addresses how to raise the organism and the effects of various teratogens on embryonic development. The site also includes chapters on: the slime mold Dictyostelium discoideum; *planarian; sea urchin; the fruit fly* Drosophila melanogaster; *chick; and amphibian.*

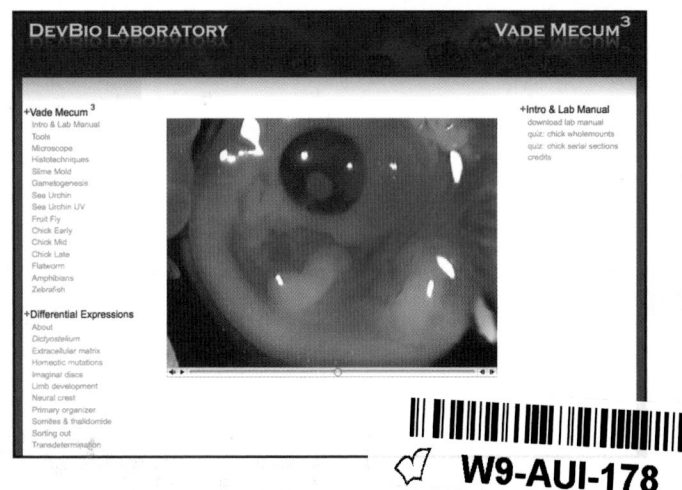

W9-AUI-178

ADDITIONAL FEATURES INCLUDE:

- **Movie excerpts** from *Differential Expressions*[2]: Short video excerpts about key concepts in development

- **PowerPoint® slides** of chick wholemounts and serial sections for self-quizzing and creating tests

- **Full video instruction** on histological techniques

- **Glossary:** Every chapter of the Laboratory Manual includes an extensive glossary.

- **Laboratory Manual:** The Third Edition of Mary S. Tyler's laboratory manual, *Developmental Biology: A Guide for Experimental Study*, designed for use with the multimedia chapters of *DevBio Laboratory: Vade Mecum³*

- **Website:** www.devbio.net augments *DevBio Laboratory*, allowing convenient access to recipes from the lab book, a searchable glossary, a module on laboratory safety, and more.

- **Study Questions**

- **Laboratory Skills Guides**

← To access *DevBio Laboratory: Vade Mecum³* go to **labs.devbio.com**, click the "**Register**" link, and follow the instructions to enter your registration code (see left) and create a username and password

Developmental Biology

TENTH EDITION

Developmental Biology

TENTH EDITION

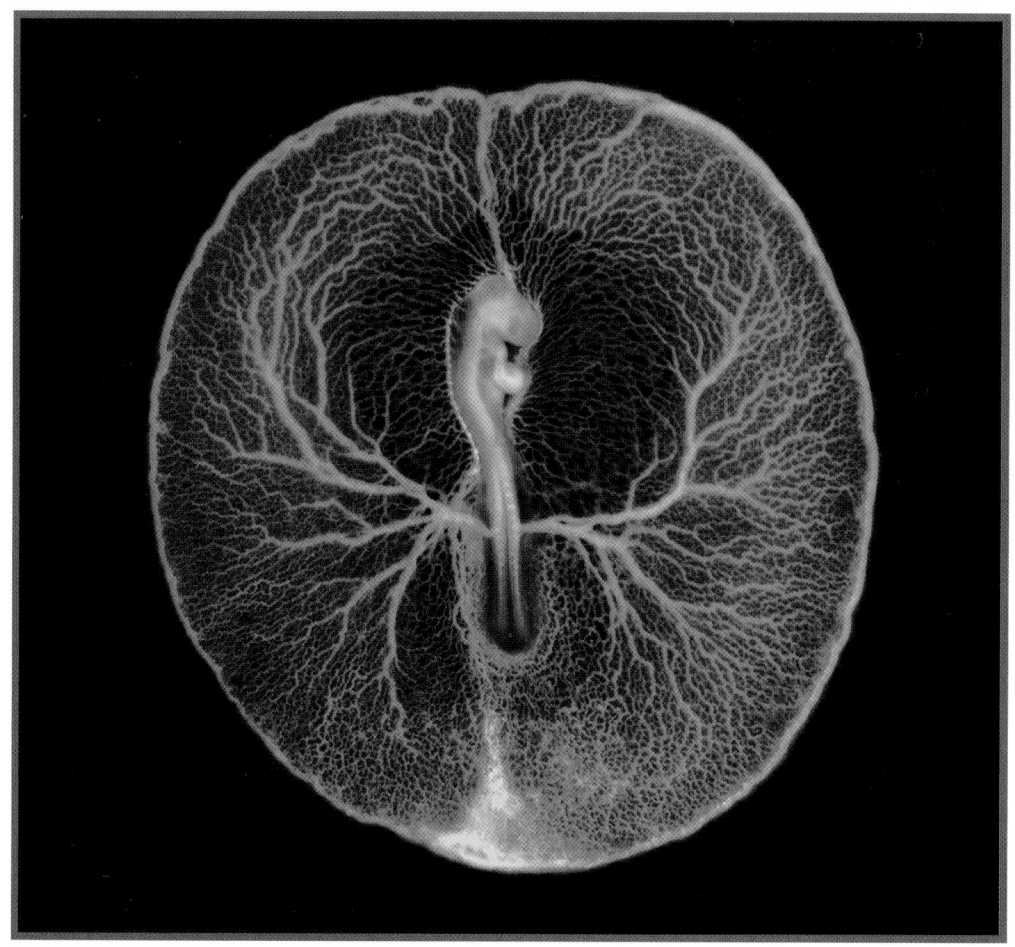

SCOTT F. GILBERT

Swarthmore College and the University of Helsinki

 **Sinauer Associates, Inc.** Publishers • Sunderland, MA USA

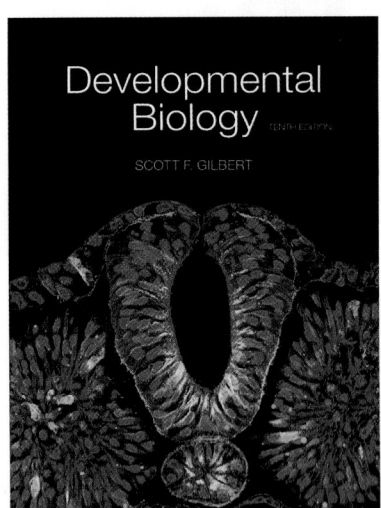

The Cover

Cross-section of neural tube formation in a 2-day chick embryo. This confocal micrograph shows the neural tube (which will become the animal's central nervous system) closing at its most dorsal (top) region. The blocks of cells on either side are the somites, which will form muscles, vertebrae, and dermis. The upper cells become the epidermis (outer skin). Cell nuclei are stained blue, the microtubules of the cytoskeleton are red, and the vitronectin of the extracellular matrix appears green. Photograph by M. Angeles Rabadán and Elisa Martí Gorostiza, Instituto de Biología Molecular de Barcelona-CSIC.

The Title Page

Fluorescence micrograph of a late 2-day chick embryo (about 45 hours after the egg was laid, at which point the heart has begun to beat). The vascular system of this embryo was revealed by injecting fluorescent beads into the circulation. The three-dimensional effect was achieved by superimposing two separate images. Photograph by Vincent Pasque. Used with permission of The Wellcome Institute.

Developmental Biology, 10th Edition

Copyright © 2014 by Sinauer Associates, Inc. All rights reserved.
This book may not be reproduced in whole or in part without permission from the publisher.

Sinauer Associates, Inc.
P.O. Box 407
Sunderland, MA 01375 USA
FAX 413-549-1118
email: orders@sinauer.com; publish@sinauer.com

About the Book

Publisher: Andrew D. Sinauer
Editor: Azelie Aquadro
Project Editor: Carol Wigg
Copy Editor: Elizabeth C. Pierson
Production Manager: Christopher Small
Book and Cover Design: Joan Gemme
Illustration Program: Dragonfly Media Group, Precision Graphics
Indexes: Grant Hackett, Carrie Crompton
Book and Cover Manufacture: Courier Companies, Inc.

Library of Congress Cataloging-in-Publication Data

Gilbert, Scott F., 1949-
  Developmental biology / Scott F. Gilbert, Swarthmore College and the University of Helsinki. -- Tenth edition.
      pages cm
  Includes index.
  ISBN 978-0-87893-978-7 (casebound)
  1. Embryology--Textbooks. 2. Developmental biology--Textbooks. I. Title.
  QL955.G48 2014
  612.6′4--dc23   2013011029

7 6 5 4 3 2

To Daniel, Sarah, David, and Natalia

Brief Contents

Contents

PART THREE   THE STEM CELL CONCEPT
Introducing Organogenesis 319

CHAPTER 10
Emergence of the Ectoderm
Central Nervous System and Epidermis 333

CHAPTER 11
Neural Crest Cells and Axonal
Specificity 375

PART FOUR SYSTEMS BIOLOGY
Expanding Developmental Biology to Medicine, Ecology, and Evolution 627

Preface

Change is the law of life. And those who look only to the past or present are certain to miss the future.

JOHN FITZGERALD KENNEDY (1963)

Metamorphosis is a time of redefinition and dramatic developmental transition. Caterpillars go from crawling and eating to flying and mating; tadpoles go from swimming to hopping, developing new muscles as their old ones decay. Today there's a feeling in the air that developmental biology is about to undergo another metamorphic molt. Don't panic!

I don't know what caterpillars experience as their new molt gives them wings. I don't know how a tadpole feels when it is about to become a frog. But I do know how developmental biology feels when it is about to undergo a major change, and this is how it feels today. It felt this way as I first wrote this textbook between 1980 and 1985. There were hints back then that developmental biology was about to change in a big way. Recombinant DNA was coming of age. It was becoming possible to identify and add new genes into an organism, and to take genes away. In 1985, the first edition of *Developmental Biology* had no mention of transcription factors or paracrine factors—but it mentioned the research that led to our discovery of them. Transcription factors entered the book's second edition three years later, and paracrine factors appeared in the next edition, three years after that. Between 1985 and 1991 the form of developmental biology changed from experimental embryology to developmental genetics. Our metamorphosis had occurred. Even journals changed their names.

My sense today is that the present metamorphosis will transform developmental genetics into an as yet unnamed developmental science that will integrate anatomy, physiology, genetics, cell biology, systems theory, genomics, and structural biology. Development has always been a science of syntheses and relationships, and these will be major themes for *all* science in the twenty-first century. Developmental biology will become a "biology without borders." The new developmental biology may be simultaneously molecular, ecological, evolutionary, and physiological. In fact, I would be surprised if it were not. As Mark Lewandoski told me at a recent meeting, "Developmental biology is dead. Long live developmental biology!"

Metamorphic change is in the nature of science. From within the larva, the rudiment of a new organism emerges. I

A plate from *Metamorphosis Insectorum Surinamensium* (1705) by the artist and scientist Maria Merian. She was one of the first scientists to document that caterpillars, pupae, and butterflies were different life cycle stages of a single organism.

hope this book helps to bridge those epistemes, the older and the newer way of approaching development. For me personally, the metamorphosis can hardly have occurred at a more appropriate time. Thirty years ago, as I started writing the first edition, I had a 2-year old child and a new job. Now that child has a 2-year old child and a new job. Revolution is nothing less than the turning of the cycle, and metamorphosis is the act of renewal. Long live developmental biology!

Acknowledgments

In book publishing, as in animal development, there are two types of relationships, the instructive and the permissive. Instructive relationships change the information in the product. They shape the phenotype. The reviewers listed on the following page have played this instructive role to great effect, letting me know what research is coming to the fore, reassuring me when my interpretations of data were sound, and correcting me when they were not. Such input both stabilizes and alters the content of the book, and to these reviewers I owe a huge debt of gratitude. If my book has some "OMG, how did he know that?" moments, the reviewers are the ones to credit. If there are any "WTF, why is this still being cited?" moments, I'm most likely the one to blame.

Permissive relationships don't change the information content, but enable that information to come into being. These are the transcription, translation, scaffolding and patterning apparatuses. Since the early editions of this book, my words, sentences, and paragraphs have been queried, clarified, rearranged, reordered, and realigned by Carol Wigg of Sinauer Associates. Carol's work on important biology textbooks deserves some kind of award for making biology accessible to the reader. It's an honor to have her work on my book. I have also been incredibly fortunate to have Elizabeth Pierson as copyeditor. She has an eagle eye, a sense of humor, and an uncanny ability to point out discrepancies hundreds of pages apart and make you smile at your own mistakes.

This is a beautiful book, and I can say that because it is not my doing. It is due in large part to the many wonderful photographs my generous colleagues continue to supply. It is also due to Chris Small and his production staff at Sinauer, especially designer Joan Gemme. Also at Sinauer, Johannah Walkowicz coordinated the academic reviews and course input for this revision, while Marie Scavotto and Nancy Asai of marketing made sure everyone knew the new edition was coming along.

The original vision of what this book could be and its evolution from its instar in 1985 has been directed by Andy Sinauer in collaboration with myriad reviewers and editors (and of course the author). Andy's many longstanding relationships with leading scientists (Andy doesn't "lose" authors) testify to his integrity and his standards for excellence, which have set a high bar few others can reach.

Like developmental biology, textbook publishing is undergoing a metamorphosis. I like to think this book helped usher in the age of electronic publishing when, back in the mid-1990s, we offered embedded websites in the text. The original website was a Gopher Wiki, and I had a map on my wall of all the biology websites then in existence. Today this book has its own large website, www.devbio.com, maintained by Jason Dirks, Nate Nolet, and the ever-growing number of Associates at Sinauer who work in multimedia and electronic formats. Among many other things, this expansive site allows the reader immediate access via links to PubMed to most of the original references cited in the book.

Also online is the wonderful Vade Mecum[3], created as a laboratory resource by Mary Tyler and Ron Kozlowski. This is the low-cost "electronic laboratory manual" for developmental biology laboratories, and it should also be looked at for its remarkable movie footage of the development of model organisms.

The administration of Swarthmore College and the University of Helsinki went above and beyond the call of duty in providing me with the facilities to write this book. Both venues remain full of stimulating people and great opportunities to continue learning. I have also been blessed with remarkable students who have never been shy about asking questions. Many of the footnotes in this book began as answers to questions posed by savvy students. Other footnotes attempt to highlight connections between developmental biology and other fields of study, particularly the perspectives offered by the humanities, which demand that we think about the context of what we do.

This is the tenth edition of *Developmental Biology*, and "what a long strange trip it's been." Either there is much more material to integrate than there used to be, or I'm getting older and slower (I'm sure it must be the former). The book is done, and I am looking forward to seeing my friends and family again! My wife, Anne Raunio, has been incredibly patient, and there seem to be all sorts of things around the house that need fixing.

SCOTT GILBERT

Reviewers of the Tenth Edition

It is no longer possible (if it ever was) for one person to comprehend this entire field. As Bob Seger so aptly sings, "I've got so much more to think about … what to leave in, what to leave out." The people who help me leave in and take out the right things are the reviewers. Their expertise in particular areas has become increasingly valuable to me. Their comments were made on early versions of each chapter, and they should not be held accountable for any errors that may appear.

Ehab Abouheif, *McGill University*

Arkhat Abzhanov, *Harvard University*

William Anderson, *Harvard University*

Jonathan Bard, *University of Edinburgh*

Julie Brill, *University of Toronto*

Marianne Bronner, *California Institute of Technology*

Donald Brown, *Carnegie Institution for Science*

Kimberly Cooper, *Harvard University*

J. K. Dale, *University of Dundee*

Diana Darnell, *University of Arizona School of Medicine*

Bradley Davidson, *Swarthmore College*

Jamie Davies, *University of Edinburgh*

Susan Ernst, *Tufts University*

Ben Ewen-Campen, *Harvard University*

Peter Koopman, *University of Queensland*

Catherine Krull, *University of Michigan*

Anthony-Samuel LaMantia, *George Washington University Medical Center*

Laura Lee, *Vanderbilt University*

Bluma Lesch, *Massachusetts Institute of Technology*

Kersti Linask, *University of South Florida College of Medicine*

Mary Mullins, *University of Pennsylvania*

Peter Rabinovitch, *University of Washington*

Peter Reddien, *Massachusetts Institute of Technology*

Janet Rossant, *The Hospital for Sick Children*

Joshua Sanes, *Harvard University*

Ana Soto, *Tufts University*

Kathleen Sulik, *University of North Carolina*

Nicole Theodosiou, *Union College*

Steven Vokes, *University of Texas*

Pablo Visconti, *University of Massachusetts*

Media and Supplements
to accompany **Developmental Biology,** Tenth Edition

eBook

Developmental Biology, Tenth Edition is available as an eBook in several formats, all at a substantial discount from the price of the printed textbook. For details on available ebook formats, please visit the Sinauer Associates website at **www.sinauer. com**, or contact your Sinauer sales representative.

For the Student

Companion Website

devbio.com

Available free of charge, this website is intended to supplement and enrich courses in developmental biology. It provides more information for advanced students as well as historical, philosophical, and ethical perspectives on issues in developmental biology. Included are articles, movies, interviews, opinions, Web links, updates, and more. References to specific website topics are included in the printed textbook throughout each chapter, as well as at the end of each chapter.

DevBio Laboratory: Vade Mecum³: An Interactive Guide to Developmental Biology

labs.devbio.com

MARY S. TYLER and RONALD N. KOZLOWSKI

Access to *DevBio Laboratory: Vade Mecum³* is included with every new copy of the textbook. (See the inside front cover for details.) *DevBio Laboratory: Vade Mecum³* is a rich multimedia learning tool that helps students understand the development of the organisms discussed in lecture and prepares them for laboratory exercises. It also includes excerpts from the *Differential Expressions²* series of videos, highlighting some major concepts in developmental biology, famous experiments, and the scientists who performed them.

Developmental Biology: A Guide for Experimental Study, Third Edition

MARY S. TYLER

(Included in *DevBio Laboratory: Vade Mecum³*)
This lab manual teaches the student to work as an independent investigator on problems in development and provides extensive background information and instructions for each experiment. It emphasizes the study of living material, intermixing developmental anatomy in an enjoyable balance, and allows students to make choices in their work.

For the Instructor

(Available to qualified adopters)

Instructor's Resource Library

The *Developmental Biology,* Tenth Edition Instructor's Resource Library includes a rich collection of visual resources for use in preparing lectures and other course materials. The IRL includes:

- All textbook figures and tables in JPEG format (both high- and low-resolution versions), formatted for optimal projection quality
- PowerPoint® presentations of all figures and tables
- Chick embryo cross-sections and chick embryo whole-mounts from *DevBio Laboratory: Vade Mecum³* (PowerPoint® format)
- Three collections of videos:
  - Developmental biology video collection
  - Video segments from *DevBio Laboratory: Vade Mecum³*
  - Video segments from *Differential Expressions²*
- Instructor's Reference Guide for *Differential Expressions²*

Also Available

The following titles are available for purchase separately or, in some cases, bundled with the textbook. Please contact Sinauer Associates for more information.

■ *Ecological Developmental Biology:*
Integrating Epigenetics, Medicine, and Evolution
Scott F. Gilbert and David Epel
Paper, 460 pages • ISBN 978-0-87893-299-3

■ *Bioethics and the New Embryology:*
Springboards for Debate
Scott F. Gilbert, Anna Tyler, and Emily Zackin
Paper, 261 pages • ISBN 978-0-7167-7345-0

■ *Differential Expressions[2]:*
Key Experiments in Developmental Biology
Mary S. Tyler, Ronald N. Kozlowski, and Scott F. Gilbert
2-DVD Set • UPC 855038001020

■ *A Dozen Eggs:*
Time-Lapse Microscopy of Normal Development
Rachel Fink
DVD • ISBN 978-0-87893-329-7

■ *Fly Cycle[2]*
Mary S. Tyler and Ronald N. Kozlowski
DVD • ISBN 978-0-87893-849-0

■ *From Egg to Tadpole:*
Early Morphogenesis in Xenopus
Jeremy D. Pickett-Heaps and Julianne Pickett-Heaps
DVD • ISBN 978-0-97752-224-8

PART ONE

QUESTIONS
Introducing Developmental Biology

Between fertilization and birth, the developing organism is known as an embryo. The concept of an embryo is a staggering one, and forming an embryo is the hardest thing you will ever do. To become an embryo, you had to build yourself from a single cell. You had to respire before you had lungs, digest before you had a gut, build bones when you were pulpy, and form orderly arrays of neurons before you knew how to think. One of the critical differences between you and a machine is that a machine is never required to function until after it is built. Every animal has to function even as it builds itself.

For animals, fungi, and plants, the sole way of getting from egg to adult is by developing an embryo. The embryo mediates between genotype and phenotype, between the inherited genes and the adult organism. Whereas most fields of biology study adult structure and function, developmental biology finds the study of the transient stages leading up to the adult to be more interesting. Developmental biology studies the initiation and construction of organisms rather than their maintenance. It is a science of becoming, a science of process.

This development, this formation of an orderly body from relatively homogeneous material, provokes profound and fundamental questions that *Homo sapiens* have been asking since the dawn of self-awareness. How does the body form with its head always above its shoulders? Why is the heart on the left side of our body? Why do we typically have five digits on each hand and not more—or fewer? Why can't we regenerate limbs? How do the sexes develop their different anatomies? Why can only females have babies?

Our answers to these questions must respect the complexity of the inquiry and must form a coherent causal network from gene through functional organ. To say that XX mammals are usually females and that XY mammals are usually males does not explain sex determination to a developmental biologist, who wants to know *how* the XX genotype produces a female and *how* the XY genotype produces a male. Similarly, a geneticist might ask how globin genes are transmitted from one generation to the next, and a physiologist might ask about the function of globin proteins in the body. But the developmental biologist asks how it is that the globin genes come to be expressed only in red blood cells, and how these genes become active only at specific times in development. (We don't know the answers yet.) Each field of biology is defined by the questions it asks. *Welcome to a wonderful set of important questions!*

The Questions of Developmental Biology

Development accomplishes two major objectives. First, it generates cellular diversity and order within the individual organism; secondly, it ensures the continuity of life from one generation to the next. Put another way, there are two fundamental questions in developmental biology. How does the fertilized egg give rise to the adult body? And how does that adult body produce yet another body? These two huge questions can be subdivided into seven general categories of questions scrutinized by developmental biologists:

- **The question of differentiation.** A single cell, the fertilized egg, gives rise to hundreds of different cell types—muscle cells, epidermal cells, neurons, lens cells, lymphocytes, blood cells, fat cells, and so on. The generation of this cellular diversity is called *differentiation*. Since every cell of the body (with very few exceptions) contains the same set of genes, how can this identical set of genetic instructions produce different types of cells? How can a single cell, the fertilized egg, generate so many different cell types?*

- **The question of morphogenesis.** How can the cells in our body organize themselves into functional structures? Our differentiated cells are not randomly distributed. Rather, they become organized into intricate tissues and organs. During development, cells divide, migrate, and die; tissues fold and separate. Our fingers are always at the tips of our hands, never in the middle; our eyes are always in our head, not in our toes or gut. This creation of ordered form is called *morphogenesis*, and it involves coordinating cell growth, cell migration, and cell death.

- **The question of growth.** If each cell in our face were to undergo just one more cell division, we would be considered horribly malformed. If each cell in our arms underwent just one more round of cell division, we could tie our shoelaces without bending over. How do

our cells know when to stop dividing? Our arms are generally the same size on both sides of the body. How is cell division so tightly regulated?

- **The question of reproduction.** The sperm and egg are very specialized cells, and only they can transmit the instructions for making an organism from one generation to the next. How are these germ cells set apart from the cells that are constructing the physical structures of the embryo, and what are the instructions in the nucleus and cytoplasm that allow them to form the next generation?

- **The question of regeneration.** Some organisms can regenerate their entire body. Some salamanders regenerate their eyes and legs, and many reptiles can regenerate their tail. Mammals are generally poor at regeneration, and yet there are some cells in our bodies—*stem cells*—that are able to form new structures even in adults. How do stem cells retain this capacity, and can we harness it to cure debilitating diseases?

- **The question of evolution.** Evolution involves inherited changes in development. When we say that today's one-toed horse had a five-toed ancestor, we are saying that changes in the development of cartilage and muscles occurred over many generations in the embryos of the horse's ancestors. How do changes in development create new body forms? Which heritable changes are possible, given the constraints imposed by the necessity that the organism survive as it develops?

- **The question of environmental integration.** The development of many (perhaps all) organisms is influenced by cues from the environment that surrounds the embryo or larvae. The sex of many species of turtles, for instance, depends on the temperature the embryo experiences while in the egg. The formation of the reproductive system in some insects depends on bacteria that are transmitted inside the egg. Moreover, certain chemicals in the environment can disrupt normal development, causing malformations in the adult. How is the development of an organism integrated into the larger context of its habitat?

The study of development has become essential for understanding all other areas of biology. Indeed, the questions asked by developmental biologists have also become critical in molecular biology, physiology, cell biology, genetics, anatomy, cancer research, neurobiology, immunology, ecology, and evolutionary biology. In turn, the many advances of molecular biology, along with new techniques of cell imaging, have finally made these questions answerable. This makes developmental biologists extremely happy, for as the

---

*Biologists recognize more than 210 different cell types in the adult human, but this number has little or no significance when studying development because there are many transient cell types that are formed during development but are not seen in the adult. Some of these embryonic cells are transitional stages or precursors of adult cell types. Other embryonic cell types perform particular functions in constructing an organ and then undergo programmed cell death after completing their tasks.

Nobel Prize-winning developmental biologist Hans Spemann stated in 1927,

We stand in the presence of riddles, but not without the hope of solving them. And riddles with the hope of solution—what more can a scientist desire?

So, like the man in the cartoon, I come bearing questions. They are questions bequeathed to us by earlier generations of biologists, philosophers, and parents. They are questions with their own history, questions discussed on an anatomical level by people such as Aristotle, William Harvey, St. Albertus Magnus, and Charles Darwin. More recently, these questions have been addressed on the cellular and molecular levels by men and women throughout the world, each of whom brings to the laboratory his or her own perspectives and training, for there is no one way to become a developmental biologist. The field has benefitted by having researchers trained in cell biology, genetics, biochemistry, immunology, and even anthropology, engineering, physics, history, and art. You are invited to become part of a community of question-askers for whom the embryo is a source of both wonder and the most interesting questions in the world.

The next three chapters will outline some of the critical framework needed to answer these questions. Chapter 1 discusses *organismal* concepts, including life cycles, the three germ layers that form the organs, and the migration of cells during development. Chapter 2 concentrates on the *genetic* approach to cell differentiation and outlines the principle of differential gene expression (which explains how different proteins can be made in different cells from the same set of inherited genes). Chapter 3 focuses on the *cellular* approach to morphogenesis, showing how communication between cells is critical for their formation into tissues and organs. Thus, you will be introduced to development at the organismal, genetic, and cellular levels, and much of the textbook thereafter will show how these levels are integrated to produce the remarkable panoply of animal development.

1

Comprehending Development
Generating New Cells and Organs

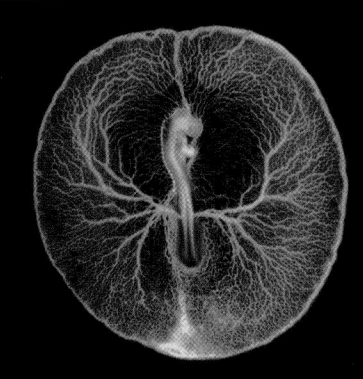

ACCORDING TO ARISTOTLE, THE FIRST EMBRYOLOGIST known to history, science begins with wonder: "It is owing to wonder that people began to philosophize, and wonder remains the beginning of knowledge" (Aristotle, *Metaphysics*, ca. 350 BCE). The development of an animal from an egg has been a source of wonder throughout history. The simple procedure of cracking open a chick egg on each successive day of its 3-week incubation period provides a remarkable experience as a thin band of cells is seen to give rise to an entire bird. Aristotle performed this procedure and noted the formation of the major organs. Anyone can wonder at this remarkable—yet commonplace—phenomenon, but it is the scientist who seeks to discover how development actually occurs. And rather than dissipating wonder, new understanding increases it.

Multicellular organisms do not spring forth fully formed. Rather, they arise by a relatively slow process of progressive change that we call **development**. In nearly all cases, the development of a multicellular organism begins with a single cell—the fertilized egg, or **zygote**, which divides mitotically to produce all the cells of the body. The study of animal development has traditionally been called **embryology**, after that phase of an organism that exists between fertilization and birth. But development does not stop at birth, or even at adulthood. Most organisms never stop developing. Each day we replace more than a gram of skin cells (the older cells being sloughed off as we move), and our bone marrow sustains the development of millions of new red blood cells every minute of our lives. In addition, some animals can regenerate severed parts, and many species undergo metamorphosis (such as the transformation of a tadpole into a frog, or a caterpillar into a butterfly). Therefore, in recent years it has become customary to speak of **developmental biology** as the discipline that studies embryonic and other developmental processes.

As the introduction to Part One noted, a scientific field is defined by the questions it seeks to answer. Most of the questions in developmental biology have been provided to it by its embryological heritage. We can identify three major approaches to studying embryology:

- Anatomical approaches
- Experimental approaches
- Genetic approaches

Each of these traditions has predominated during a different era. However, although it is true that anatomical approaches gave rise to experimental approaches, and that genetic approaches built on the foundations of the earlier two approaches, all three traditions persist to this day and continue to play a major role in developmental biology. The basis of all research in developmental biology is the changing anatomy of the organism. Today the anatomical approach to development is continually expanded and enhanced by revolutions in microscopy, computer-aided graphical reconstructions of three-dimensional objects, and methods of applying mathematics to biology. Many of the beautiful photographs in this book reflect this increasingly important component of embryology.

It is a most beautiful thing to study the different changes of life, from the microscopic changes of conception to the more apparent ones of maturity and old age.
FRANKLIN MALL
(CA. 1890)

The greatest progressive minds of embryology have not looked for hypotheses; they have looked at embryos.
JANE OPPENHEIMER
(1955)

The Cycle of Life

One of the major triumphs of descriptive embryology was the idea of a generalizable animal life cycle. Each animal, whether earthworm or eagle, termite or beagle, passes through similar stages of development. The stages of development between fertilization and hatching (or birth) are collectively called **embryogenesis**.

Throughout the animal kingdom, an incredible variety of embryonic types exist, but most patterns of embryogenesis are variations on six fundamental processes: fertilization, cleavage, gastrulation, organogenesis, metamorphosis, and gametogenesis.

1. **Fertilization** involves the fusion of the mature sex cells, the sperm and egg, which are collectively called the **gametes**. The fusion of the gamete cells stimulates the egg to begin development and initiates a new individual. The subsequent fusion of the gamete nuclei (both of which have only half the normal number of chromosomes characteristic for the species) gives the embryo its **genome**, the collection of genes that helps instruct the embryo to develop in a manner very similar to that of it parents.

2. **Cleavage** is a series of extremely rapid mitotic divisions that immediately follow fertilization. During cleavage, the enormous volume of zygote cytoplasm is divided into numerous smaller cells called **blastomeres**. By the end of cleavage, the blastomeres have usually formed a sphere, known as a **blastula**.

3. After the rate of mitotic division slows down, the blastomeres undergo dramatic movements and change their positions relative to one another. This series of extensive cell rearrangements is called **gastrulation**, and the embryo is said to be in the **gastrula** stage. As a result of gastrulation, the embryo contains three **germ layers** (**endoderm**, **ectoderm**, and **mesoderm**) that will interact to generate the organs of the body.

4. Once the germ layers are established, the cells interact with one another and rearrange themselves to produce tissues and organs. This process is called **organogenesis**. Chemical signals are exchanged between the cells of the germ layers, resulting in the formation of specific organs at specific sites. Certain cells will undergo long migrations from their place of origin to their final location. These migrating cells include the precursors of blood cells, lymph cells, pigment cells, and gametes (eggs and sperm).

5. In many species, the organism that hatches from the egg or is born into the world is not sexually mature. Rather, the organism needs to undergo **metamorphosis** to become a sexually mature adult. In most animals, the young organism is a called a **larva**, and it may look significantly different from the adult. In many species, the larval stage is the one that lasts the longest, and is used for feeding or dispersal. In such species, the adult is a brief stage whose sole purpose is to reproduce. In silkworm moths, for instance, the adults do not have mouthparts and cannot feed; the larva must eat enough so that the adult has the stored energy to survive and mate. Indeed, most female moths mate as soon as they eclose from the pupa, and they fly only once—to lay their eggs. Then they die.

6. In many species, a group of cells is set aside to produce the next generation (rather than forming the current embryo). These cells are the precursors of the gametes. The gametes and their precursor cells are collectively called **germ cells**, and they are set aside for reproductive function. All other cells of the body are called **somatic cells**. This separation of somatic cells (which give rise to the individual body) and germ cells (which contribute to the formation of a new generation) is often one of the first differentiations to occur during animal development. The germ cells eventually migrate to the gonads, where they differentiate into gametes. The development of gametes, called **gametogenesis**, is usually not completed until the organism has become physically mature. At maturity, the gametes may be released and participate in fertilization to begin a new embryo. The adult organism eventually undergoes senescence and dies, its nutrients often supporting the early embryogenesis of its offspring and its absence allowing less competition. Thus, the cycle of life is renewed.

A Frog's Life

All animal life cycles are modifications of the generalized one described above. **FIGURE 1.1** shows the development of the leopard frog *Rana pipiens* and provides a good starting point for a more detailed discussion of a representative life cycle.

Gametogenesis and fertilization

The end of one life cycle and the beginning of the next are often intricately intertwined. Life cycles are often controlled by environmental factors (tadpoles wouldn't survive if they hatched in the fall, when their food is dying), so in most frogs, gametogenesis and fertilization are seasonal events. A combination of photoperiod (hours of daylight) and temperature informs the pituitary gland of the mature female frog that it is spring. The pituitary then secretes hormones that stimulate her ovary to make the hormone estrogen. Estrogen then instructs the liver to make and secrete yolk proteins, which are then transported through the blood into the enlarging eggs in the ovary. The yolk is transported into the bottom portion of the egg, called the **vegetal hemisphere**, where it will serve as food for the developing embryo (**FIGURE 1.2A**). The upper half of the egg is called

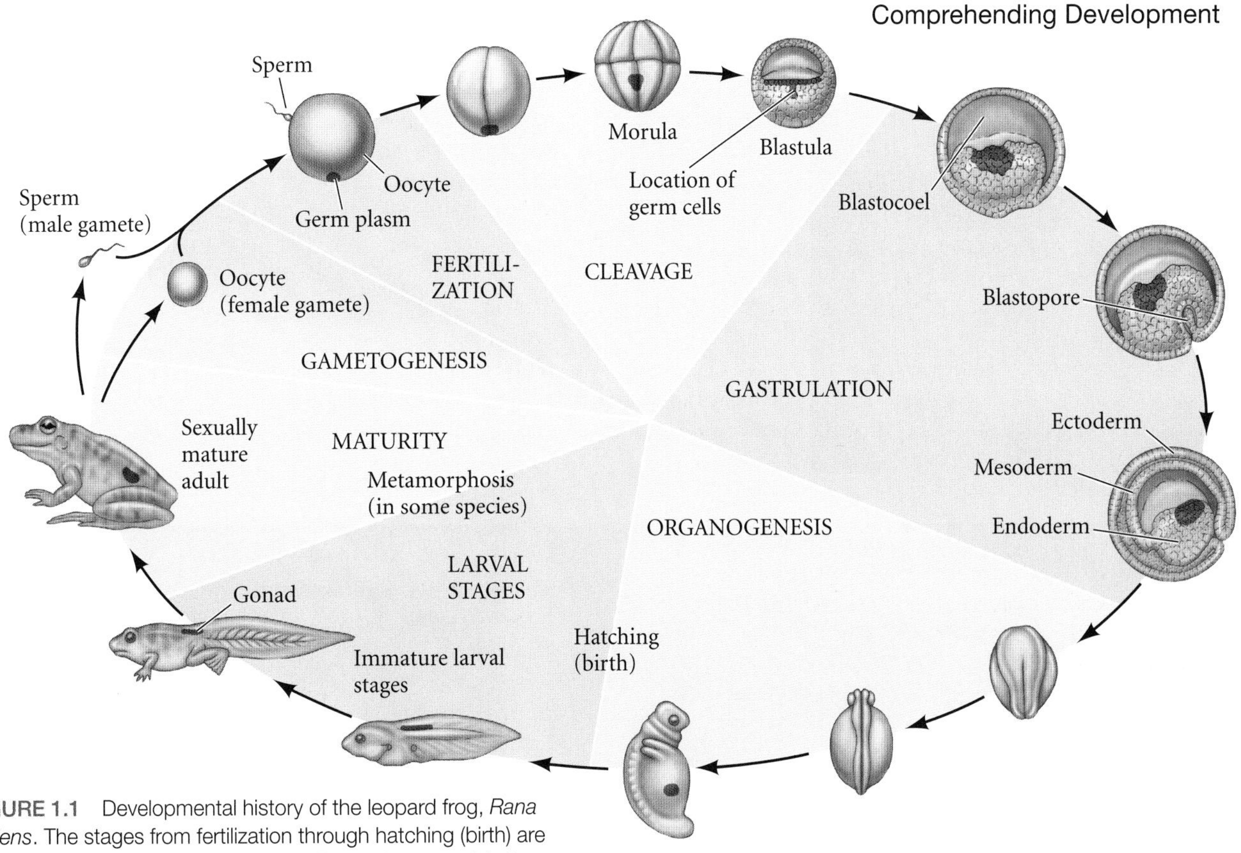

FIGURE 1.1 Developmental history of the leopard frog, *Rana pipiens*. The stages from fertilization through hatching (birth) are known collectively as embryogenesis. The region set aside for producing germ cells is shown in purple. Gametogenesis, which is completed in the sexually mature adult, begins at different times during development, depending on the species. (The sizes of the varicolored wedges shown here are arbitrary and do not correspond to the proportion of the life cycle spent in each stage.)

● **See VADE MECUM**   The amphibian life cycle

the **animal hemisphere**.* Sperm formation also occurs on a seasonal basis. Male leopard frogs make sperm during the summer, and by the time they begin hibernation in the fall they have produced all the sperm that will be available for the following spring's breeding season.

In most species of frogs, fertilization is external. The male frog grabs the female's back and fertilizes the eggs as the female releases them (**FIGURE 1.2B**). Some species lay their eggs in pond vegetation, and the egg jelly adheres to the plants and anchors the eggs (**FIGURE 1.2C**). Other species float their eggs into the center of the pond without any support. So the first important thing to remember about life cycles is that they are often intimately involved with environmental factors.

Fertilization accomplishes several things. First, it allows the haploid nucleus of the egg (the **female pronucleus**) to

merge with the haploid nucleus of the sperm (the **male pronucleus**) to form the diploid **zygote nucleus**. Second, fertilization causes the cytoplasm of the egg to move such that different parts of the cytoplasm find themselves in new locations (**FIGURE 1.2D**). This cytoplasmic migration will be important in determining the three body axes of the frog: **anterior-posterior** (head-tail), **dorsal-ventral** (back-belly), and **right-left**. Third, fertilization activates those molecules necessary to begin cell cleavage and gastrulation (Rugh 1950).

Cleavage and gastrulation

During cleavage, the volume of the frog egg stays the same, but it is divided into tens of thousands of cells (**FIGURE 1.2E–H**). The cells in the animal hemisphere of the egg divide faster than those in the vegetal hemisphere, and the cells of the vegetal hemisphere become progressively larger the more vegetal the cytoplasm. Meanwhile, a fluid-filled cavity, the **blastocoel**, forms in the animal hemisphere (**FIGURE 1.2I**). This cavity will be important for allowing cell movements to occur during gastrulation.

Gastrulation in the frog begins at a point on the embryo surface roughly 180° opposite the point of sperm entry with the formation of a dimple, called the **blastopore**. This dimple (which will mark the future dorsal side of the embryo) expands to become a ring, and cells migrating through the blastopore become the mesoderm (**FIGURE 1.3A–C**). The cells remaining on the outside become the ectoderm, and this outer layer expands to enclose the entire embryo. The large, yolky cells that remain in the vegetal hemisphere (until

---

*The use of the terms *animal* and *vegetal* for the upper and lower hemispheres of the early frog embryo reflect the division rates of the cells. The upper cells divide rapidly and become actively mobile (hence "animated"), whereas the yolk-filled cells of the lower half were seen as being immobile (hence like plants, or "vegetal").

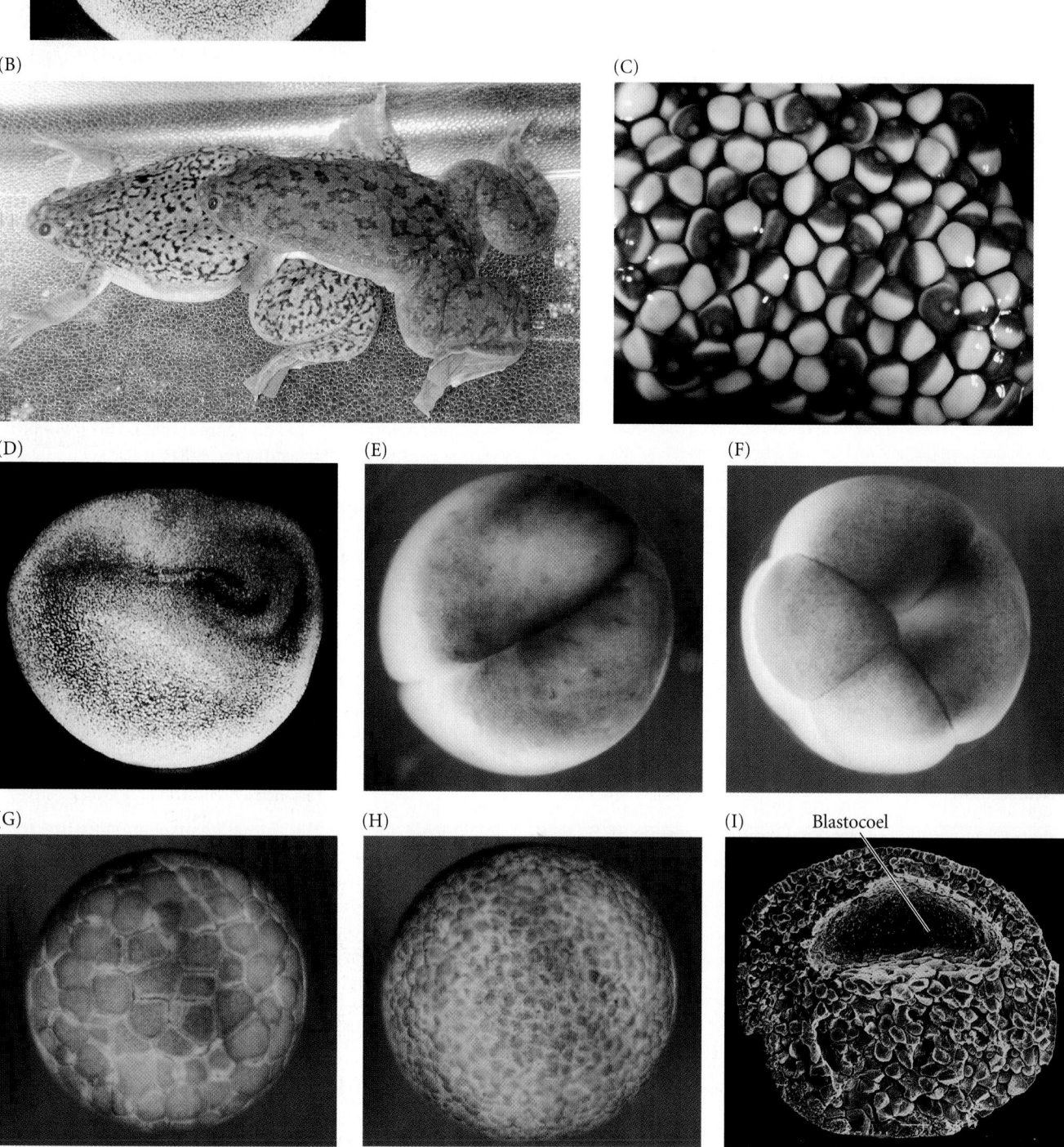

FIGURE 1.2 Early development of the frog *Xenopus laevis*. (A) As the egg matures, it accumulates yolk (here stained green) in the vegetal cytoplasm. (B) Frogs mate by amplexus, the male grasping the female around the belly and fertilizing the eggs as they are released. (C) A newly laid clutch of eggs. The brown area of each egg is the pigmented animal hemisphere. The white spot in the middle of the pigment is where the egg's nucleus resides. (D) Cytoplasm rearrangement seen during first cleavage. Compare with the initial stage seen in (A). (E) A 2-cell embryo near the end of its first cleavage. (F) An 8-cell embryo. (G) Early blastula. Note that the cells get smaller, but the volume of the egg remains the same. (H) Late blastula. (I) Cross section of a late blastula, showing the blastocoel (cavity). (A–H courtesy of Michael Danilchik and Kimberly Ray; I courtesy of J. Heasman.)

(A)

(B)

(C)

(D)

(E)

(F)

(G)

(H)

(I) Blastocoel

FIGURE 1.3 Continued development of *Xenopus laevis*. (A) Gastrulation begins with an invagination, or slit, in the future dorsal (top) side of the embryo. (B) This slit, the dorsal blastopore lip, as seen from the ventral surface (bottom) of the embryo. (C) The slit becomes a circle, the blastopore. Future mesoderm cells migrate into the interior of the embryo along the blastopore edges, and the ectoderm (future epidermis and nerves) migrates down the outside of the embryo. The remaining part, the yolk-filled endoderm, is eventually encircled. (D) Neural folds begin to form on the dorsal surface. (E) A groove can be seen where the bottom of the neural tube will be. (F) The neural folds come together at the dorsal midline, creating a neural tube. (G) Cross section of the *Xenopus* embryo at the neurula stage. (H) A pre-hatching tadpole, as the protrusions of the forebrain begin to induce eyes to form. (I) A mature tadpole, having swum away from the egg mass and feeding independently. (Courtesy of Michael Danilchik and Kimberly Ray.)

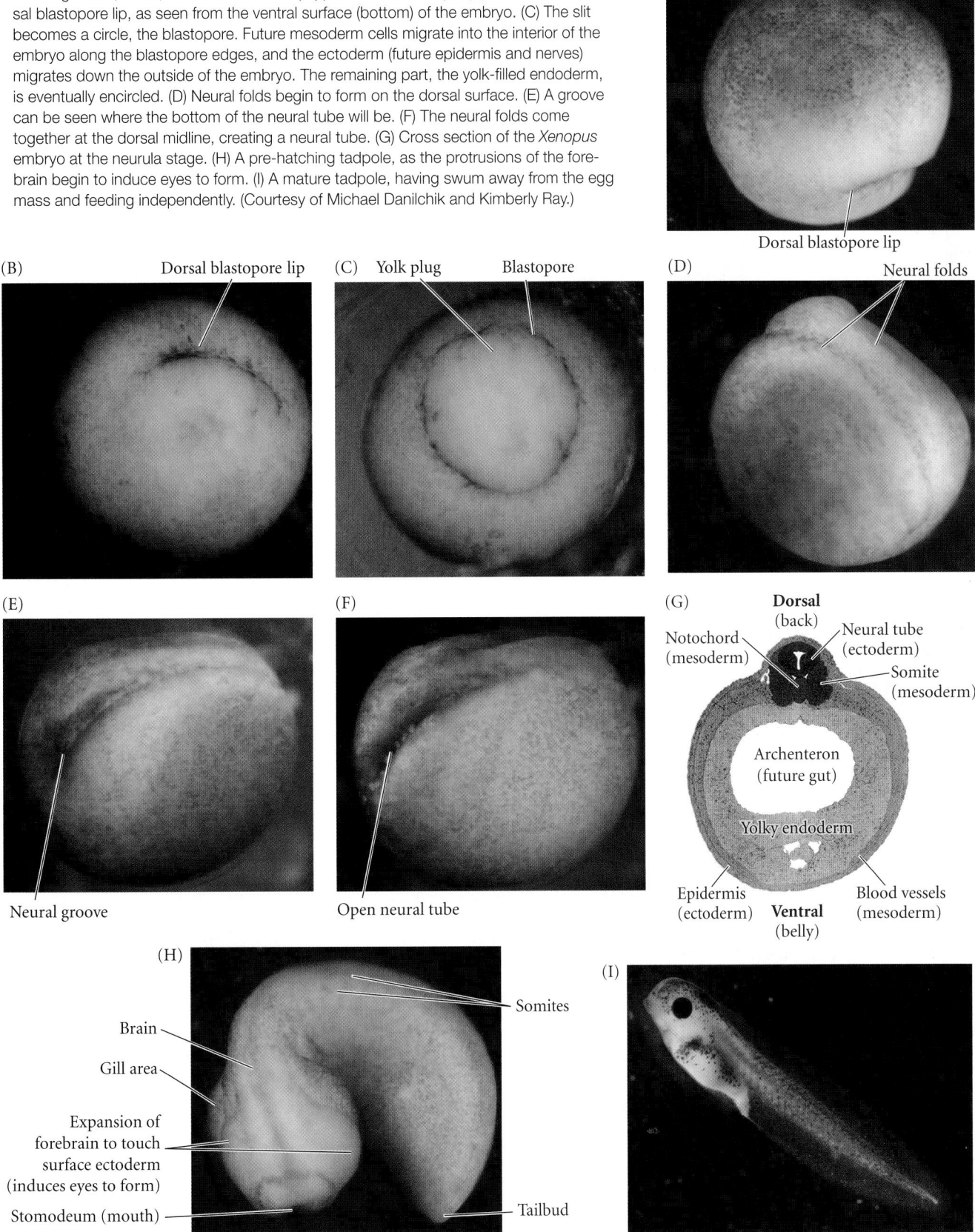

(A)

(B)

(C)

(D)

(E)

(F)

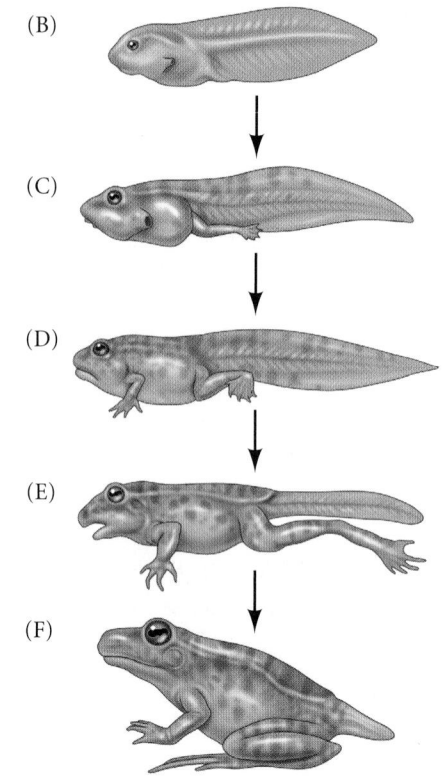

FIGURE 1.4 Metamorphosis of the frog. (A) Huge changes are obvious when one contrasts the tadpole and the adult bullfrog. Note especially the differences in jaw structure and limbs. (B) Premetamorphic tadpole. (C) Prometamorphic tadpole, showing hindlimb growth. (D) Onset of metamorphic climax as forelimbs emerge. (E,F) Climax stages. (A © Patrice Ceisel/Visuals Unlimited.)

they are encircled by the expanding ectoderm) become the endoderm. Thus, at the end of gastrulation, the ectoderm (precursor of the epidermis, brain, and nerves) is on the outside of the embryo, the endoderm (precursor of the gut and respiratory systems) is on the inside of the embryo, and the mesoderm (precursor of the connective tissue, blood, heart, skeleton, gonads, and kidneys) is between them.

Organogenesis

Organogenesis in the frog begins when the **notochord**—a rod of mesodermal cells in the most dorsal portion of the embryo*—signals the ectodermal cells above it that they are not going to become epidermis. Instead, these dorsal ectoderm cells form a tube and become the nervous system. At this stage, the embryo is called a **neurula**. The neural precursor cells elongate, stretch, and fold into the embryo, forming

---

*The notochord consists of cells such as those mentioned on page 2—i.e., cells that are important for constructing the embryo but which, having performed their tasks, die. Although adult vertebrates do not have notochords, this embryonic organ is critical for establishing the fates of the ectodermal cells above it, as we shall see in Chapters 8–10.

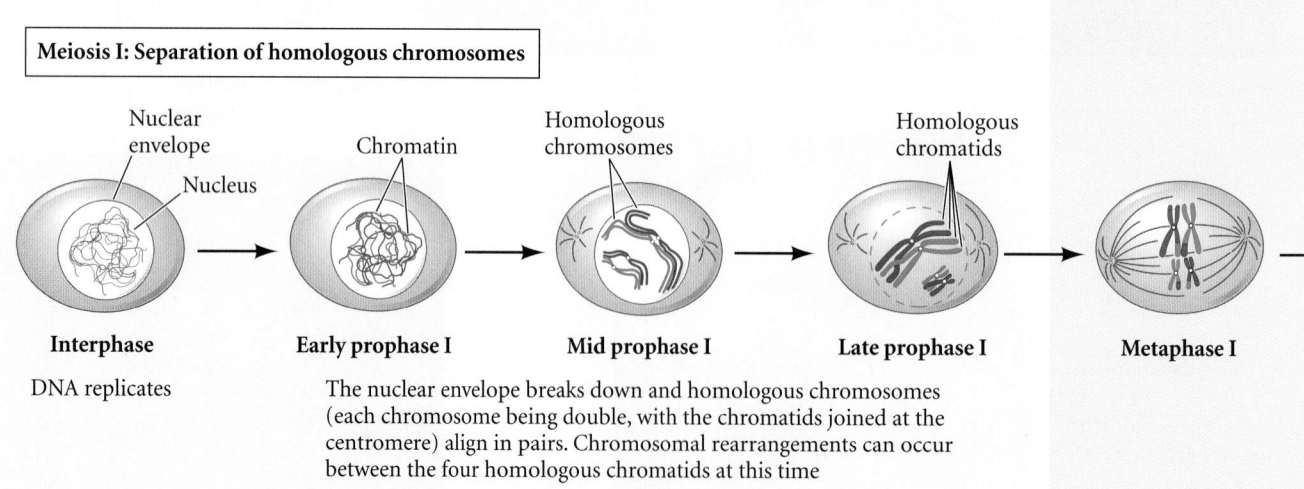

Meiosis I: Separation of homologous chromosomes

Nuclear envelope

Nucleus

Chromatin

Homologous chromosomes

Homologous chromatids

Interphase

DNA replicates

Early prophase I

Mid prophase I

Late prophase I

Metaphase I

The nuclear envelope breaks down and homologous chromosomes (each chromosome being double, with the chromatids joined at the centromere) align in pairs. Chromosomal rearrangements can occur between the four homologous chromatids at this time

the **neural tube** (FIGURE 1.3D–F); the future epidermal cells of the back cover the neural tube.

Once the neural tube has formed, it and the notochord induce changes in their neighbors, and organogenesis continues. The mesodermal tissue adjacent to the neural tube and notochord becomes segmented into **somites** (FIGURE 1.3G,H), the precursors of the frog's back muscles, spinal vertebrae, and dermis (the inner portion of the skin). The embryo develops a mouth and an anus, and it elongates into the familiar tadpole structure (FIGURE 1.3I). The neurons make their connections to the muscles and to other neurons, the gills form, and the larva is ready to hatch from its egg jelly. The hatched tadpole will feed for itself as soon as the yolk supplied by its mother is exhausted.

Metamorphosis and gametogenesis

Metamorphosis of the fully aquatic tadpole larva into an adult frog that can live on land is one of the most striking transformations in all of biology. In amphibians, metamorphosis is initiated by hormones from the tadpole's thyroid gland. (The mechanisms by which thyroid hormones accomplish these changes will be discussed in Chapter 16.) In frogs, almost every organ is subject to modification, and the resulting changes in form are striking and very obvious (FIGURE 1.4). The hindlimbs and forelimbs the adult will use for locomotion differentiate as the tadpole's paddle tail recedes. The cartilaginous tadpole skull is replaced by the predominantly bony skull of the young frog. The horny teeth the tadpole uses to tear up pond plants disappear as the mouth and jaw take a new shape, and the fly-catching tongue muscle of the frog develops. Meanwhile, the tadpole's lengthy intestine—a characteristic of herbivores—shortens to suit the more carnivorous diet of the adult frog. The gills regress and the lungs enlarge. The speed of metamorphosis is carefully keyed to environmental pressures. In temperate regions, for instance, *Rana* metamorphosis must occur before ponds freeze in winter. An adult leopard frog can burrow into the mud and survive the winter; its tadpole cannot.

As metamorphosis ends, the development of the germ cells begins. Gametogenesis can take a long time. In *Rana pipiens*, it takes 3 years for the eggs to mature in the female's ovaries. (Sperm take less time; *Rana* males are often fertile soon after metamorphosis.) To become mature, the germ cells must be competent to complete **meiosis**.

Meiosis (FIGURE 1.5) is one of the most important evolutionary processes characteristic of eukaryotic organisms. It makes fertilization possible and results in recombination of the genes from the two parents, generating unique genomes that are the raw material of biodiversity. Genetics, development, and evolution throughout the eukaryotic kingdoms are predicated on meiosis. We will discuss meiosis more thoroughly in Chapter 17, but the most important things to remember from the start are:

1. The chromosomes replicate *prior* to cell division, so that each gene is represented *four* times.

2. The replicated chromosomes (each called a **chromatid**) are held together by **centromeres**, and the four homologous chromatids pair together.

3. The first meiotic division separates the chromatid pairs from one another.

4. The second meiotic division splits the centromere such that each chromatid becomes a chromosome.

FIGURE 1.5 Summary of meiosis. The DNA replicates during interphase. During first meiotic prophase, the nuclear envelope breaks down and the homologous chromosomes (each chromosome is double, with its two chromatids joined at the centromere) align together. Chromosome rearrangements ("crossing over") can occur at this stage. After the first metaphase, the centromere remains unsplit and the pairs of homologous chromosomes are sorted into different cells. During the second meiotic division, the centromere splits and the sister chromatids are moved into separate cells, each with a haploid set of chromosomes.

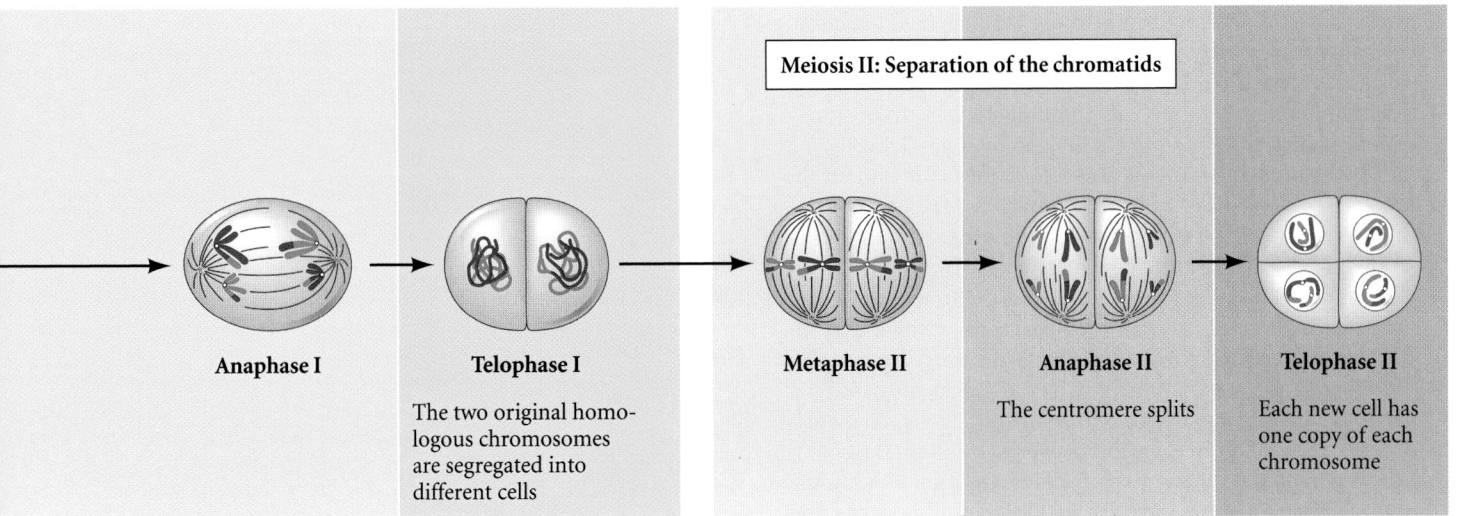

Meiosis II: Separation of the chromatids

Anaphase I

Telophase I

The two original homologous chromosomes are segregated into different cells

Metaphase II

Anaphase II

The centromere splits

Telophase II

Each new cell has one copy of each chromosome

5. Meiosis thus results in four germ cells, each with a haploid nucleus.

Having undergone meiosis, the mature sperm and egg nuclei can unite in fertilization, restoring the diploid chromosome number and initiating the events that lead to development and the continuation of the circle of life.

"How Are You?"

The fertilized egg has no heart. It has no eye. No limb is found in the zygote. So how did we become what we are? What part of the embryo forms the heart? How do the cells that form the eye's retina migrate the proper distance from the cells that form the lens? How do the tissues that form a bird's wing relate to the tissues that form fish fins or the human hand? What organs are affected by mutations in particular genes? These are the types of questions asked by developmental anatomists.

Several strands weave together to form the anatomical approaches to development. The first strand is **comparative embryology**, the study of how anatomy changes during the development of different organisms. The second strand, based on the first, is **evolutionary embryology**, the study of how changes in development may cause evolutionary change and of how an organism's ancestry may constrain the types of changes that are possible. The third strand of the anatomical approach to developmental biology is **teratology**, the study of birth defects.

● **See WEBSITE 1.1**   When does human personhood begin?

Comparative embryology

The first known study of comparative developmental anatomy was undertaken by Aristotle in the fourth century BCE. In *The Generation of Animals* (ca. 350 BCE), he noted some of the variations on the life cycle themes: some animals are born from eggs (**oviparity**, as in birds, frogs, and most invertebrates); some by live birth (**viviparity**, as in placental mammals); and some by producing an egg that hatches inside the body (**ovoviviparity**, as in certain reptiles and sharks). Aristotle also identified the two major cell division patterns by which embryos are formed: the **holoblastic** pattern of cleavage (in which the entire egg is divided into smaller cells, as it is in frogs and mammals) and the **meroblastic** pattern of cleavage (as in chicks, wherein only part of the egg is destined to become the embryo, while the other portion—the yolk—serves as nutrition for the embryo). And should anyone want to know who first figured out the functions of the mammalian placenta and the umbilical cord, it was Aristotle.

There was remarkably little progress in embryology for the two thousand years following Aristotle. It was only in 1651 that William Harvey concluded that all animals—even mammals—originate from eggs. *Ex ovo omnia* ("All from the egg") was the motto on the frontispiece of Harvey's *On the Generation of Living Creatures*, and this precluded the spontaneous generation of animals from mud or excrement. This statement was not made lightly, for Harvey knew that it went against the views of Aristotle, whom Harvey still venerated. (Aristotle had thought that menstrual fluid formed the material of the embryo, while the semen gave it form and animation.) Harvey also was the first to see the blastoderm of the chick embryo (the small region of the egg containing the yolk-free cytoplasm that gives rise to the embryo), and he was the first to notice that "islands" of blood tissue form before the heart does. Harvey also suggested that the amniotic fluid might function as a "shock absorber" for the embryo.

As might be expected, embryology remained little but speculation until the invention of the microscope allowed detailed observations (**FIGURE 1.6**). Marcello Malpighi published the first microscopic account of chick development in 1672. Here, for the first time, the neural groove (precursor of the neural tube), the muscle-forming somites, and the first circulation of the arteries and veins—to and from the yolk—were identified.

Epigenesis and preformation

With Malpighi began one of the great debates in embryology: the controversy over whether the organs of the embryo are formed de novo ("from scratch") at each generation, or whether the organs are already present, in miniature form, within the egg or sperm. The first view, called **epigenesis**, was supported by Aristotle and Harvey. The second view, called **preformation**, was reinvigorated with Malpighi's support. Malpighi showed that the unincubated* chick egg already had a great deal of structure, and this observation provided him with reasons to question epigenesis. According to the preformationist view, all the organs of the adult were prefigured in miniature within the sperm or (more usually) the egg. Organisms were not seen to be "constructed" but rather "unrolled" or "unfurled."

The preformationist hypothesis had the backing of eighteenth-century science, religion, and philosophy (Gould 1977; Roe 1981; Churchill 1991; Pinto-Correia 1997). First, if all organs were prefigured, embryonic development merely required the growth of existing structures, not the formation of new ones. No extra mysterious force was needed for embryonic development. Second, just as the adult organism was prefigured in the germ cells, another generation already existed in a prefigured state within the germ cells of the first prefigured generation. This corollary, called *emboîtment* (encapsulation), ensured that the species would remain constant. Although certain microscopists claimed to see fully formed human miniatures within the sperm or egg, the major proponents of this hypothesis—Albrecht von Haller and Charles Bonnet—knew that organ systems develop at

---

* As pointed out by Maître-Jan in 1722, the eggs Malpighi examined may technically be called "unincubated," but as they were left sitting in the Bolognese sun in August, they were not unheated. Such eggs would be expected to have developed into chicks.

(A)

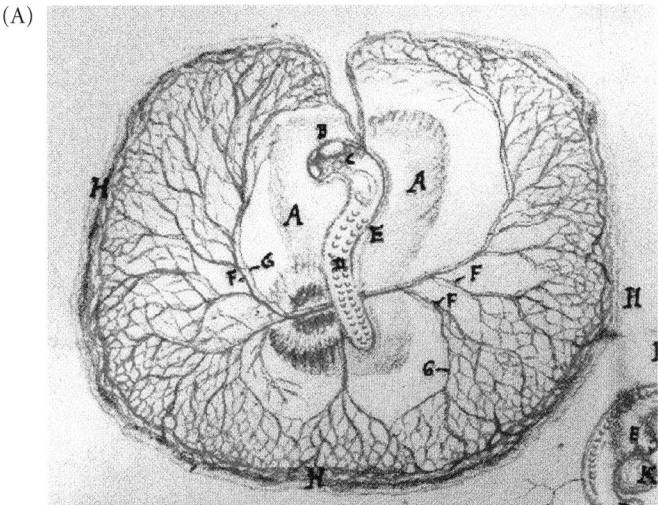

(B)

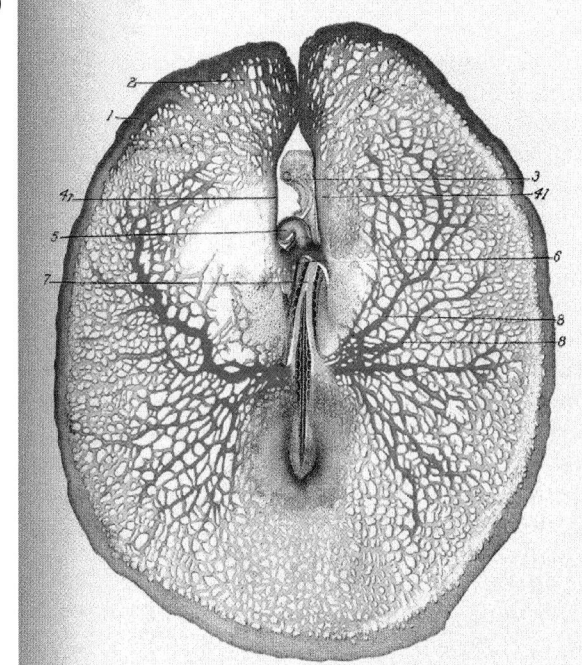

(C)

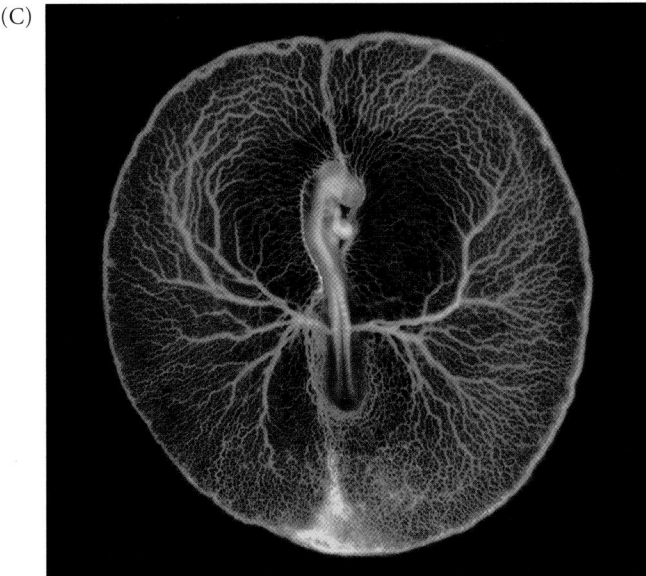

FIGURE 1.6 Depictions of chick developmental anatomy. (A) Dorsal view (looking "down" at what will become the back) of a 2-day chick embryo, as depicted by Marcello Malpighi in 1672. (B) Ventral view (looking "up" at the prospective belly) of a chick embryo at a similar stage, seen through a dissecting microscope and rendered by F. R. Lillie in 1908. (C) Dorsal view of a late 2-day chick embryo, about 45 hours after the egg was laid. The heart starts beating during day 2. The vascular system of this embryo was revealed by injecting fluorescent beads into the circulatory system. The three-dimensionality is achieved by superimposing two separate images. (A from Malpighi 1672; B from Lillie 1908; C from V. Pasque 2011, with permission of the Wellcome Institute.)

different rates, and that structures need not be in the same place in the embryo as they are in the newborn.

The preformationists had no cell theory to provide a lower limit to the size of their preformed organisms (the cell theory did not arise until the mid-1800s). Rather, said Bonnet (1764), "Nature works as small as it wishes." This view was in accord with the best science of its time, conforming to the French mathematician-philosopher René Descartes' principle of the infinite divisibility of a mechanical nature initiated, but not interfered with, by God.

Preformationism's principal failure was its inability to account for the intergenerational variations revealed by even the limited genetic evidence of the time. It was known, for instance, that the children of a white and a black parent would have intermediate skin color—an impossibility if inheritance and development were solely through either

the sperm or the egg. In more scientific studies, the German botanist Joseph Kölreuter (1766) produced hybrid tobacco plants having the characteristics of both species. Moreover, by mating the hybrid to either the male or female parent, Kölreuter was able to "revert" the hybrid back to one or the other parental type after several generations. Thus, inheritance seemed to arise from a mixture of parental components, in refutation of preformationism.

The embryological case for epigenesis was revived at the same time by Kaspar Friedrich Wolff, a German embryologist working in St. Petersburg. By carefully observing the development of chick embryos, Wolff demonstrated that the embryonic parts develop from tissues that have no counterpart in the adult organism. The heart and blood vessels (which, according to preformationism, must be present from the beginning) could be seen to develop anew in each embryo. Similarly, the intestinal tube was seen to arise by the folding of an originally flat tissue. This latter observation was explicitly detailed by Wolff, who proclaimed in 1767 that "when the formation of the intestine in this manner has been duly weighed, almost no doubt can remain, I believe, of the truth of epigenesis." To explain how an organism is created anew each generation, however, Wolff had to postulate an unknown force—the *vis essentialis* ("essential force")—which, acting according to natural laws in the same way as gravity or magnetism, would organize embryonic development.

A reconciliation between preformationism and epigenesis was attempted by the German philosopher Immanuel Kant (1724–1804) and his colleague, biologist Johann Friedrich Blumenbach (1752–1840). Attempting to construct a scientific theory of racial descent, Blumenbach postulated a mechanical, goal-directed force he called *Bildungstrieb* ("developmental force"). Such a force, he said, was not theoretical, but could be shown to exist by experimentation. A hydra, when cut, regenerates its amputated parts by rearranging existing elements (see Chapter 16). Some purposeful organizing force could be observed in operation, and this *Bildungstrieb* was a property of the organism itself, thought to be inherited through the germ cells. Thus, development could proceed through a predetermined force inherent in the matter of the embryo (Cassirer 1950; Lenoir 1980). Moreover, this force was believed to be susceptible to change, as demonstrated by the left-handed variant of snail coiling (where left-coiled snails

can produce right-coiled progeny). In this hypothesis, wherein epigenetic development is directed by preformed instructions, we are not far from the view held by modern biologists that most of the instructions for forming the organism are already present in the fertilized egg.*

Naming the parts: The primary germ layers and early organs

The end of preformationism did not come until the 1820s, when a combination of new staining techniques, improved microscopes, and institutional reforms in German universities created a revolution in descriptive embryology. The new techniques enabled microscopists to document the epigenesis of anatomical structures, and the institutional reforms provided audiences for these reports and students to carry on the work of their teachers. Among the most talented of this new group of microscopically inclined investigators were three friends, born within a year of each other, all of whom came from the Baltic region and studied in northern Germany. The work of Christian Pander, Karl Ernst von Baer, and Heinrich Rathke transformed embryology into a specialized branch of science.

---

*But not *all* the instructions are there. Later in this book, we will see that temperature, diet, predators, symbionts, crowding, and other environmental agents normally regulate gene expression in the embryo and can cause particular phenotypes to occur.

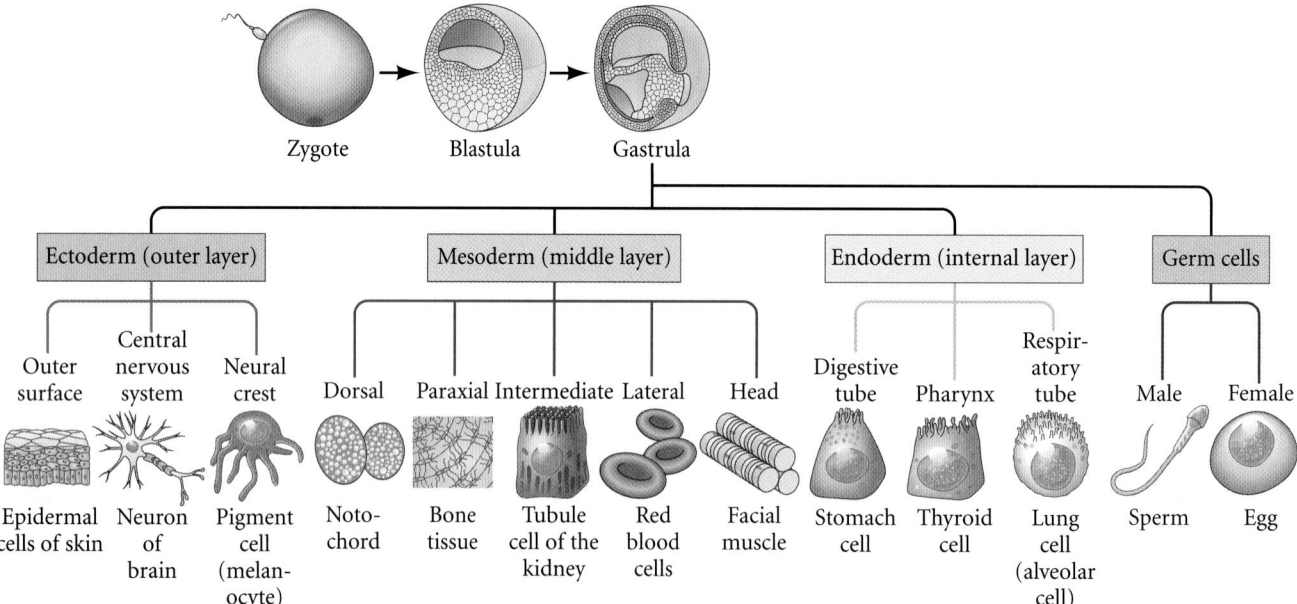

FIGURE 1.7 The dividing cells of the fertilized egg form three distinct embryonic germ layers. Each of the germ layers gives rise to myriad differentiated cell types (only a few representatives are shown here) and distinct organ systems. The germ cells (precursors of the sperm and egg) are set aside early in development and do not arise from any particular germ layer.

Pander studied the chick embryo for less than 2 years (before becoming a paleontologist), but in those 15 months, he discovered the **germ layers***—three distinct regions of the embryo that give rise to the differentiated cells types and specific organ systems (**FIGURE 1.7**).

- The **ectoderm** generates the outer layer of the embryo. It produces the surface layer (epidermis) of the skin and forms the brain and nervous system.
- The **endoderm** becomes the innermost layer of the embryo and produces the epithelium of the digestive tube and its associated organs (including the lungs).
- The **mesoderm** becomes sandwiched between the ectoderm and endoderm. It generates the blood, heart, kidney, gonads, bones, muscles, and connective tissues.

These three layers are found in the embryos of most animal phyla, as we will describe at the start of Chapter 5.

Pander and Rathke also made observations that weighted the balance in favor of epigenesis. Rathke followed the intricate development of the vertebrate skull, excretory systems, and respiratory systems, showing that these became

---

*From the same root as "germination," the Latin *germen* means "sprout" or "bud." The names of the three germ layers are from the Greek: ectoderm from *ektos* ("outside") plus *derma* ("skin"); mesoderm from *mesos* ("middle"); and endoderm from *endon* ("within").

increasingly complex. He also showed that their complexity took on different trajectories in different classes of vertebrates. For instance, Rathke was the first to identify the **pharyngeal arches** (**FIGURE 1.8**). He showed that these same embryonic structures became gill supports in fish and the jaws and ears (among other things) in mammals. Pander demonstrated that the germ layers did not form their respective organs autonomously (Pander 1817). Rather, each germ layer "is not yet independent enough to indicate what it truly is; it still needs the help of its sister travelers, and therefore, although already designated for different ends, all three influence each other collectively until each has reached an appropriate level." Pander had discovered the tissue interactions that we now call induction. No tissue is able to construct organs by itself; it must interact with other tissues, as we will describe in Chapter 3. Thus, Pander showed that preformation could not be true, since the organs come into being through interactions between simpler structures.

The four principles of Karl Ernst von Baer

Karl Ernst von Baer extended Pander's studies of the chick embryo. He recognized that there is a common pattern to all vertebrate development—that each of the three germ layers generally gives rise to the same organs, whether the organism is a fish, a frog, or a chick. He discovered the notochord, the rod of dorsalmost mesoderm that separates the embryo into right and left halves and that instructs the ectoderm

(A)

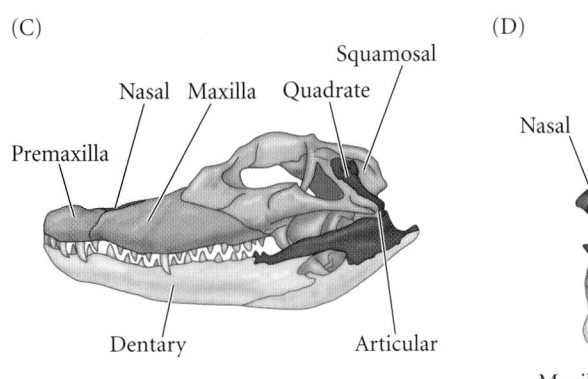

(B)

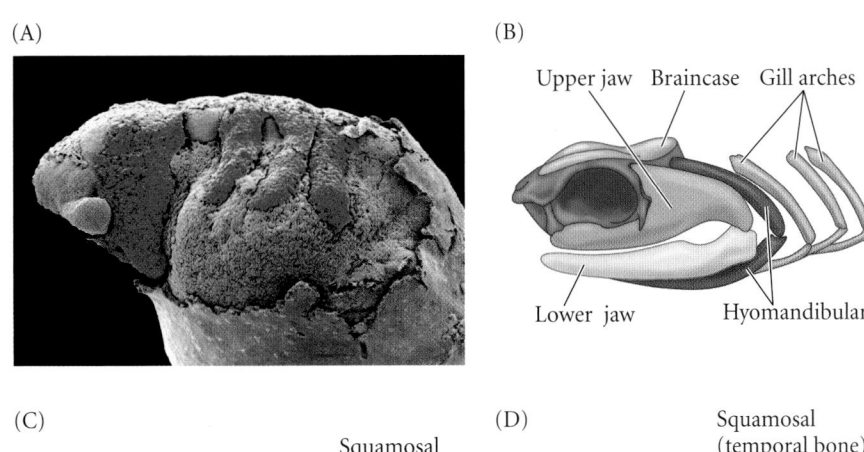

Upper jaw   Braincase   Gill arches

Lower jaw   Hyomandibular

(C)

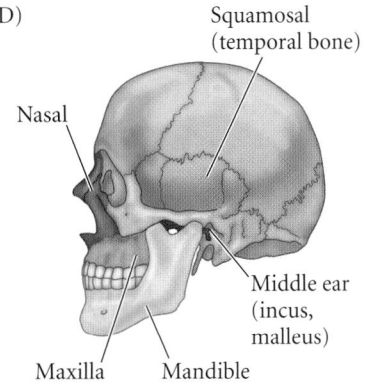

Squamosal

Nasal   Maxilla   Quadrate

Premaxilla

Dentary   Articular

(D)

Squamosal (temporal bone)

Nasal

Middle ear (incus, malleus)

Maxilla   Mandible

FIGURE 1.8 Evolution of pharyngeal arch structures in the vertebrate head. (A) Pharyngeal arches (also called branchial arches) in the embryo of the salamander *Ambystoma mexicanum*. The surface ectoderm has been removed to permit visualization of the arches (highlighted in color) as they form. (B) In adult fish, pharyngeal arch cells form the hyomandibular jaws and gill arches. (C) In amphibians, birds, and reptiles (a crocodile is shown here), these same cells form the quadrate bone of the upper jaw and the articular bone of the lower jaw. (D) In mammals, the quadrate has become internalized and forms the incus of the middle ear. The articular bone retains its contact with the quadrate, becoming the malleus of the middle ear. Thus, the cells that form gill supports in fish form the middle ear bones in mammals. (A courtesy of P. Falck and L. Olsson; B–D after Zangerl and Williams 1975.)

FIGURE 1.9 Two types of microscopy are used to visualize the notochord and its separation of vertebrate embryos (in this case a chick) into right and left halves. The notochord instructs the ectoderm above it to become the nervous system (the neural tube at this stage of development). To either side of the notochord and the neural tube are the mesodermal masses called somites, which will form vertebrae, ribs, and skeletal muscles. (A) Fluorescence micrograph stained with different dyes to highlight nuclear DNA (blue), cytoskeletal microtubules (red, yellow), and the extracellular matrix (green). (B) Scanning electron micrograph of the same stage, highlighting the three-dimensional relationship of the structures. (A courtesy of M. Angeles Rabadán and E. Martí Gorostiza; B courtesy of K. Tosney and G. Schoenwolf.)

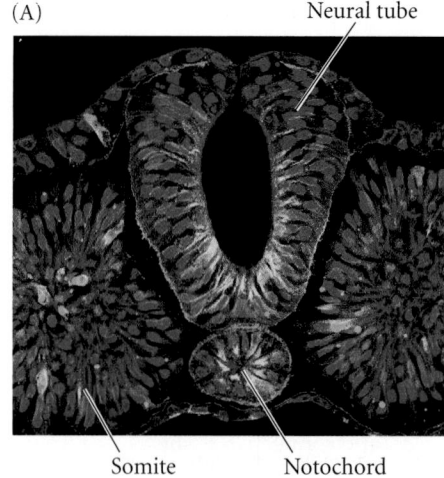

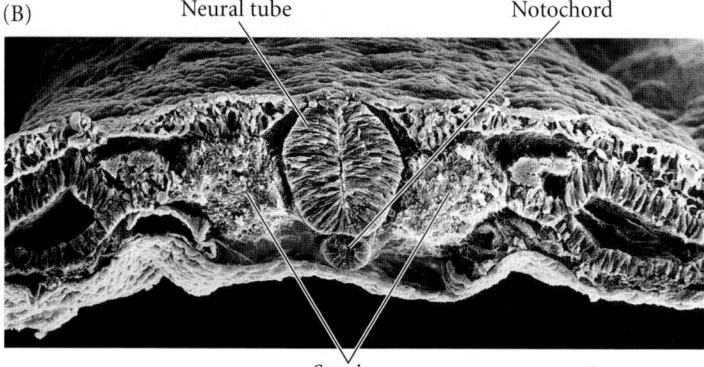

above it to become the nervous system (**FIGURE 1.9**). He also discovered the mammalian egg, that minuscule, long-sought cell that everyone believed existed but no one before von Baer had ever seen.*

In 1828, von Baer reported, "I have two small embryos preserved in alcohol, that I forgot to label. At present I am unable to determine the genus to which they belong. They may be lizards, small birds, or even mammals." Drawings of such early-stage embryos allow us to appreciate his quandary (**FIGURE 1.10**). From his detailed study of chick development and his comparison of chick embryos with the embryos of other vertebrates, von Baer derived four generalizations. Now often referred to as "von Baer's laws," they are stated here with some vertebrate examples.

1. *The general features of a large group of animals appear earlier in development than do the specialized features of a smaller group.* Although each vertebrate group may start off with different patterns of cleavage and gastrulation, they converge at a very similar structure when they begin forming their neural tube. All developing vertebrates appear very similar right after gastrulation. All vertebrate embryos have gill arches, a notochord, a spinal cord, and primitive kidneys. The structure in Figure 1.9—a notochord below a neural tube, flanked by somites—is seen in every vertebrate embryo. It is only later in development that the distinctive features of class, order, and finally species emerge.

2. *Less general characters develop from the more general, until finally the most specialized appear.* All vertebrates initially have the same type of skin. Only later does the skin

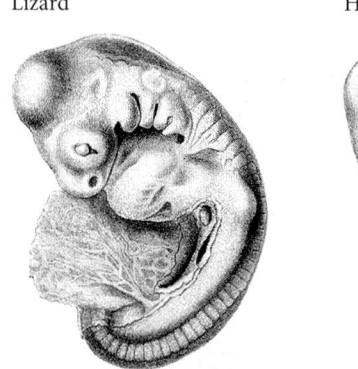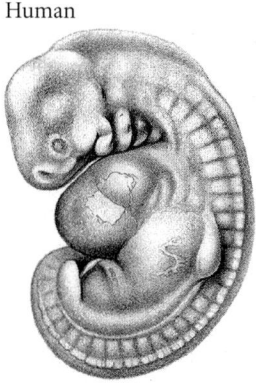

FIGURE 1.10 The eggs of birds, mammals, reptiles, fish, and amphibians start development very differently because of the enormous differences in the sizes of their eggs. By the beginning of neurulation, however, all vertebrate embryos have converged on a common structure. Here, a lizard embryo is shown next to a human embryo at a similar stage. As they develop beyond the neurula stage, the embryos of the different vertebrate groups become less and less like each other. (From Keibel 1904, 1908; see Galis and Sinervo 2002.)

*Von Baer could hardly believe that he had at last found what so many others—Harvey, de Graaf, von Haller, Prevost, Dumas, and even Purkinje—had searched for and failed to find. "I recoiled as if struck by lightening … I had to try to relax a while before I could work up enough courage to look again, as I was afraid I had been deluded by a phantom. Is it not strange that a sight which is expected, and indeed hoped for, should be frightening when it eventually materializes?"

develop fish scales, reptilian scales, bird feathers, or the hair, claws, and nails of mammals. Similarly, the early development of limbs is essentially the same in all vertebrates. Only later do the differences between legs, wings, and arms become apparent.

3. *The embryo of a given species, instead of passing through the adult stages of lower animals, departs more and more from them.* For example, as shown in Figure 1.8, the pharyngeal arches start off the same in all vertebrates. But the arch that becomes the jaw support in fish becomes part of the skull of reptiles and becomes part of the middle ear bones of mammals. Mammals never go through a fishlike stage (Reichert 1837; Rieppel 2011).

4. *Therefore, the early embryo of a higher animal is never like a lower animal, but only like its early embryo.* Human embryos never pass through a stage equivalent to an adult fish or bird. Rather, human embryos initially share characteristics in common with fish and avian embryos. Later, the mammalian and other embryos diverge, none of them passing through the stages of the others.

Recent research has confirmed von Baer's view that there is a "phylotypic stage" at which the embryos of the different classes of vertebrates all have a similar physical structure, such as the stage seen in Figure 1.10. Interestingly, at this same stage there also appears to be the least amount of difference among the genes expressed by the different vertebrate groups (Irie and Kuratani 2011).

Comparative embryonic anatomy remains an active field of research today, although it is now done in an evolutionary context. What embryonic interactions, for instance, cover the squirrel's tail with fur but provide scales on the rat's tail? The author's own research concerns how turtles get their shells—a skeletal feature generally composed of more than 50 bones that no other vertebrate possesses. What is the relationship of these new bones to the bones found in alligators and prehistoric marine reptiles? What changes in the "typical" development of the vertebrate skeleton allowed these unique bones to form? Paleontologists have noted the similarities between the developmental anatomy of chick and dinosaur embryos and have found that the embryonic chick, unlike the dinosaur, regresses its tail. They are now collaborating with developmental biologists to see if blocking this regression produces chick that more closely resembles its dinosaur ancestors (Horner and Gorman 2009).

Keeping Track of Moving Cells: Fate Maps and Cell Lineages

By the late 1800s, it had been conclusively demonstrated that the cell is the basic unit of all anatomy and physiology. Embryologists, too, began to base their field on the

cell. But unlike those who studied the adult, developmental anatomists found that cells in the embryo do not "stay put." Indeed, one of the most important conclusions of developmental anatomists is that embryonic cells do not remain in one place, nor do they keep the same shape (Larsen and McLaughlin 1987).

Early embryologists recognized that there are two major types of cells in the embryo: **epithelial cells**, which are tightly connected to one another in sheets or tubes; and **mesenchymal cells**, which are unconnected to one another and operate as independent units. Morphogenesis is brought about through a limited repertoire of variations in cellular processes within these two types of arrangements (**TABLE 1.1**):

- *Direction and number of cell divisions.* Think of the faces of two dog breeds—say, a German shepherd and a poodle. The faces are made from the same cell types, but the number and orientation of the cell divisions are different (Schoenebeck et al. 2012). Think also of the legs of a German shepherd compared with those of a dachshund. The skeleton-forming cells of the dachshund have undergone fewer cell divisions than those of taller dogs (see Figure 1.21).

- *Cell shape changes.* Cell shape change is a critical feature of development. Changing the shapes of epithelial cells often creates tubes out of sheets (as when the neural tube forms), and a shape change from epithelial to mesenchymal is critical when individual cells migrate away from the epithelial sheet (as when muscle cells are formed). (As we will see in Chapter 18, this same type of epithelial-to-mesenchymal change operates in cancer, allowing cancer cells to migrate and spread from the primary tumor to new sites.)

- *Cell migration.* Cells have to move in order to get to their appropriate locations. The germ cells have to migrate into the developing gonad, and the primordial heart cells meet in the middle of the vertebrate neck and then migrate to the left part of the chest.

- *Cell growth.* Cells can change in size. This is most apparent in the germ cells: the sperm eliminates most of its cytoplasm and becomes smaller, whereas the developing egg conserves and adds cytoplasm, becoming comparatively huge. Many cells undergo an "asymmetric" cell division that produces one big cell and one small cell, each of which may have a completely different fate.

- *Cell death.* Death is a critical part of life. The embryonic cells that constitute the webbing between our toes and fingers die before we are born. So do the cells of our tails. The orifices of our mouth, anus, and reproductive glands all form through *apoptosis*—the programmed death of certain cells at particular times and places.

- *Changes in the composition of the cell membrane or secreted products.* Cell membranes and secreted cell products influence the behavior of neighboring cells. For instance, extracellular matrices secreted by one set of cells will

---

*Von Baer formulated these generalizations prior to Darwin's theory of evolution. "Lower animals" would be those having simpler anatomies.

TABLE 1.1   Major morphogenetic processes regulated by mesenchymal and epithelial cells

Process	Action	Morphology	Example
MESENCHYMAL CELLS			
Condensation	Mesenchyme becomes epithelium		Cartilage mesenchyme
Cell division	Mitosis produces more cells (hyperplasia)		Limb mesenchyme
Cell death	Cells die (apoptosis)		Interdigital mesenchyme
Migration	Cells move at particular times and places		Heart mesenchyme
Matrix secretion and degradation	Synthesis or removal of extracellular layer		Cartilage mesenchyme
Growth	Cells get larger (hypertrophy)		Fat cells
EPITHELIAL CELLS			
Dispersal	Epithelium becomes mesenchyme (entire structure)		Müllerian duct degeneration
Delamination	Epithelium becomes mesenchyme (part of structure)		Chick hypoblast
Shape change or growth	Cells remain attached as morphology is altered		Neurulation
Cell migration (intercalation)	Rows of epithelia merge to form fewer rows		Vertebrate gastrulation
Cell division	Mitosis within row or column		Vertebrate gastrulation
Matrix secretion and degradation	Synthesis or removal of extracellular matrix		Vertebrate organ formation
Migration	Formation of free edges		Chick ectoderm

allow the migration of their neighboring cells. Extracellular matrices made by other cell types will *prohibit* the migration of the same set of cells. In this way, "paths and guiderails" are established for migrating cells.

Fate maps

Given such a dynamic situation, one of the most important programs of descriptive embryology became the tracing of **cell lineages**: following individual cells to see what those cells become. In many organisms, resolution of individual cells is not possible, but one can label groups of embryonic cells to see what that area becomes in the adult organism. By bringing such studies together, one can construct a **fate map**. These diagrams "map" larval or adult structures onto the region of the embryo from which they arose. Fate maps constitute an important foundation for experimental embryology, providing researchers with information on which portions of the embryo normally become which larval or adult structures. FIGURE 1.11 shows fate maps of some vertebrate embryos at the early gastrula stage.

Fate maps can be generated in several ways, and the technology has changed greatly over the past few years. The ability to follow cells with molecular dyes and computer imaging has altered our understanding of the origins of several cell types. Even our views of where heart cells originate has been changed (Lane and Sheets 2006; Camp et al. 2012). Mammalian embryos are among the most difficult to map (since they develop inside another organism), and researchers are actively constructing, refining, and arguing about the fate maps of mammalian embryos.

Direct observation of living embryos

Some embryos have relatively few cells, and the cytoplasm in each of the early blastomeres has a different colored pigment. In such fortunate cases, it is actually possible to look through the microscope and trace the descendants of a particular cell into the organs they generate. E. G. Conklin patiently followed the fates of each early cell of the tunicate (sea squirt) *Styela partita* (FIGURE 1.12; Conklin 1905). The muscle-forming cells of the *Styela* embryo always had a yellow color, derived from a region of cytoplasm found in the B4.1 blastomere. Conklin's fate map was confirmed by cell-removal experiments. Removal of the B4.1 cell (which according to Conklin's map should produce all the tail musculature) in fact resulted in a larva with no tail muscles (Reverberi and Minganti 1946).

● See **WEBSITE 1.2** Conklin's art and science

● See **VADE MECUM** The compound microscope

Dye marking

Most embryos are not so accommodating as to have cells of different colors. In the early years of the twentieth century, Vogt (1929) traced the fates of different areas of amphibian eggs by applying **vital dyes** to the region of interest. Vital dyes stain cells but do not kill them. Vogt mixed such dyes with agar and spread the agar on a microscope slide

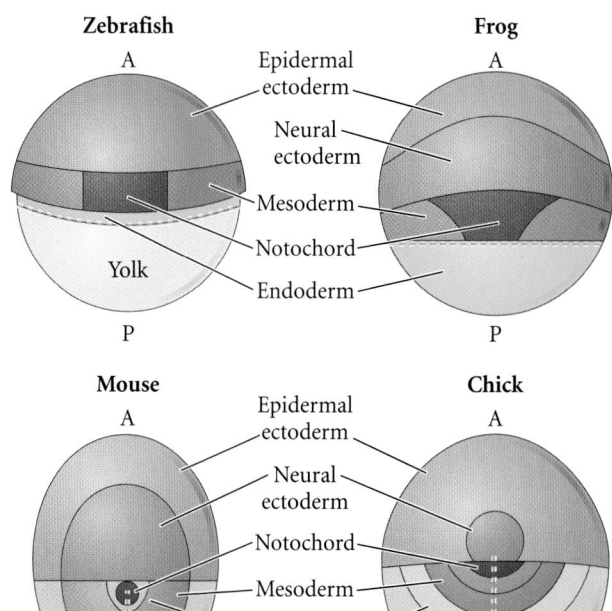

FIGURE 1.11 Fate maps of vertebrates at the early gastrula stage. All are dorsal surface views (looking "down" on the embryo at what will become its back). Despite the different appearances of the adult animals, fate maps of these four vertebrates show numerous similarities among the embryos. The cells that will form the notochord occupy a central dorsal position, while the precursors of the neural system lie immediately anterior to it. The neural ectoderm is surrounded by less dorsal ectoderm, which will form the epidermis of the skin. A indicates the anterior end of the embryo, P the posterior end. The dashed green lines indicate the site of ingression—the path cells will follow as they migrate from the exterior to the interior of the embryo.

to dry. The ends of the dyed agar were very thin. Vogt cut chips from these ends and placed them on a frog embryo. After the dye stained the cells, he removed the agar chips and could follow the stained cells' movements within the embryo (FIGURE 1.13).

One problem with vital dyes is that as they become more diluted with each cell division, they become difficult to detect. One way around this is to use **fluorescent dyes** that are so intense that once injected into individual cells, they can still be detected in the progeny of these cells many divisions later. Fluorescein-conjugated dextran, for example, can be injected into a single cell of an early embryo, and the descendants of that cell can be seen by examining the embryo under ultraviolet light (FIGURE 1.14).

● See **VADE MECUM** Histotechniques

Genetic labeling

One way of permanently marking cells and following their fates is to create embryos in which the same organism

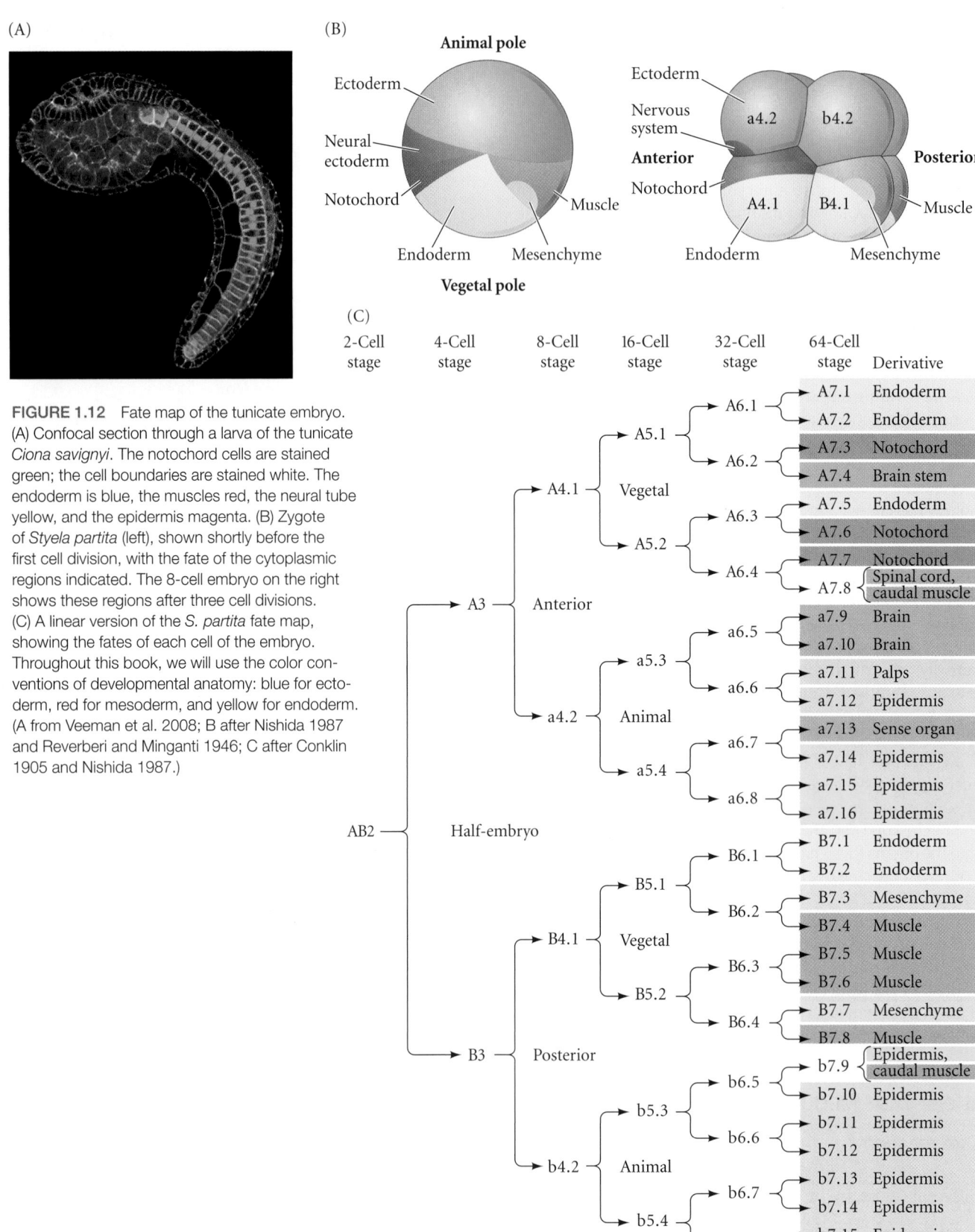

(A)

(B)

Animal pole

Ectoderm

Neural ectoderm

Notochord

Endoderm — Mesenchyme

Muscle

Vegetal pole

Ectoderm

Nervous system

Anterior

Notochord

Endoderm

a4.2   b4.2

A4.1   B4.1

Posterior

Muscle

Mesenchyme

(C)

FIGURE 1.12 Fate map of the tunicate embryo. (A) Confocal section through a larva of the tunicate *Ciona savignyi*. The notochord cells are stained green; the cell boundaries are stained white. The endoderm is blue, the muscles red, the neural tube yellow, and the epidermis magenta. (B) Zygote of *Styela partita* (left), shown shortly before the first cell division, with the fate of the cytoplasmic regions indicated. The 8-cell embryo on the right shows these regions after three cell divisions. (C) A linear version of the *S. partita* fate map, showing the fates of each cell of the embryo. Throughout this book, we will use the color conventions of developmental anatomy: blue for ectoderm, red for mesoderm, and yellow for endoderm. (A from Veeman et al. 2008; B after Nishida 1987 and Reverberi and Minganti 1946; C after Conklin 1905 and Nishida 1987.)

2-Cell stage	4-Cell stage	8-Cell stage	16-Cell stage	32-Cell stage	64-Cell stage	Derivative
					A7.1	Endoderm
				A6.1	A7.2	Endoderm
			A5.1		A7.3	Notochord
				A6.2	A7.4	Brain stem
		A4.1	Vegetal		A7.5	Endoderm
				A6.3	A7.6	Notochord
			A5.2		A7.7	Notochord
	A3			A6.4	A7.8	Spinal cord, caudal muscle
	Anterior				a7.9	Brain
				a6.5	a7.10	Brain
			a5.3		a7.11	Palps
				a6.6	a7.12	Epidermis
		a4.2	Animal		a7.13	Sense organ
				a6.7	a7.14	Epidermis
			a5.4		a7.15	Epidermis
AB2	Half-embryo			a6.8	a7.16	Epidermis
					B7.1	Endoderm
				B6.1	B7.2	Endoderm
			B5.1		B7.3	Mesenchyme
				B6.2	B7.4	Muscle
		B4.1	Vegetal		B7.5	Muscle
				B6.3	B7.6	Muscle
			B5.2		B7.7	Mesenchyme
	B3			B6.4	B7.8	Muscle
	Posterior				b7.9	Epidermis, caudal muscle
				b6.5	b7.10	Epidermis
			b5.3		b7.11	Epidermis
				b6.6	b7.12	Epidermis
		b4.2	Animal		b7.13	Epidermis
				b6.7	b7.14	Epidermis
			b5.4		b7.15	Epidermis
				b6.8	b7.16	Epidermis

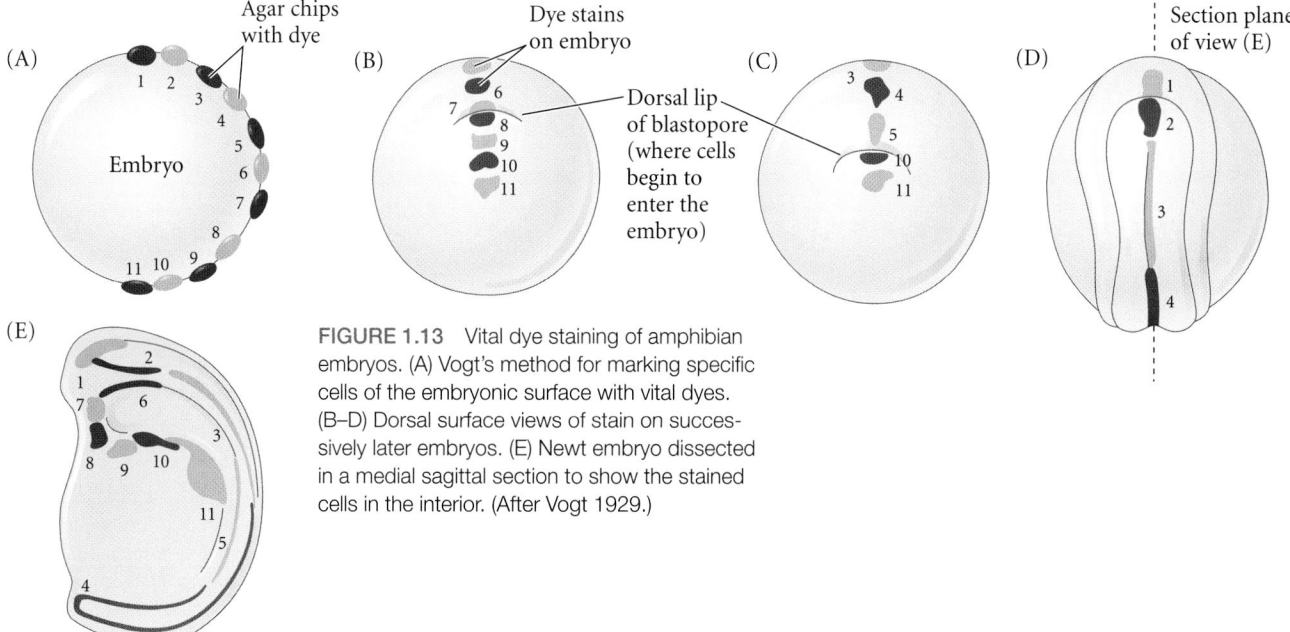

FIGURE 1.13 Vital dye staining of amphibian embryos. (A) Vogt's method for marking specific cells of the embryonic surface with vital dyes. (B–D) Dorsal surface views of stain on successively later embryos. (E) Newt embryo dissected in a medial sagittal section to show the stained cells in the interior. (After Vogt 1929.)

contains cells with different genetic constitutions. In the 1920s, the German embryologists Hilde Mangold and Hans Spemann performed some of the most important experiments in the history of embryology when they transplanted embryonic tissues from one species of newt into the embryo of a different newt species. These **chimeric embryos** —embryos made from tissues of more than one genetic source—enabled Mangold and Spemann to tell which structures arose from donor tissue and which from host tissue (see Figures 8.17 and 8.18).

One of the best examples of this technique is the construction of chimeric embryos by grafting quail cells inside a chick embryo while the chick is still in the egg. Chicks and quail embryos develop in a similar manner (especially during the early stages), and the grafted quail cells become integrated into the chick embryo and participate in the construction of the various organs. The chick that hatches will have quail cells in particular sites, depending on where the graft was placed. Quail cells differ from chick cells in two important ways. First, the quail nucleus has condensed DNA

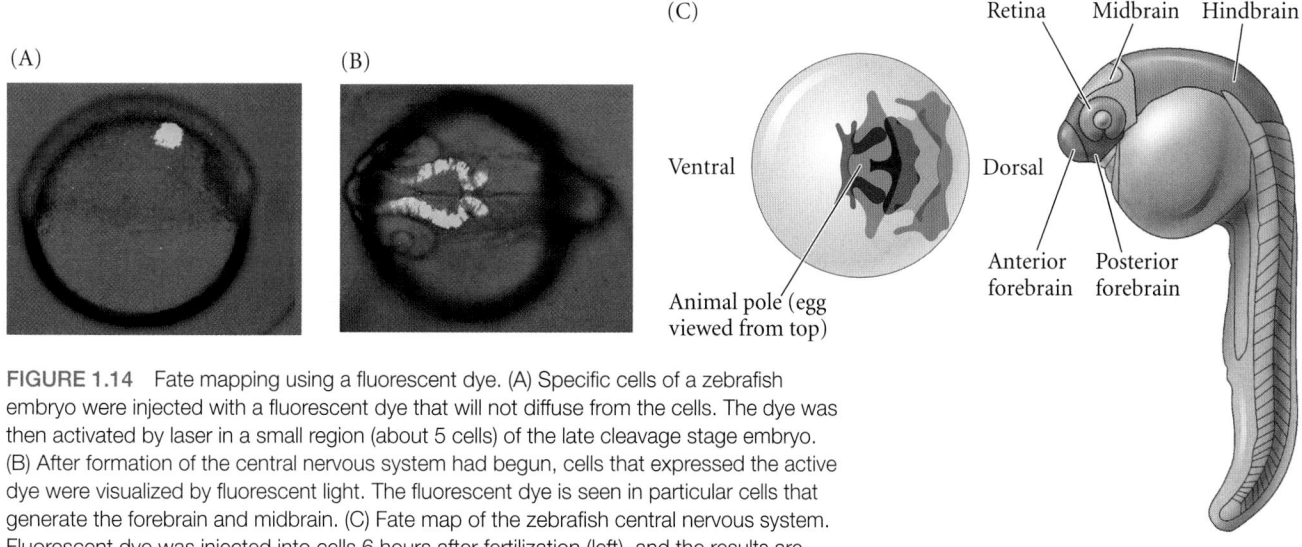

FIGURE 1.14 Fate mapping using a fluorescent dye. (A) Specific cells of a zebrafish embryo were injected with a fluorescent dye that will not diffuse from the cells. The dye was then activated by laser in a small region (about 5 cells) of the late cleavage stage embryo. (B) After formation of the central nervous system had begun, cells that expressed the active dye were visualized by fluorescent light. The fluorescent dye is seen in particular cells that generate the forebrain and midbrain. (C) Fate map of the zebrafish central nervous system. Fluorescent dye was injected into cells 6 hours after fertilization (left), and the results are color-coded onto the hatched fish (right). Overlapping colors indicate that cells from these regions of the 6-hour embryo contribute to two or more regions. (A,B from Kozlowski et al. 1998, photographs courtesy of E. Weinberg; C after Woo and Fraser 1995.)

(A)

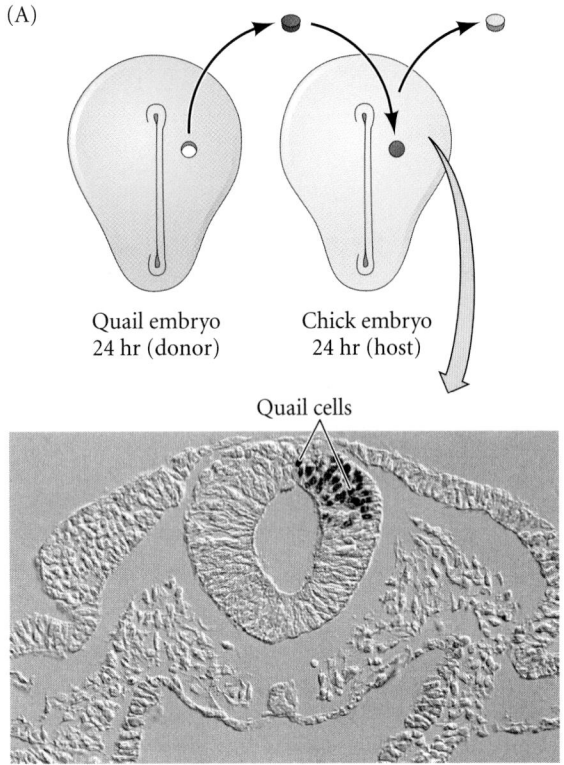

Quail embryo
24 hr (donor)

Chick embryo
24 hr (host)

Quail cells

(B)

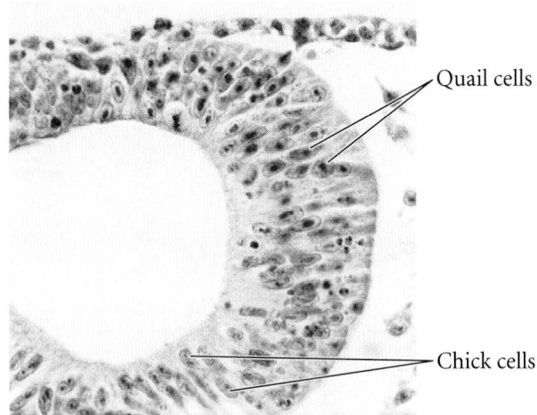

Quail cells

Chick cells

FIGURE 1.15 Genetic markers as cell lineage tracers. (A) Grafting experiment wherein the cells from a particular region of a 1-day quail embryo have been placed into a similar region of a 1-day chick embryo. After several days, the quail cells can be seen by using an antibody to quail-specific proteins (photograph at left). This region of the 3-day embryo produces cells that populate the neural tube. (B) Chick and quail cells can also be distinguished by the heterochromatin of their nuclei. The quail cells have a single large nucleolus (dense purple), distinguishing them from the diffuse nuclei of the chick. (From Darnell and Schoenwolf 1997, courtesy of the authors.)

(*heterochromatin*) concentrated around the nucleoli, making quail nuclei easily distinguishable from chick nuclei. Second, cell-specific antigens that are quail-specific can be used to find individual quail cells, even if they are "hidden" within a large population of chick cells (**FIGURE 1.15**). In this way, fine-structure maps of the chick brain and skeletal system have been produced (Le Douarin 1969; Le Douarin and Teillet 1973).

Chick-quail chimeras dramatically confirmed the extensive migrations of the neural crest cells during vertebrate development. Mary Rawles (1940) showed that the pigment cells (**melanocytes**) of the chick originate in the **neural crest**, a transient band of cells that joins the neural tube to the epidermis. When she transplanted small regions of neural crest-containing tissue from a pigmented strain of chickens into a similar position in an embryo from an unpigmented strain of chickens, the migrating pigment cells entered the epidermis and later entered the feathers (**FIGURE 1.16**). Ris (1941) used similar techniques to show that, although almost all of the external pigment of the chick embryo came from the migrating neural crest cells, the pigment of the retina formed in the retina itself and was not dependent on migrating neural crest cells. This pattern was confirmed in chick-quail hybrids, in which the quail neural crest cells produced their own pigment and pattern in the chick feathers.

● See **VADE MECUM** Chick-quail chimeras

---

* Green fluorescent protein occurs naturally in certain jellyfish. It emits bright green fluorescence when exposed to ultraviolet light and is widely used as a transgenic label for cells in developmental and other research. GFP labeling will be seen in many of the photographs throughout this book.

FIGURE 1.16 Chick resulting from transplantation of a trunk neural crest region from an embryo of a pigmented strain of chickens into the same region of an embryo of an unpigmented strain. The neural crest cells that gave rise to the pigment migrated into the wing epidermis and feathers. (From the archives of B. H. Willier.)

(A)

(B)

(C) Neural tube

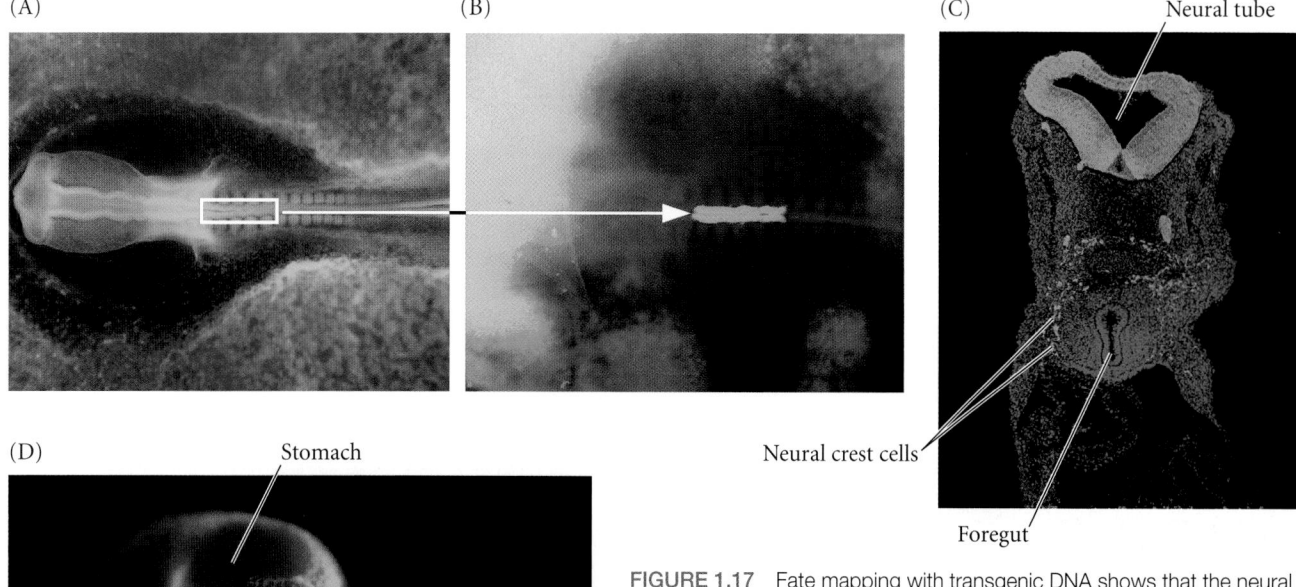

(D) Stomach

Esophagus   Midgut   Hindgut

Neural crest cells

Foregut

FIGURE 1.17 Fate mapping with transgenic DNA shows that the neural crest is critical in making the gut neurons. (A) A chick embryo containing an active gene for green fluorescent protein expresses GFP in every cell. The brain is forming on the left side of the embryo, and the bulges from the forebrain (which will become the retinas) are contacting the head ectoderm to initiate eye formation. (B) The region of the neural tube and neural crest in the presumptive neck region (rectangle in A) is excised and transplanted into a similar position in an unlabeled wild-type embryo. One can see it by its green fluorescence. (C) A day later, one can see the neural crest cells migrating from the neural tube to the stomach region. (D) In 4 more days, the neural crest cells have spread in the gut from the esophagus to the anterior end of the hindgut. (From Freem et al. 2012; photographs courtesy of A. Burns.)

Transgenic DNA chimeras

In most animals, it is difficult to meld a chimera from two species. One way of circumventing this problem is to transplant cells from a genetically modified organism. In such a technique, the genetic modification can then be traced only to those cells that express it. One version is to infect the cells of an embryo with a virus whose genes have been altered such that they express the gene for a fluorescently active protein such as **green fluorescent protein**, or **GFP**.* (This type of altered gene is called a transgene, because it contains DNA from another species.) When the infected embryonic cells are transplanted into a wild-type host, only the donor cells will express GFP and emit a visible green glow.

Freem and colleagues (2012) have used this technique to study the migration of neural crest cells to the gut of chick embryos. These neural crest cells are critical here, because they form the neurons that coordinate the muscular contractions of the gut (peristalsis) to eliminate solid waste. The parents of the GFP-labeled chick embryo had been infected with a replication-deficient virus that carried an active gene for GFP. This virus was inherited by the chick embryo and was expressed in every cell. In this way, Freem and colleagues

generated embryos in which every cell glowed green when placed under ultraviolet light (**FIGURE 1.17A**). They then transplanted the neural tube and neural crest of a GFP-transgenic embryo into a similar region of a normal chick embryo (**FIGURE 1.17B**). A day later, they could see GFP-labeled cells migrating into the stomach region (**FIGURE1.17C**), and by 7 days, the entire gut showed GFP staining up to the anterior region of the hindgut (**FIGURE 1.17D**).

Evolutionary Embryology

Charles Darwin's theory of evolution restructured comparative embryology and gave it a new focus. After reading Johannes Müller's summary of von Baer's laws in 1842, Darwin saw that embryonic resemblances would be a strong argument in favor of the genetic connectedness of different animal groups. "Community of embryonic structure reveals community of descent," he would conclude in *On the Origin of Species* in 1859. Darwin's evolutionary interpretation of von Baer's laws established a paradigm that was to be followed for many decades—namely, that relationships between groups can be established by finding common embryonic or larval forms.

(A) Barnacle

(B) Shrimp

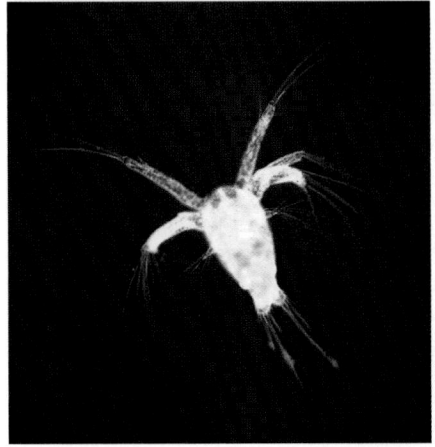

FIGURE 1.18 Larval stages reveal the common ancestry of two crustacean arthropods. (A) Barnacle. (B) Shrimp. Barnacles and shrimp both exhibit a distinctive larval stage (the nauplius) that underscores their common ancestry as crustacean arthropods, even though adult barnacles—once classified as molluscs—are sedentary, differing in body form and lifestyle from the free-swimming adult shrimp. (A © Wim van Egmond/Visuals Unlimited and © Barrie Watts/OSF/Getty; B courtesy of U.S. National Oceanic and Atmospheric Administration and © Kim Taylor/Naturepl.com.)

Even before Darwin, larval forms had been used for taxonomic classification. J. V. Thompson, for instance, demonstrated in the 1830s that larval barnacles were almost identical to larval shrimp, and therefore (correctly) counted barnacles as arthropods rather than molluscs (**FIGURE 1.18**; Winsor 1969). Darwin, an expert on barnacle taxonomy, celebrated this finding: "Even the illustrious Cuvier did not perceive that a barnacle is a crustacean, but a glance at the larva shows this in an unmistakable manner." Alexander Kowalevski (1871) made the similar type of discovery (publicized in Darwin's *The Descent of Man*, 1874) that tunicate larvae had a notochord and pharyngeal pouches, which came from the same germ layers as those same structures in fish and chicks. Thus, the tunicate (an invertebrate) was related to the vertebrates, and the two great domains of the animal kingdom—invertebrates and vertebrates— were thereby united through larval structures. Darwin (1874) was thrilled, writing, "Thus, if we may rely on embryology, ever the safest guide in classification, it seems that we have at last gained a clue to the source whence the Vertebrata were derived."

Darwin also noted that embryonic organisms sometimes make structures that are inappropriate for their adult form but that show their relatedness to other animals. He pointed out the existence of eyes in embryonic moles, pelvic bone rudiments in embryonic snakes, and teeth in baleen whale embryos.

Darwin also argued that adaptations that depart from the "type" and allow an organism to survive in its particular environment develop late in the embryo.* He noted that

---

*Moreover, as first noted by Weismann (1875), larvae must have their own adaptations to help them survive. The adult viceroy butterfly mimics the monarch butterfly, but the viceroy caterpillar does not resemble the beautiful larva of the monarch. Rather, the viceroy larva escapes detection by resembling bird droppings (Begon et al. 1986).

the differences among species within genera become greater as development persists, as predicted by von Baer's laws. Thus, Darwin recognized two ways of looking at "descent with modification." One could emphasize the common descent by pointing out embryonic similarities between two or more groups of animals, or one could emphasize the modifications by showing how development was altered to produce structures that enabled animals to adapt to particular conditions.

Embryonic homologies

One of the most important distinctions made by evolutionary embryologists was the difference between analogy and homology. Both terms refer to structures that appear to be similar. **Homologous** structures are those organs whose underlying similarity arises from their being derived from a common ancestral structure. For example, the wing of a bird and the arm of a human are homologous, both having evolved from the forelimb bones of a common ancestor. Moreover, their respective parts are homologous (**FIGURE 1.19**).

Analogous structures are those whose similarity comes from their performing a similar function rather than their arising from a common ancestor. For example, the wing of a butterfly and the wing of a bird are analogous; the two share a common function (and thus both are called wings), but the bird wing and insect wing did not arise from a common ancestral structure that became modified through evolution into bird wings and butterfly wings.*

As we will see in Chapter 20, evolutionary change is based on developmental change. The bat wing, for instance, is made in part by (1) maintaining a rapid growth rate in the cartilage that forms the fingers and (2) preventing the cell death that normally occurs in the webbing between the fingers. As seen in **FIGURE 1.20**, mice start off with webbing between their digits (as do humans and most other mammals). This webbing is important for creating the anatomical distinctions between the fingers (see Figure 14.27). Once the webbing has served that function, genetic signals cause its cells to die, leaving free digits that can grasp and manipulate. Bats, however, use their fingers for flight—a feat accomplished by changing the genes that are active in the webbing. The genes activated in embryonic bat webbing encode proteins that *prevent* cell death as well as proteins that accelerate finger elongation (Cretekos et al. 2005; Sears et al. 2006; Weatherbee et al. 2006). Thus, homologous anatomical structures can differentiate by altering development, and such changes in development can provide the variations needed for evolutionary change.

---

*Homologies must always refer to the level of organization being compared. For instance, bird and bat wings are homologous as forelimbs but not as wings. In other words, they share an underlying structure of forelimb bones because birds and mammals share a common ancestor that possessed such bones. Bats, however, descended from a long line of non-winged mammals, whereas bird wings evolved independently, from the forelimbs of ancestral reptiles. As we will see, the structure of a bat's wing is markedly different from that of a bird's wing.

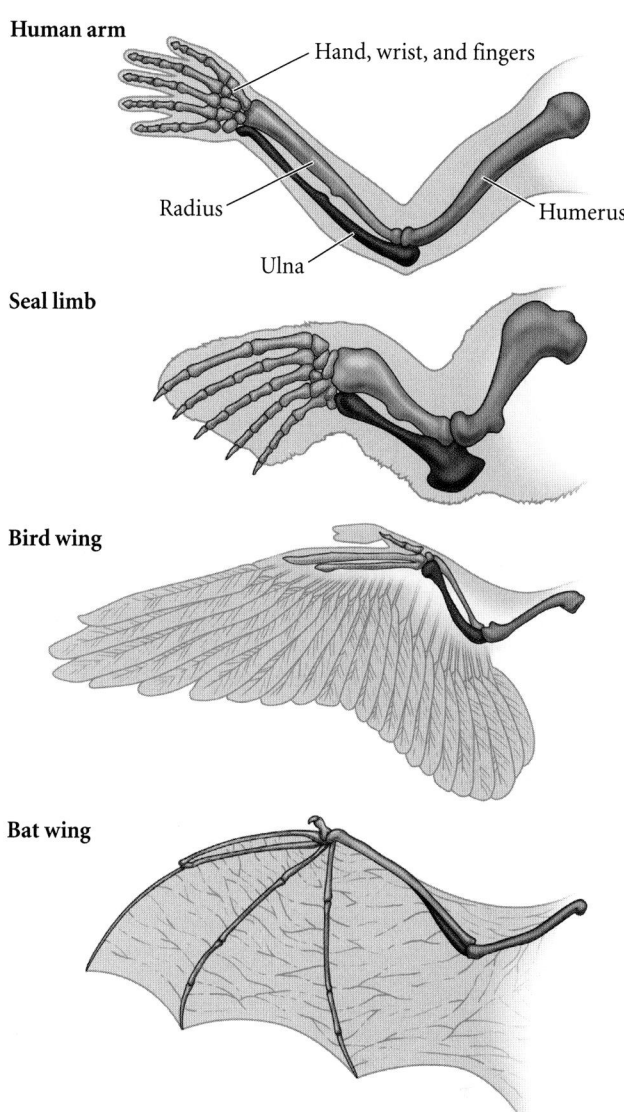

FIGURE 1.19 Homologies of structure among a human arm, a seal forelimb, a bird wing, and a bat wing; homologous supporting structures are shown in the same color. All four were derived from a common tetrapod ancestor and thus are homologous as forelimbs. The adaptations of bird and bat forelimbs to flight, however, evolved independently of each other, long after the two lineages diverged from their common ancestor. Therefore, as wings they are not homologous, but analogous.

Charles Darwin observed artificial selection in pigeon and dog breeds, and these examples remain valuable resources for studying selectable variation. For instance, the short legs of dachshunds (**FIGURE 1.21A**) were selected by breeders who wanted to use these dogs to hunt badgers (German *Dachs*, "badger" + *Hund*, "dog") in their underground burrows. The mutation that causes the dachshund's short legs involves an extra copy of the gene *Fgf4*, which tells the cartilage precursor cells that they have divided enough and can start differentiating. With this extra copy of *Fgf4*, the cartilage cells are told too early that they should stop dividing, so the legs stop

(A)

(B)

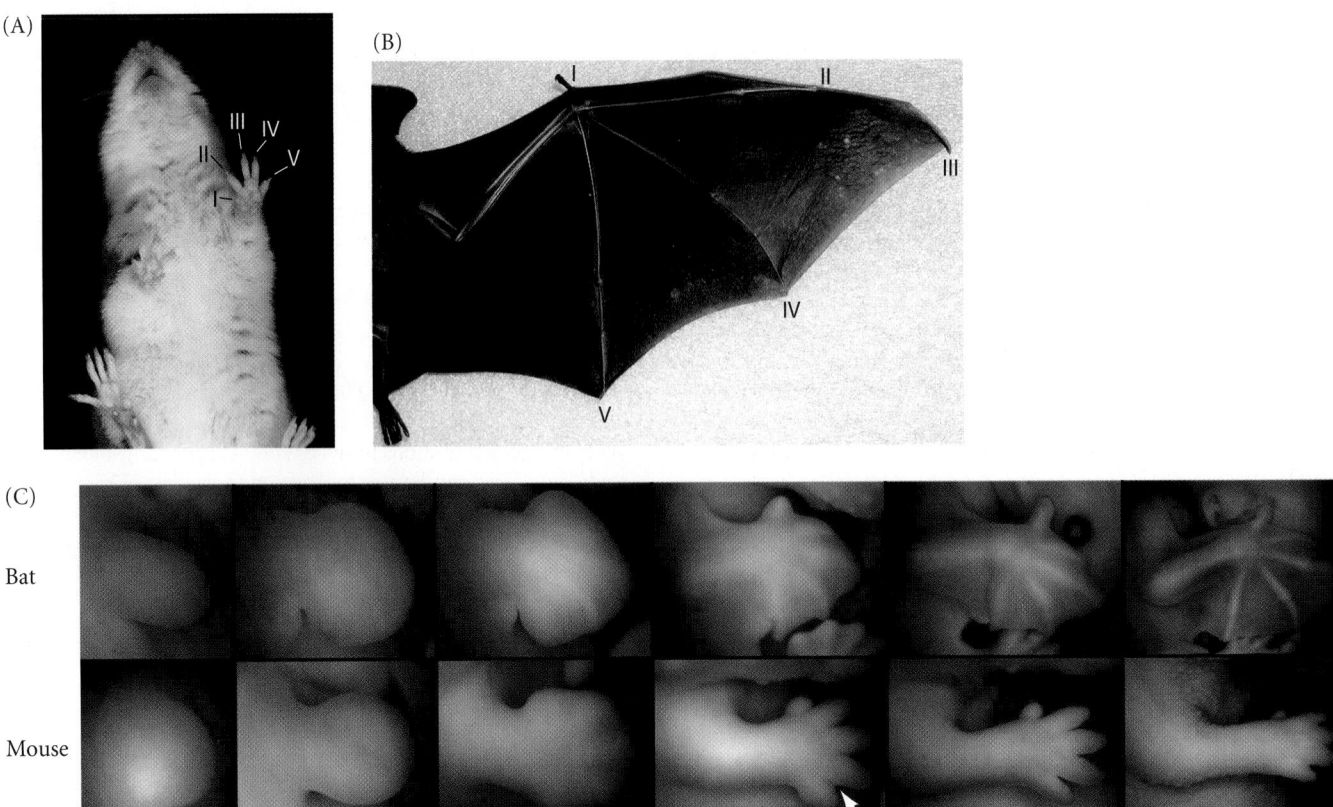

(C)

Bat

Mouse

FIGURE 1.20 Development of bat and mouse forelimbs. (A,B) Mouse and bat torsos, showing the mouse forelimb and the elongated fingers and prominent webbing in the bat wing. The digits are numbered on both animals (I, thumb; V, "pinky"). (C) Comparison of mouse and bat forelimb morphogenesis. Both limbs start as webbed appendages, but the webbing between the mouse's digits dies at embryonic day 14 (arrow). The webbing in the bat forelimb does not die and is sustained as the fingers grow. (A courtesy of D. McIntyre; B,C from Cretekos et al. 2008, courtesy of C. J. Cretekos.)

(A)

(B)

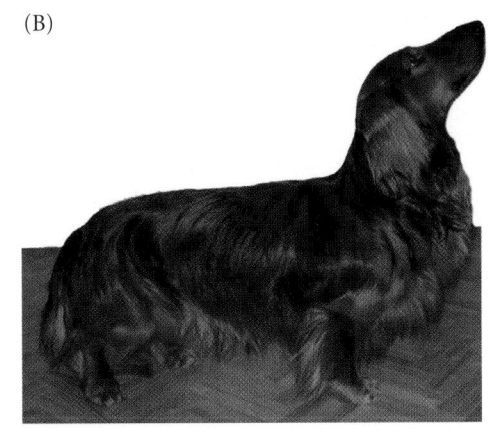

FIGURE 1.21 Selectable variation through mutations of genes that work during development. (A) The dachshund has been selected by breeders for its small legs, which enable it to seek badgers in their tunnels. The small legs are a result of premature cessation of cell division in the limb cartilage precursor cells. This premature end to cell division is caused by early activation of the cartilage FGF receptor protein, because the dachshund genome has an extra copy of the *Fgf4* gene. (B) Longhaired dachshunds have an additional mutation, a truncated *Fgf5* gene, which alters the hair follicle cycle, thereby allowing hair growth beyond the wild-type levels. (A © Alex Potemkin/istockphoto.com; B courtesy of K. Lilleväli.)

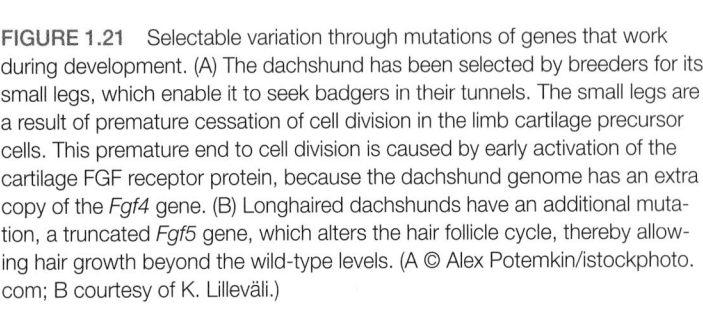

growing (Parker et al. 2009). Similarly, long-haired dachshunds (**FIGURE 1.21B**) differ from their short-haired relatives in having a mutation in the *Fgf5* gene* (Cadieu et al. 2009). This gene is involved in hair production and allows each follicle to make a longer hair shaft (Ota et al. 2002). Thus, mutations in genes controlling developmental processes can generate selectable variation.

Medical Embryology and Teratology

While embryologists could look at embryos to describe the evolution of life and how different animals form their organs, physicians became interested in embryos for more practical reasons. Between 2% and 5% of human infants are born with a readily observable anatomical abnormality (Winter 1996; Thorogood 1997). These abnormalities may include missing limbs, missing or extra digits, cleft palate, eyes that lack certain parts, hearts that lack valves, and so forth. Some birth defects are produced by mutant genes or chromosomes, and some are produced by environmental factors that impede development. Physicians need to know the causes of specific birth defects in order to counsel parents as to the risk of having another malformed infant. In addition, the study of birth defects can tell us how the human body is normally formed. In the absence of experimental data on human embryos, nature's "experiments" sometimes offer important insights into how the human body becomes organized.†

Genetic malformations and syndromes

Abnormalities caused by genetic events (gene mutations, chromosomal aneuploidies, and translocations) are called **malformations**. Developmental biologists and clinical geneticists often study human malformations (and determine their causes) using **animal models**—that is, animals that display the same abnormal phenotype. For instance, we just discussed the short limbs of the dachshund being caused by an extra copy of the *Fgf4* gene. The Fgf4 protein binds to receptor proteins on

---

*The FGF genes will be discussed throughout this book, as they regulate construction of numerous organs. Independently acquired mutations in the *Fgf5* gene are also responsible for the long-haired phenotype of Persian cats (Drögemüller et al. 2007; Kehler et al. 2007). However, *Fgf5* is not considered a good candidate to explain the woolliness of mammoths: the sequence of the *Fgf5* gene extracted from the DNA of extinct woolly mammoths appears virtually identical to that of the gene in modern elephants (Roca et al. 2009).

†The word *monster*, used frequently in textbooks prior to the mid-twentieth century to describe malformed infants, comes from the Latin *monstrare*, "to show or point out." This is also the root of the English word *demonstrate*. In the 1830s, J. F. Meckel realized that syndromes of congenital anomalies demonstrated certain principles about normal development. Parts of the body that were affected together must have some common developmental origin or mechanism that was being affected. It should also be noted that a condition considered a developmental anomaly in one situation may be considered advantageous in another. The short legs of dachshunds is only one such example.

the membranes of cartilage cells and instructs these cells *not* to divide. In humans, achondroplastic dysplasia, the most common form of dwarfism, is the result of a mutation in the receptor protein activated by Fgf4. In fact, this mutation (the most common dominant mutation in humans) makes the receptor protein active even when Fgf4 is not present. As a result, the cartilage cells get the signal to stop dividing even though Fgf4, which usually activates the signal, is absent (see Chapter 14).

Similarly, both humans and mice can be born with a condition called Hirschprung disease, in which the lower regions of the gut are unable to perform peristalsis (involuntary muscle contractions), with the resulting inability to properly eliminate feces from the body. Hirschprung disease can be lethal, although in infants today it is usually correctable by surgery. By studying gene mutations in mice with Hirschprung-like bowel malformations, candidates for the genes causing Hirschprung disease in humans were discovered. It turns out, as we will see in Chapter 11, that this condition develops when the normal migration of neural crest cells to the gut during organ development is incomplete. Studies on mutant mice revealed that this failure to complete migration can be the result of either (1) mutations in the genes encoding the proteins made by the developing gut that guide the essential neural crest cells to the intestine, or (2) mutations in the genes on the neural crest cells that encode the *receptors* for the guiding proteins. Thus, as we saw in the dwarfism example above, similar phenotypes are produced when either the signal protein or the receptor for that signal protein is mutated.

Disruptions and teratogens

Developmental abnormalities caused by exogenous agents (certain chemicals or viruses, radiation, or hyperthermia) are called **disruptions**. The agents responsible for these disruptions are called **teratogens** (Greek, "monster-formers"), and the study of how environmental agents disrupt normal development is called teratology. Teratogens were brought to the attention of the public in the early 1960s. In 1961, Lenz and McBride independently accumulated evidence that the drug thalidomide, prescribed as a mild sedative to many pregnant women, caused an enormous increase in a previously rare syndrome of congenital anomalies. The most noticeable of these anomalies was phocomelia, a condition in which the long bones of the limbs are deficient or absent (**FIGURE 1.22A**). More than 7000 affected infants were born to women who took thalidomide, and a woman need only have taken one tablet for her child to be born with all four limbs deformed (Lenz 1962, 1966; Toms 1962). Other abnormalities induced by ingesting this drug included heart defects, absence of the external ears, and malformed intestines. Nowack (1965) documented the period of susceptibility during which thalidomide caused these abnormalities (**FIGURE 1.22B**). The drug was found to be teratogenic only during days 34–50 after the last menstruation (i.e., 20–36 days postconception). From days 34 to 38, no limb abnormalities are seen, but during this period thalidomide can cause the absence or deficiency of ear components. Malformations

(A)

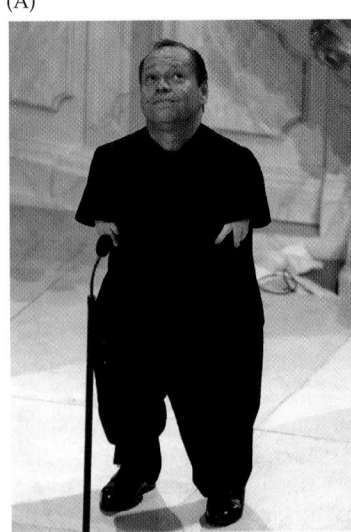

(B)

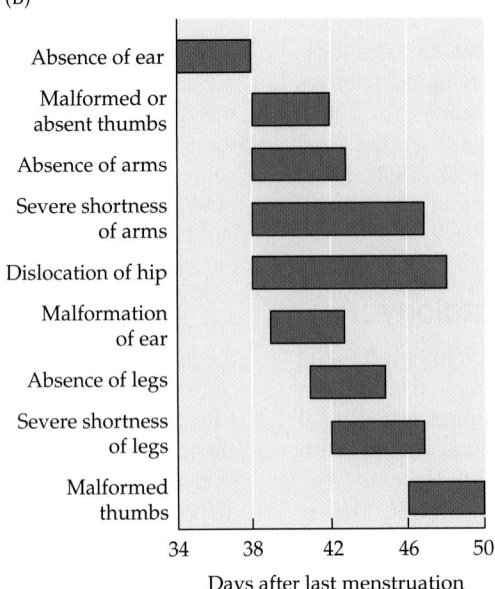

Absence of ear
Malformed or absent thumbs
Absence of arms
Severe shortness of arms
Dislocation of hip
Malformation of ear
Absence of legs
Severe shortness of legs
Malformed thumbs

34   38   42   46   50

Days after last menstruation

FIGURE 1.22 Developmental anomalies caused by environmental agents. (A) Phocomelia, the lack of proper limb development, was the most visible of the birth defects that occurred in many children born in the early 1960s whose mothers took the drug thalidomide during pregnancy. These children are now middle-aged adults; this photograph is of Grammy-nominated German singer Thomas Quasthoff. (B) Thalidomide disrupts different structures at different times of human development. (A © AP Photo.)

of the upper limbs are seen before those of the lower limbs because the developing arms form slightly before the legs.

We still do not know for certain the mechanisms by which thalidomide causes human developmental disruptions (although it may work by blocking certain molecules from the developing mesoderm, thus preventing blood vessel development). The drug was withdrawn from the market in November 1961, but is once more being prescribed (although not to pregnant women) as a potential anti-tumor and anti-autoimmunity drug (Raje and Anderson 1999).

Substances that can cause birth defects include relatively common substances such as alcohol and retinoic acid (often used to treat acne), as well many chemicals used in manufacturing and released into the environment. Heavy metals (e.g., mercury, lead, selenium) can alter brain development. For example, many lakes in the western United States have been contaminated with heavy metals from mining operations, and pregnant women are advised not to eat fish from those lakes. Teratology will be discussed extensively in Chapter 18.

The integration of anatomical information about congenital malformations with our new knowledge of the genes responsible for development has resulted in an ongoing restructuring of medicine. This integrated information is allowing us to discover the genes responsible for inherited malformations, and to identify exactly which steps in development are disrupted by specific teratogens. We will see examples of this integration throughout this text.

SNAPSHOT SUMMARY: Comprehending Development

1. The life cycle can be considered a central unit in biology; the adult form need not be paramount. The basic animal life cycle consists of fertilization, cleavage, gastrulation, germ layer formation, organogenesis, metamorphosis, adulthood, and senescence.

2. In gametogenesis, the germ cells (i.e., those cells that will become sperm or eggs) undergo meiosis. Eventually, usually after adulthood is reached, the mature gametes are released to unite during fertilization. The resulting new generation then begins development.

3. Epigenesis happens. New organisms are created de novo each generation from the relatively disordered cytoplasm of the egg.

4. Preformation is not found in the anatomical structures themselves, but in the genetic instructions that instruct their formation. The inheritance of the fertilized egg includes the genetic potentials of the organism. These preformed nuclear instructions include the ability to respond to environmental stimuli in specific ways.

5. The three germ layers give rise to specific organ systems. The ectoderm gives rise to the epidermis, nervous system, and pigment cells; the mesoderm generates the kidneys, gonads, muscles, bones, heart, and blood cells; and the endoderm forms the lining of the digestive tube and the respiratory system.

6. Karl von Baer's principles state that the general features of a large group of animals appear earlier in the embryo than do the specialized features of a smaller group. As each embryo of a given species develops, it diverges from the adult forms of other species. The early embryo of a "higher" animal species is not like the adult of a "lower" animal.

7. Labeling cells with dyes shows that some cells differentiate where they form, whereas others migrate from their original sites and differentiate in their new locations. Migratory cells include neural crest cells and the precursors of germ cells and blood cells.

8. "Community of embryonic structure reveals community of descent" (Charles Darwin, *On the Origin of Species*).

9. Homologous structures in different species are those organs whose similarity is due to sharing a common ancestral structure. Analogous structures are those organs whose similarity comes from serving a similar function (but which are not derived from a common ancestral structure).

10. Congenital anomalies can be caused by genetic factors (mutations, aneuploidies, translocations) or by environmental agents (certain chemicals, certain viruses, radiation).

11. Teratogens—environmental compounds that can alter development—act at specific times when certain organs are being formed. Similar genetic malformations can occur when communication between cells is interrupted or eliminated. The molecular signal and its receptor on the responding cell are both critical.

For Further Reading

Complete bibliographical citations for all literature cited in this chapter can be found at the free-access website **www.devbio.com**

Cadieu, E. and 19 others. 2009. Coat variation in the domestic dog is governed by variants in three genes. *Science* 326: 150–153.

Cebra-Thomas, J. A., E. Betters, M. Yin, C. Plafkin, K. McDow and S. F. Gilbert. 2007. Evidence that a late-emerging population of trunk neural crest cells forms the plastron bones in the turtle *Trachemys scripta. Evol. Dev.* 9: 267–277.

Larsen, E. and H. McLaughlin. 1987. The morphogenetic alphabet: Lessons for simple-minded genes. *BioEssays* 7: 130–132.

Le Douarin, N. M. and M.-A. Teillet. 1973. The migration of neural crest cells to the wall of the digestive tract in the avian embryo. *J. Embryol. Exp. Morphol.* 30: 31–48.

Nishida, H. 1987. Cell lineage analysis in ascidian embryos by intracellular injection of a tracer enzyme. III. Up to the tissue-restricted stage. *Dev. Biol.* 121: 526–541.

Pinto-Correia, C. 1997. *The Ovary of Eve: Egg and Sperm and Preformation.* University of Chicago Press, Chicago.

Weatherbee, S. D., R. R. Behringer, J. J. Rasweiler 4th and L. A. Niswander. 2006. Interdigital webbing retention in bat wings illustrates genetic changes underlying amniote limb diversification. *Proc. Natl. Acad. Sci. USA* 103: 15103–15107.

Winter, R. M. 1996. Analyzing human developmental abnormalities. *BioEssays* 18: 965–971.

Woo, K. and S. E. Fraser. 1995. Order and coherence in the fate map of the zebrafish embryo. *Development* 121: 2595–2609.

Go Online

WEBSITE 1.1 When does human personhood begin?
Scientists have proposed different answers to this question, depending on the stages of the human life cycle. Fertilization, gastrulation, brain function, and the time around birth each have its supporters.

WEBSITE 1.2 Conklin's art and science.
The plates from Conklin's remarkable 1905 paper are online. Looking at them, one can see the precision of his observations and how he constructed his fate map of the tunicate embryo.

Vade Mecum

The amphibian life cycle. Frogs have some of the most dramatic life cycles seen among the vertebrates. Moreover, these life cycles can be studied nearly any place in the world.

The compound microscope. The compound microscope has been the critical tool of developmental anatomists. Mastery of microscopic techniques allows one to enter an entire world of form and pattern.

Histotechniques. Most cells must be stained in order to see them; different dyes stain different types of molecules. Instructions on staining cells to observe particular structures (such as the nucleus) are given here.

Chick-quail chimeras. We are fortunate to present here a movie made by Dr. Nicole Le Douarin of her chick-quail grafts. You will be able to see how these grafts are actually done.

Outside Sites

The Embryo Project site maintained at Arizona State University is still in gestation. It contains information about the history, technology, and people who made modern embryology possible. **http://embryo.asu.edu/**.

The University of New South Wales Human Embryology Site, curated by Dr. Mark Hill, contains exceptionally well organized material on human embryology **php.med.unsw. edu.au/embryology/index.php?title=Main_Page**.

2

Differential Gene Expression in Development

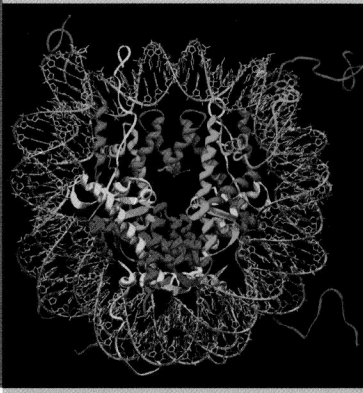

CYTOLOGICAL STUDIES DONE at the turn of the twentieth century established that the chromosomes in each cell of an organism's body are the mitotic descendants of the chromosomes established at fertilization (Wilson 1896; Boveri 1904). In other words, each somatic cell nucleus has the same chromosomes—and therefore the same set of genes—as all other somatic cell nuclei. This fundamental concept is known as **genomic equivalence**. Given this concept, biologists of the early twentieth century were faced with the question of how nuclear genes can direct development when these genes are exactly the same in every cell type (Harrison 1937; Just 1939). If every cell in the body contains the genes for hemoglobin and insulin, why are hemoglobin proteins made only in red blood cells and insulin proteins only in certain pancreas cells?

Based on the embryological evidence for genomic equivalence (as well as on bacterial models of gene regulation), a consensus emerged in the 1960s that the answer to this question lies in **differential gene expression**. The three postulates of differential gene expression are:

- Every somatic cell nucleus contains the complete genome established in the fertilized egg. In molecular terms, the DNAs of all differentiated cells are identical.

- The unused genes in differentiated cells are neither destroyed nor mutated; they retain the potential for being expressed.

- Only a small percentage of the genome is expressed in each cell, and a portion of the RNA synthesized in each cell is specific for that cell type.

By the late 1980s, it was established that gene expression can be regulated at several levels such that different cell types synthesize different sets of proteins:

- *Differential gene transcription* regulates which of the nuclear genes are transcribed into nuclear RNA.

- *Selective nuclear RNA processing* regulates which of the transcribed RNAs (or which parts of such a nuclear RNA) are able to enter into the cytoplasm and become messenger RNAs.

- *Selective messenger RNA translation* regulates which of the mRNAs in the cytoplasm are translated into proteins.

- *Differential protein modification* regulates which proteins are allowed to remain and/or function in the cell.

Some genes (such as those coding for the globin protein subunits of hemoglobin) are regulated at all these levels.

We have entered the cell, the mansion of our birth, and have started the inventory of our acquired wealth.

ALBERT CLAUDE (1974)

Embryonic development is an enormous informational transaction, in which DNA sequence data generate and guide the system-wide deployment of specific cellular functions.

E. H. DAVIDSON (2010)

● See **WEBSITE 2.1**   Does the genome or the cytoplasm direct development?

● See **WEBSITE 2.2**   The origins of developmental genetics

Evidence for Genomic Equivalence

Until the mid-twentieth century, genomic equivalence was not so much proved as it was assumed (because every cell is the mitotic descendant of the fertilized egg). One of the first tasks of developmental genetics was to determine whether every cell of an organism indeed does have the same **genome**—that is, the same set of genes—as every other cell.

Evidence that every cell in the body has the same genome originally came from the analysis of *Drosophila* chromosomes, in which the DNA of certain larval tissues undergoes numerous rounds of DNA replication without separation, such that the structure of the chromosomes can be seen. In these **polytene** (Greek, "many strands") **chromosomes**, no structural differences were seen between cells; but different regions were "puffed up" at different times and in different cell types, suggesting that these areas were actively making RNA (Beermann 1952). When Giemsa dyes allowed such observations to be made in mammalian chromosomes, it was also found that no chromosomal regions were lost in most cells. These

observations, in turn, were confirmed by nucleic acid hybridization studies, which (for instance) found globin genes in pancreatic tissue, which does not make globin proteins.

But the ultimate test of whether the nucleus of a differentiated cell has undergone irreversible functional restriction is to have that nucleus generate every other type of differentiated cell in the body. If each cell's nucleus is identical to the zygote nucleus, then each cell's nucleus should also be capable of directing the entire development of the organism when transplanted into an activated enucleated egg. Although such experiments had been proposed in the 1930s, the first demonstration that a nucleus from an adult mammalian somatic cell could direct the development of an entire animal didn't come until 1997, when Dolly the sheep was cloned.

Ian Wilmut and his colleagues took cells from the mammary gland of a 6-year-old pregnant ewe and placed them in culture (**FIGURE 2.1A**; Wilmut et al. 1997). The culture medium was formulated to keep the cell nuclei at the intact diploid stage (G1) of the cell cycle; this cell-cycle stage turned out to be critical. The researchers then obtained oocytes from a different strain of sheep and removed their nuclei. These oocytes had to be in the second meiotic metaphase—the stage at which they are usually fertilized. The donor cell and the enucleated oocyte were brought together and electric

SIDELIGHTS & SPECULATIONS

The Basic Tools of Developmental Genetics

DNA analysis

Embryologist Theodor Boveri (1904) wrote that to discover the mechanisms of development, it was "not cell nuclei, not even individual chromosomes, but certain parts of certain chromosomes from certain cells that must be isolated and collected in enormous quantities for analysis." This analysis was finally made possible by the techniques of gene cloning, DNA sequencing, Southern blotting, gene knockouts, and enhancer traps. In addition, techniques for showing which enhancers and promoters are methylated and which are unmethylated have become more important, as investigations of differential gene transcription have focused on these elements.

● *For discussions of these techniques, see Website 2.3*

RNA analysis

Differential gene transcription is critical in development. In order to know the specific time and place of gene expression, one needs to use procedures that actually locate a particular type of messenger RNA. These techniques include northern blots, RT-PCR, in situ hybridization, and array technology. To ascertain the function of these mRNAs once they are located, new techniques have been formulated, including antisense and RNA interference (which destroy mRNA messages), Cre-lox analysis (which allows the message to be made or destroyed in particular cell types), and chromatin immunoprecipitation-sequencing (ChIP-Seq) techniques (which enable one to identify proteins bound to specific DNA sequences and to identify active chromatin). In addition, "high-throughput" RNA analysis

by micro- and macroarrays enables researchers to compare thousands of mRNAs, and computer-aided synthetic techniques can predict interactions between proteins and mRNAs.

● *For discussions of these techniques, see Website 2.4*

Bioinformatics

Modern developmental genetics often involves comparing DNA sequences (especially regulatory units such as enhancers and 3′ UTRs) and looking at specific genomes to determine how genes are being regulated. Various free websites enable researchers to use the tools that allow such comparisons. Other sites are organism- or organ-specific and are used by researchers studying that particular organ or organism.

● *For more about these sites and links to them, see Website 2.5*

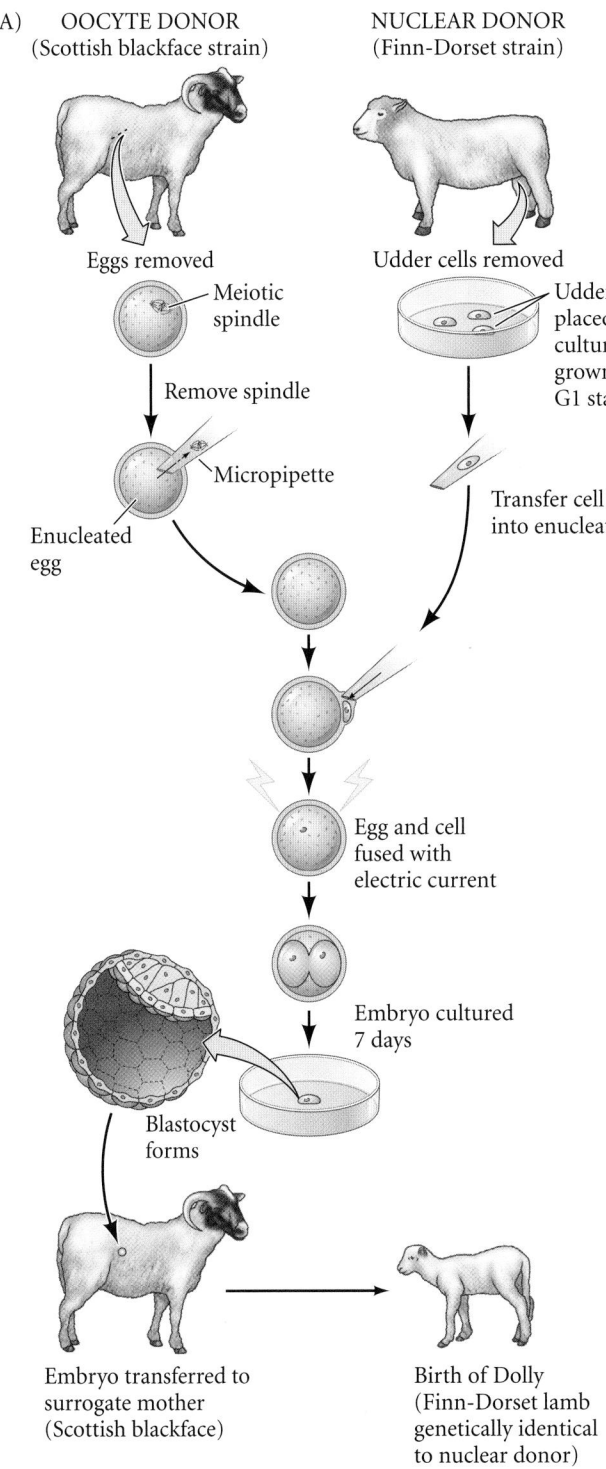

FIGURE 2.1 Cloning a mammal using nuclei from adult somatic cells. (A) Procedure used for cloning sheep. (B) Dolly, the adult sheep on the left, was derived by fusing a mammary gland cell nucleus with an enucleated oocyte, which was then implanted in a surrogate mother (of a different breed of sheep) that gave birth to Dolly. Dolly later gave birth to a lamb (Bonnie, at right) by normal reproduction. (A after Wilmut et al. 2000; B, photograph by Roddy Field © Roslin Institute.)

Of the 434 sheep oocytes originally used in this experiment, only one survived: Dolly* (**FIGURE 2.1B**). DNA analysis confirmed that the nuclei of Dolly's cells were derived from the strain of sheep from which the donor nucleus was taken (Ashworth et al. 1998; Signer et al. 1998). Cloning of adult mammals has been confirmed in guinea pigs, rabbits, rats, mice, dogs, cats, horses, and cows. In 2003, a cloned mule became the first sterile animal to be so reproduced (Woods et al. 2003). Thus, it appears that the nuclei of vertebrate adult somatic cells contain all the genes needed to generate an adult organism. No genes necessary for development have been lost or mutated in the somatic cells.†

pulses were sent through them, thereby destabilizing the cell membranes and allowing the cells to fuse. The same electric pulses that fused the cells activated the egg to begin development. The resulting embryos were eventually transferred into the uteri of pregnant sheep.

● **See WEBSITE 2.6** The 2012 Nobel Prize for Physiology or Medicine: Cloning and nuclear equivalence

*The creation of Dolly was the result of a combination of scientific and social circumstances. These circumstances involved job security, people with different areas of expertise meeting each other, children's school holidays, international politics, and who sits near whom in a pub. The complex interconnections giving rise to Dolly are told in *The Second Creation* (Wilmut et al. 2000), a book that should be read by anyone who wants to know how contemporary science actually works. As Wilmut acknowledged (p. 36), "The story may seem a bit messy, but that's because life is messy, and science is a slice of life."

†Although all the organs were properly formed in the cloned animals, many of the clones developed debilitating diseases as they matured (Humphreys et al. 2001; Jaenisch and Wilmut 2001; Kolata 2001). As we will shortly see, this problem is due in large part to the differences in methylation between the chromatin of the zygote and the differentiated cell.

Differential Gene Transcription

So how does the same genome give rise to different cell types? To understand this, one needs to understand the anatomy of the genes. One of the fundamental differences distinguishing most eukaryotic genes from prokaryotic genes is that eukaryotic genes are contained within a complex of DNA and protein called **chromatin**. The protein component constitutes about half the weight of chromatin and is composed largely of **histones**. The **nucleosome** is the basic unit of chromatin structure (**FIGURE 2.2A,B**). It is composed of an octamer of histone proteins (two molecules each of histones H2A, H2B, H3, and H4) wrapped with two loops containing approximately 147 base pairs of DNA (Kornberg and Thomas 1974). Histone H1 is bound to the 60–80 or so base pairs of "linker" DNA between the nucleosomes (Weintraub 1984,

FIGURE 2.2 Nucleosome and chromatin structure. (A) Model of nucleosome structure as seen by X-ray crystallography at a resolution of 1.9 Å. Histones H2A and H2B are yellow and red, respectively; H3 is purple and H4 is green. The DNA helix (gray) winds around the protein core. The histone "tails" that extend from the core are the sites of acetylation and methylation, which may disrupt or stabilize, respectively, the formation of nucleosome assemblages. (B) Histone H1 can draw nucleosomes together into compact forms. About 147 base pairs of DNA encircle each histone octamer, and about 60–80 base pairs of DNA link the nucleosomes together. (C) Model for the arrangement of nucleosomes in the highly compacted solenoidal chromatin structure. Histone tails protruding from the nucleosome subunits allow for the attachment of chemical groups. (D) Methyl groups condense nucleosomes more tightly, preventing access to promoter sites and thus preventing gene transcription. Acetylation loosens nucleosome packing, exposing the DNA to RNA polymerase II and transcription factors that will activate the genes. (A after Davey et al. 2002.)

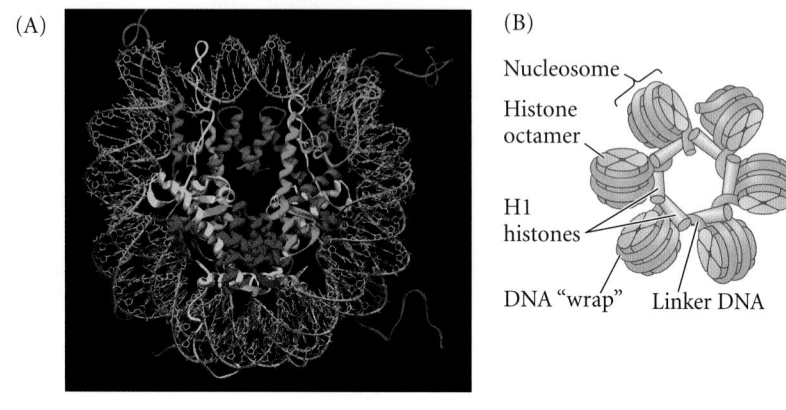

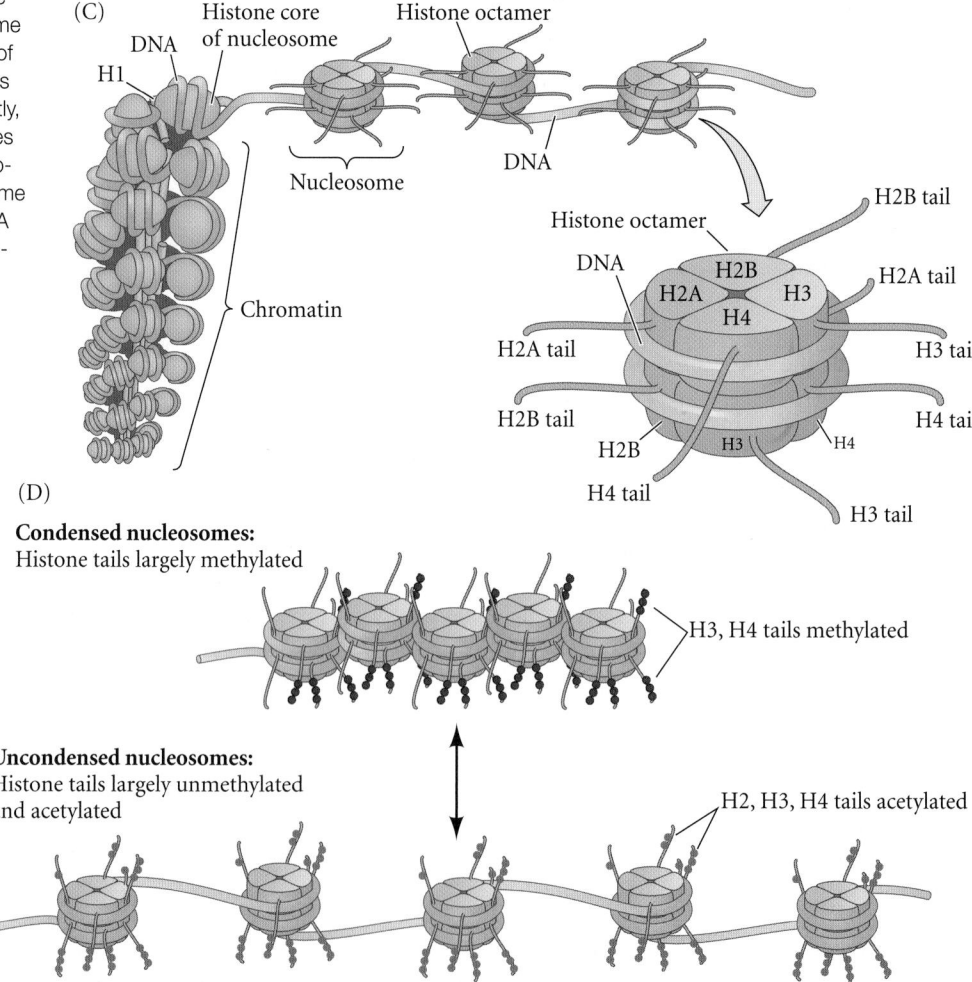

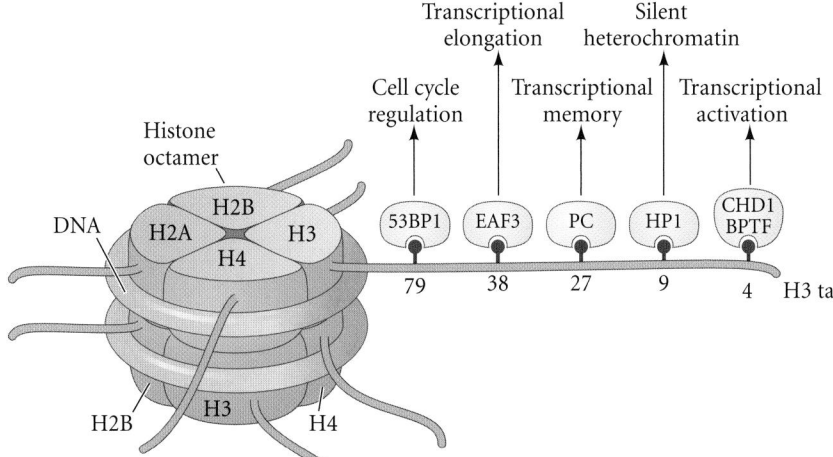

FIGURE 2.3 Histone methylations on histone H3. The tail of histone H3 (its amino-most sequence, at the beginning of the protein) sticks out from the nucleosome and is capable of being methylated or acetylated. Here, lysines can be methylated and recognized by particular proteins. Methylated lysine residues at positions 4, 38, and 79 are associated with gene activation, whereas methylated lysines at positions 9 and 27 are associated with repression. The proteins binding these sites (not shown to scale) are represented above the methyl group. (After Kouzarides and Berger 2007.)

1985). There are 14 points of contact between the DNA and the histones (Luger et al. 1997; Bartke et al. 2010).

Whereas classical geneticists have likened genes to "beads on a string," molecular geneticists liken genes to "string on the beads," an image in which the beads are nucleosomes. Much of the time, the nucleosomes appear to be wound into tight structures called **solenoids** that are stabilized by histone H1 (**FIGURE 2.2C**). This H1-dependent conformation of nucleosomes inhibits the transcription of genes in somatic cells by packing adjacent nucleosomes together into tight arrays that prevent transcription factors and RNA polymerases from gaining access to the genes (Thoma et al. 1979; Schlissel and Brown 1984).

Anatomy of the gene: Active and repressed chromatin

HISTONES AS AN ON/OFF SWITCH Histones are critical because they appear to be responsible for either facilitating or forbidding gene expression (**FIGURE 2.2D**). Repression and activation are controlled to a large extent by modifying the "tails" of histones H3 and H4 with two small organic groups: methyl (CH_3) and acetyl ($COCH_3$) residues. In general, **histone acetylation**—the addition of negatively charged acetyl groups to histones—neutralizes the basic charge of lysine and loosens the histones. This activates transcription. Enzymes known as **histone acetyltransferases** place acetyl groups on histones (especially on lysines in H3 and H4), destabilizing the nucleosomes so that they come apart easily. As might be expected, then, enzymes that *remove* acetyl groups—**histone deacetylases**—stabilize the nucleosomes and prevent transcription.

Histone methylation, the addition of methyl groups to histones by **histone methyltransferases**, can either activate or further repress transcription, depending on the amino acid being methylated and the presence of other methyl or acetyl groups in the vicinity (see Strahl and Allis 2000; Cosgrove et al. 2004). For instance, acetylation of the tails of H3 and H4 along with the addition of three methyl groups on the lysine at position 4 of H3 (i.e., H3K4me3; remember that

K is the abbreviation for lysine) is usually associated with actively transcribed chromatin. In contrast, a combined lack of acetylation of the H3 and H4 tails and methylation of the lysine in the ninth position of H3 (H3K9) is usually associated with highly repressed chromatin (Norma et al. 2001). Indeed, lysine methylations at H3K9, H3K27, and H4K20 are often associated with highly repressed chromatin. **FIGURE 2.3** depicts a nucleosome with residues on its H3 tail. Modifications of such residues regulate transcription.

If methyl groups at specific places on histones repress transcription, then getting rid of these methyl moieties should be expected to permit transcription. This has been shown to be the case in the activation of the Hox genes, a family of genes that are critical in giving cells their identities along the anterior-posterior axis. In early development, Hox genes are repressed by H3K27 trimethylation (the lysine at position 27 on histone 3 has three methyl groups: H3K27me3). However, in differentiated cells, a demethylase specific for H3K27me3 is recruited to these promoters, eliminating the methyl goups and enabling the gene to be transcribed (Agger et al. 2007; Lan et al. 2007).

The effects of methylation in controlling gene transcription are extensive. So far, we have documented transcriptional regulation by *histone* methylation. Later in this chapter we will discuss the exciting research on the control of transcription by *DNA* methylation.

Anatomy of the gene: Exons and introns

Another fundamental difference between prokaryotic and eukaryotic genes (along with the fact that eukaryotic genes are contained within chromatin) is that eukaryotic genes are not co-linear with their peptide products. Rather, the single nucleic acid strand of eukaryotic mRNA comes from noncontiguous regions on the chromosome. Between **exons**—the regions of DNA that code for a protein*—are intervening sequences called **introns** that have nothing whatsoever to

---

*The term *exon* refers to a nucleotide sequence whose RNA "exits" the nucleus. It has taken on the functional definition of a protein-encoding nucleotide sequence. Leader sequences and 3′ UTR sequences are also derived from exons, even though they are not translated into protein.

(A)

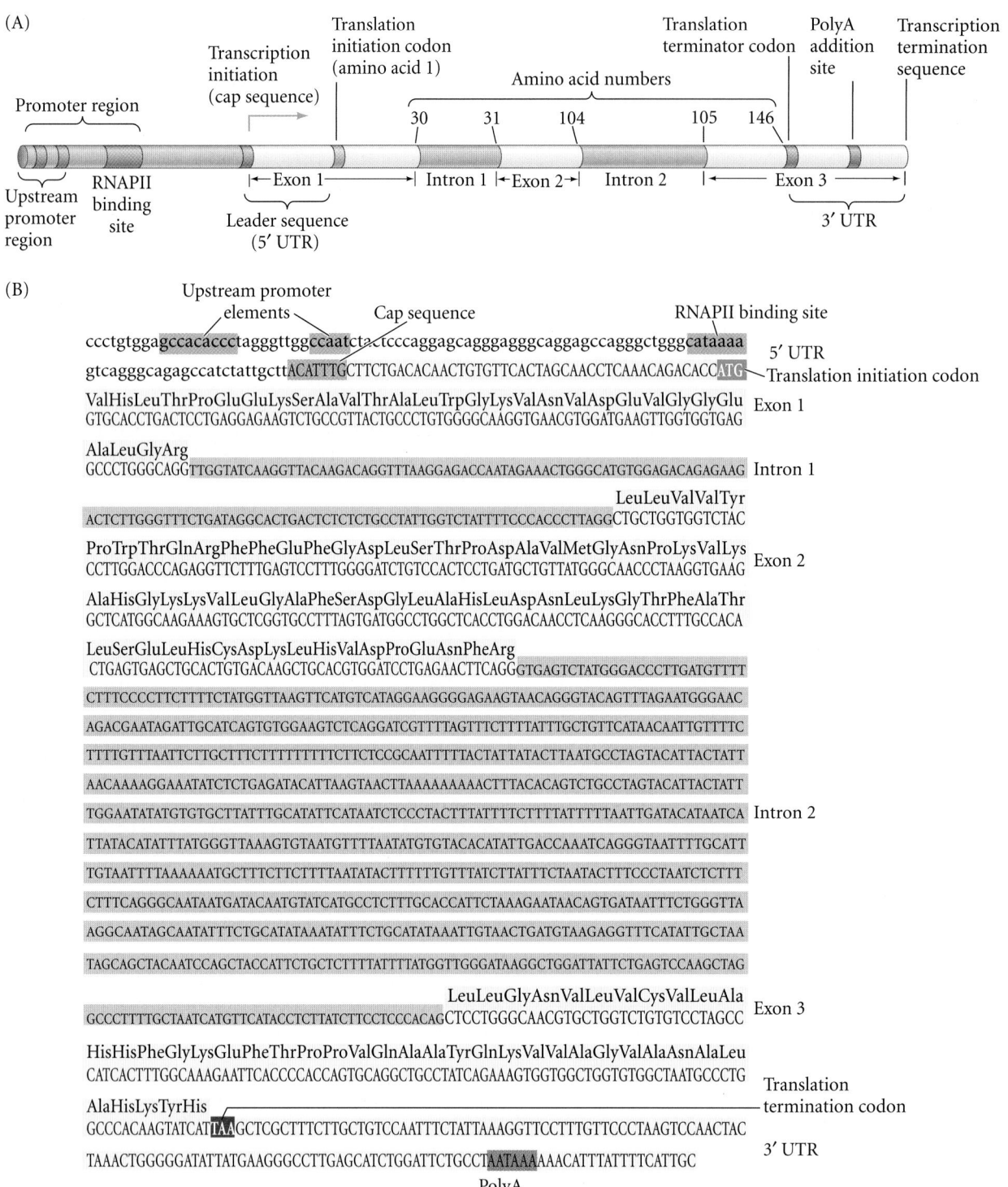

(B)

FIGURE 2.4   Nucleotide sequence of the human β-globin gene. (A) Schematic representation of the locations of the promoter region, transcription initiation site (cap sequence), 5′ untranslated region (leader sequence), exons, introns, and 3′ untranslated region. Exons are shown in color; the numbers flanking them indicate the amino acid positions each exon encodes in β-globin. (B) The nucleotide sequence shown from the 5′ end to the 3′ end of the RNA. The colors correspond to their diagrammatic representation in (A). The promoter sequences are boxed, as are the translation initiation and termination codes ATG and TAA. The large capital letters boxed in color are the bases of the exons, with the amino acids for which they code abbreviated above them. Smaller capital letters indicate the intron bases. The codons after the translation termination site exist in β-globin mRNA but are not translated into proteins. Within this group is the sequence thought to be needed for polyadenylation. By convention, only the RNA-like strand of the DNA double helix is shown. (B after Lawn et al. 1980.)

do with the amino acid sequence of the protein. The structure of a typical eukaryotic gene can be illustrated by the human β-globin gene (**FIGURE 2.4**). This gene, which encodes part of the hemoglobin protein of the red blood cells, consists of the following elements:

- A **promoter region**, which is responsible for the binding of RNA polymerase II and for the subsequent initiation of transcription. The promoter region of the human β-globin gene has three distinct units and extends from 95 to 26 base pairs before ("upstream from")* the transcription initiation site (i.e., from –95 to –26). The TBP site binds the basal transcription factor (TBP) that helps anchor RNA polymerase II to the promoter.

- The **transcription initiation site**, which for human β-globin is ACATTTG. This site is often called the **cap sequence** because it represents the 5′ end of the RNA, which will receive a "cap" of modified nucleotides soon after it is transcribed. The specific cap sequence varies among genes.

- The **5′ untranslated region (5′ UTR)**, often called the **leader sequence**. This is the sequence of 50 base pairs intervening between the initiation points of transcription and translation. The 5′ UTR can determine the rate at which translation is initiated.

- The **translation initiation site**, ATG. This codon (which becomes AUG in mRNA) is located 50 base pairs after the transcription initiation site in the human β-globin gene (although this distance differs greatly among different genes). This begins the first exon.

- The protein-encoding portion of the first exon, which contains 90 base pairs coding for amino acids 1–30 of human β-globin protein.

- An intron containing 130 base pairs with no coding sequences for β-globin. However, the structure of this intron is important in enabling the RNA to be processed into mRNA and exit from the nucleus.

- An exon containing 222 base pairs coding for amino acids 31–104.

- A large intron—850 base pairs—having nothing to do with β-globin protein structure.

- An exon containing 126 base pairs coding for amino acids 105–146 of the protein.

- A **translation termination codon**, TAA. This becomes UAA in the mRNA. When a ribosome encounters this codon, the ribosome dissociates and the protein is released.

- A **3′ untranslated region (3′ UTR)** that, although transcribed, is not translated into protein. This region includes the sequence AATAAA, which is needed for **polyadenylation**, the insertion of a "tail" of some

200–300 adenylate residues on the RNA transcript, about 20 bases downstream of the AAUAAA sequence. This polyA tail (1) confers stability on the mRNA, (2) allows the mRNA to exit the nucleus, and (3) permits the mRNA to be translated into protein.

- A **transcription termination sequence**. Transcription continues beyond the AATAAA site for about 1000 nucleotides before being terminated.

The original transcription product is called **nuclear RNA** (**nRNA**), sometimes called *heterogeneous nuclear RNA* (hnRNA) or *pre-messenger RNA* (pre-mRNA). Nuclear RNA contains the cap sequence, the 5′ UTR, exons, introns, and the 3′ UTR. Both ends of these transcripts are modified before these RNAs leave the nucleus. A cap consisting of methylated guanosine is placed on the 5′ end of the RNA in opposite polarity to the RNA itself. This means there is no free 5′ phosphate group on the nRNA. The 5′ cap is necessary for the binding of mRNA to the ribosome and for subsequent translation (Shatkin 1976). The 3′ terminus is usually modified in the nucleus by the addition of a polyA tail. The adenylate residues in this tail are put together enzymatically and are added to the transcript; they are not part of the gene sequence. Both the 5′ and 3′ modifications may protect the mRNA from exonucleases that would otherwise digest it (Sheiness and Darnell 1973; Gedamu and Dixon 1978). The modifications thus stabilize the message and its precursor.

As the nRNA leaves the nucleus, its introns are removed and the remaining exons spliced together. In this way the coding regions of the mRNA—that is, the exons—are brought together to form a single transcript, and this transcript is translated into a protein. The protein can be further modified to make it functional (**FIGURE 2.5**).

Anatomy of the gene: Promoters and enhancers

In addition to the protein-encoding region of the gene, there are regulatory sequences that can be located on either end of the gene (or even within it). These sequences—the promoters and enhancers—are necessary for controlling where and when a particular gene is transcribed.

Promoters are sites where RNA polymerase II binds to the DNA sequence to initiate transcription. Promoters of genes that synthesize messenger RNAs (i.e., those genes that encode proteins†) are typically located immediately upstream from the site where RNA polymerase II initiates transcription. Most of these promoters contain a stretch of

---

*By convention, upstream, downstream, 5′, and 3′ directions are specified in relation to the RNA. Thus, the promoter is upstream of the gene, near to and "before" its 5′ end.

†In the case of protein-encoding genes, RNA polymerase II is used for transcription. There are several types of RNA that do *not* encode proteins. These include the ribosomal RNAs and transfer RNAs (which are used in protein synthesis) and the small nuclear RNAs (which are used in RNA processing). In addition, there are regulatory RNAs (such as the microRNAs and long noncoding RNAs that we will discuss later in this chapter), which are involved in regulating gene expression and are not translated into peptides. These often are transcribed by other RNA polymerases.

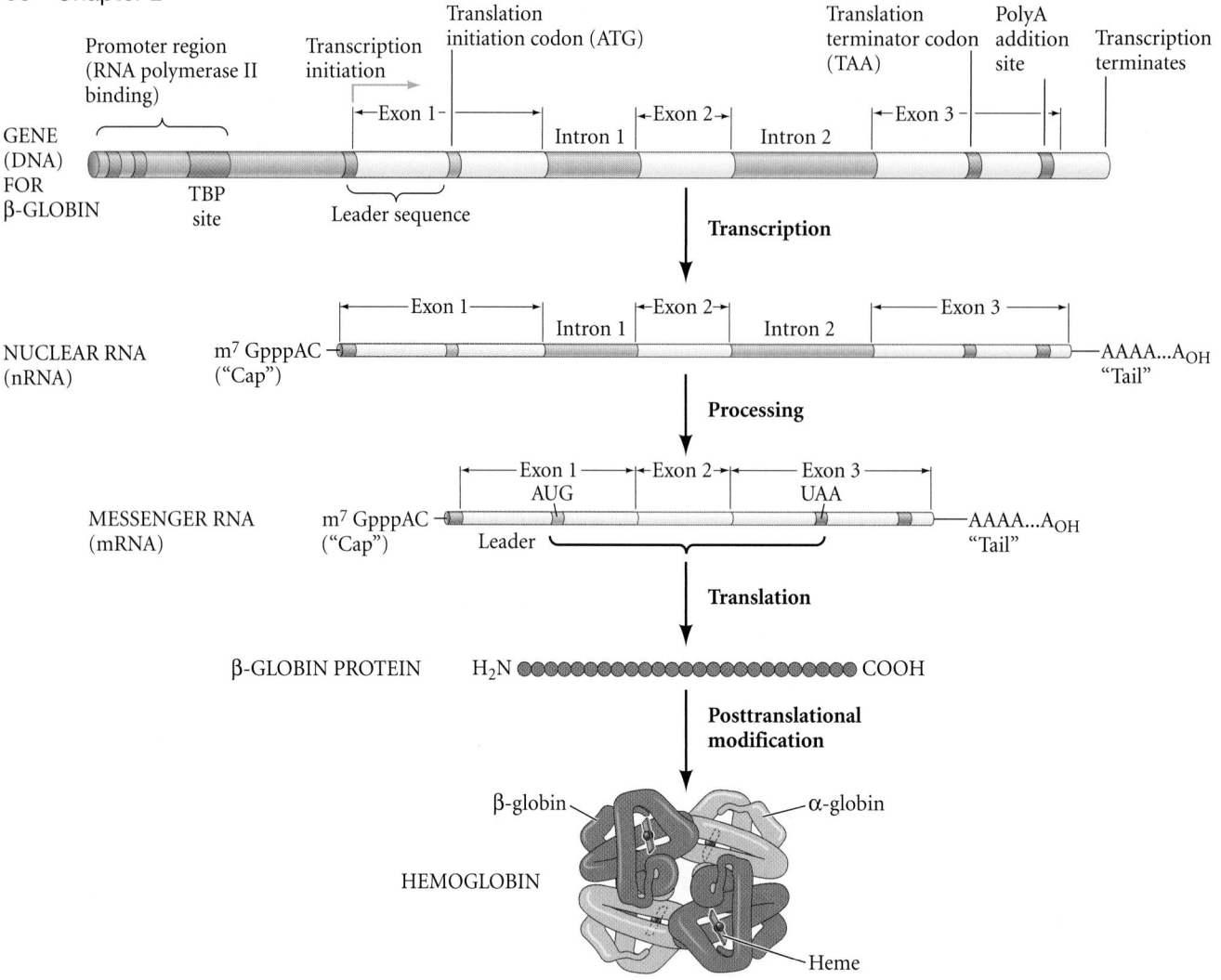

FIGURE 2.5   Steps in the production of β-globin and hemoglobin. Transcription of the β-globin gene creates a nuclear RNA containing exons and introns, as well as the cap, tail, and 3′ and 5′ untranslated regions. Processing the nuclear RNA into messenger RNA removes the introns. Translation on ribosomes uses the mRNA to encode a protein. The β-globin protein is inactive until it is modified and complexed with α-globin and heme to become active hemoglobin (bottom).

about 1000 base pairs that is rich in the sequence CpG (a C and a G connected through the normal phosphate bond). These regions are called **CpG islands** (Down and Hubbard 2002; Deaton and Bird 2011). The reason transcription is initiated near CpG islands is thought to involve proteins called **basal transcription factors** (such as TBP), which are present in every cell. These proteins bind to the CpG-rich sites and form a "saddle" that can recruit RNA polymerase II and position it so the polymerase can begin transcription (Kostrewa et al. 2009).

But RNA polymerase II does not bind to every promoter in the genome at the same time. Rather, it is recruited to and stabilized on the promoters by DNA sequences called **enhancers** that signal where and when a promoter can be used and how much gene product to make. In other words, enhancers control the efficiency and rate of transcription from a specific promoter (see Ong and Corces 2011). Enhancers bind specific **transcription factors**, proteins that activate the gene by (1) recruiting

enzymes (such as histone acetyltransferases) that break up the nucleosomes in the area or (2) stabilizing the transcription initiation complex as described above. Thus, transcription factors usually work in two nonexclusive ways:

1. Once bound, transcription factors can bind cofactors that recruit nucleosome-modifying proteins (such as histone methyltransferases and acetyltransferases) that make that area of the genome accessible for RNA polymerase II to bind and enable the chromatin in that vicinity to be unwound and transcribed.

2. Transcription factors can form bridges, looping the chromatin such that the transcription factors (and their histone-modifying enzymes) on enhancers can be brought into the vicinity of the promoter. In the activation of mammalian β-globin genes, such a bridge uniting the promoter and enhancer is formed by proteins that bind to transcription factors on both the enhancer and promoter

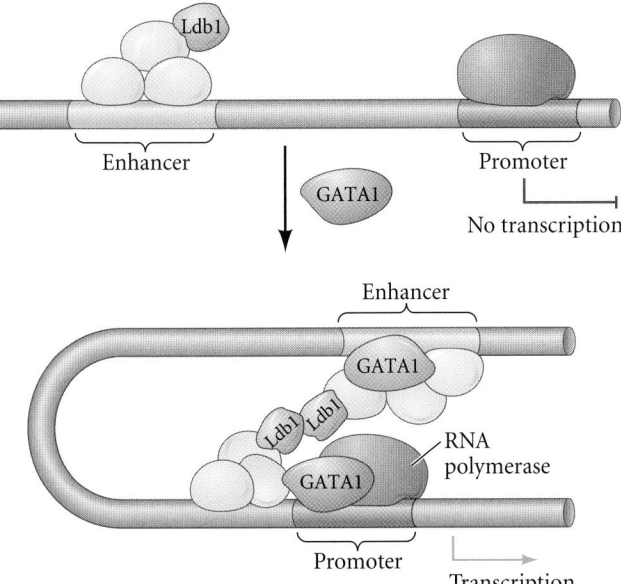

FIGURE 2.6 The bridge between enhancer and promoter can be made by transcription factors. Certain transcription factors bind to DNA on the promoter (where RNA polymerase II will initiate transcription), while other transcription factors bind to the enhancer (which regulates when and where transcription can occur). Other transcription factors do not bind to the DNA, but rather link the transcription factors that have bound to the enhancer and promoter sequences. In this way, the chromatin loops to bring the enhancer to the promoter. The example shown here is the mouse β-globin gene. (A) Transcription factors assemble on the enhancer, but the promoter is not used until the GATA1 transcription factor binds to the promoter. (B) GATA1 can recruit several other factors, including Ldb1, which forms a link uniting the enhancer-bound factors to the promoter-bound factors. (After Deng et al. 2012.)

sequences. These proteins recruit the nucleosome-modifying enzymes and TAFs that stabilize RNA polymerase II (**FIGURE 2.6**; Deng et al. 2012; Noordermeer and Duboule 2013).

THE MEDIATOR COMPLEX: LINKING ENHANCER AND PROMOTER   In many genes, a bridge between enhancer and promoter is made by a large, multimeric complex called the **Mediator**, whose nearly 30 protein subunits connect RNA polymerase II to enhancer regions that relay developmental signals (Malik and Roeder 2010). This forms the **pre-initiation complex** at the promoter. Therefore, the Mediator initiates a chromatin loop, bringing the enhancers to the promoter. This chromatin loop is stabilized by the protein cohesin, which becomes associated with the Mediator after the Mediator is bound by transcription factors (**FIGURE 2.7**).

Although the Mediator may help bring the RNA polymerase II to the promoter, in order for transcription to take place, the connection between the Mediator and the RNA polymerase II has to be broken, and RNA polymerase II must be released from the promoter. The release of RNA polymerase II is accomplished by a **transcription elongation complex** (**TEC**) made up of several transcription factors. This release coincides with the capping of the transcript and the phosphorylation of the polymerase. The enhancer-bound

Mediator complex can presumably recruit new RNA polymerases to the promoter, maintaining transcriptional activity there. However, in some instances (discussed later in the chapter), the RNA polymerase II either does not dissociate from the Mediator, or it dissociates but only transcribes a few nucleotides before it pauses. In the latter case, a transcription elongation suppressor (such as NELF) appears to prevent the transcription elongation complex from associating with the polymerase, and the RNA polymerase II is paused, held in readiness for a new developmental signal.

Enhancer functioning

One of the principal methods of identifying enhancer sequences is to clone DNA sequences flanking the gene of interest and fuse them to reporter genes whose products are both readily identifiable and not usually made in the organism being studied. Researchers can insert constructs of possible enhancers and reporter genes into embryos and then monitor the expression of the reporter gene (such as *GFP*; **FIGURE 2.8A**). If the sequence contains an enhancer, the reporter gene should become active at particular times and places. For instance, the *E. coli* gene for β-galactosidase (the *lacZ* gene) can be used as a reporter gene and fused to (1) a promoter that can be activated in any cell and (2) an enhancer that directs expression of a particular gene (*Myf5*) only in mouse muscles. When the resulting transgene is injected into a newly fertilized mouse egg and becomes incorporated into its DNA, β-galactosidase protein reveals the expression pattern of that muscle-specific gene (**FIGURE 2.8B**). More recently, genomic techniques such as ChIP-Seq (discussed later in the chapter) have enabled researchers to identify enhancer elements by sequencing the DNA regions bound by transcription factors.

Enhancers generally activate only *cis*-linked promoters (i.e., promoters on the same chromosome); therefore, they are sometimes called ***cis*-regulatory elements**.* However, because of DNA folding, enhancers can regulate genes at great distances (some as great as a million bases away) from the promoter (Visel et al. 2009). Moreover, enhancers do not need to be on the 5′ (upstream) side of the gene; they can be at the 3′ end, and are frequently in the introns (Maniatis et al. 1987). As we will see in Chapter 14, an important enhancer for a gene involved in specifying the "pinky" of each of our

---

* *Cis*- and *trans*-regulatory elements are so named by analogy with *E. coli* genetics and organic chemistry. Therefore, *cis*-elements are regulatory elements that reside on the same chromosome (*cis*-, "on the same side as"), whereas *trans*-elements are those that could be supplied from another chromosome (*trans*-, "on the other side of"). The term *cis*-regulatory elements now refers to those DNA sequences that regulate a gene on the same stretch of DNA (i.e., the promoters and enhancers). *Trans*-regulatory factors are soluble molecules whose genes are located elsewhere in the genome and which bind to the *cis*-regulatory elements. They are usually transcription factors or microRNAs. Some evidence points to the ability of an enhancer to activate a *trans* promoter (i.e., a promoter on another chromosome), but these appear to be exceptional and rare events (Noordermeer et al. 2011).

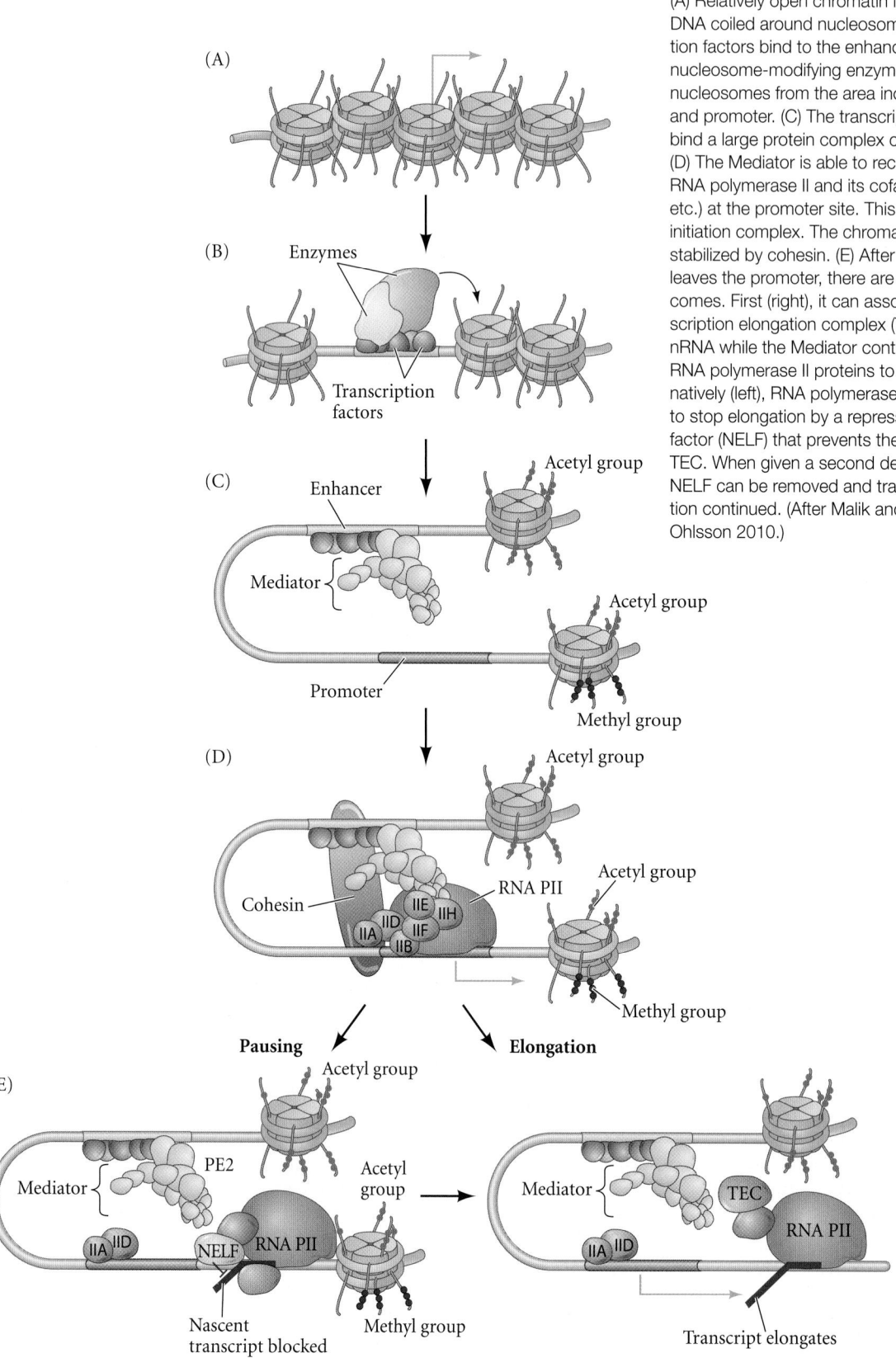

(A)

(B) Enzymes

Transcription factors

(C) Enhancer

Mediator

Promoter

Acetyl group

Acetyl group

Methyl group

(D) Acetyl group

Cohesin

IIE IIH

IIA IID IIF

IIB

RNA PII

Acetyl group

Methyl group

Pausing

Elongation

(E)

Acetyl group

Mediator

PE2

IIA IID

NELF

RNA PII

Acetyl group

Nascent transcript blocked

Methyl group

Mediator

IIA IID

TEC

RNA PII

Transcript elongates

FIGURE 2.7 The role of the Mediator complex in forming the transcription pre-initiation complex. (A) Relatively open chromatin is composed on DNA coiled around nucleosomes. (B) Transcription factors bind to the enhancer and bind nucleosome-modifying enzymes that remove nucleosomes from the area including the enhancer and promoter. (C) The transcription factors also bind a large protein complex called the Mediator. (D) The Mediator is able to recruit and stabilize RNA polymerase II and its cofactors (TAFs IIA, IIB, etc.) at the promoter site. This is called the pre-initiation complex. The chromatin looping is further stabilized by cohesin. (E) After RNA polymerase II leaves the promoter, there are generally two outcomes. First (right), it can associate with the transcription elongation complex (TEC) to elongate the nRNA while the Mediator continues to recruit new RNA polymerase II proteins to the complex. Alternatively (left), RNA polymerase II can be instructed to stop elongation by a repressive transcription factor (NELF) that prevents the assembly of the TEC. When given a second developmental signal, NELF can be removed and transcription elongation continued. (After Malik and Roeder 2010; Ohlsson 2010.)

(A)

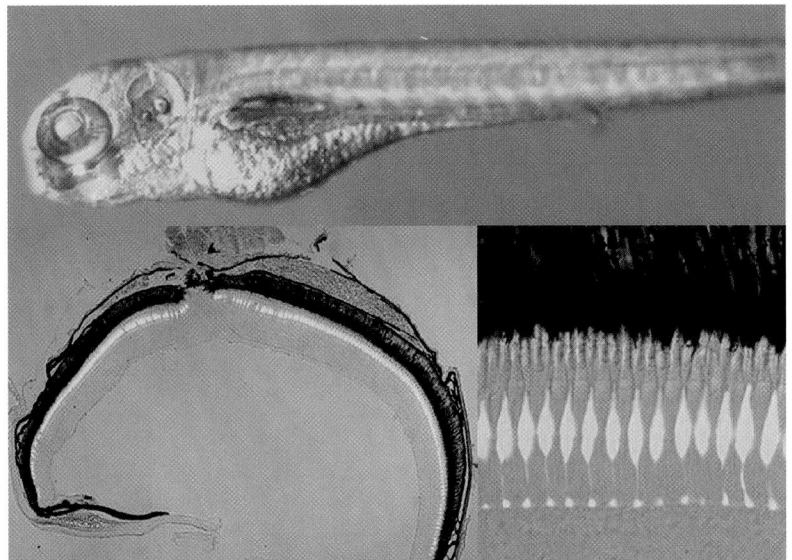

(B)

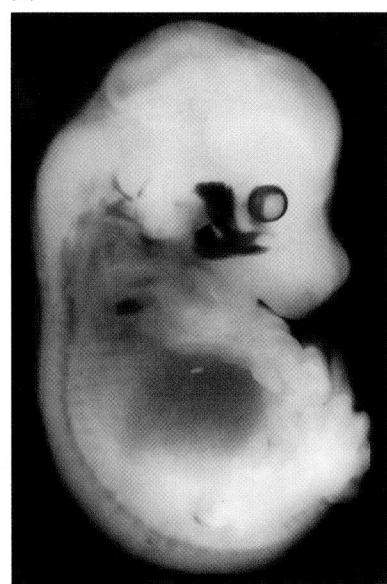

FIGURE 2.8 The genetic elements regulating tissue-specific transcription can be identified by fusing reporter genes to suspected enhancer regions of the genes expressed in particular cell types. (A) The *GFP* gene is fused to a zebrafish gene that is active only in certain cells of the retina. The result is expression of green fluorescent protein in the larval retina (below left), specifically in the cone cells (below right). (B) The enhancer region of the gene for the muscle-specific protein Myf5 is fused to a β-galactosidase reporter gene and incorporated into a mouse embryo. When stained for β-galactosidase activity (darkly staining region), the 13.5-day mouse embryo shows that the reporter gene is expressed in the muscles of the eye, face, neck, and forelimb and in the segmented myotomes (which give rise to the back musculature). (A from Takechi et al. 2003, courtesy of S. Kawamura, T. Hamaoka and M. Takechi; B courtesy of A. Patapoutian and B. Wold.)

limbs is found in an intron of *another* gene, some million base pairs away from its promoter (Lettice et al. 2008). In each cell, the enhancer becomes associated with particular transcription factors, binds nucleosome regulators and the Mediator complex, and engages with the promoter to transcribe the gene in that particular type of cell (**FIGURE 2.9A**).

ENHANCER MODULARITY The enhancer sequences on the DNA are the same in every cell type; what differs is the combination of transcription factor proteins that the enhancers experience. Once bound to enhancers, transcription factors are able to enhance or suppress the ability of RNA polymerase II to initiate transcription. Enhancers can bind several transcription factors, and it is the specific *combination* of transcription factors present that allows a gene to be active in a particular cell type. That is, the same transcription factor, in conjunction with different other factors, will activate different promoters in different cells. Moreover, the same gene can have several enhancers, with each enhancer binding transcription factors that enable that same gene to be expressed in different cell types.

The mouse *Pax6* gene (which is expressed in the lens, cornea, and retina of the eye, in the neural tube, and in the pancreas) has several enhancers (**FIGURE 2.9B,C**). The 5′ regulatory regions of the mouse *Pax6* gene were discovered by taking regions from its 5′ flanking sequence and introns and fusing them to a *lacZ* reporter gene. Each of these transgenes was then microinjected into newly fertilized mouse pronuclei,

and the resulting embryos were stained for β-galactosidase (**FIGURE 2.9D**; Kammandel et al. 1998; Williams et al. 1998). Analysis of the results revealed that the enhancer farthest upstream from the promoter contains the regions necessary for *Pax6* expression in the pancreas, while a second enhancer activates *Pax6* expression in surface ectoderm (lens, cornea, and conjunctiva). A third enhancer resides in the leader sequence; it contains the sequences that direct *Pax6* expression in the neural tube. A fourth enhancer, located in an intron shortly downstream of the translation initiation site, determines the expression of *Pax6* in the retina. The *Pax6* gene illustrates the principle of enhancer modularity, wherein having multiple, separate enhancers allows a protein to be expressed in several different tissues while not being expressed at all in others.

COMBINATORIAL ASSOCIATION While there is modularity *between* enhancers, there are codependent units *within* each enhancer. Enhancers contain regions of DNA that bind transcription factors, and it is this *combination* of transcription factors that activates the gene. For instance, the pancreas-specific enhancer of the *Pax6* gene has binding sites for the Pbx1 and Meis transcription factors (see Figure 2.9C). Both need to be present in order for the enhancer to activate *Pax6* in the pancreas cells (Zhang et al. 2006).

Moreover, the product of the *Pax6* gene encodes a transcription factor that works in combinatorial partnerships with other transcription factors. Figure 2.10 shows two gene

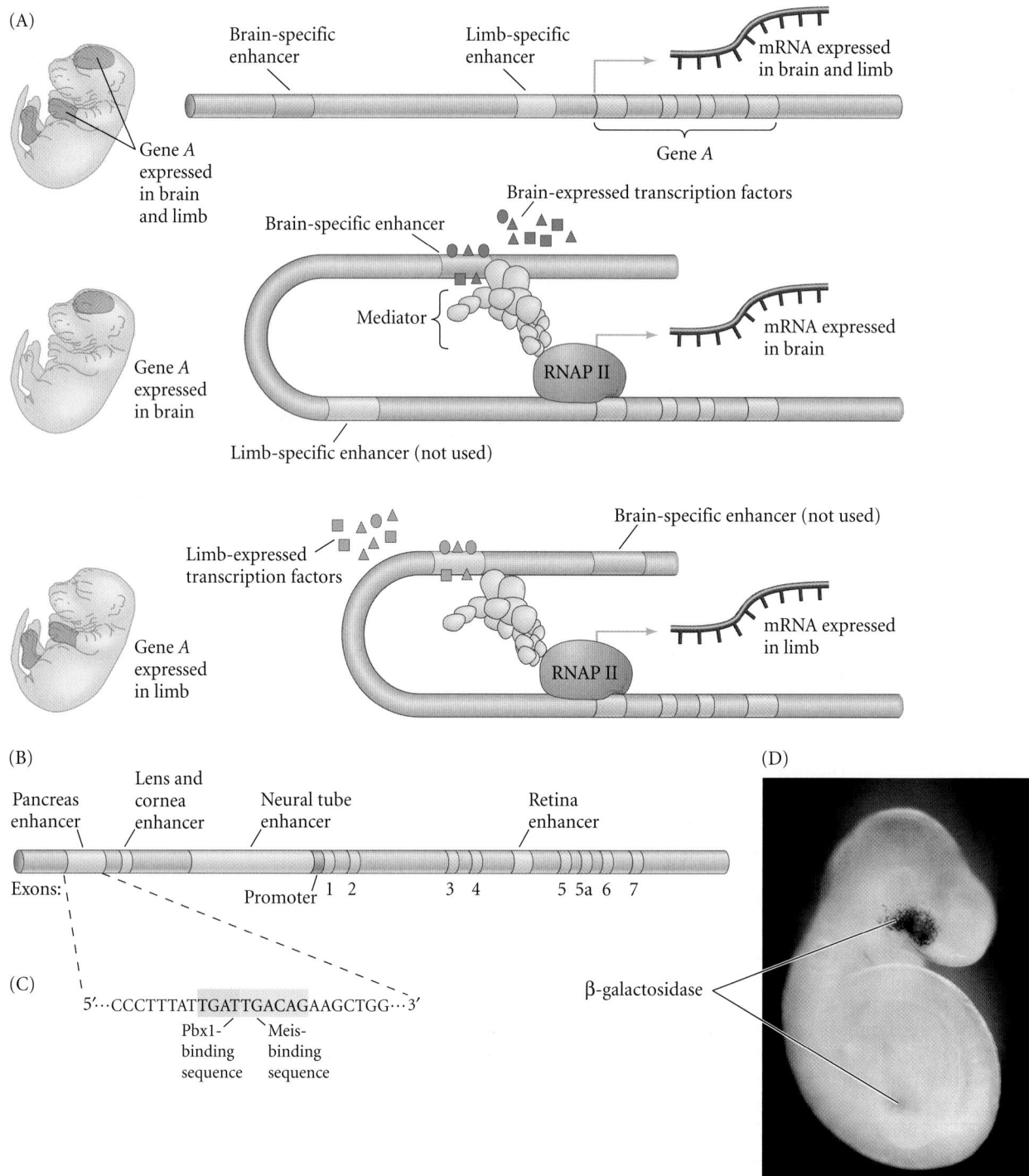

FIGURE 2.9 Enhancer region modularity. (A) Model for gene regulation by enhancers. (i) The top diagram shows the exons, introns, promoter, and enhancers of a hypothetical gene A. In situ hybridization (left) shows that gene A is expressed in limb and brain cells. (ii) In developing brain cells, brain-specific transcription factors bind to the brain enhancer, causing it to bind to the Mediator, stabilize RNA polymerase II at the promoter, and modify the nucleosomes in the region of the promoter. The gene is transcribed in the brain cells only; the limb enhancer does not function. (iii) An analogous process allows for transcription of the same gene in the cells of the limbs. The gene is not transcribed in any cell type whose transcription factors the enhancers can't bind. (B) The Pax6 protein is critical in the development of several widely different tissues. Enhancers direct *Pax6* gene expression (yellow exons 1–7) differentially in the pancreas, the lens and cornea of the eye, the retina, and the neural tube. (C) A portion of the DNA sequence of the pancreas-specific enhancer element. This sequence has binding sites for the Pbx1 and Meis transcription factors; both must be present in order to activate *Pax6* in the pancreas. (D) When the β-galactosidase reporter gene is fused to the *Pax6* enhancers for expression in the pancreas and lens/cornea, the enzyme is seen in those tissues. (A after Visel et al. 2009; D from Williams et al. 1998, courtesy of R. A. Lang.)

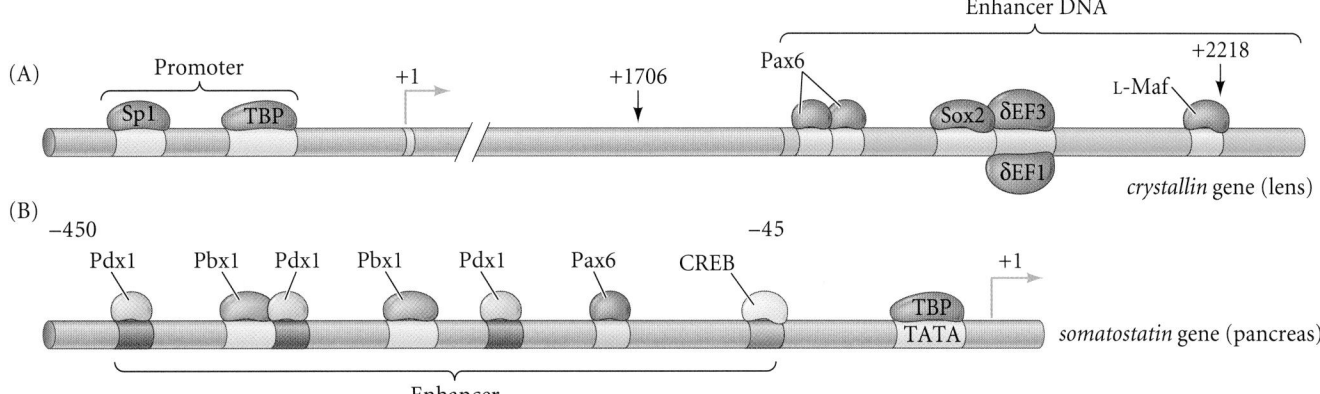

FIGURE 2.10 Modular transcriptional regulatory regions using Pax6 as an activator. (A) Promoter and enhancer of the chick δ1 lens *crystallin* gene. Pax6 interacts with two other transcription factors, Sox2 and L-Maf, to activate this gene. The protein δEF3 binds factors that permit this interaction; δEF1 binds factors that inhibit it. (B) Promoter and enhancer of the rat *somatostatin* gene. Pax6 activates this gene by cooperating with the Pbx1 and Pdx1 transcription factors. (A after Cvekl and Piatigorsky 1996; B after Andersen et al. 1999.)

enhancer regions that bind Pax6. The first is that of the chick δ1 lens *crystallin* gene (**FIGURE 2.10A**; Cvekl and Piatigorsky 1996; Muta et al. 2002). This gene encodes crystallin, a lens protein that is transparent and allows light to reach the retina. A promoter in the *crystallin* gene contains binding sites for TBP and Sp1 (basal transcriptional factors that recruit RNA polymerase II to the DNA). The gene also has an enhancer in its third intron that controls the time and place of crystallin expression. This enhancer has two Pax6-binding sites. The Pax6 protein works with the Sox2 and L-Maf transcription factors to activate the *crystallin* gene only in those head cells that are going to become lens. As we will see in Chapter 10, this means the cell (1) must be head ectoderm (which expresses Pax6), (2) must be in the region of the ectoderm capable of forming eyes (expressing L-Maf), and (3) must be in contact with the future retinal cells (which induce Sox2 expression; see Kamachi et al. 1998).

Meanwhile, Pax6 also regulates the transcription of the genes encoding insulin, glucagon, and somatostatin in the pancreas (**FIGURE 2.10B**). Here, Pax6 works in cooperation with other transcription factors such as Pdx1 (specific for the pancreatic region of the endoderm) and Pbx1 (Andersen et al. 1999; Hussain and Habener 1999). So in the absence of Pax6 the eye fails to form and the endocrine cells of the pancreas do not develop properly; these improperly developed endocrine cells produce deficient amounts of their hormones (Sander et al. 1997; Zhang et al. 2002).

Other genes are activated by Pax6 binding, and one of them is the *Pax6* gene itself. Pax6 protein can bind to a *cis*-regulatory element of the *Pax6* gene (Plaza et al. 1993). This means that once the *Pax6* gene is turned on, it will continue to be expressed, even if the signal that originally activated it is no longer present.

ENHANCERS: SUMMARY Enhancers enable genes for specific proteins to use numerous transcription factors in various combinations. Thus, *enhancers are modular* such that, for example, the *Pax6* gene is regulated by enhancers that enable it to be expressed in the eye, pancreas, and nervous system, as seen in Figure 2.9B; this is the Boolean "or" function. But *within each* cis-*regulatory module, transcription factors work in a combinatorial fashion* such that Pax6, L-Maf, and Sox2 proteins are all needed for the transcription of crystallin in the lens (see Figure 2.10A); this is the Boolean "and" function. The combinatorial association of transcription factors on enhancers leads to the spatiotemporal output of any particular gene (see Davidson 2006; Zinzen et al. 2009). This "and" function may be extremely important in activating entire groups of genes sumultaneously.

Transcription factor function

FAMILIES AND OTHER ASSOCIATIONS The science journalist Natalie Angier (1992) has written that "a series of new discoveries suggests that DNA is more like a certain type of politician, surrounded by a flock of protein handlers and advisers that must vigorously massage it, twist it, and on occasion, reinvent it before the grand blueprint of the body can make any sense at all." These "handlers and advisers" are the transcription factors. These factors can be grouped together in families based on similarities in structure (**TABLE 2.1**). The transcription factors in such a family share a common framework in their DNA-binding sites, and slight differences in the amino acids at the binding site can cause the binding site to recognize different DNA sequences.

As we have already seen, enhancers function by binding transcription factors, and each enhancer can have binding sites for several transcription factors. Transcription factors bind to the enhancer DNA with one part of the protein and use other sites on the protein to interact with one another to recruit histone-modifying enzymes.

For example, the association of the Pax6, Sox2, and L-Maf transcription factors in lens cells recruits a histone acetyltransferase that can transfer acetyl groups to the histones and dissociate the nucleosomes in that area (Yang et al. 2006). Similarly, when MITF, a transcription factor essential for ear development

TABLE 2.1 Some major transcription factor families and subfamilies

Family	Representative transcription factors	Some functions
Homeodomain:		
Hox	Hoxa1, Hoxb2, etc.	Axis formation
POU	Pit1, Unc-86, Oct-2	Pituitary development; neural fate
Lim	Lim1, Forkhead	Head development
Pax	Pax1, 2, 3, 6, etc.	Neural specification; eye development
Basic helix-loop-helix (bHLH)	MyoD, MITF, daughterless	Muscle and nerve specification; *Drosophila* sex determination; pigmentation
Basic leucine zipper (bZip)	C/EBP, AP1	Liver differentiation; fat cell specification
Zinc-finger:		
Standard	WT1, Krüppel, Engrailed	Kidney, gonad, and macrophage development; *Drosophila* segmentation
Nuclear hormone receptors	Glucocorticoid receptor, estrogen receptor, testosterone receptor, retinoic acid receptors	Secondary sex determination; craniofacial development; limb development
Sry-Sox	Sry, SoxD, Sox2	Bend DNA; mammalian primary sex determination; ectoderm differentiation

and pigment production, binds to its specific DNA sequence, it also binds a (different) histone acetyltransferase that also facilitates the dissociation of nucleosomes (Ogryzko et al. 1996; Price et al. 1998). And the Pax7 transcription factor that activates muscle-specific genes binds to the enhancer region of these genes within the muscle precursor cells. Pax7 then recruits a histone methyltransferase that methylates the lysine in the fourth position of histone H3 (H3K4), resulting in the trimethylation of this lysine and the activation of transcription (McKinnell et al. 2008). The displacement of nucleosomes along the DNA makes it possible for other transcription factors to find their binding sites (Adkins et al. 2004; Li et al. 2007).

In addition to recruiting histone modifying enzymes, transcription factors can also work by stabilizing the transcription pre-initiation complex that enables RNA polymerase II to bind to the promoter (see Figures 2.6 and 2.7). For instance, MyoD, a transcription factor that is critical for muscle cell development, stabilizes TFIIB, which supports RNA polymerase II at the promoter site (Heller and Bengal 1998). Indeed, MyoD plays several roles in activating gene expression, since it also can bind histone acetyltransferases that initiate nucleosome remodeling and dissociation (Cao et al. 2006).

One of the important consequences of the combinatorial association of transcription factors is **coordinated gene expression**. The simultaneous expression of many cell-specific genes can be explained by the binding of transcription factors by the enhancer elements. For example, many genes that are specifically activated in the lens contain an enhancer that binds Pax6. This means that all the other transcription factors might be assembled at the enhancer, but until Pax6 binds, they cannot activate the gene. Similarly, many of the coexpressed muscle-specific genes contain enhancers that bind the Mef2 transcription factor; and the enhancers on genes encoding pigment-producing enzymes bind MITF

(see Davidson 2006). In some instances, entire ensembles of transcription factors appear to direct simultaneous gene transcription. Junion and colleagues have shown, for example, that a particular ensemble of five transcription factors is bound on hundreds of enhancers that are active in the developing *Drosophila* heart muscle cells (Junion et al. 2012).

TRANSCRIPTION FACTOR DOMAINS Transcription factors have three major domains. The first is a **DNA-binding domain** that recognizes a particular DNA sequence in the enhancer. For instance, MITF, a transcription factor involved in ear and pigment cell development, recognizes the DNA sequence CATGTG, whereas Pax6 binds to a longer site, CAATTAGTCACGCTTGA (Askan and Goding 1998; Wolf et al. 2009). The DNA-binding domain of MITF is close to the amino-terminal end of the protein and contains numerous basic amino acids that make contact with the DNA (Hemesath et al. 1994; Steingrímsson et al. 1994). This assignment was confirmed by the discovery of various human and mouse mutations that map within the DNA-binding site for MITF and that prevent the attachment of the MITF protein to the DNA. Sequences for MITF binding have been found in the regulatory regions of genes encoding several pigment-cell-specific enzymes of the tyrosinase family (Bentley et al. 1994; Yasumoto et al. 1997). Without MITF, these proteins are not synthesized properly, and melanin pigment is not made. These *cis*-regulatory regions all contain the same 11-base-pair sequence, including the core sequence (CATGTG) that is recognized by MITF.

The second domain is a **trans-activating domain** that activates or suppresses the transcription of the gene whose promoter or enhancer it has bound. Usually, this *trans*-activating domain enables the transcription factor to interact with the proteins involved in binding RNA polymerase

II (such as TFIIB or TFIIE; see Sauer et al. 1995) or with enzymes that modify histones. MITF contains such a domain of amino acids in the center of the protein. When the MITF dimer is bound to its target sequence in the enhancer, the *trans*-activating region is able to bind a transcription-associated factor (TAF), p300/CBP. The p300/CPB protein is a histone acetyltransferase enzyme that can transfer acetyl groups to each histone in the nucleosomes (Ogryzko et al. 1996; Price et al. 1998). Acetylation of the nucleosomes destabilizes them and allows the genes for pigment-forming enzymes to be expressed.

Finally, there is usually a **protein-protein interaction domain** that allows the transcription factor's activity to be modulated by TAFs or other transcription factors. MITF has a protein-protein interaction domain that enables it to dimerize with another MITF protein (Ferré-D'Amaré et al. 1993). The resulting homodimer (i.e., two identical protein molecules bound together) is the functional protein that can bind to enhancer DNA and activate the transcription of certain genes (**FIGURE 2.11**).

Keeping the right genes on or off

The modifications of histones can also signal the recruitment of the proteins that can retain the memory of transcriptional state from generation to generation through mitosis. These are the proteins of the **Trithorax** and **Polycomb** families. When bound to the nucleosomes of active genes, Trithorax proteins keep these genes active, whereas Polycomb proteins, which bind to condensed nucleosomes, keep the genes in a repressed state.

The Polycomb proteins fall into two categories that act sequentially in repression. The first set has histone methyltransferase activities that methylate lysines H3K27 and H3K9 to repress gene activity. In many organisms, this repressed state is stabilized by the activity of a second set of Polycomb factors, which bind to the methylated tails of histone 3 and keep the methylation active and also methylate adjacent nucleosomes, thereby forming tightly packed repressive complexes (Grossniklaus and Paro 2007; Margueron et al. 2009).

The Trithorax proteins help retain the memory of activation; they act to counter the effect of the Polycomb proteins. Trithorax proteins can modify the nucleosomes or alter their positions on the chromatin, allowing transcription factors to bind to the DNA previously covered by them. Other Trithorax proteins keep the H3K4 lysine trimethylated (preventing its demethylation into a dimethylated, repressed state; Tan et al. 2008).

PIONEER TRANSCRIPTION FACTORS: BREAKING THE SILENCE Finding a promoter is not easy because the DNA is usually so wound up that the promoter sites are not accessible. Indeed, more than *6 feet* of DNA is packaged into the chromosomes of each human cell nucleus (Schones and Zhao 2008).

How can a transcription factor find its binding site, given that the enhancer might be covered by nucleosomes? This is the job of certain transcription factors that penetrate

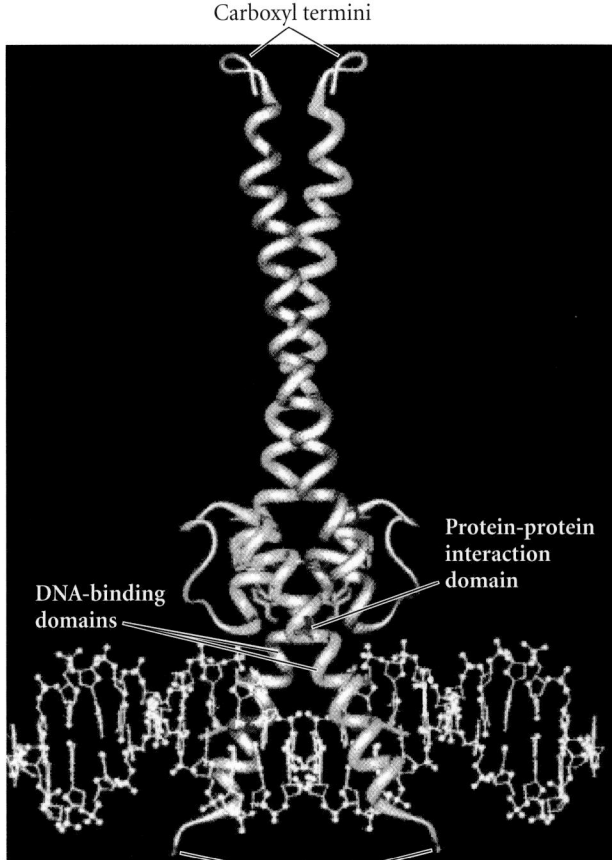

Carboxyl termini

Protein-protein interaction domain

DNA-binding domains

Amino termini

FIGURE 2.11 Three-dimensional model of the homodimeric transcription factor MITF (one protein shown in red, the other in blue) binding to a promoter element in DNA (white). The amino termini are located at the bottom of the figure and form the DNA-binding domains that recognize an 11-base-pair sequence of DNA having the core sequence CATGTG. The protein-protein interaction domain is located immediately above. MITF has the basic helix-loop-helix structure found in many transcription factors. The carboxyl end of the molecule is thought to be the *trans*-activating domains that bind the p300/CBP transcription-associated factor (TAF). (From Steingrímsson et al. 1994, courtesy of N. Jenkins.)

repressed chromatin and bind to their enhancer DNA sequences (Cirillo et al. 2002; Berkes et al. 2004). They have been called "pioneer" transcription factors, and they appear to be critical in establishing certain cell lineages. One of these transcription factors is FoxA1, which binds to certain enhancers and opens up the chromatin to allow other transcription factors access to the promoter (Lupien et al. 2008; Smale 2010). FoxA1 is extremely important in specifying liver cells, remaining bound to the DNA during mitosis and providing a mechanism to reestablish normal transcription in presumptive liver cells (Zaret et al. 2008). Another pioneer transcription factor is the Pax7 protein mentioned above. It activates muscle-specific gene transcription in a population of muscle stem cells by binding to its DNA recognition sequence and being stabilized there by dimethylated H3K4

Reprogramming Cells: Changing Cell Differentiation through Embryonic Transcription Factors

In some of the original cloning experiments, Briggs and King (1952) and John Gurdon (1962) were able to reprogram the nuclei of larval frog fibroblast or gut cells to become an entire frog. However, they did not know what proteins were doing this reprogramming. As we will see later, transcription factors are the agents responsible for giving cells their identities. By changing the expression of certain transcription factors, an entirely new network of gene expression can be effected.

Zhou and colleagues (2008) used three transcription factors to convert *exocrine* pancreatic cells (which make amylase, chymotrypsin, and other digestive enzymes) into insulin-secreting *endocrine* pancreatic β cells. The researchers infected the pancreases of living 2-month-old mice with harmless viruses containing the genes for three transcription factors: Pdx1, Ngn3, and MafA.

The Pdx1 protein stimulates the outgrowth of the digestive tube that results in the pancreatic buds. This protein is found throughout the pancreas and is critical in specifying that organ's endocrine cells, as well as in activating genes that encode endocrine proteins. Ngn3 is a transcription factor found in endocrine, but not exocrine, pancreatic cells. MafA, a transcription factor regulated by glucose levels, is found only in pancreatic β cells (i.e., those cells that make insulin) and can activate transcription of the insulin gene. MafA is also present in lens cells, but in the pancreas it is specific for insulin-secreting β-cells.

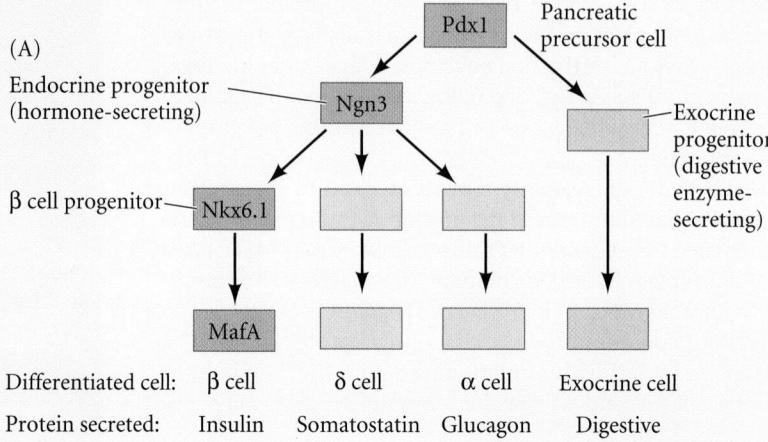

FIGURE 2.12 Pancreatic lineage and transcription factors. (A) Simplified depiction of the role of transcription factors in pancreatic islet β cell development. Pdx1 protein is critical for specifying a certain group of endoderm cells as pancreas precursors (dark purple lineage). Those descendants of Pdx1-expressing cells that express Ngn3 become the endocrine (hormone-secreting) lineages (shades of purple), whereas those that do not express Ngn3 become the exocrine (digestive enzyme-secreting) lineage of the pancreas (gold). Types of hormone-secreting cells in the pancreatic islets include the insulin-secreting β cells, the somatostatin-secreting δ cells, and the glucagon-secreting α cells. Those cells destined to become β cells express the Nkx6.1 transcription factor, which in turn will activate the gene for the MafA transcription factor found in the insulin-producing β cells. (B) New pancreatic β cells arise in adult mouse pancreas in vivo after viral delivery of three transcription factors (Ngn3, Pdx1, and MafA). Virally infected exocrine cells are detected by their expression of nuclear green fluorescent protein. Newly induced β cells are detected by insulin staining (red). Their overlap produces yellow. The nuclei of all pancreatic cells are stained blue. (A after Matsuoka et al. 2004; B courtesy of D. Melton.)

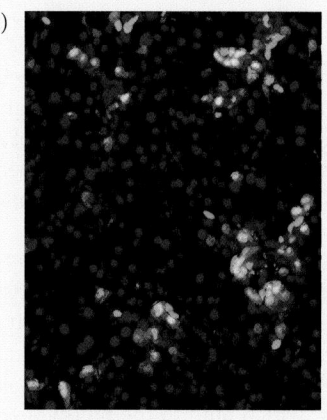

---

on the nucleosomes. It then recruits the histone methyltransferase that converts the dimethylated H3K4 into the trimethylated H3K4 associated with active transcription (McKinnell et al. 2008).

SILENCERS Silencers are DNA regulatory elements that actively repress the transcription of a particular gene. They can be viewed as "negative enhancers," and they can silence spatially (in particular cell types) or temporally (at particular times). For instance, in the mouse there is a DNA sequence that prevents a promoter's activation in any tissue *except* neurons. This sequence, given the name **neural restrictive silencer element** (**NRSE**), has been found in several mouse genes whose expression is limited to the nervous system: those encoding synapsin I, sodium channel type II, brain-derived neurotrophic factor, Ng-CAM, and L1. The protein that binds to the NRSE is a zinc-finger transcription factor called **neural restrictive silencer factor** (**NRSF**, sometimes called REST). NRSF appears to be expressed in every cell that is *not* a mature neuron (Chong et al. 1995; Schoenherr and

SIDELIGHTS & SPECULATIONS (continued)

Pdx1, Ngn3, and MafA activate other transcription factors that work in concert to turn a pancreatic endodermal cell into an insulin-secreting β cell. Zhou and colleagues found that, of all the transcription factor genes tested, these three were the only ones that were crucial for the conversion (FIGURE 2.12). Converted pancreas cells looked identical to normal β cells, and like normal β cells, they were capable of secreting insulin and curing an experimentally induced diabetes in mice.

This study has opened the door to an entirely new field of regenerative medicine, illustrating the possibilities of changing one adult cell type into another by using the transcription

factors that had made the new cell type in the embryo. In some instances, the developmental histories of the cells can be very distant. For instance, adult mouse skin fibroblasts (the mesodermally derived connective tissue of the skin) can be transformed into endodermal hepatocyte-like cells by the addition of only two liver transcription factors (Hnf4α and FoxA1). These induced hepatocytes make several liver-specific proteins and are able to substitute for liver cells in adult mice (Sekiya and Suzuki 2011). Indeed, several laboratories (Caiazzo et al. 2011; Pfeister et al. 2011; Qiang et al. 2011) been able to "reprogram" adult human and mouse fibroblasts into functional dopaminergic

neurons (i.e., the type of nerve cell that degenerates in Parkinson disease) by the addition of three particular activated transcription factor genes to adult skin cells. Other laboratories (Son et al. 2011) have used a different mix of transcription factors to convert adult human fibroblasts into functional spinal motor neurons (of the type that degenerate in Lou Gehrig syndrome). These "induced neurons" had the electrophysiological signatures of spinal nerves and formed synapses with muscle cells.

● **See VADE MECUM**
Transdetermination in *Drosophila*

Anderson 1995). When NRSE is deleted from particular neural genes, these genes are expressed in non-neural cells (FIGURE 2.13; Kallunki et al. 1995, 1997). Thus, neural-specific genes are actively repressed in non-neural cells.

A recently discovered "temporal silencer" may play a role in regulating the human globin genes. In most people, a fetal

(A)

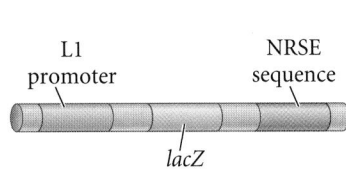
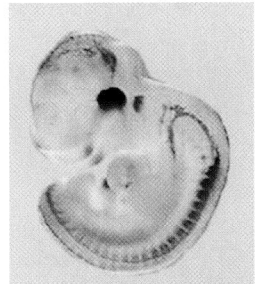

(B)

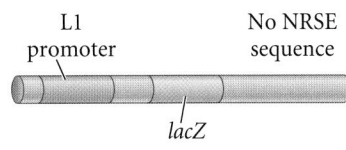
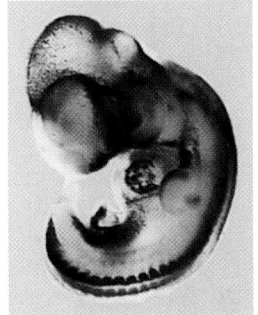

FIGURE 2.13 A silencer represses gene transcription. (A) Mouse embryo containing a transgene composed of the L1 promoter, a portion of the neuron-specific *L1* gene, and a *lacZ* gene fused to the *L1* second exon, which contains the NRSE sequence. (B) Same-stage embryo with a similar transgene but lacking the NRSE sequence. Dark areas reveal the presence of β-galactosidase (the *lacZ* product). (Photographs from Kallunki et al. 1997.)

globin gene is active from about week 12 until birth. Then, around the time of birth, the fetal globin gene is turned off and the adult globin gene is activated. However, some families show a hereditary persistence of fetal hemoglobin, with the fetal globin genes remaining active in the adults. Some of these families have a mutation in a region of DNA that usually silences the fetal globin gene at birth. In most people, this silencer contains binding sites for the transcription factors GATA1 and BCL11A, whose combination on the DNA recruits histone modification enzymes. This causes the formation of deacetylated and repressive (H3K27me3-containing) nucleosomes (Sankaran et al. 2011).

INSULATORS The boundaries of gene expression appear to be set by DNA sequences called **insulators**. Insulator sequences limit the range in which an enhancer can activate gene expression. They thereby "insulate" a promoter from being activated by another gene's enhancers. Insulator DNA binds a transcription factor called CTCF. While the mechanism by which CTCF prevents promoter activation remains elusive, recent studies have shown that bound CTCF forms a complex with cohesin (Wendt et al. 2008; Wood et al. 2010). This CTCF-cohesin complex may bind to the enhancer-bound Mediator, thereby preventing the enhancer from activating the adjacent promoter.

Mechanisms of Differential Gene Transcription

During the twentieth century we found the actors in the drama of gene transcription, but it wasn't until the twenty-first century that we found their scripts. The ability to identify protein-bound enhancers using ChIP-Seq technology showed that there are different types of promoters, and that they use different scripts to transcribe their genes.

Finding DNA regulatory sequences

How does one locate the places on the gene where a particular transcription factor binds, or where nucleosomes with specific modifications are localized? How does one determine the "regulatory architecture" of individual genes and of the entire genome? Although many techniques have been used, one of the newest and most important methods is called **ChIP-Seq**, for *Ch*romatin *Immunoprecipitation-Sequencing* (Johnson et al. 2007; Jothi et al. 2008). ChIP-Seq is based on two highly specific interactions: (1) the binding of a transcription factor or a modified nucleosome to very particular sequences of DNA (such as enhancer elements), and (2) the binding of antibody molecules specifically to the transcription factor or modified histone being studied (**FIGURE 2.14**; Liu et al. 2010).

In the first step of ChIP-Seq, chromatin is isolated and the proteins are crosslinked (usually by glutaraldehyde or formaldehyde) to the DNA to which they are bound. This prevents the nucleosome or transcription factors from dissociating from the DNA. After crosslinking, the DNA is fragmented (usually by sonication, but sometimes by enzymes) into pieces about 500 nucleotides long. The next step is to bind these proteins with an antibody that recognizes only that particular protein. Indeed, these antibodies are so specific that an antibody that recognizes histone 3 when it is dimethylated at position 4 will not recognize trimethylated histone 3 at that same position. The antibodies can be precipitated out of solution (often with magnetic beads that bind to antibodies), and they will bring down to the bottom of the test tube any DNA fragments bound by the protein of interest. These DNA fragments, once separated from the proteins, are amplified and can be sequenced and mapped to the entire genome. In this way, the DNA sequences bound specifically by particular transcription factors or nucleosomes containing modified histones can be identified very precisely.

Differentiated proteins from high and low CpG-content promoters

ChIP-Seq has overturned many of our hypotheses concerning the mechanisms by which promoters and enhancers regulate differential gene expression. It turns out that not all promoters are the same. Rather, there are two general classes of promoters that use different methods for controlling transcription. These promoter types are catalogued as having either a relatively high or a relatively low number of CpG sequences at which DNA methylation can occur.

- **High CpG-content promoters** (**HCPs**) are usually found in "developmental control genes," where they regulate synthesis of the transcription factors and other developmental regulatory proteins used in the *construction* of the organism (Zeitinger and Stark 2010; Zhou et al. 2011). The default state of these promoters is "on," and they have to be actively repressed by *histone* methylation (**FIGURE 2.15A**).

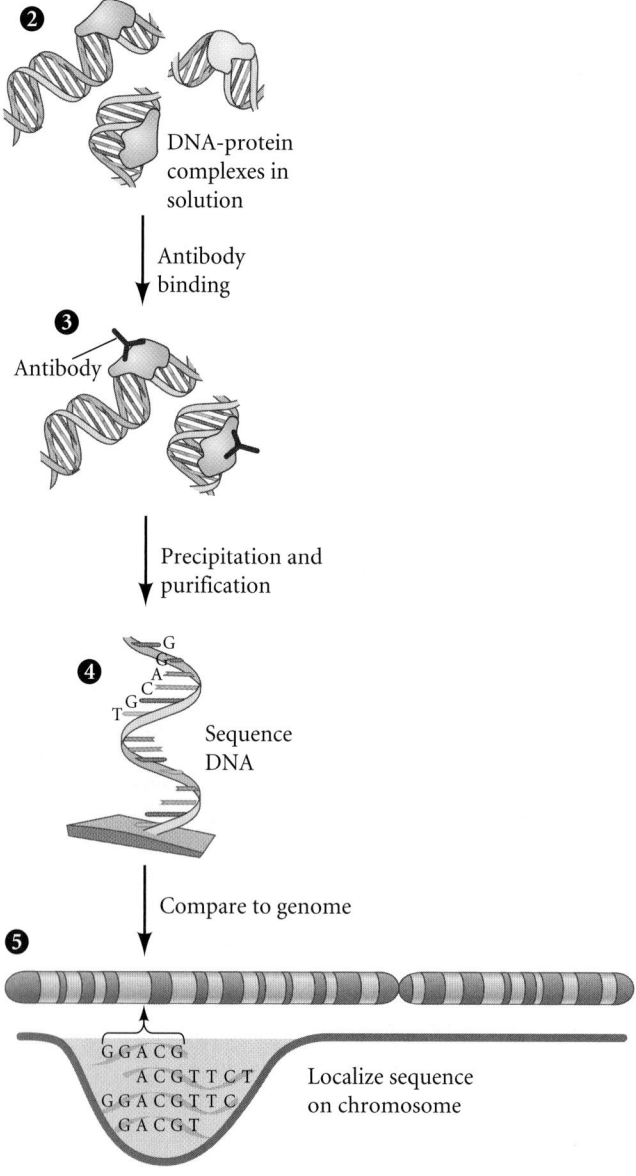

FIGURE 2.14 Chromatin immunoprecipitation-sequencing (ChIP-Seq). Chromatin is isolated from the cell nuclei. The chromatin proteins are crosslinked to their DNA binding sites, and the DNA is fragmented into small pieces. Antibodies bind to specific chromatin proteins, and the antibodies—with whatever is bound to them—are precipitated out of solution. The DNA fragments associated with the precipitated complexes are purified from the proteins and sequenced. These sequences can be compared with the genome maps to give a precise localization of what genes these proteins may be regulating. (After Szalkowski and Schmid 2011.)

❶ Chromatin

DNA with interacting proteins

Crosslink and shear

❷ DNA-protein complexes in solution

Antibody binding

❸ Antibody

Precipitation and purification

❹ Sequence DNA

Compare to genome

❺

GGACG
ACGTTCT
GGACGTTC
GACGT

Localize sequence on chromosome

(A) High CpG-content promoters (HCPs)

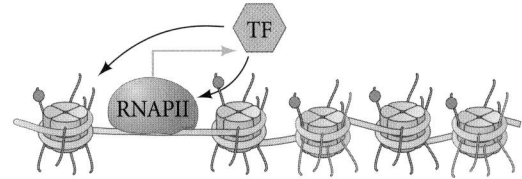

Active

'Open' chromatin RNAPII initiation (default)

Poised (intermediate state)

Bivalent chromatin modifications

Inactive

Repressed by histone modification

- H3K4me3   ○ H3K4me2   ● H3K27me3

(B) Low CpG-content promoters (LCPs)

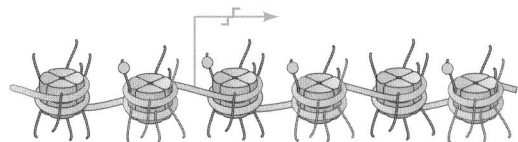

Selective use

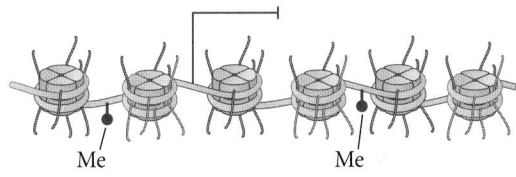

H3K4me2 chromatin modifications

Me   Me

DNA methylation, no transcription (default)

FIGURE 2.15 Chromatin regulation in HCPs and LCPs. Promoters with high and low CpG content have different modes of regulation. (A) HCPs are typically in an *active* state, with unmethylated DNA and nucleosomes rich in H3K4me3. The open chromatin allows RNA polymerase II to bind. The *poised* state of HCPs is bivalent, having both activating (H3K4me3) and repressive (H3K27me3) modifications of the nucleosomes. RNA polymerase II can bind but not transcribe. The *repressed* state is characterized by repressive histone modification, but not by extensive DNA methylation. (B) *Active* LCPs, like HCPs, have nucleosomes rich in H3K4me3 and low methylation but require stimulation by transcription factors (TF). *Poised* LCPs are capable of being activated by transcription factors have relatively unmethylated DNA and nucleosomes enriched in H3K4me2. In their usual state, LCPs are *repressed* by methylated DNA nucleosomes rich in H3K27me3. (After Zhou et al. 2011.)

- **Low CpG-content promoters** (**LCPs**) are usually found in those genes whose products characterize mature cells (e.g., the globins of red blood cells, the hormones of pancreatic cells, and the enzymes that carry out the normal maintenance functions of the cell). The default state of these promoters is "off," but they can be activated by transcription factors (**FIGURE 2.15B**). The nucleosomes on these promoters have relatively few modified histones in the repressed state. Rather, their CpG sites are usually methylated, and this methylation is critical for preventing transcription. When the DNA becomes unmethylated, the histones become modified with H3K4me3 and disperse so that RNA polymerase II can bind.

DNA methylation, another key on/off switch of transcription

Earlier in this chapter we discussed *histone* methylation and its importance for transcription. Now we look at how the *DNA itself* can be methylated to regulate transcription. Generally speaking, the promoters of inactive genes are methylated at certain cytosine residues, and the resulting methylcytosine stabilizes nucleosomes and prevents transcription factors from binding. This is especially important in the LCP promoters.

It is often assumed that a gene contains exactly the same nucleotides whether it is active or inactive; that is, a β-globin gene that is activated in a red blood cell precursor has the same nucleotides as the inactive β-globin gene in a fibroblast or retinal cell of the same animal. However, it turns out there is in fact a subtle difference. In 1948, R. D. Hotchkiss discovered a "fifth base" in DNA, 5-methylcytosine. In vertebrates, this base is made enzymatically after DNA is replicated. At this time, about 5% of the cytosines in mammalian DNA are converted to 5-methylcytosine (**FIGURE 2.16A**). This conversion can occur only when the cytosine residue is followed by a guanosine—in other words, at a CpG sequence (as we will soon see, this restriction is important). Numerous studies have shown that the degree to which the cytosines of a gene are methylated can control the level of the gene's transcription. Cytosine methylation appears to be a major mechanism

(A)

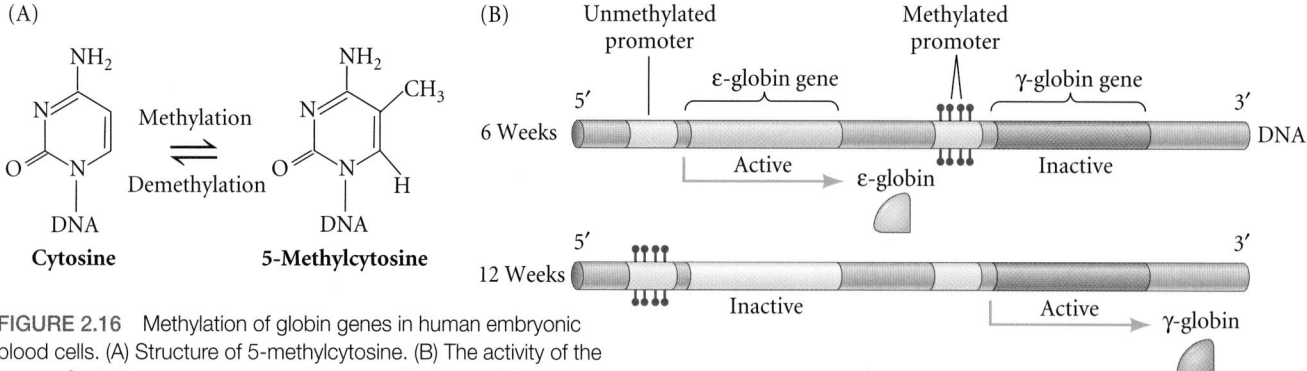

FIGURE 2.16 Methylation of globin genes in human embryonic blood cells. (A) Structure of 5-methylcytosine. (B) The activity of the human β-globin genes correlates inversely with the methylation of their promoters. (After Mavilio et al. 1983.)

of transcriptional regulation among vertebrates; however, some other species (*Drosophila* and nematodes among them) do not methylate their DNA.

In vertebrates, the presence of methylated cytosines in a gene's promoter correlates with the repression of transcription from that gene. In developing human and chick red blood cells, for example, the DNA of the globin gene promoters is almost completely unmethylated, whereas the same promoters are highly methylated in cells that do not produce globins. Moreover, the methylation pattern changes during development (FIGURE 2.16B). The cells that produce hemoglobin in the human embryo have unmethylated promoters in the genes encoding the ε-globins ("embryonic globin chains") of embryonic hemoglobin. These promoters become methylated in the fetal tissue, as the genes for fetal-specific γ-globin (rather than the embryonic chains) become activated (van der Ploeg and Flavell 1980; Groudine and Weintraub 1981; Mavilio et al. 1983). Similarly, when fetal globin gives way to adult (β) globin, promoters of the fetal (γ) globin genes become methylated.

MECHANISMS BY WHICH DNA METHYLATION BLOCKS TRANSCRIPTION DNA methylation appears to act in two ways to repress gene expression. First, it can block the binding of transcription factors to enhancers. Several transcription factors can bind to a particular sequence of unmethylated DNA, but they cannot bind to that DNA if one of its cytosines is

methylated (FIGURE 2.17). Second, a methylated cytosine can recruit the binding of proteins that facilitate the methylation or deacetylation of histones, thereby stabilizing the nucleosomes. For instance, methylated cytosines in DNA can bind particular proteins such as MeCP2. Once connected to a methylated cytosine, MeCP2 binds to histone deacetylases and histone methyltransferases, which, respectively, remove acetyl groups (FIGURE 2.18A) and add methyl groups (FIGURE 2.18B) on the histones. As a result, the nucleosomes form tight complexes with the DNA and don't allow other transcription factors and RNA polymerases to find the genes. Other proteins, such as HP1 and histone H1, will bind and aggregate methylated histones (Fuks 2005; Rupp and Becker 2005). In this way, repressed chromatin becomes associated with regions where there are methylated cytosines.

INHERITANCE AND STABILIZATION OF DNA METHYLATION PATTERNS Another enzyme recruited to the chromatin by MeCP2 is DNA methyltransferase-3 (Dnmt3). This enzyme methylates previously unmethylated cytosines on the DNA. In this way, a relatively large region can be repressed. The newly established methylation pattern is then transmitted to the next generation by DNA methyltransferase-1 (Dnmt1). This enzyme recognizes methyl cytosines on one strand of DNA and places methyl groups on the newly synthesized strand opposite it (FIGURE 2.19; see Bird 2002; Burdge et al. 2007). This is why it is necessary for the C to be next to a G in the sequence. Thus, in each cell division, the pattern of DNA methylation can be maintained. The newly synthesized (unmethylated) strand will become properly methylated when Dnmt1 binds to a methyl C on the old CpG sequence and methylates the cytosine of the CpG sequence on the complementary strand. In this way, once the

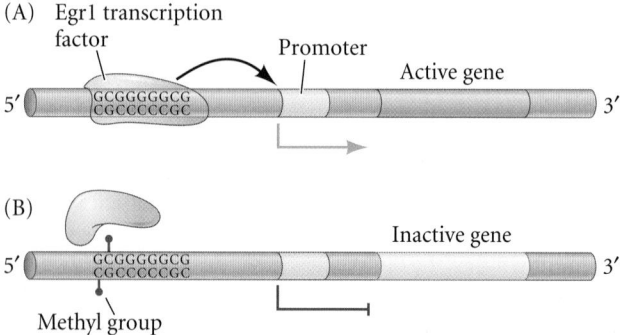

FIGURE 2.17 DNA methylation can block transcription by preventing transcription factors from binding to the enhancer region. (A) The Egr1 transcription factor can bind to specific DNA sequences such as 5'...GCGGGGGCG...3', helping activate transcription of those genes. (B) If the first cytosine residue is methylated, however, Egr1 will not bind and the gene will remain repressed. (After Weaver et al. 2005.)

(A)

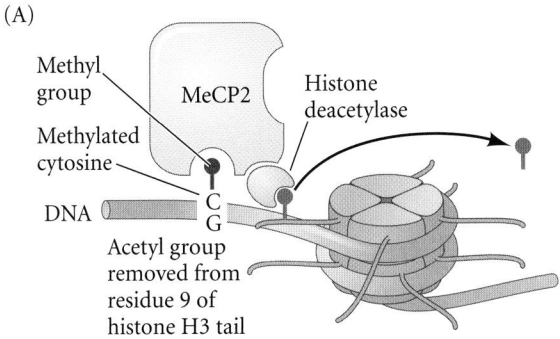

(B)

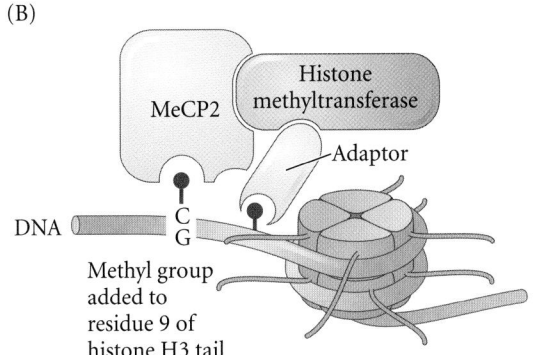

FIGURE 2.18 Modifying nucleosomes through methylated DNA. MeCP2 recognizes the methylated cytosines of DNA. It binds to the DNA and is thereby able to recruit (A) histone deacetylases (which take acetyl groups off the histones) or (B) histone methyltransferases (which add methyl groups to the histones). Both modifications promote the stability of the nucleosome and the tight packing of DNA, thereby repressing gene expression in these regions of DNA methylation. (After Fuks 2005.)

Dnmt3L. It actually has lost its enzymatic activity, but it can still bind avidly to the amino end of histone H3. However, if the lysine at H3K4 is methylated, it will not bind. Once bound, however, it recruits and/or activates a Dnmt3 methyltransferase to methylate the cytosines on nearby CG pairs (Fan et al. 2007; Ooi et al. 2007).

"Poised" chromatin

Promoters can exist in three major states: an active state, a repressed state, and an intermediate, or "poised" state (see Figure 2.15). This poised chromatin state allows for a rapid response to developmental signals, and it characterizes the high CpG-content promoters (HCPs) that regulate the transcription of developmental control genes. The DNA of HCPs is relatively unmethylated, and nucleosomes tend to be enriched with "activating" H3K4me3. As a result, RNA polymerase II is usually already present on HCPs (Hon et al. 2009; Ernst and Kellis 2010). Indeed, there is often a small, truncated transcript of nRNA already initiated (but not completed) at these promoters (see Figure 2.15). DNA methylation does not appear to play a major role in HCP regulation. Rather, HCPs can be repressed by modifying the histone 3 to H3K27me3, which recruits Polycomb repressive complex 2 (Peng et al. 2009; Li et al. 2010), a complex that appears to inhibit further RNA polymerase II binding as well as preventing elongation of the existing nRNA transcripts.

HCPs become poised for activation by having nucleosomes containing both H3K4me3 (activating) and H3K27me3 (repressive) histones (this is sometimes called a bivalent state). Thus, the rate-limiting step of RNA transcription from HCPs is not the *initiation* of transcription (as it is in the LCPs), but RNA *elongation*. This "poised for activation" state may be predominant during early development (Muse et al. 2007; Zeitlinger 2007). The genes may be put into an active state by specific transcription factors that activate the elongation of RNA transcripts (Peterlin and Price 2006), and they may be repressed later in development (Hargreaves et al. 2009; Ramirez-Carrozzi et al. 2009; Rahl et al. 2010).

These transcription factors may act on several levels to promote RNA elongation. In mammalian cells, where about 30% of the genes have promoters that already contain RNA polymerase II and nascent RNA chains (Core and Lis 2008), transcription factors appear to act through the Mediator complex. Here transcription is paused because the RNA polymerase II remains tethered to TFIID, which remains bound to the promoter sequence of the gene. This tethering is accomplished by the Mediator complex, especially by the Mediator protein Med26, which binds the Mediator to

DNA methylation pattern is established in a cell, it can be stably inherited by all the progeny of that cell.

Reinforcement between repressive chromatin and repressive DNA has also been observed. Just as methylated DNA is able to attract proteins that deacetylate histones and attract H1 linker histones (both of which will stabilize nucleosomes), so repressed states of chromatin are able to recruit enzymes that methylate DNA. DNA methylation patterns during gametogenesis depend in part on the DNA methyltransferase

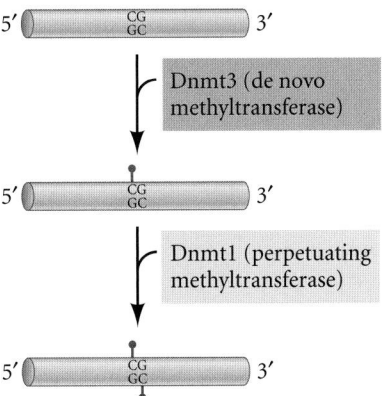

FIGURE 2.19 Two DNA methyltransferases are critically important in modifying DNA. The "de novo" methyltransferase Dnmt3 can place a methyl group on unmethylated cytosines. The "perpetuating" methyltransferase, Dnmt1, recognizes methylated Cs on one strand and methylates the C on the CG pair on the opposite strand.

Genomic Imprinting and DNA Methylation

DNA methylation has explained at least one very puzzling phenomenon, that of genomic imprinting (Ferguson-Smith 2011). It is usually assumed that the genes one inherits from one's father and the genes one inherits from one's mother are equivalent. In fact, the basis for Mendelian ratios (and the Punnett square analyses used to teach them) is that it does not matter whether the genes came from the sperm or from the egg. But in mammals, there are about 100 genes for which it *does* matter (International Human Epigenome Consortium).* In these cases, the chromosomes from the male and the female are not equivalent; only the sperm-derived or only the egg-derived allele of the gene is expressed. This means that a severe or lethal condition arises if a mutant allele is derived from one parent, but that the same mutant allele will have no deleterious effects if inherited from the other parent. In some of these cases, the nonfunctioning gene has been rendered inactive by DNA methylation. (This means that a mammal must have both a male parent and a female parent. Unlike sea urchins, flies, and even some turkeys, mammals cannot experience parthenogenesis, or "virgin birth.") The methyl groups are placed on the DNA during spermatogenesis and oogenesis by a series of enzymes that first take the existing methyl groups off the chromatin and then place new sex-specific ones on the DNA (Ciccone et al. 2009; Gu et al. 2011).

As described in this chapter, methylated DNA is associated with stable DNA silencing, either (1) by interfering with the binding of gene-activating transcription factors or (2) by recruiting repressor proteins that stabilize nucleosomes in a restrictive manner along the gene. The presence of a methyl group in the minor groove of DNA can prevent certain

*A list of imprinted mouse genes is maintained at www.har.mrc.ac.uk/research/genomic_imprinting/introduction.html.

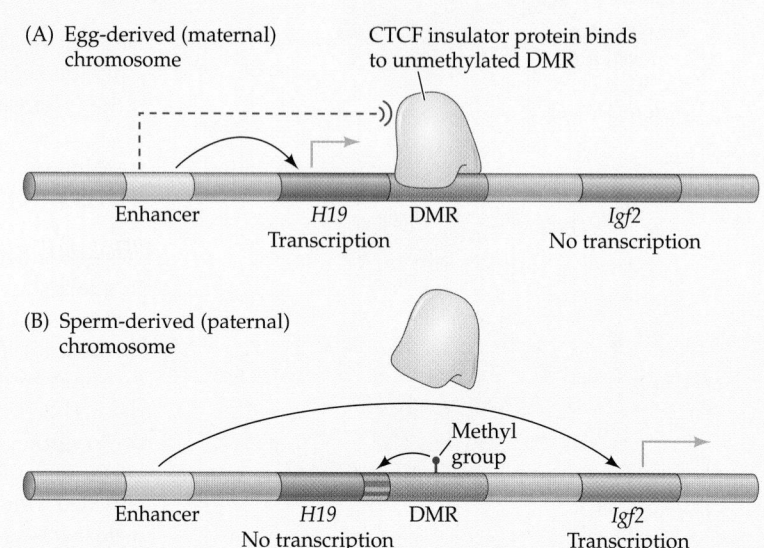

FIGURE 2.20 Regulation of the imprinted *Igf2* gene in the mouse. This gene is activated by an enhancer element it shares with the *H19* gene. The differentially methylated region (DMR) is a sequence located between the enhancer and the *Igf2* gene, and is found on both sperm- and egg-derived chromosomes. (A) In the egg-derived chromosome, the DMR is unmethylated. The CTCF insulator protein binds to the DMR and blocks the enhancer signal. (B) In the sperm-derived chromosome, the DMR is methylated. The CTCF insulator protein cannot bind to the methylated sequence, and the signal from the enhancer is able to activate *Igf2* transcription.

transcription factors from binding to the DNA, thereby preventing the gene from being activated (Watt and Molloy 1988).

For example, during early embryonic development in mice, the *Igf2* gene (for insulin-like growth factor) is transcribed only from the sperm-derived (paternal) chromosome 7. The egg-derived (maternal) *Igf2* gene does not function during embryonic development. This is because the CTCF protein is an inhibitor that can block the promoter from getting activation signals from enhancers. It binds to a region near the *Igf2* gene in females because this region is not methylated. Once bound, it prevents the maternally derived *Igf2* gene from functioning. In the paternally derived chromosome 7, the region where CTCF would bind is methylated. CTCF cannot bind and the gene is not inhibited from

functioning (**FIGURE 2.20**; Bartolomei et al. 1993; Ferguson-Smith et al. 1993; Bell and Felsenfeld 2000).

In humans, misregulation of *IGF2* methylation causes Beckwith-Wiedemann growth syndrome. Although DNA methylation is the mechanism for imprinting this gene in both mice and humans, the mechanisms responsible for the differential *Igf2* methylation between sperm and egg appear to be very different in the two species (Ferguson-Smith et al. 2003; Walter and Paulsen 2003). Differential methylation is one of the most important mechanisms of epigenetic changes and is a reminder that an organism cannot be explained solely by its genes. One needs knowledge of developmental parameters (such as whether the gene was modified by the gamete transmitting it) as well as genetic ones.

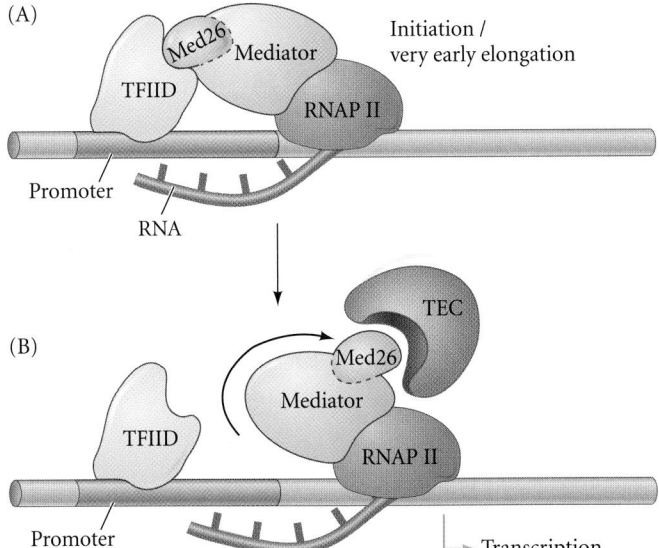

(A)

Med26 Mediator

TFIID

RNAP II

Initiation /
very early elongation

Promoter

RNA

(B)

TEC

Med26

Mediator

TFIID

RNAP II

Promoter

Transcription

FIGURE 2.21 Model for the regulation of RNA elongation by the Mediator protein Med26. In the initiation and early elongation phase of transcription, the Mediator tethers RNA polymerase II to TFIID at the promoter through its Med26 protein. The Med26 protein can also bind to the transcription elongation complex (TEC). Transcription elongation can be reactivated by transcription factors promoting the binding of Med26 to the TEC rather than to TFIID. (After Takahashi et al. 2011.)

● **See WEBSITE 2.7** Chromatin diminution
● **See WEBSITE 2.8** The nuclear envelope's roles in gene regulation

Differential RNA Processing

The regulation of gene expression is not confined to the differential transcription of DNA. Even if a particular RNA transcript is synthesized, there is no guarantee that it will create a functional protein in the cell. To become an active protein, the nuclear RNA must be (1) processed into messenger RNA by the removal of introns, (2) translocated from the nucleus to the cytoplasm, and (3) translated by the protein-synthesizing apparatus. In some cases, the synthesized protein is not in its mature form and must be (4) posttranslationally modified to become active. Regulation during development can occur at any of these steps.

The essence of differentiation is the production of different sets of proteins in different types of cells. In bacteria, differential gene expression can be effected at the levels of transcription, translation, and protein modification. In eukaryotes, however, another possible level of regulation exists—namely, control at the level of RNA processing and transport. Differential RNA processing is the *splicing* of mRNA precursors into messages that specify different proteins by using different combinations of potential exons. If an mRNA precursor had five potential exons, one cell type might use exons 1, 2, 4, and 5; a different cell type might use exons 1, 2, and 3; and yet another cell type might use all five (**FIGURE 2.22**). Thus, a single gene can produce an entire family of proteins. The different proteins encoded by the same gene are called **splicing isoforms** of the protein.

TFIID (**FIGURE 2.21**). In order to elongate the RNA, a signal must enable the multiprotein transcription elongation complex (TEC) to compete with TFIID for the favors of Med26. Once the transcription elongation complex frees the RNA polymerase II from TFIID, RNA polymerase II can become phosphorylated and travel along the DNA to transcribe the gene (Takahashi et al. 2011).

Drosophila may use a slightly different mechanism to pause the transcription from HCPs. In many of these genes, there appears to be a DNA sequence in the proximal promoter (i.e., the sequences of the promoter closest to the exons) that acts as a "pause button" (Hendrix et al. 2008). About 1500 genes in *Drosophila* embryos have RNA polymerase II already on their promoters, and these genes are primarily those active in regulating early development (Muse et al. 2007; Zeitlinger et al. 2007). It is possible that these "pause button" sequences may be more difficult to unwind, and the presence of the polymerase may enable elongation inhibitory factors to assemble there (Levine 2011).

But how does the release of a single transcript influence the synthesis of that protein? It certainly takes more than one transcript to produce significant amounts of gene product. In some cases, it appears that the transcript of the paused polymerase can recruit histone-activating proteins, enabling further transcription to occur as soon as elongation commences (Petesch and Lis 2008). It is also possible that paused RNA polymerase II prevents the assembly of new nucleosomes on the promoter, keeping the gene in an open configuration (Gilchrist et al. 2010; Nechaev et al. 2010).

Thus, both high and low CpG-content promoters regulate RNA synthesis, but they do so in different manners. LCPs are usually turned off, requiring transcription factors to enable gene expression by promoting access of RNA polymerase II to the DNA. HCPs already have initiated transcription, but transcription is not completed. Here, developmental signals allow the elongation of the nascent nRNA. Both LCPs and HCPs have repressed states that prevent transcription, as well as poised states that enable the genes to be transcribed immediately when the appropriate signal is received.

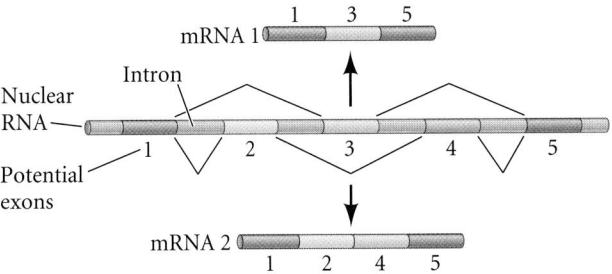

mRNA 1

1 3 5

Nuclear RNA

Intron

Potential exons

1 2 3 4 5

mRNA 2

1 2 4 5

FIGURE 2.22 Differential RNA processing. By convention, splicing paths are shown by fine V-shaped lines. Differential splicing can process the same nuclear RNA into different mRNAs by selectively using different exons.

SIDELIGHTS & SPECULATIONS

X Chromosome Inactivation: Noncoding RNAs in Transcriptional Gene Regulation

In *Drosophila*, nematodes, and mammals, females have two X chromosomes per cell, while males have a single X chromosome per cell. Unlike the Y chromosome, the X chromosome contains thousands of genes that are essential for cell activity. Yet despite the female's cells having double the number of X chromosomes, male and female cells contain approximately equal amounts of X chromosome-encoded gene products. This equalization phenomenon is called **dosage compensation**, and it can be accomplished in three ways (Migeon 2002). In *Drosophila*, the transcription rate of the male X chromosomes is doubled so that the single male X chromosome makes the same amount of transcript as the two female X chromosomes (Lucchesi and Manning 1987). This hyperactivation of the male X chromosome is accomplished by enhanced transcriptional elongation and by acetylation of the nucleosomes throughout the male's X chromosomes (Akhtar et al. 2000; Larschan et al. 2011). In *C. elegans*, both X chromosomes are partially repressed (Chu et al. 2002) so that the products of the X chromosomes are equalized between the sexes.

In mammals, dosage compensation occurs through the inactivation of one X chromosome in each female cell. Thus, each mammalian somatic cell, whether male or female, has only one functioning X chromosome. This phenomenon is called **X chromosome inactivation**. The chromatin of the inactive X chromosome is converted into **heterochromatin**—chromatin that remains condensed throughout most of the cell cycle and replicates later than most of the other chromatin (the **euchromatin**) of the nucleus. This was first shown by Mary Lyon (1961), who observed coat color patterns in mice. If a mouse is heterozygous for an autosomal gene controlling hair pigmentation, then it resembles one of its two parents, or has a color intermediate between the two. In either case, the mouse is a single color. But if a female mouse is heterozygous for

FIGURE 2.23 The kitten "CC" (left) was the first household pet to be successfully cloned using somatic nuclear transfer from "Rainbow" (right), a female calico cat. However, these two genetically identical cats are not identical because calico coloration is the result of random X chromosome inactivation. In random pigment cells, the "orange" allele is inactivated, while in other pigment cells, the "black" allele is randomly inactivated. (Courtesy of the College of Veterinary Medicine, Texas A&M University.)

a pigmentation gene *on the X chromosome*, a different result is seen: patches of one parental color alternate with patches of the other parental color. Calico and tortoiseshell cats* are normally female; their coat color alleles (black and orange) are on the X chromosome (**FIGURE 2.23**; Centerwall and Benirscke 1973). To account for these results, Lyon proposed the following hypothesis:

- Very early in the development of female mammals, both X chromosomes are active. As development proceeds, one X chromosome is inactivated in each cell (**FIGURE 2.24A**).

- This inactivation is random. In some cells, the paternally derived X chromosome is inactivated; in other cells, the maternally derived X chromosome is shut down.

*Although the terms *calico* and *tortoiseshell* are sometimes used synonymously, tortoiseshell coats are a patchwork of black and orange only; calico cats usually have white patches—that is, patches with no pigment—as well (see Figure 2.23).

- This process is irreversible. Once a particular X chromosome (either the one derived from the mother or the one derived from the father) has been inactivated in a cell, the same X chromosome is inactivated in all of that cell's progeny (**FIGURE 2.24B,C**). Because X inactivation happens relatively early in development, an entire region of cells derived from a single cell may have the same X chromosome inactivated. Thus, all tissues in female mammals are mosaics of two cell types.

The inactivation of the X chromosome is complicated; indeed, it is a bottleneck that many female embryos do not get through (Migeon 2007). The mechanisms of X chromosome inactivation appear to differ between mammalian groups (Migeon 2007; Okamoto et al. 2011), but these mechanisms each use a noncoding RNA called *Xist* (*X-i*nactivation specific *t*ranscript) to inactivate one of the two X chromosomes in each cell. *Xist* RNA is initially transcribed at low levels in both X chromosomes, but it becomes actively transcribed on only

(A)

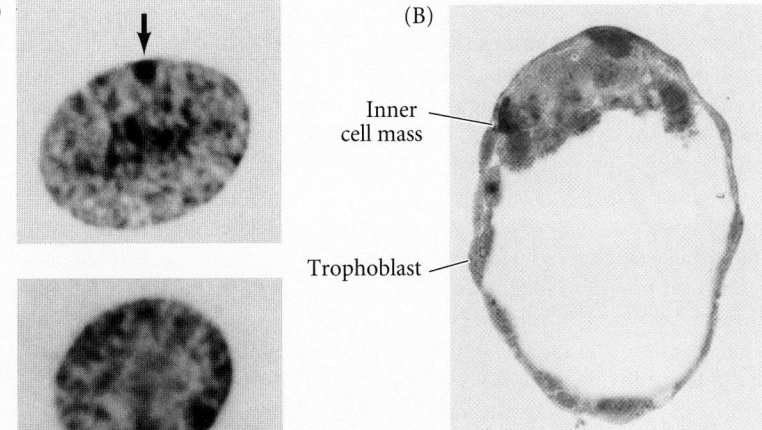

(B)

Inner
cell mass

Trophoblast

(C)

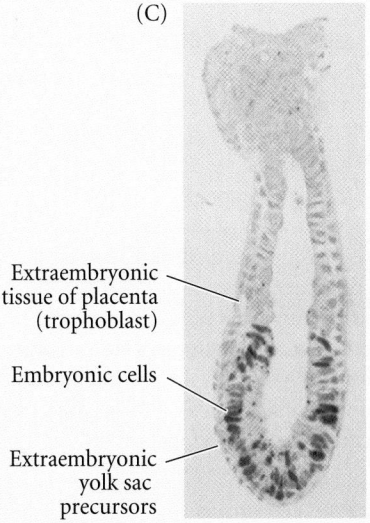

Extraembryonic
tissue of placenta
(trophoblast)

Embryonic cells

Extraembryonic
yolk sac
precursors

FIGURE 2.24 X chromosome inactivation. (A) Inactivated X chromosomes, or Barr bodies, in the nuclei of human oral epithelial cells. The top cell is from a normal XX female and has a single Barr body (arrow). In the lower cell, from a female with three X chromosomes, two Barr bodies can be seen. In both cases, only one X chromosome per cell is active. (B,C) The paternally derived X chromosome of this mouse embryo contained a *lacZ* transgene. Cells in which the paternal chromosome is active make β-galactosidase and stain blue. (B) In the early blastocyst stage (day 4), both X chromosomes are active in all cells. (C) At day 6, random inactivation of one of the chromosomes occurs. Embryonic cells in which the maternal X is active appear pink, while those with an active paternal X are blue. In mouse (but not human) trophoblasts, the paternally derived X chromosome is preferentially inactivated, so the trophoblast cells are uniformly pink. (A courtesy of M. L. Barr; B,C from Sugimoto et al. 2000, courtesy of N. Takagi.)

one X chromosome in each cell—but *which* one is random. Although *Xist* is initially present only near the site of its transcription, *Xist* transcripts spread out, eventually covering the entire chromosome and recruiting Polycomb repressive complexes (PRCs) 1 and 2 to that particular X chromosome. These repressive complexes modify the nucleosomes to prevent transcription (see Wutz 2011). The PRC-*Xist* complexes also recruit DNA methyltransferases that further stabilize the repressive state by methylating the gene promoters. In mice and humans, the promoter regions of numerous genes become methylated on the inactive X chromosome but are unmethylated on the active X chromosome (Wolf et al. 1984; Keith et al. 1986; Migeon et al. 1991). The memory of this "X inactivation" is transmitted to the progeny of the cells by

successive DNA methylation through Dnmt1 (see pp. 50–51).

Although *Xist* RNA transcription appears to be the major step in X chromosome inactivation, different mammals may regulate it in different ways. In mice, it appears that the *inactive* X chromosome is being selected. Both X chromosomes transcribe two factors that inhibit *Xist* transcription, and the local concentration of these inhibitors determines the levels of *Xist* and which X chromosome becomes inactive (Gontan et al. 2012).

In humans, however, the autosomes may provide an inhibitor of *Xist*. Humans with an extra X chromosome (X trisomy) have only one active X chromosome. On the other hand, triploid humans have an *entire* extra set of haploid chromosomes and often die around the time of birth. Some of these rare fetuses have an extra

X chromosome (a condition in which two X-bearing sperm enter the egg and the embryo has 66 autosomes and 3 X chromosomes) and often have two active X chromosomes per cell. This suggests that the *Xist* is being constitutively synthesized and that an *autosomal* inhibitor of human *Xist* transcription (i.e., an X chromosome activator) might exist (Migeon et al. 2008; Migeon 2011).

Xist is a member of a newly discovered group of transcriptional regulators called the **long noncoding RNAs (lncRNAs)**. These regulators are often used to silence genes on one of the two chromosome. The *Airn* lncRNA, for instance, silences the *Igf2R* promoter on the paternal (but not the maternal) mouse chromosome 17. (The Igf2R protein is the receptor for the Igf2 insulin growth factor mentioned earlier). The *paternal* promoter for *Airn* is unmethylated and active in the early embryo (Seidl et al. 2006; Latos et al. 2012).

● **See WEBSITE 2.9** Small noncoding RNAs that repress transcription

Creating families of proteins through differential nRNA splicing

Alternative nRNA splicing is a means of producing a wide variety of proteins from the same gene, and most vertebrate genes make nRNAs that are alternatively spliced* (Wang et al. 2008; Nilsen and Graveley 2010). The average vertebrate nRNA consists of several relatively short exons (averaging about 140 bases) separated by introns that are usually much longer. Most mammalian nRNAs contain numerous exons. By splicing together different sets of exons, different cells can make different types of mRNAs, and hence, different proteins. Recognizing a sequence of nRNA as either an exon or an intron is a crucial step in gene regulation.

Alternative nRNA splicing is based on the determination of which sequences will be spliced out as introns. This can occur in several ways. Most genes contain "consensus

---

*Mutations can generate species-specific splicing events, and tissue-specific differences in nRNA splicing between vertebrate species occur 10–100 times more frequently than changes in gene transcription (Barbosa-Morais et al. 2012; Merkin et al. 2012).

sequences" at the 5′ and 3′ ends of the introns. These sequences are the "splice sites" of the intron. The splicing of nRNA is mediated through complexes known as **spliceosomes** that bind to the splice sites. Spliceosomes are made up of small nuclear RNAs (snRNAs) and proteins called **splicing factors** that bind to splice sites or to the areas adjacent to them. By their production of specific splicing factors, cells can differ in their ability to recognize a sequence as an intron. That is to say, a sequence that is an *exon* in one cell type may be an *intron* in another (**FIGURE 2.25A,B**). In other instances, the factors in one cell might recognize different 5′ sites (at the beginning of the intron) or different 3′ sites (at the end of the intron; **FIGURE 2.25C,D**).

The 5′ splice site is normally recognized by small nuclear RNA U1 (U1 snRNA) and splicing factor 2 (SF2; also known as alternative splicing factor). The choice of alternative 3′ splice sites is often controlled by which splice site can best bind a protein called U2AF. The spliceosome forms when the proteins that accumulate at the 5′ splice site contact those proteins bound to the 3′ splice site. Once the 5′ and 3′ ends are brought together, the intervening intron is excised and the two exons are ligated together.

(A) Cassette exon: Type II procollagen

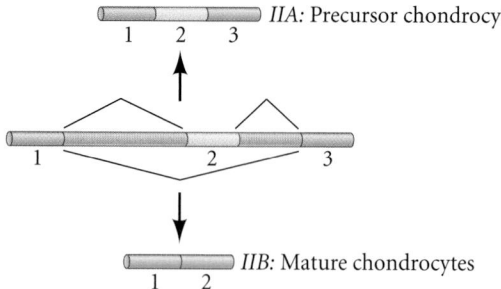

(B) Mutually exclusive exons: FgfR2

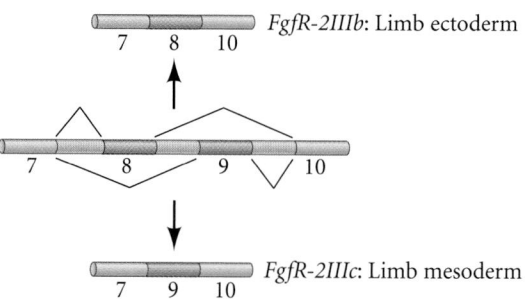

(C) Alternative 5′ splice site: Bcl-x

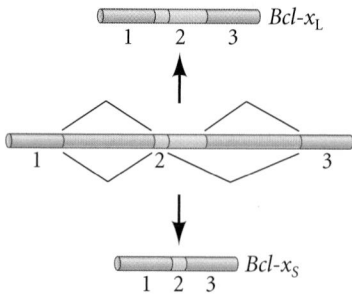

(D) Alternative 3′ splice site: Chordin

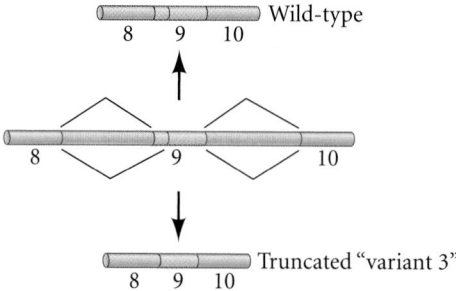

FIGURE 2.25 Some examples of alternative RNA splicing. Blue and colored portions of the bars represent exons; gray represents introns. Alternative splicing patterns are shown with V-shaped lines. (A) A "cassette" (yellow) that can be used as exon or removed as an intron distinguishes the type II collagen types of chondrocyte precursors and mature chondrocytes (cartilage cells). (B) Mutually exclusive exons distinguish fibroblast growth factor receptors found in the limb ectoderm from those found in the limb mesoderm. (C) Alternative 5′ splice site selection, such as that used to create the large and small isoforms of the protein Bcl-X. (D) Alternative 3′ splice sites are used to form the normal and truncated forms of Chordin. (After McAlinden et al. 2004.)

In some instances, alternatively spliced RNAs yield proteins that play similar yet distinguishable roles in the same cell. Different isoforms of the WT1 protein perform different functions in the development of the gonads and kidneys. The isoform without the extra exon functions as a transcription factor during kidney development, whereas the isoform containing the extra exon appears to be critical in testis development (Hammes et al. 2001; Hastie 2001).

The *Bcl-x* gene provides a good example of how alternative nRNA splicing can make a huge difference in a protein's function. If a particular DNA sequence is used as an exon, the "large Bcl-X protein," or Bcl-X_L, is made (see Figure 2.25C). This protein inhibits programmed cell death. However, if this sequence is seen as an intron, the "small Bcl-X protein" (Bcl-X_S) is made, and this protein *induces* cell death. Many tumors have a higher than normal amount of Bcl-X_L.

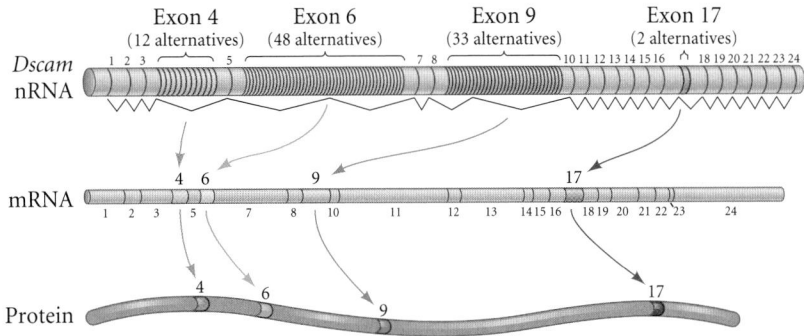

FIGURE 2.26 The *Dscam* gene of *Drosophila* can produce 38,016 different types of proteins by alternative nRNA splicing. The gene contains 24 exons. Exons 4, 6, 9, and 17 are encoded by sets of mutually exclusive possible sequences. Each messenger RNA will contain one of the 12 possible exon 4 sequences, one of the 48 possible exon 6 alternatives, one of the 33 possible exon 9 alternatives, and one of the 2 possible exon 17 sequences. The *Drosophila Dscam* gene is homologous to a DNA sequence on human chromosome 21 that is expressed in the nervous system. Disturbances of this gene in humans may lead to the neurological defects of Down syndrome. (After Yamakawa et al. 1998; Saito et al. 2000.)

If you get the impression from this discussion that a gene with dozens of introns could create literally thousands of different, related proteins through differential splicing, you are probably correct. The current champion at making multiple proteins from the same gene is the *Drosophila Dscam* gene. This gene encodes a membrane receptor protein involved in preventing dendrites from the same neuron from binding to one another (Wu et al. 2012). *Dscam* contains 115 exons. Moreover, a dozen different adjacent DNA sequences can be selected to be exon 4, and more than 30 mutually exclusive adjacent DNA sequences can become exons 6 and 9, respectively (**FIGURE 2.26**; Schmucker et al. 2000). If all possible combinations of exons are used, this one gene can produce 38,016 different proteins, and random searches for these combinations indicate that a large fraction of them are in fact made. The nRNA of *Dscam* has been found to be alternatively spliced in different axons, and when two dendrites from the same axon touch each other, they are repelled. This causes the extensive branching of the dendrites. It appears that the thousands of splicing isoforms are needed to ensure that each neuron acquires a unique identity (Schmucker 2007; Millard and Zipursky 2008; t al. 2009). The *Drosophila* genome is thought to contain only 14,000 genes, but here is a single gene that encodes three times that number of proteins!

About 92% of human genes are thought to produce multiple types of mRNA. Therefore, even though the human genome may contain 20,000 genes, its **proteome**—the number and type of proteins encoded by the genome—is far more complex. "Human genes are multitaskers," notes Christopher Burge, one of the scientists who calculated this figure (Ledford 2008). This explains an important paradox. *Homo sapiens*

has around 20,000 genes in each nucleus; so does the nematode *Caenorhabditis elegans*, a tubular creature with only 959 cells. We have more cells and cell types in the shaft of a hair than *C. elegans* has in its entire body. What's this worm doing with approximately the same number of genes as us? The answer is that *C. elegans* genes rarely make isoforms. Each gene in the worm makes but one protein, whereas in humans the same number of genes produces an enormous array of different proteins.

● See **WEBSITE 2.10** Control of early development by nuclear RNA selection

● See **WEBSITE 2.11** So you think you know what a gene is

Splicing enhancers and recognition factors

The mechanisms of differential RNA processing involve both *cis*-acting sequences on the nRNA and *trans*-acting protein factors that bind to these regions (Black 2003). The *cis*-acting sequences on nRNA are usually close to their potential 5′ or 3′ splice sites. These sequences are called **splicing enhancers**, since they promote the assembly of spliceosomes at RNA cleavage sites. (Conversely, these same sequences can be "splicing silencers" if they act to exclude exons from an mRNA sequence.) These sequences are recognized by *trans*-acting proteins, most of which can recruit spliceosomes to that area. However, some *trans*-acting proteins, such as the polyprimidine tract-binding proteins (PTPs), repress spliceosome formation where they bind. Indeed, different PTPs can control the splicing of batteries of nRNAs. For example, PTPb prevents the adult neuron-specific splicing of the neural nRNAs controlling cell fate, cell proliferation, and actin cytoskeleton, thereby keeping the neuronal precursors in a proliferating, immature state (Licatolosi et al. 2012).

The selection of particular exons is determined not only by the spliceosome-binding consensus sequences but also by numerous sequence elements that are recognized by regulatory factors that can regulate spliceosome binding (Ke and Chasin 2011). The DNA sequences that regulate whether a spliceosome can form on a particular splicing consensus sequence are splicing enhancers. As might be expected, some splicing enhancers appear to be specific for certain tissues. Muscle-specific splicing enhancers have been found around those exons characterizing muscle cell messages. These are recognized by certain proteins that are found in the muscle cells early in their development (Ryan and Cooper 1996; Charlet-B et al. 2002). Their presence is able to compete with the PTP that would otherwise prevent the inclusion of the muscle-specific exon into the mature message. In this way, an entire battery of muscle-specific isoforms can be generated. However, the context dependency of splicing is too complex to delineate by merely comparing sequences. Computational studies—in which the computer is asked to identify (1) the combination of sequence elements, (2) the proximity of these sequences to the splice junctions, and (3) the differences of splicing outcomes in different cell types—are providing our first look at a "splicing code" that may allow us to predict which exons will persist in one cell and not in others (Barash et al. 2010).

Mutations in the splicing sites can lead to alternative developmental phenotypes. Most splice site mutations lead to nonfunctional proteins and serious diseases. For instance, a single base change at the 5′ end of intron 2 in the human β-globin gene prevents splicing from occurring and generates a nonfunctional mRNA (Baird et al. 1981). This causes the absence of any β-globin from this gene, and thus a severe (and often life-threatening) type of anemia. Similarly, a mutation in the *Dystrophin* gene at a particular splice site causes the skipping of that exon and a severe form of muscular dystrophy (Sironi et al. 2001). In at least one such case of aberrant splicing, the splice site mutation was not dangerous and actually gave the patient greater strength. In this case, Schuelke and colleagues (2004) described a family in which individuals in four generations had a splice site mutation in the *myostatin* gene. Among the family members were professional athletes and a 4-year-old toddler who was able to hold two 3-kg dumbbells with his arms fully extended. The product of the normal *myostatin* gene is a factor that tells muscle precursor cells to stop dividing—that is, it is a negative regulator. In mammals (including humans and mice) with the mutation, the factor is nonfunctional and the muscle precursors are not told to differentiate until they have undergone many more rounds of cell division; the result is larger muscles (**FIGURE 2.27**).

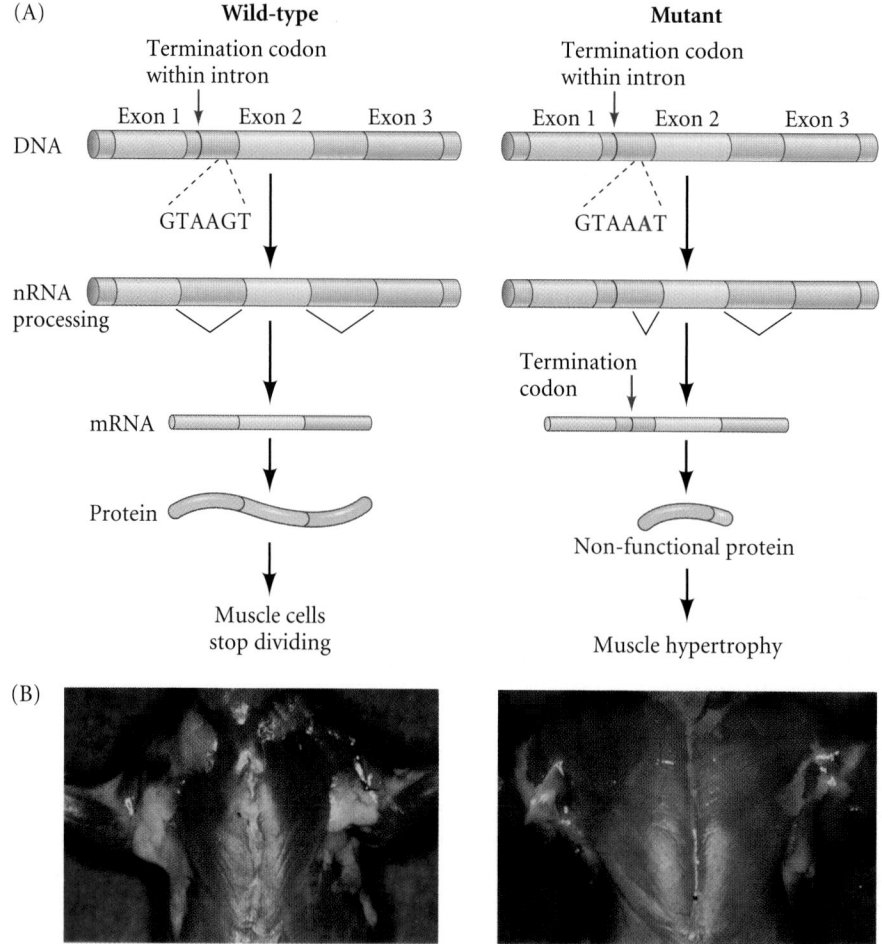

FIGURE 2.27 Muscle hypertrophy through mispliced RNA. This mutation results in a deficiency of the negative growth regulator myostatin in the muscle cells. (A) Molecular analysis of the mutation. There is no mutation in the coding sequence of the gene, but in the first intron, a mutation from a G to an A creates a new (and widely used) splicing site. This causes aberrant nRNA splicing and the inclusion of an early protein synthesis termination codon into the mRNA. Thus, proteins made from that message are short and nonfunctional. (B) Pectoral musculature of a "mighty mouse" with the mutation (right) compared with the muscles of a wild-type mouse (left). (A after Schuelke et al. 2004; B from McPherron et al. 1997, courtesy of A. C. McPherron.)

Control of Gene Expression at the Level of Translation

The splicing of nuclear RNA is intimately connected with its export through the nuclear pores and into the cytoplasm. As the introns are removed, specific proteins bind to the spliceosome and attach the spliceosome-RNA complex to nuclear pores (Luo et al. 2001; Strässer and Hurt 2001). The proteins coating the 5′ and 3′ ends of the RNA also change. The nuclear cap binding protein at the 5′ end is replaced by *eukaryotic translation initiation factor eIF4E,* and the polyA tail becomes bound by the cytoplasmic polyA binding protein. Although both of these changes facilitate the initiation of translation, there is no guarantee that the RNA will be translated once it reaches the cytoplasm. The control of gene expression at the level of translation can occur by many means; some of the most important of these are described below.

Differential mRNA longevity

The longer an mRNA persists, the more protein can be translated from it. If a message with a relatively short half-life were selectively stabilized in certain cells at certain times, it would make large amounts of its particular protein only at those times and places.

The stability of a message often depends on the length of its polyA tail. This, in turn, depends largely on sequences in the 3′ untranslated region, certain of which allow longer polyA tails than others. If these 3′ UTRs are experimentally traded, the half-lives of the resulting mRNAs are altered: long-lived messages will decay rapidly, while normally short-lived mRNAs will remain around longer (Shaw and Kamen 1986; Wilson and Treisman 1988; Decker and Parker 1995).

In some instances, mRNAs are selectively stabilized at specific times in specific cells. The mRNA for casein, the major protein of milk, has a half-life of 1.1 hours in rat mammary gland tissue. However, during periods of lactation, the presence of the hormone prolactin increases this half-life to 28.5 hours (**FIGURE 2.28**; Guyette et al. 1979). In the development of the nervous system, a set of proteins called HuD proteins stabilizes two groups of mRNAs that would otherwise perish quickly. One group of mRNAs encodes proteins that stop neuronal precursor cells from dividing, while the second group of mRNAs encodes proteins that initiate neuronal differentiation (Okano and Darnell 1997; Deschênes-Furry et al. 2006, 2007). Thus, once the HuD proteins are made, the neuronal precursor cells can become neurons.

Stored oocyte mRNAs: Selective inhibition of mRNA translation

Some of the most remarkable cases of translational regulation of gene expression occur in the oocyte. The oocyte often makes and stores mRNAs that will be used only after fertilization occurs. These messages stay in a dormant state until they are activated by ion signals (discussed in Chapter 4) that spread through the egg during ovulation or fertilization.

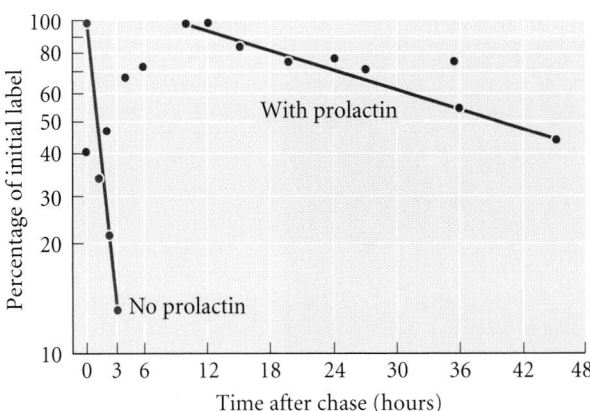

FIGURE 2.28 Degradation of casein mRNA in the presence and absence of prolactin. Cultured rat mammary cells were given radioactive RNA precursors (pulse) and, after a given time, were washed and given nonradioactive precursors (chase). This procedure labeled the casein mRNA synthesized during the pulse time. Casein mRNA was then isolated at different times following the chase and its radioactive label measured. In the absence of prolactin, the labeled (i.e., newly synthesized) casein mRNA decayed rapidly, with a half-life of 1.1 hours. When the same experiment was done in a medium containing prolactin, the half-life was extended to 28.5 hours. (After Guyette et al. 1979.)

Some of these stored mRNAs encode proteins that will be needed during cleavage, when the embryo makes enormous amounts of chromatin, cell membranes, and cytoskeletal components. These include the messages for histone proteins, the messages for the actin and tubulin proteins of the cytoskeleton, and the mRNAs for the cyclin proteins that regulate the timing of early cell division (Raff et al. 1972; Rosenthal et al. 1980; Standart et al. 1986). Indeed, in many species (including sea urchins and *Drosophila*), maintenance of the normal rate and pattern of early cell divisions does not require a nucleus; rather, it requires continued protein synthesis from stored maternal mRNAs (Wagenaar and Mazia 1978; Edgar et al. 1994). Other stored messages encode proteins that determine the fates of cells. These include the *bicoid, caudal,* and *nanos* messages that provide information in the *Drosophila* embryo for the production of its head, thorax, and abdomen.

Most translational regulation in oocytes is negative, as the "default state" of the mRNA is to be available for translation. Therefore, there must be inhibitors preventing the translation of these mRNAs in the oocyte, and these inhibitors must somehow be removed at the appropriate times around fertilization. The 5′ cap and the 3′ UTR seem especially important in regulating the accessibility of mRNA to ribosomes. If the 5′ cap is not made or if the 3′ UTR lacks a polyA tail, the message probably will not be translated. The oocytes of many species have "used these ends as means" to regulate the translation of their mRNAs.

It is important to realize that, unlike the usual representations of mRNA, most mRNAs probably form circles, with

(A) Circularized mRNA

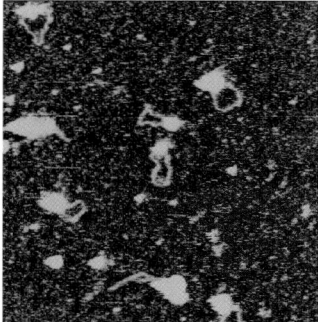

FIGURE 2.29 Translational regulation in oocytes. (A) Messenger RNAs are often found as circles, where the 5′ end and the 3′ end contact one another. Here, a yeast mRNA seen by atomic force microscopy is circularized by eIF4E and eIF4G (5′ end) and the polyA binding protein (3′ end). (B) In *Xenopus* oocytes, the 3′ and 5′ ends of the mRNA are brought together by maskin, a protein that binds CPEB on the 3′ end and eukaryotic initiation factor 4E (eIF4E) on the 5′ end. Maskin blocks the initiation of translation by preventing eIF4E from binding eIF4G. (C) When stimulated by progesterone during ovulation, a kinase phosphorylates CPEB, which can then bind CPSF. CPSF can bind polyA polymerase and initiate growth of the polyA tail. PolyA binding protein (PABP) can bind to this tail and then bind eIF4G in a stable manner. This initiation factor can then bind eIF4E and, through its association with eIF3, position a 40S ribosomal subunit on the mRNA. (A from Wells et al. 1998; B,C after Mendez and Richter 2001.)

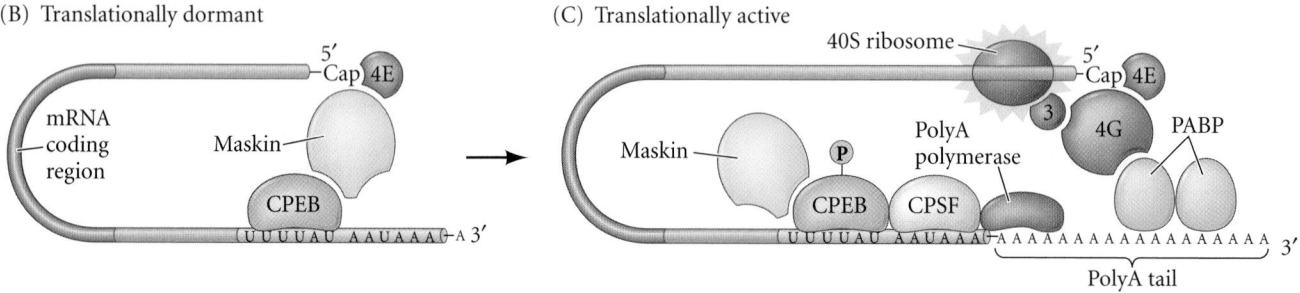

(B) Translationally dormant (C) Translationally active

their 3′ end being brought to their 5′ end by various **eukaryotic initiation factors** (**eIFs**; **FIGURE 2.29A**). The 5′ cap is bound by eukaryotic initiation factor 4E (eIF4E), a protein that is also bound to eIF4A (a helicase that unwinds double-stranded regions of RNA) and to eIF4G, a scaffold protein that allows the mRNA to bind to the ribosome through its interaction with eIF4E (Wells et al. 1998; Gross et al. 2003). The polyA binding protein, which sits on the polyA tail of the mRNA, also binds to the eIF4G protein. This brings the 3′ end of the message next to the 5′ end and allows the messenger RNA to be recognized by the ribosome. Thus, the 5′ cap is critical for translation, and some animals' oocytes have used this as a direct means of translational control. For instance, the oocyte of the tobacco hornworm moth makes some of its mRNAs without their methylated 5′ caps. In this state, they cannot be efficiently translated. However, at fertilization, a methyltransferase completes the formation of the caps, and these mRNAs can be translated (Kastern et al. 1982).

In amphibian oocytes, the 5′ and 3′ ends of many mRNAs are brought together by a protein called **maskin** (Stebbins-Boaz et al. 1999; Mendez and Richter 2001). Maskin links the 5′ and 3′ ends into a circle by binding to two other proteins, each at opposite ends of the message. First, it binds to the **cytoplasmic polyadenylation-element-binding protein** (**CPEB**) attached to the UUUUAU sequence in the 3′ UTR; second, maskin also binds to the eIF4E factor that is attached to the cap sequence. In this configuration, the mRNA cannot be translated (**FIGURE 2.29B**). The binding of eIF4E to maskin is thought to prevent the binding of eIF4E to eIF4G, a critically important translation initiation factor that brings the small ribosomal subunit to the mRNA.

Mendez and Richter (2001) proposed an intricate scenario to explain how mRNAs bound together by maskin become translated at about the time of fertilization. At ovulation (when the hormone progesterone stimulates the last meiotic divisions of the oocyte and the oocyte is released for fertilization), a kinase activated by progesterone phosphorylates the CPEB protein. The phosphorylated CPEB can now bind to CPSF, the cleavage and polyadenylation specificity factor (Mendez et al. 2000; Hodgman et al. 2001). The bound CPSF protein sits on a particular sequence of the 3′ UTR that has been shown to be critical for polyadenylation, and it complexes with a polymerase that elongates the polyA tail of the mRNA. In oocytes, a message having a short polyA tail is not degraded; however, such messages are not translated.

Once the tail is extended, molecules of the polyA binding protein (PABP) can attach to the growing tail. PABP stabilizes eIF4G, allowing it to outcompete maskin for the binding site on the eIF4E protein at the 5′ end of the mRNA. The eIF4G protein can then bind eIF3, which can position the small ribosomal subunit onto the mRNA. The small (40S) ribosomal subunit will then find the initiator tRNA, complex with the large ribosomal subunit, and initiate translation (**FIGURE 2.29C**).

In the *Drosophila* oocyte, Bicoid protein initiates head and thorax formation. Bicoid can act both as a transcription factor (activating genes such as *hunchback* that are necessary form forming the fly anterior) and as a translational inhibitor of those genes such as *caudal* that are critical for making the fly posterior (see Chapter 6). Caudal protein is important in activating those genes that specify the cells to be abdomen precursors. Bicoid inhibits *caudal* mRNA translation by binding

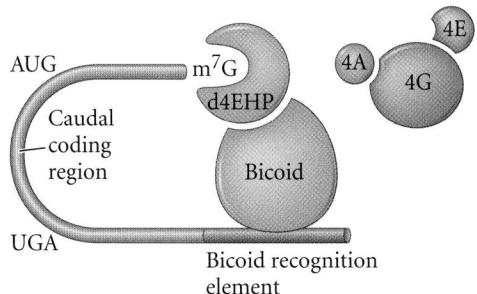

FIGURE 2.30 Protein binding in *Drosophila* oocytes. Bicoid protein binds to a recognition element in the 3′ UTR of the *caudal* message. Bicoid can bind to d4EHP, which prevents the binding of eIF4E to the cap structure. Without eIF4E, the eIF4G cannot bind and initiate translation. (After Cho et al. 2005.)

to a "bicoid recognition element," a series of nucleotides in the 3′ UTR of the *caudal* message (**FIGURE 2.30**). Once there, Bicoid can bind with and recruit another protein, d4EHP, which can compete with eIF4E protein for the cap. Without eIF4E, there is no association with eIF4G and *caudal* mRNA becomes untranslatable. As a result, the *caudal* message is not translated in the anterior of the embryo (where Bicoid is abundant) but is active in the posterior portion of the embryo.

Ribosomal selectivity: Selective activation of mRNA translation

It has long been assumed that ribosomes do not show favoritism toward translating certain mRNAs. After all, eukaryotic messages can be translated even by *E. coli* ribosomes, and ribosomes from immature red blood cells have long been used to translate mRNAs for any source. But recent evidence has shown that ribosomal proteins are not the same in all cells, and that some ribosomal proteins are necessary for translating certain messages. When Kondrashov and colleagues (2011) mapped the gene that caused numerous axial skeleton deformities in mice, they found that the mutation was not in one of the well-known genes that control skeletal polarity. Rather, it was in ribosomal protein Rpl38. When this protein is mutated, the ribosomes can still translate most messages, but the ribosomes in the skeletal precursors cannot translate the mRNA from a specific subset of Hox genes. These Hox genes, as we will see in Chapter 9, specify the type of vertebrae at each particular axial level (ribbed thoracic vertebrae, unribbed abdominal vertebrae, etc.). Without functioning Rpl38, vertebral cells are unable to form the initiation complex with mRNA from the appropriate Hox genes, and the skeleton is deformed (**FIGURE 2.31**). Mutations in other ribosomal proteins have also been found to produce deficient phenotypes (Terzian and Box 2013; Watkins-Chow et al. 2013).

microRNAs: Specific regulation of mRNA translation and transcription

If proteins can bind to specific nucleic acid sequences to block transcription or translation, you would think that

RNA would do the job even better. After all, RNA can be made specifically to bind a particular sequence. Indeed, one of the most efficient means of regulating the translation of a specific message is to make a small RNA complementary to a portion of a particular mRNA. Such a naturally occurring antisense RNA was first seen in *C. elegans* (Lee et al. 1993; Wightman et al. 1993). Here, the *lin-4* gene was found to encode a 21-nucleotide RNA that bound to multiple sites in the 3′ UTR of the *lin-14* mRNA (**FIGURE 2.32**). The *lin-14* gene encodes a transcription factor, LIN-14, that is important during the first larval phase of *C. elegans* development. It is not needed afterward, and *C. elegans* is able to inhibit synthesis of LIN-14 from these messages by activating the *lin-4* gene. The binding of *lin-4* transcripts to the *lin-14* mRNA 3′ UTR causes degradation of the *lin-14* message (Bagga et al. 2005).

The *lin-4* RNA is now thought to be the "founding member" of a very large group of **microRNAs** (**miRNAs**). Computer analysis of the human genome predicts that we have more than 1000 miRNA loci, and that these miRNAs probably modulate 50% of the protein-encoding genes in our bodies (Berezikov and Plasterk 2005; Friedman et al. 2009). These miRNAs usually contain only 22 nucleotides and are made from longer precursors. These precursors can be in independent transcription units (the *lin-4* gene is far apart from the *lin-14* gene), or they can reside in the introns of other genes (Aravin et al. 2003; Lagos-Quintana et al. 2003). The initial RNA transcript (which may contain several repeats of the miRNA sequence) forms hairpin loops wherein the RNA finds complementary structures within its strand. These stem-loop structures are processed by a set of RNases (Drosha and Dicer) to make single-stranded microRNA (**FIGURE 2.33**). The microRNA is then packaged with a series of proteins to make an **RNA-induced silencing complex** (**RISC**).

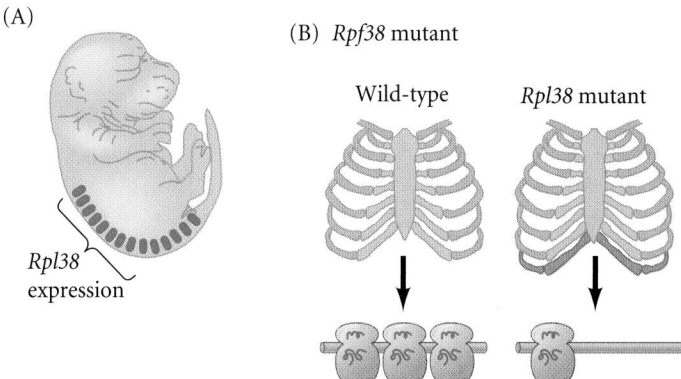

FIGURE 2.31 Model of ribosomal heterogeneity in mice. (A) Ribosomes have slightly different proteins depending on the tissue in which they reside. Ribosomal protein Rpl38 (i.e., protein 38 of the large ribosomal subunit) is concentrated in those ribosomes found in the somites that give rise to the vertebrae. (B) A wild-type embryo (left) has normal vertebrae and normal Hox gene translation. Mice deficient in Rpl38 have an extra pair of vertebrae, tail deformities, and reduced Hox gene translation. (After Kondrashov et al. 2011.)

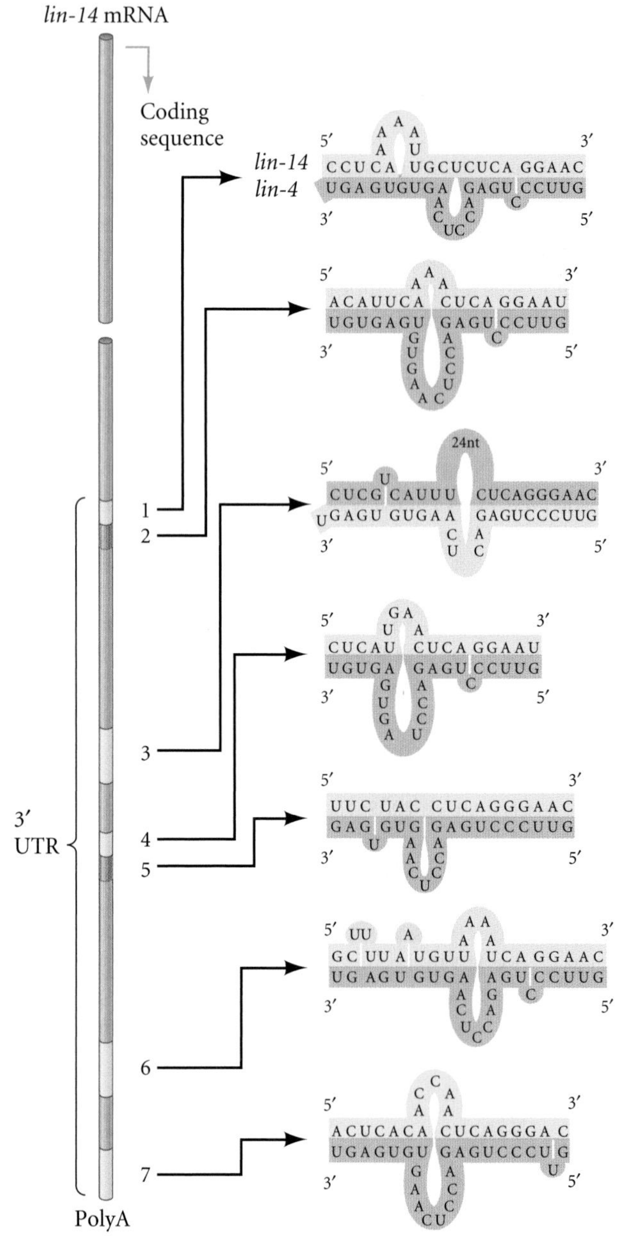

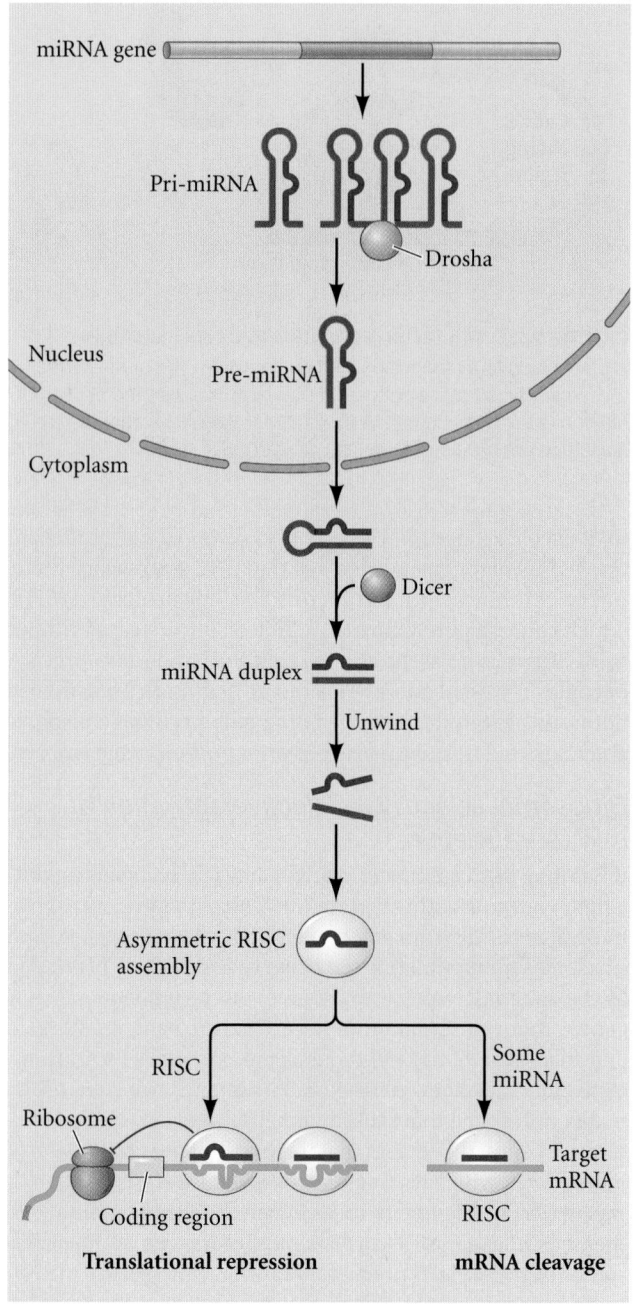

FIGURE 2.32 Hypothetical model of the regulation of *lin-14* mRNA translation by *lin-4* RNAs. The *lin-4* gene does not produce an mRNA. Rather, it produces small RNAs that are complementary to a repeated sequence in the 3′ UTR of the *lin-14* mRNA, which bind to it and prevent its translation. (After Wickens and Takayama 1995.)

FIGURE 2.33 Model for the formation and use of microRNAs. The miRNA gene encodes a pri-miRNA that often has several hairpin regions where the RNA finds nearby complementary bases with which to pair. The pri-miRNA is processed into individual pre-miRNA "hairpins" by the Drosha RNAase, and these are exported from the nucleus. Once in the cytoplasm, another RNAase, Dicer, eliminates the non-base-paired loop. Dicer also acts as a helicase to separate the strands of the double-stranded miRNA. One strand (probably recognized by placement of Dicer) is packaged with proteins into the RNA-induced silencing complex (RISC), which subsequently binds to the 3′ UTRs to effect translational suppression or cleavage, depending (at least in part) on the strength of the complementarity between the miRNA and its target. (After He and Hannon 2004.)

Proteins of the Argonaute family are particularly important members of this complex. Such small regulatory RNAs can bind to the 3′ UTR of messages and inhibit their translation. In some cases (especially when the binding of the miRNA to the 3′ UTR is perfect), the RNA is cleaved. More usually, however, several RISCs attach to sites on the 3′ UTR and prevent the message from being translated (see Bartel 2004; He and Hannon 2004).

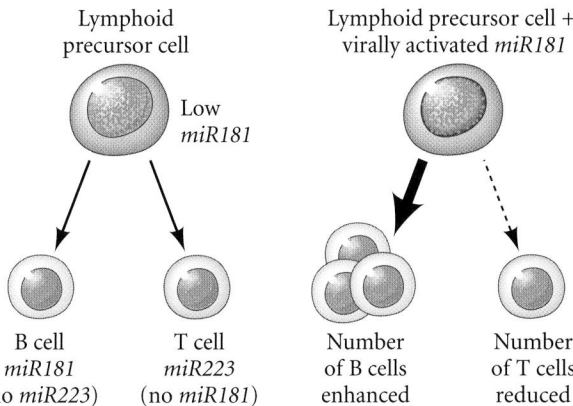

FIGURE 2.34 Lymphoid precursor cells can generate either B cells (lymphocytes that make antibodies) or T cells (lymphocytes that kill virally infected cells), depending on the organ in which the precursor resides. The regulation of the lineage pathway is controlled in part by levels of the microRNA *miR181*. The lymphocyte precursor cell has little *miR181*. A B cell has high levels of *miR181*, whereas T cells do not appear to have any. If lymphocyte precursor cells are virally transfected with *miR181*, they preferentially generate B cells at the expense of T cells.

The abundance of microRNAs and their apparent conservation among different groups—including flies, nematodes, vertebrates, and even plants—suggest that such RNA regulation is a previously unrecognized but potentially important means of regulating gene expression. This hidden layer of gene regulation parallels the better known protein-level gene control mechanisms, and it may be just as important in regulating cell fate. Indeed, computer analyses of miRNA sequences and their potential targets suggest that about half of our genes are subject to control by microRNAs (Friedman et al. 2009).

Recent studies have shown that microRNAs are involved in mammalian heart and blood cell differentiation. During mouse heart development, the microRNA *miR1* can repress the messages encoding the Hand2 transcription factor (Zhao et al. 2005). This transcription factor is critical in the proliferation of ventricle heart muscle cells, and *miR1* may control the balance between ventricle growth and differentiation. The *miR181* miRNA is essential for committing progenitor cells to differentiate into B lymphocytes, and ectopic expression of *miR181* in mice causes a preponderance of B lymphocytes (**FIGURE 2.34**; Chen et al. 2004).

MicroRNAs are also used to "clean up" and fine-tune the level of gene products. We mentioned those maternal RNAs in the oocyte that allow early development to occur. How does the embryo get rid of maternal RNAs once they have been used and the embryonic cells are making their own mRNAs? In zebrafish, this cleanup operation is assigned to microRNAs such as *miR430*. This is one of the first genes transcribed by the fish embryonic cells, and there are about 90 copies of this gene in the zebrafish genome. So the level of *miR430* goes up very rapidly. This microRNA has hundreds of targets (about 40% of the maternal RNA types), and when it binds to the 3′ UTR of these target mRNAs, these mRNAs lose their polyA tails and are degraded (Giraldez et al. 2006). Slightly later in development, this same microRNA is used in

the fish embryo to fine-tune the expression of *Nodal* mRNA (Choi et al. 2007). The consequence of this latter use of *miR430* is the determination of how many cells become committed to the endoderm and how many become committed to be mesoderm.

Although the microRNA is usually 22 bases long, it recognizes its target primarily through a "seed" region of about 5 bases in the 5′ end of the microRNA (usually at positions 2–7). This seed region recognizes targets in the 3′ UTR of the message. What happens, then, if an mRNA has a mutated 3′ UTR? Such a mutation appears to have given rise to the Texel sheep, a breed with a large and well-defined musculature that is the dominant meat-producing sheep in Europe. Genetic techniques mapped the basis of the sheep's meaty phenotype to the *myostatin* gene. We have already seen that a mutation in the *myostatin* gene that prevents the proper splicing of the nRNA can produce a large-muscled phenotype (see Figure 2.27). Another way of reducing the levels of myostatin involves a mutation in its 3′ UTR sequence. In the Texel breed, there has been a G-to-A transition in the 3′ UTR of the gene for myostatin, creating a target for the *mir1* and *mir206* microRNAs that are abundant in skeletal muscle (Clop et al. 2006). This mutation causes the depletion of *myostatin* messages and the increase in muscle mass characteristic of these sheep.

The binding of microRNAs and their associated RISCs to the 3′ UTR can regulate translation in two ways (**FIGURE 2.35**; Filipowicz et al. 2008). First, this binding can block initiation of translation, preventing the binding of initiation factors or ribosomes. The Argonaute proteins, for instance, have been found to bind directly to the methylated guanosine cap at the 5′ end of the mRNA message (Djuranovic et al. 2010, 2011). Second, this binding can recruit endonucleases that

(A) Initiation block

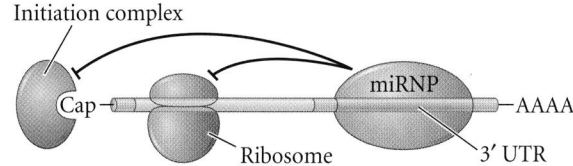

(B) Endonuclease digestion (deadenylation)

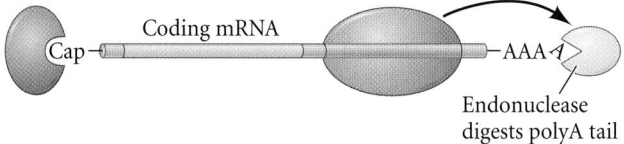

FIGURE 2.35 The miRNA complex, including numerous proteins that bind to the miRNA (miRNP), can block translation in two major ways. (A) One way is by blocking the binding of the mRNA to initiation factors or ribosomes. (B) The other way is by recruiting endonucleases to chew away the polyA tail of the mRNA, thereby causing its destruction. (After Filipowicz et al. 2008.)

digest the mRNA, usually starting with the polyA tail (Guo et al. 2010). The later seems to be commonly used in mammalian cells.

Control of RNA expression by cytoplasmic localization

Not only is the timing of mRNA translation regulated, but so is the place of RNA expression. A majority of mRNAs (about 70% in *Drosophila* embryos) are localized to specific places in the cell (Lécuyer et al. 2007). Just like the selective repression of mRNA translation, the selective localization of messages is often accomplished through their 3′ UTRs. There are three major mechanisms for the localization of an mRNA (see Palacios 2007):

- *Diffusion and local anchoring.* Messenger RNAs such as *nanos* diffuse freely in the cytoplasm. However, when they diffuse to the posterior pole of the *Drosophila* oocyte. However, they are trapped there by proteins that reside particularly in these regions. These proteins also activate the mRNA, allowing it to be translated (**FIGURE 2.36A**).

- *Localized protection.* Messenger RNAs such as those encoding the *Drosophila* heat shock protein hsp83 (which helps protect the embryos from thermal extremes) also float freely in the cytoplasm. Like *nanos* mRNA, *hsp83* accumulates at the posterior pole, but its mechanism for getting there is different. Throughout the embryo, the mRNA is degraded. However, proteins at the posterior pole protect the *hsp83* mRNA from being destroyed (**FIGURE 2.36B**).

- *Active transport along the cytoskeleton.* This is probably the most widely used mechanism for mRNA localization. Here, the 3′ UTR of the mRNA is recognized by proteins that can bind these messages to "motor proteins" that travel along the cytoskeleton to their final destination (**FIGURE 2.36C**). These motor proteins are usually ATPases such as dynein or kinesin that split ATP for their motive force. We will see in Chapters 6 and 17 that this is very important for localizing transcription factor mRNAs into different regions of the *Drosophila* oocyte.

● **See WEBSITE 2.12** Stored mRNA in brain cells

FIGURE 2.36 Localization of mRNAs. (A) Diffusion and local anchoring. *Nanos* mRNA diffuses through the *Drosophila* egg and is bound (in part by the Oskar protein) at the posterior end of the oocyte. This anchoring allows the *nanos* mRNA to be translated. (B) Localized protection. The mRNA for *Drosophila* heat shock protein (hsp83) will be degraded unless it binds to a protector protein (in this case, also at the posterior terminal of the oocyte). (C) Active transport on the cytoskeleton, causing the accumulation of mRNA at a particular site. Here, *bicoid* mRNA is transported to the anterior of the oocyte by dynein and kinesin motor proteins. Meanwhile, *oskar* mRNA is brought to the posterior pole by transport along microtubules by kinesin ATPases. (After Palacios 2007.)

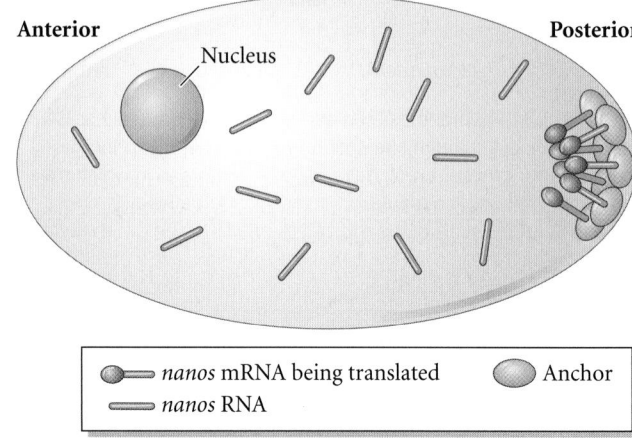

(A) Diffusion and local anchoring

Anterior **Posterior**

Nucleus

● — *nanos* mRNA being translated   ◖ Anchor
— *nanos* RNA

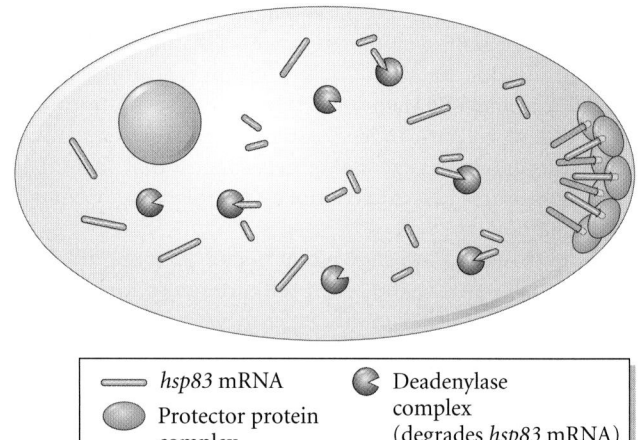

(B) Localized protection

— *hsp83* mRNA   ◖ Deadenylase complex (degrades *hsp83* mRNA)
◖ Protector protein complex

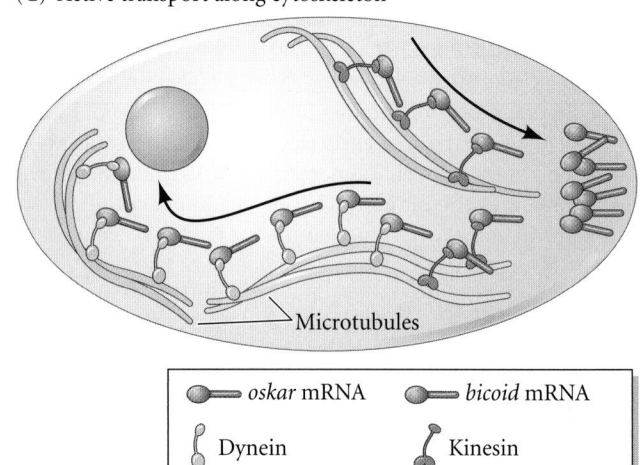

(C) Active transport along cytoskeleton

Microtubules

●— *oskar* mRNA   ●— *bicoid* mRNA
Dynein   Kinesin

Posttranslational Regulation of Gene Expression

The story is not over when a protein is synthesized. Once a protein is made, it becomes part of a larger level of organization. For instance, it may become part of the structural framework of the cell, or it may become involved in one of the many enzymatic pathways for the synthesis or breakdown of cellular metabolites. In any case, the individual protein is now part of a complex "ecosystem" that integrates it into a relationship with numerous other proteins. Several changes can still take place that determine whether or not the protein will be active.

Some newly synthesized proteins remain inactive until certain inhibitory sections are cleaved away. This is what happens when insulin is made from its larger protein precursor. Some proteins must be "addressed" to their specific intracellular destinations in order to function. Proteins are often sequestered in certain regions of the cell, such as membranes, lysosomes, nuclei, or mitochondria. Some proteins need to assemble with other proteins in order to form a functional unit. The hemoglobin protein, the microtubule, and the ribosome are all examples of multiple proteins joining together to form a functional unit. And some proteins are not active unless they bind an ion (such as Ca^{2+}) or are modified by the covalent addition of a phosphate or acetate group. The importance of this last type of protein modification will become obvious in Chapter 3, since many of the critical proteins in embryonic cells just sit there until some signal activates them.

Coda

The emergence of the physical organism is orchestrated by its inherited genome. This process does not require "decoding" that genome so much as interpreting it, much as an orchestra interprets a musical score. It is important to remember that all the processes we have discussed in this chapter—the interacting transcription factors, the binding of RNA polymerase II to the promoter, the initiation of transcription, the elongation of the mRNAs, the kinetics of RNA splicing, and the half-lives of mRNAs in the cytoplasm—are all stochastic events. Each event depends on the different concentrations of the interacting proteins (Cacace et al. 2012; Murugan and Kreiman 2012; Costa et al. 2013; Neuert et al. 2013). Thus, each organism is a unique "performance" coordinated by interactions that tell the individual cell which genes are to be expressed in that cell and which genes are to remain silent. Chapter 3 will detail the mechanisms by which cells tell one another to activate only a certain subset of the genome.

SNAPSHOT SUMMARY: Differential Gene Expression in Development

1. Evidence from molecular biology, cell biology, and somatic cell nuclear cloning has shown that each cell of the body (with very few exceptions) carries the same nuclear genome.

2. Differential gene expression from genetically identical nuclei creates different cell types. Differential gene expression can occur at the levels of gene transcription, nuclear RNA processing, mRNA translation, and protein modification. Notice that RNA processing and export can occur while the RNA is still being transcribed from the gene.

3. Chromatin is made of DNA and proteins. The histone proteins form nucleosomes, and the methylation and acetylation of specific histone residues can activate or repress gene transcription.

4. Histone methylation is often used to silence gene expression. Histones can be methylated by histone methyltransferases and demethylated by histone demethylases.

5. Acetylated histones are often associated with active gene expression. Histone acetyltransferases add acetyl groups to histones, while histone deacetylases remove them.

6. Eukaryotic genes contain promoter sequences to which RNA polymerase II can bind to initiate transcription. To accomplish this, the eukaryotic RNA polymerases are bound by a series of proteins called transcription-associated factors, including TFIID and TFIIB.

7. Eukaryotic genes expressed in specific cell types contain enhancer sequences that regulate their transcription in time and space. Enhancers activate genes on the same chromosome. Enhancer sequences can be within introns or the 3′UTR; they can even be millions of base pairs away from the gene they activate. Enhancers can also act as silencers to suppress the transcription of a gene in inappropriate cell types.

8. Specific transcription factors can recognize specific sequences of DNA in the promoter and enhancer regions. These proteins activate or repress transcription from the genes to which they have bound.

9. Enhancers work in a combinatorial fashion. The binding of several transcription factors can act to promote or inhibit transcription from a certain promoter. In some cases, transcription is activated only if both factor A and factor B are present; in other cases, transcription is activated if either factor A or factor B is present.

10. Enhancers work in a modular fashion. A gene can contain several enhancers, each directing the gene's expression in a particular cell type.

11. A gene encoding a transcription factor can maintain itself in the activated state if the transcription factor it encodes also activates its own promoter. Thus, a transcription factor gene can have one set of enhancer sequences to initiate its activation and a second set of enhancer sequences

(which bind the encoded transcription factor) to maintain its activation.

12. Transcription factors act in different ways to regulate RNA synthesis. Some transcription factors stabilize RNA polymerase II binding to the DNA; some disrupt nucleosomes, increasing the efficiency of transcription.

13. The Mediator complex often serves as the bridge between the enhancer and promoter.

14. Transcription elongation complexes enable the RNA polymerase II to be released from the pre-initiation complex and continue transcribing the DNA.

15. A transcription factor usually has three domains: a sequence-specific DNA binding domain; a *trans*-activating domain that enables the transcription factor to recruit histone remodeling enzymes; and a protein-protein interaction domain that enables it to interact with other proteins on the enhancer or promoter.

16. Even differentiated cells can be converted into another cell type by the activation of a different set of transcription factors.

17. In low CpG-content promoters, transcription correlates with a lack of methylation on the promoter and enhancer regions of genes.

18. In high CpG-content promoters, the nucleosomes often allow transcription to start but do not permit the elongation of the nRNA.

19. Differences in DNA methylation can account for genomic imprinting, wherein a gene transmitted through the sperm is expressed differently than the same gene transmitted through the egg. Some genes are active only if inherited from the sperm or the egg. The imprinting marks appear to be CpG sites that are methylated on either the maternally inherited or paternally inherited locus.

20. Maintaining active gene expression is often accomplished by Trithorax proteins, while active repression is maintained by Polycomb protein complexes that contain histone methyltransferases.

21. Insulators are DNA sequences that bind CTCF protein. Insulators limit the range over which an enhancer can activate a promoter.

22. DNA methylation can block transcription by preventing the binding of certain transcription factors or by recruiting histone methyltransferases or histone deacetylases to the chromatin.

23. Some chromatin is "poised" to act quickly to developmental signals. The mRNA of poised chromatin has begun to be transcribed and its histones have both active and repressive marks.

24. Dosage compensation enables the X chromosome-derived products of males (which have one X chromosome per cell in fruit flies, nematodes, and mammals) to equal the X chromosome-derived products of females (which have two X chromosomes per cell). This compensation is accomplished at the level of transcription, either by doubling the transcription rate from the lone X chromosome in males (*Drosophila*), decreasing the level of transcription from each X chromosome by 50% (*C. elegans*), or by inactivating a large portion of one of the two X chromosomes in females (mammals).

25. Long non-coding RNA (lncRNA) can suppress gene expression from nearby promoters. They can bind to the promoter region and can recruit histone-modifying enzymes that repress translation. *Xist* and *Airn* are two such lcRNAs.

26. Differential RNA splicing can create a family of related proteins by causing different regions of the nRNA to be read as exons or introns. What is an exon in one set of circumstances may be an intron in another.

27. Alternative RNA splicing can create several different proteins from the same pre-mRNA transcript. These proteins (splicing isoforms) can play different roles.

28. Alternative pre-mRNA slicing is accomplished by splicing site recognition factors that can be different in different cell types. Mutations in splice sites can lead to alternative phenotypes and disease.

29. Some messages are translated only at certain times. The oocyte, in particular, uses translational regulation to set aside certain messages that are transcribed during egg development but used only after the egg is fertilized. This activation is often accomplished either by the removal of inhibitory proteins or by the polyadenylation of the message.

30. MicroRNAs can act as translational inhibitors, binding to the 3′ UTR of the RNA. The microRNA recuits an RNA-induced silencing complex that either prevents translation or leads to the degradation of the mRNA.

31. Many mRNAs are localized to particular regions of the oocyte or other cells. This localization appears to be regulated by the 3′ UTR of the mRNA.

32. Ribosomes can differ in different cell types, and ribosomes in one cell may be more efficient at translating certain mRNAs than ribosomes in other cells.

33. Differential gene expression is more like interpreting a musical score than decoding a codescript. It is a stochastic phenomenon in which there are numerous events that have to take place, each having numerous interactions between component parts.

For Further Reading

Complete bibliographical citations for all literature cited in this chapter can be found at the free-access **website www.devbio.com**

Clop, A. and 16 others. 2006. A mutation creating a potential illegitimate microRNA target site in the myostatin gene affects muscularity in sheep. *Nature Genet.* 38: 813–818.

Core, L. J. and J. T. Lis. 2008. Transcriptional regulation through promoter-proximal pausing of RNA polymerase II. *Science* 319: 1791–1792.

Davidson, E. H. 2006. *The Regulatory Genome.* Academic Press, New York.

Jothi, R., S. Cuddapah, A. Barski, K. Cui and K. Zhao. 2008. Genome-wide identification of the in vivo protein-DNA binding sites from ChIP-Seq data. *Nucl. Acids. Res.* 36: 5221–5231.

Migeon, B. R. 2007. *Females Are Mosaics: X Inactivation and Sex Differences in Disease.* Oxford University Press, New York.

Muse, G. W. and 7 others. 2007. RNA polymerase is poised for activation across the genome. *Nature Genet.* 39: 1507–1511.

Ong, T.-C. and V. G. Corces. 2011. Enhancer function: New insights into the regulation of tissue-specific gene expression. *Nature Rev. Genet.* 12: 283–293.

Palacios, I. M. 2007. How does an mRNA find its way? Intracellular localization of transcripts. *Sem. Cell Dev. Biol.* 163–170.

Wilmut, I., K. Campbell and C. Tudge. 2001. *The Second Creation: Dolly and the Age of Biological Control.* Harvard University Press, Cambridge, MA.

Zhou, Q., J. Brown, A. Kanarek, J. Rajagopal and D. A. Melton. 2008. In vivo reprogramming of adult pancreatic exocrine cells to β cells. *Nature* 455: 627–632.

Zhou, V. W., A. Goren and B. E. Bernstein. 2011. Charting histone modifications and the functional organization of mammalian genomes. *Nature Rev. Genet.* 12: 7–18.

Zinzen, R. P., C. Girardot, J. Gagneur, M. Braun and E. E. Furlong. 2009. Combinatorial binding predicts spatio-temporal *cis*-regulatory activity. *Nature* 462: 65–70.

Go Online

WEBSITE 2.1 Does the genome or the cytoplasm direct development? The geneticists versus the embryologists. Geneticists were certain that genes controlled development, whereas embryologists generally favored the cytoplasm. Both sides had excellent evidence for their positions.

WEBSITE 2.2 The origins of developmental genetics. The first hypotheses for differential gene expression came from C. H. Waddington, Salome Glueksohn-Waelsch, and other scientists who understood both embryology and genetics.

WEBSITE 2.3 Techniques of DNA analysis. The entries of this website describe crucial laboratory skills, including gene cloning, DNA sequencing, Southern blotting, knockouts of specific genes, enhancer traps, and identification of methylated sites.

WEBSITE 2.4 Techniques of RNA analysis. Techniques described here include northern blots, RT-PCR, in situ hybridization, microarray technology, antisense RNA, interference RNA, Cre-lox analysis, and ChIP-Seq.

WEBSITE 2.5 Bioinformatics. This entry provides links to various free websites with tools that enable researchers to compare DNA sequences and specific genomes, with the aim of further illuminating the various mechanisms of gene regulation.

WEBSITE 2.6 The 2012 Nobel Prize for Physiology or Medicine: Cloning and nuclear equivalence. The final "proof" of genomic equivalence was the demonstration that the nuclei of differentiated somatic cells could generate any cell type in the body.

WEBSITE 2.7 Chromatin diminution. The inactivation or elimination of entire chromosomes is not uncommon among invertebrates and is sometimes used as a mechanism of sex determination. In some organisms, portions of the chromosomes condense and break off such that only the germ cells have the full chromatin complement.

WEBSITE 2.8 The nuclear envelope's roles in gene regulation. There is evidence that many genes are regulated by enzymes that are localized to the nuclear envelope. The inner portion of the nuclear envelope (the nuclear lamina) may be critical in activating and silencing transcription.

WEBSITE 2.9 Small noncoding RNAs that repress transcription. In some instances, microRNAs can bind to RNA as soon as it is transcribed, and recruit enzymes that repress further transcription. This may be the way that some silencers act.

WEBSITE 2.10 Control of early development by nuclear RNA selection. In addition to alternative nRNA splicing, the nuclear RNA to mRNA stage can also be regulated by RNA "censorship"—selecting which nuclear transcripts are processed into cytoplasmic messages. Different cells select different nuclear transcripts to be processed and sent to the cytoplasm as messenger RNA.

WEBSITE 2.11 So you think you know what a gene is. Different scientists have different definitions, and nature has given us some problematic examples of DNA sequences that may or may not be considered genes.

WEBSITE 2.12 Stored mRNA in brain cells. One of the most important areas of local translational regulation may be in the brain. The storage of long-term memory requires new protein synthesis, and the local translation of mRNAs in the dendrites of brain neurons has been proposed as a control point for increasing the strength of synaptic connections

Vade Mecum

Transdetermination in *Drosophila*. These movies describe Ernst Hadorn's discovery of transdetermination and Walter Gehring's pioneering study of homeotic mutants, changing body parts into eyes through transcription factors.

3
Cell-Cell Communication in Development

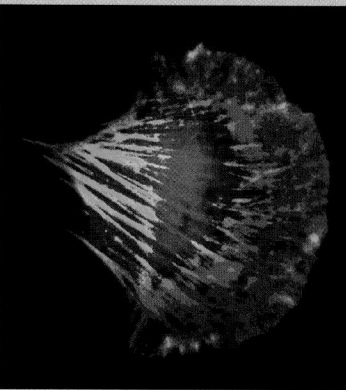

DEVELOPMENT IS MORE THAN JUST DIFFERENTIATION. The different cell types don't exist as random arrays. Rather, they form organized structures such as limbs and hearts. Moreover, the types of cells that constitute our fingers—bones, cartilage, neurons, blood cells, and others—are the same cell types that make up our pelvis and legs. Somehow, the cells must be ordered to take different shapes and make different connections. This formation of organized form is called **morphogenesis**, and it has been one of the great sources of wonder for humankind.

The twelfth-century rabbi and physician Maimonides framed the question of morphogenesis beautifully when he noted that the pious men of his day (around 1190 CE) believed that an angel of God had to enter the womb to form the organs of the embryo. This, the men said, was a miracle. How much more powerful a miracle would life be, Maimonides asked, if the Deity had made matter such that it could generate this remarkable order without a matter-molding angel having to intervene in every pregnancy? The problem addressed today is the secular version of Maimonides' question: How can matter alone construct itself into the organized tissues of the embryo? This huge question can be divided into at least five smaller questions that confront modern embryologists who study morphogenesis:

1. *How are separate tissues formed from populations of cells?* For example, how do bone cells stick to other bone cells to create a bone rather than merging with adjacent capillary cells or muscle cells? How do some bones form a pelvis while other bones form fingers? What keeps the mesoderm separate from the ectoderm such that the skin has both a dermis and an epidermis?

2. *How are organs constructed from tissues?* How are the ventricles formed in the developing heart? How are they distinguished from the atria, and how does the left ventricle lead into the aorta?

3. *How do organs form in particular locations, and how do migrating cells reach their destinations?* Why do eyes only form in the head? What causes there to be two—and usually only two—kidneys, and why are they so close to the genital regions of the body? How do some cells—such as the precursors of our pigment cells, germ cells, and blood cells—travel long distances to reach their final destinations?

4. *How do organs and their component cells grow, and how is their growth coordinated throughout development?* How can some organs, such as the skin and intestine, constantly regenerate themselves through stem cell divisions while other tissues, such as brain and bones, are unable to regenerate? What regulates the size of the lungs such that the left lung grows less than the right lung, thus allowing the heart to fit into the left side of the thorax?

5. *How do organs achieve polarity?* How is it that the retina of the eye is always within the eye, while the lens faces outward? Why do fingers never form in the middle of the hand, and how does one end of the hand form an opposable thumb while the other end forms a pinky?

The behaviour of a cell in an embryo depends on the extent to which it listens to its mother or its neighbourhood. The size and nature of the noise, the way in which it is heard, and the response are unpredictable and can only be discovered by experimentation.
JONATHAN BARD (1997)

All that you touch You Change. All that you Change Changes you. The only lasting truth Is Change.
OCTAVIA BUTLER (1998)

Answers to these questions came slowly and are still coming. In the 1850s, Robert Remak (1852, 1855) formulated the cell theory and showed that all the different cells in the body are the mitotic products of a single cell, the fertilized egg. In the mid-twentieth century, E. E. Just (1939) and Johannes Holtfreter (Townes and Holtfreter 1955) predicted that embryonic cells could have differences in their cell membrane components that would enable the formation of organs. In the late twentieth century, these membrane components—the molecules by which embryonic cells are able to adhere to, migrate over, and induce gene expression in neighboring cells—began to be discovered and described. Today these pathways and networks are being modeled, and we are beginning to understand how the cell integrates the information from its nucleus and from its surroundings to take its place in the community of cells.

As we discussed in Chapter 1, the cells of an embryo are either epithelial or mesenchymal (see Table 1.1). Epithelial cells adhere to one another and can form sheets and tubes, whereas mesenchymal cells often migrate individually and form extensive extracellular matrices that keep the individual cells separate. There appear to be only a few processes through which cells create structured organs (Newman and Bhat 2008), and all of these processes involve the cell surface. This chapter will concentrate on three behaviors requiring cell-cell communication via the cell surface: cell adhesion, cell migration, and cell signaling.

Cell Adhesion

Differential cell affinity

Just as cellular differentiation focuses on the nucleus and differential gene expression, morphogenesis focuses on the cell membrane and its interactions with the peripheral cytoplasm. Biologists once thought that the membrane was identical in all cells, but observations of fertilization and early embryonic development made by E. E. Just (1939) suggested that the cell membrane differed among cell types. The experimental analysis of morphogenesis, however, began with the experiments of Townes and Holtfreter in 1955. Taking advantage of the discovery that amphibian tissues become dissociated into single cells when placed in alkaline solutions, they prepared single-cell suspensions from each of the three germ layers of amphibian embryos soon after the neural tube had formed. Two or more of these single-cell suspensions could be combined in various ways. When the pH of the solution was normalized, the cells adhered to one another, forming aggregates on agar-coated petri dishes. By using embryos from species having cells of different sizes and colors, Townes and Holtfreter were able to follow the behavior of the recombined cells.

The results of their experiments were striking. First, they found that reaggregated cells become spatially segregated. That is, instead of two cell types remaining mixed, each type sorts out into its own region. Thus, when epidermal (ectodermal) and mesodermal cells are brought together in a mixed aggregate, the epidermal cells move to the periphery of the aggregate and the mesodermal cells move to the inside (**FIGURE 3.1**).

Second, the researchers found that the final positions of the reaggregated cells reflect their respective positions in the embryo. The reaggregated mesoderm migrates centrally with respect to the epidermis, adhering to the inner epidermal surface (**FIGURE 3.2A**). The mesoderm also migrates centrally with respect to the gut or endoderm (**FIGURE 3.2B**). However, when the three germ layers are mixed together, the

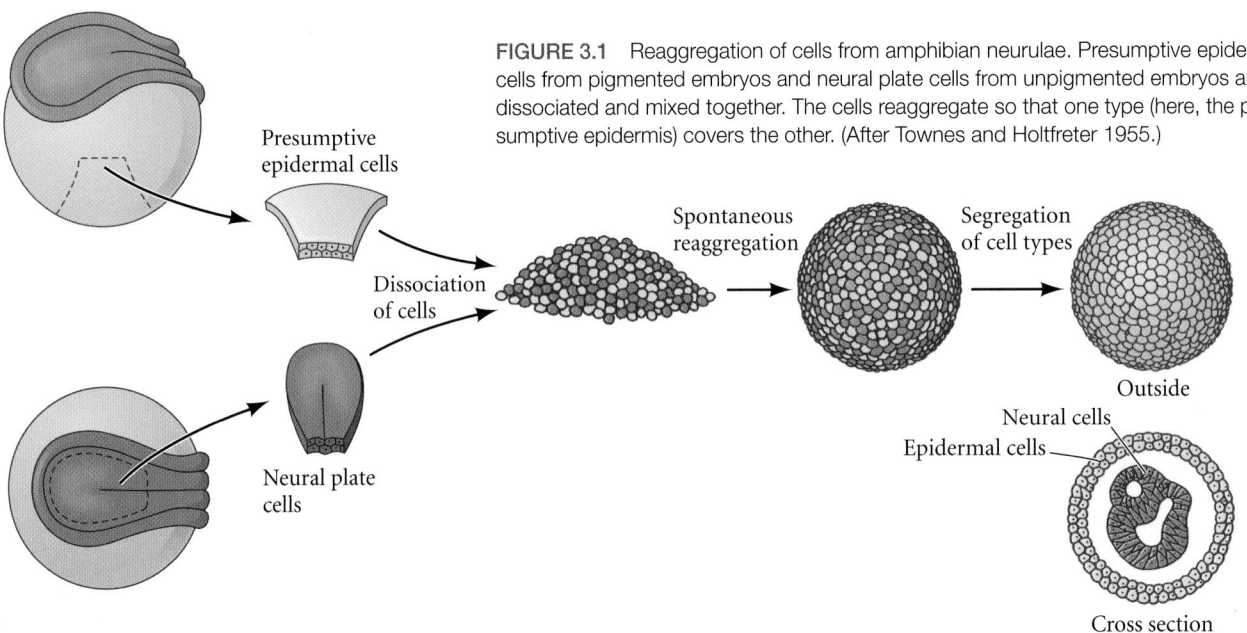

FIGURE 3.1 Reaggregation of cells from amphibian neurulae. Presumptive epidermal cells from pigmented embryos and neural plate cells from unpigmented embryos are dissociated and mixed together. The cells reaggregate so that one type (here, the presumptive epidermis) covers the other. (After Townes and Holtfreter 1955.)

Presumptive epidermal cells

Neural plate cells

Dissociation of cells

Spontaneous reaggregation

Segregation of cell types

Outside

Neural cells

Epidermal cells

Cross section

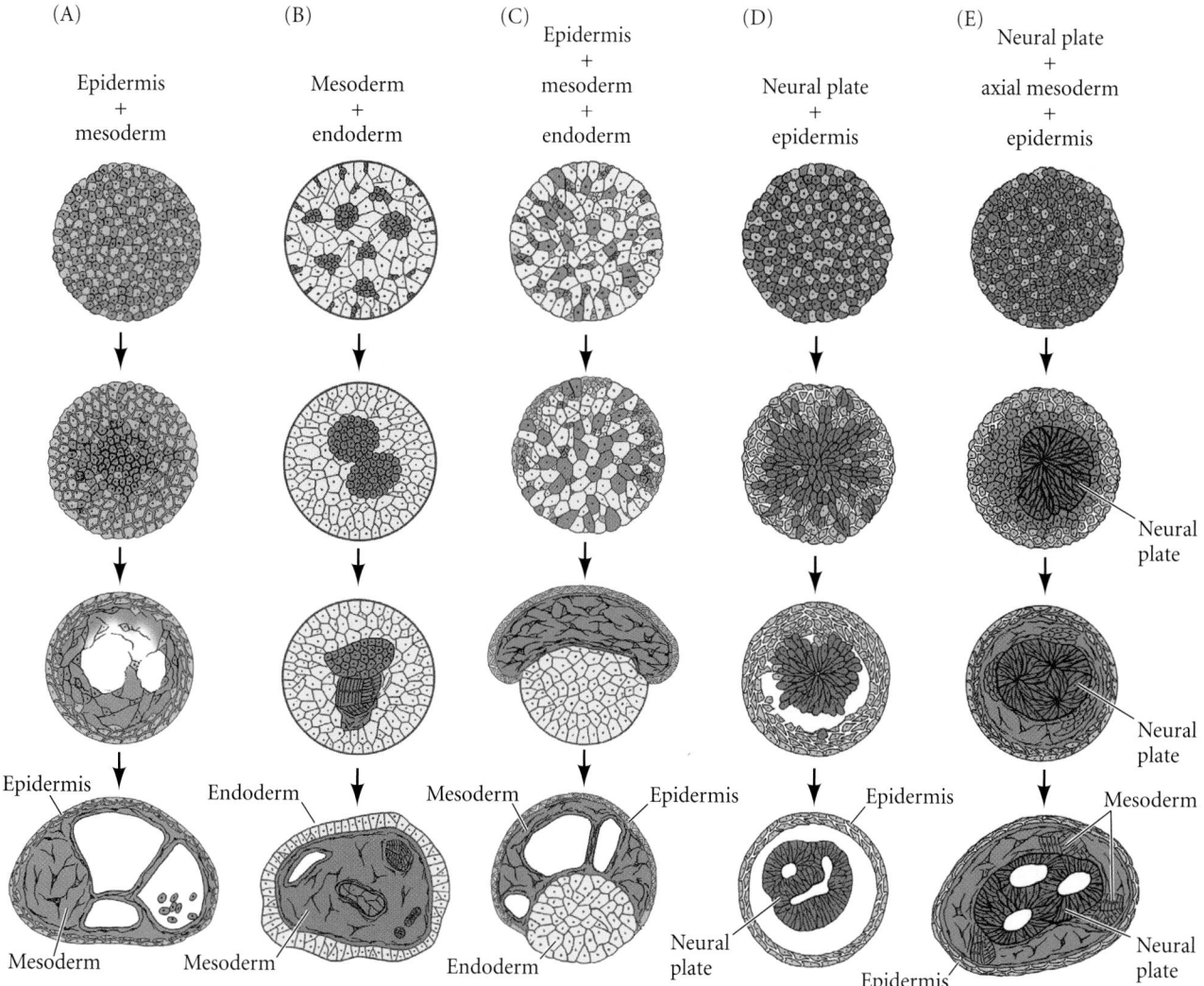

(A) Epidermis + mesoderm

(B) Mesoderm + endoderm

(C) Epidermis + mesoderm + endoderm

(D) Neural plate + epidermis

(E) Neural plate + axial mesoderm + epidermis

Neural plate

Neural plate

Epidermis

Mesoderm   Mesoderm   Endoderm

Epidermis   Mesoderm   Neural plate

Mesoderm   Endoderm

Epidermis   Neural plate   Epidermis

FIGURE 3.2 Sorting out and reconstruction of spatial relationships in aggregates of embryonic amphibian cells. (After Townes and Holtfreter 1955.)

endoderm separates from the ectoderm and mesoderm and is then enveloped by them (**FIGURE 3.2C**). In the final configuration, the ectoderm is on the periphery, the endoderm is internal, and the mesoderm lies in the region between them.

Holtfreter interpreted this finding in terms of **selective affinity**. The inner surface of the ectoderm has a positive affinity for mesodermal cells and a negative affinity for the endoderm, while the mesoderm has positive affinities for both ectodermal and endodermal cells. Mimicry of normal embryonic structure by cell aggregates is also seen in the recombination of epidermis and neural plate cells (**FIGURE 3.2D**; also see Figure 3.1). The presumptive epidermal cells migrate to the periphery as before; the neural plate cells migrate inward, forming a structure reminiscent of the neural tube. When axial mesoderm (notochord) cells are added to a suspension of presumptive epidermal and presumptive neural cells, cell segregation results in an external

epidermal layer, a centrally located neural tissue, and a layer of mesodermal tissue between them (**FIGURE 3.2E**). Somehow, the cells are able to sort out into their proper embryonic positions.

The third conclusion of Holtfreter and his colleagues was that selective affinities change during development. Such changes should be expected, because embryonic cells do not retain a single stable relationship with other cell types. For development to occur, cells must interact differently with other cell populations at specific times. Such changes in cell affinity are extremely important in the processes of morphogenesis. When tissues from later-stage mammalian and chick embryos were made into single cell suspensions (using the enzyme trypsin, which split the proteins connecting the cells together), the cells reaggregated to form tissuelike arrangements (Moscona 1961; Giudice and Just 1962).

The thermodynamic model of cell interactions

Cells, then, do not sort randomly, but can actively move to create tissue organization. What forces direct cell movement

(A)

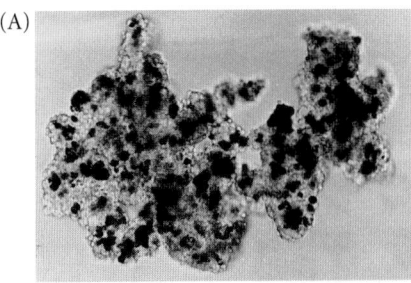

(B)

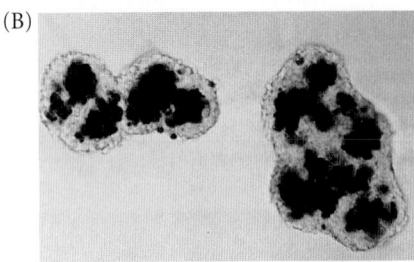

(C)

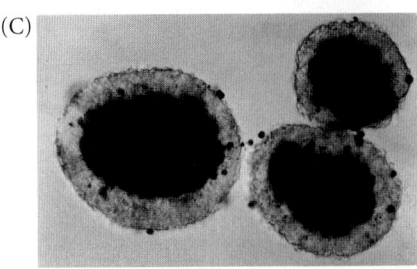

FIGURE 3.3 Aggregates formed by mixing 7-day chick embryo neural retina (unpigmented) cells with pigmented retina cells. (A) Five hours after the single-cell suspensions are mixed, aggregates of randomly distributed cells are seen. (B) At 19 hours, the pigmented retina cells are no longer seen on the periphery. (C) At 2 days, a great majority of the pigmented retina cells are located in a central internal mass, surrounded by the neural retina cells. (The scattered pigmented cells are probably dead cells.) (From Armstrong 1989, courtesy of P. B. Armstrong.)

during morphogenesis? In 1964, Malcolm Steinberg proposed the **differential adhesion hypothesis**, a model that sought to explain patterns of cell sorting based on thermodynamic principles. Using cells derived from trypsinized embryonic tissues, Steinberg showed that certain cell types migrate centrally when combined with some cell types, but migrate peripherally when combined with others. **FIGURE 3.3** illustrates the interactions between pigmented retina cells and neural retina cells. When single-cell suspensions of these two cell types are mixed together, they form aggregates of randomly arranged cells. However, after several hours, no pigmented retina cells are seen on the periphery of the aggregates, and after 2 days, two distinct layers are seen, with the pigmented retina cells lying internal to the neural retina cells. Moreover, such interactions form a hierarchy (Steinberg 1970). If the final position of cell type A is internal to a second cell type B, and the final position of B is internal to a third cell type C, then the final position of A will always be internal to

C. For example, pigmented retina cells migrate internally to neural retina cells, and heart cells migrate internally to pigmented retina cells. Therefore, heart cells migrate internally to neural retina cells.

This observation led Steinberg to propose that cells interact so as to form an aggregate with the smallest interfacial free energy. In other words, the cells rearrange themselves into the most thermodynamically stable pattern. If cell types A and B have different strengths of adhesion, and if the strength of A-A connections is greater than the strength of A-B or B-B connections, sorting will occur, with the A cells becoming central. However, if the strength of A-A connections is less than or equal to the strength of A B connections, then the aggregate will remain as a random mix of cells. Finally, if the strength of A-A connections is far greater than the strength of A-B connections—in other words, if A and B cells show essentially no adhesivity toward one another—then A cells and B cells will form separate aggregates. According to this hypothesis, the early embryo can be viewed as existing in an equilibrium state until some change in gene activity changes the cell surface molecules. The movements that result seek to restore the cells to a new equilibrium configuration. All that is required for sorting to occur is that cell types differ in the strengths of their adhesion.

This was found to be the case. In several meticulous experiments using numerous tissue types, researchers showed that those cell types that had greater surface cohesion migrated centrally compared with cells that had less surface tension (**FIGURE 3.4**; Foty et al. 1996; Krens and Heisenberg 2011). In the simplest form of this model, all cells could have the same type of "glue" on the cell surface. The amount of this "glue," or the cellular architecture that allows such a substance to be differentially distributed across the surface, could create a difference in the number of stable contacts made between cell types. In a more specific version of this model, the thermodynamic differences could be caused by different types of adhesion molecules (see Moscona 1974). When Holtfreter's studies were revisited using modern techniques, Davis and colleagues (1997) found that the tissue surface tensions of the individual germ layers were precisely those required for the sorting patterns observed both in vitro and in vivo.

● **See VADE MECUM** The differential adhesion hypothesis

Cadherins and cell adhesion

Recent evidence shows that boundaries between tissues can indeed be created by different cell types having both different types and different amounts of cell adhesion molecules. Several classes of molecules can mediate cell adhesion, but the major cell adhesion molecules appear to be the cadherins.

As their name suggests, **cadherins** are *ca*lcium-*de*pendent *adhe*sion molecules. They are critical for establishing and maintaining intercellular connections, and they appear

Tissue	Surface tension (dyne/cm)	Equilibrium configuration
Limb bud (green)	20.1	
Pigmented epithelium (red)	12.6	
Heart (yellow)	8.5	
Liver (blue)	4.6	
Neural retina (orange)	1.6	

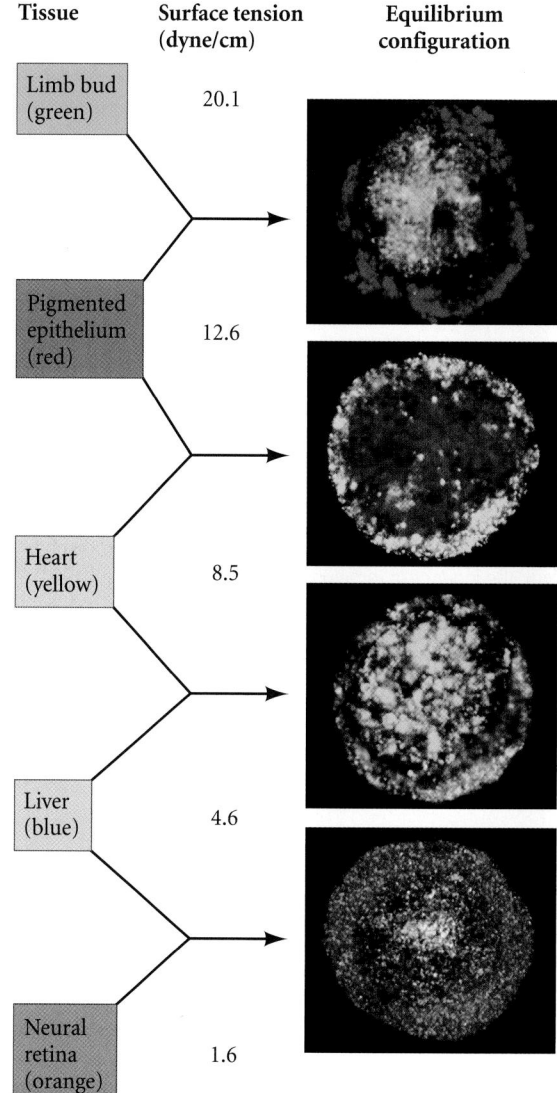

FIGURE 3.4 Hierarchy of cell sorting in order of decreasing surface tensions. The equilibrium configuration reflects the strength of cell cohesion, with the cell types having the greater cell cohesion segregating inside the cells with less cohesion. The images were obtained by sectioning the aggregates and assigning colors to the cell types by computer. The black areas represent cells whose signal was edited out in the program of image optimization. (From Foty et al. 1996, courtesy of M. S. Steinberg and R. A. Foty.)

to be crucial to the spatial segregation of cell types and to the organization of animal form (Takeichi 1987). Cadherins are transmembrane proteins that interact with other cadherins on adjacent cells. The cadherins are anchored inside the cell by a complex of proteins called **catenins** (**FIGURE 3.5A**), and the cadherin-catenin complex forms the classic adherens junctions that help hold epithelial cells together. Moreover, since the cadherins and the catenins bind to the actin (microfilament) cytoskeleton of the cell, they integrate the epithelial cells into a mechanical unit. Blocking cadherin

function (by antibodies that bind and inactivate cadherin) or blocking cadherin *synthesis* (with antisense RNA that binds cadherin messages and prevents their translation) can prevent the formation of epithelial tissues and cause the cells to disaggregate (**FIGURE 3.5B**; Takeichi et al. 1979).

Cadherins perform several related functions. First, their external domains serve to adhere cells together. Second, cadherins link to and help assemble the actin cytoskeleton, thereby providing the mechanical forces for forming sheets and tubes. Third, cadherins can serve as signaling molecules that change a cell's gene expression.

In vertebrate embryos, several major cadherin types have been identified. **E-cadherin** is expressed on all early mammalian embryonic cells, even at the zygote stage. Later in development, this molecule is restricted to epithelial tissues of embryos and adults. **P-cadherin** is found predominantly on the placenta, where it helps the placenta stick to the uterus (Nose and Takeichi 1986; Kadokawa et al. 1989). **N-cadherin** becomes highly expressed on the cells of the developing central nervous system (Hatta and Takeichi 1986), and it may play roles in mediating neural signals. **R-cadherin** is critical in retina formation (Babb et al. 2005). A class of cadherins called **protocadherins** (Sano et al. 1993) lack the attachment to the actin skeleton through catenins. Expressing similar protocadherins is an important means of keeping migrating epithelial cells together; and expressing dissimilar protocadherins is an important way of separating tissues (as when the mesoderm forming the notochord separates from the surrounding mesoderm that will form somites).

Differences in cell surface tension and the tendency of cells to bind together depend on the strength of cadherin interactions (Duguay et al. 2003). This strength can be achieved quantitatively (the more cadherins on the apposing cell surfaces, the tighter the adhesion) or qualitatively (some cadherins will bind to different cadherin types, whereas other cadherins will not bind to different types).

QUANTITY AND COHESION The ability of cells to sort themselves based on the *amount* of cadherin expression was first shown when Steinberg and Takeichi (1994) collaborated on an experiment using two cell lines that were identical except that they synthesized different amounts of P-cadherin. When these two groups of cells, each expressing a different amount of cadherin, were mixed, the cells that expressed more P-cadherin had a higher surface cohesion and migrated internally to the lower-expressing group of cells. Foty and Steinberg (2005) demonstrated that this quantitative cadherin-dependent sorting directly correlated with surface tension (**FIGURE 3.6**). The surface tensions of these aggregates are linearly related to the amount of cadherin they express on the cell surface. The cell sorting hierarchy is strictly dependent on the cadherin interactions between the cells.

Moreover, the energetic value of cadherin-cadherin binding is remarkably strong—about 3400 kcal/mole, or some 200 times stronger than most metabolic protein-protein interactions. This free energy change associated with cadherin

(A)

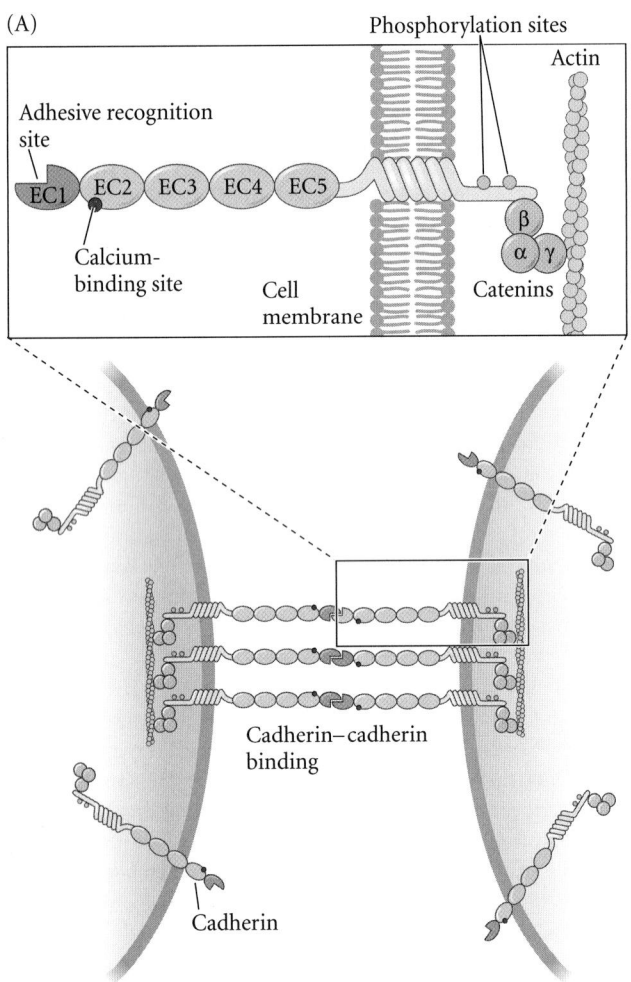

(B)

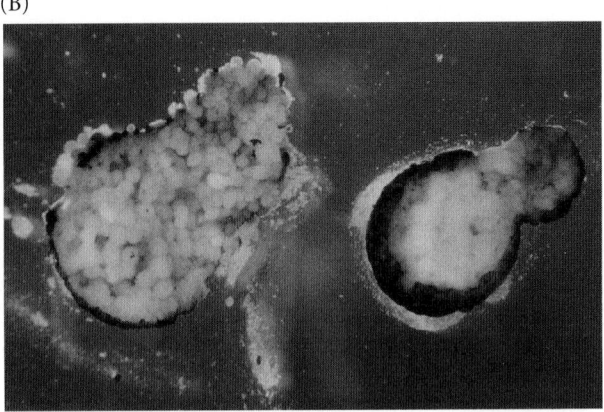

FIGURE 3.5 Cadherin-mediated cell adhesion. (A) Simplified scheme of cadherin linkage to the cytoskeleton via catenins. (B) When an oocyte is injected with an antisense oligonucleotide against a maternally inherited cadherin mRNA (thus preventing synthesis of the cadherin), the inner cells of the resulting embryo disperse when the animal cap is removed (left). In control embryos (right), the inner cells remain together. (A after Takeichi 1991; B from Heasman et al. 1994, courtesy of J. Heasman.)

(A)

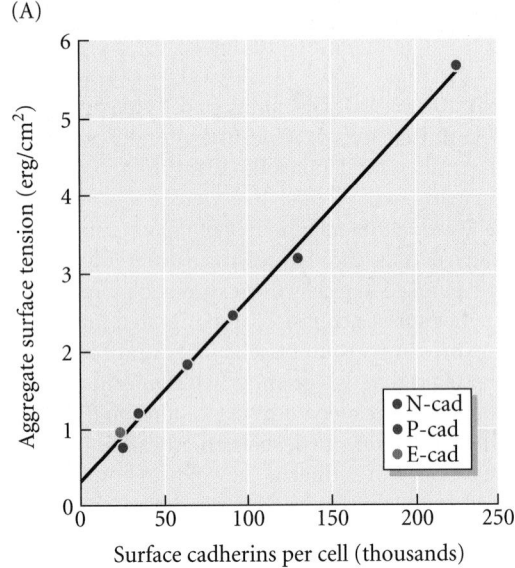

(B)

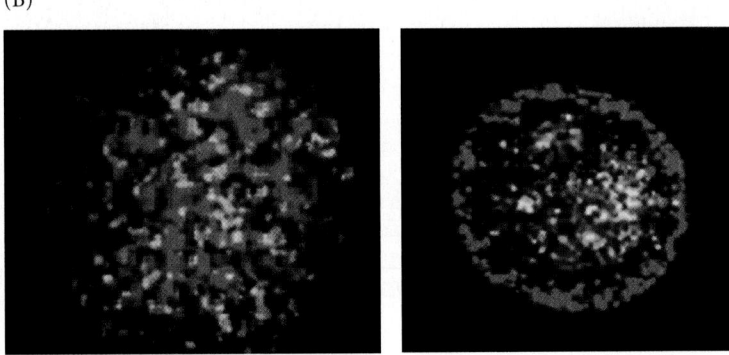

FIGURE 3.6 Importance of the amount of cadherin for correct morphogenesis. (A) Aggregate surface tension correlates with the number of cadherin molecules on the cell membranes. (B) Sorting out of two subclones having different amounts of cadherin on their cell surfaces. The green-stained cells had 2.4 times as many N-cadherin molecules in their membrane as did the other cells. (These cells had no normal cadherin genes being expressed.) At 4 hours of incubation (left), the cells are randomly distributed, but after 24 hours of incubation (right), the red cells (with a surface tension of about 2.4 erg/cm^2) have formed an envelope around the more tightly cohering (5.6 erg/cm^2) green cells. (After Foty and Steinberg 2005, photographs courtesy of R. Foty.)

function could be dissipated by depolymerizing the actin skeleton. The underlying actin cytoskeleton appears to be crucial in organizing the cadherins in a manner that allows them to form remarkably stable linkages between cells (Foty and Steinberg 2004).

TYPE, TIMING, AND GETTING ALONG WITH OTHERS The quantitative effects of cadherins are crucial, but *qualitative* interactions—that is, the *type* of cadherin expressed—also can be important. Duguay and colleagues (2003) showed, for instance, that R-cadherin and B-cadherin do *not* bind well to each other. In another example, the expression of N-cadherin is important in separating the precursors of the neural cells from the precursors of the epidermal cells. All early embryonic cells originally contain E-cadherin, but those cells destined to become the neural tube lose E-cadherin and gain N-cadherin. If epidermal cells are experimentally made to express N-cadherin, or if N-cadherin synthesis is blocked in prospective neural cells, the border between the skin and the nervous system fails to form properly (**FIGURE 3.7**; Kintner et al. 1992).

The timing of particular developmental events can also depend on cadherin expression. For instance, N-cadherin appears in the mesenchymal cells of the developing chick leg just before these cells condense and form nodules of cartilage (which are the precursors of the limb skeleton). N-cadherin is not seen prior to condensation, nor is it seen afterward. If the limbs are injected just prior to condensation with antibodies that block N-cadherin, the mesenchyme cells fail to condense and cartilage fails to form (Oberlender and Tuan 1994). It therefore appears that the signal to begin cartilage formation in the chick limb is the appearance of N-cadherin.

During development, the many cadherins also work with other adhesion systems. For example, one of the most critical times in a mammal's life occurs soon after conception, as the embryo passes from the oviduct and enters the uterus. If development is to continue, the embryo must adhere to and embed itself in the uterine wall. That is why the first differentiation event in mammalian development distinguishes the trophoblast cells (the outer cells that bind to the uterus) from the inner cell mass (those cells that will generate the embryo and eventually the mature organism; see Chapter 9). This differentiation process occurs as the embryo travels from the upper regions of the oviduct on its way to the uterus. Trophoblast cells are endowed with several adhesion molecules that anchor the embryo to the uterine wall. They contain both E- and P-cadherins (Kadokawa et al. 1989), and these two molecule types recognize similar cadherins on the uterine cells. They also have receptors (integrin proteins) for the collagen and the heparan sulfate glycoproteins of the uterine wall (Farach et al. 1987; Carson et al. 1988, 1993; Cross et al. 1994). For something as important as the implantation of the mammalian embryo, it is not surprising that multiple cell adhesion systems appear to work together.

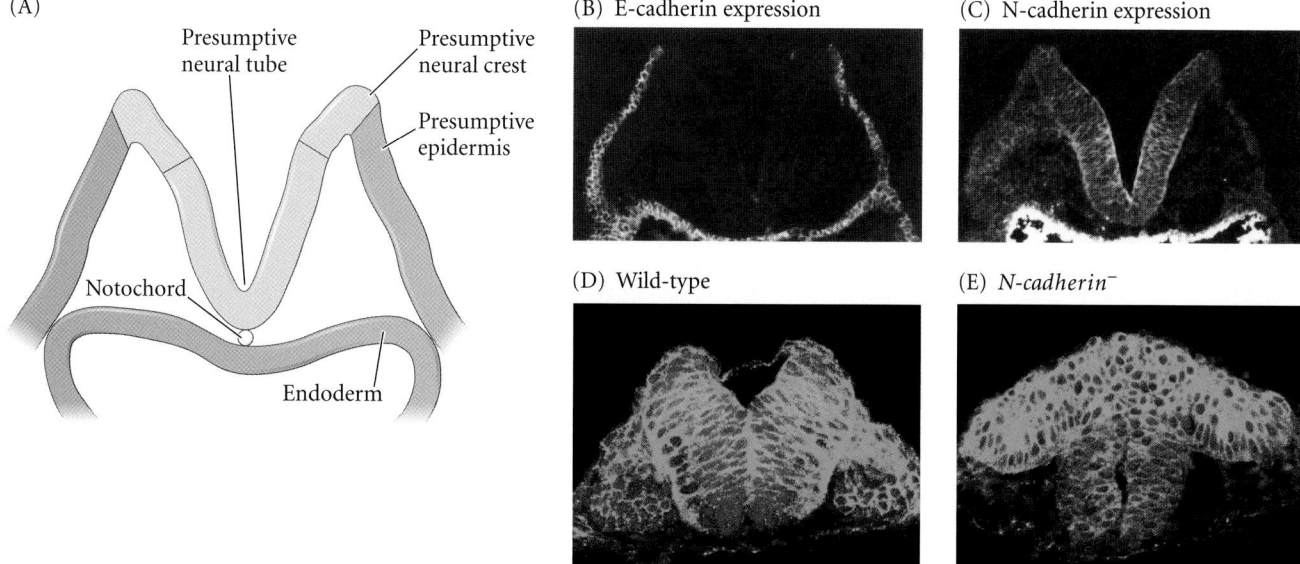

FIGURE 3.7 Importance of the types of cadherin for correct morphogenesis. (A–C) Neural and epidermal tissues in a cross section of a mouse embryo showing the domains of E-cadherin expression (B) and N-cadherin expression (C). N-cadherin is critical for separation of presumptive epidermal and neural tissues during organogenesis. (D,E) The neural tube separates cleanly from surface epidermis in wild-type zebrafish embryos (D) but not in mutant embryos where N-cadherin fails to be made (E). In these images, the cell outlines are stained green with antibodies to β-catenin, while the cell interiors are stained blue. (B,C, photographs by K. Shimamura and H. Matsunami, courtesy of M. Takeichi; D,E from Hong and Brewster 2006, courtesy of R. Brewster.)

Shape Change and Epithelial Morphogenesis: "The Force Is Strong in You"

Epithelial cells form sheets and tubes. Their ability to form such structures often depends on cell shape changes that usually involve cadherins and the actin cytoskeleton. The extracellular domains of cadherins bind groups of cells together, while the intracellular domains of the cadherins alter the actin cytoskeleton. The proteins mediating this cadherin-dependent remodeling of the cytoskeleton are usually (1) non-muscle myosin, which provides the energy for actin contraction, and (2) the Rho family of GTPases, which convert soluble actin into fibrous actin cables that anchor at the cadherins. These Rho GTPases are generally divided into three groups that have different but overlapping functions. **RhoA** primarily organizes stress fibers, such as those that transiently anchor cells during cell migration. **Rac1** is involved in producing lamellipodia (the broad, membranous sheet extending from migrating cells). **Cdc42** is used primarily in constructing filopodia (the thin extensions of cytoplasm used to sense the cells' environment).

Two examples of cadherin-dependent remodeling of the cytoskeleton are the formation of the neural tube in vertebrates and the internalization of the mesoderm in *Drosophila*. In both cases, the cells (neural ectoderm in vertebrates, mesoderm in *Drosophila*) are on the outside of the embryo, and it is critical that they migrate to the inside.

Involution of the frog neural tube

In the early frog embryo, each cell membrane can contain several types of cadherins. Each cell of the gastrula is covered with C-cadherin. However, the presumptive *neural tube* ectoderm cell membranes also contain N-cadherin concentrated in their apical (upper) regions; the presumptive *epidermal* cells of the ectoderm express E-cadherin on their lateral

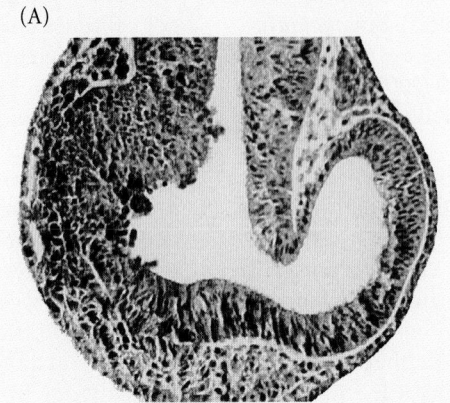
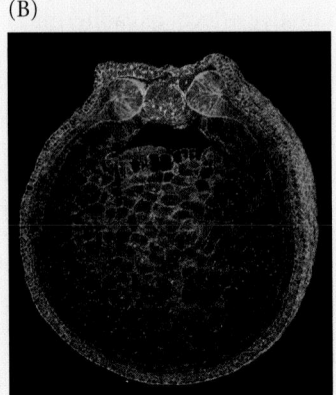

FIGURE 3.8 Importance of cadherin in cell adhesion and morphogenetic movements. (A) Frog gastrula injected with a nonfunctioning N-cadherin gene on one side. The uninjected side (right) develops normally; on the injected side (left), the epidermis and neural tissue fail to separate. (B) Cross section of a *Xenopus* neurula stained for F-actin (green). The red-staining cells of the neural plate (uppermost cells) and notochord (the mesodermal rod beneath them) have in them a morpholino oligonucleotide that prevents *N-cadherin* mRNA from being translated. The neural plate cells fail to invaginate into the embryo or to form a neural tube because of the loss of the N-cadherin-based actin assembly in their apical cytoplasm. (A from Kintner et al. 1992, courtesy of C. Kintner; B from Nandadasa et al. 2009, courtesy of C. Wylie.)

and basal (lower) surfaces. The actin organized in the apical region of the neural cells causes them to change shape and enter the internal region of the embryo as a neural tube. The actin organized on the lateral sides of the epidermal cells enables the migratory movements of the epidermal (skin) cells over the surface of the embryo. If N-cadherin is experimentally removed from a frog gastrula, the cells still adhere (thanks to the C-cadherin that is still present), but the actin (and the activated myosin that binds to it) fail to assemble apically and there is no neurulation: the presumptive neural cells do not enter the embryo, and no neural tube forms (**FIGURE 3.8**; Nandadasa et al. 2009).

Drosophila mesoderm formation

In *Drosophila*, the mesoderm is formed from epithelial cells on the ventral side of the embryo. These cells form a furrow and then migrate inside the embryo (**FIGURE 3.9**). To create

this furrow, the cube-shaped cells become wedge-shaped, constricting at their apical surfaces. This transition creates a force that pushes the ventral cells inside the embryo. What creates this force? The apical constriction is brought about by the rearrangement of actin microfilaments and myosin II (a "non-skeletal myosin") to the apical end of the cell (**FIGURE 3.10**). Actin microfilaments are part of the cytoskeleton and are often found on the periphery of the cell. (Indeed, they are critical for producing the cleavage furrows of cell division.) The instructions for this apical constriction appear to emanate from the *Twist* gene, which is only expressed in the nuclei of the ventralmost cells (Kölsch et al. 2007). The Twist protein activates other genes, whose protein products cause the actin cytoskeleton to build up on the apical side of the cell. This build-up is accomplished by the binding of a Rho GTPase and β-catenin to E-cadherin on the apical portion of

(A)

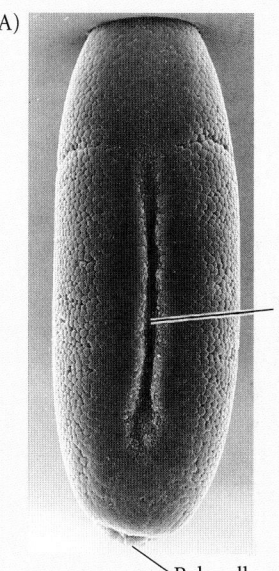

Ventral
furrow

Pole cells

(B)

FIGURE 3.9 Ventral furrow formation during *Drosophila* gastrulation internalizes the cells that will become the mesoderm. (A) Ventral furrow in the *Drosophila* embryo seen by scanning electron microscopy. (B) Cross section through the center of such an embryo, demonstrating cell shape changes and the redistribution of protein products along the dorsal-ventral axis. Apical constrictions can be observed, mediated by G proteins in the furrow. (A courtesy of F. R. Turner; B courtesy of V. Kölsch and M. Leptin.)

the membrane in the most ventral cells. Once stabilized, the actin-myosin complex in the apical cortex constricts like the drawstring of a purse, causing the cells to change shape, buckle inward, and enter the embryo to form the mesoderm (Dawes-Hoang et al. 2005).

External signals: Insect trachea

In the above cases, the instructions for folding come from inside the cell. Instructions for cell shape change can also arise outside the cell. For instance, the tracheal (respiratory) system in *Drosophila* embryos develops from epithelial sacs. The approximately 80 cells in each of these sacs become reorganized into primary, secondary, and tertiary branches without any cell division or cell death (Ghabrial and Krasnow 2006). This reorganization is initiated when nearby cells secrete a protein called Branchless, which acts as a

(A)   Dorsal

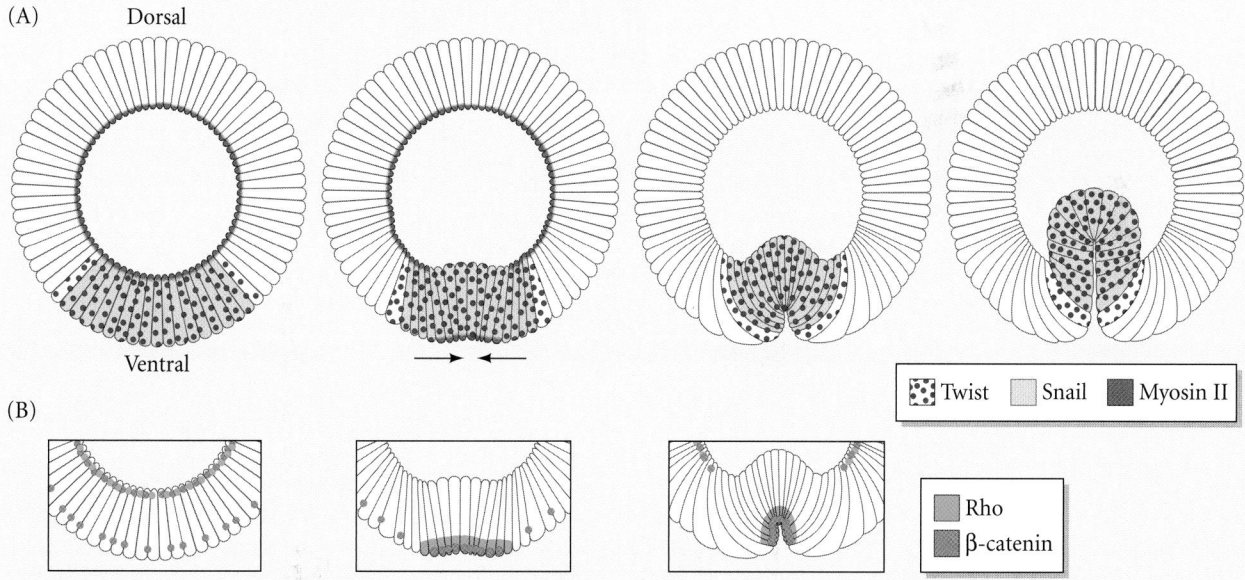

Ventral

(B)

| ⠿ Twist | ☐ Snail | ■ Myosin II |

| ▨ Rho |
| ▨ β-catenin |

FIGURE 3.10 Getting mesodermal cells inside the embryo during *Drosophila* gastrulation by regulation of the cytoskeleton. (A) Schematic representation of ventral furrow formation shown as cross sections through the *Drosophila* embryo, progressing in time from left to right. The ventral cells are defined by the expression of transcription factors Twist and Snail. These cells accumulate myosin II at their apical surfaces. When myosin II interacts with actin already present, the cells begin to constrict apically

and thus invaginate. (B) Close-ups of the ventral domain. Before the initiation of ventral furrow formation, Rho GTPase (green) and β-catenin (orange) both reside along the basal surface (facing the interior of the embryo) of the ventral cells. β-catenin is also found in a subapical region in all cells. Formation of the ventral furrow begins with the relocalization of Rho and β-catenin, which move from the basal surface to accumulate apically, at the opposite end of the cell. (From Kölsch et al. 2007.)

chemoattractant.* Branchless binds to a receptor on the cell membranes of the epithelial cells. The cells receiving the most Branchless protein lead the rest, while the followers (connected to each other by cadherins) receive a signal from the leading cells to form the tracheal tube (FIGURE 3.11). It is the lead cell that will change its shape (by rearranging its actin-myosin cytoskeleton via a Rho GTPase-mediated process, just like the mesodermal cells) to migrate and to form the secondary branches. During this migration, cadherin proteins are regulated such that the epithelial cells can migrate over one another to form a tube while keeping their integrity as an epithelium (Cela and Llimagas 2006).

But another external force is also at work. The dorsalmost secondary branches of the sacs move along a groove that forms between the developing muscles. These tertiary cell migrations cause the trachea to become segmented around the musculature (Franch-Marro and Casanova 2000). In this way, the respiratory tubes are placed close to the larval musculature.

*Chemoattractants are usually diffusible molecules that attract a cell to migrate along an increasing concentration gradient toward the cells secreting the factor. There are also *chemorepulsive* factors that send the migrating cells in an opposite direction. Generally speaking, chemotactic *factors*—soluble factors that cause cells to move in a particular direction—are assumed to be chemoattractive unless otherwise described.

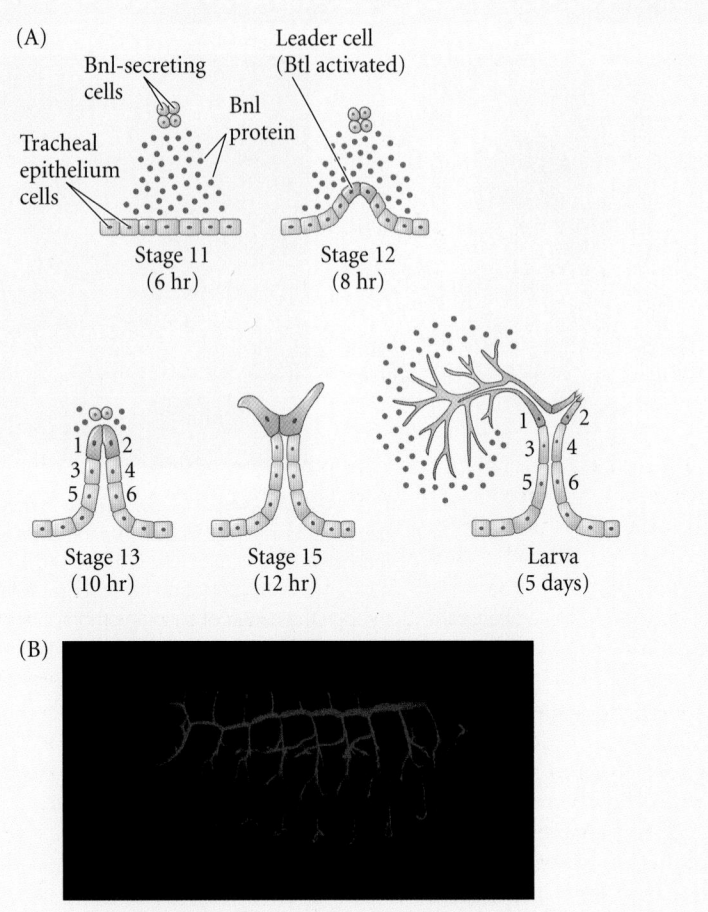

FIGURE 3.11 Tracheal development in *Drosophila*. (A) Diagram of dorsal tracheal branch budding from tracheal epithelium. Nearby cells secrete Branchless protein (Bnl; blue dots), which activates Breathless protein (Btl) on tracheal cells. The activated Btl induces migration of the leader cells and tube formation; the dorsal branch cells are numbered 1–6. Branchless also induces unicellular secondary branches (stage 15). (B) Larval *Drosophila* tracheal system visualized with a fluorescent red antibody. Note the intercalated branching pattern. (A after Ghabrial and Krasnow 2006; B from Casanova 2007.)

Cell Migration

Cell migration is a common feature of both epithelial and mesenchymal cells (Kurosaka and Kashina 2008). For example, the cells of the embryo move extensively during gastrulation to form the three germ layers; the neural tube folds into the vertebrate embryo; the mesoderm folds into the fly embryo; and the precursors of the germ cells, blood cells, and pigment cells undergo individual and extensive migrations. Cell migration is a combination of motility and guidance. As Rørth (2011) noted, motility is the legwork, but it can go anywhere and thus requires guidance. The nose smells food, and this sensation directs our movements. Both the signal and the movement are needed.

In epithelia, the motive force for migration is usually provided by the cells at the edge of the sheet, and the rest of the cells follow passively. In mesenchymal cell migration, individual cells become polarized and migrate through the extracellular milieu. In both cases, there is a broad reorganization of the actin cytoskeleton. The first stage of migration is **polarization**, wherein a cell defines its front and its back (FIGURE 3.12A). Polarization can be directed by diffusing signals (such as a chemotactic protein) or by signals from the extracellular matrix. These signals reorganize the cytoskeleton so that the cell has a front and a back, and so that the front part of the cell becomes structurally different from the back of the cell (Rodionov and Borisy 1997; Malikov et al. 2005).

(A)

(B)

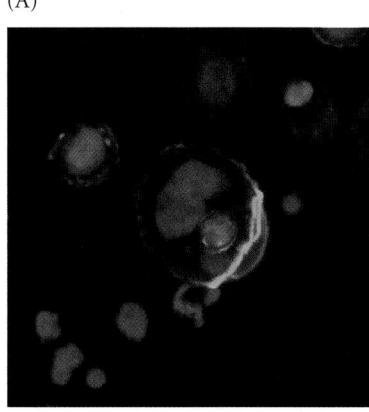
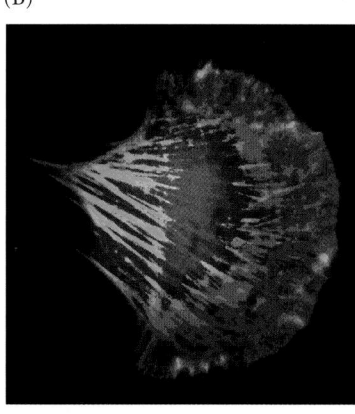

FIGURE 3.12   Cell migration. (A) Polarization of a migrating cell. Cell interiors are stained blue. The actin (stained green with antibodies to β-catenin) becomes redistributed by interstitial flow to the leading edge of the cell. (B) In the lamellipodium of a migrating mesenchymal cell, the ratio of filamentous actin to globular actin is visualized by different colors, blue being the highest filamentous actin level, red being the lowest. (A from Shields et al. 2007, courtesy of M. A. Swartz; B from Cramer et al. 2002, courtesy of L. Cramer.)

The second stage of migration is the protrusion of the cell's leading edge. The mechanical force for this is the polymerization of the actin microfilaments at the cell membrane, creating long parallel bundles (forming filopodia) or broad sheets (forming lamellipodia; FIGURE 3.12B). The membrane-bound Rho GTPase activate the WASP-N proteins to nucleate actin and connect it to cadherins and the cell membrane (Co et al. 2007).

The third stage of migration involves the adhesion of the cell to its extracellular substrate. The moving cell, needing something to push off on, attaches to the surrounding matrix. The key molecules in this process (as we will detail later in this chapter) are proteins called **integrins**. Integrins span the cell membrane, connecting the extracellular matrix outside the cell to the actin cytoskeleton on the inside of the cell. These connections of actin to integrin form **focal adhesions** on the cell membrane where the membrane contacts the extracellular matrix. Myosin and its regulators provide the motive force along these actin microfilaments, and they are linked with the lamellipodial actin at the sites of adhesion (Giannone et al. 2007).

The fourth stage of cell migration concerns the release of adhesions in the rear, allowing the cell to migrate in the forward direction. It is probable that stretch-sensitive calcium channels are opened and that the released calcium ions activate proteases that destroy the focal adhesion sites.

Cell Signaling

Induction and competence

From the earliest stages of development through the adult, cell behaviors such as adhesion, migration, differentiation, and division are regulated by signals from one cell being received by another cell. Indeed, these interactions (which are often reciprocal, as we will describe later) are what allow organs to be constructed. The development of the vertebrate eye is a classic example used to describe the modus operandi of tissue organization via intercellular interactions.

In the vertebrate eye, light is transmitted through the transparent corneal tissue and focused by the lens tissue (the diameter of which is controlled by muscle tissue), eventually impinging on the tissue of the neural retina. The precise arrangement of tissues in the eye cannot be disturbed without impairing its function. Such coordination in the construction of organs is accomplished by one group of cells changing the behavior of an adjacent set of cells, thereby causing them to change their shape, mitotic rate, or cell fate. This kind of interaction at close range between two or more cells or tissues of different histories and properties is called **induction**.

There are at least two components to every inductive interaction. The first component is the **inducer**, the tissue that produces a signal (or signals) that changes the cellular behavior of the other tissue. Often this signal is a secreted protein called a paracrine factor. **Paracrine factors** are proteins made by a cell or a group of cells that alter the behavior or differentiation of adjacent cells. In contrast to endocrine factors (hormones), which travel through the blood and exert their effects on cells and tissues far away,* paracrine factors are secreted into the extracellular space and influence their close neighbors. (The Branchless protein is such a factor; see Figure 3.11.) The second component, the **responder**, is the cell or tissue being induced. Cells of the responding tissue must have both a receptor protein for the inducing factor (e.g., the receptor for Branchless is the Breathless protein) and the *ability* to respond to the signal. The ability to respond to a specific inductive signal is called **competence** (Waddington 1940).

For instance, in the initiation of the vertebrate eyes, paired regions of the brain bulge out and approach the surface ectoderm of the head. The head ectoderm is competent to respond to the paracrine factors made by these brain bulges (the **optic vesicles**), and the head ectoderm receiving these

---

*Some endocrine factors are active in development; these include the hormones estrogen, testosterone, and thyroxine. They affect many tissues simultaneously and often coordinate development throughout the body (as in metamorphosis or the morphogenesis of sexual phenotypes). These hormones work directly, binding to a dormant transcription factor ("receptor"), thereby activating the transcription factor and allowing it to enter the nucleus and bind to DNA.

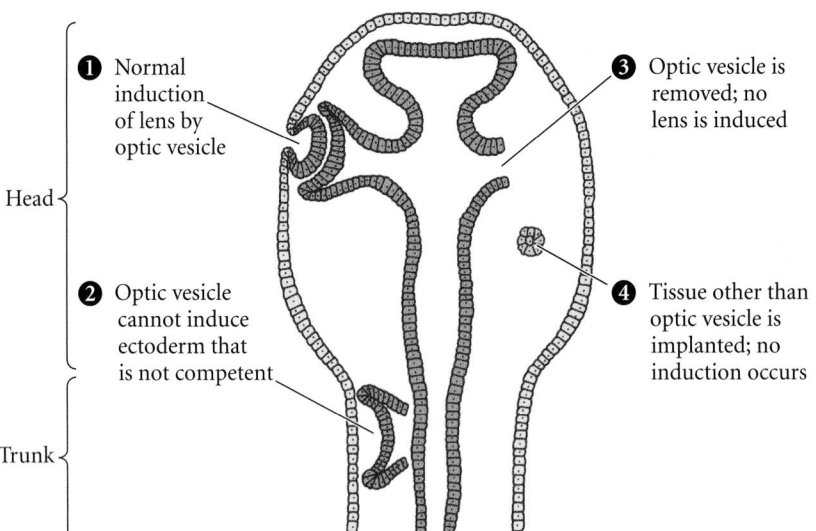

FIGURE 3.13 Ectodermal competence and the ability to respond to the optic vesicle inducer in *Xenopus*. The optic vesicle is able to induce lens formation in the anterior portion of the ectoderm (1) but not in the presumptive trunk and abdomen (2). If the optic vesicle is removed (3), the surface ectoderm forms either an abnormal lens or no lens at all. Most other tissues are not able to substitute for the optic vesicle (4).

paracrine factors is induced to form the lens of the eye. The genes for lens proteins become induced in the head ectoderm cells and are expressed in these cells. The Rho-family GTPases are activated to control the elongation and curvature of the lens fibers (see Chapter 10; Maddala et al. 2008). Moreover, the prospective lens cells secrete paracrine factors that instruct the optic vesicle to form the retina. Thus, the two major parts of the eye co-construct each other, and the eye forms from reciprocal paracrine interactions. The head ectoderm is the only region capable of responding to the optic vesicle. If an optic vesicle from a *Xenopus laevis* embryo is placed underneath head ectoderm in a different part of the head from where the frog's optic vesicle normally occurs, the vesicle will induce that ectoderm to form lens tissue; trunk ectoderm, however, will not respond to the optic vesicle (**FIGURE 3.13**; Saha et al. 1989; Grainger 1992). Only head

ectoderm is *competent* to respond to the signals from the optic vesicle by producing a lens.

Often, one induction will give a tissue the competence to respond to another inducer. Studies on amphibians suggest that the first inducers of the lens may be the foregut endoderm and heart-forming mesoderm that underlie the lens-forming ectoderm during the early and mid gastrula stages (Jacobson 1963, 1966). The anterior neural plate may produce the next signals, including a signal that promotes the synthesis of Pax6 transcription factor in the anterior ectoderm (**FIGURE 3.14**; Zygar et al. 1998). Pax6 is important in providing the competence for the ectoderm to respond to the inducers from the optic cup (Fujiwara et al. 1994). Thus, although the optic vesicle appears to be *the* inducer, the anterior ectoderm has already been induced by at least two other tissues. (The optic vesicle's situation is like that of the player who kicks the

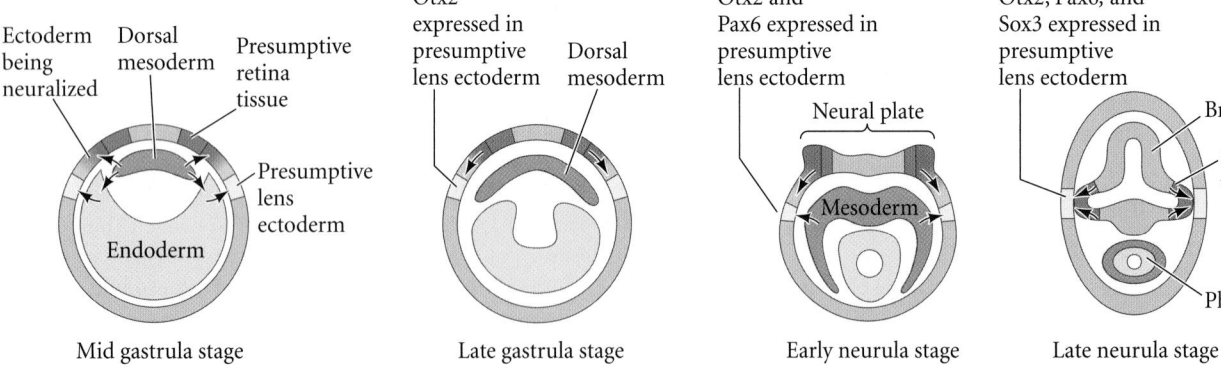

FIGURE 3.14 Sequence of amphibian lens induction postulated by experiments on embryos of the frog *Xenopus laevis*. Unidentified inducers (possibly from the foregut endoderm and cardiac mesoderm) cause the synthesis of the Otx2 transcription factor in the head ectoderm during the late gastrula stage. As the neural folds rise, inducers from the anterior neural plate (including the region that will form the retina) induce Pax6 expression in the anterior

ectoderm that can form lens tissue. Expression of Pax6 protein may constitute the competence of the surface ectoderm to respond to the optic vesicle during the late neurula stage. The optic vesicle secretes BMP and FGF family paracrine factors that induce the synthesis of the Sox transcription factors and initiate observable lens formation. (After Grainger 1992.)

"winning" goal in a soccer match.) The optic vesicle appears to secrete two paracrine factors, one of which is BMP4 (Furuta and Hogan 1998), a protein that is received by the lens cells and induces the production of the Sox transcription factors. The other paracrine factor is Fgf8, a signal that induces the appearance of the L-Maf transcription factor (Ogino and Yasuda 1998; Vogel-Höpker et al. 2000). As we saw in Chapter 2, the combination of Pax6, Sox2, and L-Maf in the ectoderm is needed for the production of the lens and the activation of lens-specific genes such as δ-*crystallin*.

Reciprocal induction

Another feature of induction is the reciprocal nature of many inductive interactions. To continue the above example, once

the lens has formed, it induces other tissues. One of these responding tissues is the optic vesicle itself; thus, the inducer becomes the induced. Under the influence of factors secreted by the lens, the optic vesicle becomes the optic cup, and the wall of the optic cup differentiates into two layers, the pigmented retina and the neural retina (**FIGURE 3.15**; Cvekl and Piatigorsky 1996; Strickler et al. 2007). Such interactions are called **reciprocal inductions**.

Another principle can be seen in such reciprocal inductions: a structure does not need to be fully differentiated in order to have a function. As we will detail in Chapter 10, the optic vesicle induces the surface ectoderm to become a lens before the optic vesicle has become the retina. Similarly, the developing lens reciprocates by inducing the optic vesicle

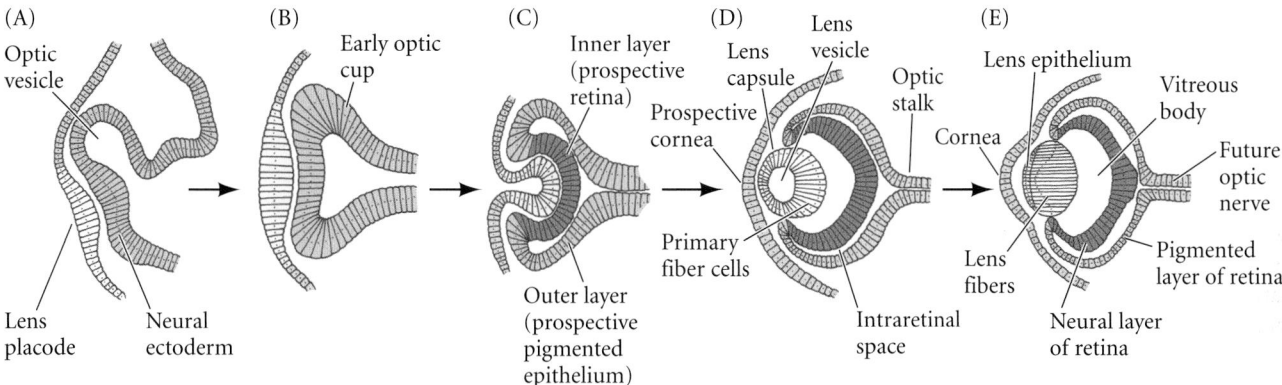

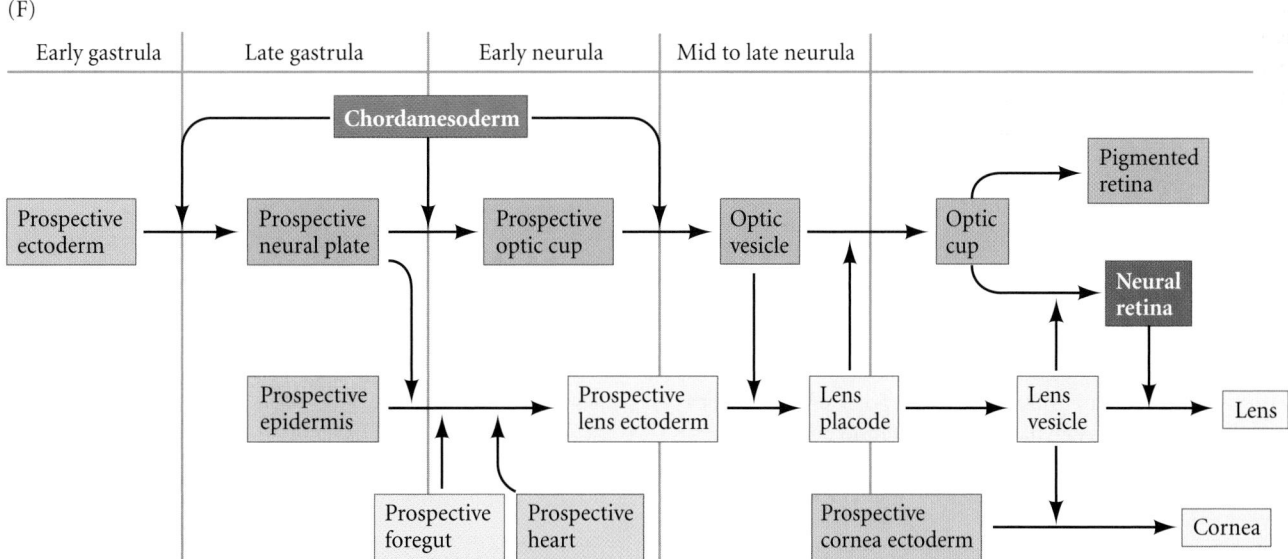

FIGURE 3.15 Schematic diagram of induction of the mouse lens. (A) At embryonic day 9, the optic vesicle extends toward the surface ectoderm from the forebrain. The lens placode (the prospective lens) appears as a local thickening of the surface ectoderm near the optic vesicle. (B) By the middle of day 9, the lens placode has enlarged and the optic vesicle has formed an optic cup. (C) By the middle of day 10, the central portion of the lens-forming ectoderm

invaginates while the two layers of the retina become distinguished. (D) By the middle of day 11, the lens vesicle has formed. (E) By day 13, the lens consists of anterior cuboidal epithelial cells and elongating posterior fiber cells. The cornea develops in front of the lens. (F) Summary of some of the inductive interactions during eye development. (A–E after Cvekl and Piatigorsky 1996.)

FIGURE 3.16 Feather induction in the chick. (A) Feather tracts on the dorsum of a 9-day chick embryo. Note that each feather primordium is located between the primordia of adjacent rows. (B) In situ hybridization of a 10-day chick embryo shows Sonic hedgehog expression (dark spots) in the ectoderm of the developing feathers and scales. (A courtesy of P. Sengal; B courtesy of W.-S. Kim and J. F. Fallon.)

(A)

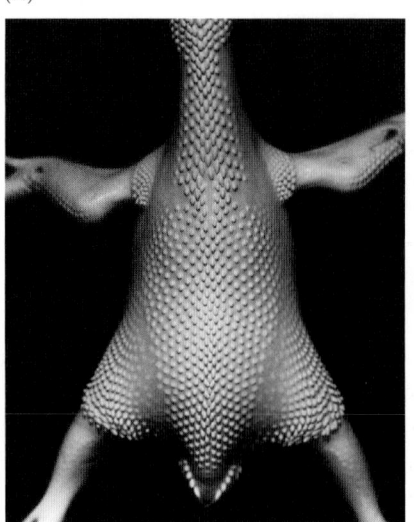

(B)

before the lens forms its characteristic fibers. Thus, before a tissue has its "adult" functions, it has critically important transient functions in building the organs of the embryo.

INSTRUCTIVE AND PERMISSIVE INTERACTIONS Howard Holtzer (1968) distinguished two major modes of inductive interaction. In **instructive interaction**, a signal from the inducing cell is necessary for initiating new gene expression in the responding cell. Without the inducing cell, the responding cell is not capable of differentiating in that particular way. For example, when a *Xenopus* optic vesicle is experimentally placed under a new region of head ectoderm and causes that region of the ectoderm to form a lens, that is an instructive interaction.

The second type of inductive interaction is **permissive interaction**. Here, the responding tissue has already been specified and needs only an environment that allows the expression of these traits. For instance, many tissues need an extracellular matrix in order to develop. The extracellular matrix does not alter the type of cell that is produced, but it enables what has already been determined to be expressed.*

Epithelial-mesenchymal interactions

Some of the best-studied cases of induction involve the interactions of sheets of epithelial cells with adjacent mesenchymal cells. All organs consist of an epithelium and an associated mesenchyme, so these interactions are among the most important phenomena in nature. Some examples are listed in **TABLE 3.1**.

---

* It is easy to distinguish permissive and instructive interactions by an analogy with a more familiar situation. This textbook is made possible by both permissive and instructive interactions. A reviewer can convince me to change the material in the chapters. This is an instructive interaction, as the information expressed in the book is changed from what it would have been. However, the information in the book could not be expressed at all without permissive interactions with the publisher and printer.

REGIONAL SPECIFICITY OF INDUCTION Using the induction of cutaneous (skin) structures as our examples, we will look at the properties of epithelial-mesenchymal interactions. The first of these properties is the regional specificity of induction. Skin is composed of two main tissues: an outer epidermis (an epithelial tissue derived from ectoderm) and a dermis (a mesenchymal tissue derived from mesoderm). The chick epidermis secretes proteins that signal the underlying dermal cells to form condensations, and the condensed dermal mesenchyme responds by secreting factors that cause the epidermis to form regionally specific cutaneous structures (**FIGURE 3.16**; Nohno et al. 1995; Ting-Berreth and Chuong 1996). These structures can be the broad feathers of the wing,

TABLE 3.1 Some epithelial-mesenchymal interactions

Organ	Mesenchymal component	Epithelial component
Cutaneous structures (hair, feathers, sweat glands, mammary glands)	Epidermis (ectoderm)	Dermis (mesoderm)
Limb	Epidermis (ectoderm)	Mesenchyme (mesoderm)
Gut organs (liver, pancreas, salivary glands)	Epithelium (endoderm)	Mesenchyme (mesoderm)
Foregut and respiratory-associated organs (lungs, thymus, thyroid)	Epithelium (endoderm)	Mesenchyme (mesoderm)
Kidney	Ureteric bud (mesoderm)	Mesenchyme epithelium (mesoderm)
Tooth	Jaw epithelium (ectoderm)	Mesenchyme (neural crest)

the narrow feathers of the thigh, or the scales and claws of the feet.

As **FIGURE 3.17** demonstrates, the dermal mesenchyme is responsible for the regional specificity of induction in the competent epidermal epithelium. Researchers can separate the embryonic epithelium and mesenchyme from each other and recombine them in different ways (Saunders et al. 1957). The same epithelium develops cutaneous structures according to the region from which the mesenchyme was taken. Here, the mesenchyme plays an instructive role, calling into play different sets of genes in the responding epithelial cells.

GENETIC SPECIFICITY OF INDUCTION The second property of epithelial-mesenchymal interactions is the genetic specificity of induction. Whereas the mesenchyme may instruct the epithelium as to what sets of genes to activate, the responding epithelium can comply with these instructions only so far as its genome permits. This property was discovered through experiments involving the transplantation of tissues from one species to another.

In one of the most dramatic examples of interspecific induction, Hans Spemann and Oscar Schotté (1932) transplanted flank ectoderm from an early *frog* gastrula to the region of a *newt* gastrula destined to become parts of the mouth. Similarly, they placed presumptive flank ectodermal tissue from a *newt* gastrula into the presumptive oral regions of *frog* embryos. The structures of the mouth region differ greatly between salamander and frog larvae. The salamander larva has club-shaped balancers beneath its mouth, whereas

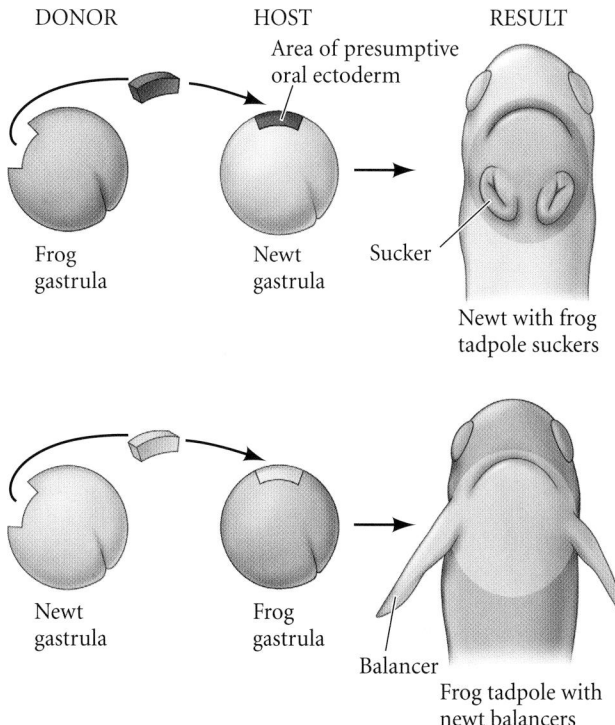

FIGURE 3.18   Genetic specificity of induction in amphibians. Reciprocal transplantation between the presumptive oral ectoderm regions of salamander and frog gastrulae leads to newts with tadpole suckers and tadpoles with newt balancers. (After Hamburgh 1970.)

the frog tadpole produces mucus-secreting glands and suckers. The frog tadpole also has a horny jaw without teeth, whereas the salamander has a set of calcareous teeth in its jaw. The larvae resulting from the transplants were chimeras. The salamander larvae had froglike mouths, and the frog tadpoles had salamander teeth and balancers (**FIGURE 3.18**). In other words, the mesenchymal cells instructed the ectoderm to make a mouth, but the ectoderm responded by making the only kind of mouth it "knew" how to make, no matter how inappropriate.*

Thus, the instructions sent by the mesenchymal tissue can cross species barriers. Salamanders respond to frog inducers, and chick tissue responds to mammalian inducers. The response of the epithelium, however, is species-specific. So, whereas organ-type specificity (e.g., feather or claw) is usually controlled by the mesenchyme, species specificity is usually controlled by the responding epithelium. As we will see in Chapter 20, major evolutionary changes in the phenotype can be brought about by changing the response to a particular inducer.

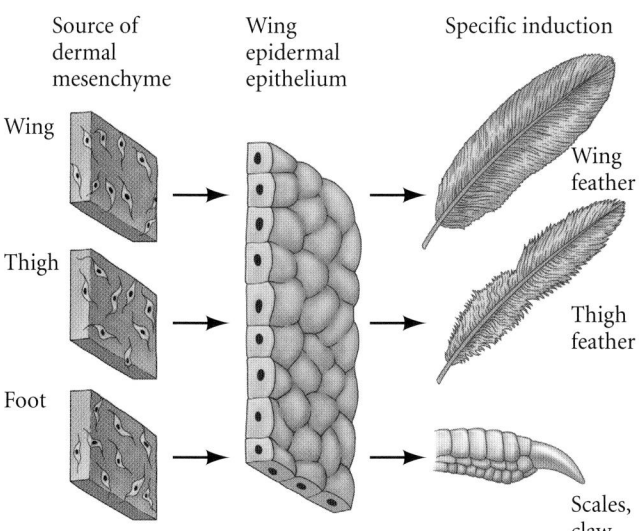

FIGURE 3.17   Regional specificity of induction in the chick. When cells from different regions of the dermis (mesenchyme) are recombined with the epidermis (epithelium), the type of cutaneous structure made by the epidermal epithelium is determined by the original source of the mesenchyme. (After Saunders 1980.)

---

*Spemann is reported to have put it this way: "The ectoderm says to the inducer, 'you tell me to make a mouth; all right, I'll do so, but I can't make your kind of mouth; I can make my own and I'll do that.'" (Quoted in Harrison 1933.)

Paracrine Factors: Inducer Molecules

How are the signals between inducer and responder transmitted? While studying the mechanisms of induction that produce the kidney tubules and teeth, Grobstein (1956) and others (Saxén et al. 1976; Slavkin and Bringas 1976) found that some inductive events could occur despite a filter separating the epithelial and mesenchymal cells. Other inductions, however, were blocked by the filter. The researchers therefore concluded that some of the inducers were soluble molecules that could pass through the small pores of the filter, and that other inductive events required physical contact between the epithelial and mesenchymal cells.

When membrane proteins on one cell surface interact with receptor proteins on adjacent cell surfaces, these events are called **juxtacrine interactions** (since the cell membranes are *juxtaposed*). When proteins synthesized by one cell can diffuse over small distances to induce changes in neighboring cells, the event is called a **paracrine interaction**. Paracrine factors are diffusible molecules that work in a range of about 15 cell diameters, or about 40–200 μm (Bollenbach et al. 2008; Harvey and Smith 2009).

A specific type of paracrine interaction is the **autocrine interaction**. Autocrine interactions occur when the same cells that secrete paracrine factors also respond to them. In other words, the cell synthesizes a molecule for which it has its own receptor. Although autocrine regulation is not common, it is seen in placental cytotrophoblast cells; these cells synthesize and secrete platelet-derived growth factor, whose receptor is on the cytotrophoblast cell membrane (Goustin et al. 1985). The result is the explosive proliferation of that tissue.

Signal transduction cascades: The response to inducers

The induction of numerous organs is effected by a relatively small set of paracrine factors. The embryo inherits a rather compact genetic "tool kit" and uses many of the same proteins to construct the heart, kidneys, teeth, eyes, and other organs. Moreover, the same proteins are used throughout the animal kingdom—the factors active in creating the *Drosophila* eye or heart are very similar to those used in generating mammalian organs. Many paracrine factors can be grouped into one of four major families on the basis of their structure:

1. The fibroblast growth factor (FGF) family
2. The Hedgehog family
3. The Wnt family
4. The TGF-β superfamily, encompassing the TGF-β family, the activin family, the bone morphogenetic proteins (BMPs), the Nodal proteins, the Vg1 family, and several other related proteins

Paracrine factors function by binding to a receptor that initiates a series of enzymatic reactions within the cell. These enzymatic reactions have as their end point either the

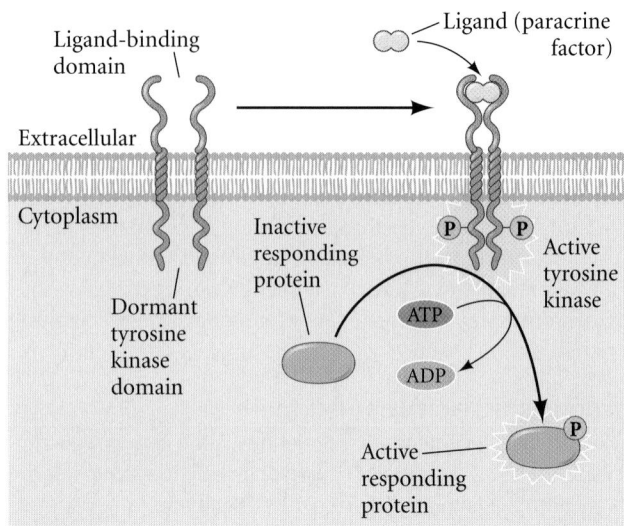

FIGURE 3.19 Structure and function of a receptor tyrosine kinase. The binding of a paracrine factor (such as Fgf8) by the extracellular portion of the receptor protein activates the dormant tyrosine kinase, whose enzyme activity phosphorylates specific tyrosine residues of certain proteins.

regulation of transcription factors (such that different genes are expressed in the cells reacting to these paracrine factors) and/or the regulation of the cytoskeleton (such that the cells responding to the paracrine factors alter their shape or are permitted to migrate). These pathways of responses to the paracrine factor often have several end points and are called **signal transduction cascades**.

The major signal transduction pathways all appear to be variations on a common and rather elegant theme, exemplified in **FIGURE 3.19**. Each receptor spans the cell membrane and has an extracellular region, a transmembrane region, and a cytoplasmic region. When a paracrine factor binds to its receptor's extracellular domain, the paracrine factor induces a conformational change in the receptor's structure. This shape change is transmitted through the membrane and alters the shape of the receptor's cytoplasmic domain, giving that domain the ability to activate cytoplasmic proteins. Often such a conformational change confers enzymatic activity on the domain—usually a kinase activity that can use ATP to phosphorylate specific tyrosine residues of particular proteins. Thus, this type of receptor is often called a **receptor tyrosine kinase** (**RTK**). The active receptor can now catalyze reactions that phosphorylate other proteins, and this phosphorylation in turn activates their latent activities. Eventually, the cascade of phosphorylation activates a dormant transcription factor or a set of cytoskeletal proteins.

Fibroblast growth factors and the RTK pathway

The **fibroblast growth factor** (**FGF**) family of paracrine factors comprises nearly two dozen structurally related members, and the FGF genes can generate hundreds of protein

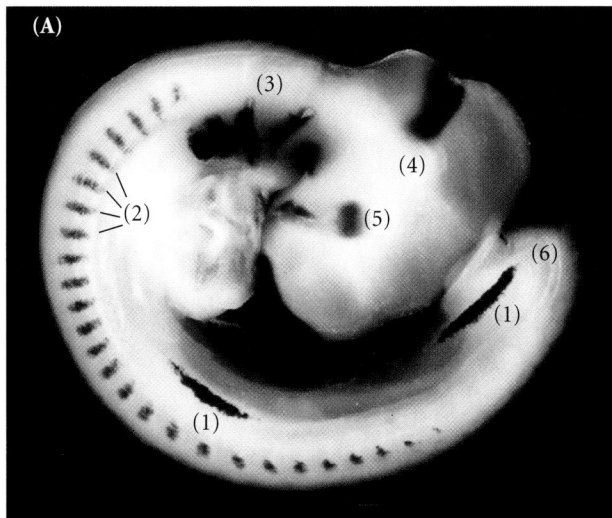

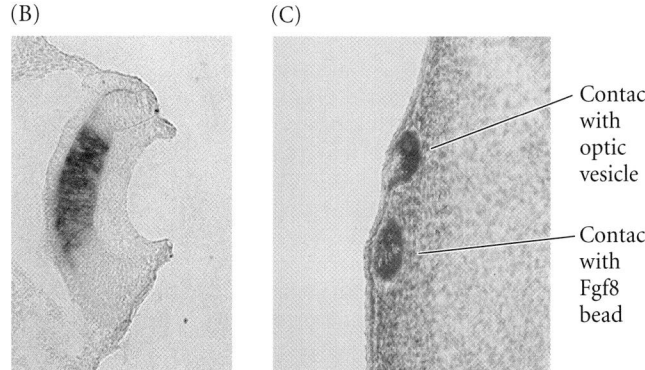

FIGURE 3.20 Fgf8 in the developing chick. (A) *Fgf8* gene expression pattern in the 3-day chick embryo, shown by in situ hybridization. Fgf8 protein (dark areas) is seen in the distalmost limb bud ectoderm (1); in the somitic mesoderm (the segmented blocks of cells along the anterior-posterior axis (2); in the branchial arches of the neck (3); at the boundary between the midbrain and hindbrain (4); in the optic vesicle of the developing eye (5); and in the tail (6). (B) In situ hybridization of *Fgf8* in the optic vesicle. The *Fgf8* mRNA (purple) is localized to the presumptive neural retina of the optic cup and is in direct contact with the outer ectoderm cells that will become the lens. (C) Ectopic expression of L-Maf in competent ectoderm can be induced by the optic vesicle (above) and by an Fgf8-containing bead (below). (A courtesy of E. Laufer, C.-Y. Yeo, and C. Tabin; B,C courtesy of A. Vogel-Höpker.)

isoforms by varying their RNA splicing or initiation codons in different tissues (Lappi 1995). Fgf1 protein is also known as acidic FGF and appears to be important during regeneration (Yang et al. 2005); Fgf2 is sometimes called basic FGF and is very important in blood vessel formation; and Fgf7 sometimes goes by the name of keratinocyte growth factor and is critical in skin development. Although FGFs can often substitute for one another, the expression patterns of the FGFs and their receptors give them separate functions. In *Drosophila*, Breathless is an FGF protein.

One member of this family, Fgf8, is especially important during limb development and lens induction. Fgf8 is usually made by the optic vesicle that contacts the outer ectoderm of the head (**FIGURE 3.20**; Vogel-Höpker et al. 2000). After contact with the outer ectoderm occurs, *Fgf8* gene expression becomes concentrated in the region of the presumptive neural retina—the tissue directly apposed to the presumptive lens. Moreover, if Fgf8-containing beads are placed adjacent to head ectoderm, this ectopic Fgf8 will induce this ectoderm to produce ectopic lenses and to express the lens-associated transcription factor L-Maf (see Figure 3.20C). FGFs often work by activating a set of receptor tyrosine kinases called the **fibroblast growth factor receptors** (**FGFRs**). The Branchless protein is an FGFR in *Drosophila*.

When an FGFR binds an FGF (and only when it binds an FGF), the dormant kinase is activated and phosphorylates certain proteins (including other FGFRs) within the responding cell. These proteins, once activated, can perform new functions. The **RTK pathway** was one of the first signal transduction pathways to unite various areas of developmental biology (**FIGURE 3.21**). Researchers studying *Drosophila* eyes, nematode vulvae, and human cancers found that they were all studying the same genes.

The pathway begins at the cell surface, where an RTK binds its specific paracrine factor. Paracrine factor ligands that bind to RTKs include the fibroblast growth factors, epidermal growth factors, platelet-derived growth factors, and stem cell factor. Each RTK can bind only one (or one small set) of these ligands.*

The RTK spans the cell membrane, and when it binds its protein it undergoes a conformational change that enables it to dimerize with another RTK. This conformational change stimulates the latent kinase activity of each RTK, and these receptors phosphorylate each other on particular tyrosine residues (see Figure 3.19). Thus, the binding of the paracrine factor to its receptor causes autophosphorylation of the cytoplasmic domain of the receptor.

The phosphorylated tyrosine on the receptor is then recognized by an adaptor protein that serves as a bridge linking the phosphorylated RTK to a powerful intracellular signaling system. While binding to the phosphorylated RTK through one of the RTK's cytoplasmic domains, the adaptor protein also activates a G protein, such as Ras. Normally, the G protein is in an inactive, GDP-bound state. The activated receptor stimulates the adaptor protein to activate the **GTP exchange factor** (**GEF**, also called **guanine nucleotide releasing factor**, or **GNRP**). GEF exchanges a phosphate from a GTP to transform the bound GDP into GTP. The GTP-bound G protein is an active form that transmits the signal to the next molecule. After the signal is delivered, the GTP on the G protein is hydrolyzed back into GDP. This catalysis is

---

* *Ligand* is a general term used for an agent that binds to a protein; it often designates a signaling molecule, such as a paracrine factor.

FIGURE 3.21 The widely used RTK signal transduction pathway. The receptor tyrosine kinase is dimerized by the ligand (a paracrine factor). which causes the autophosphorylation of the receptor. The adaptor protein recognizes the phosphorylated tyrosines on the RTK and activates an intermediate protein, GEF, which activates the Ras G protein by allowing phosphorylation of the GDP-bound Ras. At the same time, the GAP protein stimulates hydrolysis of this phosphate bond, returning Ras to its inactive state. The active Ras activates the Raf protein kinase C (PKC), which in turn phosphorylates a series of kinases (such as MEK). Eventually, the activated kinase ERK alters gene expression in the nucleus of the responding cell by phosphorylating certain transcription factors (which can then enter the nucleus to change the types of genes transcribed) and certain translation factors (which alter the level of protein synthesis). In many cases, this pathway is reinforced by the release of calcium ions. A simplified version of the pathway is shown on the left.

Ligand
↓
RTK
↓
GEF
↓
RAS
↓
RAF
↓
MEK
↓
ERK
↓
Transcription factor
↓
Transcription

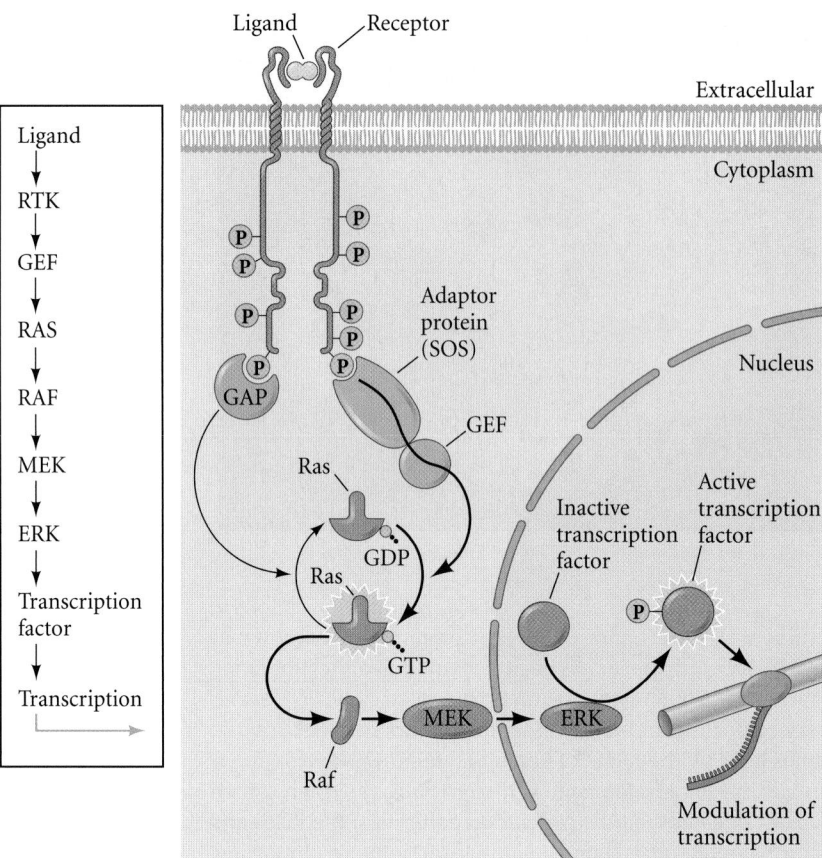

greatly stimulated by the complexing of the Ras protein with the **GTPase-activating protein** (**GAP**). In this way, the G protein is returned to its inactive state, where it can await further signaling. Without the GAP protein, Ras protein cannot efficiently catalyze GTP, and so remains in its active configuration for a longer time (Cales et al. 1988; McCormick 1989). Mutations in the *RAS* gene account for a large proportion of cancerous human tumors (Shih and Weinberg 1982), and the mutations of *RAS* that make it oncogenic all inhibit the binding of the GAP protein.

The active Ras G protein associates with a kinase called Raf. The G protein recruits the inactive Raf kinase to the cell membrane, where it becomes active (Leevers et al. 1994; Stokoe et al. 1994). Raf kinase activates the MEK protein by phosphorylating it. MEK is itself a kinase, which activates the ERK protein by phosphorylation. In turn, ERK is a kinase that enters the nucleus and phosphorylates certain transcription factors.

For instance, to effect the activation of MITF (discussed in Chapter 2), a paracrine factor called **stem cell factor** binds to an RTK called Kit. This binding dimerizes the Kit protein, causing it to phosphorylate itself. Phosphorylated Kit activates the pathway whereby phosphorylated ERK is able to phosphorylate the MITF transcription factor (Hsu et al. 1997; Hemesath et al. 1998). Only the phosphorylated form of MITF is able to bind the p300/CBP histone acetyltransferase

protein that enables it to activate transcription of the genes encoding tyrosinase and other proteins of the melanin-formation pathway (**FIGURE 3.22**; Price et al. 1998).

The JAK-STAT pathway

Fibroblast growth factors can also activate the JAK-STAT cascade. This pathway is extremely important in the differentiation of blood cells, the growth of limbs, and the activation of the casein gene during milk production (**FIGURE 3.23**; Briscoe et al. 1994; Groner and Gouilleux 1995). The cascade starts when a paracrine factor is bound by the extracellular domain of a receptors that spans the cell membrane, with the cytoplasmic domain of the receptor being linked to **JAK** (*J*anus *k*inase) proteins. The binding of paracrine factor to the receptor activates the JAK kinases and causes them to phosphorylate the **STAT** (*s*ignal *t*ransducers and *a*ctivators of *t*ranscription) family of transcription factors (Ihle 1996, 2001). The phosphorylated STAT is a transcription factor that can now enter into the nucleus and bind to its enhancers.

The JAK-STAT pathway is critically important in regulating human fetal bone growth. Mutations that prematurely activate the STAT pathway have been implicated in some severe forms of dwarfism, such as the lethal condition thanatophoric dysplasia, in which the growth plates of the rib and limb bones fail to proliferate. The short-limbed newborn dies because its ribs cannot support breathing. The genetic lesion

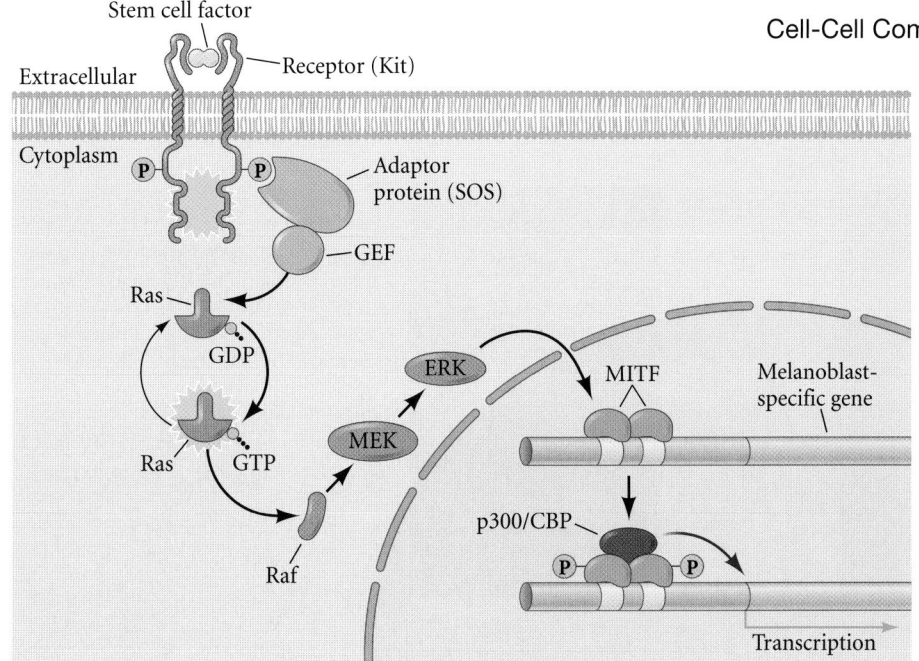

FIGURE 3.22 Activation of MITF transcription factor through the binding of stem cell factor by the Kit receptor tyrosine kinase. The information received at the cell membrane is sent to the nucleus by the RTK signal transduction pathway. When the receptor domain of the Kit RTK protein binds the stem cell factor, Kit dimerizes and becomes phosphorylated. This phosphorylation is used to activate the Ras G protein, which activates the chain of kinases that will phosphorylate the MITF protein. Once phosphorylated, MITF can bind the cofactor p300/CBP, acetylate the nucleosome histones, and initiate transcription of the genes needed for melanocyte development. (After Price et al. 1998.)

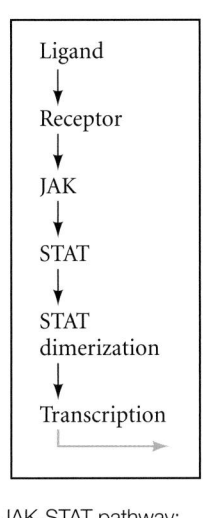

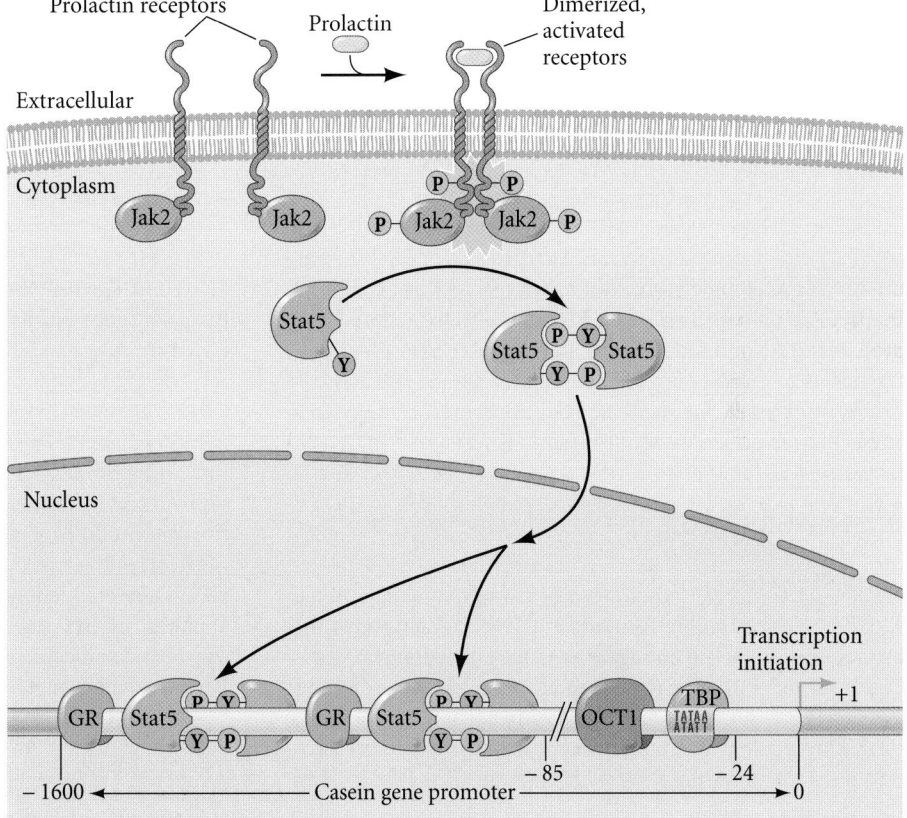

FIGURE 3.23 A JAK-STAT pathway: the casein gene activation pathway activated by prolactin. The casein gene is activated during the last (lactogenic) phase of mammary gland development, and its signal is the secretion of the hormone prolactin from the anterior pituitary gland. Prolactin causes the dimerization of prolactin receptors in the mammary duct epithelial cells. A particular JAK protein (Jak2) is "hitched" to the cytoplasmic domain of these receptors. When the receptors bind prolactin and dimerize, the JAK proteins phosphorylate each other and the dimerized receptors, activating the dormant kinase activity of the receptors. The activated receptors add a phosphate group to a tyrosine residue (Y) of a particular STAT protein—in this case, Stat5. This allows Stat5 to dimerize, be translocated into the nucleus, and bind to particular regions of DNA. In combination with other transcription factors (which presumably have been waiting for its arrival), the Stat5 protein activates transcription of the casein gene. GR is the glucocorticoid receptor, OCT1 is a general transcription factor, and TBP is the major promoter-binding protein that anchors RNA polymerase II (see Chapter 2) responsible for binding RNA polymerase II. A simplified diagram is shown to the right. (For details, see Groner and Gouilleux 1995.)

Premature activation
of FgfR3 kinase

FGF
receptor 3
(FgfR3)

Mutation
site

Phosphorylated
Stat1

Kinase
domains

Cartilage growth
stops before birth

Narrow chest,
extremely
short limbs

Thanatophoric
dysplasia

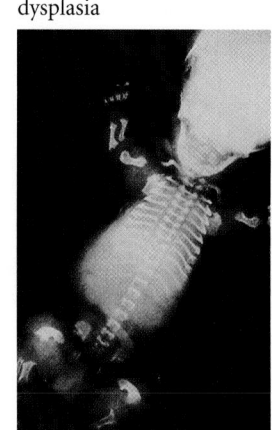

FIGURE 3.24 A mutation in the gene for FgfR3 causes the premature constitutive activation of the STAT pathway and the production of phosphorylated Stat1 protein. This transcription factor activates genes that cause the premature termination of chondrocyte cell division. The result is thanatophoric dysplasia, a condition of failed bone growth that results in the death of the newborn infant because the thoracic cage cannot expand to allow breathing. (After Gilbert-Barness and Opitz 1996.)

responsible is in *FGFR3*, the gene encoding fibroblast growth factor receptor 3 (**FIGURE 3.24**; Rousseau et al. 1994; Shiang et al. 1994). *FGFR3* is expressed in the cartilage precursor cells (chondrocytes) in the growth plates of the long bones. Normally, the FgfR3 protein (a receptor tyrosine kinase) is activated by a fibroblast growth factor and signals the chondrocytes to stop dividing and begin differentiating into cartilage. This signal is mediated by the Stat1 protein, which is phosphorylated by activated FgfR3 and then translocated into the nucleus. Inside the nucleus, Stat1 activates the genes encoding a cell cycle inhibitor, the p21 protein (Su et al. 1997). Thus, the mutations causing thanatophoric dwarfism result from a gain-of-function mutation in the *FGFR3* gene. The mutant receptor gene is active constitutively—that is, without the need to be activated by an FGF signal (Deng et al. 1996; Webster and Donoghue 1996). Chondrocytes stop proliferating shortly after they are formed and the bones fail to grow. Other mutations that activate *FGFR3* prematurely but to a lesser degree produce achondroplasic (short-limbed) dwarfism (Legeai-Mallet et al. 2004).

● **See WEBSITE 3.1** FGF receptor mutations

The Hedgehog family

The proteins of the **Hedgehog family** of paracrine factors are often used by the embryo to induce particular cell types and to create boundaries between tissues. Hedgehog proteins are processed such that only the amino-terminal two-thirds of the molecule is secreted; once this takes place, the protein must become complexed with a molecule of cholesterol in order to function. Cholesterol is critical in enabling the Hedgehog protein to anchor to its receptor cell's cell membrane—a critical step in binding the paracrine factor to its receptor (Grover et al. 2011).

Vertebrates have at least three homologues of the *Drosophila hedgehog* gene: *sonic hedgehog* (*shh*), *desert hedgehog* (*dhh*), and *indian hedgehog* (*ihh*). The Desert hedgehog protein is found in the Sertoli cells of the testes, and mice homozygous for a null allele of *Dhh* exhibit defective spermatogenesis.

Indian hedgehog is expressed in the gut and cartilage and is important in postnatal bone growth (Bitgood and McMahon 1995; Bitgood et al. 1996). Sonic hedgehog* has the greatest number of functions of the three vertebrate Hedgehog homologues. Among other important functions, Sonic hedgehog is responsible for assuring that motor neurons come only from the ventral portion of the neural tube (see Chapter 11), that a portion of each somite forms the vertebrae (see Chapter 12), that the feathers of the chick form in their proper places (see Figure 3.16), and that our pinkies are always our most posterior digits (see Chapter 14). Sonic hedgehog often works with other paracrine factors, such as Wnt and FGF proteins.

THE HEDGEHOG PATHWAY Proteins of the Hedgehog family function by binding to a receptor called Patched. Patched, however, is not a signal transducer. Rather, the Patched protein prevents the Smoothened protein from functioning. In the absence of Hedgehog binding to Patched, Smoothened is inactive, and the Cubitus interruptus (Ci) protein in *Drosophila* (or the homologous Gli protein in vertebrates) is tethered to the microtubules of the responding cell. While on the microtubules, Gli is cleaved in such a way that a portion of it enters the nucleus and acts as a transcriptional repressor. When Hedgehog binds to Patched, the Patched protein's shape is altered such that it no longer inhibits Smoothened.

---

*Yes, it is named after the Sega Genesis character. The original *hedgehog* gene was found in *Drosophila*, in which genes are named after their mutant phenotypes—the loss-of-function *hedgehog* mutation causes the fly embryo to be covered with pointy denticles on its cuticle, so it looks like a hedgehog. The vertebrate Hedgehog genes were discovered by searching vertebrate gene libraries (chick, rat, zebrafish) with probes that would find sequences similar to that of the fruit fly *hedgehog* gene. Riddle and colleagues (1993) discovered three genes homologous to *Drosophila hedgehog*. Two were named after existing species of hedgehog; the third was named after the animated character. Two other Hedgehog genes, found only in fish, were originally named *echidna hedgehog* (possibly after Sonic's cartoon friend) and *Tiggywinkle hedgehog* (after Beatrix Potter's fictional hedgehog), but they are now referred to as *ihh-b* and *shh-b*, respectively.

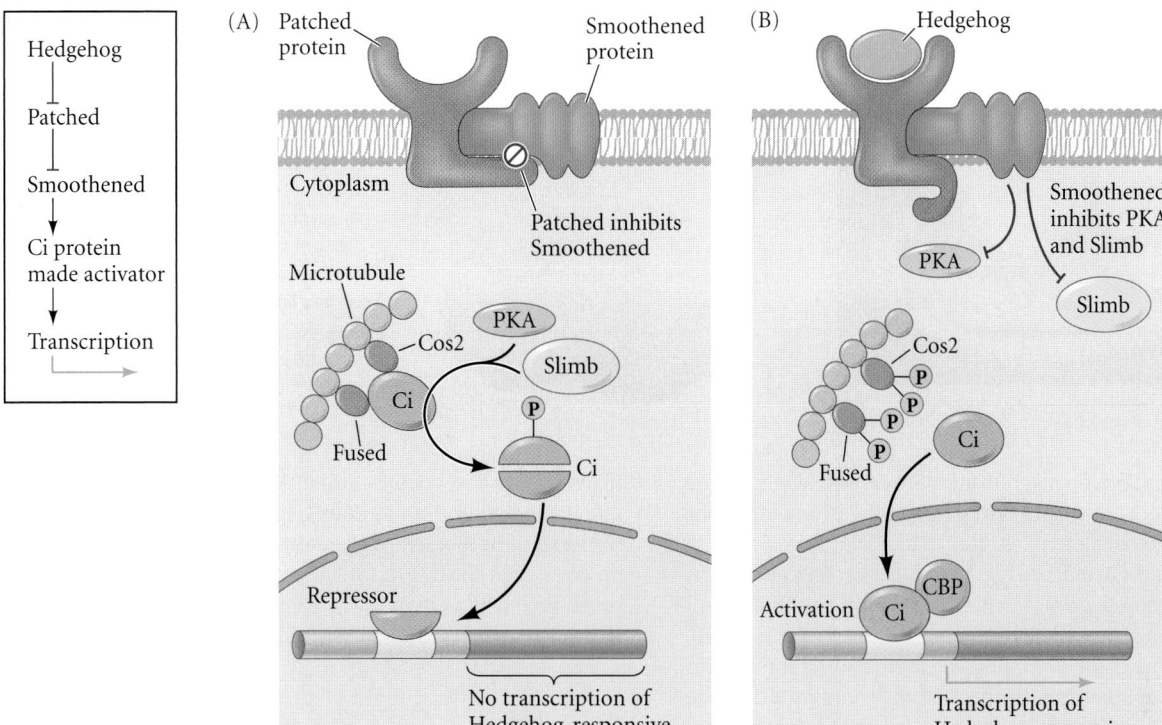

FIGURE 3.25 Hedgehog signal transduction pathway. Patched protein in the cell membrane is an inhibitor of the Smoothened protein. (A) In the absence of Hedgehog binding to Patched, the Ci protein is tethered to the microtubules by the Cos2 and Fused proteins. This binding allows the PKA and Slimb proteins to cleave Ci into a transcriptional repressor that blocks the transcription of particular genes. (B) When Hedgehog binds to Patched, its conformation changes, releasing the inhibition of the Smoothened protein. Smoothened then releases Ci from the microtubules (probably by adding more phosphates to the Cos2 and Fused proteins) and inactivates the cleavage proteins PKA and Slimb. The Ci protein enters the nucleus, binds a CBP protein, and acts as a transcriptional activator of particular genes. (After Johnson and Scott 1998.)

Smoothened acts (probably by phosphorylation) to release Ci from the microtubules and to prevent its being cleaved. The intact Ci protein can now enter the nucleus, where it acts as a transcriptional *activator* of the same genes it used to repress (**FIGURE 3.25**; Aza-Blanc et al. 1997; Lum and Beachy 2004).

The Hedgehog pathway is extremely important in vertebrate limb development, neural differentiation, and facial morphogenesis (**FIGURE 3.26A**; McMahon et al. 2003). When mice were made homozygous for a mutant allele of *Sonic hedgehog*, they had major limb and facial abnormalities. The midline of the face was severely reduced and a single eye formed in the center of the forehead, a condition known as *cyclopia* after the one-eyed Cyclops of Homer's *Odyssey* (**FIGURE 3.26B**; Chiang et al. 1996). In later development, Sonic hedgehog is critical for feather formation in the chick embryo and for hair formation in mammals (Harris et al. 2002; Michino et al. 2003).

While mutations that inactivate the Hedgehog pathway can cause malformations, mutations that activate the pathway ectopically can cause cancers. If the Patched protein is mutated in somatic tissues such that it can no longer inhibit Smoothened, it can cause tumors of the basal cell layer of the epidermis (basal cell carcinomas). Heritable mutations of the *patched* gene cause basal cell nevus syndrome, a rare autosomal dominant condition characterized by both developmental anomalies (fused fingers; rib and facial abnormalities) and multiple malignant tumors such as basal cell carcinoma (Hahn et al. 1996; Johnson et al. 1996).

One remarkable feature of the Hedgehog signal transduction pathway is the important role of cholesterol. Cholesterol is critical for the catalytic cleavage of Sonic hedgehog protein. Only the amino-terminal portion of the protein is functional and secreted. Cholesterol also binds to the active N-terminus of Sonic hedgehog, which is then able to diffuse over a range of a few hundred μm (about 30 cell diameters in the mouse limb). Without this cholesterol modification, Shh diffuses too quickly and dissipates into the surrounding space. Indeed, Hedgehog proteins probably do not diffuse as single molecules but are linked together through their cholesterol-containing regions into lipoprotein packets (Breitling 2007; Guerrero and Chiang 2007). In addition, cholesterol is involved in the binding of Sonic hedgehog to the Patched receptor protein.

(A)

(B)

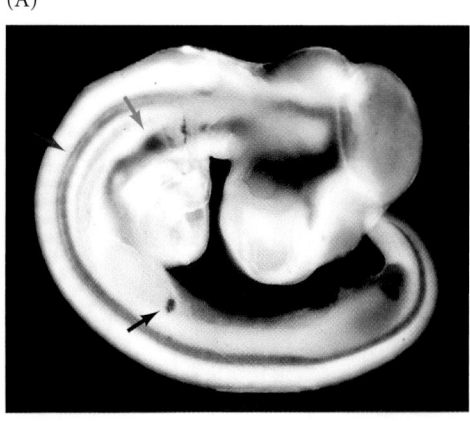

FIGURE 3.26 (A) Sonic hedgehog is shown by in situ hybridization to be expressed in the nervous system (red arrow), gut (blue arrow), and limb bud (black arrow) of a 3-day chick embryo. (B) Head of a cyclopic lamb born of a ewe that ate *Veratrum californicum* early in pregnancy. The cerebral hemispheres fused, resulting in the formation of a single, central eye and no pituitary gland. The jervine alkaloid made by this plant inhibits cholesterol synthesis, which is needed for Hedgehog production and reception. (A courtesy of C. Tabin; B courtesy of L. James and USDA Poisonous Plant Laboratory.)

Some human cyclopia syndromes are caused by mutations in genes that encode either Sonic hedgehog or the enzymes that synthesize cholesterol (Kelley et al. 1996; Roessler et al. 1996; Opitz and Furtado 2013). Moreover, certain chemicals that induce cyclopia do so by interfering with the cholesterol biosynthetic enzymes (Beachy et al. 1997; Cooper et al. 1998). Two teratogens known to cause cyclopia in vertebrates are jervine and cyclopamine. Both substances are found in the plant *Veratrum californicum*, and both block the synthesis of cholesterol (see Figure 3.26B; Keeler and Binns 1968).

● **See VADE MECUM** Cyclopia induced in zebrafish

The Wnt family

The Wnts are a family of cysteine-rich glycoproteins, and there are at least 15 members of this gene family in vertebrates* (Nusse and Varmus 2012). Their name is a fusion of the name of the *Drosophila* segment polarity gene *wingless* with the name of one of its vertebrate homologues, *integrated*. In flies, mutations in *wingless* prevent the formation of the wing. Wnt proteins also are critical in establishing the polarity of insect and vertebrate limbs, in promoting the proliferation of stem cells, and in several steps of the mammalian urogenital system development (FIGURE 3.27).

THE CANONICAL WNT PATHWAY Members of the Wnt family usually interact with a pair of transmembrane receptor proteins, one from the Frizzled family and one large

---

* A summary of all the Wnt proteins and Wnt signaling components can be found at **http://www.stanford.edu/~rnusse/wntwindow.htm.**

transmembrane protein called LRP5/6 (Logan and Nusse 2004; MacDonald et al. 2009). In the absence of Wnts, the transcriptional cofactor β-catenin is constantly being degraded by a protein degradation complex containing several proteins (such as axin and APC), as well as **glycogen synthase kinase 3** (**GSK3**). GSK3 phosphorylates β-catenin so

(A)

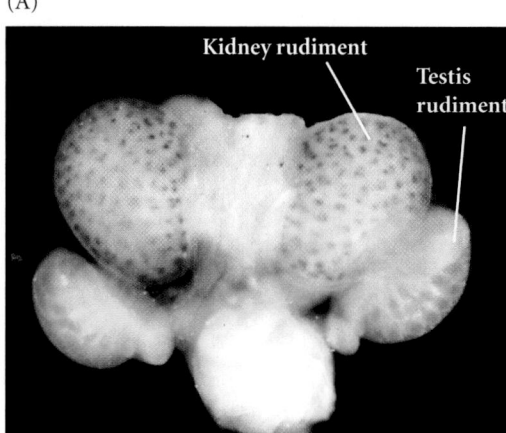

(B) (C)

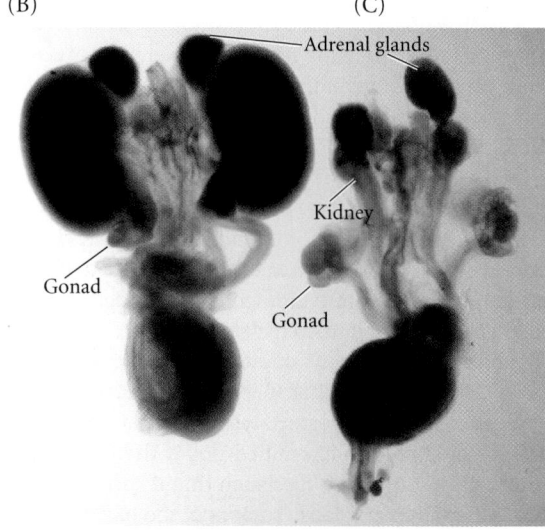

FIGURE 3.27 Wnt4 is necessary for kidney development and for female sex determination. (A) Wnt4 expression in a 14-day mouse embryonic male urogenital rudiment. Expression (purple staining) is seen in the mesenchyme that condenses to form the kidney's nephrons. (B) Urogenital rudiment of a wild-type newborn female mouse. (C) Urogenital rudiment of a newborn female mouse with targeted knockout of *Wnt4* shows that the kidney fails to develop. In addition, the ovary starts synthesizing testosterone and becomes surrounded by a modified male duct system. (Courtesy of J. Perasaari and S. Vainio.)

(A)

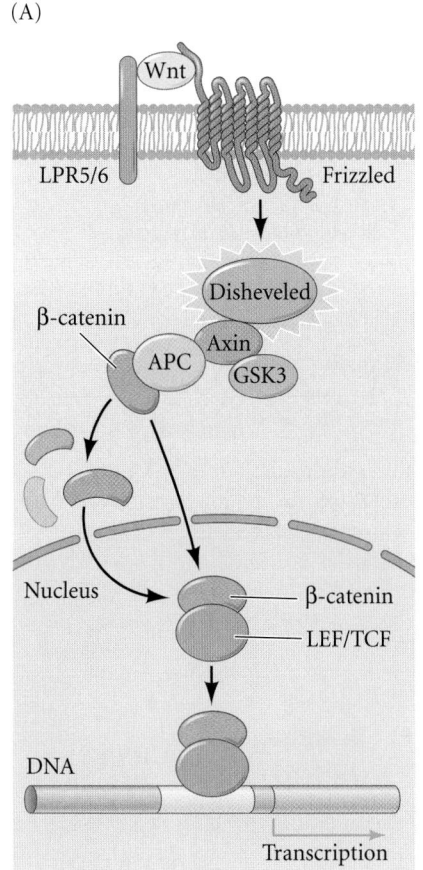

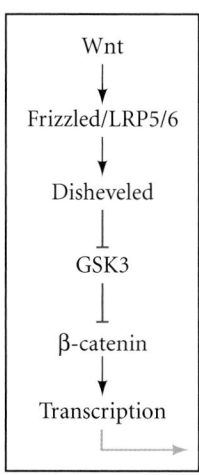

(B)

(C)

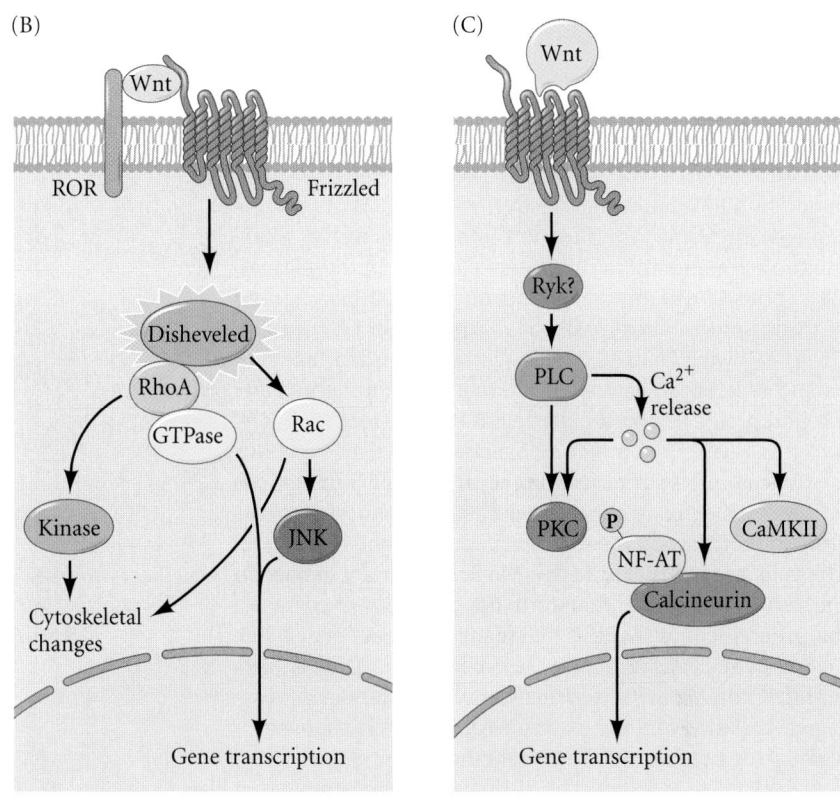

FIGURE 3.28 Wnt signal transduction pathways. (A) The canonical Wnt pathway. The Wnt protein binds to its receptor, a member of the Frizzled family. In the case of certain Wnt proteins, Frizzled then activates Disheveled, allowing Disheveled to become an inhibitor of glycogen synthase kinase 3 (GSK3). GSK3, if it were active, would prevent the dissociation of β-catenin from the APC protein. So, by inhibiting GSK3, the Wnt signal frees β-catenin to associate with an LEF or TCF protein and become an active transcription factor. (B) In a pathway that regulates cell morphology, division, and movement, certain Wnt proteins activate Frizzled in a way that causes Frizzled to activate Disheveled (which had been tethered to the cell membrane through the Pickle protein). Here, Disheveled activates Rac and RhoA proteins that coordinate the cytoskeleton and that can also regulate gene expression. (C) In a third pathway, certain Wnt proteins activate Frizzled receptors in a way that releases calcium ions and can result in Ca^{2+}-dependent gene expression. (After MacDonald et al. 2009.)

that it will be recognized and degraded by proteosoms. Wnt-responsive genes are repressed by the LEF/TCF transcription factor, which binds histone deacetylases.

When Wnt proteins come into contact with a cell, they bring together the Frizzled and LRP5/6 receptors. This linkage causes LRP5/6 to bind both axin and GSK3, and it allows the Frizzled protein to bind Disheveled. Disheveled stabilizes the axin, keeping GSK3 bound to the cell membrane and thereby preventing GSK3 from phosphorylating β-catenin;

so the β-catenin accumulates and enters the nucleus. There, it binds to the LEF/TCF transcription factor, displaces the histone deacetylase, and activates transcription. The binding of β-catenin to the LEF/TCF protein converts the repressor into a transcriptional activator, thereby activating Wnt-responsive genes (**FIGURE 3.28A**; Cadigan and Nusse 1997; Niehrs 2012).

This model is undoubtedly an oversimplification, because different cells use the pathway in different ways (see McEwen and Peifer 2001; Nusse 2012). Natural inhibitors of the Wnt pathway (such as Dickkopf) often bind to the LRP5/6 protein, blocking Wnt binding. One overriding principle is readily evident in both the Wnt pathway and the Hedgehog pathway: *activation is often accomplished by inhibiting an inhibitor*. Thus, in the Wnt pathway, the GSK3 protein is an inhibitor that is itself repressed by the Wnt signal.

THE NONCANONICAL WNT PATHWAYS The pathway described above is often called the "canonical" Wnt pathway because it was the first one to be discovered. However, in addition to sending signals to the nucleus, Wnt proteins can also affect calcium transport into cells as well as altering the actin and microtubule cytoskeleton. In these cases, Wnt activates alternative, "noncanonical" pathways. One of these noncanonical pathways is the planar cell polarity (PCP) pathway (**FIGURE 3.28B**). Certain Wnts (such as Wnt 5a and Wnt11) can activate Disheveled by binding to a different receptor (Frizzled paired with ROR instead of LPR5), and this receptor complex phosphorylates Disheveled in a way that allows it to interact with a Rho GTPase. This GTPase can activate the kinases that phosphorylate cytoskeletal proteins and thereby alter cell shape and movement. The cells are instructed to divide such that they remain in the same plane (rather than form upper and lower compartments) and to move within that plane (Shulman et al. 1998; Winter et al. 2001; Witte et al. 2010; Ho et al. 2012). In vertebrates, this polarity information is very important during gastrulation and the extension of the anterior-posterior axis during neurulation. Recent studies (Ciruna et al. 2006; Sepich et al. 2011; Habib et al. 2013) have demonstrated that this Wnt pathway regulates the cytoskeleton to control the plane of mitotic division during this time.

A third Wnt pathway diverges earlier than Disheveled. Here, the receptor protein (possibly Ryk) activates a phospholipase (PLC) that synthesizes a compound that releases calcium ions from the endoplasmic reticulum (**FIGURE 3.28C**). The released calcium can activate enzymes, transcription factors, and translation factors. In zebrafish, Ryk deficiency impairs Wnt-directed calcium influx into cells as well as directional cell movement (Lin et al. 2010).

The TGF-β superfamily

There are more than 30 structurally related members of the **TGF-β superfamily**,* and they regulate some of the most important interactions in development (**FIGURE 3.29**). The TGF-β superfamily includes the TGF-β family, the Nodal and activin families, the bone morphogenetic proteins (BMPs), the Vg1 family, and other proteins, including glial-derived neurotrophic factor (GDNF; necessary for kidney and enteric neuron differentiation) and anti-Müllerian hormone (AMH), a paracrine factor involved in mammalian sex determination.

- Among members of the **TGF-β family**, TGF-β1, 2, 3, and 5 are important in regulating the formation of the extracellular matrix between cells and for regulating

---

*TGF stands for *transforming growth factor*. The designation "superfamily" is often applied when each of the different classes of molecules constitutes a "family." The members of a superfamily all have similar structures but are not as similar as the molecules within each "family" are to one another.

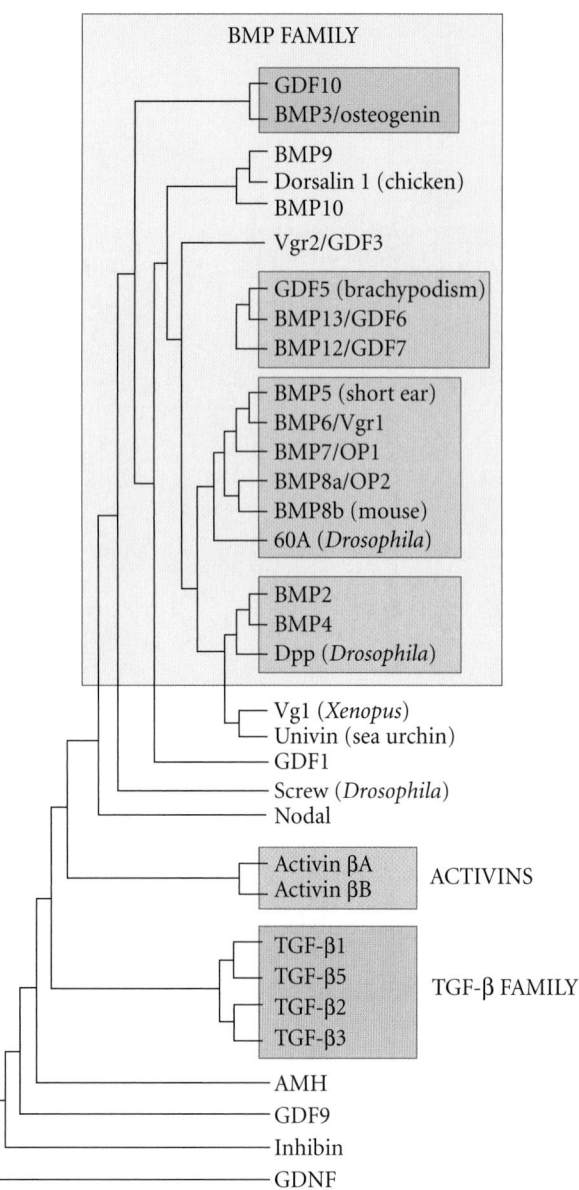

FIGURE 3.29 Relationships among members of the TGF-β superfamily. (After Hogan 1996.)

cell division (both positively and negatively). TGF-β1 increases the amount of extracellular matrix that epithelial cells make (both by stimulating collagen and fibronectin synthesis and by inhibiting matrix degradation). TGF-β proteins may be critical in controlling where and when epithelia branch to form the ducts of kidneys, lungs, and salivary glands (Daniel 1989; Hardman et al. 1994; Ritvos et al. 1995). The effects of the individual TGF-β family members are difficult to sort out because members of the TGF-β family appear to function similarly and can compensate for losses of the others when expressed together.

- The members of the **BMP family** can be distinguished from other members of the TGF-β superfamily by having seven (rather than nine) conserved cysteines in the mature polypeptide. Because they were originally

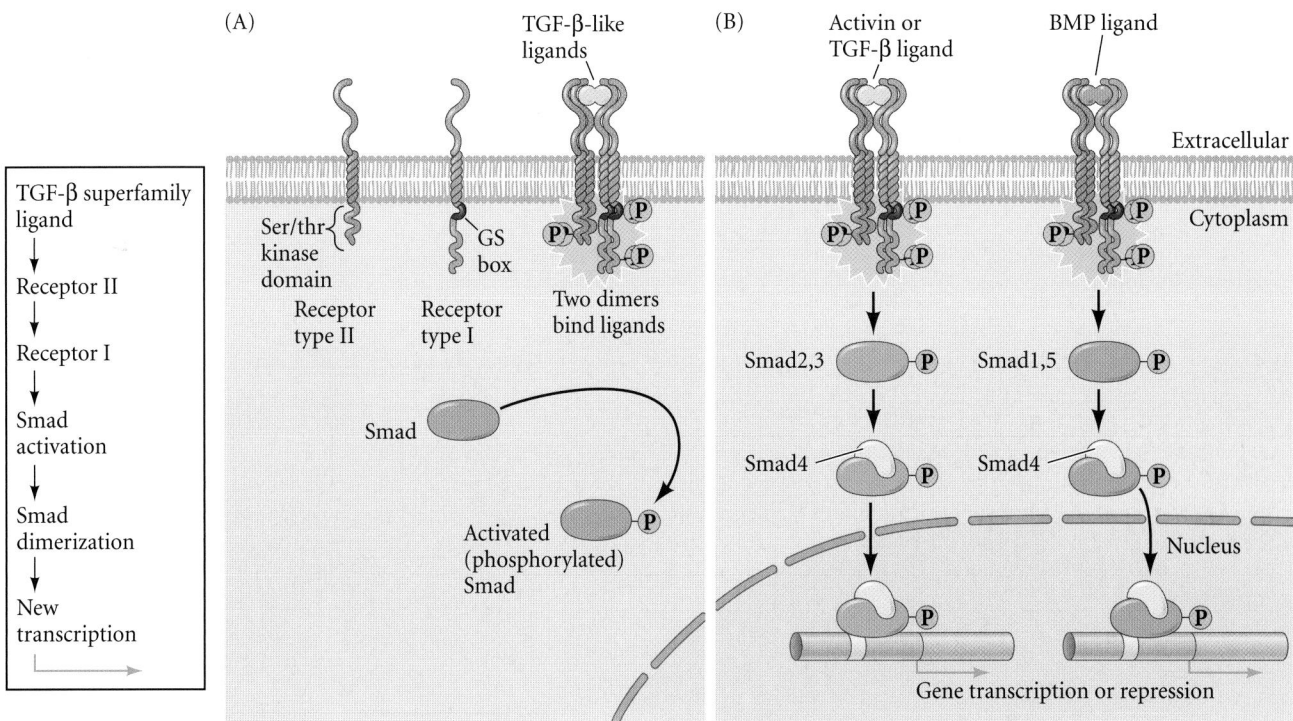

FIGURE 3.30 The Smad pathway is activated by TGF-β superfamily ligands. (A) An activation complex is formed by the binding of the ligand by the type I and type II receptors. This allows the type II receptor to phosphorylate the type I receptor on particular serine or threonine residues. The phosphorylated type I receptor protein can now phosphorylate the Smad proteins. (B) Those receptors that bind TGF-β family proteins or members of the activin family phosphorylate Smads 2 and 3. Those receptors that bind to BMP family proteins phosphorylate Smads 1 and 5. These Smads can complex with Smad4 to form active transcription factors. A simplified version of the pathway is shown at the left.

discovered by their ability to induce bone formation, they were given the name **bone morphogenetic proteins**. But it turns out that bone formation is only one of their many functions; the BMPs are extremely multifunctional.* They have been found to regulate cell division, apoptosis (programmed cell death; see Sidelights & Speculations), cell migration, and differentiation (Hogan 1996). They include proteins such as BMP4 (which in some tissues causes bone formation, in other tissues causes cell death, and in other instances specifies the epidermis) and BMP7 (which is important in neural tube polarity, kidney development, and sperm formation). The BMP4 homologue in *Drosophila* is critically involved in forming appendages, including the limbs, wings, genitalia, and antennae. Indeed, the malformations of 15 such structures have given this homologue the name Decapentaplegic (DPP). As it (rather oddly) turns out, BMP1 is not a member of the BMP family at all; it is a protease. BMPs

are thought to work by diffusion from the cells producing them (Ohkawara et al. 2002).

- The **Nodal** and **activin** proteins are extremely important in specifying the different regions of the mesoderm and for distinguishing the left and right sides of the vertebrate body axis.

THE SMAD PATHWAY Members of the TGF-β superfamily activate members of the **Smad family** of transcription factors (Heldin et al. 1997; Shi and Massagué 2003). The TGF-β ligand binds to a type II TGF-β receptor, which allows that receptor to bind to a type I TGF-β receptor. Once the two receptors are in close contact, the type II receptor phosphorylates a serine or threonine on the type I receptor, thereby activating it. The activated type I receptor can now phosphorylate the Smad† proteins (**FIGURE 3.30A**). Smads 1 and 5 are activated by the BMP family of TGF-β factors, while the receptors binding activin, Nodal, and the TGF-β family phosphorylate Smads 2 and 3. These phosphorylated Smads bind to Smad4 and form the transcription factor complex that will enter the nucleus (**FIGURE 3.30B**).

---

* One of the many reasons why humans don't seem to need an enormous genome is that the gene products—proteins—involved in our construction and development often have many functions. Many of the proteins we are familiar with in adults (such as hemoglobin, keratins, insulin, and the like) *do* have only one function, which led to the erroneous conclusion that this is the norm.

---

†Researchers named the Smad proteins by merging the names of the first identified members of this family: the *C. elegans* SMA protein and the *Drosophila* Mad protein

Cell Death Pathways

"To be, or not to be: that is the question." While we all are poised at life-or-death decisions, this existential dichotomy is exceptionally stark for embryonic cells. Programmed cell death, or **apoptosis**,* is a normal part of development (see Fuchs and Steller 2011). In the nematode *C. elegans*, in which we can count the number of cells as the animal develops, exactly 131 cells die according to the normal developmental pattern. All the cells of this nematode are programmed to die unless they are actively told not to undergo apoptosis. In an adult human, as many as 10^{11} cells die each day and are replaced by other cells. (Indeed, the mass of cells we lose each year through normal cell death is close to our entire body weight!) During embryonic development, we were constantly making and destroying cells, and we generated about three times as many neurons as we eventually ended up with when we were born. Lewis Thomas (1992) aptly noted,

> By the time I was born, more of me had died than survived. It was no wonder I cannot remember; during that time I went through brain after brain for nine months, finally contriving the one model that could be human, equipped for language.

Apoptosis is necessary not only for the proper spacing and orientation of neurons, but also for generating the middle ear space, the vaginal

*The term *apoptosis* (both *p*s are pronounced) comes from the Greek word for the natural process of leaves falling from trees or petals falling from flowers. Apoptosis is an active process that can be subject to evolutionary selection. A second type of cell death, necrosis, is a pathological death caused by external factors such as inflammation or toxic injury. The "cell death pathway" discussed here can also be used for other functions. Low levels of caspase-3, for instance, can cleave particular proteins, and these cleaved proteins can regulate cell development. In this way, the cell death pathway can be converted into a cell differentiation pathway (Basu et al. 2012).

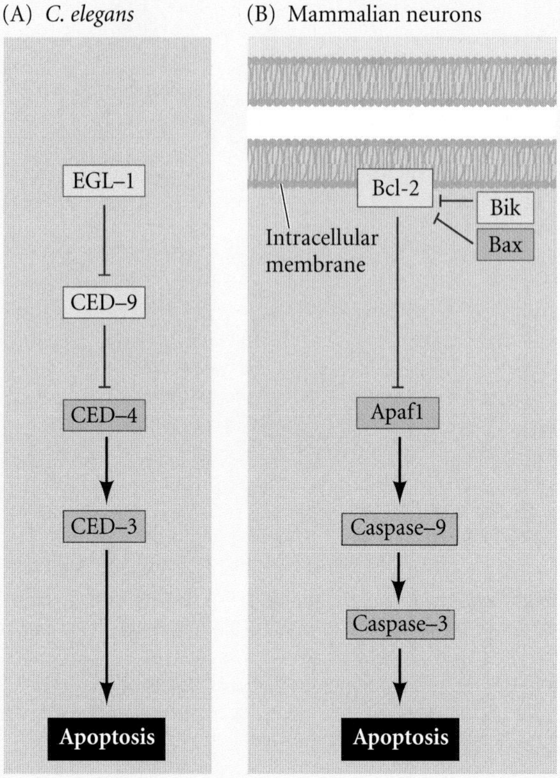

FIGURE 3.31 Apoptosis pathways in nematodes and mammals. (A) In *C. elegans*, the CED-4 protein is a protease-activating factor that can activate the CED-3 protease. The CED-3 protease initiates the cell destruction events. CED-9 can inhibit CED-4 (and CED-9 can be inhibited upstream by EGL-1). (B) In mammals, a similar pathway exists, and appears to function in a similar manner. In this hypothetical scheme for the regulation of apoptosis in mammalian neurons, Bcl-X$_L$ (a member of the Bcl-2 family) binds Apaf1 and prevents it from activating the precursor of caspase-9. The signal for apoptosis allows another protein (here, Bik) to inhibit the binding of Apaf1 to Bcl-X$_L$. Apaf1 is now able to bind to the caspase-9 precursor and cleave it. Caspase-9 dimerizes and activates caspase-3, which initiates apoptosis. The same colors are used to represent homologous proteins. (After Adams and Cory 1998.)

opening, and the spaces between our fingers and toes (Saunders and Fallon 1966; Rodriguez et al. 1997; Roberts and Miller 1998). Apoptosis prunes unneeded structures (frog tails, male mammary tissue), controls the number of cells in particular tissues (neurons in vertebrates and flies), and sculpts complex organs (palate, retina, digits, and heart).

Different tissues use different signals for apoptosis. One of the signals often used in vertebrates is bone morphogenetic protein 4. BMP4 is an excellent example of an important principle in developmental biology: What a paracrine factor does depends on the competence of its targets.† Some tissues, such as connective tissue, respond to BMP4

†As we will see throughout the book, "targets" are rarely the passive entities the metaphor would imply.

by differentiating into bone. Others, such as the frog gastrula ectoderm, respond to BMP4 by differentiating into skin. Still others, such as the cells in the webbing between our developing fingers and toes, respond to BMP by dying. In these cases, BMP is said to activate the "cell death pathway."

The pathways for apoptosis were delineated primarily through genetic studies of *C. elegans*. Indeed, the importance of these pathways was recognized by awarding a Nobel Prize to Sydney Brenner, Robert Horvitz, and Jonathan Sulston in 2002. It was found that the proteins encoded by the *ced-3* and *ced-4* genes were essential for apoptosis, and that in the cells that did not undergo apoptosis, those genes were turned off by the product of the *ced-9* gene (FIGURE 3.31A; Hengartner et al. 1992). The CED-4 protein is a protease-activating factor that activates the gene for CED-3, a

protease that initiates destruction of the cell. CED-9 can bind to and inactivate CED-4. Mutations that inactivate the gene for CED-9 cause numerous cells that would normally survive to activate their *ced-3* and *ced-4* genes and die, leading to the death of the entire embryo. Conversely, gain-of-function mutations in the *ced-9* gene cause its protein to be made in cells that would normally die, resulting in those cells surviving. Thus, the *ced-9* gene appears to be a binary switch that regulates the choice between life and death on the cellular level. It is possible that every cell in the nematode embryo is poised to die, with those cells that survive being rescued by the activation of the *ced-9* gene.

The CED-3 and CED-4 proteins are at the center of the apoptosis pathway that is common to all animals studied. The trigger for apoptosis can be a developmental cue such as

a particular molecule (e.g., BMP4 or glucocorticoids) or the loss of adhesion to a matrix. Either type of cue can activate CED-3 or CED-4 proteins or inactivate CED-9 molecules. In mammals, the homologues of the CED-9 protein are members of the Bcl-2 family (which includes Bcl-2, Bcl-X, and similar proteins; FIGURE 3.31B). The functional similarities are so strong that if an active human *BCL-2* gene is placed in *C. elegans* embryos, it prevents normally occurring cell death (Vaux et al. 1992).

The mammalian homologue of CED-4 is Apaf1 (apoptotic protease activating factor 1), and it participates in the cytochrome *c*-dependent activation of the mammalian CED-3 homologues, the proteases caspase-9 and caspase-3 (see Figure 3.31B; Shaham and Horvitz 1996; Cecconi et al. 1998; Yoshida et al. 1998). Activation of the caspase proteins results in autodigestion—caspases are strong proteases that digest the cell from within, cleaving cellular proteins and fragmenting the DNA.

While apoptosis-deficient nematodes deficient for CED-4 are viable (despite having 15% more cells than wild-type worms), mice with loss-of-function mutations for either caspase-3 or caspase-9 die around birth from massive cell overgrowth in the nervous system (FIGURE 3.32; Kuida et al. 1996, 1998; Jacobson et al. 1997). Mice homozygous for targeted deletions of *Apaf1* have similarly severe craniofacial abnormalities, brain overgrowth, and webbing between their toes.

● See WEBSITE 3.2 The uses of apoptosis

(A) *caspase-9*$^{+/+}$ (Wild-type)   (B) *caspase-9*$^{-/-}$ (Knockout)

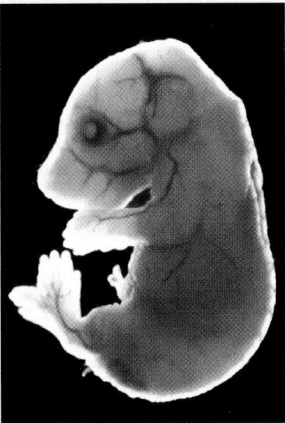 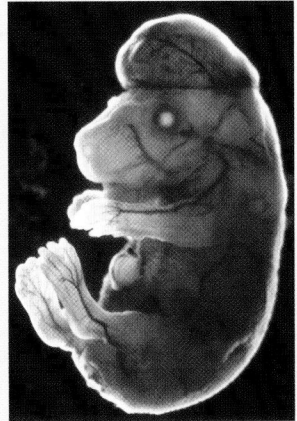

(C) Wild-type   (D) Knockout

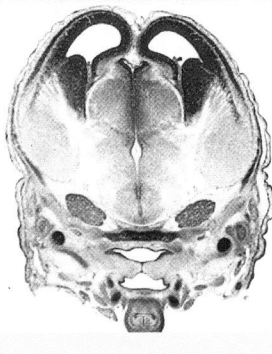 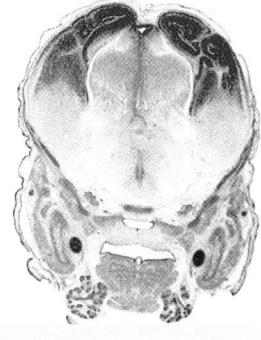

FIGURE 3.32 Disruption of normal brain development by blocking apoptosis. In mice in which the genes for caspase-9 have been knocked out, normal neural apoptosis fails to occur, and the overproliferation of brain neurons is obvious. (A) A 6-day embryonic wild-type mouse. (B) A *caspase-9* knockout mouse of the same age. The enlarged brain protrudes above the face, and the limbs are still webbed. (C,D) This effect is confirmed by cross sections through the forebrain at day 13.5. The knockout exhibits thickened ventricle walls and the near-obliteration of the ventricles. (From Kuida et al. 1998.)

Other paracrine factors

Although most paracrine factors are members of one of the four families described above (FGF, Hedgehog, Wnt, or the TGF-β superfamily), some paracrine factors have few or no close relatives. Epidermal growth factor, hepatocyte growth factor, neurotrophins, and stem cell factor are not included among these four groups, but each plays important roles during development. In addition, there are numerous paracrine factors involved almost exclusively with developing blood cells: erythropoietin, the cytokines, and the interleukins. We will discuss these factors when we detail blood cell formation in Chapter 13.

Juxtacrine Signaling

In juxtacrine interactions, proteins from the inducing cell interact with receptor proteins of adjacent responding cells without diffusing from the cell producing it. Two of the most widely used families of juxtacrine factors are the **Notch proteins** (which bind to a family of ligands exemplified by the Delta protein) and the **eph receptors** and their **ephrin ligands**. When the ephrin on one cell binds with the eph receptor on an adjacent cell, signals are sent to each of the two cells (Davy et al. 2004; Davy and Soriano 2005). These signals are often those of either attraction or repulsion, and ephrins are often seen where cells are being told where to migrate or where boundaries are forming. We will see the ephrins and the eph receptors functioning in the formation of blood vessels, neurons, and somites. For the moment, we will look at the Notch proteins and their ligands.

The Notch pathway: Juxtaposed ligands and receptors

While most known regulators of induction are diffusible proteins, some inducing proteins remain bound to the inducing cell surface. In one such pathway, cells expressing the Delta, Jagged, or Serrate proteins in their cell membranes activate neighboring cells that contain Notch protein in their cell membranes (see Artavanis-Tsakakonas and Muskavitch 2010). Notch extends through the cell membrane, and its external surface contacts Delta, Jagged, or Serrate proteins extending out from an adjacent cell. When complexed to one of these ligands, Notch undergoes a conformational change that enables a part of its cytoplasmic domain to be cut off by the presenilin-1 protease. The cleaved portion enters the nucleus and binds to a dormant transcription factor of the CSL family. When bound to the Notch protein, the CSL transcription factors activate their target genes (**FIGURE 3.33**; Lecourtois and Schweisguth 1998; Schroeder et al. 1998; Struhl and Adachi 1998). This activation is thought to involve the recruitment of histone acetyltransferases (Wallberg et al. 2002). Thus, Notch can be considered as a transcription factor tethered to the cell membrane. When the attachment is broken, Notch (or a piece of it) can detach from the cell membrane and enter the nucleus (Kopan 2002).

Notch proteins are involved in the formation of numerous vertebrate organs—kidney, pancreas, and heart—and they are extremely important receptors in the nervous system. In both the vertebrate and *Drosophila* nervous systems, the binding of Delta to Notch tells the receiving cell not to become neural (Chitnis et al. 1995; Wang et al. 1998). In the

(A)

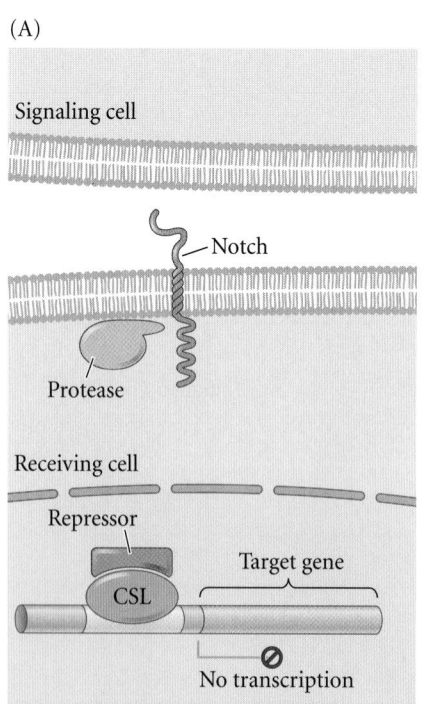

(B)

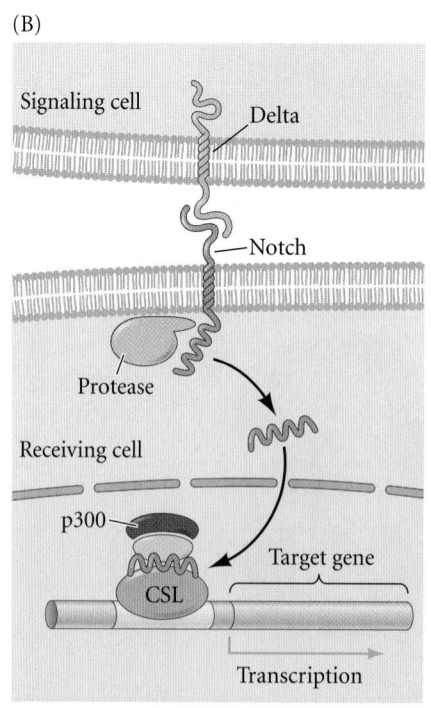

FIGURE 3.33 Mechanism of Notch activity. (A) Prior to Notch signaling, a CSL transcription factor (such as Suppressor of hairless or CBF1) is on the enhancer of Notch-regulated genes. The CSL binds repressors of transcription. (B) Model for the activation of Notch. A ligand (Delta, Jagged, or Serrate protein) on one cell binds to the extracellular domain of the Notch protein on an adjacent cell. This binding causes a shape change in the intracellular domain of Notch, which activates a protease. The protease cleaves Notch and allows the intracellular region of the Notch protein to enter the nucleus and bind the CSL transcription factor. This intercellular region of Notch displaces the repressor proteins and binds activators of transcription, including the histone acetyltransferase p300. The activated CSL can then transcribe its target genes. (After K. Koziol-Dube, Pers. Comm.)

SIDELIGHTS & SPECULATIONS

Juxtacrine Signaling and Cell Patterning

nduction does indeed occur on the cell-to-cell level, and one of the best examples is the formation of the vulva in the nematode worm *Caenorhabditis elegans*. Remarkably, the signal transduction pathways involved turn out to be the same as those used in the formation of retinal receptors in *Drosophila*; only the targeted transcription factors are different. In both cases, an epidermal growth factor-like inducer activates the RTK pathway.

Vulval induction in *C. elegans*

Most *C. elegans* individuals are hermaphrodites. In their early development, they are male and the gonad produces sperm, which is stored for later use. As they grow older,

they develop ovaries. The eggs "roll" through the region of sperm storage, are fertilized inside the nematode, and then pass out of the body through the vulva (see Figure 5.20).

The formation of the vulva in *C. elegans* represents a case in which one inductive signal generates a variety of cell types. This organ forms during the larval stage from six cells called the **vulval precursor cells** (**VPCs**). The cell connecting the overlying gonad to the vulval precursor cells is called the **anchor cell** (FIGURE 3.34). The anchor cell secretes LIN-3 protein, a paracrine factor (similar to mammalian epidermal growth factor, or EGF) that activates the RTK pathway (Hill and Sternberg 1992). If the anchor

cell is destroyed (or if the *lin-3* gene is mutated), the VPCs will not form a vulva, and instead become part of the hypodermis (skin) (Kimble 1981).

The six VPCs influenced by the anchor cell form an **equivalence group**. Each member of this group is competent to become induced by the anchor cell and can assume any of three fates, depending on its proximity to the anchor cell. The cell directly beneath the anchor cell divides to form the central vulval cells. The two cells flanking that central cell divide to become the lateral vulval cells, while the three cells farther away from the anchor cell generate hypodermal cells. If the anchor cell is destroyed, all six cells of the equivalence group

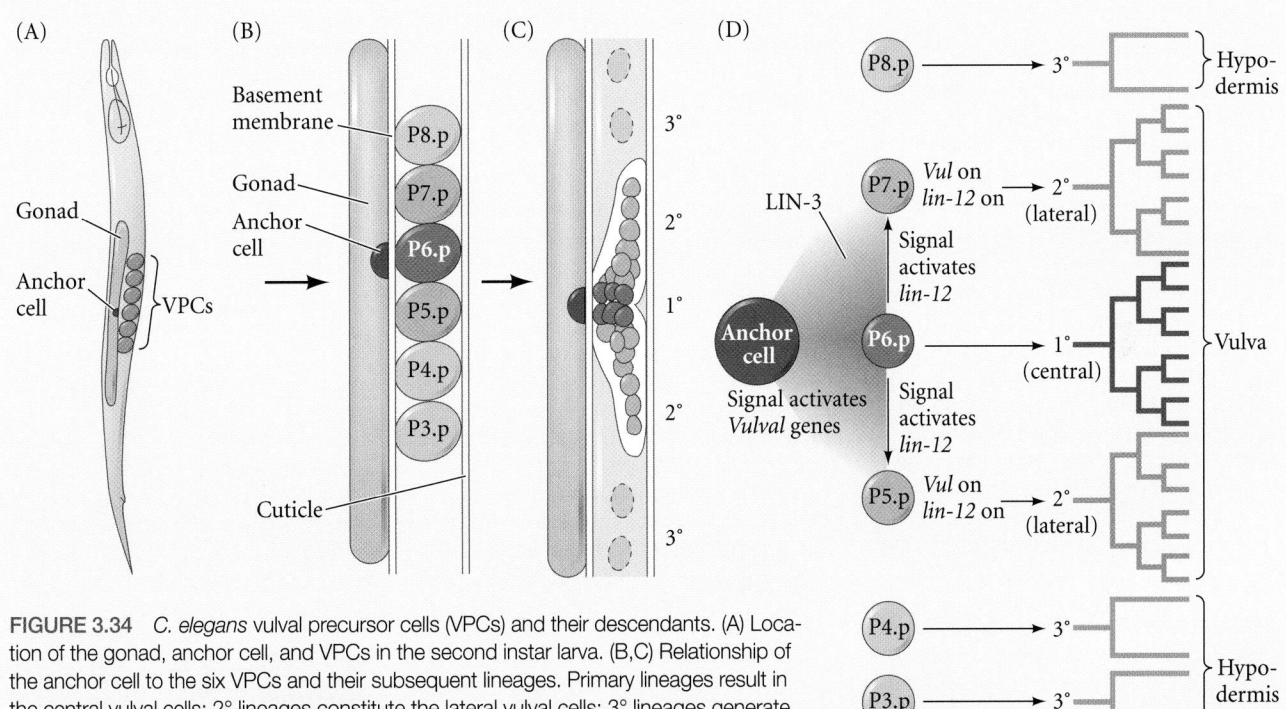

FIGURE 3.34 *C. elegans* vulval precursor cells (VPCs) and their descendants. (A) Location of the gonad, anchor cell, and VPCs in the second instar larva. (B,C) Relationship of the anchor cell to the six VPCs and their subsequent lineages. Primary lineages result in the central vulval cells; 2° lineages constitute the lateral vulval cells; 3° lineages generate hypodermal cells. (C) Outline of the vulva in the fourth instar larva. The circles represent the positions of the nuclei. (D) Model for the determination of vulval cell lineages in *C. elegans*. The LIN-3 signal from the anchor cell causes the determination of the P6.p cell to generate the central vulval lineage (dark purple). Lower concentrations of LIN-3 cause the P5.p and P7.p cells to form the lateral vulval lineages. The P6.p (central lineage) cell also secretes a short-range juxtacrine signal that induces the neighboring cells to activate the LIN-12 (Notch) protein. This signal prevents the P5.p and P7.p cells from generating the 1°, central vulval cell lineage. (After Katz and Sternberg 1996.)

divide once and contribute to the hypodermal tissue. If the three central VPCs are destroyed, the three outer cells, which normally form hypodermis, generate vulval cells instead.

The LIN-3 protein is received by the LET-23 receptor tyrosine kinase on the VPCs, and the signal is transferred to the nucleus through the RTK pathway. The target of the kinase cascade is LIN-31 (Tan et al. 1998). When this protein is phosphorylated in the nucleus, it loses its inhibitory protein partner and is able to function as a transcription factor, promoting vulval cell fates. Two mechanisms coordinate the formation of the vulva through this induction, as shown in Figure 3.34 (Katz and Sternberg 1996; Félix 2007):

- LIN-3 forms a concentration gradient in which the VPC closest to the anchor cell (i.e., the P6.p cell) receives the highest concentration of LIN-3 and generates the central vulval cells. The two adjacent VPCs (P5.p and P7.p) receive lower amounts of LIN-3 and become the lateral vulval cells. VPCs farther away from the anchor cell do not receive enough LIN-3 to have an effect, so they become hypodermis (Katz et al. 1995).

- In addition to forming the central vulval lineage, the VPC closest to the anchor cell also signals laterally to the two adjacent (P5.p and P7.p) cells and instructs them not to generate the central vulval lineages. The P5.p and P7.p cells receive the signal through the LIN-12 (Notch) proteins on their cell membranes. The Notch signal activates a microRNA, *mir-61*, which represses the gene that would specify central vulval fate,

as well as promoting those genes that are involved in forming the lateral vulval cells (Sternberg 1988; Yoo et al. 2005). The lateral cells do not instruct the peripheral VPCs to do anything, so they become hypodermis (Koga and Ohshima 1995; Simske and Kim 1995).

Cell-cell interactions and chance in the determination of cell types

Vuval development in *C. elegans* offers several examples of induction on the cellular level (Barkoulas et al. 2013). We have already discussed the reception of the EGF-like LIN-3 signal by the cells of the equivalence group that forms the vulva. But before this induction occurs, an earlier interaction has formed the anchor cell. The formation of the anchor cell is mediated by *lin-12*, the *C. elegans* homologue of the *Notch* gene. In wild-type *C. elegans* hermaphrodites, two adjacent cells, Z1.ppp and Z4.aaa, have the potential to become the anchor cell. They interact in a manner that causes one of them to become the anchor cell while the other one becomes the precursor of the uterine tissue. In loss-of-function *lin-12* mutants, both cells become anchor cells, whereas in gain-of-function mutations, both cells become uterine precursors (Greenwald et al. 1983). Studies using genetic mosaics and cell ablations have shown that this decision is made in the second larval stage, and that the *lin-12* gene needs to function only in that cell destined to become the uterine precursor cell. The presumptive anchor cell does not need it. Seydoux and Greenwald (1989) speculate that these two cells

originally synthesize both the signal for uterine differentiation (the LAG-2 protein, homologous to Delta) and the receptor for this molecule (the LIN-12 protein, homologous to Notch; Wilkinson et al. 1994).

During a particular time in larval development, the cell that, by chance, is secreting more LAG-2 causes its neighbor to cease its production of this differentiation signal and to increase its production of LIN-12. The cell secreting LAG-2 becomes the gonadal anchor cell, while the cell receiving the signal through its LIN-12 protein becomes the ventral uterine precursor cell (**FIGURE 3.35**). Thus, the two cells are thought to determine each other prior to their respective differentiation events. When LIN-12 is used again during vulva formation, it is activated by the primary vulval lineage to stop the lateral vulval cells from forming the central vulval phenotype (see Figure 3.34). Thus, the anchor cell/ventral uterine precursor decision illustrates two important aspects of determination in two originally equivalent cells. First, the initial difference between the two cells is created by chance. Second, this initial difference is reinforced by feedback.

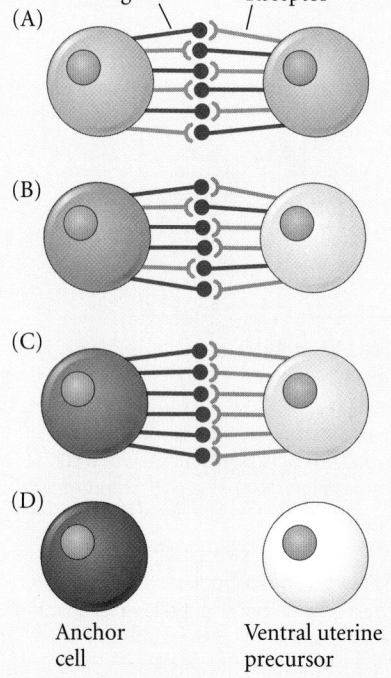

FIGURE 3.35 Model for the generation of two cell types (anchor cell and ventral uterine precursor cell) from two equivalent cells (Z1.ppp and Z4.aaa) in *C. elegans*. (A) The cells start off as equivalent, producing fluctuating amounts of signal and receptor (inverted arrow). The *lag-2* gene is thought to encode the signal; the *lin-12* gene is thought to encode the receptor. Reception of the signal turns down LAG-2 (Delta) production and upregulates LIN-12 (Notch). (B) A stochastic (chance) event causes one cell to produce more LAG-2 than the other cell at some particular critical time. This stimulates more LIN-12 production in the neighboring cell. (C) This difference is amplified, since the cell producing more LIN-12 produces less LAG-2. Eventually, just one cell is delivering the LAG-2 signal, and the other cell is receiving it. (D) The signaling cell becomes the anchor cell; the receiving cell becomes the ventral uterine precursor cell. (After Greenwald and Rubin 1992.)

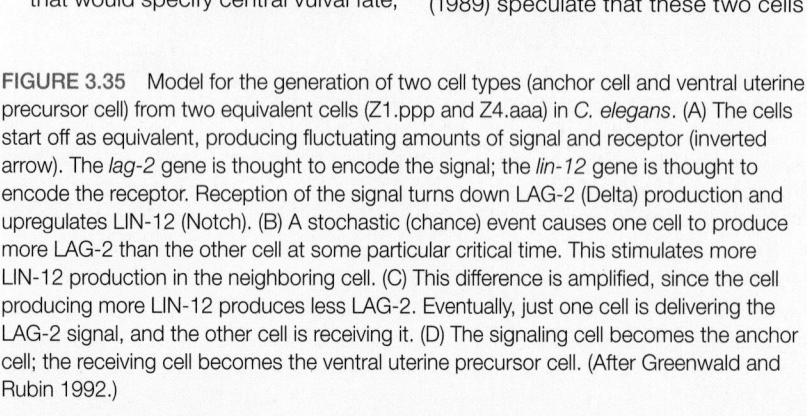

vertebrate eye, the interactions between Notch and its ligands regulate which cells become optic neurons and which become glial cells (Dorsky et al. 1997; Wang et al. 1998). Notch proteins are also important in the patterning of the nematode vulva. The vulval precursor cell closest to the anchor cell becomes the central vulva cell, and this cell is able to prevent its neighbors from becoming central vulval cells by signaling to them through its Notch homologue, the LIN-12 receptor (Berset et al. 2001).

● See WEBSITE 3.3 Notch mutations

The Extracellular Matrix as a Source of Developmental Signals

The **extracellular matrix** is an insoluble network consisting of macromolecules secreted by cells into their immediate environment. These macromolecules form a region of noncellular material in the interstices between the cells. The extracellular matrix is a critical region for much of animal development. Cell adhesion, cell migration, and the formation of epithelial sheets and tubes all depend on the ability of cells to form attachments to extracellular matrices. In some cases, as in the formation of epithelia, these attachments have to be extremely strong. In other instances, as when cells migrate, attachments have to be made, broken, and made again. In some cases, the extracellular matrix merely serves as a permissive substrate to which cells can adhere, or on which they can migrate. In other cases, it provides the directions for cell movement or the signal for a developmental event. Extracellular matrices are made up of the matrix protein collagen, proteoglycans, and a variety of specialized glycoprotein molecules such as fibronectin and laminin.

Proteoglycans play critically important roles in the delivery of the paracrine factors. These large molecules consist of core proteins (such as syndecan) with covalently attached glycosaminoglycan polysaccharide side chains. Two of the most widespread proteoglycans are heparan sulfate and chondroitin sulfate. Heparan sulfate can bind many members of the TGF-β, Wnt, and FGF families, and it appears to be essential for presenting the paracrine factor in high concentrations to its receptors. In *Drosophila, C. elegans,* and mice, mutations that prevent proteoglycan protein or carbohydrate synthesis block normal cell migration, morphogenesis, and differentiation (García-García and Anderson 2003; Hwang et al. 2003; Kirn-Safran et al. 2004).

The large glycoproteins are responsible for organizing the matrix and the cells into an ordered structure. **Fibronectin** is a very large (460 kDa) glycoprotein dimer synthesized by numerous cell types. One function of fibronectin is to serve as a general adhesive molecule, linking cells to one another and to other substrates such as collagen and proteoglycans. Fibronectin has several distinct binding sites, and their interaction with the appropriate molecules results in the proper alignment of cells with their extracellular matrix (**FIGURE 3.36A**). Fibronectin also has an important role in cell migration, since the "roads" over which certain migrating cells travel are paved with this protein. Fibronectin paths lead germ cells to the gonads and heart cells to the midline of the embryo. If chick embryos are injected with antibodies to fibronectin, the heart-forming cells fail to reach the midline, and two separate hearts develop (Heasman et al. 1981; Linask and Lash 1988).

Laminin (another large glycoprotein) and **type IV collagen** are major components of a type of extracellular matrix called the **basal lamina**. The basal lamina is characterized by closely knit sheets that surround epithelial tissue (**FIGURE 3.36B**).

(A)

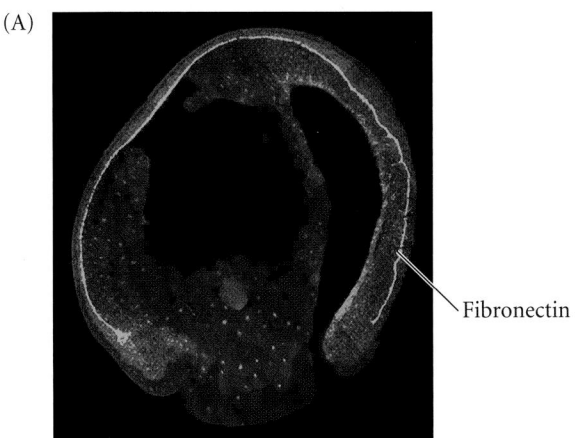

Fibronectin

(B)

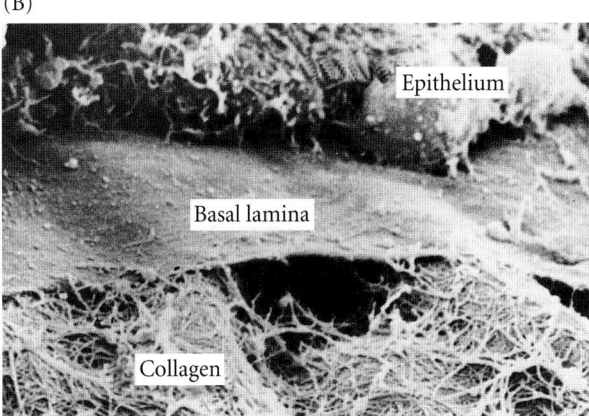

Epithelium

Basal lamina

Collagen

FIGURE 3.36 Extracellular matrices in the developing embryo. (A) Fluorescent antibodies to fibronectin show fibronectin deposition as a green band in the *Xenopus* embryo during gastrulation. The fibronectin will orient the movements of the mesoderm cells. (B) Fibronectin links together migrating cells, collagen, heparan sulfate, and other extracellular matrix proteins. This scanning electron micrograph shows the extracellular matrix at the junction of the epithelial cells (above) and mesenchymal cells (below). The epithelial cells synthesize a tight, laminin-based basal lamina, whereas the mesenchymal cells secrete a loose reticular lamina made primarily of collagen. (A courtesy of M. Marsden and D. W. DeSimone; B courtesy of R. L. Trelsted.)

The adhesion of epithelial cells to laminin (on which they sit) is much greater than the affinity of mesenchymal cells for fibronectin (to which they must bind and release if they are to migrate). Like fibronectin, laminin plays a role in assembling the extracellular matrix, promoting cell adhesion and growth, changing cell shape, and permitting cell migration (Hakamori et al. 1984; Morris et al. 2003).

● **See VADE MECUM** Elements of the extracellular matrix

● **See WEBSITE 3.4** Maintaining cell differentiation

Integrins: Receptors for extracellular matrix molecules

The ability of a cell to bind to adhesive glycoproteins such as laminin or fibronectin depends on its expressing membrane receptors for the cell-binding sites of these large molecules. The main fibronectin receptors were identified by using antibodies that block the attachment of cells to fibronectin (Chen et al. 1985; Knudsen et al. 1985). The main fibronectin receptor was found to be an extremely large protein that could bind fibronectin on the outside of the cell, span the membrane, and bind cytoskeletal proteins on the inside of the cell. Thus, the fibronectin receptor complex appears to span the cell membrane and to unite the extracellular and intracellular matrices (**FIGURE 3.37**).

This family of receptor proteins are called integrins because they *integrate* the extracellular and intracellular scaffolds, allowing them to work together (Horwitz et al.

1986; Tamkun et al. 1986). On the extracellular side, integrins bind to the amino acid sequence arginine-glycine-aspartate (RGD), found in several extracellular matrix adhesive proteins, including fibronectin, vitronectin (found in the basal lamina of the eye), and laminin (Ruoslahti and Pierschbacher 1987). On the cytoplasmic side, integrins bind to talin and α-actinin, two proteins that connect to actin microfilaments. This dual binding enables the cell to move by contracting the actin microfilaments against the fixed extracellular matrix.

Integrins can also signal from the outside of the cell to the inside of the cell, altering gene expression (Walker et al. 2002). Bissell and her colleagues (1982; Martins-Green and Bissell 1995) have shown that integrin is critical for inducing specific gene expression in developing tissues, especially those of the liver, testis, and mammary gland. In the mammary gland, extracellular laminin is able to signal the expression of estrogen receptor and casein protein genes through the integrin proteins (Streuli et al. 1991; Notenboom et al. 1996; Muschler et al. 1999; Novaro et al. 2003).

The presence of bound integrin prevents the activation of genes that promote apoptosis (Montgomery et al. 1994; Frisch and Ruoslahti 1997). For instance, the chondrocytes that produce the cartilage of our vertebrae and limbs can survive and differentiate only if they are surrounded by an extracellular matrix and are joined to that matrix through their integrins (Hirsch et al. 1997). If chondrocytes from the developing chick sternum are incubated with antibodies that block the binding of integrins to the extracellular matrix, they shrivel up and die. Indeed, when focal adhesions linking an epithelial cell to its extracellular matrix are broken, the caspase-dependent apoptosis pathway is activated and the cell dies. Such "death-upon-detachment" is a special type of apoptosis called **anoikis**, and it appears to be a major weapon against cancer (Frisch and Francis 1994; Chiarugi and Giannoni 2008).

While the mechanisms by which bound integrins inhibit apoptosis remain controversial, the extracellular matrix is obviously an important source of signals that can be transduced into the nucleus to produce specific gene expression. Some of the genes induced by matrix attachment are being identified. When plated onto tissue culture plastic, mouse mammary gland cells will divide (**FIGURE 3.38**). Indeed, genes for cell division (*c-myc*, *cyclinD1*) are expressed, while genes for differentiated products of the mammary gland (casein, lactoferrin, whey acidic protein) are not expressed. If the same cells are plated onto plastic coated with a basal lamina, the cells stop dividing and the genes of differentiated mammary gland cells are expressed. This happens only

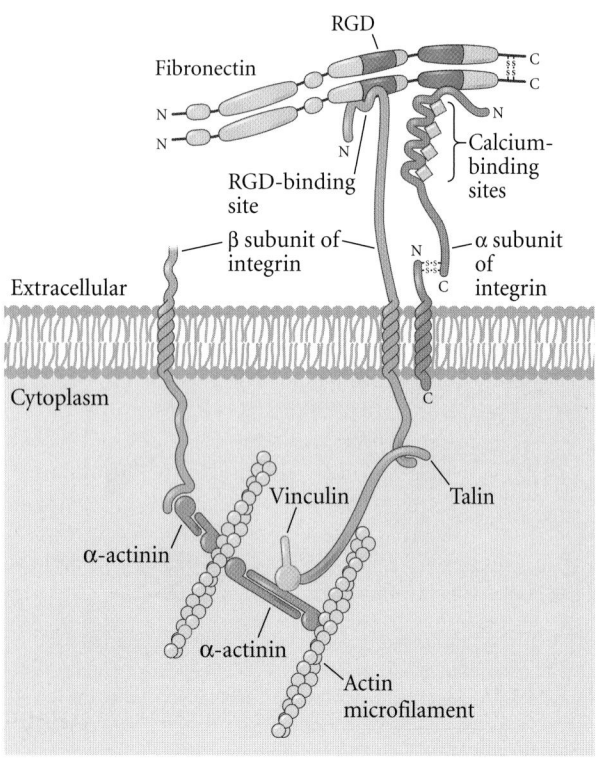

FIGURE 3.37 Simplified diagram of the fibronectin receptor complex. The integrins of the complex are membrane-spanning receptor proteins that bind fibronectin on the outside of the cell while binding cytoskeletal proteins on the inside of the cell. (After Luna and Hitt 1992.)

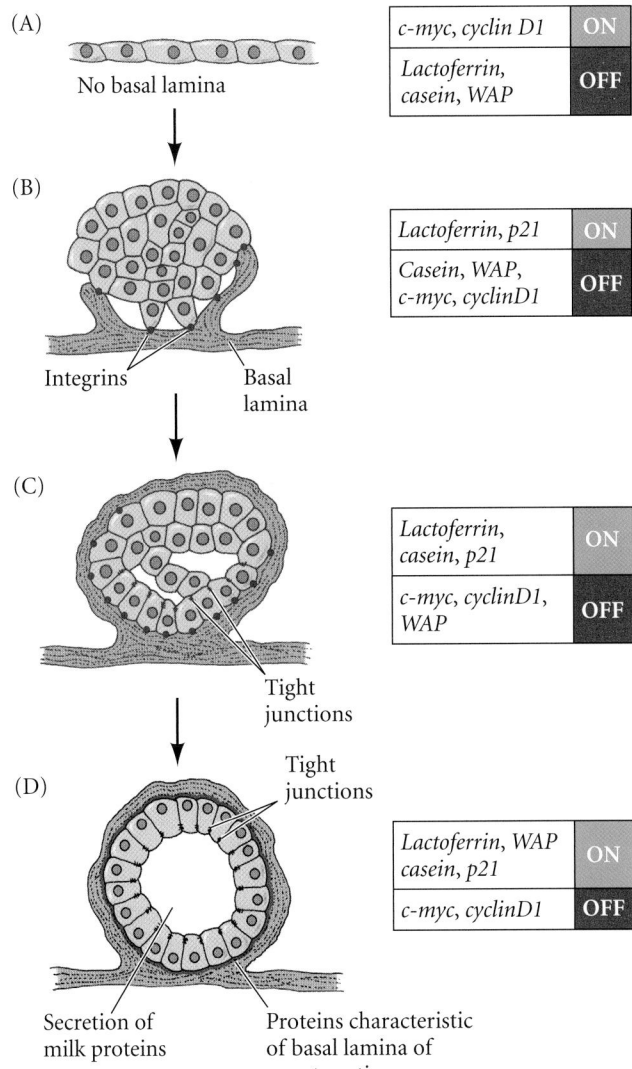

(A) No basal lamina

| c-myc, cyclin D1 | ON |
| Lactoferrin, casein, WAP | OFF |

(B) Integrins   Basal lamina

| Lactoferrin, p21 | ON |
| Casein, WAP, c-myc, cyclinD1 | OFF |

(C) Tight junctions

| Lactoferrin, casein, p21 | ON |
| c-myc, cyclinD1, WAP | OFF |

(D) Tight junctions

| Lactoferrin, WAP casein, p21 | ON |
| c-myc, cyclinD1 | OFF |

Secretion of milk proteins   Proteins characteristic of basal lamina of secretory tissue

FIGURE 3.38 Basal lamina-directed gene expression in mammary gland tissue. (A) Mouse mammary gland tissue divides when placed on tissue culture plastic. The genes encoding cell division proteins are on, and the genes capable of synthesizing the differentiated products of the mammary gland—lactoferrin, casein, and whey acidic protein (WAP)—are off. (B) When these cells are placed on a basal lamina, the genes for cell division proteins are turned off, while the genes encoding inhibitors of cell division (such as p21) and the gene for lactoferrin are turned on. (C,D) The mammary gland cells wrap the basal lamina around them, forming a secretory epithelium. The genes for casein and WAP are sequentially activated. (After Bissell et al. 2003.)

after the integrins of the mammary gland cells bind to the laminin of the basal lamina. Then the gene for lactoferrin is expressed, as is the gene for p21, a cell division inhibitor. The *c-myc* and *cyclinD1* genes become silent. Eventually, all the genes for the developmental products of the mammary gland are expressed, and the cell division genes remain

turned off. By this time, the mammary gland cells have enveloped themselves in a basal lamina, forming a secretory epithelium reminiscent of the mammary gland tissue. The binding of integrins to laminin is essential for transcription of the casein gene, and the integrins act in concert with prolactin (see Figure 3.23) to activate that gene's expression (Roskelley et al. 1994; Muschler et al. 1999).

Several studies have shown that the binding of integrins to an extracellular matrix can stimulate the RTK pathway. When an integrin on the cell membrane of one cell binds to fibronectin or collagen secreted by a neighboring cell, the integrin can activate the RTK cascade through an adaptor protein-like complex that connects the integrin to the Ras G protein (Wary et al. 1998).

The Epithelial-Mesenchymal Transition

One important developmental phenomenon, the **epithelial-mesenchymal transition**, or **EMT**, integrates all the processes we have discussed in this chapter. EMT is an orderly series of events whereby epithelial cells are transformed into mesenchymal cells. In this transition, a polarized stationary epithelial cell, which normally interacts with basal lamina through its basal surface, becomes a migratory mesenchymal cell that can invade tissues and form organs in new places (**FIGURE 3.39A**; see Sleepman and Thiery 2011). EMT is usually initiated when paracrine factors from neighboring cells activate gene expression in the target cells, instructing the target cells to downregulate their cadherins, release their attachment to laminin and other basal lamina components, rearrange their actin cytoskeleton, and secrete new extracellular matrix molecules characteristic of mesenchymal cells.

Epithelial-mesenchymal transition is critical during development (**FIGURE 3.39B,C**). Examples of developmental processes in which this transition is active include (1) the formation of neural crest cells from the dorsalmost region of the neural tube; (2) the formation of mesoderm in chick embryos, wherein cells that had been part of an epithelial layer become mesodermal and migrate into the embryo; and (3) the formation of vertebrae precursor cells from the somites, wherein these cells detach from the somite and migrate around the developing spinal cord. EMT is also important in adults, in whom it is needed for wound healing. However, the most critical adult form of EMT is seen in cancer metastasis, wherein cells that have been part of a solid tumor mass leave that tumor to invade other tissues and form secondary tumors elsewhere in the body. It appears that in metastasis, the processes that generated the cellular transition in the embryo are reactivated, allowing cancer cells to migrate and become invasive. Cadherins are downregulated, the actin cytoskeleton is reorganized, and the cells secrete mesenchymal extracellular matrix while undergoing cell division (Acloque et al. 2009; Kalluri and Weinberg 2009).

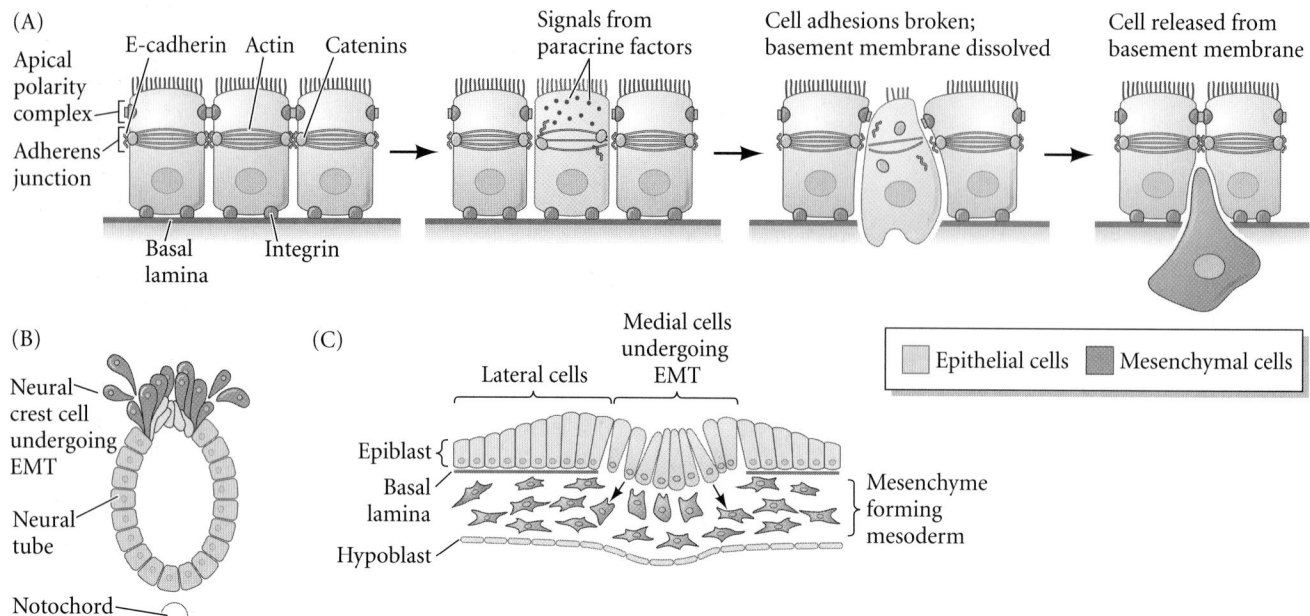

FIGURE 3.39 Epithelial-mesenchymal transition, or EMT. (A) Normal epithelial cells are attached to one another through adherens junctions containing cadherin, catenins, and actin rings. They are attached to the basal lamina through integrins. Paracrine factors can repress the expression of genes that encode these cellular components, causing the cell to lose polarity, lose attachment to the basal lamina, and lose cohesion with other epithelial cells. Cytoskeletal remodeling occurs, as well as the secretion of proteases and extracellular matrix molecules that enable the migration of the newly formed mesenchymal cell. (B,C) EMT is seen in vertebrate embryos during the normal formation of neural crest from the dorsal region of the neural tube (B), and during the formation of the mesoderm by mesenchymal cells delaminating from the epiblast (C).

The Cell Biology of Paracrine Signaling

We have been discussing cell membrane dynamics and cell signaling as if they were two separate entities; but their functioning is closely related. Paracrine factors can rearrange the cell surface, and the cell surface is critical in regulating paracrine factor synthesis, flow, and function. The actions of paracrine signals often change the composition of the cell membrane.

ENDOSOMAL SIGNALING IN THE WNT PATHWAY When Wnt binds to its receptors, the β-catenin destruction complex binds to the receptor, and the entire complex (including the receptor and its bound Wnt) is internalized into the cell in membrane-bound vesicles called **endosomes** (FIGURE 3.40; Taelman et al. 2010; Niehrs 2012). This removes the complex, targets it for degradation, and enables the survival of β-catenin. The internalization of the signaling complex appears to be critical for the accumulation of β-catenin, and proteins that aid in this endocytosis (such as R-spondins) make the Wnt pathway more efficient (Ohkawara et al. 2011).

DIFFUSION OF PARACRINE FACTORS Paracrine factors do not flow freely through the extracellular space. Rather, the factor can be bound by the cell membranes and extracellular matrices of the tissues. In some cases, such binding can impede the spread of a paracrine morphogen and even target the paracrine factor for degradation (Capurro et al. 2008; Schwank et al. 2011). Wnt proteins, for instance, do not diffuse far from the cells secreting them unless helped by other proteins. Thus, the range of Wnt factors is significantly extended when the nearby cells secrete proteins that bind to the paracrine factor and prevent it from binding prematurely to the target tissue (FIGURE 3.41; Mulligan et al. 2012). Similarly, heparan sulfate proteogycans on the extracellular matrix often promote the stability of FGF, BMP, and Wnt proteins, perhaps shielding them from protein-digesting enzymes and thus affording them more time to diffuse (Akiyama et al. 2008; Yan and Lin 2009; Christian 2011; Nahmad and Lander 2011). Cell surface proteoglycans can also store the paracrine factors so that they can be presented to their receptors at a high density (Berendsen et al. 2011; Müller and Schier 2011). Thus, the target tissue is not passive. It can promote diffusion, retard diffusion, or degrade the paracrine factor.

CILIA AND LAMELLIPODIA AS SIGNAL RECEPTION CENTERS In many cases, the reception of paracrine factors is not uniform throughout the cell membrane. For instance, the reception of Hedgehog proteins in vertebrates occurs on cilia, extensions of the cell membrane made by microtubules (Huangfu et al. 2003; Goetz and Anderson 2010). In unstimulated cells, the Patched protein (the Hedgehog receptor; see Figure 3.25) is located on the cilia, while the Smoothened

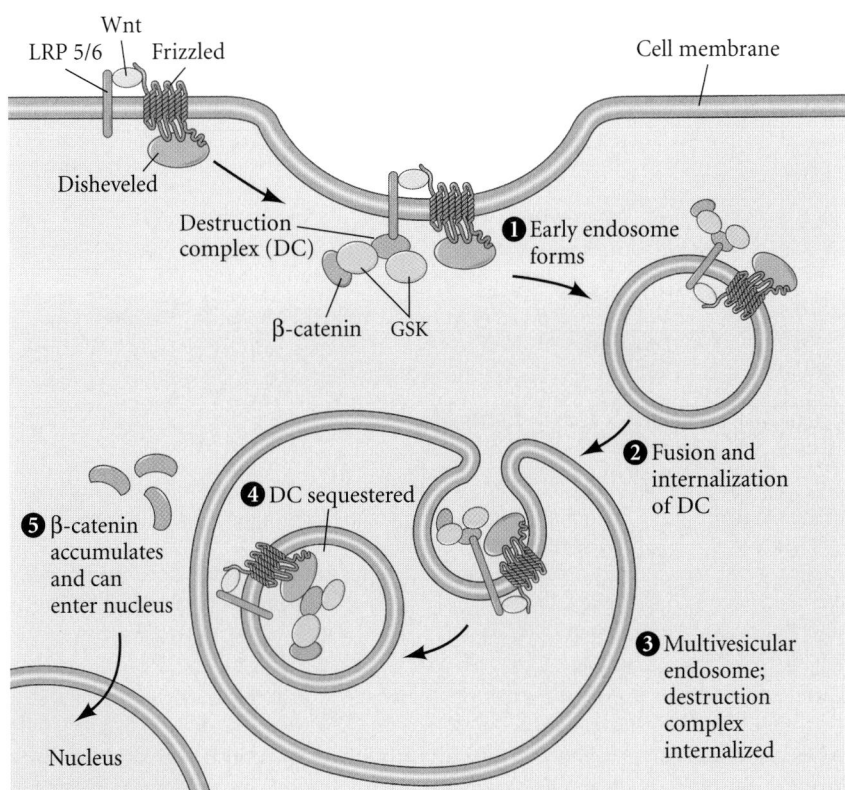

FIGURE 3.40   The Wnt pathway, showing the packaging of the β-catenin destruction apparatus into endosomes as a consequence of Wnt signaling. A major mechanism for separating β-catenin from the enzymes that would destroy it is to package the complex in membrane-bound vesicles called endosomes. When Wnt binds to Frizzled, Frizzled can bind the destruction complex and the entire complex (including the bound Wnt and its receptor) is internalized, allowing β-catenin to accumulate rather than being degraded. (After Taelman et al. 2010.)

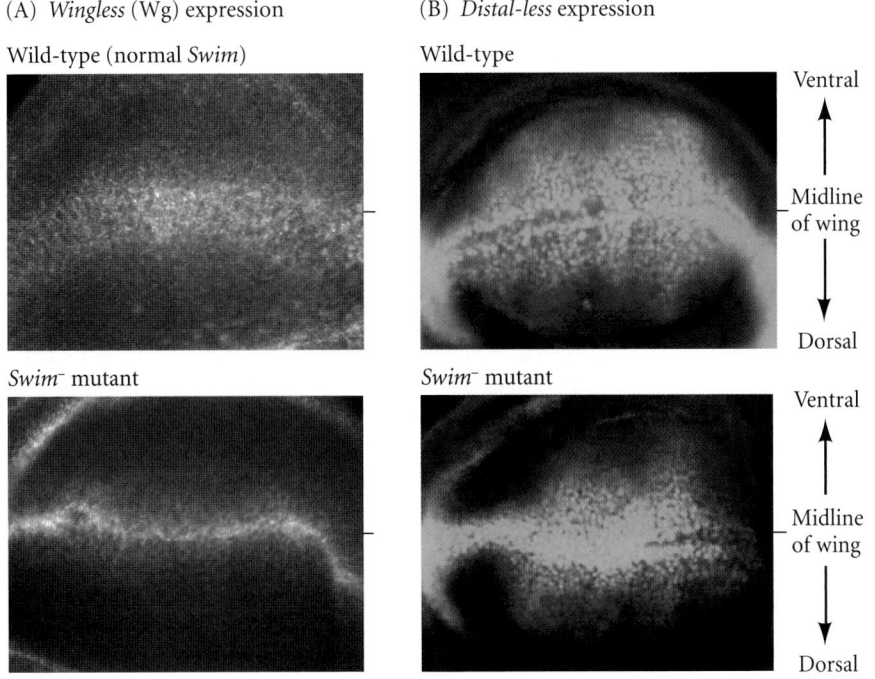

FIGURE 3.41   Wnt paracrine factor diffusion is affected by other proteins. (A) Diffusion of Wingless (Wg, a Wnt paracrine factor) throughout the developing *Drosophila* wing (above) is enhanced by Swim, a protein that stabilizes Wg and that is made by some of the wing cells. When Swim is not present, as in the mutant below, Wg does not disperse but is confined to the narrow band of *Wg*-expressing cells. (B) Similarly, Wingless usually activates the *Distal-less* gene (green) in much of the wild-type wing (seen above). However, in *Swim*-mutant flies, the range of *Distal-less* expression is confined to those areas near the band of *Wg*-expressing cells. (From Mulligan et al. 2012.)

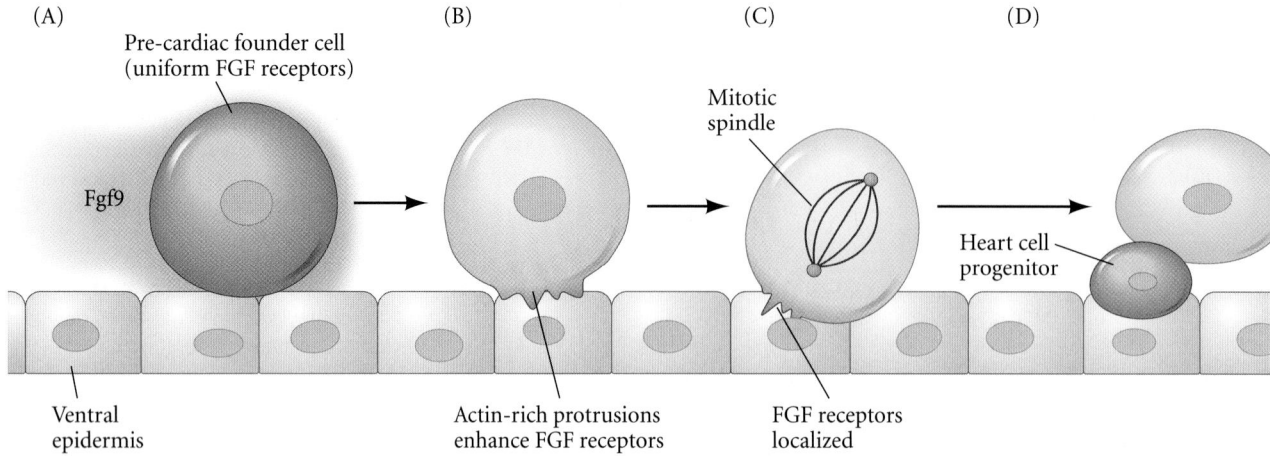

(A)

Pre-cardiac founder cell
(uniform FGF receptors)

Fgf9

(B)

(C)

Mitotic
spindle

(D)

Heart cell
progenitor

Ventral
epidermis

Actin-rich protrusions
enhance FGF receptors

FGF receptors
localized

FIGURE 3.42 Model for differential specification of the tunicate heart progenitor lineage. (A) Uniform exposure to Fgf9 leads to uniform FGF receptor occupancy on all parts of the founder cell membrane. (B) Actin-rich protrusions on the ventral-anterior membrane of the cell are associated with high FGF receptor activation. (C) As the progenitor cell enter mitosis, invasive protrusions of the ventral-anterior cell membrane facilitate restrict FGF receptors to this region. (D) Following asymmetric cell division, the FGF-activated MAPK pathway is restricted to the ventral daughter cell, leading to differential expression of heart progenitor genes. (After Cooley et al. 2011.)

protein is in the cell membrane (or membrane vesicle) close to the cilia. Patched prevents Smoothened from entering the cilia (Milenkovic et al. 2009; Wang et al. 2009). When Hedgehog binds to Patched, Smoothened is allowed to join it on the ciliary cell membrane, where it inhibits the PKA and Slimb proteins that make the repressive form of the Gli transcription factor (Gli is the vertebrate homologue of the *Drosophila* Ci protein shown in Figure 3.25). The microtubules of vertebrate cilia help transport activated Gli protein into the cytoplasm, and mutations that knock out cilia formation in vertebrates also preclude Hedgehog signaling.

In tunicates, an asymmetric division of a pre-cardiac founder cell gives rise to the heart progenitors. Although both daughter cells are exposed to the inductive signal Fgf9, only the smaller of the two responds to generate the heart progenitor lineage. During asymmetric division, localized protrusions (lamellopodia) form on the anterior ventral side of the founder cell (Cooley et al. 2011). These protrusions are actin-rich (unlike the microtubule-rich cilia) and result from the polarized localization of a Rho GTPase (Cdc42) in this region. It is possible that the underlying extracellular matrix of the ventral epidermis stimulates this localization. At the same time, FGF receptor activity becomes concentrated in these protrusions. When the cell divides, the smaller daughter inherits these localized, activated FGF receptors, leading to differential activation of the genes that will form heart muscle (**FIGURE 3.42**). It is possible that, in the case of both the Hedgehog-receiving cilia and the FGF-receiving lamellipodia, the paracrine factor receptors become clustered to a high density and that the signal transducing proteins are brought into contact with each other.

There is thus an intimate connection, both anatomically and physiologically, among the cell membrane, the internal cytoskeleton, and signal transduction pathways. These components of the cellular periphery will be used in many ways to generate the variety of developing tissues and organs.

Coda

Communication among cells is critical for building tissues and organs. We now have a mechanism that explains how the molelcular products produced by a group of cells can change gene expression and cytoskeletal arrangement in neighboring cells. Cells communicate through diffusible paracrine factors and membrane-associated juxtacrine factors. The receptors for these factors span the cell membrane, and binding the factor alters the shape of the receptor. This shape change on the outside of the cell affects the shape of the receptor inside the cell, and this latter change can give the intracellular portion of the receptor a new property. It now has the ability to activate the enzymatic reactions that constitute the signal transduction pathways. These enzymes can activate transcription factors in the nucleus, thereby changing gene expression, and they can alter cytoskeletal proteins, thereby regulating cell movement. Thus, changes occurring on the cell membrane can be communicated both to the cell nucleus and to the cytoskeleton. These pathways uniting paracrine and juxtacrine binding on the cell membrane to gene expression changes in the nucleus are the fundamental modules of animal development.

SNAPSHOT SUMMARY: Cell-Cell Communication in Development

1. The sorting out of one cell type from another results from differences in the cell membrane.

2. The membrane structures responsible for cell sorting out are often cadherin proteins that change the surface tension properties of the cells.

3. Cadherin proteins can cause cells to sort out by both quantitative differences (different amounts of cadherin) or qualitative differences (different types of cadherin). Cadherins appear to be critical during certain morphological changes.

4. Migration occurs through changes in the actin cytoskeleton. These changes can be directed by internal instructions (from the nucleus) or by external instructions (from the extracellular matrix or chemoattractant molecules).

5. Inductive interactions involve inducing and responding tissues.

6. The ability to respond to inductive signals depends on the competence of the responding cells.

7. Reciprocal induction occurs when the two interacting tissues are both inducers and are competent to respond to each other's signals.

8. Cascades of inductive events are responsible for organ formation.

9. Regionally specific inductions can generate different structures from the same responding tissue.

10. The specific response to an inducer is determined by the genome of the responding tissue.

11. Programmed cell death is one possible response to inductive stimuli. Apoptosis is a critical part of life.

12. Paracrine interactions occur when a cell or tissue secretes proteins that induce changes in neighboring cells. Juxtacrine interactions are inductive interactions that take place between the cell membranes of adjacent cells or between a cell membrane and an extracellular matrix secreted by another cell.

13. Paracrine factors are proteins secreted by inducing cells. These factors bind to cell membrane receptors in competent responding cells.

14. Competent cells respond to paracrine factors through signal transduction pathways. Competence is the ability to bind and to respond to inducers, and it is often the result of a prior induction.

15. Signal transduction pathways begin with a paracrine or juxtacrine factor causing a conformational change in its cell membrane receptor. The new shape can result in enzymatic activity in the cytoplasmic domain of the receptor protein. This activity allows the receptor to phosphorylate other cytoplasmic proteins. Eventually, a cascade of such reactions activates a transcription factor (or set of factors) that activates or represses specific gene activity.

16. The differentiated state can be maintained by positive feedback loops involving transcription factors, autocrine factors, or paracrine factors.

17. The extracellular matrix is a source of signals for the differentiating cells and plays critical roles in cell migration.

18. Cells can convert from being epithelial to being mesenchymal and vice versa. The epithelial-mesenchymal transition is a series of transformations involved in the dispersion of neural crest cells and the creation of vertebrae from somitic cells. In adults, EMT is involved in wound healing and cancer metastasis.

19. The cell surface is intimately involved with cell signaling. Proteoglycans and other membrane components can expand or restrict the diffusion of paracrine factors.

20. Specializations of the cell surface, including cilia and lamellipodia, may concentrate receptors for paracrine and extracellular matrix proteins. The reception of Hedgehog proteins occurs on cilia.

For Further Reading

Complete bibliographical citations for all literature cited in this chapter can be found at the free-access website **www.devbio.com**

Ananthakrishnan, R. and A. Ehrlicher. 2007. The forces behind cell movement. *Int. J. Biol. Sci.* 3: 303–317.

Briscoe, J., P. A. Lawrence and J.-P. Vincent (eds.). 2010. *Generation and Interpretation of Morphogen Gradients.* Cold Spring Harbor Press, NY.

Cadigan, K. M. and R. Nusse. 1997. Wnt signaling: A common theme in animal development. *Genes Dev.* 24: 3286–3306.

Casanova, J. 2007. The emergence of shape: notions from the study of the *Drosophila* tracheal system. *EMBO Rep.* 4: 335–339.

Cooper, M. K., J. A. Porter, K. E. Young and P. A. Beachy. 1998. Teratogen-mediated inhibition of target tissue response to hedgehog signaling. *Science* 280: 1603–1607.

Engler, A., S. Sen, H. Sweeney and D. Discher. 2006. Matrix elasticity directs stem cell lineage specification. *Cell* 126: 677–689.

Foty, R. A. and M. S. Steinberg. 2005. The differential adhesion hypothesis: a direct evaluation. *Dev. Biol.* 278: 255–263.

Harvey, S. A. and J. C. Smith. 2008. Visualisation and quantification of morphogen gradient formation in the zebrafish. *PLoS Biol.* 7(5):e1000101.

Heldin, C.-H., K. Miyazono and P. ten Dijke. 1997. TGF-b signaling from cell membrane to nucleus through SMAD proteins. *Nature* 390: 465–471.

Huangfu, D., A. Liu, A. S. Rakeman, N. S. Murcia, L. Niswander and K. V. Anderson. 2003. Hedgehog signalling in the mouse requires intraflagellar transport proteins. *Nature* 426: 83–87.

Müller P., and A. F. Schier. 2011. Extracellular movement of signaling molecules. *Dev. Cell* 21: 145–158.

Nahmad, M. and A. D. Lander. 2011. Spatiotemporal mechanisms of morphogen gradient interpretation. *Curr. Opin. Genet. Dev.* 21: 726–731.

Rousseau, F. and 7 others. 1994. Mutations in the gene encoding fibroblast growth factor receptor-3 in achondroplasia. *Nature* 371: 252–254.

Steinberg, M. S. and S. F. Gilbert. 2004. Townes and Holtfreter (1955): Directed movements and selective adhesion of embryonic amphibian cells. *J. Exp. Zool. A: Comp Exp. Biol.* 301: 701–706.

Go Online

WEBSITE 3.1 FGF receptor mutations. Mutations of human FGF receptors have been associated with several skeletal malformation syndromes, including syndromes wherein skull cartilage, rib cartilage, or limb cartilage fail to grow or differentiate.

WEBSITE 3.2 The uses of apoptosis. Apoptosis is used for numerous processes throughout development. This website explores the role of apoptosis in such phenomena as *Drosophila* germ cell development and the eyes of blind cave fish.

WEBSITE 3.3 Notch mutations. Mutations in the genes that encode Notch proteins can cause nervous system abnormalities in humans. Humans have more than one Notch gene and more than one ligand. Their interactions may be critical in neural development. Moreover, the association of Notch with the presenilin protease suggests that disruption of Notch functioning might lead to Alzheimer disease.

WEBSITE 3.4 Maintaining cell differentiation. Paracrine factors and transcription factors can establish positive feedback loops such that a cell remains differentiated even after the signal initiating differentiation has ceased.

Vade Mecum

The differential adhesion hypothesis. These movies show the pioneering work of Townes and Holtfreter and Malcolm Steinberg. These experiments demonstrated the phenomenon of cell sorting and how cell surface adhesion molecules can direct sorting behaviors.

Cyclopia induced in zebrafish. As seen in the segment on zebrafish development, alcohol can act as a teratogen and induce cyclopia in these embryos.

Elements of the extracellular matrix. Movies review the molecular components of the extracellular matrix, how cells are influenced by them, and the work of Elizabeth Hay, who was among the first scientists to show the importance of ECM to tissue differentiation.

PART TWO

SPECIFICATION
Introducing Cell Commitment and Early Embryonic Development

The cartoon above illustrates the fundamental phenomenon and mystery of development: a complex and exquisitely ordered body is generated from the relatively unorganized fertilized egg. In 1883, one of America's first embryologists, William Keith Brooks, wrote of "the greatest of all wonders of the material universe: the existence, in a simple, unorganized egg, of a power to produce a definite adult animal." Brooks reflected that this property is so complex that "we may fairly ask what hope there is of discovering its solution, of reaching its true meaning, its hidden laws and causes." Biologists are now piecing together these "hidden laws and causes" by finding the ways that cells interpret the genome that is the same in every embryonic cell. How are certain genes activated and repressed in one group of cells to turn them into mesoderm while a different set of genes is regulated to instruct cells to become endoderm? Are there any principles or strategies that characterize the origins of different cell types?

Levels of Commitment

The generation of specialized cell types is called **differentiation**. But differentiation is only the last, overt stage in a series of events that commit a particular blastomere to become a particular cell type (TABLE P2.1). A red blood cell obviously differs radically in its protein composition and cell structure from a lens cell in the eye or a neuron in the brain. But these changes in cellular biochemistry and function are preceded by a process resulting in the **commitment** of the cell to a certain fate. During the course of commitment, the cell might not look differentiated, but its developmental fate has become restricted.

The process of commitment can be divided into two stages (Harrison 1933; Slack 1991). The first stage is a labile phase called **specification**. The fate of a cell or a tissue is said to be *specified* when it is capable of differentiating autonomously (i.e., by itself) when placed in a petri dish or test tube—that is, in an environment that is neutral with respect to the developmental pathway. At the stage of specification, cell commitment is still capable of being reversed.

TABLE P2.1 Some differentiated cell types and their major products

Type of cell	Differentiated cell product	Specialized function
Keratinocyte (epidermal cell)	Keratin	Protection against abrasion, desiccation
Erythrocyte (red blood cell)	Hemoglobin	Transport of oxygen
Lens cell	Crystallins	Transmission of light
B lymphocyte	Immunoglobulins	Synthesis of antibodies
T lymphocyte	Cytokines	Destruction of foreign cells; regulation of immune response
Melanocyte	Melanin	Pigment production
Pancreatic islet (β) cell	Insulin	Regulation of carbohydrate metabolism
Leydig cell (♂)	Testosterone	Male sexual characteristics
Chondrocyte (cartilage cell)	Chondroitin sulfate; type II collagen	Tendons and ligaments
Osteoblast (bone-forming cell)	Bone matrix	Skeletal support
Myocyte (muscle cell)	Actin and myosin	Muscle contraction
Hepatocyte (liver cell)	Serum albumin; numerous enzymes	Production of serum proteins and numerous enzymatic functions
Neurons	Neurotransmitters (acetylcholine, serotonin, etc.)	Transmission of communication signals in the nervous system
Tubule cell (♀) of hen oviduct	Ovalbumin	Egg white proteins for nutrition and protection of the embryo
Follicle cell (♀) of insect ovary	Chorion proteins	Eggshell proteins for protection of embryo

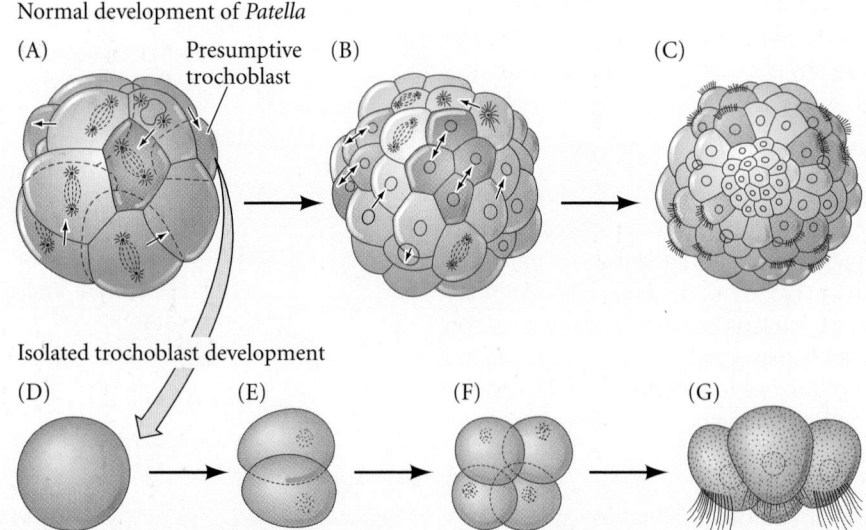

Normal development of *Patella*

(A) Presumptive trochoblast (B) (C)

Isolated trochoblast development

(D) (E) (F) (G)

FIGURE P2.1 Autonomous specification. (A–C) Differentiation of trochoblast (ciliated) cells of the snail *Patella*. (A) 16-Cell stage seen from the side; the presumptive trochoblast cells are shown in pink. (B) 48-Cell stage. (C) Ciliated larval stage, seen from the animal pole. (D–G) Differentiation of a *Patella* trochoblast cell isolated from the 16-cell stage and cultured in vitro. Even in isolated culture, the cells divide and become ciliated at the correct time. (After Wilson 1904.)

The second stage of commitment is **determination**. A cell or tissue is said to be *determined* when it is capable of differentiating autonomously even when placed into another region of the embryo—a decidedly non-neutral environment. If a cell or tissue type is able to differentiate according to its specified fate even under these circumstances, it is assumed that commitment is irreversible.

In addition to the *stages* of commitment (specification, determination, differentiation), there are three major *strategies* of specification: autonomous, conditional, and syncytial. Embryos of different species use different combinations of these strategies. All three strategies are based on the apportioning of certain sets of transcription factors to different cells.

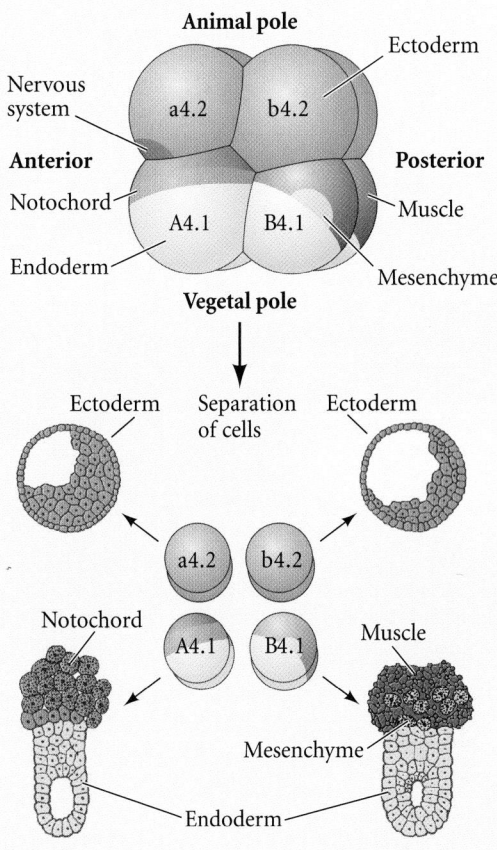

FIGURE P2.2 Autonomous specification in the early tunicate embryo. When the four blastomere pairs of the 8-cell embryo are dissociated, each forms the structures it would have formed had it remained in the embryo. The tunicate nervous system, however, is conditionally specified (see Chapter 7). The fate map shows that the left and right sides of the tunicate embryo produce identical cell lineages. Here the muscle-forming yellow cytoplasm is colored red to conform with its being associated with mesoderm. (After Reverberi and Minganti 1946.)

Autonomous Specification

The first strategy of commitment is **autonomous specification**. Here, the blastomere inherits a set of transcription factors from the egg cytoplasm. These transcription factors regulate gene expression, directing the cell into a particular path of development. In other words, the egg cytoplasm is not homogeneous; rather, different regions of the egg contain different **morphogenetic determinants**—transcription factors or their mRNAs—that will influence the cell's development. *In autonomous specification, the cell "knows" what it is to become very early and without interacting with other cells* (FIGURE P2.1).

For instance, in tunicate (sea squirt) embryos (see Figure 1.12), each blastomere will form most of its respective cell types even when separated from the remainder of the embryo. In the 8-cell tunicate embryo, the two blastomeres that are going to generate tail muscles (the B4.1 blastomeres; FIGURE P2.2) contain a yellow-pigmented cytoplasm that has within it the mRNA for a muscle-specific transcription factor called Macho, and those blastomeres that acquire this region of yellow cytoplasm will give rise to muscle cells. Indeed, if the *macho*-containing yellow cytoplasm is placed into other cells, those cells will form tail muscles (FIGURE P2.3; Whittaker

1973; Nishida and Sawada 2001). Conversely, if the cells normally containing this cytoplasm are removed, the embryo will not form tail muscles. Thus, the tail muscles of tunicates are formed autonomously by acquiring the mRNA for a transcription factor from the egg cytoplasm.

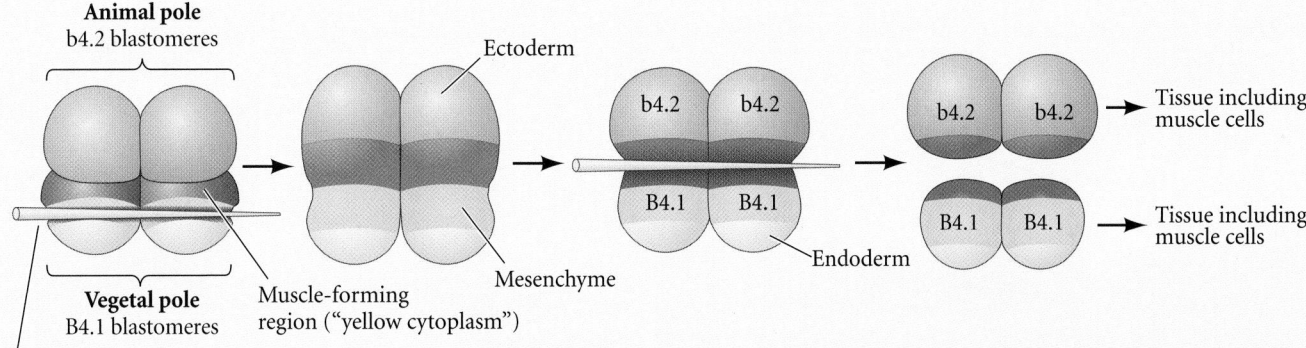

Needle pushes muscle-forming cytoplasm into animal cells

FIGURE P2.3 Microsurgery on tunicate eggs forces some of the cytoplasm of the muscle-forming B4.1 blastomeres to enter the b4.2 (epidermis- and nerve-producing) blastomere pair. Pressing the B4.1 blastomeres with a glass needle causes the regression of the cleavage furrow. The furrow re-forms at a more vegetal position where the cells are cut with a needle. The new furrow thereby separates the cells in such a way that the b4.2 blastomeres receive some of the muscle-forming B4.1 yellow cytoplasm. These modified b4.2 cells then produce muscle cells as well as their normal ectodermal progeny. As we will see in Chapter 7, the yellow cytoplasm contains a transcription factor (Macho) that activates muscle-specific genes. (After Whittaker 1982.)

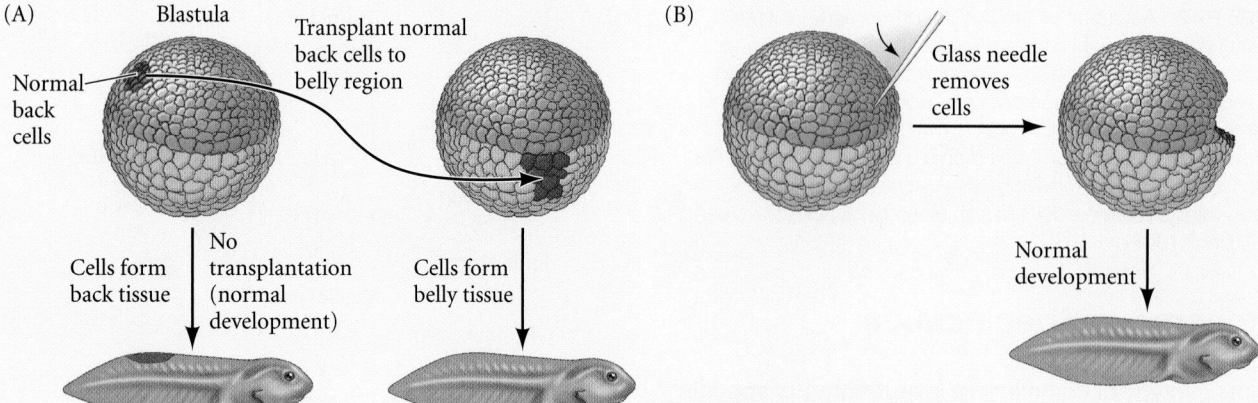

(A) Blastula

Normal back cells

Transplant normal back cells to belly region

Cells form back tissue

No transplantation (normal development)

Cells form belly tissue

(B)

Glass needle removes cells

Normal development

FIGURE P2.4   Conditional specification. (A) What a cell becomes depends on its position in the embryo. Its fate is determined by interactions with neighboring cells. (B) If cells are removed from the embryo, the remaining cells can regulate and compensate for the missing part.

When most of the cells of an early embryo are determined by autonomous specification, it gives the appearance that the animal is fully specified this way. This is not the case, and even in tunicate embryos the nervous system arises conditionally (i.e., via cell-cell interactions). However, embryologists have traditionally called such embryos *mosaic embryos*, since they develop like a mosaic of individually laid tiles, with each cell receiving its instructions independently of any interaction with other cells.

Conditional Specification

Conditional specification is the ability of cells to achieve their respective fates by interacting with other cells (**FIGURE P2.4**). Here, what a cell becomes is in large measure specified by paracrine factors secreted by its neighbors (see Chapter 3). In one of the ironies of research, conditional specification was demonstrated by attempts to disprove it. In 1888, August Weismann proposed the first testable model of cell specification, the **germ plasm theory**, in which

each cell of the embryo would develop autonomously. He boldly proposed that the sperm and egg provided equal chromosomal contributions, both quantitatively and qualitatively, to the new organism. Moreover, he postulated that the chromosomes carried the inherited potentials of this new organism.* However, not all the determinants on the chromosomes were thought to enter every cell of the embryo.

---

*Note that embryologists were thinking in terms of chromosomal mechanisms of inheritance some 15 years before the rediscovery of Mendel's work. Weismann (1892, 1893) also speculated that these nuclear determinants of inheritance functioned by elaborating substances that became active in the cytoplasm!

FIGURE P2.5   Weismann's germ plasm theory of inheritance. The germ cell (blue) gives rise to the differentiating somatic cells of the body, as well as to new germ cells. Weismann hypothesized that only the germ cells contained all the inherited determinants. The somatic cells were each thought to contain a subset of the determinants, and the types of determinants found in a somatic cell's nucleus would determine its differentiated type. (After Wilson 1896.)

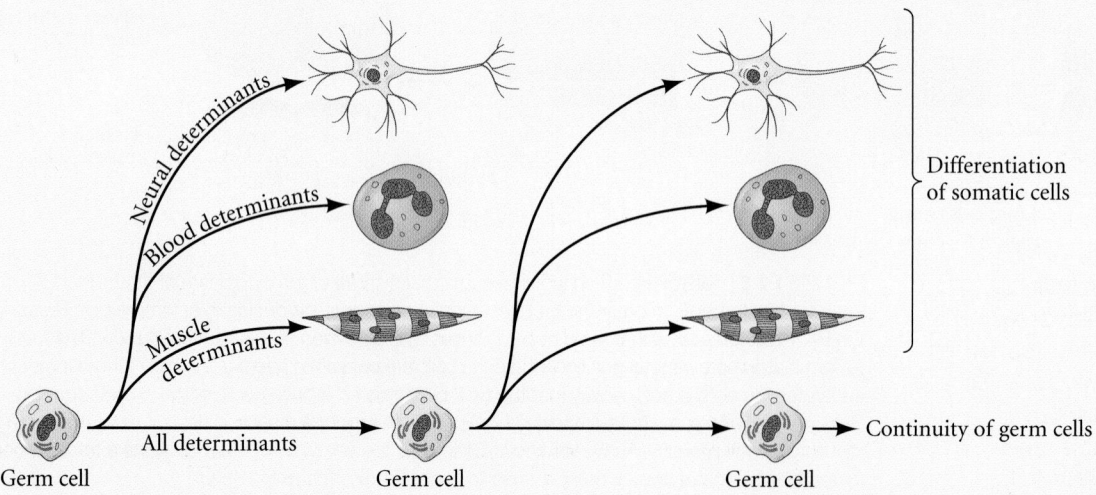

Neural determinants

Blood determinants

Muscle determinants

Differentiation of somatic cells

Germ cell

All determinants

Germ cell

Germ cell

Continuity of germ cells

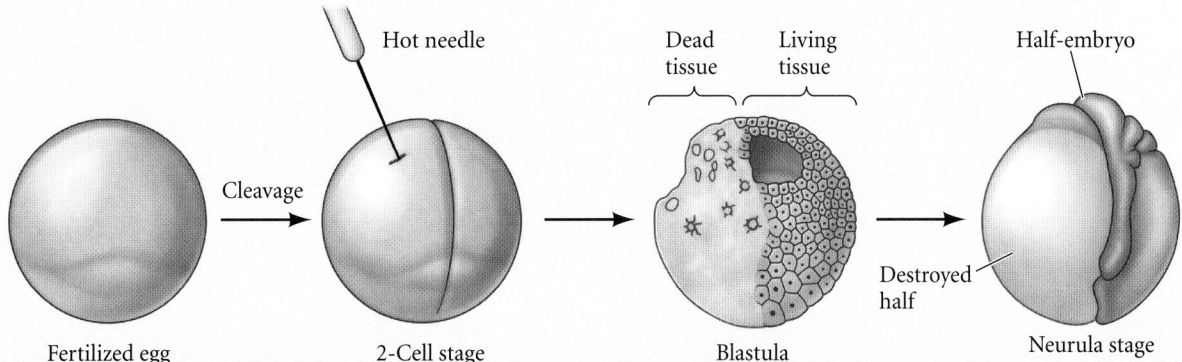

Fertilized egg   2-Cell stage   Blastula   Neurula stage

Hot needle

Cleavage

Dead tissue   Living tissue

Half-embryo

Destroyed half

FIGURE P2.6   Roux's attempt to demonstrate autonomous specification. Destroying (but not removing) one cell of a 2-cell frog embryo resulted in the development of only one half of the embryo.

Instead of dividing equally, the chromosomes were hypothesized to divide in such a way that different determinants entered different cells. Whereas the fertilized egg would carry the full complement of determinants, certain somatic cells would retain the "blood-forming" determinants while others retained the "muscle-forming" determinants, and so forth (**FIGURE P2.5**). Only the nuclei in those cells destined to become germ cells (gametes) were postulated to contain all the different types of determinants.

In postulating his germ plasm model, Weismann proposed a hypothesis of development that could be tested immediately. Based on the fate map of the frog embryo, Weismann claimed that when the first cleavage division separated the future right half of the embryo from the future left half, there would be a separation of "right" determinants from "left" determinants in the resulting blastomeres. Wilhelm Roux tested Weismann's hypothesis by using a hot needle to kill one of the cells in a 2-cell frog embryo—with the result that only the right or left half of a larva developed (**FIGURE P2.6**). Based on this result, Roux claimed that specification was autonomous, and that all the instructions for normal development were present inside each cell.

Roux's colleague Hans Dreisch, however, obtained opposite results. Whereas Roux's studies were **defect experiments** that answered the question of how the embryo would develop when a subset of blastomeres was destroyed, Driesch (1892) sought to extend this research by performing **isolation experiments** (**FIGURE P2.7**).

He separated sea urchin blastomeres from each other by vigorous shaking (or later, by placing them in calcium-free seawater). To Driesch's surprise, each of the blastomeres from a 2-cell embryo developed into a complete larva. Similarly, when Driesch separated the blastomeres of 4- and 8-cell embryos, some of the isolated cells produced entire

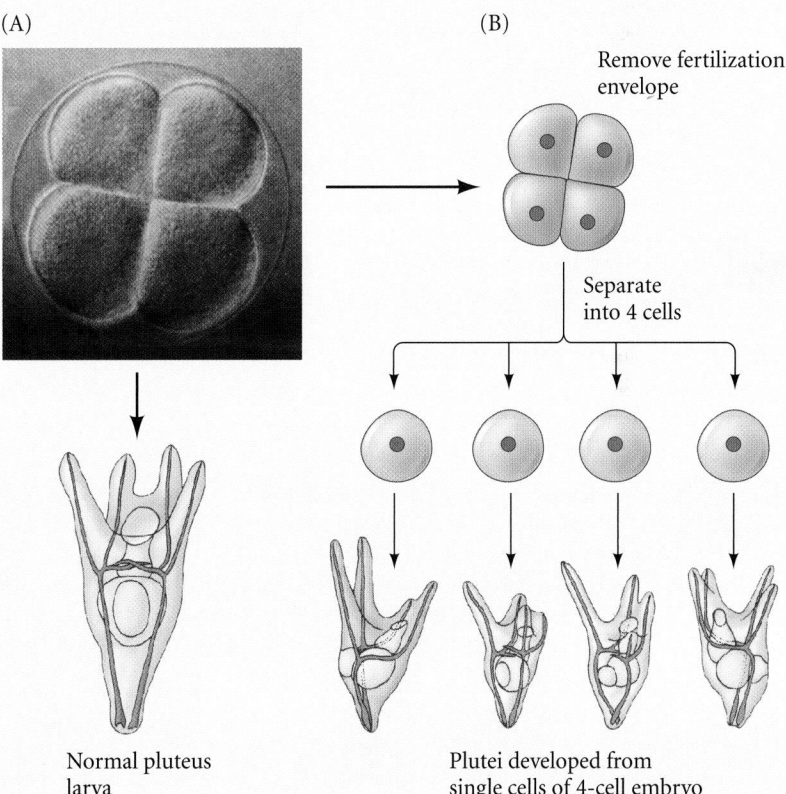

(A)

(B)

Remove fertilization envelope

Separate into 4 cells

Normal pluteus larva

Plutei developed from single cells of 4-cell embryo

FIGURE P2.7   Driesch's demonstration of conditional (regulative) specification. (A) An intact 4-cell sea urchin embryo generates a normal pluteus larva. (B) When one removes the 4-cell embryo from its fertilization envelope and isolates each of the four cells, each cell can form a smaller, but normal, pluteus larva. (All larvae are drawn to the same scale.) Note that the four larvae derived in this way are not identical, despite their ability to generate all the necessary cell types. Such variation is also seen in adult sea urchins formed in this way (see Marcus 1979). (A, photograph courtesy of G. Watchmaker.)

(A) Normal cleavage

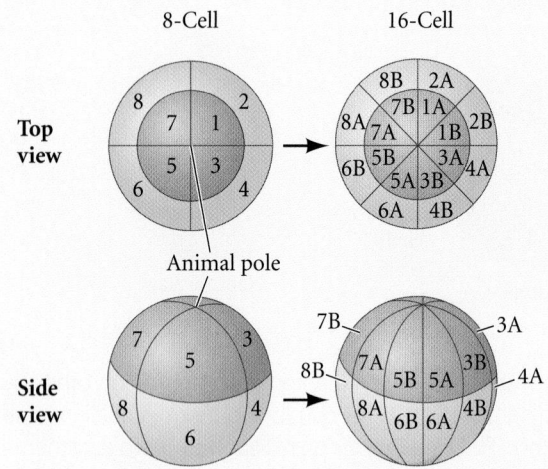

(B) Cleavage under pressure

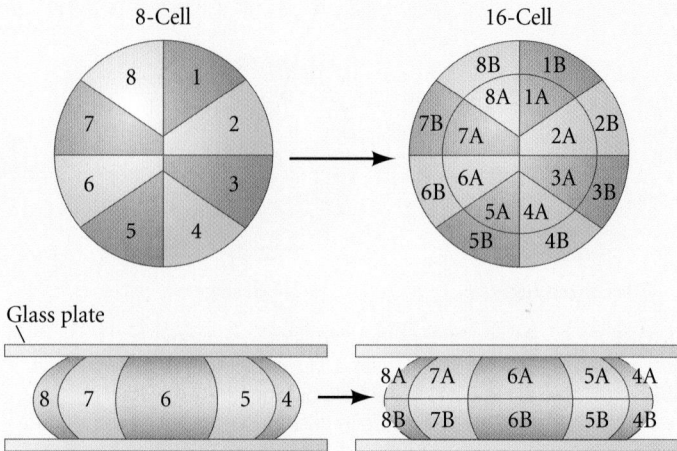

FIGURE P2.8   Driesch's pressure-plate experiment for altering the distribution of nuclei. (A) Normal cleavage in 8- to 16-cell sea urchin embryos, seen from the animal pole (upper sequence) and from the side (lower sequence). The nuclei are numbered. (B) Abnormal cleavage planes formed under pressure, as seen from the animal pole and from the side. (After Huxley and de Beer 1934.)

pluteus larvae. Here was a result drastically different from the predictions of Weismann and Roux. Rather than self-differentiating into its future embryonic part, each isolated blastomere *regulated* its development to produce a complete organism. These experiments provided the first experimentally observable evidence that a cell's fate depends on that cell's neighbors. *In conditional specification, interactions between cells determine their fates,* rather than some cytoplasmic factor particular to that type of cell.

Driesch confirmed conditional development in sea urchin embryos by performing an intricate **recombination experiment**. In sea urchin eggs, the first two cleavage planes are normally meridional, passing through both the animal and vegetal poles, whereas the third division is equatorial, dividing the embryo into four upper and four lower cells (**FIGURE P2.8A**). Driesch (1893) changed the direction of the third cleavage by gently compressing early embryos between two glass plates, thus causing the third division to be meridional like the preceding two. After he released the pressure, the fourth division was equatorial. This procedure reshuffled the nuclei, placing nuclei that normally would have been in the region destined to form endoderm into the presumptive ectoderm region. In other words, some nuclei that would normally have produced ventral structures were now found in the dorsal cells (**FIGURE P2.8B**). If segregation of nuclear determinants had occurred (as proposed by Weismann and Roux), the resulting embryo should have been strangely disordered. However, Driesch obtained normal larvae from these embryos. He thus concluded that "the relative position

of a blastomere within the whole will probably in a general way determine what shall come from it."

The consequences of these experiments were momentous, both for embryology and for Driesch personally.* First, Driesch had demonstrated that the prospective potency of an isolated blastomere (i.e., those cell types it was possible for it to form) is greater than the blastomere's actual prospective fate (those cell types it would normally give rise to over the unaltered course of its development). According to Weismann and Roux, the prospective potency and the prospective fate of a blastomere should have been identical. Second, Driesch concluded that the sea urchin embryo is a "harmonious equipotential system" because all of its potentially independent parts interacted together to form a single organism. Driesch's experiment implies that cell interaction is critical for normal development. Moreover, if each early blastomere can form all the embryonic cells when isolated, then it follows that, in normal development, the community of cells must prevent it from doing so (Hamburger 1997).

Third, Driesch concluded that the fate of a nucleus depended solely on its location in the embryo. Interactions between cells determined their fates. We now know (and will see in Chapters 7 and 8) that sea urchins and frogs alike use

---

*This idea of nuclear equivalence and the ability of cells to interact eventually caused Driesch to abandon science. Driesch, who thought the embryo was like a machine, could not explain how the embryo could make its missing parts or how a cell could change its fate to become another cell type. Harking back to Aristotle, he invoked a vital force, *entelechy* ("internal goal-directed force"), to explain how development

proceeds. Essentially, he believed that the embryo was imbued with an internal psyche and the wisdom to accomplish its goals despite the obstacles embryologists placed in its path. However, others, especially Oscar Hertwig (1894), were able to incorporate Driesch's experiments into a more sophisticated experimental embryology (which will be discussed in the introduction to Part Four of this book).

both autonomous and conditional specification of their early embryonic cells. Moreover, both animal groups use a similar strategy and even similar molecules during early development. In the 16-cell sea urchin embryo, a group of cells called the micromeres inherits a set of transcription factors from the egg cytoplasm. These transcription factors cause the micromeres to develop *autonomously* into the larval skeleton—but these same factors also activate genes for paracrine and juxtacrine factors that are then secreted by the micromeres and *conditionally* specify the cells around them.

Embryos (especially vertebrate embryos) in which most of the early blastomeres are conditionally specified have traditionally been called *regulative embryos*. But as we become more cognizant of the manner in which both autonomous and conditional specification are used in each embryo, the notions of "mosaic" and "regulative" embryos appear less and less tenable. Indeed, attempts to get rid of these distinctions were begun by the embryologist Edmund B. Wilson (1894, 1904) more than a century ago.

Morphogen Gradients and Conditional Specification

One of the most important mechanisms of conditional specification involves paracrine factors and morphogen gradients. A **morphogen** (Greek, "form-giver") is a diffusable biochemical molecule that can determine the fate of a cell by its concentration.* That is, cells exposed to high levels of a morphogen activate different genes than those cells exposed to lower levels. Morphogens can be transcription factors produced within cells (as in the *Drosophila* embryos described in the next section). They can also be paracrine factors that are produced in one group of cells and then travel to another population of cells, specifying the target cells to have similar or different fates according to the concentration of morphogen. Uncommitted cells exposed to high concentrations of the morphogen (nearest its source of production) are specified as one cell type. When the morphogen's concentration drops below a certain threshold, a different cell fate is specified. When the concentration falls even lower, a cell that initially was of the same uncommitted type is specified in yet a third distinct manner (**FIGURE P2.9**).

Regulation by gradients of paracrine factor concentration was elegantly demonstrated by the specification of different mesodermal cell types in the frog *Xenopus laevis* by activin, a paracrine factor of the TGF-β family (**FIGURE P2.10**; Green and Smith 1990; Gurdon et al. 1994).

---

*Although there is overlap in the terminology, a *morphogen* specifies cells in a quantitative ("more or less") manner, whereas a *morphogenetic determinant* specifies cells in a qualitative ("present or absent") way. Morphogens are analog; morphogenetic determinants are digital.

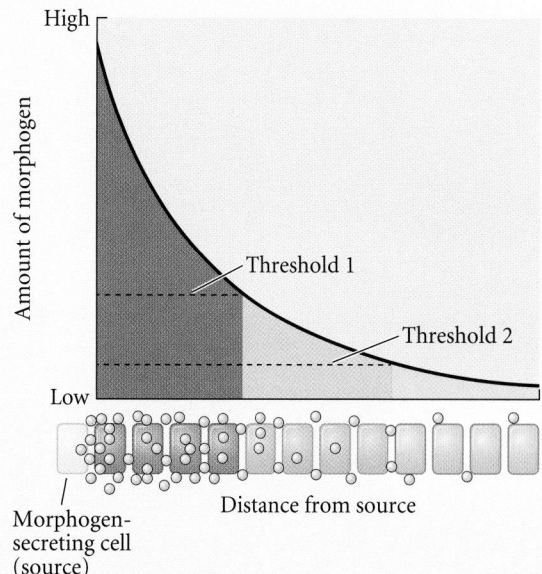

FIGURE P2.9 Specification of uniform cells into three cell types by a morphogen gradient. A morphogenetic paracrine factor is secreted from source cells (gold) and forms a concentration gradient within the responsive tissue. Cells exposed to morphogen concentrations above threshold 1 activate certain genes (dark red). Cells exposed to intermediate concentrations (between thresholds 1 and 2) activate a different set of genes (pink) and also inhibit the genes induced at the higher concentrations. Those cells encountering low concentrations of morphogen (below threshold 2) activate a third set of genes (blue). (After Rogers and Schier 2011.)

Activin-secreting beads were placed on unspecified cells from an early *Xenopus* embryo. The activin then diffused from the beads. At high concentrations (about 300 molecules/cell), activin induced expression of the *goosecoid* gene, whose product is a transcription factor that specifies the frog's dorsal-most structures. At slightly lower concentrations of activin (about 100 molecules per cell), the same tissue activated the *Xbra* gene and was specified to become muscle. At still lower concentrations, these genes were not activated, and the "default" gene expression instructed the cells to become blood vessels and heart (Dyson and Gurdon 1998).

As mentioned in Chapter 3, the range of a paracrine factor (and thus the shape of its morphogen gradient) depends on several aspects of that factor's synthesis, transport, and degradation. In some cases, cell surface molecules stabilize the paracrine factor and aid in its diffusion, whereas in other cases cell surface moieties retard diffusion and enhance degradation. Such diffusion-regulating interactions between morphogens and extracellular matrix factors are very important in coordinating organ growth and shape (Ben Zvi et al. 2010, 2011).

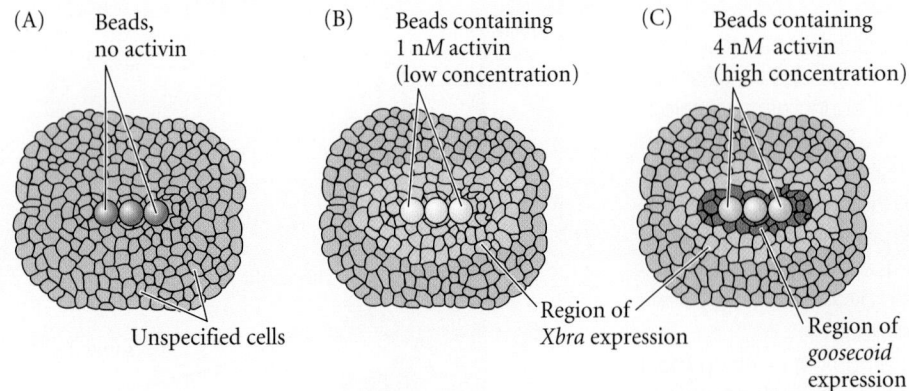

(A) Beads, no activin

Unspecified cells

(B) Beads containing 1 n*M* activin (low concentration)

Region of *Xbra* expression

(C) Beads containing 4 n*M* activin (high concentration)

Region of *goosecoid* expression

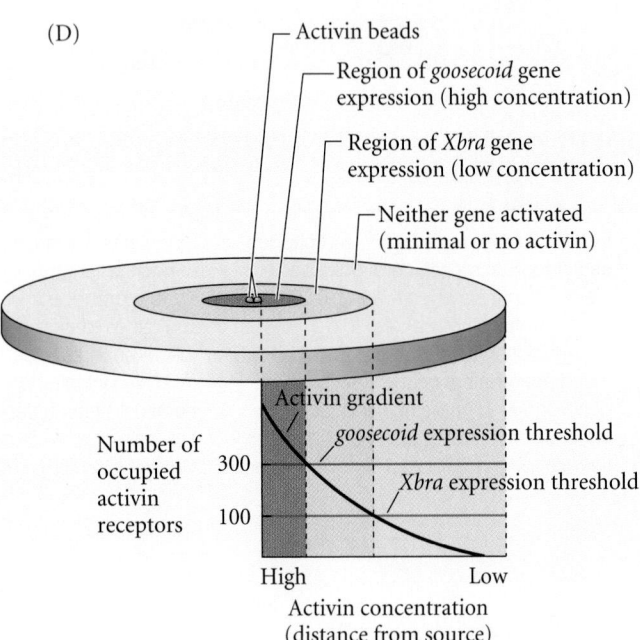

(D)

— Activin beads

— Region of *goosecoid* gene expression (high concentration)

— Region of *Xbra* gene expression (low concentration)

— Neither gene activated (minimal or no activin)

Activin gradient

goosecoid expression threshold

Xbra expression threshold

Number of occupied activin receptors

300

100

High — Low

Activin concentration (distance from source)

FIGURE P2.10 A gradient of the paracrine factor activin, a morphogen, causes concentration-dependent expression differences of two genes in unspecified amphibian cells. (A) Beads containing no activin did not elicit expression (i.e., transcription of mRNA) of either the *Xbra* or *goosecoid* gene. (B) Beads containing 1 n*M* activin elicited *Xbra* expression in nearby cells. (C) Beads containing 4 n*M* activin elicited *Xbra* expression, but only at a distance of several cell diameters from the beads. However, *goosecoid* expression is seen near the source bead. Thus, it appears that *Xbra* is induced at particular concentrations of activin and that *goosecoid* is induced at higher concentrations. (D) Interpretation of the *Xenopus* activin gradient. High concentrations of activin activate *goosecoid*, whereas lower concentrations activate *Xbra*. A threshold value appears to exist that determines whether a cell will express *goosecoid*, *Xbra*, or neither gene. In addition, Brachyury (the *Xbra* protein product in *Xenopus*) inhibits the expression of *goosecoid*, thereby creating a distinct boundary. This pattern correlates with the number of activin receptors occupied on individual cells. (After Gurdon et al. 1994; Dyson and Gurdon 1998.)

Syncytial Specification

In addition to autonomous and conditional specification, there is a third strategy that uses elements of both. In early embryos of insects, nuclei divide within the egg; but the cell does not divide. In other words, many nuclei are formed within one common cytoplasm. A cytoplasm that contains many nuclei is called a **syncytium**, and the specification of presumptive cells within such a syncytium is called **syncytial specification**. As in the other eggs we have mentioned, the insect egg cytoplasm is not uniform. Nuclei in the anterior part of the cell are exposed to cytoplasmic transcription factors that are not present in the posterior part of the cell, and vice versa. The interactions of nuclei and transcription factors that eventually result in cell specification take place in a common cytoplasm.

Each nucleus in *Drosophila* is given positional information (i.e., whether that nucleus is to become part of the anterior, posterior, or midsection of the body) by transcription factors acting as morphogens. We have just discussed gradients of paracrine factor morphogens *between* cells. In syncytial specification, there are morphogen gradients of transcription factors *within* a cell. These intracellular proteins are made in specific sites in the embryo, diffuse over long distances, and form gradients in which the highest concentration is at the point of synthesis and gets lower as the morphogen diffuses away from its source and degrades over time. The concentration of specific morphogens at a particular site tells the nuclei where they are in relation to the source of the morphogens. As we will detail in Chapter 6, the anteriormost portion of the *Drosophila* embryo produces a transcription factor called Bicoid with a concentration that is highest in the anterior and declines toward the posterior. The posteriormost portion of the egg forms a posterior-to-anterior gradient of the transcription factor Caudal. Thus, the long axis of the *Drosophila* egg is spanned by opposing gradients—Bicoid from the anterior and Caudal from the posterior (**FIGURE P2.11**). Different concentrations and ratios of the Bicoid and Caudal proteins activate different sets of genes in the syncytial nuclei. Those nuclei in regions containing high amounts of Bicoid and little Caudal are instructed to activate the genes that produce the head. Nuclei in regions with slightly less Bicoid and

FIGURE P2.11 Syncytial specification in *Drosophila melanogaster*. Anterior-posterior specification originates from morphogen gradients in the egg cytoplasm, specifically of the transcription factors Bicoid and Caudal. The concentrations and ratios of these two proteins distinguish each position along the axis from any other position. When nuclear division occurs, the amounts of each morphogen transcribed differentially activate transcription of the various nuclear genes that specify the segment identities of the larval and the adult fly. (As we will see in Chapter 6, the Caudal gradient is itself constructed by interactions between constituents of the egg cytoplasm.)

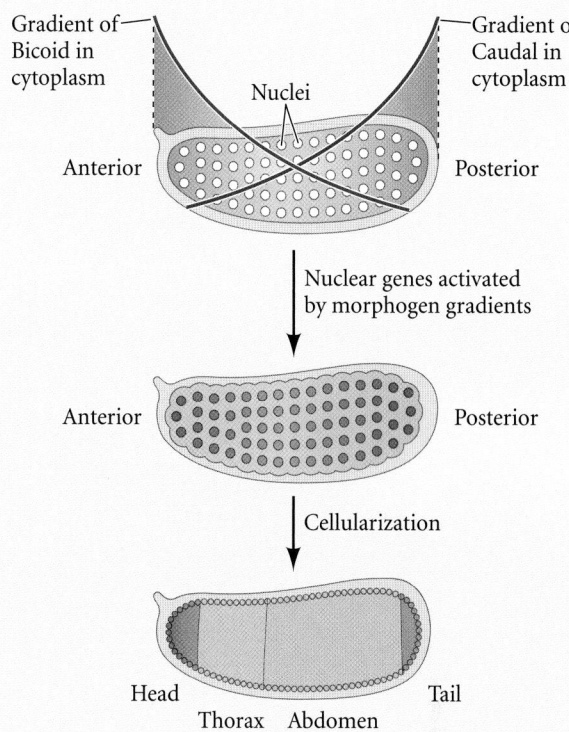

a small amount of Caudal are instructed to activate genes that generate the thorax. In regions with little or no Bicoid but plenty of Caudal, the activated genes form abdominal structures (Nüsslein-Volhard et al. 1987). Thus, when the syncytial nuclei are eventually incorporated into cells, these cells will have their *general* fate specified. Afterward, the *specific* fate of each cell will become determined both autonomously (from transcription factors acquired by the cell's nucleus from the egg cytoplasm) and conditionally (by interactions between the cell and its neighbors).

Summary

Each of the three major strategies of cell specification (summarized in TABLE P2.2) offers a different way of providing each embryonic cell with a set of transcription factors that will activate specific genes and cause that cell to differentiate into a particular cell type. In autonomous specification, the transcription factors (or their mRNAs) are already present in the cytoplasm of the cell. In conditional specification, the set of transcription factors is determined by paracrine and juxtacrine interactions with neighboring cells. In syncytial specification, specification is achieved via interactions not between cells, but between regions of the egg cytoplasm. These regional interactions give each cell a different ratio of particular transcription factors, and these ratios determine which genes are on and which are off.

The chapters that follow will describe the early development of several well-studied organisms. In these chapters, we will see how the mechanisms of fertilization, cleavage, and gastrulation use the three modes of specification to produce committed cell types and to organize the early embryo.

TABLE P2.2 Modes of cell type specification

AUTONOMOUS SPECIFICATION

Predominates in most invertebrates.

Specification by differential acquisition of certain cytoplasmic molecules present in the egg.

Invariant cleavages produce the same lineages in each embryo of the species; blastomere fates are generally invariant.

Cell type specification precedes any large-scale embryonic cell migration.

Results in "mosaic" development: cells cannot change fate if a blastomere is lost.

CONDITIONAL SPECIFICATION

Predominates in vertebrates and a few invertebrates.

Specification by interactions among cells. Positions of cells relative to each other are key.

Variable cleavages, no invariant fate assignment to cells.

Massive cell rearrangements and migrations precede or accompany specification.

Capacity for "regulative" development allows cells to acquire different functions as a result of interactions with neighboring cells.

SYNCYTIAL SPECIFICATION

Predominates in most insect classes.

Specification of body regions by interactions between cytoplasmic regions prior to cellularization of the blastoderm.

Variable cleavage produces no rigid cell fates for particular nuclei.

After cellularization, both autonomous and conditional specification are seen.

Source: After Davidson 1991.

4

Fertilization
Beginning a New Organism

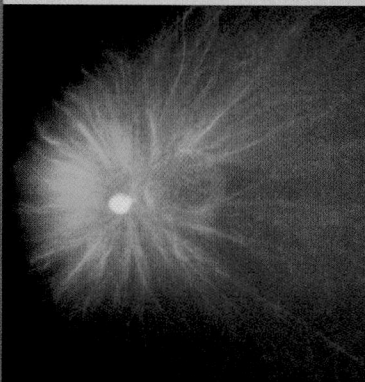

FERTILIZATION IS THE PROCESS whereby the sperm and the egg—collectively called the **gametes**—fuse together to begin the creation of a new individual whose genome is derived from both parents. Fertilization accomplishes two separate ends: sex (the combining of genes derived from two parents) and reproduction (the generation of a new organism). Thus, the first function of fertilization is to transmit genes from parent to offspring, and the second is to initiate in the egg cytoplasm those reactions that permit development to proceed.

Although the details of fertilization vary from species to species, conception generally consists of four major events:

1. Contact and recognition between sperm and egg. In most cases, this ensures that the sperm and egg are of the same species.

2. Regulation of sperm entry into the egg. Only one sperm nucleus can ultimately unite with the egg nucleus. This is usually accomplished by allowing only one sperm to enter the egg and actively inhibiting any others from entering.

3. Fusion of the genetic material of sperm and egg.

4. Activation of egg metabolism to start development.

Structure of the Gametes

A complex dialogue exists between egg and sperm. The egg activates the sperm metabolism that is essential for fertilization, and the sperm reciprocates by activating the egg metabolism needed for the onset of development. But before we investigate these aspects of fertilization, we need to consider the structures of the sperm and egg—the two cell types specialized for fertilization.*

Sperm

It is only within the past 135 years that the sperm's role in fertilization has been known. Anton van Leeuwenhoek, the Dutch microscopist who co-discovered sperm in the 1670s, first believed them to be parasitic animals living within the semen (hence the term *spermatozoa*, meaning "seed animals"). Although he originally assumed that they had nothing to do with reproducing the organism in which they were found, he later came to believe that each sperm contained a preformed embryo. Leeuwenhoek (1685) wrote that sperm were seeds (both *sperma* and *semen* mean "seed") and that the female merely provided the nutrient soil in which the seeds were planted. In this, he was

---

*Many developmental biology courses begin with gametogenesis and meiosis. This author believes that meiosis and gametogenesis are the culminating processes in development, and that they cannot be properly appreciated without first understanding somatic organogenesis and differentiation. Also, having gonad formation and gametogenesis in the last lectures completes a circle.

Urge and urge and urge,
Always the procreant urge
of the world.

Out of the dimness opposite equals advance,

Always substance and increase, always sex,

Always a knit of identity, always distinction,
WALT WHITMAN
(1855)

The final aim of all love intrigues, be they comic or tragic, is really of more importance than all other ends in human life. What it turns upon is nothing less than the composition of the next generation.
A. SCHOPENHAUER
(QUOTED BY
C. DARWIN,
1871)

returning to a notion of procreation promulgated by Aristotle 2000 years earlier.

Try as he might, Leeuwenhoek was continually disappointed in his attempts to find preformed embryos within spermatozoa. Nicolas Hartsoeker, the other co-discoverer of sperm, drew a picture of what he hoped to find: a miniscule human ("homunculus") within the sperm (**FIGURE 4.1**). This belief that the sperm contained the entire embryonic organism never gained much acceptance, as it implied an enormous waste of potential life. Most investigators regarded the sperm as unimportant.*

● See **WEBSITE 4.1** Leeuwenhoek and images of homunculi

The first evidence suggesting the importance of sperm in reproduction came from a series of experiments performed by Lazzaro Spallanzani in the late 1700s. Spallanzani induced male toads to ejaculate into taffeta breeches and found toad semen so filtered to be devoid of sperm; such semen did not fertilize eggs. He even showed that semen had to touch the eggs in order to be functional. However, Spallanzani (like many others) felt that the spermatic "animals" were parasites in the fluid; he thought the embryo was contained within the egg and needed spermatic fluid to activate it (see Pinto-Correia 1997).

The combination of better microscopic lenses and the elucidation of the cell theory (i.e., that all life is cellular, and all cells come from preexisting cells) led to a new appreciation of sperm function. In 1824, J. L. Prevost and J. B. Dumas claimed that sperm were not parasites, but rather the active agents of fertilization. They noted the universal existence of sperm in sexually mature males and their absence in immature and aged individuals. These observations, coupled with the known absence of sperm in the sterile mule, convinced Prevost and Dumas that "there exists an intimate relation between their presence in the organs and the fecundating capacity of the animal." They proposed that the sperm entered the egg and contributed materially to the next generation.

These claims were largely disregarded until the 1840s, when A. von Kolliker described the formation of sperm from cells in the adult testes. He ridiculed the idea that the semen could be normal and yet support such an enormous number of parasites. Even so, von Kolliker denied there was any physical contact between sperm and egg. He believed that the sperm excited the egg to develop in much the same way a magnet communicates its presence to iron. It was not until 1876 that Oscar Hertwig and Herman Fol independently demonstrated sperm entry into the egg and the union of the two cells' nuclei. Hertwig had been seeking an organism suitable for detailed microscopic observations, and he found the Mediterranean sea urchin (*Paracentrotus lividus*) to be perfect

FIGURE 4.1 The human infant preformed in the sperm, as depicted by Nicolas Hartsoeker (1694).

for this purpose. Not only was it common throughout the region and sexually mature throughout most of the year, but its eggs were available in large numbers and were transparent even at high magnifications.

When Hertwig mixed suspensions of sperm together with egg suspensions, he repeatedly observed sperm entering the eggs and saw sperm and egg nuclei unite. He also noted that *only one sperm was seen to enter each egg, and that all the nuclei of the resulting embryo were derived mitotically from the nucleus created at fertilization.* Fol made similar observations and also detailed the mechanism of sperm entry.† Fertilization was at last recognized as the union of sperm and egg, and the union of sea urchin gametes remains one of the best-studied examples of fertilization.

● See **WEBSITE 4.2** The origins of fertilization research

Sperm anatomy

Each sperm cell consists of a haploid nucleus, a propulsion system to move the nucleus, and a sac of enzymes that enable the nucleus to enter the egg. In most species, almost all of the cell's cytoplasm is eliminated during sperm maturation, leaving only certain organelles that are modified for spermatic function (**FIGURE 4.2**). During the course of maturation, the sperm's haploid nucleus becomes very streamlined and its DNA becomes tightly compressed. In front of this compressed haploid nucleus lies the **acrosomal vesicle**, or **acrosome**. The acrosome is derived from the cell's Golgi apparatus and contains enzymes that digest proteins and complex sugars; thus, the acrosome can be considered a modified secretory vesicle. The enzymes stored in the acrosome can digest a path through the outer coverings of the egg. In many species, a region of globular actin proteins lies between the sperm nucleus and the acrosomal vesicle. These proteins are used to extend a fingerlike **acrosomal process** from the sperm during the early stages of fertilization. In sea urchins

---

*Indeed, sperm was discovered around 1676, whereas the events of fertilization were not elucidated until 1876. Thus, for some 200 years people had no idea what the sperm actually did. See Pinto-Correia 1997 for details of this remarkable story.

†Hertwig and Fol were not the first scientists to report fertilization in the sea urchin. At least three other astute observers—Adolphe Dufossé, Karl Ernst von Baer, and Alphonse Derbés—observed sperm-egg contact in 1847. Briggs and Wessel (2006) suggest that the convulsions of Europe during 1848, the low opinion German scientists had of French biology, and the tenuousness of these results (given poor microscopy and the lack of a theory in which to place them) may have confined these papers to obscurity.

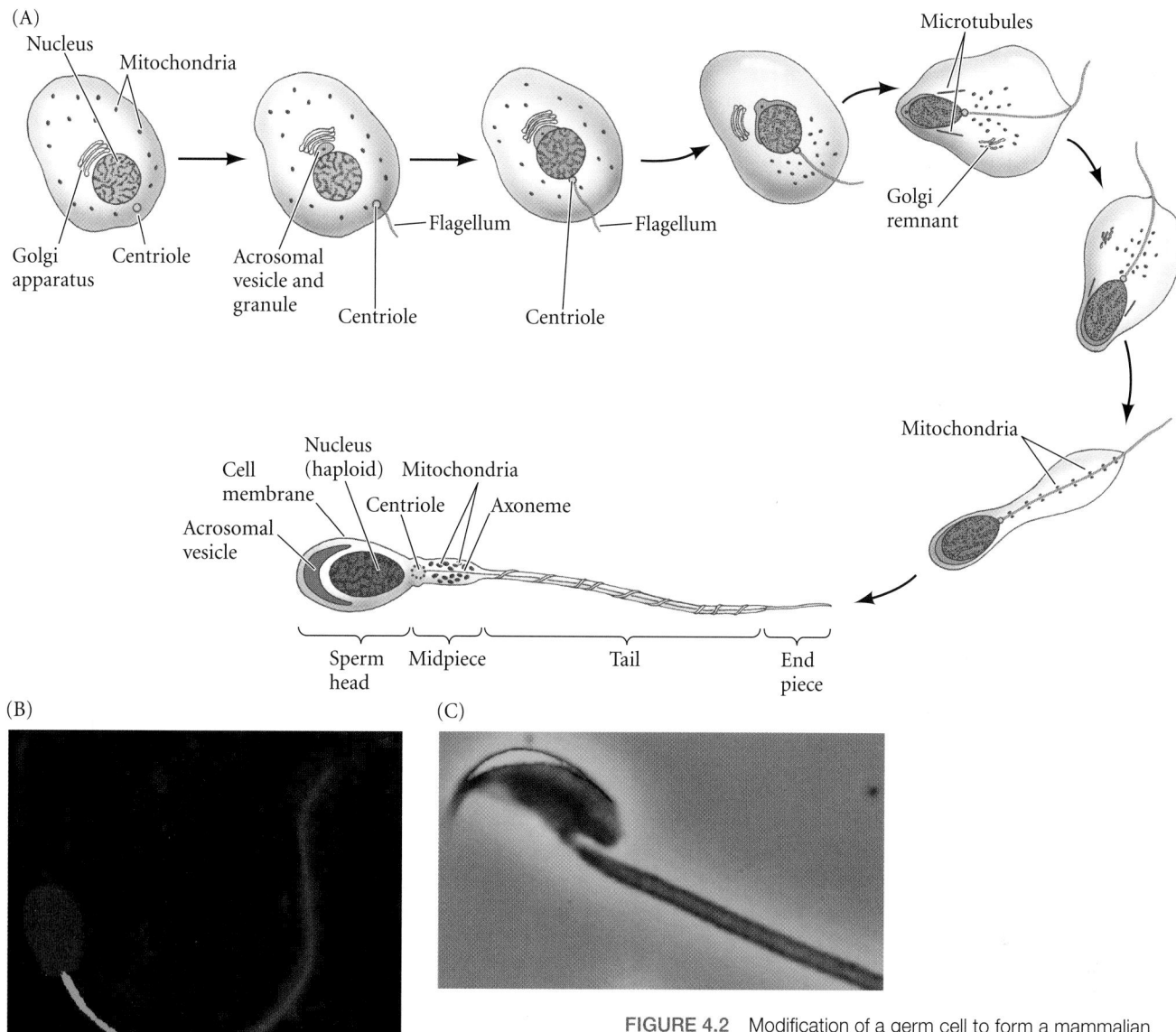

(A) Nucleus Mitochondria

Golgi apparatus Centriole

Acrosomal vesicle and granule Flagellum Centriole

Flagellum Centriole

Microtubules

Golgi remnant

Mitochondria

Cell membrane Nucleus (haploid) Mitochondria

Acrosomal vesicle Centriole Axoneme

Sperm head Midpiece Tail End piece

(B)

(C)

FIGURE 4.2 Modification of a germ cell to form a mammalian sperm. (A) The centriole produces a long flagellum at what will be the posterior end of the sperm. The Golgi apparatus forms the acrosomal vesicle at the future anterior end. Mitochondria collect around the flagellum near the base of the haploid nucleus and become incorporated into the midpiece ("neck") of the sperm. The remaining cytoplasm is jettisoned, and the nucleus condenses. The size of the mature sperm has been enlarged relative to the other stages. (B) Mature bull sperm. The DNA is stained blue, mitochondria are stained green, and the tubulin of the flagellum is stained red. (C) The acrosomal vesicle of this mouse sperm is stained green by the fusion of proacrosin with green fluorescent protein (GFP). (A after Clermont and Leblond 1955; B from Sutovsky et al. 1996, courtesy of G. Schatten; C courtesy of K.-S. Kim and G. L. Gerton.)

and several other species, recognition between sperm and egg involves molecules on the acrosomal process. Together, the acrosome and nucleus constitute the **sperm head**.

The means by which sperm are propelled vary according to how the species has adapted to environmental conditions. In most species, an individual sperm is able to travel by whipping its **flagellum**. The major motor portion of the flagellum is the **axoneme**, a structure formed by microtubules emanating from the centriole at the base of the sperm nucleus. The core of the axoneme consists of two central microtubules surrounded by a row of 9 doublet microtubules (**FIGURE 4.3A,B**). Actually, only one microtubule of each doublet is completely circular, having 13 protofilaments; the other is C-shaped and has only 11 protofilaments (**FIGURE 4.3C**). The interconnected protofilaments are made exclusively of the dimeric protein **tubulin**.

Although tubulin is the basis for the structure of the flagellum, other proteins are also critical for flagellar function. The force for sperm propulsion is provided by **dynein**, a protein attached to the microtubules (see Figure 4.3B). Dynein is an ATPase—an enzyme that hydrolyzes ATP, converting the released chemical energy into mechanical energy to propel the sperm. This energy allows the active sliding of the outer doublet microtubules, causing the flagellum to bend (Ogawa et al. 1977; Shinyoji et al. 1998). The importance of dynein can be seen in individuals with a genetic syndrome known as the Kartagener triad. These individuals lack functional dynein in all their ciliated and flagellated cells, rendering these structures immotile (Afzelius 1976). Thus, males with Kartagener triad are sterile (immotile sperm). Both men and women affected by the syndrome are susceptible to bronchial infections (immotile respiratory cilia) and have a 50% chance of having the heart on the right side of the body (a condition known as *situs inversus*, caused by immotile cilia in the center of the embryo; see Chapter 9).

The "9 + 2" microtubule arrangement with dynein arms has been conserved in axonemes throughout the eukaryotic kingdoms, suggesting that this arrangement is extremely well suited for transmitting energy for movement. The ATP needed to move the flagellum and propel the sperm comes from rings of mitochondria located in the **midpiece** of the sperm. In many species (notably mammals), a layer of dense fibers has interposed itself between the mitochondrial sheath and the cell membrane. This fiber layer stiffens the sperm tail. Because the thickness of this layer decreases toward the tip, the fibers probably prevent the sperm head from being whipped around too suddenly. Thus, the sperm cell has undergone extensive modification for the transport of its nucleus to the egg.

In mammals, the differentiation of sperm is not completed in the testes. Although they are able to move, the sperm released during ejaculation do not yet have the capacity to bind to and fertilize an egg. The final stages of sperm maturation, cumulatively referred to as **capacitation**, do not occur in mammals until the sperm has been inside the female reproductive tract for a certain period of time.

The egg

CYTOPLASM AND NUCLEUS   All the material necessary for the beginning of growth and development must be stored in the egg, or **ovum**. Whereas the sperm eliminates most of its cytoplasm as it matures, the developing egg (called the **oocyte** before it reaches the stage of meiosis at which it is

(A) Sperm

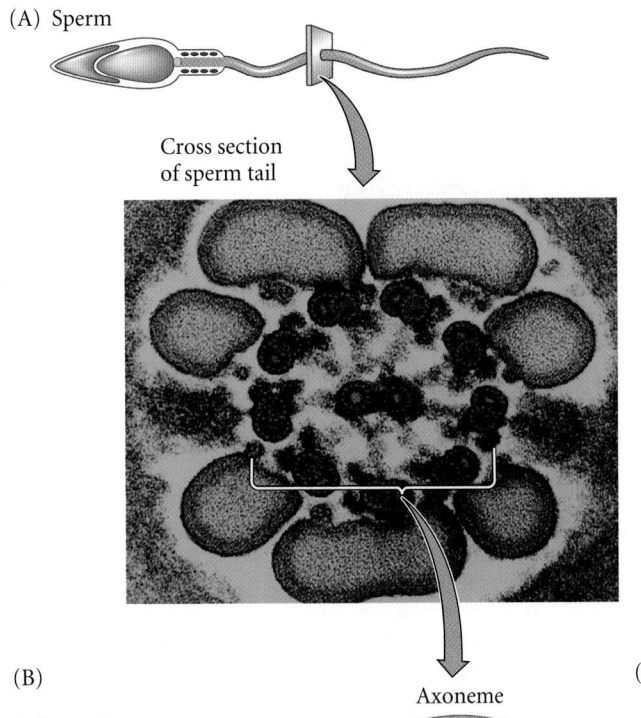

Cross section
of sperm tail

FIGURE 4.3   Motile apparatus of the sperm. (A) Cross section of the flagellum of a mammalian spermatozoon, showing the central axoneme and external fibers. (B) Interpretive diagram of the axoneme, showing the "9 + 2" arrangement of the microtubules and other flagellar components. The dynein arms contain the ATPases that provide the energy for flagellar movement. (C) Association of tubulin protofilaments into a microtubule doublet. One portion of the doublet is a fully circular microtubule comprising 13 protofilaments. The second portion of the doublet contains only 11 (occasionally 10) protofilaments. (A © D. M. Phillips/Photo Researchers, Inc.)

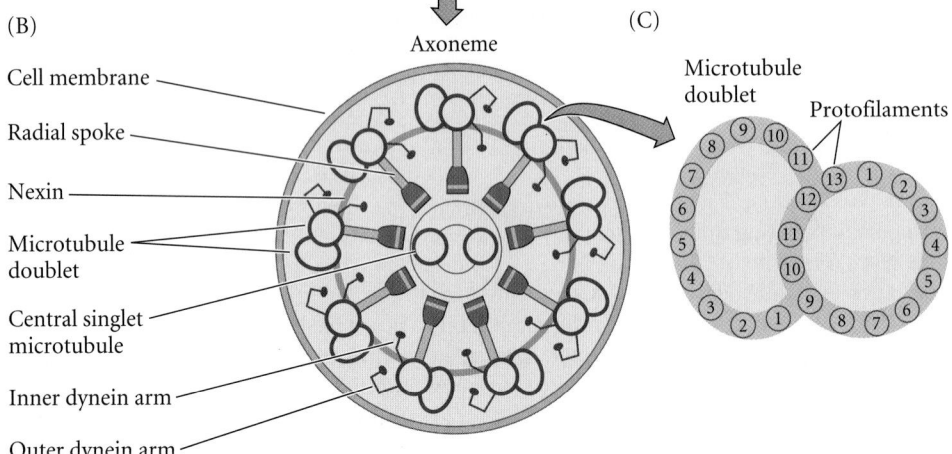

(B)

Axoneme

Cell membrane

Radial spoke

Nexin

Microtubule
doublet

Central singlet
microtubule

Inner dynein arm

Outer dynein arm

(C)

Microtubule
doublet

Protofilaments

fertilized*) not only conserves the material it has, but actively accumulates more. The meiotic divisions that form the oocyte conserve its cytoplasm rather than giving half of it away; at the same time, the oocyte either synthesizes or absorbs proteins such as yolk that act as food reservoirs for the developing embryo. Birds' eggs are enormous single cells, swollen with accumulated yolk. Even eggs with relatively sparse yolk are large compared with sperm. The volume of a sea urchin egg is about 200 picoliters (2×10^{-4} mm^3), more than 10,000 times the volume of sea urchin sperm (**FIGURE 4.4**). So even though sperm and egg have equal, haploid *nuclear* components, the egg also accumulates a remarkable cytoplasmic storehouse during its maturation. This cytoplasmic trove includes the following:

- **Nutritive proteins**. The early embryonic cells must have a supply of energy and amino acids. In many species, this is accomplished by accumulating yolk proteins in the egg. Many of these yolk proteins are made in other organs (e.g., liver, fat bodies) and travel through the maternal blood to the oocyte.

- **Ribosomes and tRNA**. The early embryo must make many of its own structural proteins and enzymes, and in some species there is a burst of protein synthesis soon after fertilization. Protein synthesis is accomplished by ribosomes and tRNA that exist in the egg. The developing egg has special mechanisms for synthesizing ribosomes; certain amphibian oocytes produce as many as 10^{12} ribosomes during their meiotic prophase.

- **Messenger RNAs**. The oocyte not only accumulates proteins, it also accumulates mRNAs that encode proteins for the early stages of development. It is estimated that sea urchin eggs contain thousands of different types of mRNA that remain repressed until after fertilization (see Chapter 2).

- **Morphogenetic factors**. Molecules that direct the differentiation of cells into certain cell types are present in the egg. These include transcription factors and paracrine factors. In many species, they are localized in different regions of the egg and become segregated into different cells during cleavage.

- **Protective chemicals**. The embryo cannot run away from predators or move to a safer environment, so it must be equipped to deal with threats. Many eggs

---

*Eggs over easy: The terminology of eggs is confusing. In general, an *egg* is a female gamete capable of binding sperm and being fertilized. An *oocyte* is a developing egg that cannot yet bind sperm or be fertilized (Wessel 2009). The problems in terminology come from the fact that the eggs of different species are in different stages of meiosis (see Figure 4.5). The human egg, for example, is in second meiotic metaphase when it binds sperm, whereas the sea urchin egg has completed all of its meiotic divisions when it binds sperm. The contents of the egg also vary greatly from species to species. The synthesis and placement of these materials will be addressed in Chapter 17, where we discuss the differentiation of the germ cells.

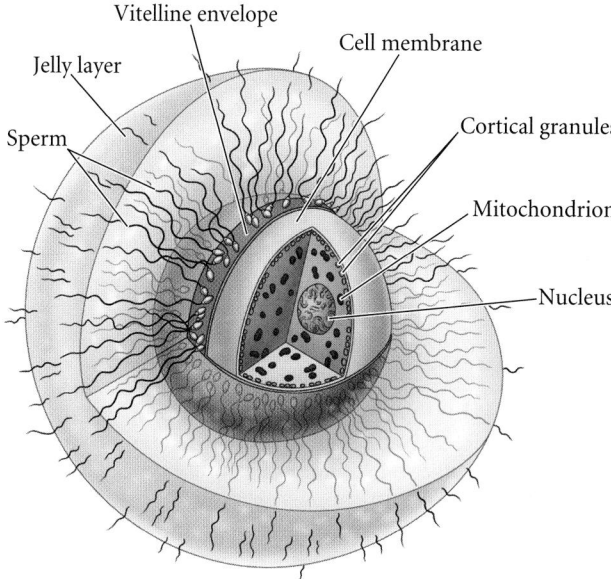

FIGURE 4.4   Structure of the sea urchin egg at fertilization. The drawing shows the relative sizes of egg and sperm. (After Epel 1977.)

contain ultraviolet filters and DNA repair enzymes that protect them from sunlight, and some eggs contain molecules that potential predators find distasteful. The yolk of bird eggs contains antibodies.

Within the enormous volume of egg cytoplasm resides a large nucleus. In a few species (such as sea urchins), this **female pronucleus** is already haploid at the time of fertilization. In other species (including many worms and most mammals), the egg nucleus is still diploid—the sperm enters before the egg's meiotic divisions are completed (**FIGURE 4.5**). In these species, the final stages of egg meiosis will take place after the sperm's nuclear material—the **male pronucleus**—is already inside the egg cytoplasm.

● See **WEBSITE 4.3** The egg and its environment

CELL MEMBRANE AND EXTRACELLULAR ENVELOPE   The cell membrane enclosing the egg cytoplasm regulates the flow of specific ions during fertilization and must be capable of fusing with the sperm cell membrane. Outside this egg cell membrane is an extracellular matrix that forms a fibrous mat around the egg and is often involved in sperm-egg recognition (Correia and Carroll 1997). In invertebrates, this structure is usually called the **vitelline envelope** (**FIGURE 4.6A**). The vitelline envelope contains several different glycoproteins. It is supplemented by extensions of membrane glycoproteins from the cell membrane and by proteinaceous "posts" that adhere the vitelline envelope to the cell membrane (Mozingo and Chandler 1991). The vitelline envelope is essential for the species-specific binding of sperm. Many types of eggs also have a layer of **egg jelly** outside the vitelline envelope (see Figure 4.4). This glycoprotein meshwork

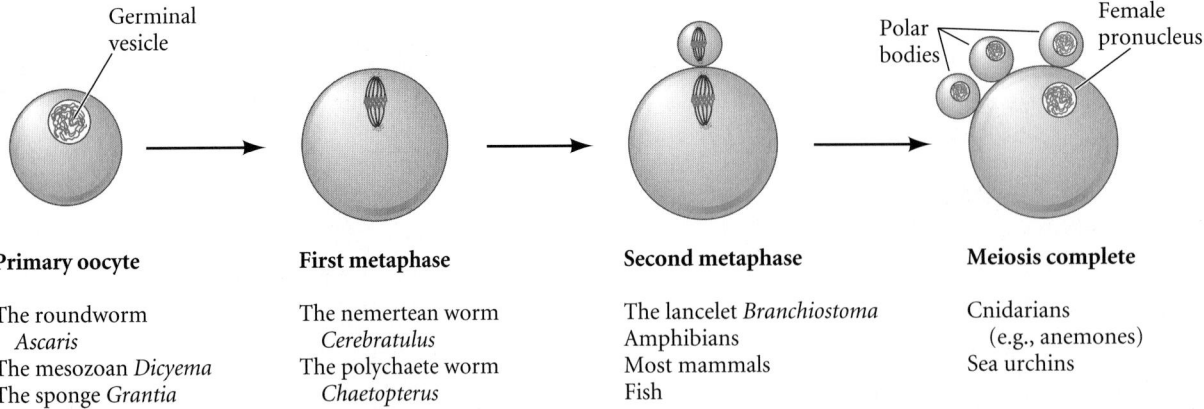

Primary oocyte

The roundworm
  Ascaris
The mesozoan *Dicyema*
The sponge *Grantia*
The polychaete worm
  Myzostoma
The clam worm *Nereis*
The clam *Spisula*
The echiuroid worm
  Urechis
Dogs and foxes

First metaphase

The nemertean worm
  Cerebratulus
The polychaete worm
  Chaetopterus
The mollusc *Dentalium*
The core worm
  Pectinaria
Many insects
Starfish

Second metaphase

The lancelet *Branchiostoma*
Amphibians
Most mammals
Fish

Meiosis complete

Cnidarians
  (e.g., anemones)
Sea urchins

FIGURE 4.5   Stages of egg maturation at the time of sperm entry in different animal species. Note that in most species, sperm entry occurs before the egg nucleus has completed meiosis. The germinal vesicle is the name given to the large diploid nucleus of the primary oocyte. The polar bodies are nonfunctional cells produced by meiosis (see Chapter 17). (After Austin 1965.)

can have numerous functions, but most commonly it is used either to attract or to activate sperm. The egg, then, is a cell specialized for receiving sperm and initiating development.

Lying immediately beneath the cell membrane of most eggs is a thin layer (about 5 μm) of gel-like cytoplasm called the **cortex**. The cytoplasm in this region is stiffer than the internal cytoplasm and contains high concentrations of globular actin molecules. During fertilization, these actin molecules polymerize to form long cables of actin known as **microfilaments**. Microfilaments are necessary for cell division. They are also used to extend the egg surface into small projections called **microvilli**, which may aid sperm entry into the cell (**FIGURE 4.6B**; see also Figure 4.15).

Also within the cortex are the **cortical granules** (see Figures 4.4 and 4.6B). These membrane-bound, Golgi-derived structures contain proteolytic enzymes and are thus homologous to the acrosomal vesicle of the sperm. However, whereas a sea urchin sperm contains just one acrosomal vesicle, each sea urchin egg contains approximately 15,000 cortical granules. Moreover, in addition to digestive enzymes, the cortical granules contain mucopolysaccharides, adhesive glycoproteins, and hyalin protein. As we will soon describe, the enzymes and mucopolysaccharides help prevent polyspermy—that is, they prevent additional sperm from entering the egg after the first sperm has entered—while hyalin

(A)

(B)

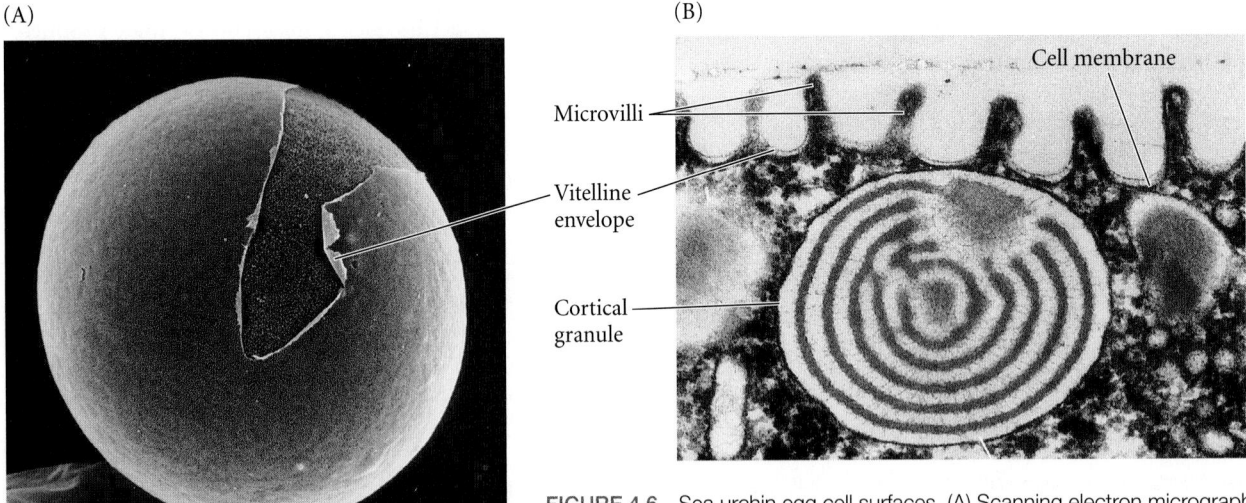

FIGURE 4.6   Sea urchin egg cell surfaces. (A) Scanning electron micrograph of an egg before fertilization. The cell membrane is exposed where the vitelline envelope has been torn. (B) Transmission electron micrograph of an unfertilized egg, showing microvilli and cell membrane, which are closely covered by the vitelline envelope. A cortical granule lies directly beneath the cell membrane. (From Schroeder 1979, courtesy of T. E. Schroeder.)

(A)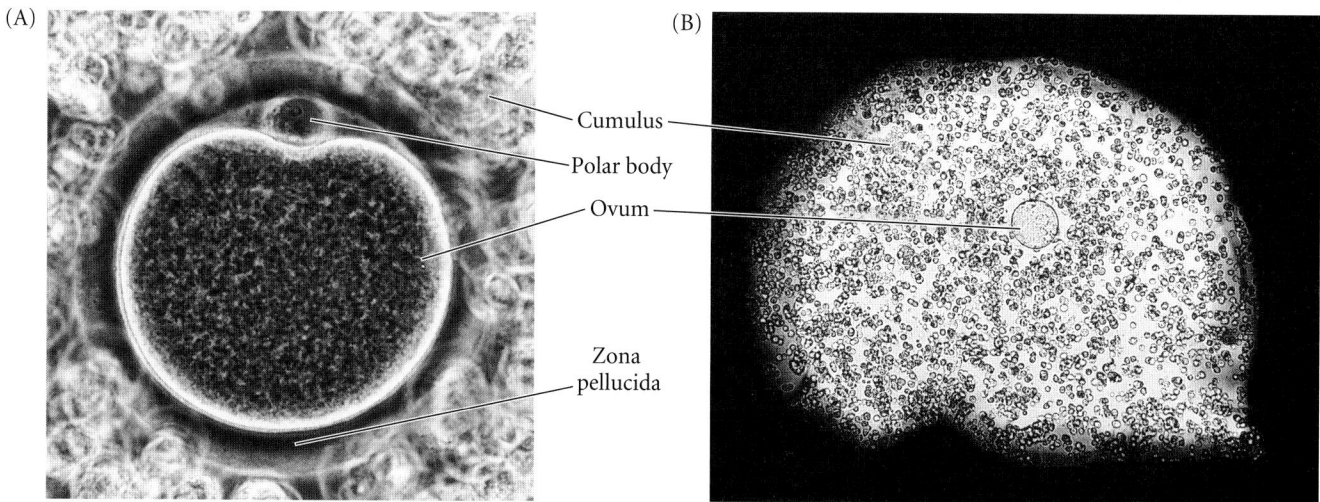

Cumulus
Polar body
Ovum

Zona
pellucida

FIGURE 4.7 Mammalian eggs immediately before fertilization. (A) The hamster egg, or ovum, is encased in the zona pellucida, which in turn is surrounded by the cells of the cumulus. A polar body cell, produced during meiosis, is visible within the zona pellucida. (B) At lower magnification, a mouse oocyte is shown surrounded by the cumulus. Colloidal carbon particles (India ink, seen here as the black background) are excluded by the hyaluronidate matrix. (Courtesy of R. Yanagimachi.)

and the adhesive glycoproteins surround the early embryo, providing support for cleavage-stage blastomeres.

In mammalian eggs, the extracellular envelope is a separate, thick matrix called the **zona pellucida**. The mammalian egg is also surrounded by a layer of cells called the **cumulus** (**FIGURE 4.7**), which is made up of the ovarian follicular cells that were nurturing the egg at the time of its release from the ovary. Mammalian sperm have to get past these cells to fertilize the egg. The innermost layer of cumulus cells, immediately adjacent to the zona pellucida, is called the **corona radiata**.

● **See VADE MECUM** Gametogenesis

Recognition of egg and sperm

The interaction of sperm and egg generally proceeds according to five steps (**FIGURE 4.8**; Vacquier 1998):

1. Chemoattraction of the sperm to the egg by soluble molecules secreted by the egg

2. Exocytosis of the sperm acrosomal vesicle and release of its enzymes

3. Binding of the sperm to the extracellular matrix (vitelline envelope or zona pellucida) of the egg

4. Passage of the sperm through this extracellular matrix

5. Fusion of the egg and sperm cell membranes

After these steps are accomplished, the haploid sperm and egg nuclei can meet and the reactions that initiate development can begin. In this chapter, we will focus on these events in sea urchins, which undergo external fertilization, and in mice, which undergo internal fertilization. Some variations of fertilization events will be described

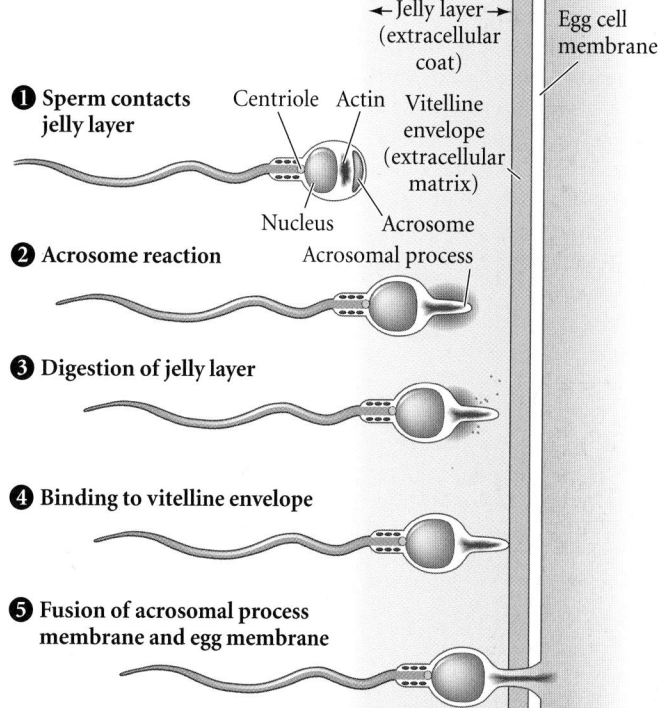

❶ **Sperm contacts jelly layer**

Jelly layer (extracellular coat)
Egg cell membrane
Centriole Actin Vitelline envelope (extracellular matrix)
Nucleus Acrosome

❷ **Acrosome reaction**
Acrosomal process

❸ **Digestion of jelly layer**

❹ **Binding to vitelline envelope**

❺ **Fusion of acrosomal process membrane and egg membrane**

FIGURE 4.8 Summary of events leading to the fusion of egg and sperm cell membranes in sea urchin fertilization, which is external. (1) The sperm is chemotactically attracted to and activated by the egg. (2, 3) Contact with the egg jelly triggers the acrosome reaction, allowing the acrosomal process to form and release proteolytic enzymes. (4) The sperm adheres to the vitelline envelope and lyses a hole in it. (5) The sperm adheres to the egg cell membrane and fuses with it. The sperm pronucleus can now enter the egg cytoplasm.

in subsequent chapters as we study the development of particular organisms.

External Fertilization in Sea Urchins

In many species, the meeting of sperm and egg is not a simple matter. Many marine organisms release their gametes into the environment. That environment may be as small as a tide pool or as large as an ocean. Moreover, this environment is shared with other species that may shed their gametes at the same time. Such organisms are faced with two problems: How can sperm and eggs meet in such a dilute concentration, and how can sperm be prevented from attempting to fertilize eggs of another species? In addition to simply producing enormous numbers of gametes, two major mechanisms have evolved to solve these problems: species-specific sperm attraction and species-specific sperm activation. Here we describe these events as they occur in sea urchins.

Sperm attraction: Action at a distance

Species-specific sperm attraction has been documented in numerous species, including cnidarians, molluscs, echinoderms, amphibians, and urochordates (Miller 1985; Yoshida et al. 1993; Burnett et al. 2008). In many species, sperm are attracted toward eggs of their species by **chemotaxis**—that is, by following a gradient of a chemical secreted by the egg. Miller (1978) demonstrated that the eggs of the cnidarian *Orthopyxis caliculata* not only secrete a chemotactic factor but also regulate the timing of its release. Developing oocytes at various stages in their maturation were fixed on microscope slides, and sperm were released at a certain distance from the eggs. Miller found that when sperm were added to oocytes that had not yet completed their second meiotic division, there was no attraction of sperm to eggs. However, after the second meiotic division was finished and the eggs were ready to be fertilized, the sperm migrated toward them. Thus, these oocytes control not only the type of sperm they attract, but also the time at which they attract them.

The mechanisms of chemotaxis differ among species (see Metz 1978; Eisenbach 2004), and the chemotactic molecules are different even in closely related species. In sea urchins,

sperm motility is acquired when the sperm are spawned into seawater. As long as sperm cells are in the testes, they cannot move because their internal pH is kept low (about pH 7.2) by the high concentrations of CO_2 in the gonad. However, once sperm are spawned into seawater, their pH is elevated to about 7.6, resulting in the activation of the dynein ATPase. The splitting of ATP provides the energy for the flagella to wave, and the sperm begin swimming vigorously (Christen et al. 1982).

But the ability to move does not provide the sperm with direction. In echinoderms, direction is provided by small chemotactic peptides called **sperm-activating peptides** (**SAPs**). One such SAP is resact, a 14-amino acid peptide that has been isolated from the egg jelly of the sea urchin *Arbacia punctulata* (Ward et al. 1985). Resact diffuses readily from the egg jelly into seawater and has a profound effect at very low concentrations when added to a suspension of *Arbacia* sperm. When a drop of seawater containing *Arbacia* sperm is placed on a microscope slide, the sperm generally swim in circles about 50 μm in diameter. Within seconds after a small amount of resact is injected, sperm migrate into the region of the injection and congregate there (**FIGURE 4.9**). As resact diffuses from the area of injection, more sperm are recruited into the growing cluster.

Resact is specific for *A. punctulata* and does not attract sperm of other species. (An analogous compound, speract, has been isolated from the purple sea urchin, *Strongylocentrotus purpuratus*.) *A. punctulata* sperm have receptors in their cell membranes that bind resact (Ramarao and Garbers 1985; Bentley et al. 1986). When the extracellular side of the receptor binds resact, it activates latent guanylyl cyclase in the cytoplasmic side of the receptor (**FIGURE 4.10**). Active guanylyl cyclase causes the sperm cell to produce more cyclic GMP

FIGURE 4.9 Sperm chemotaxis in the sea urchin *Arbacia punctulata*. One nanoliter of a 10-n*M* solution of resact is injected into a 20-microliter drop of sperm suspension. (A) A 1-second photographic exposure showing sperm swimming in tight circles before the addition of resact. The position of the injection pipette is shown by the white lines. (B–D) Similar 1-second exposures showing migration of sperm to the center of the resact gradient 20, 40, and 90 seconds after injection. (From Ward et al. 1985, courtesy of V. D. Vacquier.)

(A)   (B)   (C)   (D)

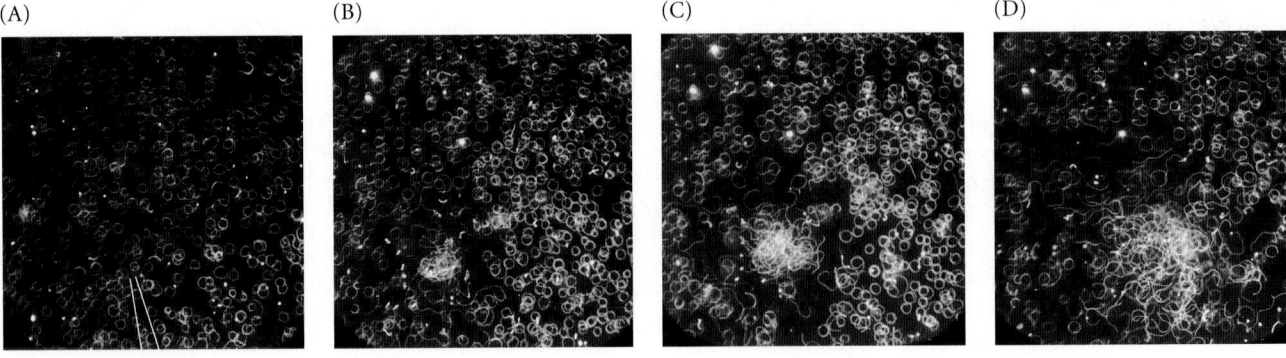

(A)

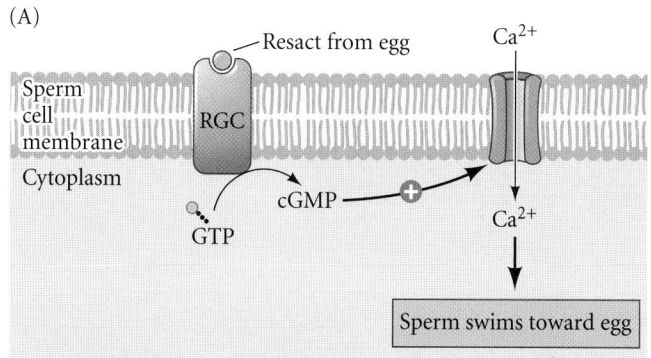

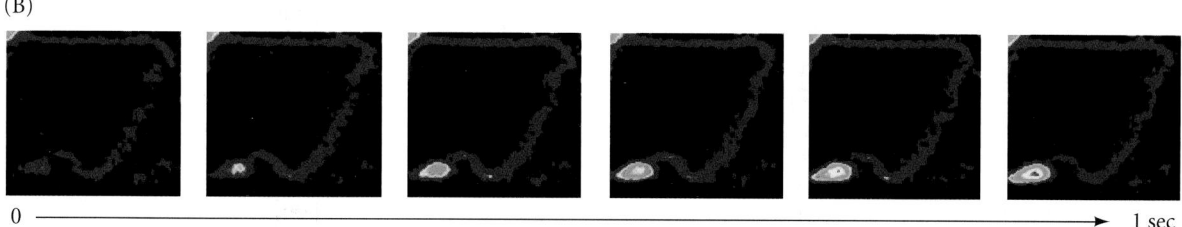

FIGURE 4.10   Model for chemotactic peptides in sea urchin sperm. (A) Resact from *Arbacia* egg jelly binds to its receptor on the sperm. This activates the receptor's guanylyl cyclase (RGC) activity, forming intracellular cGMP in the sperm. The cGMP opens calcium channels in the sperm cell membrane, allowing Ca^{2+} to enter the sperm. The influx of Ca^{2+} activates sperm motility, and the sperm swims up the resact gradient toward the egg. (B) Ca^{2+} levels in different regions of *Strongylocentrotus purpuratus* sperm after exposure to 125 n*M* speract (this species' analog of resact). Red indicates the highest level of Ca^{2+}, blue the lowest. The sperm head reaches its peak Ca^{2+} levels within 1 second. (A after Kirkman-Brown et al. 2003; B from Wood et al. 2003, courtesy of M. Whitaker.)

(cGMP), a compound that activates a calcium channel in the cell membrane of the sperm tail, allowing the influx of calcium ions (Ca^{2+}) from the seawater into the tail (Nishigaki et al. 2000; Wood et al. 2005). The sperm sense the gradient of SAP by curving their tails, interspersing straight swimming with a "turn" to sense the environment (Guerrero et al. 2010). The binding of a single resact molecule may be enough to provide direction for the sperm, which swim up a concentration gradient of this compound until they reach the egg (Kaupp et al. 2003; Kirkman-Brown et al. 2003). Thus, resact functions as a sperm-*attracting* peptide as well as a sperm-activating peptide. (In some organisms, the functions of sperm attraction and sperm activation are performed by different compounds.)

Resact (and various sperm-activating proteins found in the egg jellies of other species) act by causing dramatic and immediate increases in mitochondrial respiration and sperm motility (Hardy et al. 1994; Inamdar et al. 2007). The increases in cGMP and Ca^{2+} activate both the mitochondrial ATP-generating apparatus and the dynein ATPase that stimulates flagellar movement in the sperm (Shimomura et al. 1986; Cook and Babcock 1993). Thus, upon meeting resact, *Arbacia* sperm are instructed where to go and are given the motive force to get there.

The acrosome reaction

A second interaction between sperm and egg jelly results in the **acrosome reaction**. In most marine invertebrates, the acrosome reaction has two components: the fusion of the acrosomal vesicle with the sperm cell membrane (an exocytosis that results in the release of the contents of the acrosomal vesicle), and the extension of the acrosomal process (Dan 1952; Colwin and Colwin 1963). The acrosome reaction in sea urchins is initiated by contact of the sperm with the egg jelly. Contact causes

the exocytosis of the sperm's acrosomal vesicle. The proteolytic enzymes and proteasomes (protein-digesting complexes) thus released digest a path through the jelly coat to the egg cell surface (Dan 1967; Franklin 1970). Once the sperm reaches the egg surface, the acrosomal process adheres to the vitelline envelope and tethers the sperm to the egg. It is possible that proteasomes from the acrosome coat the acrosomal process, allowing it to digest the vitelline envelope at the point of attachment and proceed toward the egg (Yokota and Sawada 2007).

In sea urchins, sulfate-containing polysaccharides in the egg jelly bind to specific receptors located directly above the acrosomal vesicle on the sperm cell membrane. These polysaccharides are often highly species-specific, and egg jelly factors from one species of sea urchin generally fail to activate the acrosome reaction even in closely related species (**FIGURE 4.11**; Hirohashi and Vacquier 2002; Hirohashi et al. 2002; Vilela-Silva et al. 2008). Thus, activation of the acrosome reaction serves as a barrier to interspecies (and thus unviable) fertilizations. This is important when numerous species inhabit the same habitat and when their spawning seasons overlap.

In *Stongylocentrotus purpuratus*, the acrosome reaction is initiated by a repeating polymer of fucose sulfate. When this sulfated polysaccharide binds to its receptor on the sperm, the receptor activates three sperm membrane proteins: (1) a calcium transport channel that allows Ca^{2+} to enter the sperm head; (2) a sodium-hydrogen exchanger that pumps sodium ions (Na^+) into the sperm as it pumps hydrogen ions (H^+) out; and (3) a phospholipase enzyme that makes another second messenger, the phospholipid inositol trisphosphate (IP_3, of which we will hear much more later in the chapter). IP_3 is able to release Ca^{2+} from *inside* the sperm, probably from within the acrosome itself (Domino and Garbers 1988; Domino et

(A)

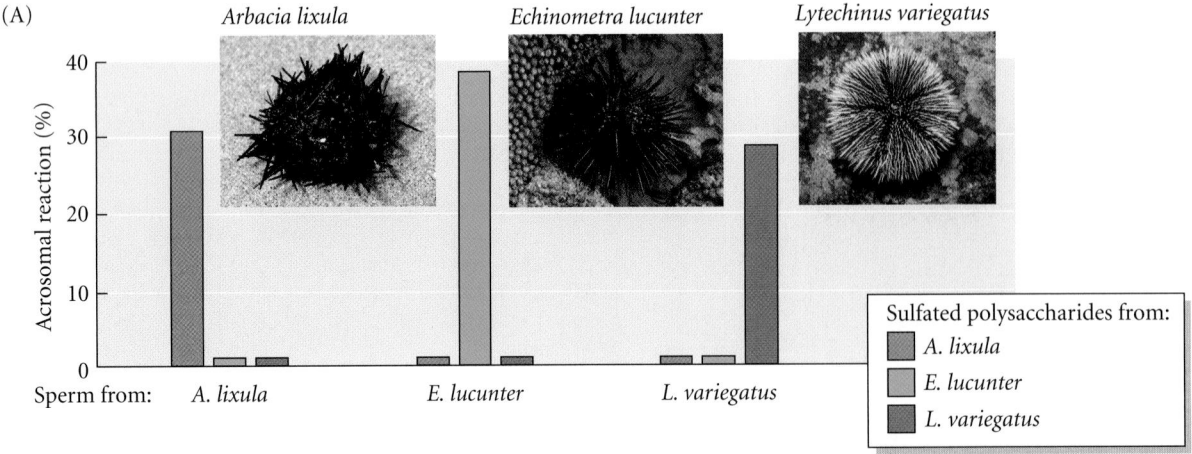

Arbacia lixula   Echinometra lucunter   Lytechinus variegatus

Acrosomal reaction (%)

Sperm from: A. lixula   E. lucunter   L. variegatus

Sulfated polysaccharides from:
- A. lixula
- E. lucunter
- L. variegatus

(B)

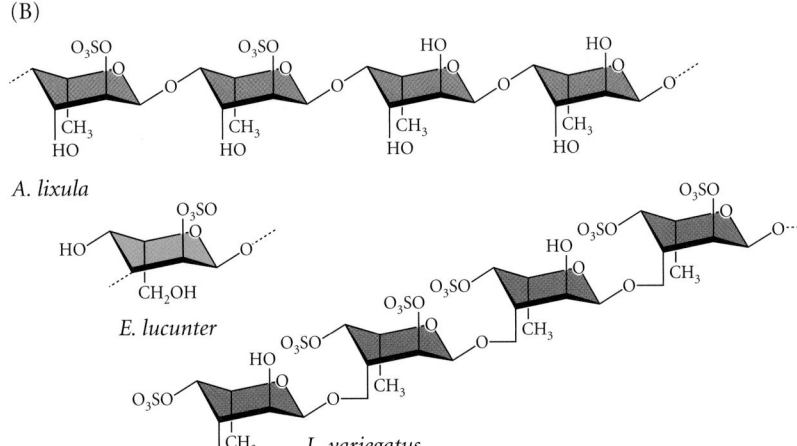

A. lixula

E. lucunter

L. variegatus

FIGURE 4.11 Species-specific induction of the acrosome reaction by sulfated polysaccharides characterizing the egg jelly coats of three species of sea urchins that co-inhabit the intertidal around Rio de Janeiro. (A) The histograms compare the ability of each polysaccharide to induce the acrosome reaction in the different species of sperm. (B) Chemical structures of the acrosome reaction-inducing sulfated polysaccharides reveal their species-specificity. (After Vilela-Silva et al. 2008; photographs left to right © Interfoto/Alamy; © FLPA/AGE Fotostock; © Water Frame/Alamy.)

al. 1989; Hirohashi and Vacquier 2003). The elevated Ca^{2+} level in a relatively basic cytoplasm triggers the fusion of the acrosomal membrane with the adjacent sperm cell membrane (**FIGURE 4.12A–C**), releasing enzymes that can lyse a path through the egg jelly to the vitelline envelope.

The second part of the acrosome reaction involves the extension of the acrosomal process (**FIGURE 4.12D**). This protrusion arises through the polymerization of globular actin molecules into actin filaments (Tilney et al. 1978). The influx of Ca^{2+} is thought to activate the protein RhoB in the acrosomal region and midpiece of sea urchin sperm (Castellano et al. 1997; de la Sancha 2007). This GTP-binding protein helps organize the actin cytoskeleton in many types of cells and is thought to be active in polymerizing actin to make the acrosomal process.

Recognition of the egg's extracellular coat

The sea urchin sperm's contact with an egg's jelly coat provides the first set of species-specific recognition events (i.e., sperm attraction, activation, and acrosome reaction). Another

critical species-specific binding event must occur once the sperm has penetrated the egg jelly and its acrosomal process contacts the surface of the egg (**FIGURE 4.13A**). The acrosomal protein mediating this recognition in sea urchins is an insoluble, 30,500-Da protein called **bindin**. In 1977, Vacquier and co-workers isolated bindin from the acrosome of *Strongylocentrotus purpuratus* and found it to be capable of binding to dejellied eggs of the same species. Further, sperm bindin, like egg jelly polysaccharides, is usually species-specific: bindin isolated from the acrosomes of *S. purpuratus* binds to its own dejellied eggs but not to those of *S. franciscanus* (**FIGURE 4.13B**; Glabe and Vacquier 1977; Glabe and Lennarz 1979).

Biochemical studies have confirmed that the bindins of closely related sea urchin species have different protein sequences. This finding implies the existence of species-specific bindin *receptors* on the egg vitelline envelope. Such receptors were suggested by the experiments of Vacquier and Payne (1973), who saturated sea urchin eggs with sperm. As seen in **FIGURE 4.14A**, sperm binding does not occur over the entire egg surface. Even at saturating numbers of sperm

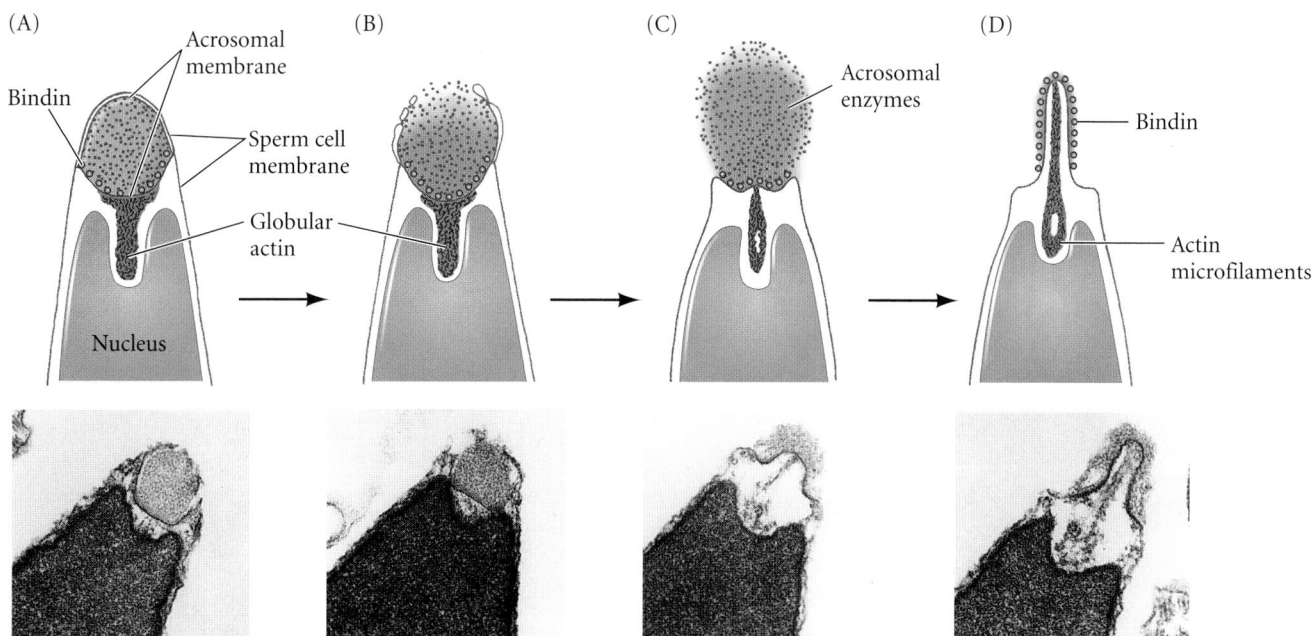

FIGURE 4.12 Acrosome reaction in sea urchin sperm. (A–C) The portion of the acrosomal membrane lying directly beneath the sperm cell membrane fuses with the cell membrane to release the contents of the acrosomal vesicle. (D) The actin molecules assemble to produce microfilaments, extending the acrosomal process outward. Actual photographs of the acrosome reaction in sea urchin sperm are shown below the diagrams. (After Summers and Hylander 1974; photographs courtesy of G. L. Decker and W. J. Lennarz.)

(approximately 1500), there appears to be room on the ovum for more sperm heads, implying a limiting number of sperm-binding sites. EBR1, a 350-kDa glycoprotein that displays the properties expected of a bindin receptor, has been isolated from sea urchin eggs (FIGURE 4.14B; Kamei and Glabe 2003).

These bindin receptors are thought to be aggregated into complexes on the vitelline envelope, and hundreds of such complexes may be needed to tether the sperm to the egg. The receptor for sperm bindin on the egg vitelline envelope appears to recognize the protein portion of bindin on the acrosome (FIGURE 4.14C) in a species-specific manner. Closely related species of sea urchins (i.e., different species in the same genus) have divergent bindin receptors, and eggs will adhere only to the bindin of their own species (FIGURE 4.14D). Thus, species-specific recognition of sea urchin gametes can occur at the levels of sperm attraction, sperm activation, the acrosome reaction, and sperm adhesion to the egg surface.

FIGURE 4.13 Species-specific binding of acrosomal process to egg surface in sea urchins. (A) Actual contact of a sea urchin sperm acrosomal process with an egg microvillus. (B) In vitro model of species-specific binding. The agglutination of dejellied eggs by bindin was measured by adding bindin particles to a plastic well containing a suspension of eggs. After 2–5 minutes of gentle shaking, the wells were photographed. Each bindin bound to and agglutinated only eggs from its own species. (A from Epel 1977, courtesy of F. D. Collins and D. Epel; B based on photographs in Glabe and Vacquier 1977.)

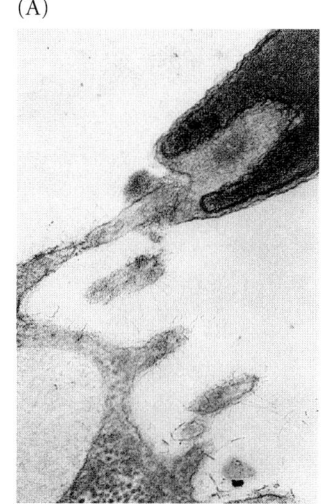

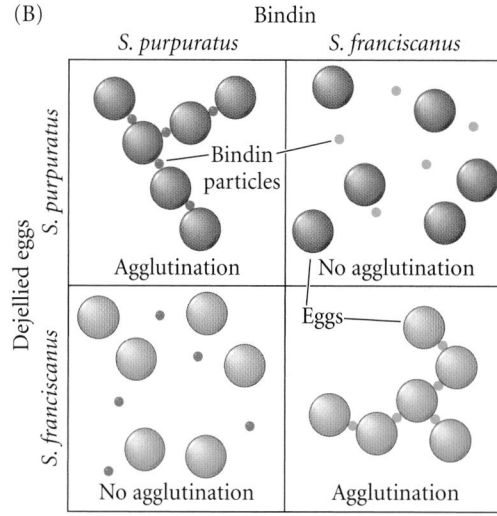

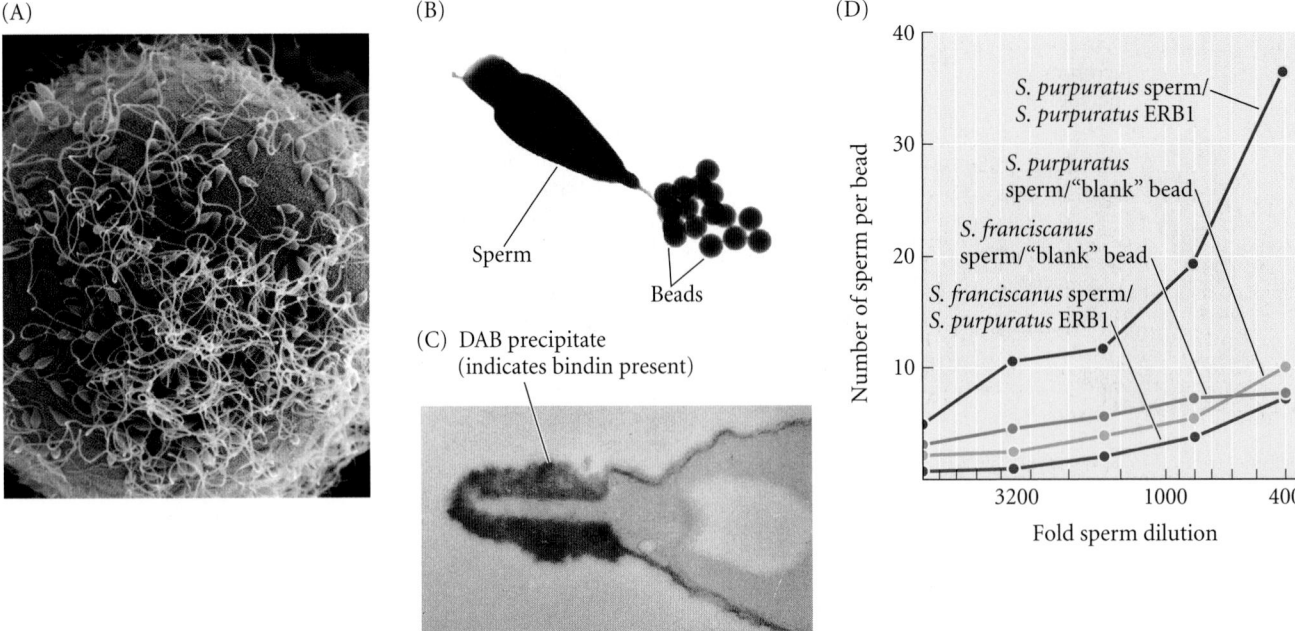

(A)

(B)

Sperm

Beads

(C) DAB precipitate (indicates bindin present)

(D)

FIGURE 4.14 Bindin receptors on the sea urchin egg. (A) Scanning electron micrograph of sea urchin sperm bound to the vitelline envelope of an egg. Although this egg is saturated with sperm, there appears to be room on the surface for more sperm, implying the existence of a limited number of bindin receptors. (B) *Strongylocentrotus purpuratus* sperm bind to polystyrene beads that have been coated with purified bindin receptor protein. (C) Immunochemically labeled bindin (the label manifests as a dark precipitate of diaminobenzidine, DAB) is seen to be localized to the acrosomal process after the acrosome reaction. (D) Species-specific binding of sea urchin sperm to ERB1. *S. purpuratus* sperm bound to beads coated with EBR1 bindin receptor purified from *S. purpuratus* eggs, but *S. franciscanus* sperm did not. Neither sperm bound to uncoated "blank" beads. (A © Mia Tegner/SPL/Photo Researchers, Inc.; B from Foltz et al. 1993; C from Moy and Vacquier 1979, courtesy of V. Vacquier; D after Kamei and Glabe 2003.)

Bindin and other gamete recognition proteins are among the fastest evolving proteins known (Metz and Palumbi 1996; Swanson and Vacquier 2002). Even when closely related urchin species have near-identity of every other protein, their bindins and bindin receptors may have diverged significantly. The evolution of gamete recognition proteins shows the hallmarks of sexual selection and coevolved genes. It is thought that coevolution of the genes encoding male and female gamete recognition proteins can lead to reproductive barriers that have the potential to drive speciation by dividing a population into different mating groups (see Clark et al. 2009; Palumbi 2009).

Fusion of the egg and sperm cell membranes

Once the sperm has traveled to the egg and undergone the acrosome reaction, the fusion of the sperm cell membrane with the egg cell membrane can begin. The entry of a sperm into a sea urchin egg is illustrated in **FIGURE 4.15**. Sperm-egg fusion appears to cause the polymerization of actin in the egg to form a **fertilization cone** (Summers et al. 1975). Homology between the egg and the sperm is again demonstrated, since the sperm's acrosomal process also appears to be formed by the polymerization of actin. The actin from the gametes forms a connection that widens the cytoplasmic bridge between the egg and the sperm. The sperm nucleus and tail pass through this bridge.

Fusion is an active process, often mediated by specific "fusogenic" proteins. Indeed, sea urchin sperm bindin plays a second role as a fusogenic protein. In addition to recognizing the egg, bindin contains a long stretch of hydrophobic amino acids near its amino terminus, and this region is able to fuse phospholipid vesicles in vitro (Ulrich et al. 1999; Gage et al. 2004). Under the ionic conditions present in the mature unfertilized egg, bindin can cause the sperm and egg membranes to fuse.

One egg, one sperm

As soon as one sperm has entered the egg, the fusibility of the egg membrane—which was necessary to get the sperm inside the egg—becomes a dangerous liability. In the normal case—**monospermy**—only one sperm enters the egg, and the haploid sperm nucleus combines with the haploid egg nucleus to form the diploid nucleus of the fertilized egg (zygote), thus restoring the chromosome number appropriate for the species. During cleavage, the centriole provided by the sperm divides to form the two poles of the mitotic spindle while the egg-derived centriole is degraded.

In most animals, any sperm that enters the egg can provide a haploid nucleus and a centriole. The entrance of multiple sperm—**polyspermy**—leads to disastrous consequences in most organisms. In sea urchins, fertilization by two sperm results in a triploid nucleus, in which each

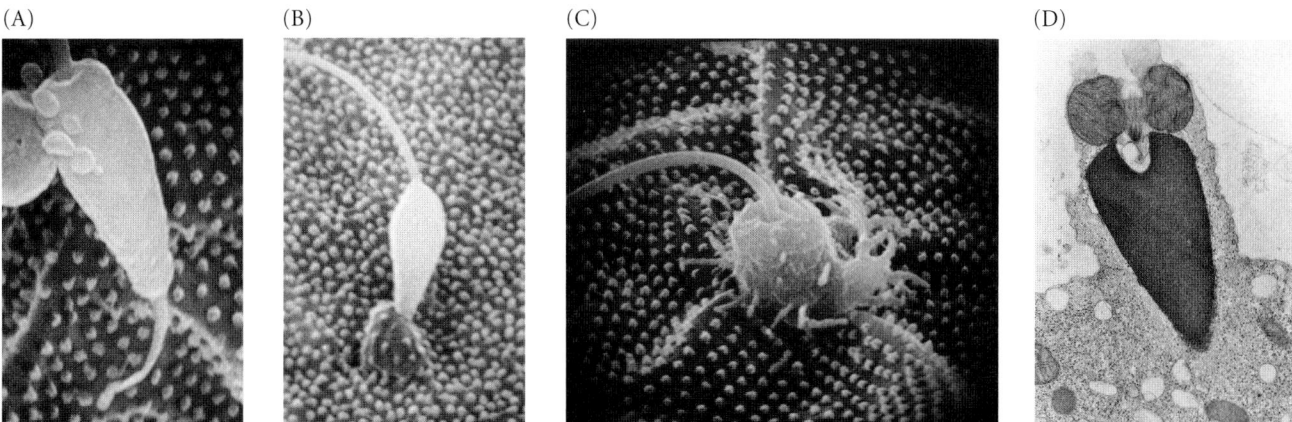

FIGURE 4.15   Scanning electron micrographs of the entry of sperm into sea urchin eggs. (A) Contact of sperm head with egg microvillus through the acrosomal process. (B) Formation of fertilization cone. (C) Internalization of sperm within the egg. (D) Transmission electron micrograph of sperm internalization through the fertilization cone. (A–C from Schatten and Mazia 1976, courtesy of G. Schatten; D courtesy of F. J. Longo.)

chromosome is represented three times rather than twice. Worse, each sperm's centriole divides to form the two poles of a mitotic apparatus, so instead of a bipolar mitotic spindle separating the chromosomes into two cells, the triploid chromosomes may be divided into as many as four cells, with some cells receiving extra copies of certain chromosomes while other cells lack them (**FIGURE 4.16**). Theodor Boveri demonstrated in 1902 that such cells either die or develop abnormally.

The fast block to polyspermy

The most straightforward way to prevent the union of more than two haploid nuclei is to prevent more than one sperm from entering the egg. Different mechanisms to prevent polyspermy have evolved, two of which are seen in the sea urchin egg. An initial, fast reaction, accomplished by an electric change in the egg cell membrane, is followed by a slower reaction caused by the exocytosis of the cortical granules (Just 1919).

The **fast block to polyspermy** is achieved by changing the electric potential of the egg cell membrane immediately upon the entry of a sperm. By selectively prohibiting the entry of sodium ions (Na^+) into the egg, the egg cell membrane maintains an electrical voltage gap between the egg and its environment. This **resting membrane potential** is generally about 70 mV—usually expressed as –70 mV because the inside of the cell is negatively charged with respect to the exterior. Within 1–3 seconds after the binding of the first sperm, the membrane potential shifts to a *positive* level with respect to the exterior, about +20 mV (**FIGURE 4.17A**; Jaffe 1980; Longo et al. 1986). Sperm can fuse with membranes having a resting potential of –70 mV but cannot fuse with membranes having a positive resting potential, so no more sperm can fuse to the egg. The shift from negative to positive is the result of a small influx of Na^+ into the egg through

newly opened sodium channels. It is not known whether the increased sodium permeability of the egg is due to the *binding* of the first sperm, or to the *fusion* of the first sperm with the egg (Gould and Stephano 1987, 1991; McCulloh and Chambers 1992).

The importance of Na^+ and the change in resting potential from negative to positive was demonstrated by Laurinda Jaffe and colleagues. They found that polyspermy can be induced if an electric current is applied to artificially keep the sea urchin egg membrane potential negative. Conversely, fertilization can be prevented entirely by artificially keeping the membrane potential of eggs positive (Jaffe 1976). The fast block to polyspermy can also be circumvented by lowering the concentration of Na^+ in the surrounding water (**FIGURE 4.17B–D**). If the supply of sodium ions is not sufficient to cause the positive shift in membrane potential, polyspermy occurs (Gould-Somero et al. 1979; Jaffe 1980).

It is not known how the change in membrane potential acts on the sperm to block secondary fertilization. Most likely, the sperm carry a voltage-sensitive component (possibly a positively charged fusogenic protein), and the insertion of this component into the egg cell membrane could be regulated by the electric charge across the membrane (Iwao and Jaffe 1989). There are data to suggest that the fusogenic region of bindin will not function on a positively charged surface (Rocha et al. 2007). An electric block to polyspermy also occurs in frogs* (Cross and Elinson 1980), but probably not in most mammals (Jaffe and Cross 1983).

● See **WEBSITE 4.4** Blocks to polyspermy
● See **VADE MECUM** E. E. Just

---

*You might ask, as did one student, how amphibians can have a fast block to polyspermy, since their eggs are fertilized in pond water, which lacks the high Na^+ concentrations found in seawater. It turns out that the ion channels that open in frog egg membranes at fertilization are chloride channels instead of sodium channels as in sea urchin eggs. The concentration of Cl^- inside the frog egg is much higher than that of pond water. Thus, when chloride channels open at fertilization, the negatively charged chloride ions flow *out* of the cytoplasm, leaving the inside of the egg at a positive potential (see Jaffe and Schlicter 1985; Glahn and Nuccitelli 2003).

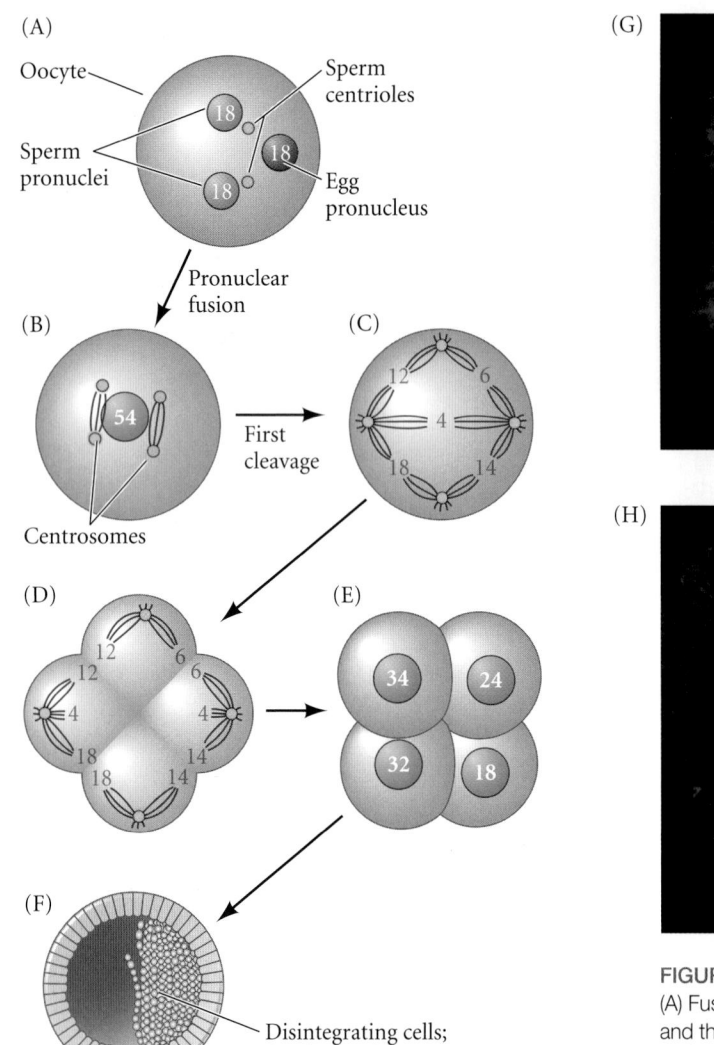

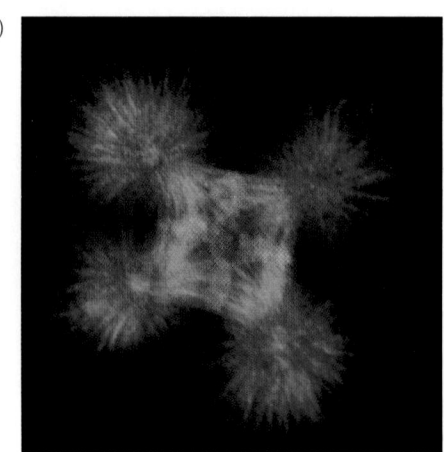

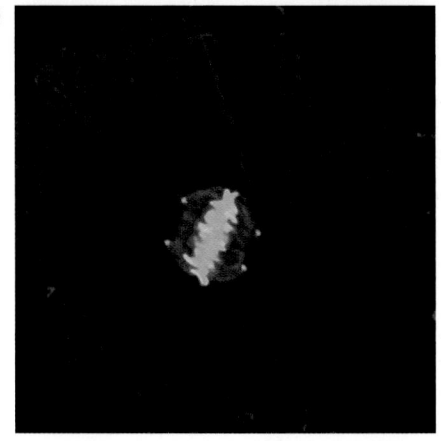

FIGURE 4.16 Aberrant development in a dispermic sea urchin egg. (A) Fusion of three haploid nuclei, each containing 18 chromosomes, and the division of the two sperm centrioles to form four centrosomes (mitotic poles). (B,C) The 54 chromosomes randomly assort on the four spindles. (D) At anaphase of the first division, the duplicated chromosomes are pulled to the four poles. (E) Four cells containing different numbers and types of chromosomes are formed, thereby causing (F) the early death of the embryo. (G) First metaphase of a dispermic sea urchin egg akin to (D). The microtubules are stained green; the DNA stain appears orange. The triploid DNA is being split into four chromosomally unbalanced cells instead of the normal two cells with equal chromosome complements. (H) Human dispermic egg at first mitosis. The four centrioles are stained yellow, while the microtubules of the spindle apparatus (and of the two sperm tails) are stained red. The three sets of chromosomes divided by these four poles are stained blue. (A–F after Boveri 1907; G courtesy of J. Holy; H from Simerly et al. 1999, courtesy of G. Schatten.)

The slow block to polyspermy

The fast block to polyspermy is transient, since the membrane potential of the sea urchin egg remains positive for only about a minute. This brief potential shift is not sufficient to prevent polyspermy permanently, and polyspermy can still occur if the sperm bound to the vitelline envelope are not somehow removed (Carroll and Epel 1975). This sperm removal is accomplished by the **cortical granule reaction**, also known as the **slow block to polyspermy**. This slower, mechanical block to polyspermy becomes active about a minute after the first successful sperm-egg fusion (Just 1919). This reaction is found in many animal species, including sea urchins and most mammals.

Directly beneath the sea urchin egg cell membrane are about 15,000 cortical granules, each about 1 μm in diameter (see Figure 4.6B). Upon sperm entry, cortical granules fuse with the egg cell membrane and release their contents into the space between the cell membrane and the fibrous mat of vitelline envelope proteins. Several proteins are released by this cortical granule exocytosis, one of which is *cortical granule serine*

protease. This enzyme cleaves the protein posts that connect the vitelline envelope proteins to the egg cell membrane; it also clips off the bindin receptors and any sperm attached to them (Vacquier et al. 1973; Glabe and Vacquier 1978; Haley and Wessel 1999, 2004).

The components of the cortical granules bind to the vitelline envelope to form a **fertilization envelope**. The

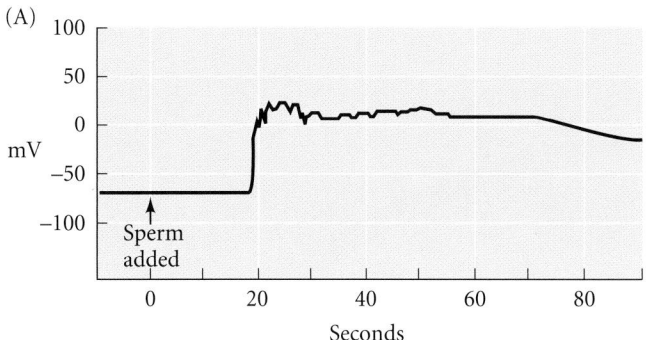

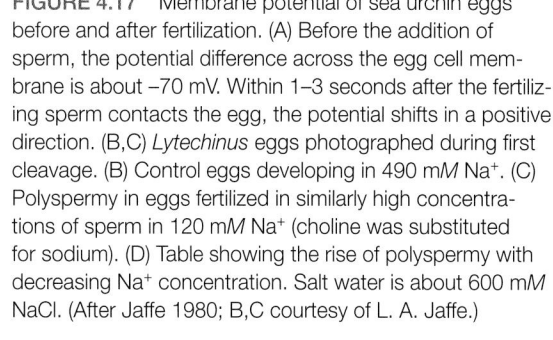

FIGURE 4.17 Membrane potential of sea urchin eggs before and after fertilization. (A) Before the addition of sperm, the potential difference across the egg cell membrane is about –70 mV. Within 1–3 seconds after the fertilizing sperm contacts the egg, the potential shifts in a positive direction. (B,C) *Lytechinus* eggs photographed during first cleavage. (B) Control eggs developing in 490 m*M* Na⁺. (C) Polyspermy in eggs fertilized in similarly high concentrations of sperm in 120 m*M* Na⁺ (choline was substituted for sodium). (D) Table showing the rise of polyspermy with decreasing Na⁺ concentration. Salt water is about 600 m*M* NaCl. (After Jaffe 1980; B,C courtesy of L. A. Jaffe.)

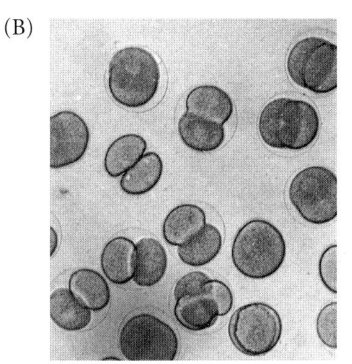

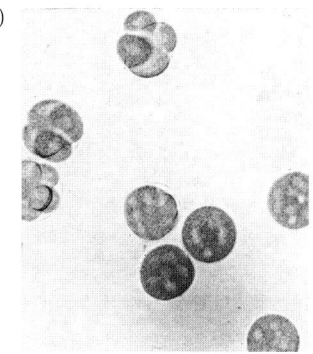

(D)

Na⁺ (mM)	Polyspermic eggs (%)
490	22
360	26
120	97
50	100

fertilization envelope starts to form at the site of sperm entry and continues its expansion around the egg. This process starts about 20 seconds after sperm attachment and is complete by the end of the first minute of fertilization (**FIGURE 4.18**; Wong and Wessel 2004, 2008).

The fertilization envelope is elevated from the cell membrane by *mucopolysaccharides* released by the cortical granules. These viscous compounds absorb water to expand the space between the cell membrane and the fertilization envelope, so that the envelope moves radially away from the egg. The fertilization envelope is then stabilized by crosslinking adjacent proteins through egg-specific *peroxidase enzymes* and a *transglutaminase* released from the cortical granules (**FIGURE 4.19**; Foerder and Shapiro 1977; Wong et al. 2004; Wong and Wessel 2009). This crosslinking allows the egg and early embryo to resist the shear forces of the ocean's intertidal waves. As this is happening, a fourth set of cortical granule proteins, including *hyalin,* forms a coating around the egg (Hylander and Summers 1982). The egg extends elongated microvilli whose tips attach to this **hyaline layer**, which provides support for the blastomeres during cleavage.

● See **WEBSITE 4.5** Building the egg's extracellular matrix

● See **VADE MECUM** Sea urchin fertilization

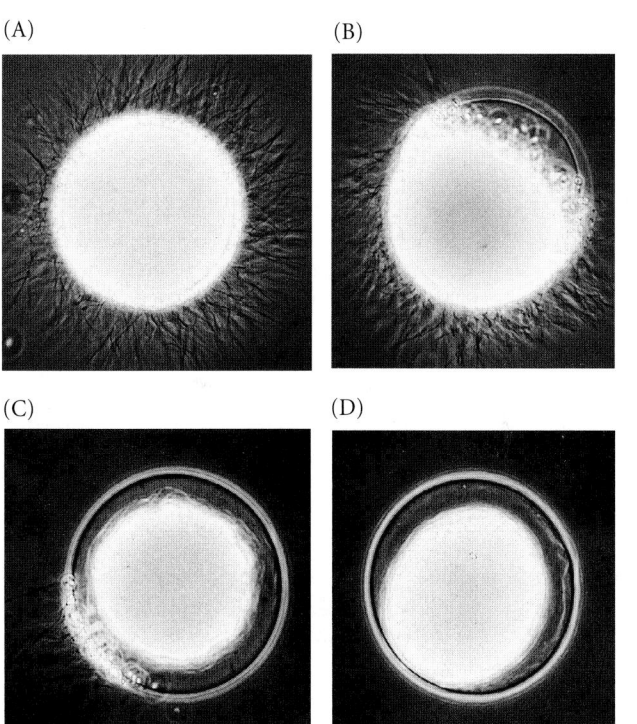

FIGURE 4.18 Formation of the fertilization envelope and removal of excess sperm. To create these photographs, sperm were added to sea urchin eggs, and the suspension was then fixed in formaldehyde to prevent further reactions. (A) At 10 seconds after sperm addition, sperm surround the egg. (B,C) At 25 and 35 seconds after insemination, respectively, a fertilization envelope is forming around the egg, starting at the point of sperm entry. (D) The fertilization envelope is complete, and excess sperm have been removed. (From Vacquier and Payne 1973, courtesy of V. D. Vacquier.)

(A)

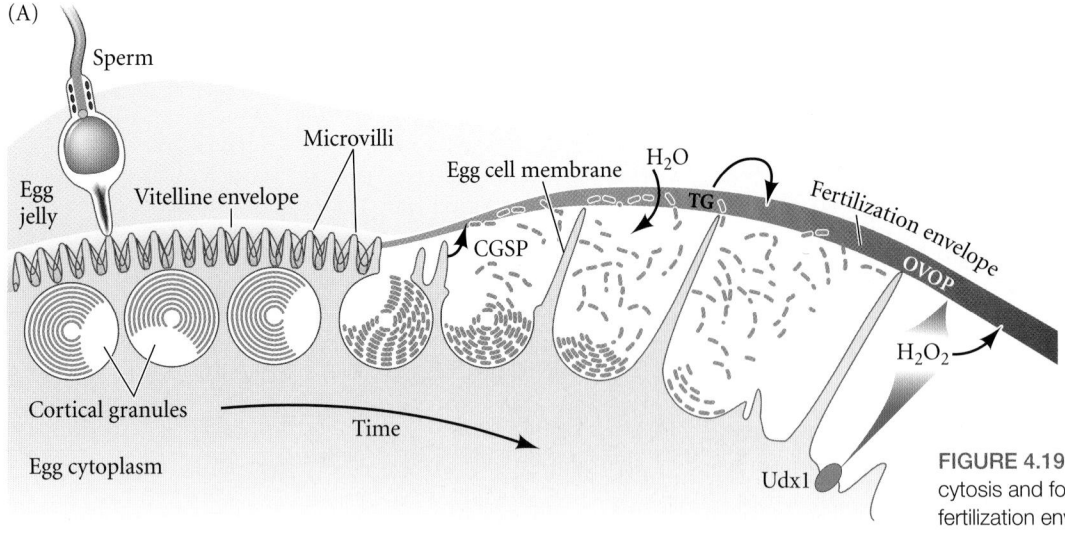

Sperm

Egg jelly

Vitelline envelope

Microvilli

Egg cell membrane

H_2O

CGSP

TG

Fertilization envelope

OVOP

H_2O_2

Cortical granules

Time

Egg cytoplasm

Udx1

(B) Unfertilized

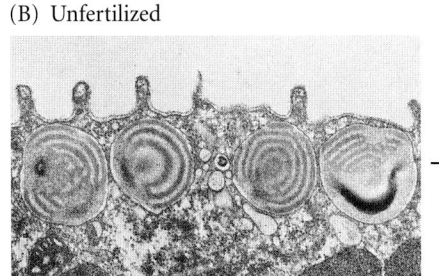

(C) Recently fertilized

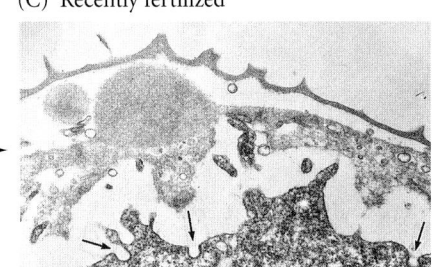

FIGURE 4.19 Cortical granule exocytosis and formation of the sea urchin fertilization envelope. (A) Schematic diagram of events leading to the formation of the fertilization envelope. As cortical granules undergo exocytosis, they release cortical granule serine protease (CGSP), an enzyme that cleaves the proteins linking the vitelline envelope to the cell membrane. Mucopolysaccharides released by the cortical granules form an osmotic gradient, causing water to enter and swell the space between the vitelline envelope and the cell membrane. The enzyme Udx1 in the former cortical granule membrane catalyzes the formation of hydrogen peroxide (H_2O_2), the substrate for soluble ovoperoxidase (OVOP). OVOP and transglutaminases (TG) harden the vitelline envelope, now called the fertilization envelope. (B,C) Transmission electron micrographs of the cortex of an unfertilized sea urchin egg and the same region of a recently fertilized egg. The raised fertilization envelope and the points at which the cortical granules have fused with the egg cell membrane of the egg (arrows) are visible in (C). (A after Wong et al. 2008; B,C from Chandler and Heuser 1979, courtesy of D. E. Chandler.)

Calcium as the initiator of the cortical granule reaction

The mechanism of cortical granule exocytosis is similar to that of the exocytosis of the acrosome, and it may involve many of the same molecules.* Upon fertilization, the concentration of free Ca^{2+} in the egg cytoplasm increases greatly. In this high-calcium environment, the cortical granule membranes fuse with the egg cell membrane, releasing their contents (see Figure 4.19A). Once the fusion of the cortical granules begins near the point of sperm entry, a wave of cortical granule exocytosis propagates around the cortex to the opposite side of the egg.

---

*Exocytotic reactions like the cortical granule reaction and the acrosome reaction are also seen in the release of insulin from pancreatic cells and in the release of neurotransmitters from synaptic terminals. In all cases, there is Ca^{2+}-mediated fusion of the secretory vesicle and the cell membrane. Indeed, the similarity of acrosomal vesicle exocytosis and synaptic vesicle exocytosis may be quite deep. Studies of acrosome reactions in sea urchins and mammals suggest that when the receptors for the sperm-activating ligands bind these molecules, the resulting depolarization of the membrane opens voltage-dependent Ca^{2+} channels in a manner reminiscent of synaptic transmission (González-Martínez et al. 1992; Tulsani and Abou-Haila 2004). The proteins that dock the cortical granules of the egg to the cell membrane also appear to be homologous to those used in the axon terminal. The synaptic granules of the neurons, the acrosomal vesicle of the sperm, and the cortical granules of the egg all appear to use synaptotagmin to bind calcium and initiate fusion of the vesicle with the cell membrane (Bi et al. 1995; Leguia et al. 2006; Roggero et al. 2007).

In sea urchins and mammals, the rise in Ca^{2+} concentration responsible for the cortical granule reaction is not due to an influx of calcium into the egg, but comes from within the egg itself. The release of calcium from intracellular storage can be monitored visually using calcium-activated luminescent dyes such as aequorin (a protein that, like GFP, is isolated from luminescent jellyfish) or fluorescent dyes such as fura-2. These dyes emit light when they bind free Ca^{2+}. When a sea urchin egg is injected with dye and then fertilized, a striking wave of calcium release propagates across the egg and is visualized as a band of light that starts at the point of sperm entry and proceeds actively to the other end of the cell (**FIGURE 4.20**; Steinhardt et al. 1977; Hafner et al. 1988). The entire release of Ca^{2+} is complete within roughly 30 seconds, and free Ca^{2+} is re-sequestered shortly after being released.

Several experiments have demonstrated that Ca^{2+} is directly responsible for propagating the cortical granule reaction, and that these ions are stored within the egg itself. The drug A23187 is a calcium ionophore (a compound that allows

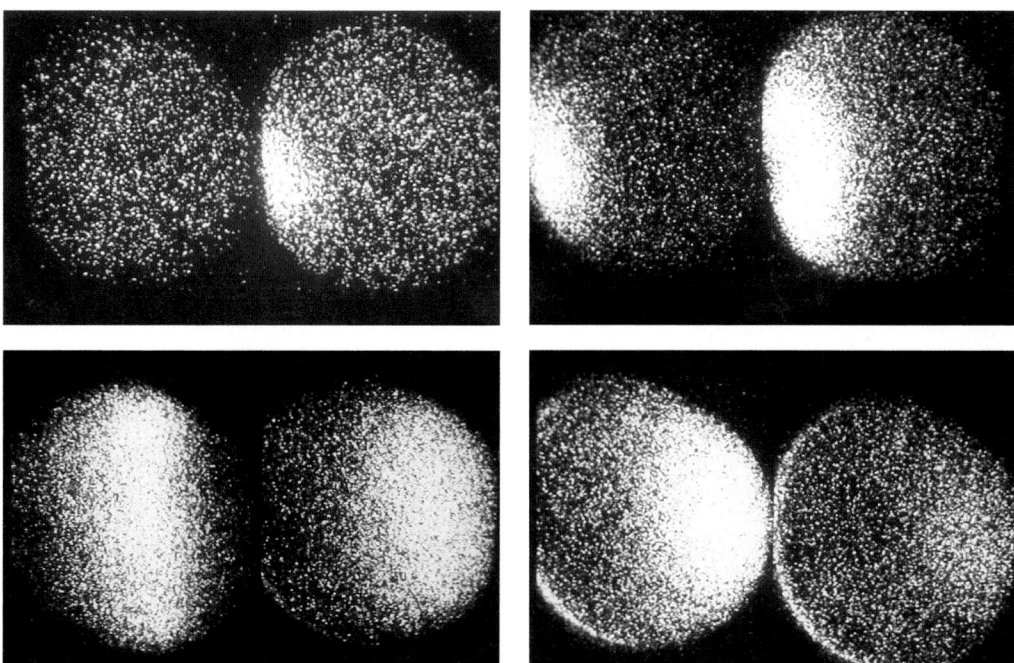

FIGURE 4.20 Ca^{2+} release across a sea urchin egg during fertilization. The egg is pre-loaded with a dye that fluoresces when it binds Ca^{2+}. When a sperm fuses with the egg, a wave of calcium release is seen, beginning at the site of sperm entry and propagating across the egg. The wave does not simply diffuse but travels actively, taking about 30 seconds to traverse the egg. (Courtesy of G. Schatten.)

the diffusion of ions such as Ca^{2+} across lipid membranes, permitting them to traverse otherwise impermeable barriers). Placing unfertilized sea urchin eggs into seawater containing A23187 causes the cortical granule reaction and the elevation of the fertilization envelope. Moreover, this reaction occurs in the absence of any Ca^{2+} in the surrounding water, and thus

the A23187 must be stimulating the release of Ca^{2+} already sequestered in organelles within the egg (Chambers et al. 1974; Steinhardt and Epel 1974).

In sea urchins and vertebrates (but not snails and worms), the Ca^{2+} responsible for the cortical granule reaction is stored in the endoplasmic reticulum of the egg (Eisen and Reynolds 1985; Terasaki and Sardet 1991). In sea urchins and frogs, this reticulum is pronounced in the cortex and surrounds the cortical granules (FIGURE 4.21; Gardiner and Grey 1983; Luttmer and Longo 1985). The cortical granules are themselves tethered to the cell membrane by a series of integral membrane proteins that facilitate calcium-mediated exocytosis (Conner

(A)

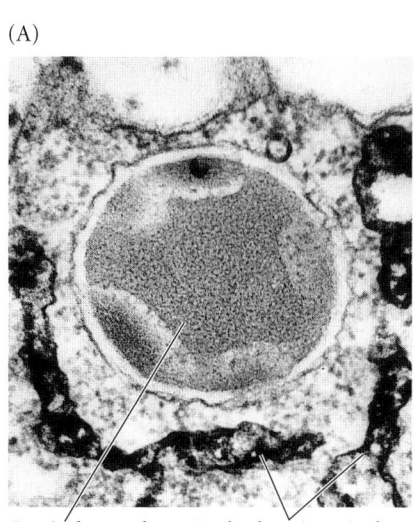

Cortical granule   Endoplasmic reticulum

(B)

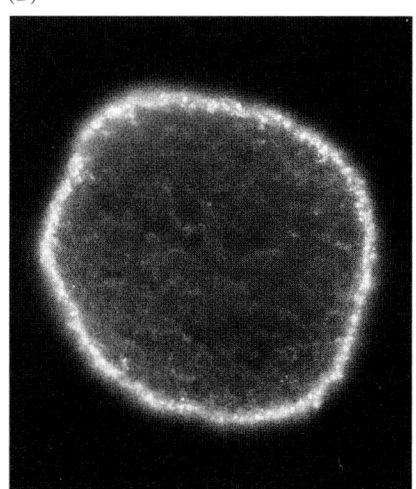

FIGURE 4.21   Endoplasmic reticulum surrounding cortical granules in sea urchin eggs. (A) The endoplasmic reticulum has been stained to allow visualization by transmission electron microscopy. The cortical granule is seen to be surrounded by dark-stained endoplasmic reticulum. (B) An entire egg stained with fluorescent antibodies to calcium-dependent calcium release channels. The antibodies show these channels in the cortical endoplasmic reticulum. (A from Luttmer and Longo 1985, courtesy of S. Luttmer; B from McPherson et al. 1992, courtesy of F. J. Longo.)

et al. 1997; Conner and Wessel 1998). Thus, as soon as Ca^{2+} is released from the endoplasmic reticulum, the cortical granules fuse with the cell membrane above them. Once initiated, the release of calcium is self-propagating. Free calcium is able to release sequestered calcium from its storage sites, thus causing a wave of Ca^{2+} release and cortical granule exocytosis.

Activation of Egg Metabolism in Sea Urchins

Although fertilization is often depicted as nothing more than the means to merge two haploid nuclei, it has an equally important role in initiating the processes that begin development. These events happen in the cytoplasm and occur without the involvement of the parental nuclei.* In addition to initiating the slow block to polyspermy (through cortical granule exocytosis), the release of Ca^{2+} that occurs when the sperm enters the egg is critical for activating the egg's metabolism and initiating development. Calcium ions release the inhibitors from maternally stored messages, allowing these mRNAs to be translated; they also release the inhibition of nuclear division, thereby allowing cleavage to occur. Indeed,

---

*In certain salamanders, this developmental function of fertilization has been totally divorced from the genetic function. The silver salamander (*Ambystoma platineum*) is a hybrid subspecies consisting solely of females. Each female produces an egg with an unreduced chromosome number. This egg, however, cannot develop on its own, so the silver salamander mates with a male Jefferson salamander (*A. jeffersonianum*). The sperm from the male Jefferson salamander only stimulates the egg's development; it does not contribute genetic material (Uzzell 1964). For details of this complex mechanism of procreation, see Bogart et al. 1989.

throughout the animal kingdom, calcium ions are used to activate development during fertilization.

Release of intracellular calcium ions

The way Ca^{2+} is released varies from species to species (see Parrington et al. 2007). One way, first proposed by Jacques Loeb (1899, 1902), is that a soluble factor from the sperm is introduced into the egg at the time of cell fusion, and this substance activates the egg by changing the ionic composition of the cytoplasm (**FIGURE 4.22A**). This mechanism, as we will see later, probably works in mammals. The other mechanism, proposed by Loeb's rival Frank Lillie (1913), is that the sperm binds to receptors on the egg cell surface and changes their conformation, thus initiating reactions within the cytoplasm that activate the egg (**FIGURE 4.22B**). This is probably what happens in sea urchins.

IP_3: THE RELEASER OF CA^{2+} If Ca^{2+} from the egg's endoplasmic reticulum is responsible for the cortical granule reaction and the reactivation of development, what releases Ca^{2+}? Throughout the animal kingdom, it has been found that **inositol 1,4,5-trisphosphate (IP_3)** is the primary mechanism for releasing Ca^{2+} from intracellular storage.

The IP_3 pathway is shown in **FIGURE 4.23**. The membrane phospholipid **phosphatidylinositol 4,5-bisphosphate (PIP_2)** is split by the enzyme **phospholipase C (PLC)** to yield two active compounds: IP_3 and **diacylglycerol (DAG)**. IP_3 is able to release Ca^{2+} into the cytoplasm by opening the calcium channels of the endoplasmic reticulum. DAG activates protein kinase C, which in turn activates a protein that exchanges sodium ions for hydrogen ions, raising the pH of the egg (Nishizuka 1986; Swann and Whitaker 1986). This Na^+-H^+ exchange pump also requires Ca^{2+}. The result of PLC activation, therefore, is the liberation of Ca^{2+} and the alkalinization

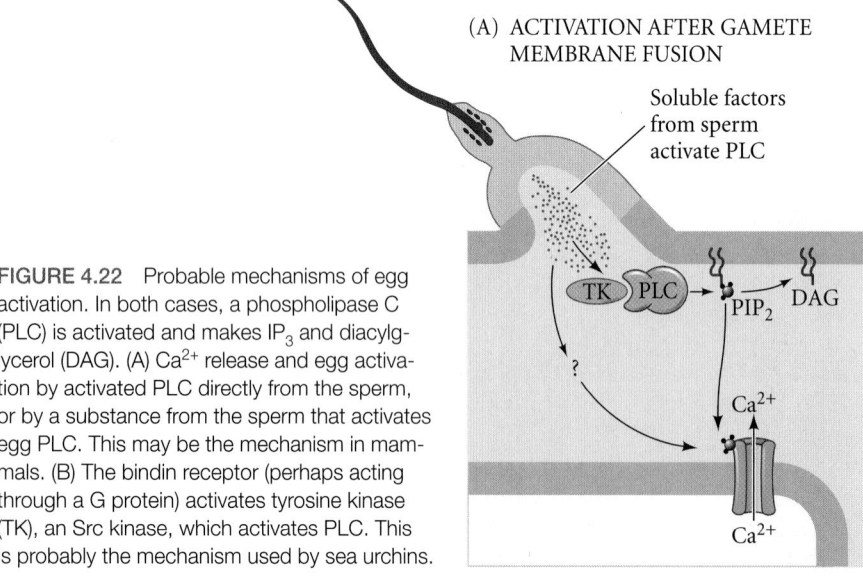

(A) ACTIVATION AFTER GAMETE MEMBRANE FUSION

Soluble factors from sperm activate PLC

TK — PLC → PIP₂ → DAG

?

Ca²⁺

Ca²⁺

FIGURE 4.22 Probable mechanisms of egg activation. In both cases, a phospholipase C (PLC) is activated and makes IP_3 and diacylglycerol (DAG). (A) Ca^{2+} release and egg activation by activated PLC directly from the sperm, or by a substance from the sperm that activates egg PLC. This may be the mechanism in mammals. (B) The bindin receptor (perhaps acting through a G protein) activates tyrosine kinase (TK), an Src kinase, which activates PLC. This is probably the mechanism used by sea urchins.

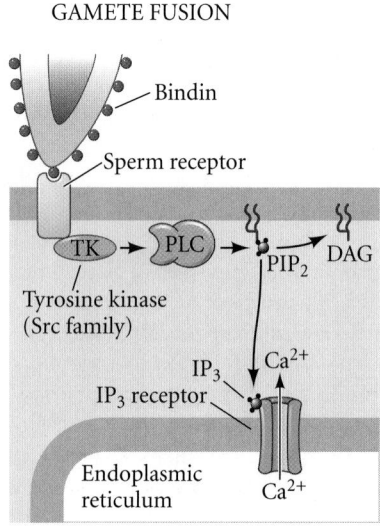

(B) ACTIVATION PRIOR TO GAMETE FUSION

Bindin

Sperm receptor

TK → PLC → PIP₂ → DAG

Tyrosine kinase (Src family)

IP_3

IP_3 receptor

Ca²⁺

Endoplasmic reticulum

Ca²⁺

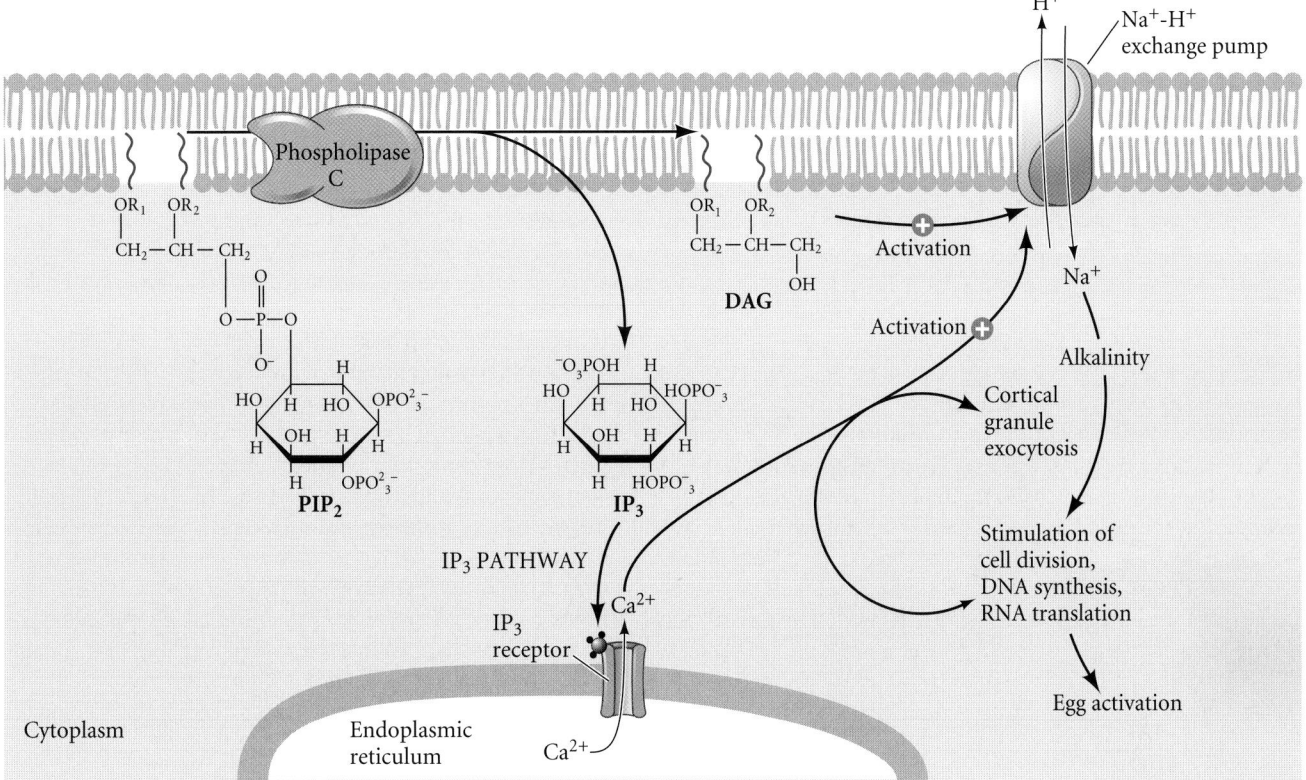

FIGURE 4.23 Roles of inositol phosphates in releasing calcium from the endoplasmic reticulum and the initiation of development. Phospholipase C splits PIP$_2$ into IP$_3$ and DAG. IP$_3$ releases calcium from the endoplasmic reticulum, and DAG, with assistance from the released Ca^{2+}, activates the sodium-hydrogen exchange pump in the membrane.

of the egg, and both of the compounds this activation creates—IP$_3$ and DAG—are involved in the initiation of development.

In sea urchin eggs, IP$_3$ is formed initially at the site of sperm entry and can be detected within seconds of sperm-egg attachment. The inhibition of IP$_3$ synthesis prevents Ca^{2+} release (Lee and Shen 1998; Carroll et al. 2000), whereas injected IP$_3$ can release sequestered Ca^{2+}, leading to cortical granule exocytosis (Whitaker and Irvine 1984; Busa et al. 1985). Moreover, these IP$_3$-mediated effects can be thwarted by preinjecting the egg with calcium-chelating agents (Turner et al. 1986).

IP$_3$-responsive calcium channels have been found in the egg endoplasmic reticulum. The IP$_3$ formed at the site of sperm entry is thought to bind to IP$_3$ receptors in these calcium channels, effecting a local release of Ca^{2+} (Ferris et al. 1989; Furuichi et al. 1989). Once released, Ca^{2+} can diffuse directly, or it can facilitate the release of more Ca^{2+} by binding to *calcium-triggered calcium-release receptors*, also located in the cortical endoplasmic reticulum (McPherson et al. 1992). These receptors release stored Ca^{2+} when they bind Ca^{2+}, so binding Ca^{2+} releases more Ca^{2+}, which binds to more receptors, and so on. The resulting wave of calcium release is propagated throughout the cell, starting at the point of sperm

entry (see Figure 4.20). The cortical granules, which fuse with the cell membrane in the presence of high Ca^{2+} concentrations, respond with a wave of exocytosis that follows the calcium wave. Mohri and colleagues (1995) have shown that IP$_3$-released Ca^{2+} is both necessary and sufficient for initiating the wave of calcium release.

PHOSPHOLIPASE C: THE GENERATOR OF IP$_3$ If IP$_3$ is necessary for Ca^{2+} release and phospholipase C is required in order to generate IP$_3$, the question then becomes, What activates PLC? This question has not been easy to address since (1) there are numerous types of PLC that (2) can be activated through different pathways, and (3) different species use different mechanisms to activate PLC. Results from studies of sea urchin eggs suggest that the active PLC in echinoderms is a member of the γ (gamma) family of PLCs (Carroll et al. 1997, 1999; Shearer et al. 1999). Inhibitors that specifically block PLCγ inhibit IP$_3$ production as well as Ca^{2+} release. Moreover, these inhibitors can be circumvented by microinjecting IP$_3$ into the egg.

SRC KINASES: A LINK BETWEEN SPERM AND PLCγ The finding that the γ class of PLCs was responsible for generating IP$_3$ during echinoderm fertilization spurred investigators to look at exactly which proteins activate this particular class of phospholipases. Their work soon came to focus on the **Src family** of protein kinases. Src proteins are found in the cortical cytoplasm of sea urchin and starfish eggs, where they can

SIDELIGHTS & SPECULATIONS

Rules of Evidence: "Find It, Lose It, Move It"

Biology, like all science, does not deal with Facts; rather, it deals with evidence. Several types of evidence are presented in this book, and they are not equivalent in strength. As an example, we use here the analysis of the role of Ca^{2+} in egg activation.

Correlative evidence

The first, and weakest, type of evidence is **correlative evidence**. Here, we find correlations between two or more events and then make the inference that one event causes the other. For example, upon the meeting of sea urchin sperm and egg, a wave of free Ca^{2+} spreads across the egg (see Figure 4.20), and this wave of Ca^{2+} is thought to activate the egg. This chain of events has been shown in several ways, most convincingly by aequorin fluorescence (Steinhardt et al. 1977; Shimomura 1995; Steinhardt 2006).

Although one might infer that the meeting of egg and sperm caused the Ca^{2+} wave, and that this Ca^{2+} wave caused egg activation, such a correlation of events with one another does not necessarily demonstrate a causal relationship. It is possible the meeting of gametes first caused the flow of Ca^{2+} across the egg and then, separately and by some other mechanism, activated the egg. It is also conceivable that some aspect of egg activation caused the Ca^{2+} release. The correlated occurrences of these events could even be coincidental and have no relationship to one another.*

*In a tongue-in-cheek letter spoofing such correlative evidence, Sies (1988) demonstrated a remarkably good correlation between the number of storks seen in West Germany from 1965 to 1980 and the number of babies born during those same years. Any cause-and-effect scenario between storks and babies, however, would certainly fly in the face of the evidence presented in this chapter.

Correlative evidence provides a starting point for many investigations, but one cannot say that one event causes another based solely on correlation.

Functional evidence

The next type of evidence is called **loss-of-function evidence**, also known as **negative inference evidence**. Here, the absence of the postulated cause is associated with the absence of the effect. While stronger than correlative evidence, loss-of-function evidence still does not exclude other explanations. For instance, when calcium chelators such as EDTA were injected into the egg prior to fertilization, released Ca^{2+} failed to activate the egg. This would imply that Ca^{2+} is necessary for egg activation. However, data from such inhibitory studies (including studies from loss-of-function mutations) always leave open the possibility that the inhibitor suppresses more than just the process being studied. For instance, when protease inhibitors caused the failure of mammalian fertilization, it was assumed that these inhibitors were blocking the action of proteases released from the acrosome. As a result, biologists thought that the mammalian acrosome releases soluble proteases that digest the zona. Later experiments, however, demonstrated that the protease inhibitors inhibited the acrosome reaction itself so that the proteases were never released (Llanos et al. 1993). Thus, one couldn't tell whether soluble proteases played any role in mammalian fertilization.

The strongest type of evidence is **gain-of-function evidence**. Here the initiation of the first event causes the second event to happen even in instances where or when neither event usually occurs. Thus, when calcium ionophores (which can shuttle Ca^{2+} across membranes) were added to

unfertilized eggs, Ca^{2+} was released from intracellular storage and the eggs became activated even without fertilization (Steinhardt and Epel 1974).

Progression of evidence and coherence

Correlative ("find it"), loss-of-function ("lose it"), and gain-of-function ("move it") evidence must consistently support each other to establish and solidify a conclusion. This progression of "find it, lose it, move it" evidence is at the core of nearly all studies of developmental mechanism (Adams 2000). Sometimes it can be found in a single paper, and sometimes, as the case above illustrates, the evidence comes from many laboratories. "Every scientist," writes Fleck (1979), "knows just how little a single experiment can prove or convince." Rather, "an entire system of experiments and controls is needed." Science is a communal endeavor, and it is doubtful that any great discovery is the achievement of a single experiment, or of any individual.

Science also accepts evidence better when it fits into a system of other findings. This is often called **coherence**. For instance, the ability of calcium to activate the egg became a standard part of fertilization physiology when Ca^{2+} was shown to cause both the resumption of cell division and the initiation of translation—two separate components of egg activation. Also, once the sperm was found to activate phospholipase C—the enzyme that synthesizes IP_3—and IP_3 was found to activate intracellular calcium release in numerous cells, the release of Ca^{2+} became understood as being the central element of sea urchin egg activation. It fit into a much wider picture of physiological calcium release, and the mechanisms for its synthesis and its effects all fit together.

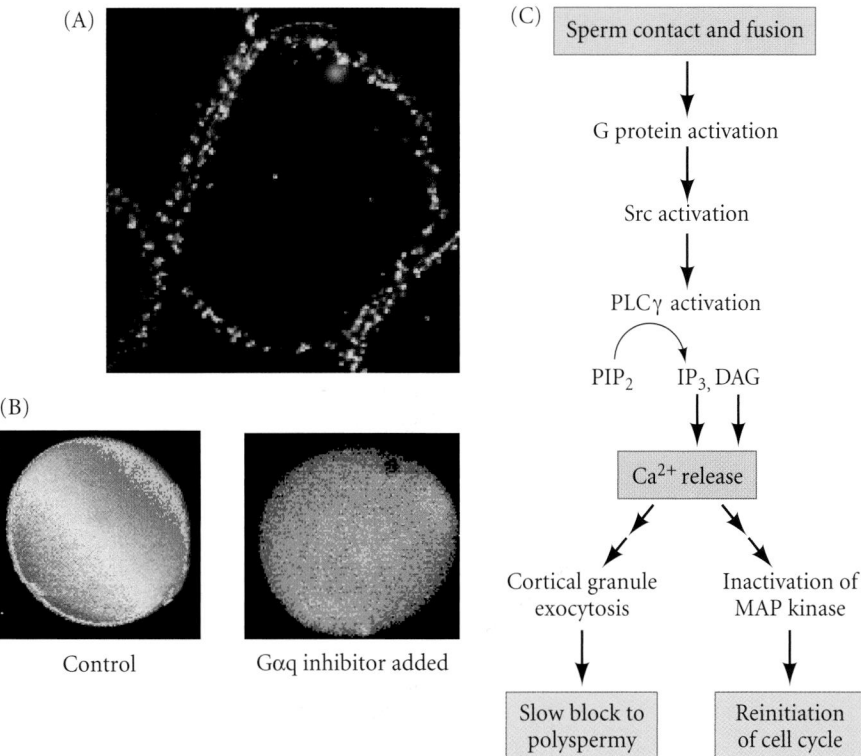

(A)

(B)

Control   Gαq inhibitor added

(C)

Sperm contact and fusion

↓

G protein activation

↓

Src activation

↓

PLCγ activation

PIP₂   IP₃, DAG

Ca²⁺ release

Cortical granule exocytosis   Inactivation of MAP kinase

Slow block to polyspermy   Reinitiation of cell cycle

FIGURE 4.24 G protein involvement in Ca^{2+} entry into sea urchin eggs. (A) Mature sea urchin egg immunologically labeled for the cortical granule protein hyaline (red) and the G protein Gαq (green). (The overlap of signals produces the yellow color.) Gαq is localized to the cortex. (B) A wave of Ca^{2+} appears in the control egg (computer-enhanced to show relative intensities, with red being the highest), but not in the egg injected with an inhibitor of the Gαq protein. (C) Possible model for egg activation by the influx of Ca^{2+}. (After Voronina and Wessel 2003; photographs courtesy of G. M. Wessel.)

form a complex with PLCγ. Inhibition of Src kinases lowered and delayed the amount of Ca^{2+} released (Kinsey and Shen 2000; Giusti et al. 2003; Townley et al. 2009).

So what activates Src kinase activity? One possibility is heterotrimeric G proteins in the cortex of the egg (**FIGURE 4.24**). Such G proteins are known to activate Src kinases in mammalian somatic cells, so the cortical G proteins of sea urchin eggs seem like good candidates, and blocking these G proteins prevented Ca^{2+} release (Voronina and Wessel 2003, 2004). It is also possible that these G proteins activate PLC directly. Indeed, there may be more than one pathway and more than one way to activate Ca^{2+} release.

Thus, in sea urchins, it is thought that the binding of sperm to the egg (or possibly the fusion of sperm and egg) activates PLCγ through G proteins and Src kinases. The IP₃ thus generated opens calcium channels in the nearby cortical endoplasmic reticulum, allowing the initial and local outflow of Ca^{2+}. This first efflux of ions opens calcium-gated calcium release channels, causing a wave of Ca^{2+} that flows across the egg from the point of sperm entry to the opposite side of the egg. In so doing, some of the Ca^{2+} initiates cortical granule exocytosis, fusing the cortical granule with the egg cell membrane. Other Ca^{2+} would be bound by proteins such as calmodulin, which is activated by Ca^{2+} and can regulate numerous functions.

NAADP It is possible that diffusible compounds from the sperm also release calcium directly to the endoplasmic

reticulum. There is evidence that *nicotinic acid adenine dinucleotide phosphate* (NAADP), a linear dinucleotide derived from NADP, serves as a sperm-borne Ca^{2+} releaser. NAADP frees stored Ca^{2+} from membrane vesicles during muscle contraction, insulin secretion, and neurotransmitter release (Lee 2001). Upon contact with egg jelly, NAADP concentration in sea uchin sperm increases tenfold, reaching levels that appear to be more than sufficient to release stored Ca^{2+} within the egg (Churchill et al. 2003; Morgan and Galione 2007). Thus, among sea urchins, two pathways may have evolved for the release of stored calcium.

Effects of calcium release

The flux of calcium across the egg activates a preprogrammed set of metabolic events. The responses of the sea urchin egg to the sperm can be divided into "early" responses, which occur within seconds of the cortical granule reaction, and "late" responses, which take place several minutes after fertilization begins (**TABLE 4.1**).

EARLY RESPONSES As we have seen, contact or fusion of a sea urchin sperm and egg activates two major blocks to polyspermy: the fast block, initiated by sodium influx into the cell; and the cortical granule reaction, or slow block, initiated by the intracellular release of Ca^{2+}. The same release of Ca^{2+} responsible for the cortical granule reaction is also responsible for the re-entry of the egg into the cell cycle

TABLE 4.1 Events of sea urchin fertilization

Event	Approximate time postinsemination[a]
EARLY RESPONSES	
Sperm-egg binding	0 sec
Fertilization potential rise (fast block to polyspermy)	within 1 sec
Sperm–egg membrane fusion	within 1 sec
Calcium increase first detected	10 sec
Cortical granule exocytosis (slow block to polyspermy)	15–60 sec
LATE RESPONSES	
Activation of NAD kinase	starts at 1 min
Increase in NADP$^+$ and NADPH	starts at 1 min
Increase in O_2 consumption	starts at 1 min
Sperm entry	1–2 min
Acid efflux	1–5 min
Increase in pH (remains high)	1–5 min
Sperm chromatin decondensation	2–12 min
Sperm nucleus migration to egg center	2–12 min
Egg nucleus migration to sperm nucleus	5–10 min
Activation of protein synthesis	starts at 5–10 min
Activation of amino acid transport	starts at 5–10 min
Initiation of DNA synthesis	20–40 min
Mitosis	60–80 min
First cleavage	85–95 min

Main sources: Whitaker and Steinhardt 1985; Mohri et al. 1995.

[a]Approximate times based on data from *S. purpuratus* (15–17°C), *L. pictus* (16–18°C), *A. punctulata* (18–20°C), and *L. variegatus* (22–24°C). The timing of events within the first minute is best known for *L. variegatus*, so times are listed for that species.

and the reactivation of egg protein synthesis. Ca^{2+} levels in the egg increase from 0.05 to between 1 and 5 μ*M*, and in almost all species this occurs as a wave or succession of waves that sweep across the egg beginning at the site of sperm-egg fusion (Jaffe 1983; Terasaki and Sardet 1991; Stricker 1999; see Figure 4.20).

Calcium release activates a series of metabolic reactions that initiate embryonic development (**FIGURE 4.25**). One of these is the activation of the enzyme NAD$^+$ kinase, which converts NAD$^+$ to NADP$^+$ (Epel et al. 1981). Since NADP$^+$ (but not NAD$^+$) can be used as a coenzyme for lipid biosynthesis, such a conversion has important consequences for lipid metabolism and thus may be important in the construction of the many new cell membranes required during cleavage. NADP$^+$ is also used to make NAADP, which appears to boost calcium release even further (see the preceding page).

Calcium release also affects oxygen consumption. A burst of oxygen reduction (to hydrogen peroxide) is seen during fertilization, and much of this "respiratory burst" is used to crosslink the fertilization envelope (see Figure 4.19). Udx1, the enzyme responsible for this reduction of oxygen, is also NADPH-dependent (Heinecke and Shapiro 1989; Wong et al. 2004). Lastly, NADPH helps regenerate glutathione and ovothiols, molecules that may be crucial scavengers of free radicals that could otherwise damage the DNA of the egg and early embryo (Mead and Epel 1995).

LATE RESPONSES: RESUMPTION OF PROTEIN AND DNA SYNTHESIS The late responses of fertilization include the activation of a new burst of DNA and protein synthesis. In sea urchins, the fusion of egg and sperm causes the intracellular pH to increase. This rise in intracellular pH begins with a second influx of sodium ions (Na$^+$) from seawater, which causes a 1:1 exchange between the sodium ions and hydrogen ions (H$^+$) from inside the egg. This loss of H$^+$ causes the pH within the egg to rise (Shen and Steinhardt 1978; Michael and Walt 1999).

It is thought that pH increase and Ca^{2+} elevation act together to stimulate new DNA and protein synthesis (Winkler et al. 1980; Whitaker and Steinhardt 1982; Rees et al. 1995). If one experimentally elevates the pH of an unfertilized egg to a level similar to that of a fertilized egg, DNA synthesis and nuclear envelope breakdown ensue, just as if the egg were fertilized (Miller and Epel 1999). Calcium ions are also critical to new DNA synthesis. The wave of free Ca^{2+} inactivates the enzyme MAP kinase, converting it from a phosphorylated (active) to an unphosphorylated (inactive) form, thus removing an inhibition on DNA synthesis (Carroll et al. 2000). DNA synthesis can then resume.

In sea urchins, a burst of protein synthesis usually occurs within several minutes after sperm entry. This protein synthesis does not depend on the synthesis of new messenger RNA; rather, it uses mRNAs already present in the oocyte cytoplasm (**FIGURE 4.26**). These mRNAs encode proteins such as histones, tubulins, actins, and morphogenetic factors that are used during early development. Such a burst of protein synthesis can be induced by artificially raising the pH of the cytoplasm using ammonium ions (Winkler et al. 1980).

One mechanism for this global rise in the translation of messages stored in the oocyte appears to be the release of inhibitors from the mRNA. In Chapter 2 we discussed maskin, an inhibitor of translation in the unfertilized amphibian oocyte. In sea urchins, a similar inhibitor binds translation initiation factor eIF4E at the 5′ end of several mRNAs and prevents these mRNAs from being translated. Upon fertilization, however, this inhibitor—the eIF4E-binding protein—becomes phosphorylated and is degraded, thus allowing eIF4E to complex with other translation factors and permit protein synthesis from the stored sea urchin mRNAs (Cormier et al. 2001;

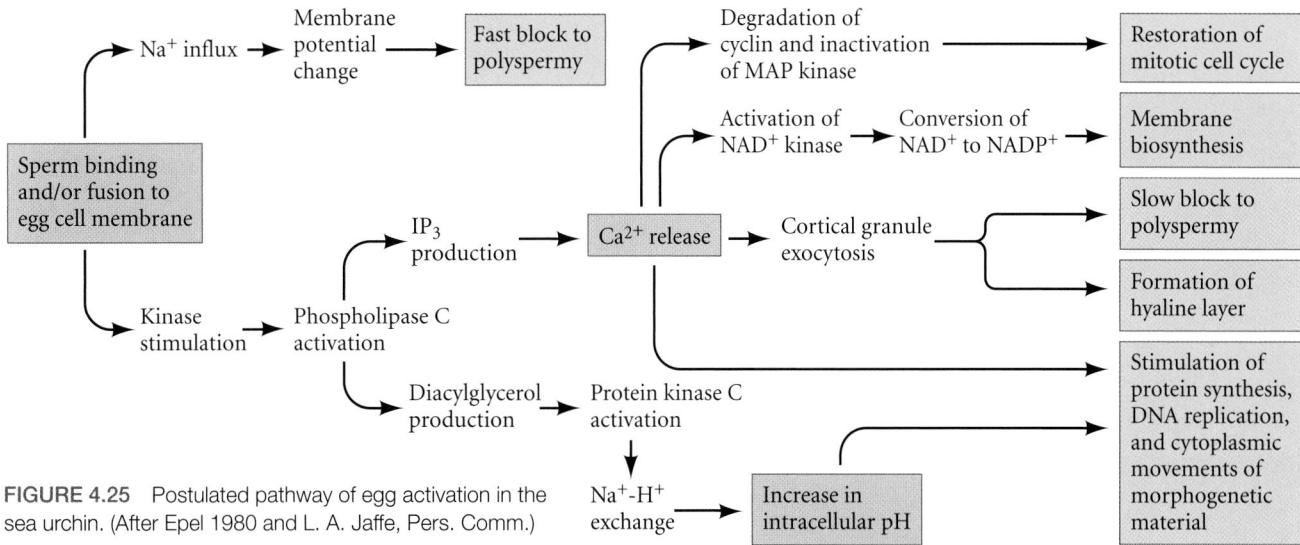

FIGURE 4.25   Postulated pathway of egg activation in the sea urchin. (After Epel 1980 and L. A. Jaffe, Pers. Comm.)

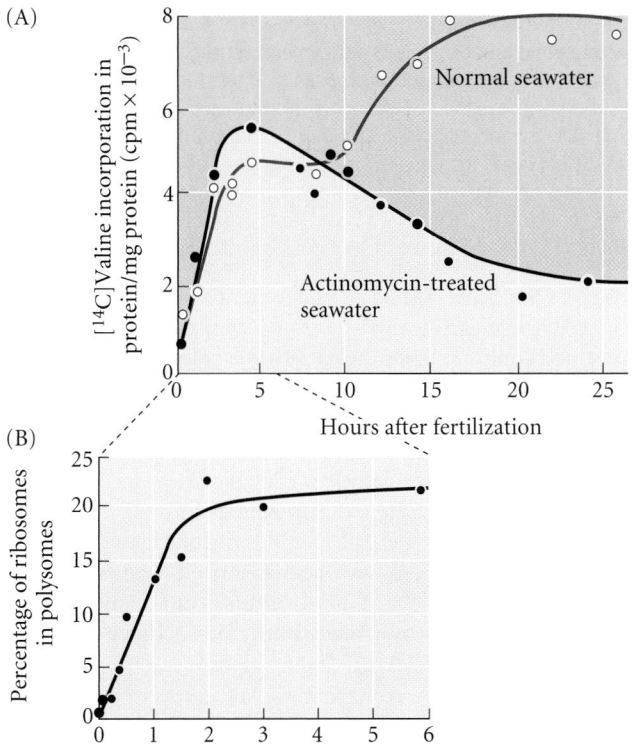

FIGURE 4.26   A burst of protein synthesis at fertilization uses mRNAs stored in the oocyte cytoplasm. (A) Protein synthesis in embryos of the sea urchin *Arbacia punctulata* fertilized in the presence or absence of actinomycin D, an inhibitor of transcription. For the first few hours, protein synthesis occurs with very little new transcription from the zygote or embryo nuclei. A second burst of protein synthesis occurs during the mid-blastula stage, at about 12 hours after fertilization. This burst represents translation of newly transcribed messages and therefore is not seen in embryos growing in actinomycin. (B) The percentage of ribosomes recruited into polysomes increases during the first hours of sea urchin development, especially during the first cell cycle. (A after Gross et al. 1964; B after Humphreys 1971.)

Oulhen et al. 2007). One of the mRNAs "freed" by the degradation of eIF4E-binding protein is the message encoding cyclin B (Salaun et al. 2003, 2004). The cyclin B protein combines with Cdk1 cyclin to create **mitosis-promoting factor** (**MPF**), which is required to initiate cell division.

Fusion of Genetic Material in Sea Urchins

After the sperm and egg cell membranes fuse, the sperm nucleus and its centriole separate from the mitochondria and flagellum. The mitochondria and the flagellum disintegrate inside the egg, so very few, if any, sperm-derived mitochondria are found in developing or adult organisms. Thus, although each gamete contributes a haploid genome to the zygote, the *mitochondrial* genome is transmitted primarily by the maternal parent. Conversely, in almost all animals studied (the mouse being the major exception), the centrosome needed to produce the mitotic spindle of the subsequent divisions is derived from the sperm centriole (see Figure 4.16; Sluder et al. 1989, 1993).

Fertilization in sea urchin eggs occurs after the second meiotic division, so there is already a haploid female pronucleus present when the sperm enters the egg cytoplasm. Once inside the egg, the sperm nucleus undergoes a dramatic transformation as it decondenses to form the haploid male pronucleus. First, the nuclear envelope vesiculates into small packets, exposing the compact sperm chromatin to the egg cytoplasm (Longo and Kunkle 1978; Poccia and Collas 1997). Then proteins holding the sperm chromatin in its condensed, inactive state are exchanged for other proteins derived from the egg cytoplasm. This exchange permits the decondensation of the sperm chromatin. Once decondensed, the DNA adheres to the nuclear envelope, where DNA polymerase can initiate replication (Infante et al. 1973).

Sperm chromosome decondensation appears to be initiated by the phosphorylation of the nuclear envelope lamin protein and the phosphorylation of two sperm-specific histones that bind tightly to the DNA. The process begins when

(A)

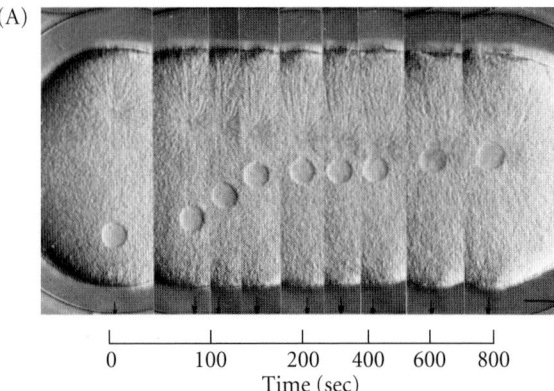

```
|_____|_____|_____|____|____|____|
0    100   200   400  600  800
          Time (sec)
```

(B)

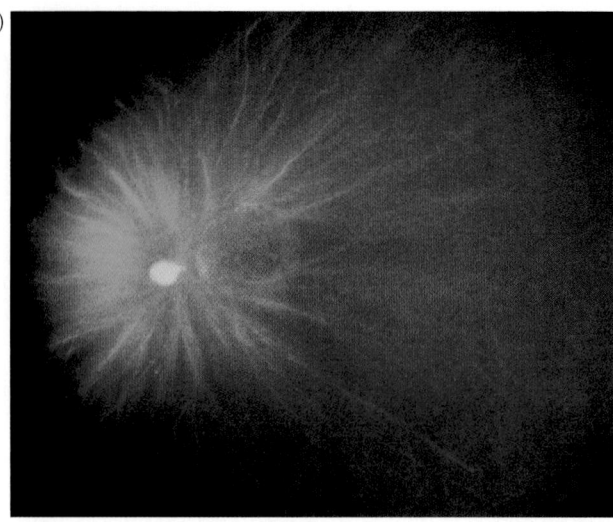

(C)

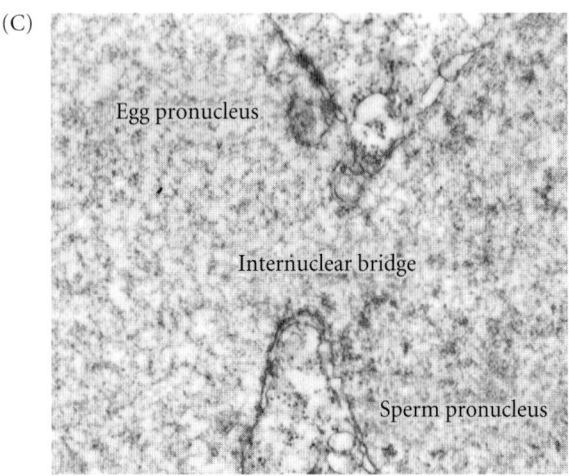

Egg pronucleus

Internuclear bridge

Sperm pronucleus

FIGURE 4.27 Nuclear events in the fertilization of the sea urchin. (A) Sequential photographs showing the migration of the egg pronucleus and the sperm pronucleus toward each other in an egg of *Clypeaster japonicus*. The sperm pronucleus is surrounded by its aster of microtubules. (B) The two pronuclei migrate toward each other on these microtubular processes. (The pronuclear DNA is stained blue by Hoechst dye.) The microtubules (stained green with fluorescent antibodies to tubulin) radiate from the centrosome associated with the (smaller) male pronucleus and reach toward the female pronucleus. (C) Fusion of pronuclei in the sea urchin egg. (A from Hamaguchi and Hiramoto 1980, courtesy of the authors; B from Holy and Schatten 1991, courtesy of J. Holy; C courtesy of F. J. Longo.)

sperm comes into contact with a certain glycoprotein in the egg jelly that elevates the level of cAMP-dependent protein kinase activity. These protein kinases phosphorylate several of the basic residues of the sperm-specific histones and thereby interfere with their binding to DNA (Garbers et al. 1980; Porter and Vacquier 1986; Stephens et al. 2002). This loosening is thought to facilitate the replacement of the sperm-specific histones with other histones that have been stored in the oocyte cytoplasm (Green and Poccia 1985).

But how do the sperm and egg pronuclei find each other? After the sea urchin sperm enters the egg cytoplasm, the male pronucleus separates from the tail and rotates 180 degrees so that the sperm centriole is between the sperm pronucleus and the egg pronucleus. The sperm centriole then acts as a microtubule organizing center, extending its own microtubules and integrating them with egg microtubules to form an aster.* Microtubules extend throughout the egg and contact the female pronucleus, at which point the two

pronuclei migrate toward each other. Their fusion forms the diploid zygote nucleus (**FIGURE 4.27**). DNA synthesis can begin either in the pronuclear stage or after the formation of the zygote nucleus, and depends on the level of Ca^{2+} released earlier in fertilization (Jaffe et al. 2001).

At this point, the diploid nucleus has formed. DNA synthesis and protein synthesis have commenced, and the inhibitions to cell division have been removed. The sea urchin can now begin to form a multicellular organism. We will describe the means by which sea urchins achieve multicellularity in Chapter 7.

Internal Fertilization in Mammals

It is very difficult to study any interactions between the mammalian sperm and egg that take place prior to these gametes making contact. One obvious reason for this is that mammalian fertilization occurs inside the oviducts of the female: although it is relatively easy to mimic the conditions surrounding sea urchin fertilization using natural or artificial seawater, we do not yet know the components of the various natural environments that mammalian sperm encounter as they travel to the egg.

A second reason why it is difficult to study mammalian fertilization is that the sperm population ejaculated into the female is probably heterogeneous, containing spermatozoa

---

*When Oscar Hertwig observed this radial array of sperm asters forming in his newly fertilized sea urchin eggs, he called it "the sun in the egg" and thought it was the happy indication of a successful fertilization (Hertwig 1877). More recently, Simerly and co-workers (1999) found that certain types of human male infertility are due to defects in the centriole's ability to form these microtubular asters. This deficiency results in the failure of pronuclear migration and the cessation of further development.

at different stages of maturation. Out of the 280×10^6 human sperm normally ejaculated during coitus, only about 200 reach the vicinity of the egg (Ralt et al. 1991). Thus, since fewer than 1 in 10,000 sperm even gets close to the egg, it is difficult to assay those molecules that might enable the sperm to swim toward the egg and become activated.

A third reason why it has been difficult to elucidate the details of mammalian fertilization is the recent discovery (see Clark 2011 and below) that there may be multiple mechanisms by which mammalian sperm can undergo the acrosome reaction and bind to the zona pellucida.

Getting the gametes into the oviduct: Translocation and capacitation

The female reproductive tract is not a passive conduit through which sperm race, but a highly specialized set of tissues that actively regulate the transport and maturity of both gametes. Both the male and female gametes use a combination of small-scale biochemical interactions and large-scale physical propulsion to get to the **ampulla**, the region of the oviduct where fertilization takes place.

TRANSLOCATION   A mammalian oocyte just released from the ovary is surrounded by a matrix containing cumulus cells. (Cumulus cells are the cells of the ovarian follicle to which the developing oocyte was attached; see Figures 4.7 and 17.31). If this matrix is experimentally removed or significantly altered, the fimbriae of the oviduct will not "pick up" the oocyte-cumulus complex (see Figure 9.16), nor will the complex be able to enter the oviduct (Talbot et al. 1999). Once it is picked up, a combination of ciliary beating and muscle contractions transport the oocyte-cumulus complex to the appropriate position for its fertilization in the oviduct.

The translocation of sperm from the vagina to the oviduct involves many processes that work at different times and places. Sperm motility (i.e., flagellar action) is probably a minor factor in getting the sperm into the oviduct, although such motility is required for mouse sperm to travel through the cervical mucus, and for sperm to encounter the egg once they are in the oviduct. Sperm are found in the oviducts of mice, hamsters, guinea pigs, cows, and humans within 30 minutes of sperm deposition in the vagina—a time "too short to have been attained by even the most Olympian sperm relying on their own flagellar power" (Storey 1995). Rather, sperm appear to be transported to the oviduct by the muscular activity of the uterus. Recent studies of sperm motility have led to several conclusions, including the following:

1. Uterine muscle contractions are critical in getting the sperm into the oviduct.

2. The region of the oviduct before the ampulla may slow down sperm and release them slowly.

3. Sperm (flagellar) motility is important once sperm arrive within the oviduct; sperm become hyperactive in the vicinity of the oocyte.

4. Sperm may receive directional cues from temperature gradients between the regions of the oviduct and from chemical cues derived from the oocyte or cumulus.

5. During this trek from the vagina to the ampullary region of the oviduct, the sperm matures such that it has the capacity to fertilize the egg when the two finally meet.

CAPACITATION   Newly ejaculated mammalian sperm are unable to undergo the acrosome reaction or fertilize an egg until they have resided for some time in the female reproductive tract (Chang 1951; Austin 1952). The set of physiological changes by which sperm become competent to fertilize the egg is called **capacitation**. Sperm that are not capacitated are "held up" in the cumulus matrix and are unable to reach the egg (Austin 1960; Corselli and Talbot 1987). Capacitation can be accomplished in vitro by incubating sperm in a tissue culture medium (such media contain calcium ions, bicarbonate, and serum albumin) or in fluid taken from the oviducts.

Contrary to popular belief, "the race is not always to the swiftest." Wilcox and colleagues (1995) found that nearly all human pregnancies result from sexual intercourse during a 6-day period ending on the day of ovulation. This means that the fertilizing sperm could have taken as long as 6 days to make the journey to the oviduct. Although some human sperm reach the ampulla of the oviduct within a half-hour after intercourse, "speedy" sperm may have little chance of fertilizing the egg because they have not undergone capacitation.

Eisenbach (1995) has proposed a hypothesis wherein capacitation is a transient event, and sperm are given a relatively brief window of competence during which they can successfully fertilize the egg. As the sperm reach the ampulla, they acquire competence—but they lose it if they stay around too long. By binding and capacitating sperm, the oviduct releases "packets" of capacitated sperm at various intervals, thereby prolonging the time that fertilization can be successful.

The molecular processes of capacitation prepare the sperm for the acrosome reaction and enable the motility of the sperm to become hyperactive (**FIGURE 4.28**). Capacitation is a suite of sequential events (occurring at different time periods) and spatially regulated events (some of them occurring in the tail and others in the head). Although the details of these processes still await description (they are notoriously difficult to study), five sets of molecular changes are considered to be important:

1. The sperm cell membrane is altered by the removal of cholesterol by albumin proteins in the female reproductive tract (Cross 1998). The cholesterol efflux from the sperm cell membrane is thought to change the location of "lipid rafts," isolated regions of the cell membrane that often contain receptor proteins. Originally located throughout the sperm cell membrane, lipid rafts now cluster over the anterior sperm head. These lipid microdomains contain proteins that can bind the zona pellucida and participate in the acrosome reaction (Bou Khalil et al. 2006; Gadella et al. 2008).

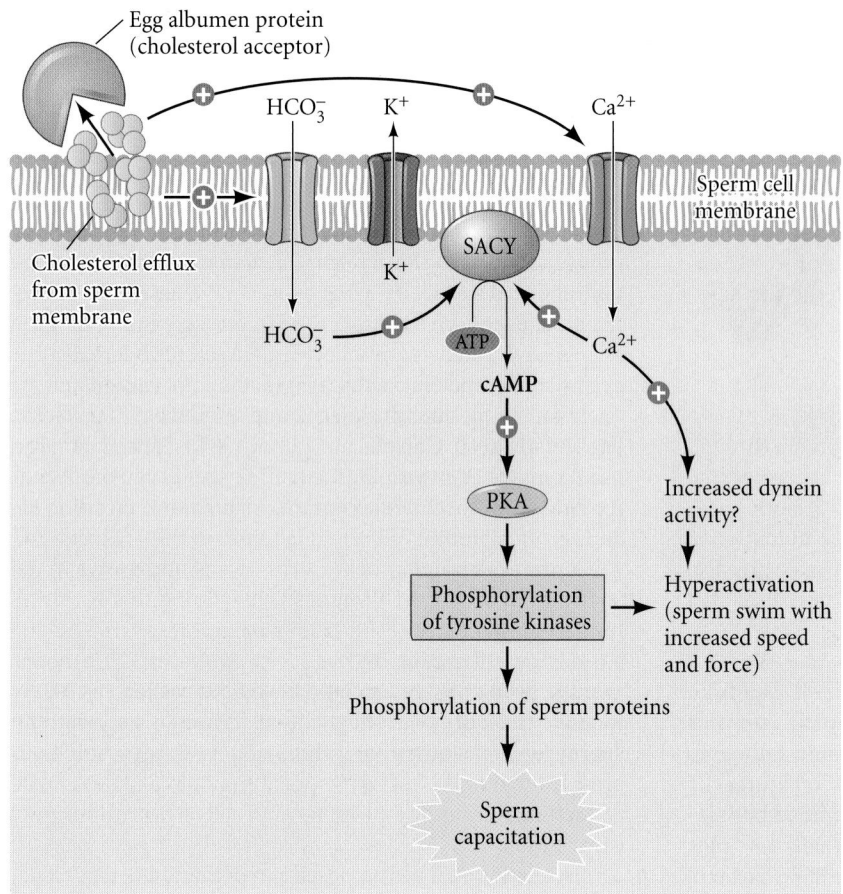

FIGURE 4.28 Hypothetical model for mammalian sperm capacitation. The pathway is modulated by the removal of cholesterol from the sperm cell membrane, which allows the influx of bicarbonate ions (HCO_3^-) and calcium ions (Ca^{2+}). These ions activate adenylate kinase (SACY), thereby elevating cAMP concentrations. The high cAMP levels then activate protein kinase A (PKA). Active PKA phosphorylates several tyrosine kinases, which in turn phosphorylate several sperm proteins, leading to capacitation. Increased intracellular Ca^{2+} also activates the phosphorylation of these proteins, as well as contributing to the hyperactivation of the sperm. (After Visconti et al. 2011.)

2. Particular proteins or carbohydrates on the sperm surface are lost during capacitation (Lopez et al. 1985; Wilson and Oliphant 1987). It is possible that these compounds block the recognition sites for the sperm proteins that bind to the zona pellucida. It has been suggested that the unmasking of these sites might be one of the effects of cholesterol depletion (Benoff 1993).

3. The membrane potential of the sperm cell membrane becomes more negative as potassium ions leave the sperm. This change in membrane potential may allow calcium channels to be opened and permit calcium to enter the sperm. Calcium and bicarbonate ions may be critical in activating cAMP production and in facilitating the membrane fusion events of the acrosome reaction (Visconti et al. 1995; Arnoult et al. 1999). The influx of bicarbonate ions and possibly other ions causes alkalinization of the sperm, raising its pH. This will be critical in the subsequent activation of calcium channels (Navarro et al. 2007).

4. Protein phosphorylation occurs (Galantino-Homer et al. 1997; Arcelay et al. 2008). In particular, two chaperone (heat-shock) proteins migrate to the surface of the sperm head when they are phosphorylated. One of these chaperone proteins is Izumo, which is critical in sperm-egg fusion (see below; Baker et al. 2010).

5. The outer acrosomal membrane changes and comes into contact with the sperm cell membrane in a way that prepares it for fusion (Tulsiani and Abou-Haila 2004).

There may be an important connection between sperm translocation and capacitation. Smith (1998) and Suarez (1998) have documented that before entering the ampulla of the oviduct, the uncapacitated sperm bind actively to the membranes of the oviduct cells in the narrow passage (the isthmus) preceding it (**FIGURE 4.29**; see also Figure 9.16). This binding is temporary and appears to be broken when the sperm become capacitated. Moreover, the life span of the sperm is significantly lengthened by this binding, and its capacitation is slowed down. This restriction of sperm entry into the ampulla, the slowing down of capacitation, and the expansion of sperm life span may have important consequences (Töpfer-Petersen et al. 2002; Gwathmey et al. 2003). The binding action may function as a block to polyspermy by preventing many sperm from reaching the egg at the same time (if the oviduct isthmus is excised in cows, a much higher rate of polyspermy results). In addition, slowing the rate of sperm capacitation and extending the active life of sperm may maximize the probability that sperm will still be available to meet the egg in the ampulla.

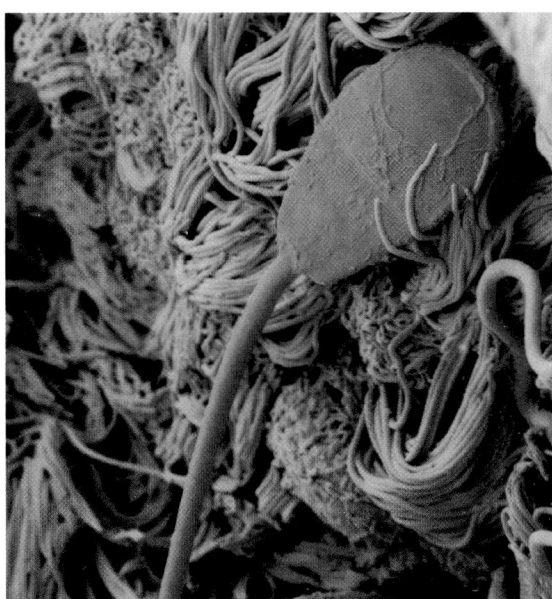

FIGURE 4.29 Scanning electron micrograph (artificially colored) showing bull sperm as it adheres to the membranes of epithelial cells in the oviduct of a cow prior to entering the ampulla. (From Lefebvre et al. 1995, courtesy of S. Suarez.)

In the vicinity of the oocyte: Hyperactivation, thermotaxis, and chemotaxis

Different regions of the female reproductive tract may secrete different, regionally specific molecules, and these molecules may influence sperm motility as well as capacitation. During capacitation, sperm become hyperactivated—they swim at higher velocities and generate greater force. This hyperactivation appears to be mediated through the opening of a sperm-specific calcium channel located in the sperm tail (Ren et al. 2001; Quill et al. 2003; see Figure 4.28). The symmetric beating of the flagellum is changed into a rapid asynchronous beat with a higher degree of bending. The power of the beat and the direction of sperm head movement are thought to release the sperm from their binding with the oviduct epithelial cells. Indeed, only hyperactivated sperm are seen to detach and continue their journey to the egg (Suarez 2008a,b).

Hyperactivation may enable sperm to sense a chemotactic gradient (Armon and Eisenbach 2011). In addition, Suarez and co-workers (1991) have shown that although this behavior is not conducive to travel through low-viscosity fluids, it appears to be extremely well suited for linear sperm movement in the viscous fluid that sperm might encounter in the oviduct. Hyperactivation, along with a hyaluronidase enzyme on the outside of the sperm cell membrane, enables the sperm to digest a path through the extracellular matrix of the cumulus cells (Lin et al. 1994; Kimura et al. 2009).

An old joke claims that the reason a man has to release so many sperm at each ejaculation is that none of these male gametes is willing to ask for directions. So what *does* provide the sperm with directions? Heat is one cue that capacitated sperm can use to find the egg. There is a thermal gradient of 2°C between the isthmus of the oviduct and the warmer ampullary region (Bahat et al. 2003, 2006). Capacitated mammalian sperm can sense thermal differences as small as 0.014°C over a millimeter and tend to migrate toward the higher temperature (Bahat et al. 2012). This ability to sense the difference and preferentially swim from cooler to warmer sites (thermotaxis) is found only in capacitated sperm.

Once in the ampullary region, a second sensing mechanism, chemotaxis, may come into play. It appears that the oocyte and its accompanying cumulus cells secrete molecules that attract capacitated (and only capacitated) sperm toward the egg during the last stages of sperm migration (Ralt et al. 1991; Cohen-Dayag et al. 1995; Eisenbach and Tur-Kaspa 1999; Wang et al. 2001). The identity of these chemotactic compounds is being investigated, but one of them appears to be the hormone **progesterone**. Guidobaldi and colleagues (2008) have shown that progesterone secreted from the cumulus cells surrounding the rabbit oocyte is bound by capacitated sperm and is used as a directional cue by the sperm. In humans, progesterone has been shown to bind to a receptor that activates the Ca^{2+} channels in the cell membrane of the sperm tail, leading to sperm hyperactivity (Lishko et al. 2011; Strünker et al. 2011). This activation takes place only after the sperm's intracellular pH has increased, which may help explain why capacitation is needed in order for sperm to reach and fertilize the egg (Navarro et al. 2007).

Thus it appears that, just as in the case of sperm-activating peptides in sea urchins, progesterone both provides direction and activates motility. Moreover, as in certain invertebrate eggs, it appears that the human egg secretes a chemotactic factor only when it is capable of being fertilized, and that sperm are attracted to such a compound only when they are capable of fertilizing the egg.

The acrosome reaction and recognition at the zona pellucida

Before the mammalian sperm can bind to the oocyte, it must first bind to and penetrate the egg's zona pellucida. The zona pellucida in mammals plays a role analogous to that of the vitelline envelope in invertebrates; the zona, however, is a far thicker and denser structure than the vitelline envelope. The mouse zona pellucida is made of three major glycoproteins—**ZP1**, **ZP2**, and **ZP3** (**zona proteins 1**, **2**, and **3**)—along with accessory proteins that bind to the zona's integral structure. The binding of sperm to the zona is relatively, but not absolutely, species-specific.

For more than 30 years the prevailing model has been that the capacitated mouse sperm bind to ZP3 on the zona pellucida, and that ZP3 causes the acrosome reaction to occur. Thus, the enzymes from the acrosome would be concentrated at the site where the sperm needs to digest a path through the zona pellucida. ZP3-binding proteins were indeed found on the acrosomal membrane of mammalian sperm. However, new technologies allowing more sensitive detection of

(A)

(B)

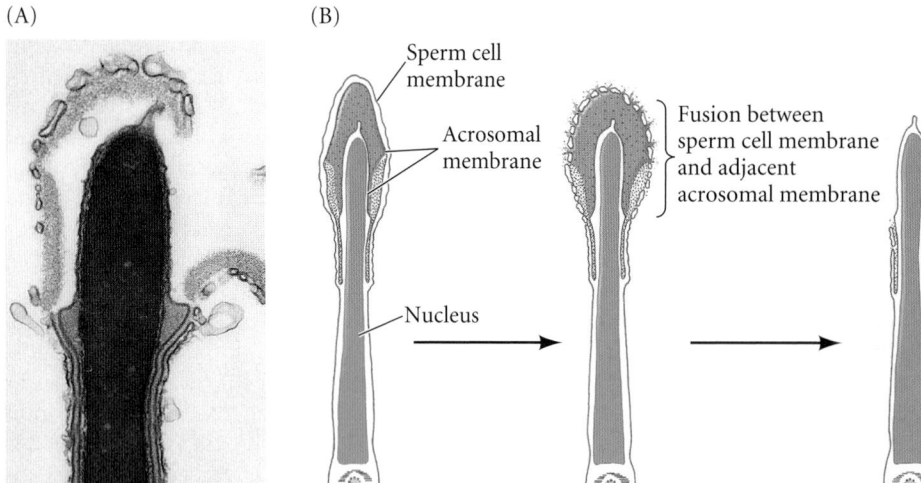

Sperm cell
membrane

Acrosomal
membrane

Fusion between
sperm cell membrane
and adjacent
acrosomal membrane

Nucleus

Centriole

FIGURE 4.30 Acrosome reaction in hamster sperm. (A) Transmission electron micrograph of hamster sperm undergoing the acrosome reaction. The acrosomal membrane can be seen to form vesicles. (B) Interpretive diagram of electron micrographs showing the fusion of the acrosomal and cell membranes in the sperm head. (A from Meizel 1984, courtesy of S. Meizel; B after Yanagimachi and Noda 1970.)

acrosome-reacted sperm have shown that most sperm that bind to the zona have already undergone the acrosome reaction. Earlier investigations (Huang et al. 1981; Yanagamachi and Phillips 1984), looking at fertilization in hamsters and guinea pigs (which have large acrosomes; **FIGURE 4.30**), had suggested that the mammalian acrosome reaction might occur in the cumulus. But these observations were eclipsed by the mouse data for intact sperm binding to the zona and undergoing their acrosome reaction on ZP3. However, the "early acrosome reaction" model was revitalized when Gahlay and

colleagues (2010) provided evidence that the zona may not actually be able to induce the acrosome reaction. The eggs of mice having mutations in ZP3 such that it could not induce the acrosome reaction were still fertilized. This would suggest that the fertilizing sperm underwent the acrosome reaction either before binding to the zona or while passing through it.

Evidence for the acrosome reaction occurring prior to zona binding came from studies by Jin and colleagues (2011), who used video recording to look at mouse sperm entering the cumulus-oocyte complex (**FIGURE 4.31A**). They placed a single cumulus-oocyte complex beneath a small coverslip and added capacitated sperm with GFP label in the acrosomes. Sperm whose heads glowed green under fluorescent light had intact acrosomes, while those that did not fluoresce had undergone the acrosome reaction. (The presence of GFP in the acrosome did not damage the sperm or hamper its abilities; Nakanishi et al. 1999.) By examining the recorded images, Jin and colleagues found that "successful" sperm—those that actually fertilized

(A)

0.00   4.30   5.37   6.50
Seconds

(B)

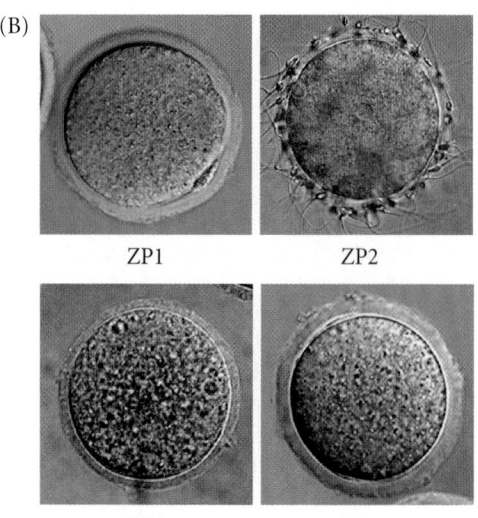

ZP1   ZP2

ZP3   ZP4

FIGURE 4.31 Acrosome-reacted mouse sperm bind to the zona and are successful at fertilizing the egg. Sperm were labeled such that the intact acrosome fluoresced green; all sperm tails had red fluorescent markers. When the labeled sperm were allowed to interact with the mouse egg and cumulus, the resulting video revealed that the fertilizing sperm (arrowhead at 4.30 seconds) showed no green fluorescence when it reached the surface of the zona pellucida at 6.20 seconds—meaning it had undergone the acrosome reaction before that time. An adajacent sperm did fluoresce green, meaning its acrosome remained intact. Such acrosome-intact sperm remained bound to the zona without undergoing the acrosome reaction or progressing to the egg cell membrane. (B) Gain-of-function experiment demonstrating that human sperm bind to ZP2. The human zona pellucida contains four zona proteins. Of these, ZP4 is not found in the mouse zona. Transgenic mouse oocytes were constructed that expressed the three normal mouse zona proteins and also one of the four human zona proteins. When human sperm were added to the mouse oocytes, they bound only to those transgenic oocytes that expressed human ZP2. Human sperm did not bind to cells expressing human ZP1, ZP3, or ZP4. (A from Jin et al. 2011, courtesy of N. Hirohashi; B from Baibakov et al. 2012.)

an egg—had already undergone the acrosome reaction by the time they were first seen in the cumulus. Those sperm that underwent the acrosome reaction on the zona were almost always unsuccessful. Thus it appears that most sperm undergo the acrosome reaction in the cumulus, and that they probably bind to ZP2. In a gain-of-function experiment (Baibakov et al. 2012), ZP2 was shown to be critical for human sperm-egg binding. Baibakov and colleges added the different human zona proteins separately to the zona of mouse eggs. Only the mouse eggs with the *human* ZP2 bound human sperm (**FIG-URE 4.31B**). It is possible that the binding of acrosome-intact sperm to ZP3 may be a process used by sperm that arrive late to the egg, perhaps after the cumulus has dispersed (see Clark 2011; Yanagimachi 2011). If this is indeed the case, we still do not know the agent(s) that normally initiate the mammalian acrosome reaction.

Gamete fusion and the prevention of polyspermy

In mammals, it is not the tip of the sperm head that makes contact with the egg (as happens in the perpendicular entry of sea urchin sperm) but the side of the sperm head (**FIGURE** 4.32A,B). The acrosome reaction, in addition to expelling the enzymatic contents of the acrosome, also exposes the inner acrosomal membrane to the outside. The junction between this inner acrosomal membrane and the sperm cell membrane is called the **equatorial region**, and this is where membrane fusion between sperm and egg begins (**FIGURE** 4.32C,D). As in sea urchin gamete fusion, the sperm is bound to regions of the egg where actin polymerizes to extend microvilli to the sperm (Yanagimachi and Noda 1970).

The mechanism of mammalian gamete fusion is still controversial (see Ikawa et al. 2008; Lefèvre et al. 2010). Gene knockout experiments suggest that mammalian gamete fusion may depend on interaction between a sperm protein and integrin-associated CD9 protein on the egg (Le Naour et al. 2000; Miyado et al. 2000; Evans 2001). CD9 protein has been localized to the membranes of the egg microvilli, and female mice with the *CD9* gene knocked out are infertile because their eggs fail to fuse with sperm (Kaji et al. 2002; Runge et al. 2006). This infertility can be reversed by the microinjection of mRNA encoding either mouse or human CD9 protein. It is not known exactly how these proteins

(A)

(B)

(D)

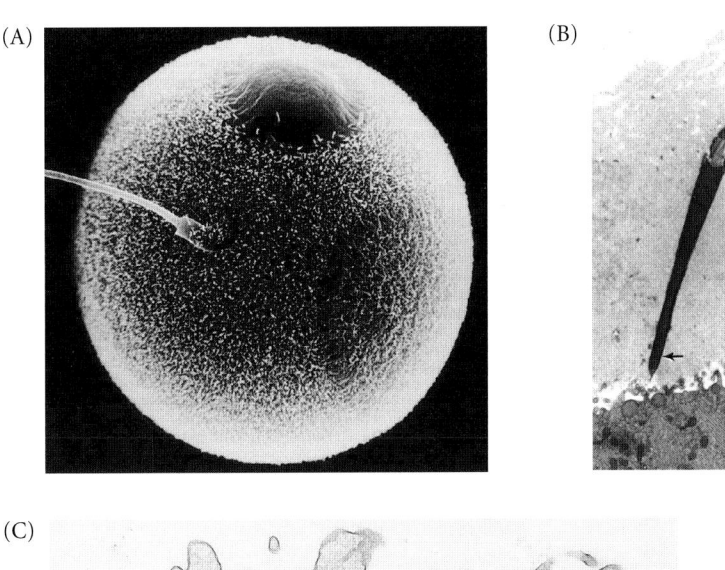

(C)

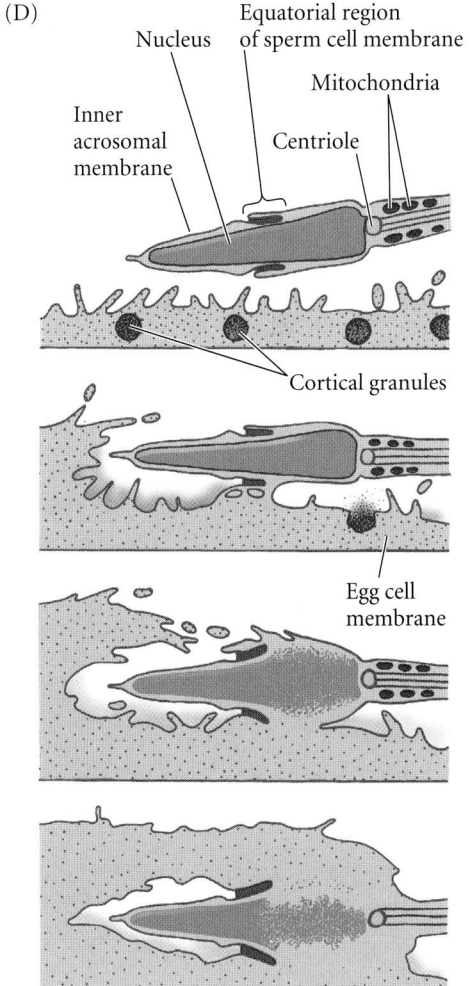

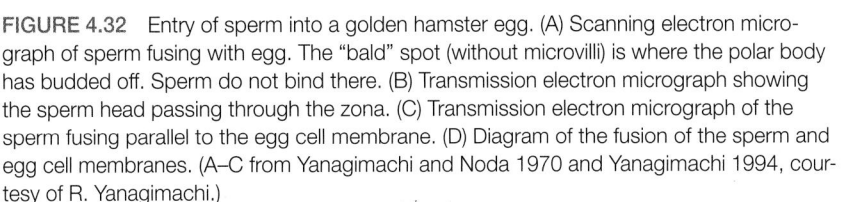

FIGURE 4.32 Entry of sperm into a golden hamster egg. (A) Scanning electron micrograph of sperm fusing with egg. The "bald" spot (without microvilli) is where the polar body has budded off. Sperm do not bind there. (B) Transmission electron micrograph showing the sperm head passing through the zona. (C) Transmission electron micrograph of the sperm fusing parallel to the egg cell membrane. (D) Diagram of the fusion of the sperm and egg cell membranes. (A–C from Yanagimachi and Noda 1970 and Yanagimachi 1994, courtesy of R. Yanagimachi.)

facilitate membrane fusion, but CD9 is also known to be critical for the fusion of myocytes (muscle cell precursors) to form striated muscle (Tachibana and Hemler 1999).

On the sperm side of the mammalian fusion process, Inoue and colleagues (2005) have implicated an immunoglobulin-like protein, named Izumo after a Japanese shrine dedicated to marriage. Sperm from mice carrying loss-of-function mutations in the *Izumo* gene are able to bind and penetrate the zona pellucida, but are not able to fuse with the egg cell membrane. Human sperm also contain Izumo protein, and antibodies directed against Izumo prevent sperm-egg fusion in humans as well. There are other candidates for sperm fusion proteins; indeed, there may be several sperm-egg binding systems operating, each of which may be necessary but not sufficient to ensure proper gamete binding and fusion. It is not yet known whether Izumo on the sperm and CD9 on the egg bind one another.

Polyspermy is a problem for mammals as well as for sea urchins. In mammals, an electrical "fast" block to polyspermy has not yet been detected; it may not even be needed, given the limited number of sperm that reach the ovulated egg (Gardner and Evans 2006). However, a *slow* block to polyspermy in mammals occurs when enzymes released by the cortical granules modify the zona pellucida sperm receptor proteins such that they can no longer bind sperm (Bleil and Wassarman 1980). ZP2 is clipped by the cortical granule protease ovastacin and loses its ability to bind sperm (Moller and Wassarman 1989). This protease, ovastacin, is found in the cortical granules of unfertilized eggs and is released during cortical granule fusion. Indeed, if a mouse egg bears a mutant ZP2 that cannot be cleaved, polyspermy will occur (**FIGURE 4.33**; Gahlay et al. 2010; Burkart et al. 2012). Thus, once one sperm has entered the egg, other sperm can no longer initiate or maintain their binding to the zona pellucida and are rapidly shed.

Fusion of genetic material

As in sea urchins, the mammalian sperm that finally enters the egg carries its genetic contribution in a haploid pronucleus. In mammals, however, the process of pronuclear migration takes about 12 hours, compared with less than 1 hour in the sea urchin. The DNA of the sperm pronucleus is bound by **protamines**—basic proteins that are tightly compacted through disulfide bonds. Glutathione in the egg cytoplasm reduces these disulfide bonds and allows the sperm chromatin to uncoil (Calvin and Bedford 1971; Kvist et al. 1980; Perreault et al. 1988).

The mammalian sperm enters the oocyte while the oocyte nucleus is "arrested" in metaphase of its second meiotic division (**FIGURE 4.34A,B**; see also Figure 4.5). As described for the sea urchin, the calcium oscillations brought about by sperm entry inactivate MAP kinase and allow DNA synthesis. But unlike the sea urchin egg, which is already in a

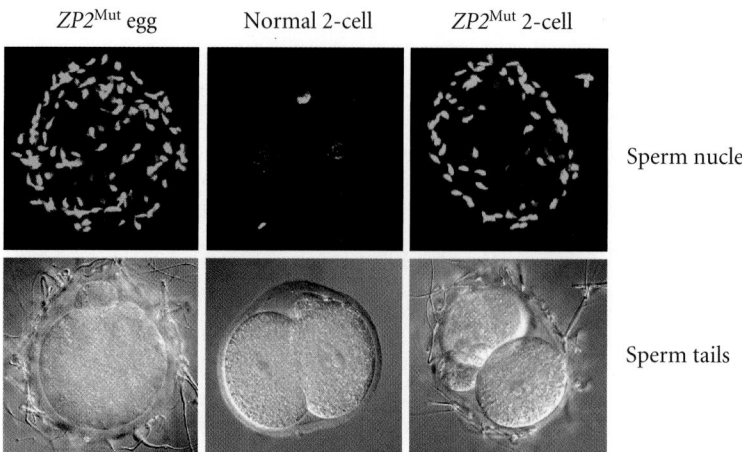

FIGURE 4.33 Cleaved ZP2 is necessary for the block to polyspermy in mammals. Eggs and embryos were visualized by fluorescence microscopy (to see sperm nuclei; top row) and brightfield microscopy (differential interference contrast, to see sperm tails; bottom row). Sperm bound normally to eggs containing a mutant ZP2 that could not be cleaved. However, the egg with normal (i.e., cleavable) ZP2 got rid of sperm by the 2-cell stage, whereas the egg with the mutant (uncleavable) ZP2 retained sperm. (From Gahlay et al. 2010, photograph courtesy of J. Dean.)

haploid state, the chromosomes of the mammalian oocyte are still in the middle of meiotic metaphase. Oscillations in the level of Ca^{2+} activate another kinase that leads to the proteolysis of cyclin (thus allowing the cell cycle to continue) and securin (the protein holding the metaphase chromosomes together), eventually resulting in a haploid female pronucleus (Watanabe et al. 1991; Johnson et al. 1998).

DNA synthesis occurs separately in the male and female pronuclei. The centrosome (new centriole) accompanying the male pronucleus produces its asters (largely from proteins stored in the oocyte). The microtubules join the two pronuclei and enable them to migrate toward one another. Upon meeting, the two nuclear envelopes break down. However, instead of producing a common zygote nucleus (as in sea urchins), the chromatin condenses into chromosomes that orient themselves on a common mitotic spindle (**FIGURE 4.34C,D**). Thus, in mammals a true diploid nucleus is first seen not in the zygote, but at the 2-cell stage.

Each sperm brings into the egg not only its pronucleus but also its mitochondria, its centriole, and a tiny amount of cytoplasm. The sperm mitochondria and their DNA are degraded in the egg cytoplasm, so all of the new individual's mitochondria are derived from its mother. The egg and embryo appear to get rid of the paternal mitochondria both by dilution and by actively targeting them for destruction (Cummins et al. 1998; Shitara et al. 1998; Schwartz and Vissing 2002). In most mammals, however, the sperm centriole not only survives but serves as the organizing agent for making the new mitotic spindle.

Activation of the mammalian egg

As in every other animal studied, a transient rise in cytoplasmic Ca^{2+} is necessary for egg activation in mammals. The

(A) 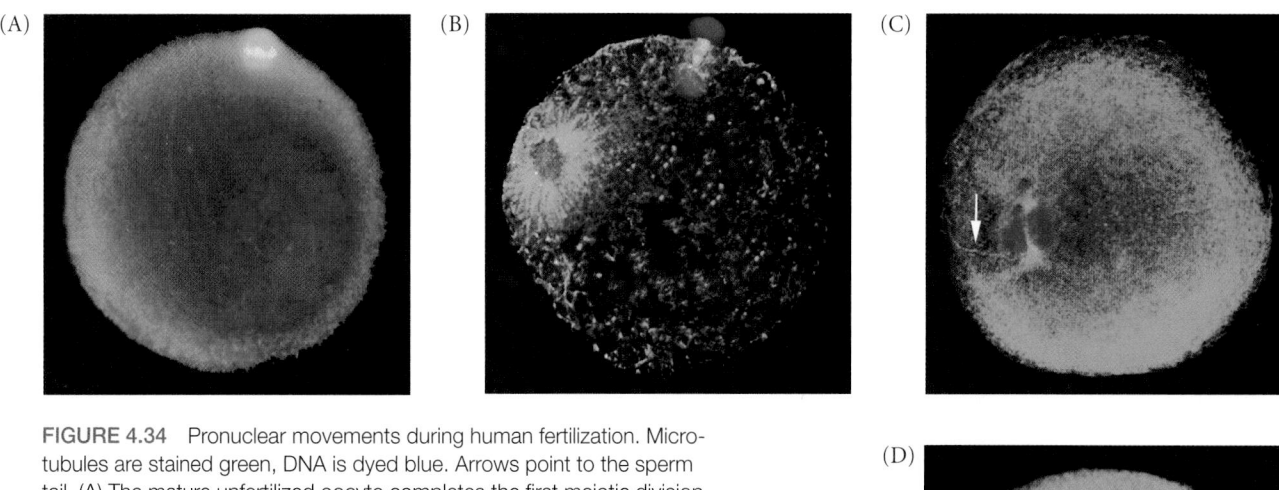 (B)

(C)

(D)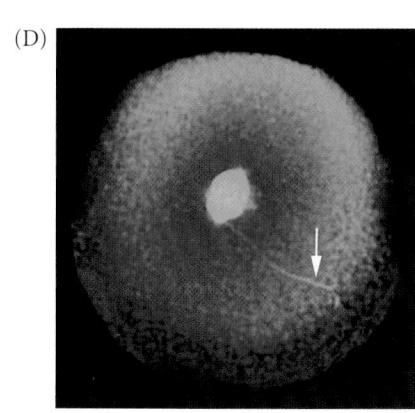

FIGURE 4.34 Pronuclear movements during human fertilization. Microtubules are stained green, DNA is dyed blue. Arrows point to the sperm tail. (A) The mature unfertilized oocyte completes the first meiotic division, budding off a polar body. (B) As the sperm enters the oocyte (left side), microtubules condense around it as the oocyte completes its second meiotic division at the periphery. (C) By 15 hours after fertilization, the two pronuclei have come together, and the centrosome splits to organize a bipolar microtubule array. The sperm tail is still seen (arrow). (D) At prometaphase, chromosomes from the sperm and egg intermix on the metaphase equator and a mitotic spindle initiates the first mitotic division. The sperm tail can still be seen. (From Simerly et al. 1995, courtesy of G. Schatten.)

sperm induces a series of Ca^{2+} waves that can last for hours, terminating in egg activation (i.e., resumption of meiosis, cortical granule exocytosis, and release of the inhibition on maternal mRNAs) and the formation of the male and female pronuclei. And, again as in sea urchins, fertilization triggers intracellular Ca^{2+} release through the production of IP_3 by the enzyme phospholipase C (Swann et al. 2006; Igarashi et al. 2007).

However, the mammalian PLC responsible for egg activation and pronucleus formation may in fact come from the sperm rather than from the egg. Some of the first observations for a sperm-derived PLC came from studies of intracytoplasmic sperm injection (ICSI), an experimental treatment for curing infertility. Here, sperm are directly injected into oocyte cytoplasm, bypassing any interaction with the egg cell membrane. To the surprise of many biologists (who had assumed that sperm *binding* to some egg receptor protein was critical for egg activation), this treatment worked. The human egg was activated and pronuclei formed. Injecting mouse sperm into mouse eggs will also induce fertilization-like Ca^{2+} oscillations in the egg and lead to complete development (Kimura and Yanagimachi 1995).

It appeared that an activator of Ca^{2+} release was stored in the sperm head (see Figure 4.23A). This activator turned out to be a soluble sperm PLC enzyme, **PLCζ** (zeta), that is delivered to the egg by gamete fusion. In mice, expression of PLCζ mRNA in the egg produces Ca^{2+} oscillations, and removing PLCζ from mouse sperm (by antibodies or RNAi) abolishes the sperm's calcium-inducing activity (Saunders et al. 2002;

Yoda et al. 2004; Knott et al. 2005). Human sperm that are unsuccessful in ICSI have been shown to have little or no functional PLCζ. In fact, normal human sperm can activate Ca^{2+} oscillations when injected into mouse eggs, but sperm lacking PLCζ do not (Yoon et al. 2008).

Whereas sea urchin eggs usually are activated as a single wave of Ca^{2+} crosses from the point of sperm entry, the mammalian egg is traversed by numerous waves of calcium ions (Miyazaki et al. 1992; Ajduk et al. 2008; Ducibella and Fissore 2008). The extent (amplitude, duration, and number) of these Ca^{2+} oscillations appears to regulate the timing of mammalian egg activation (Ducibella et al. 2002; Ozil et al. 2005; Toth et al. 2006). Thus, cortical granule exocytosis occurs just before the resumption of meiosis and much before the translation of maternal mRNAs.

In mammals, the Ca^{2+} released by IP_3 binds to a series of proteins including calmodulin-activated protein kinase (which will be important in eliminating the inhibitors of mRNA translation), MAP kinase (which allows the resumption of meiosis), and synaptotagmin (which helps initiate cortical granule fusion). Unused Ca^{2+} is pumped back into the endoplasmic reticulum, and additional Ca^{2+} is acquired from outside the cell. This recruitment of extracellular Ca^{2+} appears to be necessary for the egg to complete meiosis. If Ca^{2+} influx is blocked, the second polar body does not form; instead, the result is two nonviable (triploid) egg pronuclei (Maio et al. 2012).

The Nonequivalence of Mammalian Pronuclei

It is generally assumed that males and females carry equivalent haploid genomes. Indeed, one of the fundamental tenets of Mendelian genetics is that genes derived from the sperm are functionally equivalent to those derived from the egg. However, as we saw in Chapter 2, genomic imprinting can occur in mammals such that the sperm-derived genome and the egg-derived genome may be functionally different and play complementary roles during certain stages of development. Genomic imprinting is thought to be caused by the different patterns of cytosine methylation.

The first evidence for nonequivalence came from studies of a human tumor called a **hydatidiform mole**, which resembles placental tissue. A majority of such moles have been shown to arise when a haploid sperm fertilizes an egg in which the female pronucleus is absent. After entering the egg, the sperm chromosomes duplicate themselves, thereby restoring the diploid chromosome number. However, the entire genome is derived from the sperm (Jacobs et al. 1980; Ohama et al. 1981). The cells survive, divide, and have a normal chromosome number, but development is abnormal. Instead of forming an embryo, the egg becomes a mass of placenta-like cells.

Normal development does not occur when the entire genome comes from the male parent. Normal mammalian development also does not occur when the genome is derived totally from the egg.* Placing mouse oocytes in a culture medium that artificially activates the oocyte while suppressing the formation of the second polar body produces diploid mouse eggs whose genes are derived exclusively from the oocyte (Kaufman et al. 1977). These eggs divide to form embryos with spinal cords, muscles, skeletons, and organs, including beating hearts. However, development does not continue, and by day 10 or 11

*The eggs of many invertebrates and some vertebrates are capable of producing a normal embryo in the absence of any spermatic contribution, an ability known as *parthenogenesis* (Greek, "virgin birth"). Mammals, however, do not exhibit parthenogenesis.

TABLE 4.2 Pronuclear transplantation experiments

Class of reconstructed zygotes	Operation	Number of successful transplants	Number of surviving progeny
Bimaternal		339	0
Bipaternal		328	0
Control		348	18

Source: McGrath and Solter 1984.

(halfway through the mouse's gestation), these parthenogenetic embryos deteriorate. Neither human nor mouse development can be completed solely with egg-derived chromosomes.

That male and female pronuclei are both needed for normal development was also shown by pronuclear transplantation experiments (Surani and Barton 1983; McGrath and Solter 1984; Surani et al. 1986). Either male or female pronuclei can be removed from recently fertilized mouse eggs and added to other recently fertilized eggs. (The two pronuclei can be distinguished at this stage because the female pronucleus is the one beneath the polar bodies.) Thus, zygotes with two male or two female pronuclei can be constructed. Although these eggs will form diploid cells that undergo normal cleavage, eggs whose genes are derived solely from sperm nuclei or solely from oocyte nuclei do not develop to birth. Control eggs undergoing such transplantation (i.e., eggs containing one male and one female pronucleus taken from different zygotes) can develop normally (**TABLE 4.2**). Thus, for mammalian development to occur, both the sperm-derived and the egg-derived pronuclei are critical.

The reason for the necessity of having both sperm-derived and egg-derived genomes is that there are approximately 100 mammalian genes that are expressed only from the sperm-derived chromosomes or only from the egg-derived chromosomes. Such genes are called **imprinted genes**. For example, the human *DLK1* gene, encoding a signaling protein that inhibits fat cell formation, is only expressed from the sperm-derived chromosome 14. The *UBE3A* gene on chromosome 15 is only expressed from the maternally derived chromosome, where its protein product will help regulate brain cell mitoses (Singhmar and Kumar 2011). Thus a *UBE3A* mutation on the maternal chromosome cannot be compensated for by a wild-type allele on the paternal chromosome. The regulator of these transcriptional differences is DNA methylation. **Differentially methylated regions (DMRs)** are found near genes that regulate transcription. The DMRs for maternal-specific methylation (i.e., gene expression from the paternal copy) are usually located at promoters of these genes, whereas the DMRs for paternal-specific methylation (i.e., gene expression from the maternal copy) are usually located in intergenic regions (such as that controlling the CTCF-dependent insulator element discussed in Chapter 2). These methylation patterns are erased during germ cell migration and established during gametogenesis (see Bartolomei and Ferguson-Smith 2011).

The importance of DNA methylation in this block to parthenogenesis was demonstrated when Kono and colleagues (2004) generated a female mouse whose genes came exclusively from two oocytes. To accomplish this feat, they had to mutate the DNA methylation system in one of the oocyte genomes to make it more like that of a male mouse, and then they had to perform two rounds of nuclear transfer. "Men," as one reviewer remarked, "do not need to fear becoming redundant any time soon" (Vogel 2004).

Coda

Fertilization is not a moment or an event, but a process of carefully orchestrated and coordinated events including the contact and fusion of gametes, the fusion of nuclei, and the activation of development. It is a process whereby two cells, each at the verge of death, unite to create a new organism that will have numerous cell types and organs. It is just the beginning of a series of cell-cell interactions that characterize animal development.

SNAPSHOT SUMMARY: Fertilization

1. Fertilization accomplishes two separate activities: sex (the combining of genes derived from two parents) and reproduction (the creation of a new organism).

2. The events of fertilization usually include (1) contact and recognition between sperm and egg; (2) regulation of sperm entry into the egg; (3) fusion of genetic material from the two gametes; and (4) activation of egg metabolism to start development.

3. The sperm head consists of a haploid nucleus and an acrosome. The acrosome is derived from the Golgi apparatus and contains enzymes needed to digest extracellular coats surrounding the egg. The midpiece of the sperm contain mitochondria and the centriole that generates the microtubules of the flagellum. Energy for flagellar motion comes from mitochondrial ATP and a dynein ATPase in the flagellum.

4. The female gamete can be an egg (with a haploid nucleus, as in sea urchins) or an oocyte (in an earlier stage of development, as in mammals). The egg (or oocyte) has a large mass of cytoplasm storing ribosomes and nutritive proteins. Some mRNAs and proteins that will be used as morphogenetic factors are also stored in the egg. Many eggs also contain protective agents needed for survival in their particular environment.

5. Surrounding the egg cell membrane is an extracellular layer often used in sperm recognition. In most animals, this extracellular layer is the vitelline envelope. In mammals, it is the much thicker zona pellucida. Cortical granules lie beneath the egg's cell membrane.

6. Neither the egg nor the sperm is the "active" or "passive" partner. The sperm is activated by the egg, and the egg is activated by the sperm. Both activations involve calcium ions and membrane fusions.

7. In many organisms, eggs secrete diffusible molecules that attract and activate the sperm.

8. Species-specific chemotactic molecules secreted by the egg can attract sperm that are capable of fertilizing it. In sea urchins, the chemotactic peptides resact and speract have been shown to increase sperm motility and provide direction toward an egg of the correct species.

9. The acrosome reaction releases enzymes exocytotically. These proteolytic enzymes digest the egg's protective coating, allowing the sperm to reach and fuse with the egg cell membrane. In sea urchins, this reaction in the sperm is initiated by compounds in the egg jelly. Globular actin polymerizes to extend the acrosomal process. Bindin on the acrosomal process is recognized by a protein complex on the sea urchin egg surface.

10. Fusion between sperm and egg is probably mediated by protein molecules whose hydrophobic groups can merge the sperm and egg cell membranes. In sea urchins, bindin may mediate gamete recognition and fusion.

11. Polyspermy results when two or more sperm fertilize an egg. It is usually lethal, since it results in blastomeres with different numbers and types of chromosomes.

12. Many species have two blocks to polyspermy. The fast block is immediate and causes the egg membrane resting potential to rise. Sperm can no longer fuse with the egg. In sea urchins this is mediated by the influx of sodium ions. The slow block, or cortical granule reaction, is physical and is mediated by calcium ions. A wave of Ca^{2+} propagates from the point of sperm entry, causing the cortical granules to fuse with the egg cell membrane. The released contents of these granules cause the vitelline envelope to rise and harden into the fertilization envelope.

13. The fusion of sperm and egg results in the activation of crucial metabolic reactions in the egg. These reactions include reinitiation of the egg's cell cycle and subsequent mitotic division, and the resumption of DNA and protein synthesis.

14. In all species studied, free Ca^{2+}, supported by the alkalinization of the egg, activates egg metabolism, protein synthesis, and DNA synthesis. Inositol trisphosphate (IP_3) is responsible for releasing Ca^{2+} from storage in the endoplasmic reticulum. DAG (diacylglycerol) is thought to initiate the rise in egg pH.

15. IP_3 is generated by phospholipases. Different species may use different mechanisms to activate the phospholipases.

16. Genetic material is carried in a male and a female pronucleus, which migrate toward each other. In sea urchins, the male and female pronuclei merge and a diploid zygote nucleus is formed. DNA replication occurs after pronuclear fusion.

17. Mammalian fertilization takes place internally, within the female reproductive tract. The cells and tissues of the female reproductive tract actively regulate the transport and maturity of both the male and female gametes.

18. The translocation of sperm from the vagina to the egg is regulated by the muscular activity of the uterus, by the

binding of sperm in the isthmus of the oviduct, and by directional cues from the oocyte and/or the cumulus cells surrounding it.

19. Mammalian sperm must be capacitated in the female reproductive tract before they are capable of fertilizing the egg. Capacitation is the result of biochemical changes in the sperm cell membrane and the alkalinization of its cytoplasm. Capacitated mammalian sperm can penetrate the cumulus and bind the zona pellucida.

20. In one model of sperm-zona binding, the acrosome-intact sperm bind to ZP3 on the zona, and ZP3 induces the sperm to undergo the acrosome reaction on the zona pellucida. In another, more recent, model, the acrosome reaction is induced in the cumulus, and the acrosome-reacted sperm bind to ZP2.

21. In mammals, blocks to polyspermy include modification of the zona proteins by the contents of the cortical granules so that sperm can no longer bind to the zona.

22. The rise in intracellular free Ca^{2+} at fertilization in amphibians and mammals causes the degradation of cyclin and the inactivation of MAP kinase, allowing the second meiotic metaphase to be completed and the formation of the haploid female pronucleus.

23. In mammals, DNA replication takes place as the pronuclei are traveling toward each other. The pronuclear membranes disintegrate as the pronuclei approach each other, and their chromosomes gather around a common metaphase plate.

24. The male and female pronuclei of mammals are not equivalent. If the zygote's genetic material is derived solely from one parent or the other, normal development will not take place. This difference in the male and female genomes is thought to be the result of different methylation patterns on the genes.

For Further Reading

Complete bibliographical citations for all literature cited in this chapter can be found at the free-access website **www.devbio.com**

Bartolomei, M. S. and A. C. Ferguson-Smith. 2011. Mammalian genomic imprinting. *Cold Spring Harbor Persp. Biol.* 2011; 3:a002592.

Boveri, T. 1902. On multipolar mitosis as a means of analysis of the cell nucleus. [Translated by S. Gluecksohn-Waelsch.] In B. H. Willier and J. M. Oppenheimer (eds.), *Foundations of Experimental Embryology*. Hafner, New York, 1974.

Briggs, E. and G. M. Wessel. 2006. In the beginning…: Animal fertilization and sea urchin development. *Dev. Biol.* 300: 15–26.

Gahlay, G., L. Gauthier, B. Baibakov, O. Epifano and J. Dean. 2010. Gamete recognition in mice depends on the cleavage status of an egg's zona pellucida protein. *Science* 329: 216–219.

Glabe, C. G. and V. D. Vacquier. 1978. Egg surface glycoprotein receptor for sea urchin sperm bindin. *Proc. Natl. Acad. Sci. USA* 75: 881–885.

Jaffe, L. A. 1976. Fast block to polyspermy in sea urchins is electrically mediated. *Nature* 261: 68–71.

Jin, M. and 7 others. 2011. Most fertilizing mouse spermatozoa begin their acrosome reaction before contact with the zona pellucida during in vitro fertilization. *Proc. Natl. Acad. Sci. USA* 108: 4892–4896.

Just, E. E. 1919. The fertilization reaction in *Echinarachinus parma. Biol. Bull.* 36: 1–10.

Knott, J. G., M. Kurokawa, R. A. Fissore, R. M. Schultz and C. J. Williams. 2005. Transgenic RNA interference reveals role for mouse sperm phospholipase Cζ in triggering Ca^{2+} oscillations during fertilization. *Biol Reprod.* 72: 992–996.

Parrington, J., L. C. Davis, A. Galione and G. Wessel. 2007. Flipping the switch: How a sperm activates the egg at fertilization. *Dev. Dyn.* 236: 2027–2038.

Vacquier, V. D. and G. W. Moy. 1977. Isolation of bindin: The protein responsible for adhesion of sperm to sea urchin eggs. *Proc. Natl. Acad. Sci. USA* 74: 2456–2460.

Go Online

WEBSITE 4.1 Leeuwenhoek and images of homunculi. Scholars in the 1600s thought that either the sperm or the egg carried the rudiments of the adult body. Moreover, these views became distorted by contemporary commentators and later historians.

WEBSITE 4.2 The origins of fertilization research. Studies by Hertwig, Fol, Boveri, and Auerbach investigated fertilization by integrating cytology with genetics. The debates over meiosis and nuclear structure were critical in these investigations of fertilization.

WEBSITE 4.3 The egg and its environment. Most eggs are not found in laboratories. Eggs have evolved remarkable ways to protect themselves in specific natural environments.

WEBSITE 4.4 Blocks to polyspermy. Theodore Boveri's analysis of polyspermy is a classic of experimental and descriptive biology. E. E. Just's delineation of the fast and slow blocks was a critical paper in embryology. Both papers are reprinted here, along with commentaries.

WEBSITE 4.5 Building the egg's extracellular matrix. In sea urchins, the cortical granules secrete not only hyalin but a number of proteins that construct the extracellular matrix of the embryo. This highly coordinated process results in sequential layers.

Vade Mecum

Gametogenesis. Stained sections of testis and ovary illustrate the process of gametogenesis, the streamlining of developing sperm, and the remarkable growth of the egg as it stores nutrients for its long journey. You can see this in movies and labeled photographs that take you at each step deeper into the mammalian gonad.

E. E. Just. Blocks to polyspermy were discovered in the early 1900s by the African-American embryologist Ernest Just, who became one of the few embryologists ever to be honored on a postage stamp. The Sea Urchin segment contains videos of Just's work on sea urchin fertilization.

Sea urchin fertilization. The remarkable reactions that prevent polyspermy in a fertilized sea urchin egg can be seen in the raising of the fertilization envelope. The Sea Urchin segment contains movies of this event shown in real time.

Outside Sites

YouTube videos worth a look include one of a calcium wave traveling across several sea urchin eggs:
youtube.com/watch?v=BH06WgFua_4
and an amazing video of human in vitro fertilization:
youtube.com/watch?v=gtPd4Yn_18c&feature=related

5

Early Development
Rapid Specification in Snails and Nematodes

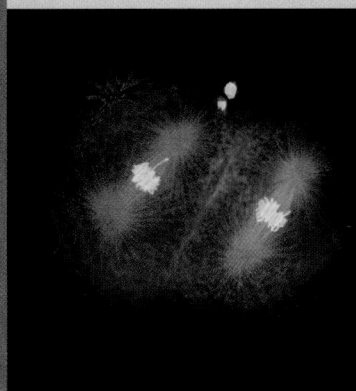

"THE GENERAL POLARITY, SYMMETRY, AND PATTERN of the embryo are egg characters which were determined before fertilization," wrote Edwin Grant Conklin in 1920. Indeed, as we will see, the egg cytoplasm plays a major role in determining patterns of cleavage, gastrulation, and cell specification. It does so by interacting with the nuclear genome established at fertilization. Fertilization gave the organism a new genome and rearranged its cytoplasm. Now the zygote begins the production of a multicellular organism.

During *cleavage*, rapid cell divisions divide the cytoplasm of the fertilized egg into numerous cells. These cells undergo dramatic displacements during *gastrulation*, a process whereby they move to different parts of the embryo and acquire new neighbors. During cleavage and gastrulation, the major axes of the embryo are determined and the embryonic cells begin to acquire their respective fates. Three body axes must be specified: the anterior-posterior (head-tail) axis; the dorsal-ventral (back-belly) axis; and the left-right axis. Different species specify these axes at different times, using different mechanisms. Cleavage always precedes gastrulation, but in some species axis formation begins as early as oocyte formation. It can be completed during cleavage (as in *Drosophila*) or extend all the way through gastrulation (as in *Xenopus*).

This and the remaining chapters in this unit will look at some of the different ways these universal processes take place. This chapter will present an overview of developmental phenomena common to all animals, after which we will detail the development of two invertebrate groups, the gastropod molluscs (represented by snails) and the nematodes (represented by *Caenorhabditis elegans*.) We will begin, however, by taking a brief look at evolution and classification as seen through the lens of animal development.

Developmental Patterns among the Metazoa

The classification of organisms depends on the patterns of their development. To be a **eukaryotic organism** means that the cell contains a nucleus and several distinct chromosomes that undergo mitosis. To be a **multicellular eukaryotic organism** (plant, fungus, or animal) means that the cells formed by mitosis remain together as a functional whole and that subsequent generations form the same coherent individuals composed of several cells. To be a **metazoan** means to be an animal, and to be an animal means to undergo gastrulation. All animals gastrulate, and nothing gastrulates that isn't an animal. Most of the remainder of this book concerns the development of metazoans.

When we say that there are 35 metazoan phyla, we are stating that there are 35 surviving patterns of animal development (see Davidson and Erwin 2009). The most striking pattern is that life has not evolved in a straight line, but in branching pathways. FIGURE 5.1 shows the four major branches of metazoans: the sponges, diploblasts, protostomes, and deuterostomes.

Hence, studying the period of cleavage, we approach the source whence emerge the progressively branched streams of differentiation that end finally in almost quiet pools, the individual cells of the complex adult organism.

E. E. JUST (1939)

It is not birth, marriage, or death, but gastrulation, which is truly the most important time in your life.

LEWIS WOLPERT (1986)

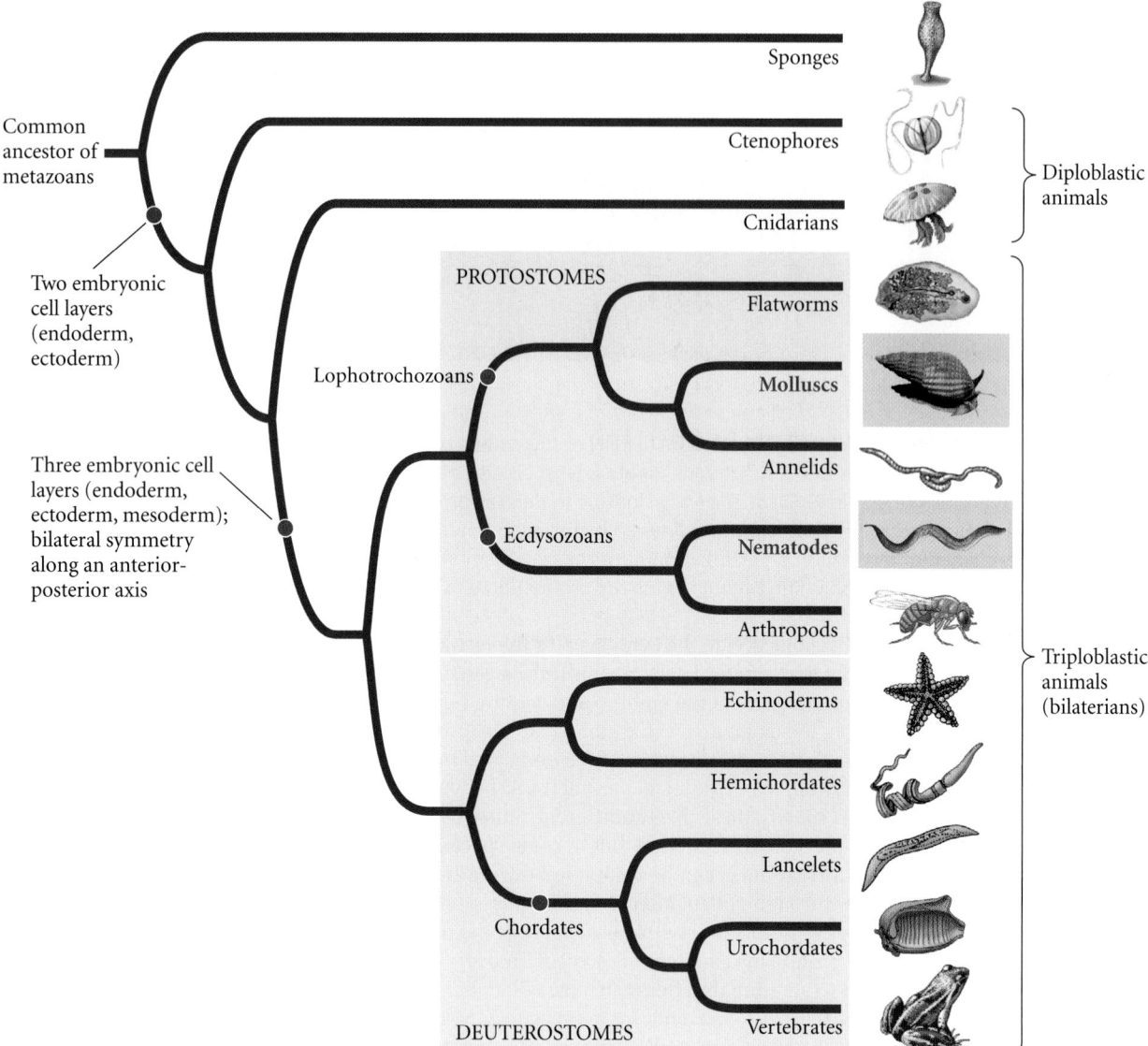

FIGURE 5.1 A simple rendition of the Tree of Life showing the four broad divisions of the animals (metazoans) as defined by their developmental patterns: the developmentally unique sponges, the diploblastic animals, the protostomes, and the deuterostomes. That the evolutionary relationships among the metazoan groups are based on their development has been confirmed by DNA analysis. Photographs of a gastropod mollusc (snail) and the nematode *Caenorhabditis elegans* represent the organisms whose development is covered in this chapter. (Sources include Bourlat et al. 2006; Delsuc et al. 2006; Schierwater et al. 2009; Hejnol 2012; Hillis et al. 2013.)

Sponges and the diploblastic animals

SPONGES The sponges (Porifera) develop so differently from any other animal group that some taxonomists do not consider them metazoans, calling them *parazoans* instead. Most recent taxonomies, however, do place the sponges among the metazoans (see Degnan et al. 2005). A sponge has three major types of somatic cells, one of which, the **archeocyte**, can differentiate into all the other cell types in the body. Individual cells of a sponge that is passed through a sieve can reaggregate to form new sponges. Moreover, in some instances, such reaggregation is species-specific: if individual sponge cells from two different species are mixed together, each of the reaggregated sponges contains cells from only one species (Wilson 1907). In these cases, it is thought that the motile archeocytes collect cells from their own species and not from others (Turner 1978).

Sponges contain no mesoderm, so the Porifera have no true organ systems, nor do they have a digestive tube, circulatory system, nerves, or muscles. Thus, even though they undergo gastrulation and pass through an embryonic and a larval stage, sponges are very unlike most metazoans. However, sponges do share many features of development (including gene regulatory proteins and signaling cascades) with all other metazoan phyla, suggesting that they share a common origin (Fell 1997; Coutinho et al. 1998; King 2004).

DIPLOBLASTS Animals that have two germ layers—ectoderm and endoderm but little or no mesoderm—are referred to as **diploblasts**. The diploblasts have traditionally included the cnidarians (jellyfish and hydras) and the ctenophores (comb jellies). It has long been thought that the members of these phyla have radial symmetry and no mesoderm, whereas the triploblast phyla (all other animals) have bilateral symmetry and a third, mesodermal, germ layer. However, this clear-cut demarcation is now being questioned in regard to the cnidarians. Although certain cnidarians (such as *Hydra*) have no true mesoderm, others do seem to have some mesoderm, and some display bilateral symmetry at parts of their life cycle (Martindale et al. 2004; Martindale 2005). However, the mesoderm of cnidarians may have evolved independently of the mesoderm found in the protostomes and deuterostomes. We now are aware that jellyfish possess striated muscle (necessary for that fascinating propulsion movement), but their muscles do not seem related either molecularly or developmentally to the mesodermally derived muscles of vertebrates or insects (Steinmetz et al. 2012). This independent generation of contractile cells appears to represent a remarkable case of evolutionary convergence.

The triploblastic animals: Protostomes and deuterostomes

The vast majority of metazoan species have three germ layers and are thus **triploblasts**. The evolution of the mesoderm enabled greater mobility and larger bodies because it became the animal's musculature and circulatory system. Triploblastic animals are also called **bilaterians** because they have bilateral symmetry—that is, they have right and left sides. Bilaterians are further classified as either protostomes or deuterostomes.

PROTOSTOMES Protostomes (Greek, "mouth first"), which include the mollusc, arthropod, and worm phyla, are so called because the mouth is formed first, at or near the opening to the gut that is produced during gastrulation. The anus forms later, at a different location. The protostome **coelom**, or body cavity, forms from the hollowing out of a previously solid cord of mesodermal cells.

There are two major branches of protostomes. The **Ecdysozoans** (Greek *ecdysis*, "get out of" or "shed") are those animals that molt their exterior skeletons. The most prominent ecdysozoan group is Arthropoda, the arthropods, a well-studied phylum that includes the insects, arachnids, mites, crustaceans, and millipedes. Molecular analysis has also placed another molting group, the nematodes, in this clade. Members of the second major protostome group, the **Lophotrochozoans**, are characterized by a common type of cleavage (spiral) and a common larval form (the trochophore). The trochophore (Greek *trochos*, "wheel") is a planktonic (free-swimming) larval form with characteristic bands of locomotive cilia. Adults of some lophotrochozoan species have a distinctive feeding apparatus, the lophophore. Lophotrochozoan phyla include the flatworms, annelids, and molluscs.

DEUTEROSTOMES The major **deuterostome** lineages are the chordates (including the vertebrates) and the echinoderms. Although it may seem strange to classify humans, fish, and frogs in the same broad group as starfish and sea urchins, certain embryological features stress this kinship. First, in deuterostomes ("mouth second"), the oral opening is formed after the anal opening. Also, whereas protostomes generally form their body cavity by hollowing out a solid block of mesoderm (**schizocoelous** formation of the body cavity), most deuterostomes form their body cavity from mesodermal pouches extending from the gut (**enterocoelous** formation of the body cavity). It should be mentioned that there are many exceptions to these generalizations (see Martín-Durán et al. 2012).

The lancelets (Cephalochordata) and the tunicates (Urochordata) are invertebrates—they have no backbone. However, the *larvae* of these organisms have a notochord and pharyngeal arches (head structures), indicating that they are chordates. (The "chord" in "chordates" refers to the notochord.) This discovery, made by Alexander Kowalevsky (1867, 1868), was a milestone in biology. The developmental stages of these organisms united the invertebrates and vertebrates into a single "animal kingdom." Darwin (1874) rejoiced, noting that vertebrates probably arose from a group of animals that resembled larval tunicates. Indeed, Urochordata is now considered to be the group most closely related to the vertebrates. This relationship has been demonstrated both by developmental affinities and by molecular analysis (Bourlat et al. 2006; Delsuc et al. 2006), reversing a previous view that cephalochordates were the sister group to vertebrates.

EARLY DEVELOPMENTAL PROCESSES: AN OVERVIEW

Cleavage

Once fertilization is complete, the development of a multicellular organism continues by a process called **cleavage**, a series of mitotic divisions whereby the enormous volume of egg cytoplasm is divided into numerous smaller, nucleated cells. These cleavage-stage cells are called **blastomeres**.* In most species (mammals being the chief exception), both the initial rate of cell division and the placement of the blastomeres with respect to one another are under the control of

---

* We will be using an entire "blast" vocabulary. A **blastomere** is a cell derived from cleavage in an early embryo. A **blastula** is an embryonic stage composed of blastomeres; a mammalian blastula is called a **blastocyst** (see Chapter 9). The cavity within the blastula is the **blastocoel**. (A blastula that lacks a blastocoel is called a **stereoblastula**.) The invagination where gastrulation begins is the **blastopore**.

the proteins and mRNAs stored in the oocyte. Only later do the rates of cell division and the placement of cells come under the control of the newly formed genome.

During the initial phase of development, when cleavage rhythms are controlled by maternal factors, the cytoplasmic volume does not increase. Rather, the zygote cytoplasm is divided into increasingly smaller cells. The zygote is divided first in half, then quarters, then eighths, and so forth. Cleavage occurs very rapidly in most invertebrates, probably as an adaptation to generate a large number of cells quickly and to restore the somatic ratio of nuclear volume to cytoplasmic volume. The embryo often accomplishes this by abolishing the gap periods of the cell cycle (the G1 and G2 phases), when growth can occur. A frog egg, for example, can divide into 37,000 cells in just 43 hours. Mitosis in cleavage-stage *Drosophila* embryos occurs every 10 minutes for more than 2 hours, and forms some 50,000 cells in just 12 hours.

From fertilization to cleavage

As we saw in Chapter 4, fertilization activates protein synthesis, DNA synthesis, and the cell cycle. One of the most important events in this transition from fertilization to cleavage is the activation of **mitosis-promoting factor**, or **MPF**. MPF was first discovered as the major factor responsible for the resumption of meiotic cell divisions in the ovulated frog egg. It continues to play a role after fertilization, regulating the cell cycle of early blastomeres.

Blastomeres generally progress through a biphasic cell cycle consisting of just two steps: M (mitosis) and S (DNA synthesis)

(FIGURE 5.2). The MPF activity of early blastomeres is highest during M and undetectable during S. The shift between the M and S phases in blastomeres is driven solely by the gain and loss of MPF activity. When MPF is made in these cells, they enter into M. Their nuclear envelope breaks down and their chromatin condenses into chromosomes. After an hour, MPF is degraded and the chromosomes return to S phase (Gerhart et al. 1984; Newport and Kirschner 1984).

What causes this cyclical activity of MPF? Mitosis-promoting factor consists of two subunits. The larger subunit, **cyclin B**, displays the cyclical behavior that is key to mitotic regulation, accumulating during S and being degraded after the cells have reached M (Evans et al. 1983; Swenson et al. 1986). Cyclin B is often encoded by mRNAs stored in the oocyte cytoplasm, and if the translation of this message is specifically inhibited, the cell will not enter mitosis (Minshull et al. 1989).

Cyclin B regulates the small subunit of MPF, the **cyclin-dependent kinase (CDK)**. This kinase activates mitosis by phosphorylating several target proteins, including histones, the nuclear envelope lamin proteins, and the regulatory subunit of cytoplasmic myosin. It is the actions of this small kinase subunit that bring about chromatin condensation, nuclear envelope depolymerization, and the organization of the mitotic spindle. Without the cyclin B subunit, however, the cyclin-dependent kinase subunit of MPF will not function.

The presence of cyclin B is controlled by several proteins that ensure its periodic synthesis and degradation. In most

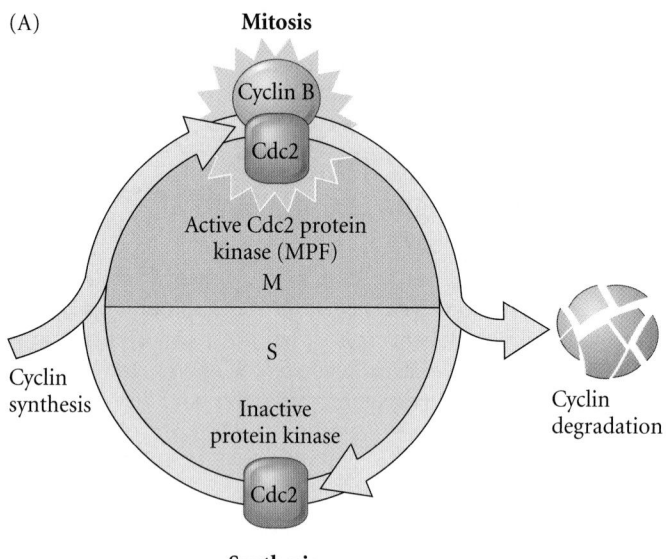

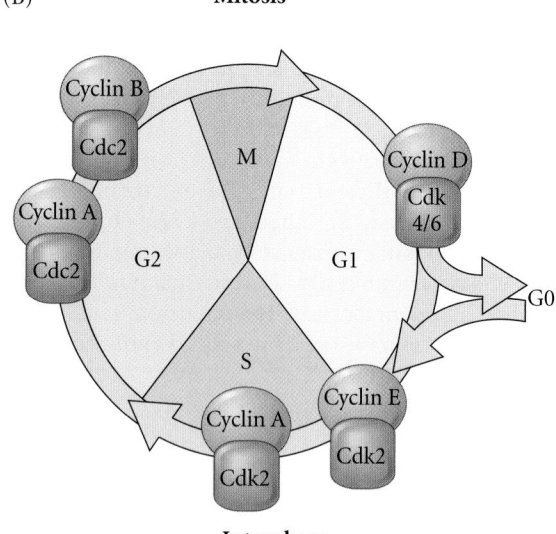

FIGURE 5.2 Cell cycles of somatic cells and early blastomeres. (A) The biphasic cell cycle of early amphibian blastomeres has only two states, S and M. Cyclin B synthesis allows progression to M (mitosis), whereas degradation of cyclin B allows cells to pass into S (synthesis) phase. (B) The complete cell cycle of a typical somatic cell. Mitosis (M) is followed by an interphase stage. Interphase is subdivided into G1, S (synthesis), and G2 phases. Cells that are differentiating are usually taken "out" of the cell cycle and are in an extended G1 phase called G0. The cyclins responsible for the progression through the cell cycle and their respective kinases are shown at their point of cell cycle regulation. (B after Nigg 1995.)

Rapid Specification in Snails and Nematodes

(A)

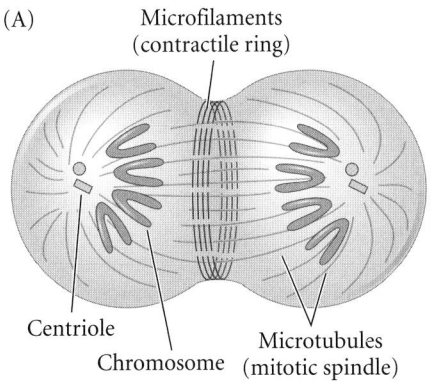

Microfilaments
(contractile ring)

Centriole

Chromosome

Microtubules
(mitotic spindle)

(B)

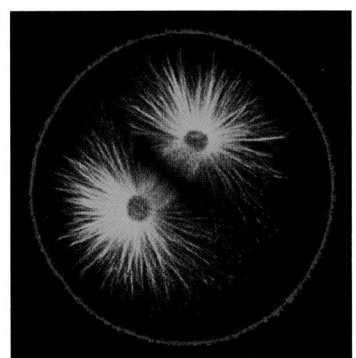

FIGURE 5.3 Roles of microtubules and microfilaments in cell division. (A) Diagram of first-cleavage telophase in a sea urchin egg. The chromosomes are drawn to the centrioles by microtubules while the cytoplasm is being pinched in by the contraction of microfilaments. (B) Confocal fluorescent image of an echinoderm embryo undergoing first cleavage (early anaphase). Microtubules are stained green, actin microfilaments are stained red. (C) Confocal fluorescent image of a sea urchin embryo at the very end of first cleavage. Microtubules are orange; actin proteins (both unpolymerized and in microfilaments) are blue. (B,C courtesy of G. von Dassow and the Center for Cell Dynamics.)

(C)

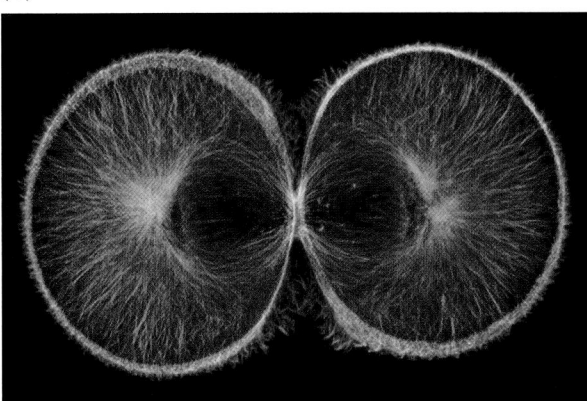

of MPF. After numerous synchronous rounds of mitosis, the cells begin to "go their own way." Third, new mRNAs are transcribed. Many of these messages encode proteins that will become necessary for gastrulation. In several species, if transcription is blocked cell division will still occur at normal rates and times, but the embryo will not be able to initiate gastrulation. Many of these new messenger RNAs are also used for cell specification. As we will see when we look at sea urchin embryos in Chapter 7, the new mRNA expression patterns of the mid-blastula transition map out territories where specific types of cells will later differentiate.

The cytoskeletal mechanisms of mitosis

Cleavage is the result of two coordinated processes. The first of these is **karyokinesis,** the mitotic division of the cell's nucleus. The mechanical agent of karyokinesis is the **mitotic spindle**, with its microtubules composed of tubulin (the same type of protein that makes up the sperm flagellum). The second process is **cytokinesis**, the division of the cell itself. The mechanical agent of cytokinesis is a **contractile ring** of microfilaments made of actin (the same type of protein that extends the egg microvilli and the sperm acrosomal process). TABLE 5.1 compares these agents of cell division. The relationship and coordination between these two systems during cleavage are depicted in FIGURE 5.3A, in which a sea urchin egg is shown undergoing first cleavage. The mitotic spindle and contractile ring are perpendicular to each other, and

species studied, the regulators of cyclin B (and thus of MPF) are stored in the egg cytoplasm. Therefore, the cell cycle remains independent of the nuclear genome for a number of cell divisions. These early divisions tend to be rapid and synchronous. However, as the cytoplasmic components are used up, the nucleus begins to synthesize them. In several species, the embryo now enters a **mid-blastula transition** (**MBT**), in which several new properties are added to the biphasic cell divisions of the embryo. First, the "gap" stages (G1 and G2) are added to the cell cycle (see Figure 5.2B). *Xenopus* embryos add G1 and G2 phases to the cell cycle shortly after the twelfth cleavage. *Drosophila* adds G2 during cycle 14 and G1 during cycle 17 (Newport and Kirschner 1982; Edgar et al. 1986). Second, the synchronicity of cell division is lost, because different cells synthesize different regulators

TABLE 5.1 Karyokinesis and cytokinesis

Process	Mechanical agent	Major protein composition	Location	Major disruptive drug
Karyokinesis	Mitotic spindle	Tubulin microtubules	Central cytoplasm	Colchicine, nocodazole[a]
Cytokinesis	Contractile ring	Actin microfilaments	Cortical cytoplasm	Cytochalasin B

[a]Because colchicine has been found to independently inhibit several membrane functions, including osmoregulation and the transport of ions and nucleosides, nocodazole has become the major drug used to inhibit microtubule-mediated processes (see Hardin 1987).

the spindle is internal to the contractile ring. The contractile ring creates a **cleavage furrow**, which eventually bisects the plane of mitosis, thereby creating two genetically equivalent blastomeres.

The actin microfilaments are found in the cortex (outer cytoplasm) of the egg rather than in the central cytoplasm. Under the confocal microscope, the ring of microfilaments can be seen forming a distinct cortical band 0.1 μm wide (**FIGURE 5.3B,C**). This contractile ring exists only during cleavage and extends 8–10 μm into the center of the egg. It is responsible for exerting the force that splits the zygote into blastomeres; if the ring is disrupted, cytokinesis stops. Schroeder (1973) likened the contractile ring to an "intracellular purse-string," tightening about the egg as cleavage continues. This tightening of the microfilamentous ring creates the cleavage furrow. Microtubules are also seen near the cleavage furrow (in addition to their role in creating the mitotic spindle), since they are needed to bring membrane material to the site of membrane addition (Danilchik et al. 1998).

Although karyokinesis and cytokinesis are usually coordinated, they are sometimes modified by natural or experimental conditions. The placement of the centrioles is critical in orienting the mitotic spindle, and thus the division plane of the blastomeres. Depending on the placement of the centrioles, the blastomeres can separate either into dorsal and ventral daughter cells, anterior and posterior daughter cells, or left and right daughter cells. The spindle can even be at an angle such that one daughter cell is clockwise or counterclockwise to the other. As we will discuss in Chapter 6, during the cleavage of insect eggs karyokinesis occurs several times before cytokinesis takes place, so that numerous nuclei exist within the same cell. The outer membrane of that one large cell eventually indents, separating the nuclei and forming individual cells.

Patterns of embryonic cleavage

In 1923, the embryologist E. B. Wilson reflected on how little we knew about cleavage: "To our limited intelligence, it would seem a simple task to divide a nucleus into equal parts. The cell, manifestly, entertains a very different opinion." Indeed, different organisms undergo cleavage in distinctly different ways. The pattern of embryonic cleavage peculiar to a species is determined by two major parameters: (1) the amount and distribution of yolk protein within the cytoplasm, and (2) factors in the egg cytoplasm that influence the angle of the mitotic spindle and the timing of its formation.

The amount and distribution of yolk determine where cleavage can occur and the relative size of the blastomeres. In general, yolk inhibits cleavage. When one pole of the egg is relatively yolk-free, cellular divisions occur there at a faster rate than at the opposite pole. The yolk-rich pole is referred to as the **vegetal pole**; the yolk concentration in the **animal pole** is relatively low. The zygote nucleus is frequently displaced toward the animal pole. **FIGURE 5.4** provides a classification of cleavage types and shows the influence of yolk on cleavage symmetry and pattern.

At one extreme are the eggs of sea urchins, mammals, and snails. These eggs have sparse, equally distributed yolk and are thus **isolecithal** (Greek, "equal yolk"). In these species, cleavage is **holoblastic** (Greek *holos*, "complete"), meaning that the cleavage furrow extends through the entire egg. With little yolk, these embryos must have some other way of obtaining food. Most will generate a voracious larval form, while mammals will obtain their nutrition from the maternal placenta.

At the other extreme are the eggs of insects, fish, reptiles, and birds. Most of their cell volumes are made up of yolk. The yolk must be sufficient to nourish these animals throughout embryonic development. Zygotes containing large accumulations of yolk undergo **meroblastic cleavage** (Greek *meros*, "part"), wherein only a portion of the cytoplasm is cleaved. The cleavage furrow does not penetrate the yolky portion of the cytoplasm because the yolk platelets impede membrane formation there. Insect eggs have yolk in the center (i.e., they are **centrolecithal**), and the divisions of the cytoplasm occur only in the rim of cytoplasm, around the periphery of the cell (i.e., **superficial cleavage**). The eggs of birds and fish have only one small area of the egg that is free of yolk (**telolecithal** eggs), and therefore the cell divisions occur only in this small disc of cytoplasm, giving rise to **discoidal cleavage**. These are general rules, however, and even closely related species have evolved different patterns of cleavage in different environments.

Yolk is just one factor influencing a species' pattern of cleavage. There are also, as Conklin had intuited, inherited patterns of cell division superimposed on the constraints of the yolk. The importance of this inheritance can readily be seen in isolecithal eggs. In the absence of a large concentration of yolk, **holoblastic cleavage** takes place. Four major patterns of this cleavage type can be observed: *radial*, *spiral*, *bilateral*, and *rotational* holoblastic cleavage. In this chapter we will see examples of two of these cleavage patterns—spiral cleavage in snails and rotational cleavage in nematodes.

Gastrulation and Axis Formation

The blastula consists of numerous cells, the positions of which were established during cleavage. During **gastrulation**, these cells are given new positions and new neighbors, and the multilayered body plan of the organism is established. The cells that will form the endodermal and mesodermal organs are brought to the inside of the embryo, while the cells that will form the skin and nervous system are spread over its outside surface. Thus, the three germ layers—outer ectoderm, inner endoderm, and interstitial mesoderm—are first produced during gastrulation. In addition, the stage is set for the interactions of these newly positioned tissues.

Gastrulation usually proceeds by some combination of several types of movements. These movements involve the

I. HOLOBLASTIC (COMPLETE) CLEAVAGE

  A. **Isolecithal**
    (Sparce, evenly distributed yolk)

    1. Radial cleavage
       Echinoderms, amphioxus

    2. Spiral cleavage
       Annelids, molluscs,
       flatworms

    3. Bilateral cleavage
       Tunicates

    4. Rotational cleavage
       Mammals, nematodes

  B. **Mesolecithal**
    (Moderate vegetal yolk disposition)

    Displaced radial cleavage
    Amphibians

II. MEROBLASTIC (INCOMPLETE) CLEAVAGE

  A. **Telolecithal**
    (Dense yolk throughout most of cell)

    1. Bilateral cleavage
       Cephalopod molluscs

    2. Discoidal cleavage
       Fish, reptiles, birds

  B. **Centrolecithal**
    (Yolk in center of egg)

    Superficial cleavage
    Most insects

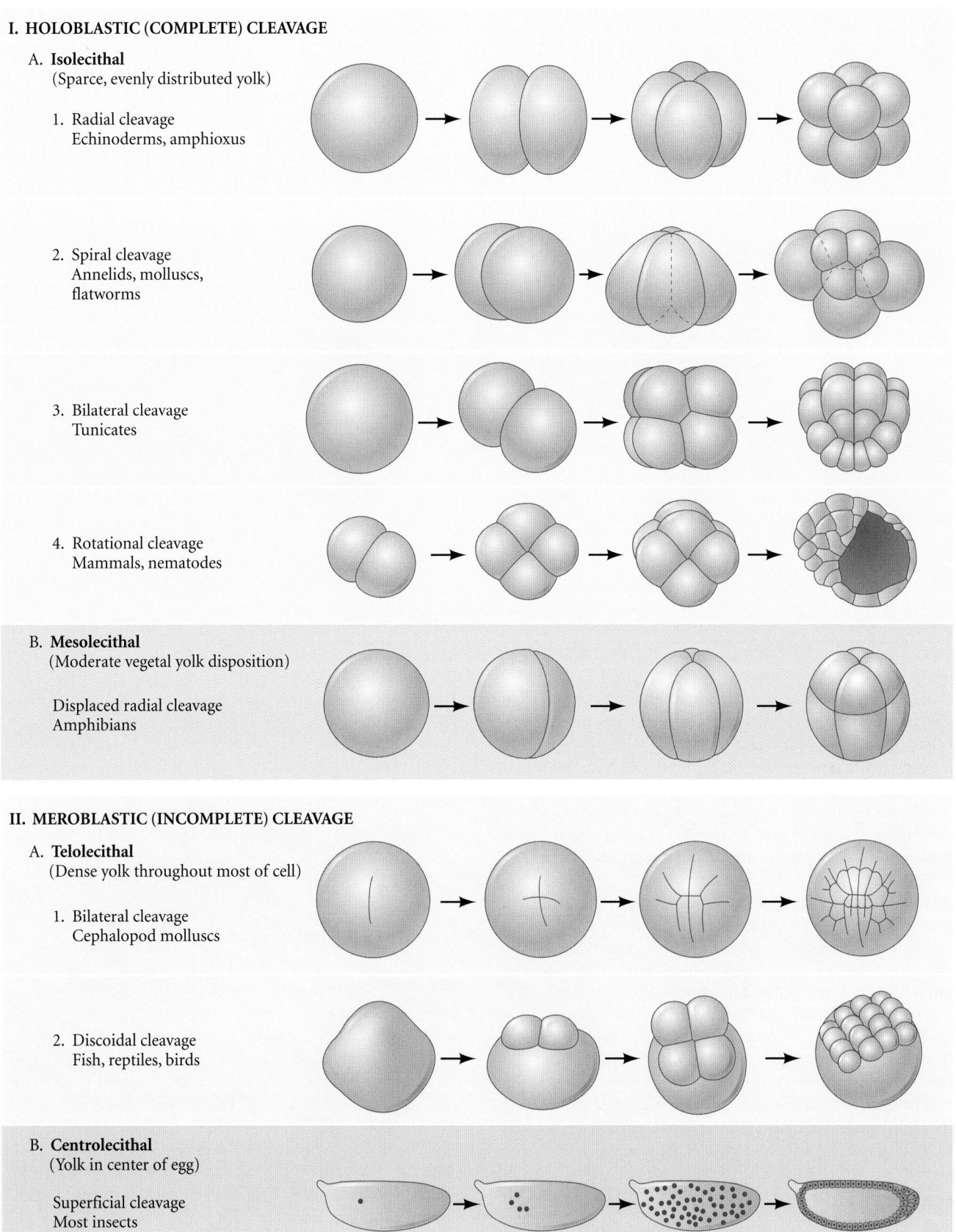

FIGURE 5.4   Summary of the main patterns of cleavage.

TABLE 5.2   Types of cell movement during gastrulation[a]

Type of movement	Description	Illustration	Example
Invagination	Infolding of a sheet (epithelium) of cells, much like the indention of a soft rubber ball when it is poked.		Sea urchin endoderm
Involution	Inward movement of an expanding outer layer so that it spreads over the internal surface of the remaining external cells.		Amphibian mesoderm
Ingression	Migration of individual cells from the surface into the embryo's interior. Individual cells become mesenchymal (i.e., separate from one another) and migrate independently.		Sea urchin mesoderm, *Drosophila* neuroblasts
Delamination	Splitting of one cellular sheet into two more or less parallel sheets. While on a cellular basis it resembles ingression, the result is the formation of a new (additional) epithelial sheet of cells.		Hypoblast formation in birds and mammals
Epiboly	Movement of epithelial sheets (usually ectodermal cells) spreading as a unit (rather than individually) to enclose deeper layers of the embryo. Can occur by cells dividing, by cells changing their shape, or by several layers of cells intercalating into fewer layers; often, all three mechanisms are used.		Ectoderm formation in sea urchins, tunicates, and amphibians

[a]The gastrulation of any particular organism is an ensemble of several of these movements.

entire embryo, and cell migrations in one part of the gastrulating embryo must be intimately coordinated with other movements that are taking place simultaneously. Although patterns of gastrulation vary enormously throughout the animal kingdom, all of the patterns are different combinations of the five basic types of cell movements—**invagination**, **involution**, **ingression**, **delamination**, and **epiboly**—described in **TABLE 5.2**.

In addition to establishing which cells will be in which germ layer, embryos must develop three crucial axes that are the foundation of the body: the anterior-posterior axis, the dorsal-ventral axis, and the right-left axis (**FIGURE 5.5**). The **anterior-posterior (AP** or **anteroposterior) axis** is the line extending from head to tail (or mouth to anus in those organisms that lack a head and tail). The **dorsal-ventral (DV** or **dorsoventral) axis** is the line extending from back (dorsum) to belly (ventrum). The **right-left axis** is a line between the two lateral sides of the body. Although humans (for example) may look symmetrical, recall that in most of us, the heart and

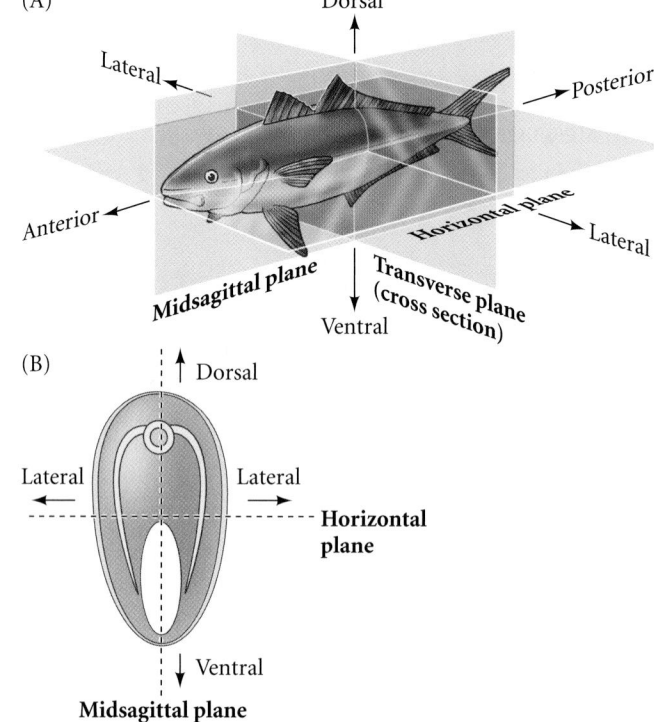

FIGURE 5.5   Axes of a bilaterally symmetrical animal. (A) A single plane, the midsagittal plane, divides the animal into left and right halves. (B) Cross sections bisecting the anterior-posterior axis.

liver are in the left half of the body. Somehow, the embryo knows that some organs belong on one side and other organs go on the other.

The remainder of this chapter and Chapters 6–9 will look at how representative species in several groups undergo cleavage, gastrulation, axis specification, and cell fate determination. With the exception of the human examples in Chapter 9, virtually all of the species and groups described (including fruit flies, snails, sea urchins, nematodes, zebrafish, chicks, and mice) have been important **model systems** for developmental biologists. In other words, these species are easily maintained in the laboratory and have special properties* that allow their mechanisms of development to be readily observed. The model systems represent a wide variety of cleavage types, patterns of gastrulation, and ways of specifying axes and cell fates.

We will now turn to a detailed description of early development in snails (shelled gastropod molluscs) and *C. elegans* (an extremely well-studied species of nematode worm). Despite their differences, the embryos of these two invertebrate groups are both characterized by what Davidson (2001) has called "Type I embryogenesis." This type of embryogenesis includes:

- Immediate activation of the zygotic genes
- Rapid specification of the blastomeres by the products of the zygotic genes and by maternally active genes
- A relatively small number of cells (a few hundred or fewer) at the start of gastrulation

EARLY DEVELOPMENT IN SNAILS

Snails have a long history as model organisms in developmental biology. They are abundant along the shores of all continents, they grow well in the laboratory, and they show variations in their development that can be correlated with their environmental needs. Snails also have large eggs and develop rapidly, specifying cell types very readily. Since the late 1800s, snails have been used as examples of autonomous (mosaic) development, where the loss of an early blastomere causes the loss of an entire structure. Indeed, in snail embryos, the cells responsible for certain organs can be localized to a remarkable degree. The results of experimental embryology can now be extended (and explained) by molecular analyses, leading to fascinating syntheses of development and evolution (Conklin 1897; Henry et al. 2010a).

---

*These properties include quick generation time, large litters, amenability to genetic and surgical manipulation, and the ability to develop under laboratory conditions. However, this very ability to develop in the laboratory sometimes precludes our asking certain questions concerning the relationship of development to an organism's natural habitat. These questions will be addressed in Chapter 19.

Cleavage in Snail Embryos

"As for the rest, the spiral is the fundamental theme of the molluscan organism. They are animals that twisted over themselves" (Flusser 2011). Indeed, the shells of snails are spirals, their larvae undergo a 180° torsion that brings their anus anteriorly above their head, and (most importantly) the cleavage of their early embryos is spiral. **Spiral holoblastic cleavage** is characteristic of several animal groups, including annelid worms, polyhelminth flatworms, and most molluscs (see Lambert 2010). The cleavage planes of spirally cleaving embryos are not parallel or perpendicular to the animal-vegetal axis of the egg; rather, cleavage is at oblique angles, forming a "spiral" arrangement of daughter blastomeres. The cells are in intimate contact with each other, producing the most thermodynamically stable packing orientation, much like adjacent soap bubbles. Moreover, spirally cleaving embryos usually undergo relatively fewer divisions before they begin gastrulation, making it possible to follow the fate of each cell of the blastula. When the fates of the individual blastomeres from annelid, flatworm, and mollusc embryos were compared, many of the same cells were seen in the same places, and their general fates were identical (Wilson 1898; Hejnol et al. 2010). Blastulae produced by spiral cleavage have no blastocoel and are called **stereoblastulae**.

FIGURE 5.6 depicts the cleavage pattern typical of many molluscan embryos. The first two cleavages are nearly meridional, producing four large macromeres (labeled A, B, C, and D). In many species, these four blastomeres are different sizes (D being the largest), a characteristic that allows them to be individually identified. In each successive cleavage, each **macromere** buds off a small **micromere** at its animal pole. Each successive quartet of micromeres is displaced to the right or to the left of its sister macromere, creating the characteristic spiral pattern. Looking down on the embryo from the animal pole, the upper ends of the mitotic spindles appear to alternate clockwise and counterclockwise (**FIGURE 5.7**). This arrangement causes alternate micromeres to form obliquely to the left and to the right of their macromeres.

At the third cleavage, the A macromere gives rise to two daughter cells, macromere 1A and micromere 1a. The B, C, and D cells behave similarly, producing the first quartet of micromeres. In most species, these micromeres are to the *right* of their macromeres (looking down on the animal pole). At the fourth cleavage, macromere 1A divides to form macromere 2A and micromere 2a, and micromere 1a divides to form two more micromeres, $1a^1$ and $1a^2$ (see Figure 5.6). The micromeres of this second quartet are to the left of the macromeres. Further cleavage yields blastomeres 3A and 3a from macromere 2A, and micromere $1a^2$ divides to produce cells $1a^{21}$ and $1a^{22}$. In normal development, the first-quartet micromeres form the head structures, while the second-quartet micromeres form the statocyst (balance organ) and shell. These fates are specified both by cytoplasmic localization and by induction (Cather 1967; Clement 1967; Render 1991; Sweet 1998).

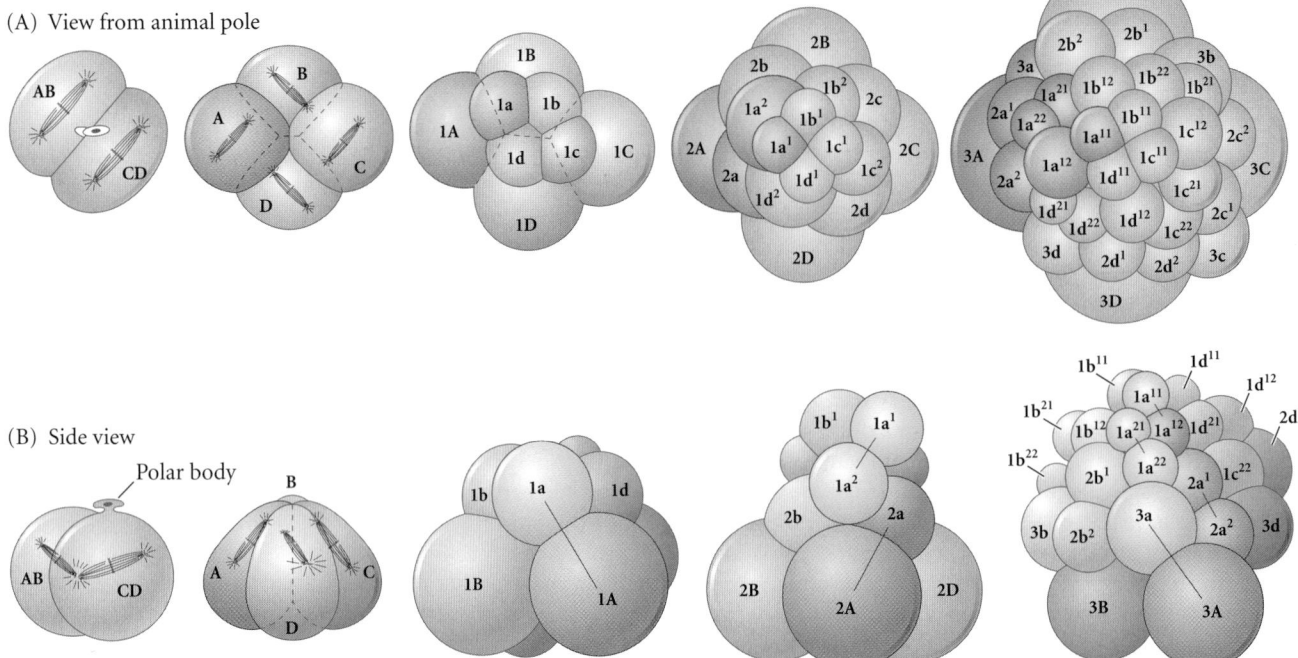

(A) View from animal pole

(B) Side view

Polar body

FIGURE 5.6 Spiral cleavage of the mollusc *Trochus* viewed from the animal pole (A) and from one side (B). Cells derived from the A blastomere are shown in color. The mitotic spindles, sketched in the early stages, divide the cells unequally and at an angle to the vertical and horizontal axes. Each successive quartet of micromeres (lowercase letters) is displaced to the right or to the left of its sister macromere (uppercase letters), creating the characteristic spiral pattern.

(A)

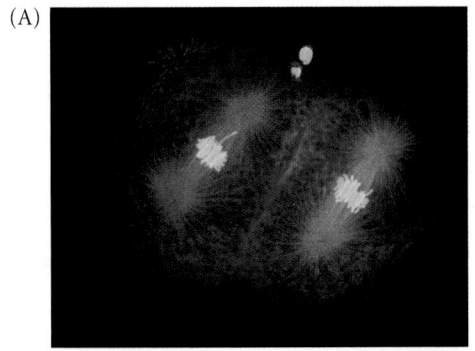

FIGURE 5.7 Spiral cleavage in molluscs. (A) The spiral nature of third cleavage can be seen in the confocal fluorescence micrograph of the 4-cell embryo of the clam *Acila castrenis*. Microtubules stain red, RNA stains green, and DNA stains yellow. Two cells and a portion of a third cell are visible; a polar body can be seen at the top of the micrograph. (B–D) Cleavage in the mud snail *Ilyanassa obsoleta*. The D blastomere is larger than the others, allowing the identification of each cell. Cleavage is dextral. (B) 8-cell stage. PB, polar body (a remnant of meiosis). (C) Mid-fourth cleavage (12-cell embryo). The macromeres have already divided into large and small spirally oriented cells; 1a–d have not divided yet. (D) 32-cell embryo. (A courtesy of G. von Dassow and the Center for Cell Dynamics; B–D from Craig and Morrill 1986, courtesy of the authors.)

(B)   (C)   (D)

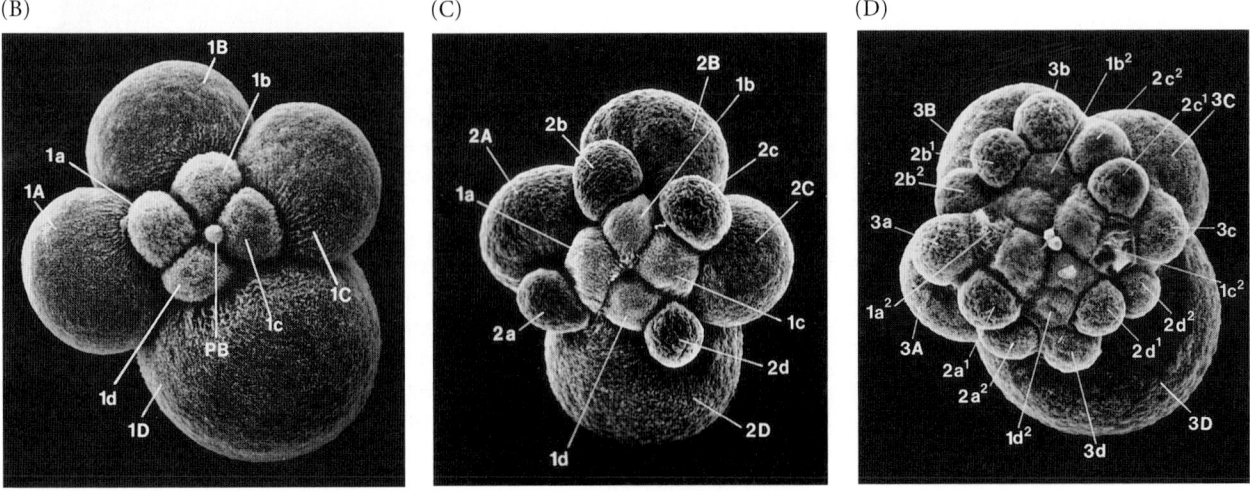

The orientation of the cleavage plane to the left or to the right is controlled by cytoplasmic factors in the oocyte. This was discovered by analyzing mutations of snail coiling. Some snails have their coils opening to the right of their shells (**dextral coiling**), whereas the coils of other snails open to the left (**sinistral coiling**). Usually the direction of coiling is the same for all members of a given species, but occasional mutants are found (i.e., in a population of right-coiling snails, a few individuals will be found with coils that open on the left). Crampton (1894) analyzed the embryos of such aberrant snails and found that their early cleavage differed from the norm. The orientation of the cells after the second cleavage was different in the sinistrally coiling snails as a result of a different orientation of the mitotic apparatus (**FIGURE 5.8**). You can see in Figure 5.8 that the position of the 4d blastomere is different in the right-coiling and left-coiling snail embryos. This 4d blastomere is rather special. It is often called the **mesentoblast**, since its progeny include most of the mesodermal organs (heart and larval muscle) and endodermal organs (gut tube).

In snails such as *Lymnaea*, the direction of snail shell coiling is controlled by a single pair of genes (Sturtevant 1923; Boycott et al. 1930; Shibazaki 2004). In *Lymnaea peregra*, rare mutants exhibiting sinistral coiling were found and mated with wild-type, dextrally coiling snails. These matings showed that the right-coiling allele, *D*, is dominant to the left-coiling allele, *d*. However, the direction of cleavage is determined not by the genotype of the developing snail but by the genotype of the snail's *mother*. A *dd* female snail can produce only sinistrally coiling offspring, even if the offspring's genotype is *Dd*. A *Dd* individual will coil either left or right, depending on the genotype of its mother. Such matings produce a chart like this:

		Genotype	Phenotype
DD ♀ × *dd* ♂	→	*Dd*	All right-coiling
DD ♂ × *dd* ♀	→	*Dd*	All left-coiling
Dd × *Dd*	→	1*DD*:2*Dd*:1*dd*	All right-coiling

The genetic factors involved in snail coiling are brought to the embryo by the oocyte cytoplasm. It is the genotype of the ovary in which the oocyte develops that determines which orientation cleavage will take. When Freeman and Lundelius (1982) injected a small amount of cytoplasm from dextrally coiling snails into the eggs of *dd* mothers, the resulting embryos coiled to the right. Cytoplasm from sinistrally coiling snails did not affect right-coiling embryos. These findings confirmed that the wild-type mothers were placing a factor into their eggs that was absent or defective in the *dd* mothers. We still have not identified the molecule produced by the gene designated *D* (or *d*). One candidate gene for *D* was *Par6*, an evolutionarily conserved gene that helps position the mitotic spindle. However, the sequence and localization of the PAR-6 protein were identical in the right- and left-coiling variants of *Lymnaea* (Homma et al. 2011).

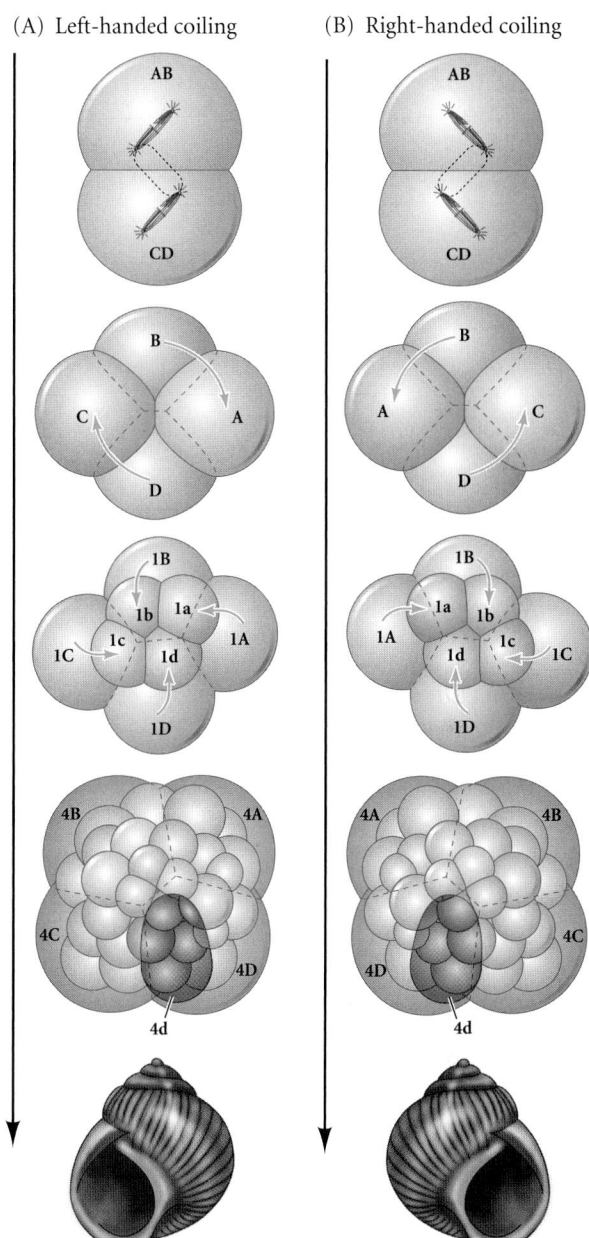

FIGURE 5.8 Looking down on the animal pole of left-coiling (A) and right-coiling (B) snails. The origin of sinistral and dextral coiling can be traced to the orientation of the mitotic spindle at the second cleavage. Left- and right-coiling snails develop as mirror images of each other. (After Morgan 1927.)

Some results, however, have come from analyzing how the right-left axis comes into being. As also happens in sea urchins and in vertebrates, the right-left axis in snails is defined by the Nodal family of paracrine factors. In snails, Nodal protein activates genes on the right side of dextrally coiling embryos and on the left side of sinistrally coiling embryos. Using glass needles to change the direction of cleavage at the 8-cell stage changes the location of *Nodal* gene expression (Grande and Patel 2009; Kuroda et al. 2009). Nodal appears to be expressed in the C-quadrant micromere lineages (which give rise to the ectoderm) and induces expression of the gene for the Pitx transcription factor (a target of Nodal in vertebrate axis formation) in the neighboring D-quadrant blastomeres.

● See **WEBSITE 5.1** Alfred Sturtevant and the genetics of snail coiling

The snail fate map

The fate maps of *Ilyanassa obsoleta* and *Crepidula fornicata* were constructed by injecting large polymers conjugated to fluorescent dyes into specific micromeres (Render 1997; Hejnol et al. 2007). The fluorescence is maintained over the period of embryogenesis and can be seen in the larval tissue derived from the injected cells. The results of the *Crepidula* studies, shown in **FIGURE 5.9**, indicated that the second-quartet micromeres (2a–d) generally contribute to the shell-forming mantle, the velum (the ciliated border of the larva), the mouth, and the heart. The third-quartet micromeres (3a–d) generate large regions of the foot, velum, esophagus, and heart. The 4d cell—the mesentoblast—contributes to the larval kidney, heart, retractor muscles, and intestine.

Autonomous cell specification and the polar lobe

Molluscs provide some of the most impressive examples of autonomous development, in which the blastomeres are specified by morphogenetic determinants located in specific regions of the oocyte. Autonomous specification of early blastomeres is especially prominent in those groups of animals having spiral cleavage, all of which initiate gastrulation at the future anterior end after only a few cell divisions. In molluscs, the mRNAs for some transcription factors and paracrine factors are placed in particular cells by associating them with certain centrosomes (**FIGURE 5.10**; Lambert and Nagy 2002; Kingsley et al. 2007; Henry et al. 2010b,c). This association allows the mRNA to enter specifically into one of the two daughter cells. In many instances, the mRNAs that get transported together into a particular tier of blastomeres have very similar 3′ tails, suggesting that the identity of the micromere tiers may be controlled largely by the 3′ untranslated regions (UTRs) of the mRNAs that attach to the centrosomes at each division (**FIGURE 5.11**; Rabinowitz and Lambert 2010). In other cases, the patterning molecules appear to be bound to a certain region of the egg that will form a unique structure called the polar lobe.

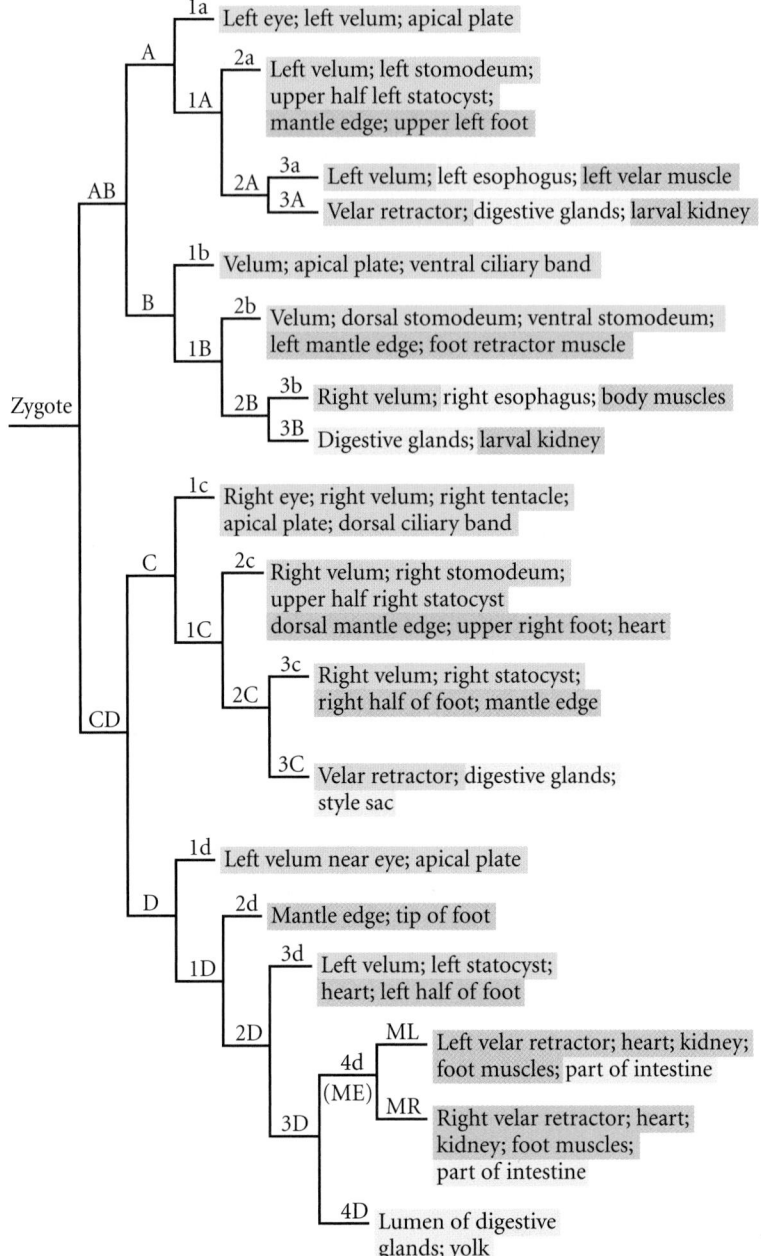

FIGURE 5.9   Fate map of the slipper snail *Crepidula fornicata*. Beads containing fluorescent dyes were injected into individual blastomeres. When the embryos developed into larvae, the descendants of each blastomere could be identified by their fluorescences. (After Render 1997 and Henry et al. 2010.)

THE POLAR LOBE   E. B. Wilson and his student H. E. Crampton observed that certain spirally cleaving embryos (mostly in molluscs and annelids) extrude a bulb of cytoplasm—the **polar lobe**—immediately before first cleavage. In some species of snails, the region uniting the polar lobe to the rest of the egg becomes a fine tube. The first cleavage splits the zygote asymmetrically, so that the polar lobe is connected only to the CD blastomere (**FIGURE 5.12A**). In several species,

(A) (B) (C)

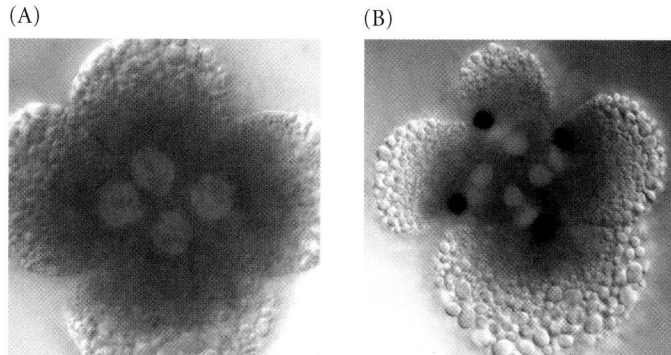

FIGURE 5.10 Association of *decapentaplegic* (*dpp*) mRNA with specific centrosomes of *Ilyanassa*. (A) In situ hybridization of the mRNA for Dpp in the 4-cell snail embryo shows no Dpp accumulation. (B) At prophase of the 4- to 8-cell stage, *dpp* mRNA (black) accumulates at one centrosome of the pair forming the mitotic spindle. (C) As mitosis continues, *dpp* mRNA is seen to attend the centrosome in the macromere rather than the centrosome in the micromere of each cell. The BMP-like paracrine factor encoded by *dpp* is critical to molluscan development. (From Lambert and Nagy 2002, courtesy of L. Nagy.)

nearly one-third of the total cytoplasmic volume is contained in this anucleate lobe, giving it the appearance of another cell (**FIGURE 5.12B**). The resulting three-lobed structure is often referred to as the **trefoil stage** embryo (**FIGURE 5.12C**).

Crampton (1896) showed that if one removes the polar lobe at the trefoil stage, the remaining cells divide normally. However, the resulting larva is incomplete (**FIGURE 5.13**), wholly lacking its endoderm (intestine) and mesodermal organs (such as the heart), as well as some ectodermal organs (such as eyes). Moreover, Crampton demonstrated that the same type of abnormal larva can be produced by removing the D blastomere from the 4-cell embryo. Crampton thus concluded that the polar lobe cytoplasm contains the endodermal and mesodermal determinants, and that these determinants give the D blastomere its endomesoderm-forming capacity. Crampton also showed that the localization of the mesodermal determinants is established shortly after fertilization, thereby demonstrating that a specific cytoplasmic region of the egg, destined

for inclusion in the D blastomere, contains whatever factors are necessary for the special cleavage rhythms of the D blastomere and for the differentiation of the mesoderm.

Centrifugation studies have demonstrated that the morphogenetic determinants sequestered in the polar lobe are probably located in the lobe's cytoskeleton or cortex, not in the its diffusible cytoplasm (Clement 1968). Van den Biggelaar (1977) obtained similar results when he removed the cytoplasm from the polar lobe with a micropipette. Cytoplasm from other regions of the cell flowed into the polar lobe, replacing the portion he removed, and subsequent development of these embryos was normal. In addition, when he added the diffusible polar lobe cytoplasm to the B blastomere, no duplicated structures were seen (Verdonk and Cather 1983). Therefore, the diffusible part of the polar lobe cytoplasm does not contain the morphogenetic determinants; they probably reside in the nonfluid cortical cytoplasm or on the cytoskeleton.

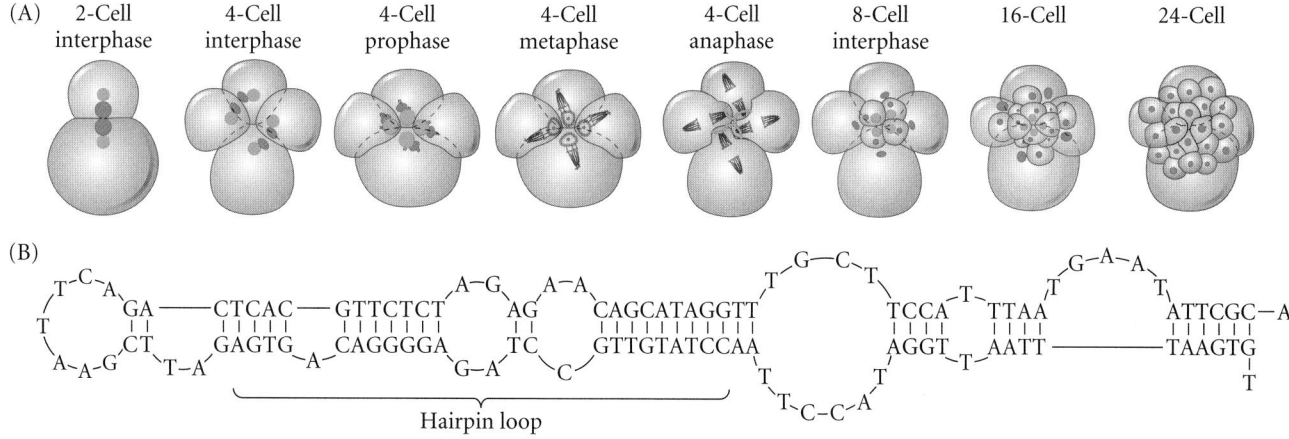

FIGURE 5.11 Importance of the 3' UTR for association of mRNAs with specific centrosomes. In *Ilyanassa*, the *R5LE* message is usually segregated into the first tier of micromeres. The message binds to one side of the centrosome complex (the side that will be in the small micromere.) (A) Normal *R5LE* mRNA distribution from the 2-cell through the 24-cell stage. The mRNA (green) associates with the centrosomic region (blue) that will generate the micromere tier and becomes localized to particular blastomeres by the 24-cell stage. (B) Hairpin loop of the 3' UTR of the *R5LE* message. (After Rabinowitz and Lambert 2010.)

(A)

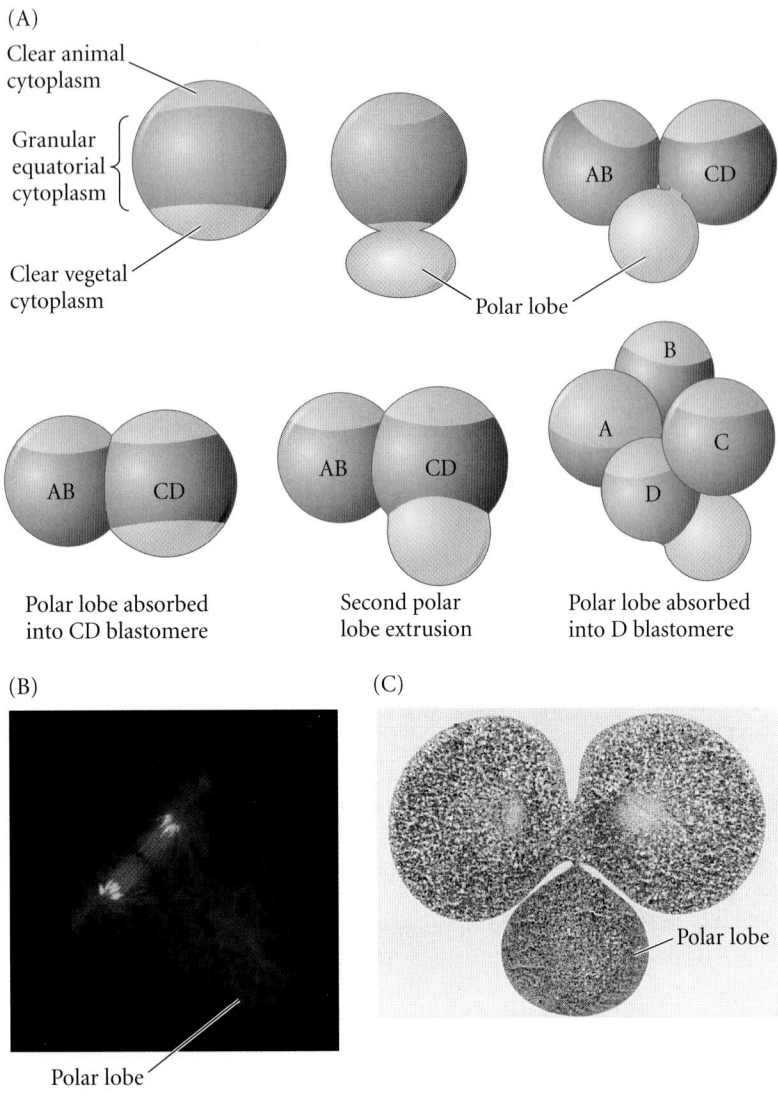

Clear animal cytoplasm

Granular equatorial cytoplasm

Clear vegetal cytoplasm

Polar lobe

AB

CD

Polar lobe absorbed into CD blastomere

Second polar lobe extrusion

Polar lobe absorbed into D blastomere

B

A

C

D

(B)

Polar lobe

(C)

Polar lobe

FIGURE 5.12 Polar lobe formation. (A) During cleavage, extrusion and reincorporation of the polar lobe occur twice. The CD blastomere absorbs the polar lobe material but extrudes it again prior to second cleavage. After this division, the polar lobe is attached only to the D blastomere, which absorbs its material. From this point on, no polar lobe is formed. (B) Late in the first division of a scallop embryo, the anucleate polar lobe (lower right) contains nearly one-third of the cytoplasmic volume. Microtubules are stained red, RNA is green, and the chromosomal DNA appears yellow. (C) Section through first-cleavage, or trefoil-stage, embryo of *Dentalium*. (A after Wilson 1904; B courtesy of G. von Dassow and the Center for Cell Dynamics; C courtesy of M. R. Dohmen.)

THE D BLASTOMERES Clement (1962) also analyzed the further development of the D blastomere in order to observe the further appropriation of these determinants. The development of the D blastomere can be traced in Figure 5.7B–D. This macromere, having received the contents of the polar lobe, is larger than the other three. When one removes the D blastomere or its first or second macromere derivatives (1D or 2D), one obtains an incomplete larva, lacking heart, intestine, velum, shell gland, eyes, and foot. This is essentially the same phenotype one gets when one removes the polar lobe. Since the D blastomeres do not directly contribute cells to many of these structures, it appears that the D-quadrant macromeres are involved in inducing other cells to have these fates.

When one removes the 3D blastomere shortly after the division of the 2D cell to form the 3D and 3d blastomeres, the larva produced looks similar to those formed by the removal of the D, 1D, or 2D macromeres. However, ablation of the 3D blastomere at a later time produces an almost normal larva, with eyes, foot, velum, and some shell gland, but no heart or intestine (see Figure 5.13). After the 4d cell is given off (by the division of the 3D blastomere), removal of the D derivative (the 4D cell) produces no qualitative difference in development. In fact, all the essential determinants for

(A)

(B)

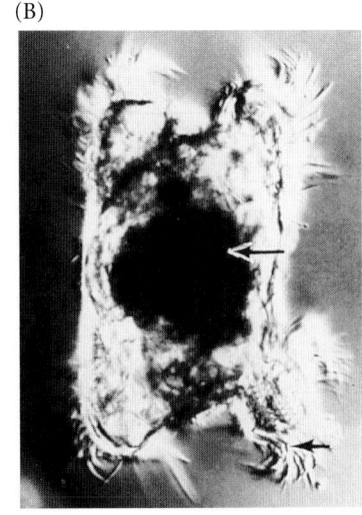

Y

E

V   VC

DV

V

ST

F

S

FIGURE 5.13 Importance of the polar lobe in the development of *Ilyanassa*. (A) Normal trochophore larva. (B) Abnormal larva, typical of those produced when the polar lobe of the D blastomere is removed. (E, eye; F, foot; S, shell; ST, statocyst; V, velum; VC, velar cilia; Y, residual yolk; ES, everted stomodeum; DV, disorganized velum.) (From Newrock and Raff 1975, courtesy of K. Newrock.)

SIDELIGHTS  SPECULATIONS

Adaptation by Modifying Embryonic Cleavage

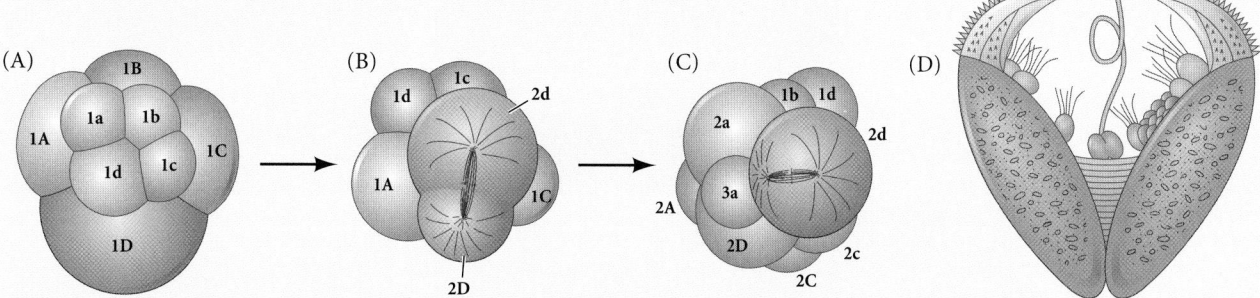

FIGURE 5.14 Formation of a glochidium larva by the modification of spiral cleavage. After the 8-cell embryo is formed (A), the placement of the mitotic spindle causes most of the D cytoplasm to enter the 2d blastomere (B). This large 2d blastomere divides (C), eventually giving rise to the large "bear-trap" shell of the larva (D). (After Raff and Kaufman 1983.)

Evolution is often the result of the hereditary alterations of embryonic development. Sometimes we are able to identify a specific modification of embryogenesis that has enabled the organism to survive in an otherwise inhospitable environment. One such modification, discovered by Frank Lillie in 1898, is brought about by an alteration of the typical pattern of molluscan spiral cleavage in the unionid family of clams.

Unlike most clams, *Unio* and its relatives live in swift-flowing streams. Streams create a problem for the dispersal of larvae: because the adults are sedentary, free-swimming larvae would always be carried downstream by the current. *Unio* clams have adapted to this environment via two modifications of their development. The first is an alteration in embryonic cleavage. In typical molluscan cleavage, either all the macromeres are equal in size or the 2D blastomere is the largest cell at that embryonic stage. However, cell division in *Unio* is such that the *2d* blastomere gets the largest amount of cytoplasm (**FIGURE 5.14**). This cell divides to produce most of the larval structures, including a gland capable of producing a large shell. The resulting larva is called a **glochidium** and resembles a tiny bear trap. Glochidia have sensitive hairs that cause the valves of the shell to snap shut when they are touched by the gills or fins of a wandering fish. The larvae can thus attach themselves to the fish and "hitchhike" until they are ready to drop off and metamorphose into adult clams. In this manner, they can spread upstream as well as downstream.

In some unionid species, glochidia are released from the female's brood pouch (marsupium) and then wait passively for a fish to swim by. Some other species, such as *Lampsilis altilis*, have increased the chances of their larvae finding a fish by yet another developmental modification. Many clams develop a thin mantle that flaps around the shell and surrounds the brood pouch. In some unionids, the shape of the brood pouch and the undulations of the mantle mimic the shape and swimming behavior of a minnow (Welsh 1969). To make the deception even better, the clams develop a black "eyespot" on one end and a flaring "tail" on the other (**FIGURE 5.15**). When a predatory fish is lured within range of this "prey," the clam discharges the glochidia from the brood pouch and the larvae attach to the fish's gills. Thus, the modification of existing developmental patterns has permitted unionid clams to survive in challenging environments.

FIGURE 5.15 Phony fish atop the unionid clam *Lampsilis altilis.* The "fish" is actually the brood pouch and mantle of the clam. The "eyes" and flaring "tail" attract predatory fish, and the glochidium larvae attach to the fish's gills. (Courtesy of Wendell R. Haag/USDA Forest Service.)

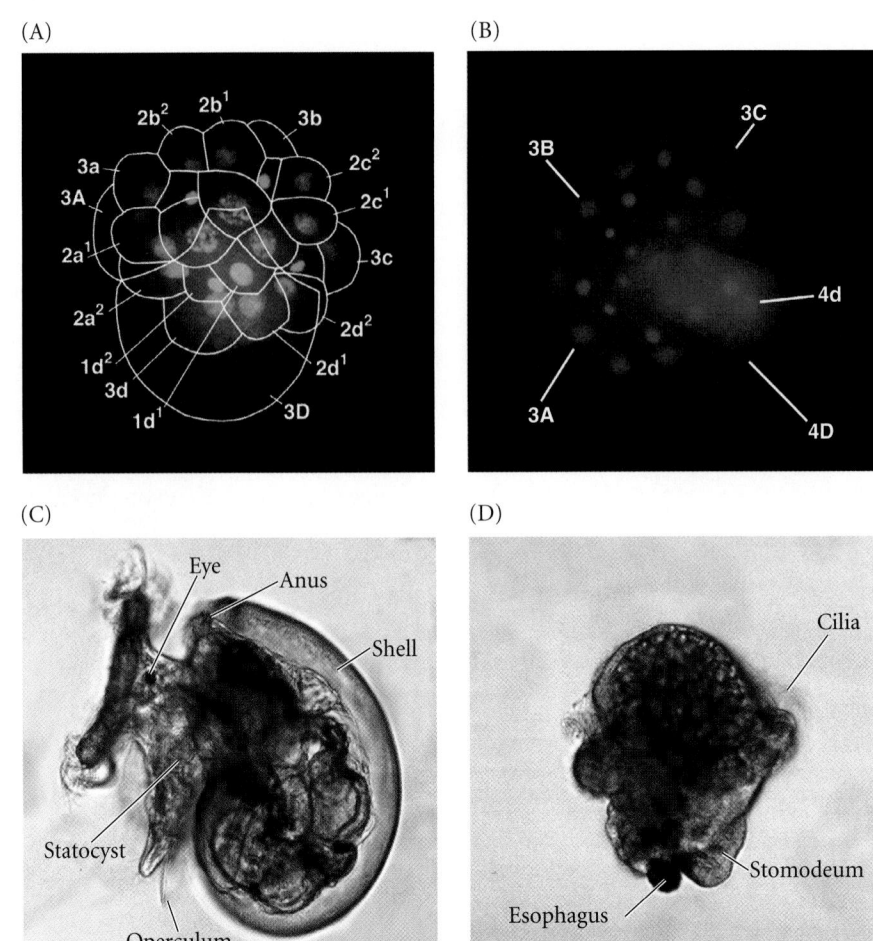

FIGURE 5.16 MAP kinase activity activated by D-quadrant snail blastomeres. (A) The 3D blastomere activates MAP kinase activity in adjacent *Ilyanassa* micromeres. Activated MAP kinase (blue stain) can be seen in the 3D macromere of *Ilyanassa* and in the micromeres above it ($1a–d^1$, $1d^2$, $2d^1$, $2d^2$, 3d). The nuclei are counterstained green, and the cell boundaries have been superimposed on the photographic image. Staining was done 30 minutes after the formation of the 3D macromere. (B) The presence of MAP kinase activity in the 4d blastomere (but not 4D) of *Crepidula*. (C) Control larva grown to veliger stage. (The veliger larval structure (with its ciliated "sail") is characteristic of gastropod and bivalve molluscs.) (D) Veliger larva treated with MAP kinase inhibitor 15 minutes after 3D blastomere formed. The shell, eye, statocyst, and operculum have not developed. (A,C,D from Lambert and Nagy 2001; B from Henry and Perry 2008, courtesy of the authors.)

heart and intestine formation are now in the 4d blastomere (also called the mesentoblast, as mentioned earlier), and removal of that cell results in a heartless and gutless larva (Clement 1986). The 4d blastomere is responsible for forming (at its next division) the two bilaterally paired blastomeres that give rise to both the mesodermal (heart) and endodermal (intestine) organs.

The mesodermal and endodermal determinants of the 3D macromere, then, are transferred to the 4d blastomere, while the inductive ability of the 3D blastomere (to induce eyes and shell gland, for instance) is needed during the time the 3D cell is formed but is not required afterward. The 3D cell appears to activate the MAP kinase signaling pathway in the micromeres above it (Lambert and Nagy 2001). In cells stained for activated MAP kinase, the stain is seen in those cells that require the signal from the 3D macromere for normal differentiation (**FIGURE 5.16**). Removal of 3D prevents MAP kinase signaling, and if the MAP kinase signaling is blocked by specific inhibitors, the resulting larvae look precisely like those formed by the deletion of the D blastomeres. Thus, the 3D macromere appears to activate the MAP kinase cascade in the ectodermal (eye- and shell gland-forming) micromeres above it. After the 3D divides to produce the 4d and 4D blastomeres, MAP kinase activity persists only in the 4d cell. If the MAP kinase is inhibited then, the structures (such as the heart) produced by the 4d derivatives fail to form (Henry and Perry 2008; Lambert 2008).

The 4d blastomere has at least two morphogenetic determinants not found in other cells. First, the cell appears to be specified by the presence of β-catenin, which becomes selectively stabilized in the 4d mesentoblast and its immediate progeny (**FIGURE 5.17A**; Henry et al. 2008; Rabinowitz et al. 2008). When translational inhibitors prevented β-catenin protein synthesis in the 4d cell, the cell did not differentiate into heart, muscles, or gut, and gastrulation failed to occur in those embryos (Henry et al. 2010b). Indeed, β-catenin may have an evolutionarily conserved role in mediating autonomous specification throughout the animal kingdom; in subsequent chapters we will see a similar role for this protein in both sea urchin and frog embryos.

The 4d blastomere also contains the *Nanos* mRNA and protein (**FIGURE 5.17B**). As with β-catenin, blocking translation of *Nanos* mRNA prevents formation of the larval muscles, heart, and intestine, completely disrupting the strict linear pattern of 4d cleavage (Rabinowitz et al. 2008). In addition, the germline cells (sperm and egg progenitors) do not form. As we will see throughout the book, the Nanos protein is often involved in specification of germ cell progenitors. Spiral-cleaving embryos often contain Nanos and another germline-associated protein, Vasa, in the progeny of the 4d cells. However, *which* progeny of 4d become the somatic cells (muscle, heart, gut) and which become germline cells is probably not autonomously specified, but appears to be induced by neighboring cell groups (Kranz et al. 2010).

(A)

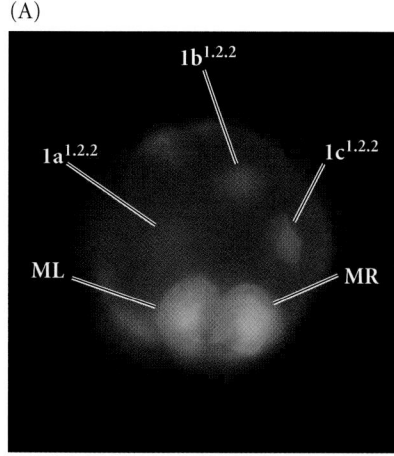

(B)

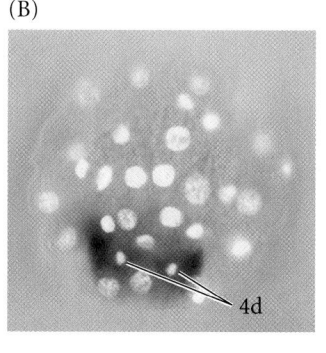 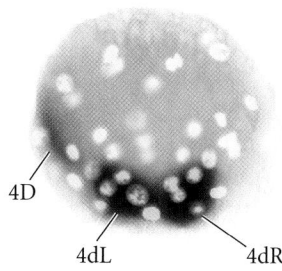

FIGURE 5.17 Morphogenetic determinants in the 4d snail blastomere. (A) β-Catenin expression in the progeny of the 4d blastomere of *Crepidula*. (B) *Nanos* mRNA localization (purple) in the dividing 4d blastomere and in its right and left progeny, 4dL and 4dR, of *Ilyanassa*. (A from Henry et al. 2010; B from Rabinowitz et al. 2008.)

SUMMARY Experiments have demonstrated that the non-diffusible polar lobe cytoplasm that is localized to the D blastomere is extremely important in normal molluscan development for several reasons:

- It contains the determinants for the proper cleavage rhythm and the cleavage orientation of the D blastomere.
- It contains certain determinants (those entering the 4d blastomere and hence leading to the mesentoblasts) for autonomous mesodermal and intestinal differentiation.
- It is responsible for permitting the inductive interactions (through the material entering the 3D blastomere) leading to the formation of the shell gland and eye.

In addition to these roles in cell differentiation, the material in the polar lobe is responsible for specifying the dorsal-ventral polarity of the embryo. When polar lobe material is forced to pass into the AB blastomere as well as into the CD blastomere, twin larvae form that are joined at their ventral surfaces (Guerrier et al. 1978; Henry and Martindale 1987).

The polar lobe is clearly important in normal snail development, but we still do not know the mechanisms for most of its effects. One possible clue has been provided by Atkinson (1987), who observed differentiated cells of the velum,

digestive system, and shell gland in lobeless embryos. But even though lobeless embryos can produce these cells, they appear unable to organize them into functional tissues and organs. Tissues of the digestive tract can be found but are not connected; individual muscle cells are scattered around the lobeless larva but are not organized into a functional muscle tissue. Thus, the developmental functions of the polar lobe are probably very complex and may be essential for axis formation.

● See WEBSITE 5.2 Modifications of cell fate in spiralian eggs

Gastrulation in Snails

The snail stereoblastula is relatively small, and its cell fates have already been determined by the D series of macromeres. Gastrulation is accomplished primarily by epiboly, wherein the micromeres at the animal cap multiply and "overgrow" the vegetal macromeres (Collier 1997; van den Biggelaar and Dictus 2004). Eventually, the micromeres cover the entire embryo, leaving a small blastopore slit at the vegetal pole (**FIGURE 5.18**). Molluscs are protostomes, forming their mouth regions from the blastopore; thus, this slit will become the mouth.

(A)

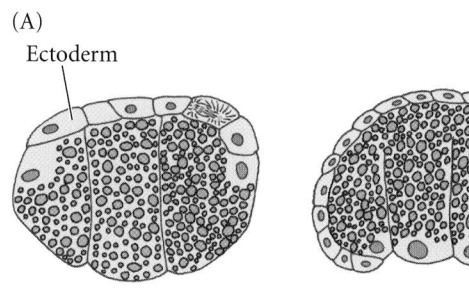

(B)

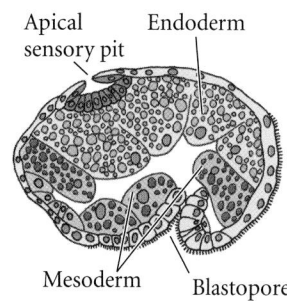

FIGURE 5.18 Gastrulation in molluscs. (A) Gastrulation in the snail *Crepidula*. The ectoderm undergoes epiboly from the animal pole and envelops the other cells of the embryo. (B) Late gastrula of the clam *Acila*, stained for actin microfilaments (orange) and nucleic acid (blue). (A after Conklin 1897; B courtesy of G. von Dassow and the Center for Cell Dynamics.)

THE NEMATODE *C. ELEGANS*

Unlike the snail, with its long embryological pedigree, the nematode *Caenorhabditis elegans* (usually referred to as *C. elegans*) is a thoroughly modern model system, uniting developmental biology with molecular genetics. In the 1970s, Sydney Brenner and his students sought an organism wherein it might be possible to identify each gene involved in development as well as to trace the lineage of each and every cell (Brenner 1974). Nematode roundworms seemed like a good group to start with because embryologists such as Richard Goldschmidt and Theodor Boveri had already shown that several nematode species have a relatively small number of chromosomes and a small number of cells with invariant cell lineages.

Brenner and his colleagues eventually settled on *C. elegans*, a small (1 mm long), free-living (i.e., nonparasitic) soil nematode with relatively few cell types. *C. elegans* has a rapid period of embryogenesis (about 16 hours), which it can accomplish in a petri dish (**FIGURE 5.19A**). Moreover, its predominant adult form is hermaphroditic, with each individual producing both eggs and sperm. These roundworms can reproduce either by self-fertilization or by cross-fertilization with the infrequently occurring males.

The body of an adult *C. elegans* hermaphrodite contains exactly 959 somatic cells, and the entire cell lineage has been traced through its transparent cuticle (**FIGURE 5.19B**; Sulston and Horvitz 1977; Kimble and Hirsh 1979). For each cell in the embryo, we can say where it came from (i.e., which cells

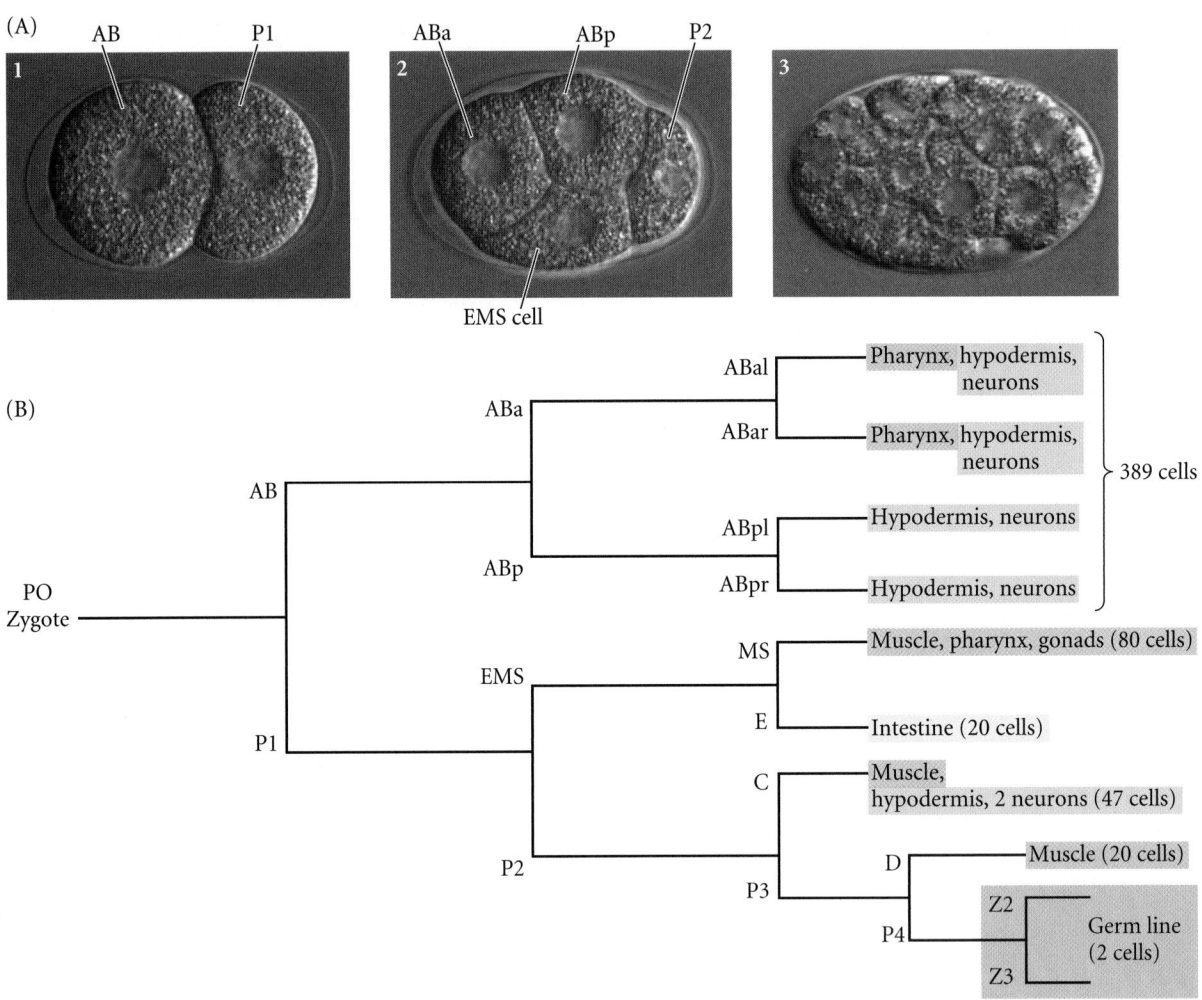

FIGURE 5.19 Development in the nematode *Caenorhabditis elegans* is rapid and results in an adult with exactly 959 somatic cells. Individual cell lineages have been traced through the course of the animal's development. (A) Scanning electron micrographs of the cleaving embryo. (1) The AB cell (left) and the P1 cell (right) are the result of the first asymmetric division. Each will give rise to a different cell lineage. (2) The 4-cell embryo shows ABa, ABp, P2, and EMS cells. (3) Gastrulation is initiated by the movement of E-derived cells toward the center of the embryo. (B) Abbreviated cell lineage chart. The germ line segregates into the posterior portion of the most posterior (P) cell. The first three cell divisions produce the AB, C, MS, and E lineages. The number of derived cells (in parentheses) refers to the 558 cells present in the newly hatched larva. Some of these continue to divide to produce the 959 somatic cells of the adult. (A courtesy of D. G. Morton and K. Kemphues; B after Pines 1992, based on Sulston and Horvitz 1977 and Sulston et al. 1983.)

in earlier embryonic stages were its progenitors) and which tissues it will contribute to forming. Furthermore, unlike vertebrate cell lineages, the *C. elegans* lineage is almost entirely invariant from one individual to the next; there is little room for randomness (Sulston et al. 1983). It also has a very compact genome. The *C. elegans* genome was the first complete sequence ever obtained for a multicellular organism (*C. elegans* Sequencing Consortium 1999). Although it has about the same number of genes as humans (18,000–20,000 genes whereas *Homo sapiens* has 20,000–25,000), the nematode has only about 3% the number of nucleotides in its genome (Hodgkin 1998, 2001).*

C. elegans displays the rudiments of nearly all the major types of bodily systems (feeding, nervous, reproductive, etc.—although there is no skeleton), and it exhibits an aging phenotype before it dies. Neurobiologists celebrate its minimal nervous system (302 neurons), and each of its 7,600 synapses have been identified (White et al. 1986; Seifert et al. 2006). In addition, *C. elegans* is particularly friendly to molecular biologists. DNA injected into *C. elegans* cells is readily incorporated into their nuclei, and *C. elegans* can take up antisense RNA from its culture medium.

Cleavage and Axis Formation in *C. elegans*

Fertilization in *C. elegans* is a not your typical sperm-meets-egg story. As most *C. elegans* individuals are hermaphrodites, they produce both sperm and eggs, and fertilization occurs within the adult individual. The egg becomes fertilized by rolling through a region of the embryo (the spermatheca) containing mature sperm (**FIGURE 5.20A,B**). The sperm are not typical long-tailed, streamlined cells, but are small, round, unflagellated cells that travel slowly by amoeboid motion. When a sperm fuses with the egg cell membrane, polyspermy is prevented by the rapid synthesis of chitin (the protein comprising the cuticle) by the newly fertilized egg (Johnston et al. 2010). The fertilized egg undergoes early divisions and is extruded through the vulva.

---

*This similarity in gene number is rather surprising, to say the least. "What does a worm want with 20,000 genes?" wrote Jonathan Hodgkin, curator of the *C. elegans* gene map. Humans have trillions of cells, a four-chambered heart, an incredibly regionalized brain, intricate limbs, and remarkable vascular networks. *C. elegans*, on the other hand, has no hands. Nor does the nematode have chambers in its heart, or any head to speak of. Thousands of these organisms would fit under our fingernails (which *C. elegans* also lacks). Hodgkin (2001) notes that, whereas human developmental proteins often have many functions, each *C. elegans* protein appears to have only a single function. In addition, nematode genes do not have nearly the capacity for producing alternatively spliced RNAs that human genes do, so humans get more types of protein per gene. In addition, *C. elegans* may have duplicated many of its genes, thus inflating its gene number. Whatever the case, the mere *number* of genes does not seem to be responsible for the huge physical differences between worms and humans.

Rotational cleavage of the egg

The zygote of *C. elegans* exhibits rotational holoblastic cleavage (**FIGURE 5.20C**). During early cleavage, each asymmetrical division produces one founder cell (denoted AB, E, MS, C, and D) that produces differentiated descendants; and one stem cell (the P1–P4 lineage). The anterior-posterior axis is determined before the first cell division, and the cleavage furrow is located asymmetrically along this axis of the egg, closer to what will be the posterior pole. The first cleavage forms an anterior founder cell (AB) and a posterior stem cell (P1). The dorsal-ventral axis is determined during the second division. The founder cell (AB) divides equatorially (longitudinally, 90° to the anterior-posterior axis), while the P1 cell divides meridionally (transversely) to produce another founder cell (EMS) and a posterior stem cell (P2). The EMS cell marks the ventral region of the developing embryo. The stem cell lineage always undergoes meridional division to produce (1) an anterior founder cell and (2) a posterior cell that will continue the stem cell lineage. The right-left axis is seen at the transition between the 4- and 8-cell stage. Here, the locations of two "granddaughters" of the AB cell (ABal and ABpl) are on the left side, while two others (ABar and ABpr) are on the right (see Figure 5.20C).

Anterior-posterior axis formation

The decision as to which end of the egg will become the anterior and which the posterior seems to reside with the position of the sperm pronucleus (**FIGURE 5.21**). When the sperm pronucleus enters the oocyte cytoplasm, the oocyte has no polarity. However, the oocyte does have a distinct arrangement of "partitioning-defective," or **PAR proteins**,† in its cytoplasm. PAR-3 and PAR-6, interacting with the protein kinase PKC-3 (mutations of which cause defective partitioning), are uniformly distributed in the cortical cytoplasm. PKC-3 restricts PAR-1 and PAR-2 to the internal cytoplasm by phosphorylating them. The sperm centrosome (microtubule-organizing center) contacts the cortical cytoplasm through its microtubules and initiates cytoplasmic movements that push the male pronucleus to the nearest end of the oblong oocyte. That end becomes the posterior pole (Goldstein and Hird 1996). Moreover, these microtubules locally protect PAR-2 from phosphorylation, thereby allowing PAR-2 (and its binding partner, PAR-1) into the cortex nearest the centrosome. Once PAR-1 is in the cortical cytoplasm, it phosphorylates PAR-3, causing PAR-3 (and its binding partner, PKC-3) to leave the cortex. At the same

---

†Although originally discovered in *C. elegans*, many species use the PAR proteins in establishing cell polarity. They are critical for forming the anterior and posterior regions of *Drosophila* oocytes, and they distinguish the basal and apical ends of *Drosophila* epithelial cells. *Drosophila* PAR proteins are also important in distinguishing which product of a neural stem cell division becomes the neuron and which remains a stem cell. PAR-1 homologues in mammals also appear to be critical in neural polarity (Goldstein and Macara 2007; Nance and Zallen 2011).

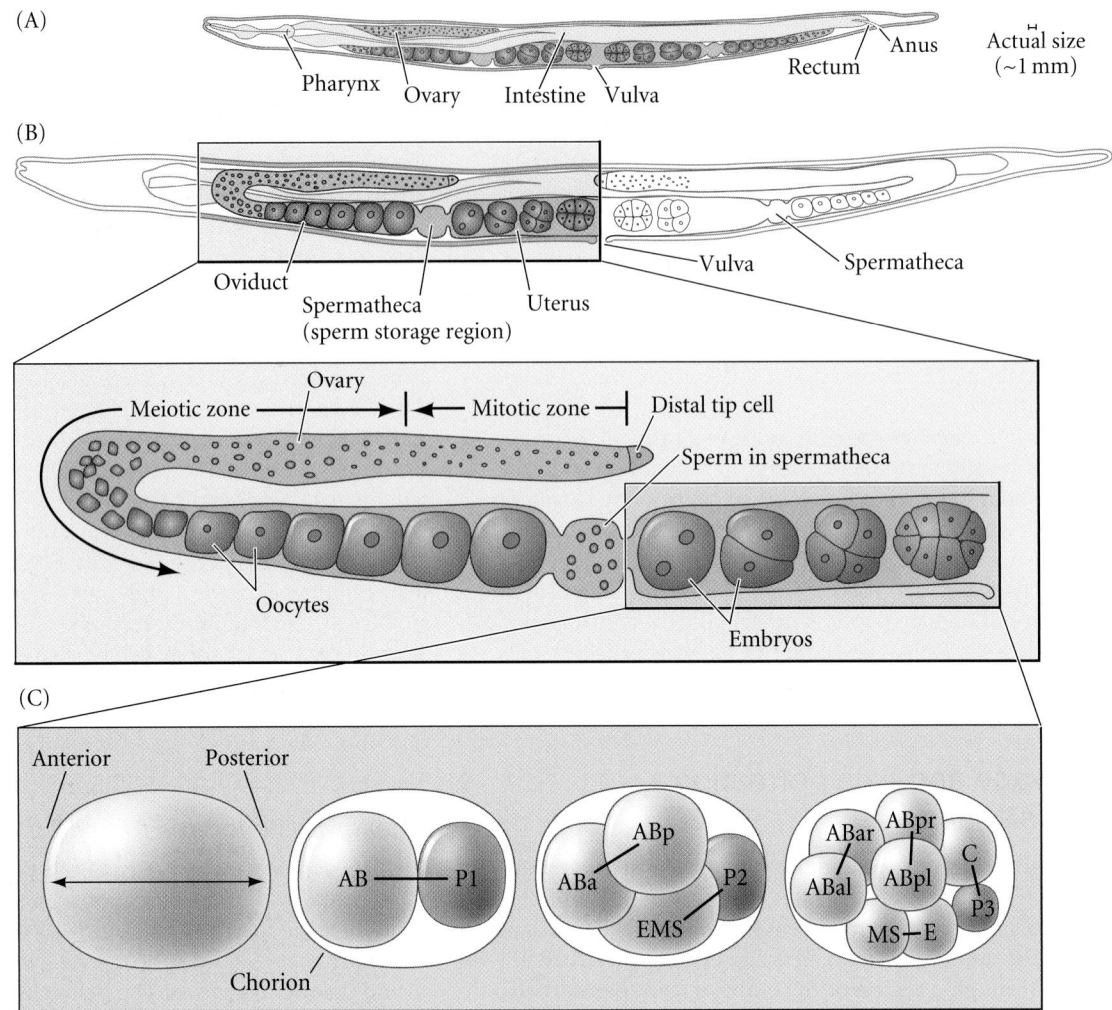

FIGURE 5.20 Fertilization and early cleavages in *C. elegans*. (A) Side view of adult hermaphrodite. Sperm are stored such that a mature egg must pass through the sperm on its way to the vulva. (B) The germ cells undergo mitosis near the distal tip of the gonad. As they move farther from the distal tip, they enter meiosis. Early meioses form sperm, which are stored in the spermatheca. Later meioses form eggs, which are fertilized as they roll through the spermatheca. (C) Early development occurs as the egg is fertilized and moves toward the vulva. The P lineage consists of stem cells that will eventually form the germ cells. (After Pines 1992, based on Sulston and Horvitz 1977 and Sulston et al. 1983.)

time, the sperm microtubules induce the contraction of the actin-myosin cytoskeleton toward the anterior, thereby clearing PAR-3, PAR-6, and PKC-3 from the posterior of the 1-cell embryo. During first cleavage, the metaphase plate is closer to the posterior, and the fertilized egg is divided into two cells, one having the anterior PARs (PAR-6 and PAR-3) and one having the posterior PARs (PAR-2 and PAR-1) (Goehring et al. 2011; Motegi et al. 2011).

Dorsal-ventral and right-left axis formation

The dorsal-ventral axis of *C. elegans* is established in the division of the AB cell. As the cell divides, it becomes longer than the eggshell is wide. This squeezing causes the daughter cells to slide, one becoming anterior and one posterior (hence their respective names, ABa and ABp; see Figure 5.20C). The squeezing also causes the ABp cell to take a position above the EMS cell that results from the division of the P1 blastomere.

The ABp cell thus defines the future dorsal side of the embryo, while the EMS cell—the precursor of the muscle and gut cells—marks the future ventral surface of the embryo.

The left-right axis is not readily seen until the 12-cell stage, when the MS blastomere (from the division of the EMS cell) contacts half the "granddaughters" of the ABa cell, distinguishing the right side of the body from the left side (Evans et al. 1994). This asymmetric signaling sets the stage for several other inductive events that make the right side of the larva differ from the left (Hutter and Schnabel 1995). Indeed, even the different neuronal fates seen on the left and right sides of the *C. elegans* brain can be traced back to that single change at the 12-cell stage (Poole and Hobert 2006). Although readily seen at the 12-cell stage, the first indication of left-right symmetry probably occurs at the zygote stage. Just prior to first cleavage, the embryo rotates 120° inside its vitelline envelope. This rotation is always in the same

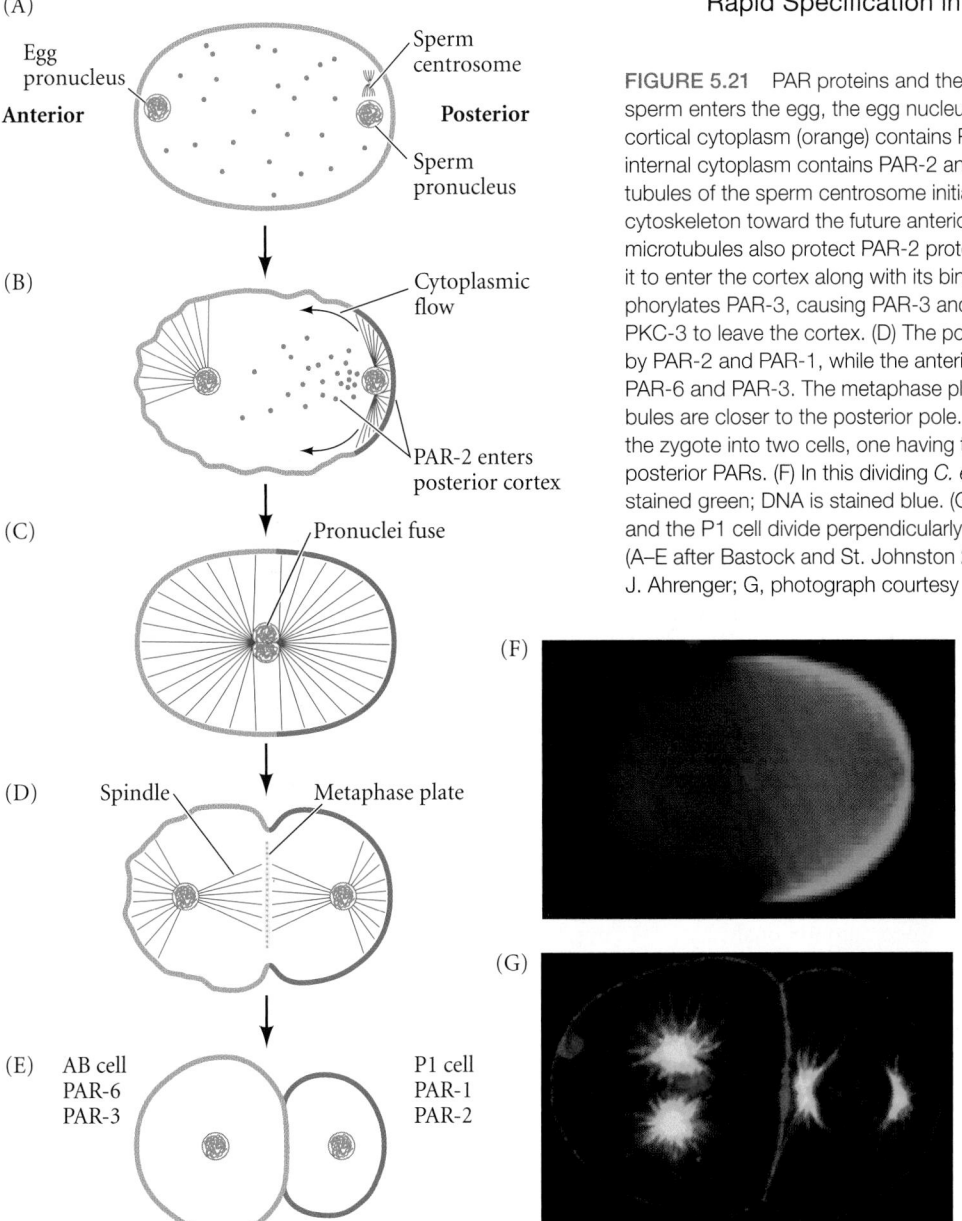

(A)

Egg
pronucleus

Anterior

Sperm
centrosome

Posterior

Sperm
pronucleus

(B)

Cytoplasmic
flow

PAR-2 enters
posterior cortex

(C)

Pronuclei fuse

(D)

Spindle

Metaphase plate

(E)

AB cell
PAR-6
PAR-3

P1 cell
PAR-1
PAR-2

(F)

(G)

FIGURE 5.21 PAR proteins and the establishment of polarity. (A) When sperm enters the egg, the egg nucleus is undergoing meiosis (left). The cortical cytoplasm (orange) contains PAR-3, PAR-6, and PKC-3, and the internal cytoplasm contains PAR-2 and PAR-1 (purple dots). (B,C) Microtubules of the sperm centrosome initiate contraction of the actin-based cytoskeleton toward the future anterior side of the embryo. These sperm microtubules also protect PAR-2 protein from phosphorylation, allowing it to enter the cortex along with its binding partner, PAR-1. PAR-1 phosphorylates PAR-3, causing PAR-3 and its binding partners PAR-6 and PKC-3 to leave the cortex. (D) The posterior of the cell becomes defined by PAR-2 and PAR-1, while the anterior of the cell becomes defined by PAR-6 and PAR-3. The metaphase plate is asymmetric, as the microtubules are closer to the posterior pole. (E) The metaphase plate separates the zygote into two cells, one having the anterior PARs and one the posterior PARs. (F) In this dividing *C. elegans* zygote, PAR-2 protein is stained green; DNA is stained blue. (G) In second division, the AB cell and the P1 cell divide perpendicularly (90° differently from each other). (A–E after Bastock and St. Johnston 2011; F, photograph courtesy of J. Ahrenger; G, photograph courtesy of J. White.)

direction relative to the already established anterior-posterior axis, indicating that the embryo already has a left-right chirality. If cytoskeleton proteins or the PAR proteins are inhibited, the direction of the rotation and subsequent chirality become random (Wood and Schonegg 2005; Pohl 2011).

Control of blastomere identity

C. elegans demonstrates both the conditional and autonomous modes of cell specification. Both modes can be seen if the first two blastomeres are experimentally separated (Priess and Thomson 1987). The P1 cell develops autonomously without the presence of AB, generating all the cells it would normally make, and the result is the posterior half of an embryo. However, the AB cell in isolation makes only a small fraction of the cell types it would normally make. For instance, the resulting ABa blastomere fails to make the anterior pharyngeal muscles that it would have made in an intact embryo.

Therefore, the specification of the AB blastomere is conditional, and it needs to interact with the descendants of the P1 cell in order to develop normally.

AUTONOMOUS SPECIFICATION The determination of the P1 lineages appears to be autonomous, with cell fates determined by internal cytoplasmic factors rather than by interactions with neighboring cells (see Maduro 2006). The SKN-1, PAL-1, and PIE-1 proteins encode transcription factors that act intrinsically to determine the fates of cells derived from the four P1-derived somatic founder cells (MS, E, C, and D).

The **SKN-1** protein is a maternally expressed transcription factor that controls the fate of the EMS blastomere, the cell that generates the posterior pharynx. After first cleavage, only the posterior blastomere—P1—has the ability to produce pharyngeal cells when isolated. After P1 divides, only EMS is able to generate pharyngeal muscle cells in isolation

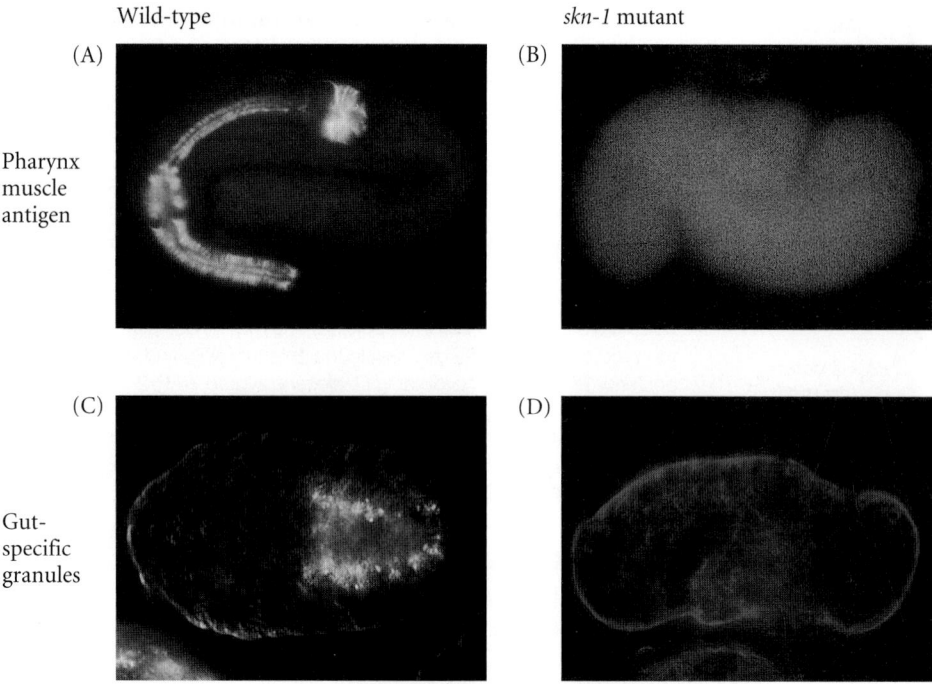

Wild-type   skn-1 mutant

(A)   (B)

Pharynx
muscle
antigen

(C)   (D)

Gut-
specific
granules

FIGURE 5.22 Deficiencies of intestine and pharynx in *skn-1* mutants of *C. elegans*. Embryos derived from wild-type females (A,C) and from females homozygous for mutant *skn-1* (B,D) were tested for the presence of pharyngeal muscles (A,B) and gut-specific granules (C,D). A pharyngeal muscle-specific antibody labels the pharynx musculature of those embryos derived from wild-type females (A) but does not bind to any structure in the embryos from *skn-1* mutant females (B). Similarly, the gut granules characteristic of embryonic intestines (C) are absent from embryos derived from the *skn-1* mutant females (D). (From Bowerman et al. 1992a, courtesy of B. Bowerman.)

(Priess and Thomson 1987). Similarly, when the EMS cell divides, only one of its progeny, MS, has the intrinsic ability to generate pharyngeal tissue. These findings suggest that pharyngeal cell fate may be determined autonomously, by maternal factors residing in the cytoplasm that are parceled out to these particular cells.

Bowerman and co-workers (1992a,b, 1993) found maternal effect mutants lacking pharyngeal cells and were able to isolate a mutation in the *skn-1* gene. Embryos from homozygous *skn-1*-deficient mothers lack both pharyngeal mesoderm and endoderm derivatives of EMS (**FIGURE 5.22**). Instead of making the normal intestinal and pharyngeal structures, these embryos seem to make extra hypodermal (skin) and body wall tissue where their intestine and pharynx should be. In other words, the EMS blastomere appears to be respecified as C. Only those cells destined to form pharynx or intestine are affected by this mutation. Moreover, the protein encoded by the *skn-1* gene has a DNA-binding site motif similar to that seen in the bZip family of transcription factors (Blackwell et al. 1994). SKN-1 is a maternal transcription factor that activates the transcription of at least two genes, *med-1* and *med-2*, whose products are also transcription factors. The MED transcription factors appear to specify the entire fate of the EMS cell, since expression of the *med* genes in other cells can cause non-EMS cells to become EMS even if SKN-1 is absent (Maduro et al. 2001).

Another transcription factor, **PAL-1**, is also required for the differentiation of the P1 lineage. PAL-1 activity is needed for the normal development of the *somatic* (but not the germline) descendants of the P2 blastomere, where it specifies muscle production. Embryos lacking PAL-1 have no somatic cell types derived from the C and D stem cells (Hunter and

Kenyon 1996). PAL-1 is regulated by the MEX-3 protein, an RNA-binding protein that appears to inhibit the translation of *pal-1* mRNA. Wherever MEX-3 is expressed, PAL-1 is absent. Thus, in *mex-3*-deficient mutants, PAL-1 is seen in every blastomere. SKN-1 also inhibits PAL-1 (thereby preventing it from becoming active in the EMS cell). But what keeps *pal-1* from functioning in the prospective germ cells and turning them into muscles? In the germ line, PAL-1 synthesis is prevented by the PUF-8 protein, which binds to the 3′UTR of *pal-1* mRNA and blocks its translation (Mainpal et al. 2011).

A third transcription factor, **PIE-1**, is necessary for germline cell fate. PIE-1 is placed into the P blastomeres through the action of the PAR-1 protein (**FIGURE 5.23**), and it appears to inhibit both SKN-1 and PAL-1 function in the P2 and subsequent germline cells (Hunter and Kenyon 1996). Mutations of the maternal *pie-1* gene result in germline blastomeres adopting somatic fates, with the P2 cell behaving similarly to a wild-type EMS blastomere. The localization and the genetic properties of PIE-1 suggest that it represses the establishment of somatic cell fate and preserves the totipotency of the germ cell lineage (Mello et al. 1996; Seydoux et al. 1996).

CONDITIONAL SPECIFICATION As mentioned earlier, the *C. elegans* embryo uses both autonomous and conditional modes of specification. Conditional specification can be seen in the development of the endoderm cell lineage. At the 4-cell stage, the EMS cell requires a signal from its neighbor (and sister cell), the P2 blastomere. Usually, the EMS cell divides into an MS cell (which produces mesodermal muscles) and an E cell (which produces the intestinal endoderm). If the P2 cell is removed at the early 4-cell stage, the EMS cell will divide into two MS cells, and no endoderm will be produced. If the EMS

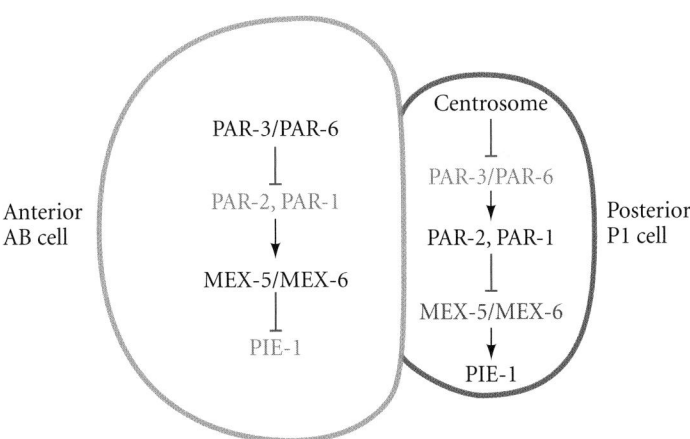

FIGURE 5.23 Segregation of PIE-1 determinant into the P1 blastomere at the 2-cell stage. The sperm centrosome inhibits the presence of the PAR-3/PAR-6 complex in the posterior of the egg. This allows the function of PAR-2 and PAR-1, which inhibit the MEX-5 and MEX-6 proteins that would degrade PIE-1. So while PIE-1 is degraded in the resulting anterior AB cell, it is preserved in the posterior P1 cell. (After Gönczy and Rose 2005.)

cell is recombined with the P2 blastomere, however, it will form endoderm; it will not do so, however, when combined with ABa, ABp, or both AB derivatives (Goldstein 1992). Specification of the MS cell begins with maternal SKN-1 activating the genes encoding transcription factors such as MED-1 and MED-2. The POP-1 signal blocks the pathway to the E (endodermal) fate in the prospective MS cell to become MS by blocking the ability of MED-1 and MED-2 to activate the *tbx-35* gene (**FIGURE 5.24**; Broitman-Maduro et al. 2006; Maduro 2009). Throughout the animal kingdom, TBX proteins are known to be active in mesoderm formation; TBX-35 acts to activate the mesodermal genes in the pharynx (*pha-4*) and muscles (*hlh-1*) of *C. elegans*.

The P2 cell produces a signal that interacts with the EMS cell and instructs the EMS daughter next to it to become the

E cell. This message is transmitted through the Wnt signaling cascade (**FIGURE 5.25**; Rocheleau et al. 1997; Thorpe et al. 1997; Walston et al. 2004). The P2 cell produces the MOM-2 protein, a *C. elegans* Wnt protein. MOM-2 is received in the EMS cell by the MOM-5 protein, a *C. elegans* version of the Wnt receptor protein Frizzled. The result of this signaling cascade is to downregulate the expression of the *pop-1* gene in the EMS daughter destined to become the E cell. In *pop-1*-deficient embryos, both EMS daughter cells become E cells (Lin et al. 1995; Park et al. 2004).

The P2 cell is also critical in giving the signal that distinguishes ABp from its sister, ABa (see Figure 5.25). ABa gives rise to neurons, hypodermis, and the anterior pharynx cells, while ABp makes only neurons and hypodermal cells. However, if one experimentally reverses the positions of these two cells, their fates are similarly reversed and a normal embryo forms. In other words, ABa and ABp are equivalent cells whose fates are determined by their positions in the embryo (Priess and Thomson 1987). Transplantation and genetic studies have shown that ABp becomes different from ABa through its interaction with the P2 cell. In an unperturbed embryo, both ABa and ABp contact the EMS blastomere, but only ABp contacts the P2 cell. If the P2 cell is killed at the early 4-cell stage, the ABp cell does not generate its normal complement of cells (Bowerman et al. 1992a,b). Contact between ABp and P2 is essential for the specification of ABp cell fates, and the ABa cell can be made into an ABp-type cell if it is forced into contact with P2 (Hutter and Schnabel 1994; Mello et al. 1994).

This interaction is mediated by the GLP-1 protein on the ABp cell and the APX-1 (anterior pharynx excess) protein on the P2 blastomere. In embryos whose mothers have mutant *glp-1*, ABp is transformed into an ABa cell (Hutter and Schnabel 1994; Mello et al. 1994). The GLP-1 protein is a member of a widely conserved family called the Notch proteins, which serve as cell membrane receptors in many cell-cell

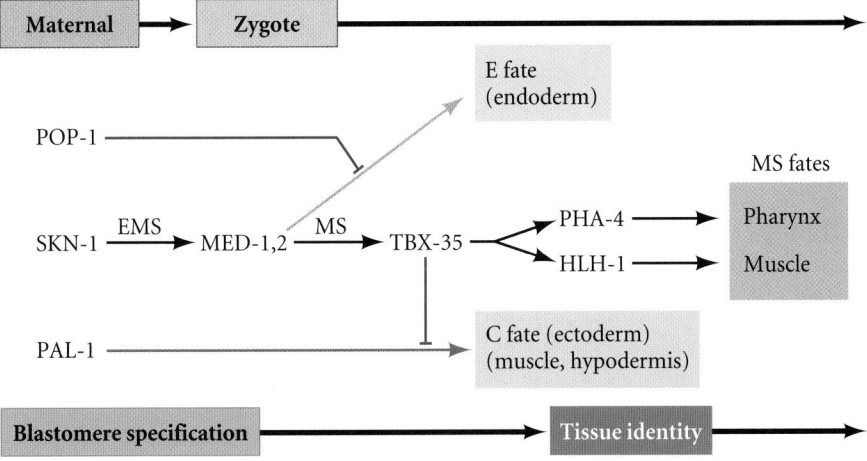

FIGURE 5.24 Model for specification of the MS blastomere. Maternal SKN-1 activates GATA transcription factors MED-1 and MED-2 in the EMS cell. The POP-1 signal prevents these proteins from activating the endodermal transcription factors (such as END-1) and instead activates the *tbx-35* gene. The TBX-35 transcription factor activates mesodermal genes in the MS cell, including *pha-4* in the pharynx lineage and *hlh-1* (which encodes a myogenic transcription factor) in muscles. TBX-35 also inhibits *pal-1* gene expression, thereby preventing the MS cell from acquiring the C-blastomere fates. (After Broitman-Maduro et al. 2006.)

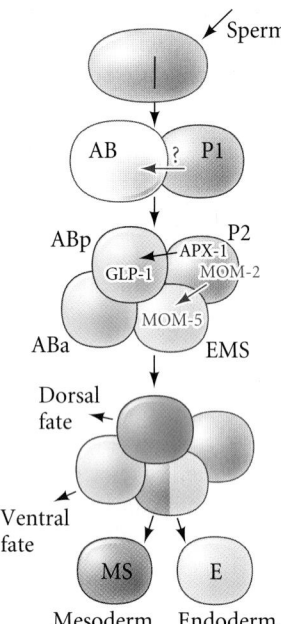

FIGURE 5.25 Cell-cell signaling in the 4-cell embryo of *C. elegans*. The P2 cell produces two signals: (1) the juxtacrine protein APX-1 (a Delta homologue), which is bound by GLP-1 (Notch) on the ABp cell, and (2) the paracrine protein MOM-2 (Wnt), which is bound by the MOM-5 (Frizzled) protein on the EMS cell. (After Han 1998.)

interactions; it is seen on both the ABa and ABp cells (Evans et al. 1994).* One of the most important ligands for Notch proteins such as GLP-1 is another cell surface protein called Delta. In *C. elegans*, the Delta-like protein is APX-1, and it is found on the P2 cell (Mango et al. 1994a; Mello et al. 1994). This APX-1 signal breaks the symmetry between ABa and ABp, since it stimulates the GLP-1 protein solely on the AB descendant that it touches—namely, the ABp blastomere. In doing this, the P2 cell establishes the dorsal-ventral axis of *C. elegans* and confers on the ABp blastomere a fate different from that of its sister cell.

INTEGRATION OF AUTONOMOUS AND CONDITIONAL SPECI-FICATION: DIFFERENTIATION OF THE *C. ELEGANS* PHARYNX It should become apparent from the above discussion that the pharynx is generated by two sets of cells. One group of pharyngeal precursors comes from the EMS cell and is dependent on the maternal *skn-1* gene. The second group of pharyngeal precursors comes from the ABa blastomere and

---

*GLP-1 protein is localized in the ABa and ABp blastomeres, but the maternally encoded *glp-1* mRNA is found throughout the embryo. Evans and colleagues (1994) have postulated that there might be some translational determinant in the AB blastomere that enables the *glp-1* message to be translated in its descendants. The *glp-1* gene is also active in regulating postembryonic cell-cell interactions. It is used later by the distal tip cell of the gonad to control the number of germ cells entering meiosis; hence the name GLP, for "germ line proliferation."

is dependent on GLP-1 signaling from the EMS cell. In both cases, the pharyngeal precursor cells (and only these cells) are instructed to activate the *pha-4* gene (Mango et al. 1994b). The *pha-4* gene encodes a transcription factor that resembles the mammalian HNF3β protein. Microarray studies by Gaudet and Mango (2002) revealed that the PHA-4 transcription factor activates almost all of the pharynx-specific genes. It appears that the PHA-4 transcription factor may be the node that takes the maternal inputs and transforms them into a signal that transcribes the zygotic genes necessary for pharynx development.

Gastrulation in *C. elegans*

Gastrulation in *C. elegans* starts extremely early, just after the generation of the P4 cell in the 26-cell embryo (**FIGURE 5.26**; Skiba and Schierenberg 1992). At this time, the two daughters of the E cell (Ea and Ep) migrate from the ventral side into the center of the embryo. There they divide to form a gut consisting of 20 cells. There is a very small and transient blastocoel prior to the movement of the Ea and Ep cells, and their inward migration creates a tiny blastopore. The next cell to migrate through this blastopore is the P4 cell, the precursor of the germ cells. It migrates to a position beneath the gut primordium. The mesodermal cells move in next: the descendants of the MS cell migrate inward from the anterior side of the blastopore, and the C- and D-derived muscle precursors enter from the posterior side. These cells flank the gut tube on the left and right sides (Schierenberg 1997). Finally, about 6 hours after

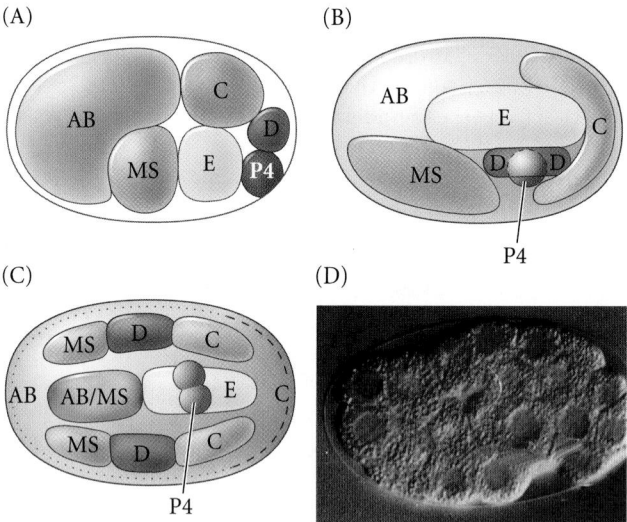

FIGURE 5.26 Gastrulation in *C. elegans*. (A) Positions of founder cells and their descendants at the 26-cell stage, at the start of gastrulation. (B) 102-cell stage, after the migration of the E, P4, and D descendants. (C) Positions of the cells near the end of gastrulation. The dotted and dashed lines represent regions of the hypodermis contributed by AB and C, respectively. (D) Early gastrulation, as the two E cells start moving inward. (After Schierenberg 1997; photograph courtesy of E. Schierenberg.)

fertilization, the AB-derived cells that contribute to the pharynx are brought inside, while the hypoblast cells (precursors of the hypodermal skin cells) move ventrally by epiboly, eventually closing the blastopore. The two sides of the hypodermis are sealed by E-cadherin on the tips of the leading cells that meet at the ventral midline (Raich et al. 1999).

During the next 6 hours, the cells move and develop into organs, while the ball-shaped embryo stretches out to become a worm with 556 somatic cells and 2 germline stem cells (see Priess and Hirsh 1986; Schierenberg 1997). There is evidence (Schnabel et al. 2006) that, although these gastrulation movements provide a good first approximation of the final form, an additional "cell focusing" is used to move cells into functional arrangements. Here cells of the same fate sort out along the anterior-posterior axis. Other modeling takes place as well; an additional 115 cells undergo apoptosis (programmed cell death; see Chapter 3). After four molts, the worm is a sexually mature, hermaphroditic adult, containing exactly 959 somatic cells as well as numerous sperm and eggs.

One characteristic of *C. elegans* that distinguishes it from most other well-studied developing organisms is the prevalence of cell fusion. During *C. elegans* gastrulation, about one-third of all the cells fuse together to form syncytial cells containing many nuclei. The 186 cells that comprise the hypodermis (skin) of the nematode fuse into 8 syncytial cells, and cell fusion is also seen in the vulva, uterus, and pharynx. The functions of these fusion events can be determined by observing mutations that prevent syncytia from forming (Shemer and Podbilewicz 2000, 2003). The fusion prevents individual cells from migrating beyond their normal borders. In the vulva, cell fusion prevents hypodermis cells from adopting a vulval fate and making an ectopic (and nonfunctional) vulva.

The *C. elegans* research program integrates genetics, cell biology, embryology, and even ecology to provide an understanding of the networks that govern cell differentiation and morphogenesis. In addition to providing some remarkable insights into how gene expression can change during development, studies of *C. elegans* have also humbled us by demonstrating how complex these networks are. Even in an organism as "simple" as *C. elegans*, with only a few genes and cell types, the right side of the body is made in a different manner from the left. The identification of the genes mentioned above is just the beginning of our effort to understand the complex interacting systems of development.

● See WEBSITE 5.3 Heterochronic genes and the control of larval stages

---

SNAPSHOT SUMMARY: Early Development in Snails and Nematodes

1. During cleavage, embryos do not usually grow. Rather, the volume of the oocyte is cleaved into numerous cells. The major exceptions to this rule are the mammals.

2. The blastomere cell cycle is governed by the synthesis and degradation of cyclin B. Cyclin B synthesis promotes the formation of mitosis-promoting factor. Degradation of cyclin B brings the cell back to the S phase. The G phases are added at the mid-blastula transition.

3. The movements of gastrulation include invagination, involution, ingression, delamination, and epiboly.

4. Three axes form the foundations of the bilateral body structure: the anterior-posterior axis (head to tail, or mouth to anus); the dorsal-ventral axis (back to belly); and the right-left axis (the two lateral sides of the body).

5. Body axes are established in different ways in different species. In some species the axes are established at fertilization through determinants in the egg cytoplasm. In others, such as nematodes and snails, the axes are established by cell interactions later in development.

6. Both snails and nematodes have holoblastic cleavage. In snails, cleavage is spiral; in nematodes, it is rotational.

7. In snails and *C. elegans*, gastrulation begins when there are relatively few cells. The blastopore becomes the mouth (the protostome mode of gastrulation).

8. Spiral cleavage in snails results in stereoblastulae (i.e., blastulae with no blastocoels). The direction of the cleavage spirals is regulated by a factor encoded by the mother and placed in the oocyte. Spiral cleavage can be modified by evolution, and adaptations of spiral cleavage have allowed some molluscs to survive in otherwise harsh environments.

9. The polar lobe of certain molluscs contains the morphogenetic determinants for mesoderm and endoderm. These determinants enter the D blastomere.

10. The soil nematode *Caenorhabditis elegans* was chosen as a model organism because it has a small number of cells, has a small genome, is easily bred and maintained, has a short life span, can be genetically manipulated, and has a cuticle through which one can see cell movements.

11. In the early divisions of the *C. elegans* zygote, one daughter cell becomes a founder cell (producing differentiated descendants) and the other becomes a stem cell (producing other founder cells and the germ line).

12. Blastomere identity in *C. elegans* is regulated by both autonomous and conditional specification.

For Further Reading

Complete bibliographical citations for all literature cited in this chapter can be found at the free-access website **www.devbio.com**

Lambert, J. D. 2010. Developmental patterns in spiralian embryos. *Curr. Biol.* 20: 272–277.

Resnick T. D., K. A. McCulloch and A. E. Rougvie. 2010. miRNAs give worms the time of their lives: Small RNAs and temporal control in *Caenorhabditis elegans. Dev. Dyn.* 239:1477–1489.

Shibazaki, Y., M. Shimizu and R. Kuroda. 2004. Body handedness is directed by genetically determined cytoskeletal dynamics in the early embryo. *Curr. Biol.* 14: 1462–1467.

Sulston, J. E., J. Schierenberg, J. White and N. Thomson. 1983. The embryonic cell lineage of the nematode *Caenorhabditis elegans. Dev. Biol.* 100: 64–119.

Go Online

WEBSITE 5.1 Alfred Sturtevant and the genetics of snail coiling. By a masterful thought experiment, Sturtevant demonstrated the power of applying genetics to embryology. To do this, he applied Mendelian genetics to the study of snail coiling.

WEBSITE 5.2 Modifications of cell fate in spiralian eggs. Within the gastropods, differences in the timing of cell fate result in significantly different body plans. Furthermore, in the leeches and nemerteans, the spiralian cleavage pattern has been modified to produce new types of body plans.

WEBSITE 5.3 Heterochronic genes and the control of larval stages. *C. elegans* undergoes four larval stages before becoming an adult. These stages are regulated by microRNAs that control the translation of particular messages.

Outside Sites

The *C. elegans* community has been especially cognizant of the need for excellent online sources of information. Wormbase, at **http://www.wormbase.org/**, is the one-stop-shopping site for anything Caenorhabitic. The Goldstein laboratory has excellent movies of *C. elegans* development at **http://www.bio.unc.edu/faculty/goldstein/lab/movies.html**. WormBook, **http://www.wormbook.org/**, is a superb web-based textbook.

6

The Genetics of Axis Specification in *Drosophila*

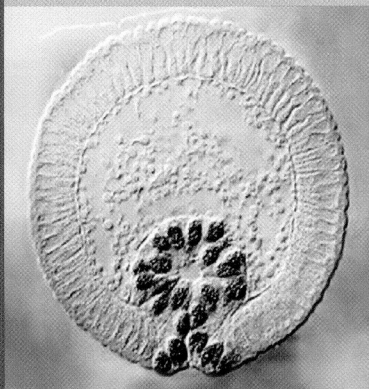

THANKS LARGELY TO STUDIES spearheaded by Thomas Hunt Morgan's laboratory during the first two decades of the twentieth century, we know more about the genetics of *Drosophila melanogaster* than that of any other multicellular organism (**FIGURE 6.1**). The reasons have to do with both the flies themselves and with the people who first studied them. *Drosophila* is easy to breed, hardy, prolific, tolerant of diverse conditions, and the polytene chromosomes of its larvae (see Figure 16.18) are readily identified. The progress of *Drosophila* genetics was aided by the relatively free access to the mutant strains and breeding techniques of other researchers. Mutants were considered the property of the entire scientific community, and Morgan's laboratory established a database and exchange network whereby anyone could obtain them.

Undergraduates (starting with Calvin Bridges and Alfred Sturtevant) played important roles in *Drosophila* research, which achieved its original popularity as a source of undergraduate research projects. As historian Robert Kohler noted (1994), "Departments of biology were cash poor but rich in one resource: cheap, eager, renewable student labor." The *Drosophila* genetics program was "designed by young persons to be a young person's game," and the students set the rules for *Drosophila* research: "No trade secrets, no monopolies, no poaching, no ambushes."

But *Drosophila* was a difficult organism on which to study embryology. Although Jack Schultz (originally in Morgan's laboratory) and others following him attempted to relate the genetics of *Drosophila* to its development, the fly embryos proved complex and intractable, being neither large enough to manipulate experimentally nor transparent enough to observe microscopically. It was not until the techniques of molecular biology allowed researchers to identify and manipulate the insect's genes and RNAs that its genetics could be related to its development. And when that happened, a revolution occurred in the field of biology. This revolution is continuing, in large part because of the complete sequencing of the *Drosophila* genome and the ability to generate transgenic flies at high frequency (Pfeiffer et al. 2009; del Valle Rodríguez et al. 2011). Researchers are now able to identify developmental interactions taking place in very small regions of the embryo, and to identify enhancers and other regions of the genome that control developmental processes. The merging of our knowledge of the molecular aspects of *Drosophila* genetics with our knowledge of the fly's development built the foundations on which the current sciences of developmental genetics and evolutionary developmental biology are based.

EARLY *DROSOPHILA* DEVELOPMENT

We have already discussed the specification of early embryonic cells by cytoplasmic determinants stored in the oocyte. The cell membranes that form during cleavage establish the region of cytoplasm incorporated into each new blastomere, and the morphogenetic determinants in the incorporated cytoplasm then direct differential gene expression in each cell. But in *Drosophila* development, cell membranes do not form until after the thirteenth nuclear division. Prior to this time, the dividing nuclei all share

Those of us who are at work on Drosophila *find a particular point to the question. For the genetic material available is all that could be desired, and even embryological experiments can be done.... It is for us to make use of these opportunities. We have a complete story to unravel, because we can work things from both ends at once.*

JACK SCHULTZ (1935)

The chief advantage of Drosophila *initially was one that historians have overlooked: it was an excellent organism for student projects.*

ROBERT E. KOHLER (1994)

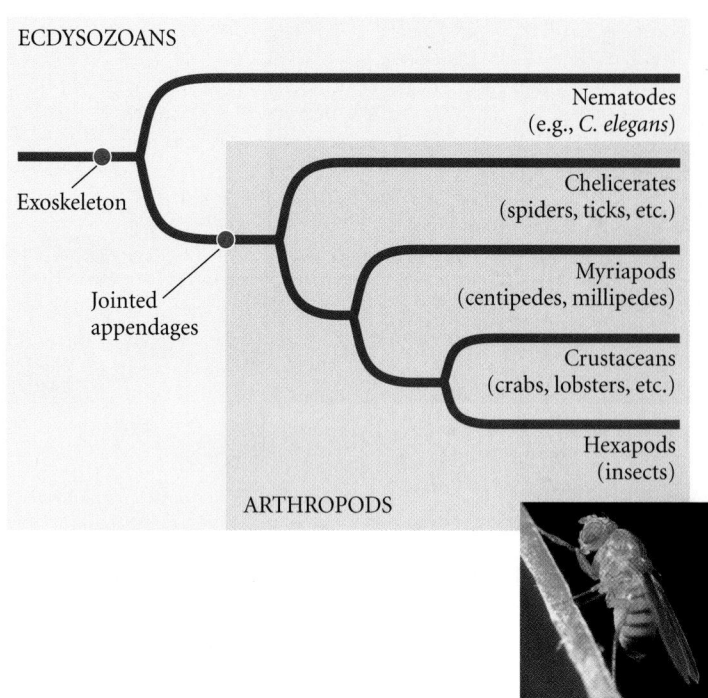

ECDYSOZOANS

Exoskeleton

Jointed appendages

Nematodes (e.g., *C. elegans*)

Chelicerates (spiders, ticks, etc.)

Myriapods (centipedes, millipedes)

Crustaceans (crabs, lobsters, etc.)

Hexapods (insects)

ARTHROPODS

FIGURE 6.1 Like the nematode *C. elegans* discussed in Chapter 5, members of the "superclade" Arthropoda are ecdysozoans. Adult arthropods have jointed appendages, a characteristic they share with the tetrapod vertebrates. The fruit fly *Drosophila melanogaster* represents the insects, the animal clade with the greatest number of species. Long a favored model organism of geneticists, in the last three decades the developmental biology of *Drosophila* also has been extensively studied, with results that have illuminated the entire field.

a common cytoplasm and material can diffuse throughout the whole embryo. The specification of cell types along the anterior-posterior and dorsal-ventral axes is accomplished by the interactions of components *within* the single multinucleated cell. Moreover, these axial differences are initiated at an earlier developmental stage by the position of the egg within the mother's egg chamber. Whereas the sperm entry site may fix the axes in nematodes and tunicates, the fly's anterior-posterior and dorsal-ventral axes are specified by interactions between the egg and its surrounding follicle cells prior to fertilization.

Fertilization

Drosophila fertilization is a remarkable series of events and is quite different from fertilizations we've described previously.

- *The sperm enters an egg that is already activated.* Egg activation in *Drosophila* is accomplished at ovulation, a few minutes *before* fertilization begins. As the *Drosophila* oocyte squeezes through a narrow orifice, calcium channels open and Ca²⁺ flows in. The oocyte nucleus then resumes its meiotic divisions and the cytoplasmic mRNAs become translated without fertilization (Mahowald et al. 1983; Fitch and Wakimoto 1998; Heifetz et al. 2001; Horner and Wolfner 2008).

- *There is only one site where the sperm can enter the egg.* This is the **micropyle**, a tunnel in the chorion (eggshell) located at the future dorsal anterior region of the embryo. The micropyle allows sperm to pass through it one at a time and probably prevents polyspermy in

Drosophila. There are no cortical granules to block polyspermy, although cortical changes are seen.

- *By the time the sperm enters the egg, the egg already has begun to specify the body axes;* thus the sperm enters an egg that is already organizing itself as an embryo.

In addition, there is competition between sperm. In some species of *Drosophila*, the sperm tail is longer than the adult fly; in *D. melanogaster*, it is 1.8 mm long—some 300 times longer than a human sperm. This huge tail is thought to block other sperm from entering the egg. The entire sperm (supersized tail and all) gets incorporated into the oocyte cytoplasm, and the sperm cell membrane does not break down until after it is fully inside the oocyte (Snook and Karr 1998; Clark et al. 1999).

● See WEBSITE 6.1 *Drosophila* fertilization

Cleavage

Most insect eggs undergo **superficial cleavage**, wherein a large mass of centrally located yolk confines cleavage to the cytoplasmic rim of the egg. One of the fascinating features of this cleavage pattern is that cells do not form until after the nuclei have divided several times. In the *Drosophila* egg, karyokinesis (nuclear division) occurs without cytokinesis (cell division) so as to create a **syncytium**, a single cell with many nuclei residing in a common cytoplasm (**FIGURE 6.2**). The zygote nucleus undergoes several mitotic divisions within the central portion of the egg; 256 nuclei are produced by a series of eight nuclear divisions averaging 8 minutes each (**FIGURE 6.3A,B**). This rapid rate of division is accomplished by repeated rounds of alternating S (DNA replication) and M (mitosis) phases in the absence of the gap (G) phases of the cell cycle. During the ninth division cycle, approximately five nuclei reach the surface of the posterior pole of the embryo. These nuclei become enclosed by cell membranes and generate the **pole cells** that give rise to the gametes of the adult. At cycle 10, the other nuclei migrate to the cortex (periphery) of the egg and the mitoses continue, albeit at a progressively slower rate (**FIGURE 6.3C,D**; Foe et al. 2000). During these stages of nuclear division,

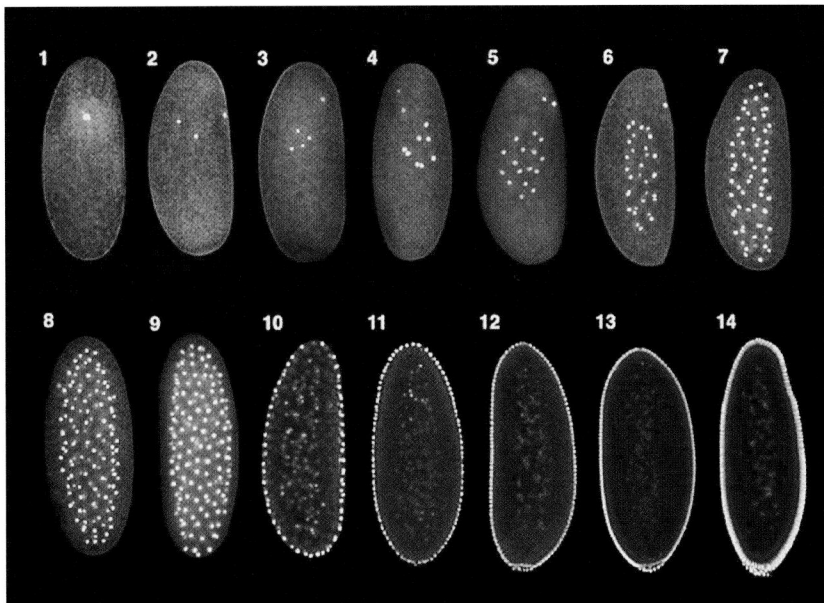

FIGURE 6.2 Laser confocal micrographs of stained chromatin showing syncytial nuclear divisions and superficial cleavage in a series of *Drosophila* embryos. The future anterior end is positioned upward; numbers refer to the cell division cycle. The early nuclear divisions occur centrally within a syncytium. Later, the nuclei and their cytoplasmic islands (energids) migrate to the periphery of the cell. This creates the syncytial blastoderm. After cycle 13, the cellular blastoderm forms by ingression of cell membranes between nuclei. The pole cells (germ cell precursors) form in the posterior. (Courtesy of D. Daily and W. Sullivan.)

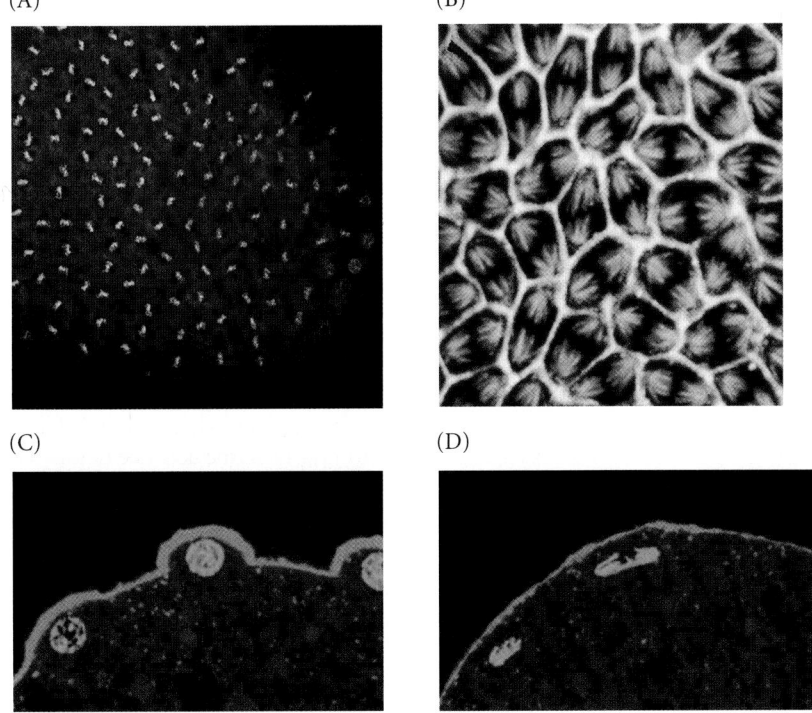

FIGURE 6.3 Nuclear and cell division in *Drosophila* embryos. (A) Nuclear division (but not cell division) can be seen in a syncytial *Drosophila* embryo using a dye that stains DNA. The first region to cellularize, the pole region, can be seen forming the cells in the posterior region of the embryo that will eventually become the germ cells (sperm or eggs) of the fly. (B) Chromosomes dividing at the cortex of a syncytial blastoderm. Although there are no cell boundaries, actin (green) can be seen forming regions within which each nucleus divides. The microtubules of the mitotic apparatus are stained red with antibodies to tubulin. (C,D) Cross section of a part of a cycle 10 *Drosophila* embryo showing nuclei (green) at the cortex of the syncytial cell, adjacent to a layer of actin microfilaments (red). (C) Interphase nuclei. (D) Nuclei in anaphase, dividing parallel to the cortex and enabling the nuclei to stay in the cell periphery. (A from Bonnefoy et al. 2007; B from Sullivan et al. 1993, courtesy of W. Theurkauf and W. Sullivan; C,D from Foe 2000, courtesy of V. Foe.)

(A)

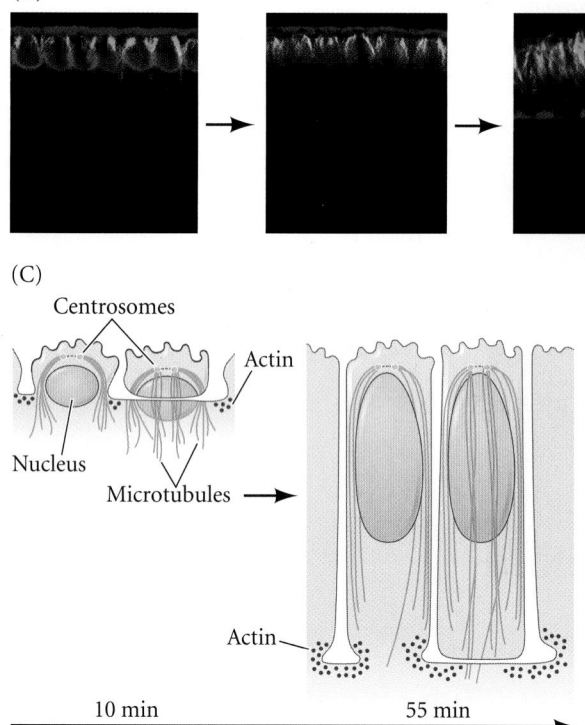

(B)

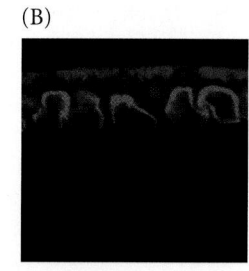

(C)

Centrosomes

Actin

Nucleus

Microtubules →

Actin

10 min   55 min

Cellularization

FIGURE 6.4 Formation of the cellular blastoderm in *Drosophila*. Nuclear shape change and cellularization are coordinated through the cytoskeleton. (A) Cellularization and nuclear shape change shown by staining the embryo for microtubules (green), microfilaments (blue), and nuclei (red). The red stain in the nuclei is due to the presence of the Kugelkern protein, one of the earliest proteins made from the zygotic nuclei. It is essential for nuclear elongation. (B) This embryo was treated with nocadozole to disrupt microtubules. The nuclei fail to elongate, and cellularization is prevented. (C) Diagrammatic representation of cell formation and nuclear elongation. (After Brandt et al. 2006; photographs courtesy of J. Grosshans and A. Brandt.)

the embryo is called a **syncytial blastoderm**, since no cell membranes exist other than that of the egg itself.

The nuclei divide within a common cytoplasm, but this does not mean the cytoplasm is itself uniform. Karr and Alberts (1986) have shown that each nucleus within the syncytial blastoderm is contained within its own little territory of cytoskeletal proteins. When the nuclei reach the periphery of the egg during the tenth cleavage cycle, each nucleus becomes surrounded by microtubules and microfilaments. The nuclei and their associated cytoplasmic islands are called **energids**. Following division cycle 13, the cell membrane (which had covered the egg) folds inward between the nuclei, eventually partitioning off each energid into a single cell. This process creates the **cellular blastoderm**, in which all the cells are arranged in a single-layered jacket around the yolky core of the egg (Turner and Mahowald 1977; Foe and Alberts 1983; Mavrakis et al. 2009).

Like any other cell formation, the formation of the cellular blastoderm involves a delicate interplay between microtubules and microfilaments (**FIGURE 6.4**). The membrane movements, nuclear elongation, and actin polymerization all appear to be coordinated by the microtubules (Riparbelli et al. 2007). The first phase of blastoderm cellularization is characterized by the invagination of cell membranes between the nuclei to form furrow canals. This process can be inhibited by drugs that block microtubules. After the furrow canals have passed the level of the nuclei, the second phase of cellularization occurs. The rate of invagination increases, and

the actin-membrane complex begins to constrict at what will be the basal end of the cell (Foe et al. 1993; Schejter and Wieschaus 1993; Mazumdar and Mazumdar 2002). In *Drosophila*, the cellular blastoderm consists of approximately 6000 cells and is formed within 4 hours of fertilization.

The mid-blastula transition

After the nuclei reach the periphery, the time required to complete each of the next four divisions becomes progressively longer. Whereas cycles 1–10 average 8 minutes each, cycle 13—the last cycle in the syncytial blastoderm—takes 25 minutes to complete. Cycle 14, in which the *Drosophila* embryo forms cells (i.e., after 13 divisions), is asynchronous. Some groups of cells complete this cycle in 75 minutes, other groups take 175 minutes (Foe 1989). Zygotic gene transcription (which begins around cycle 11) is greatly enhanced at this stage. This slowdown of nuclear division, cellularization, and concomitant increase in new RNA transcription is often referred to as the **mid-blastula transition** (see Chapter 5). It is at this stage that the maternally provided mRNAs are degraded and control of development is handed over to the zygotic genome (Brandt et al. 2006; De Renzis et al. 2007; Benoit et al. 2009). Such a **maternal-to-zygotic transition** is seen in the embryos of numerous vertebrate and invertebrate phyla.

In *Drosophila*, the coordination of the mid-blastula transition and the maternal-to-zygotic transition is controlled by several factors, including (1) the ratio of chromatin to cytoplasm; (2) Smaug protein; and (3) the Zelda transcription factor. The ratio of chromatin to cytoplasm is a consequence of the increasing amount of DNA while the cytoplasm remains constant (Newport and Kirschner 1982; Edgar et al. 1986a). Edgar and his colleagues compared the early development of wild-type *Drosophila* embryos with that of haploid mutants. The haploid *Drosophila* embryos had half the wild-type quantity of chromatin at each cell division. Hence, a haploid embryo at cell division cycle 8 had the same amount of chromatin that a wild-type embryo had at cycle 7. The investigators found that, whereas wild-type embryos formed a cellular blastoderm immediately after the thirteenth DNA division, haploid embryos underwent an extra, fourteenth DNA division before cellularization. Moreover, the lengths of cycles 11–14 in wild-type embryos corresponded to those of cycles 12–15 in the haploid embryos. Thus, the haploid embryos followed a pattern similar to that of the wild-type embryos—but they lagged by one cell cycle.

Smaug (yes, it's named after the dragon in *The Hobbit*) is an RNA-binding protein with known roles in translational repression. During the mid-blastula transition, however, it targets the maternal mRNAs for destruction (Tadros et al. 2007; Benoit et al. 2009). Embryos produced by *Smaug* mutant females show disruption of the slowing down of DNA division, a block in cellularization, and a failure to increase zygotic (nuclear) genome transcription. Smaug is encoded by a maternal mRNA, and Smaug protein levels increase during the early cleavage divisions. These levels peak when the zygotic genome begins efficient transcription. Moreover, if Smaug is artificially added to the anterior of an early *Drosophila* embryo, there results a concomitant gradient in the timing of maternal transcript destruction, cleavage cell cycle delays, zygotic gene transcription, cellularization, and gastrulation. Thus, Smaug accumulation appears to regulate the progression from maternal to zygotic control of development and coordinates this progression with the mid-blastula transition.

In addition to its involvement in the decline of maternal mRNAs, the maternal-to-zygotic transition involves the activation of the zygotic genes. This activation appears to be regulated by the transcription factor Zelda (Liang et al. 2008), a name that stands for "zinc-finger early *Drosophila* activator." Zelda is encoded by a maternal mRNA and binds to a CAGGTAG motif found in the promoters of the earliest transcribed zygotic genes. Many of the genes that initiate the pathways of sex determination, dorsal-ventral polarity, and anterior-posterior polarity are initiated by Zelda, and if this transcription factor is absent, these genes are not turned on at the correct times or places. It is possible that as concentrations of Zelda increase during the first few hours of *Drosophila* development, the genes with the highest affinity for Zelda get activated first, and those with lower affinities may be activated later (Harrison et al. 2011; Nien et al. 2011).

- ● See VADE MECUM *Drosophila* development
- ● See WEBSITE 6.2 The early development of other insects

Gastrulation

The general body plan of *Drosophila* is the same in the embryo, the larva, and the adult, each of which has a distinct head end and a distinct tail end, between which are repeating segmental units (FIGURE 6.5). Three of these segments form the thorax, while another eight segments

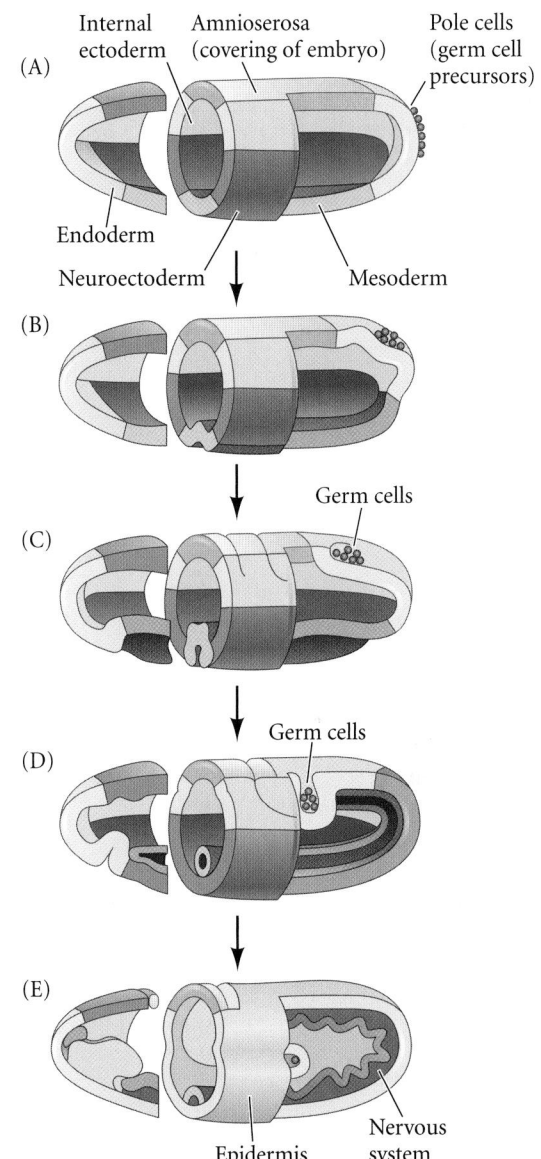

FIGURE 6.5 Schematic representation of gastrulation in *Drosophila*. Anterior is to the left; dorsal is facing upward. (A,B) Surface and cutaway views showing the fates of the tissues immediately prior to gastrulation. (C) The beginning of gastrulation as the ventral mesoderm invaginates into the embryo. (D) This view corresponds to Figure 6.6A, while (E) corresponds to Figure 6.6B,C. In (E), the neuroectoderm is largely differentiated into the nervous system and the epidermis. (After Campos-Ortega and Hartenstein 1985.)

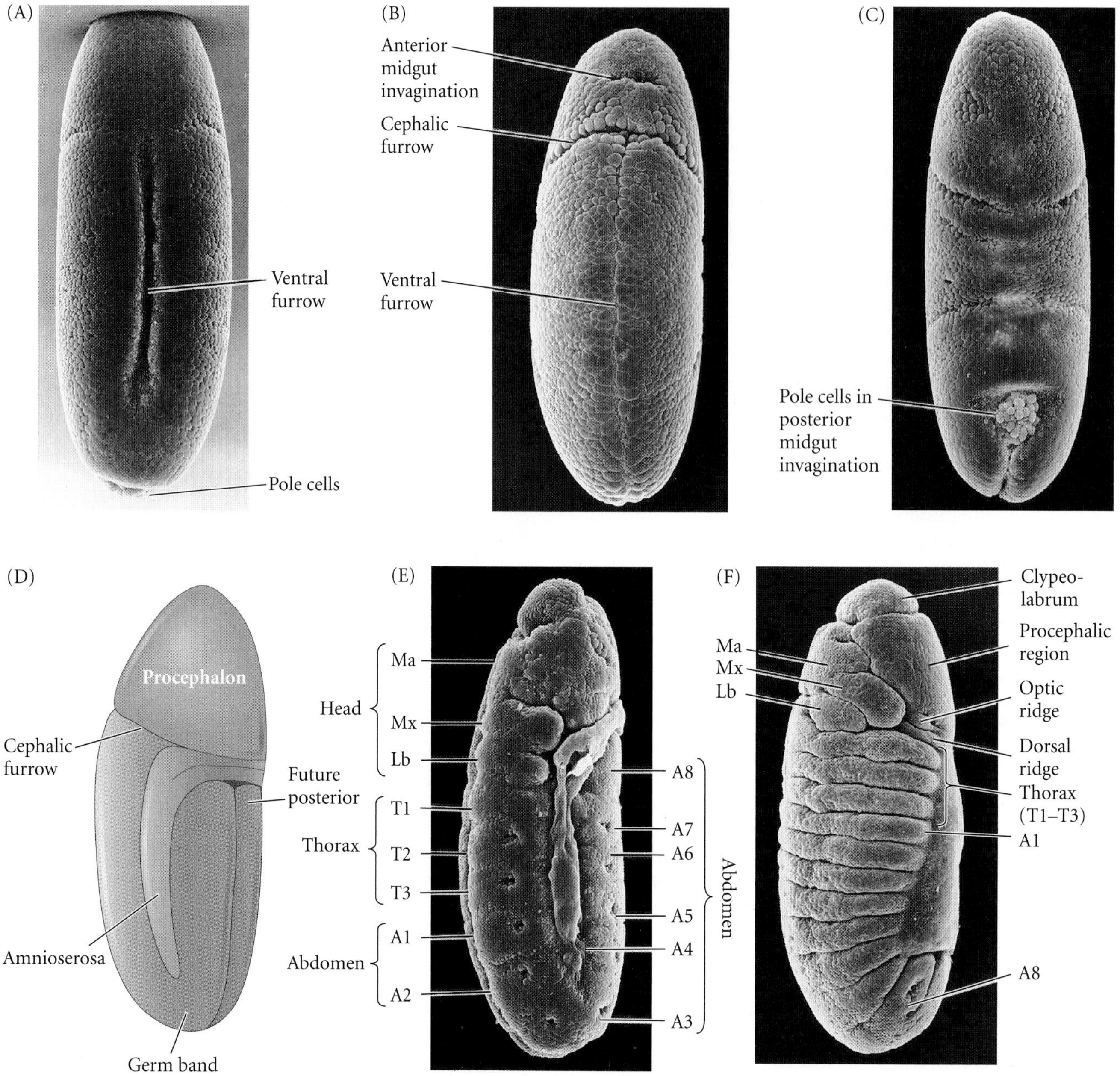

(A)

Ventral
furrow

Pole cells

(B)

Anterior
midgut
invagination

Cephalic
furrow

Ventral
furrow

(C)

Pole cells in
posterior
midgut
invagination

(D)

Procephalon

Cephalic
furrow

Future
posterior

Amnioserosa

Germ band

(E)

Head { Ma, Mx, Lb

Thorax { T1, T2, T3

Abdomen { A1, A2

A8
A7
A6
A5
A4
A3

Abdomen

(F)

Ma
Mx
Lb

Clypeo-
labrum

Procephalic
region

Optic
ridge

Dorsal
ridge

Thorax
(T1–T3)

A1

A8

FIGURE 6.6 Gastrulation in *Drosophila*. The anterior of each gastrulating embryo points upward in this series of scanning electron micrographs. (A) Ventral furrow beginning to form as cells flanking the ventral midline invaginate. (B) Closing of ventral furrow, with mesodermal cells placed internally and surface ectoderm flanking the ventral midline. (C) Dorsal view of a slightly older embryo, showing the pole cells and posterior endoderm sinking into the embryo. (D) Schematic representation showing dorsolateral view of an embryo at fullest germ band extension, just prior to segmentation. The cephalic furrow separates the future head region (procephalon) from the germ band, which will form the thorax and abdomen. (E) Lateral view, showing fullest extension of the germ band and the beginnings of segmentation. Subtle indentations mark the incipient segments along the germ band. Ma, Mx, and Lb correspond to the mandibular, maxillary, and labial head segments; T1–T3 are the thoracic segments; and A1–A8 are the abdominal segments. (F) Germ band reversing direction. The true segments are now visible, as well as the other territories of the dorsal head, such as the clypeolabrum, procephalic region, optic ridge, and dorsal ridge. (G) Newly hatched first instar larva. (Photographs courtesy of F. R. Turner; D after Campos-Ortega and Hartenstein 1985.)

Anterior segment

(G)

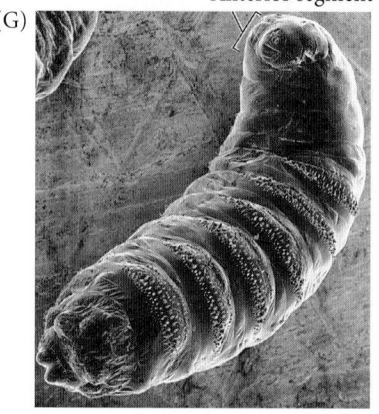

form the abdomen. Each segment of the adult fly has its own identity. The first thoracic segment, for example, has only legs; the second thoracic segment has legs and wings; and the third thoracic segment has legs and halteres (balancing organs). Thoracic and abdominal segments can also be distinguished from each other by differences in the cuticle of the newly hatched first instar larvae.

Gastrulation begins shortly after the mid-blastula transition. The first movements of *Drosophila* gastrulation segregate the presumptive mesoderm, endoderm, and ectoderm. The prospective mesoderm—about 1000 cells constituting the ventral midline of the embryo—folds inward to produce the **ventral furrow** (**FIGURE 6.6A**). This furrow eventually pinches off from the surface to become a ventral tube within the embryo. The prospective endoderm invaginates to form two pockets at the anterior and posterior ends of the ventral furrow. The pole cells are internalized along with the endoderm (**FIGURE 6.6B,C**). At this time, the embryo bends to form the **cephalic furrow**.

The ectodermal cells on the surface and the mesoderm undergo convergence and extension, migrating toward the ventral midline to form the **germ band**, a collection of cells along the ventral midline that includes all the cells that will form the trunk of the embryo. The germ band extends posteriorly and, perhaps because of the egg case, wraps around the top (dorsal) surface of the embryo (**FIGURE 6.6D**). Thus, at the end of germ band formation, the cells destined to form the most posterior larval structures are located immediately behind the future head region (**FIGURE 6.6E**). At this time, the body segments begin to appear, dividing the ectoderm and mesoderm. The germ band then retracts, placing the presumptive posterior segments at the posterior tip of the embryo (**FIGURE 6.6F**). At the dorsal surface, the two sides of the epidermis are brought together in a process called **dorsal closure**. The amnioserosa (the extraembryonic layer that surrounds the embryo), which had been the most dorsal structure, interacts with the epidermal cells to stimulate their migration (reviewed in Panfilio 2007; Heisenberg 2009).

While the germ band is in its extended position, several key morphogenetic processes occur: organogenesis, segmentation (**FIGURE 6.7A**), and segregation of the imaginal discs.* The nervous system forms from two regions of ventral ectoderm. Neuroblasts differentiate from this neurogenic ectoderm within each segment (and also from the nonsegmented

---

*Imaginal discs are those cells set aside to produce the adult structures. The details of imaginal disc differentiation will be discussed in Chapter 16. For more information on *Drosophila* developmental anatomy, see Bate and Martinez-Arias 1993; Tyler and Schetzer 1996; and Schwalm 1997.

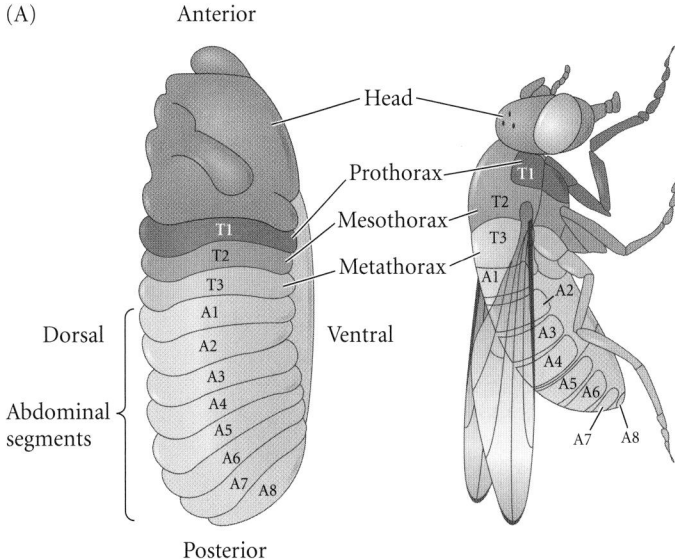

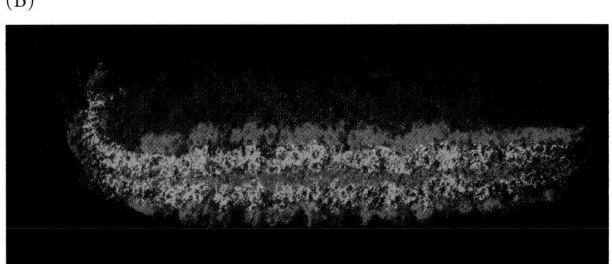

FIGURE 6.7 Axis formation in *Drosophila*. (A) Comparison of larval (left) and adult (right) segmentation. In the adult, the three thoracic segments can be distinguished by their appendages: T1 (prothorax) has legs only; T2 (mesothorax) has wings and legs; T3 (metathorax) has halteres (not visible) and legs. (B) During gastrulation, the mesodermal cells in the most ventral region enter the embryo, and the neurogenic cells expressing *Short gastrulation* (*Sog*) become the ventralmost cells of the embryo. *Sog*, blue; *ventral nervous system defective*, green; *intermediate neuroblast defective*, red. (B courtesy of E. Bier.)

region of the head ectoderm). Therefore, in insects such as *Drosophila*, the nervous system is located ventrally, rather than being derived from a dorsal neural tube as in vertebrates (**FIGURE 6.7B**).

GENES THAT PATTERN THE *DROSOPHILA* BODY PLAN

Most of the genes involved in shaping the larval and adult forms of *Drosophila* were identified in the early 1990s using a powerful "forward genetics" approach. The basic strategy was to randomly mutagenize flies and then screen for mutations that disrupted the normal formation of the body plan. Some of these mutations were quite fantastic, including embryos and adult flies in which specific body structures were either missing or in the wrong place. These mutant collections were distributed to many different laboratories. The genes involved in the mutant phenotypes were cloned and then characterized with respect to their expression patterns and their functions. This combined effort has led to a molecular understanding of body plan development in *Drosophila* that is unparalleled in all of biology, and in 1995 the work resulted in a Nobel Prize for Edward Lewis, Christiane Nüsslein-Volhard, and Eric Wieschaus.

The rest of this chapter details the genetics of *Drosophila* development as we have come to understand it over the past two decades. First we will examine how the dorsal-ventral and anterior-posterior axes of the embryo are established by interactions between the developing oocyte and its surrounding follicle cells. Next we will see how dorsal-ventral patterning gradients are formed within the embryo, and how these gradients specify different tissue types. The third part of the discussion will examine how segments are formed along the anterior-posterior axis, and how the different segments become specialized. Finally, we will briefly show how the positioning of embryonic tissues along the two primary axes specifies these tissues to become particular organs.

Primary Axis Formation during Oogenesis

The processes of embryogenesis may begin officially at fertilization, but many of the molecular events critical for *Drosophila* embryogenesis actually occur during oogenesis. Each oocyte is descended from a single female germ cell—the **oogonium**. Before oogenesis begins, the oogonium divides four times with incomplete cytokinesis, giving rise to 16 interconnected cells: 15 **nurse cells** and the single oocyte precursor (**FIGURE 6.8A**). These 16 germline cells plus a surrounding epithelial layer of somatic follicle cells constitute the **egg chamber** in which the oocyte will develop. As the oocyte precursor develops at the posterior end of the egg chamber, numerous mRNAs made in the nurse cells are transported along microtubules through the cellular interconnections into the enlarging oocyte.

Anterior-posterior polarity in the oocyte

The follicular epithelium surrounding the developing oocyte is initially uniform with respect to cell fate, but this uniformity is broken by two signals organized by the oocyte nucleus. Interestingly, both of these signals involve the same gene, *gurken*. The *gurken* message appears to be synthesized in the nurse cells, but it becomes transported specifically to the oocyte cytoplasm. Here it is localized between the nucleus and the cell membrane and is translated into Gurken protein (Cáceres and Nilson 2005). At this time the oocyte nucleus is very near the posterior tip of the egg chamber, and the Gurken signal is received by

the follicle cells at that position through a receptor protein encoded by the *torpedo* gene* (see Figure 6.8A). This signal results in the "posteriorization" of these follicle cells (**FIGURE 6.8B**). The posterior follicle cells send a signal back into the oocyte. The identity of this signal is not yet known, but it recruits the Par-1 protein to the posterior edge of the oocyte cytoplasm (**FIGURE 6.8C**; Doerflinger et al. 2006). Par-1 protein organizes microtubules specifically with their minus (cap) and plus (growing) ends at the anterior and posterior ends of the oocyte, respectively (Gonzalez-Reyes et al. 1995; Roth et al. 1995; Januschke et al. 2006).

The orientation of the microtubules is critical, because different microtubule motor proteins will transport their mRNA or protein cargoes in different directions. The motor protein kinesin, for instance, is an ATPase that will use the energy of ATP to transport material to the plus end of the microtubule. Dynein, however, is a "minus-directed" motor protein that will transport its cargo the opposite way. One of the messages transported by kinesin along the microtubules to the posterior end of the oocyte is *oskar* mRNA (Zimyanin et al. 2008). The *oskar* mRNA is not able to be translated until it reaches the posterior cortex, at which time it generates the Oskar protein. Oskar recruits more Par-1 protein, thereby stabilizing the microtubule orientation and allowing more material to be recruited to the posterior pole of the oocyte (Doerflinger et al. 2006; Zimyanin et al. 2007). The posterior pole will thereby have its own distinctive cytoplasm, called **pole plasm**, which contains the determinants for producing the abdomen and the germ cells.

This cytoskeletal rearrangement in the oocyte is accompanied by an increase in oocyte volume, owing to transfer of cytoplasmic components from the nurse cells. These components include maternal messengers such as the *bicoid* and *nanos* mRNAs. These mRNAs are carried by motor proteins along the microtubules to the anterior and posterior ends of the oocyte, respectively (**FIGURE 6.8D–F**). As we shall soon see, the protein products encoded by *bicoid* and *nanos* are critical for establishing the anterior-posterior polarity of the embryo.

Dorsal-ventral patterning in the oocyte

As oocyte volume increases, the oocyte nucleus is pushed by the growing microtubules to an anterior dorsal position where a second major signaling event takes place (Zhao et al. 2012). Here the *gurken* message becomes localized in a crescent between the oocyte nucleus and the oocyte cell membrane, and its protein product forms an anterior-posterior gradient along the dorsal surface of the oocyte (**FIGURE 6.9**; Neuman-Silberberg and Schüpbach 1993). Since it can diffuse only a short distance, Gurken protein reaches only

---

*Molecular analysis has established that *gurken* encodes a homologue of the vertebrate epidermal growth factor (EGF), whereas *torpedo* encodes a homologue of the vertebrate EGF receptor (Price et al. 1989; Neuman-Silberberg and Schüpbach 1993).

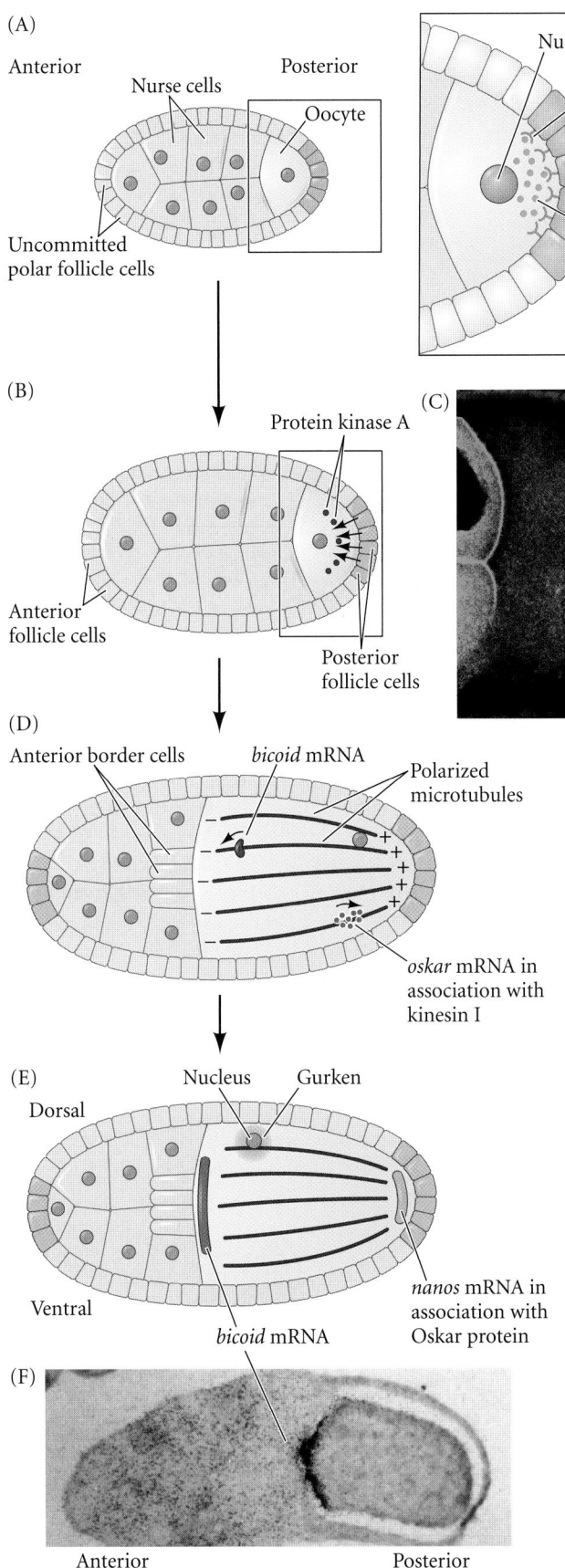

(A)

Anterior
Posterior

Nurse cells
Oocyte

Uncommitted
polar follicle cells

Nucleus
Torpedo
(Gurken
receptor)
Terminal
follicle
cells
Gurken
protein

(B)

Protein kinase A

Anterior
follicle cells

Posterior
follicle cells

(C)

(D)

Anterior border cells
bicoid mRNA

Polarized
microtubules

oskar mRNA in
association with
kinesin I

(E)

Nucleus   Gurken

Dorsal

Ventral

bicoid mRNA

nanos mRNA in
association with
Oskar protein

(F)

Anterior   Posterior

FIGURE 6.8 The anterior-posterior axis is specified during oogenesis. (A) The oocyte moves into the posterior region of the egg chamber, while nurse cells fill the anterior portion. The oocyte nucleus moves toward the terminal follicle cells and synthesizes Gurken protein (green). The terminal follicle cells express Torpedo, the receptor for Gurken. (B) When Gurken binds to Torpedo, the terminal follicle cells differentiate into posterior follicle cells and synthesize a molecule that activates protein kinase A in the egg. Protein kinase A orients the microtubules such that the growing (plus) ends are at the posterior (depicted in panel D). (C) Par-1 protein (green) localizes to the cortical cytoplasm of nurse cells and to the posterior pole of the oocyte. (The Staufen protein marking the posterior pole is labeled red; the red and green signals combine to fluoresce yellow.) (D) *bicoid* mRNA binds to dynein, a "minus-directed" motor protein associated with the non-growing end of microtubules; dynein moves the *bicoid* mRNA to the anterior end of the egg. *oskar* mRNA becomes complexed to kinesin I, a "plus-directed" motor protein that moves it toward the growing end of the microtubules at the posterior region, where Oskar protein can bind *nanos* mRNA. (E) The nucleus (with its associated Gurken protein) migrates along the microtubules to the dorsal anterior region of the oocyte and induces the adjacent follicle cells to become the dorsal follicle cells. (F) Photomicrograph of *bicoid* mRNA (stained black) passing from the nurse cells and localizing to the anterior end of the oocyte during oogenesis. (C courtesy of H. Doerflinger; F from Stephanson et al. 1988, courtesy of the authors.)

those follicle cells closest to the oocyte nucleus, and it signals those cells to become the more columnar **dorsal follicle cells** (Montell et al. 1991; Schüpbach et al. 1991; see Figure 6.8E). This establishes the dorsal-ventral polarity in the follicle cell layer that surrounds the growing oocyte.

Maternal deficiencies of either the *gurken* or the *torpedo* gene cause ventralization of the embryo. However, *gurken* is active only in the oocyte, whereas *torpedo* is active only in the somatic follicle cells. This fact was revealed by experiments with germline/somatic chimeras. In one such experiment, Trudi Schüpbach (1987) transplanted germ cell precursors from wild-type embryos into embryos whose mothers carried the *torpedo* mutation. Conversely, she transplanted the germ cell precursors from *torpedo* mutants into wild-type embryos (**FIGURE 6.10**). The wild-type eggs produced mutant, ventralized embryos when they developed in a *torpedo* mutant mother's egg chamber. The mutant eggs were able to produce normal embryos if they developed in a wild-type ovary. Thus, unlike Gurken, Torpedo is needed in the follicle cells, but not

(A)

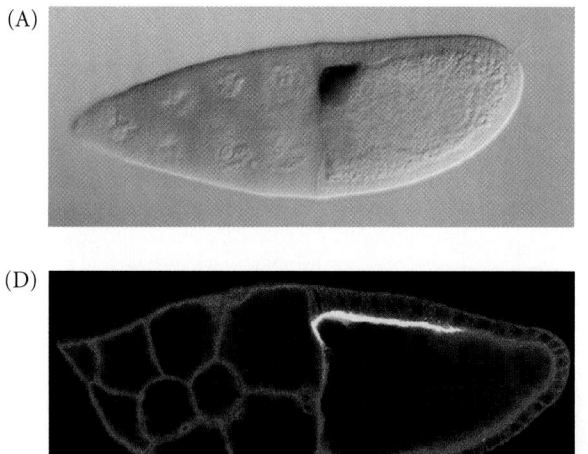

(D)

(B)

(C)

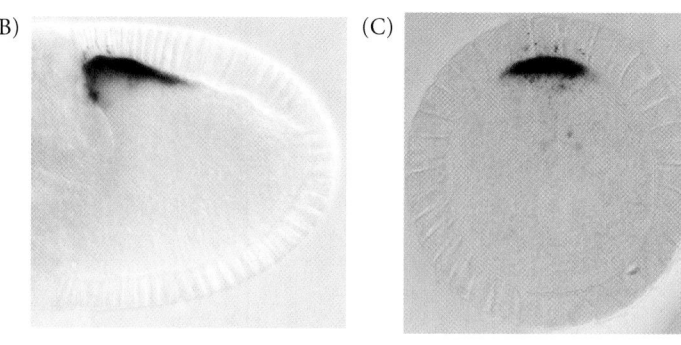

FIGURE 6.9 Expression of Gurken between the oocyte nucleus and the dorsal anterior cell membrane. (A) The *gurken* mRNA is localized between the oocyte nucleus and the dorsal follicle cells of the ovary. Anterior is to the left; dorsal faces upward. (B) Gurken protein is similarly located (shown here in a younger-stage oocyte than A). (C) Cross section of the egg through the region of *gurken* expression. (D) A more mature oocyte, showing Gurken protein (yellow) across the dorsal region. The actin is stained red, showing cell boundaries. As the oocyte grows, follicle cells migrate across the top of the oocyte, becoming exposed to Gurken. (A from Ray and Schüpbach 1996, courtesy of T. Schüpbach; B,C from Peri et al. 1999, courtesy of S. Roth; D courtesy of C. van Buskirk and T. Schüpbach.)

in the egg itself, for normal dorsal-ventral patterning of the embryo.

The Gurken-Torpedo signal that specifies dorsalized follicle cells initiates a cascade of gene activities that create the dorsal-ventral axis of the embryo. The activated Torpedo receptor protein activates Mirror, a transcription factor that represses expression of the *pipe* gene (Andreu et al. 2012; Fuchs et al. 2012). As a result, Pipe is made only in the ventral follicle cells (**FIGURE 6.11A**; Sen et al. 1998; Amiri and Stein 2002). Pipe protein modifies the ventral vitelline envelope by sulfating its proteins. This allows the Gastrulation-defective protein to bind to the vitelline envelope (only

in the ventral region) and to recruit other proteins to make a complex that will cleave the Easter protein into its active protease form (**FIGURE 6.11B**; Cho et al. 2010, 2012). Easter then cleaves the Spätzle protein (Chasan et al. 1992; Hong and Hashimoto 1995; LeMosy et al. 2001), and the cleaved Spätzle protein is the ligand that binds to and activates the

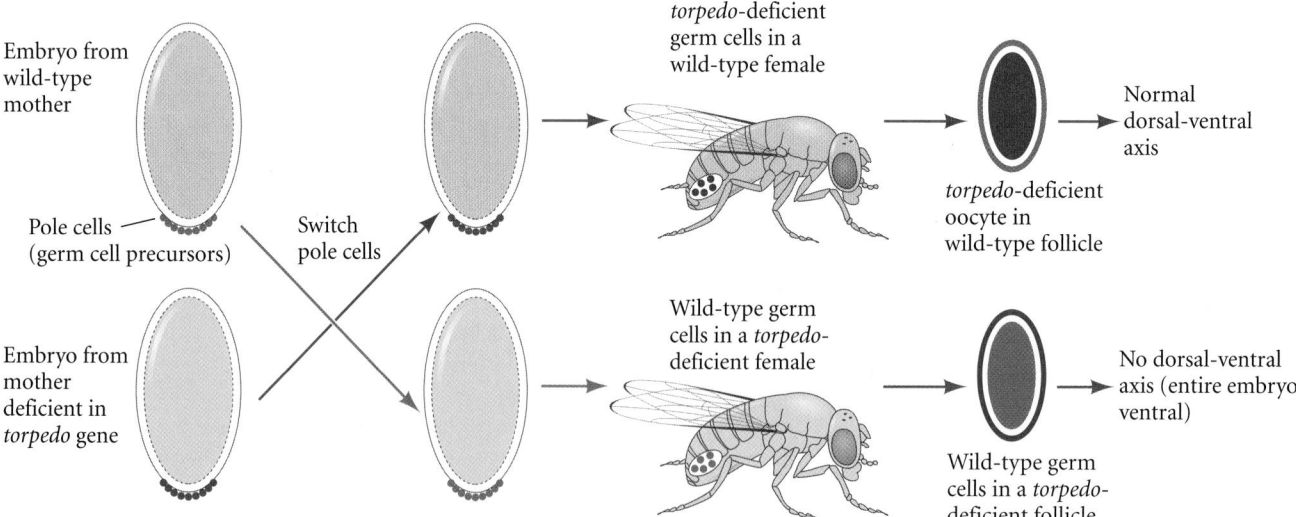

FIGURE 6.10 The Torpedo protein is required in follicle cells, but not in the oocyte, for normal dorsal-ventral patterning of the embryo. Germline chimeras made by interchanging pole cells (germ cell precursors) between wild-type embryos and embryos from mothers homozygous for a mutation of the *torpedo* gene. These transplants produced wild-type females whose eggs came from

mutant mothers, and *torpedo*-deficient females that laid wild-type eggs. The *torpedo*-deficient eggs produced normal embryos when they developed in the wild-type ovary, whereas the wild-type eggs produced ventralized embryos when they developed in the mutant mother's ovary.

1. Oocyte nucleus travels to anterior dorsal side of oocyte where it localizes *gurken* mRNA.

2. Translated Gurken is received by Torpedo proteins.

3a. Torpedo signal causes follicle cells to differentiate to a dorsal morphology.

3b. Pipe synthesis is inhibited in dorsal follicle cells.

4. Gurken does not diffuse to ventral follicle cells.

5a. Ventral follicle cells synthesize Pipe.

5b. Pipe signal sulfates ventral vitelline proteins.

6. Sulfated vitelline membrane proteins bind Gastrulation-defective (GD).

7a. GD cleaves Snake to its active form and forms a complex with Snake and uncleaved Easter proteins.

7b. Easter protein is cleaved into its active form.

8. Cleaved Easter binds to and cleaves Spätzle; activated Spätzle binds to Toll receptor protein.

9. Toll activation activates Tube and Pelle, which phosphorylate the Cactus protein. Cactus is degraded, releasing it from Dorsal.

10. Dorsal protein enters the nucleus and ventralizes the cell.

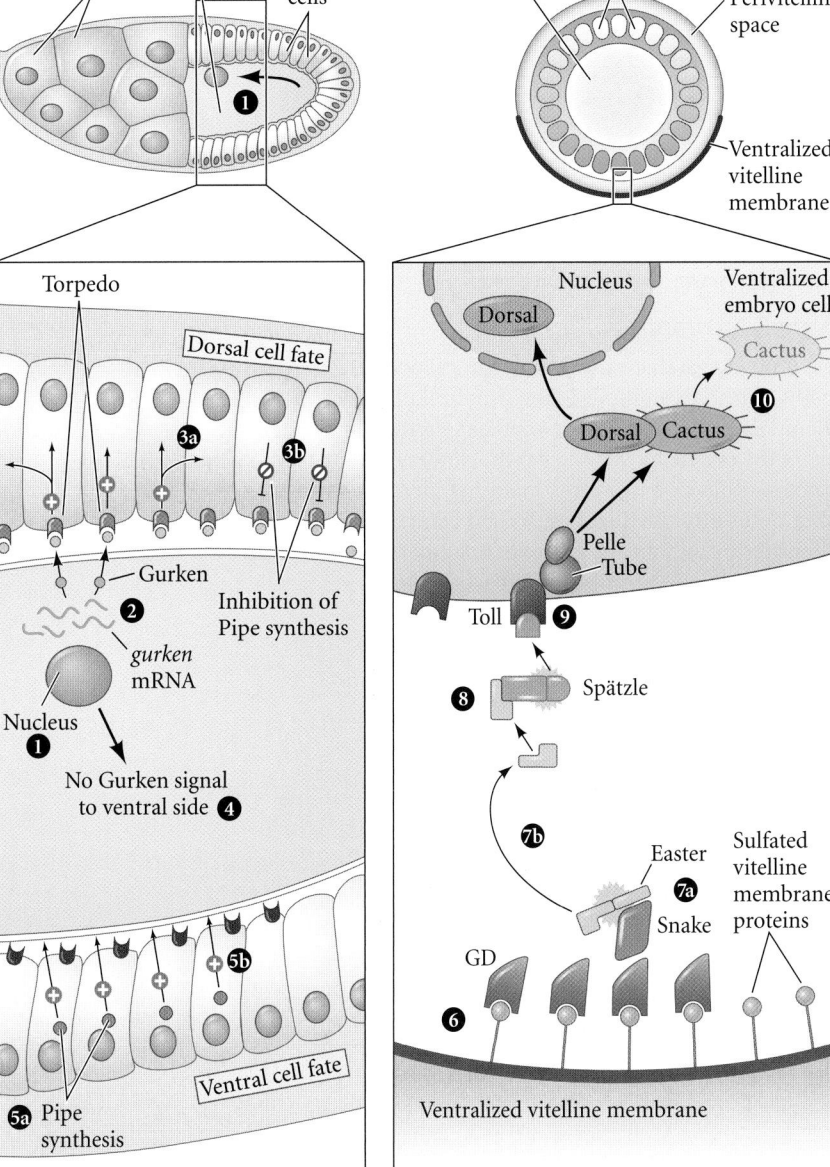

FIGURE 6.11 Generating dorsal-ventral polarity in *Drosophila*. (A) The nucleus of the oocyte travels to what will become the dorsal side of the embryo. The *gurken* genes of the oocyte synthesize mRNA that becomes localized between the oocyte nucleus and the cell membrane, where it is translated into Gurken protein. The Gurken signal is received by the Torpedo receptor protein made by the follicle cells (see Figure 6.8). Given the short diffusibility of the signal, only the follicle cells closest to the oocyte nucleus (i.e., the dorsal follicle cells) receive the Gurken signal, which causes the follicle cells to take on a characteristic dorsal follicle morphology and inhibits the synthesis of Pipe protein. Therefore, Pipe protein is made only by the *ventral* follicle cells. (B) The ventral region, at a slightly later stage of development. Sulfated proteins on the ventral region of the vitelline envelope recruit Gastrulation-defective (GD), which in turn complexes with other proteins, initiating a cascade that results in the cleaved Spätzle protein binding to the Toll receptor. The resulting cascade ventralizes the cell. (After van Eeden and St. Johnston 1999; Cho et al. 2010.)

Toll receptor. It is obviously important that the cleavage of Spätzle be limited to the most ventral portion of the embryo. This is accomplished by the secretion of a protease inhibitor from the follicle cells of the ovary (Hashimoto et al. 2003; Ligoxygakis et al. 2003).

Toll protein is a maternal product that is evenly distributed throughout the cell membrane of the egg (Hashimoto et al. 1988, 1991), but it becomes activated only by binding Spätzle, which is produced only on the ventral side of the egg. The ventral Toll receptors bind the mature Spätzle protein, and the membrane containing the activated Toll protein undergoes endocytosis. Signaling from the Toll receptor is believed to occur in these cytoplasmic endosomes rather than on the cell surface (Lund et al. 2010). Therefore, the Toll receptors on the ventral side of the egg are transducing a signal into the egg, whereas the Toll receptors on the dorsal side are not. This localized activation establishes the dorsal-ventral polarity of the oocyte.

Generating the Dorsal-Ventral Pattern in the Embryo

Dorsal, the ventral morphogen

The protein that distinguishes dorsum (back) from ventrum (belly) in the fly embryo is the product of the *dorsal* gene. The mRNA transcript of the mother's *dorsal* gene is deposited in the oocyte by the nurse cells. However, Dorsal protein is not synthesized from this maternal message until about 90 minutes after fertilization. When Dorsal is translated, it is found throughout the embryo, not just on the ventral or dorsal side. How can this protein act as a morphogen if it is located everywhere in the embryo?

The answer to this question was unexpected (Roth et al. 1989; Rushlow et al. 1989; Steward 1989). Although Dorsal protein is found throughout the syncytial blastoderm of the early *Drosophila* embryo, it is translocated into nuclei only in the ventral part of the embryo. In the nucleus, Dorsal protein acts as a transcription factor, binding to certain genes to activate or repress their transcription. If Dorsal does not enter the nucleus, the genes responsible for specifying ventral cell types are not transcribed, the genes responsible for specifying dorsal cell types are not repressed, and all the cells of the embryo become specified as dorsal cells.*

This model of dorsal-ventral axis formation in *Drosophila* is supported by analyses of maternal effect mutations that give rise to an entirely dorsalized or an entirely ventralized phenotype (**FIGURE 6.12**; Anderson and Nüsslein-Volhard 1984). In mutants in which all the cells are dorsalized (evident from their dorsal-specific exoskeleton), Dorsal does not enter the nucleus of any cell. Conversely, in mutants in which all cells have a ventral phenotype, Dorsal protein is found in every cell nucleus.

Establishing a nuclear Dorsal gradient

So how does the Dorsal protein enter into the nuclei only of the ventral cells? When Dorsal is first produced, it is complexed with a protein called Cactus in the cytoplasm of the syncytial blastoderm. As long as Cactus is bound to it, Dorsal remains in the cytoplasm. Dorsal enters ventral nuclei in response to a signaling pathway that frees it from Cactus (see Figure 6.11B). This separation of Dorsal from Cactus is initiated by the ventral activation of the Toll receptor. When Spätzle binds

---

*Remember that a gene in *Drosophila* is usually named after its mutant phenotype. Thus, the product of the *dorsal* gene is necessary for the differentiation of ventral cells. That is, in the absence of *dorsal*, the ventral cells become dorsalized.

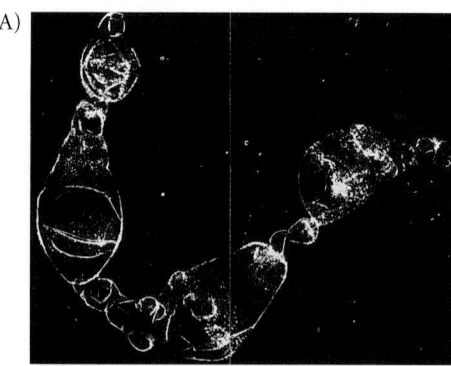

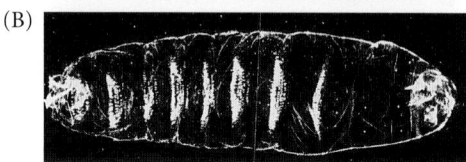

FIGURE 6.12 Mutations affecting distribution of the Dorsal protein, as seen in the exoskeleton (cuticle) patterns of larvae. (A) Deformed larva consisting entirely of dorsal cells. Larvae like these developed from the eggs of a female homozygous for a mutation of the *snake* gene, one of the maternal effect genes involved in the signaling cascade that establishes a Dorsal gradient. (B) Larva that developed from *snake* mutant eggs that received injections of mRNA from wild-type eggs. Larvae like this one have a wild-type appearance. (From Anderson and Nüsslein-Volhard 1984, courtesy of C. Nüsslein-Volhard.)

to and activates the Toll protein, Toll activates a protein kinase called Pelle. Another protein, Tube, is probably necessary for bringing Pelle to the cell membrane, where it can be activated (Galindo et al. 1995). The activated Pelle protein kinase (probably through an intermediate) can phosphorylate Cactus. Once phosphorylated, Cactus is degraded and Dorsal can enter the nucleus (Kidd 1992; Shelton and Wasserman 1993; Whalen and Steward 1993; Reach et al. 1996). Since Toll is activated by a gradient of Spätzle protein that is highest in the most ventral region, there is a corresponding gradient of Dorsal translocation in the ventral cells of the embryo, with the highest concentrations of Dorsal in the most ventral cell nuclei.†

---

†Recall that maternal effect mutations (as in the coiling mutant in snails discussed in Chapter 5) involve those genes that are active in the female and provide materials for the oocyte cytoplasm. The process described for the translocation of Dorsal protein into the nucleus is very similar to the process for the translocation of the NF-κB transcription factor into the nucleus of mammalian lymphocytes. In fact, there is substantial homology between NF-κB and Dorsal, between I-B and Cactus, between Toll and the interleukin 1 receptor, between Pelle and an IL1-associated protein kinase, and between the DNA sequences recognized by Dorsal and by NF-κB (González-Crespo and Levine 1994; Cao et al. 1996). Thus, the biochemical pathway used to specify dorsal-ventral polarity in *Drosophila* appears to be homologous to that used to differentiate lymphocytes in mammals.

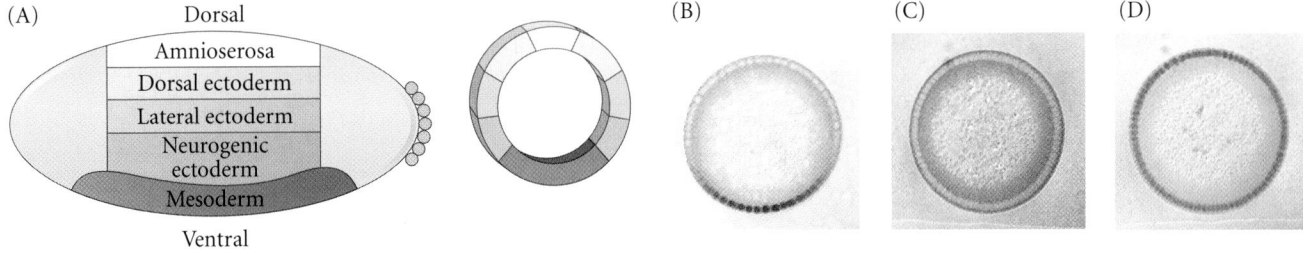

(A)

Dorsal

Amnioserosa

Dorsal ectoderm

Lateral ectoderm

Neurogenic ectoderm

Mesoderm

Ventral

LATERAL VIEW   TRANSVERSE SECTION

(B)   (C)   (D)

FIGURE 6.13 Specification of cell fate by the gradient of Dorsal protein. The translocation of Dorsal protein into ventral, but not lateral or dorsal, nuclei produces a gradient whereby the ventral cells with the most Dorsal protein become mesoderm precursors. (A) Fate map of a lateral cross section through the *Drosophila* embryo at division cycle 14. The most ventral part becomes the mesoderm; the next higher portion becomes the neurogenic (ventral) ectoderm. The lateral and dorsal ectoderm can be distinguished in the cuticle, and the dorsalmost region becomes the amnioserosa, the extraembryonic layer that surrounds the embryo. (B–D) Transverse sections of embryos stained with antibody to show the presence of Dorsal protein (dark-stained area). (B) A wild-type embryo, showing Dorsal protein in the ventralmost nuclei. (C) A dorsalized mutant, showing no localization of Dorsal protein in any nucleus. (D) A ventralized mutant, in which Dorsal protein has entered the nucleus of every cell. (A after Rushlow et al. 1989; B–D from Roth et al. 1989, courtesy of the authors.)

● See **WEBSITE 6.3** Evidence for gradients in insect development

Effects of the Dorsal protein gradient

What does the Dorsal protein do once it is located in the nuclei of the ventral cells? A look at the fate map of a cross section through the *Drosophila* embryo at the division cycle 14 shows that the 16 cells with the highest concentration of Dorsal are those that generate the mesoderm (**FIGURE 6.13**). The next cell up from this region generates the specialized glial and neural cells of the midline. The next two cells give rise to the ventrolateral epidermis and ventral nerve cord, and the nine cells above them produce the dorsal epidermis. The most dorsal group of six cells generates the amnioserosal covering of the embryo (Ferguson and Anderson 1991). This fate map is generated by the gradient of Dorsal protein in the nuclei. Large amounts of Dorsal instruct the cells to become mesoderm, whereas lesser amounts instruct the cells to become glial or ectodermal tissue (Jiang and Levine 1993; Hong et al. 2008).

The first morphogenetic event of *Drosophila* gastrulation is the invagination of the 16 ventralmost cells of the embryo to create the ventral furrow (**FIGURE 6.14**, also see Figure 6.6A). All of the body muscles, fat bodies, and gonads derive from these mesodermal cells (Foe 1989).

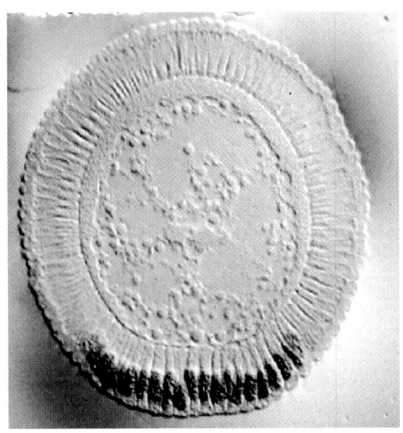
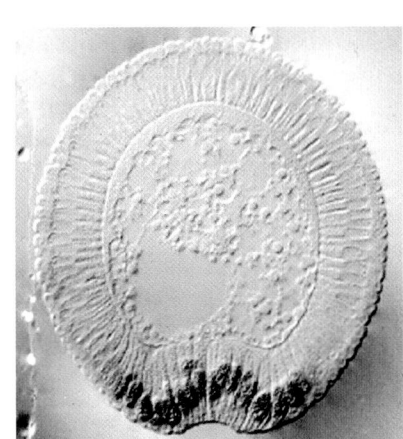
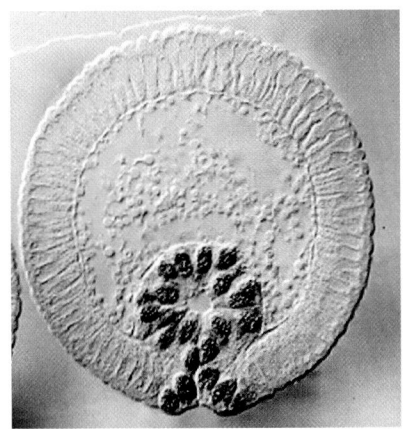
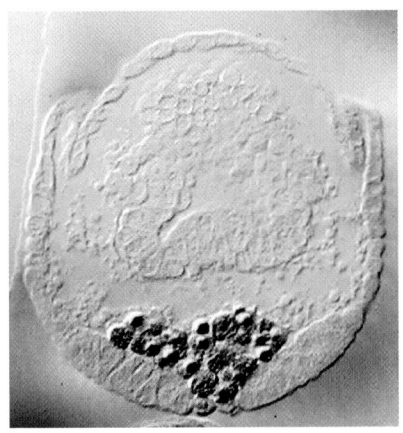

FIGURE 6.14 Gastrulation in *Drosophila*. In this cross section, the mesodermal cells at the ventral portion of the embryo buckle inward, forming the ventral furrow (see Figure 6.6A,B). This furrow becomes a tube that invaginates into the embryo and then flattens and generates the mesodermal organs. The nuclei are stained with antibody to the Twist protein, a marker for the mesoderm. (From Leptin 1991a, courtesy of M. Leptin.)

(A) Dorsal patterning

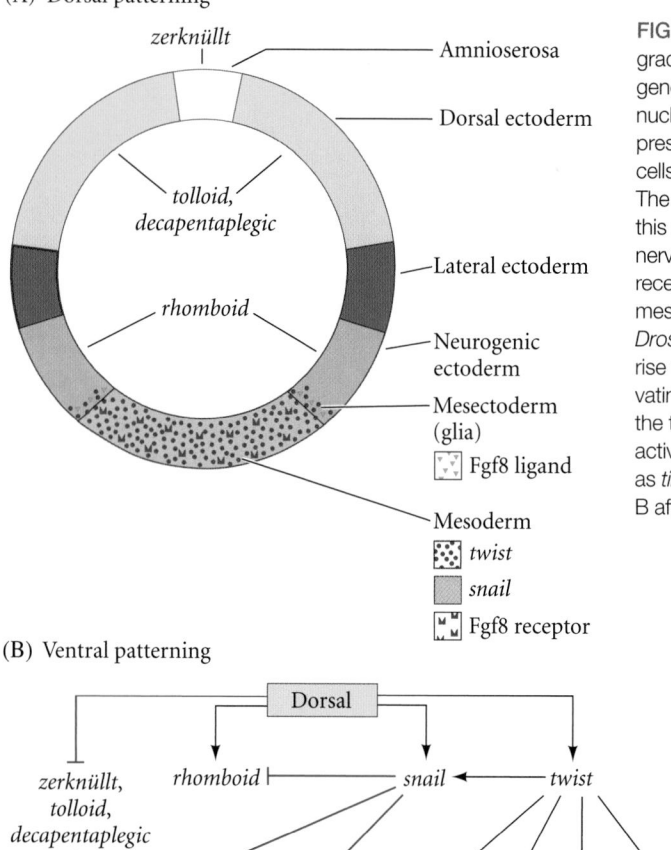

(B) Ventral patterning

FIGURE 6.15 Subdivision of the *Drosophila* dorsal-ventral axis by the gradient of Dorsal protein in the nuclei. (A) Dorsal activates the zygotic genes *rhomboid*, *twist*, *fgf8*, *fgf8 receptor*, and *snail*, depending on its nuclear concentration. The mesoderm forms where Twist and Snail are present, and the glial cells form where Twist and Rhomboid interact. Those cells with Rhomboid, but no Snail or Twist, form the neurogenic ectoderm. The Fgf receptor is expressed in the mesoderm, and the Fgf8 ligands for this receptor are expressed in the mesectoderm (glia and midline central nervous system), adjacent to the mesoderm. The binding of Fgf8 to its receptor triggers the cell movements required for the ingression of the mesoderm. (B) Interactions in the specification of the ventral portion of the *Drosophila* embryo. Dorsal protein inhibits those genes that would give rise to dorsal structures (*tolloid*, *decapentaplegic*, and *zerknüllt*) while activating the three ventral genes. Snail protein, formed most ventrally, inhibits the transcription of *rhomboid* and prevents ectoderm formation. Twist activates *dMet2* and *bagpipe* (which activate muscle differentiation) as well as *tinman* (heart muscle development). (A after Steward and Govind 1993; B after Furlong et al. 2001 and Leptin and Affolter 2004.)

enables the future mesoderm to respond to random wavelike contraction events by invaginating into the embryo (Pouille et al. 2009).

The *rhomboid* and *fgf8* genes are interesting because they are activated by Dorsal but repressed by Snail. Thus, *rhomboid* and *fgf8* are not expressed in the most ventral cells (i.e., the mesodermal precursors) but are expressed in the cells adjacent to the mesoderm. These *rhomboid*- and *fgf8*-expressing cells will become the mesectoderm. The mesectoderm tissue is fated to become the ventral midline, once the mesoderm invaginates and brings these ventrolateral regions together. This mesectoderm gives rise to glial cells and to the midline structures of the central nervous system. Unlike the neurogenic ectoderm adjacent to it, the mesectoderm cells never form typical neuroblasts, never form epidermis, and are not a stem cell population (see Figure 6.15).

The high concentration of Twist protein in the nuclei of the ventralmost cells activates the gene for the Fgf8 receptor (the product of the *heartless* gene) in the presumptive mesoderm (Jiang and Levine 1993; Gryzik and Müller 2004; Strathopoulos et al. 2004). The expression and secretion of Fgf8 by the presumptive neural ectoderm is received by its receptor on the mesoderm cells, causing these mesoderm cells to invaginate into the embryo and flatten against the ectoderm (see Figure 6.14).

Meanwhile, *intermediate* levels of nuclear Dorsal activate transcription of the *Short gastrulation (Sog)* gene in two lateral stripes that flank the ventral *twist* expression domain, each 12–14 cells wide (François et al. 1994; Srinivasan et al. 2002). *Sog* encodes a protein that prevents the ectoderm in this region from becoming epidermis and begins the processes of neural differentiation (**FIGURE 6.16A**). At gastrulation, when

Dorsal protein specifies these cells to become mesoderm in two ways. First, the protein activates specific genes that create the mesodermal phenotype. Five of the target genes for the Dorsal protein are *twist*, *snail*, *fgf8*, *fgf8 receptor*, and *rhomboid* (**FIGURE 6.15**). These genes are transcribed only in nuclei that have received high concentrations of Dorsal, since their enhancers do not bind Dorsal with a very high affinity (Thisse et al. 1988, 1991; Jiang et al. 1991; Pan et al. 1991). Both Snail and Twist are also needed for the complete mesodermal phenotype and proper gastrulation (Leptin et al. 1991b). The Twist protein activates mesodermal genes, whereas the Snail protein represses particular non-mesodermal genes that might otherwise be active. The combination of Snail and Twist transcription factors in the future mesoderm cells also induces myosin contractile proteins to accumulate at the apical ends of the mesoderm cells. This localization

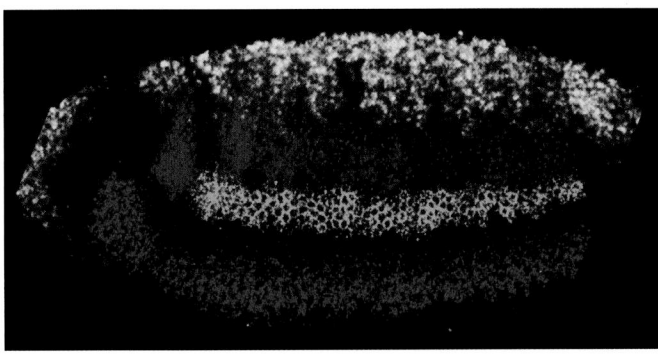

FIGURE 6.16 Dorsal-ventral patterning in *Drosophila*. The readout of the Dorsal gradient can be seen in the trunk region of a whole-mount stained embryo. The expression of the most ventral gene, *ventral nervous system defective* (blue), is from the neurogenic ectoderm. The *intermediate neuroblast defective* gene (green) is expressed in lateral ectoderm. Red represents the *muscle-specific homeobox* gene, expressed in the mesoderm above the intermediate neuroblasts. The dorsalmost tissue expresses *decapentaplegic* (yellow). (From Kosman et al. 2004, courtesy of D. Kosman and E. Bier.)

the mesoderm (at the most ventral region) invaginates into the embryo, the *Sog*-expressing cells become the most ventral cells (**FIGURE 6.16B**).

Dorsal protein also determines the mesoderm indirectly. In addition to activating the mesoderm-stimulating genes (*twist* and *snail*), it directly inhibits the dorsalizing genes *zerknüllt* (*zen*) and *decapentaplegic* (*dpp*). Thus, in the same cells Dorsal can act as an activator of some genes and a repressor of others. Whether Dorsal activates or represses a given gene depends on the structure of the gene's enhancers. The *zen* enhancer has a silencer region that contains a binding site for Dorsal as well as a second binding site for two other DNA-binding proteins. These two other proteins enable Dorsal to bind a transcriptional repressor protein (Groucho) and bring it to the DNA (Valentine et al. 1998). Mutants of *Dorsal* express *dpp* and *zen* genes throughout the embryo (Rushlow et al. 1987), and embryos deficient in *dpp* and *zen* fail to form dorsal structures (Irish and Gelbart 1987). Thus, in wild-type embryos, the mesodermal precursors

SIDELIGHTS & SPECULATIONS

The Left-Right and Inside-Out Axes

Although *Drosophila* may look bilaterally symmetric, there are left-right asymmetries both in the embryonic hindgut (which loops to the left) and in the adult hindgut and gonads. The mechanism that produces this asymmetry is different from that known to produce left-right asymmetry in vertebrates. Whereas vertebrate asymmetry appears to be regulated by microtubules, asymmetry in *Drosophila* appears to be regulated by microfilaments (Hozumi et al. 2006; Spéder et al. 2006). Very little is known about the formation of the left-right axis in *Drosophila*. If the actin microfilaments are disrupted in the *Drosophila* embryo, many defects occur, and the left-right pattern is randomized. The dextral or sinistral orientation of the gut depends on the activity of Myosin-1, a protein that interacts with these actin microfilaments. Loss-of-function mutations of the gene encoding Myosin-1 (or failure to activate this gene) reverses the body axis (**FIGURE 6.17**; see Coutelis et al. 2013).

Moreover, because the insect embryo is built along the cortex of a

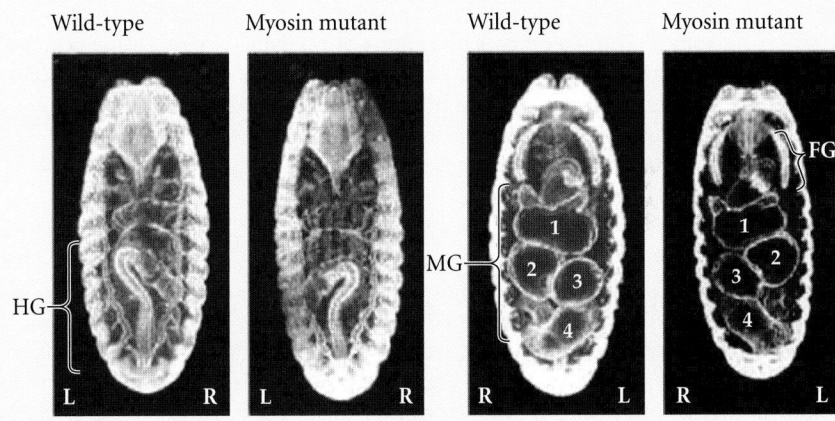

FIGURE 6.17 Left-right axis formation in *Drosophila* involves the microfilament cytoskeleton. Mutations in the myosin gene *Myo31DF* can reverse the insect's left-right asymmetry. Here the embryonic gut is seen in dorsal and ventral perspectives, showing that the asymmetry of the gut is reversed in the *myosin* mutant larva. HG, hindgut; MG, midgut; FG, foregut. (From Hozumi et al. 2006, courtesy of K. Matsuno.)

yolk-filled cytoplasm, it also has an "inside-out" axis. The outer (apical) part of each blastoderm cell is made from the egg cell membrane, while its inner (basal) membrane contacts the yolk. The *zerknüllt* (*zen*) gene is necessary for the dorsal closure

of the insect. In beetles and bugs, using RNAi to eliminate *zen* function causes the embryos to evert, resulting in "inside-out" embryos with their leg rudiments protruding into the yolk rather than outside the embryo (van der Zee et al. 2005; Panfilio 2009).

express *twist* and *snail* (but not *zen* or *dpp*); precursors of the dorsal epidermis and amnioserosa express *zen* and *dpp* (but not *twist* or *snail*). Glial (mesectoderm) precursors express *twist* and *rhomboid*, while the lateral neural ectodermal precursors do not express any of these five genes (Kosman et al. 1991; Ray and Schüpbach 1996). By the cellular responses to the Dorsal protein gradient, the embryo becomes subdivided from the ventral to dorsal regions into mesoderm, neurogenic ectoderm, epidermis (from the lateral and dorsal ectoderm), and amnioserosa (see Figure 6.13A).

Segmentation and the Anterior-Posterior Body Plan

The genetic screens pioneered by Nüsslein-Volhard and Wieschaus identified a hierarchy of genes that (1) establish anterior-posterior polarity and (2) divide the embryo into a specific number of segments, each with a different identity (FIGURE 6.18). This hierarchy is initiated by **maternal effect genes** that produce messenger RNAs localized to different regions of the egg. These mRNAs encode transcriptional and translational regulatory proteins that diffuse through the syncytial blastoderm and activate or repress the expression of certain zygotic genes.

The first such zygotic genes to be expressed are called **gap genes** (because mutations in them cause gaps in the segmentation pattern). These genes are expressed in certain broad (about three segments wide), partially overlapping domains. These gap genes encode transcription factors, and differing combinations and concentrations of the gap gene proteins regulate the transcription of **pair-rule genes**, which divide the embryo into periodic units. The transcription of the different pair-rule genes results in a striped pattern of seven transverse bands perpendicular to the anterior-posterior axis. The proteins encoded by the pair-rule genes are transcription factors that activate the **segment polarity genes**, whose mRNA and protein products divide the embryo into 14-segment-wide units, establishing the periodicity of the embryo. At the same time, the protein products of the gap, pair-rule, and segment polarity genes interact to regulate another class of genes, the **homeotic selector genes**, whose transcription determines the developmental fate of each segment.

Maternal gradients: Polarity regulation by oocyte cytoplasm

Classical experiments demonstrated that there are at least two "organizing centers" in the insect egg, one in the anterior of the egg and one in its posterior. Klaus Sander (1975) found that if he ligated the egg early in development, separating the anterior half from the posterior half, one half developed into an anterior embryo and one half developed into a posterior embryo, but neither contained the middle segments of the embryo. The later in development the ligature was made, the fewer middle segments were missing.

Thus, it appeared that there were indeed morphogenetic gradients emanating from the two poles during cleavage, and that these gradients interacted to produce the positional information determining the identity of each segment.

Moreover, when the RNA in the anterior of insect eggs was destroyed (by either ultraviolet light or RNase), the resulting embryos lacked a head and thorax. Instead, these embryos developed two abdomens and two **telsons** (tails) with mirror-image symmetry: telson-abdomen-abdomen-telson (FIGURE 6.19; Kalthoff and Sander 1968; Kandler-Singer and Kalthoff 1976). Sander's laboratory postulated the existence of a gradient at both ends of the egg, and hypothesized that the egg sequesters an mRNA that generates a gradient of anterior-forming material.

The molecular model: Protein gradients in the early embryo

In the late 1980s, the gradient hypothesis was united with a genetic approach to the study of *Drosophila* embryogenesis. If there were gradients, what were the morphogens whose concentrations changed over space? What were the genes that shaped these gradients? And did these morphogens act by activating or inhibiting certain genes in the areas where they were concentrated? Christiane Nüsslein-Volhard led a research program that addressed these questions. The researchers found that one set of genes encoded morphogens for the anterior part of the embryo, another set of genes encoded morphogens responsible for organizing the posterior region of the embryo, and a third set of genes encoded proteins that produced the terminal regions at both ends of the embryo (TABLE 6.1).

● See WEBSITE 6.4 Christiane Nüsslein-Volhard and the molecular approach to development

Two maternal messenger RNAs, *bicoid* and *nanos*, are most critical to the formation of the anterior-posterior axis. The *bicoid* mRNAs are located near the anterior tip of the unfertilized egg, and *nanos* messages are located at the posterior tip. These distributions occur as a result of the dramatic polarization of the microtubule networks in the developing oocyte (see Figure 6.8). After ovulation and fertilization, the *bicoid* and *nanos* mRNAs are translated into proteins that can diffuse in the syncytial blastoderm, forming gradients that are critical for anterior-posterior patterning (FIGURE 6.20; see also Figure 6.18B).

BICOID AS THE ANTERIOR MORPHOGEN That Bicoid was the head morphogen of *Drosophila* was demonstrated by the "find it, lose it, move it" experimentation scheme. Christiane Nüsslein-Volhard, Wolfgang Driever, and their colleagues (Driever and Nüsslein-Volhard 1988a,b; Driever et al. 1990) showed that (1) Bicoid protein was found in a gradient, highest in the anterior (head-forming) region; (2) embryos lacking Bicoid could not form a head; and (3) when *bicoid* mRNA

FIGURE 6.18 Generalized model of *Drosophila* anterior-posterior pattern formation. Anterior is to the left; the dorsal surface faces upward. (A) The pattern is established by maternal effect genes that form gradients and regions of morphogenetic proteins. These proteins are transcription factors that activate the gap genes, which define broad territories of the embryo. The gap genes enable the expression of the pair-rule genes, each of which divides the embryo into regions about two segments wide. The segment polarity genes then divide the embryo into segment-sized units along the anterior-posterior axis. Together, the actions of these genes define the spatial domains of the homeotic genes that define the identities of each of the segments. In this way, periodicity is generated from nonperiodicity, and each segment is given a unique identity. (B) Maternal effect genes. The anterior axis is specified by the gradient of Bicoid protein (yellow through red; yellow being the highest concentration). (C) Gap gene protein expression and overlap. The domain of Hunchback protein (orange) and the domain of Krüppel protein (green) overlap to form a region containing both transcription factors (yellow). (D) Products of the *fushi tarazu* pair-rule gene form seven bands across the blastoderm of the embryo. (E) Products of the segment polarity gene *engrailed*, seen here at the extended germ band stage. (B courtesy of C. Nüsslein-Volhard; C courtesy of C. Rushlow and M. Levine; D courtesy of D. W. Knowles; E courtesy of S. Carroll and S. Paddock.)

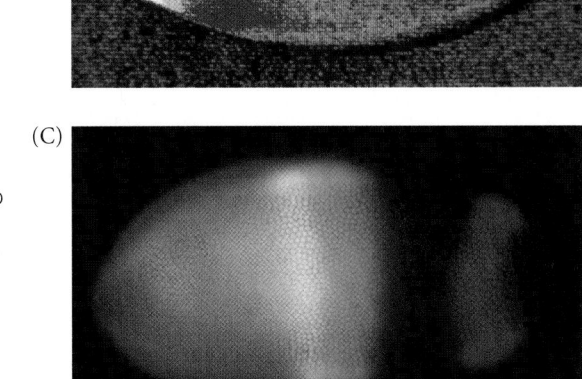

(B)

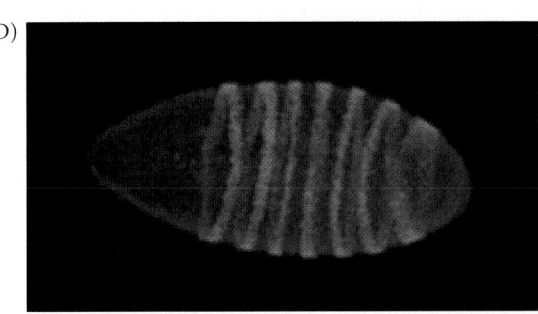

(C)

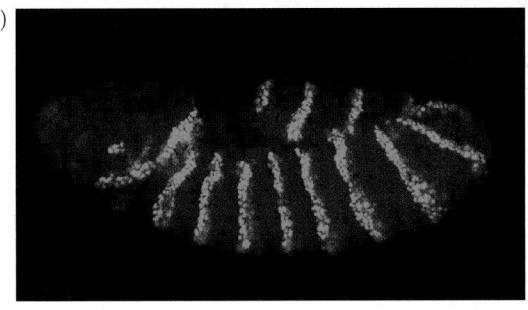

(D)

(E)

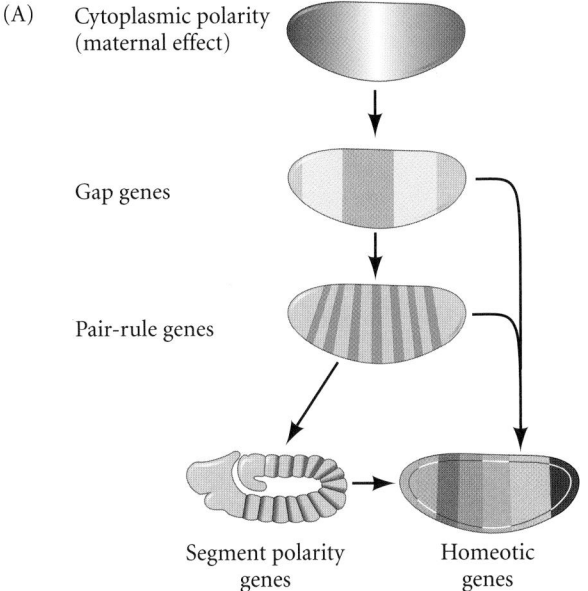

(A) Cytoplasmic polarity (maternal effect)

Gap genes

Pair-rule genes

Segment polarity genes

Homeotic genes

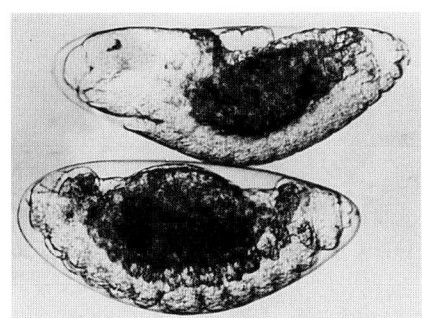

FIGURE 6.19 Normal and irradiated embryos of the midge *Smittia*. The normal embryo (top) shows a head on the left and abdominal segments on the right. The UV-irradiated embryo (bottom) has no head region but has abdominal and tail segments at both ends. (From Kalthoff 1969, courtesy of K. Kalthoff.)

TABLE 6.1 Maternal effect genes that establish the anterior-posterior polarity of the *Drosophila* embryo

Gene	Mutant phenotype	Proposed function
ANTERIOR GROUP		
bicoid (bcd)	Head and thorax deleted, replaced by inverted telson	Graded anterior morphogen; contains homeodomain; represses *caudal* mRNA
exuperantia (exu)	Anterior head structures deleted	Anchors *bicoid* mRNA
swallow (swa)	Anterior head structures deleted	Anchors *bicoid* mRNA
POSTERIOR GROUP		
nanos (nos)	No abdomen	Posterior morphogen; represses *hunchback* mRNA
tudor (tud)	No abdomen, no pole cells	Localization of Nanos protein
oskar (osk)	No abdomen, no pole cells	Localization of Nanos protein
vasa (vas)	No abdomen, no pole cells; oogenesis defective	Localization of Nanos protein
valois (val)	No abdomen, no pole cells; cellularization defective	Stabilization of the Nanos localization complex
pumilio (pum)	No abdomen	Helps Nanos protein bind *hunchback* message
caudal (cad)	No abdomen	Activates posterior terminal genes
TERMINAL GROUP		
torso (tor)	No termini	Possible morphogen for termini
trunk (trk)	No termini	Transmits Torsolike signal to Torso
fs(1)Nasrat[fs(1)N]	No termini; collapsed eggs	Transmits Torsolike signal to Torso
fs(1)polehole[fs(1)ph]	No termini; collapsed eggs	Transmits Torsolike signal to Torso

Source: After Anderson 1989.

was added to Bicoid-deficient embryos in different places, the place where *bicoid* mRNA was injected became the head (**FIGURE 6.21**). Moreover, the areas around the site of Bicoid injection became the thorax, as expected from a concentration-dependent signal. When injected into the anterior of *bicoid*-deficient embryos (whose mothers lacked *bicoid* genes), the *bicoid* mRNA "rescued" the embryos and they developed normal anterior-posterior polarity. Moreover, any location in an embryo where the *bicoid* message was injected became the head. If *bicoid* mRNA was injected into the center of an

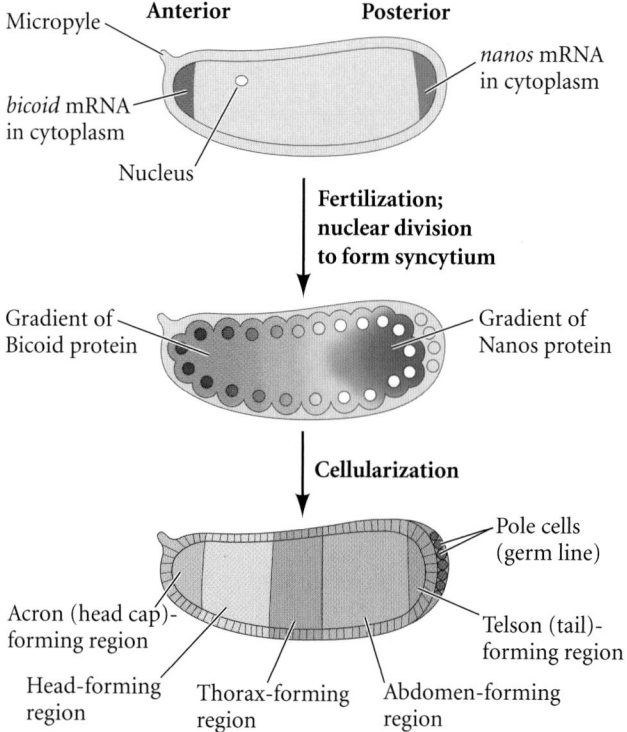

FIGURE 6.20 Syncytial specification in *Drosophila*. Anterior-posterior specification originates from morphogen gradients in the egg cytoplasm. *bicoid* mRNA is stabilized in the most anterior portion of the egg, while *nanos* mRNA is tethered to the posterior end. (The anterior can be recognized by the micropyle on the shell; this structure permits sperm to enter.) When the egg is laid and fertilized, these two mRNAs are translated into proteins. The Bicoid protein forms a gradient that is highest at the anterior end, and the Nanos protein forms a gradient that is highest at the posterior end. These two proteins form a coordinate system based on their ratios. Each position along the axis is thus distinguished from any other position. When the nuclei form, each nucleus is given its positional information by the ratio of these proteins. The proteins forming these gradients activate the transcription of the genes specifying the segmental identities of the larva and the adult fly.

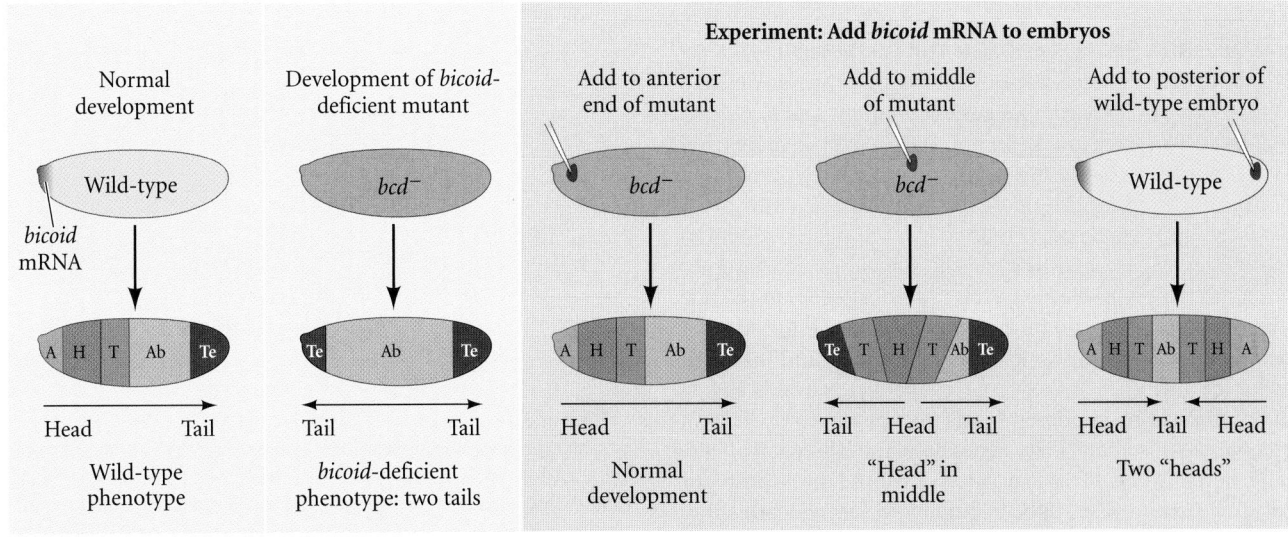

Experiment: Add *bicoid* mRNA to embryos

| Normal development | Development of *bicoid*-deficient mutant | Add to anterior end of mutant | Add to middle of mutant | Add to posterior of wild-type embryo |

Wild-type / *bicoid* mRNA

Wild-type phenotype

bicoid-deficient phenotype: two tails

Normal development

"Head" in middle

Two "heads"

| A | Acron | H | Head | T | Thorax | Ab | Abdomen | Te | Telson |

FIGURE 6.21 Schematic representation of experiments demonstrating that the *bicoid* gene encodes the morphogen responsible for head structures in *Drosophila*. The phenotypes of *bicoid*-deficient and wild-type embryos are shown at left. When *bicoid*-deficient embryos are injected with *bicoid* mRNA, the point of injection forms the head structures. When the posterior pole of an early-cleavage wild-type embryo is injected with *bicoid* mRNA, head structures form at both poles. (After Driever et al. 1990.)

embryo, then that middle region became the head, with the regions on either side of it becoming thorax structures. If a large amount of *bicoid* mRNA was injected into the posterior end of a wild-type embryo (with its own endogenous *bicoid* message in its anterior pole), two heads emerged, one at either end (Driever et al. 1990).

BICOID mRNA LOCALIZATION IN THE ANTERIOR POLE OF THE OOCYTE

The 3′untranslated region (UTR) of *bicoid* mRNA contains sequences that are critical for its localization at the anterior pole (**FIGURE 6.22**; Ferrandon et al. 1997; Macdonald and Kerr 1998; Spirov et al. 2009). These sequences interact with the Exuperantia and Swallow proteins while the messages are still in the nurse cells of the egg chamber (Schnorrer et al. 2000). Experiments in which fluorescently labeled *bicoid* mRNA was microinjected into living egg chambers of wild-type or mutant flies indicate that Exuperantia must be present in the nurse cells for anterior localization. But Exuperantia alone is not sufficient to bring the *bicoid* message into the oocyte (Cha et al. 2001; Reichmann and Ephrussi 2005). The *bicoid*-Exuparentia complex is transported out of the nurse cells and into the oocyte via microtubules, seeming to ride on a kinesin

ATPase (Arn et al. 2003). Once inside the oocyte, the *bicoid* mRNA attaches to dynein proteins that are maintained at the microtubule organizing center (the "minus end") at the anterior of the oocyte (see Figure 6.8; Cha et al. 2001). About 90% of the *bicoid* mRNA is localized to the apical 20% of the oocyte, with its concentration peaking at 0.7% of the oocyte length (Little et al. 2011).

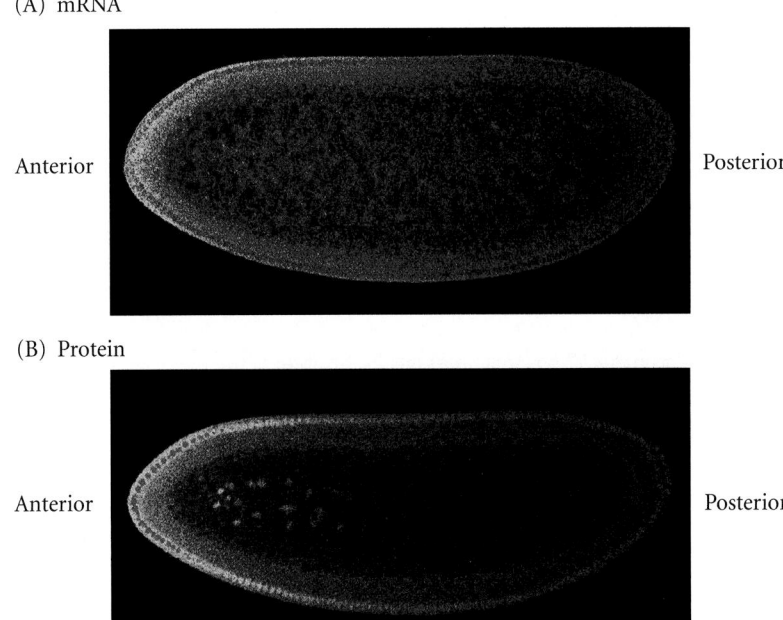

(A) mRNA

Anterior — Posterior

(B) Protein

Anterior — Posterior

FIGURE 6.22 The *bicoid* mRNA and protein gradients shown by in situ hybridization and confocal microscopy. (A) *bicoid* mRNA shows a steep gradient across the anterior portion of the oocyte. (B) When the mRNA is translated, the Bicoid protein gradient can be seen in the anterior nuclei. Anterior is to the left; the dorsal surface is upward. (After Spirov et al., courtesy of S. Baumgartner.)

NANOS mRNA LOCALIZATION IN THE POSTERIOR POLE OF THE OOCYTE The posterior organizing center is defined by the activities of the *nanos* gene (Lehmann and Nüsslein-Volhard 1991; Wang and Lehmann 1991; Wharton and Struhl 1991). While the *bicoid* message is actively transported and bound to the anterior end of the microtubules, the *nanos* message appears to get "trapped" in the posterior end of the oocyte by passive diffusion. The *nanos* message becomes bound to the cytoskeleton in the posterior region of the egg through its 3′UTR and its association with the products of several other genes (*oskar, valois, vasa, staufen,* and *tudor*).* If *nanos* (or any other of these maternal effect genes) is absent in the mother, no abdomen forms in the embryo (Lehmann and Nüsslein-Volhard 1986; Schüpbach and Wieschaus 1986). But before the *nanos* message can be localized in the posterior cortex, a *nanos* mRNA-specific trap has to be made; this trap is the Oskar protein (Ephrussi et al. 1991). The *oskar* message and the Staufen protein are transported to the posterior end of the oocyte by the kinesin motor protein (see Figure 6.8). There they become bound to the actin microfilaments of the cortex. Staufen allows the translation of the *oskar* message, and the resulting Oskar protein is capable of binding the *nanos* message (Brendza et al. 2000; Hatchet and Ephrussi 2004).

Most *nanos* mRNA, however, is not trapped. Rather, it is bound in the cytoplasm by the translation inhibitors Smaug and CUP. Smaug binds to the 3′UTR of *nanos* mRNA and recruits the CUP protein that prevents the association of the message with the ribosome as well as recruiting other proteins that deadenylate the message and target it for degradation (Rouget et al. 2010). If the *nanos*-Smaug-CUP complex reaches the posterior pole, however, Oskar can dissociate CUP from Smaug, allowing the mRNA to be bound at the posterior and ready for translation (Forrest et al. 2004; Nelson et al. 2004).

Thus, at the completion of oogenesis, the *bicoid* message is anchored at the anterior end of the oocyte and the *nanos* message is tethered to the posterior end (Frigerio et al. 1986; Berleth et al. 1988; Gavis and Lehmann 1992; Little et al. 2011). These two mRNAs are dormant until ovulation and fertilization, at which time they are translated. Since the Bicoid and Nanos *protein products* are not bound to the cytoskeleton, they diffuse toward the middle regions of the early embryo, creating the two opposing gradients that establish the anterior-posterior polarity of the embryo. Mathematical

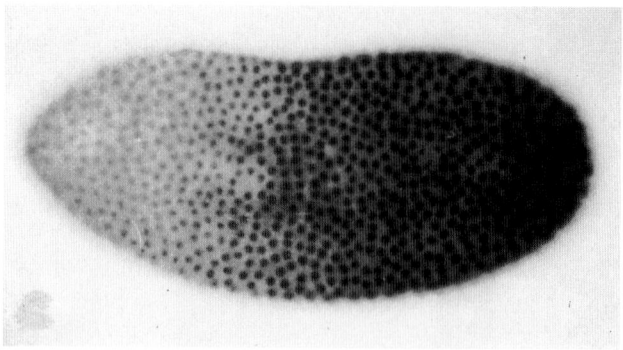

FIGURE 6.23 Caudal protein gradient of a wild-type *Drosophila* embryo at the syncytial blastoderm stage. Anterior is to the left. The protein (stained darkly) enters the nuclei and helps specify posterior fates. Compare with the complementary gradient of Bicoid protein in Figure 6.22. (From Macdonald and Struhl 1986, courtesy of G. Struhl.)

models indicate that these gradients are established by protein diffusion as well as by the active degradation of the proteins (Little et al. 2011; Liu and Ma 2011).

Two other maternally provided mRNAs (*hunchback, hb;* and *caudal, cad*) are critical for patterning the anterior and posterior regions of the body plan, respectively (Lehmann et al. 1987; Wu and Lengyel 1998). These two mRNAs are synthesized by the nurse cells of the ovary and transported to the oocyte, where they are distributed ubiquitously throughout the syncytial blastoderm. But if they are not localized, how do they mediate their localized patterning activities? It turns out that translation of the *hb* and *cad* mRNAs is repressed by the diffusion gradients of Nanos and Bicoid proteins, respectively.

GRADIENTS OF SPECIFIC TRANSLATIONAL INHIBITORS In the anterior region, Bicoid protein prevents translation of the *hunchback* message. Bicoid binds to a specific region of *caudal*'s 3′UTR. Here, it binds Bin3, a protein that stabilizes an inhibitory complex that prevents the binding of the mRNA 5′ cap to the ribosome. By recruiting this translational inhibitor, Bicoid prevents translation of *caudal* in the anterior of the embryo (**FIGURE 6.23**; Rivera-Pomar et al. 1996; Cho et al. 2006; Signh et al. 2011). This suppression is necessary; if Caudal protein is made in the embryo's anterior, the head and thorax do not form properly. Caudal activates the genes responsible for the invagination of the hindgut and thus is critical in specifying the posterior domains of the embryo.

In the posterior region, Nanos protein prevents translation of the *hunchback* message. Nanos in the posterior of the embryo forms a complex with several other ubiquitous proteins, including Pumilio and Brat. This complex binds to the 3′UTR of the *hunchback* message, where it recruits d4EHP and prevents the *hunchback* message from attaching to ribosomes (Tautz 1988; Cho et al. 2006).

---

*Like the placement of the *bicoid* message, localization of the *nanos* message is determined by its 3′UTR. If the *bicoid* 3′UTR is experimentally transferred to the protein-encoding region of *nanos* mRNA, the *nanos* message gets localized in the anterior of the egg. When this chimeric mRNA is translated, Nanos protein inhibits translation of *hunchback* and *bicoid* mRNAs and the embryo forms two abdomens—one in the anterior of the embryo and one in the posterior (Gavis and Lehmann 1992). We will see these posterior-forming proteins in later chapters when we discuss germ cell formation.

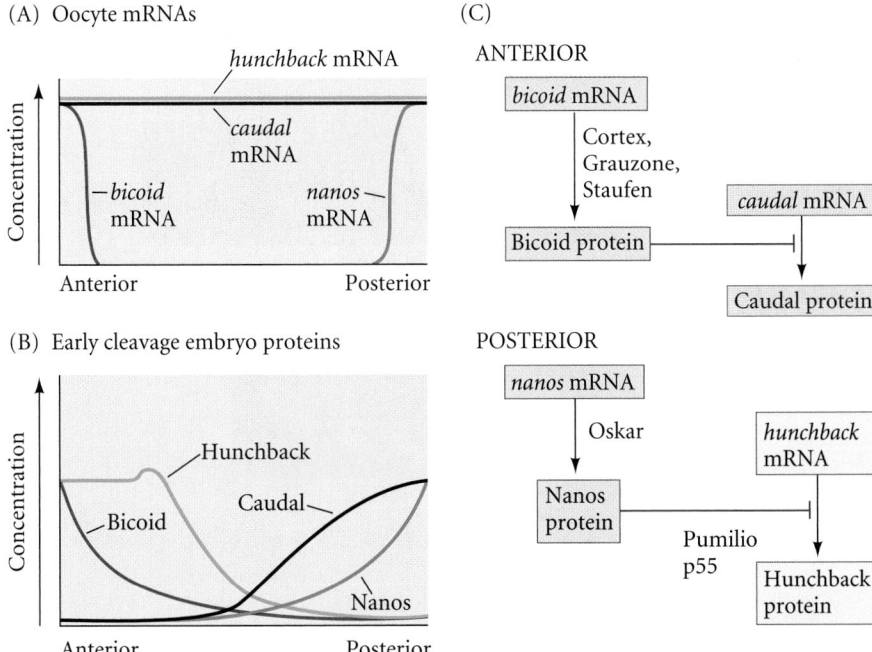

FIGURE 6.24 Model of anterior-posterior pattern generation by *Drosophila* maternal effect genes. (A) The *bicoid*, *nanos*, *hunchback*, and *caudal* mRNAs are deposited in the oocyte by the ovarian nurse cells. The *bicoid* message is sequestered anteriorly; the *nanos* message is localized to the posterior pole. (B) Upon translation, the Bicoid protein gradient extends from anterior to posterior, while the Nanos protein gradient extends from posterior to anterior. Nanos inhibits the translation of the *hunchback* message (in the posterior), while Bicoid prevents the translation of the *caudal* message (in the anterior). This inhibition results in opposing Caudal and Hunchback gradients. The Hunchback gradient is secondarily strengthened by transcription of the *hunchback* gene in the anterior nuclei (since Bicoid acts as a transcription factor to activate *hunchback* transcription). (C) Parallel interactions whereby translational gene regulation establishes the anterior-posterior patterning of the *Drosophila* embryo. (C after Macdonald and Smibert 1996.)

The result of these interactions is the creation of four maternal protein gradients in the early embryo (**FIGURE 6.24**):

- An anterior-to-posterior gradient of Bicoid protein
- An anterior-to-posterior gradient of Hunchback protein
- A posterior-to-anterior gradient of Nanos protein
- A posterior-to-anterior gradient of Caudal protein

The stage is now set for the activation of zygotic genes in the insect's nuclei, which were busy dividing while these four protein gradients were being established.

The anterior organizing center: The Bicoid and Hunchback gradients

In *Drosophila*, the phenotype of *bicoid* mutants provides valuable information about the function of morphogenetic gradients (**FIGURE 6.25A–C**). Instead of having anterior structures (acron, head, and thorax) followed by abdominal structures and a telson, the structure of a *bicoid* mutant is telson-abdomen-abdomen-telson (**FIGURE 6.25D**). It would appear that these embryos lack whatever substances are needed for the formation of the anterior structures. Moreover, one could hypothesize that the substance these mutants lack is the one postulated by Sander and Kalthoff to turn on genes for the anterior structures and turn off genes for the telson structures (compare Figures 6.19 and 6.25D).

Several studies support the view that the product of the wild-type *bicoid* gene is the morphogen that controls anterior development. The first type of evidence came from experiments that altered the shape of the Bicoid protein gradient. As we have seen, the *exuperantia* and *swallow* genes are responsible for keeping the *bicoid* message at the anterior pole of the egg. In their absence, *bicoid* mRNA diffuses farther into the posterior of the egg and the protein gradient is less steep (Driever and Nüsslein-Volhard 1988a). The phenotype produced by *exuperantia* and *swallow* mutants is similar to that of *bicoid*-deficient embryos but is less severe; these embryos lack their most anterior structures and have an extended mouth and thoracic region. Furthermore, by adding extra copies of the *bicoid* gene, the Bicoid protein gradient can be extended into more posterior regions, causing anterior structures such as the cephalic furrow to be expressed in a more posterior position (Driever and Nüsslein-Volhard 1988a; Struhl et al. 1989). Thus, altering the Bicoid gradient correspondingly alters the fate of specific embryonic regions.

(A)

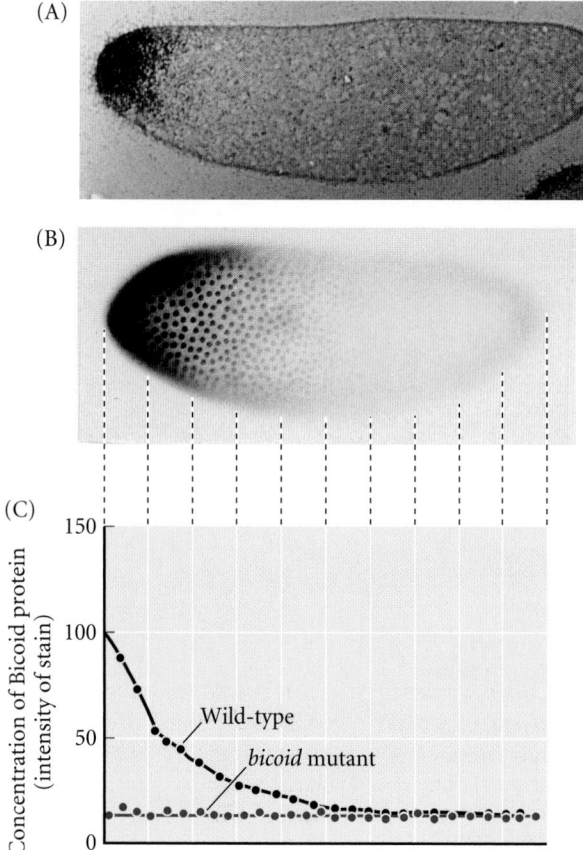

(B)

(C)

(D)

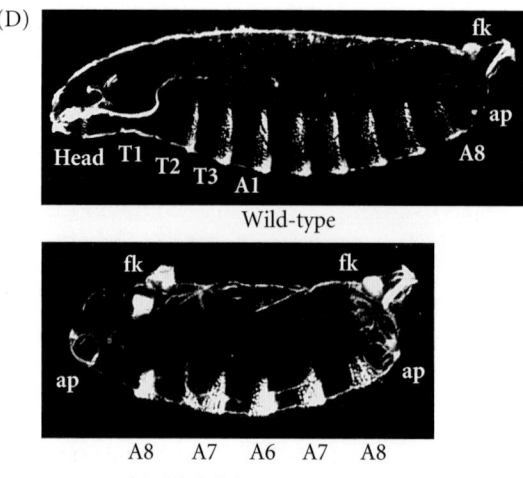

FIGURE 6.25 Bicoid protein gradient in the early *Drosophila* embryo. (A) Localization of *bicoid* mRNA to the anterior tip of the embryo in a steep gradient. (B) Bicoid protein gradient shortly after fertilization. Note that the concentration is greatest anteriorly and trails off posteriorly. Notice also that Bicoid is concentrated in the nuclei. (C) Densitometric scan of the Bicoid protein gradient. The upper curve (black) represents the Bicoid gradient in wild-type embryos. The lower curve (red) represents Bicoid in embryos of *bicoid* mutant mothers. (D) Phenotype of cuticle from a strongly affected embryo produced by a female fly deficient in the *bicoid* gene compared with the wild-type cuticle pattern. The head and thorax of the *bicoid* mutant have been replaced by a second set of posterior telson structures, abbreviated fk (filzkörper neurons) and ap (anal plates). (A from Kaufman et al. 1990; B,C from Driever and Nüsslein-Volhard 1988b; D from Driever et al. 1990, courtesy of the authors.)

It had been thought that, once the *bicoid* message was translated, the gradient of Bicoid protein would be generated simply by diffusion of the protein; the reality is a bit more complicated. In 2007, Thomas Gregor and his colleagues demonstrated that the speed of diffusion cannot account for the rapid deployment of the Bicoid protein gradient. Shortly thereafter, using highly sensitive confocal microscopy, Weil and colleagues (2008) showed that the anteriorly localized *bicoid* message became dispersed by egg activation (at ovulation), and Spirov and collaborators (2009) showed that the *bicoid* mRNA was transported along microtubules to form a gradient that prefigured the gradient of its protein (see Figures 6.22 and 6.25A,B). The *bicoid* mRNA gradient is established at nuclear cycle 10 (the beginning of the syncytial blastoderm stage), persists through nuclear division 13, and disappears as the mRNA is degraded during the initial stages of cycle 14 (when the blastoderm becomes cellular).

Whether the gradient of Bicoid protein arises from diffusion from a single source or from localized synthesis, it appears to act as a morphogen. (As described in the Part Two opening essay, morphogens are substances that differentially specify the fates of cells by different concentrations.) High concentrations of Bicoid produce anterior head structures.

Slightly less Bicoid tells the cells to become jaws. A moderate concentration of Bicoid is responsible for instructing cells to become the thorax, whereas the abdomen is characterized as lacking Bicoid.

How might a gradient of Bicoid protein control the determination of the anterior-posterior axis? As discussed earlier (see Figure P2.11), Bicoid protein inhibits the translation of *caudal*, and *caudal*'s protein product is critical for the specification of the fly posterior. However, Bicoid's primary function is to act as a transcription factor that activates the expression of target genes in the anterior part of the embryo.*

The first target of Bicoid to be discovered was the *hunchback* (*hb*) gene. In the late 1980s, two laboratories independently demonstrated that Bicoid binds to and activates *hb* (Driever and Nüsslein-Volhard 1989; Struhl et al. 1989). Bicoid-dependent transcription of *hb* is seen only in the anterior half of the embryo—the region where Bicoid is

---

**bicoid appears to be a relatively "new" gene that evolved in the Dipteran (two-winged insects such as flies) lineage; it has not been found in other insect lineages. The anterior determinant in other insect groups includes the Orthodenticle and Hunchback proteins, both of which are induced in the *Drosophila* anterior by Bicoid (Wilson and Dearden 2011).

found. This transcription reinforces the gradient of maternal Hunchback protein produced by Nanos-dependent translational repression. Mutants deficient in maternal and zygotic *hb* genes lack mouthparts and thorax structures. Therefore, both maternal and zygotic Hunchback contribute to the anterior patterning of the embryo.

Based on two pieces of evidence, Driever and co-workers (1989) predicted that Bicoid must activate at least one other anterior gene besides *hb*. First, deletions of *hb* produced only some of the defects seen in the *bicoid* mutant phenotype. Second, the *swallow* and *exuperantia* experiments showed that only moderate levels of Bicoid protein are needed to activate thorax formation (i.e., *hunchback* gene expression), but head formation requires higher Bicoid concentrations. Since then, a large number of Bicoid target genes have been identified. These include the head gap genes *buttonhead*, *empty spiracles*, and *orthodenticle*, which are expressed in specific subregions of the anterior part of the embryo (Cohen and Jürgens 1990; Finkelstein and Perrimon 1990; Grossniklaus et al. 1994).

Driever and co-workers (1989) predicted that the promoters of such a head-specific gap gene would have low-affinity binding sites for Bicoid, causing them to be activated only at extremely high concentrations of Bicoid—that is, near the anterior tip of the embryo. In addition to needing high Bicoid levels for activation, transcription of these genes also requires the presence of Hunchback protein (Simpson-Brose et al. 1994; Reinitz et al. 1995). Bicoid and Hunchback act synergistically at the enhancers of these "head genes" to promote their transcription in a feedforward manner.

In the posterior half of the embryo, the Caudal protein gradient also activates a number of zygotic genes, including the gap genes *knirps* (*kni*) and *giant* (*gt*), which are critical for abdominal development (Rivera-Pomar et al. 1995; Schulz and Tautz 1995).

The terminal gene group

In addition to the anterior and posterior morphogens, there is a third set of maternal genes whose proteins generate the unsegmented extremities of the anterior-posterior axis: the **acron** (the terminal portion of the head that includes the brain) and the telson (tail). Mutations in these terminal genes result in the loss of the acron and the most anterior head segments as well as the telson and the most posterior abdominal segments (Degelmann et al. 1986; Klingler et al. 1988). A critical gene here appears to be **torso**, a gene encoding a receptor tyrosine kinase (RTK; see Chapter 3). The embryos of mothers with mutations of *torso* have neither acron nor telson, suggesting that the two termini of the embryo are formed through the same pathway. The *torso* mRNA is synthesized by the ovarian cells, deposited in the oocyte, and translated after fertilization. The transmembrane Torso protein is not spatially restricted to the ends of the egg but is evenly distributed throughout the cell membrane (Casanova and Struhl 1989). Indeed, a gain-of-function mutation of *torso*, which imparts constitutive activity to the receptor, converts the entire anterior half of the embryo into an acron and the entire posterior half into a telson. Thus, Torso must normally be activated only at the ends of the egg.

Stevens and her colleagues (1990) have shown that this is the case. Torso protein is activated by the follicle cells only at the two poles of the oocyte. Two pieces of evidence suggest that the activator of Torso is probably the Torso-like protein: first, loss-of-function mutations in the *torsolike* gene create a phenotype almost identical to that produced by *torso* mutants; and second, ectopic expression of Torso-like protein activates Torso in the new location. The *torsolike* gene is usually expressed only in the anterior and posterior follicle cells, and secreted Torsolike protein can cross the perivitelline space to activate Torso in the egg membrane (Martin et al. 1994; Furriols et al. 1998). In this manner, Torsolike activates Torso in the anterior and posterior regions of the oocyte membrane.

The end products of the RTK cascade activated by Torso diffuse into the cytoplasm at both ends of the embryo (**FIGURE 6.26**; Gabay et al. 1997). These kinases are thought to inactivate the Capicua protein, a transcriptional repressor of the *tailless* and *huckebein* gap genes (Ajuria et al. 2011); it is these two gap genes that specify the termini of the embryo. The distinction between the anterior and posterior termini depends on the presence of Bicoid. If *tailless* and *huckebein* act alone, the terminal region differentiates into a telson. However, if Bicoid is also present (at the anterior end), the terminal region forms an acron (Pignoni et al. 1992).

Summarizing early anterior-posterior axis specification in Drosophila

The anterior-posterior axis of the *Drosophila* embryo is specified by three sets of genes:

1. **Genes that define the anterior organizing center.** Located at the anterior end of the embryo, the anterior organizing center acts through a gradient of Bicoid protein. Bicoid functions both as a *transcription factor* to activate anterior-specific gap genes and as a *translational repressor* to suppress posterior-specific gap genes.

2. **Genes that define the posterior organizing center**. The posterior organizing center is located at the posterior pole. This center acts *translationally* through the Nanos protein to inhibit anterior formation, and *transcriptionally* through the Caudal protein to activate those genes that form the abdomen.

3. **Genes that define the terminal boundary regions**. The boundaries of the acron and telson are defined by the product of the *torso* gene, which is activated at the tips of the embryo.

The next step in development will be to use these gradients of transcription factors to activate specific genes along the anterior-posterior axis.

(A)

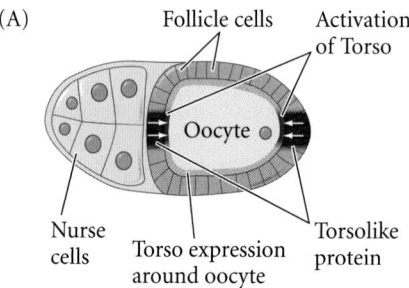

(B)

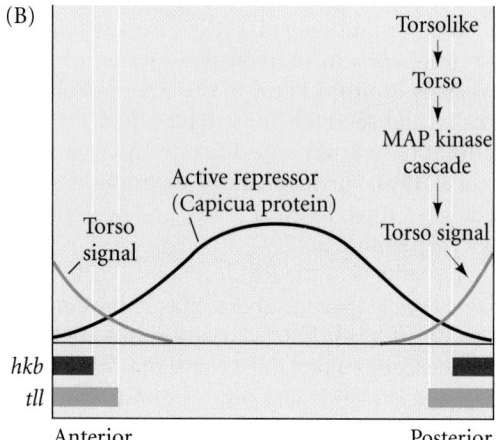

FIGURE 6.26 Formation of the unsegmented extremities by Torso signaling. (A) Torsolike protein is expressed by the follicle cells at the poles of the oocyte. Torso protein is uniformly distributed throughout the cell membrane of the oocyte. Torsolike activates Torso at the poles (see Casanova et al. 1995). (B) Inactivation of the transcriptional repression of *huckebein* (*hkb*) and *tailless* (*tll*) genes. The Torso signal antagonizes the Capicua protein. Capicua represses *tailless* and *huckebein* expression. The distinction between whether an acron/head or a tail is formed depends on the presence of Bicoid. (A after Gabay et al. 1997; B after Paroush et al. 1997.)

Segmentation Genes

Cell fate commitment in *Drosophila* appears to have two steps: specification and determination (Slack 1983). Early in development, the fate of a cell depends on cues provided by protein gradients. This specification of cell fate is flexible and can still be altered in response to signals from other cells. Eventually, however, the cells undergo a transition from this loose type of commitment to an irreversible determination. At this point, the fate of a cell becomes cell-intrinsic.*

---

*Aficionados of information theory will recognize that the process by which the anterior-posterior information in morphogenetic gradients is transferred to discrete and different parasegments represents a transition from analog to digital specification. Specification is analog, determination digital. This process enables the transient information of the gradients in the syncytial blastoderm to be stabilized so that it can be used much later in development (Baumgartner and Noll 1990).

TABLE 6.2 Major genes affecting segmentation pattern in *Drosophila*

Category	Gene name
Gap genes	*Krüppel (Kr)*
	knirps (kni)
	hunchback (hb)
	giant (gt)
	tailless (tll)
	huckebein (hkb)
	buttonhead (btd)
	empty spiracles (ems)
	orthodenticle (otd)
Pair-rule genes (primary)	*hairy (h)*
	even-skipped (eve)
	runt (run)
Pair-rule genes (secondary)	*fushi tarazu (ftz)*
	odd-paired (opa)
	odd-skipped (odd)
	sloppy-paired (slp)
	paired (prd)
Segment polarity genes	*engrailed (en)*
	wingless (wg)
	cubitus interruptusD (ciD)
	hedgehog (hh)
	fused (fu)
	armadillo (arm)
	patched (ptc)
	gooseberry (gsb)
	pangolin (pan)

The transition from specification to determination in *Drosophila* is mediated by **segmentation genes** that divide the early embryo into a repeating series of segmental primordia along the anterior-posterior axis. Segmentation genes were originally defined by zygotic mutations that disrupted the body plan, and these genes were divided into three groups based on their mutant phenotypes (**TABLE 6.2**; Nüsslein-Volhard and Wieschaus 1980):

1. *Gap mutants* lack large regions of the body (several contiguous segments; **FIGURE 6.27A**).
2. *Pair-rule mutants* lack portions of every other segment (**FIGURE 6.27B**).
3. *Segment polarity mutants* show defects (deletions, duplications, polarity reversals) in every segment (**FIGURE 6.27C**).

Segments and parasegments

Mutations in segmentation genes result in *Drosophila* embryos that lack certain segments or parts of segments.

(A) Gap: *Krüppel* (as an example)

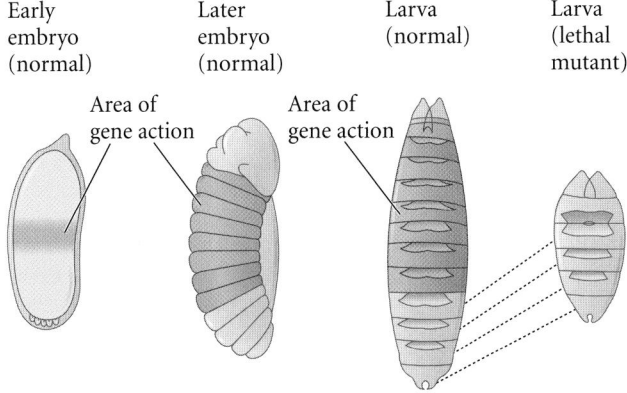

Early embryo (normal) Later embryo (normal) Larva (normal) Larva (lethal mutant)

Area of gene action   Area of gene action

(B) Pair-rule: *fushi tarazu* (as an example)

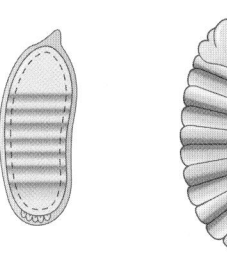 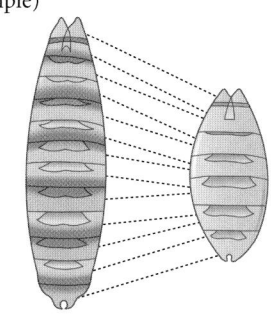

(C) Segment polarity: *engrailed* (as an example)

 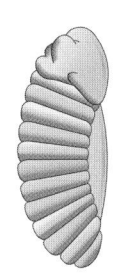 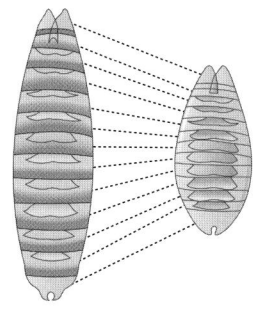

FIGURE 6.27 Three types of segmentation gene mutations. The left side shows the early-cleavage embryo (yellow), with the region where the particular gene is normally transcribed in wild-type embryos shown in blue. These areas are deleted as the mutants develop into late-stage embryos.

However, early researchers found a surprising aspect of these mutations: many of them did not affect actual segments. Rather, they affected the posterior compartment of one segment and the anterior compartment of the immediately posterior segment (**FIGURE 6.28**). These "transegmental" units were named **parasegments** (Martinez-Arias and Lawrence 1985).

Once the means to detect gene expression patterns were available, it was discovered that the expression patterns in the early embryo are delineated by parasegmental boundaries, not by the boundaries of the segments. Thus, the parasegment appears to be the fundamental unit of *embryonic* gene expression. Although parasegmental organization is also seen in the nerve cord of adult *Drosophila*, it is not seen in the adult epidermis (the most obvious manifestation of segmentation), nor is it found in the adult musculature. These adult structures are organized along the segmental pattern. In *Drosophila*, segmental grooves appear in the epidermis when the germ band is retracted; the muscle-forming mesoderm becomes segmental later in development.

One can think about the segmental and parasegmental organization schemes as representing different ways of organizing the compartments along the anterior-posterior axis of the embryo. The cells of one compartment do not mix with cells of neighboring compartments, and parasegments and segments are out of phase by one compartment.*

The gap genes

The gap genes are activated or repressed by the maternal effect genes, and are expressed in one or two broad domains along the anterior-posterior axis. These expression patterns correlate quite well with the regions of the embryo that are missing in gap mutations. For example, *Krüppel* is expressed primarily in parasegments 4–6, in the center of the embryo (see Figures 6.27A and 6.18C); in the absence of the Krüppel protein, the embryo lacks segments from these and the immediately adjacent regions.

Deletions caused by mutations in three gap genes—*hunchback*, *Krüppel*, and *knirps*—span the entire segmented region of the *Drosophila* embryo. The gap gene *giant* overlaps with these three, and the gap genes *tailless* and *huckebein* are expressed in domains near the anterior and posterior ends of the embryo.

The expression patterns of the gap genes are highly dynamic. These genes usually show low levels of transcriptional activity across the entire embryo that become consolidated into discrete regions of high activity as cleavage continues (Jäckle et al. 1986). The Hunchback gradient is particularly important in establishing the initial gap gene expression patterns. By the end of nuclear division cycle 12, Hunchback is found at high levels across the anterior part of the embryo. Hunchback then forms a steep gradient through about 15 nuclei near the middle of the embryo (see Figures 6.18C and 6.24). The last third of the embryo has undetectable Hunchback levels at this time. The transcription patterns of the anterior gap genes are initiated by the different concentrations of the Hunchback and

---

*The two modes of segmentation may be required for the coordination of movement in the adult fly. In arthropods, the ganglia of the ventral nerve cord are organized by parasegments, but the cuticle grooves and musculature are segmental. This shift in frame by one compartment allows the muscles on both sides of any particular epidermal segment to be coordinated by the same ganglion. This, in turn, allows rapid and coordinated muscle contractions for locomotion (Deutsch 2004). A similar situation occurs in vertebrates, where the posterior portion of the anterior somite combines with the anterior portion of the next somite.

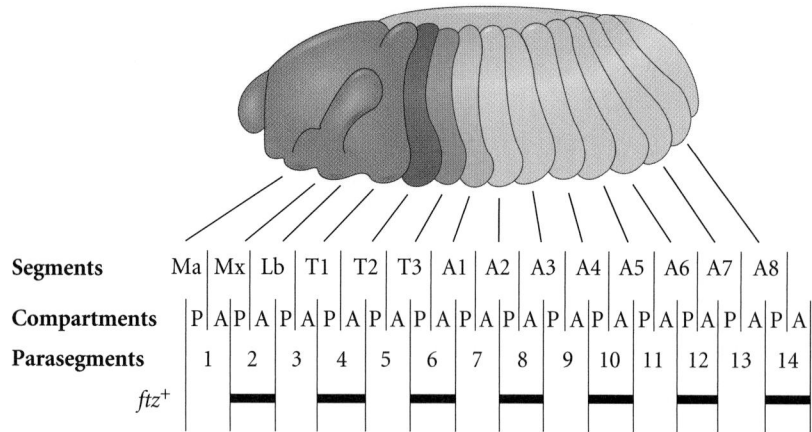

FIGURE 6.28 Overlap of segments and parasegments. Parasegments in the *Drosophila* embryo are shifted one compartment forward in relation to the segments. Ma, Mx, and Lb are the mandibular, maxillary, and labial head segments; T1–T3 are the thoracic segments; and A1–A8 are abdominal segments. Each segment has an anterior (A) and a posterior (P) compartment. Each parasegment (numbered 1–14) consists of the posterior compartment of one segment and the anterior compartment of the segment in the next posterior position. Black bars indicate the boundaries of gene expression observed in the *fushi tarazu* (*ftz*) mutant (see Figure 6.27B). (After Martinez-Arias and Lawrence 1985.)

Bicoid proteins. High levels of Bicoid and Hunchback induce the expression of *giant*, while the *Krüppel* transcript appears over the region where Hunchback begins to decline. High levels of Hunchback also prevent the transcription of the posterior gap genes (such as *knirps* and *giant*) in the embryo's anterior (Struhl et al. 1992). It is thought that a gradient of the Caudal protein, highest at the posterior pole, is responsible for activating the abdominal gap genes *knirps* and *giant* in the posterior part of the embryo. The *giant* gene thus has two methods for its activation, one for its anterior expression band and one for its posterior expression band (Rivera-Pomar 1995; Schulz and Tautz 1995).

After the initial gap gene expression patterns have been established by the maternal effect gradients and Hunchback, they are stabilized and maintained by repressive interactions between the different gap gene products themselves.* These boundary-forming inhibitions are thought to be directly mediated by the gap gene products, because all four major gap genes (*hunchback, giant, Krüppel,* and *knirps*) encode DNA-binding proteins (Knipple et al. 1985; Gaul and Jäckle 1990; Capovilla et al. 1992). One such model, established by genetic experiments, biochemical analyses, and mathematical modeling, is presented in **FIGURE 6.29A** (Papatsenko and Levine 2011). The model depicts a network with three major toggle switches (**FIGURE 6.29B–D**). Two of these switches are the strong mutual inhibition between Hunchback and Knirps, and the strong mutual inhibition between Giant and Krüppel (Jaeger et al. 2004). The third is the concentration-dependent interaction between Hunchback and Krüppel. At high doses, Hunchback inhibits the production of Krüppel protein, but at moderate doses (at about 50% of the embryo length) Hunchback promotes Krüppel formation (see Figure 6.29C).

The end result of these repressive interactions is the creation of a precise system of overlapping mRNA expression patterns. Each domain serves as a source for diffusion of gap proteins into adjacent embryonic regions. This creates

a significant overlap (at least eight nuclei, which accounts for about two segment primordia) between adjacent gap protein domains. This was demonstrated in a striking manner by Stanojević and co-workers (1989). They fixed cellularizing blastoderms (see Figure 6.2), stained Hunchback protein with an antibody carrying a red dye, and simultaneously stained Krüppel protein with an antibody carrying a green dye. Cellularizing regions that contained both proteins bound both antibodies and stained bright yellow (see Figure 6.18C). Krüppel overlaps with Knirps in a similar manner in the posterior region of the embryo (Pankratz et al. 1990). The precision of these patterns is maintained by having redundant enhancers; if one of these enhancers fails to work, there is a high probability that the other will still function (Perry et al. 2011).

The pair-rule genes

The first indication of segmentation in the fly embryo comes when the pair-rule genes are expressed during nuclear division cycle 13, as the cells begin to form at the periphery of the embryo. The transcription patterns of these genes divide the embryo into regions that are precursors of the segmental body plan. As can be seen in **FIGURE 6.30** and (in Figure 6.18D), one vertical band of nuclei (the cells are just beginning to form) expresses a pair-rule gene, the next band of nuclei does not express it, and then the next band expresses it again. The result is a "zebra stripe" pattern along the anterior-posterior axis, dividing the embryo into 15 subunits (Hafen et al. 1984). Eight genes are currently known to be capable of dividing the early embryo in this fashion, and they overlap one another so as to give each cell in the parasegment a specific set of transcription factors (see Table 6.2).

The primary pair-rule genes include *hairy, even-skipped,* and *runt,* each of which is expressed in seven stripes. All three build their striped patterns from scratch, using distinct enhancers and regulatory mechanisms for each stripe. These enhancers are often modular: control over expression in each stripe is located in a discrete region of the DNA, and these DNA regions often contain binding sites recognized by gap proteins. Thus, it is thought that the different concentrations

---

*The interactions between these genes and gene products are facilitated by the fact that these reactions occur within a syncytium, in which the cell membranes have not yet formed.

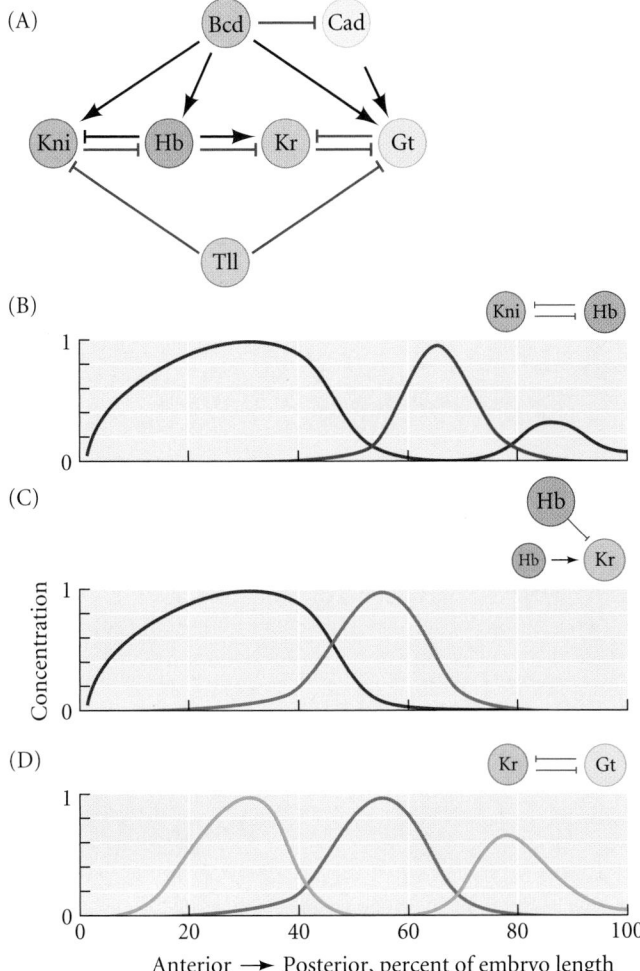

(A)

(B)

(C)

Concentration

(D)

Anterior ⟶ Posterior, percent of embryo length

FIGURE 6.29 Architecture of the gap gene network. These interactions are supported by mathematical modeling, genetic data, and biochemical analyses. (A) The anterior-posterior gradient of Bicoid (Bcd) and Caudal (Cad) regulates expression of Knirps (Kni), Hunchback (Hb), Krüppel (Kr; weakly activated by both Bicoid and Caudal proteins), and Giant (Gt). Tailless (Tll) prevents these patterning pathways at the terminal ends of the embryo. (B–D) The three "toggle switches" activated along the anterior-posterior axis to establish gap gene domains. (B) The mutual inhibition of Knirps and Hunchback positions the Knirps protein domain at around 60–80% along the anterior-posterior axis. (C) Hunchback inhibits Krüppel expression at high concentrations but promotes it at intermediate concentrations. (D) Krüppel and Giant mutually inhibit each other's synthesis. (After Papatsenko and Levine 2011.)

separate stripe or a pair of stripes. For instance, *even-skipped* stripe 2 is controlled by a 500-bp region that is activated by Bicoid and Hunchback and repressed by both Giant and Krüppel proteins (**FIGURE 6.32**; Small et al. 1991, 1992; Stanojevíc et al. 1991; Janssens et al. 2006). The anterior border is maintained by repressive influences from Giant, while the posterior border is maintained by Krüppel. DNase I footprinting showed that the minimal enhancer region for this stripe contains five binding sites for Bicoid, one for Hunchback, three for Krüppel, and three for Giant. Thus, this region is thought to act as a switch that can directly sense the concentrations of these proteins and make on/off transcriptional decisions.

The importance of these enhancer elements can be shown by both genetic and biochemical means. First, a mutation in a particular enhancer can delete its particular stripe and no other. Second, if a reporter gene (such as *lacZ*, which encodes β-galactosidase) is fused to one of the enhancers, the reporter gene is expressed only in that particular stripe (see Figure 6.31; Fujioka et al. 1999). Third, placement of the stripes can be altered by deleting the gap genes that regulate them. Thus, stripe placement is a result of (1) the modular *cis*-regulatory enhancer elements of the pair-rule genes and (2) the *trans*-regulatory gap gene and maternal gene proteins that bind to these enhancer sites.

Once initiated by the gap gene proteins, the transcription pattern of the primary pair-rule genes becomes stabilized by interactions among their products (Levine and Harding 1989). The primary pair-rule genes also form the context that allows or inhibits expression of the later-acting secondary pair-rule genes. One such gene is *fushi tarazu* (*ftz*), which means "too few segments" in Japanese (**FIGURE 6.33**). Early in division cycle 14, *ftz* mRNA and its protein are seen throughout the segmented portion of the embryo. However, as the proteins from the primary pair-rule genes begin to interact with the *ftz* enhancer, the *ftz* gene is repressed in certain bands of nuclei to create interstripe regions. Meanwhile, the Ftz protein interacts with its own promoter to stimulate more transcription of *ftz* where it is already present (Edgar et al. 1986b; Karr and Kornberg 1989; Schier and Gehring 1992).

of gap proteins determine whether or not a pair-rule gene is transcribed.

One of the best-studied primary pair-rule genes is *even-skipped* (**FIGURE 6.31**). Its enhancer region is composed of modular units arranged such that each enhancer regulates a

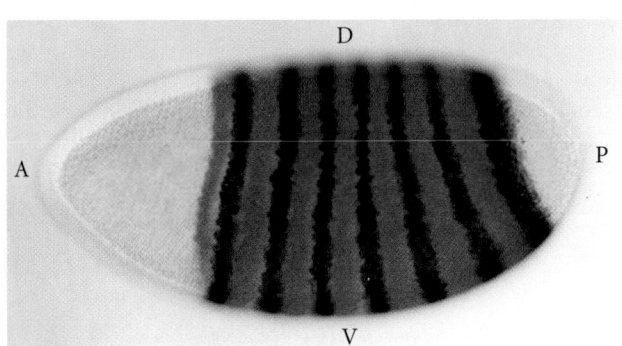

FIGURE 6.30 Messenger RNA expression patterns of two pair-rule genes, *even-skipped* (red) and *fushi tarazu* (black) in the *Drosophila* blastoderm. Each gene is expressed as a series of seven stripes. Anterior is to the left, dorsal is up. (Courtesy of S. Small.)

(A)

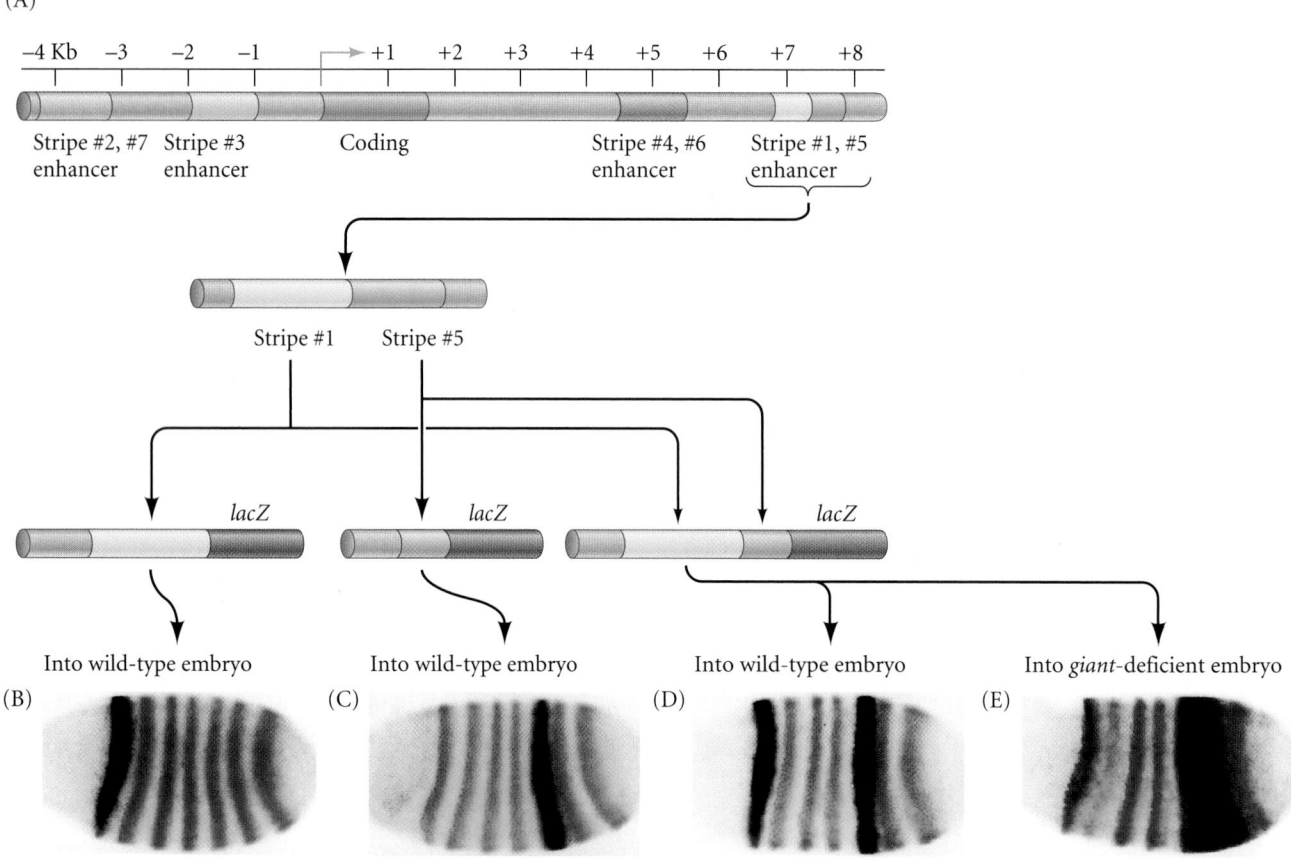

Into wild-type embryo Into wild-type embryo Into wild-type embryo Into *giant*-deficient embryo

FIGURE 6.31 Specific promoter regions of the *even-skipped* (*eve*) gene control specific transcription bands in the embryo. (A) Partial map of the *eve* promoter, showing the regions responsible for the various stripes. (B–E) A reporter β-galactosidase gene (*lacZ*) was fused to different regions of the *eve* promoter and injected into fly embryos. The resulting embryos were stained (orange bands) for the presence of Even-skipped protein. (B–D) Wild-type embryos that were injected with *lacZ* transgenes containing the enhancer region specific for stripe 1 (B), stripe 5 (C), or both regions (D). (E) The enhancer region for stripes 1 and 5 was injected into an embryo deficient in *giant*. Here the posterior border of stripe 5 is missing. (After Fujioka et al. 1999 and Sackerson et al. 1999; photographs courtesy of M. Fujioka and J. B. Jaynes.)

The eight known pair-rule genes are all expressed in striped patterns, but the patterns are not coincident with each other. Rather, each row of nuclei within a parasegment has its own array of pair-rule products that distinguishes it from any other row. These products activate the next level of segmentation genes, the segment polarity genes.

The segment polarity genes

So far our discussion has described interactions between molecules within the syncytial embryo. But once cells form, interactions take place between the cells. These interactions are mediated by the segment polarity genes, and they accomplish two important tasks. First, they reinforce the

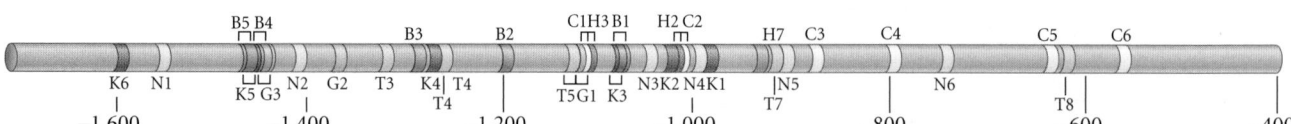

FIGURE 6.32 Model for formation of the second stripe of transcription from the *even-skipped* gene. The enhancer element for stripe 2 regulation contains binding sequences for several maternal and gap gene proteins. Activators (e.g., Bicoid and Hunchback) are noted above the line; repressors (e.g., Krüppel and Giant) are shown below. Note that nearly every activator site is closely linked to a repressor site, suggesting competitive interactions at these positions. (Moreover, a protein that is a repressor for stripe 2 may be an activator for stripe 5; it depends on which proteins bind next to them.) B, Bicoid; C, Caudal; G, Giant; H, Hunchback; K, Krüppel; N, Knirps; T, Tailless. (After Janssens et al. 2006.)

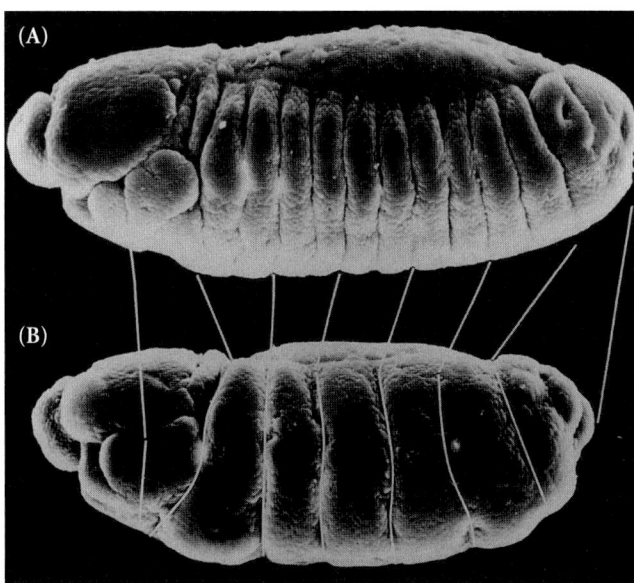

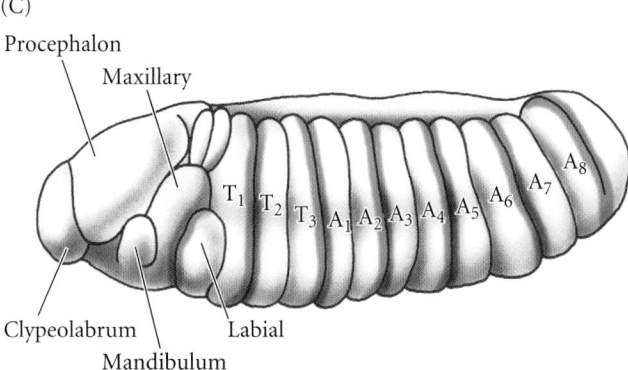

Procephalon
Maxillary

T₁ T₂ T₃ A₁ A₂ A₃ A₄ A₅ A₆ A₇ A₈

Clypeolabrum   Labial
Mandibulum

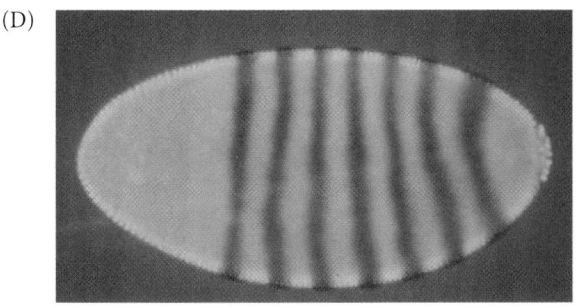

FIGURE 6.33 Defects seen in the *fushi tarazu* mutant. Anterior is to the left; dorsal surface faces upward. (A) Scanning electron micrograph of a wild-type embryo, seen in lateral view. (B) A *fushi tarazu*–mutant embryo at the same stage. The white lines connect the homologous portions of the segmented germ band. (C) Diagram of wild-type embryonic segmentation. The areas shaded in purple show the parasegments of the germ band that are missing in the mutant embryo. (D) Transcription pattern of the *fushi tarazu* gene. (After Kaufman et al. 1990; A,B courtesy of T. Kaufman; D courtesy of T. Karr.)

Wingless protein. (Wingless is the *Drosophila* Wnt protein.) The key to this pattern is the activation of the *engrailed* (*en*) gene in those cells that are going to express Hedgehog. The *engrailed* gene is activated in cells that have high levels of the Even-skipped, Fushi tarazu, or Paired transcription factors; *engrailed* is repressed in those cells with high levels of Odd-skipped, Runt, or Sloppy-paired proteins. As a result, the Engrailed protein is found in 14 stripes across the anterior-posterior axis of the embryo (see Figure 6.18E). (Indeed, in *ftz*-deficient embryos, only seven bands of *engrailed* are expressed.)

These stripes of *engrailed* transcription mark the anterior compartment of each parasegment (and the posterior compartment of each segment). The *wingless* (*wg*) gene is activated in those bands of cells that receive little or no Even-skipped or Fushi tarazu protein, but which do contain Sloppy-paired. This pattern causes *wingless* to be transcribed solely in the row of cells directly anterior to the cells where *engrailed* is transcribed (FIGURE 6.34A).

Once *wingless* and *engrailed* expression patterns are established in adjacent cells, this pattern must be maintained to retain the parasegmental periodicity of the body plan. It should be remembered that the mRNAs and proteins involved in initiating these patterns are short-lived, and that the patterns must be maintained after their initiators are no longer being synthesized. The maintenance of these patterns is regulated by reciprocal interaction between neighboring cells: cells secreting Hedgehog activate *wingless* expression in their neighbors, and the Wingless protein signal, which is received by the cells that secreted Hedgehog, serves to maintain *hedgehog* (*hh*) expression (FIGURE 6.34B). Wingless protein also acts in an autocrine fashion, maintaining its own expression (Sánchez et al. 2008).

In the cells transcribing the *wingless* gene, *wingless* mRNA is translocated by its 3′UTR to the apex of the cell (Simmonds et al. 2001; Wilkie and Davis 2001). At the apex, the *wingless* message is translated and secreted from the cell. The cells expressing *engrailed* can bind this protein because they contain Frizzled, which is the *Drosophila* membrane receptor protein for Wingless (Bhanot et al. 1996). Binding of Wingless to Frizzled activates the Wnt signal transduction pathway, resulting in the continued expression of *engrailed* (Siegfried et al. 1994).

This activation starts another portion of this reciprocal pathway. The Engrailed protein activates the transcription

parasegmental periodicity established by the earlier transcription factors. Second, through this cell-to-cell signaling, cell fates are established within each parasegment.

The segment polarity genes encode proteins that are constituents of the Wnt and Hedgehog signaling pathways (see Chapter 3). Mutations in these genes lead to defects in segmentation and in gene expression pattern across each parasegment. The development of the normal pattern relies on the fact that only one row of cells in each parasegment is permitted to express the Hedgehog protein, and only one row of cells in each parasegment is permitted to express the

FIGURE 6.34 Model for transcription of the segment polarity genes *engrailed* (*en*) and *wingless* (*wg*). (A) Expression of *wg* and *en* is initiated by pair-rule genes. The *en* gene is expressed in cells that contain high concentrations of either Even-skipped or Fushi tarazu proteins. The *wg* gene is transcribed when neither *eve* nor *ftz* genes are active, but when a third gene (probably *sloppy-paired*) is expressed. (B) The continued expression of *wg* and *en* is maintained by interactions between the Engrailed- and Wingless-expressing cells. Wingless protein is secreted and diffuses to the surrounding cells. In those cells competent to express Engrailed (i.e., those having Eve or Ftz proteins), Wingless protein is bound by the Frizzled and Lrp6 receptor proteins, which enables the activation of the *en* gene via the Wnt signal transduction pathway. (Armadillo is the *Drosophila* name for β-catenin.) Engrailed protein activates the transcription of the *hedgehog* gene and also activates its own (*en*) gene transcription. Hedgehog protein diffuses from these cells and binds to the Patched receptor protein on neighboring cells. The Hedgehog signal enables the transcription of the *wg* gene and the subsequent secretion of the Wingless protein. For a more complex view, see Sánchez et al. 2008.

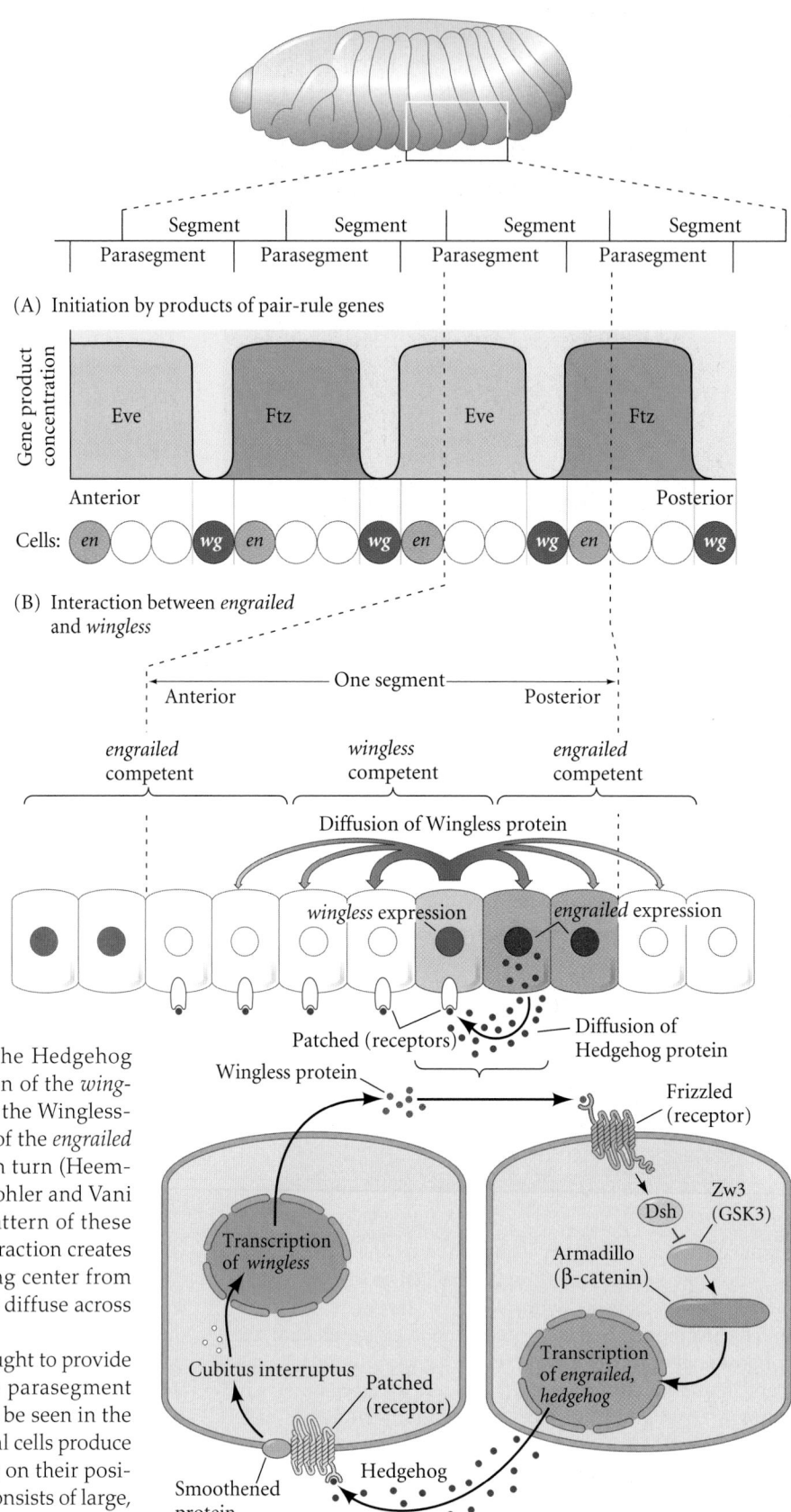

of the *hedgehog* gene in the *engrailed*-expressing cells. Hedgehog protein can bind to its receptor protein (Patched) on neighboring cells. When it binds to the adjacent posterior cells, it stimulates the expression of the *wingless* gene. The result is a reciprocal loop wherein the Engrailed-synthesizing cells secrete the Hedgehog protein, which maintains the expression of the *wingless* gene in the neighboring cells, while the Wingless-secreting cells maintain the expression of the *engrailed* and *hedgehog* genes in their neighbors in turn (Heemskerk et al. 1991; Ingham et al. 1991; Mohler and Vani 1992). In this way, the transcription pattern of these two types of cells is stabilized. This interaction creates a stable boundary, as well as a signaling center from which Hedgehog and Wingless proteins diffuse across the parasegment.

The diffusion of these proteins is thought to provide the gradients by which the cells of the parasegment acquire their identities. This process can be seen in the dorsal epidermis, where the rows of larval cells produce different cuticular structures depending on their position in the segment. The 1° row of cells consists of large, pigmented spikes called denticles. Posterior to these

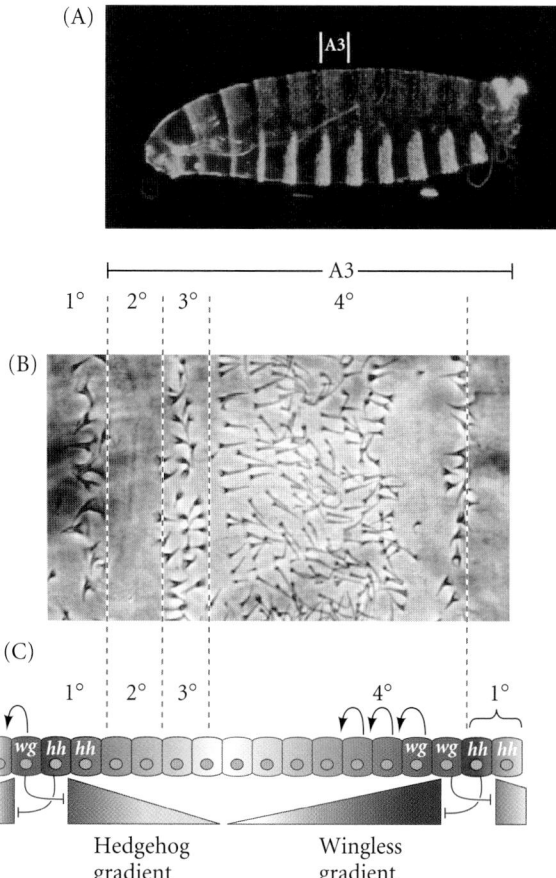

FIGURE 6.35 Cell specification by the Wingless/Hedgehog signaling center. (A) Bright-field photograph of wild-type *Drosophila* embryo, showing the position of the third abdominal segment. Anterior is to the left; the dorsal surface faces upward. (B) Close-up of the dorsal area of the A3 segment, showing the different cuticular structures made by the 1°, 2°, 3°, and 4° rows of cells. (C) A model for the roles of Wingless and Hedgehog. Each signal is responsible for roughly half the pattern. Either each signal acts in a graded manner (shown here as gradients decreasing with distance from their respective sources) to specify the fates of cells at a distance from these sources, or each signal acts locally on the neighboring cells to initiate a cascade of inductions (shown here as sequential arrows). (After Heemskerk and DiNardo 1994; A,B courtesy of the authors.)

cells, the 2° row produces a smooth epidermal cuticle. The next two cell rows have a 3° fate, making small, thick hairs; they are followed by several rows of cells that adopt the 4° fate, producing fine hairs (**FIGURE 6.35**).

The fates of the cells can be altered by experimentally increasing or decreasing the levels of Hedgehog or Wingless (Heemskerk and DiNardo 1994; Bokor and DiNardo 1996; Porter et al. 1996). These two proteins thus appear to be necessary for elaborating the entire pattern of cell types across the parasegment. Gradients of Hedgehog and Wingless are interpreted by a second series of protein gradients within the cells. This second set of gradients provides certain cells with

the receptors for Hedgehog and (often) with the receptor for Wingless (Casal et al. 2002; Lander et al. 2002). The resulting pattern of cell fates also changes the focus of patterning from parasegment to segment. There are now external markers, as the *engrailed*-expressing cells become the most posterior cells of each segment.

● See **WEBSITE 6.5** Asymmetrical spread of morphogens

● See **WEBSITE 6.6** Getting a head in the fly

The Homeotic Selector Genes

After the segmental boundaries are set, the pair-rule and gap genes interact to regulate the homeotic selector genes, which specify the characteristic structures of each segment (Lewis 1978). By the end of the cellular blastoderm stage, each segment primordium has been given an individual identity by its unique constellation of gap, pair-rule, and homeotic gene products (Levine and Harding 1989). Two regions of *Drosophila* chromosome III contain most of these homeotic genes (**FIGURE 6.36**). The first region, known as the **Antennapedia complex**, contains the homeotic genes *labial* (*lab*), *Antennapedia* (*Antp*), *sex combs reduced* (*scr*), *deformed* (*dfd*), and *proboscipedia* (*pb*). The *labial* and *deformed* genes specify the head segments, while *sex combs reduced* and *Antennapedia* contribute to giving the thoracic segments their identities. The *proboscipedia* gene appears to act only in adults, but in its absence, the labial palps of the mouth are transformed into legs (Wakimoto et al. 1984; Kaufman et al. 1990; Maeda and Karch 2009).

The second region of homeotic genes is the **bithorax complex** (Lewis 1978; Maeda and Karch 2009). Three protein-coding genes are found in this complex: *Ultrabithorax* (*Ubx*), which is required for the identity of the third thoracic segment; and the *abdominal A* (*abdA*) and *Abdominal B* (*AbdB*) genes, which are responsible for the segmental identities of the abdominal segments (Sánchez-Herrero et al. 1985). The chromosome region containing both the Antennapedia complex and the bithorax complex is often referred to as the **homeotic complex** (**Hom-C**).

Because the homeotic selector genes are responsible for the specification of fly body parts, mutations in them lead to bizarre phenotypes. In 1894, William Bateson called these organisms **homeotic mutants**, and they have fascinated developmental biologists for decades.* For example, the

---

* *Homeo*, from the Greek, means "similar." *Homeotic mutants* are mutants in which one structure is replaced by another (as where an antenna is replaced by a leg). *Homeotic genes* are those genes whose mutation can cause such transformations; thus, homeotic genes are genes that specify the identity of a particular body segment. The *homeobox* is a conserved DNA sequence of about 180 base pairs that is shared by many homeotic genes. This sequence encodes the 60-amino acid *homeodomain*, which recognizes specific DNA sequences. The homeodomain is an important region of the transcription factors encoded by homeotic genes. However, not all genes containing homeoboxes are homeotic genes.

(A)

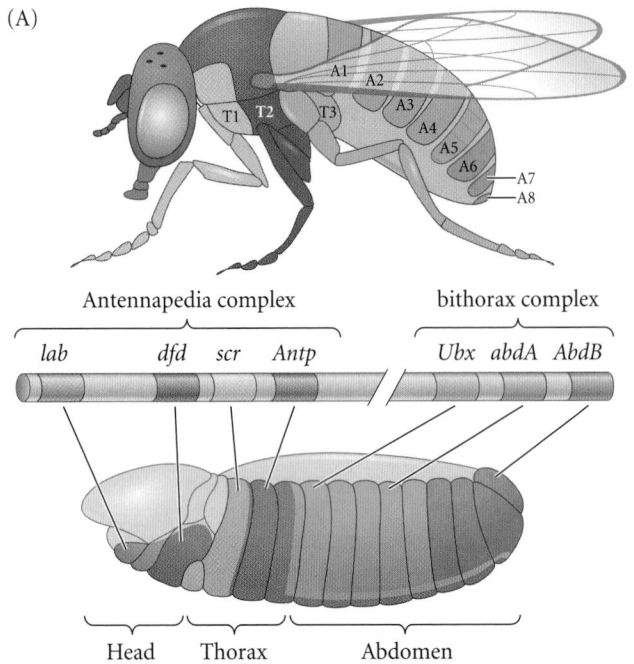

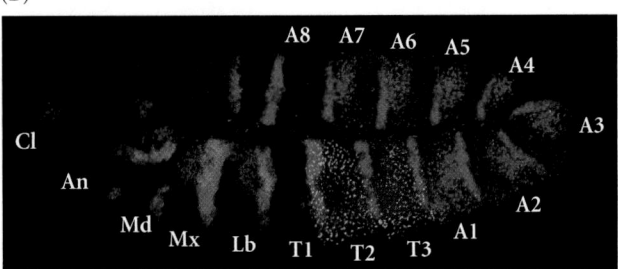

(B)

FIGURE 6.36 Homeotic gene expression in *Drosophila*.
(A) Expression map of the homeotic genes. In the center are the
genes of the Antennapedia and bithorax complexes and their functional domains. Below and above the gene map, the regions of
homeotic gene expression (both mRNA and protein) in the blastoderm of the *Drosophila* embryo and the regions that form from them
in the adult fly are shown. (B) In situ hybridization for four genes at
a slightly later stage (the extended germ band). The *engrailed* (blue)
expression pattern separates the body into segments; *Antennapedia* (green) and *Ultrabithorax* (purple) separate the thoracic and
abdominal regions; *Distal-less* (red) shows the placement of jaws
and the beginnings of limbs. (A after Kaufman et al. 1990 and Des-
sain et al. 1992; B courtesy of D. Kosman.)

FIGURE 6.37 A four-winged fruit fly constructed by putting
together three mutations in *cis*-regulators of the *Ultrabithorax*
gene. These mutations effectively transform the third thoracic segment
into another second thoracic segment (i.e., halteres into wings).
(Courtesy of E. B. Lewis.)

body of the normal adult fly contains three thoracic segments, each of which produces a pair of legs. The first thoracic segment does not produce any other appendages, but
the second thoracic segment produces a pair of wings in
addition to its legs. The third thoracic segment produces a
pair of wings and a pair of balancers known as **halteres**. In
homeotic mutants, these specific segmental identities can be
changed. When the *Ultrabithorax* gene is deleted, the third

thoracic segment (characterized by halteres) is transformed
into another second thoracic segment. The result is a fly
with four wings (**FIGURE 6.37**)—an embarrassing situation
for a classic dipteran.*

Similarly, Antennapedia protein usually specifies the second thoracic segment of the fly. But when flies have a mutation wherein the *Antennapedia* gene is expressed in the head
(as well as in the thorax), legs rather than antennae grow out
of the head sockets (**FIGURE 6.38**). This is partly because, in
addition to promoting the formation of thoracic structures,
the Antennapedia protein binds to and represses the enhancers of at least two genes, *homothorax* and *eyeless*, which
encode transcription factors that are critical for antenna and
eye formation, respectively (Casares and Mann 1998; Plaza
et al. 2001). Therefore, one of Antennapedia's functions is to
repress the genes that would trigger antenna and eye development. In the recessive mutant of *Antennapedia*, the gene
fails to be expressed in the second thoracic segment, and
antennae sprout in the leg positions (Struhl 1981; Frischer et
al. 1986; Schneuwly et al. 1987).

The major homeotic selector genes have been cloned and
their expression analyzed by in situ hybridization (Harding
et al. 1985; Akam 1987). Transcripts from each gene can
be detected in specific regions of the embryo (see Figure
6.36B) and are especially prominent in the central nervous
system.

---

* Dipterans (two-winged insects such as flies) are thought to
have evolved from four-winged insects; it is possible that this
change arose via alterations in the bithorax complex. Chapter
20 includes more speculation on the relationship between the
homeotic complex and evolution.

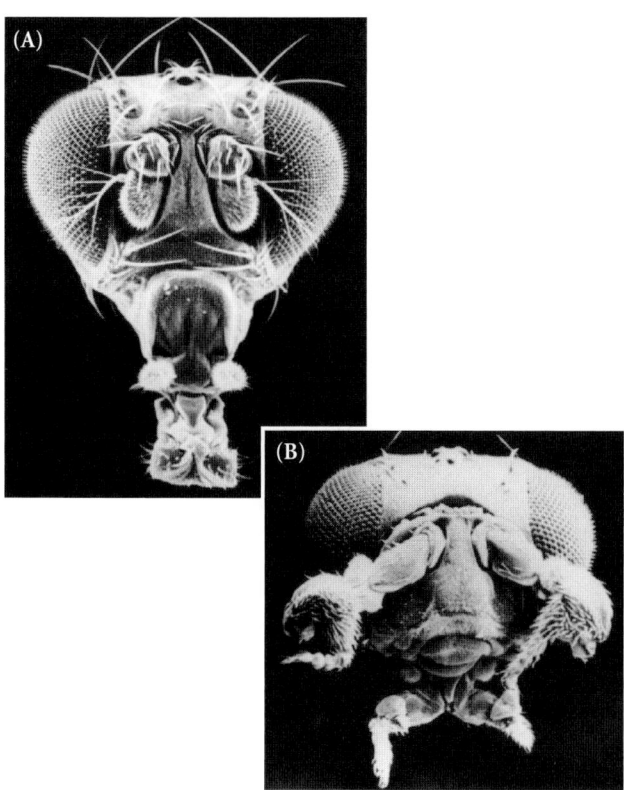

FIGURE 6.38 (A) Head of a wild-type fruit fly. (B) Head of a fly containing the *Antennapedia* mutation that converts antennae into legs. (From Kaufman et al. 1990, courtesy of T. C. Kaufman.)

Initiating and maintaining the patterns of homeotic gene expression

The initial domains of homeotic gene expression are influenced by the gap genes and pair-rule genes. For instance, expression of the *abdA* and *AbdB* genes is repressed by the gap gene proteins Hunchback and Krüppel. This inhibition prevents these abdomen-specifying genes from being expressed in the head and thorax (Casares and Sánchez-Herrero 1995). Conversely, the *Antennapedia* gene is activated by particular levels of Hunchback (needing both the maternal and the zygotically transcribed messages), so *Antennapedia* is originally transcribed in parasegment 4, specifying the mesothoracic (T2) segment (Wu et al. 2001).

The expression of homeotic genes is a dynamic process. The *Antennapedia* gene, for instance, although initially expressed in presumptive parasegment 4, soon appears in parasegment 5. As the germ band expands, *Antp* expression is seen in the presumptive neural tube as far posterior as parasegment 12. During further development, the domain of *Antp* expression contracts again, and *Antp* transcripts are localized strongly to parasegments 4 and 5. Like that of other homeotic genes, *Antp* expression is negatively regulated by all

the homeotic gene products expressed posterior to it (Levine and Harding 1989; González-Reyes and Morata 1990). In other words, each of the bithorax complex genes represses the expression of *Antp*. If the *Ultrabithorax* gene is deleted, *Antp* activity extends through the region that would normally have expressed *Ubx* and stops where the *Abd* region begins. (This allows the third thoracic segment to form wings like the second thoracic segment, as in Figure 6.37.) If the entire bithorax complex is deleted, *Antp* expression extends throughout the abdomen. (Such a larva does not survive, but the cuticle pattern throughout the abdomen is that of the second thoracic segment.)

As we have seen, the proteins encoded by the gap and pair-rule genes are transient; however, in order for differentiation to occur, the identities of the segments must be stabilized. So, once the transcription patterns of the homeotic genes have become stabilized, they are "locked" into place by alteration of the chromatin conformation in these genes. The repression of homeotic genes is maintained by the **Polycomb** family of proteins, while the active chromatin conformation appears to be maintained by the **Trithorax** proteins (Ingham and Whittle 1980; McKeon and Brock 1991; Simon et al. 1992).

Realisator genes

Homeotic genes don't do the work alone. In fact, they appear to regulate the action from up in the "executive suite," while the actual business of making an organ is done by other genes on the "factory floor." In this scenario, the homeotic genes work by activating or repressing a group of "realisator genes" that are the targets of the homeotic gene proteins and that function to form the specified tissue or organ primordia (Garcia-Bellido 1975).

Such a pathway for one simple structure—the posterior spiracle—is well on its way to being elucidated. This organ is a simple tube connecting to the trachea and a protuberance called the Filzkörper (see Figure 6.25D). The posterior spiracle is made in the eighth abdominal segment and is under the control of the Hox gene *AbdB*. Lovegrove and colleagues (2006) have found that the AbdB protein controls four genes that are necessary for posterior spiracle formation: *Spalt* (*Sal*), *Cut* (*Ct*), *Empty spiracles* (*Ems*), and *Unpaired* (*Upd*). The first three encode transcription factors; the fourth encodes a paracrine factor. None of them are transcribed without AbdB. Moreover, if these genes are independently activated in the absence of AbdB, a posterior spiracle will form.

Controlled by AbdB, these four regulator genes in turn control the expression of the realisator genes that control cell structure and function. *Spalt* and *Cut* encode proteins that activate the cadherin genes necessary for cell adhesion and the invagination of the spiracle. *Empty spiracles* and *Unpaired* encode proteins that control the small G proteins (such as Gef64C) that organize the actin cytoskeleton and the cell polarizing proteins that control the elongation of the spiracle (**FIGURE 6.39**).

FIGURE 6.39 Developmental control of posterior spiracle formation through AbdB. The homeotic selector protein AbdB (with the interaction of cofactors) activates the transcription of four genes encoding "intermediate" regulators. The proteins encoded by these genes—Spalt (Sal), Cut (Ct), Empty spiracles (Ems), and Unpaired (Upd)—are necessary and sufficient for specifying posterior spiracle development. They control (directly or indirectly) the local expression of a battery of realisator genes that influence morphogenetic processes such as cell adhesion (cadherins), cell polarity (Crumbs), and cytoskeletal organization (small G proteins). (After Lohmann 2006; Lovegrove et al. 2006.)

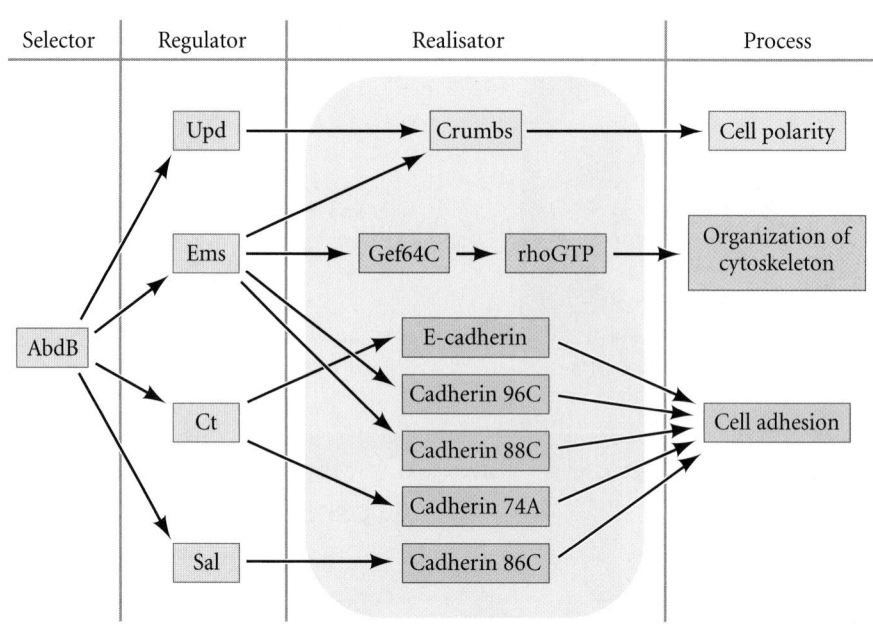

Axes and Organ Primordia: The Cartesian Coordinate Model

The anterior-posterior and dorsal-ventral axes of *Drosophila* embryos form a coordinate system that can be used to specify positions within the embryo (**FIGURE 6.40A**). Theoretically, cells that are initially equivalent in developmental potential can respond to their position by expressing different sets of genes. This type of specification has been demonstrated in the formation of the salivary gland rudiments (Panzer et al. 1992; Bradley et al. 2001; Zhou et al. 2001).

Drosophila salivary glands form only in the strip of cells defined by the activity of the *sex combs reduced* (*scr*) gene along the anterior-posterior axis (parasegment 2). No salivary glands form in *scr*-deficient mutants. Moreover, if *scr* is experimentally expressed throughout the embryo, salivary gland primordia form in a ventrolateral stripe along most of the length of the embryo. The formation of salivary glands along the dorsal-ventral axis is repressed by both Decapentaplegic and Dorsal proteins, which inhibit salivary gland formation both dorsally and ventrally. Thus, the salivary glands form at the intersection of the vertical *scr* expression band (parasegment 2) and the horizontal region in the middle of the embryo's circumference that has neither Decapentaplegic nor Dorsal (**FIGURE 6.40B**). The cells that form the salivary glands are directed to do so by the intersecting gene activities along the anterior-posterior and dorsal-ventral axes.

A similar situation is seen with tissues that are found in every segment of the fly. Neuroblasts arise from 10 clusters of 4 to 6 cells each that form on each side in every segment

(A)

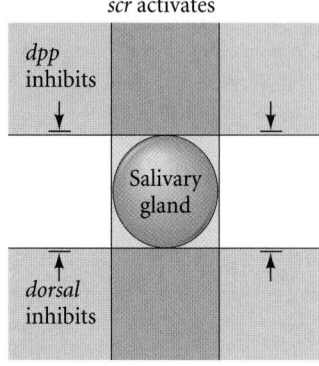

(B)

FIGURE 6.40 Cartesian coordinate system mapped out by gene expression patterns. (A) A grid (ventral view, looking "up" at the embryo) formed by the expression of *short-gastrulation* (red), *intermediate neuroblast defective* (green), and *muscle segment homeobox* (magenta) along the dorsal-ventral axis, and by the expression of *wingless* (yellow) and *engrailed* (green) transcripts along the anterior-posterior axis. (B) Coordinates for the expression of genes giving rise to *Drosophila* salivary glands. These genes are activated by the protein product of the *sex combs reduced* (*scr*) homeotic gene in a narrow band along the anterior-posterior axis, and they are inhibited in the regions marked by *decapentaplegic* (*dpp*) and *dorsal* gene products along the dorsal-ventral axis. This pattern allows salivary glands to form in the midline of the embryo in the second parasegment. (A courtesy of D. Kosman; B after Panzer et al. 1992.)

in the strip of neural ectoderm at the midline of the embryo (Skeath and Carroll 1992). The cells in each cluster interact (via the Notch pathway discussed in Chapter 3) to generate a single neural cell from each cluster. Skeath and colleagues (1993) have shown that the pattern of neural gene transcription is imposed by a coordinate system. Their expression is repressed by the Decapentaplegic and Snail proteins along the dorsal-ventral axis while positive enhancement by pair-rule genes along the anterior-posterior axis causes their repetition in each half-segment. It is very likely, then, that the positions of organ primordia in the fly are specified via a two-dimensional coordinate system based on the intersection of the anterior-posterior and dorsal-ventral axes.

Coda

Genetic studies of the *Drosophila* embryo have uncovered numerous genes that are responsible for specification of the anterior-posterior and dorsal-ventral axes. Mutations of *Drosophila* genes have given us our first glimpses of the multiple levels of pattern regulation in a complex organism and have enabled us to isolate these genes and their products. Most importantly, however, as we will see in forthcoming chapters, the insights arising from work on *Drosophila* genes have been pivotal in helping us understand the general mechanism of pattern formation used not only by insects but throughout the animal kingdom.

SNAPSHOT SUMMARY: *Drosophila* Development and Axis Specification

1. *Drosophila* cleavage is superficial. The nuclei divide 13 times before forming cells. Before cell formation, the nuclei reside in a syncytial blastoderm. Each nucleus is surrounded by actin-filled cytoplasm.

2. When the cells form, the *Drosophila* embryo undergoes a mid-blastula transition, wherein the cleavages become asynchronous and new mRNA is made. At this time, there is a transfer from maternal to zygotic control of development.

3. Gastrulation begins with the invagination of the most ventral region (the presumptive mesoderm), which causes the formation of a ventral furrow. The germ band expands such that the future posterior segments curl just behind the presumptive head.

4. The genes regulating pattern formation in *Drosophila* operate according to certain principles:

   • There are *morphogens*—such as Bicoid and Dorsal—whose gradients determine the specification of different cell types. These morphogens can be transcription factors.

   • There is a *temporal order* wherein different classes of genes are transcribed, and the products of one gene often regulate the expression of another gene.

   • *Boundaries* of gene expression can be created by the interaction between transcription factors and their gene targets. Here, the transcription factors transcribed earlier regulate the expression of the next set of genes.

   • *Translational control* is extremely important in the early embryo, and localized mRNAs are critical in patterning the embryo.

   • *Individual cell fates* are not defined immediately. Rather, there is a stepwise specification wherein a given field is divided and subdivided, eventually regulating individual cell fates.

5. Maternal effect genes are responsible for the initiation of anterior-posterior polarity. *bicoid* mRNA is bound by its 3′ UTR to the cytoskeleton in the future anterior pole; *nanos* mRNA is sequestered by its 3′ UTR in the future posterior pole. *hunchback* and *caudal* messages are seen throughout the embryo.

6. Dorsal-ventral polarity is regulated by the entry of Dorsal protein into the nucleus. Dorsal-ventral polarity is initiated when the nucleus moves to the dorsal-anterior of the oocyte and sequesters the *gurken* message, enabling it to synthesize proteins in the dorsal side of the egg.

7. Gurken protein is secreted from the oocyte and binds to its receptor (Torpedo) on the follicle cells. This binding dorsalizes the follicle cells, preventing them from synthesizing Pipe.

8. Pipe sulfates the vitelline membrane proteins beneath the ventral side of the embryo, allowing them to bind Gastrulation-defection (GD). GD initiates a complex that allows the activation of the Easter protease, which cleaves the Spätzle zymogen into its active form, which is the ligand for the Toll protein on the ventral embryonic cells.

9. Activated Toll protein initiates a cascade that phosphorylates the Cactus protein, which has been bound to Dorsal. Phosphorylated Cactus is degraded, allowing Dorsal to enter the nucleus. Once in the nucleus, Dorsal activates the genes responsible for the ventral cell fates and represses those genes whose proteins would specify dorsal cell fates.

10. Dorsal protein forms a gradient as it enters the various nuclei. Those nuclei at the most ventral surface incorporate the most Dorsal protein and become mesoderm; those more lateral become neurogenic ectoderm.

11. Bicoid and Hunchback proteins activate the genes responsible for the anterior portion of the fly; Caudal activates genes responsible for posterior development.

12. The unsegmented anterior and posterior extremities are regulated by the activation of Torso protein at the anterior and posterior poles of the egg.

13. The gap genes respond to concentrations of the maternal effect gene proteins. Their protein products interact with each other such that each gap gene protein defines specific regions of the embryo.

14. The gap gene proteins activate and repress the pair-rule genes. The pair-rule genes have modular promoters such that they become activated in seven "stripes." Their boundaries of transcription are defined by the gap genes. The pair-rule genes form seven bands of transcription along the anterior-posterior axis, each one comprising two parasegments.

15. The pair-rule gene products activate *engrailed* and *wingless* expression in adjacent cells. The *engrailed*-expressing cells form the anterior boundary of each parasegment. These cells form a signaling center that organizes the cuticle formation and segmental structure of the embryo.

16. Homeotic selector genes are found in two complexes on chromosome III of *Drosophila*. Together, these regions are called Hom-C, the homeotic gene complex. The genes are arranged in the same order as their transcriptional expression. Genes of the Hom-C specify the individual segments, and mutations in these genes are capable of transforming one segment into another.

17. Expression of each homeotic selector gene is regulated by the gap and pair-rule genes. Expression of gap and pair-rule genes is refined and maintained by interactions whereby their protein products prevent the transcription of neighboring Hom-C genes.

18. The targets of the Hom-C proteins are the realisator genes responsible for constructing the specific structure.

19. Organs form at the intersection of dorsal-ventral and anterior-posterior regions of gene expression.

For Further Reading

Complete bibliographical citations for all literature cited in this chapter can be found at the free-access website **www.devbio.com**

Driever, W. and C. Nüsslein-Volhard. 1988a. The Bicoid protein determines position in the *Drosophila* embryo in a concentration-dependent manner. *Cell* 54: 95–104.

Fujioka, M., Y. Emi-Sarker, G. L. Yusibova, T. Goto and J. B. Jaynes. 1999. Analysis of an *even-skipped* rescue transgene reveals both composite and discrete neuronal and early blastoderm enhancers, and multi-stripe positioning by gap gene repressor gradients. *Development* 126: 2527–2538.

Lehmann, R. and C. Nüsslein-Volhard. 1991. The maternal gene *nanos* has a central role in posterior pattern formation of the *Drosophila* embryo. *Development* 112: 679–691.

Leptin, M. 1991. *twist* and *snail* as positive and negative regulators during *Drosophila* mesoderm development. *Genes Dev.* 5: 1568–1576.

Lewis, E. B. 1978. A gene complex controlling segmentation in *Drosophila*. *Nature* 276: 565–570.

Maeda, R. K. and F. Karch. 2009. The Bithorax complex of *Drosophila*: An exceptional Hox cluster. *Curr. Top. Dev. Biol.* 88: 1–33.

Martinez-Arias, A. and P. A. Lawrence. 1985. Parasegments and compartments in the *Drosophila* embryo. *Nature* 313: 639–642.

Panfilio, K. A. 2007. Extraembryonic development in insects and the acrobatics of blastokinesis. *Dev. Biol.* 313: 471–491.

Pankratz, M. J., E. Seifert, N. Gerwin, B. Billi, U. Nauber and H. Jäckle. 1990. Gradients of *Krüppel* and *knirps* gene products direct pair-rule gene stripe patterning in the posterior region of the *Drosophila* embryo. *Cell* 61: 309–317.

Pfeiffer, B. D. and 16 others. 2009. Tools for neuroanatomy and neurogenetics in *Drosophila*. *Proc. Natl. Acad. Sci. USA* 105: 9715–9720.

Roth, S., D. Stein and C. Nüsslein-Volhard. 1989. A gradient of nuclear localization of the dorsal protein determines dorsoventral pattern in the *Drosophila* embryo. *Cell* 59: 1189–1202.

Schüpbach, T. 1987. Germ line and soma cooperate during oogenesis to establish the dorsoventral pattern of egg shell and embryo in *Drosophila melanogaster*. *Cell* 49: 699–707.

Struhl, G. 1981. A homeotic mutation transforming leg to antenna in *Drosophila*. *Nature* 292: 635–638.

Wang, C. and R. Lehman. 1991. Nanos is the localized posterior determinate in *Drosophila*. *Cell* 66: 637–647.

Go Online

WEBSITE 6.1 *Drosophila* fertilization. Fertilization of *Drosophila* can only occur in the region of the oocyte that will become the anterior of the embryo. Moreover, the sperm tail appears to stay in this region.

WEBSITE 6.2 The early development of other insects. *Drosophila* is a highly derived species. There are other insect species that develop in ways very different from the "standard" fruit fly.

WEBSITE 6.3 Evidence for gradients in insect development. The original evidence for gradients in insect development came from studies providing evidence for two "organization centers" in the egg, one located anteriorly and one located posteriorly.

WEBSITE 6.4 Christiane Nüsslein-Volhard and the molecular approach to development. The research that revolutionized developmental biology had to wait for someone to synthesize molecular biology, embryology, and *Drosophila* genetics.

WEBSITE 6.5 Asymmetrical spread of morphogens. It is unlikely that morphogens such as Wingless spread by free diffusion. The asymmetry of Wingless diffusion suggests that neighboring cells play a crucial role in moving this protein.

WEBSITE 6.6 Getting a head in the fly. The segment polarity genes may act differently in the head than in the trunk. Indeed, the formation of the *Drosophila* head may differ significantly from the way the rest of the body is formed.

Vade Mecum

***Drosophila* development.** The Vade Mecum sites have remarkable time-lapse sequences of *Drosophila* development, including cleavage and gastrulation. This segment also provides access to the fly life cycle. The color coding superimposed on the germ layers allows you to readily understand tissue movements.

Outside Sites

"The Interactive Fly," compiled by Thomas Brody, provides an index to the major *Drosophila* websites worldwide. It is hosted by the Society for Developmental Biology (SDB) at **http://www.sdbonline.org/fly/aimain/1aahome.htm**. Two notable entries accessible through the site are "Atlas of Fly Development" by Voker Hartenstein (**http://www. sdbonline.org/fly/atlas/00atlas.htm**) and "Stages in Fly Development: The Movies" (**http://www.sdbonline.org/fly/ aimain/2stages.htm**).

7

Sea Urchins and Tunicates
Deuterostome Invertebrates

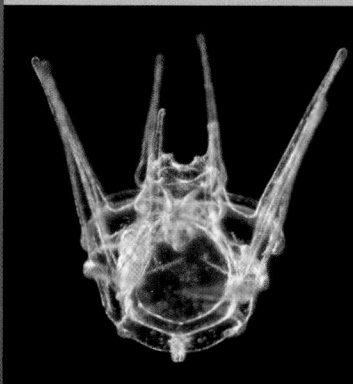

CHAPTERS 5 AND 6 DESCRIBED IN DETAIL the processes of protostome development in representative species from three groups, the molluscs, nematodes, and insects. There are many fewer species of deuterostomes than there are of protostomes, but among the deuterostome groups are all the vertebrates, including fish, amphibians, reptiles, birds, and mammals. Besides these vertebrates, several invertebrate groups follow the deuterostome pattern of development. These include the hemichordates (acorn worms), cephalochordates (amphioxus), echinoderms (sea urchins, sea stars, sea cucumbers, etc.), and urochordates (tunicates, also called sea squirts). The echinoderms (especially sea urchins) and the tunicates have been the subjects of critically important studies in developmental biology (FIGURE 7.1). Indeed, conditional specification ("regulative development") was first discovered in sea urchins, and tunicates provided the first evidence for autonomous specification ("mosaic development"). As we will see, it turns out that both groups of organisms use both modes of specification.

EARLY DEVELOPMENT IN SEA URCHINS

Sea urchins have been exceptionally important organisms in studying how genes regulate the formation of the body. Indeed, sea urchin embryos provided the first evidence that chromosomes were needed for development, that DNA and RNA were present in each animal cell, that messenger RNAs directed protein synthesis, that stored messenger RNA provided the proteins for early embryonic development, that cyclins controlled cell division, and that enhancers were modular (Ernst 2011; McClay 2011). The first cloned eukaryotic gene encoded a sea urchin histone protein (Kedes et al. 1975), and the first evidence for chromatin remodeling concerned histone alterations during sea urchin development (Newrock et al. 1978). With the advent of new genetic techniques, sea urchin embryos continue to be critically important organisms for delineating the mechanisms by which genetic interactions specify different cell fates.

Sea Urchin Cleavage

Sea urchins exhibit **radial holoblastic cleavage** (FIGURES 7.2 and 7.3). Recall from Chapter 5 that this type of cleavage occurs in eggs with sparse yolk, and that holoblastic cleavage furrows extend through the entire egg (see Figure 5.4). In sea urchins, the first seven cleavage divisions are stereotypic in that the same pattern is followed in every individual of the same species. The first and second cleavages are both meridional and are perpendicular to each other (that is to say, the cleavage furrows pass through the animal and vegetal poles). The third cleavage is equatorial, perpendicular to the first two cleavage planes, and separates the animal and vegetal hemispheres from each other (see Figure 7.2A, top row, and Figure 7.3A–C). The fourth cleavage, however, is very different. The four cells of the animal tier divide meridionally into eight blastomeres, each with the same volume. These eight cells are called **mesomeres**. The vegetal tier,

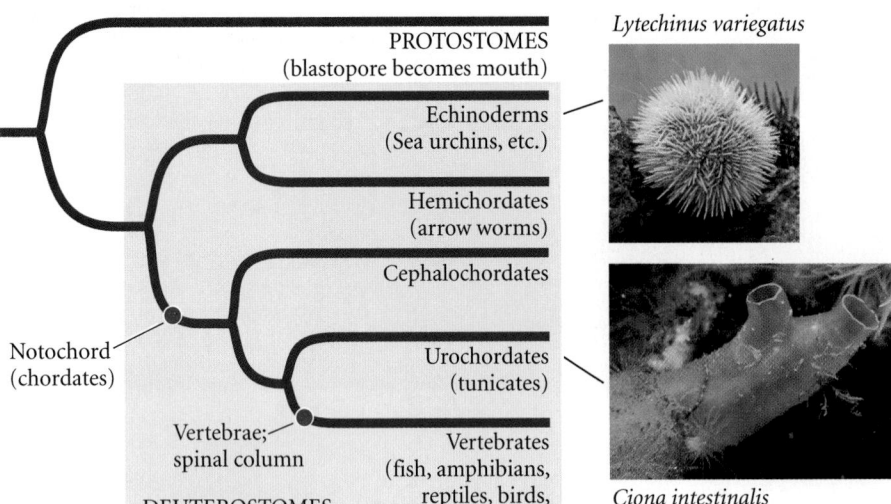

FIGURE 7.1 The echinoderms and tunicates represent deuterostome invertebrates. The tunicates, however, are classified as chordates because their larvae possess a notochord, dorsal neural tube, and pharyngeal arches. The tunicates are referred to as urochordates, a name that emphasizes their affinity with the other chordate groups. The green sea urchin *Lytechinus variegatus* and the tunicate *Ciona intestinalis* are two widely studied model organisms. (*L. variegatus* photograph courtesy of David McIntyre; *C. intestinalis* photograph © Nature Picture Library/Alamy.)

however, undergoes an unequal equatorial cleavage (see Figure 7.2B) to produce four large cells—the **macromeres**—and four smaller **micromeres** at the vegetal pole. As the 16-cell embryo cleaves, the eight "animal" mesomeres divide equatorially to produce two tiers, an_1 and an_2, one staggered above the other. The macromeres divide meridionally, forming a tier of eight cells below an_2 (see Figure 7.2A, bottom row). Somewhat later, the micromeres divide unequally, producing a cluster of four small micromeres at the tip of the vegetal pole, beneath a tier of four large micromeres. The small micromeres divide once more, then stop dividing until the larval stage. At the sixth division, the animal hemisphere cells divide meridionally while the vegetal cells divide equatorially; this pattern is reversed in the seventh division (see Figure 7.2A, bottom row). At that time, the embryo is a 120-cell* blastula in which the cells form a **blastocoel**: a hollow

FIGURE 7.2 Cleavage in the sea urchin. (A) Planes of cleavage in the first three divisions, and the formation of tiers of cells in divisions 3–6. (B) Confocal fluorescence micrograph of the unequal cell division that initiates the 16-cell stage (asterisk in A), highlighting the unequal equatorial cleavage of the vegetal blastomeres to produce the micromeres and macromeres. (B courtesy of G. van Dassow and the Center for Cell Dynamics.)

---

*You might have been expecting a 128-cell embryo, but remember that the small micromeres stopped dividing

(A)

(B)

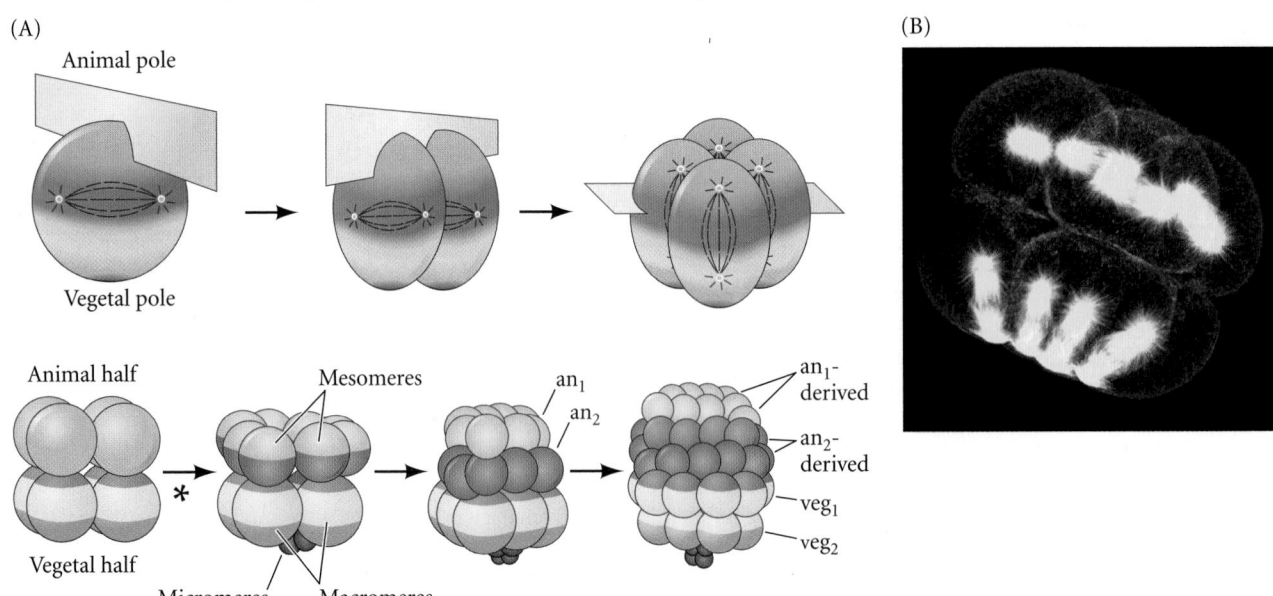

(A)

(B) Fertilization envelope

(C)

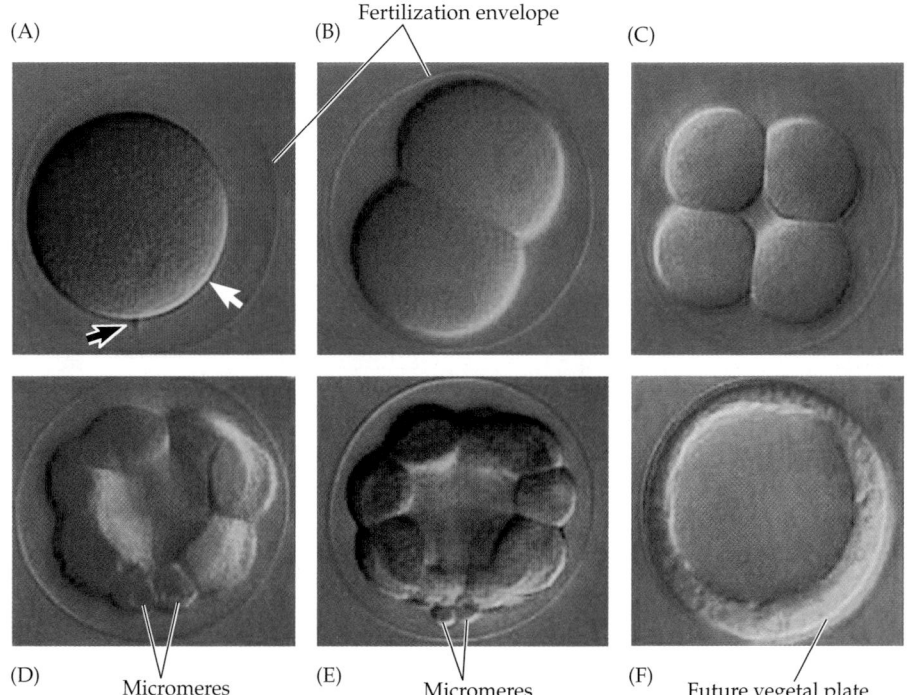

(D) Micromeres

(E) Micromeres

(F) Future vegetal plate

FIGURE 7.3 Micrographs of cleavage in live embryos of the sea urchin *Lytechinus variegatus*, seen from the side. (A) The 1-cell embryo (zygote). The site of sperm entry is marked with a black arrow; a white arrow marks the vegetal pole. The fertilization envelope surrounding the embryo is clearly visible. (B) 2-Cell stage. (C) 8-Cell stage. (D) 16-Cell stage. Micromeres have formed at the vegetal pole. (E) 32-Cell stage. (F) The blastula has hatched from the fertilization envelope. The vegetal plate is beginning to thicken. (Courtesy of J. Hardin.)

sphere surrounding a central cavity (see Figure 7.3F). From here on, the pattern of divisions becomes less regular.

Blastula formation

By the blastula stage, all the cells of the developing sea urchin are the same size, the micromeres having slowed down their cell divisions. Every cell is in contact with the proteinaceous fluid of the blastocoel on the inside and with the hyaline layer on the outside. Tight junctions unite the once loosely connected blastomeres into a seamless epithelial sheet that completely encircles the blastocoel. As the cells continue to divide, the blastula remains one cell layer thick, thinning out as it expands. This is accomplished by the adhesion of the blastomeres to the hyaline layer and by an influx of water that expands the blastocoel (Dan 1960; Wolpert and Gustafson 1961; Ettensohn and Ingersoll 1992).

These rapid and invariant cell cleavages last through the ninth or tenth division, depending on the species. By this time, the fates of the cells have become specified (discussed in the next section), and each cell becomes ciliated on the region of the cell membrane farthest from the blastocoel. Thus, there is apical (outside)-basal (inside) polarity in each of the embryonic cells, and there is evidence that PAR proteins are involved in distinguishing the basal cell membranes (Alford et al. 2009). This ciliated blastula begins to rotate within the fertilization envelope. Soon afterward, differences are seen in the cells. The cells at the vegetal pole of the blastula begin to thicken, forming a **vegetal plate** (see Figure 7.3F). The cells of the animal hemisphere synthesize and secrete a hatching enzyme that digests the fertilization envelope (Lepage et al. 1992). The embryo is now a free-swimming **hatched blastula**.

Fate maps and the determination of sea urchin blastomeres

The first fate maps of the sea urchin embryo followed the descendants of each of the 16-cell-stage blastomeres. More recent investigations have refined these maps by following the fates of individual cells that have been injected with fluorescent dyes that glow in the injected cells' progeny for many cell divisions (see Chapter 1). Such studies have shown that by the 60-cell stage, most of the embryonic cell fates are specified, but the cells are not irreversibly committed. In other words, particular blastomeres consistently produce the same cell types in each embryo, but these cells remain pluripotent and can give rise to other cell types if experimentally placed in a different part of the embryo.

A fate map of the 60-cell sea urchin embryo is shown in **FIGURE 7.4**. The animal half of the embryo consistently gives rise to the ectoderm—the larval skin and its neurons. The veg_1 layer produces cells that can enter into either the larval ectodermal or the endodermal organs. The veg_2 layer gives rise to cells that can populate three different structures—the endoderm, the coelom (internal mesodermal body wall), and the **non-skeletogenic mesenchyme** (sometimes called secondary mesenchyme), which generates pigment cells, immunocytes, and muscle cells. The first tier of micromeres (the large micromeres) produces the **skeletogenic mesenchyme** (also called primary mesenchyme), which forms the larval skeleton. The second-tier micromeres (i.e., the small micromeres) play no role in embryonic development. Rather, they contribute cells to the larval coelom from which the tissues of the adult are derived during metamorphosis (Logan and McClay 1997, 1999; Wray 1999). These second-tier

(A)

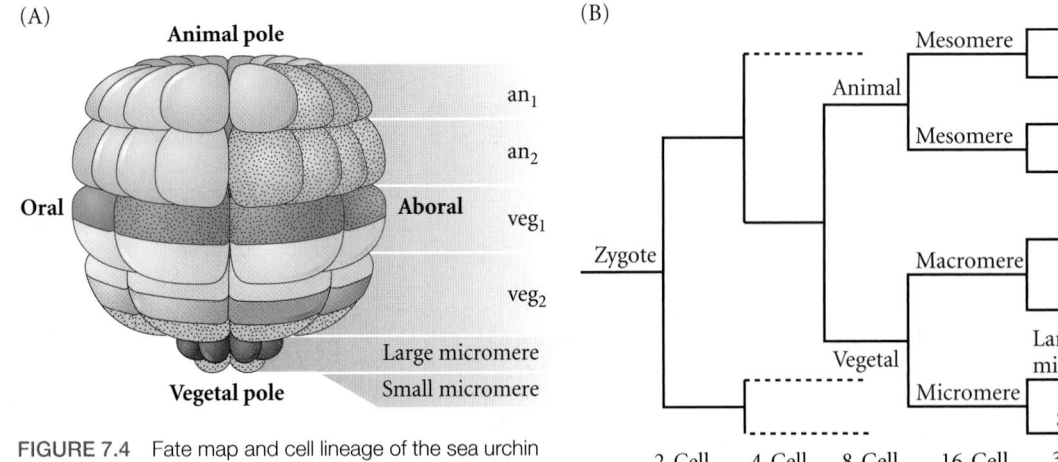

(B)

FIGURE 7.4 Fate map and cell lineage of the sea urchin *Strongylocentrotus purpuratus*. (A) The 60-cell embryo is shown, with the left side facing the viewer. Blastomere fates are segregated along the animal-vegetal axis of the egg. (B) Cell lineage map of the embryo. For simplicity, only one of the four embryonic cells is shown beyond second cleavage (solid lines). The veg_1 tier gives rise to both ectodermal and endodermal lineages. The body cavity (coelom) will form in the larva from two sources: the second tier of micromeres and some veg_2 cells. (After Logan and McClay 1999, Wray 1999.)

micromeres also contribute to producing the germline cells (Yajima and Wessel 2011.)

The fates of the different cell layers are determined in a two-step process:

- First, the large micromeres are *autonomously* specified. They inherit maternal determinants that had been deposited at the vegetal pole of the egg and that become incorporated into the four micromeres at fourth cleavage. These cells are thus determined to become skeletogenic mesenchyme cells that will leave the blastula epithelium to enter the blastocoel, migrate to particular positions along the blastocoel wall, and then differentiate into the larval skeleton.

- Second, the autonomously specified large micromeres are now able to produce paracrine and juxtacrine factors that *conditionally* specify the fates of their neighbors. The micromeres produce a signal that tells the cells above them to become endoderm and induces them to invaginate into the embryo.

The ability of the micromeres to produce signals that change the fates of the neighboring cells is so pronounced that if micromeres are removed from the embryo and placed on

top of an isolated **animal cap**—that is, the top two animal tiers that usually become ectoderm—the animal cap cells will generate endoderm and a more or less normal larva will develop (**FIGURE 7.5**; Hörstadius 1939).

These skeletogenic micromeres are the first cells whose fates are specified autonomously. If micromeres are isolated

FIGURE 7.5 Ability of micromeres to induce presumptive ectodermal cells to acquire other fates. (A) Normal development of the 64-cell sea urchin embryo, showing the fates of the different layers. (B) An isolated animal hemisphere becomes a ciliated ball of undifferentiated ectodermal cells called a *Dauerblastula*. (C) When an isolated animal hemisphere is combined with isolated micromeres, a recognizable pluteus larva is formed, with all the endoderm derived from the animal hemisphere. (After Hörstadius 1939.)

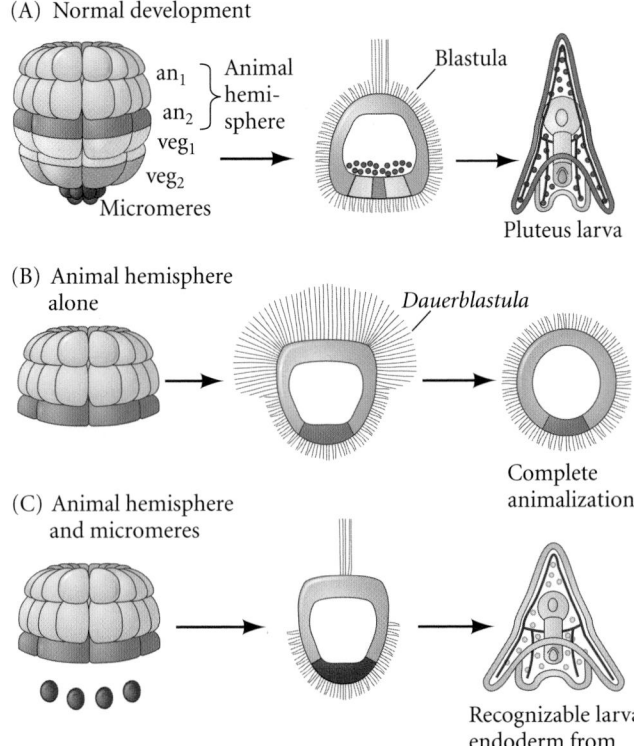

from the 16-cell embryo and placed in petri dishes, they will divide the appropriate number of times and produce the skeletal spicules (Okazaki 1975). Thus, the isolated micromeres do not need any other signals to generate their skeletal fates. Moreover, if skeletogenic micromeres are transplanted into the animal region of the blastula, not only will their descendants form skeletal spicules, but the transplanted micromeres will alter the fates of nearby cells by inducing a secondary site for gastrulation. Cells that would normally have produced ectodermal skin cells will be respecified as endoderm and will produce a secondary gut (**FIGURE 7.6**; Hörstadius 1973; Ransick and Davidson 1993). Therefore, the inducing ability of the micromeres is also established autonomously.

Gene regulatory networks and skeletogenic mesenchyme specification

Heredity, according to the embryologist E. B. Wilson, is the transmission from generation to generation of a particular

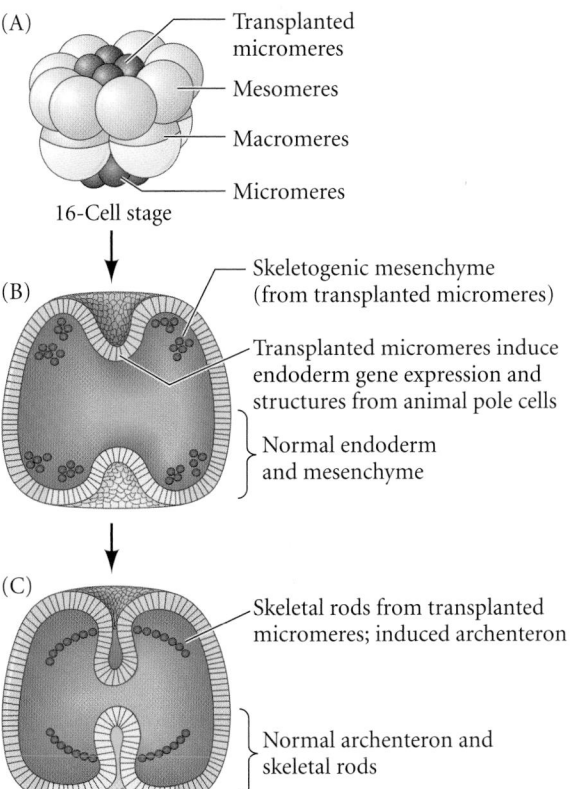

FIGURE 7.6 Ability of micromeres to induce a secondary axis in sea urchin embryos. (A) Micromeres are transplanted from the vegetal pole of a 16-cell embryo into the animal pole of a host 16-cell embryo. (B) The transplanted micromeres invaginate into the blastocoel to create a new set of skeletogenic mesenchyme cells, and they induce the animal cells next to them to become vegetal plate endoderm cells. (C) The transplanted micromeres differentiate into skeletal rods while the induced animal cap cells form a secondary archenteron. Meanwhile, gastrulation proceeds normally from the original vegetal plate of the host. (After Ransick and Davidson 1993.)

pattern of development, and evolution is the hereditary alteration of such a plan. Wilson was probably the first scientist to write (in 1895) that the instructions for development were somehow stored in chromosomal DNA and were transmitted by the chromosomes at fertilization. However, he had no way of knowing how the chromosomal information organized matter into forming an embryo.

Studies from the sea urchin developmental biology community have begun to demonstrate how DNA can be regulated to specify the cells and direct the morphogenesis of the developing organism. Eric Davidson's group has pioneered a network approach to development in which they envision *cis*-regulatory elements (such as promoters and enhancers) in a logic circuit connected to each other by transcription factors (see http://sugp.caltech.edu/endomes; Davidson and Levine 2008; Oliveri et al. 2008). The network receives its first inputs from transcription factors in the egg cytoplasm; from then on, the network self-assembles from (1) the ability of the maternal transcription factors to recognize *cis*-regulatory elements of particular genes that encode other transcription factors, and (2) the ability of this new set of transcription factors to activate paracrine signaling pathways that activate specific transcription factors in neighboring cells. The studies show the regulatory logic by which the genes of the sea urchin interact to specify and generate characteristic cell types. The researchers refer to such a set of interconnections among cell-type specifying genes as a **gene regulatory network** (**GRN**).

Here we will focus on the earliest part of one such GRN: the reactions by which the skeletogenic mesenchyme cells of the sea urchin embryo receive their developmental fate and interactive properties. Skeletogenic mesenchyme cells are those cells (descended from the micromeres) that are autonomously specified to ingress into the blastocoel and become the skeleton of the sea urchin larva. And, as we have seen, they are also the cells that induce their neighbors to become endoderm (gut) and non-skeletogenic mesenchyme (pigment; coelom) cells (see Figure 7.6). Nearly all the genes of this particular GRN have been identified and found to act as described.

DISHEVELED AND β-CATENIN: SPECIFYING THE MICROMERES

The specification of the micromere lineage (and hence the rest of the embryo) begins inside the undivided egg. The initial regulatory inputs are two transcription regulators, Disheveled and β-catenin, both of which are found in the cytoplasm and are inherited by the micromeres as soon as they are formed, at the fourth cleavage. During oogenesis, Disheveled becomes located in the vegetal cortex of the egg (**FIGURE 7.7A**; Weitzel et al. 2004; Leonard and Ettensohn 2007), where it prevents the degradation of β-catenin in the micromere and veg_2-tier macromere cells. The β-catenin then enters the nucleus, where it combines with the TCF transcription factor to activate gene expression from specific promoters.

Several pieces of evidence suggest that β-catenin specifies the micromeres. First, during normal sea urchin

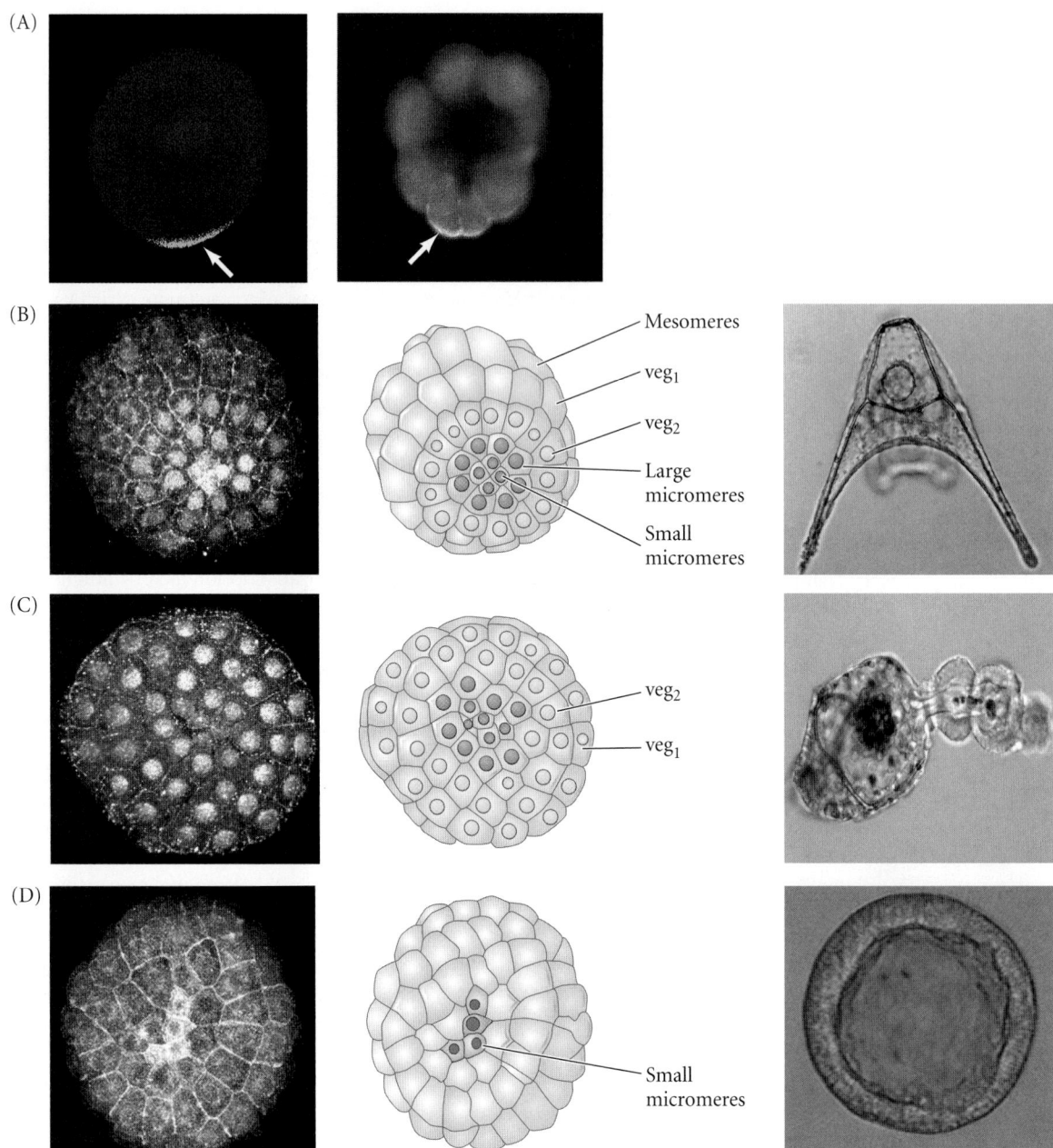

FIGURE 7.7 Role of the Disheveled and β-catenin proteins in specifying the vegetal cells of the sea urchin embryo. (A) Localization of Disheveled (arrows) in the vegetal cortex of the sea urchin oocyte before fertilization (left) and in the region of a 16-cell embryo about to become the micromeres (right). (B) During normal development, β-catenin accumulates predominantly in the micromeres and somewhat less in the veg₂ tier cells. (C) In embryos treated with lithium chloride, β-catenin accumulates in the nuclei of all blastula cells (probably by LiCl blocking the GSK3 enzyme of the Wnt pathway), and the animal cells become specified as endoderm and mesoderm. (D) When β-catenin is prevented from entering the nuclei (i.e., it remains in the cytoplasm), the vegetal cell fates are not specified, and the entire embryo develops as a ciliated ectodermal ball. (A,B from Weitzel et al. 2004, courtesy of C. Ettensohn; B–D from Logan et al. 1998, courtesy of D. McClay.)

development, β-catenin accumulates in the nuclei of those cells fated to become endoderm and mesoderm (**FIGURE 7.7B**). This accumulation is autonomous and can occur even if the micromere precursors are separated from the rest of the embryo. Second, this nuclear accumulation appears to

be responsible for specifying the vegetal half of the embryo. It is possible that levels of nuclear β-catenin accumulation help determine the mesodermal and endodermal fates of the vegetal cells (Kenny et al. 2003). Treating sea urchin embryos with lithium chloride allows β-catenin to accumulate in

(A)

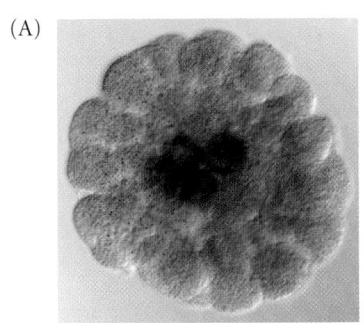

FIGURE 7.8 Simplified illustration of the double-negative gated "circuit" for micromere specification. (A) In situ hybridization reveals the accumulation of *Pmar1* mRNA (dark purple) in the micromeres. (B) Otx and β-catenin from the maternal cytoplasm are concentrated at the vegetal pole of the egg. These transcriptional regulators are inherited by the micromeres and activate the *Pmar1* gene. *Pmar1* encodes a repressor of *HesC*, which in turn encodes a repressor (hence the "double-negative") of several genes involved in micromere specification (e.g., *Alx1*, *Tbr*, and *Ets*). Genes encoding signaling proteins (e.g., *Delta*) are also under the control of HesC. In the micromeres, where activated Pmar1 protein represses the *HesC* repressor, the micromere specification and signaling genes are active. In the veg₂ cells, *Pmar1* is not activated and the HesC gene product shuts down the skeletogenic genes; however, those cells containing Notch can respond to the Delta signal from the skeletogenic mesenchyme. The gene expression patterns are seen below. U represents ubiquitous activating transcription factors. (After Oliveri et al. 2008; photograph courtesy of P. Oliveri.)

(B)

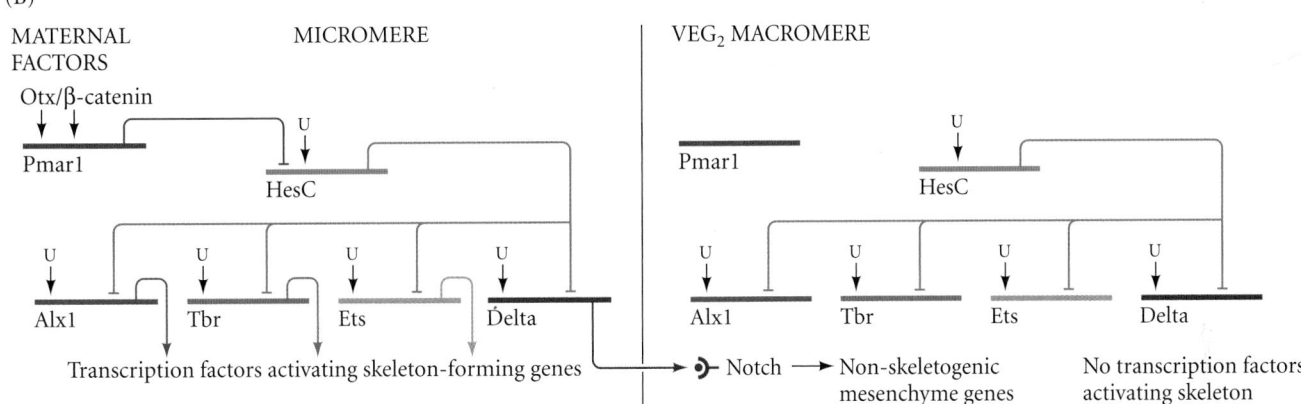

every cell of the embryo and transforms presumptive ectoderm into endoderm (**FIGURE 7.7C**). Conversely, experimental procedures that inhibit β-catenin accumulation in the vegetal cell nuclei prevent the formation of endoderm and mesoderm (**FIGURE 7.7D**; Logan et al. 1998; Wikramanayake et al. 1998).

PMAR1 AND HESC: A DOUBLE-NEGATIVE GATE The next micromere regulatory input comes from the Otx transcription factor, which is also enriched in the micromere cytoplasm. Otx interacts with the β-catenin/TCF complex at the enhancer of the *Pmar1* gene to activate *Pmar1* gene transcription in the micromeres shortly after their formation (**FIGURE 7.8A**; Oliveri et al. 2008). Pmar1 protein is a repressor of *HesC*, a gene that encodes another repressive transcription factor. *HesC* is expressed in every nucleus of the sea urchin embryo *except* those of the micromeres.* In the micromeres, where *Pmar1* is activated, the *HesC* gene is repressed. This mechanism, whereby a repressor locks the genes of specification and these genes can be unlocked by

the repressor of that repressor (in other words, when activation occurs by the repression of a repressor), is called a **double-negative gate** (**FIGURES 7.8B** and **7.9A**). Such a gate allows for tight regulation of fate specification: it promotes the expression of these genes where the input occurs, and it represses the same genes in every other cell type (Oliveri et al. 2008).

The genes repressed by HesC are those involved in micromere specification and differentiation: *Alx1*, *Ets1*, *Tbr*, *Tel*, and *SoxC*. Each of these genes can be activated by ubiquitous transcription factors, but these positive transcription factors cannot work while HesC repressor protein binds to their respective enhancers. When the Pmar1 protein is present, it represses *HesC*, and all these genes become active (Revilla-i-Domingo et al. 2007). The newly activated genes synthesize transcription factors that activate another set of genes, most of which are genes that activate skeletal determinants. These transcription factors also activate each other's genes, so that once one factor is activated, it maintains the activity of the other skeletogenic genes. This stabilizes the regulatory state of the skeletogenic mesenchyme cells.

Another way micromeres retain their specification is to secrete an autocrine factor, Wnt8 (Angerer and Angerer 2000; Wikramanayake et al. 2004). As soon as the micromeres form, maternal β-catenin and Otx activate the *Blimp1* gene, whose product (in conjunction with more β-catenin) activates the *Wnt8* gene. Wnt8 protein is then received by the same micromeres that made it (i.e., autocrine regulation), activating

---

*This is an oversimplification of a very complex process. In the pathway to micromere specification, other transcription factors must be expressed, and the maternal transcription factor SoxB1 has to be eliminated from the vegetal pole or it will inhibit the activation of *Pmar1*. In addition, the cytoskeletal processes partitioning the cells and anchoring certain factors are not considered here. For complete details of the model, see the continually updated website http://sugp.caltech.edu/endomes/.

(A)

(B)

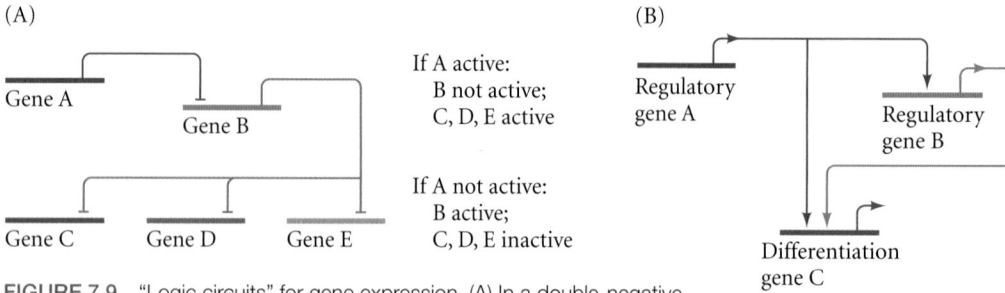

If A active:
  B not active;
  C, D, E active

If A not active:
  B active;
  C, D, E inactive

Regulatory gene A

Regulatory gene B

Differentiation gene C

FIGURE 7.9 "Logic circuits" for gene expression. (A) In a double-negative gate, a single gene encodes a repressor of an entire battery of genes. When this repressor gene is repressed, the battery of genes is expressed. (B) In a feedforward circuit, gene product A activates both gene B and gene C, and gene B also activates gene C. Feedforward circuits provide an efficient way to amplify a signal in one direction.

the micromeres' own genes for β-catenin. Because β-catenin activates *Blimp1*, this autocrine regulation sets up a positive feedback loop between Blimp1 and Wnt8 that establishes a source of β-catenin for the micromere nuclei.

In contrast to the double-negative gate that *specifies* the micromeres, control of the genes that *differentiate* the cells of the sea urchin skeleton operates on a feedforward process (**FIGURE 7.9B**). Here, regulatory gene A produces a transcription factor that is needed for differentiation gene C and also activates regulatory gene B, which produces a transcription factor also needed for differentiation gene C. This feedforward process is often used to stabilize active gene expression and makes the resulting cell type irreversible.

EVOLUTION BY SUBROUTINE COOPTION The skeletogenic portion of the micromere gene regulatory network shown in Figure 7.8B appears to have arisen from the recruitment of a network "subroutine" that in most echinoderms (including sea urchins) is used for making the adult skeleton (Gao and Davidson 2008). The cooption of subroutines by a new lineage is one of the ways evolution occurs. It happens that the GRN of the micromeres in sea urchin embryos is very different from that in other echinoderms. Only in the micromeres of sea urchins has the skeletogenic subroutine (which in all other echinoderms is activated late in development) come under the control of the genes that specify cells to the micromere lineage. The most important evolutionary events were those placing the skeletogenic genes *Alx1* and *Ets1* (necessary for adult skeletal development) and *Tbr* (used in later larval skeleton formation) under the regulation of the *Pmar1-HesC* double-negative gate. This occurred through mutations in the *cis*-regulatory regions of these genes. Thus, the skeletogenic properties that distinguish the sea urchin micromeres appear to have arisen through the recruitment of a preexisting skeletogenic regulatory system by the micromere lineage gene regulatory system.

Specification of the vegetal cells

The skeletogenic micromeres also produce signals that can induce changes in other tissues. One of these signals is the TGF-β family paracrine factor activin. Expression of the gene for activin is also under the control of the *Pmar1-HesC* double-negative gate, and activin secretion appears to be critical for endoderm formation (Sethi et al. 2009). Indeed, if *Pmar1* mRNA is injected into an animal cell, that animal cell will develop into a skeletogenic mesenchyme cell, and the cells adjacent to it will start developing like a macromere (Oliveri et al. 2003). If the activin signal is blocked, the adjacent cells do not become endoderm* (Ransick and Davidson 1995; Sherwood and McClay 1999; Sweet et al. 1999).

Another cell-specifying signal from the micromeres is the juxtacrine protein Delta, another factor that is controlled by the double-negative gate. Delta functions by activating Notch proteins on the adjacent veg$_2$ cells and later will act on the adjacent small micromeres. Delta causes these cells to become the non-skeletogenic mesenchyme cells by activating the Gcm transcription factor and repressing the FoxA transcription factor (which activates the endoderm-specific genes). The upper veg$_2$ cells, since they do not receive the Delta signal, retain FoxA expression, and this pushes them in the direction of becoming endodermal cells (Croce and McClay 2010).

In sum, the genes of the sea urchin micromeres specify their cell fates autonomously and also specify the fates of their

---

*Recall the experiments in Figure 7.6, which demonstrated that the micromeres are able to induce a second embryonic axis when transplanted to the animal hemisphere. However, micromeres in which β-catenin is prevented from entering the nucleus are unable to induce the animal cells to form endoderm, and no second axis forms (Logan et al. 1998). β-Catenin also accumulates in macromeres, but by a different means, and the *Pmar1* gene is not activated in macromeres (possibly due to the presence of SoxB1; see Kenny et al. 2003 and Lhomond et al. 2012).

neighbors conditionally. The original inputs come from the maternal cytoplasm and activate genes that unlock repressors of a specific cell fate. Once the maternal cytoplasmic factors accomplish their functions, the nuclear genome takes over.

Axis specification

In the sea urchin blastula, the general cell fates (ectoderm, endoderm, skeletogenic mesenchyme, etc.) line up along the animal-vegetal axis—an axis that is established in the egg cytoplasm prior to fertilization. The animal-vegetal axis also appears to structure the future anterior-posterior axis, with the vegetal region sequestering those maternal components necessary for posterior development (Boveri 1901; Maruyama et al. 1985).

In most sea urchins, the dorsal-ventral and left-right axes are specified after fertilization, but the manner of their specification is just now beginning to be understood. Lineage tracer dye injected into one blastomere at the 2-cell stage demonstrated that, in nearly all cases, the oral pole of the future oral-aboral (mouth-anus; ventral-dorsal) axis lies 45° clockwise from the first cleavage plane as viewed from the animal pole (Cameron et al. 1989). The oral-aboral axis appears to form through the activation of the *Nodal* gene in the oral (but not in the aboral) ectoderm during gastrulation (**FIGURE 7.10A**; Duboc et al. 2004; Flowers et al. 2004). *Nodal* transcription appears to be initiated by a small difference in the redox state of the ectoderm: the prospective oral side of the embryo has a higher mitochondrial respiration rate than the prospective aboral side (Coffman and Davidson 2001; Coffman et al. 2009). This difference activates different sets of transcription factors that in turn activate the *Nodal* gene in the oral region (Nam et al. 2007).

The role of *Nodal* was discovered through the classic "find it, lose it, move it" mode of experimentation described in Chapter 4 (Duboc et al. 2004). Researchers cloned a sea urchin *Nodal* gene and, using in situ hybridization, demonstrated that Nodal protein becomes expressed in the presumptive oral ectoderm at about the 60-cell stage. Nodal then becomes prominent on one side of the blastula, and this side becomes the oral (mouth) side of the gastrula. When the researchers prevented translation of the *Nodal* message, development was normal until the mesenchyme blastula stage—but the larvae never obtained bilateral symmetry, the archenteron did not bend to one side to form the mouth, and the skeletogenic mesenchyme did not separate into the two sets of spicule-forming skeleton cells. Moreover, genes usually activated in the oral ectoderm were not expressed. Conversely, when researchers induced ectopic expression of the *Nodal* gene throughout the ectoderm, all of the ectoderm appeared to become oral. Thus, Nodal appears to be crucial in establishing the oral ectoderm.

While *Nodal* expression during gastrulation establishes the oral-aboral axis of the larva, the right-left axis of the larva appears to be established by Nodal signaling *after* gastrulation, in the early larval stages (Duboc et al. 2005). At that time, *Nodal* expression moves to the future right side of the larva* (**FIGURE 7.10B**). Importantly, it is expressed in the right coelomic pouch where it appears to restrict the pouch's growth, thus allowing the adult sea urchin to arise solely from the left coelomic pouch. Inhibiting Nodal signaling results in the formation of an imaginal rudiment (a pouch whose cells form the adult urchin; see Figure 7.19) on both sides of the larva, leading to twin urchins, whereas forcing *Nodal* expression on both sides of the embryo prevents either pouch from developing into an imaginal rudiment. Thus, although there is no right-left partitioning of the *adult* sea urchin body, the distinguishing of left and right sides is critical for sea urchin development.

● **See WEBSITE 7.1** Sea urchin cell specification

Sea Urchin Gastrulation

The late sea urchin blastula consists of a single layer of about 1000 epithelial cells that form a hollow ball, somewhat flattened at the vegetal end. The blastomeres are derived from different regions of the zygote and have different sizes and properties. **FIGURES 7.11** and **7.12** illustrate development of the blastula through gastrulation to the **pluteus larva** stage that is characteristic of sea urchins. The drawings show the fate of each cell layer during gastrulation. As can be seen, the cells that are destined to become the endoderm (gut)

(A)   (B)

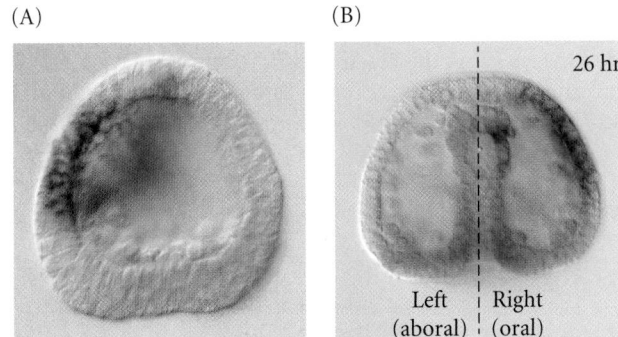

FIGURE 7.10 *Nodal* expression in the developing sea urchin. (A) In the mesenchyme blastula stage, as the micromeres enter the blastocoel, *Nodal* is expressed on the cells that will become the oral ectoderm (purple). (B) Later in development, after the endoderm reaches the oral area, *Nodal* expression shifts to the right half of the embryo and can be seen in all germ layers. This view is rotated 90° from that of (A). (From Duboc et al. 2004, 2005; photographs courtesy of T. Lepage.)

---

*As we will see in subsequent chapters, Nodal initiates a pathway that determines the left-right axis in vertebrates as well. However, in vertebrates, Nodal specifies the left side rather than the right side as in echinoderms.

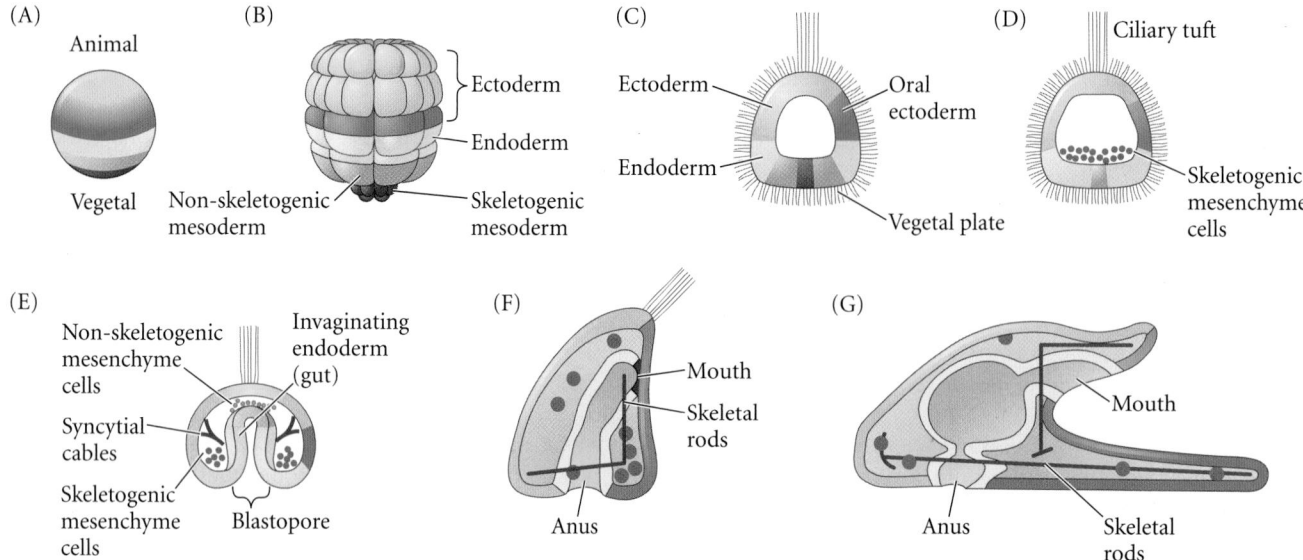

FIGURE 7.11 Normal sea urchin development, following the fate of the cellular layers of the blastula. (A,B) Fate maps of the zygote and the 64-cell embryo. (C) Blastula with ciliary tuft and flattened vegetal plate. (D) Late blastula with skeletogenic mesenchyme. (E) Gastrula with syncytial cables and non-skeletogenic mesenchyme. (F) Prism-stage larva. (G) Pluteus larva. (Courtesy of D. McClay.)

and mesoderm (skeleton) are still on the outside (see Figure 7.11A–C) and need to be brought inside the embryo through gastrulation.

Ingression of the skeletogenic mesenchyme

Shortly after the blastula hatches from its fertilization envelope, the descendants of the large micromeres undergo an **epithelial-mesenchymal transition** (see Chapter 3). The epithelial cells change their shape, lose their adhesions to their neighboring cells, and break away from the epithelium to enter the blastocoel as skeletogenic mesenchyme cells (Figure 7.12, 9–10 hours). The skeletogenic mesenchyme cells

then begin extending and contracting long, thin (250 nm in diameter and 25 μm long) processes called **filopodia**. At first the cells appear to move randomly along the inner blastocoel surface, actively making and breaking filopodial connections to the wall of the blastocoel. Eventually, however, they become localized within the prospective ventrolateral region of the blastocoel. Here they fuse into syncytial cables that will form the axis of the calcium carbonate spicules of the larval skeletal rods (see Figure 7.11D–F).

THE IMPORTANCE OF EXTRACELLULAR LAMINA INSIDE THE BLASTOCOEL The ingression of the large micromere descendants into the blastocoel is a result of their losing their affinity for their neighbors and for the hyaline membrane; instead these cells acquire a strong affinity for a group of proteins that line the blastocoel. This model of mesenchymal migration was first proposed by Gustafson and Wolpert (1967) and was confirmed in 1985, when Rachel Fink and

TABLE 7.1 Affinities of mesenchymal and nonmesenchymal cells for cellular and extracellular components

Cell type	Dislodgment force (in dynes)[a]		
	Hyaline	Gastrula cell monolayers	Basal lamina
16-Cell-stage micromeres	5.8×10^{-5}	6.8×10^{-5}	4.8×10^{-7}
Migratory-stage mesenchyme cells	1.2×10^{-7}	1.2×10^{-7}	1.5×10^{-5}
Gastrula ectoderm and endoderm	5.0×10^{-5}	5.0×10^{-5}	5.0×10^{-7}

Source: After Fink and McClay 1985.

[a]Tested cells were allowed to adhere to plates containing hyaline, extracellular basal lamina, or cell monolayers. The plates were inverted and centrifuged at various strengths to dislodge the cells. The dislodgement force is calculated from the centrifugal force needed to remove the test cells from the substrate.

David McClay measured the strength of sea urchin blastomere adhesion to the hyaline layer, to the basal lamina lining the blastocoel, and to other blastomeres.

Initially, all the cells of the blastula are connected on their outer surface to the hyaline layer, and on their inner surface to a basal lamina secreted by the cells. On their lateral surfaces, each cell has another cell for a neighbor. Fink and McClay found that the prospective ectoderm and endoderm cells (descendants of the mesomeres and macromeres, respectively) bind tightly to one another and to the hyaline layer, but adhere only loosely to the basal lamina (TABLE 7.1). The micromeres initially display a similar pattern of binding. However, the micromere pattern changes at gastrulation. Whereas the other cells retain their tight binding to the hyaline layer and to their neighbors, the skeletogenic mesenchyme precursors lose their affinities for these structures (which drop to about 2% of their original value), while their affinity for components of the basal lamina and extracellular matrix increases a hundredfold. These changes have been correlated with changes in cell surface molecules that occur during this time (Wessel and McClay 1985), and proteins such as fibronectin, integrin, laminin, and cadherins have been shown to be involved in cellular ingression.

These changes in affinity cause the skeletogenic mesenchyme precursors to release their attachments to the external hyaline layer and to their neighboring cells and, drawn in by the basal lamina, to migrate up into the blastocoel (FIGURE 7.13A). There is a heavy concentration of extracellular material around the ingressing mesenchyme cells (FIGURE 7.13B,C). Once inside the blastocoel, these cells appear to migrate along the extracellular matrix of the blastocoel wall, extending their filopodia in front of them (Galileo and Morrill 1985; Karp and Solursh 1985; Cherr et al. 1992). Several proteins (including a fibronectin-like protein and a particular sulfated glycoprotein) are necessary to initiate and maintain this migration (Wessel et al. 1984; Lane and Solursh 1991; Berg et al. 1996).

At two sites near the future ventral side of the larva, many skeletogenic mesenchyme cells cluster together, fuse with one another, and initiate spicule formation (FIGURE 7.14; Hodor and Ettensohn 1998). If a labeled micromere from another embryo is injected into the blastocoel of a gastrulating sea urchin embryo, it migrates to the correct location and contributes to the formation of the embryonic spicules (Ettensohn 1990; Peterson and McClay 2003). It is thought that the

9 hr

9.5 hr

10 hr

10.5 hr

11 hr

11.5 hr   Blastopore

12 hr

13 hr

13.5 hr   Syncytial cables

15 hr

17 hr

18 hr   Blastopore
Syncytial cables

FIGURE 7.12 *Entire sequence of gastrulation in* Lytechinus variegatus. *Times show the length of development at 25°C. Courtesy of J. Morrill; pluteus larva courtesy of G. Watchmaker.)*

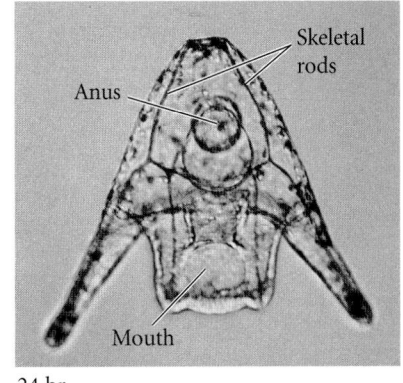

Skeletal rods

Anus

Mouth

24 hr

(A)

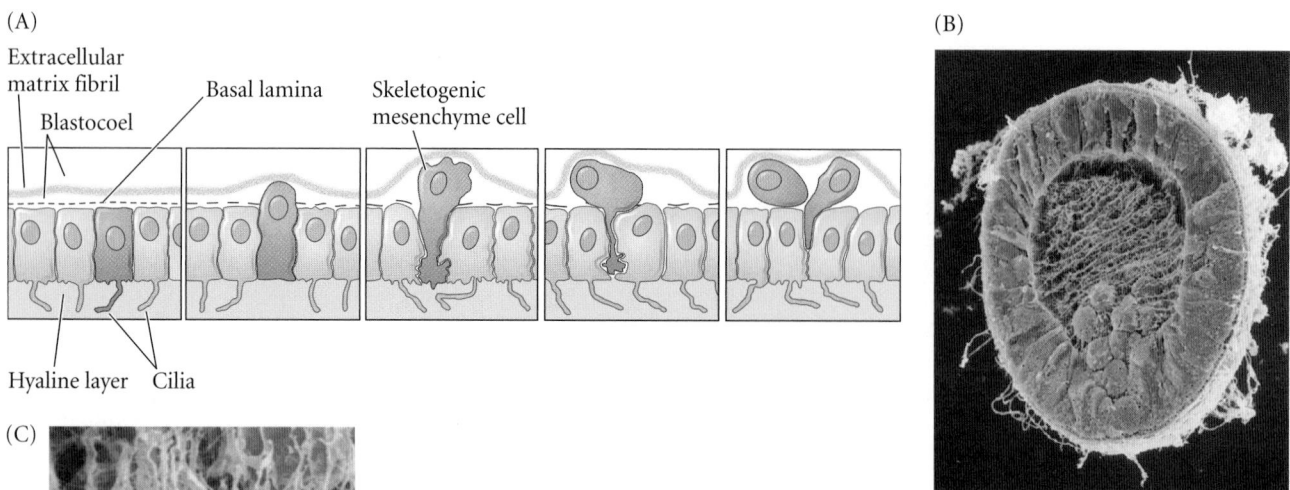

Extracellular matrix fibril

Basal lamina

Blastocoel

Skeletogenic mesenchyme cell

Hyaline layer   Cilia

(B)

(C)

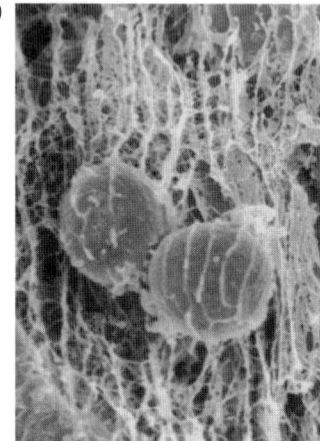

FIGURE 7.13   Ingression of skeletogenic mesenchyme cells. (A) Interpretative depiction of changes in the adhesive affinities of the presumptive skeletogenic mesenchyme cells (pink). These cells lose their affinities for hyalin and for their neighboring blastomeres while gaining an affinity for the proteins of the basal lamina. Nonmesenchymal blastomeres retain their original high affinities for the hyaline layer and neighboring cells. (B) Scanning electron micrograph of skeletogenic mesenchyme cells enmeshed in the extracellular matrix of an early *Strongylocentrotus* gastrula. (C) Gastrula-stage mesenchyme cell migration. The extracellular matrix fibrils of the blastocoel lie parallel to the animal-vegetal axis and are intimately associated with the skeletogenic mesenchyme cells. (B,C from Cherr et al. 1992, courtesy of the authors.)

necessary positional information is provided by the prospective ectodermal cells and their basal laminae (**FIGURE 7.15A**; Harkey and Whiteley 1980; Armstrong et al. 1993; Malinda and Ettensohn 1994). Only the skeletogenic mesenchyme cells (and not other cell types or latex beads) are capable of responding to these patterning cues (Ettensohn and McClay 1986). The extremely fine filopodia on the skeletogenic mesenchyme cells explore and sense the blastocoel wall and appear to be sensing dorsal-ventral and animal-vegetal patterning cues from the ectoderm (**FIGURE 7.15B**; Malinda et al. 1995; Miller et al. 1995).

Two signals secreted by the blastula wall appear to be critical for this migration. VEGF paracrine factors are emitted from two small regions of the ectoderm where the skeletogenic mesenchyme cells will congregate (Duloquin et al. 2007), and a fibroblast growth factor (FGF) paracrine factor is made in the equatorial belt between endoderm and ectoderm, becoming defined into the lateral domains where the skeletogenic mesenchyme cells collect (**FIGURE 7.15C**; Röttinger et al. 2008). The skeletogenic mesenchyme cells migrate to these points of VEGF and FGF synthesis and arrange themselves in a ring along the animal-vegetal axis.

(A)

(B)

FIGURE 7.14   Formation of syncytial cables by skeletogenic mesenchyme cells of the sea urchin. (A) Skeletogenic mesenchyme cells in the early gastrula align and fuse to lay down the matrix of the calcium carbonate spicule (arrows). (B) Scanning electron micrograph of spicules formed by the fusing of skeletogenic mesenchyme cells into syncytial cables. (A from Ettensohn 1990; B from Morrill and Santos 1985.)

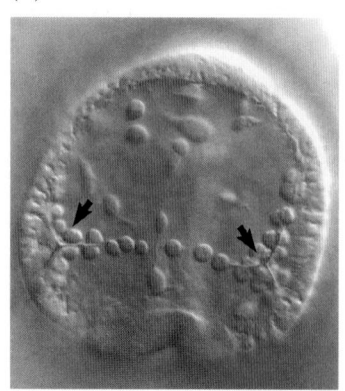

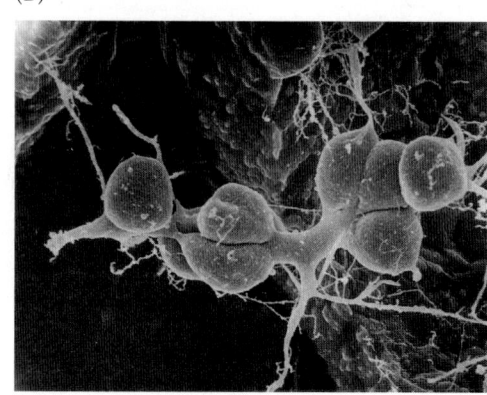

(A)

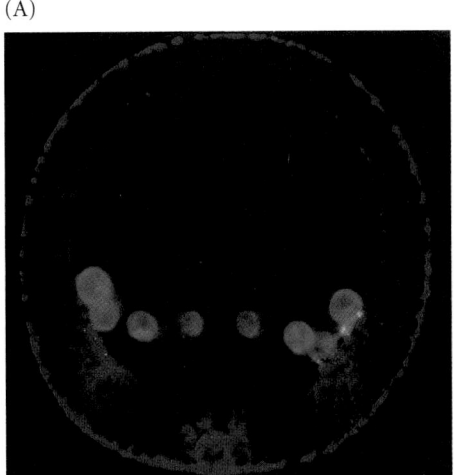

(B)

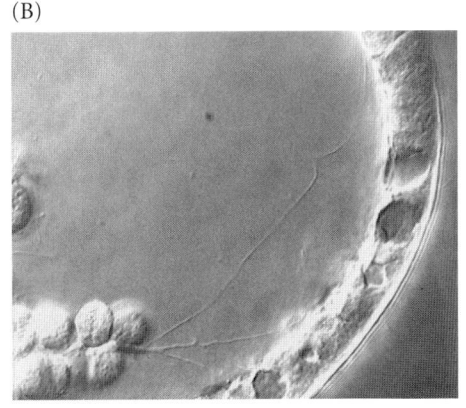

(C)

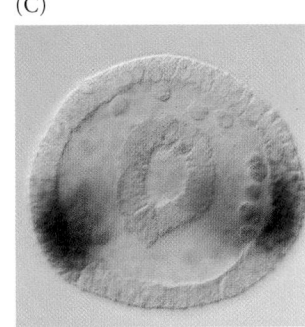

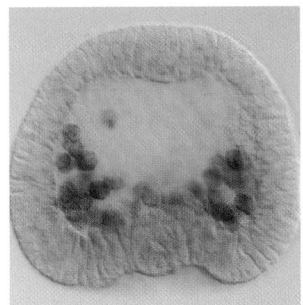

FIGURE 7.15 Localization of skeletogenic mesenchyme cells in the sea urchin. (A) Localization of the micromeres to form the calcium carbonate skeleton is determined by the ectodermal cells. Skeletogenic mesenchyme cells are stained green; β-catenin is red; skeletogenic mesenchyme cells appear to accumulate in those regions characterized by high β-catenin concentrations. (B) Nomarski videomicrograph showing a long, thin filopodium extending from a skeletogenic mesenchyme cell to the ectodermal wall of the gastrula, as well as a shorter filopodium extending inward from the ectoderm. Mesenchymal filopodia extend through the extracellular matrix and directly contact the cell membrane of the ectodermal cells. (C) Seen in cross section through the archenteron (top), the surface ectoderm expresses FGF in the particular locations where skeletogenic micromeres congregate. Moreover, the ingressing skeletal micromeres (bottom; longitudinal section) express the FGF receptor. When FGF signaling is suppressed, the skeleton does not form properly. (B from Miller et al. 1995, photographs courtesy of J. R. Miller and D. McClay; C from Röttinger et al. 2008, photographs courtesy of T. Lepage.)

INTEGRATION OF THE GENE REGULATORY NETWORK AND MORPHOGENESIS The skeletogenic mesenchyme cell receptor for VEGF is under the control of the micromere GRN network and thereby connects morphogenesis to cell specification. As we have seen, skeletogenic mesenchyme cells ingress into the blastocoel at a particular time. Moreover, this ingression is a "cell-autonomous" property (i.e., not controlled by neighboring cells), and skeletogenic mesenchyme cells will ingress into the blastocoels even if transplanted into younger or older embryos (Peterson and McClay 2003). The preparation for cell ingression involves endocytosis of the original micromere cell membrane and its replacement with a new one (Wu et al. 2007). This new membrane lacks the cadherins necessary for cell-cell adhesion (and presumably has altered receptors for hyalin and basal lamina proteins as well).

Endocytosis and the downregulation of cadherins are controlled by the transcription factor Snail, and the gene encoding Snail is activated by the Alx1 transcription factor, which in turn is regulated by the double-negative gate of the gene regulatory network. The use of Snail to downregulate cadherins and prepare the cell for its epithelial-mesenchymal transition is a morphogenetic subroutine that exists throughout the animal kingdom. In the case of sea urchins, it has been connected to the specification program of the micromere lineage. In this manner, the micromere cells alter their cell membranes and leave the blastula epithelium by ingressing into the blastocoel.

Invagination of the archenteron

FIRST STAGE OF ARCHENTERON INVAGINATION As the skeletogenic mesenchyme cells leave the vegetal region of the spherical embryo, important changes are occurring in the cells that remain there. These cells thicken and flatten to form a vegetal plate, changing the shape of the blastula (see Figure 7.12, 9 hours). The vegetal plate cells remain bound to one another and to the hyaline layer of the egg, and they move to fill the gaps caused by the ingression of the skeletogenic mesenchyme. The vegetal plate involutes inward by altering its cell shape, then invaginates about one-fourth to one-half of the way into the blastocoel (**FIGURE 7.16A**; see also Figure 7.12, 10.5–11.5 hours) before invagination suddenly ceases. The invaginated region is called the **archenteron** (primitive gut), and the opening of the archenteron at the vegetal pole is called the **blastopore**.

The movement of the vegetal plate into the blastocoel appears to be initiated by shape changes in the vegetal plate cells and in the extracellular matrix underlying them (see Kominami and Takata 2004). Actin microfilaments collect in the apical ends of the vegetal cells, causing these ends to constrict, forming bottle-shaped vegetal cells that pucker inward (Kimberly and Hardin 1998; Beane et al. 2006). Destroying

(A)

(B)

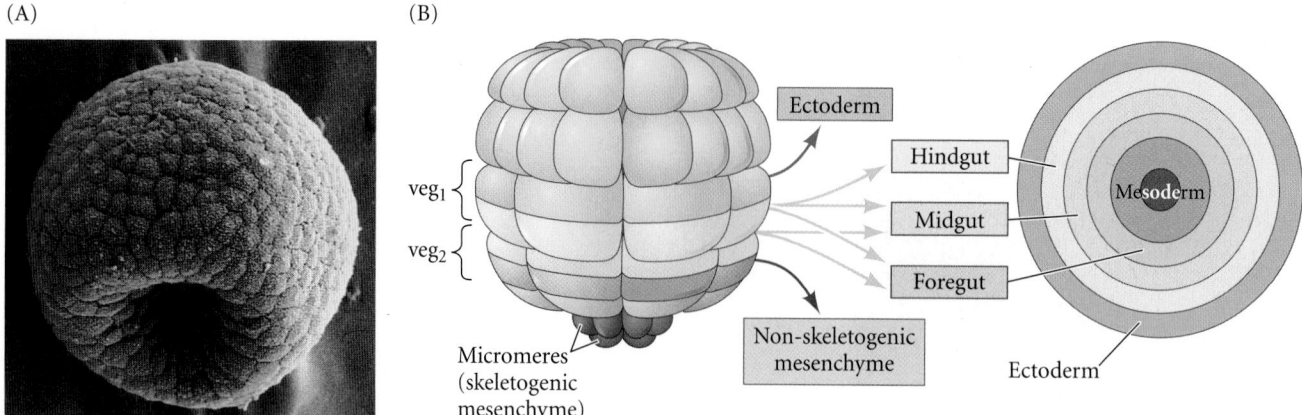

FIGURE 7.16 Invagination of the vegetal plate. (A) Vegetal plate invagination in *Lytechinus variegatus*, seen by scanning electron microscopy of the external surface of the early gastrula. The blastopore is clearly visible. (B) Fate map of the vegetal plate of the sea urchin embryo, looking "upward" at the vegetal surface. The central portion becomes the non-skeletogenic mesenchyme cells, while the concentric layers around it become the foregut, midgut, and hindgut, respectively. The boundary where the endoderm meets the ectoderm marks the anus. The non-skeletogenic mesenchyme and foregut come from the veg_2 layer, the midgut comes from veg_1 and veg_2 cells, and the hindgut (and the ectoderm in contact with it) comes from the veg_1 layer. (A from Morrill and Santos 1985, courtesy of J. B. Morrill; B after Logan and McClay 1999.)

these cells with lasers retards gastrulation. In addition, the hyaline layer at the vegetal plate buckles inward due to changes in its composition, directed by the vegetal plate cells (Lane et al. 1993).

At the stage when the skeletogenic mesenchyme cells begin ingressing into the blastocoel, the fates of the vegetal plate cells have already been specified (Ruffins and Ettensohn 1996). The non-skeletogenic mesenchyme is the first group of cells to invaginate, forming the tip of the archenteron and leading the way into the blastocoel. The non-skeletogenic mesenchyme will form the pigment cells, the musculature around the gut, and contribute to the coelomic pouches. The endodermal cells adjacent to the micromere-derived mesenchyme become foregut, migrating the farthest distance into the blastocoel. The next layer of endodermal cells becomes midgut, and the last circumferential row to invaginate forms the hindgut and anus (FIGURE 7.16B).

SECOND AND THIRD STAGES OF ARCHENTERON INVAGINATION After a brief pause following initial invagination, the second phase of archenteron formation begins. The archenteron extends dramatically, sometimes tripling in length. In this process of extension, the wide, short gut rudiment is transformed into a long, thin tube (FIGURE 7.17; see also Figure 7.12, 12 hours). To accomplish this extension, numerous cellular phenomena work together. First, the endoderm cells proliferate as they enter into the embryo. Second, the clones derived from these cells slide past one another, like the extension of a telescope. And lastly, the cells rearrange themselves by intercalating between one another, like lanes of traffic merging (Ettensohn 1985; Hardin and Cheng 1986; Martins et al. 1998; Martik et al. 2012). This phenomenon, where cells intercalate to narrow the tissue and at the same time move it forward, is called **convergent extension**.

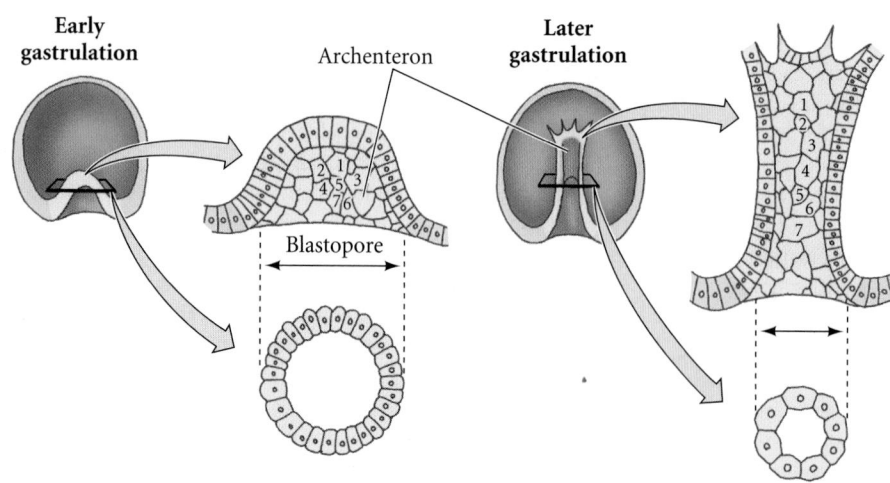

FIGURE 7.17 Cell rearrangement during extension of the archenteron in sea urchin embryos. In this species, the early archenteron has 20–30 cells around its circumference. Later in gastrulation, the archenteron has a circumference made by only 6–8 cells. (After Hardin 1990.)

(A)

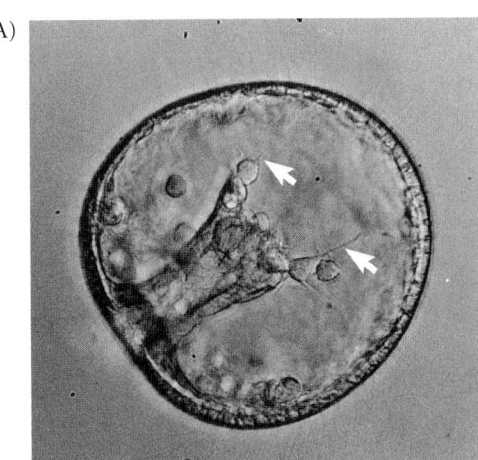

(B)

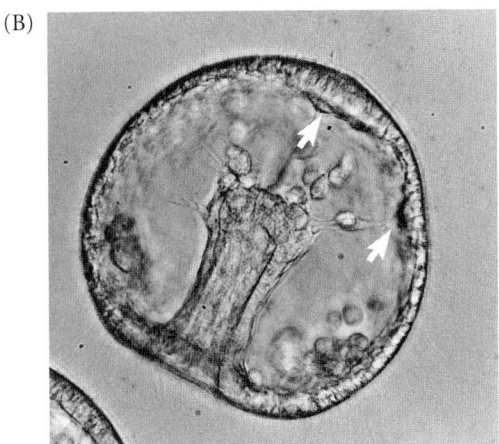

FIGURE 7.18 Mid-gastrula stage of *Lytechinus pictus*, showing filopodial extensions of non-skeletogenic mesenchyme. (A) Non-skeletogenic mesenchyme cells extend filopodia (arrows) from the tip of the archenteron. (B) Filopodial cables connect the blastocoel wall to the archenteron tip. The tension of the cables can be seen as they pull on the blastocoel wall at the point of attachment (arrows). (Courtesy of C. Ettensohn.)

The final phase of archenteron elongation is initiated by the tension provided by non-skeletogenic mesenchyme cells, which form at the tip of the archenteron and remain there. These cells extend filopodia through the blastocoel fluid to contact the inner surface of the blastocoel wall (Dan and Okazaki 1956; Schroeder 1981). The filopodia attach to the wall at the junctions between the blastomeres and then shorten, pulling up the archenteron (**FIGURE 7.18**; see also Figure 7.12, 12 and 13 hours). Hardin (1988) ablated non-skeletogenic mesenchyme cells of *Lytechinus pictus* gastrulae with a laser, with the result that the archenteron could elongate only to about two-thirds of the normal length. If a few non-skeletogenic mesenchyme cells were left, elongation continued, although at a slower rate. Thus, in this species the non-skeletogenic mesenchyme cells play an essential role in pulling the archenteron upward to the blastocoel wall during the last phase of invagination.

But can the filopodia of non-skeletogenic mesenchyme cells attach to any part of the blastocoel wall, or is there a specific target in the animal hemisphere that must be present for attachment to take place? Is there a region of the blastocoel wall that is already committed to becoming the ventral side of the larva? Studies by Hardin and McClay (1990) show that there is indeed a specific target site for filopodia that differs from other regions of the animal hemisphere. The filopodia extend, touch the blastocoel wall at random sites, and then retract. However, when filopodia contact a particular region of the wall, they remain attached and flatten out against this region, pulling the archenteron toward it. When Hardin and McClay poked in the other side of the blastocoel wall so that contacts were made most readily with that region, the filopodia continued to extend and retract after touching it. Only when the filopodia found their target tissue did they cease these movements. If the gastrula was constricted so that the filopodia never reached the target area, the non-skeletogenic mesenchyme cells continued to explore until they eventually moved off the archenteron and found the target as freely migrating cells. There appears, then, to be a target region on what is to become the ventral side of the larva that is recognized by the non-skeletogenic mesenchyme cells, and which positions the archenteron near the region where the mouth will form. Thus, as is characteristic of deuterostomes, the blastopore marks the position of the anus.

THE PLUTEUS LARVA AND IMAGINAL RUDIMENT As the top of the archenteron meets the blastocoel wall in the target region, many of the non-skeletogenic mesenchyme cells disperse into the blastocoel, where they proliferate to form the mesodermal organs (see Figure 7.12, 13.5 hours). Where the archenteron contacts the wall, a mouth eventually forms. The mouth fuses with the archenteron to create a continuous digestive tube of the pluteus larva.

As the pluteus elongates, the coelomic cavities form from non-skeletogenic mesenchyme and veg$_2$ cells near the tip of the archenteron. Under the influence of Nodal protein, the right coelomic sac remains rudimentary while the left coelomic sac undergoes extensive development to form many of the structures of the adult sea urchin (see Figure 7.10 and its discussion). This growth involves BMP signaling on the left side, by the veg$_2$ cells. The BMP signal is necessary for the specification and organization of the left coelomic pouch. The small micromeres, which give rise to the germ cells, are preferentially retained by the left coelomic pouch (Luo and Su 2012; Warner et al. 2012).

The left sac eventually splits into three smaller sacs. An invagination from the ectoderm fuses with the middle sac to form the **imaginal rudiment**. This rudiment develops a fivefold (pentaradial) symmetry (**FIGURE 7.19**), and skeletogenic mesenchyme cells enter the rudiment to synthesize the first skeletal plates of the shell. The left side of the pluteus becomes, in effect, the future oral surface of the adult sea urchin (Bury 1895; Aihara and Amemiya 2001). During metamorphosis, the imaginal rudiment separates from the larva, which then degenerates. While the imaginal rudiment (now called a juvenile) is re-forming its digestive tract and settling

(A)

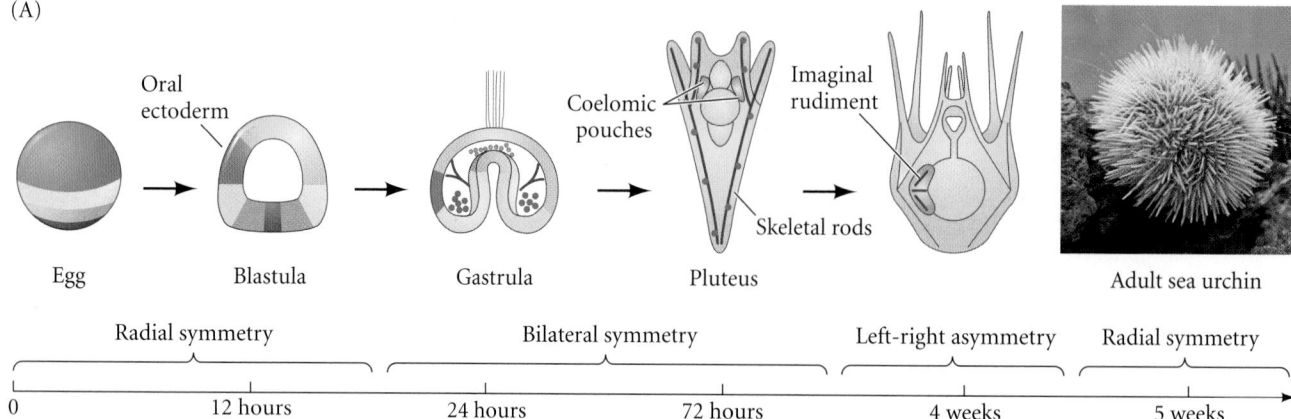

Egg — Blastula — Gastrula — Pluteus — Adult sea urchin

| Radial symmetry | Bilateral symmetry | Left-right asymmetry | Radial symmetry |

0 — 12 hours — 24 hours — 72 hours — 4 weeks — 5 weeks

FIGURE 7.19 Left-right specification is crucial for the asymmetric formation of the imaginal rudiment on the left side of the bilateral pluteus larva. The adult echinoderm emerges from this imaginal rudiment through metamorphosis. (A) The archenteron forms bilateral coelomic pouches, but only the left pouch proliferates and differentiates to form the rudiment. In addition to these veg_2 cells, some small micromeres enter the rudiment and are thought to contribute to the germline. (B) This imaginal rudiment growing in the left side of a pluteus larva will become the adult sea urchin; the larval stage will be jettisoned. The fivefold (pentaradial) symmetry of the rudiment is obvious. (A after Bessodes et al. 2012; B courtesy of G. Wray.)

(B)

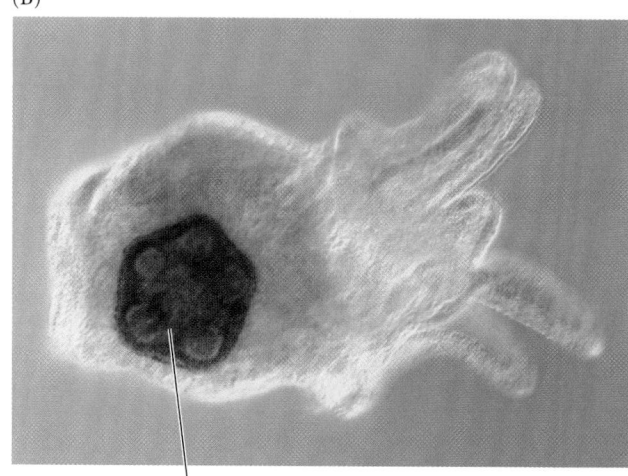

Imaginal rudiment

on the ocean floor, it is dependent on the nutrition it received from the jettisoned larva.

The pentaradial symmetry of adult echinoderms is unique and distinguishes them from the many bilaterally symmetrical animals. Note, however, that pluteus larvae *are* bilaterally symmetrical, evidence that echinoderms share a common ancestor with the bilaterally symmetrical chordates (see Figure 7.1). The next section will discuss the tunicates, an invertebrate group whose inclusion among the chordates is likewise the result of understanding these animals' early development.

● **See VADE MECUM** Sea urchin development

EARLY DEVELOPMENT IN TUNICATES

Tunicate Cleavage

Tunicates (also known as ascidians or "sea squirts") are fascinating animals for several reasons, but the foremost is that they are invertebrate chordates. As Lemaire (2009) has written, "looking at an adult ascidian, it is difficult, and slightly degrading, to imagine that we are close cousins to these creatures." But even though tunicates lack vertebrae at all stages of their life cycles, the free-swimming tunicate larva, or "tadpole," has a notochord and a dorsal nerve cord, making these animals chordates (see Figure 7.1). When the tadpole undergoes metamorphosis, its nerve cord and notochord

degenerate, and it secretes a cellulose tunic that is the source of the tunicates' name.

Tunicates have **bilateral holoblastic cleavage** (**FIGURE 7.20**). The most striking feature of this type of cleavage is that the first cleavage plane establishes the earliest axis of symmetry in the embryo, separating the embryo into its future right and left sides. Each successive division orients itself to this plane of symmetry, and the half-embryo formed on one side of the first cleavage plane is the mirror image of the half-embryo on the other side. The second cleavage is meridional, like the first, but unlike the first division, it does not pass through the center of the egg. Rather, it creates two large anterior cells (the A and a blastomeres) and two smaller posterior cells (the B and b blastomeres). Each side now has a large and a small blastomere.

Indeed, from the 8- through the 64-cell stages, every cell division is asymmetrical, such that the posterior blastomeres are always smaller than the anterior blastomeres (Nishida 2005; Sardet et al. 2007). Prior to each of these unequal cleavages, the posterior centrosome in the blastomere migrates toward the **centrosome-attracting body** (**CAB**), a macroscopic subcellular structure composed of endoplasmic

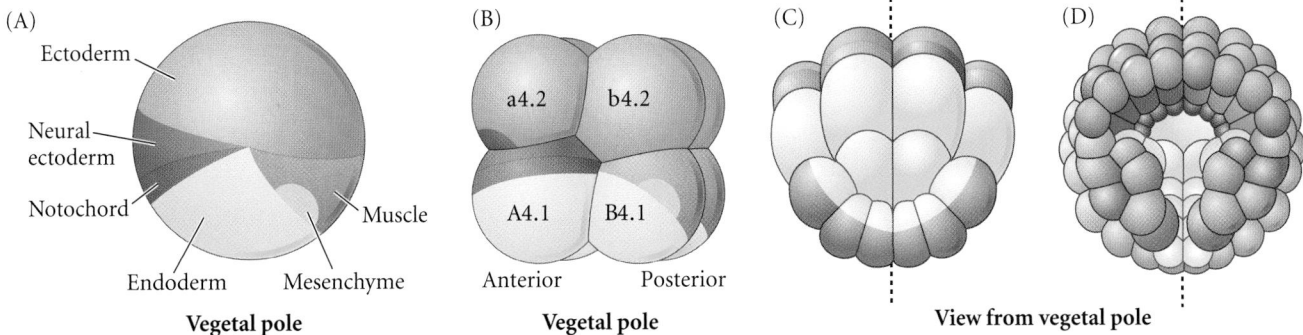

FIGURE 7.20 Bilateral symmetry in the egg of the ascidian tunicate *Styela partita*. (The cell lineages of *Styela* are shown in Figure 1.12C.) (A) Uncleaved egg. The regions of cytoplasm destined to form particular organs are labeled here and coded by color throughout the diagrams. (B) 8-Cell embryo, showing the blastomeres and

the fates of various cells. The embryo can be viewed as two 4-cell halves; from here on, each division on the right side of the embryo has a mirror-image division on the left. (C,D) Views of later embryos from the vegetal pole. The dashed line shows the plane of bilateral symmetry. (A after Balinsky 1981.)

reticulum. The CAB connects to the cell membrane through a network of PAR proteins that position the centrosomes asymmetrically in the cell (as seen for *C. elegans* in Figure 5.21), resulting in one large and one small cell at each of these three divisions. The CAB also attracts particular mRNAs in such a way that these messengers are placed in the posteriormost (i.e., smaller) cell of each division (Hibino et al. 1998; Nishikata et al. 1999; Patalano et al. 2006). In this way, the CAB integrates cell patterning with cell determination.* At the 64-cell stage, a small blastocoel is formed and gastrulation begins from the vegetal pole.

The tunicate fate map

Figure 7.20 shows the fate map and cell lineages of the tunicate *Styela partita* (see also Figure 1.12). Most early tunicate

---

*This description should sound like the discussion of the posterior cytoplasm of *Drosophila* eggs (Chapter 6). Indeed, mRNAs are localized in the CAB by their 3′ UTRs, the CAB is enriched with vesicles, and some of the mRNAs of the CAB become partitioned into the germ cells while others help construct the anterior-posterior axis (Makabe and Nishida 2012).

blastomeres are specified autonomously, each cell acquiring a specific type of cytoplasm that will determine its fate. In many tunicate species, the different regions of cytoplasm have distinct pigmentation, and the cell fates can easily be seen to correspond to the type of cytoplasm taken up by each cell.

These cytoplasmic regions are apportioned to the egg during fertilization. In the unfertilized egg of *S. partita*, a central gray cytoplasm is enveloped by a cortical layer containing yellow lipid inclusions (**FIGURE 7.21A**). During meiosis, the breakdown of the nucleus releases a clear substance that accumulates in the animal hemisphere of the egg. Within 5 minutes of sperm entry, the inner clear and

FIGURE 7.21 Cytoplasmic rearrangement in the fertilized egg of *Styela partita*. (A) Before fertilization, yellow cortical cytoplasm surrounds the gray yolky inner cytoplasm. (B) After sperm entry (in the vegetal hemisphere of the oocyte), the yellow cortical cytoplasm and the clear cytoplasm derived from the breakdown of the oocyte nucleus contract vegetally toward the sperm. (C) As the sperm pronucleus migrates animally toward the newly formed egg pronucleus, the yellow and clear cytoplasms move with it. (D) The final position of the yellow cytoplasm marks the location where cells give rise to tail muscles. (After Conklin 1905.)

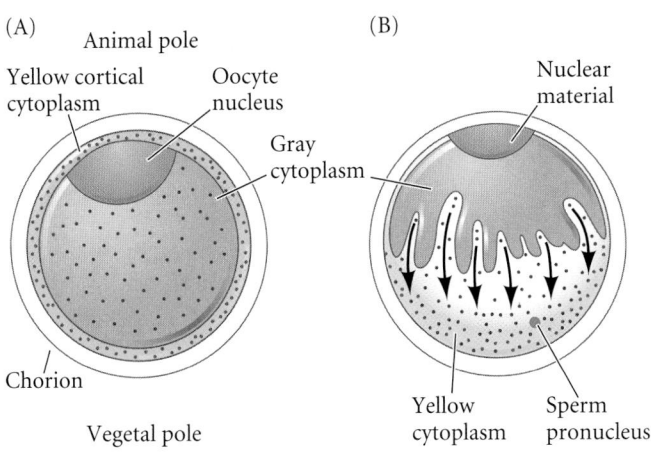

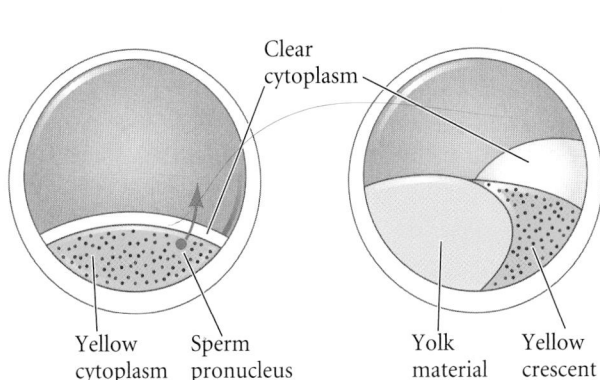

cortical yellow cytoplasms contract into the vegetal (lower) hemisphere of the egg (Prodon et al. 2005, 2008; Sardet et al. 2005). As the male pronucleus migrates from the vegetal pole to the equator of the cell along the future posterior side of the embryo, the yellow lipid inclusions migrate with it. This migration forms the **yellow crescent**, extending from the vegetal pole to the equator (FIGURE 7.21B–D); this region will produce most of the tail muscles of the tunicate larva. The movement of these cytoplasmic regions depends on microtubules that are generated by the sperm centriole and on a wave of calcium ions that contracts the animal pole cytoplasm (Sawada and Schatten 1989; Speksnijder et al. 1990; Roegiers et al. 1995).

Edwin Conklin (1905) took advantage of the differing coloration of these regions of cytoplasm to follow each of the cells of the tunicate embryo to its fate in the larva (see Figure 1.12C). He found that cells

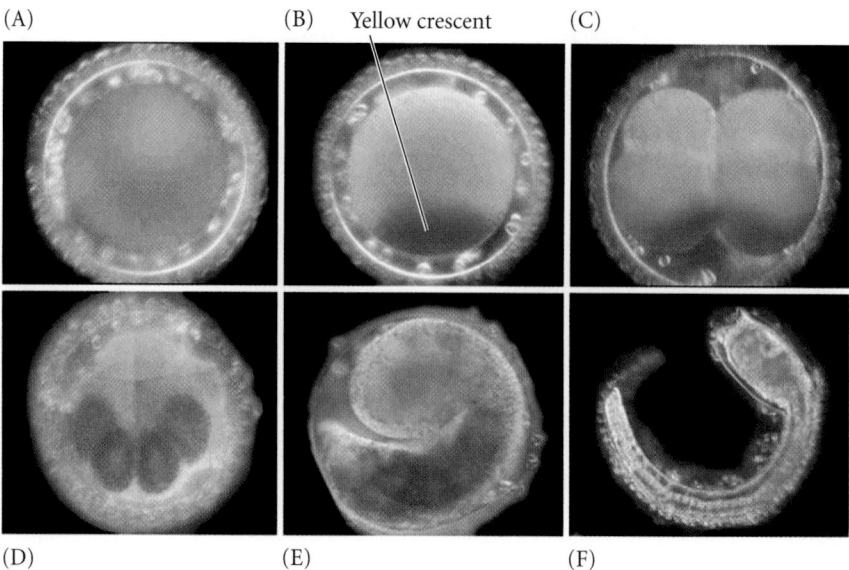

FIGURE 7.22 Cytoplasmic segregation in the egg of *Boltenia villosa*. The yellow crescent, originally seen in the vegetal pole, becomes segregated into the B4.1 blastomere pair and thence into the muscle cells. (From Swalla 2004, courtesy of B. Swalla, K. Zigler, and M. Baltzley.)

receiving clear cytoplasm become ectoderm; those containing yellow cytoplasm give rise to mesodermal cells; those incorporating slate gray inclusions become endoderm; and light gray cells become the neural tube and notochord. The cytoplasmic regions are localized bilaterally around the plane of symmetry, so they are bisected by the first cleavage furrow into the right and left halves of the embryo. The second cleavage causes the prospective mesoderm to lie in the two posterior cells, while the prospective neural ectoderm and chordamesoderm (notochord) will be formed from the two anterior cells (see Figure 7.20). The third division further partitions these cytoplasmic regions such that the mesoderm-forming cells are confined to the two vegetal posterior blastomeres, while the chordamesoderm cells are restricted to the two vegetal anterior cells.

● See WEBSITE 7.2 The experimental analysis of tunicate cell specification

Autonomous and conditional specification of tunicate blastomeres

The autonomous specification of tunicate blastomeres was one of the first observations in the field of experimental embryology (Chabry 1888). Cohen and Berrill (1936) confirmed Chaby's and Conklin's results, and by counting the number of notochord and muscle cells, they demonstrated that larvae derived from only one of the first two blastomeres (the right or the left) had half the expected number of cells. When the 8-cell embryo is separated into its four doublets

(the right and left sides being equivalent), both autonomous and conditional specification are seen (Reverberi and Minganti 1946; see Figure P2.2). Autonomous specification is seen in the gut endoderm, muscle mesoderm, and skin ectoderm (see Lemaire 2009). Conditional specification (by induction) is seen in the formation of the brain, notochord, heart, and mesenchyme cells. Indeed, a majority of the cell lineages involve some inductions.

AUTONOMOUS SPECIFICATION OF THE MYOPLASM: THE YELLOW CRESCENT AND MACHO-1 From the cell lineage studies of Conklin and others, it was known that only one pair of blastomeres (posterior vegetal; B4.1) in the 8-cell embryo is capable of producing tail muscle tissue (Whittaker 1982). These cells contain the yellow crescent cytoplasm. When yellow crescent cytoplasm is transferred from the B4.1 (muscle-forming) blastomere to the b4.2 or a4.2 (ectoderm-forming) blastomeres of an 8-cell tunicate embryo, the ectoderm-forming blastomeres generate muscle cells as well as its normal ectodermal progeny (FIGURE 7.22; see also Figure P2.3). Cytoplasm from the yellow crescent area of the fertilized egg can also cause the a4.2 blastomere to express muscle-specific proteins (Nishida 1992a). We can conclude, then, that the determination of the muscle-forming descendants of the B4.1 blastomere and the activation of the muscle-specific genes in these cells are controlled by the spatial localization of the morphogenetic determinants within the egg cytoplasm.

But what *are* these muscle-promoting determinants? Using RNA hybridization techniques, Nishida and Sawada

(A) 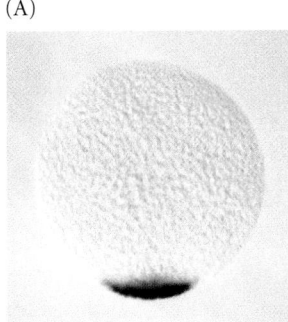 (B) 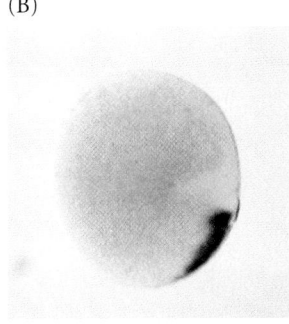 (C)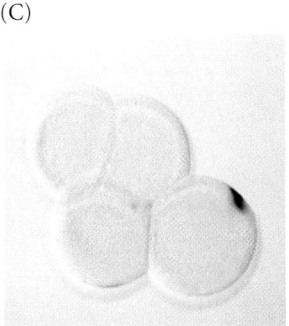

FIGURE 7.23 Autonomous specification by a morphogenetic factor. The *macho-1* mRNA message is localized to the muscle-forming tunicate cytoplasm. In situ hybridization shows the *macho-1* message found first in the vegetal pole cytoplasm (A), then migrating up the presumptive posterior surface of the egg (B) and becoming localized in the B4.1 blastomere (C). (From Nishida and Sawada 2001, courtesy of H. Nishida and N. Satoh.)

(2001) found particular mRNAs to be highly enriched in the vegetal hemisphere of the tunicate *Halocynthia roretzi*. One of these RNA messages encodes a zinc-finger transcription factor called Macho-1. *Macho-1* mRNA is concentrated in the vegetal hemisphere of the unfertilized egg and remains present during early fertilization. It appears to migrate with the yellow crescent cytoplasm into the posterior vegetal region of the egg during the second half of the first cell cycle. By the 8-cell stage, *macho-1* is found only in the B4.1 blastomeres. At the 16- and 32-cell stages, it is seen in those blastomeres that give rise to the muscle cells* (**FIGURE 7.23**).

When antisense oligonucleotides to deplete *macho-1* mRNA were injected into unfertilized eggs, the resulting larvae lacked all the muscles usually formed by the descendants of the B4.1 blastomere. The tails of these larvae were severely shortened (although the larvae did have the secondary muscles that are generated through the interactions of A4.1 and b4.2 blastomeres), but the other regions of the tadpoles appeared structurally and biochemically normal. Moreover, B4.1 blastomeres isolated from *macho-1*-depleted embryos failed to produce muscle tissue. Nishida and Sawada then injected *macho-1* mRNA into cells that would not normally form muscle and found that these ectoderm or endoderm precursors did generate muscle cells when given *macho-1* mRNA.

Macho-1 protein turns out to be a transcription factor that is required for the activation of several mesodermal genes, including *muscle actin*, *myosin*, *tbx6*, and *snail* (Sawada et al. 2005; Yagi et al. 2004). Of these gene products, only the Tbx6 protein produced muscle differentiation (as Macho-1 did) when expressed in cells ectopically. Macho-1 thus appears to directly activate a set of *tbx6* genes, and Tbx6 proteins activate the rest of muscle development (Yagi et al. 2005). Thus, the *macho-1* message is found at the right place and at the right time, and these experiments suggest that Macho-1 protein is both necessary and sufficient to promote muscle differentiation in certain tunicate cells.

Macho-1 and Tbx6 also appear to activate (possibly in a feedforward manner) the muscle-specific gene *snail*. Snail protein is important in preventing *Brachyury* expression in presumptive muscle cells, and thereby prevents the muscle precursors from becoming notochord cells.† It appears, then, that Macho-1 is a critical transcription factor of the tunicate yellow crescent, muscle-forming cytoplasm. Macho-1 activates a transcription factor cascade that promotes muscle differentiation while at the same time inhibiting notochord specification.

AUTONOMOUS SPECIFICATION OF ENDODERM: β-CATENIN
Presumptive endoderm originates from the vegetal A4.1 and B4.1 blastomeres. The specification of these cells coincides with the localization of β-catenin, discussed earlier in regard to sea urchin endoderm specification. Inhibition of β-catenin results in the loss of endoderm and its replacement by ectoderm in the tunicate embryo (**FIGURE 7.24**; Imai et al. 2000). Conversely, increasing β-catenin synthesis causes an increase in the endoderm at the expense of the ectoderm (just as in sea urchins). The β-catenin transcription factor appears to function by activating the synthesis of the homeobox transcription factor Lhx3. Inhibition of the *lhx3* message prohibits the differentiation of endoderm (Satou et al. 2001).

CONDITIONAL SPECIFICATION OF MESENCHYME AND NOTOCHORD BY THE ENDODERM While most tunicate muscles are specified autonomously from the yellow crescent cytoplasm, the most posterior muscle cells form through conditional specification by interactions with the descendants of the A4.1 and b4.2 blastomeres (Nishida 1987, 1992a,b). Moreover, the notochord, brain, heart, and mesenchyme also form through inductive interactions. In fact, the notochord and mesenchyme appear to be induced by an FGF secreted by the endoderm cells (Nakatani et al. 1996; Kim et al. 2000; Imai et al. 2002).

Macho-1 mRNA is also localized in the cells that become the mesenchyme. However, FGF signals from the endoderm prevent these mesenchyme precursors from developing into muscle cells, as we will see later.

---

†We will see the importance of *Brachyury* in vertebrate notochord formation as well. Indeed, the notochord is the "cord" that links the tunicates with the vertebrates, and *Brachyury* appears to be the gene that specifies the notochord (Satoh et al. 2012). As we will also see, *Tbx6* (which is closely related to *Brachyury*) is important in forming vertebrate musculature.

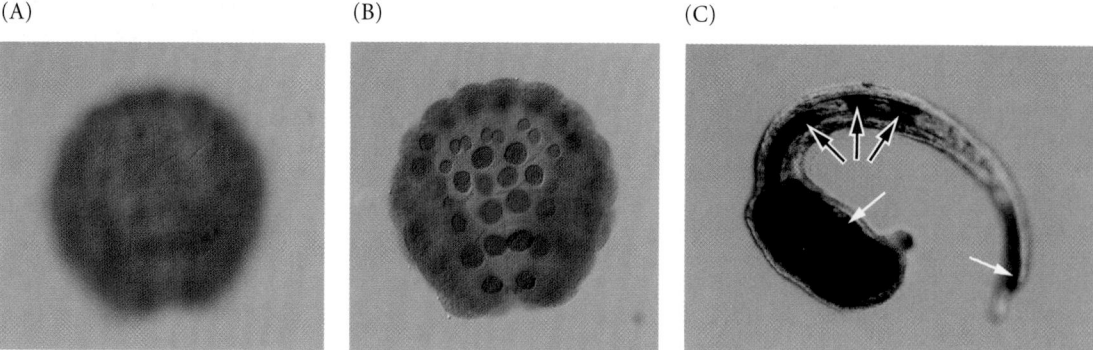

(A) (B) (C)

FIGURE 7.24 Antibody staining of β-catenin protein shows its involvement with endoderm formation. (A) No β-catenin is seen in the animal pole nuclei of a 110-cell *Ciona* embryo. (B) In contrast, nuclear β-catenin is readily seen in the nuclei in the vegetal endoderm precursors at the 110-cell stage. (C) When β-catenin is expressed in notochordal precursor cells, those cells will become endoderm and express endodermal markers such as alkaline phosphatase. The white arrows show normal endoderm; the black arrows show notochordal cells that are expressing endodermal enzymes. (From Imai et al. 2000, courtesy of H. Nishida and N. Satoh.)

The presence of Macho-1 in the posterior vegetal cytoplasm causes those posterior cells that will become mesenchyme to respond differently to the FGF signal than do the cells that will form neural structures (**FIGURE 7.25**; Kobayashi et al. 2003). Macho-1 prevents notochord induction in the mesenchymal cell precursors by activating the *snail* gene (which will in turn suppress the activation of *Brachyury*). Thus, Macho-1 is not only a muscle-activating determinant, it is also a factor that distinguishes cell response to the FGF signal. These FGF-responding cells do not become muscle because FGF activates cascades that block muscle formation (a role for these factors that is conserved in vertebrates). As seen in Figure 7.25, the presence of Macho-1 changes the responses to endodermal FGFs, causing the anterior cells to form notochord while the posterior cells become mesenchyme.

Specification of the embryonic axes

The axes of the tunicate larva are among its earliest commitments. Indeed, all of its embryonic axes are determined by the cytoplasm of the zygote *prior* to first cleavage (Sardet et al. 2007). The first axis to be determined is the dorsal-ventral axis, which is defined by the cap of cytoplasm at the vegetal pole. This vegetal cap is enriched for mitochondria, endoplasmic reticum components, and specific maternal mRNAs (such as *macho-1*) that will be involved in cell specification. This vegetal cap prefigures the future dorsal side of the larva and the site where gastrulation is initiated (Bates and Jeffery 1988). When small regions of vegetal pole cytoplasm were removed from zygotes between the first and second waves of zygote cytoplasmic movement, those zygotes neither gastrulated nor formed a dorsal-ventral axis.

The anterior-posterior axis is the second axis to appear and is determined during the migration of the oocyte cytoplasm during fertilization. Microtubules originating from the sperm centrosome, followed by cortical actin microfilaments, cause the vegetal cap to become repositioned to what will be the posterior region of the embryo. This can be followed readily, since the yellow crescent forms in the region of the egg that will become the posterior side of the larva (see Figures 7.21 and 7.22). When roughly 10% of the cytoplasm from this posterior vegetal region of the egg was removed after the second wave of cytoplasmic movement, most of the embryos failed to form an anterior-posterior axis. Rather,

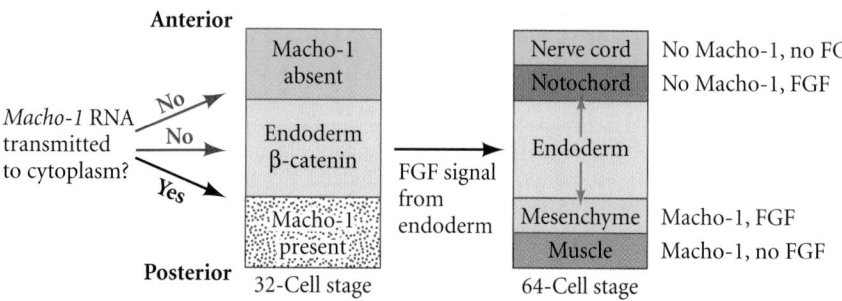

FIGURE 7.25 The two-step process for specifying the marginal cells of the tunicate embryo. The first step involves the acquisition (or nonacquisition) by the cells of the Macho-1 transcription factor. The second step involves the reception (or nonreception) of the FGF signal from the endoderm. (After Kobayashi et al. 2003.)

these embryos developed into radially symmetrical larvae with anterior fates. This posterior vegetal cytoplasm (PVC) is "dominant" to other cytoplasms in that when it was transplanted into the anterior vegetal region of zygotes that had their own PVC removed, the anterior of the cell became the new posterior, and the axis was reversed (Nishida 1994).

The specification of the left-right axis appears to involve the *Nodal* gene, just as it does in snails, sea urchins, and vertebrates (Morokuma et al. 2002; Yoshida and Saiga 2008). Although the first cleavage divides the tunicate embryo into symmetrical right and left halves, this symmetry is broken before the embryo hatches from its chorion. In the tunicate neurula, the *Nodal* gene is expressed only on the left side, and the brain vesicle and sensory pigment cells are located only on the right side of the brain. The asymmetric expression of *Nodal* appears to result from the rotation of the neurula-stage embryo within its vitelline envelope (Nishide et al. 2012).

When seen from the posterior, the tunicate neurula rotates along the anterior-posterior axis in a counterclockwise direction. When the rotation ceases, the future left portion of the embryo is downward, where the left epidermis touches the vitelline envelope. This contact causes *Nodal* expression in the left (but not the right) epidermis. When neurulae were sandwiched between two pieces of vitelline envelope such that both the right and left sides of the neurula touched the envelope, both sides showed *Nodal* expression and the right-left asymmetry of the brain structures was randomized. It therefore appears that juxtacrine signals from the vitelline envelope induce *Nodal* expression in the neurula, thereby creating left-right asymmetries.

Gastrulation in Tunicates

Tunicates, like sea urchins and vertebrates, follow the deuterostome pattern of gastrulation in which the blastopore becomes the anus. Tunicate gastrulation is characterized by the invagination of the endoderm, the involution of the mesoderm, and the epiboly of the ectoderm. About 4–5 hours after fertilization, the vegetal (endoderm) cells assume a wedge shape, expanding their apical margins and contracting near their vegetal margins (**FIGURE 7.26**). The A8.1 and B8.1 blastomere pairs appear to lead this invagination into the center of the embryo. The invagination forms a blastopore whose lips will become the mesodermal cells. The presumptive notochord cells are now on the anterior portion of the blastopore lip, while the presumptive tail muscle cells (from the yellow crescent) are on the posterior lip. The lateral lips comprise those cells that will become mesenchyme.

The second step of gastrulation involves the involution of the mesoderm. The presumptive mesoderm cells involute over the lips of the blastopore and, by migrating over the basal surfaces of the ectodermal cells, move inside the embryo. The ectodermal cells flatten and epibolize over the mesoderm and endoderm, eventually covering the embryo. After gastrulation

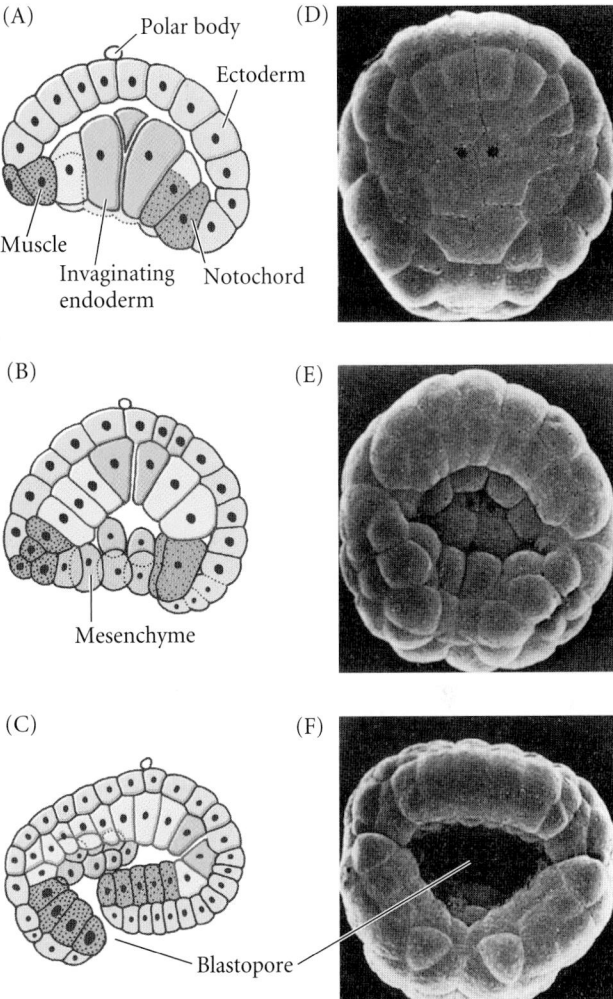

FIGURE 7.26 Gastrulation in the tunicate. Cross sections (A–C) and scanning electron micrographs viewed from the vegetal pole (D–F) illustrate the invagination of the endoderm (A,D), the involution of the mesoderm (B,E), and the epiboly of the ectoderm (C,F). Cell fates are color-coded as in Figure 7.20. (From Satoh 1978 and Jeffery and Swalla 1997, courtesy of N. Satoh.)

is complete, the embryo elongates along its anterior-posterior axis. The dorsal ectodermal cells that are the precursors of the neural tube invaginate into the embryo and are enclosed by neural folds. This process forms the neural tube, which will form a brain anteriorly and a spinal chord posteriorly. Meanwhile, the presumptive notochord cells on the right and left sides of the embryo migrate to the midline and interdigitate to form the notochord. The 40 cells of the notochord rearrange themselves from a 4 × 10 sheet of cells into a single row of 40 cells, extending the body axis along the anterior-posterior dimension (**FIGURE 7.27**; Jiang et al. 2005). This intercalation and migration of notochord cells is another example of convergent extension, a phenomenon seen throughout development.

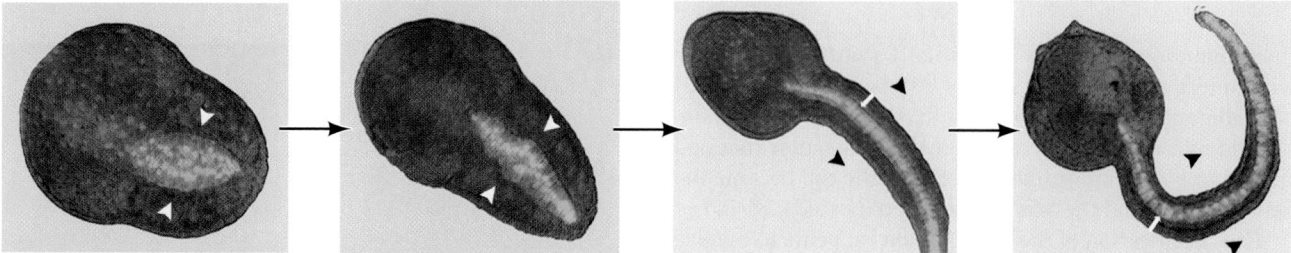

FIGURE 7.27 Convergent extension of the tunicate notochord. The notochord is visualized by a green fluorescent protein (GFP) probe fused to a promoter of the *Brachyury* gene, which is usually expressed in the notochord. The notochordal precursor cells converge and extend the notochord down the length of the animal's tail. (From Deschet et al. 2003, courtesy of the authors.)

Indeed, the convergent extension of notochordal precursor cells is characteristic of all chordates.

The muscle cells of the tunicate tail differentiate on either side of the neural tube and notochord (Jeffery and Swalla 1997). This forms the tadpole-like body of the larva (see the final panel of Figure 7.27). At the 110-cell stage, the B7.5 blastomere pairs express the conserved heart transcription factor Mesp. The anterior daughters of these B7.5 blastomeres respond to FGF signals to activate the cytoskeletal genes responsible for migration as well as the genes responsible for heart differentiation. These two cells migrate to form two regions of cardiac mesoderm on the left and right ventral sides of the tadpole, just anterior to the tail. Like the heart precursor cells of vertebrate embryos, these two cell clusters migrate to meet at the ventral midline of the larva (Davidson and Levine 2003; Satou et al. 2004; Christiaen et al. 2008). After metamorphosis, they will form the functional heart of the adult. During this metamorphosis, the tail and brain degenerate and the tunicate no longer moves.*

Coda

Chapters 5, 6, and 7 have described the events of early development—cleavage, gastrulation, and formation of the body axes—in several invertebrate species that have long served as models for developmental biologists. Next we will turn to these events as they occur in representative species of several vertebrate groups, including fish, amphibians, birds, and mammals (including humans).

In addition, we saw in the tunicate the formation of an anterior head and a notochord that divides the organism into right and left sides. This makes the organism a chordate, even though these structures degenerate in the adult. In the next two chapters, we will see development of various chordate groups—fish, amphibians, birds, and mammals.

---

*Such a process, according to neurobiologist Rodolfo Llinás (1987), is "paralleled by some human academics upon obtaining university tenure."

SNAPSHOT SUMMARY: Early Development in Sea Urchins and Tunicates

1. In both sea urchins and tunicates, the blastopore becomes the anus and the mouth is formed elsewhere; this deuterostome mode of gastrulation is also characteristic of chordates (including vertebrates).

2. Sea urchin cleavage is radial and holoblastic. At fourth cleavage, however, the vegetal tier divides into large macromeres and small micromeres. The animal pole divides to form the mesomeres.

3. Sea urchin cell fates are determined both by autonomous and conditional modes of specification. The micromeres are specified autonomously and become a major signaling center for conditional specification of other lineages. Maternal β-catenin is important for the autonomous specification of the micromeres.

4. Differential cell adhesion is important in regulating sea urchin gastrulation. The micromeres detach first from the vegetal plate and move into the blastocoel. They form the skeletogenic mesenchyme, which becomes the skeletal rods of the pluteus larva. The vegetal plate invaginates to form the endodermal archenteron, with a tip of non-skeletogenic mesenchyme cells. The archenteron elongates by convergent extension and is guided to the future mouth region by the non-skeletogenic mesenchyme.

5. The large micromeres become the skeleton of the larva; the small micromeres become the germ cells of the adult.

6. The micromeres regulate the fates of their neighboring cells through juxtacrine and paracrine pathways. They can convert animal cells into endoderm.

7. Gene regulatory networks are modules that work by logic circuits to integrate inputs into coherent cellular outputs. The micromeres integrate maternal components such that the placement of Disheveled at the vegetal pole enables the formation of β-catenin to help activate the *Pmar1* gene whose products inhibits the *HesC* gene. The product of

HesC inhibits skeletogenic genes. Thus, by locally inhibiting the inhibitor, the most vegetal cells become committed to skeleton production. This is called a double-negative gate.

8. The Nodal protein determines the oral-aboral axis of the sea urchin pluteus larva during early gastrulation. Later, during late gastrulation and the early larval period, it specifies the right half of the larva.

9. The ingression of the skeletogenic mesenchyme is accomplished through an epithelial-mesenchymal transition in which these cells lose cadherins and gain affinity to adhere to the matrix within the blastocoel.

10. Archenteron invagination and growth are coordinated by cell shape changes, cell proliferation, and convergent extension. In the final stage in invagination, the tip of the archenteron is actively pulled to the blastocoel roof by the non-skeletogenic mesenchyme cells.

11. The adult rudiment, called the imaginal rudiment, forms from the left coelomic pouch under the influence of BMPs.

12. The tunicate embryo divides holoblastically and bilaterally.

13. The tunicate fate map is identical on its right and left sides. The yellow cytoplasm contains muscle-forming determinants that act autonomously. The heart and nervous system are formed conditionally by signaling interactions between blastomeres.

14. Macho-1 is a tunicate muscle determinant (a transcription factor that activates muscle-specifying genes). The notochord and mesenchyme are generated conditionally by paracrine factors such as FGF.

15. The transcription factor Nodal appears to specify the left-right axis in tunicates, in which it is expressed solely on the left side of the body (as is also the case in snails and vertebrates).

16. Tunicate gastrulation brings the mesoderm and endoderm cells inside the body. The notochord elongates by convergent extension, and the muscle cells of the tail differentiate on either side of the notochord to form the tunicate tadpole.

For Further Reading

Complete bibliographical citations for all literature cited in this chapter can be found at the free-access website **www.devbio.com**

Lemaire, P. 2009. Unfolding a chordate developmental program, one cell at a time: Invariant lineages, short-range inductions, and evolutionary plasticity in ascidians. *Dev. Biol.* 332: 48–60.

Nishida, H. and K. Sawada. 2001. *Macho-1* encodes a localized mRNA in ascidian eggs that specifies muscle fate during embryogenesis. *Nature* 409: 724–729.

Revilla-i-Domingo, R., P. Oliveri and E. H. Davidson. 2007. A missing link in the sea urchin embryo gene regulatory network:

hesC and the double-negative specification of micromeres. *Proc. Natl. Acad. Sci. USA* 104: 12383–12388.

Wu, S. Y., M. Ferkowicz and D. R. McClay. 2007. Ingression of primary mesenchyme cells of the sea urchin embryo: A precisely timed epithelial mesenchymal transition. *Birth Def. Res. C Embryol. Today* 81: 241–252.

Go Online

WEBSITE 7.1 Sea urchin cell specification. The specification of sea urchin cells was one of the first major research projects in experimental embryology and remains a fascinating area of research. It appears that the initial signaling parses the blastula into domains characterized by the expression of specific transcription factors.

WEBSITE 7.2 The experimental analysis of tunicate cell specification. Researchers analyzing tunicate development are using biochemical and molecular probes to find the morphogenetic determinants that are segregated to different regions of the egg cytoplasm. You may also want to look back at Website 1.2, which includes E. G. Conklin's remarkable 1905 fate map of the tunicate embryo. Conklin's studies have been the basis for all subsequent research on the autonomous specification of tunicates.

Vade Mecum

Sea urchin development. This is an excellent review of sea urchin development, as well as questions on the fundamentals of echinoderm cleavage and gastrulation.

Outside Sites

Stanford University hosts two valuable and freely accessible websites on sea urchin development. (1) A set of interactive tutorials developed in conjunction with the National Science Foundation is found at **virtualurchin.stanford.edu/**. (2) Numerous ways of studying sea urchin development in the laboratory are described at **stanford.edu/group/Urchin/contents.html**.

Other laboratory protocols can be found at **swarthmore.edu/NatSci/sgilber1/DB_lab/Urchin/urchin_protocols.html**.

At **sugp.caltech.edu/endomes/**, the Davidson Laboratory Sea Urchin Specification Project provides updated systems diagrams showing an hour-by-hour account of the specification of the sea urchin cell types during the early cleavage stages.

The genome of the sea urchin *Strongylocentrotus purpuratus* was sequenced in 2006. The urchin gene database can be accessed at **spbase.org**.

The Four-Dimensional Ascidian Body Atlas website at **chordate.bpni.bio.keio.ac.jp/faba2/2.0/top.html** provides confocal images of developing tunicates.

The French site Aniseed (**www.aniseed.cnrs.fr**) has become a major resource for tunicate developmental biologists.

The site **readbookonline.net/read/5019/15171/** contains Scottish folklorist and poet Andrew Lang's beautiful piece of academic doggerel on tunicates.

8

Early Development in Vertebrates
Amphibians and Fish

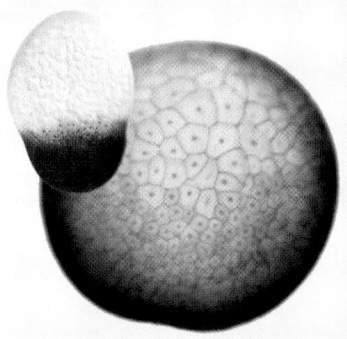

DESPITE VAST DIFFERENCES IN THEIR ADULT MORPHOLOGY, the early development of each of the vertebrate groups is very similar. Fish and amphibians are two of the most easily studied vertebrates. In both cases, hundreds of eggs are laid externally and fertilized simultaneously. Fish and amphibians are **anamniotic** vertebrates—they do not form the amnion that permits embryonic development to take place on dry land (**FIGURE 8.1**). But developing amphibians and fish generate body axes and organs employing many of the same processes and genes used by other vertebrates, including humans.

Amphibians were the first vertebrates to serve as models of development. In recent years, studies on the zebrafish (*Danio rerio*) have extended our knowledge to another vertebrate group and have reinforced the findings of early amphibian embryologists. Together, the amphibian (especially the frog *Xenopus*) and zebrafish research communities have established the general rules of vertebrate development, which (as we will see in Chapter 9) are maintained among the amniote vertebrates (reptiles, birds, and mammals). We start here with the amphibians because it was these organisms that provided us with the "backbone" of questions and tentative conclusions that have enabled vertebrate developmental biology to flourish.

EARLY AMPHIBIAN DEVELOPMENT

Amphibian embryos once dominated the field of experimental embryology. With their large cells and rapid development, salamander and frog embryos were excellently suited for transplantation experiments. However, amphibian embryos fell out of favor during the early days of developmental genetics, in part because these animals undergo a long period of growth before they become fertile and because their chromosomes are often found in several copies, precluding easy mutagenesis.* But with the advent of molecular techniques such as in situ hybridization, antisense oligonucleotides, chromatin immunoprecipitation, and dominant negative proteins, researchers have returned to the study of amphibian embryos and have been able to integrate their molecular analyses with earlier experimental findings. The results have been spectacular, revealing new vistas of how vertebrate bodies are patterned and structured.

---

*In the 1960s, the African clawed frog *Xenopus laevis* largely replaced both salamanders and frogs of the widespread genus *Rana* as the model organism of choice. *Xenopus* has the rare property of not having a breeding season, and unlike most amphibians, it can generate embryos all year. Unfortunately, *X. laevis* has four copies of each chromosome rather than the more usual two and takes 1–2 years to reach sexual maturity. Another *Xenopus* species, *X. tropicalis*, is often studied now; it has all the advantages of *X. laevis*, plus it is diploid and reaches sexual maturity in a mere 6 months (Gurdon and Hopwood 2000; Hirsch et al. 2002).

We are standing and walking with parts of our body which could have been used for thinking had they developed in another part of the embryo.
HANS SPEMANN (1943)

Theories come and theories go. The frog remains.
JEAN ROSTAND (1960)

(A)

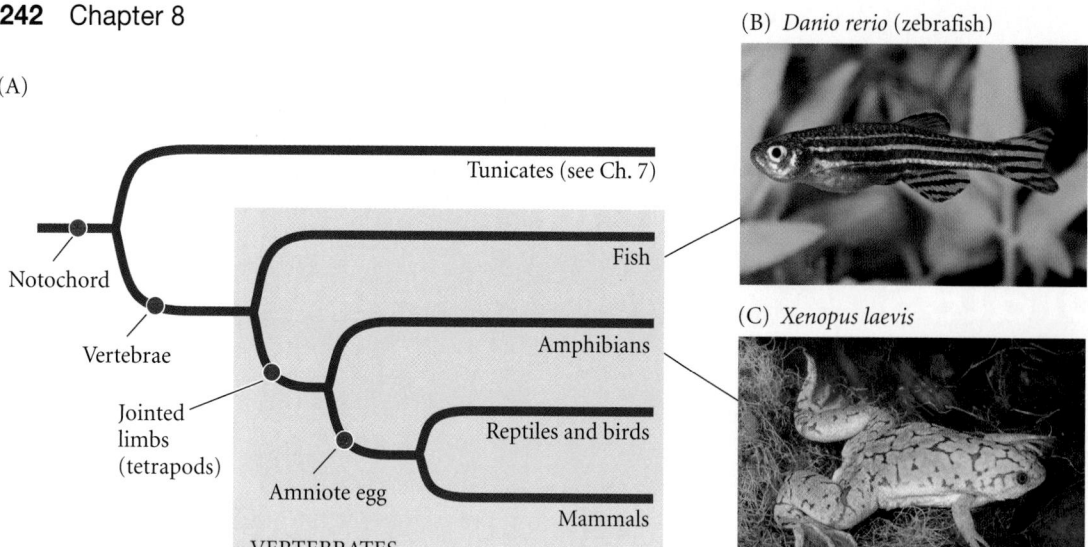

(B) *Danio rerio* (zebrafish)

(C) *Xenopus laevis*

FIGURE 8.1 (A) Phylogenetic tree of the chordates showing the relationship of the vertebrate groups. The embryonic development of fish and amphibians must be carried out in moist environments. The evolution of the shelled amniote egg permitted development to proceed on dry land for the reptiles and their descendants, as we will see in Chapter 9. (B) The zebrafish (*Danio rerio*) has become a popular model organism for the study of development. It is the first vertebrate species to be subjected to mutagenesis studies similar to those that have been carried out on *Drosophila*. (C) *Xenopus laevis*, the African clawed frog, is one of the most studied amphibians because it has the rare property of not having a breeding season and thus can generate embryos year-round.

Fertilization, Cortical Rotation, and Cleavage

Frogs usually have external fertilization, with the male fertilizing the eggs as the female is laying them. Even before fertilization, the frog egg has polarity, in that the dense yolk is at the vegetal (bottom) end, whereas the animal part of the egg (the upper half) has very little yolk. As we will also see, certain proteins and mRNAs are already localized in specific regions of the unfertilized egg.

Fertilization can occur anywhere in the animal hemisphere of the amphibian egg. The point of sperm entry is important because it determines dorsal-ventral polarity. The point of sperm entry marks the ventral (belly) side of the embryo, while the site 180° opposite the point of sperm entry will mark the dorsal (spinal) side.* The sperm centriole, which enters the egg with the sperm nucleus, organizes the microtubules of the egg into parallel tracks in the vegetal cytoplasm, separating the outer cortical cytoplasm from the yolky internal cytoplasm (**FIGURE 8.2A,B**). These microtubular tracks allow the cortical cytoplasm to rotate with respect to the inner cytoplasm. Indeed, these parallel arrays are first seen immediately before rotation starts, and they disappear when rotation ceases (Elinson and Rowning 1988; Houliston and Elinson 1991).

In the 1-cell embryo, the cortical cytoplasm rotates about 30° with respect to the internal cytoplasm (**FIGURE 8.2C**). In some eggs, this exposes gray-colored inner cytoplasm directly opposite the sperm entry point (**FIGURE 8.2D**; Roux 1887; Ancel and Vintenberger 1948). This region, the **gray crescent**, is where gastrulation will begin. Even in *Xenopus* eggs, which do not expose a gray crescent, cortical rotation occurs and cytoplasmic movements can be seen (Manes and Elinson 1980; Vincent et al. 1986). Gastrulation begins at the part of the egg opposite the point of sperm entry, and this region will become the dorsal portion of the embryo. Thus, the microtubular arrays organized by the sperm centriole at fertilization will become extremely important in initiating the dorsal-ventral axis of the larva.

Unequal radial holoblastic cleavage

Cleavage in most frog and salamander embryos is radially symmetrical and holoblastic, just like echinoderm cleavage. The amphibian egg, however, contains much more yolk. This yolk, which is concentrated in the vegetal hemisphere, is an impediment to cleavage. Thus, the first division begins at the animal pole and slowly extends down into the vegetal region (**FIGURE 8.3A**). In those species having a gray crescent (especially salamanders and frogs of the genus *Rana*), the first cleavage division usually bisects the gray crescent.

While the first cleavage furrow is still cleaving the yolky cytoplasm of the vegetal hemisphere, the second cleavage has already started near the animal pole. This cleavage is at right angles to the first one and is also meridional (**FIGURE 8.4A,B**). The third cleavage is equatorial. However, because of the vegetally placed yolk, the third cleavage furrow is not at the equator but is displaced toward the animal pole (Valles

---

* The axis between the point of sperm entry and the dorsal side approximates, but does not exactly correspond to, the actual ventral-dorsal axis of the amphibian larva (Lane and Smith 1999; Lane and Sheets 2002). However, as the literature in the field has traditionally equated these two axes, we will use the classical terminology here.

(A) 0.50

(B) 0.70

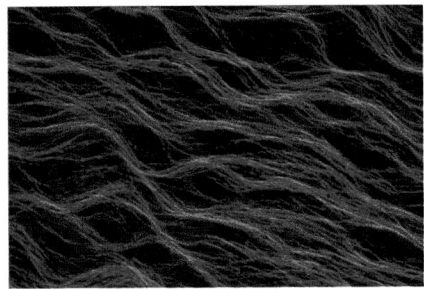

FIGURE 8.2 Reorganization of the cytoplasm and cortical rotation produce the gray crescent in frog eggs. (A,B) Parallel arrays of microtubules (visualized here using fluorescent antibodies to tubulin) form in the vegetal hemisphere of the egg, along the future dorsal-ventral axis. (A) With 50% of the first cell cycle complete, microtubules are present, but they lack polarity. (B) By 70% completion of the cycle, the vegetal shear zone is characterized by a parallel array of microtubules; cortical rotation begins at this time. At the end of rotation, the microtubules will depolymerize. (C) Schematic cross section of cortical rotation. At left, the egg is shown about midway through the first cell cycle. It has radial symmetry around the animal-vegetal axis. The sperm nucleus has entered at one side and is migrating inward. At right, 80% into first cleavage, the cortical cytoplasm has rotated 30° relative to the internal cytoplasm. Gastrulation will begin in the gray crescent—the region opposite the point of sperm entry, where the greatest displacement of cytoplasm occurs. (D) Gray crescent of *Rana pipiens*. Immediately after cortical rotation (left), lighter gray pigmentation is exposed beneath the heavily pigmented cortical cytoplasm. The first cleavage furrow (right) bisects this gray crescent. (A,B from Cha and Gard 1999, courtesy of the authors; C after Gerhart et al. 1989; D courtesy of R. P. Elinson.)

(C)

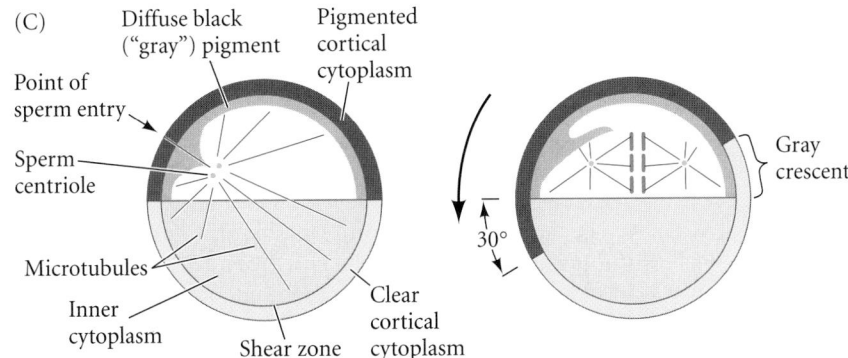

(D)

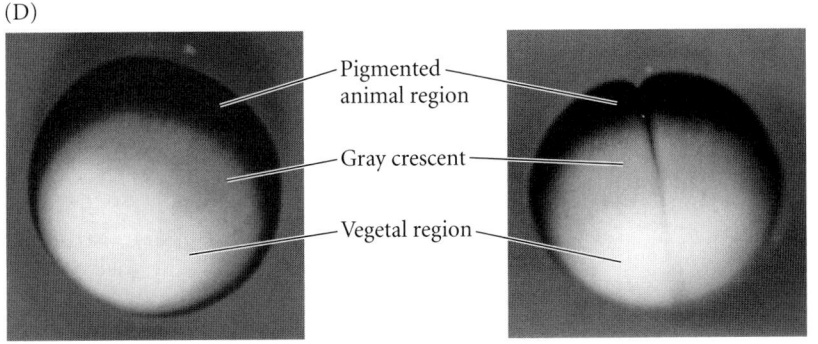

(A)

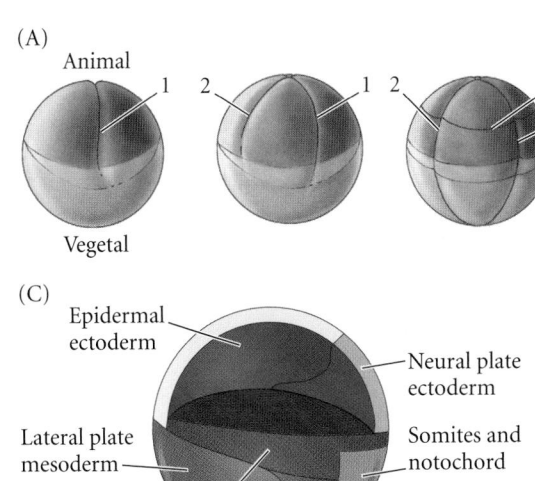

(B)

(C)

FIGURE 8.3 Cleavage of a *Xenopus* egg. (A) The first three cleavage furrows, numbered in order of appearance. Because the vegetal yolk impedes cleavage, the second division begins in the animal region of the egg before the first division has divided the vegetal cytoplasm completely. The third division is displaced toward the animal pole. (B) As cleavage progresses, the vegetal hemisphere ultimately contains larger and fewer blastomeres than the animal hemisphere. The final drawing shows a cross section through a mid-blastula stage embryo. (C) Fate map of the *Xenopus* embryo superimposed on the mid-blastula stage. (A,B after Carlson 1981; C after Lane and Smith 1999 and Newman and Kreig 1999.)

(A)

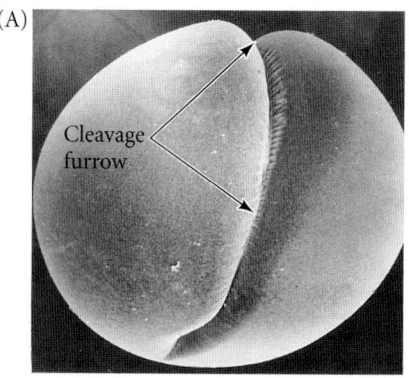

Cleavage furrow

(B)

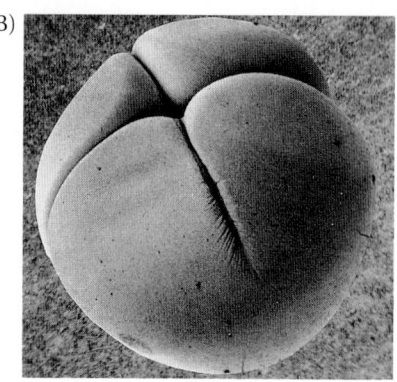

(C)

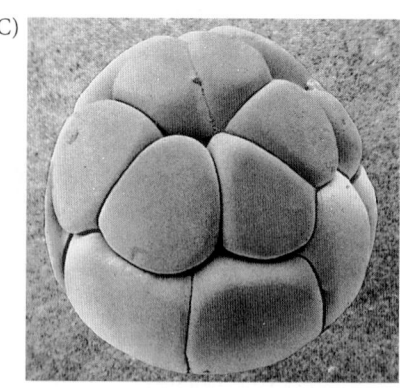

FIGURE 8.4 Scanning electron micrographs of frog egg cleavage. (A) First cleavage. (B) Second cleavage (4 cells). (C) Fourth cleavage (16 cells), showing the size discrepancy between the animal and vegetal cells after the third division. (A from Beams and Kessel 1976, courtesy of the authors; B,C courtesy of L. Biedler.)

et al. 2002). It divides the amphibian embryo into four small animal blastomeres (micromeres) and four large blastomeres (macromeres) in the vegetal region (**FIGURE 8.4C**). Despite their unequal sizes, the blastomeres continue to divide at the same rate until the twelfth cell cycle (with only a small delay of the vegetal cleavages). As cleavage progresses, the animal region becomes packed with numerous small cells, while the vegetal region contains a relatively small number of large, yolk-laden macromeres. An amphibian embryo containing 16 to 64 cells is commonly called a **morula** (plural *morulae*; Latin, "mulberry," whose shape it vaguely resembles). At the 128-cell stage, the blastocoel becomes apparent, and the embryo is considered a blastula (**FIGURE 8.3B**).

Numerous cell adhesion molecules keep the cleaving blastomeres together. One of the most important of these is EP-cadherin. The mRNA for this protein is supplied in the oocyte cytoplasm. If this message is destroyed by antisense oligonucleotides so that no EP-cadherin is made, the adhesion between blastomeres is dramatically reduced, resulting in the obliteration of the blastocoel (**FIGURE 8.5**; Heasman et al. 1994a,b).

Although amphibian development differs from species to species (see Hurtado and De Robertis 2007), in general the animal hemisphere cells will give rise to the ectoderm, the vegetal cells will give rise to the endoderm, and the cells beneath the blastocoel cavity will become mesoderm (**FIGURE 8.3C**). The cells opposite the point of sperm entry will become the neural ectoderm, the notochord mesoderm, and the pharyngeal (head) endoderm (Keller 1975, 1976; Landstrom and Løvtrup 1979).

The amphibian blastocoel serves two major functions. First, it can change its shape such that cell migration can occur during gastrulation; and second, it prevents the cells beneath it from interacting prematurely with the cells above it. When Nieuwkoop (1973) took embryonic newt cells from the roof of the blastocoel in the animal hemisphere (a region often called the **animal cap**) and placed them next to the yolky vegetal cells from the base of the blastocoel, the animal cap cells differentiated into mesodermal tissue instead of ectoderm. Thus, the blastocoel prevents the premature contact of the vegetal cells with the animal cap cells, and keeps the animal cap cells undifferentiated.

(A)

(B)

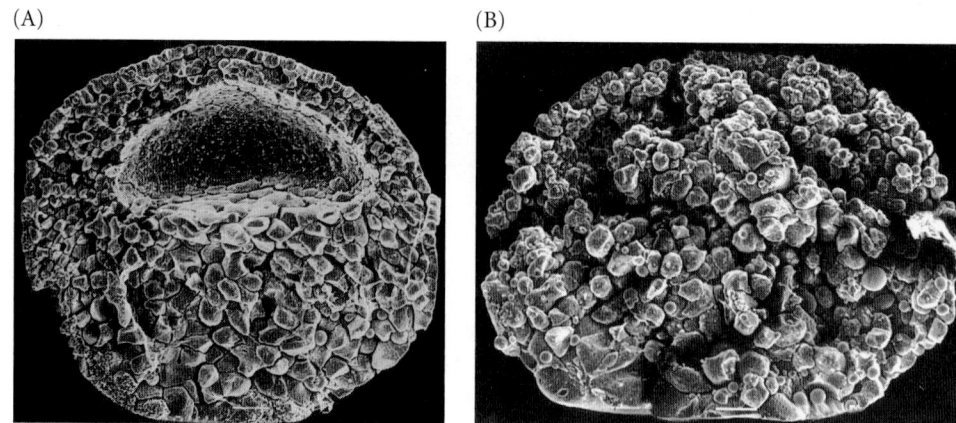

FIGURE 8.5 Lack of EP-cadherin in *Xenopus* results in the failure of blastomere adhesion. (A) Control embryo. (B) Embryo in which the mRNA for EP-cadherin has been depleted. The blastomeres have failed to adhere to one another, and the blastocoel has failed to form. (From Heasman et al. 1994b, courtesy of J. Heasman.)

The mid-blastula transition: Preparing for gastrulation

An important precondition for gastrulation is the activation of the zygotic genome (that is, the genes within each nucleus of the embryo). In *Xenopus laevis*, only a few genes appear to be transcribed during early cleavage. For the most part, nuclear genes are not activated until late in the twelfth cell cycle (Newport and Kirschner 1982a,b; Yang et al. 2002). At that time, the embryo experiences a **mid-blastula transition** (**MBT**; see Chapters 5 and 6). Different genes begin to be transcribed in different cells, the cell cycle acquires gap phases, and the blastomeres acquire the capacity to become motile. It is thought that some factor in the egg is being absorbed by the newly made chromatin because (as in *Drosophila*) the time of this transition can be changed experimentally by altering the ratio of chromatin to cytoplasm in the cell (Newport and Kirschner 1982a,b).

Some of the events that trigger the mid-blastula transition involve chromatin modification. First, certain promoters are demethylated, allowing transcription of these genes. During the late blastula stages, there is a loss of methylation on the promoters of genes that are activated at MBT. This demethylation is not seen on promoters that are not activated at MBT, nor is it observed in the coding regions of MBT-activated genes. The methylation of lysine-4 on histone H3 (forming a trimethylated lysine associated with active transcription) is also seen on the 5′ ends of many genes during the MBT. It appears, then, that modification of certain promoters and their associated nucleosomes may play a pivotal role in regulating the timing of gene expression at the mid-blastula transition (Stancheva et al. 2002; Akkers et al. 2009).

It is thought that once the chromatin at the promoters has been remodeled, various transcription factors (such as the VegT protein, formed in the vegetal cytoplasm from localized maternal mRNA) bind to the promoters and initiate new transcription. For instance, the vegetal cells (under the direction of the VegT protein) become the endoderm and begin secreting factors that induce the cells above them to become the mesoderm (see Figure 8.14).

● See **WEBSITE 8.1** Movies of amphibian development

Amphibian Gastrulation

The study of amphibian gastrulation is both one of the oldest and one of the newest areas of experimental embryology (see Beetschen 2001; Braukmann and Gilbert 2005). Even though amphibian gastrulation has been studied extensively for the past century, most of our theories concerning its mechanisms have been revised over the past decade. The study of these developmental movements has been complicated by the fact that there is no single way amphibians gastrulate; different species employ different means to achieve the same goal. In recent years, the most intensive investigations have focused

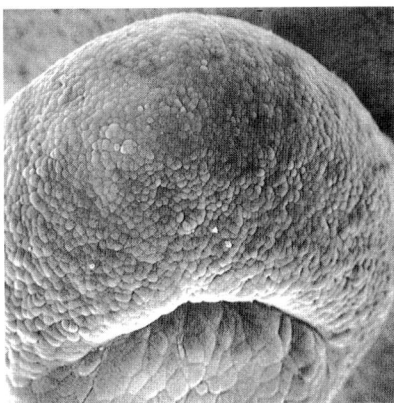

FIGURE 8.6 Surface view of an early dorsal blastopore lip of *Xenopus*. The size difference between the animal and vegetal blastomeres is readily apparent. (Courtesy of C. Phillips.)

on *Xenopus laevis*, so we will concentrate on the mechanisms of gastrulation in that species.

Vegetal rotation and the invagination of the bottle cells

Amphibian blastulae are faced with the same tasks as the invertebrate blastulae we followed in Chapters 5 through 7—namely, to bring inside the embryo those areas destined to form the endodermal organs; to surround the embryo with cells capable of forming the ectoderm; and to place the mesodermal cells in the proper positions between the ectoderm and the endoderm. The cell movements of gastrulation that will accomplish this are initiated on the future dorsal side of the embryo, just below the equator, in the region of the gray crescent (i.e., the region opposite the point of sperm entry; see Figure 8.2). Here the cells invaginate to form the slitlike blastopore (**FIGURE 8.6**). These **bottle cells** change their shape dramatically. The main body of each cell is displaced toward the inside of the embryo while maintaining contact with the outside surface by way of a slender neck. As in sea urchins, the bottle cells will initiate the formation of the archenteron (primitive gut).* However, unlike sea urchins, gastrulation in the frog begins not in the most vegetal region but

---

* Ray Keller and his students showed that the peculiar shape change of the bottle cells is needed to *initiate* gastrulation in *Xenopus*. It is the constriction of these cells that forms the blastopore, and it brings subsurface marginal cells into contact with the basal region of the surface blastomeres. Once this contact is made, the marginal cells begin to migrate along the extracellular matrix on the basal region of these surface cells. When such involution movements are underway, the bottle cells are no longer essential. At this point, they have done their job and can be removed without stopping gastrulation (Keller 1981; Hardin and Keller 1988). Thus, in *Xenopus*, the major factor in the movement of cells into the embryo appears to be the involution of the subsurface cells rather than the invagination of superficial marginal bottle cells.

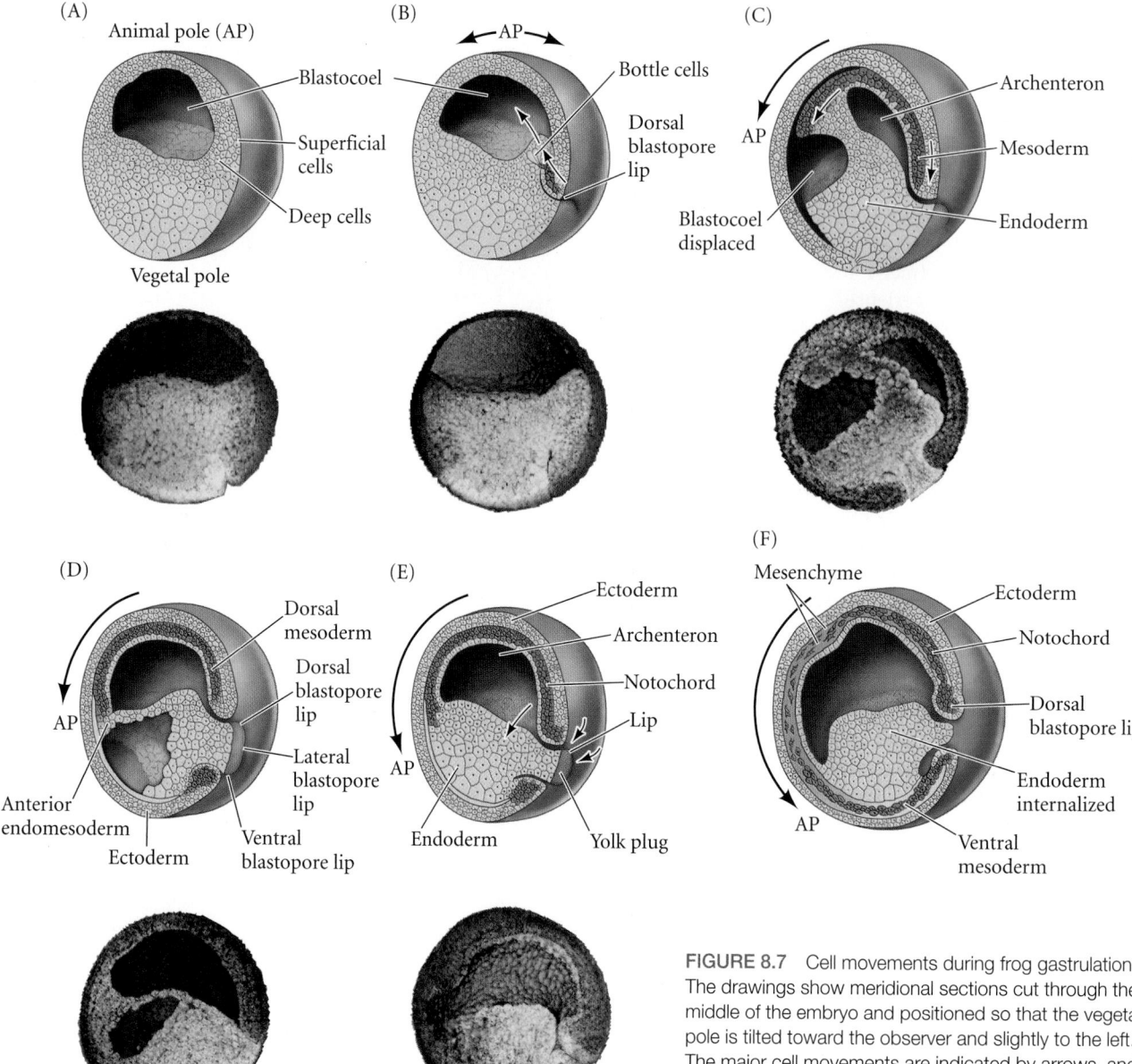

FIGURE 8.7 Cell movements during frog gastrulation. The drawings show meridional sections cut through the middle of the embryo and positioned so that the vegetal pole is tilted toward the observer and slightly to the left. The major cell movements are indicated by arrows, and the superficial animal hemisphere cells are colored so that their movements can be followed. Below the drawings are corresponding micrographs imaged with a surface imaging microscope (see Ewald et al. 2002). (A,B) Early gastrulation. The bottle cells of the margin move inward to form the dorsal lip of the blastopore, and the mesodermal precursors involute under the roof of the blastocoel. AP marks the position of the animal pole, which will change as gastrulation continues. (C,D) Mid-gastrulation. The archenteron forms and displaces the blastocoel, and cells migrate from the lateral and ventral lips of the blastopore into the embryo. The cells of the animal hemisphere migrate down toward the vegetal region, moving the blastopore to the region near the vegetal pole. (E,F) Toward the end of gastrulation, the blastocoel is obliterated, the endoderm has been internalized, and the mesodermal cells have been positioned between the ectoderm and endoderm. (Drawings after Keller 1986; micrographs courtesy of Andrew Ewald and Scott Fraser.)

in the **marginal zone**—the region surrounding the equator of the blastula, where the animal and vegetal hemispheres meet (**FIGURE 8.7A,B**). Here the endodermal cells are not as large or as yolky as the most vegetal blastomeres.

But cell involution is not a passive event. At least 2 hours before the bottle cells form, internal cell rearrangements propel the cells of the dorsal floor of the blastocoel toward the animal cap. This **vegetal rotation** places the prospective pharyngeal endoderm cells adjacent to the blastocoel and immediately above the involuting mesoderm (see Figure 8.8). These cells then migrate along the basal surface of the blastocoel roof, traveling toward the future anterior of the embryo (**FIGURE 8.7C–E**; Nieuwkoop and Florschütz 1950; Winklbauer and Schürfeld 1999; Ibrahim and Winklbauer

2001). The superficial layer of marginal cells is pulled inward to form the endodermal lining of the archenteron, merely because it is attached to the actively migrating deep cells. Although experimentally removing the bottle cells does not affect the involution of the deep or superficial marginal zone cells into the embryo, removal of the deep **involuting marginal zone (IMZ)** cells stops archenteron formation.

● See VADE MECUM Amphibian development

INVOLUTION AT THE BLASTOPORE LIP After the bottle cells have brought the involuting marginal zone into contact with the blastocoel wall, the IMZ cells involute into the embryo. As the migrating marginal cells reach the lip of the blastopore, they turn inward and travel along the inner surface of the outer animal hemisphere cells (i.e., the blastocoel roof; **FIGURE 8.7D–F**). The order of the march into the embryo is determined by the vegetal rotation that abuts the prospective pharyngeal endoderm against the inside of the animal cap tissue (**FIGURE 8.8**; Winklebauer and Damm 2011). Meanwhile, the animals cells undergo epiboly, producing a stream of cells that converge at and become the **dorsal blastopore lip**.

The first cells to compose the dorsal blastopore lip and enter the embryo are the prospective pharyngeal endoderm of the foregut. These cells begin to transcribe the *hhex* gene, which encodes a transcription factor that is critical for forming the head and heart (Rankin et al. 2011). As these first cells pass into the interior of the embryo, the dorsal blastopore lip becomes composed of cells that involute into the embryo to become the **prechordal plate** (the precursor of the head mesoderm). These cells transcribe the *goosecoid* gene, whose product is a transcription factor that activates numerous genes controlling head formation. The next cells involuting through the dorsal blastopore lip are the **chordamesoderm** cells. These cells will form the **notochord**, the transient mesodermal rod that

plays an important role in inducing and patterning the nervous system. Chordamesoderm cells express the *Xbra* (*Brachyury*) gene, whose product is a transcription factor critical for spinal cord formation. Thus, the cells constituting the dorsal blastopore lip are constantly changing as the original cells migrate into the embryo and are replaced by cells migrating downward, inward, and upward.

As the new cells enter the embryo, the blastocoel is displaced to the side opposite the dorsal lip. Meanwhile, the lip expands laterally and ventrally as bottle cell formation and involution continue around the blastopore. The widening blastopore "crescent" develops lateral lips and, finally, a ventral lip over which additional mesodermal and endodermal precursor cells pass (**FIGURE 8.9**). With the formation of the ventral lip, the blastopore has formed a ring around the large endodermal cells that remain exposed on the vegetal surface. This remaining patch of endoderm is called the **yolk plug**; it, too, is eventually internalized. At that point, all the endodermal precursors have been brought into the interior of the embryo, the ectoderm has encircled the surface, and the mesoderm has been brought between them. The first cells into the blastopore become the most anterior.

CONVERGENT EXTENSION OF THE DORSAL MESODERM Involution begins dorsally, led by the pharyngeal endoderm and the head mesoderm. These tissues will migrate most anteriorly beneath the surface ectoderm.* Meanwhile, as the lip of the blastopore expands to have dorsolateral, lateral,

---

* The pharyngeal endoderm and head mesoderm cannot be separated experimentally at this stage, so they are sometimes referred to collectively as the head endomesoderm. The notochord is the basic unit of the dorsal mesoderm, but it is thought that the dorsal portion of the somites may have similar properties to the notochord.

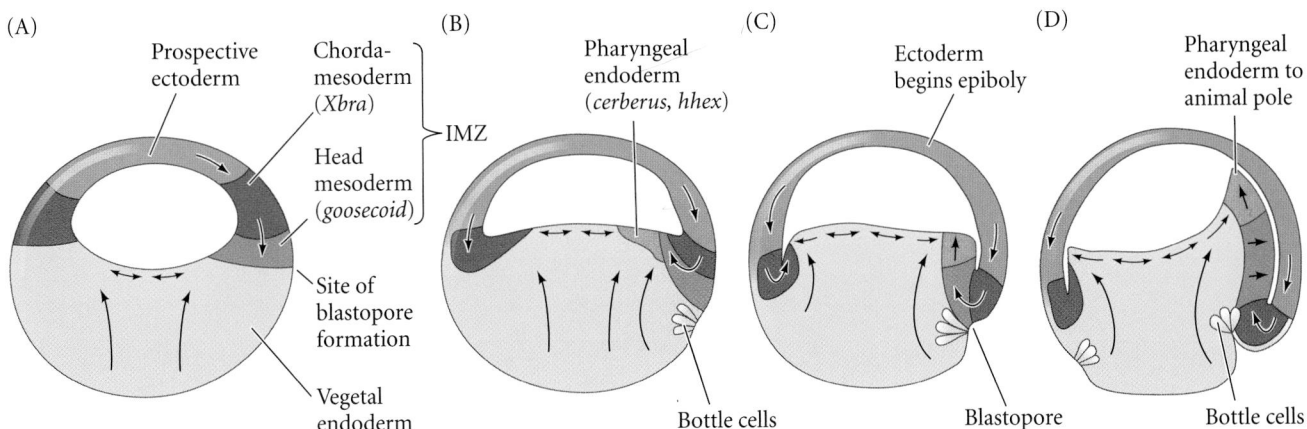

FIGURE 8.8 Early movements of *Xenopus* gastrulation. (A) At the beginning of gastrulation, the involuting marginal zone (IMZ) forms. Pink represents the prospective head mesoderm (*goosecoid* expression). Chordamesoderm (*Xbra* expression) is red. (B) Vegetal rotation (arrows) pushes the prospective pharyngeal endoderm (orange; specified by *hhex* and *cerberus* expression) to the side of the blastocoel. (C,D) The vegetal endoderm (yellow) movements push the pharyngeal endoderm forward, driving the mesoderm passively into the embryo and toward the animal pole. The ectoderm (blue) begins epiboly. (After Winklbauer and Schürfeld 1999.)

(A)

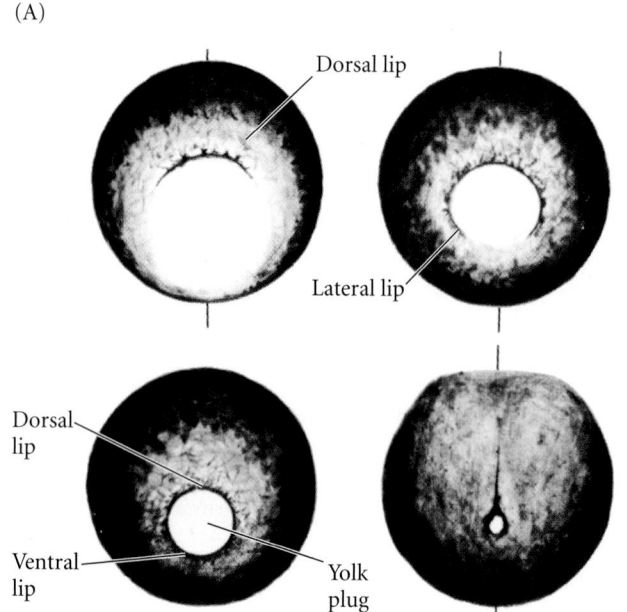

(B)

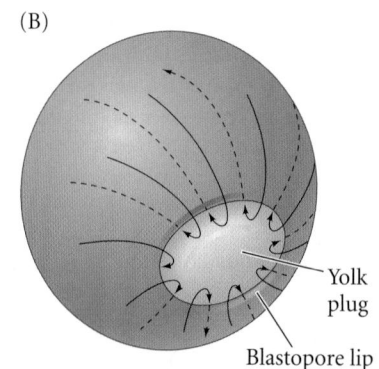

FIGURE 8.9 Epiboly of the ectoderm. (A) Changes in the region around the blastopore as the dorsal, lateral, and ventral lips are formed in succession. When the ventral lip completes the circle, the endoderm becomes progressively internalized. The four stages shown here correspond to Figure 8.7B–E. (B) Summary of epiboly of the ectoderm and involution of the mesodermal cells migrating into the blastopore and then under the surface. The endoderm beneath the blastopore lip (the yolk plug) is not mobile and is enclosed by these movements. (A from Balinsky 1975, courtesy of B. I. Balinsky.)

and ventral sides, the prospective heart, kidney, and ventral mesodermal precursor cells enter into the embryo.

FIGURE 8.10 depicts the behavior of the involuting marginal zone cells at successive stages of *Xenopus* gastrulation (Keller and Schoenwolf 1977; Hardin and Keller 1988). The IMZ is originally several layers thick. Shortly before their involution through the blastopore lip, the several layers of deep IMZ cells intercalate radially to form one thin, broad layer. This intercalation further extends the IMZ vegetally

(see Figure 8.10A). At the same time, the superficial cells spread out by dividing and flattening. When the deep cells reach the blastopore lip, they involute into the embryo and initiate a second type of intercalation. This intercalation causes a convergent extension along the mediolateral axis that integrates several mesodermal streams to form a long, narrow band (see Figure 8.10B). The anterior part of this band migrates toward the animal cap. Thus, the mesodermal stream continues to migrate toward the animal pole, and the

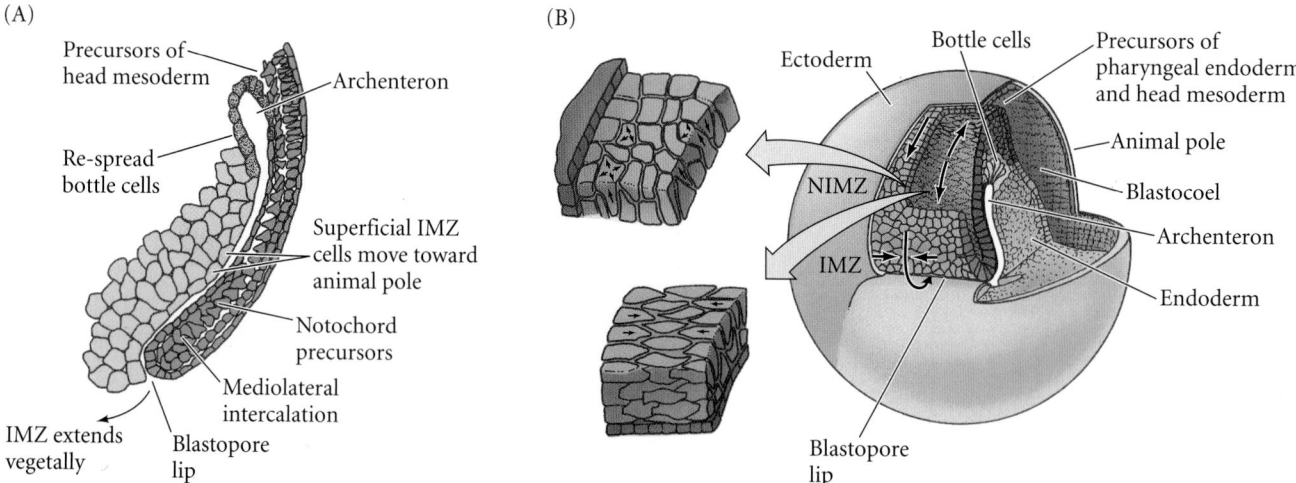

FIGURE 8.10 *Xenopus* gastrulation continues. (A) The deep marginal cells flatten, and the formerly superficial cells form the wall of the archenteron. (B) Radial intercalation, looking down at the dorsal blastopore lip from the dorsal surface. In the noninvoluting marginal zone (NIMZ) and the upper portion of the IMZ, deep (mesodermal) cells are intercalating radially to make a thin band of flattened cells.

This thinning of several layers into a few causes convergent extension (white arrows) toward the blastopore lip. Just above the lip, mediolateral intercalation of the cells produces stresses that pull the IMZ over the lip. After involuting over the lip, mediolateral intercalation continues, elongating and narrowing the axial mesoderm. (After Wilson and Keller 1991 and Winklbauer and Schürfeld 1999.)

overlying layer of superficial cells (including the bottle cells) is passively pulled toward the animal pole, thereby forming the endodermal roof of the archenteron (see Figure 8.10A). The radial and mediolateral intercalations of the deep layer of cells appear to be responsible for the continued movement of mesoderm into the embryo.

Several forces appear to drive convergent extension. The first force is a polarized cell cohesion, wherein the involuted mesodermal cells send out protrusions to contact one another. These "reachings out" are not random, but occur toward the midline of the embryo and require an extracellular fibronectin matrix (Goto et al. 2005; Davidson et al. 2008). The second force is differential cell cohesion. During gastrulation, the genes encoding adhesion proteins **paraxial protocadherin** and **axial protocadherin** become expressed specifically in the paraxial (somite-forming) mesoderm and the notochord, respectively (**FIGURE 8.11**). An experimental dominant negative form of axial protocadherin prevents the presumptive notochord cells from sorting out from the paraxial mesoderm and blocks normal axis formation. A dominant negative paraxial protocadherin (which is secreted instead of being bound to the cell membrane) prevents convergent extension* (Kim et al. 1998; Kuroda et al. 2002). Moreover, the expression domain of paraxial protocadherin characterizes the trunk mesodermal cells, which undergo convergent extension, distinguishing them from the head mesodermal cells, which do not undergo convergent extension.

The third factor regulating convergent extension is calcium flux. Wallingford and colleagues (2001) found that dramatic waves of calcium ions (Ca^{2+}) surge across the dorsal tissues undergoing convergent extension, causing waves of contraction within the tissue. Ca^{2+} is released from intracellular stores and is required for convergent extension. If Ca^{2+} release is blocked, normal cell specification still occurs, but the dorsal mesoderm neither converges nor extends.

This calcium flux is thought to regulate the contraction of the actin microfilaments, which may help explain why the head region does not undergo convergent extension. The head region, but not the trunk, expresses the *Otx* gene (the vertebrate homologue of the *Drosophila orthodenticle* gene), which encodes a transcription factor expressed in the most anterior region of the embryo. In addition to activating several genes necessary for forebrain development, the Otx2 protein activates the *calponin* gene (Morgan et al. 1999). Calponin protein binds to actin and myosin and prevents actin microfilaments from contracting during the calcium flux. These findings support a model wherein regulatory proteins cause changes in the outer surface of the tissue and generate mechanical traction forces that either

---

* Dominant negative proteins are mutated forms of the wild-type protein that interfere with the normal functioning of the wild-type protein. Thus, a dominant negative protein will have an effect similar to a loss-of-function mutation in the gene encoding the particular protein.

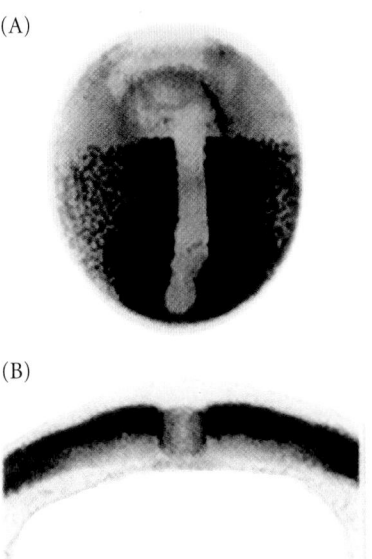

FIGURE 8.11 Protocadherin expression separates axial and paraxial mesoderm. (A) The expression pattern of paraxial protocadherin during late gastrulation (dark areas) shows the distinct downregulation in the notochord and the absence of expression in the head region. (B) Double-stained cross section through a late *Xenopus* gastrula shows the separation of notochord (reddish-brown staining for Chordin) and the paraxial mesoderm (midnight blue staining for paraxial protocadherin). (Courtesy of E. M. De Robertis.)

prevent or encourage cell migration (Beloussov et al. 2006; Davidson et al. 2008).

As mesodermal movement progresses, convergent extension continues to narrow and lengthen the involuting marginal zone. The involuting cells contain the prospective endodermal roof of the archenteron in its outer layer and the prospective mesodermal cells, including those of the notochord, adjacent to the ectodermal roof. Toward the end of gastrulation, the centrally located notochord separates from the somitic mesoderm on either side of it, and the notochord elongates separately as its cells continue to intercalate (Wilson and Keller 1991). This may in part be a consequence of the different adhesion molecules in the axial and paraxial mesoderms (see Figure 8.11; Kim et al. 1998; Park et al. 2011). This convergent extension of the mesoderm appears to be autonomous—the movements of these cells occur even if this region of the embryo is experimentally isolated from the rest of the embryo (Keller 1986).

Those mesodermal cells entering through the dorsal lip of the blastopore give rise to the central dorsal mesoderm (notochord and somites), while the remainder of the body mesoderm (which forms the heart, kidneys, bones, and parts of several other organs) enters through the ventral and lateral blastopore lips to create the **mesodermal mantle**. The endoderm is derived from the superficial cells of the involuting marginal zone that form the lining of the archenteron roof and from the subblastoporal vegetal cells that become

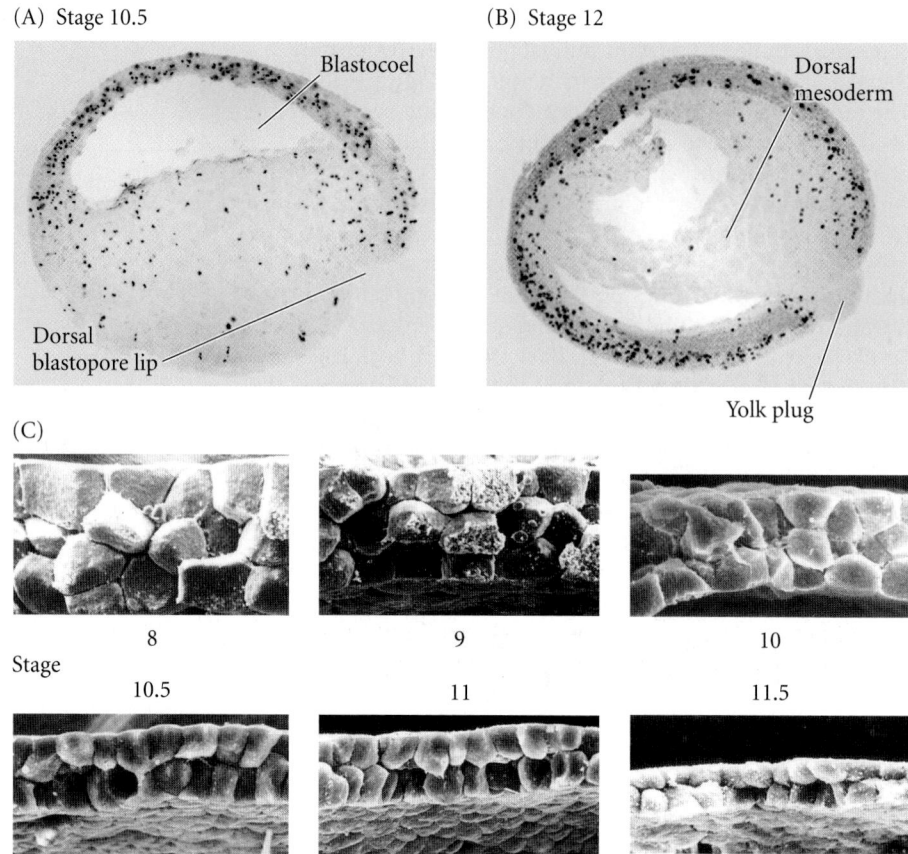

(A) Stage 10.5

Blastocoel

Dorsal
blastopore lip

(B) Stage 12

Dorsal
mesoderm

Yolk plug

(C)

Stage   8   9   10

10.5   11   11.5

FIGURE 8.12 Epiboly of the ectoderm is accomplished by cell division and intercalation. (A,B) Cell division in the presumptive ectoderm shown by staining for phosphorylated histone H3, a marker of mitosis. Stained nuclei appear black. (A) In early gastrulae, most cell division occurs in the presumptive ectoderm of the animal hemisphere. (B) In late gastrulae, cell division can be seen throughout the ectodermal layer. (Interestingly, the dorsal mesoderm shows no cell division). (C) Scanning electron micrographs of the *Xenopus* blastocoel roof, showing the changes in cell shape and arrangement. Stages 8 and 9 are blastulae; stages 10–11.5 represent progressively later gastrulae. (A,B after Saka and Smith 2001, photographs courtesy of the authors; C from Keller 1980, courtesy of R. E. Keller.)

the archenteron floor (Keller 1986). The remnant of the blastopore—where the endoderm meets the ectoderm—now becomes the anus.*

● **See WEBSITE 8.2** Migration of the mesodermal mantle

Epiboly of the prospective ectoderm

During gastrulation, the animal cap and noninvoluting marginal zone (NIMZ) cells expand by epiboly to cover the entire embryo (see Figure 8.10B). These cells will form the surface ectoderm. One important mechanism of epiboly in *Xenopus* gastrulation appears to be an increase in cell number (through division) coupled with a concurrent integration of several deep layers into one (**FIGURE 8.12**; Keller and

---

* As gastrulation expert Ray Keller famously remarked, "Gastrulation is the time when a vertebrate takes its head out of its anus."

Schoenwolf 1977; Keller and Danilchik 1988; Saka and Smith 2001). A second mechanism of *Xenopus* epiboly involves the assembly of fibronectin into fibrils by the blastocoel roof. This fibrillar fibronectin is critical in allowing the vegetal migration of the animal cap cells and enclosure of the embryo (Rozario et al. 2009). In *Xenopus* and many other amphibians, it appears that the involuting mesodermal precursors migrate toward the animal pole by traveling on an extracellular lattice of fibronectin secreted by the presumptive ectoderm cells of the blastocoel roof (**FIGURE 8.13A,B**).

Confirmation of fibronectin's importance for the involuting mesoderm came from experiments with a chemically synthesized peptide fragment that was able to compete with fibronectin for the binding sites of embryonic cells (Boucaut et al. 1984). If fibronectin were essential for cell migration, then cells binding this synthesized peptide fragment instead of extracellular fibronectin should stop migrating. Unable to find their "road," these prospective mesodermal cells should cease involution. That is precisely what happened, and the mesodermal

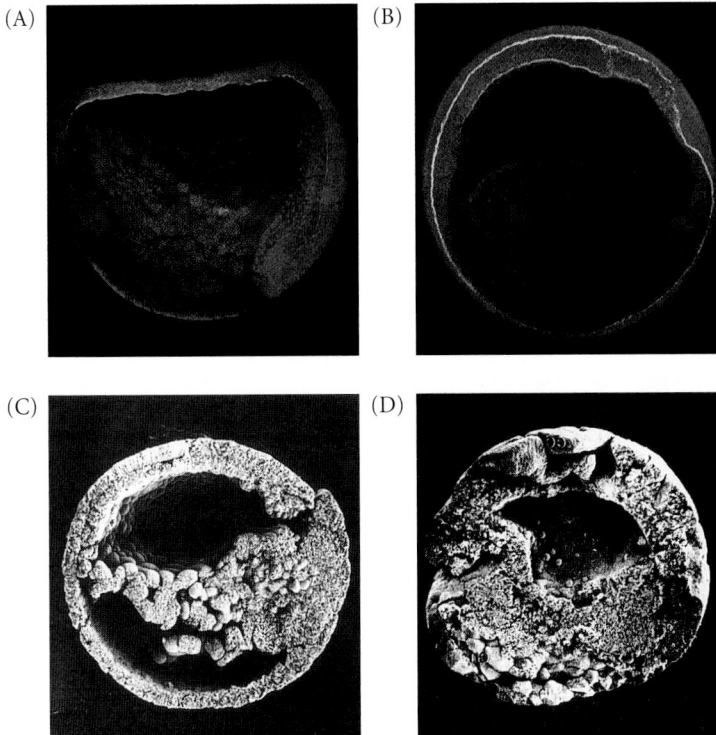

FIGURE 8.13 Fibronectin and amphibian gastrulation. (A,B) Sagittal section of *Xenopus* embryos at early (A) and late (B) gastrulation. The fibronectin lattice on the blastocoel roof is identified by fluorescent antibody labeling (yellow), while the embryonic cells are counterstained red. (C) Scanning electron micrograph of a normal salamander embryo injected with a control solution at the blastula stage. (D) Salamander embryo at the same stage but injected with a synthesized cell-binding fragment that competes with fibronectin. The archenteron has failed to form, and the mesodermal precursors have not undergone involution and remain on the embryo's surface. (A,B from Marsden and DeSimone 2001, photographs courtesy of the authors; C,D from Boucaut et al. 1984, courtesy of J.-C. Boucaut and J.-P. Thiery.)

precursors remained outside the embryo, forming a convoluted cell mass (**FIGURE 8.13C,D**). Thus, the fibronectin-containing extracellular matrix appears to provide both a substrate for adhesion as well as cues for the direction of cell migration.

Progressive Determination of the Amphibian Axes

Specification of the germ layers

As we have seen, the unfertilized amphibian egg has polarity along the animal-vegetal axis. Thus, the germ layers can be mapped onto the oocyte even before fertilization. The animal hemisphere blastomeres will become the cells of the ectoderm (skin and nerves); the vegetal hemisphere cells will form the cells of the gut and associated organs (endoderm); and the mesodermal cells will form from the internal cytoplasm around the equator. This general fate map is thought to be imposed on the embryo by the vegetal cells. The vegetal cells have two major functions: (1) to differentiate into endoderm and (2) to induce the cells immediately above them to become mesoderm.

The mechanism for this "bottom-up" specification of the frog embryo resides in a set of mRNAs that are tethered to the vegetal cortex. This includes the mRNA encoding the transcription factor **VegT**, which becomes apportioned to the vegetal cells during cleavage. VegT is critical in generating both the endodermal and mesodermal lineages. When VegT transcripts are destroyed by antisense oligonucleotides, the entire embryo becomes epidermis, with no mesodermal or endodermal components (Zhang et al. 1998; Taverner et al. 2005). *VegT* mRNA is translated shortly after fertilization. Its product activates a set of genes prior to the mid-blastula transition. One of the genes activated by this VegT protein encodes the Sox17 transcription factor. Sox17, in turn, is critical for activating the genes that specify cells to be endoderm. Thus, the vegetal cells become the endodermal cells.

Another set of early genes activated by VegT encodes Nodal paracrine factors that instruct the cell layers *above* them to become mesoderm (see Figure 8.24C; Skirkanich et al. 2011). Nodal secreted from the vegetal cells in the nascent endoderm and signal the cells above them to express phosphorylated Smad2. Smad2 helps activate the *eomesodermin* and *Brachyury* (*Xbra*) genes in those cells, causing the cells to become specified as mesoderm. The Eomesodermin and Smad2 proteins working together can activate the zygotic genes for the VegT proteins, thus creating a positive feedforward loop that is critical in sustaining the mesoderm (**FIGURE 8.14**). In the absence of such induction, cells become ectoderm (Fukuda et al. 2010).

In addition, the *Vg1* mRNA that has been stored in the vegetal cytoplasm is also translated. The production of Vg1 (another Nodal-like protein) is needed to activate other genes in the dorsal mesoderm. If either Nodal or Vg1 signaling is blocked, there is little or no mesoderm induction (Kofron et al. 1999; Agius et al. 2000; Birsoy et al. 2006). Thus, by the late blastula stage, the fundamental germ layers are becoming specified. The vegetal cells are specified as endoderm through transcription factors such as Sox17. The equatorial cells are specified as mesoderm by transcription factors such as Eomesodermin. And the animal cap—which has not begun receiving signals yet—becomes specified as ectoderm (see Figure 8.14). The critical factor in this partition of the embryo into the three germ-layer regions appears to be Nodal-like paracrine factors, whose actions are stimulated by T-box transcription factors such as VegT and Eomesodermin.

The dorsal-ventral and anterior-posterior axes

Although animal-vegetal polarity begins the specification of which cells will form each germ layer, the anterior-posterior, dorsal-ventral, and left-right axes are specified by events triggered at fertilization but not realized until gastrulation. In *Xenopus* (and in other amphibians), the formation of the

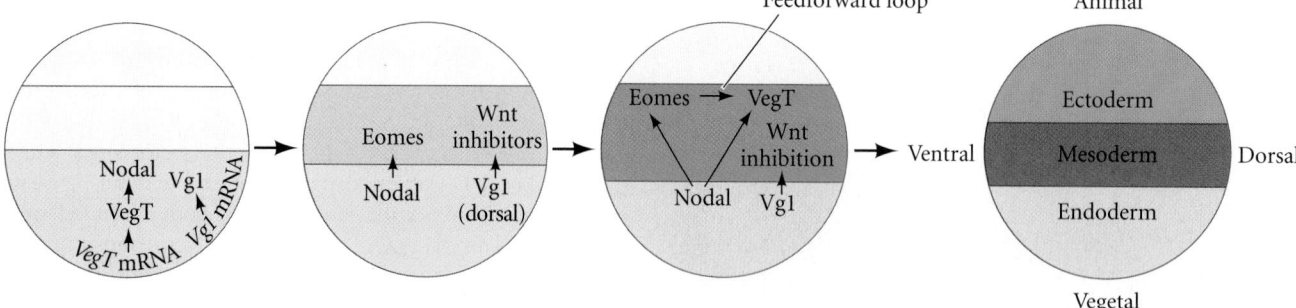

FIGURE 8.14 Model for the specification of the mesoderm. The vegetal region of the oocyte has accumulated mRNA for the transcription factor VegT and (in the future dorsal region) mRNA for the Nodal paracrine factor Vg1. At the late blastula stage, the *Vg1* mRNA is translated and Vg1 induces the future dorsal mesoderm to transcribe the genes for several Wnt antagonists (such as Dickkopf). The *VegT* message is also translated, and VegT activates nuclear genes encoding Nodal proteins. These TGF-β family members activate the expression of the transcription factor Eomesodermin in the presumptive mesoderm. Eomesodermin, with the help of activated Smad2 from the Nodal proteins, activates nuclear genes encoding VegT. In this way, VegT expression has gone from maternal mRNAs in the presumptive endoderm to nuclear expression in the presumptive mesoderm. (After Fukuda et al. 2010.)

anterior-posterior axis is inextricably linked to the formation of the dorsal-ventral axis. This, as we will see, is predicated on fertilization events that will place the transcription factor β-catenin in the region of the egg opposite the point of sperm entry and will specify that region of the egg to be the dorsal region of the embryo.

Once β-catenin is localized in this region of the egg, the cells containing β-catenin will induce expression of certain genes and thus initiate the movement of the involuting mesoderm. This movement of mesoderm will establish the anterior-posterior axis of the embryo. The first mesodermal cells to migrate over the dorsal blastopore lip will induce the ectoderm above them to produce anterior structures, such as the forebrain; mesoderm that involutes later will signal the ectoderm to form more posterior structures, such as the hindbrain and spinal cord. This process, whereby the central nervous system forms through interactions with the underlying mesoderm, has been called *primary embryonic induction* and is one of the principal ways that the vertebrate body becomes organized. Indeed, as we will now see, its discoverers called the dorsal blastopore lip and its descendants "the organizer," and found that this region is different from all the other parts of the embryo. In the early twentieth century, experiments by Hans Spemann and his students at the University of Freiburg, Germany, framed the questions that experimental embryologists would continue to ask for most of the rest of the century and resulted in a Nobel Prize for Spemann in 1935 (see Hamburger 1988; De Robertis and Aréchaga 2001; Sander and Fässler 2001). In recent times, the ongoing saga of discovery in identifying the molecules associated with these inductive processes has provided some of the most exciting moments in contemporary science.

The Work of Hans Spemann and Hilde Mangold

Autonomous specification versus inductive interactions

The experiment that began the Spemann laboratory's research program was performed in 1903, when Spemann demonstrated that early newt blastomeres have identical nuclei, each capable of producing an entire larva. His procedure was ingenious: Shortly after fertilizing a newt egg, Spemann used a baby's hair (taken from his infant daughter) to "lasso" the zygote in the plane of the first cleavage. He then partially constricted the egg, causing all the nuclear divisions to remain on one side of the constriction. Eventually—often as late as the 16-cell stage—a nucleus would escape across the constriction into the non-nucleated side. Cleavage then began on this side too, whereupon Spemann tightened the lasso until the two halves were completely separated. Twin larvae developed, one slightly more advanced than the other (**FIGURE 8.15**). Spemann concluded from this experiment that early amphibian nuclei were genetically identical and that each cell was capable of giving rise to an entire organism.

However, when Spemann performed a similar experiment with the constriction still longitudinal, but perpendicular to the plane of the first cleavage (i.e., separating the future dorsal and ventral regions rather than the right and left sides), he obtained a different result altogether. The nuclei continued to divide on both sides of the constriction, but only one side—the future dorsal side of the embryo—gave rise to a normal larva. The other side produced an unorganized tissue mass of ventral cells, which Spemann called the *Bauchstück*—the belly piece. This tissue mass was a ball of epidermal cells (ectoderm) containing blood and mesenchyme (mesoderm) and gut cells (endoderm), but it contained no dorsal structures such as nervous system, notochord, or somites.

Why should these two experiments give such different results? One possibility was that when the egg was divided perpendicular to the first cleavage plane, some *cytoplasmic* substance was not equally distributed into the two halves. Fortunately, the salamander egg was a good place to test that hypothesis. As we saw earlier in this chapter (see Figure 8.2), there are dramatic movements in the cytoplasm following the fertilization of amphibian eggs, and in some amphibians these movements expose a gray, crescent-shaped area of cytoplasm in the region directly opposite the point

FIGURE 8.15 Spemann's demonstration of nuclear equivalence in newt cleavage. (A) When the fertilized egg of the newt *Triturus taeniatus* was constricted by a ligature, the nucleus was restricted to one half of the embryo. The cleavage on that side of the embryo reached the 8-cell stage, while the other side remained undivided. (B) At the 16-cell stage, a single nucleus entered the as-yet undivided half, and the ligature was further constricted to complete the separation of the two halves. (C) After 14 days, each side had developed into a normal embryo. (After Spemann 1938.)

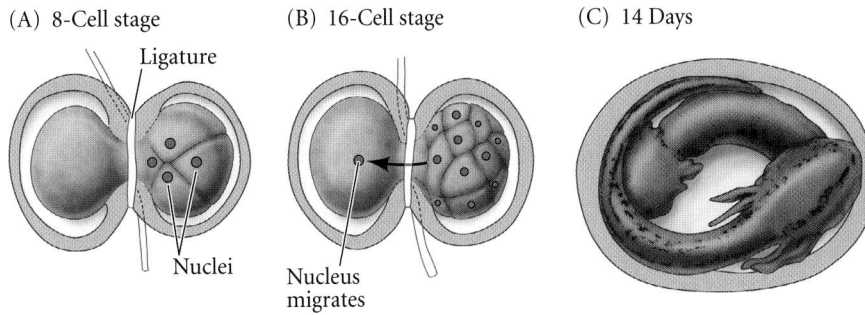

(A) 8-Cell stage   Ligature   Nuclei

(B) 16-Cell stage   Nucleus migrates

(C) 14 Days

of sperm entry. The first cleavage plane normally splits this gray crescent equally between the two blastomeres (see Figure 8.2D). If these cells are then separated, two complete larvae develop (**FIGURE 8.16A**). However, should this cleavage plane be aberrant (either in the rare natural event or in an experiment), the gray crescent material passes into only one of the two blastomeres. Spemann's work revealed that when two blastomeres are separated such that only one of the two cells contains the crescent, only the blastomere containing the gray crescent develops normally (**FIGURE 8.16B**).

It appeared, then, that *something in the region of the gray crescent was essential for proper embryonic development.* But how did it function? What role did it play in normal development? The most important clue came from fate maps, which showed that the gray crescent region gives rise to those cells that form the dorsal lip of the blastopore. These dorsal lip cells are committed to invaginate into the blastula, initiating gastrulation and the formation of the head endomesoderm and notochord. Because all future amphibian development depends on the interaction of cells that are rearranged during gastrulation, Spemann speculated that the importance of the gray crescent material lies in its ability to initiate gastrulation, and that crucial changes in cell potency occur during gastrulation. In 1918, he performed experiments that showed both statements to be true. He found that the cells of the *early* gastrula were uncommitted, but that the fates of *late* gastrula cells were determined.

Spemann's demonstration involved exchanging tissues between the gastrulae of two species of newts whose embryos were differently pigmented—the darkly pigmented *Triturus taeniatus* and the nonpigmented *T. cristatus*. When a region of prospective epidermal cells from an early gastrula of one species was transplanted into an area in an early gastrula of the other species and placed in a region where neural tissue normally formed, the transplanted cells gave rise to neural tissue. When prospective neural tissue from early gastrulae was transplanted to the region fated to become belly skin, the neural tissue became epidermal (**FIGURE 8.17A; TABLE 8.1**). Thus, cells of the *early* newt gastrula exhibit *conditional* (i.e., regulative, induction-dependent) development because their ultimate fate depends on their location in the embryo.

However, when the same interspecies transplantation experiments were performed on *late* gastrulae, Spemann obtained completely different results. Rather than differentiating in accordance with their new location, the transplanted cells exhibited *autonomous* (mosaic, independent)

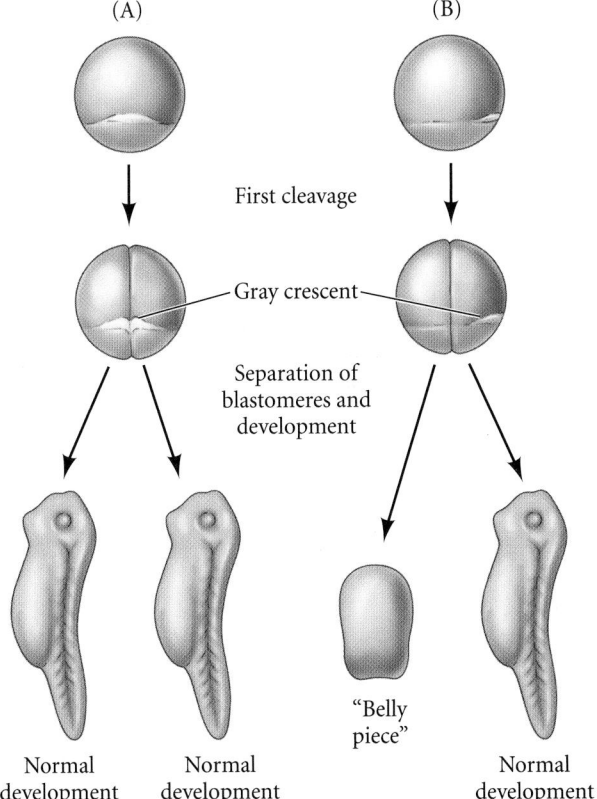

(A)   (B)

First cleavage

Gray crescent

Separation of blastomeres and development

Normal development   Normal development   "Belly piece"   Normal development

FIGURE 8.16 Asymmetry in the amphibian egg. (A) When the egg is divided along the plane of first cleavage into two blastomeres, each of which gets half of the gray crescent, each experimentally separated cell develops into a normal embryo. (B) When only one of the two blastomeres receives the entire gray crescent, it alone forms a normal embryo. The other blastomere produces a mass of unorganized tissue lacking dorsal structures. (After Spemann 1938.)

TABLE 8.1 Results of tissue transplantation during early- and late-stage newt gastrulae

Donor region	Host region	Differentiation of donor tissue	Conclusion
EARLY GASTRULA			
Prospective neurons	Prospective epidermis	Epidermis	Conditional development
Prospective epidermis	Prospective neurons	Neurons	Conditional development
LATE GASTRULA			
Prospective neurons	Prospective epidermis	Neurons	Autonomous development
Prospective epidermis	Prospective neurons	Epidermis	Autonomous development

development. Their prospective fate was *determined*, and the cells developed independently of their new embryonic location. Specifically, prospective neural cells now developed into brain tissue even when placed in the region of prospective epidermis (**FIGURE 8.17B**), and prospective epidermis formed

skin even in the region of the prospective neural tube. Within the time separating early and late gastrulation, the potencies of these groups of cells had become restricted to their eventual paths of differentiation. Something was causing them to become committed to epidermal and neural fates. What was happening?

Primary embryonic induction

The most spectacular transplantation experiments were published by Spemann and his doctoral student Hilde Mangold in 1924.* They showed that, of all the tissues in the early gastrula, only one has its fate autonomously determined. This self-determining tissue is the dorsal lip of the blastopore—the tissue derived from the gray crescent cytoplasm. When this tissue was transplanted into the presumptive belly skin region of another gastrula, it not only continued to be dorsal blastopore lip but also initiated gastrulation and embryogenesis in the surrounding tissue!

In these experiments, Spemann and Mangold once again used the differently pigmented embryos of *Triturus taeniatus* and *T. cristatus* so they could identify host and donor tissues on the basis of color. When the dorsal lip of an early *T. taeniatus* gastrula was removed and implanted into the region of an early *T. cristatus* gastrula fated to become ventral epidermis (belly skin), the dorsal lip tissue invaginated just as it would normally have done (showing self-determination) and disappeared beneath the vegetal cells (**FIGURE 8.18A**). The pigmented donor tissue then continued to self-differentiate into the chordamesoderm (notochord) and other mesodermal structures that normally form from the dorsal lip (**FIGURE 8.18B**). As the donor-derived mesodermal cells moved forward, host cells began to participate in the production of a new embryo, becoming organs that normally they never

(A) Transplantation in early gastrula

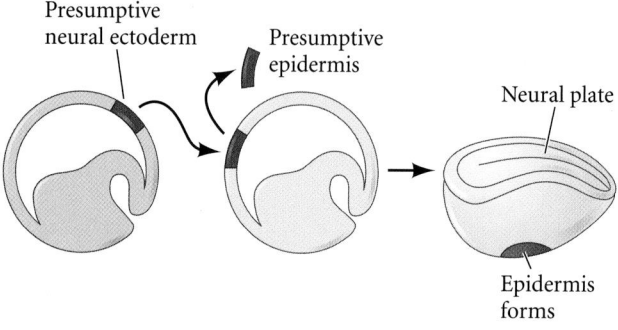

Presumptive neural ectoderm → Presumptive epidermis → Neural plate / Epidermis forms

(B) Transplantation in late gastrula

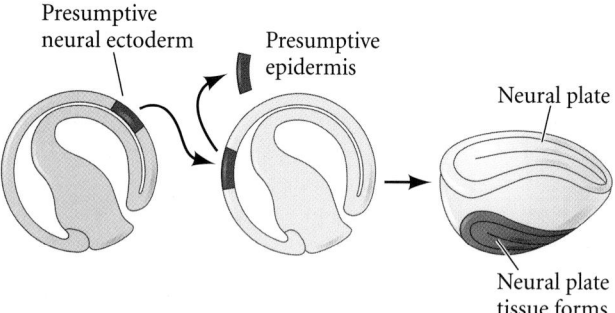

Presumptive neural ectoderm → Presumptive epidermis → Neural plate / Neural plate tissue forms

FIGURE 8.17 Determination of ectoderm during newt gastrulation. Presumptive neural ectoderm from one newt embryo is transplanted into a region in another embryo that normally becomes epidermis. (A) When the tissues are transferred between early gastrulae, the presumptive neural tissue develops into epidermis, and only one neural plate is seen. (B) When the same experiment is performed using late-gastrula tissues, the presumptive neural cells form neural tissue, thereby causing two neural plates to form on the host. (After Saxén and Toivonen 1962.)

*Hilde Proescholdt Mangold died in a tragic accident in 1924, when her kitchen's gasoline heater exploded. She was 26 years old, and her paper was just about to be published. Hers is one of the very few doctoral theses in biology that have directly resulted in the awarding of a Nobel Prize. For more information about Hilde Mangold, her times, and the experiments that identified the organizer, see Hamburger 1984, 1988, and Fässler and Sander 1996.

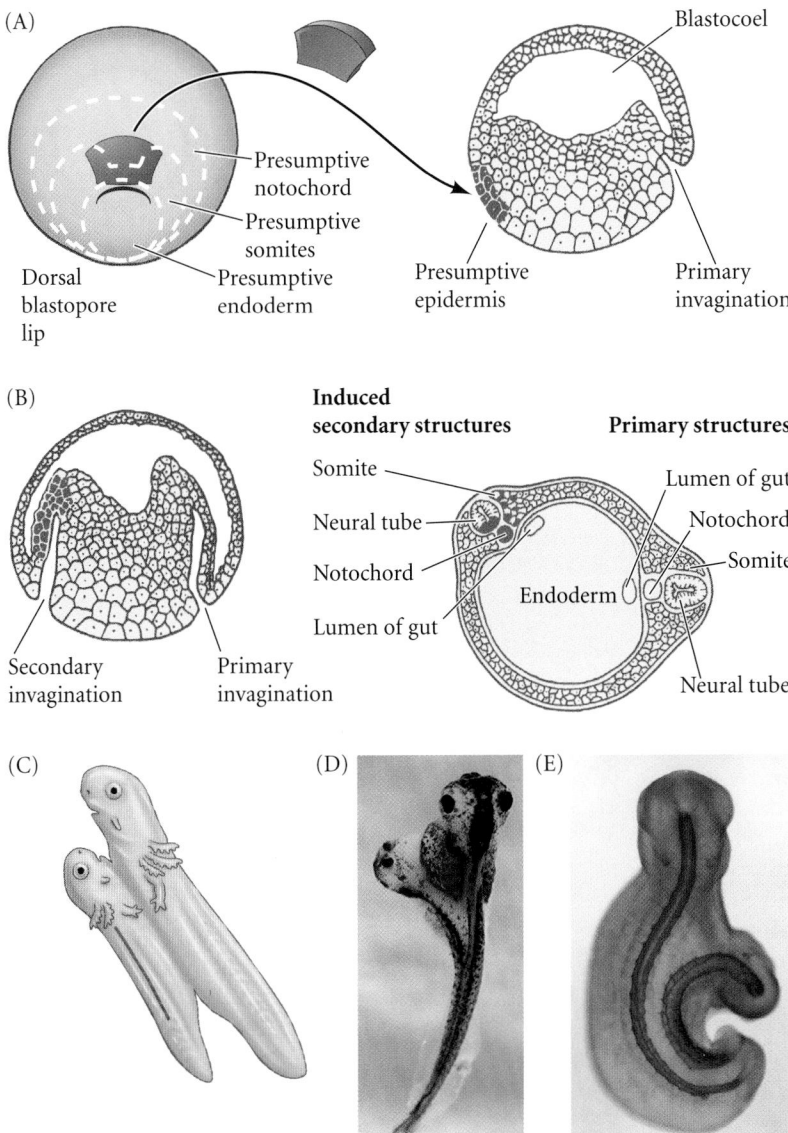

FIGURE 8.18 Organization of a secondary axis by dorsal blastopore lip tissue. (A–C) Spemann and Mangold's 1924 experiments visualized the process by using differently pigmented newt embryos. (A) Dorsal lip tissue from an early *T. taeniatus* gastrula is transplanted into a *T. cristatus* gastrula in the region that normally becomes ventral epidermis. (B) The donor tissue invaginates and forms a second archenteron, and then a second embryonic axis. Both donor and host tissues are seen in the new neural tube, notochord, and somites. (C) Eventually, a second embryo forms, joined to the host. (D) Live twinned *Xenopus* larvae generated by transplanting a dorsal blastopore lip into the ventral region of an early-gastrula host embryo. (E) Similar twinned larvae are seen from below and stained for notochord; the original and secondary notochords can be seen. (A–C after Hamburger 1988; D,E photographs by A. Wills, courtesy of R. Harland.)

would have formed. In this secondary embryo, a somite could be seen containing both pigmented (donor) and nonpigmented (host) tissue. Even more spectacularly, the dorsal lip cells were able to interact with the host tissues to form a complete neural plate from host ectoderm. Eventually, a secondary embryo formed, conjoined face to face with its host (**FIGURE 8.18C**). The results of these technically difficult experiments have been confirmed many times and in many amphibian species, including *Xenopus* (**FIGURE 8.18D,E**; Capuron 1968; Smith and Slack 1983; Recanzone and Harris 1985).

Spemann referred to the dorsal lip cells and their derivatives (notochord and head endomesoderm) as the **organizer** because (1) they induced the host's ventral tissues to change their fates to form a neural tube and dorsal mesodermal tissue (such as somites), and (2) they organized host and donor tissues into a secondary embryo with clear anterior-posterior and dorsal-ventral axes. He proposed that during normal development, these cells "organize" the dorsal ectoderm

into a neural tube and transform the flanking mesoderm into the anterior-posterior body axis (Spemann 1938). It is now known (thanks largely to Spemann and his students) that the interaction of the chordamesoderm and ectoderm is not sufficient to organize the entire embryo. Rather, it initiates a series of sequential inductive events. Because there are numerous inductions during embryonic development, this key induction—in which the progeny of dorsal lip cells induce the dorsal axis and the neural tube—is traditionally called the **primary embryonic induction**.*

● **See WEBSITE 8.3** Spemann, Mangold, and the organizer

---

*This classical term has been a source of confusion because the induction of the neural tube by the notochord is no longer considered the first inductive process in the embryo. We will soon discuss inductive events that precede this "primary" induction.

Molecular Mechanisms of Amphibian Axis Formation

The experiments of Spemann and Mangold showed that the dorsal lip of the blastopore, along with the dorsal mesoderm and pharyngeal endoderm that form from it, constituted an "organizer" able to instruct the formation of embryonic axes. But the mechanisms by which the organizer itself was constructed and through which it operated remained a mystery. Indeed, it is said that Spemann and Mangold's landmark paper posed more questions than it answered. Among those questions were:

- How did the organizer get its properties? What caused the dorsal blastopore lip to differ from any other region of the embryo?

- What factors were being secreted from the organizer to cause the formation of the neural tube and to create the anterior-posterior, dorsal-ventral, and left-right axes?

- How did the different parts of the neural tube become established, with the most anterior becoming the sensory organs and forebrain and the most posterior becoming spinal cord?

Spemann and Mangold's description of the organizer was the starting point of one of the first truly international scientific research programs: the search for the organizer molecules (see Gilbert and Saxén 1993; Armon 2012). Researchers from Britain, Germany, France, the United States, Belgium, Finland, Japan, and the Soviet Union all tried to find these remarkable substances. R. G. Harrison referred to the amphibian gastrula as the "new Yukon to which eager miners were now rushing to dig for gold around the blastopore" (see Twitty 1966, p. 39). Unfortunately, their early picks and shovels proved too blunt to uncover the molecules involved. The proteins responsible for induction were present in concentrations too small for biochemical analyses, and the large quantity of yolk and lipids in the amphibian egg further interfered with protein purification (Grunz 1997). The analysis of organizer molecules had to wait until recombinant DNA technologies enabled investigators to make cDNA clones from blastopore lip mRNA and to see which of these clones encoded factors that could dorsalize the embryo. We are now able to take up each of the above four questions in turn.

How does the organizer form?

Why are the dozen or so initial cells of the organizer positioned opposite the point of sperm entry, and what determines their fate so early? Recent evidence provides an unexpected answer: these cells are in the right place at the right time, at a point where two signals converge. The first signal tells the cells that they are dorsal. The second signal says that these cells are mesoderm. These signals interact to create a polarity within the mesoderm that is the basis for specifying the organizer and for creating dorsal-ventral polarity.

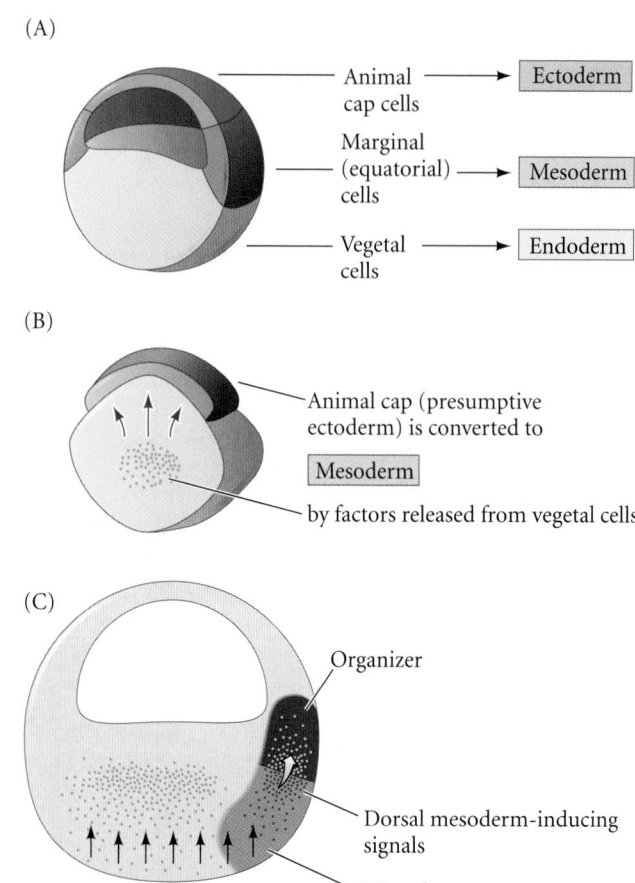

FIGURE 8.19 Summary of experiments by Nieuwkoop and by Nakamura and Takasaki, showing mesodermal induction by vegetal endoderm. (A) Isolated animal cap cells become a mass of ciliated ectoderm, isolated equatorial (marginal zone) cells become mesoderm, and isolated vegetal cells generate gutlike tissue. (B) If animal cap cells are combined with vegetal cap cells, many of the animal cells generate mesodermal tissue. (C) Simplified model for mesoderm induction in *Xenopus*. A ventral signal (probably a complex set of signals from activin-like TGF-β factors and FGFs) is released throughout the vegetal region of the embryo. This signal induces the marginal cells to become mesoderm. On the dorsal side (away from the point of sperm entry), a signal is released by the vegetal cells of the Nieuwkoop center. This dorsal signal induces the formation of the Spemann organizer in the overlying marginal zone cells. The possible identity of this signal will be discussed later in this chapter. (C after De Robertis et al. 1992.)

THE DORSAL SIGNAL: β-CATENIN It turns out that one of the reasons the organizer cells are special is that these mesodermal cells reside above a special group of vegetal cells. One of the major clues in determining how the dorsal blastopore lip obtains its properties came from the experiments of Pieter Nieuwkoop (1969, 1973, 1977) and Osamu Nakamura. These studies showed that the organizer receives its special properties from signals coming from the prospective endoderm beneath it.

Nakamura and Takasaki (1970) showed that the mesoderm arises from the marginal (equatorial) cells at the border between the animal and vegetal poles. The Nakamura and Nieuwkoop laboratories then demonstrated that the properties of this newly formed mesoderm were induced by the vegetal (presumptive endoderm) cells underlying them. Nieuwkoop removed the equatorial cells (i.e., presumptive mesoderm) from a blastula and showed that neither the animal cap (presumptive ectoderm) nor the vegetal cap (presumptive endoderm) produced any mesodermal tissue. However, when the two caps were recombined, the animal cap cells were induced to form mesodermal structures such as notochord, muscles, kidney cells, and blood cells (**FIGURE 8.19**). The polarity of this induction (i.e., whether the animal cells formed dorsal mesoderm or ventral mesoderm) depended on whether the endodermal (vegetal) fragment was taken from the dorsal or the ventral side: ventral and lateral vegetal cells (those closer to the site of sperm entry) induced ventral (mesenchyme, blood)

and intermediate (kidney) mesoderm, while the dorsalmost vegetal cells specified dorsal mesoderm components (somites, notochord)—including those having the properties of the organizer. These dorsalmost vegetal cells of the blastula, which are capable of inducing the organizer, have been called the **Nieuwkoop center** (Gerhart et al. 1989).

The Nieuwkoop center was demonstrated in the *Xenopus* embryo by transplantation and recombination experiments. First, Gimlich and Gerhart (Gimlich and Gerhart 1984; Gimlich 1985, 1986) performed an experiment analogous to the Spemann and Mangold studies, except that they used early *Xenopus* blastulae rather than newt gastrulae. When they transplanted the dorsalmost vegetal blastomere from one blastula into the ventral vegetal side of another blastula, two embryonic axes formed (**FIGURE 8.20A**). Second, Dale and Slack (1987) recombined single vegetal blastomeres from a 32-cell *Xenopus* embryo with the uppermost animal tier of a fluorescently labeled embryo of the same stage. The

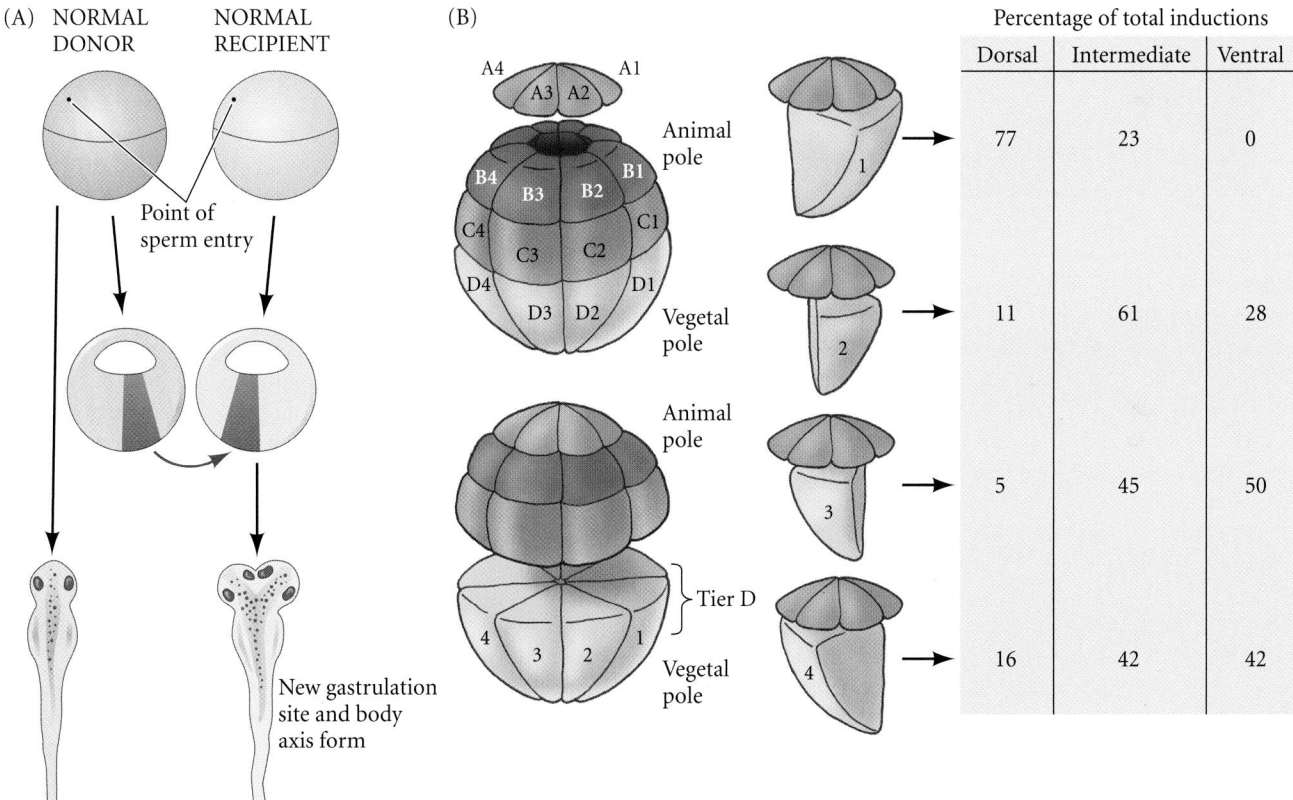

FIGURE 8.20   Transplantation and recombination experiments on *Xenopus* embryos demonstrate that the vegetal cells underlying the prospective dorsal blastopore lip region are responsible for initiating gastrulation. (A) Formation of a new gastrulation site and body axis by the transplantation of the dorsalmost vegetal cells of a 64-cell embryo into the ventralmost vegetal region of another embryo. (B) The regional specificity of mesoderm induction demonstrated by recombining blastomeres of 32-cell *Xenopus* embryos. Animal pole cells were labeled with fluorescent polymers so their descendants could be identified, then combined with individual vegetal blastomeres. The inductions resulting from these recombinations are summarized at the right. D1, the dorsalmost vegetal blastomere, was the most likely to induce the animal pole cells to form dorsal mesoderm. These dorsalmost vegetal cells constitute the Nieuwkoop center. (A after Gimlich and Gerhart 1984; B after Dale and Slack 1987.)

(A)

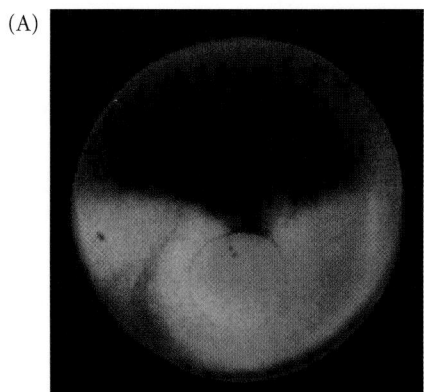

(B)

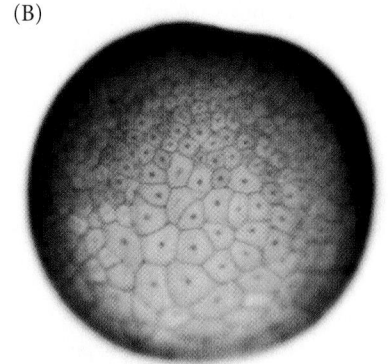

(C)

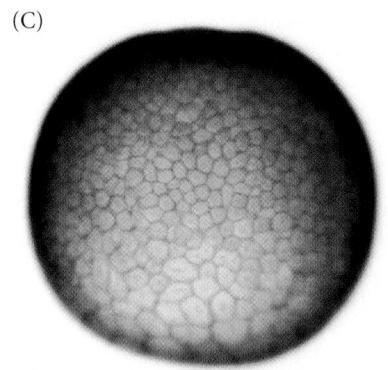

(D)

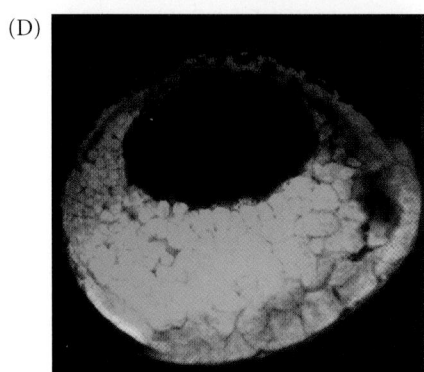

FIGURE 8.21 Role of Wnt pathway proteins in dorsal-ventral axis specification. (A–D) Differential translocation of β-catenin into *Xenopus* blastomere nuclei. (A) Early 2-cell stage, showing β-catenin (orange) predominantly at the dorsal surface. (B) Presumptive dorsal side of a blastula stained for β-catenin shows nuclear localization. (C) Such nuclear localization is not seen on the ventral side of the same embryo. (D) Dorsal localization of β-catenin persists through the gastrula stage. (A,D courtesy of R. T. Moon; B,C from Schneider et al. 1996, courtesy of P. Hausen.)

dorsalmost vegetal cell, as expected, induced the animal pole cells to become dorsal mesoderm. The remaining vegetal cells usually induced the animal cells to produce either intermediate or ventral mesodermal tissues (**FIGURE 8.20B**). Holowacz and Elinson (1993) found that cortical cytoplasm from the dorsal vegetal cells of the 16-cell *Xenopus* embryo was able to induce the formation of secondary axes when injected into ventral vegetal cells. Thus, dorsal vegetal cells can induce animal cells to become dorsal mesodermal tissue.

So one important question became, What gives the dorsalmost vegetal cells their special properties? The major candidate for the factor that forms the Nieuwkoop center in these vegetal cells is **β-catenin**. We have encountered this multifunctional protein several times already. It can act as an anchor for cell membrane cadherins (see Chapter 3) or as a nuclear transcription factor (induced by the Wnt pathway). As we saw in Chapter 7, β-catenin is responsible for specifying the micromeres of the sea urchin embryo. β-catenin is a key player in the formation of the dorsal amphibian tissues, and experimental depletion of this molecule results in the lack of dorsal structures (Heasman et al. 1994a). Moreover, injection of exogenous β-catenin into the ventral side of an embryo produces a secondary axis (Funayama et al. 1995; Guger and Gumbiner 1995).

In *Xenopus* embryos, β-catenin is initially synthesized throughout the embryo from maternal mRNA (Yost et al. 1996; Larabell et al. 1997). It begins to accumulate in the dorsal region of the egg during the cytoplasmic movements of

fertilization and continues to accumulate preferentially at the dorsal side throughout early cleavage. This accumulation is seen in the nuclei of the dorsal cells and appears to cover both the Nieuwkoop center and organizer regions (**FIGURE 8.21**; Schneider et al. 1996; Larabell et al. 1997).

If β-catenin is originally found throughout the embryo, how does it become localized specifically to the side opposite sperm entry? The answer appears to reside in the localizations of proteins in the egg cortical cytoplasm. Three proteins—Wnt11, GSK3-binding protein (GBP), and Disheveled (Dsh)—are translocated from the vegetal pole of the egg to the future dorsal side of the embryo during fertilization. From research done on the Wnt pathway, we have learned that β-catenin is targeted for destruction by glycogen synthase kinase 3 (GSK3; see Chapter 3). Indeed, activated GSK3 destroys β-catenin and blocks axis formation when added to the egg, and if endogenous GSK3 is knocked out by a dominant negative form of GSK3 in the ventral cells of the early embryo, a second axis forms (see Figure 8.22F; He et al. 1995; Pierce and Kimelman 1995; Yost et al. 1996).

GSK3 can be inactivated by GBP and Disheveled. These two proteins release GSK3 from the degradation complex and prevent it from binding β-catenin and targeting it for destruction. During the first cell cycle, when the microtubules form parallel tracts in the vegetal portion of the egg, GBP travels along the microtubules by binding to kinesin, an ATPase motor protein that travels on microtubules. Kinesin always migrates toward the growing end of the microtubules, and in this case, that means moving to the point opposite sperm entry, i.e., the future dorsal side (**FIGURE 8.22A–C**). Disheveled, which is originally found in the vegetal pole cortex, grabs onto the GPB, and it too becomes translocated along the microtubular monorail (Miller et al. 1999; Weaver et al. 2003). The cortical rotation is probably important in orienting

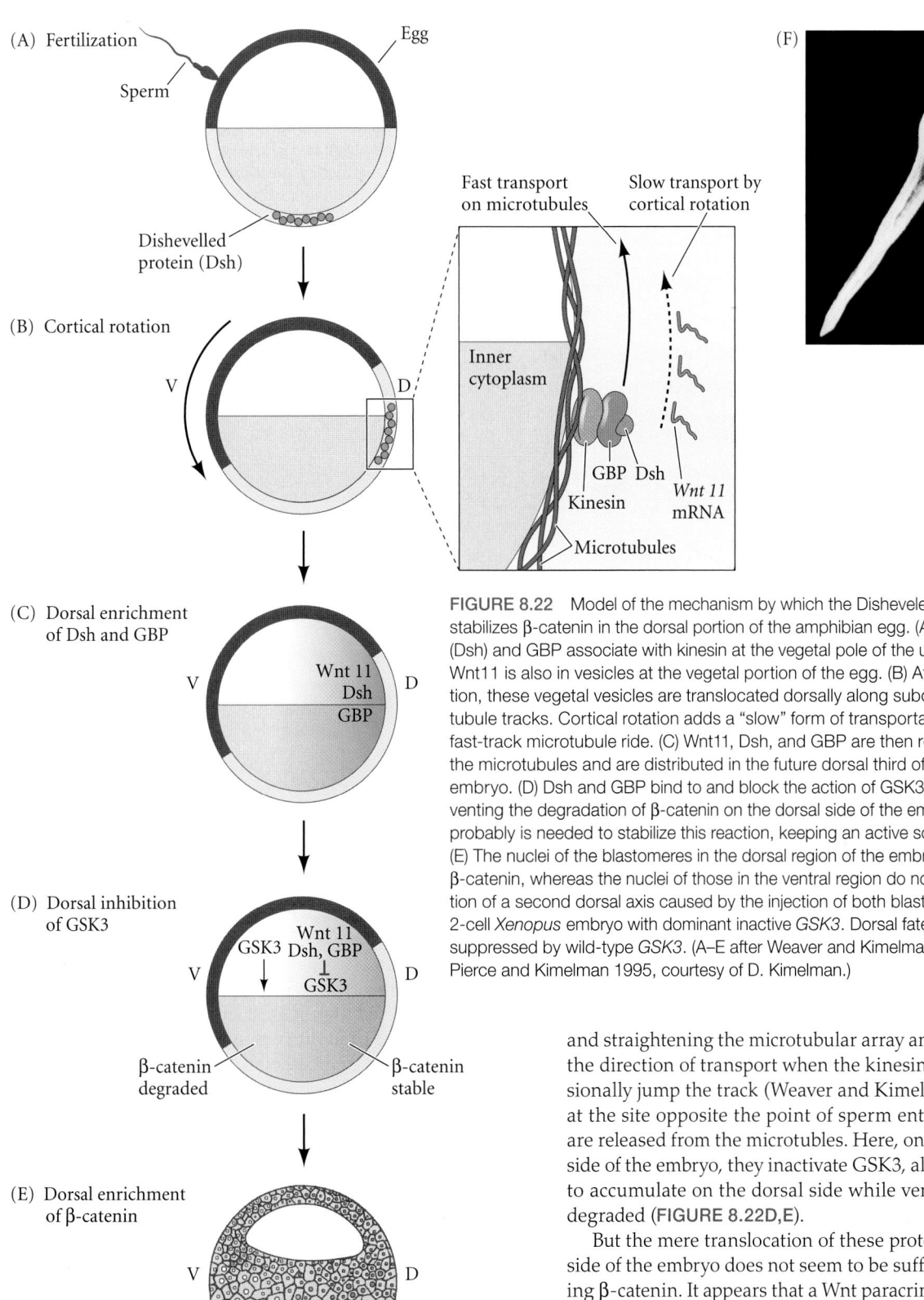

(A) Fertilization

Sperm

Egg

Dishevelled protein (Dsh)

(B) Cortical rotation

V

D

Fast transport on microtubules

Slow transport by cortical rotation

Inner cytoplasm

GBP Dsh

Kinesin

Wnt 11 mRNA

Microtubules

(C) Dorsal enrichment of Dsh and GBP

V

Wnt 11
Dsh
GBP

D

(D) Dorsal inhibition of GSK3

V

Wnt 11
GSK3 Dsh, GBP
⊥
GSK3

D

β-catenin degraded

β-catenin stable

(E) Dorsal enrichment of β-catenin

V

D

No β-catenin in ventral nuclei

β-catenin in dorsal nuclei

(F)

FIGURE 8.22 Model of the mechanism by which the Disheveled protein stabilizes β-catenin in the dorsal portion of the amphibian egg. (A) Disheveled (Dsh) and GBP associate with kinesin at the vegetal pole of the unfertilized egg. Wnt11 is also in vesicles at the vegetal portion of the egg. (B) After fertilization, these vegetal vesicles are translocated dorsally along subcortical microtubule tracks. Cortical rotation adds a "slow" form of transportation to the fast-track microtubule ride. (C) Wnt11, Dsh, and GBP are then released from the microtubules and are distributed in the future dorsal third of the 1-cell embryo. (D) Dsh and GBP bind to and block the action of GSK3, thereby preventing the degradation of β-catenin on the dorsal side of the embryo. Wnt11 probably is needed to stabilize this reaction, keeping an active source of Dsh. (E) The nuclei of the blastomeres in the dorsal region of the embryo receive β-catenin, whereas the nuclei of those in the ventral region do not. (F) Formation of a second dorsal axis caused by the injection of both blastomeres of a 2-cell *Xenopus* embryo with dominant inactive *GSK3*. Dorsal fate is actively suppressed by wild-type *GSK3*. (A–E after Weaver and Kimelman 2004; F from Pierce and Kimelman 1995, courtesy of D. Kimelman.)

and straightening the microtubular array and in maintaining the direction of transport when the kinesin complexes occasionally jump the track (Weaver and Kimelman 2004). Once at the site opposite the point of sperm entry, GBP and Dsh are released from the microtubules. Here, on the future dorsal side of the embryo, they inactivate GSK3, allowing β-catenin to accumulate on the dorsal side while ventral β-catenin is degraded (**FIGURE 8.22D,E**).

But the mere translocation of these proteins to the dorsal side of the embryo does not seem to be sufficient for protecting β-catenin. It appears that a Wnt paracrine factor has to be secreted there to activate the β-catenin protection pathway; this is accomplished by Wnt11 (see Figure 8.22). If Wnt11 synthesis is suppressed (by the injection of antisense Wnt11 oligonucleotides into the oocytes), the organizer fails to form. Furthermore, *Wnt11* mRNA is localized to the vegetal cortex

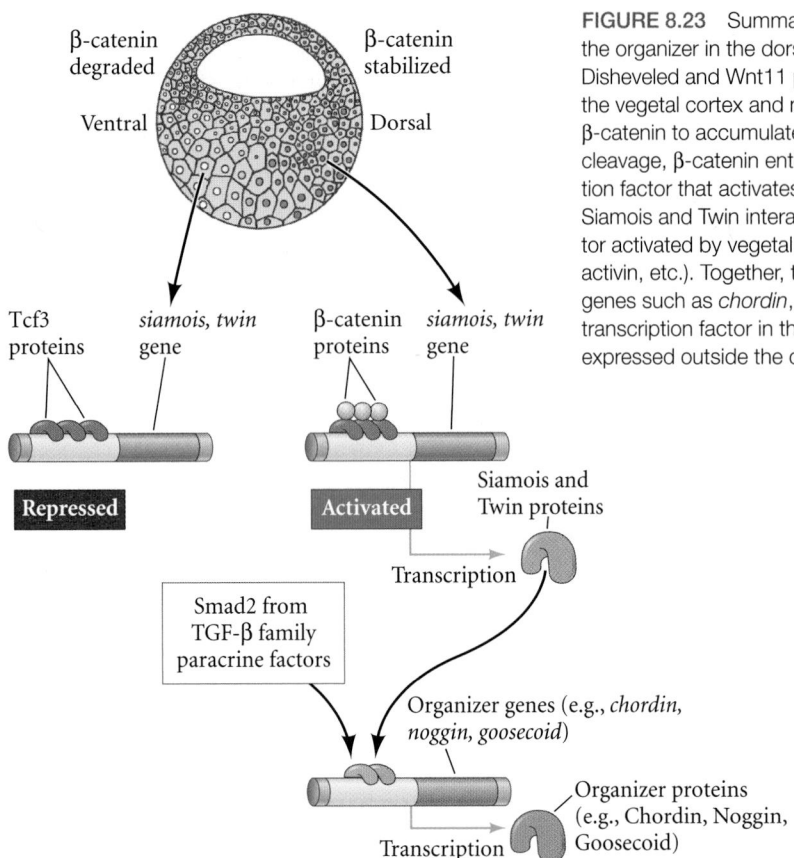

β-catenin
degraded

β-catenin
stabilized

Ventral

Dorsal

Tcf3
proteins

siamois, twin
gene

β-catenin
proteins

siamois, twin
gene

Repressed

Activated

Siamois and
Twin proteins

Transcription

Smad2 from
TGF-β family
paracrine factors

Organizer genes (e.g., *chordin,
noggin, goosecoid*)

Organizer proteins
(e.g., Chordin, Noggin,
Goosecoid)

Transcription

FIGURE 8.23 Summary of events hypothesized to bring about induction of the organizer in the dorsal mesoderm. Microtubules allow the translocation of Disheveled and Wnt11 proteins to the dorsal side of the embryo. Dsh (from the vegetal cortex and newly made by Wnt11) binds GSK3, thereby allowing β-catenin to accumulate in the future dorsal portion of the embryo. During cleavage, β-catenin enters the nuclei and binds with Tcf3 to form a transcription factor that activates genes encoding proteins such as Siamois and Twin. Siamois and Twin interact in the organizer with the Smad2 transcription factor activated by vegetal TGF-β family members (Nodal-related proteins, Vg1, activin, etc.). Together, these three transcription factors activate the "organizer" genes such as *chordin, noggin,* and *goosecoid*. The presence of the VegT transcription factor in the endoderm prevents the organizer genes from being expressed outside the organizer area. (After Moon and Kimelman 1998.)

during oogenesis and is translocated to the future dorsal portion of the embryo by the cortical rotation of the egg cytoplasm (Tao et al. 2005; Cuykendall and Houston 2009). Here it is translated into a protein that becomes concentrated in and secreted on the dorsal side of the embryo (Ku and Melton 1993; Schroeder et al. 1999; White and Heasman 2008).

Thus, during first cleavage, GBP, Dsh, and Wnt11 are brought into the future dorsal section of the embryo where GBP and Dsh can *initiate* the inactivation of GSK3 and the consequent protection of β-catenin. The signal from Wnt11 amplifies the signal and *stabilizes* GBP and Dsh and organizes them to protect β-catenin; β-catenin can associate with other transcription factors, giving these factors new properties. It is known, for example, that *Xenopus* β-catenin can combine with a ubiquitous transcription factor known as Tcf3, converting the Tcf3 repressor into an activator of transcription. Expression of a mutant form of Tcf3 that lacks the β-catenin binding domain results in embryos without dorsal structures (Molenaar et al. 1996).

The β-catenin/Tcf3 complex binds to the promoters of several genes whose activity is critical for axis formation. Two of these genes, *twin* and *siamois*, encode homeodomain transcription factors and are expressed in the organizer region immediately following the mid-blastula transition. If these genes are ectopically expressed in the ventral cells, a secondary axis emerges on the former ventral side of the embryo; and if cortical microtubular polymerization is prevented,

siamois expression is eliminated (Lemaire et al. 1995; Brannon and Kimelman 1996). The Tcf3 protein is thought to inhibit *siamois* and *twin* transcription when it binds to those genes' promoters in the absence of β-catenin. However, when β-catenin binds to Tcf3, the repressor is converted into an activator, and *twin* and *siamois* are activated (**FIGURE 8.23**).

Siamois and Twin proteins bind to the enhancers of several genes involved in organizer function (Fan and Sokol 1997; Bae et al 2011). These include genes encoding the transcription factors Goosecoid and Xlim1 (which are critical in specifying the dorsal mesoderm) and the paracrine factor antagonists Noggin, Chordin, Frzb, and Cerberus (which specify the ectoderm to become neural; Laurent et al. 1997; Engleka and Kessler 2001). In the vegetal cells, Siamois and Twin appear to combine with vegetal transcription factors to help activate endodermal genes (Lemaire et al. 1998). Thus, one could expect that if the dorsal side of the embryo contained β catenin, β catenin would allow this region to express Twin and Siamois, which in turn would initiate formation of the organizer.

THE VEGETAL NODAL-RELATED SIGNAL Yet another factor appears to be critical in activating the genes that characterize the organizer cells. This other factor is the phosphorylated Smad2 transcription factor (discussed earlier), which is essential in forming the mesoderm. Smad2 is activated in the mesodermal cells when it becomes phosphorylated by

Nodal-related paracrine factors secreted by the vegetal cells beneath the mesoderm (Brannon and Kimelman 1996; Engle-ka and Kessler 2001). Activated Smad2 usually binds with a partner to form a complex that acts as a transcription factor.

At the late blastula stages, there is a *gradient* of Nodal-related proteins across the endoderm, with low concentrations ventrally and high concentrations dorsally (Onuma et al. 2002; Rex et al. 2002; Chea et al. 2005). Because Vg1 and the Nodal-related proteins act through the same pathway (i.e., by activating the Smad2 transcription factor), we would expect them to produce an additive signal (Agius et al. 2000). Indeed, this appears to be the case.

The Nodal-related gradient is produced in large part by β-catenin. Higher levels of β-catenin activate more Nodal-related genes than do low concentrations (**FIGURE 8.24**). In the most dorsal (Nieuwkoop center) blastomeres, β-catenin cooperates with the VegT transcription factor to activate the *Xenopus nodal-related 1, 5,* and *6* (*Xnr1, 5,* and *6*) genes even before the mid-blastula transition. The more ventral blastomeres in the endoderm lack the expression of these Nodal-related genes. In the region that will become the most anterior portion of the organizer—the pharyngeal endoderm—higher levels of Nodal-related proteins produce higher concentrations of activated Smad2. Smad2 can bind to the promoter of the *hhex* gene, and in concert with Twin

and Siamois (induced by β-catenin), Hhex activates genes that specify pharyngeal endoderm cells to become foregut endoderm and to induce anterior brain development (Smithers and Jones 2002; Rankin et al. 2011). Slightly lower levels of Smad2 are believed to activate *goosecoid* expression in the cells that will become the prechordal mesoderm and noto-chord. Even lower amounts of Smad2 result in the formation of lateral and ventral mesoderm.

In summary, then, the formation of the dorsal mesoderm and the organizer originates through the activation of critical transcription factors by intersecting pathways. The first pathway is the Wnt/β-catenin pathway that activates genes encoding the Siamois and Twin transcription factors. The second pathway is the vegetal pathway that activates the expression of Nodal-related paracrine factors, which in turn activate the Smad2 transcription factor in the mesodermal cells above them. The high levels of Smad2 and Siamois/Twin transcription factor proteins work within the dorsal mesoderm cells and activate the genes that give these cells their "organizer" properties (Germain et al. 2000; Cho 2012; review Figures 8.22–8.24).

Functions of the organizer

While the Nieuwkoop center cells remain endodermal, the cells of the organizer become the dorsal mesoderm and

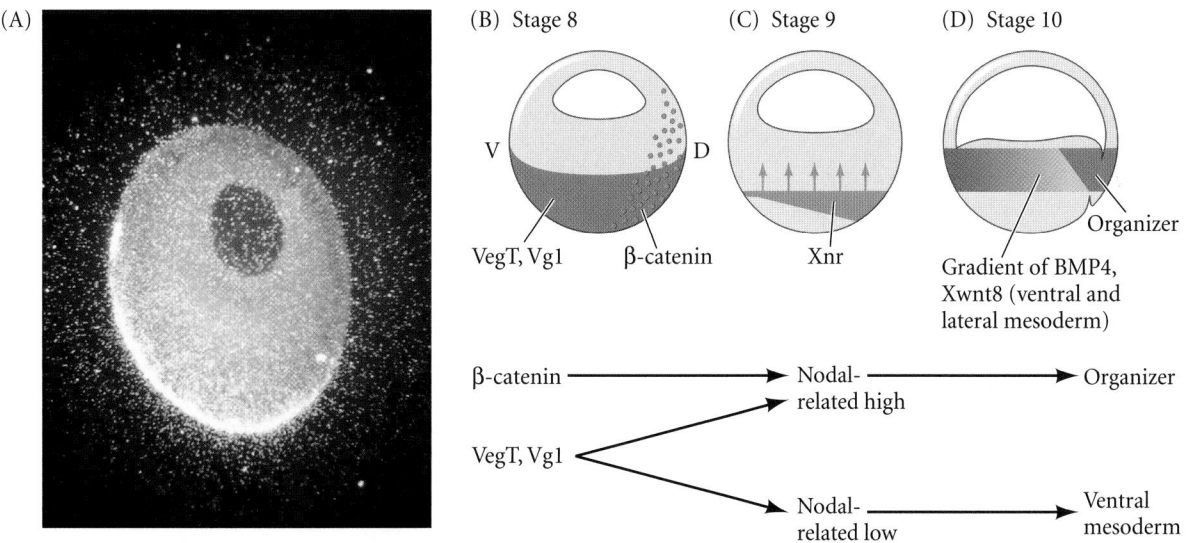

FIGURE 8.24 Vegetal induction of mesoderm. (A) The maternal RNA encoding Vg1 (bright white crescent) is tethered to the vegetal cortex of a *Xenopus* oocyte. The message (along with the maternal VegT message) will be translated at fertilization. Both proteins appear to be crucial for the ability of vegetal cells to induce cells above them to become mesoderm. (B–D) Model for mesoderm induction and organizer formation by the interaction of β-catenin and TGF-β proteins. (B) At late blastula stages, Vg1 and VegT are found in the vegetal hemisphere; β-catenin is located in the dorsal region. (C) β-catenin acts synergistically with Vg1 and VegT to activate the *Xenopus nodal-related* (*Xnr*) genes. This creates a gradient of Xnr proteins across the endoderm, highest in the dorsal region. (D) The mesoderm is specified by the Xnr gradient. Mesodermal regions with little or no Xnr have high levels of BMP4 and Xwnt8; they become ventral mesoderm. Those having intermediate concentrations of Xnr become lateral mesoderm. Where there is a high concentration of Xnr, *goosecoid* and other dorsal mesodermal genes are activated and the mesodermal tissue becomes the organizer. (A courtesy of D. Melton; B–D after Agius et al. 2000.)

migrate underneath the dorsal ectoderm. It is now thought that the cells of the organizer ultimately contribute to four cell types: pharyngeal endoderm, head mesoderm (prechordal plate), dorsal mesoderm (primarily the notochord), and the dorsal blastopore lip (Keller 1976; Gont et al. 1993). The pharyngeal endoderm and prechordal plate lead the migration of the organizer tissue and induce the forebrain and midbrain. The dorsal mesoderm induces the hindbrain and trunk. The dorsal blastopore lip remaining at the end of gastrulation eventually becomes the chordaneural hinge that induces the tip of the tail.

The properties of the organizer tissue can be divided into four major functions:

1. The ability to self-differentiate into dorsal mesoderm (prechordal plate, chordamesoderm, etc.)

2. The ability to dorsalize the surrounding mesoderm into paraxial (somite-forming) mesoderm when it otherwise would form ventral mesoderm

3. The ability to dorsalize the ectoderm and induce formation of the neural tube

4. The ability to initiate the movements of gastrulation

As we have just seen, the Smad2 and β-catenin transcription factors cooperate to activate several genes. Many of these genes encode secreted proteins that will act to organize the embryo (**TABLE 8.2**).

● **See WEBSITE 8.4** Early attempts to locate the organizer molecules

Induction of neural ectoderm and dorsal mesoderm: BMP inhibitors

Evidence from experimental embryology showed that one of the most critical properties of the organizer was its production of soluble factors. The evidence for such diffusible signals from the organizer came from several sources. First, Hans Holtfreter (1933) showed that if the notochord fails to migrate beneath the ectoderm, the ectoderm will not become neural tissue (and will become epidermis). More definitive evidence for the importance of soluble factors came later from the transfilter studies of Finnish investigators (Saxén 1961; Toivonen et al. 1975; Toivonen and Wartiovaara 1976). Here, newt dorsal lip tissue was placed on one side of a filter fine enough so that no processes could fit through the pores, and competent gastrula ectoderm was placed on the other side. After several hours, neural structures were observed in the ectodermal tissue (**FIGURE 8.25**). The identities of the factors diffusing from the organizer, however, took another quarter of a century to find.

It turned out that scientists were looking for the wrong mechanism. They were searching for a molecule secreted by the organizer and received by the ectoderm that would then convert the ectoderm into neural tissue. However, molecular studies led to a remarkable and non-obvious conclusion: *it is the epidermis that is induced to form, not the neural tissue.* The

TABLE 8.2 Proteins expressed solely or almost exclusively in the organizer (partial list)

Nuclear proteins	Secreted proteins
Twin	Chordin
Siamois	Dickkopf
Xlim1	ADMP
Xnot	Frzb
Otx2	Noggin
XFD1	Follistatin
XANF1	Sonic hedgehog
Goosecoid	Cerberus
HNF3β	Nodal-related proteins (several)
Hhex	ILGF

ectoderm is induced to become epidermal tissue by binding **bone morphogenetic proteins** (**BMPs**), whereas the nervous system forms from that region of the ectoderm that is *protected* from epidermal induction by BMP-inhibiting molecules (Hemmati-Brivanlou and Melton 1994, 1997). In other words, (1) the "default fate" of the ectoderm is to become neural tissue; (2) certain parts of the embryo induce the ectoderm to become epidermal tissue by secreting BMPs; and (3) the organizer tissue acts by secreting molecules that block BMPs, thereby allowing the ectoderm "protected" by these BMP inhibitors to become neural tissue.

Three of the major BMP inhibitors secreted by the organizer are Noggin, Chordin, and Follistatin. The genes encoding these proteins are some of the most critical genes activated by Smad2 and Siamois/Twin (Carnac et al. 1996; Fan and Sokol 1997; Kessler 1997). In addition, a fourth BMP inhibitor,

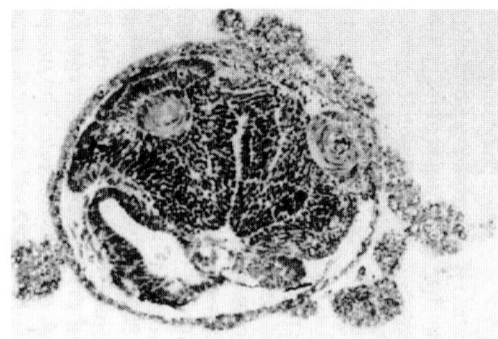

FIGURE 8.25 Neural structures induced in presumptive ectoderm by newt dorsal lip tissue, separated from the ectoderm by a nucleopore filter with an average pore diameter of 0.05 mm. Anterior neural tissues are evident, including some induced eyes. (From Toivonen 1979, courtesy of L. Saxén.)

(A)

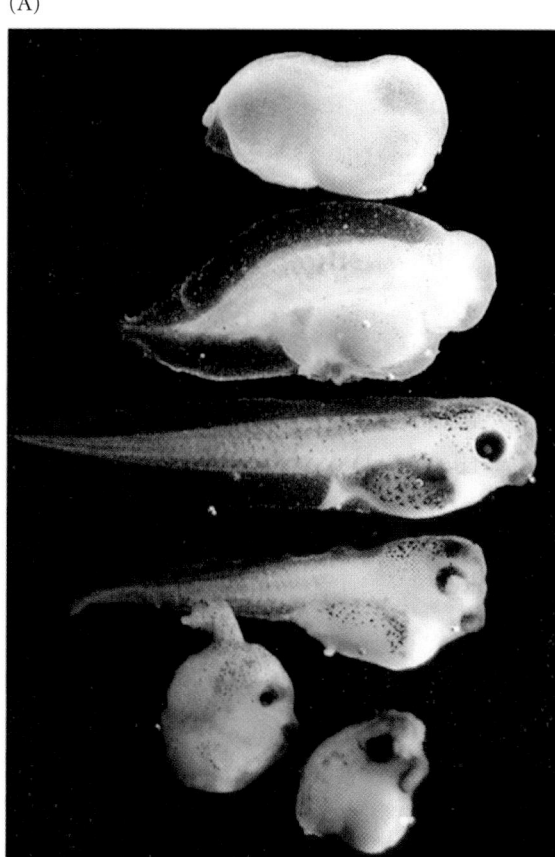

(B)

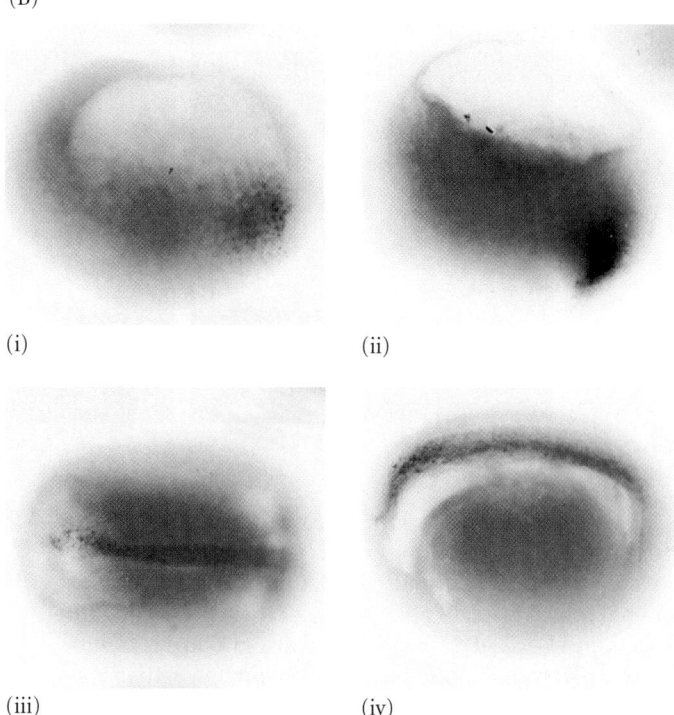

(i)　　　　　(ii)

(iii)　　　　　(iv)

FIGURE 8.26 The soluble protein Noggin dorsalizes the amphibian embryo. (A) Rescue of dorsal structures by Noggin protein. When *Xenopus* eggs are exposed to ultraviolet radiation, cortical rotation fails to occur, and the embryos lack dorsal structures (top). If such an embryo is injected with *noggin* mRNA, it develops dorsal structures in a dosage-related fashion (top to bottom). If too much *noggin* message is injected, the embryo produces dorsal anterior tissue at the expense of ventral and posterior tissue, becoming little more than a head (bottom). (B) Localization of *noggin* mRNA in the organizer tissue, shown by in situ hybridization. At gastrulation (i), *noggin* mRNA (dark areas) accumulates in the dorsal marginal zone. When cells involute (ii), *noggin* mRNA is seen in the dorsal blastopore lip. During convergent extension (iii), *noggin* is expressed in the precursors of the notochord, prechordal plate, and pharyngeal endoderm, which (iv) extend beneath the ectoderm in the center of the embryo. (Courtesy of R. M. Harland.)

Norrin, appears to be stored in the animal pole of the oocyte and functions to block BMPs in the dorsal ectoderm.

NOGGIN In 1992, Smith and Harland constructed a cDNA plasmid library from dorsalized (lithium chloride-treated) gastrulae. Messenger RNAs synthesized from sets of these plasmids were injected into ventralized embryos (having no neural tube) produced by irradiating early embryos with ultraviolet light. Those plasmid sets whose mRNAs rescued dorsal structures in these embryos were split into smaller sets, and so on, until single-plasmid clones were isolated whose mRNAs were able to restore the dorsal tissue in such embryos. One of these clones contained the gene for the protein Noggin (**FIGURE 8.26A**). Injection of *noggin* mRNA into 1-cell, UV-irradiated embryos completely rescued dorsal development and allowed the formation of a complete embryo.

Noggin is a secreted protein that is able to accomplish two of the major functions of the organizer: it induces dorsal ectoderm to form neural tissue, and it dorsalizes mesoderm cells that would otherwise contribute to the ventral mesoderm (Smith et al. 1993). Smith and Harland showed that newly transcribed *noggin* mRNA is first localized in the dorsal blastopore lip region and then becomes expressed in the notochord (**FIGURE 8.26B**). Noggin binds to BMP4 and BMP2 and inhibits their binding to receptors (Zimmerman et al. 1996).

CHORDIN Chordin protein was isolated from clones of cDNA whose mRNAs were present in dorsalized, but not in ventralized, embryos (Sasai et al. 1994). These cDNA clones were tested by injecting them into ventral blastomeres and seeing whether they induced secondary axes. One of the clones capable of inducing a secondary neural tube contained the *chordin* gene; *chordin* mRNA was found to be localized in the dorsal blastopore lip and later in the notochord (**FIGURE 8.27**). Morpholino antisense oligomers directed against the *chordin* message blocked the ability of an organizer graft to induce a secondary central nervous system (Oelgeschläger et al. 2003). Of all organizer genes observed, *chordin* is the one most acutely activated by β-catenin (Wessely et al. 2004). Like Noggin, Chordin binds directly to BMP4 and BMP2 and prevents their complexing with their receptors (Piccolo et al. 1996).

(A)   (B)   (C)

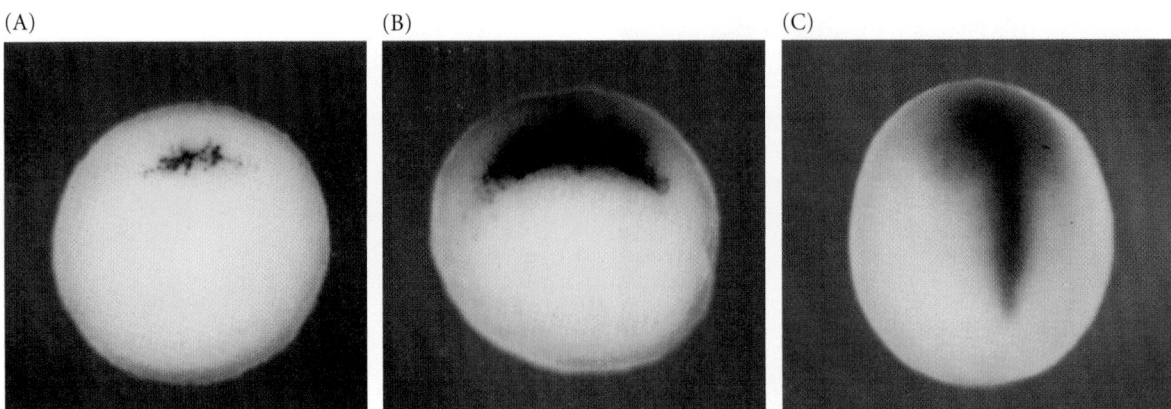

FIGURE 8.27 Localization of *chordin* mRNA. (A) Whole-mount in situ hybridization shows that just prior to gastrulation, *chordin* mRNA (dark area) is expressed in the region that will become the dorsal blastopore lip. (B) As gastrulation begins, *chordin* is expressed at the dorsal blastopore lip. (C) In later stages of gastrulation, the *chordin* message is seen in the organizer tissues. (From Sasai et al. 1994, courtesy of E. De Robertis.)

FOLLISTATIN The mRNA for a third organizer-secreted protein, Follistatin, is also transcribed in the dorsal blastopore lip and notochord. Follistatin was found in the organizer as an unexpected result of an experiment that was looking for something else. Ali Hemmati-Brivanlou and Douglas Melton (1992, 1994) wanted to see whether the protein activin was required for mesoderm induction. In searching for the mesoderm inducer, they found that Follistatin, an inhibitor of both activin and BMPs, caused ectoderm to become neural tissue. They then proposed that under normal conditions, ectoderm becomes neural unless induced to become epidermal by the BMPs. This model was supported by, and explained, certain cell dissociation experiments that had also produced odd results. Three 1989 studies—by Grunz and Tacke, Sato and Sargent, and Godsave and Slack—showed that when whole embryos or their animal caps were dissociated, they formed neural tissue. This result would be explainable if the "default state" of the ectoderm was not epidermal, but neural, and tissue had to be induced to have an epidermal phenotype. Thus, we conclude that the organizer blocks this epidermalizing induction by inactivating BMPs.

NORRIN For several decades, it has been thought that the dorsal ectoderm is somehow biased to become neural (see Savage and Philips 1989). A fourth BMP inhibitor, Norrin, has been discovered that, unlike Chordin, Noggin, and Follistatin, is already present in the animal portion of the oocyte. In the gastrula, Norrin becomes incorporated into the dorsal animal cells that will become the anterior dorsal (i.e., head) ectoderm. Here, adjacent to the organizer, Norrin appears to block BMP in a cell-autonomous manner (Kuroda et al. 2004; Xu et al. 2012) and thus may predispose this area of the ectoderm to become neural tissue.

Epidermal inducers: The BMPs

In *Xenopus*, the major epidermal inducers are the bone morphogenetic proteins, especially BMP4 and its close relatives BMP2 and BMP7. There is an antagonistic relationship between these BMPs and the organizer. If the mRNA for BMP4 is injected into *Xenopus* eggs, all the mesoderm in the embryo becomes ventrolateral mesoderm. Involution is delayed and, when it does occur, has a ventral rather than a dorsal character (Dale et al. 1992; Jones et al. 1992). Conversely, overexpression of a dominant negative BMP4 receptor results in the formation of twinned axes (Graff et al. 1994; Suzuki et al. 1994). In 1995, Wilson and Hemmati-Brivanlou demonstrated that BMP4 induces ectodermal cells to become epidermal. By 1996, several laboratories had demonstrated that Noggin, Chordin, and Follistatin are all secreted by the organizer, and that each of them prevents BMP from binding to and inducing the ectoderm and mesoderm cells near the organizer (Piccolo et al. 1996; Zimmerman et al. 1996; Iemura et al. 1998).

BMP4 is expressed initially throughout the ectodermal and mesodermal regions of the late blastula. However, during gastrulation, *bmp4* transcripts become restricted to the ventrolateral marginal zone. This is because the Goosecoid protein (as well as some other transcription factors) is induced by the Siamois/Twin and Smad2 interactions in the dorsal (organizer) mesoderm starting at the beginning of gastrulation (Blitz and Cho 1995; Yao and Kessler 2001). These transcription factors repress *bmp4* and *wnt8* transcription (Hemmati-Brivanlou and Thomsen 1995; Northrop et al. 1995; Steinbeisser et al. 1995; Glavic et al. 2001). In the ectoderm, BMPs repress the genes (such as *neurogenin*) involved in forming neural tissue, while activating other genes involved in epidermal specification (Lee et al. 1995). In the mesoderm, it appears that graded levels of BMP4 activate different sets of mesodermal genes: an absence of BMP4 specifies the dorsal mesoderm; a low amount specifies the intermediate mesoderm; and a high amount specifies the lateral mesoderm (**FIGURE 8.28**; Gawantka et al. 1995; Hemmati-Brivanlou and Thomsen 1995; Dosch et al. 1997).

(A)

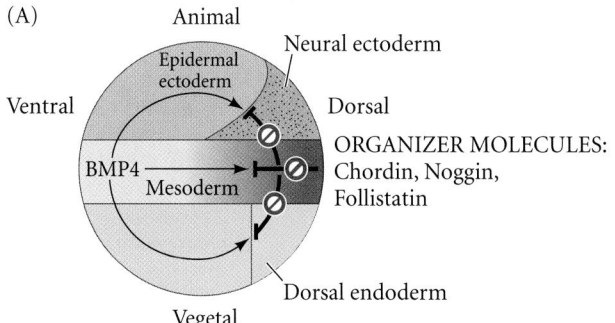

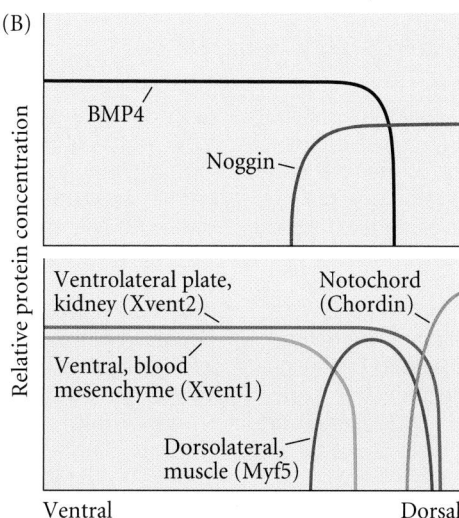

FIGURE 8.28 Model for the action of the organizer. (A) BMP4 (along with certain other molecules) is a powerful ventralizing factor. Organizer proteins such as Chordin, Noggin, and Follistatin block the action of BMP4; their inhibitory effects can be seen in all three germ layers. (B) BMP4 may elicit the expression of different genes in a concentration-dependent fashion. Thus, in the regions of *noggin* and *chordin* expression, BMP4 is totally prevented from binding, and these tissues become notochord (organizer) tissue. Slightly farther away from the organizer, the *myf5* gene is activated, producing a marker for the dorsolateral muscles. As more and more BMP4 molecules are allowed to bind to the cells, the *Xvent2* (ventrolateral) and *Xvent1* (ventral) genes become expressed. (After Dosch et al. 1997; De Robertis et al. 2000.)

In 2005, two important sets of experiments confirmed the default model and the importance of blocking BMPs to specify the nervous system. First, Khokha and colleagues (2005) used antisense morpholinos to eliminate three BMP antagonists (Noggin, Chordin, and Follistatin) in *Xenopus*. The resulting embryos had catastrophic failure of dorsal development and lacked neural plates and dorsal mesoderm (**FIGURE 8.29A,B**). Second, Reversade and colleagues blocked BMP activity with antisense morpholinos (Reversade et al. 2005; Reversade and De Robertis 2005). When they simultaneously blocked the formation of BMPs 2, 4, and 7, the neural tube became greatly expanded, taking over a much larger region of the ectoderm (**FIGURE 8.29C**). When they did a quadruple inactivation of the three BMPs *and* ADMP (another protein of the BMP family), the entire ectoderm became neural

(**FIGURE 8.29D**). Thus, the epidermis is instructed by BMP signaling, and the organizer specifies the ectoderm above it to become neural by blocking that BMP signal from reaching the adjacent ectoderm.

In the absence of BMP signaling, the FoxD5 transcription factor becomes expressed in the presumptive neural ectoderm. It initiates a pathway that leads to the stabilization of neural identity in most of the induced ectodermal cells, while allowing the formation of an immature, stem cell-like state in other induced cells (see Rogers et al 2009).

FIGURE 8.29 Control of neural specification by levels of BMPs. (A,B) Lack of dorsal structures in *Xenopus* embryos whose BMP-inhibitor genes *chordin*, *noggin*, and *follistatin* were eliminated by antisense morpholino oligonucleotides. (A) Control embryo with neural folds stained for the expression of the neural gene *Sox2*. (B) Lack of neural tube and *Sox2* expression in an embryo treated with the morpholinos against three BMP inhibitors. (C,D) Expanded neural development. (C) The neural tube, visualized by *Sox2* staining, is greatly enlarged in embryos treated with antisense morpholinos that destroy BMPs 2, 4, and 7. (D) Complete transformation of the entire ectoderm into neural ectoderm (and loss of the dorsal-ventral axis) by inactivation of ADMP as well as BMPs 2, 4, and 7. (A,B from Khokha et al. 2005, courtesy of R. Harland; C,D from Reversade and De Robertis 2005.)

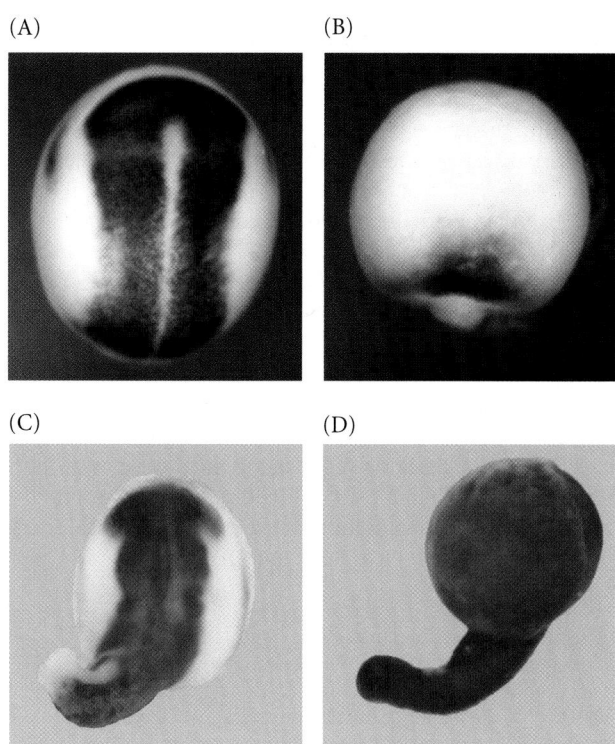

SIDELIGHTS & SPECULATIONS

BMP4 and Geoffroy's Lobster

The hypothesis that the organizer secretes proteins that block BMPs received further credence from an unexpected source—the emerging field of evolutionary developmental biology (see Chapter 20). Researchers have discovered that the same Chordin-BMP4 interaction that instructs the formation of the neural tube in vertebrates also forms neural tissue in fruit flies (Holley et al. 1995; Schmidt et al. 1995; De Robertis and Sasai 1996). The dorsal neural tube of the vertebrate and the ventral neural cord of the fly appear to be generated by the same set of instructions.

The *Drosophila* homologue of the *bmp4* gene is *decapentaplegic* (*dpp*). As discussed in Chapter 6, Dpp protein is responsible for patterning the fly's dorsal-ventral axis; it is present in the dorsal portion of the fly embryo and diffuses ventrally. Dpp is opposed by a protein called Short-gastrulation

(Sog), which is the *Drosophila* homologue of Chordin. These insect homologues not only appear to be similar to their vertebrate counterparts, they can actually substitute for each other. When *sog* mRNA is injected into ventral regions of *Xenopus* embryos, it induces the amphibian notochord and neural tube. Injecting *chordin* mRNA into *Drosophila* embryos produces ventral nervous tissue.

Although Chordin dorsalizes the *Xenopus* embryo, it ventralizes *Drosophila*. In *Drosophila*, Dpp is made dorsally; in *Xenopus*, BMP4 is made ventrally. In both cases, Sog/Chordin helps specify neural tissue by blocking the effects of Dpp/BMP4. In *Drosophila*, Sog interacts with Tolloid and several other proteins to create a gradient of Sog proteins. In *Xenopus*, the homologues of the same proteins act to create a gradient of Chordin (see Figure 20.9; Hawley et al. 1995;

Holley et al. 1995; De Robertis et al. 2000).

In 1822, the French anatomist Étienne Geoffroy Saint-Hilaire provoked one of the most heated and critical confrontations in biology when he proposed that the lobster was but a vertebrate upside down. He claimed that the ventral side of the lobster (with its nerve cord) was homologous to the dorsal side of the vertebrate (Appel 1987). It seems that he was correct on the molecular level, if not on the anatomical level. The instructions for producing a nervous system in fact may have evolved only once, and the myriad animal lineages may all have used this same set of instructions—just in different places. The BMP4 (Dpp)/Chordin (Sog) interaction is an example of "homologous processes," suggesting a unity of developmental principles among all animals (Gilbert and Bolker 2001).

The Regional Specificity of Neural Induction

One of the most important phenomena in neural induction is the regional specificity of the neural structures that are produced. Forebrain, hindbrain, and spinocaudal regions of the neural tube must be properly organized in an anterior-to-posterior direction. The organizer tissue not only induces the neural tube, it also specifies the regions of the neural tube. This region-specific induction was demonstrated by Hilde Mangold's husband, Otto Mangold, in 1933. He transplanted four successive regions of the archenteron roof of late-gastrula newt embryos into the blastocoels of early-gastrula embryos. The most anterior portion of the archenteron roof (containing head mesoderm) induced balancers and portions of the oral apparatus; the next most anterior section induced the formation of various head structures, including nose, eyes, balancers, and otic vesicles; the third section (including the notochord) induced the hindbrain structures; and the most posterior section induced the formation of dorsal trunk and tail mesoderm* (FIGURE 8.30A–D).

---

*The induction of dorsal mesoderm—rather than the dorsal ectoderm of the nervous system—by the posterior end of the notochord was confirmed by Bijtel (1931) and Spofford (1945), who showed that the posterior fifth of the neural plate gives rise to tail somites and the posterior portions of the pronephric kidney duct.

In further experiments, Mangold demonstrated that when dorsal blastopore lips from early salamander gastrulae were transplanted into other *early* salamander gastrulae, they formed secondary heads. When dorsal lips from *later* gastrulas were transplanted into early salamander gastrulae, however, they induced the formation of secondary tails (FIGURE 8.30E,F; Mangold 1933). These results show that the first cells of the organizer to enter the embryo induce the formation of brains and heads, while those cells that form the dorsal lip of later-stage embryos induce the cells above them to become spinal cords and tails.

The question then became, What are the molecules being secreted by the organizer in a regional fashion such that the first cells involuting through the blastopore lip (the endomesoderm) induce head structures, whereas the next portion of involuting mesoderm (notochord) produces trunk and tail structures? FIGURE 8.31 shows a possible model for these inductions, the elements of which we will now describe in detail.

The head inducer: Wnt antagonists

One of the unifying features of animal development has been the varied roles of Wnt proteins in specifying the anterior-posterior axis. In every vertebrate studied (and in many invertebrates, including echinoderms, cnidarians, and flatworms), a gradient of Wnt signals is highest in the posterior and lowest in the head. In most cases, Wnts are made in the posterior regions and Wnt antagonists in the

REGIONAL SPECIFICITY OF INDUCTION

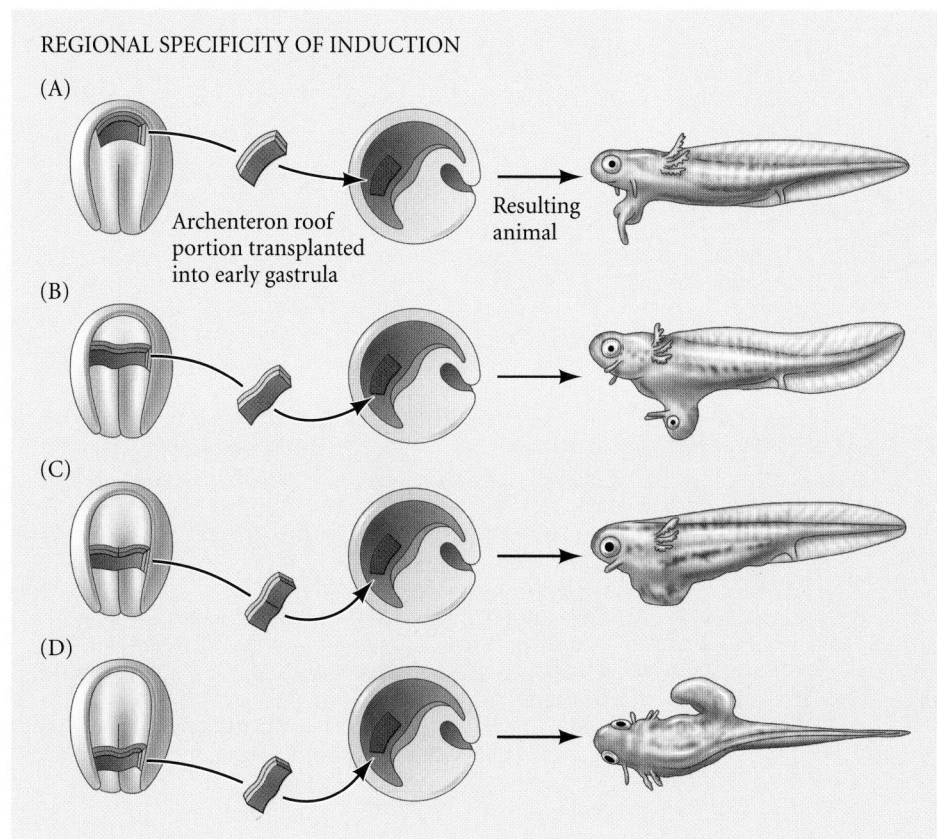

(A)

Archenteron roof portion transplanted into early gastrula

Resulting animal

(B)

(C)

(D)

FIGURE 8.30 Regional and temporal specificity of induction. (A–D) Regional specificity of structural induction can be demonstrated by implanting different regions (color) of the archenteron roof into early *Triturus* gastrulae. The resulting embryos develop secondary dorsal structures. (A) Head with balancers. (B) Head with balancers, eyes, and forebrain. (C) Posterior part of head, diencephalon, and otic vesicles. (D) Trunk-tail segment. (E,F) Temporal specificity of inducing ability. (E) Young dorsal lips (which will form the anterior portion of the organizer) induce anterior dorsal structures when transplanted into early newt gastrulae. (F) Older dorsal lips transplanted into early newt gastrulae produce more posterior dorsal structures. (A–D after Mangold 1933; E,F after Saxén and Toivonen 1962.)

TEMPORAL SPECIFICITY OF INDUCTION

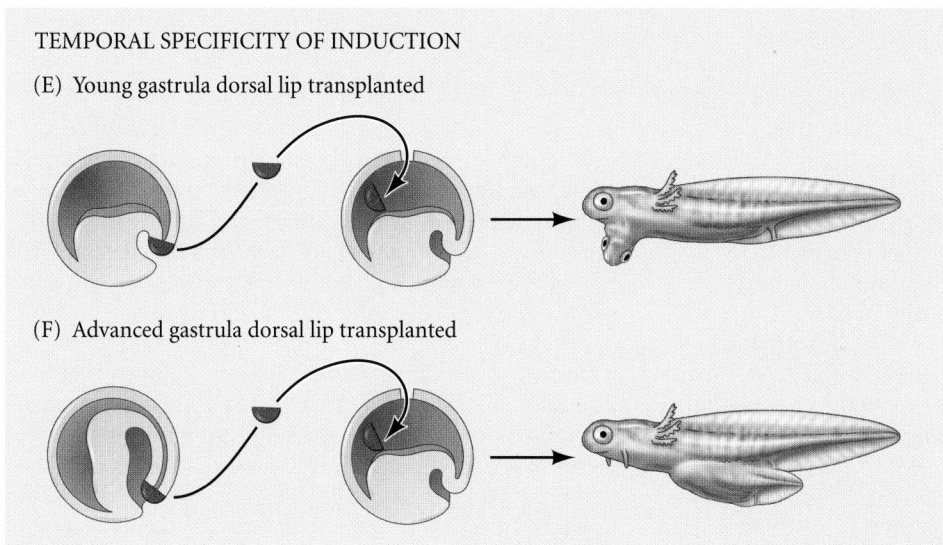

(E) Young gastrula dorsal lip transplanted

(F) Advanced gastrula dorsal lip transplanted

anterior (Petersen and Reddien 2009). The most anterior regions of the head and brain are underlain not by notochord but by pharyngeal endoderm and head (prechordal) mesoderm (see Figures 8.7C,D and 8.31A). This endomesodermal tissue constitutes the leading edge of the dorsal blastopore lip. Recent studies have shown that these cells not only induce the most anterior head structures, but that they do it by blocking the Wnt pathway as well as by blocking BMPs. The Wnt antagonists appear to be induced by the high levels of phosphorylated Smad2 coming from Nodal

and Vg1 secreted by the vegetal cells (Agius et al. 2000; Bisroy et al. 2006).

CERBERUS In 1996, Bouwmeester and colleagues showed that the induction of the most anterior head structures could be accomplished by a secreted protein called Cerberus (named after the three-headed dog that guarded the entrance to Hades in Greek mythology). Unlike the other proteins secreted by the organizer, Cerberus promotes the formation of the cement gland (the most anterior region of

(A)

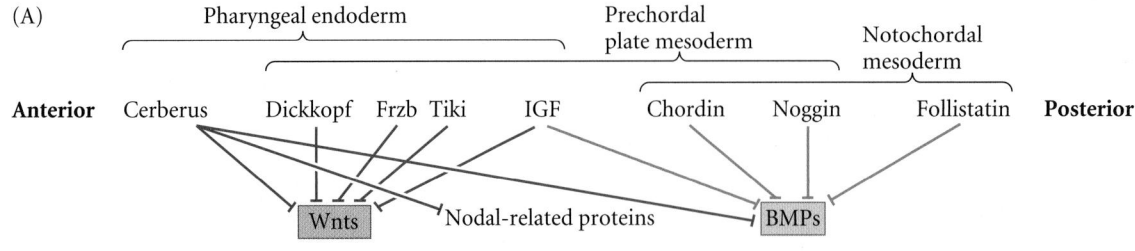

(B)

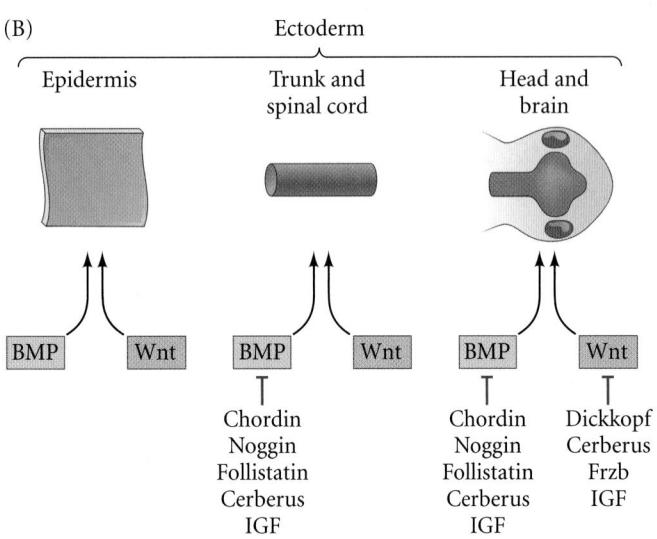

FIGURE 8.31 Paracrine factor antagonists from the organizer are able to block specific paracrine factors to distinguish head from tail. (A) The pharyngeal endoderm that underlies the head secretes Dickkopf, Frzb, and Cerberus. Dickkopf and Frzb block Wnt proteins; Cerberus blocks Wnts, Nodal-related proteins, and BMPs. The prechordal plate secretes the Wnt blockers Dickkopf and Frzb, as well as BMP blockers Chordin and Noggin. The notochord contains the BMP blockers Chordin, Noggin, and Follistatin but does not secrete Wnt blockers. Insulin-like growth factor (IGF) from the head endomesoderm probably acts at the junction of the notochord and prechordal mesoderm. (B) Summary of paracrine antagonist function in the ectoderm. Brain formation requires inhibiting both the Wnt and BMP pathways. Spinal cord neurons are produced when Wnt functions without the presence of BMPs. Epidermis is formed when both the Wnt and BMP pathways are operating.

tadpole ectoderm), eyes, and olfactory (nasal) placodes. When *cerberus* mRNA was injected into a vegetal ventral *Xenopus* blastomere at the 32-cell stage, ectopic head structures were formed (**FIGURE 8.32A**). These head structures arose from the injected cell as well as from neighboring cells.

The *cerberus* gene is expressed in the pharyngeal endomesoderm cells that arise from the deep cells of the early dorsal lip. Cerberus protein can bind BMPs, Nodal-related proteins, and Xwnt8 (see Figure 8.31A; Piccolo et al. 1999). When Cerberus synthesis is blocked, the levels of BMP, Nodal-related proteins, and Wnts all rise in the anterior of the embryo, and the ability of the anterior endomesoderm to induce a head is severely diminished (Silva et al. 2003).

FRZB, DICKKOPF, AND TIKI Shortly after the attributes of Cerberus were demonstrated, two other proteins, Frzb and Dickkopf, were discovered to be synthesized in the involuting endomesoderm. Frzb (pronounced "frisbee") is a small, soluble form of Frizzled (the Wnt receptor), and it is capable of binding Wnt proteins in solution (**FIGURE 8.32B,C**; Leyns et al. 1997; Wang et al. 1997). Frzb is synthesized predominantly in the endomesoderm cells beneath the prospective brain (**FIGURE 8.32D**). If embryos are made to synthesize excess Frzb, Wnt signaling fails to occur throughout the embryo; such embryos lack ventral posterior structures and become "all head." The Dickkopf protein (German, "thick head," "stubborn") also appears to interact directly with the Wnt receptors, preventing Wnt signaling (Mao et al. 2001, 2002). Injection of antibodies against Dickkopf causes the resulting embryos to have small,

deformed heads with no forebrain (Glinka et al. 1998). Therefore, the induction of trunk structures may be caused by the blockade of BMP signaling from the notochord, while Wnt signals are allowed to proceed. However, to produce a head, both the BMP signal and the Wnt signal must be blocked. This Wnt blockade comes from the endomesoderm, the most anterior portion of the organizer (Glinka et al. 1997).

A third organizer protein, Tiki, has recently been found to bind to Wnt proteins during gastrulation. Tiki not only prevents the binding of Wnts to their receptors, it cleaves the protein to render it nonfunctional. Tiki is synthesized primarily in the anterior regions of the organizer, and morpholino knockdown experiments show that Tiki is crucial for head formation in *Xenopus* (Zhang et al. 2012.)

INSULIN-LIKE GROWTH FACTORS In addition to those proteins that block BMPs and Wnts by physically binding to these paracrine factors, the head region contains yet another set of proteins that prevent BMP and Wnt signals from reaching the nucleus. Pera and colleagues (2001) showed that **insulin-like growth factors** (**IGFs**) are required for the formation of the anterior neural tube, including the brain and sensory placodes. IGFs accumulate in the dorsal midline and are especially prominent in the anterior neural tube. When injected into ventral mesodermal blastomeres, mRNA from IGFs causes the formation of ectopic heads, while blocking IGF receptors results in the lack of head formation.

IGFs appear to work by initiating a receptor tyrosine kinase (RTK) signal transduction cascade (see Chapter 3) that

(A)

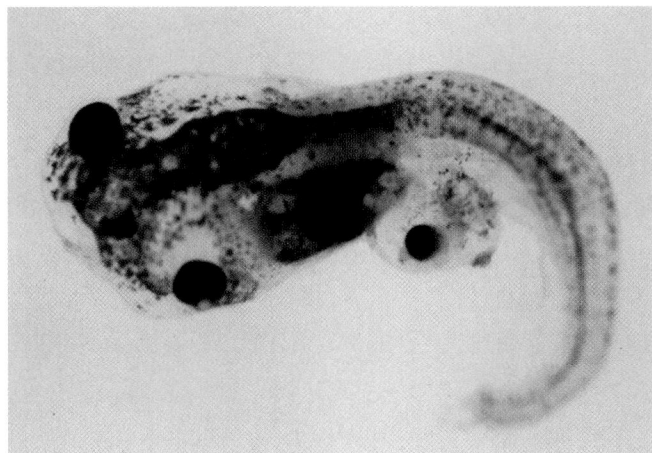

(B)

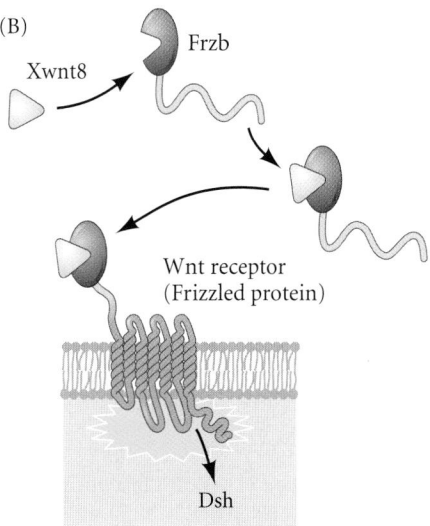

(C)

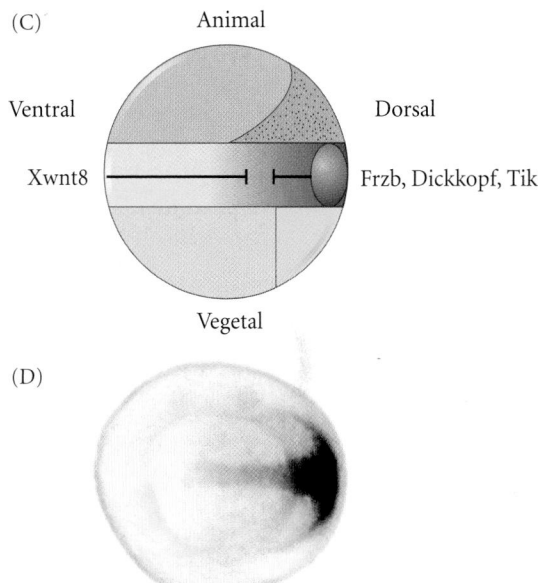

FIGURE 8.32 Inhibiting Wnt signaling enables head formation. (A) Injecting *cerberus* mRNA into a single D4 (ventral vegetal) blastomere of a 32-cell *Xenopus* embryo induces head structures as well as a duplicated heart and liver. The secondary eye (a single cyclopic eye) and olfactory placode can be readily seen. Xwnt8 is capable of ventralizing the mesoderm and preventing anterior head formation in the ectoderm. (B) Frzb protein is secreted by the anterior region of the organizer. It must bind to Xwnt8 before that inducer can bind to its receptor. Frzb resembles the Wnt–binding domain of the Wnt receptor (Frizzled protein), but Frzb is a soluble molecule. (C) Xwnt8 is made throughout the marginal zone. (D) Double in situ hybridization localizing Frzb (dark stain) and Chordin (reddish stain) messages. The *frzb* mRNA is seen to be transcribed in the head endomesoderm of the organizer, but not in the notochord (where *chordin* is expressed). (A from Bouwmeester et al. 1996; D from Leyns et al. 1997; photographs courtesy of E. M. De Robertis.)

interferes with the signal transduction pathways of both BMPs and Wnts (Richard-Parpaillon et al. 2002; Pera et al. 2003).

Trunk patterning: Wnt signals and retinoic acid

Toivonen and Saxén provided evidence for a gradient of a posteriorizing factor that would act to specify the trunk and tail tissues of the amphibian embryo* (Toivonen and Saxén 1955, 1968; reviewed in Saxén 2001). This factor's activity would be highest in the posterior of the embryo and weakened anteriorly. Recent studies have extended this model and have proposed candidates for posteriorizing molecules. The primary protein involved in posteriorizing the neural tube is

---

*The tail inducer was initially thought to be part of the trunk inducer, since transplantation of the late dorsal blastopore lip into the blastocoel often produced larvae with extra tails. However, it appears that tails are normally formed by interactions between the neural plate and the posterior mesoderm during the neurula stage (and thus are generated outside the organizer). Here, Wnt, BMPs, and Nodal signaling all seem to be required (Tucker and Slack 1995; Niehrs 2004). Interestingly, all three of these signaling pathways have to be inactivated if the head is to form.

thought to be a member of the Wnt family of paracrine factors, most likely Xwnt8 (Domingos et al. 2001; Kiecker and Niehrs 2001).

It appears that a gradient of Wnt proteins is necessary for specifying the posterior region of the neural plate (the trunk and tail; Hoppler et al. 1996; Niehrs 2004). In *Xenopus*, an endogenous gradient of Wnt signaling and β-catenin is highest in the posterior and absent in the anterior (**FIGURE 8.33A**). Moreover, if Xwnt8 is added to developing embryos, spinal cord-like neurons are seen more anteriorly in the embryo, and the most anterior markers of the forebrain are absent. Conversely, suppressing Wnt signaling (by adding Frzb or Dickkopf to the developing embryo) leads to the expression of the anteriormost markers in more posterior neural cells. Therefore, there appear to be two major gradients in the amphibian gastrula—a BMP gradient that

FIGURE 8.33 Signaling gradients and axis specification. (A) A Wnt signaling pathway posteriorizes the neural tube. Gastrulating embryos were stained for β-catenin and the density of the stain compared between regions of the ectodermal cells, revealing a gradient of β-catenin in the presumptive neural plate. (B) The Wnt gradient specifies posterior-anterior polarity and the BMP gradient specifies dorsal-ventral polarity. This double-gradient interaction was discovered in amphibians but has now been shown to be characteristic of animal development. (After Saxén and Toivonen 1962; Kiecker and Niehrs 2001 and Niehrs 2004.)

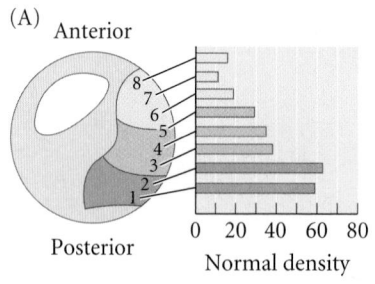
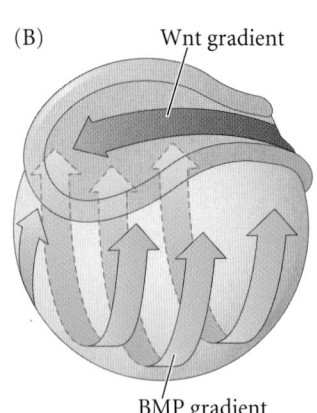

specifies the dorsal-ventral axis and a Wnt gradient specifying the anterior-posterior axis (**FIGURE 8.33B**). It must be remembered, too, that both of these axes are established by the initial axes of Nodal-like TGF-β paracrine factors and β-catenin across the vegetal cells. The basic model of neural induction, then, looks like the diagram in **FIGURE 8.34**.

While the Wnt proteins probably play a major role in specifying the anterior-posterior axis, they are probably not the only agents involved. Fibroblast growth factors appear to be critical in allowing the cells to respond to the Wnt signal (Holowacz and Sokol 1999; Domingos et al. 2001). Retinoic acid (RA) also is seen to have a gradient highest at the posterior end of the neural plate, and RA can also posteriorize the neural tube in a concentration-dependent manner (Cho and De Robertis 1990; Sive and Cheng 1991; Chen et al. 1994). RA signaling appears to be especially important in patterning the hindbrain and appears to interact with FGF signals to activate the posterior Hox genes (Kolm et al. 1997; Dupé and Lumsden 2001; Shiotsugu et al. 2004). Together, the posterior-to-anterior Wnt, FGF, and RA gradients function to determine the boundaries of the Hox genes along the anterior-posterior axis (Wacker et al. 2004; Durston et al. 2010a,b).

- See **WEBSITE 8.5** Gradients and Hox gene expression

- See **VADE MECUM** The primary organizer and the double-gradient hypothesis

Specifying the Left-Right Axis

Although the developing tadpole looks symmetrical from the outside, several internal organs, such as the heart and the gut tube, are not evenly balanced on the right and left sides. In other words, in addition to its dorsal-ventral and anterior-posterior axes, the embryo has a left-right axis. In all vertebrates studied so far, the crucial event in left-right axis formation is the expression of a *nodal* gene in the lateral plate mesoderm on the left side of the embryo. In *Xenopus*, this gene is *Xnr1* (*Xenopus nodal-related 1*). If *Xnr1* expression is permitted to occur on the right-hand side, the position of the heart (normally found on the left) and the coiling of the gut are randomized.

But what limits *Xnr1* expression solely to the left-hand side? As in other vertebrates (as we will see in Chapter 9), the concentration of a Nodal protein to the left side is caused by the clockwise rotation of cilia found in the organizer region. In *Xenopus*, these specific cilia are formed during the later stages of gastrulation (after the original specification of the mesoderm) at the dorsal blastopore lip (Schweickert

FIGURE 8.34 Model of organizer function and axis specification in the *Xenopus* gastrula. (1) BMP inhibitors from organizer tissue (dorsal mesoderm and pharyngeal mesendoderm) block the formation of epidermis, ventrolateral mesoderm, and ventrolateral endoderm. (2) Wnt inhibitors in the anterior of the organizer (pharyngeal endomesoderm) allow the induction of head structures. (3) A gradient of caudalizing factors (Wnts, FGFs, and retinoic acid) results in the regional expression of Hox genes, which specify the regions of the neural tube.

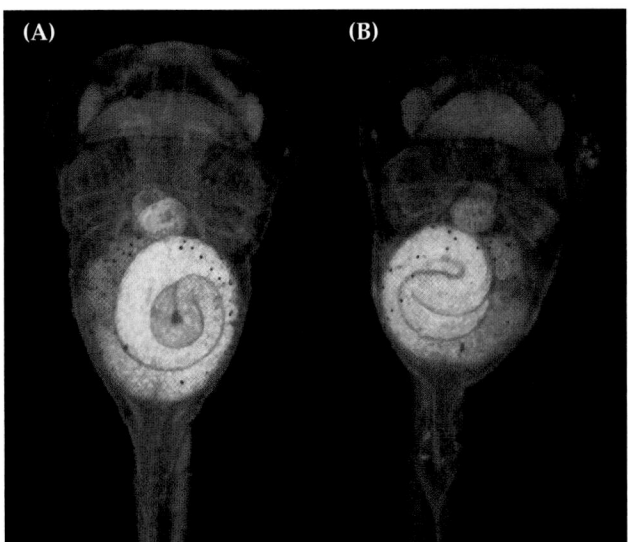

FIGURE 8.35 Pitx2 determines the direction of heart looping and gut coiling. (A) Wild-type *Xenopus* tadpole viewed from the ventral side, showing rightward heart looping and counterclockwise gut coiling. (B) If an embryo is injected with Pitx2 so that this protein is present in the mesoderm of both the right and left sides (instead of just the left side), heart looping and gut coiling are random with respect to each other. Sometimes this treatment results in complete reversals, as in this embryo, in which the heart loops to the left and the gut coils in a clockwise manner. (From Ryan et al. 1998, courtesy of J. C. Izpisúa-Belmonte.)

et al. 2007; Blum et al. 2009). That is, they are located in the posterior region of the embryo, at the site where the archenteron is still forming. If rotation of these cilia is blocked, *Xnr1* expression fails to occur in the mesoderm and laterality defects result.*

The pathway by which Xnr1 protein instructs the heart and gut to fold properly is unknown, but one of the key genes activated by Xnr1 appears to be *pitx2*, which normally is expressed only on the left side of the embryo. Pitx2 protein persists on the left side as the heart and gut develop, controlling their respective positions; if Pitx2 is injected into the right side of an embryo, heart placement and gut coiling are randomized (**FIGURE 8.35**; Ryan et al. 1998). As we will see, the pathway through which Nodal protein establishes left-right polarity by activating *pitx2* on the left side of the embryo is conserved throughout all vertebrate lineages.

EARLY ZEBRAFISH DEVELOPMENT

In recent years, the teleost (bony) fish *Danio rerio*, commonly known as the zebrafish, has joined *Xenopus* as a widely

---

*The rotation of the cilia is based on microtubules and dynein. There is evidence that dynein and the microtubular proteins play an even earlier role, asymmetrically distributing cytoplasm to the right and left sides of the egg during the first cleavage division (see Vanderberg and Levin 2010; Lobikin et al. 2012; Schweikert et al. 2012). This asymmetry may involve the production of serotonin, which may create the ciliary flow (Beyer et al. 2012).

studied model of vertebrate development (see Figure 8.1B). Despite differences in their cleavage patterns (*Xenopus* eggs are holoblastic, dividing the entire egg, whereas the yolky zebrafish egg is meroblastic), *Xenopus* and *Danio* form their body axes and specify their cells in very similar ways.

Zebrafish have large broods, breed all year, are easily maintained, have transparent embryos that develop outside the mother (an important feature for microscopy), and can be raised so that mutants can be readily discovered and propagated in the laboratory. In addition, these fish develop rapidly. By 24 hours after fertilization, the embryo has already formed most of its organ primordia and displays a characteristic tadpole-like form (**FIGURE 8.36**; see Granato and Nüsslein-Volhard 1996; Langeland and Kimmel 1997). Furthermore, the ability to microinject fluorescent dyes into single blastomeres and to generate transgenes driving cell-type-specific fluorescent protein expression has allowed scientists to follow individual living cells as an organ develops.

The zebrafish is the first vertebrate that has been studied by intensive mutagenesis screens. By treating parents with mutagens and selectively breeding their progeny, scientists have found thousands of mutations whose normally functioning genes are critical for zebrafish development. The traditional method of genetic screening (modeled after large-scale screens in *Drosophila*) begins when the male parental fish are treated with a chemical mutagen that will cause random mutations in their germ cells. Each mutagenized male is then mated with a wild-type female fish to generate F_1 fish. Individuals in the F_1 generation carry the mutations inherited from their father. If the mutation is dominant, it will be expressed in the F_1 generation. If the mutation is recessive, the F_1 fish will not show a mutant phenotype, since the wild-type dominant allele will mask the mutation. The F_1 fish are then mated with wild-type fish to produce an F_2 generation that includes both males and females that carry the mutant allele. When two F_2 parents carry the same recessive mutation, there is a 25% chance that their offspring will show the mutant phenotype (**FIGURE 8.37**). Since zebrafish development occurs in the open (as opposed to within an opaque shell or inside the mother's body), abnormal developmental stages can be readily observed, and the defects in development can often be traced to changes in a particular group of cells (Driever et al. 1996; Haffter et al. 1996). Recently, high throughput methods of gene analysis and targeted exonucleases have propelled the analysis of zebrafish development, enabling mutations in particular genes to be rapidly generated, identified, and bred (see Lawson and Wolfe 2011; Huang et al. 2012).

Like *Xenopus* embryos, zebrafish embryos are susceptible to morpholino antisense molecules (Zhong et al. 2001), and researchers can use this method to test whether a particular gene is required for a particular function. Furthermore, the green fluorescent protein (GFP) reporter gene can be fused with specific zebrafish promoters and enhancers and inserted into the fish embryos. The resulting transgenic fish express

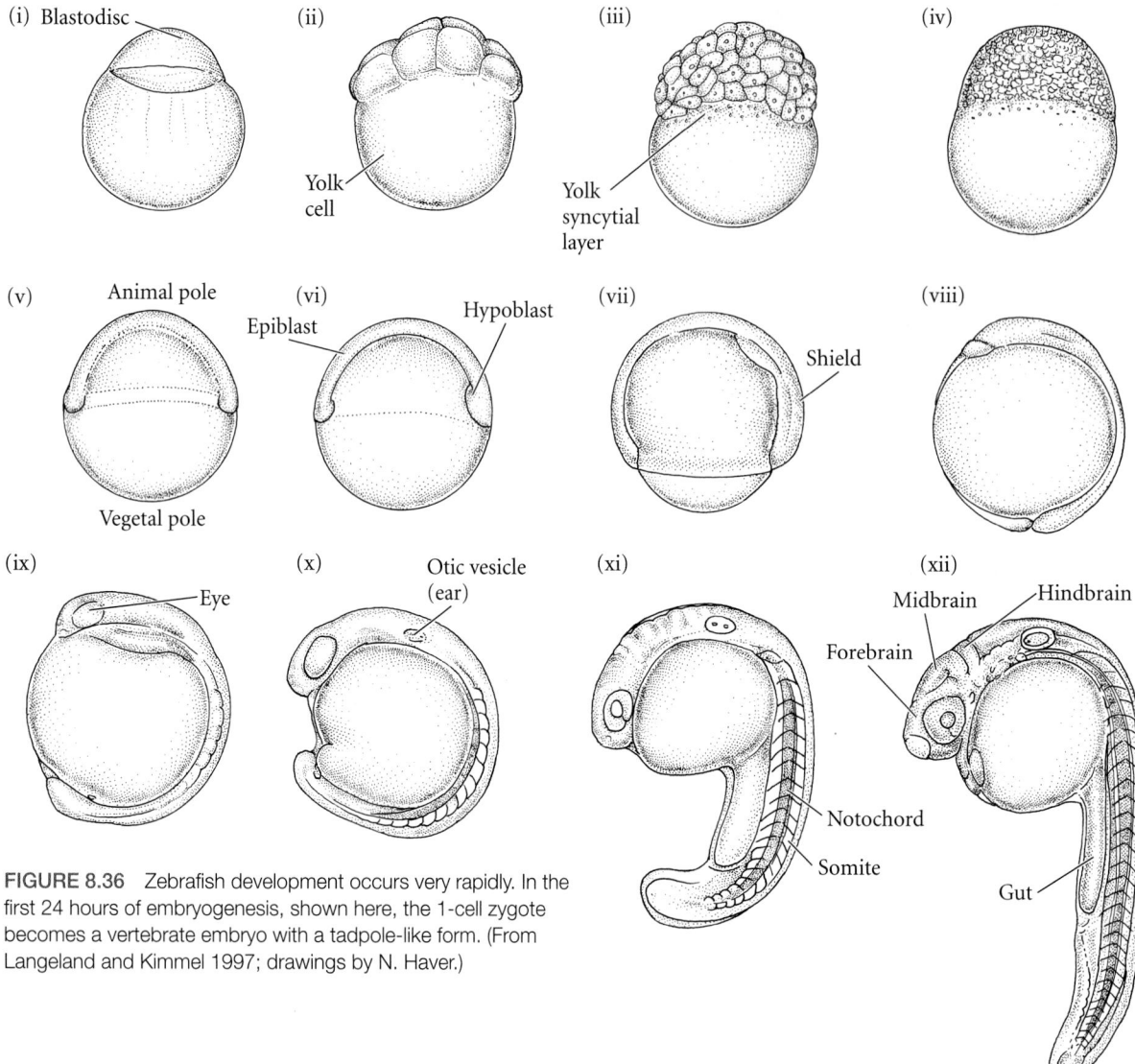

FIGURE 8.36 Zebrafish development occurs very rapidly. In the first 24 hours of embryogenesis, shown here, the 1-cell zygote becomes a vertebrate embryo with a tadpole-like form. (From Langeland and Kimmel 1997; drawings by N. Haver.)

GFP at the same times and places as they express the proteins controlled by these regulatory sequences. The amazing thing is that one can observe the reporter protein in living transparent embryos* (**FIGURE 8.38**).

The similarity of developmental mechanisms among all vertebrates and the ability of *Danio rerio* to be genetically manipulated has given this small fish an important role in investigating the genes that operate during human development (Mudbhary and Sadler 2011). When developmental biologists screened zebrafish mutants for cystic kidney disease, they found 12 different genes. Two of these genes were known to cause human cystic kidney disease, and the other

10 were as-yet unknown genes that were found to interact with the first two in a common pathway. Moreover, that pathway, which involves the synthesis of cilia, was not what had been expected. Thus, the zebrafish studies disclosed an important and previously unknown pathway to explore human birth defects (Sun et al. 2004). Zebrafish embryos are also permeable to small molecules placed in the water—a property that allows us to test drugs that may be deleterious to vertebrate development. For instance, zebrafish development can be altered by the addition of ethanol or retinoic acid, both of which produce malformations in the fish that resemble human developmental syndromes known to be caused by these molecules (Blader and Strähle 1998). As one zebrafish researcher joked, "Fish really are just little people with fins" (Bradbury 2004).

---

*GFP-transgenic zebrafish were the first commercially available transgenic pets, marketed as "Glo-Fish."

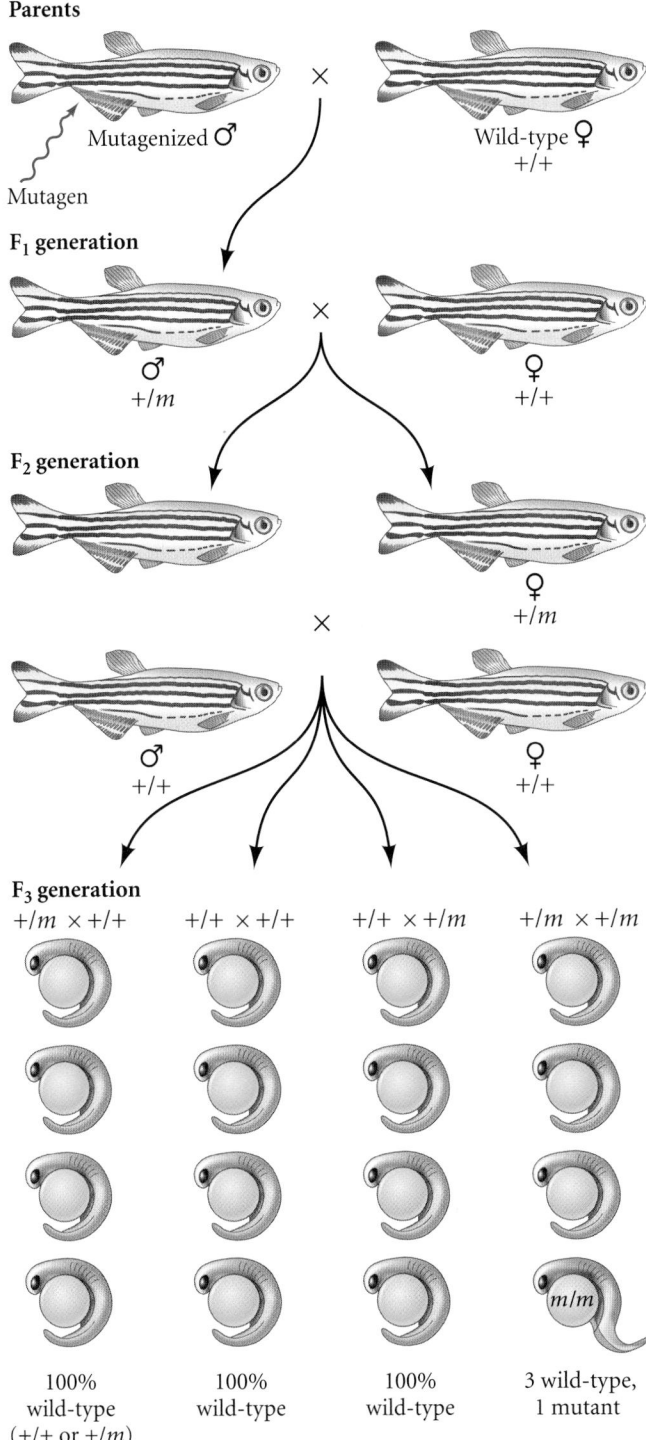

Parents

Mutagenized ♂

Mutagen

Wild-type ♀
+/+

F₁ generation

♂
+/m

♀
+/+

F₂ generation

♀
+/m

♂
+/+

♀
+/+

F₃ generation

+/m × +/+   +/+ × +/+   +/+ × +/m   +/m × +/m

100%
wild-type
(+/+ or +/m)

100%
wild-type

100%
wild-type

3 wild-type,
1 mutant

m/m

FIGURE 8.37 Screening protocol for identifying mutations of zebrafish development. The male parent is mutagenized and mated with a wild-type (+/+) female. If some of the male's sperm carry a recessive mutant allele (m), then some of the F₁ progeny of the mating will inherit that allele. F₁ individuals (here shown as a male carrying the mutant allele m) are then mated with wild-type partners. This creates an F₂ generation wherein some males and some females carry the recessive mutant allele. When the F₂ fish are mated, approximately 25% of their progeny will show the mutant phenotype. (After Haffter et al. 1996.)

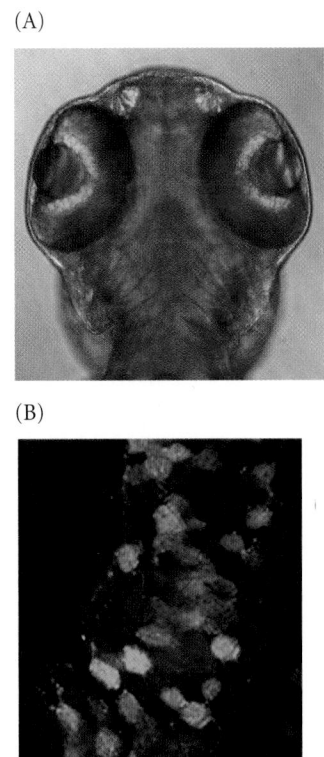

(A)

(B)

FIGURE 8.38 The gene for green fluorescent protein (GFP) was fused to the regulatory region of a zebrafish *sonic hedgehog* gene. As a result, GFP was synthesized wherever the Hedgehog protein is normally expressed in the fish embryo. (A) In the head of a zebrafish embryo, GFP is seen in the developing retina and nasal placodes. (B) Because GFP is expressed by individual cells, scientists can see precisely which cells make GFP, and thus which cells normally transcribe the gene of interest (in this case, *sonic hedgehog* in the retina). (Photographs courtesy of U. Strahle and C. Neumann.)

Cleavage

The eggs of most bony fish are **telolecithal**, meaning that most of the egg cell cytoplasm is occupied by yolk. Cleavage can take place only in the **blastodisc**, a thin region of yolk-free cytoplasm at the animal pole of the egg. The cell divisions do not completely divide the egg, so this type of cleavage is called **meroblastic** (Greek *meros*, "part"). Since only the blastodisc becomes the embryo, this type of meroblastic cleavage is referred to as **discoidal**.

Scanning electron micrographs show beautifully the incomplete nature of discoidal meroblastic cleavage in fish eggs (**FIGURE 8.39**). The calcium waves initiated at fertilization stimulate the contraction of the actin cytoskeleton to squeeze non-yolky cytoplasm into the animal pole of the egg. This process converts the spherical egg into a pear-shaped structure with an apical blastodisc (Leung et al. 1998, 2000). In fish, there are many waves of calcium release, and they orchestrate the processes of cell division. The calcium ions

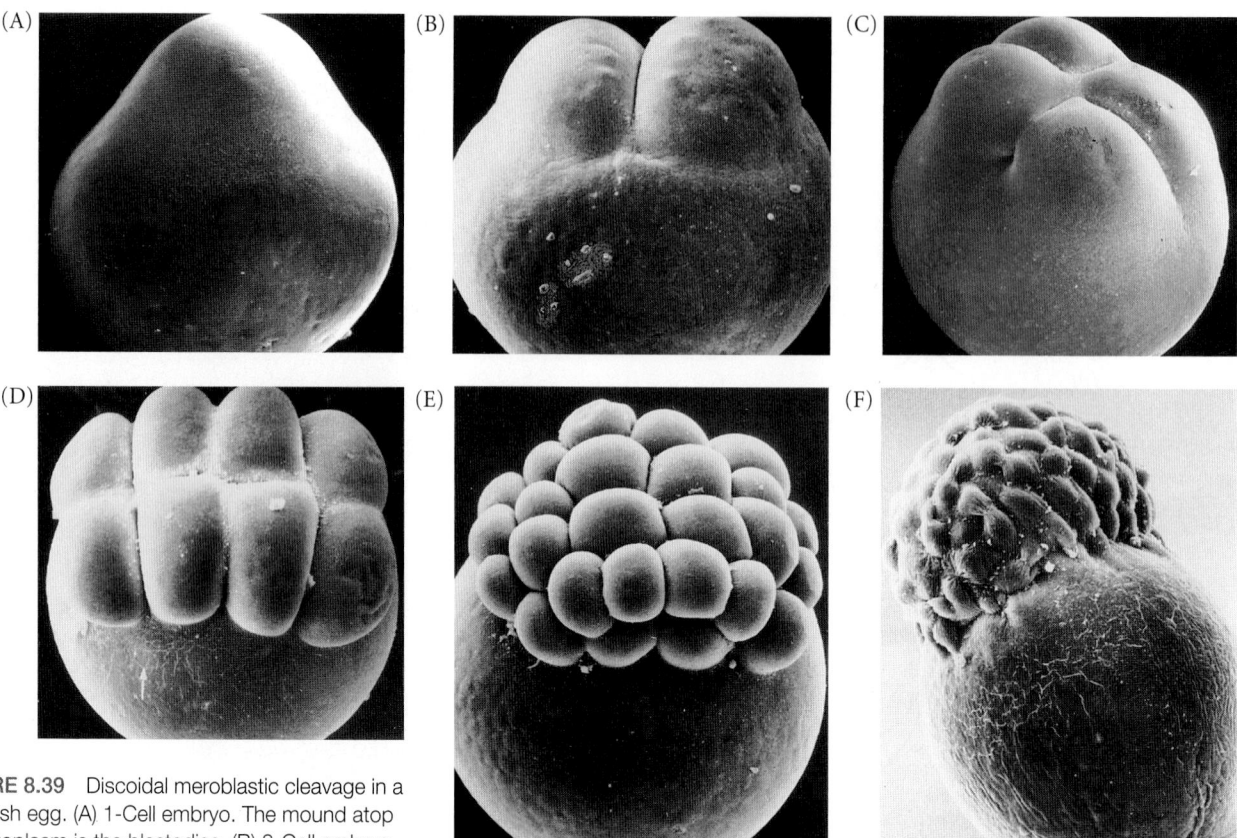

FIGURE 8.39 Discoidal meroblastic cleavage in a zebrafish egg. (A) 1-Cell embryo. The mound atop the cytoplasm is the blastodisc. (B) 2-Cell embryo. (C) 4-Cell embryo. (D) 8-Cell embryo, wherein two rows of four cells are formed. (E) 32-Cell embryo. (F) 64-Cell embryo, wherein the blastodisc can be seen atop the yolk cell. (From Beams and Kessel 1976, courtesy of the authors.)

coordinate the mitotic apparatus with the actin cytoskeleton, help propagate cell division across the cell surface, help deepen the cleavage furrow, and heal the membrane after the separation of the blastomeres (Lee et al. 2003).

The first cell divisions follow a highly reproducible pattern of meridional and equatorial cleavages. These divisions are rapid, taking only about 15 minutes each. The first 10 divisions occur synchronously, forming a mound of cells that sits at the animal pole of a large **yolk cell**. This mound of cells constitutes the **blastoderm**. Initially, all the cells maintain some open connection with one another and with the underlying yolk cell, so that moderately sized (17-kDa) molecules can pass freely from one blastomere to the next (Kane and Kimmel 1993; Kimmel and Law 1985). Remarkably, as the daughter cells migrate away from one another, they often retain these bridges through long tunnels connecting the cells (Caneparo et al. 2011).

Maternal effect mutations have shown the importance of oocyte proteins and mRNAs in embryonic polarity, cell division, and axis formation (Dosch et al. 2004; Langdon and Mullins 2011). As in frogs, the microtubules are important roads upon which morphogenetic determinants travel, and maternal mutants affecting the formation of the microtubal cytoskeleton prevent the normal positioning of the cleavage furrow and of mRNAs in the early embryo (Kishimoto et al. 2004).

Fish embryos, like those of many other embryos, undergo a mid-blastula transition (seen around the tenth cell division in zebrafish) when zygotic gene transcription begins, cell divisions slow, and cell movements become evident (Kane and Kimmel 1993). At this time, three distinct cell populations can be distinguished. The first of these is the **yolk syncytial layer**, or **YSL** (Agassiz and Whitman 1884; Carvalho and Heisenberg 2010). The YSL is formed at the ninth or tenth cell cycle, when the cells at the vegetal edge of the blastoderm fuse with the underlying yolk cell. This fusion produces a ring of nuclei in the part of the yolk cell cytoplasm that sits just beneath the blastoderm. Later, as the blastoderm expands vegetally to surround the yolk cell, some of the yolk syncytial nuclei will move under the blastoderm to form the **internal YSL** (**iYSL**), and others will move vegetally, staying ahead of the blastoderm margin, to form the **external YSL** (**eYSL**; FIGURE 8.40A,B). The YSL will be important for directing some of the cell movements of gastrulation.

The second cell population distinguished at the mid-blastula transition is the **enveloping layer** (**EVL**). It is made up of the most superficial cells from the blastoderm, which form an epithelial sheet a single cell layer thick. The EVL is a protective covering derived from the outer epidermis and is sloughed off after 2 weeks. It allows the embryo to develop

(A)

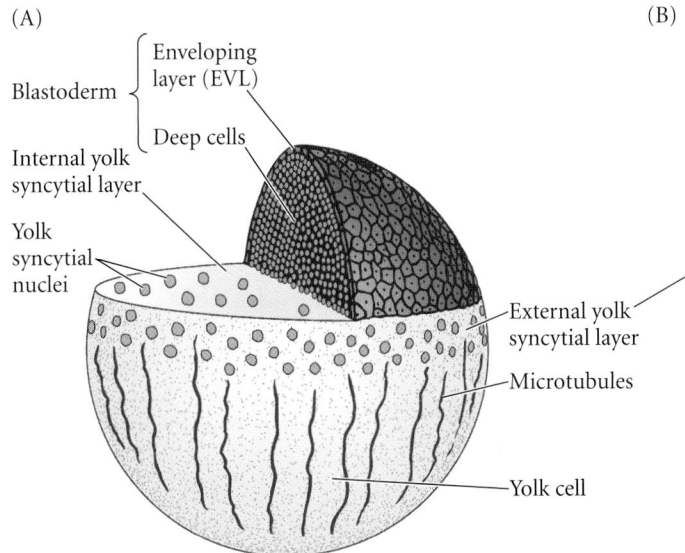

Blastoderm { Enveloping layer (EVL)

Deep cells

Internal yolk syncytial layer

Yolk syncytial nuclei

External yolk syncytial layer

Microtubules

Yolk cell

(C)

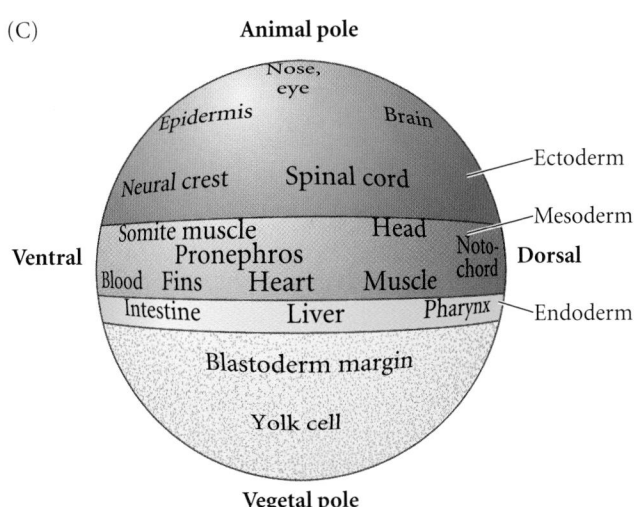

Animal pole

Nose, eye

Epidermis

Brain

Neural crest

Spinal cord

Ectoderm

Somite muscle

Head

Mesoderm

Ventral

Pronephros

Noto-chord

Dorsal

Blood Fins

Heart

Muscle

Intestine

Liver

Pharynx

Endoderm

Blastoderm margin

Yolk cell

Vegetal pole

(B)

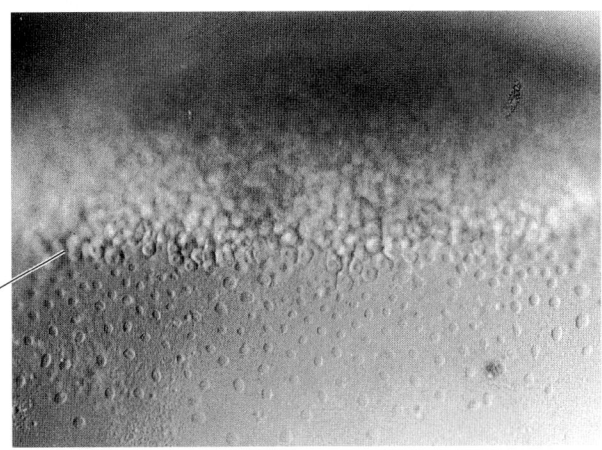

FIGURE 8.40 Fish blastula. (A) Prior to gastrulation, the deep cells are surrounded by the enveloping layer (EVL). The animal surface of the yolk cell is flat and contains the nuclei of the yolk syncytial layer (YSL). Microtubules extend through the yolky cytoplasm and the external YSL. (B) Late-blastula-stage embryo of the minnow *Fundulus*, showing the external YSL. The nuclei of these cells were derived from cells at the margin of the blastoderm, which released their nuclei into the yolky cytoplasm. (C) Fate map of the deep cells after cell mixing has stopped. This is a lateral view; for the sake of clarity, not all organ fates are labeled. (A,C after Langeland and Kimmel 1997; B from Trinkaus 1993, courtesy of J. P. Trinkaus.)

Gastrulation and Formation of the Germ Layers

All three layers of the zebrafish blastoderm undergo epiboly. The first cell movement of fish gastrulation is the epiboly of the blastoderm cells over the yolk, and this is thought to be controlled both by maternal proteins (such as Eomesodermin) and by new proteins transcribed from the YSL nuclei (Du et al. 2012). In the initial phase of this movement, the deep cells of the blastoderm move outward to intercalate with the more superficial cells, and the yolk cell (with its syncytial nuclei) pushes upward (Warga and Kimmel 1990). This intercalation of cells causes a flattening of the "dome" of the blastoderm cells (**FIGURE 8.41A**).

PROGRESSION OF EPIBOLY When about half the yolk is covered, a new set of movements is initiated. The YSL nuclei divide such that some nuclei (constituting the eYSL) remain in the upper cortex of the yolk cell, while the iYSL lies beneath the blastoderm. The enveloping layer is tightly joined to the iYSL by E-cadherin and tight junctions (Shimizu et al. 2005a; Siddiqui et al. 2010) and is dragged ventrally as the iYSL nuclei migrate. That the vegetal migration of the blastoderm margin is dependent on the epiboly of the YSL can be demonstrated by severing the attachments between the YSL and the EVL. When this is done, the EVL and the deep cells spring back to the top of the yolk, while the YSL continues its expansion around the yolk cell (Trinkaus 1984, 1992).

in a hypotonic solution that would otherwise burst the cells (Fukazawa et al. 2010). Between the EVL and the YSL is the third set of blastomeres, the **deep cells**. These are the cells that give rise to the embryo proper.

The fates of the early blastoderm cells are not determined, and cell lineage studies (in which a nondiffusible fluorescent dye is injected into a cell so that its descendants can be followed) show that there is much cell mixing during cleavage. Moreover, any one of these early blastomeres can give rise to an unpredictable variety of tissue descendants (Kimmel and Warga 1987; Helde et al. 1994). A fate map of the blastoderm cells can be made shortly before gastrulation begins. At this time, cells in specific regions of the embryo give rise to certain tissues in a highly predictable manner (**FIGURE 8.40C**; see also Figure 1.11), although they remain plastic, and cell fates can change if tissue is grafted to a new site.

● See **VADE MECUM** Zebrafish development

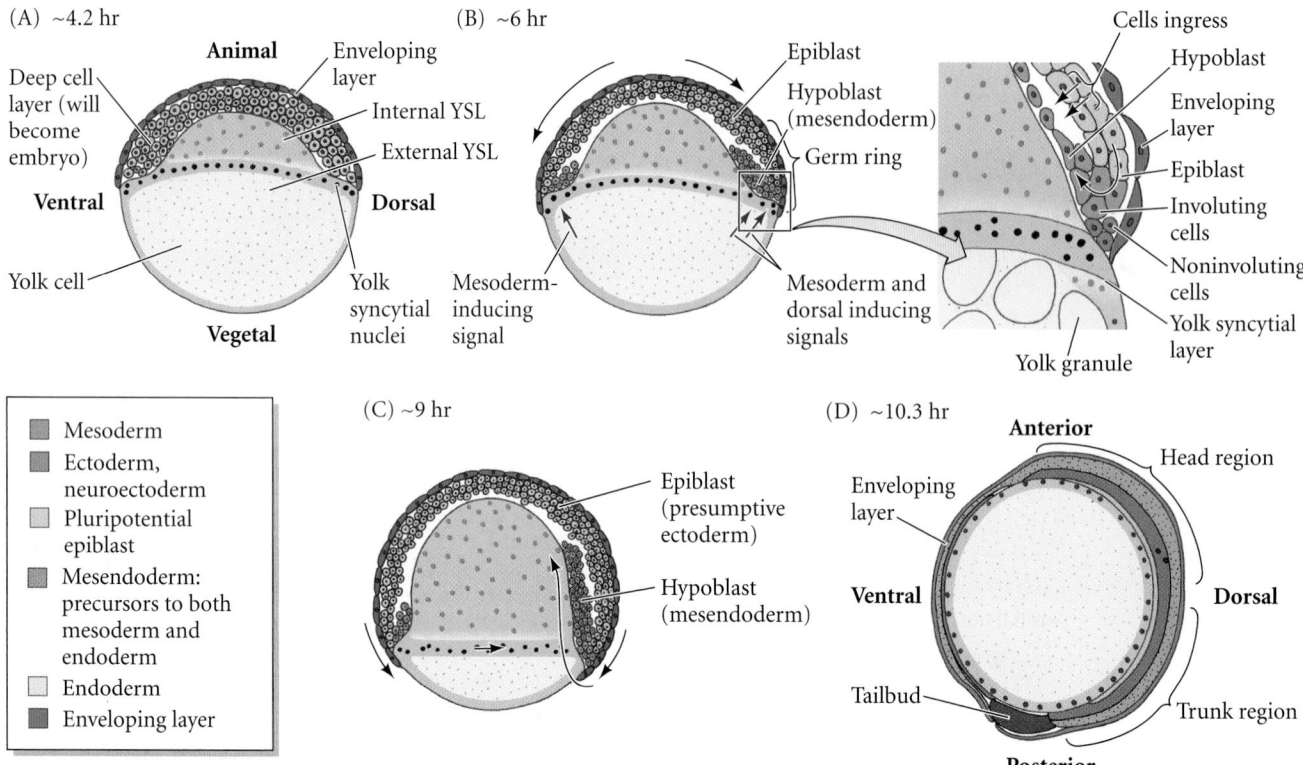

FIGURE 8.41 Cell movements during zebrafish gastrulation. (A) The blastoderm at 30% completion of epiboly (about 4.7 hr). (B) Formation of the hypoblast, either by involution of cells at the margin of the epibolizing blastoderm or by delamination and ingression of cells from the epiblast (6 hr). A close-up of the marginal region is at the right. (C) As ectodermal epiboly nears completion, the hypoblast, carrying the mesoderm and endoderm precursors, begins to cover the yolk. (D) Completion of gastrulation (10.3 hr). The germ layers (yellow endoderm, blue ectoderm, red mesoderm) are present. (After Driever 1995; Langeland and Kimmel 1997; Carvalho and Heisenberg 2010; Lepage and Bruce 2010.)

The migration of the YSL ventrally depends partially on the expansion of this layer by cell division and intercalation, and partly on the cytoskeletal network within the yolk cell (see Lepage and Bruce 2010). An actomyosin band forms in the eYSL, at the boundary between the YSL and the EVL. This pulls down the YSL/EVL at its vegetal connection by means of contraction and friction (Behrndt et al. 2012). Meanwhile, the eYSL nuclei appear to migrate along the microtubles aligned along the animal-vegetal axis of the yolk cell, presumably pulling the iYSL and its accompanying EVL over the yolk cell. (Radiation or drugs that block the polymerization of tubulin slow epiboly; Strahle and Jesuthasan 1993; Solnica-Krezel and Driever 1994.) At the end of gastrulation, the entire yolk cell is covered by the blastoderm.

INTERNALIZATION OF THE HYPOBLAST After the blastoderm cells have covered about half the zebrafish yolk cell, a thickening occurs throughout the margin of the deep cells. This thickening, called the **germ ring**, is composed of a superficial layer, the epiblast (which will become the ectoderm); and an inner layer, the **hypoblast** (which will become endoderm and mesoderm). The hypoblast forms in a synchronous "wave" of internalization (Keller et al. 2008) that has some characteristics of ingression (especially in the dorsal region; see Carmany-Rampey and Schier 2001) and some elements of involution (especially in the future ventral regions). Thus, as the cells of the blastoderm undergo epiboly around the yolk, they are also internalizing cells at the blastoderm margin to form the hypoblast. The epiblast cells (presumptive ectoderm) do not involute, whereas the deep cells (the future mesoderm and endoderm) do. As the hypoblast cells internalize, the future mesoderm cells (the majority of the hypoblast cells) initially migrate vegetally while proliferating to make new mesoderm cells. Later, they alter their direction and proceed toward the animal pole. The endodermal precursors, however, appear to move randomly over the yolk (Pézeron et al. 2008).

THE EMBRYONIC SHIELD AND THE NEURAL KEEL Once the hypoblast has formed, cells of the epiblast and hypoblast intercalate on the future dorsal side of the embryo to form a localized thickening, the **embryonic shield** (Schmitz and Campos-Ortega 1994). Here, the cells converge and extend anteriorly, eventually narrowing along the dorsal midline (**FIGURE 8.42A**). This convergent extension in the hypoblast

(A)

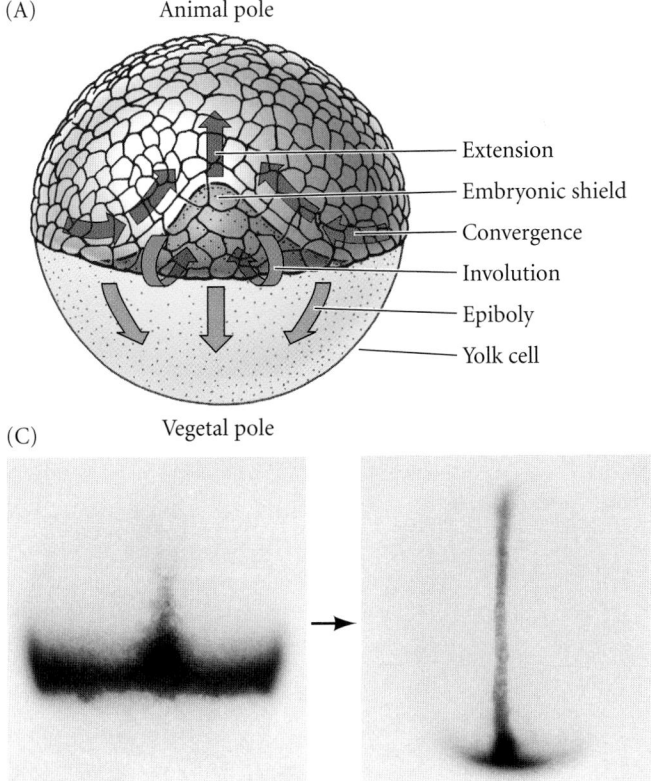

Animal pole

— Extension
— Embryonic shield
— Convergence
— Involution
— Epiboly
— Yolk cell

Vegetal pole

(B)

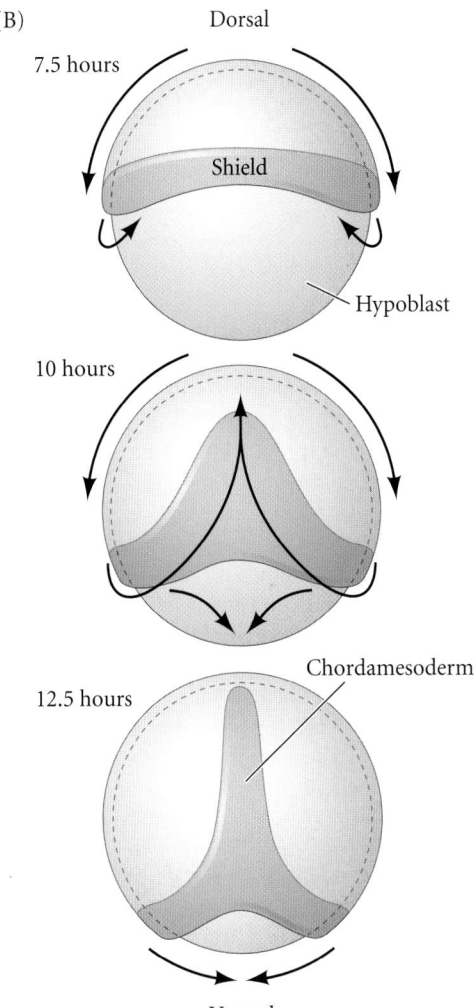

Dorsal

7.5 hours

Shield

Hypoblast

10 hours

12.5 hours

Chordamesoderm

Ventral

(C)

FIGURE 8.42 Convergence and extension in the zebrafish gastrula. (A) Dorsal view of convergence and extension movements during zebrafish gastrulation. Epiboly spreads the blastoderm over the yolk; involution and ingression generate the hypoblast; convergence and extension bring the hypoblast and epiblast cells to the dorsal side to form the embryonic shield. Within the shield, intercalation extends the chordamesoderm toward the animal pole. (B) Model of mesendoderm (hypoblast) formation. Numbers indicate hours after fertilization. On the future dorsal side, the internalized cells undergo convergent extension to form the chordamesoderm (notochord) and the paraxial (somitic) mesoderm adjacent to it. On the ventral side, the hypoblast cells migrate with the epibolizing epiblast toward the vegetal pole, eventually converging there. (C) Convergent extension of the chordamesoderm of the hypoblast cells. These cells are marked by their expression of the *no-tail* gene (dark areas) encoding a T-box transcription factor. (A,C from Langeland and Kimmel 1997, courtesy of the authors; B after Keller et al. 2008.)

forms the chordamesoderm, the precursor of the notochord (Trinkaus 1992; **FIGURE 8.42B,C**). As we will see, the embryonic shield is functionally equivalent to the dorsal blastopore lip of amphibians, since it can organize a secondary embryonic axis when transplanted to a host embryo (Oppenheimer 1936; Ho 1992).

The cells adjacent to the chordamesoderm—the paraxial mesoderm cells—are the precursors of the mesodermal somites. Concomitant convergence and extension in the epiblast bring presumptive neural cells from the epiblast into the dorsal midline, where they form the **neural keel**. The neural keel, a band of neural precursors that extends over the axial and paraxial mesoderm, eventually develops a slit-like lumen to become the neural tube and to enter into the embryo. Those cells remaining in the epiblast become the epidermis. On the ventral side (see Figure 8.42B), the hypoblast ring moves toward the vegetal pole, migrating directly beneath the epiblast that is epibolizing itself over the yolk cell. Eventually, the ring closes at the vegetal pole, completing the internalization of those cells that are going to become the mesoderm and endoderm (Keller et al. 2008).

● **See WEBSITE 8.6** GFP zebrafish embryos: Movies and photographs

Axis Formation

By different mechanisms, the *Xenopus* egg and zebrafish egg have reached the same state: they have become multicellular; they have undergone gastrulation; and they have positioned their germ layers such that the ectoderm is on the outside, the endoderm is on the inside, and the mesoderm lies between them. We will now see that zebrafish form their body axes in ways very similar to those of *Xenopus*, and using very similar molecules.

Dorsal-ventral axis formation

As mentioned above, the embryonic shield of fish is homologous to the dorsal blastopore lip of amphibians, and it is critical in establishing the dorsal-ventral axis. Shield tissue can convert lateral and ventral mesoderm (blood and connective tissue precursors) into dorsal mesoderm (notochord and somites), and it can cause the ectoderm to become neural rather than epidermal. This transformative capacity was shown by transplantation experiments in which the embryonic shield of an early-gastrula embryo was transplanted to the ventral side of another (**FIGURE 8.43**; Oppenheimer 1936; Koshida et al. 1998). Two axes formed, sharing a common yolk cell. Although the prechordal plate and notochord were derived from the donor embryonic shield, the other organs of the secondary axis came from host tissues that would normally form ventral structures. The new axis had been induced by the donor cells.

Like the amphibian blastopore lip, the embryonic shield forms the prechordal plate and the notochord of the developing embryo. The prechordal plate cells are the first to involute, and they migrate toward the animal pole (Dumortier et al. 2012). The presumptive prechordal plate and notochord are responsible for inducing ectoderm to become neural ectoderm, and they appear to do this in a manner very much like the homologous structures in amphibians.* Like amphibians, fish specify the epidermis by using bone morphogenetic proteins (BMPs) and certain Wnt proteins to induce the ectoderm (see Tucker et al. 2008). These BMPs and Wnts are made in the ventral and lateral regions of the embryo. The notochords of both fish and amphibians secrete factors that block this induction, thereby allowing the ectoderm to become neural. In fish, BMP2B induces embryonic cells to acquire ventral and lateral fates, and Wnt8 ventralizes, lateralizes, and posteriorizes the embryonic tissues (see Schier 2001). The protein secreted by the chordamesoderm that binds with and inactivates BMP2B is the zebrafish homologue of Chordin (**FIGURE 8.44**; Kishimoto et al. 1997; Schulte-Merker et al. 1997). If the *chordin* gene is mutated, the neural tube fails to form properly; and if *chordin*, *noggin*, and *follistatin* homologues are all inactivated, the neural tube fails to form at all, just like in *Xenopus* (Dal-Pra et al. 2006).

Extracellular BMP antagonists such as Chordin and Noggin are not the only important factors for specifying the neural plate. FGFs made in the dorsal side of the embryo also

---

* Another similarity between the amphibian and fish organizers is that they can be duplicated by rotating the egg and changing the orientation of the microtubules (Fluck et al. 1998). One difference in the axial development of these groups is that in amphibians, the prechordal plate is necessary for inducing the anterior brain to form. In zebrafish, although the prechordal plate appears to be necessary for forming ventral neural structures, the anterior regions of the brain can form in its absence (Schier et al. 1997; Schier and Talbot 1998).

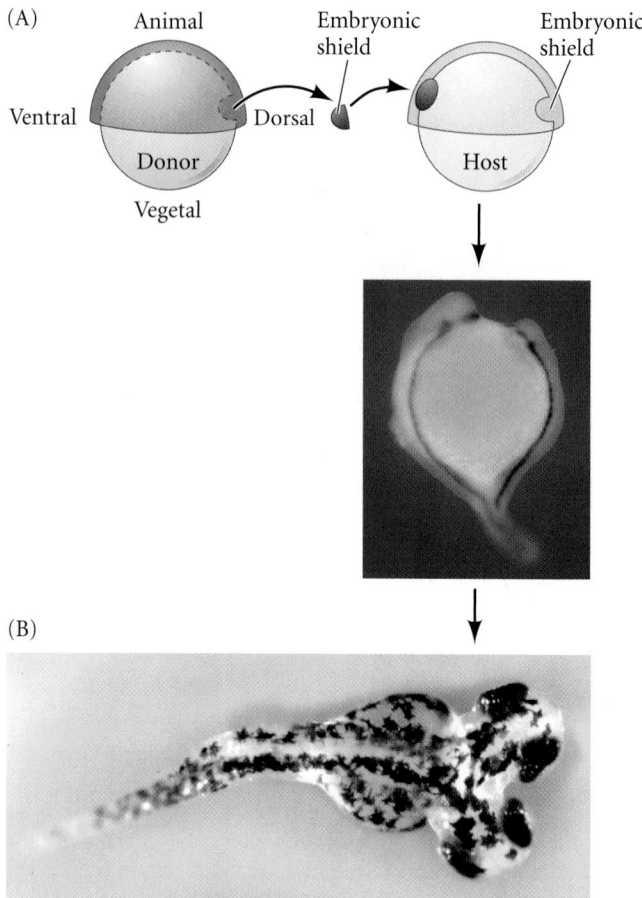

FIGURE 8.43   The embryonic shield as organizer in the fish embryo. (A) A donor embryonic shield (about 100 cells from a stained embryo) is transplanted into a host embryo at the same early-gastrula stage. The result is two embryonic axes joined to the host's yolk cell. In the photograph, both axes have been stained for *sonic hedgehog* mRNA, which is expressed in the ventral midline. (The embryo to the right is the secondary axis.) (B) The same effect can be achieved by activating nuclear catenin in embryos at sites opposite where the embryonic shield will form. (A after Shinya et al. 1999, photograph courtesy of the authors; B courtesy of J. C. Izpisúa-Belmonte.)

inhibit BMP gene expresson (Fürthauer et al. 2004; Tsang et al. 2004; Little and Mullins 2006). In the caudal region of the embryo, FGF signaling is the predominant neural specifier (Kudoh et al. 2004). And as in *Xenopus*, insulin-like growth factors (IGFs) also play a role in the production of the anterior neural plate. Zebrafish IGFs appear to upregulate *chordin* and *goosecoid* while restricting the expression of *bmp2b*. Although IGFs appear to be made throughout the embryo, during gastrulation the IGF *receptors* are found predominantly in the anterior portion of the embryo (Eivers et al. 2004). Also, Wnt inhibitors appear to play roles in head formation, as antisense morpholinos that downregulate Wnt3a and Wnt8 lead to the anteriorization of the neural plate at the expense of trunk structures (Shimizu et al. 2005b).

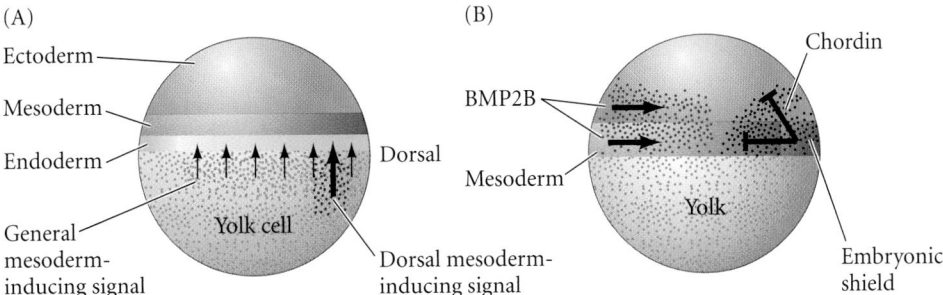

FIGURE 8.44 Axis formation in the zebrafish embryo. (A) Prior to gastrulation, the zebrafish blastoderm is arranged with the presumptive ectoderm near the animal pole, the presumptive mesoderm beneath it, and the presumptive endoderm sitting atop the yolk cell. The yolk syncytial layer (and possibly the endoderm) sends two signals to the presumptive mesoderm. One signal (lighter arrows) induces the mesoderm; a second signal (heavy arrow) specifically induces an area of mesoderm to become the dorsal mesoderm (embryonic shield). (B) Formation of the dorsal-ventral axis. During gastrulation, the ventral mesoderm secretes BMP2B (arrows) to induce the ventral and lateral mesodermal and epidermal differentiation. The dorsal mesoderm secretes factors (such as Chordin) that block BMP2B and dorsalize the mesoderm and ectoderm (converting the latter into neural tissue). (After Schier and Talbot 1998.)

The fish Nieuwkoop center

It thus appears that zebrafish have an "organizer" region that dorsalizes the region through BMP and Wnt inhibitors related to similar proteins seen in *Xenopus*. It also turns out that both zebrafish and *Xenopus* use β-catenin and Nodal-related proteins to form the dorsal mesoderm and enable this mesoderm to express the organizer genes. Both species use BMPs and Wnts to lateralize and posteriorize the embryo, and in both the organizer genes encode proteins such as Chordin, Noggin, and Dickkopf that inhibit BMPs and Wnts. Furthermore, later in development both zebrafish and *Xenopus* use a particular Wnt protein to posteriorize the ectoderm, forming the trunk neural tube. In some instances, fish and amphibians use these proteins in different ways, but in both groups the result is a structure that is definitely recognizable as a vertebrate embryo.

The zebrafish embryonic shield appears to acquire its organizing ability in much the same way as its amphibian counterparts do. In amphibians, the endodermal cells beneath the dorsal blastopore lip (i.e., the Nieuwkoop center) accumulate β-catenin synthesized from maternal messages, enabling the amphibian endoderm to induce the cells above it to become the dorsal blastopore lip. In zebrafish, the nuclei in that part of the yolk syncytial layer lying beneath the cells express Nodal-related proteins, and while β-catenin proteins are also made throughout the YSL, they become stabilized through Wnt signaling on the presumptive dorsal side of the embryo, the embryonic shield (see Langdon and Mullins 2011). Indeed, the presence of β-catenin distinguishes dorsal YSL from the lateral and ventral YSL regions* (**FIGURE 8.45A**; Schneider et al. 1996). Inducing β-catenin accumulation on the ventral side of the egg results in dorsalization and a second embryonic axis (Kelly et al. 1995).

β-catenin in the embryonic shield combines with the zebrafish homologue of Tcf3 to form a transcription factor that activates genes encoding several mesoderm-patterning proteins. One of them is an FGF that represses BMP gene expression on the dorsal side of the embryo (Fürthauer et al. 2004; Tsang et al. 2004). Other patterning proteins induced by β-catenin include Squint and Bozozok, which are very similar to the proteins that pattern the amphibian mesoderm. Squint is a Nodal-related paracrine factor, while Bozozok† is a homeodomain protein similar to the amphibian Nieuwkoop center protein Siamois.

Bozokok works in several ways (see Langdon and Mullins 2011). First, acting alone, it can repress BMP and Wnt genes that would promote ventral functions (Solnica-Krezel and Driever 2001). Second, it suppresses a transcriptional inhibitor (the *vega1* gene), allowing the organizer genes to function (Kawahara et al. 2000). Third, Bozozok and Squint act individually to activate the *chordin* gene, and they also act synergistically to activate other organizer genes such as *goosecoid*, *noggin*, and *dickkopf* (**FIGURE 8.45B**; Sampath et al. 1998; Gritsman et al. 2000; Schier and Talbot 2001). These genes encode proteins that block BMPs and Wnts and allow the specification of the dorsal mesoderm and neural ectoderm. Thus, the embryonic shield is considered equivalent to the amphibian organizer, and the dorsal part of the yolk cell, together with the dorsal marginal blastomeres (the precursors of the Kupffer cells; see below) can be thought of as the Nieuwkoop center of the teleost fish embryo.

---

* Some of the endodermal cells that accumulate β-catenin will become the precursors of the ciliated cells of Kupffer's vesicle (Cooper and D'Amico 1996). As we will discuss in the final section of this chapter, these cells are critical in determining the left-right axis of the embryo.

---

† *Bozozok* is Japanese slang for an arrogant youth on a motorcycle. The gene's name was derived from the severe loss-of-function phenotype wherein a single-eyed embryo curves ventrally over the yolk cell (i.e., resembling a rider on a fast motorcycle). However, this gene is also known as *dharma* (after a famous Buddhist priest) because embryos with gain of function—too much of this protein, the result of experimentally injecting its mRNA into the embryo—develop huge eyes and head, but no trunk or tail; they thus resemble Japanese Dharma dolls (Yamanaka et al. 1998; Fekany et al. 1999).

(A)

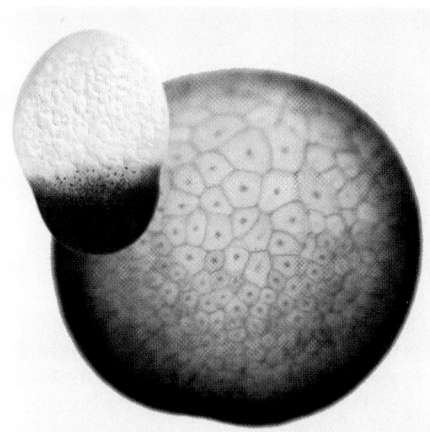

(B)

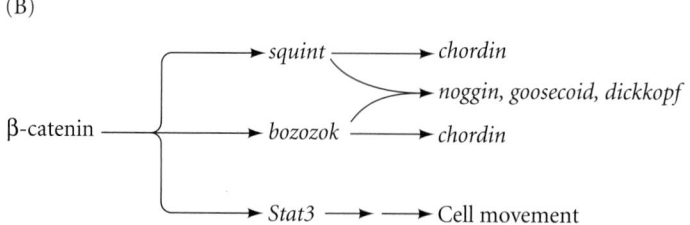

FIGURE 8.45 β-catenin activates organizer genes in the zebrafish. (A) Nuclear localization of β-catenin marks the dorsal side of the *Xenopus* blastula (larger image) and helps form its Nieuwkoop center beneath the organizer. In the zebrafish late blastula (smaller image), nuclear localization of β-catenin is seen in the yolk syncytial layer nuclei beneath the future embryonic shield. (B) β-catenin activates *squint* and *bozozok*, whose proteins activate organizer-specific genes as well as the *Stat3* gene whose product is necessary for gastrulation movements. (A courtesy of S. Schneider; B after Schier and Talbot 2001.)

Anterior-posterior axis formation

The patterning of the neural ectoderm along the anterior-posterior axis in the zebrafish appears to be the result of the interplay of FGFs, Wnts, and retinoic acid, similar to that seen in *Xenopus*. In fish embryos, there seem to be two separate processes. First a Wnt signal represses the expression of anterior genes; then Wnts, retinoic acid, and FGFs are required to activate the posterior genes.

This regulation of anterior-posterior identity appears to be coordinated by **retinoic acid-4-hydroxylase**, an enzyme that degrades RA (Kudoh et al. 2002; Dobbs-McAuliffe et al. 2004). The gene encoding this enzyme, *cyp26*, is expressed specifically in the region of the embryo destined to become the anterior end. Indeed, this gene's expression is first seen during the late blastula stage, and by the time of gastrulation it defines the presumptive anterior neural plate. Retinoic acid-4-hydroxylase prevents the accumulation of RA at the embryo's anterior end, blocking the expression of the posterior genes there. This inhibition is reciprocated, since the posteriorly expressed FGFs and Wnts inhibit the expression of the *cyp26* gene, as well as inhibiting the expression of the head-specifying gene *Otx2*. This mutual inhibition creates a border between the zone of posterior gene expression and the zone of anterior gene expression. As epiboly continues, more and more of the body axis is specified to become posterior (**FIGURE 8.46A**).

Retinoic acid acts as a morphogen, regulating cell properties depending on its concentration. Cells receiving very little RA express anterior genes; cells receiving high levels of RA express posterior genes; and those cells receiving intermediate levels of RA express genes characteristic of cells between the anterior and posterior regions. This morphogenesis is extremely important in the hindbrain, where different levels of RA specify different types of cells along the anterior-posterior axis (White et al. 2007). FGFs positively regulate RA accumulation both by inhibiting the expression of *cyp26* (which encodes the enzyme that degrades RA) and by enhancing the synthesis of the enzyme that synthesizes RA from vitamin A (**FIGURE 8.46B**). The result is a gradient of Cyp26, which in turn generates an RA gradient that is stable and can be sustained throughout the time the embryo is elongating.

● **See VADE MECUM** Retinoic acid as a teratogen

Left-right axis formation

In all vertebrates studied, the right and left sides differ both anatomically and developmentally. In fish, the heart is on the left side and there are different structures in the left and right regions of the brain (**FIGURE 8.47**). Moreover, as in other vertebrates, the cells on the left side of the body are given that information by Notch and Nodal signaling and by the Pitx2 transcription factor, whereas the cells on the right side of the body are exposed to FGF signaling. The ways the different vertebrate classes accomplish this asymmetry differ, but recent evidence suggests that the currents produced by cilia in the node may be responsible for left-right axis formation in all the vertebrate classes (Okada et al. 2005); when the gene for a dynein subunit of cilia was cloned, it was found to be expressed in the ventral portion of the node (or organizer) in mouse, chick, *Xenopus*, and zebrafish embryos (Essner et al. 2002, 2005). Using extremely rapid (500 frames/sec) photography, Okada and colleagues (2005) showed that the clockwise rotational motion of cilia and the leftward flow of particles were conserved in all these vertebrate groups, despite the different types and shapes of their nodal structures.

In zebrafish, the nodal structure housing the cilia that control left-right asymmetry is a transient fluid-filled organ called **Kupffer's vesicle**. As mentioned earlier, Kupffer's vesicle arises from a group of dorsal cells near the embryonic shield shortly after gastrulation. Essner and colleagues (2002, 2005) were able to inject small beads into Kupffer's vesicle and see their translocation from one side of the vesicle to the other. Blocking ciliary function by preventing the synthesis of dynein or by ablating the precursors of the ciliated cells prevented normal left-right axis formation. The cilia activate the Nodal-related genes on the left side of the embryo. These genes are critically important in initiating the cascades of paracrine factors that are specific to the left side of the body (Rebagliati et al. 1998; Long et al. 2003).

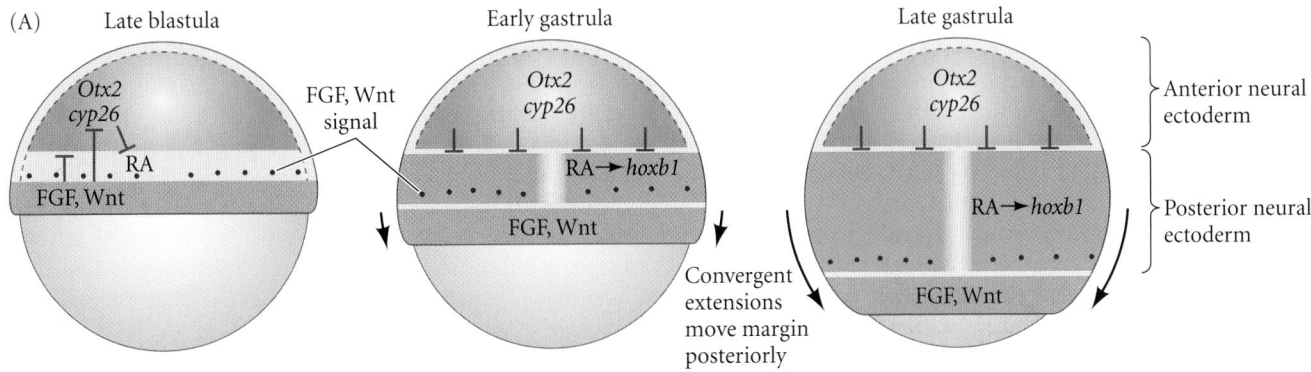

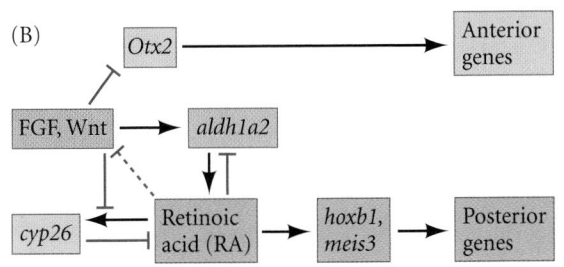

FIGURE 8.46 Model for specification of zebrafish neural ectoderm by FGF, Wnt, and retinoic acid (RA) signaling. The *cyp26* gene encodes retinoic acid-4-hydroxylase, a cytochrome family enzyme that degrades RA. (A) Temporal sequence of the posteriorization process. At the late blastula stage, the *cyp26* gene is confined to the anterior region by FGFs and Wnts from the margin. After the start of gastrulation, convergent extension takes the margin farther from the anterior. RA accumulates in this region, activating genes associated with the posterior neural ectoderm. (B) Pathway through which a boundary can form between anterior (*cyp26*-, *Otx2*-expressing) and posterior (*hoxb1*-, *meis3*-expressing) neural ectoderm. In the posterior region, FGFs and/or Wnt suppress anterior genes such as *Otx2*. The FGF/Wnt signal in the posterior also suppresses expression of *cyp26* (which encodes the enzyme that *degrades* RA) and enhances the expression of *aldh1a2* (which encodes the enzyme that *synthesizes* RA), enabling RA to accumulate and activate the posterior genes. (After Kudoh et al. 2002 and White et al. 2007.)

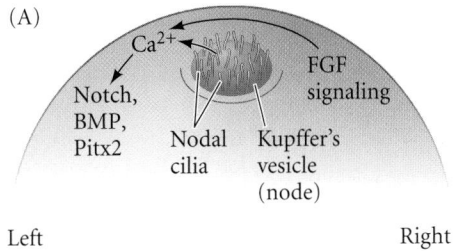

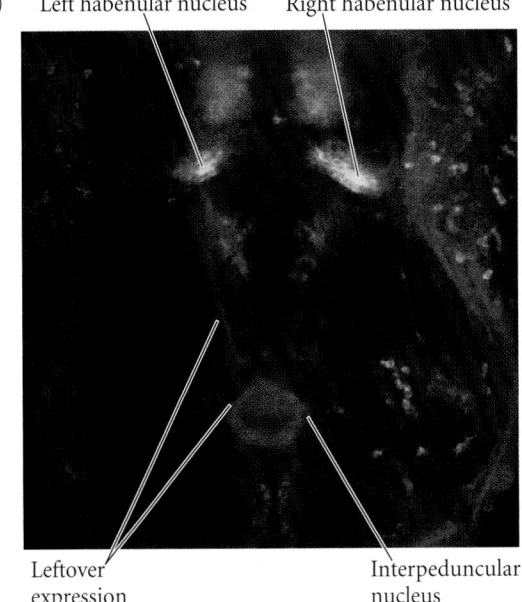

FIGURE 8.47 Left-right asymmetry in the zebrafish embryo. (A) Model for asymmetric gene expression. As in other vertebrates, Nodal-related genes are expressed solely on the embryo's left side: Nodal cilia in Kupffer's vesicle create a current that releases Ca^{2+}, stimulating the Notch and BMP pathways and activating the Pitx2 transcription factor in the left-hand mesoderm (blue). By contrast, FGF expression is seen predominantly on the right-hand side (red). (B) Brain asymmetry in zebrafish. Antibody staining of the Leftover (red) and Right-on (green) proteins in neurons of the habenular nucleus (a behavior-controlling region of the zebrafish forebrain) and the axonal projections to their midbrain target (the interpeduncular nucleus) reveals marked asymmetry. Most Leftover-positive axons emerge from the left habenula to innervate the target. (A after Okada et al. 2005; B from Gamse et al. 2005, courtesy of M. Halpern.)

Coda

Researchers analyzing the development of *Xenopus laevis* and *Danio rerio* are finally putting names to the "agents" and "soluble factors" postulated by the early experimental embryologists. We are also delineating the intercellular pathways of paracrine and transcription factors that constitute the first steps in organogenesis. The international research program initiated by Spemann's laboratory in the 1920s is reaching fruition, and this research has revealed layers of complexity beyond anything Spemann could have conceived. Just as Spemann's experiments told us how much we didn't know, the answers we have found to these older questions have generated an entirely new set of questions. The ability to perform live imaging of cells, the ability to delete and add new genes to particular cells, and the ability to produce mutant lines of zebrafish have all helped us understand the mechanisms by which order is generated in the embryo. The remarkable conservation of mechanisms and molecules between model amphibian and fish species, despite the initial differences in their cleavage, shows very strong evolutionary constraints on how this order comes into being.

SNAPSHOT SUMMARY: Early Development of Amphibians and Fish

1. Amphibian cleavage is holoblastic, but it is unequal because of the presence of yolk in the vegetal hemisphere.

2. Amphibian gastrulation begins with the invagination of the bottle cells, followed by the coordinated involution of the mesoderm and the epiboly of the ectoderm. Vegetal rotation plays a significant role in directing the involution.

3. The driving forces for ectodermal epiboly and the convergent extension of the mesoderm are the intercalation events in which several tissue layers merge. Fibronectin plays a critical role in enabling the mesodermal cells to migrate into the embryo.

4. The dorsal lip of the blastopore forms the organizer tissue of the amphibian gastrula. This tissue dorsalizes the ectoderm, transforming it into neural tissue, and it transforms ventral mesoderm into lateral and dorsal mesoderm.

5. The organizer consists of pharyngeal endoderm, head mesoderm, notochord, and dorsal blastopore lip tissues. The organizer functions by secreting proteins (Noggin, Chordin, and Follistatin) that block the BMP signal that would otherwise ventralize the mesoderm and activate the epidermal genes in the ectoderm.

6. Dorsal-ventral specification begins with maternal messages and proteins stored in the vegetal cytoplasm. These include Nodal-like paracrine factors, transcription factors (such as VegT), and agents that protect β-catenin from degradation.

7. The organizer is itself induced by the Nieuwkoop center, located in the dorsalmost vegetal cells. This center is formed by the translocation of the Disheveled protein and Wnt11 to the dorsal side of the egg to stabilize β-catenin in the dorsal cells of the embryo.

8. The Nieuwkoop center is formed by the accumulation of β-catenin, which can complex with Tcf3 to form a transcription factor complex that can activate the transcription of the *siamois* and *twin* genes on the dorsal side of the embryo.

9. The Siamois and Twin proteins collaborate with activated Smad2 transcription factors generated by the TGF-β pathway (Nodal, Vg1) to activate genes encoding BMP inhibitors. These inhibitors include the secreted factors Noggin, Chordin, and Follistatin, as well as the transcription factor Goosecoid.

10. In the presence of BMP inhibitors, ectodermal cells form neural tissue. The action of BMP on ectodermal cells causes them to become epidermis.

11. In the head region, an additional set of proteins (Cerberus, Frzb, Dickkopf, Tiki) blocks the Wnt signal from the ventral and lateral mesoderm.

12. Wnt signaling causes a gradient of β-catenin along the anterior-posterior axis of the neural plate. This graded signaling appears to specify the regionalization of the neural tube.

13. Insulin-like growth factors (IGFs) help transform the neural tube into anterior (forebrain) tissue.

14. The left-right axis appears to be initiated by the activation of a Nodal protein solely on the left side of the embryo. In *Xenopus*, as in other vertebrates, Nodal protein activates expression of *pitx2*, which is critical in distinguishing left-sidedness from right-sidedness in the heart and gut tubes.

15. Cleavage in fish is meroblastic. The deep cells of the blastoderm form between the yolk syncytial layer and the enveloping layer. These deep cells migrate over the top of the yolk, forming the hypoblast and epiblast layers.

16. On the future dorsal side, the hypoblast and epiblast intercalate to form the embryonic shield, a structure homologous to the amphibian organizer. Transplantation of the embryonic shield into the ventral side of another embryo will cause a second embryonic axis to form.

17. In both amphibians and fish, neural ectoderm is permitted to form where the BMP-mediated induction of epidermal tissue is prevented. The fish embryonic shield, like the amphibian dorsal blastopore lip, secretes the BMP antagonists. Like the amphibian organizer, the shield receives its abilities by being induced by β-catenin and by underlying endodermal cells expressing Nodal-related paracrine factors.

For Further Reading

Complete bibliographical citations for all literature cited in this chapter can be found at the free-access website **www.devbio.com**

Cho, K. W. Y., B. Blumberg, H. Steinbeisser and E. De Robertis. 1991. Molecular nature of Spemann's organizer: The role of the *Xenopus* homeobox gene *goosecoid*. *Cell* 67: 1111–1120.

De Robertis, E. M. 2006. Spemann's organizer and self-regulation in amphibian embryos. *Nature Rev. Mol. Cell Biol.* 7: 296–302.

Essner, J. J., J. D. Amack, M. K. Nyholm, E. B. Harris and H. J. Yost. 2005. Kupffer's vesicle is a ciliated organ of asymmetry in the zebrafish embryo that initiates left-right development of the brain, heart and gut. *Development* 132: 1247–1260.

Khokha, M. K., J. Yeh, T. C. Grammer and R. M. Harland. 2005. Depletion of three BMP antagonists from Spemann's organizer leads to catastrophic loss of dorsal structures. *Dev. Cell* 8: 401–411.

Langdon, Y. G. and M. C. Mullins. 2011. Maternal and zygotic control of zebrafish dorsoventral axial patterning. *Annu. Rev. Genet.* 45: 357–377.

Larabell, C. A. and 7 others. 1997. Establishment of the dorsal-ventral axis in *Xenopus* embryos is presaged by early asymmetries in β-catenin which are modulated by the Wnt signaling pathway. *J. Cell Biol.* 136: 1123–1136.

Lepage, E. S. and A. E. Bruce. 2010. Zebrafish epiboly: Mechanics and mechanisms. *Int. J. Dev. Biol.* 54: 1213–1228.

Niehrs, C. 2004. Regionally specific induction by the Spemann-Mangold organizer. *Nature Rev. Genet.* 5: 425–434.

Piccolo, S., E. Agius, L. Leyns, S. Bhattacharyya, H. Grunz, T. Bouwmeester and E. M. DeRobertis. 1999. The head inducer Cerberus is a multifunctional antagonist of Nodal, BMP, and Wnt signals. *Nature* 397: 707–710.

Reversade, B., H. Kuroda, H. Lee, A. Mays, and E. M. De Robertis. 2005. Deletion of BMP2, BMP4, and BMP7 and Spemann organizer signals induces massive brain formation in *Xenopus* embryos. *Development* 132: 3381–3392.

Spemann, H. and H. Mangold. 1924. Induction of embryonic primordia by implantation of organizers from a different species. (Trans. V. Hamburger.) In B. H. Willier and J. M. Oppenheimer (eds.), *Foundations of Experimental Embryology*. Hafner, New York, pp. 144–184. Reprinted in *Int. J. Dev. Biol.* 45: 13–38.

Winklbauer, R. and E. W. Damm. 2011. Internalizing the vegetal cell mass before and during amphibian gastrulation: Vegetal rotation and related movements. *WIREs Dev Biol.* Doi:10.1002/wdev.26.

Go Online

WEBSITE 8.1 Movies of amphibian development. A compilation of amphibian development movies.

WEBSITE 8.2 Migration of the mesodermal mantle. Different growth rates coupled with the intercalation of cell layers allows the mesoderm to expand in a tightly coordinated fashion.

WEBSITE 8.3 Spemann, Mangold, and the organizer. Spemann did not see the importance of this work the first time he and Mangold did it. This website provides a more detailed account of why Spemann and Mangold did this experiment.

WEBSITE 8.4 Early attempts to locate the organizer molecules. Although Spemann did not believe that molecules alone could organize the embryo, his students began a long quest for these factors.

WEBSITE 8.5 Gradients and Hox gene expression. The mechanisms by which the gradients of Wnt, RA, and FGFs specify the Hox genes in the neural ectoderm are still not agreed upon.

WEBSITE 8.6 GFP zebrafish embryos: Movies and photographs. The ability to photograph and film living embryos expressing the GFP reporter gene driven by promoters of specific genes has opened a new dimension in developmental biology and has allowed us to link gene structure to developmental anatomy.

Vade Mecum

Amphibian development. The events of cleavage and gastrulation are difficult to envision without three-dimensional models. You can see movies of such 3-D models, as well as footage of a living *Xenopus* embryo.

The primary organizer and the double-gradient hypothesis. These movies explain Spemann and Mangold's discovery of the primary organizer, Holtfreter's discovery of the "dead organizer," and Lauri Saxén's work on the double-gradient hypothesis.

Zebrafish development. This full account of zebrafish development includes time-lapse movies of the beautiful and rapid development of this organism.

Retinoic acid as a teratogen. One feature in the zebrafish development segment is a visualization of the teratogenic effects of retinoic acid on development.

Outside Sites

A *Xenopus* web resource called Xenbase chronicles research and discoveries in the genetics, development, and cell biology of *Xenopus* and is updated frequently. **http://www.xenbase.org**

The zebrafish model organism database Zfin contains protocols, atlases, and books concerning zebrafish. **http://zfin.org**

9

Early Development in Vertebrates
Birds and Mammals

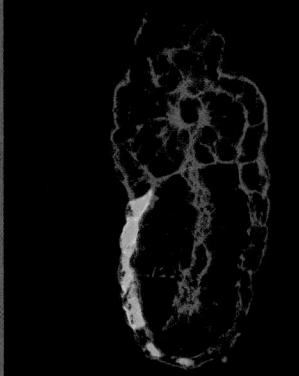

THIS FINAL CHAPTER ON THE PROCESSES OF EARLY DEVELOPMENT extends our survey of vertebrate development to include the **amniotes**—those vertebrates whose embryos form an amnion, or water sac (i.e., the reptiles, birds, and mammals). Birds and reptiles follow a very similar pattern of development (Gilland and Burke 2004; Coolen et al. 2008), and birds are considered by modern taxonomists to be a reptilian clade (**FIGURE 9.1A**).

The **amniote egg** is characterized by a set of membranes that together enable the embryo to survive on land (**FIGURE 9.1B**). First, the **amnion**, for which the amniote egg is named, is formed early in embryonic development and enables the embryo to float in a fluid environment that protects it from dessication. Another cell layer derived from the embryo, the **yolk sac**, enables nutrient uptake and the development of the circulatory system. The **allantois**, developing at the posterior of the embryo, stores waste products. The **chorion** contains blood vessels that exchange gases with the outside environment. In birds and most reptiles, the embryo and its membranes are enclosed in a hard or leathery shell within which the embryo develops outside the mother's body. Cleavage in bird and reptile eggs, like that of the bony fishes described in the last chapter, is meroblastic, with only a small portion of the egg cytoplasm being used to make the cells of the embryo. The vast majority of the large egg is composed of yolk that will nourish the growing embryo.

In most mammals, holoblastic cleavage is modified to accommodate the formation of a **placenta**, which enables the embryo to develop inside another organism. The placenta is an organ containing tissues and blood vessels from both the embryo and the mother. Gas exchange, nutrient uptake, and waste elimination take place through the placenta.

Most research on the development of birds has been done on the domestic chicken (*Gallus gallus*). Ever since Aristotle first observed and recorded the details of its 3-week-long development, the chick has been a favorite organism for embryological studies. It is accessible year-round and is easily maintained. Moreover, at any particular temperature, its developmental stage can be accurately predicted, so large numbers of same-stage embryos can be obtained and manipulated. Chick organ formation is accomplished by genes and cell movements similar to those of mammalian organ formation, and the chick is one of the few organisms whose embryos are amenable to both surgical and genetic manipulations (Stern 2005a). Thus, the chick embryo has often served as a surrogate for human embryos, as has the ubiquitous laboratory mouse.

The mouse is the mammalian model organism of choice and is the subject of many studies involving genetic and surgical manipulation. In addition, the mouse genome was the first to be sequenced, and when it was first published many scientists felt it was more valuable than knowing the human genome sequence. Their reasoning was that "working on mouse models allows the manipulation of each and every gene to determine their functions" (Gunter and Dhand 2002). One cannot do that with humans. Human development is a subject of medical as well as general scientific interest, however, and the latter

My dear fellow … life is infinitely stranger than anything which the mind of man could invent. We would not dare to conceive the things which are really mere commonplaces of existence.

A. CONAN DOYLE (1891)

Between the fifth and tenth days the lump of stem cells differentiates into the overall building plan of the [mouse] embryo and its organs. It is a bit like a lump of iron turning into the space shuttle. In fact it is the profoundest wonder we can still imagine and accept, and at the same time so usual that we have to force ourselves to wonder about the wondrousness of this wonder.

MIROSLAV HOLUB (1990)

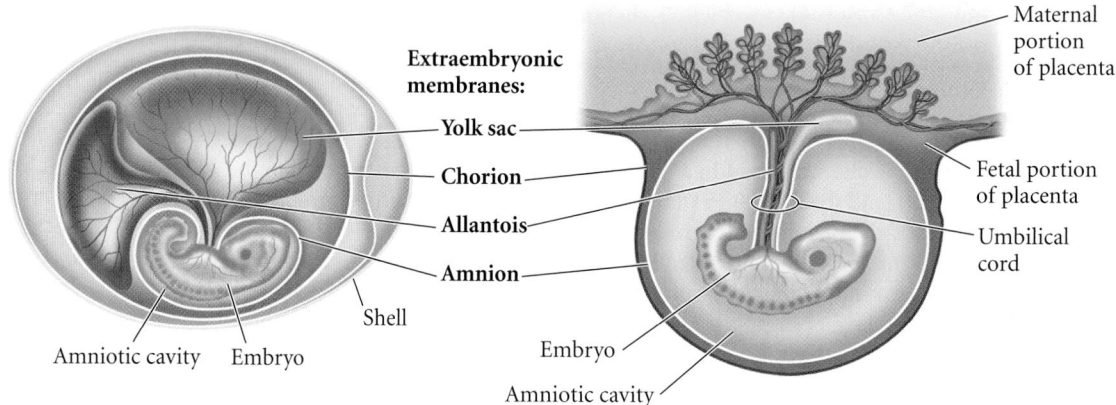

(A)
AMNIOTE VERTEBRATES

Reptiles

Gallus gallus

Feathers, flight

Birds

Amniote egg

Monotremes (egg-laying mammals: platypus, echidna)

Mammary glands

Placenta

Placental mammals

Mus musculus

(B)

Extraembryonic membranes:

Yolk sac

Chorion

Allantois

Amnion

Shell

Amniotic cavity   Embryo

Maternal portion of placenta

Fetal portion of placenta

Umbilical cord

Embryo

Amniotic cavity

FIGURE 9.1 The membranes of the amniote egg characterize reptiles, birds, and mammals. (A) Phylogenetic relationships of the amniotes. Note that birds are considered reptiles by most modern taxonomists, but for physiological studies they are often treated as separate taxa. (Other flying and feathered reptile groups have not survived to the present day.) The domestic chicken (*Gallus gallus*) is the most widely studied bird species. Among mammals, the development of the laboratory mouse *Mus musculus* is the most widely studied. Both avian and mouse studies contribute to our understanding of human development. (B) The shelled amniote egg (as exemplified by the chicken egg on the left) permitted animals to develop away from bodies of water. The amnion provides a "water sac" in which the embryo develops; the allantois stores wastes; and the blood vessels of the chorion exchange gases and the nutrients from the yolk sac. In mammals (right), this arrangement is modified such that the blood vessels acquire nutrients and exchange gases via a placenta joined to the mother's uterus rather than from the yolk sac. (B after Sadava et al. 2013.)

● **See WEBSITE 9.1** The extraembryonic membranes

sections of this chapter will cover early human development, illustrating the application of many of the principles we have described in model organisms.

EARLY DEVELOPMENT IN BIRDS

Cleavage

Fertilization of the chick egg occurs in the hen's oviduct, before the albumin and shell are secreted to cover it. Like the egg of the zebrafish, the chick egg is telolecithal, with a small disc of cytoplasm—the **blastodisc**—sitting on top of a large yolk. Like fish eggs, the yolky eggs of birds undergo **discoidal meroblastic cleavage**. Cleavage occurs only in the blastodisc, which is about 2–3 mm in diameter and is located at the animal pole of the egg (**FIGURE 9.2A**). The first cleavage furrow appears centrally in the blastodisc; other cleavages follow to create a single-layered **blastoderm** (**FIGURE 9.2B–F**). As in the fish embryo, the cleavages do not extend into the yolky cytoplasm, so the early-cleavage cells are continuous with one another and with the yolk at their bases. Thereafter, equatorial and vertical cleavages divide the blastoderm into a tissue 5–6 cell layers thick, and the cells become linked together by tight junctions (Bellairs et al. 1978; Eyal-Giladi 1991).

Between the blastoderm and the yolk of avian eggs is a space called the **subgerminal cavity**, which is created when the blastoderm cells absorb water from the albumin ("egg white") and secrete the fluid between themselves and the yolk (New 1956). At this stage, the deep cells in the center of the blastoderm appear to be shed and die, leaving behind a

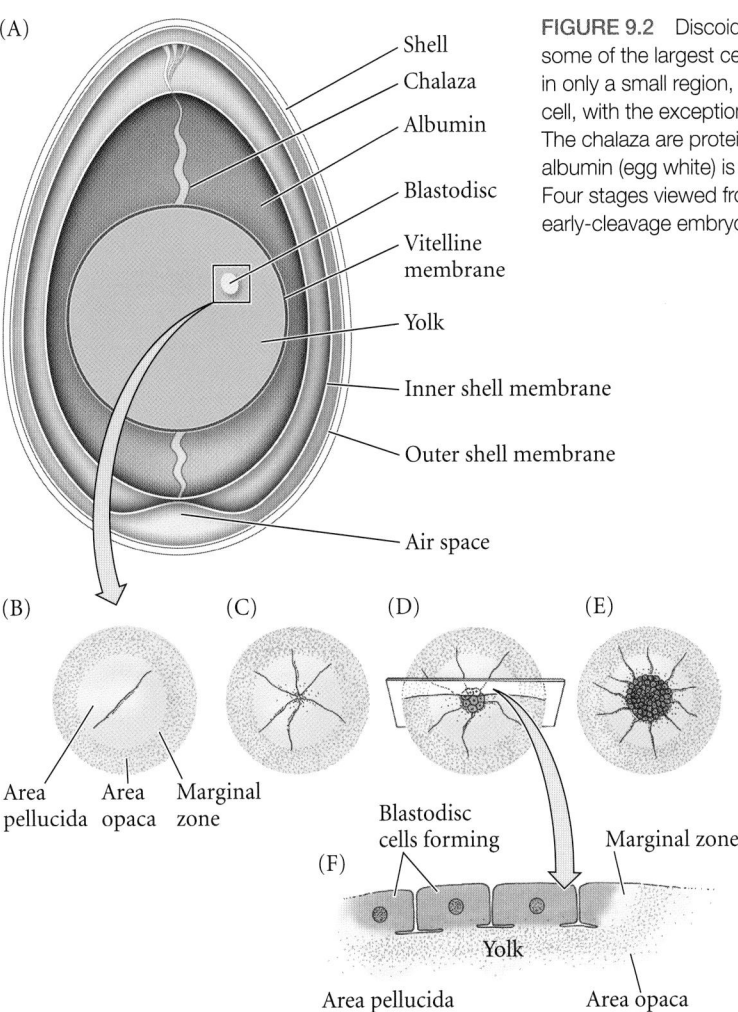

(A)

Shell
Chalaza
Albumin
Blastodisc
Vitelline membrane
Yolk
Inner shell membrane
Outer shell membrane
Air space

(B)

(C)

(D)

(E)

Area pellucida   Area opaca   Marginal zone

Blastodisc cells forming

Marginal zone

(F)

Yolk

Area pellucida   Area opaca

FIGURE 9.2 Discoidal meroblastic cleavage in a chick egg. (A) Avian eggs include some of the largest cells known to science (inches across), but cleavage takes place in only a small region, the blastodisc. The yolk fills up the entire cytoplasm of the egg cell, with the exception of a small blastodisc in which development will take place. The chalaza are protein strings that keep the yolky egg cell centered in the shell. The albumin (egg white) is secreted onto the egg in its passage out of the oviduct. (B–E) Four stages viewed from the animal pole (the future dorsal side of the embryo). (F) An early-cleavage embryo viewed from the side. (After Bellairs et al. 1978.)

posterior edge of the area pellucida. In between the area opaca and Koller's sickle is a beltlike region called the **posterior marginal zone** (**PMZ**). A sheet of cells derived from Koller's sickle migrates anteriorly beneath the surface. Meanwhile, cells in more anterior regions of the epiblast have delaminated and stay attached to the epiblast, to form hypoblast "islands," an archipelago of disconnected clusters of 5–20 cells each that migrate and become the **primary hypoblast** (**FIGURE 9.3A,B**). The sheet of cells that grows anteriorly from Koller's sickle combines with the primary hypoblast to form the complete hypoblast layer, also called the **secondary hypoblast** or **endoblast** (**FIGURE 9.3C–E**; Eyal-Giladi et al. 1992; Bertocchini and Stern 2002; Khaner 2007a,b). The resulting two-layered blastoderm (epiblast and hypoblast) is joined together at the marginal zone of the area opaca, and the space between the layers forms a blastocoel. Thus, although the shape and formation of the avian blastodisc differs from those of the amphibian, fish, or echinoderm blastula, the overall spatial relationships are retained.

The avian embryo comes entirely from the epiblast; the hypoblast does not contribute any cells to the developing embryo (Rosenquist 1966, 1972). Rather, the hypoblast cells form portions of the extraembryonic membranes (see Figure 9.1B), especially the yolk sac and the stalk linking the yolk mass to the endodermal digestive tube. Hypoblast cells also provide chemical signals that specify the migration of epiblast cells. However, the three germ layers of the embryo proper (plus the amnion, chorion, and allantois extraembryonic membranes) are formed solely from the epiblast (Schoenwolf 1991).

1-cell-thick **area pellucida**; this part of the blastoderm forms most of the actual embryo. The peripheral ring of blastoderm cells that have not shed their deep cells constitutes the **area opaca**. Between the area pellucida and the area opaca is a thin layer of cells called the **marginal zone*** (Eyal-Giladi 1997; Arendt and Nübler-Jung 1999). Some marginal zone cells become very important in determining cell fate during early chick development.

Gastrulation of the Avian Embryo

The hypoblast

By the time a hen has laid an egg, the blastoderm contains some 50,000 cells. At this time, most of the cells of the area pellucida remain at the surface, forming an "upper layer" called the **epiblast**. Shortly after the egg is laid, a local thickening of the epiblast, called **Koller's sickle**, is formed at the

The primitive streak

Avian, reptilian, and mammalian gastrulation takes place through the **primitive streak**, the equivalent of the blastopore lip of amphibian embryos. Dye-marking experiments and time-lapse cinemicrography indicate that the primitive streak first arises from Koller's sickle and the epiblast above it (Bachvarova et al. 1998; Lawson and Schoenwolf 2001a,b; Voiculescu et al. 2007).

As cells converge to form the primitive streak, a depression forms within the streak. This depression is called the **primitive groove**, and it serves as an opening through which migrating cells pass into the deep layers of the

---

*Arendt and Nübler-Jung (1999) have argued that the region should be called the marginal *belt* to distinguish it from the marginal *zone* of amphibians. Here we will continue to use the earlier nomenclature.

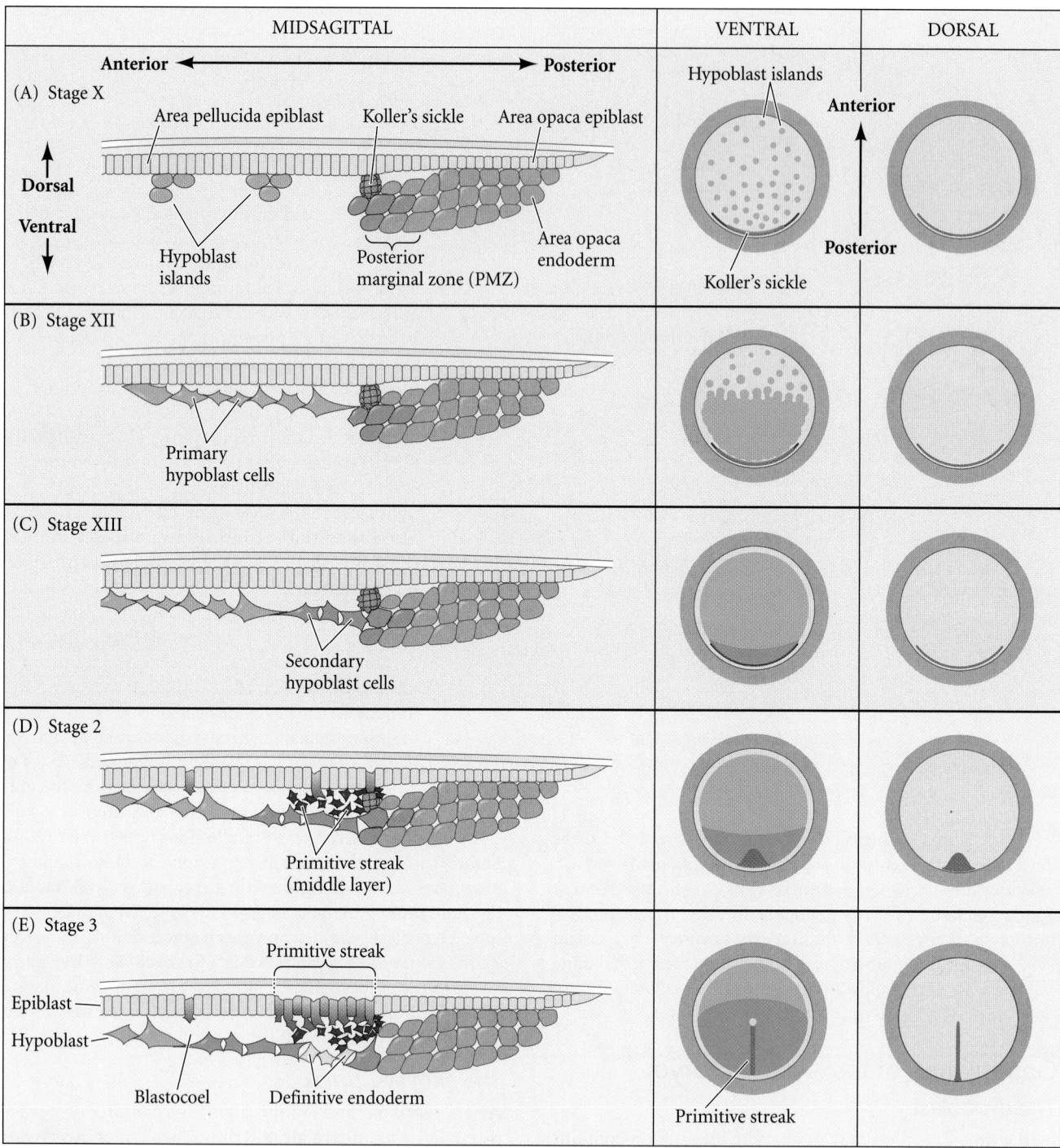

FIGURE 9.3 Formation of the chick blastoderm. The left column is a diagrammatic midsagittal section through part of the blastoderm. The middle column depicts the entire embryo viewed from the ventral side, showing the migration of the primary hypoblast and the secondary hypoblast (endoblast) cells. The right column shows the entire embryo seen from the dorsal side. (A–C) Events prior to laying of the shelled egg. (A) Stage X embryo, where islands of hypoblast cells can be seen, as well as a congregation of hypoblast cells around Koller's sickle. (B) By stage XII, a sheet of cells that grows anteriorly from Koller's sickle combines with the hypoblast islands to form the complete hypoblast layer. (C) By stage XIII, just prior to primitive streak formation, the formation of the hypoblast just been completed. (D) By stage 2 (12–14 hours after the egg is laid), the primitive streak cells form a third layer that lies between the hypoblast and epiblast cells. (E) By stage 3 (15–17 hours post laying), the primitive streak has become a definitive region of the epiblast, with cells migrating through it to become the mesoderm and endoderm. (After Stern 2004.)

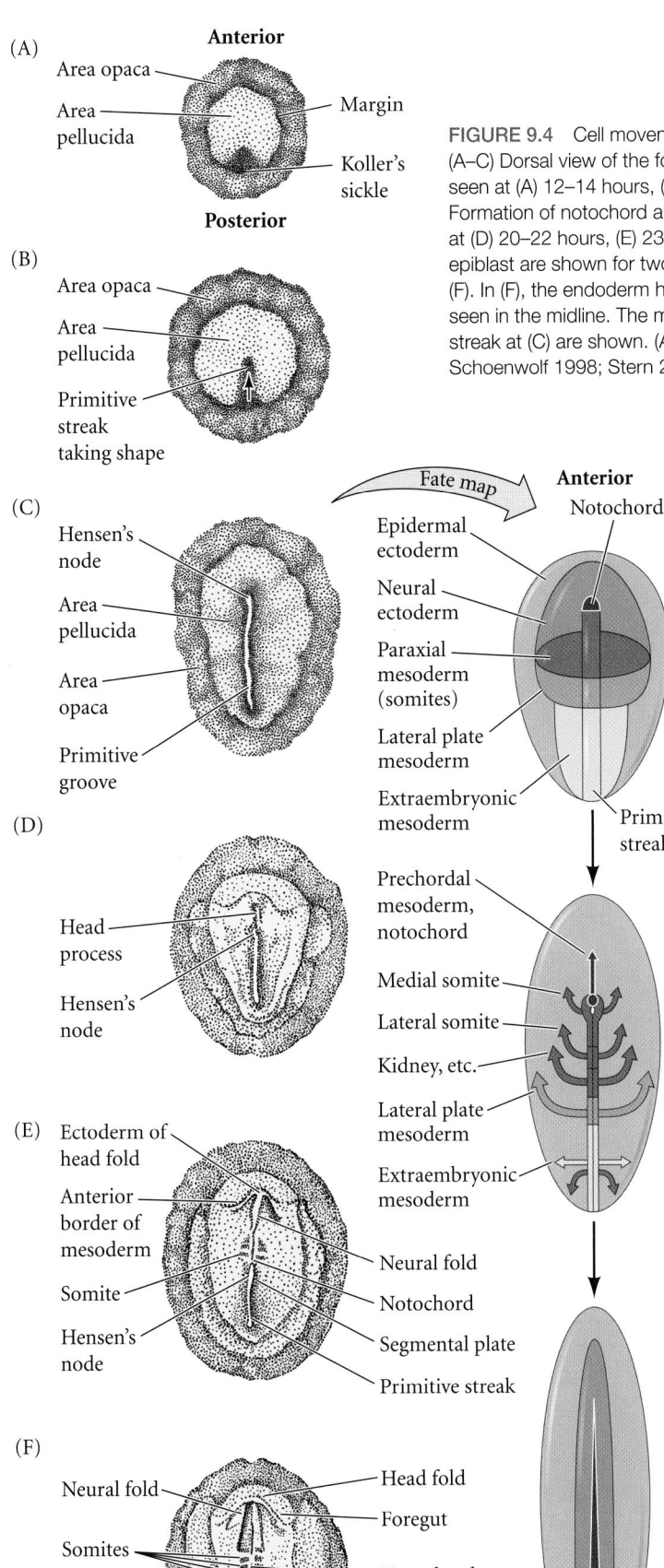

(A)

Anterior

Area opaca

Area pellucida

Margin

Koller's sickle

Posterior

(B)

Area opaca

Area pellucida

Primitive streak taking shape

(C)

Hensen's node

Area pellucida

Area opaca

Primitive groove

Fate map

Anterior

Notochord

Epidermal ectoderm

Neural ectoderm

Paraxial mesoderm (somites)

Lateral plate mesoderm

Extraembryonic mesoderm

Primitive streak

(D)

Head process

Hensen's node

Prechordal mesoderm, notochord

Medial somite

Lateral somite

Kidney, etc.

Lateral plate mesoderm

Extraembryonic mesoderm

(E)

Ectoderm of head fold

Anterior border of mesoderm

Somite

Hensen's node

Neural fold

Notochord

Segmental plate

Primitive streak

(F)

Neural fold

Somites

Segmental plate

Primitive streak

Head fold

Foregut

Notochord

FIGURE 9.4 Cell movements of the primitive streak and fate map of the chick embryo. (A–C) Dorsal view of the formation and elongation of the primitive streak. The blastoderm is seen at (A) 12–14 hours, (B) 15–17 hours, and (C) 18–20 hours after the egg is laid. (D–F) Formation of notochord and mesodermal somites as the primitive streak regresses, shown at (D) 20–22 hours, (E) 23–25 hours, and (F) the four-somite stage. Fate maps of the chick epiblast are shown for two stages, the definitive primitive streak stage (C) and neurulation (F). In (F), the endoderm has ingressed beneath the epiblast, and convergent extension is seen in the midline. The movements of the mesodermal precursors through the primitive streak at (C) are shown. (Adapted from several sources, especially Spratt 1946; Smith and Schoenwolf 1998; Stern 2005a,b.)

embryo (**FIGURE 9.4**). Thus, the primitive groove is homologous to the amphibian blastopore, and the primitive streak is homologous to the blastopore lip. At the anterior end of the primitive streak is a regional thickening of cells called **Hensen's node** (also known as the *primitive knot*; see Figure 9.4C). The center of Hensen's node contains a funnel-shaped depression (sometimes called the **primitive pit**) through which cells can enter the embryo to form the notochord and prechordal plate. Hensen's node is the functional equivalent of the dorsal lip of the amphibian blastopore (i.e., the organizer)* and the fish embryonic shield (Boettger et al. 2001).

The primitive streak defines the major body axes of the avian embryo. It extends from posterior to anterior; migrating cells enter through its dorsal side and move to its ventral side; and it separates the left portion of the embryo from the right. The axis of the streak is equivalent to the dorsal-ventral axis of amphibians. The anterior end of the streak—Hensen's node—gives rise to the prechordal mesoderm, notochord, and medial part of the somites. Cells that ingress through the middle of the streak give rise to the lateral part of the somites and to the heart and kidneys. Cells in the posterior portion of the streak make the lateral plate and extraembryonic mesoderm (Psychoyos and Stern 1996). After the ingression of the mesoderm cells, epiblast cells remaining outside, but close to, the streak will form medial (dorsal) structures such as the neural plate, while those epiblast cells farther from the streak will become epidermis (see Figure 9.4, right-hand panels).

ELONGATION OF THE PRIMITIVE STREAK As cells enter the primitive streak, they undergo an epithelial-mesenchymal transformation, and the basal lamina beneath them breaks down.

*Frank M. Balfour proposed the homology of the amphibian blastopore and the chick primitive streak in 1873, while he was still an undergraduate (Hall 2003). August Rauber (1876) provided further evidence for their homology.

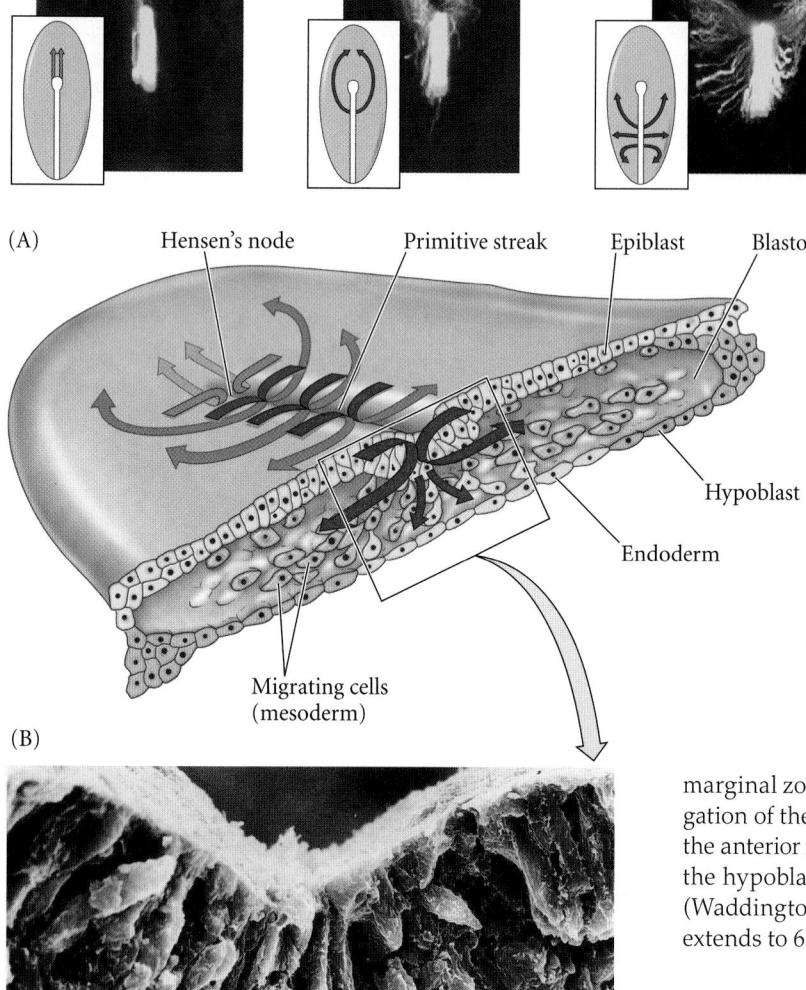

(A)

Hensen's node Primitive streak Epiblast Blastocoel

Hypoblast

Endoderm

Migrating cells
(mesoderm)

(B)

FIGURE 9.5 Migration of endodermal and mesodermal cells through the primitive streak. (A) Stereogram of a gastrulating chick embryo, showing the relationship of the primitive streak, the migrating cells, and the hypoblast and epiblast of the blastoderm. The lower layer becomes a mosaic of hypoblast and endodermal cells; the hypoblast cells eventually sort out to form a layer beneath the endoderm and contribute to the yolk sac. Above each region of the stereogram are micrographs showing the tracks of GFP-labeled cells at that position in the primitive streak. Cells migrating through Hensen's node travel anteriorly to form the prechordal plate and notochord; those migrating through the next anterior region of the streak travel laterally but converge near the midline to make notochord and somites; those from the middle of the streak form intermediate mesoderm and lateral plate mesoderm (see the fate maps in Figure 9.4). Farther posterior, the cells migrating through the primitive streak make the extraembryonic mesoderm (not shown). (B) This scanning electron micrograph shows epiblast cells passing into the blastocoel and extending their apical ends to become bottle cells. (A after Balinsky 1975, photographs from Yang et al. 2002; B from Solursh and Revel 1978, courtesy of M. Solursh and C. J. Weijer.)

The streak elongates toward the future head region as more anterior cells migrate toward the center of the embryo. Convergent extension is responsible for the progression of the streak—a doubling in streak length is accompanied by a concomitant halving of its width (see Figure 9.4B). Cell division adds to the length produced by convergent extension, and some of the cells from the anterior portion of the epiblast contribute to the formation of Hensen's node (Streit et al. 2000; Lawson and Schoenwolf 2001b).

At the same time, the secondary hypoblast (endoblast) cells continue to migrate anteriorly from the posterior

marginal zone of the blastoderm (see Figure 9.3E). The elongation of the primitive streak appears to be coextensive with the anterior migration of these secondary hypoblast cells, and the hypoblast directs the movement of the primitive streak (Waddington 1933; Foley et al. 2000). The streak eventually extends to 60–75% of the length of the area pellucida.

FORMATION OF ENDODERM AND MESODERM The basic rule of amniote cell specification is that germ layer identity (ectoderm, mesoderm, or endoderm) is established before gastrulation starts (see Chapman et al. 2007), but the specification of cell type is controlled by inductive influences during and after migration through the primitive streak. As soon as the primitive streak has formed, epiblast cells begin to migrate through it and into the blastocoel. The streak thus has a continually changing cell population. Cells migrating through the anterior end pass down into the blastocoel and migrate anteriorly, forming the endoderm, head mesoderm, and notochord; cells passing through the more posterior portions of the primitive streak give rise to the majority of mesodermal tissues (**FIGURE 9.5**; Rosenquist et al. 1966; Schoenwolf et al. 1992).

The first cells to migrate through Hensen's node are those destined to become the pharyngeal endoderm of the foregut. Once deep within the embryo, these endodermal cells migrate anteriorly and eventually displace the hypoblast cells, causing the hypoblast cells to be confined to a region in the anterior portion of the area pellucida. This anterior region, the **germinal crescent**, does not form any embryonic structures, but it does contain the precursors of the germ cells, which later migrate through the blood vessels to the gonads (see Chapter 17).

The next cells entering through Hensen's node also move anteriorly, but they do not travel as far ventrally as the presumptive foregut endodermal cells. Rather, they remain between the endoderm and the epiblast to form the **prechordal plate mesoderm** (Psychoyos and Stern 1996). Thus, the head of the avian embryo forms anterior (rostral) to Hensen's node.

The next cells passing through Hensen's node become the **chordamesoderm**. The chordamesoderm has two components: the head process and the notochord. The most anterior part, the **head process**, is formed by central mesoderm cells migrating anteriorly, behind the prechordal plate mesoderm and toward the rostral tip of the embryo (see Figures 9.4 and 9.5). The head process will underlie those cells that will form the forebrain and midbrain. As the primitive streak regresses (see below), the cells deposited by the regressing Hensen's node will become the notochord.

Molecular mechanisms of migration through the primitive streak

FORMATION OF THE PRIMITIVE STREAK   The migration of chick epiblast cells to form the primitive streak was first analyzed by Ludwig Gräper, who in 1926 made time-lapse movies of labeled cells under the microscope. He wrote that these movements reminded him of the Polonaise, a courtly dance in which men and women move in parallel rows along the sides of the room, and the man and woman at the "posterior end" leave their respective lines to dance forward through the center. The mechanism for the cellular "dance" was revealed by Voiculescu and colleagues (2007), who used a modern version of cinemicrography (specifically, multiphoton time-lapse microscopy) that identified individual moving cells. They found that cells came down the sides of the epiblast to undergo a medially directed intercalation of cells in the posterior margin where the primitive streak was forming (**FIGURE 9.6**). And although the movement may look like a dance from far away, "at high power, it looks like a rush hour" (Stern 2007).

This rush to the center is mediated by the activation of the Wnt planar cell polarity pathway (see Chapter 3) in the epiblast next to Koller's sickle, at the posterior edge of the embryo. If this pathway is blocked, the mesoderm and endoderm form peripherally instead of centrally. The Wnt pathway in turn appears to be activated by **fibroblast growth factors** (**FGFs**) produced by the hypoblast. If the hypoblast is rotated, the orientation of the primitive streak follows it. Moreover, if FGF signaling is activated in the margin of the epiblast, Wnt signaling will occur there and the orientation of the primitive streak will change, as if the hypoblast had been placed there. The cell migrations that form the primitive streak thus appear to be regulated by

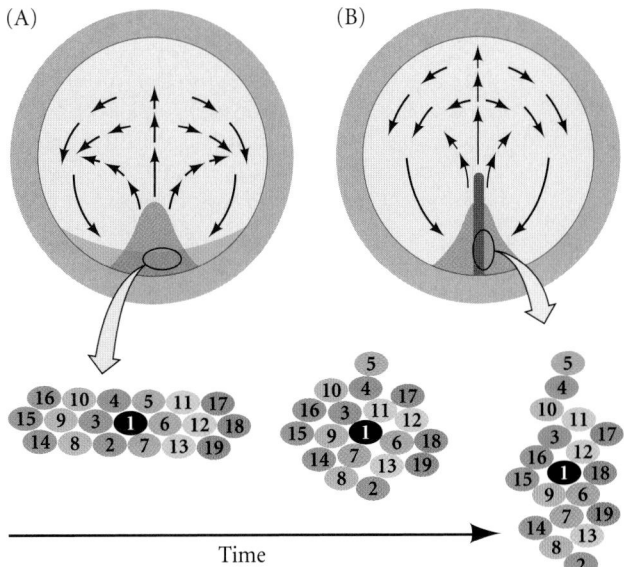

FIGURE 9.6   Mediolateral intercalation in the formation of the primitive streak. Chick embryos at (A) stage XIII (immediately prior to primitive streak formation) and (B) stage 2 (shortly after primitive streak formation). Arrows show cell displacement toward the streak and in front of it. The red area represents the streak-forming region; in (A), the original location of this region is shown in green. The circled areas are represented in the lower row. Each colored disc represents an individual cell, and the cells become mediolaterally intercalated as the primitive streak forms. (After Voiculescu et al. 2007.)

FGFs coming from the hypoblast, which activates the Wnt planar cell polarity pathway in the epiblast.

MIGRATION THROUGH THE PRIMITIVE STREAK   Cells migrate to the primitive streak, and as they enter the embryo, the cells separate into two layers. The deep layer joins the hypoblast along its midline, displacing the hypoblast cells to the sides. These deep-moving cells give rise to the endodermal organs of the embryo, as well as to most of the extraembryonic membranes (the hypoblast and peripheral cells of the area opaca form the rest). The second migrating layer spreads to form a loose layer of cells between the endoderm and the epiblast. This middle layer of cells generates the mesodermal portions of the embryo and the mesoderm lining the extraembryonic membranes.

The migration of mesodermal cells through the anterior primitive streak and their condensation to form the chordamesoderm also appear to be controlled by FGF and Wnt signaling. Fgf8 is expressed in the primitive streak and repels migrating cells away from the streak. Yang and colleagues (2002) were able to follow the trajectories of cells as they migrated through the primitive streak (see Figure 9.5) and were able to deflect these normal trajectories by using beads that released Fgf8.

Once cells migrate away from the streak, further movement of the mesodermal precursors appears to be regulated by Wnt proteins. In the more posterior regions, Wnt5a is unopposed and directs the cells to migrate broadly and become lateral plate mesoderm (Chapter 13). In the more anterior regions of the streak, however, Wnt5a is opposed by Wnt3a, which inhibits migration and causes the cells to form paraxial mesoderm (Chapter 12). Indeed, the addition of Wnt3a-secreting pellets to the posterior primitive streak suppresses lateral migration and prevents the formation of lateral plate mesoderm (Sweetman et al. 2008). By 22 hours of incubation, most of the presumptive endodermal cells are in the interior of the embryo, although presumptive mesodermal cells continue to migrate inward for a longer time.

Regression of the primitive streak and epiboly of the ectoderm

Now a new phase of development begins. As mesodermal ingression continues, the primitive streak starts to regress, moving Hensen's node from near the center of the area pellucida to a more posterior position (**FIGURE 9.7**). The regressing streak leaves in its wake the dorsal axis of the embryo,

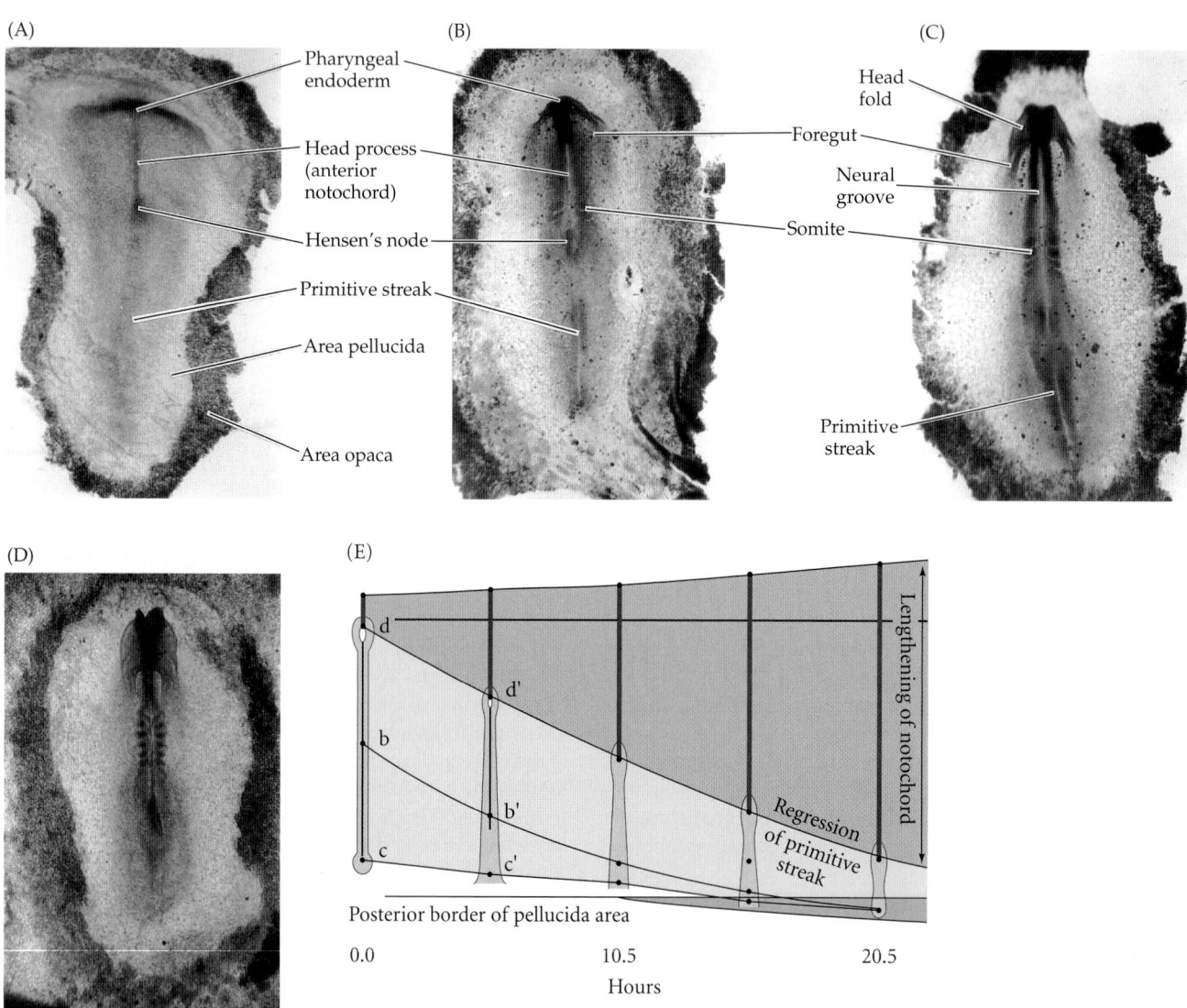

FIGURE 9.7 Chick gastrulation 24–28 hours after fertilization. (A) The primitive streak at full extension (24 hours). The head process (anterior notochord) can be seen extending from Hensen's node. (B) Two-somite stage (25 hours). Pharyngeal endoderm is seen anteriorly, while the anterior notochord pushes up the head process beneath it. The primitive streak is regressing. (C) Four-somite stage (27 hours). (D) At 28 hours, the primitive streak has regressed to the caudal portion of the embryo. (E) Regression of the primitive streak, leaving the notochord in its wake. Various points of the streak (represented by letters) were followed after it achieved its maximum length. The x axis (time) represents hours after achieving maximum length (the reference line is about 18 hours of incubation). (A–D courtesy of K. Linask; E after Spratt 1947.)

including the notochord. The notochord is laid down in a head-to-tail direction, starting at the level where the ears and hindbrain form and extending caudally to the tailbud. As in the frog, the pharyngeal endoderm and head mesoendoderm will induce the anterior parts of the brain, while the notochord will induce the hindbrain and spinal cord. By this time, all the presumptive endodermal and mesodermal cells have entered the embryo and the epiblast is composed entirely of presumptive ectodermal cells.

While the presumptive mesodermal and endodermal cells are moving inward, the ectodermal precursors proliferate and migrate to surround the yolk by epiboly. The enclosure of the yolk by the ectoderm (again, reminiscent of the epiboly of the amphibian ectoderm) is a Herculean task that takes the greater part of 4 days to complete. It involves the continuous production of new cellular material and the migration of the presumptive ectodermal cells along the underside of the vitelline envelope (New 1959; Spratt 1963). Interestingly, only the cells of the outer margin of the area opaca attach firmly to the vitelline envelope. These cells are inherently different from the other blastoderm cells, as they can extend enormous (500 μm) cytoplasmic processes onto the vitelline envelope. These elongated filopodia are believed to be the locomotor apparatus of the marginal cells, by which the marginal cells pull other ectodermal cells around the yolk (Schlesinger 1958). The filopodia bind to fibronectin, a laminar protein that is a component of the chick vitelline envelope. If the contact between the marginal cells and the fibronectin is experimentally broken by adding a soluble polypeptide similar to fibronectin, the filopodia retract and ectodermal migration ceases (Lash et al. 1990).

Thus, as avian gastrulation draws to a close, the ectoderm has surrounded the embryo, the endoderm has replaced the hypoblast, and the mesoderm has positioned itself between these two regions. Although we have identified many of the processes involved in avian gastrulation, we are only beginning to understand the molecular mechanisms by which some of these processes are carried out.

● See **WEBSITE 9.2** Epiblast cell heterogeneity

● See **VADE MECUM** Chick development

Axis Specification and the Avian "Organizer"

As a consequence of the sequence in which the head endomesoderm and notochord are established, avian (and mammalian, reptilian, and teleost fish) embryos exhibit a distinct anterior-to-posterior gradient of developmental maturity. While cells of the posterior portions of the embryo are still part of a primitive streak and colonizing the mesoderm, cells at the anterior end are already starting to form organs (see Darnell et al. 1999). For the next several days, the anterior end of the embryo is more advanced in its development (having had a "head start," if you will) than the posterior end. Although the formation of the chick body axes is accomplished during gastrulation, axis specification begins earlier, during the cleavage stage.

The role of gravity and the PMZ

The conversion of the radially symmetrical blastoderm into a bilaterally symmetrical structure appears to be determined by gravity. As the ovum passes through the hen's reproductive tract, it is rotated for about 20 hours in the shell gland. This spinning, at a rate of 15 revolutions per hour, shifts the yolk such that its lighter components (probably containing stored maternal determinants for development) lie beneath one side of the blastoderm. This imbalance tips up one end of the blastoderm, and that end becomes the posterior marginal zone, where primitive streak formation begins (**FIGURE 9.8**; Kochav and Eyal-Giladi 1971; Callebaut et al. 2004).

It is not known what interactions cause this specific portion of the blastoderm to become the PMZ. Early on, the ability to initiate a primitive streak is found throughout the marginal zone; if the blastoderm is separated into parts, each with its own marginal zone, each part will form its own primitive streak (Spratt and Haas 1960; Bertocchini and Stern 2012). However, once the PMZ has formed, it controls the other regions of the margin. Not only do the cells of the PMZ initiate gastrulation, they also prevent other regions of the margin from forming their own primitive streaks (Khaner and Eyal-Giladi 1989; Eyal-Giladi et al. 1992; Bertocchini et al. 2004).

It now seems apparent that the PMZ contains cells that act as the equivalent of the amphibian Nieuwkoop center. When

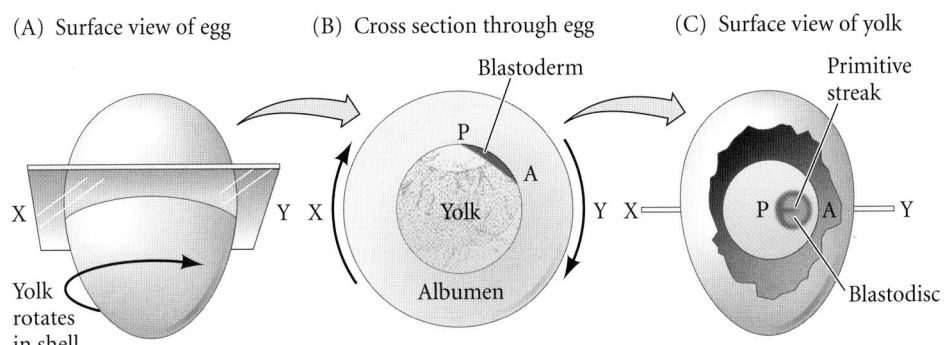

(A) Surface view of egg (B) Cross section through egg (C) Surface view of yolk

FIGURE 9.8 Specification of the chick anterior-posterior axis by gravity. (A) Rotation in the shell gland results in (B) the lighter components of the yolk pushing up one side of the blastoderm. (C) This more elevated region becomes the posterior of the embryo. (After Wolpert et al. 1998.)

(A)

Anterior

Hypoblast

Epiblast

Koller's
sickle

Posterior
marginal
zone (PMZ)

Marginal
zone

Area
opaca

Yolk

Posterior

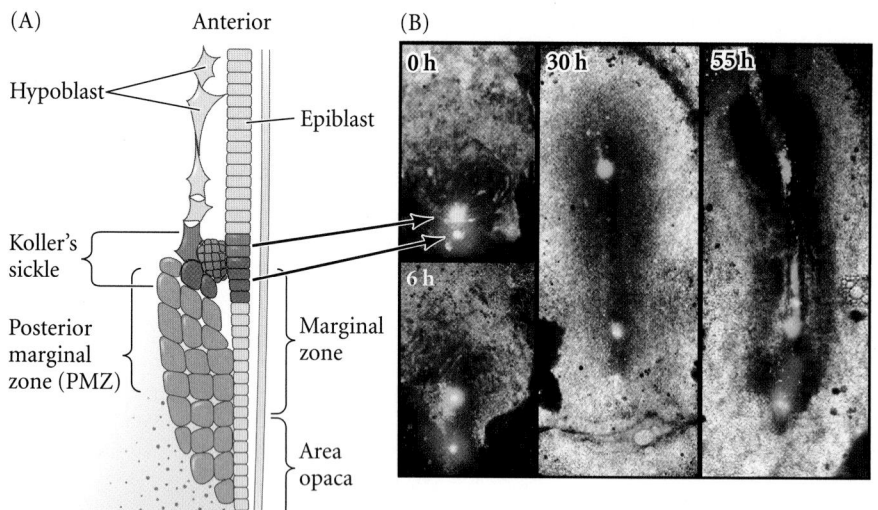

(B)

0 h   30 h   55 h

6 h

FIGURE 9.9 Formation of Hensen's node from Koller's sickle. (A) Diagram of posterior end of an early (pre-streak) embryo, showing the cells labeled with fluorescent dyes in the photographs. (B) Just before gastrulation, cells in the anterior end of Koller's sickle (the epiblast and middle layer) were labeled with green dye. Cells of the posterior portion of Koller's sickle were labeled with red dye. As the cells migrated, the anterior cells formed Hensen's node and its notochord derivatives. The posterior cells formed the posterior region of the primitive streak. The time after dye injection is labeled on each photograph. (A after Bachvarova et al. 1998; B courtesy of R. F. Bachvarova.)

placed in the anterior region of the marginal zone, a graft of PMZ tissue (posterior to and including Koller's sickle) is able to induce a primitive streak and Hensen's node without contributing cells to either structure (Bachvarova et al. 1998; Khaner 1998). Current evidence suggests that the entire marginal zone produces Wnt8c (capable of inducing β-catenin) and that, like the amphibian Nieuwkoop center, the PMZ cells secrete Vg1, a member of the TGF-β family (Mitrani et al. 1990; Hume and Dodd 1993; Seleiro et al. 1996).

Wnt8c and Vg1 act together to induce expression of Nodal (another secreted TGF-β protein) in the future embryonic epiblast next to Koller's sickle and the PMZ (Skromne and Stern 2002). Thus, the pattern appears similar to that of amphibian embryos. Recent studies suggest that Nodal activity is needed to initiate the primitive streak, and that it is the secretion of Cerberus—an antagonist of Nodal—by the primary hypoblast cells that prevents primitive streak formation (Bertocchini and Stern 2002; Bertocchini et al. 2004). As the primary hypoblast cells move away from the PMZ, Cerberus protein is no longer present, allowing Nodal activity and therefore formation of the primitive streak in the posterior epiblast. Once formed, however, the streak secretes its own Nodal antagonist (the Lefty protein), thereby preventing any further primitive streaks from forming. Eventually, the Cerberus-secreting hypoblast cells are pushed to the future anterior of the embryo, where they contribute to ensuring that neural cells in this region become forebrain rather than more posterior structures of the nervous system.*

* Conjoined twins may be formed by having by having two sources of *Nodal* expression within the same blastodisc. Although this has not been shown in mammals, experimentation with chick embryos can produce two axes in the same blastodisc by circumventing the usual inhibition of *Nodal* by the Vg1-secreting posterior cells (Bertocchini et al. 2004).

The chick "organizer" and the role of fibroblast growth factors

The "organizer" of the chick embryo forms from cells initially located just anterior to the posterior marginal zone. The epiblast and middle layer cells in the anterior portion of Koller's sickle become Hensen's node, as described earlier. The posterior portions of Koller's sickle contribute to the posterior portion of the primitive streak (**FIGURE 9.9**). Hensen's node has long been known to be the avian equivalent of the amphibian dorsal blastopore lip, since (1) it is the region whose cells become the prechordal plate and chordamesoderm, (2) it is the region whose cells can both induce and pattern a second embryonic axis when transplanted into other locations of the gastrula (**FIGURE 9.10**; Waddington 1933,1934; Gallera 1966; Nicolet 1970), and (3) it expresses the same marker genes as Spemann's organizer in amphibians and the embryonic shield of teleost fishes, such as the transcription factor Goosecoid (Izpisua-Belmonte et al. 1993). Moreover, Hensen's node can induce neural tissue when grafted into fish, amphibian, or mammalian embryos (Waddington 1936; Kintner and Dodd 1991; Hatta and Takahashi 1996).

As is the case in all vertebrates, the dorsal mesoderm is able to induce the formation of the central nervous system in the ectoderm overlying it. The cells of Hensen's node and its derivatives act like the amphibian organizer, and they secrete BMP inhibitors such as Chordin, Noggin, and Nodal. These proteins repress BMP signaling and dorsalize the ectoderm and mesoderm (**FIGURE 9.11**). However, repression of BMP signals by these antagonists does not appear to be sufficient for neural induction (see Stern 2005b). Fibroblast growth factors synthesized in the hypoblast and in Hensen's node precursor cells just prior to gastrulation appear to be critical for preparing the epiblast to generate neuronal phenotypes. FGFs can block BMP signaling, but this fact alone does not account for the ability of FGFs to induce a transient expression of pre-neural genes in the epiblast (Streit et al. 1998, 2000). These neural genes do not stay active unless they are

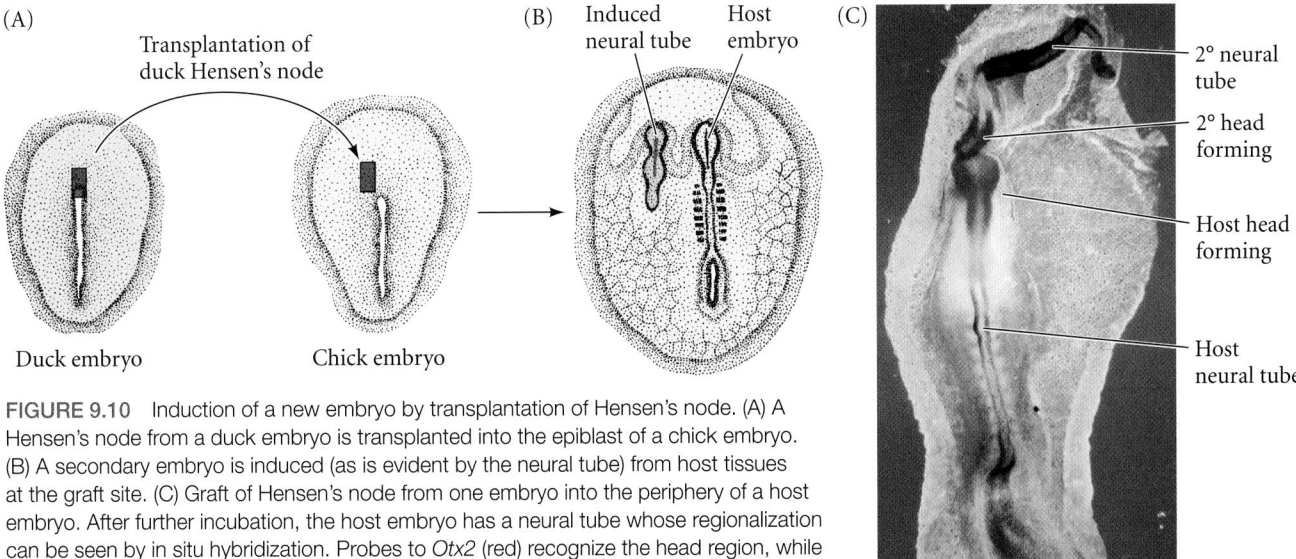

FIGURE 9.10 Induction of a new embryo by transplantation of Hensen's node. (A) A Hensen's node from a duck embryo is transplanted into the epiblast of a chick embryo. (B) A secondary embryo is induced (as is evident by the neural tube) from host tissues at the graft site. (C) Graft of Hensen's node from one embryo into the periphery of a host embryo. After further incubation, the host embryo has a neural tube whose regionalization can be seen by in situ hybridization. Probes to *Otx2* (red) recognize the head region, while probes to *Hoxb1* (blue) recognize the trunk neural tube. The donor node has induced the formation of a secondary axis, complete with head and trunk regions. (A,B after Waddington 1933; C from Boettger et al. 2001.)

supported by BMP antagonists (Streit et al. 1998, 2000; Albazerchi and Stern 2006). Thus, FGF signaling inhibits BMPs from inducing the genes that specify ectoderm to become epidermis, and they activate the genes that specify ectoderm to become neural.

Indeed, fibroblast growth factors play four fundamental roles in cell specification during gastrulation:

1. As in all vertebrates, FGFs are responsible for specifying the mesoderm. FGFs from the hypoblast (in collaboration with Nodal from the posterior marginal zone) accomplish this specification by activating the *Brachyury* and *Tbx6* genes in the cells passing through the primitive streak (**FIGURE 9.12**; Sheng et al. 2003).

2. FGFs separate mesoderm formation from neurulation. The mechanism by which FGFs help end mesoderm ingression and stabilize the epiblast appears to be due to a gene called *Churchill*. While FGFs are rapidly inducing the mesoderm, they are also slowly inducing activation of *Churchill* in the ectoderm. The Churchill protein (so named because the protein's two zinc fingers extend like the British prime minister's famous "V for Victory" symbol) can activate the Smad-interacting protein SIP1. SIP1 controls the genes whose expression is required for ingression of cells through the primitive streak. Thus, once activated, SIP1 helps prospective neural plate cells remain in the epiblast.

3. FGFs help bring about neurulation in the central ectodermal cells. SIP1, probably through its interaction with

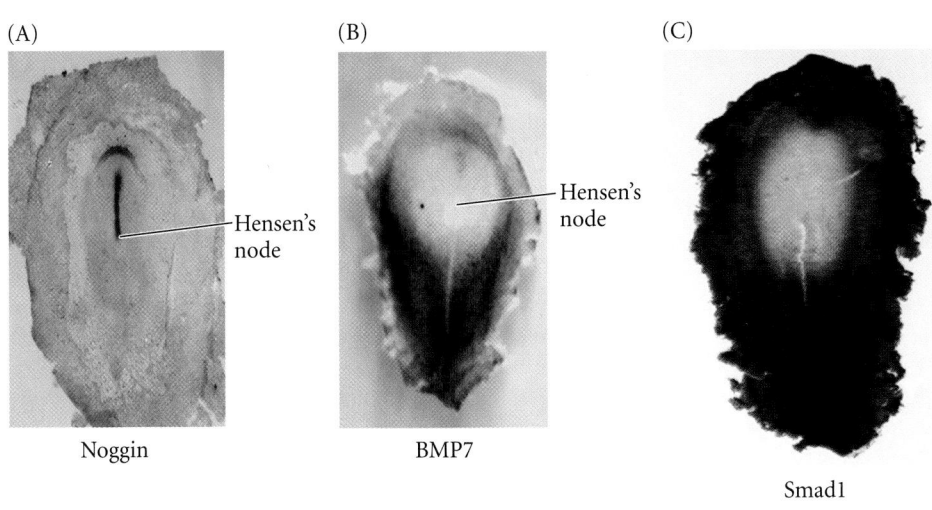

FIGURE 9.11 Possible contribution to chick neural induction by the inhibition of BMP signaling. (A) In a neurulating embryo, Noggin protein (purple) is expressed in the notochord and the pharyngeal endoderm. (B) BMP7 expression (dark purple), which had encompassed the entire epiblast, becomes restricted to the non-neural regions of the ectoderm. (C) Similarly, the product of BMP signaling, the phosphorylated form of Smad1 (recognized by antibodies to the phosphorylated form of the protein; dark brown) is not seen in the neural plate. (From Faure et al. 2002, courtesy of the authors.)

(A)

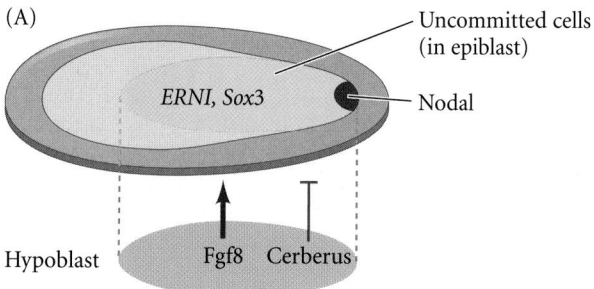

(B)

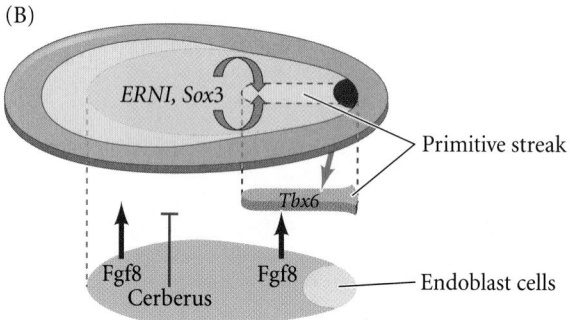

(C)

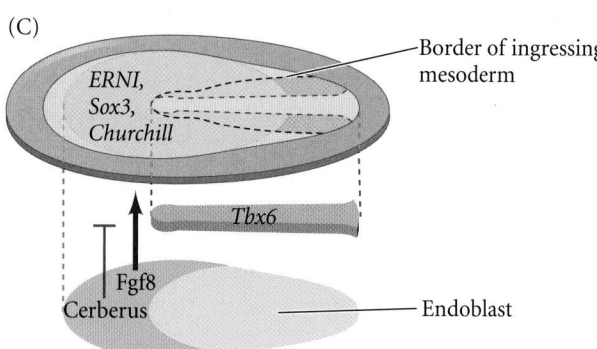

(D)

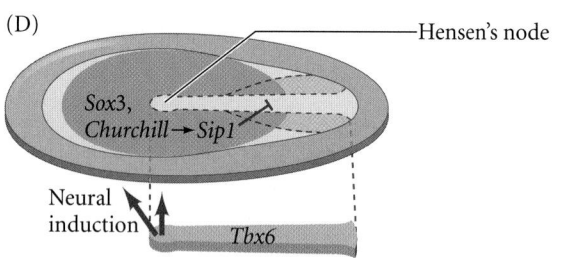

FIGURE 9.12 *Model by which FGFs regulate mesoderm forma-tion and neurulation. (A) Stage XI, where the hypoblast (green) secretes Fgf8, which induces pre-neural genes* ERNI *and* Sox3 *(blue) in the epiblast. The cells in this domain, however, remain uncommitted. Nodal, expressed in the posterior epiblast, cannot function; it is inhibited by the Cerberus protein secreted by the hypoblast. (B) At around stage 1, the hypoblast is displaced from the posterior edge by the endoblast (secondary hypoblast; gold), allowing Nodal to function. Nodal plus Fgf8 induces* Brachyury *and* Tbx6 *expression to specify the mesoderm and initiate the ingression of mesoderm cells through the primitive streak (red). (C) At stage 4, continued Fgf8 expression activates* Churchill *in the epiblast. (D) By end of stage 4, Churchill protein induces Sip1, which blocks* Brachyury *and* Tbx6, *preventing further ingression of epiblast cells through the streak. The remaining epiblast cells can now become sensitized to neural inducers from Hensen's node (purple). (After Sheng et al. 2003.)*

Anterior-posterior patterning

The patterning of the definitive anterior-posterior axis occurs differently for the mesoderm and neural ectoderm, but in both cases the process involves timing (the sequential generation of cells from a zone of undifferentiated proliferating cells) and the influence of caudalizing molecules. In the ectoderm, most of the initial neural plate corresponds to the future head region (from forebrain to the level of the future ear vesicle, which lies adjacent to Hensen's node at full primitive streak stage). A small region of neural ectoderm just lateral and posterior to the node (sometimes called the caudal lateral epiblast) will give rise to the rest of the nervous system, including the posterior hindbrain and all of the spinal cord. As the primitive streak regresses, this latter region regresses with Hensen's node and adds cells to the caudal end of the elongating neural plate. It appears that FGF signaling in the streak and paraxial (future somite) mesoderm keeps this region "young" and undifferentiated as it regresses, and that this is antagonized by retinoic acid (RA) activity as cells leave this zone (**FIGURE 9.13**; Diez del Corral et al. 2003).

While they are still in the epiblast, but close to the primitive streak, the mesoderm cells appear to receive instructions that tell them exactly where they are along the anterior-posterior axis. The entire length of the notochord at the midline is derived from cells that are present in Hensen's node by the full primitive streak stage. A population of progenitor cells remains in the node; their descendants gradually leave as the node regresses, laying down the chordamesoderm and the ventral midline of the neural tube (the future floor plate of the spinal cord) (Selleck and Stern 1991; Psychoyos and Stern 1996; Tzouanacou et al. 2009). Therefore, anterior-posterior identities along the axis from the hindbrain to the tail are specified as a function of the time of emergence from the primitive streak and Hensen's node.

It has been proposed that the length of time cells are resident in the primitive streak region determines which Hox genes are expressed by the cells. This pattern of Hox gene expression can also be under the influence of the FGF and

Smad1, may make the prospective neural plate cells less sensitive to BMP.

4. FGFs induce *ERNI* and *Sox3*, two pre-neural genes that initiate the signaling cascade leading to the production of neural tissue.

Thus, FGFs appear to be critically important regulators of cell fate in the early chick embryo (Streit et al. 2000; Sheng et al. 2003; Albazerchi and Stern 2006).

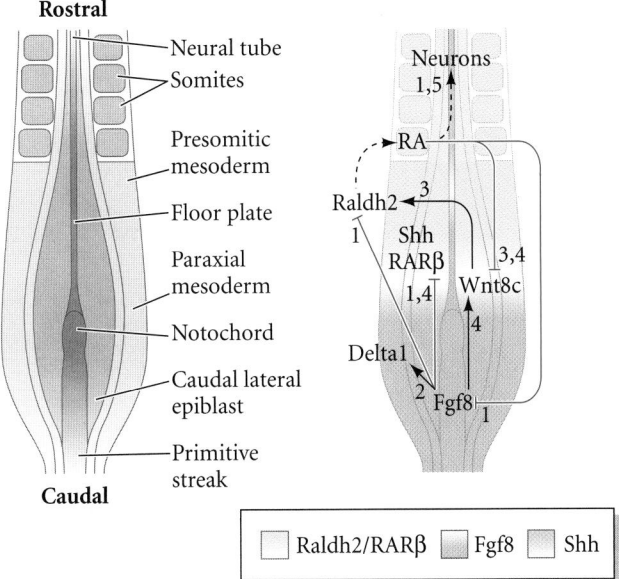

Rostral

Neural tube
Somites
Presomitic mesoderm
Floor plate
Paraxial mesoderm
Notochord
Caudal lateral epiblast
Primitive streak

Caudal

Neurons
1,5
RA
Raldh2   3
Shh
RARβ   3,4
1,4   Wnt8c
4
Delta1
2   Fgf8   1

☐ Raldh2/RARβ   ☐ Fgf8   ☐ Shh

FIGURE 9.13 Signals that regulate axis extension in chick embryos. In the stage 10 chick embryo, Fgf8 inhibits expression of the retinoic acid (RA) synthesizing enzyme Raldh2 in the presomitic mesoderm (1) and the expression of the RA receptor RARβ in the neural ectoderm (4), thus preventing RA from triggering differentiation in the caudal-lateral epiblast cells (those cells adjacent to the node/streak border and which give rise to lateral and dorsal neural tube) and the caudalmost paraxial mesoderm (1,5). In addition, Fgf8 inhibits Sonic hedgehog (Shh) expression in the neural tube floorplate, controlling the onset of ventral patterning genes (1). FGF signaling is also required for expression of Delta1 in the medial portion of the caudal-lateral epiblast cells (2) and promotes expression of Wnt8c (4). As Fgf8 decays in the caudal paraxial mesoderm, Wnt signaling, most likely provided by Wnt8c, now acts to promote Raldh2 in the adjacent paraxial mesoderm (4). RA produced by Raldh2 activity represses Fgf8 (1) and Wnt8c (3,4). (After Wilson et al. 2009.)

retinoic acid gradients (Gaunt 1991; Wilson et al. 2009). As we will detail later in the chapter, Hox genes are the vertebrate homologues of the homeotic (Hom-C) genes of *Drosophila*. And just as in *Drosophila* embryos, vertebrate Hox genes specify the identity of cells along the anterior-posterior axis. In the case of vertebrates, however, there are four gene clusters (*HoxA*, *HoxB*, *HoxC*, and *HoxD*) instead of just one, and rather than individual Hox genes appearing at particular segmental levels, there is a nested set of Hox gene expression. For example, the mesodermal precursor cells are patterned along the anterior-posterior axis by the *HoxB* genes, which appear to inform the cells when to leave the epiblast

and ingress into the primitive streak. "Anterior" Hox genes (which are identified with lower numbers, e.g., *Hoxb4*) are expressed early and extend farther anterior than genes such as *Hoxb9*, which is expressed later and does not extend as far into the embryo's anterior (Iimura and Pourquié 2006; **FIGURE 9.14**). Thus, the more posterior cells express more Hox genes than the more anterior cells do.

Left-right axis formation

The vertebrate body has distinct right and left sides. The heart and spleen, for instance, are generally on the left side of the body, whereas the liver is usually on the right. The distinction between the sides is primarily regulated by the left-sided expression of two proteins: the paracrine factor Nodal and the transcription factor Pitx2. However, the mechanism by which *Nodal* gene expression is activated in the left side of the body differs among the vertebrate classes. The ease with which chick embryos can be manipulated has allowed scientists to elucidate the pathways of left-right axis determination in birds more readily than in other vertebrates.

As the primitive streak reaches its maximum length, transcription of the *Sonic hedgehog* (*Shh*) gene becomes restricted to the left side of the embryo, controlled by activin and its receptor (**FIGURE 9.15A**). Activin signaling, along with BMP4, appears to block the expression of Sonic hedgehog protein and to

(A) *Hoxb4*

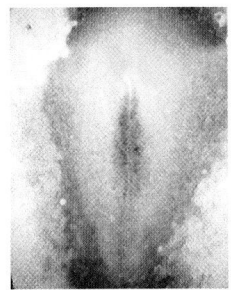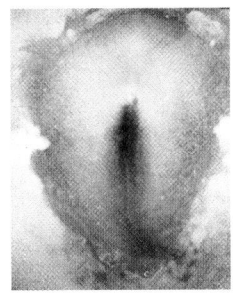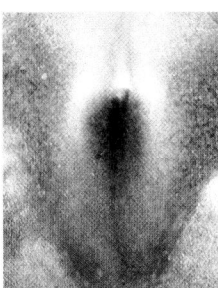

(B) *Hoxb9*

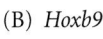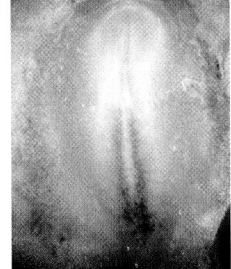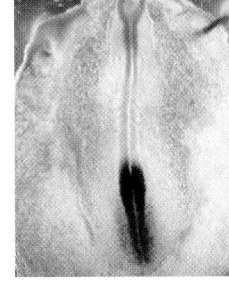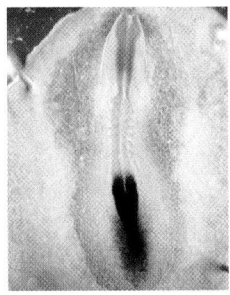

FIGURE 9.14 Hox gene activation begins when the mesodermal precursor cells are still in the epiblast. These genes are activated in an anterior-to-posterior fashion. Migration into the primitive streak is regulated by the Hox gene expression pattern (the most posterior Hox gene having preference). (A) *Hoxb4* expression in stage-4, -5, and -6 chick embryos. (B) *Hoxb9* expression in stage-7, -8, and -8+ chick embryos. Note that the anterior border of expression is more posterior than the anterior border of *Hoxb4* expression. (From Iimura and Pourquié 2006, courtesy of O. Pourquié.)

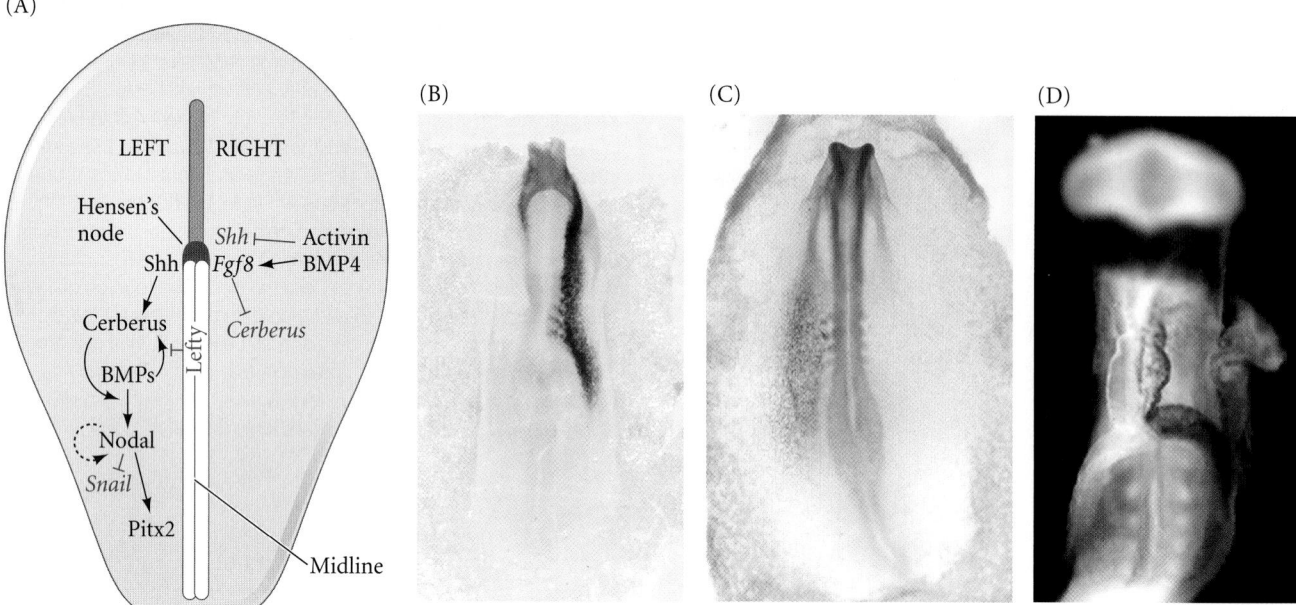

FIGURE 9.15 Model for generating left-right asymmetry in the chick embryo. (A) On the left side of Hensen's node, Sonic hedgehog (Shh) activates Cerberus, which stimulates BMPs to induce the expression of Nodal. In the presence of Nodal, the *Pitx2* gene is activated. Pitx2 protein is active in the various organ primordia and specifies which side will be the left. On the right side of the embryo, activin is expressed, along with activin receptor IIa. This activates Fgf8, a protein that blocks expression of the gene for Cerberus. In the absence of Cerberus, Nodal is not activated and thus Pitx2 is not expressed. (B) Whole-mount in situ hybridization of *Cerberus* mRNA. This view is from the ventral surface ("from below," so the expression seems to be on the right). Dorsally, the expression pattern would be on the left. (C) Whole-mount in situ hybridization using probes for the chick *Nodal* message (stained purple) shows its expression in the lateral plate mesoderm only on the left side of the embryo. This view is from the dorsal side. (D) Similar in situ hybridization, using the probe for *Pitx2* at a later stage of development. The embryo is seen from its ventral surface. At this stage, the heart is forming, and *Pitx2* expression can be seen on the left side of the heart tube (as well as symmetrically in more anterior tissues. (A after Raya and Izpisua-Belmonte 2004; B from Rodriguez-Esteban et al. 1999, courtesy of J. Izpisúa-Belmonte; C courtesy of C. Stern; D from Logan et al. 1998, courtesy of C. Tabin.)

activate expression of Fgf8 protein on the right side of the embryo. Fgf8 blocks expression of the paracrine factor Cerberus on the right-hand side; it may also activate a signaling cascade that instructs the mesoderm to have right-sided capacities (Schlueter and Brand 2009).

Meanwhile, on the left side of the body, Shh protein activates Cerberus (**FIGURE 9.15B**), which in this case acts with BMP to stimulate the synthesis of Nodal protein (Yu et al. 2008). Nodal activates the *Pitx2* gene while repressing *Snail*. In addition, Lefty1 in the ventral midline prevents the Cerberus signal from passing to the right side of the embryo (**FIGURE 9.15C,D**). As in *Xenopus*, Pitx2 is crucial in directing the asymmetry of the embryonic structures. Experimentally induced expression of either Nodal or Pitx2 on the right side of the chick embryo reverses the asymmetry or causes randomization of asymmetry on the right or left sides* (Levin et al. 1995; Logan et al. 1998; Ryan et al. 1998).

---

* In humans, homozygous loss of *PITX2* causes Rieger's syndrome, a condition characterized by asymmetry anomalies. A similar condition is caused by knocking out the *Pitx2* gene in mice (Fu et al. 1998; Lin et al. 1999).

The real mystery is, What processes create the original asymmetry of Shh and Fgf8? One important observation is that the first asymmetry seen during the formation of Hensen's node in chicks involves Fgf8- and Shh-expressing cells rearranging themselves to converge on the right-hand side of the node (Cui et al. 2009; Gros et al. 2009). Therefore, the differences in gene expression can be traced back to differences in cell migration to the right and left sides of the embryo. What establishes this initial asymmetry is still unknown, but it may be a physical displacement of cells around the node (Tsikolia et al. 2012).

EARLY MAMMALIAN DEVELOPMENT

Cleavage

Mammalian eggs are among the smallest in the animal kingdom, making them hard to manipulate experimentally. The human zygote, for instance, is only 100 μm in diameter—barely visible to the eye and less than one-thousandth the volume of a *Xenopus laevis* egg. Also, mammalian zygotes are not produced in numbers comparable to sea urchin or

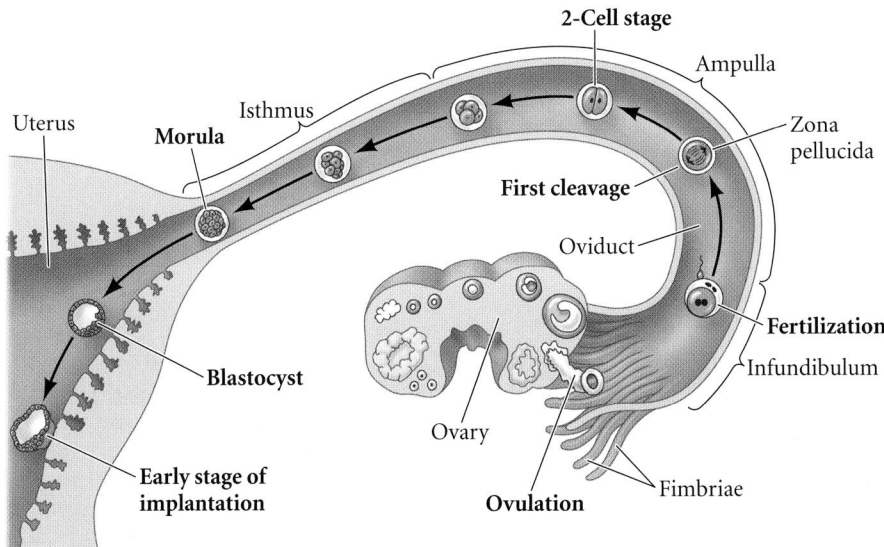

FIGURE 9.16 Development of a human embryo from fertilization to implantation. Compaction of the human embryo occurs on day 4, at the 10-cell stage. The embryo "hatches" from the zona pellucida upon reaching the uterus. During its migration to the uterus, the zona prevents the embryo from prematurely adhering to the oviduct rather than traveling to the uterus.

frog zygotes; a female mammal usually ovulates fewer than 10 eggs at a given time, so it is difficult to obtain enough material for biochemical studies. As a final hurdle, the development of mammalian embryos is accomplished inside another organism rather than in the external environment (although early embryos prior to implantation can be cultured and observed in vitro). Most research on mammalian development has focused on the mouse, since mice are relatively easy to breed, have large litters, and are easily housed in laboratories.

The unique nature of mammalian cleavage

Prior to fertilization, the mammalian oocyte, wrapped in cumulus cells, is released from the ovary and swept by the fimbriae into the oviduct (**FIGURE 9.16**). Fertilization occurs in the **ampulla** of the oviduct, a region close to the ovary. Meiosis is completed after sperm entry, and the first cleavage begins about a day later (see Figure 4.34). The positioning of the first cleavage plane may depend on the point of sperm entry (Piotrowska and Zernicka-Goetz 2001), and in mice, a sperm-borne microRNA (miRNA-34c) is required to initiate this first cell division. This microRNA appears to bind and inhibit Bcl-2, a protein that prevents the entry of the cell cycle into S phase (Liu et al. 2012). The two nuclei produced by this cleavage are the first nuclei to contain the entire genome, since the haploid pronculei enter cell division upon meeting (see Chapter 4).

Cleavages in mammalian eggs are among the slowest in the animal kingdom, taking place some 12–24 hours apart. Meanwhile, the cilia in the oviduct push the embryo toward the uterus; the first cleavages occur along this journey. In addition to the slowness of cell division, several other features distinguish mammalian cleavage, including the unique orientation of mammalian blastomeres relative to one another. In many but not all mammalian embryos, the first cleavage is a normal meridional division; however, in the second cleavage, one of the two blastomeres divides meridionally and the other divides equatorially (**FIGURE 9.17**). This is called **rotational cleavage** (Gulyas 1975).

Another major difference between mammalian cleavage and that of most other embryos is the marked asynchrony of early cell division. Mammalian blastomeres do not all divide at the same time. Thus, mammalian embryos do not increase exponentially from 2 to 4 to 8 cells, but frequently contain odd numbers of cells. Furthermore, the mammalian genome, unlike the genomes of rapidly developing animals, is activated during early cleavage and zygotically transcribed proteins are necessary for cleavage and development. Maternally encoded proteins can persist through most of the cleavage

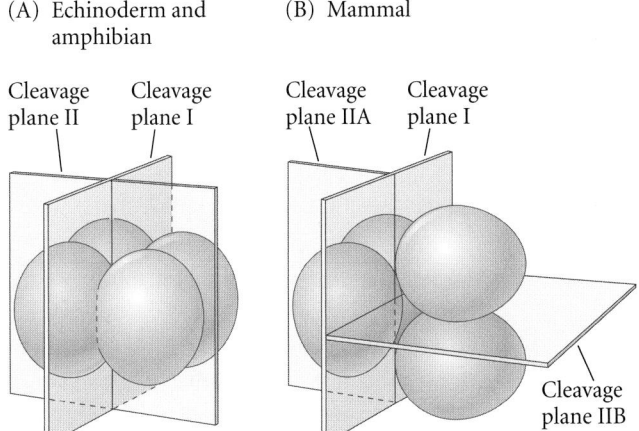

FIGURE 9.17 Comparison of early cleavage in (A) echinoderms and amphibians (radial cleavage) and (B) mammals (rotational cleavage). Nematodes also have a rotational form of cleavage, but they do not form the blastocyst structure characteristic of mammals. (After Gulyas 1975.)

SIDELIGHTS & SPECULATIONS

The Decisions That Really Shape Your Life

The philosopher and theologian Søren Kierkegaard wrote that we define ourselves by the choices we make. It seems that the embryo already knows this. The decision to become either trophoblast or inner cell mass (ICM) is the first binary decision in mammalian life. Later in development, embryonic cells must loose their pluripotency and decide on what it is they are going to grow up to be. In the first decision, Oct4 mutually represses Cdx2 expression, enabling some cells to be trophoblast and other cells to become the pluripotent cells of the ICM. In the second decision, each of the cells of the ICM expresses either Nanog or Gata6, thereby retaining its pluripotency (Nanog) or becoming primitive endoderm (Gata6).

Trophoblast or ICM: The First Decision of the Rest of Your Life

Prior to blastocyst formation, each embryonic blastomere expresses both the Cdx2 and the Oct4 transcription factors (Niwa et al. 2005; Dietrch and Hiiragi 2007; Ralston and Rossant 2008) and appears to be capable of becoming either ICM or trophoblast (Hiiragi and Solter 2004; Motosugi et al. 2005; Kurotaki et al. 2007). However, once the decision to become either trophoblast or ICM is made, the cell expresses a set of genes specific to each region. The pluripotency of the ICM is maintained by a core of three transcription factors, Oct4, Sox2, and Nanog. These proteins bind to the enhancers of their own genes to maintain their expression while at the same time activating one another's enhancers (FIGURE 9.18). Thus, when one of these genes is

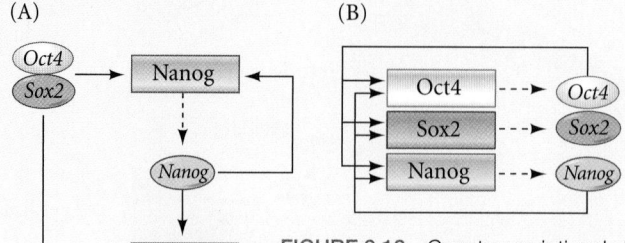

FIGURE 9.18 Core transcriptional circuitry for the pluripotency of ES cells (A) Feedforward circuit in which Oct4/Sox2 dimers activate *Nanog** genes. Nanog protein then activates its own gene as well as genes promoting pluripotency. (B) The interconnected regulatory circuit whereby Oct4, Sox2, and Nanog each activate themselves and each other's synthesis. (After Boyer et al. 2005.)

activated, the other ones are too. Acting in concert, Sox2 and Oct4 form a dimer and often reside on enhancers adjacent to Nanog, activating those genes required to maintain pluripotency in embryonic stem (ES) cells and repressing those genes whose products would lead to differentiation (Marson et al. 2008; Young 2011). These transcription factors appear to work by recruiting RNA polymerase II to the promoters of those genes being activated while recruiting histone methyltransferases to those genes being repressed (Kagey et al. 2010; Adamo et al. 2011).

Only trophoblast cells synthesize the transcription factor Cdx2, which downregulates Oct4 and Nanog (Strumpf et al. 2005). The activation of the *Cdx2* gene in the trophoblast cells appears to be regulated by the Yap protein, which in turn is a co-factor for the transcription factor Tead4 (FIGURE 9.19A). Tead4 is found in the nuclei of both the inner and outer cells of the blastocyst, but it is activated by Yap only in the outer compartment. That is because Yap can enter the nucleus in the outer cells and thereby allow Tead4 to transcribe trophoblast-specifying genes such as

Cdx2 and *eomesodermin* (*Eomes*). In contrast, the inner cells, with each of their surfaces surrounded by other cells, activate the gene for Lats, a protein kinase that phosphorylates Yap (FIGURE 9.19B). Phosphorylated Yap cannot enter the nucleus and is subsequently degraded (Nishioka et al. 2009). Therefore, in the inner cells, Tead4 cannot function and *Cdx2* remains untranscribed. Cdx2 blocks the expression of Oct4, and Oct4 blocks the expression of Cdx2. In this way, the two lineages become separated.

The End of Pluripotency: More Choices

As important as pluripotency is, leaving pluripotency behind is just as important. The embryonic stem cells

---

* The research leading to the discovery of Nanog was motivated in part by the desire to convert human somatic cells into pluripotent stem cell lines. The name derives from Tir Nanog, the mythical Celtic land of perpetual youth.

---

stages and play important roles in early development. In the mouse and goat, the activation of zygotic (i.e., nuclear) genes begins in the late zygote and continues through the 2-cell stage (Zeng and Schultz 2005; Rother et al. 2011). In humans, the zygotic genes are activated slightly later, around the 8-cell stage (Piko and Clegg 1982; Braude et al. 1988; Dobson et al. 2004).

In order for the zygotic genes to be activated, the parental chromatin undergoes many changes. New histones are

placed on the DNA during the early cell divisions, and the gamete-specific DNA methyl groups are removed (except for those on imprinted genes; see Chapter 2). New DNA methylation patterns characteristic of totipotent and pluripotent cells are established (Abdalla et al. 2009; Hales et al. 2011). Thus, by the 16-cell stage, the genome of each cell is hypomethylated and each of these 16 cells appears to be pluripotent (Tarkowski et al. 2010). The stage is now set for cell differentiation to take place.

SIDELIGHTS SPECULATIONS (continued)

(A)

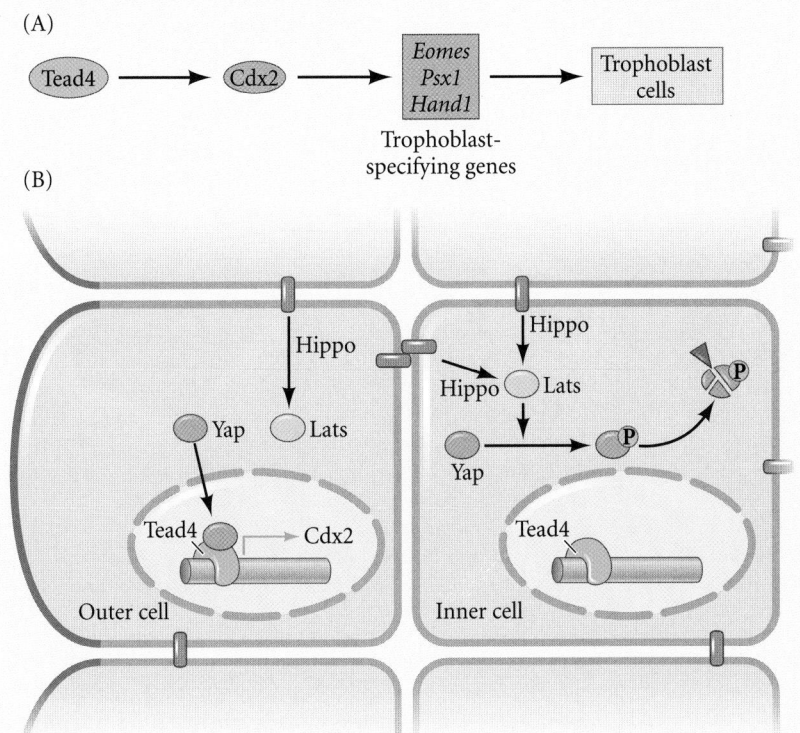

(B)

FIGURE 9.19 Possible pathway initiating the distinction between inner cell mass and trophoblast. (A) The Tead4 transcription factor, when active, promotes transcription of the *Cdx2* gene. Together, the Tead4 and Cdx2 transcription factors activate the genes that specify the outer cells to become the trophoblast. (B) Model for Tead4 activation. In the outer cells, the lack of cells surrounding the embryo sends a signal (as yet unknown) that blocks the Hippo pathway from activating the Lats protein. In the absence of functional Lats, the Yap transcriptional co-factor can bind with Tead4 to activate the *Cdx2* gene. In the inner cells, the Hippo pathway is active and the Lats kinase phosphorylates the Yap transcriptional co-activator. The phosphorylated form of Yap does not enter the nucleus and is targeted for degradation. (C) Mouse blastocyst in which the Oct4 protein in the ICM is stained orange. The extracellular lineages (trophoblast and hypoblast) are stained green. (A,B after Nishioka et al. 2009; C courtesy of J. Rossant.)

(C)

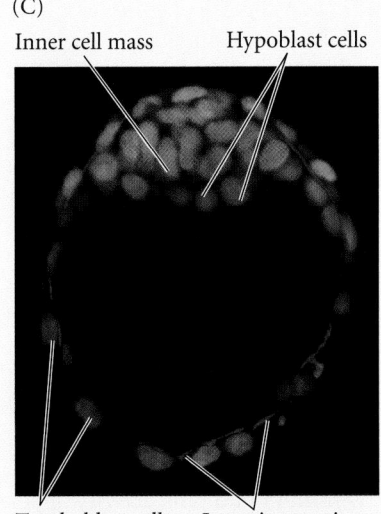

must retain their pluripotency, but they must also produce progeny that differentiate into particular cell types.

Pluripotency ends at gastrulation. Only the primordial germ cells, set aside in the extraembryonic tissues at the posterior end of the embryo, remain pluripotent, capable of engendering the next generation of individual organisms. The cells that are to become *this* generation's body have to make choices. If the cell can no longer be an embryonic stem cell, what sort of cell will it become? The answer may depend on the paracrine milieu experienced by the cell, and these paracrine factors appear to act by regulating the ratio of Oct4 and Sox2 transcription factors present in the cells of the

ICM (Gabut et al. 2011; Thompson et al. 2011). Without Oct4, Sox2 will activate genes for the synthesis of retinoic acid and FGFs that instruct the cells to become ectoderm and will repress those genes that would cause the cells to become the mesoderm or endoderm cells that migrate through the primitive streak. Without Sox2, Oct4 will activate the genes (such as those for activin and Wnts) promoting mesodermal and endodermal specification and will repress those genes instructing cells to become ectoderm (specifically, neural ectoderm, although this can be altered by exposure to BMPs to generate epidermal ectoderm). The first decisions have been made, and the grand migrations of gastrulation can begin.

Compaction

One of the most crucial events of mammalian cleavage is **compaction**. Mouse blastomeres through the 8-cell stage form a loose arrangement (**FIGURE 9.20A–C**). Following the third cleavage, however, the blastomeres undergo a spectacular change in their behavior. Cell adhesion proteins such as E-cadherin become expressed, and the blastomeres gradually huddle together and form a compact ball of cells (**FIGURE 9.20D**; Peyrieras et al. 1983; Fleming et al. 2001). This tightly

packed arrangement is stabilized by tight junctions that form between the outside cells of the ball, sealing off the inside of the sphere. The cells within the sphere form gap junctions, thereby enabling small molecules and ions to pass between them.

The cells of the compacted 8-cell embryo divide to produce a 16-cell **morula** (**FIGURE 9.20E**). The morula consists of a small group of internal cells surrounded by a larger group of external cells (Barlow et al. 1972). Most of the descendants

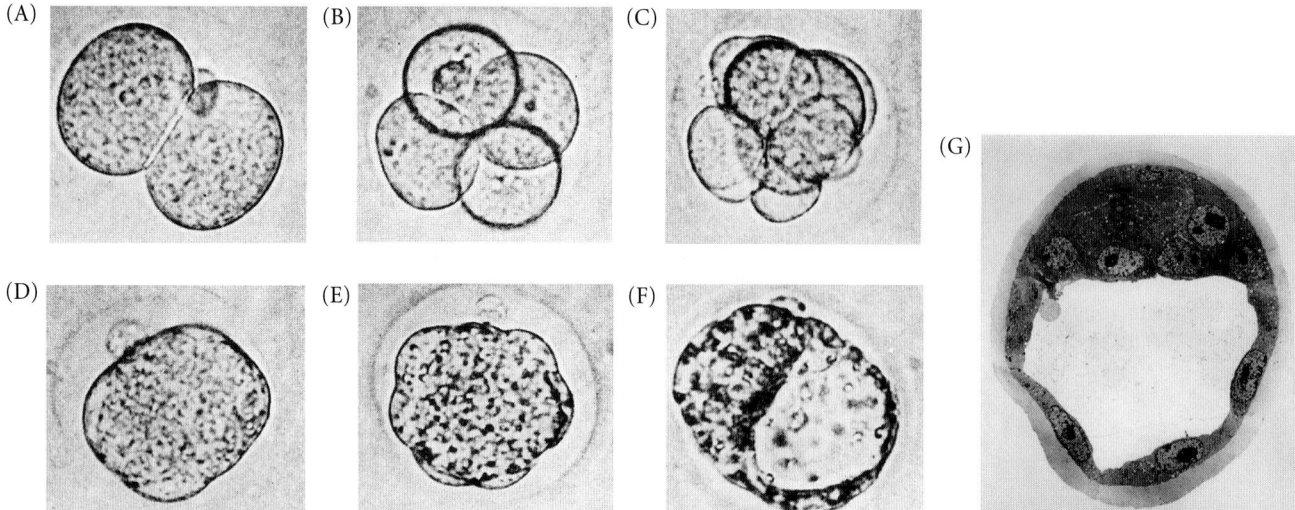

FIGURE 9.20 Cleavage of a single mouse embryo in vitro. (A) 2-Cell stage. (B) 4-Cell stage. (C) Early 8-cell stage. (D) Compacted 8-cell stage. (E) Morula. (F) Blastocyst. (G) Electron micrograph through the center of a mouse blastocyst. (A–F from Mulnard 1967, courtesy of J. G. Mulnard; G from Ducibella et al. 1975, courtesy of T. Ducibella.)

of the external cells become **trophoblast** (trophectoderm) cells, whereas the internal cells give rise to the **inner cell mass** (**ICM**). The inner cell mass, which will give rise to the embryo, becomes positioned on one side of the ring of trophoblast cells; the resulting **blastocyst** is another hallmark of mammalian cleavage (**FIGURE 9.20F,G**).

The trophoblast cells produce no embryonic structures, but rather form the tissues of the chorion, the extraembryonic membrane and portion of the placenta that enables the fetus to get oxygen and nourishment from the mother. The chorion also secretes hormones that cause the mother's uterus to retain the fetus and produces regulators of the immune response so that the mother will not reject the embryo (as she would an organ graft).

It is important to remember that a crucial outcome of these first divisions is the generation of cells that attach the embryo to the uterus. Thus, formation of the trophectoderm is the first differentiation event in mammalian development. The earliest blastomeres (such as each blastomere of a 2-cell embryo) can form both trophoblast cells and the embryo precursor cells of the ICM. These very early cells are said to be **totipotent** (Latin, "capable of everything"). The inner cell mass is said to be **pluripotent** (Latin, "capable of many things"). That is, each cell of the ICM can generate any cell type in the body but is no longer able to form the trophoblast.

● **See WEBSITE 9.3** Mechanisms of compaction and formation of the inner cell mass

● **See WEBSITE 9.4** Human cleavage and compaction

In mice, the embryo proper is derived from the inner cell mass of the 16-cell stage, supplemented by cells dividing from the outer cells of the morula during the transition to the 32-cell stage (Pedersen et al. 1986; Fleming 1987; McDole et al. 2011). The cells of the ICM give rise to the embryo and its associated yolk sac, allantois, and amnion. By the 64-cell stage, the ICM (comprising approximately 13 cells at that stage) and the trophoblast cells have become separate cell layers, neither of which contributes cells to the other group (Dyce et al. 1987; Fleming 1987). The ICM actively supports the trophoblast, secreting proteins that stimulate the trophoblast cells to divide (Tanaka et al. 1998).

Initially, the morula does not have an internal cavity. However, during a process called **cavitation**, the trophoblast cells secrete fluid into the morula to create a blastocoel. The membranes of trophoblast cells contain sodium pumps (an Na^+-K^+ ATPase and an Na^+-H^+ exchanger) that pump Na^+ into the central cavity. The subsequent accumulation of Na^+ draws in water osmotically, creating and enlarging the blastocoel (Borland 1977; Ekkert et al. 2004; Kawagishi et al. 2004). Interestingly, this sodium pumping activity appears to be stimulated by the oviduct cells on which the embryo is traveling toward the uterus (Xu et al. 2004). As the blastocoel expands, the inner cell mass becomes positioned on one side of the ring of trophoblast cells, resulting in the distinctive mammalian blastocyst.*

---

*Although the mammalian blastocyst was discovered by Rauber in 1881, its first public display was probably in Gustav Klimt's 1908 painting *Danae*, in which blastocyst-like patterns are featured on the heroine's robe as she becomes impregnated by Zeus (Gilbert and Braukmann 2011).

(A)

(B)

(C) Trophoblast

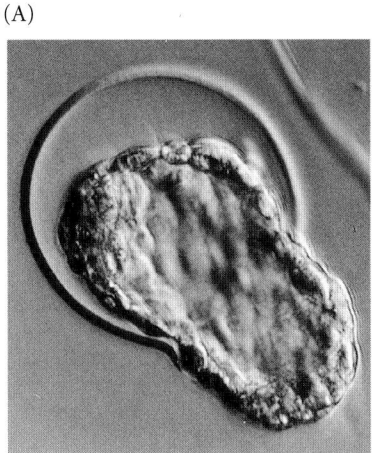
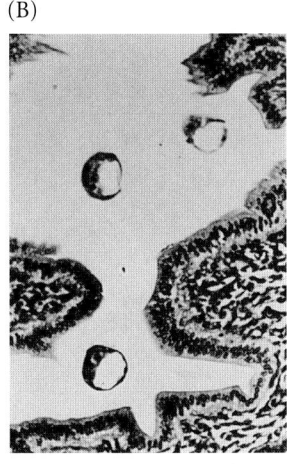
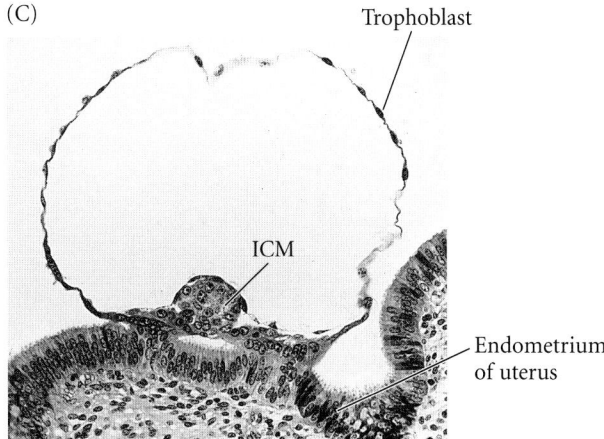

ICM

Endometrium
of uterus

FIGURE 9.21 Hatching from the zona and implantation of the mammalian blastocyst in the uterus. (A) Mouse blastocyst hatching from the zona pellucida. (B) Mouse blastocysts entering the uterus. (C) Initial implantation of a rhesus monkey blastocyst. (A from Mark et al. 1985, courtesy of E. Lacy; B from Rugh 1967; C, Carnegie Institution of Washington, Chester Reather, photographer.)

Escape from the zona pellucida

While the embryo is moving through the oviduct en route to the uterus, the blastocyst expands within the zona pellucida (the extracellular matrix of the egg that was essential for sperm binding during fertilization; see Chapter 4). During this time, the zona pellucida prevents the blastocyst from adhering to the oviduct walls. (If this happens—as it sometimes does in humans—it is called an ectopic, or "tubal," pregnancy, a dangerous condition because an embryo implanted in the oviduct can cause a life-threatening hemorrhage when it begins to grow.) When the embryo reaches the uterus, it must "hatch" from the zona so that it can adhere to the uterine wall.

The mouse blastocyst hatches from the zona pellucida by digesting a small hole in it and squeezing through the hole as the blastocyst expands (**FIGURE 9.21A**). A trypsin-like protease secreted by the trophoblast seems responsible for hatching the blastocyst from the zona (Perona and Wassarman 1986; O'Sullivan et al. 2001). Once outside the zona, the blastocyst can make direct contact with the uterus (**FIGURE 9.21B,C**). The **endometrium**—the epithelial lining of the uterus—has been altered by estrogen and progesterone hormones and has made an extensive extracellular matrix that "catches" the blastocyst. This extracellular matrix is composed of complex sugars, collagen, laminin, fibronectin, cadherins, hyaluronic acid, and heparan sulfate receptors (see Wang and Dey 2006; Ramathal et al. 2011).

After the initial binding, several other adhesion systems appear to coordinate their efforts to keep the blastocyst tightly bound to the uterine lining. The trophoblast cells synthesize integrins that bind to the uterine collagen, fibronectin,

and laminin, and they synthesize heparan sulfate proteoglycan precisely prior to implantation (see Carson et al. 1993). P-cadherins (see Chapter 3) on the trophoblast and uterine endometrium also help dock the embryo to the uterus Once in contact with the endometrium, Wnt proteins (from the trophoblast, the endometrium, or from both) instruct the trophoblast to secrete a set of proteases, including collagenase, stromelysin, and plasminogen activator. These protein-digesting enzymes digest the extracellular matrix of the uterine tissue, enabling the blastocyst to bury itself within the uterine wall (Strickland et al. 1976; Brenner et al. 1989; Pollheimer et al. 2006).

Mammalian Gastrulation

Birds and mammals are both descendants of reptilian species (albeit different reptilian species). It is not surprising, therefore, that mammalian development parallels that of reptiles and birds. What *is* surprising is that the gastrulation movements of reptilian and avian embryos, which evolved as an adaptation to yolky eggs, are retained in the mammalian embryo even in the absence of large amounts of yolk. The mammalian inner cell mass can be envisioned as sitting atop an imaginary ball of yolk, following instructions that seem more appropriate to its reptilian ancestors.

Modifications for development inside another organism

The mammalian embryo obtains nutrients directly from its mother and does not rely on stored yolk. This adaptation has entailed a dramatic restructuring of the maternal anatomy (such as expansion of the oviduct to form the uterus) as well as the development of a fetal organ capable of absorbing maternal nutrients. The origins of early mammalian tissues

Twins and Chimeras

The early cells of the mammalian embryo can replace each other and compensate for a missing cell. This regulative ability was first demonstrated in 1952, when Seidel destroyed one cell of a 2-cell rabbit embryo and the remaining cell produced an entire embryo. The regulative capacity of the early embryo is also seen in humans. Human twins are classified into two major groups: **monozygotic** (one-egg; identical) twins and **dizygotic** (two-egg; fraternal) twins. Fraternal twins are the result of two separate fertilization events, whereas identical twins are formed from a single embryo whose cells somehow become dissociated from one another.

Identical twins, which occur in roughly 1 in 400 human births, may be produced by the separation of early blastomeres, or even by the separation of the inner cell mass into two regions within the same blastocyst. About 33% of identical twins have two complete and separate chorions, indicating that separation occurred before the formation of the trophoblast tissue at day 5 (FIGURE 9.22A). Other identical twins share a common chorion, suggesting that the split occurred within the inner cell mass after the trophoblast formed. By day 9, the human embryo has completed the construction of another extraembryonic layer, the lining of the amnion. If separation of the embryo comes after the formation of the chorion on day 5 but before the formation of the amnion on day 9, then the resulting embryos should have one chorion and two amnions (FIGURE 9.22B). This happens in about two-thirds of human identical twins. A small percentage of identical twins are born within a single chorion and amnion (FIGURE 9.22C), meaning the division of the embryo came after day 9.

The ability to produce an entire embryo from cells that normally would have contributed to only a portion of the embryo is seen in the ability of two or more early embryos to form one

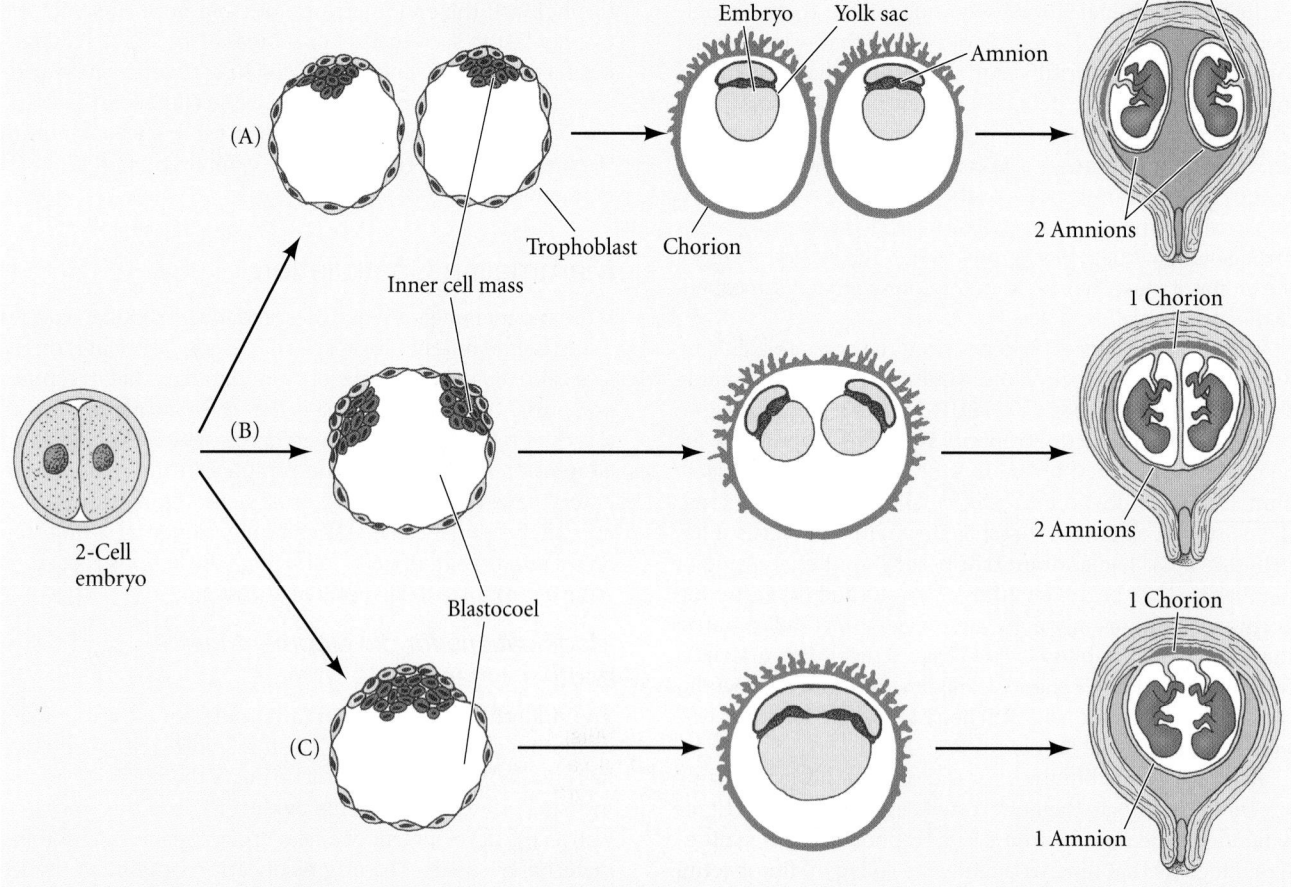

FIGURE 9.22 The timing of human monozygotic twinning with relation to extraembryonic membranes. (A) Splitting occurs before formation of the trophoblast, so each twin has its own chorion and amnion. (B) Splitting occurs after trophoblast formation but before amnion formation, resulting in twins having individual amniotic sacs but sharing one chorion. (C) Splitting after amnion formation leads to twins in one amniotic sac and a single chorion. (After Langman 1981.)

chimeric individual rather than twins, triplets, or a multiheaded individual. Chimeric mice can be produced by artificially aggregating two or more early-cleavage (usually 4- or 8-cell) embryos to form a composite embryo (FIGURE 9.23A). Markert and Petters (1978) have shown that three early 4-cell embryos can unite to form a common compacted morula and that the resulting mouse can have the coat colors of the three different strains. Moreover, they showed that each of the three embryos gave rise to precursors of the gametes. When a chimeric (black/brown/white) female mouse was mated to a white-furred (recessive) male, offspring of each of the three colors were produced (FIGURE 9.23B).

There is even evidence that human embryos can form chimeras (de la Chappelle et al. 1974; Mayr et al. 1979). Some individuals have two genetically different cell types (XX and XY) within the same body, each with its own set of genetically defined characteristics. The simplest explanation for such a phenomenon is that these individuals resulted from the aggregation of two embryos, one male and one female, that were developing at the same time. If this explanation is correct, then two fraternal twins have fused to create the individual* (see Yu et al. 2002).

According to these studies of twins and chimeras, each cell of the inner cell mass should be able to produce any cell of the body. This hypothesis has been confirmed, and it has important consequences for the study of mammalian development. When ICM

*There are other explanations, at least for some chimeras. Souter and colleagues (2007) have shown that in at least one XX/XY human the maternal alleles were identical and the paternal alleles differed. This would be expected if the egg underwent parthenogenic activation (meiosis without sperm activation) and each of the meiotic cells was then fertilized by a different sperm (one bearing an X chromosome, one a Y). The intermingling of the cells would produce the chimera, which in the case recounted by Souter et al. was a true hermaphrodite, having both male and female sex organs (see Chapter 15). We still do not know the mechanisms through which human twins and chimeras form.

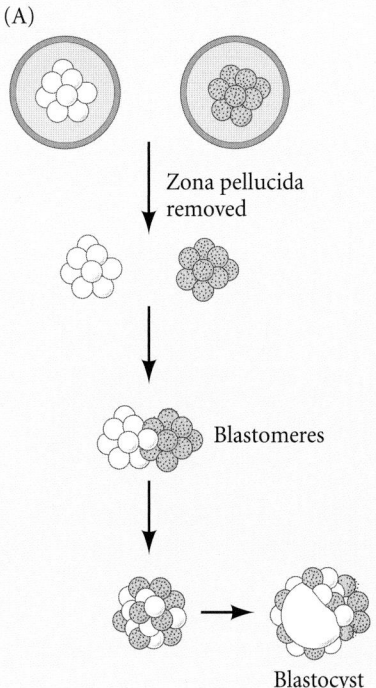
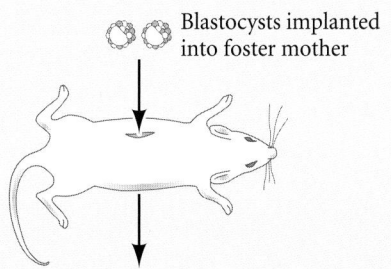
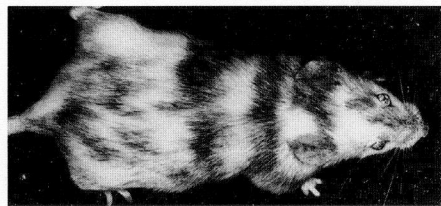

FIGURE 9.23 Production of chimeric mice. (A) Experimental procedures used to produce chimeric mice. Early 4-cell embryos of genetically distinct mice (here, with coat color differences) are isolated from mouse oviducts and brought together after their zonae are removed by proteolytic enzymes. The cells form a composite blastocyst, which is implanted into the uterus of a foster mother. The photograph shows one of the actual chimeric mice produced in this manner. (B) An adult female chimeric mouse (bottom) produced from the fusion of three 4-cell embryos: one from two white-furred parents, one from two black-furred parents, and one from two brown-furred parents. The resulting mouse has coat colors from all three embryos. Moreover, each embryo contributed germline cells, as is evidenced by the three colors of offspring (above) produced when this chimeric female was mated with recessive (white-furred) males. (A courtesy of B. Mintz; B from Markert and Petters 1978, courtesy of C. Markert.)

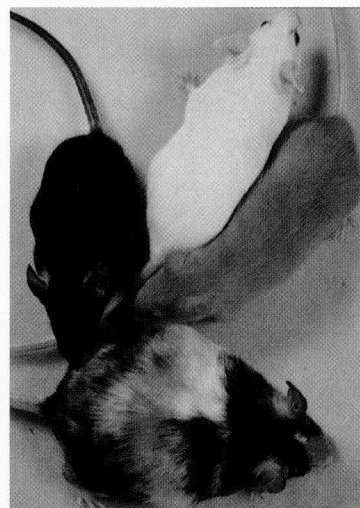

cells are isolated and grown under certain conditions, they remain undifferentiated and continue to divide in culture (Evans and Kaufman 1981; Martin 1981). Such cells are **embryonic stem cells** (**ES cells**). When ES cells are injected into a mouse blastocyst, they can integrate into the host inner cell mass. The resulting embryo has cells from both the host

and the donor tissue. This technique has become extremely important in determining the function of genes during mammalian development.

● See WEBSITE 9.5
Nonidentical monozygotic twins

● See WEBSITE 9.6
Conjoined twins

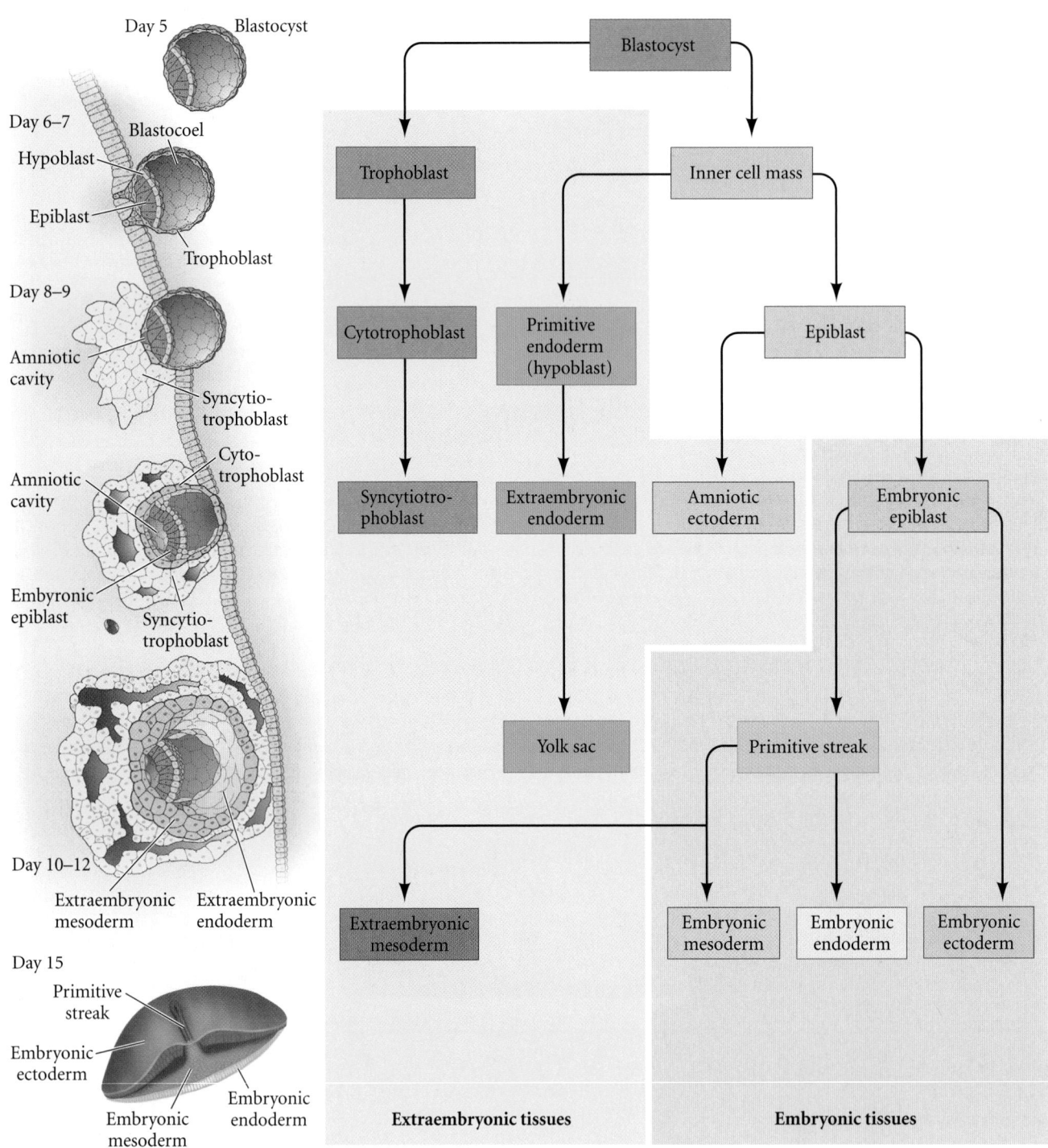

FIGURE 9.24 Tissue and germ layer formation in the early human embryo. Days 5–9: Implantation of the blastocyst. The inner cell mass delaminates hypoblast cells that line the blastocoel, forming the extraembryonic endoderm of the primitive yolk sac and a bilayered (epiblast and hypoblast) blastodisc. Days 10–12: The trophoblast divides into the cytotrophoblast, which will form the villi, and the syncytiotrophoblast, which will ingress into the uterine tissue to form the chorion. Days 12–15: Gastrulation and formation of primitive streak. Meanwhile, the epiblast splits into the amniotic ectoderm (which encircles the amniotic cavity) and the embryonic epiblast. The adult mammal (ectoderm, endoderm, mesoderm, and germ cells) forms from the cells of the embryonic epiblast. The extraembryonic endoderm forms the yolk sac. The actual size of the embryo at this stage is about that of the period at the end of this sentence.

are summarized in **FIGURE 9.24**. As we saw above, the first distinction is that between inner cell mass and trophoblast. The trophoblast develops through several stages, eventually becoming the chorion, the embryonically derived portion of the placenta. Trophoblast cells also induce the mother's uterine cells to form the maternal portion of the placenta, the **decidua**. The decidua becomes rich in the blood vessels that will provide oxygen and nutrients to the embryo. The inner cell mass gives rise to the epiblast and the hypoblast (primitive endoderm). The hypoblast will generate yolk sac cells, while the epiblast will generate the embryo, the amnion, and the allantois.

THE PRIMITIVE ENDODERM: THE MAMMALIAN HYPOBLAST

The first segregation of cells within the inner cell mass forms two layers. The lower layer, in contact with the blastoel, is called the **primitive endoderm**, and it is homologous to the hypoblast of the chick embryo. The remaining inner cell mass tissue above it is the epiblast. The primitive endoderm will form the yolk sac of the embryo, and like the chick hypoblast, will be used for positioning the site of gastrulation, regulating cell movements in the epiblast, and promoting the maturation of blood cells. Moreover, the primitive endoderm, like the chick hypoblast, is an extraembryonic layer and does not provide cells to the actual embryo (see Stern and Downs 2012).

Whether a mouse ICM cell becomes epiblast or primitive endoderm may depend on *when* the cell became part of the ICM (Bruce and Zernicka-Goetz 2010; Morris et al. 2010). Cells that become internalized in the division from 8 to 16 cells appear biased to become pluripotent epiblast cells, while the future primitive endoderm may be generated by cells entering the ICM during the division from 16 to 32 cells (**FIGURE 9.25**). At that stage, the blastomeres of the ICM are a mosaic of future epiblast cells (expressing Nanog transcription factor,

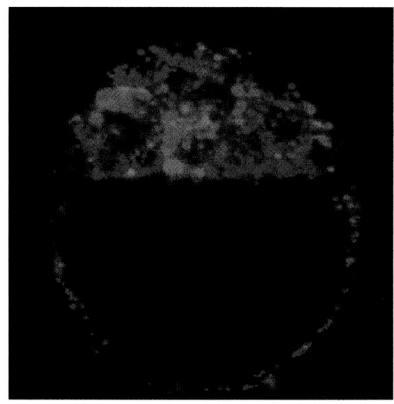

FIGURE 9.25 Mouse embryo at day 3.5 (early blastocyst), showing the random expression of Nanog (blue, for the epiblast) and Gata6 (red, for the visceral endoderm) in the inner cell mass. In another 24 hours, the cells will sort out: the hypoblast cells will abut the blastocoel, and the epiblast cells will be between the hypoblast cells and the trophoblast (as in Figure 9.19C). (F courtesy of J. Rossant.)

which promotes pluripotency; see Sidelights & Speculations, p. 300) and primitive endoderm cells (expressing Gata6 transcription factor) a full day before the layers segregate at day 4.5 (Chazaud et al. 2006). Levels of FGF signaling within the ICM determine the final identity of epiblast or primitive endoderm, with cells receiving higher levels of FGF becoming primitive endoderm (Yamanaka et al. 2010).

The epiblast and primitive endoderm form a structure called the **bilaminar germ disc** (**FIGURE 9.26A**). The primitive endoderm cells expand to line the blastocoel cavity, where they give rise to the yolk sac. The primitive endoderm cells contacting the epiblast are the **visceral endoderm**, while those yolk sac cells

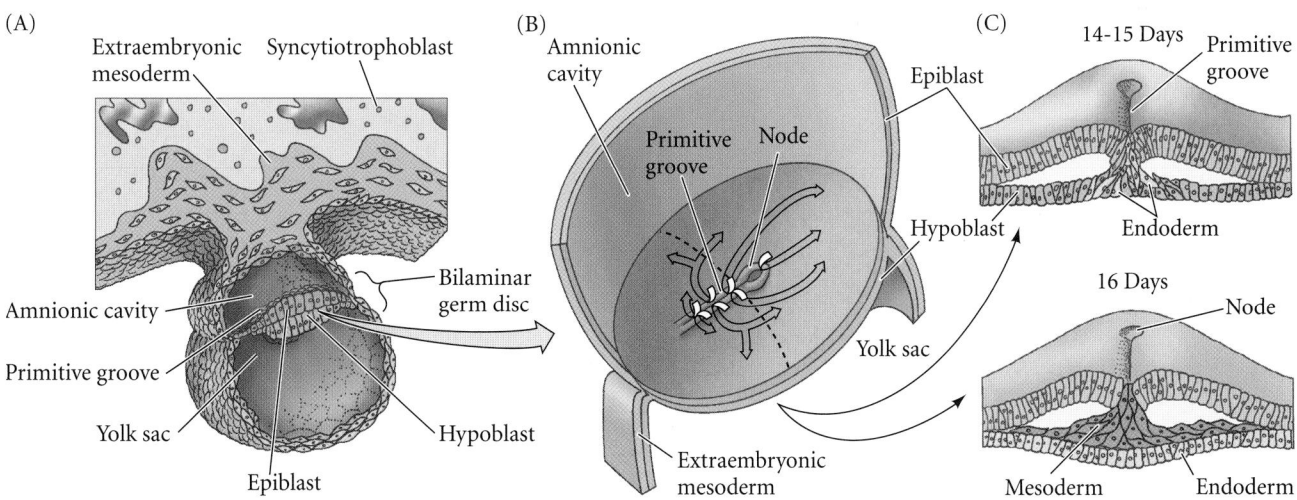

FIGURE 9.26 Amnion structure and cell movements during human gastrulation. (A,B) Human embryo and uterine connections at day 15 of gestation. (A) Sagittal section through the midline. (B) View looking down on the dorsal surface of the embryo. Movements of the epiblast cells through the primitive streak and the node and underneath the epiblast are superimposed on the dorsal surface view. (C) At days 14 and 15 the ingressing epiblast cells are thought to replace the hypoblast cells (which contribute to the yolk sac lining), and at day 16 the ingressing cells fan out to form the mesodermal layer. (After Larsen 1993.)

contacting the trophoblast are the **parietal endoderm**. The epiblast cell layer is split by small clefts that eventually coalesce to separate the embryonic epiblast from the other epiblast cells that form the amnion. Once the amnion is completed, the amniotic cavity fills with **amniotic fluid**, a secretion that serves as a shock absorber as well as preventing the developing embryo from drying out. The embryonic epiblast is thought to contain all the cells that will generate the actual embryo and is similar in many ways to the avian epiblast.

By labeling individual cells of the epiblast with horseradish peroxidase, Kirstie Lawson and her colleagues (1991) were able to construct a detailed fate map of the mouse epiblast (see Figure 1.11). Gastrulation begins at the posterior end of the embryo, and this is where the cells of the **node*** arise (**FIGURE 9.26B,C**). Like the chick epiblast cells, mammalian mesoderm and endoderm cells originate in the epiblast, undergo epithelial-mesenchymal transition, lose E-cadherin, and migrate through a primitive streak as individual mesenchymal cells (Burdsal et al. 1993). Those cells arising from the node give rise to the notochord. However, in contrast to notochord formation in the chick, the cells that form the mouse notochord are thought to become integrated into the endoderm of the primitive gut (Jurand 1974; Sulik et al. 1994). These cells can be seen as a band of small, ciliated cells extending rostrally from the node. They form the notochord by converging medially and "budding" off in a dorsal direction from the roof of the gut. The timing of these developmental events varies enormously in mammals. In humans, the migration of cells forming the mesoderm doesn't start until day 16—around the time that a mouse embryo is almost ready to be born (see Figure 9.26C; Larsen 1993).

Cell migration and specification appear to be coordinated by fibroblast growth factors. The cells of the primitive streak appear to be capable of both synthesizing and responding to FGFs (Sun et al. 1999; Ciruna and Rossant 2001). In embryos that are homozygous for the loss of the *Fgf8* gene or its receptor, cells fail to emigrate from the primitive streak, and neither mesoderm nor endoderm are formed. Fgf8 (and perhaps other FGFs) probably control cell movement into the primitive streak by downregulating the E-cadherin that holds the epiblast cells together. Fgf8 may also control cell specification by regulating *snail*, *Brachyury*, and *Tbx6*, three genes that are essential (as they are in the chick embryo) for mesodermal migration, specification, and patterning.

The ectodermal precursors are located anterior and lateral to the fully extended primitive streak, as in the chick epiblast; however, in some instances (also as in the chick embryo), a single cell gives rise to descendants in more than one germ layer, or to both embryonic and extraembryonic derivatives. Thus, at the epiblast stage these lineages have not become fully separate from one another. Indeed, in mice, some of the visceral endoderm, which had been extraembryonic, is able to intercalate with the definitive endoderm and become part of the gut (Kwon et al. 2008).

Formation of the chorion

While the embryonic epiblast is undergoing cell movements reminiscent of those seen in reptilian or avian gastrulation, the extraembryonic cells are forming the placenta, a distinctly mammalian set of tissues that enable the embryo to survive within the maternal uterus. Although the initial trophoblast cells of mice and humans divide like most other cells of the body, they give rise to a population of cells in which nuclear division occurs in the absence of cytokinesis. The original trophoblast cells constitute a layer called the **cytotrophoblast**, whereas the multinucleated cell type forms the **syncytiotrophoblast**. The cytotrophoblast initially adheres to the endometrium through a series of adhesion molecules, as we saw above. Moreover, cytotrophoblasts contain proteolytic enzymes that enable them to enter the uterine wall and remodel the uterine blood vessels so that the maternal blood bathes fetal blood vessels. The syncytiotrophoblast tissue is thought to further the progression of the embryo into the uterine wall by digesting uterine tissue. The cytotrophoblast secretes paracrine factors that attract maternal blood vessels and gradually displace their vascular tissue such that the vessels become lined with trophoblast cells (Fisher et al. 1989; Knöfler 2010). Shortly thereafter, mesodermal tissue extends outward from the gastrulating embryo (see Figure 9.26C). Studies of human and rhesus monkey embryos have suggested that the yolk sac as well as primitive streak-derived cells contribute this extraembryonic mesoderm (Bianchi et al. 1993).

The extraembryonic mesoderm joins the trophoblastic extensions and gives rise to the blood vessels that carry nutrients from the mother to the embryo. The narrow connecting stalk of extraembryonic mesoderm that links the embryo to the trophoblast eventually forms the vessels of the umbilical cord. The fully developed extraembryonic organ, consisting of trophoblast tissue and blood vessel-containing mesoderm, is the chorion, and it fuses with the uterine wall decidua to create the placenta. Thus the placenta is an organ derived from two genetically different organisms.

The chorion may be very closely apposed to maternal tissues while still being readily separable from them (as in the contact placenta of the pig), or it may be so intimately integrated with maternal tissues that the two cannot be separated without damage to both the mother and the developing fetus (as in the deciduous placenta of most mammals, including humans).† Beyond their role in nourishing the

---

* In mammalian development, Hensen's node is usually just called "the node," despite the fact that Hensen discovered this structure in rabbit and guinea pig embryos.

---

†There are numerous types of placentas, and the extraembryonic membranes form differently in different orders of mammals (see Cruz and Pedersen 1991). Although mice and humans gastrulate and implant in a similar fashion, their extraembryonic structures are distinctive. It is very risky to extrapolate developmental phenomena from one group of mammals to another. Even Leonardo da Vinci got caught (Renfree 1982). His remarkable drawing of the human fetus inside the placenta is stunning art but poor science: the placenta is that of a cow.

(A)

(B)

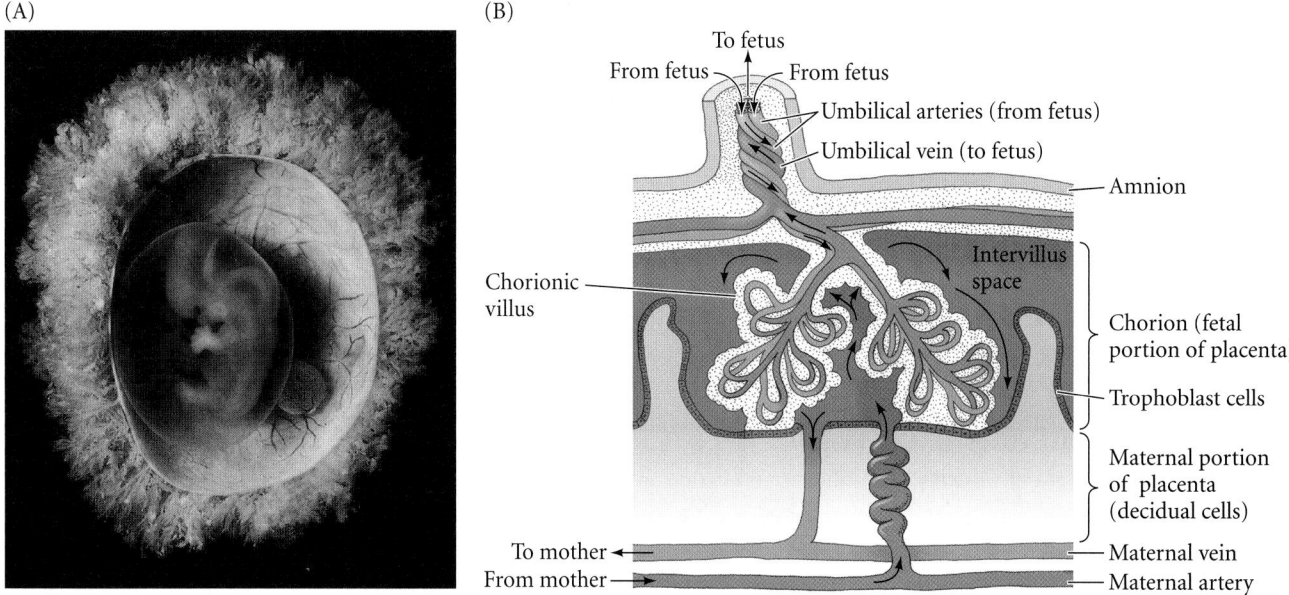

FIGURE 9.27   (A) Human embryo and placenta after 6 weeks of gestation. The embryo lies within the amnion, and its blood vessels can be seen extending into the chorionic villi. The small sphere to the right of the embryo is the yolk sac. (B) Relationship of the chorionic villi to the maternal blood supply in the primate uterus. In the umbilicus, there are two arteries and a single vein. (A from Carnegie Institution of Washington, courtesy of Chester F. Reather.)

embryo, placentas are endocrine and immunological organs. They produce hormones (such as progesterone) that enable the uterus to retain the pregnancy and accelerate mammary gland development. Recent studies also suggest that the placenta generates several mechanisms that block the mother's immune system from attacking foreign substances produced by the embryo or fetus (Warning et al. 2011; Rowe et al. 2012).

FIGURE 9.27A shows the relationships between the embryonic and extraembryonic tissues of a 6.5-week human embryo. The embryo is seen encased in the amnion and is further shielded by the chorion. The blood vessels extending to and from the chorion are readily observable, as are the villi that project from the outer surface of the chorion. These villi contain the blood vessels and allow the chorion to have a large area exposed to the maternal blood. Although fetal and maternal circulatory systems normally never merge, diffusion of soluble substances can occur through the villi (FIGURE 9.27B). In this manner, the mother provides the fetus with nutrients and oxygen, and the fetus sends its waste products (mainly carbon dioxide and urea) into the maternal circulation. The maternal and fetal blood cells usually do not mix, although a small number of fetal red blood cells are seen in the maternal blood circulation.

● See WEBSITE 9.7 Placental functions

Mammalian Axis Formation

The anterior-posterior axis: Two signaling centers

The formation of the mammalian anterior-posterior axis has been studied most extensively in mice. The structure of the mouse epiblast, however, differs from that of humans in that it is cup-shaped rather than disc-shaped. Whereas the human embryo looks very much like the chick embryo, the mouse embryo "drops" such that it looks like a droplet enclosed by the primitive endoderm (FIGURE 9.28A).

The mammalian embryo appears to have two signaling centers: one in the node (equivalent to Hensen's node and the trunk portion of the amphibian organizer) and one in the **anterior visceral endoderm** (**AVE**; Beddington and Robertson 1999; Foley et al. 2000). The node appears to be responsible for the patterning of most of the body axis, and the two signaling centers work together to pattern the anterior region of the embryo (Bachiller et al. 2000).

The signals that initiate the primitive streak appear to come from interactions between the trophoblast-derived extraembryonic ectoderm and the epiblast. BMP4 originating from the extraembryonic ectoderm instructs the adjacent epiblast cells to make Wnt3a and Nodal. However, the AVE prevents Wnt3a and Nodal from having an effect on the anterior side of the embryo, by secreting antagonists of these paracrine factors, Lefty-1 and Cerberus (FIGURE 9.28B; Brennan et al. 2001; Perea-Gomez et al. 2001; Yamamoto et al. 2004). As a result, Wnt3a activates the *Brachyury* gene in cells of the posterior but not the anterior epiblast, generating mesoderm cells (Bertocchini et al. 2002; Perea-Gomez et al. 2002).

Once formed, the node secretes Chordin; the head process and notochord will later add Noggin. These two BMP

(A)

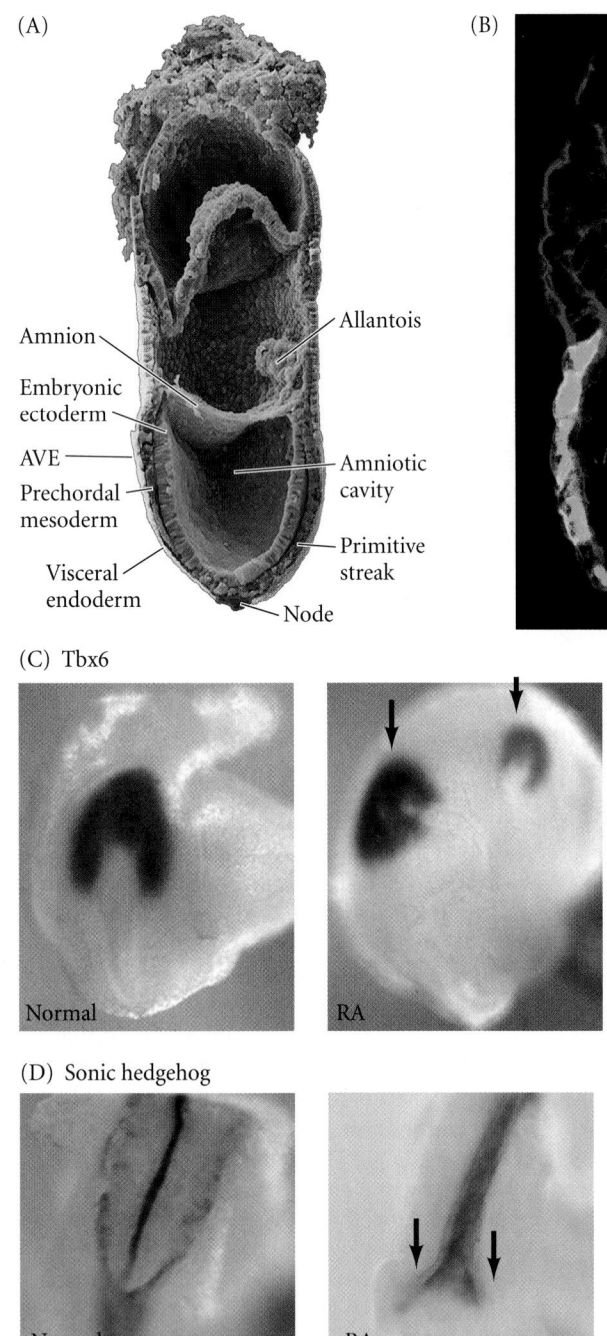

Amnion

Embryonic
ectoderm

AVE

Prechordal
mesoderm

Visceral
endoderm

Allantois

Amniotic
cavity

Primitive
streak

Node

(B)

(C) Tbx6

Normal   RA

(D) Sonic hedgehog

Normal   RA

FIGURE 9.28 Axis and notochord formation in the mouse. (A) In the 7-day mouse embryo, the dorsal surface of the epiblast (embryonic ectoderm) is in contact with the amniotic cavity. The ventral surface of the epiblast contacts the newly formed mesoderm. In this cuplike arrangement, the endoderm covers the surface of the embryo. The node is at the bottom of the cup, and it has generated chordamesoderm. The two signaling centers, the node and the anterior visceral endoderm (AVE), are located on opposite sides of the cup. Eventually, the notochord will link them. The caudal side of the embryo is marked by the presence of the allantois. (B) Confocal fluorescence image of *Cerberus* gene expression, with the *Cerberus* gene fused to a gene for GFP. At this stage, the Cerberus-synthesizing cells are migrating to the most anterior region of the visceral endoderm. (C,D) Retinoic acid (RA) appears to inhibit the migration of the AVE precursors. In normal mouse embryos (left column), Tbx6 is expressed in the anterior primitive streak and in the epiblast surrounding the node (C), while Sonic hedgehog is expressed in the notochord (D). In embryos treated with RA during early gastrulation (right column), there are often two axes. Here, two areas of Tbx6 are observed and the notochord is bifurcated (arrows). (A courtesy of K. Sulik; B courtesy of J. Belo; C,D from Liao and Collins 2008, courtesy of K. Sulik and G. Schoenwolf.)

Anterior-posterior patterning by FGF and retinoic acid gradients

The head region of the mammalian embryo is devoid of Nodal signaling, and BMPs, FGFs, and Wnts are also inhibited. The posterior region is characterized by Nodal, BMPs, Wnts, FGFs, and retinoic acid. There appears to be a gradient of Wnt, BMP, and FGF proteins that is highest in the posterior and drops off strongly near the anterior region. Moreover, in the anterior half of the embryo, starting at the node, there is a high concentration of antagonists that prevent BMPs and Wnts from acting (**FIGURE 9.29A**). The Fgf8 gradient is created by the decay of mRNA: *Fgf8* is expressed at the growing posterior tip of the embryo, but the *Fgf8* message is slowly degraded in the newly formed tissues. Thus there is a gradient of *Fgf8* mRNA across the posterior of the embryo, which is converted into an Fgf8 protein gradient (**FIGURE 9.29B**; Dubrulle and Pourquié 2004).

In addition to FGFs, the late-stage gastrula has a gradient of retinoic acid, with RA levels high in the posterior regions and low in the anterior of the embryo. This gradient (like that of chick, frog, and fish embryos) appears to be controlled by the expression of RA-synthesizing enzymes in the embryo's posterior and RA-degrading enzymes in the anterior parts of the embryo (Sakai et al. 2001; Oosterveen et al. 2004).

antagonists are not expressed in the AVE. Whereas knockouts of either the *Chordin* or the *Noggin* gene do not affect development, mice missing *both* genes lack a forebrain, nose, and other facial structures. Dickkopf is expressed in both the AVE and in the node, but only the Dickkopf from the node is critical for head development (Mukhopadhyay et al. 2001). This regulation can be altered by retinoic acid (RA), which appears to inhibit the migration of the AVE precursors. As a result, more than one body axis can form (**FIGURE 9.28C,D**; Liao and Collins 2008).

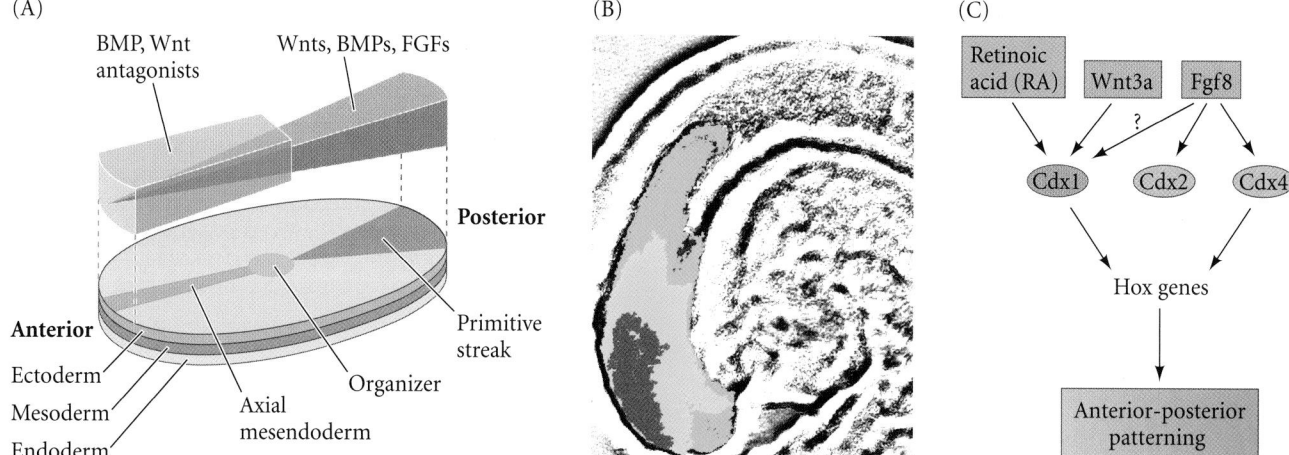

FIGURE 9.29 Anterior-posterior patterning in the mouse embryo. (A) Concentration gradients of BMPs, Wnts, and FGFs in the late-gastrula mouse embryo (depicted as a flattened disc). The primitive streak and other posterior tissues are the sources of Wnt and BMP proteins, whereas the organizer and its derivatives (such as the notochord) produce antagonists. Fgf8 is expressed in the posterior tip of the gastrula and continues to be made in the tailbud. Its mRNA decays, creating a gradient across the posterior portion of the embryo. (B) Fgf8 gradient in the tailbud region of a 9-day mouse embryo. The highest amount of Fgf8 (red) is found near the tip. The gradient was determined by in situ hybridization of an Fgf8 probe and staining for increasing amounts of time. (C) Retinoic acid, Wnt3a, and Fgf8 each contribute to posterior patterning, but they are integrated by the Cdx family of proteins that regulates the activity of the Hox genes. (A after Robb and Tam 2004; B from Dubrulle and Pourquié 2004, courtesy of O. Pourquié; C after Lohnes 2003.)

The FGF gradient patterns the posterior portion of the embryo by working through the Cdx family of caudal-related genes (**FIGURE 9.29C**; Lohnes 2003). The Cdx genes, in turn, integrate the various posteriorization signals and activate particular Hox genes.

Anterior-posterior patterning: The Hox code hypothesis

In all vertebrates, anterior-posterior polarity becomes specified by the expression of Hox genes. Vertebrate Hox genes are homologous to the homeotic selector genes (Hom-C genes) of the fruit fly (see Chapter 6). The *Drosophila* homeotic gene complex on chromosome 3 contains the *Antennapedia* and *bithorax* clusters (see Figure 6.36) and can be seen as a single functional unit. (Indeed, in some other insects, such as the flour beetle *Tribolium*, it *is* a single physical unit.) All of the known mammalian genomes contain four copies of the Hox complex per haploid set, located on four different chromosomes (*Hoxa* through *Hoxd* in the mouse, *HOXA* through *HOXD* in humans; see Boncinelli et al. 1988; McGinnis and Krumlauf 1992; Scott 1992).

The order of these genes on their respective chromosomes is remarkably similar between insects and humans, as is the expression pattern of these genes. Those mammalian genes homologous to the *Drosophila labial, proboscipedia,* and *deformed* genes are expressed anteriorly and early, whereas those genes homologous to the *Drosophila AbdB* gene are expressed posteriorly and later. As in *Drosophila*, a separate set of genes in mice encodes the transcription factors that regulate head formation. In *Drosophila*, these are the *orthodenticle* and *empty spiracles* genes. In mice, the midbrain and forebrain are made through the expression of genes homologous to these—*Otx2* and *Emx* (see Kurokawa et al. 2004; Simeone 2004).

The mammalian Hox/HOX genes are numbered from 1 to 13, starting from that end of each complex that is expressed most anteriorly. **FIGURE 9.30** shows the relationships between the *Drosophila* and mouse homeotic gene sets. The equivalent genes in each mouse complex (such as *Hoxa4, b4, c4,* and *d4*) are **paralogues**—that is, it is thought that the four mammalian Hox complexes were formed by chromosome duplications. Because the correspondence between the *Drosophila* Hom-C genes and mouse Hox genes is not one-to-one, it is likely that independent gene duplications and deletions have occurred since these two animal groups diverged (Hunt and Krumlauf 1992). Indeed, the most posterior mouse Hox gene (equivalent to *Drosophila AbdB*) underwent its own set of duplications in some mammalian chromosomes.

Hox gene expression can be seen along the mammalian body axis (in the neural tube, neural crest, paraxial mesoderm, and surface ectoderm) from the anterior boundary of the hindbrain through the tail. The regions of expression are not in register, but the 3′ Hox genes (homologous to *labial, proboscopedia,* and *deformed* of the fly) are expressed more anteriorly than the 5′ Hox genes (homologous to *Ubx, abdA,* and *AbdB*). Thus, one generally finds the genes of paralogous group 4 expressed anteriorly to those of paralogous group 5, and so forth (see Figure 9.30; Wilkinson et al. 1989; Keynes and Lumsden 1990). Mutations in the Hox genes suggest that the regional identity along the anterior-posterior axis is determined primarily by the most posterior Hox gene expressed in that region.

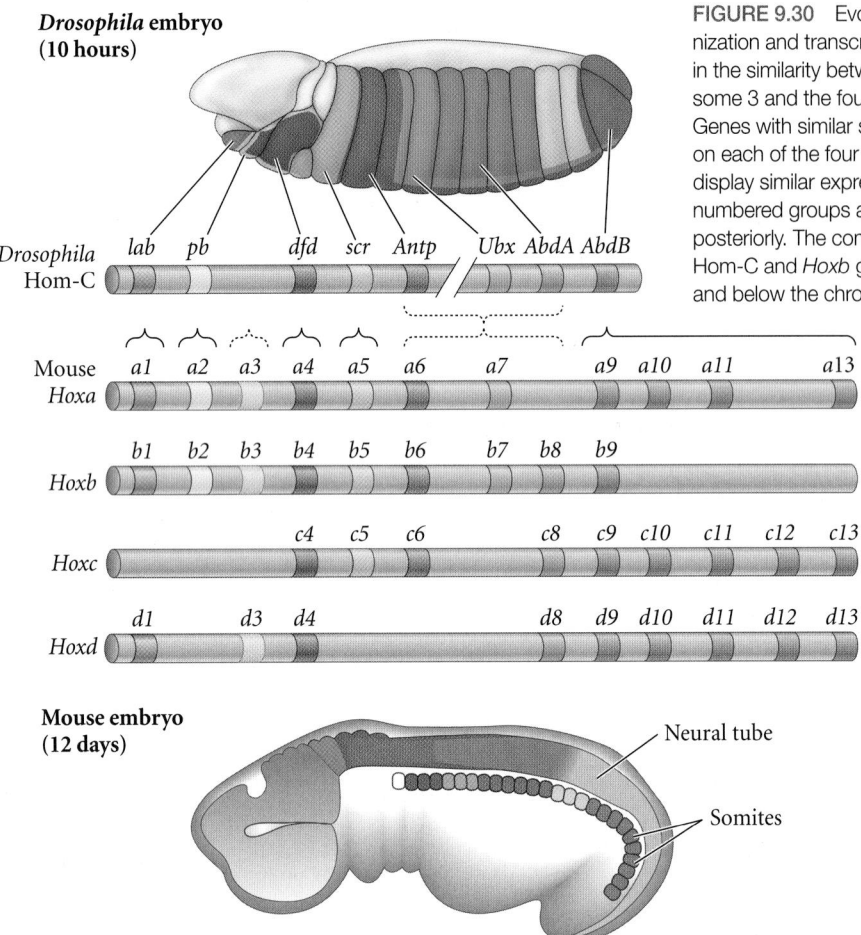

Drosophila embryo (10 hours)

Drosophila Hom-C

lab   pb   dfd   scr   Antp   Ubx AbdA AbdB

Mouse Hoxa
a1   a2   a3   a4   a5   a6   a7   a9   a10   a11   a13

Hoxb
b1   b2   b3   b4   b5   b6   b7   b8   b9

Hoxc
c4   c5   c6   c8   c9   c10   c11   c12   c13

Hoxd
d1   d3   d4   d8   d9   d10   d11   d12   d13

Mouse embryo (12 days)

Neural tube

Somites

FIGURE 9.30 Evolutionary conservation of homeotic gene organization and transcriptional expression in fruit flies and mice is seen in the similarity between the Hom-C cluster on *Drosophila* chromosome 3 and the four Hox gene clusters in the mouse genome. Genes with similar structures occupy the same relative positions on each of the four chromosomes, and paralogous gene groups display similar expression patterns. The mouse genes in the higher-numbered groups are expressed later in development and more posteriorly. The comparison of the transcription patterns of the Hom-C and *Hoxb* genes of *Drosophila* and mice are shown above and below the chromosomes, respectively. (After Carroll 1995.)

EXPERIMENTAL ANALYSIS OF THE HOX CODE The expression patterns of mouse Hox genes suggest a code whereby certain combinations of Hox genes specify a particular region of the anterior-posterior axis (Hunt and Krumlauf 1991). Particular sets of paralogous genes provide segmental identity along the anterior-posterior axis of the body. Evidence for such a code comes from three sources:

- Comparative anatomy, in which the types of vertebrae in different vertebrate species are correlated with the constellation of Hox gene expression

- Gene targeting or "knockout" experiments in which mice are constructed that lack both copies of one or more Hox genes

- Retinoic acid teratogenesis, in which mouse embryos exposed to RA show an atypical pattern of Hox gene expression along the anterior-posterior axis and abnormal differentiation of their axial structures

COMPARATIVE ANATOMY AND HOX GENE EXPRESSION A new type of comparative embryology is emerging based on the comparison of gene expression patterns that produce the phenotypes of different species. Gaunt (1994) and Burke and

her collaborators (1995) have compared the vertebrae of the mouse and the chick (FIGURE 9.31A). Although the mouse and chick have a similar number of vertebrae, they apportion them differently. Mice (like all mammals, be they giraffes or whales) have 7 cervical (neck) vertebrae. These are followed by 13 thoracic (rib) vertebrae, 6 lumbar (abdominal) vertebrae, 4 sacral (hip) vertebrae, and a variable (20+) number of caudal (tail) vertebrae. The chick, by contrast, has 14 cervical vertebrae, 7 thoracic vertebrae, 12 or 13 (depending on the strain) lumbosacral vertebrae, and 5 coccygeal (fused tail) vertebrae. The researchers asked, Does the constellation of Hox gene expression correlate with the type of vertebra formed (e.g., cervical or thoracic) or with the relative position of the vertebrae (e.g., number 8 or 9)?

The answer is that the constellation of Hox gene expression predicts the type of vertebra formed. In the mouse, the transition between cervical and thoracic vertebrae is between vertebrae 7 and 8; in the chick, it is between vertebrae 14 and 15 (FIGURE 9.31B). In both cases, the *Hox5* paralogues are expressed in the last cervical vertebra, while the anterior boundary of the *Hox6* paralogues extends to the first thoracic vertebra. Similarly, in both animals, the thoracic-lumbar transition is seen at the boundary between the *Hox9* and *Hox10* paralogous groups. It appears there is a code of differing Hox gene expression along the anterior-posterior axis, and that code determines the type of vertebra formed.

GENE TARGETING As noted above, there is a specific pattern to the number and type of vertebrae in mice, and the Hox gene expression pattern dictates which vertebral type will form (FIGURE 9.32A). This was demonstrated when all six copies of the *Hox10* paralogous group (i.e., *Hoxa10, c10,* and *d10* in Figure 9.30) were knocked out and no lumbar vertebrae developed. Instead, the presumptive lumbar

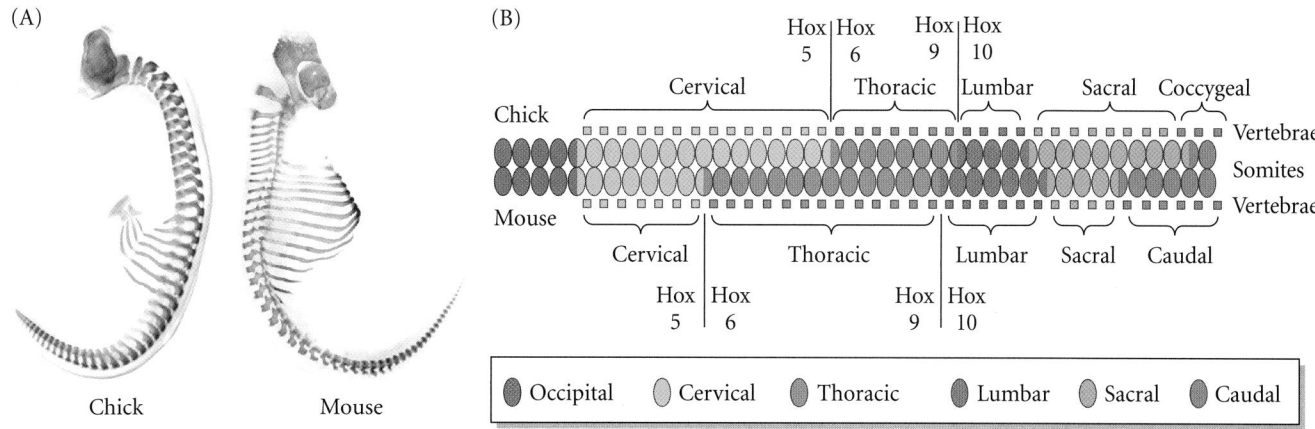

FIGURE 9.31 Schematic representation of the chick and mouse vertebral pattern along the anterior-posterior axis. (A) Axial skeletons stained with alcian blue at comparable stages of development. The chick has twice as many cervical vertebrae as the mouse. (B) The boundaries of expression of certain Hox gene paralogous groups (*Hox5/6* and *Hox9/10*) have been mapped onto the vertebral type domains. (A from Kmita and Duboule 2003, courtesy of M. Kmita and D. Duboule; B after Burke et al. 1995.)

vertebrae formed ribs and other characteristics similar to those of thoracic vertebrae (**FIGURE 9.32B**). This was a homeotic transformation comparable to those seen in insects; however, the redundancy of genes in the mouse made it much more difficult to produce, because the existence of even one copy of the *Hox10* group genes prevented the transformation (Wellik and Capecchi 2003; Wellik 2009). Similarly, when all six copies of the *Hox11* group were knocked out, the thoracic and lumbar vertebrae were normal, but the sacral vertebrae failed to form and were replaced by lumbar vertebrae (**FIGURE 9.32C**).

More recently, a *Hoxb6* gene was placed on a Delta enhancer, causing it to be expressed in each somite. The result was a "snakelike" mouse where each somite had formed a rib-bearing thoracic vertebra (see Figure 12.11C).

RETINOIC ACID TERATOGENESIS Homeotic changes are also seen when mouse embryos are exposed to teratogenic doses of retinoic acid, a derivative of vitamin A. As we saw earlier, by day 7 of development, an RA gradient has been established that is high in the posterior regions and low in the anterior portions of the embryo. This gradient appears to be controlled by the differential synthesis or degradation of RA in the different parts of the embryo. Hox genes are responsive

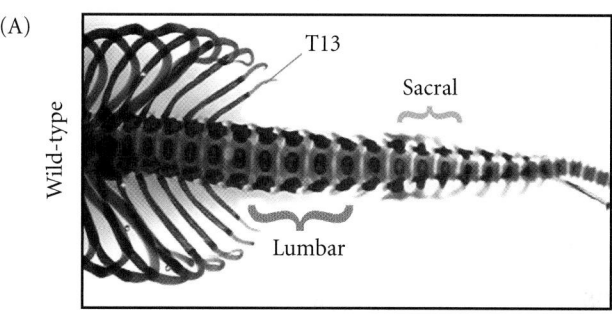
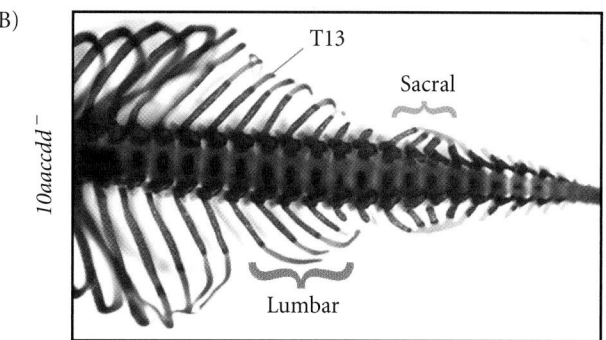
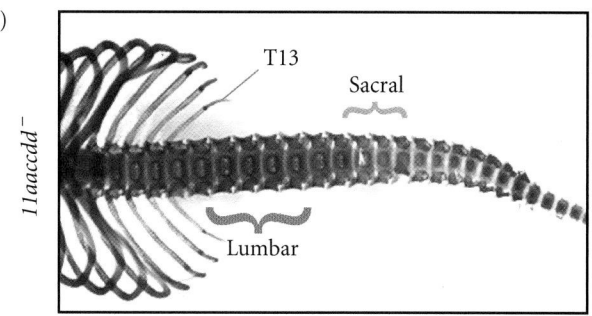

FIGURE 9.32 Axial skeletons of mice in gene knockout experiments. Each photograph is of an 18.5-day embryo, looking upward at the ventral region from the middle of the thorax toward the tail. (A) Wild-type mouse. (B) Complete knockout of *Hox10* paralogues (*Hox10aaccdd*) converts lumbar vertebrae (after the thirteenth thoracic vertebrae) into ribbed thoracic vertebrae. (C) Complete knockout of *Hox11* paralogues (*Hox11aaccdd*) transforms the sacral vertebrae into copies of lumbar vertebrae. (After Wellik and Capecchi 2003, courtesy of M. Capecchi.)

to retinoic acid either by virtue of having RA receptor sites in their enhancers or by being responsive to *Cdx* (the mammalian homologue of the *Drosophila caudal* gene), which is activated by RA (Conlon and Rossant 1992; Kessel 1992; Sakai et al. 2001; Lohnes 2003).

Exogenous retinoic acid can mimic the RA concentrations normally encountered only by the posterior cells, and high doses of RA can activate Hox genes in more anterior locations along the anterior-posterior axis (Kessel and Gruss 1991; Allan et al. 2001). Thus, when excess retinoic acid is administered to mouse embryos on day 8 of gestation, shifts in Hox gene expression occur such that the last cervical vertebra is turned into a thoracic (ribbed) vertebra. Conversely, impairment of retinoic acid function causes Hox gene expression to become more posterior, and the first thoracic vertebra becomes a copy of the cervical vertebrae.

The dorsal-ventral and left-right axes

THE DORSAL-VENTRAL AXIS Very little is known about the mechanisms of dorsal-ventral axis formation in mammals. After the fifth cell division in the mouse embryo, the blastocyst cavity begins to form, and the inner cell mass resides on one side of this cavity. This axis is probably created by the ellipsoidal shape of the zona pellucida (Kurotaki et al. 2007). In mice and humans, the hypoblast forms on the side of the inner cell mass that is exposed to the blastocyst fluid, while the dorsal axis forms from those ICM cells that are in contact with the trophoblast and amniotic cavity. Thus, the dorsal-ventral axis of the embryo is defined, in part, by the embryonic-abembryonic axis of the blastocyst. The embryonic region contains the ICM, while the abembryonic region is that part of the blastocyst opposite the ICM. The first dorsal-ventral polarity is seen at the blastocyst stage, and as development proceeds, the primitive streak maintains this polarity by causing migration ventrally from the dorsal surface of the embryo (Goulding et al. 1993).

THE LEFT-RIGHT AXIS The internal organs of the mammalian body are not symmetric, as the placement of the spleen, heart, and liver is determined along a right-left axis (**FIGURE 9.33A–C**). As in the chick embryo, the left-right axis appears to be due to the activation of Nodal proteins and the Pitx2 transcription factor on the left side of the lateral plate mesoderm, while Cerberus, an inhibitor of Nodal protein, is expressed on the right (see Figure 9.15; Collignon et al. 1996; Lowe et al. 1996; Meno et al. 1996). In mammals, the distinction between left and right sides begins in the ciliary cells of the node (**FIGURE 9.33D**). The cilia cause fluid in the node to flow from right to left (clockwise when viewed from the ventral side). When Nonaka and colleagues (1998) knocked out a mouse gene encoding the ciliary motor protein dynein (see Chapter 4), the nodal cilia did not move and the situs (lateral position) of each asymmetrical organ was randomized. (This helped explain the clinical observation that humans with a dynein deficiency have immotile cilia and a random chance of having their heart on the left or right side of the body; Afzelius

1976). Moreover, when Nonaka and colleagues (2002) cultured early mouse embryos under an artificial flow of medium from left to right, they obtained a reversal of the left-right axis.

The mechanism for this rotation appears to be the placement of the basal body of the cilium on each of the 200 or so monociliated node cells. The basal body giving rise to each cilium is at the posterior of each cell and extends out the ventral surface. Thus, the placement of cilia integrates information concerning the anterior-posterior and dorsal-ventral axes to construct the right-left axis (Guirao et al. 2010; Hashimoto et al. 2010). The placement of the cilia is governed by the planar cell polarity (PCP) pathway, possibly directed by a Wnt. Mutations in PCP pathway signaling molecules can randomize localization of cilia in these cells, also causing randomization of the left-right axis.

But how does rotation generate a body axis? It appears that the cells neighboring the node, the **crown cells,** are responsible for sensing the flow. Crown cells have immobile cilia, and theses cilia are affected by the movement of fluids. The fluid movement activates the Pkd2 protein on their cilia. A cascade initiated by Pdk2 (in a manner still unknown) appears to suppress the synthesis of Cerberus and thereby activate the expression of Nodal (Kawasumi et al. 2011; Yoshiba et al. 2012). Nodal is thought to bind in an autocrine manner to crown cells to maintain its own transcription; so Cerberus (which is made by the crown cells on the right side) would inhibit the maitenance of Nodal expression. In this way, Nodal expression is maintained on the left side, where it can activate *Pitx1,* which determines the left- and right-sidedness of the tissue.

Coda

Variations on the important themes of development have evolved in the different vertebrate groups (**FIGURE 9.34**). The major themes of vertebrate gastrulation include:

1. Internalization of the endoderm and mesoderm
2. Epiboly of the ectoderm around the entire embryo
3. Convergence of the internal cells to the midline
4. Extension of the body along the anterior-posterior axis

Although fish, amphibian, avian, and mammalian embryos have different patterns of cleavage and gastrulation, they use many of the same molecules to accomplish the same goals. Each group uses gradients of Nodal proteins to establish polarity along the dorsal-ventral axis. In *Xenopus* and zebrafish, maternal factors induce Nodal proteins in the vegetal hemisphere or marginal zone. In the chick, Nodal expression is induced by Wnt and Vg1 emanating from the posterior marginal zone, while elsewhere Nodal activity is suppressed by the hypoblast. In the mouse, the hypoblast similarly restricts Nodal activity, although the source of its ability to do so remains uncertain.

Each of these vertebrate groups uses BMP inhibitors to specify the dorsal axis; however, they use them in different ways. Similarly, Wnt inhibition and Otx2 expression are

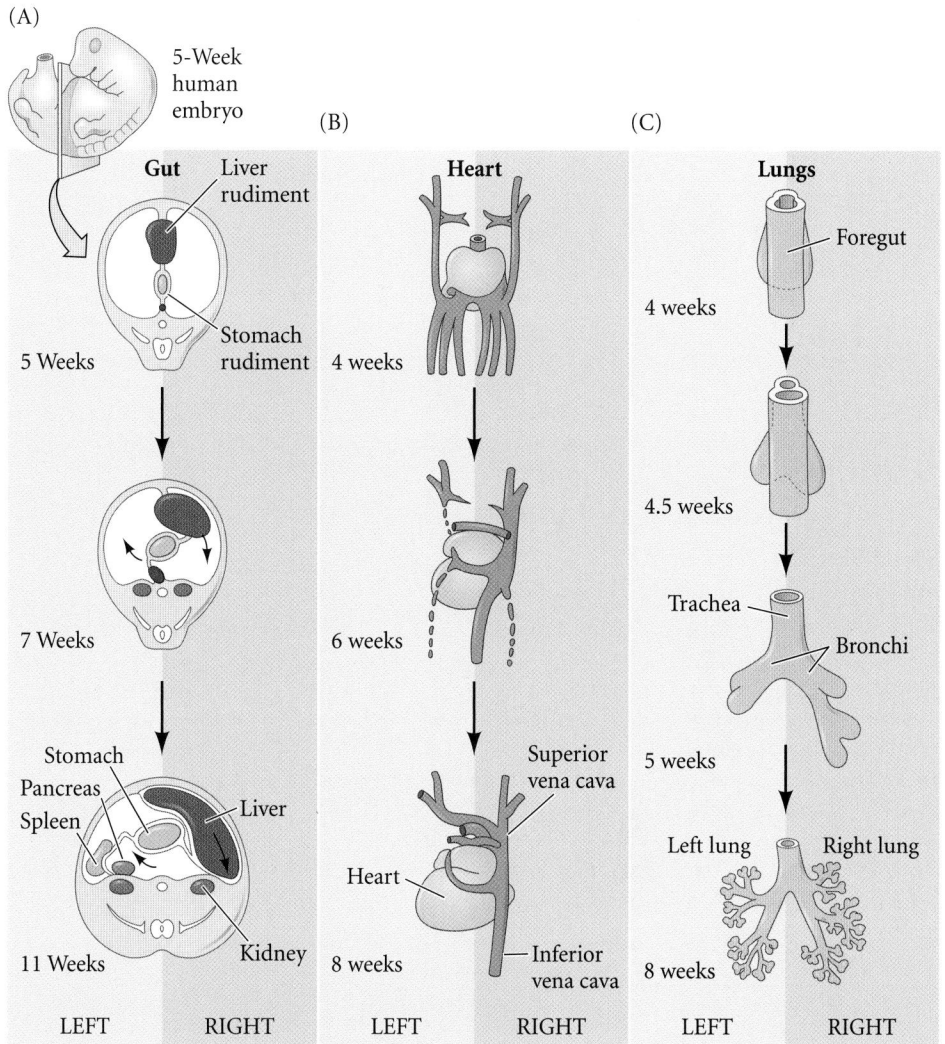

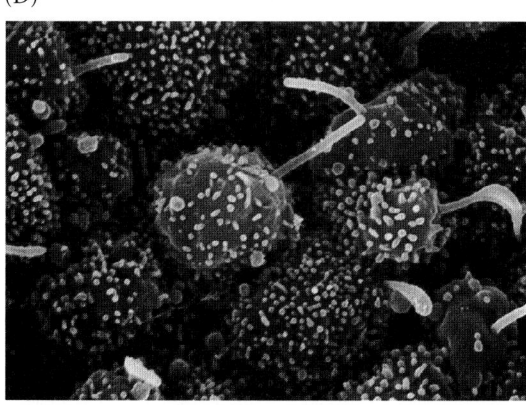

FIGURE 9.33 Left-right asymmetry in the developing human. (A) Abdominal cross sections show that the originally symmetrical organ rudiments acquire asymmetric positions by week 11. The liver moves to the right and the spleen moves to the left. (B) Not only does the heart move to the left side of the body, but the originally symmetrical veins of the heart regress differentially to form the superior and inferior venae cavae, which connect only to the right side of the heart. (C) The right lung branches into three lobes, while the left lung (near the heart) forms only two lobes. In human males, the scrotum also forms asymmetrically. (D) Ciliated cells of the mouse node, each with a cilium extending from the posterior ventral region of the cell. (A–C After Kosaki and Casey 1998; D courtesy of K. Sulik and G. C. Schoenwolf.)

important in specifying the anterior regions of the embryo, but different groups of cells may be expressing these proteins. In all cases, the region of the body from the hindbrain to the tail is specified by Hox genes. Finally, the left-right axis is established through the expression of Nodal on the left-hand side of the embryo. Nodal activates Pitx2, leading to the differences between the left and right sides of the embryo. How Nodal becomes expressed on the left side appears to differ among the vertebrate groups. But overall, despite their initial differences in cleavage and gastrulation, the vertebrates have maintained very similar ways of establishing the three body axes.

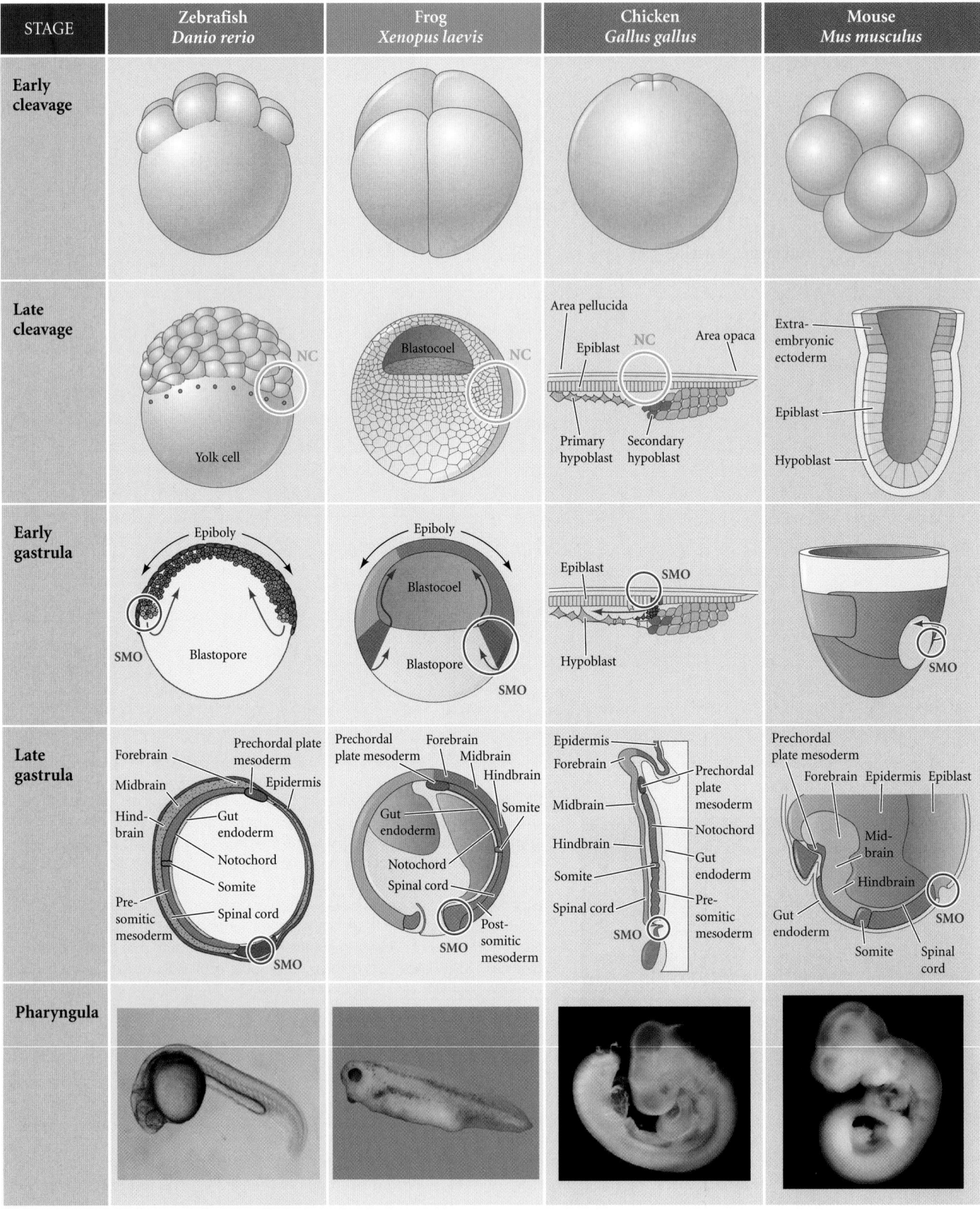

STAGE	Zebrafish *Danio rerio*	Frog *Xenopus laevis*	Chicken *Gallus gallus*	Mouse *Mus musculus*
Early cleavage				
Late cleavage				
Early gastrula				
Late gastrula				
Pharyngula				

◀ **FIGURE 9.34** Early development in four vertebrates. Cleavage differs greatly among the four groups. Zebrafish and chicks have meroblastic discoidal cleavage; frogs have unequal holoblastic cleavage; and mammals have equal holoblastic cleavage. These cleavage patterns form different structures, but there are many conserved features, such as the Nieuwkoop center (NC; green circles). As gastrulation begins, each of the groups has cells equivalent to the Spemann-Mangold organizer (SMO; red circles). The SMO marks the beginning of the blastopore region, and the remainder of the blastopore is indicated by the red arrows extending from the organizer. By the late gastrula stage, the endoderm (yellow) is inside the embryo, the ectoderm (blue, purple) surrounds the embryo, and the mesoderm (red) is between the endoderm and ectoderm. The regionalization of the mesoderm has also begun. The bottom row shows the pharyngula stage that immediately follows gastrulation. This stage, with a pharynx, a central neural tube and notochord flanked by somites, and a sensory cephalic (head) region, characterizes the vertebrates. (After Solnica-Krezel 2005.)

SNAPSHOT SUMMARY: Early Development of Birds and Mammals

1. Reptiles and birds, like fish, undergo discoidal meroblastic cleavage, wherein the early cell divisions do not cut through the yolk of the egg. These early cells form a blastoderm.

2. In each class of vertebrates, neural ectoderm is permitted to form where the BMP-mediated induction of epidermal tissue is prevented.

3. In chick embryos, early cleavage forms an area opaca and an area pellucida. The region between them is the marginal zone. Gastrulation begins in the area pellucida next to the posterior marginal zone, as the hypoblast and primitive streak both start there.

4. The primitive streak is derived from anterior epiblast cells and the central cells of Koller's sickle. As the primitive streak extends rostrally, Hensen's node is formed. Cells migrating out of Hensen's node become prechordal mesendoderm and are followed by the head process and notochord cells.

5. The prechordal plate helps induce formation of the forebrain; the chordamesoderm induces formation of the midbrain, hindbrain, and spinal cord. The first cells migrating laterally through the primitive streak become endoderm, displacing the hypoblast. The mesoderm cells then migrate through the primitive streak. Meanwhile, the surface ectoderm undergoes epiboly around the yolk.

6. In birds, gravity helps determine the position of the primitive streak (which defines the future anterior-posterior axis). The left-right axis is formed by the expression of Nodal protein on the left side of the embryo, which signals Pitx2 expression on the left side of developing organs.

7. The hypoblast helps determine the body axes of the embryo, and its migration determines the anterior-posterior axis.

8. Mammals undergo a variation of holoblastic rotational cleavage that is characterized by a slow rate of cell division, a unique cleavage orientation, lack of divisional synchrony, and formation of a blastocyst.

9. The blastocyst forms after the blastomeres undergo compaction. It contains outer cells—the trophoblast cells—that become the chorion, and an inner cell mass that becomes the amnion and the embryo.

10. The inner cell mass cells are pluripotent and can be cultured as embryonic stem cells. They give rise to the epiblast and to the visceral endoderm (hypoblast.)

11. The chorion forms the fetal portion of the placenta, which functions to provide oxygen and nutrition to the embryo, to provide hormones for the maintenance of pregnancy, and to block the potential immune response of the mother to the developing fetus.

12. Mammalian gastrulation is not unlike that of birds. There appear to be two signaling centers, one in the node and one in the anterior visceral endoderm. The latter center is critical for head development, while the former is critical in inducing the nervous system and in patterning axial structures caudally from the midbrain.

13. Hox genes pattern the anterior-posterior axis and help specify positions along that axis. If Hox genes are knocked out, segment-specific malformations can arise. Similarly, causing the ectopic expression of Hox genes can alter the body axis.

14. The homology of gene structure and the similarity of expression patterns between *Drosophila* and mammalian Hox genes suggest that this patterning mechanism is extremely ancient.

15. The mammalian left-right axis is specified similarly to that of the chick, but with some significant differences in the roles of certain genes.

16. In amniote gastrulation, the pluripotent epithelium, or epiblast, produces the mesoderm and endoderm (which migrate through the primitive streak), and the precursors of the ectoderm, which remain on the surface. By the end of gastrulation, the head and anterior trunk structures are formed. Elongation of the embryo continues through precursor cells in the caudal epiblast surrounding the posteriorized Hensen's node.

For Further Reading

Complete bibliographical citations for all literature cited in this chapter can be found at the free-access website **www.devbio.com**

Beddington, R. S. P. and E. J. Robertson. 1999. Axis development and early asymmetry in mammals. *Cell* 96: 195–209.

Bertocchini, F. and C. D. Stern. 2002. The hypoblast of the chick embryo positions the primitive streak by antagonizing Nodal signaling. *Dev. Cell* 3: 735–744.

Boyer, L. A. and 13 others. 2005. Core transcriptional regulatory circuitry in human embryonic stem cells. *Cell* 122: 947–956.

Burke, A. C., A. C. Nelson, B. A. Morgan and C. Tabin. 1995. Hox genes and the evolution of vertebrate axial morphology. *Development* 121: 333–346.

Markert, C. L. and R. M. Petters. 1978. Manufactured hexaparental mice show that adults are derived from three embryonic cells. *Science* 202: 56–58.

Stern, C. D. and K. M. Downs. 2012. The hypoblast (visceral endoderm): An evo-devo perspective. *Development* 139: 1059–1069.

Strumpf, D., C.-A. Mao, Y. Yamanaka, A. Ralston, K. Chawengsaksophak, F. Beck and J. Rossant. 2005. Cdx2 is required for correct cell fate specification and differentiation of trophectoderm in the mouse blastocyst. *Development* 132: 2093–2102.

Go Online

WEBSITE 9.1 The extraembryonic membranes. The amniote embryo is supported by a variety of extraembryonic membranes that provide nourishment, protection, and waste disposal services.

WEBSITE 9.2 Epiblast cell heterogeneity. Although the early epiblast appears uniform, different cells have different molecules on their cell surfaces. This variability allows some of them to remain in the epiblast while others migrate into the embryo.

WEBSITE 9.3 Mechanisms of compaction and formation of the inner cell mass. What determines whether a cell is to become a trophoblast cell or a member of the inner cell mass? It may just be a matter of chance. However, once the decision is made, different genes are switched on.

WEBSITE 9.4 Human cleavage and compaction. XY blastomeres have a slight growth advantage that may have had profound effects on in vitro fertility operations.

WEBSITE 9.5 Nonidentical monozygotic twins. Although monozygotic twins have the same genome, random developmental factors or the uterine environment may give them dramatically different phenotypes.

WEBSITE 9.6 Conjoined twins. There are rare events in which more than one set of axes is induced in the same embryo. These events can produce conjoined twins—twins that share some parts of their bodies. The medical and social issues raised by conjoined twins provide a fascinating look at what people throughout history have considered "individuality."

WEBSITE 9.7 Placental functions. Placentas are nutritional, endocrine, and immunological organs. They provide hormones that enable the uterus to retain the pregnancy and also accelerate mammary gland development. Placentas also block the potential immune response of the mother against the developing fetus. Recent studies suggest that the placenta uses several mechanisms to block the mother's immune response.

Vade Mecum

Chick development. Viewing these movies of 3-D models of chick cleavage and gastrulation will help you understand these phenomena.

Outside Sites

The University of North Carolina School of Medicine has developed an excellent site for medical embryology. (**http://www.med.unc.edu/embryo_images/**)

The Edinburgh Mouse Atlas Project is a remarkable database for both anatomy and gene expression of Burns's "wee little beastie." (**http://genex.hgu.mrc.ac.uk/**)

The University of New South Wales Embryology Course, constructed by Mark Hill, serves as a reference for human developmental anatomy, with connections to websites for other organisms. (**http://embryology.med.unsw.edu.au/**)

The Visible Embryo shows the progress of human development, using the Carnegie Institution embryos, which have become the standard for staging. (**http://www.visembryo.com/**)

PART THREE

THE STEM CELL CONCEPT
Introducing Organogenesis

"REMEMBER WHEN YOU SAID STEM CELL RESEARCH WAS ALL HYPE?"

We have now explored the development of the embryo from fertilization through gastrulation. In the next chapters, we will see how the three germ layers interact with each other to begin **organogenesis**—the processes of organ formation. Early organogenesis is a symphony of interactions between different parts of the embryo, and some of these interactions create privileged sites called *stem cell niches*. These niches provide a milieu of extracellular matrices, juxtacrine factors, and paracrine factors that allow cells residing within them to remain relatively undifferentiated. These relatively undifferentiated cells become **stem cells**, and their presence has become central to our vision of organogenesis and critical to the field of modern medicine.

The Stem Cell Concept

A stem cell can be defined as a relatively undifferentiated cell that when it divides produces (1) at least one of two daughter cells that retains its undifferentiated character (a process called self-renewal); and (2) a daughter cell that can undergo further differentiation. Thus, a stem cell has the potential to renew itself at each division (so that there is always a supply of stem cells) while also producing a daughter cell capable of responding to its environment by differentiating in a particular manner. This strategy, in which two types of cells (the remaining stem cell and the developmentally committed cell) are produced at each division, is called the *single stem cell asymmetry* mode and is seen in many types of stem cells (**FIGURE P3.1A**). An alternative (but not mutually exclusive) mode of retaining cell homeostasis is the *population asymmetry* mode of stem cell division. Here, some stem cells are more prone to produce differentiated progeny, and this is compensated for by another set of stem cells that divide symmetrically to produce more stem cells (**FIGURE P3.1B**; Watt and Hogan 2000; Simons and Clevers 2011).

In several organs (such as adult bone marrow, as we will describe below) there are subpopulations of stem cells. Often there is a population of *multipotent stem cells*, whose progeny include another multipotent stem cell as well as a stem cell that is committed to a particular cell lineage (**FIGURE P3.1C**). These *committed stem*

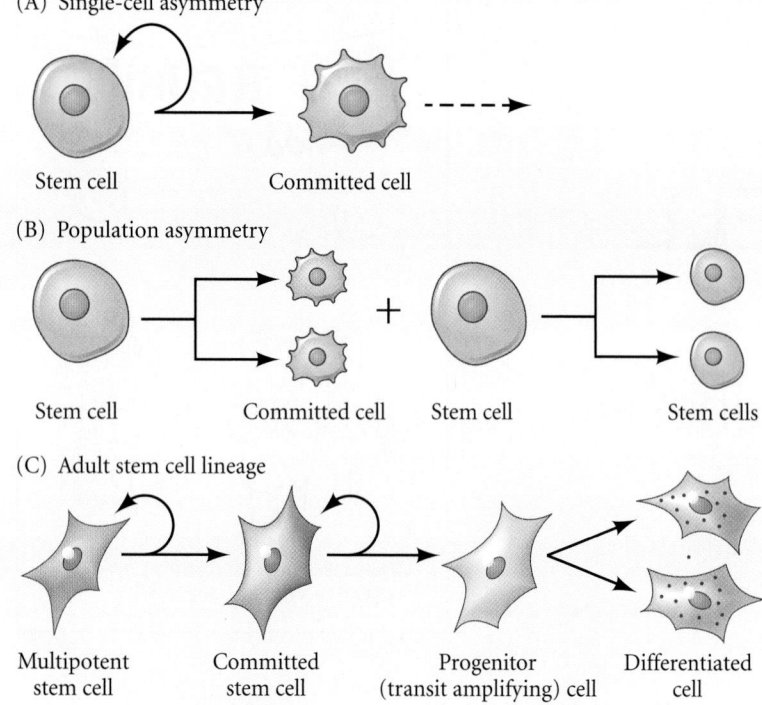

(A) Single-cell asymmetry

Stem cell   Committed cell

(B) Population asymmetry

Stem cell   Committed cell   Stem cell   Stem cells

(C) Adult stem cell lineage

Multipotent stem cell   Committed stem cell   Progenitor (transit amplifying) cell   Differentiated cell

FIGURE P3.1 The stem cell concept. (A) The fundamental notion of a stem cell is that it can make more stem cells while also producing cells committed to undergoing differentiation. This is called asymmetric cell division. (B) A population of stem cells can also be maintained through population asymmetry. Here, some stem cells are more prone to have committed progeny. This is balanced by other stem cells dividing symmetrically to produce two more stem cells. (C) In many organs, stem cell lineages pass from a multipotent stem cell (capable of forming numerous types of cells) to a committed stem cell that makes one or very few types of cells, to a progenitor (transit amplifying) cell that proliferates and is committed to becoming a particular cell type, to a particular type of differentiated cell.

cells produce other committed stem cells as well as *progenitor cells* (also known as *transit amplifying cells*) that are committed to a particular cell fate and that divide to produce many such cells. Progenitor cells are not stem cells, as they can only undergo a few rounds of cell division.

In some organs, such as the gut, epidermis, and bone marrow, stem cells regularly divide to replace worn-out cells and repair damaged tissues. In others, such as the prostate and heart, stem cells divide only under special physiological conditions, usually in response to stress or the need to repair the organ. We will describe various such stem cell lineages throughout the rest of the book.

PROOF OF CONCEPT: THE HEMATOPOIETIC STEM CELL In the mid-twentieth century, biologists thought that

cell specification and differentiation occurred in the early embryo and that after this stage there was only the growth of the existing parts. In the 1960s, however, biologists studying blood cell development began to ponder a remarkable phenomenon. All blood cells—red blood cells (erythrocytes), white blood cells (granulocytes), platelets, and even lymphocytes—are constantly being produced in the bone marrow. Billions of blood cells are destroyed by the spleen each hour, but an equal number of blood cells are generated to replace them.

In an elegant series of experiments, Ernest McCulloch and James Till demonstrated that there was a common generative cell—the "hematopoietic [blood-forming] stem cell"—that gave rise to all the different types of blood cells* (FIGURE P3.2A). In a 1961 experiment, Till and McCulloch injected bone marrow cells from a donor mouse into lethally irradiated mice† of the same genetic strain. Some of the individual donor cells produced discrete nodules on the spleens of the host animals (FIGURE P3.2B), and these nodules contained red blood cells, white blood cells, and platelet precursor cells. Later experiments in which the donor marrow cells were irradiated to genetically mark each cell with random chromosome breaks confirmed that each of these different cell types in a nodule arose from a single cell, and that there were even lymphocytes present in some of these nodules (Becker et al. 1963). One cell could generate numerous different cell types.

For this "colony-forming cell" to be a true stem cell, however, it had to produce not only the differentiated blood cells but also more colony-forming cells. This was shown to be the case by taking the nodule derived from a single genetically marked colony-forming cell and injecting cells of the nodule into another irradiated mouse. Many spleen colonies emerged, each of them having the same chromosomal

---

*The concept that the different blood cell types were continuously generated by a hematopoietic stem cell was first proposed in 1909 by the Russian histologist Alexander Maximov. He is credited with coining the word *Stammzelle* ("stem-cell") to refer to the regenerative capacities of these cells. Maximov had been a student of Oskar Hertwig, one of the leading German embryologists. He worked in St. Petersburg as a professor of embryology and histology, but his studies were curtailed by the Russian Revolution. He fled the Soviet Union in 1922, having bribed a guard at the Finnish border with a bottle of laboratory ethanol (Konstantinov 2000).

†The mice were "lethally irradiated" in that X-rays were used to destroy their immune and blood cell precursors. Thus, the mice survived only if there were stem cells in the transplant that could regenerate blood cells and lymphocytes to replace the destroyed cells.

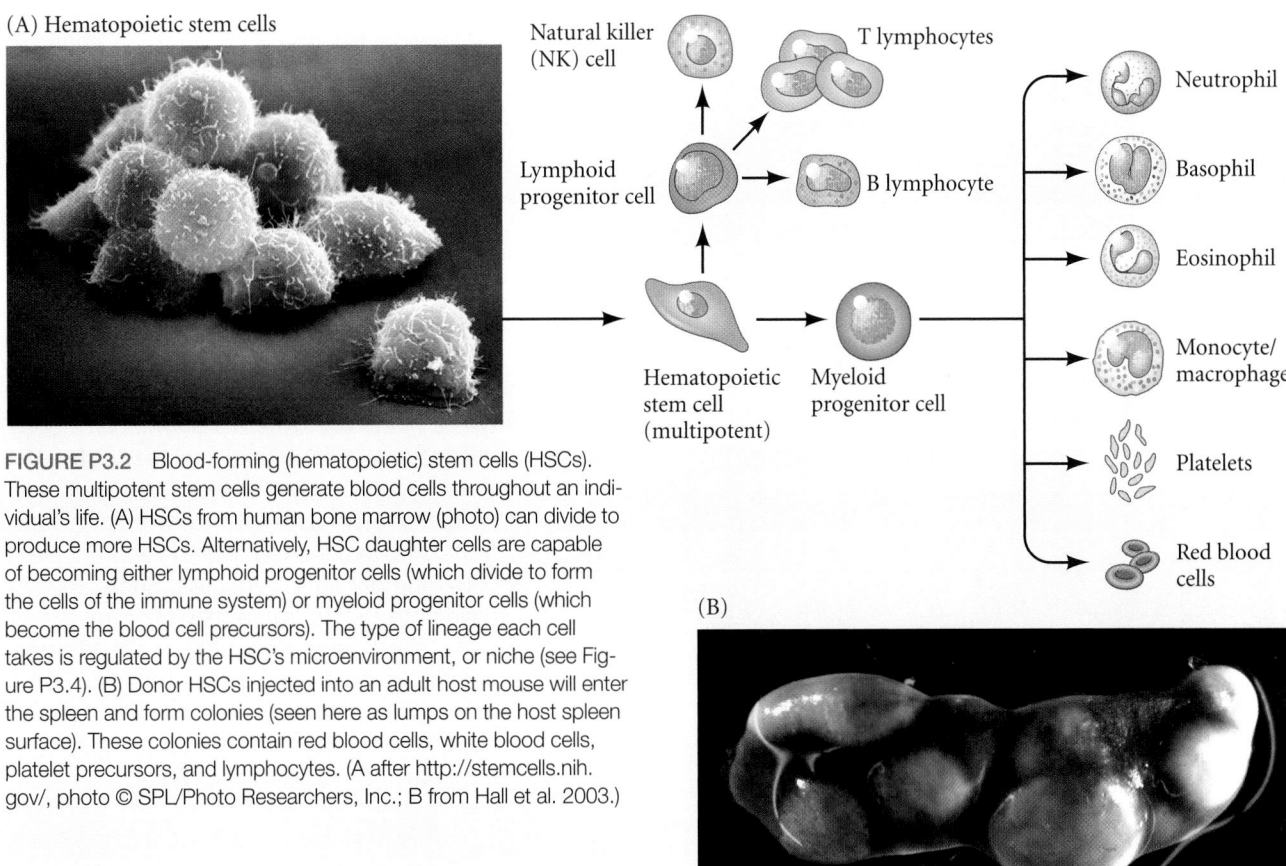

(A) Hematopoietic stem cells

Natural killer (NK) cell

T lymphocytes

Neutrophil

Lymphoid progenitor cell

B lymphocyte

Basophil

Eosinophil

Hematopoietic stem cell (multipotent)

Myeloid progenitor cell

Monocyte/macrophage

Platelets

Red blood cells

FIGURE P3.2   Blood-forming (hematopoietic) stem cells (HSCs). These multipotent stem cells generate blood cells throughout an individual's life. (A) HSCs from human bone marrow (photo) can divide to produce more HSCs. Alternatively, HSC daughter cells are capable of becoming either lymphoid progenitor cells (which divide to form the cells of the immune system) or myeloid progenitor cells (which become the blood cell precursors). The type of lineage each cell takes is regulated by the HSC's microenvironment, or niche (see Figure P3.4). (B) Donor HSCs injected into an adult host mouse will enter the spleen and form colonies (seen here as lumps on the host spleen surface). These colonies contain red blood cells, white blood cells, platelet precursors, and lymphocytes. (A after http://stemcells.nih.gov/, photo © SPL/Photo Researchers, Inc.; B from Hall et al. 2003.)

(B)

arrangement as the original colony (Till et al. 1964; Jurásková and Tkadlécek 1965; Humphries et al. 1979). Thus we see that a single marrow cell can form numerous different cell types and can also undergo self-renewal. But these blood cell precursors were still thought of as an exception. It wasn't until cell imaging and molecular biology had matured that the stem cell concept could be applied throughout development.

Stem Cell Vocabulary

Numerous terms are used to describe stem cells, and there is only now beginning to be general agreement on how these terms are used. The names of the two major divisions of stem cells are based on their sources. **Embryonic stem cells** are derived from the inner cell mass of mammalian blastocysts (see Chapter 9) or from fetal gamete progenitor cells (germ cells; see Chapter 17). These cells are capable of producing all the cells of the embryo (i.e., a complete organism). **Adult stem cells** are found in the tissues of organs after the organ has matured. These stem cells, which are usually involved in replacing and repairing tissues of that particular organ, can form only a subset of cell types.

STEM CELL POTENCY   The ability of a particular stem cell to generate numerous different types of differentiated cells is its *potency* (**FIGURE P3.3**). In mammals, **totipotent cells** are capable of forming every cell in the embryo and, in addition, the trophoblast cells of the placenta. The only totipotent cells are the zygote and (probably) the first 4–8 blastomeres that form prior to compaction (see Chapter 9). **Pluripotent stem cells** have the ability to become all the cell types of the embryo body but cannot generate the trophoblast. Usually pluripotent stem cells are derived from the inner cell mass of the mammalian blastocyst. However, undifferentiated germ cells (and germ cell tumors such as teratocarcinomas; see Chapter 17) can also form pluripotent stem cells.

Multipotent stem cells can be in either the embryo or the adult, and their commitment is limited to a relatively small subset of all the possible cells of the body (see Figure P3.1C). The hematopoietic stem cell, for instance, is a multipotent cell that can form the white blood cell, lymphocyte, and red blood cell lineages. **Unipotent stem cells** are found in particular tissues and are involved in regenerating a particular type of cell. Spermatogonia, for example, are stem cells that give rise only to sperm. Whereas pluripotent

Potential		Cell	Source
Totipotent	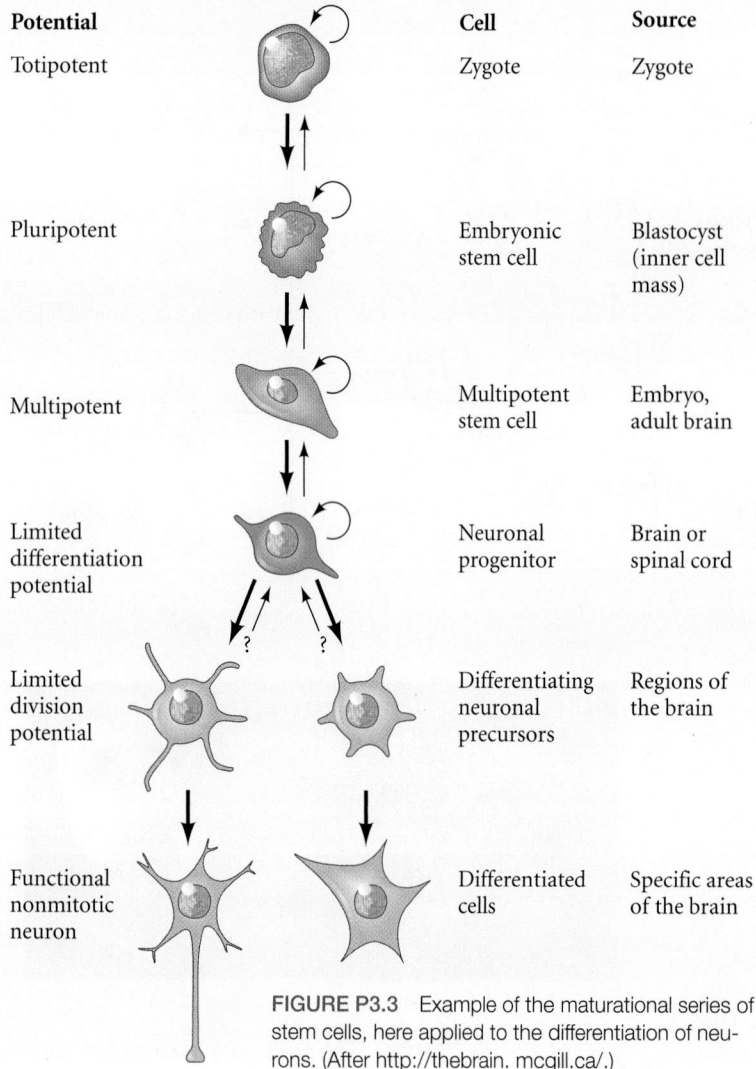	Zygote	Zygote
Pluripotent		Embryonic stem cell	Blastocyst (inner cell mass)
Multipotent		Multipotent stem cell	Embryo, adult brain
Limited differentiation potential		Neuronal progenitor	Brain or spinal cord
Limited division potential		Differentiating neuronal precursors	Regions of the brain
Functional nonmitotic neuron		Differentiated cells	Specific areas of the brain

FIGURE P3.3 Example of the maturational series of stem cells, here applied to the differentiation of neurons. (After http://thebrain. mcgill.ca/.)

stem cells can produce cells of all three germ layers (as well as the germ cells), multipotent and unipotent stem cells are often grouped together as **committed stem cells**, as they have the potential to become only a few cell types.

PROGENITOR CELLS Although they are related to stem cells, **progenitor cells** are not capable of unlimited self-renewal; they have the capacity to divide only a few times before differentiating (Seaberg and van der Kooy 2003). They are sometimes called "transit amplifying cells," since they usually divide while migrating away from the stem cell niche. Both unipotent stem cells and progenitor cells have been called *lineage-restricted cells*, but unipotent stem cells have the capacity for self-renewal, whereas progenitor cells do not. Progenitor cells are usually more differentiated than stem cells, having become committed to become a particular type of cell. In many instances, stem cell division generates progeny that become progenitor cells, as is seen in the formation of the blood cells, sperm cells, and the nervous

system (see Figures P3.2 and P3.3). Yet another term, *precursor cell* (or simply *precursors*) is widely used in developmental biology to denote any ancestral cell type (either stem cell or progenitor cell) of a particular lineage. Thus it is often used when such distinctions do not matter or are not known (see Tajbakhsh 2009)

Adult Stem Cells

Numerous adult organs contain stem cells that can give rise to a limited set of cell and tissue types. In addition to the well-known hematopoietic stem cells, developmental biologists have discovered epidermal stem cells, neural stem cells, hair stem cells, melanocyte stem cells, muscle stem cells, tooth stem cells, gut stem cells, and germline stem cells. Such cells are difficult to isolate, since they often represent fewer than 1 out of every 1000 cells in an organ. In addition, many stem cells appear to have a relatively low rate of cell division and do not proliferate readily. However, neither of these facts precludes their usefulness. Each year some 40,000 bone marrow transplant procedures are performed in which hematopoietic stem cells are transferred from one person to another. These multipotent stem cells are rare (about 1 in every 15,000 bone marrow cells), but such transplantation treatment works well for people suffering from red blood cell deficiencies or leukemias.

Techniques to selectively allow the growth and isolation of adult stem cells may allow some organ deficiencies to be treated in the same way as blood cell deficiencies—by administering a source of committed stem cells. In mice, very few blood stem cells (perhaps even one) can reconstitute the mouse's blood and immune systems (Osawa et al. 1996), and a single mammary stem cell can generate an entire mammary gland (including epithelium, muscles, and stroma; Shackleton et al. 2006). Carvey and colleagues have shown that when neural stem cells from the midbrain of adult rats are cultured in a mixture of paracrine factors, they will differentiate into dopaminergic neurons that can ameliorate the rodent version of Parkinson disease (Carvey et al. 2001; see Hall et al. 2007).

Stem Cell Niches

Many adult tissues and organs contain stem cells that undergo continual renewal; these include the mammalian epidermis, hair follicles, intestinal villi, blood cells, and sperm cells, as well as *Drosophila* intestine, sperm, and egg cells. Such adult stem cells must maintain the long-term ability to divide,

TABLE P3.1   Some human stem cell niches

Stem cell type	Niche location	Cellular components of niche
HIGH TURNOVER[a]		
Mesenchymal stem cells (MSCs)	Bone marrow, adipose tissue, heart, placenta, umbilical cord	Probably blood vessel epithelium
Hematopoietic (blood-forming) stem cells (HSCs)	Bone marrow	Macrophages, T_{reg} cells, osteoblasts, pericytes, glia, neurons, MSCs
Intestine	Base of small intestinal crypts	Paneth cells, MSCs
Epidermis (skin)	Basal layer of epidermis	Dermal fibroblasts
Hair follicle	Bulge[b]	Dermal papillae, adipocyte precursors, subcutaneous fat, keratin[b]
Sperm (♂)	Testes	Sertoli cells[c]
LOW TURNOVER[a]		
Brain (neurons and glia)	Subventricular zone, subgranular zone	Ependymal cells, blood vessel epithelium
Skeletal muscle	Between basal lamina and muscle fibers	Muscle fiber cells

Sources: After Kolf et al. 2007; Hsu and Fuchs 2012.

[a]Niches with high rates of cell turnover are constantly producing new cells for bodily maintenance. Niches with low turnover have stem cells for repair, slow growth, and (in the case of neurons) learning.

[b]See Figures 10.40 and 10.42

[c]See Figure 17.28

producing some daughter cells that are differentiated and others that remain stem cells. The ability of a cell to become an adult stem cell is determined in large part by where it resides. The continuously proliferating stem cells are housed in regulatory microenvironments called **stem cell niches** (Schofield 1978). Stem cell niches are particular locations that allow the controlled self-renewal and survival of the stem cells within the niche and the controlled differentiation of those stem cell progeny that leave the niche (TABLE P3.1).

Stem cell microenvironments usually regulate proliferation and differentiation via paracrine or juxtacrine factors produced by the cells that make up the niche (Moore and Lemischka 2006; Jones and Wagers 2008). In many cases, paracrine and juxtacrine factors within the niche maintain the stem cells in an uncommitted state. Once stem cells leave the niche, however, these factors cannot reach them and the cells begin differentiating. For example, the distal tip cell of the *C. elegans* gonad provides the niche for the nematode's germline stem cells (see Chapter 17). Notch signaling in the distal tip cell suppresses differentiation of the germline stem cells. Those cells leaving the range of Notch signaling start differentiating into germ cells.

The regulation of stem cell division versus differentiation is critical. Too much differentiation depletes the stem cell population and promotes the phenotypes of aging or decay; too much stem cell proliferation can cause cancers to arise. This essential regulation appears to be effected by a balance of antagonistic paracrine factors. In the mammalian hair stem cells niche, for instance, a high BMP-to-Wnt ratio promotes stem cell self-renewal, whereas a high Wnt-to-BMP ratio signals hair follicle differentiation (Kandyba et al. 2013.)

THE HEMATOPOIETIC STEM CELL NICHE   One of the most important stem cell niches is the mammalian hematopoietic stem cell (HSC) niche that produces the many different types of blood cells described earlier. The HSC niche is found in the hollow cavities of trabecular bones (such as the femur) where the bone marrow resides (FIGURE P3.4). Here, the HSCs are in close proximity to the bone cells (osteocytes) and the endothelial cells that line the blood vessels. A complex cocktail of paracrine factors, including Wnts, angiopoietin, and stem cell factor, combines with cell surface signals from Notch and integrin to regulate stem cell proliferation and differentiation. Hormonal signals and pressure from the blood vessels, as well as neurotransmitters from adjacent axons, also help regulate hematopoiesis (Spiegel et al. 2008; Malhotra and Kincade 2009). In this way, the HSC niche can regulate for the production of more white blood cells (as in the case of infections) or more red blood cells (as when one climbs to high altitudes).

In the blood and many other organs (including the skin and intestines), there can be two or more interacting populations of stem cells in a niche, some that are rapidly cycling and others that are more quiescent (Jaenisch et al. 2011; Takeda et al. 2011). Wilson and colleagues (2008, 2009) have demonstrated that there are two subpopulations of HSCs in the blood. One subpopulation can divide rapidly in response to immediate needs, while a quiescent population is held in reserve and

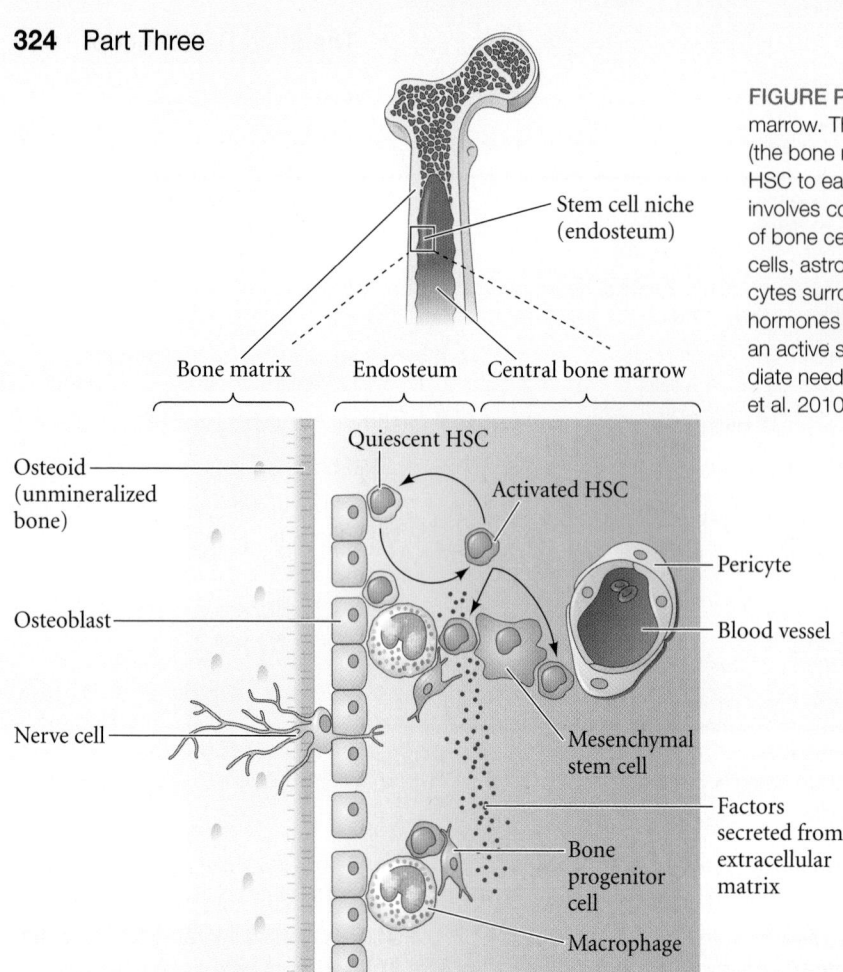

Stem cell niche (endosteum)

Bone matrix | Endosteum | Central bone marrow

Osteoid (unmineralized bone)

Quiescent HSC

Activated HSC

Pericyte

Osteoblast

Blood vessel

Nerve cell

Mesenchymal stem cell

Factors secreted from extracellular matrix

Bone progenitor cell

Macrophage

FIGURE P3.4 The hematopoietic stem cell niche in bone marrow. The HSC-generating niche is in the endosteum (the bone marrow close to the bone). The designation of an HSC to each particular blood cell lineage (see Figure P3.2A) involves contact between the stem cells and the matrices of bone cells, paracrine and juxtacrine factors from stromal cells, astrocytes, macrophages, osteoblasts, and the pericytes surrounding the blood vessels, as well as systemic hormones and neural signals. The stem cells cycle between an active state in which they can divide in response to immediate needs and a quiescent, "reserve" state. (After Lévesque et al. 2010.)

appears to consist of those cells having the greatest potential for self-renewal. Depending on physiological conditions, cells from one stem cell population can enter the other group.

Myeloproliferative disease is a cancer of blood stem cells and their non-lymphocytic derivatives. It results from a failure of the niche to provide the signals for proper blood cell differentiation (Walkley et al. 2007a,b). When immature osteoblasts of the bone marrow matrix were specifically prevented from functioning, HSCs proliferated rapidly without differentiating. The resulting overproduction of stem cells in turn led to myeloproliferative disease (Raaijmakers et al. 2010, 2012).

GERM CELL NICHES IN *DROSOPHILA* Stem cell niches in the testes of male *Drosophila* illustrate the importance of proximity and asymmetric cell division. The stem cells for sperm reside in a regulatory microenvironment called the **hub** (FIGURE P3.5). The hub consists of about a dozen somatic testes cells and is surrounded by 5–9 germ stem cells. The division of the sperm stem cell is asymmetric, always producing one cell that remains attached to the hub and one unattached cell. The daughter cell attached to the hub is maintained as a stem cell, while the cell that is not touching the hub becomes a **gonialblast**—a committed cell that will divide to become the precursors of the sperm cells. The somatic cells of the hub create this asymmetric

proliferation by secreting the paracrine factor Unpaired onto the cells attached to them. Unpaired protein activates the JAK-STAT pathway in the adjacent germ stem cells to specify their self-renewal. Cells that are distant from the paracrine factor do not receive this signal and begin their differentiation into the sperm cell lineage (Kiger et al. 2001; Tulina and Matunis 2001).

Physically, this asymmetric division involves the interactions between the sperm stem cells and the somatic cells. In the division of the stem cell, one centrosome remains attached to the cortex at the contact site between the stem cell and the somatic cells. The other centrosome moves to the opposite side, thus establishing a mitotic spindle that will produce one daughter cell attached to the hub and one daughter cell away from it (Yamashita et al. 2003). (We will see a similar inheritance of centrosomes in the division of mammalian neural stem cells.) The cell adhesion molecules linking the hub and stem cells together are probably involved in retaining one of the centrosomes in the region where the two cells touch. Here we see stem cell production using asymmetric cell division.

THE INTESTINAL STEM CELL NICHE The epithelium of the mammalian intestine constitutes a stem cell niche. Several stem cells reside at the base of each crypt in the mouse small intestine, and each of these stem cells divides once a day (Lander et al. 2012). Some daughter cells stay in the crypts as stem cells, while others become progenitor cells and divide rapidly. These transit amplifying cells differentiate into the cells characteristic of the small intestine—enterocytes, goblet cells, and enteroendocrine cells—as they move up each villus. Upon reaching the tip of the intestinal villus, they undergo apoptosis.*

---

*This is highly reminiscent of *Hydra* growth (see Chapter 16), where each cell is formed at the animal's base, migrates and becomes part of the differentiated body, and is eventually shed from the tips of the arms.

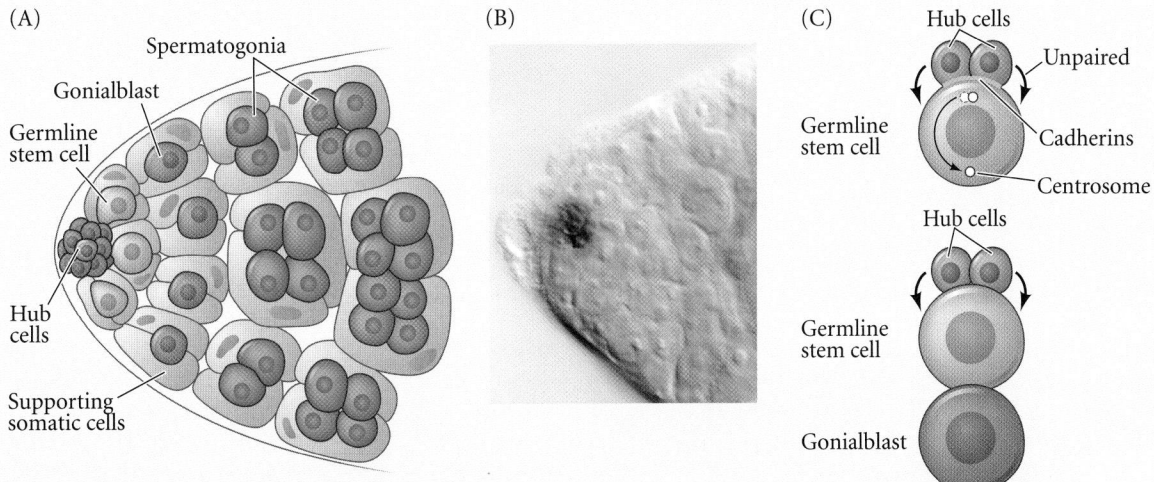

FIGURE P3.5 Stem cell niche in *Drosophila* testes. (A) The apical hub consists of about 12 somatic cells, to which are attached 5–9 germ stem cells. The germ stem cells divide asymmetrically to form another germ stem cell (which remains attached to the somatic hub cells) and a gonialblast that will divide to form the sperm precursors (the spermatogonia and the spermatocyte cysts where meiosis is initiated). (B) Reporter β-galactosidase inserted into the gene for Unpaired reveals that this protein is transcribed in the somatic hub cells. (C) Cell division pattern of the germline stem cells, wherein one of the centrosomes remains in the cortical cytoplasm near the site of hub cell adhesion, while the other centrosome migrates to the opposite pole of the germline stem cell. This results in one cell remaining attached to the hub and the other cell detaching from it and differentiating. (After Tulina and Matunis 2001; B courtesy of E. Matunis.)

Lineage-tracing studies (Barker et al. 2007; Sato et al. 2011) have shown that intestinal stem cells (expressing the Lgr5 protein) can generate all the differentiated cells of the intestine, and that a single such stem cell grown in culture can produce a "mini-gut" containing all the intestinal cell types. One of the ways it does this is to generate its own stem cell niche, a feat that involves a specific differentiated cell type, the Paneth cell (**FIGURE P3.6**).

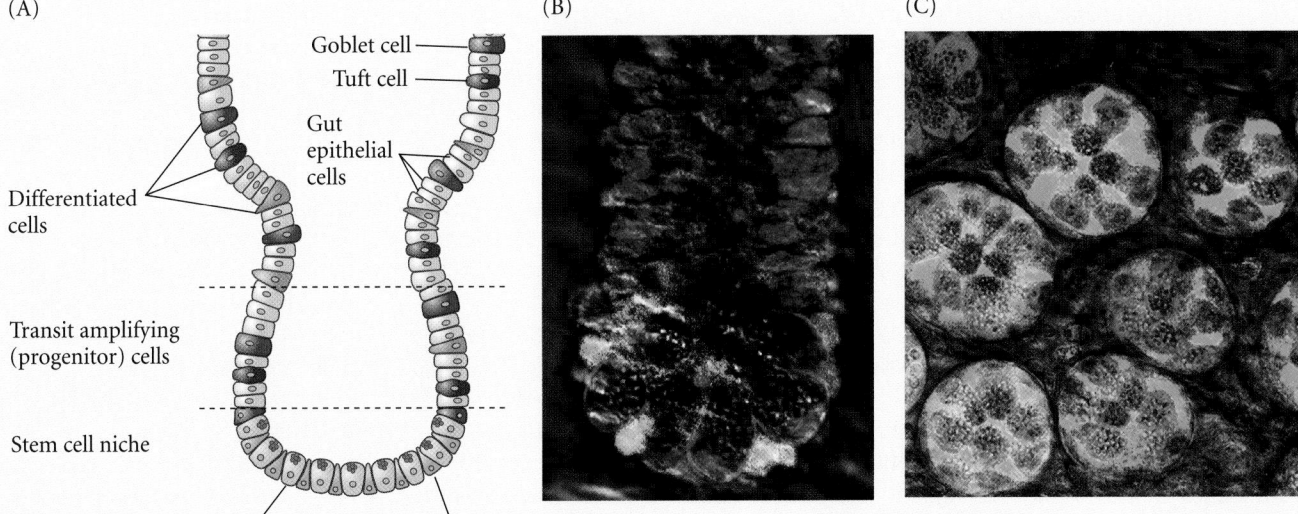

FIGURE P3.6 Intestinal stem cell niche and differentiation. (A) Diagram depicting a small intestine crypt and cell differentiation. Stem cells generate transit amplifying cells that migrate out of a niche in the crypt's base and travel up the villus as they begin to differentiate into the different cell types of the intestinal tract. Paneth cells, however, remain in the niche after they differentiate, intercalated among the stem cells. (B) Longitudinal section of an intestinal crypt. Paneth cells are stained red. Stem cells have been labeled to express GFP under the control of the *Lgr5* gene promoter and are seen as green. Nuclear DNA is counterstained blue. (C) Confocal micrograph looking down on the bottom of an isolated intestinal crypt, showing the Paneth cells (red) in intimate association with the stem cells (green). (A after Noah et al. 2011; B,C from Sato et al. 2011, photographs courtesy of H. Clevers and T. Sato.)

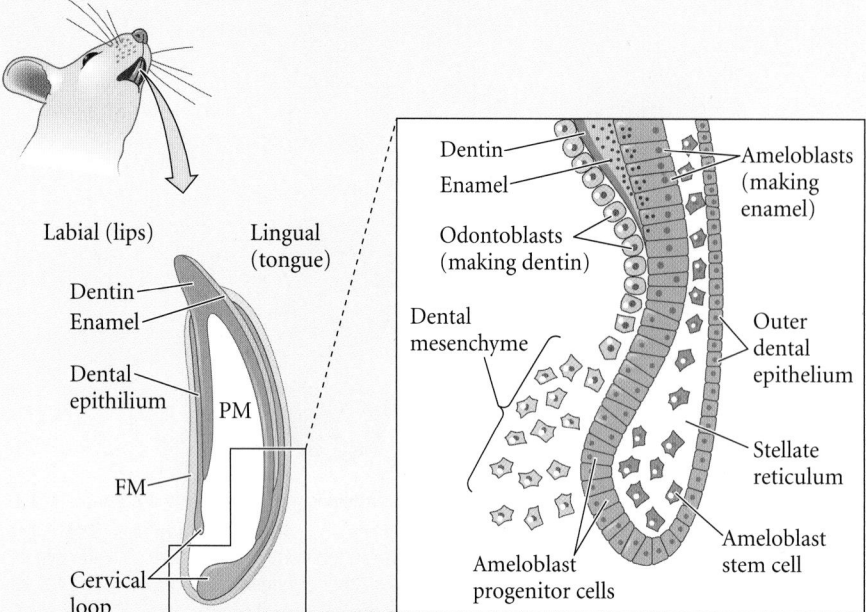

FIGURE P3.7 The cervical loop of the mouse incisor is a stem cell niche for the enamel-secreting ameloblast cells. These cells migrate from the base of the stellate reticulum into the enamel layer, allowing the teeth to keep growing. (After Wang et al. 2007.)

Like other differentiated intestinal cells, Paneth cells are produced by the progeny of the adult intestinal stem cells. Paneth cells settle at the base of the crypt (see Figure P3.6) and express several paracrine and juxtacrine factors, including Wnt3a, epidermal growth factor, and Delta-like4, an activator of Notch (Sato et al. 2009). Deleting the Paneth cells destroys the ability of the stem cells to generate other tissues. Each niche contains about 15 Paneth cells and an equal number of intestinal stem cells. In this case, the transit amplifying cells do not appear to be generated by asymmetric cell division; rather, the stem cells double their number each day by symmetric divisions, as in Figure P3.1B. After this doubling, they adopt stem cell or transit amplifying cell fates by competing for the Paneth cells' surfaces (Snippert et al. 2010; Sato et al. 2011). Only half (a random half) can bind to the Paneth cells, and these cells remain stem cells.

TOOTH REGENERATION AND LOST NICHES There are instances of some animals having lost a stem cell niche while related animals retained them. Rodent incisors, for instance, differ from human incisors (and those of most other mammalian groups) in that they continue to grow throughout the lifetime of the animal. In the much-studied mouse, each incisor has two stem cell niches, one on the "inside," facing into the mouth, and one on the "outside," facing the lips (**FIGURE P3.7**). Most other mammals lack these incisor stem cell niches, and thus their teeth do not regenerate.

Again, a balance of paracrine factors is at work regulating proliferation and differentiation in the mouse incisor niche. Stem cells are kept in a proliferative and non-differentiated state by an integrated network of paracrine factors that include FGF proteins (and BMP inhibitors), while BMP proteins (and other FGF inhibitors) promote differentiation into enamel-forming ameloblast cells (Tummers and Thesleff 2009).

Mesenchymal Stem Cells: Multipotent Adult Stem Cells

Most (if not all) adult stem cells are restricted to forming only a few cell types (Wagers et al. 2002). For example, when hematopoietic stem cells marked with green fluorescent protein were placed in mice, their labeled descendants were found throughout the animals' blood but not in any other tissue. Some adult stem cells, however, appear to have a surprisingly large degree of plasticity. These multipotent cells are called **mesenchymal stem cells**, or **MSCs** (sometimes called **bone marrow-derived stem cells**, or **BMDCs**), and their potency remains a controversial subject.

Originally found in bone marrow (Friedenstein et al. 1968; Caplan 1991), multipotent MSCs have also been found in adult tissue such as fat, muscle, thymus, and dental pulp, as well as in the umbilical cord and placenta (see Kuhn and Tuan 2010; Nazarov et al. 2012). Indeed, the finding that human umbilical cords and deciduous ("baby") teeth contain MSCs (Gronthos et al. 2000; Hirata et al. 2004; Traggiai et al. 2004; Perry et al. 2008) has led some physicians to propose that parents freeze cells from their child's umbilical cord or teeth so that these cells will be available for transplantation later in life.* However, the crucial test for pluripotency—the ability of a mouse stem cell to generate cells of all germ layers when inserted into a blastocyst—has not yet been achieved.

Mesenchymal stem cells *are* able to give rise to numerous bone, cartilage, muscle, and fat lineages (Pittenger et al. 1999; Dezawa et al. 2005; Jackson et al. 2010). The differentiation of MSCs is predicated on both paracrine

---

*Another argument for saving umbilical cord cells is that they contain hematopoietic stem cells that might be transplanted into the child should she or he later develop leukemia (see Goessling et al. 2011).

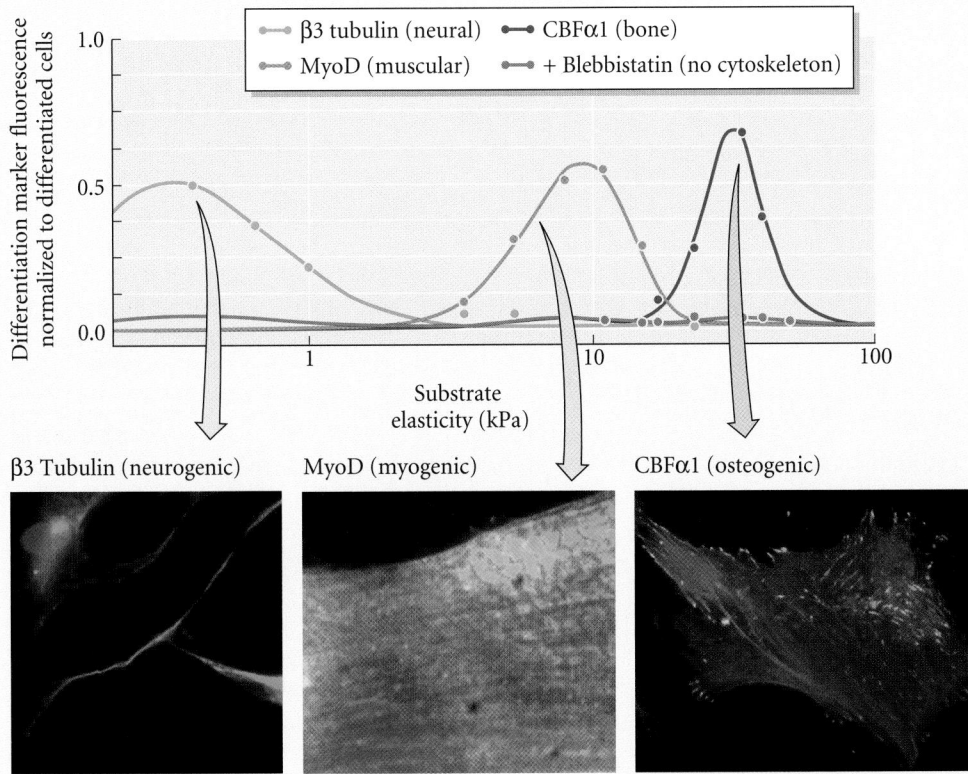

FIGURE P3.8 Mesenchymal stem cell differentiation is influenced by the elasticity of the matrices upon which the cells sit. On collagen-coated gels having elasticity similar to that of the brain (about 01–1 kPa), human MSCs differentiated into cells containing neural markers (such as β3-tubulin) but not into cells containing muscle cell markers (MyoD) or bone cell markers (CBFα1). As the gels became stiffer, the MSCs generated cells exhibiting muscle-specific proteins, and even stiffer matrices elicited the differentiation of cells with bone markers. Differentiation of the MSC on any matrix could be abolished with blebbistatin, which inhibits microfilament assembly at the cell membrane. (After Engler et al. 2006; photographs courtesy of J. Shields.)

factors and cell matrix molecules in the stem cell niche. Certain cell matrix components, especially laminin, have been implicated in keeping MSCs in a state of undifferentiated "stemness" (Kuhn and Tuan 2010). Certain paracrine factors appear to direct development into specific lineages. In one study (Ng et al. 2008), platelet-derived growth factor was critical for chondrogenesis and fat formation, TGF-β signaling was important for chondrogenesis, and FGF signaling was crucial for the differentiation of MSCs into bone cells.

In addition to paracrine factors, the surfaces on which the stem cells reside may also enhance the repertoire of cell types derivable from mesenchymal stem cells. For example, if human MSCs are placed on soft matrices of collagen-coated polyacrilamide, they differentiate into neurons. A moderately elastic matrix of the same materials causes the same stem cells to become muscle cells, whereas harder matrices cause the MSCs to produce bone cells (**FIGURE P3.8**; Engler et al. 2006). It is not yet known whether this range of potency is found normally in the body.

Mesenchymal stem cells have been linked to normal growth and repair conditions in the human body. Indeed, one premature aging syndrome, Hutchinson-Gilford progeria, appears to be caused by the inability of MSCs to differentiate into certain cell types, such as fat cells (Scaffidi and Misteli 2008). These findings lead to speculation that the loss of MSC ability to differentiate may be a component of the normal aging syndrome. Moreover, MSCs may work in ways other than differentiating into needed cell types. They may produce paracrine factors that aid other, more specific stem cells to divide and repair tissues (Gnecchi et al. 2009).

Pluripotent Embryonic Stem Cells

Pluripotent embryonic cells are a special case because these stem cells can generate all 220 cell types of the adult mammalian body (see Shevde 2012). In the laboratory, pluripotent embryonic cells are derived from two major sources (**FIGURE P3.9**). The first source is the inner cell mass of the early embryo, that part of the mammalian embryo that gives rise to the embryo's body (as opposed to the outer cells that generate the placenta). The inner cell mass gives rise to the **embryonic stem cells (ESCs)**. The second source is primordial germ cells that have not yet differentiated into the sperm or egg. When isolated from the embryo, these cells divide, remain diploid, and are sometimes called **embryonic germ cells (EGCs)**. In both instances, the cells demonstrate pluripotency, the ability to generate all the cells of the body (Shamblott et al. 1998; Thomson et al. 1998).

The pluripotency of embryonic cells is maintained by a core of three transcription factors: Oct4, Sox2, and Nanog. Acting in concert, these factors activate the genes required to maintain pluripotency and repress those genes whose

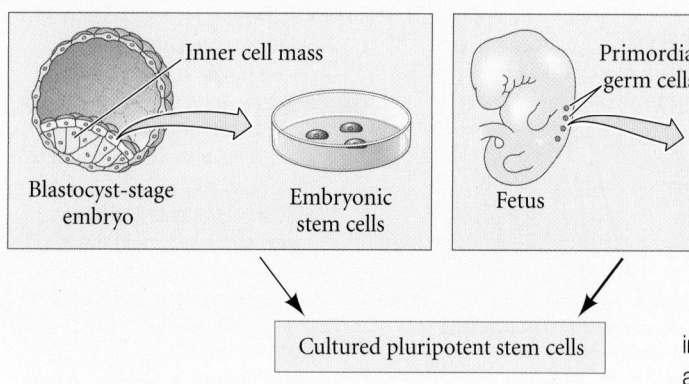

FIGURE P3.9   Major sources of pluripotent stem cells. Embryonic stem cells (ES cells) arise from culturing the inner cell mass of the early embryo. Embryonic germ cells (EG cells) are derived from primordial germ cells that have not yet reached the gonads.

products would lead to differentiation (Marson et al. 2008; Young 2011). Moreover, these pluripotent stem cells can respond to paracrine factors and differentiate as they would in the body. By placing them into media containing different paracrine factors, embryonic stem cells can be directed into particular paths of differentiation (**FIGURE P3.10**).

The power of embryonic stem cells is that they often retain their ability to differentiate into any cell type, even when placed into an adult body. For instance, Kerr and colleagues (2003) found that human EGCs were able to cure motor neuron

injuries in adult rats, both by differentiating into new neurons and by producing paracrine factors (BDNF and TGF-α) that prevent the death of existing neurons. Similarly, ES cells from monkey blastocysts were seen to cure a Parkinson-like condition in adult monkeys whose dopaminergic neurons had been destroyed (Takagi and Takahashi, 2005).

Stem Cell Therapy

This ability to generate all the cell types of the body offers the possibility of stem cell therapy for numerous human conditions in which adult cells degenerate (such as Alzheimer disease, Parkinson disease, diabetes, and cirrhosis). However, there are two major problems with using human embryonic stem cells therapeutically. One is that transplanted ES cells come from another individual and thus are not the same genotype as the patient; ES cells can be rejected by the patient's immune system, just like any other organ transplant.* The second problem involves the several social and ethical issues raised by the fact that the ES cells are taken from human embryos (Gilbert et al. 2005; Siegel 2008; NSF 2012). Both problems could be circumvented (at least theoretically) by transforming somatic cells (such as skin fibroblasts, which are adult dermal cells that grow readily in culture) into pluripotent cells.

INDUCED PLURIPOTENT STEM CELLS: A WAVE OF THE FUTURE?   Although we know that the nuclei of differentiated somatic cells retain copies of an individual's entire genome, biologists have long thought that potency was like going down a steep hill. Once differentiated, we believed, a cell could not be restored to a more potent state. However, our newfound knowledge of the transcription factors needed to maintain pluripotency has illuminated a startlingly easy way to generate stem cells from somatic cells.

In 2006, Kazutoshi Takahashi and Shinya Yamanaka of Kyoto University demonstrated that by inserting activated copies of four genes that encoded some of these critical

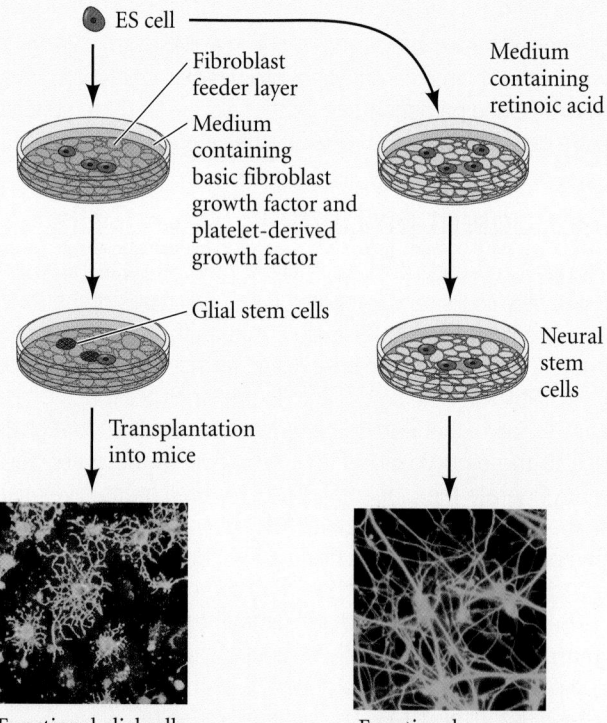

FIGURE P3.10   Inducing stem cell differentiation. The differentiation of mouse embryonic stem cells into lineage-restricted (neuronal and glial) stem cells can be accomplished by altering the media in which the ES cells grow. (Photographs from Brüstle et al. 1999 and Wickelgren 1999, courtesy of O. Brüstle and J. W. McDonald.)

---

*One reason that diseases of the brain are being targeted is that the brain and the eyes are among the few places where immune rejection is not a big problem. The blood-brain barrier of the brain's endothelial cells keeps them shielded from the immune system.

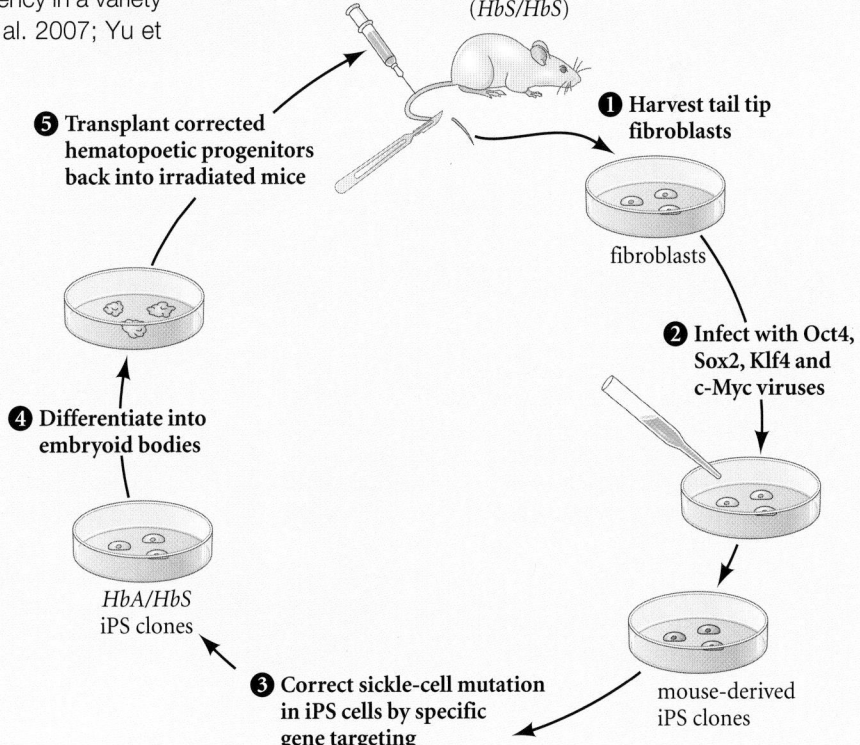

ES cell

Differentiation

?

Reprogramming

Differentiation

Somatic cell

iPS cell

Mouse gene combinations for iPS induction:
- *Oct3/4, Sox2, c-Myc, Klf4*
- *Oct3/4, Sox2, N-Myc, Klf4*
- *Oct3/4, Sox2, Klf4*
- *Oct3/4, Sox2, Lin28, Nanog*

FIGURE P3.11 Pluripotent ES cells derived from the early embryo can generate all the types of somatic cells. Differentiated somatic cells can be reprogrammed to induced pluripotent stem (iPS) cells by the introduction of a defined and limited set of transcription factors. Sox2 and Oct4 are critical for the induction of pluripotency, and they can work in combination with other transcription factors (such as c-Myc and Klf4). Under certain culture conditions, the potency of iPS cells appears to be identical to that of ES cells. Both iPS cells and ES cells are capable of generating entire embryos. (After Nishikawa et al. 2008.)

transcription factors, nearly any cell in the adult mouse body could be made into an **induced pluripotent stem cell (iPS cell)** with the pluripotency of an embryonic stem cell.* These genes were *Sox2* and *Oct4* (which activated Nanog and other transcription factors that established pluripotency and blocked differentiation), c-*Myc* (which opened up chromatin and made the genes accessible to Sox2, Oct4, and Nanog), and *Klf4*, which prevents cell death (**FIGURE P3.11**; see also Figure 9.18).

Within 6 months of the publication of this work, three groups of scientists reported findings that the same or similar transcription factors had induced pluripotency in a variety of human differentiated cells (Takahashi et al. 2007; Yu et

al. 2007; Park et al. 2008). By 2012, modifications of the culture techniques made it possible for the gene expression of mouse iPS cells to become identical to that of mouse embryonic stem cells (Stadtfeld et al. 2012). Entire mouse embryos could be generated from single iPS cells, showing complete pluripotency.

The importance of iPS cells is not that they can make entire embryos, however, but that they can make cells that are able to repair damaged organs. Thus, in addition to allowing us to understand organogesis, embryonic stem cells and induced pluripotent cells have opened up an entirely new field of therapy, often called **regenerative medicine** (Wu and Hochelinger 2011; Robinton and Daley 2012). The therapeutic potential of iPS cells was demonstrated by the ability of iPS-derived hematopoietic stem cells to correct a sickle-cell anemia phenotype in mice (**FIGURE P3.12**; Hanna

---

*This achievement earned Shinya Yamanaka the 2012 Nobel Prize in Physiology or Medicine.

FIGURE P3.12 Protocol for curing a "human" disease in a mouse using iPS cells plus recombinant genetics. (1) Tail-tip fibroblasts are taken from a mouse whose genome contains the human alleles for sickle-cell anemia (*HbS*) and no mouse genes for this protein. (2) The cells are cultured and infected with viruses containing the four transcription factors known to induce pluripotency. (3) The iPS cells are identified by their distinctive shapes and are given DNA containing the wild-type allele of human globin (*HbA*). (4) The embryos are allowed to differentiate in culture. They form "embryoid bodies" that contain blood-forming stem cells. (5) Hematopoietic progenitor and stem cells from these embryoid bodies are injected into the original mouse and cure its sickle-cell anemia. (After Hanna et al. 2007.)

Humanized sickle cell anemia mouse model (*HbS/HbS*)

❶ Harvest tail tip fibroblasts

fibroblasts

❷ Infect with Oct4, Sox2, Klf4 and c-Myc viruses

mouse-derived iPS clones

❸ Correct sickle-cell mutation in iPS cells by specific gene targeting

HbA/HbS iPS clones

❹ Differentiate into embryoid bodies

❺ Transplant corrected hematopoetic progenitors back into irradiated mice

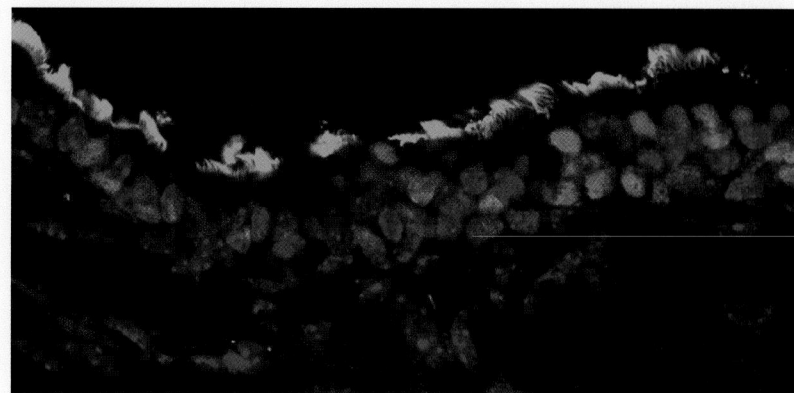

FIGURE P3.13 Lung epithelium derived from mouse iPS cells. The lung transcription factor Nxk2.1 is stained red, indicating that the iPS cells cells have become lung epithelia. The tubulin of the epithelial cilia, whose functions are disturbed in patients with cystic fibrosis, is stained green. Nuclei are blue. (Courtesy of J. Rajagopal.)

et al. 2007). Ongoing studies are attempting to determine if such therapy could cure human conditions such as diabetes, macular degeneration, Parkinson disease, and Alzheimer disease, as well as liver and heart disease.

Using iPS cells also allows medical researchers to experiment on diseased human tissue, and thus to study conditions in ways that were previously problematic. Mice, for instance, do not get the same type of cystic fibrosis that humans get, and experimentation on human subjects is fraught with ethical issues. Mou and colleagues (2012) circumvented these obstacles by first determining that mouse iPS cells could be turned into lung epithelia by placing them sequentially into BMP-, FGF-, and Wnt-containing media, mimicking the signals such cells would receive during development (**FIGURE P3.13**). The researchers then made iPS cells from a person with cystic fibrosis and used the same protocol to turn these into lung epithelium. These cells formed respiratory epithelium that had the characteristics of

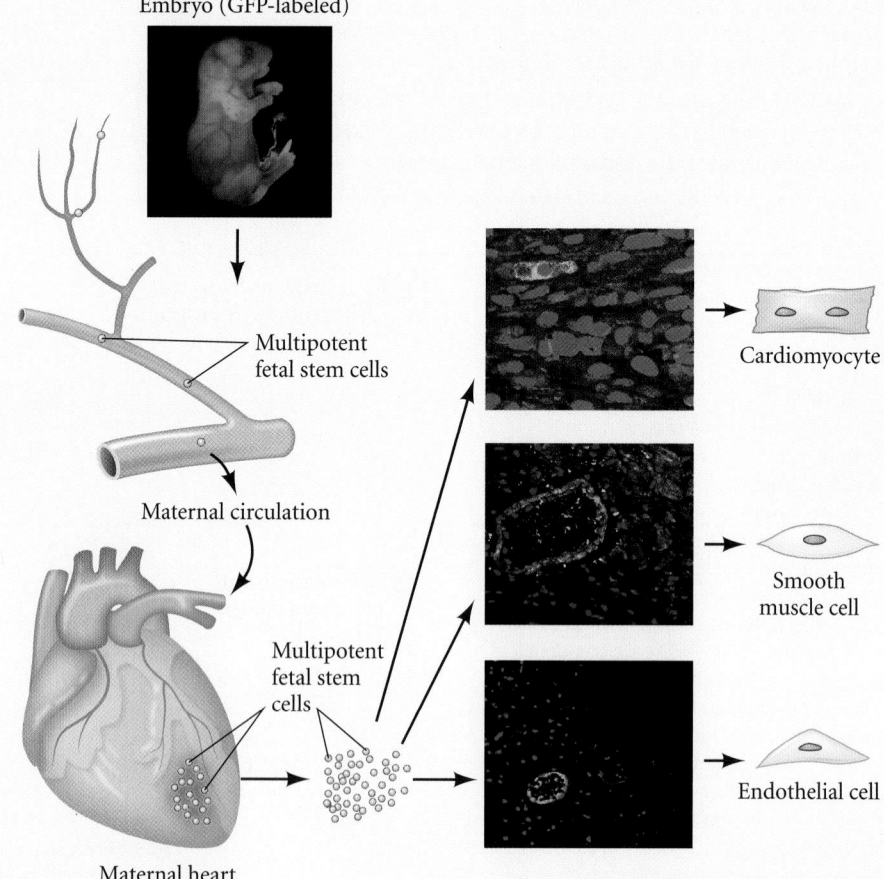

FIGURE P3.14 Possible pathway for the trafficking of cells from the fetus across the placenta and into maternal circulation. Multipotent stem cells from a mouse fetus were labeled to express GFP. The labeled fetal stem cells were seen to integrate into the mother's heart (which had been surgically damaged) and to generate diverse cardiac lineages, including cardiomyocytes, smooth muscle cells, and endothelial cells. Cardiomyocytes stain green (GFP) and red (sarcomeric actin); smooth muscle cells stain green (GFP) and red (smooth muscle actin); endothelial cells stain green (GFP) and red (CD31). (After Kara et al. 2012, courtesy of H. Chaudry and R. Kara.)

human cystic fibrosis. This type of iPS research may allow the discovery of drugs that can ameliorate the human disease phenotype.

Further, iPS cells have been induced to form numerous cell types, many of which are functional when the differentiated cells are transplanted back into the organism from which they were derived. Even sperm cells and oocytes have been generated by mouse iPS cells. Here, the iPS cells are induced to form primordial germ cells (PGCs) by placing them into mixtures of paracrine factors. When these induced PGC cells (which were derived from skin fibroblasts) were aggregated with gonadal tissues, the cells proceded through meiosis and became functional gametes (Hayashi et al. 2011, 2012). This work could become significant in circumventing many types of infertility, as well as allowing scientists to study the details of meiosis.

NATURAL STEM CELL THERAPY: COOPERATION BETWEEN FETUS AND MOTHER   Pluripotent and multipotent stem cells have been found in an unexpected place: in the blood of pregnant mice and women. It appears that stem cells from the fetus enter the maternal circulation (Khosrotehrani et al. 2004, 2005). In humans, these cells often leave the mother's blood and integrate into her existing organs, where they express the differentiation markers associated with the adult tissue into which they have integrated. It further appears that these embryo-derived stem cells may integrate preferentially into any damaged or diseased organs in the mother (**FIGURE P3.14**; Kara et al. 2012). Such fetal stem cells have been seen even decades after the pregnancy. Thus, women who have borne children can have an additional set of stem cells or progenitor cells

in their organs (Bianchi et al. 1996). It is possible that such fetal stem cells help the mother during pregnancy and the immediate stresses of labor and delivery.

A New Perspective on Organogenesis

The ability to induce, isolate, and manipulate stem cells offers a vision of regenerative medicine wherein patients can have their diseased organs regrown and replaced by their own stem cells. Stem cells also offer fascinating avenues for the treatment of numerous diseases. Indeed, when one thinks about the mechanisms of aging, the replacement of diseased body tissues, and even the enhancement of bodily abilities, the line between medicine and science fiction becomes very tenuous. Developmental biologists have to consider not only the biology of stem cells, but also the ethics, medical economics, and theories of justice with which they interact (see Faden et al. 2003; Dresser 2010; Buchanan 2011).

Several stem cell therapy protocols are even now being tested on human patients (Normile 2012; Cyranoski 2013), and stem cell research may be the beginning of a revolution that will be as important for medicine (and as transformative for society) as the research on infectious microbes was a century ago. But beyond the potential medical applications, stem cells tell us a great many facts about how the body is constructed and how it maintains its structure. Organs often form by the regulation of stem cells, and we will see that the skin, hair, blood, and parts of the nervous system routinely use stem cells in their construction. Stem cells certainly give credence to the view that "development never stops."

10
Emergence of the Ectoderm
Central Nervous System and Epidermis

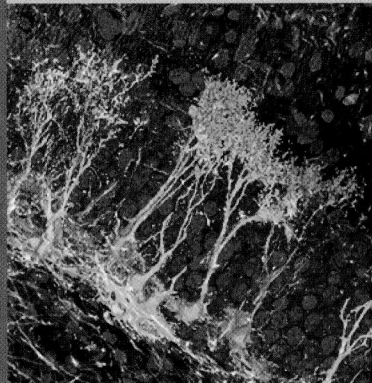

THE VERTEBRATE ECTODERM, the outer germ layer covering the late-stage gastrula, has three major responsibilities (FIGURE 10.1):

1. One part of the ectoderm will become the **neural plate**, the presumptive neural tissue induced by the prechordal plate and notochord during gastrulation. The neural plate involutes into the body to form the **neural tube**, the precursor of the **central nervous system** (the brain and spinal cord).

2. Another part of this germ layer will become the outer layer of the skin, the **epidermis**—the largest organ of the vertebrate body. The epidermis forms an elastic, waterproof, and constantly regenerating barrier between the organism and the outside world.

3. Between the compartments forming the epidermis and the central nervous system lies the presumptive **neural crest**. The cells of the neural crest migrate away from the dorsal center of the embryo (between the neural tube and epidermis) to generate, among other things, the peripheral nervous system and pigment cells (melanocytes).

The processes by which the three ectodermal regions are made physically and functionally distinct from one another is called **neurulation**, and an embryo undergoing these processes is called a **neurula** (FIGURE 10.2). As we saw in the preceding chapters, the specification of the ectoderm is accomplished primarily by regulating the levels of BMP experienced by the ectodermal cells. High levels of BMP specify the cells to become epidermis. Very low levels specify the cells to become neural plate. Intermediate levels effect the formation of the neural crest cells. Neurulation follows directly upon gastrulation.

The specification, migration, and differentiation of the neural crest cells will be described in Chapter 11. This chapter follows the story of the neural and epidermal precursor cells. We will begin with the neural precursors, those cells that form first the neural plate and then all the nerve and glial cells of the vertebrate central nervous system.

NEURAL PLATE AND CENTRAL NERVOUS SYSTEM

The cells of the neural plate are characterized by expressing the Sox family of transcription factors (Sox1, 2, and 3)—factors that (1) activate the genes that specify cells to be neural plate and (2) inhibit the formation of epidermis and neural crest by blocking the transcription and signaling of BMPs (Archer et al. 2010). In this, we see once again an important principle of development: *the signals promoting the specification of one cell type often also block the specification of an alternative cell type.* The expression of Sox transcription factors establishes the neural plate cells as neural precursors that can form all the cell types of the central nervous system (see Figure P3.3; Wilson and Edlund 2001).

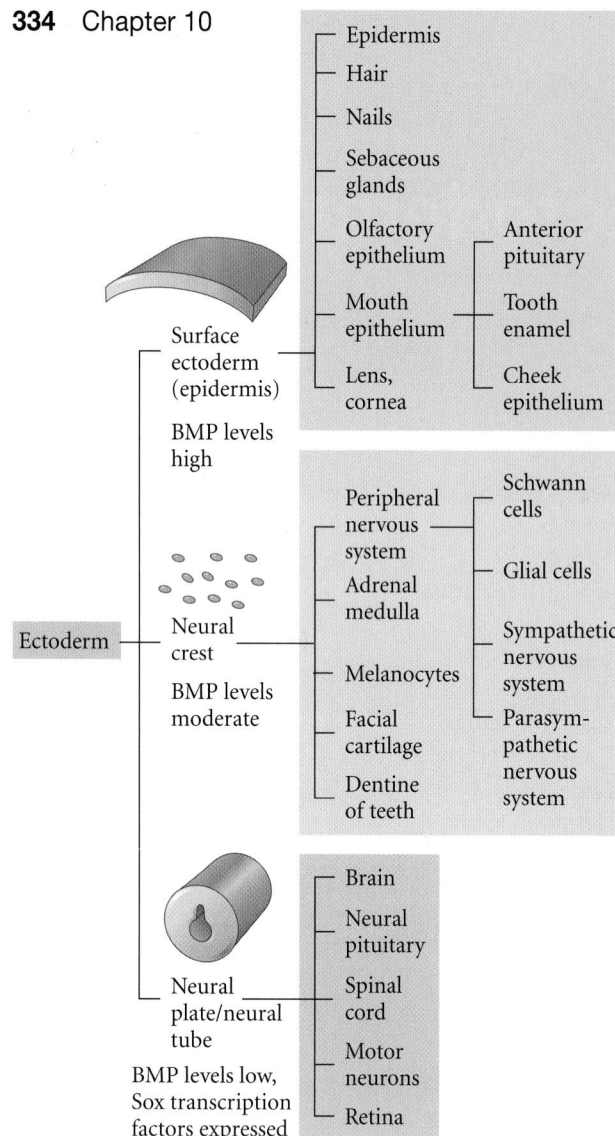

FIGURE 10.1 Major derivatives of the ectoderm germ layer. The ectoderm is divided into three major domains: the surface ectoderm (primarily epidermis), the neural crest (peripheral neurons, pigment, facial cartilage), and the neural tube (brain and spinal cord).

Formation of the Neural Tube

The neural plate lies on the surface of the embryo. But the nervous system does not lie on the outside of the body. Somehow the neural plate has to move inside the embryo and form a neural tube. This is accomplished in two major steps. In **primary neurulation**, the cells surrounding the neural plate direct the neural plate cells to proliferate, invaginate into the body, and separate from the surface to form a hollow tube. In **secondary neurulation**, the neural tube arises from the coalescence of mesenchyme cells into a solid cord that subsequently forms cavities that coalesce to create a hollow tube. In general, the *anterior* portion of the neural tube is made by primary neurulation, while the *posterior* portion of the neural tube is the product of secondary neurulation. The complete

neural tube forms by joining these two separately formed tubes together (Harrington et al. 2009).

In birds, primary neurulation generates the neural tube anterior to the hindlimbs (at the twenty-eighth somite pair; Pasteels 1937; Catala et al. 1996). In mammals, secondary neurulation begins at the level of the sacral vertebrae of the tail (around the thirty-fifth somite pair; Schoenwolf 1984; Nievelstein et al. 1993). In fish and amphibians (e.g., zebrafish and *Xenopus*), only the tail neural tube is derived from secondary neurulation (Gont et al. 1993; Lowery and Sive 2004).

Primary neurulation

The process of primary neurulation appears to be similar in all vertebrates. **FIGURE 10.3** illustrates the process in amphibians (Gallera 1971). Shortly after the neural plate has formed, its edges thicken and move upward to form the **neural folds**, while a U-shaped **neural groove** appears in the center of the plate, dividing the future right and left sides of the embryo. The neural folds on the lateral sides of the neural plate migrate toward the midline of the embryo, eventually fusing to form the neural tube beneath the overlying ectoderm. In some species, the cells at the dorsalmost portion of the neural tube (i.e., the cells that had been the neural folds and contacting the presumptive epidermal cells) become the neural crest cells.

Primary neurulation can be divided into four distinct but spatially and temporally overlapping stages:

1. *Extension and folding of the neural plate.* Divisions of the neural plate cells are preferentially in the anterior-posterior direction (often referred to as the **rostral-caudal**, or beak-tail, direction). These events occur even if the neural plate tissue is isolated from the rest of the embryo. However, in order to roll into a neural tube, the presumptive epidermis is also needed (**FIGURE 10.4A**; Jacobson and Moury 1995; Moury and Schoenwolf 1995; Sausedo et al. 1997).

2. *Bending of the neural plate.* The bending of the neural plate involves the formation of hinge regions where the neural plate contacts surrounding tissues. In birds and mammals, the cells at the midline of the neural plate form the **medial hinge point** (**MHP**; Schoenwolf 1991a,b; Catala et al. 1996). MHP cells become anchored to the notochord beneath them and form a hinge, which forms a furrow at the dorsal midline. The notochord induces the MHP cells to decrease in height and to become wedge-shaped (**FIGURE 10.4B,C**; van Straaten et al. 1988; Smith and Schoenwolf 1989).

3. *Convergence of the neural folds.* Shortly thereafter, two **dorsolateral hinge points** (**DLHPs**) are induced by the non-neural ectoderm and are anchored to the surface (epidermal) ectoderm. Noggin from the surface ectoderm blocks BMPs in the lateral neural plate and allows the cytoskeleton to elongate the DLHP cells and make them wedge-shaped. Sonic hedgehog from the notochord

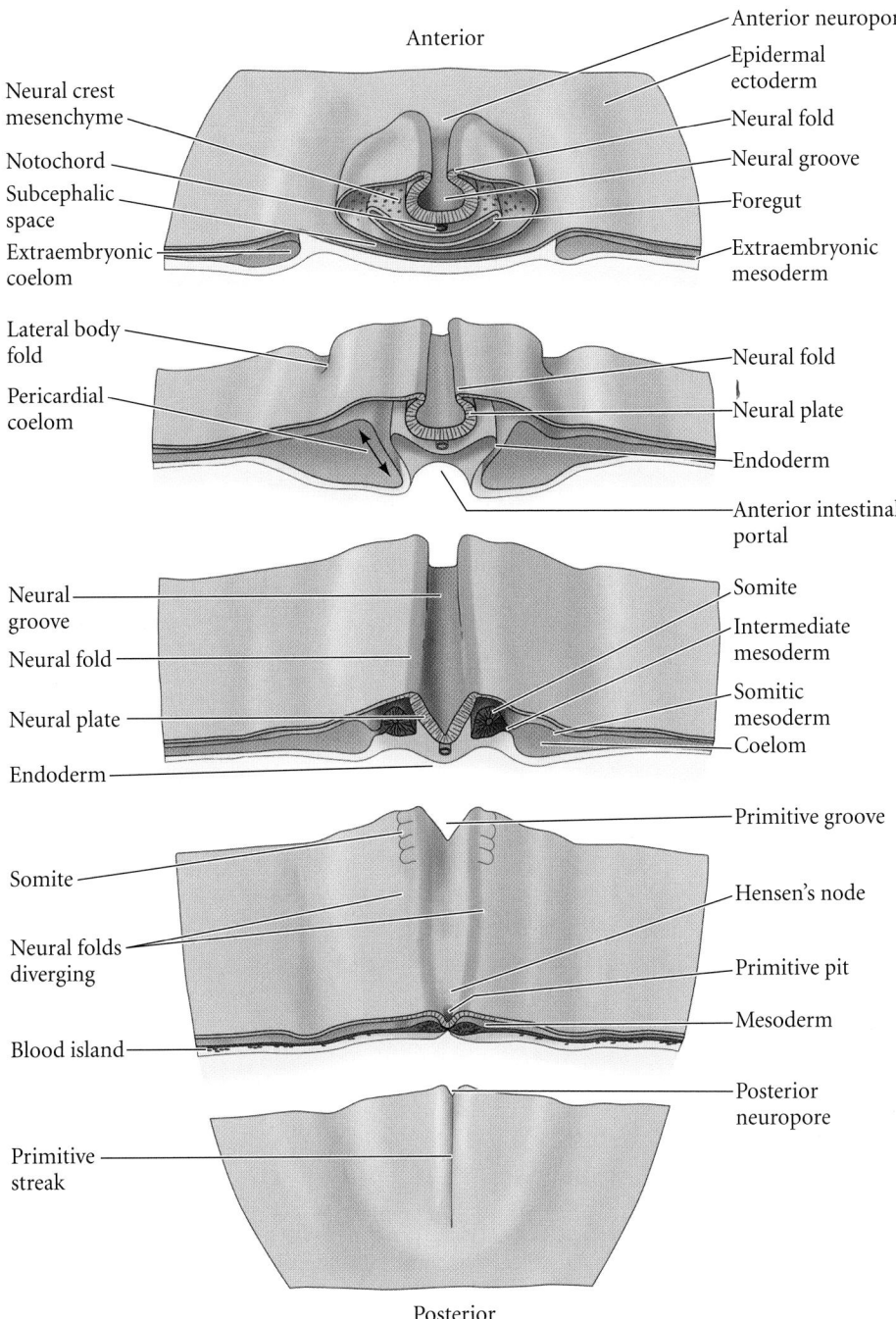

Anterior

Anterior neuropore
Epidermal ectoderm
Neural fold
Neural groove
Foregut
Extraembryonic mesoderm

Neural crest mesenchyme
Notochord
Subcephalic space
Extraembryonic coelom

Lateral body fold
Pericardial coelom

Neural fold
Neural plate
Endoderm
Anterior intestinal portal

Neural groove
Neural fold
Neural plate
Endoderm

Somite
Intermediate mesoderm
Somitic mesoderm
Coelom

Somite
Neural folds diverging
Blood island

Primitive groove
Hensen's node
Primitive pit
Mesoderm
Posterior neuropore

Primitive streak

Posterior

FIGURE 10.2 The neurulating chick embryo (dorsal view) at about 24 hours. The cephalic (head) region has undergone neurulation, while the caudal (tail) region is still undergoing gastrulation. (After Patten 1971.)

blocks Noggin in the more medial neural plate cells (Ybot-Gonzalez et al. 2007). After the initial furrowing of the neural plate, the plate bends around the hinge regions. Each hinge acts as a pivot that directs the rotation of the cells around it (Smith and Schoenwolf 1991). The surface ectoderm pushes toward the midline of the embryo, providing another motive force for bending the neural plate, causing the neural folds to converge toward each other (FIGURE 10.4D; Alvarez and Schoenwolf 1992;

Lawson et al. 2001). This movement of the presumptive epidermis and the anchoring of the neural plate to the underlying mesoderm may also be important for ensuring that the neural tube invaginates inward, or into the embryo and not outward (Schoenwolf 1991a).

4. *Closure of the neural tube.* The neural tube closes as the paired neural folds are brought together at the dorsal midline. The folds adhere to each other, and the cells from the two folds merge (FIGURE 10.4E).

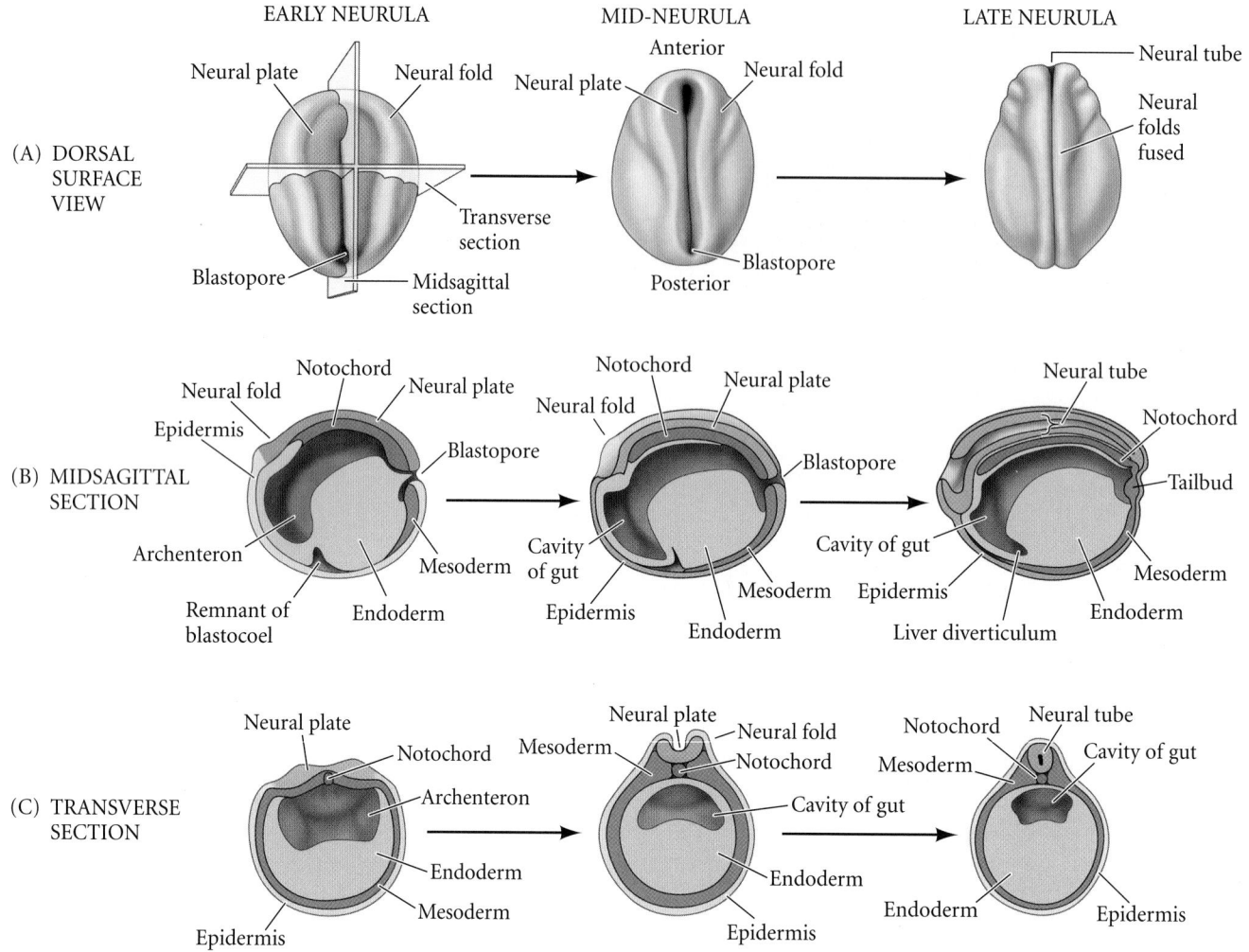

FIGURE 10.3 Three views of primary neurulation in an amphibian embryo, showing early (left), middle (center), and late (right) neurulae in each case. (A) Looking down on the dorsal surface of the whole embryo. (B) Midsagittal section through the medial plane of the embryo. (C) Transverse section through the center of the embryo. (After Balinsky 1975.)

EVENTS OF NEURAL TUBE CLOSURE   Closure of the neural tube does not occur simultaneously throughout the ectoderm. This phenomenon is best seen in amniote vertebrates (reptiles, birds, and mammals) whose body axis is elongated prior to neurulation. In amniotes, induction occurs in an anterior-to-posterior fashion; so in the 24-hour chick embryo, neurulation in the **cephalic** (head) region is well advanced, while the **caudal** (tail) region of the embryo is still undergoing gastrulation (see Figure 10.2). The two open ends of the neural tube are called the **anterior neuropore** and the **posterior neuropore**.

In chicks, neural tube closure is initiated at the level of the future midbrain and "zips up" in both directions. By contrast, in mammals neural tube closure is initiated at several places along the anterior-posterior axis (**FIGURE 10.5**). In humans there are probably five sites of neural tube closure (see Figure 10.5B; Nakatsu et al. 2000; O'Rahilly and Muller 2002),

FIGURE 10.4   Primary neurulation: neural tube formation in the chick embryo. (A,1) Cells of the neural plate can be distinguished as elongated cells in the dorsal region of the ectoderm. Folding begins as the medial hinge point (MHP) cells anchor to the notochord and change their shape, while the presumptive epidermal cells move toward the dorsal midline. (B,2) The neural folds are elevated as the presumptive epidermis continues to move toward the dorsal midline. (C) Elevated neural folds stained to show the extracellular matrix (green) and the actin cytoskeleton (red) concentrated in the apical portions of the neural plate cells. (D,3) Convergence of the neural folds occurs as the cells at the dorsolateral hinge point (DLHP) become wedge-shaped and the epidermal cells push toward the center. (E,4) The neural folds are brought into contact with one another, and the neural crest cells link the neural tube with the epidermis. The neural crest cells disperse, leaving the neural tube separate from the epidermis. (Scanning electron micrographs courtesy of K. Tosney and G. Schoenwolf, drawings after Smith and Schoenwolf 1997; C courtesy of E. Marti Gorostiza and M. Angeles Rabadán.)

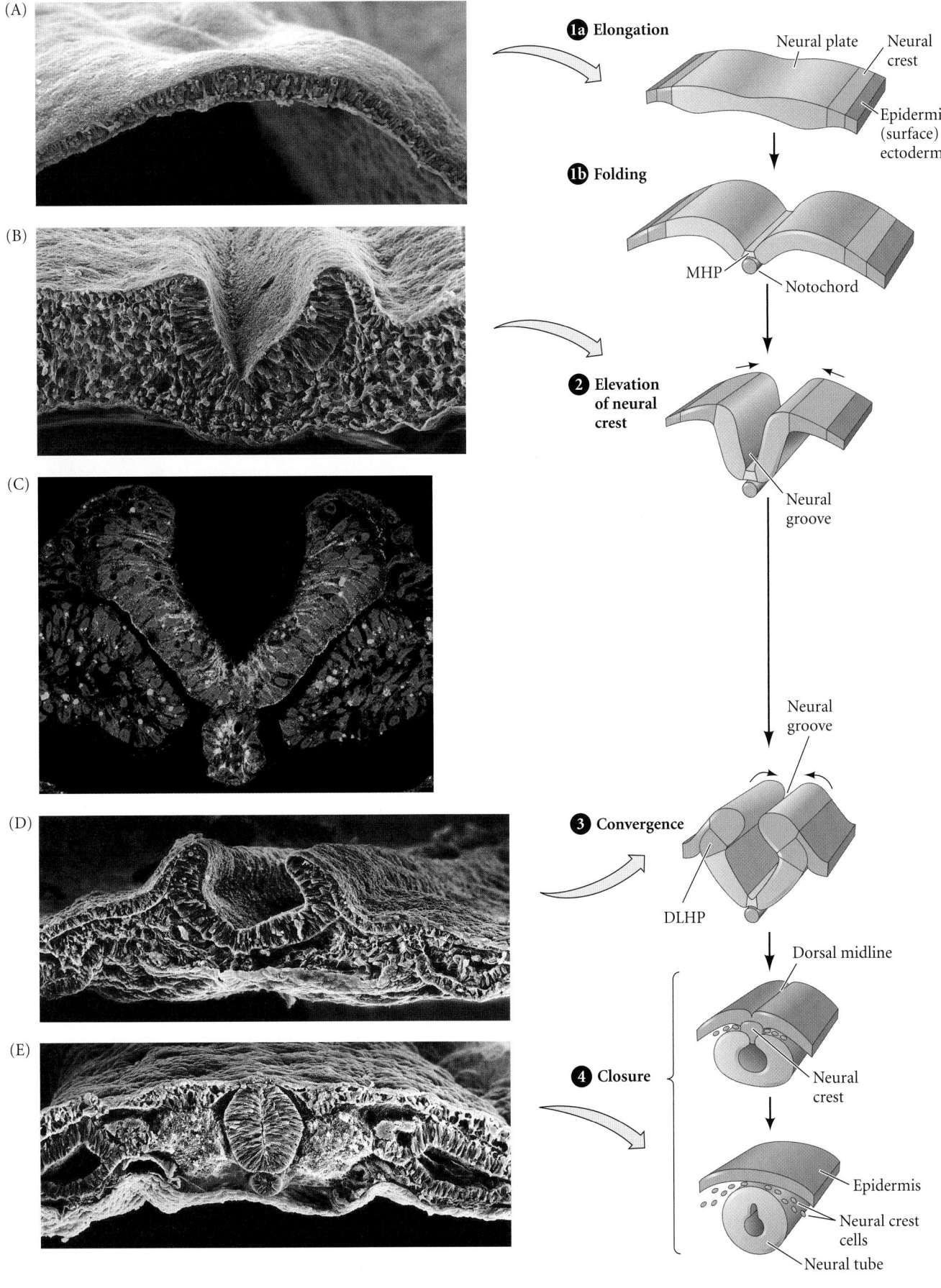

(A)

(B)

(C)

(D)

(E)

1a Elongation

Neural plate   Neural crest

Epidermis (surface) ectoderm

1b Folding

MHP   Notochord

2 Elevation of neural crest

Neural groove

Neural groove

3 Convergence

DLHP

Dorsal midline

Neural crest

4 Closure

Epidermis

Neural crest cells

Neural tube

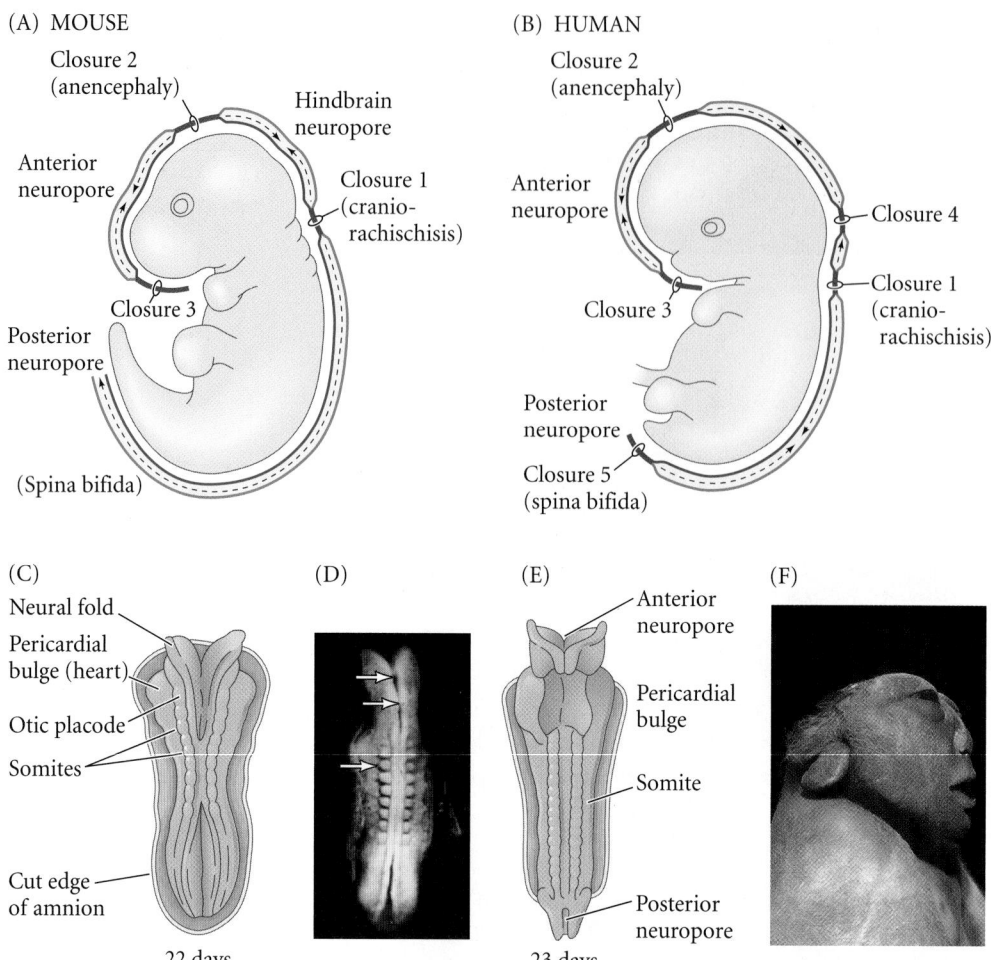

(A) MOUSE

Closure 2 (anencephaly)

Anterior neuropore

Hindbrain neuropore

Closure 1 (cranio-rachischisis)

Closure 3

Posterior neuropore

(Spina bifida)

(B) HUMAN

Closure 2 (anencephaly)

Anterior neuropore

Closure 3

Closure 4

Closure 1 (cranio-rachischisis)

Posterior neuropore

Closure 5 (spina bifida)

(C)

Neural fold

Pericardial bulge (heart)

Otic placode

Somites

Cut edge of amnion

22 days

(D)

(E)

Anterior neuropore

Pericardial bulge

Somite

Posterior neuropore

23 days

(F)

FIGURE 10.5 Neurulation in the mammalian embryo. (A,B) Initiation sites for neural tube closure of mouse (A) and human (B) embryos. In addition to the three sites found in mice, neural tube closure in humans also initiates at the posterior end of the hindbrain and in the lumbar region. (C) Dorsal view of a 22-day (8-somite) human embryo initiating neurulation. Both anterior and posterior neuropores are open to the amniotic fluid. (D) A 10-somite human embryo showing some of the major sites of neural tube closure (arrows). (E) Dorsal view of a 23-day neurulating human embryo with only its neuropores open. (F) Photograph of a stillborn infant with anencephaly. (A,B after Bassuk and Kibar 2009; D from Nakatsu et al. 2000; F courtesy of National March of Dimes.)

and the closure mechanism may differ at each site (Rifat et al. 2010). The rostral closure site (closure site 1) is located at the junction of the spinal cord and hindbrain and appears to close, as does the chick neural tube, by a "zippering" of the dorsal neural tube. However, at closure site 2, located at the midbrain/forebrain boundary, it is the dorsalmost epithelial cells that appear to do the job. They extend long filopodial extensions toward each other and then bring the neural cells into contact (**FIGURE 10.6**; Pyrgaki et al. 2010). At closure site 3 (the rostral forebrain), the dorsolateral hinge points appear to be fully responsible for the neural tube closure.

The neural tube eventually forms a closed cylinder that separates from the surface ectoderm. This separation appears to be mediated by the expression of different cell adhesion molecules. Although the cells that will become the neural tube originally express E-cadherin, they stop producing this protein as the neural tube forms, and instead synthesize N-cadherin (**FIGURE 10.7A**). As a result, the surface ectoderm and neural tube tissues no longer adhere to each other. If the surface ectoderm is experimentally made to express N-cadherin (by injecting N-cadherin mRNA into one cell of a 2-cell *Xenopus* embryo), the separation of the neural tube from the presumptive epidermis is dramatically impeded (**FIGURE 10.7B**; Detrick et al. 1990; Fujimori et al. 1990). The Grainyhead transcription factors are especially important in this process (Rifat et al. 2010; Werth et al. 2010; Pyrgaki et al. 2011). Grainyhead-like2, for instance, controls a battery of cell adhesion molecules and downregulates E-caderin synthesis in the neural folds. Mice with mutant *Grainyhead-2* genes have severe neural tube defects (including a split face).

● **See VADE MECUM** Chick neurulation

(A)

(B)

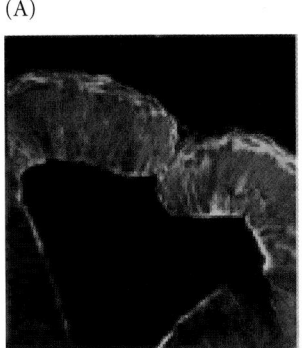
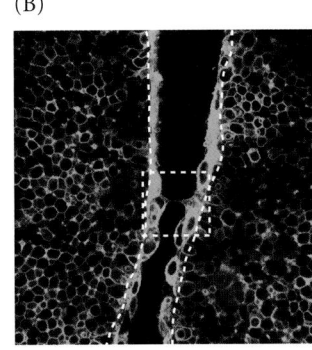
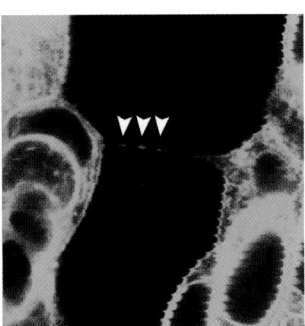

FIGURE 10.6 Neural tube closure at mouse site 2. (A) Optical section through a mouse embryo as the neural folds are close but not yet closed. The single layer of non-neural surface ectoderm (large, flattened cells; stained green) has wrapped itself around the neural ectoderm (stained blue) at the edge of the closing neural folds. (B) Dotted lines show the border between neural and non-neural ectoderm. Cellular bridges from the non-neural ectoderm connect the two juxtaposed neural folds. A close-up of one of these bridges is seen at the right (arrowheads). (From Pyrgaki et al. 2011; photographs courtesy of L. Niswander.)

NEURAL TUBE CLOSURE DEFECTS In humans, neural tube closure defects occur in about 1 in every 1000 live births. Failure to close the posterior neuropore (closure site 5) around day 27 of development results in a condition called **spina bifida**, the severity of which depends on how much of the spinal cord remains exposed. Failure to close site 2 or site 3 in the rostral neural tube keeps the anterior neuropore open, resulting in a lethal condition called **anencephaly** in which the forebrain remains in contact with the amniotic fluid and subsequently degenerates. The fetal forebrain ceases development, and the vault of the skull fails to form (see Figure 10.5F). The failure of the entire neural tube to close over the body axis is called **craniorachischisis**.

Failure to close the neural tube can result from both genetic and environmental causes (Fournier-Thibault et al.

2009; Harris and Juriloff 2010). Mutations (first found in mice) in genes such as *Pax3, Sonic hedgehog, Grainyhead, AP-2,* and *Openbrain* are essential for the formation of the mammalian neural tube, but loss-of-function mutations of these genes are very rare in humans. However, dietary factors, such as cholesterol and folate (also known as folic acid or vitamin B$_9$), also appear to be vital for human neural tube closure. Cholesterol probably functions as a cofactor in the critical Sonic hedgehog signaling pathway. Although the exact role of folate remains unknown, it may be the limiting factor in regulating DNA synthesis during cell division in the brain (Anderson et al. 2013). Whatever its mechanism, it has been estimated that more than half of all human neural tube birth defects can be prevented by pregnant women taking supplemental folate. Therefore, the U.S. Public Health Service recommends that women of childbearing age take 0.4 milligrams of folate daily (Milunsky et al. 1989; Centers for Disease Control 1992; Czeizel and Dudas 1992).

Secondary neurulation

Secondary neurulation involves the production of mesenchyme cells from the prospective ectoderm and endoderm, followed by the condensation of these cells into a **medullary cord** beneath the surface ectoderm (**FIGURE 10.8A,B**). After

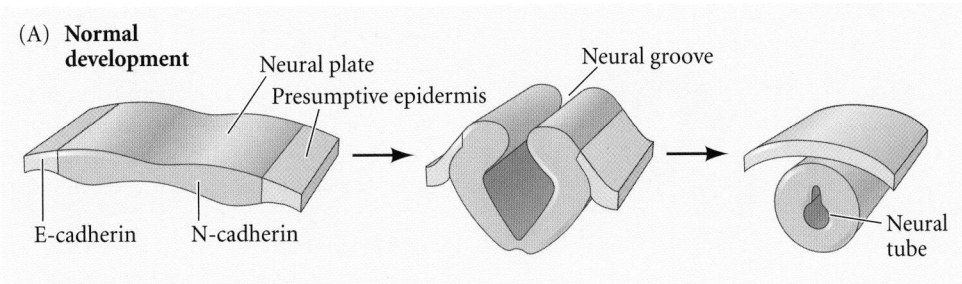
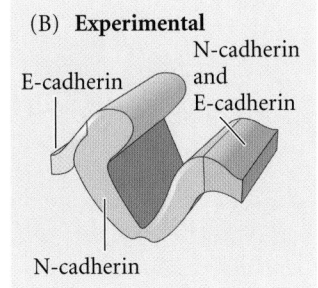

FIGURE 10.7 Expression of N- and E-cadherin adhesion proteins during neurulation in *Xenopus*. (A) Normal development. In the neural plate stage, N-cadherin is seen in the neural plate, while E-cadherin is seen on the presumptive epidermis. Eventually, the N-cadherin-bearing neural cells separate from the E-cadherin-containing epidermal cells. (The neural crest cells express neither N- nor E-cadherin, and they disperse.) (B) No separation of the neural tube occurs when one side of the frog embryo is injected with N-cadherin mRNA, so that N-cadherin is expressed in the epidermal cells as well as in the presumptive neural tube.

FIGURE 10.8 Secondary neurulation in the caudal region of a 25-somite chick embryo. (A) Mesenchymal cells condense to form the medullary cord at the most caudal end of the chick tailbud. (B) The medullary cord at a slightly more anterior position in the tailbud. (C) The neural tube is cavitating and the notochord forming; note the presence of separate lumens. (D) The lumens coalesce to form the central canal of the neural tube. (From Catala et al. 1995; photographs courtesy of N. M. Le Douarin.)

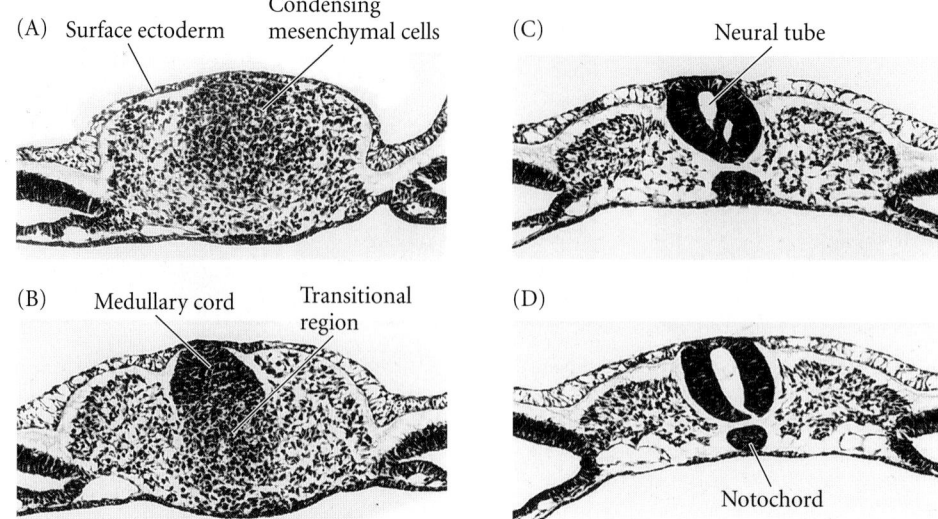

this mesenchymal-epithelial transition, the central portion of this cord undergoes cavitation to form several hollow spaces, or **lumens** (**FIGURE 10.8C**); the lumens then coalesce into a single central cavity (**FIGURE 10.8D**; Schoenwolf and Delongo 1980).

We have seen that after Hensen's node has migrated to the posterior end of the embryo, the caudal region of the epiblast contains a precursor cell population that gives rise to both neural ectoderm and paraxial (somite) mesoderm as the embryo's trunk elongates (Tzouanacou et al. 2009). The ectodermal cells that will form the posterior (secondary) neural tube express the *Sox2* gene, whereas the ingressing mesodermal cells (which no longer encounter high levels of BMPs as they migrate beneath the epiblast) do not express *Sox2*. Rather, they express *Tbx6* and form somites (Shimokita and Takahashi 2010; Takemoto et al. 2011). The ability of the Tbx6 transcription factor to repress neural-inducing *Sox2* expression explains the bizarre phenotype of homozygous *Tbx6* mouse mutants, which have three neural tubes (**FIGURE 10.9**; Chapman and Papaioannou 1998; Takemoto et al. 2011). In these mutants, the two rods of paraxial mesoderm have become neural tubes. Thus, the epiblast surrounding the rostral primitive streak (the caudal lateral epiblast; see Chapter 9) contains a common precursor pool for paraxial mesoderm and for the neural plate that forms the caudal hindbrain and spinal cord (Cambray and Wilson 2007; Wilson et al. 2009).

In human and chick embryos, there appears to be a transitional region at the junction of the anterior (primary) and posterior (secondary) neural tubes. In human embryos, coalescing cavities are seen in the transitional region, but the neural tube also forms by the bending of neural plate cells. Some posterior neural tube anomalies result when the two regions of the neural tube fail to coalesce (Saitsu et al. 2007). Given the prevalence of human posterior spinal cord malformations, further understanding of the mechanisms of secondary neurulation may have important clinical implications.

(A) Wild-type

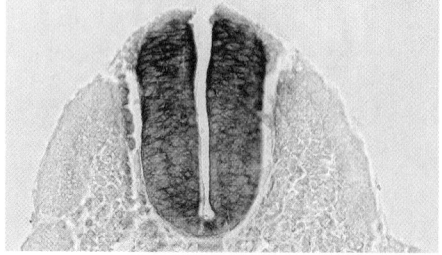

(B) *Tbx6* mutant

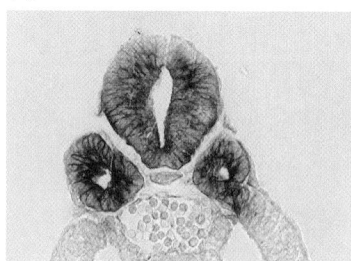

FIGURE 10.9 Homozygous *Tbx6* mutant mice have three neural tubes posterior to somite 6 level. (A) Wild-type mouse with a single Sox2-expressing neural tube flanked by mesodermal somites. (B) In mutant mice lacking the *Tbx6* gene, the ingressing cells remain neural (i.e., they continue to express Sox2) and form two paraxial neural tubes as well as the normal central neural tube. (From Takemoto et al. 2011; photographs courtesy of H. Kondoh.)

Building the Brain

The construction of an organ that perceives, thinks, loves, hates, remembers, changes, deceives itself, and coordinates all of our conscious and unconscious bodily processes is undoubtedly the most challenging of all developmental enigmas. A combination of genetic, cellular, and organismal approaches is now giving us a very preliminary understanding of how the basic anatomy of the brain becomes ordered.

Differentiation of the neural tube into the various regions of the brain and spinal cord occurs simultaneously in three different ways. On the gross anatomical level,

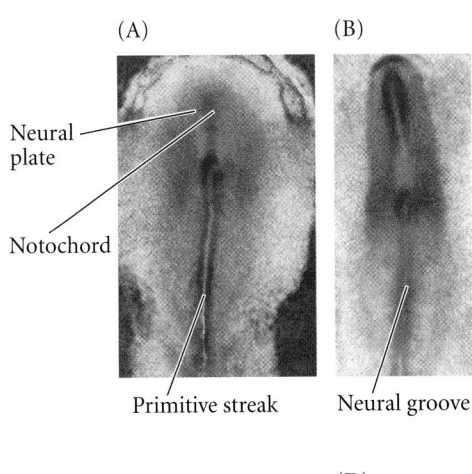

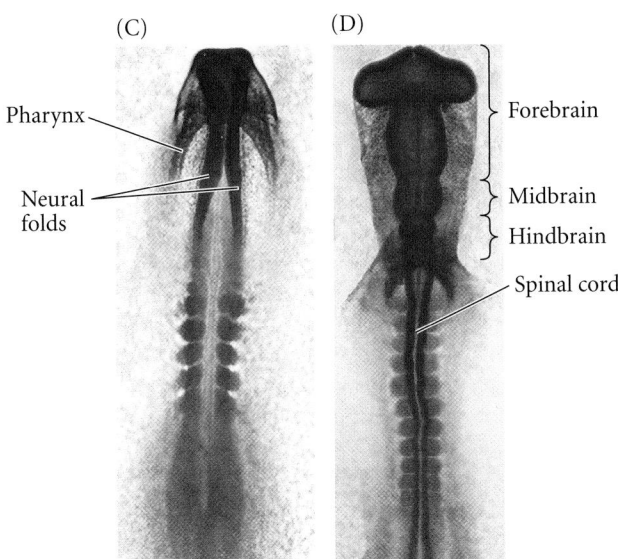

the neural tube and its lumen bulge and constrict to form the chambers of the brain and spinal cord. At the tissue level, the cell populations in the wall of the neural tube arrange themselves into the different functional regions of the brain and spinal cord. Finally, on the cellular level, the neuroepithelial cells themselves differentiate into the numerous types of nerve cells (**neurons**) and associative cells (**glia**) present in the body.

The anterior-posterior axis

The early development of most vertebrate brains is similar (**FIGURE 10.10A–D**). But because the human brain may be the most organized piece of matter in the solar system and is arguably the most interesting organ in the animal kingdom, we will concentrate on the development that is supposed to make *Homo* sapient.

The early mammalian neural tube is a straight structure. However, even before the posterior portion of the tube has formed, the most anterior portion of the tube is undergoing

FIGURE 10.10 Early brain development and formation of the first brain chambers. (A–D) Chick brain development. (A) Flat neural plate with underlying notochord (head process). (B) Neural groove. (C) Neural folds begin closing at the dorsalmost region, forming the incipient neural tube. (D) Neural tube, showing the three brain regions and the spinal cord. The neural tube remains open at the anterior end, and the optic bulges (which become the retinas) have extended to the lateral margins of the head. (E) In humans, the three primary brain vesicles become further subdivided as development continues. At the right is a list of the adult derivatives formed by the walls and cavities of the brain, along with some of their functions. (A–D courtesy of G. C. Schoenwolf; E after Moore and Persaud 1993.)

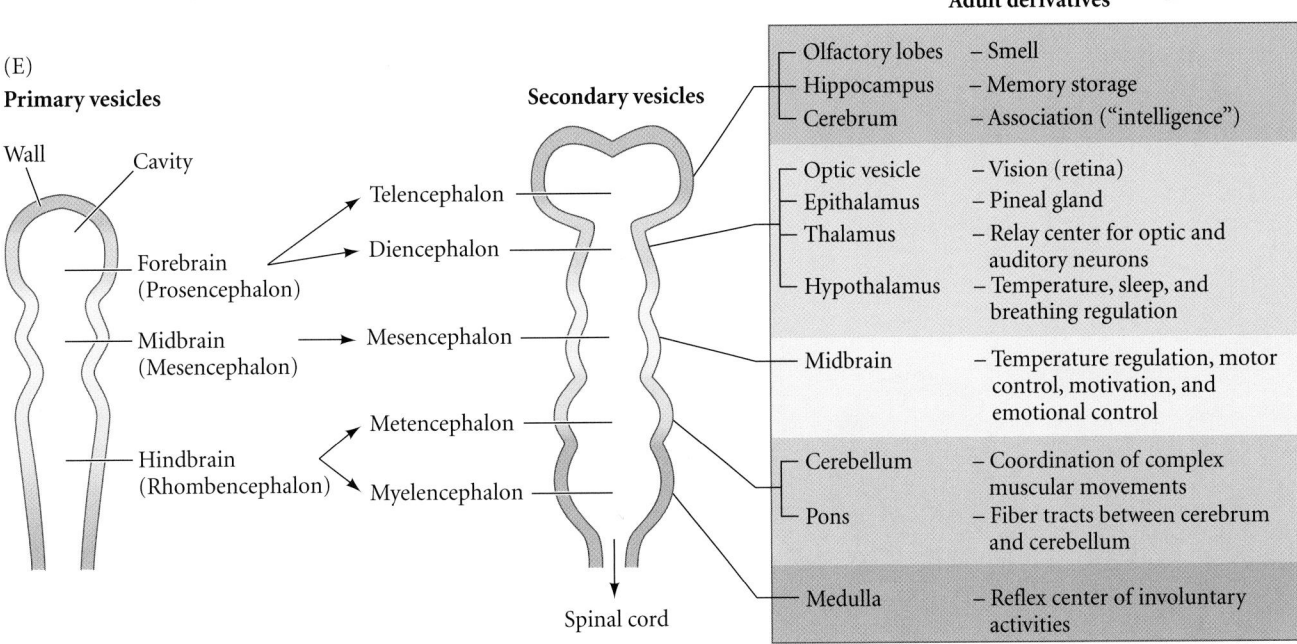

Adult derivatives

Olfactory lobes	– Smell
Hippocampus	– Memory storage
Cerebrum	– Association ("intelligence")
Optic vesicle	– Vision (retina)
Epithalamus	– Pineal gland
Thalamus	– Relay center for optic and auditory neurons
Hypothalamus	– Temperature, sleep, and breathing regulation
Midbrain	– Temperature regulation, motor control, motivation, and emotional control
Cerebellum	– Coordination of complex muscular movements
Pons	– Fiber tracts between cerebrum and cerebellum
Medulla	– Reflex center of involuntary activities

(A)   (B)

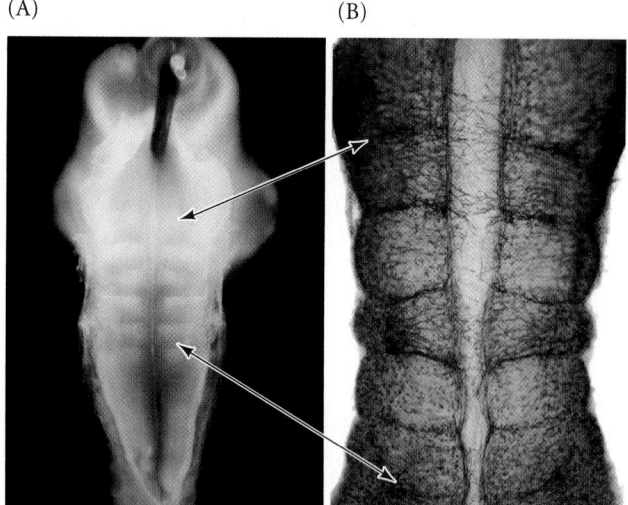

FIGURE 10.11   Rhombomeres of the chick hindbrain. (A) Hindbrain of a 3-day chick embryo. The roof plate has been removed so that the segmented morphology of the neural epithelium can be seen. The r1/r2 boundary is at the upper arrow, and the r6/r7 boundary is at the lower arrow. (B) A chick hindbrain at a similar stage stained with antibody to a neurofilament subunit. The rhombomere boundaries are emphasized because they serve as channels for neurons crossing from one side of the brain to the other. (From Lumsden 2004, courtesy of A. Lumsden.)

drastic changes. In the anterior region, the neural tube balloons into the three primary vesicles: the forebrain (**prosencephalon**), which forms the cerebral hemispheres; the midbrain (**mesencephalon**), whose neurons are involved in motivation, movement, and depression (Niwa et al. 2013; Tye et al. 2013); and the hindbrain (**rhombencephalon**),

which becomes the cerebellum and the medulla oblongata (the most primitive area of the brain and the center of involuntary activities such as breathing) (**FIGURE 10.10E**). By the time the posterior end of the neural tube closes, secondary vesicles have formed. The forebrain becomes the telencephalon (which forms the cerebral hemispheres) and the diencephalon (which will form the optic vesicle that initiates eye development).

The rhombencephalon develops a segmental pattern that specifies the places where certain nerves originate. Periodic swellings called **rhombomeres** divide the rhombencephalon into smaller compartments. The rhombomeres represent separate "territories" in that the cells within each rhombomere mix freely within it but do not mix with cells from adjacent rhombomeres (Guthrie and Lumsden 1991; Lumsden 2004). Each rhombomere expresses a unique combination of transcription factors, thereby generating rhombomere-specific patterns of neuronal differentiation. Thus, each rhombomere produces neurons with different fates. As we will see in Chapter 11, the neural crest cells derived from the rhombomeres will form **ganglia**—clusters of neuronal cell bodies whose axons form a nerve. Each of these rhombomeric ganglia produces a different type of nerve. The generation of the cranial nerves from the rhombomeres has been studied most extensively in the chick, in which the first neurons appear in the even-numbered rhombomeres, r2, r4, and r6 (**FIGURE 10.11**; Lumsden and Keynes 1989). Neurons originating from r2 ganglia form the fifth (trigeminal) cranial nerve; those from r4 form the seventh (facial) and eighth (vestibuloacoustic) cranial nerves; and those from r6 form the ninth (glossopharyngeal) cranial nerve.

● See **WEBSITE 10.1** Specifying the brain boundaries

(A)   (B)   (D)

(C)

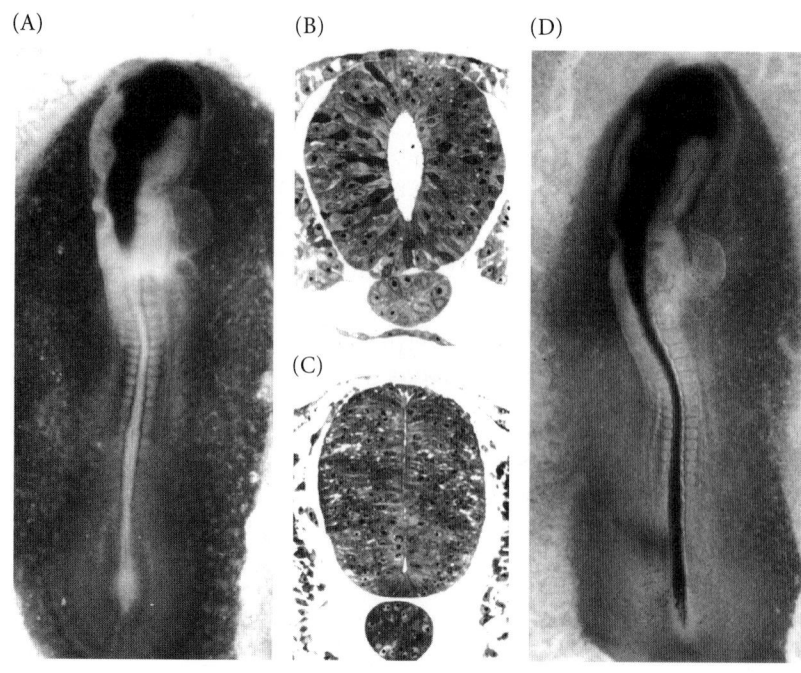

FIGURE 10.12   Occlusion of the neural tube allows expansion of the future brain region. (A) Dye injected into the anterior portion of a 3-day chick neural tube fills the brain region but does not pass into the spinal region. (B,C) Sections of the chick neural tube at the base of the brain before occlusion (B) and during occlusion (C). (D) Reopening of the occlusion after initial brain enlargement allows dye to pass from the brain region into the spinal cord region. (Courtesy of M. Desmond.)

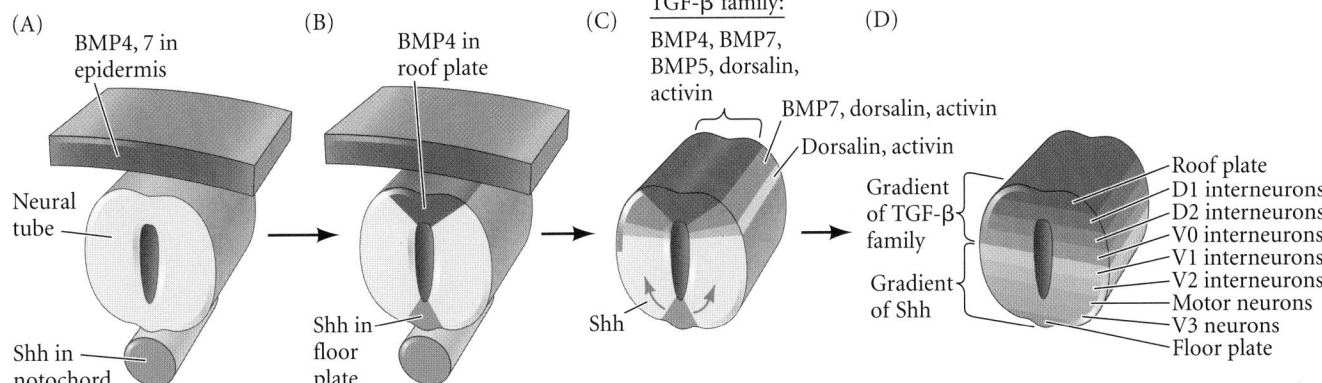

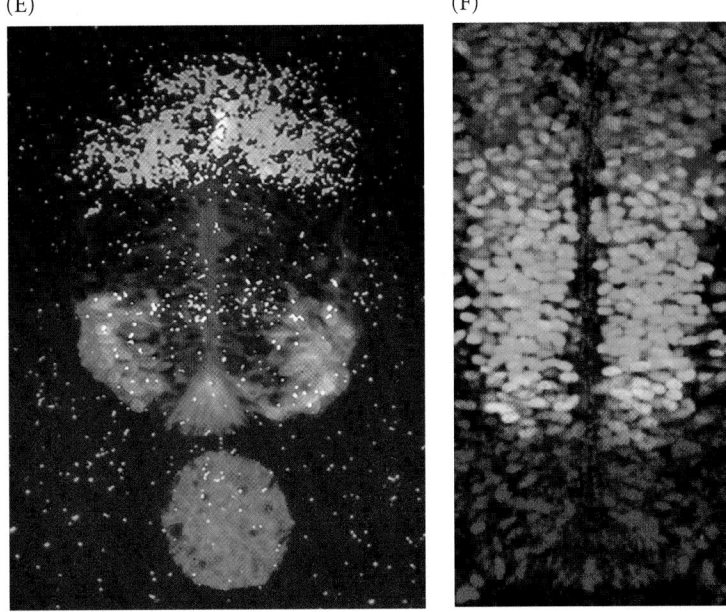

FIGURE 10.13 Dorsal-ventral specification of the neural tube. (A) The newly formed neural tube is influenced by two signaling centers. The roof of the neural tube is exposed to BMP4 and BMP7 from the epidermis, and the floor of the neural tube is exposed to Sonic hedgehog protein (Shh) from the notochord. (B) Secondary signaling centers are established in the neural tube. BMP4 is expressed and secreted from the roof plate cells; Sonic hedgehog is expressed and secreted from the floor plate cells. (C) BMP4 establishes a nested cascade of TGF-β factors spreading ventrally into the neural tube from the roof plate. Sonic hedgehog diffuses dorsally as a gradient from the floor plate cells. (D) The neurons of the spinal cord are given their identities by their exposure to these gradients of paracrine factors. The amounts and types of paracrine factors present cause different transcription factors to be activated in the nuclei of these cells, depending on their position in the neural tube. (E) Chick neural tube showing areas of Sonic hedgehog (green) and the expression domain of the protein dorsalin (blue; dorsalin is a member of the TGF-β superfamily). Motor neurons induced by a particular concentration of Sonic hedgehog are stained orange/yellow. (F) In situ hybridization for three other transcription factors: Pax7 (blue, characteristic of the dorsal neural tube cells), Pax6 (green), and Nkx6.1 (red). Where Nkx6.1 and Pax6 overlap (yellow), the motor neurons become specified. (E from Jessell 2000, courtesy of T. M. Jessell; F courtesy of J. Briscoe.)

The physical division of the prospective brain from the prospective spinal cord is done by occluding the lumen of the neural tube at the boundary between these regions. As the neural folds close in the region between the presumptive brain and spinal cord, the surrounding dorsal tissues push inward to constrict the neural tube at the base of the brain (**FIGURE 10.12**; Desmond 1982; Desmond and Schoenwolf 1986). The prospective brain cells then secrete cerebrospinal fluid into the lumen, expanding the embryonic brain's volume enormously. In chicks and zebrafish, if the fluid pressure in the anterior portion of an occluded neural tube is experimentally decreased, the brain forms abnormally and contains many fewer cells than normal (Desmond and Levitan 2002; Lowery and Sive 2005). The occluded region of the neural tube reopens after the initial rapid enlargement of the brain ventricles. The anterior-posterior patterning of the hindbrain and spinal cord is controlled by a series of genes that include the Hox gene complexes.

The dorsal-ventral axis

The neural tube is polarized along its dorsal-ventral axis. In the spinal cord, for instance, the dorsal region is the place where the spinal neurons receive input from sensory neurons, whereas the ventral region is where the motor neurons reside. In the middle are numerous interneurons that relay information between the sensory and motor neurons.

The dorsal-ventral polarity of the neural tube is induced by signals coming from its immediate environment. The ventral pattern is imposed by the notochord, while the dorsal pattern is induced by the overlying epidermis (**FIGURE 10.13A–D**). Specification of the axis is initiated by two major paracrine factors: Sonic hedgehog protein (Shh) originating from the notochord, and TGF-β proteins originating in the dorsal ectoderm (**FIGURE 10.13E,F**). In both cases, these factors induce a second signaling center within the neural tube itself. Sonic hedgehog is secreted from the notochord and induces the medial hinge point cells to become the **floor plate** of the neural tube. The floor plate cells also secrete Sonic hedgehog, and this paracrine factor from the floor plate cells forms a gradient that is highest at the most ventral

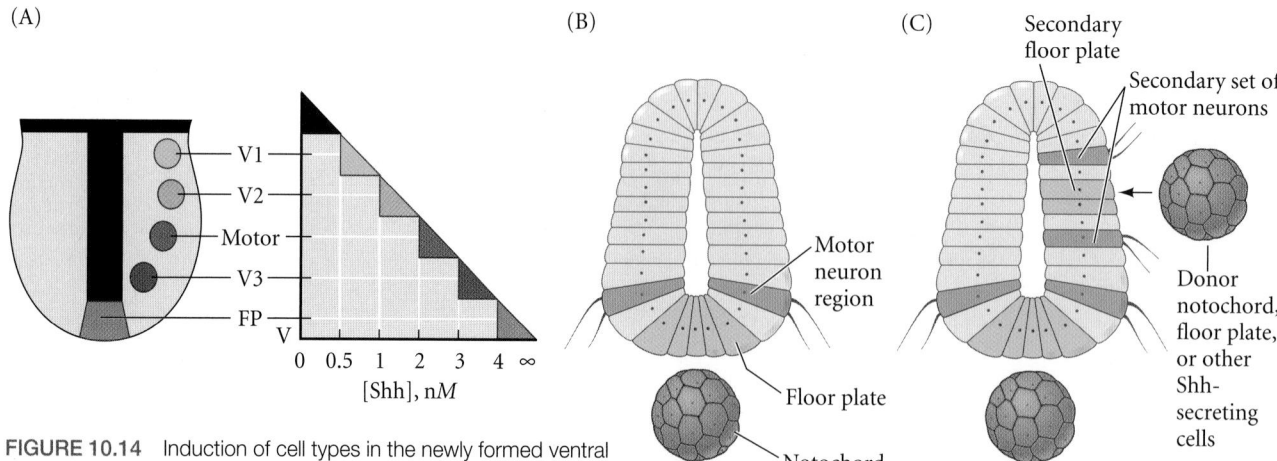

(A)

(B)

(C)

Secondary floor plate

Secondary set of motor neurons

Motor neuron region

Floor plate

Notochord

Donor notochord, floor plate, or other Shh-secreting cells

FIGURE 10.14 Induction of cell types in the newly formed ventral neural tube. (A) Relationship between Sonic hedgehog concentrations, the generation of particular neuronal types in vitro, and distance from the notochord. (B) Cells closest to the notochord become the floor plate neurons; motor neurons emerge on the ventrolateral sides. (C) If a second notochord, floor plate, or any other Sonic hedgehog-secreting cell is placed adjacent to the neural tube, it induces a second set of floor plate neurons, as well as two other sets of motor neurons. (A after Briscoe et al. 1999; B,C after Placzek et al. 1990.)

portion of the neural tube (**FIGURE 10.14A**; Roelink et al. 1995; Briscoe et al. 1999).

The importance of Sonic hedgehog in patterning the ventral portion of the neural tube has been confirmed by experiments that again demonstrate the principles of "find it, lose it, move it" (see p. 136). If a piece of notochord is removed from an embryo, the neural tube adjacent to the deleted region will have no floor plate cells (Placzek et al. 1990). Moreover, if notochord fragments are taken from one embryo and transplanted to the lateral side of a host neural tube, the host neural tube will form another set of floor plate cells at its sides

(**FIGURE 10.14B,C**). These ectopic floor plate cells induce the formation of motor neurons on either side of them. The same results can be obtained if the notochord fragments are replaced by pellets of cultured cells secreting Sonic hedgehog (Echelard et al. 1993).

The dorsal fates of the neural tube are established by proteins of the TGF-β superfamily, especially BMPs 4 and 7, dorsalin, and activin (Liem et al. 1995, 1997, 2000). Initially, BMP4 and BMP7 are found in the epidermis. Just as the notochord establishes a secondary signaling center—the floor plate cells—on the ventral side of the neural tube, the epidermis establishes a secondary signaling center by inducing BMP4 expression in the **roof plate** cells of the neural tube. The BMP4 protein from the roof plate induces a cascade of TGF-β proteins in adjacent cells (see Figure 10.13C). Dorsal sets of cells are thus exposed to higher concentrations of TGF-β proteins, and at earlier times, when compared with the more ventral neural cells. The

(A)

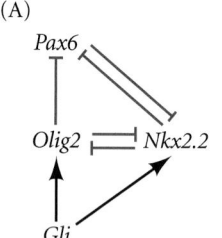

(B)

FIGURE 10.15 Model for interpreting the Shh morphogen gradient. (A) The gene regulatory network (GRN) combining the Gli (activated by the Hedgehog pathway), Olig2, Nkx2.2, and Pax6 transcription factors. (B) Signal-mediated patterning of the ventral portion of the neural tube. At the earliest time of induction (t_0–t_1), Shh from the notochord (green triangles) induces Gli (purple) in the floor plate cells. This is not sufficient to activate Olig2 or Nkx2.2, or to repress Pax6. As development ensues, Gli is able to induce Olig2, which inhibits Nkx2.2 and Pax6. As the most ventral cells experience higher concentrations of Shh for longer periods, Nkx2.2 is activated and suppresses Olig2. This pattern can be retained even when Gli levels decrease. (After Balaskas et al. 2012.)

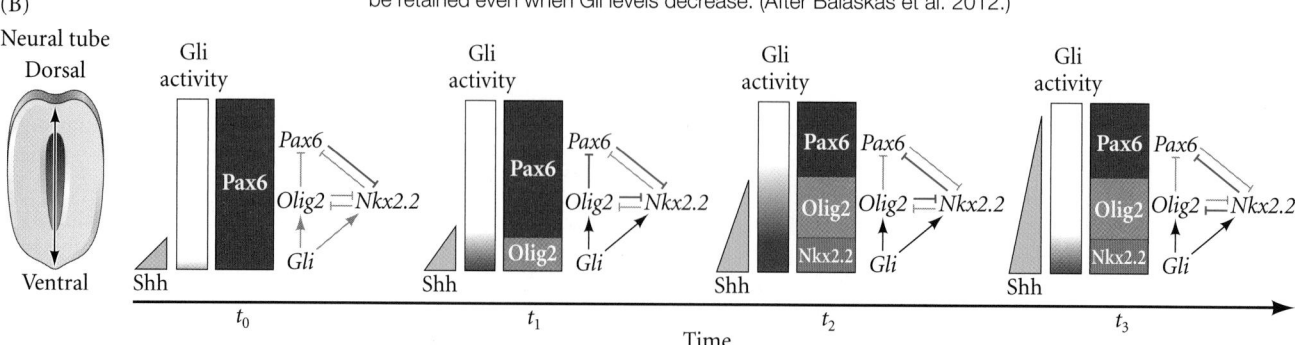

importance of the TGF-β superfamily factors in patterning the dorsal portion of the neural tube was demonstrated by the phenotypes of zebrafish mutants. Those mutants deficient in certain BMPs lacked dorsal and intermediate types of neurons (Nguyen et al. 2000).

The fate of a particular cell in the neural tube depends on its distance from these signaling centers. Cells adjacent to the floor plate that receive high concentrations of Sonic hedgehog synthesize the Nkx6.1 and Nkx2.2 transcription factors and become the ventral (V3) neurons. The cells dorsal to these, exposed to slightly less Sonic hedgehog (and slightly more TGF-β factors), produce Pax6 and Olig2 and become the motor neurons. The next two groups of cells, receiving progressively less Sonic hedgehog, express Pax6 alone and become the V2 and V1 interneurons (see Figure 10.14; Lee and Pfaff 2001; Muhr et al. 2001).

It had been thought that the intersecting gradients of Shh and TGF-β signals would be sufficient to instruct the synthesis of different transcription factors. However, the regulatory network is far more complex and integrates both temporal and spatial characters. It integrates four feedforward (positive) loops and seven feedback loops into a coherent framework that enables the cells to express different transcription factors as the Shh gradient grows and stabilizes the factors once the gradient has been completed (**FIGURE 10.15**).

Differentiation of Neurons in the Brain

The human brain consists of more than 10^{11} neurons associated with perhaps 10 times that many glial cells (Noctor et al. 2007). The brain contains a wide variety of neuronal and glial cell types, from the relatively small (e.g., granule cells) to the comparatively enormous (e.g., Purkinje neurons). In generating this diversity, the multipotent neuroepithelial cells of the neural tube give rise to three main cell types.

1. They may become **ventricular (ependymal) cells** that remain integral components of the neural tube lining and that secrete the cerebrospinal fluid.

2. They may generate precursors of the *neurons* that conduct electric potentials and coordinate our bodily functions, our thoughts, and our sensations of the world.

3. They may give rise to precursors of the *glial cells* that aid in constructing the nervous system, provide insulation around the neurons, and may be important in memory storage.

The differentiation of precursor cells is believed to be largely determined by the environment they enter (Rakic and Goldman 1982). At least in some cases, a given neuroepithelial cell can give rise to both neurons and glia (Turner and Cepko 1987).

ANATOMY OF THE NEURON The fine, branching extensions of the neuron that are used to pick up electric impulses from other cells are called **dendrites** (**FIGURE 10.16**). Some neurons develop only a few dendrites, whereas others (such as the Purkinje neurons) develop extensive, branching *dendritic arbors*. Very few dendrites are found on cortical neurons at birth, and one of the amazing events of the first year of human life is the increase in the number of these receptive processes. During this year, each cortical neuron develops enough dendritic surface to accommodate as many as 100,000 connections, or **synapses**, with other neurons. The average cortical neuron connects with 10,000 other neural cells, enabling the human cortex to function as the center for learning, reasoning, and memory; to develop the capacity for symbolic expression; and to interpret and produce voluntary responses to sensory stimuli (such as light, sound, and touch).

Another important feature of a developing neuron is its **axon**. Whereas dendrites are often numerous and do not extend far from the neuronal cell body, or **soma**, axons may extend 2–3 feet (see Figure 10.16). The pain receptors on your big toe, for example, must transmit their messages all the way to your spinal cord. One of the fundamental concepts of neurobiology is that the axon is a continuous extension of the nerve cell body. At the beginning of the twentieth century, there were many competing theories of axon formation. Theodor Schwann, one of the founders of the cell theory, believed that numerous neural cells linked themselves together in a chain to form an axon. Viktor Hensen, the discoverer of the embryonic node, thought that the axon formed around preexisting cytoplasmic threads between the cells. Wilhelm His (1886) and Santiago Ramón y Cajal (1890) postulated that the

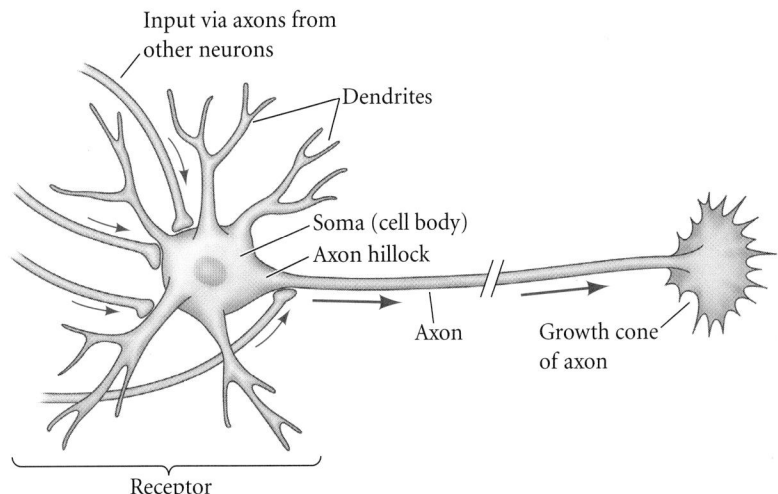

FIGURE 10.16 A motor neuron. Electric impulses (red arrows) are received by the dendrites, and the stimulated neuron transmits impulses through its axon to its target tissue. The axon (which may be 2–3 feet long) is a cellular process through which the neuron sends its signals. The growth cone of the axon is both a locomotor and a sensory apparatus that actively explores the environment and picks up directional cues telling it where to go. Eventually the growth cone will form a connection, or synapse, with the axon's target tissue.

axon was an outgrowth (albeit an extremely large one) of the neuron's soma.

AXON OUTGROWTH In 1907, Ross Harrison demonstrated the validity of the outgrowth theory in an elegant experiment that founded both the science of developmental neurobiology and the technique of tissue culture. Harrison isolated a portion of the neural tube from a 3-mm frog tadpole. (At this stage, shortly after the closure of the neural tube, there is no visible differentiation of axons.) He placed this neuroblast-containing tissue in a drop of frog lymph on a coverslip and inverted the coverslip over a depression slide so he could watch what was happening within this "hanging drop." What Harrison saw was the emergence of axons as outgrowths from the neuroblasts, elongating at about 56 mm per hour.

Nerve outgrowth is led by the tip of the axon, called the **growth cone** (FIGURE 10.17A). The growth cone does not proceed in a straight line but rather "feels" its way along the substrate. The growth cone moves by the elongation and contraction of pointed filopodia called **microspikes** (FIGURE 10.17B). These microspikes contain microfilaments, which are oriented parallel to the long axis of the axon. (This mechanism is similar to that seen in the filopodial microfilaments of secondary mesenchyme cells in echinoderms; see Chapter 7.) Treating neurons with cytochalasin B destroys the actin microspikes, inhibiting their further advance (Yamada et al. 1971; Forscher and Smith 1988). Within the axon itself, structural support is provided by microtubules, and the axon will retract if the neuron is placed in a solution of colchicine (an inhibitor of microtubule polymerization). Thus, the developing neuron displays the same mechanisms we noted in the dorsolateral hinge points of the neural tube: elongation by microtubules and apical shape changes by microfilaments.

As in most migrating cells, the exploratory microspikes of the growth cone attach to the substrate and exert a force that pulls the rest of the cell forward. Axons will not grow if the growth cone fails to advance (Lamoureux et al. 1989).

In addition to their structural role in axonal migration, the microspikes also have a sensory function. Fanning out in front of the growth cone, each microspike samples the microenvironment and sends signals back to the soma (Davenport et al. 1993). As we will see in Chapter 11, microspikes are the fundamental organelles involved in neuronal pathfinding.

NEURONAL SIGNALING Neurons transmit information via electric impulses that travel from one region of the body to another along the axons. To prevent dispersal of the electric signal and to facilitate conduction to its target cell, the axon is insulated at intervals by glial cells. Within the central nervous system, axons are insulated at intervals by processes that originate from a type of glial cell called an **oligodendrocyte**. The oligodendrocyte wraps itself around the developing axon, then produces a specialized cell membrane called a **myelin sheath**. In the peripheral nervous system, myelination is accomplished by a glial cell type called the **Schwann cell** (FIGURE 10.18). Transplantation experiments have shown that the axon, and not the glial cell, controls the thickness of the myelin sheath. Mikhailov and colleagues (2004) have demonstrated that sheath diameter is regulated by the amount of neuregulin-1 secreted by the axon.

The myelin sheath is essential for proper nerve function, and also helps keep the axons alive for decades. The demyelination of nerve fibers is associated with convulsions, paralysis, and certain debilitating afflictions such as multiple sclerosis (Emery 2010; Nave et al. 2010). There are also mouse mutants in which subsets of neurons are poorly myelinated. In the *trembler* mutant, the Schwann cells are unable to produce a particular protein component such that myelination is deficient in the peripheral nervous system but normal in the central nervous system. Conversely, in the mouse mutant *jimpy*, the central nervous system is deficient in myelin while the peripheral nerves are unaffected (Sidman et al. 1964; Henry and Sidman 1988).

A variety of different molecules, known as **neurotransmitters**, are critical in generating many action potentials. Axons are specialized for secreting specific neurotransmitters across

(A)

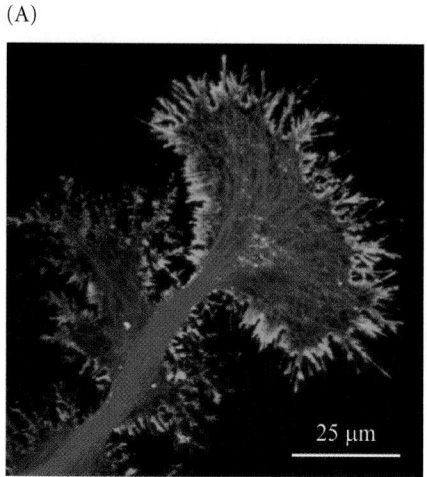

25 μm

(B)

Microspikes

Growth cone

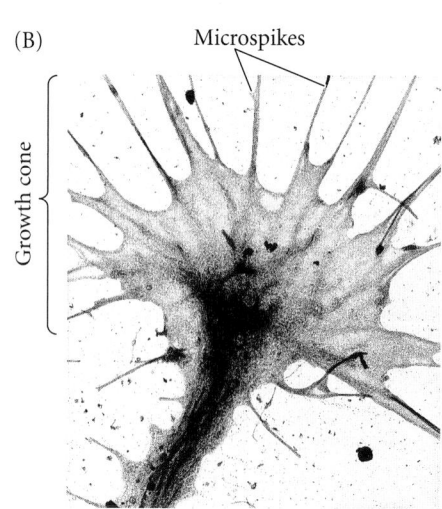

FIGURE 10.17 Axon growth cones. (A) Growth cone of the hawkmoth *Manduca sexta* during axon extension and pathfinding. The actin in the filopodia is stained green with fluorescent phalloidin, while the microtubules are stained red with a fluorescent antibody to tubulin. (B) Actin microspikes in an axon growth cone, seen by transmission electron microscopy. (A courtesy of R. B. Levin and R. Luedemanan; B from Letourneau 1979.)

(A)

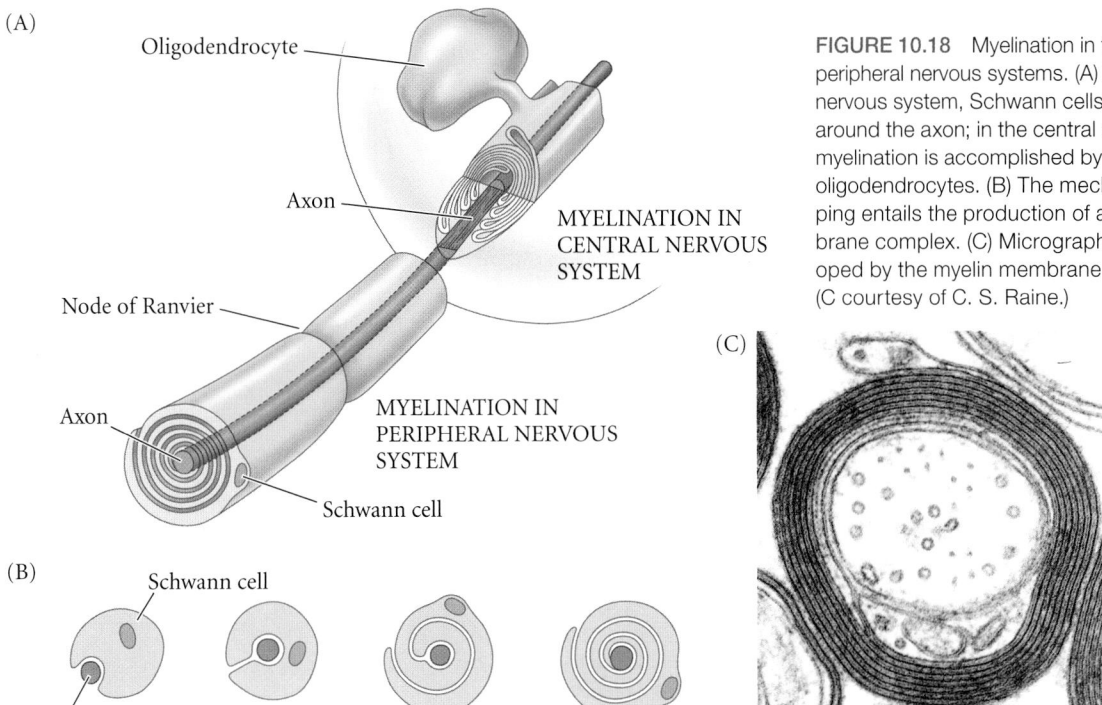

Oligodendrocyte

Axon

MYELINATION IN
CENTRAL NERVOUS
SYSTEM

Node of Ranvier

Axon

MYELINATION IN
PERIPHERAL NERVOUS
SYSTEM

Schwann cell

(B)

Schwann cell

Axon

(C)

FIGURE 10.18 Myelination in the central and peripheral nervous systems. (A) In the peripheral nervous system, Schwann cells wrap themselves around the axon; in the central nervous system, myelination is accomplished by the processes of oligodendrocytes. (B) The mechanism of this wrapping entails the production of an enormous membrane complex. (C) Micrograph of an axon enveloped by the myelin membrane of a Schwann cell. (C courtesy of C. S. Raine.)

a small gap—the **synaptic cleft**—that separates the axon of a signaling neuron from the axon or surface of its target cell. Some neurons develop the ability to synthesize and secrete acetylcholine, whereas others develop the enzymatic pathways for making and secreting epinephrine, norepinephrine, octopamine, glutamate, serotonin, γ-aminobutyric acid (GABA), or dopamine, among other neurotransmitters. Each neuron must activate those genes responsible for making the enzymes that can synthesize its neurotransmitter. Thus, neuronal development involves both structural and molecular differentiation. We will discuss the regeneration of neurons and their axons at some length in Chapter 16, but glial cells are probably very important in permitting or preventing axon regeneration.

Tissue Architecture of the Central Nervous System

The neurons of the brain are organized into layers (**laminae**) and clusters (**nuclei***), each having different functions and connections. The original neural tube is composed of a **germinal neuroepithelium**—a layer of rapidly dividing neural stem cells one cell layer thick. Sauer and colleagues (1935) showed that the cells of the germinal neuroepithelium are continuous from the luminal surface of the neural tube to the outside surface, but that the *cell* nuclei are at different

---

* In neuroanatomy, the term *nucleus* refers to an anatomically discrete collection of neurons within the brain that typically serves a specific function. Note that this is a completely distinct structure from the *cell nucleus* (a term that appears later in this same paragraph).

heights, giving the superficial impression that the neural tube has numerous cell layers (**FIGURE 10.19A**). The nuclei move within their cells as they progress through the cell cycle. DNA synthesis (S phase) occurs while the nucleus is at the outside edge of the neural tube, and the nucleus migrates toward the lumen as the cell cycle proceeds.

Neural stem cell differentiation

Mitosis occurs on the luminal side of the cell layer. When mammalian neural tube cells are labeled with radioactive thymidine during early development, 100% of them incorporate this base into their DNA (Fujita 1964). Shortly thereafter, however, certain cells stop incorporating these DNA precursors, indicating that they are no longer participating in DNA synthesis and mitosis. These cells then migrate away from the lumen of the neural tube and differentiate into neuronal and glial cells (Fujita 1966; Jacobson 1968). The division of these neural progenitor cells, or **neuroblasts**, at the lumen regulates the exposure of these cells to different paracrine environments along the anterior-posterior axis of the neural tube, enabling different regions to have varying percentages of differentiating and proliferative cells (Taverna and Huttner 2010).

If the labeled progeny of dividing cells are found in the outer cortex in the adult brain, then those neurons must have migrated to their cortical positions from the germinal neuroepithelium. When a cell of the germinal neuroepithelium is ready to generate neurons (instead of more neural stem cells), the plane of cell division shifts. Instead of having both cells attached to the luminal surface, one of the two daughter cells remains in the epithelium while the other becomes detached. The cell connected to the luminal surface usually remains

(A)

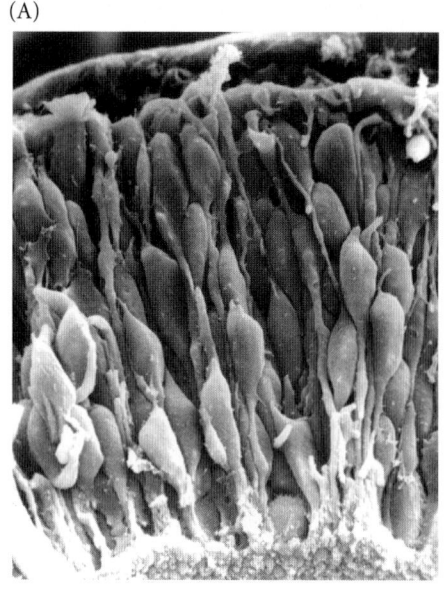

(B)

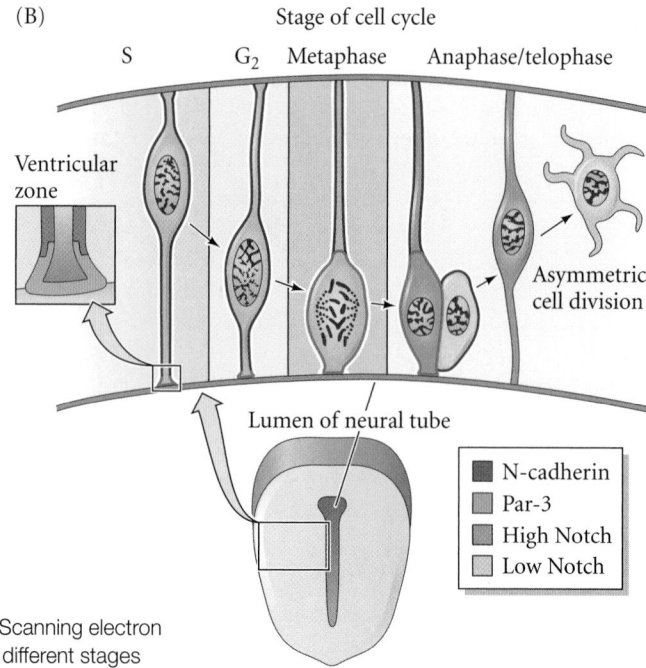

FIGURE 10.19 Neural stem cells in the germinal epithelium. (A) Scanning electron micrograph of a newly formed chick neural tube, showing cells at different stages of their cell cycles. (B) Schematic section of a chick embryo neural tube, showing the position of the nucleus and Par-3 protein in a neuroepithelial cell as a function of the cell cycle. Mitotic cells are found near the inner surface of the neural tube, adjacent to the lumen. The dynamic distribution of Par-3 protein in these luminal stem cells regulates the synthesis of Notch signaling pathway components in the cell membrane of the daughter cells. At mitosis, Par-3 (green) becomes localized primarily to one of the two daughter cells. That daughter cell will express Notch and remain a stem cell; the cell receiving less Par-3 will express less Notch and become a neuroblast (neural progenitor cell). After producing neurons, neuroblasts switch to the production of glial cells. (A courtesy of K. Tosney; B after Bultje et al. 2009 and Liu et al. 2011.)

a stem cell, while the other cell migrates and differentiates (Chenn and McConnell 1995; Hollyday 2001). The first progenitor cells leaving the lumen are neuroblasts. After a time, differentiation will lead to glioblasts (glial progenitor cells; see Kageyama et al. 2008).

This shift from *symmetric horizontal* division (yielding two neural stem cells attached to the lumen) to *asymmetric vertical* division (yielding a luminal stem cell and a neural progenitor cell; see Figure 10.19B) is mediated by interactions involving the cytoskeleton, Par proteins, and the cerebrospinal fluid. The mammalian homologue of the *C. elegans* PAR-3 protein is responsible (as in the nematode) for maintaining apical-basal polarity of the cells. In the developing brain, Par-3 recruits a complex in the apical portion of the cell that contains receptors for cell division molecules (such as insulin-like growth factors, FGFs, and Sonic hedgehog) that are present in the cerebrospinal fluid (Lehtinen et al. 2011). This promotes cell proliferation. Par-3 also regulates the placement of the Notch signaling proteins to different parts of the stem cell. When the cell divides at the lumen, one daughter cell receives more Par-3 protein than the other daughter cell (**FIGURE 10.19B**). The daughter cell receiving more Par-3 develops high Notch

signaling activity and remains a stem cell. The other daughter cell expresses high amounts of the Delta protein (Delta being the Notch receptor) and becomes primed for neuronal differentiation (Bultje et al. 2009).

This vertical division is the last time the migrating cell will divide and is called that neuron's "birthday." Different types of neurons and glial cells have birthdays at different times. Labeling cells at different times during development shows that the cells with the earliest birthdays migrate the shortest distances; those with later birthdays migrate to form the more superficial regions of the brain cortex. Subsequent differentiation depends on the positions the neurons occupy once outside the germinal neuroepithelium (Letourneau 1977; Jacobson 1991).

As the neural tube matures, the progeny of the neuro-epithelial stem cells become **radial glial cells** (**RGCs**). For decades it has been known that RGCs guide neural progenitor cell migration from the inner (luminal) region to the outer zones (Rakic 1971). However, it was only in this century that cell fate studies (using time-lapse movies of cells infected with a retrovirus expressing GFP) demonstrated that RGCs are in fact neural stem cells that undergo symmetrical and asymmetrical divisions (Malatesta et al. 2000; Miyata et al. 2001; Noctor et al. 2001). Thus, the neural progenitor cells formed as the progeny of stem cells can actually use their "sister" stem cell's connection between luminal and outer surfaces to migrate to their appropriate positions. The asymmetric division takes place in the **ventricular zone** (the zone forming the ventricle and touching the cerebrospinal fluid). As the progenitor cells migrate away, they form a **subventricular zone** immediately adjacent to the ventricular zone containing the stem cells (FIGURE 10.20). In the subventricular zone, each of the progenitor cells divides symmetrically. Thus, these neural progenitors are transit amplifying cells. In mice, each of the neural progenitor cells usually undergoes only one symmetrical division to produce two neurons; but in humans and other primates, these transit amplifying cells probably undergo several more divisions (Noctor et al. 2004; Liu et al. 2011).

As the stem cells adjacent to the lumen continue to divide, the migrating cells form a second layer around the original neural tube. This layer becomes progressively thicker as more cells are added to it from the germinal neuroepithelium. This new layer is called the **cortical mantle** (or **intermediate**) **zone**. The mantle zone cells differentiate into both neurons and glia. The neurons make connections among themselves and send forth axons away from the lumen, thereby creating a cell-poor **marginal zone**. Eventually glial cells cover many of the axons in the marginal zone in myelin sheaths, giving them a whitish appearance. Hence, the axonal marginal layer is often called **white matter**, while the mantle zone, containing the neuronal cell bodies, is referred to as **gray matter**. The germinal epithelium of the ventricular zone will later shrink to become the **ependyma** that lines the brain cavity.

Spinal cord and medulla organization

In the spinal cord and medulla, the basic three-zone pattern of ventricular (ependymal), mantle, and marginal layers is retained throughout development. When viewed in

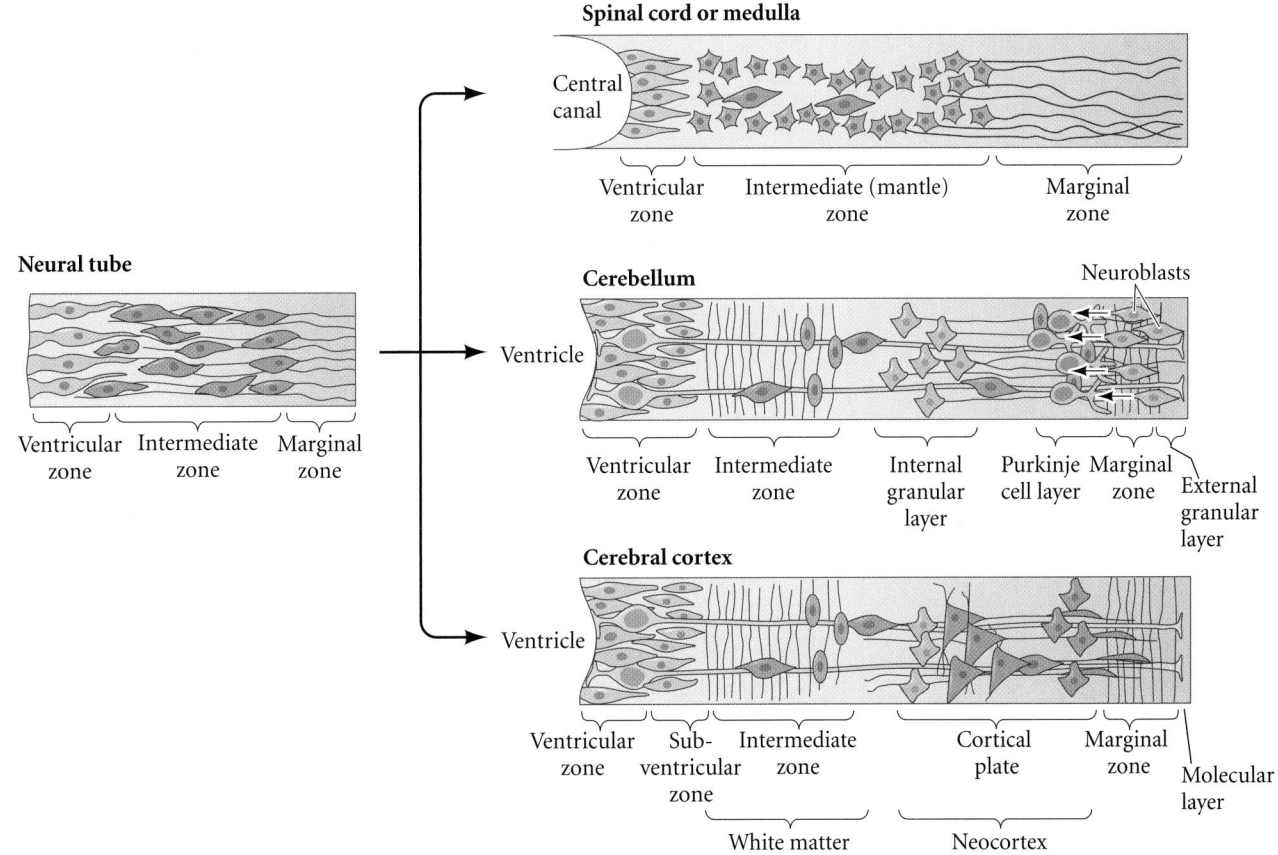

FIGURE 10.20 Differentiation of the walls of the neural tube. A section of a 5-week human neural tube (left) reveals three zones: ventricular (ependymal), intermediate (mantle), and marginal. In the spinal cord and medulla (top row), the ventricular zone remains the sole source of neurons and glial cells. In the cerebellum (middle row), a second mitotic layer, the external granular layer, forms at the region farthest removed from the ventricular zone. Neuroblasts from this layer migrate back into the intermediate zone to form the internal granular layer. In the cerebral cortex (bottom row), the migrating neuroblasts and glioblasts form a cortical plate containing six layers. (After Jacobson 1991.)

cross section, the gray matter (mantle) gradually becomes a butterfly-shaped structure surrounded by white matter, and both become encased in connective tissue. As the neural tube matures, a longitudinal groove—the **sulcus limitans**—divides it into dorsal and ventral halves. The dorsal portion receives input from sensory neurons, whereas the ventral portion is involved in effecting various motor functions (**FIGURE 10.21**). This developmental anatomy generates the basis of medullary and spinal cord physiology (such as the reflex arch).

Cerebellar organization

In the brain, cell migration, differential neuronal proliferation, and selective cell death produce modifications of the three-zone pattern seen in Figure 10.20. Cerebellar development results in a highly folded cortex composed of Purkinje neurons and granule neurons integrated into "nuclei" that control balance functions and relay information from the cerebellar cortex to other brain regions. In the development of the cerebellum, the critical event appears to be the migration of the neural progenitor cells (i.e., neuroblasts) to the outer surface of the developing cerebellum. Here, they form a new germinal zone, the **external granular layer** (see Figure 10.20), near the outer boundary of the neural tube.

At the outer boundary of the external granular layer, which is 1–2 cells thick, neuroblasts proliferate and come into contact with cells that secrete BMPs. The BMPs specify the postmitotic products of these neuroblasts to become a type of neuron

called **granule cells** (Alder et al. 1999). Granule cells migrate back toward the ventricular (ependymal) zone, where they produce a region called the **internal granular layer**. Meanwhile, the original ventricular zone of the cerebellum generates a wide variety of neurons and glial cells, including the distinctive and large **Purkinje neurons**, the major cell type of the cerebellum (**FIGURE 10.22**). Purkinje neurons secrete Sonic hedgehog, which sustains the division of granule cell precursors in the external granular layer (Wallace 1999). Each Purkinje neuron has an enormous **dendritic arbor** that spreads like a tree above a bulblike cell body (see Figure 10.22B). A typical Purkinje neuron may form as many as 100,000 synapses with other neurons—more connections than any other type of neuron studied. Each Purkinje neuron also sends out a slender axon, which connects to neurons in the deep cerebellar nuclei.

Purkinje neurons are critical in the electrical pathway of the cerebellum. All electric impulses eventually regulate the activity of these neurons, which are the only output neurons of the cerebellar cortex. For this to happen, the proper cells must differentiate at the appropriate place and time. How is this accomplished?

One mechanism thought to be important for positioning young neurons in the developing mammalian brain is **glial guidance** (Rakic 1972; Hatten 1990). Throughout the cortex, neurons are seen to ride a "glial monorail" to their respective destinations. In the cerebellum, the granule cell precursors travel on the long processes of the **Bergmann glia**, a type of glial cell that extends a thin process throughout the

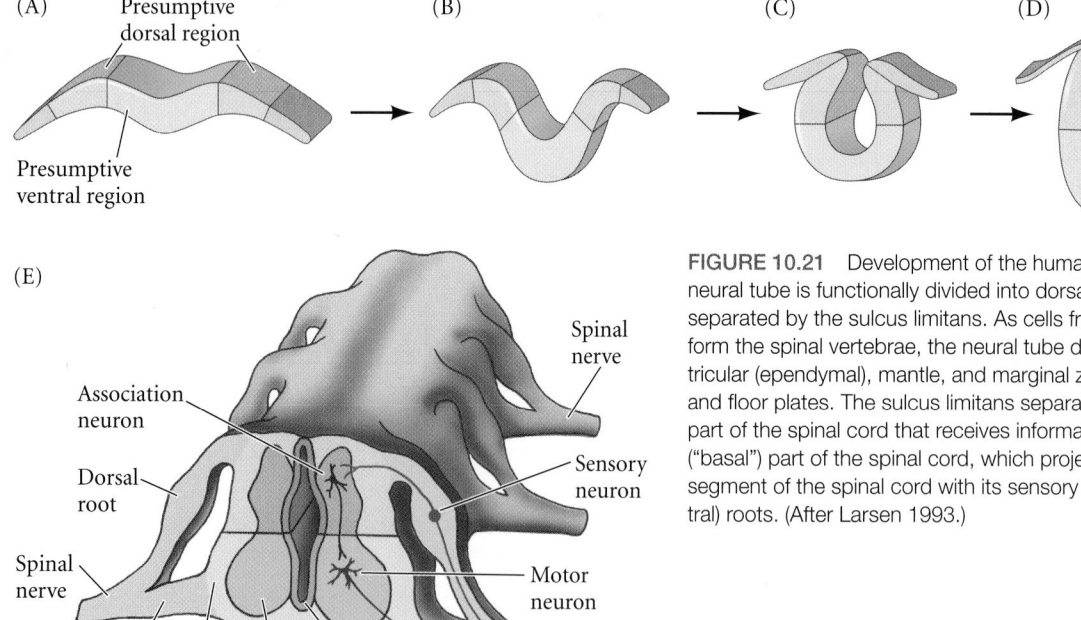

FIGURE 10.21   Development of the human spinal cord. (A–D) The neural tube is functionally divided into dorsal and ventral regions, separated by the sulcus limitans. As cells from the adjacent somites form the spinal vertebrae, the neural tube differentiates into the ventricular (ependymal), mantle, and marginal zones, as well as the roof and floor plates. The sulcus limitans separates the dorsal ("alar") part of the spinal cord that receives information from the ventral ("basal") part of the spinal cord, which projects motor neurns. (E) A segment of the spinal cord with its sensory (dorsal) and motor (ventral) roots. (After Larsen 1993.)

(A)

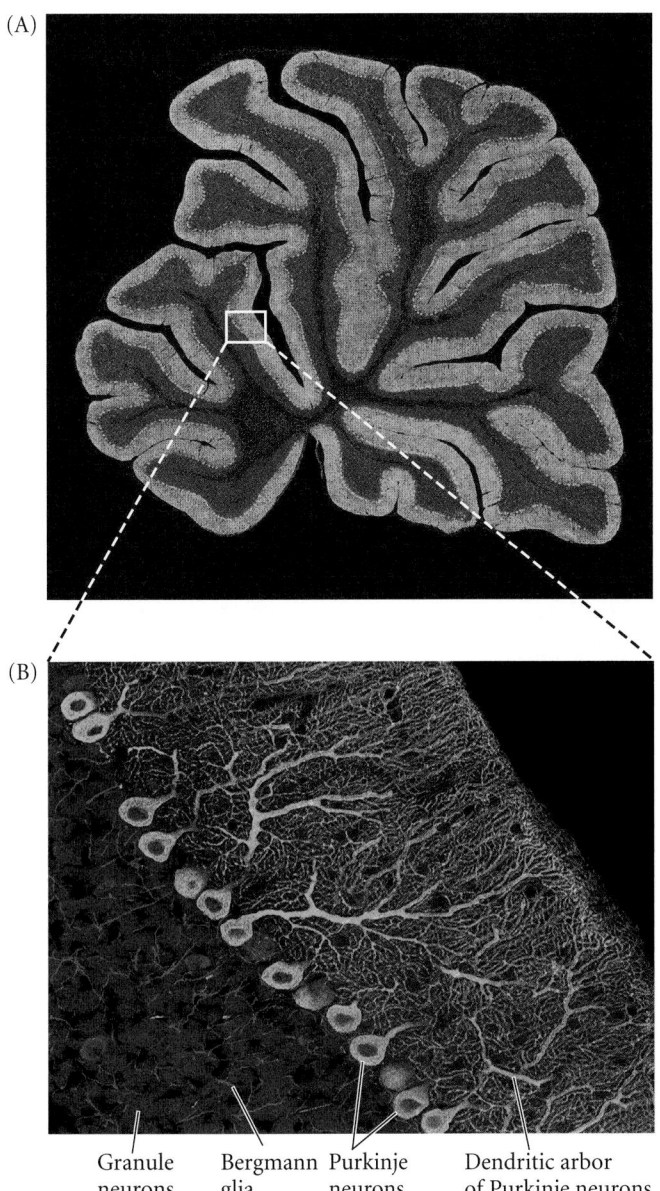

(B)

| Granule neurons | Bergmann glia | Purkinje neurons | Dendritic arbor of Purkinje neurons |

FIGURE 10.22 Cerebellar organization. (A) Sagittal section of a fluorescently labeled rat cerebellum photographed using dual-photon confocal microscopy. (B) A vast enlargement of one area of (A). Purkinje neurons are light blue with bright green processes, Bergmann glia are red, and granule cells are dark blue. This close-up illustrates the highly structured organization of neurons and glial cells. (Courtesy of T. Deerinck and M. Ellisman, University of California, San Diego.)

germinative neuroepithelium (see Figure 10.22B; Rakic and Sidman 1973; Rakic 1975). As **FIGURE 10.23** illustrates, this neuron-glia interaction is a complex and fascinating series of events, involving reciprocal recognition between glia and neuroblasts (Hatten 1990; Komuro and Rakic 1992). It appears that migration of the newborn neuroblasts involves

the loss of adhesion molecules linking the neuroblast to the germinal layer cells and the acquisition of a set of adhesion molecules that attach it to the glia (Famulski et al. 2010). The molecules involved in this adhesion were discovered through a number of mouse mutants that could not keep their balance (and were given names such as *reeler, staggerer*, and *weaver* that reflected their movement problems.) In *reeler* brains, glial cells lack the extracellular matrix protein, reelin, that permits the neuroblasts to bind them. Another adhesion protein, astrotactin, is needed by the granule cell neurons to maintain their adhesion to the glial process. If the astrotactin on a neuron is masked by antibodies to that protein, the neuron will fail to adhere to the glial processes (Edmondson et al. 1988; Fishell and Hatten 1991). The direction of this migration appears to be regulated by a complex series of events orchestrated by brain-derived neurotrophic factor (BDNF), a paracrine factor made by the internal granular layer (Zhou et al. 2007).

Cerebral organization

The three-zone arrangement of the neural tube is also modified in the cerebrum. The cerebrum is organized in two distinct ways. First, like the cerebellum, it is organized vertically into layers that interact with one another. Certain neuroblasts from the mantle zone migrate on glial processes that extend through the white matter to generate a second zone of neurons at the outer surface of the brain. This new layer of gray matter will become the **neocortex**. The specification of the neocortex is accomplished largely through the Lhx2 transcription factor, which activates numerous other cerebral genes. In *Lhx2*-deficient mice, the cerebral cortex fails to form (Mangale et al. 2008; Chou et al. 2009).

The neocortex eventually stratifies into six layers of neuronal cell bodies; the adult forms of these layers are not completed until the middle of childhood. Each layer of the neocortex differs from the others in its functional properties, the types of neurons found there, and the sets of connections they make. For instance, neurons in layer 4 receive their major input from the thalamus (a region that forms from the diencephalon), whereas neurons in layer 6 send their major output back to the thalamus.

In addition to the six vertical layers, the cerebral cortex is organized horizontally into more than 40 regions that regulate anatomically and functionally distinct processes. For instance, neurons of the visual cortex in layer 6 project axons to the lateral geniculate nucleus of the thalamus, which is involved in vision, while neurons of the auditory cortex of layer 6 (located more anteriorly than the visual cortex) project axons to the medial geniculate nucleus of the thalamus, which functions in hearing. One of the major questions in developmental neurobiology is whether the different functional regions of the cerebral cortex are already specified in the ventricular region, or if specification is accomplished much later by the synaptic connections between the regions. Evidence that specification is early (and that there might be

(A)

Leading process
of neuron

Migrating
neuron

Process of
glial cell

(B)

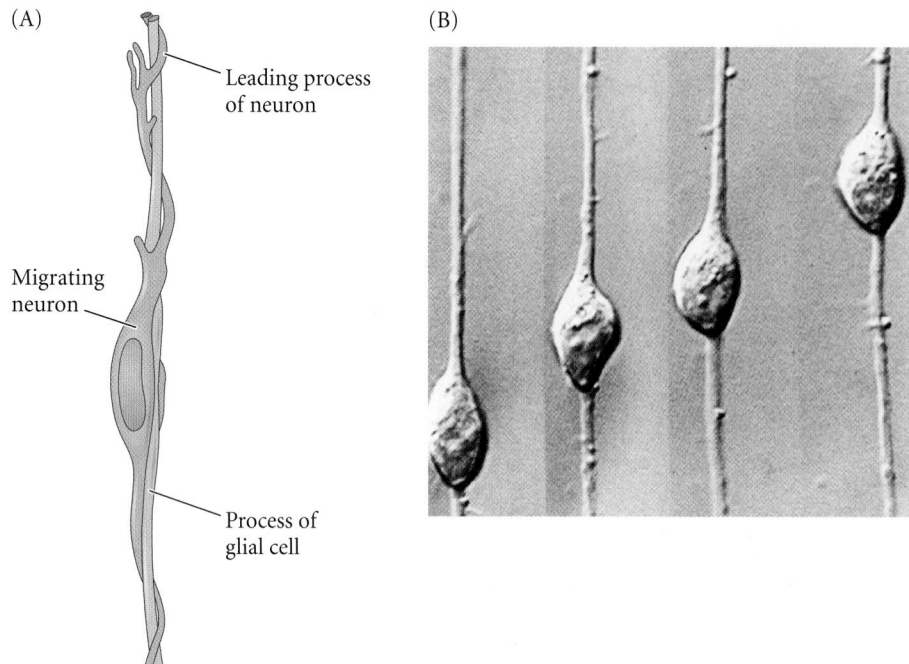

FIGURE 10.23 Neuron-glia interaction in the mouse. (A) Diagram of a cortical neuron migrating on a glial cell process. (B) Sequential photographs of a neuron migrating on a cerebellar glial process. The leading process has several filopodial extensions. The neuron can reach speeds of about 40 mm per hour as it travels. (A after Rakic 1975; B from Hatten 1990, photograph courtesy of M. Hatten.)

some "proto-map" of the cerebral cortex) is suggested by certain human mutations that destroy the layering and functional abilities in only one part of the cortex, leaving the other regions intact (Piao et al. 2004).

Indeed, most of the neuroblasts generated in the ventricular zone migrate outward along radial glial processes to form the **cortical plate** at the outer surface of the brain. As in the rest of the brain, those neurons with the earliest birthdays form the layer closest to the ventricle. Subsequent neurons travel greater distances to form the more superficial layers of the cortex. This process forms an "inside-out" gradient of development (Rakic 1974). McConnell and Kaznowski (1991) have shown that the determination of laminar identity (i.e., which layer a cell migrates to) is made during the final cell division. Newly generated neuronal precursors transplanted after this last division from young brains (where they would form layer 6) into older brains whose migratory neurons are forming layer 2 are committed to their fate, and migrate only to layer 6. However, if these cells are transplanted prior to their final division (during mid-S phase), they are uncommitted and can migrate to layer 2 (**FIGURE 10.24**). The fates of neuronal precursors from older brains are more fixed. The neuronal precursor cells formed early in development have the potential to become any neuron (at layers 2 or 6, for instance); later precursor cells give rise only to upper-level

(layer 2) neurons (Frantz and McConnell 1996). Once the cells arrive at their final destination, it is thought that they produce particular adhesion molecules that organize them together as brain nuclei (Matsunami and Takeichi 1995).

As summarized by Gaiano (2008):

The construction of the mammalian neocortex is perhaps the most complex biological process that occurs in nature. A pool of seemingly homogeneous stem cells first undergoes proliferative expansion and diversification and then initiates the production of successive waves of neurons. As these neurons are generated, they take up residence in the nascent cortical plate where they integrate into the developing neocortical circuitry. The spatial and temporal coordination of neuronal generation, migration, and differentiation is tightly regulated and of paramount importance to the creation of a mature brain capable of processing and reacting to sensory input from the environment and of conscious thought.

And—as if this weren't enough—after neurogenesis subsides, the stem cells of the vertebrate cortex start making glia.

Retroviral tracers and time-lapse microcinematography have shown that individual precursor cells make both neurons and glia (Walsh and Reid 1995; Qian et al. 2000; Shen et al. 2006). In mice, neurons are formed from embryonic days 12 through 18. Then, at embryonic day 18, the same precursor cells generate glia. When cortical progenitor cells (such as radial glia and their more committed descendants) are cultured on embryonic day 12, they differentiate into neurons for the first few days but after multiple days switch and start making glia (Götz and Barde 2005). Both internal and external factors regulate this transition. Cortical precursor cells form neurons when cultured on embryonic cortical slices, but they become glia when placed on older

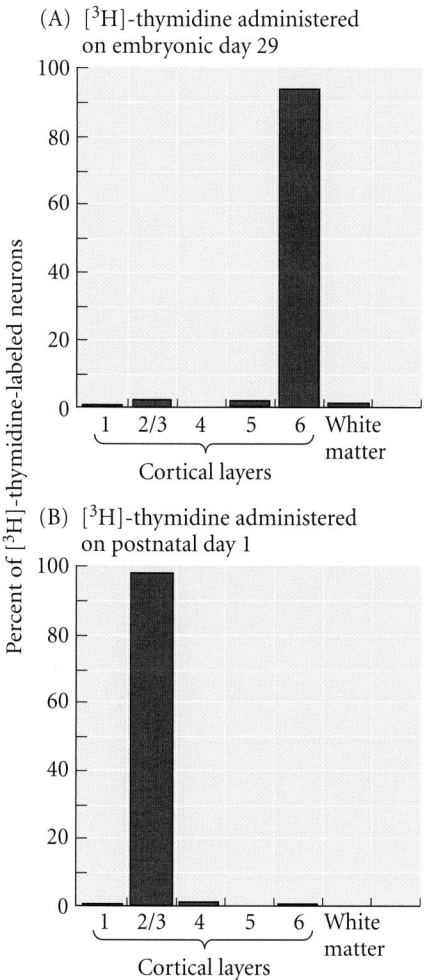

(A) [³H]-thymidine administered on embryonic day 29

(B) [³H]-thymidine administered on postnatal day 1

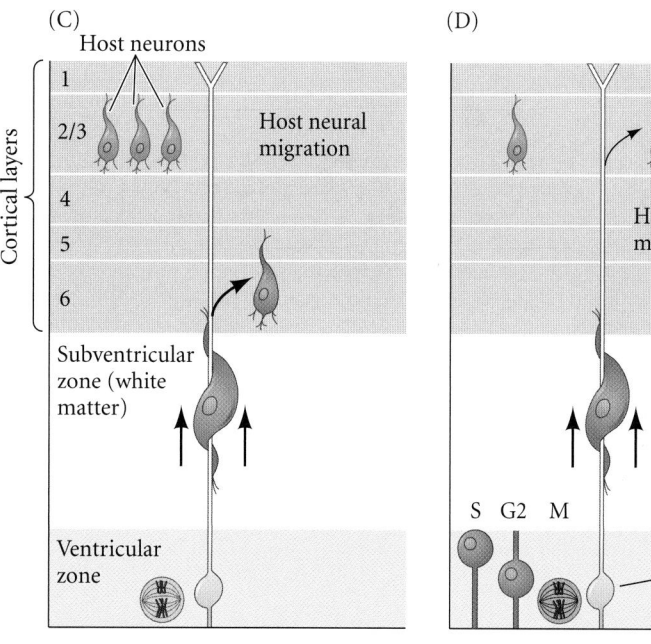

FIGURE 10.24   Determination of cortical laminar identity in the ferret cerebrum. (A) "Early" neuronal precursors (birthdays on embryonic day 29) migrate to layer 6. (B) "Late" neuronal precursors (birthdays on postnatal day 1) migrate farther, into layers 2 and 3. (C) When early neuronal precursors (dark blue) are transplanted into older ventricular zones after their last mitotic S phase, the neurons they form migrate to layer 6. (D) If these precursors are transplanted before or during their last S phase, however, they migrate (with the host neurons) to layer 2. (After McConnell and Kaznowski 1991.)

cortical slices (Morrow et al. 2001). One of the main factors involved in this environmental regulation is CT-1, a paracrine factor that activates the JAK-STAT pathway. The STAT transcription factors activate glial-specific genes. Moreover, this transition must be done in the absence of neurotrophic factors, as these activate the RTK-MAPK pathway, resulting in neuron-specific gene expression (see Miller and Gauthier 2007).

STEM AND PRECURSOR CELLS OF THE NEOCORTEX   The stem cells of the neocortex are originally generated by symmetrical division of the neural tube epithelium before neurogenesis (neuron formation) actually occurs. Initially, the proliferative layer of the mouse cortex forms the ventricular zone (VZ). Shortly thereafter (around day 13), it divides to give rise to a subventricular zone (SVZ) directly outside it. Together, these zones form the germinal strata that generate the neuroblasts (neuronal precursor cells) that migrate into the cortical plate and form the layers of neurons. The VZ will form the lower (deeper) layers of neurons, while the SVZ will give rise to those cells that form the upper layer of neurons (Frantz et al. 1994).

There are two major progenitor cells in the germinal strata: radial glial cells (RGCs) and intermediate progenitor cells (IPCs; **FIGURE 10.25**). Radial glial cells (which are both stem cells and can act as glial support cells) are found in the ventricular zone. At each division, they generate another radial glial cell and a more committed cell type. Interestingly, at each division, the cell receiving the "old" centriole (which contains different proteins than the newly made centriole) stays in the VZ, while the cell receiving the "young" centriole leaves to differentiate (Wang et al. 2009). The more committed cells can be either neuroblasts (which divide to generate neurons) or IPCs, which migrate to the SVZ, where they generate neuroblasts. A single stem cell in the ventricular layer can give rise to neurons (and glial cells) in any of the cortical layers (Walsh and Cepko 1988).

The switch from progenitor cell to a migrating neuroblast is thought to be mediated by the phosphorylation of the DISC1 protein. In its unphosphorylated form, DISC1 promotes cell proliferation. Once phosphorylated, however, it blocks cell division and switches the cytoskeleton into that of migrating cells (Ishizuka et al. 2011). Progenitor cells containing variants of the DISC1 protein that cannot

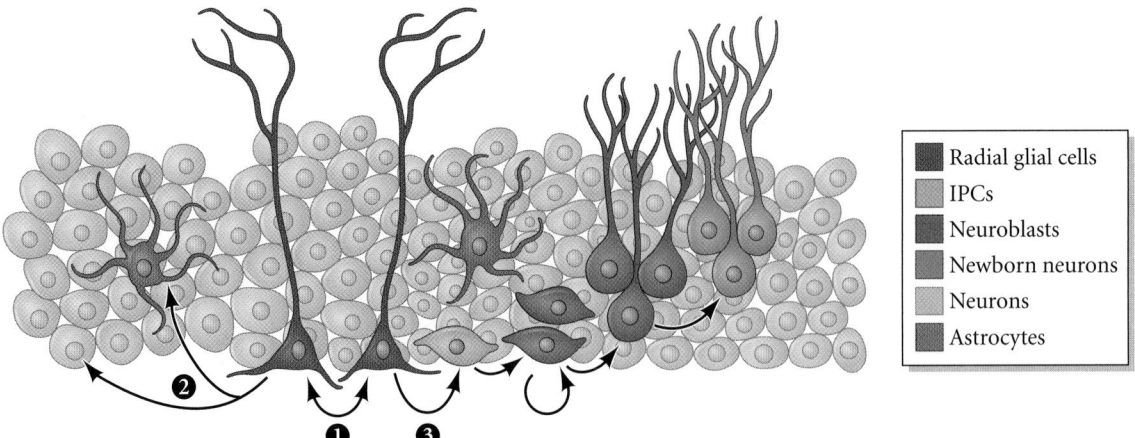

FIGURE 10.25 Possible model for neurogenesis and gliogenesis from hippocampal stem cells (radial glial cells) in the hippocampus. (1) The radial glial cell can be self-renewing and divide symmetrically to produce more stem cells. (2) Alternatively, an RGC can generate both adult neurons and adult glial cells (astrocytes). One labeled cell can create a clone containing both neurons and glia. (3) The production of neurons may require an intermediate progenitor cell (IPC) that gives rise to the transit amplifying neuroblasts. (After Taylor 2011.)

be phosphorylated are able to proliferate but cannot migrate, and aberrations in the DISC1 protein have been associated with a range of mental illnesses.*

● **See WEBSITE 10.2** Horizontal and vertical specification of the cerebrum

Adult Neural Stem Cells

Until recently, it was generally believed that once the mammalian nervous system was mature, no new neurons were "born"—in other words, the neurons formed in utero and during the first few years of life were all we could ever expect to have. The good news from recent studies, however, is that the adult brain *is* capable of producing new neurons, and environmental stimulation can increase the number of these new neurons.

In these experiments, researchers injected adult mice, rats, and marmosets with bromodeoxyuridine (BrdU), a nucleoside that resembles thymidine. BrdU is incorporated into a cell's DNA only if the cell is undergoing DNA replication; therefore, any cell labeled with BrdU must have been undergoing DNA synthesis during the time it was exposed to BrdU. This labeling technique revealed that adult mice produce *thousands* of new neurons each day. Moreover, these new brain cells integrated with other cells of the brain, had normal neuronal morphology, and exhibited action potentials (**FIGURE 10.26A,B**; van Praag et al. 2002). These new cells turn out to be radial glial cells similar to the cells that produced the brain during embryogenesis,

and like the original radial glial stem cells, they are self-renewing and multipotent (i.e., capable of forming both neurons and glia; Bonagudi et al. 2011).

Injecting humans with BrdU is usually unethical, since large doses of BrdU are often lethal. However, in certain cancer patients, the progress of chemotherapy is monitored by transfusing the patient with a small amount of BrdU. Gage and colleagues took postmortem samples from the brains of five such patients who died between 16 and 781 days after the BrdU infusion (see Erikkson et al. 1998). In all five subjects, they saw labeled (new) neurons in the granular cell layer of the hippocampal dentate gyrus (a part of the brain where memories may be formed). The BrdU-labeled cells also stained for neuron-specific markers (**FIGURE 10.26C**). Thus, although the rate of new neuron formation in adulthood may be relatively low, the human brain is not an anatomical *fait accompli* at birth, or even after childhood.

Production of neurons in adults appears to be limited to (1) the subventricular zone (adjacent to the ventricular zone) along the walls of the lateral ventricles and (2) certain regions of the hippocampus (Kempermann et al. 1997a,b; Kornack and Rakic 1999; van Praag et al. 1999; Ihrie and Alvarez-Buylla 2011). In the subventricular zone, the type of neuron produced is determined by the paracrine factors that are secreted by other cells in the neighborhood. Sonic hedgehog, produced by a small group of neurons in the ventral forebrain, is an especially important paracrine factor in producing ventral neuronal types (such as those whose defects cause Parkinson disease). If these Shh-secreting neurons are ablated, no new ventral neurons form; and if Shh is ectopically placed near the dorsal portions of the subventricular zone, those new neuroblasts are transformed into ventral neuroblasts (Ihrie et al. 2011).

---

*DISC1 mutations are not the cause of most cases of these diseases, but they are seen in the rare cases where these behavioral diseases are transmitted as Mendelian traits (see Marx 2007).

(A) (B)

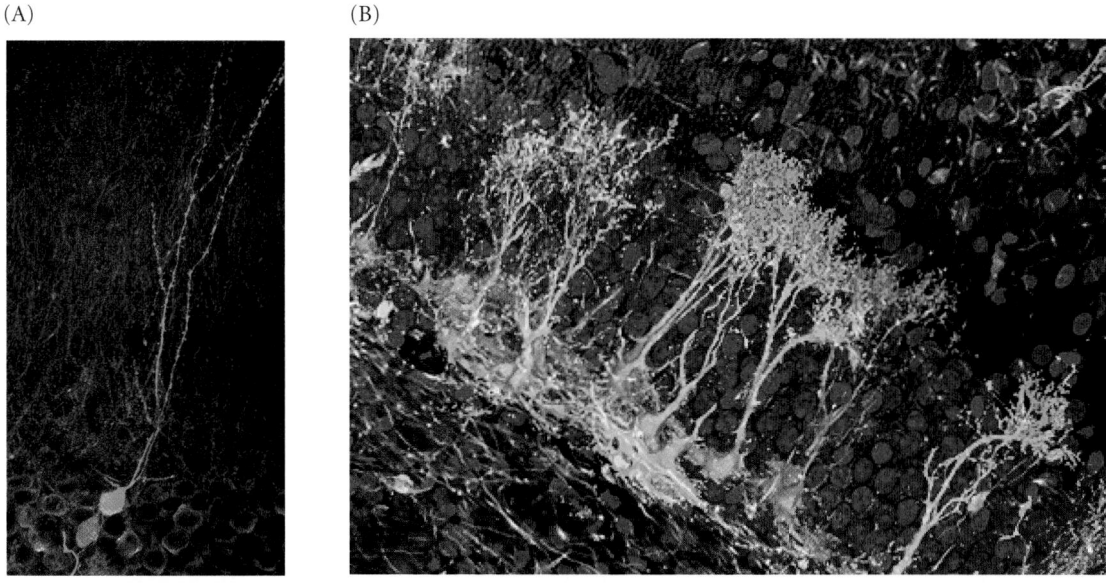

(C)

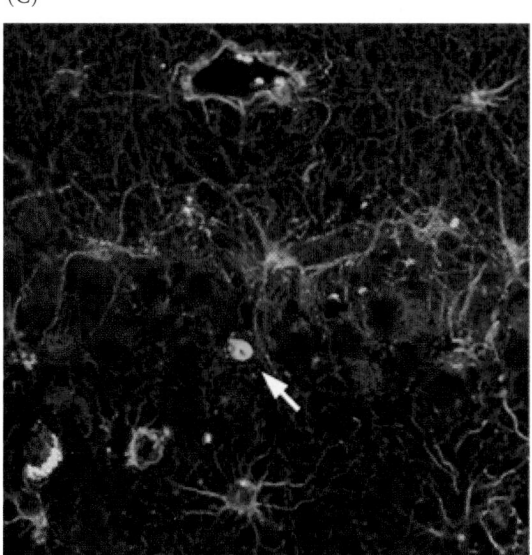

FIGURE 10.26 Evidence of adult neural stem cells. The green staining, which indicates newly divided cells, is from a fluorescent antibody against BrdU (a thymidine analogue that is taken up only during the S phase of the cell cycle). (A) Newly generated adult mouse neurons (green cells) have a normal morphology and receive synaptic inputs. The red spots are synaptophysin, a protein found on the dendrites at the synapses of axons from other neurons. (B) In the adult mouse hippocampus, the soma of neural stem cells (green) reside below their progeny, the granule neurons (red). The long axons of the stem cells can be seen extending through this region of neurons. (C) A newly generated neuron (arrow) in the adult human brain (specifically, in the dentate gyrus of the hippocampus). The red fluorescence is from an antibody that stains only neural cells. Yellow indicates the overlap of red and green. Glial cells are stained purple. (A from van Praag et al. 2002; B courtesy of E. Enikolopov and A.-S. Chiang; C from Erikksson et al. 1998, photograph courtesy of F. H. Gage.)

Adult neural stem cells represent only about 0.3% of the ventricle wall cell population, but they can be distinguished from more differentiated cells by their cell surface proteins* (Rietze et al. 2001). In the adult mouse, thousands of new neuroblasts are generated each day, migrating from the lateral subventricular zone to the olfactory bulb, where they differentiate into several different types of neurons. Recent evidence suggests that these stem cells are not multipotent (becoming specified only when they reach the olfactory bulb) but instead are a population of heterogeneous neuroblasts that are already committed to becoming certain neuronal types (Merkle et al. 2007). These adult neural stem cells proliferate in response to exercise, learning, and stress (Zhang et al. 2008).

Before they become neurons, neural stem cells are characterized by the expression of the NRSE translational inhibitor that prevents neuronal differentiation by binding to a silencer region of DNA (see Chapter 2). When neural stem cells begin to differentiate, they synthesize a small, double-stranded RNA that has the same sequence as the silencer and which might bind NRSE and thereby permit neuronal differentiation (Kuwabara et al. 2004). The use of cultured neuronal stem cells to regenerate or repair parts of the adult brain will be considered in Chapter 18.

---

*These neural stem cells may have particular physiological roles as well. During pregnancy, the hormone prolactin stimulates production of neuronal progenitor cells in the subventricular zone of the adult mouse forebrain. These progenitor cells migrate to produce olfactory neurons that may be important for maternal behavior of rearing offspring (Shingo et al. 2003).

The Unique Development of the Human Brain

There are many differences between humans and our closest relatives, the chimpanzees. These include our hairless, sweaty skin and striding bipedal posture. Male humans also lack the penile bone and the keratinous penile spines that characterize the external genitalia of other male primates. However, the most striking and significant differences occur in brain development. The enormous growth and asymmetry of our neocortex and our ability to reason, remember, plan for the future, and learn language and cultural skills make humans unique in the animal kingdom (Varki et al. 2008). The development of the human neocortex is strikingly plastic and is an almost constant work in progress. Several developmental phenomena have been identified that distinguish the development of the human brain from that of other species, including other primates:

- Retention of the fetal neuronal growth rate after birth
- Activity of human-specific RNA genes
- High levels of transcription
- Human-specific alleles of developmental regulatory genes
- Continuation of brain maturation into adulthood

Fetal neuronal growth rate after birth

If there is one developmental trait that distinguishes humans from the rest of the animal kingdom, it is our retention of the fetal neuronal growth rate. Both human and ape brains have a high growth rate before birth. After birth, however, this rate slows greatly in the apes, whereas human brain growth continues at a rapid rate for about 2 years (**FIGURE 10.27A**; Martin 1990; see Leigh 2004). Portmann (1941), Montagu (1962), and Gould (1977) have each made the claim that we are essentially "extrauterine fetuses" for the first year of life.

During early postnatal development, we add approximately 250,000 neurons per minute (Purves and Lichtman 1985). The ratio of brain weight to body weight at birth is similar for great apes and humans, but by adulthood the ratio for humans is literally "off the chart" when compared with that of other primates (**FIGURE 10.27B**; Bogin 1997). Indeed, if one follows the charts of ape maturity, human gestation should be 21 months. Our "premature" birth is an evolutionary compromise based on maternal pelvic width, fetal head circumference, and fetal lung maturity. The mechanism for retaining the fetal neuronal growth rate beyond birth has been called *hypermorphosis*—the extension of development beyond its ancestral state (Vrba 1996; Vinicius and Lahr 2003).

In addition to the neurons made after birth, the number of synapses increases by an astronomical number. At the cellular level, no fewer than 30,000 synapses per cm² of cortex are formed *every second* during the first few years of human life (Rose 1998; Barinaga 2003). It is speculated that these new neurons and rapidly proliferating neural connections enable plasticity and learning, create an enormous storage potential for memories, and enable us to develop skills such as language, humor, and music—that is, those things that help make us human.

● See **WEBSITE 10.3** Neuronal growth and the invention of childhood

Genes for neuronal growth

Which genes distinguish us from our closest relatives, the chimpanzees? Humans and chimpanzees have remarkably similar genomes. When protein-encoding DNAs are compared, they are around 99% identical. However, protein-coding regions comprise only around 2% of either genome. In 1975, King and Wilson concluded from

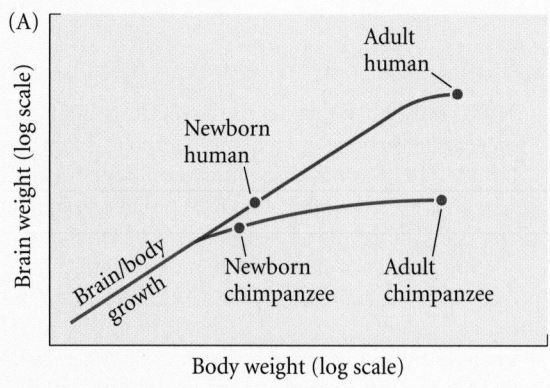

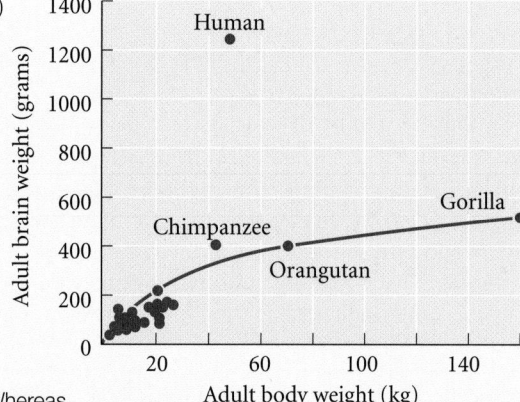

FIGURE 10.27 Retention of fetal neuronal growth rate in humans. (A) Whereas other primates (e.g., chimpanzees) complete neuron proliferation around the time of birth, in newborn humans neurons continue to proliferate at the same rate as the fetal brain neurons. (B) The brain/body weight ratio (encephalization index) of humans is about 3.5 times higher than that of apes. (After Bogin 1997.)

their studies of human and chimpanzee proteins that "The organismal differences between chimpanzees and humans would then result chiefly from genetic changes in a few regulatory systems, while amino acid substitutions in general would rarely be a key factor in major adaptive shifts." This was one of the first suggestions that evolution can occur through changes in developmental regulatory genes. When the total genomes are compared, humans and chimpanzees differ at about 4% of their sequences, most of them in noncoding regions (see Varki et al. 2008).

Although there are some brain growth genes (e.g., *ASPM* and *microcephalin*) whose sequences differ between humans and apes, these differences have not been correlated with the huge growth of human brains. Rather, the critical differences appear to reside in the DNA sequences that control these genes. These could be in enhancer regions of the DNA or in the DNA that produces "noncoding RNAs." These noncoding RNAs are highly expressed in the developing brain, and while not producing any protein product themselves, they may regulate the transcription or translation of neuronal transcription factors. Computer analysis comparing different mammalian genomes may have found such genes (Pollard et al. 2006a,b; Prabhakar et al. 2006). First, these studies identified a relatively small group of noncoding DNA regions that were conserved among non-human mammals. This represents about 2% of the genome, and it was assumed that if these regions were conserved throughout mammalian evolution, they were important.

The studies then compared these sequences to their human homologues to see if any of these regions have changed between humans and other mammals. About 50 such regions were found where the sequence is highly conserved among mammals but has diverged rapidly between humans and chimpanzees. The most rapid divergence is seen in the sequence *HAR1* (*human accelerated region-1*), where 18 sequence changes were seen between

chimpanzees and humans. *HAR1* is expressed in the developing brains of humans and apes, especially in the *Reelin*-expressing Cajal-Retzius neurons that are known to be responsible for directing neuronal migration during the formation of the six-layered neocortex. Research is ongoing to discover the function of *HAR1* and the other HAR genes that are in the conserved noncoding region of the genome.

A similar search for human-specific DNA deletions in primate genomes found some fascinating candidates. Recalling that the loss of an inhibitor is equivalent to the gain of an activator (think about the Wnt pathway or the double-negative gate in sea urchin blastomeres), McLean and colleagues (2011) uncovered 510 sequences that are present in the genomes of chimpanzees and other mammals but not in humans. One of these deletions is in the forebrain enhancer of the *GADD45G* gene. This gene encodes a growth suppressor that is normally expressed in the ventral forebrain region of chimpanzees and mice, but not humans (**FIGURE 10.28**). When a marker gene is placed onto a chimpanzee *GADD45G* enhancer, it is expressed in the mouse brain. If placed on a human *GADD45G* enhancer, it will not be expressed there.

Another particularly human gene expressed in the brain cortex is an inhibitor of the gene for the GTPase SRGAP2, which downregulates neuronal migration and regulates dendrite growth. Most animals, including the other apes, have one copy of this gene per haploid chromosome set. Humans, however, have duplicated this gene, and one of the duplicated copies produces a protein that binds to and inhibits the original protein. When this duplicated human gene is placed into mice, the mouse neocortex receives more neurons, and the dendrites on these neurons have an increased density of longer spines. Intriguingly, molecular evidence suggests that this duplication arose about 2.4 million years ago—right about the time our Anthropithecene

Preoptic area   Septum

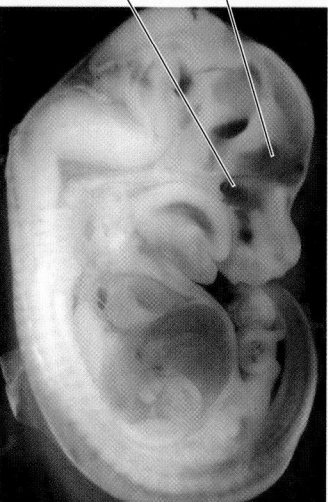
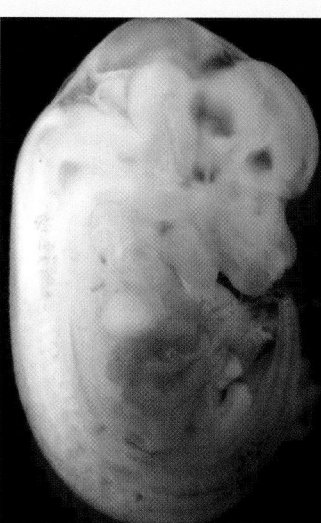

Chimp *GADD45G* enhancer   Mouse *GADD45G* enhancer

FIGURE 10.28 An "absentee candidate" for accelerated brain growth. The growth-suppressing gene *GADD45G* is not expressed in the human brain because the human *GADD45G* enhancer lacks a specific DNA sequence found in other mammals; this sequence activates the gene, suppressing growth in certain brain regions. When this DNA sequence is placed into a gene encoding β-galactosidase and inserted into mouse embryos, brain-specific expression is directed by both chimpanzee (left) and mouse (right) sequences. (From McLean et al. 2011; photograph courtesy of D. M. Kingsley.)

ancestors started growing larger brains and using tools (Dennis et al. 2012; Charrier et al. 2012).

High transcriptional activity

In the 1970s, A. C. Wilson suggested that the difference between humans and chimpanzees might reside in the *amount* of proteins made from their genes (see Gibbons 1998), and there is now evidence supporting this hypothesis. Using microarrays to study global patterns of gene expression, several recent investigations have found that, although the quantities and types of genes expressed in human and chimpanzee livers and blood are indeed extremely similar, human *brains* produce more than five times as much mRNA as chimpanzee brains (Enard et al. 2002a; Preuss et al. 2004). In humans, transcription of some genes (such as *SPTLC1*, a gene whose defect causes sensory nerve damage) was elevated 18-fold over the same genes' expression in the chimpanzee cortex, while other genes (such as *DDX17*, whose product is involved in RNA processing) are expressed 10 times less in human than in chimpanzee cortices.

Speech, language, and the *FOXP2* gene

Spoken language is a characteristically human trait and is presumed to be the prerequisite for the evolution of cultures. Speech entails the fine-scale control of the larynx (voice box) and

mouth. Individuals who are heterozygous for mutations at the *FOXP2* locus have severe problems with language articulation and with forming sentences (Vargha-Khadem et al. 1995; Lai et al. 2001). This observation has provided genetic anthropologists with an interesting gene to study. Enard and colleagues (2002b) have shown that, although the *FOXP2* gene is conserved throughout most of mammalian evolution, it has a unique form in humans, having accumulated at least two amino acid-changing mutations just since our divergence from the common ancestor of humans and chimpanzees. These differences are significant, since human and chimpanzee forms of the Foxp2 protein differentially regulate more than 100 genes (Konopka et al. 2009).

In the mouse, the *Foxp2* gene is expressed in the developing brain, but its major site of expression is the lung (Shu et al. 2001). In humans, *FOXP2* is predominantly expressed in those brain regions that coordinate speech (i.e., the caudate nucleus and inferior olive nuclei); these sites are abnormal in patients with *FOXP2* deficiency (Lai et al. 2003). In the cortical regions regulating language and speech, the human-specific FoxP2 appears to promote the expression of these specific transcripts during brain development (Lambert et al. 2011). In birds, the Foxp2 protein is associated with song learning, and the experimental downregulation of *Foxp2*

expression in certain areas of the brain prevents young male birds from imitating their species-specific song (Teramitsu and White 2006; Haesler et al. 2007). Although it is not certain that *FOXP2* is the most critical gene for human language acquisition, it seems to be very important for allowing the orofacial movements and grammar characteristic of human speech.

Teenage brains: Wired and unchained

Until recently, most scientists thought that after the initial growth of neurons during fetal development and early childhood, rapid neural proliferation ceased. However, magnetic resonance imaging (MRI) studies have shown that the brain keeps developing until around puberty, and that not all areas of the brain mature simultaneously (Giedd et al. 1999; Sowell et al. 1999). Soon after puberty, neuronal growth ceases and pruning occurs. The time of this pruning correlates with the time when language acquisition becomes difficult (which may be why children learn language more readily than adults). There is also a wave of myelin production ("white matter" from the glial cells that surround neuronal axons) at this time. Myelination is critical for proper neural functioning, and although myelination continues throughout adulthood (Lebel and Beaulieu 2011), the greatest differences between brains in early puberty and those in early adulthood involve the

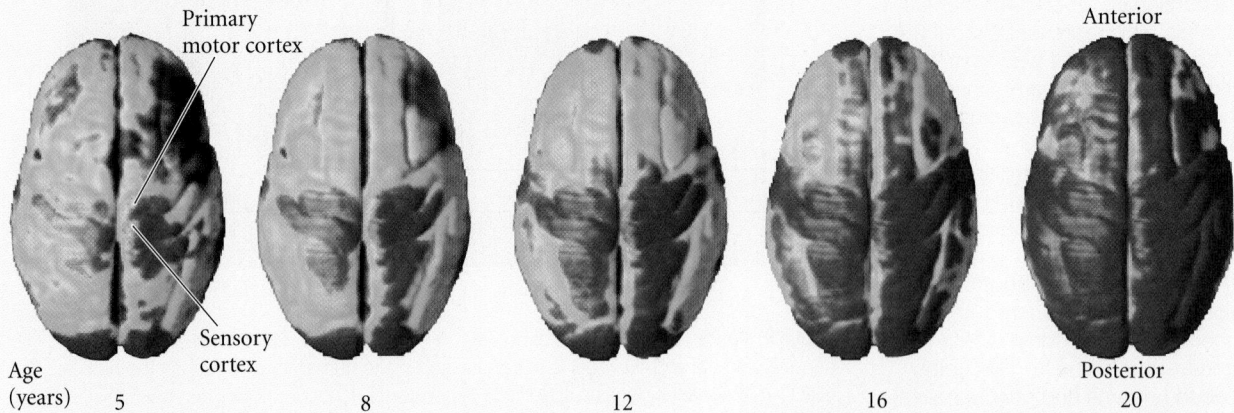

FIGURE 10.29 Dorsal view of the human brain showing the progression of myelination ("white matter") over the cortical surface during adolescence. (Images courtesy of N. Gogtay.)

SIDELIGHTS **SPECULATIONS** (continued)

frontal cortex (**FIGURE 10.29**; Sowell et al. 1999; Gogtay et al. 2004). These differences in brain development may explain the extreme responses teenagers have to certain stimuli, as well as their ability to learn certain tasks.

In tests using functional MRI to scan subjects' brains while emotion-charged pictures flashed on a computer screen, the brains of young teens showed activity in the amygdala, which mediates fear and strong emotions. When older teens were shown the same pictures, most of their brain activity was centered in the frontal lobe, an area involved in more reasoned perceptions (Baird et al. 1999; Luna et al. 2001). As of now, data on groups shows these differences best. However, improving technology is beginning to allow the maturity of a single individual's brain to be assessed (Dosenbach et al. 2010). The teenage brain is a complicated and dynamic entity that is not easily understood (as any parent knows). But once through these years, the resulting adult brain is usually capable of making reasoned decisions, even in the onslaught of emotional situations.

THE DYNAMICS OF OPTIC DEVELOPMENT: THE VERTEBRATE EYE

An individual gains knowledge of its environment through its sensory organs. The major sensory organs of the head develop from interactions of the neural tube with a series of epidermal thickenings called the **cranial ectodermal placodes*** (discussed in more detail in Chapter 11). Most of these placodes form neurons and sensory epithelia. The two **olfactory placodes** form the nasal epithelium and its embedded olfactory (smell) receptors as well as the ganglia for the olfactory nerves. Similarly, the two **otic placodes** invaginate to form the inner ear labyrinth, whose neurons form the acoustic ganglia that enable us to hear. In this section, we will focus on the development of the eye from the **lens placode**.

The lens placode does not form neurons. Rather, it forms the transparent lens that allows light to impinge on the retina. The retina develops from a bulge in the forebrain. The interactions between the cells of the lens placode and the presumptive retina structure the eye via a cascade of reciprocal changes that enable the construction of an intricately complex organ.

We first described the interactive induction of the vertebrate eye in Chapter 3. At gastrulation, the involuting prechordal plate and foregut endoderm interact with the adjacent prospective head ectoderm to give the head ectoderm a lens-forming bias (Saha et al. 1989). In mammals, this induces the Pax6 transcription factor in the ectoderm, which is critical in the competence of the ectoderm to respond to subsequent signals. But not all parts of the head ectoderm eventually form lenses, and the lens must have a precise spatial relationship with the retina. The activation of the head ectoderm's latent lens-forming ability and the positioning of the lens in relation to the retina are accomplished by the **optic vesicles** that extend from the diencephalon of the forebrain.

FIGURE 10.30 shows the development of the vertebrate eye. Where the optic vesicle contacts the head

---

* The term *placode* refers to a thickened portion of germ layer (usually ectoderm) tissue from which an organ develops.

(A) 4-mm embryo

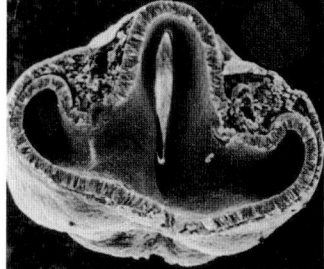

(B) 4.5-mm embryo

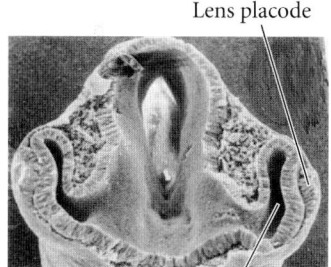

Lens placode

Optic vesicle

(C) 5-mm embryo

Lens vesicle

(D) 7-mm embryo

Retina   Lens

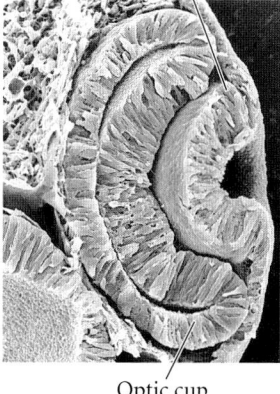

Optic cup

Cornea

FIGURE 10.30 Development of the vertebrate eye (see also Figure 3.15A). (A) The optic vesicle evaginates from the brain and contacts the overlying ectoderm, inducing a lens placode. (B,C) The overlying ectoderm differentiates into lens cells as the optic vesicle folds in on itself, and the lens placode becomes the lens vesicle. (C) The optic vesicle becomes the neural and pigmented retina as the lens is internalized. (D) The lens vesicle induces the overlying ectoderm to become the cornea. (A–C from Hilfer and Yang 1980, courtesy of S. R. Hilfer; D courtesy of K. Tosney.)

ectoderm, it induces the ectoderm to lengthen, forming the lens placode. The optic vesicle then bends to form the two-layered **optic cup**, and in so doing draws the developing lens into the embryo. This invagination is accomplished by the cells of the lens placode extending adhesive filopodia to contact the optic vesicle (Chauhan et al. 2009). As the optic vesicle becomes the optic cup, its two layers differentiate. The cells of the *outer layer* produce melanin pigment (being one of the few tissues other than the neural crest cells that can form this pigment) and ultimately become the **pigmented retina**. The cells of the *inner layer* proliferate rapidly and generate a variety of glial cells, ganglion cells, interneurons, and light-sensitive photoreceptor neurons that collectively constitute the **neural retina**. The retinal ganglion cells are neurons whose axons send electric impulses to the brain. Their axons meet at the base of the eye and travel down the optic stalk, which is then called the **optic nerve**. The inner cells of the optic cup (which will become the neural retina) induce the lens placode to become the **lens vesicle**, which will eventually differentiate into the lens cells of the eye.

Formation of the Eye Field: The Beginnings of the Retina

The details of eye development tell us how the eye comes to be made only in the head and why only two eyes normally form. These details show that the precise arrangement of the eye is the result of multiple layers of inductive events involving gene expression differences in both time and place. The story begins with formation of the **eye field** during specification of the neural tube. The anterior portion of the neural tube, where both BMP and Wnt pathways are inhibited, is specified by the *Otx2* gene expression. Noggin is especially important, as it not only blocks BMPs, thus allowing *Otx2* expression, but also inhibits expression of the transcription factor ET, one of the first proteins expressed in the eye field. However, once Otx2 protein accumulates in the ventral head region, it blocks Noggin's ability to inhibit the *ET* gene, so ET protein is produced.

One of the genes controlled by ET is *Rx*, whose product helps specify the retina. Rx (for "retinal homeobox") is a transcription factor that acts first by inhibiting *Otx2* (since it has done its jobs and can now get in the way), and second by activating *Pax6*, the major gene in forming the eye field in the anterior neural plate (**FIGURE 10.31A–C**; Zuber et al. 2003; Zuber 2010). Pax6 protein is especially important in the specification of the lens and retina; indeed, it appears to be a common denominator for specifying photoreceptive cells in all phyla, vertebrate and invertebrate (Halder et al. 1995).

Humans and mice heterozygous for loss-of-function mutations in *Pax6* have small eyes, whereas homozygotic mice and humans (and *Drosophila*) lack eyes altogether (**FIGURE 10.31D**; Jordan et al. 1992; Glaser et al. 1994; Quiring

FIGURE 10.31 Dynamic formation of the eye field in the anterior neural plate. (A) Formation of the eye field. Light blue represents the neural plate; moderate blue indicates the area of *Otx2* expression (forebrain); and dark blue indicates the region of the eye field as it forms in the forebrain. (B) Dynamic expression of transcription factors leading to specification of the eye field. Prior to stage 10, Noggin inhibits *ET* expression but promotes expression of *Otx2*. Otx2 protein then represses the inhibition of *ET* by Noggin signaling. The ET transcription factor activates *Rx*, which encodes a transcription factor that blocks *Otx2* and promotes *Pax6* expression. Pax6 protein initiates the cascade of gene expression constituting the eye field (at right). (C) Location of the transcription factors in the nascent eye field of stage 12.5 (early neurula) and stage 15 (mid-neurula) *Xenopus* embryos, showing a concentric organization of transcription factors having domains of decreasing size: Six3 > Pax6 > Rx > Lhx2 > ET. (D) Eye development in a normal mouse embryo (left) and lack of eyes in a mouse embryo whose *Rx* gene has been knocked out (right). (E) Expression pattern of the *Xenopus Xrx1* gene in the single eye field of the early neurula (left) and in the two developing retinas (as well as in the pineal, an organ that has a presumptive retina-like set of photoreceptors) of a newly hatched tadpole (right). (A–C after Zuber et al. 2003; D,E after Bailey et al. 2004, photographs courtesy of M. Jamrich.)

et al. 1994). In both flies and vertebrates, Pax6 initiates a cascade of transcription factors (such as Six3, Rx, and Sox2) with overlapping functions. These factors mutually activate one another to generate a single large, eye-forming field in the center of the ventral forebrain (**FIGURE 10.31E**; Tétreault et al. 2009; Fuhrmann 2010). The final result, however, is two eyes that are more lateral in the head. The main player in separating the single vertebrate eye field into two bilateral fields is our old friend Sonic hedgehog.

Shh from the prechordal plate suppresses *Pax6* expression in the center of the neural tube, dividing the field in two (**FIGURE 10.32**). If the mouse *Shh* gene is mutated, or if the processing of this protein is inhibited, the single median eye field does not split. The result is **cyclopia**—a single eye in the center of the face, usually below the nose (see Figure 10.32C and Figure 3.26B; Chiang et al. 1996; Kelley et al. 1996; Roessler et al. 1996). Conversely, if too much Shh is synthesized by the prechordal plate, *Pax6* is suppressed in too large an area and the eyes fail to form at all. This phenomenon may explain why cave-dwelling fish are blind. Yamamoto and colleagues (2004) demonstrated that the difference between surface populations of the Mexican tetra fish (*Astyanax mexicanus*) and eyeless cave-dwelling populations of the same species is the amount of Shh secreted from the prechordal plate. Elevated Shh was probably selected in cave-dwelling species because it resulted in heightened oral sensing and larger jaws (Yamamoto et al. 2009). However, Shh also downregulates *Pax6*, resulting in the disruption of optic cup development, apoptosis of lens cells, and arrested eye development (**FIGURE 10.33**).

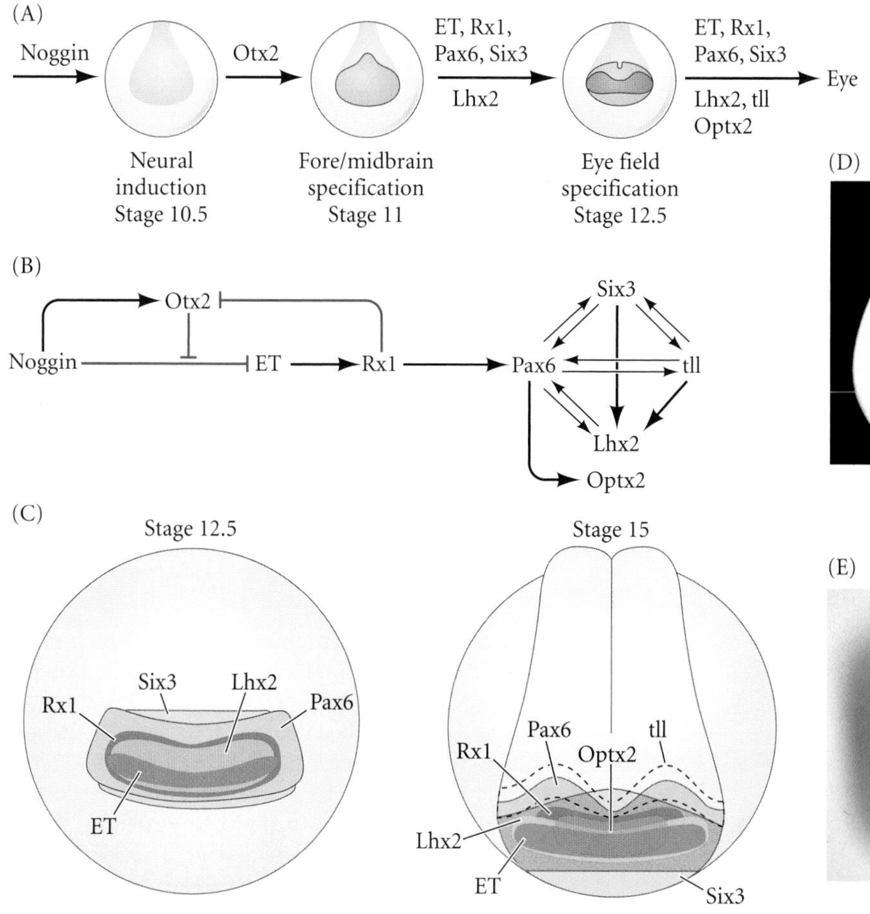

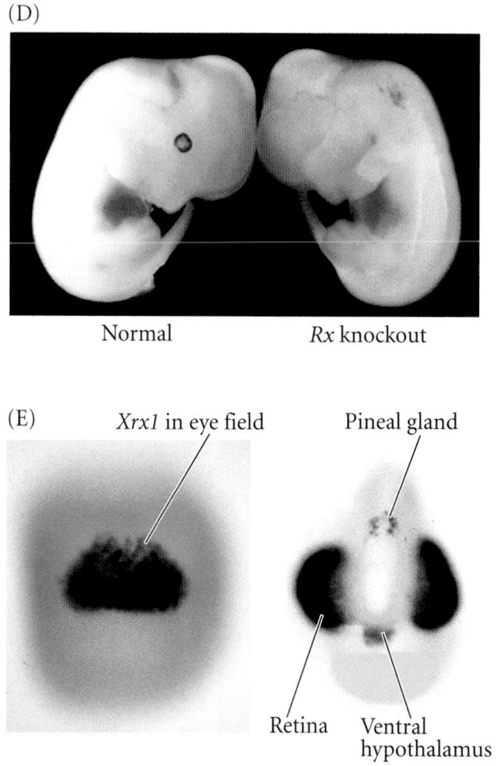

(A)

Noggin → Neural induction Stage 10.5 → Otx2 → Fore/midbrain specification Stage 11 → ET, Rx1, Pax6, Six3 Lhx2 → Eye field specification Stage 12.5 → ET, Rx1, Pax6, Six3 Lhx2, tll Optx2 → Eye

(B)

Noggin → Otx2 ⊣ ET → Rx1 → Pax6
Six3, tll, Lhx2, Optx2

(C)

Stage 12.5 — Rx1, Six3, Lhx2, Pax6, ET

Stage 15 — Rx1, Pax6, Optx2, tll, Lhx2, ET, Six3

(D) Normal   *Rx* knockout

(E) *Xrx1* in eye field   Pineal gland   Retina   Ventral hypothalamus

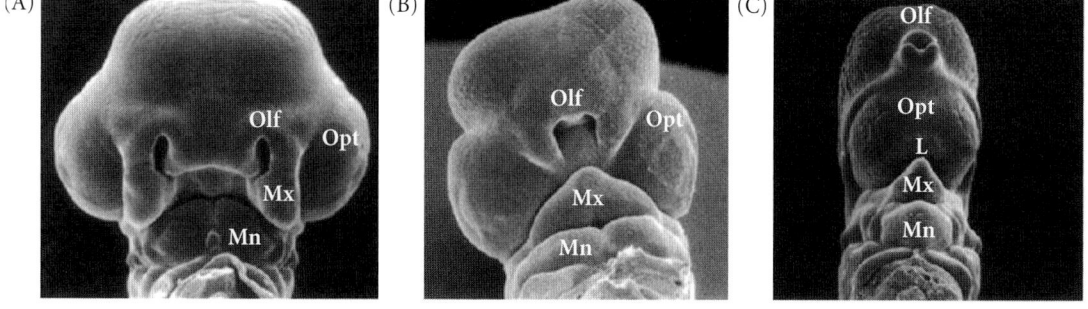

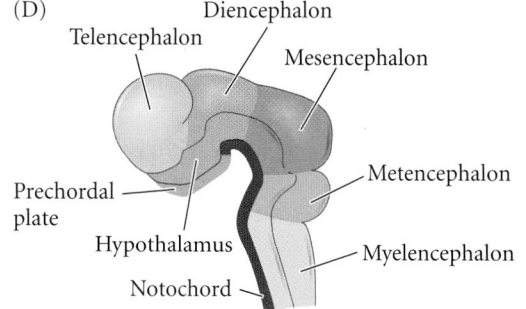

(A) Olf, Opt, Mx, Mn

(B) Olf, Opt, Mx, Mn

(C) Olf, Opt, L, Mx, Mn

(D) Diencephalon, Telencephalon, Mesencephalon, Metencephalon, Myelencephalon, Notochord, Hypothalamus, Prechordal plate

FIGURE 10.32 Sonic hedgehog separates the eye field into bilateral fields. Jervine, an alkaloid found in certain plants, inhibits endogenous *Shh* signaling. (A) Scanning electron micrograph showing the external facial features of a normal mouse embryo. (B) Mouse embryos exposed to 10 μM jervine had variable loss of midline tissue and resulting fusion of the paired, lateral olfactory processes (Olf), optic vesicles (Opt), and maxillary (Mx) and mandibular (Mn) processes of the jaw. (C) Complete fusion of the mouse optic vesicles and lenses (L) resulted in cyclopia. (D) Drawing showing the location of the prechordal plate (the source of Shh) in the 12-day mouse embryo. (A–C from Cooper et al. 1998, courtesy of P. A. Beachy.)

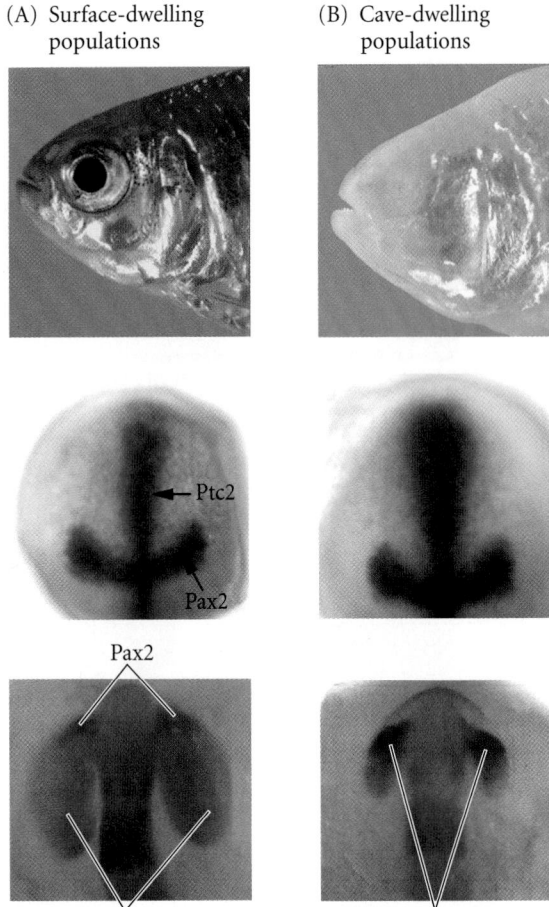

(A) Surface-dwelling populations

(B) Cave-dwelling populations

Ptc2

Pax2

Pax2

Pax6

Pax2

FIGURE 10.33 Surface-dwelling (A) and cave-dwelling (B) Mexican tetras (*Astyanax mexicanus*). The eye fails to form in the population that has lived in caves for more than 10,000 years (top right). Two genes that respond to Shh proteins, *Ptc2* and *Pax2*, are expressed in broader domains in the cavefish embryos than in those of surface dwellers (center). The embryonic optic vesicles (bottom) of surface-dwelling fish are normal size and have small domains of *Pax2* expression (specifying the optic stalk). The optic vesicles of the cave-dwelling fishes' embryos (where *Pax6* is usually expressed) are much smaller, and the *Pax2*-expressing region has grown at the expense of the *Pax6* region. (From Yamamoto et al. 2004; photographs courtesy of W. Jeffery.)

The Lens-Retina Induction Cascade

Once the eye field split is accomplished, how do the two fields form the eyes? Modern studies of vertebrate eye formation were initiated by Hans Spemann (1901), who found that when he destroyed the anterior neural plate on one side of the embryo, no lens formed on the operated side. Something in the neural plate was necessary for the lens to form. Soon afterward, Warren Lewis (1904) found that when he placed anterior neural tube under a different part of the head epidermis, the neural tube became retina and the skin became a lens. More recent studies (see Grainger 1992; Ogino et al.

2012) have shown that although the story is more complicated, the eye field of the anterior neural plate induces the epidermis above it to become the lens; and as each lens begins to form, it induces the eye field to become the retina. The development of the eye is a beautiful example of reciprocal embryonic induction (**FIGURE 10.34**).

Before epidermal tissue has the ability to be induced to form lens by the presumptive retinal cells, it has to become competent to respond to the signals sent by those cells. Jacobson (1963, 1966) showed that the epidermis that can respond to the eye field is first conditioned by passing over the pharyngeal endoderm and heart-forming mesoderm during gastrulation. These developing organs probably supply the area with antagonists that block the BMP and Wnt pathways, and these immature organs may be critical for inducing *Pax6* and other genes specific for the anterior ectoderm (Donner et al. 2006). Meanwhile, in the brain, the bilateral ventral forebrain eye fields evaginate as the Rx protein activates *Nlcam*, a gene whose cell-surface product regulates the evagination of the retinal precursor cells from the ventral forebrain (Brown et al. 2010). These evaginations become the optic vesicles. When the cells of the optic vesicles touch surface ectoderm, both tissues are changed. The optic vesicle cells flatten against the surface ectoderm and produce BMP4, Fgf8, and Delta (Ogino et al. 2012). These inducers instruct the cells of the surface ectoderm to elongate and become lens placode cells. As these surface ectoderm cells become the lens placode, they secrete FGFs that instruct the adjacent cells of the optic vesicle to activate the *Vsx2* gene that characterizes the neural retina. The dermal mesenchyme surrounding the optic vesicle instructs most of the outer optic vesicle cells to activate the *Mitf* gene, which will instruct them to form the pigment melanin (Burmeister et al.1996; Nguyen and Arnheiter 2000). Thus, the most distal part of the optic vesicle (those cells touching the surface ectoderm) is instructed to become neural retina, while the cells adjacent to this region are instructed to become *pigmented* retina (see Fuhrmann 2010).

Once the eye fields are specified and divided, much of the development into the optic cup is remarkably autonomous. In mice, a single homogeneous embryonic stem cell population, when placed on a three-dimensional extracellular matrix in the presence of appropriate paracrine factors, can generate an optic vesicle. It will first create an ectodermal sphere, which then produces a "bud" with an inner and an outer wall. These interact such that the outer wall secretes Wnts and become characterized by the MITF transcription factor and melanin pigment. That is, it will become pigmented retina cells. Simultaneously, the inner layer becomes specified as neural retina, characterized by such transcription factors as Six3 and Chx10. Moreover, this optic vesicle, without any external pressure, will invaginate to become an optic cup, and the inner portion will differentiate into a retina-like structure that contains each of the major types of retinal neuron, including the photoreceptors. This indicates that once the eye field is formed, the neural retina and pigmented retina will segregate from one another (patterning), the folding will occur by intrinsic cell

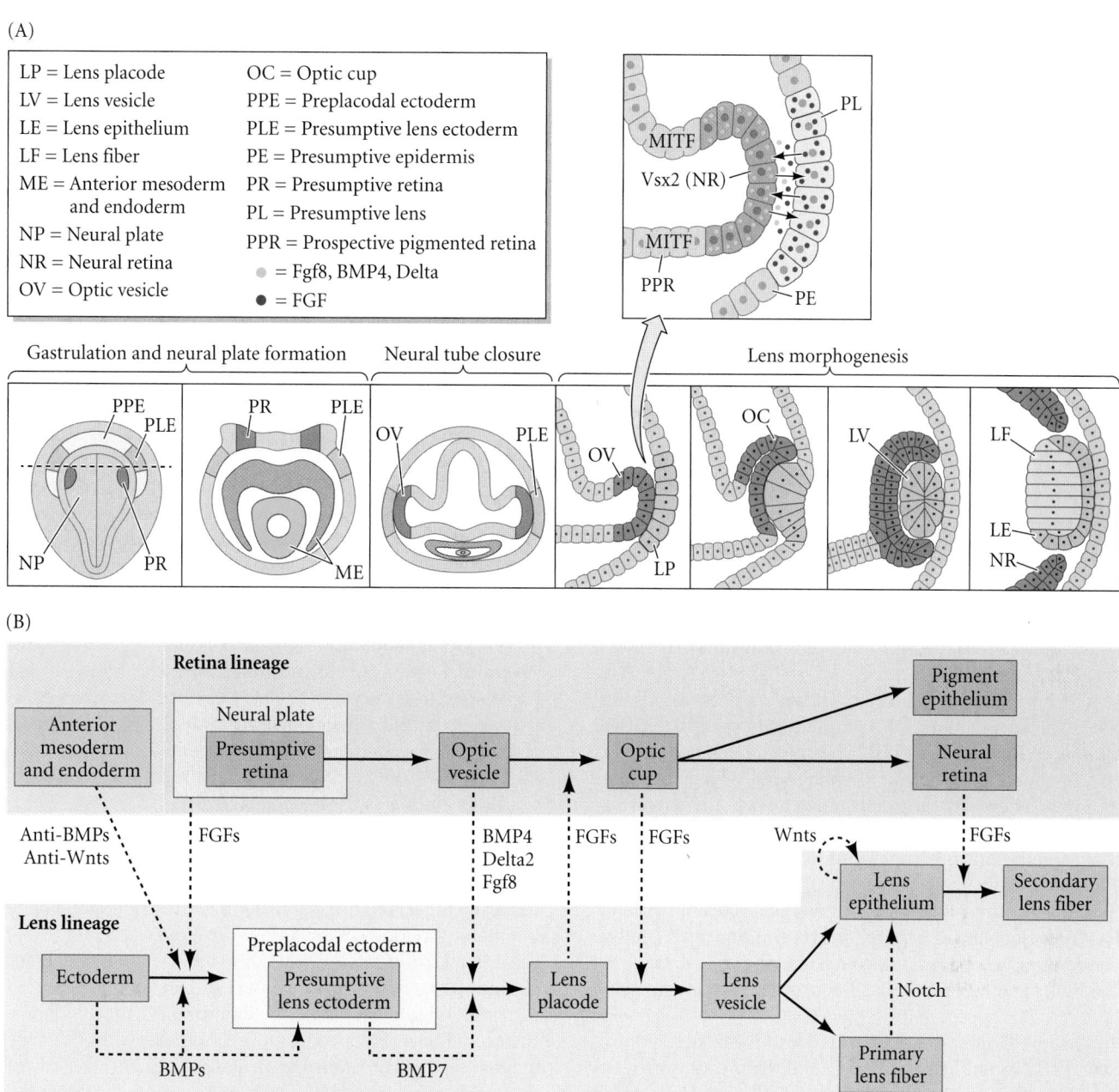

FIGURE 10.34 Reciprocal embryonic interactions between the developing lens placode and the optic vesicle from the brain. (A) Diagrammatic rendering of the major anatomical changes from gastrulation through lens morphogenesis. These interactions start with the presumptive lens ectoderm being influenced by the neural plate, cardiac mesoderm, and pharyngeal mesoderm. Later, the optic vesicle—a bulge from the diencephalon—touches the presumptive lens ectoderm, triggering a series of interactions that turns the optic vesicle into a two-layered optic cup, converts the inner layer of the optic cup into the neural retina, and causes the lens placode to involute and form the lens vesicle. The expanded inset (upper right) shows the critical interactions between the optic vesicle and the presumptive lens cells of the lens placode. (B) Some of the paracrine factors involved in lens development. At different stages and in different tissues, the FGFs and their receptors may be different The three arrows below show some of the genes that become expressed in the presumptive lens during the indicated timeframes. (After Ogino et al. 2012.)

shape changes (morphogenesis), and the neural retina cells will differentiate into different neuronal types (differentiation) in the correct spatial arrangement (Eiraku et al. 2011; Sasai et al. 2013).

The optic vesicle then adheres to the lens placode and changes its shape to form the optic cup. The presumptive neural retina adheres to the lens placode, drawing it into the embryo, while the outer wall of the optic cup becomes the pigmented retina. FGF from the optic cup activates a new set of genes in the lens placode, transforming it into the lens vesicle, which will form the cells of the lens.

Lens and cornea differentiation

The differentiation of the lens tissue into a transparent membrane capable of directing light onto the retina involves changes in cell structure and shape as well as the synthesis of transparent, lens-specific proteins called **crystallins**. As the lens cells continue to grow, they synthesize crystallins, which eventually fill up the cell and cause the extrusion of the nucleus. The lens cells must curve properly, and this curvature is caused by balancing the Rho-generated apical *constriction* of microfilaments with Rac-generated actin polymerization that *extends* the microfilaments along the apical-basal axis (Chauhan et al. 2011).*

The crystallin-containing cellular fibers eventually fill the space between the two layers of the lens vesicle (**FIGURE 10.35A,B**; Piatigorsky 1981). The anterior cells of the lens vesicle constitute a germinal epithelium, which continues dividing. These dividing cells move toward the equator of the vesicle, and as they pass through the equatorial region, they, too, begin to elongate into cellular fibers (**FIGURE 10.35C,D**). Thus, the lens contains three regions: an anterior zone of dividing epithelial cells, an equatorial zone of cellular elongation, and a posterior and central zone of crystallin-containing fiber cells. This arrangement persists throughout the lifetime of the animal as fibers are continuously being laid down (Papaconstantinou 1967).

The initial differentiation of lens-forming tissues requires contact between the optic vesicle and the presumptive lens ectoderm. In *Xenopus*, for instance, Delta proteins on the optic vesicle activate the Notch receptors on the presumptive lens ectoderm (Ogino et al. 2008). The Notch intracellular domain binds to an enhancer element of the *Lens1* gene, and in the presence of the Otx2 transcription factor (which is expressed throughout the entire head region), *Lens1* is activated. The Lens1 protein is itself a transcription factor that is essential for epithelial cell proliferation (making and growing the lens placode) and eventually for closing the lens vesicle. In this interaction we see a principle that is observed

throughout development—namely, that some transcription factors (such as Otx2) specify a particular field and provide competence for cells to respond to a more specific induction (such as Notch) within the field.

Paracrine factors from the optic vesicle also induce lens-specific transcription factors. Regulation of the *crystallin* genes is under the control of Pax6, Sox2, and L-Maf (**FIGURE 10.35E**). Like Otx2, Pax6 appears in the head ectoderm before the lens is formed, and Sox2 is induced in the lens placode by BMP4 secreted from the optic vesicle. Coexpression of Pax6 and Sox2 in the same cells initiates lens differentiation and activates *crystallin* genes. Appearing later than Sox2, L-Maf is induced by Fgf8 secreted by the optic vesicle and is needed for the maintenance of *crystallin* gene expression and the completion of lens fiber differentiation (Kondoh et al. 2004; Reza et al. 2007).

Relatively little is known about the development of the cornea. Shortly after the lens vesicle has detached from the surface ectoderm, the lens vesicle stimulates the overlying ectoderm to secrete layers of collagen into which neural crest cells migrate and make new cell layers while secreting a corneal-specific extracellular matrix (Meier and Hay 1974; Johnston et al. 1979; Kanakubo et al. 2006). These cells condense to form several flat layers of cells, eventually becoming the corneal precursor cells (see Figure 10.35A; Cvekl and Tamm 2004). As these cells mature, they dehydrate and form tight junctions among the cells, uniting with the surface ectoderm (Kurpakus et al. 1994; Gage et al. 2005) to become the cornea. Intraocular fluid pressure (from the aqueous humor) is necessary for the correct curvature of the cornea, allowing light to be focused on the retina (Coulombre 1956, 1965).

Repair and regeneration are critical to the cornea since, like the epidermis, it is exposed to the outside world. The main problem for the cornea is reactive oxygen species (ROS; see Chapter 16) that damage DNA and proteins. The major sources of ROS are the amniotic fluid (as an embryo) and ultraviolet light (as an adult). One protective mechanism is the production of the iron-binding protein ferritin (Linsenmeyer et al. 2005; Beazley et al. 2009). The second mode of protection is a layer of basal cells that continually renew the corneal cells throughout the life of the individual. Long-lived stem cells found at the edge of the cornea contribute to corneal repair and can regenerate the cornea in humans (Cotsarelis et al. 1989; Tsai et al. 2000; Majo et al. 2008).

Neural retina differentiation

Like the cerebral and cerebellar cortices, the neural retina develops into a layered array of different neuronal types. These layers include the light- and color-sensitive photoreceptor cells (**rods** and **cones**); the cell bodies of the ganglion cells; and **bipolar interneurons** that transmit electric stimuli from the rods and cones to the ganglion cells (**FIGURE 10.36**). In addition, the retina contains numerous **Müller glial cells** that maintain its integrity, **amacrine neurons** (which lack large axons), and **horizontal neurons** that transmit electric impulses in the plane of the retina.

---

*As you may recall from Chapter 3, Rac and Rho are the two Rho-family GTPases that regulate cell shape and motility by reorganizing the cytoskeletal subunits. Rho is often involved in contractility, while Rac specializes in growth and spreading. We'll see these antagonistic cytoskeletal masons numerous times in the next few chapters.

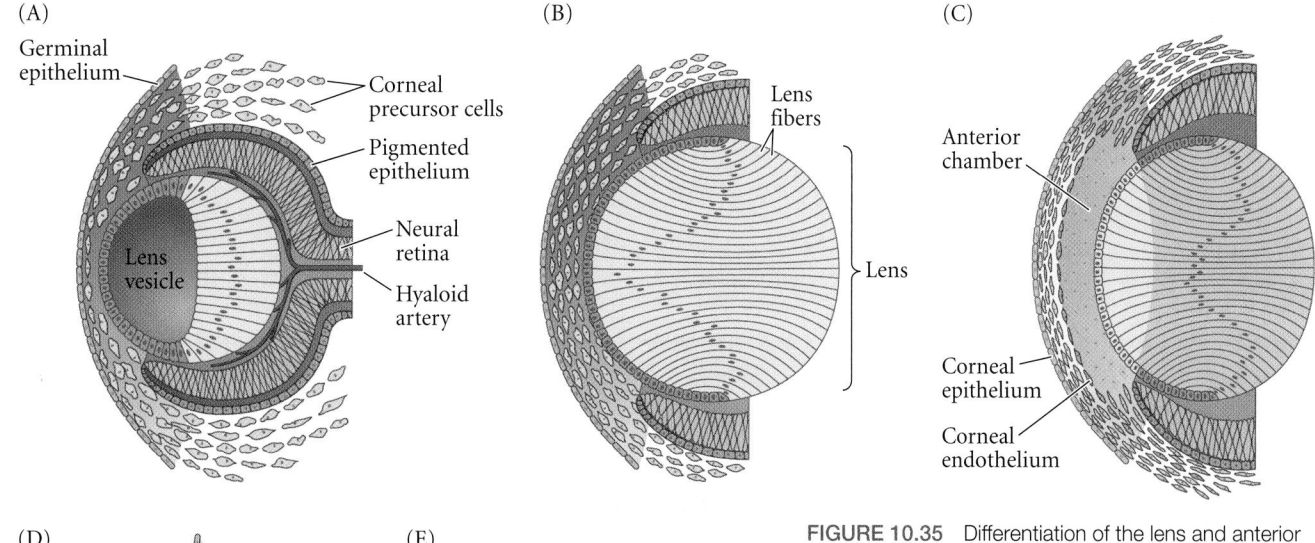

FIGURE 10.35 Differentiation of the lens and anterior portion of the mouse eye. (A) At embryonic day 13, the lens vesicle detaches from the surface ectoderm and invaginates into the optic cup. Corneal precursors (mesenchymal cells) from the neural crest migrate into this space. The elongation of the interior lens cells begins, producing primary lens fibers. (B) At day 14, the lens is filled with crystallin-synthesizing fibers. The neural crest-derived mesenchyme cells between the lens and surface condense to form several layers. (C) At day 15, the lens detaches from the corneal layers, generating an anterior cavity. (D) The surface ectoderm at the anterior side becomes the corneal epithelium, and at day 15.5, corneal layers differentiate and begin to become transparent. The anterior edge of the optic cup enlarges to form a non-neural region containing the iris muscles and the ciliary body. New lens cells are derived from the anterior lens epithelium. As the lens grows, the nuclei of the primary lens cells degenerate and new lens fibers grow from the epithelium on the lateral sides. (E) Close binding of the Sox2 and Pax6 transcription factors on a small region of the δ-*crystallin* enhancer. (A–D after Cvekl and Tamm 2004; E after Kondoh et al. 2004.)

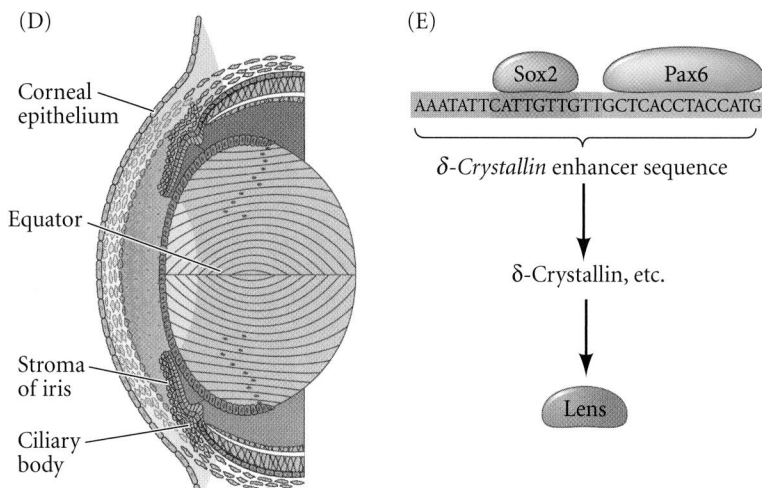

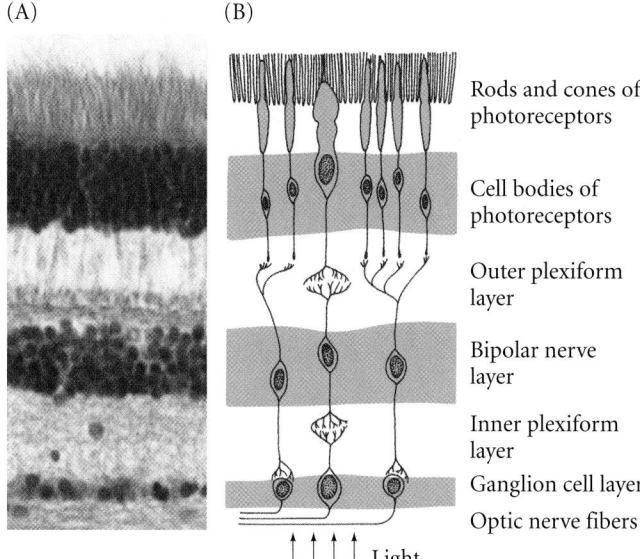

FIGURE 10.36 Retinal neurons sort out into functional layers during development. (A) The three layers of neurons in the adult retina and the synapses between them. (B) A functional depiction of the major neuronal pathway in the retina. Light traverses the layers until it is received by the photoreceptors. The axons of the photoreceptors synapse with bipolar neurons, which transmit electric signals to the ganglion cells. The axons of the ganglion cells join to form the optic nerve, which enters the brain. (A courtesy of G. Grunwald.)

The neuroblasts of the retina are competent to generate all seven retinal cell types. For instance, if one injects an individual retinal neuroblast with a genetic marker, that marker will be seen in a strip that can include all the different cell types of the retina (Turner and Cepko 1987; Yang 2004). In amphibians, the type of neuron produced from the multipotent retinal stem cell appears to depend on the *timing* of gene *translation,* not the *location* of gene *transcription.* Photoreceptor neurons, for instance, are specified by expression of the *Xotx5b* gene, while expression of *Xotx2* and *Xvsx1* are critical for specifying the bipolar neurons.

Interestingly, these three genes are *transcribed* in all retinal cells, but they are *translated* differently (**FIGURE 10.37A**). Those neurons whose birthday is at stage 30 translate the *Xotx5b* mRNA and become photoreceptors, while those neurons forming later (birthdays at stage 35) translate the *Xotx2* and *Xvsx1* messages and become bipolar interneurons (Decembrini et al. 2006, 2009).

This time-dependent regulation of translation is mediated by four microRNAs: *miRNA-129, miRNA-155, miRNA-214,* and *miRNA-222.* These microRNAs are highly expressed in early retinal progenitor cells and bind to the 3′ UTRs of

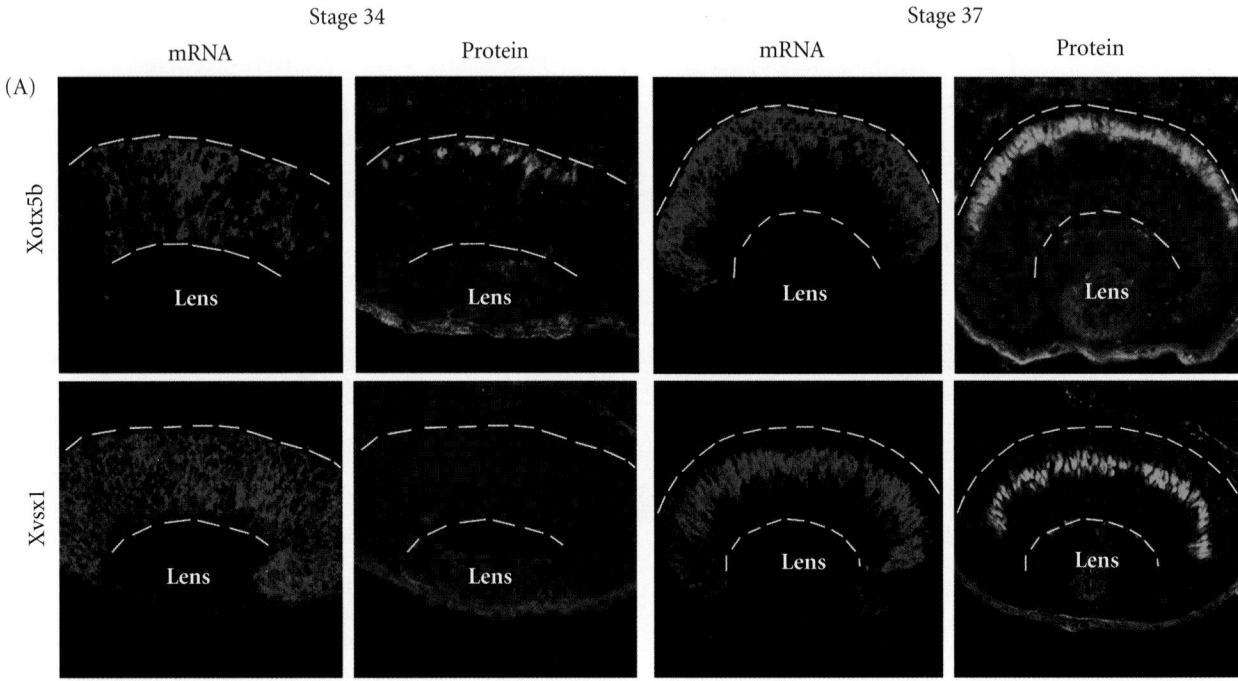

FIGURE 10.37 Timing of retinal neurogenesis by the translation of mRNAs encoding transcription factors Xotx5b and Xvsx1. (A) Comparison of *Xotx5b* (photoreceptor) and *Xvsx1* (bipolar neuron) mRNAs and proteins at stages 34 (early retinal neurogenesis) and 37 (late retinal neurogenesis). The mRNAs for both proteins are synthesized, but at stage 34 only the *Xotx5b* message has been translated. At stage 37, both messages have been translated. (B) The diminishing expression of *miRNA-222* (pink) from stage 30 through stage 42. (From Decembrini et al. 2006, 2010.)

the *Xotx2* and *Xvsx1* mRNAs, inhibiting their translation and thus preventing early-differentiating retinal cells from becoming bipolar neurons. Expression of these microRNAs diminishes with time (**FIGURE 10.37B**; Decembrini et al. 2009), so that later-forming neurons are no longer blocked from acquiring bipolar interneuron characteristics.

Not all the cells of the optic cup become neural tissue. The tips of the optic cup on either side of the lens develop into a pigmented ring of muscular tissue called the **iris**. The iris muscles control the size of the pupil (and give an individual his or her characteristic eye color). At the junction between the neural retina and the iris, the optic cup forms the **ciliary body**. This tissue secretes the **aqueous humor**, a fluid needed for the nutrition of the lens and for forming the pressure needed to stabilize the curvature of the eye and the constant distance between the lens and the cornea.

● See **WEBSITE 10.4** Why babies don't see well

THE EPIDERMIS AND ITS CUTANEOUS APPENDAGES

The skin—a tough, elastic, water-impermeable membrane—is the largest organ in our bodies. Moreover, skin is constantly being renewed. This regenerative ability is due to a population of epidermal stem cells that last the lifetime of our bodies. Mammalian skin has three major components: (1) a stratified epidermis; (2) an underlying dermis composed of loosely packed fibroblasts; and (3) neural crest-derived melanocytes that reside in the basal epidermis and hair follicles. It is the melanocytes (discussed in Chapter 11) that provide the skin's pigmentation. In addition, a subcutaneous ("below the skin") fat layer is present beneath the dermis.

Origin of the Epidermis

The epidermis originates from the ectodermal* cells covering the embryo after neurulation. As detailed in Chapter 8, this surface ectoderm is induced to form epidermis rather than neural tissue by the actions of BMPs. The BMPs promote epidermal specification and at the same time induce transcription factors that block the neural pathway (see Bakkers et al. 2002). Again, we see the principle that the

---

*To review the vocabulary, *epidermis* is the outer layer of skin. The *ectoderm* is the germ layer that forms the epidermis, neural tube, and neural crest. *Epithelial* refers to a sheet of cells that are held together tightly (as opposed to the loosely connected *mesenchymal* cells; see Chapter 3). Epithelia can be produced by any germ layer. The epidermis and the neural tube both happen to be ectodermal epithelia; the lining of the gut is an *endodermal* epithelium.

specification of one tissue also involves blocking the specification of an alternative tissue.

The epidermis is only one cell layer thick to start with, but in most vertebrates it soon becomes a two-layered structure. The outer layer gives rise to the **periderm**, a temporary covering that is shed once the inner layer differentiates to form a true epidermis. The inner layer, called the **basal layer** or **stratum germinativum**, contains epidermal stem cells† attached to a basal lamina that the stem cells themselves help to make (**FIGURE 10.38**). Like the neural stem cells of the ependymal layer, the epidermal stem cells divide asymmetrically: the daughter cell that remains attached to the basal lamina remains a

---

†In addition to these interfollicular stem cells (those in the basal skin between hair follicles), it is probable that there are also epidermal stem cells in the permanent portion of the hair (the "bulge," discussed in the next section) that may be used in emergencies when more epidermis is needed (Blanpain et al. 2007; Janich et al. 2011).

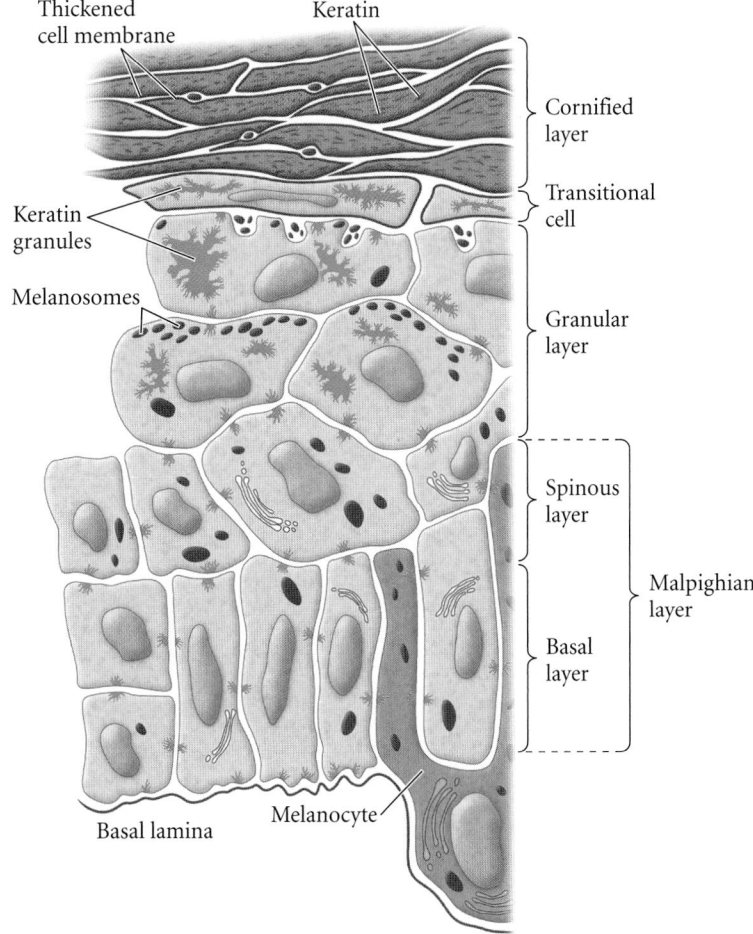

FIGURE 10.38 Layers of the human epidermis. The basal cells are mitotically active, whereas the fully keratinized cells characteristic of external skin are dead and are continually shed. The keratinocytes obtain their pigment through the transfer of melanosomes from the processes of melanocytes that reside in the basal layer. (After Montagna and Parakkal 1974.)

stem cell, while the cell that leaves the basal layer migrates outward and starts differentiating. Just as in neural stem cells, this differentiation is positively regulated by the Notch pathway (Nguyen et al. 2006; Aguirre et al. 2010). In the absence of Notch signaling, there is hyperproliferation of the dividing cells (Ezratty et al. 2011). This Notch signal promotes the synthesis of the keratins characteristic of skin and joins them into dense intermediate filaments (Lechler and Fuchs 2005; Williams et al. 2011). It is likely that all the basal layer cells of the adult mouse skin have stem cell properties (Clayton et al. 2007).

Cell division from the basal layer produces younger cells and pushes the older cells to the border of the skin. This is unlike the "inside-out" patterning of the neural tube, where newly generated neurons migrate through the layers of older cells on their way to the periphery. (But the neural cells don't form anew each day like the epidermal cells.) After the synthesis of the differentiated products, the cells cease transcriptional and metabolic activities. These differentiated epidermal cells, the **keratinocytes**, are bound tightly together and produce a water-impermeable seal of lipid and protein.

As they reach the surface, the cells are dead, flattened sacs of keratin protein, and their nuclei are pushed to one edge of the cell. These cells constitute the **cornified layer**, or **stratum corneum**. Throughout life, the dead keratinocytes of the cornified layer are shed—humans lose about 1.5 grams of these cells every day*—and are replaced by new cells. In mice, the journey from the basal layer to the sloughed cell takes about 2 weeks. Human epidermis turns over a bit more slowly. The proliferative ability of the basal layer is remarkable in that it can supply the cellular material to continuously replace 1–2 m² of skin for several decades throughout adult life.

Several factors stimulate development of the epidermis. BMPs help initiate epidermal production by inducing the p63 transcription factor in the basal layer. This transcription factor's multiple roles may depend in part on different splicing isoforms of p63 that are expressed in the epidermis. The p63 protein is required for keratinocyte proliferation and differentiation (Truong and Khavari 2007); it also appears to stimulate the production of the Notch ligand Jagged. Jagged is a juxtracrine protein in the basal cells that activates the Notch protein on the cells above them, activating the keratinocyte differentiation pathway and preventing further cell divisions (see Mack et al. 2005; Blanpain and Fuchs 2009).

The Cutaneous Appendages: Mammalian Hair

The epidermis and dermis interact at specific sites to create the sweat glands and the **cutaneous appendages**: hairs, scales, scutes (such as the coverings of turtle shells), or feathers, depending on the species. The formation of these appendages requires a series of reciprocal inductive interactions between the dermal mesenchyme and the ectodermal epithelium, resulting in the formation of epidermal placodes (thickenings) that are the precursors of hair follicles. Epidermal cells in the regions capable of forming these placodes secrete Wnt protein, and Wnt signaling is critical for the initiation of follicle development (see Reddy et al. 2001; Andl et al. 2002).

Hair follicles and hair generation in mammals

In mammals, the first indication that a hair follicle placode will form at a particular place is an aggregation of cells in the basal layer of the epidermis (**FIGURE 10.39A,B**). This aggregation is directed by the underlying dermal fibroblast cells and occurs at different times and different places in the embryo. The basal epidermal cells elongate, divide, and sink into the dermis. The dermal fibroblasts respond to this ingression of epidermal cells by forming a small node (the **dermal papilla**) beneath the hair germ (**FIGURE 10.39C**). The dermal papilla then pushes up on the basal stem cells and stimulates them to divide more rapidly. The basal cells respond by producing postmitotic cells that will differentiate into the keratinized hair shaft (**FIGURE 10.39D–G**; see Hardy 1992; Miller et al. 1993; Duverger and Morasso 2009).

Wnt signaling is critical in placode formation. Forced expression of the Wnt-inhibitor Dickkopf, or deletion of epithelial β-catenin, precludes placode formation, whereas mutation of epithelial β-catenin to a constitutively active form results in adoption of hairlike fate by the entire surface ectoderm. This indicates that ectodermal β-catenin signaling determines hair follicle versus epidermal fate (Gat et al. 1998; Andl et al. 2002; Närhi et al. 2008; Zhang et al. 2008, 2009). The patterning of the surface ectoderm into regions of hair follicles and interfollicular epidermis is thought to rely on a reaction-diffusion mechanism based on competition between placode promoting-factors such as Wnts and secreted inhibitors, including the Dickkopf family of Wnt inhibitors and several BMP family members (**FIGURE 10.40**; Jiang et al. 2004; Mou et al. 2006; Sick et al. 2006; Bazzi et al. 2007).

In one such regulatory loop, Wnt activates expression of its own inhibitor, Dickkopf (Dkk), in placodes. As Dkk is thought to be more diffusible than Wnt ligands, it may suppress placode fate in cells adjacent to the placode, thereby contributing to patterning (Sick et al. 2006; Bazzi et al. 2007). Expression of Dkk is additionally enhanced by another signaling pathway that is crucial for placode formation: the ectodysplasin (EDA) pathway† (Fliniaux et al. 2008; Zhang et al. 2009).

Once placodes are established, signals from each placode induce clumping of underlying dermal fibroblasts, forming a dermal condensate (see Figure 10.39B). Reciprocal signaling between the condensed dermis and the epithelial placode causes the placodal tissue to proliferate and extend downward into the dermis. Eventually, the epithelium surrounds the dermal condensate, which develops into the hair follicle dermal

---

*Most of this skin becomes "house dust" on furniture and floors. If you doubt this, burn some dust; it will smell like singed skin.

†Interestingly, such Wnt signaling interactions occur during healing of large wounds in adult mouse skin, resulting in regeneration of follicles within the wound (Ito et al. 2007).

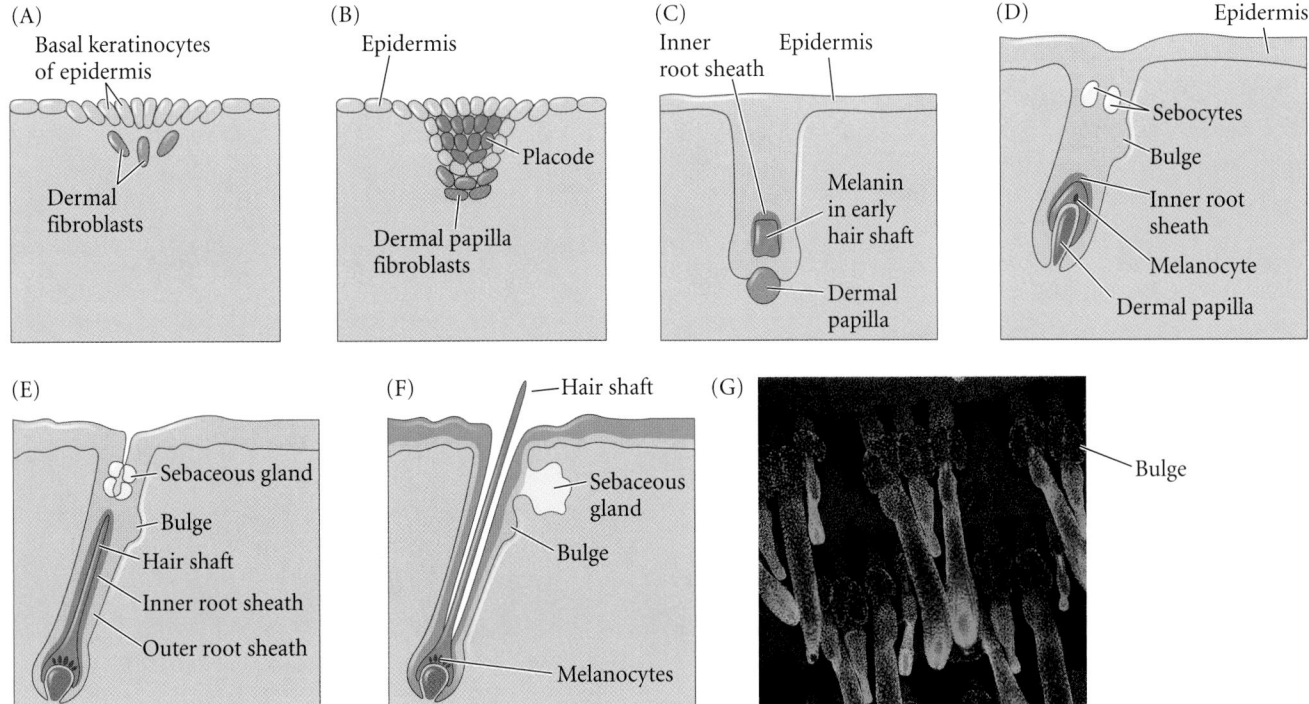

FIGURE 10.39 Early development of the hair follicle and hair shaft. (A) Signals initiate local proliferation of the basal keratinocytes in the epidermis. (B) Proliferation of epidermal cells results in formation of the hair follicle placode, which signals the dermal mesenchymal cells to aggregate beneath it into a dermal papilla. (C) The papilla signals the proliferation of the hair germ, making it into a primitive hair shaft ("hair peg"). (D) The hair shaft engulfs the dermal papilla and forms the inner hair root directly above the papilla. Sebaceous cells (sebocytes) and the bulge appear as melanin granules enter into the cortex. (E) Sebaceous glands form, and the hair canal is made. The hair shaft differentiates the inner root sheath of epidermal cells. (F) The sebaceous gland is localized on the lateral wall of the follicle, while the hair shaft extends into the hair canal and out past the skin. (G) Confocal image taken through mouse hair follicles. The bulge is near the top (dark blue) and is connected to the epidermis by the arrector pili muscles (red). The keratin-producing cells of the hair shaft are green; nuclei are stained blue. (A–F after Philpott and Paus 1998; G courtesy of I. Smyth.)

papilla, an important signaling center in the mature hair follicle. Further proliferation and differentiation of the epithelial cells result in the formation of the inner root sheath and hair shaft of the mature follicle, processes that are likely to require lateral communication between epithelial cells (see Hardy 1992; Millar 2002). As the hair follicle matures, an epithelial swelling begins to form on the periphery of the hair germ and eventually develops into the **sebaceous gland** (see Figure 10.39E,F). Sebaceous glands produce an oily secretion known as **sebum** that functions to lubricate the hair follicles and skin.

The first hairs in the human embryo are of a thin, closely spaced type called **lanugo**. This type of hair is usually shed before birth and is replaced (at least partially by new follicles) by the short and silky **vellus hair**. Vellus hair remains on many parts of the human body that are usually considered hairless, such as the forehead and eyelids. In other areas of the body, vellus hair gives way to longer, thicker "terminal" hair. During a person's life, some of the same follicles that first produce vellus hair later form terminal hair and still later revert to vellus hair production. Hair follicles in the armpits, for example, grow vellus hair until adolescence, when they begin producing terminal shafts in response to androgens.

Conversely, in male pattern baldness, scalp hair follicles revert to producing unpigmented and very fine vellus hair (Montagna and Parakkal 1974).

Adult stem cells and hair regeneration

There appear to be three stem cell populations involved in producing epidermal structures. One, discussed earlier, is found in the germinal layer of the epidermis and generates the keratinocytes that characterize the interfollicular epidermis. A second group of stem cells is critical for forming the sebaceous gland of each hair shaft; and a third group is critical for regenerating the hair shaft itself. Interestingly, it seems that there is also a primitive stem cell that can form all the others (Snippert et al. 2010), and members of each stem cell group can be recruited to any of the other three pools if needed (Levy et al. 2007; Fuchs et al. 2008).

Hair is one structure that mammals are able to regenerate. Throughout life, hair follicles undergo cycles of growth (known as **anagen**), regression (**catagen**), rest (**telogen**), and regrowth. Hair length is determined by the amount of time the hair follicle spends in the anagen phase. Human scalp hair can spend several years in anagen, whereas arm hair grows

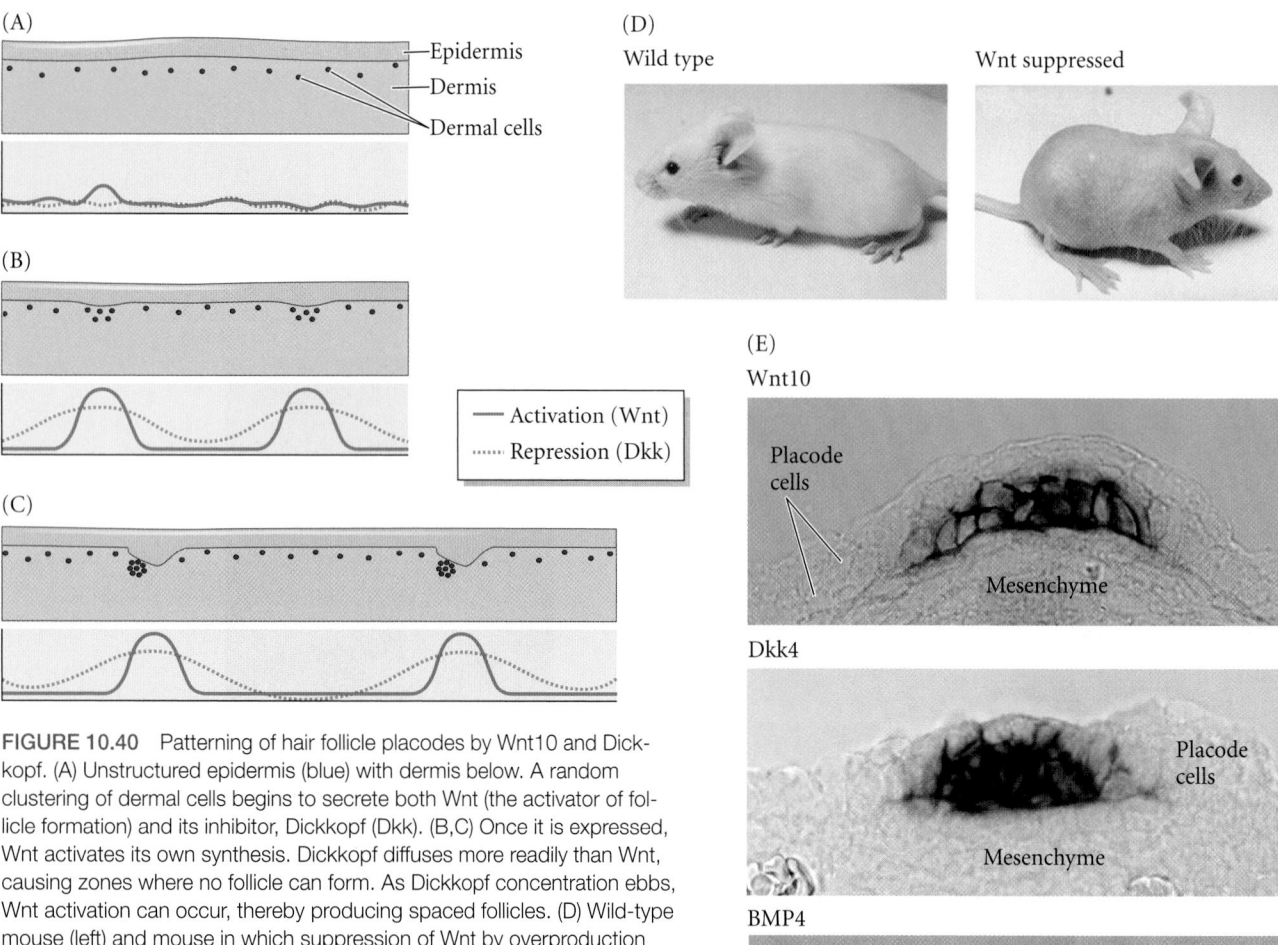

FIGURE 10.40 Patterning of hair follicle placodes by Wnt10 and Dick-kopf. (A) Unstructured epidermis (blue) with dermis below. A random clustering of dermal cells begins to secrete both Wnt (the activator of follicle formation) and its inhibitor, Dickkopf (Dkk). (B,C) Once it is expressed, Wnt activates its own synthesis. Dickkopf diffuses more readily than Wnt, causing zones where no follicle can form. As Dickkopf concentration ebbs, Wnt activation can occur, thereby producing spaced follicles. (D) Wild-type mouse (left) and mouse in which suppression of Wnt by overproduction of Dickkopf has severely lowered the density of hair follicles (right). (E) Wnt and Dickkopf are both synthesized in the placode cells as they aggregate. BMP4 is made in the mesenchymal cells beneath them. (A–C after Schlake and Sick 2007; D from Sick et al. 2006, photographs courtesy of S. Sick; E from Bazzi et al. 2007.)

for only 6–12 weeks in each cycle. The ability of hair follicles to regenerate depends on the existence of a population of epithelial stem cells that forms in the permanent **bulge** region of the follicle late in embryogenesis. When Philipp Stöhr drew the histology of the human hair for his 1903 textbook, he showed this bulge (*Wulst*) as the attachment site for the arrector pili muscles (which give a person "goosebumps" when they contract). Research carried out during the 1990s suggested that the bulge houses populations of at least two remarkable adult stem cells: the **hair follicle stem cells** (HFSCs), which gives rise to the hair shaft and sheath (Cotsarelis et al. 1990; Morris and Potten 1999; Taylor et al. 2000); and the **melanocyte stem cells**, which give rise to the pigment of the skin and hair (see Chapter 11; Nishimura et al. 2002). The bulge appears to be a niche that allows adult cells to retain the quality of "stemness." Stem cells in the bulge can regenerate all the epithelial cell types of the hair, and their selective destruction causes the loss of the follicle.

The entire skin organ seems to be involved in hair cycling (**FIGURE 10.41**). The HFSCs reside in the outer layer of the bulge. The inner bulge cells are the progeny of the HFSCs, and they secrete BMP5 and Fgf18, two repressors of HFSC proliferation. In addition, the dermal fibroblasts and the subcutaneous fat cells also make growth-suppressive BMPs. The HFSCs are activated at the beginning of the anagen phase by signals from the dermal papilla of condensed mesenchyme. These signals are FGFs, Wnts, and BMP antagonists, and they direct the epidermal stem cells to migrate out of the bulge. There, the epidermal stem cells produce transit amplifying cells that proliferate downward and generate the seven concentric columns of cells that form the outer root sheath from bulge to matrix.

This activation of the dermal papilla appears to be regulated by the microenvironment of the dermis. The underlying

(A) Telogen
(Stem cell quiescence)

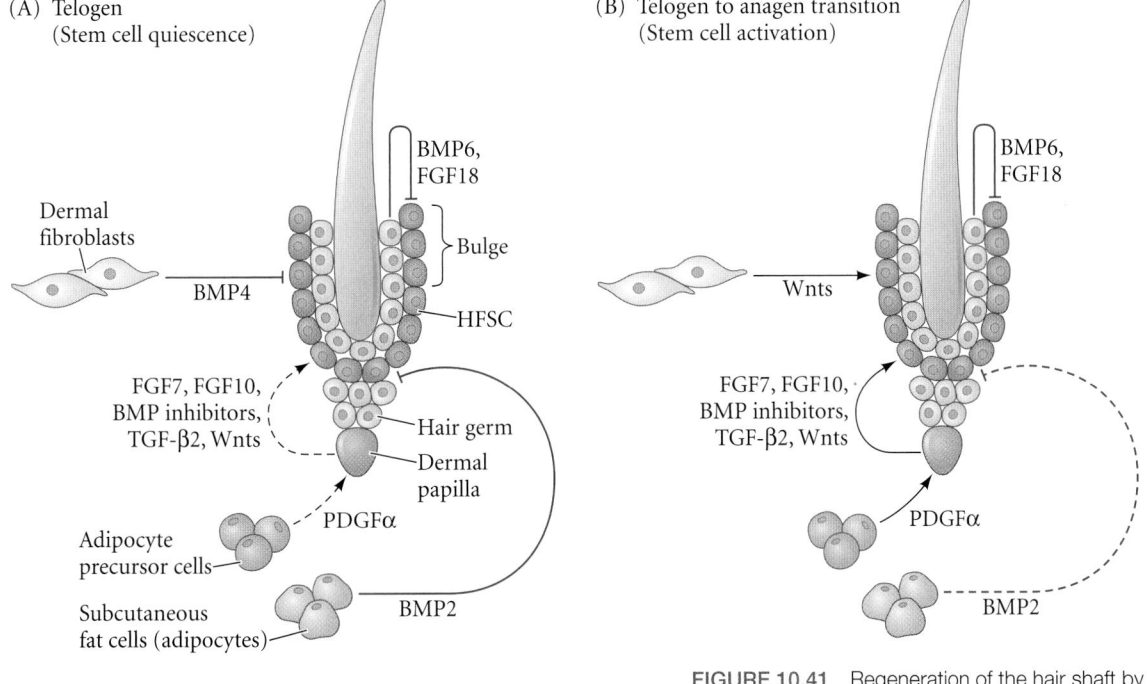

(B) Telogen to anagen transition
(Stem cell activation)

(C) Anagen

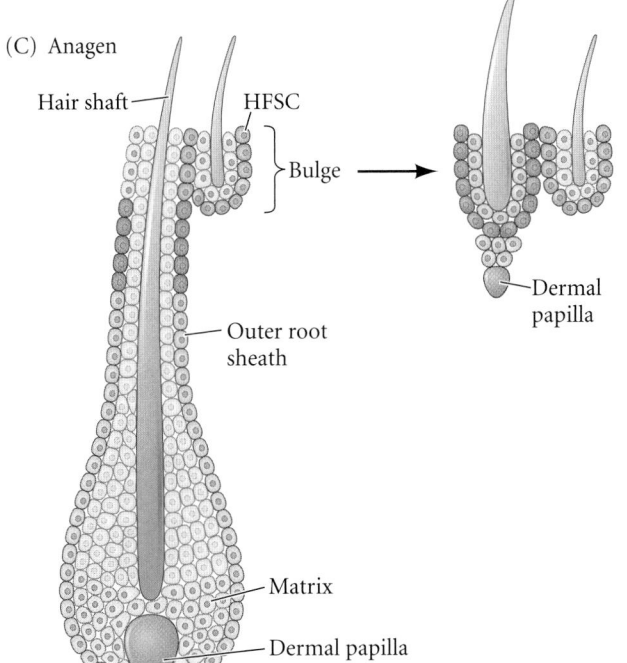

FIGURE 10.41 Regeneration of the hair shaft by the bulge hair follicle stem cells (HFSCs). (A) During quiescence (telogen), the condensed mesenchyme of the dermal papilla contacts the stem cells (blue) in the outer layer of the bulge. The HFSCs are quiescent because of BMP6 and Fgf18 produced by the inner bulge layer, whose cells descend from the HFSC of the outer bulge, and because of other BMPs produced by dermal mesoderm (fibroblasts) and fat cells (adipocytes). (B) At the transition from telogen to anagen (growth), the dermal papilla is induced by the mesenchyme cells to produce activators of hair growth (FGFs and Wnts) as well as antagonists of BMP signals. This causes the proliferation and differentiation of the hair follicle. (C) Contacting the dermal papilla at the base, the cells divide rapidly to generate the hair shaft and its channel. The cells close to the earlier bulge become the outer layer of the bulge and have stem cell properties. An inner layer, several cell layers thick, is also derived from the HFSCs, and it eventually inhibits HFSC proliferation. At catagen (not shown), most cells undergo apoptosis, but those remaining stem cells survive in the bulge region. An epithelial strand then brings up the dermal papilla to the region of the bulge, and the interaction between them appears to generate the hair germ of the next generation of hair. (After Hsu and Fuchs 2012.)

dermis appears to make more Wnts and less BMPs, and the adipocyte precursor cells make more of the paracrine factor PDGF, which stimulates the dermal papilla (Närhi et al. 2008; Plikus et al. 2008; Rendl et al. 2008; Hsu and Fuchs 2012). As the dermal papilla is moved farther away by the growth of the sheath cells, its signals are not received by the hair germ, and the hair germ return to quiescence. During the latter part of anagen, prostaglandins appear to prevent the production of transit amplifying cells. During catagen, most of the hair shaft cells undergo apoptosis. However, upper follicular stem cells remain. The outer root sheath cells close to the old bulge contain HFSCs and become the outer layer, while those nearer to the matrix differentiate to become the inner layer of the bulge. The apoptosis causes the outer cells to be in contact with the dermal papilla in readiness for the next cycle of activation (Hsu et al. 2011).

Male pattern baldness, which is characterized by a decrease in hair follicle size, appears to be caused by the progressive inability of the HFSCs to generate the transit amplifying progenitor cells. This cessation of progenitor cell production appears to be due to the prolonged synthesis of prostaglandin PGD2, which is normally used to stop hair growth at the end of the anagen phase. Bald men have higher levels of this factor, and transgenic mice overexpressing the enzymes leading to prostaglandin PGD2 synthesis have hair loss. Moreover, the genes encoding the enzymes producing this prostaglandin are upregulated by testosterone* (Garza et al. 2011, 2012).

---

*Testosterone and (more importantly, as we will see in Chapter 15) its derivative hydrotestosterone are critical in producing male pattern baldness. Ancient civilizations noted that eunuchs (castrated males) did not become naturally bald. Testosterone does not appear to play roles in female hair thinning (Kaufman 2002).

● See **WEBSITE 10.5** Normal variation in human hair production

● See **WEBSITE 10.6** Ectodysplasin pathway and other mutations of hair development

Coda

We have now reached the point where the vertebrate ectoderm has formed two of its major derivatives, the nervous system and epidermis. The body is covered by an epidermis, and there is an internalized central nervous system. Both are forming regionalized structures. In the next chapter we will define further regions of the developing ectoderm, placodes, and neural crest cells, and we will look at how the neurons are led to their targets.

SNAPSHOT SUMMARY: Neural and Epidermal Ectoderm

1. The neural tube forms from the shaping and folding of the neural plate. In primary neurulation, the surface ectoderm folds into a tube that separates from the surface. In secondary neurulation, the ectoderm forms a cord, then forms a cavity within the cord.

2. Primary neurulation is regulated both by intrinsic and extrinsic forces. Intrinsic wedging occurs within cells of the hinge regions, bending the neural plate. Extrinsic forces include the migration of the surface ectoderm toward the center of the embryo.

3. Neural tube closure is also the result of extrinsic and intrinsic forces. In humans, congenital anomalies can result if the neural tube fails to close. Folate is important in mediating neural tube closure.

4. After the node has reached the posterior of the epiblast, certain cells contribute to both the paraxial mesoderm and the neural tube.

5. The neural crest cells arise at the borders of the neural tube and surface ectoderm. They become located between the neural tube and surface ectoderm, and they migrate away from this region to become peripheral neural, glial, and pigment cells.

6. There is a gradient of maturity in many embryos (especially those of amniotes) so that the anterior develops earlier than the posterior.

7. The brain forms three primary vesicles: prosencephalon (forebrain), mesencephalon (midbrain), and rhombencephalon (hindbrain). The prosencephalon and rhombencephalon become subdivided.

8. The brain expands as fluid secretion puts positive pressure on the vesicles.

9. The dorsal-ventral patterning of the neural tube is accomplished by proteins of the TGF-β superfamily secreted from the surface ectoderm and the roof plate of the neural tube, and by Sonic hedgehog protein secreted from the notochord and floor plate cells. Temporal and spatial gradients of Shh trigger the synthesis of particular transcription factors that specify the neuroepithelium.

10. Dendrites receive signals from other neurons, axons transmit signals to other neurons. The gap between cells where signals are transferred from one neuron to another (through the release of neurotransmitters) is called a synapse.

11. Axons grow from the nerve cell body, or soma. They are led by the growth cone.

12. Neural stem cells have been observed in the adult human brain. Humans can continue making neurons throughout life, although at nowhere near the fetal rate.

13. The neurons of the brain are organized into laminae (layers) and nuclei (clusters).

14. New neurons are formed by the division of neural stem cells in the wall of the neural tube (called the ventricular zone). The resulting neural precursors, or neuroblasts, can migrate away from the ventricular zone and form a new layer, called the mantle zone (gray matter). Neurons forming later have to migrate through the existing layers. This process forms the cortical layers.

15. In the cerebellum, migrating neurons form a second germinal zone, called the external granular layer. Other neurons

migrate out of the ventricular zone on the processes of glial cells.

16. The cerebral cortex in humans, called the neocortex, has six layers. Cell fates are often fixed as they undergo their last division. Neurons derived from the same stem cell may end up in different functional regions of the brain.

17. The specification of one type of cell is often done in parallel with blocking the specification of an alternative cell type. Thus, specifying a cell to become epidermis also involves preventing the cell from becoming neural.

18. Human brains appear to differ from those of other primates by their retention of the fetal neuronal growth rate during early childhood, the high transcriptional activity, the presence of human-specific alleles of developmental regulatory genes, including the loss of transcriptional regulators.

19. The vertebrate retina forms from an optic vesicle that extends from the brain. Pax6 plays a major role in eye formation, and the downregulation of Pax6 by Sonic hedgehog (Shh) in the center of the brain splits the eye-forming

region of the brain in half. If Shh is not expressed there, a single medial eye results.

20. The photoreceptor cells of the retina gather light and transmit an electric impulse through interneurons to the retinal ganglion cells. The axons of the retinal ganglion cells form the optic nerve.

21. The lens and cornea form from the surface ectoderm. Both must become transparent.

22. Reciprocal induction is critical in the specification and differentiation of the retina and lens. The cells that form the organs have two "lives." In the embryonic life, they construct the organs; in the adult life, they function as part of an organ. The body is constructed by cells that are not performing their adult roles.

23. The basal layer of the surface ectoderm becomes the germinal layer of the skin. Epidermal stem cells divide to produce differentiated keratinocytes and more stem cells.

24. The hair follicular stem cells, which regenerate hair follicles during periods of cyclical growth, reside in the bulge of the hair follicle.

For Further Reading

Complete bibliographical citations for all literature cited in this chapter can be found at the free-access website **www.devbio.com**

Balaskas, N. and 7 others. 2012. Gene regulatory logic for reading the Sonic hedgehog signaling gradient in the vertebrate neural tube. *Cell* 14: 273–284.

Catala, M., M.-A. Teillet, E. M. De Robertis and N. M. Le Douarin. 1996. A spinal cord fate map in the avian embryo: While regressing, Hensen's node lays down the notochord and floor plate thus joining the spinal cord lateral walls. *Development* 122: 2599–2610.

Erikksson, P. S., E. Perfiliea, T. Björn-Erikksson, A.-M. Alborn, C. Nordberg, D. A. Peterson and F. H. Gage. 1998. Neurogenesis in the adult human hippocampus. *Nature Med.* 4: 1313–1317.

Halder, G., P. Callaerts and W. J. Gehring. 1995. Induction of ectopic eyes by targeted expression of the *eyeless* gene in *Drosophila. Science* 267: 1788–1792.

Hatten, M. E. 1990. Riding the glial monorail: A common mechanism for glial-guided neuronal migration in different regions of the mammalian brain. *Trends Neurosci.* 13: 179–184.

Hsu, Y. C., H. A. Pasolli and E. Fuchs. 2011. Dynamics between stem cells, niche, and progeny in the hair follicle. *Cell* 144: 92-105.

Jessell, T. M. 2000. Neuronal specification in the spinal cord: Inductive signals and transcriptional codes. *Nature Rev. Genet.* 1: 20–210.

Lawson, A., H. Anderson and G. C. Schoenwolf. 2001. Cellular mechanisms of neural fold formation and morphogenesis in the chick embryo. *Anat. Rec.* 262: 153–168.

McLean, C. Y. and 12 others. 2011. Human-specific loss of regulatory DNA and the evolution of human-specific traits. *Nature* 471: 216–219.

Milunsky, A., H. Jick, S. S. Jick, C. L. Bruell, D. S. Maclaughlen, K. J. Rothman and W. Willett. 1989. Multivitamin folic acid supplementation in early pregnancy reduces the prevalence of neural tube defects. *J. Am. Med. Assoc.* 262: 2847–2852.

Ogino, H., H. Ochi, H. M. Reza and K. Yasuda. 2012. Transcription factors involved in lens development from the preplacodal ectoderm. *Dev. Biol.* 363: 333–347.

Sick, S., S. Reinker, J. Timmer and T. Schlake. 2006. WNT and DKK determine hair follicle spacing through a reaction-diffusion mechanism. *Science* 314: 1447–1450.

Turner, D. L. and C. L. Cepko. 1987. A common progenitor for neurons and glia persists in rat retina late in development. *Nature* 328: 131–136.

Varki, A., D. H. Geschwind and E. E. Eichler. 2008. Explaining human uniqueness: Genome interactions with environment, behaviour, and culture. *Nature Rev. Genet.* 9: 749–763.

Go Online

WEBSITE 10.1 Specifying the brain boundaries. The Pax transcription factors and the paracrine factor Fgf8 are critical in establishing the boundaries of the forebrain, midbrain, and hindbrain.

WEBSITE 10.2 Horizontal and vertical specification of the cerebrum. Neither the vertical not horizontal organization of the cerebral cortex is clonally specified. Cell migration is critical, and paracrine factors from neighboring cells play major roles in migration and specification.

WEBSITE 10.3 Neuronal growth and the invention of childhood. An interesting hypothesis claims that the caloric requirements of this brain growth necessitated a new stage of the human life cycle—childhood—during which the child is actively fed by adults.

WEBSITE 10.4 Why babies don't see well. The retinal photoreceptors are not fully developed at birth. As the child grows older, the density of photoreceptors increases, allowing far better discrimination and nearly 350 times the light-absorbing capacity that is present at birth.

WEBSITE 10.5 Normal variation in human hair production. The human hair has a complex life cycle. Moreover, some hairs (such as those of our eyelashes) grow short while other hairs (such as those of our scalp) grow long. The pattern of hair size and thickness (or lack thereof) is determined by paracrine and endocrine factors.

WEBSITE 10.6 Ectodysplasin pathway and other mutations of hair development. Some people are born without the ability to grow hair, whereas others develop hair over their entire bodies. Other conditions prevent the growth of hair, teeth, and mammary glands. These genetic conditions give us insights into the mechanisms of normal hair growth.

Vade Mecum

Chick neurulation. By 33 hours of incubation, neurulation in the chick embryo is well underway. Both whole mounts and a complete set of serial cross sections through a 33-hour chick embryo are included in the "Chick-Mid" segment so you can see this amazing event. The serial sections can be displayed either as a continuum in movie format or individually, along with labels and color-coding that designates germ layers.

Other Resources

The University of New South Wales Embryology site has excellent programs for understanding human development, especially medically relevant areas of human embryology. **php.med.unsw.edu.au/embryology/index.php?title=Main_Page**

The National Institutes of Health public website on neural tube disease has vetted the websites of several agencies specializing in different topics. **www.nlm.nih.gov/medlineplus/neuraltubedefects.html**

The Embryo Project has several articles on special topics and historical topics and presents them in an integrated format. **embryo.asu.edu/**

11

Neural Crest Cells and Axonal Specificity

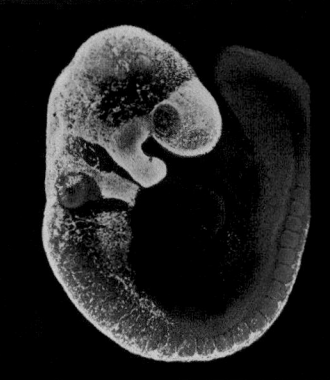

THIS CHAPTER CONTINUES THE DISCUSSION of ectodermal development, focusing on two remarkable entities: (1) the neural crest cells, which are responsible for generating the facial skeleton, pigment cells, and peripheral nervous system; and (2) the nerve axons, whose growth cones guide them to their destinations. Neural crest cells and axon growth cones share the property of having to migrate far from their source of origin to specific places in the embryo; both must recognize and respond to signals that guide them along specific routes to their final destination. Moreover, neural crest cells and axonal growth cones recognize many of the same signals.

THE NEURAL CREST

Although it is derived from the ectoderm, the neural crest is so important that it has sometimes been called the "fourth germ layer" (see Hall 2009). It has even been said, somewhat hyperbolically, that "the only interesting thing about vertebrates is the neural crest" (quoted in Thorogood 1989). Certainly, the emergence of the neural crest is one of the pivotal events of animal evolution, as it led to the jaws, face, skull, and sensory ganglia of the vertebrates (Northcutt and Gans 1983). The neural crest is a transient structure. Adults do not have a neural crest, nor do late-stage vertebrate embryos. Rather, the cells of the neural crest undergo an epithelial-mesenchymal transition from the dorsal neural tube, after which they migrate extensively to generate a prodigious number of differentiated cell types (FIGURE 11.1A,B; TABLE 11.1).

The neural crest is a population of cells that can produce tissues as diverse as (1) the neurons and glial cells of the sensory, sympathetic, and parasympathetic nervous systems; (2) the epinephrine-producing (medulla) cells of the adrenal gland; (3) the pigment-containing cells of the epidermis; and (4) many of the skeletal and connective tissue components of the head. It remains uncertain whether the majority of the individual cells that leave the neural crest are multipotent or whether most are already restricted to certain fates. Bronner-Fraser and Fraser (1988, 1989) provided evidence that many individual trunk neural crest cells are multipotent as they leave the crest. They injected fluorescent dextran molecules into individual chick neural crest cells while the cells were still within the neural tube, and then looked to see what types of cells their descendants became after migration. The progeny of a single neural crest cell could become sensory neurons, melanocytes (pigment-forming cells), glia (including Schwann cells), and adrenomedullary cells (FIGURE 11.1C). Henion and Weston (1997) also found that the initial avian trunk neural crest population was a heterogeneous mixture of precursor cells, and that nearly half of the cells that emerge from the neural crest only generate a single cell type. Recent studies of the cranial region of the chick neural crest provide evidence that a large majority of early migrating neural crest cells can generate nearly all the numerous cells types (Calloni et al. 2009).

Reviewing the results of numerous laboratories as well as their own studies, Nicole Le Douarin and others proposed a model whereby an original multipotent neural crest cell

(A)

(B)

(C)

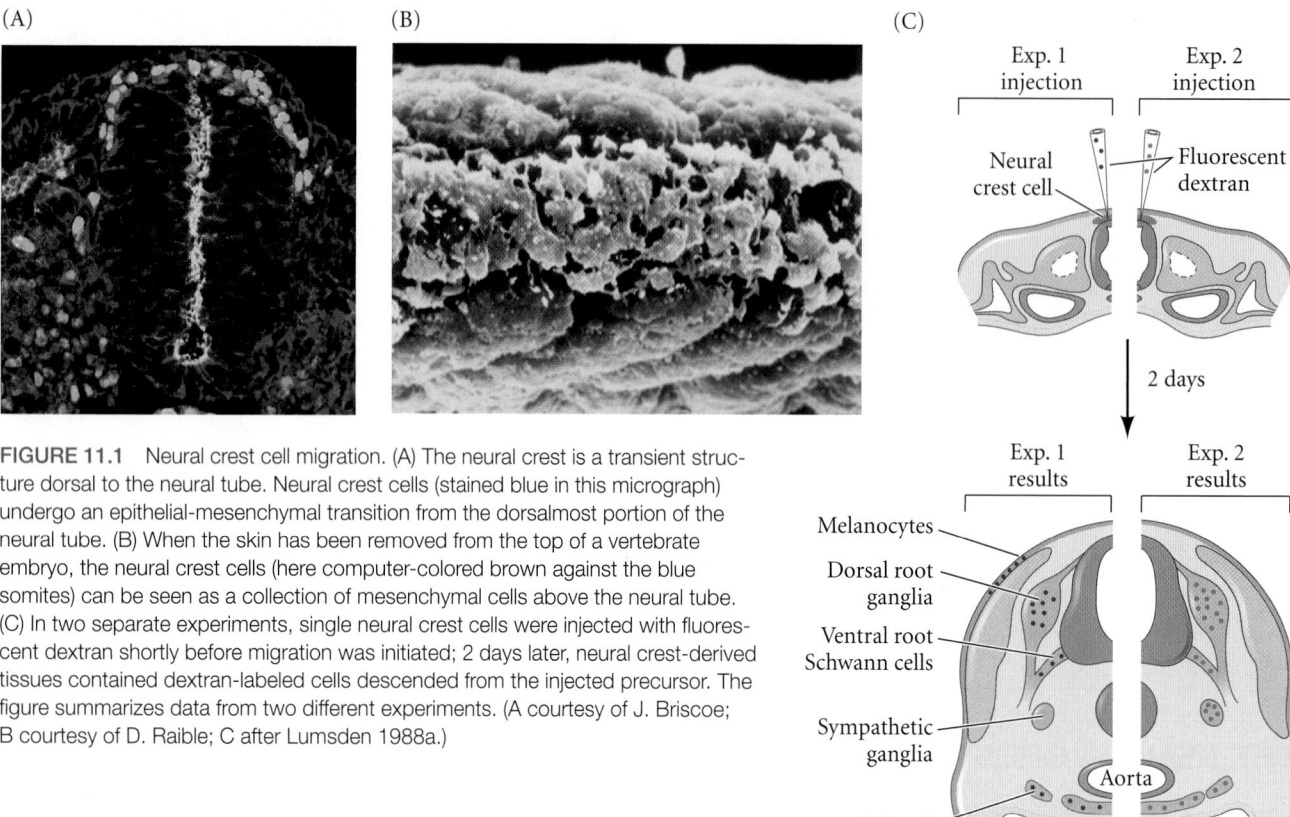

FIGURE 11.1 Neural crest cell migration. (A) The neural crest is a transient structure dorsal to the neural tube. Neural crest cells (stained blue in this micrograph) undergo an epithelial-mesenchymal transition from the dorsalmost portion of the neural tube. (B) When the skin has been removed from the top of a vertebrate embryo, the neural crest cells (here computer-colored brown against the blue somites) can be seen as a collection of mesenchymal cells above the neural tube. (C) In two separate experiments, single neural crest cells were injected with fluorescent dextran shortly before migration was initiated; 2 days later, neural crest-derived tissues contained dextran-labeled cells descended from the injected precursor. The figure summarizes data from two different experiments. (A courtesy of J. Briscoe; B courtesy of D. Raible; C after Lumsden 1988a.)

TABLE 11.1 Some derivatives of the neural crest

Derivative	Cell type or structure derived
Peripheral nervous system (PNS)	Neurons, including sensory ganglia, sympathetic and parasympathetic ganglia, and plexuses NeuroGlial cells Schwann cells and other glial cells
Endocrine and paraendocrine derivatives	Adrenal medulla Calcitonin-secreting cells Carotid body type I cells
Pigment cells	Epidermal pigment cells
Facial cartilage and bones	Facial and anterior ventral skull cartilage and bones
Connective tissue	Corneal endothelium and stroma Tooth papillae Dermis, smooth muscle, and adipose tissue of skin, head, and neck Connective tissue of salivary, lachrymal, thymus, thyroid, and pituitary glands Connective tissue and smooth muscle in arteries of aortic arch origin

Source: After Jacobson 1991, based on multiple sources.

divides and progressively refines its developmental potentials (**FIGURE 11.2**; see Creuzet et al. 2004; Martinez-Morales et al. 2007; Le Douarin et al. 2008). Thus, the individual cells of the neural crest may differ greatly in their developmental potential and commitments. While some appear to be multipotent precursor cells, others may be more restricted in their developmental fates. Stemple and Anderson (1992) showed that neural crest cells have the ability to self-renew, at least for a limited time, suggesting that neural crest cells have stem cell-like properties.

Specification of Neural Crest Cells

Although neural crest cells are not apparent until they emigrate from the neural tube, induction of these cells first occurs during early gastrulation, at the border between the presumptive epidermis and the presumptive neural plate (the region that will form the central nervous system). The specification of the neural crest at the neural plate-epidermis boundary is a multistep process (see Huang and Saint-Jeannet 2004; Meulemans and Bronner-Fraser 2004). The first step appears to be the specification of the **neural plate border**. The cells in this border between the neural plate and

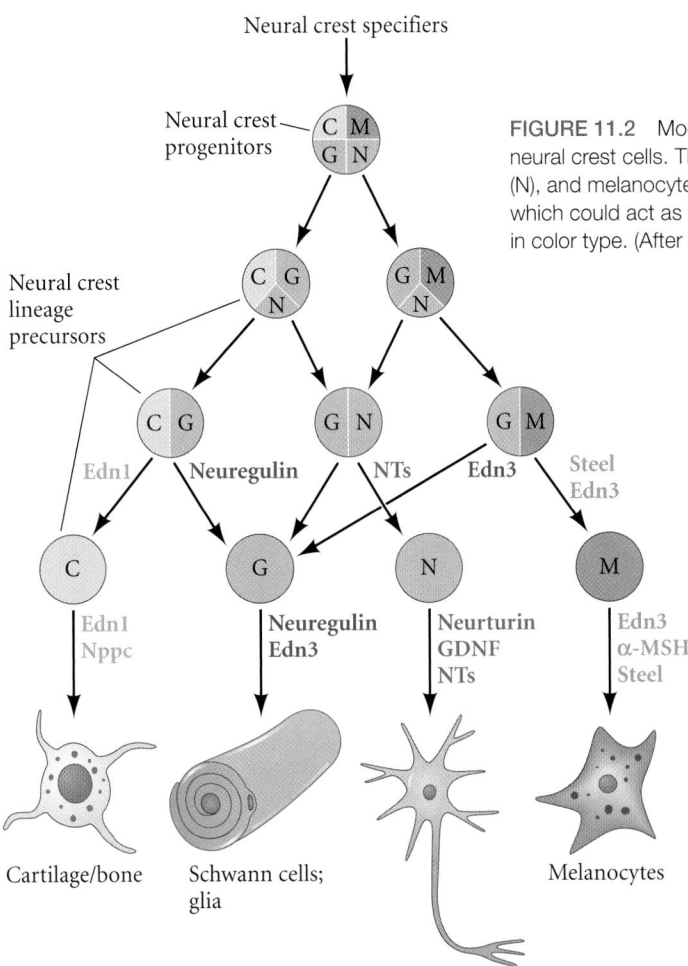

Neural crest specifiers

Neural crest progenitors

Neural crest lineage precursors

Edn1 **Neuregulin** NTs **Edn3** Steel Edn3

Edn1 Nppc **Neuregulin Edn3** **Neurturin GDNF NTs** Edn3 α-MSH Steel

Cartilage/bone Schwann cells; glia Melanocytes

Neurons

FIGURE 11.2 Model for neural crest lineage segregation and the heterogeneity of neural crest cells. The committed precursors of cartilage/bone (C), glia (G), neurons (N), and melanocytes (M) are derived from intermediate progenitor cells, some of which could act as stem cells. The paracrine factors regulating these steps are shown in color type. (After Martinez-Morales et al. 2007.)

the epidermis will become the neural crest and (in the anterior) the **placodes** that generate the eye, ear, nose, and other sensory organs. In amphibians, this border appears to be specified by the interplay between a number of **neural plate inductive signals**, including BMPs, Wnts, and FGFs. Indeed, in the 1940s Raven and Kloos (1945) showed that although the presumptive notochord could induce both the amphibian neural plate and neural crest tissue (presumably blocking nearly all BMPs), the somite and lateral plate mesoderm could induce only the neural crest. In chick embryos, neural crest specification occurs during gastrulation, when the borders between the neural and non-neural ectoderm are still forming (Basch et al. 2006; Schmidt et al. 2007; Ezin et al. 2009). Here, the neural plate inductive signals (especially BMPs and Wnts) secreted from the ventral ectoderm and paraxial mesoderm interact to specify the boundaries.

In the anterior region, the timing of BMP and Wnt expression is critical for discriminating between neural plate, epidermis, placode, and neural crest cell tissues (**FIGURE 11.3A**). As we saw in Chapters 8 and 9, if both BMP and Wnt signaling are continuous, the fate of the ectoderm is epidermal; but if BMP antagonists (e.g., Noggin or FGFs) block BMP signaling, the ectoderm becomes neural. Studies by Patthey and colleagues

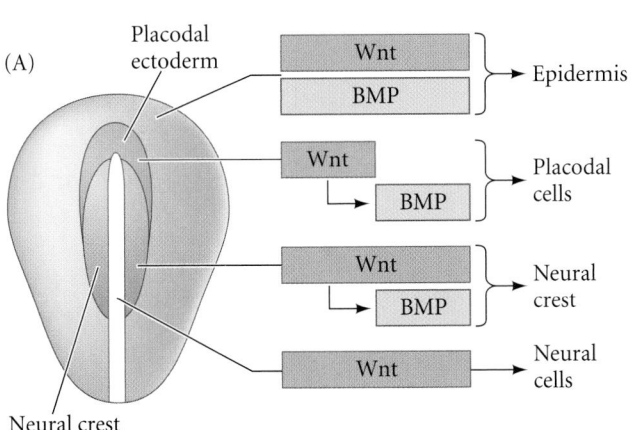

(A)

Placodal ectoderm

Wnt / BMP → Epidermis

Wnt → BMP → Placodal cells

Wnt → BMP → Neural crest

Wnt → Neural cells

Neural crest

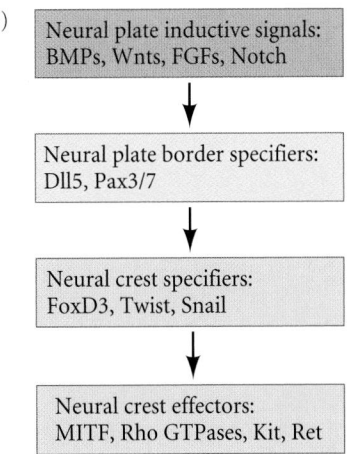

(B)

Neural plate inductive signals: BMPs, Wnts, FGFs, Notch

↓

Neural plate border specifiers: Dll5, Pax3/7

↓

Neural crest specifiers: FoxD3, Twist, Snail

↓

Neural crest effectors: MITF, Rho GTPases, Kit, Ret

FIGURE 11.3 Specification of neural crest cells. (A) The neural plate is bordered by neural crest anteriorly and caudally, and by placodal ectoderm anteriorly. If the ectodermal cells receive both BMP and Wnt for an extended period of time, they become epidermis. If Wnt induces BMPs and is then downregulated, the cells become placodal cells (expressing the placode specifier genes *Six1*, *Six4*, and *Eya2*). If Wnt induces BMP but remains active, these

border cells between the neural plate and epidermis become neural crest (expressing neural crest specifier genes *Pax7*, *Snail2*, and *Sox9*). If they receive Wnt only (because the BMP signal is blocked by Noggin or FGF), the ectodermal cells become neural cells. (B) Stages in the specification of neural crest cells. (A after Patthey et al. 2009; B after Nikitina and Bronner-Fraser 2009.)

(2008, 2009) have shown that if Wnts induce BMPs and then Wnt signaling is turned *off*, the cells become committed to be anterior placodes, whereas if the Wnt signaling induces BMPs but stays *on*, the cells become capable of becoming neural crest. These include the dorsal portion of the neural tube, where cells can migrate to the dorsal midline and emigrate as neural crest cells (McKinney et al. 2013).

The neural plate induces in these border cells a set of transcription factors called **neural plate border specifiers** (FIGURE 11.3B). These factors, including Pax3, Pax7, and Distalless, collectively confer upon the border region the ability to form neural crest as well as dorsal neural tube cell types. The border-specifying transcription factors then induce a second set of more specific transcription factors, the **neural crest specifiers**, in those cells that are to become the neural crest. These include genes encoding the transcription factors Foxd3, Sox9, Sox10, Id, Twist, and Snail.

When Foxd3, Snail, Sox9, and Sox10 are experimentally expressed in the lateral neural tube, these lateral neuroepithelial cells become neural crest-like, undergo epithelial-mesenchymal transition, and delaminate from the neuroepithelium. Sox9 and Snail together are sufficient to induce such a transition in neuroepithelial cells. Sox9 is also required for the survival of trunk neural crest cells after delamination, since in the absence of Sox9, neural crest cells undergo apoptosis as soon as they delaminate. Foxd3 may play many roles.

It is needed for the expression of the cell surface proteins needed for cell migration, and also appears to be critical for the specification of ectodermal cells as neural crest. Inhibiting expression of the *Foxd3* gene inhibits neural crest differentiation. Conversely, when *Foxd3* is expressed ectopically by electroporating the active gene into neural plate cells, those neural plate cells express proteins characteristic of the neural crest (Nieto et al. 1994; Taneyhill et al. 2007; Teng et al. 2008).

The *Sox10* gene appears to be one of the most critical regulators of neural crest specification. It is crucial not only for the delamination of neural crest cells from the neural tube, but also for the differentiation of the numerous neural crest lineages (Kelsh 2006; Betancur et al. 2010). Sox10 protein binds to the enhancers of numerous target genes. These target genes encode the neural crest effectors, which include some transcription factors (such as MITF in the melanocyte lineage that forms pigment cells); small G proteins (such as Rho GTPases) that allow cells to change shape and migrate; and cell surface receptors (such as receptor tyrosine kinases and the endothelin receptor such as EDNRB2) that allow the neural crest cells to respond to patterning and chemotactic proteins in their environments (FIGURE 11.4).

● **See VADE MECUM**   Nicole Le Douarin and the neural crest

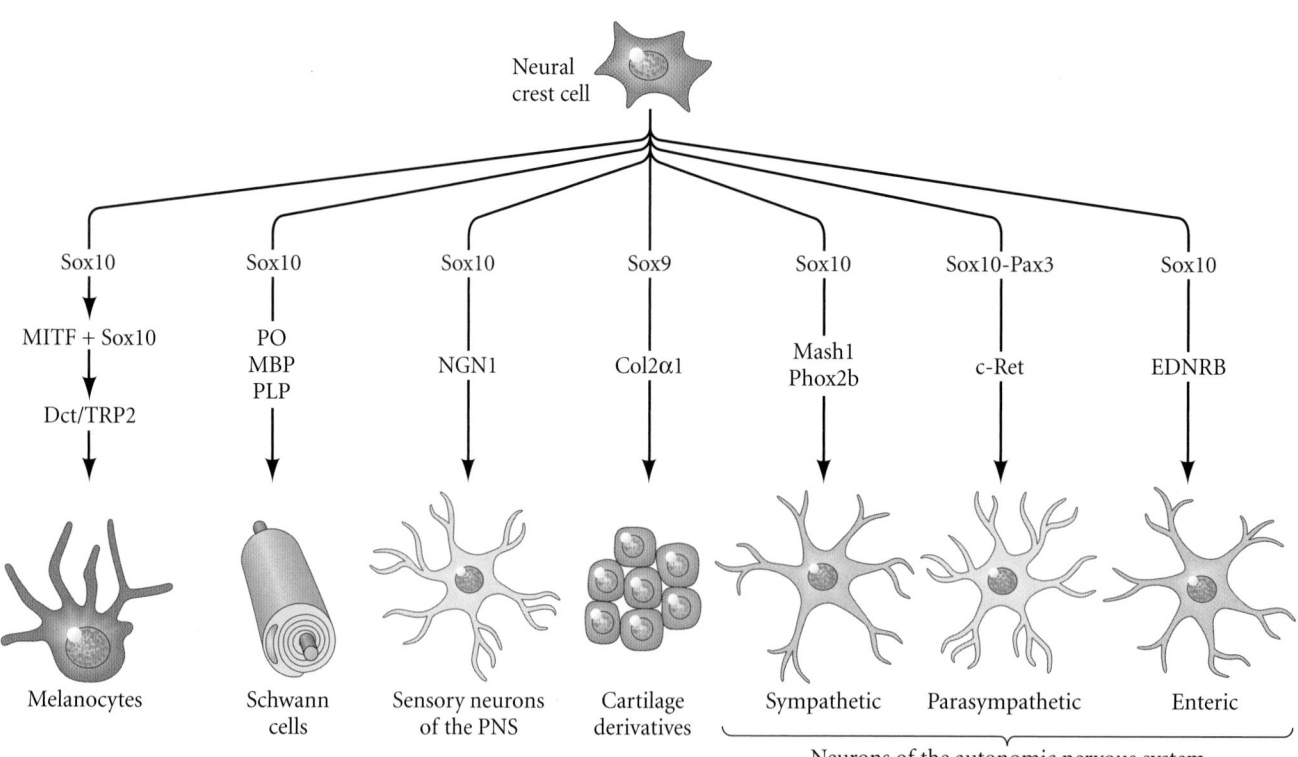

FIGURE 11.4   Sox10 transcription factor proteins mediate neural crest cell differentiation. In the final differentiation processes of neural crest cells, Sox10 binds to enhancers adjacent to genes involved in this terminal differentiation. In the cartilage lineage, a related protein, Sox9, appears to substitute for Sox10. (After Sauka-Spengler and Bronner 2010.)

Regionalization of the Neural Crest

The neural crest is a transient structure whose cells undergo an epithelial-mesenchymal transition to disperse throughout the body (see Figure 11.1A). But at different levels along the anterior-posterior axis, these cells enter different tissues and form different cell types. The crest can be divided into four main (but overlapping) anatomical regions, each with characteristic derivatives and functions (**FIGURE 11.5**):

• **Cranial (cephalic) neural crest cells** migrate to produce the craniofacial mesenchyme, which differentiates into the cartilage, bone, cranial neurons, glia, pigment cells, and connective tissues of the face. These cells also enter the pharyngeal arches and pouches to give rise to thymic cells, the odontoblasts of the tooth primordia, and the bones of the middle ear and jaw.*

• The **cardiac neural crest** is a subregion of the cranial neural crest and extends from the otic (ear) placodes to the third somites (Kirby 1987; Kirby and Waldo 1990). Cardiac neural crest cells develop into melanocytes, neurons, cartilage, and connective tissue (of the third, fourth, and sixth pharyngeal arches). This region of the neural crest also produces the entire muscular-connective tissue wall of the large arteries (the "outflow tracts") as they arise from the heart, as well as contributing to the septum that separates pulmonary circulation from the aorta (Le Lièvre and Le Douarin 1975; Sizarov et al. 2012).

• **Trunk neural crest** cells take one of two major pathways. One migratory pathway takes trunk neural crest cells ventrolaterally through the anterior half of each somitic sclerotome. **Sclerotomes**, derived from the somites, are blocks of mesodermal cells that will differentiate into the vertebral cartilage of the spine. Trunk neural crest cells that remain in the sclerotomes form the **dorsal root ganglia** containing the sensory neurons. Cells that continue traveling more ventrally form the sympathetic ganglia, the adrenal medulla, and the nerve clusters surrounding the aorta. The second major migratory path for trunk neural crest cells proceeds dorsolaterally, allowing the precursors of melanocytes to move through the dermis from the dorsum to the belly (Harris and Erickson 2007).

• The **vagal** and **sacral neural crest** cells generate the **parasympathetic (enteric) ganglia** of the gut (Le Douarin and Teillet 1973; Pomeranz et al. 1991). The vagal (neck) neural crest overlaps the cranial/trunk crest boundary, lying opposite chick somites 1–7, while the sacral neural crest lies posterior to somite 28. Failure of neural crest cell migration from these regions to the colon results in the absence of enteric ganglia and thus to the absence of peristaltic movement in the bowels (Hirschsprung disease; see Chapter 1 and p. 384).

Trunk neural crest cells and cranial neural crest cells are not equivalent. Cranial crest cells can form cartilage, muscle, and bone, whereas trunk crest cells cannot. When trunk neural crest cells are transplanted into the head region, they can migrate to the sites of cartilage and cornea formation, but they make neither cartilage nor cornea (Noden 1978; Nakamura and Ayer-Le Lievre 1982; Lwigale et al. 2004). However, both cranial and trunk neural crest cells can generate neurons, melanocytes, and glia. The cranial neural crest cells that normally migrate into the eye region to become cartilage cells can form sensory ganglion neurons, adrenomedullary cells, glia, and Schwann cells if the cranial region is transplanted into the trunk region (Noden 1978; Schweizer et al. 1983).

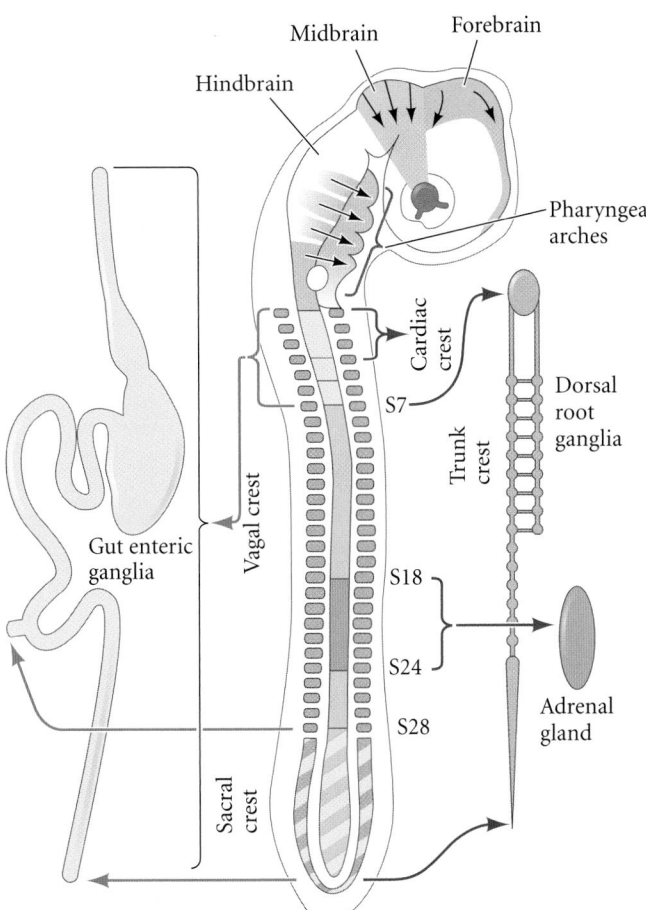

FIGURE 11.5 Regions of the chick neural crest. The cranial neural crest migrates into the pharyngeal arches and the face to form the bones and cartilage of the face and neck. It also produces the cranial nerves. The vagal neural crest (near somites 1–7) and the sacral neural crest (posterior to somite 28) form the parasympathetic nerves of the gut. The cardiac neural crest cells arise near somites 1–3; they are critical in making the division between the aorta and the pulmonary artery. Neural crest cells of the trunk (from about somite 6 through the tail) make sympathetic neurons and pigment cells (melanocytes), and a subset of these (at the level of somites 18–24) forms the medulla portion of the adrenal gland. (After Le Douarin 1982.)

---

*The pharyngeal (branchial) arches (see Figure 1.8A) are outpocketings of the head and neck region into which cranial neural crest cells migrate. The pharyngeal pouches form between these arches and become the thyroid, parathyroid, and thymus.

The inability of the trunk neural crest to form skeleton is most likely due to the expression of Hox genes in the trunk neural crest. If Hox genes are expressed in the cranial neural crest, these cells fail to make skeletal tissue; if trunk crest cells lose Hox gene expression, they can form skeleton. Moreover, if transplanted into the trunk region, cranial crest cells participate in forming trunk cartilage that normally does not arise from neural crest components. This ability to form bone may have been a primitive property of the neural crest and may have been critical for forming the bony armor found in several extinct fish species (Smith and Hall 1993). In other words, the trunk crest has apparently lost the ability to form bone, rather than the cranial crest acquiring this ability. McGonnell and Graham (2002) have shown that bone-forming capacity may still be latent in the trunk neural crest: if cultured in certain hormones and vitamins, the trunk cells become capable of forming bone and cartilage when placed into the head region. Moreover, Abzhanov and colleagues (2003) have shown that the trunk crest cells can act like cranial crest cells (and make skeletal tissue) if the trunk cells are cultured in conditions that cause them to lose the expression of their Hox genes.

So, even though the cells of the cranial neural crest and trunk neural crest are multipotent (a cranial crest cell can form neurons, cartilage, bone, and muscles; a trunk neural crest cell can form glia, pigment cells, and neurons), they have different repertoires of cell types that they can generate under normal conditions.

Trunk Neural Crest

Migration pathways of trunk neural crest cells

Cells migrating from the trunk-level neural crest follow one of two major pathways (**FIGURE 11.6A**). Many cells that leave early follow a **ventral pathway** away from the neural tube. Fate mapping experiments show that these cells become sensory (dorsal root) and autonomic neurons, adrenomedullary cells, and Schwann cells and other glial cells (Weston 1963; Le Douarin and Teillet 1974). In birds and mammals (but not fish and frogs*), these cells migrate ventrally through the anterior, but not the posterior, section of the sclerotomes (**FIGURE 11.6B–D**; Rickmann et al. 1985; Bronner-Fraser 1986; Loring and Erickson 1987; Teillet et al. 1987).

Trunk crest cells that emigrate via the second pathway—the **dorsolateral pathway**—become melanocytes, the melanin-forming pigment cells. These cells travel between the epidermis and the dermis, entering the ectoderm through minute holes in the basal lamina (which they themselves may create). Once in the ectoderm, they colonize the skin and hair follicles (Mayer 1973; Erickson et al. 1992). The dorsolateral pathway was demonstrated in a series of classic experiments by Mary Rawles (1948), who transplanted the neural tube

---

*In the migration of fish neural crest cells, the sclerotome is not important; rather, the myotome appears to guide the migration of the crest cells ventrally (Morin-Kensicki and Eisen 1997).

and neural crest from a pigmented strain of chickens into the neural tube of an albino chick embryo and saw pigmented feathers on otherwise white wings (see Figure 1.16).

By transplanting quail neural tubes or neural folds into chick embryos, Teillet and co-workers (1987) were able to mark neural crest cells both genetically and immunologically. The antibody marker recognized and labeled neural crest cells of both species; the genetic marker enabled the investigators to distinguish between quail and chick cells (see Figure 1.15A). These studies showed that neural crest cells initially located opposite the posterior region of a somite migrate anteriorly or posteriorly along the neural tube, and then enter the anterior region of their own or an adjacent somite. These cells join with the neural crest cells that initially were opposite the anterior portion of the somite, and they form the same structures. Thus, each dorsal root ganglion is composed of neural crest cell populations forming adjacent to three somites: one from the neural crest opposite the anterior portion of the somite, and one each from the two neural crest regions opposite the posterior portions of its own and the neighboring somites.

Mechanisms of trunk neural crest migration

Any analysis of migration (be it of birds, butterflies, or neural crest cells) has to ask these questions:

1. What signals initiate migration?
2. When does the migratory agent become competent to respond to these signals?
3. How do the migratory agents know the route to travel?
4. What signals indicate that the destination has been reached?

The timing of neural crest cell migration is controlled by the neural tube's environment. The trigger for the epithelial-mesenchymal transition (EMT) appears to be the activation of the Wnt genes by BMPs. The BMPs (which can be produced by the dorsal region of the neural tube; see Chapter 10) are held in check by Noggin produced in the somites (see Chapter 12). When the somites cease making Noggin, the neural tube BMPs can function, and they activate EMT in the dorsal neural tube cells (Burstyn-Cohen et al. 2004).

Wnt and BMP signals help promote expression of proteins from the Snail, Foxd, and Rho families in those cells destined to become emigrating neural crest (**FIGURE 11.7**; Nieto et al. 1994; Mancilla and Mayor 1996; LaBonne and Bronner-Fraser 1998; Liu and Jessell 1998). Noncanonical Wnt signaling is critical for activating the small Rho GTPases. RhoA, for instance, helps organize actin into filopodial projections, and RhoA and Rac1 (another Rho GTPase) facilitate the expression of *Foxd3* and the Snail family genes. RhoB is involved in establishing the cytoskeletal conditions for migration by promoting actin polymerization into microfilaments and the attachment of these microfilaments to the cell membrane (Hall 1998; De Calisto et al. 2005).

The crest cells cannot leave the neural tube as long as they are tightly connected to one another, and one of the

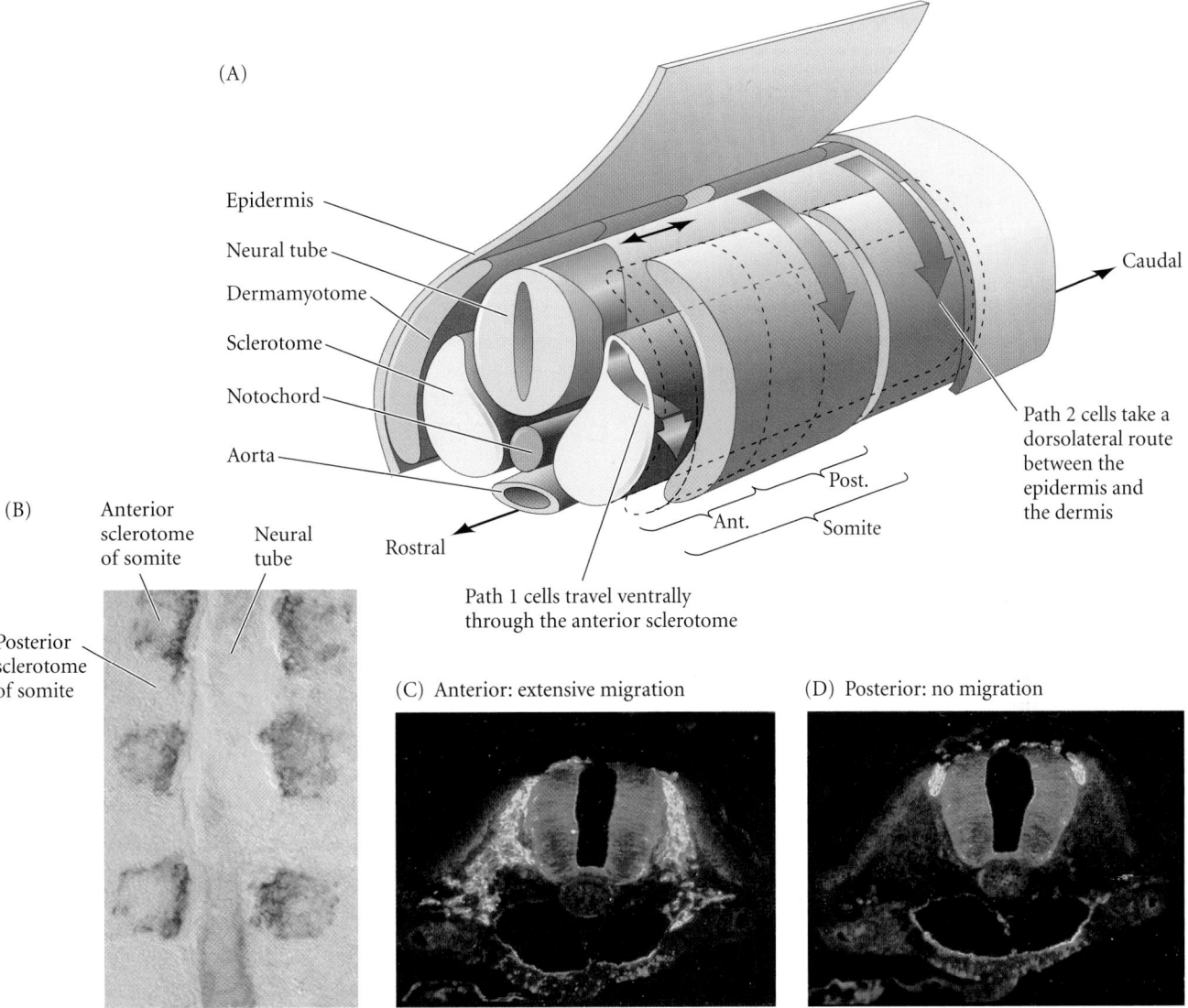

(A)

Epidermis

Neural tube

Dermamyotome

Sclerotome

Notochord

Aorta

Caudal

Path 2 cells take a dorsolateral route between the epidermis and the dermis

Post.

Ant.

Somite

Rostral

Path 1 cells travel ventrally through the anterior sclerotome

(B)

Anterior sclerotome of somite

Neural tube

Posterior sclerotome of somite

(C) Anterior: extensive migration

(D) Posterior: no migration

FIGURE 11.6 Neural crest cell migration in the trunk of the chick embryo. (A) Schematic diagram of trunk neural crest cell migration. Cells taking path 1 (the ventral pathway) travel ventrally through the anterior of the sclerotome (that portion of the somite that generates vertebral cartilage). Those cells initially opposite the posterior portion of a sclerotome migrate along the neural tube until they come to an anterior region. These cells contribute to the sympathetic and parasympathetic ganglia as well as to the adrenomedullary cells and dorsal root ganglia. Other trunk neural crest cells enter path 2 (the dorsolateral pathway) somewhat later. These cells travel along a dorsolateral route beneath the ectoderm and become pigment-producing melanocytes. (Migration pathways are shown on only one side of the embryo.) (B) These fluorescence photomicrographs of longitudinal sections of a 2-day chick embryo are stained red with antibody to HNK-1, which selectively recognizes neural crest cells. Extensive staining is seen in the anterior, but not in the posterior, half of each sclerotome. (C,D) Cross sections through these areas, showing extensive migration through the anterior portion of the sclerotome (C), but no migration through the posterior portion (D). Here, the antibodies to HNK-1 are stained green. (B from Wang and Anderson 1997; C,D from Bronner-Fraser 1986, courtesy of the author.)

functions of the Snail proteins is to activate factors that dissociate the cadherins that bind these cells together. Two cell adhesion proteins, cadherin 6B and N-cadherin, are present on cells of the neuroepithelium but are downregulated in the dorsal neural tube cells that become neural crest. Snail downregulates cadherin 6B and the tight junction proteins that bind epithelial cells together.* Later, after migration, the migrating neural crest cells re-express cadherins as they aggregate to form the dorsal root and sympathetic ganglia (Takeichi 1988; Akitaya and Bronner-Fraser 1992; Coles et al. 2007).

---

*This loss of cadherin expression, activation of Rho GTPases, and gain of the cell's ability to migrate on extracellular matrices are all characteristics not only of neural crest cells but of metastasizing tumor cells. It was recently found that cancer cells use the same pathway to reorganize their cytoskeleton as neural crest cells do (Murphy et al. 2011; Powell et al. 2013).

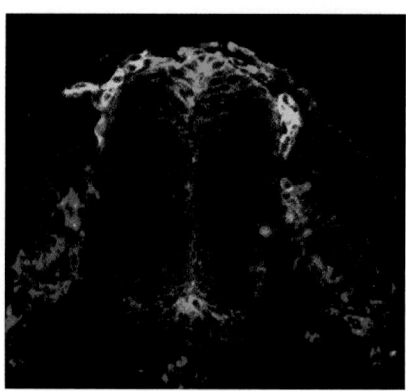

FIGURE 11.7 Migrating neural crest cells are stained red by HNK-1 antibody, which recognizes a cell surface carbohydrate involved in neural crest cell migration. RhoB protein (green stain) is expressed in cells as they leave the neural crest. Cells expressing both HNK-1 and RhoB appear yellow. (From Liu and Jessell 1998, courtesy of T. M. Jessell.)

The pushing out of neural crest cells from the dorsal neural tube appears to be facilitated by fellow neural crest cells (Carmona-Fontaine et al. 2008). When migrating neural crest cells meet, they stop migration in that direction and extend protrusions from the opposite side of the cell. This **contact inhibition** of locomotion results in forward migration of the leading edge of the cells. Contact between two neural crest cells (but not contact of a neural crest cell with another cell type) causes the reassortment of noncanonical Wnt pathway proteins to the site of contact, where Wnts activate the RhoA protein, which in turn disaggregates the cytoskeletons of the lamellipodia responsible for migration, allowing them to form on the opposite side of the cell.

The ventral pathway

The choice between the dorsolateral versus the ventral trunk pathway is made at the dorsal neural tube shortly after neural crest cell specification (Harris and Erickson 2007). The earliest migrating cells are inhibited from entering the dorsolateral pathway by chondroitin sulfate proteoglycans, ephrins, Slit proteins, and probably several other molecules. Because they are so inhibited, these cells turn around and migrate ventrally and there give rise to the neurons and glial cells of the peripheral nervous system.

The next choice concerns whether these ventrally migrating cells migrate *between* the somites (to form the sympathetic ganglia of the aorta) or *through* the somites (Schwarz et al. 2009). In the mouse embryo, the first few neural crest cells that form go between the somites, but this pathway is soon blocked by **semaphorin-3F**, a protein that repels neural crest cells; thus, most neural crest cells traveling ventrally migrate through the somites. These cells migrate through the anterior portion of each sclerotome (the sclerotome is the portion of the somite that gives rise to the cartilage of the spine) and associate with proteins of the extracellular matrices such as fibronectin and laminin which are permissive for migration (Newgreen and Gooday 1985; Newgreen et al. 1986).

The extracellular matrices of the sclerotome differ in the anterior and posterior region of each somite, and only the extracellular matrix of the *anterior* sclerotome allows neural crest cell migration. Like the extracellular matrix molecules that prevented neural crest cells from migrating dorsolaterally, the extracellular matrix of the posterior portion of each sclerotome contains proteins that actively exclude neural crest cells. Besides semaphorin-3F, these proteins include the **ephrins** (**FIGURE 11.8A**). The ephrin on the posterior sclerotome is recognized by its receptor, Eph, on the neural crest cells. Similarly, semaphorin-3F on the posterior sclerotome cells is recognized by its receptor, neuropilin-2, on the migrating neural crest cells. When neural crest cells are plated on a culture dish containing stripes of immobilized cell membrane proteins alternately with and without ephrins, the cells leave the ephrin-containing stripes and move along the stripes that lack ephrin (**FIGURE 11.8B**; Krull et al. 1997; Wang and Anderson 1997; Davy and Soriano 2007). Similarly, neural crest cells fail to migrate on substrates containing semaphorin-3F; and mutant mice deficient in either semaphorin-3F or neuropilin-2 have severe neural crest migration abnormalities throughout the trunk, since the neural crest cells migrate through both anterior and posterior halves of the somite. This patterning of neural crest cell migration generates the overall segmental character of the peripheral nervous system, reflected in the positioning of the dorsal root ganglia and other neural crest-derived structures.

CELL DIFFERENTIATION IN THE VENTRAL PATHWAY The neural crest cells entering the somites differentiate to become two major types of neurons, depending on their location. Those cells that differentiate within the sclerotome give rise to the dorsal root ganglia. These contain the sensory neurons that relay information regarding touch, pain, and temperature back to the central nervous system.* It is likely that, as they begin to migrate ventrally, the neural crest cells produce progeny that express different receptors. Migrating cells that have receptors for neurotrophin and Wnt respond to those proteins (which are produced by the dorsal neural tube) and differentiate close to the neural tube into the glia and neurons of the dorsal root ganglia (Weston 1963). Within the dorsal root ganglia, those cells having more Notch become the glia, while those cells having more Delta (the Notch ligand) become the neurons (Wakamatsu et al. 2000; Harris and Erickson 2007).

The neural crest cells that lack Wnt and neurotrophin receptors continue migrating. They migrate through the anterior portion of the sclerotome and continue ventrally until they reach the dorsal aorta (but stopping before they enter the gut) and become the sympathetic ganglia (Vogel and Weston 1990). At trunk levels, they contribute to the epinephrine-secreting

---

*These are *afferent* neurons, since they carry information from sensory cells *to* the spinal cord and brain. *Efferent* neurons carry information *away* from the central nervous system; these are the motor neurons generated in the ventral region of the neural tube (as discussed in Chapter 10).

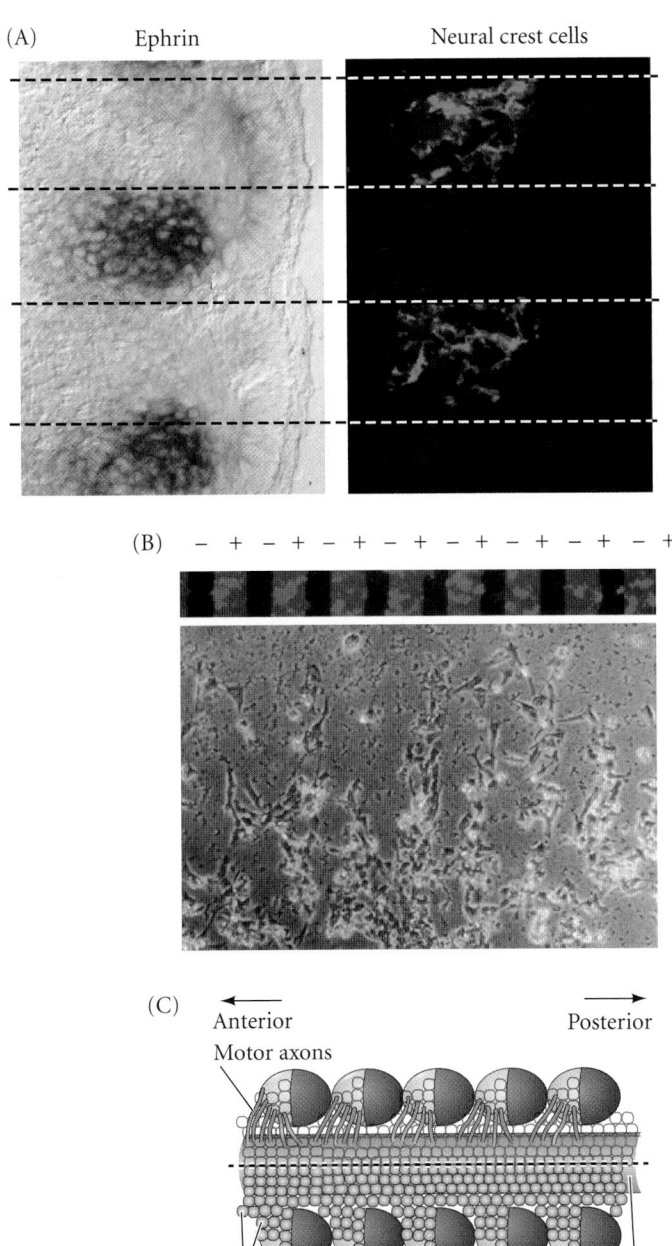

(A) Ephrin   Neural crest cells

(B) − + − + − + − + − + − + − +

(C)

← Anterior   Posterior →

Motor axons

Neural crest cells   Sclerotome of somites   Spinal cord

FIGURE 11.8 Segmental restriction of neural crest cells and motor neurons by the ephrin proteins of the sclerotome. (A) Negative correlation between regions of ephrin in the sclerotome (dark blue stain, left) and the presence of neural crest cells (green HNK-1 stain, right). (B) When neural crest cells are plated on fibronectin-containing matrices with alternating stripes of ephrin, they bind to those regions lacking ephrin. (C) Composite scheme showing the migration of spinal cord neural crest cells and motor neurons through the ephrin-deficient anterior regions of the sclerotomes. (For clarity, the neural crest cells and motor neurons are each depicted on only one side of the spinal cord.) (A,B from Krull et al. 1997; C after O'Leary and Wilkinson 1999.)

sympathetic (adrenergic, "flight or fight") neurons of the autonomic nervous system, as well as to the adrenal medulla (**FIGURE 11.9A**). At cardiac and vagal axial levels, they become acetylcholine-secreting parasympathetic (cholinergic, "rest and digest") neurons, including the enteric neurons of the gut. These cell lineages may each arise from a mutipotent neural crest progenitor cell, and the restriction of fate into these three lineages may come relatively late (Sieber-Blum 1989). BMPs from the aorta appear to convert neural crest cells into the sympathetic and adrenal lineages, whereas glucocorticoids from the adrenal cortex block neuron formation, directing the neural crest cells near them to become adrenomedullary cells (Unsicker et al. 1978; Doupe et al. 1985; Anderson and Axel 1986; Vogel and Weston 1990). The cells destined to become the epinephrine-secreting cells of the adrenal medulla retain their responsiveness to BMP and migrate toward the BMP4-secreting adrenal cortical cells. The remaining cells become the sympathetic ganglia surrounding the aorta (Saito et al. 2012).

Moreover, when chick vagal and thoracic neural crests are reciprocally transplanted, the former thoracic crest produces the cholinergic neurons of the parasympathetic ganglia, and the former vagal crest forms adrenergic neurons in the sympathetic ganglia (Le Douarin et al. 1975). Kahn and co-workers (1980) found that premigratory neural crest cells from both the thoracic and the vagal regions contain enzymes for synthesizing both acetylcholine and norepinephrine. Thus, there is good evidence that, although some neural crest cells are committed soon after their formation, the differentiation of the ventrally migrating neural crest cells depends on the pathway they follow and their final location.

GOING FOR THE GUT The next choice involves which neural crest cells are able to colonize the gut and which cannot. This distinction involves both extracellular matrix components and soluble paracrine factors. Neural crest cells from the vagal and sacral regions form the enteric ganglia of the gut tube and control intestinal peristalsis. Cells from the vagal neural crest, once past the somites, enter into the foregut and spread to most of the digestive tube, while the sacral neural crest cells colonize the hindgut (**FIGURE 11.9B**). Various inhibitory extracellular matrix proteins (including the Slit proteins) block the more ventral migration of trunk neural crest cells into the gut, but these inhibitory proteins are absent around the vagal and sacral crest, allowing these neural crest cells to reach the gut tissue. Once in the vicinity of the developing gut, these crest cells are attracted to the digestive tube by **glial-derived neurotrophic factor** (**GDNF**), a paracrine factor produced by the gut mesenchyme (Young et al. 2001; Natarajan et al. 2002). GDNF from the gut mesenchyme binds to its receptor, Ret, on the neural crest cells. The vagal neural crest cells have more Ret in their cell membranes than do the sacral cells, and this makes them more invasive (Delalande et al. 2008).

(A)

(B)

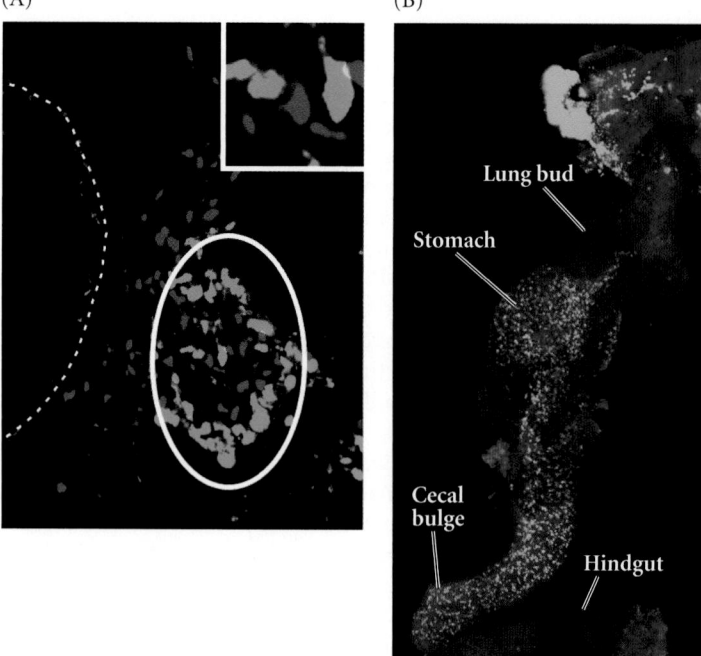

Lung bud

Stomach

Cecal bulge

Hindgut

FIGURE 11.9 Entry of neural crest cells into the gut and adrenal gland. (A) Migrating neural crest cells (stained red for the Sox8 transcription factor) migrating toward the adrenal cortical cells (stained green for SF1). The limits of the adrenal gland are circled; the dorsal aorta boundary is shown by a dotted line. (B) Neural crest cells form the enteric (gut) ganglia necessary for peristalsis. Confocal image (200× magnification) of an 11.5-day mouse gut showing the migration of neural crest cells (stained for Phox2b) through the foregut and into the cecal bulge of the intestine. (A from Reiprich et al. 2008; B from Corpening et al. 2008.)

the anterior gut and the ratio of cell motility to gut growth. These results were not obvious, and they show the power of combining experimental and mathematical approaches to development.

The dorsolateral migration pathway

It appears that the cells that take the dorsolateral pathway have already become specified as melanoblasts—pigment cell progenitors—and that they are led along the dorsolateral route by chemotactic factors and cell matrix glycoproteins (**FIGURE 11.10**). In the chick (but not in the mouse), the first neural crest cells to migrate enter the ventral pathway, whereas cells that migrate later enter the dorsolateral pathway (see Harris and Erickson 2007). These late-migrating cells remain above the neural tube in what is often called the "staging area," and it is these cells that become specified as melanoblasts (Weston and Butler 1966; Tosney 2004). The switch between glial/neural precursor and melanoblast precursor seems to be controlled by the Foxd3 transcription factor. If Foxd3 is present, it represses expression of **MITF**, a transcription factor necessary for melanoblast specification and pigment production. If *Foxd3* expression is downregulated, MITF is expressed and the cells become melanoblasts. MITF is involved in three signaling cascades. The first cascade activates those genes responsible for pigment production; the second allows these neural crest cells to travel along the dorsolateral pathway into the skin; and the third prevents apoptosis in the migrating cells (Kos et al. 2001; McGill et al. 2002; Thomas and Erickson 2009). In humans, heterozygosity for *MITF* causes there to be fewer pigment cells reaching the center of the body, and a hypopigmented streak may go through the hair. In some animals, however, heterozygosity for *Mitf* causes a random death of melanoblasts, as in certain breeds of dogs and horses (**FIGURE 11.11**).

GDNF activates cell division, directs cell migration into the gut mesoderm, and induces neural differentiation (Mwizerwa et al. 2011). If either GDNF or Ret is deficient in mice or humans, the pup or child suffers from Hirschsprung disease, a syndrome wherein the intestine cannot properly void solid wastes. In humans, this is most often due to the failure of the vagal neural crest cells to complete their colonization of the hindgut, thus leaving a section of the lower intestine without the ability to undergo peristalsis. By combining the experimental analysis of crest cell migration with mathematical modeling, Landman and colleagues (2007) modeled the migration of the vagal crest cells and explained the genetic deficiencies that cause Hirschsprung disease. In this model, the vagal crest cells normally do not migrate in a directed manner once they are in the anterior portion of the gut. Rather, they proliferate until all the niches in that region of the intestine are saturated, after which the migrating front moves posteriorly. Meanwhile, the gut itself continues to elongate. Whether or not the colonization is complete depends on the initial number of vagal crest cells entering

Once specified, the melanoblasts in the staging area upregulate the ephrin receptor (Eph B2) and the endothelin receptor (EDNRB2). This allows them to migrate along extracellular matrices containing ephrin and endothelin-3 (see Figure 11.10B; Harris et al. 2008). Indeed, the melanocyte lineage migrates on exactly those same molecules that *repelled* the glial/neural lineage of crest cells. Ephrin expressed along the dorsolateral migration pathway stimulates the migration of melanocyte precursor cells. Ephrin activates its receptor, Eph B2, on the neural crest cell membrane, and this Eph signaling appears to be critical for promoting this migration. Disruption of Eph signaling in late-migrating neural crest cells prevents their dorsolateral migration (Santiago and Erickson 2002).

(A)

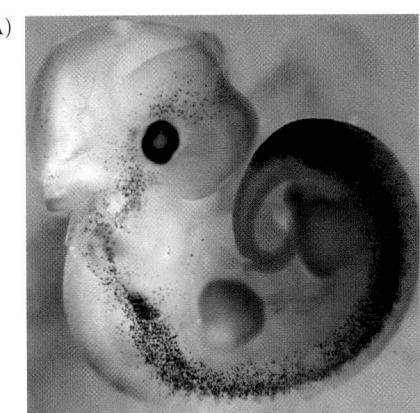

(B)

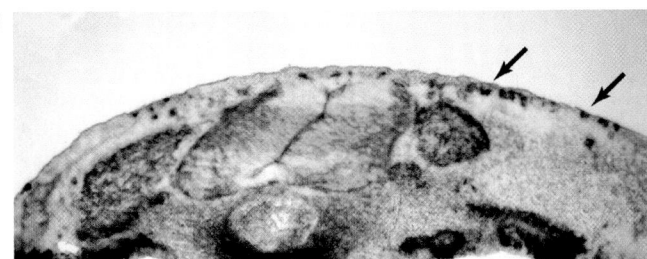

FIGURE 11.10 Neural crest cell migration in the dorsolateral pathway through the skin. (A) Whole mount in situ hybridization of day 11 mouse embryo staining neural crest-derived melanoblasts (purple). (B) Stage 18 chick embryo seen in cross section through the trunk. Melanoblasts (arrows) can be seen moving through the dermis, from the neural crest region toward the periphery. (A from Baxter and Pavan 2003; B from Santiago and Erickson 2002, courtesy of C. Erickson.)

In mammals (but not in chicks), the Kit receptor protein is critical in causing the committed melanoblast precurors to migrate on the dorsolateral pathway. This protein is found on those mouse neural crest cells that also express MITF—that is, the presumptive neuroblasts. Kit protein binds **stem cell factor** (**SCF**), which is made by the dermal cells. When bound to SCF, Kit prevents apoptosis and stimulates cell division among the melanoblast precursors. If mice or humans do not make sufficient amounts of Kit, the neural crest cells do not proliferate enough to cover the entire skin (**FIGURE 11.11C,D**; Spritz et al. 1992). Moreover, SCF is critical for dorsolateral migration. If SCF is experimentally secreted from tissues (such as cheek epithelium or the footpads) that do not usually synthesize this protein (and do not usually have melanocytes), neural crest cells will enter those regions and become melanocytes (Kunisada et al. 1998; Wilson et al. 2004).

(A)

(C)

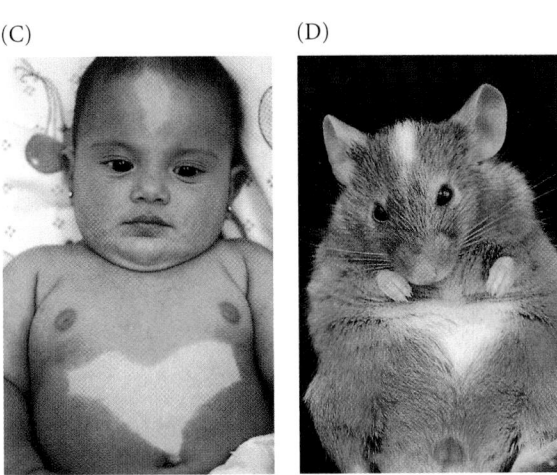

(B)

FIGURE 11.11 Variable melanoblast migration, caused by different mutations. (A,B) In several animals, the random death of melanoblasts provides spotted pigmentation. Migrating melanoblasts induce the blood vessels to form in the inner ear, and without these vessels, the cochlea degenerates, and the animal cannot hear in that ear. This is often the case with (A) Dalmation dogs, which are heterozygous for *Mitf*, and (B) American Paint horses, which are thought to be heterozygous for endothelin receptor B. (C) Piebaldism in a human infant, where pigment fails to form in regions of the body, is caused by a mutation in the *KIT* gene. Kit protein is essential for the proliferation and migration of neural crest cells, germ cell precursors, and blood cell precursors. (D) Mice can also have a *Kit* mutation, and provide important models for piebaldism and for melanoblast migration. (A © Robert Pickett/Corbis; B © M. J. Barrett/Alamy; C,D courtesy of R. A. Fleischman.)

In vertebrates, all pigment cells except those of the pigmented retina are derived from the neural crest. Eventually, the melanoblasts enter the epidermis, where they rapidly enter the developing hair follicles or feather primordia and take up residence at the base of the follicle bulge. In mice, all of the melanoblasts go into the hair follicle (the mouse epidermis is transparent), whereas in birds and humans, both the epidermis and its cutaneous appendage (hair, feathers) become pigmented. In the feather or hair follicle, the melanoblasts become **melanocyte stem cells** (Mayer 1973; Nishimura et al. 2002). In the hair shaft, they reside in the bulge, along with the hair follicle stem cells (see Figure 10.39). The melanocyte stem cells appear to be supported by the hair follicle stem cells, which make a cell-membrane collagen (collagen 17) that is needed for the self-renewal of both hair follicle and melanocyte stem cells (Tanimura et al. 2011). The melanocyte stem cells are thought to produce transit amplifying cells that migrate outside the bulge at the beginning of each hair development cycle to differentiate into mature melanocytes and provide pigmentation to the hair shaft. The epithelial cells secrete Fgf2, which stimulates melanocytes to transfer packets of pigment into the epithelial cells (Weiner et al. 2007). Nishimura and colleagues (2005) have documented that the reason the hair of mice and humans grays with age is that melanocyte stem cells become depleted from the bulge.

Thus, the differentiation of the trunk neural crest is accomplished by (1) autonomous factors (such as the Hox genes distinguishing trunk and cranial neural crest cells, or MITF committing cells to a melanocyte lineage), (2) specific conditions of the environment (such as the adrenal cortex inducing adjacent neural crest cells into adrenomedullary cells), or (3) a combination of the two (as when cells migrating through the sclerotome respond to Wnt signals depending on their types of receptors). The fate of an individual neural crest cell is determined both by its starting position (anterior-posterior along the neural tube) and by its migratory path.

Cranial Neural Crest

The head, comprising the face and the skull, is the most anatomically sophisticated portion of the vertebrate body (Northcutt and Gans 1983; Wilkie and Morriss-Kay 2001). The head is largely the product of the cranial neural crest, and the evolution of jaws, teeth, and facial cartilage occurs through changes in the placement of these cells (see Chapter 20).

Like the trunk neural crest, the cranial crest can form pigment cells, glial cells, and peripheral neurons; but in addition, it can generate bone, cartilage, and connective tissue. The cranial neural crest is a mixed population of cells in different stages of commitment, and about 10% of the population is made up of multipotent progenitor cells that can differentiate to become neurons, glia, melanocytes, muscle cells, cartilage, and bone (Calloni et al. 2009). In mice and humans, the cranial neural crest cells migrate from the neural folds even before they have fused together (Nichols 1981; Betters et al. 2010). Subsequent

FIGURE 11.12 Cranial neural crest cell migration in the mammalian head. (A) Migration of GFP-labeled neural crest cells in a day 9.5 mouse embryo, emphasizing the colonization of the pharyngeal arches and frontonasal process. (B) Migrational pathways from the cranial neural crest into the pharyngeal arches (p1–4) and frontonasal process. (C) Continued migration of the cranial neural crest to produce the human face. The frontonasal process contributes to the forehead, nose, philtrum of the upper lip (the area between the lip and nose), and primary palate. The lateral nasal process generates the sides of the nose. The maxillomandibular process gives rise to the lower jaw, much of the upper jaw, and the sides of the middle and lower regions of the face. (D) Structures formed in the human face by the mesenchymal cells of the neural crest. The cartilaginous elements of the pharyngeal arches are indicated by colors, and the darker pink region indicates the facial skeleton produced by anterior regions of the cranial neural crest. (A courtesy of P. Trainor and A. Barlow; B after Le Douarin 2004; C after Helms et al. 2005; D after Carlson 1999.)

migration of these cells is directed by an underlying segmentation of the hindbrain. As mentioned in Chapter 10, the hindbrain is segmented along the anterior-posterior axis into compartments called rhombomeres. The cranial neural crest cells migrate ventrally from those regions anterior to rhombomere 8 into the pharyngeal arches and the frontonasal process that forms the face. The final destination of these crest cells will determine their eventual fate (**FIGURE 11.12A,B; TABLE 11.2**).

The cranial crest cells follow one of three major streams:

1. Neural crest cells from the midbrain and rhombomeres 1 and 2 of the hindbrain migrate to the first pharyngeal arch (the mandibular arch), forming the jawbones as well as the incus and malleus bones of the middle ear. These cells will also differentiate into neurons of the trigeminal ganglion—the cranial nerve that innervates the teeth and jaw—and will contribute to the ciliary ganglion that innervates the ciliary muscle of the eye. These neural crest cells are also pulled by the expanding epidermis to generate the **frontonasal process**, the bone-forming region that becomes the forehead, the middle of the nose, and the primary palate. Thus, the cranial neural crest cells generate much of the facial skeleton (**FIGURE 11.12C**; Le Douarin and Kalcheim 1999; Wada et al. 2011).

2. Neural crest cells from rhombomere 4 populate the second pharyngeal arch, forming the upper portion of the hyoid cartilage of the neck as well as the stapes bone of the middle ear (**FIGURE 11.12D**). These cells will also contribute neurons of the facial nerve. The hyoid cartilage eventually ossifies to provide the bone in the neck that attaches the muscles of the larynx and tongue.

3. Neural crest cells from rhombomeres 6–8 migrate into the third and fourth pharyngeal arches and pouches to form the lower portion of the hyoid cartilage as well as contributing cells to the thymus, parathyroid, and thyroid glands (see Figure 11.12D; Serbedzija et al. 1992; Creuzet et al. 2005). These neural crest cells also go to the region of the

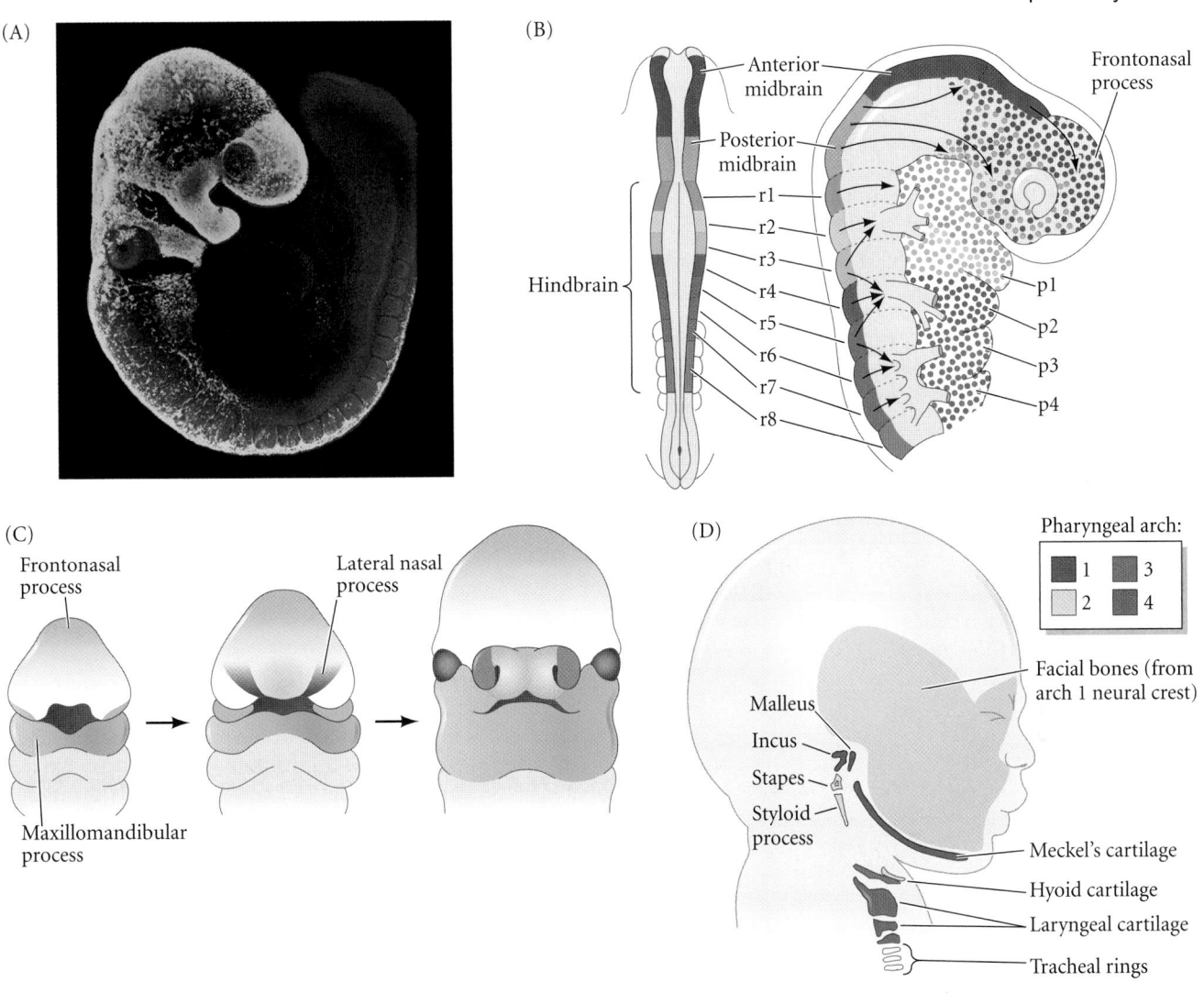

TABLE 11.2 Some derivatives of the pharyngeal arches

Pharyngeal arch	Skeletal elements (neural crest plus mesoderm)	Arches, arteries (mesoderm)	Muscles (mesoderm)	Cranial nerves (neural tube)
1	Incus and malleus (from neural crest); mandible, maxilla, and temporal bone regions (from neural crest)	Maxillary branch of the carotid artery (to the ear, nose, and jaw)	Jaw muscles; floor of mouth; muscles of the ear and soft palate	Maxillary and mandibular divisions of trigeminal nerve (V)
2	Stapes bone of the middle ear; styloid process of temporal bone; part of hyoid bone of neck (all from neural crest cartilage)	Arteries to the ear region: corticotympanic artery (adult); stapedial artery (embryo)	Muscles of facial expression; jaw and upper neck muscles	Facial nerve (VII)
3	Lower rim and greater horns of hyoid bone (from neural crest)	Common carotid artery; root of internal carotid	Stylopharyngeus (to elevate the phaynx)	Glossopharyngeal nerve (IX)
4	Laryngeal cartilages (from lateral plate mesoderm)	Arch of aorta; right subclavian artery;original spouts of pulmonary arteries	Constrictors of pharynx and vocal cords	Superior laryngeal branch of vagus nerve (X)
6[a]	Laryngeal cartilages (from lateral plate mesoderm)	Ductus arteriosus; roots of definitive pulmonary arteries	Intrinsic muscles of larynx	Recurrent laryngeal branch of vagus nerve (X)

Source: Adapted from Larsen 1993.

[a] The fifth arch degenerates in humans.

developing heart, where they help construct the outflow tracts (i.e., the aorta and pulmonary artery). If the neural crest is removed from those regions, these structures fail to form (Bockman and Kirby 1984). Some of these cells migrate caudally to the clavicle (collarbone), where they settle at the sites that will be used for the attachment of certain neck muscles (McGonnell et al. 2001).

Relatively fewer neural crest cells are produced by rhombomeres 3 and 5, and they do not migrate laterally, but rather join the even-numbered streams anterior and posterior to these odd-numbered rhombomeres.

The three streams of cells are kept from dispersing through interactions of the cells with their environment and with one another. In frog embryos, there is evidence that the separate streams are kept apart by ephrins. Blocking the activity of the Eph receptors causes cells from the different streams to mix together (Smith et al. 1997; Helbling et al. 1998). Observations of labeled neural crest cells from chick hindbrain, wherein individually marked cells were followed by cameras focused through a Teflon membrane window in the egg, showed that the migrating cells were "kept in line" not only by restrictions provided by neighboring cells, but also by the lead cells passing material to those cells behind them. It appears that cranial neural crest cells extend long, slender bridges that temporarily connect cells and influence the migration of the later cells to "follow the leader"* (Kulesa and Fraser 2000; McKinney et al. 2011).

Intramembranous ossification: Neural crest-derived head skeleton

Bones form in two major ways. In one way, mesenchyme becomes cartilage and the cartilage is replaced by bone; this is called **endochondral ossification**, which we will describe in detail in Chapter 12. The other type of bone formation, where mesenchyme forms bones directly, is called **intrambran-ous ossification** (**FIGURE 11.13**). Both mesodermal and ectodermal (i.e., neural crest-derived) mesenchyme undergo intramem-branous ossification in forming the face and skull.

The pathway from neural crest to intramembranous bone begins when cranial neural crest cells, under the influence of BMPs from the head epidermis, proliferate and condense into compact nodules (**FIGURE 11.14**). High levels of BMPs induce these nodules to become cartilage, whereas lower levels of BMPs induce them to become pre-osteoblast progenitor cells that express the Runx2 transcription factor and the mRNA for collagens II and IX. Later, these cells downregulate Runx2 and begin expressing the *osteopontin* gene, giving them a phenotype similar to that of a developing chondrocyte (cartilage cell); thus, this stage is called a **chondrocyte-like osteoblast**. Under the influence of Indian hedgehog (which it secretes and probably receives in an autocrine fashion), the chondrocyte-like osteoblast becomes a mature **osteoblast**—a committed bone precursor cell (Abzhanov et al. 2007). The osteoblasts secrete a collagen-proteoglycan **osteoid matrix** that is able to bind calcium. Osteoblasts that are embedded in the calcified matrix become **osteocytes** (bone cells). As calcification proceeds, bony spicules radiate out from the region

FIGURE 11.13 Intramembranous ossification. (A) Mouse mesenchyme cells condense and change shape to produce osteoblasts, which deposit osteoid matrix. Osteoblasts then become arrayed along the calcified region of the matrix. Osteoblasts that are embedded within the calcified matrix become osteocytes. (B) Bone formation in chick embryo heads (days 9–13) as revealed with alizarin red (which stains bone matrix). (A from Komori et al. 1997; B from Abzhanov et al. 2007, courtesy of P. Abzhanov.)

---

*We have seen similar phenomena before, as in the cells of the chick neural folds (some of which probably become neural crest cells), early zebrafish blastomeres, and the extensions of sea urchin micromeres.

(A)

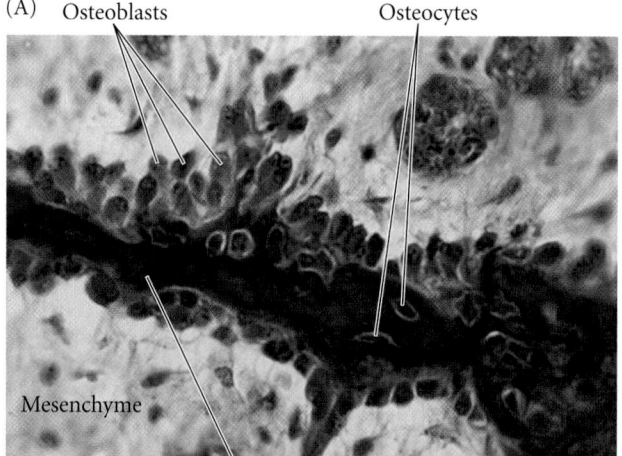

Osteoblasts   Osteocytes

Mesenchyme

Spicule of bone

(B)

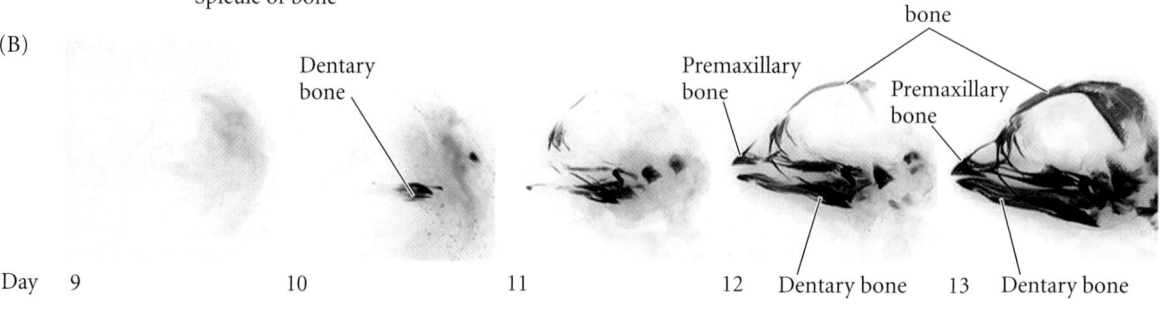

Dentary bone

Premaxillary bone

Frontal bone

Premaxillary bone

Dentary bone   Dentary bone

Day   9   10   11   12   13

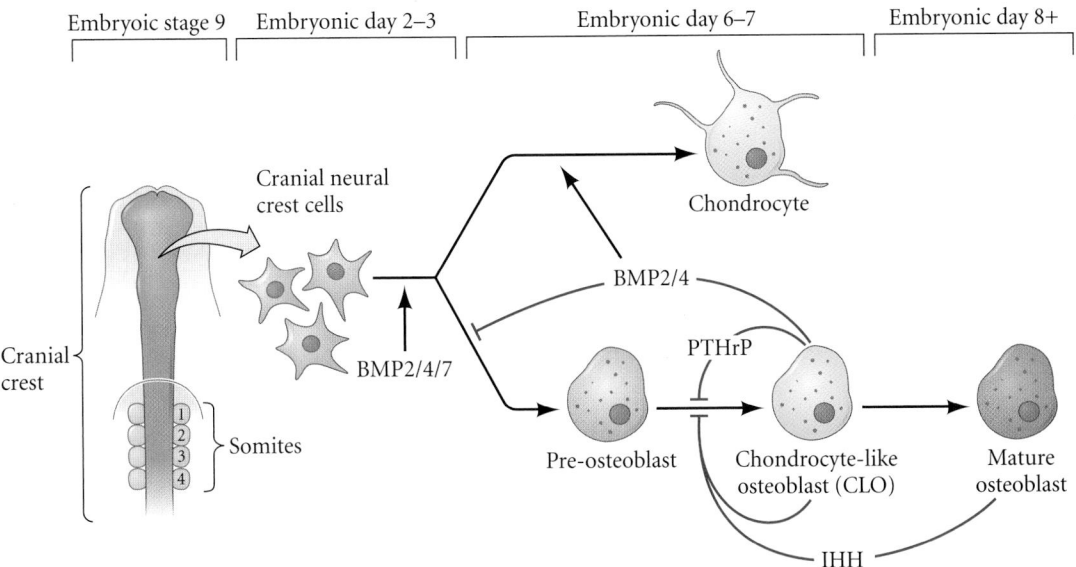

| Embryoic stage 9 | Embryonic day 2–3 | Embryonic day 6–7 | Embryonic day 8+ |

FIGURE 11.14 During craniofacial development, chick mesencephalic cranial neural crest cells migrate to become the mesenchyme of the future face and much of the skull. Cells of the early cranial skeletogenic condensations depend on BMP2/4/7 activities to form pre-osteoblastic progenitors (which eventually become bone), whereas high levels of BMP2 and/or BMP4 alone induce a chondrogenic (cartilage) fate. Differentiation into the chondrocyte-like osteoblasts is negatively regulated by both Indian hedgehog (IHH) and parathyroid hormone-related protein (PTHrP) activities that are probably autocrine. (After Abzhanov et al. 2007.)

where ossification began. Furthermore, the entire region of calcified spicules becomes surrounded by compact mesenchymal cells that form the **periosteum** (a membrane of cells that surrounds bone). The cells on the inner surface of the periosteum also become osteoblasts and deposit matrix parallel to the existing spicules. In this manner, many layers of bone are formed.

The vertebrate skull, or **cranium**, is composed of the **neurocranium** (skull vault and base) and the **viscerocranium** (jaws and other pharyngeal arch derivatives). Skull bones are derived from both the neural crest and the head mesoderm (Le Lièvre 1978; Noden 1978; Evans and Noden 2006). While the neural crest origin of the viscerocranium has been well documented, the contributions of cranial neural crest cells to the skull vault are more controversial. In 2002, Jiang and colleagues constructed transgenic mice that expressed β-galactosidase only in their cranial neural crest cells.* When the embryonic mice were stained for β-galactosidase, the cells forming the anterior portion of the head—the nasal, frontal, alisphenoid, and squamosal bones—turned blue; the parietal bone of the skull did not (**FIGURE 11.15A,B**). The boundary between neural crest-forming head bone and mesoderm-forming head bone is between the frontal and parietal bones (**FIGURE 11.15C**; Yoshida et al. 2008). Although the specifics may vary among the vertebrate groups, in general the front of the head is derived from the neural crest while the back of the skull is derived from a combination of neural crest-derived and mesodermal bones. The neural crest contribution to facial muscle mixes with the cells of the cranial mesoderm, such that facial muscles probably also have dual origins (Grenier et al. 2009).

Given that the neural crest forms our facial skeleton, it follows that even small variations in the rate and direction of cranial neural crest cell divisions will determine what we look like. Moreover, since we look more like our biological parents than our friends do (at least, we hope this is true), such small variations must be hereditary. The regulation of our facial features is probably coordinated in large part by numerous paracrine growth factors. BMPs (especially BMP3) and Wnt signaling cause the protrusion of the frontonasal and maxillary processes, giving shape to the face (Brugmann et al. 2006; Schoenebeck et al. 2012). FGFs from the pharyngeal endoderm are responsible for the attraction of the cranial neural crest cells into the arches as well as for patterning the skeletal elements within the arches. Fgf8 is both a survival factor for the cranial crest cells and is critical for the proliferation cells forming the facial skeleton (Trocovic et al. 2003, 2005; Creuzet et al. 2004, 2005). The FGFs work in concert with BMPs, sometimes activating them and sometimes repressing them (Lee et al. 2001; Holleville et al. 2003; Das and Crump 2012).

---

*These experiments were done using the Cre-lox technique. The mice were heterozygous for both (1) a β-galactosidase allele that could be expressed only when Cre-recombinase was activated in that cell, and (2) a Cre-recombinase allele fused to a *Wnt1* gene promoter. Thus, the gene for β-galactosidase was activated (blue stain) only in those cells expressing Wnt1—a protein that is activated in the cranial neural crest and in certain brain cells.

(A) *Wnt1-Cre*: Neural crest-derived bone

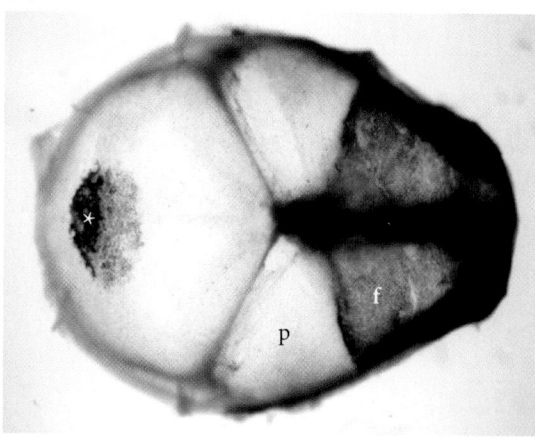

FIGURE 11.15   Cranial neural crest cells in embryonic mice, stained for β-galactosidase expression. (A) In the *Wnt1-Cre* strain, the β-galactosidase gene is expressed wherever Wnt1 (a neural crest marker) would be expressed. This dorsal view of a 17.5-day embryonic mouse shows staining in the frontal bone (f) and interparietal bone (asterisk) but not in the parietal bone (p). (B) The *Mesp-Cre* strain of mice expresses β-galactosidease in those cells derived from the mesoderm. Here, a reciprocal pattern of staining is seen, and the parietal bone is blue. (C) Summary diagram of results from mapping with Sox9 and Wnt1 markers (Als, alisphenoid; Bs, basisphenoid; Ex, exoccipital; Max, Maxilla; Nc, nasal capsule; Os, orbitosphenoid; So, supraoccipital; Sq, squamosal). (A,B from Yoshida 2008, courtesy of G. Morriss-Kay; C from several sources, including Noden and Schneider 2006 and Lee and Saint-Jeannet 2011.)

(B) *Mesp-Cre*: Mesoderm-derived bone

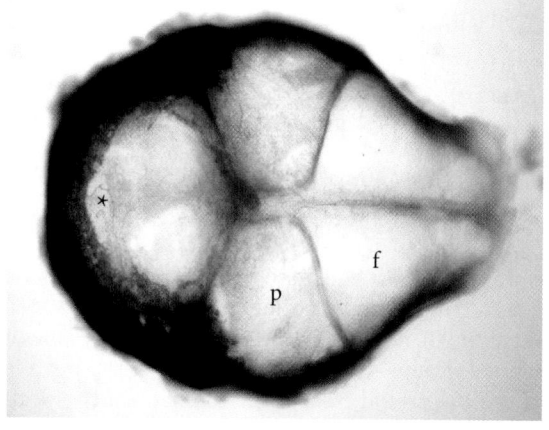

(C)

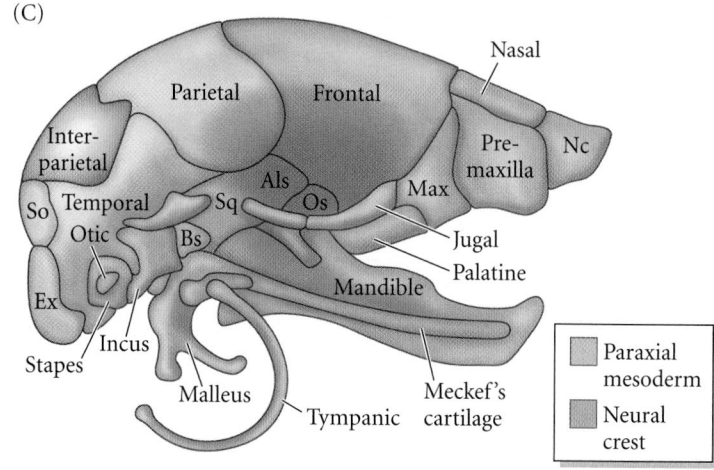

FGFs also work in concert with Sonic hedgehog (Shh; Haworth et al. 2008). We saw in Chapter 10 that Shh is critical for the proper growth of the facial midline (see Figure 10.32), and Shh is also crucial for shaping the neural crest derivatives of the head. The epithelia (both neural and epidermal) of the dorsal part of the frontonasal process secrete Fgf8, while the ventral epithelia of the frontonasal process secrete Shh. The crest-derived mesenchyme between the epithelia receives both signals. Where these signals meet is where a bird's beak cartilage grows out; if the region of frontonasal process containing the FGF/Shh boundary is inverted in the chick, an upside-down beak forms (Hu et al. 2003; Abzhanov and Tabin 2004). Variants in some regulators and targets of paracrine pathways have been shown to cause normal facial variation, as have variants of the *Pax3* gene, which is expressed in the cranial neural crest cells (Liu et al. 2012).

Coordination of face and brain growth

It is a generalization in clinical genetics that "the face reflects the brain." Although this is not always the case, physicians are aware that children with facial anomalies may have brain malformations as well. The coordination between facial form and brain growth was highlighted in studies by Le Douarin and her colleagues (2007). First they found that the region of cranial neural crest that forms the facial skeleton is also critical for the growth of the anterior brain (FIGURE 11.16). When that region of chick neural crest was removed, not only did the bird's face fail to form, but the telencephalon failed to grow as well. Next they found that the forebrain development could be rescued by adding Fgf8-containing beads to the anterior neural ridge (the neural folds of the anterior neuropore). This finding was strange, however, because cranial neural crest cells do not make or secrete Fgf8; the anterior neural ridge usually does. It seemed that removing the cranial neural crest cells prevented the anterior neural ridge from making the Fgf8 necessary for forebrain proliferation.

Looking at the effects of activated genes added to the anterior neural ridge region, Le Douarin and colleagues hypothesized that the BMP4 from the surface ectoderm was capable of blocking Fgf8. The cranial neural crest cells secreted Noggin and Gremlin, two extracellular proteins that bind to and inactivate BMP4. This allows the synthesis of Fgf8 in the anterior neural ridge and the development of the forebrain structures. Thus, the cranial neural crest cells not only provide the cells that build the facial skeleton and

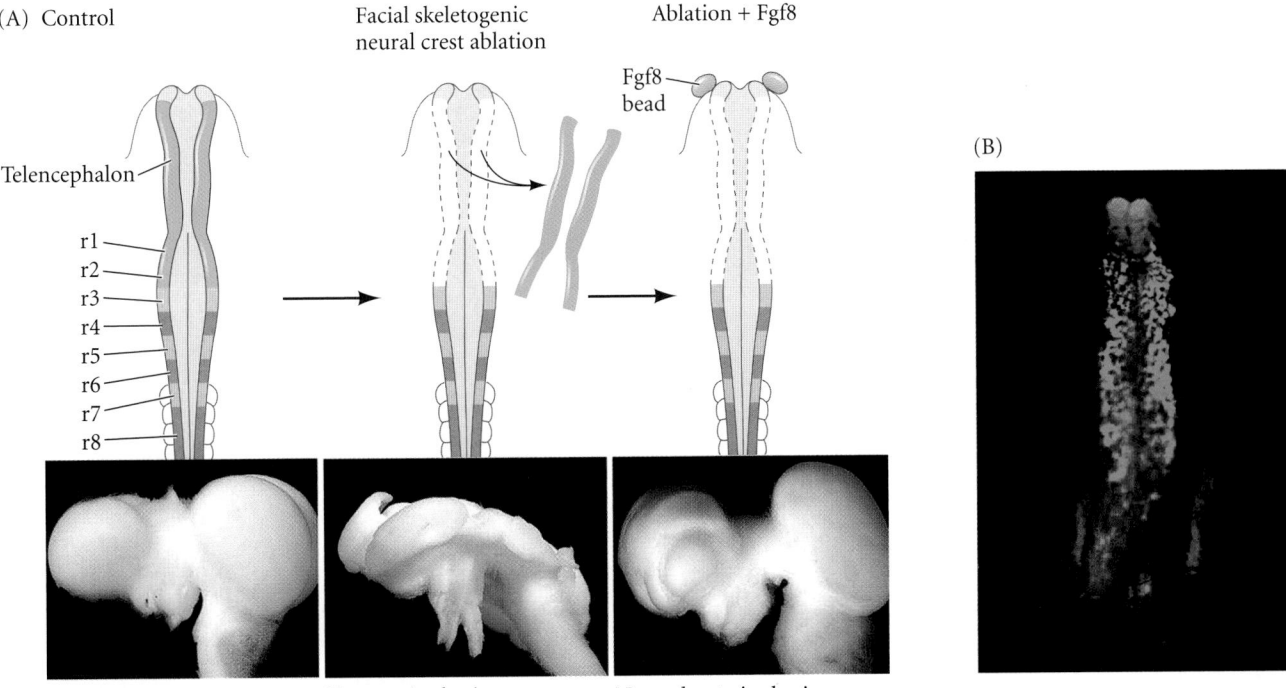

(A) Control

Facial skeletogenic
neural crest ablation

Ablation + Fgf8

Fgf8
bead

Telencephalon

r1
r2
r3
r4
r5
r6
r7
r8

Control   No anterior brain   Normal anterior brain

(B)

FIGURE 11.16 The cranial neural crest that forms the facial skeleton is also critical for the growth of the anterior region of the brain. (A) Removal of the facial skeleton-forming neural crest cells from a 6-somite-stage chick embryo stops the telencephalon from forming, as well as inhibiting formation of the facial skeleton. However, telencephalon development can be rescued by adding Fgf8-containing beads to the anterior neural ridge. (B) Embryo stained with HNK-1 (which labels neural crest cells green). Fgf8 appears pink in this micrograph. (After Creuzet et al. 2006, 2009; photographs courtesy of N. Le Douarin.)

connective tissues, they also regulate the production of Fgf8 in the anterior neural ridge, thereby allowing development of the midbrain and forebrain.

Tooth formation

The cranial neural crest cells of the first pharyngeal arch also form the interior, dentin-secreting odontoblasts of the teeth. The jaw epithelium becomes the outer, enamel-secreting ameloblasts. The mouse tooth is specified by either Fgf8 or BMP4. Fgf8 induces the expression of the Barx1 transcription factor and molds the tooth bud to form molars, whereas BMP4 induces the expression of the Msx1 and Msx2 transcription factors that mold the tooth to form incisors (Mina and Kollar 1987; Lumsden 1988b; Tucker et al. 1998).

The signaling center of the tooth is the **enamel knot**, a group of cells induced in the epithelium by the neural crest-derived mesenchyme (Jernvall et al. 1994; Vaahtokari et al. 1996a,b; Tummers and Thesleff 2009). The enamel knot secretes a cocktail of paracrine factors, including Shh, BMPs 2, 4, and 7, and Fgf4. Shh and Fgf4 induce the proliferation of cells to form a cusp, while the BMPs inhibit the formation of new enamel knots. These proteins are thought to act through

a reaction-diffusion mechanism to pattern the cusps of the tooth (as described in Chapter 20).

● See **WEBSITE 11.1** Why birds don't have teeth

Cardiac Neural Crest

The heart originally forms in the neck region, directly beneath the pharyngeal arches, so it should not be surprising that it acquires cells from the neural crest. The pharyngeal ectoderm and endoderm both secrete Fgf8, which acts as a chemotactic factor to draw neural crest cells into the area. Indeed, if beads containing large amounts of Fgf8 are placed dorsally to the chick pharynx, the cardiac neural crest cells will migrate there, instead (Sato et al. 2011.) The caudal region of the cranial neural crest is sometimes called the cardiac neural crest, since its cells (and only those particular neural crest cells) generate the endothelium of the aortic arch arteries and the septum between the aorta and the pulmonary artery (**FIGURE 11.17**; Kirby 1989; Waldo et al. 1998). Cardiac crest cells also enter pharyngeal arches 3, 4, and 6 to become portions of other neck structures such as the thyroid, parathyroid, and thymus glands. These cells are often referred to as the circumpharyngeal crest (Kuratani and Kirby 1991, 1992). In the thymus, neural crest-derived cells are especially important in one of the most critical functions of adaptive immunity: regulating the exit of mature T cells from the thymus and into the circulation (Zachariah and Cyster 2010). It is also likely that the carotid body, which monitors oxygen in the blood and regulates respiration accordingly, is derived from the cardiac neural crest (see Pardal et al. 2007).

In mice, cardiac neural crest cells are peculiar in that they express the transcription factor Pax3. Mutations of the *Pax3*

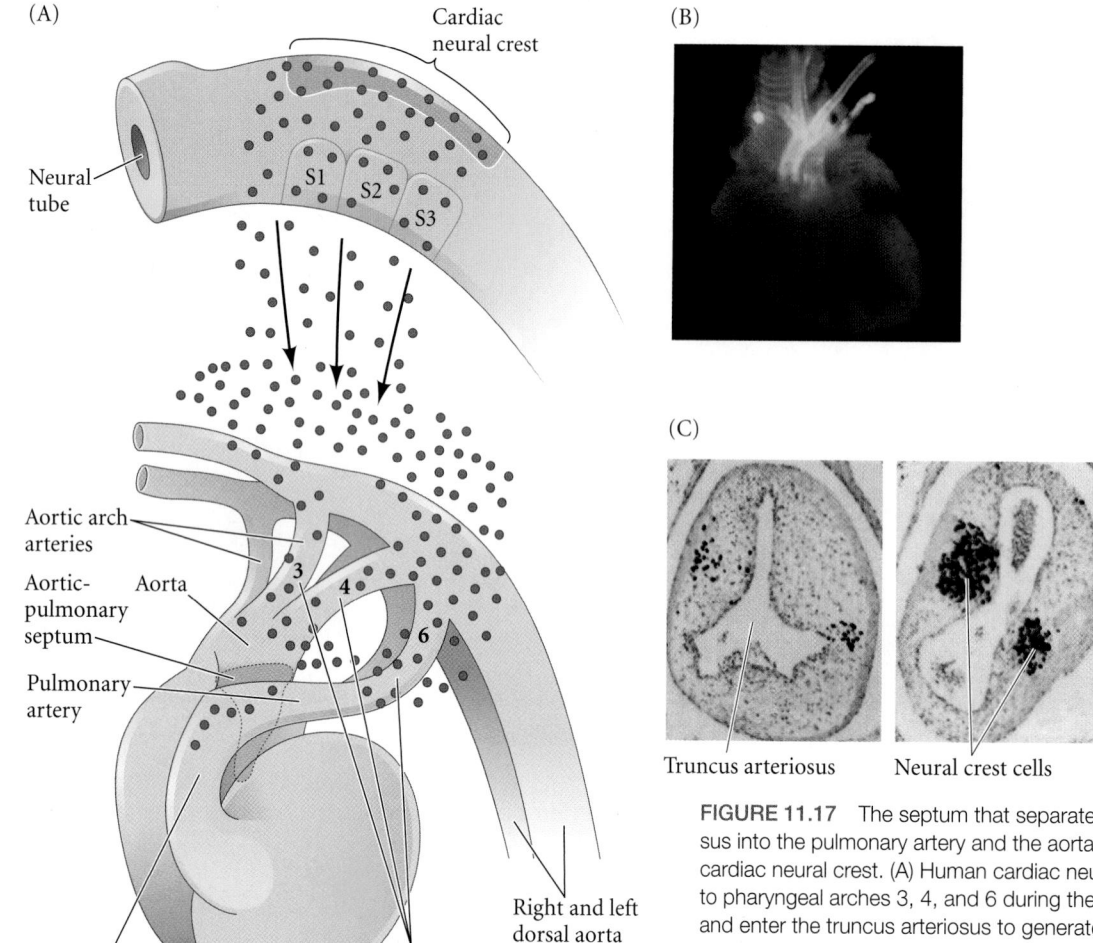

(A)

Cardiac neural crest

Neural tube

S1 S2 S3

Aortic arch arteries

Aortic-pulmonary septum

Aorta

3 4

6

Pulmonary artery

Outflow tract

Pharyngeal arches

Right and left dorsal aorta

(B)

Neural crest cells in septum

Pulmonary artery

(C)

Truncus arteriosus   Neural crest cells   Aorta

FIGURE 11.17 The septum that separates the truncus arteriosus into the pulmonary artery and the aorta forms from cells of the cardiac neural crest. (A) Human cardiac neural crest cells migrate to pharyngeal arches 3, 4, and 6 during the fifth week of gestation and enter the truncus arteriosus to generate the septum. (B) In a transgenic mouse where the fluorescent green protein is expressed only in cells having the Pax3 cardiac neural crest marker, the outflow regions of the heart become labeled. (C) Quail cardiac neural crest cells were transplanted into the analogous region of a chick embryo, and the embryos were allowed to develop. Quail cardiac neural crest cells are visualized by a quail-specific antibody, which stains them darkly. In the heart, these cells can be seen separating the truncus arteriosus into the pulmonary artery and the aorta. (A after Hutson and Kirby 2007; B from Stoller and Epstein 2005; C from Waldo et al. 1998, courtesy of K. Waldo and M. Kirby.)

gene result in fewer cardiac neural crest cells, which in turn leads to persistent truncus arteriosus (failure of the aorta and pulmonary artery to separate), as well as to defects in the thymus, thyroid, and parathyroid glands (Conway et al. 1997, 2000). The path from the dorsal neural tube to the heart appears to involve the coordination between the attractive cues provided by semaphorin-3C and the repulsive signals provided by semaphorin-6 (Toyofuku et al. 2008). Congenital heart defects in humans and mice often occur along with defects in the parathyroid, thyroid, or thymus glands. It would not be surprising to find that all these problems are linked to defects in the migration of cells from the neural crest (Hutson and Kirby 2007).

Cranial Placodes

In addition to forming the cranial neural crest cells, the anterior borders between the epidermal and neural ectoderm form the **cranial ectodermal placodes**, which are local and transient thickenings of the ectoderm in the head and neck. (The cranial neural crest and the cranial placodes may have originated from the same cell population during early vertebrate evolution; see Northcutt and Gans 1983; Baker and

Bronner-Fraser 1997.) The most anterior placode—the olfactory placode—is initially found within the anterior neural tube and subsequently moves laterally. Other placodes are found adjacent to the neural tube, between the epidermis and cranial neural crest. With some contributions from the cranial neural crest, the cranial ectodermal placodes generate most of the peripheral neurons of the head, associated with hearing, balance, taste, and smell (**FIGURE 11.18**; also see Figure 11.3), whereas the cranial neural crest contributes all of the glia. In the case of the trigeminal ganglion, the proximal neurons are formed from neural crest cells (Baker and Bronner-Fraser 2001) and the distal ones from the trigeminal placode. For example, the olfactory placode gives rise to the sensory neurons involved in smell, as well as to migratory neurons that will travel into the brain and secrete gonadotropin releasing hormone. The otic placode gives rise to the

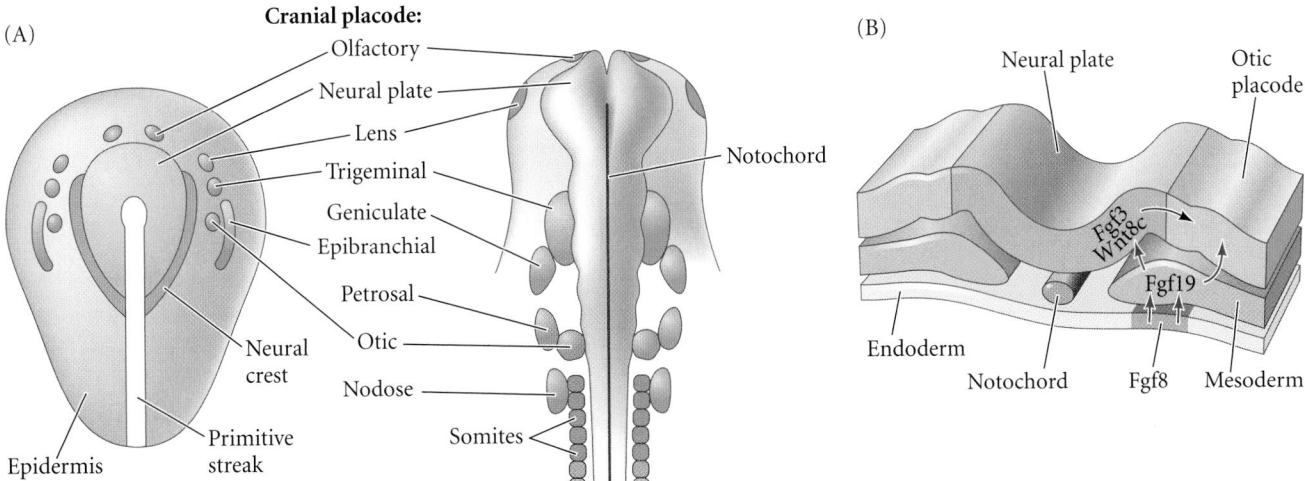

FIGURE 11.18 Cranial placodes form sensory neurons. (A) Fate map of the cranial placodes in the developing chick embryo at the neural plate (left) and 8-somite (right) stages. (B) Induction of the otic (inner ear) placode in the chick embryo. A portion of the pharyngeal endoderm secretes Fgf8, which induces the mesoderm overlying it to secrete Fgf19. Fgf19 is received by both the prospective otic placode and the adjacent neural plate. Fgf19 instructs the neural plate to secrete Wnt8c and Fgf3, two paracrine factors that work synergistically to induce *Pax2* and other genes that allow the cells to produce the otic placode and become sensory cells. (After Schlosser 2010.)

sensory epithelium of the ear and to neurons that help form the cochlear-vestibular ganglion. The lens placode is the only placode that does not form neurons.

The placodes are induced by their neighboring tissues, and there is evidence that the different placodes are each a small portion of what had earlier been a common "pan-placodal" territory (see Figure 11.3A; Streit 2004; Schlosser 2005). Histological evidence shows that the anterior neural plate is surrounded by a single thickening during the early neurula stages, and the cranial placodes may arise from a common set of inductive interactions between the pharyngeal endoderm and head mesoderm (Platt 1896; Jacobson 1966). The anterior-posterior and lateral boundaries are set by retinoic acid, working through Fgf8 (Janesick et al. 2012). Detailed fate mapping studies have confirmed that during the neurula stages, all the placodal precursors are located in a horseshoe-shaped domain that surrounds the anterior neural plate and cranial neural folds (see Figure 11.18A; Pieper et al. 2011). Jacobson (1963) also showed that the pre-placodal cells adjacent to the anterior neural tube are competent to give rise to any placode. This columnar pre-placodal epithelium contains the transcription factors Six1, Six4, and Eya2. These proteins are maintained in all the placodes and are downregulated in the interplacodal regions (Bhattacharyya et al. 2004; Schlosser and Ahrens 2004). Later, specific interactions define the individual placodes. For instance, the chick otic placode, which develops into the sensory cells of the inner ear, is induced by a combination of FGF and Wnt signals (Ladher et al. 2000, 2005). Here, Fgf19 from the underlying cranial paraxial mesoderm is received by both the presumptive otic vesicle and

the adjacent neural plate. Fgf19 induces the neural plate to secrete Wnt8c and Fgf3, which in turn act synergistically to induce formation of the otic placode. The localization of Fgf19 to the specific region of the mesoderm is controlled by Fgf8 secreted in the endodermal region beneath it (see Figure 11.18B).

The epibranchial placodes form dorsally to the point at which the pharyngeal pouches contact the epidermis. The epibranchial placodes split to form the geniculate, petrosal, and nodose placodes, which give rise to the sensory neurons of the facial, glossopharyngeal, and vagal nerves, respectively. The connections made by these placodal neurons are critical in that they enable taste and other pharyngeal sensations to be appreciated. But how do these neurons find their way into the hindbrain? Late-migrating cranial neural crest cells do not travel ventrally to enter the pharyngeal arches; rather, they migrate dorsally to generate glial cells (Weston and Butler 1966; Baker et al. 1997). These glia form tracks that guide neurons from the epibranchial placodes to the hindbrain. Indeed, if the hindbrain is removed before the neural crest cells emigrate, neurons leaving the placodes enter a crest-free environment and fail to migrate into the hindbrain (Begbie and Graham 2001). Therefore, glial cells from the late-migrating cranial neural crest cells are critical in organizing the innervation of the hindbrain. Throughout the head, the neural crest and the sensory placodes—those structures between the epidermis and neural plate—provide the sensory neurons of our ears, nose, tongue, facial skin, and balance system.

● See WEBSITE 11.2 Kallmann syndrome

● See WEBSITE 11.3 The human cranial nerves

NEURON SPECIFICATION AND AXON SPECIFICITY

Unlike most cells, neurons are not confined to their immediate space but can produce axons that may extend for meters. As we saw in Chapter 10, the axon has its own locomotory apparatus that resides in the axonal growth cone (see Figures 10.16 and 10.17). The growth cone has been called "a neural crest cell on a leash" because, like neural crest cells, it migrates and senses the environment. Moreover, it can respond to the same types of signals that migrating cells sense. The cues for axonal migration, moreover, may be even more specific than those used to guide specific cell types to particular areas. Each of the 10^{11} neurons in the human brain has the potential to interact specifically with thousands of other neurons (**FIGURE 11.19**). A large neuron (such as a Purkinje cell or motor neuron) can receive input from more than 10^5 other neurons (Gershon et al. 1985). Understanding the generation of this stunningly ordered complexity is one of the greatest challenges to modern science.

Goodman and Doe (1993) list eight stages of neurogenesis:

1. Induction and patterning of a neuron-forming (neurogenic) region
2. Birth and migration of neurons and glia
3. Specification of cell fates
4. Guidance of axonal growth cones to specific targets
5. Formation of synaptic connections
6. Binding of trophic factors for survival and differentiation

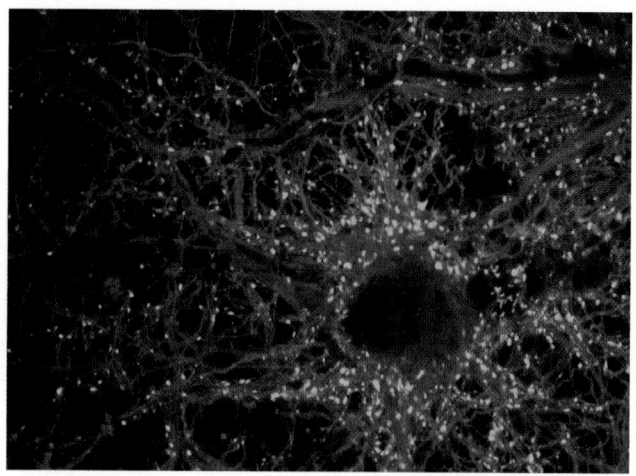

FIGURE 11.19 Connections of axons to a cultured rat hippocampal neuron. The neuron has been stained red with fluorescent antibodies to tubulin. The neuron appears to be outlined by the synaptic protein synapsin (green), which is present in the terminals of axons that contact it. (Photograph courtesy of R. Fitzsimmons and PerkinElmer Life Sciences.)

7. Competitive rearrangement of functional synapses
8. Continued synaptic plasticity during the organism's lifetime

We described the first two stages in Chapter 10. Here we continue our investigation of the processes of neural development.

● See WEBSITE 11.4 The evolution of developmental neurobiology

SIDELIGHTS & SPECULATIONS

The Growth Cone

The navigation of axons to their appropriate targets depends on guidance molecules in the extracellular environment of the **growth cone**, a specialized structure at the tip of the extending axon. It is the growth cone that turns or doesn't turn in response to guidance cues as the axon seeks to make appropriate synaptic connection. Such differential responsiveness is due to disparities in the expression of receptors on the growth cone cell membrane. Growth cones are simultaneously sensory and motile, and they enable axons to navigate certain paths and to recognize specific targets. Growth cones have the ability to sense the environment and translate the extracellular signals into a directed movement. This use of directional cues to facilitate specific migration is accomplished by altering the cytoskeleton, changing membrane growth, and coordinating cell adhesion and cell movement (Vitriol and Zheng 2012).

Growth cones were discovered by the Spanish biologist, pathologist, and artist Santiago Ramón y Cajal (1890), who believed that this region of the axon responded to external stimuli to find neuronal targets. Ross G. Harrison (1910), a pioneering American developmental biologist, was able to culture neuroblasts and see the growth cone in live cells. The growth cone was found to have two major compartments. The peripheral region contains two types of actin-associated membrane protrusions. The **lamellipodia**, broad membranous sheets containing short branched actin networks, act as the migratory network of the growth cone. The **filopodia**, membranes extended by long bundles of filamentous actin, act as the sensory network. The central region of the growth cone contains microtubules that extend the axon shaft and support mitochondria and other organelles (**FIGURE 11.20A**). A transition zone between the central and

(A)

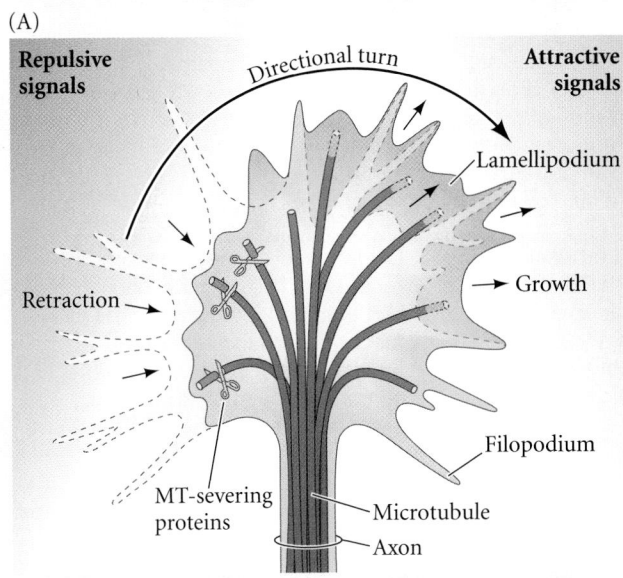

FIGURE 11.20 Growth cone response to attractive and repulsive forces. (A) The periphery of the growth cone contains lamellipodia and filopodia. The lamellipodia are the major motile apparatus and are seen in the regions that are turning toward a stimulus. The filopodia are sensory. Both contain actin microfilaments. There is also a central region of microtubules, some of which extend outward and join the filopodia. The transition region between them contains regulatory proteins that can extend or retract the cytoskeleton. The microtubules entering the peripheral area can be lengthened or shortened by proteins activated by the attractive or repulsive stimuli. During attraction, proteins bind to the plus ends of the microtubules, stabilizing and lengthening them. On the side opposite the attractive cue, microtubules are removed from the periphery. (B) The four major ligands providing cues to the growth cone (ephrins, netrins, slit proteins, and semaphorins) bind to receptors that stabilize or destabilize actin microfilaments. The Rho family of GTPases (RhoA, Rac1, and Cdc42) act as mediators between the receptors and the agents carrying out the cyoskeletal changes. (A after Sanes et al. 2006; B after Lowery and Van Vactor 2009.)

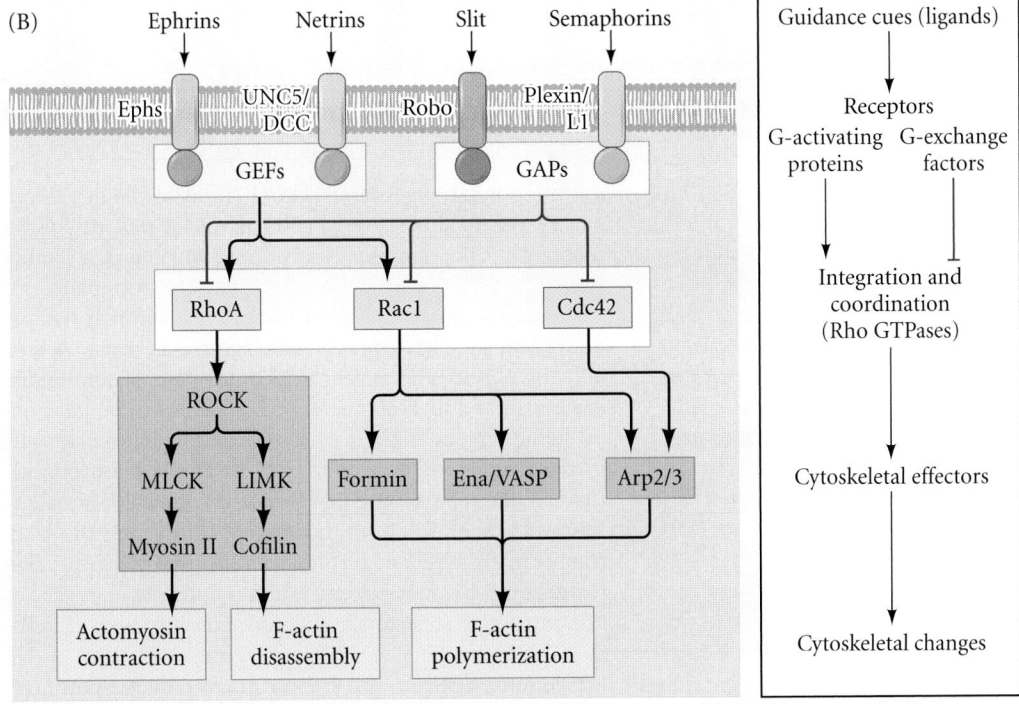

peripheral regions may coordinate the actin and tubulin growth (Rodriguez et al. 2003; Lowery and Van Vactor 2009). The actin-based membrane protrusions coupled with selective adhesion and membrane recycling provide the force for axon movement and directionality.

The regulation of actin polymerization drives growth cone movement and thus is the target of many molecular guidance pathways. Rho GTPases regulate the growth of actin microfilaments. These GTPases can be activated or repressed by receptors binding ephrins, netrins, Slit proteins, or semaphorins (**FIGURE 11.20B**). Similarly, the regulation of tubulin polymerization into microtubules is important, as tubulin is encouraged to polymerize on the side of the growth cone receiving attractant stimuli, and it is inhibited from polymerizing (indeed, the tubulin is depolymerized and recycled) on the side opposite the attractive stimuli (Vitriol and Zheng 2012).

Adhesion is thought to provide the "clutch" for directional movement. Visualize actin as being linked to the cell membrane. Now contemplate that the dynamics of actin assembly and disassembly cause a retrograde flow of actin—that is, actin is moving *away from* the tip of the growth cone and *toward* the cell body. But if the cell membrane is anchored to an external adhesion molecule (through its integrins or cadherins), then the membrane is propelled forward (Bard et al. 2008; Chan and Odde 2008). If there is no such anchoring adhesion, there is no net movement. But if the adhesion is too stable, the growth cone also stops moving. Thus, adhesions have to be made and broken for the growth cone

to progress. These transitory adhesive complexes are referred to as **focal adhesions**, and they bind actin internally and the extracellular environment externally. Focal adhesions may have as many as 100 different protein components (Geiger and Yamada 2011). One of these components, focal adhesion kinase (FAK), appears to be critical for the assembly, stabilization, and degradation of the focal adhesions (Mitra et al. 2005; Chacon and Fazzari 2011), and appears to be able to recognize both attractive and repulsive stimuli. The investigation of the focal adhesion components is just beginning to delineate the mechanisms by which traction is coordinated with cytoskeletal growth and membrane turnover.

Because the membranes turn over, the growth cone grows by the exocytosis of vesicles and the incorporation of their membranes into the cell membrane. These vesicles (sometimes called "enlargeosomes") are constructed in the neuron cell body

and travel on the microtubules to the center of the growth cone (Pfenniger et al. 2003; Rachetti et al. 2010). Most of these membrane sacs are involved in the constitutive growth of the axon, not directional growth of the growth. However, some of these vesicles are transported from the central region into the periphery in response to Ca^{2+} signals coming from membrane receptors. The vesicles then integrate into the tip of the growth cone and function (perhaps exclusively) in turning the growth cone toward attractive stimuli (Tojima et al. 2007). Repulsive cues have been found to initiate endocytosis (the formation of vesicles from cell membranes) in those areas they contact, and this would have the effect of both removing the receptor and of diminishing the amount of cell membrane in that area (Tojima et al. 2010; Hines et al. 2010). Thus, by cytoskeleton assembly, cell adhesion, and membrane turnover, the growth cone leads a migrating axon toward its appropriate target.

The Generation of Neuronal Diversity

Neurons are specified in a hierarchical manner. The first decision is whether the ectodermal epithelium will become neural epithelium, epidermal epithelium, or a neural plate border region (neural crest and placodes). If the choice is neural epithelium, the next decision (as we saw in Chapter 10) is to become either a glial cell or a neuron. For a cell that is to become a neuron, the next decision is, what type of neuron? Will it become a motor neuron, a sensory neuron, a commissural neuron, or some other type? After this fate is determined, still another decision gives the neuron a specific target. To illustrate this process of progressive specification, we will focus on the motor neurons of vertebrates (Bonanomi and Pfaff 2010). These are the neurons in the ventral portion of the spinal cord and brain stem that extend into the periphery to innervate muscle cells and give them commands for contraction.

Vertebrates form a dorsal neural tube by blocking a BMP signal, and the specification of neural (as opposed to glial or epidermal) fate is accomplished through the Notch-Delta pathway (see Chapter 3). The specification of the *type* of neuron appears to be controlled by the position of the neuronal precursor within the neural tube and by its birthday. As described in Chapter 10, neurons at the ventrolateral margin of the vertebrate neural tube become motor neurons, whereas interneurons are derived from cells in the dorsal region of the neural tube. Grafting floor plate or notochord cells (which

secrete Sonic hedgehog) to lateral areas of the neural tube can respecify dorsolateral cells as motor neurons, which means that the specification of neuron type is probably a function of the cell's position relative to the floor plate.

Erickson and colleagues (1996) have shown that two periods of Shh signaling are needed to specify the motor neurons: an early period wherein the cells of the ventrolateral margin are instructed to become ventral neurons, and a later period (which includes the S phase of its last cell division) that instructs a ventral neuron to become a motor neuron rather than an interneuron. The first decision is probably regulated by the secretion of Shh from the notochord, whereas the second is more likely regulated by Shh from the floor plate cells. Sonic hedgehog appears to specify motor neurons by inducing certain transcription factors at different concentrations (Erickson et al. 1992; Tanabe et al. 1998; see Figure 10.14).

The next decision involves target specificity. Once a cell becomes a motor neuron, will that motor neuron innervate the thigh, the forelimb, or the tongue? The anterior-posterior specification of the neural tube is regulated primarily by Hox genes from the hindbrain through the spinal cord, and by specific head genes (such as *Otx*) in the brain (Dasen et al. 2005; Jung et al. 2010). Within a region of the body along the anterior-posterior axis, motor neuron specificity is regulated by the neuron's age at its final cell division, which in turn involves activation of the Foxp1 transcription factor in the motor neuron precursor cells. In the absence of Foxp1, motor

(A)

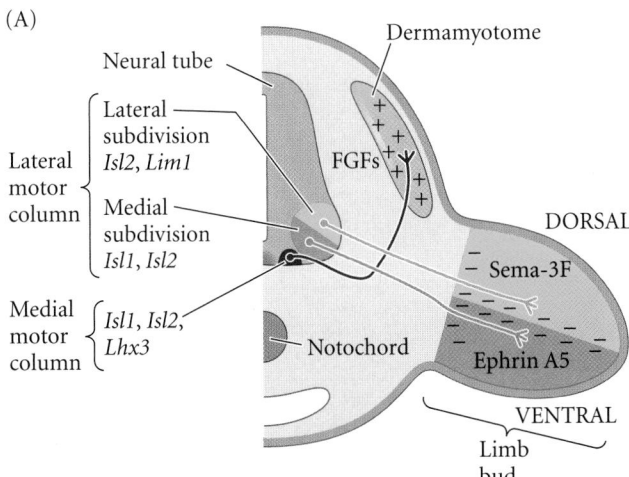

(B)

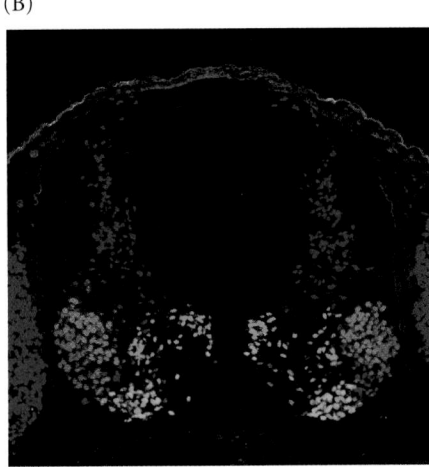

FIGURE 11.21 Motor neuron organization and Lim specification in the spinal cord innervating the chick hindlimb. (A) Neurons in each of three different columns express specific sets of Lim family genes (including *Isl1* and *Isl2*), and neurons within each column make similar pathfinding decisions. Neurons of the medial motor column are attracted to the axial muscles by FGFs secreted by the dermamyotome. Neurons of the lateral motor column send axons to the limb musculature. Where these columns are subdivided, medial subdivisions project to ventral positions because they are repelled by semaphorin-3F in the dorsal limb bud; and lateral subdivisions send axons to dorsal regions of the limb bud, as they are repelled by ephrin A5 synthesized in the ventral half. (B) Motor neurons in different regions of the chick spinal cord express different transcription factors (visualized here using various stains), giving them different cell surface receptors that affect axonal migration. (A after Polleux et al. 2007; B courtesy of J. S. Dasen.)

neurons at any given section along the axis choose their targets randomly (Dasen et al. 2008). Foxp1 apparently enables the Hox genes to act in concert with those genes whose activation depends on the level of the spinal cord in which the motor neuron precursor resides. As discussed in Chapter 10, a neuron's birthday determines which layer of the cortex it will enter. As younger cells migrate to the periphery, they must pass through neurons that differentiated earlier in development. Similarly, as younger motor neurons migrate through the region of older motor neurons in the intermediate zone of the spinal cord, they express new transcription factors as a result of a retinoic acid (or other retinoid) signal secreted by the early-born motor neurons (Sockanathan and Jessell 1998). These transcription factors are encoded by the Lim genes and are structurally related to those proteins encoded by the Hox genes. The combination of Hox (anterior-posterior axis) and Lim (proximal-distal axis) genes expressed by a neuron provides the information specifying which targets that neuron seeks out.

As a result of their differing birthdays and migration patterns, motor neurons form three major groups: the columns of Terni (CT), and the lateral and medial motor columns

(LMC and MMC). The cell bodies of the motor neurons projecting to a single muscle are pooled in a longitudinal column of the spinal cord. This pooling is performed by different cadherins that become expressed on these different populations of cells (Landmesser 1978; Hollyday 1980; Price et al. 2002). The pools are grouped into the CT, LMC, and MMC, and neurons in similar places have similar targets. For instance, in the chick hindlimb, muscles are innervated by the LMC axons, with lateral neurons entering the dorsal musculature, while the motor neurons of the MMC innervate ventral limb musculature (**FIGURE 11.21**; Tosney et al. 1995; Polleux et al. 2007). This arrangement of motor neurons is consistent throughout the vertebrates.

The targets of motor neurons are specified before their axons extend into the periphery. This was shown by Lance-Jones and Landmesser (1980), who reversed segments of the chick spinal cord so that the motor neurons were placed in new locations. The axons went to their original targets, not to the ones expected from their new positions (**FIGURE 11.22**). The molecular basis for this target specificity resides in the members of the Hox and Lim protein families that are induced during neuronal development (Tsushida et al. 1994; Sharma et al. 2000; Price and Briscoe 2004; Bonanomi and Pfaff 2010). For instance, all motor neurons express Islet1 and (slightly later) Islet2. If no other Lim protein is expressed, the neurons project to the ventral limb muscles. This is because the axons (just like the trunk neural crest cells) synthesize neuropilin-2, the receptor for the chemorepellant semaphorin-3F, which is made in the dorsal part of the limb bud. However, if Lim1 protein is also synthesized, the motor neurons project dorsally to the dorsal limb muscles because Lim1 induces the expression of Eph A4, the receptor for the chemorepellent protein ephrin A5, which is made in the ventral part of the limb bud. Thus, the innervation of the limb by motor neurons depends on repulsive signals. The motor neurons entering the axial muscles of the body wall, however, are brought there by chemoattraction. (Indeed, these axons make an abrupt turn to get to the developing musculature.) This is because these motor

(A)

(B) 2–5 Days

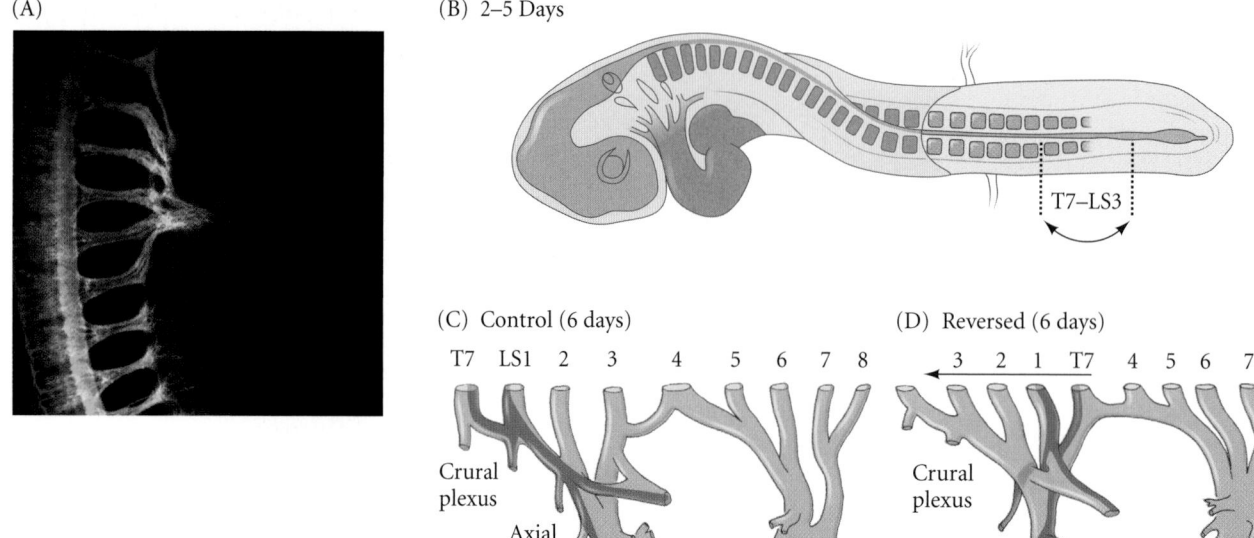

(C) Control (6 days)

(D) Reversed (6 days)

FIGURE 11.22 Compensation for small dislocations of axonal initiation position in the chick embryo. (A) Axons from motor neurons and sensory neurons group together (fasciculate) before finding their muscle targets Here, motor nerves (stained green with GFP) and sensory neurons (stained red with antibodies) fasciculate before entering the limb bud of a 10.5-day mouse embryo. (B) A length of spinal cord comprising segments T7–LS3 (seventh thoracic to third lumbosacral segments) is reversed in a 2.5-day embryo. (C) Normal pattern of axon projection to the forelimb muscles at 6 days. (D) Projection of axons from the reversed segment at 6 days. The ectopically placed neurons eventually found their proper neural pathways and innervated the appropriate muscles. (A from Huettl et al. 2011, courtesy of Dr. A. Huber-Brösamle; B–D after Lance-Jones and Landmesser 1980.)

neurons express Lhx3, which induces the expression of a receptor for FGFs such as those secreted by the dermamyotome (the somatic region that contains muscle precursor cells).

Motor neurons seek their targets through intrinsic "programs" that assign different motor neurons different cell surface molecules that determine the responsiveness of the axon growth cones to guidance cues in their path and on their targets. However, this is not the case with sensory neurons. It has long been known that sensory neurons need motor neurons to make the appropriate connections (Hamburger 1929; Landmesser et al. 1983; Honig et al. 1986). It appears that the subtypes of motor neurons produce specific compounds (such as Ephs) that cause the sensory neurons to follow them (see Figure 11.23; Huettl et al. 2011; Wang et al. 2011). In this way, a close connection forms between motor neurons and their associated sensory neurons (see Figure 11.22A).

Pattern Generation in the Nervous System

How does the neuronal axon "know" how to traverse numerous potential target cells to make its specific connection? Ross G. Harrison (1910) first suggested that the specificity of axonal growth is due to **pioneer nerve fibers**, axons that go ahead of other axons and serve as guides for them.* This observation simplified, but did not solve, the problem of how neurons form appropriate patterns of interconnection. Harrison also noted, however, that axons must grow on a solid substrate, and he speculated that differences among embryonic surfaces might allow axons to travel in certain specified directions. The final connections would occur by complementary interactions on the target cell surface:

> *That it must be a sort of a surface reaction between each kind of nerve fiber and the particular structure to be innervated seems clear from the fact that sensory and motor fibers, though running close together in the same bundle, nevertheless form proper peripheral connections, the one with the epidermis and the other with the muscle. … The foregoing facts suggest that there may be a certain analogy here with the union of egg and sperm cell.*

---

*The growth cones of pioneer neurons migrate to their target tissue while embryonic distances are still short and the intervening embryonic tissue is still relatively uncomplicated. Later in development, other neurons bind to pioneer neurons and thereby enter the target tissue. Klose and Bentley (1989) have shown that in some cases, pioneer neurons die after the "follow-up" neurons reach their destination. Yet if the pioneer neurons are prevented from differentiating, the other axons do not reach their target tissue.

Research on the specificity of neuronal connections has focused on two major systems: motor neurons, whose axons travel from the spinal cord to a specific muscle; and the optic system, where axons originating in the retina find their way back into the brain. In both cases, the specificity of axonal connections unfolds in three steps (Goodman and Shatz 1993):

1. *Pathway selection*. The axons travel along a route that leads them to a particular region of the embryo.

2. *Target selection*. The axons, once they reach the correct area, recognize and bind to a set of cells with which they may form stable connections.

3. *Address selection*. The initial patterns are refined such that each axon binds to a small subset (sometimes only one) of its possible targets.

The first two processes are independent of neuronal activity. The third process involves interactions between several active neurons and converts the overlapping projections into a fine-tuned pattern of connections.

It has been known since the 1930s that the axons of motor neurons can find their appropriate muscles even if the neural activity of the axons is blocked. Twitty (who was Harrison's student) and his colleagues found that embryos of the newt *Taricha torosa* secrete tetrodotoxin (TTX), a toxin that blocks neural transmission in other species. By grafting pieces of *T. torosa* embryos onto embryos of other salamander species, they were able to paralyze the host embryos for days while development occurred. Normal neuronal connections were made, even though no neural activity could occur. At about the time the tadpoles were ready to feed, the neurotoxin wore off, and the young salamanders swam and fed normally (Twitty and Johnson 1934; Twitty 1937). More recent experiments using zebrafish mutants with nonfunctional neurotransmitter receptors similarly demonstrated that motor neurons can establish their normal patterns of innervation in the absence of neuronal activity (Westerfield et al. 1990). But the question remains: How are the neurons' axons instructed where to go?

Cell adhesion and contact guidance by attraction and repulsion

The initial pathway an axonal growth cone follows is determined by the environment the growth cone experiences. The polarity of a neuron—that is, which part of the cell will extend the axon—is determined largely by the neuron's response to cell adhesive cues in its immediate environment. Integrins and N-cadherins serve as receptors to orient the neuron in accordance with cues from the extracellular matrices and membranes of surrounding cells (Myers et al. 2011; Randlett et al. 2011; Gärtner et al. 2012). These receptors recruit actin, which forms microfilaments in the specified area. The microfilaments transport the motor protein dynein, which in turn recruits microtubules, which can extend the axon (Ligon et al. 2001).

Once an axon begins to form, its growth cone encounters different substrates. The growth cone adheres to certain substrates and moves in their direction. Other substrates cause the growth cone to retract, preventing its axon from growing in that direction. Growth cones prefer to migrate on surfaces that are more adhesive than their surroundings, and a track of adhesive molecules (such as laminin) can direct them to their targets (Letourneau 1979; Akers et al. 1981; Gundersen 1987).

In addition to general extracellular matrix cues, there are protein guidance cues specific to certain groups of neurons. Axons in the developing nervous system respond to attractive and repulsive signals from four major protein families: ephrins, semaphorins, netrins, and the Slit proteins—the same proteins we saw regulating neural crest cell migration (see Kolodkin and Tessier-Lavigne 2010). We have already seen that neural crest cells are patterned by their recognition of ephrin, and that what is an attractive cue to one set of cells (such as the presumptive melanocytes going through the dermis) can be a repulsive signal to other cells (such as the presumptive sympathetic ganglia). We will see that whether a guidance signal is attractive or repulsive can depend on (1) the type of cell receiving that signal and (2) the time when a cell receives the signal.

Membrane proteins: The ephrins and semaphorins

Members of two membrane protein families, the ephrins and the semaphorins, are involved in neural patterning. Just as neural crest cells are inhibited from migrating across the posterior portion of a sclerotome, the axons from the dorsal root ganglia and motor neurons also pass only through the anterior portion of each sclerotome and avoid migrating through the posterior portion (**FIGURE 11.23A**; also see Figure 11.8). Davies and colleagues (1990) showed that membranes isolated from the posterior portion of a somite cause the growth cones of these neurons to collapse (**FIGURE 11.23B,C**). These growth cones contain Eph receptors (which bind ephrins) and neuropilin receptors (which bind semaphorins) and are thus responsive to ephrins and semaphorins on the posterior sclerotome cells (Wang and Anderson 1997; Krull et al. 1999; Kuan et al. 2004). In this way, the same signals that pattern neural crest cell migration also pattern the spinal neuronal outgrowths.

EPHRINS   As any psychologist knows, the line between attraction and repulsion is often thin, and at the base of both phenomena is some sort of recognition event. Context is critical. Ephrins can either repel or attract a neuron, depending on the neuron's history and which ephrin receptor (the Eph proteins) is being used. Most ephrin-Eph interactions result in repulsion. However, when Eph A7 recognizes ephrin A5 in the mouse neural tube, the result is attraction, and the interaction between these two proteins is critical for the closure of the neural tube. The Eph A7 and ephrin A5 proteins cause adhesion of the neural plate cells, and deletion of either gene results in a condition resembling anencephaly (see Figure 10.5). The change from repulsion to adhesion is caused by

FIGURE 11.23 Repulsion of dorsal root ganglion growth cones. (A) Motor axons migrating through the rostral (anterior), but not the caudal (posterior), compartments of each sclerotome. (B) In vitro assay, wherein ephrin stripes were placed on a background surface of laminin. Motor axons grew only where the ephrin was absent. (C) Inhibition of growth cones by ephrin after 10 minutes of incubation. The left-hand photograph shows a control axon subjected to a similar (but not inhibitory) compound; the axon on the right was exposed to an ephrin found in the posterior somite. (From Wang and Anderson 1997, courtesy of the authors.)

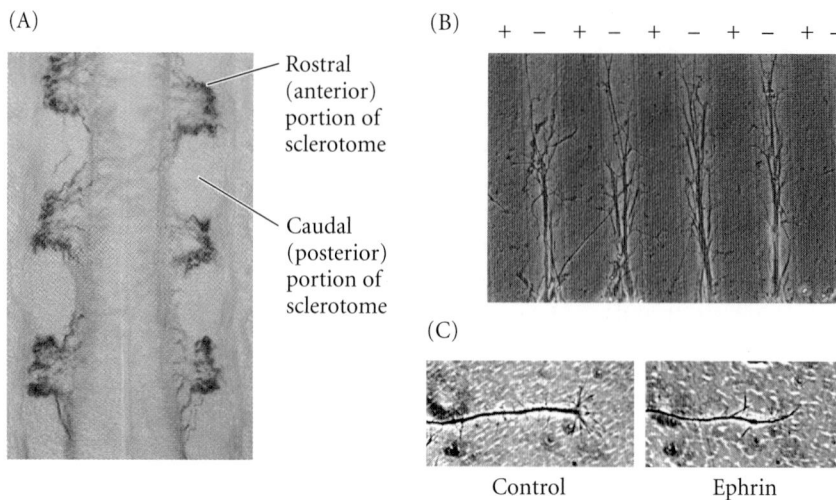

Rostral (anterior) portion of sclerotome

Caudal (posterior) portion of sclerotome

Control   Ephrin

alternative RNA processing. By using a different splice site, the mouse neural plate produces Eph A7 lacking the tyrosine kinase domain that transmits the repulsive signal (Holmberg et al. 2000). The result is that the cells recognize each other through these proteins, and no repulsion occurs.

SEMAPHORINS   Found throughout the animal kingdom, the semaphorins usually guide growth cones by selective repulsion. They are especially important in forcing "turns" when an axon must change direction. Semaphorin-1, for example, is a transmembrane protein that is expressed in a band of epithelial cells in the developing insect limb. This protein appears to inhibit the growth cones of the Ti1 sensory neurons from moving forward, thus causing them to turn (**FIGURE 11.24**; Kolodkin et al. 1992, 1993). In *Drosophila*, semaphorin-2 is secreted by a single large thoracic muscle. In this way, the thoracic muscle prevents itself from being innervated by inappropriate axons (Matthes et al. 1995).

The proteins of the semaphorin-3 family, also known as collapsins, are found in mammals and birds. These secreted proteins collapse the growth cones of axons originating in the dorsal root ganglia (Luo et al. 1993). There are several types of neurons in the dorsal root ganglia whose axons enter the dorsal spinal cord. Most of these axons are prevented from traveling farther and entering the ventral spinal cord; however, a subset of them does travel ventrally through the other neural cells (**FIGURE 11.25**). These particular axons are not inhibited by semaphorin-3, whereas those of the other neurons are (Messersmith et al. 1995). This finding suggests that semaphorin-3 patterns sensory projections from the dorsal root ganglia by selectively repelling certain axons so that they terminate dorsally. A similar scheme is seen in the brain, where semaphorin made in one region of the brain prevents the entry of neurons that originated in another region (Marín et al. 2001).

In some circumstances, semaphorins can also be attractants. Semaphorin-3A is a classic chemorepellant for axons coming from pyramidal neurons in the mammalian cortex.

However, it is a chemoattractant for *dendrites* of the same cells. In this way, a target can "reach out" to the dendrites of these cells without attracting their axons as well (Polleux et al. 2000).

Guidance by diffusible molecules

NETRINS AND THEIR RECEPTORS   The idea that chemotactic cues guide axons in the developing nervous system was first proposed by Santiago Ramón y Cajal (1892). He suggested that diffusible molecules might signal the commissural neurons of the spinal cord to send axons from their dorsal positions in the neural tube to the ventral floor plate. Commissural neurons coordinate right and left motor activities,

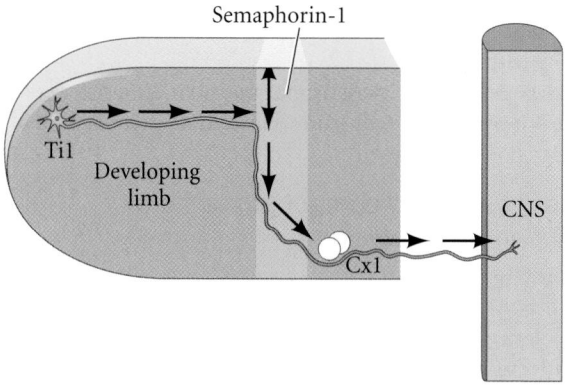

Semaphorin-1

Ti1

Developing limb

Cx1

CNS

Ventral nerve cord

FIGURE 11.24 Action of semaphorin-1 in the developing grasshopper limb. The axon of sensory neuron Ti1 projects toward the central nervous system. (The arrows represent sequential steps en route.) When it reaches a band of semaphorin-1-expressing epithelial cells, the axon reorients its growth cone and extends ventrally along the distal boundary of the semaphorin-1-expressing cells. When its filopodia connect to the Cx1 pair of cells, the growth cone crosses the boundary and projects into the CNS. When semaphorin-1 is blocked by antibodies, the growth cone searches randomly for the Cx1 cells. (After Kolodkin et al. 1993.)

(A)

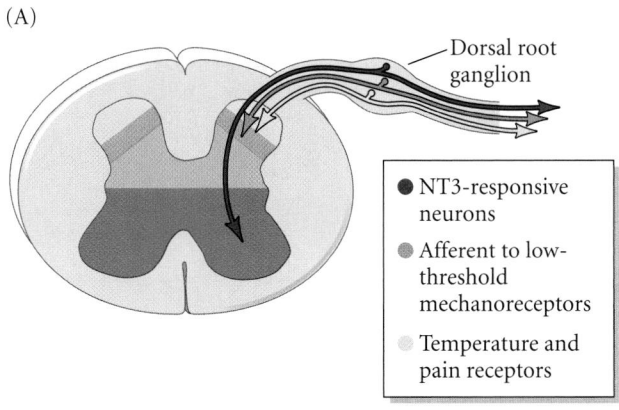

Dorsal root ganglion

- ● NT3-responsive neurons
- ● Afferent to low-threshold mechanoreceptors
- ○ Temperature and pain receptors

FIGURE 11.25 Semaphorin-3 as a selective inhibitor of axonal projections into the ventral spinal cord. (A) Trajectory of axons in relation to semaphorin-3 expression in the spinal cord of a 14-day embryonic rat. Neurons that are responsive to neurotrophin 3 (NT3) can travel to the ventral region of the spinal cord, but the afferent axons for the mechanoreceptors and for temperature and pain receptor neurons terminate dorsally. (B) Transgenic chick fibroblast cells that secrete semaphorin-3 inhibit the outgrowth of mechanoreceptor axons. These axons are growing in medium treated with nerve growth factor (NGF), which stimulates their growth, but are still inhibited from growing toward the source of semaphorin-3. (C) Neurons that are responsive to NT3 for growth are not inhibited from extending toward the source of semaphorin-3 when grown with NT3. (A after Marx 1995; B,C from Messersmith et al. 1995, courtesy of A. Kolodkin.)

(B)

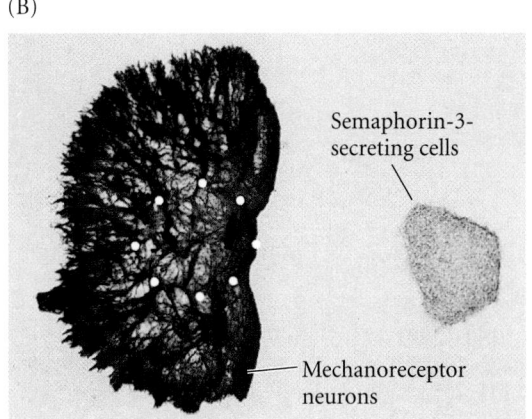

Semaphorin-3-secreting cells

Mechanoreceptor neurons

(C)

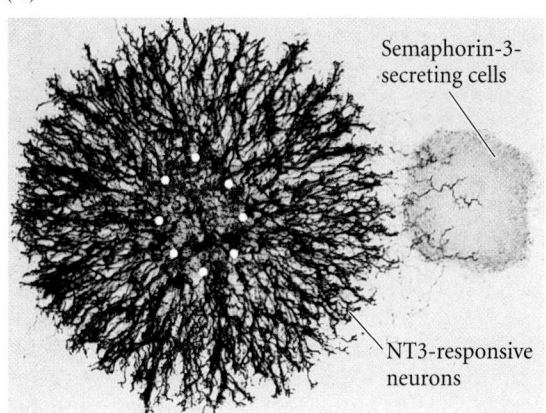

Semaphorin-3-secreting cells

NT3-responsive neurons

and to accomplish this they somehow must migrate to (and through) the ventral midline. The axons of commissural neurons begin growing ventrally down the side of the neural tube. However, about two-thirds of the way down, the axons change direction and project through the ventrolateral (motor) neuron area of the neural tube toward the floor plate cells (**FIGURE 11.26A**).

There appear to be two systems involved in attracting the commissural neurons to the ventral midline. The first is the Sonic hedgehog protein made in the floor plate and which starts the commissural neurons on their ventral migrations (see Figure 10.14; Charron et al. 2003). If Shh is inhibited by cyclopamine or conditionally knocked out of the floor plate cells, the commissural axons have difficulty getting to the ventral midline. However, a Sonic hedgehog gradient does not provide a full explanation of the migration; some other factor is also involved. In 1994, Serafini and colleagues developed an assay that allowed them to screen for the presence of a presumptive diffusible molecule that might be guiding the commissural neurons. When dorsal spinal cord explants from chick embryos were explanted onto collagen gels, the presence of floor plate cells near them promoted the outgrowth of commissural axons. Serafini and co-workers took fractions of embryonic chick brain homogenate and tested them to see if any of the proteins therein mimicked explant activity.

This research resulted in the identification of two proteins, **netrin-1** and **netrin-2**. Netrin-1 is made by and secreted from the floor plate cells, whereas netrin-2 is synthesized in the lower region of the spinal cord, but not in the floor plate (**FIGURE 11.26B**). It is possible that the commissural neurons first encounter a gradient of netrin-2 and Shh, which brings them into the domain of the steeper netrin-1 gradient. The netrins are recognized by the receptors DCC and DSCAM, found in axon growth cones (Liu et al. 2009).

Although they are soluble molecules, both netrins become associated with the extracellular matrix.* Such associations can play important roles and may change the effect of the

---

*Nature doesn't necessarily conform to the categories we humans create. The binding of a soluble factor to the extracellular matrix makes for an interesting ambiguity between *chemotaxis* (movement toward a specific chemical substance) and *haptotaxis* (migration along a preferred substrate). There is also some confusion between the terms *neurotropic* and *neurotrophic*. Neuro*tropic* (Latin, *tropicus*, "a turning movement") means that something attracts the neuron. Neuro*trophic* (Greek, *trophikos*, "nursing" or "nourishing") refers to a factor's ability to keep the neuron alive, usually by supplying growth factors. Since many agents have both properties, they are alternatively called *neurotropins* and *neurotrophins*. In the recent literature, *neurotrophin* appears to be more widely used.

(A)

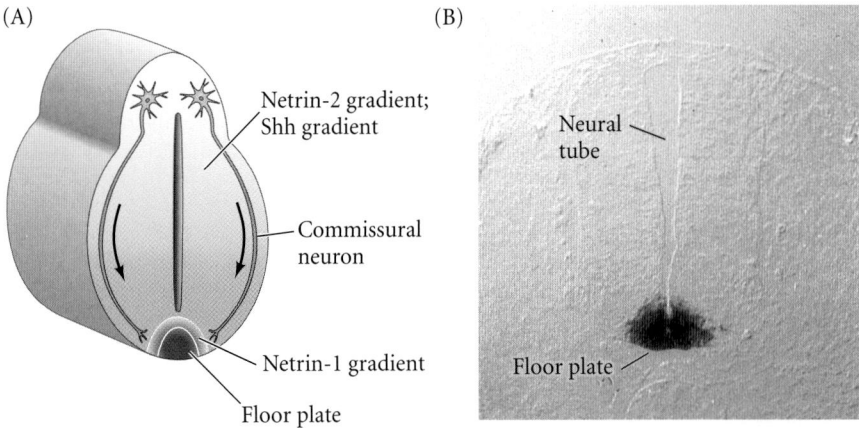

Netrin-2 gradient; Shh gradient

Commissural neuron

Netrin-1 gradient

Floor plate

(B)

Neural tube

Floor plate

FIGURE 11.26 Trajectory of commissural axons in the rat spinal cord. (A) Schematic drawing of a model wherein commissural neurons first experience a gradient of Sonic hedgehog and netrin-2, and then a steeper gradient of netrin-1. The commissural axons are chemotactically guided ventrally down the lateral margin of the spinal cord toward the floor plate. Upon reaching the floor plate, contact guidance from the floor plate cells causes the axons to change direction. (B) Autoradiographic localization of *netrin-1* mRNA by in situ hybridization of anti-sense RNA to the hindbrain of a young rat embryo. *Netrin-1* mRNA (dark area) is concentrated in the floor plate neurons. (B from Kennedy et al. 1994, courtesy of M. Tessier-Lavigne.)

netrin from attractive to repulsive. The growth cones of *Xenopus* retinal neurons, for example, are attracted to netrin-1 and are guided to the head of the optic nerve by this diffusible factor. Once there, however, the combination of netrin-1 and laminin *prevents* the axons from leaving the optic nerve. It appears that the laminin of the extracellular matrix surrounding the optic nerve converts the netrin from an attractive molecule to a repulsive one (Höpker et al. 1999).

The structures of the netrin proteins have numerous regions of homology with UNC-6, a protein implicated in directing the migration of axons around the body wall of *C. elegans*. In the wild-type nematode, UNC-6 induces axons from certain centrally located sensory neurons to move ventrally while inducing ventrally placed motor neurons to extend axons dorsally. In *unc-6* loss-of-function mutations, neither of these migrations occurs (Hedgecock et al. 1990; Ishii et al. 1992; Hamelin et al. 1993). Mutations of the *unc-40* gene disrupt ventral (but not dorsal) axon migration, whereas mutations of the *unc-5* gene prevent only dorsal migration (**FIGURE 11.27**). Genetic and biochemical evidence suggests that UNC-5 and UNC-40 are portions of the UNC-6 receptor complex, and that UNC-5 can convert a UNC-40-mediated attraction into a repulsion (Leonardo et al. 1997; Hong et al. 1999; Chang 2004).

There is reciprocity in science. Just as research on vertebrate netrin genes led to the discovery of their *C. elegans* homologues, research on the nematode *unc-5* gene led to the discovery of the gene encoding the mammalian netrin receptor. This turns out to be a gene whose mutation in mice causes a disease called rostral cerebellar malformation (Ackerman et al. 1997; Leonardo et al. 1997).

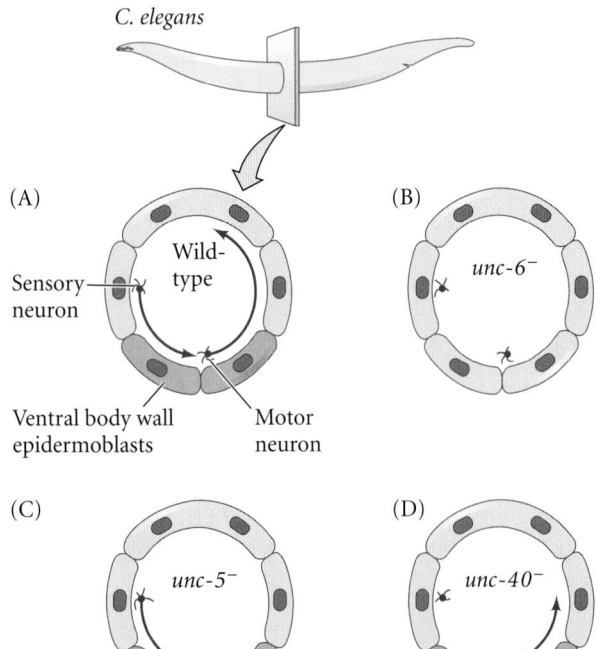

C. elegans

(A)

Sensory neuron

Wild-type

Ventral body wall epidermoblasts

Motor neuron

(B)

unc-6⁻

(C)

unc-5⁻

(D)

unc-40⁻

FIGURE 11.27 UNC expression and function in axonal guidance. (A) In the body of the wild-type *C. elegans* embryo, sensory neurons project ventrally and motor neurons project dorsally. The ventral body wall epidermoblasts expressing UNC-6 are darkly shaded. (B) In the *unc-6* mutant embryo, neither of these migrations occurs. (C) The *unc-5* loss-of-function mutation affects only the dorsal movements of the motor neurons. (D) The *unc-40* loss-of-function mutation affects only the ventral migration of the sensory growth cones. (After Goodman 1994.)

(A) Slit protein

(B) Robo protein

(C) Wild-type

(D) *Slit*$^{-/-}$

(E)

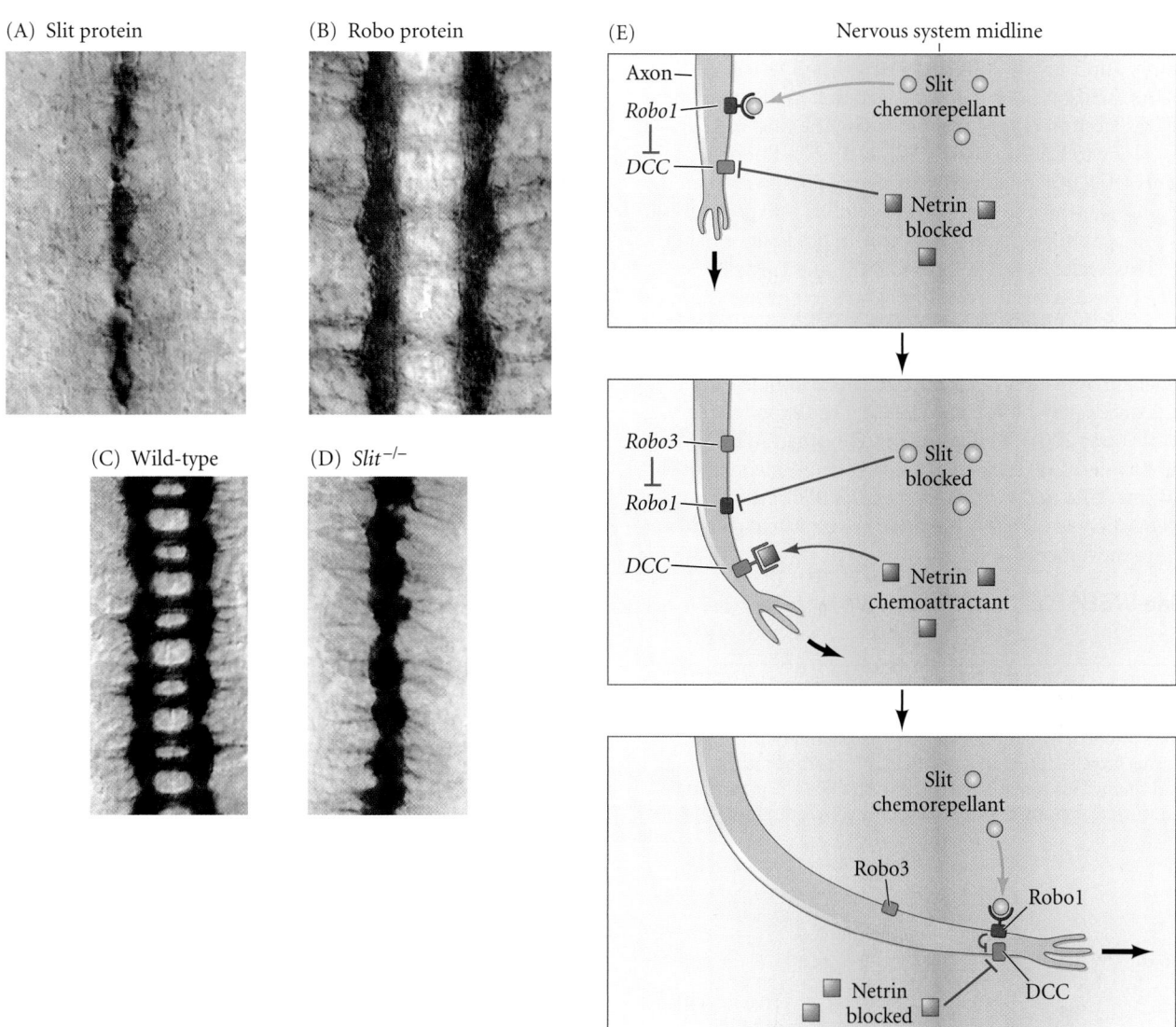

Nervous system midline

FIGURE 11.28 Robo/Slit regulation of midline crossing by neurons. (A–D) Robo and Slit in the *Drosophila* central nervous system. (A) Antibody staining reveals Slit protein in the midline glial cells. (B) Robo protein appears along the neurons of the longitudinal tracts of the CNS axon scaffold. (C) Wild-type CNS axon scaffold shows the ladderlike arrangement of neurons crossing the midline. (D) Staining of the CNS axon scaffold with antibodies to all CNS neurons in a *Slit* loss-of-function mutant shows axons entering but failing to leave the midline (instead of running alongside it). (E) Regulation by Slit and Robo in vertebrates. Neurons are prevented from crossing the midline by Slit, which activates the *Robo1* gene. Robo1 then blocks DCC from binding to the netrin proteins. When the axon gets near the midline, Robo3 is expressed and blocks these functions of Robo1, thus enabling netrin to bind to DCC and turning the axon growth cone toward the midline. After the cone crosses the midline, Robo1 is upregulated and Robo3 is downregulated. This allows Slit to act as a chemorepellent, forcing the growth cone away from the midline. (A–D from Kidd et al. 1999, courtesy of C. S. Goodman; E after Woods 2004.)

SLIT AND ROBO Diffusible proteins also provide guidance by repulsion. One important chemorepulsive molecule are the Slit proteins. In *Drosophila*, Slit is secreted by the neural cells in the midline and acts to prevent most neurons from crossing the midline from either side. The growth cones of *Drosophila* neurons express Roundabout (Robo) protein, which is the receptor for Slit. In this way, most *Drosophila* neurons are prevented from migrating across the midline. However, the commissural neurons that traverse the embryo from side to side avoid this repulsion by downregulating Robo protein as they approach the midline. Once the growth cone is across the middle of the embryo, the neurons re-express Robo and become sensitive again to the midline inhibitory actions of Slit (**FIGURE 11.28A**; Brose et al. 1999; Kidd et al. 1999; Orgogozo et al. 2004).

In vertebrates, the Slit/Robo system cooperates with the netrin/DCC system to permit the commissural neurons to cross the midline; there are several vertebrate Robo and Slit proteins, and they do different tasks (Mambetisaeva et al. 2005). As the axon extends toward its target in the developing brain, the neuron is kept from crossing the midline by Slit, which binds to Robo1 (**FIGURE 11.28B**). The Robo1 protein prevents DCC from binding to the netrin proteins. When the axon gets near the midline, however, Robo3 is expressed and blocks Robo1. Robo1 is no longer able to bind Slit or to block DCC. This enables netrin to bind DCC and turns the axon growth cone toward the midline. The axon grows through the midline, but after crossing it, Robo3 is downregulated while Robo1 is upregulated. This allows Slit to act as a chemorepellent, forcing the growth cone away from the midline. DCC is once again blocked, preventing the axon from going back* (Woods 2004). Mutations in the human *ROBO3* gene disrupt the normal crossing of axons from one side of the brain's medulla to the other (Jen et al. 2004). Among other problems, people with this mutation are unable to coordinate their eye movements.

● See WEBSITE 11.5 The early evidence for chemotaxis

---

*In the mammalian forebrain, Slit also acts to orient guidepost neurons in the midline to permit the movement of thalamic neurons into a new mammalian-specific tract. In reptiles and birds, this area is closed and the thalamic neurons must migrate around them. This new tract may have enabled the expansion of the neocortex characteristic of mammals (Bielle et al. 2011).

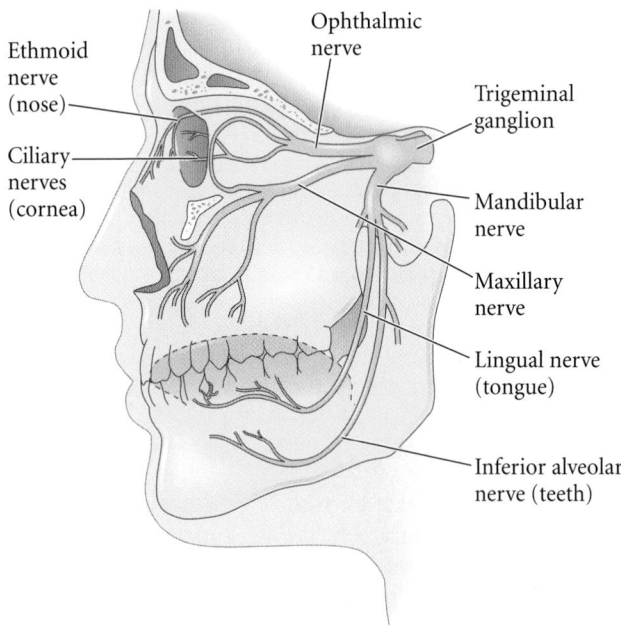

Ethmoid nerve (nose)

Ophthalmic nerve

Ciliary nerves (cornea)

Trigeminal ganglion

Mandibular nerve

Maxillary nerve

Lingual nerve (tongue)

Inferior alveolar nerve (teeth)

FIGURE 11.29 The trigeminal ganglion has three main branches: the ophthalmic (to the eyes), the maxillary (to the upper jaw), and the mandibular (to the lower jaw). The growth of these nerves is regulated by BMP4 from the target tissues combined with differential receptors for the BMPs on the neurons.

Target selection

In some cases, nerves in the same ganglion may have several different targets. How do the different neurons know where to go? It seems they use the same strategy as the neural crest cells. In some cases, different neurons in the same ganglion have different receptors and can therefore respond to certain cues and not to others. For instance, some neurons in the superior cervical ganglia (the biggest ganglia in the neck) go toward the carotid artery, while other neurons from these ganglia do not. It appears that those axons extending to the carotid artery follow the blood vessels that lead there. These blood vessels secrete small peptides called **endothelins**. In addition to their adult roles constricting blood vessels, endothelins appear to have an embryonic role, as they are able to direct the migration of certain neural crest cells (such as those entering the gut) and of certain sympathetic axons that have endothelin receptors on their membranes (Makita et al. 2008). Similarly, the trigeminal ganglion has three peripheral axon bundles that innervate the eye regions, the upper jaw, and the lower jaw (**FIGURE 11.29**); these include the neurons dentists "put to sleep" with novocaine while filling cavities. BMP4 from the target organs causes the differential growth and differentiation of these neurons, but intrinsic differences in transcription factors enable them to respond differently to this signal and allow their axons to migrate in their particular ways (Hodge et al. 2007).

Once a neuron reaches a group of cells in which lie its potential targets, it is responsive to various proteins produced by the target cells.* Both attractive and repulsive forces steer the axons to their appropriate destinations. As we will see, the amount of repulsive proteins (such as ephrins) can be critical in getting particular neurons to particular targets (Gosse et al. 2008).

In addition to the ephrin, Slit, semaphorin, and netrin proteins already mentioned, some target cells produce a set of chemotactic proteins collectively called **neurotrophins**. The neurotrophins include **nerve growth factor** (**NGF**), **brain-derived neurotrophic factor** (**BDNF**), **conserved dopamine neurotrophic factor** (**CDNF**), and **neurotrophins 3 and 4/5** (**NT3, NT4/5**). These proteins are released from potential target tissues and work at short ranges as either chemotactic factors or chemorepulsive factors (Paves and Saarma 1997). Each can promote and attract the growth of some axons to its source while inhibiting other axons. For instance, sensory neurons from the rat dorsal root ganglia are attracted to sources of NT3 (**FIGURE 11.30**) but are inhibited by BDNF. Neurotrophins are probably transported from the axon growth cone to the soma of the neuron. For instance, NGF derived from the hippocampus of the brain binds to receptors on the axons of basal forebrain neurons and is endocytosed into these neurons. It is then transported back to the nerve cell body, where it stimulates gene expression. Increased expression of *App* (a

---

*As will be seen throughout developmental biology, the metaphor of a "target" is problematic. Here, the "target" is not a passive entity, but an importantly active one.

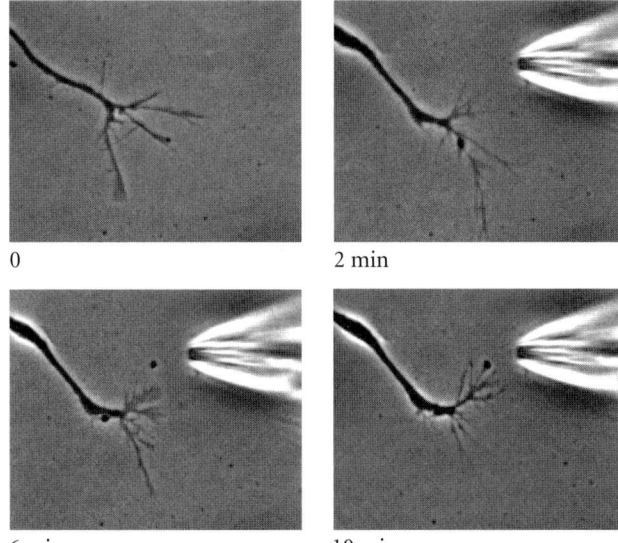

0   2 min

6 min   10 min

FIGURE 11.30 Embryonic axon from a rat dorsal root ganglion turning in response to a source of NT3. The photographs document the growth cone's turn over a 10-minute period. The same growth cone was insensitive to other neurotrophins. (From Paves and Saarma 1997, courtesy of M. Saarma.)

gene on chromosome 21, encoding ameloid precursor protein) is seen in people with Down syndrome and Alzheimer disease. Increased App protein blocks the retrograde transport of NGF from the axon to the cell body and affects the sensitivity and localization of receptor for NGF on the cell membrane (Salehi et al. 2006; Matrone et al. 2011). Thus, the NGF pathway is being studied for possible roles in the treatment of impaired cognition.

The attachment of an axon to its target can be either "digital" or "analogue." In "analogue" mode, different axons recognize the same molecule on the target, but the *amount* of the molecule on the target appears to be critical to the connections that form. This may be the case in the attachment of retinal neurons to the tectum in the fish brain (Gosse et al. 2008). In other cases, there may be extremely molecule-specific qualitative ("digital") binding such that certain connections are neuron-specific. This may be the case for retinal neurons in *Drosophila*. Dscam protein (mentioned in Chapter 2) has several thousand splicing isoforms, and this variety might enable highly specific recognition of a given neuron with its target neurons (Millard et al. 2010; Zipursky and Sanes 2010). Given the complexity of neural connections, it is probable that both qualitative and quantitative cues are used. Growth cones do not rely on a single type of molecule to recognize their target, but integrate the simultaneously presented attractive and repulsive cues, selecting their targets based on the combined input of these multiple signals (Winberg et al. 1998).

Synapse formation: Activity-dependent development

When an axon contacts its target (usually either a muscle cell or another neuron), it forms a specialized junction called a **synapse**. The axon terminal of the **presynaptic neuron**

(i.e., the neuron transmitting the signal) releases chemical neurotransmitters that depolarize or hyperpolarize the membrane of the target cell (the **postsynaptic cell**). The neurotransmitters are released into the synaptic cleft between the two cells, where they bind to receptors in the target cell.

The construction of a synapse involves several steps (Burden 1998). When motor neurons in the spinal cord extend axons to muscles, the growth cones that contact newly formed muscle cells migrate over their surfaces. When a growth cone first adheres to the cell membrane of a muscle fiber, no specializations can be seen in either membrane. However, the axon terminal soon begins to accumulate neurotransmitter-containing synaptic vesicles, the membranes of both cells thicken at the region of contact, and the synaptic cleft between the cells fills with an extracellular matrix that includes a specific form of laminin (**FIGURE 11.31A–C**). This muscle-derived laminin specifically binds the growth cones of motor neurons and may act as a "stop signal" for axonal growth (Martin et al. 1995; Noakes et al. 1995). In at least some neuron-to-neuron synapses, the synapse is stabilized by N-cadherin. The activity of the synapse releases N-cadherin from storage vesicles in the growth cone (Tanaka et al. 2000).

In muscles, after the first contact is made, growth cones from other axons converge at the site to form additional synapses. During mammalian development, all muscles that have been studied are innervated by at least two axons. However, this *polyneuronal innervation* is transient; soon after birth, all but one of the axon branches are retracted (**FIGURE 11.31D–F**). This "address selection" is based on competition between the axons (Purves and Lichtman 1980; Thompson 1983; Colman et al. 1997). When one of the motor neurons is active, it suppresses the synapses of the other neurons, possibly through a nitric oxide-dependent mechanism (Dan and Poo 1992; Wang et al. 1995). Eventually, the less active synapses are eliminated. The remaining axon terminal expands and is ensheathed by a Schwann cell (see Figure 11.31E).

Differential survival after innervation: The role of neuroptrophins

One of the most puzzling phenomena in the development of the nervous system is neuronal cell death. In many parts of the vertebrate central and peripheral nervous systems, more than half the neurons die during the normal course of development. Moreover, there does not seem to be great correlation in apoptosis patterns across species. For example, about 80% of a cat's retinal ganglion cells die, whereas in the chick retina this figure is only 40%. In fish and amphibian retinas, no ganglion cells appear to die (Patterson 1992).

The apoptotic death of a neuron is not caused by any obvious defect in the neuron itself. Indeed, before dying, these neurons have differentiated and successfully extended axons to their targets. Rather, it appears that the target tissue regulates the number of axons innervating it by limiting the supply of neurotrophin. In addition to their roles as chemotrophic factors described in the previous section, neurotrophins

(A) Growth cone contacts myotube

(B) Neural agrin induces ACh receptors to cluster

(C) Synaptic basal lamina forms

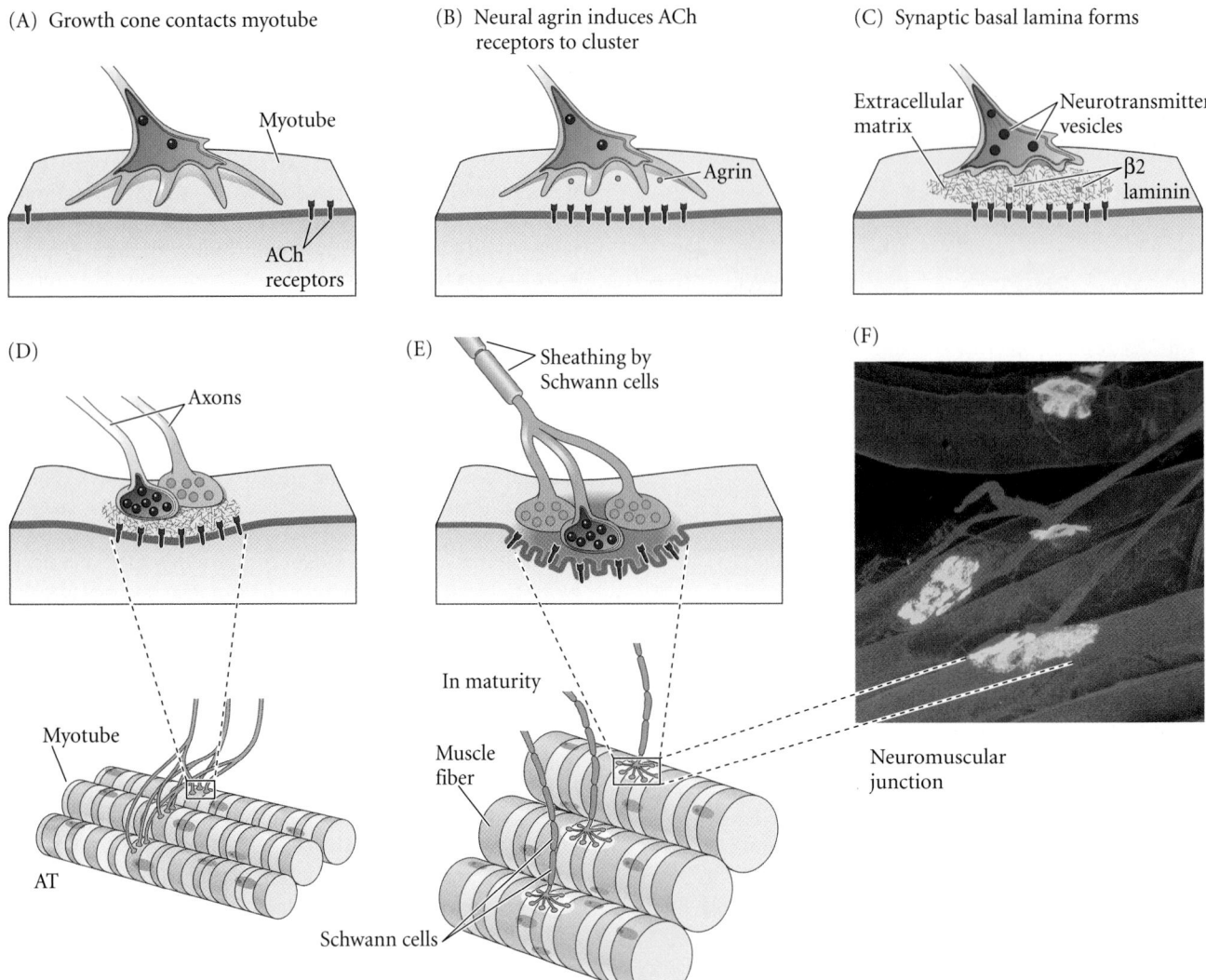

(D)

(E)

(F)

FIGURE 11.31 Differentiation of a motor neuron synapse with a muscle in mammals. (A) A growth cone approaches a developing muscle cell. (B) The axon stops and forms an unspecialized contact on the muscle surface. Agrin, a protein released by the neuron, causes acetylcholine (ACh) receptors to cluster near the axon. (C) Neurotransmitter vesicles enter the axon terminal (AT), and an extracellular matrix connects the axon terminal to the muscle cell as the synapse widens. This matrix contains a nerve-specific laminin. (D) Other axons converge on the same synaptic site. The wider view (below) shows muscle innervation by several axons (seen in mammals at birth). (E) All but one of the axons are eliminated. The remaining axon can branch to form a complex neuromuscular junction with the muscle fiber. Each axon terminal is sheathed by a Schwann cell process, and folds form in the muscle cell membrane. The overview shows muscle innervation several weeks after birth. (F) Whole mount view of mature neuromuscular junction in a mouse. (A–E after Hall and Sanes 1993; Purves 1994; Hall 1995; F courtesy of M. A. Ruegg.)

regulate the survival of different subsets of neurons (**FIGURE 11.32**). NGF, for example, is necessary for the survival of sympathetic and sensory neurons. Treating mouse embryos with anti-NGF antibodies reduces the number of trigeminal sympathetic and dorsal root ganglion neurons to 20% of their control numbers (Levi-Montalcini and Booker 1960; Pearson et al. 1983). Furthermore, removing these neurons' target tissues results in the death of the neurons that would have innervated them, and there is a good correlation between the amount of NGF secreted and the survival of the neurons

that innervate these tissues (Korsching and Thoenen 1983; Harper and Davies 1990).

BDNF does not affect sympathetic or sensory neurons, but it can rescue fetal motor neurons in vivo from normally occurring cell death, and from induced cell death following the removal of their target tissue. The results of these in vitro studies have been corroborated by gene knockout experiments, wherein the deletion of particular neurotrophic factors causes the loss of only certain subsets of neurons (Crowley et al. 1994; Jones et al. 1994).

FIGURE 11.32 Effects of NGF (top row) and BDNF (bottom row) on axonal outgrowths from (A) sympathetic ganglia, (B) dorsal root ganglia, and (C) nodose (taste perception) ganglia. While both NGF and BDNF had a mild stimulatory effect on dorsal root ganglia axonal outgrowth, the sympathetic ganglia responded dramatically to NGF and hardly at all to BDNF; the converse was true of the nodose ganglia. (From Ibáñez et al. 1991.)

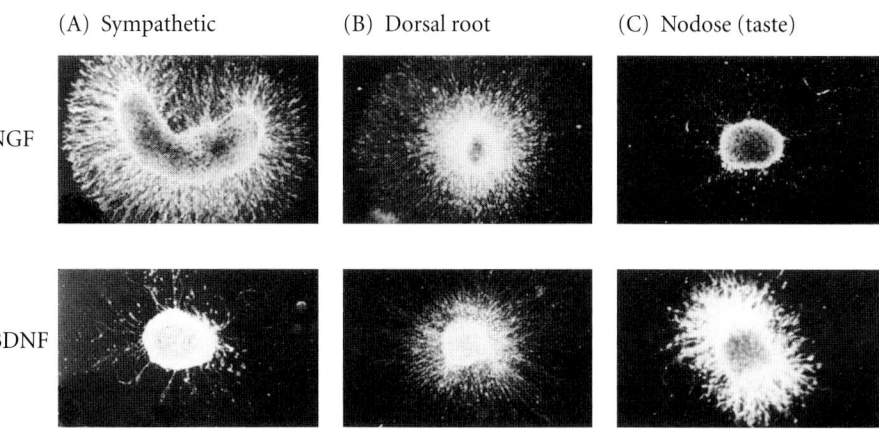

Neurotrophic factors are produced continuously in adults, and their loss may produce debilitating diseases. BDNF is required for the survival of a particular subset of neurons in the striatum (a region of the brain involved in modulating the intensity of coordinated muscle activity such as movement, balance, and walking) and enables these neurons to differentiate and synthesize the receptor for dopamine. BDNF in this region of the brain is upregulated by Huntingtin, a protein that is mutated in Huntington disease. Patients with Huntington disease have decreased production of BDNF, which leads to the death of striatal neurons (Guillin et al. 2001; Zuccato et al. 2001). The result is a series of cognitive abnormalities, involuntary muscle movements, and eventual death. Two other neurotrophins—glial-derived neurotrophic factor (GDNF, which we discussed earlier in terms of neural crest migration) and conserved dopamine neurotrophic factor (CDNF)—enhance the survival of the midbrain dopaminergic neurons, whose destruction characterizes Parkinson disease (Lin et al. 1993; Lindholm et al. 2007). The midbrain dopaminergic neurons send axons to the cells of the striatum, whose ability to respond to dopamine signals is dependent on BDNF. Drugs that activate the neurotrophic factors are being tested for the ability to cure Parkinson and Alzheimer diseases (Youdim 2013).

● **See WEBSITE 11.6** The development of behaviors: Constancy and plasticity

Paths to Glory: The Travels of the Retinal Ganglion Axons

Nearly all the mechanisms for neuronal specification and axon specificity mentioned in this chapter can be seen in the ways individual retinal neurons send axons to the visual processing areas of the brain. Although here we will describe the events as they occur in non-mammalian vertebrates, the principles and overall processes apply to mammals as well.

GROWTH OF THE RETINAL GANGLION AXON TO THE OPTIC NERVE The first steps in getting retinal ganglion cell (RGC) axons to their specific regions of the optic tectum take place within the retina. As the RGCs differentiate, their position in the inner margin of the retina is determined by cadherin molecules (N-cadherin and retina-specific R-cadherin) on their cell membranes (Matsunaga et al. 1988; van Horck et al. 2004). The RGC axons grow along the inner surface of the retina toward the optic disc (the head of the optic nerve). The mature human optic nerve will contain more than a million retinal ganglion axons.

The adhesion and growth of the retinal ganglion axons along the inner surface of the retina may be governed by its laminin-containing basal lamina. The embryonic lens and the periphery of the retina secrete inhibitory factors (probably chondroitin sulfate proteoglycans) that repel the ganglial cell axons, thereby preventing them from traveling in the wrong direction (**FIGURE 11.33**; Hynes and Lander 1992; Ohta et al. 1999). Moreover, N-CAM may also be especially important here, since the directional migration of the retinal ganglion growth cones depends on the N-CAM-expressing glial endfeet at the inner retinal surface (Stier and Schlosshauer 1995). The secretion of netrin-1 by the cells of the optic disc (where the axons are assembled to form the optic nerve) plays a role in this migration as well. Mice lacking the genes for either netrin-1 or for the netrin receptor (found in the retinal ganglion axons) have poorly formed optic nerves, as many of the axons fail to leave the eye and grow randomly around the disc (Deiner et al. 1997). The role of netrin may change in different parts of the eye. At the entrance to the optic nerve, netrin-1 is coexpressed with laminin on the surface of the retina. Laminin converts netrin from having an attractive signal to having a repulsive signal. This repulsion might "push" the growth cone away from the retinal surface and into the head of the optic nerve, where netrin is expressed without laminin (Mann et al. 2004).

(A)

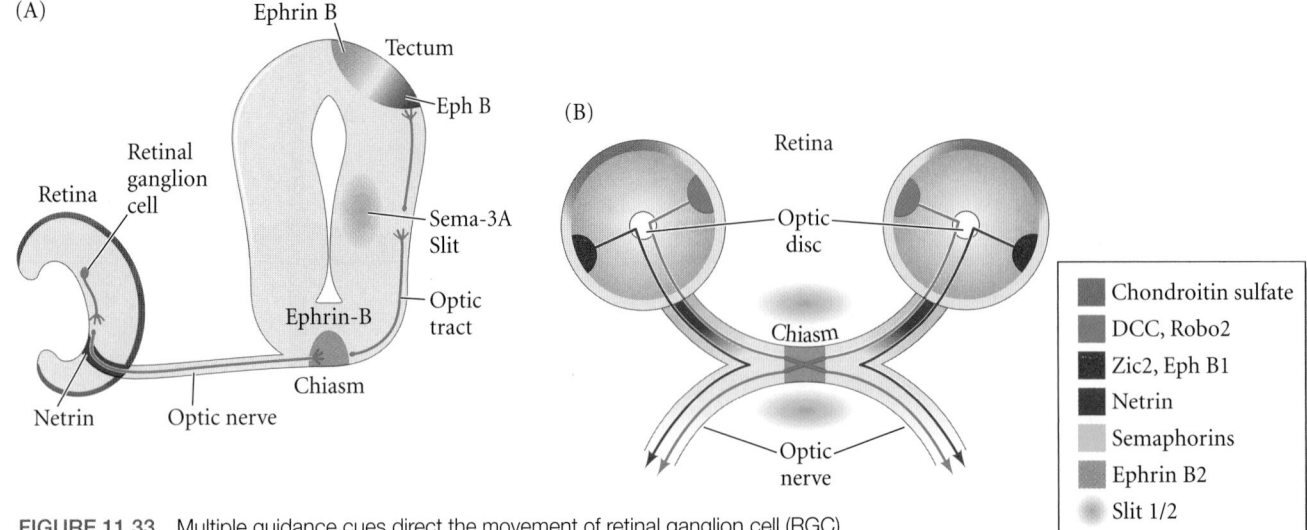

FIGURE 11.33 Multiple guidance cues direct the movement of retinal ganglion cell (RGC) axons to the optic tectum. Guidance molecules belonging to the netrin, Slit, semaphorin, and ephrin families are expressed in discrete regions at several sites along the pathway to direct the RGC growth cones. RGC axons are repelled from the retinal periphery, probably by chondroitin sulfate. At the optic disc, the axons exit the retina and enter the optic nerve, guided by netrin-/DCC-mediated attraction. Once in the optic nerve, the axons are kept within the pathway by inhibitory interactions. Slit proteins in the optic chiasm create zones of inhibition. Zic2-expressing ganglia in the ventrotemporal retina project Eph B1-expressing axons, which are repelled at the chiasm by ephrin B2, thus terminating at ipsilateral (same-side) targets. Neurons from the medial portions of the retina do not express Eph B1 and proceed to the opposite (contralateral) side. (A) Cross section. (B) Dorsal view. Not all cues are shown. (A after van Horck et al. 2004; B after Harada et al. 2007.)

Upon their arrival at the optic nerve, the migrating axons fasciculate (bundle) with axons already present there. N-CAM and L1 cell adhesion molecules are critical to this fasciculation, and antibodies against L1 or N-CAM cause the axons to enter the optic nerve in a disorderly fashion, which in turn causes them to emerge at the wrong positions in the tectum (Thanos et al. 1984; Brittis et al. 1995; Yin et al. 1995).

GROWTH OF THE RETINAL GANGLION AXON THROUGH THE OPTIC CHIASM When the axons enter the optic nerve, they grow on glial cells toward the midbrain. In non-mammalian vertebrates, the axons will go to a portion of the brain called the optic tectum. (Mammalian axons go to the lateral geniculate nuclei). After leaving the eye, the retinal axons appear to grow on surfaces of netrin, surrounded on all sides by semaphorins keeping them on track by providing repulsive cues (see Harada et al. 2007). Upon entering the brain, mammalian retinal ganglion axons reach the optic chiasm, where they have to "decide" if they are to continue straight or if they are to turn 90° and enter the other side of the brain. In the optic chiasm area, semaphorins are no longer present, but Slit proteins take over their function in establishing corridors through which the axons can travel. It appears that those axons not destined to cross to the opposite side of the brain are repulsed from doing so when they enter the optic chiasm (Godement et al. 1990). The basis of this repulsion

appears to be the synthesis of ephrin on the neurons in the chiasm and of ephrin receptors (Ephs) on the retinal axons (Cheng et al. 1995; Marcus et al. 2000).

In the mouse eye, the Eph B1 receptor is expressed on those temporal axons that are repelled by the optic chiasm's ephrin B2, and those axons project to the side of the tectum on the same side as their eye (the ipsilateral projections); Eph B1 is nearly absent on axons that are allowed to cross over. Mice lacking the *Eph B1* gene show hardly any ipsilateral projections. This pattern of *Eph B1* expression appears to be regulated by the Zic2 transcription factor found on those retinal axons that do form ipsilateral projections (Herrera et al. 2003; Williams et al. 2003; Pak et al. 2004).

Ephrin appears to play a similar role in the retinotectal mapping in the frog. In the developing frog, the ventral regions express the Eph B receptor, while the dorsal axons do not. Before metamorphosis, both axons cross the optic chiasm. However, when the frog nervous system is being remodeled during metamorphosis, the chiasm expresses ephrin B, which causes a subpopulation of ventral cells to be repulsed and project to the same side rather than cross the chiasm (Mann et al. 2002). (This allows the frog to have binocular vision, which is very good if one is trying to catch flies with one's tongue.)

Laminin appears to promote crossing of the optic chiasm. On their way to the optic tectum, the axons of non-mammalian vertebrates travel on a pathway (the optic tract) over glial cells whose surfaces are coated with laminin. Very few areas of the brain contain laminin, and the laminin in this pathway exists only when the optic nerve fibers are growing on it (Cohen et al. 1987).

The Brainbow

One of the difficulties encountered by developmental biologists is trying to follow any one of the thousands of axons from its source to its specific destination. (Think about how difficult it is to trace wires from a computer to a surge protector; then multiply by several hundred thousand.) But now there is a way that scientists can color-code axons using fluorescent proteins and Cre-lox recombination techniques. Livet and colleagues (2007) made a transgene for genes encoding four proteins, each fluorescing at a different wavelength (orange, yellow, magenta, and blue). At the termination of each fluorescent protein gene is a 3′ UTR, and in front of each is a variant sequence encoding a specific lox recombination site (**FIGURE 11.34A**). At the beginning of the entire transgene is a constitutively active promoter and three lox recombination sites (corresponding to the three lox recombination sites in front of each gene). When Cre is introduced into the cell, random recombination (by Cre bringing two similar Lox proteins together) can cause the promoter to be brought next to any of the fluorescent protein genes. Moreover, if more than one transgene enters the neuron, combinatorial expression can yield nearly 100 different colors (**FIGURE 11.34B**). The gene expression activated by the Cre recombinase is cell autonomous, and different cells, adjacent to one another, will almost certainly have different colors. This ability to uniquely label and trace individual axons through the embryo has been expanded for use in tracing the development of neural circuits during the development of many species as well as in regenerating neurons (Lang et al. 2012).

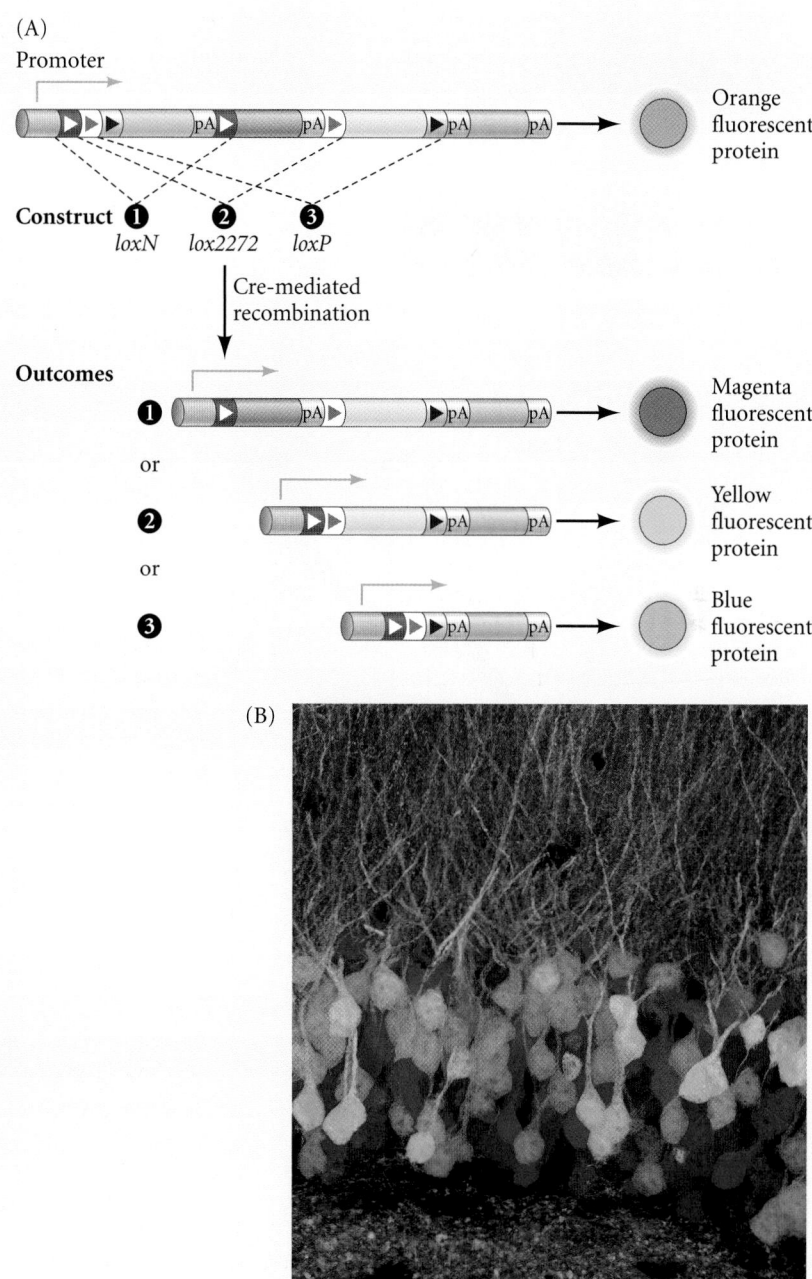

FIGURE 11.34 Making a "brainbow." (A) Experimentally manipulated mice contain a region of DNA constructed such that lox sites at the promoter randomly find homologous lox sites adjacent to one of four genes, each encoding a different fluorescent protein. Depending on the rearrangement induced by the Cre protein, different fluorescent proteins are expressed in different neurons. (B) The result is that adjacent neurons fluoresce different colors, allowing researchers to trace axonal routes and connections through the embryo. (A after Livet et al. 2007; B courtesy of J. Lichtman.)

(A)

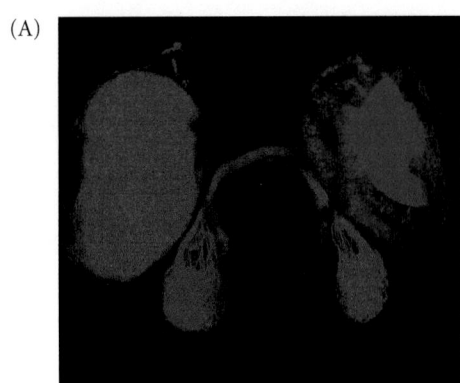

FIGURE 11.35   Retinotectal projections. (A) Confocal micrograph of axons entering the tecta of a 5-day zebrafish embryo. Fluorescent dyes were injected into the eyes of zebrafish embryos mounted in agarose. The dyes diffused down the axons and into each tectum, showing the retinal axons from the right eye going to the left tectum and vice versa. (B) Map of the normal retinotectal projection in adult *Xenopus*. The right eye innervates the left tectum, and the left eye innervates the right tectum. The dorsal (D) portion of the retina innervates the lateral (L) regions of the tectum. The nasal (anterior) region of the retina projects to the caudal (C) region of the tectum. (A courtesy of M. Wilson; B after Holt 2002, courtesy of C. Holt.)

(B)

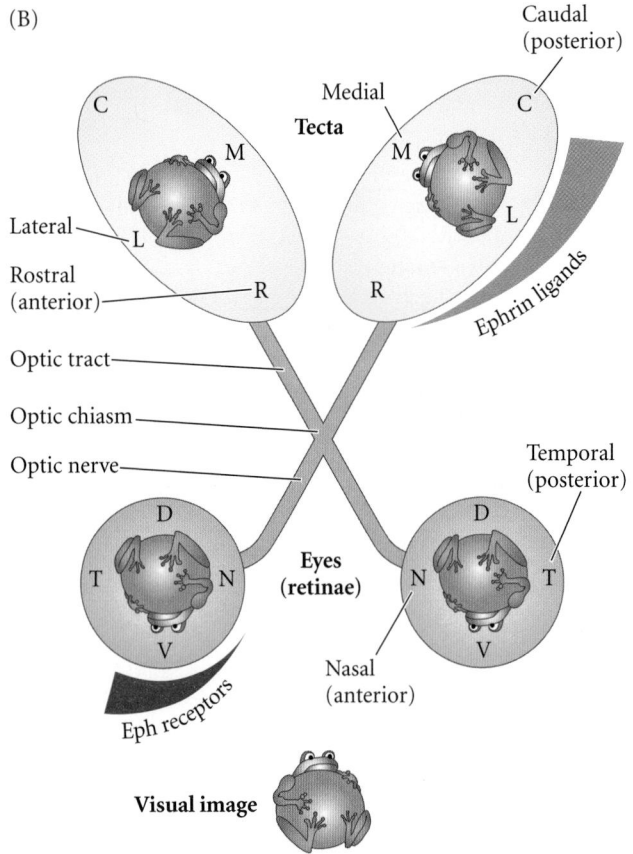

The map of retinal connections to the frog optic tectum (the **retinotectal projection**) was detailed by Marcus Jacobson (1967). Jacobson created this map by shining a narrow beam of light on a small, limited region of the retina and noting, by means of a recording electrode in the tectum, which tectal cells were being stimulated. The retinotectal projection of *Xenopus laevis* is shown in **FIGURE 11.35B**. Light illuminating the ventral part of the retina stimulates cells on the lateral surface of the tectum. Similarly, light focused on the temporal (posterior) part of the retina stimulates cells in the caudal portion of the tectum. These studies demonstrate a point-for-point correspondence between the cells of the retina and the cells of the tectum. When a group of retinal cells is activated, a very small and specific group of tectal cells is stimulated. Furthermore, the points form a continuum; in other words, adjacent points on the retina project onto adjacent points on the tectum. This arrangement enables the frog to see an unbroken image. This intricate specificity caused Sperry (1965) to put forward the **chemoaffinity hypothesis**:

> *The complicated nerve fiber circuits of the brain grow, assemble, and organize themselves through the use of intricate chemical codes under genetic control. Early in development, the nerve cells, numbering in the millions, acquire and retain thereafter, individual identification tags, chemical in nature, by which they can be distinguished and recognized from one another.*

Current theories do not propose a point-for-point specificity between each axon and the neuron that it contacts. Rather, evidence now demonstrates that gradients of adhesivity (especially those involving repulsion) play a role in defining the territories that the axons enter, and that activity-driven competition between these neurons determines the final connection of each axon.

ADHESIVE SPECIFICITIES IN DIFFERENT REGIONS OF THE OPTIC TECTUM   There is good evidence that retinal ganglion cells can distinguish between regions of the optic tectum. Cells taken from the ventral half of the chick neural retina preferentially adhere to dorsal (medial) halves of the tectum, and vice versa (Gottlieb et al. 1976; Roth and Marchase 1976; Halfter et al. 1981). Retinal ganglion cells are specified along the dorsal-ventral axis by a gradient of transcription factors. Dorsal retinal cells are characterized by high concentrations

TARGET SELECTION   When the axons come to the end of the laminin-lined optic tract, they spread out and find their specific targets in the optic tectum. Studies on frogs and fish (in which retinal neurons from each eye project to the opposite side of the brain) have indicated that each retinal ganglion axon sends its impulse to one specific site (a cell or small group of cells) within the optic tectum (**FIGURE 11.35A**; Sperry 1951). There are two optic tecta in the frog brain. The axons from the right eye form synapses with the left optic tectum, and those from the left eye form synapses in the right optic tectum.

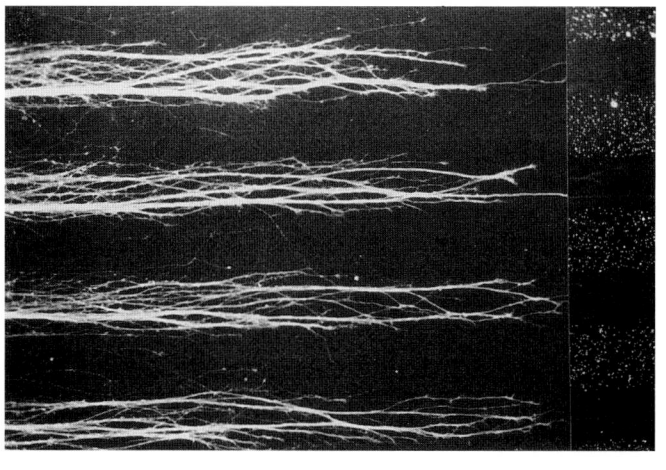

Tectal membranes

Anterior

Posterior

Anterior

Posterior

Anterior

Posterior

Anterior

FIGURE 11.36 Differential repulsion of temporal retinal ganglion axons on tectal membranes. Alternating stripes of anterior and posterior tectal membranes were absorbed onto filter paper. When axons from temporal (posterior) retinal ganglion cells were grown on such alternating carpets, they preferentially extended axons on the anterior tectal membranes. (From Walter et al. 1987.)

of Tbx5 transcription factor, whereas ventral cells have high levels of Pax2. These transcription factors are induced by paracrine factors (BMP4 and retinoic acid, respectively) from nearby tissues (Koshiba-Takeuchi et al. 2000). Misexpression of Tbx5 in the early chick retina results in marked abnormalities of the retinotectal projection. Therefore, the retinal ganglion cells are specified according to their location.

One gradient that has been identified functionally is a gradient of repulsion that is highest in the posterior tectum and weakest in the anterior tectum. Bonhoeffer and colleagues (Walter et al. 1987; Baier and Bonhoeffer 1992) prepared a "carpet" of tectal membranes with alternating "stripes" of membrane derived from the posterior and the anterior tecta. They then let cells from the nasal (anterior) or temporal (posterior) regions of the retina extend axons into the carpet. The nasal ganglion cells extended axons equally well on both the anterior and posterior tectal membranes. The neurons from the temporal side of the retina, however, extended axons only on the anterior tectal membranes (**FIGURE 11.36**). When the growth cone of a temporal retinal ganglion axon contacted a posterior tectal cell membrane, the growth cone's filopodia withdrew, and the cone collapsed and retracted (Cox et al. 1990).

The basis for this specificity appears to be two sets of gradients along the tectum and retina. The first gradient set consists of ephrin proteins and their receptors. In the optic tectum, ephrin proteins (especially ephrins A2 and A5) are found in gradients that are highest in the posterior (caudal) tectum and decline anteriorly (rostrally) (**FIGURE 11.37A**). Moreover, cloned ephrin proteins have the ability to repulse axons, and ectopically expressed ephrin will prohibit axons from the temporal (but not from the nasal) regions of the retina from projecting to where it is expressed (Drescher et al. 1995; Nakamoto et al. 1996). The complementary Eph receptors have been found on chick retinal ganglion cells, expressed in a temporal-to-nasal gradient along the retinal ganglion axons (Cheng et al. 1995). This gradient appears to be due to a spatially and temporally regulated expression of retinoic acid (Sen et al. 2005).

Ephrins appears to be remarkably pliable molecules. Concentration differences in ephrin A in the tectum can account for the smooth topographic map (wherein the position of neurons in the retina maps continuously onto the targets). Hansen and colleagues (2004) have shown that ephrin A can be an attractive as well as a repulsive signal for retinal axons. Moreover, their quantitative assay for axon growth showed that the origin of the axon determined whether it was attracted or repulsed by ephrins. Axon growth is promoted by low ephrin A concentrations that are anterior to the proper target and is inhibited by higher concentrations posterior to the correct target (**FIGURE 11.37B**). Each axon is thus led to the appropriate place and then told to go no farther. At that equilibrium point, there would be no growth and no inhibition, and the synapses with the target tectal neurons could be made.

The second set of gradients parallels the ephrins and Ephs. The tectum has a gradient of Wnt3 that is highest at the medial region and lowest laterally (like the ephrin gradient). In the retina, a gradient of Wnt receptor is highest ventrally (like the Eph proteins). The two sets of gradients are both required to specify the position of the axon in the tectum (Schmitt et al. 2006).

We have learned that guidance molecules can determine the path that neurons take, and that quantitative, rather than qualitative, information is critical in reaching the target. Also, we see that most of the pathway and target finding is independent of neural activity; but neural activity plays a critical role in the precise localization and size of the arrachments of retinal ganglia neurons to the optic tectum. This research provides an outline for the detailed analysis of what is happening. We know very little, for instance, about how the neurons find the different layers of the cortex such that some axons enter the layers devoted to brightness intensity, whereas others enter layers that interpret color or movement (see Joste and Huberman 2010). New computer-enabled visual analyses will be critical in uncovering these mechanisms and integrating them into a unified vision of the development of how vision is generated.

(A)

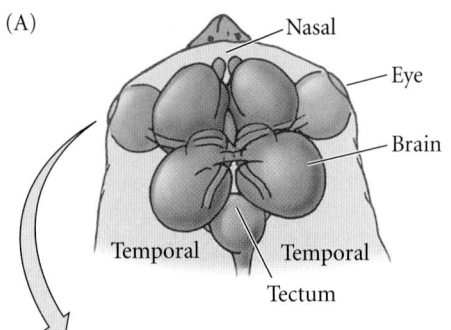

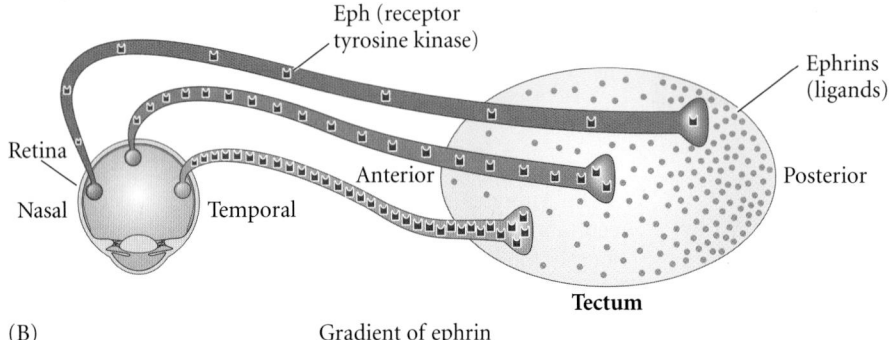

FIGURE 11.37 Differential retinotectal adhesion is guided by gradients of Eph receptors and their ligands. (A) Representation of the dual gradients of Eph receptor tyrosine kinase in the retina and its ligands (ephrin A2 and A5) in the tectum. (B) Experiment showing that temporal, but not nasal, retinal ganglion axons respond to a gradient of ephrin ligand in tectal membranes by turning away or slowing down. An equilibrium of attractive and repulsive forces inherent in the gradient may lead specific axons to their targets. (After Barinaga 1995; Hansen et al. 2004.)

(B)

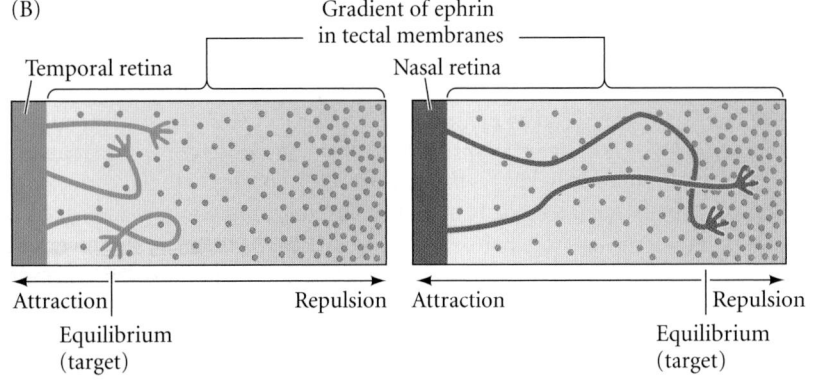

SNAPSHOT SUMMARY: Neural Crest Cells and Axonal Specificity

1. The neural crest is a transitory structure. Its cells migrate to become numerous different cell types.

2. Trunk neural crest cells can migrate dorsolaterally into the ectoderm, where they become melanocytes and dorsal root ganglia cells. They can also migrate ventrally, to become sympathetic and parasympathetic neurons and adrenomedullary cells.

3. Cranial neural crest cells enter the pharyngeal arches to become the cartilage of the jaw and the bones of the middle ear. They also form the bones of the frontonasal process, the papillae of the teeth, and the cranial nerves.

4. Cardiac neural crest enters the heart and forms the septum (separating wall) between the pulmonary artery and aorta.

5. The formation of the neural crest depends on interactions between the prospective epidermis and the neural plate.

Paracrine factors from these regions induce the formation of transcription factors that enable neural crest cells to emigrate.

6. The path a neural crest cell takes depends on the extracellular environment it meets.

7. Trunk neural crest cells will migrate through the anterior portion of each sclerotome, but not through the posterior portion of a sclerotome. Semaphorin and ephrin proteins expressed in the posterior portion of each sclerotome can prevent neural crest cell migration.

8. Some neural crest cells appear to be capable of forming a large repertoire of cell types. Other neural crest cells may be restricted even before they migrate. The final destination of the neural crest cell can sometimes change its specification.

9. The fates of the cranial neural crest cells are influenced by Hox genes. They can acquire their Hox gene expression pattern through interaction with neighboring cells.

10. Motor neurons are specified according to their position in the neural tube. The Lim family of transcription factors plays an important role in this specification.

11. The targets of motor neurons are specified before their axons extend into the periphery.

12. The growth cone is the locomotor organelle of the neuron, and it senses environmental cues. Axons can find their targets without neuronal activity.

13. Some proteins are generally permissive to neuron adhesion and provide substrates on which axons can migrate. Other substances prohibit migration.

14. The growth cone integrates cues from the external environment and uses them to alter cytoskeletal formation, membrane deposition, and adhesive connections to the substrate.

15. Some growth cones recognize molecules that are present in very specific areas and are guided by these molecules to their respective targets.

16. Some neurons are "kept in line" by repulsive molecules. If the neurons wander off the path to their target, these molecules send them back. Some molecules, such as the semaphorins, are selectively repulsive to particular sets of neurons.

17. Some neurons sense gradients of a protein and are brought to their target by following these gradients. The netrins may work in this fashion.

18. Target selection can be brought about by neurotrophins, proteins that are made by the target tissue and that stimulate the particular set of axons able to innervate it. In some cases, the target makes only enough of these factors to support a single axon. Neurotrophins also play a role in the apoptosis of many neurons.

19. Synapse formation has an activity-dependent component. An active neuron can suppress synapse formation by other neurons on the same target.

20. Retinal ganglion cells in frogs and chicks send axons that bind to specific regions of the optic tectum. This process is mediated by numerous interactions, and target selection appears to be mediated through ephrins.

For Further Reading

Complete bibliographical citations for all literature cited in this chapter can be found at the free-access website **www.devbio.com**

Bard, L., C. Boscher, M. Lambert, R. M. Mège, D. Choquet and O. Thoumine. 2008. A molecular clutch between the actin flow and N-cadherin adhesions drives growth cone migration. *J. Neurosci.* 28: 5879–5890.

Bronner-Fraser, M. and S. E. Fraser. 1988. Cell lineage analysis reveals multipotency of some avian neural crest cells. *Nature* 335: 161–164.

Hall, B. K. 2000. The neural crest as a fourth germ layer and vertebrates as quadroblastic, not triploblastic. *Evol. Dev.* 2: 3–5.

Harada, T., C. Harada and L. F. Parada. 2007. Molecular regulation of visual system development: More than meets the eye. *Genes Dev.* 21: 367–378.

Harris, M. L. and C. A. Erickson. 2007. Lineage specification in neural crest cell pathfinding. *Dev. Dyn.* 236: 1–19.

Kolodkin, A. L. and M. Tessier-Lavigne. 2011. Mechanisms and molecules of neuronal wiring: A primer. *Cold Spring Harbor Persp. Biol.* 3(6): pii: a001727.

Le Douarin, N. M. 2004. The avian embryo as a model to study the development of the neural crest: A long and still ongoing study. *Mech. Dev.* 121: 1089–1102.

Livet, J. and 7 others. 2007. Transgenic strategies for combinatorial expression of fluorescent proteins in the nervous system. *Nature* 450: 56–62.

Lowery, L. A. and V. Van Vactor. 2009. The trip of the tip: Understanding the growth cone machinery. *Nature Rev. Mol. Cell Biol.* 10: 332–343.

Mambetisaeva, E., T. W. Andrews, L. Camurri, A. Annan and V. Sundaresan. 2005. Robo family of proteins exhibit differential expression in mouse spinal cord and Robo-Slit interaction is required for midline crossing in vertebrate spinal cord. *Dev. Dynam.* 233: 41–51.

Schlosser, G. 2005. Evolutionary origins of vertebrate placodes: Insights from developmental studies and from comparisons with other deuterostomes. *J. Exp. Zool.* 304B: 347–399.

Teillet, M.-A., C. Kalcheim and N. M. Le Douarin. 1987. Formation of the dorsal root ganglia in the avian embryo: Segmental origin and migratory behavior of neural crest progenitor cells. *Dev. Biol.* 120: 329–347.

Tojima, T., H. Akiyama, R. Itofusa, Y. Li, H. Katayama, A. Miyawaki and H. Kamiguchi. 2007. Attractive axon guidance involves asymmetric membrane transport and exocytosis in the growth cone. *Nature Neurosci.* 10: 58–66.

Tosney, K. W. 2004. Long-distance cue from emerging dermis stimulates neural crest melanoblast migration. *Dev. Dynam.* 229: 99–108.

Vaahtokari, A., T. Aberg, J. Jernvall, S. Keränen and I. Thesleff. 1996. The enamel knot as a signalling center in the developing mouse tooth. *Mech. Dev.* 54: 39–43.

Waldo, K., S. Miyagawa-Tomita, D. Kumiski and M. L. Kirby. 1998. Cardiac neural crest cells provide new insight into septation of the cardiac outflow tract: Aortic sac to ventricular septal closure. *Dev. Biol.* 196: 129–144.

Walter, J., S. Henke-Fahle and F. Bonhoeffer. 1987. Avoidance of posterior tectal membranes by temporal retinal axons. *Development* 101: 909–913.

Yoshida, T., P. Vivatbutsiri, G. Morriss-Kay, Y. Saga and S. Iseki. 2008. Cell lineage in mammalian craniofacial mesenchyme. *Mech. Dev.* 125: 797–808.

Go Online

WEBSITE 11.1 Why birds don't have teeth. The formation of the face and jaw is coordinated by a series of cranial neural crest cell migrations and by the interactions of these neural crest cells with surrounding tissues.

WEBSITE 11.2 Kallmann syndrome. Some infertile men have no sense of smell. The relationship between sense of smell and male fertility was elusive until the gene for Kallmann syndrome was identified. The gene produces a protein that is necessary for the proper migration of both olfactory axons and hormone-secreting neurons from the olfactory placode.

WEBSITE 11.3 The human cranial nerves. Much of our perception of the outside world is mediated by our 12 cranial nerves. These nerves are made up of elements derived from the neural crest and from the sensory placodes.

WEBSITE 11.4 The evolution of developmental neurobiology. Santiago Ramón y Cajal, Viktor Hamburger, and Rita Levi-Montalcini helped bring order to the study of neural development by identifying some of the important questions that still occupy us today.

WEBSITE 11.5 The early evidence for chemotaxis. Before molecular techniques, investigators used transplantation experiments and ingenuity to reveal evidence that chemotactic molecules were being released by target tissues.

WEBSITE 11.6 The development of behaviors: Constancy and plasticity. The correlation of certain neuronal connections with specific behaviors is one of the fascinating aspects of developmental neurobiology.

Vade Mecum

Nicole Le Douarin and the neural crest. The segment on Dr. Le Douarin's work shows original footage of the experimental techniques and results of her work on neural crest cell regionalization, migration, and differentiation.

Outside Sites

There is an animation of the pharyngeal arch and pouch formation in humans at **youtube.com/ watch?v=tsa4uZRKbu8**

12
Paraxial and Intermediate Mesoderm

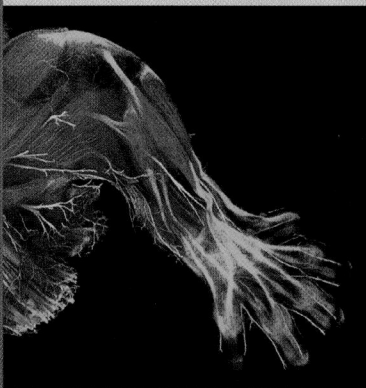

CHAPTERS 10 AND 11 FOLLOWED THE FORMATION of various tissues from the vertebrate ectoderm. In this chapter and the next, we will follow the development of the mesodermal and endodermal germ layers. We will see that the endoderm forms the epithelial cell lining of the digestive and respiratory tubes, with their associated organs. The mesoderm generates almost all of the organs between the ectodermal wall and the endodermal tissues. These include the musculoskeletal tissues, the cardiovascular tissues, and the urogenital tissues, as well as the mesenchymal cells that associate with the ectodermal and endodermal epithelia, helping them to form their own tissues.

The trunk mesoderm of a neurula-stage embryo can be subdivided into four regions (FIGURE 12.1A):

1. The central region of trunk mesoderm is the **chordamesoderm**. This tissue forms the notochord, a transient organ whose major functions include inducing and patterning the neural tube and establishing the anterior-posterior body axis. The formation of the notochord on the future dorsal side of the embryo was discussed in Chapters 8 and 9.

2. Flanking the notochord on both sides is the **paraxial**, or **somitic**, **mesoderm**. The tissues developing from this region will be located in the back of the embryo, surrounding the spinal cord. The cells in this region will form somites—blocks of mesodermal cells on either side of the neural tube—which will produce muscle and many of the connective tissues of the back (dermis, muscle, and skeletal elements such as the vertebrae and ribs). The anteriormost paraxial mesoderm does not segment; it becomes the **head mesoderm**, which (along with the neural crest) forms the skeleton, muscles, and connective tissue of the face and skull.

3. The **intermediate mesoderm** forms the urogenital system, consisting of the kidneys, the gonads, and their associated ducts. The outer (cortical) portion of the adrenal gland also derives from this region.

4. Farthest away from the notochord, the **lateral plate mesoderm** gives rise to the heart, blood vessels, and blood cells of the circulatory system, as well as to the lining of the body cavities. It gives rise to the pelvic and limb skeleton (but not the limb muscles, which are somitic in origin). Lateral plate mesoderm also helps form a series of extraembryonic membranes that are important for transporting nutrients to the embryo.

These four subdivisions are thought to be specified along the mediolateral (center-to-side) axis by increasing amounts of BMPs (Pourquié et al. 1996; Tonegawa et al. 1997). The more lateral mesoderm of the chick embryo expresses higher levels of BMP4 than do the midline areas, and one can change the identity of the mesodermal tissue by altering BMP expression. While it is not known how this patterning is accomplished, it is thought that the different BMP concentrations may cause differential expression of the Forkhead (Fox) family of transcription factors. *Foxf1* is transcribed in those regions that will become the lateral

Of physiology from top to toe I sing,

Not physiognomy alone or brain alone

is worthy for the Muse,

I say the form complete is worthier far,

The Female equally with the Male I sing.
  WALT WHITMAN (1867)

Built of 206 bones, the skeleton is a living cathedral of ivory vaults, ribs, and buttresses—a structure at once light and strong.
  NATALIE ANGIER (1994)

(A)

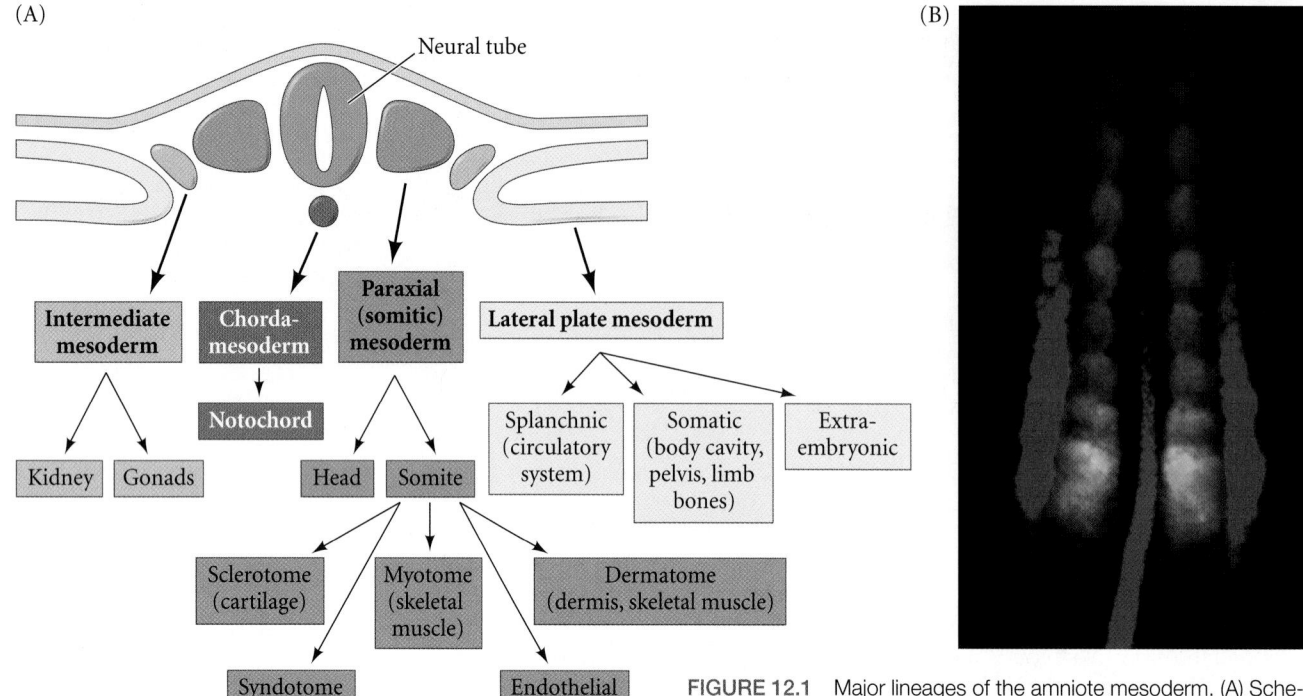

(B)

FIGURE 12.1 Major lineages of the amniote mesoderm. (A) Schematic of the mesodermal compartments of the amniote embryo. (B) Staining for the medial mesodermal compartments in the trunk of a 12-somite chick embryo (about 33 hours). In situ hybridization was performed with probes binding to *Chordin* mRNA (blue) in the notochord, *Paraxis* mRNA (green) in the somites, and *Pax2* mRNA (red) in the intermediate mesoderm. (B from Denkers et al. 2004, courtesy of T. J. Mauch.)

plate and extraembryonic mesoderm, whereas *Foxc1* and *Foxc2* are expressed in the paraxial mesoderm that will form the somites (Wilm et al. 2004). If *Foxc1* and *Foxc2* are both deleted from the mouse genome, the paraxial mesoderm is respecified as intermediate mesoderm and initiates the expression of *Pax2*, which encodes a major transcription factor of the intermediate mesoderm (**FIGURE 12.1B**).

Anterior to the trunk mesoderm is a fifth mesodermal region, the head mesoderm, consisting of the unsegmented paraxial mesoderm and prechordal mesoderm. This region provides the head mesenchyme that forms much of the connective tissues and musculature of the head (Evans and Noden 2006). The muscles derived from the head mesoderm form differently than those formed from the somites. Not only do they have their own set of transcription factors, but the head and trunk muscles are affected by different types of muscular dystrophies (Emery 2002; Bothe and Dietrich 2006; Harel et al. 2009).

PARAXIAL MESODERM: THE SOMITES AND THEIR DERIVATIVES

One of the major tasks of gastrulation is to create a mesodermal layer between the endoderm and the ectoderm. As seen in **FIGURE 12.2**, the formation of mesodermal and endodermal tissues is not subsequent to neural tube formation but occurs synchronously. The notochord extends beneath the neural tube, from the base of the head into the tail. On either side of the neural tube lie thick bands of mesodermal cells.

These bands of paraxial mesoderm are referred to as the **presomitic mesoderm**. As the primitive streak regresses and the neural folds begin to gather at the center of the embryo, the cells of the presomitic mesoderm will form **somites**.

The paraxial mesoderm appears to be specified by the antagonism of BMP signaling by the Noggin protein. Noggin is usually synthesized by the early presomitic mesoderm, and if Noggin-expressing cells are placed into presumptive lateral plate mesoderm, the lateral plate tissue will be respecified into somite-forming paraxial mesoderm (**FIGURE 12.3**; Tonegawa and Takahashi 1998; Gerhart et al. 2011).

Mature somites contain three major compartments: the **sclerotome**, which forms the vertebrae and rib cartilage; the **myotome**, which forms the musculature of the back, rib cage, and ventral body wall; and the **dermamyotome**, which also contains skeletal muscle progenitor cells (including muscle progenitors that migrate into the limbs), as well as the cells that generate the dermis of the back. Additional, smaller compartments are formed from these three. The *syndetome* arising from the most dorsal sclerotome cells generates the tendons, while the most internal cells of the sclerotome (sometimes called the *arthrotome*) become the vertebral joints, the outer portion of the intervertebral discs, and the proximal portion of the ribs (Mittapalli et al. 2005; Christ et al. 2007). Moreover, an as-yet unnamed group of cells in the posterior

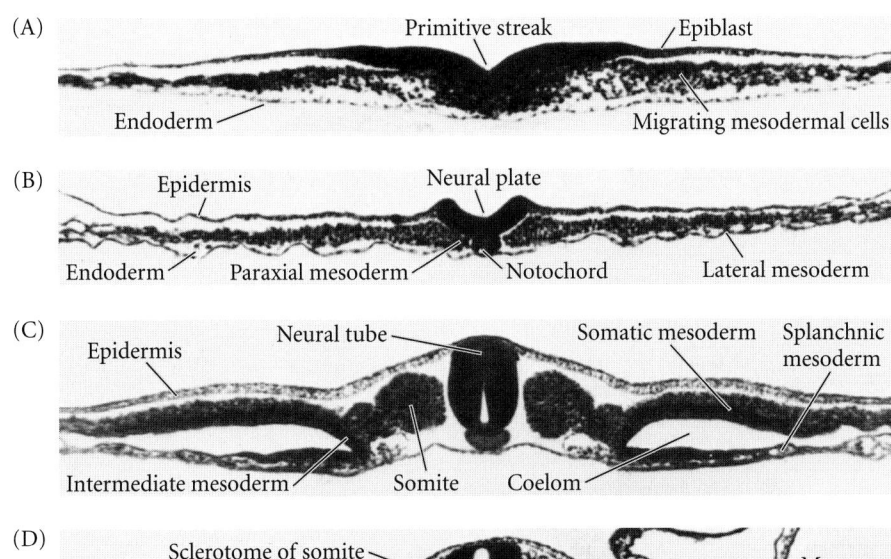

(A) Primitive streak / Epiblast

Endoderm / Migrating mesodermal cells

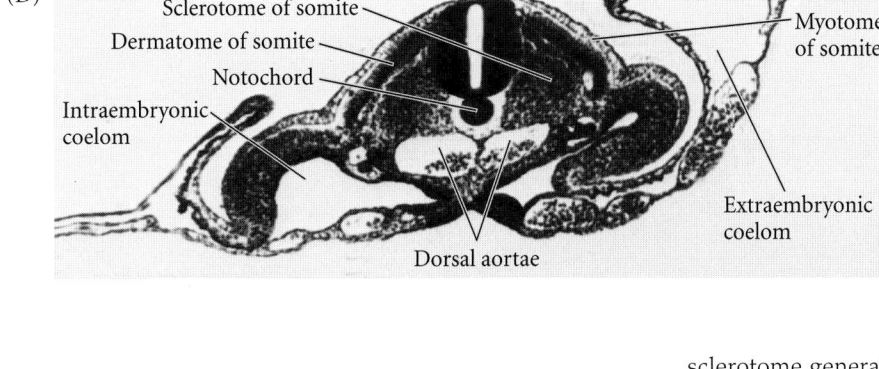

(B) Epidermis / Neural plate

Endoderm / Paraxial mesoderm / Notochord / Lateral mesoderm

(C) Epidermis / Neural tube / Somatic mesoderm / Splanchnic mesoderm

Intermediate mesoderm / Somite / Coelom

(D) Sclerotome of somite / Dermatome of somite / Notochord / Intraembryonic coelom / Myotome of somite / Extraembryonic coelom / Dorsal aortae

FIGURE 12.2 Gastrulation and neurulation in the chick embryo, focusing on the mesodermal component. (A) Primitive streak region, showing migrating mesodermal and endodermal precursors. (B) Formation of the notochord and paraxial mesoderm. (C,D) Differentiation of the somites, coelom, and the two aortae (which will eventually fuse). A–C, 24-hour embryos; D, 48-hour embryo.

sclerotome generates vascular cells of the dorsal aorta and intervertebral blood vessels (**TABLE 12.1**; Pardanaud et al. 1996; Sato et al. 2008; Ohata et al. 2009).

● See **VADE MECUM** Mesoderm in the vertebrate embryo

Formation of the Somites

The important components of somite formation, or **somitogenesis**, are (1) fissure formation (separation of the somites), (2) periodicity, (3) epithelialization, (4) specification, and (5) differentiation. The first somites appear in the anterior portion of the trunk, and new somites "bud off" from the rostral end of the presomitic mesoderm at regular intervals (**FIGURE 12.4**). The anterior mesoderm arises from cells that have migrated through the primitive streak. The posterior of amniote embryos (as we saw in Chapter 10) contains a precursor population whose cells become either central neural tube (*Sox2*-expressing) or paraxial mesoderm (*Tbx6*-expressing). The nascent paraxial mesoderm cells are released from this

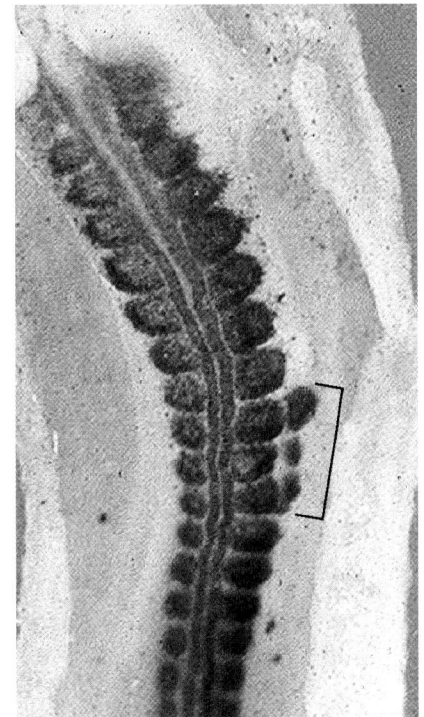

FIGURE 12.3 Specification of somites. Placing Noggin-secreting cells into a prospective region of chick lateral plate mesoderm will respecify that mesoderm into somite-forming paraxial mesoderm. Induced somites (bracketed) were detected by in situ hybridization with Pax3. (From Tonegawa and Takahashi 1998, courtesy of Y. Takahashi.)

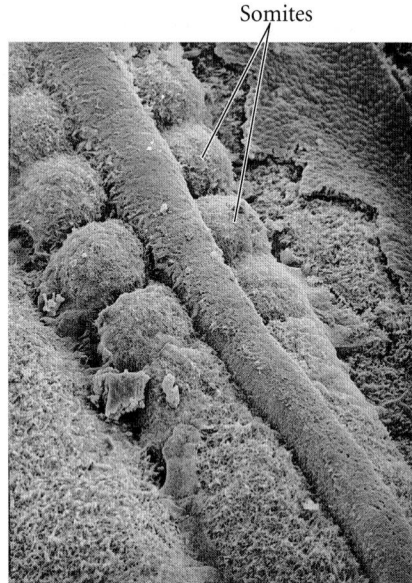

Somites

FIGURE 12.4 Neural tube and somites seen by scanning electron microscopy. When the surface ectoderm is peeled away, well-formed somites are revealed, as well as paraxial mesoderm (bottom right) that has not yet separated into distinct somites. A rounding of the paraxial mesoderm into a somitomere is seen at the lower left, and neural crest cells can be seen migrating ventrally from the roof of the neural tube. (Courtesy of K. W. Tosney.)

TABLE 12.1 Derivatives of the somite

Traditional view	Current view
DERMAMYOTOME	
Myotome forms skeletal muscles	Lateral edges generate primary myotome that forms muscle
Dermatome forms back dermis	Central region forms muscle, muscle stem cells, dermis, brown fat cells
SCLEROTOME	
Forms vertebral and rib cartilage	Forms vertebral and rib cartilage
	Dorsal region forms tendons (syndetome)
	Medial region forms blood vessels and meninges
	Central mesenchymal region forms joints (arthrotome)
	Forms smooth muscle cells of dorsal aorta

pool and become positioned in the tail end of the rods of presomitic mesoderm. These cells are originally mesenchymal, but as they mature they become organized into whorls of cells, the somite precursors that are sometimes called **somitomeres** (Meier 1979).

The somitomeres become compacted and split apart as fissures separate them into discrete, immature somites. The mesenchymal cells making up the immature somite now change, with the outer cells joining into an epithelium while the inner cells remain mesenchymal. Because individual embryos in any species can develop at slightly different rates (as when chick embryos are incubated at slightly different temperatures), the number of somites present is usually the best indicator of how far development has proceeded. The number of somites in an adult individual is species-specific and is set at the initial stages of presomitic mesoderm formation. Chicks have about 50 somites, mice have 65 (with many in the tail), and humans generally have 33. Some snakes have as many as 500. When *Xenopus* or mouse embryos are experimentally or genetically reduced in size, the number of somites generated remains the same but each somite is smaller than normal (Tam 1981).

Where somites form: The Notch pathway

One of the key agents in determining where somites form is the Notch signaling pathway (see Aulehla and Pourquié 2008). When a small group of cells from a region constituting the posterior border at the presumptive somite boundary is

transplanted into a region of presomitic mesoderm that would not ordinarily be part of the boundary area, a new boundary is created. The transplanted boundary cells instruct the cells anterior to them to epithelialize and separate. Nonboundary cells will not induce border formation when transplanted to a nonborder area. However, these cells can acquire boundary-forming ability if an activated Notch protein is electroporated into them (**FIGURE 12.5**; Sato et al. 2002). Morimoto and colleagues (2005) have been able to visualize the endogenous level of Notch activity in mouse embryos and have shown that it oscillates in a segmentally defined pattern. The somite boundaries were formed at the interface between the Notch-expressing and Notch-nonexpressing areas. The expression pattern of *Hairy1*—a segmentation gene regulated by Notch activity—at the anterior of the presomitic mesoderm at the end of each cycle of expression was exactly where the transplantation experiments showed Notch expression to be important (see Figure 12.8).

Another critical gene in somite formation is *Mesp2/c-Meso1* (the first name refers to the mouse homologue, the second to the chick homologue; for convenience in this chapter we will refer to simply *Mesp*). *Mesp* is activated by Notch, and its protein product, a transcription factor, initiates the reactions that *suppress* Notch activity. This cycle of alternating activation and suppression causes *Mesp* expression to oscillate in time and space. It is initially expressed in a somite-wide domain as a result of Notch activity. It is then repressed in the posterior half of this domain but is maintained in the anterior half where it in turn represses Notch activity. Wherever *Mesp* expression is maintained, that site is the most anterior group of cells in the next somite, and the boundary forms immediately anterior to those cells (**FIGURE 12.6**).

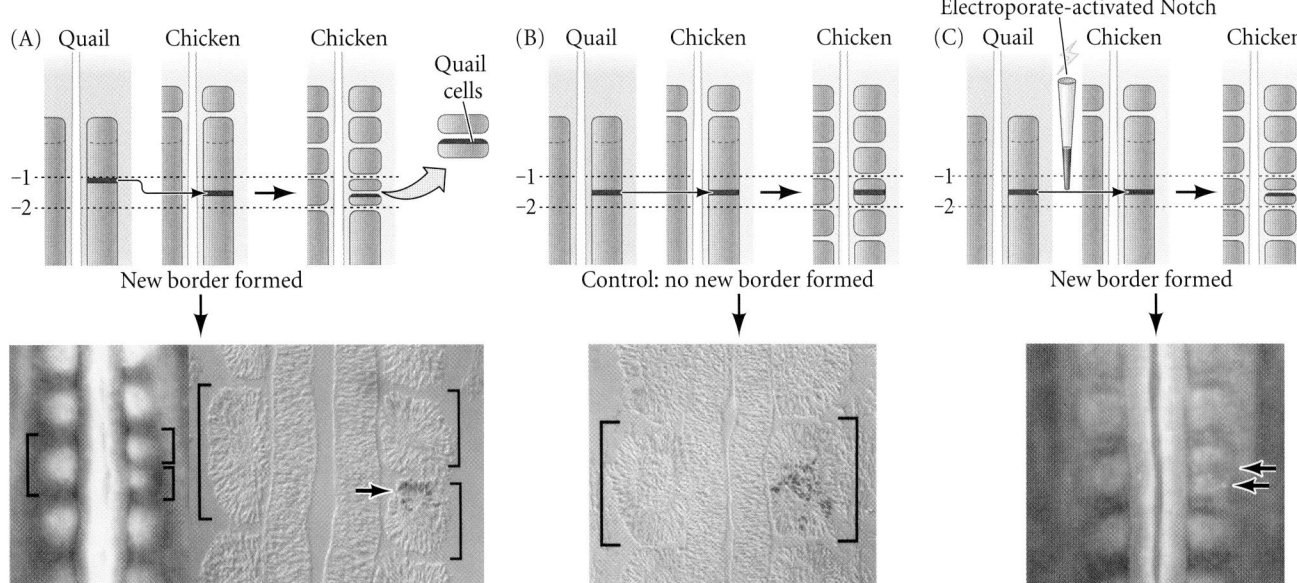

FIGURE 12.5 Notch signaling and somite formation. (A) Transplantation of a prospective somite boundary region into a nonboundary region creates a new boundary and a new somite. The transplanted quail cells can be identified by staining for a quail-specific protein. (B) Transplantation of nonboundary cells into a nonboundary region does not create a new boundary or a new somite. (C) Transplantation of a nonboundary region that has had Notch activated will cause a new somite boundary to occur. (After Sato et al. 2002, photographs courtesy of Y. Takahashi.)

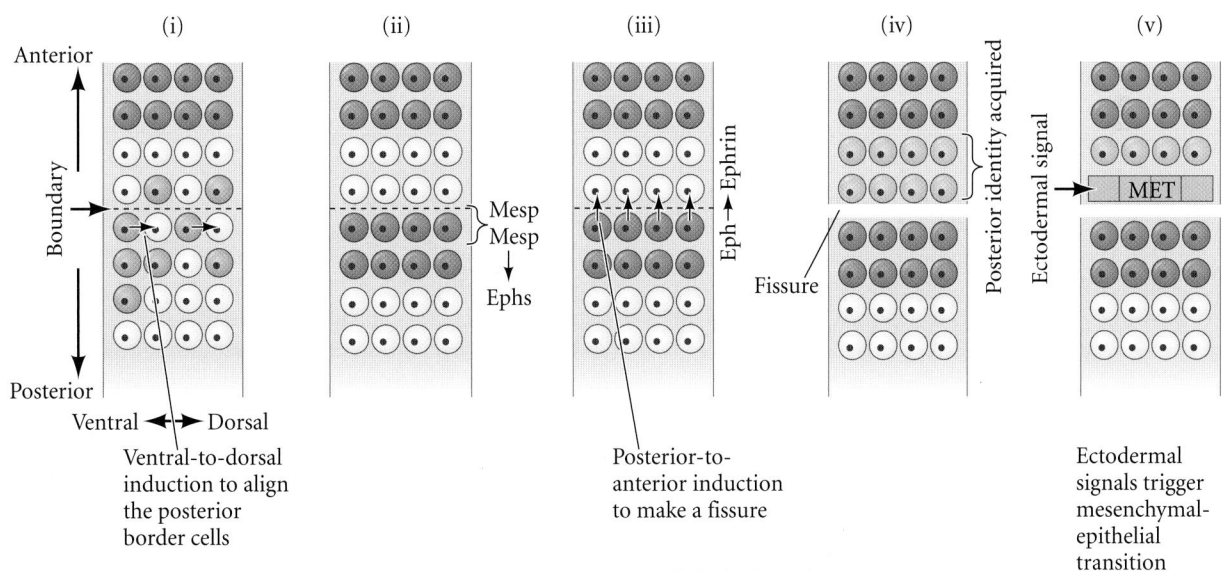

FIGURE 12.6 Sequential molecular and cellular events in somitogenesis. (i) At the boundary (determined by Notch signaling), a ventral-to-dorsal signal aligns the posterior border cells (i.e., those cells immediately posterior to the presumptive border). (ii, iii) Mesp induces Ephs in the posterior border cells, and the Ephs induce ephrin in the cells across the border. This creates the fissure. (iv) As the fissure forms, a separate signal aligns the cells that will form the posterior boundary of the somite. (v) Ectodermal signals act on GTPases to coordinate the transition from mesenchymal to epithelial cell, completing the somite. (After Takahashi and Sato 2008.)

(A)

Somites:
Anterior
Posterior

Eph A4
Ephrin B2

Presomitic
mesoderm

(B)

Somite number

1 2 3 4 5 7 8 9

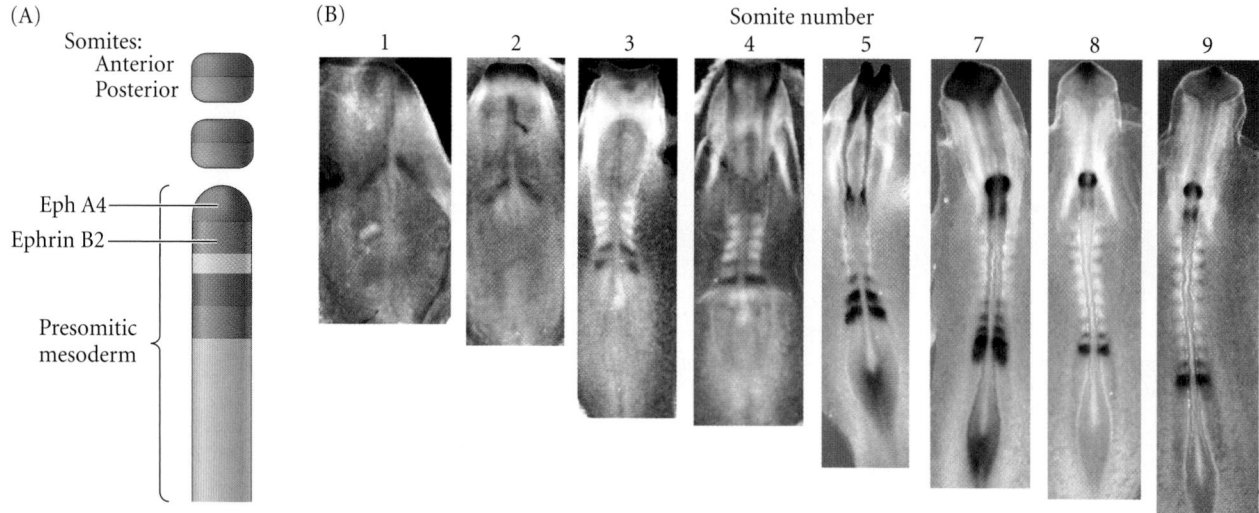

FIGURE 12.7 Ephrin and its receptor constitute a possible fissure site for somite formation. (A) Expression pattern of the receptor tyrosine kinase Eph A4 (blue) and its ligand, ephrin B2 (red), as somites develop. The somite boundary forms at the junction between the region of ephrin expression on the posterior of the last somite formed and the region of Eph A4 expression on the anterior of the next somite to form. In the presomitic mesoderm, the pattern is created anew as each somite buds off. The posteriormost region of the next somite to form does not express ephrin until that somite is ready to separate. (B) In situ hybridization showing Eph A4 (dark blue) expression as new somites form in the chick embryo. (A after Durbin et al. 1998; B courtesy of J. Kastner.)

Within that anterior domain of the next prospective somite, *Mesp* induces Eph A4 (one of the compounds whose repulsive interaction separates the somites) (Saga et al. 1997; Watanabe et al. 2005). In the caudal (posterior) half of the prospective somite, where *Mesp* is not expressed, Notch activity induces the expression of the transcription factor Uncx4.1, which specifies the somite's posterior identity (Takahashi et al. 2000; Saga 2007). In this way, the somite boundary is determined, and the somite is given anterior/posterior polarity at the same time. This division into anterior and posterior somite compartments, as we saw in Chapter 11, is critical for the migration patterns of neural crest cells and neurons.

Separation of somites from the presomitic mesoderm

Two proteins whose roles appear to be critical for fissure formation and somite separation are the Eph tyrosine kinases and their ligands, the ephrin proteins. We saw in Chapter 11 that the Eph tyrosine kinase receptors and their ephrin ligands are able to elicit cell-cell repulsion between the posterior somite and migrating neural crest cells. The separation of the somite from the anterior end of the presomitic mesoderm occurs at the ephrin B2/Eph A4 border. In the zebrafish, the boundary between the most recently separated somite and the presomitic mesoderm forms between ephrin B2 in the

posterior of the somite and Eph A4 in the most anterior portion of the presomitic mesoderm (**FIGURE 12.7**; Durbin et al. 1998). Eph A4 is restricted to the boundary area in chick embryos as well. Interfering with this signaling (by injecting embryos with mRNA encoding dominant negative Ephs) leads to the formation of abnormal somite boundaries.

In addition to the posterior-to-anterior induction of the fissure (from Eph proteins to ephrin proteins on their neighboring cells), a second signal originates from the ventral posterior cells of the somite, putting all the cells in register so that the cut is clean from the ventral to the doral aspects of the somite (Sato and Takahashi 2005).

Periodicity of somite formation

Somites appear at exactly the same time on both sides of the embryo. Even when isolated from the rest of the body, presomitic mesoderm will segment at the appropriate time and in the right direction (Palmeirim et al. 1997). This synchronized somite formation depends on a "clock-and-wave" mechanism in which an oscillating signal generated by the Notch pathway sets the periodicity of the process (i.e., the "clock"), and a rostral-to-caudal gradient provides a moving "wave" of FGF activity. Each oscillation of FGF organizes groups of presomitic cells that will then segment together (see Maroto et al. 2012). In the chick embryo, a new somite is formed about every 90 minutes. In mouse embryos, this time frame is more variable but is closer to 2 hours (Tam 1981).

Although the mechanisms controlling the temporal periodicity of somite formation involve the Notch pathway, different species regulate Notch expression in different ways (Eckalbar et al. 2012). Evidence that the Notch pathway is involved in regulating this periodicity comes from research showing that targets of this pathway show a dynamic and cyclical expression profile across the presomitic mesoderm. The *Hairy1* gene was the first Notch target found to show this pattern of expression, appearing as a wave that traverses

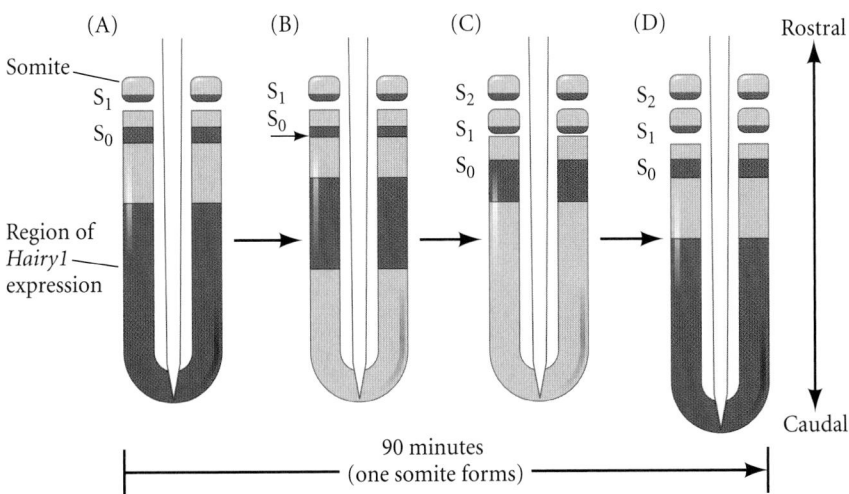

FIGURE 12.8 Somite formation correlates with the wavelike expression of the *Hairy1* gene in the chick. (A) In the posterior portion of a chick embryo somite, S_1 has just budded off the presomitic mesoderm. Expression of the *Hairy1* gene (purple) is seen in the caudal half of this somite, as well as in the posterior portion of the presomitic mesoderm and in a thin band that will form the caudal half of the next somite (S_0). (B) A caudal fissure (small arrow) begins to separate the new somite from the presomitic mesoderm. The posterior region of *Hairy1* expression extends anteriorly. (C) The newly formed somite is now referred to as S_1; it retains the expression of *Hairy1* in its caudal half, as the posterior domain of *Hairy1* expression moves farther anteriorly and shortens. The former S_1 somite, now called S_2, undergoes differentiation. (D) The formation of somite S_1 is complete, and the anterior region of what had been the posterior *Hairy1* expression pattern is now the anterior expression pattern. It will become the caudal domain of the next somite. The entire process takes 90 minutes.

the presomitic mesoderm in a caudal-to-rostral (tail-to-head) direction (**FIGURE 12.8**). The *Hairy1* gene is expressed first in a broad domain at the caudal end of the presomitic mesoderm. This expression domain moves anteriorly while narrowing until it reaches the rostral end, at which time a new wave of expression begins in the caudal end. The time it takes for a wave of expression to cross the presomitic mesoderm is 90 minutes in the chick—which is, not coincidentally, the

time it takes to from a pair of somites in this species. This dynamic expression is not due to cell movement but to cells turning on and off the gene in different regions of the tissue. In addition to *Hairy1*, a variety of other Notch target genes, as well as some targets of the Wnt pathway, are transcribed in a cyclic fashion and function as an autonomous segmentation "clock" in the presomitic mesoderm (Palmeirim et al. 1997; Jouve et al. 2000, 2002).

SIDELIGHTS & SPECULATIONS

Coordinating Waves and Clocks in Somite Formation

A cyclic activation of Notch appears to be critical for forming the somites, but what controls Notch activation? The predominant model of somite formation is the clock-and-wave model proposed by Cooke and Zeeman (1976). Although there are interspecies differences in exactly which gene products oscillate, in all

vertebrate species the clock involves a negative feedback loop of the Notch signaling pathway. Thus, in all vertebrates at least one of the Notch target genes having dynamic oscillating expression in the presomitic mesoderm is also able to inhibit the gene for Notch; this establishes a negative feedback loop. These inhibitory

proteins are unstable, and when the inhibitor is degraded, Notch becomes active again. Such feedback creates a cycle (the "clock") in which the *Notch* gene is turned on and off by a protein that Notch itself induces. These off-and-on oscillations could provide the molecular basis for the periodicity of somite segmentation (Holley and Nüsslein-Volhard 2000; Jiang et al. 2000; Dale et al. 2003). The Mesp2 protein may be the critical link regulated by the Notch clock, since the output from Mesp2 controls the ephrins that mediate the separation of the block of cells that form the somite (Morimoto et al. 2005).

The "wave" regulating the competence of cells to respond to the clock appears to be the Fgf8 gradient, which becomes caudally displaced as more cells are added to the posterior presomitic mesoderm and more somites bud off the anterior presomitic mesoderm. As long as the presomitic mesenchyme is in a region of relatively high Fgf8 concentration, the clock will not function. At least in zebrafish, this appears to be due to the repression of Delta, the major ligand of Notch. The binding of Fgf8 to its receptor enables the expression of the Her13.2 protein, which is necessary to inhibit transcription of Delta (see Dequéant and Pourquié 2008). Interestingly, the Fgf8 signal can be perturbed by the right-left laterality signals being given

at the same time, and the presence of retinoic acid insulates the somite wavefront from the laterality signals of Nodal proteins. If no retinoic acid is made, somitogenesis becomes asymmetric (Kawakami et al. 2005; Brend and Holley 2009).

In chicks and mice, the situation appears to be far more complicated. The Notch pathway still provides the clock, but the clock appears to be sensitive not only to FGFs but also to the Wnt signaling pathway (which also exhibits a posterior-to-anterior gradient) and to the retinoic acid gradient (which extends in the opposite direction). FGFs (especially Fgf8) are made predominantly in the tailbud, which serves as a source of new cells. It has recently been proposed (Niwa et al. 2011; Pourquié 2011) that Fgf8 synthesis also cycles, but at a different frequency than the Notch ligands. FGF signals are needed to get cells to migrate anteriorly out of the tailbud, but as long as FGFs activate the ERK transcription factors, the cells remain unresponsive to Notch ligands. But

because of its own cycling (probably by synthesizing its own inhibitors and by inhibition by retinoic acid), FGF signaling is downregulated at certain areas in the paraxial mesoderm, and cells in these regions become competent to respond to Notch signals (**FIGURE 12.9**). Thus, FGFs establish the *placement* (i.e., the wavefront) of cells that are competent to respond to oscillating Notch signal (the clock), which can induce Mesp2.*

Notch oscillation appears to be regulated differently across the vertebrates (Krol et al. 2011; Eckalbar et al. 2012). The Notch-induced (and Notch-inhibiting) *Hairy/Hes* genes appear to be conserved in all vertebrate somitogenesis systems. Interestingly, *Hes* genes also have been found to be cyclically expressed in the formation of arthropod body segments (Chipman and Akam 2008; Pueyo et al. 2008), suggesting that *Hairy/Hes* control of the Notch pathway is at the core of a system for body segmentation that has been modified in both vertebrates and invertebrates.

---

* Pourquié (2011) has noted that this situation seems a lot like the one that exists in the limb bud (see Chapter 14), wherein a pool of newly formed cells that has been maintained by FGFs in a relatively undifferentiated and migratory state becomes differentiated into periodic elements (limb cartilage) by interacting gradients of FGFs and retinoic acid.

FIGURE 12.9 Possible model of "clock and wavefront" somite specification. (A–C) Formation of a bi-stable zone where somites can be formed at the junction of the retinoic acid (anterior) and FGF (posterior) domains. (A) Fgf8 expression in the posterior part of the embryo. (B) RNA for Raldh2 (retinoic acid-synthesizing enzyme) in the central part of the embryo. (C) Mesp showing region where the somite formation will occur later. Asterisks show the last formed somite. The dotted line approximates the region where somites are being determined. (D) In each panel, the FGF-induced transcription factor (pERK) and the somite-specifying transcription factor (Mesp) are shown on the left, and the Notch-generated transcription factor (NICD) and the Hes7 inhibitory protein induced by NICD are on the right. (A–C after Pourquié 2011; D after Niwa et al. 2011.)

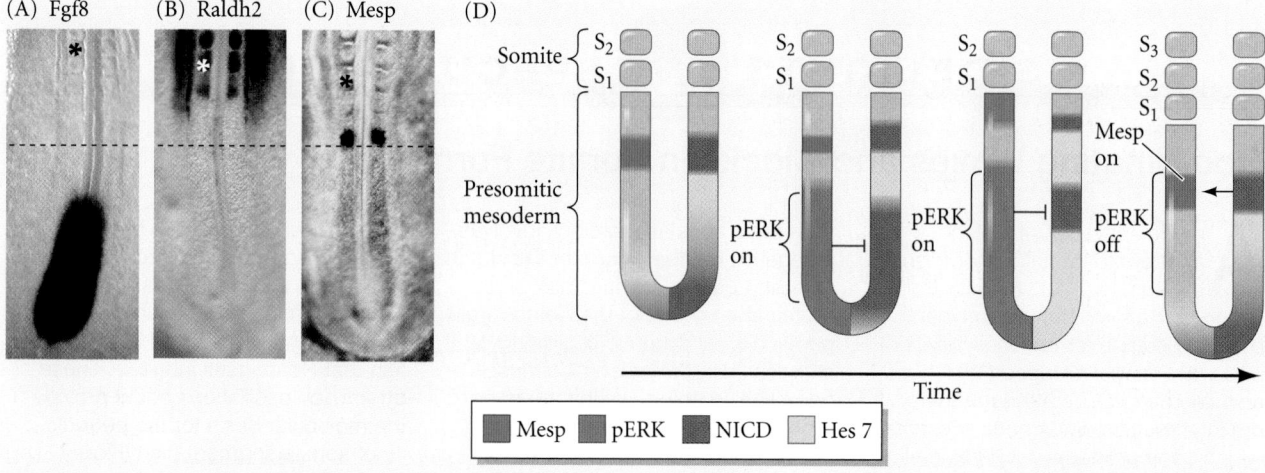

(A)   Epithelial cells

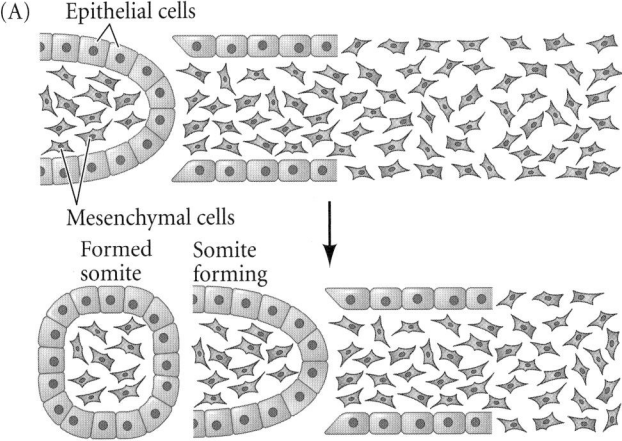

Mesenchymal cells

Formed   Somite
somite   forming

(B)

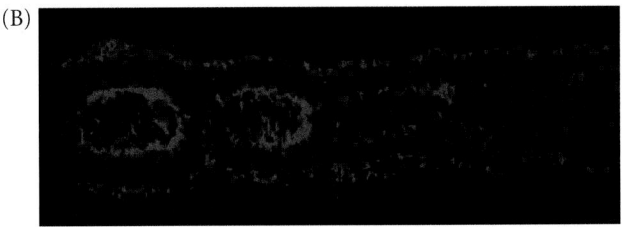

FIGURE 12.10 Epithelialization and de-epithelialization in somites of a chick embryo. (A) Changes in cell shape from mesenchymal (pink) to epithelial (gray) cells when a somite forms from presomitic mesenchyme. A formed somite is surrounded by epithelial cells, with mesenchymal cells remaining inside. In chickens, epithelialization occurs first at the posterior edge of the somite, with the anterior edge becoming epithelial later. (B) Changes in cell polarity as somites form are revealed by staining that visualizes F-actin accumulation (red). (A after Nakaya et al. 2004; B courtesy of Y. Takahashi.)

Epithelialization of the somites

Several studies in the chick have shown that epithelialization occurs immediately after somite fission occurs. As seen in Figure 12.4, the cells of a newly formed somite are randomly organized as a mesenchymal mass. These cells have to be compacted into an outer epithelium and an internal mesenchyme (**FIGURE 12.10**). Ectodermal signals appear to cause the peripheral presomitic cells to undergo mesenchymal-epithelial transition by lowering the cells' Cdc42 levels. Low Cdc42 levels alter the cytoskeleton, allowing epithelial cells to form a box around the remaining mesenchymal cells, which have a higher level of Cdc42. Another GTPase, Rac1, must be present at a certain level that allows it to activate Paraxis, another transcription factor involved in epithelialization (Burgess et al. 1995; Barnes et al. 1997; Nakaya et al. 2004).

The epithelialization of each somite is stabilized by synthesis of the extracellular matrix protein fibronectin and the adhesion protein N-cadherin (Lash and Yamada 1986; Hatta et al. 1987; Saga et al. 1997; Linask et al. 1998). N-cadherin links the adjoining cells into an epithelium, while the

fibronectin matrix acts alongside the ephrin and Eph to promote the separation of the somites from each other (Martins et al. 2009).

Specification of the somites along the anterior-posterior axis

Although all somites look identical, they will form different structures. For instance, the somites that form the cervical vertebrae of the neck and lumbar vertebrae of the abdomen are not capable of forming ribs; ribs are generated only by the somites that form the thoracic vertebrae, and this specification of thoracic vertebrae occurs very early in development. The presomitic mesoderm is determined by its position along the anterior-posterior axis before somitogenesis. If one isolates the region of chick presomitic mesoderm that will give rise to a thoracic somite and transplants this mesoderm into the cervical (neck) region of a younger embryo, the host embryo will develop ribs in its neck—but only on the side where the thoracic mesoderm has been transplanted (**FIGURE 12.11A**; Kieny et al. 1972; Nowicki and Burke 2000).

The vertebral column therefore exhibits regionalization along the anterior-posterior axis. The somites are specified according to the Hox genes they express (see Chapter 9). The *Hoxb* genes are first activated in a temporal and co-linear fashion in the paraxial mesoderm precursor cells while these cells still reside in the epiblast, lateral to the primitive streak. It is probable that *Hoxb* activation regulates the migration of these cells from the epiblast into the streak (Iimura and Pourquié 2006). If these cells are transplanted prior to this migration, the profile of Hox gene expression can be altered (McGrew 2008).

By contrast, once the paraxial mesoderm cells have taken up position in the presomitic mesoderm, Hox gene expression appears more fixed. Indeed, once established, each somite retains its pattern of Hox gene expression, even if that somite is transplanted into another region of the embryo (Nowicki and Burke 2000). If the Hox gene pattern is altered, so is the specification of the mesoderm. For instance, if the entire presomitic mesoderm ectopically expresses *Hoxb6*, ribs will form from every vertebra (**FIGURE 12.11B**). Moreover, if Fgf8 protein levels are manipulated to create extra (albeit smaller) somites, the appropriate Hox gene expression will be activated in the appropriately numbered somite, even if it is in a different position along the anterior-posterior axis. The regulation of Hox genes by the segmentation clock presumably allows coordination between the formation and the specification of the new segments. Mutations that affect the autonomous segmentation clock also affect activation of the appropriate Hox genes (Dubrulle et al. 2001; Zákány et al. 2001).

Differentiation of the somites

In contrast to the early commitment of the positional identity of the presomitic mesoderm along the anterior-posterior body axis, the commitment of the cells *within* a somite toward a particular cell fate occurs relatively late, after the somite has

(A)

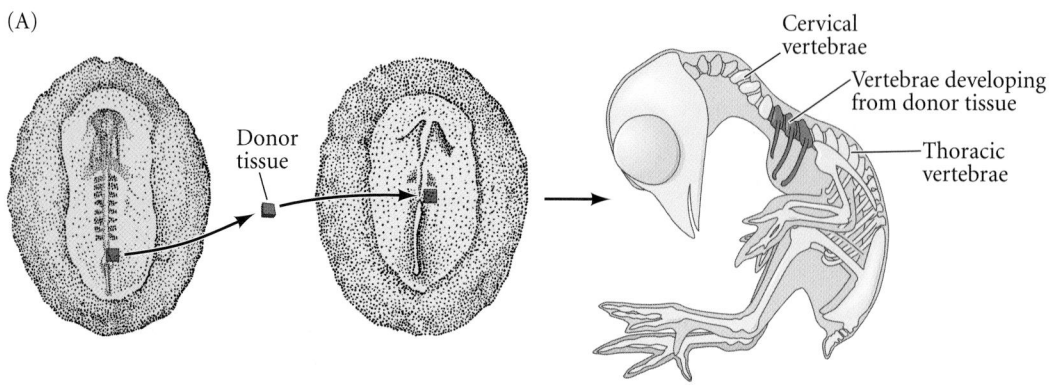

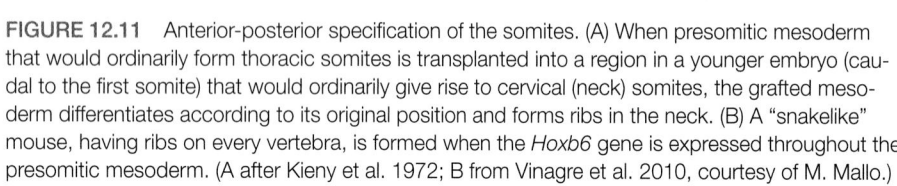

(B)

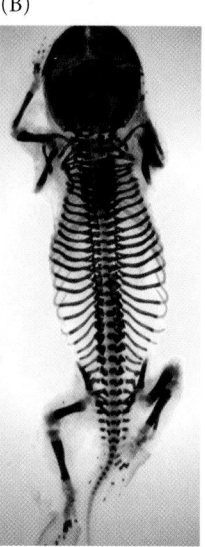

FIGURE 12.11 Anterior-posterior specification of the somites. (A) When presomitic mesoderm that would ordinarily form thoracic somites is transplanted into a region in a younger embryo (caudal to the first somite) that would ordinarily give rise to cervical (neck) somites, the grafted mesoderm differentiates according to its original position and forms ribs in the neck. (B) A "snakelike" mouse, having ribs on every vertebra, is formed when the *Hoxb6* gene is expressed throughout the presomitic mesoderm. (A after Kieny et al. 1972; B from Vinagre et al. 2010, courtesy of M. Mallo.)

already formed. When the somite is first separated from the presomitic mesoderm, any of its cells can become any of the somite-derived structures. These structures include:

- The cartilage of the vertebrae and ribs
- The muscles of the rib cage, limbs, abdominal wall, back, and tongue
- The tendons that connect the muscles to the bones
- The dermis of the dorsal skin
- Vascular cells that contribute to the formation of the aorta and the intervertebral blood vessels
- The cellular sheaths, or *meninges*, of the spinal cord that protect the central nervous system

Thus, the somite contains a population of multipotent cells whose specification depends on their location within the somite and their subsequent response to paracrine factors from the surrounding tissues (neural tube, notochord, epidermis, and intermediate mesoderm).

As the somite matures, its various regions become committed to forming only certain cell types. First, the ventral-medial cells of the somite (those cells closest to the neural tube and notochord) undergo mitosis, lose their round epithelial characteristics, and become mesenchymal cells again. This portion of the somite is the sclerotome, and these mesenchymal cells ultimately become the cartilage cells (chondrocytes) of the vertebrae and a major part of each rib (FIGURE 12.12A,B,E).

The remaining epithelial portion of the somite is the dermamyotome. Fate mapping with chick-quail chimeras (Ordahl and Le Douarin 1992; Brand-Saberi et al. 1996; Kato and Aoyama 1998) has revealed that the dermamyotome is arranged into three regions (FIGURE 12.12C,D,F). The cells in the two lateral portions of this epithelium (i.e.,

the dorsomedial and ventrolateral lips closest to and farthest from the neural tube, respectively) are the myotomes and will form muscle cells. The central region, the dermatome, will form back dermis and several other derivatives. In the lateral myotomes, muscle precursor cells will migrate beneath the dermamyotome to produce a lower layer of muscle precursor cells, the **myoblasts**. Those myoblasts in the myotome closest to the neural tube form the centrally located **primaxial muscles**,* which include the intercostal musculature between the ribs and the deep muscles of the back; those myoblasts formed in the region farthest from the neural tube produce the **abaxial muscles** of the body wall, limbs, and tongue (FIGURE 12.13A–C). The boundary between the primaxial and abaxial muscles and between the somite-derived and lateral plate-derived dermis is called the **lateral somitic frontier** (Christ and Ordahl 1995; Burke and Nowicki 2003; Nowicki et al. 2003). Various transcription factors distinguish the primaxial and abaxial muscles (FIGURE 12.13D).

The central portion of the dermamyotome has traditionally been called the **dermatome**, since its major product is the precursors of the dermis of the back. (The dermis of the ventral portion of the body is derived from the lateral plate, and the dermis of the head and neck comes, at least in part, from the cranial neural crest.) However, recent studies have shown that this central region of the dermamyotome also gives rise to a third population of muscle cells (Gros et al. 2005; Relaix

---

*As used here, the terms *primaxial* and *abaxial* designate the muscles from the medial and lateral portions of the somite, respectively. The terms *epaxial* and *hypaxial* are commonly used, but these terms are derived from secondary modifications of the adult anatomy (the hypaxial muscles being innervated by the ventral regions of the spinal cord) rather than from the somitic myotome lineages (see Nowicki 2003).

(A) 2-Day embryo

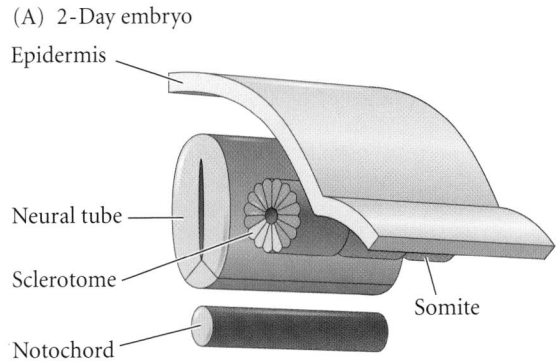

Epidermis

Neural tube

Sclerotome

Notochord

Somite

(B) 3-Day embryo

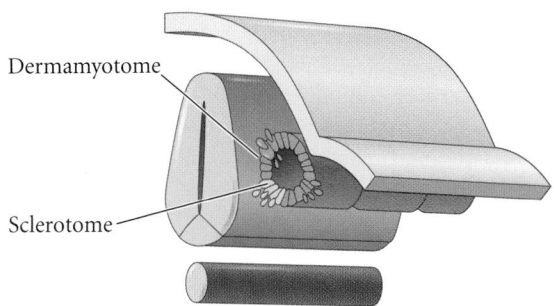

Dermamyotome

Sclerotome

(C) 4-Day embryo

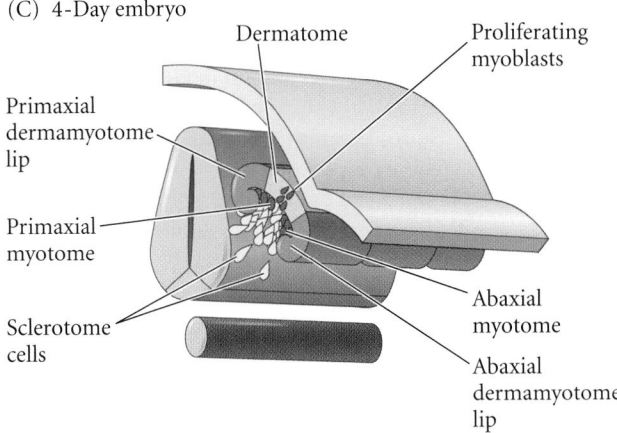

Dermatome

Proliferating myoblasts

Primaxial dermamyotome lip

Primaxial myotome

Sclerotome cells

Abaxial myotome

Abaxial dermamyotome lip

(D) Late 4-Day embryo

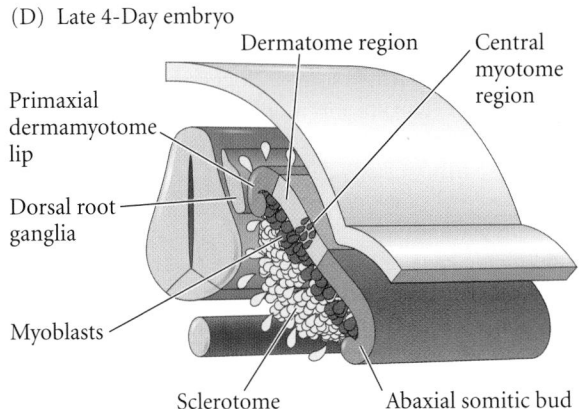

Dermatome region

Central myotome region

Primaxial dermamyotome lip

Dorsal root ganglia

Myoblasts

Sclerotome

Abaxial somitic bud

(E)

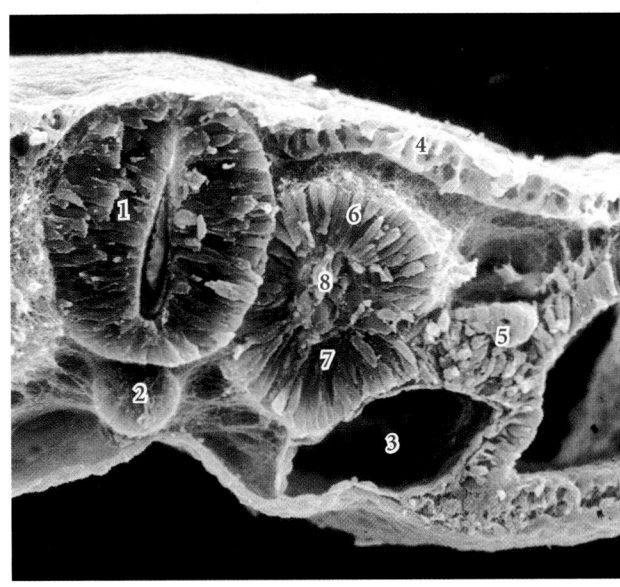

FIGURE 12.12 Transverse section through the trunk of a chick embryo on days 2–4. (A) In the 2-day somite, the sclerotome cells can be distinguished from the rest of the somite. (B) On day 3, the sclerotome cells lose their adhesion to one another and migrate toward the neural tube. (C) On day 4, the remaining cells divide. The medial cells form a primaxial myotome beneath the dermamyotome, while the lateral cells form an abaxial myotome. (D) A layer of muscle cell precursors (the myotome) forms beneath the epithelial dermamyotome. (E,F) Scanning electron micrographs correspond to (A) and (D), respectively; 1, neural tube; 2, notochord; 3, dorsal aorta; 4, surface ectoderm; 5, intermediate mesoderm; 6, dorsal half of somite; 7, ventral half of somite; 8, somitocoel/arthrotome; 9, central sclerotome; 10, ventral sclerotome; 11, lateral sclerotome; 12, dorsal sclerotome; 13, dermamyotome. (A,B after Langman 1981; C,D after Ordahl 1993; E,F from Christ et al. 2007, courtesy of H. J. Jacob and B. Christ.)

(F)

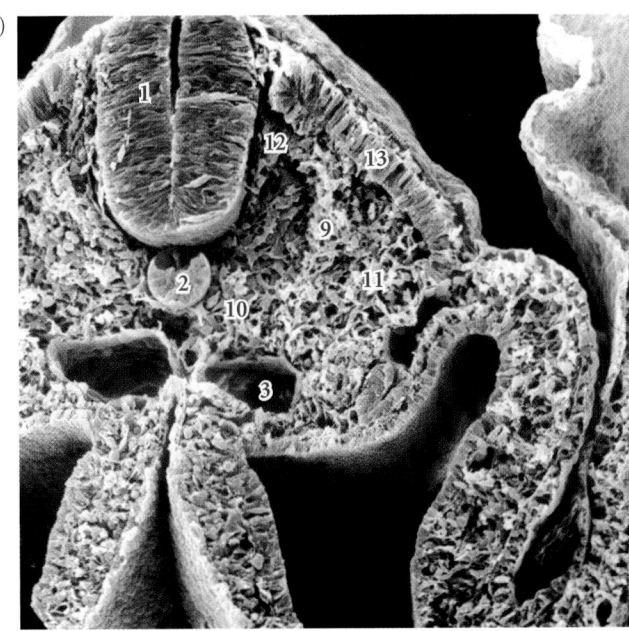

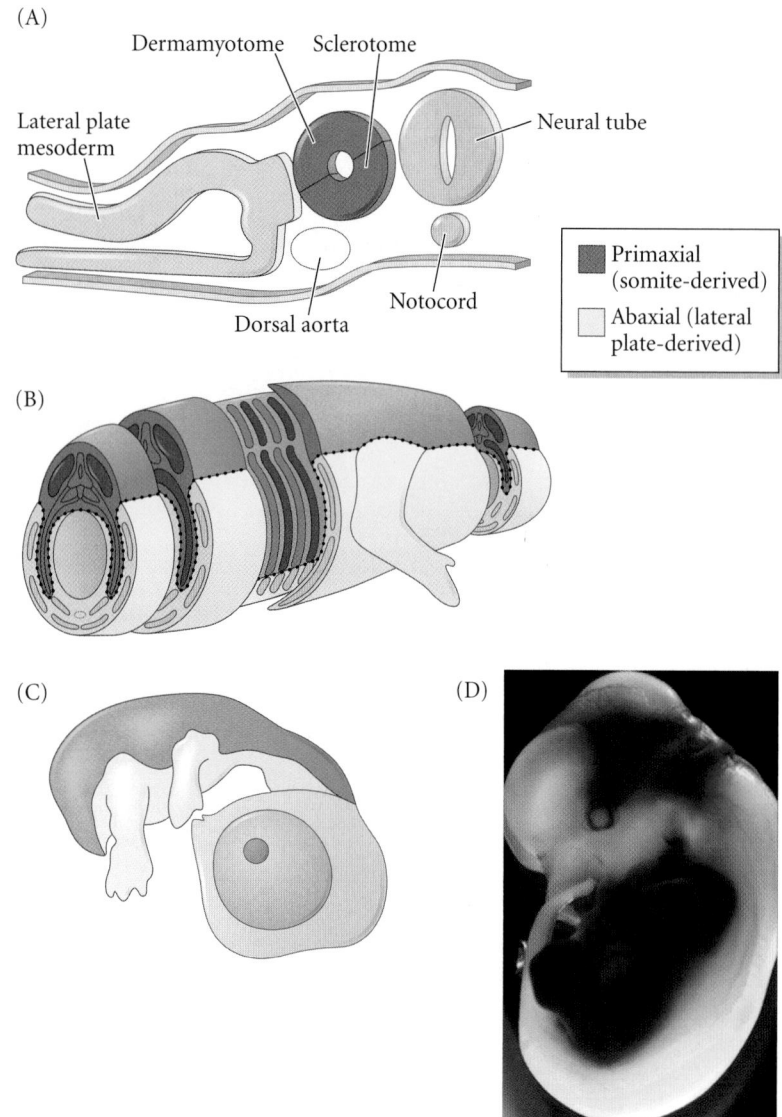

FIGURE 12.13 Primaxial and abaxial domains of vertebrate mesoderm. (A) Mesoderm (red) differentiation in an early-stage chick embryo. (B,C) Two views of a day-9 chick embryo. Regionalization of the mesoderm is apparent. (D) Day-13 chick in which *Prx1* gene expression is revealed by a dark stain. In the chick trunk, *Prox1* is expressed in the abaxial region. The boundary between the stained and unstained regions is the lateral somitic frontier. (A–C after Winslow et al. 2007; D from Durland et al. 2008, courtesy of A. C. Burke.)

The muscle precursor cells that delaminate from the epithelial plate to join the primary myotome cells remain undifferentiated, and they proliferate rapidly to account for most of the myoblast cells. While most of these progenitor cells differentiate to form muscles, some remain undifferentiated and surround the mature muscle cells. These undifferentiated cells become the *satellite cells* responsible for postnatal muscle growth and muscle repair.

Determination of the sclerotome

Like the proverbial piece of real estate, the destiny of a particular region of the somite depends on three things: location, location, and location. As shown in FIGURE 12.14, the locations of the somitic regions place them close to different signaling centers, such as the notochord (source of Sonic hedgehog and Noggin), the neural tube (source of Wnts and BMPs), and the surface epithelium (also a source of Wnts and BMPs).

The specification of a somite is accomplished by the interaction of several tissues. The ventromedial portion of the somite is induced to become the sclerotome by paracrine factors (especially Sonic hedgehog) secreted from the notochord (Fan and Tessier-Lavigne 1994; Johnson et al. 1994). If portions of the chick notochord are transplanted next to other regions of the somite, those regions will also become sclerotome cells. The notochord and somites also secrete Noggin and Gremlin, two BMP antagonists. The absence of BMPs is critical in permitting Sonic hedgehog to induce cartilage expression, and if either of these inhibitors is deficient, the sclerotome fails to form and the chicks lack normal vertebrae. Paracrine factors induce the presumptive sclerotome cells to express the transcription factor Pax1, which is required for their epithelial-mesenchymal transition and subsequent differentiation into cartilage (Smith and Tuan 1996). In this EMT, the epithelial cells lose N-cadherin expression and become motile mesenchyme (Sosic et al. 1997). Sclerotome cells also express inhibitors of the myogenic (muscle-forming) transcription factors

et al. 2005). Therefore, some researchers (Christ and Ordahl 1995; Christ et al. 2007) prefer to retain the term *dermamyotome* (or *central dermamyotome*) for this epithelial region. Soon, however, this part of the somite also undergoes an epithelial-mesenchymal transition (EMT). FGF signals from the myotome activate the transcription of the *Snail2* gene in the central dermamyotome cells, and the Snail2 protein is a well-known regulator of EMT (see Chapters 3 and 11; Delfini et al. 2009). During EMT, the mitotic spindles of the epithelial cells are realigned so that cell division takes place along the dorsal-ventral axis. The ventral daughter cell joins the other myoblasts from the myotomes, while the other daughter cell locates dorsally, becoming a precursor of the dermis. The N-cadherin holding the cells together is downregulated and the two daughter cells go their separate ways, with the remaining N-cadherin found only on those cells entering the myotome (Ben-Yair and Kalcheim 2005).

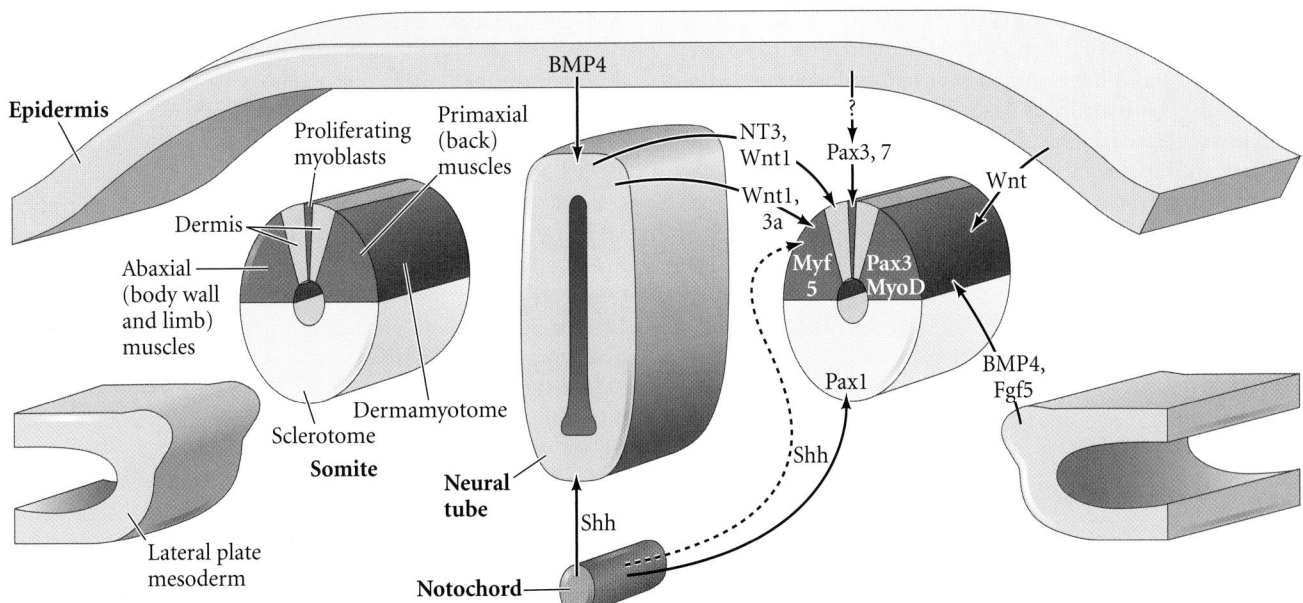

FIGURE 12.14 Model of major postulated interactions in the patterning of the somite. The sclerotome is white, while the dermamyotome regions are red and pink. A combination of Wnts (probably Wnt1 and Wnt3a) is induced by BMP4 in the dorsal neural tube. These Wnt proteins, in combination with low concentrations of Sonic hedgehog from the notochord and floor plate, induce the primaxial myotome, which synthesizes the myogenic transcription factor Myf5. High concentrations of Shh from the notochord and neural tube floorplate induce Pax1 expression in those cells fated to become the sclerotome. Certain concentrations of neurotrophin-3 (NT3) from the dorsal neural tube appear to specify the dermatome, while Wnt proteins from the epidermis, in conjunction with BMP4 and Fgf5 from the lateral plate mesoderm, are thought to induce the primaxial myotome. The proliferating myoblasts are characterized by Pax 3 and Pax7 and are induced by Wnts n the epidermis. (After Cossu et al. 1996b.)

(myogenic regulatory factors, or MRFs, of which we'll soon hear more) (Chen et al. 1996).

The sclerotome contains several regions, each of which becomes specified according to its location. While most sclerotome cells become the prescursors of the vertebral and rib cartilage, the medial sclerotome cells closest to the neural tube generate the meninges (coverings) of the spinal cord as well as giving rise to blood vessels that will provide the spinal cord with nutrients and oxygen (Halata et al. 1990; Nimmagadda et al. 2007). The cells in the center of the somite (which remain mesenchymal) also contribute to the sclerotome, becoming the vertebral joints, the cartilagenous discs between the vertebrae, and the portions of the ribs closest to the vertebrae (Mittapalli et al. 2005; Christ et al. 2007). This region of the somite has been called the **arthrotome**.

The notochord, with its secretion of Sonic hedgehog, is thus critical for sclerotome formation; we will also see that notochord-generated compounds direct the migration of sclerotome cells to the center of the embryo, where they form the vertebrae. But what happens to the notochord, that central mesodermal structure that induced the nervous system and caused the sclerotome to form? After it has provided the axial integrity of the early embryo and has patterned the ventral neural tube, most of it degenerates by apoptosis. This apoptosis is probably signaled by mechanical forces.

Wherever the sclerotome cells have formed a vertebral body, the notochordal cells die. However, between the vertebrae, the notochordal cells form the central part of the intervertebral discs, the **nuclei pulposi** (Aszódi et al. 1998; Guehring et al. 2009; Risbud and Shapiro 2011). These are the spinal discs that "slip" in certain back injuries.

Determination of the central dermamyotome

The central dermamyotome generates muscle precursors as well as the dermal cells that constitute the connective tissue layer of the *dorsal* skin. (The dermis of the *ventral* and *lateral* sides of the body is derived from the lateral plate mesoderm that forms the body wall.) There is a sharp demarcation between the somite- and lateral plate-derived dermis.

The maintenance of the central dermamyotome depends on Wnt6 coming from the epidermis (Christ et al. 2007), and its epithelial-mesenchymal transition appears to be regulated by neurotrophin 3 (NT3) and Wnt1, two factors secreted by the neural tube. Antibodies that block NT3 activity prevent the conversion of epithelial dermatome into the loose dermal mesenchyme that migrates beneath the epidermis (Brill et al. 1995). Removing or rotating the neural tube prevents this dermis from forming (Takahashi et al. 1992; Olivera-Martinez et al. 2002). The Wnt signals from the epidermis promote the differentiation of the dorsally migrating central dermamyotome cells into dermis (Atit et al. 2006).

But muscle precursor cells and dermal cells are not the only derivatives of the central dermamyotome. Atit and her colleagues (2006) have shown that **brown adipose cells** ("brown fat") are also somite-derived and appear to come from the central dermamyotome. Brown fat plays active roles in energy utilization by burning fat (unlike the better known adipose tissue, or "white fat," which stores fat). Tseng and colleagues (2008) have found that skeletal muscle and brown fat cells share the same somitic precursor that originally expresses myogenic regulatory factors. In brown fat precursor cells, the transcription factor PRDM16 is induced (probably by BMP7); PRDM16 appears to be critical for the conversion of myoblasts to brown fat cells, as it activates a battery of genes that are specific for the fat-burning metabolism of brown adipocytes (Kajimura et al. 2009).

Determination of the myotome

All the skeletal musculature in the vertebrate body with the exception of the head muscles comes from the dermamyotome of the somite. The myotome forms from the lateral edges of the dermamyotome and folds to form a layer between the more peripheral dermamyotome and the more medial sclerotome. The major transcription factors associated with (and causing) muscle development are the **myogenic regulatory factors** (**MRFs**; sometimes called the myogenic bHLH proteins). This family of transcription factors include MyoD, Myf5, myogenin, and MRF4. Each member of this family can activate the genes of the other family members, leading to positive feedback regulation so powerful that the activation of an MRF in nearly any cell in the body converts that cell into muscle.* The MRFs bind to and activate genes that are critical for muscle function. For instance, the MyoD protein appears to directly activate the muscle-specific creatine phosphokinase gene by binding to the DNA immediately upstream from it (Lassar et al. 1989). There are also two MyoD-binding sites on the DNA adjacent to the genes encoding a subunit of the chicken muscle acetylcholine receptor (Piette et al. 1990). MyoD also directly activates its own gene. Therefore, once the *myoD* gene is activated, its protein product binds to the DNA immediately upstream of *myoD* and keeps this gene active. Many of the MRFs are active only if they associate with a muscle-specific co-factor from the Mef2 (myocyte enhancer factor-2) family of proteins. MyoD can activate the *Mef2* gene, and thereby regulate the differential timing of muscle gene expression.

As discussed above, the myotome is induced in the somite at two different places by at least two distinct signals (see Punch et al. 2009). Studies using transplantation

---

* A general rule of development, as in the U.S. constitution, is that powerful entities must be powerfully regulated. As a result of their power to convert any cell into muscle, the MRFs are among the most powerfully controlled entities of the genome. They are controlled at several points in transcription, as well as in processing, translation, and posttranslational modification (see Sartorelli and Juan 2011; Ling et al. 2012).

and knockout mice indicate that the *primaxial* myoblasts from the medial portion of the somite are induced by factors from the neural tube—probably Wnt1 and Wnt3a from the dorsal region and low levels of Sonic hedgehog from the floor plate of the neural tube (Münsterberg et al. 1995; Stern et al. 1995; Borycki et al. 2000). These induce the Pax3-containing cells of the somite to activate the *Myf5* gene in the primaxial myotome. Myf5 (in concert with Mef2 and either Six1 or Six4) activates the myogenin and MRF4 genes, whose proteins activate the muscle-specific batteries of genes (**FIGURE 12.15A**; Buckingham et al. 2006). The cells of the primaxial myotome appear to be originally confined by the laminin extracellular matrix that outlines the dermamyotome and myotome. However, as the myoblasts mature, this matrix dissolves, and the primaxial myoblasts migrate along fibronectin cables. Eventually they align, fuse, and elongate to become the deep muscles of the back, connecting to the developing vertebrae and ribs (Deries et al. 2010, 2012).

The abaxial myoblasts that form the limb and ventral body wall musculature arise from the lateral edge of the somite. Two conditions appear necessary to produce these muscle precursors: (1) the presence of Wnt signals and (2) the absence of BMPs (**FIGURE 12.15B**; Marcelle et al. 1997; Reshef et al. 1998). Wnt proteins (especially Wnt7a) are made from the epidermis (see Figure 12.14; Cossu et al. 1996a; Pourquié et al. 1996; Dietrich et al. 1998), but the BMP4 made by the adjacent lateral plate mesoderm would normally prevent muscles from forming.

What, then, is inhibiting BMP activity? Recent studies (Gerhart et al. 2006, 2011) have found that the somites have attached at their tips a population of cells that secrete the BMP inhibitor Noggin. These Noggin-secreting cells arise in the blastocyst, become part of the epiblast cells, and distinguish themselves by expressing the mRNA for MyoD but do not translate this mRNA into protein. These particular cells migrate to become paraxial mesoderm, specifically sorting out to the dorsomedial and ventrolateral lips of the dermamyotome. There they synthesize and secrete Noggin, thus promoting differentiation of myoblasts. If Noggin-secreting cells are removed from the epiblast, there is a decrease in the skeletal musculature throughout the body, and the ventral body wall is so weak that the heart and abdominal organs often are herniated through it (**FIGURE 12.16**). (This defect can be prevented by implanting Noggin-releasing beads into somites lacking these cells.) Once BMP is inhibited, Wnt7 can induce MyoD in the competent dermamyotome cells, which activates the battery of MRF proteins that generate the muscle precursor cells.

Another population of cells—the migrating neural crest cells—also affects myotome development in the somite. Recall from Chapter 11 that trunk neural crest cells migrate through the early somite. As they begin to migrate, the Delta-expressing neural crest cells touch the Notch-containing membranes of the primaxial myotome cells, which helps induce *Myf5* expression in those cells (Rios et al. 2011). Moreover, as they migrate through the sclerotome, the neural crest cells secrete neuregulin-1, a paracrine factor that prevents the

(A) Paraxial myotome

(B) Abaxial myotome

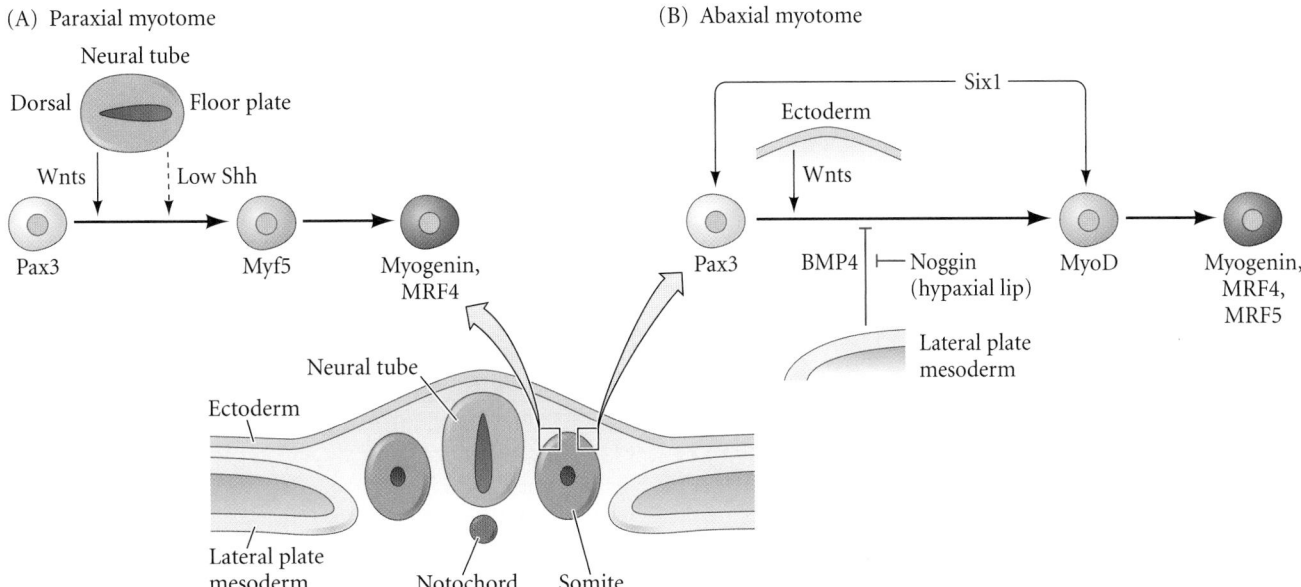

FIGURE 12.15 Differential gene expression in myotome. (A) The primaxial myotome is thought to be specified by a combination of Wnts (probably Wnt1 and Wnt3a) from the dorsal neural tube, in combination with low concentrations of Sonic hedgehog from the floor plate of the neural tube. Pax3 in somatic cells allows expression of Myf5 in response to paracrine factors. This allows the cells to synthesize the myogenic transcription factor Myf5. In combination with a Six protein and Mef2, Myf5 activates the genes responsible for activating myogenin and MRF4. (B) BMP4 is inhibited by Noggin produced by cells that migrate specifically to the lips of the somite. In the absence of BMP4, Wnt proteins from the epidermis are thought to induce the abaxial myotome. (After Punch et al. 2009.)

(A) Control   (B) Ablated

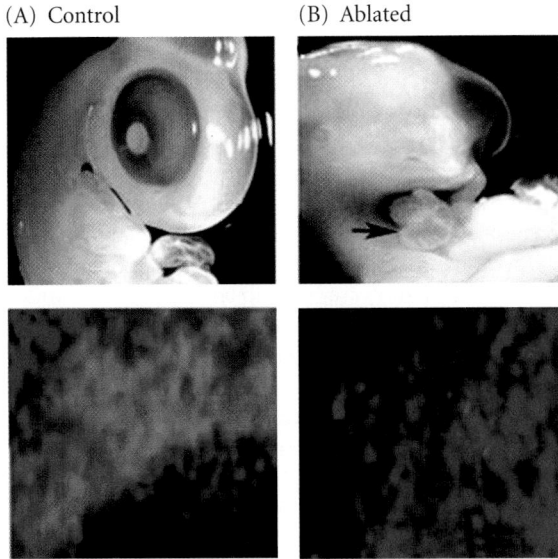

FIGURE 12.16 Ablating Noggin-secreting epiblast cells results in severe muscle defects. Noggin-secreting epiblast cells were ablated in stage 2 chick embryos using antibodies against G8. (A) The control embryo has normal morphology and abundant staining of myosin (lower photograph) in the muscles. (B) Embryos whose Noggin-secreting epiblast cells are ablated have severe eye defects, severely reduced somatic musculature, and herniation of abdominal organs through the thin abdominal wall (arrow). Severely reduced musculature (sparse myosin in lower photograph) is characteristic of these embryos. (From Gerhart et al. 2006, courtesy of J. Gerhart and M. George-Weinstein.)

premature differentiation of myoblasts into muscle cells (Ho et al. 2011). Thus, neural crest cells have an embryonic function as well as an adult function. Their adult function is to serve as sensory neurons, glial, or adrenal cells; their embryonic function includes regulating myotome development.*

● See WEBSITE 12.1 Cranial paraxial mesoderm

Myoblasts and myofibers

The cells producing the myogenic regulatory factors are the myoblasts—committed muscle cell precursors. But unlike most cells of the body, muscle cells do not function as "individuals." Rather, several myoblasts align together and fuse their cell membranes to form a **myofiber**, a single large cell with several nuclei that is characteristic of muscle tissue (Konigsberg 1963; Mintz and Baker 1967; Richardson et al. 2008). Studies on mouse embryos show that by the time a mouse is born, it has the adult number of myofibers, and that these multinucleated myofibers grow during the first week after birth by the fusion of singly nucleated myoblasts (Ontell

---

*We see here another example of the reciprocal induction that is so critical in eye development (see Chapters 3 and 10) and which will feature prominently in the discussion of kidney development later in this chapter. The dorsal neural tube (through BMPs and Wnts) helps pattern the somites; the somites (through paracrine factors and extracellular matrices) help initiate and pattern neural crest cell migration; and the migrating neural crest cells regulate the production of the myotome in the somite (see Kalcheim 2011).

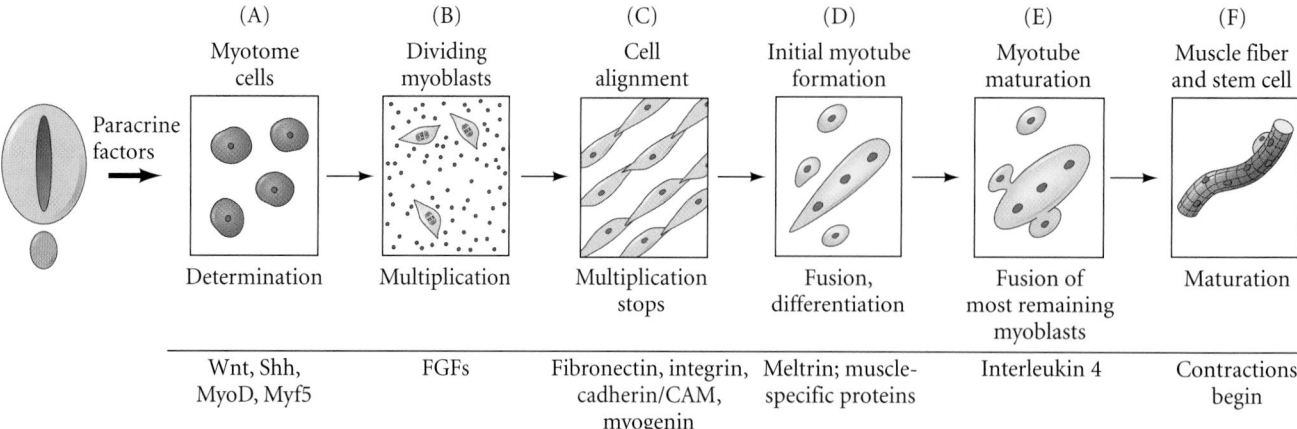

	(A) Myotome cells	(B) Dividing myoblasts	(C) Cell alignment	(D) Initial myotube formation	(E) Myotube maturation	(F) Muscle fiber and stem cell
	Determination	Multiplication	Multiplication stops	Fusion, differentiation	Fusion of most remaining myoblasts	Maturation
	Wnt, Shh, MyoD, Myf5	FGFs	Fibronectin, integrin, cadherin/CAM, myogenin	Meltrin; muscle- specific proteins	Interleukin 4	Contractions begin

Paracrine factors

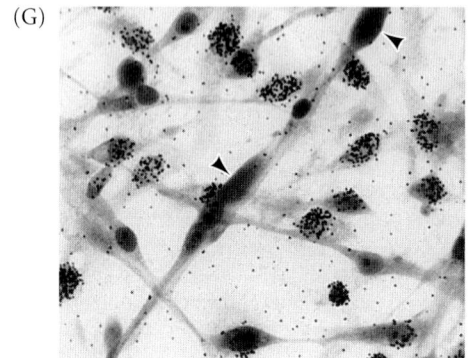

(G)

FIGURE 12.17 Conversion of myoblasts into muscles in culture. (A) Determination of myotome cells by paracrine factors. (B) Committed myoblasts divide in the presence of growth factors (primarily FGFs) but show no obvious muscle-specific proteins. (C–E) When the growth factors are used up, the myoblasts cease dividing, align, and fuse into myotubes. (F) The myotubes become organized into muscle fibers that spontaneously contract. (G) Autoradiograph showing DNA synthesis in myoblasts and the exit of fusing cells from the cell cycle. Phospholipase C can "freeze" myoblasts after they have aligned with other myoblasts but before their membranes fuse. Cultured myoblasts were treated with phospholipase C and then exposed to radioactive thymidine. Unattached myoblasts continued to divide and thus incorporated the radioactive thymidine into their DNA. Aligned (but not yet fused) cells (arrowheads) did not incorporate the label. (A–F after Wolpert 1998; G from Nameroff and Munar 1976, courtesy of M. Nameroff.)

et al. 1988; Abmayr and Pavlath 2012). After the first week, muscle cells can continue to grow by the fusion of satellite cells into existing myofiber and by an increase in contractile proteins within the myofibers.

MYOBLAST FUSION The first step in fusion requires the myoblasts to exit the cell cycle, and this is thought to be accomplished by Hedgehog signals (Osborn et al. 2011). Next, the myoblasts secrete fibronectin and other proteins onto their extracellular matrices and bind to it through α5β1 integrin, a major receptor for these extracellular matrix components (Menko and Boettiger 1987; Boettiger et al. 1995; Sunadome et al. 2011). If this adhesion is experimentally blocked, no further muscle development ensues, so it appears that the signal from the integrin-fibronectin attachment is critical for instructing myoblasts to differentiate into muscle cells (**FIGURE 12.17**).

The third step is the alignment of the myoblasts into chains. This step is mediated by cell membrane glycoproteins, including several cadherins (Knudsen 1985; Knudsen et al. 1990). Recognition and alignment between cells take place only if the cells are myoblasts. Fusion can occur even between chick and rat myoblasts in culture (Yaffe and Feldman 1965); the identity of the species is not critical. The internal cytoplasm is rearranged in preparation for the fusion, with actin regulating the regions of contact between the cells (Duan and Gallagher 2009).

The fourth step is the cell fusion event itself. As in most membrane fusions, calcium ions are critical, and fusion can

be activated by calcium transporters, such as A23187, that carry Ca²⁺ across cell membranes (Shainberg et al. 1969; David et al. 1981). Fusion appears to be mediated by a set of metalloproteinases called **meltrins**. Meltrins were discovered during a search for myoblast proteins that would be homologous to fertilin, a protein implicated in sperm-egg membrane fusion. Yagami-Hiromasa and colleagues (1995) found that one of these, meltrin-α, is expressed in myoblasts at about the same time that fusion begins, and that antisense RNA to the meltrin-α message inhibited fusion when added to myoblasts. As the myoblasts become capable of fusing, another myogenic regulatory factor—**myogenin**—becomes active. Myogenin binds to the regulatory region of several muscle-specific genes and activates their expression. Thus, whereas MyoD and Myf5 are active in the lineage specification of muscle cells, myogenin appears to mediate muscle cell differentiation (Bergstrom and Tapscott 2001).

Cell fusion ends with the re-sealing ("healing") of the newly apposed membranes. This is accomplished by proteins such as myoferlin and dysferlin, which stabilize the membrane phospholipids (Doherty et al. 2005). These proteins are similar to those that re-seal the membranes at axon nerve synapses after membrane vesicle fusion releases neurotransmitters.

MYOFIBER GROWTH After the original fusion of myoblasts into a myofiber, the myofiber secretes the paracrine factor interleukin 4 (IL4). Although IL4 was originally thought to work exclusively in the adult immune system, Horsely and

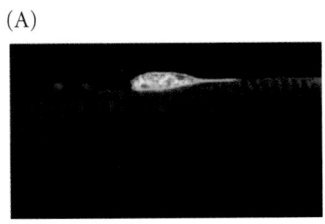 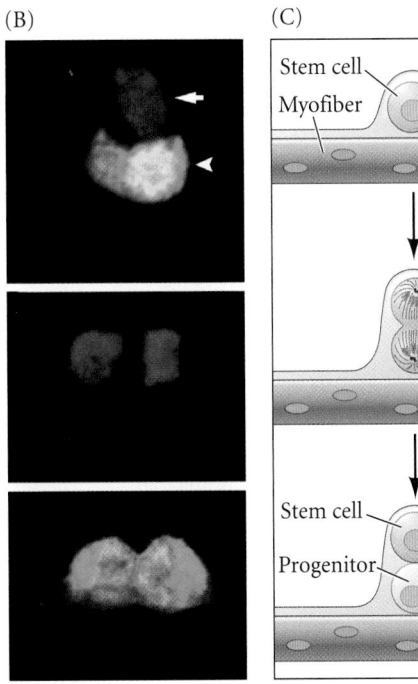 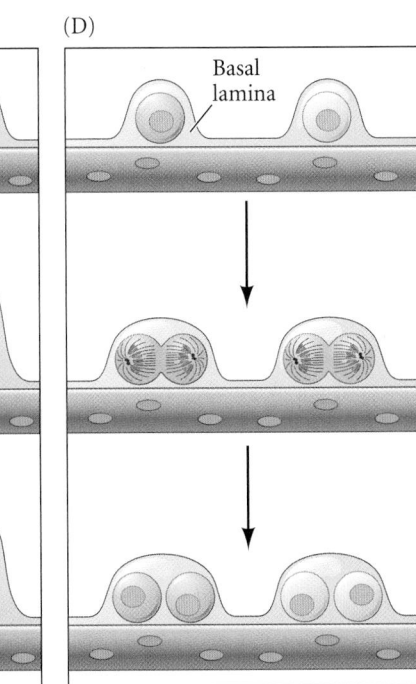

FIGURE 12.18 Satellite cells and muscle growth. (A) Satellite cells (stained with antibodies to the Pax7 protein) reside between the myofiber cell membrane and the basal lamina. (B) The top photograph shows asymmetric cell division of a satellite stem cell and the distinction between the daughter cell keeping Pax7 (stem cell; red) and the daughter cell downregulating Pax7 and expressing Myf5 (progenitor cell; green). This corresponds to the diagrammatic representation in (C). The bottom two photographs show symmetric division, where stem cells and progenitor cells make more stem and progenitor cells, respectively, as panel (D) shows in diagrammatic form. (After Bentzinger et al. 2012; photographs courtesy of F. Bentzinger and M. A. Rudnicki.)

colleagues (2003) found that IL4 secreted by new myofibers recruits other myoblasts to fuse with the tube, thereby forming the mature myofiber (see Figure 12.17).

The number of muscle fibers in the embryo and the growth of these fibers after birth appear to be negatively regulated by **myostatin**, a member of the TGF-β family (McPherron et al. 1997; Lee 2004). Myostatin is made by developing and adult skeletal muscle and most probably works in an autocrine fashion. As mentioned in Chapter 2, *myostatin* loss-of-function mutations allow both hyperplasia (more fibers) and hypertrophy (larger fibers) of the muscle. These changes give rise to Herculean phenotypes in dogs,* cattle, mice, and humans (see Figure 2.27).

Satellite cells: Unfused muscle progenitor cells

Any dancer, athlete, or sports fan knows that (1) adult muscles grow larger when they are exercised, and (2) muscles are capable of limited regeneration following injury. The growth and regeneration of muscles both arise from **satellite cells**, populations of stem cells and progenitor cells that reside alongside the adult muscle fibers. Satellite cells respond to injury or exercise by proliferating into myogenic cells that fuse and form new muscle fibers. Lineage tracing using chick-quail chimeras indicates that satellite cells are somite-derived myoblasts that have not fused and remain potentially available throughout adult life (Armand et al. 1983).

* A loss-of-function mutation in the *myostatin* gene has found its way into whippets bred for dog racing. In these dogs, the homozygous loss-of-function condition is not advantageous, but heterozygotes have more muscle power and are significantly overrepresented among the top racers (Mosher et al. 2007).

The source of mouse and chick satellite cells appears to be the central part of the dermamyotome (Ben-Yair and Kalcheim 2005; Gros et al. 2005; Kassar-Duchossoy et al. 2005; Relaix et al. 2005). While the myoblast-forming cells of the dermamyotome form at the lips and express Myf5 and MyoD, the cells that enter the myotome from the central region usually express Pax3 and Pax7 as well as microRNAs miRNA-489 and miRNA-31. The combination of Pax3 and Pax7 appears to inhibit MyoD expression (and thus muscle differentiation) in these cells; Pax7 also protects the satellite cells against apoptosis (Olguin and Olwin 2004; Kassar-Duchossoy et al. 2005; Buckingham et al. 2006). The two microRNAs appear to prevent the translation of factors such as Myf5 that would promote muscle cell differentiation (Cheung et al. 2012; Crist et al. 2012).

Recent experiments have shown that satellite cells are not a homogeneous population but contain both stem cells and progenitor cells. The stem cells represent only about 10% of the satellite cells and are found, with the other satellite cells, between the cell membrane and the extracellular basal lamina of mature myofibers. Satellite stem cells express Pax7 but not Myf5 (this is designated Pax7$^+$/Myf5$^-$) and can divide asynchronously to produce two types of cells: another Pax7$^+$/Myf5$^-$ stem cell and a Pax7$^+$/Myf5$^+$ satellite progenitor cell (**FIGURE 12.18**). This progenitor cell differentiates into muscle. The Pax7$^+$/Myf5$^-$ cells, when transplanted into other muscles, contribute to the stem cell population there (Kuang et al. 2007). The factor responsible for the asymmetry of this division appears to be miRNA-489, which is found in quiescent stem cells. Upon division, miRNA-489 remains in the daughter that remains a stem cell, but it is absent in the cell that becomes part of the muscle. MiRNA-489 inhibits

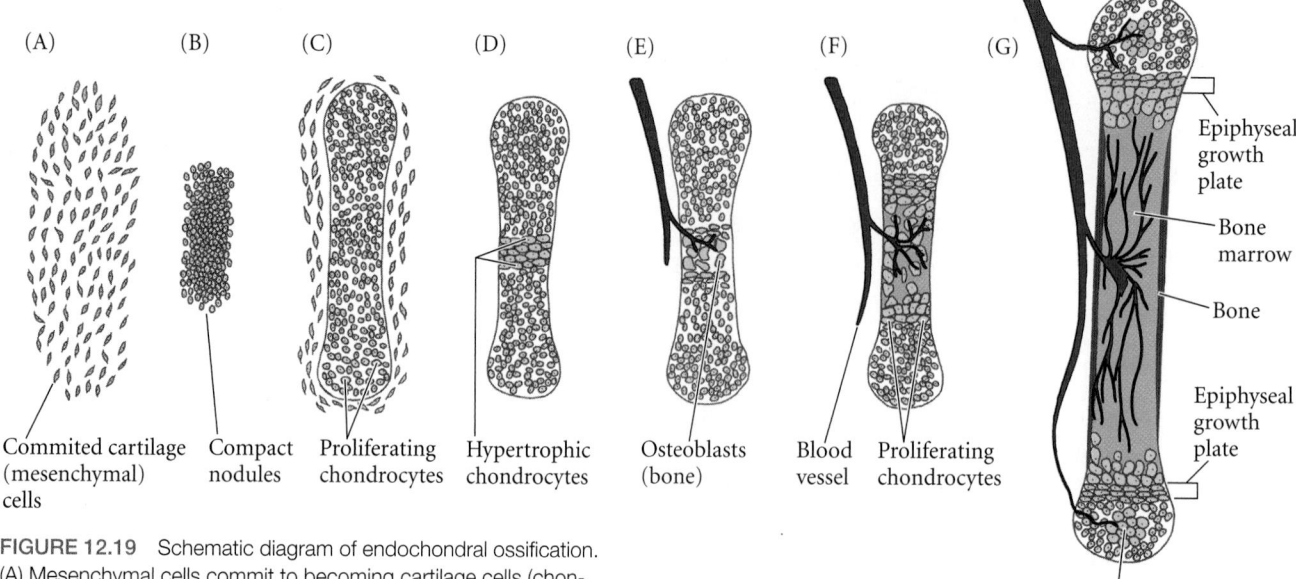

Commited cartilage (mesenchymal) cells | Compact nodules | Proliferating chondrocytes | Hypertrophic chondrocytes | Osteoblasts (bone) | Blood vessel | Proliferating chondrocytes

Epiphyseal growth plate
Bone marrow
Bone
Epiphyseal growth plate

Secondary ossification center

FIGURE 12.19 Schematic diagram of endochondral ossification. (A) Mesenchymal cells commit to becoming cartilage cells (chondrocytes). (B) Committed mesenchyme condenses into compact nodules. (C) Nodules differentiate into chondrocytes and proliferate to form the cartilage model of bone. (D) Chondrocytes undergo hypertrophy and apoptosis while they change and mineralize their extracellular matrix. (E) Apoptosis of chondrocytes allows blood vessels to enter. (F) Blood vessels bring in osteoblasts, which bind to the degenerating cartilaginous matrix and deposit bone matrix. (G) Bone formation and growth consist of ordered arrays of proliferating, hypertrophic, and mineralizing chondrocytes. Secondary ossification centers also form as blood vessels enter near the tips of the bone. (After Horton 1990.)

the translation of the message for Dek which becomes translated in the daughter cell that differentiates. Dek encodes a chromatin protein that promotes the transient proliferation of the progenitor cells (Cheung et al. 2012). Thus, miRNA-489 maintains the quiescent state of an adult muscle stem cell population.

● **See WEBSITE 12.2** Muscle formation

Osteogenesis: The Development of Bones

Three distinct lineages generate the skeleton. The paraxial mesoderm generates the vertebral and craniofacial bones, the lateral plate mesoderm generates the limb skeleton, and the cranial neural crest gives rise to some of the craniofacial bones and cartilage. There are two major modes of bone formation, or **osteogenesis**, and both involve the transformation of preexisting mesenchymal tissue into bone tissue. The direct conversion of mesenchyme into bone is called intramembranous ossification and was discussed in Chapter 11. In other cases, the mesenchymal cells differentiate into cartilage, which is later replaced by bone in a process called **endochondral ossification**.

Endochondral ossification

Endochondral ossification involves the formation of cartilage tissue from aggregated mesenchymal cells and the subsequent replacement of cartilage tissue by bone (Horton 1990). This is the type of bone formation characteristic of the vertebrae, ribs, and limbs. The vertebrae and ribs form from the somites, whereas the limb bones (to be discussed in Chapter 14) form from the lateral plate mesoderm.

The process of endochondral ossification can be divided into five stages, as seen in **FIGURE 12.19**.

PHASES 1 AND 2: COMMITMENT AND COMPACTION First, the mesenchymal cells commit to becoming cartilage (**FIGURE 12.19A**). This commitment is stimulated by Sonic hedgehog, which induces nearby sclerotome cells to express the Pax1 transcription factor (Cserjesi et al. 1995; Sosic et al. 1997). Pax1 initiates a cascade that is dependent on external paracrine factors and internal transcription factors.

During the second phase of endochondral ossification, the committed mesenchyme cells condense into compact nodules (**FIGURE 12.19B**). These inner cells become committed to generating cartilage, while the outer cells become committed to becoming bone. BMPs appear to be critical in this stage. They are responsible for inducing the expression of the adhesion molecules N-cadherin and N-CAM and the transcription factor Sox9. N-cadherin appears to be important in the initiation of these condensations, and N-CAM seems to be critical for maintaining them (Oberlender and Tuan 1994; Hall and Miyake 1995). Sox9 activates other transcription factors as well as a suite of genes, including those encoding collagen II and aggrecan, which are critical in cartilage function. In humans, mutations of the *SOX9* gene cause camptomelic dysplasia, a rare disorder of skeletal development that results in deformities of most of the bones of the body. Most affected babies die from respiratory failure

(A)

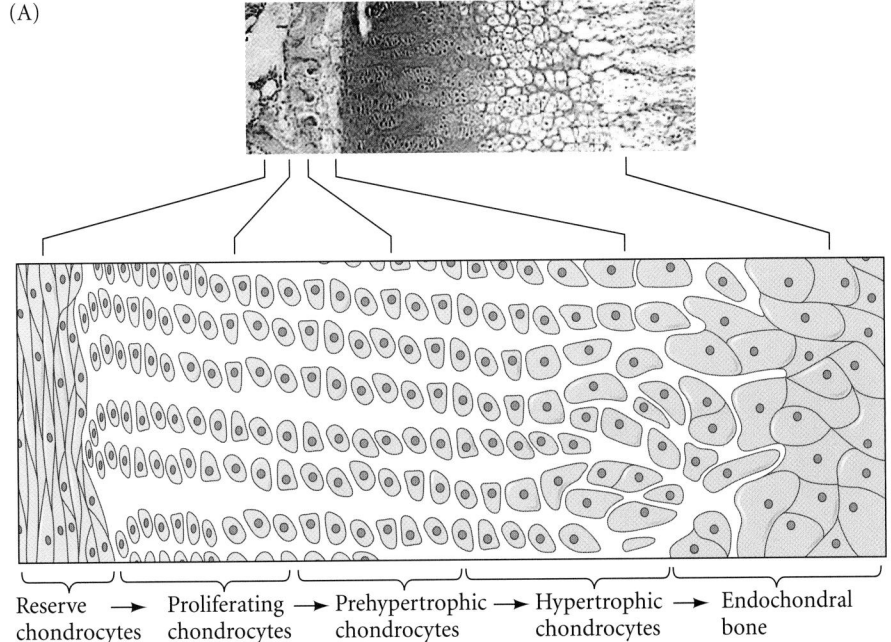

Reserve → Proliferating → Prehypertrophic → Hypertrophic → Endochondral
chondrocytes   chondrocytes   chondrocytes   chondrocytes   bone

FIGURE 12.20 Endochondral ossification. (A) Long bone undergoing endochondral ossification. The cartilage is stained with alcian blue, and the bone is stained with alizarin red. Below is a diagram of the transition zone wherein the chondrocytes (cartilage cells) divide, enlarge, die, and are replaced by osteocytes (mature bone cells). (B) Paracrine and transcription factors active in the transition of cartilage to bone. The mesenchymal sclerotome cell can become a chondrocyte (characterized by the Sox9 transcription factor) or an osteocyte (characterized by the Osterix transcription factor), depending on the types of paracrine factors it experiences. The paracrine factor Indian hedgehog (Ihh), secreted by the growing chondrocytes, appears to repress Twist, an inhibitor of Runx2. Runx2 is critical for directing cell fate into the bone pathway; it activates Osterix, which in turn activates the bone-specific proteins. (After Long 2012.)

(B)

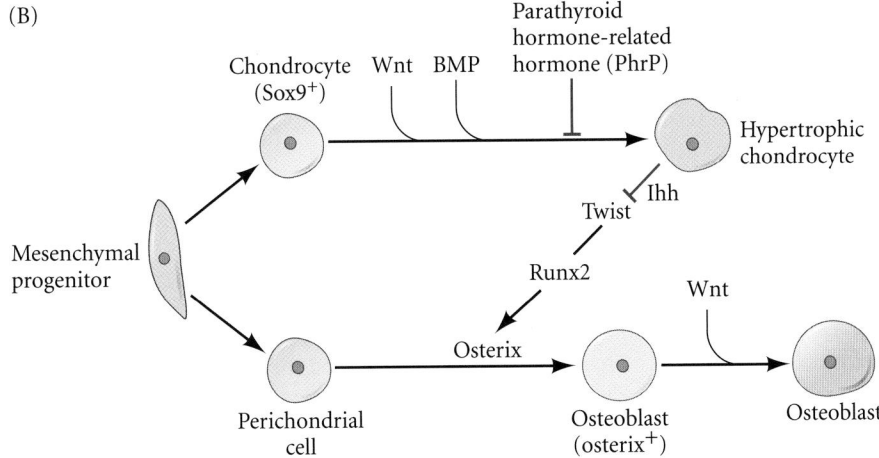

due to poorly formed tracheal and rib cartilage (Wright et al. 1995).

PHASES 3 AND 4: PROLIFERATION AND GROWTH During the third phase of endochondral ossification, the chondrocytes proliferate rapidly to form a cartilaginous model for the bone (**FIGURE 12.19C**). As they divide, the chondrocytes secrete a cartilage-specific extracellular matrix. The outermost cells become the **perichondrium** that ensheaths the cartilage.

In the fourth phase, the chondrocytes stop dividing and increase their volume dramatically, becoming **hypertrophic chondrocytes** (**FIGURE 12.19D**; **FIGURE 12.20**). This step appears to be mediated by the transcription factor Runx2 (also called CBFα1), which is necessary for the development of both intramembranous and endochondral bone. *Runx2* expression is regulated by histone deacetylase-4 (HDAC4), a form of chromatin-restructuring enzyme that is expressed solely in the prehypertrophic cartilage. If HDAC4 is

overexpressed in the cartilaginous ribs or limbs, ossification is seriously delayed; if the *Hdac4* gene is knocked out of the mouse genome, the limbs and ribs ossify prematurely (Vega et al. 2004). Hypertrophic cartilage is exceptionally important in regulating the final size of the long bone (see Figure 14.33). Indeed, the greatest contribution to the growth rate in mammals is the relative size of the hypertrophic cartilage (Cooper et al. 2013). The swelling of this cartilage determines the elongation rate of each skeletal element and is responsible for the differences in the growth rates between different skeletal elements both within an organism (hands versus legs, for instance) and between related organisms (the legs of a mouse versus the legs of a jerboa).

These large chondrocytes alter the matrix they produce (by adding collagen X and more fibronectin) to enable it to become mineralized (calcified) by calcium phosphate. These hypertrophic cartilage cells also secrete two factors that will be critical for the transformation of cartilage into bone. First,

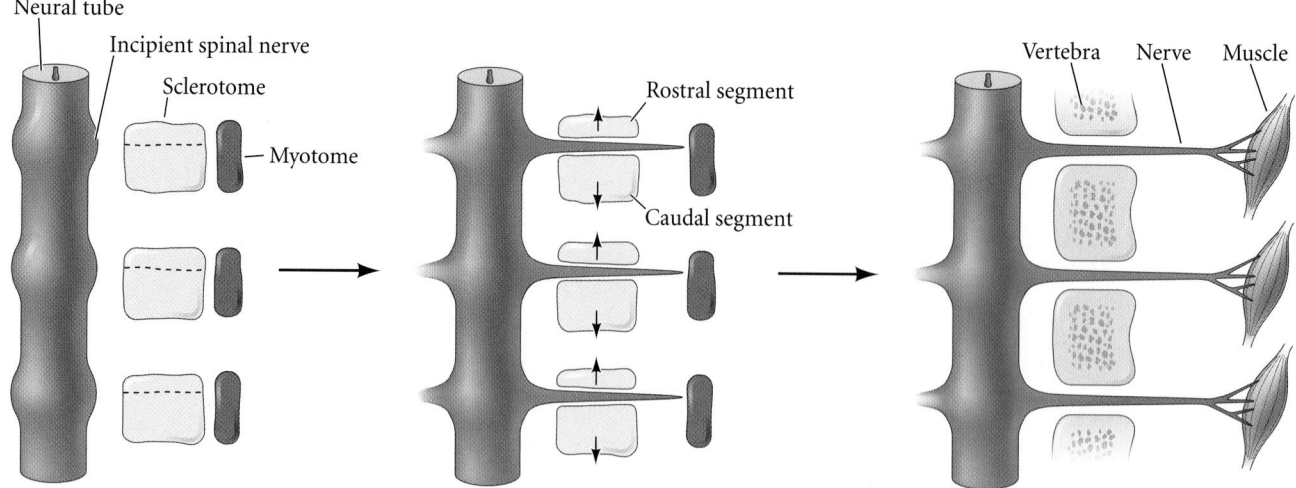

they secrete the angiogenesis factor VEGF, which can transform mesodermal mesenchyme cells into blood vessels (see Chapter 13; Gerber et al. 1999; Haigh et al. 2000). Second, they secrete Indian hedgehog, a member of the Hedgehog family (a close cousin of Sonic) that activates *Runx2* transcription in the perichondrial cells surrounding the cartilage primordium. This initiates the differentiation of those cells into bone-forming **osteoblasts**. Mice lacking the *Indian hedgehog* gene completely lack the osteoblasts of the endochondral skeleton (trunk and limbs), although the osteoblasts formed in the head and face by intramembranous ossification form normally (St-Jacques et al. 1999).

PHASE 5: CHONDROCYTE DEATH AND BONE CELL GENERATION   In the fifth phase, the hypertrophic chondrocytes die by apoptosis (Hatori et al. 1995; Rajpurohit et al. 1999). The hypertrophic cartilage is replaced by bone cells both on the outside and inside, and blood vessels invade the cartilage model (**FIGURE 12.19E–G**). On the *outside*, the osteoblasts begin forming bone matrix, constructing a bone collar around the calcified and partially degraded cartilage matrix (Bruder and Caplan 1989; Hatori et al. 1995; St-Jacques et al. 1999). The osteoblasts become responsive to Wnt signals that upregulate Osterix, a transcription factor that instructs the osteoblasts to become mature bone cells, or **osteocytes** (Nakashima et al. 2002; Hu et al. 2005).

New bone material is added peripherally from the *internal* surface of the **periosteum**, a fibrous sheath covering the developing bone. The periosteum contains connective tissue, capillaries, and bone progenitor cells (Long et al. 2012). At the same time, there is a hollowing out of the internal region of the bone to form the bone marrow cavity. As cartilage cells die, they alter the extracellular matrix, releasing the VEGF. VEGF stimulates the formation of blood vessels around the dying cartilage. If the blood vessels are inhibited from forming, bone development is significantly delayed (Karsenty and Wagner 2002; Yin et al. 2002). The blood vessels bring in both osteoblasts and **osteoclasts**—multinucleated cells that eat the debris of the apoptotic chondrocytes, thus creating the hollow bone marrow cavity (Kahn and Simmons 1975;

FIGURE 12.21   Re-specification of the sclerotome to form vertebrae. Each sclerotome splits into a rostral and a caudal segment. As the spinal neurons grow outward to innervate the muscles from the myotome, the rostral segment of each sclerotome combines with the caudal segment of the next anterior sclerotome to form a vertebral rudiment. (After Larson 1998.)

Manolagas and Jilka 1995). Osteoclasts are not derived from the somite; rather, they are derived from a blood cell lineage (in the lateral plate mesoderm) and come from the same precursors as macrophage blood cells (Ash et al. 1980; Blair et al. 1986).

● **See WEBSITE 12.3** Paracrine factors, their receptors, and human bone growth

Vertebrae formation

The notochord appears to induce its surrounding mesenchyme cells to secrete epimorphin. Epimorphin then attracts sclerotome cells to the region around the notochord and neural tube, where they begin to condense and differentiate into cartilage (Oka et al. 2006). However, before the sclerotome-derived cells form a vertebra, each sclerotome must split into a rostral (anterior) and a caudal segment (**FIGURE 12.21**). As motor neurons from the neural tube grow laterally to innervate the newly forming muscles, the rostral segment of each sclerotome recombines with the caudal segment of the next anterior sclerotome to form the vertebral rudiment (Remak 1850; Aoyama and Asamoto 2000; Morin-Kensicki et al. 2002). This **resegmentation** enables the muscles to coordinate the movement of the skeleton, permitting the body to move laterally.

The resegmentation of the sclerotome to allow coordinated movement is reminiscent of the strategy used by insects when constructing segments out of parasegments (see Chapter 6). The bending and twisting movements of the spine are permitted by the intervertebral (synovial) joints that form from the arthrotome region of the sclerotome. Removal of sclerotome cells from the arthrotome leads to the failure of synovial joints to form and to the fusion of adjacent vertebrae (Mittapalli et al. 2005).

SIDELIGHTS & SPECULATIONS

Mechanotransduction in the Musculoskeletal System

The ability of cells to sense their environment and convert mechanical forces into molecular signals is called **mechanotransduction**, and the importance of mechanotransduction to development is just beginning to be recognized. We saw this importance in the discussion of how extracellular mechanical signals changed the differentiation of stem cells, and mechanical forces appear to be significant in the formation of bones, muscles, and tendons, and perhaps also for their repair and regeneration in the adult. However, very little is known about how mechanical stress is sensed, quantified, and transmitted as a change in cytoplasmic chemicals.

Exercise and increased muscle mass

We know that physical forces generated by exercise cause muscles to enlarge. Exercise stimulates protein synthesis in the muscle cells, and each nucleus in the multinucleate fiber appears to have a region around it where protein synthesis is regulated (Lai et al. 2004; Quaisar et al. 2012). If physical stress continues, the force appears to cause the muscle satellite cells to proliferate and to fuse with the existing muscle fibers. Indeed, endurance exercise has been shown to increase the number of satellite cells in the elderly (Shefer et al. 2010). Insulin-like growth factor, acting as an autocrine secretion from muscle cells, is a candidate for causing such muscle growth (Yang 1996; Goldspink 2004; Sculthorpe et al. 2012), but how this factor or any other is induced by stress remains unknown.

Also, in a way that is not yet understood, the tension produced by weight-bearing loads activates production of TGF-β2 and 3 in the tendon cells (Maeda et al. 2011). Indeed, mice lacking these genes completely lack tendons. The TGF-β pathway (through the Smad2/3 transcription factors) continues to activate the gene for the transcription factor scleraxis after the initial FGF signaling; in turn, scleraxis

activates the genes responsible for forming the extracellular matrix. Moreover, TGF-β produced by the developing tendon may recruit cells from the cartilage and muscle to make the bridge between these three tissues (Blitz et al. 2009; Pryce et al. 2009).

Vertebrate bone development

Vertebrate skeletal bone development also depends on mechanotransduction. Tension and stress forces activate the gene for Indian hedgehog (Ihh), a paracrine factor that activates the bone morphogenetic proteins (BMPs; Wu et al. 2001). In the chick, several bones do not form if embryonic movement in the egg is suppressed. One of these bones is the fibular crest, which connects the tibia directly to the fibula (FIGURE 12.22A,B). This direct connection is believed to be important in the evolution of birds, and the fibular crest is a universal feature of the bird hindlimb (Müller and Steicher 1989; Müller 2003).

The jaws of cichlid fish differ enormously, depending on the food they eat (FIGURE 12.22C; Meyer 1987). Similarly, normal primate jaw development may be predicated on how much tension is produced by grinding food: mechanical tension appears to stimulate *Indian hedgehog* expression in mammalian mandibular cartilage (Tang et al. 2004). If an infant monkey is given soft food, its lower jaw is smaller than normal. Corruccini and Beecher (1982, 1984) and Varrela (1992) have shown that people in cultures where infants are fed hard food have jaws that "fit" better, and these researchers speculate that soft baby food may explain why so many children in Western societies need braces. Indeed, the notion that mechanical tension can change jaw size and shape is the basis of the functional hypothesis of modern orthodontics (Moss 1962, 1997).

In mammals, muscle force within the embryo is critical for the normal shaping of bone and the development of load-bearing capacity (Sharir et al. 2011). After birth, the patella (knee cap) is formed by pressure on

the skeleton, and it is thought that the aberrant skeletal development seen in persons with cerebral palsy is caused by the absence of pressure on these bones (Shefelbine and Carter 2004; Ward et al. 2006).

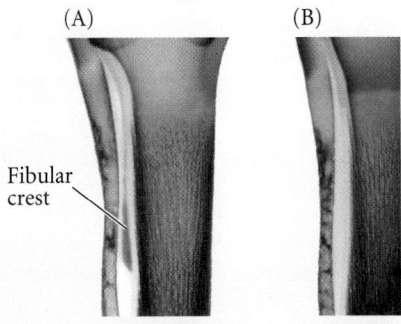

Fibular crest

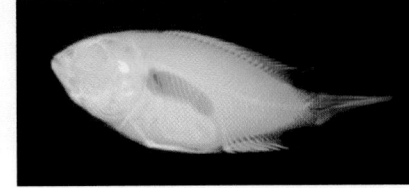

FIGURE 12.22 Phenotypes can be produced by stress force on muscular and skeletal tissues. (A,B) The avian fibular crest (syndesmosis tibiofibularis) connects the tibia directly to the fibula. The fibular crest is formed when the movement of the active embryo inside the egg puts physical stress on the tibia. (A) Fibular crest forming in the connective tissue of a 13-day chick embryo. (B) Absence of fibular crest in the connective tissue of a 13-day embryo whose movement was inhibited. The blue dye stains cartilage; the red dye stains the bone elements. (C) The jaws of cichlid fish are shaped by the hardness of the food they eat. Different diets give different jaw structures. (A,B from Müller 2003, courtesy of G. Müller; C from Meyer 1987, courtesy of A. Meyer.)

(A)

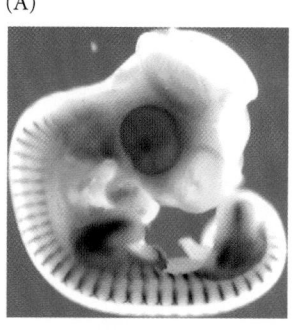

(B)

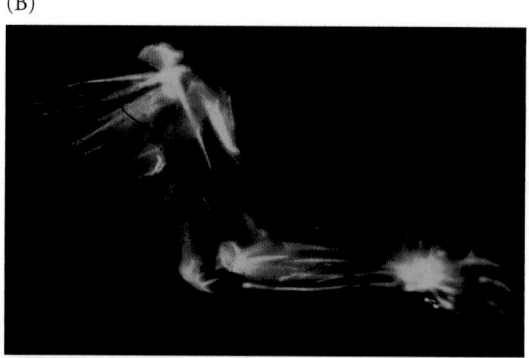

(C)

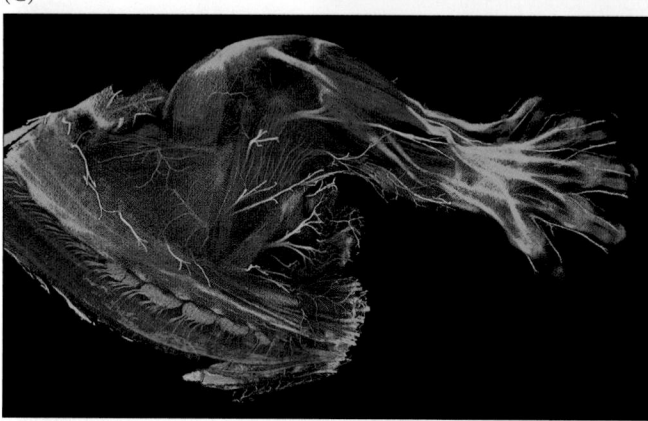

FIGURE 12.23 Scleraxis is expressed in the progenitors of the tendons. (A) In situ hybridization showing scleraxis pattern in the developing chick embryo. (B) Areas of scleraxis expression in the tendons of a newborn mouse forelimb (ventral view). The *GFP* gene had been fused on the scleraxis enhancer. (C) Wrist and digit of a newborn mouse, showing scleraxis (GFP) in the tendons (green) connecting muscles (stained red with antibodies to myosin) to the digit and wrist. The neurons have been stained blue by antibodies to neurofilament proteins. (A,B from Schweitzer et al. 2001, courtesy of R. Schweitzer; C courtesy of A. K. Lewis and G. Kardon.)

Tendon Formation: The Syndetome

The most dorsal part of the sclerotome will become the fourth compartment of the somite, the **syndetome**. The tendon-forming cells of the syndetome can be visualized by their expression of the *scleraxis* gene (**FIGURE 12.23**; Schweitzer et al. 2001; Brent et al. 2003). Because there is no obvious morphological distinction between the sclerotome and syndetome cells (they are both mesenchymal), our knowledge of this somitic compartment had to wait until we had molecular markers (Pax1 for the sclerotome, scleraxis for the syndetome) that could distinguish between them and allow researchers to follow the cells' fates.

Because the tendons connect muscles to bones, it is not surprising that the syndetome (Greek *syn*, "connected") is derived from the most dorsal portion of the sclerotome—that is, from sclerotome cells adjacent to the muscle-forming myotome (**FIGURE 12.24A**). The syndetome is made from the myotome's secretion of Fgf8 onto the immediately subjacent row of sclerotome cells (Brent et al. 2003; Brent and Tabin 2004). Other transcription factors limit the expression of scleraxis to the anterior and posterior portions of the syndetome, causing two stripes of scleraxis expression. Meanwhile, the developing cartilage cells, under the influence of Sonic hedgehog from the notochord and floorplate, synthesize Sox5 and Sox6—transcription factors that block scleraxis transcription while activating the cartilage-promoting factor Sox9

(**FIGURE 12.24B**; Yamashita et al. 2012). In this way, the cartilage protects itself from any spread of the Fgf8 signal. The tendons then associate with the muscles directly above them and with the skeleton (including the ribs) on either side of them (**FIGURE 12.24C**; Brent et al. 2005).

Formation of the Dorsal Aorta

Most of the circulatory system of an early-stage amniote embryo is directed outside the embryo, its job being to obtain nutrients from the yolk or placenta. The *intraembryonic* circulatory system begins with the formation of the dorsal aorta. The dorsal aorta is composed of two cell layers: an internal lining of endothelial cells surrounded concentrically by a layer of smooth muscle cells. Elsewhere in the body, these two layers of blood vessels are usually derived from the lateral plate mesoderm, as will be detailed in Chapter 13. However, the posterior sclerotome provides the endothelial cells and smooth muscle cells for the dorsal aorta and intervertebral blood vessels (Pardanaud et al. 1996; Wiegreffe et al. 2007). The presumptive endothelial cells are induced by Notch signaling and are instructed to migrate ventrally by a chemoattractant that has been made by the primary dorsal aorta, a transitory structure made by the lateral plate mesoderm. Eventually, the endothelial cells from the sclerotome replace the cells of the primary dorsal aorta (which will become part of the blood stem cell population) (**FIGURE 12.25**; Pouget et al. 2008; Sato et al. 2008; Ohata et al. 2009).

INTERMEDIATE MESODERM: THE UROGENITAL SYSTEM

The intermediate mesoderm generates the urogenital system—the kidneys, the gonads, and their respective duct systems. Reserving the gonads for our discussion of sex determination in Chapter 15, we will concentrate here on the development of the mammalian kidney.

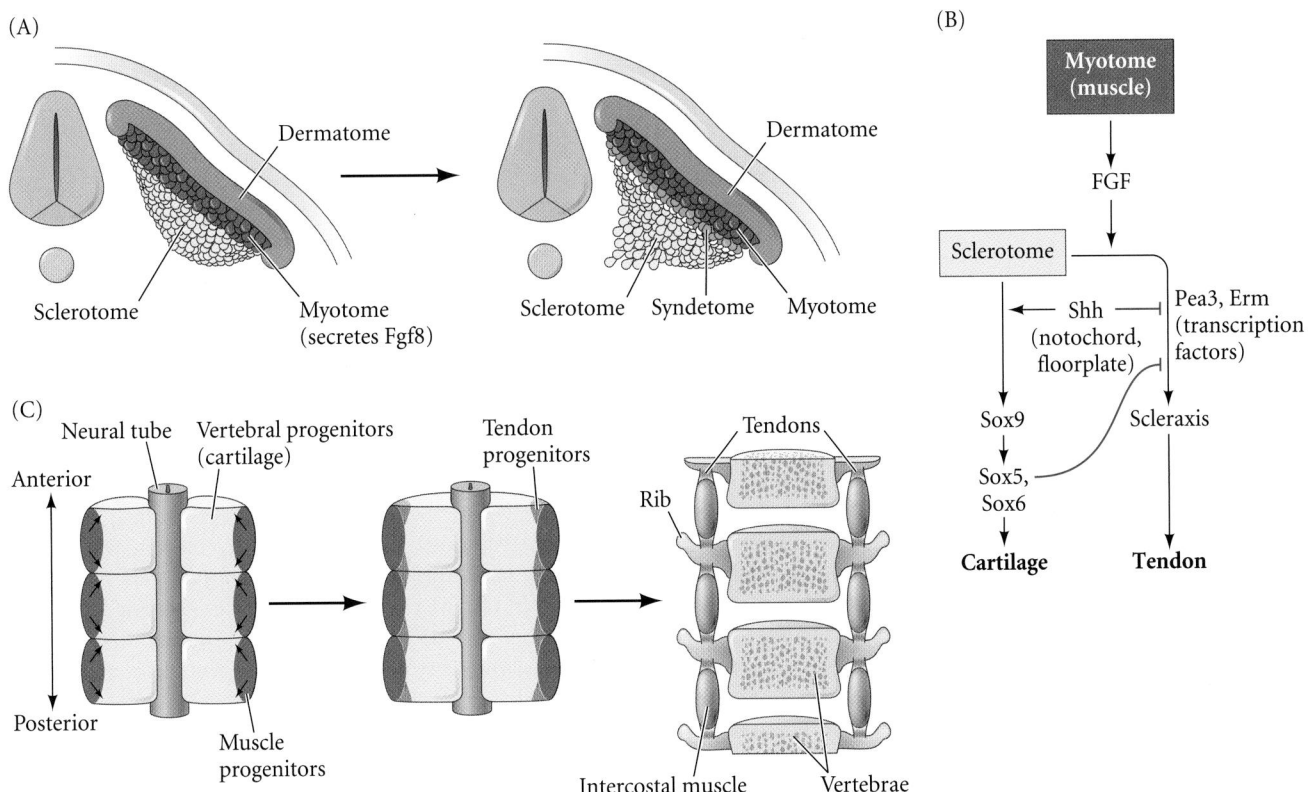

FIGURE 12.24 Induction of scleraxis in the chick sclerotome by Fgf8 from the myotome. (A) The dermatome, myotome, and sclerotome are established before the tendon precursors are specified. Tendon precursors (syndetome) are specified in the dorsalmost tier of sclerotome cells by Fgf8 received from the myotome. (B) Pathway by which Fgf8 signals from the muscle precursor cells induce the subjacent sclerotome cells to become tendons. (C) Syndetome cells migrate (arrows) along the developing vertebrae. They differentiate into tendons that connect the ribs to the intercostal muscles beloved by devotees of spareribs. (A,C after Brent et al. 2003.)

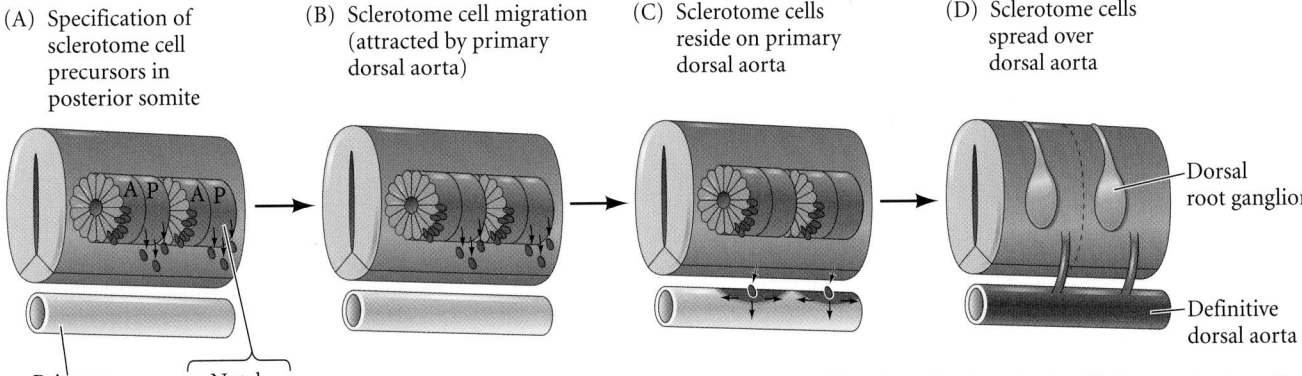

FIGURE 12.25 Model showing contribution of somitic cells to the dorsal aorta. (A) At an early stage, the primary dorsal aorta is of lateral plate origin (pink). A subpopulation of sclerotome cells becomes specified by Notch in the posterior half of somites as endothelial precursors. (B) Chemoattractants made in the primary dorsal aorta cause these cells to migrate through the somite to the aorta. (C) The sclerotome cells take up residence in the dorsal region of the vessel. (D) These cells then spread along both the anterior-posterior and dorsal-ventral axes, ultimately occupying the entire region of the aorta. The primary aortic endothelial cells become blood cell precursors. (After Sato et al. 2008.)

The Progression of Kidney Types

The physiologist and philosopher of science Homer Smith noted in 1953 that "our kidneys constitute the major foundation of our philosophical freedom. Only because they work the way they do has it become possible for us to have bone, muscles, glands, and brains." While this statement may smack of hyperbole, the human kidney is a remarkably intricate organ whose importance cannot be overestimated. Its functional unit, the **nephron**, contains more than 10,000 cells and at least 12 different cell types, each cell type having a specific function and being located in a particular place in relation to the others along the length of the nephron.

Mammalian kidney development progresses through three major stages. The first two stages are transient; only the third and last persists as a functional kidney. Early in development (day 22 in humans; day 8 in mice), the **pronephric duct** arises in the intermediate mesoderm just ventral to the anterior somites. The cells of this duct migrate caudally, and the anterior region of the duct induces the adjacent mesenchyme to form the **pronephros**, or tubules of the initial kidney (FIGURE 12.26A). The pronephric tubules form functioning kidneys in fish and in amphibian larvae, but they are not believed to be active in amniotes. In mammals, the pronephric tubules and the anterior portion of the pronephric duct

degenerate, but the more caudal portions of the pronephric duct persist and serve as the central component of the excretory system throughout development (Toivonen 1945; Saxén 1987). This remaining duct is often referred to as the **nephric**, or **Wolffian**, **duct**.

As the pronephric tubules degenerate, the middle portion of the nephric duct induces a new set of kidney tubules in the adjacent mesenchyme. This set of tubules constitutes the **mesonephros**, or mesonephric kidney (FIGURE 12.26B; Sainio and Raatikainen-Ahokas 1999). In some mammalian species, the mesonephros functions briefly in urine filtration, but in mice and rats it does not function as a working kidney. In humans, about 30 mesonephric tubules form, beginning around day 25. As more tubules are induced caudally, the anterior mesonephric tubules begin to regress through apoptosis (although in mice, the anterior tubules remain while the posterior ones regress; FIGURE 12.26C,D). While it remains unknown whether the human mesonephros actually filters blood and makes urine, it does provide important developmental functions during its brief existence. First, as we will see in Chapter 13, it is one of the main sources of the hematopoietic stem cells necessary for blood cell development (Medvinsky and Dzierzak 1996; Wintour et al. 1996). Second, in male mammals, some of the mesonephric tubules persist to become the tubes that transport the sperm from

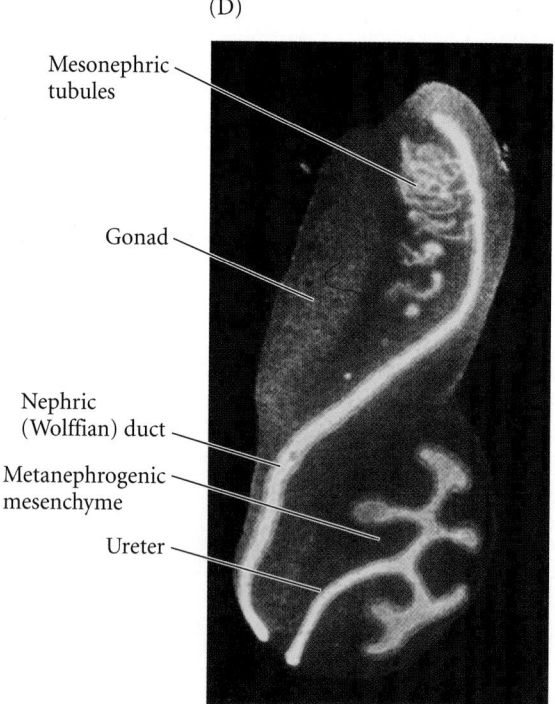

FIGURE 12.26 General scheme of development in the vertebrate kidney. (A) The original tubules, constituting the pronephros, are induced from the nephrogenic mesenchyme by the pronephric duct as it migrates caudally. (B) As the pronephros degenerates, the mesonephric tubules form. (C) The final mammalian kidney, the metanephros, is induced by the ureteric bud, which branches from

the nephric duct. (D) Intermediate mesoderm of a 13-day mouse embryo showing initiation of the metanephric kidney (bottom) while the mesonephros is still apparent. The duct tissue is stained with a fluorescent antibody to a cytokeratin found in the pronephric duct and its derivatives. (A–C after Saxén 1987; D courtesy of S. Vainio.)

(A)

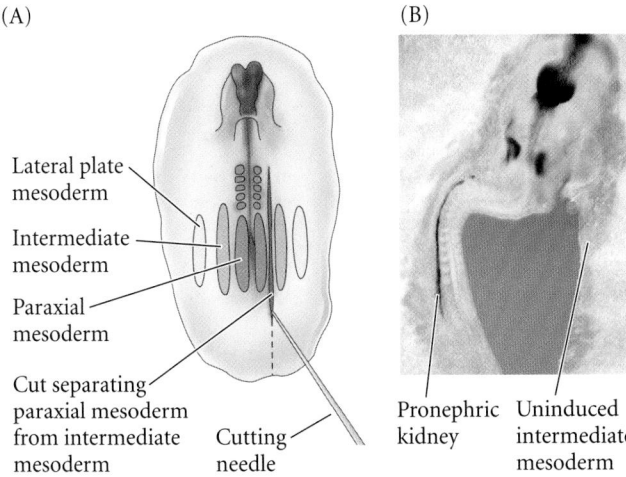

Lateral plate mesoderm

Intermediate mesoderm

Paraxial mesoderm

Cut separating paraxial mesoderm from intermediate mesoderm

Cutting needle

(B)

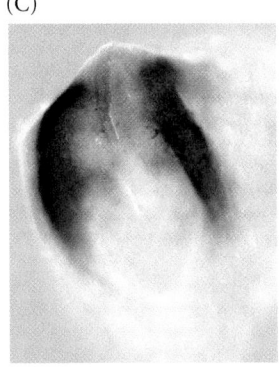

Pronephric kidney   Uninduced intermediate mesoderm

(C)

FIGURE 12.27   Signals from the paraxial mesoderm induce pronephros formation in the intermediate mesoderm of the chick embryo. (A) The paraxial mesoderm was surgically separated from the intermediate mesoderm on the right side of the body. (B) As a result, a pronephric kidney (Pax2-staining duct) developed only on the left side. (C) Lim1 expression in an 8-day mouse embryo, showing the prospective intermediate mesoderm. (A,B after Mauch et al. 2000; B courtesy of T. J. Mauch and G. C. Schoenwolf; C courtesy of K. Sainio and M. Hytönen.)

the testes to the urethra (the epididymis and vas deferens; see Chapter 15).

The permanent kidney of amniotes, the **metanephros**, is generated by some of the same components as the earlier, transient kidney types (see Figure 12.26C). It is thought to originate through a complex set of interactions between epithelial and mesenchymal components of the intermediate mesoderm (reviewed in Costantini and Kopan 2010). In the first steps, the **metanephrogenic mesenchyme** is committed and forms in the posterior regions of the intermediate mesoderm, where it induces the formation of a branch from each of the paired nephric ducts. These epithelial branches are called the **ureteric buds**. These buds eventually grow out from the nephric duct to become the collecting ducts and ureters that take the urine to the bladder. When the ureteric buds emerge from the nephric duct, they enter the metanephrogenic mesenchyme. The ureteric buds induce this mesenchymal tissue to condense around them and to differentiate into the nephrons of the mammalian kidney. As this mesenchyme differentiates, it tells the ureteric bud to branch and grow.

Specification of the Intermediate Mesoderm: Pax2/8 and Lim1

The intermediate mesoderm of the chick embryo acquires its ability to form kidneys through its interactions with the paraxial mesoderm. While its bias to become intermediate mesoderm is probably established through a BMP gradient, specification appears to become stabilized through signals from the paraxial mesoderm. Mauch and her colleagues (2000) showed that signals from the paraxial mesoderm induced primitive kidney formation in the intermediate mesoderm of the chick embryo. They cut developing embryos such that the intermediate mesoderm could not contact the paraxial mesoderm on one side of the body. That side of the body (where contact with the paraxial mesoderm was abolished) did not form kidneys, but the undisturbed side was able to form kidneys (**FIGURE 12.27A,B**). The paraxial mesoderm appears to

be both necessary and sufficient for inducing kidney-forming ability in the intermediate mesoderm, since co-culturing lateral plate mesoderm with paraxial mesoderm causes pronephric tubules to form in the lateral plate mesoderm, and no other cell type can accomplish this.

These interactions induce the expression of a set of homeodomain transcription factors—including Lim1 (sometimes called Lhx1), Pax2, and Pax8—that cause the intermediate mesoderm to form the kidney (**FIGURE 12.27C**; Karavanov et al. 1998; Kobayashi et al. 2005; Cirio et al. 2011). In *Xenopus*, Pax8 and Lim1 have overlapping boundaries, and kidney development originates from cells that express both. Ectopic co-expression of Pax8 and Lim1 will produce kidneys in other tissues as well (Carroll and Vize 1999).

In the chick embryo, Pax2 and Lim1 are expressed in the intermediate mesoderm, starting at the level of the sixth somite (i.e., only in the trunk, not in the head). If Pax2 is experimentally induced in the presomitic mesoderm, it converts that paraxial mesoderm into intermediate mesoderm, causing it to express Lim1 and form kidneys (Mauch et al. 2000; Suetsugu et al. 2005). Similarly, in mouse embryos with knockouts of both the *Pax2* and *Pax8* genes, the mesenchyme-epithelial transition necessary to form the kidney duct fails, the cells undergo apoptosis, and no nephric structures form (Bouchard et al. 2002). Moreover, in the mouse, Lim1 and Pax2 proteins appear to induce one another.

Lim1 plays several roles in the formation of the mouse kidney. First it is needed for converting the intermediate mesenchyme into the kidney duct (Tsang et al. 2000), and later it is required for the formation of the ureteric bud and the tubular structure both in mesonephric and metanephric mesenchyme (Shawlot and Behringer 1995; Karavanov et al. 1998; Kobayashi et al. 2005).

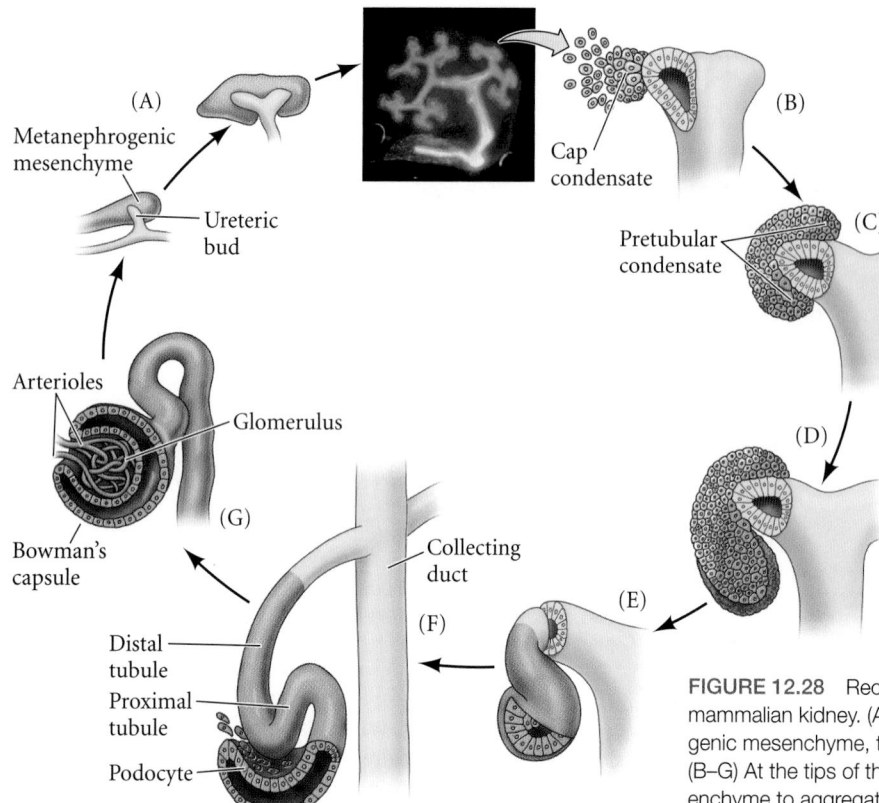

FIGURE 12.28 Reciprocal induction in the development of the mammalian kidney. (A) As the ureteric bud enters the metanephrogenic mesenchyme, the mesenchyme induces the bud to branch. (B–G) At the tips of the branches, the epithelium induces the mesenchyme to aggregate and cavitate to form the renal tubules and glomeruli (where the blood from the arteriole is filtered). When the mesenchyme has condensed into an epithelium, it digests the basal lamina of the ureteric bud cells that induced it and connects to the ureteric bud epithelium. A portion of the aggregated mesenchyme (the pretubular condensate) becomes the nephron (renal tubules and Bowman's capsule), while the ureteric bud becomes the collecting duct for the urine. (After Saxén 1987; Sariola 2002.)

The anterior border of the Lim1- and Pax2-expressing cells appears to be established by the cells above a certain region losing their competence to respond to activin, a TGF-βΠ family paracrine factor secreted by the neural tube. This competence is established by the transcription factor Hoxb4, which is not expressed in the anteriormost region of the intermediate mesoderm. The anterior boundary of Hoxb4 is established by a retinoic acid gradient, and adding activin locally will allow the kidney to extend anteriorly (Barak et al. 2005; Preger-Ben Noon et al. 2009).

Reciprocal Interactions of Developing Kidney Tissues

The two intermediate mesodermal tissues—the ureteric bud and the metanephrogenic mesenchyme—interact and reciprocally induce each other to form the kidney (FIGURE 12.28). The metanephrogenic mesenchyme causes the ureteric bud to elongate and branch. The tips of these branches induce the loose mesenchyme cells to form epithelial aggregates. Each aggregated nodule of about 20 cells proliferates and differentiates into the intricate structure of a renal nephron. Each nodule first elongates into a "comma" shape, then forms a characteristic S-shaped tube. Soon afterward, the cells of this epithelial structure begin to differentiate into regionally specific cell types, including the capsule cells, the podocytes, and the proximal and distal tubule cells. While

this transformation is happening, the cells of the S-shaped nodule closest to the ureter ducts break down the basal lamina of the ureteric bud ducts and migrate into the duct region.* This creates an open connection between the ureteric bud and the newly formed tubule, allowing material to pass from one into the other (Bard et al. 2001; Kao et al. 2012). These tubules derived from the mesenchyme form the secretory nephrons of the functioning kidney, and the branched ureteric bud gives rise to the renal collecting ducts and to the ureter, which drains the urine from the kidney.

Clifford Grobstein (1955, 1956) documented this reciprocal induction in vitro. He separated the ureteric bud from the metanephrogenic mesenchyme and cultured them

---

*The intricate coordination of nephron development with the blood capillaries that the nephrons filter is accomplished by the secretion of VEGF from the podocytes. VEGF, as we will see in Chapter 13, is a powerful inducer of blood vessels, causing endothelial cells from the dorsal aorta to form the capillary loops of the glomerular filtration apparatus (Aitkenhead et al. 1998; Klanke et al. 2002).

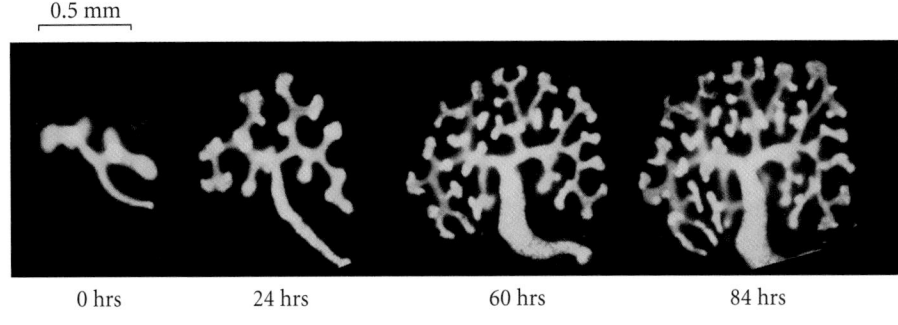

0.5 mm

0 hrs   24 hrs   60 hrs   84 hrs

FIGURE 12.29 Kidney induction observed in vitro. (A) A kidney rudiment from an 11.5-day mouse embryo was placed into culture. This transgenic mouse had a *GFP* gene fused to a Hoxb7 promoter, so it expressed green fluorescent protein in the nephric (Wolffian) duct and in the ureteric buds. Since GFP can be photographed in living tissues, the kidney could be followed as it developed. (Srinivas et al. 1999, courtesy of F. Costantini.)

either individually or together. In the absence of mesenchyme, the ureteric bud does not branch. In the absence of the ureteric bud, the mesenchyme soon dies. But when they are placed together, the ureteric bud grows and branches, and nephrons form throughout the mesenchyme. This has been confirmed by experiments using GFP-labeled proteins to monitor cell division and branching (**FIGURE 12.29**; Srinvas et al. 1999).

Mechanisms of reciprocal induction

The induction of the metanephros can be viewed as a dialogue between the ureteric bud and the metanephrogenic mesenchyme. As the dialogue continues, both tissues are altered. We will eavesdrop on this dialogue more intently than we have done for other organs, in part because the kidney has become a model for organogenesis. There are many reasons for this. First, early kidney development has only two major components. Second, the identity and roles of many of the paracrine and transcription factors produced during this dialogue have been discovered from studies of knockout mice. Third, the absence of many of these transcription factors is associated with serious pathologies characterized by absent or rudimentary kidneys. While there are several simultaneous dialogues between different groups of kidney cells, there appear to be at least nine critical sets of signals operating in the reciprocal induction of the metanephros (see Costantini 2012).

STEP 1: FORMATION OF METANEPHROGENIC MESENCHYME: FOXES, HOXES, AND WT1   Only the metanephrogenic mesenchyme has the competence to respond to the ureteric bud and form kidney tubules; indeed, this mesenchyme cannot become any tissue other than renal tissues (nephrons and their supporting stroma). If induced by non-nephric tissues (such as embryonic salivary gland or neural tube tissue), metanephric mesenchyme responds by forming renal tissue and no other structures (Saxén 1970; Sariola et al. 1982).

The positional specification of the metanephrogenic mesenchyme is positively regulated by Hoxb4 and negatively regulated by the transcription factors Foxc1 and Foxc2. *Foxc1⁻/2⁻* double-mutant mice have an expanded metanephric area that induces extra ureters and kidneys (Kume et al.

2000). Next, the permanent kidney-forming metanephrogenic mesenchyme is specified by genes of the Hox11 paralogue group. When Hox11 genes are knocked out of mouse embryos, mesenchymal differentiation is arrested and the mesenchyme cannot induce the ureteric bud to form (Patterson et al. 2001; Wellik et al. 2002). The competence to respond to ureteric bud inducers is regulated by a tumor suppressor gene, *WT1*. Without *WT1*, the metanephric mesenchyme cells remain uninduced and die (Kreidberg et al. 1993). In situ hybridization shows that, normally, *WT1* is first expressed in the intermediate mesoderm prior to kidney formation and is then expressed in the developing kidney, gonad, and mesothelium (Pritchard-Jones et al. 1990; van Heyningen et al. 1990; Armstrong et al. 1992).

STEP 2: THE METANEPHROGENIC MESENCHYME SECRETES RETINOIC ACID AND GDNF TO INDUCE AND DIRECT THE URETERIC BUD   So what are all these factors doing in the metanephrogenic mesenchyme? It seems that they are setting the stage for the secretion of paracrine factors that can induce the ureteric buds to emerge. The first of these paracrine factors is retinoic acid, and this factor directs the nearby nephric duct to express the Ret receptor (Rosselot et al. 2010). Ret is the receptor for **glial-derived neurotrophic factor** (**GDNF**), and GDNF is the second of these paracrine factors.*

GDNF becomes expressed through a complex network initiated by Pax2 and Hox11 (Xu et al. 1999; Wellik et al. 2002). Pax2 and Hox11 (in concert with other transcription factors that permit this interaction) activate GDNF expression in the metanephrogenic mesenchyme (see Brodbeck and Englert 2004). If GDNF were secreted throughout the metanephrogenic mesenchyme, numerous epithelial buds would sprout from the nephric duct; thus, GDNF expression must be limited to the posterior region of this mesenchymal tissue. This restriction is accomplished by the proteins Sprouty1 and Robo2 (Robo2 being the same protein that helps deflect

---

*This is the same compound that we saw in Chapter 11 as critical for the induction of dopaminergic neurons in the mammalian brain and for the entry of neural crest cells into the gut. GDNF is one busy protein; we will meet it again when we discuss sperm cell production in Chapter 17.

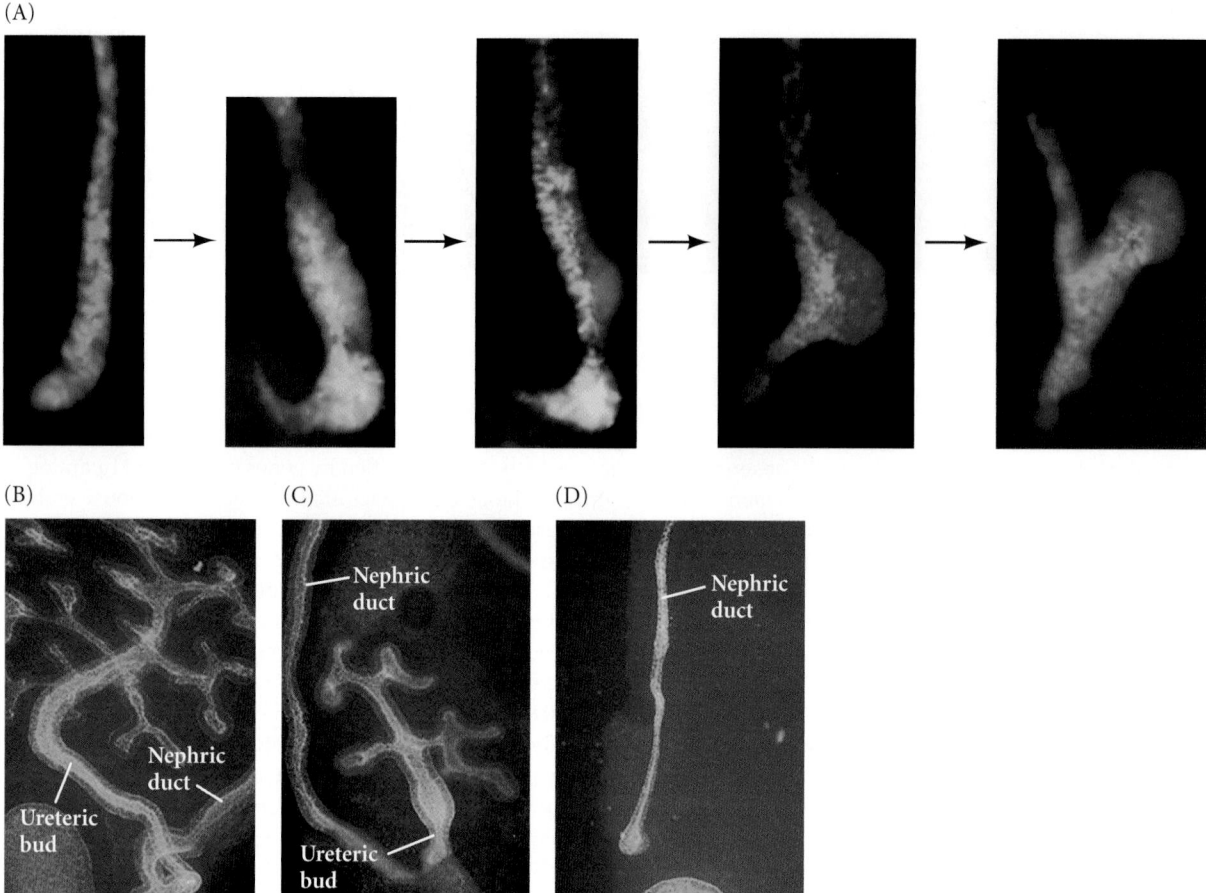

FIGURE 12.30 Ureteric bud growth is dependent on GDNF and its receptors. (A) When mice are constructed from Ret-deficient cells (green) and Ret-expressing cells (blue), the cells expressing Ret migrate to form the tips of the ureteric bud. (B) The ureteric bud from an 11.5-day wild-type mouse kidney cultured for 72 hours has a characteristic branching pattern. (C) In embryonic mice hetero-zygous for a mutation of the gene encoding GDNF, both the size of the ureteric bud and the number and length of its branches are reduced. (D) In mouse embryos missing both copies of the *Gdnf* gene, the ureteric bud does not form. (A from Chi et al. 2009, cour-tesy of F. Costantini; B–D from Pichel et al. 1996, courtesy of J. G. Pichel and H. Sariola.)

the ureter, those cells expressing the Ret receptor migrate to positions closest to the source of GDNF (**FIGURE 12.30A**; Chi et al. 2009). Mice whose *Gdnf* or GDNF receptor genes are knocked out die soon after birth from renal agenesis—lack of kidneys (**FIGURE 12.30B–D**; Moore et al. 1996; Pichel et al. 1996; Sánchez et al. 1996). The ability of nephric duct cells to proliferate appears to be suppressed by activin, and the major mechanism of GDNF action may be to locally suppress the activities of inhibitory activin. When activin was experi-mentally inhibited, GDNF induced numerous ureteric buds (Maeshima et al. 2006).

axon growth cones; see Chapter 11). If the *Robo2* gene is mutated, both the anterior and posterior metanephrogenic mesenchyme express GDNF, and the nephric duct sends out an anterior and a posterior ureteric bud (Grieshammer et al. 2004). Sprouty1 is an inhibitor of FGF signaling. In mice whose *Sprouty1* gene is knocked out, GDNF is produced throughout the metanephrogenic mesenchyme, initiating numerous ureteric buds (Basson et al. 2005).

The receptors for GDNF (the Ret tyrosine kinase receptor and the GFRα1 co-receptor) are synthesized in the nephric ducts and later become concentrated in the growing ureteric buds (Schuchardt et al. 1996). Indeed, during formation of

STEP 3: THE URETERIC BUD SECRETES FGF2 AND BMP7 TO PREVENT MESENCHYMAL APOPTOSIS The third signal in kidney development is sent from the ureteric bud to the metanephrogenic mesenchyme, and it alters the fate of the mesenchyme cells. If left uninduced by the ureteric bud, the mesenchyme cells undergo apoptosis (Grobstein 1955; Koseki et al. 1992). However, if induced by the ureteric bud, the mes-enchyme cells are rescued from the precipice of death and are converted into proliferating stem cells (Bard and Ross 1991; Bard et al. 1996). The factors secreted from the ure-teric bud include Fgf2 and BMP7. Fgf2 has three modes of action in that it inhibits apoptosis, promotes the condensation of mesenchyme cells, and maintains the synthesis of WT1 (Perantoni et al. 1995). BMP7 has similar effects (Dudley et al. 1995; Luo et al. 1995).

STEP 4: WNT9B AND WNT6 FROM THE URETERIC BUD INDUCE MESENCHYME CELLS TO AGGREGATE The ureteric bud causes dramatic changes in the behavior of the metanephrogenic mesenchyme cells, converting them into an epithelium. The newly induced mesenchyme synthesizes E-cadherin, which causes the mesenchyme cells to clump together. These aggregated nodes of mesenchyme now synthesize an epithelial basal lamina containing type IV collagen and laminin. At the same time, the mesenchyme cells synthesize receptors for laminin, allowing the aggregated cells to become an epithelium (Ekblom et al. 1994; Müller et al. 1997). The cytoskeleton also changes from one characteristic of mesenchyme cells to one typical of epithelial cells (Ekblom et al. 1983; Lehtonen et al. 1985).

The transition from mesenchymal to epithelial organization may be mediated by several molecules, including the expression of Pax2 in the newly induced mesenchyme cells. When antisense RNA to *Pax2* prevents the translation of the *Pax2* mRNA that is transcribed as a response to induction, the mesenchyme cells of cultured kidney rudiments fail to condense (Rothenpieler and Dressler 1993). Thus, Pax2 may play several roles during kidney formation.

In the mouse, Wnt6 and Wnt9b from the lateral sides of the ureteric bud (but not from the tip) are critical for transforming the metanephrogenic mesenchyme cells into tubular epithelium. The mesenchyme has receptors for these Wnts. Wnt6 appears to promote the condensation of mesenchyme in an FGF-independent way (Itäranta et al. 2002), and induces Wnt4 in the mesenchyme. As we will see, Wnt4 is very important for the formation of the nephron, and mice deficient in Wnt9b lack kidneys* (**FIGURE 12.31**; see also Figure 3.27).

STEP 5: WNT4 CONVERTS AGGREGATED MESENCHYME CELLS INTO A NEPHRON Once induced, and after it has started to condense, the mesenchyme begins to secrete Wnt4, which acts in an autocrine fashion to complete the transition from mesenchymal mass to epithelium (Stark et al. 1994; Kispert et al. 1998). Wnt4 expression is found in the condensing mesenchyme cells, in the resulting S-shaped tubules, and in the region where the newly epithelialized cells fuse with the ureteric bud tips. In mice lacking the *Wnt4* gene, the mesenchyme becomes condensed but does not form epithelia. Therefore, the ureteric bud induces the changes in the

---

*Wnt9b appears to be critical for inducing mesenchyme-epithelium transitions throughout the intermediate mesoderm. This includes the formation of the nephric (Wolffian) and Müllerian ducts, as well as the kidney tubules. In addition to having no kidneys, *Wnt9b*-deficient mice have no uterus (if they are female) and no vas deferens (if male). In rats, leukemia inhibitory factor (LIF) appears to substitute for Wnt6 and Wnt9 (Barasch et al. 1999; Carroll et al. 2005).

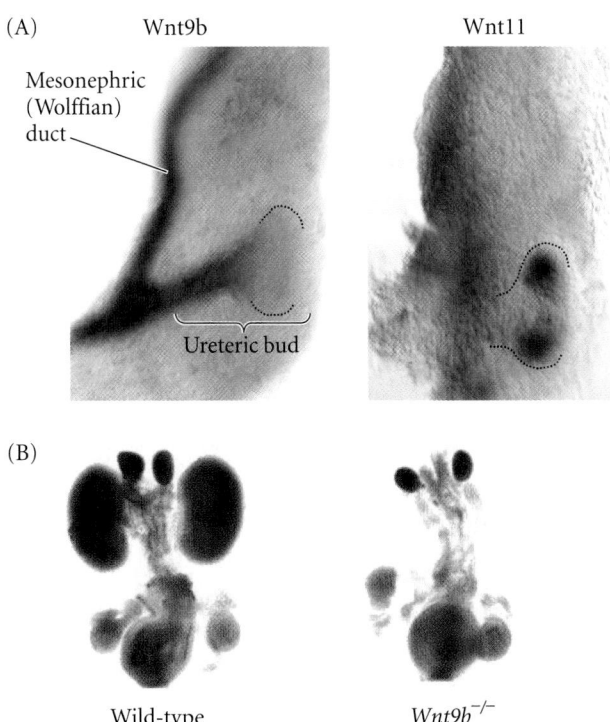

(A) Wnt9b   Wnt11

Mesonephric (Wolffian) duct

Ureteric bud

(B)

Wild-type   *Wnt9b*⁻/⁻

FIGURE 12.31 Wnts are critical for kidney development. (A) In the 11-day mouse kidney, Wnt9b is found on the stalk of the ureteric bud, while Wnt11 is found at the tips. Wnt9b induces the metanephrogenic mesenchyme to condense; Wnt11 will partition the metanephrogenic mesoderm to induce branching of the ureteric bud. Borders of the bud are indicated by a dashed line. (B) A wild-type 18.5-day male mouse (left) has normal kidneys, adrenal glands, and ureters. In a mouse deficient for *Wnt9b* (right), the kidneys are absent. (From Carroll et al. 2005.)

metanephrogenic mesenchyme by secreting FGFs, Wnt9b, and Wnt6; but these changes are mediated by the effects of the mesenchyme's secretion of Wnt4 on itself.

One molecule that may be involved in the transition from aggregated mesenchyme to nephrons is the Lim1 homeodomain transcription factor (Karavanov et al. 1998; Kobayashi et al. 2005). This protein is found in the mesenchyme cells after they have condensed around the ureteric bud, and its expression persists in the developing nephron (**FIGURE 12.32**). Two other proteins that may be critical for the conversion of the aggregated cells into a nephron are polycystins 1 and 2. These proteins are the products of the genes whose loss-of-function alleles give rise to human polycystic kidney disease. Mice deficient in these genes have abnormal, swollen nephrons (Ward et al. 1996; van Adelsberg et al. 1997).

STEP 6: SIGNALS FROM THE MESENCHYME INDUCE THE BRANCHING OF THE URETERIC BUD Recent evidence has implicated several paracrine factors in the branching of the ureteric bud, and these factors probably work as pushes and pulls. Some factors may preserve the extracellular matrix

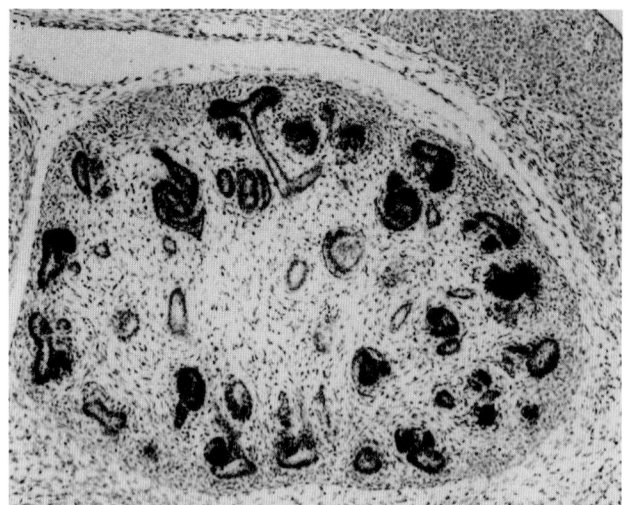

FIGURE 12.32 Lim1 expression (dark stain) in a 19-day embryonic mouse kidney. In situ hybridization shows high levels of Lim1 protein in the newly epithelialized comma-shaped and S-shaped bodies that will become nephrons. Compare with earlier Lim1 expression shown in Figure 12.26C. (From Karavanov et al. 1998, courtesy of A. A. Karavanov.)

surrounding the epithelium, thereby preventing branching from taking place (the "push"), while other factors may cause the digestion of this extracellular matrix, permitting branching to occur (the "pull").

The first protein regulating ureteric bud branching is GDNF (Sainio et al. 1997). GDNF from the mesenchyme not only induces the initial ureteric bud from the nephric duct, but it can also induce secondary buds from the ureteric bud once the bud has entered the mesenchyme (**FIGURE 12.33**). GDNF from the mesenchyme promotes cell division in the Ret-expressing cells at the tip of the ureteric bud (Shakya

et al. 2005). GDNF also appears to induce Wnt11 synthesis in these responsive cells at the tip of the bud (see Figure 12.31A), and Wnt11 reciprocates by regulating GDNF levels. Wnt11 may also be important in inducing the differentiation of the metanephrogenic mesenchyme (Kuure et al. 2007). The cooperation between the GDNF/Ret pathway and the Wnt pathway appears to coordinate the balance between branching and metanephrogenic cell division such that continued metanephric development is ensured (Majumdar et al. 2003).

The second candidate branch-regulating molecule is transforming growth factor-β1 (TGF-β1). When exogenous TGF-β1 is added to cultured kidneys, it prevents the epithelium from branching (**FIGURE 12.34A,B**; Ritvos et al. 1995). TGF-β1 is known to promote the synthesis of extracellular matrix proteins and to inhibit the metalloproteinases that can digest these matrices (Penttinen et al. 1988; Nakamura et al. 1990). Thus, it is possible that TGF-β1 stabilizes branches once they form.

A third molecule that may be important in epithelial branching is BMP4 (Miyazaki et al. 2000). BMP4 is found in the mesenchymal cells surrounding the nephric duct, and BMP4 receptors are found in the epithelial tissue of the duct. Because BMPs antagonize branching signals, BMP4 restricts the branching of the duct to the appropriate sites. When the BMP4 signaling cascade is activated ectopically in embryonic mouse kidney rudiments, it severely distorts the normal branching pattern (**FIGURE 12.34C**). BMPs are also critical in regulating the number of nephrons (Di Giovanni et al. 2011).

The fourth molecule involved in branching is collagen XVIII, which is part of the extracellular matrix induced by the mesenchyme. It may provide the specificity for the branching pattern (Lin et al. 2001). Collagen XVIII is found on the branches of the kidney epithelium but not at the tips; in the developing lung, the reciprocal pattern is seen. This pattern is generated in part by GDNF, which downregulates collagen XVIII expression in the tips of the ureteric bud branches.

(A)

(B)

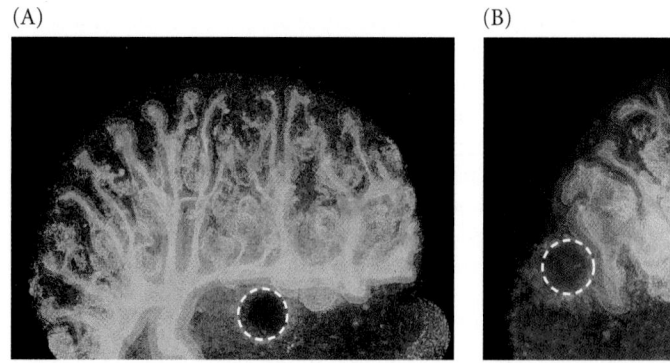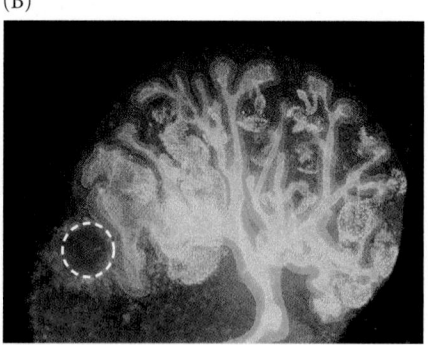

FIGURE 12.33 Effect of GDNF on branching of the ureteric epithelium. The ureteric bud and its branches are stained orange (with antibodies to cytokeratin 18), while the nephrons are stained green (with antibodies to nephron brush border antigens). (A) A 13-day embryonic mouse kidney cultured 2 days with a control bead (circle) has a normal branching pattern. (B) A similar kidney cultured 2 days with a GDNF-soaked bead shows a distorted pattern, as new branches are induced in the vicinity of the bead. (From Sainio et al. 1997, courtesy of K. Sainio.)

(A) (B)

(C) (D)

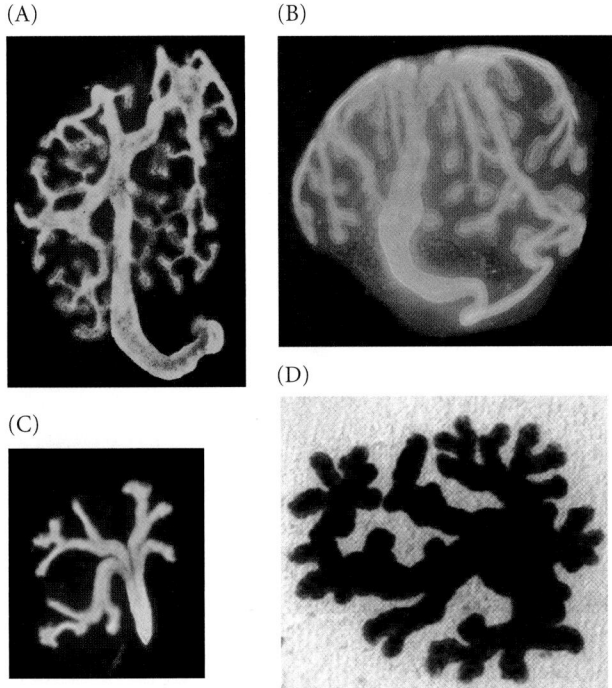

FIGURE 12.34 Signaling molecules and branching of the ureteric epithelium. (A) An 11-day embryonic mouse kidney cultured for 4 days in control medium has a normal branching pattern. (B) An 11-day mouse kidney cultured in TGF-β1 shows no branching until reaching the periphery of the mesenchyme, and the branches formed are elongated. (C) An 11-day mouse kidney cultured in activin (which activates the same receptor as BMP4) shows a marked distortion of branching. (D) Epithelial branching of kidney cells grown in lung mesenchyme takes on an appearance similar to that of lung epithelium. (A–C from Ritvos et al. 1995; D from Lin et al. 2001. Photographs courtesy of Y. Lin and S. Vainio.)

When ureteric duct epithelium is incubated in lung mesenchyme, the collagen XVIII expression pattern seen is typical of that of the lung, and the branching pattern resembles that of the lung epithelium (FIGURE 12.34D).

STEPS 7 AND 8: DIFFERENTIATION OF THE NEPHRON AND GROWTH OF THE URETERIC BUD The interactions we have described to this point create a *cap condensate* of metanephrogenic mesenchyme cells that covers the tips of the ureteric bud branches. The cap condensate becomes a multipotent, self-renewing stem cell population that can make all the different cell types of the nephron. Fgf9 and Fgf20 from the ureteric bud promote the survival and "stemness" of this mesenchyme (Mugford et al. 2009; Barak et al. 2012). The cap condensate reciprocates by converting vitamin A to retinoic acid, and this RA functions to retain the expression of Ret (one of the GDNF receptors) in the ureteric bud (Batourina et al. 2002). Ret, along with Fgf7 (also secreted by the mesenchymal cap) is essential for generating an appropriate number of nephrons in the kidney (Qiao et al. 1999; Chi et al. 2004).

STEP 9: INSERTING THE URETER INTO THE BLADDER The branching epithelium becomes the collecting system of the kidney. This epithelium collects the filtered urine from the nephron and secretes antidiuretic hormone for the resorption of water (a process that, not so incidentally, makes life on land possible). The rest of the ureteric bud becomes the ureter, the tube that carries urine into the bladder. The junction between the ureter and bladder is extremely important, and hydronephrosis, a birth defect involving renal filtration, occurs when this junction is so tightly formed that urine cannot enter the bladder. The ureter is made into a watertight connecting duct by the condensation of mesenchymal cells around it (but not around the collecting ducts). These mesenchymal cells become smooth muscle cells capable of wave-like contractions (peristalsis) that allow the urine to move into the bladder; these cells also secrete BMP4 (Cebrian et al. 2004). BMP4 upregulates genes for uroplakin, a cell membrane protein that causes differentiation of this region of the ureteric bud into the ureter. BMP inhibitors protect the region of the ureteric bud that forms the collecting ducts from this differentiation.

The bladder develops from a portion of the cloaca (FIGURE 12.35A,B). The cloaca* is an endodermally lined chamber at the caudal end of the embryo that will become the waste receptacle for both the intestine and the kidney. Amphibians, reptiles, and birds still have this organ and use it to void both liquid and solid wastes. In mammals, the cloaca becomes divided by a septum into the urogenital sinus and the rectum. Part of the urogenital sinus becomes the bladder, while another part becomes the urethra (which will carry the urine out of the body). The ureteric bud originally empties into the bladder via the nephric (Wolffian) duct (which grows toward the bladder by an as-yet-unknown mechanism). Once at the bladder, the urogenital sinus cells of the bladder wrap themselves around both the ureter and the nephric duct. Then the nephric ducts migrate ventrally, opening into the urethra rather than into the bladder and the nephric duct (FIGURE 12.35C–F). The caudal end of the nephric duct appears to undergo apoptosis, allowing the ureter to separate from the nephric duct. Expansion of the bladder then moves the ureter to its final position at the neck of the bladder (Batourina et al. 2002; Mendelsohn 2009). In females, the entire nephric duct degenerates, while the Müllerian duct opens into the vagina (see Chapter 15). In males, however, the nephric duct also forms the sperm outflow track, so males expel sperm and urine through the same opening.

The Hox13 paralogue group appears to be important in specifying the distal ureter. In human *HOXA13* defects, there are abnormalities of the cloaca, the male and female reproductive tracts, and the urethra (Pinsky 1974; Mortlock and Innis 1997). Because *HOXA13* is also involved in digit specification (see Chapter 14), the person's fingers and toes are also

---

*The term *cloaca* is Latin for "sewer"—a bad joke on the part of the early European anatomists.

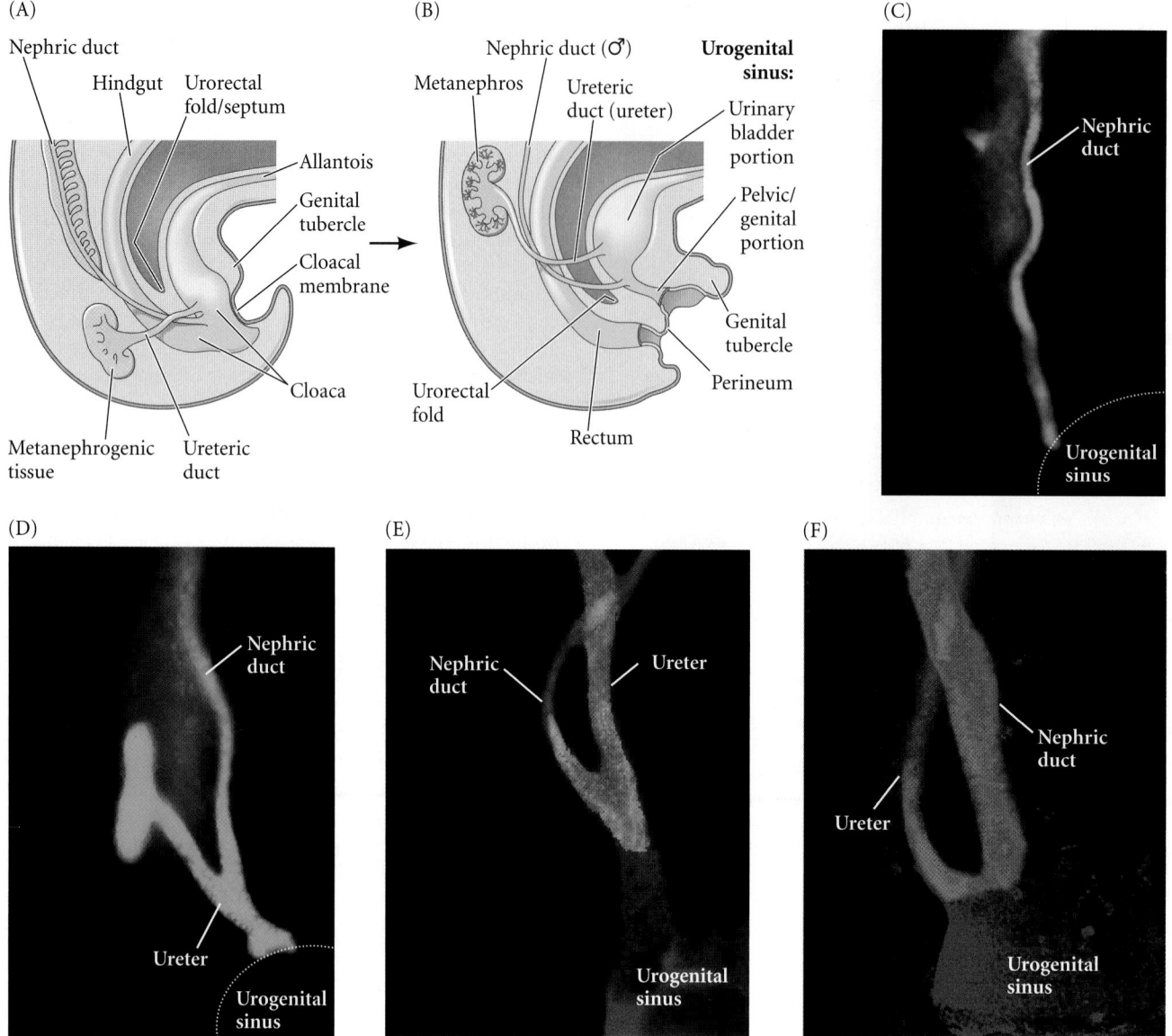

FIGURE 12.35 Development of the bladder and its connection to the kidney via the ureter. (A) The cloaca originates as an endodermal collecting area that opens into the allantois. (B) The urogenital septum divides the cloaca into the future rectum and the urogenital sinus. The bladder forms from the anterior portion of that sinus, and the urethra develops from the posterior region of the sinus. The space between the rectal opening and the urinary opening is the perineum. (C–F) Insertion of the ureter into the embryonic mouse bladder. (C) Day-10 mouse urogenital tract. The nephric duct is stained with GFP attached to a Hoxb7 enhancer. (D) Urogenital tract from a day-11 embryo, after ureteric bud outgrowth. (E) Whole mount urogenital tract from a day-12 embryo. The ducts are stained green and the urogenital sinus red. (F) The ureter separates from the nephric duct and forms a separate opening into the bladder. (A,B after Cochard 2002; C–F from Batourina et al. 2002, courtesy of C. Mendelsohn.)

malformed; thus, the syndrome has been called "hand-foot-genital syndrome."

FURTHER STEPS The next stages of kidney development occur in the separate groups. The glomerulus, for instance, forms from the epithelialized descendants of the metanephrogenic mesenchyme located distally, while the more proximal descendants, closer to the collecting ducts, become the distal and proximal convoluted tubules (see Massa et al. 2013). The developing kidney epitomizes the reciprocal interactions needed to form an organ. It also shows us that we have only begun to understand how organs form.

Coda

The central mesodermal compartments—the notochord, paraxial, and intermediate mesoderm—form the core of our musculoskeletal and reproductive systems. We have seen how reciprocal induction functions both in the relationship between tendons and muscles and in kidney formation. We have also seen how some cells (such as the somite cells) are multipotent and can generate numerous cell types, while others (such as the muscle satellite cells) can produce but

one type. Probably the most important phenomenon of the central mesodermal compartments is the rich tapestry of integrated co-construction of organs, which emerges from interactions *within* the mesodermal lineages and *between* the mesoderm and its surrounding ectoderm and endoderm. We will now proceed to the lateral mesodermal compartments and their interactions among themselves and with the endoderm to provide the basis for circulation and nutrient absorption in the developing body.

SNAPSHOT SUMMARY: Paraxial and Intermediate Mesoderm

1. The paraxial mesoderm forms blocks of tissue called somites. Somites give rise to three major divisions: the sclerotome, the myotome, and the central dermamyotome.

2. Somites are formed from unsegmented presomitic mesoderm by the interactions of several proteins. The Notch pathway is extremely important in this process, and Eph receptor systems may be involved in the separation of the somites from the unsegmented paraxial mesoderm. N-cadherin, fibronectin, and Rac1 appear to be important in causing these cells to become epithelial.

3. The sclerotome forms the vertebral cartilage. In thoracic vertebrae, the sclerotome cells also form the ribs. The intervertebral joints as well as the meninges and dorsal aortic cells also come from the sclerotome.

4. The primaxial myotome forms the back musculature. The abaxial myotome forms the muscles of the body wall, limb, diaphragm, and tongue.

5. The central dermamyotome forms the dermis of the back, as well as forming precursors of muscle and brown fat cells.

6. The somite regions are specified by paracrine factors secreted by neighboring tissues. The sclerotome is specified to a large degree by Sonic hedgehog, which is secreted by the notochord and floor plate cells. The two myotome regions are specified by different factors, and in both instances, myogenic regulatory factors are induced in the cells that will become muscles.

7. To form muscles, the myoblasts stop dividing, align themselves into myotubes, and fuse.

8. The major lineages that form the skeleton are the somites (axial skeleton), lateral plate mesoderm (appendages), and neural crest and head mesoderm (skull and face).

9. There are two major types of ossification. In intramembranous ossification, which occurs primarily in the skull and facial bones, neural crest and head mesenchyme is converted directly into bone. In endochondral ossification,

mesenchyme cells become cartilage. These cartilaginous models are later replaced by bone cells.

10. Hypertrophic cartilage cells make Indian hedgehog and VEGF. Ihh initiates osteoblast differentiation into bone, and VEGF induces the construction of capillaries that enable bone cells to be brought into the degenerating cartilage.

11. Osteoclasts continually remodel bone throughout a person's lifetime. The hollowing out of bone for the bone marrow is accomplished by osteoclasts.

12. Tendons are formed through the conversion of the dorsalmost layer of sclerotome cells into syndetome cells by FGFs secreted by the myotome.

13. The intermediate mesoderm is specified through interactions with the paraxial mesoderm. It generates the kidneys and gonads. This specification requires Pax2, Pax8, and Lim1.

14. The metanephric kidney of mammals is formed by the reciprocal interactions of the metanephrogenic mesenchyme and a branch of the nephric duct called the ureteric bud.

15. The metanephrogenic mesenchyme becomes competent to form nephrons by expressing WT1, and it starts to secrete GDNF. GDNF is secreted by the mesoderm and induces the formation of the ureteric bud.

16. The ureteric bud secretes Fgf2 and BMP7 to prevent apoptosis in the metanephrogenic mesenchyme. Without these factors, this kidney-forming mesenchyme dies.

17. The ureteric bud secretes Wnt9b and Wnt6, which induce the competent metanephrogenic mesenchyme to form epithelial tubules. As they form these tubules, the cells secrete Wnt4, which promotes and maintains their epithelialization.

18. The condensing mesenchyme secretes paracrine factors that mediate the branching of the ureteric bud. These factors include GDNF, BMP4, and TGF-β1. The branching also depends on the extracellular matrix of the epithelium.

For Further Reading

Complete bibliographical citations for all literature cited in this chapter can be found at the free-access website **www.devbio.com**

Brent, A. E., R. Schweitzer and C. J. Tabin. 2003. A somitic compartment of tendon precursors. *Cell* 113: 235–248.

Christ, B., R. Huang and M. Scaal. 2007. Amniote somite derivatives. *Dev. Dyn.* 236: 2382–2396.

Costantini, F. and R. Kopan. 2010. Patterning a complex organ: Branching morphogenesis and nephron segmentation in kidney development. *Dev. Cell* 18: 698–712.

Gerber, H. P., T. H. Vu, A. M. Ryan, J. Kowalski and Z. Werb. 1999. VEGF couples hypertrophic cartilage remodeling, ossification, and angiogenesis during endochondral bone formation. *Nature Med.* 5: 623–628.

Kuang, S., K. Kuroda, F. Le Grand and M. A. Rudnicki. 2007. Asymmetric self-renewal and commitment of satellite stem cells in muscle. *Cell* 129: 999–1010.

Mauch, T. J., G. Yang, M. Wright, D. Smith and G. C. Schoenwolf. 2000. Signals from trunk paraxial mesoderm induce pronephros formation in chick intermediate mesoderm. *Dev. Biol.* 220: 62–75.

Ordahl, C. P. and N. Le Douarin. 1992. Two myogenic lineages within the developing somite. *Development* 114: 339–353.

Palmeirim, I., D. Henrique, D. Ish-Horowicz and O. Pourquié. 1997. Avian hairy gene expression identifies a molecular clock linked to vertebrate segmentation and somitogenesis. *Cell* 91: 639–648.

Pichel, J. G. and 11 others. 1996. Defects in enteric innervation and kidney development in mice lacking GDNF. *Nature* 382: 73–76.

Go Online

WEBSITE 12.1 Cranial paraxial mesoderm. Most of the head musculature does not come from somites. Rather, it comes from the cranial paraxial (prechordal plate) mesoderm. These cells originate adjacent to the sides of the brain, and they migrate to their respective destinations.

WEBSITE 12.2 Muscle formation. Research on chimeric mice has shown that skeletal muscle becomes multinucleate by the fusion of cells, whereas heart muscle becomes multinucleate by nuclear divisions within a cell.

WEBSITE 12.3 Paracrine factors, their receptors, and human bone growth. Mutations in the genes encoding paracrine factors and their receptors cause numerous skeletal anomalies in humans and mice.

Vade Mecum

Mesoderm in the vertebrate embryo. The organization of the mesoderm in the neurula stage is similar for all vertebrates. You can see this organization by viewing serial sections of the chick embryo in the Chick-Mid segment.

13

Lateral Plate Mesoderm and the Endoderm

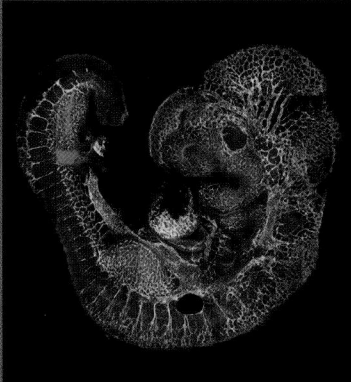

IN THE CHAOS OF THE ENGLISH CIVIL WARS, William Harvey, physician to the king and discoverer of the chick blastoderm, was comforted by viewing the heart as the undisputed ruler of the body, through whose divinely ordained powers the lawful growth of the organism was assured. Later embryologists saw the heart as more of a servant than a ruler, the chamberlain of the household who assured that nutrients reached the apically located brain and the peripherally located muscles. In either metaphor, the heart, circulation, and digestive system were seen to be absolutely critical for development. As Harvey persuasively argued in 1651, the chick embryo must form its own blood without any help from the hen, and this blood is crucial in embryonic growth. How this happened was a mystery to him. "What artificer," he wrote, could create blood "when there is yet no liver in being?" The nutrition provided by the egg was also paramount to Harvey. His conclusion about the nutritive value of the yolk and albumen was "The egge is, as it were, an exposed womb; wherein there is a substance concluded as the Representative and Substitute, or Vicar of the breasts."

The heart and circulatory system that so intrigued Harvey arise from the vertebrate embryo's lateral plate mesoderm. The lateral plate mesoderm resides on the lateral side of each of the two bands of intermediate mesoderm (see Figures 12.1 and 12.2). Each of these lateral plates splits horizontally into two layers. The dorsal layer is the **somatic (parietal) mesoderm**, which underlies the ectoderm and, together with the ectoderm, forms the **somatopleure**. The ventral layer is the **splanchnic (visceral) mesoderm**, which overlies the endoderm and, together with the endoderm, forms the **splanchnopleure** (FIGURE 13.1A). The space between these two layers becomes the body cavity—the **coelom**—which stretches from the future neck region to the posterior of the body.

During later development, the right-side and left-side coeloms fuse, and folds of tissue extend from the somatic mesoderm, dividing the coelom into separate cavities. In mammals, these mesodermal folds subdivide the coelom into the **pleural**, **pericardial**, and **peritoneal cavities**, enveloping the thorax, heart, and abdomen, respectively. The mechanism for creating the linings of these body cavities from the lateral plate mesoderm has changed little throughout vertebrate evolution, and the development of the amniote mesoderm can be compared with similar stages of frog embryos (FIGURE 13.1B,C).

This chapter outlines the mechanisms by which the circulatory system, the respiratory system, and the digestive system emerge in the amniote embryo. We will first discuss the formation of the heart and then proceed to the mechanisms whereby the blood vessels and blood cells develop. At the end of the chapter, we will briefly follow the development of the gut tube and its associated organs. As we will see, the cells and organs of all these systems arise from the interaction of the lateral plate mesoderm and the endoderm.

The Heart of Creatures is the Foundation of Life, the Prince of All, the Sun of the Microcosm, on which all Vegetation doth depend, from whence all Vigor and Strength doth flow.
  WILLIAM HARVEY (1628)

Blut is ein ganz besonderer Saft. [Blood is a very special juice.]
  WOLFGANG GOETHE (1805)

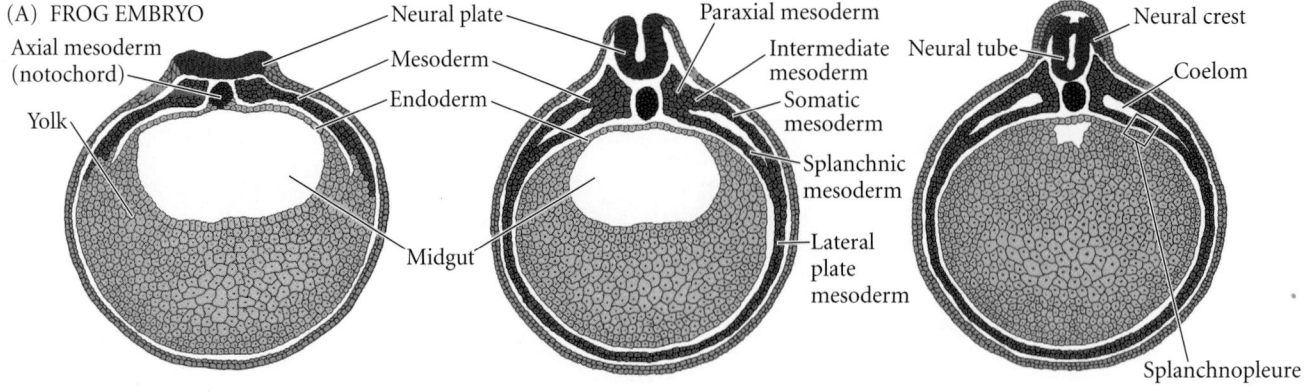

(A) FROG EMBRYO

Axial mesoderm (notochord)
Yolk
Neural plate
Mesoderm
Endoderm
Midgut
Paraxial mesoderm
Intermediate mesoderm
Somatic mesoderm
Splanchnic mesoderm
Lateral plate mesoderm
Neural tube
Neural crest
Coelom
Splanchnopleure

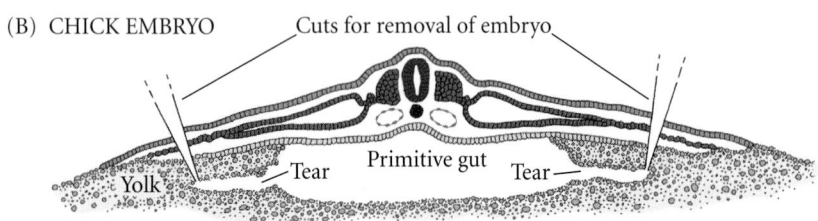

(B) CHICK EMBRYO
Cuts for removal of embryo
Yolk
Tear
Primitive gut
Tear

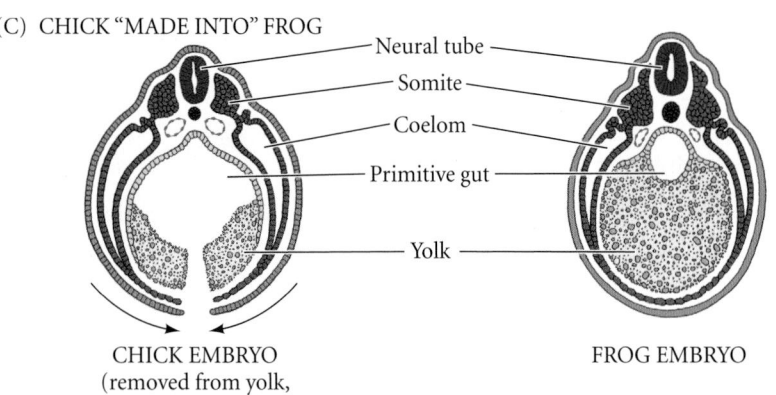

(C) CHICK "MADE INTO" FROG
Neural tube
Somite
Coelom
Primitive gut
Yolk

CHICK EMBRYO
(removed from yolk, edges pulled together)

FROG EMBRYO

FIGURE 13.1 Mesodermal development in frog and chick embryos. (A) Neurula-stage frog embryos, showing progressive development of the mesoderm and coelom. (B) Transverse section of a chick embryo. (C) When the chick embryo is separated from its enormous yolk mass, it resembles the amphibian neurula at a similar stage. (A after Rugh 1951; B,C after Patten 1951.)

● See WEBSITE 13.1 Coelom formation

FORMATION AND DEVELOPMENT OF THE CIRCULATORY SYSTEM

Consisting of a heart, blood cells, and an intricate system of blood vessels, the circulatory system provides nourishment to the developing vertebrate embryo and is the embryo's first functional unit. Few events in developmental biology are as exciting and as accessible as watching the heart beating in a two-day chick embryo, pumping the first blood cells into vessels that have not even formed valves yet. The development of the circulatory system provides excellent examples of induction, specification, cell migration, organ formation, and the role of stem cells in both embryonic development and adult tissue regeneration.

Heart Development

The circulatory system is the first working unit in the developing embryo, and the heart is the first functional organ.

Like other organs, the heart arises through the specification of precursor cells, the migration of these precursor cells to the organ-forming region, the specification of cell types through signaling interactions within and between tissues, and the coordination of morphogenesis, growth, and cell differentiation.

Formation of the heart fields

The vertebrate heart arises from two regions of splanchnic mesoderm—one on each side of the body—that interact with adjacent tissue to become specified for heart development. In the early amniote gastrula, the heart progenitor cells (about 50 of them in mice) are located in two small patches, one on each side of the epiblast, close to the rostral portion of the primitive streak. These cells migrate together through the streak and form two groups of lateral plate mesoderm cells, positioned anteriorly at the level of the node (Tam et al. 1997; Colas et al. 2000). The general specification of a **heart field**,

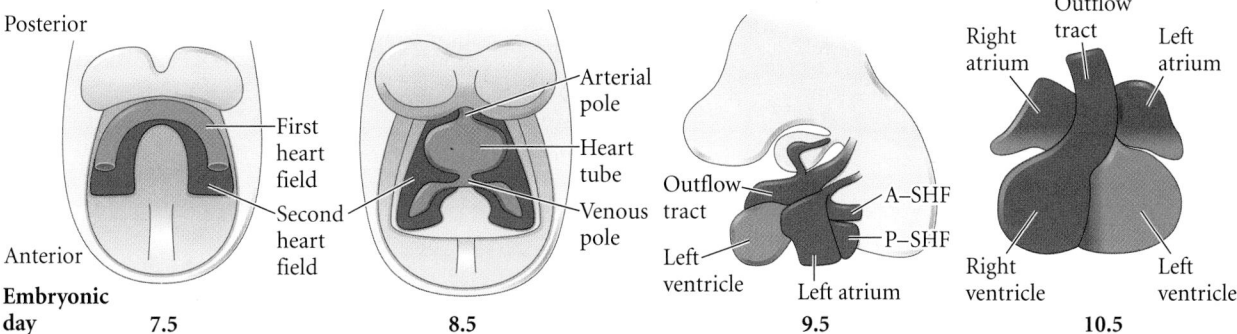

Posterior

First heart field

Second heart field

Anterior

Arterial pole

Heart tube

Venous pole

Outflow tract

Left ventricle

A–SHF

P–SHF

Left atrium

Right atrium

Outflow tract

Left atrium

Right ventricle

Left ventricle

Embryonic day   7.5   8.5   9.5   10.5

FIGURE 13.2 First and second heart fields in the mouse. On embryonic day 7.5, the heart fields from each side of the body have joined into a common cardiac crescent that contains the first and second heart fields. The first heart field contributes primarily to the left ventricle. By day 10.5, the second heart field contributes to the other three chambers—the right ventricle and the left and right atria—as well as to the outflow tract that originally includes the aorta and the pulmonary artery. A-SHF, P-SHF anterior and posterior components of the second heart field. (After Kelly 2012.)

also known as **cardiogenic mesoderm**, has already started during this cellular migration. Labeling experiments by Stalberg and DeHann (1969) and Abu-Issa and Kirby (2008) have shown that the progenitor cells of the heart field migrate such that the medial-lateral (center-to-side) arrangement of these early cells will become the anterior-posterior (rostral-caudal) axis of a linear **heart tube**.

The vertebrate heart field is divided into two regions (**FIGURE 13.2**). The first heart field appears to form the scaffold of the developing heart. The progenitor cells of the first field fuse at the midline to form the heart tube. However, these cells have limited proliferative ability and will generate only the left ventricle of the adult heart (i.e., the chamber that pumps blood into the aorta). The progenitors of the second heart field add cells to both ends of the heart tube. On one end, these cells will produce the atria, as well as the **outflow tract** (the **conus arteriosus** and **truncus arteriosus**), which becomes the base of the aorta and pulmonary arteries (de la Cruz 1989; Kelly 2012). On the other end, the second heart field will generate the right ventricle as well as the inflow region, where the pulmonary vein and the vena cava enter the heart (van den Berg et al. 2009; van den Berg and Moorman 2011). All the cell types of the heart—the **cardiomyocytes** that form the muscular layers, the **endocardium** that forms the internal layer, the **endocardial cushions** of the valves, the **epicardium** that forms the coronary blood vessels that feed the heart, and the **Purkinje fibers*** that coordinate the heartbeat—are generated from these heart fields

* Note that these specialized myocardial nerve fibers are not the same thing as the Purkinje *neurons* of the cerebellum mentioned in Chapter 10. Both were named for the nineteenth-century Czech anatomist and histologist Jan Purkinje.

(Mikawa 1999; van Wijk et al. 2009). Moreover, as we will discuss later, it appears that each progenitor cell is capable of becoming any of the differentiated heart cell types. The cardiac precursor cells will be supplemented by cardiac neural crest cells; the latter help make the outflow tract and the septum that separates the aorta from the pulmonary trunk (see Figure 11.17; Porras and Brown 2008).

Specification of the cardiogenic mesoderm

The cardiogenic mesoderm cells are specified by their interactions with the pharyngeal endoderm and notochord. The heart does not form if this anterior endoderm is removed, and posterior endoderm cannot induce heart cells to form (Nascone and Mercola 1995; Schultheiss et al. 1995). BMPs (especially BMP2) from the anterior endoderm promote both heart and blood development. Endodermal BMPs also induce Fgf8 synthesis in the endoderm directly beneath the cardiogenic mesoderm, and Fgf8 appears to be critical for the expression of cardiac proteins (Alsan and Schultheiss 2002).

Inhibitory signals prevent heart structures from forming where they should not be made. First, the notochord secretes Noggin and Chordin, blocking BMP signaling in the center of the embryo, and specific Noggin-secreting cells in the myotome prevent heart cell specification of the somites. Second, Wnt proteins from the neural tube, especially Wnt3a and Wnt8, *inhibit* heart formation but *promote* blood formation. The anterior endoderm, moreover, produces Wnt inhibitors such as Cerberus, Dickkopf, and Crescent, which prevent Wnts from binding to their receptors. In this way, cardiac precursor cells are specified in those places where BMPs (from the lateral mesoderm and endoderm) and Wnt antagonists (from the anterior endoderm) coincide (**FIGURE 13.3A**; Marvin et al. 2001; Schneider and Mercola 2001; Tzahor and Lassar 2001; Gerhart et al. 2011).

In the absence of Wnt signals, BMPs activate *nkx2-5* and **mesp1**, two genes that are critical in the regulatory network that specifies the heart cells (**FIGURE 13.3B**). The *nkx2-5* gene has functions in heart development that are conserved across species (Komuro and Izumo 1993; Lints et al. 1993; Sugi and Lough 1994; Schultheiss et al. 1995; Andrée et al. 1998). In *Drosophila*, the *Nkx2-5* homologue is called "tinman," as

(A)

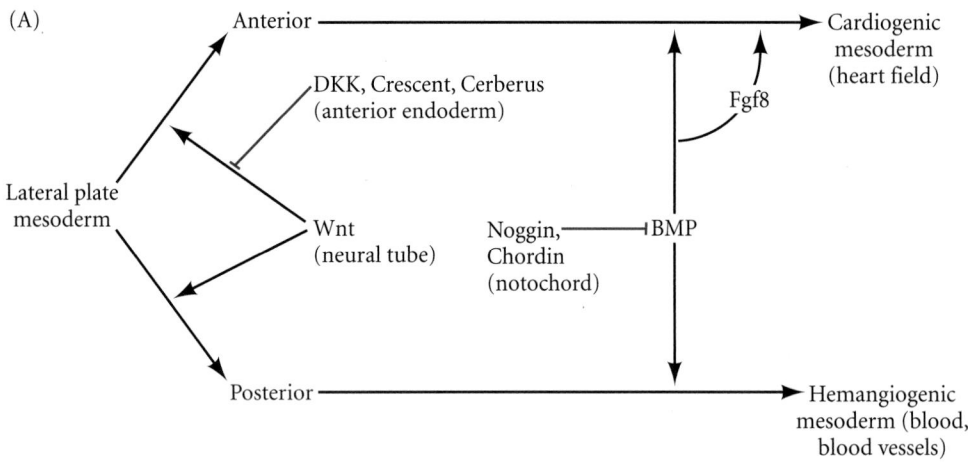

(B)

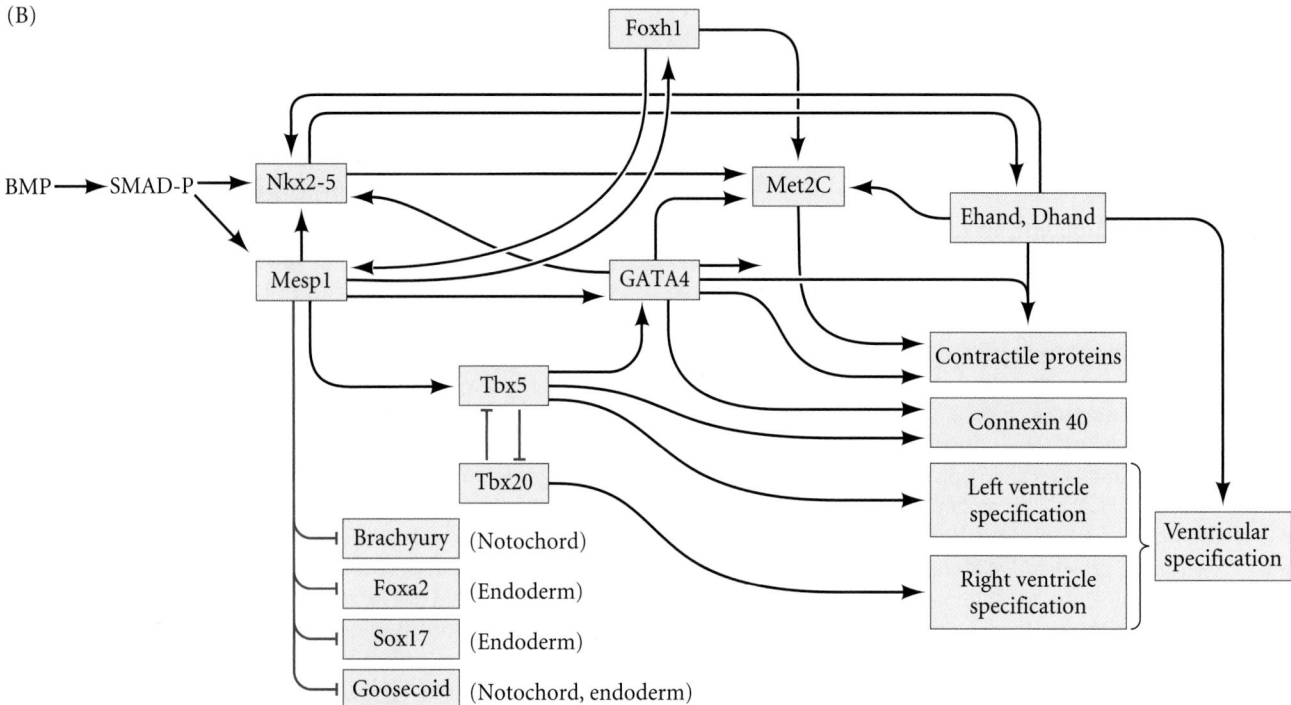

FIGURE 13.3 Model of inductive interactions involving the BMP and Wnt pathways that form the boundaries of the cardiogenic mesoderm. (A) Wnt signals from the neural tube instruct lateral plate mesoderm to become precursors of the blood and blood vessels. In the anterior portion of the body, Wnt inhibitors (Dickkopf, Crescent, Cerberus) from the pharyngeal endoderm prevent Wnt from functioning, allowing later signals (BMP, Fgf8) to convert lateral plate mesoderm into cardiogenic mesoderm. BMP signals will also be important for the differentiation of hemangiogenic (blood, blood vessel) mesoderm. In the center of the embryo, Noggin and Chordin signals from the notochord block BMPs. Thus, the cardiac and blood-forming fields do not form in the center of the embryo. (B) Model gene regulatory network for the vertebrate heart initiated by BMP signals. BMP signaling activates the pivotal switches Nkx2-5 and Mesp1. These transcription factors act in concert to activate numerous heart-forming genes. Mesp1 has also been shown to repress genes that would otherwise specify the cell into other fates. The antagonism between Tbx20 (right side) and Tbx5 (left side) can also be seen. This model is provisional, as new ChIP-Seq techniques have identified thousands of promoters activated at different stages of heart development. (A after Davidson 2006; B after May et al. 2012.)

loss-of-function mutants lack a heart. Nkx2-5 can also down-regulate BMPs, and in early heart cell development it limits the number of heart cell precursors that can form the heart fields. If the *nkx2-5* gene is specifically knocked out in those cells destined to become ventricles, these chambers express BMP10, resulting in massive overgrowth of the ventricles such that the ventricular chambers fill with muscle cells (Pashmforoush et al. 2004; Prall et al. 2007).

The other gene activated by BMPs is *mesp1*.* Mesp1 protein cooperates with Nkx2-5 to activate the genes that specify the heart. Mesp1 also acts to prevent heart progenitors from being respecified as some other type of mesoderm. First, it activates the *dickkopf* gene in these cells (David et al. 2008), thereby preventing Wnts from transforming them into vascular cells. Second, Mesp1 represses *brachyury*, *sox17*, and *goosecoid* so the cardiac precursor cells will not become endoderm, somite, or notochord (Bondue and Blanpain 2010). Mesp1 also promotes the expression of those genes whose products allow cell migration, and once these cardiogenic precursor cells are committed to become a heart, the cells migrate to the midline to form the heart tube (Lazic and Scott 2011).

The study of the gene expression networks in the heart is proceeding extremely rapidly since the availability of ChIP-Seq techniques (see Chapter 2) to identify and sequence enhancer regions active in any given stage of heart development. This approach has already identified thousands of enhancer sequences from fetal and adult tissues, and these sequences are usually able to direct the transcription of marker genes in the heart (May et al. 2012).

Migration of the cardiac precursor cells

As the presumptive heart cells move anteriorly between the ectoderm and endoderm toward the middle of the embryo, they remain in close contact with the endoderm surface (Linask and Lash 1986). In the chick, the directionality of this migration appears to be provided by the foregut endoderm. If the cardiac region endoderm is rotated with respect to the rest of the embryo, migration of the cardiogenic mesoderm cells is reversed. It is thought that the endodermal component responsible for this movement is an anterior-to-posterior concentration gradient of fibronectin. Antibodies against fibronectin stop the migration, whereas antibodies against other extracellular matrix components do not (Linask and Lash 1988).

This movement produces two populations of migrating cardiac precursor cells, on the right and left sides of the embryo. Each side has its own first and second heart fields, and each of these populations starts to form its own heart tube. In the chick, the fields are brought together around the 7-somite stage, when the foregut is formed by

---

* *Mesp1* is a close relative of *Mesp2*, which directs somitogenesis (Chaper 12). The tunicate (Chapter 7) has only one *Mesp* gene, and it specifies heart development through the activation of *Nkx* and *Hand* genes, just as occurs in vertebrates (Satou et al. 2004). BMPs may be activating *Mesp1* indirectly by inducing the expression of the Eomesodermin transcription factor. Eomeosdermin is important for both endoderm and mesodermal lineages. In the early mouse epiblast (which has low amounts of Nodal), Eomesodermin activates *Mesp1*. Later, as the primitive streak elongates, Eomesodermin acts with Nodal to activate genes for the Sox17 and Foxa2 transcription factors that specify the definitive endoderm (Costelo et al. 2011).

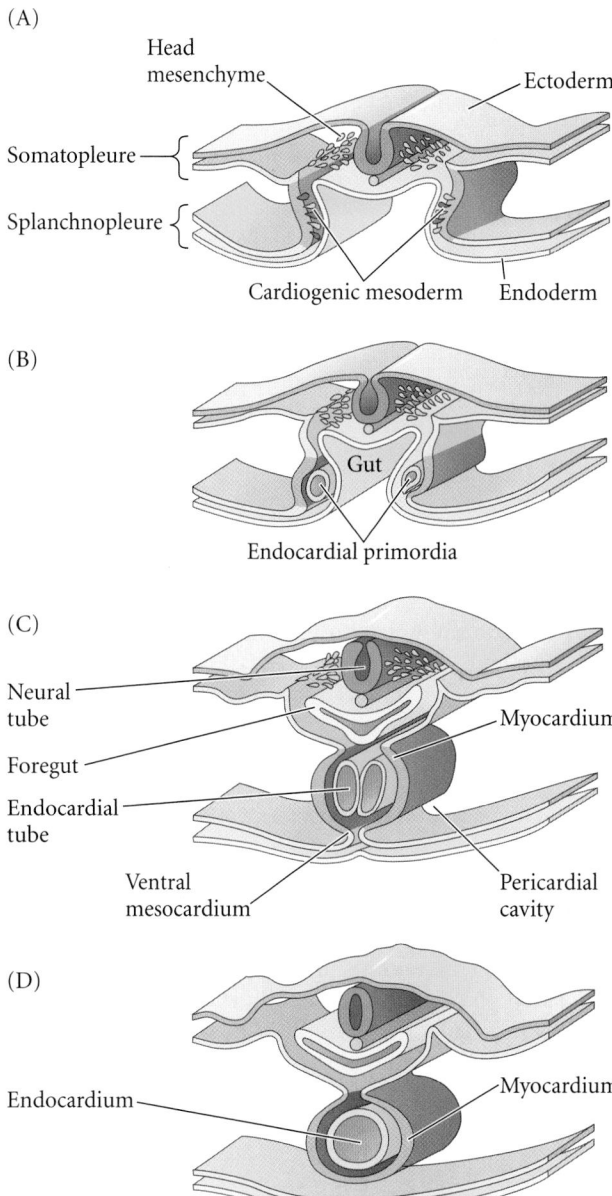

(A)
Head mesenchyme
Ectoderm
Somatopleure
Splanchnopleure
Cardiogenic mesoderm   Endoderm

(B)
Gut
Endocardial primordia

(C)
Neural tube
Foregut
Endocardial tube
Ventral mesocardium
Myocardium
Pericardial cavity

(D)
Endocardium
Myocardium

FIGURE 13.4 Formation of chick heart from splanchnic lateral plate mesoderm. The endocardium forms the inner lining of the heart, the myocardium forms the heart muscles, and the epicardium will eventually cover the heart. Transverse sections through the heart-forming region of the chick embryo are shown at (A) 25 hours, (B) 26 hours, (C) 28 hours, and (D) 29 hours. (After Carlson 1981.)

the inward folding of the splanchnopleure (**FIGURE 13.4**). This movement places the two cardiac tubes together (Varner and Taber 2012). The two endocardial tubes lie within the common tube for a short time, but eventually these two tubes also fuse. The bilateral origin of the heart can be demonstrated by surgically preventing the merger of the lateral

FIGURE 13.5 Migration of heart primordia. (A) Cardia bifida (two hearts) in a chick embryo, induced by surgically cutting the ventral midline, thereby preventing the two heart primordia from fusing. (B) Wild-type zebrafish and (C) *miles apart* mutant, stained with probes for the cardiac myosin light chain. There is a lack of migration in the *miles apart* mutant. (D) Mouse heart stained with antisense RNA probe to ventricular myosin shows fusion of the heart primordia in a wild-type 13.5-day embryo. (E) Cardia bifida in a *Foxp4*-deficient mouse embryo. Interestingly, each of these hearts has ventricles and atria, and they both loop and form all four chambers with normal left-right asymmetry. (A courtesy of R. L. DeHaan; B,C from Kupperman et al. 2000, courtesy of Y. R. Didier; D,E from Li et al. 2004, courtesy of E. E. Morrisey.)

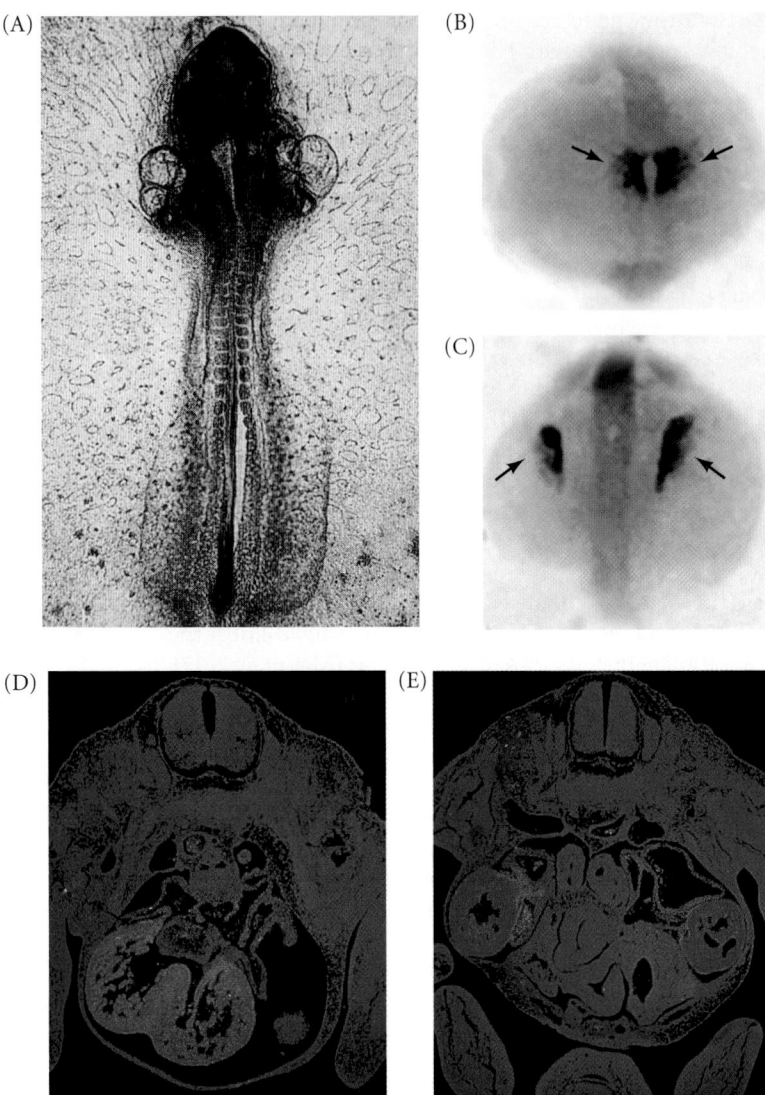

plate mesoderm (Gräper 1907; DeHaan 1959). This manipulation results in a condition called **cardia bifida**, in which two separate hearts form, one on each side of the body (**FIGURE 13.5A**). Thus, endoderm specifies heart progenitors, gives directionality to their migration, and mechanically pulls the two heart fields together.

Although the chick is an excellent model for surgical manipulation, mouse and zebrafish embryos have been more tractable genetically. In the zebrafish, heart precursor cells migrate actively from the lateral edges toward the midline. Several mutations affecting endoderm differentiation disrupt this process, indicating that, as in the chick, the endoderm is critical for cardiac precursor specification and migration. The *faust* gene, which encodes the GATA5 protein, is expressed in the endoderm and is required for the migration of cardiac precursor cells to the midline and also for their division and specification. It appears to be important in the pathway leading to activation of the zebrafish *nkx2-5* gene in the cardiac precursor cells (Reiter et al. 1999). Another particularly interesting zebrafish mutation is *miles apart*. Its phenotype is limited to cardiac precursor migration to the midline and resembles the cardia bifida seen in experimentally manipulated chick embryos (**FIGURE 13.5B,C**). Differentiation is not affected, and the fish form two normal heart tubes; the tubes are not connected properly to the blood vessels, however, and thus cannot support circulation. The *miles apart* gene encodes a protein that regulates the interactions of the cardiac cells with fibronectin, and it is expressed in the endoderm on either side of the midline (Kupperman et al. 2000; Matsui et al. 2007).

In mice, cardia bifida can also be produced by mutations of genes that are expressed in the endoderm. One of these genes, *Foxp4*, encodes a transcription factor expressed in the early foregut cells, along the pathway the cardiogenic

precursors travel toward the midline. In these mutants, each heart primordia develops *separately*, and the embryonic mouse contains two hearts, one on each side of the body (**FIGURE 13.5D,E**; Li et al. 2004).

However, as the cells of the first heart field migrate along the endoderm to form the heart tube, the cells of the second heart field remain in contact with the pharyngeal endoderm. Here they are kept in a state of proliferation by a combination of paracrine factors (probably Sonic hedgehog, Fgf8, and Wnts) (Chen et al. 2007; Lin et al. 2007). The cells of the second heart field can be distinguished by their expression of the Islet1 transcription factor. These cells will also begin to synthesize and secrete Fgf8, which acts in an autocrine manner to stimulate the cells to migrate and add themselves to the anterior and posterior portions of the heart tube formed by the first heart field progenitor cells (Park et al. 2008). The anterior region of the second heart field contributes to the right ventricle and the outflow tract, whereas its posterior

(A) Chick, stage 8

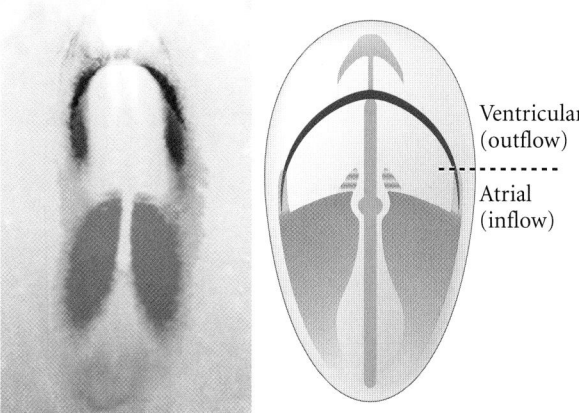

Ventricular (outflow)

Atrial (inflow)

(B) Mouse, 8 days

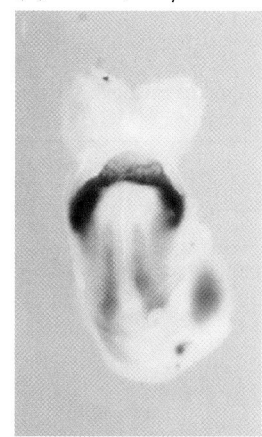

FIGURE 13.6 Double in situ hybridization for the expression of *RADH2* (orange), which encodes the retinoic acid-synthesizing enzyme retinaldehyde dehydrogenase-2; and *Tbx5* (purple), a marker for the early heart fields. In the developmental stages seen here, the heart precursor cells are exposed to progressively increasing amounts of retinoic acid. (A) Chick, stage 8 (26–29 hours). (B) Mouse, 8 days. (From Simões-Costa et al. 2005, courtesy of J. Xavier-Neto.)

region generates the atria (Zaffran et al. 2004; Verzi et al. 2005; Galli et al. 2008).

As the second heart field precursor cells migrate, the posterior region becomes exposed to increasingly higher concentrations of retinoic acid (RA) produced by the posterior mesoderm. RA is critical in specifying these posterior precursor cells to become the inflow, or "venous," portions of the heart—the sinus venosus and atria. Originally, these fates are not fixed, as transplantation or rotation experiments show that these precursor cells can regulate and differentiate in accordance with a new environment. But once the posterior cardiac precursors enter the realm of active RA synthesis, they express the gene for retinaldehyde dehydrogenase; they then can produce their own RA, and their posterior fate becomes committed (**FIGURE 13.6**; Simões-Costa et al. 2005).

Retinoic acid regulates the expression of Hox genes (especially *Hoxa1*, *Hoxb1*, and *Hoxa3*), which appear to promote different regional identities in the second heart field precursors (Bertrand et al. 2011). In mice, the outflow tract region, as well as the cardiac neural crest cells that enter this region of the second heart field, display differential Hox gene expression based on their exposure to RA (Diman et al. 2011). This ability of RA to specify and commit heart precursor to become atria explains its teratogenic effects on heart development, wherein exposure of vertebrate embryos to RA can cause expansion of atrial tissues at the expense of ventricular tissues (Stainier and Fishman 1992; Hochgreb et al. 2003).

● See VADE MECUM Early heart development

Heart cell differentiation

One of the most important recent discoveries of cardiac development has been the demonstration that all the different cells of the heart—the ventricular myocytes, the atrial myocytes, the smooth muscles that generate the venous and arterial vasculature, the endothelial lining of the heart and valves, and the epicardium that forms an envelope for the heart—are derived from the same progenitor cell type (Kattman et al. 2006; Moretti et al. 2006; Wu et al. 2006).

The heart fields contain multipotent progenitor cells. Indeed, there appears to be an early progenitor cell population that bears responsibility for forming the entire circulatory system. Under one set of influences, its descendants become **hemangioblasts**, those cells that form blood vessels and blood cells; under the conditions in the heart fields, its descendants form **multipotent cardiac precursor cells** (**FIGURE 13.7**; Anton et al. 2007). Several investigators have proposed slightly different pathways for generating these cells, but the differences may be caused by the ability of the heart precursor cells to differentiate according to their microenvironment (Linask 2003).*

INITIAL CELL DIFFERENTIATION Several genes are expressed very early during heart development (see Figure 13.3B). Nkx2-5 and Mesp1 are critical in initiating a self-sustaining gene regulatory network. One of the genes active in this network encodes the GATA4 transcription factor, which is first seen in the precardiac cells of chicks and mice when these cells emerge from the primitive streak. GATA4 is necessary for activating numerous heart-specific genes as well as for activating expression of the gene for N-cadherin, a protein that is critical for both the formation of the cardiac epithelium and the fusion of the two heart rudiments into one tube (Linask 1992; Zhang et al. 2003).

In addition to activating a group of core heart-forming genes, Mesp1 also helps activate different patterns of protein synthesis in the heart fields on each side of the embryo. Mesp1 and Nkx2-5 instruct the cells of the second heart field to express the *Foxh1* gene, which commits these heart

---

*The existence of multipotent cardiac progenitor cells is exciting because it might provide a way to restore heart tissue in adults suffering from heart attacks. Mammals are one of the relatively few groups of animals that cannot regenerate their hearts; fish and salamanders are particularly good at it. Although fewer than 1% of human heart cells divide in any year, it appears that there are multipotent cardiac progenitor cells in the epicardial layer surrounding the heart muscle (Bergmann et al. 2009; Smart et al. 2011). Such progenitors might be expanded into a population of cells that can restore heart function.

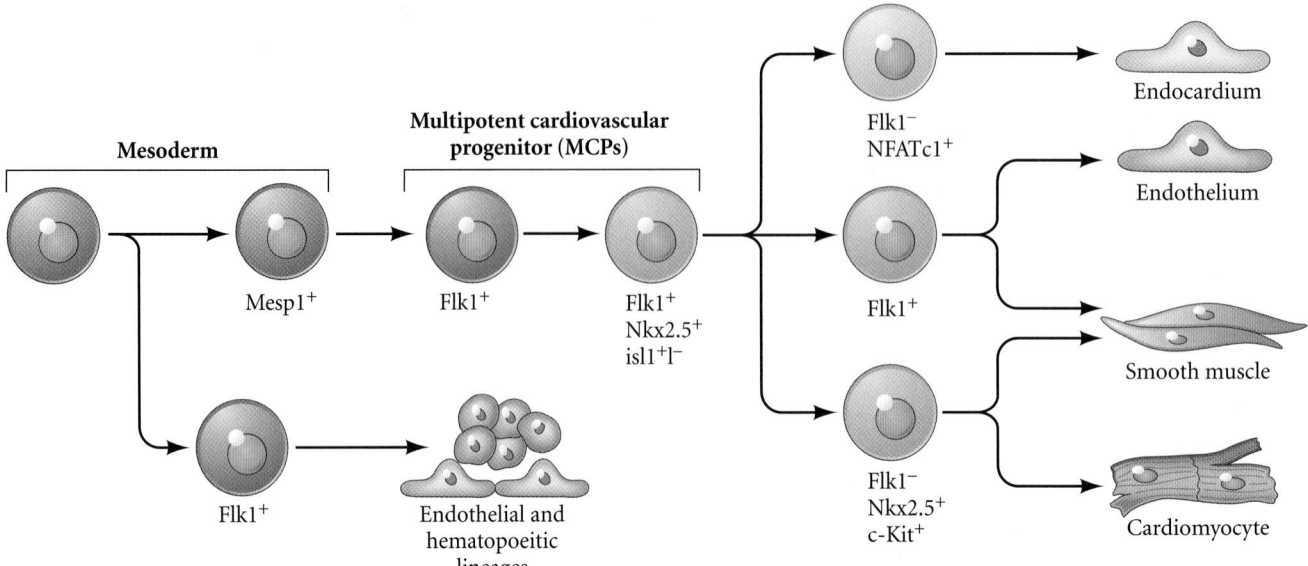

FIGURE 13.7 Model for early cardiovascular lineages. The splanchnic mesoderm gives rise to two lineages, both of which have Flk1 (a VEGF receptor) on their cell membranes. The earlier population gives rise to the hemangioblasts (precursors to blood cells and blood vessels), whereas the later population gives rise to the cardiac (heart) precursor cells. This latter population in turn gives rise to a variety of cell types whose relationships are still obscure; however, all the cell types of the heart can be traced back to the cardiac precursor cells. (After Anton 2007 and DeLaughter et al. 2011.)

precursor cells to become the right ventricle and outflow tract (von Both et al. 2004). In the first heart field, Mesp1 activates the *Tbx5* gene, whose product is critical for heart tube and left ventricle development (see Figure 13.6; Koshiba-Takeuchi et al. 2009). In these early cells, Tbx5 acts with GATA4 and Nkx2-5 to activate numerous genes involved in heart specification. Later, Tbx5 protein becomes restricted to the atria and *left* ventricle. The ventricular septum (the wall separating the left and right ventricles) is formed at the boundary between those cells expressing *Tbx5* and those that do not. Tbx5 protein works antagonistically to Tbx20, which becomes expressed in the *right* ventricle. When the *Tbx5* expression domain is ectopically expanded, the location of the ventricular septum shifts to this new location. Moreover, a conditional knockout of the mouse *Tbx5* gene—specifically inactivating it during ventricular development—leads to the formation of a lizardlike ventricle that lacks any septation (Takeuchi et al. 2003; Koshiba-Takeuchi et al. 2009). Thus, Tbx5 is extremely important in separating the left and right ventricles. Mutations in human *TBX5* cause Holt-Oram syndrome (Bruneau et al. 1999), characterized by abnormalities of the heart and upper limbs.

Fusion of the heart rudiments and initial heartbeats

Cell differentiation occurs independently in the right and left heart-forming primordia (each containing first and second heart fields). As they migrate toward each other, the ventral splanchnic mesoderm cells of the primordia begin to express N-cadherin on their apices, sort out from the somatic (parietal) mesoderm cells, and join together to form an epithelial layer. This joining of the somatic mesoderm will lead to the formation of the pericardial cavity, the sac in which the heart is formed (Linask 1992). A small population of splanchnic mesoderm then downregulates N-cadherin and delaminates from the epithelium to form the **endocardium**, the lining of the heart that is continuous with the blood vessels (see Figure 13.4C,D). Indeed, these cells may already have been specified as endocardial precursors in the second heart field (Milgrom-Hoffman et al. 2011). The endocardial cells produce many of the heart valves, secrete the proteins that regulate myocardial growth, and regulate the placement of nervous tissue in the heart. The epithelial layer of splanchnic mesoderm forms the **myocardium** (Manasek 1968; Linask et al. 1997), which will give rise to the cardiac muscles that will pump for the lifetime of the organism.

Pulsations of the chick heart begin while the paired primordia are still fusing. Heart muscle cells develop an inherent ability to contract, and isolated heart cells from 7-day rat or chick embryos will continue to beat when placed in petri dishes (Harary and Farley 1963; DeHaan 1967; Imanaka-Yoshida et al. 1998). The pulsations are made possible by the appearance of the sodium-calcium exchange pump in the muscle cell membrane; inhibiting this channel's function prevents the heartbeat from starting (Wakimoto et al. 2000; Linask et al. 2001). Eventually, the rhythmicity of the heartbeat becomes coordinated by the sinus venosus. The electric impulses generated here initiate waves of muscle contraction through the tubular heart. In this way, the heart can pump blood even before its intricate system of valves has been completed.

SIDELIGHTS & SPECULATIONS

A Minimalist Heart

Both the chick and the mammalian heart are complex, rather baroque structures. Even the two-chambered heart of zebrafish (one atrium and one ventricle) displays a complex organogenesis, with a first and second heart field (Zhou et al. 2011). For a simpler model, we can look to the tunicates, the closest invertebrate relative of the vertebrates (see Chapter 7). Early tunicate development is accomplished with very few cells, and the larval heart rudiment is a single chamber consisting of about two dozen cells. Despite this remarkable simplicity, tunicate heart development displays numerous ancestral characters also found in vertebrate cardiogenesis. As in vertebrate embryos, tunicate cardiac precursor cells form bilateral cell clusters that migrate anteriorly and ventrally along the endoderm and fuse at the ventral midline (Davidson et al. 2005). The few cells that form the tunicate heart appear to have the same basic pattern of transcription factors that we see in the mouse and chick heart lineages.

In the gastrulating tunicate embryo, the heart lineage is represented only by two pairs of mesodermal cells near the vegetal pole. Each side of the embryo contains a pair of B8.9 and B8.10 blastomeres that express the MesP transcription factor, just like the cardiac precursor cells of vertebrates (**FIGURE 13.8**; Davidson et al. 2006). During neurulation, each of these four **cardiac founder cells** divides asymmetrically to produce a small cell that generates the cardiac precursors and a larger cell that generates anterior muscle precursors. The anterior tail cells do not migrate, but they express retinaldehyde dehydrogenase and initiate a retinoic acid gradient that specifies cells as in vertebrate embryos (see Figure 13.6).

Moreover, as in the vertebrate cardiac precursors, it appears that FGF signaling is critical for the production of heart cells, with FGF signals combining with Mesp to induce the expression of Nkx2-5, GATA, and Hand-family transcription factors (Davidson and Levine 2003; Simões-Costa et al. 2005). The large cells and fixed cell lineages of tunicate embryos enable one to study in detail the mechanism of FGF signaling (see Figure 3.42).

There may even be a second heart field in tunicates. The B7.5 blastomeres, which give rise to the B8.9 and B8.10 cells, also give rise to the lineage that forms the atrial siphon muscles. Interestingly, these cells express the same transcription factors as the second heart field precursors (Stolfi et al. 2010) and may be homologous to the vertebrate head musculature (which also shares a cellular origin with the second heart field.) Thus, the tunicate, with its small number of cells, fixed cell lineages, and rapid development, may become an excellent model organism in which to study the complex organogenesis seen in the vertebrate heart.

FIGURE 13.8 Heart development in the tunicate *Ciona*. (A) In the tailbud-stage embryo, transgenic *MesP-GFP* glows in regions where *MesP* is activated by Tbx6 in the B8.9 and B8.10 blastomeres. (B) At a slightly later stage, the heart precursors (green) migrate into the head region. (C) Ventrolateral view in which both left and right heart and muscle precursors can be observed. Cell divisions are forming both the heart (left) and the anterior muscles. (From Davidson et al. 2006.)

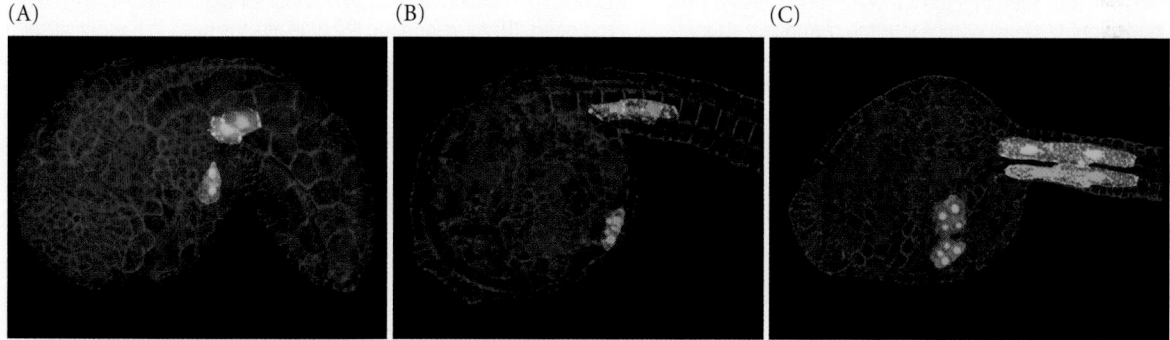

(A)   (B)   (C)

Eventually, the rhythm of the heartbeat will be governed by the "pacemaker," the sinoatrial node of the right atrium. The pacemaker cells appear to be cardiac myocytes that have been transformed into electrical conductive tissue by Tbx3. Tbx3 appears to (1) repress those genes involved in atrial myocyte differentiation and those genes (such as those encoding gap junction proteins) that enabled the earlier contractions, and (2) activate those genes whose products increase the resting potential of the sinoatrial cells and their conducting network throughout the heart (Hoogaars et al. 2007; Bakker et al. 2012).

Looping and formation of heart chambers

In 3-day chick embryos and 4-week human embryos, the heart is a two-chambered tube, with an atrium to receive blood and a ventricle to pump blood out. (In the chick

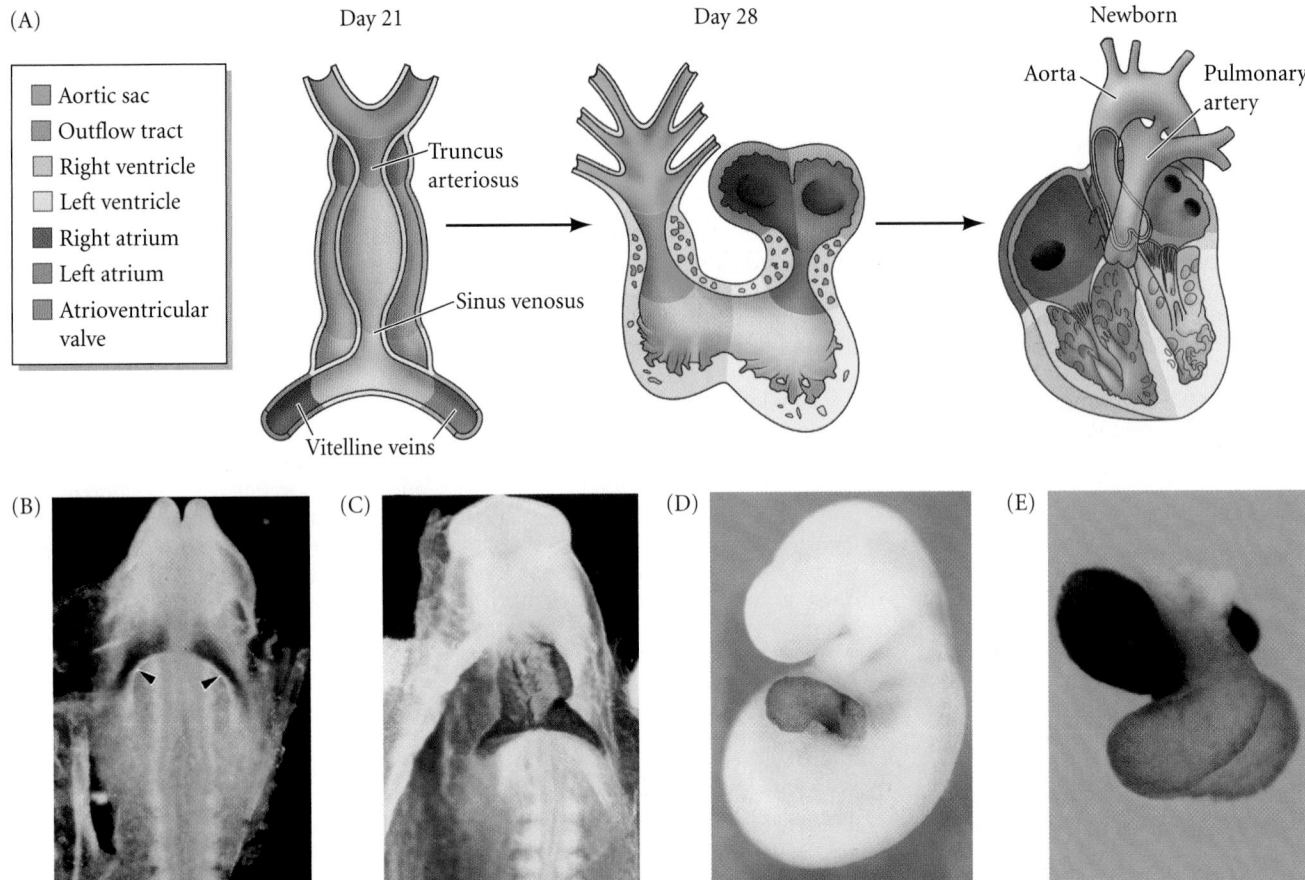

FIGURE 13.9 Cardiac looping and chamber formation. (A) Schematic diagram of cardiac morphogenesis in humans. On day 21, the heart is a single-chambered tube. Regional specification of the tube is shown by the different colors. By day 28, cardiac looping has occurred, placing the presumptive atria anterior to the presumptive ventricles. In the newborn, the valves and chambers establish circulatory routes such that the left ventrical pumps into the aorta, while the right ventrical pumps into the pulmonary artery to the lungs. (B,C) *Xin* expression in the fusion of left and right heart primordia of a chick. The cells fated to form the myocardium are shown by staining for the *Xin* message, whose protein product is essential for the looping of the heart tube. (B) Stage-9 chick neurula, in which expression of *Xin* (purple) is seen in the two symmetrical heart-forming fields (arrows). (C) Stage-10 chick embryo, showing fusion of the two heart-forming regions prior to looping.

(D,E) Specification of the atria and ventricles occurs even before heart looping. The atria and ventricles of the mouse embryo have separate types of myosin proteins, which allows them to be differentially stained. In these photographs, atrial myosin is stained blue and ventricular myosin is stained orange. (D) In the tubular heart (prior to looping), the two myosins (and their respective stains) overlap at the atrioventricular channel joining the future regions of the heart. (E) After looping, the blue stain is seen in the definitive atria and inflow tract, while the orange stain is seen in the ventricles. The unstained region above the ventricles is the truncus arteriosus. Derived primarily from the neural crest, the truncus arteriosus becomes separated into the aorta and pulmonary arteries. (A after Srivastava and Olson 2000; B,C from Wang et al. 1999, courtesy of J. J.-C. Lin; D,E from Xavier-Neto et al. 1999, courtesy of N. Rosenthal.)

embryo, the unaided eye can see the remarkable cycle of blood entering the lower chamber and being pumped out through the aorta.) Looping of the heart converts the original anterior-posterior polarity of the heart tube into the right-left polarity seen in the adult organism. When this looping is completed, the portion of the heart tube destined to become the atria lies anterior to the portion that will become the ventricles (**FIGURE 13.9**).

The direction of heart looping is dependent on the left-right patterning proteins (Nodal and Pitx2) discussed in Chapter 8. Within the heart primordium, Nkx2-5 regulates the Hand1 and Hand2 transcription factors. Both Hand proteins appear to be synthesized throughout the early heart tube, but as looping commences, Hand1 becomes restricted to the future left ventricle and Hand2 to the right. Without these proteins, looping is abnormal, since the ventricles fail to form properly* (Srivastava et al. 1995; Biben and Harvey 1997).

---

*Zebrafish, with only one ventricle, have only one type of Hand protein. When the gene encoding this protein is mutated, the entire ventricular portion of the heart fails to form (Srivastava and Olson 2000). Nongenetic agents are also critical in normal zebrafish heart formation. In the absence of high-pressure blood flow, heart looping, chamber formation, and valve development are impaired (Hove et al. 2003).

(A) Day 33

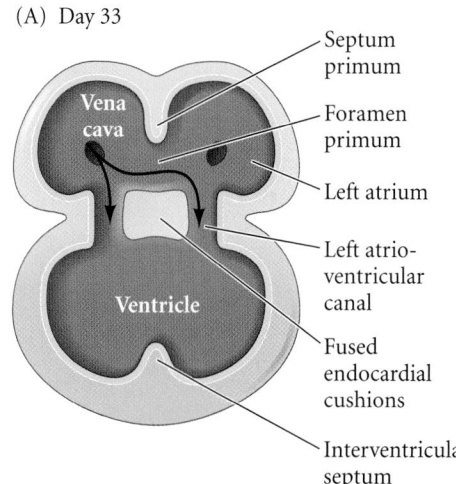

Septum primum

Foramen primum

Left atrium

Left atrio-ventricular canal

Fused endocardial cushions

Interventricular septum

Vena cava

Ventricle

(B) Third month

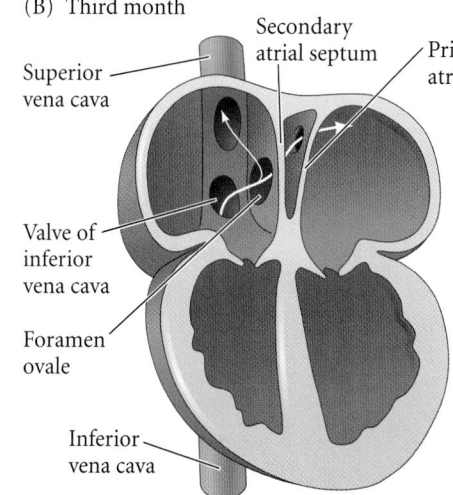

Secondary atrial septum

Primary atrial septum

Superior vena cava

Valve of inferior vena cava

Foramen ovale

Inferior vena cava

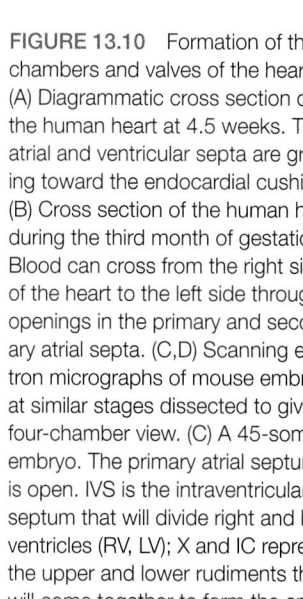

FIGURE 13.10 Formation of the chambers and valves of the heart. (A) Diagrammatic cross section of the human heart at 4.5 weeks. The atrial and ventricular septa are growing toward the endocardial cushions. (B) Cross section of the human heart during the third month of gestation. Blood can cross from the right side of the heart to the left side through openings in the primary and secondary atrial septa. (C,D) Scanning electron micrographs of mouse embryos at similar stages dissected to give a four-chamber view. (C) A 45-somite embryo. The primary atrial septum is open. IVS is the intraventricular septum that will divide right and left ventricles (RV, LV); X and IC represent the upper and lower rudiments that will come together to form the endocardial cushions separating atria from ventricles. (D) A 48-somite embryo, when the muscular ventricles are forming and the four chambers are beginning to form their boundaries. (A,B after Larsen 1993; C,D from Webb et al. 1998.)

(C)

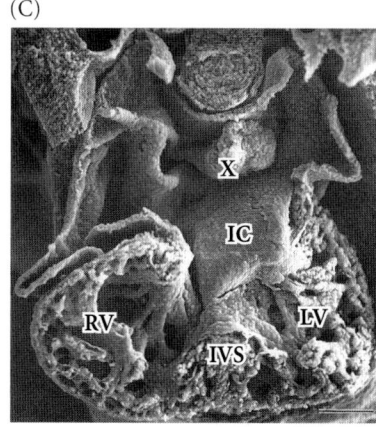

(D)

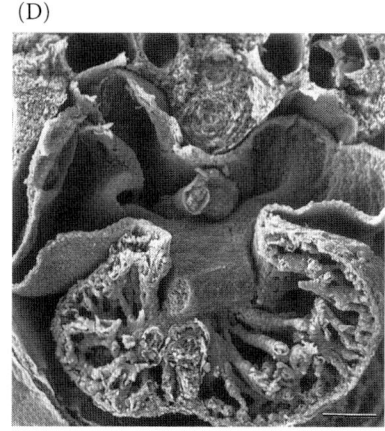

Cytoskeletal proteins and their adhesion to the extracellular matrices are very important in these turning events, suggesting that biophysical forces (mechanotransduction; see Sidelights & Speculations, p. 435) also are important for the looping process (Hove et al. 2003; Garita et al. 2011). First, the myosin-like protein **flectin** regulates the physical tension of the heart tissues differently on different sides of the heart (Tsuda et al. 1996; Lu et al. 2008). Second, the transcription factors Nkx2-5 and Mef2C activate the *Xin* (Chinese for "heart") gene, whose product binds to actin microfilaments and mediates cytoskeletal changes that are essential for heart looping (Wang et al. 1999; Grosskurth et al. 2008). Third, metalloproteinases—especially metalloproteinase-2 (MMP2)—are critical for remodeling the cytoskeleton. If *MMP2* gene expression is blocked, the extracellular matrix fails to change, the asymmetric cell divisions (which cause the left dorsal side to grow faster than the right) fail to occur, and heart looping stops (Linask et al. 2005).

The formation of the heart valves—four leaflike flaps that must open and shut without failure once each second for the duration of an individual's life—is just starting to be understood. In mammals, **endocardial cushions** form from the

endocardium and divide the tube into right and left atrioventricular channels. Meanwhile, the primitive atrium is partitioned by two **septa** that grow ventrally toward the endocardial cushions (**FIGURE 13.10A**). The cushions are thought to serve as the "glue" for the normal septation of the mammalian heart into four chambers (Webb et al. 1998; Sizarov et al. 2012). Moreover, they are thought to serve as valves during early heart development (Lamers et al. 1988). As the heart enlarges, however, specialized valves develop to prevent the return of blood into the atria and to prevent the mixing of bloods from the two sides of the heart (**FIGURE 13.10B**). These valves begin to form when cells from the myocardium produce a factor that causes cells from the adjacent endocardium to detach and enter the hyaluronate-rich "cardiac jelly" extracellular matrix between the two layers (Markwald et al. 1977; Potts et al. 1991). In mammals, BMP2 appears to be necessary for inducing this epithelial-mesenchymal transition (EMT) and for forming the endocardial cushions from cardiac myocytes (Ma et al. 2005; Rivera-Feliciano and Tabin 2006). BMP2 induces the transcription factors Tbx2 and Tbx3, which promote EMT by activating the genes encoding TGF-β2, Twist, and the enzymes that synthesize the

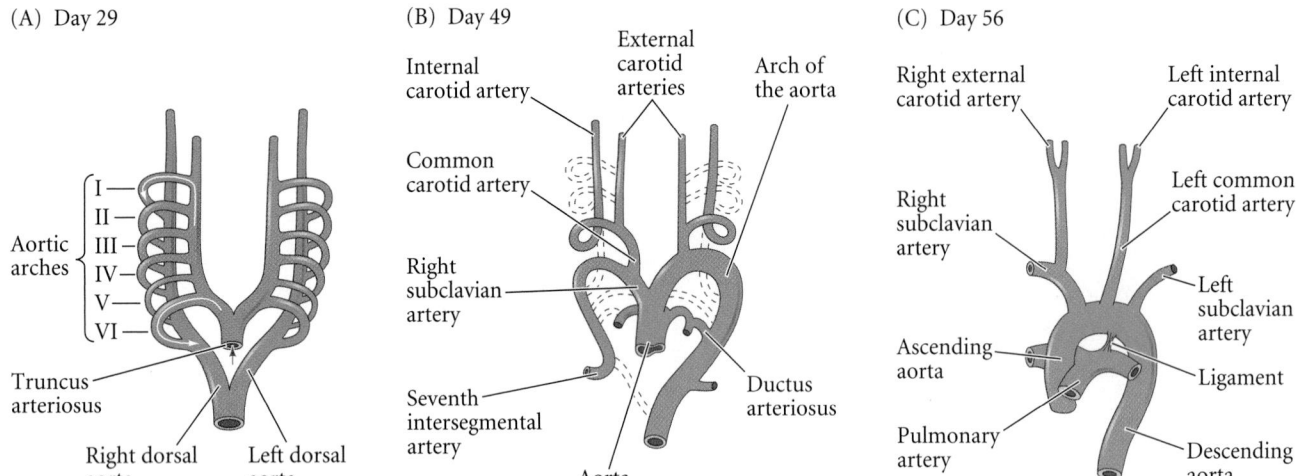

(A) Day 29

Aortic arches { I — II — III — IV — V — VI — }

Truncus arteriosus

Right dorsal aorta   Left dorsal aorta

(B) Day 49

Internal carotid artery

External carotid arteries

Arch of the aorta

Common carotid artery

Right subclavian artery

Seventh intersegmental artery

Aorta

Ductus arteriosus

(C) Day 56

Right external carotid artery

Left internal carotid artery

Right subclavian artery

Left common carotid artery

Left subclavian artery

Ascending aorta

Ligament

Pulmonary artery

Descending aorta

FIGURE 13.11 Aortic arches of the human embryo. (A) Originally, the truncus arteriosus pumps blood into the aorta, which branches on either side of the foregut. The six aortic arches take blood from the truncus arteriosus and allow it to flow into the dorsal aorta. (B) As development proceeds, arches begin to disintegrate or become modified (the dotted lines indicate degenerating structures). The first two pairs of arches completely disappear. The third arches become the internal carotid arteries. The right fourth arch becomes the right subclavian artery, and the left fourth arch becomes the major arch of the aorta. The fifth arches disintegrate. The right sixth arch disappears, but the left sixth arch gives rise to the pulmonary arteries and the ductus arteriosus (which will start closing on the first breath). (C) Eventually, the remaining arches are modified and the adult arterial system is formed. However, numerous variations of this scheme are found in the human population.

hyaluronic acid that separates the cells and becomes a major part of the cardiac jelly (Shirai et al. 2009).

With the formation of the septa (which usually occurs in the seventh week of human development), the heart is a four-chambered structure with the pulmonary artery connected to the right ventricle and the aorta connected to the left ventricle (**FIGURE 13.10C,D**). The septa between the fetal atria have openings in them, however, so blood can still cross from one side of the heart into the other. This crossing of blood is needed for the survival of the fetus before circulation to functional lungs has been established. Upon the infant's first breath, the septal openings close and the left and right circulatory loops are established.

Blood Vessel Formation

Although the heart is the first functional organ of the body, it does not begin to pump until the vascular system of the embryo has established its first circulatory loops. Rather than sprouting from the heart, the blood vessels form independently, linking up to the heart soon afterward. Everyone's circulatory system is different, since the genome cannot encode the intricate series of connections between the arteries and veins. Indeed, chance plays a major role in establishing the microanatomy of the circulatory system. However, all circulatory systems in a given species look very much alike because the development of the circulatory system is severely constrained by physiological, evolutionary, and physical parameters.

Constraints on the construction of blood vessels

The first constraint on vascular development is *physiological*. Unlike new machines, which do not need to function until they leave the assembly line, new organisms have to function even as they develop. Embryonic cells must obtain nourishment before there is an intestine, use oxygen before there are lungs, and excrete wastes before there are kidneys. All these functions are mediated through the embryonic circulatory system. Therefore, the circulatory physiology of the developing embryo must differ from that of the adult organism. Food is absorbed not through the intestine, but from either the yolk or the placenta, and respiration is conducted not through the gills or lungs, but through the chorionic or allantoic membranes (see Figure 9.1). The major embryonic blood vessels must be constructed to serve these extraembryonic structures.

The second constraint is *evolutionary*. The mammalian embryo extends blood vessels to the yolk sac even though there is no yolk inside (see Figure 13.16B). Moreover, the blood leaving the heart via the truncus arteriosus passes through vessels that loop over the foregut to reach the dorsal aorta. Six pairs of these aortic arches loop over the pharynx (**FIGURE 13.11**). In primitive fish, these arches persist and enable the gills to oxygenate the blood through the gills. In adult birds and mammals, in which lungs oxygenate the blood, such a system makes little sense—but all six pairs of aortic arches are formed in mammalian and avian embryos before the system eventually becomes simplified into a single aortic arch. Thus, even though our physiology does not

SIDELIGHTS & SPECULATIONS

Redirecting Blood Flow in the Newborn Mammal

Although the developing fetus shares with the adult the need to get oxygen and nutrients to its tissues, the physiology of the fetus differs dramatically from that of the adult. Chief among the differences is the fetus's lack of functional lungs and intestines. All of its oxygen and nutrients must come from the placenta. This observation raises two questions. First, how does the fetus obtain oxygen and nutrients from maternal blood? And second, how is blood circulation redirected to the lungs once the umbilical cord is cut and breathing becomes necessary?

Human embryonic circulation

The human embryonic circulatory system is a modification of that used in birds and reptiles. (The circulatory system to and from the chick embryo and yolk sac is shown in Figure 1.16.) Blood pumped through the dorsal aorta passes over the aortic arches and down into the embryo. Some of this blood leaves the embryo through the vitelline arteries and enters the yolk sac. Nutrients and oxygen are absorbed from the yolk, and blood returns through the vitelline veins to re-enter the heart through the sinus venosus.

Mammalian embryos obtain food and oxygen from the placenta. Thus, although the embryo has vessels homologous to the vitelline veins, the main supply of food and oxygen comes from the umbilical vein, which unites the embryo with the placenta (FIGURE 13.12). This vein, which brings oxygenated, food-laden blood into the embryo, is derived from what would be the right vitelline vein in birds. The umbilical artery, carrying wastes to the placenta, is derived from what would have become the allantoic artery of the chick. It extends from the caudal portion of the aorta and proceeds along the allantois and then out to the placenta.

Fetal hemoglobin

The solution to the fetus's problem of getting oxygen from its mother's blood involves the development of a specialized fetal hemoglobin. The hemoglobin in fetal red blood cells differs slightly from that in adult corpuscles. Two of the four peptides—the alpha (α) chains—that make up fetal and adult hemoglobin chains are identical, but adult hemoglobin has two beta (β) chains, whereas the fetus has two gamma (γ) chains (FIGURE 13.13). The β chains bind the natural regulator diphosphoglycerate, which assists in the unloading of oxygen. The γ chain proteins do not bind diphosphoglycerate as well, and therefore have a higher affinity for oxygen. In the low-oxygen environment of the placenta, oxygen is released from adult hemoglobin. In this same environment, fetal hemoglobin does not release oxygen,

but binds it. This small difference in oxygen affinity mediates the transfer of oxygen from the mother to the fetus. In the fetus, the myoglobin of the fetal muscles has an even higher affinity for oxygen, so oxygen molecules pass from fetal hemoglobin to the fetal muscles.

Fetal hemoglobin is not deleterious to the newborn, and in humans, the replacement of fetal hemoglobin-containing blood cells with adult hemoglobin-containing blood cells is not complete until about 6 months after birth (the molecular basis for this switch in globins was discussed in Chapter 2).

From fetal to newborn circulation

Once the fetus is no longer obtaining its oxygen from its mother, how does it restructure its circulation to get oxygen from its own lungs? During fetal

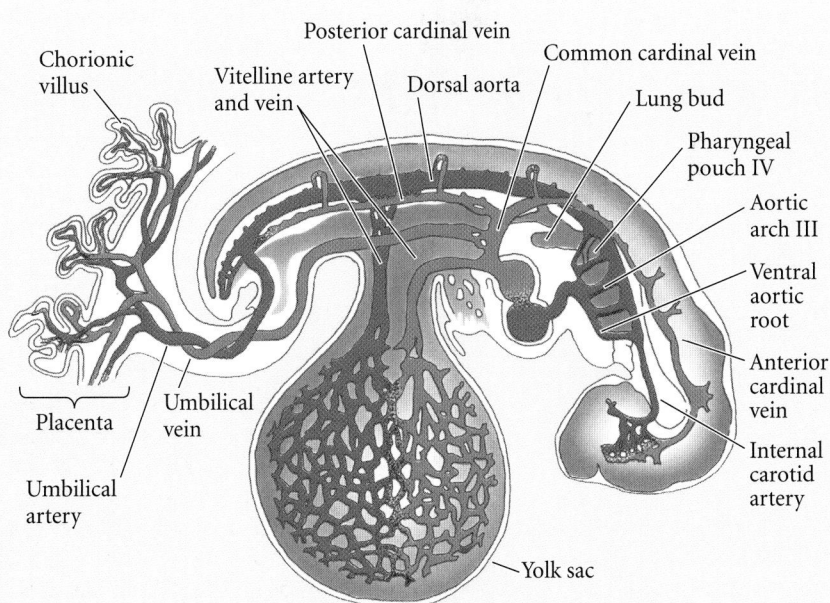

FIGURE 13.12 Circulatory system of a 4-week human embryo. Although at this stage all the major blood vessels are paired left and right, only the right vessels are shown here. Arteries are shown in red, veins in blue. (After Carlson 1981.)

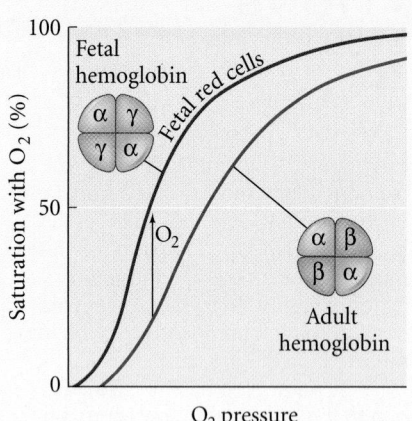

FIGURE 13.13 Adult and fetal hemoglobin molecules differ in their globin subunits. The fetal γ chain binds diphosphoglycerate less avidly than does the adult β chain. Consequently, fetal hemoglobin can bind oxygen more efficiently than can adult hemoglobin. In the placenta, there is a net flow (arrow) of oxygen from the mother's blood (which gives up oxygen to the tissues at lower oxygen pressures) to the fetal blood (which at the same pressure is still taking up oxygen).

development, an opening—the **ductus arteriosus**—diverts blood from the pulmonary artery into the aorta (and thus to the placenta). Because blood does not return from the pulmonary vein in the fetus, the developing mammal has to have some other way of getting blood into the left ventricle to be pumped. This is accomplished by the **foramen ovale**, an opening in the septum separating the right and left atria. Blood can enter the right atrium, pass through the foramen into the left atrium, and then enter the left ventricle (**FIGURE 13.14**). When the first breath is drawn, blood pressure in the left side of the heart increases. This pressure closes the septa over the foramen ovale, thereby separating the pulmonary and systemic circulations. Moreover, the decrease in prostaglandins experienced by the newborn causes the muscles surrounding the ductus arteriosus to close that opening as well (Nguyen et al. 1997). Thus, when breathing begins, the respiratory circulation is shunted from the placenta to the lungs.

In some infants, the septa fail to close and the foramen ovale is left open. Indeed, atrial and ventricular septum defects are among the most common congenital anomalies. Usually the atrial opening is so small that there are no physical symptoms, and the foramen eventually closes. If it does not close completely, however, and the secondary septum fails to form, the atrial septal opening may cause enlargement of the right side of the heart, which can lead to heart failure in early adulthood. This fine-tuning of septal growth is controlled by miRNA1-2, a microRNA that regulates translation of several proteins involved in cardiac muscle growth and electrical conduction (Zhao et al. 2007).

FIGURE 13.14 Redirection of human blood flow at birth. The expansion of air into the lungs causes pressure changes that redirect the flow of blood in the newborn infant. The ductus arteriosus squeezes shut, breaking off the connection between the aorta and the pulmonary artery, and the foramen ovale, a passageway between the left and right atria, also closes. In this way, pulmonary circulation is separated from systemic circulation.

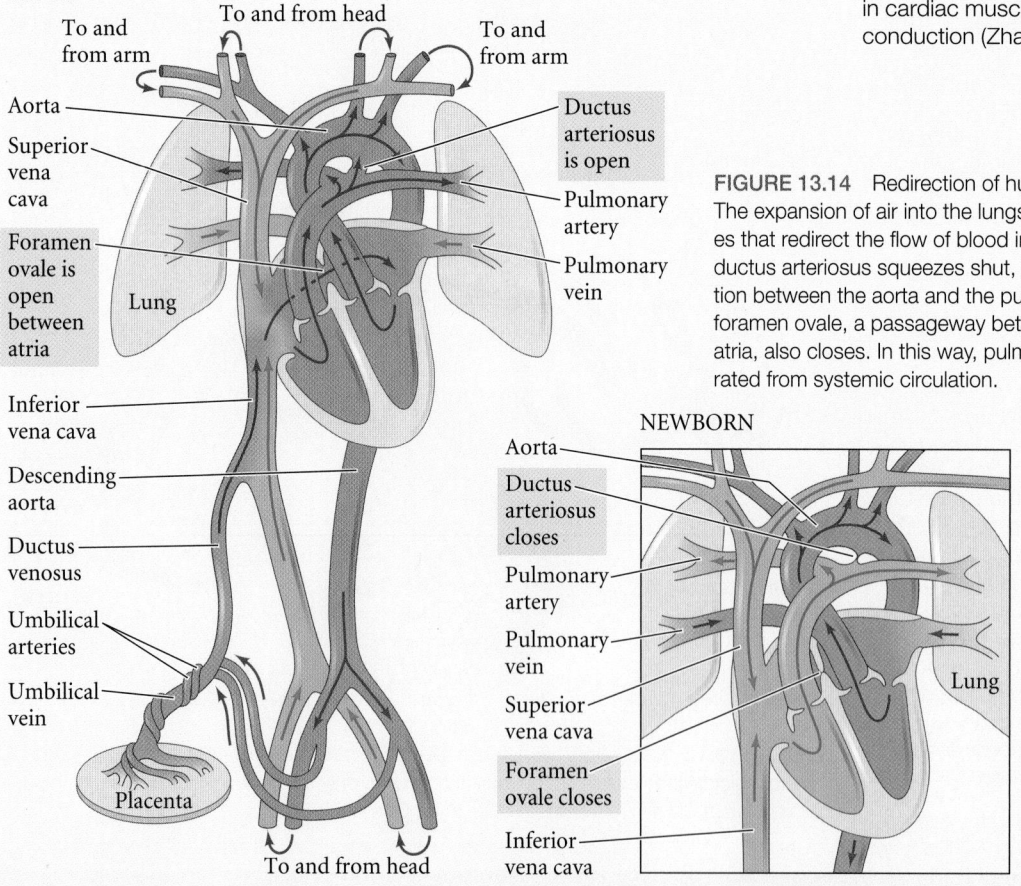

require such a structure, our embryonic condition reflects our evolutionary history.

The third set of constraints is *physical*. According to the laws of fluid movement, the most effective transport of fluids is performed by large tubes. As the radius of a blood vessel gets smaller, resistance to flow increases as r^{-4} (Poiseuille's law). A blood vessel that is half as wide as another has a resistance to flow 16 times greater. However, diffusion of nutrients can take place only when blood flows slowly and has access to cell membranes. So here is a paradox: the constraints of diffusion mandate that vessels be small, whereas the laws of hydraulics mandate that vessels be large. This paradox has been solved by the evolution of circulatory systems with a hierarchy of vessel sizes (LaBarbera 1990). In dogs, for example, blood in the large vessels (aorta and vena cava) flows more than 100 times faster than it does in the capillaries. With a system of large vessels specialized for transport and small vessels specialized for diffusion (where the blood spends most of its time), nutrients and oxygen can reach the individual cells of the growing organism. This hierarchy is seen very early in development; it is already well established in the 3-day chick embryo.

But this is not the entire story. If fluid under constant pressure moves directly from a large-diameter tube into a small-diameter tube (as in a hose nozzle), the fluid velocity increases. The evolutionary solution to this problem was the emergence of many smaller vessels branching out from a larger one, making the collective cross-sectional area of all the smaller vessels greater than that of the larger vessel. Circulatory systems show a relationship (known as Murray's law) in which the cube of the radius of the parent vessel approximates the sum of the cubes of the radii of the smaller vessels. Computer models of blood vessel formation must take into account not only gene expression patterns but also the fluid dynamics of blood flow (Gödde and Kurz 2001). The construction of any circulatory system negotiates among all of these physical, physiological, and evolutionary constraints.

Vasculogenesis: The initial formation of blood vessels

The development of blood vessels occurs by two temporally separate processes: **vasculogenesis** and **angiogenesis** (**FIGURE 13.15**). During vasculogenesis, a network of blood

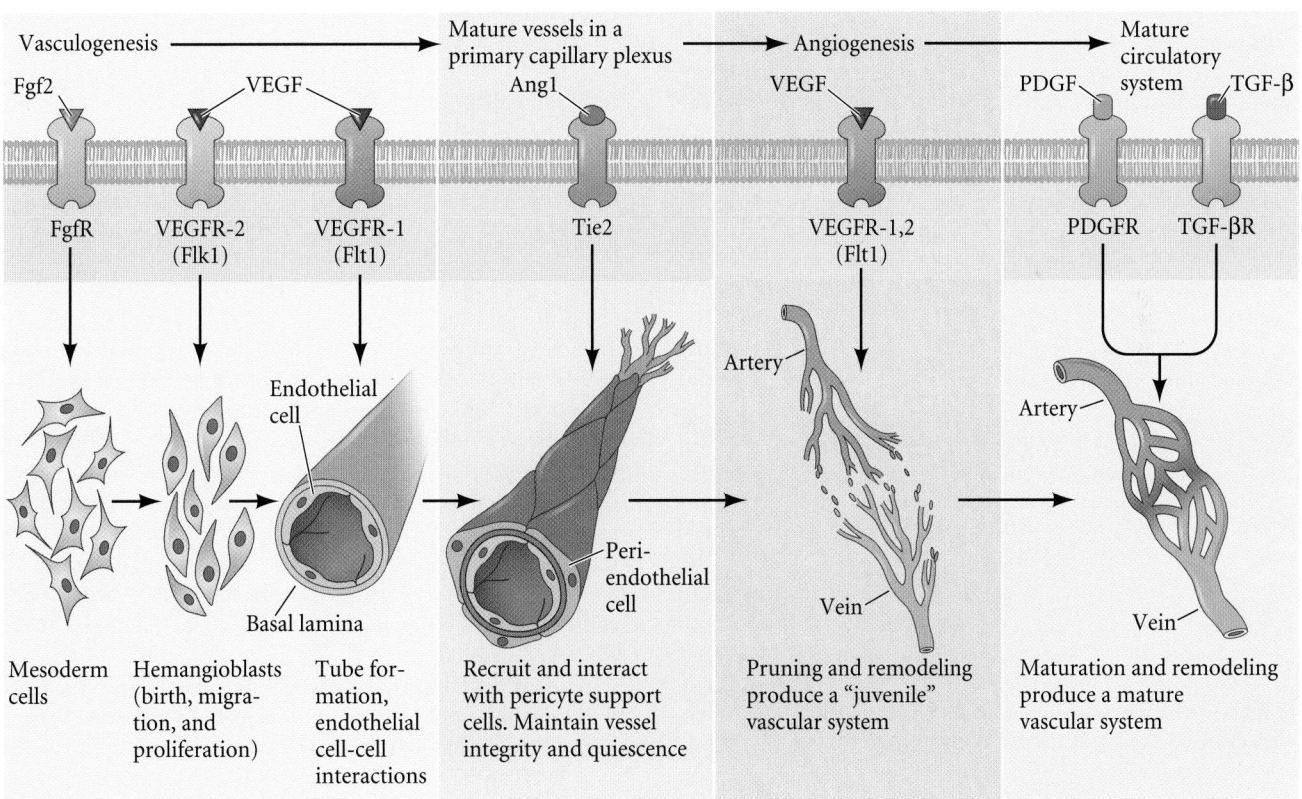

FIGURE 13.15 Vasculogenesis and angiogenesis. Vasculogenesis involves the formation of blood islands and the construction of capillary networks from them. Angiogenesis involves the formation of new blood vessels by remodeling and building on older ones. Angiogenesis finishes the circulatory connections begun by vasculogenesis. The major paracrine factors involved in each step are shown at the top of the diagram, and their receptors (on the vessel-forming cells) are shown beneath them. (After Hanahan 1997 and Risau 1997.)

vessels is created de novo from the lateral plate mesoderm. During angiogenesis, this primary network is remodeled and pruned into a distinct capillary bed, arteries, and veins.

In the first phase of vasculogenesis, cells leaving the primitive streak in the posterior of the embryo become hemangioblasts,* the precursors of both blood cells and blood vessels. Labeling zebrafish embryos with fluorescent probes to make single-cell fate maps confirms that hemangioblasts are the common progenitor for both the hematopoietic (blood) and endothelial (vascular) lineages in zebrafish (Paik and Zon 2010). This population of bipotential progenitor cells is found only in the ventral portion of the aorta, the region that had been known to produce these two cell types. The pathway permitting such aortic cells to differentiate into hemangioblasts appears to be induced by the *Cdx4* gene, while the determination of whether the hemangioblast becomes a blood cell precursor or a blood vessel precursor is regulated by the Notch signaling pathway. Notch signaling increases the conversion of hemangioblasts into blood cell precursors, whereas reduced amounts of Notch cause hemangioblasts to become endothelial (Vogeli et al. 2006; Hart et al. 2007; Lee et al. 2009). Notch signaling activates the expression of the Runx1 transcription factor, which, as we shall shortly see, appears to be conserved throughout vertebrates in inducing the conversion of endothelial cells to blood stem cells (Burns et al. 2005, 2009).

In amniotes, however, there is much more controversy about hemangioblasts, and there may be more than one pathway by which to generate blood vessels and blood cells (Ueno and Weissman 2006; Weng et al. 2007). Hemangioblasts in the splanchnic mesoderm condense into aggregations that are often called **blood islands**† (Shalaby et al. 1997; Huber et al. 2004). It is generally thought that the inner cells of these blood islands become blood progenitor cells, while the outer cells become **angioblasts**, the progenitors of blood vessels. In the second phase of vasculogenesis, the angioblasts multiply and differentiate into **endothelial cells**, which form the

---

* A combination of BMP, Wnt, and Notch signals activate the Etv2 transcription factor in these lateral plate mesoderm cells, converting them into hemangioblasts. The prefixes *hem-* and *hemato-* refer to blood (as in hemoglobin). Similarly, the prefix *angio-* refers to blood vessels. The suffix *-blast* denotes a rapidly dividing cell, usually a stem cell. The suffixes *-poiesis* and *-poietic* refer to generation or formation (*poeisis* is also the root of the word *poetry*). Thus, hematopoietic stem cells are those cells that generate the different types of blood cells. The Latin suffix *-genesis* (as in *angiogenesis*) means the same as the Greek *-poiesis*. The names of zebrafish hematopoietic mutants can be very poetic. Most are named after wines, and one of the genes producing a bloodless phenotype is named *vlad tepes,* after the historic Vlad Dracula.

† Again, the endoderm plays a major role in lateral plate mesoderm specification. Here, the visceral endoderm of the splanchnopleure interacts with the yolk sac mesoderm to induce the blood islands. The endoderm is probably secreting Indian hedgehog, a paracrine factor that activates BMP4 expression in the mesoderm. BMP4 expression feeds back on the mesoderm itself, causing it to form hemangioblasts (Baron 2001).

lining of the blood vessels. In the third phase, the endothelial cells form tubes and connect to form the **primary capillary plexus**, a network of capillaries.

SITES OF VASCULOGENESIS In amniotes, formation of the primary vascular networks occurs in two distinct and independent regions. First, **extraembryonic vasculogenesis** occurs in the blood islands of the yolk sac. These are the blood islands formed by the hemangioblasts, and they give rise to the early vasculature needed to feed the embryo and also to a red blood cell population that functions in the early embryo (**FIGURE 13.16A**). Second, **intraembryonic vasculogenesis** forms the dorsal aorta, and vessels from this large vessel connect with capillary networks that form from mesodermal cells within each organ.

The aggregation of endothelium-forming cells in the yolk sac is a critical step in amniote development, for the blood islands that line the yolk sac produce the veins that bring nutrients to the embryo and transport gases to and from the sites of respiratory exchange (**FIGURE 13.16B**). In birds, these vessels are called the vitelline veins; in mammals, they are called the **omphalomesenteric**, or more usually, **umbilical veins**. In the chick, blood islands are first seen in the area opaca, when the primitive streak is at its fullest extent (Pardanaud et al. 1987). They form cords of hemangioblasts, which soon become hollowed out and become the flat endothelial cells lining the vessels (while the central cells give rise to blood cells). As the blood islands grow, they eventually merge to form the capillary network draining into the two vitelline veins, which bring food and blood cells to the newly formed heart.

High-resolution time-lapse movies of zebrafish vascular vessel formation in culture and in transparent embryos have shown that the cells get together in groups and form the lumen of the blood vessel by the fusion of intracellular vacuoles (Ellertsdóttir et al. 2010). It appears that there are two major mechanisms used in forming the lumen. First, the endothelial cells aggregate; then fluid-filled vacuoles (formed by endocytosis) fuse with the cell membrane at those sites where the cells meet vertically. The result is a fluid-filled lumen between the cells (**FIGURE 13.17A**; Blum et al. 2008). In a second manner, vessels can be formed by intracellular endocytosis and the organization of a channel *within* the cell. The fusion of these cells at the tips of each cell would result in formation of the vessel (**FIGURE 13.17B**; Kamei et al. 2006).

The intraembryonic vascular networks usually arise from individual angioblast progenitor cells in the mesoderm surrounding a developing organ. These cells do not appear to be associated with blood cell formation (Noden 1989; Pardanaud et al. 1989; Risau 1995). It is important to realize that these intraembryonic capillary networks arise in or around the organ itself and are not extensions of larger vessels. Indeed, in some cases the developing organ produces paracrine factors that induce blood vessels to form only in its own mesenchyme (Auerbach et al. 1985; LeCouter et al. 2001). This

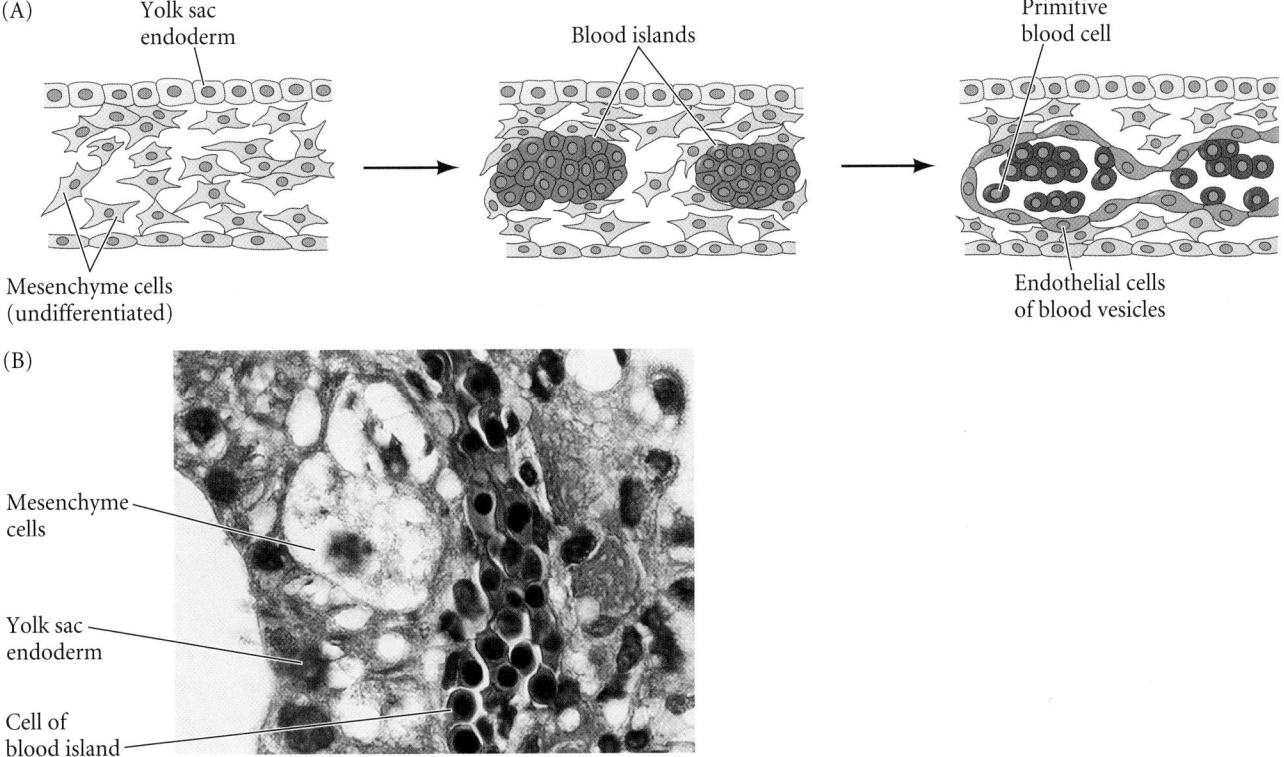

FIGURE 13.16 Vasculogenesis. (A) Blood vessel formation is first seen in the wall of the yolk sac, where undifferentiated mesenchyme cells cluster to form blood islands. The centers of these clusters form the blood cells, and the outsides of the clusters develop into blood vessel endothelial cells. (B) A human blood island in the mesoderm surrounding the yolk sac. (The photomicrograph is from a tubal pregnancy—an embryo that had to be removed because it implanted in an oviduct rather than in the uterus.) (A after Langman 1981; B from Katayama and Kayano 1999, courtesy of the authors.)

allows each capillary network to have its own specific properties. For instance, the capillary network that forms in the brain is modified by Wnt proteins to produce the extracellular matrices of the blood-brain barrier and to express the glucose transporter proteins that enable the brain to consume 25% of the body's oxygen (Stenman et al. 2008). In limbs, the chondrogenic nodules that form the bones produce VEGF paracrine factors to generate blood vessels in the surrounding mesenchyme (Eshkar-Oren et al. 2009).

GROWTH FACTORS AND VASCULOGENESIS Three growth factors are critically responsible for initiating vasculogenesis (see Figure 13.15). One of these, **basic fibroblast growth factor (Fgf2)**, is required for the generation of hemangioblasts from the splanchnic mesoderm. When cells from quail blastodiscs are dissociated in culture, they do not form blood islands or endothelial cells. However, when these cells are cultured with Fgf2 protein, blood islands emerge and form endothelial cells (Flamme and Risau 1992). Fgf2 is synthesized in the chick embryonic chorioallantoic membrane and is responsible for the vascularization of this tissue (Ribatti et al. 1995).

The second family of proteins involved in vasculogenesis is the **vascular endothelial growth factors (VEGFs)**. This family includes several VEGFs, as well as placental growth factor (PlGF), which directs the expansive growth of blood vessels in the placenta. Each VEGF appears to enable the differentiation of the angioblasts and their multiplication to form endothelial tubes. The most important VEGF in normal development, VEGF-A, is secreted by the mesenchymal cells near the blood islands, and hemangioblasts and angioblasts have receptors for this VEGF* (Millauer et al. 1993). If mouse embryos lack the genes encoding either VEGF-A or

*VEGF needs to be regulated very carefully in adults, and studies indicate that it can be affected by diet. The consumption of green tea has been associated with lower incidences of human cancer and the inhibition of tumor cell growth in laboratory animals. Cao and Cao (1999) have shown that green tea and one of its components, epigallocatechin-3-gallate (EGCG), prevent angiogenesis by inhibiting VEGF. Moreover, in mice given green tea instead of water (at levels similar to humans after drinking 2–3 cups of tea), the ability of VEGF to stimulate new blood vessel formation was reduced by more than 50%.

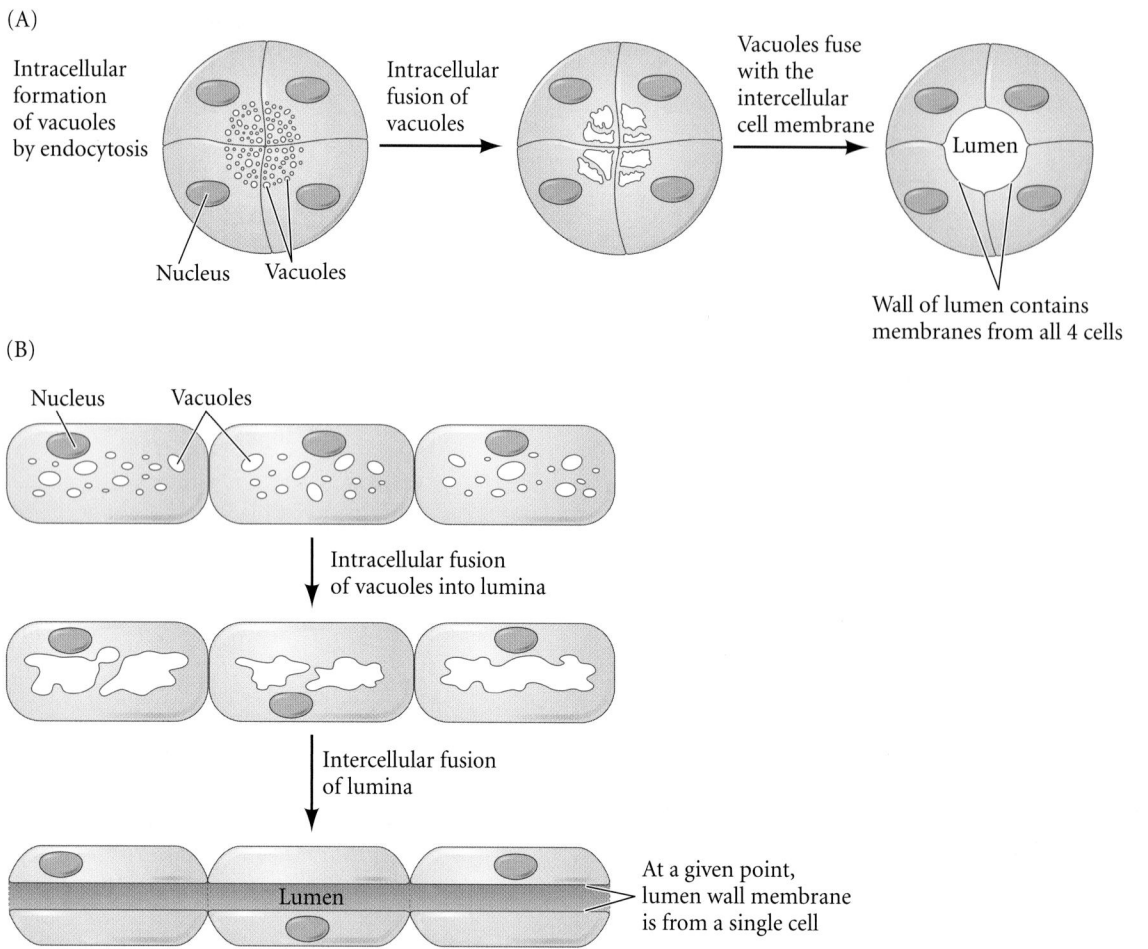

(A)

Intracellular formation of vacuoles by endocytosis

Nucleus   Vacuoles

Intracellular fusion of vacuoles

Vacuoles fuse with the intercellular cell membrane

Lumen

Wall of lumen contains membranes from all 4 cells

(B)

Nucleus   Vacuoles

Intracellular fusion of vacuoles into lumina

Intercellular fusion of lumina

Lumen

At a given point, lumen wall membrane is from a single cell

FIGURE 13.17 Forming a vascular tube. The lumen, or central space, within vascular tubes can be formed by the fusion of intracellular vacuoles. (A) In one scenario, vacuoles form within the endothelial cells by endocytosis: vacuoles merge with other vacuoles to form larger vacuoles. Large vacuoles then fuse with the cell membrane at the point where the cells come together, forming the lumen. (B) Alternatively, vacuoles may form intracellular lumina within each cell; these individual lumina then fuse such that in any single portion of the final lumen, the lining is made up of membranes from the same cell. (A after Mostov and Martin-Belmonte 2006; B after Kamei et al. 2006.)

its major receptor (the Flk1 receptor tyrosine kinase), yolk sac blood islands fail to appear and vasculogenesis fails to take place (**FIGURE 13.18A**; Ferrara et al. 1996). Mice lacking genes for the Flk1 receptor protein have blood islands and differentiated endothelial cells, but these cells are not organized into blood vessels (Fong et al. 1995; Shalaby et al. 1995). As we saw in Chapter 12, VEGF-A is also important in forming blood vessels to the developing bone and kidney.

A third set of proteins, the **angiopoietins**, mediate the interaction between the endothelial cells and the **pericytes**— smooth muscle-like cells that the endothelial cells recruit to cover them. Mutations of either the angiopoietins or their receptor protein, Tie2, lead to malformed blood vessels deficient in the smooth muscles that normally surround them (Davis et al. 1996; Suri et al. 1996; Vikkula et al. 1996; Moyon 2001).

Angiogenesis: Sprouting of blood vessels and remodeling of vascular beds

After an initial phase of vasculogenesis, angiogenesis begins. By this process, the primary capillary networks are remodeled and veins and arteries are made (see Figure 13.15). The critical factor for angiogenesis is VEGF-A (Adams and Alitalo 2007). In many cases, an organ will secrete VEGF-A in order to induce the migration of endothelial cells from existing blood vessels into the organ and to cause them to form capillary networks there. Other factors, including hypoxia (low oxygen levels), can also induce the secretion of VEGF-A and induce blood vessel formation.

During angiogenesis, some endothelial cells in the existing blood vessel can respond to the VEGF signal and begin "sprouting" to form a new vessel. These cells are known as the **tip cells**, and they differ from the other vessel cells. (If all the endothelial cells responded equally, then the original

(A)

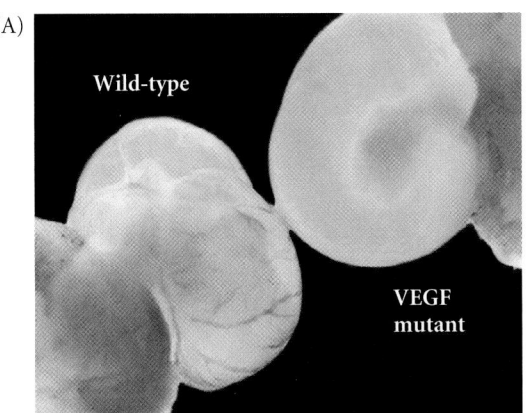

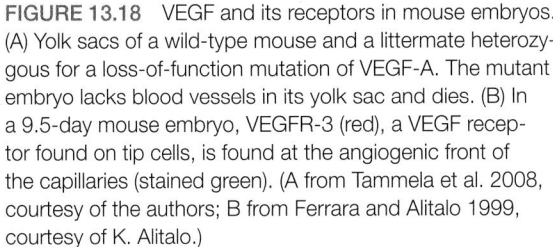

(B)

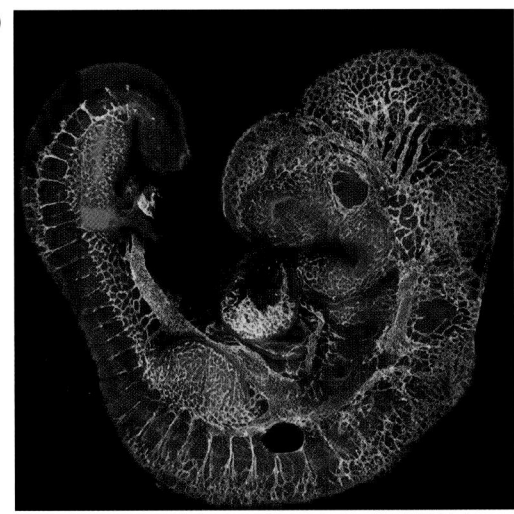

FIGURE 13.18 VEGF and its receptors in mouse embryos. (A) Yolk sacs of a wild-type mouse and a littermate heterozygous for a loss-of-function mutation of VEGF-A. The mutant embryo lacks blood vessels in its yolk sac and dies. (B) In a 9.5-day mouse embryo, VEGFR-3 (red), a VEGF receptor found on tip cells, is found at the angiogenic front of the capillaries (stained green). (A from Tammela et al. 2008, courtesy of the authors; B from Ferrara and Alitalo 1999, courtesy of K. Alitalo.)

blood vessel would fall apart.) The tip cells express the Notch ligand Delta-like-4 (Dll4) on their cell surfaces. This protein activates Notch signaling in the adjacent cells, preventing them from responding to VEGF-A (Noguera-Troise et al. 2006; Ridgway et al. 2006; Hellström et al. 2007). If the expression of Dll4 is experimentally reduced, tip cells form along a large portion of the blood vessel in response to VEGF-A.

The tip cells produce filopodia that are densely packed with VEGFR-2 (VEGF receptor-2) on their cell surfaces. They also express another VEGF receptor, VEGFR-3, and blocking VEGFR-3 greatly suppresses sprouting (**FIGURE 13.18B**; Tammela et al. 2008). These receptors enable the tip cell to extend toward the source of VEGF, and when the cell divides, the division is along the gradient of VEGFs. Indeed, the filopodia of the tip cells act just like the filopodia of neural crest cells and neural growth cones, and they respond to similar cues (Carmeliet and Tessier-Lavigne 2005; Eichmann et al. 2005). Semaphorins, netrins, neuropilins, and split proteins have roles in directing the sprouting tip cells to the source of VEGF.

Arterial and venous differentiation

Arteries and veins differ substantially from one another even though they are made from the same endothelial precursor cells. Arteries have an extensive coating of smooth muscle and a rich and elastic extracellular matrix. Veins have less extensive musculature and are characterized by valves that direct the flow of blood. A key to our understanding of the mechanism by which veins and arteries form was the discovery that the primary capillary plexus in mice actually contains two types of endothelial cells. The precursors of the arteries contain ephrin B2 in their cell membranes, and the precursors of the veins contain one of the receptors for this molecule, Eph B4 tyrosine kinase, in their cell membranes (Wang et al. 1998). If ephrin B2 is knocked out in mice, vasculogenesis occurs but angiogenesis does not. It is thought that during angiogenesis Eph B4 interacts with its ligand, ephrin B2, in two ways. First, at the borders of the venous and arterial capillaries, it ensures that arterial capillaries connect only to venous ones. Second, in nonborder areas, it ensures that the fusion of capillaries to make larger vessels occurs only between the same type of vessel (**FIGURE 13.19**). As we saw before, the same proteins involved in neural patterning are involved in endothelial patterning.

In zebrafish, the separation of arterial cells and venous cells occurs very early in development. The angioblasts develop in the posterior part of the lateral plate mesoderm, and they migrate to the midline of the embryo, where they coalesce to form the aorta (artery) and the cardinal vein beneath it (**FIGURE 13.20**). Zhong and colleagues (2001) followed individual angioblasts and found that, contrary to expectations, all the progeny of a single angioblast formed either veins or arteries, never both. In other words, each angioblast was already specified as to whether it would form aorta or cardinal vein. This specification appears to be controlled by the Notch signaling pathway* (Lawson et al. 2001, 2002). Repression of Notch signaling resulted in the loss of ephrin B2-expressing arteries and their replacement by veins. Conversely, activation of Notch signaling suppressed venous development, causing more arterial cells to form. Activation of the Notch proteins in the membranes of the presumptive angioblasts causes the activation of the transcription factor Gridlock. Gridlock in turn activates ephrin B2 and other arterial markers, while those

---

*The coordinated use of Notch and Eph signaling pathways is also seen to regulate the production of neuroblasts and somites.

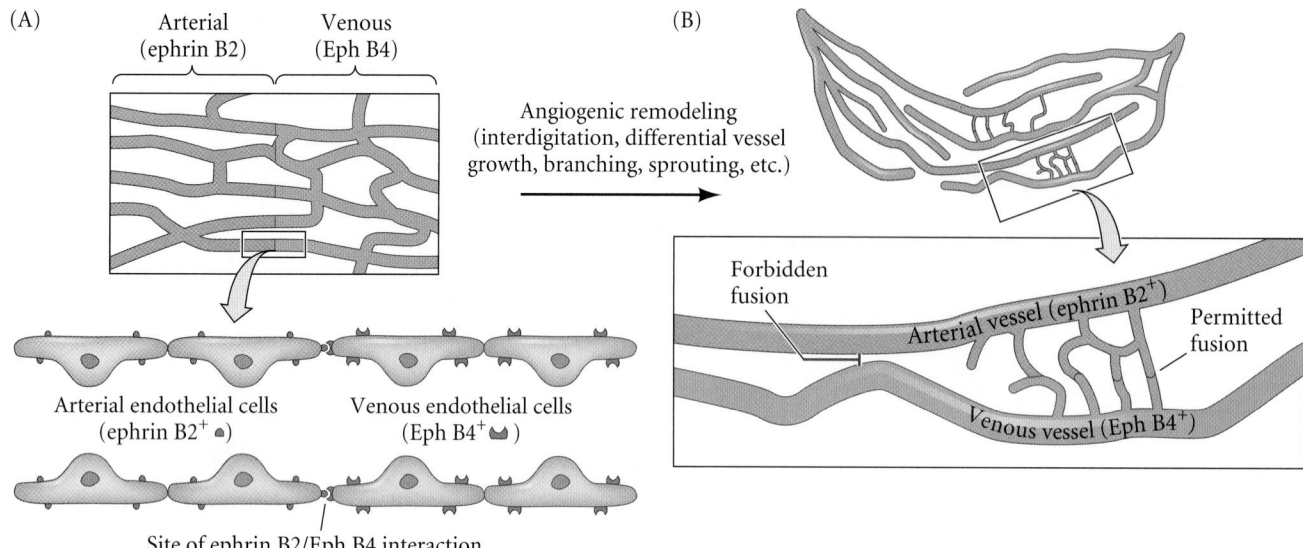

FIGURE 13.19 Roles of ephrin and Eph receptors during angiogenesis. (A) Primary capillary plexus produced by vasculogenesis. The arterial and venous endothelial cells have sorted themselves out by the presence of ephrin B2 or Eph B4 in their respective cell membranes. (B) A maturing vascular network wherein the ephrin-Eph interaction mediates the joining of small branches (future capillaries) and may prevent fusion laterally. (After Yancopopoulos et al. 1998.)

angioblasts with low amounts of Gridlock became Eph B4-expressing vein cells.

Weinstein and Lawson (2003) speculate that vascular beds are formed in a two-step process. First, new arteries form in response to VEGF. Second, these arteries then induce neighboring angioblasts (possibly through the ephrin-Eph interactions) to form the venous vessels that will provide the return for the arterial blood (**FIGURE 13.21A**). This speculation fits well with the detailed observations of chick vascular development done (and exquisitely drawn) by Popoff (1894) and

Isida (1956). These researchers found that the vitelline arteries appeared first within the capillary network and that these capillaries appeared to induce veins on either side of them (**FIGURE 13.21B–D**).

Organ-specific capillary formation

As mentioned for the cases of the brain and the placenta, several organs induce vasculogenesis and angiogenesis in their own mesenchyme. One of the main inducers of VEGF proteins is hypoxia (low oxygen). The HIF-1α transcription

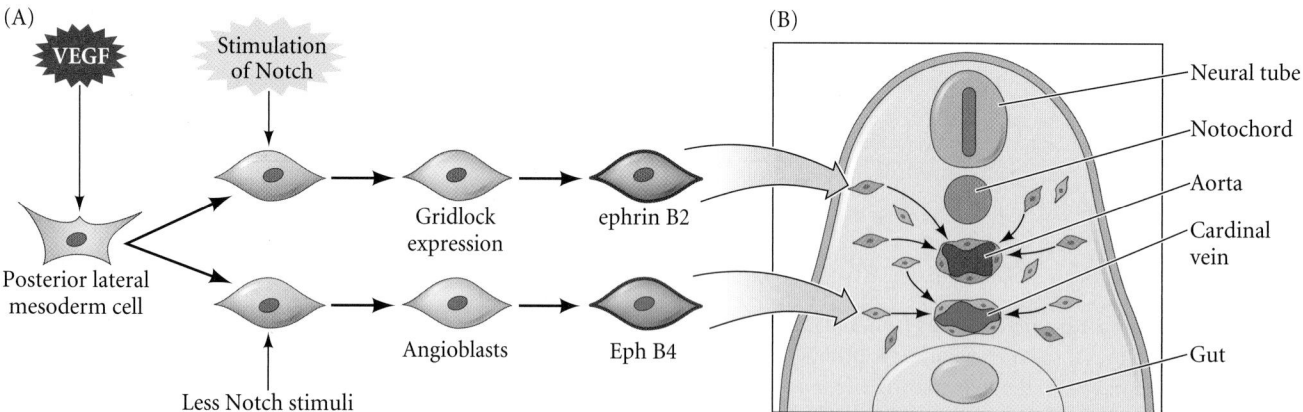

FIGURE 13.20 Blood vessel specification in the zebrafish embryo. (A) Angioblasts experiencing activation of Notch upregulate the Gridlock transcription factor. These cells express ephrin B2 and become aorta cells. Those angioblasts experiencing significantly less Notch activation do not express Gridlock, and they become Eph B4-expressing cells of the cardinal vein. (B) Once committed to forming veins or arteries, the cells migrate toward the midline of the embryo and contribute to forming the aorta or cardinal vein.

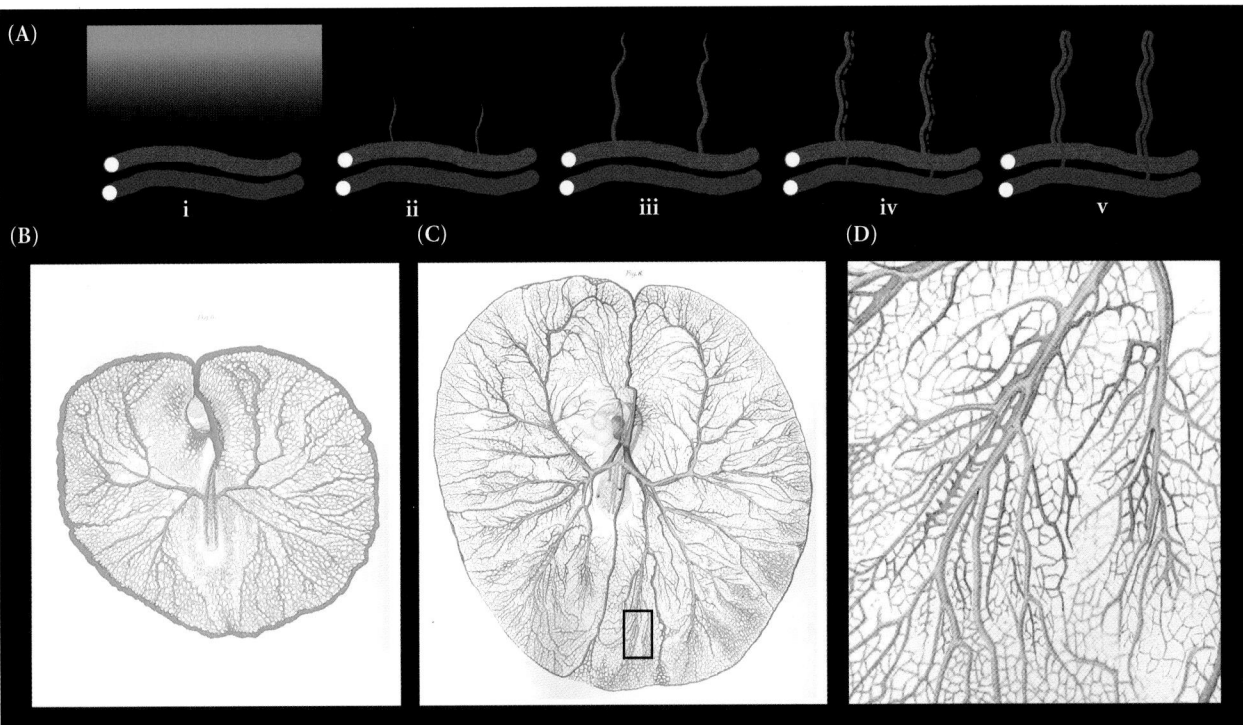

FIGURE 13.21 Blood vessel formation in the chick blastoderm. (A) In response to VEGF (green gradient), endothelial cells are induced to become arteries (red), and these arteries induce veins (blue) to form adjacent to them. New arterial vessels sprout from the arteries and then induce venous vessels adjacent to them. (B) In the chick embryo, a complex branched venous network emerges in the vascular region, with venous drainage at the periphery, via the marginal vein. (C) At later stages, collateral veins emerge adjacent to the arteries. (D) Higher magnification of the boxed region of (C). (After Weinstein and Lawson 2003; B–D modified from Popoff 1894, courtesy of N. D. Weinstein.)

factor that activates the *VEGF-A* gene (among others) is functional only at lower oxygen levels (Cramer et al. 2004). The competence of the mesenchyme cells to respond to this signal is governed by their extracellular matrices. Some extracellular matrices can stress the cell membranes, activating certain GTPases. These GTPases can activate transcription factors (such as GATA2) that activate the genes encoding VEGF receptors, while inactivating the transcription factors that inhibit VEGF receptors* (Mammoto et al. 2009).

---

*This mechanotransduction of VEGF receptors helps explain one of the great conundrums of developmental anatomy: why the aortic arches are asymmetrical. The sixth aortic arch develops only on the left side, whereas the right aortic arch degenerates (see Figure 13.11). The Nodal-induced Pitx2 signal (see Chapter 8) causes rotation of the outflow tract, producing an asymmetric blood flow into the left arch. The left side gets the blood, and the shear force from the blood flow activates the *VEGFR-2* gene on that side only. Without VEGFR-2, the endothelial cells on the right side degenerate. Thus, the sixth aortic arch forms only on the left side (Yashiro et al. 2007). A related event in zebrafish is regulated by flow-induced microRNAs (Nicoli et al. 2010).

The kidney vasculature is mainly derived from the sprouting of endothelial cells from the dorsal aorta during the initial steps of nephrogenesis. The developing nephrons secrete VEGF, thus allowing the blood vessels to enter the developing kidney and form the capillary loops of the glomerular apparatus (Kitamoto et al. 2002). Another striking example of organ-specific angiogenesis is produced by the developing peripheral nerves. Anatomists have known for decades that blood vessels follow peripheral nerves (see Greenberg and Jin 2005). Their proximity allows the nerves to obtain oxygen and allows hormones in the blood to regulate vasoconstriction and vasodilation.

The mechanism allowing the nerves and blood vessels to become adjacent is a reciprocal induction: the nerves secrete an angiogenesis factor, and the blood vessels secrete a nerve growth factor. Mukouyama and colleagues (2002, 2005) have demonstrated that arteries become associated with peripheral nerves, although veins do not (**FIGURE 13.22**). Moreover, peripheral nerves induce arteries to form near them. If the peripheral nerves in the skin fail to form (because of mutations that specifically target peripheral neurons), the arteries likewise fail to form properly. If other mutations cause

(A) Arteries   Nerves

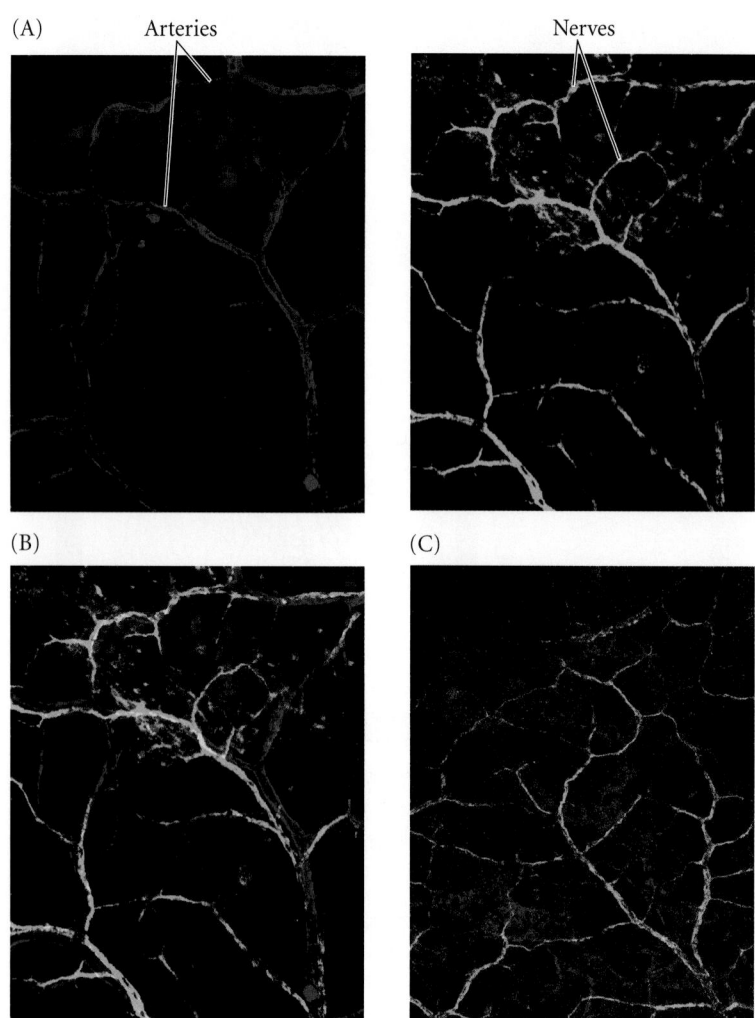

(B)   (C)

FIGURE 13.22 Arteries are specifically aligned with peripheral nerves in mouse limb skin. (A) Antibody staining of arteries (red; left) and nerves (green; right). (B) Placing the photographs together reveals that the arteries and nerves coincide. (C) Doing the same operation with stained nerves and veins reveals that the veins and nerves do not follow one another. (From Mukouyama et al. 2002, courtesy of Y. Mukouyama.)

redifferentiate as arteries, while the endothelial cells on the surface redifferentiate as veins. The mechanisms responsible for this change of fate are not yet known (Red-Horse et al. 2010.)

Anti-angiogenesis in normal and abnormal development

Like any powerful process in development, angiogenesis has to be powerfully regulated. Blood vessel formation must be told to cease, and in some tissues blood vessel formation must be prevented. For example, the cornea of most mammals is avascular.* This absence of blood vessels allows the transparency of the cornea and optical acuity. The cornea appears to have evolved two ways of keeping blood vessels out of the cornea. The first mechanism involves preventing the release of VEGF from the extracellular matrix in which it is stored (Seo et al. 2012). In addition, Ambati and colleagues (2006) have shown that the cornea secretes a soluble form of the VEGF receptor that "traps" VEGF and prevents angiogenesis in the cornea.

Soluble VEGF receptor also appears to be part of the normal mechanisms for regulating the increased formation of vasculature in the uterus during pregnancy. However, if too much soluble VEGF receptor is produced during pregnancy, there can be a dramatic reduction of normal angiogenesis. The spiral arteries that supply the fetus with nutrition fail to form, and the capillary bed of the kidneys is also reduced. This is thought to be a major cause of **preeclampsia**, a condition of pregnant women characterized by hypertension and poor renal filtration (both of which are kidney problems) and fetal distress. Preeclampsia is the leading cause of premature birth and a major cause of maternal and fetal deaths (Levine et al. 2006; Mutter and Karumanchi et al. 2008).

the peripheral neurons to form haphazardly, the arteries will follow them. This property is due to the secretion of VEGF by the neurons and their associated Schwann cells, which is necessary for arterial formation. Thus, the peripheral nerves appear to provide a template for organ-specific angiogenesis through their ability to secrete VEGF. This interaction is not a one-way street; in some instances, the blood vessels are formed in an area first, before the neurons enter. In those cases, the vascular smooth muscle cells can secrete a compound (most likely GDNF) that allows the neuron to grow alongside it. In this way, neurons can reach their destinations by following the blood vessels (Honma et al. 2002; Li et al. 2013).

One of the most interesting exceptions to the rules of blood vessel formation involves the coronary arteries— those arteries that feed the heart and whose malfunctions cause heart attacks that cause more than 7 million deaths in the United States each year. The coronary arteries arise from the venous endothelial cells of the inflow tract (the sinus venosis). These cells dedifferentiate and migrate across the heart. Those vessels that go into the myocardium then

---

*The manatee is the only mammal known to have a vascularized cornea, and it turns out that this exception proves the rule: the cornea of the manatee does not express the soluble VEGF receptor. The manatee's closest relatives (dugongs and elephants) do express it, and their corneas are avascular (Ambati et al. 2006). This morphological distinction among related taxa provides further evidence of the importance of soluble VEGF in preventing corneal vascularization.

(A)

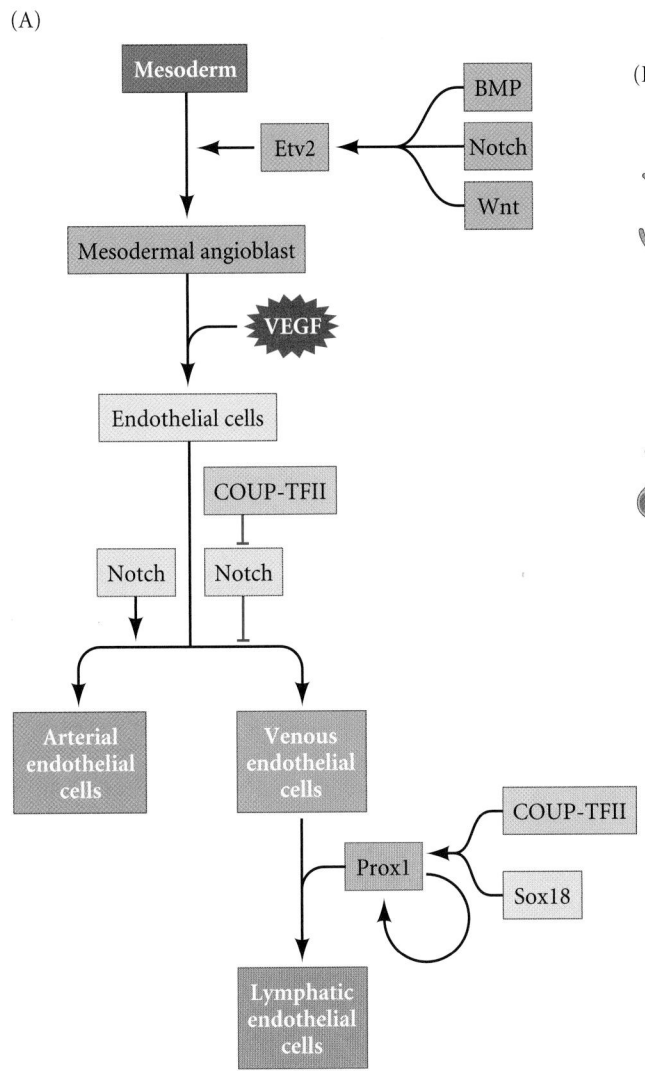

(B)

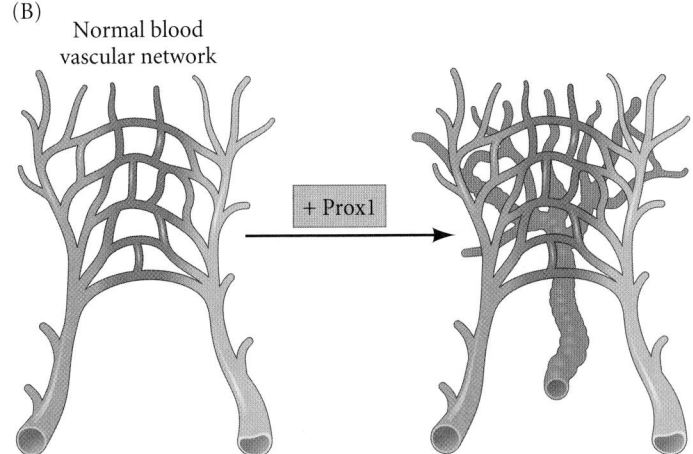

Normal blood
vascular network

+ Prox1

FIGURE 13.23 Origin of the lymphatic vessels. (A) The endothelial precursor (angioblast) cells are determined to be arterial or venous depending on their response to Notch signals. In veins, the COUP-TFII transcription factor inhibits Notch signaling. The venous endothelial cells can be further transformed into lymphatic vessels by the expression of *Prox1*. The *Sox18* gene appears to be needed prior to *Prox1* expression. (B) Upon Prox1 signaling in the venous endothelial cells, lymphatic vessels emerge. (After Adams and Alitalo 2007; Oliver and Srinivasan 2010.)

Too much VEGF can also be dangerous. Abnormal blood vessel formation occurs in solid tumors and in the retina of patients with diabetes. This vascularization results in the growth and spread of tumor cells and blindness, respectively. By targeting the VEGF receptors and the Notch pathway involved in regulating them, researchers are seeking ways to block angiogenesis and prevent cancer cells and the retina from becoming vascularized (Miller et al 2013; Wilson et al. 2013).

The lymphatic vessels

In addition to the blood vessels, there is a second circulatory system, the **lymphatic vasculature**. The lymphatic vasculature forms a separate system of vessels that is essential for draining fluid and transporting lymphocytes. The development of the lymphatic system commences when a subset of endothelial cells from the jugular vein (in the neck) sprout to form the lymphatic sacs. After the formation of these sacs, the peripheral lymphatic vessels are generated by further sprouting (Sabin 1902; van der Putte 1975). The lymphatic vessels lack both the pericytes and the extracellular matrix that surround the blood vessels, making the lymphatic circulation much more permeable to interstitial fluid (Wang and Oliver 2010).

Commitment to the lymphatic lineage appears to be mediated through the **Prox1** transcription factor, which downregulates blood vessel-specific genes and upregulates genes involved in forming lymphatic vessels (Wigle and Oliver 1999; Wigle et al. 2002; Françoise et al. 2008). Indeed, it appears that the ground state is to make venous blood vessel endothelial cells, and that upon *Prox1* expression, these vessels acquire the lymphatic fate (Wigle et al. 2002; Srinivasan et al. 2007). Moreover, if *Prox1* expression is downregulated, the lymphatic endothelial cells return to a venous-like condition (Johnson et al. 2008). This makes the lymphatic vessel cell one of the few cell types whose maintenance is dependent on continued expression of a particular gene that initiated its differentiation (**FIGURE 13.23**).

One of the genes upregulated by Prox1 is *VEGFR-3*, which encodes the receptor for the paracrine factor VEGF-C. As important as VEGF-A is for blood vessel development, VEGF-C is equally necessary for proper lymphatic development (**FIGURE 13.24**; Karkkainen et al. 2004; Alitalo et al. 2005). VEGF-C produced in the area of the jugular vein attracts Prox1-positive endothelial cells out from the vein and then promotes their proliferation and development into the lymphatic sacs (see Adams and Alitalo 2007; Hosking and Makinen 2007).

(A) Wild type   VEGF-C-deficient

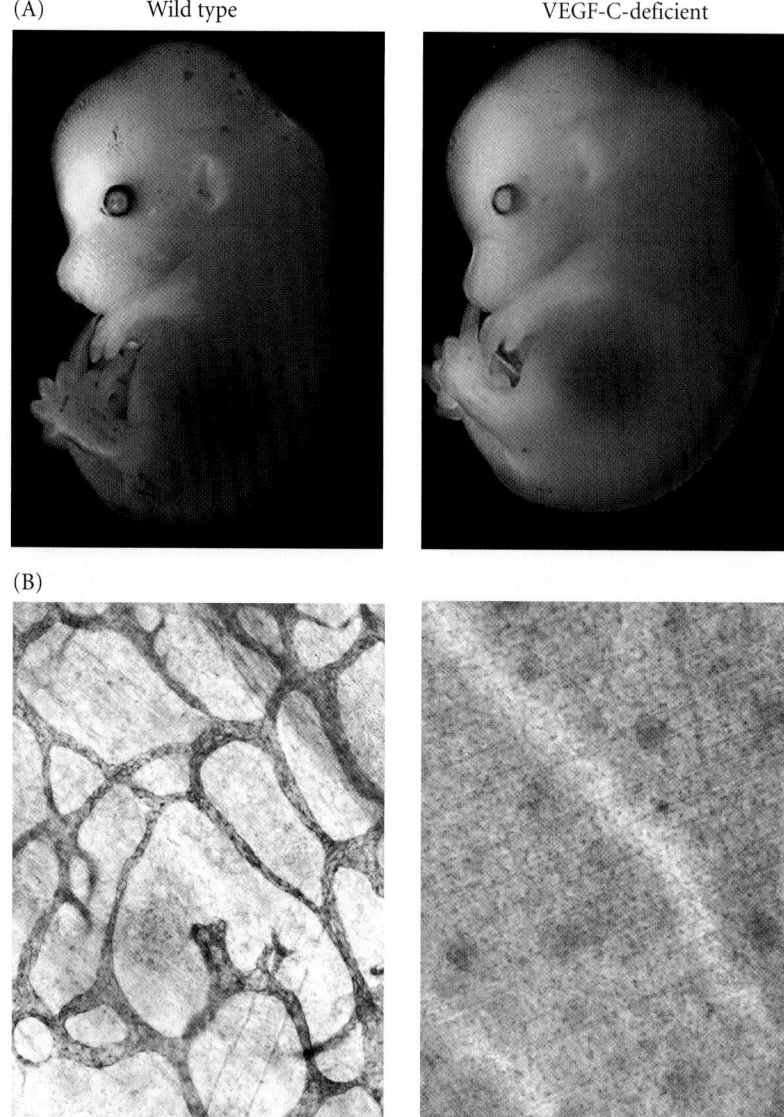

(B)

FIGURE 13.24 VEGF-C is critical for the formation of lymphatic vessels. (A) Compared with the wild-type control, a 15.5-day mouse embryo heterozygous for a *VEGF-C* deficiency suffers from severe edema (bloating with excess fluid). (B) 16.5-day mouse embryos stained for lymphatic vasculature. The lack of lymphatics in the skin of the *VEGF-C* mutant (right) is obvious when compared with that of the wild-type embryo (left). (From Karkkainen et al. 2004, courtesy of K. Alitalo.)

Hematopoiesis: The Stem Cell Concept

Each day we lose and replace about 10^{11} red blood cells. As red blood cells are destroyed in the spleen, their replacements come from populations of stem cells. As we mentioned in the introduction to Part Three, a stem cell is capable of extensive proliferation, creating both more stem cells (self-renewal) and differentiated cell progeny (see Figures P3.1B and P3.3). In the case of **hematopoiesis**—the generation of blood cells—stem cells divide to produce (1) more stem cells and (2) progenitor cells that can respond to the environment around

them to differentiate into mature blood cell types. The critical stem cell in hematopoiesis is the pluripotent hematopoietic stem cell, or simply the **hematopoietic stem cell (HSC)**, which is capable of producing all the blood cells and lymphocytes of the body. It achieves this by generating a series of intermediate progenitor cells whose potency is restricted to certain lineages.

Sites of hematopoiesis

Where do hematopoietic stem cells come from? In the early 1900s, numerous investigators (looking at many different vertebrate species, including mongooses, bats and humans) observed the emergence of blood cells from the ventral endothelium of the aorta (Adamo and Garcia-Cardeña 2012). In the 1960s, however, experiments on mice concluded that all hematopoietic stem cells are derived from cells originating in the extraembryonic mesoderm surrounding the yolk sac. The aortic hematopoiesis was thought of as an intermediate stop that the stem cells had on their way to the spleen and bone marrow (the sites of adult hematopoiesis in the mice).

However, in 1975 Dieterlen-Lièvre performed a transplantation of early chick yolk sacs onto 2-day (precirculation) quail embryos. Chick and quail blood cells can be readily distinguished under the microscope, and the chimeric animal survives. Dieterlen-Lièvre's analysis indicated that all the blood cells of the late quail embryo originated from the quail host and not from the transplanted chick yolk sac. Moreover, hematopoietic activity within the embryo was restricted to one major site: the ventral portion of the aorta (Dieterlen-Lièvre and Martin 1981). The grafting of splanchopleure from this **aorta-gonad-mesonephros (AGM) region** from one genetically variant mouse to another confirmed that in mammals, too, definitive hematopoiesis takes place from inside the embryo (Godin et al. 1993; Medvinsky et al. 1993). Soon afterward, hematopoietic stem cells were identified in clusters of cells that were observed on the ventral region of the 10.5-day embryonic mouse aorta (Cumano et al. 1996; Medvinsky and Dziermak 1996).

While there is evidence that some yolk sac hematopoietic stem cells persist in the adult mouse (see Samokhvalov et al. 2007), it is generally thought that the yolk sac hematopoietic stem cells in mammals produce blood cells that will allow oxygen to be transported to the early embryo, but that nearly all the stem cells found in the adult are those from

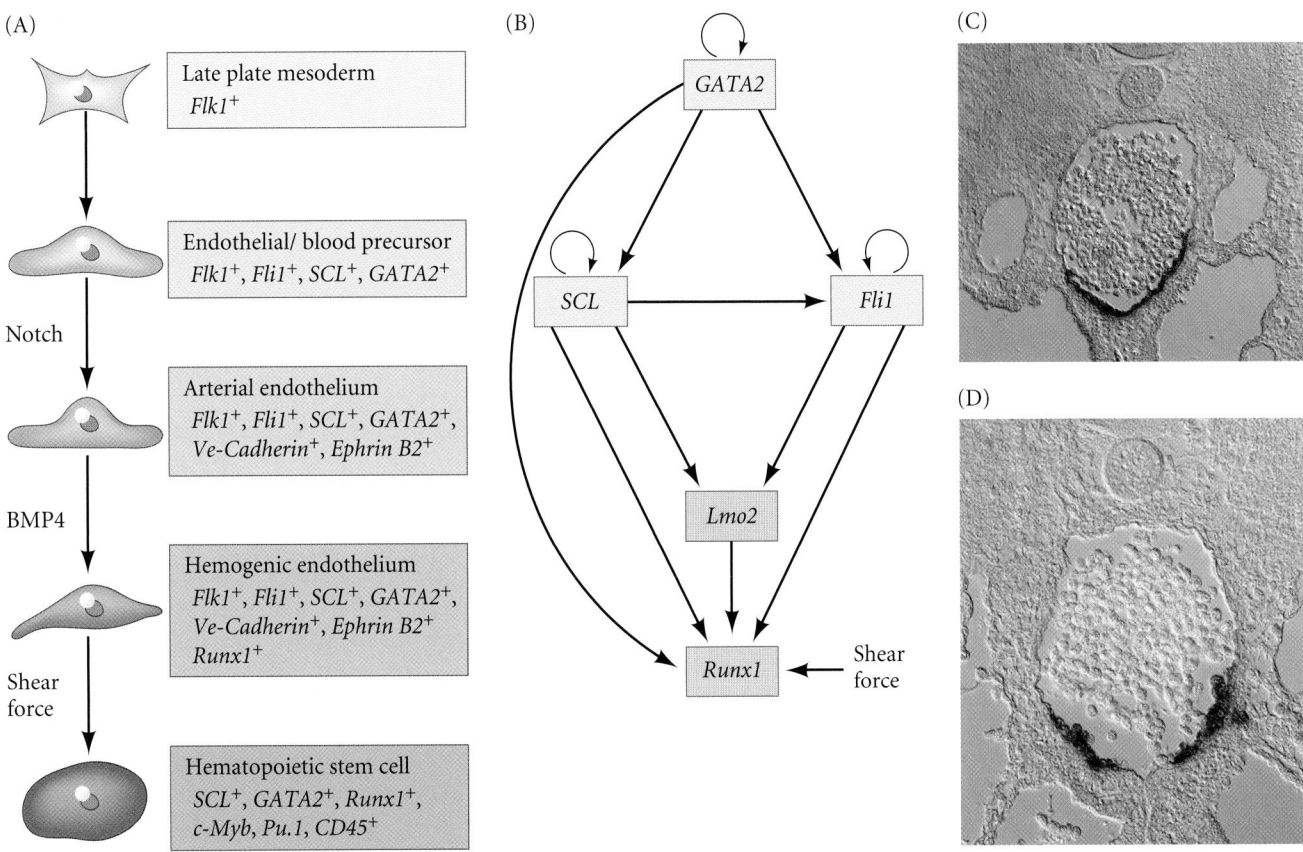

FIGURE 13.25 Pathway for hematopoietic stem cell formation. (A) In the developing mouse, hematopoietic stem cells arise from the hemogenic endothelium of the aorta. Runx1 is critical for this conversion of endothelial cells into blood stem cells. The lateral plate mesodermal heritage of the hemogenic endothelium is seen, as well as the juxtacrine and paracrine factors that brought it to its fate. The transcription factors associated with each stage are shown at the right. (B) A simplified view of the factors activating *Runx1* in the gene regulatory network establishing the hematopoietic stem cell in the mouse. The GATA2, Fli1, and Scl transcription factors bind together at adjacent sites on a single enhancer 23 base pairs downstream from the *Runx1* transcription initiation site. Scl is critical for preventing blood and vascular cells from becoming heart muscle. The mechanism by which shear force is mechanotransduced to help activate *Runx1* remains unknown. (C) *Runx1* expression (purple) in the stage-19 chick embryo, when the *Runx1*-expressing cells have become part of the blood vessel. (D) *Runx1*-expressing cells at stage 21 in the chick embryo. Hematopoietic colonies are now visible. (A after Swiers et al. 2010; B after Pimanda and Göttgens 2010; C,D from Jaffredo et al. 2010.)

the AGM that have migrated to the bone marrow (Jaffredo et al. 2010).

In 2009, several laboratories proposed a new mechanism for blood cell production. This new hypothesis was based on the discovery of a new cell type, the **hemogenic endothelial cell**.* Recall that in our discussion of somites (Chapter 12), we noted that the sclerotome produces angioblasts that migrate to the dorsal aorta and replace most of the primary dorsal aorta cells. Before their replacement, the remaining primary, lateral plate-derived, endothelial cells of the dorsal

aorta, now in the ventral area of the blood vessel, give rise to blood-forming stem cells. These blood vessel-derived hematopoietic stem cells are the critical source of adult blood stem cells. By analyzing the types of cells made by the blood vessel endothelium, researchers were able to isolate the hemogenic endothelial cells and showed that they produce the hematopoietic stem cells that migrate to the liver and bone marrow (Eilken et al. 2009; Lancrin et al. 2009). Furthermore, the transition from endothelial cell to hematopoietic stem cell was seen to be mediated by the activation of the Runx1 transcription factor (**FIGURE 13.25**). In mice lacking Runx1, the blood stem cells failed to form in the yolk sac, umbilical arteries, dorsal aorta, and placental vessels (Chen et al. 2009).

The *Runx1* gene appears to be regulated by a complex and dynamic circuitry. Moreover, *Runx1* expression is not initiated until after the heart starts beating. If cardiac mutations

---

*The relationship of the hemogenic endothelial cell and the hemangioblast is controversial. In general, it is thought that hemangioblasts generate the hemogenic endothelial cells (see Ueno and Weissman 2010) and that the hemangioblast is a precursor to the hemogenic endothelium.

prevent fluid flow through the aorta, *Runx1* is not expressed. Rather, shear forces (that is, friction) from the fluid flow are required to activate *Runx1* in the ventral endothelium of the dorsal aorta (Adamo et al. 2009; North et al. 2009).* The shear forces appear to elevate levels of nitric oxide (NO) in the endothelium, and NO, in turn, activates (perhaps through cGMP) *Runx1* and other genes known to be critical for blood cell formation. The transition from hemogenic endothelial cell to HSC does not appear to be caused by an asymmetric cell division. Rather, there is a rearrangement of cytoskeleton and tight junctions that resembles an epithelial-mesenchymal transition such as those seen in the sclerotome or dermamyotome (Yue et al. 2012).

In non-amniote vertebrates, the splanchnopleure is also the source of the hematopoietic stem cells, and BMPs are crucial in inducing the blood-forming cells in all vertebrates studied. In *Xenopus*, the ventral mesoderm forms a large blood island that is the first site of hematopoiesis. Ectopic BMP2 and BMP4 can induce blood cell and blood vessel formation in *Xenopus*, and interference with BMP signaling prevents blood formation (Maeno et al. 1994; Hemmati-Brivanlou and Thomsen 1995). In the zebrafish, both yolk sac and aortic hematopoiesis are seen. As in the mammalian embryo, the second wave of hematopoiesis is from the aorta. Hematopoietic stem cells can be seen arising from the ventral aortic endothelium (Bertrand et al. 2010; Kissa and Herbomel 2010), and the same genetic pathways leading to *Runx1* expression (including the BMPs) regulate this second, definitive wave of hematopoiesis (Mullins et al. 1996; Paik and Zon 2010).

The bone marrow HSC niche

The hematopoietic stem cells of the aorta generate HSCs that come to reside first in the liver and then in the bone marrow (Coskun and Hirschi 2010). In humans, the aorta generates blood cells around days 27 through 40 (Tavian and Péault 2005). The bone marrow HSC is a remarkable cell, in that it is the common precursor of red blood cells (erythrocytes), white blood cells (granulocytes, neutrophils, and platelets), monocytes (macrophages and osteoclasts), and lymphocytes (see Figure P3.2A). When transplanted into inbred, irradiated mice (that are genetically identical to the donor cells and whose own stem cells have been eliminated by radiation), HSCs can repopulate the mouse with all the blood and lymphoid cell types. It is estimated that only about 1 in every 10,000 blood cells is a pluripotent HSC (Berardi et al. 1995). In humans, "bone marrow transplants" are used to transfer healthy HSCs into people whose lymphocytes, red blood cells, or white blood cells have been wiped out by disease, drugs, or radiation. In recent years, more than 50,000 such transplantations have been performed annually (Gratwohl et al. 2010).

The maintenance of the HSC depends on the stem cell niche, and especially on the ability of the HSC to receive the paracrine factor **SCF**. SCF stands for **stem cell factor**, and it binds to the Kit receptor protein. (This binding is critical for sperm and pigment stem cells as well.) Because it was important to determine which cells of the stem cell niche were supplying SCF, Ding and colleagues (2012) constructed genetically recombined mice in which the gene for SCF was replaced by the gene for green fluorescent protein in all or in selected cell types. When all the cell types of the niche expressed GFP instead of SCF, the HSCs died. When they deleted SCF production in just certain cell types, they found that replacing SCF with GFP in blood cells, bone cells, or marrow mesenchymal cells did not block HSC maintenance. However, when they got rid of SCF expression in either the endothelial cells or the perivascular cells surrounding the endothelial cells, many fewer HSCs survived. And when SCF synthesis was turned off in both of these cell types (but not in the others), all the HSCs perished. It appears, then, that the SCF needed for HSC survival is made primarily by the perivascular cells, with some contribution from the endothelial cells (**FIGURE 13.26**).

These perivascular cells may be important in another manner: the hematopoietic stem cell niche is possibly derived from its own stem cell. Sacchetti and colleagues (2007) have shown that the perivascular cells of the human hematopoietic microenvironment, a relatively minor cell population in the bone marrow (see Figure P3.4) and having no known function there, are actually hematopoietic niche stem cells. When such cells are implanted into the skin of immunosuppressed mice, the cells divide and differentiate into small hematopoietic microenvironments, complete with subendothelial cells and miniature bones. Mouse hematopoietic stem cells migrate into these regions and start forming blood. Moreover, when those new perivascular cells are transferred to yet other mice, they too develop a new hematopoietic microenvironment.†

SCF is not the only paracrine factor that HSCs require; there are numerous others, and these probably render the HSCs competent to respond to the paracrine and juxtacrine factors that will direct cell differentiation. As mentioned in the Part Three opener, stem cell niches often contain long-term, quiescent HSCs that are used to generate progenitor cells on a continual basis, as well as shorter-acting HSCs that can respond to immediate physiological needs. Wnts that activate the noncanonical pathways are secreted by niche osteoblasts to maintain the quiescent HSCs, whereas the

---

*Shear stress from blood flow is a major player in development. Recall that it is also required for normal heart development (Mironov et al. 2005) and for the correct patterning of blood vessels (Lucitti et al. 2007; Yashiro et al. 2007). It is also needed for the fragmentation of the platelet precursor cell—the megakaryocyte—into platelets. The megakaryocyte in the bone marrow inserts small processes into the blood vessels surrounding the stem cell niche, and the shear force there fragments these processes into platelets (Junt et al. 2007).

---

†Ian Wilmut, who directed the cloning program that produced Dolly, famously remarked, "Although the story is complicated, it is biology, not physics: that is to say, it is not *weird*" (2001, p. 17). One sometimes has to remember this dictum.

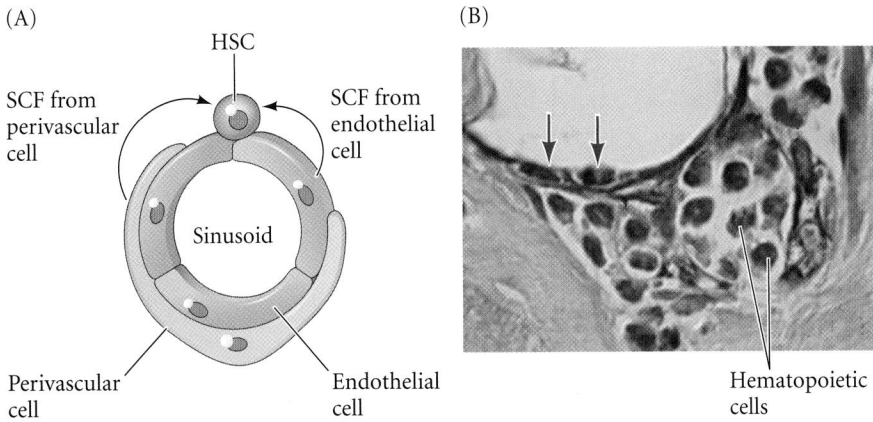

(A)

HSC

SCF from
perivascular
cell

SCF from
endothelial
cell

Sinusoid

Perivascular
cell

Endothelial
cell

(B)

Hematopoietic
cells

FIGURE 13.26 The home for HSCs appears to be a niche where stem cell factor (SCF) can be made by the perivascular (subendothelial) cells as well as by the endothelial cells of the bone marrow sinusoids. (A) Simplified diagram of the sinusoid with its endothelial cells and a surrounding perivascular cell. (B) Development of a stem cell niche when human perivascular cells (stained brown with antibodies to the subendothelial cell marker CD146) implanted into a mouse. At 8 weeks, processes from perivascular cells establish contacts with hematopoietic stem cells (as in human bone marrow). The red arrows show hematopoietic cells between endothelial and perivascular cells. (A after Shestopalov and Zon 2012; B froom Sacchetti et al. 2007, courtesy of P. Bianco.)

canonical Wnt pathway may be critical for inducing them to become the rapidly proliferating HSCs (Reya et al. 2003; Sugimura et al. 2012).

Hematopoietic inductive microenvironments

Endocrine, paracrine, and juxtacrine factors are thought to push blood cell differentiation down one path or another (**FIGURE 13.27A**). One of the major *endocrine* factors (i.e., hormones) is **erythropoietin**, which appears to cause the the common myeloid precursor cell (CMP) to make more megakaryocyte/erythroid precursor cells (MEPs) and biases the MEPs to make more erythrocytes (Lu et al. 2008; Klimchenko et al. 2009). The *paracrine* factors involved in blood cell and lymphocyte formation are the **cytokines**. Cytokines can be made by several cell types, but they are collected and concentrated by the extracellular matrix of the stromal (mesenchymal) cells at the sites of hematopoiesis (Hunt et al. 1987; Whitlock et al. 1987). For instance, granulocyte-macrophage colony-stimulating factor (GM-CSF) and the multilineage growth factor interleukin 3 (IL3) both bind to the heparan sulfate glycosaminoglycan of the bone marrow stroma (Gordon et al. 1987; Roberts et al. 1988). The extracellular matrix is then able to present these paracrine factors to the stem cells in concentrations high enough to bind to their respective receptors. At different stages of maturation, the stem cells become competent to respond to different factors.

The developmental path taken by a descendant of a pluripotent HSC depends on which growth factors it meets, and

is therefore determined by the stromal cells. Wolf and Trentin (1968) demonstrated that short-range interactions between stromal cells and stem cells determine the developmental fates of the stem cells' progeny. These investigators placed plugs of bone marrow in a spleen and then injected stem cells into it. Those CMPs that came to reside in the spleen formed colonies that were predominantly erythroid, whereas those that came to reside in the bone marrow formed colonies that were predominantly granulocytic. Colonies that straddled the borders of the two tissue types were predominantly erythroid in the spleen and granulocytic in the marrow. Such regions of determination are referred to as **hematopoietic inductive microenvironments** (**HIMs**). As expected, the HIMs induce different sets of transcription factors in these cells, and these transcription factors specify the fate of the particular cells (see Kluger et al. 2004).

The transcription factors of the HIM may act by pulling the equilibrium of the stem cells transcription network in different directions (**FIGURE 13.27B**; Krumsiek et al. 2011; Wontakai et al. 2012). By decomposing the interactions into negative feedback loops and feedforward loops (of both activation and repression), there appear to be only four stable configurations that this network can have. Such stable configurations are called "attractor states" in systems theory, and these attractor states correspond to four cell types. Moreover, certain mutations will make certain attractor states impossible, and these are the mutations that block the differentiation of certain cell types.

In conclusion

We ask a lot of our circulatory system. We require a flawless flow of blood through the valves each second of our lives; we demand fine-tuned coordination between our brain, heart, bone marrow, and hormones such that the cardiac muscle contractions adapt to our physiological needs; and we demand that the production of our blood cells—cells made by precursors that formed in our embryo—be so precise that we get neither cancer nor anemia. Given all this, it is not surprising that blood cell differentiation, heart development, and vessel formation are now among the most important fields of study in medical

(A)

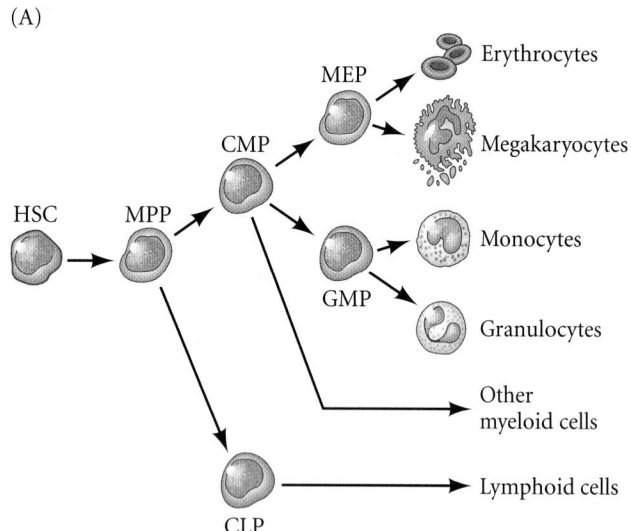

(B)

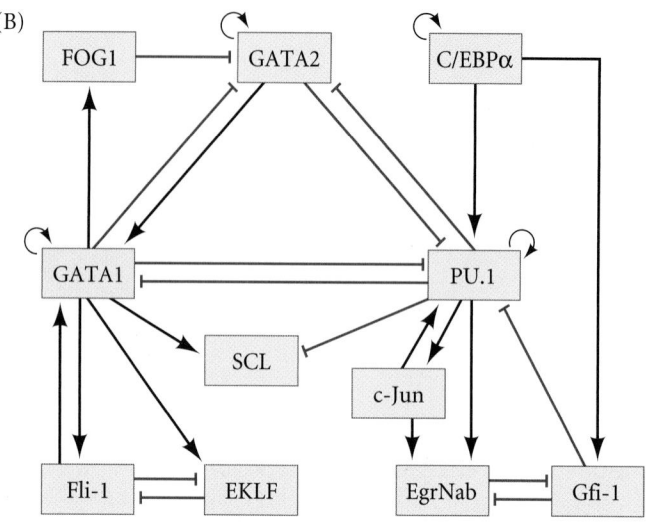

FIGURE 13.27 Hematopoetic stem cell differentiation occurs in a stepwise fashion with numerous binary decisions. (A) Simplified lineage map of HSC progeny. The HSC can divide to generate a multipotent progenitor (MPP) as well as another HSC. Depending on the paracrine, juxtacrine, and endocrine factors in its environment, the MPP can produce a blood precursor cell (the common myeloid precursor, or CMP) or a lymphocyte precursor cell, the common lymphoid progenitor (CLP). The CMP generates the megakaryocyte-erythroid precursor (MEP) which, depending on the extracellular environment, can generate the erythroid lineage (to red blood cells, the erythrocytes) or the megakaryocyte lineage (which becomes the platelets). The CMP, under other conditions, can give rise to the granulocyte-monocyte precursor (GMP) lineage that becomes the white blood cells (neutrophils, basophils, and eosinophils) or the monocytes (which become macrophages). (B) Regulatory interactions among 11 transcription factor genes (and their products) in the CMP that are critical for the decisions that make the blood cell lineages. (After Krumsiek et al. 2012.)

science. As we will see in Chapter 18, controlling blood cell differentiation and stem cell proliferation is at the root of leukemia research, and regulating angiogenesis holds promise for preventing tumor formation. Congenital heart defects are among the most prevalent types of birth defects, and cardiovascular disease is the most common cause of death in industrialized nations. The questions of cardiogenesis, angiogenesis, and hematopoiesis that engaged Aristotle and Harvey still excite major research programs today.

ENDODERM

In the adult amniote, the endoderm is essential for the uptake of oxygen into the body and for the digestion and absorption of foods. In the embryo, the endoderm's first major function is inducing the formation of several mesodermal organs. As we have seen in this and earlier chapters, the endoderm is critical for instructing the formation of the notochord, the heart, the blood vessels, and even the mesodermal germ layer.* The second embryonic function is to construct the linings

---

* Some scientists think that the mesoderm and endoderm were originally a single germ layer, the "mesendoderm," that accomplished all these functions. Recall from Chapter 5 that what in vertebrates is a broad territory of embryonic cells is the progeny of a single mesentoblast in many invertebrates. The signals that regulate the mesentoblast and the entire mesodermal and endodermal territory in vertebrates may be very similar (Henry and Melton 1998; Maduro et al. 2001; Rodaway and Patient 2001).

of two tubes within the vertebrate body. The **digestive tube** extends the length of the body. Buds from the digestive tube form the liver, gallbladder, and pancreas. The **respiratory tube** forms as an outgrowth of the digestive tube and eventually bifurcates into the two lungs. The region of the digestive tube anterior to the point where the respiratory tube branches off is the **pharynx**. Epithelial outpockets of the pharynx give rise to the tonsils, to the thyroid, thymus, and parathyroid glands, and eventually to the respiratory tube itself.

In 1769, Caspar Friedrich Wolff, an embryologist in St. Petersburg, demonstrated that the gut tube formed by the curving of an initially flat sheet. His *De Formatione Intestinorum* was the first microscopic evidence for *epigenesis*—the view that the embryo constructed itself "from scratch" and that no small, preformed individual resided within the sperm or egg (see Chapter 1). In addition to confirming this, modern work has shown that gut tube development begins at two sites that migrate toward each other and fuse in the center (Lawson et al. 1986; Franklin et al. 2008). In the foregut, cells from the lateral portions of the anterior endoderm move ventrally to form the tube of the **anterior intestinal portal** (**AIP**); the **caudal intestinal portal** (**CIP**) forms from the posterior endoderm. The AIP and CIP migrate toward each other and come together to form the midgut (**FIGURE 13.28**).

There is an ectodermal entrance at either end of the gut tube. At first, the oral end is blocked by a region of ectoderm called the **oral plate**, or **stomodeum**. Eventually (at about 22 days in human embryos), the stomodeum breaks,

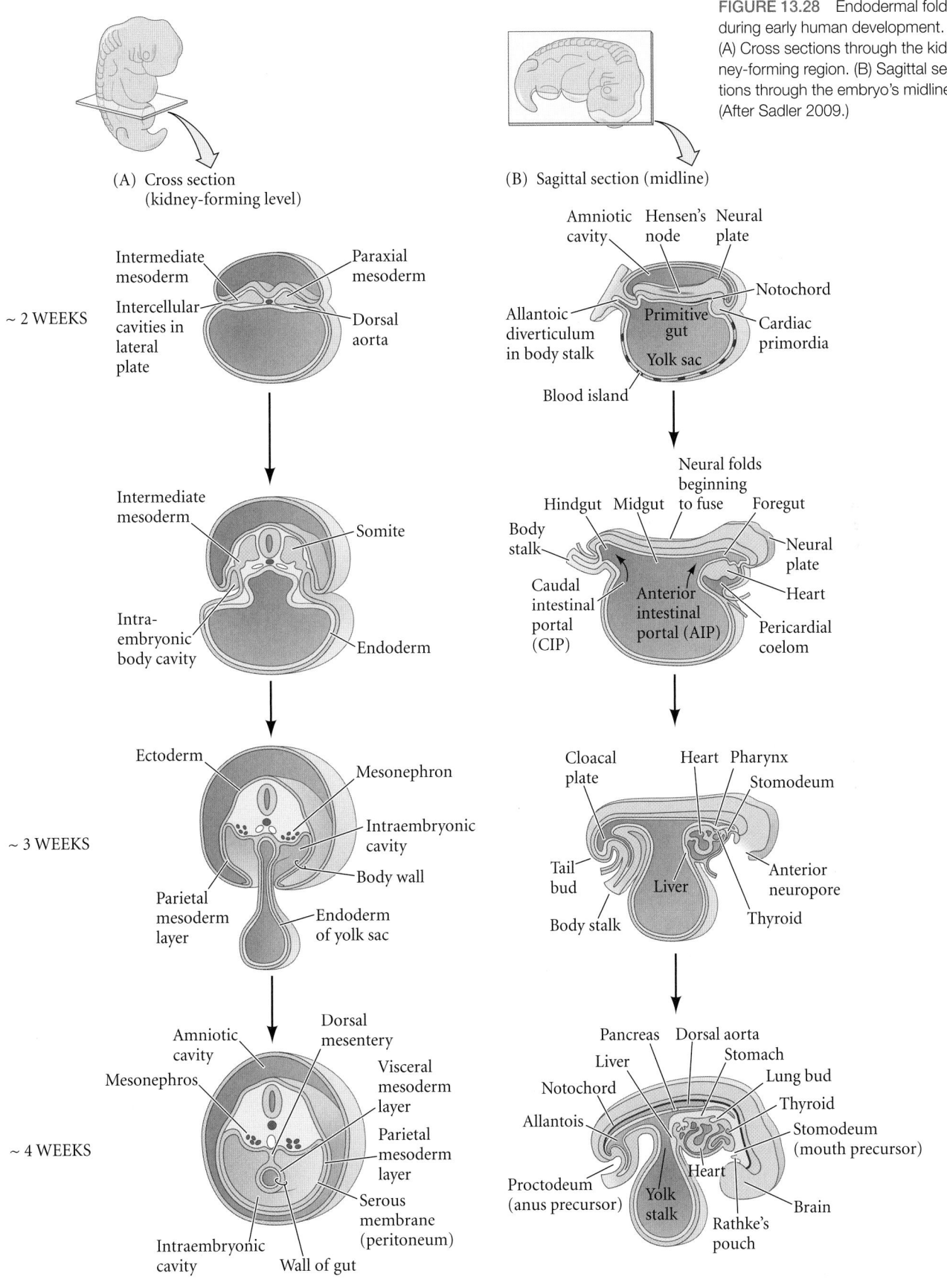

FIGURE 13.28 Endodermal folding during early human development. (A) Cross sections through the kidney-forming region. (B) Sagittal sections through the embryo's midline. (After Sadler 2009.)

(A) Cross section (kidney-forming level)

(B) Sagittal section (midline)

~ 2 WEEKS

Intermediate mesoderm
Intercellular cavities in lateral plate
Paraxial mesoderm
Dorsal aorta

Amniotic cavity
Hensen's node
Neural plate
Notochord
Allantoic diverticulum in body stalk
Primitive gut
Cardiac primordia
Yolk sac
Blood island

Intermediate mesoderm
Somite
Intra-embryonic body cavity
Endoderm

Neural folds beginning to fuse
Hindgut
Midgut
Foregut
Body stalk
Neural plate
Caudal intestinal portal (CIP)
Anterior intestinal portal (AIP)
Heart
Pericardial coelom

~ 3 WEEKS

Ectoderm
Mesonephron
Intraembryonic cavity
Body wall
Parietal mesoderm layer
Endoderm of yolk sac

Cloacal plate
Heart
Pharynx
Stomodeum
Tail bud
Body stalk
Liver
Anterior neuropore
Thyroid

~ 4 WEEKS

Amniotic cavity
Dorsal mesentery
Mesonephros
Visceral mesoderm layer
Parietal mesoderm layer
Serous membrane (peritoneum)
Intraembryonic cavity
Wall of gut

Pancreas
Dorsal aorta
Liver
Stomach
Notochord
Lung bud
Allantois
Thyroid
Stomodeum (mouth precursor)
Proctodeum (anus precursor)
Yolk stalk
Heart
Brain
Rathke's pouch

creating the oral opening of the digestive tube. The opening itself is lined by ectodermal cells. This arrangement creates an interesting situation, because the oral plate ectoderm is in contact with the brain ectoderm, which has curved around toward the ventral portion of the embryo. These two ectodermal regions interact with each other, with the roof of the oral region forming Rathke's pouch and becoming the glandular portion of the pituitary gland. The neural tissue on the floor of the diencephalon gives rise to the infundibulum, which becomes the neural portion of the pituitary. Thus, the pituitary gland has a dual origin, which is reflected in its adult functions. There is a similar meeting of endoderm and ectoderm at the anus; this is called the **anorectal junction**.

The Pharynx

The anterior endodermal portion of the digestive and respiratory tubes begins in the pharynx. Using a reporter gene (*Sox17*, which we'll discuss later) that gets activated only in the endoderm, Rothova and colleagues (2012) found that there is a dividing line between the mouth ectoderm and endoderm. In mammals, the teeth and the major salivary glands are from the ectoderm. The anterior taste buds are ectodermal, but the posterior taste buds, as well as some of the posterior salivary and mucus glands, are derived from the endoderm.

The embryonic pharynx in mammals contains four pairs of endodermally derived **pharyngeal pouches**. Between these pouches are the four **pharyngeal arches** (FIGURE 13.29). The first pair of pharyngeal pouches become the auditory cavities of the middle ear and the associated eustachian tubes. The second pair of pouches give rise to the walls of the tonsils. The thymus is derived from the third pair of pharyngeal pouches; it will direct the differentiation of T lymphocytes during later stages of development. One pair of parathyroid glands is also derived from the third pair of pharyngeal pouches, while the other pair is derived from the fourth pair of pouches. In addition to these paired pouches, a small, central diverticulum is formed between the second pharyngeal pouches on the floor of the pharynx. This pocket of endoderm and mesenchyme will bud off from the pharynx and migrate down the neck to become the thyroid gland. The respiratory tube sprouts from the pharyngeal floor (between the fourth pair of pharyngeal pouches) to form the lungs, as we will see below.

The pharynx is where the endoderm meets the ectoderm, and the endoderm plays a critical role in determining which pouches develop. Sonic hedgehog from the endoderm appears to act as a survival factor, preventing apoptosis of the neural crest cells (Moore-Scott and Manley 2005). In zebrafish, genetic analysis combined with transplantation studies has shown that FGFs (mainly Fgf3 and Fgf8) from the ectoderm and mesoderm also are important not only for the migration and survival of neural crest cells, but also for the formation of the pouches themselves. Mice deficient in both *Fgf8* and *Fgf3* genes lack all the pharyngeal pouches, even when endoderm is present. Instead of migrating laterally and ventrally to form pouches, the endoderm remains in the anterior pharynx and does not spread out (Crump et al. 2004).

The Digestive Tube and Its Derivatives

Posterior to the pharynx, the digestive tube constricts to form the esophagus, which is followed in sequence by the

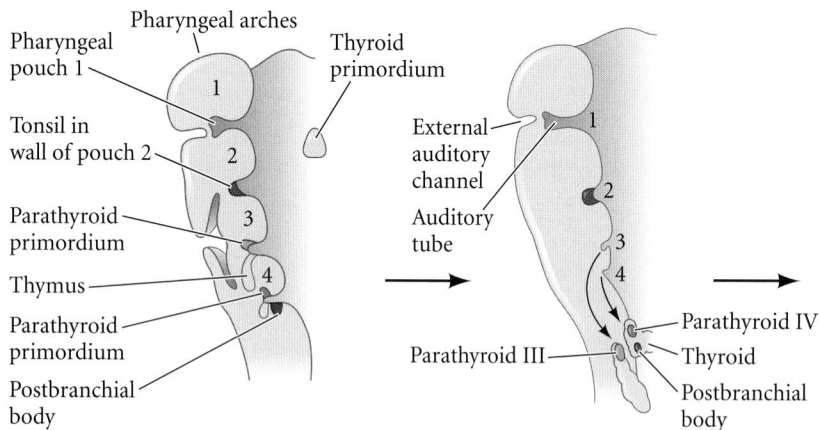
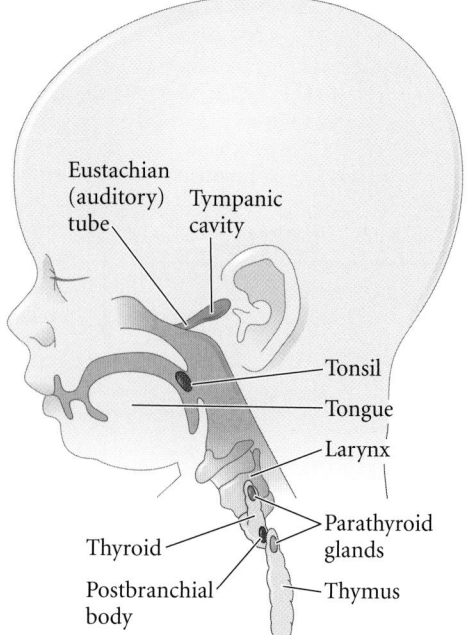

FIGURE 13.29 Formation of glandular primordia from the pharyngeal pouches. The end of each of the first pharyngeal pouches becomes the tympanic cavity of the middle ear and the eustachian tube. The second pouches receive aggregates of lymphoid tissue and become the tonsils. The dorsal portion of the third pharyngeal pouches forms part of the parathyroid gland, while the ventral portion forms the thymus. Both migrate caudally and meet with the tissue from the fourth pharyngeal pouches to form the rest of the parathyroid and the postbranchial body. The thyroid, which had originated in the midline of the pharynx, also migrates caudally into the neck region. (After Carlson 1981.)

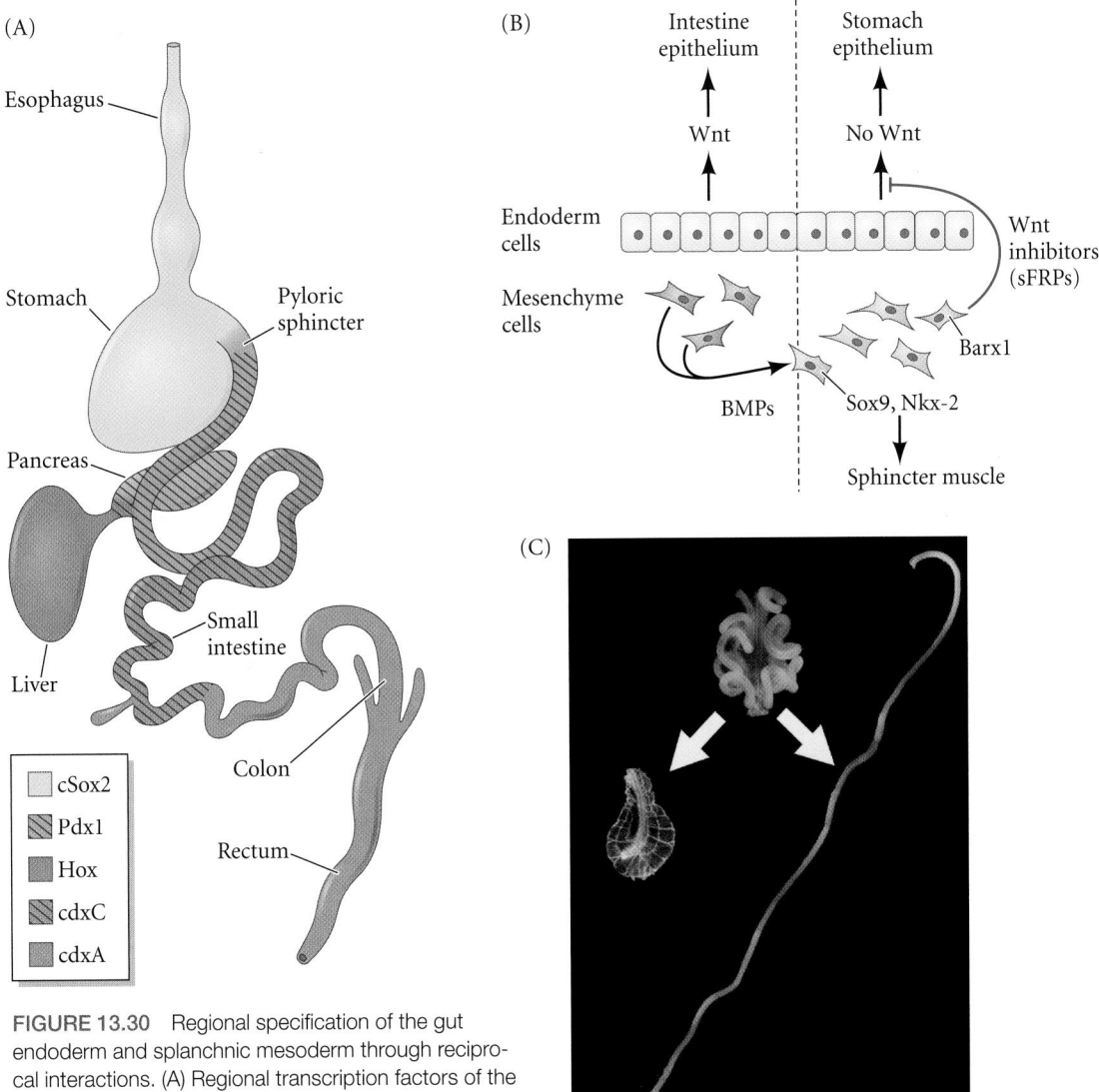

(A)

Esophagus

Stomach

Pyloric sphincter

Pancreas

Liver

Small intestine

Colon

Rectum

	cSox2
	Pdx1
	Hox
	cdxC
	cdxA

(B)

Intestine epithelium

Stomach epithelium

Wnt

No Wnt

Endoderm cells

Mesenchyme cells

Wnt inhibitors (sFRPs)

Barx1

BMPs

Sox9, Nkx-2

Sphincter muscle

(C)

FIGURE 13.30 Regional specification of the gut endoderm and splanchnic mesoderm through reciprocal interactions. (A) Regional transcription factors of the (mature) chick gut endoderm. These factors are seen prior to interactions with the mesoderm, but they are not stabilized. (B) Possible mechanism by which mesenchymal cells may induce endoderm to become either intestine or stomach, depending on the region. (C) Surgically separating the chick endodermal gut tube from the dorsal mesentery at embryonic day 12 causes the mesentery to shrivel up and the gut tube to straighten. The original gut-mesentery association (top) holds the digestive tube in place. When the two parts are separated on embryonic day 20, the mesentery (left) shrivels up, while the gut tube (right) straightens. (A after Grapin-Botton et al. 2001; C after Savin et al. 2011.)

stomach, small intestine, and large intestine. The endodermal cells generate only the lining of the digestive tube and its glands; mesenchyme cells from the splanchnic portion of the lateral plate mesoderm will surround the tube to provide the muscles for peristalsis.

As **FIGURE 13.30A** shows, the stomach develops as a dilated region of the gut close to the pharynx. The intestines develop more caudally, and the connection between the intestine and yolk sac is eventually severed. The intestine originally ends in the endodermal cloaca, but after the cloaca separates into the bladder and rectal regions (see Chapter 12), the intestine joins with the rectum. At the caudal end of the rectum, a depression forms where the endoderm meets the overlying ectoderm, and a thin **cloacal membrane** separates the two tissues. When the cloacal membrane eventually ruptures, the resultant opening becomes the anus.

Specification of the gut tissue

The production of endoderm is one of the first decisions made by the embryo. Indeed, in amphibian embryos, endoderm is specified autonomously by the presence of the **Sox17** transcription factor. Dominant negative forms of Sox17 (having repressive instead of activator subunits) block endoderm formation in the vegetal blastomeres, while the overexpression of the wild-type form expands the endodermal domain (Hudson et al. 1997; Henry and Melton 1998). In zebrafish and mice, *Sox17* also appears to be specifying endoderm, and mice whose *Sox17* genes have been knocked out have deficient gut endoderm (Kanai-Azuma et al. 2002). Indeed, when *Sox17*-containing adenoviruses are induced to express *Sox17* in embryonic stem cells, the embryonic stem cells produce endodermal derivatives (Takayama et al. 2011).

Although Sox17 helps specify the digestive tube, it does not give it its remarkable polarity. The digestive tube proceeds from the pharynx to the anus, differentiating along the way into the esophagus, stomach, duodenum, and intestines, and putting out branches that become (among other things) the thyroid, thymus, pancreas, and liver. What tells the endodermal tube to become these tissues at particular places? Why do we never see a mouth opening directly into a stomach? The endoderm and the splanchnic lateral plate mesoderm undergo a complicated set of interactions, and the signals for generating the different gut tissues appear to be conserved throughout the vertebrate classes (see Wallace and Pack 2003). One possible model for the polarity of the gut tube starts off with the specification of the pharynx and then follows with the specification of the remainder of the gut tube. Studies on chick embryos using beads containing either retinoic acid or inhibitors of its synthesis (Bayha et al. 2009) show that the pharynx can develop only in areas containing little or no RA, whereas the RA gradient patterns the pharyngeal arch endoderm in a graded manner. This is probably accomplished by activating and repressing particular sets of transcription factor genes.

The second phase of gut specification is thought to involve signals from the splanchnic mesoderm-derived mesenchyme surrounding the endodermal tube. As the digestive tube meets different mesenchymes, the mesenchyme cells instruct the endoderm to differentiate into esophagus, stomach, small intestine, and colon (Okada 1960; Gumpel-Pinot et al. 1978; Fukumachi and Takayama 1980; Kedinger et al. 1990). Wnt signals are thought to be especially important (Sherwood et al. 2011). The initial ("default") specification of the entire gut tube is thought to be anterior (i.e., stomach/esophagus). However, graded Wnt signaling from the posterior mesoderm (instructed by RA and FGF gradients) provides a signal that induces in the gut endoderm the posteriorizing transcription factors Cdx1 and Cdx2, as well as the paracrine factor Indian hedgehog. At high concentration, the Cdx transcription factors induce the formation of the large intestine, whereas at lower concentrations they induce the formation of the small intestine. Indeed, when β-catenin is artificially expressed in the *foregut* tissue, Cdx2 is activated and the anterior endoderm tissue is transformed into the more posterior intestinal type of tissue (Sherwood et al. 2011).

The molecular signals by which Wnt signals from the mesenchyme influence the gut tube are just becoming known (**FIGURE 13.30B**). Cdx2, for instance, suppresses genes such as *Hhex* and thereby prevents the stomach, liver, and pancreas from forming in the posterior (Bossard and Zaret 2000; McLin et al. 2007). In the anterior regions of the gut tube (which form the thymus, pancreas, stomach, and liver), Wnt signaling is blocked. In the stomach-forming domain, the mesenchyme lining the gut tube expresses the transcription factor Barx1, which activates production of Frzb-like Wnt-blocking proteins (sFRP1 and sFRP2). These Wnt antagonists block Wnt signaling in the vicinity of the stomach but not around the intestine. (Indeed, *Barx1*-deficient mice do not develop stomachs and express intestinal markers in that tissue; Kim et al. 2005.)

Wnt-based polarity may be transient and needs to be strengthened and refined by further interactions between the endoderm and the surrounding mesenchymes. Roberts and colleagues (1995, 1998) have implicated Sonic hedgehog (Shh) in endodermal specification. Shh is thought to be made by the endoderm and secreted in different concentrations at different sites. Its targets appear to be the mesodermal cells surrounding the gut tube. The secretion of Shh by the hindgut endoderm induces a nested pattern of "posterior" Hox gene expression in the mesoderm. As in the vertebrae (see Chapter 9), the anterior borders of Hox gene expression delineate the morphological boundaries of the regions that will form the cloaca, large intestine, cecum, mid-cecum (at the midgut-hindgut border), and posterior portion of the midgut (Roberts et al. 1995; Yokouchi et al. 1995). When Hox-expressing viruses cause misexpression of Hox genes in the mesoderm, the mesodermal cells alter the differentiation of the adjacent endoderm (Roberts et al. 1998). The Hox genes are thought to specify the mesoderm so that it can further interact with the endodermal tube and more finely specify its regions.

Once the boundaries of the transcription factors are established, differentiation can begin. The regional differentiation of the mesoderm (into smooth muscle types) and the regional differentiation of the endoderm (into different functional units such as the stomach, duodenum, and small intestine) are synchronized. For instance, in certain regions the intestinal mesenchyme secretes BMP4, which instructs the mesoderm anterior to it to express the Sox9 and Nkx2-5 transcription factors. These factors tell the mesoderm to become the muscles of the pyloric sphincter rather than the smooth muscle that normally lines the stomach and intestine (Theodosiou and Tabin 2005).

The interaction between the splanchnic mesoderm and the endoderm continues long after the specification stage of development. One derivative of the splanchnic mesoderm is the **dorsal mesentery**. This is a fibrous membrane that connects the endoderm to the body wall. The looping of the intestine is driven by a combination of growth intrinsic to the endoderm coupled with the connection of that tube to the dorsal mesentery (Savin et al. 2011). If the connection is

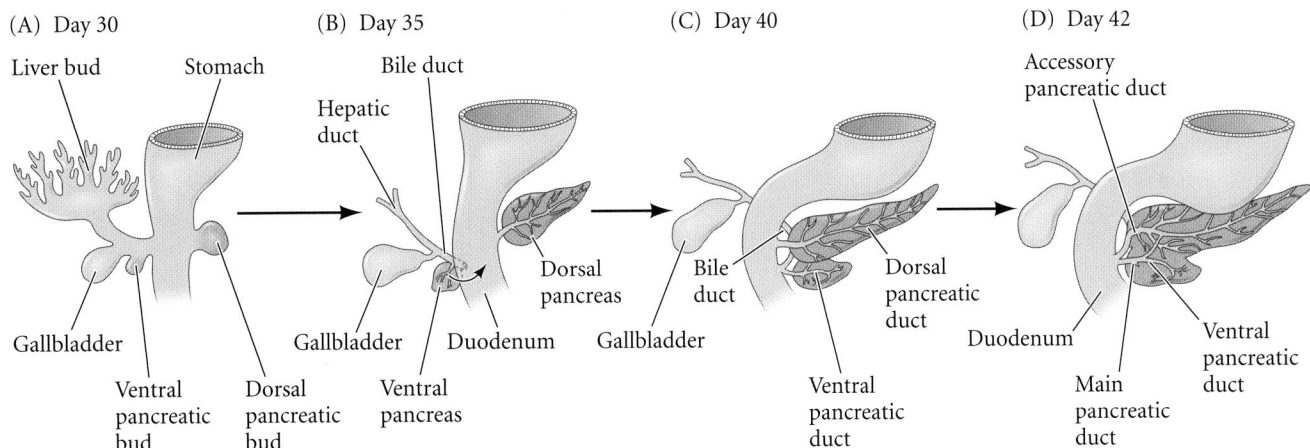

(A) Day 30

Liver bud   Stomach

Gallbladder

Ventral
pancreatic
bud

Dorsal
pancreatic
bud

(B) Day 35

Bile duct

Hepatic
duct

Gallbladder   Ventral
pancreas

Duodenum

Dorsal
pancreas

(C) Day 40

Bile
duct

Gallbladder

Ventral
pancreatic
duct

Dorsal
pancreatic
duct

(D) Day 42

Accessory
pancreatic duct

Duodenum

Main
pancreatic
duct

Ventral
pancreatic
duct

FIGURE 13.31 Pancreatic development in humans. (A) At 30 days, the ventral pancreatic bud is close to the liver primordium. (B) By 35 days, it begins migrating posteriorly, and (C) comes into contact with the dorsal pancreatic bud during the sixth week of development. (D) In most individuals, the dorsal pancreatic bud loses its duct into the duodenum; however, in about 10% of the population, the dual duct system persists.

severed, the mesentery shrinks and the gut becomes a long, thin, unfolded tube (**FIGURE 13.30C**).

The further development of the intestine involves (1) the differentiation of the Paneth cells and progeny of intestinal stem cells (see Figure P3.6) and (2) the interactions between the gut epithelium and symbiotic bacteria to complete the differentiation of the cell types. This will be discussed in Chapter 19.

Liver, pancreas, and gallbladder

The endoderm also forms the lining of three accessory organs that develop immediately caudal to the stomach: the liver, pancreas, and gallbladder. The **hepatic diverticulum** is a bud of endoderm that extends out from the foregut into the surrounding mesenchyme. The endoderm of this bud comes from two populations of cells—a lateral group that exclusively forms liver cells, and ventral-medial endoderm cells that form several midgut regions, including the liver (Tremblay and Zaret 2005). The mesenchyme induces this endoderm to proliferate, branch, and form the glandular epithelium of the liver. A portion of the hepatic diverticulum (the region closest to the digestive tube) continues to function as the drainage duct of the liver, and a branch from this duct produces the gallbladder (**FIGURE 13.31**).

SIDELIGHTS & SPECULATIONS

Specification of Liver and Pancreas

There is an intimate relationship between the splanchnic lateral plate mesoderm and the foregut endoderm. Just as the foregut endoderm is critical in specifying the cardiogenic mesoderm, the blood vessel endothelial cells induce the endodermal tube to produce the liver primordium and the pancreatic rudiments. Moreover, the chromatin of the multipotent cells of the ventral foregut endoderm may be primed for their differential activation. The genes involved in forming the liver progenitors have their genes silenced in a different manner than those genes involved in

forming the pancreatic progenitors. Thus, a single signal may be able to de-repress an entire battery of specification genes (Xu et al. 2011).

Liver formation

The expression of liver-specific genes (such as the genes for α-fetoprotein and albumin) can occur in any region of the gut tube that is exposed to cardiogenic mesoderm. However, this induction can occur only if the notochord is removed. If the notochord is placed by the portion of the endoderm normally induced by the cardiogenic mesoderm to become liver, the

endoderm will not form liver (hepatic) tissue. Therefore, the developing heart appears to induce the liver to form, while the presence of the notochord inhibits liver formation (**FIGURE 13.32**). This induction is probably due to FGFs secreted by the developing heart cells (Le Douarin 1975; Gualdi et al. 1996; Jung et al. 1999).

However, Matsumoto and colleagues (2001) found that the heart is not the only mesodermal derivative needed to form the liver. Blood vessel endothelial cells are also critical. If endothelial cells are not present in the area around the hepatic region of

FIGURE 13.32 Positive and negative signaling in the formation of the hepatic (liver) endoderm. The ectoderm and the notochord block the ability of the endoderm to express liver-specific genes. The cardiogenic mesoderm, probably through Fgf1 or Fgf2, promotes liver-specific gene transcription by blocking the inhibitory factors induced by the surrounding tissue. (After Gualdi et al. 1996.)

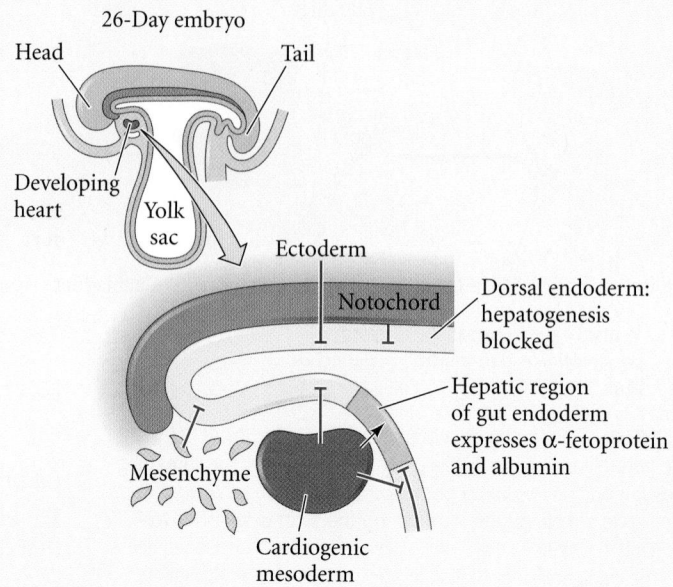

the gut tube, the liver buds fail to form. This induction occurs even before the endothelial cells have formed tubes, so it does not have anything to do with getting nutrients or oxygen into the region. Thus, the heart endothelial cells have a developmental function in addition to their circulatory roles: they induce the formation of the liver bud by secreting FGFs.

But in order to respond to the FGF signal, the endoderm has to become competent. This competence is given to the foregut endoderm by the **Forkhead** transcription factors. Forkhead transcription factors Foxa1 and Foxa2 are required to open the chromatin surrounding the liver-specific genes. These pioneer transcription factors displace nucleosomes from the regulatory regions surrounding these genes and are required before the FGF signal is given (Lee et al. 2005; Hirai et al. 2010). Mouse embryos lacking *Foxa1* and *Foxa2* expression in their endoderm fail to produce a liver bud or to express liver-specific enzymes.

Once the signal is given, other Forkhead transcription factors, such as HNF4α, become critical. HNF4α is essential for the morphological and biochemical differentiation of the

hepatic bud into liver tissue (Parviz et al. 2003). When conditional mutants of HNF4α were made such that this factor was absent only in the developing liver, neither the tissue architecture, cellular structure, or liver-specific enzymes formed in the liver bud cells. Meanwhile, Odom and colleagues (2004) found that Forkhead factors were also critical for the differentiation of the endocrine islands of the pancreas. HNF4α bound to the regulatory regions of almost half the actively transcribed pancreas-specific genes in these tissues, including those involved in insulin secretion. A link between HNF4α mutations and late-onset type 2 diabetes has been observed (see Kulkarni and Kahn 2004), confirming the importance of this transcription factor in pancreatic as well as hepatic development.

Pancreas formation

The formation of the pancreas may be the flip side of liver formation. Whereas the heart cells promote and the notochord prevents liver formation, the notochord may actively promote pancreas formation, while the heart may block the pancreas from forming. It seems that this particular region of

the digestive tube has the ability to become either pancreas or liver. One set of conditions (presence of heart, absence of notochord) induces the liver, while another set of conditions (presence of notochord, absence of heart) causes the pancreas to form.

The notochord activates pancreas development by repressing *shh* expression in the endoderm (Apelqvist et al. 1997; Hebrok et al. 1998). (This was a surprising finding, since we saw in Chapter 11 that the notochord is a source of Shh protein and an inducer of further *shh* gene expression in ectodermal tissues.) Sonic hedgehog is expressed throughout the gut endoderm, *except* in the region that will form the pancreas. The notochord in this region secretes Fgf2 and activin, which are able to downregulate *shh* expression. If *shh* is experimentally expressed in this region, the tissue reverts to being intestinal (Jonnson et al. 1994; Ahlgren et al. 1996; Offield et al. 1996).

The lack of Shh in the pancreas-forming region of the gut seems to enable it to respond to signals coming from the blood vessel endothelium. Indeed, pancreatic development is initiated at precisely those

SIDELIGHTS & SPECULATIONS (continued)

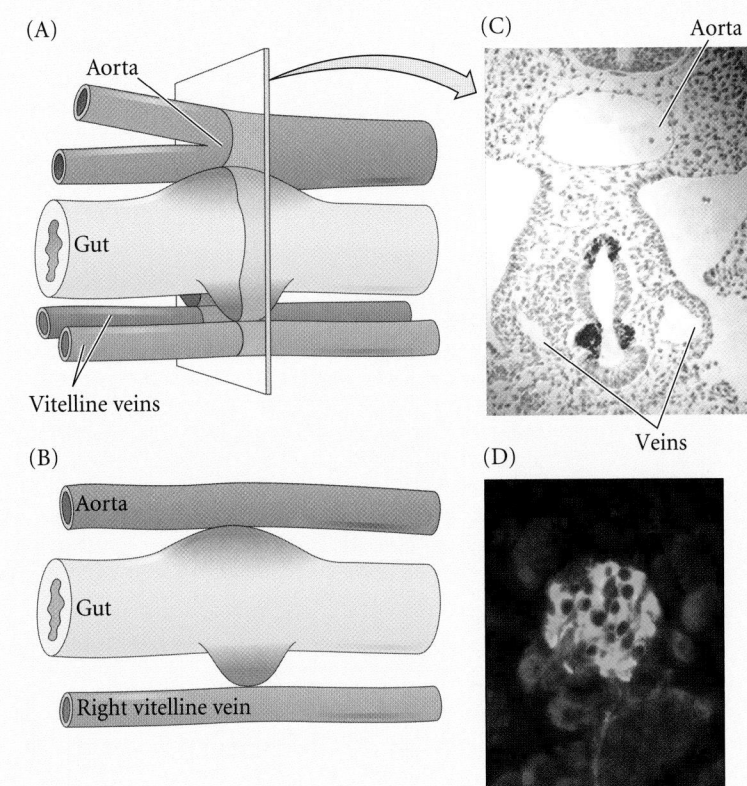

(A)

Aorta

Gut

Vitelline veins

(B)

Aorta

Gut

Right vitelline vein

(C)

Aorta

Veins

(D)

FIGURE 13.33 Induction of *Pdx1* gene expression in the gut epithelium. (A) In the chick embryo, *Pdx1* (purple) is expressed in the gut tube and is induced by contact with the aorta and vitelline veins. The regions of *Pdx1* gene expression create the dorsal and ventral rudiments of the pancreas. (B) In the mouse embryo, only the right vitelline vein survives, and it contacts the gut endothelium. *Pdx1* gene expression is seen only on this side, and only one ventral pancreatic bud emerges. (C) In situ hybridization of *Pdx1* mRNA in a section through the region of contact between the blood vessels and the gut tube of a mouse embryo. The regions of *Pdx1* expression show as deep blue. (D) Blood vessels (stained red) direct islets (stained green with antibodies to insulin) of chick embryo to differentiate. The nuclei are stained deep blue.

three locations where the foregut endoderm contacts the endothelium of the major blood vessels. It is at these points—where the endodermal tube meets the aorta and the vitelline veins—that the transcription factors Pdx1 and Ptf1a are expressed (**FIGURE 13.33A–C**; Lammert et al. 2001; Yoshitomi and Zaret 2004). If the blood vessels are removed from this area, the Pdx1 and Ptf1a expression regions fail to form, and the pancreatic endoderm fails to bud; if more blood vessels form in this area, more of the endodermal tube becomes pancreatic tissues.

The association of the pancreatic tissues with blood vessels is critical in the formation of the insulin-secreting cells of the pancreas. Pdx1 appears to act in concert with other such transcription factors to form the endocrine cells of the pancreas, the islets of Langerhans (Odom et al. 2004; Burlison et al. 2008; Dong et

al. 2008). The exocrine cells (which produce digestive enzymes such as chymotrypsin) and the endocrine cells appear to have the same progenitor (Fishman and Melton 2002), and the level of Ptf1a appears to regulate the proportion of cells in these lineages. The exocrine pancreatic cells have higher amounts of Ptf1a (Dong et al. 2008). The islet cells secrete VEGF to attract blood vessels, and these vessels surround the developing islet (**FIGURE 13.33D**).

The endocrine progenitor cells can form two populations. One is the progenitor of the β and δ cells of the islets of Langerhans. The other is the progenitor of the α and PP cells (which produce pancreatic polypeptide, a hormone that regulates gut endocrine secretion). The progenitor of the β and δ cells expresses the transcription factor Pax4, while the α/PP progenitor cell expresses Arx. These are mutually exclusive states, so the cell becomes

one cell type or the other. If the cell expresses Pax4 (becoming a β/δ progenitor), it still has a further choice. If it expresses MafA, it becomes a β cell (secreting insulin). If it doesn't express this gene, it becomes a δ cell (**FIGURE 13.34**).

The hierarchical dichotomous system shown in Figure 13.34 resembles the scheme of blood production from a single hematopoietic stem cell (see Figure 13.27). Modeling by Zhou and colleagues (2011) suggests that the different cell types can be seen as attractor states that result from the possible interactions of a common set of transcription factors. Indeed, Dhawan and colleagues (2011) have found that when the gene for Dnmt1 methyltransferase is knocked out in insulin-producing β cell, the methylation patterns change such that the *Arx* promoter is de-repressed and the *Arx* gene activated, converting β cells into glucagon-producing α cells.

Pancreatic β cell identity is thereby maintained by the methylation-mediated repression of *Arx*.

These transcription factor networks may enable the reprogramming of one cell type into another. Horb and colleagues (2003) have shown that Pdx1 can respecify developing liver tissue into pancreas. When *Xenopus* tadpoles were given a *pdx1* gene attached to a promoter active in liver cells, Pdx1 was made in the liver, and the liver was converted into a pancreas containing both exocrine and endocrine cells. Thus, Pdx1 appears to be the critical factor in distinguishing the liver from the pancreatic mode of development. As mentioned in Chapter 2, the expression of Ngn3, Pdx1, and MafA reprograms differentiated exocrine pancreatic cells of adult mice into functional β cells (see Figure 2.12; Zhou et al. 2007). This may become a very important tool in treating diabetes, which is caused by the destruction of pancreatic β cells.

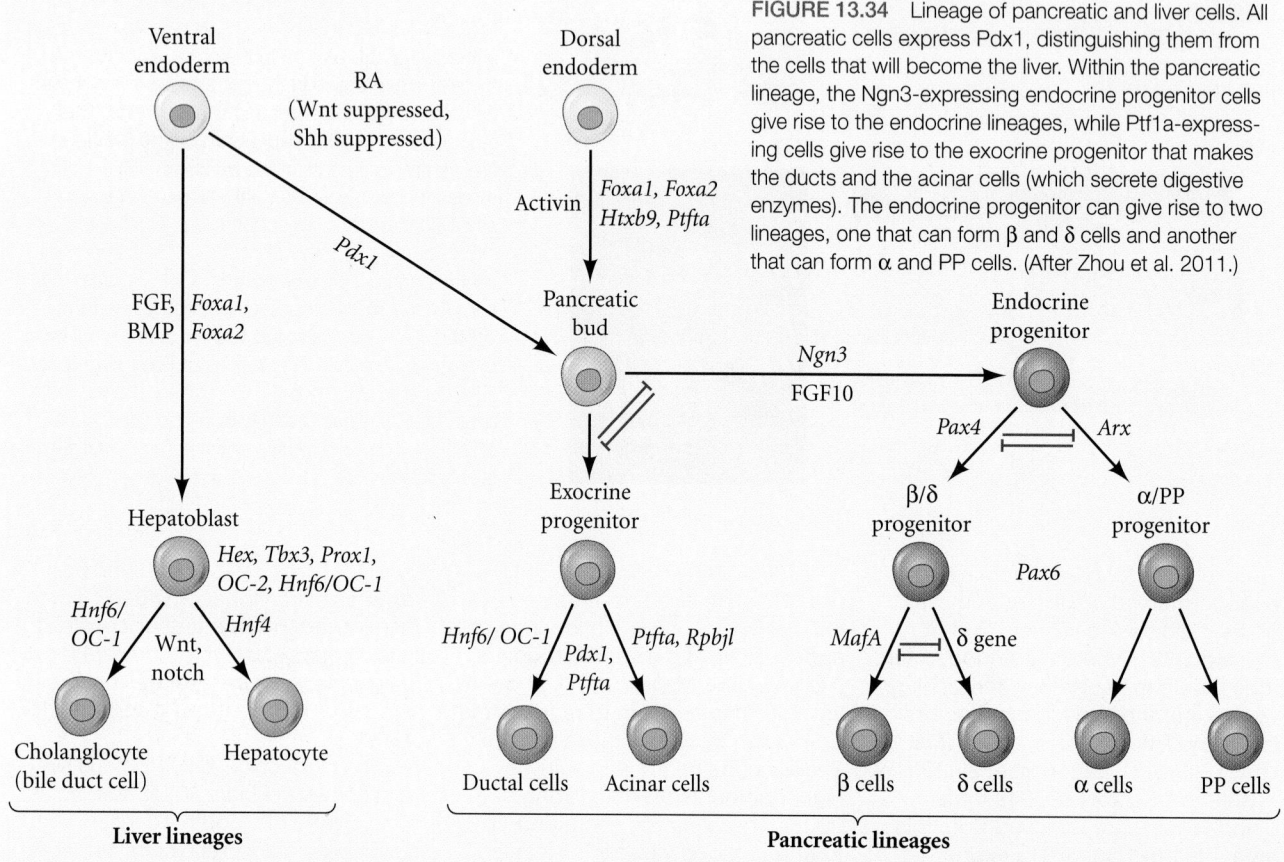

FIGURE 13.34 Lineage of pancreatic and liver cells. All pancreatic cells express Pdx1, distinguishing them from the cells that will become the liver. Within the pancreatic lineage, the Ngn3-expressing endocrine progenitor cells give rise to the endocrine lineages, while Ptf1a-expressing cells give rise to the exocrine progenitor that makes the ducts and the acinar cells (which secrete digestive enzymes). The endocrine progenitor can give rise to two lineages, one that can form β and δ cells and another that can form α and PP cells. (After Zhou et al. 2011.)

The pancreas develops from the fusion of distinct dorsal and ventral diverticula. As they grow, they come closer together and eventually fuse. In humans, only the ventral duct survives to carry digestive enzymes into the intestine. In other species (such as the dog), both dorsal and ventral ducts empty into the intestine.

The Respiratory Tube

The lungs are a derivative of the digestive tube, even though they serve no role in digestion. In the center of the pharyngeal floor, between the fourth pair of pharyngeal pouches, the **laryngotracheal groove** extends ventrally (**FIGURE 13.35A–C**). This groove then bifurcates into the branches that form the paired bronchi and lungs. The laryngotracheal endoderm becomes the lining of the trachea, the two bronchi, and the air sacs (alveoli) of the lungs. Sometimes this separation is not complete and a baby is born with a connection between the two tubes. This digestive and respiratory condition is called a **tracheal-esophageal fistula** and must be surgically repaired so the baby can breathe and swallow properly.

The formation of the esophagus from the gut tube shows again the interactions between the endoderm and specific mesenchyme. This time (and it is later in development), the difference is between dorsal and ventral regions of the body. Wnt signals from the mesenchyme cause the accumulation of β-catenin in the region of the gut tube that will become the lung and trachea. Without these signals, the separation

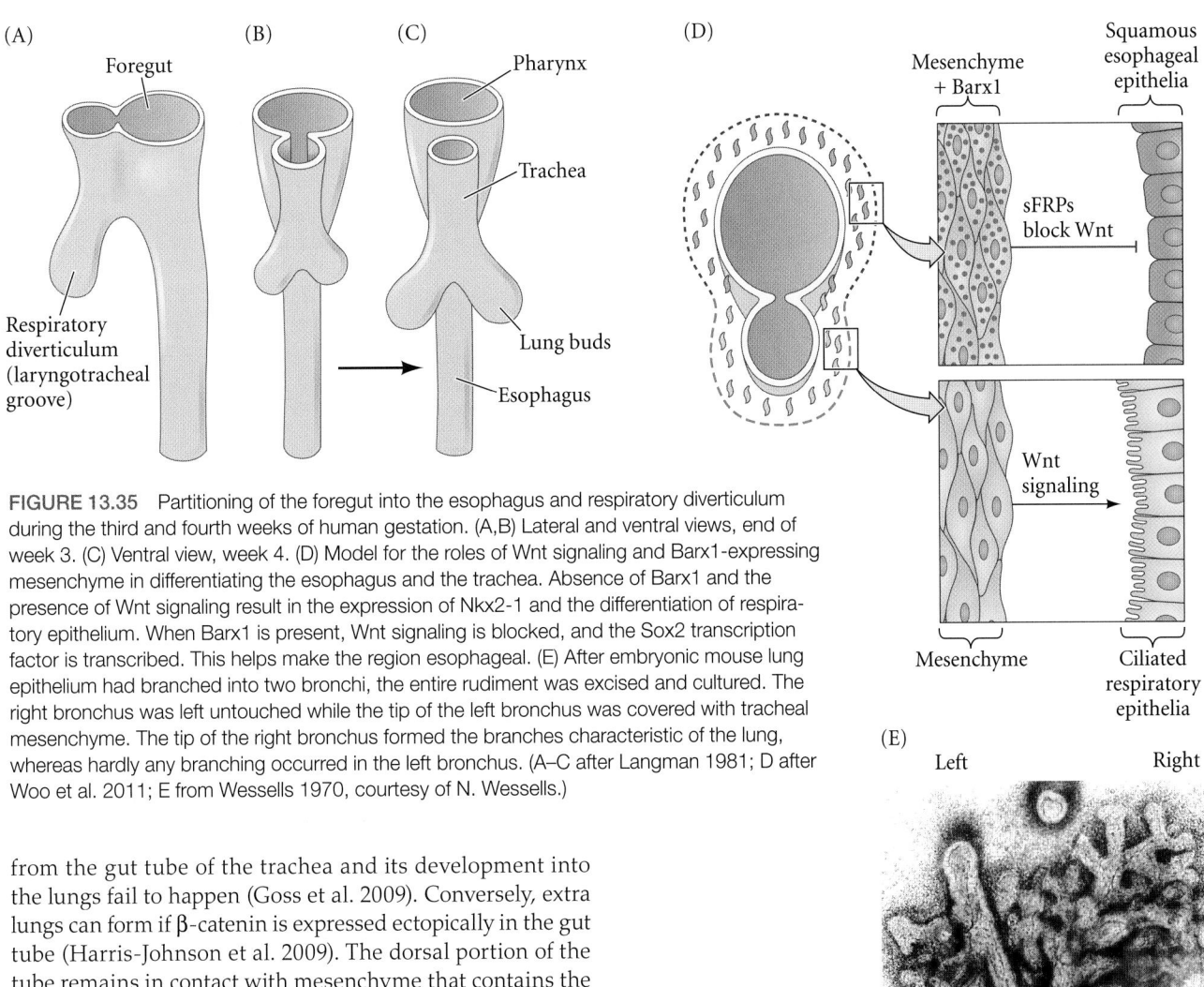

FIGURE 13.35 Partitioning of the foregut into the esophagus and respiratory diverticulum during the third and fourth weeks of human gestation. (A,B) Lateral and ventral views, end of week 3. (C) Ventral view, week 4. (D) Model for the roles of Wnt signaling and Barx1-expressing mesenchyme in differentiating the esophagus and the trachea. Absence of Barx1 and the presence of Wnt signaling result in the expression of Nkx2-1 and the differentiation of respiratory epithelium. When Barx1 is present, Wnt signaling is blocked, and the Sox2 transcription factor is transcribed. This helps make the region esophageal. (E) After embryonic mouse lung epithelium had branched into two bronchi, the entire rudiment was excised and cultured. The right bronchus was left untouched while the tip of the left bronchus was covered with tracheal mesenchyme. The tip of the right bronchus formed the branches characteristic of the lung, whereas hardly any branching occurred in the left bronchus. (A–C after Langman 1981; D after Woo et al. 2011; E from Wessells 1970, courtesy of N. Wessells.)

from the gut tube of the trachea and its development into the lungs fail to happen (Goss et al. 2009). Conversely, extra lungs can form if β-catenin is expressed ectopically in the gut tube (Harris-Johnson et al. 2009). The dorsal portion of the tube remains in contact with mesenchyme that contains the Barx1 transcription factor and which is producing the Wnt-blocking sFRPs. (The sFRP are soluble Frizzled-related proteins similar to Frzb (Chapter 8), which can to bind Wnts and prevent their reaching their cell membrane receptors). Thus, sFRPs block Wnt activity and help specify the epithelium of the esophagus. The ventral portion of the tube, however, comes into contact with a mesenchyme that does not produce sFRPs. The Wnt signals (which had been blocked earlier) convert the tube into the ciliated respiratory epithelium of the trachea (**FIGURE 13.35D**; Woo et al. 2011).

As in the digestive tube, the regional specificity of the mesenchyme determines the differentiation of the developing respiratory tube. In the developing mammal, the respiratory epithelium responds in two distinct fashions. In the region of the neck, it grows straight, forming the trachea. After entering the thorax, it branches, forming the two bronchi and then the lungs. The respiratory epithelium of an embryonic mouse can be isolated soon after it has split into two bronchi, and the two sides can be treated differently. **FIGURE 13.35E** shows the result when the right bronchial epithelium was allowed to retain its lung mesenchyme while the left bronchus was surrounded with tracheal mesenchyme (Wessells

1970). The right bronchus proliferated and branched under the influence of the lung mesenchyme, whereas the left bronchus continued to grow in an unbranched manner. Moreover, the differentiation of the respiratory epithelia into trachea cells or lung cells depends on the mesenchyme it encounters (Shannon et al. 1998).

● **See WEBSITE 13.2** Induction of the lung

The lungs are among the last of the mammalian organs to fully differentiate. The lungs must be able to draw in oxygen at the newborn's first breath. To accomplish this, the alveolar cells secrete a surfactant into the fluid bathing the lungs. This surfactant, consisting of specific proteins and phospholipids such as sphingomyelin and lecithin, is secreted very late in gestation, and it usually reaches physiologically useful levels at about week 34 of human gestation. The surfactant

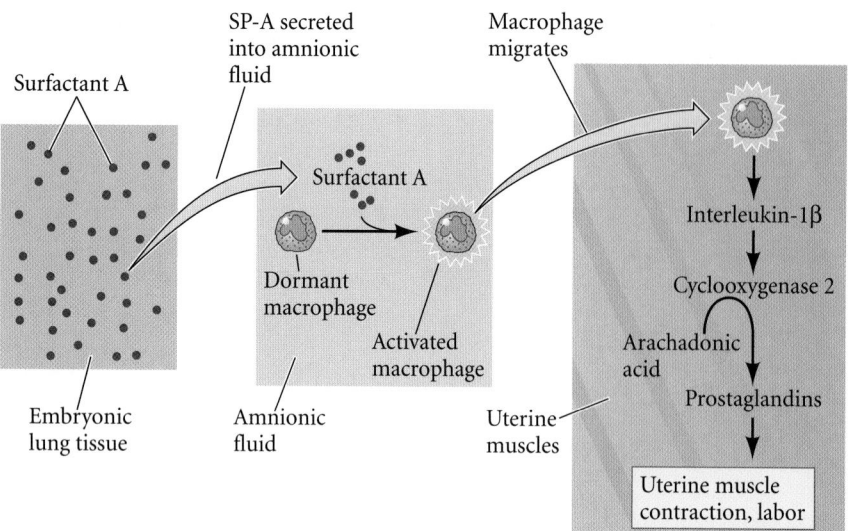

FIGURE 13.36 The immune system relays a signal from the embryonic lung. Surfactant protein A (SP-A) activates macrophages in the amniotic fluid to migrate into the uterine muscles, where the macrophages secrete IL1β. IL1β stimulates production of cyclooxygenase-2, an enzyme that in turn triggers the production of the prostaglandin hormones responsible for initiating uterine muscle contractions and birth.

enables the alveolar cells to touch one another without sticking together. Thus, infants born prematurely often have difficulty breathing and have to be placed on respirators until their surfactant-producing cells mature.

Mammalian birth occurs very soon after lung maturation. Some evidence suggests that the embryonic lung may actually signal the mother to start delivery. Condon and colleagues (2004) have shown that surfactant protein A—one of the final products produced by the embryonic mouse lung—activates macrophages in the amniotic fluid. These macrophages migrate from the amnion into the uterine muscle, where they produce immune system proteins such as interleukin 1β (IL1β). IL1β initiates the contractions of labor, both by activating cyclooxygenase-2 (which stimulates production of the prostaglandins that contract the uterine muscle cells) and by antagonizing the progesterone receptor (**FIGURE 13.36**). Mice deficient in surfacant proteins have a significant delay in the onset of labor, and surfactant-stimulated macrophages

injected into the uteri of female mice induce early labor (Montalbano et al. 2013). Thus, the signal for birth may be transmitted to the mother via her immune system.

Coda

So far the chapters of Parts Two and Three have described how the germ layers of the embryo undergo the processes of cell and tissue specification, followed by organogenesis and resulting in the formation of an organism. The remaining chapters in this section will explore the unique formation of vertebrate limbs, the genetics of sex determination, and various aspects of postembryonic development. We will end Part Three back where Part Two began—with the setting aside and specification of the cells of the germ line that are brought together in fertilization to start the processes anew in the ongoing circle of life.

SNAPSHOT SUMMARY: Lateral Plate Mesoderm and the Endoderm

1. The lateral plate mesoderm splits into two layers. The dorsal layer is the somatic (parietal) mesoderm, which underlies the ectoderm and forms the somatopleure. The ventral layer is the splanchnic (visceral) mesoderm, which overlies the endoderm and forms the splanchnopleure.

2. The space between the two layers of lateral plate mesoderm forms the body cavity, or coelom.

3. The heart arises from splanchnic mesoderm on both sides of the body. This region of cells is called the heart field, or cardiogenic mesoderm. The cardiogenic mesoderm is specified by BMPs in the absence of Wnt signals.

4. The Nkx2-5, Mesp1, and GATA transcription factors are important in committing the cardiogenic mesoderm to

become heart cells. These cardiac precursor cells migrate from the sides to the midline of the embryo, in the neck region.

5. There are two major heart fields one each side of the body. Each heart field has two regions: The first heart field forms the scaffold of the heart tube and will form the left ventricle. The rest of the heart is made largely by the second heart field.

6. A cardiac precursor cell can form each of the major lineages of the heart. The cardiogenic mesoderm forms the endocardium (which is continuous with the blood vessels) and the myocardium (the muscular component of the heart).

7. The endocardial tubes form separately and then fuse. The looping of the heart transforms the original anterior-posterior polarity of the heart tube into a right-left polarity.

8. Retinoic acid is important in determining the anterior-posterior polarity of the heart.

9. Tbx transcription factors are critical for specifying the heart chambers and for establishing the electrical circuitry of the heart.

10. Coronary arteries and lymphatic vessels both come from the reprogramming of veins.

11. In mammals, fetal circulation differs dramatically from adult circulation. When the infant takes its first breath, changes in air pressure close the foramen ovale through which blood had passed from the right to the left atrium. At that time, the lungs, rather than the placenta, become the source of oxygen.

12. Blood vessel formation is constrained by physiological, evolutionary, and physical parameters. The subdividing of a large vessel into numerous smaller ones allows rapid transport of the blood to regions of gas and nutrient diffusion.

13. Blood vessels are constructed by two processes, vasculogenesis and angiogenesis. Vasculogenesis involves the condensing of splanchnic mesoderm cells to form blood islands. The outer cells of these islands become endothelial (blood vessel) cells. Angiogenesis involves the remodeling of existing blood vessels.

14. Numerous paracrine factors are essential in blood vessel formation. Fgf2 is needed for specifying the angioblasts. VEGF-A is essential for the differentiation of the angioblasts. Angiopoietins allow the smooth muscle cells (and smooth muscle-like pericytes) to cover the vessels. Ephrin ligands and Eph receptor tyrosine kinases are critical for capillary bed formation.

15. The pluripotent hematopoietic stem cell (HSC) generates other pluripotent stem cells, as well as lineage-restricted stem cells. It gives rise to both blood cells and lymphocytes.

16. In vertebrates, HSCs are thought to originate from hemogenic endothelial cells that characterize the blood islands, the dorsal aorta, and the placental vessels. The definitive HSC appears to be derived from the ventral portion of the aorta.

17. The common myeloid precursor (CMP) is a blood stem cell that can generate the more committed stem cells for the different blood lineages. Hematopoietic inductive microenvironments (HIMs) determine the blood cell differentiation.

18. The HSC depends on stem cell factor, which is provided to it by the perivascular cells of the sinusoids contained in the stem cell niche.

19. The endoderm is specified in part by the Sox17 transcription factor. In vertebrates, the endoderm constructs the digestive tube and the respiratory tube.

20. Four pairs of pharyngeal pouches become the endodermal lining of the eustacian tubes, the tonsils, the thymus, and the parathyroid glands. The thyroid also forms in this region of endoderm.

21. The gut tissue forms by reciprocal interactions between the endoderm and the mesoderm. Wnt signals from the mesoderm and Sonic hedgehog from the endoderm appear to play a role in inducing a nested pattern of Hox gene expression in the mesoderm surrounding the gut. The regionalized mesoderm then instructs the endodermal tube to become the different organs of the digestive tract.

22. The endoderm helps specify the splanchnic mesoderm; the splanchnic mesoderm, especially the heart and the blood vessels, helps specify the endoderm.

23. The pancreas forms in a region of endoderm that lacks *sonic hedgehog* expression. The Pdx1 and Ptf1a transcription factors are expressed in this region.

24. The endocrine and exocrine cells of the pancreas have a common origin. The Ngn3 transcription factor probably decides endocrine fate.

25. The respiratory tube is derived as an outpocketing of the digestive tube. The regional specificity of the mesenchyme it meets determines whether the tube remains straight (as in the trachea) or branches (as in the bronchi and alveoli).

For Further Reading

Complete bibliographical citations for all literature cited in this chapter can be found at the free-access website **www.devbio.com**

Adamo, L. and G. García-Cardeña. 2012. The vascular origin of hematopoietic cells. *Dev. Biol.* 362: 1–10.

Bakker, M. L. and 12 others. 2012. T-box transcription factor TBX3 reprograms mature cardiac myocytes into pacemaker-like cells. *Cardiovasc. Res.* 94: 439-49.

Cooley, J., S. Whitaker, S. Sweeney, S. Fraser and B. Davidson. 2011. Cytoskeletal polarity mediates localized induction of the heart progenitor lineage. *Nat. Cell Biol.* 13: 952–957.

Ding, L., T. L. Saunders, G. Enikolopov and S. J. Morrison. 2012. Endothelial and perivascular cells maintain haematopoietic stem cells. *Nature* 481: 457–462.

Fishman, M. P. and D. A. Melton. 2002. Pancreatic lineage analysis using a retroviral vector in embryonic mice demonstrates a common progenitor for endocrine and exocrine cells. *Int. J. Dev. Biol.* 46: 201–207.

Karkkainenen, M. J. and 11 others. 2004. Vascular endothelial growth factor C is required for sprouting of the first lymphatic vessels from embryonic aveins. *Nature Immunol.* 5: 74–80.

Kim, B.-M., G. Buchner, I. Miletich, P. T. Sharpe and R. A. Shivdasani. 2005. The stomach mesenchymal transcription factor Barx1 specifies gastric epithelial identity through inhibition of transient Wnt signaling. *Dev. Cell* 8: 611–622.

Odom, D. T. and 12 others. 2004. Control of pancreas and liver gene expression by HNF transcription factors. *Science* 303: 1378–1381.

Oliver, G. and R. S. Srinivasan. 2010. Endothelial cell plasticity: how to become and remain a lymphatic endothelial cell. *Development* 137: 363–372.

Sizarov, A., J. Ya, B. A. de Boer, W. H. Lamers, V. M. Christoffels and A. F. Moorman. 2011. Formation of the building plan of the human heart: Morphogenesis, growth, and differentiation. *Circulation* 123:1125–1135.

Wang, H. U., Z.-F. Chen and D. J. Anderson. 1998. Molecular distinction and angiogenic interaction between embryonic arteries and veins revealed by ephrin-B2 and its receptor Eph-B4. *Cell* 93: 741–753.

Go Online

WEBSITE 13.1 Coelom formation. Coelom formation is readily visualized by animations. The animation presented here shows the expansion of the mesoderm during chick development.

WEBSITE 13.2 Induction of the lung. The induction of the lung involves interplay between FGFs and Shh. However, it appears to be different from the induction of either the pancreas or the liver.

Vade Mecum

Early heart development. The vertebrate heart begins to function early in its development. You can see this in movies of the living chick embryo at early stages when the heart is little more than a looped tube.

Outside Sites

The University of New South Wales embryology course has excellent graphics for studying human heart and blood vessel development. **php.med.unsw.edu.au/embryology/ index.php?title=Cardiovascular_System_Development**

14

Development of the Tetrapod Limb

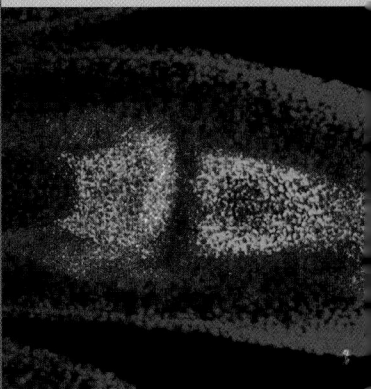

CONSIDER YOUR LIMB. First, consider its polarity. It has fingers or toes at one end, a humerus or femur at the other. You won't find anyone with fingers in the middle of their arm. Next, consider the subtle but obvious differences between your hands and your feet. If your fingers were replaced by toes, you would know it. But then consider how *similar* the bones of your feet are to the bones of your hand. It's easy to see that they share a common pattern. And finally, consider the size and length of your limbs. Both hands are remarkably similar in size; so are both your feet. And after about 20 years of growth, each of your feet turns out, independently, to be the same length. These commonplace phenomena present fascinating questions to the developmental biologist. How is it that we have four limbs and not six or eight? Why do fingers form at one end of the limb and nowhere else? How is it that the little finger develops at one edge of the limb and the thumb at the other? How does the forelimb grow differently than the hindlimb? How can limb growth be so precisely regulated?

All of these questions are really about **pattern formation**: the set of processes by which embryonic cells form ordered spatial arrangements of differentiated tissues. Pattern formation is one of the most dramatic properties of a developing organism, and one that has provoked awe in scientists and laypeople alike. It is one thing to differentiate the chondroblasts and osteoblasts that synthesize the cartilage and bone matrices, respectively; it is another thing to produce those cells in a temporal-spatial orientation that generates a functional bone. It is still another thing to make that bone a humerus and not a pelvis or a femur. The tissues of the finger—bone, cartilage, blood vessels, nerves, dermis, and epidermis—are the same in the toe and thigh; it is their arrangement that differs. The ability of limb cells to sense their relative positions and to differentiate with regard to those positions has been—and still is—the subject of intense study, experimentation, and debate.

The bones of any tetrapod limb, be it arm or leg, wing or flipper, consist of a proximal **stylopod** (humerus/femur) adjacent to the body wall; a **zeugopod** (radius-ulna/tibia-fibula) in the middle region; and a distal **autopod** (carpals-fingers/tarsals-toes) (FIGURE 14.1). The positional information needed to construct a limb has to function in a three-dimensional* coordinate system:

- The first dimension is the *proximal-distal axis* ("close-far"; that is, shoulder-finger or hip-toe). The bones of the limb are formed by endochondral ossification. They are initially cartilaginous, but eventually most of the cartilage is replaced by bone. Somehow the limb cells develop differently at early stages of development (when they make the stylopod) than at later stages (when they make the autopod).

- The second dimension is the *anterior-posterior axis* (thumb-pinkie). Our little fingers or toes mark the *posterior* end, while our thumbs or big toes are at the *anterior*

---

*Actually, it is a *four*-dimensional system, in which time is the fourth axis. Developmental biologists get used to seeing nature in four dimensions.

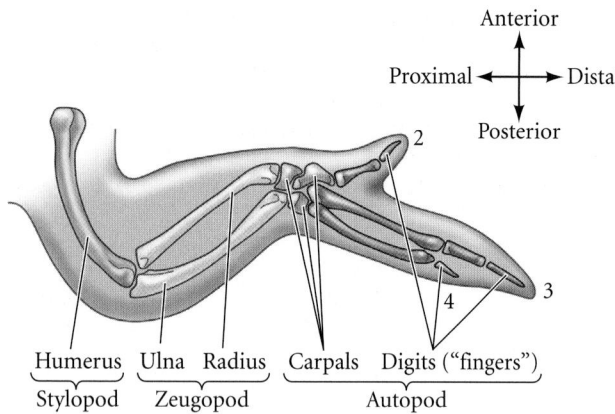

FIGURE 14.1 Skeletal pattern of the chick wing. According to convention, the digits are numbered 2, 3, and 4. The cartilage condensations forming the digits appear similar to those forming digits 2, 3, and 4 of mice and humans; however, new evidence (discussed in Website 14.2) suggests that the correct designation may be 1, 2, and 3.

end. In humans, it is obvious that each hand develops as a mirror image of the other. One can imagine other arrangements to exist—such as the thumb developing on the left side of both hands—but these patterns do not occur.

- Finally, limbs have a *dorsal-ventral axis*: our palms (ventral) are readily distinguishable from our knuckles (dorsal).

During the past decade, proteins have been identified that play critical roles in the formation of each of the three limb axes. The proximal-distal axis (shoulder-finger; hip-toe) is regulated in large part by proteins of the fibroblast growth factor (FGF) family. The anterior-posterior axis (thumb-pinkie) is primarily controlled by Sonic hedgehog. The dorsal-ventral axis (knuckle-palm) is regulated, at least in part, by Wnt7a. The interactions of these proteins mutually support one another and determine the differentiation of cell types.

The fundamental problem of morphogenesis—how specific structures arise in particular places—is exemplified in limb development. Because the limbs, unlike the heart or brain, are not essential for embryonic or fetal life, one can experimentally remove or transplant parts of the developing limb, or create limb-specific mutants, without interfering with the vital processes of the organism. Such experiments have shown that the basic "morphogenetic rules" for forming a limb appear to be the same in all tetrapods. Grafted pieces of reptile or mammalian limb buds can direct the formation of chick limbs, and regions of frog limb buds can direct the patterning of salamander limbs (Fallon and Crosby 1977; Sessions et al. 1989; see Hinchliffe 1991). Moreover, as will be detailed in Chapter 16, *regenerating* salamander limbs appear to follow many of the same rules as developing limbs (Muneoka and Bryant 1982). But what are these morphogenetic rules?

Formation of the Limb Bud

Specification of the limb fields

Limbs do not form just anywhere along the body axis but are generated at discrete positions. The mesodermal cells that give rise to a vertebrate limb can be identified by (1) removing certain groups of cells and observing that a limb does not develop in their absence (Detwiler 1918; Harrison 1918); (2) transplanting groups of cells to a new location and observing that they form a limb in this new place (Hertwig 1925); and (3) marking groups of cells with dyes or radioactive precursors and observing that their descendants partake in limb development (Rosenquist 1971).

FIGURE 14.2 shows the prospective forelimb areas in the tailbud stage of a salamander embryo. The fate-mapping studies on salamanders, pioneered by Ross Granville Harrison's laboratory (see Harrison 1918, 1969), showed that the center of this disc of cells in the somatic region of the lateral plate mesoderm normally gives rise to the limb itself. Adjacent to it are the cells that will form the peribrachial flank tissue and the shoulder girdle. However, if all these cells are extirpated from the embryo, a limb will still form (albeit somewhat later) from an additional ring of cells that surrounds this area but would not normally form a limb. If this surrounding ring of cells is included in the extirpated tissue, no limb will develop. This larger region, representing all the cells in the area capable of forming a limb on their own, is the **limb field**.

Limb development begins when mesenchyme cells migrate from the limb fields of the lateral plate mesoderm (to form the limb *skeletal* precursor cells) and from the somites (the limb *muscle* precursor cells) at the same level. These mesenchymal cells accumulate under the ectodermal tissue to create a circular bulge called a **limb bud** (FIGURE 14.3A).

Vertebrates have no more than four limb buds per embryo, and limb buds are always paired opposite each other with

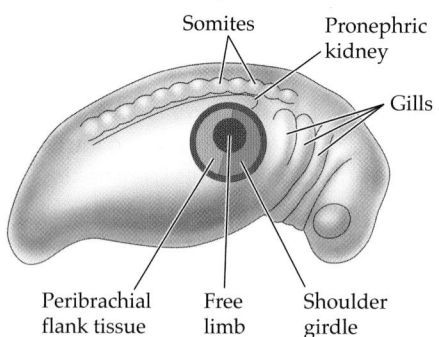

FIGURE 14.2 Prospective forelimb field of the salamander *Ambystoma maculatum*. The central area contains cells destined to form the limb per se (the free limb). The cells surrounding the free limb give rise to the peribrachial flank tissue and the shoulder girdle. The ring of cells outside these regions usually is not included in the limb, but can form a limb if the more central tissues are extirpated. (After Stocum and Fallon 1982.)

(A)

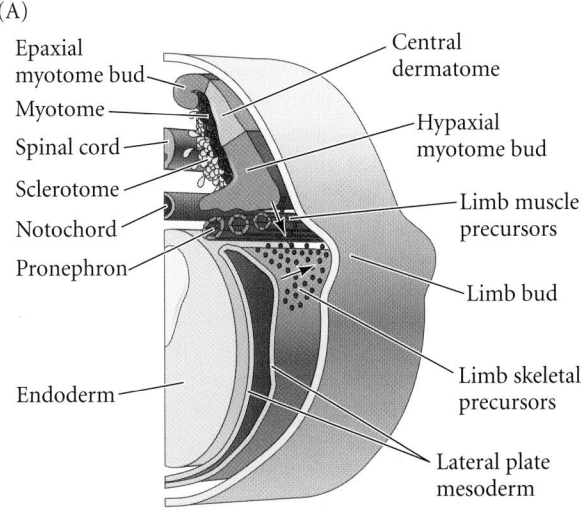

Epaxial
myotome bud
Myotome
Spinal cord
Sclerotome
Notochord
Pronephron

Endoderm

Central
dermatome

Hypaxial
myotome bud

Limb muscle
precursors

Limb bud

Limb skeletal
precursors

Lateral plate
mesoderm

FIGURE 14.3 Emergence of the limb bud. (A) Proliferation of mesenchymal cells (arrows) from the somatic region of the lateral plate mesoderm causes the limb bud in the amphibian embryo to bulge outward. These cells generate the skeletal elements of the limb. Contributions of myoblasts from the lateral myotome provide the limb's musculature. (B) Entry of myoblasts (purple) into the limb bud. This computer stereogram was created from sections of an in situ hybridization to the *myf5* mRNA found in developing muscle cells. If you can cross your eyes (or try focusing "past" the page, looking through it to your toes), the three-dimensionality of the stereogram will become apparent. (B courtesy of J. Streicher and G. Müller.)

(B)

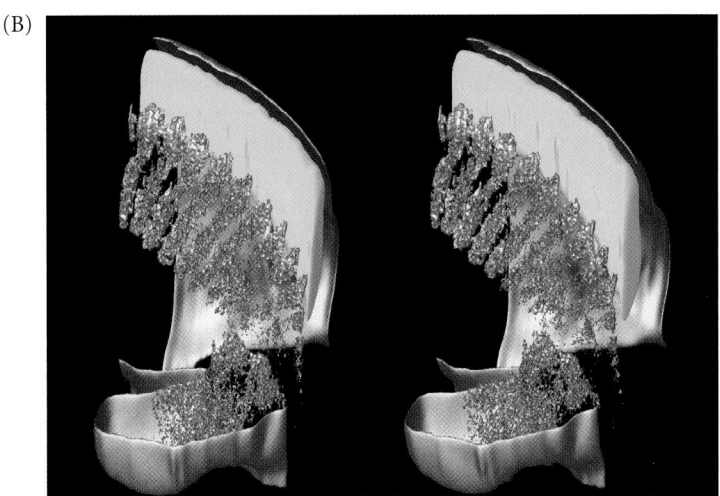

respect to the midline. Although the limbs of different vertebrates differ with respect to the somite level at which they arise, their position is constant with respect to the level of Hox gene expression along the anterior-posterior axis (see Chapter 9). For instance, in fish (in which the pectoral and pelvic fins correspond to the anterior and posterior limbs, respectively), amphibians, birds, and mammals, the forelimb buds are found at the most anterior expression region of *Hoxc6*, the position of the first thoracic vertebra* (Oliver et al. 1988; Molven et al. 1990; Burke et al. 1995). The lateral plate mesoderm in the four limb fields is also unique in that it induces myoblasts to migrate out from the somites and enter the limb bud to become the limb musculature (**FIGURE 14.3B**; Hayashi and Ozawa 1995). It is probable that positional information from

*Interestingly, Hox gene expression in at least some snakes (such as *Python*) creates a pattern in which each somite is specified to become a thoracic (ribbed) vertebra. The patterns of Hox gene expression associated with limb-forming regions are not seen (Cohn and Tickle 1999; see Chapter 13).

the Hox gene expression domains causes the paraxial mesoderm in the limb-forming regions to be different from all other paraxial mesoderm. Transplantation experiments in which paraxial mesoderm from different locations is placed adjacent to flank lateral plate shows that the paraxial mesoderm from limb-forming regions promotes limb bud formation, whereas the paraxial mesoderm from limbless flank actively represses limb formation (Noro et al. 2011).

When it first forms, the limb field has the ability to regulate for lost or added parts. In the tailbud stage of *Ambystoma*, any half of the limb disc is able to generate an entire limb when grafted to a new site (Harrison 1918). This potency can also be shown by splitting the limb disc vertically into two or more segments and placing thin barriers between the segments to prevent their reunion. When this is done, each segment develops into a full limb. Thus, like an early sea urchin embryo, the limb field represents a "harmonious equipotent system" wherein a cell can be instructed to form any part of the limb. The regulative ability of the limb bud has recently been highlighted by a remarkable experiment of nature. In

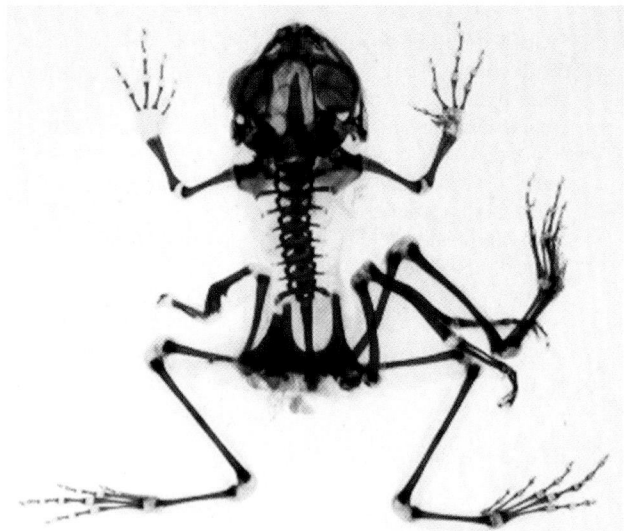

FIGURE 14.4 Multilimbed Pacific tree frog (*Hyla regilla*), the result of infestation of the tadpole-stage developing limb buds by trematode cysts. The parasitic cysts apparently split the developing limb buds in several places, resulting in extra limbs. In this adult frog's skeleton, the cartilage is stained blue; the bones are stained red. (Courtesy of S. Sessions.)

numerous ponds in the United States, multilegged frogs and salamanders have been found (**FIGURE 14.4**). The presence of these extra appendages has been linked to the infestation of the larval abdomen by parasitic trematode worms. The eggs of these worms apparently split the developing tadpole limb buds in several places, and the limb bud fragments develop as multiple limbs (Sessions and Ruth 1990; Sessions et al. 1999).

Induction of the early limb buds: Wnt proteins and fibroblast growth factors

Molecular studies on the earliest stages of limb formation have shown that the signal for limb bud formation comes from the lateral plate mesoderm cells that will become the prospective limb skeleton. These cells secrete the paracrine factor Fgf10, which is capable of initiating the limb-forming

interactions between ectoderm and mesoderm. If beads containing Fgf10 are placed ectopically beneath the flank ectoderm, extra limbs emerge (**FIGURE 14.5**; Ohuchi et al. 1997; Sekine et al. 1999). The Fgf10-containing beads appear to stimulate Wnt proteins in the surface ectoderm, and these Wnt proteins instruct the surface ectoderm to make Fgf8. This is interesting, because once Fgf8 is made in the surface ectoderm, the surface ectoderm elongates to become the **apical ectodermal ridge** (**AER**), a structure that is the source of FGFs for limb outgrowth and patterning. One of the main functions of the AER is to tell the mesenchyme cells directly beneath it to continue making Fgf10. In this way, a positive feedback loop is created wherein mesodermal Fgf10 tells the surface ectoderm to make Fgf8, and the surface ectoderm tells the underlying mesoderm to make Fgf10 (see Figure 14.6). Moreover, the somites that are not adjacent to the limb bud-forming regions may prevent the spread of the signals by preventing cells from responding to Fgf8 (Noro et al. 2011).

Specification of forelimb or hindlimb

The mechanisms by which instructions from the Hox code enable the production of Fgf10 are being investigated (**FIGURE 14.6**). Specific Hox proteins may be permitting the synthesis of retinoic acid in certain regions of the lateral plate mesoderm. RA then induces the expression of particular Tbx transcription factors (Agarwal et al. 2003; Grandel and Brand 2011). The Tbx factors then activate Fgf10. In this manner, the question of how Hox gene expression causes Fgf10 expression may be intimately linked to the question of how the forelimbs and hindlimbs are distinguished. In mice, the gene encoding Tbx5 is transcribed in the anterior lateral plate mesoderm and in the forelimbs, while the genes encoding Tbx4 and Pitx1 are expressed in the posterior lateral plate mesoderm and in the hindlimbs* (Chapman et al. 1996; Gibson-Brown et al. 1996; Takeuchi et al. 1999).

---

*"Tbx" stands for T-box, a specific DNA-binding domain. The *T* (*Brachyury*) gene and its relatives have a sequence that encodes this domain. We discussed Tbx5 in the context of heart ventricle development in Chapter 13.

FIGURE 14.5 Fgf10 expression and action in the developing chick limb. (A) Fgf10 becomes expressed in the lateral plate mesoderm in precisely those positions (arrows) where limbs normally form. (B) When transgenic cells that secrete Fgf10 are placed in the flanks of a chick embryo, the Fgf10 can cause the formation of an ectopic limb (arrow). (From Ohuchi et al. 1997, courtesy of S. Noji.)

(A)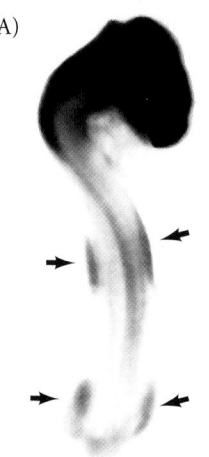

(B)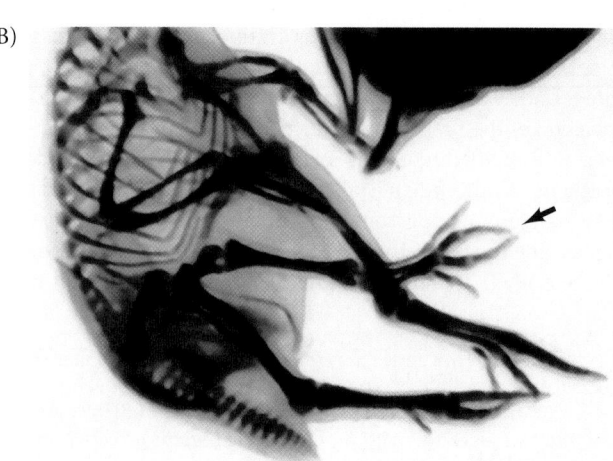

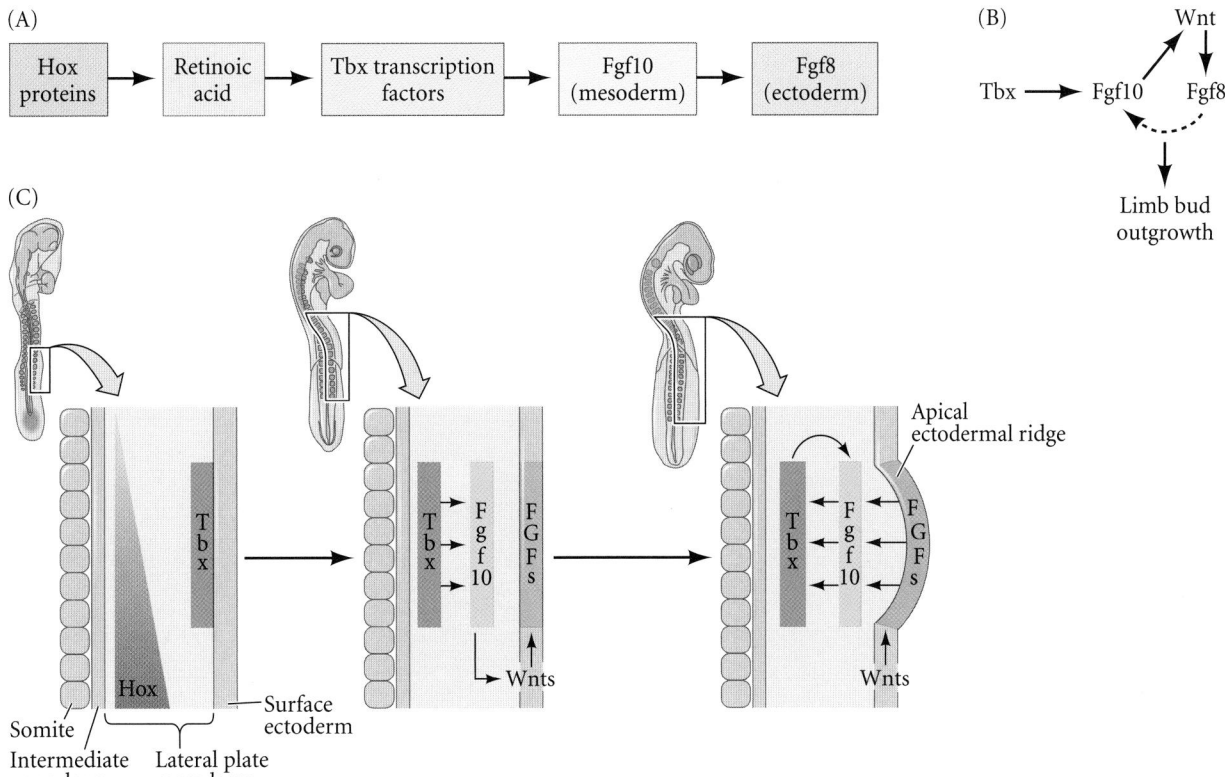

FIGURE 14.6 Model for initiation of the forelimb bud in mouse embryos. (A) Hox proteins establish conditions for the synthesis of retinoic acid in the lateral plate mesoderm. RA causes the induction of transcription factors Tbx5 (forelimb) and Tbx4 (hindlimb). The Tbx transcription factors induce Fgf10 in the lateral plate mesoderm, and this induces (via Wnt signaling) *Fgf8* expression in the ectoderm. The positive feedback loop between Fgf10 in the mesoderm and Fgf8 in the ectoderm initiates the outgrowth of the limb. (B) Schematic representation whereby Fgf10 expression depends on earlier expression of a Tbx gene, whose expression depends on Hox genes. (C) Geographic representation of the pathway whereby Hox gene expression in the lateral plate mesoderm activates Tbx in a particular region of that germ layer. Tbx induces the expression of Fgf10 from the lateral plate mesoderm cells, and Fgf10 in turn induces Wnts in the surface ectoderm. These Wnts cause the expression of FGFs in the surface ectoderm, causing the ectoderm to become the apical ectodermal ridge (AER). The FGFs and Wnts from the AER stimulate the continued production of Fgf10 in the mesoderm beneath it. (After Agarwal et al. 2003.)

Several laboratories (Logan et al. 1998; Ohuchi et al. 1998; Rodriguez-Esteban et al. 1999; Takeuchi et al. 1999, among others) have provided gain-of-function evidence that Tbx4 and Tbx5 are critical in the specification of hindlimbs and forelimbs, respectively. Before limb formation, there are normally regions of Tbx4 in the posterior portion of the lateral plate mesoderm (including the region that will form the hindlimbs) and regions of Tbx5 in the anterior portion (including the region that will form the forelimbs). When FGF-secreting beads were used to induce an ectopic limb between the chick hindlimb and forelimb buds, the type of limb produced was determined by which Tbx protein was expressed. Limb buds induced by placing FGF beads close to the hindlimb (opposite somite 25) expressed Tbx4 and became hindlimbs. Buds induced close to the forelimb (opposite somite 17) expressed Tbx5 and developed as forelimbs (wings). Limb buds induced in the center of the flank tissue expressed Tbx5 in the anterior portion of the limb and *Tbx4*

in the posterior portion; these limbs developed as chimeric structures, with the anterior resembling a forelimb and the posterior resembling a hindlimb (**FIGURE 14.7**). Moreover, when a chick embryo was made to express Tbx4 throughout the flank tissue (by infecting the tissue with a virus that expressed Tbx4), limbs induced in the anterior region of the flank often became legs instead of wings.

However, the Tbx genes are not the complete story of limb specification, and mice may differ from chicks. In mice, *Tbx4* and *Tbx5* are responsible for the outgrowth of the limbs but not their specification (Agarwal et al. 2003). When either *Tbx5* or *Tbx4* is knocked out, the lateral plate mesoderm fails to make Fgf10 in the forelimb or hindlimb regions. Therefore, it seems likely that Hox proteins may activate Tbx transcription factors and that Tbx factors activate Fgf10 to initiate limb bud formation (see Figure 14.6). Moreover, Tbx4 can be replaced by Tbx5 without changing the limb from forelimb to hindlimb. In mice, it appears to be Pitx1 that is

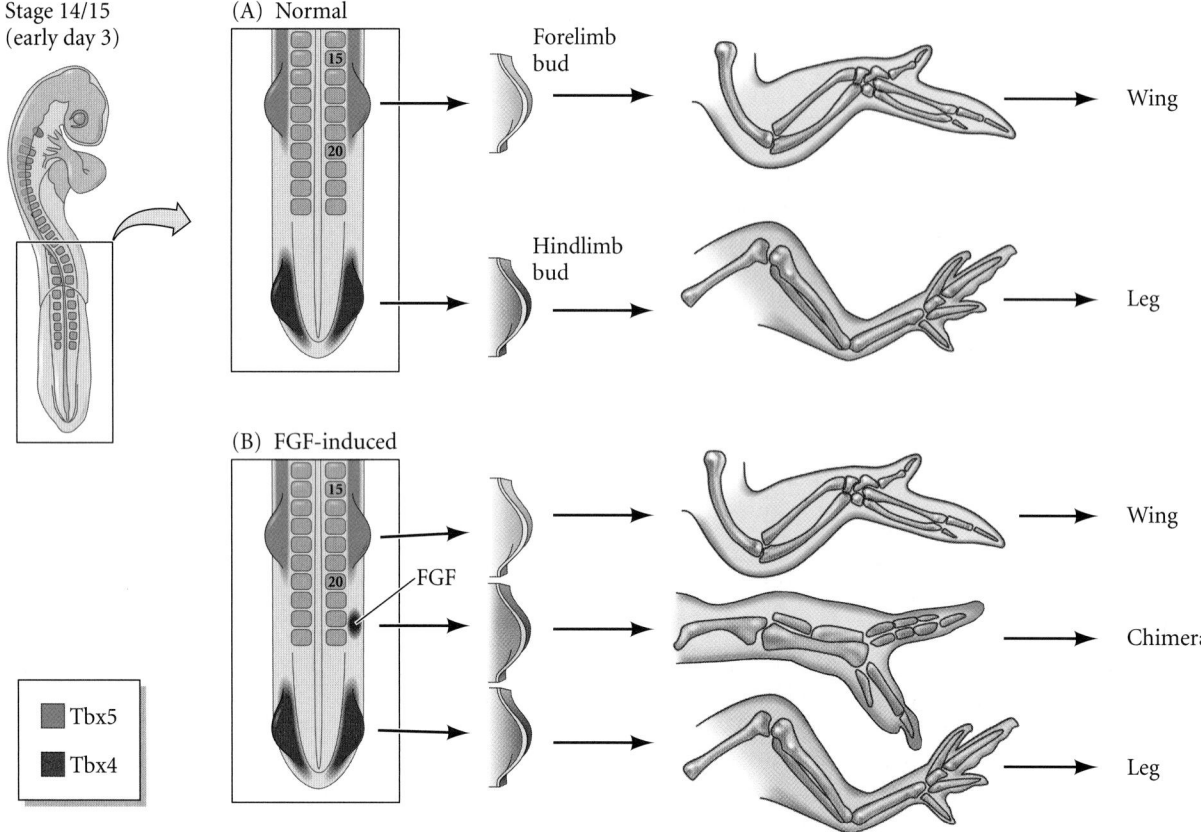

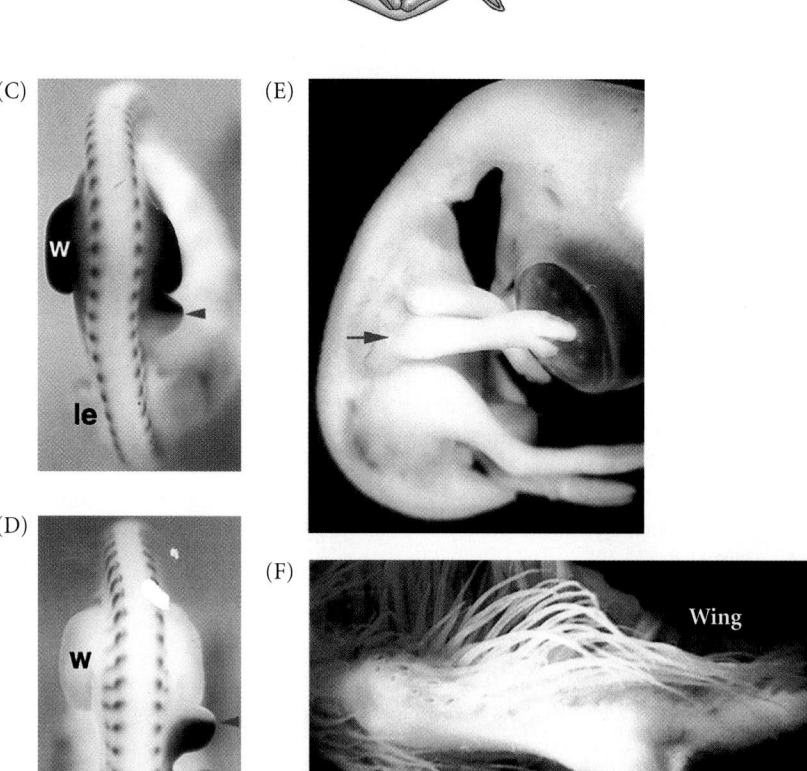

FIGURE 14.7 Specification of limb type in the chick by Tbx4 and Tbx5. (A) In situ hybridizations show that during normal chick development, Tbx5 (blue) is found in the anterior lateral plate mesoderm, while Tbx4 (red) is found in the posterior lateral plate mesoderm. Tbx5-containing limb buds produce wings, whereas Tbx4-containing limb buds generate legs. (B) If a new limb bud is induced with an FGF-secreting bead, the type of limb formed depends on which Tbx gene is expressed in the limb bud. If placed between the regions of *Tbx4* and *Tbx5* expression, the bead will induce the expression of *Tbx4* posteriorly and *Tbx5* anteriorly. The resulting limb bud will also express *Tbx5* anteriorly and *Tbx4* posteriorly and will generate a chimeric limb. (C) Expression of *Tbx5* in the forelimb (w, wing) buds and in the anterior portion of a limb bud induced by an FGF-secreting bead. (The somite level can be determined by staining for *Mrf4* mRNA, which is localized to the myotomes.) (D) Expression of *Tbx4* in the hindlimb (le, leg) buds and in the posterior portion of an FGF-induced limb bud. (E,F) A chimeric limb induced by an FGF bead contains anterior wing structures (feathers) and posterior leg structures (scales). (F) is at a later developmental stage, after feathers form. (A,B after Ohuchi et al. 1998, Ohuchi and Noji 1999; C–F courtesy of S. Noji.)

critical in constructing the hindlimbs. Indeed, misexpression of *Pitx1* in the mouse forelimb results in the transformation and translocation of the muscles, bones, and tendons, such that the forelimb acquires a hindlimb-like morphology (Minguillon et al. 2005; DeLaurier et al. 2006; Ouimette et al. 2009). *Tbx4* expressed in the mouse forelimb will not do this. Moreover, Pitx1 activates hindlimb-specific genes in the forelimb, including *Hoxc10* and *Tbx4*. Ongoing research is seeking to discover the mechanisms by which the forelimbs and hindlimbs become distinct from one another and how each grows to a specific length.

Generating the Proximal-Distal Axis of the Limb

The apical ectodermal ridge

When mesenchyme cells enter the limb field, they secrete Fgf10 that induces the overlying ectoderm to form the apical ectodermal ridge (FIGURE 14.8). The AER runs along the distal margin of the limb bud and will become a major signaling center for the developing limb (Saunders 1948; Kieny 1960; Saunders and Reuss 1974; Fernandez-Teran and Ros 2008). Its roles include (1) maintaining the mesenchyme beneath it in a plastic, proliferating state that enables the linear (proximal-distal, or shoulder-finger) growth of the limb; (2) maintaining the expression of those molecules that generate the anterior-posterior (thumb-pinkie) axis; and (3) interacting with the proteins specifying the anterior-posterior and dorsal-ventral (knuckle-palm) axes so that each cell is given instructions on how to differentiate.

The proximal-distal growth and differentiation of the limb bud are made possible by a series of interactions between the AER and the limb bud mesenchyme directly

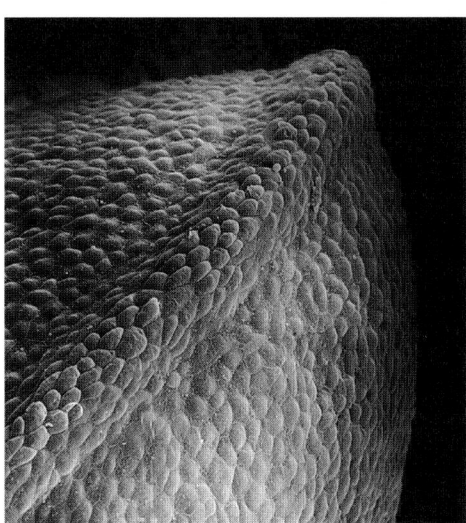

FIGURE 14.8 Scanning electron micrograph of an early chick forelimb bud, with the apical ectodermal ridge in the foreground. (Courtesy of K. W. Tosney.)

(200 mm) beneath it. This distal mesenchyme is called the **progress zone** (**PZ**) mesenchyme (and sometimes the **undifferentiated zone**), since its proliferative activity extends the limb bud (Harrison 1918; Saunders 1948; Tabin and Wolpert 2007). These interactions were demonstrated by the results of several experiments on chick embryos (FIGURE 14.9):

1. If the AER is removed at any time during limb development, further development of distal limb skeletal elements ceases.

2. If an extra AER is grafted onto an existing limb bud, supernumerary structures are formed, usually toward the distal end of the limb.

3. If leg mesenchyme is placed directly beneath the wing AER, distal hindlimb structures (toes) develop at the end of the limb. (However, if this mesenchyme is placed farther from the AER, the hindlimb [leg] mesenchyme becomes integrated into wing structures.)

4. If limb mesenchyme is replaced by nonlimb mesenchyme beneath the AER, the AER regresses and limb development ceases.

Thus, although the mesenchyme cells induce and sustain the AER and determine the type of limb to be formed, the AER is responsible for the sustained outgrowth and development of the limb (Zwilling 1955; Saunders et al. 1957; Saunders 1972; Krabbenhoft and Fallon 1989). The AER keeps the mesenchyme cells directly beneath it in a state of mitotic proliferation and prevents them from forming cartilage (see ten Berge et al. 2008). Hurle and co-workers (1989) found that if they cut away a small portion of the AER in a region that would normally fall between the digits of the chick leg, an extra digit emerged at that position.*

FGF and Wnt proteins also regulate the shape and growth of the early limb bud. The initial limb mesenchyme cells are not randomly organized. Rather, they show a polarity such that they extend their long axes perpendicular to the ectoderm (Gros et al. 2010; Sato et al. 2010). Wnt signaling appears to determine this orientation and also promotes cell division in the plane of the ectoderm, thereby exending the limb bud outward. FGF signals increase the velocity of cell migration, causing cells to migrate distally. As a result, the limb bud flattens in the anterior-posterior axis while growing distally (i.e., toward the source of Fgf8).

● See **WEBSITE 14.1** Induction of the AER

● See **VADE MECUM** Patterning the limb

---

*When referring to the hand, one has an orderly set of names to specify each digit (*digitus pollicis, d. indicis, d. medius, d. annularis,* and *d. minimus,* respectively, from thumb to little finger). No such nomenclature exists for the pedal digits, but the plan proposed by Phillips (1991) has much merit. The pedal digits, from hallux to small toe, would be named *porcellus fori, p. domi, p. carnivorus, p. non voratus,* and *p. plorans domi,* respectively.

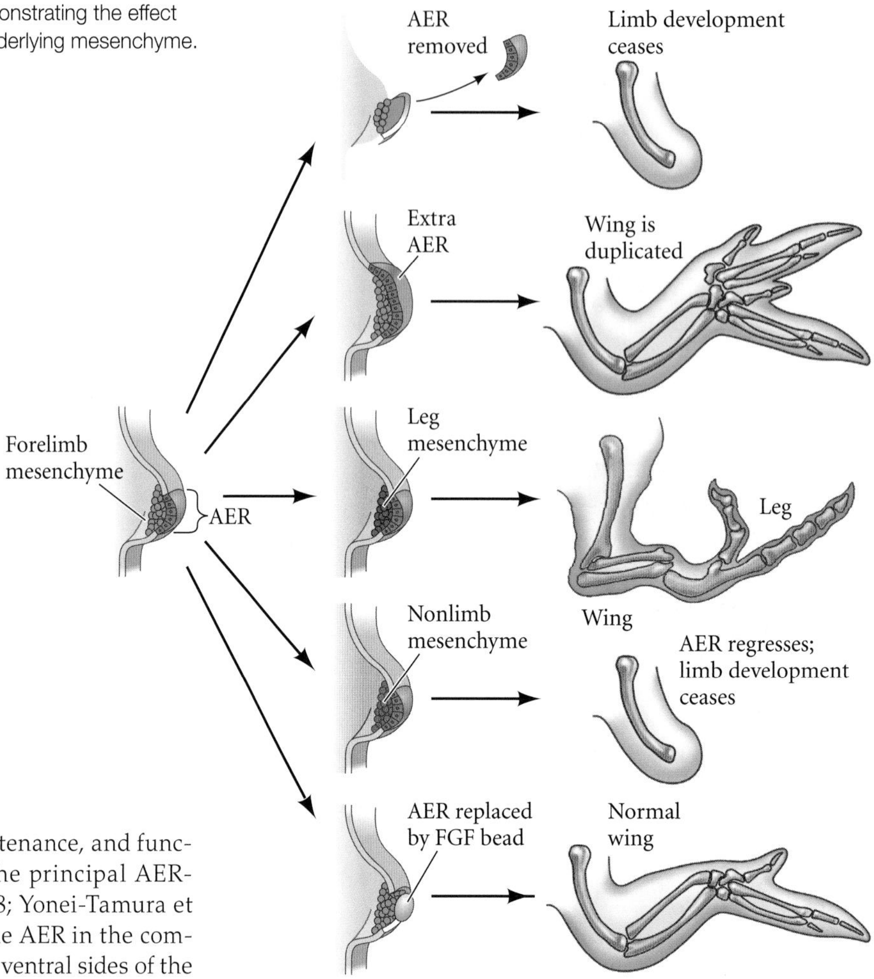

FIGURE 14.9 Summary of experiments demonstrating the effect of the apical ectodermal ridge (AER) on the underlying mesenchyme. (After Wessells 1977.)

AER removed → Limb development ceases

Extra AER → Wing is duplicated

Leg mesenchyme → Leg

Forelimb mesenchyme → AER

Nonlimb mesenchyme → AER regresses; limb development ceases

Wing

AER replaced by FGF bead → Normal wing

FGFs in the induction of the AER

FGFs are critical for the induction, maintenance, and function of the apical ectodermal ridge. The principal AER-inducing protein is Fgf10 (Xu et al. 1998; Yonei-Tamura et al. 1999). Fgf10 is capable of inducing the AER in the competent ectoderm between the dorsal and ventral sides of the embryo. The boundary where dorsal and ventral ectoderm meet is critical to the placement of the AER. In mutants in which the limb bud is dorsalized and there is no dorsal-ventral junction (as in the chick mutant *limbless*), the AER fails to form and limb development ceases (Carrington

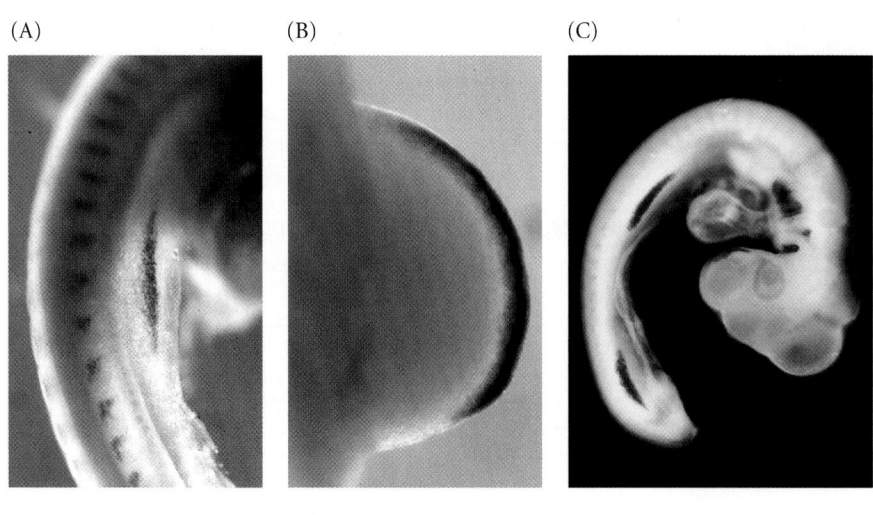

(A)   (B)   (C)

FIGURE 14.10 Fgf8 in the apical ecto-dermal ridge. (A) In situ hybridization showing expression of *Fgf8* message (stained dark purple) in the ectoderm as the limb bud begins to form. (B) Expression of *Fgf8* RNA in the AER, the source of mitotic signals to the underlying mesoderm. (C) In the normal 3-day chick embryo, *Fgf8* is expressed in the AER of both the forelimb and hindlimb buds. It is also expressed in several other places in the embryo, including the pharyngeal arches. (A,B courtesy of J. C. Izpisúa-Belmonte; C courtesy of A. López-Martínez and J. F. Fallon.)

(A)

Graft locations

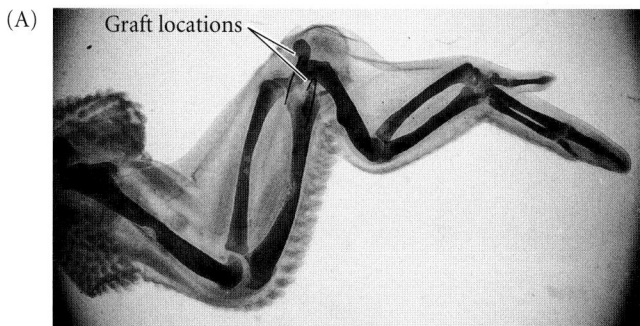

(B)

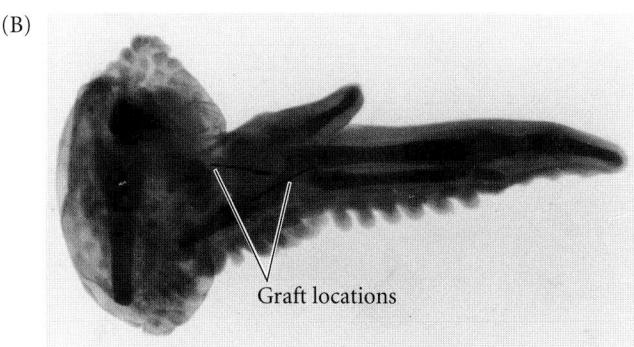

Graft locations

FIGURE 14.11 Control of proximal-distal specification by the progress zone mesenchyme. (A) An extra set of ulna and radius formed when an early wing-bud progress zone was transplanted to a late wing bud that had already formed ulna and radius. (B) Lack of intermediate structures seen when a late wing-bud progress zone was transplanted to an early wing bud. (From Summerbell and Lewis 1975, courtesy of D. Summerbell.)

and Fallon 1988; Laufer et al. 1997a; Rodriguez-Esteban et al. 1997; Tanaka et al. 1997). Fgf10 from the lateral plate mesoderm acts by inducing a Wnt protein (Wnt3a in chicks; Wnt3 in humans and mice) in the prospective limb bud ectoderm. The Wnt protein acts through the canonical β-catenin pathway to induce Fgf8 expression and form the AER (**FIGURE 14.10**; see also Figure 14.6; Fernandez-Teran and Ros 2008).

Fgf8 is the major active factor in the AER, and Fgf8-secreting beads can substitute for the AER functions in inducing limb growth. There are other FGFs (such as Fgf4) made by the AER, and these probably add to the general "FGF effect" (Lewandoski et al. 2000; Boulet et al. 2004). Fgf8 stimulates mitosis in the mesenchyme cells beneath it, causing these cells to keep expressing Fgf10 (see Figure 14.6). Thus, a positive feedback loop is established: Fgf10 in the mesenchyme induces Fgf8 in the AER, and Fgf8 in the AER maintains Fgf10 expression in the mesenchyme. Each FGF activates the synthesis of the other (Mahmood et al. 1995; Crossley et al. 1996; Vogel et al. 1996; Ohuchi et al. 1997; Kawakami et al. 2001). The continued expression of FGFs maintains mitosis in the mesenchyme beneath the AER.

Specifying the limb mesoderm: Determining the proximal-distal polarity of the limb skeleton

In 1948, John Saunders made a simple and profound observation: if the AER is removed from an early-stage wing bud, only a humerus forms. If the AER is removed slightly later, humerus, radius, and ulna form (Saunders 1948; Iten 1982; Rowe and Fallon 1982). Explaining how this happens has not been easy. First it had to be determined whether the positional information for proximal-distal polarity resided in the AER or in the progress zone mesenchyme. Through a series of reciprocal transplantations, this specificity was found to reside in the mesenchyme. If the *AER* had provided the positional information—somehow instructing the undifferentiated mesoderm beneath it as to what structures to make—then older AERs combined with younger mesoderm should have produced limbs with deletions in the middle, whereas younger AERs combined with older mesoderm should have produced duplications of structures. This was not found to be the case; rather, normal limbs formed in both experiments (Rubin and Saunders 1972). But when the entire progress zone (including both the mesoderm and the AER) from an early embryo was placed on the limb bud of a later-stage embryo, new proximal structures were produced beyond those already present. Conversely, when old progress zones were added to young limb buds, distal structures developed immediately, so that digits were seen to emerge from the humerus without an intervening ulna and radius (**FIGURE 14.11**; Summerbell and Lewis 1975).

The dual-gradient model of limb patterning

But how does the mesenchyme specify the proximal-distal axis? In 2010, evidence on chick limb patterning converged on a model of two opposing gradients: a gradient of FGFs and Wnts from the distal AER, and a gradient of retinoic acid from the proximal flank tissue (**FIGURE 14.12A**). Such a two-gradient explanation had been proposed earlier for amphibian limb regeneration (see Maden 1985; Crawford and Stocum 1988) and had even been hypothesized for embryonic limb patterning (see Mercader et al. 2000). Actual evidence for the model eventually came from mesenchyme transplantation experiments by Cooper and her colleagues in the United States and by Roselló-Díez and his colleagues in Spain (Cooper et al. 2010; Roselló-Díez et al. 2010). In both laboratories, researchers took undifferentiated limb bud mesenchyme cells and "repacked" them into the ectodermal hull of a young limb bud. As expected, the age of the mesenchyme determined the type of bones formed. However, the type of bone formed became more proximal (in the stylopod direction) if young limb bud mesenchyme had been treated with RA in the presence of Wnt and FGF; and it became more distal (toward the autopod) if the mesenchyme had been treated with only FGFs and Wnts (**FIGURE 14.12B,C**). Moreover, if the actions of FGFs were inhibited, the bones

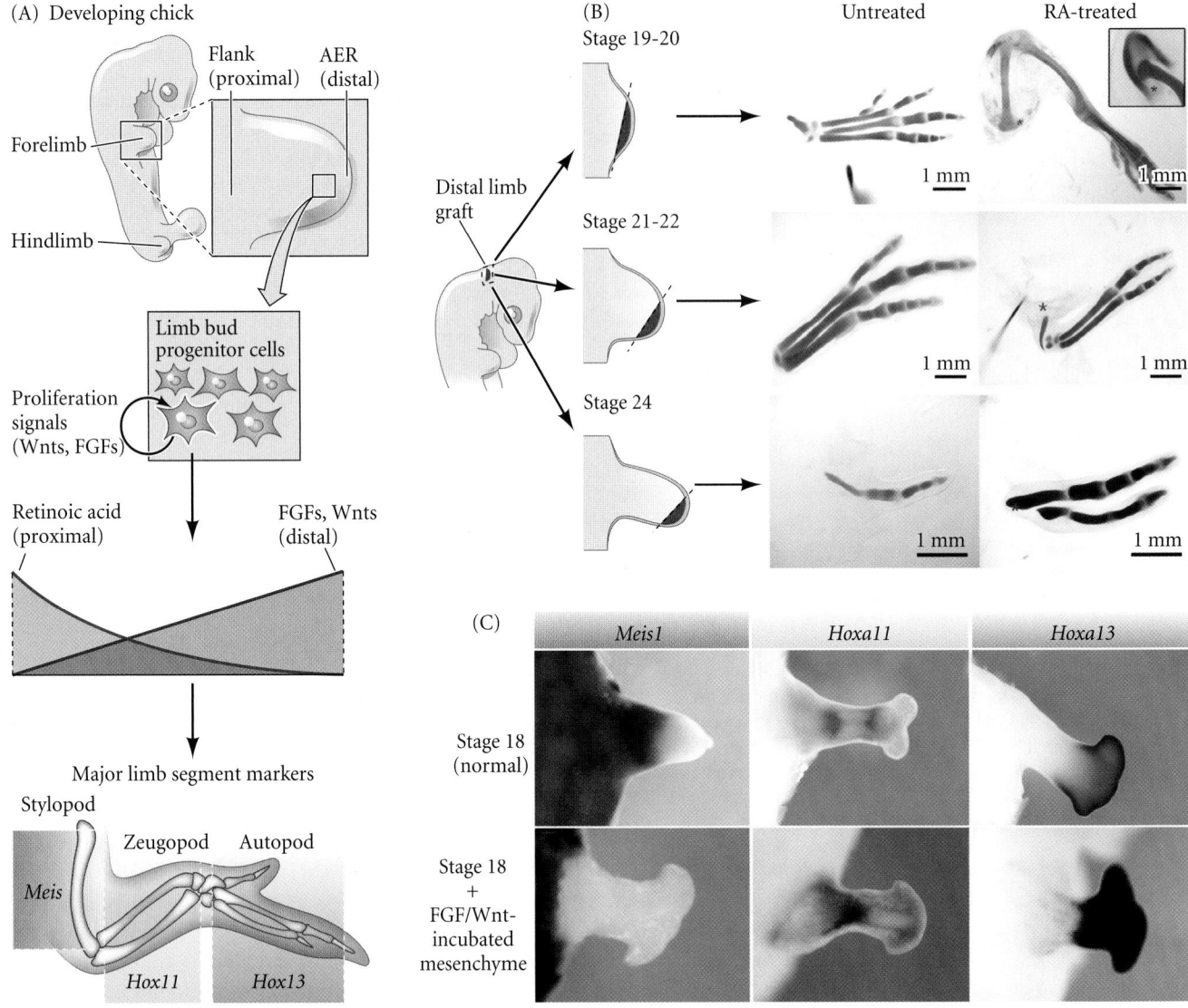

FIGURE 14.12 (A) Model for limb patterning, whereby the proximal-distal axis is generated by opposing gradients of retinoic acid (RA) from the proximal flank and of FGFs and Wnts from the distal AER. (B) Grafting procedure showing that RA proximalizes the bones forming from the transplanted mesenchyme. Untreated mesenchyme from the tip of the limb bud (shaded) generated specific limb cartilage depending on age. However, when treated with 1 mg/ml RA, the skeleton that formed became more proximal. (C) Transcription factors change when FGFs and Wnts are applied to the grafted mesenchyme from limb bids of different ages. *Meis1* is specific for the stylopod; *Hoxa11* for the zeugopod; and Hoxa13 for the autopod. In the earliest limb bud (stage 18), the mesenchyme will form all three types of cartilage. However, when the mesenchyme is first incubated in Fgf8 and Wnt3a, the autopod transcription factor (Hoxa13) is greatly expressed, whereas the stylopod marker (*Meis*) is drastically reduced. The addition of RA in culture is required to maintain competence to express *Meis1* proximally. (A after Macken and Lewandoski 2011; B after Roselló-Díez et al. 2010; C after Cooper et al. 2010, photographs courtesy of the authors.)

became more proximal; and if RA synthesis was inhibited, the bones became more distal. Thus, there appears to be a balance between the proximalizing of the bones by RA from the flank and the distalizing of bones by the FGFs and Wnts of the AER. The opposing gradients may accomplish this

balance by laying down a segmental pattern of different transcription factors in the mesenchyme. Such opposing gradients are probably a common mechanism for cell specification, as we've already seen in the early *Drosophila* embryo (see Chapter 6).

SIDELIGHTS & SPECULATIONS

A Reaction-Diffusion Mechanism for Building the Limb Skeletal Pattern

Genes and proteins don't produce a skeleton. Cells do. The cell types of the stylopod and the autopod are identical; it's only how they are arranged in space that differs. Why is only one cartilage element formed in the stylopod while two are formed in the zeugopod and several in the autopod? How do the gradients tell the cells how to form different parts of the skeleton in different places? Why are the fingers and toes always at the distal end of the limb? The answers may come from a model that involves the diffusion of two or more paracrine factors. This **reaction-diffusion mechanism** for developmental patterning was formulated by Alan Turing (1952), and it is a way of generating complex chemical patterns out of substances that are initially homogenously distributed. Turing realized that this patterning would not occur in the presence of just a single morphogen, but that it *could* be achieved by two homogeneously distributed substances ("substance P" and "substance S") if the rates of production of each substance depended on the other.

He went on to show that the dynamics of such a network could produce stable patterns that could be used to drive developmental change.

In Turing's model, substance P promotes the production of more substance P as well as substance S. Substance S, however, inhibits the production of substance P. Turing's mathematics show that if S diffuses more readily than P, sharp waves of concentration differences will be generated for substance P (**FIGURE 14.13**). The reaction-diffusion mechanism predicts alternating areas of high and low concentrations of some substance. When the concentration of the substance is above a certain threshold level, a cell (or group of cells) can be instructed to differentiate in a certain way. An important feature of Turing's model is that particular chemical wavelengths will be amplified while all others will be suppressed. As local concentrations of P increase, the values of S form a peak centering on the P peak, but becoming broader and shallower because substance S diffuses more rapidly. These S

peaks inhibit other P peaks from forming. But which of the many P peaks will survive? That depends on the size and shape of the tissues in which the oscillating reaction is occurring. This pattern is analogous to the harmonics of vibrating strings, as in a guitar: only certain resonance vibrations are permitted. The wavelength comes from the constants, particularly the ratio of the diffusion constants. The mathematics describing which particular wavelengths are selected consist of complex polynomial equations (which are now solved computationally).

Turing's model has produced fascinating results when applied to limb development. It appears that reaction-diffusion kinetics can tell us how the limb bud acquires its proximal-distal polarity and also how the number of digits is regulated at the distal tip of the limb. First, the AER is seen as dividing the limb into three domains (**FIGURE 14.14**).The area immediately adjacent to the AER is prevented from forming cartilage by Wnt inhibition and thus is called the inhibitory domain.

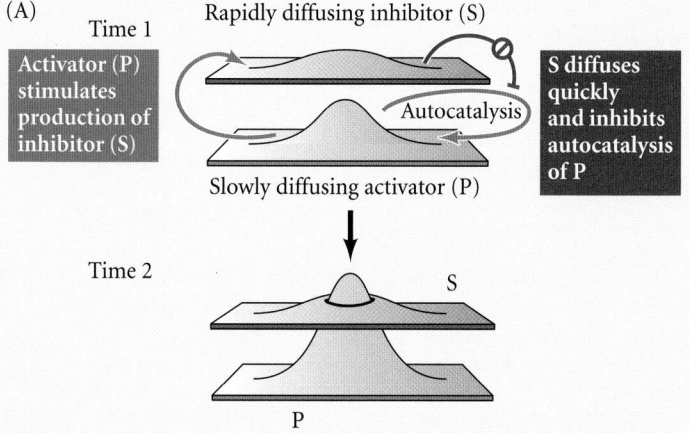

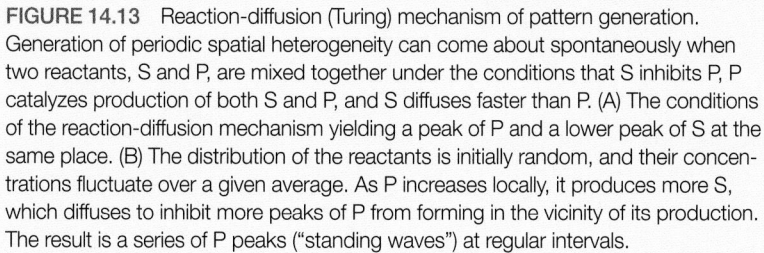

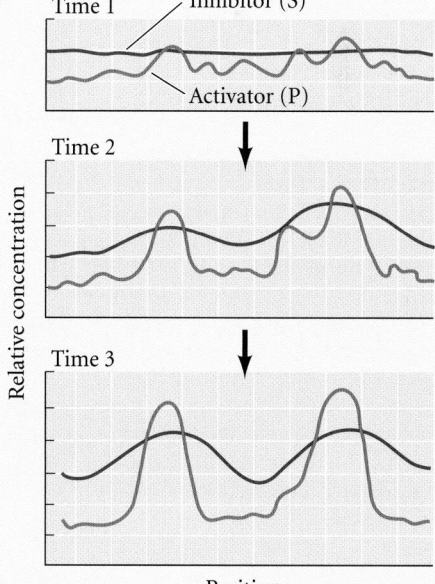

FIGURE 14.13 Reaction-diffusion (Turing) mechanism of pattern generation. Generation of periodic spatial heterogeneity can come about spontaneously when two reactants, S and P, are mixed together under the conditions that S inhibits P, P catalyzes production of both S and P, and S diffuses faster than P. (A) The conditions of the reaction-diffusion mechanism yielding a peak of P and a lower peak of S at the same place. (B) The distribution of the reactants is initially random, and their concentrations fluctuate over a given average. As P increases locally, it produces more S, which diffuses to inhibit more peaks of P from forming in the vicinity of its production. The result is a series of P peaks ("standing waves") at regular intervals.

SIDELIGHTS & SPECULATIONS (continued)

Behind the inhibitory domain is the morphogenetically active domain, where cartilage-forming nodules coalesce. Behind this is the "frozen" domain, where the actual skeleton begins to take shape. As mentioned earlier, Gros and colleagues (2010) found that the Wnt and FGF proteins secreted from the AER induce specific patterns of cell division and growth. Stuart Newman's laboratory (Newman and Bhat 2007; Zhu et al. 2011) has demonstrated that a reaction-diffusion mechanism can pattern this limb mesenchyme, and that size and shape matter. First, the limb mesenchyme cells can synthesize activators of cartilage nodule formation. These activators include BMPs, activins, and certain cell-binding proteins called galectins. They can induce the formation of certain cell adhesion molecules and extracellular matrix proteins that cause cells to aggregate together to form the cartilaginous skeleton. However, these same cells also synthesize *inhibitors* of aggregation, such as Noggin and inhibitory galectins. As a result, what were once cartilage-forming aggregates inhibit the areas surrounding them from forming more such aggregates. Broader space allows more aggregates to form.

At different sizes of the limb, different numbers of pre-cartilaginous condensations can form. First, a single condensation can fit (humerus), then two (ulna and radius), then several (wrist, digits). In this reaction-diffusion hypothesis, the aggregations of pre-cartilage mesenchyme actively recruit more cells from the surrounding area and laterally inhibit the formation of other foci of condensation. The number of these, then, depends on the geometry of the tissue and the strength of the lateral inhibition. Once formed, the aggregates of mesenchyme interact with each other not only to recruit more cells but also to express the transcription factors (Sox9) and extracellular matrix (collagen 2) characteristic of cartilage (Lorda-Diez et al. 2011).

This model predicts that slight size changes of the distal limb bud will alter the number of digits. This is indeed found to be the case, and it may be a very simple way of gaining or losing digits during evolution.* According to

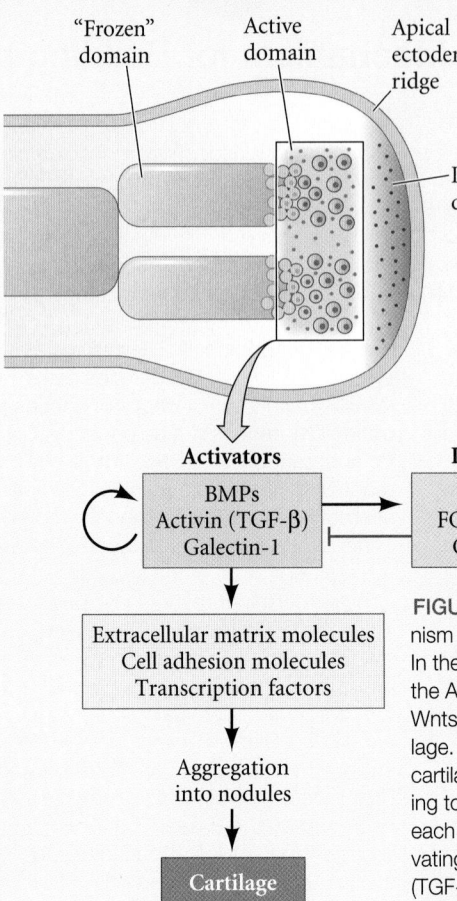

FIGURE 14.14 Reaction-diffusion mechanism for proximal-distal limb specification. In the inhibitory domain immediately outside the AER, cells are kept dividing by FGFs and Wnts and are prevented from forming cartilage. Behind this area, in the active domain, cartilaginous nodules actively form according to a reaction-diffusion mechanism. Here each cell secretes and can respond to activating paracrine factors of the TGF-β family (TGF-β, BMPs, activin) and cell adhesion factors such as galectin-1. These factors stimulate their own synthesis as well as that of the extracellular matrix and cell adhesion proteins that promote aggregation. The activating cells also stimulate the synthesis of diffusible inhibitors of aggregation (including Noggin and galectin-8), preventing cell adhesion in neighboring regions. The places where nodules can form are governed by the geometry of the limb bud (i.e., the geometry decides how many "waves" of activator will be allowed). In the "frozen" domain, the aggregated nodules can now differentiate into cartilage, thus "freezing" the configuration. (After Zhu et al. 2010.)

the model, waves of synthesis and inhibition would form the original pattern of the limb. By placing such constraints as geometry, diffusibility, and the rates of synthesis and degradation of each activator and inhibitor, Zhu and colleagues have been able to model the types of skeleton formed as the limb bud grows. First, the computer model accurately mimicked the normal patterning of the limb (FIGURE 14.15A). Next, it modeled the aberrant skeletons formed as a result of manipulations (FIGURE 14.15B) and mutation (FIGURE 14.15C). Altering the geometries

also could yield patterns seen in fossil limbs (FIGURE 14.15D).

*Thus, dogs with larger limb buds may generate enough cells such that an additional cartilage condensation can fit into the autopod (Alberch 1985). Such appears to be the case with St. Bernard and Great Pyrenees breeds. Fondon and Garner (2004) have shown that one allele of the *Alx-4* gene was found to be homozygous in only one breed of dogs, the Great Pyrenees. These dogs are characterized by polydactyly (an extra toe, the dew claw). Here, there is apparently more growth of the limb bud, such that another condensation can emerge in the autopod.

SIDELIGHTS SPECULATIONS   (continued)

(A)

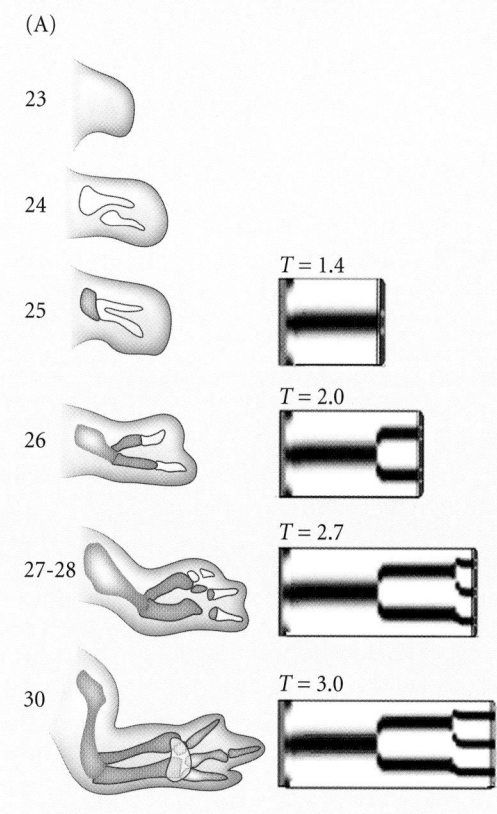

(B)

(C)

(D)

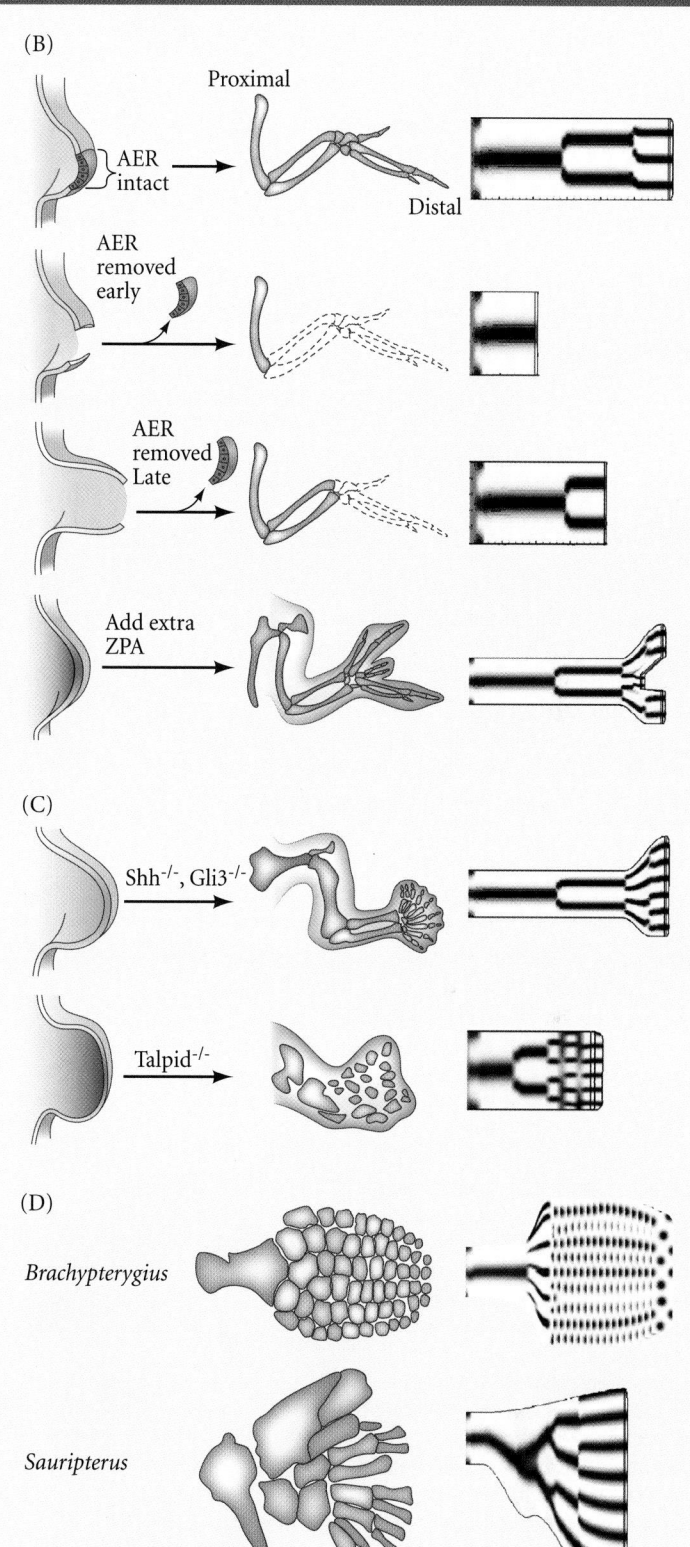

FIGURE 14.15 Computer simulations of limb development can be made after providing experimentally derived parameters of limb bud geometry and of rates of diffusion, synthesis, and degradation of activators and inhibitors. (A) This model predicts the sequential formation of the single stylopod, the double zeugopod, and the multiple cartilaginous rods of the autopod. Stages of chick embryos are on the left. *T* (time) is in arbitrary units for comparing the computer stages. (B,C) The chick model also mimics the observed patterns of skeleton aberration formed by (B) experimental manipulations and (C) mutations. (D) Interestingly, changing the parameters slightly will also generate the observed limb skeletons found in fossils such as the fish-like aquatic reptile *Brachypterygius* (whose forelimb had a paddlelike shape) and *Sauripterus*, one of the first land-dwelling reptiles. (After Zhu et al. 2010.)

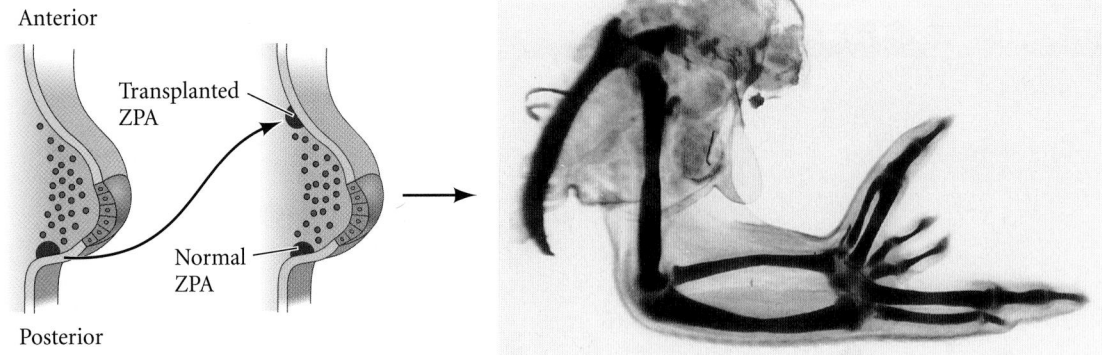

FIGURE 14.16 When a ZPA is grafted to anterior limb bud mesoderm, duplicated digits emerge as a mirror image of the normal digits. (From Honig and Summerbell 1985, photograph courtesy of D. Summerbell.)

Specifying the Anterior-Posterior Axis

The zone of polarizing activity

The specification of the anterior-posterior axis of the limb is the earliest restriction in limb bud cell potency from the pluripotent condition. In the chick, this axis is specified shortly before a limb bud is recognizable. Hamburger (1938) showed that as early as the 16-somite stage, prospective wing mesoderm transplanted to the flank area develops into a limb with the anterior-posterior and dorsal-ventral polarities of the donor graft, not those of the host tissue.

Several experiments (Saunders and Gasseling 1968; Tickle et al. 1975) suggest that the anterior-posterior axis is specified by a small block of mesodermal tissue near the posterior junction of the young limb bud and the body wall. When tissue from this region is taken from a young limb bud and transplanted to a position on the anterior side of another limb bud, the number of digits on the resulting wing is doubled (**FIGURE 14.16**). Moreover, the structures of the extra set of

digits are mirror images of the normally produced structures. Polarity has been maintained, but the information is now coming from both an anterior and a posterior direction. Thus, this region of the mesoderm has been called the **zone of polarizing activity** (**ZPA**).

Sonic hedgehog defines the ZPA

The search for the molecule(s) that confer polarizing activity on the ZPA became one of the most intensive quests in developmental biology. In 1993, Riddle and colleagues showed by in situ hybridization that *sonic hedgehog* (*shh*), a vertebrate homologue of the *Drosophila hedgehog* gene, was expressed specifically in that region of the limb bud known to be the ZPA (**FIGURE 14.17A**). As evidence that this association between the ZPA and *sonic hedgehog* was more than

(B) Stage 19–23 embryo

(A)

FIGURE 14.17 Sonic hedgehog protein is expressed in the ZPA. (A) In situ hybridization showing the sites of *sonic hedgehog* expression (arrows) in the posterior mesoderm of the chick limb buds. These are precisely the regions that transplantation experiments defined as the ZPA. (B) Assay for polarizing activity of Sonic hedgehog protein. The *shh* gene was inserted adjacent to an active promoter of a chicken virus, and the recombinant virus was placed into cultured chick embryo fibroblast cells. The virally transfected cells were pelleted and implanted in the anterior margin of a limb bud of a chick embryo. The resulting limbs produced mirror-image digits, showing that the secreted protein had polarizing activity. (A courtesy of R. D. Riddle; B after Riddle et al. 1993.)

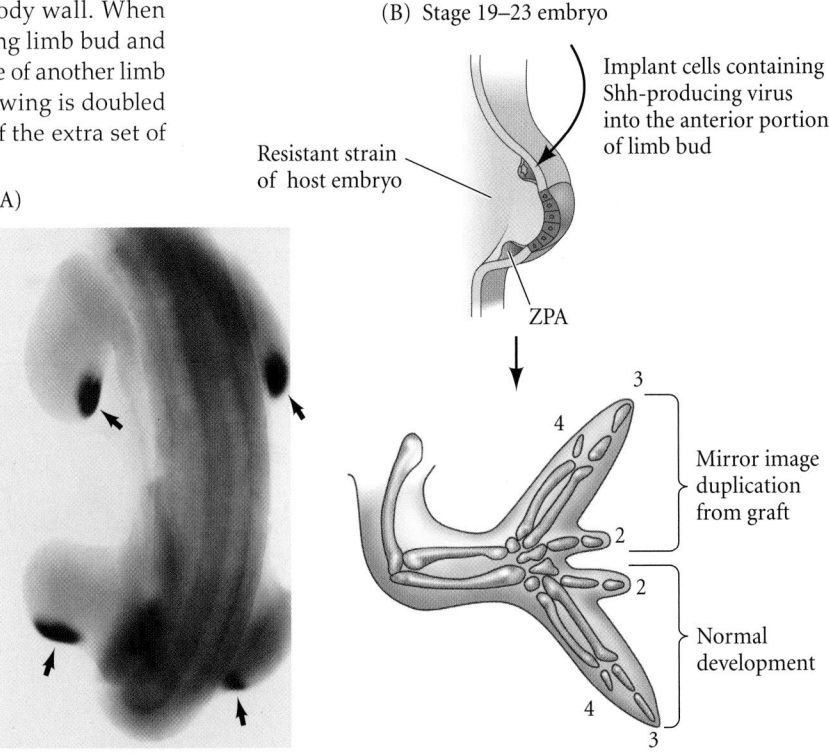

(A) Thumb region

(B)

Extra digits

Thumb region

(C) Wild type

GCTTGTTTTTTTTGCCACT**G**ATGATCCATAA-

(D) *Hx* mutant

GCTTGTTTTTTTTGCCACT**A**ATGATCCATAA-

(E)

(F)

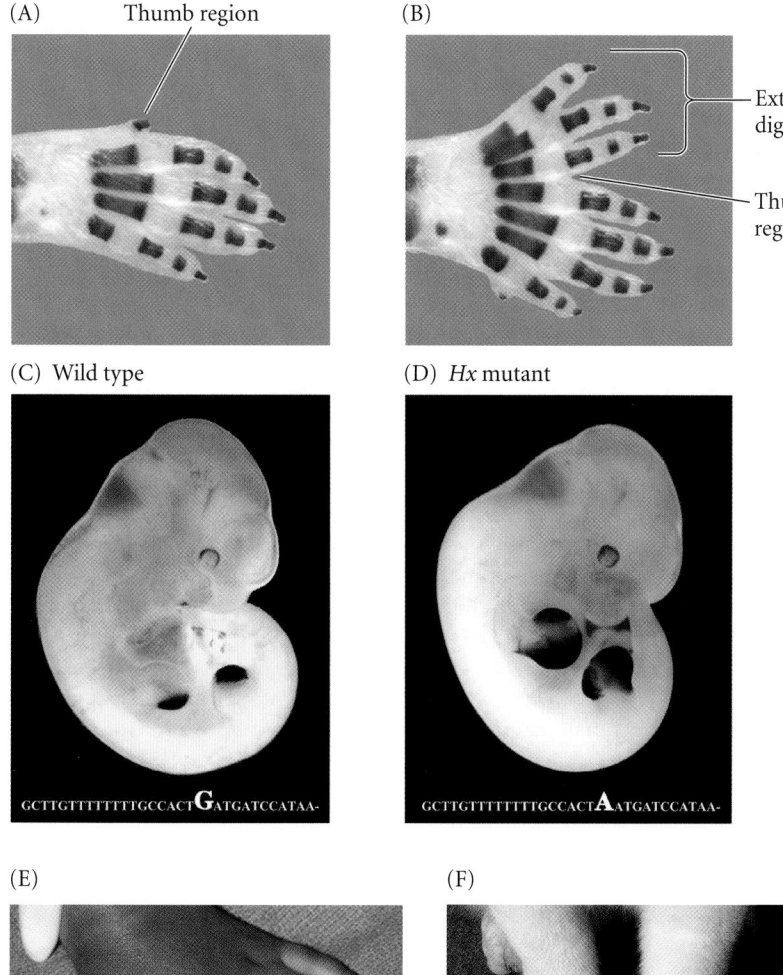

FIGURE 14.18 Ectopic expression of *Sonic hedgehog* in the anterior limb causes extra digit formation. (A) Wild-type mouse paw. The bones are stained with alizarin red. (B) *Hx* (*hemimelic extra-toes*) mutant mouse paws, showing the extra digits associated with the anterior ("thumb") region. (The small extra nodule of posterior bone is peculiar to the *Hx* phenotype on the genetic background used and is not seen on other genetic backgrounds.) (C) Reporter constructs from wild-type *Shh* limb enhancer direct transcription solely to the posterior part of each mouse limb bud (i.e., in the ZPA). (D) Reporter constructs from the *hx* mutant direct transcription to both the anterior and posterior regions of each mouse limb bud. The wild-type and mutant *Shh* limb-specific enhancer region DNA sequences are shown below and highlight the single G-to-A nucleotide substitution that differentiates the two. (E) A similar mutation in the human long-range enhancer for *SHH* causes similar mirror-image hand duplications. (F) Descendants of Ernest Hemingway's polydactylous pet cats still inhabit the Hemingway home in Key West, Florida, and display a mutation in this long-range enhancer. (A–D From Maas and Fallon 2005, courtesy of B. Robert, Y. Lallemand, S. A. Maas, and J. F. Fallon; E from Yang and Kozin 2009, courtesy of S. Kozin; F, photograph by the author.)

just a correlation, Riddle and co-workers (1993) demonstrated that the secretion of Sonic hedgehog protein is sufficient for polarizing activity. They transfected embryonic chick fibroblasts (which normally would never synthesize Shh) with a viral vector containing the *shh* gene (**FIGURE 14.17B**). The gene became expressed, translated, and secreted in these fibroblasts, which were then inserted under the anterior ectoderm of an early chick limb bud. Mirror-image digit duplications like those induced by ZPA transplants were the result. Moreover, beads containing Sonic hedgehog protein were shown to cause the same duplications (López-Martínez et al. 1995; Yang et al. 1997). Thus, Sonic hedgehog appears to be the active agent of the ZPA.

This fact was confirmed by a remarkable gain-of-function mutation. The *hemimelic extra-toes* (*hx*) mutant of mice has

extra digits on the thumb side of the paws (**FIGURE 14.18A,B**). This phenotype is associated with a single base-pair difference in the limb-specific *Shh* enhancer, a highly conserved region located a long distance (about 1 million base pairs) upstream from the *Shh* gene itself (Lettice et al. 2003; Sagai et al. 2005). Maas and Fallon (2005) made a reporter construct by fusing a β-galactosidase gene to this long-range limb enhancer region from both wild-type and *hx*-mutant genes. They injected these reporter constructs into the pronuclei of newly fertilized mouse eggs to obtain transgenic embryos. In the transgenic mouse embryos carrying the reporter gene with wild-type limb enhancer, staining for β-galactosidase activity revealed a single patch of expression in the posterior mesoderm of each limb bud (i.e., in the ZPA; **FIGURE 14.18C**). However, the mice carrying the mutant *Hx* reporter construct showed β-galactosidase activity in *both* the anterior and posterior regions of the limb bud (**FIGURE 14.18D**). It appears that this enhancer has both positive and negative functions, and that in the anterior region of the limb bud, some factor represses the ability of this enhancer to activate *Shh* transcription. The inhibitors probably cannot bind to the mutated enhancer, and thus *Shh* expression is seen in both the anterior and posterior regions of the limb bud. This anterior *Shh* expression, in turn, causes extra digits to develop in the mutant mice. Similar mutations in the long-range limb enhancer of Sonic hedgehog produce polydactylous phenotypes in humans and other mammals, such as the polydactylous cats bred by author Ernest Hemingway (**FIGURE 14.18E,F**; Gurnett et al. 2007; Lettice et al. 2008; Sun et al. 2008).

Specifying digit identity by Sonic hedgehog

How does Sonic hedgehog specify the identities of the digits? Surprisingly, when scientists were able to perform fine-scale fate-mapping experiments on the Shh-secreting cells of the ZPA (using recombinase to express a label only in those cells expressing *shh*), they found that cells that expressed *shh* at any time did not undergo apoptosis in the way that the AER does after it finishes its job. Rather, the descendants of Shh-secreting cells become the bone and muscle of the posterior limb (Ahn and Joyner 2004; Harfe et al. 2004). Indeed, digits 5 and 4 (and part of digit 3) of the mouse hindlimb are formed from the descendants of Shh-secreting cells (**FIGURE 14.19**).

It seems that specification of the digits is primarily dependent on the amount of time the *shh* gene is expressed and only a little bit on the concentration of Shh protein that other cells receive (see Tabin and McMahon 2008). The difference between digits 4 and 5 is that the cells of the more posterior digit 5 express *shh* longer and are exposed to Shh (in an autocrine manner) for a longer time. Digit 3 is made up of some cells that secreted Shh for a shorter period of time than those of digit 4, and they also depend on Shh diffusion from the ZPA (indicated by the fact that digit 4 is lost when Shh is modified such that it cannot diffuse away from cells). Digit 2 is dependent entirely on Shh diffusion for its specification, and digit 1 is specified independently of Shh. Indeed, in a naturally occurring chick mutant that lacks *Shh* expression in the limb, the only digit that forms is digit 1. Furthermore, when the genes for Shh and Gli3 are conditionally knocked out of the mouse limb, the resulting limbs have numerous digits, but the digits have no obvious specificity (Litingtung et al. 2002; Ros et al. 2003; Scherz et al. 2007). Vargas and Fallon (2005) propose that digit 1 is specified by *Hoxd13* in the absence of *Hoxd12*. Forced expression of *Hoxd12* throughout the digital primordia leads to the transformation of digit 1 into a more posterior digit (Knezevic et al. 1997).

By using conditional knockouts of *Shh*, where researchers could stop *Shh* expression at different times during mouse development, Zhu and Mackem (2011) found that Sonic hedgehog works by two temporally distinct mechanisms. The

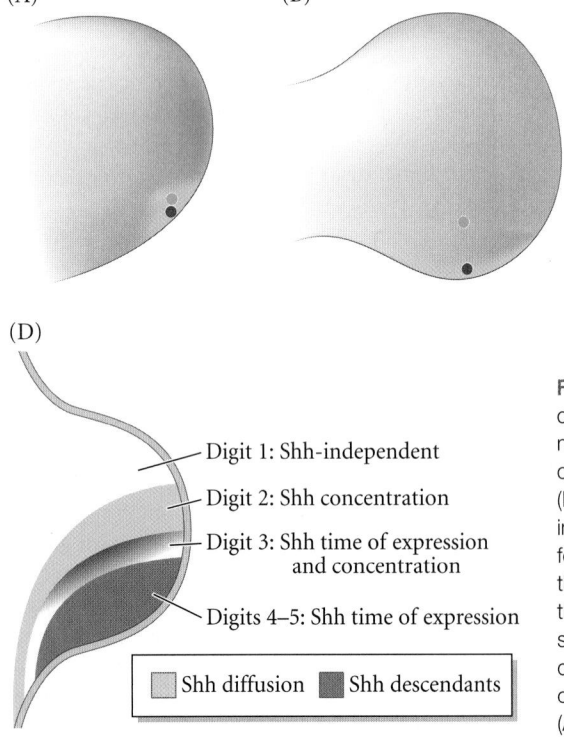

(A) (B) (C)

1
2
3
4
5

(D)

Digit 1: Shh-independent
Digit 2: Shh concentration
Digit 3: Shh time of expression and concentration
Digits 4–5: Shh time of expression

Shh diffusion   Shh descendants

FIGURE 14.19 The descendants of Shh-secreting cells form digits 4 and 5 and contribute to the specification of digits 2 and 3 in the mouse limb. (A) In the early mouse hindlimb bud, the progenitors of digit 4 (green dot) and the progenitors of digit 5 (red dot) are both in the ZPA and express *Sonic hedgehog* (light green). (B) At later stages of limb development, the cells forming digit 5 are still expressing *Shh* in the ZPA, but the cells forming digit 4 no longer do. (C) When the digits form, the cells in digit 5 will have seen high levels of Shh protein for a longer time than the cells in digit 4. (D) Schematic by which digits 4 and 5 are specified by the amount of time they were exposed to Shh in an autocrine fashion; digit 3 is specified by the amount of time the cells were exposed to Shh both in an autocrine and paracrine fashion. Digit 2 is specified by the concentration of Shh its cells received by paracrine diffusion, and digit 1 is specified independently of Shh. (After Harfe et al. 2004.)

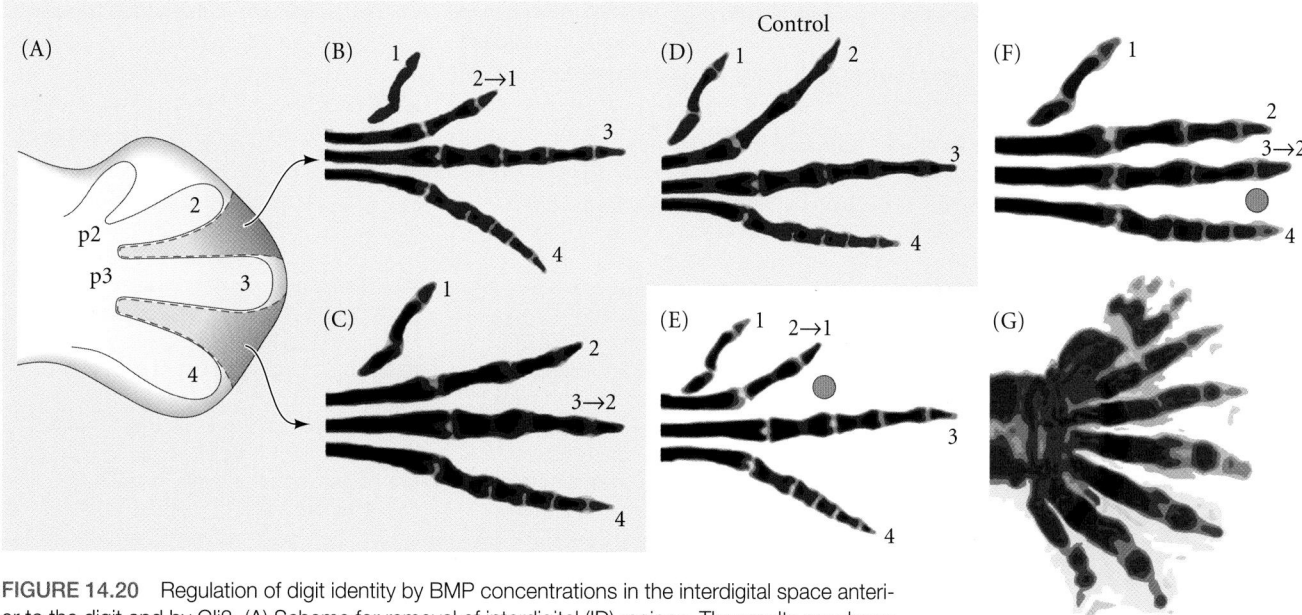

FIGURE 14.20 Regulation of digit identity by BMP concentrations in the interdigital space anterior to the digit and by Gli3. (A) Scheme for removal of interdigital (ID) regions. The results are shown in (B) and (C), respectively. (B) Removal of ID region 2 between digit primordia 2 (p2) and 3 (p3) causes digit 2 to change to the structure of digit 1. (C) Removing ID region 3 (between digit primordia 3 and 4) causes digit 3 to form the structures of digit 2. (D) Control digits and their ID spaces. (E,F) The same transformations as in (B) and (C) can be obtained by adding beads containing the BMP inhibitor Noggin to the ID regions. (E) When a Noggin-containing bead (green dot) is placed in ID region 2, digit 2 is transformed into a copy of digit 1. (F) When the Noggin bead is placed in ID region 3, digit 3 is transformed into a copy of digit 2. (G) The forelimb of a mouse homozygous for deletions of both *Gli3* and *Shh* is characterized by extra digits of no specific type. (After Dahn and Fallon 2000; Litingtung et al. 2002; B–G, photographs courtesy of R. D. Dahn and J. F. Fallon.)

first phase involves the specification of digit identity (from the posterior pinky to the anterior thumb). In this phase, Shh acts as a morphogen, with the digit identities being specified first by the concentration of Shh in that region of the limb bud, and then by the duration of exposure to Shh. In the second phase, Shh works as a mitogen to stimulate the proliferation and expansion of the limb bud mesenchyme, thus helping shape the limb bud.

The mechanism by which Sonic hedgehog establishes a digit's identity may involve cell cycle regulation and the BMP pathway The concentrations of Shh produce the Gli3 transcriptional activator, which regulates expression of Shh target genes. These targets include the genes for the BMP antagonist Gremlin, the cell cycle regulator Cdk6, and the genes that synthesize hyaluronic acid (a component of cell adhesion). Shh (through Gli3) restricts the proliferation of cartilage progenitor cells (by downregulating Cdk6) and promotes their BMP-stimulated differentiation into cartilage by inhibiting the BMP antagonist Gremlin and by upregulating hyaluronic acid synthase 2 (Vokes et al. 2008; Liu et al. 2012; Lopez-Rios et al. 2012).

Shh initiates and sustains a gradient of BMP proteins across the limb bud, and this BMP gradient can specify the digits (Laufer et al. 1994; Kawakami et al. 1996; Drossopoulou et al. 2000). Identity is not specified directly in each digit primordium, however. Rather, the identity of each digit is determined by the *interdigital* mesoderm—that is, by the webbing between the digits (the region of mesenchyme that will shortly undergo apoptosis).

The interdigital tissue specifies the identity of the digit forming anteriorly to it (i.e., toward the thumb or big toe). Thus, when Dahn and Fallon (2000) removed the webbing between the cartilaginous condensations forming chick hindlimb digits 2 and 3, the second digit was changed into a copy of digit 1. Similarly, when the webbing on the other side of digit 3 was removed, the third digit formed a copy of digit 2 (**FIGURE 14.20A–C**). Moreover, the positional value of the webbing could be altered by changing the BMP level. When beads containing BMP antagonists such as Noggin were placed in the webbing between digits 3 and 4, digit 3 was anteriorly transformed into digit 2 (**FIGURE 14.20D–F**). Each digit has a characteristic array of nodules that form the digit skeleton, and Suzuki and colleagues (2008) have shown that the different levels of BMP signaling in the interdigital webbing regulate the recruitment of progress zone mesenchymal cells into the nodules that make the digits.

SIDELIGHTS & SPECULATIONS

Hox Gene Changes during Limb Development

We have seen that the Hox genes play a role in specifying where the limbs will form. Now we will see that they play a second role in specifying whether a particular mesenchymal cell will become stylopod, zeugopod, or autopod, and a third role in specifying both the ZPA and the identity of the digits.

Hox specification code for limb skeleton identity

The 5′ (*AbdB*-like) portions (paralogues 9–13) of the Hoxa and Hoxd gene complexes appear to be active in the limb buds of mice. Based on the expression patterns of these genes, and on naturally occurring and gene knockout mutations, Mario Capecchi's laboratory (Davis et al. 1995) proposed a model wherein these Hox genes specify the identity of a limb region (**FIGURE 14.21A,B**). Here, Hox9 and Hox10 paralogues specify the stylopod, Hox11 paralogues specify the zeugopod, and Hox12 and Hox13 paralogues specify the autopod. This scenario has been confirmed by numerous recent experiments. For instance, when Wellik and Capecchi (2003) knocked out all six Hox10 paralogues (*Hox10aaccdd*) from mouse embryos, the resulting mice not only had severe axial skeletal defects, they also had no femur or patella. (These mice did have humeruses, because the Hox9 paralogues are expressed in the forelimb stylopod but not in the hindlimb stylopod.) When all six Hox11 paralogues were knocked out, the resulting hindlimbs had femurs but neither tibias nor fibulas (and the forelimbs lacked the ulnas and radii). Thus, the Hox11

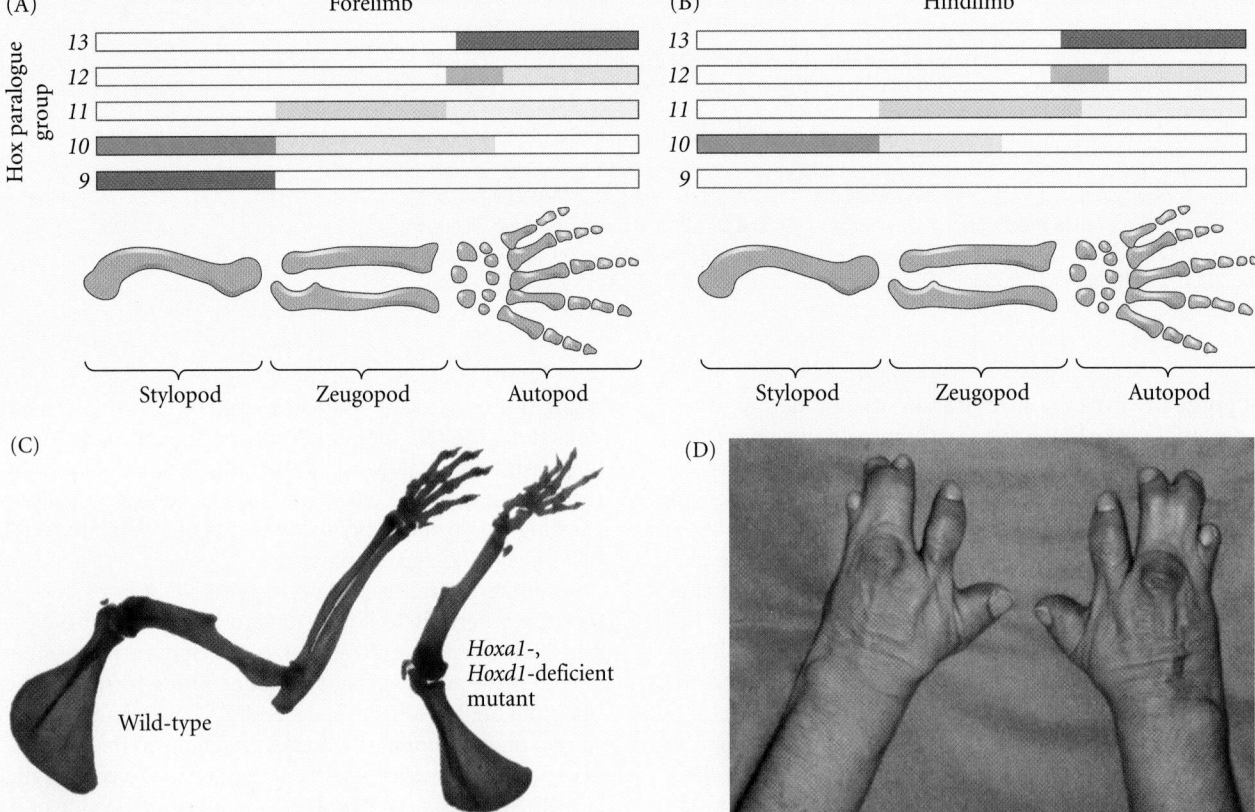

FIGURE 14.21 Deletion of limb bone elements by the deletion of paralogous Hox genes. (A) 5′ Hox gene patterning of the forelimb. *Hox9* and *Hox10* paralogues specify the humerus (stylopod). *Hox10* paralogues are expressed to a lesser extent in the radius and ulna (zeugopod). *Hox11* paralogues are chiefly responsible for patterning the zeugopod. *Hox12* and *Hox13* paralogues function in the autopod, with *Hox12* paralogues functioning primarily in the wrist and to a lesser extent in the digits. (B) Similar but somewhat differing pattern is seen in the hindlimb. (C) Forelimb of a wild-type mouse (left) and of a double-mutant mouse that lacks functional *Hoxa11* and *Hoxd11* genes (right). The ulna and radius are severely reduced or absent in the mutant. (D) Human polysyndactyly ("many fingers joined together") syndrome resulting from a homozygous mutation at the *HOXD13* loci. This syndrome includes malformations of the urogenital system, which also expresses *HOXD13*. (A,B after Wellik and Capecchi 2003; C from Davis et al. 1995, courtesy of M. Capecchi; D from Muragaki et al. 1996, courtesy of B. Olsen.)

knockouts got rid of the zeugopods (**FIGURE 14.21C**). Similarly, knocking out all four *Hoxa13* and *Hoxd13* loci resulted in loss of the autopod (Fromental-Ramain et al. 1996). Humans homozygous for a *HOXD13* mutation show abnormalities of the hands and feet wherein the digits fuse (**FIGURE 14.21D**), and humans with homozygous mutant alleles of *HOXA13* also have deformities of their autopods (Muragaki et al. 1996; Mortlock and Innis 1997). In both mice and humans, the autopod (the most distal portion of the limb) is affected by the loss of function of the most 5′ Hox genes.

Hox specification of the digits

Hox gene expression—especially that of the *Hoxd* cluster—functions in two phases (Zakany et al. 2004; Tarchini and Duboule 2006; also see Abbasi 2011). The first phase is important for the specification of the stylopod and zeugopod, discussed above. The later phase of *Hoxd* expression helps specify the autopod. There are two major "early" regulatory regions. These regions are composed of numerous enhancers that work together to activate the *Hoxd* genes in a specific temporal and spatial array. The major early

regulatory region, called ELCR (*early limb control regulatory region*), activates transcription in a time-dependent manner: the closer the gene is to the ELCR, the earlier it is activated. The second early regulatory region, POST ("posterior restriction"), imposes spatial restrictions on the 5′ *Hoxd* genes (*d10–13*), such that the genes closest to this region have the most restricted position, starting from the posterior margin of the limb bud (**FIGURE 14.22A,B**). This pattern of nested Hoxd proteins is essential for activating the long-range enhancer (ZRS) of

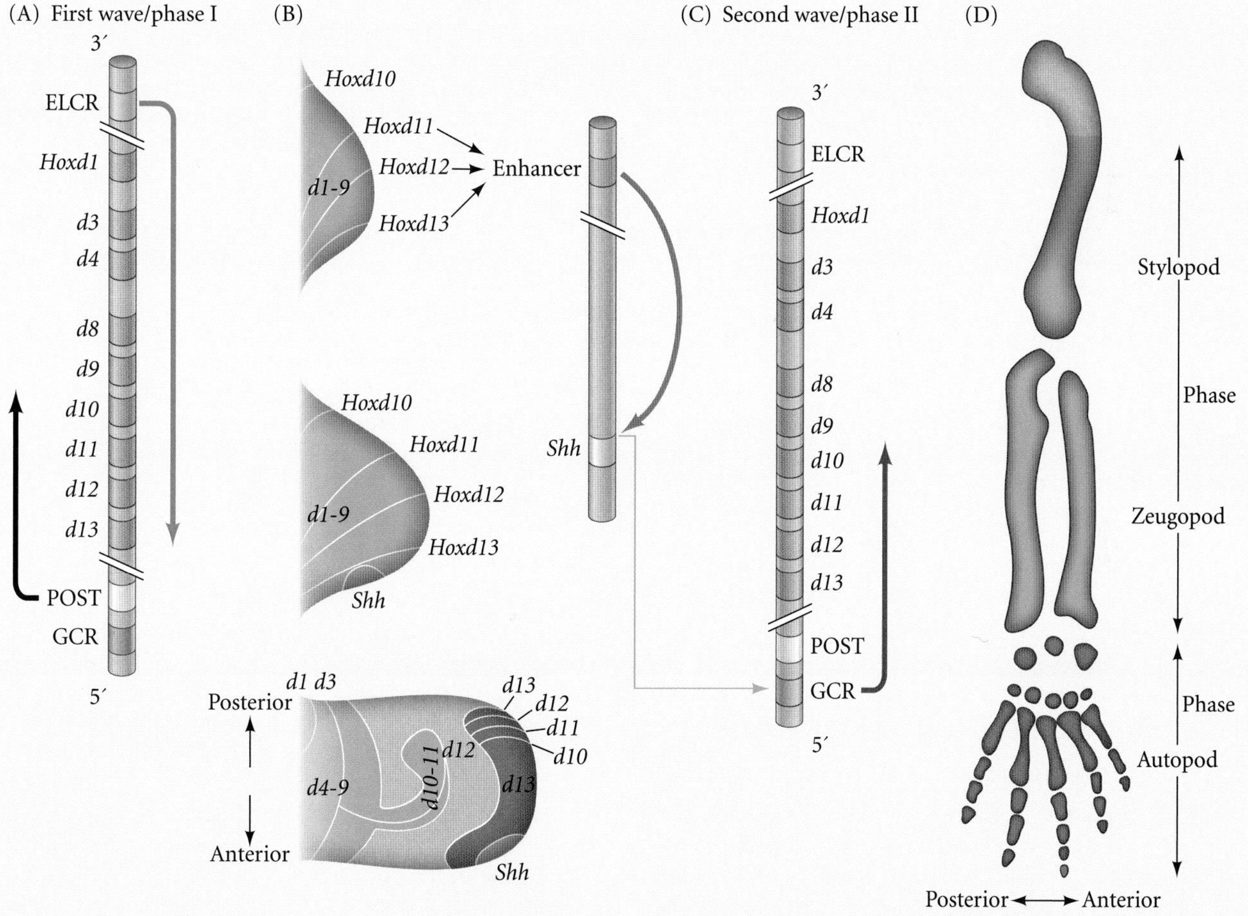

FIGURE 14.22 Changes in *Hoxd* gene expression regulate patterning of the tetrapod limb in two independent phases. (A) The first phase of *Hoxd* expression is initiated as the limb bud forms. The ELCR regulatory element activates those genes closest to it earlier than those genes away from it, while the POST regulatory element acts negatively to restrict the anterior expression of these genes in the opposite direction. (B) This results in expression domains such that *Hoxd13* is confined to the most posterior region, while *Hoxd12* is allowed to expand more anteriorly. The 5′*Hoxd* genes activate the long-range Shh enhancer (ZRS), thereby creating the ZPA in the posterior limb mesoderm. (C) In the second phase, Shh activates the GCR regulatory locus, which inverts the *Hoxd* expression pattern such that *Hoxd13* is located more anteriorly than the other *Hoxd* genes (B, bottom red). (D) Skeletal elements specified by the early (blue) and later (red) phases. The dotted lines in the limb bud show the boundaries of gene expression. (After Abbasi 2011.)

the *Sonic hedgehog* gene, thereby activating *Shh* expression in the posterior limb bud mesoderm and forming the ZPA (Tarchini et al. 2006; Galli et al. 2010). In addition, the presence of Hoxb8 in the mesenchyme appears to help define the anterior/posterior posterior boundary of the forelimb bud. If Hoxb8 is expressed ectopically in the anterior compartment of the mouse forelimb bud, a ZPA will also form there (Charite et al. 1994; Hornstein et al. 2005).

The ZPA now acts to alter the Hoxd expression patterns. Sonic hedgehog expressed from the posterior margin activates a second set of enhancers called the GCR (global control region; Spitz et al. 2003; Montavon et al. 2011.) The *Shh* genes closest to the GCR are expressed most broadly. This expression inverts the original pattern of *Hoxd10–13* expression, such that *Hoxd13* is expressed at the highest level and extends most anteriorly. *Hoxd12, d11,* and *d10* are expressed in slightly narrower domains, so that the most anterior digit (e.g., the thumb) expresses *Hoxd13* but no other Hox gene (Montavon et al. 2008). Thus, the first phase of Hoxd gene expression helps specify the ZPA, while in the second phase of Hoxd expression, the ZPA instructs the expression patterns, and these patterns define the identities of the digits.

Moreover, transplantation of either the ZPA or other Shh-secreting cells to the anterior margin of the limb bud at this stage leads to the formation of

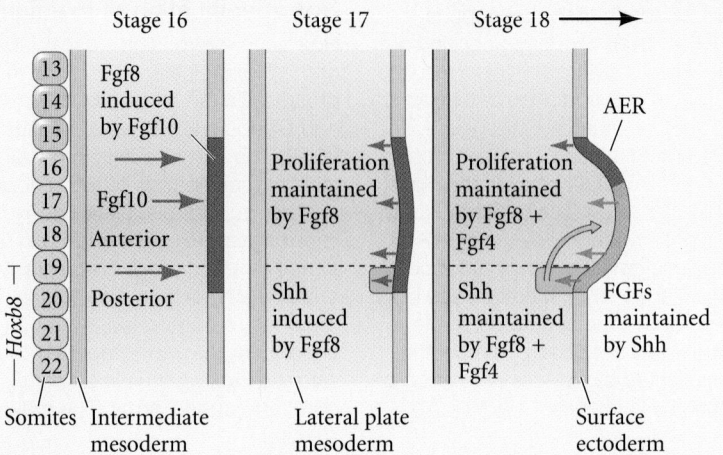

FIGURE 14.23 Feedback between the AER and ZPA in the chick forelimb bud. Stage 16 is about 54 hours gestation, and the other two stages are about 5 hours apart. At stage 17, the newly induced AER secretes Fgf8 into the underlying mesenchyme. The mesenchyme expressing *Hoxb8* is induced to express *shh*, thereby forming the ZPA in the posterior region of the forelimb bud. At stage 18, Shh protein maintains FGF expression in the AER, and these FGFs maintain *shh* expression.

mirror-image patterns of Hoxd gene expression and results in mirror-image digit patterns (Izpisúa-Belmonte et al. 1991; Nohno et al. 1991; Riddle et al. 1993). In normal embryos, the *shh* gene is kept active by the FGFs of the limb bud, while FGF genes in the AER are kept active by Shh protein. Thus, once the AER and ZPA are established, they mutually support one another (**FIGURE 14.23**).

But what genes are these Hox proteins regulating? Some clues come from the analysis of mutations of the Hox13 series of genes. As mentioned

above, people with mutations in the *HOXD13* gene have portions of their autopods that fuse together rather than separate. Ectopic expression of the chicken *Hoxa13* gene (usually expressed in the distal ends of developing chick limbs) appears to make the cells expressing it "stickier" as well as making the "wavelength" between cartilage and web smaller. These properties might cause the cartilaginous nodules to condense in specific ways (Yokouchi et al. 1995; Newman 1996; Sheth et al. 2012).

Generation of the Dorsal-Ventral Axis

The third axis of the limb distinguishes the dorsal half of the limb (knuckles, nails) from the ventral half (pads, soles). In 1974, MacCabe and co-workers demonstrated that the dorsal-ventral polarity of the limb bud is determined by the ectoderm encasing it. If the ectoderm is rotated 180° with respect to the limb bud mesenchyme, the dorsal-ventral axis is partially reversed; the distal elements (digits) are "upside-down." This suggests that the late specification of the dorsal-ventral axis of the limb is regulated by its ectodermal component.

One molecule that appears to be particularly important in specifying dorsal-ventral polarity is **Wnt7a**. The *Wnt7a* gene is expressed in the dorsal (but not the ventral) ectoderm of chick and mouse limb buds (Dealy 1993; Parr et al. 1993). When Parr and McMahon (1995) knocked out the

Wnt7a gene, the resulting mouse embryos had ventral footpads on both surfaces of their paws, showing that Wnt7a is needed for the dorsal patterning of the limb (**FIGURE 14.24**).

Wnt7a is the first known dorsal-ventral axis gene expressed in limb development. It induces activation of the *Lim1* gene in the dorsal mesenchyme. *Lim1* encodes a transcription factor that appears to be essential for specifying dorsal cell fates in the limb. If Lim1 protein is expressed in the ventral mesenchyme cells, those cells develop a dorsal phenotype (Riddle et al. 1995; Vogel et al. 1995; Altabef and Tickle 2002). Human and mouse *lim1* mutants also reveal this gene's importance for specifying dorsal limb fates. *Lim1* knockouts in mice produce a syndrome in which the dorsal limb phenotype is lacking, and loss-of-function mutations in humans result in nail-patella syndrome (no nails on the digits, no kneecaps), in which the dorsal sides of the limbs have

(A)

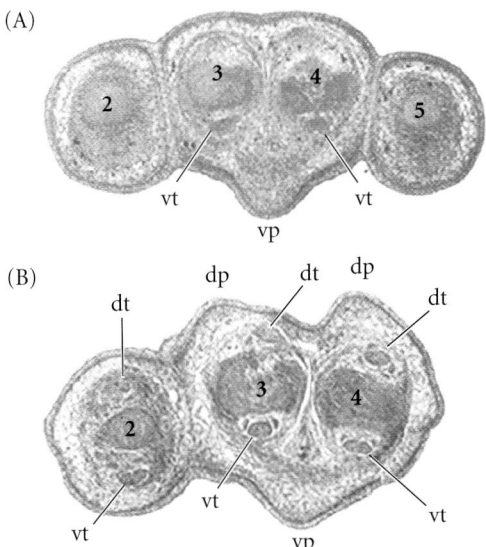

(B)

FIGURE 14.24 Dorsal-to-ventral transformations of limb regions in mice deficient for both *Wnt7a* genes. (A) Histological section (stained with hemotoxylin and eosin) of wild-type 15.5-day embryonic mouse forelimb paw. The ventral tendons (vt) and ventral footpads (vp) are readily seen. Numbers indicate digit identity. (B) Same section through a mutant embryo deficient in *Wnt7a*. Ventral tendons and footpads are duplicated on what would normally be the dorsal surface of the paw—that is, the dorsal tendons (dt) and dorsal footpads (dp). (From Parr and McMahon 1995; photographs courtesy of the authors.)

(A)   (B)

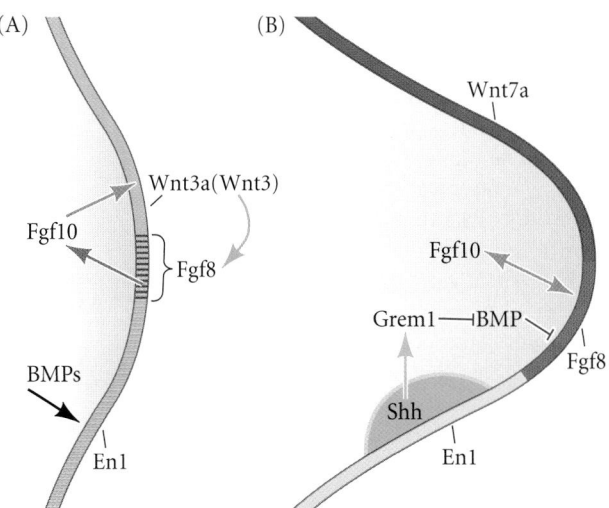

FIGURE 14.25 Early interactions between the AER and limb bud mesenchyme. (A) In the limb bud, Fgf10 from mesenchyme generated by the lateral plate mesoderm activates a Wnt (Wnt3a in chicks; Wnt3 in mice and humans) in the ectoderm. Wnt activates the β-catenin pathway, which induces synthesis of Fgf8 in the region near the AER. Fgf8 activates Fgf10, causing a positive feedback loop. BMPs expressed in the ventral ectoderm cause expression of Engrailed-1, which represses Wnt7a expression. (B) As the limb bud grows, Sonic hedgehog in the posterior mesenchyme creates a new signaling center that induces posterior-anterior polarity, and it also activates Gremlin (Grem1) to prevent mesenchymal BMPs from blocking FGF synthesis in the AER. (After Fernandez-Teran and Ros 2008.)

been ventralized (Chen et al. 1998; Dreyer et al. 1998). The Lim1 protein probably specifies the cells to differentiate in a dorsal manner, and this is critical, as we saw in Chapter 11, for the innervation of motor neurons (whose growth cones recognize inhibitory factors made differentially in the dorsal and ventral compartments of the limb bud).

Conversely, the transcription factor Engrailed-1 marks the ventral ectoderm of the limb bud and is induced by BMPs. If BMPs are knocked out in the early limb bud, Engrailed-1 is not expressed and Wnt7a is expressed in both dorsal and ventral ectoderm. The result is a malformed limb that is dorsal on both sides (Ahn et al. 2001; Pizette et al. 2001.)

Coordinating the Three Axes

The mechanisms specifying the three axes of the tetrapod limb are interrelated and coordinated at both the genetic and cellular levels (see Rabinowitz and Vokes 2012). At first, when the limb bud is relatively small, an initial positive feedback loop between Fgf10 and Fgf8 is established (**FIGURE 14.25A**). As the limb bud grows, the ZPA is established and another regulatory loop is created (**FIGURE 14.25B**). Sonic hedgehog in the ZPA activates Gremlin, which inhibits BMPs. BMPs in the mesoderm would downregulate FGFs in the AER were in not for the Shh-dependent expression of a BMP inhibitor, Gremlin (Niswander et al. 1994; Zúñiga et al. 1999; Scherz et al. 2004; Vokes et al. 2008). After this, the interactivity gets complicated.

Depending on the levels of FGFs in the AER, the ZPA can be either activated or shut down; two feedback loops have been demonstrated (**FIGURE 14.26A**; Verheyden and Sun 2008; Bénazet et al. 2009). At first, relatively low levels of AER

FGFs activate Shh and keep the ZPA functioning (**FIGURE 14.26B**; also see Figure 14.25B). The FGF signals appear to repress the proteins Etv4 and Etv5, which are repressors of Sonic hedgehog transcription (Mao et al. 2009; Zhang et al. 2009). Thus, the AER and ZPA mutually support each other through the positive loop of Sonic hedgehog and FGFs (Todt and Fallon 1987; Laufer et al. 1994; Niswander et al. 1994). In this phase of limb development, levels of Gremlin (a powerful BMP antagonist) are high, and the positive FGF/Shh loop sustains the limb growth. However, as FGF levels rise (**FIGURE 14.26C**), they *block* Gremlin expression in the distal mesenchyme. This repression of Gremlin synthesis activates the inhibitory feedback loop and creates a zone of cells in the distal mesenchyme that do not secrete Gremlin.

As long as the Gremlin signal can diffuse to the AER, FGFs will be made and the AER maintained. It is thought that as the limb bud expands, the Gremlin signal eventually fails to reach the ectoderm. At that time, the BMPs abrogate FGF synthesis, the AER collapses, and the ZPA (with no FGFs to support it) is terminated. The embryonic phase of limb development ends (**FIGURE 14.26D**).

The dorsal-ventral axis is also coordinated with these other two axes. Indeed, the *Wnt7a*-deficient mice described earlier lacked not only dorsal limb structures but also posterior digits, suggesting that Wnt7a is also needed for the anterior-posterior axis (Parr and McMahon 1995). Yang and Niswander (1995) made a similar set of observations in chick embryos. These investigators removed the dorsal ectoderm from developing limbs and found that this resulted in the loss of posterior skeletal elements from the limbs. The reason these limbs lacked posterior digits was that *Shh* expression was greatly reduced. Viral-induced expression of *Wnt7a* was able to substitute for the dorsal ectoderm signal and restore *Shh* expression and posterior phenotypes. These findings showed that Sonic hedgehog synthesis is stimulated by the combination of Fgf4 and Wnt7a proteins. Conversely, overactive Wnt signaling in the ventral ectoderm causes an overgrowth of the AER and extra digits, indicating that the proximal-distal patterning is not independent of dorsal-ventral patterning either (Loomis et al. 1998; Adamska et al. 2004).

Thus, at the end of limb patterning, BMPs are responsible for simultaneously shutting down the AER, indirectly shutting down the ZPA, and inhibiting the Wnt7a signal along the dorsal-ventral axis (Pizette et al. 2001). The BMP signal eliminates growth and patterning along all three axes. When exogenous BMP is applied to the AER, the elongated epithelium of the AER reverts to a cuboidal epithelium and ceases to produce FGFs; and when BMPs are inhibited by Noggin, the AER continues to persist days after it would normally have regressed (Gañan et al. 1998; Pizette and Niswander 1999).

Cell Death and the Formation of Digits and Joints

Sculpting the autopod

Cell death plays a role in sculpting the tetrapod limb. Indeed, cell death is essential if our joints are to form and if our fingers are to become separate (Zaleske 1985; Zuzarte-Luis and Hurle 2005). The death (or lack of death) of specific cells in the vertebrate limb is genetically programmed and has been selected for over the course of evolution. The difference between a chicken's foot and the webbed foot of a duck is the presence or absence of cell death between the digits (**FIGURE 14.27**). Saunders and co-workers have shown that after a certain stage, chick cells between the digit cartilage are destined to die, and will do so even if transplanted to another region of the embryo or placed in culture (Saunders et al. 1962; Saunders and Fallon 1966). Before that time, however, transplantation to a duck limb will save them. Between the time when the cell's death is determined and when death actually takes place, levels of DNA, RNA, and protein synthesis in the cell decrease dramatically (Pollak and Fallon 1976).

In addition to the **interdigital necrotic zone**, three other regions of the limb are "sculpted" by cell death. The ulna and radius are separated from each other by an **interior necrotic zone**, and two other regions, the **anterior** and **posterior necrotic zones**, further shape the end of the limb (see Figure 14.27B; Saunders and Fallon 1966). Although these zones are referred to as "necrotic," this term is a holdover from the days when no distinction was made between necrotic cell death and apoptotic cell death (see Chapter 3). These cells die by apoptosis, and the death of the interdigital tissue is associated with the fragmentation of their DNA (Mori et al. 1995).

The signal for apoptosis in the autopod is provided by the BMP proteins. BMP2, BMP4, and BMP7 are each expressed in the interdigital mesenchyme, and blocking BMP signaling (by infecting progress zone cells with retroviruses carrying dominant negative BMP receptors) prevents interdigital apoptosis (Zou and Niswander 1996; Yokouchi et al. 1996; Abara-Buis et al. 2011). Because these BMPs are expressed throughout the progress zone mesenchyme, it is thought that cell death would be the default state unless there were active suppression of the BMPs. This suppression may come from the Noggin protein, which is made in the developing cartilage of the digits and in the perichondrial cells surrounding it (Capdevila and Johnson 1998; Merino et al. 1998). If Noggin is expressed throughout the limb bud, no apoptosis is seen.

Forming the joints

The function first ascribed to BMPs was the formation, not the destruction, of bone and cartilage tissue. In the developing limb, BMPs induce the mesenchymal cells either to undergo apoptosis or to become cartilage-producing chondrocytes—depending on the stage of development. The same BMPs can induce death or differentiation, depending on the responding cell's history. This **context dependency** of signal

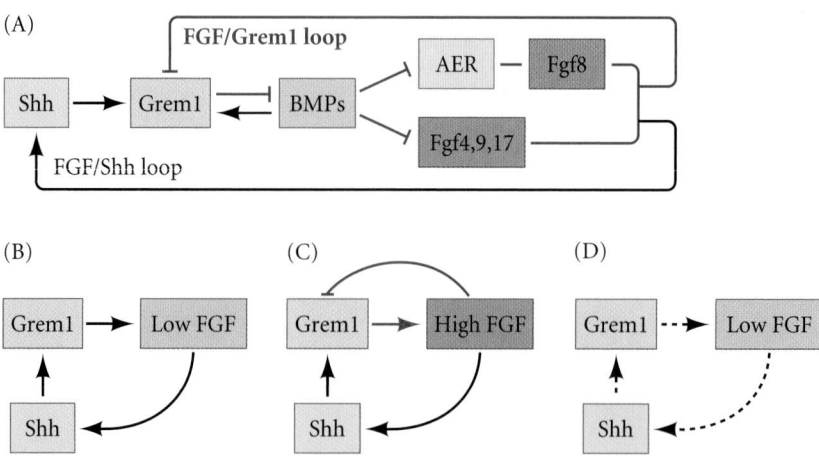

FIGURE 14.26 One of several models to explain the coordinated functions of the AER and ZPA in terminating limb outgrowth. (A) Two feedback loops link the AER and ZPA. In the positive feedback loop (black arrow; below), FGFs 4, 9, and 17 from the AER activate Shh, stabilizing the ZPA. In the reciprocal inhibitory loop (red; above), Shh from the ZPA activates Gremlin (Grem1), which blocks BMPs, thus preventing BMP-mediated inactivation of FGFs in the AER. (B) The feedback loops create a mutual accelerated synthesis of Shh (ZPA) and FGFs (AER). (C) As FGF concentration climbs, it eventually reaches a threshold where it inhibits Gremlin, thus allowing the BMPs to begin repressing the AER FGFs. As more cells multiply in the area not expressing Gremlin, the Gremlin signal near the AER is too weak to prevent the BMPs from repressing the FGFs. (D) At that point, the AER disappears, removing the signals that stabilize the ZPA. The ZPA then also disappears. (After Verheyden and Sun 2008.)

(A) DUCK LEG PRIMORDIUM
Minimal cell death

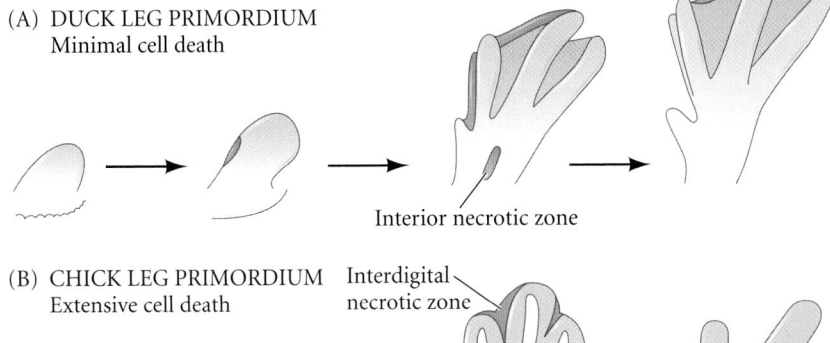

Interior necrotic zone

(B) CHICK LEG PRIMORDIUM
Extensive cell death

Interdigital necrotic zone

Anterior necrotic zone

Posterior necrotic zone

Interior necrotic zone

FIGURE 14.27 Patterns of cell death in leg primordia of (A) duck and (B) chick embryos. Shading indicates areas of cell death. In the duck, the regions of cell death are very small, whereas there are extensive regions of cell death in the interdigital tissue of the chick leg. (After Saunders and Fallon 1966.)

action is a critical concept in developmental biology. It is also critical for the formation of joints. Macias and colleagues (1997) have shown that during early limb bud stages (before cartilage condensation), beads secreting BMP2 or BMP7 cause apoptosis. Two days later, the same beads cause the limb bud cells to form cartilage.

In the normally developing limb, BMPs use both of these properties to form joints. Several BMPs are made in the perichondrial cells surrounding the condensing chondrocytes and promote further cartilage formation* (**FIGURE 14.28A**). Another BMP, GDF5, is expressed at the regions between the bones, where joints will form (**FIGURE 14.28B**; Macias et al. 1997; Brunet et al. 1998). Mouse mutations have suggested that the function of this protein in joint formation is critical. Mutations of *Gdf5* produce brachypodia, a condition characterized by the lack of limb joints (Storm and Kingsley 1999). In mice homozygous for loss-of-function of the BMP antagonist *Noggin*, no joints form. Rather, BMP7 in these *Noggin*-defective embryos appears to recruit nearly all the surrounding mesenchyme into the digits (**FIGURE 14.28C**).

---

*In bats, the amount of BMPs synthesized is extraordinarily high and recruits more mesenchymal cells into the cartilage, thereby extending the digits to a much greater length than in most other mammals (Cooper et al. 2012).

FIGURE 14.28 Possible involvement of BMPs in stabilizing cartilage and apoptosis. (A) Model for the dual role of BMP signals in limb mesodermal cells. BMP can be received in the presence of FGFs (to produce apoptosis) or Wnts (to induce bone). When FGFs from the AER are present, Dickkopf (Dkk) is activated. This protein mediates apoptosis and at the same time inhibits Wnt from aiding in skeleton formation. (B,C) Effects of Noggin. (B) A 16.5-day autopod from a wild-type mouse, showing *Gdf5* expression (dark blue) at the joints. (C) A 16.5-day *Noggin*-deficient mutant mouse autopod, showing neither joints nor *Gdf5* expression. Presumably, in the absence of Noggin, BMP7 was able to convert nearly all the mesenchyme into cartilage. (A after Grotewold and Rüther 2002; B,C from Brunet et al. 1998, courtesy of A. P. McMahon.)

(A)

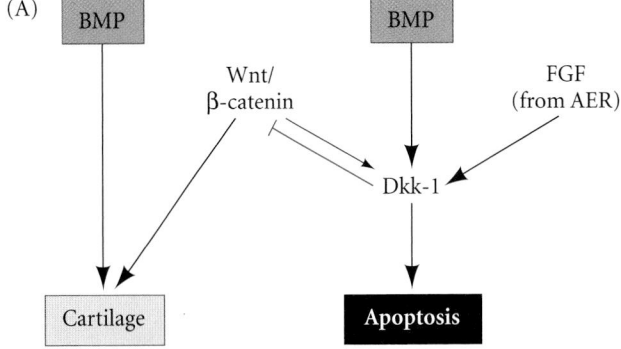

(B)

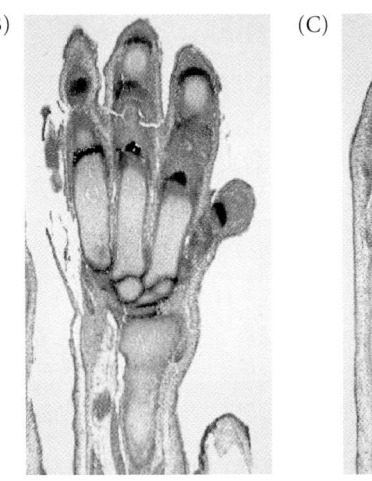

(C)

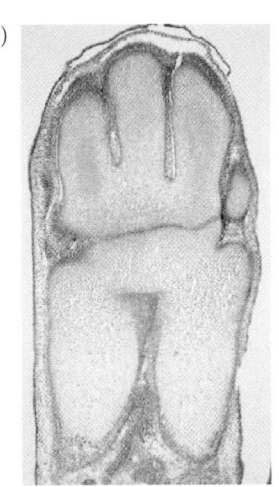

SIDELIGHTS & SPECULATIONS

Evolution by Altering Limb Signaling Centers

In the quotation from *On the Origin of Species* that opens this chapter, Charles Darwin noted that the differences between horse legs, porpoise flippers, and human hands are underlain by a similar pattern of bone formation. Darwin proposed that these bones evolved from a common ancestor, but he did not know how. C. H. Waddington pointed out that such evolution was predicated on changing the development of the limb in ways that it could be selected. In other words, he proposed that changes in development caused the arrival of new variations. These variations could then be tested by natural selection. We have now discovered several ways by which "tinkering" with limb signaling molecules can generate new limb morphologies. In these ways, limb evolution can be caused by developmental changes.*

Web-footed friends

We can start with the sculpting of the autopod. The regulation of BMPs is critical in creating the webbed feet of ducks (Laufer et al. 1997b; Merino et al. 1999). The interdigital regions of duck feet exhibit the same pattern of BMP expression as the webbing of chick feet. However, whereas the interdigital regions of the chick feet appear to undergo BMP-mediated apoptosis, developing duck feet synthesize the BMP inhibitor Gremlin and block this regional cell death (**FIGURE 14.29**). Moreover, the webbing of chick feet can be preserved if Gremlin-soaked beads are placed in the interdigital regions (**FIGURE 14.30**). Thus, the evolution of web-footed birds probably involved the inhibition of BMP-mediated apoptosis in the interdigital regions. In Chapter 20 we will find that the bat embryo uses a similar mechanism to acquire its wings.

Tinkering with the signaling centers: Making whales

Numerous transition fossils attest to the evolution of modern cetaceans (whales, dolphins, porpoises) from hoofed land mammals (Gingrich et al. 1994; Thewissen et al. 2007, 2009). Numerous changes in the anatomy were made, but few are as striking as the conversion of a forelimb into a flipper and the elimination of the hindlimb altogether. These events

*Earlier in this chapter, we noted that developmental biologists get used to thinking in four dimensions. Evolutionary developmental biologists have to think in *five* dimensions—the three standard dimensions of space, the dimension of developmental time (hours or days), and the dimension of evolutionary time (millions of years).

FIGURE 14.29 Autopods of chicken feet (upper row) and duck feet (lower row) are shown at similar stages. Both show BMP4 expression (dark blue) in the interdigital webbing; BMP4 induces apoptosis. The duck foot (but not the chicken foot) expresses the BMP4-inhibitory protein Gremlin (dark brown; arrows) in the interdigital webbing. Thus, the chicken foot undergoes interdigital apoptosis (as seen by neutral red dye accumulation in the dying cells), but the duck foot does not. (Courtesy of J. Hurle and E. Laufer.)

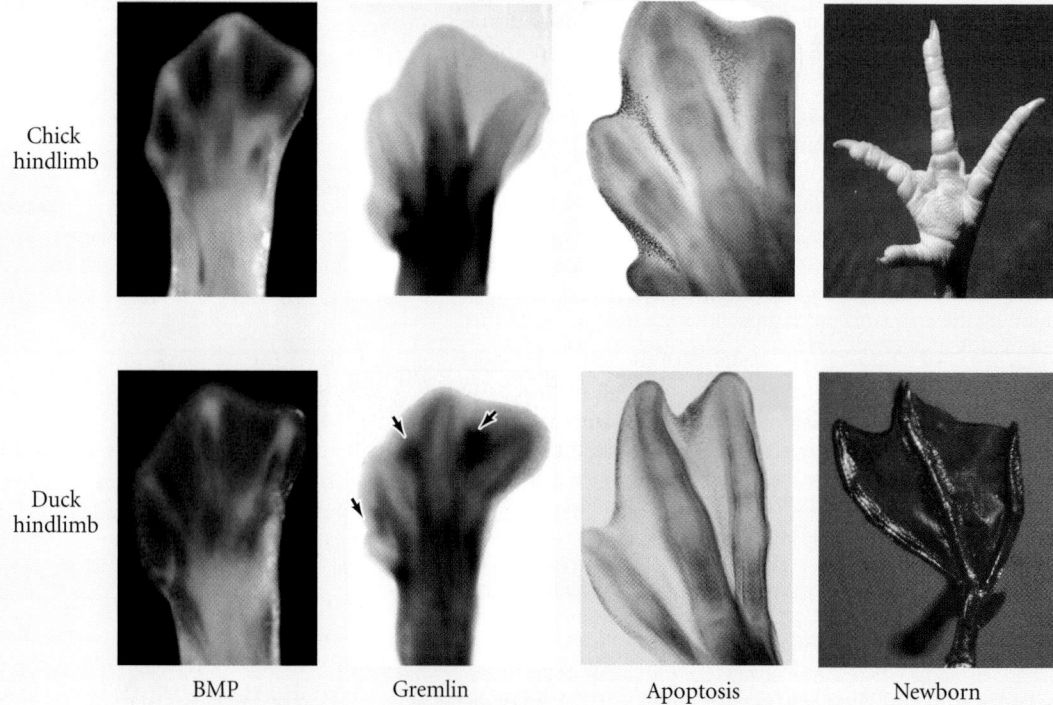

Chick hindlimb

Duck hindlimb

BMP Gremlin Apoptosis Newborn

(A) (B)

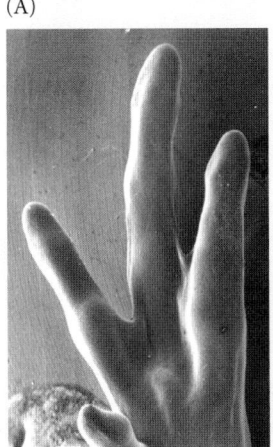
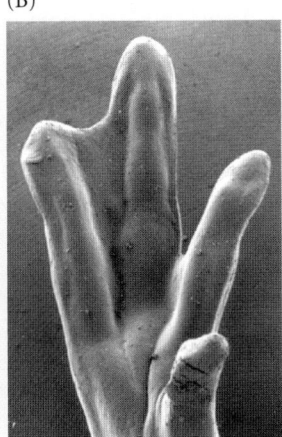

FIGURE 14.30 Inhibition of cell death by inhibiting BMPs. (A) Control chick limb shows extensive apoptosis in the space between the digits, leading to the absence of webbing. (B) When beads soaked with Gremlin protein are placed in the interdigital mesoderm, the webbing persists and generates a ducklike pattern. (From Merino et al. 1999, courtesy of E. Hurle.)

were accomplished by modifying the signaling centers of the ancestral cetacean limb buds in three ways. First the FGF signaling of the forelimb AER was preserved for a much longer duration. This caused the formation of longer fingers by the continual addition of phalanges. Second, interdigital apoptosis was prevented by blocking BMP activity in a manner similar to that described above for duck feet. Third, the Sonic hedgehog signal from the hindlimb ZPA ceased early in development. Once the ZPA signal diminished, the AER could not be sustained and the hindlimb ceased to develop (Thewissen et al. 2006). **FIGURE 14.31** shows the elongated flipper phalanges and truncated hindlimb of an embryonic dolphin. Thus, despite creationists claiming there is no way whales could have evolved from land mammals (see Gish 1985), in fact the combination of developmental biology and paleontology explains the phenomenon extremely well.

● **See WEBSITE 14.2** Dinosaurs and chicken fingers

Remnants of pelvic girdle

Hyperphalangy in digit 2 and 3

FIGURE 14.31 A 110-day embryo of a pantropical spotted dolphin (*Stella attenuata*), stained to show bones (red) and cartilage (blue). Hyperphalangy (extremely long fingers) is seen in the forelimb (correlating with continued Fgf8 expression in the AER), while a rudimentary hindlimb is seen (correlating with the reduction of AER signaling following the elimination of Shh from the ZPA). (From Cooper 2009; courtesy of L. N. Cooper and the Thewissen laboratory.)

Wnt proteins and blood vessels also appear to be critical in joint formation. The conversion of mesenchyme cells into nodules of cartilage-forming tissue establishes the bone boundaries. The mesenchyme will not form such nodules in the presence of blood vessels, and one of the first indications of cartilage formation is the regression of blood vessels in the region wherein the nodule will form (Yin and Pacifici 2001). Wnt proteins are critical in sustaining transcription of *Gdf5*, and β-catenin produced by the Wnts is able to suppress the *Sox9* and *collagen-2* genes that characterize pre-cartilage cells (Hartmann and Tabin 2001; Tufan and Tuan 2001).

Joints are not merely just "absences" of bone. Rather, they are complex structures that incorporate a lubrication system, an immune system, and a ligament system, all joined to the proper articulation of the skeleton. One of the critical elements of joint formation that allows this differentiation is muscle contraction. In normal joint formation, the cells that will form the joint lose their chondrocyte characteristics (such as expression of collagen 2 and Sox9) and instead begin to express GDF5, Wnt4, Wnt9a, and Ext1 (a protein necessary for synthesizing heparan sulfate). These cells will form the articulate cartilage and the synovium (which secretes lubricating synovial fluid) (Koyama et al. 2008; Mundy et al. 2011). Kahn and colleagues (2009) have shown that the movement of the bones is necessary for maintaining this commitment to form joints. In mutant mice where the muscles do not form or are paralyzed, the joint cells revert back to a cartilaginous phenotype.

The discovery of the fossil *Tiktaalik roseae*, the "fish with fingers," highlights the importance of joint development in limb evolution. Fish fins, including those of some of the most primitive species, develop using the same three Hox gene expression phases as tetrapods use to form their limbs (Davis et al. 2007; Ahn and Ho 2008). What may have allowed the

(A)

(B)

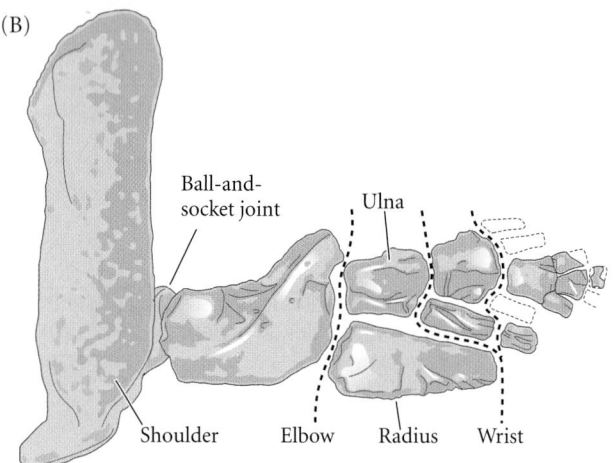

Ball-and-
socket joint

Ulna

Shoulder Elbow Radius Wrist

(C)

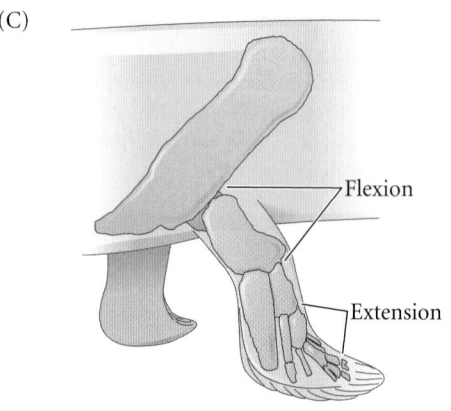

Flexion

Extension

FIGURE 14.32 *Tiktaalik*, a fish with wrists and fingers, lived in shallow waters about 375 million years ago. (A) This reconstruction shows *Tiktaalik's* fishlike gills, fins, scales, and (lack of) neck. The external nostrils on its snout, however, indicate that it could breath air. (B) Fossilized *Tiktaalik* bones reveal the beginnings of digits, wrists, elbows, and shoulders, and suggest that this amphibian-like fish could propel itself on stream bottoms and perhaps live on land for short durations. The joints of the fin included a ball-and-socket joint in the shoulder and a planar joint that allowed the wrist to bend. Other joints allowed the animal to get purchase on its substrate. (C) Resistant contact with a substrate would allow flexion at the proximal joints (shoulder and elbow) and extension at the distal ones (wrist and digits). (A, model and photograph © Tyler Keillor; B,C after Shubin et al. 2006.)

independent modification of fin bones into limb bones are the joints. The joints of *Tiktaalik's* pectoral fins are very similar to those of amphibians and indicate that *Tiktaalik* had mobile wrists and a substrate-supported stance in which the elbow and shoulder could flex (**FIGURE 14.32**; Shubin et al. 2006; Shubin 2008). In addition, the presence of wristlike structures and the loss of dermal scales in these regions suggest that this Devonian fish was able to propel itself on moist substrates. So *Tiktaalik* is thought to be a transition between fish and amphibians—a "fishapod" (as one of its discoverers, Neil Shubin, called it), "capable of doing push-ups."

Continued Limb Growth: Epiphyseal Plates

The three axes of human limbs are specified in highly asymmetrical fashion. However, their growth—for the next 16 years—is so symmetrical that the length of one arm matches the other to within 0.2% (Ballock and O'Keefe 2003; Wolpert 2010). If all of our cartilage were turned into bone before birth, we could not grow any larger, and our bones would be only as large as the original cartilaginous model. However, as we saw in Chapter 12, growth zones (epiphyseal plates) form at the proximal and distal ends of each developing long bone. At the portion of the epiphysial plate farthest from the new bone is a germinal region of cartilaginous stem cells, followed by regions of proliferating cartilage cells, mature cartilage cells (chondrocytes), and hypertrophic cartilage cells. The hypertrophic cartilage cells (which increase their size five- to tenfold) undergo apoptosis and are replaced by bone cells (osteocytes).

In the long bones of many mammals (including humans), this endochondral ossification spreads outward in both directions from the center of the bone (**FIGURE 14.33A**; see also Figures 12.19 and 12.20). Although more than 10,000 new cartilage cells may be made daily, the number appears to be identical in each arm. Cartilage proliferation in humans is high through about 3 years of age. However, the major factor of differential growth (between the arms and legs of a human, or between the legs of a human versus the legs of a dog) is probably the swelling of the hypertrophic cartilage. Not only can different numbers of cartilage cells be made, but those cells can enlarge to different sizes (Cooper et al. 2013). After a growth spurt at puberty, the epiphyseal growth plates fuse and there are no longer any stem cells for growth. As long as the epiphyseal plates are able to produce chondrocytes, the bone continues to grow.

Fibroblast growth factor receptors: Dwarfism

The rate of growth appears to be intrinsic to each bone. Each growth plate is controlled locally (probably by differences in the sensitivity to growth factors), but the coordinated growth of the entire skeleton is maintained by circulating factors. Thus, when transplantations are made between growth plates of old and young mammals, the growth rate of the transplanted growth plate depends on the age of the donor animal, not that of the host animal (Wolpert 2010). However, discoveries of human and mouse mutations resulting in abnormal skeletal development have provided remarkable insights into how hormones and paracrine factors can control the eventual size of the limbs.

Fibroblast growth factors are critically important in halting the growth of the epiphyseal plates, telling the cells to differentiate rather than divide (Deng et al. 1996; Webster and Donoghue 1996). In humans, mutations of the receptors for FGFs can cause these receptors to become active prior to receiving the normal FGF signal. Such mutations give rise to the major types of human dwarfism. **Achondroplasia** is a dominant condition caused by mutations in the transmembrane region of FGF receptor 3 (FgfR3). Roughly 95% of achondroplastic dwarfs have the same mutation of FgfR3, a base-pair substitution that converts glycine to arginine at position 380 in the transmembrane region of the protein. In addition, mutations in the extracellular portion of the FgfR3 protein or in the tyrosine kinase intracellular domain may result in thanatophoric dysplasia (see Figure 3.24), a lethal form of dwarfism that resembles homozygous achondroplasia (Bellus et al. 1995; Tavormina et al. 1995).

As mentioned in Chapter 1, dachshunds have an achondroplastic mutation, but its cause is slightly different than that of the human form. Dachshunds have an extra copy of the *Fgf4* gene, which is also expressed in the developing limb (see Figure 1.21). This causes excess production of Fgf4, activating FgfR3 and accelerating the pathway that stops the growth of chondroblasts and hastens their differentiation. The same extra copy of *Fgf4* has been found in other short-limbed dogs, such as corgis and basset hounds (Parker et al. 2009).

Growth hormone and estrogen receptors

One of the major hormones involved in growth control is, of course, growth hormone (GH). Growth hormone usually acts indirectly, instructing the liver to synthesize insulin-like growth factor-I (IGF-I). IGF-I stimulates the production and division of the cartilage cells. In most giants, there is an increase in the production of growth hormone. Among African Pygmies, a people noted for short stature, there appears to be a severe reduction in the number of growth hormone receptors, resulting in a reduction of liver IGF-I and small limbs (Bozzola et al. 2009). Similarly, certain small breeds of dogs (such as Pomeranians and toy poodles) are characterized by a reduced-function allele of *Igf1*, whereas large breeds (such as Newfoundlands and St. Bernards) are characterized by different allele of this gene (Sytter et al. 2007; Parker et al. 2010).

The pubertal growth spurt and subsequent cessation of growth are induced by increased growth hormone production regulated by the gonadal steroid hormones (Kaplan and Grumbach 1990). Estrogen appears to be important for bone growth in both men and women. Relatively low levels of estrogen stimulate skeletal growth in men and women, whereas the higher levels that occur at the end of puberty induce apoptosis in the hypertrophic chondrocytes and stimulate the invasion of bone-forming osteoblasts into the growth plate. There have been several documented cases of

(A)

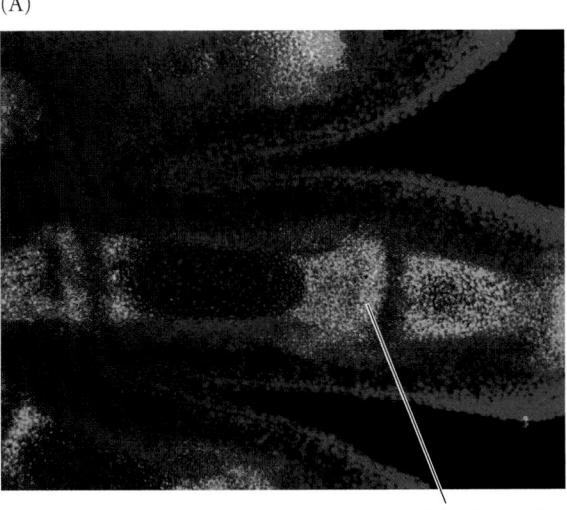

Joint formation

(B)

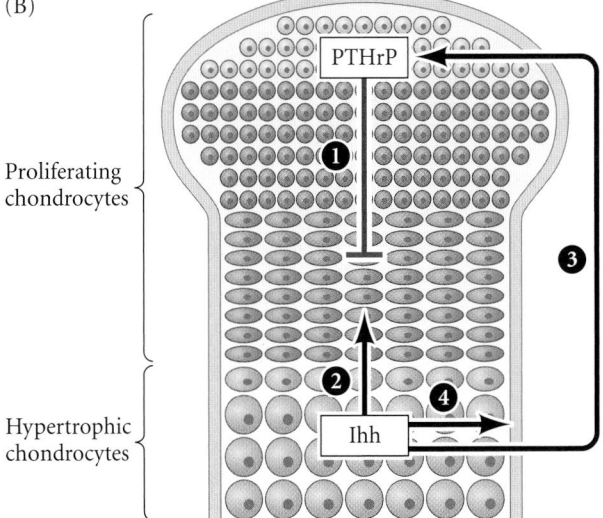

FIGURE 14.33 Model for endochondral ossification in the limb. (A) In situ hybridization in the limb of a 14.5-day mouse embryo shows *collagen-2* mRNA (green) in the proliferative growth plate cartilage and *Gdf5* mRNA (red) in the region destined to become a joint. Nuclei are stained blue. (B) Regulation of cartilage cell proliferation by PTHrP and Indian hedgehog. (1) PTHrP acts on receptors on proliferating chondrocytes to keep them dividing and thereby delays the production of Ihh. (2) When the source of PTHrP production becomes sufficiently far away, then Ihh is synthesized. Indian hedgehog acts on its receptor on chondrocytes to increase the rate of proliferation and (3) to stimulate the production of PTHrP at the ends of bones. (4) Ihh also acts on the perichondrial cells enveloping the cartilage to convert them into the osteoblasts of the bone collar. (A courtesy of Dr. P. Tylzanowski; B after Kronenberg 2003.)

men who do not produce estrogen (Smith 1994; Juul 2001). These men continue to grow even in adulthood and may approach 7 feet in height. Their epiphyseal growth plates do not mature and thus remain full of dividing chondroblasts. In most cases, this deficiency is due to a loss-of-function mutation in the gene encoding aromatase, the enzyme that converts testosterone into estrogen; in such instances, therapy with estrogens is able to stop the excessive growth.

Estrogen receptors are found in the cells that regulate growth hormone production, as well as in all the cells of the human growth plate. Thus, the effects of estrogen on growth may involve both the regulation of growth hormone and more local effects on the growth plate itself (Juul 2001; Karimian and Sävendahl 2011). Both growth hormone and estrogen antagonists (such as aromatase) can increase final height of young boys and girls, but there are often severe side-effects (Karimian and Sävendahl 2011; Mauras 2011). It is even possible that estrogen ensures that all the growth plates of the body close at about the same time (Nilsson and Baron 2004).

Parathyroid hormone-related peptide and Indian hedgehog

The growth plate is a rapidly changing population of cells. In young rats, for instance, there are about 40 cells in a column of chondrocytes. During peak growth, about eight cells per day die and are replaced by bone, and another eight chondrocytes replace them (Wilsman et al. 1998). The length of these proliferating chondrocyte columns is controlled largely by parathyroid hormone-related peptide (PTHrP) and Indian hedgehog (Ihh). Ihh maintains chondrocyte cell division directly (by activating cyclin D1 synthesis) and prevents the chondrocytes' immediate hypertrophy. Ihh is made by the cartilage at the "bottom" of the column as the chondrocytes begin to hypertrophy. This Ihh diffuses "upward," stimulating cell division and increasing the number of new chondrocytes in the column. PTHrP, which is stimulated by Ihh, prevents premature hypertrophic differentiation in the young chondrocytes at the base of the growth plate. As the chondrocytes leave the influence of PTHrP, they begin to hypertrophy, die, and are replaced by bone. In this way, Ihh and PTHrP regulate the length of the chondrocyte columns (**FIGURE 14.33B**; Vortkamp et al. 1996; Kobayashi et al. 2005). In humans, mild loss-of-function mutations in the protein encoding PTHrP result in short stature and short fingers (Provot and Schipani 2005).

Coda

FGF proteins generate the proximal-distal axis of the vertebrate limb; Shh and BMPs generate the anterior-posterior axis; and Wnts appear to mediate formation of the dorsal-ventral axis. Thus, all the major paracrine factor families act in coordination to build the limb. While many of the "executives" of the limb bud formation have been identified, we are just starting to discern how the activity of these paracrine factors influences where the cartilaginous condensations will form, how the skeletal elements are sculpted, how each digit becomes specified, and where the tendons and muscles will insert onto the skeletal elements. Limb development is therefore a meeting place for developmental biology, evolutionary biology, and medicine. Within the next decade, we can expect to know the bases for numerous congenital diseases of limb formation, and perhaps we will understand how limbs are modified into flippers, wings, hands, and legs.

SNAPSHOT SUMMARY: The Tetrapod Limb

1. The positions where limbs emerge from the body axis depend on Hox gene expression.

2. The proximal-distal axis of the developing limb is initiated by the induction of the ectoderm at the dorsal-ventral boundary by Fgf10 from the mesenchyme. This induction forms the apical ectodermal ridge (AER). The AER secretes Fgf8, which keeps the underlying mesenchyme proliferative and undifferentiated. This area of mesenchyme is called the progress zone.

3. Fgf and Wnt proteins regulate the shape and growth of the limb bud by controlling the plane rapidity of cell division.

4. Two opposing gradients, one of Fgfs and Wnts from the AER, the other of retinoic acid from the flank, pattern the limb.

5. As the limb grows outward, the stylopod forms first, then the zeugopod, and last, the autopod. Each phase of limb development is characterized by a specific pattern of Hox gene expression. The evolution of the autopod involved a duplication and reversal of Hox gene expression that distinguishes fish fins from tetrapod limbs.

6. Computer models suggest that a reaction-diffusion mechanism can explain the constant pattern of stylopod-zeugopod-autopd seen in tetrapod limbs.

7. The anterior-posterior axis is defined by the expression of Sonic hedgehog in the zone of polarizing activity a region in the posterior mesoderm of the limb bud. If ZPA tissue (or Sonic hedgehog-secreting cells or beads) is placed in the anterior margin of a limb bud, a second, mirror-image pattern of Hox gene expression occurs and there is a corresponding mirror-image duplication of the digits.

8. The ZPA is maintained by the interaction of FGFs from the AER with mesenchyme made competent to express Sonic

hedgehog by its expression of particular Hox genes. Sonic hedgehog acts in turn, probably in an indirect manner, and probably through the Gli factors, to change the expression of the Hox genes in the limb bud.

9. Sonic hedgehog specifies digits in at least two ways. It works through BMP inhibition in the interdigital mesenchyme, and it also regulates the proliferation of digit cartilage.

10. Mutations in the long-range enhancer for Shh can cause polydactyly by creating a second ZPA in the anterior margin of the limb bud.

11. The dorsal-ventral axis is formed in part by the expression of Wnt7a in the dorsal portion of the limb ectoderm. Wnt7a also maintains the expression level of Sonic hedgehog in the ZPA and of Fgf4 in the posterior AER. Fgf4 and Sonic hedgehog reciprocally maintain each other's expression.

12. The levels of FGFs in the AER can either support or inhibit the production of Shh by the ZPA. As the limb bud grows and more FGFs are produced in the AER, inhibition of Shh occurs. This causes the lowering of FGFs, and eventually proximodistal outgrowth ceases.

13. Cell death in the limb is mediated by BMPs and is necessary for the formation of digits and joints. Differences

between the unwebbed chicken foot and the webbed duck foot can be explained by differences in the expression of Gremlin, a protein that antagonizes BMPs.

14. The ends of the long bones of humans and other mammals contain cartilaginous regions called epiphyseal growth plates. The cartilage in these regions proliferates, allowing the resulting bone to grow larger. Eventually the cartilage is replaced by bone and growth stops.

15. The differences in bone growth between different parts of the body and between different species is due primarily to the expansion of hypertrophic cartilage (which will later be replaced by bone).

16. By modifying paracrine factor secretion, different limb morphologies can form, initiating the development of webbed feet, flippers, or hands. By eliminating the synthesis of certain paracrine factors, limbs can be prevented from forming (as in whales and snakes).

17. The BMPs are involved both in inducing apoptosis and in differentiating the mesenchymal cells into cartilage. The regulation of BMP effects by the Noggin and Gremlin proteins is critical in forming the joints between the bones of the limb and in regulating proximal-distal outgrowth.

For Further Reading

Complete bibliographical citations for all literature cited in this chapter can be found at the free-access website **www.devbio.com**

Cooper, K. L., J. K. Hu, D. ten Berge, M. Fernandez-Teran, M. A. Ros and C. J. Tabin. 2011. Initiation of proximal-distal patterning in the vertebrate limb by signals and growth. *Science* 332: 1083–1086.

Mahmood, R. and 9 others. 1995. A role for Fgf-8 in the initiation and maintenance of vertebrate limb outgrowth. *Curr. Biol.* 5: 797–806.

Merino, R., J. Rodriguez-Leon, D. Macias, Y. Gañan, A. N. Economides and J. M. Hurle. 1999. The BMP antagonist Gremlin regulates outgrowth, chondrogenesis and programmed cell death in the developing limb. *Development* 126: 5515–5522.

Niswander, L., S. Jeffrey, G. R. Martin and C. Tickle. 1994. A positive feedback loop coordinates growth and patterning in the vertebrate limb. *Nature* 371: 609–612.

Rabinowitz, A. H. and S. A. Vokes. 2012. Integration of the transcription networks regulating limb morphogenesis. *Dev. Biol.* 368: 165–180.

Riddle, R. D., R. L. Johnson, E. Laufer and C. Tabin. 1993. Sonic hedgehog mediates the polarizing activity of the ZPA. *Cell* 75: 1401–1416.

Roselló-Díez, A., M. A. Ros and M. Torres. 2011. Diffusible signals, not autonomous mechanisms, determine the main proximodistal limb subdivision. *Science* 332:1086–1088.

Saunders, J. W., Jr. 1948. The proximodistal sequence of origin of the parts of the chick wing and the role of the ectoderm. Republished in *J. Exper. Biol.* 282 (1998): 626–668.

Saunders, J. W., Jr. and M. T. Gasseling. 1968. Ectodermal-mesodermal interactions in the origin of limb symmetry. In Epithelial-Mesenchymal Interactions in Development (ed. R. Fleishmajer), pp. 78–79. Baltimore, MD: Williams and Wilkins.

Scherz, P. J., E. McGlinn, S. Nissim and C. J. Tabin. 2007. Extended exposure to Sonic hedgehog is required for patterning the posterior digits of the vertebrate limb. *Dev. Biol.* 308: 343–354.

Sekine, K. and 10 others. 1999. Fgf10 is essential for limb and lung formation. *Nature Genet.* 21: 138–141.

Todt ,W. L. and J. F. Fallon. 1987. Posterior apical ectodermal ridge removal in the chick wing bud triggers a series of events resulting in defective anterior pattern formation. *Development* 101: 501–515.

Verheyden, J. M. and X. Sun. 2008. An Fgf–Gremlin inhibitory feedback loop triggers termination of limb bud outgrowth. *Nature* 454: 638–641.

Vortkamp, A., K. Lee, B. Lanske, G. V. Segre, H. M. Kronenberg and C. J. Tabin. 1996. Regulation of rate of cartilage differentiation by Indian hedgehog and PTH-related protein. *Science* 273: 613–622.

WEBSITE 14.1 Induction of the AER. The induction of the AER is a complex event involving the interaction between the dorsal and ventral compartments of the ectoderm. The Notch pathway may be critical in this process. Misexpression of the genes in this pathway can result in the absence or duplication of limbs.

WEBSITE 14.2 Dinosaurs and chicken fingers. One piece of evidence that birds were descended from dinosaurs is that they both retain three digits. But to some, it seems that birds have digits 2, 3, and 4, whereas dinosaurs had digits 1, 2, and 3. Recent evidence has uncovered a change in gene transcription that might explain this conundrum.

Vade Mecum

Patterning the limb. An interview with John Saunders contains movies of his work on limb development, which identified the AER and the ZPA as two of the major signaling centers in limb formation. His transplantation studies provided the framework for the molecular characterization of the mechanisms of limb formation.

15

Sex Determination

HOW AN INDIVIDUAL'S SEX IS DETERMINED has been one of the great questions of natural philosophy since antiquity. It is also a topic where social beliefs about what is "normal" behavior in men and women have played major roles in interpreting science. In 1890, Geddes and Thomson claimed that females were more passive than males and that this was reflected in their germ cells (the active sperm, the passive egg). Thus, they concluded, those environmental factors favoring the storage of energy and nutrients predisposed one to have female offspring, whereas factors favoring the utilization of energy and nutrients influenced one to have male offspring. As we saw in Chapter 4, however, egg and sperm are both active participants in fertilization. And as we will see in this chapter, male and female offspring are generated by equivalent, equally active, gene-directed processes, and neither is "higher" or "lower" than the other. We also know, however, that even though environmental factors do not determine sex in either mammals or flies, there are many animal species whose sex *is* determined by environmental factors such as temperature.

Indeed, the environmental view of sex determination remained the only major scientific theory of mammalian sex determination until the identification of the X and Y chromosomes and the correlation of the female sex with an XX karyotype (chromosome complement) and the male sex with either XY or XO karyotypes (Stevens 1905; Wilson 1905; see Gilbert 1978). This correlation suggested strongly that a specific nuclear component was responsible for directing the development of the sexual phenotype.

● See WEBSITE 15.1 Social critique of sex determination research

CHROMOSOMAL SEX DETERMINATION

There are several ways chromosomes can determine the sex of an embryo. In mammals, the presence of either a second X chromosome or a Y chromosome determines whether the embryo is to be female (XX) or male (XY). In birds, the situation is reversed (Smith and Sinclair 2001): the male has the two similar sex chromosomes (ZZ), and the female has the unmatched pair (ZW). In flies, the Y chromosome plays no role in sex determination, but the number of X chromosomes appears to determine the sexual phenotype. In other insects (especially hymenopterans such as bees, wasps, and ants), fertilized, diploid eggs develop into females, whereas the unfertilized, haploid eggs become males (Beukeboom 1995; Gempe et al. 2009). This chapter will discuss only two of the many chromosomal modes of sex determination: sex determination in placental mammals and sex determination in *Drosophila*.

The Mammalian Pattern

Mammalian sex determination is governed by the gonad-forming genes and by the hormones elaborated by the gonads. **Primary sex determination** is the determination

*Sexual reproduction is …
the masterpiece of nature.*
ERASMUS DARWIN
(1791)

*It is quaint to notice that
the number of specula-
tions connected with the
nature of sex have well-
nigh doubled since Drelincourt, in the eighteenth
century, brought together
two hundred and sixty-two
"groundless hypotheses,"
and since Blumenbach
caustically remarked that
nothing was more certain
than that Drelincourt's
own theory formed the two
hundred and sixty-third.*

J. A. THOMSON (1926)

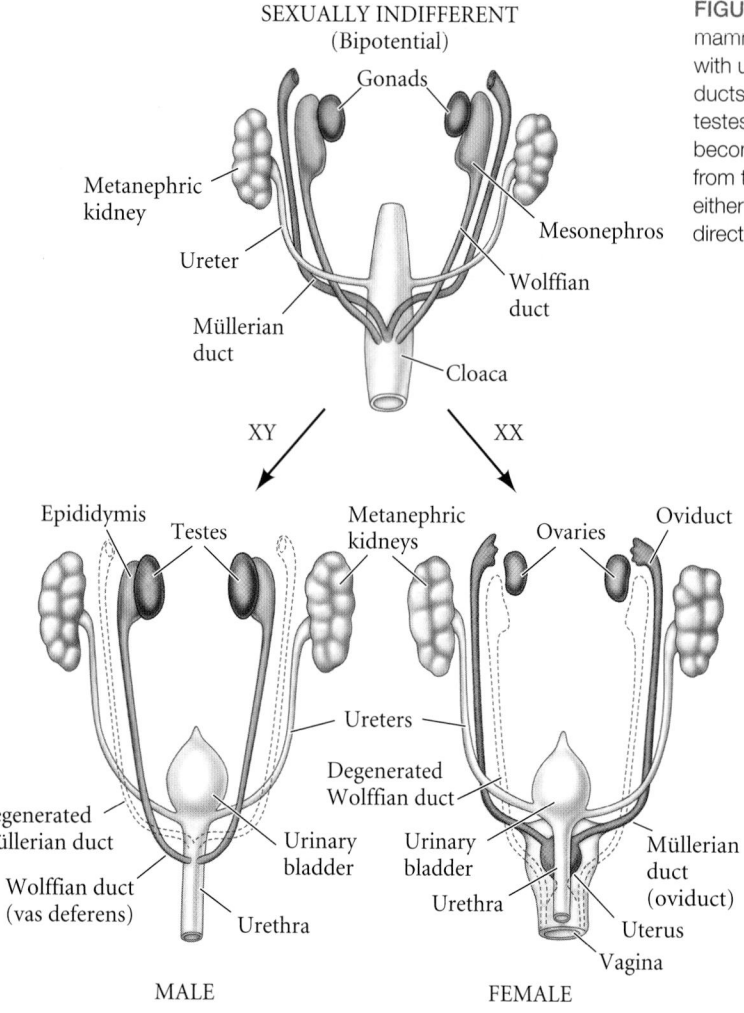

SEXUALLY INDIFFERENT
(Bipotential)

Gonads

Metanephric
kidney

Mesonephros

Ureter

Wolffian
duct

Müllerian
duct

Cloaca

XY   XX

Epididymis   Metanephric   Oviduct
Testes   kidneys   Ovaries

Ureters

Degenerated
Wolffian duct

Degenerated
Müllerian duct   Urinary   Urinary   Müllerian
bladder   bladder   duct
(oviduct)
Wolffian duct   Urethra
(vas deferens)   Urethra   Uterus
Vagina

MALE   FEMALE

FIGURE 15.1 Development of gonads and their ducts in mammals. Originally, a bipotential (indifferent) gonad develops, with undifferentiated Müllerian ducts (female) and Wolffian ducts (male) ducts both present. If XY, the gonads becomes testes and the Wolffian duct persists. If XX, the gonads become ovaries and the Müllerian duct persists. Hormones from the gonads will cause the external genitalia to develop either in the male direction (penis and scrotum) or the female direction (clitoris, labia majora).

GONADS		
Gonadal type	Testis	Ovary
Germ cell location	Inside testis cords (in medulla of testis)	Inside follicles of ovarian cortex
DUCTS		
Remaining duct	Wolffian	Müllerian
Duct differentiation	Vas deferens, epididymis, seminal vesicle	Oviduct, uterus, cervix, upper portion of vagina
UROGENITAL SINUS	Prostate	Skene's glands
LABIOSCROTAL FOLDS	Scrotum	Labia majora
GENITAL TUBERCLE	Penis	Clitoris

chromosomal. The formation both of ovaries and of testes is an active, gene-directed process. Moreover, as we shall see, both the male and female gonads diverge from a common precursor, the **bipotential**, or **indifferent**, gonad (**FIGURE 15.1**).

Primary sex determination in mammals is determined by whether an organism has an XX or an XY karyotype. In most cases, the female's karyotype is XX and the male's is XY. Every individual must carry at least one X chromosome. Since the diploid female is XX, each of her haploid eggs has a single X chromosome. The male, being XY, generates two populations of haploid sperm: half will bear an X chromosome, half a Y. If at fertilization the egg receives a second X chromosome from the sperm, the resulting individual is XX, forms ovaries, and is female; if the egg receives a Y chromosome from the sperm, the individual is XY, forms testes, and is male (**FIGURE 15.2A**).

The Y chromosome carries a gene that encodes a **testis-determining factor** that organizes the bipotential gonad into a testis. This was demonstrated in 1959 when karyotyping showed that XXY individuals were male (despite having two X chromosomes) and that individuals having only one X chromosome were female (Ford et al. 1959; Jacobs and Strong 1959). XXY men (Klinefelter syndrome) have functioning testes. Women with a single X chromosome and no second X or Y (i.e., XO, sometimes called Turner syndrome) begin making ovaries, but the ovarian follicles cannot be maintained without the second X chromosome. Thus, a second X chromosome completes the ovaries, whereas the presence of a Y chromosome (even when multiple X chromosomes are present) initiates the development of testes.

The reason the Y chromosome is able to direct testis formation even if more than one X chromosome is present may

of the gonads—the egg-forming ovaries or sperm-forming testes. **Secondary sex determination** is the determination of the male or female phenotype by the hormones produced by the gonads. In mammals, primary sex determination is

(A)

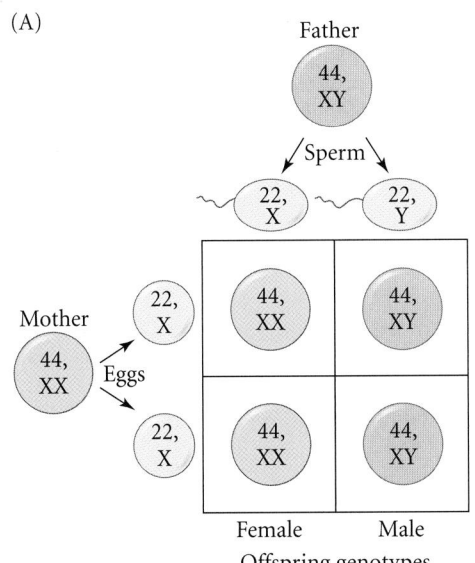

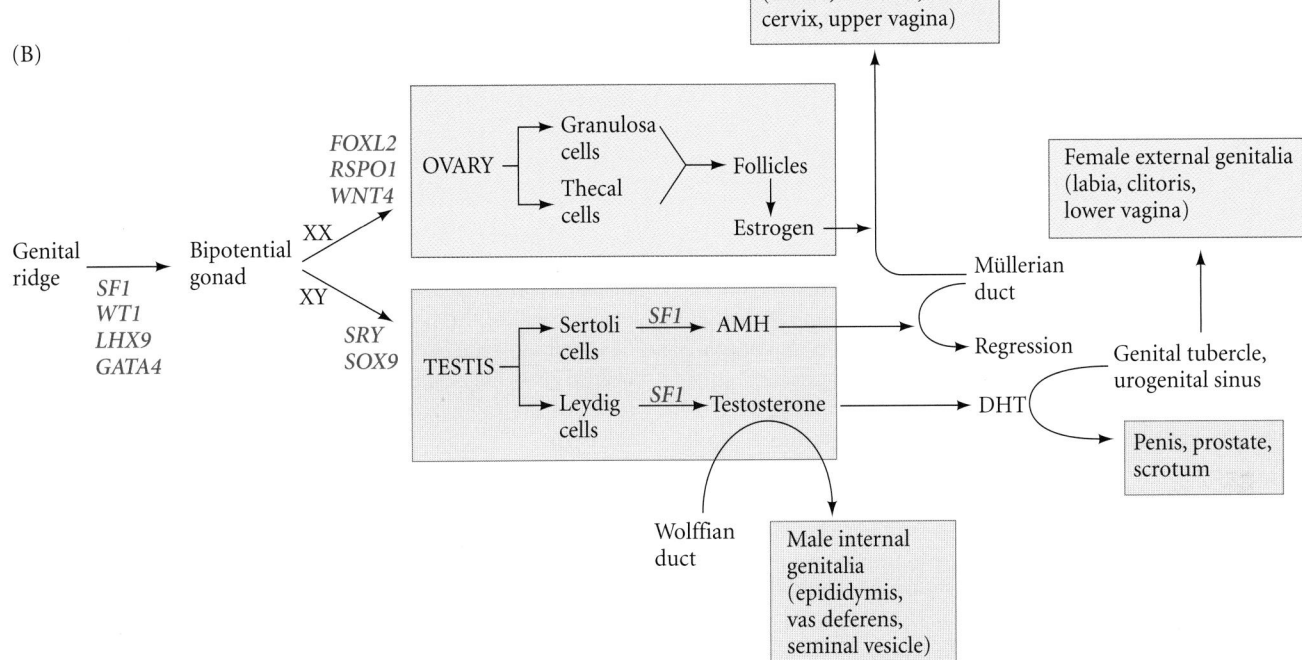

FIGURE 15.2 Sex determination in placental mammals. (A) Mammalian chromosomal sex determination results in approximately equal numbers of male and female offspring. (B) Postulated cascades leading to male and female phenotypes in mammals. The conversion of the genital ridge into the bipotential gonad requires, among others, the *Sf1*, *Wt1*, and *Lhx9* genes; mice lacking any of these genes lack gonads. The bipotential gonad appears to be moved into the female pathway (ovary development) by the *Foxl2*, *Wnt4,* and *Rspo1* genes and into the male pathway (testis development) by the *Sry* gene (on the Y chromosome), which triggers the activity of *Sox9*. (Lower levels of Wnt4 are also present in the male gonad.) The ovary makes thecal cells and granulosa cells, which together are capable of synthesizing estrogen. Under the influence of estrogen (first from the mother, then from the fetal gonads), the Müllerian duct differentiates into the female reproductive tract, the internal and external genitalia develop, and the offspring develops the secondary sex characteristics of a female. The testis makes two major hormones involved in sex determination. The first, anti-Müllerian hormone (AMH), causes the Müllerian duct to regress. The second, testosterone, causes differentiation of the Wolffian duct into the male internal genitalia. In the urogenital region, testosterone is converted into dihydrotestosterone (DHT), which causes the morphogenesis of the penis and prostate gland. (B after Marx 1995; Birk et al. 2000.)

be a matter of timing. It appears there is a crucial window of opportunity during gonad development during which the testis-determining factor (now known to be the product of the *Sry* gene) can function. If the *Sry* gene is present, it usually acts during this duration to promote testis formation and to inhibit ovary formation. If the *Sry* gene is not present (or if it fails to act at the appropriate time), the ovary-forming genes are the ones that will function (**FIGURE 15.2B**; Hiramatsu et al. 2009; Kashimada and Koopman 2010).

Once primary (chromosomal) determination has established the gonads, the gonads begin to produce the hormones and paracrine factors that govern secondary sex

determination—development of the sexual phenotype outside the gonads. This includes the male or female duct systems and the external genitalia. A male mammal has a penis, scrotum (testicle sac), seminal vesicles, and prostate gland. A female mammal has a uterus, oviducts, cervix, vagina, clitoris, labia, and mammary glands.* In many species, each sex also has a sex-specific body size, vocal cartilage, and musculature. Secondary sex characteristics

---

*The naturalist Carolus Linnaeus named the mammals after this female secondary sexual trait in the seventeenth century. The politics of this decision is discussed in Schiebinger 1993.

are usually determined by hormones and paracrine factors secreted from the gonads. In the absence of gonads, it appears the female phenotype is generated. When Jost (1947, 1953) removed fetal rabbit gonads before they had differentiated, the resulting rabbits had a female phenotype, regardless of whether their genotype was XX or XY.

The general scheme of primary sex determination is shown in Figure 15.2B. If the embryonic cells have two X chromosomes and no Y chromosome, the gonadal primordia develop into ovaries. The ovaries produce **estrogen**, a hormone that enables the development of the **Müllerian duct** into the uterus, oviducts, cervix, and upper portion of the vagina (Fisher et al. 1998; Couse et al. 1999; Couse and Korach 2001). If embryonic cells contain both an X and a Y chromosome, testes form and secrete two major factors. The first is a TGF-β family paracrine factor called **anti-Müllerian hormone** (**AMH**; sometimes called Müllerian inhibiting factor, MIF). AMH destroys the Müllerian duct, thus preventing formation of the uterus and oviducts. The second factor is the steroid hormone **testosterone**. Testosterone masculinizes the fetus, stimulating formation of the penis, male duct system, scrotum, and other portions of the male anatomy, as well as inhibiting development of the breast primordia.

Primary Sex Determination in Mammals

The developing gonads

Mammalian gonads embody a unique embryological situation. All other organ rudiments normally can differentiate into only one type of organ. A lung rudiment can only become a lung, and a liver rudiment can develop only into a liver. The gonadal rudiment, however, has two options: it can develop into either an ovary or a testis. These organs have very different tissue architectures. The path of differentiation taken by this rudiment is dictated by the genotype and determines the future sexual development of the organism (Lillie 1917). But before this decision is made, the mammalian gonad first develops through a bipotential, or indifferent, stage, during which time it has neither female nor male characteristics (see Figure 15.1).

In humans, the bipotential gonadal rudiments appear during week 4 and remain sexually indifferent until week 7. The gonadal rudiments are paired regions of the intermediate mesoderm; they form adjacent to the developing kidneys. The ventral surface of these rudiments comprises the coelomic epithelium. During the indifferent stage, this epithelium and its associated mesenchyme proliferate, forming the two **genital ridges** (FIGURE 15.3A,B). The paired genital ridges are the somatic (i.e., non-germ cell) components of the gonads. The germ cells migrate into the gonad during week 6 and are surrounded by the somatic cells.

MALE GONAD DEVELOPMENT If the fetus is XY, the somatic cells continue to proliferate through the eighth week, and then a subset of these cells initiate their differentiation into

Sertoli cells. During week 8, the developing Sertoli cells surround the incoming germ cells and organize themselves into the **testis cords**. These cords form loops in the medullary (central) region of the developing testis and are connected to a network of thin canals, called the **rete testis**, located near the mesonephric duct (FIGURE 15.3C,D). Eventually, the region containing the testis cords and germ cells is enclosed by a thick extracellular matrix, the **tunica albuginea**. Thus, when the germ cells enter the male gonads, they will develop within the testis cords, inside the organ.

The Sertoli cells of the fetal testis cords secrete the AMH that blocks development of the female ducts, and these same Sertoli epithelial cells will later support the development of sperm throughout the lifetime of the male mammal. Meanwhile, during fetal development, the interstitial mesenchyme cells of the testes differentiate into **Leydig cells**, which make testosterone.

Later in development (at puberty in humans; shortly after birth in mice, which procreate much faster), the testis cords mature to form the **seminiferous tubules**. The germ cells migrate to the periphery of these tubules, where they establish the spermatogonial stem cell population that produces sperm throughout the lifetime of the male (see Figure 17.28). In the mature seminiferous tubule, sperm are transported from the inside of the testis through the rete testis, which joins the **efferent ducts**. These efferent ducts are the remodeled tubules of the mesonephric kidney. They link the seminiferous tubules to the **Wolffian duct** (also called the **mesonephric duct**), which used to be the collecting tube of the mesonephric kidney* (see Chapter 12). During male development, the Wolffian duct differentiates to become the **epididymis** (adjacent to the testis) and the **vas deferens**, the tube through which sperm pass into the urethra and out of the body.

FEMALE GONAD DEVELOPMENT In females, the Müllerian duct remains intact and differentiates into the oviducts, uterus, cervix, and upper vagina. In the absence of adequate testosterone, the Wolffian duct degenerates (see Figure 15.2). The germ cells accumulate near the outer surface of the gonad, interspersed with the gonadal somatic cells. Near the time of birth, each individual germ cell is surrounded by somatic cells (FIGURE 15.3E,F). The germ cells will become the ova, and the surrounding cortical epithelial cells will differentiate into **granulosa** cells. As will be detailed in Chapter 17, the mesenchyme cells of the ovary differentiate into **thecal** cells. Together, the thecal and granulosa cells form **follicles** that envelop the germ cells and secrete steroid hormones. Each follicle contains a single germ cell—an **oogonium** (egg precursor)—which will enter

---

* As discussed in Chapter 12, the mesonephric kidney is one of the three major kidney stages seen during mammalian development, but it does not function as a kidney in most mammals.

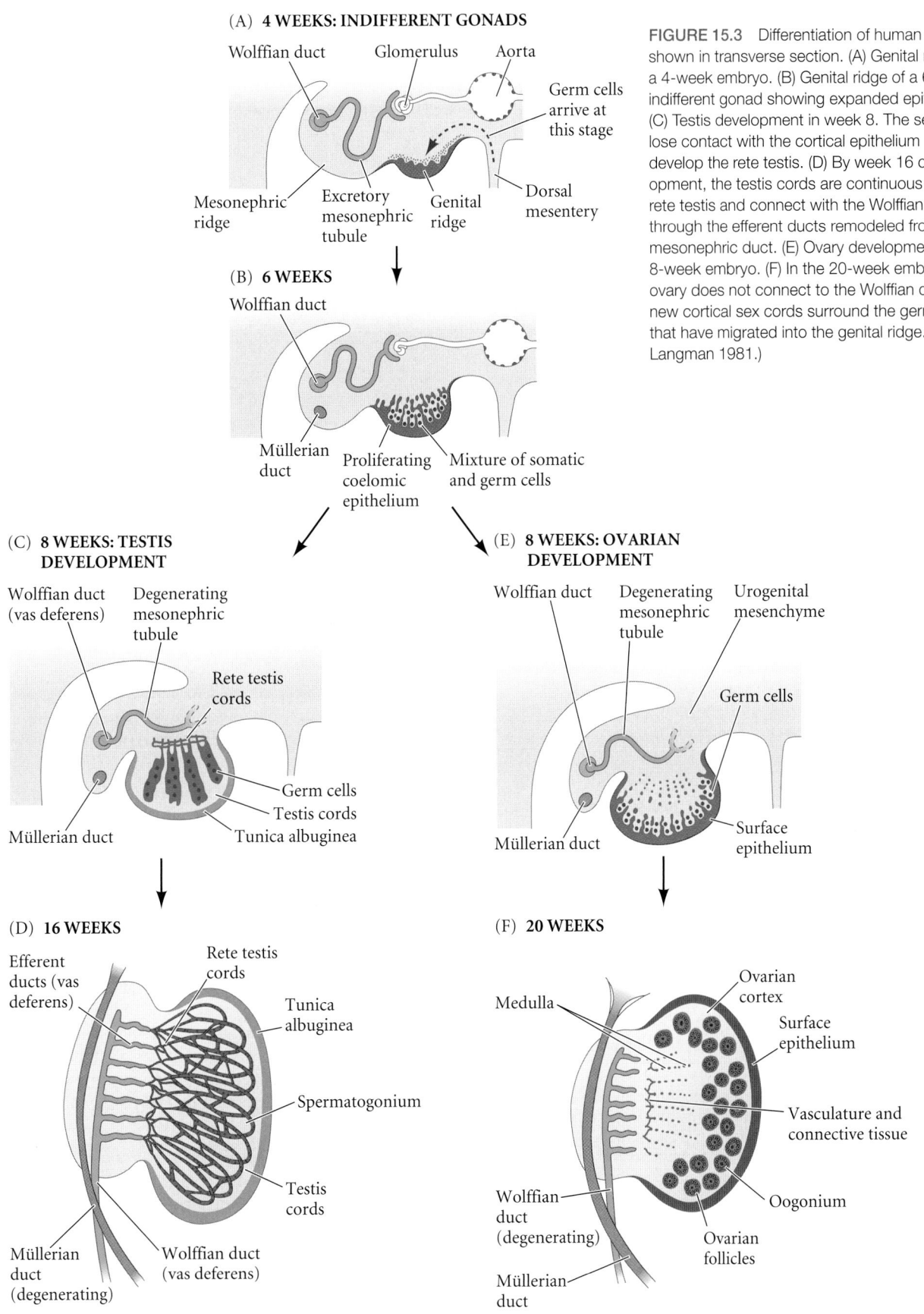

(A) 4 WEEKS: INDIFFERENT GONADS

Wolffian duct · Glomerulus · Aorta · Germ cells arrive at this stage · Dorsal mesentery · Mesonephric ridge · Excretory mesonephric tubule · Genital ridge

(B) 6 WEEKS

Wolffian duct · Müllerian duct · Proliferating coelomic epithelium · Mixture of somatic and germ cells

(C) 8 WEEKS: TESTIS DEVELOPMENT

Wolffian duct (vas deferens) · Degenerating mesonephric tubule · Rete testis cords · Germ cells · Testis cords · Tunica albuginea · Müllerian duct

(D) 16 WEEKS

Efferent ducts (vas deferens) · Rete testis cords · Tunica albuginea · Spermatogonium · Testis cords · Müllerian duct (degenerating) · Wolffian duct (vas deferens)

(E) 8 WEEKS: OVARIAN DEVELOPMENT

Wolffian duct · Degenerating mesonephric tubule · Urogenital mesenchyme · Germ cells · Surface epithelium · Müllerian duct

(F) 20 WEEKS

Medulla · Ovarian cortex · Surface epithelium · Vasculature and connective tissue · Oogonium · Ovarian follicles · Wolffian duct (degenerating) · Müllerian duct

FIGURE 15.3 Differentiation of human gonads shown in transverse section. (A) Genital ridge of a 4-week embryo. (B) Genital ridge of a 6-week indifferent gonad showing expanded epithelium. (C) Testis development in week 8. The sex cords lose contact with the cortical epithelium and develop the rete testis. (D) By week 16 of development, the testis cords are continuous with the rete testis and connect with the Wolffian duct through the efferent ducts remodeled from the mesonephric duct. (E) Ovary development in an 8-week embryo. (F) In the 20-week embryo, the ovary does not connect to the Wolffian duct, and new cortical sex cords surround the germ cells that have migrated into the genital ridge. (After Langman 1981.)

meiosis at this time.* These germ cells are required for the gonadal cells to complete their differentiation into ovarian tissue (McLaren 1991).

Genetic mechanisms of primary sex determination: Making decisions

Several human genes have been identified whose function is necessary for normal sexual differentiation. Because the phenotype of mutations in sex-determining genes is often sterility, clinical infertility studies have been useful in identifying those genes that are active in determining whether humans become male or female. Experimental manipulations to confirm the functions of these genes can then be done in mice. Although the story unfolded in the following paragraphs demonstrates the remarkable progress that has been made in recent years, we still do not fully understand how all these gonad-determining genes interact. The problem of primary sex determination remains (as it has since prehistory) one of biology's great mysteries.

The story starts in the genital ridge, which can become either type of gonad. Here, the genes for Wt1, Wnt4, Lhx9, GATA4, and Sf1 are expressed, and the loss of function of any one of them will prevent the normal development of either male or female gonads.

- If no Y chromosome is present, these factors are thought to activate further expression of Wnt4 (already expressed at low levels in the genital epithelium) and of a small soluble protein called **R-spondin1 (Rspo1)**. Rspo1 binds to its cell membrane receptor and further stimulates the Disheveled protein of the Wnt pathway, making the Wnt pathway more efficient at producing β-catenin. One of the functions of β-catenin in gonadal cells is to further

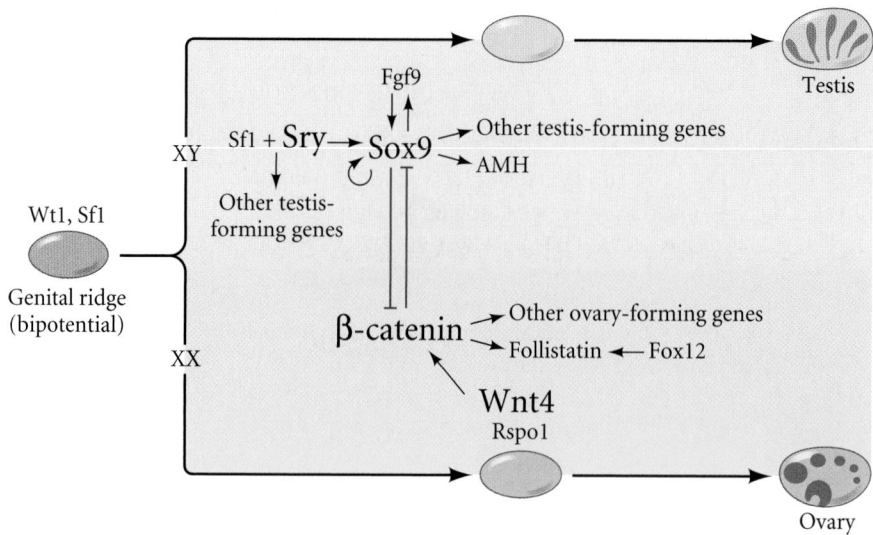

FIGURE 15.4 Possible mechanism for the initiation of primary sex determination in mammals. While we do not know the specific interactions involved, this model attempts to organize the data into a coherent sequence. If Sry is *not present* (pink region), the interactions between paracrine and transcription factors in the developing genital ridge activate *Wnt4* and *Rspo1*. Wnt4 activates the canonical Wnt pathway, which is made more efficient by Rspo1. The Wnt pathway causes the accumulation of β-catenin, and large accumulation of β-catenin stimulates further Wnt4 activity. This continual production of β-catenin both induces the transcription of ovary-producing genes and blocks the testis-determining pathway by interfering with *Sox9* activity. If Sry is *present* (blue region), it may block β-catenin signaling (thus halting ovary generation) and, along with Sf1, activate the *Sox9* gene. Sox9 activates Fgf9 synthesis, which stimulates testis development and promotes further Sox9 synthesis. Sox9 also prevents β-catenin's activation of ovary-producing genes. Sry may also activate other genes (such as *TCF21* and *NT3*) that help generate Sertoli cells. In summary, then, a Wnt4/β-catenin loop specifies the ovaries, whereas a Sox9/Fgf9 loop specifies the testes. One of the targets of the Wnt pathway is the *follistatin* gene, whose product organizes the granulosa cells of the ovary. Transcription factor Foxl2, which is activated (in a still unknown way) in the ovary, is also involved in inducing follistatin synthesis. The XY pathway appears to have an earlier initiation; if it does not function, the XX pathway takes over. (After Sekido and Lovell-Badge 2009; McClelland et al. 2012.)

activate Rspo1 and Wnt4, creating a positive feedback loop between these two proteins. A second role of β-catenin is to initiate the ovarian pathway of development; a third role is to prevent the accumulation of Sox9, a protein crucial for testis determination (Maatouk et al. 2008; Bernard et al. 2012).

- If a Y chromosome is present, the same set of factors in the genital ridge activates the *Sry* gene on the Y chromosome. Sry protein binds to the enhancer of the *Sox9* gene and elevates expression of this key gene in the testis-determining pathway (Bradford et al. 2009b; Sekido and Lovell-Badge 2009). Sox9 and Sry may also act (either directly or indirectly) to block the ovary-forming pathway, possibly by blocking β-catenin (Bernard et al. 2008; Lau and Li 2009).

One possible model of how sex determination can be initiated is shown in **FIGURE 15.4**. Here we see an important rule of animal development: a pathway for cell specification often has two components. One branch of the pathway says "Make A," while the other branch says "and don't make B." In the case of the gonads, the male pathway says "Make testes and don't make ovaries," while the female pathway says "Make ovaries and don't make testes."

---

*There is a reciprocal relationship between the germ cells and the gonadal somatic cells. The germ cells are originally bipotential and can become either sperm or eggs. Once in the male or female sex cords, however, they are instructed to either begin (and remain in) meiosis and become eggs, or to remain meiotically dormant and become spermatogonia (McLaren 1995; Brennan and Capel 2004). In XX gonads, germ cells are essential for the maintenance of ovarian follicles. Without germ cells, the follicles degenerate into cordlike structures and express male-specific markers. In XY gonads, the germ cells help support the differentiation of Sertoli cells, although testis cords will form without the germ cells, albeit a bit later.

The ovary pathway: Wnt4 and R-spondin1

In mice, the paracrine factor Wnt4 is expressed in the genital ridges of both sexes, but its expression later becomes undetectable in XY gonads (which become testes), whereas it is maintained in XX gonads as they begin to form ovaries. In transgenic XX mice that lack the *Wnt4* gene, the ovary fails to form properly, and the cells transiently express testis-specific markers, including Sox9, testosterone-producing enzymes, and AMH (Vainio et al. 1999; Heikkilä et al. 2005). Thus, Wnt4 appears to be an important factor in ovary formation, although it is not the only determining factor.

R-spondin1 (Rspo1) is also critical in ovary formation, since in human case studies several XX individuals with *RSPO1* gene mutations became males (Parma et al. 2006). Rspo1 acts in synergy with Wnt4 to produce β-catenin, which appears to be critical both in activating further ovarian development and in blocking the synthesis of Sox9. Sox9 (as we will see below) is necessary in testis determination (Maatouk et al. 2008; Jameson et al. 2012). In XY individuals with a duplication of the region on chromosome 1 that contains both the *WNT4* and *RSPO1* genes, the pathways that make β-catenin override the male pathway, resulting in a male-to-female sex reversal. Similarly, if XY mice are induced to overexpress β-catenin in their gonadal rudiments, they will form ovaries rather than testes. Indeed, β-catenin appears to be a key "pro-ovarian/anti-testis" signaling molecule in all vertebrate groups, as it is seen in the female gonads (but not in the male gonads) of birds, mammals, and turtles. These groups have very different modes of sex determination, yet Rspo1 and β-catenin are made in the ovaries of each of them (**FIGURE 15.5**; Maatouk et al. 2008; Cool and Capel 2009; Smith et al. 2009).

Certain transcription factors that appear to be activated by β-catenin are found exclusively in the ovaries. One possible target for β-catenin is the gene encoding TAFII 105 (Freiman et al. 2002). This transcription factor subunit (which helps bind RNA polymerase to promoters) is seen only in ovarian follicle cells. Female mice lacking this subunit have small ovaries with few, if any, mature follicles. The transcription factor **Foxl2** is another protein that is strongly upregulated in ovaries, and XX mice homozygous for mutant *Foxl2* alleles develop male-like gonad structure and upregulate Sox9 expression and testosterone production. Both the Foxl2 and the Wnt4 pathway are critical for activation of the *Follistatin* gene (Ottolenghi et al. 2005; Kashimada et al. 2011; Pisarka et al. 2011). Follistatin, an inhibitor of TGF-β family paracrine factors, is thought to be the protein responsible for organizing the epithelium into the granulosa cells of the ovary (Yao et al. 2004). XX mice lacking Follistatin in the developing gonad undergo a partial sex reversal, forming testicle-like structures. We are just beginning to figure out how the components of the ovary-forming pathway are integrated together.

The testis pathway

SRY: THE Y CHROMOSOME SEX DETERMINANT In humans, the major gene for testis determination resides on the short

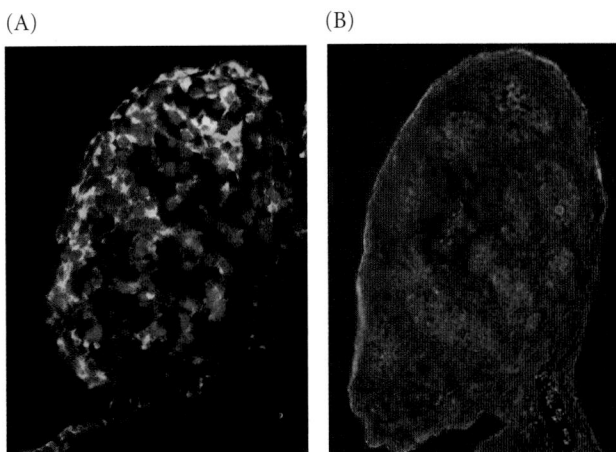

(A) (B)

FIGURE 15.5 Localization of Rspo1 protein in embryonic day 14.5 mouse gonads. Immunofluorescent probes were used to identify Rspo1 (green) and the meiotic germ cell marker, Scp3 (red). (A) Rspo1 was found on somatic cells and at the germ cell surface of the ovaries. (B) These antibodies revealed neither Rspo1 nor Scp3 in the developing testes. (The germ cells in the male gonads have not entered meiosis at this point in development, whereas the ovarian germ cells have.) (From Smith et al. 2008; photograph courtesy of C. Smith.)

arm of the Y chromosome. Individuals born with the short but not the long arm of the Y chromosome are male, whereas individuals born with the long but not the short arm of the Y chromosome are female. By analyzing the DNA of rare XX men and XY women, the position of the testis-determining gene was narrowed down to a 35,000-base-pair region of the Y chromosome found near the tip of the short arm. In this region, Sinclair and colleagues (1990) found a male-specific DNA sequence that encodes a peptide of 223 amino acids. This gene is called **Sry** (**sex-determining region of the Y chromosome**), and there is extensive evidence that it is indeed the gene that encodes the human testis-determining factor.

SRY is found in normal XY males and also in the rare XX males; it is absent from normal XX females and from many XY females. Approximately 15% of XY females have the *SRY* gene, but their copies of the gene contain point or frameshift mutations that prevent Sry protein from binding to DNA (Pontiggia et al. 1994; Werner et al. 1995). If the *SRY* gene actually does encode the major testis-determining factor, one would expect it to act in the genital ridge immediately before or during testis differentiation. This prediction has been found to be the case in studies of the homologous gene in mice. The mouse *Sry* gene also correlates with the presence of testes; it is present in XX males and absent in XY females (Gubbay et al. 1990). *Sry* is expressed in the somatic cells of the bipotential gonads of XY mice immediately before the differentiation of these cells into Sertoli cells; its expression then disappears (Koopman et al. 1990; Hacker et al. 1995; Sekido et al. 2004).

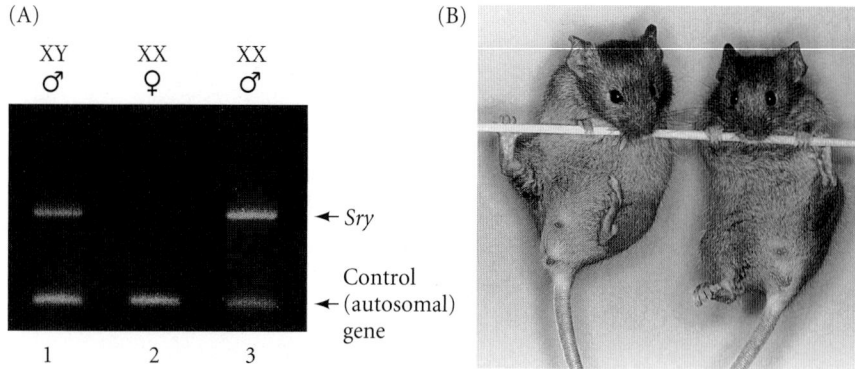

(A)

XY ♂   XX ♀   XX ♂

← Sry

Control (autosomal) gene ←

1   2   3

(B)

FIGURE 15.6 An XX mouse transgenic for *Sry* is male. (A) Polymerase chain reaction followed by electrophoresis shows the presence of the *Sry* gene in normal XY males and in a transgenic XX/*Sry* mouse. The gene is absent in a female XX littermate. (B) The external genitalia of the transgenic mouse are male (right) and are essentially the same as those in an XY male (left). (From Koopman et al. 1991; photographs courtesy of the authors.)

The most impressive evidence for *Sry* being the gene for testis-determining factor comes from transgenic mice. If *Sry* induces testis formation, then inserting *Sry* DNA into the genome of a normal XX mouse zygote should cause that XX mouse to form testes. Koopman and colleagues (1991) took the 14-kilobase region of DNA that includes the *Sry* gene (and presumably its regulatory elements) and microinjected this sequence into the pronuclei of newly fertilized mouse zygotes. In several instances, XX embryos injected with this sequence developed testes, male accessory organs, and a penis* (**FIGURE 15.6**). Therefore, we conclude that *Sry/SRY* is the major gene on the Y chromosome for testis determination in mammals.

SOX9: AN AUTOSOMAL TESTIS-DETERMINING GENE

For all its importance in sex determination, the *Sry* gene is probably active for only a few hours during gonadal development in mice. During this time, it synthesizes the Sry transcription factor, whose primary role appears to be the activation of the *Sox9* gene (Sekido and Lovell-Badge 2008; for other targets of Sry, see Website 15.2). *Sox9* is an autosomal gene involved in several developmental processes, most notably bone formation. In the genital ridge, however, *Sox9* induces testis formation. XX humans who have an extra copy of *SOX9* develop as males even if they have no *SRY* gene, and XX mice transgenic for *Sox9* develop testes (**FIGURE 15.7A–D**; Huang et al. 1999; Qin and Bishop 2005). Knocking out the *Sox9* gene in the gonads of XY mice causes complete sex reversal (Barrionuevo et al. 2006). Thus, even if *Sry* is present, mouse gonads cannot form testes if *Sox9* is absent, so it appears that *Sox9* can replace *Sry* in testis formation. This is not altogether surprising: although the *Sry* gene is found specifically in mammals, *Sox9* is found throughout the vertebrate phyla.

Indeed, *Sox9* appears to be the central male-determining gene in vertebrates. In mammals, it is activated by Sry protein; in birds, frogs, and fish, it appears to be activated by

the dosage of Dmrt1; and in those vertebrates with temperature-dependent sex determination, it is often activated (directly or indirectly) by the male-producing temperature. *Sox9* may thus be the older and more central sex determination gene (Pask and Graves 1999). Expression of the *Sox9* gene is specifically upregulated by the combined expression of Sry and Sf1 proteins in Sertoli cell precursors (**FIGURE 15.7E–H**; Sekido et al. 2004; Sekido and Lovell-Badge 2008). Thus, Sry may act merely as a "switch" operating during a very short time to activate *Sox9*, and the Sox9 protein may initiate the conserved evolutionary pathway to testis formation. So, borrowing Eric Idle's phrase, Sekido and Lovell-Badge (2009) propose that Sry initiates testis formation by "a wink and a nudge."

The Sox9 protein has several functions. First, it appears to be able to activate its own promoter, thereby allowing it to be transcribed for long periods of time. Second, it blocks the ability of β-catenin to induce ovary formation, either directly or indirectly (Wilhelm et al. 2009). Third, it binds to *cis*-regulatory regions of numerous genes necessary for testis production (Bradford et al. 2009). Fourth, Sox9 binds to the promoter site on the gene for anti-Müllerian hormone, providing a critical link in the pathway toward a male phenotype (Arango et al. 1999; de Santa Barbara et al. 2000). Fifth, Sox9 promotes the expression of the gene encoding Fgf9, a paracrine factor critical for testis development. Fgf9 is also essential for maintaining *Sox9* gene transcription, thereby establishing a positive feedback loop driving the male pathway (Kim et al. 2007).

● See **WEBSITE 15.2** Finding the male-determining genes

FIBROBLAST GROWTH FACTOR 9

When the gene for fibroblast growth factor 9 (*Fgf9*) is knocked out in mice, the homozygous mutants are almost all female. Fgf9, whose expression is dependent on Sox9 (Capel et al. 1999; Colvin et al. 2001), plays several roles in testis formation:

1. Fgf9 causes proliferation of the Sertoli cell precursors and stimulates their differentiation (Schmahl et al. 2004; Willerton et al. 2004).

---

*These embryos did not form functional sperm—but they were not expected to. The presence of two X chromosomes prevents sperm formation in XXY mice and men, and the transgenic mice lacked the rest of the Y chromosome, which contains genes needed for spermatogenesis.

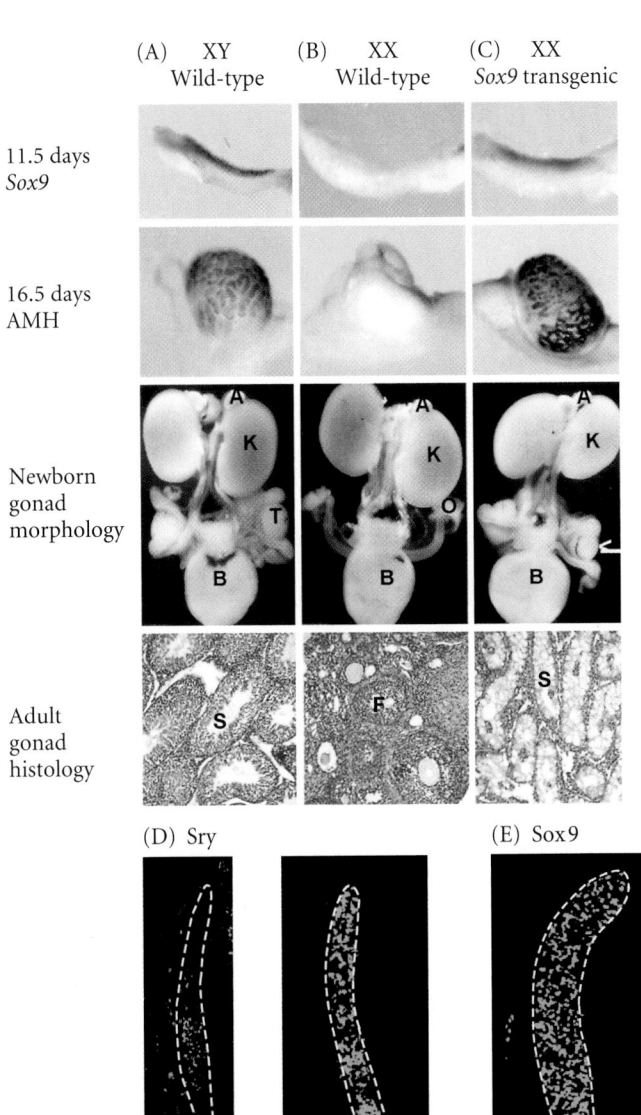

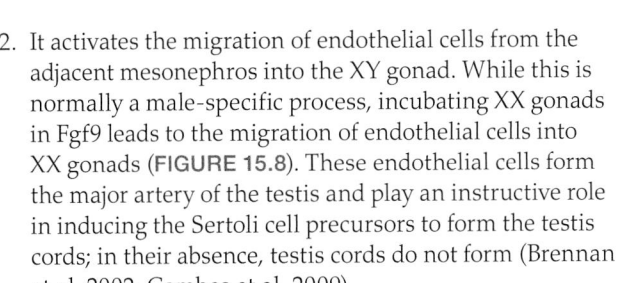

(A) XY Wild-type	(B) XX Wild-type	(C) XX *Sox9* transgenic	

11.5 days *Sox9*

16.5 days AMH

Newborn gonad morphology

Adult gonad histology

(D) Sry

(E) Sox9

11.0 days | 11.5 days | 12.0 days | 13.5 days

FIGURE 15.7 Ability of *Sox9* to generate testes. (A) A wild-type XY mouse embryo expresses *Sox9* in the genital ridge at 11.5 days postconception, anti-Müllerian hormone in the embryonic gonad Sertoli cells at 16.5 days, and eventually forms descended testes with seminiferous tubules. K, kidneys; A, adrenal glands; B, bladder; T, testis; O, ovary; S, seminiferous tubule; F, follicle cell. (B) The wild-type XX embryo shows neither *Sox9* expression nor AMH. It constructs ovaries with mature follicle cells. (C) An XX embryo with the *Sox9* transgene inserted expresses *Sox9* and has AMH in 16.5-day Sertoli cells. It has descended testes, but the seminiferous tubules lack sperm (due to the presence of two X chromosomes in the Sertoli cells). (D,E) Chronological sequence from the expression of *Sry* in the genital ridge to that of *Sox9* in the Sertoli cells. (D) *Sry* expression. At day 11.0, Sry protein (green) is seen in the center of the genital ridge. At day 11.5, the domain of *Sry* expression increases and *Sox9* expression is activated. (E) By day 12.0, Sox9 protein (green) is seen in the same cells that earlier expressed *Sry*. By day 13.5, Sox9 is seen in those cells of the testis tubule that will become Sertoli cells. (A–C from Vidal et al. 2001, photographs courtesy of A. Schedl; D,E from Kashimada and Koopman, 2010, courtesy of P. Koopman.)

2. It activates the migration of endothelial cells from the adjacent mesonephros into the XY gonad. While this is normally a male-specific process, incubating XX gonads in Fgf9 leads to the migration of endothelial cells into XX gonads (**FIGURE 15.8**). These endothelial cells form the major artery of the testis and play an instructive role in inducing the Sertoli cell precursors to form the testis cords; in their absence, testis cords do not form (Brennan et al. 2002; Combes et al. 2009).

3. It is required for maintaining *Sox9* expression in the presumptive Sertoli cells and directs their formation into tubules. Moreover, since it can act as both an autocrine and a paracrine factor, Fgf9 may coordinate Sertoli cell development by reinforcing *Sox9* expression in all the cells of the tissue (Hiramatsu et al. 2009). Such a

"community effect" may be important in achieving the integrated assembly of testis tubules (Palmer and Burgoyne 1991; Cool and Capel 2009).

4. It represses Wnt4 signaling, which would otherwise direct ovarian development (Maatouk et al. 2008; Jameson et al. 2012).

5. Finally, Fgf9 appears to help coordinate the sex determination of the gonad with that of the germ cells. As we will see in Chapter 17, those mammalian germ cells destined to become eggs enter meiosis quickly upon entering the gonad, whereas germ cells destined to become sperm delay their entry into meiosis until puberty. Fgf9 is one of the factors that blocks the immediate entry of germ cells into meiosis, thereby placing them onto the sperm-forming pathway (Barrios et al. 2010; Bowles et al. 2010).

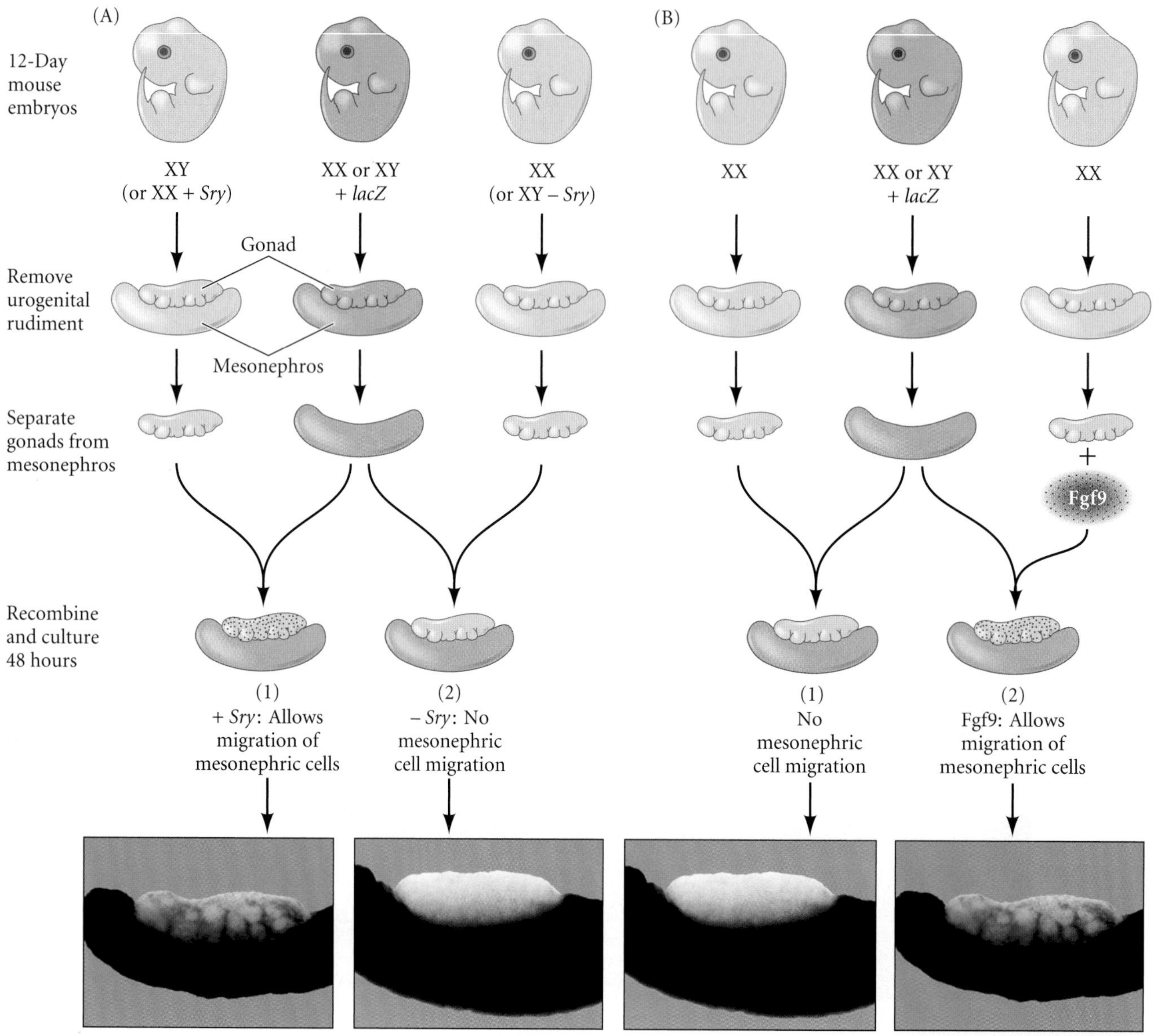

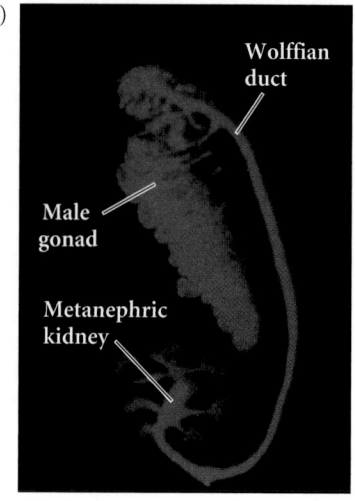

FIGURE 15.8 Migration of mesonephric endothelial cells into *Sry*+ gonadal rudiments. In the experiment diagrammed, urogenital ridges (containing both the mesonephric kidneys and bipotential gonadal rudiments) were collected from 12-day embryonic mice. Some of the mice were marked with a β-galactosidase transgene (*lacZ*) that is active in every cell. Thus, every cell of these mice turned blue when stained for β-galactosidase. The gonad and mesonephros were separated and recombined, using gonadal tissue from unlabeled mice and mesonephros from labeled mice. (A) Migration of mesonephric cells into the gonad was seen (1) when the gonadal cells were XY or when they were XX with a *Sry* transgene. No migration of mesonephric tissue into the gonad was seen (2) when the gonad contained either XX cells or XY cells in which the Y chromosome had a deletion in the *Sry* gene. The sex chromosomes of the mesonephros did not affect the migration. (B) Gonadal rudiments for XX mice could induce mesonephric cell migration if these rudiments had been incubated with Fgf9. (C) Intimate relation between the mesonephric duct and the developing gonad in a 16-day male mouse embryo. The duct tissue will form the efferent ducts of the testes and has been stained for cytokeratin-8. (A,B after Capel et al. 1999, photographs courtesy of B. Capel; C from Sariola and Saarma 1999, courtesy of H. Sariola.)

**SF1: A CRITICAL LINK BETWEEN SRY AND THE MALE DEVEL-
OPMENTAL PATHWAYS** The transcription factor **steroido-
genic factor 1** (**Sf1**) is necessary to make the bipotential
gonad. But whereas Sf1 levels decline in the genital ridge of
XX mouse embryos, they remain high in the developing testis. It is thought that Sry (either directly or indirectly) maintains *Sf1* gene expression. Sf1 protein appears to be active in
masculinizing both the Leydig and the Sertoli cells. In the
Sertoli cells, Sf1 works in collaboration with Sry to activate
Sox9 (Sekido and Lovell-Badge 2008) and then, working
with Sox9, elevates levels of anti-Müllerian hormone transcription (Shen et al. 1994; Arango et al. 1999). In the Leydig
cells, Sf1 activates genes encoding the enzymes that make
testosterone.

The right time and the right place

Having the right genes doesn't necessarily mean you'll get
the organ you expect. Studies of mice have shown that the
Sry gene of some strains of mice failed to produce testes
when bred onto a different genetic background (Eicher and
Washburn 1983; Washburn and Eicher 1989; Eicher et al.
1996). This failure can be attributed either to a delay in *Sry*
expression or to the failure of the protein to accumulate to
the critical threshold level required to trigger *Sox9* expression
and launch the male pathway. By the time *Sox9* gets turned

on, it is too late—the gonad is already following the path to
become an ovary (Bullejos and Koopman 2005; Wilhelm et
al. 2009).

The importance of timing was confirmed when Hiramatsu
and collaborators (2009) were able to place the mouse *Sry*
gene onto the regulatory sequences of a heat-sensitive gene,
allowing them to activate *Sry* at any time in mouse development by merely raising the embryo's temperature. When they
delayed *Sry* activation by as little as 6 hours, testis formation
failed and ovaries started to develop (**FIGURE 15.9**). Thus,
there appears to be a brief window during which the testis-
forming genes can function. If this window of opportunity is
missed, the ovary-forming pathway is activated.

Hermaphrodites are individuals in which both ovarian and testicular tissues exist; they have either ovotestes
(gonads containing both ovarian and testicular tissue) or an
ovary on one side and a testis on the other.* As seen in Figure 15.9, ovotestes can be generated when the *Sry* gene is

---

*This anatomical phenotype is named for Hermaphroditos, a
young man in Greek mythology whose beauty inflamed the
ardor of the water nymph Salmacis. She wished to be united
with him forever, and the gods, in their literal fashion, granted
her wish. Hermaphroditism is often considered to be one of the
"intersex" conditions discussed on p. 531.

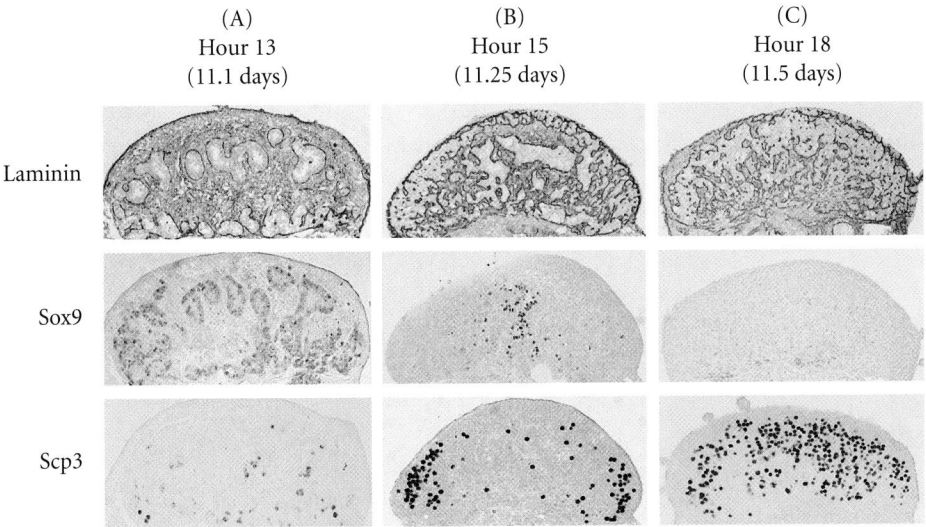

| | (A)
Hour 13
(11.1 days) | (B)
Hour 15
(11.25 days) | (C)
Hour 18
(11.5 days) |

Laminin

Sox9

Scp3

FIGURE 15.9 Experimental delay of *Sry* gene activation by 6 hours leads to failure of testis development and the initiation of ovary development. Genital ridges were removed from XX mice carrying
a heat-inducible *Sry* gene. These tissues were then heat-shocked at different times to activate *Sry*
and then allowed to mature. (A) Those genital tissues experiencing *Sry* induction at 11.1 days of
development (when Sry is normally activated) produced testes. Their laminin distribution showed
Sertoli cells, *Sox9* (a marker of testis development) was active, and *Scp3*, a marker of ovary development, was absent. (B) Three hours later, the activation of *Sry* caused a central testicular area to
form, with ovary-like structures forming peripherally. *Sox9* was present in the central testicular region,
while *Scp3* was found in the periphery. (C) If *Sry* was activated in the genital tissues 6 hours later, the
structures formed ovarian tissue, *Sox9* was absent, and *Scp3* was seen throughout the tissue. (After
Hiramatsu et al. 2009.)

activated later than normal. Hermaphrodites can also result in those very rare instances when a Y chromosome is translocated to an X chromosome. In those tissues where the translocated Y is on the active X chromosome, the Y chromosome will be active and the *SRY* gene will be transcribed; in cells where the Y chromosome is on the inactive X chromosome, the Y chromosome will also be inactive (Berkovitz et al. 1992; Margarit et al. 2000). A gonadal mosaic for expression of *SRY* can develop into a testis, an ovary, or an ovotestis, depending on the percentage of cells expressing *SRY* in the Sertoli cell precursors (see Brennan and Capel 2004; Kashinada and Koopman 2010).

Secondary Sex Determination in Mammals: Hormonal Regulation of the Sexual Phenotype

Primary sex determination is the formation of either an ovary or a testis from the bipotential gonad. This process, however, does not give the complete sexual phenotype. In mammals, secondary sex determination is the development of the female and male phenotypes in response to hormones secreted by the ovaries and testes. Both female and male secondary sex determination have two major temporal phases. The first phase occurs within the embryo during organogenesis; the second occurs at puberty.

During embryonic development, hormones and paracrine signals coordinate the development of the gonads with the development of secondary sexual organs. In females, the Müllerian ducts persist and, through the actions of estrogen, differentiate to become the uterus, cervix, oviducts, and upper vagina (see Figure 15.2). The **genital tubercle** becomes differentiated into the clitoris, and the **labioscrotal folds** become the labia majora. The Wolffian ducts require testosterone to persist, and thus they atrophy in females. In females, the portion of the **urogenital sinus** that does not become the bladder and urethra becomes Skene's glands, paired organs that makes secretions similar to those of the prostate.

The coordination of the male phenotype involves the secretion of two testicular factors. The first of these is anti-Müllerian hormone, a BMP-like paracrine factor made by the Sertoli cells. AMH causes the degeneration of the Müllerian duct. The second is the steroid hormone testosterone, an **androgen** (masculinizing substance) secreted from the fetal Leydig cells. Testosterone causes the Wolffian ducts to differentiate into sperm-carrying tubes (the epididymis and vas deferens) as well as the seminal vesicle (which emerges as an outpocketing of the vas deferens), and it causes the genital tubercle to develop into the penis and the labioscrotal folds to develop into the scrotum. In males, the urogenital sinus, in addition to forming the bladder and urethra, also forms the prostate gland.

The mechanism by which testosterone (and, as we shall see, its more powerful derivative dihydrotestosterone) masculinizes the external genitalia is thought to involve its interaction with the Wnt pathway (**FIGURE 15.10**). The Wnt pathway, which in the genital *ridge* activates the female trajectory, acts in the genital *tubercle* to activate male development (Mazahery et al. 2013). The Wnt antagonist Dickkopf is made in the urogenital swellings and can be downregulated by testosterone and upregulated by anti-androgens. This finding led to a model wherein the urogenital swellings of XX individuals make Dickkopf, thus preventing the activity of Wnt in the mesenchyme, blocking further growth and leading to the feminization of the genital tubercle by estrogens (Holdereger and Keefer 1986; Miyagawa et al. 2009). In females, then, the genital tubercle becomes the clitoris and the labioscrotal

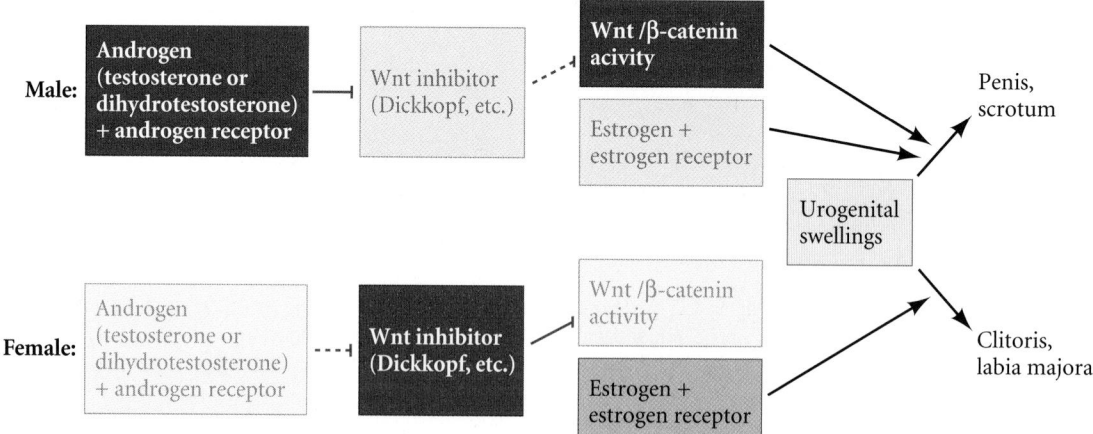

FIGURE 15.10 Model for the formation of external genitalia. In this schema, the mesenchyme in the urogenital swellings secretes inhibitors of Wnt signaling. In the absence of Wnt signaling, estrogen modifies the genital tubercle into the clitoris and the labioscrotal folds into the labia majora surrounding the vagina. In males, however, androgens (such as testosterone and dihydrotestosterone) bind to the androgen receptor in the mesenchymal cells and prevent the synthesis of the Wnt inhibitors. Wnt signaling is permitted, and it causes the genital tubercle to become the penis and the labioscrotal folds to become the scrotum. (After Miyagawa et al. 2009.)

folds become the labia majora. In males, however, testosterone and dihydrotestosterone bind to the androgen (testosterone) receptor in the mesenchyme and prevent the expression of Wnt inhibitors (thus permitting Wnt expression in the mesenchyme). With the influence of these Wnts, male urogenital swellings are converted into the penis and the scrotum.

The genetic analysis of secondary sex determination

The existence of separate and independent AMH and testosterone pathways of masculinization is demonstrated by people with **androgen insensitivity syndrome**. These XY individuals have the *SRY* gene and thus have testes that make testosterone and AMH. However, they have a mutation in the gene encoding the androgen *receptor* protein that binds testosterone and brings it into the nucleus. Therefore, these individuals cannot respond to the testosterone made by their testes (Meyer et al. 1975; Jääskeluäinen 2012). They can, however, respond to the estrogen made by their adrenal glands (which is normal for both XX and XY individuals), so they develop female external sex characteristics (FIGURE 15.11). Despite their distinctly female appearance, these individuals are chromosomally male. They have testes, and even though they cannot respond to testosterone, they produce and respond to AMH. Thus, their Müllerian ducts degenerate. Persons with androgen insensitivity syndrome develop as normal-appearing but sterile women, lacking a uterus and oviducts and having internal testes in the abdomen.

Such phenotypes, in which male and female traits are seen in the same individual, are called **intersex** conditions.*Although most people have a reasonably good correlation of their genetic and anatomical sexual phenotypes, about 0.4–1.7% of the population differs from the strictly dimorphic condition (Blackless et al. 2000; Hull 2003; Hughes et al. 2006). Androgen insensitivity syndrome is one of several intersex conditions that have traditionally been labeled **pseudohermaphroditism**. In pseudohermaphrodites, there is only one type of gonad, but the secondary sex characteristics differ from what would be expected from the gonadal sex. In humans, male pseudohermaphroditism (wherein the gonadal sex is male and the secondary sex characteristics are female) can be caused by mutations in the androgen (testosterone) receptor or by mutations affecting testosterone synthesis (Geissler et al. 1994).

---

* Chromosomal hermaphroditism (genotype XXY, discussed on p. 520) is also included among the "intersex" phenotypes. The language used to group these conditions is being debated. Some activists, physicians, and parents wish to eliminate the term "intersex" to avoid confusion of these anatomical conditions with identity issues such as homosexuality. They prefer to call these conditions "disorders of sex development." In contrast, other activists do not want to medicalize this condition and find the "disorder" category offensive to individuals who do not feel there is anything wrong with their health. For a more detailed analysis of intersexuality, see Gilbert et al. 2005, Austin et al. 2011, or the websites listed at the end of this chapter.

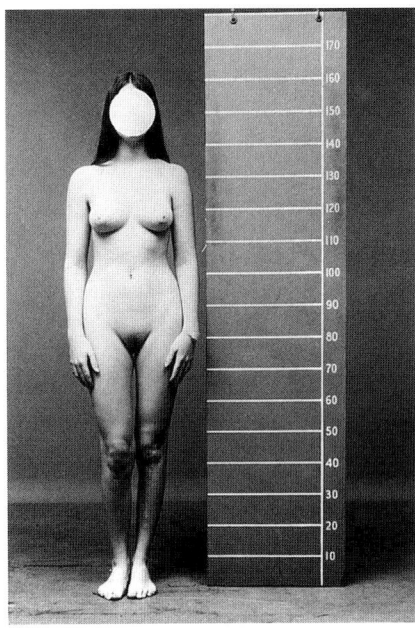

FIGURE 15.11 Androgen insensitivity syndrome. Despite having the XY karyotype, individuals with this syndrome appear female. They cannot respond to testosterone but can respond to estrogen, so they develop female secondary sex characteristics (i.e., labia and a clitoris rather than a scrotum and a penis). Internally, they lack the Müllerian duct derivatives and have undescended testes. (Courtesy of C. B. Hammond.)

Female pseudohermaphroditism, in which the gonadal sex is female but the person is outwardly male, can be the result of overproduction of androgens in the ovary or adrenal gland. The most common cause of this latter condition is **congenital adrenal hyperplasia**, in which there is a genetic deficiency of an enzyme that metabolizes cortisol steroids in the adrenal gland. In the absence of this enzyme, testosterone-like steroids accumulate and can bind to the androgen receptor, thus masculinizing the fetus (Migeon and Wisniewski 2001; Merke et al. 2002).

TESTOSTERONE AND DIHYDROTESTOSTERONE Although testosterone is one of the two primary masculinizing factors, there is evidence that it might not be the active masculinizing hormone in certain tissues. Although testosterone is responsible for promoting the formation of the male structures that develop from the Wolffian duct primordium, testosterone does not directly masculinize the urethra, prostate, penis, or scrotum. These latter functions are controlled by **5α-dihydrotestosterone**, or **DHT** (FIGURE 15.12). Siiteri and Wilson (1974) showed that testosterone is converted to DHT in the urogenital sinus and swellings but not in the Wolffian duct. DHT appears to be a more potent hormone than testosterone and is most active in prenatal and early childhood.

The importance of DHT in the early development of the male gonads was demonstrated by Imperato-McGinley

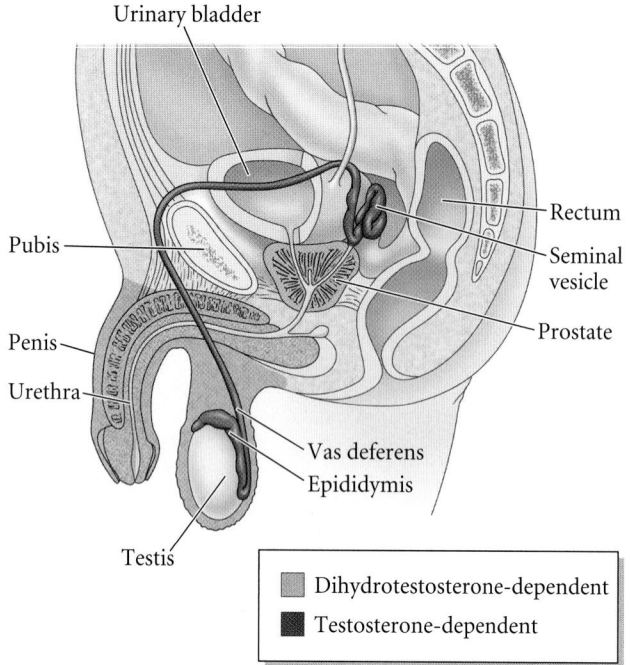

Urinary bladder

Pubis

Penis

Urethra

Rectum

Seminal vesicle

Prostate

Vas deferens

Epididymis

Testis

- Dihydrotestosterone-dependent
- Testosterone-dependent

FIGURE 15.12 Testosterone- and dihydrotestosterone-dependent regions of the human male genital system. (After Imperato-McGinley et al. 1974.)

and her colleagues (1974) when they studied a phenotypically remarkable syndrome in several inhabitants of a small community in the Dominican Republic. Individuals with this syndrome were found to lack a functional gene for the enzyme 5α-ketosteroid reductase 2—the enzyme that converts testosterone to DHT (Andersson et al. 1991; Thigpen et al. 1992). Chromosomally XY children with this syndrome have functional testes, but the testes remain inside the abdomen and do not descend before birth. These children appear to be girls and are raised as such. Their internal anatomy, however, is male: they have Wolffian duct development and Müllerian duct degeneration, along with their functional testes. At puberty, when the testes produce high levels of testosterone (which appears to compensate for the lack of DHT), their external genitalia are able to respond to the hormone and differentiate. The penis enlarges, the scrotum descends, and the person originally believed to be a girl is revealed to be a young man. Studies of this condition led to the current perception that the formation of the external genitalia is under the control of dihydrotestosterone, whereas Wolffian duct differentiation is controlled by testosterone itself.

● **See WEBSITE 15.3** Dihydrotestosterone in adult men

● **See WEBSITE 15.4** Insulin-like hormone 3

ANTI-MÜLLERIAN HORMONE Anti-Müllerian hormone, a member of the TGF-β family of growth and differentiation

factors, is secreted from the fetal Sertoli cells and causes the degeneration of the Müllerian duct (Tran et al. 1977; Cate et al. 1986). AMH is thought to bind to the mesenchyme cells surrounding the Müllerian duct, causing these cells to secrete factors that induce apoptosis in the duct's epithelium and breaks down the basal lamina surrounding the duct (Trelstad et al. 1982; Roberts et al. 1999, 2002).

ESTROGEN The steroid hormone estrogen is needed for complete postnatal development of both the Müllerian and the Wolffian ducts, and is necessary for fertility in both males and females. In females, estrogen induces the differentiation of the Müllerian duct into the uterus, oviducts, cervix, and upper vagina. In female mice whose genes for estrogen receptors are knocked out, the germ cells die in the adult, and the granulosa cells that had enveloped them start developing into Sertoli-like cells (Couse et al. 1999). Male mice with knockouts of estrogen receptor genes produce few sperm. One of the functions of the male efferent duct (vas efferens) cells is to absorb most of the water from the lumen of the rete testis. This absorption of water, which is regulated by estrogen, concentrates the sperm, giving them a longer life span and providing more sperm per ejaculate. If estrogen or its receptor is absent in male mice, water is not absorbed and the mouse is sterile (Hess et al. 1997). Although blood concentrations of estrogen are in general higher in females than in males, the concentration of estrogen in the rete testis is even higher than in female blood.

In summary, primary sex determination in mammals is regulated by the chromosomes, which results in the production of testes in XY individuals and ovaries in XX individuals. The gonads then produce the hormones that coordinate the different parts of the body to have a male or female phenotype. This secondary sex determination is thus usually coordinated with the primary sex determination.

Chromosomal Sex Determination in *Drosophila*

Although both mammals and fruit flies produce XX females and XY males, their chromosomes achieve these ends using very different means. In mammals, the Y chromosome plays a pivotal role in determining the male sex.

In *Drosophila*, the Y chromosome is not involved in determining sex. A fruit fly's sex is determined predominantly by the number of X chromosomes in each cell. If there is only one X chromosome in a diploid cell, the fly is male. If there are two X chromosomes in a diploid cell, the fly is female. Should a fly have two X chromosomes and three sets of autosomes, it is a mosaic, where some of the cells are male and some of the cells are female. In flies, the Y chromosome seems to be a collection of genes that are active in forming sperm in adults, but not in sex determination. Thus, while XO mammals are sterile females (no Y chromosome, thus no *Sry* gene), XO *Drosophila* are sterile males (one X chromosome per diploid set).

SIDELIGHTS  SPECULATIONS

Brain Sex: Secondary Sex Determination through Multiple Pathways

We have known for a long time that the brain, like other tissues, is responsive to the steroid hormones produced by the gonads. Now, recent evidence suggests that sex differences in the brain can be observed even before the gonads mature, and that the brain may experience direct regulation by the X and Y chromosomes (Arnold and Burgoyne 2004). Moreover, environmental cues can also change sexual behaviors through DNA methylation. Thus, genetic, hormonal, and environmental agents all play roles in generating sexual behaviors (McCarthy and Arnold 2011; Ngun et al. 2011).

Several areas of the mammalian brain are known to be involved in sex-specific behaviors, and these brain regions are thought to be regulated by gonadal steroid hormones. The cyclic secretion of luteinizing hormone (involved in ovulation) by the pituitary in adult female rats is dependent on a *lack* of testosterone during the first week of the animal's life. The luteinizing hormone secretion of female rats can be made noncyclic by giving them testosterone 4 days after birth. Conversely, the luteinizing hormone secretion of males can be

made cyclical by removing their testes within a day of birth (Barraclough and Gorski 1962).

In these regions of the brain, sex hormones probably act during the fetal or neonatal stage of a mammal's life to organize the nervous system in a sex-specific manner. Then, during adult life (especially puberty), the same hormones may have transitory motivational (or "activational") effects. This model of the hormonal basis of sex-specific brain development and behavior is called the **organization/activation hypothesis**. Ironically, the hormone chiefly responsible for determining the male neural pattern is **estradiol**, a form of estrogen.* Testosterone from fetal or neonatal blood can be converted into estradiol by the enzyme **aromatase** (**FIGURE 15.13A**). This conversion occurs in the hypothalamus and limbic system—two areas of the brain known to regulate hormone secretion and reproductive behavior (Reddy et al. 1974; McEwen et al. 1977). When the estrogen receptors have been knocked out in mice, male sexual behavior has been lost completely (Ogawa et al. 2000; Kudwa et al. 2005). Thus, in rodents, testosterone exerts its effects on the nervous

system by being converted into estradiol in the brain.

But the fetal environment is rich in estrogens from the gonads and placenta. What stops these estrogens from masculinizing the nervous system of a female fetus? In both male and female rats, fetal estrogen is bound by α-fetoprotein, which binds and inactivates estrogen, but not testosterone. Relationships among estradiol, aromatase, and α-fetoprotein have been analyzed by observing sexual behaviors in mice that have loss-of-function mutations for aromatase and α-fetoprotein. The brain and the behaviors of mice lacking α-fetoprotein have been defeminized, showing that α-fetoprotein prevents the female brain from receiving circulating estrogens.

Indeed, female mice whose α-fetoprotein genes have been knocked out are sterile because the brain genes controlling ovulation (such as those for gonadotropin-releasing hormone) are downregulated. However, this lack of ovulation can be reversed (and the

---

*The terms *estrogen* and *estradiol* are often used interchangeably. However, estrogen refers to a class of steroid hormones responsible (among other functions) for establishing and maintaining specific female characteristics. Estradiol is one of these hormones, and in most mammals (including humans) it is the most potent of the estrogens. The enzyme's name, *aromatase*, has nothing to do with aroma (although aromas are certainly crucial to rodent sex), but refers to the destabilization of hydrogen bonds in the steroid ring structure.

(A) Aromatase converts androgens to estrogens

Aromatase

Testosterone

Estradiol

(B)

FIGURE 15.13 Organization of brain development by hormones. (A) The enzyme aromatase converts androgens (such as testosterone) into estrogens (such as estradiol). The name of the enzyme comes from its ability to aromatize the six-carbon ring by reducing the ring-stabilizing keto group (=O) to a hydroxyl group (—OH). This biochemical change allows the hormones to bind to different receptors (dormant transcription factors) and activate different genes. (B) Female lordosis behavior (in which the female mouse bends her spine so the male can readily mate with her) remains present in mice administered the aromatase inhibitor ATD; the behavior is abolished when the genes for α-fetoprotein are knocked out (AFP-KO). However, when aromatase is blocked in female mice without α-fetoprotein (AFP-KO-ATD), lordosis behavior is restored. (After Bakker and Baum 2008.)

normal female pattern of gene expression established) if such mice are also given drugs that block aromatase. Similarly, the amount of lordosis (a swayback posture taken by female rodents that permits males to mate with them) is almost completely abrogated in female mice lacking functional α-fetoprotein genes. This behavior, too, can be restored by treating the mice prenatally with aromatase inhibitors (FIGURE 15.13B; Bakker et al. 2006; De Mees et al. 2006; Bakker and Baum 2008).

While the *prenatal* lack of estrogen and testosterone may be critical for the formation of female brains, the feminization of the rodent brain may require estrogens *after* birth. This is suggested by the behavioral phenotypes of mice whose aromatase genes have been knocked out. Their female-specific behaviors (e.g., lordosis; the ability to discriminate male pheromones) are also impaired (Bakker and Baum 2008).

Pheromones and the hormonal pathway

Pheromones—sex-specific chemicals secreted into the atmosphere—play major roles in sexual behaviors in rodents. If the vomeronasal organ (which is responsible for sensing pheromones in rodents and some other mammals, although such an organ is not present in humans) or the genes involved in pheromone recognition are removed from male mice, they fail to discriminate between males and females and attempt to mate with both. If this pheromone recognition system is removed from female mice, they lack certain female behaviors and acquire the full set of male courtship behaviors (including mounting, pelvic thrusting, and solicitation of females).

Thus, it appears that the neural circuitry for both male and female behaviors exists in every mouse brain, but the *interpretation* of pheromone signals is what distinguishes male from female brains. In females, the "feminine" pattern of behavior is activated (sexual receptivity to males, lactating behavior with pups), while the "masculine" pattern (if it's male, fight it; if it's female, mount it) is repressed. In males, the pheromones activate this "masculine"

pattern, while the "feminine" pathway is suppressed (Kimchi et al. 2007). The interpretation of pheromone signals is thought to take place in the medial preoptic area/anterior hypothalamus region of the brain, and we know this region to be sexually dimorphic as a result of prenatal estrogen exposure. Thus, the organizational abilities of testosterone may act largely to effect changes in this small area of the brain, and once this region is organized, it will interpret the pheromone signals to activate either the male or the female sets of neurons (Baum 2009).

Intrinsic sex differences

The first indication that something besides testosterone and estrogen was important in forming sexually different structures in the brain came from studies on Parkinson disease during which embryonic rat brains were dissected before the gonads matured. These studies indicated that brains from XX embryos had more epinephrine-secreting neurons than XY embryonic brains (Beyer et al. 1991). Later studies, using microarrays and PCR, demonstrated that more than 50 genes in the mouse brain are expressed in sexually dimorphic patterns *before gonad differentiation has occurred* (Dewing et al. 2003). Moreover, the mouse *Sry* gene, in addition to being expressed in the embryonic testes, is also expressed in the fetal and adult brain. The human *SRY* gene appears to be expressed in the adult brain as well (Lahr et al. 1995; Mayer et al. 1998, 2000). *SRY* is specifically active in the substantia nigra of the male hypothalamus, where it helps regulate the gene for tyrosine hydroxylase, an enzyme that is critical for the production of the neurotransmitter dopamine (Dewing et al. 2006).

Numerous regions of the brain that are not involved in sexually specific functions have been found to be sexually dimorphic, and in these instances, the sex chromosomes produce effects as important as hormones. Therefore, McCarthy and Arnold (2011) postulated that rather than there being a few hormonally dependent dimorphic structures in an otherwise "monomorphic" brain, the entire brain might be

"male" or "female" as a result of sex chromosome differences.

Stunning demonstrations that sexual dimorphism in the brain can be caused before gonadal hormone synthesis come from natural and experimental conditions in birds. One big difference between male and female finches is that large regions of the male brain are devoted to producing songs. Male finches sing; the females do not. While testosterone is important in the formation of the song centers in finches (and, when added experimentally, can cause female birds to sing), blocking those hormones in males does not prevent normal development of the song centers or singing. Genetically male birds form these brain regions even without male hormones (Mathews and Arnold 1990).

A natural experiment presented itself in the form of a bird that was half male and half female, divided down the middle (FIGURE 15.14). Such animals, where some body parts are male and others female, are called **gynandromorphs** (Greek *gynos*, "female"; *andros*, "male"; *morphos*, "form"). Agate and colleagues (2002) showed that the gynandromorph finch had ZZ (male) sex chromosomes on its right side and ZW (female) sex chromosomes on its left.* Its testes produced testosterone, and the bird sang like a male and copulated with females. However, although many brain structures were similar on both sides, some brain regions differed between the male and female halves. The song circuits on the right side had a more masculine phenotype than similar structures on the left, showing that both intrinsic and hormonal influences were important.

The roles of experience

Usually we think of DNA as controlling neural anatomy, and neural anatomy as controlling behaviors. This is the lesson that genetic mental retardation syndromes have taught us. But new

---

*Unlike mammals, sex in birds is cell-autonomous, with each cell making its own sexual decision. Circulating hormones do not usually integrate the sexual phenotype (Zhao et al. 2010).

Right (♂)   Left (♀)

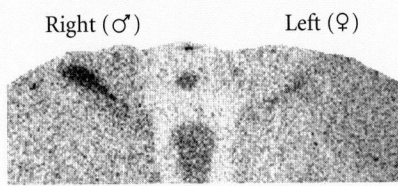

FIGURE 15.14 Gynandromorph finch with ZZ (male) cells on its right side and ZW (female) cells on its left side. Since plumage is controlled by genes on the sex chromosomes, the adult finch has male plumage on its right and female plumage on its left. Micrographs show the difference in brain regions between the right and left sides, indicated by staining of the neurons of the HVC nucleus (a neuron cluster involved in bird song production). (From Agate et al. 2002.)

research is claiming that the pathway is not one-way and that behaviors can control both gene expression and nervous system anatomy. One of the most sexually dimorphic regions of the rat central nervous is the spinal nucleus of the bulbocavernosus (SNB). This controls the pelvic thrusting muscles during mating, and it is larger in the male. The SNB is also testosterone-sensitive, and it shrinks when rats are castrated (unless the rats are given replacement testosterone). Interestingly, the size of SNB neurons changes with sexual behavior, becoming smaller as male rats mate more frequently. "It is possible," noted Breedlove (1997), "that differences in sexual behavior cause, rather than are caused by, differences in brain structure."

The roles of experience in causing changes in brain gene expression and behavior were highlighted in a series of studies involving the effects of maternal care on the behaviors of rats. Maternal care during the first week of life involves grooming and licking the young pups. Those female rat pups that experience such maternal care when young will provide such maternal care to their own offspring, whereas female pups that do not receive such maternal attention will not. The licking and grooming responses are largely regulated through the estrogen-responsive neurons of the medial preoptic area (MPOA), a sexually dimorphic region of the brain. When estrogen binds to its receptors in MPOA neurons, these neurons activate the genes that encode receptors for oxytocin, the hormone involved with nursing and grooming.

So how is this trait inherited? It turns out that the key player is the experience of being licked and groomed. Licking and grooming by the mother alters the DNA methylation pattern of brain-specific enhancers in the major estrogen receptor gene (*ERα*) in the pups (Meaney and Szyf 2005; Champagne et al. 2006). In the MPOA neurons, licking and grooming decreases the amount of DNA methylation. This enables the Stat5 transcription factor to bind and permit the estrogen receptor gene to be transcribed at high levels. This ensures the high levels of estrogen receptors needed to stimulate licking and grooming behaviors (**FIGURE 15.15**). Thus, mothers that lick and groom their offspring tend to have daughters that will lick and groom their offspring. Cross-fostering (giving the newborn pups of "high licking and grooming" mothers to "low licking and grooming mothers" and vice versa) has demonstrated that this neonatal experience does indeed cause the gene expression differences (Cameron et al. 2008).

Interestingly, in another area of the brain, the anterior paraventricular nucleus (PVN), rat pups that experience high levels of licking and grooming have a highly methylated promoter on this gene, thereby downregulating the estrogen receptor in this region.

The PVN helps regulate gonadotropins. Rats experiencing low levels of maternal licking and grooming have high levels of estrogen receptors in the PVN and correspondingly high levels of gonadotropins. As they mature, these rats are predisposed to a suite of sexual behaviors that include precocious puberty, heightened sexual activity, and lack of attention to their pups (Cameron et al. 2008). Thus, experience can create changes in gene expression and neuroanatomy. Moreover, inherited variation can come about by experience-induced changes of DNA methylation. The distinction between nature and nurture disappears in this environmental regulation of gene expression.

The human element

The mammalian examples above are exclusively from rodents. It is a very risky business extrapolating from such rodent studies to humans. Human fetuses, for instance, do not make a strong estrogen-binding protein and have a much higher level of free estrogen than do rodent embryos (see Nagel and vom Saal 2003). So although the organization/activation hypothesis explains many of the hormonal effects on rodent development, one of its fundamental assumptions—that α-fetoprotein strongly binds estrogens during prenatal development—does not in fact hold true for humans.

Human sexual behaviors differ from those of rodents in many ways, and so does brain development (see Jordan-Young 2010). Outside of physiological events such as ovulation, no sex-specific behavior has yet been identified in humans. Moreover, humans do not use pheromones as a primary sexual attractant (sight and touch being far more critical). The evidence that there are differences in brain anatomy between male homosexuals and heterosexuals has been disputed, and even so, brain anatomy can be altered by experience. No "gay gene" has been discovered, and the concordance of gender identity between identical twins is only 30%—far from the 100% expected if sexual orientation were strictly genetic (Bailey et al. 2000; CRC 2006). Moreover, behaviors that are seen as "masculine" in one culture may be considered "feminine" in another, and vice versa (see Jacklin 1981; Bleier 1984; Fausto-Sterling 1992; Kandel et al. 1995). How humans acquire gendered behaviors appears to involve a remarkable complex set of interactions between genes, hormones, nerves, and environment. As will be discussed in Chapter 19, we inherit a genome that can produce a genetically constrained range of different phenotypes. Indeed, behaviors are the "final phenotype." They have resisted explanation because the link between genotype and behavior is relatively weak, and because behavioral phenotypes are so heavily influenced by environment.

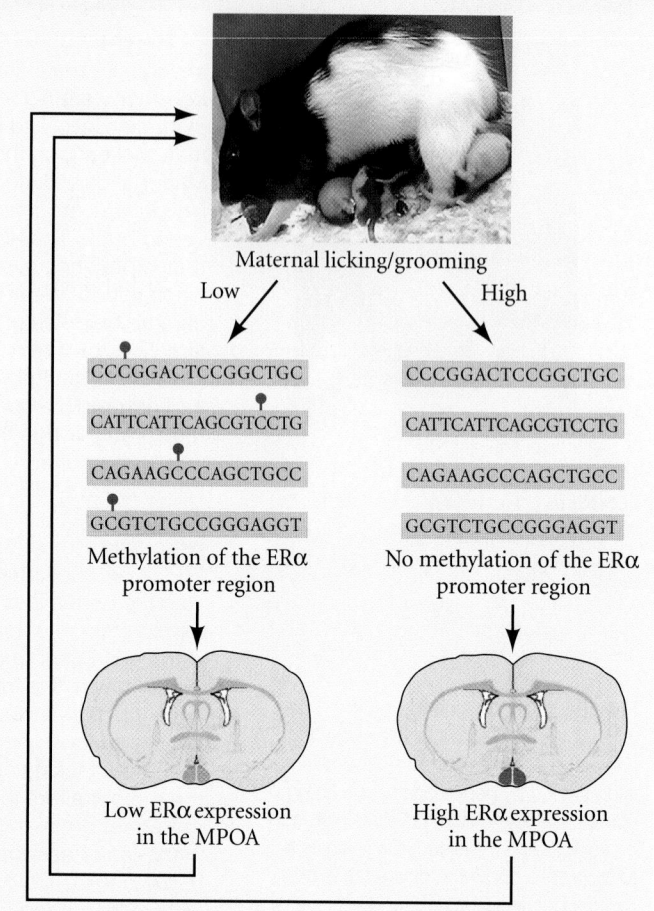

FIGURE 15.15 Environmental regulation of DNA methylation and the transmission of behavior. Female rats vary in the amount of time and effort spent licking and grooming their newborn pups. Pups that experience high levels of licking and grooming have a relatively unmethylated promoter region on the gene encoding estrogen receptor-α (ERα). This gene becomes active in the MPOA and promotes the oxytocin-based licking and grooming in females. Among pups that do not experience high levels of licking and grooming, the ERα promoter is relatively highly methylated, causing the gene to function at lower levels. The result is a decrease in oxytocin-induced licking and grooming behavior. Thus, the behavior modifies the DNA, and this in turn affects the behavior. (After Champagne 2008; photograph from Meaney and Szyf 2005.)

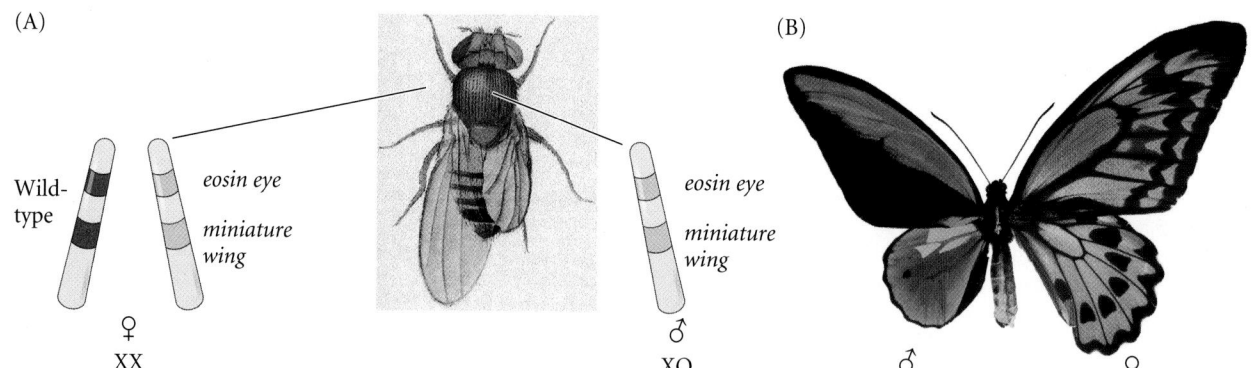

FIGURE 15.16 Gynandromorph insects. (A) *D. melanogaster* in which the left side is female (XX) and the right side is male (XO). The male side has lost an X chromosome bearing the wild-type alleles of eye color and wing shape, thereby allowing expression of the recessive alleles *eosin eye* and *miniature wing* on the remaining X chromosome. (B) Birdwing butterfly *Ornithoptera croesus*. The smaller male half is red, black, and yellow, while the female half is larger and brown. (A, drawing by Edith Wallace from Morgan and Bridges 1919; B, Montreal Insectarium, photograph by the author.)

In *Drosophila*, and in insects in general, one can observe gynandromorphs—animals in which certain regions of the body are male and other regions are female (**FIGURE 15.16**; see also Figure 15.14). Gynandromorph fruit flies result when an X chromosome is lost from one embryonic nucleus. The cells descended from that cell, instead of being XX (female), are XO (male). The XO cells display male characteristics, whereas the XX cells display female traits, suggesting that, in *Drosophila*, each cell makes its own sexual "decision." Indeed, in their classic discussion of gynandromorphs, Morgan and Bridges (1919) concluded, "Male and female parts and their sex-linked characters are strictly self-determining, each developing according to its own aspiration," and each sexual decision is "not interfered with by the aspirations of its neighbors, nor is it overruled by the action of the gonads." Although there are organs that are exceptions to this rule (notably the external genitalia), it remains a good general principle of *Drosophila* sexual development.

The Sex-lethal gene

Although it had long been thought that a fruit fly's sex was determined by the X-to-autosome (X:A) ratio (Bridges 1925), this assessment was based largely on flies with aberrant numbers of chromosomes. Recent molecular analyses suggest that X chromosome number alone is the primary sex determinant in normal diploid insects (Erickson and Quintero 2007). The main basis for this assertion is the fact that the X chromosome contains genes encoding transcription factors that activate the critical gene in *Drosophila* sex determination, the autosomal locus **Sex-lethal** (**Sxl**). The Sex-lethal protein is a splicing factor that will initiate a cascade of RNA processing events that will eventually lead to male-specific and female-specific transcription factors (**FIGURE 15.17**). These transcription factors (the Doublesex proteins) will then differentially activate the genes involved in either the male phenotype (testes, sex combs, pigmentation, etc.) or the female phenotype (ovaries, yolk proteins, pigmentation).

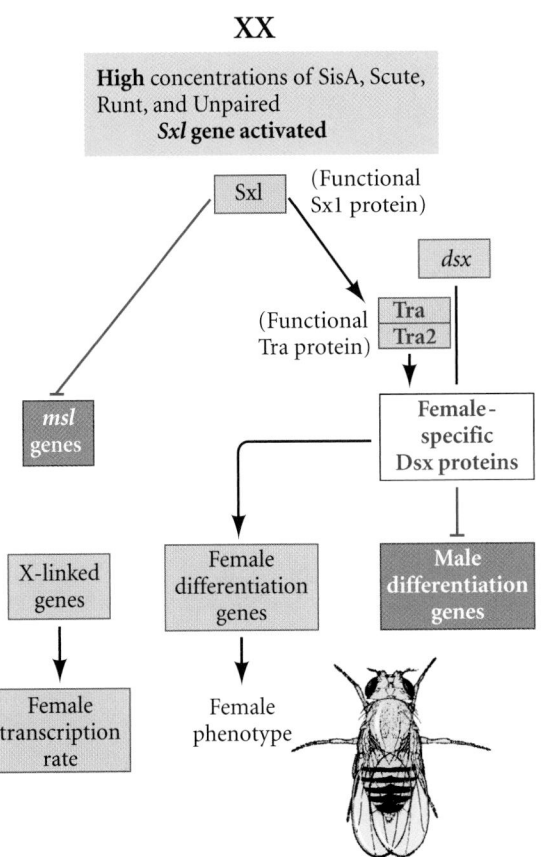

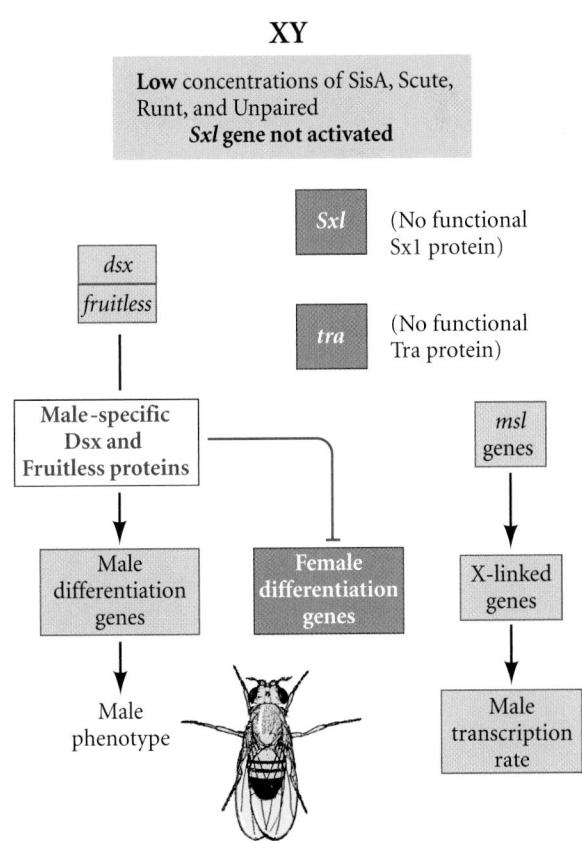

FIGURE 15.17 Proposed regulatory cascade for *Drosophila* somatic sex determination. Transcription factors from the X chromosomes activate the *Sxl* gene in females (XX) but not in males (XY). The Sex-lethal protein performs three main functions. First, it activates its own transcription, ensuring further Sxl production. Second, it represses the translation of *msl2* mRNA, a factor that facilitates transcription from the X chromosome. This equalizes the amount of transcription from the two X chromosomes in females with that of the single X chromosome in males. Third, Sxl enables the splicing of the *transformer-1* (*tra1*) pre-mRNA into functional proteins. The Tra proteins process *doublesex* (*dsx*) pre-mRNA in a female-specific manner that provides most of the female body with its sexual fate. They also process the *fruitless* pre-mRNA in a female-specific manner, giving the fly female-specific behavior. In the absence of Sxl (and thus the Tra proteins), *dsx* and *fruitless* pre-mRNAs are processed in the male-specific manner. (After Baker et al. 1987.)

ACTIVATING SEX-LETHAL The number of X chromosomes appears to have only a single function: activating or not activating the early expression of *Sex-lethal.** *Sxl* encodes an RNA splicing factor that will regulate gonad development and will also regulate the amount of gene expression from the X chromosome. The gene has two promoters. The early promoter is active only in XX cells; the later promoter is active in both XX and XY cells. The X chromosome appears to encode four protein factors that activate the early promoter of *Sxl.* Three of these proteins are transcription factors—SisA, Scute, and Runt—that bind to the early promoter to activate transcription. The fourth protein, Unpaired, is a secreted factor that reinforces the other three proteins through the JAK-STAT pathway (Sefton et al. 2000; Avila and Erickson 2007). If these factors accumulate so they are present in amounts above a certain threshold, the *Sxl* gene is activated through its early promoter (Erickson and Quintero 2007; Gonzales et al. 2008). The result is the transcription of *Sxl* early in XX embryos, during the syncytial blastoderm stage.

This *Sxl* pre-RNA transcribed from the *early* promoter of XX embryos lacks exon 3, which contains a stop codon. Thus, Sxl protein that is made early is spliced in a manner such that exon 3 is absent, and early XX embryos have complete and functional Sxl protein (**FIGURE 15.18**). In XY embryos, the early promoter of *Sxl* is not active and no functional Sry protein is present. However, later in development, as cellularization is taking place, the *late* promoter becomes active and the *Sxl* gene is transcribed in *both* males and females. In XX cells, Sxl protein made from the early promoter can bind to its own pre-mRNA and splice it in a "female" direction. In this case, Sxl binds to and blocks the splicing complex on exon 3 (Johnson et al. 2010; Salz 2011). As a result, exon 3 is skipped and is not included in the *Sxl* mRNA. Thus, early production ensures that functional full-length (354-amino acid) Sxl protein is made if the cells are XX (Bell et al. 1988; Keyes et al. 1992). In XY cells, however, the early promoter is not active (because it lacked the X-encoded transcription factors) and there is no early Sxl protein. Therefore, the *Sxl* pre-mRNA of XY cells is spliced in a manner that *includes* exon 3, with its termination codon. Protein synthesis ends at the third exon (after amino acid 48), and the Sxl is nonfunctional.

TARGETS OF SEX-LETHAL The protein made by the female-specific *Sxl* transcript contains regions that are important for binding to RNA. There appear to be three major RNA targets to which the female-specific *Sxl* transcript binds. One of these is the pre-mRNA of *Sxl* itself. Here, Sxl acts as a transcription factor to retain the activity of the late promoter of the *Sxl* gene. Another target is the *msl2* gene that controls dosage compensation (see below). Indeed, if the *Sxl*

*This gene's gory name is derived from the fact that mutations of this gene can result in aberrant dosage compensation of X-linked genes (see Chapter 2). As a result, there is inadequate transcription of genes encoded by the X chromosome, and the embryo dies.

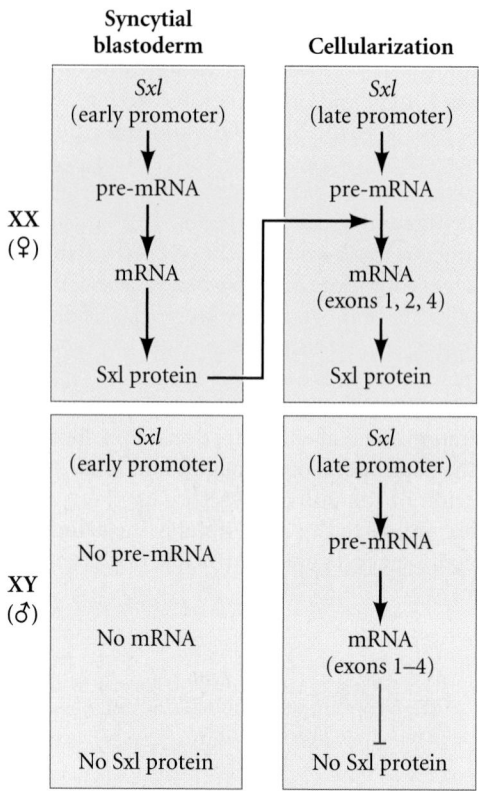

FIGURE 15.18 Differential RNA splicing and sex-specific expression of *Sex-lethal*. In the syncytial blastoderm of XX flies, transcription factors from the two X chromosomes are sufficient to activate the early promoter of the *Sxl* gene. This "early" transcript is spliced into an mRNA lacking exon 3 and makes a functional Sxl protein. The early promoter of XY flies is not activated, and males lack functional Sxl. By the cellularizing blastoderm stage, the late promoter of *Sxl* is active in both XX and XY flies. In XX flies, Sxl already present in the embryo prevents the splicing of exon 3 into mRNA and functional Sxl protein is made. Sxl then binds to its own promoter to keep it active; it also functions to splice downstream pre-mRNAs. In XY embryos, no Sxl is present and exon 3 is spliced into the mRNA. Because of the termination codon in exon 3, males do not make functional Sxl. (After Salz 2011.)

gene is nonfunctional in a cell with two X chromosomes, the dosage compensation system will not work, and the result is cell death (hence the gene's name). The third target is the pre-mRNA of *transformer* (*tra*)—the next gene in the cascade (**FIGURE 15.19**; Bell et al. 1988; Nagoshi et al. 1988).

The pre-mRNA of *transformer* (so named because loss-of-function mutations turn females into males) is spliced into a functional mRNA by Sxl protein. The *tra* pre-mRNA is made in both male and female cells; however, in the presence of Sxl protein, the *tra* transcript is alternatively spliced to create a female-specific mRNA, as well as a nonspecific mRNA that is found in both females and males. Like the male *Sxl* message, the nonspecific *tra* mRNA contains a termination codon early in the message that renders the protein nonfunctional (Boggs et al. 1987). In *tra*, the second exon of the nonspecific

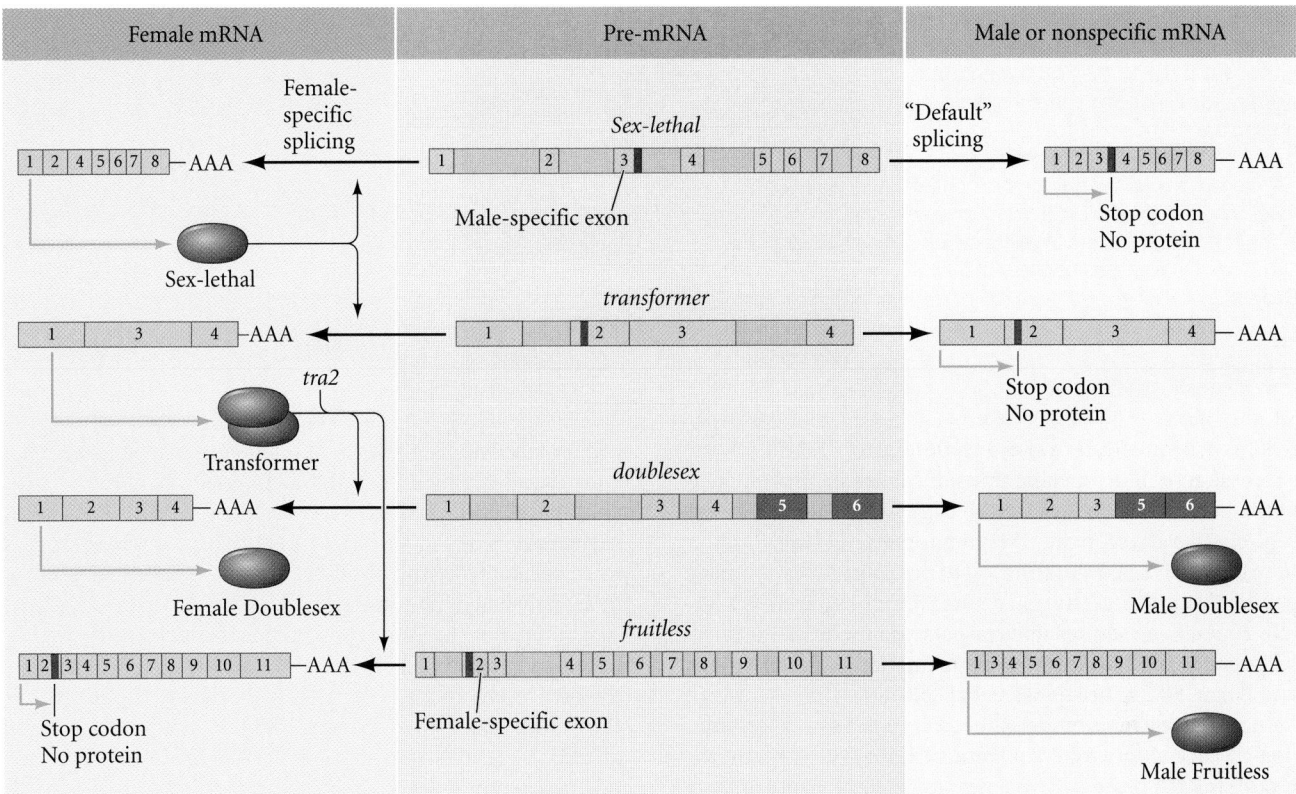

FIGURE 15.19 Sex-specific RNA splicing in four major *Drosophila* sex-determining genes. The pre-mRNAs (shown in the center of diagram) are identical in both male and female nuclei. In each case, the female-specific transcript is shown at the left, while the default transcript (whether male or nonspecific) is shown to the right. Exons are numbered, and the positions of termination codons are marked. *Sex-lethal*, *transformer*, and *doublesex* are all part of the genetic cascade of primary sex determination. The transcription pattern of *fruitless* determines the secondary characteristic of courtship behavior. (After Baker 1989; Baker et al. 2001.)

mRNA contains the termination codon and is not utilized in the female-specific message (see Figures 15.17 and 15.19).

How is it that females make a different mRNA from the males? The female-specific Sxl protein activates a 3′ splice site that causes *tra* pre-mRNA to be processed in a way that splices out the second exon. To do this, Sxl protein blocks the binding of splicing factor U2AF to the nonspecific splice site of the *tra* message by specifically binding to the polypyrimidine tract adjacent to it (Handa et al. 1999). This causes U2AF to bind to the lower-affinity (female-specific) 3′ splice site and generate a female-specific mRNA (Valcárcel et al. 1993). The female-specific Tra protein works in concert with the product of the *transformer-2* (*tra2*) gene to help generate the female phenotype by splicing the *doublesex* gene in a female-specific manner.

Doublesex: The switch gene for sex determination

The *Drosophila* **doublesex** (**dsx**) gene is active in both males and females, but its primary transcript is processed in a sex-specific manner (Baker et al. 1987). This alternative RNA processing is the result of the action of the *tra* and *tra2* gene

products on the *dsx* gene (see Figures 15.17 and 15.19). If the Tra2 and female-specific Tra proteins are both present, the *dsx* transcript is processed in a female-specific manner (Ryner and Baker 1991). The female splicing pattern produces a female-specific protein that activates female-specific genes (such as those of the yolk proteins) and inhibits male development. If no functional Tra is produced, a male-specific transcript of *dsx* is made. The male transcript encodes an active transcription factor that inhibits female traits and promotes male traits. In the embryonic gonad, Dsx regulates all known aspects of sexually dimorphic gonad cell fate.

In XX flies, the female Doublesex protein (DsxF) combines with the product of the *intersex* gene (*Ix*) to make a transcription factor complex that is responsible for promoting female-specific traits. This "Doublesex complex" activates the *Wingless* (*Wg*) gene, whose Wnt-family product promotes growth of the female portions of the genital disc. It also represses the *Fgf* genes responsible for making male accessory organs, activates the genes responsible for making yolk proteins, promotes the growth of the sperm storage duct, and modifies *bricabrac* (*bab*) gene expression to give the female-specific pigmentation profile. In contrast, the male Doublesex protein

(A)

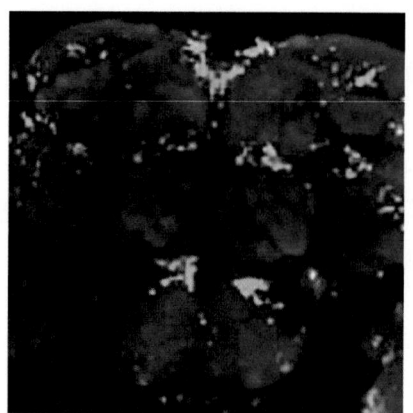

(B)

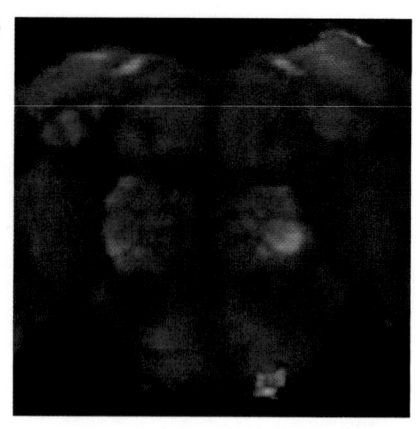

FIGURE 15.20 Sexually dimorphic neural circuits in adult male and female *Drosophila* brains. (A) In males, the Fruitless protein is expressed in subsets of neurons throughout the brain. These neurons are linked to male courtship and mating behaviors. (B) The female *fruitless* transcript is nonfunctional, and the protein is not expressed in the female brain. (From Demir and Dickson 2005, photographs courtesy of B. J. Dickson.)

(Dsx^M) acts directly as a transcription factor and directs the expression of male-specific traits. It causes the male region of the genital disc to grow at the expense of the female disc regions. It activates the BMP homologue *Decapentaplegic* (*Dpp*), as well as stimulating *Fgf* genes to produce the male genital disc and accessory structures. Dsx^M also converts certain cuticular structures into claspers and modifies the *brica-brac* gene to produce the male pigmentation pattern (Ahmad and Baker 2002; Christiansen et al. 2002).

According to this model, the result of the sex determination cascade summarized in Figure 15.17 comes down to the type of mRNA processed from the *doublesex* transcript. If there are two X chromosomes, the transcription factors activating the early promoter of *Sxl* reach a critical concentration, and *Sxl* makes a splicing factor that causes the *transformer* gene transcript to be spliced in a female-specific manner. This female-specific protein interacts with the *tra2* splicing factor, causing *dsx* pre-mRNA to be spliced in a female-specific manner. If the *dsx* transcript is not acted on in this way, it is processed in a "default" manner to make the male-specific message.

Brain sex in *Drosophila*

Our discussion of sexual dimorphism in *Drosophila* has so far been limited to nonbehavioral aspects of development. However, as in mammals, there appears to be a separate "brain sex" pathway in *Drosophila* that provides individuals of each sex with the appropriate set of courtship and aggression behaviors. Among *Drosophila*, mating takes place soon after the flies emerge from their pupal cases and does not need to be learned. Thus, the mating behaviors seem to be "hard-wired" into the insect genome.*

The behaviors of *Drosophila* courtship and mating are quite complicated. A male must first confirm that the individual he is approaching is a female. Once this is established, he must orient his body toward the female and follow a specific series of movements that include following

the female, tapping the female, playing a species-specific courtship song by vibrating his wings, licking the female, and finally, curling his abdomen so that he is in a position to mate. Each of these sex-specific courtship behaviors appear to be regulated by the products of *fruitless*, a gene expressed in certain sets of neurons involved with male sexual behaviors (**FIGURE 15.20A**). These include subsets of neurons involved in taste, hearing, smell, and touch, and in total they represent about 2% of all the neurons in the adult male (Lee et al. 2000; Billeter and Goodwin 2004; Stockinger et al. 2005). Fruitless also retains certain male-specific neural circuits; the neurons in these circuits die during female development (and in *fruitless* mutants; see Kimura et al. 2005).

As with *doublesex* pre-mRNA, the Tra and Tra2 proteins splice *fruitless* pre-mRNA into a female-specific message; the default splicing pattern is male. So the female makes Tra protein and processes the *fruitless* pre-mRNA in one way, whereas the male, lacking Tra, processes the *fruitless* message in another way. Female *fruitless* mRNA includes a termination sequence in an early exon; therefore the female does not make functional Fruitless protein (**FIGURE 5.20B**). The male fly makes an mRNA that does not contain the stop codon (Heinrichs et al. 1998), and the protein it transcribes is a zinc-finger transcription factor. Using homologous recombination to force the transcription of particular splicing forms, Demir and Dickson (2005) showed that it is Fruitless, and not the flies' anatomy, that controls their sexual behavior. When female flies were induced to make the male-specific Fruitless protein, they performed the entire male courtship ritual and tried to mate with other females.

In normal females, the courtship ritual is not as involved as in males. However, females have the ability to be receptive to a male's entreaties or to rebuff them. The product of the *retained* gene (*rtn*) is critical in this female mating behavior. Both sexes express this gene, since it is also involved in axon pathfinding. However, female flies with a loss-of-function allele of *rtn* resist male courtship and are thus rendered sterile by their own behavior (Ditch et al. 2005).

The splicing of the *fruitless* transcripts not only regulates sex-specific courtship patterns, it also regulates sex-specific aggression patterns as well. Female flies having a male

---

*This is not to say that flies don't learn; indeed, one thing they do learn is to avoid bad sexual encounters. A male that has been brushed off (quite literally) by a female because she has recently mated hesitates before starting to court another female (Siegel and Hall 1979; MacBride et al. 1999).

Fruitless protein not only tend to court females, they also will fight males and try to establish themselves at the top of a dominance hierarchy. Male flies having a mutant *fruitless* allele will show female-specific aggression against other females (Vrontou et al. 2006). It appears that the presence or absence of Fruitless proteins generate male- and female-specific neural circuits that drive these behaviors. In this manner, the same stimulus (such as a pheromone) will activate different circuits and elicit different behaviors from male and female flies (Ruta et al. 2010; Yu et al. 2010; Ito et al. 2012).

Dosage Compensation

In animals whose sex is determined by sex chromosomes, there has to be some mechanism by which the amount of X chromosome gene expression is equalized for males and females. This mechanism is known as **dosage compensation**. In Chapter 2 we discussed mammalian X-chromosome inactivation, whereby one of the X chromosomes is inactivated so that the transcription product level is the same in both XX cells and XY cells. In the worm *Caenorhabditis elegans*, dosage compensation occurs by lowering the transcription rates of *both* X chromosomes so that product levels are the same as those of XO individuals.

In *Drosophila*, the female X chromosome is not suppressed; rather, the male's single X chromosome is hyperactivated. This "hypertranscription" is accomplished at the level of translation and is mediated by the Sxl protein. Sxl (which is made by the female cells; see earlier discussion) binds to the 5' leader sequence and the 3' untranslated regions of the *msl2* message. The bound Sxl inhibits the attachment of *msl2* mRNA to the ribosome and prevents the ribosome from getting to the mRNA coding region (Beckman et al. 2005). The result is that female cells do not produce Msl2 protein (see Figure 15.17), but Msl2 *is* made in male cells, in which Sxl is not present. Mls2 is part of a protein-mRNA complex that targets the X chromosome and loosens its chromatin structure by acetylating histone 4 (see Figure 2.3). In this way, transcription factors gain access to the X chromosome at a much higher frequency in males than in females—hence, "hypertranscription."

ENVIRONMENTAL SEX DETERMINATION

Temperature-Dependent Sex Determination in Reptiles

While the sex of most snakes and lizards is determined by sex chromosomes at the time of fertilization, the sex of most turtles and all species of crocodilians is determined *after* fertilization, by the embryonic environment. In these reptiles, the temperature of the eggs during a certain period of development is the deciding factor in determining sex, and small changes in temperature can cause dramatic changes in the sex ratio (Bull 1980; Crews 2003). Often, eggs incubated at low temperatures produce one sex, whereas eggs incubated at

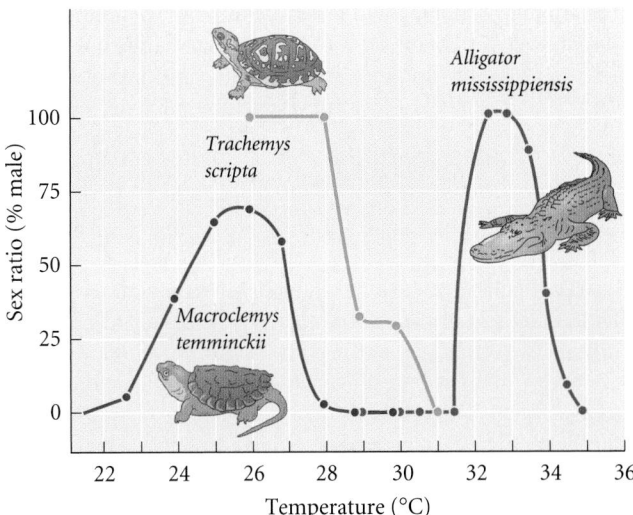

FIGURE 15.21 Temperature-dependent sex determination in three species of reptiles: the American alligator (*Alligator mississippiensis*), red-eared slider turtle (*Trachemys scripta elegans*), and alligator snapping turtle (*Macroclemys temminckii*). (After Crain and Guillette 1998.)

higher temperatures produce the other. There is only a small range of temperatures that permits both males and females to hatch from the same brood of eggs.*

FIGURE 15.21 shows the abrupt temperature-induced change in sex ratios for the red-eared slider turtle. If a brood of eggs is incubated at a temperature below 28°C, all the turtles hatching from the eggs will be male. Above 31°C, every egg gives rise to a female. At temperatures in between, the brood will give rise to individuals of both sexes. Variations on this theme also exist. The eggs of the snapping turtle *Macroclemys*, for instance, become female at either cool (22°C or lower) or hot (28°C or above) temperatures. Between these extremes, males predominate.

As we will see in Chapter 19, there can be multiple pathways of sex determination in the same individual. Under normal temperature conditions, the sex of the lizard *Bassiana duperreyi* is determined by sex chromosomes (XY males; XX females). However, at low temperatures, the environmental component overrides the genetic sex-determining mechanism, and all the offspring in cool nests are male (even if their chromosomes are XX; Radder et al. 2008).

One of the best-studied reptiles is the European pond turtle, *Emys orbicularis*. In laboratory studies, incubating *Emys* eggs at temperatures above 30°C produces all females, whereas temperatures below 25°C produce all-male broods. The threshold temperature (at which the sex ratio is even) is 28.5°C (Pieau et al. 1994). The developmental "window" during which sex determination occurs can be discovered by incubating eggs at the male-producing temperature for a certain amount of time and then shifting them to an incubator at

*The evolutionary advantages and disadvantages of temperature-dependent sex determination are discussed in Chapter 19.

the female-producing temperature (and vice versa). In *Emys*, the middle third of development appears to be the most critical for sex determination, and it is believed that the turtles cannot reverse their sex after this period.

The aromatase hypothesis for environmental sex determination

The enzyme aromatase, which converts testosterone into estrogen (see Sidelights & Speculations, p. 533), appears to be a particularly important target for environmental triggers. Unlike the situation in mammals, whose primary sex determination is a function of the X and Y chromosomes, primary sex determination in reptiles and birds is influenced by hormones, and estrogen is essential if ovaries are to develop. In reptiles, estrogen can override temperature, inducing ovarian differentiation even at masculinizing temperatures. Similarly, experimentally exposing eggs to inhibitors of estrogen synthesis produces male offspring, even if the eggs are incubated at temperatures that usually produce females (Dorizzi et al. 1994; Rhen and Lang 1994). The sensitive time for the effects of estrogens and their inhibitors coincides with the time when sex determination usually occurs (Bull et al. 1988; Gutzke and Chymiy 1988).

The estrogen-synthesis inhibitors used in the experiments mentioned above worked by blocking aromatase action, showing that experimentally low aromatase levels yield male offspring.* This correlation appears to hold under natural conditions as well. The aromatase activity of the turtle *Emys* is very low at the male-promoting temperature of 25°C. At the female-promoting temperature of 30°C, aromatase

---

*One remarkable finding is that injecting an aromatase inhibitor into the eggs of an all-female, parthenogenetic species of lizard causes the formation of males (Wibbels and Crews 1994).

activity increases dramatically during the critical period for sex determination (Desvages et al. 1993; Pieau et al. 1994; Shoemaker et al. 2007). Temperature-dependent aromatase activity is also seen in diamondback terrapins (*Malaclemys terrapin*) and in the red-eared slider turtle (*Trachemys scripta*). In both these species, inhibition of aromatase masculinizes the gonads (Jeyasuria et al. 1994; Barske and Capel 2010). In *T. scripta*, incubation at female temperature downregulates the testis-forming gene *Sox9*. Embryos treated with estrogen at the male-forming temperature also lose *Sox9* expression, and embryos treated with aromatase inhibitors (blocking estrogen synthesis) at female temperatures fail to downregulate *Sox9*. These data suggest that endogenous estrogen (produced through aromatase reactions in the early gonad) causes ovary formation in turtles while blocking *Sox9* expression. If aromatase is not synthesized, estrogen is not made and the gonad becomes male (Barske and Capel 2010).

Although there appear to be several ways that temperature can regulate sex determination, the blocking of testis development by aromatase appears to be one of the most conserved schemes (Valenzuela et al. 2013). Moreover, it is possible that aromatase expression is activated differently in different species. In some species, the aromatase *protein* may be temperature sensitive. In other species, the expression of the aromatase *gene* may be differentially activated at high temperatures (see Murdock and Wibbels 2006; Shoemaker et al. 2007; Valenzuela et al. 2013).

Estrogens, aromatase, sex reversal, and conservation biology

Over the last two decades, data have emerged showing that several of the polychlorinated biphenyl compounds (PCBs), a class of widespread pollutants introduced into the environment by humans, can act as estrogens (e.g., see Bergeron et al.

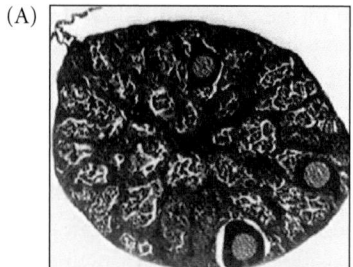 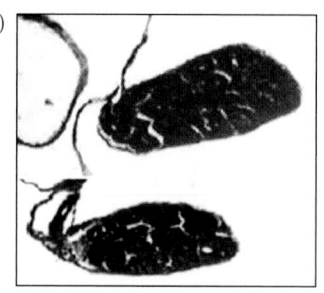 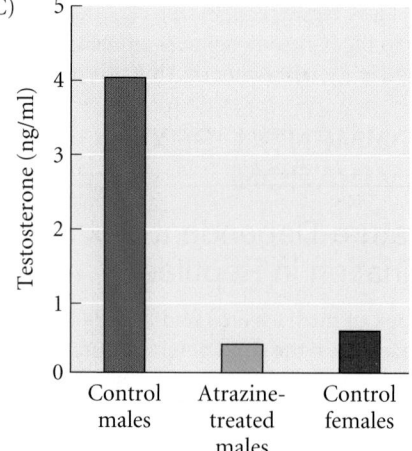

FIGURE 15.22 Demasculinization of frogs by low amounts of atrazine. (A) Testis of a frog from a natural site having 0.5 parts per billion (ppb) atrazine. The testis contains three lobules that are developing both sperm and an oocyte. (B) Two testes of a frog from a natural site containing 0.8 ppb atrazine. These organs show severe testicular dysgenesis, which characterized 28% of the frogs found at that site. (C) Effect of a 46-day exposure to 25 ppb atrazine on testosterone levels in the blood plasma of sexually mature male *Xenopus*. Levels in control males were some tenfold higher than in control females; atrazine-treated males had plasma testosterone levels at or below those of control females. (A,B after Hayes et al. 2003, photographs courtesy of T. Hayes; C after Hayes et al. 2002a.)

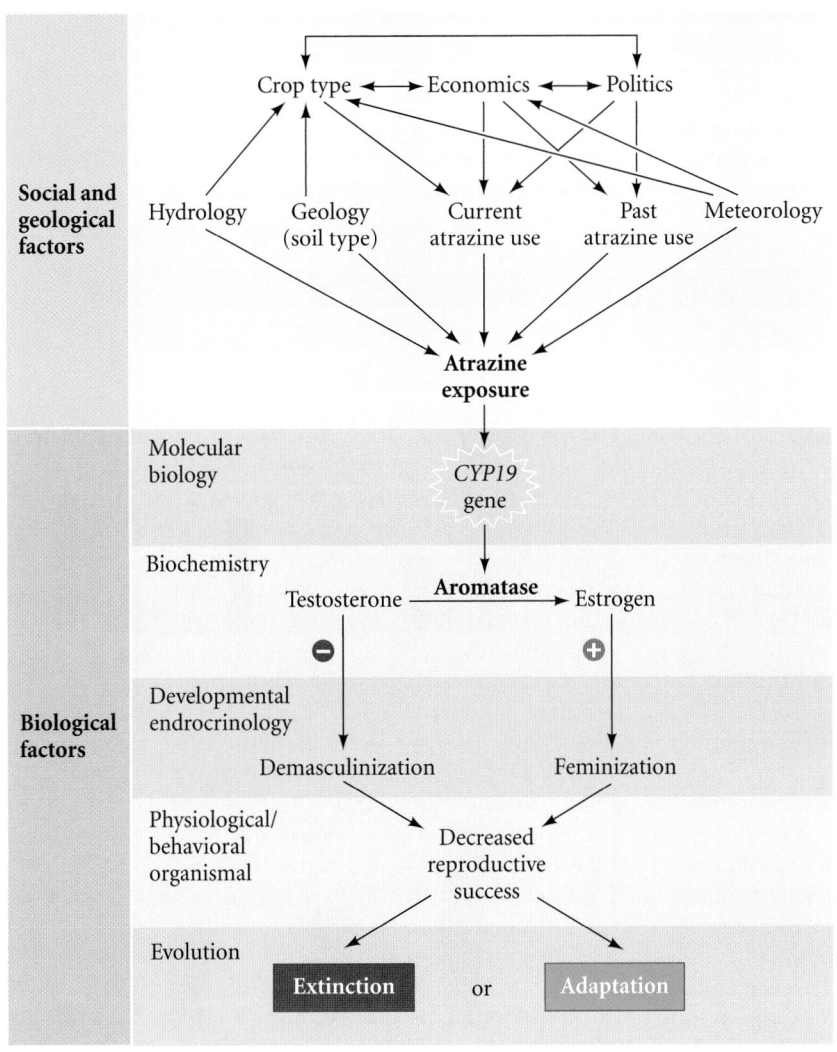

FIGURE 15.23 Possible chain of causation leading to the feminization of male frogs and the decline of frog populations in regions where atrazine has been used to control weed populations. Social, geological, and biological agents are shown. *Cyp19* is the gene encoding aromatase, and it has been shown that transcription of the human *CYP19* gene is induced by atrazine (Sanderson et al. 2000). (From Hayes 2005.)

tadpoles developed ovaries in addition to testes. At 1 ppb atrazine, the vocal sacs (which a male frog must have in order to signal and obtain a potential mate) failed to develop properly. Similar experiments in outdoor environments more similar to natural conditions (Langlois et al. 2010) also showed that male frogs (*Rana pipiens*) had been transformed into females by atrazine.

Atrazine induces aromatase, and aromatase can convert testosterone into estrogen (Crain et al.1997; Fan et al. 2007). It can feminize gonad development in larvae, and it can change gonads in adult males into ovaries. In laboratory experiments, the testosterone levels of adult male frogs were reduced by 90% (to levels of control females) when they were exposed, as sexually mature adults, to 25 ppb atrazine (Hayes et al. 2002a). This is an ecologically relevant dose, since the allowable amount of atrazine in U.S. drinking water is 3 ppb, and atrazine levels can reach 224 ppb in streams of the midwestern United States (Battaglin et al. 2000; Barbash et al. 2001). Even at doses as low as 2.5 ppb, the sexual behavior of the male frogs was severely diminished such that matings were relatively rare, and in 10% of the cases, males exposed to atrazine became functional egg-laying females (Hayes et al. 2010).

Given the amount of atrazine in the water supply and the sensitivity of frogs to this compound, the situation could be devastating to wild populations. In a field study, Hayes and his colleagues collected leopard frogs and water at eight sites across the central United States (Hayes et al. 2002b, 2003). They sent the water samples to two separate laboratories for the determination of atrazine, coding the frog specimens so that the technicians dissecting the gonads did not know which site the animals came from. The results showed that all but one site contained atrazine—and this was the only site from which the frogs had no gonadal abnormalities. At concentrations as low as 0.1 ppb, leopard frogs displayed testicular dysgenesis (stunted growth of the testes) or conversion to ovaries. In many examples, oocytes were found in the testes (see Figure 15.22A).

Concern over atrazine's apparent ability to disrupt sex hormones in both wildlife and humans has resulted in bans on the use of this herbicide by France, Germany, Italy, Norway, Sweden, and Switzerland (Dalton 2002). Many geographical and social concerns mediate atrazine use (**FIGURE 15.23**), and the company making atrazine has lobbied against the work of independent researchers whose research indicates that it may cause reproductive malfunctions or cancers in wildlife and humans

1994, 1999). PCBs can reverse the sex of turtles raised at "male" temperatures. This knowledge has important consequences in environmental conservation efforts to protect endangered species (such as turtles, amphibians, and crocodiles) in which hormones can effect changes in primary sex determination. Indeed, some reptile conservation biologists advocate using hormonal treatments to elevate the percentage of females in endangered species (www.reptileconservation.org).

Changing sex ratios may endanger certain species by generating single-sex populations (see Chapter 19). The survival of some amphibian species may be at risk from herbicides that either promote or destroy estrogens. One such case involves the development of hermaphroditic and demasculinized frogs after exposure to extremely low doses of the weed killer atrazine, the most widely used herbicide in the world (**FIGURE 15.22**). Hayes and colleagues (2002a) found that exposing tadpoles to atrazine concentrations as low as 0.1 part per billion (ppb) produced gonadal and other sexual anomalies in male frogs. At 0.1 ppb and higher, many male

(see Blumenstyk 2003). The effects of endocrine disruptors and global warming on species with temperature-dependent sex determination will be explored further in Chapter 19.

● **See WEBSITE 15.5** Forms of hermaphroditism

Location-Dependent Sex Determination

Environmental factors other than temperature can also be sex determinants in some species. For example, it has been known since the nineteenth century that the sex of the "spoon" worm *Bonellia viridis* depends on where a larva settles (Baltzer 1914). If a *Bonellia* larva lands on the ocean floor, it develops into a 10-cm-long female. If the larva is attracted to a female's proboscis, it travels along the tube until it enters the female's body. There it differentiates into a minute (1–3 mm long) male that is essentially a sperm-producing symbiont of the female (**FIGURE 15.24**).

Another species in which sex determination is affected by the location of the organism is the slipper snail *Crepidula fornicata*. In this species, individuals pile up on top of one another in a great mound. Young individuals are always male, but this phase is followed by the degeneration of the male reproductive system and a period of lability. The next phase can be either male or female, depending on the animal's position in the mound. If the snail is attached to a female, it will become male; if it is removed from its attachment, it will become female. Similarly, the presence of large numbers of males causes some of the males to become females. However, once an individual becomes female, it will not revert to being male (Coe 1936; Collin 1995; Warner et al. 1996).

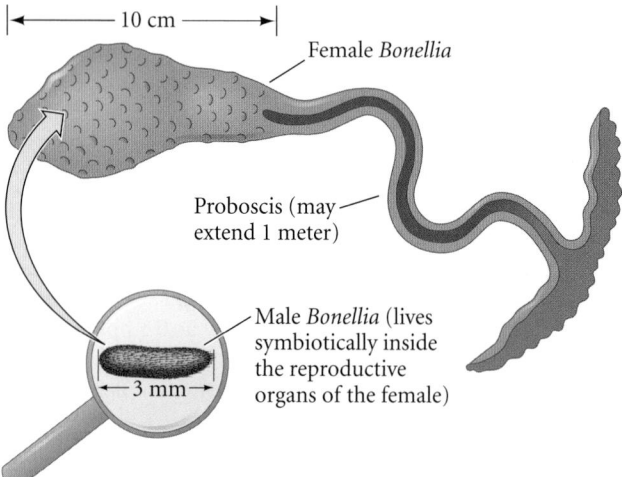

FIGURE 15.24 Sex determination in *Bonellia viridis*. Larvae that settle on the ocean floor become female. The mature female's body is about 10 cm long, with a 1-meter-long proboscis that emits chemicals that attract other *B. viridis* larvae. Larvae that land on the proboscis are taken into the female's body, where they develop and live symbiotically as tiny (1–3 mm long) males.

Many fish change their sex based on social interactions; such changes are mediated by the neuroendocrine system (Godwin et al. 2003, 2009). Interestingly, although the trigger of the sex change may be stress hormones (e.g., cortisol) that induce sex-specific neuropeptides, the effector of the change may be (once again) aromatase. There are two

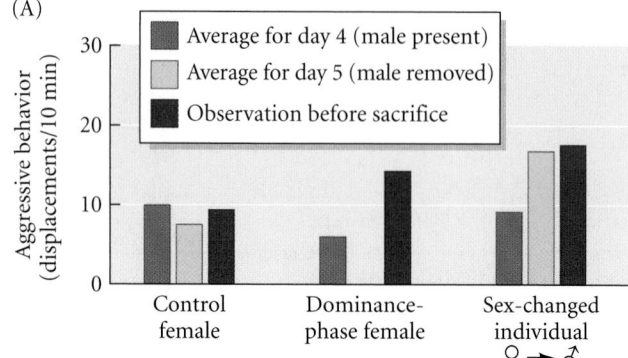

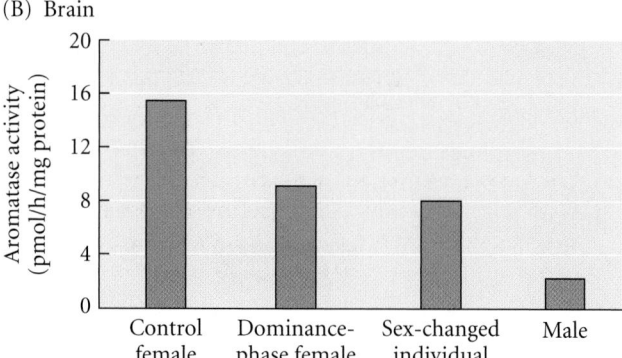

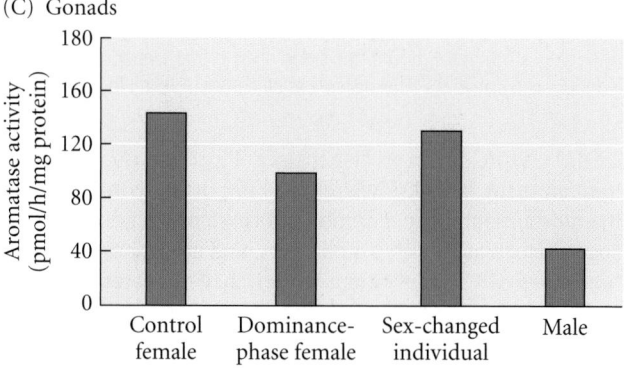

FIGURE 15.25 Aggressive behavior and aromatase activity (AA) in the brain and gonads of the goby *Lythrypnus dalli*. (A) On day 4 (prior to removing the male), there was no statistical difference in average daily displacements among the largest females. On day 5, the male was removed and dominant females increased their aggressive behavior. (Dominance-phase fish have no day 5 data because they were sacrificed on day 5 or just after.) (B,C) Brain (B) but not gonadal (C) AA was significantly lower in dominance-phase and sex-changed individuals compared with control females. Established males had lower brain AA than all other groups and lower gonadal AA than all groups except dominance-phase females. (After Black et al. 2005.)

variants of aromatase in many animals, one expressed in the brain and one in the gonads. Black and colleagues (2005) showed a striking correlation between changes in brain aromatase levels and changes in sexual behavior during the sexual transitions of gobies, in which the school typically contains only one male and multiple females. The removal of the male from a stable group caused a rapid increase (>200%) in the aggressive behavior of the largest female, which then became a male in about a week's time. This transformation may have resulted from an increase in brain testosterone levels, since within hours upon removal of the male, these dominant females developed a lower brain aromatase than the other females (**FIGURE 15.25**). The *gonadal* aromatase levels, however, stayed the same, and gonadal sex change came later. In porgy fish, aromatase inhibitors can block the natural sex change and induce male development (Lee et al. 2002). Thus, changes in the social group, perceived by the nervous system, became expressed by the hormonal system within hours, thereby changing the behavioral phenotype of the female fish. Interestingly, when it comes to behaviors in fish, sex is in the brain before it is in the gonads.

Coda

Nature has many variations on her masterpiece. In some species, including most mammals and insects, sex is determined by chromosomes; in other species, sex is a matter of environmental conditions. In yet other species, both environmental and genotypic sex determination function, often in different geographical areas. Different environmental or genetic stimuli may trigger sex determination through a series of conserved pathways. As Crews and Bull (2009) have reflected, "it is possible that the developmental decision of male versus female does not flow through a single gene but is instead determined by a 'parliamentary' system involving networks of genes that have simultaneous inputs to several components of the downstream cascade." We are finally beginning to understand the mechanisms by which this "masterpiece of nature" is created.

SNAPSHOT SUMMARY: SEX DETERMINATION

1. In mammals, primary sex determination (the determination of gonadal sex) is a function of the sex chromosomes. XX individuals are usually females, XY individuals are usually males.

2. The mammalian Y chromosome plays a key role in male sex determination. XY and XX mammals both have a bipotential gonad. In XY animals, Sertoli cells differentiate and enclose the germ cells within testis cords. The interstitial mesenchyme generates other testicular cell types, including the testosterone-secreting Leydig cells.

3. In XX mammals, the germ cells become surrounded by follicle cells in the cortex of the gonadal rudiment. The epithelium of the follicles becomes the granulosa cells; the mesenchyme generates the thecal cells.

4. In humans, the *SRY* gene is the testis-determining factor on the Y chromosome. It synthesizes a nucleic acid-binding protein that functions as a transcription factor to activate the evolutionarily conserved *SOX9* gene.

5. The *SOX9* gene product can also initiate testis formation. Functioning as a transcription factor, it binds to the gene encoding anti-Müllerian hormone and other genes. Fgf9 and Sox9 proteins have a positive feedback loop that activates testicular development and suppresses ovarian development.

6. Wnt4 and Rspo1 are involved in mammalian ovary formation. These proteins upregulate production of β-catenin; the functions of β-catenin include promoting the ovarian pathway of development while blocking the testicular pathway of development . The Foxl2 transcription factor is also required and appears to act in parallel with the Wnt4/Rspo1 pathway.

7. Secondary sex determination in mammals involves the factors produced by the developing gonads. In male mammals, the Müllerian duct is destroyed by the AMH produced by the Sertoli cells, while testosterone produced by the Leydig cells enables the Wolffian duct to differentiate into the vas deferens and seminal vesicle. In female mammals, the Wolffian duct degenerates with the lack of testosterone, whereas the Müllerian duct persists and is differentiated by estrogen into the oviducts, uterus, cervix, and upper portion of the vagina.

8. The conversion of testosterone to dihydrotestosterone in the genital rudiment and prostate gland precursor enables the differentiation of the penis, scrotum, and prostate gland.

9. Individuals with mutations of these hormones or their receptors may have a discordance between their primary and secondary sex characteristics.

10. In *Drosophila*, sex is determined by the number of X chromosomes in the cell; the Y chromosome does not play a role in sex determination. There are no sex hormones, so each cell makes a sex-determination "decision." However, paracrine factors play important roles in forming the genital structures.

11. The *Drosophila Sex-lethal* gene is activated in females (by the accumulation of proteins encoded on the X chromosomes), but the protein does not form in males because of translational termination. Sxl protein acts as an RNA

splicing factor to splice an inhibitory exon from the *transformer* (*tra*) transcript. Therefore, female flies have an active Tra protein but males do not.

12. The Tra protein also acts as an RNA splicing factor to splice exons from the *doublesex* (*dsx*) transcript. The *dsx* gene is transcribed in both XX and XY cells, but its pre-mRNA is processed to form different mRNAs, depending on whether Tra protein is present. The proteins translated from both *dsx* messages are active, and they activate or inhibit transcription of a set of genes involved in producing the sexually dimorphic traits of the fly.

13. Sex determination of the brain may have different downstream agents than in other regions of the body. *Drosophila* Tra proteins also activate the *fruitless* gene in males (but not in females); in mammals, the *Sry* gene may activate brain sexual differentiation independently from the hormonal pathways.

14. Dosage compensation is critical for the regulation of gene expression in the embryo. With the same number of autosomes, the transcription from the X chromosome must be equalized for XX females and XY males. In mammals, one X chromosome of XX females is inactivated. In *Drosophila*, the single X chromosome of XY males is hyperactivated.

15. In turtles and alligators, sex is often determined by the temperature experienced by the embryo during the time of gonad determination. Because estrogen is necessary for ovary development in these species, it is possible that differing levels of aromatase (an enzyme that can convert testosterone into estrogen) distinguish male from female patterns of gonadal differentiation.

16. Aromatase may be activated by environmental agents, causing demasculinization of the male gonads in those animals where primary sex determination can be effected by hormones.

17. In some species, such as *Bonellia* and *Crepidula*, sex is determined by the position of the individual with regard to other individuals of the same species. In fish, numerous environmental factors—especially temperature and the number of males already present in the population—can determine the sex of an individual.

For Further Reading

Complete bibliographical citations for all literature cited in this chapter can be found at the free-access website **www.devbio.com**

Bell, L. R., J. I. Horabin, P. Schedl and T. W. Cline. 1991. Positive autoregulation of *Sex-lethal* by alternative splicing maintains the female determined state in *Drosophila. Cell* 65: 229–239.

Champagne, F. A., I. C. Weaver, J. Diorio, S. Dymov, M. Szyf and M. J. Meaney. 2006. Maternal care associated with methylation of the estrogen receptor-α1b promoter and estrogen receptor-α expression in the medial preoptic area of female offspring. *Endocrinology* 147: 2909–2915.

Erickson, J. W. and J. J. Quintero. 2007. Indirect effects of ploidy suggest X chromosome dose, not the X:A ratio, signals sex in *Drosophila. PLoS Biol.* Dec. 5(12):e332.

Hayes, T. B., A. Collins, M. Lee, M. Mendoza, N. Noriega, A. Stuart and A. Vonk. 2002. Hermaphroditic, demasculinized frogs after exposure to the herbicide atrazine at low ecologically relevant doses. *Proc. Natl. Acad. Sci. USA* 99: 5476–5480.

Hiramatsu, R. and 9 others. 2009. A critical time window of *Sry* action in gonadal sex determination in mice. *Development* 136: 129–138.

Imperato-McGinley, J., L. Guerrero, T. Gautier and R. E. Peterson. 1974. Steroid 5α-reductase deficiency in man: An inherited form of male pseudohermaphroditism. *Science* 186: 1213–1215.

Jordan-Young, R. M. 2010. *Brainstorm: The Flaws in the Science of Sex Differences*. Harvard University Press, Cambridge, MA.

Koopman, P., J. Gubbay, N. Vivian, P. Goodfellow and R. Lovell-Badge. 1991. Male development of chromosomally female mice transgenic for *Sry. Nature* 351: 117–121.

Maatouk, D. M., L. DiNapoli, A. Alvers, K. L. Parker, M. M. Taketo and B. Capel. 2008. Stabilization of β-catenin in XY gonads causes male-to-female sex-reversal. *Hum. Mol. Genet.* 17: 2949–2955.

Miyamoto, Y., H. Taniguchi, F. Hamel, D. W. Silversides and R. S. Viger. 2008. GATA4/WT1 cooperation regulates transcription of genes required for mammalian sex determination and differentiation. *BMC Mol. Biol.* 29: 9–44.

Sekido, R. and R. Lovell-Badge. 2008. Sex determination involves synergistic action of Sry and Sf1 on a specific *Sox9* enhancer. *Nature* 453: 930–934.

Sekido, R. and R. Lovell-Badge. 2009. Sex determination and SRY: Down to a wink and a nudge? *Trends Genet.* 25: 19–29.

Go Online

WEBSITE 15.1 Social critique of sex determination research. In numerous cultures, women are seen as the "default state" and men are seen as having "something extra." Historians and biologists show that, until recently, such biases characterized the scientific study of human sex determination.

WEBSITE 15.2 Finding the male-determining genes. The mapping of the testis-determining factor to the Sry region of the Y chromosome took scientists more than 50 years to accomplish. Moreover, other testis-forming genes that act downstream of Sry have been found on autosomes. These include genes that are very important in gonad differentiation in other vertebrates.

WEBSITE 15.3 Dihydrotestosterone in adult men. The drug finasteride, which inhibits the conversion of testosterone to dihydrotestosterone, is being used to treat prostate growth and male pattern baldness.

WEBSITE 15.4 Insulin-like hormone 3. In addition to testosterone, the Leydig cells secrete insulin-like hormone 3 (INSL3). This hormone is required for the descent of the gonads into the scrotum. Males lacking INSL3 are infertile because the testes do not descend. In females, lack of this hormone deregulates the menstrual cycle.

WEBSITE 15.5 Forms of hermaphroditism. In *C. elegans* and many other invertebrates, hermaphroditism is the general rule. These animals may be born with both ovaries and testes, or they may develop one set of gonads first and the other later (sequential hermaphroditism). In some cases, hermaphrodites can fertilize their own eggs. This is the case for *C. elegans* and the common earthworm, as well as for the mangrove killifish *Kryptolebias marmoratus* (the only vertebrate known to self-fertilize; Mourabit et al. 2011).

Outside Sites

For discussions of disorders of sexual development, including intersex conditions, see the American Academy of Pediatrics site at **www.pediatriccareonline.org/pco/ub/view/Point-of-Care-Quick-Reference/397188/all/disorders_of_sexual_development**.

The Intersex Society of North America website, **http://isna.org**, is an excellent resource for information on human intersex conditions. The Wikipedia entry on intersexuality, **en.wikipedia.org/wiki/Intersexuality**, has a good discussion about the language used to describe these conditions. Also, see the website of Advocates for Informed Choice, **aiclegal.org/**, an organization concerned with the legal rights of children born with variations of sexual anatomy.

16

Postembryonic Development
Metamorphosis, Regeneration, and Aging

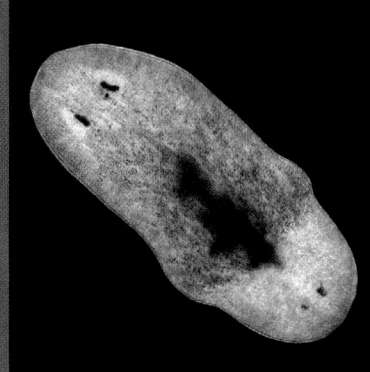

DEVELOPMENT NEVER CEASES. Throughout life, we continuously generate new blood cells, lymphocytes, keratinocytes, and digestive tract epithelium from stem cells. In addition to these continuous daily changes, there are instances in which postembryonic development is obvious—sometimes even startling. One such instance is metamorphosis, the transition from a larval stage to an adult stage. In many species that undergo metamorphosis, a large proportion of the animal's structure changes, and the larva and the adult are unrecognizable as being the same individual (see Figure 1.4). Another startling type of postembryonic development is regeneration, the creation of a new organ by an adult animal after the original one has been removed. Some adult salamanders, for instance, can regrow limbs and tails after these appendages have been amputated.

The third category of postnatal developmental change encompasses those alterations of form and function associated with aging in adult organisms. This area is controversial. Some scientists believe that the processes of age-associated degeneration are not properly part of the study of developmental biology. In this view, aging involves the random decay of normative processes. Others scientists claim that the genetically determined, species-specific patterns of aging are an important part of the life cycle and believe that **gerontology**—the scientific study of aging—is rightly part of developmental biology. As Peter Medawar (1957) noted, "That which we call 'development' when looked at from the birth and the life end becomes 'senescence' when looked at from its close." Whatever their relationship to embryonic development, metamorphosis, regeneration, and aging are critical topics for the biology of the twenty-first century.

METAMORPHOSIS: THE HORMONAL REACTIVATION OF DEVELOPMENT

Animals (including humans) whose young are essentially smaller versions of the adult are referred to as **direct developers**. Most animal species, however, are **indirect developers** whose life cycle includes a larval stage with characteristics very different from those of the adult organism, which emerges only after a period of **metamorphosis**.

Metamorphosis is both a developmental and an ecological transition. Developmentally, metamorphosis involves dramatic changes. During metamorphosis, developmental processes are reactivated by specific hormones and the entire organism changes morphologically, physiologically, and behaviorally to prepare itself for a new mode of existence. Ecologically, metamorphosis is associated with changes of habitat, food, and behaviors (Jacobs et al. 2006). Very often, larval forms are specialized for some function such as growth or dispersal, whereas the adult is specialized for reproduction. *Cecropia* moths, for example, hatch from eggs and develop as wingless juveniles—caterpillars—for several months. After metamorphosis, the adult insects spend a day or so as fully developed winged moths and must mate quickly before they die. The moths never eat, and in fact have no mouthparts during this brief reproductive phase of the life cycle. As

The earth-bound early stages built enormous digestive tracts and hauled them around on caterpillar treads. Later in the life-history these assets could be liquidated and reinvested in the construction of an entirely new organism—a flying-machine devoted to sex.

CARROLL M. WILLIAMS
(1958)

I'd give my right arm to know the secret of regeneration.

OSCAR E. SCHOTTÉ
(1950)

Death is very likely the single best invention of life.

STEVE JOBS (2005)

might be expected, the juvenile and adult forms often live in different environments.

Among indirect developers, there are two major types of larvae.* **Secondary larvae** are found among those animals whose larvae and adults possess the same basic body plan. Thus, despite the obvious differences between the caterpillar and the butterfly, these two life stages retain the same major body axes and develop by deleting and modifying old parts while adding new structures into a preexisting framework. Similarly, the frog tadpole, although specialized for an aquatic environment, is a secondary larva, organized on the same pattern as the adult will be (Jagersten 1972; Raff and Raff 2009). Larvae that represent dramatically different body plans than the adult form and that are morphologically distinct from the adult are called **primary larvae**. Sea urchin larvae, for instance, are bilaterally symmetrical organisms that float among and collect food in the plankton of the open ocean. The sea urchin adult is pentameral (i.e., has fivefold symmetry) and feeds by scraping algae from rocks on the seafloor. There is no trace of the adult form in the body plan of the juvenile (see Figure 7.19).

Metamorphosis is one of the most striking of developmental phenomena, and the extensive *morphological* changes undergone by some species have fascinated developmental anatomists for centuries (Merian 1705; Swammerdam 1737). But we know only an outline of the *molecular* bases of metamorphosis, and only for a handful of species. The following sections will detail metamorphosis in amphibians (especially *Xenopus laevis*) and insects (especially *Drosophila*).

Amphibian Metamorphosis

Amphibians are named for their ability to undergo metamorphosis, their appellation coming from the Greek *amphi* ("double") and *bios* ("life"). Amphibian metamorphosis is associated with morphological changes that prepare an aquatic organism for a primarily terrestrial existence. In **urodeles** (salamanders), these changes include the resorption of the tail fin, the destruction of the external gills, and a change in skin structure. In **anurans** (frogs and toads), the changes are more dramatic, with almost every organ subject to modification (**TABLE 16.1**; see also Figure 1.1). The changes in amphibian metamorphosis are initiated by thyroid hormones such as **thyroxine (T$_4$)** and **tri-iodothyronine (T$_3$)** that travel through the blood to reach all the organs of the larva. When the larval organs encounter these thyroid hormones, they can respond in any of four ways: growth, death, remodeling, and respecification.

Morphological changes associated with amphibian metamorphosis

GROWTH OF NEW STRUCTURES   The hormone tri-iodothyronine induces certain adult-specific organs to form. The limbs of the adult frog emerge from specific sites on the metamorphosing tadpole, and in the eye, nictitating membranes and eyelids both emerge. Moreover, T$_3$ induces the proliferation and differentiation of new neurons to serve these organs. As the limbs grow out from the body axis, new neurons proliferate and differentiate in the spinal cord. These neurons send axons to the newly formed limb musculature (Marsh-Armstrong et al. 2004). Blocking T$_3$ activity prevents these neurons from forming and causes paralysis of the limbs.

One readily observed consequence of anuran metamorphosis is the movement of the eyes to the front of the head from their originally lateral position (**FIGURE 16.1A,B**).† The lateral eyes of the tadpole are typical of preyed-upon herbivores, whereas the frontally located eyes of the frog befit its more predatory lifestyle. To catch its prey, the frog needs to see in three dimensions. That is, it has to acquire a *binocular field of vision*, where inputs from both eyes converge in the brain (see Figure 11.35B). In the tadpole, the right eye innervates the left side of the brain, and vice versa; there are no ipsilateral (same-side) projections of the retinal neurons. During metamorphosis, however, ipsilateral pathways emerge, enabling input from both eyes to reach the same area of the brain (Currie and Cowan 1974; Hoskins and Grobstein 1985a).

In *Xenopus*, these new pathways result not from the remodeling of existing neurons but from the formation of new neurons that differentiate in response to thyroid hormones (Hoskins and Grobstein 1985a,b). The ability of these axons to project ipsilaterally results from the induction of ephrin B in the optic chiasm by the thyroid hormones (Nakagawa et al. 2000). Ephrin B is also found in the optic chiasm of mammals (which have ipsilateral projections throughout life) but not in the chiasm of fish and birds (which have only contralateral projections). As shown in Chapter 11, ephrins can repel certain neurons, causing them to project in one direction rather than in another (**FIGURE 16.1C,D**).

CELL DEATH DURING METAMORPHOSIS   The hormone T$_3$ induces certain larval-specific structures to die. Thus, T$_3$ causes the degeneration of the paddle-like tail and the oxygen-procuring gills that were important for larval (but not adult) movement and respiration. While it is obvious that the tadpole's tail muscles and skin die, is this death murder or

---

* Although there is controversy on the subject, larvae probably evolved after the adult form had been established. In other words, animals evolved through direct development, and larval forms came about as specializations for feeding or dispersal during the early part of the life cycle (Jenner 2000; Rouse 2000; Raff and Raff 2009). Even so, the biphasic life cycle may be a trait characteristic of metazoans (see Degnan and Degnan 2010).

† One of the most spectacular movements of eyes during metamorphosis occurs in flatfish such as flounder. Originally, a flounder's eyes, like the lateral eyes of other fish species, are on opposite sides of its face. However, during metamorphosis, one of the eyes migrates across the head to meet the eye on the other side (Hashimoto et al. 2002; Bao et al. 2005). This allows the flatfish to dwell on the ocean bottom, looking upward.

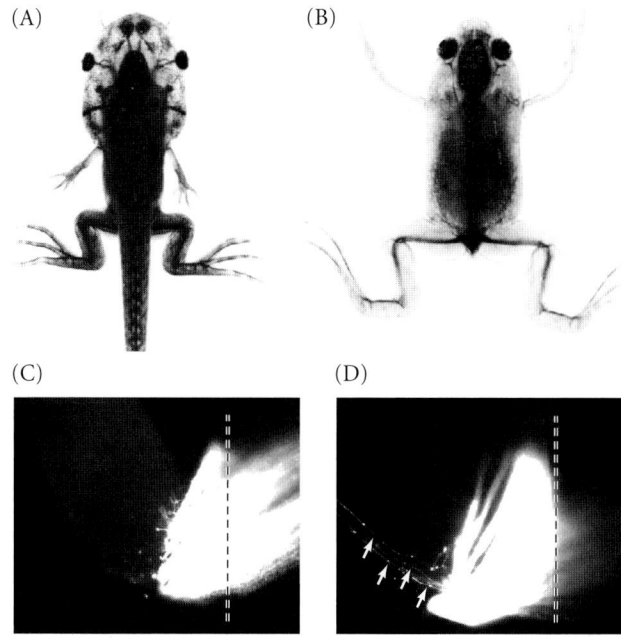

(A) (B) (C) (D)

FIGURE 16.1 Eye migration and associated neuronal changes during metamorphosis of the *Xenopus laevis* tadpole. (A) The eyes of the tadpole are laterally placed, so there is relatively little binocular field of vision. (B) The eyes migrate dorsally and rostrally during metamorphosis, creating a large binocular field for the adult frog. (C,D) Retinal projections of metamorphosing tadpole. The dye DiI was placed on a cut stump of the optic nerve to label the retinal projection. (C) In early and middle stages of metamorphosis, axons project across the midline (dashed line) from one side of the brain to the other. (D) In late metamorphosis, ephrin B is produced in the optic chiasm as certain neurons (arrows) are formed that project ipsilaterally. (A,B from Hoskins and Grobstein 1984, courtesy of P. Grobstein; C,D from Nakagawa et al. 2000, courtesy of C. E. Holt.)

suicide? In other words, is T_3 telling the cells to kill themselves, or is T_3 telling something else to kill the cells? Recent evidence suggests that the first part of tail resorption is caused by suicide, but the last remnants of the tadpole tail must be killed off by other means. When tadpole muscle cells were injected with a dominant negative T_3 receptor (and therefore could not respond to T_3), the muscle cells survived, indicating that T_3 told them to kill themselves by apoptosis (Nakajima and Yaoita 2003; Nakajima et al. 2005). This was confirmed by the demonstration that the apoptosis-inducing enzyme caspase-9 is important in causing cell death in the tadpole muscle cells (Rowe et al. 2005). However, later in metamorphosis, the tail muscles are eaten by macrophages, perhaps because the extracellular matrix that supported the muscle cells has been digested by proteases.

Death also comes to the tadpole's red blood cells. During metamorphosis, tadpole hemoglobin is replaced by adult hemoglobin, which binds oxygen more slowly and releases it more rapidly (McCutcheon 1936; Riggs 1951). The red blood cells carrying the tadpole hemoglobin have a different shape than the adult red blood cells, and these larval red blood cells are specifically digested—"eaten," if you will—by macrophages in the liver and spleen (Hasebe et al. 1999).

REMODELING DURING METAMORPHOSIS Among frogs and toads, certain larval structures are remodeled for adult needs. The larval intestine, with its numerous coils for digesting plant material, is converted into a shorter intestine for a carnivorous diet. Schrieber and his colleagues (2005) have demonstrated that the new cells of the adult intestine are derived from functioning cells of the larval intestine (instead of there being a subpopulation of stem cells that give rise to the adult intestine). The formation and differentiation of this new intestinal epithelium are probably triggered by the digestion of the old extracellular matrix by the metalloproteinase stromelysin-3, and by the new transcription of the *bmp4* and

TABLE 16.1 Some metamorphic changes in anurans

System	Larva	Adult
Locomotory	Aquatic; tail fins	Terrestrial; tailless tetrapod
Respiratory	Gills, skin, lungs; larval hemoglobins	Skin, lungs; adult hemoglobins
Circulatory	Aortic arches; aorta; anterior, posterior, and common jugular veins	Carotid arch; systemic arch; cardinal veins
Nutritional	Herbivorous: long spiral gut; intestinal symbionts; small mouth, horny jaws, labial teeth	Carnivorous: short gut; proteases; large mouth with long tongue
Nervous	Lack of nictitating membrane; porphyropsin, lateral line system, Mauthner neurons	Development of ocular muscles, nictitating membrane, rhodopsin; loss of lateral line system, degeneration of Mauthner neurons; tympanic membrane
Excretory	Largely ammonia, some urea (ammonotelic)	Largely urea; high activity of enzymes of ornithine-urea cycle (ureotelic)
Integumental	Thin, bilayered epidermis with thin dermis; no mucous glands or granular glands	Stratified squamous epidermis with adult keratins; well-developed dermis contains mucous glands and granular glands secreting antimicrobial peptides

Source: Data from Turner and Bagnara 1976 and Reilly et al. 1994.

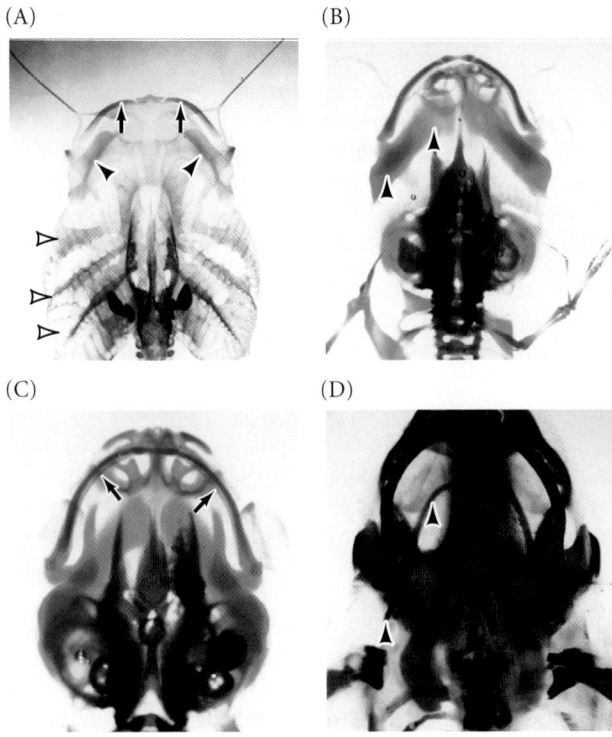

(A) (B)

(C) (D)

FIGURE 16.2 Changes in the *Xenopus* skull during metamorphosis. Whole mounts were stained with alcian blue to stain cartilage and alizarin red to stain bone. (A) Prior to metamorphosis, the pharyngeal (branchial) arch cartilage (open arrowheads) is prominent, Meckel's cartilage (arrows) is at the tip of the head, and the ceratohyal cartilage (arrowheads) is relatively wide and anteriorly placed. (B–D) As metamorphosis ensues, the pharyngeal arch cartilage disappears, Meckel's cartilage elongates, the mandible (lower jawbone) forms around Meckel's cartilage, and the ceratohyal cartilage narrows and becomes more posteriorly located. (From Berry et al. 1998, courtesy of D. D. Brown.)

sonic hedgehog genes (Stolow and Shi 1995; Ishizuya-Oka et al. 2001; Fu et al. 2005). The elimination of the original extracellular matrix probably results in the apoptosis of those epithelial cells that were attached to it.* Those remaining epithelial cells appear to de-differentiate to become intestinal stem cells that can respond to those signals (Hasabe et aal. 2013).

Much of the nervous system is remodeled as neurons grow and innervate new targets. The change in the optic nerve pathway was described earlier. Other larval neurons, such as certain motor neurons in the tadpole jaw, switch their allegiances from larval muscle to newly formed adult muscle (Alley and Barnes 1983). Still others, such as the cells

---

* Many epithelial cells are dependent on their attachment to the extracellular matrix to prevent apoptosis. The rapid apoptosis that occurs with the loss of extracellular matrix attachment has a special designation, *anoikis* (Frisch and Screaton 2001; see Chapter 3)

innervating the tongue muscle (a newly formed muscle not present in the larva), lie dormant during the tadpole stage and form their first synapses during metamorphosis (Grobstein 1987). The lateral line system of the tadpole (which allows the tadpole to sense water movement and helps it hear) degenerates, and the ears undergo further differentiation (see Fritzsch et al. 1988). The middle ear develops, as does the tympanic membrane characteristic of frog and toad outer ears. (Tadpoles experience a brief period of deafness as the neurons change targets; see Boatright-Horowitz and Simmons 1997.) Thus, the anuran nervous system undergoes enormous restructuring as some neurons die, others are born, and others change their specificity.

The shape of the anuran skull also changes significantly as practically every structural component of the head is remodeled (Trueb and Hanken 1992; Berry et al. 1998). The most obvious change is that new bone is being made. The tadpole skull is primarily neural crest-derived cartilage; the adult skull is primarily neural crest-derived bone (**FIGURE 16.2**; Gross and Hanken 2005). Another outstanding change is the formation of the lower jaw. Here, Meckel's cartilage elongates to nearly double its original length, and dermal bone forms around it. While Meckel's cartilage is growing, the gills and pharyngeal arch cartilage (which were necessary for aquatic respiration in the tadpole) degenerate. Other cartilage is extensively remodeled. Thus, as in the nervous system, some skeletal elements proliferate, some die, and some are remodeled.

BIOCHEMICAL RESPECIFICATION IN THE LIVER In addition to the obvious morphological changes, important biochemical transformations occur during metamorphosis as T_3 induces a new set of proteins in existing cells. One of the most dramatic biochemical changes occurs in the liver. Tadpoles, like most freshwater fish, are ammonotelic—that is, they excrete ammonia. Like most terrestrial vertebrates, many adult frogs (such as the genus *Rana*, although not the more aquatic *Xenopus*) are ureotelic: they excrete urea, which requires less water than ammonia excretion. During metamorphosis, the liver begins to synthesize the enzymes necessary to create urea from carbon dioxide and ammonia (**FIGURE 16.3**).

T_3 may regulate this change by inducing a set of transcription factors that specifically activates expression of the urea-cycle genes while suppressing the genes responsible for ammonia synthesis (Cohen 1970; Atkinson et al. 1996, 1998). Mukhi and colleagues (2010) showed that T_3 activates adult hepatic genes while repressing larval hepatic genes in the same cell. Moreover, for a brief time during metamorphosis, the same liver cell contains mRNAs for both larval and adult proteins.

CELL-TYPE SWITCHING IN RED BLOOD CELLS Although most of the changes during amphibian metamorphosis involve cell death or cell remodeling, the change between larval and adult blood cells appears to involve a change in cell populations. Here, T_3 causes the death of the red blood cell

(A)

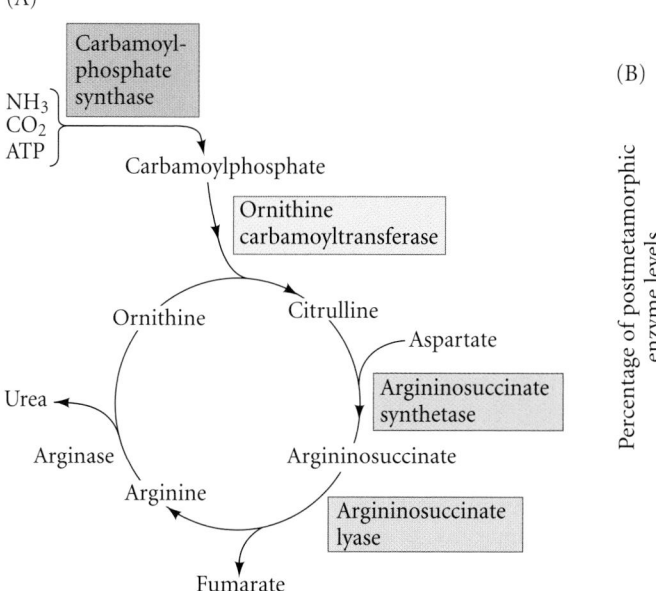

(B)

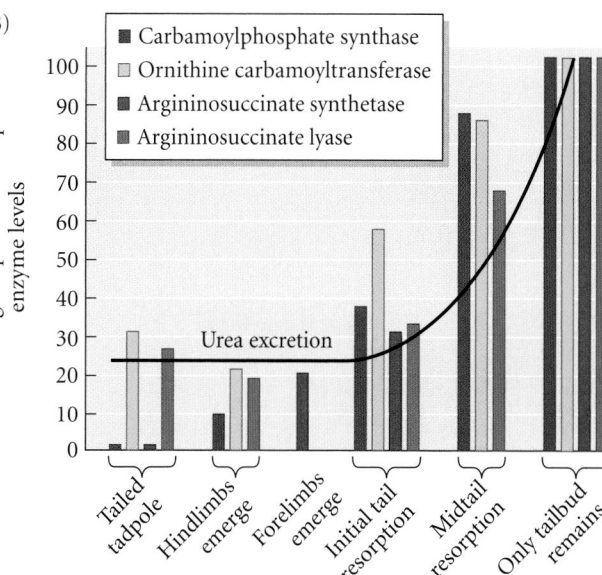

FIGURE 16.3 Development of the urea cycle during anuran metamorphosis. (A) Major features of the urea cycle, by which nitrogenous wastes are detoxified and excreted with minimal water loss. (B) The emergence of urea-cycle enzyme activities correlates with metamorphic changes in the frog *Rana catesbeiana*. (After Cohen 1970.)

precursors that make larval hemoglobin and the proliferation of those precursors that make adult hemoglobin (Nishikawa and Hayashi 1999; Mukhi et al. 2010). Larval and adult hemoglobin mRNAs are not seen in the same precursor cell (**FIGURE 16.4**).

Hormonal control of amphibian metamorphosis

The control of metamorphosis by thyroid hormones was first demonstrated in 1912 by J. F. Gudernatsch at Cornell, who

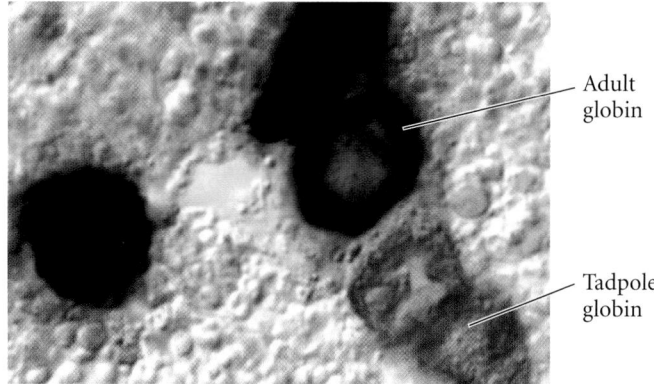

FIGURE 16.4 T_3 replaces the population of larval blood cell progenitor cells with a population of adult blood cell progenitor cells. At metamorphic climax, both populations of progenitor cells can be observed, one staining for tadpole globin mRNA (red) and one staining for adult globin mRNA (purple). No cell contains both types of mRNA. (After Mukhi et al. 2010.)

discovered that tadpoles metamorphosed prematurely when fed powdered horse thyroid glands. In a complementary study at the University of Kansas, Bennet Allen found that when he removed or destroyed the thyroid rudiment of early tadpoles, the larvae never metamorphosed but instead grew into giant tadpoles (Allen 1916). Subsequent studies showed that the sequential steps of anuran metamorphosis are regulated by increasing amounts of thyroid hormone (see Saxén et al. 1957; Kollros 1961; Hanken and Hall 1988). Some events (such as the development of limbs) occur early, when the concentration of thyroid hormones is low; other events (such as the resorption of the tail and remodeling of the intestine) occur later, after the hormones reach higher concentrations. These observations gave rise to a **threshold model**, wherein the different events of metamorphosis are triggered by different concentrations of thyroid hormones. Although the threshold model remains useful, molecular studies have shown that the timing of the events of amphibian metamorphosis is more complex than just increasing hormone concentrations.

The metamorphic changes of frog development are brought about by (1) the secretion of the hormone thyroxine (T_4) into the blood by the thyroid gland; (2) the conversion of T_4 into the more active hormone, tri-iodothyronine (T_3) by the target tissues; and (3) the degradation of T_3 in the target tissues (**FIGURE 16.5**). T_3 binds to the nuclear **thyroid hormone receptors** (**TRs**) with much higher affinity than does T_4, and causes these receptors to become transcriptional activators of gene expression. Thus, the levels of both T_3 and TRs in the target tissues are essential for producing the metamorphic response in each tissue (Kistler et al. 1977; Robinson et al. 1977; Becker et al. 1997).

The concentration of T_3 in each tissue is regulated by the concentration of T_4 in the blood and by two critical intracellular enzymes that remove iodine atoms from T_4 and T_3. **Type II deiodinase** removes an iodine atom from the outer

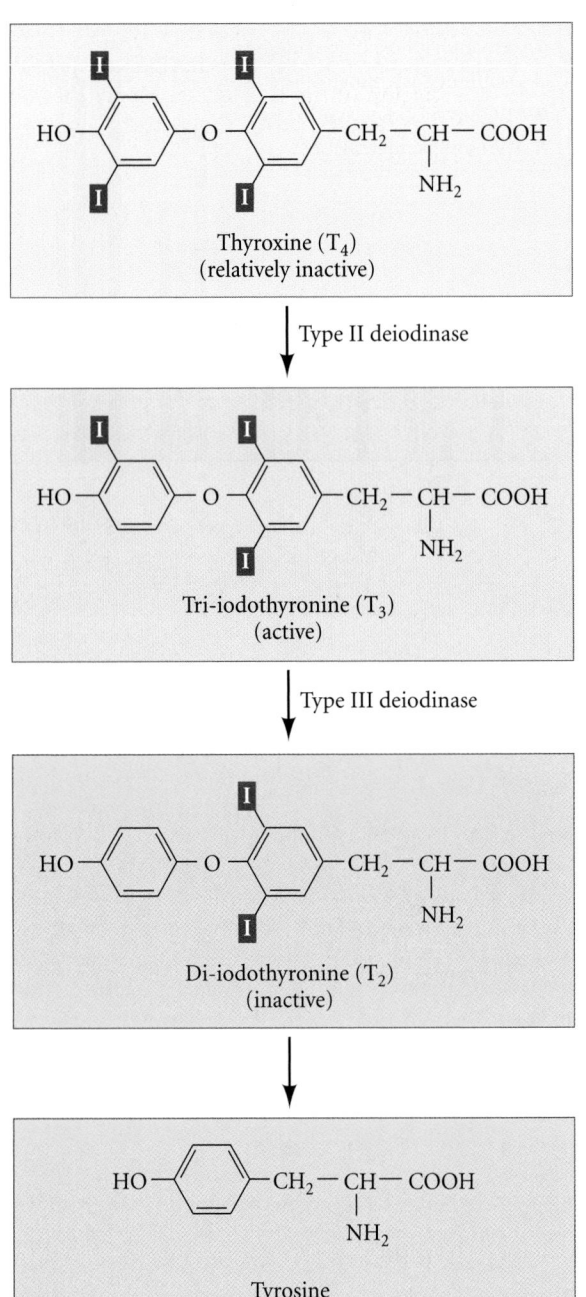

FIGURE 16.5 Metabolism of thyroxine (T_4) and tri-iodothyronine (T_3). T_4 serves as a prohormone. It is converted in the peripheral tissues to the active hormone T_3 by deiodinase II. T_3 can be inactivated by deiodinase III, which converts T_3 into di-iodothyronine and then to tyrosine.

ring of the precursor hormone (T_4) to convert it into the more active hormone T_3. **Type III deiodinase** removes an iodine atom from the inner ring of T_3 to convert it into an inactive compound that will eventually be metabolized to tyrosine (Becker et al. 1997). Tadpoles that are genetically modified to

overexpress type III deiodinase in their target tissues never complete metamorphosis (Huang et al. 1999).

There are two types of thyroid hormone receptors. In *Xenopus*, **thyroid hormone receptor α (TRα)** is widely distributed throughout all tissues and is present even before the organism has a thyroid gland. **Thyroid hormone receptor β (TRβ)**, however, is the product of a gene that is directly activated by thyroid hormones. TRβ levels are very low before the advent of metamorphosis; as the levels of thyroid hormone increase during metamorphosis, so do the intracellular levels of TRβ (Yaoita and Brown 1990; Eliceiri and Brown 1994).

The TRs do not work alone, but form dimers with the retinoid receptor RXR. TR-RXR dimers bind thyroid hormones and can then effect transcription (Mangelsdorf and Evans 1995; Wong and Shi 1995; Wolffe and Shi 1999). The TR-RXR complex appears to be physically associated with appropriate promoters and enhancers even before it binds T_3 (Grimaldi et al. 2012). In its unbound state, TR-RXR is a transcriptional repressor, recruiting histone deacetylases to its target genes. However, when T_3 is added to the complex, T_3-TR-RXR activates those same genes by recruiting histone acetyltransferases (Sachs et al. 2001; Buchholz et al. 2003; Havis et al. 2003; Paul and Shi 2003).

Metamorphosis is often divided into stages based on the concentration of thyroid hormones in the blood. During the first stage, **premetamorphosis**, the thyroid gland has begun to mature and is secreting low levels of T_4 and very low levels of T_3. T_4 secretion may be initiated by corticotropin-releasing hormone (CRH, which in mammals initiates the stress response). CRH may act directly on the frog pituitary, instructing it to release thyroid-stimulating hormone (TSH), or it may act generally to make the body cells responsive to low amounts of T_3 (Denver 1993, 2003).

The tissues that respond earliest to the thyroid hormones are those that express high levels of deiodinase II, and can thus convert T_4 directly into T_3 (Cai and Brown 2004). For instance, the limb rudiments, which have high levels of both deiodinase II and TRα, can convert T_4 into T_3 and use it immediately through the TRα receptor. Thus, during the early stage of metamorphosis, the limb rudiments are able to receive thyroid hormone and use it to start leg growth (Becker et al. 1997; Huang et al. 2001; Schreiber et al. 2001).

As the thyroid matures to the stage of **prometamorphosis**, it secretes more hormones. However, many major changes (such as tail resorption, gill resorption, and intestinal remodeling) must wait until the **metamorphic climax** stage. At that time, the concentration of T_4 rises dramatically and TRβ levels peak inside the cells. Since one of the target genes of T_3 is the *TRβ* gene, it is possible that TRβ is the principal receptor mediating the metamorphic climax. In the tail, there is only a small amount of TRα during premetamorphosis, and deiodinase II is not detectable. However, during prometamorphosis, rising levels of thyroid hormones induce higher levels of TRβ. At metamorphic climax, deiodinase II is expressed and the tail begins to be resorbed. In this way, the tail undergoes absorption only *after* the legs are functional

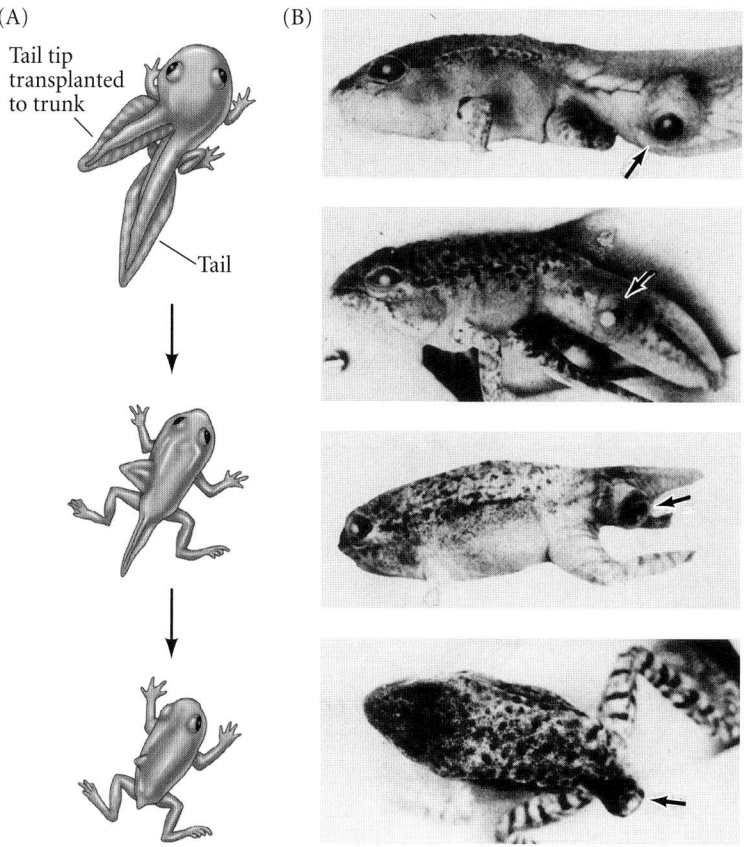

(A)

Tail tip transplanted to trunk

Tail

(B)

FIGURE 16.6 Regional specificity during frog metamorphosis. (A) Tail tips regress even when transplanted into the trunk. (B) Eye cups remain intact even when transplanted into the regressing tail. (After Schwind 1933.)

(otherwise, the poor amphibian would have no means of locomotion). The wisdom of the frog is simple: never get rid of your tail before your legs are working.

Some tissues do not seem to be responsive to thyroid hormones. For instance, thyroid hormones instruct the *ventral* retina to express ephrin B and to generate the ipsilateral neurons shown in Figure 16.1D. The *dorsal* retina, however, is not responsive to thyroid hormones and does not generate new neurons. The dorsal retina appears to be insulated from thyroid hormones by expressing deiodinase III, which degrades the T_3 produced by deiodinase II. If deiodinase III is activated in the ventral retina, neurons will not proliferate and no ipsilateral axons will be formed (Kawahara et al. 1999; Marsh-Armstrong et al. 1999).

The frog brain also undergoes changes during metamorphosis, and one of its functions is to downregulate metamorphosis once metamorphic climax has been reached. Thyroid hormones eventually induce a negative feedback loop, shutting down the pituitary cells that instruct the thyroid to secrete them (Saxén et al. 1957; Kollros 1961; White and Nicoll 1981). Huang and colleagues (2001) have shown that, at the climax of metamorphosis, deiodinase II expression is seen in those cells of the anterior pituitary that secrete

thyrotropin, the hormone that activates thyroid hormone expression. The resulting T_3 suppresses transcription of the thyrotropin gene, thereby initiating the negative feedback loop so that less thyroid hormone is made.

● **See VADE MECUM** Amphibian metamorphosis and frog calls

Regionally specific developmental programs

By regulating the amount of T_3 and TRs in their cells, the different regions of the body can respond to thyroid hormones at different times. The type of response (proliferation, apoptosis, differentiation, migration) is determined by other factors already present in the different tissues. The same stimulus causes some tissues to degenerate while stimulating others to develop and differentiate, as exemplified by the process of tail degeneration. Thus, thyroid hormone instructs the limb bud muscles to grow (they die without thyroxine) while instructing the tail muscles to undergo apoptosis (Cai et al. 2007).

The resorption of the tadpole's tail structures is brought about by apoptosis and is relatively rapid, since the bony skeleton does not extend to the tail (Wassersug 1989). After apoptosis has taken place, macrophages collect in the tail region and digest the debris with their enzymes, especially collagenases and metalloproteinases. The result is that the tail becomes a large sac of proteolytic enzymes* (Kaltenbach et al. 1979; Oofusa and Yoshizato 1991; Patterson et al. 1995). The tail epidermis acts differently than the head or trunk epidermis. During metamorphic climax, the larval skin is instructed to undergo apoptosis. The tadpole head and body are able to generate a new epidermis from epithelial stem cells. The tail epidermis, however, lacks these stem cells and fails to generate new skin (Suzuki et al. 2002).

Organ-specific responses to thyroid hormones have been dramatically demonstrated by transplanting a tail tip to the trunk region and by placing an eye cup in the tail (Schwind 1933; Geigy 1941). Tail tip tissue placed in the trunk is not protected from degeneration, but the eye cup retains its integrity despite the fact that it lies within the degenerating tail (**FIGURE 16.6**). Thus, the degeneration of the tail represents an organ-specific programmed cell death response, and only specific tissues die when the signal is given. Such programmed cell deaths are important in molding the body.

The metamorphosis of tadpoles into frogs is one of the most rapid and accessible examples of development, obvious

---

*Interestingly, the degeneration of the human tail during week 4 of gestation resembles the resorption of the tadpole tail (see Fallon and Simandl 1978).

even to the eyes of children. Yet it still presents an enormous set of enigmas. As Don Brown and Liquan Cai (2007) have asked, "What will encourage the modern generation of scientists to study the wonderful biological problems presented by amphibian metamorphosis?" Humans may have medically centered interests in preventing aging and in regenerating amputated parts; undergoing metamorphosis is not high on our to-do lists. However, the sheer wonder of these phenomena, the ability of computer-aided sequence analysis to obtain the data, and the desire to preserve these species should be enough motivation to study these phenomena.

SIDELIGHTS & SPECULATIONS

Variations on the Theme of Amphibian Metamorphosis

Many amphibians have altered their life cycle by modifying the duration of their larval stage. This phenomenon, whereby animals change the relative time of appearance and rate of development of characters present in their ancestors, is called **heterochrony**. Here we will describe three extreme types of heterochrony:

- *Neoteny* refers to the retention of the juvenile form as a result of retarded body development relative to the development of the germ cells and gonads (which achieve maturity at the normal time).

- *Progenesis* also involves the retention of the juvenile form, but in this case the gonads and germ cells develop at a faster rate than normal, becoming sexually mature while the rest of the body is still in a juvenile phase.

- In *direct development*, the embryo abandons the stages of larval development entirely and proceeds to construct a small adult.

Neoteny

In certain salamanders, the reproductive system and germ cells mature while the rest of the body retains its juvenile form throughout life. In most such species, metamorphosis fails to occur and sexual maturity takes place in a "larval" body.

The Mexican axolotl (*Ambystoma mexicanum*) does not undergo metamorphosis in nature because its pituitary gland does not release the thyrotropin (thyroid-stimulating hormone) that would activate T_4 synthesis (Prahlad and DeLanney 1965; Norris et al. 1973; Taurog et al. 1974). The axolotl does synthesize functional thyroid hormone receptors, however, and when investigators administered either thyroid hormones or thyrotropin, they found that the salamander metamorphosed into an adult form not seen in nature (**FIGURE 16.7**; Huxley 1920; Safi et al. 2004).

Other species of *Ambystoma*, such as *A. tigrinum*, metamorphose only in response to cues from the environment. In parts of its range, *A. tigrinum* is neotenic: its gonads and germ cells mature and the salamander mates successfully while the rest of the body retains its aquatic larval form. However, in other regions of its range, the larval form is transitory, leading to the land-dwelling adult tiger salamander.

(A)

(B)

FIGURE 16.7 Metamorphosis in *Ambystoma*. (A) Normal adult *Ambystoma*, with prominent gills and broad tail. (B) Metamorphosed *Ambystoma* not seen in natural populations. This individual was grown in water supplemented with thyroxine. Its gills have regressed, and its skin has changed significantly. (A © Mark Boulton/Alamy; B courtesy of K. Crawford.)

The ability to remain aquatic is highly adaptive in locations where the terrestrial environment is too dry to sustain the adult form of this salamander (Duellman and Trueb 1986).

Some salamanders are permanently neotenic, even in the laboratory. Whereas T_3 is able to produce the long-lost adult form of *A. mexicanum*, the neotenic species of *Necturus* and *Siren* remain unresponsive to thyroid hormones (Frieden 1981). Strangely, *Necturus* was recently found to have functional thyroid hormone receptors. It appears that these receptors do not bind to those genes that initiate and promote metamorphosis (Safi et al. 2006; Vlaeminck-Guillem et al. 2006).

De Beer (1940) and Gould (1977) have speculated that neoteny is a major factor in the evolution of more complex taxa. By retarding the development of somatic tissues, neoteny may give natural selection a flexible substrate. According to Gould (1977, p. 283), neoteny may "provide an escape from specialization. Animals can relinquish their highly specialized adult forms, return to the lability of youth, and prepare themselves for new evolutionary directions."

Progenesis

In progenesis, gonadal maturation is accelerated while the rest of the body develops normally to a certain stage. Progenesis has enabled some salamander species to find new ecological niches. *Bolitoglossa occidentalis* is a tropical salamander that, unlike other members of its genus, lives in trees. This salamander's webbed feet and small body size suit it for arboreal existence, the webbed feet producing suction for climbing and the small body making such traction efficient. Alberch and Alberch (1981) showed that *B. occidentalis* resembles juveniles of the related species *B. subpalmata* and *B. rostrata* (whose young are small, with digits that have not yet grown past their webbing). *B. occidentalis* reaches sexual maturity at a much smaller size than its relatives, and this appears to have given it a phenotype that made tree-dwelling possible.

Direct development

While some animals have extended their larval life stage, others have "accelerated" their development by abandoning their larval form for direct development. Thus, there are frog species that lack tadpoles and sea urchins that have no pluteus larvae.

Elinson and his colleagues (del Pino and Elinson 1983; Elinson 1987) have studied *Eleutherodactylus coqui*, a small frog that is one of the most abundant vertebrates on the island of Puerto Rico. Unlike the eggs of *Rana* and *Xenopus*, the eggs of *E. coqui* are fertilized while they are still in the female's body. Each egg is about 3.5 mm in diameter (roughly 20 times the volume of a *Xenopus* egg). After the eggs are laid, the male gently sits on the developing embryos, protecting them from predators and desiccation (Taigen et al. 1984).

Early *E. coqui* development is like that of most frogs. Cleavage is holoblastic, gastrulation is initiated at a subequatorial position, and the neural folds become elevated from the surface. However, shortly after the neural tube closes, limb buds appear on the surface (FIGURE 16.8A,B). This early emergence of limb buds is the first indication that this animal will not pass through the usual limbless tadpole stage. Moreover, the development of *E. coqui* is modified such that the modeling of most of its features—including its limbs—does not depend on thyroid hormones. Its thyroid gland does develop, however, and thyroid hormones appear to be critical for the eventual resorption of the tail (which is used as a respiratory rather than a locomotor organ), the differentiation of the skin, and the remodeling of the kidney and musculature (Lynn and Peadon 1955; Callery and Elinson 2000). It appears that the thyroid-dependent phase has been pushed back into embryonic growth (Hanken et al. 1992; Callery et al. 2001). What emerges from the egg jelly 3 weeks after fertilization is not a tadpole but a tiny frog (FIGURE 16.8C).

Direct-developing frogs do not need ponds for their larval stages and can therefore colonize habitats

(A)

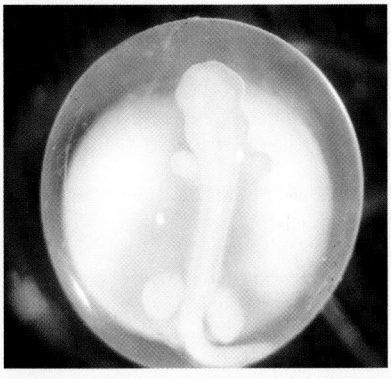

(B)

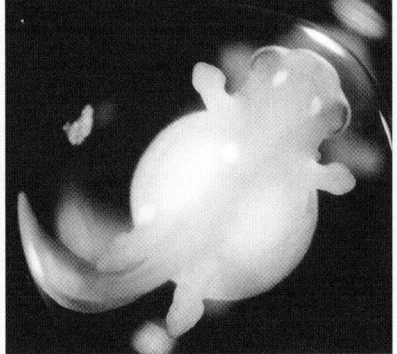

(C)

FIGURE 16.8   Direct development of the frog *Eleutherodactylus coqui*. (A) Limb buds are seen as the embryo develops on the yolk. (B) As the yolk is used up, the limb buds are easily seen. (C) Three weeks after fertilization, tiny froglets hatch. They are seen here in a petri dish and on a Canadian dime. (Courtesy of R. P. Elinson.)

that are inaccessible to other frogs. Direct development also occurs in other phyla, in which it is also correlated with a large egg. It seems that if nutrition can be provided in the egg, the life cycle need not have a food-gathering larval stage.

Tadpole-rearing behaviors

Most temperate-zone frogs do not invest time or energy in providing for their tadpoles. However, among tropical frogs, there are numerous species in which adult frogs take painstaking care of their tadpoles. An example is the poison dart frog *Dendrobates*, found in the rain forests of Central and South America. Most of the time, these highly toxic frogs live in the leaf litter of the forest floor. After the eggs are laid in a damp leaf, a parent (sometimes the male, sometimes the female, according to the species) stands guard over the eggs. If the ground gets too dry, the frog will urinate on the eggs to keep them moist. When the eggs mature into tadpoles, the guarding parent allows them to wriggle onto its back (**FIGURE 16.9A**). The parent then climbs into the canopy until it finds a bromeliad plant with a small pool of water in its leaf base. Here it deposits one of its tadpoles, then goes back for another, and so on until the entire brood has been placed into numerous small pools. The female returns each day to these pools and deposits a small number of unfertilized eggs into them, thus replenishing the tadpoles' food supply until they complete metamorphosis (Mitchell 1988; van Wijngaarden and Bolanos 1992; Brust 1993). It is not known how the female frog remembers—or is informed about—where the tadpoles have been deposited.

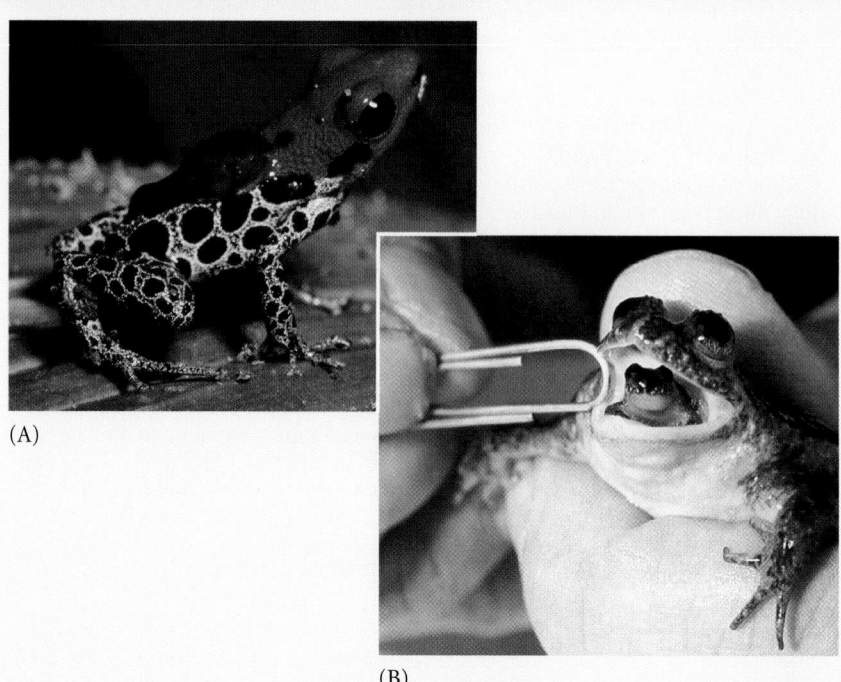

(A)

(B)

FIGURE 16.9 Parental care of tadpoles. (A) Tadpoles of the poison dart frog *Dendrobates* are carried on their parent's back to small pools of water in the Peruvian rain forest canopy. (B) This female *Rheobatrachus* of Australia brooded more than a dozen tadpoles in her stomach. They emerged after completing metamorphosis. Unfortunately, the last time anyone saw a *Rheobatrachus* frog alive was in 1985. (A © Michael Doolittle/Alamy; B courtesy of M. Tyler.)

Brooding frogs carry their developing eggs in depressions in their skin. Some species brood their tadpoles in their mouth and spit out their progeny when their tadpoles undergo metamorphosis. Even more impressive, the gastric-brooding frogs of Australia, *Rheobatrachus silus* and *R. vitellinus*, eat their eggs. The eggs develop into larvae, and the larvae undergo metamorphosis in the mother's stomach. About 8 weeks after being swallowed alive, about two dozen small frogs emerge from the female's mouth (**FIGURE 16.9B**; Corben et al. 1974; Tyler 1983). What stops the *Rheobatrachus* eggs from being digested or excreted? It appears that the eggs secrete prostaglandins that stop acid secretion and prevent peristaltic contractions in the stomach (Tyler et al. 1983). During this time, the stomach is fundamentally a uterus, and the frog does not eat. After the oral birth, the parent's stomach morphology and function return to normal. Unfortunately, both of these remarkable frog species are now believed to be extinct. No member of either *Rheobatrachus* species has been seen since the mid-1980s.

Metamorphosis in Insects

Whereas amphibian metamorphosis is largely characterized by the remodeling of existing tissues, insect metamorphosis primarily involves the destruction of larval tissues and their replacement by an entirely different population of cells. Insects grow by molting—shedding their cuticle—and forming a new cuticle as their size increases. There are three major patterns of insect development. A few insects, such as springtails, have no larval stage and undergo direct, or **ametabolous**, development (**FIGURE 16.10A**). Immediately after they hatch, ametabolous insects have a **pronymph** stage bearing the structures that enabled it to get out of the egg. But after this transitory stage, the insect looks like a small adult; it grows larger after each molt but is unchanged in form (Truman and Riddiford 1999).

Other insects, notably grasshoppers and bugs, undergo a gradual, or **hemimetabolous**, metamorphosis (**FIGURE**

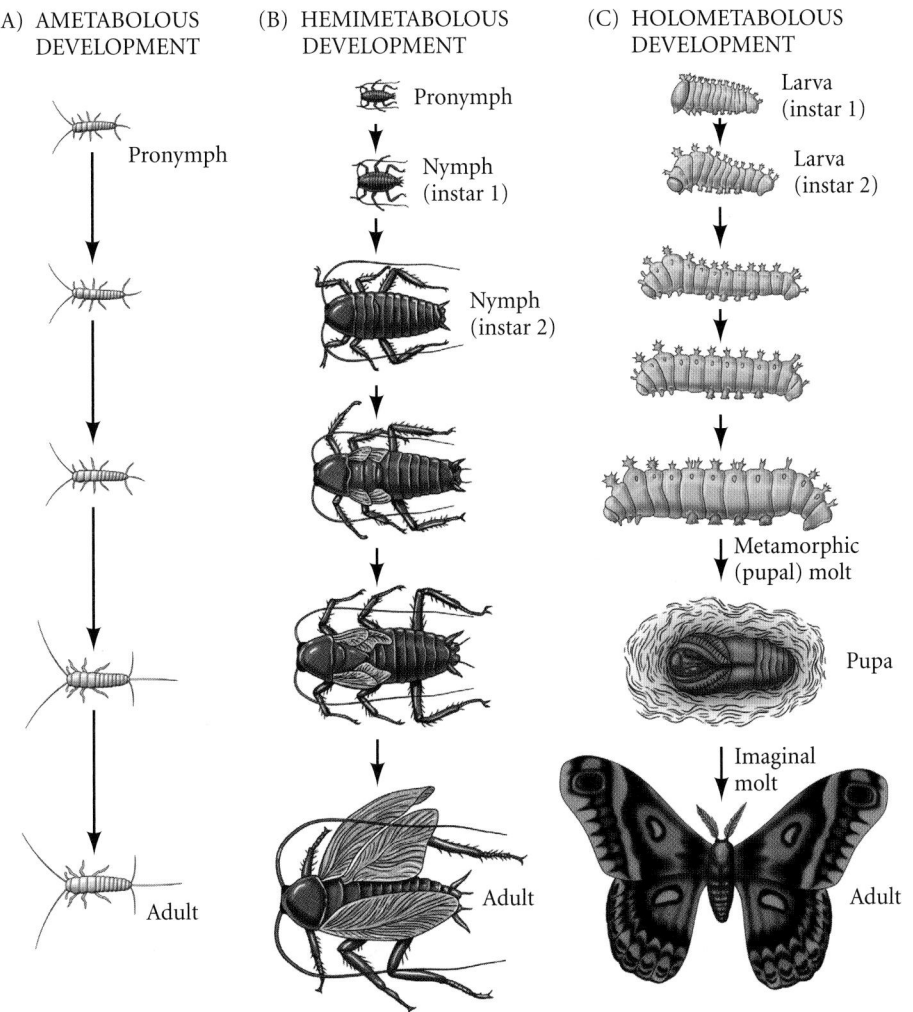

(A) AMETABOLOUS DEVELOPMENT

Pronymph

Adult

(B) HEMIMETABOLOUS DEVELOPMENT

Pronymph

Nymph (instar 1)

Nymph (instar 2)

Adult

(C) HOLOMETABOLOUS DEVELOPMENT

Larva (instar 1)

Larva (instar 2)

Metamorphic (pupal) molt

Pupa

Imaginal molt

Adult

FIGURE 16.10 Modes of insect development. Molts are represented as arrows. (A) Ametabolous (direct) development in a silverfish. After a brief pronymph stage, the insect looks like a small adult. (B) Hemimetabolous (gradual) metamorphosis in a cockroach. After a very brief pronymph phase, the insect becomes a nymph. After each molt, the next nymphal instar looks more like an adult, gradually growing wings and genital organs. (C) Holometabolous (complete) metamorphosis in a moth. After hatching as a larva, the insect undergoes successive larval molts until a metamorphic molt causes it to enter the pupal stage. Then an imaginal molt turns it into an adult.

16.10B). After spending a very brief period of time as a pronymph (whose cuticle is often shed as the insect hatches), the insect looks like an immature adult and is called a **nymph**. The rudiments of the wings, genital organs, and other adult structures are present and become progressively more mature with each molt. At the final molt, the emerging insect is a winged and sexually mature adult, or **imago**.

In the **holometabolous** development of insects such as flies, beetles, moths, and butterflies, there is no pronymph stage (**FIGURE 16.10C**). The juvenile form that hatches from the egg is called a **larva**. The larva (a caterpillar, grub, or maggot) undergoes a series of molts as it becomes larger. The stages between these larval molts are called **instars**. The number of larval molts before becoming an adult is characteristic of a species, although environmental factors can increase or decrease the number. The larval instars grow in a stepwise fashion, each instar being larger than the previous one. Finally, there is a dramatic and sudden transformation between the larval and adult stages: after the final instar, the larva undergoes a **metamorphic molt** to become a **pupa**. The pupa does not feed, and its energy must come from those foods it ingested as a larva. During

pupation, adult structures form and replace the larval structures. Eventually, an **imaginal molt** enables the adult (imago) to shed its pupal case and emerge. While the larva is said to *hatch* from an egg, the imago is said to *eclose* from the pupa.

Imaginal discs

In holometabolous insects, the transformation from juvenile into adult occurs within the pupal cuticle. Most of the larval body is systematically destroyed by programmed cell death, while new adult organs develop from relatively undifferentiated nests of **imaginal cells**. Thus, within any larva there are two distinct populations of cells: the larval cells, which are used for the functions of the juvenile insect; and thousands of imaginal cells, which lie within the larva in clusters, awaiting the signal to differentiate.

There are three main types of imaginal cells (**FIGURE 16.11**):

1. The cells of **imaginal discs** will form the cuticular structures of the adult, including the wings, legs, antennae, eyes, head, thorax, and genitalia.

Discs for:

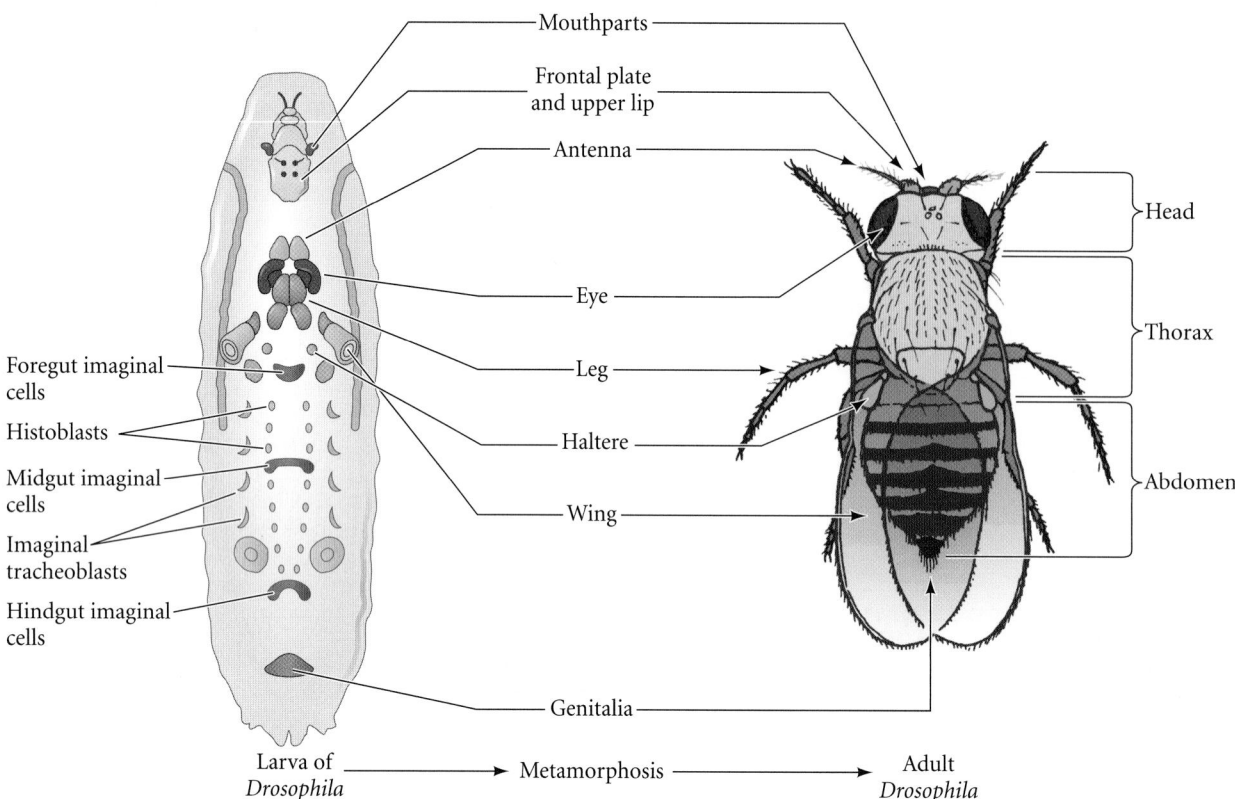

FIGURE 16.11 Locations and developmental fates of imaginal discs and imaginal tissues in the third instar larva of *Drosophila melanogaster*. (After Kalm et al. 1995.)

2. **Histoblasts** (tissue-forming cells) are imaginal cells that will form the adult abdomen.

3. In addition, there are clusters of imaginal cells within each organ that will proliferate to form the adult organ as the larval organ degenerates.

In the newly hatched larvae, the imaginal discs are visible as local thickenings of the epidermis. Each disc in the early larva has about 10–50 cells, and there are 19 such discs in *Drosophila*. The epidermis of the head, thorax, and limbs comes from nine bilateral pairs of discs, whereas the epidermis of the genitalia is derived from a single disc at the midline.

Whereas most larval cells have a very limited mitotic capacity, imaginal discs divide rapidly at specific characteristic times. As their cells proliferate, the discs form a tubular epithelium that folds in on itself in a compact spiral (**FIGURE 16.12A**). At metamorphosis, these cells proliferate even further as they differentiate, and elongate (**FIGURE 16.12B**). The fate map and elongation sequence of one of the six *Drosophila* leg discs is shown in **FIGURE 16.13**. At the end of the third instar, just before pupation, the leg disc is an epithelial sac connected by a thin stalk to the larval epidermis. On one side

of the sac, the epithelium is coiled into a series of concentric folds "reminiscent of a Danish pastry" (Kalm et al. 1995). As pupation begins, the cells at the center of the disc telescope out to become the most distal portions of the leg—the claws and the tarsus. The outer cells become the proximal structures—the coxa and the adjoining epidermis (Schubiger 1968). After differentiating, the cells of the appendages

(A)

(B)

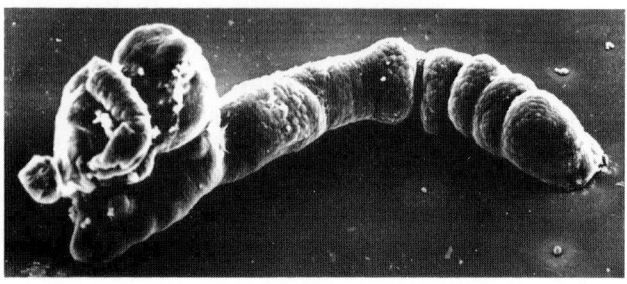

FIGURE 16.12 Imaginal disc elongation. Scanning electron micrograph of *Drosophila* third instar leg disc (A) before and (B) after elongation. (From Fristrom et al. 1977; courtesy of D. Fristrom.)

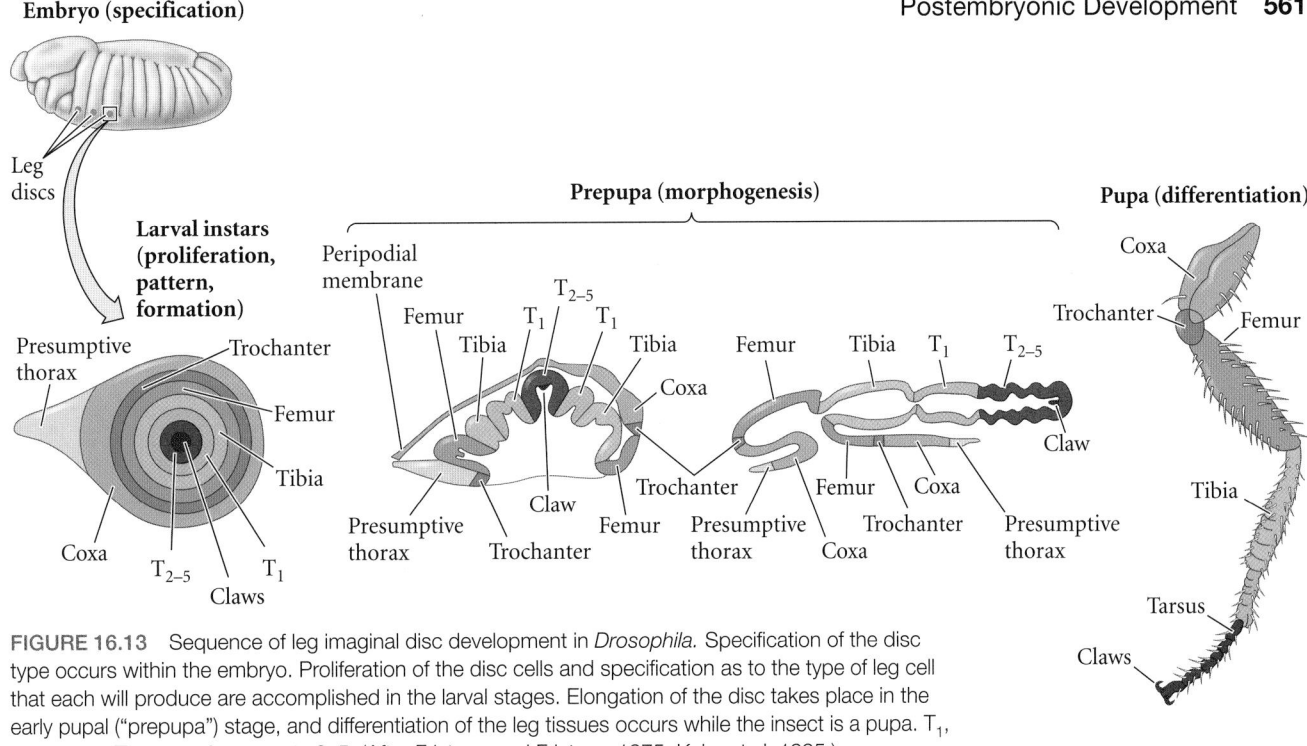

FIGURE 16.13 Sequence of leg imaginal disc development in *Drosophila*. Specification of the disc type occurs within the embryo. Proliferation of the disc cells and specification as to the type of leg cell that each will produce are accomplished in the larval stages. Elongation of the disc takes place in the early pupal ("prepupa") stage, and differentiation of the leg tissues occurs while the insect is a pupa. T_1, basitarsus; T_{2-5}, tarsal segments 2–5. (After Fristrom and Fristrom 1975; Kalm et al. 1995.)

and epidermis secrete a cuticle appropriate for each specific region. Although the disc is composed primarily of epidermal cells, a small number of **adepithelial cells** migrate into the disc early in development. During the pupal stage, these cells give rise to the muscles and nerves that serve the legs.

SPECIFICATION AND PROLIFERATION Specification of the general cell fates (i.e., that the disc is to be a leg disc and not a wing disc) occurs in the embryo and is mediated primarily by the Hox genes such as *Ultrabithorax* and *Antennapedia*. The more specific cell fates are specified in the larval stages, as the cells proliferate (Kalm et al. 1995). The type of leg structure (claw, femur, etc.) generated is determined by the interactions between several genes in the imaginal disc. **FIGURE 16.14** shows the expression of three genes involved in determining the proximal-distal axis of the fly leg. In the third instar leg disc, the center of the disc secretes the highest concentration of two morphogens, Wingless (Wg, a Wnt

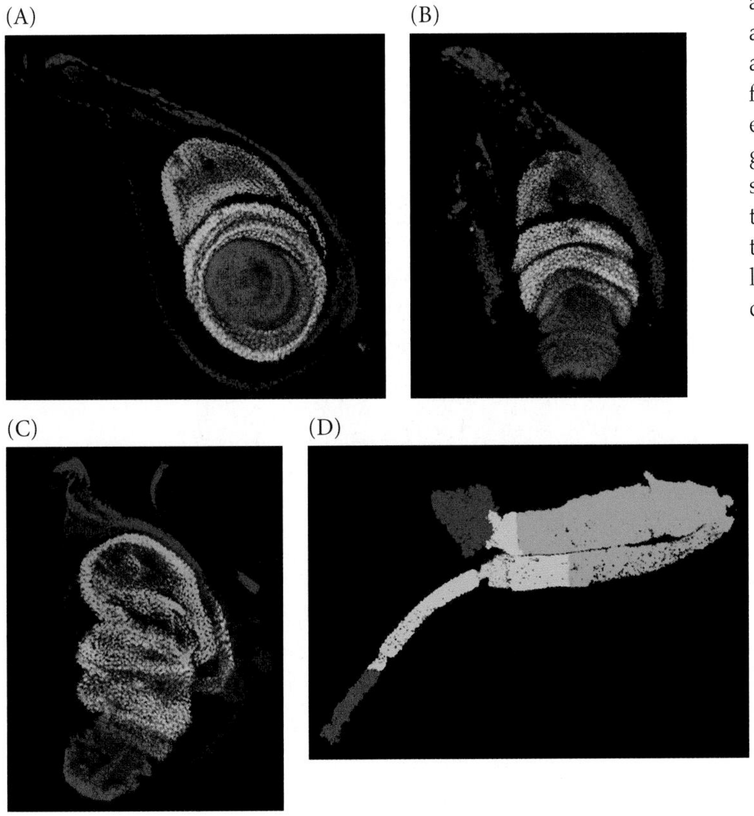

(A)

(B)

(C)

(D)

FIGURE 16.14 Expression of transcription factor genes in the *Drosophila* leg disc. At the periphery, the *homothorax* gene (purple) establishes the boundary for the coxa. The expression of the *dachshund* gene (green) locates the femur and proximal tibia. The most distal structures, the claw and distal tarsal segments, arise from the expression domain of *Distal-less* (red) in the center of the imaginal disc. The overlap of *dachshund* and *Distal-less* appears yellow and specifies the distal tibia and trochanter segments. (A–C) Gene expression at successively later stages of pupal development. (D) Localization of expression domains of the genes onto a leg immediately prior to eclosion. The areas where there is overlap between expression domains are shown in yellow, aqua, and orange. (From Abu-Shaar and Mann 1998, courtesy of R. S. Mann.)

paracrine factor) and Decapentaplegic (Dpp, a BMP paracrine factor). High concentrations of these paracrine factors cause expression of the *Distal-less* gene. Moderate concentrations cause the expression of the *dachshund* gene, and lower concentrations cause the expression of the *homothorax* gene.

Those cells expressing *Distal-less* telescope out to become the most distal structures of the leg—the claw and distal tarsal segments. Those expressing *homothorax* become the most proximal structure, the coxa. Cells expressing *dachshund* become the femur and proximal tibia. Areas where the transcription factors overlap produce the trochanter and distal tibia (Abu-Shaar and Mann 1998). These regions of gene expression are stabilized by inhibitory interactions between the protein products of these genes and of the neighboring genes. In this manner, the gradient of Wg and Dpp proteins is converted into discrete domains of gene expression that specify the different regions of the *Drosophila* leg.

EVERSION AND DIFFERENTIATION   The mature leg disc in the third instar of *Drosophila* does not look anything like the adult structure. It is determined but not yet differentiated; its differentiation requires a signal, in the form of a set of pulses of the "molting" hormone **20-hydroxyecdysone** (**20E**; see Figure 16.17A). The first pulse, occurring in the late larval stages, initiates formation of the pupa, arrests cell division in the disc, and initiates the cell shape changes that drive the eversion of the leg. Studies by Condic and her colleagues have demonstrated that the elongation of imaginal discs occurs without cell division and is due primarily to cell shape changes within the disc epithelium (Condic et

al. 1990). Using fluorescently labeled phalloidin to stain the peripheral microfilaments of leg disc cells, they showed that the cells of early third instar discs are tightly arranged along the proximal-distal axis. When the hormonal signal to differentiate is given, the cells change their shape and the leg is everted, the central cells of the disc becoming the most distal (claw) cells of the limb. The leg structures will differentiate within the pupa, so that by the time the adult fly ecloses they are fully formed and functional.

Determination of the wing imaginal discs

The largest of *Drosophila*'s imaginal discs is that of the wing, containing some 60,000 cells. (In contrast, the leg and haltere discs contain about 10,000 cells each; Fristrom 1972.) The wing discs are distinguished from the other imaginal discs by the expression of the *vestigial* gene (Kim et al. 1996). When this gene is expressed in any other imaginal disc, wing tissue emerges.

ANTERIOR AND POSTERIOR COMPARTMENTS   The axes of the wing are specified by gene expression patterns that divide the embryo into discrete but interacting compartments (FIGURE 16.15A; Meinhardt 1980; Causo et al. 1993; Tabata et al. 1995). In the first instar, expression of the *engrailed* gene distinguishes the posterior compartment of the wing from the anterior compartment. The Engrailed transcription factor is expressed only in the posterior compartment, and in those cells it activates the gene for the BMP-like paracrine factor Hedgehog. Hedgehog functions only when cells have the receptor (Patched) to receive it. In a complex manner, the

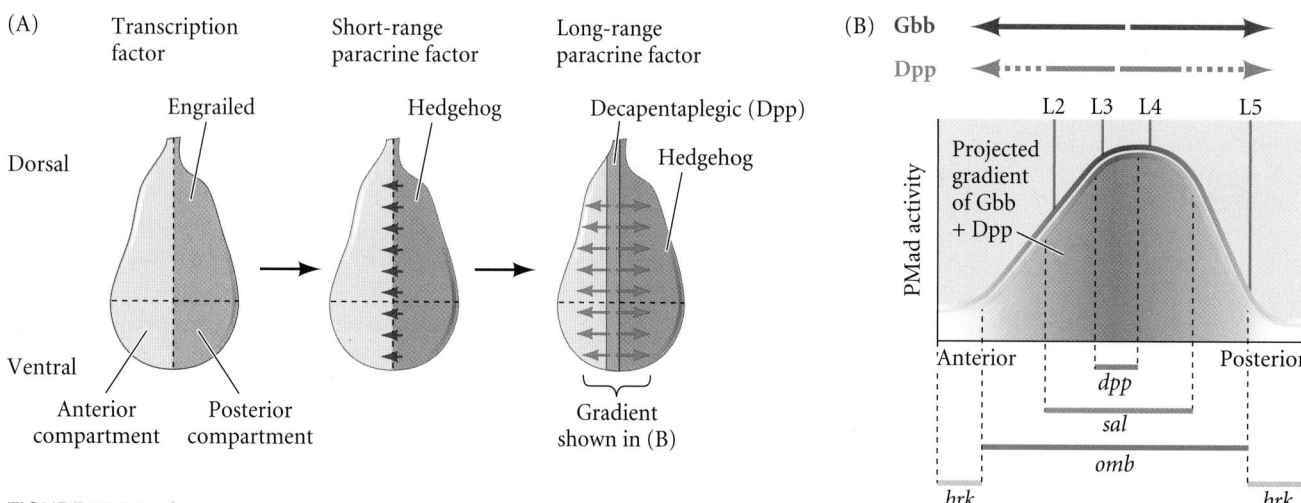

FIGURE 16.15   Compartmentalization and anterior-posterior patterning in the wing imaginal disc. (A) In the first instar larva, the anterior-posterior axis has been formed and can be recognized by the expression of the *engrailed* gene in the posterior compartment. Engrailed, a transcription factor, activates the *hedgehog* gene. Hedgehog acts as a short-range paracrine factor to activate *decapentaplegic* (*dpp*) in the anterior cells adjacent to the posterior compartment, where Dpp and a related protein, Glass-bottom boat (Gbb), act over a longer range. (B) Dpp and Gbb proteins create

a concentration gradient of BMP-like signaling, measured by the phosphorylation of Mad (pMad). High concentrations of Dpp plus Gbb near the source activate both the *spalt* (*sal*) and *optomotor blind* (*omb*) genes. Lower concentrations (near the periphery) activate *omb* but not *sal*. When Dpp plus Gbb levels drop below a certain threshold, *brinker* (*brk*) is no longer repressed. L2–L5 mark the longitudinal wing veins, with L2 being the most anterior. (After Bangi and Wharton 2006.)

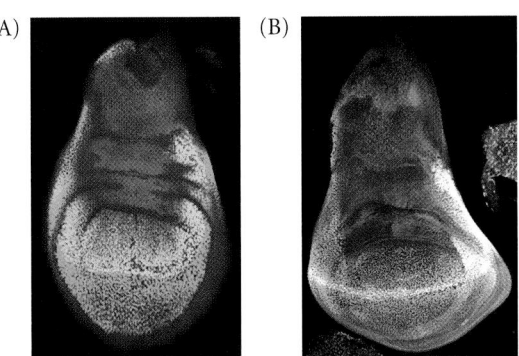

(A)   (B)

FIGURE 16.16 Determining the dorsal-ventral axis. (A) The prospective ventral surface of the wing is stained by antibodies to Vestigial protein (green), while the prospective dorsal surface is stained by antibodies to Apterous protein (red). The region of yellow illustrates where the two proteins overlap in the margin. (B) Wingless protein (purple) synthesized at the marginal juncture organizes the wing disc along the dorsal-ventral axis. The expression of Vestigial (green) is seen in cells close to those expressing Wingless. (C) The dorsal and ventral portions of the wing disc telescope out to form the two-layered wing. Gene expression patterns are indicated on the double-layered wing. (A,B courtesy of S. Carroll and S. Paddock.)

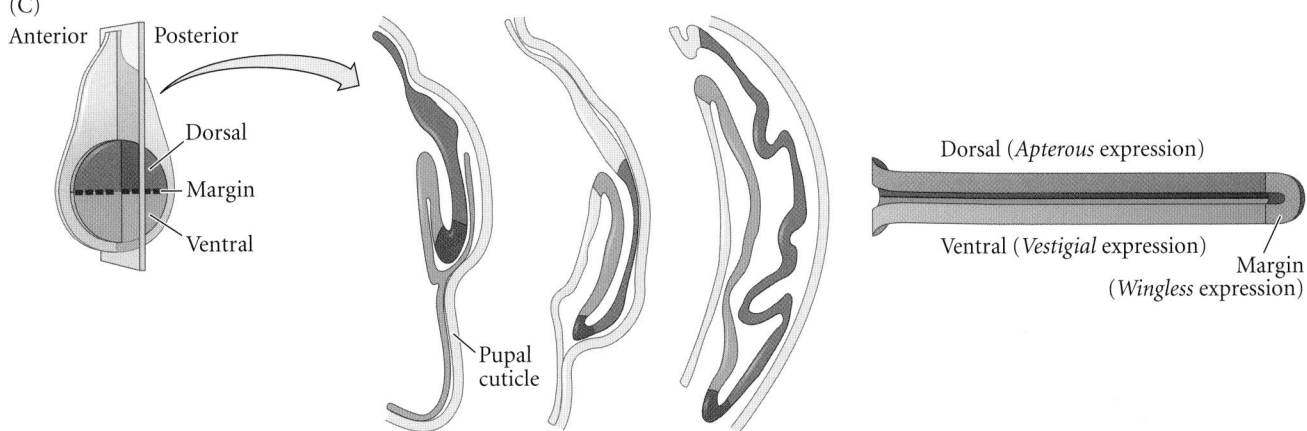

(C)

Anterior   Posterior

Dorsal

Margin

Ventral

Pupal cuticle

Dorsal (*Apterous* expression)

Ventral (*Vestigial* expression)

Margin (*Wingless* expression)

diffusion of Hedgehog activates the gene encoding Deca-pentaplegic (Dpp) in a narrow stripe of cells in the anterior region of the wing disc (Ho et al. 2005).

Dpp and a co-expressed BMP called Glass-bottom boat (Gbb) establish a gradient of BMP signaling activity (Matsuda and Shimmi 2012). BMPs activate the Mad transcription factor (a Smad protein) by phosphorylating it, so this gradient can be measured by the phosphorylation of Mad. Dpp is a short-range paracrine factor, whereas Gbb exhibits a much longer range of diffusion to create a gradient (**FIGURE 16.15B**; Bangi and Wharton 2006). This signaling gradient regulates the amount of cell proliferation in the wing regions and also specifies cell fates (Rogulja and Irvine 2005). Several transcription factor genes respond differently to activated Mad. At high levels, the *spalt* (*sal*) and *optomotor blind* (*omb*) genes are activated, whereas at low levels (where Gbb provides the primary signal), only *omb* is activated. Below a particular level of phosphorylated Mad activity, the *brinker* (*brk*) gene is no longer inhibited; thus, *brk* is expressed outside the signaling domain. Specific cell fates of the wing are specified in response to the action of these transcription factors. (For example, the fifth longitudinal vein of the wing is formed at the border of *optomotor blind* and *brinker*; see Figure 16.15B).

DORSAL-VENTRAL AND PROXIMAL-DISTAL AXES The dorsal-ventral axis of the wing is formed at the second instar stage by the expression of the *apterous* gene in the prospective dorsal cells of the wing disc (Blair 1993; Diaz-Benjumea and Cohen 1993). Here, the upper layer of the wing is distinguished from the lower layer of the wing blade (Bryant 1970; Garcia-Bellido et al. 1973). The *vestigial* gene remains "on" in the ventral portion of the wing disc (**FIGURE 16.16A**). The dorsal portion of the wing synthesizes transmembrane proteins that prevent the intermixing of the dorsal and ventral cells (Milán et al. 2005). At the boundary between the dorsal and ventral compartments, the Apterous and Vestigial transcription factors interact to activate the gene encoding the Wnt paracrine factor Wingless (**FIGURE 16.16B**; see also Figure 3.41). Neumann and Cohen (1996) showed that Wingless protein acts as a growth factor to promote the cell proliferation that extends the wing.* Wingless also helps establish the proximal-distal axis of the wing: high levels of Wingless activate the *Distal-less* gene, which specifies the most distal regions of the wing (Neumann and Cohen 1996, 1997; Zecca et al. 1996). This occurs in the central region of the disc and "telescopes" outward as the distal margin of the wing blade (**FIGURE 16.16C**).

● **See WEBSITE 16.1** The molecular biology of wing formation

● **See WEBSITE 16.2** Homologous specification

---

* As discussed in Chapter 3, the diffusion of Wingless and other paracrine factors depends on their binding to extracellular matrix proteins. The diffusion of Wingless and Hedgehog may be facilitated when these factors cluster on lipid spheres that can travel between cells without getting caught in the extracellular matrix (Glise et al. 2005; Gorfinkiel et al. 2005; Panáková et al. 2005).

(A) Juvenile hormone (JH)

Ecdysone

20-Hydroxyecdysone (20E)
("molting" hormone)

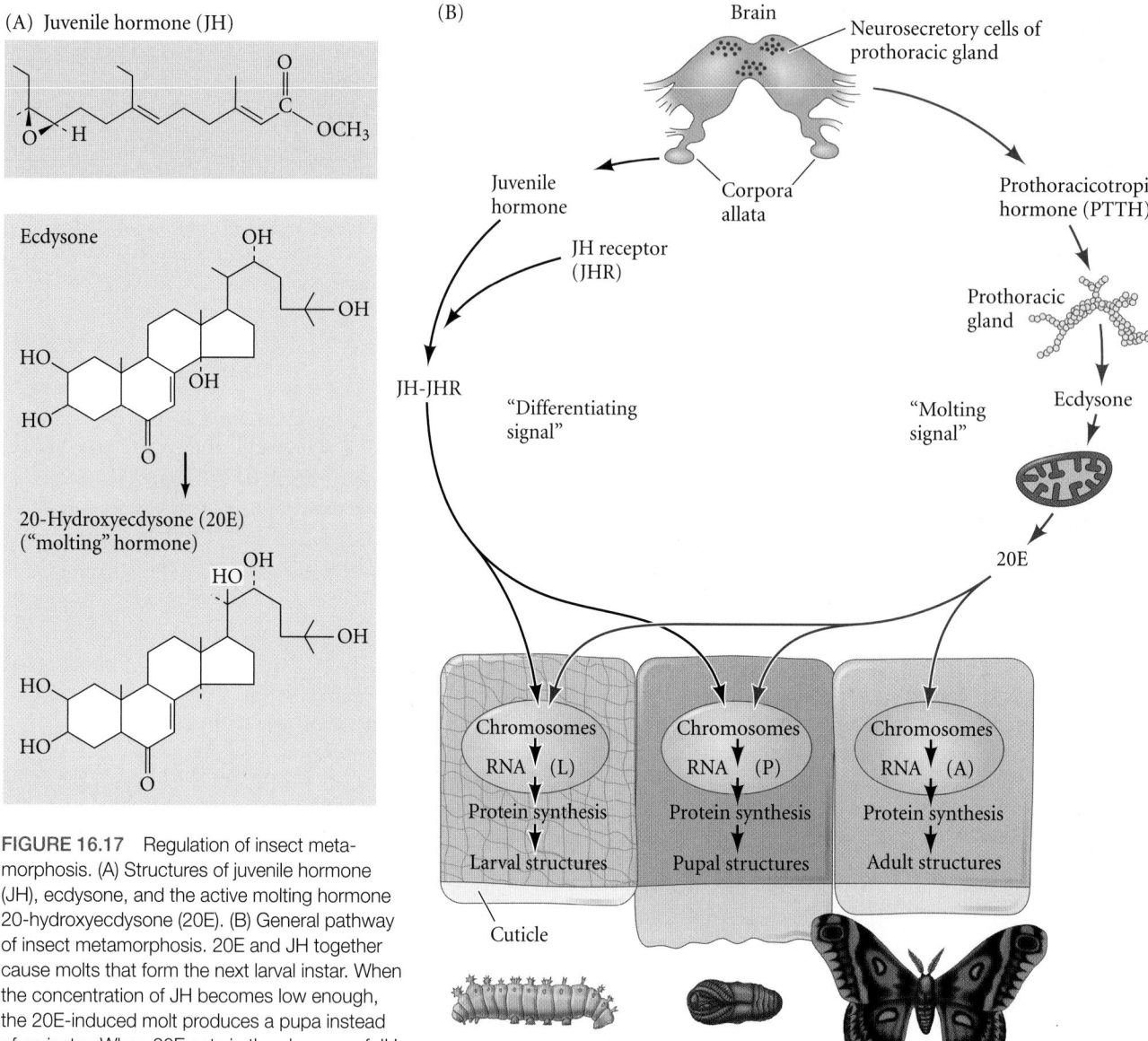

FIGURE 16.17 Regulation of insect metamorphosis. (A) Structures of juvenile hormone (JH), ecdysone, and the active molting hormone 20-hydroxyecdysone (20E). (B) General pathway of insect metamorphosis. 20E and JH together cause molts that form the next larval instar. When the concentration of JH becomes low enough, the 20E-induced molt produces a pupa instead of an instar. When 20E acts in the absence of JH, the imaginal discs differentiate and the molt gives rise to an adult (imago). (After Gilbert and Goodman 1981.)

Hormonal control of insect metamorphosis

Although the details of insect metamorphosis differ among species, the general pattern of hormonal action is very similar. Like amphibian metamorphosis, the metamorphosis of insects is regulated by systemic hormonal signals, which are controlled by neurohormones from the brain (for reviews, see Gilbert and Goodman 1981; Riddiford 1996). Insect molting and metamorphosis are controlled by two effector hormones: the steroid 20-hydroxyecdysone (20E) and the lipid **juvenile hormone** (**JH**) (**FIGURE 16.17A**). 20E initiates and coordinates each molt and regulates the changes in gene expression that occur during metamorphosis. JH prevents the ecdysone-induced changes in gene expression that are necessary for metamorphosis. Thus, its presence during a molt ensures that

the result of that molt is another larval instar, not a pupa or an adult.

The molting process is initiated in the brain, where neurosecretory cells release **prothoracicotropic hormone** (**PTTH**) in response to neural, hormonal, or environmental signals (**FIGURE 16.17B**). PTTH is a peptide hormone with a molecular weight of approximately 40,000, and it stimulates the production of **ecdysone** by the **prothoracic gland** by activating the RTK pathway in those cells (Rewitz et al. 2009; Ou et al. 2011). Ecdysone is modified in peripheral tissues to become the active molting hormone 20E. Each molt is initiated by one or more pulses of 20E. For a larval molt, the first pulse produces a small rise in the 20E concentration in the larval hemolymph (blood) and elicits a change in cellular commitment in the epidermis. A second, larger pulse of 20E initiates the differentiation events associated with molting. These

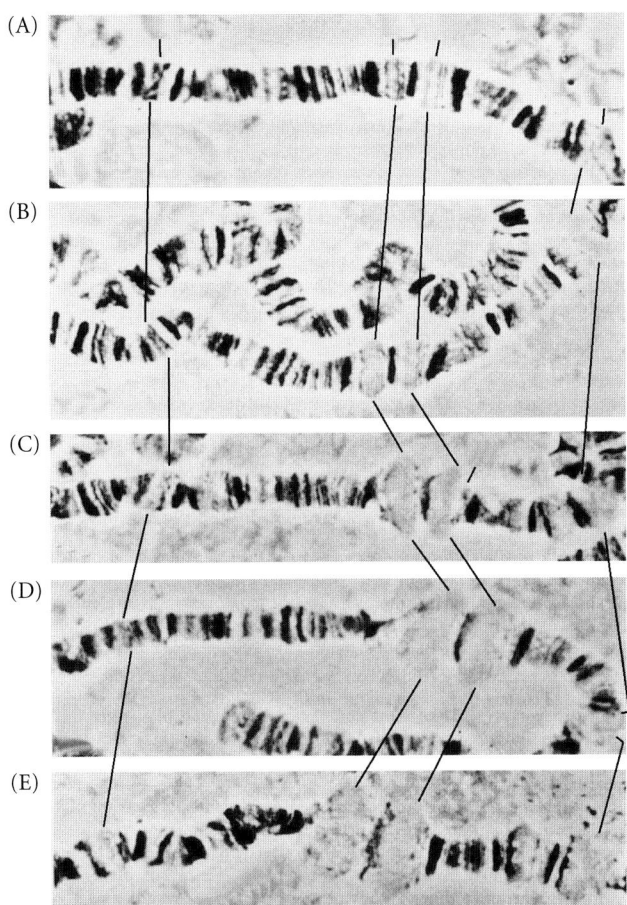

FIGURE 16.18 20E-induced puffs in cultured salivary gland cells of *D. melanogaster*. (A) Uninduced control. (B–E) 20E-stimulated chromosomes at (B) 25 minutes, (C) 1 hour, (D) 2 hours, and (E) 4 hours. (Courtesy of M. Ashburner.)

● See **VADE MECUM** Chromosome squash

pulses of 20E commit and stimulate the epidermal cells to synthesize enzymes that digest the old cuticle and synthesize a new one.

Larval-to-larval molts are produced when there are large titres of juvenile hormone. Juvenile hormone is secreted by the **corpora allata**. The secretory cells of the corpora allata are active during larval molts but inactive during the metamorphic molt and the imaginal molt. As long as JH is present, the 20E-stimulated molts result in a new larval instar. In the last larval instar, however, the medial nerve from the brain to the corpora allata inhibits these glands from producing JH, and there is a simultaneous increase in the body's ability to degrade existing JH (Safranek and Williams 1989). Both these mechanisms cause JH levels to drop below a critical threshold value, triggering the release of PTTH from the brain (Nijhout and Williams 1974; Rountree and Bollenbacher 1986). PTTH, in turn, stimulates the prothoracic gland to secrete a small amount of ecdysone. The resulting pulse of 20E, in the absence of high levels of JH, commits the epidermal cells to pupal development. Larva-specific mRNAs are not replaced, and new mRNAs are synthesized whose protein products inhibit the transcription of the larval messages.

There are two major pulses of 20E during *Drosophila* metamorphosis. The first pulse occurs in the third instar larva and triggers the "prepupal" morphogenesis of the leg and wing imaginal discs, as well as the death of the larval hindgut. The larva stops eating and migrates to find a site to begin pupation. The second 20E pulse occurs 10–12 hours later and tells the prepupa to become a pupa. The head inverts and the salivary glands degenerate (Riddiford 1982; Nijhout 1994). It appears, then, that the first pulse of 20E during the last larval instar triggers the processes that inactivate larva-specific genes and initiates the morphogenesis of imaginal disc structures. The second pulse transcribes pupa-specific genes and initiates the molt (Nijhout 1994). At the imaginal molt, when 20E acts in the absence of juvenile hormone, the imaginal discs fully differentiate and the molt gives rise to an adult.

● See **WEBSITE 16.3** Insect metamorphosis

The molecular biology of 20-hydroxyecdysone activity

ECDYSONE RECEPTORS 20-Hydroxyecdysone cannot bind to DNA by itself. Like amphibian thyroid hormones, 20E first binds to nuclear receptors. These proteins, called ecdysone receptors (EcRs), are almost identical in structure to the thyroid hormone receptors of amphibians. An EcR protein forms an active molecule by pairing with an Ultraspiracle (Usp) protein, the homologue of amphibian RXR that helps form the active thyroid hormone receptor (Koelle et al. 1991; Yao et al. 1992; Thomas et al. 1993). In the absence of the hormone-bound EcR, the Usp protein binds to the ecdysone-responsive genes and inhibits their transcription. This inhibition is converted into activation when the ecdysone receptor binds to the Usp (Schubiger and Truman 2000).

Although there is only one gene for EcR, the EcR mRNA transcript can be spliced in at least three different ways to form three distinct proteins. All three EcR proteins have the same domains for 20E and DNA binding, but they differ in their amino-terminal domains. The type of EcR present in a cell may inform that cell how to act when it receives a hormonal signal (Talbot et al. 1993; Truman et al. 1994). All cells appear to have some EcRs of each type, but the strictly larval tissues and neurons that die when exposed to 20E are characterized by their abundance of the EcR-B1 isoform of the ecdysone receptor. Imaginal discs and differentiating neurons, by contrast, show a preponderance of the EcR-A isoform. Mutations in specific codons that are found in only some of the splicing isoforms indicate that the different forms of EcR play different roles in metamorphosis and that the different receptors activate different sets of genes when they bind 20E (Davis et al. 2005).

BINDING OF 20-HYDROXYECDYSONE TO DNA During molting and metamorphosis, certain regions of the polytene chromosomes of *Drosophila* puff out in the cells of certain organs at certain times (**FIGURE 16.18**; Clever 1966; Ashburner 1972;

Ashburner and Berondes 1978). These chromosome puffs are areas where DNA is being actively transcribed. Moreover, these organ-specific patterns of chromosome puffing can be reproduced by culturing larval tissue and adding hormones to the medium, or by adding 20E to an early-stage larva. When 20E is added to larval salivary glands, certain puffs are produced and others regress. The puffing is mediated by the binding of 20E at specific places on the chromosomes; fluorescent antibodies against 20E find this hormone localized to the regions of the genome that are sensitive to it (Gronemeyer and Pongs 1980). At these sites, the ecdysone-bound receptor complex recruits a histone methyltransferase that methylates lysine-4 of histone H3, thereby loosening the nucleosomes in that area (Sedkov et al. 2003).

20E-regulated chromosome puffing occurs during the late stages of the third instar *Drosophila* larva, as it prepares to form the pupa. The puffs can be divided into three categories: "early" puffs that 20E induces rapidly; "intermolt" puffs that 20E causes to regress; and "late" puffs that are first seen several hours after 20E stimulation. For example, in the larval salivary gland, about six puffs emerge within a few minutes of hydroxyecdysone treatment. No new protein has to be made in order for these early puffs to be induced. A much larger set of puffs is induced later in development, and these late puffs do need protein synthesis to become transcribed. Ashburner (1974, 1990) hypothesized that the "early puff" genes make a protein product that is essential for the activation of the "late puff" genes and that, moreover, this early regulatory protein itself turns off the transcription of the early genes.* These insights have been confirmed by molecular analyses.

FIGURE 16.19A shows a simplified schematic for the framework of metamorphosis in *Drosophila*. First, 20E binds to the EcR/USP receptor complex. It activates the "early response genes," including *E74* and *E75* (the puffs in Figure 16.18), as well as *Broad* and the *EcR* gene itself. The transcription factors encoded by these genes activate a second series of genes, such as *E75*, *DHR4*, and *DHR3*. The products of these genes are transcription factors that work together to form the pupa. Second, the products of the second-wave genes shut off the early response genes so that they do not interfere with this second burst of 20E. Third, 20E activates the genes whose products inactivate and degrade ecdysone itself. In this way, the nucleus is cleared of the hormone so that it can respond to a second pulse. Moreover, 20E usually inhibits the gene encoding βFTZ-F1. Now this transcription factor can be synthesized, and it enables a new set of genes to respond to the second burst of 20E (Rewitz et al. 2009). Moreover, DHR4

coordinates growth and behavior in the larva. It allows the larva to stop feeding once it reaches a certain weight and to begin searching for a place to glue itself to and form a pupa (Urness and Thummel 1995; Crossgrove et al. 1996; King-Jones et al. 2005).

The effects of these two 20E pulses can be extremely different. One example of this is the ecdysone-mediated changes in the larval salivary gland. The early pulse of 20E activates the *Broad* gene, which encodes a family of transcription factors through differential RNA splicing. The targets of the Broad complex proteins include those genes that encode the salivary gland "glue proteins." The glue proteins allow the larva to adhere to a solid surface, where it becomes a pupa (Guay and Guild 1991). Here, 20E binds to the EcR-A isoform of the ecdysone receptor (**FIGURE 16.19B**). When complexed with USP, it activates the transcription of early response genes *E74*, *E75*, and *Broad*. But now a different set of targets is activated. These transcription factors activate the genes encoding the apoptosis-promoting proteins Hid and Reaper, as well as blocking the expression of the *diap2* gene (which would otherwise repress apoptosis). Thus, the first 20E pulse stimulates the function of the larval salivary gland, then the second pulse of 20E calls for the destruction of this larval organ (Buszczak and Segraves 2000; Jiang et al. 2000).

Like the ecdysone receptor gene, the *Broad* gene can generate several different transcription factor proteins through differentially initiated and spliced messages. Moreover, the variants of the ecdysone receptor may induce the synthesis of particular variants of the Broad proteins. Organs such as the larval salivary gland that are destined for death during metamorphosis express the Z1 isoform; imaginal discs destined for differentiation express the Z2 isoform; and the central nervous system (which undergoes marked remodeling during metamorphosis) expresses all isoforms, with Z3 predominating (Emery et al. 1994; Crossgrove et al. 1996).

When juvenile hormone is present, however, the *Broad* gene is repressed, and metamorphosis does not take place (Riddiford 1972; Zhou and Riddiford 2002; Hiruma and Kaneko 2013). JH maintains the "status quo" of larval-to-larval molts by binding to its nuclear receptor (the Met protein†), converting the receptor into a transcription factor. The JH-bound Met protein activates the *Kr-h1* gene, whose product, a repressive transcription factor, blocks the activation of the *Broad* gene (**FIGURE 16.19C**; Minakuchi et al. 2008; Charles et al. 2011; Li et al. 2011). Thus, in the presence of JH, the *Broad* gene is not activated and metamorphosis is blocked.

● See WEBSITE 16.4 Precocenes and synthetic JH

---

*The observation that 20E controlled the transcriptional units of chromosomes was an extremely important and exciting discovery. This was our first real glimpse of gene regulation in eukaryotic organisms. At the time when this discovery was made, the only examples of transcriptional gene regulation were in bacteria.

†Not to be confused with the unrelated Met receptor in vertebrates (which is a cell membrane receptor for hepatocyte growth factor), the Met receptor for JH was identified by its ability to bind methoprene, an insecticide that works by mimicking JH, thus preventing metamorphosis (Konopova and Jindra 2007; Charles et al. 2011).

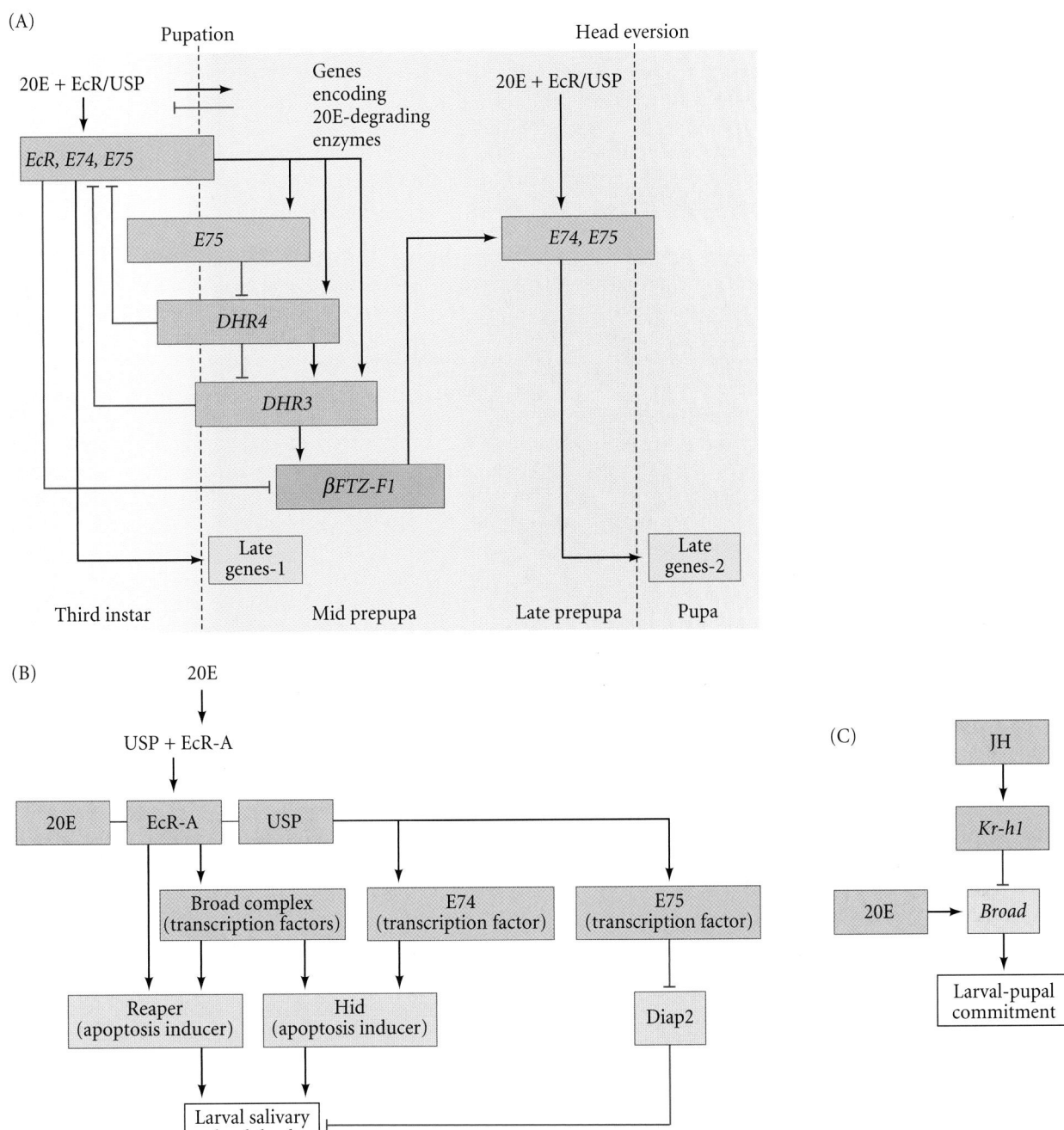

FIGURE 16.19 20-Hydroxyecdysone initiates developmental cascades. (A) Schematic of the major gene expression cascade in *Drosophila* metamorphosis. When 20E binds to the EcR/USP receptor complex, it activates the early response genes, including *E74*, *E75*, and *Broad*. Their products activate the "late genes." The activated EcR/USP complex also activates a series of genes whose products are transcription factors and which activate the βFTZ-F1 gene. The βFTZ-F1 protein modifies the chromatin so that the next 20E pulse activates a different set of late genes. The products of these genes also inhibit the early-expressed genes, including those for the EcR receptor. (B) Postulated cascade leading from ecdysone reception to death of the larval salivary gland. The 20E binds to the EcR-A isoform of the ecdysone receptor. After complexing with USP, the activated transcription factor complex stimulates transcription of the early response genes *E74A*, *E75B*, and the *Broad* complex. These promote apoptosis in the salivary gland cells. (C) When juvenile hormone binds to its receptor, Met, it activates the *Kr-h1* gene. Kr-h1 protein is a repressive transcription factor that blocks activation of the *Broad* gene by 20E. (A after King-Jones et al. 2005, Rewitz et al. 2010; B after Buszczak and Segraves 2000; C after Hiruma and Kaneko 2013.)

REGENERATION

Regeneration is the reactivation of development in postembryonic life to restore missing or damaged tissues. The ability to regenerate amputated body parts or repair nonfunctioning organs is so "unhuman" that it has been a source of fascination to humans since the beginnings of biological science. It is difficult to behold the phenomenon of limb regeneration in newts or sea stars without wondering why we cannot grow back our own arms and legs. What gives these animals an ability we so sorely lack? In fact, experimental biology was born of the efforts of eighteenth-century naturalists to answer this question. The regeneration experiments of Tremblay (*Hydra*, a cnidarian), Réaumur (crustaceans), and Spallanzani (salamanders) set the standard for experimental research and for the intelligent discussion of one's data (see Dinsmore 1991).

More than two centuries later, we are beginning to find answers to the great questions of regeneration, and at some point we may be able to alter the human body so as to permit our own limbs, nerves, and organs to regenerate. Success would mean that severed limbs could be restored, diseased organs could be removed and then regrown, and nerve cells altered by age, disease, or trauma could once again function normally. Modern medical attempts to coax human bone and neural tissue to regenerate will be discussed in Chapter 18, but to bring these treatments to humans, we must first understand how regeneration occurs in those species that already have this ability.* Our recently acquired knowledge of the roles of paracrine factors in organ formation, and our ability to clone the genes that produce those factors, have propelled what Susan Bryant (1999) has called "a regeneration renaissance." Since *renaissance* literally means "rebirth," and since regeneration can be seen as a return to the embryonic state, the term is apt in many ways.

Regeneration does in fact take place in nearly all species and can occur in four major ways:

1. *Stem-cell mediated regeneration.* Stem cells allow an organism to regrow certain organs or tissues that have been lost; examples include the regrowth of hair shafts from follicular stem cells in the hair bulge and the continual replacement of blood cells from the hematopoietic stem cells in the bone marrow.

2. *Epimorphosis.* In some species, adult structures can undergo *de*differentiation to form a relatively

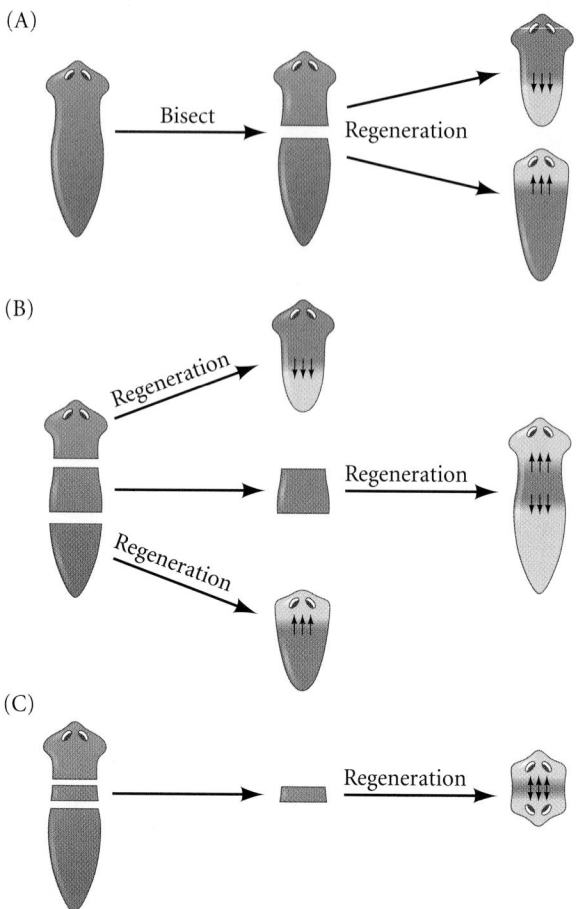

FIGURE 16.20 Flatworm regeneration and its limits. (A) If a planarian flatworm is cut in half, the anterior portion of the lower half regenerates a head while the posterior of the upper half regenerates a tail. The same tissue can generate a head (if it is at the anterior portion of the tail piece) or a tail (if it is at the posterior portion of the head piece). (B) If a flatworm is cut into three pieces, the middle piece will regenerate a head from its anterior end and a tail from its posterior end. (C) If the middle slice is too narrow, there is no discernible morphogen gradient within it, and regeneration is abnormal. (After Gosse 1969.)

---

*Mammals do have some ability to regenerate. In addition to regenerating body parts continuously through adult stem cells, rodents and humans can regenerate the tips of their digits if the animal is young enough. A regeneration blastema forms at the tip of the limb, and this is composed of progenitor cells (Fernando et al. 2011). As in the case of salamander limb regeneration, respecification does not occur (Lehoczky et al. 2011; Rinkevich et al. 2011). The new epidermis is derived from ectoderm-restricted progenitor cells, and the new bone similarly comes from osteoblast progenitor cells.

undifferentiated mass of cells that then redifferentiates to form the new structure. Such regeneration is characteristic of regenerating amphibian limbs.

3. *Morphallaxis.* Here, regeneration occurs through the repatterning of existing tissues, and there is little new growth. Such regeneration is seen in *Hydra*.

4. *Compensatory regeneration.* Here, the differentiated cells divide but maintain their differentiated functions. The new cells do not come from stem cells, nor do they come from the dedifferentiation of the adult cells. Each cell produces cells similar to itself; no mass of undifferentiated tissue forms. This type of regeneration is characteristic of the mammalian liver.

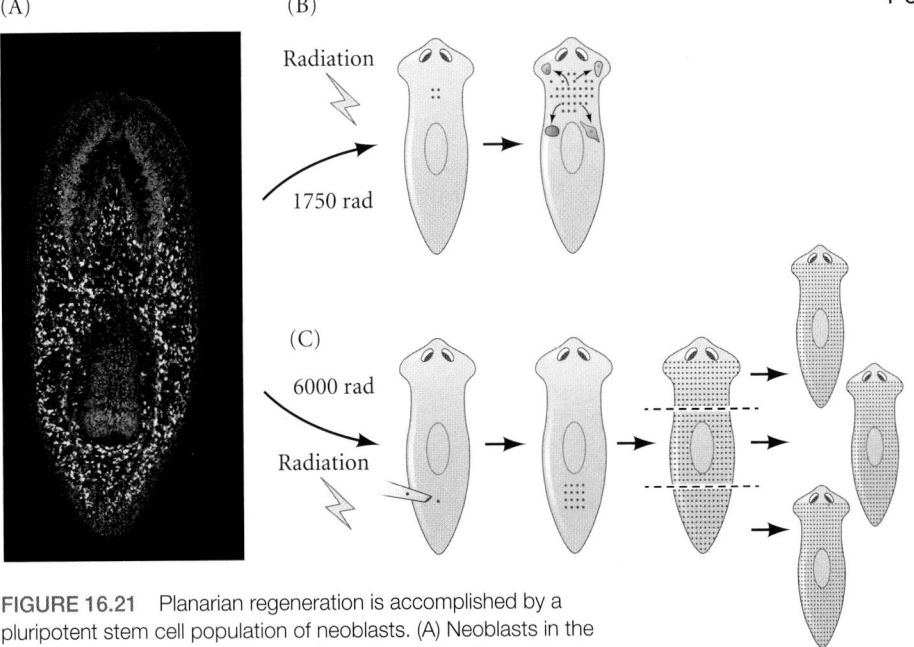

FIGURE 16.21 Planarian regeneration is accomplished by a pluripotent stem cell population of neoblasts. (A) Neoblasts in the planarian flatworm *Schmidtea mediterranea*, stained by antibodies to phosphorylated histone 3. Each pluripotent neoblast generates a colony of neoblast cells (red; nuclei are stained blue) in the flatworm. These clonogenic neoblast cells produce the differentiating cells of the regenerating flatworm. The nuclei are stained blue. Neoblasts are scattered throughout the body posterior to the eyes (although they are not present in the centrally located pharynx). (B) Irradiation with 1750 rad kills almost all neoblasts. If even one survives, a single clonogenic neoblast can divide to generate a colony of dividing cells that will ultimately produce the differentiated cells of the organs. (C) Irradiation with 6000 rad eliminates all dividing cells. Transplanting a single clonogenic neoblast from a donor strain (red) results not only in the production of all the cell types in the organism but also restores the organism's capacity for regeneration. (A courtesy of P. W. Reddien; B after Tanaka and Reddien 2011.)

Stem-Cell Mediated Regeneration in Flatworms

Planarian flatworms reproduce by binary fission, splitting their body from top to bottom and regenerating the right and left halves. It has only recently been shown that the cells capable of doing this are the same pluripotent stem cells that repair and replace body parts. In addition to their ability to regenerate during reproduction, it has been known since the 1700s that when planarians are cut in half, the head half will regenerate a tail from the wound site while the tail half will regenerate a head (**FIGURE 16.20A**; Pallas 1766). It was not until 1905, however, that Thomas Hunt Morgan (1905; Sunderland 2010) and C. M. Child (1905) realized that such polarity indicated an important principle of development.* Morgan pointed out that if both the head and the tail were cut off a flatworm, thus trisecting the animal, the medial segment would regenerate a head from the former anterior end and a tail from the former posterior end—never the reverse (**FIGURE 16.20B**). Furthermore, if the medial segment were sufficiently small, the regenerating portions would be abnormal (**FIGURE 16.20C**). Morgan and Child both postulated a gradient of anterior-producing materials concentrated in the head region. The middle segment would be told what to regenerate at both ends by the concentration gradient of these materials. If the piece were too small, however, the gradient would not be sensed within the segment.

The first question was, Which cells formed the new head or tail? For decades, it was believed that the old cells *dedifferentiated* at the cut ends of the planaria to form a **regeneration blastema**, a collection of relatively undifferentiated cells that would be organized into new structures by paracrine factors located at the wound surface (see Baguña 2012). However, in 2011 a series of experiments by Wagner and colleagues provided substantial evidence that dedifferentiation does *not* occur. Rather, the regeneration blastema forms from pluripotent stem cells called **clonogenic neoblasts**, a set of pluripotent cells in flatworms that serve as stem cells to replace the aging cells of the adult body (**FIGURE 16.21A**; Newmark and Sánchez Alvarado 2000; Pellettieri and Sánchez Alvarado 2007).

Clonogenic neoblasts can migrate to a wound site and regenerate the tissue. Wagner and colleagues were able to show that if planaria were irradiated at a dosage such that nearly all neoblasts were destroyed (dividing cells are killed more readily by radiation—a fact that is the basis of irradiating cancer sites), there would be some individuals in which a single clonogenic neoblast survived. From this neoblast, dividing progenitor cells formed, ultimately producing cell types of all germ layers and demonstrating the presence of pluripotent cells in the adult body (**FIGURE 16.21B**).

---

*Before 1910, "fly lab" maestro Thomas Hunt Morgan was well known for his research into flatworm regeneration. Indeed, it was only in 1900 that Morgan first mentioned *Drosophila*—as food for his flatworms! He was even able to "stain" the flatworms' digestive tubes by feeding them pigmented *Drosophila* eyes. Later, when he founded modern genetics, Morgan denounced flatworms as a model for heredity in favor of *Drosophila* (see Mittman and Fausto-Sterling 1992).

(A)

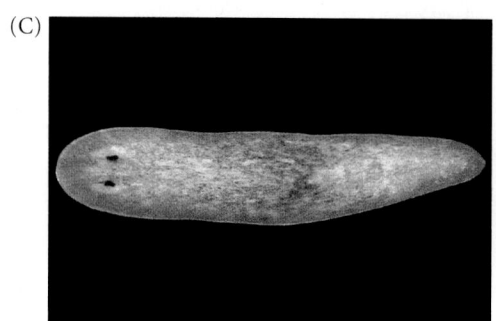

(B)

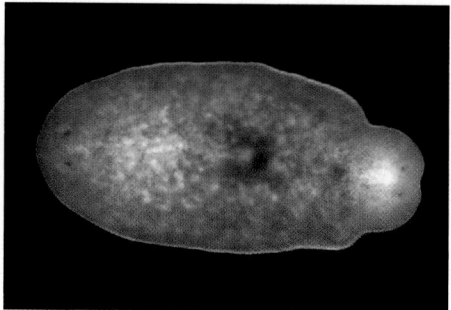

(C)

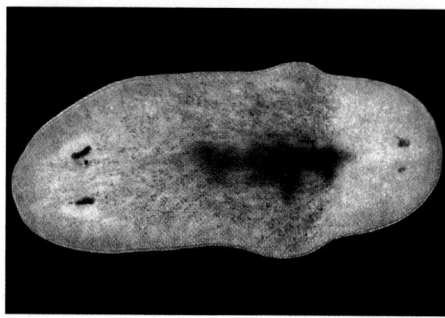

Normal

β-catenin blocked

Wnt1 blocked

FIGURE 16.22 Polarity in planarian regeneration (A) Dorsal-ventral polarity is patterned by BMP activity (as it is throughout the animal kingdom). As in insects, the source of BMP becomes the dorsal side. (B) Anterior-posterior polarity is determined by Wnt signaling, with Wnt production in the posterior and Wnt antagonists made in the anterior. (C) Normally, Wnts are produced in the posterior blastema, and the result is a tail. However, if the Wnt pathway is blocked by using RNA interference against either β-catenin or Wnt1 messages, the posterior blastema regenerates a head, thereby forming a worm with heads at both ends. (A,B after Gurley et al. 2008; C from Reddien 2011, photographs courtesy of D. Reddien.)

Next the researchers irradiated planaria so effectively that all dividing cells were destroyed (**FIGURE 16.21C**). These planaria died because of failed tissue replacement. However, transplantation of a single clonogenic neoblast into such an irradiated flatworm could, in some cases, restore all the cells of the organism. Not only did the flatworm survive, but it could split into more planaria; these new planaria were completely of the genotype of the single donor neoblast. This demonstrated that the regeneration of the flatworm was the result of the production of new cells from adult pluripotent stem cells.

This raises new questions. First, it is not yet known if all neoblasts are pluripotent; nor is it known what their embryological origin is. The hierarchy of differentiation whereby the neoblast creates the 30 or so cell types of the adult planarian is still undiscovered and is presently being investigated (see Wagner et al. 2012). A second question involves polarity. How does the flatworm tell the posterior blastema to become tail and the anterior blastema to become head? Moreover, how do the blastema-derived cells retain their dorsal-ventral polarity?

Remarkably, flatworms appear to establish these gradients using the same factors and processes known to be used by vertebrates (Petersen and Reddien 2009). Just as in vertebrates and flies, the dorsal-ventral axis of regenerating cells is regulated by BMP and its inhibitors. BMP expression defines the dorsal region in flatworms, as it does in flies (**FIGURE 16.22A**; Orii and Watanabe 2007; Gaviño and Reddien 2011; Molina et al. 2011). The anterior-posterior polarity of the regenerating flatworm appears to be regulated by Wnts and β-catenins (**FIGURE 16.22B**; Gurley et al. 2008; Petersen and Reddien 2008, 2011). β-catenin is activated (via Wnts) in the posterior-facing blastema (which generates tails). Just as in vertebrate development, repressors of Wnt signaling prevent β-catenin production in the *anterior*-facing blastema (which forms heads). If β-catenin is eliminated from the posterior (tail-forming) blastema by RNA interference, that blastema will form a head (**FIGURE 16.22C**).*

● **See VADE MECUM** Flatworm regeneration

---

*Indeed, when RNAi completely eliminates β-catenin from nonregenerating flatworms, the entire organism becomes a head, with eyes all around the periphery (Gurley et al. 2008; Iglesias et al. 2008). And if a Wnt inhibitor is knocked out, causing too much Wnt to be synthesized, the anterior-facing blastema will form a tail (Petersen and Reddien 2011).

Epimorphic Regeneration of Salamander Limbs

When an adult salamander limb is amputated, the remaining limb cells reconstruct a new limb, complete with all its differentiated cells arranged in the proper order. Remarkably, the limb regenerates only the missing structures and no more. For example, when a wrist is amputated, the salamander forms a new wrist but not a new elbow. In some way, the salamander limb "knows" where the proximal-distal axis has been severed and is able to regenerate from that point on (FIGURE 16.23).

Salamanders accomplish epimorphic regeneration by cell dedifferentiation to form a **regeneration blastema**—an aggregation of relatively undifferentiated cells derived from the originally differentiated tissue—which then proliferates and redifferentiates into the new limb parts (see Brockes and Kumar 2002; Gardiner et al. 2002). Bone, dermis, and cartilage just beneath the site of amputation contribute to the regeneration blastema, as do satellite cells from nearby muscles (Morrison et al. 2006). However, unlike the flatworm, which formed its regeneration blastema from adult pluripotent stem cells, much of the salamander limb's regeneration blastema appears to arise from the dedifferentiation of adult cells followed by cell division and the redifferentiation of those cells back into their original cell types.*

---

*There is still controversy over whether this is "true" dedifferentiation of normally postmitotic cells or whether much of the limb blastema is formed from the activation of uncommitted stem cells (see Nacu and Tanaka 2011).

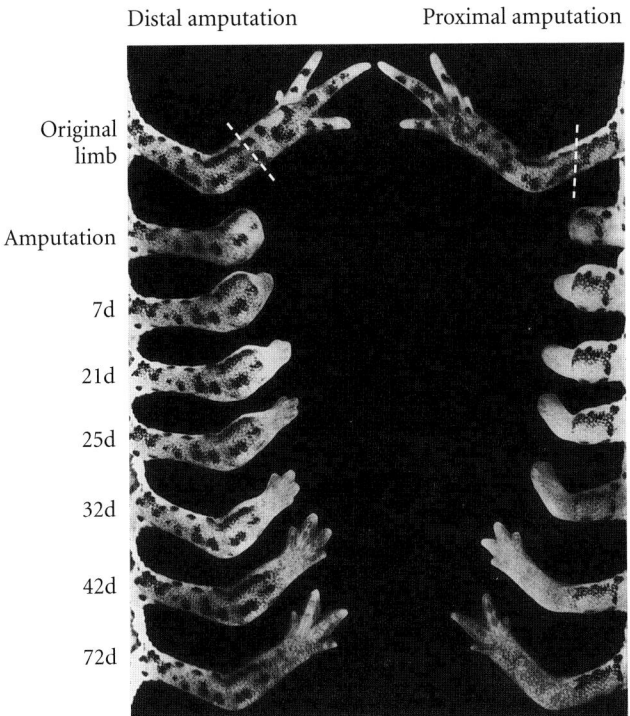

Distal amputation   Proximal amputation

Original limb

Amputation

7d

21d

25d

32d

42d

72d

Formation of the apical ectodermal cap and regeneration blastema

When a salamander limb is amputated, a plasma clot forms. Within 6–12 hours, epidermal cells from the remaining stump migrate to cover the wound surface, forming the **wound epidermis**. In contrast to wound healing in mammals, no scar forms, and the dermis does not move with the epidermis to cover the site of amputation. The nerves innervating the limb degenerate for a short distance proximal to the plane of amputation (see Chernoff and Stocum 1995).

During the next 4 days, the extracellular matrices of the tissues beneath the wound epidermis are degraded by proteases, liberating single cells that undergo dramatic dedifferentiation: bone cells, cartilage cells, fibroblasts, and myocytes all lose their differentiated characteristics. Genes that are expressed in differentiated tissues (such as the *mrf4* and *myf5* genes expressed in muscle cells) are downregulated, while there is a dramatic increase in the expression of genes such as *msx1* that are associated with the proliferating progress zone mesenchyme of the embryonic limb (Simon et al. 1995). This cell mass is the regeneration blastema, and these are the cells that will continue to proliferate, and that will eventually redifferentiate to form the new structures of the limb (FIGURE 16.24; Butler 1935). Moreover, during this time, the wound epidermis thickens to form the **apical epidermal cap** (**AEC**), which acts similarly to the apical ectodermal ridge during normal limb development (see Chapter 14; Han et al. 2001).

Thus, the previously well-structured limb region at the cut edge of the stump forms a proliferating mass of indistinguishable cells just beneath the apical ectodermal cap. One of the major questions of regeneration has been: do the cells keep a "memory" of what they had been? In other words, do new muscles arise from old muscle cells, or can any cell of the blastema become a muscle? Kragl and colleagues (2009) found that the blastema is not a collection of homogeneous, fully dedifferentiated cells. Rather, in the regenerating limbs of the axolotl salamander, muscle cells arise only from old muscle cells, dermal cells come only from old dermal cells, and cartilage can arise only from old cartilage or old dermal cells. Thus, the blastema is not a collection of unspecified multipotent progenitor cells. Rather, the cells retain their specification, and the blastema is a heterogeneous assortment of *restricted* progenitor cells.

Kragl and colleagues performed an experiment in which they transplanted limb tissue from a salamander whose cells expressed green fluorescent protein (GFP) into different

FIGURE 16.23 Regeneration of a salamander forelimb. The amputation shown on the left was made below the elbow; the amputation shown on the right cut through the humerus. In both instances, the correct positional information was respecified and a normal limb was regenerated within 72 days. (From Goss 1969, courtesy of R. J. Goss.)

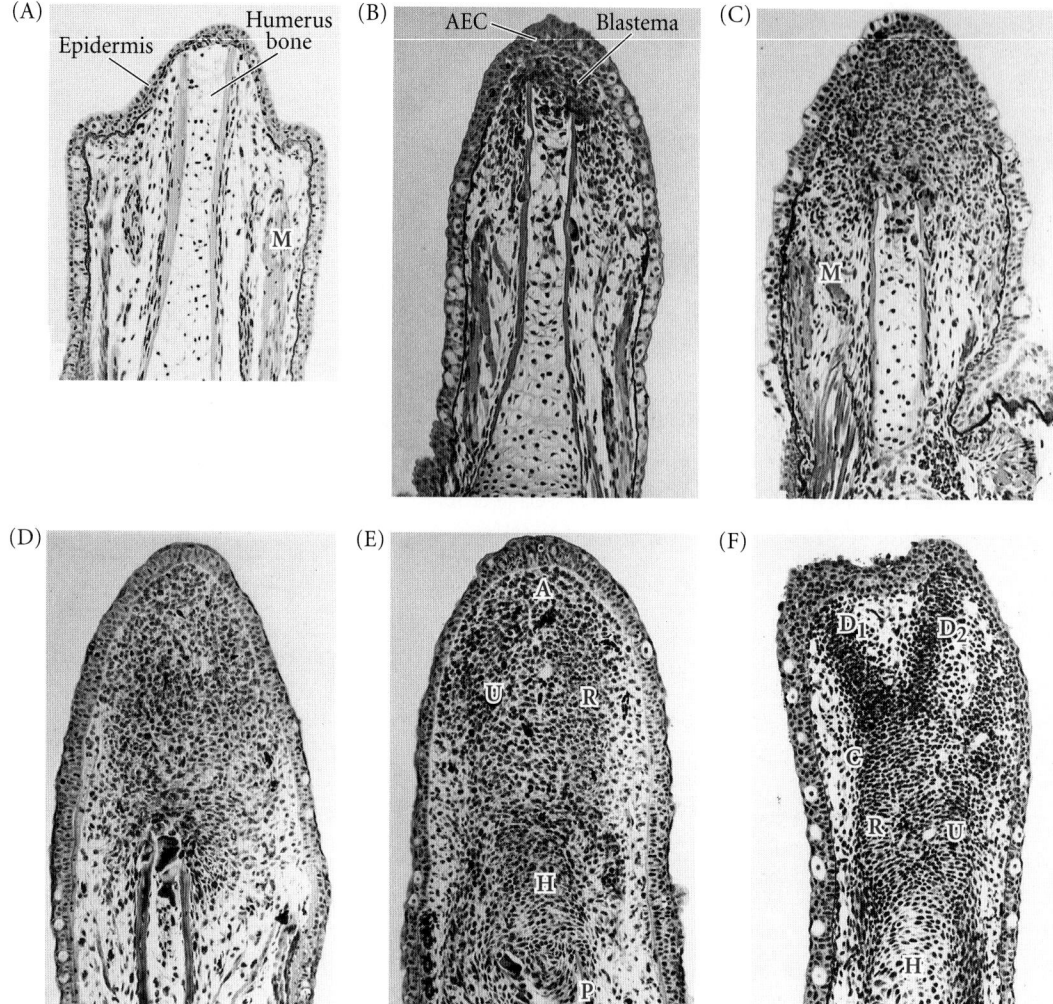

FIGURE 16.24 Regeneration in the larval forelimb of the spotted salamander *Ambystoma maculatum*. (A) Longitudinal section of the upper arm, 2 days after amputation. The skin and muscle (M) have retracted from the tip of the humerus. (B) At 5 days after amputation, a thin accumulation of blastema cells is seen beneath the thickened epidermis, where the apical ectodermal cap (AEC) forms. (C) At 7 days, a large population of mitotically active blastema cells lies distal to the humerus. (D) At 8 days, the blastema elongates by mitotic activity; much dedifferentiation has occurred.

(E) At 9 days, early redifferentiation can be seen. Chondrogenesis has begun in the proximal part of the regenerating humerus (H). The letter A marks the apical mesenchyme of the blastema, and H and R are the precartilaginous condensations that will form the ulna and radius, respectively. P represents the stump where the amputation was made. (F) At 10 days after amputation, the precartilaginous condensations for the carpal bones (ankle, C) and the first two digits (D_1, D_2) can also be seen. (From Stocum 1979, courtesy of D. L. Stocum.)

regions of limbs of normal salamanders that did not have the *GFP* transgene (**FIGURE 16.25**). If they transplanted the GFP-expressing limb cartilage into a salamander limb that did not contain the *GFP* transgene, the GFP-expressing cartilage would integrate normally into the limb skeleton. They later amputated the limb through the region containing GFP-marked cartilage cells. The blastema was found to contain GFP-expressing cells, and when the blastema differentiated, the only GFP-expressing cells found were in the limb cartilage. Similarly, GFP-marked muscle cells gave rise only to

muscle, and GFP-marked epidermal cells only produced the epidermis of the regenerated limb.

Proliferation of the blastema cells: The requirement for nerves and the AEC

The growth of the regeneration blastema depends on the presence of both the apical ectodermal cap and nerves. The AEC stimulates the growth of the blastema by secreting Fgf8 (just as the apical ectodermal ridge does in normal limb development), but the effect of the AEC is only possible if

(A)

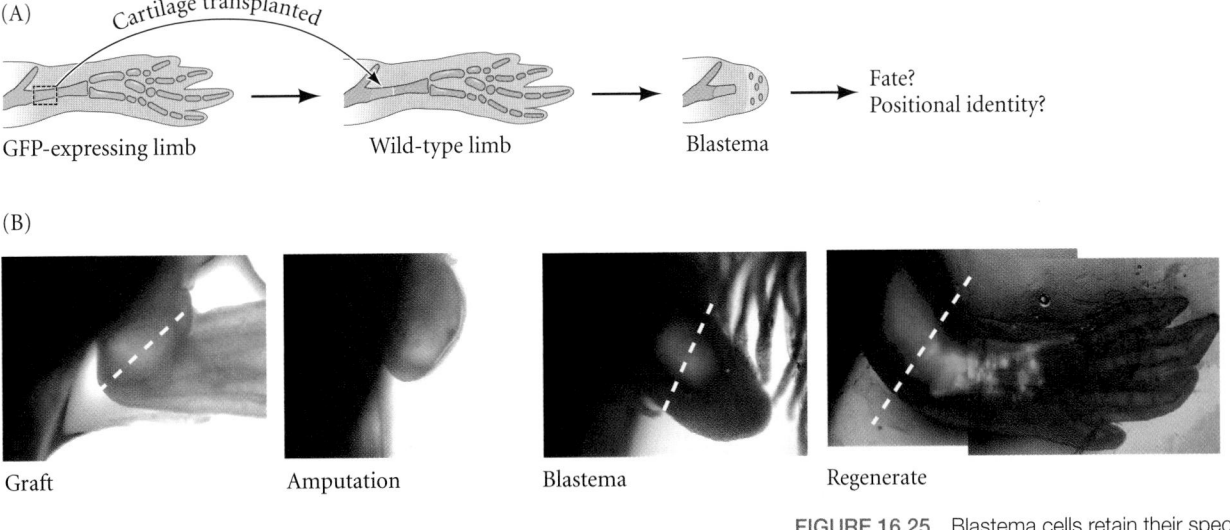

GFP-expressing limb　　　　Wild-type limb　　　　Blastema

Cartilage transplanted

Fate?
Positional identity?

(B)

Graft　　　　　　Amputation　　　　　Blastema　　　　　Regenerate

(C)

Regenerate

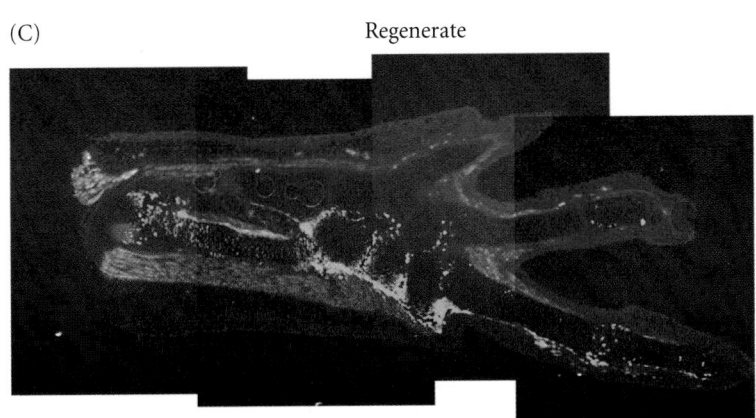

FIGURE 16.25 Blastema cells retain their specification, even though they dedifferentiate. (A,B) Schematic of the procedure wherein a particular tissue (in this case, cartilage) is transplanted from a salamander expressing a *GFP* transgene into a wild-type salamander limb. Later, the limb is amputated through the region of the limb containing GFP expression, and a blastema is formed containing GFP-expressing cells that had been cartilage precursors. The regenerated limb is then studied to see if GFP is found only in the regenerated cartilage tissues or in other tissues. (C) Longitudinal section of a regenerated limb 30 days after amputation. Muscle cells are stained red, nuclei are stained blue. The majority of GFP cells (green) were found in regenerated cartilage; no GFP was seen in the muscle. (After Kragl et al. 2009, courtesy of E. Tanaka.)

nerves are present (Mullen et al. 1996). Singer (1954) demonstrated that a minimum number of nerve fibers must be present for regeneration to take place. The neurons are also believed to release factors necessary for the proliferation of the blastema cells (Singer and Caston 1972; Mescher and Tassava 1975). There have been many candidates for such a nerve-derived blastema mitogen, but the one that is probably the best candidate is **newt anterior gradient protein (nAG)**. This protein can cause blastema cells to proliferate in culture, and it permits normal regeneration in limbs that have been denervated (**FIGURE 16.26**; Kumar et al. 2007a). If activated nAG genes are electoporated into the dedifferentiating tissues of limbs that have been denervated, the limbs are able to regenerate. If nAG is not administered, the limbs remain stumps. Moreover, nAG is only minimally expressed in normal limbs, but it is induced in the Schwann cells that surround the neurons within 5 days of amputation.

The creation of the amphibian regeneration blastema may also depend on maintaining ion currents driven through the stump; if this electric field is suppressed, the regeneration blastema fails to form (Altizer et al. 2002). Such fields have been shown to be necessary for the regeneration of tails in the frog *Xenopus laevis* (an anuran amphibian). The *Xenopus* tadpole regenerates its tail, and the notochord, muscles, and spinal cord each regenerate from the corresponding tissue in the stump (Deuchar 1975; Slack et al. 2004). In this frog, the V-ATPase proton pump is activated within 6 hours after tail amputation, changing the membrane voltage and establishing flow of protons through the blastema (Adams et al. 2007). If this proton pump is inactivated either by mutation or by drugs, depolarization of the blastema cells fails to occur and there is no regeneration.

As mentioned above, the generation of the salamander limb along the anterior-posterior axis follows rules similar to those that generate the developing limb. Indeed, the opposing retinoic acid-FGF gradients postulated for the development of the limb were first hypothesized for the regeneration of limb structures along this axis (Crawford and Stocum 1988).

● **See WEBSITE 16.5** Axes of regeneration

● **See WEBSITE 16.6** Regeneration in annelid worms

(A)

Right limb denervated | Both limbs amputated | Electroporation with nAG

Regeneration

7 days   5 days

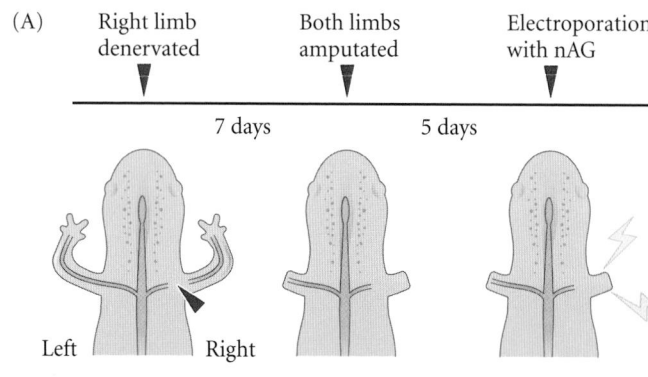

Left   Right

(B)

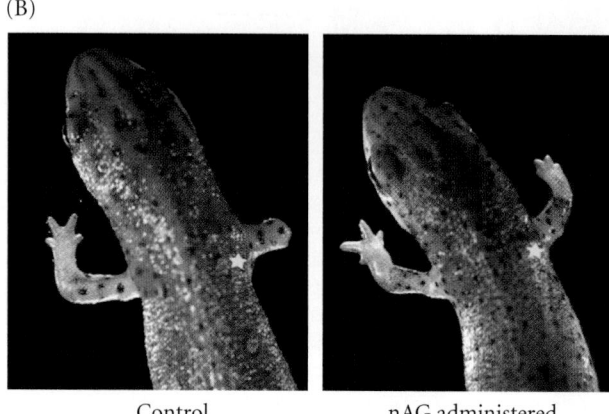

Control   nAG administered

FIGURE 16.26 Regeneration of newt limbs depends on nAG (normally supplied by the limb nerves). (A) Schematic of the procedure. The limb is denervated and 7 days later is amputated. After 5 five days, nAG is electroporated into the limb blastema. (B) Results show that in the denervated control (not given nAG), the amputated limb (yellow star) remains a stump. The limb that is given nAG regenerates tissues and proximal-distal polarity. (After Yin and Poss 2008, courtesy of K. Poss.)

Regeneration in Hydra: Morphallaxis and Epimorphosis

*Hydra** is a genus of freshwater cnidarians. Most hydras are tiny—about 0.5 cm long. A hydra has a tubular body, with a "head" at its distal end and a "foot" at its proximal end. The "foot," or **basal disc**, enables the animal to stick to rocks or the undersides of pond plants. The "head" consists of a conical **hypostome** region (containing the mouth) and a ring of tentacles (which catch food) beneath it. Hydras are diploblastic animals, having only ectoderm and endoderm (see Figure 5.1). Although these animals lack a true mesoderm, they do contain secretory cells, gametes, stinging cells (nematocytes), and neurons that are not part of the two epithelial layers. Hydras can reproduce sexually but do so only under adverse conditions (such as crowding or cold weather). They usually multiply asexually, by budding off a new individual. The buds form about two-thirds of the way down the animal's body axis (**FIGURE 16.27A**).

A hydra's body is not particularly stable. In humans and flies, for instance, a skin cell in the body's trunk is not expected to migrate and eventually be sloughed off from the face or foot. But that is precisely what does happen in hydras. The

---

*The Hydra, a many-headed serpent, is another character from Greek mythology. Whenever one of this monster's heads was chopped off, it regenerated two new ones. Hercules finally defeated the Hydra by cauterizing the stumps of its heads with fire. Hercules seems to have had a significant interest in regeneration—he also freed the bound Prometheus, thereby stopping his daily series of partial hepatectomies (see p. 578).

(A)

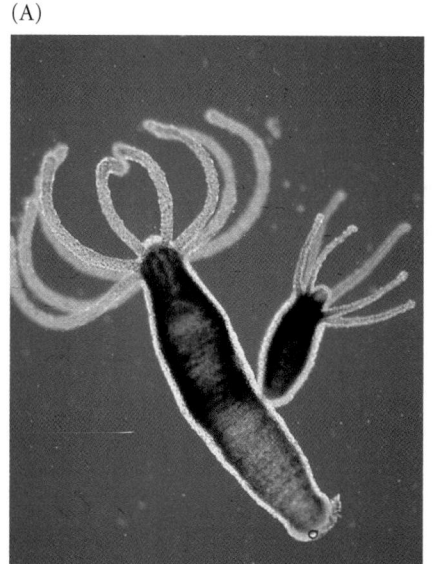

(B)

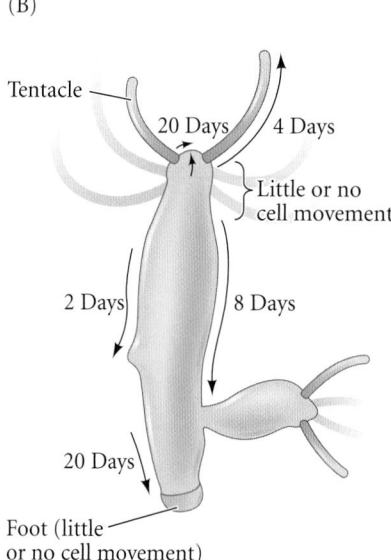

Tentacle

20 Days   4 Days

Little or no cell movement

2 Days   8 Days

20 Days

Foot (little or no cell movement)

FIGURE 16.27 Budding in *Hydra*. (A) A new individual buds about two-thirds down the side of an adult hydra. (B) Cell movements in *Hydra* were traced by following the migration of labeled tissues. The arrows indicate the starting and leaving positions of the labeled cells. The bracket indicates regions in which no net cell movement took place. Cell division takes place throughout the body column except at the tentacles and foot. (A © Biophoto/Photo Researchers Inc.; B after Steele 2002.)

cells of the body column are constantly undergoing mitosis and are eventually displaced to the extremities of the column, from which they are shed (**FIGURE 16.27B**; Campbell 1967a,b). Thus, each cell plays several roles, depending on how old it is; and the signals specifying cell fate must be active all the time. In a sense, a hydra's body is always regenerating.

This cellular replacement is generated from three cell types. Endodermal cells and ectodermal cells divide continuously, each producing more epithelia. The third cell type is a multipotent **interstitial stem cell** found within the ectodermal layer. This stem cell generates neurons, secretory cells, nematocytes, and gametes. The three cell types are all that is needed to form a hydra, and if hydra cells are separated and reaggregated, a new hydra will form (Gierer et al. 1972; Technau 2000; Bode 2011).

The head activator

Experimental embryology—indeed, experimental biology—can be said to have started with Abraham Tremblay's studies of *Hydra* regeneration.* In 1741, he reported that "the story of the Phoenix who is reborn form his own ashes, fabulous as it is, offers nothing more marvelous than the discovery of which we are going to speak." He found that if he cut the hydra into as many as 40 pieces, "there are reborn as many complete animals similar to the first." Each piece would regenerate a head at its original apical end and a foot at its original basal end. It would be like generating an entire person from their kneecap or neck.

Every portion of the hydra's body column along the apical-basal axis is potentially able to form both a head and a foot. However, the animal's polarity is coordinated by a series of morphogenetic gradients that permit the head to form only at one place and the basal disc to form only at another. Evidence for such gradients was first obtained from grafting experiments begun by Ethel Browne in the early 1900s. When hypostome tissue from one hydra is transplanted into the middle of another hydra, the transplanted tissue forms a new apical-basal axis, with the hypostome extending outward (**FIGURE 16.28A**). When a basal disc is grafted to the middle of a host hydra, a new axis also forms, but with the opposite polarity, extending a basal disc (**FIGURE 16.28B**). When tissues from both ends are transplanted simultaneously into the middle of a host, no new axis is formed, or the new axis has little polarity (**FIGURE 16.28C**; Browne 1909; Newman 1974). These experiments have been interpreted to indicate the existence of a **head activation gradient** (highest at the hypostome) and a **foot activation gradient** (highest at the basal disc). The head activation gradient can be

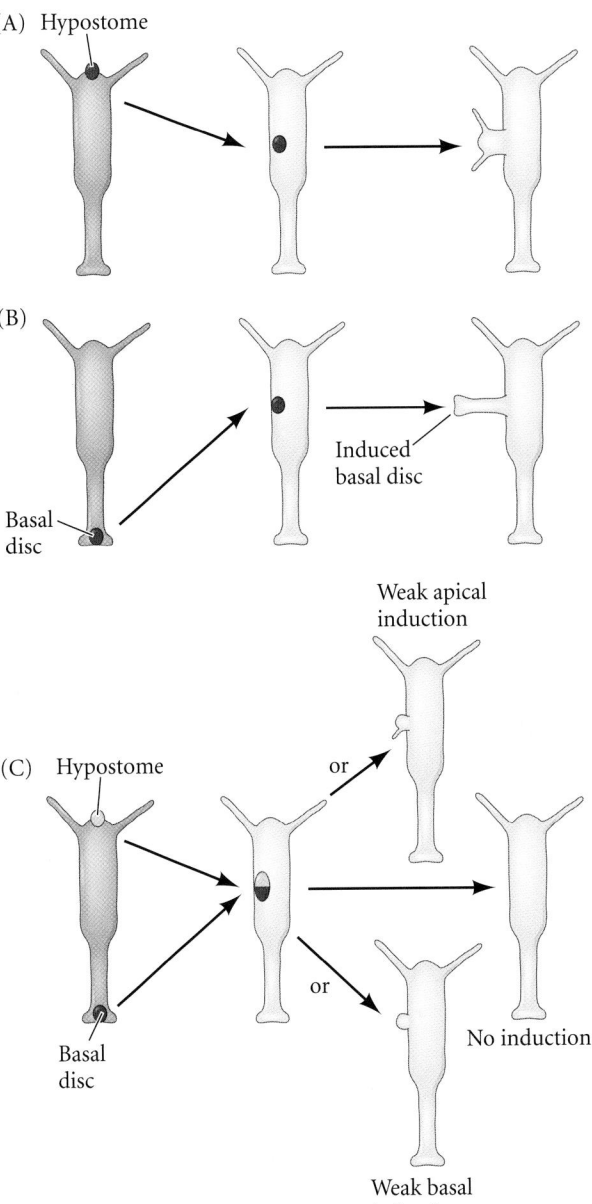

FIGURE 16.28 Grafting experiments demonstrating different morphogenetic capabilities in different regions of the *Hydra* apical-basal axis. (A) Hypostome tissue grafted onto a host trunk induces a secondary axis with an extended hypostome. (B) Basal disc tissue grafted onto a host trunk induces a secondary axis with an extended basal disc. (C) If hypostome and basal disc tissues are transplanted together, only weak (if any) inductions are seen. (After Newman 1974.)

*Tremblay's advice to researchers who would enter this new field is pertinent even today: he advises us to go directly to nature and to avoid the prejudices that our education has given us. Moreover, "one should not become disheartened by want of success, but should try anew whatever has failed. It is even good to repeat successful experiments a number of times. All that is possible to see is not discovered, and often cannot be discovered, the first time." (Quoted in Dinsmore 1991.)

measured by implanting rings of tissue from various levels of a donor hydra into a particular region of the host trunk (Wilby and Webster 1970; Herlands and Bode 1974; MacWilliams 1983b). The higher the level of head activator in the donor tissue, the greater the percentage of implants that will induce the formation of new heads. The head activation factor is concentrated in the hypostome and decreases linearly toward the basal disc.

Ethel Browne (1909; Lenhoff 1991) noted that the hypostome acted as an "organizer" of the hydra. This notion has been confirmed by Broun and Bode (2002), who demonstrated that (1) when transplanted, the hypostome can induce host tissue to form a second body axis, (2) the hypostome produces both the head activation and head inhibition signals, (3) the hypostome is the only "self-differentiating" region of the hydra, and (4) it contains a "head inhibition signal" that suppresses the formation of new organizing centers.

By inserting small pieces of hypostome tissue into a host hydra whose cells were labeled with India ink (colloidal carbon), Broun and Bode found that the hypostome induced a new body axis and that almost all of the resulting head tissue came from *host* tissue, not from the differentiation of donor tissue (**FIGURE 16.29A**). In contrast, when tissues from other regions (such as the subhypostomal region) were grafted into a host trunk, the head and apical trunk of the new hydra were made from the grafted *donor* tissue (**FIGURE 16.29B**). In other words, only the hypostome region could alter the fates of the trunk cells and cause them to become head cells. Broun and Bode also found that the signal did not have to emanate from a permanent graft. Even transient contact with the hypostome region was sufficient to induce a new axis from a host hydra. In these cases, *all* the tissue of the new axis came from the host.

The major head inducer of the hypostome organizer is a set of Wnt proteins acting through the canonical β-catenin pathway* (Hobmayer et al. 2000; Broun et al. 2005; Lengfeld et al. 2009; also see Bode 2009). These Wnt proteins are seen in the apical end of the early bud, defining the hypostome region as the bud elongates (**FIGURE 16.30**). If GSK3 is inhibited throughout the body axis, ectopic tentacles form at all levels, and each piece of the trunk has the ability to stimulate the outgrowth of new buds. Moreover, when the hypostome is brought into contact with the trunk of an adult hydra, it induces expression of the *Brachyury* gene, just as vertebrate organizers do—even though hydras lack mesoderm (Broun et al. 1999; Broun and Bode 2002). This induction is dependent on the Wnt proteins.

When a hydra is cut, the Wnt pathway is activated in the lower portion (i.e., in that portion that will form a new head). If the cut is made close to the head (just below the hypostome), Wnt3 in the epithelial cells will cause the remodeling of existing cells to form the head. No proliferation is seen in this case; hence this is morphallactic regeneration. However, if the hydra is cut at its midsection, the cells derived from the

interstitial cell (neurons, nematocytes, etc.) undergo apoptosis immediately below the cut site. But before dying, these cells produce a burst of Wnt3. The Wnt3 produced by these dying cells activates β-catenin in the interstitial cells beneath them. This β-catenin surge causes a wave of proliferation in the interstitial cells, as well as remodeling in the epithelial cells. So regeneration here is epimorphic (Chera et al. 2009). Canonical Wnt signaling is thus important both in normal budding and in head regeneration.

● See WEBSITE 16.7 Ethel Browne and the organizer

(A)

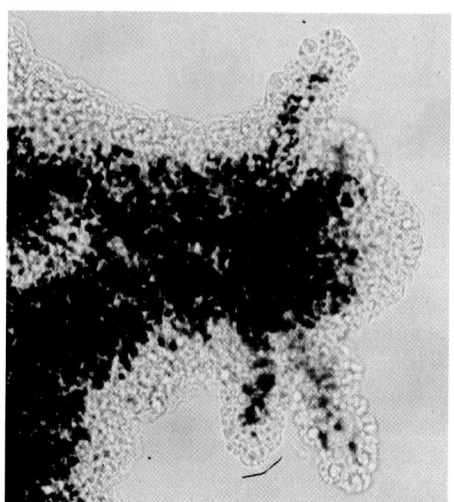

(B)

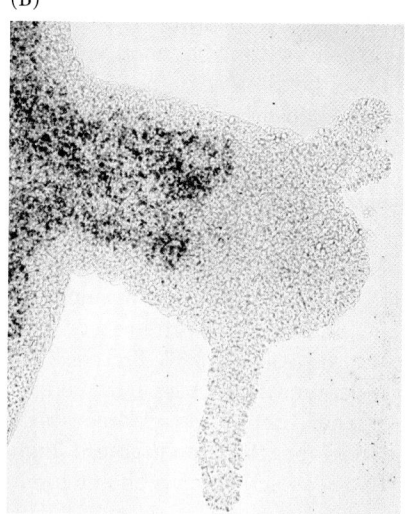

FIGURE 16.29 Formation of secondary axes following transplantation of head regions into the trunk of a hydra. The host endoderm was stained with India ink. (A) Hypostome tissue grafted onto the trunk induces the host's own trunk tissue to become tentacles and head. (B) Subhypostomal donor tissue placed on the host trunk self-differentiates into a head and upper trunk. (From Broun and Bode 2002, courtesy of H. R. Bode.)

---

*It is interesting that although Wnt signaling is used in both flatworm and hydra regeneration, its effects are different. In flatworms, Wnt signals the formation of the tail, and its inhibition is needed to form heads. In hydra, Wnt signaling appears to establish heads—or at least the part that has a mouth and resembles a head. (Just because we call it a head doesn't make it homologous to the face of bilaterians. There's still conflict about this.)

(A) (B) (C) (D)

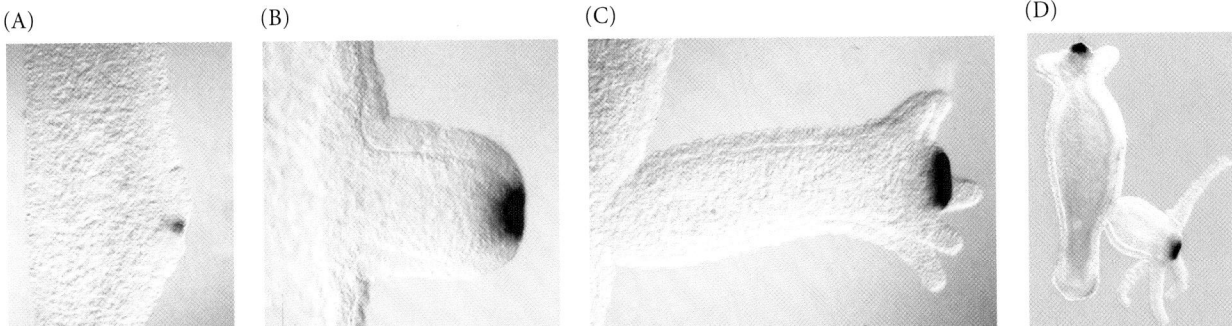

FIGURE 16.30 Wnt expression (purple) during *Hydra* budding. (A) Early bud. (B) Mid-stage bud. (C) Bud with early tentacles. (D) Adult with late bud. (From Hobmayer et al. 2000, courtesy of T. W. Holstein and B. Hobmeyer.)

The head inhibition gradients

If any tissue of the hydra body column is capable of forming a head, why doesn't it all do so? In 1926, Rand and colleagues showed that normal regeneration of the hypostome is inhibited when an intact hypostome is grafted adjacent to the amputation site. Moreover, if a graft of subhypostomal tissue (from the region just below the hypostome, where there is a relatively high concentration of head activator) is placed in the same region of a host hydra, no secondary axis forms (**FIGURE 16.31A**). The host head appears to make an inhibitor that prevents the grafted tissue from forming a head and secondary axis. However, if one grafts subhypostomal tissue to a decapitated host hydra, a second axis does form (**FIGURE 16.31B**). A gradient of this inhibitor appears to extend from the head down the body column, and can be measured by grafting subhypostomal tissue into various regions along the trunks of host hydras. This tissue will not produce a head when implanted into the apical area of an intact host hydra, but it will form a head if placed lower on the host (**FIGURE 16.31C**). The head inhibitor remains unknown, but it appears to be labile, with a half-life of only 2–3 hours (Wilby and Webster 1970; Mac-Williams 1983a). It is thought that the head inhibitor and the head activator (Wnts) are both made in the hypostome, but that the head inhibition gradient falls off more rapidly than the head activator gradient (see Bode 2011). The place where the head activator is uninhibited by the head inhibitor becomes the budding zone.

But that does not account for the bottom third of the column. What prevents cells there from becoming heads?

FIGURE 16.31 Grafting experiments provide evidence for a head inhibition gradient. (A) Subhypostomal tissue does not generate a new head when placed close to an existing host head. (B) Subhypostomal tissue generates a head if the existing host head is removed. A head also forms at the site where the host's head was amputated. (C) Subhypostomal tissue generates a new head when placed far away from an existing host head. (After Newman 1974.)

Head formation at the base appears to be prevented by the production of another substance, a foot activator (MacWilliams et al. 1970; Hicklin and Wolpert 1973; Schmidt and Schaller 1976; Meinhardt 1993; Grens et al. 1999). The inhibition gradients for the head and the foot may be important in determining where and when a bud can form. In young adult hydras, the gradients of head and foot inhibitors appear to block bud formation. However, as the hydra grows, the

(A) Intact host: No secondary axis induced

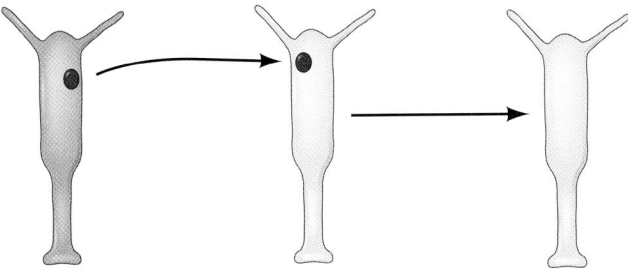

(B) Host's head removed: Secondary axis induced

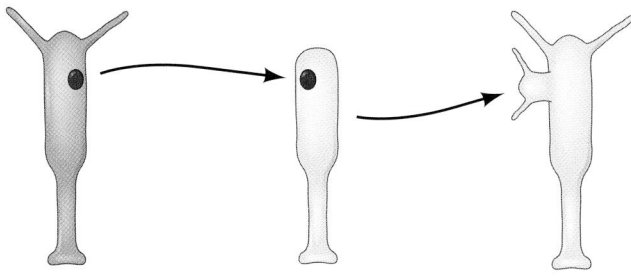

(C) Intact host: Graft away from head region induces secondary axis

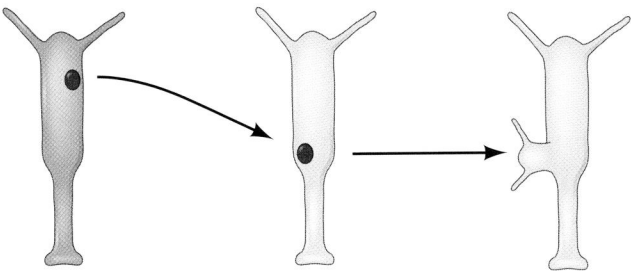

(A)

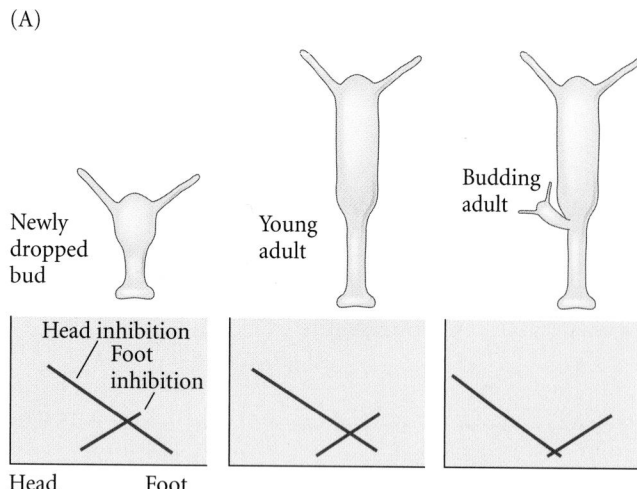

(B)

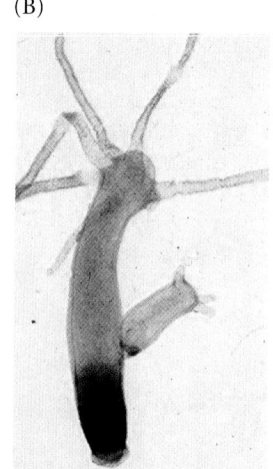

FIGURE 16.32 Bud location as a function of head and foot inhibition gradients. (A) Head inhibition (blue) and foot inhibition (red) gradients in newly dropped buds, young adults, and budding adults. (B) Expression of the Shinguard protein in a graded fashion in a budding hydra. (A after Bode and Bode 1984; B from Bridge et al. 2000.)

sources of these labile substances grow farther apart, creating a region of tissue about two-thirds down the trunk where levels of both inhibitors are minimal. This region is where the bud forms (**FIGURE 16.32A**; Shostak 1974; Bode and Bode 1984; Schiliro et al. 1999).

Certain mutants of *Hydra* have defects in their ability to form buds, and these defects can be explained by alterations of the inhibition gradients. The *L4* mutant of *Hydra magnipapillata*, for instance, forms buds very slowly, and only after reaching a size about twice as long as wild-type individuals. The amount of head inhibitor in these mutants was found to be much greater than in wild-type individuals (Takano and Sugiyama 1983).

Several small peptides have been found to activate foot formation, and researchers are just beginning to sort out the mechanisms by which these proteins arise and function (see Harafuji et al. 2001; Siebert et al. 2005). However, the specification of cells as they migrate from the basal region through the body column may be mediated by a gradient of tyrosine kinase. The product of the *shinguard* gene is a tyrosine kinase that extends in a gradient from the ectoderm just above the basal disc through the lower region of the trunk. Buds appear to form where this gradient fades (**FIGURE 16.32B**). The *shinguard* gene appears to be activated through the product of the *manacle* gene, a putative transcription factor that is expressed earlier in the basal disc ectoderm (Bridge et al. 2000).

Compensatory Regeneration in the Mammalian Liver

According to Greek mythology, Prometheus's punishment for bringing the gift of fire to humans was to be chained to

a rock and to have an eagle tear out and eat a portion of his liver each day. His liver then regenerated each night, providing a continuous food supply for the eagle and eternal punishment for Prometheus. Today the standard assay for liver regeneration is a partial hepatectomy, where specific lobes of the liver are removed (*after* anesthesia is administered, unlike the fate of Prometheus), leaving the other hepatic lobes intact. Although the removed lobe does not grow back, the remaining lobes enlarge to compensate for the loss of the missing tissue (Higgins and Anderson 1931). The amount of liver regenerated is equivalent to the amount of liver removed. Such **compensatory regeneration**—the division of differentiated cells to recover the structure and function of an injured organ—has been demonstrated in the mammalian liver and in the zebrafish heart (Poss et al. 2002).

The human liver regenerates by the proliferation of existing tissue. Surprisingly, the regenerating liver cells do not fully dedifferentiate when they re-enter the cell cycle. No regeneration blastema is formed. Rather, mammalian liver regeneration appears to have two other lines of defense, the first of which consists of normal, mature, adult hepatocytes. These mature cells, which usually are not dividing, are instructed to rejoin the cell cycle and proliferate until they have compensated for the missing part. The second line of defense, discussed below, is a population of hepatic progenitor cells that are normally quiescent but which are activated when the injury is severe and adult hepatocytes cannot regenerate well (because of senescence, alcohol, or disease).

In normal liver regeneration, the five types of liver cells—hepatocytes, duct cells, fat-storing (Ito) cells, endothelial cells, and Kupffer macrophages—all begin dividing to produce more of themselves. Each type retains its cellular identity, and the liver retains its ability to synthesize the liver-specific enzymes necessary for glucose regulation, toxin degradation, bile synthesis, albumin production, and other hepatic functions even as it regenerates itself (Michalopoulos and DeFrances 1997).

There are probably several redundant pathways that initiate liver cell proliferation (Riehle et al. 2011). Global gene profiling indicates that the end result of these pathways is to downregulate (but not totally suppress) the genes involved in the differentiated functions of liver cells while activating those genes committing the cell to mitosis (White et al. 2005). The removal or injury of the liver is sensed through the bloodstream, as some liver-specific factors are lost while others (such as bile acids and gut lipopolysaccharides) increase. These lipopolysaccharides activate some non-hepatocytes to secrete paracrine factors that allow the remaining hepatocytes to re-enter the cell cycle. The Kupffer cell secretes interleukin 6 (IL6) and tumor necrosis factor-α (which are usually involved with activating the adult immune system), while the stellate cells secrete the paracrine factors **hepatocyte growth factor** (**HGF**, or **scatter factor**) and TGF-β. The specialized blood vessels of the liver also produce HGF, as well as Wnt2 (Ding et al. 2010).

However, hepatocytes that are still connected to one another in an epithelium cannot respond to HGF. The hepatocytes activate cMet (the receptor for HGF) within an hour of partial hepatectomy, and the blocking of cMet (by RNA interference or knockout) blocks liver regeneration (Borowiak et al. 2004; Huh et al. 2004; Paranjpe et al. 2007). The trauma of partial hepatectomy may activate metalloproteinases that digest the extracellular matrix and permit the hepatocytes to separate and proliferate. These enzymes also may cleave HGF to its active form (Mars et al. 1995). Together, the factors produced by the endothelial cells, Kupffer cells, and stellate cells allow the hepatocytes to divide by preventing apoptosis, activating cyclins D and E, and repressing cyclin inhibitors such as p27 (see Taub 2004). The liver stops growing at the appropriate size; the mechanism for how this is achieved is not yet known. One clue, though, comes from parabiosis experiments, wherein the circulatory systems of two rats are surgically joined together. Partial hepatectomy in one parabiosed rat will cause the other rat's liver to enlarge (Moolten and Bucher 1967). Therefore, some factor or factors in the blood appear to be establishing the size of the liver. Huang and colleagues (2006) have proposed that these factors are bile acids that are secreted by the liver and positively regulate hepatocyte growth. Partial hepatectomy stimulates the release of bile acids into the blood. These bile acids are received by the hepatocytes and activate the Fxr transcription factor, which promotes cell division. Mice without functional Fxr cannot regenerate their livers. Therefore, bile acids (a relatively small percentage of the products secreted by the liver) appear to regulate the size of the liver and to keep it at a particular volume of cells.

Because human livers have the power to regenerate, a patient's diseased liver can be replaced by compatible liver tissue from a living donor (usually a genetically close relative, whose own liver grows back). Human livers regenerate more slowly than those of mice, but function is restored quickly (Pascher et al. 2002; Olthoff 2003). In addition, mammalian livers possess a "second line" of regenerative ability. If the hepatocytes are unable to regenerate the liver sufficiently within a certain amount of time, the **oval cells** divide to form new hepatocytes. Oval cells are a small progenitor cell population that can produce hepatocytes and bile duct cells. They appear to be kept in reserve and are used only *after* the hepatocytes have attempted to heal the liver (Fausto and Campbell 2005; Knight et al. 2005). The molecular mechanisms by which these factors interact and by which the liver is first told to begin regenerating and then to stop regenerating after reaching the appropriate size remain to be discovered.

AGING: THE BIOLOGY OF SENESCENCE

Entropy always wins. A multicellular organism is able to develop and maintain its identity for only so long before deterioration prevails over synthesis and the organism ages. **Aging** can be defined as the time-related deterioration of the physiological functions necessary for survival and fertility. The characteristics of aging—as distinguished from diseases of aging, such as cancer and heart disease—affect all the individuals of a species. The aging process has two major facets. The first is simply how long an organism lives; the second concerns the physiological deterioration, or **senescence**, that characterizes old age. These topics are often viewed as being interrelated.

Aging and senescence have both genetic and environmental components. The interplay among mutations, environmental factors, and random epigenetic change makes these phenomena both fascinating and frustrating to study. Moreover, in recent years, new molecular and stem cell technologies have allowed new approaches to these age-old questions (see Stocum 2006; Carlson 2007). If aging is a puzzle, then the pieces are beginning to come together. We will first discuss some of the pieces and then look at some attempts to integrate them into a coherent picture of senescence and longevity.

Genes and Aging

Genetic factors play roles in determining longevity both between and within species (Wilson et al. 2007). A mouse can live for 3 years; humans can live for decades. The **maximum lifespan** is the maximum number of years an individual of a given species has been known to survive and is characteristic of that species (Coles 2004). As of 2013, the maximum verified human life span stood at 122.5 years. The life spans of some tortoises and lake trout are uncertain but are estimated to extend beyond 150 years. The maximum life span of a domestic dog is about 20 years, and that of a laboratory mouse is 4.5 years; most mice in the wild do not live to celebrate even their first birthday. If a fruit fly survives to eclose (in the wild, more than 90% die as larvae), it has a maximum life span of 3 months.

(A)

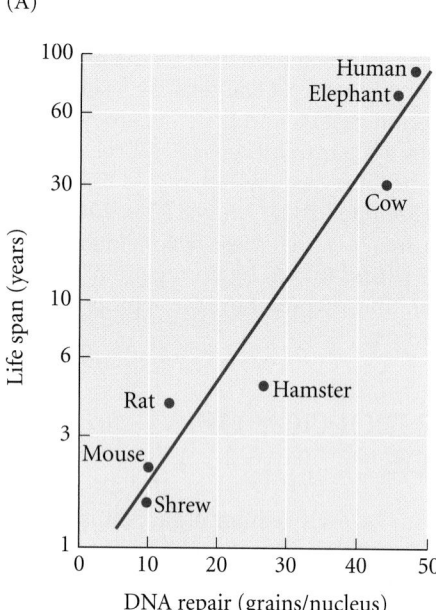

(B)

FIGURE 16.33 Life span and the aging phenotype. (A) Correlation between life span and the ability of fibroblasts to repair DNA in various mammalian species. Repair capacity is represented in autoradiography by the number of grains from radioactive thymidine per cell nucleus. Note that the y axis (life span) is a logarithmic scale. (B) Hutchinson-Gilford progeria. Although they are not yet 8 years old, these children have a phenotype similar to that of an aged person. The hair loss, fat distribution, and skin transparency are characteristic of the normal aging pattern as seen in elderly adults. The mutation causes an aberrant nuclear envelope protein that appears to prevent DNA repair (Coppede and Migliore 2010). (A after Hart and Setlow 1974; B © Associated Press.)

Most people, however, cannot expect to live 122 years. **Life expectancy**—on average, the length of time a given individual of a given species can expect to live—is not characteristic of species but of populations. It is sometimes defined as the age at which half the population still survives. A baby born in England during the 1780s could expect to live to be 35 years old. In Massachusetts during that same time, life expectancy was 28 years. These ages represent the normal range of human life expectancy for most of the human race throughout recorded history (Arking 1998). Even today, in some countries (Angola, Chad, Afghanistan, and several others) life expectancy is around 45 years. In the United States, males born at the turn of the twenty-first century have a life expectancy of about 75 years, and females can expect to live around 80 years.*

Given that in most times and places people did not live much past the age of 40, our awareness of human aging is relatively new. In 1900, 50% of Americans were dead before the age of 60; a 70-year-old person was exceptional in 1900 but is commonplace today. People in 1900 did not have the "luxury" of dying from heart attacks or cancers, because these conditions are most likely to affect people over 50. Rather, many people died (as they are still dying in large parts of the world) from microbial and viral infections. Until recently, relatively few people exhibited the general human senescent phenotype: gray hair, sagging and wrinkling skin, arthritic joints, osteoporosis (loss of bone calcium), loss of muscle fibers and muscular strength, memory loss, eyesight deterioration, and slowed sexual responsiveness. As the melancholy Jacques notes in Shakespeare's *As You Like It*, those who did survive to senescence left the world "*sans* teeth, *sans* eyes, *sans* taste, *sans* everything."

Species-specific life spans appear to be determined by genes that effect a trade-off between the energy used for early growth and reproduction (which results in somatic damage) versus the energy allocated for maintenance and repair. In other words, aging results from natural selection operating more strongly on early survival and reproduction than on having a vigorous postreproductive life. Molecular evidence (see Vijg and Campisi 2008; Kenyon 2010) indicates that certain genetic components of longevity are conserved between species—flies, worms, mammals, and even yeast all appear to use the same set of genes to promote survival and longevity. Four sets of genes are well known to be involved in aging and its prevention, and each set appears to be conserved between phyla and even kingdoms of organisms. These are the genes encoding (1) DNA repair enzymes, (2) proteins of the insulin signaling pathway, (3) proteins in the

---

*When Social Security was enacted in the United States in 1935, the average working citizen died before age 65. Thus, he (and it usually was a he) was not expected to get back as much as he had paid into the system. Similarly, marriage "until death do us part" was easier to achieve when death occurred in the third or fourth decade of life. Before antibiotics, the death rate of young women due to infections associated with childbirth was high throughout the world.

mTOR signaling pathway, a cascade that regulates translation, and (4) chromatin remodeling enzymes.

DNA repair enzymes

DNA repair enzymes appear be critically important in preventing senescence (Gornova et al. 2007). Individuals of species whose cells have more efficient DNA repair enzymes live longer (**FIGURE 16.33A**; Hart and Setlow 1974). Certain premature aging syndromes (**progerias**) in humans and mice appear to be caused by mutations that prevent the functioning of DNA repair enzymes (**FIGURE 16.33B**; Sun et al. 1998; Shen and Loeb 2001; de Boer 2002).

"Wear-and-tear" theories of aging are among the oldest hypotheses proposed to account for the human senescent phenotype (Weismann 1891; Medawar 1952). As one gets older, small traumas to the body and its genome build up. At the molecular level, the number of point mutations increases with age, and the efficiency of the enzymes encoded by our genes decreases (Singh et al. 2001; Baily et al. 2004; Rossi et al. 2007). Moreover, if mutations occur in the genes encoding transcriptional or translational proteins, the cell may make an even greater number of faulty proteins (Orgel 1963; Murray and Holliday 1981; Kamileri 2012).

REACTIVE OXYGEN SPECIES Two major sources of mutation are radiation and **reactive oxygen species** (**ROS**). The ROS produced by normal metabolism can oxidize and damage cell membranes, proteins, and nucleic acids. Some 2–3% of the oxygen atoms taken up by our mitochondria are reduced insufficiently and form ROS: superoxide ions, hydroxyl ("free") radicals, and hydrogen peroxide. Evidence that ROS molecules are critical in the aging process includes observations that fruit flies and nematodes overexpressing the enzymes that destroy ROS (catalase and superoxide dismutase) live significantly longer than do control animals (Orr and Sohal 1994; Parkes et al. 1998; Sun and Tower 1999; Feng et al. 2001). However, these correlations have not held up in some other studies, so the genetic ability to destroy free oxygen radicals may not be critically important to longevity (Pérez et al. 2009; van Raamsdonk and Hekimi 2012).

TELOMERASE AND P53 The transcription factor **p53**, one of the most important regulators of cell division, can stop the cell cycle, cause cellular senescence in rapidly dividing cells, instruct genes to initiate cellular apoptosis, and activate DNA repair enzymes. In most cells, p53 is bound to a repressor protein that keeps p53 inactive. However, ultraviolet radiation, oxidative stress, and other factors that cause DNA damage will separate p53 from its repressor, allowing it to function. The induction of apoptosis or cellular senescence by p53 can be beneficial (when destroying cancer cells) or deleterious (when destroying, say, neurons or stem cells).

SIDELIGHTS & SPECULATIONS

Exceptions to the Aging Rule

There are a few species in which aging seems to be optional, and these may hold some important clues to how animals can live longer and retain their health. Turtles, for instance, are a symbol of longevity in many cultures. Many turtle species not only live a long time, they don't seem to undergo a typical aging syndrome. Turtles seem to have "negligible senescence," in that their mortality rate does not increase with age, and their reproductive rate does not decrease with age. In these species, older females lay as many eggs (if not more) as their younger counterparts. Miller (2001) showed that a 60-year-old female three-toed box turtle (*Terrapene carolina triunguis*) lays as many eggs annually as she ever did. If turtle telomeres shorten with age, it happens (like so many turtle things)

extremely slowly (Girondot and Garcia 1998; Hatase et al. 2008). Interestingly, turtles have special adaptations against oxygen deprivation, and these enzymes also protect against ROS (Congdon et al. 2003; Lutz et al. 2003; Krivoruchko and Storey 2010).

In monarch butterflies (*Danaus plexippus*), adults that migrate to wintering grounds in the mountains of central Mexico live several months (August–March), whereas their summer counterparts live only about 2 months (May–July). The regulation of this difference appears to be juvenile hormone (Herman and Tatar 2001). The migrating butterflies are sterile because of suppressed synthesis of JH. If migrants are given JH in the laboratory, they regain fertility but lose their longevity. Conversely, when summer monarchs have their corpora

allata removed (so they no longer make JH), their longevity increases 100%. Mutations in the insulin signaling pathway of *Drosophila* likewise decrease JH synthesis (Tu et al. 2005). This decrease in JH makes the flies small, sterile, and long-lived, adding to whatever longevity-producing effect protection against ROS might have.

Finally, there may be organisms, especially cnidarians, that have actually cheated death. Hydra appear to be immortal, retaining their stem cell populations. The hydrozoan cnidarian *Turritopsis nutricula* has evolved a remarkable variation on this theme. It has both a polyp stage (like the hydra) and a medusa (jellyfish) stage in its life cycle. However, the *Turritopsis* medusa can revert to its polyp stage *after* becoming sexually mature (Bavestrello et al. 1992; Piraino et al. 1996).

One of the chief ways of activating p53 (and related proteins such as p63) is to damage the **telomeres**, the protective nucleoprotein caps on the tips of the chromosomes (like the aglets on shoelaces that keep them from unwinding). When p53 is activated by damaged telomeres, DNA replication halts, and if the repair doesn't work, apoptosis is initiated. If the cell is a stem cell or some other rapidly replicating cell, this will reduce the numbers of cells produced, and the lack of stem cells will produce an "aged" phenotype. The relationship between shortened telomeres and stem cell depletion has been seen in degenerative diseases such as mouse muscular dystrophy (Sacco et al. 2010).

There is a positive correlation in humans between telomere length and longevity (Atzmon et al. 2010), and telomeres appear to shorten with age in the stem cell compartments of mice and humans (Zhang and Ju 2010). The enzyme complex that maintains telomere integrity is **telomerase**, which acts as an antisenescence complex. Mice and humans with telomerase deficiencies age prematurely (Mitchell et al. 1999). Overexpressing telomerase or reactivating it in senescent cells extends longevity in mice without increasing cancer (Tomas-Loba et al. 2008; Jaskelioff et al. 2011; Bernardes et al. 2012).

Aging and the insulin signaling cascade

One criticism of the idea that there are genetic "programs" for aging asks how evolution could have selected for them. Once an organism has passed reproductive age and raised its offspring to sexual maturity, it becomes "an excrescence on the tree of life" (Rostand 1962); natural selection presumably cannot act on traits that affect an organism only after it has reproduced. But "How can evolution select for a way to degenerate?" may be the wrong question. Evolution probably can't select for such traits. The right question may be, "How can evolution select for phenotypes that postpone reproduction or sexual maturity?" There is often a trade-off between reproduction and maintenance, and in many species reproduction and senescence are closely linked.

Recent studies of mice, *Caenorhabditis elegans*, and *Drosophila* suggest that there is a conserved genetic pathway that regulates aging, and that it can indeed be selected for. This pathway involves the response to insulin and insulin-like growth factors. In *C. elegans*, a larva proceeds through four larval stages, after which it becomes an adult. If the nematodes are overcrowded or if there is insufficient food, the larva can enter a metabolically dormant **dauer larva** stage, a nonfeeding state of **diapause**, a condition in which development and aging are suspended. The nematode can remain in the dauer stage for up to 6 months (rather than becoming an adult that lives only a few weeks). In this state it has increased resistance to oxygen radicals that can crosslink proteins and destroy DNA. The pathway that regulates both dauer larva formation and longevity has been identified as the **insulin signaling pathway** (Kimura et al. 1997; Guarente and Kenyon 2000; Gerisch et al. 2001; Pierce et al. 2001).

In *C. elegans*, favorable environments signal activation of the insulin receptor homologue DAF-2, and this receptor stimulates the onset of adulthood (**FIGURE 16.34A**). Poor environments fail to activate the DAF-2 receptor, and dauer formation ensues. While severe loss-of-function alleles in the insulin signaling pathway cause the formation of dauer larvae in any environment, weak mutations in the pathway enable the animals to reach adulthood and live four times longer than wild-type animals.

Downregulation of the insulin signaling pathway has several other functions. First, it appears to influence metabolism, decreasing mitochondrial electron transport. Second, when the DAF-2 receptor is not active, cells increase the production of enzymes that prevent oxidative damage, as well as DNA repair enzymes (Honda and Honda 1999; Tran et al. 2002). Third, this lack of insulin signaling decreases fertility (Gems et al. 1998). This increase in DNA synthetic enzymes and in enzymes that protect against ROS is due to the Foxo/DAF-16 transcription factor. This Forkhead-type transcription factor is inhibited by the insulin receptor (DAF-2) signal. When that signal is absent, Foxo/DAF-16 can function, and this factor promotes longevity in ways not yet deciphered. It is possible that Foxo/DAF-16 activates the expression of genes involved in producing anti-stress proteins within the cell as well as lipid signals that help extend life to those cells nearby (Zhang et al. 2013). The Foxo transcription factor has been associated with longevity throughout the animal kingdom. Indeed, it has recently been shown to be one of the major drivers of stem cell renewal in immortal hydras (Boehm et al. 2012).

It is possible that this system also operates in mammals, but the mammalian insulin and insulin-like growth factor pathways are so integrated with embryonic development and adult metabolism that mutations often have numerous and deleterious effects (such as diabetes or Donahue syndrome). However, there is evidence that the insulin signaling pathway does affect life span in mammals (**FIGURE 16.34B**). Dog breeds with low levels of insulin-like growth factor 1 (IGF-1) live longer than breeds with higher levels of this factor. Mice with loss-of-function mutations of the insulin signaling pathway live longer than their wild-type littermates (see Partridge and Gems 2002; Blüher et al. 2003; Kurosu et al. 2005). Holzenberger and colleagues (2003) found that mice heterozygous for the insulin receptor IGF-1R not only lived about 30% longer than their wild-type littermates, they also had greater resistance to oxidative stress. In addition, mice lacking one copy of their IGF-1R gene lived about 25% longer than wild-type mice.

The insulin signaling pathway also appears to regulate life span in *Drosophila* (**FIGURE 16.34C**). Flies with weak loss-of-function mutations of the insulin receptor gene or genes in the insulin signaling pathway live nearly 85% longer than wild-type flies (Clancy et al. 2001; Tatar et al. 2001). These long-lived mutants are sterile, and their metabolism resembles that of flies that are in diapause (Kenyon 2001). The insulin receptor in *Drosophila* is thought to regulate a

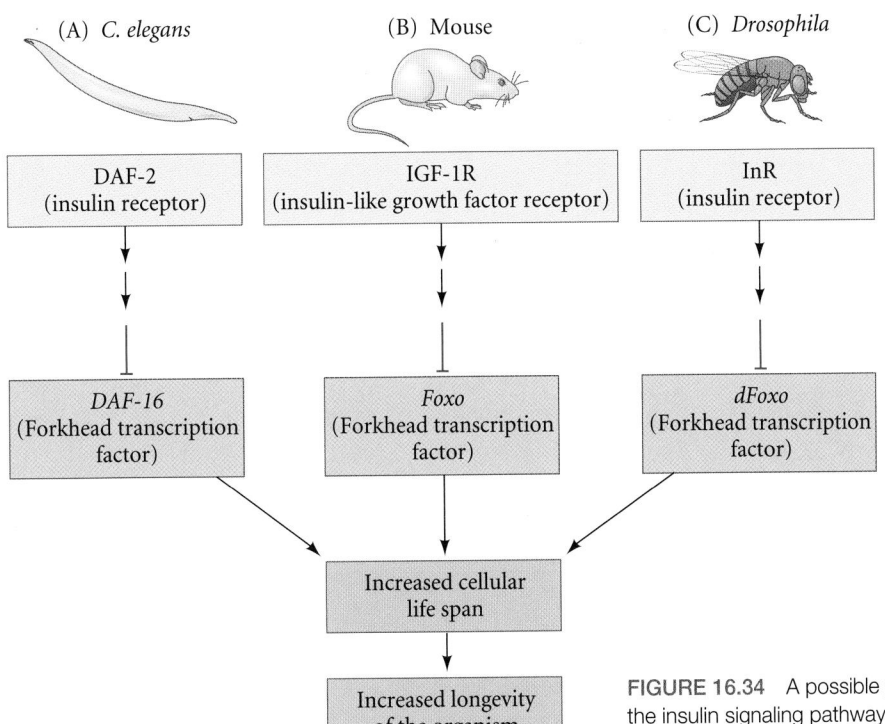

(A) *C. elegans*

(B) Mouse

(C) *Drosophila*

DAF-2
(insulin receptor)

IGF-1R
(insulin-like growth factor receptor)

InR
(insulin receptor)

DAF-16
(Forkhead transcription factor)

Foxo
(Forkhead transcription factor)

dFoxo
(Forkhead transcription factor)

Increased cellular life span

Increased longevity of the organism

FIGURE 16.34 A possible pathway for regulating longevity. In each case, the insulin signaling pathway inhibits the synthesis of the Foxo transcription factor proteins that would otherwise increase cellular longevity.

Forkhead transcription factor (dFoxo) similar to the Foxo/DAF-16 protein of *C. elegans*. When the *Drosophila dFoxo* gene is activated in the fat body, it can lengthen the fly's life span (Giannakou et al. 2004; Hwangbo et al. 2004). From an evolutionary point of view, the insulin pathway may mediate a trade-off between reproduction and survival/maintenance. Many (although not all) of the long-lived mutants have reduced fertility. Thus, it is interesting that another longevity signal originates in the gonad. When the germline cells are removed from *C. elegans*, the worms live longer. The germline stem cells produce a substance that blocks the effects of a longevity-inducing steroid hormone (Hsin and Kenyon 1999; Gerisch et al. 2001; Shen et al. 2012).

Calorie restriction is another way of downregulating the insulin pathway (Kenyon 2001; Roth et al. 2002; Holzenberger et al. 2003). Calorie restriction may reduce levels of IGF-1 (the main ligand of IGF-1R) and of circulating insulin, although other mechanisms are also being explored (e.g., Selman et al. 2009). Studies in primates (including humans) have not concluded that low calorie intake extends their longevity, although it does appear to retard the age-associated decline of heartbeat variability and motor coordination (see Colman et al. 2009; Mattison et al. 2012; Stein et al. 2012).

The mTOR pathway

One of the main ways by which the insulin signaling pathway might function to lower longevity is to activate mTORC1, a protein kinase complex that promotes translation in response to nutrients and hormones (Lamming et al. 2012; Johnson et al. 2013). Thus, the insulin signaling pathway depresses Foxo and at the same time activates mTORC1. Dietary restriction reduces mTORC1 activity, and mice with

reduced mTORC1 levels had longer lives, better protection against age-related cognitive dysfunction, and more functional stem cells than control mice (Chen et al. 2009; Harrison 2009; Halloran et al. 2012; Majumder et al. 2012; Yilmaz et al. 2012). Reducing mTORC1 also increases the amount of **autophagy**, the removal and replacement of damaged organelles and senescent cells. Many of the maladies associated with old age appear to be the result of failed autophagy and replacement (Baker et al. 2011). The mechanisms by which reduced mTORC1 accomplishes these feats are still unknown, and this pathway is an area of active study.

Chromatin modification

Chromatin modification may be very important in aging. The **sirtuin genes**, which encode histone deacetylation (chromatin-silencing) enzymes, have been found to prevent aging throughout the eukaryotic kingdoms, including in yeasts and mammals (Howitz et al. 2003; Oberdoerffer et al. 2008). Sirtuins prevent genes from being expressed at the wrong times and places, and they help repair chromatin breaks. When DNA strands break (as inevitably happens as the body ages), sirtuin proteins are called on to fix them and cannot attend to their usual functions. Thus, genes that are usually silenced become active as the cells age.

Alternatively, there are other areas of the body, such as the brain, where histone deacetylases can generate an aging phenotype. Cognitive decline, especially in the ability to recall past experiences, is a normal part of the mammalian aging syndrome. Long-term memories are stabilized by chromatin remodeling in the hippocampus and frontal lobes of the brain, a process involving DNA methylation and histone modifications (Swank et al. 2001; Korzus 2004; Miller et al.

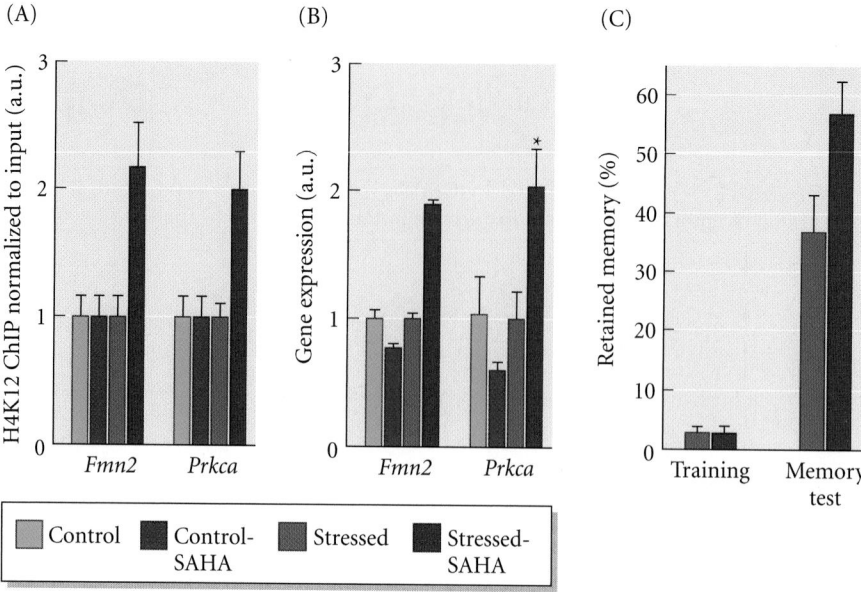

FIGURE 16.35 Age-related memory decline in mice can be reversed by inhibitors of histone deacetylases. Mice were either unstressed (control) or stressed to form a new memory. (A) H4K12 identified by chromatin immunoprecipitation (ChIP) assays in the coding regions of three genes. The stressed mice treated with the inhibitor of histone deacetylase (SAHA) had the highest level of H4K12. (B) The stressed mice treated with SAHA also had the highest levels of expression of *Fmn2* and *Prkca*, two genes that have been associated with memory formation. (C) Mice that were stressed stabilized a memory of this stress better if they had been treated with the inhibitor of histone deacetylases. (After Peleg et al. 2010.)

2008; Penner et al. 2011). Peleg and colleagues (2010) have shown that the normal transcription associated with long-term memory stabilization is disrupted as mice age, and that this lack of transcription is associated with lessened H4K12 acetylation. Indeed, this ability to store memory can be retrieved by infusing into the hippocampus an inhibitor of histone deacetylase (**FIGURE 16.35**).

Integrating the conserved aging pathways

The proteins involved in the insulin signaling pathway and the DNA repair pathway interact with one another (**FIGURE 16.36**). The p53 factor that induces cell cycle arrest also blocks the activity of the receptor for insulin-like growth factor 1. And sirtuin proteins, (which may also activate Foxo proteins), can also block p53. In some cases, the same protein is

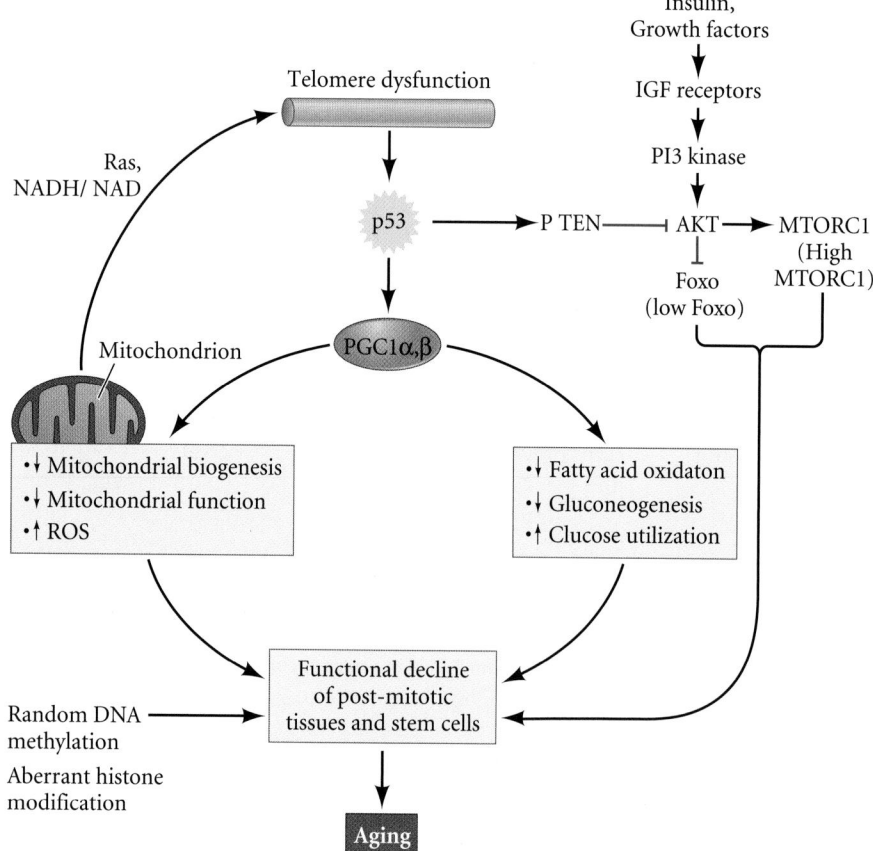

FIGURE 16.36 The pathways inducing aging are united and often reciprocal. The p53 induced by telomere dysfunction binds to and represses the expression of *PGC-1α* and *PGC-1β*. The repression of both co-activating transcription factors blocks mitochondrial production and function, leading to lower ATP generation and increased production of reactive oxygen species (ROS). PGCs also play a crucial role in energy metabolism, by regulating fatty acid oxidation, gluconeogenesis, glucose uptake, and oxidation. A decline in mitochondrial function and metabolite processing would lead to functional decline in tissue stem cells and postmitotic tissues, thereby promoting senescence. Moreover, compromised mitochondrial function can lead to more ROS and iron-sulphur clusters, which damage DNA and further activate p53. Active p53 can also block the the insulin signaling pathway that causes low levels of Foxo transcription factor and high levels of the mTORC1 complex. In addition, random epigenetic events can decrease the expression of particular proteins. (After Vijg and Campisi 2008; Sahin and DePinho 2012.)

SIDELIGHTS  SPECULATIONS

Young Blood: Serum Factors and Progenitor Cells

One of the hallmarks of aging is the declining ability of stem cells and progenitor cells to restore damaged or nonfunctioning tissues. A decline in muscle progenitor (satellite) cell activity when Notch signaling is lost results in a significant decrease in the ability to maintain muscle function. Similarly, an age-dependent decline in liver progenitor cell division impairs liver regeneration because of a decline in the transcription factor cEBPα. And the age-associated graying of mammalian hair appears to be due to the apoptosis of melanocyte stem cells in the hair bulge niche (Nishimura et al. 2005; Robinson and Fisher 2009). One of the questions, then, becomes: Is this part of the aging syndrome caused by the declining function of stem cells or by a declining ability of the stem cell niche to support them?

One way to test this is by "fusing" an old mouse to a young mouse. This can be done by a technique called **parabiosis**, wherein the animals' circulatory systems are surgically joined so that the two mice share one blood supply. If an aged and a young mouse are parabiosed, the stem cells of the old mouse are exposed to factors in young blood serum (and vice versa). This *heterochronic parabiosis* has been seen to restore the activity of old stem cells. Notch signaling of the muscle stem cells regained its youthful levels, and muscle cell regeneration was restored. Similarly, liver progenitor cells regained "young" levels of cEBPα—and their ability to regenerate (Conboy et al. 2005; 2012).

In the nervous system, however, the old appears to dominate the young. In heterochronically parabiosed mice, blood-borne factors from old mice restrict the proliferation of young mouse neural stem cells and decrease learning and memory. Moreover, when young mice were given blood plasma from other young mice, they maintained a normal young level of Doublecortin in their cortex.

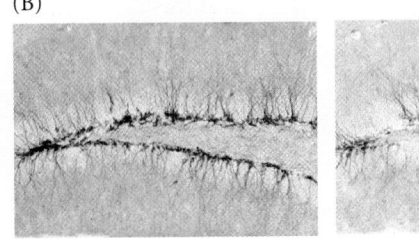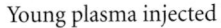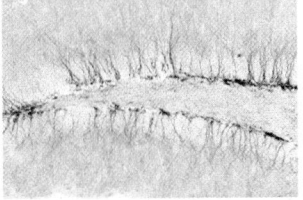

Young plasma injected   Old plasma injected

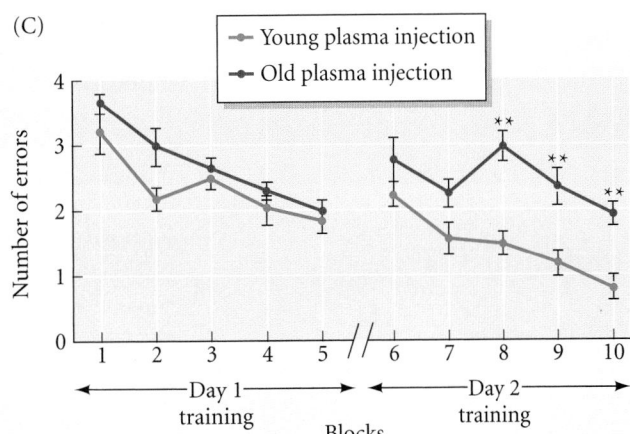

FIGURE 16.37 Factors in the plasma (the liquid portion of the blood) of old mice alter the development of new neurons and behaviors in young mice. (A) Protocol whereby plasma from young mice is injected into other young mice (left), or plasma from old mice is injected into young mice (right). (B) Young mice receiving young plasma continue to manufacture new neurons (here darkly stained with antibodies to Doublecortin). whereas the number of new neurons decreases in young mice injected with old plasma. (C) In training to do a particular task, mice receiving young or old plasma initially had the same number of errors. One day later, mice that received young plasma remembered their former mistakes and made fewer errors than mice that received old plasma. (After Villeda et al. 2011.)

Doublecortin is a protein found in newly generated neurons. However, when the injected plasma came from old mice, the number of cells containing Doublecortin dropped dramatically, as did the mouse's cognitive functions

(**FIGURE 16.37**). Villeda and colleagues (2011) identified the blood-borne agent as cytokine CC11, a protein whose levels normally rise in the plasma (and cerebrospinal fluid) of healthy aging people.

(A)

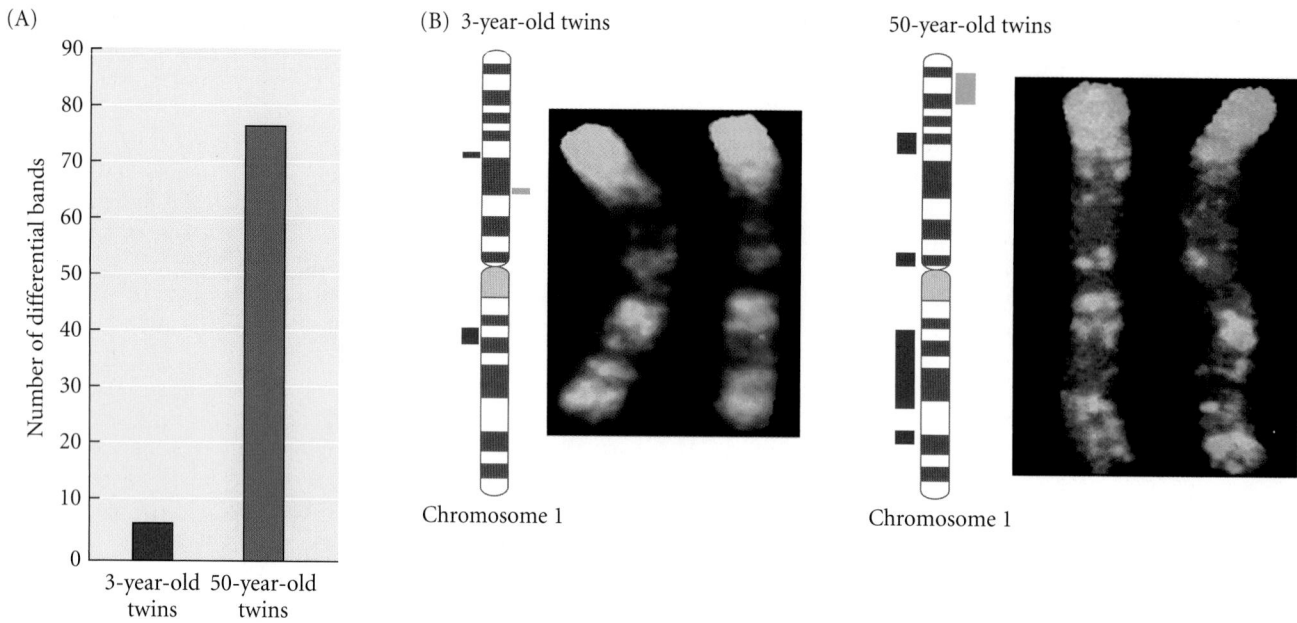

(B) 3-year-old twins

Chromosome 1

50-year-old twins

Chromosome 1

(C)

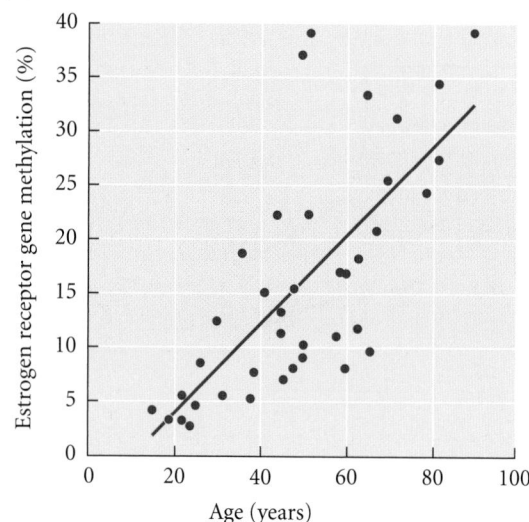

FIGURE 16.38 Differential DNA methylation patterns in aging twins. (A) In bisulfite sequence mapping, regions of DNA that are unmethylated will be cut by restriction enzymes (because bisulfite converts unmethylated cytosine to uracil), but methylated sites will not be cut. The histogram summarizes the number of differences in the resulting restriction maps of 3-year-old and 50-year-old twins. (B) A more recent technique of revealing methylation differences and similarities between twins is to mark the DNA from one twin with a red dye and that from the other with a green dye. One can then collect only the nonmethylated DNA and bind it to metaphase chromosomes. If the bands are red or green, it means the DNA from one twin bound but the DNA from the other twin did not. If the region is yellow, it means the red and green DNAs bound equally. (C) Methylation of the estrogen receptor genes occurs as a function of normal aging. (A,B after Fraga et al. 2005, photographs courtesy of M. Esteller; C after Issa et al. 1994.)

involved in both the DNA repair and insulin signaling pathways (Niedernhofer et al. 2006).

Recently, p53 was seen to link the telomere-shortening pathway of senescence with the mitochondrial and reactive oxygen theories of aging (Sahin et al. 2011; Sahin and DePinho 2012). First, DNA damage (such as telomere shortening) would activate p53, and as mentioned above, such p53-mediated apoptosis would drive functional decline in high-turnover tissues with rapidly proliferating stem cells. Such a decline of stem cells is characteristic of the central nervous system of aging mammals (Villeda et al. 2011). But p53 was also found (by analyzing transcriptomes) to bind to and profoundly repress the transcription factors PGC-1α and PGC-1β. These transcription factors are crucial for mitochondrial biogenesis (making more mitochondria) and for oxidative phosphorylation. Without proper oxidative phosphorylation, reactive oxygen radicals are more readily formed. These effects would cause the depletion of function and senescence of long-lived tissue, such as those in neurons and the heart.

Environmental and Epigenetic Causes of Aging

The general senescent phenotype is characteristic of each species. But what causes it? This question can be asked at many levels. We have just discussed how various pathways in the cell can control the amount of damage—DNA repair enzymes, Foxo proteins, telomerase, low mTORC1 levels. But what might be creating the problems? As we mentioned earlier, DNA damage by radiation and ROS may be part of the problem. Another part of the problem may be **random epigenetic drift**. Given that appropriate methylation is essential for normal development, one can immediately see that inappropriate epigenetic methylation could have dire consequences.

Recent studies have supported the idea that random epigenetic drift occurs and that inappropriate methylation is a factor in aging and cancers. Some of the evidence for this hypothesis comes from monzygotic ("identical") twins. Most identical twins start life with very few differences in appearance or behaviors, but accumulate these differences with age (FIGURE 16.38A,B). Experience counts, and both random events and lifestyles may be reflected in phenotypes. Twin pairs were nearly indistinguishable in DNA methylation and histone acetylation patterns when young, but older monozygous twins exhibited very different patterns of both these epigenetic markers (Fraga et al. 2005; Bell et al. 2012). This affected their gene expression patterns, such that older twin pairs had different patterns of gene expression, whereas younger twin pairs had very similar expression patterns.

The idea that random epigenetic drift inactivates important genes without any particular environmental cue gives rise to an entirely new hypothesis of aging. Instead of randomly accumulated mutations—which might be due to specific mutagens—we are at the mercy of chance accumulations of errors made by the DNA methylating and demethylating enzymes. Indeed, our DNA methylating enzymes, unlike the DNA polymerases, are prone to errors. DNA methyltransferases are not the most fastidious of enzymes. At each round of DNA replication, they must methylate the appropriate cytosine residues and leave the others unmethylated. This is not always done properly, and such errors accumulate as we age (Maegawa et al. 2010).

In fact, the methylation of numerous sites in the genome is linear with age and can predict the age of an individual to within 5 years (Bocklandt et al. 2011; Hannum et al. 2012). Most of these sites are found in genes known to be involved in aging-related conditions such as atherosclerosis, cancer, and neurodegenerative diseases. For instance, methylation of

the promoter region of estrogen receptor (ER) genes is known to increase with age (FIGURE 16.38C), resulting in the inactivation of ER genes in the smooth muscles of the circulatory system (Issa et al. 1994). Moreover, atherosclerotic plaques (thickened artery walls that occlude blood vessels) show greater methylation of the ER genes than does the tissue around them (Post et al. 1999; Kim et al. 2007). Thus, DNA methylation-associated inactivation of the estrogen receptors in vascular tissue may play a role in aging of the vascular system. This potentially reversible defect may provide a new target for intervention in heart disease.

Promoting Longevity

Several interacting agents may promote longevity. These include calorie restriction, protection against oxidative stress, factors activated by a suppressed insulin pathway, and factors affected by decreased mTOR signaling (e.g., decreased protein translation and augmented autophagy). It is not yet known how these factors interact—whether they are part of a single "longevity pathway" or if they act separately. Moreover, genetics and diet do not appear to be the full answer to aging. Chance, it seems, still plays a role. When clonally identical *C. elegans* are fed an identical diet, some organisms still live longer than others, and different organs deteriorate more rapidly in different individuals (Herndon et al. 2002). Mutations are randomly occurring events, and they may play a role in the aging process.

As advances in our ability to prevent and cure disease lead to increased human life expectancy, we are still left with a general aging syndrome that is characteristic of our species. Unless attention is paid to this general aging syndrome, we risk ending up like Tithonios, the miserable wretch of Greek mythology to whom the gods awarded eternal life, but not eternal youth. However, our new knowledge of regeneration is being put to use by medicine, and we may soon be able to ameliorate some of the symptoms of aging.

SNAPSHOT SUMMARY   Metamorphosis, Regeneration, and Aging

1. Amphibian metamorphosis includes both morphological and biochemical changes. Some structures are remodeled, some are replaced, and some new structures are formed.

2. The hormone responsible for amphibian metamorphosis is tri-iodothyronine (T_3). The synthesis of T_3 from thyroxine (T_4) and the degradation of T_3 by deiodinases can regulate metamorphosis in different tissues. T_3 binds to thyroid hormone receptors and acts predominantly at the transcriptional level.

3. Many changes during amphibian metamorphosis are regionally specific. The tail muscles degenerate; the trunk muscles persist. An eye will persist even if transplanted into a degenerating tail.

4. Metamorphic change in amphibians can be brought about by cell death, cell differentiation, or by cell-type switching.

5. The specific timing of metamorphic events can be orchestrated by the different events occurring at different levels of thyroid hormones. The wisdom of the frog is simple: don't regress your tail until you have constructed hindlimbs.

6. Heterochrony involves changes in the relative rates of development of different parts of the animal. In neoteny, the larval form is retained while the gonads and germ cells mature at their normal rate. In progenesis, the gonads and germ cells mature rapidly while the rest of the body matures normally. In both instances, the animal can mate while retaining its larval or juvenile form.

7. Animals with direct development do not have a larval stage. Primary larvae (such as those of sea urchins) specify their body axes differently than the adult, whereas secondary larvae (such as those of insects and amphibians) have body axes that are the same as adults of the species.

8. Ametabolous insects undergo direct development. Hemimetabolous insects pass through nymph stages wherein the immature organism is usually a smaller version of the adult.

9. In holometabolous insects, there is a dramatic metamorphosis from larva to pupa to sexually mature adult. In the stages between larval molts, the larva is called an instar. After the last instar, the larva undergoes a metamorphic molt to become a pupa. The pupa undergoes an imaginal molt to become an adult.

10. During the pupal stage, the imaginal discs and histoblasts grow and differentiate to produce the structures of the adult body.

11. The anterior-posterior, dorsal-ventral, and proximal-distal axes are sequentially specified by interactions between different compartments in the imaginal discs. The disc "telescopes out" during development, its central regions becoming distal.

12. Molting is caused by the hormone 20-hydroxyecdysone (20E). In the presence of high levels of juvenile hormone, the molt gives rise to another larval instar. In low concentrations of juvenile hormone, the molt produces a pupa; if no juvenile hormone is present, the molt is an imaginal molt.

13. The ecdysone receptor gene produces a nuclear RNA that can form at least three different proteins. The types of ecdysone receptors in a cell may influence the response of that cell to 20E. The ecdysone receptors bind to DNA to activate or repress transcription.

14. There are four major types of regeneration. In stem-cell mediated regeneration, new cells are routinely produced to replace the ones that die. In epimorphosis (such as regenerating limbs), tissues form into a regeneration blastema, divide, and redifferentiate into the new structure. In morphallaxis (characteristic of *Hydra*), there is a repatterning of existing tissue with little or no growth. In compensatory regeneration (such as in the mammalian liver), cells divide but retain their differentiated state.

15. In flatworms, regeneration occurs by forming a regeneration blastema produced by pluripotent neoblasts. Wnt signaling gradients appear to direct the anterior-posterior differentiation of these cells.

16. In regenerating limb blastemas, cells do not become multipotent. Rather, the cells retain their specification, such that neurons come from preexisting neurons and muscles come from preexisting muscle cells (that have become mononucleate) or from muscle stem cells. The mitogens, such as nAG, are provided by the AEC and the glial surrounding the limb axons.

17. Salamander limb regeneration appears to use the same pattern formation system as the developing limb.

18. *Hydra* appears to have a head activation gradient, a head inhibition gradient, a foot activation gradient, and a foot inhibition gradient. Budding occurs where these gradients are minimal.

19. The hypostome region of *Hydra* appears to be an organizer region that secretes paracrine factors to alter the fates of surrounding tissue.

20. In the mammalian liver, no regenerating blastema is formed, and the liver regenerates the same volume as it lost. Each cell appears to generate its own cell type; and a reserve population of multipotent progenitor cells divides when these tissues cannot regenerate the missing portions.

21. The maximum life span of a species is the longest time an individual of that species has been observed to survive. Life expectancy is usually defined as the age at which approximately 50% of the members of a given population still survive.

22. Aging is the time-related deterioration of the physiological functions necessary for survival and reproduction. The phenotypic changes of senescence (which affect all members of a species) are not to be confused with diseases of senescence, such as cancer and heart disease (which affect some individuals but not others).

23. Reactive oxygen species (ROS) can damage cell membranes, inactivate proteins, and mutate DNA. Mutations that alter the ability to make or degrade ROS can change the life span.

24. Proteins that regulate DNA repair and cell division (such as p53 and telomerase) may be important regulators of aging.

25. An insulin signaling pathway, involving a receptor for insulin and insulin-like proteins, may be an important component of genetically limited life spans. It may upregulate mTORC1 and downregulate Foxo transcription factors.

26. Random DNA methylation appears to repress gene expression as a cell ages. Enzymes involved with chromatin modification may be important mediators of such aging events.

27. In many cases, the aging phenotype is the result of apoptosis of stem cells or progenitor cells.

For Further Reading

Complete bibliographical citations for all literature cited in this chapter can be found at the free access website **www.devbio.com**

Baguñà, J. 2012. The planarian neoblast: The rambling history of its origin and some current black boxes. *Int. J. Dev. Biol.* 56: 19–37.

Broun, M., L. Gee, B. Reinhardt and H. R. Bode. 2005. Formation of the head organizer in hydra involves the canonical Wnt pathway. *Development* 132: 2907.

Cai, L. and D. D. Brown. 2004. Expression of type II iodothyronine deiodinase marks the time that a tissue responds to thyroid hormone-induced metamorphosis in *Xenopus laevis. Dev. Biol.* 266: 87–95.

Carlson, B. M. 2007. *Principles of Regenerative Biology.* Academic Press, New York.

Fraga, M. F. and 20 others. 2005. Epigenetic differences arise during the lifetime of monozygotic twins. *Proc. Natl. Acad. Sci. USA* 102: 10604–10609.

Hiruma. K. and Y. Kaneko. 2013. Hormonal regulation of insect metamorphosis with special reference to juvenile hormone biosynthesis. *Curr. Top. Dev. Biol.* 103: 73–100.

Jiang, C., A. F. Lamblin, H. Steller and C. S. Thummel. 2000. A steroid-triggered transcriptional hierarchy controls salivary gland cell death during *Drosophila* metamorphosis. *Mol. Cell* 5: 445–455.

Johnson, S. C., P. S. Rabinovitch and M. Kaeberlein. 2013. mTOR is a key modulator of ageing and age-related disease. *Nature* 493: 338–345.

Kenyon, C. J. 2010. The genetics of ageing. *Nature* 464: 504–512.

Kragl, M., D. Knapp, E. Nacu, S. Khattak, M. Maden, H. H. Epperlein and E. M. Tanaka. 2009. Cells keep a memory of their tissue origin during axolotl limb regeneration. *Nature* 460: 60–65.

Kumar, A., J. W. Godwin, P. B. Gates, A. A. Garza-Garcia and J. P. Brockes. 2007. Molecular basis for the nerve dependence of limb regeneration in an adult vertebrate. *Science* 318: 772–777.

Stocum, D. L. 2006. *Regenerative Biology and Medicine.* Academic Press, New York.

Taub, R. 2004. Liver regeneration: From myth to mechanism. *Nature Rev. Mol. Cell Biol.* 5: 836–847.

Wagner, D. E., J. J. Ho and P. W. Reddien. 2012. Genetic regulators of a pluripotent adult stem cell system in planarians identified by RNAi and clonal analysis. *Cell Stem Cell* 10: 299–311.

Go Online

WEBSITE 16.1 The molecular biology of wing formation. Formation of the *Drosophila* wing involves the interaction of more than 200 genes. This site discusses some of these gene interactions.

WEBSITE 16.2 Homologous specification. If a group of cells in one imaginal disc are mutated such that they give rise to a structure characteristic of another imaginal disc (for instance, cells from a leg disc giving rise to antennal structures), the regional specification of those structures will be in accordance with their position in the original disc.

WEBSITE 16.3 Insect metamorphosis. The four links on this website discuss (1) the experiments of Wigglesworth and others who identified the hormones of metamorphosis and the glands producing them; (2) the variations that *Drosophila* and other insects play on the general theme of metamorphosis; (3) the remodeling of the insect nervous system during metamorphosis; and (4) a microarray analysis of *Drosophila* metamorphosis wherein several thousand genes are simultaneously screened.

WEBSITE 16.4 Precocenes and synthetic JH. Given the voracity of insect larvae, it is amazing that any plants survive. However, many plants get revenge on their predators by making compounds that alter insect metamorphosis, thus preventing the animals from developing or reproducing.

WEBSITE 16.5 Axes of regeneration. The phenomena of epimorphic regeneration can be seen formally as events that reestablish continuity among tissues that the amputation has severed. How does the amputated stump know to start at the cut site and not start making a whole new arm from the shoulder? How does it build the same dorsal-ventral axis as the stump?

WEBSITE 16.6 Regeneration in annelid worms. An easy laboratory exercise can discover the rules by which worms regenerate their segments. This website details some of those experiments.

WEBSITE 16.7 Ethel Browne and the organizer. As detailed in Chapter 8, Spemann and Mangold's work with amphibians brought the concept of "the organizer" into embryology, and Spemann's laboratory helped make the idea a unifying principle of embryology. However, it has been argued that the concept actually had its origins in Ethel Browne's experiments on *Hydra*.

Vade Mecum

Amphibian metamorphosis and frog calls. For photographs of amphibian metamorphosis (and for the sounds of the adult frogs), check out the metamorphosis and frog call sections of the Amphibian segment.

Chromosome squash. The Fruit Fly segments contains a sequence showing how to do a chromosome squash using the *Drosophila* larval salivary gland.

Flatworm regeneration. You should see it for yourself. Flatworms are easy to obtain, and cutting the animal in half does nothing more than what the animal does to itself. Here are videos and easy instructions for experimenting with these fascinating animals.

17 The Saga of the Germ Line

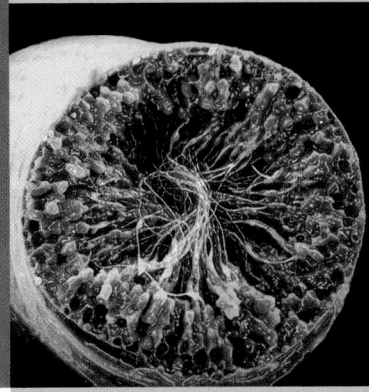

WE ARE ABOUT TO COME FULL CIRCLE. We began our analysis of animal development by discussing fertilization, and we will finish our studies of individual development by investigating **gametogenesis**, the processes by which the gametes (sperm and egg) are formed. In addition to forming its own body, an individual animal must set aside **germ cells** that will provide the material and instructions for initiating bodies in the *next* generation. In many animals, including insects, roundworms, and vertebrates, there is a clear and early separation of germ cells from somatic cell types. In several other animal phyla (and throughout the plant kingdom), this division is not as well established. In these species (which include cnidarians, flatworms, and tunicates), somatic cells can readily give rise to germ cells throughout an animal's life. The zooids, buds, and polyps of many invertebrate species testify to the ability of adult somatic cells to give rise to germ cells (Liu and Berrill 1948; Buss 1987).

In those organisms that do have an established **germ line** that separates from the somatic cells, this germ line can acquire its specification either autonomously or by induction (**FIGURE 17.1**). In autonomous germ cell specification, certain regions of the egg cytoplasm contain factors (such as Nanos, Vasa, Tudor, and Piwi) that specify the cells to be germ cells. The cytoplasmic region containing these components is often referred to as the **germ plasm**. In other groups (such as salamanders and mammals), the germ cell determinants are induced by neighboring cells (Extravour and Akam 2003; Ewen-Campen et al. 2010). Here, the germ plasm is created anew at each generation.

Germ cells provide the continuity of life between generations and thus are sometimes said to be "immortal." While this is a bit of an overstatement, the life of the germ cell persists while the body that has housed it decays and dies. August Weismann, who in 1889 popularized this idea of the mortal body temporarily housing a line of immortal germ cells, noted that the mitotic ancestors of our own germ cells once resided in the gonads of reptiles, amphibians, fish, and invertebrates.

Conserved Molecular Components

Although the mechanisms used to specify the germ cells vary enormously across the animal kingdom, the proteins expressed by germ cells once they are formed are remarkably conserved. These proteins, which include the Vasa, Nanos, Tudor, and Piwi family proteins, can be seen in the germ cells of cnidarians, flies, and mammals (Ewen-Campen et al. 2010; Leclére et al. 2012).

Vasa proteins are required for germ cells in nearly all animals studied. They are involved in binding RNA and most likely activate germ-cell-specific messages. In *Drosophila*, Vasa binds to the 3′ UTR of certain mRNAs and then binds to eukaryotic translation initiation factor 5 (eIF5), facilitating the initiation of translation for these messages (Liu et al. 2009; Parsyan et al. 2011). In chickens, experimentally induced Vasa can direct embryonic stem cells toward a germ cell fate (Lavial et al. 2009).

And the end of all our exploring
Will be to arrive where we started
And know the place for the first time.
T. S. ELIOT (1942)

When the spermatozoon enters the egg, it enters a cell system which has already achieved a certain degree of organization.
ERNST HADORN (1955)

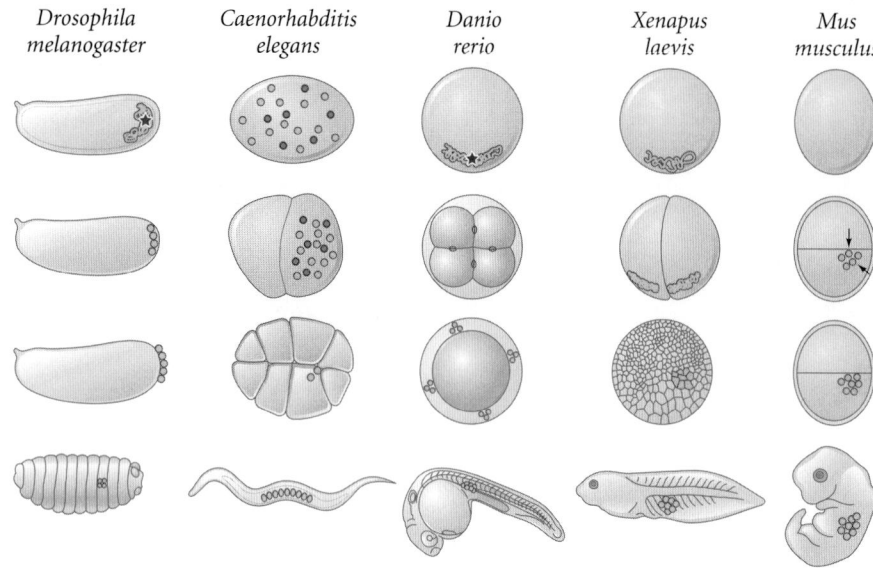

| *Drosophila melanogaster* | *Caenorhabditis elegans* | *Danio rerio* | *Xenapus laevis* | *Mus musculus* |

FIGURE 17.1 Germ plasm localization in different groups of animals. In each of the species studied, germ cell-specific products, including the mRNA and proteins for Vasa, Nanos, Piwi, and Tudor, become localized in specific cells (yellow). In many instances, these substances are already present in the egg and get organized into a particular cell group. In *Drosophila*, Oskar protein (star) organizes germ plasm in the posterior. In *C. elegans*, PIE-1 (blue) plays an important organizing role. In zebrafish (*Danio rerio*), Bucky ball (star) organizes the germ plasm. In *Xenopus* frogs, the germ plasm becomes localized near the vegetal cell membrane. In mammals such as mice, induction (arrows) causes the formation of germ plasm in the posterior epiblast. (After Ewen-Campen et al. 2010.)

Nanos proteins bind to their partner, Pumilio, to form a very potent repressive dimer. Nanos can block RNA translation, and Pumilio binds to the 3′UTRs of specific mRNAs. In *Drosophila*, Nanos and Pumilio repress the translation of numerous mRNAs, and in so doing prevent the cell from becoming part of any germ layer, prevent the cell cycle from continuing, and prevent apoptosis (Kobayashi et al. 1996; Asaoka-Taguchi et al. 1999; Hayashi et al. 2004).

Tudor proteins were discovered in *Drosophila*, in which females carrying these genes are sterile* and do not form pole cells (Boswell and Mahowald 1985). It appears that Tudor proteins interact with the **Piwi** proteins that are involved in transcriptionally silencing portions of the genome, especially active transposons.

Interestingly, whether they are formed autonomously or by induction, germ cells usually do not arise within the gonads. Rather, the gamete progenitor cells—the **primordial germ cells** (**PGCs**)—arise elsewhere and migrate into the developing gonads. The first step in gametogenesis, then, involves forming the PGCs and getting them into the developing gonad. Therefore, our discussion of gametogenesis will include:

- Formation of the germ plasm and determination of the primordial germ cells
- Migration of the PGCs into the developing gonads
- Meiosis
- Differentiation of the sperm and egg cells (gametogenesis)
- Hormonal control of gamete maturation and ovulation

---

*Tudor, like Vasa, is named after a European royal lineage that perished when the female rulers lacked heirs.

Germ Cell Specification

The specification of the germ cell lineage is a critically important event. While the characteristics of mature germ cells are very similar (as mentioned above), the mechanisms and molecules involved in specifying them can be quite different. The specification of germ cells in some major groups will be considered here.

Germ cell determination in nematodes

BOVERI'S EXPERIMENTS ON *PARASCARIS* Theodor Boveri (1862–1915) was the first person to observe an organism's chromosomes throughout its development. In so doing, he documented the existence of a germ plasm. The roundworm *Parascaris aequorum* (formerly known as *Ascaris megalocephala*) has only two chromosomes per haploid cell, thereby allowing detailed observations of its individual chromosomes. The cleavage plane of the first embryonic division is unusual in that it is equatorial, separating the animal half from the vegetal half of the zygote (**FIGURE 17.2A**). More bizarre, however, is the behavior of the chromosomes in the subsequent division of these first two blastomeres. The chromosomes in the animal blastomere fragment into dozens of pieces just before this cell divides. This phenomenon is called **chromosome diminution**, because only a portion of the original chromosomes survive. Numerous genes are lost when the chromosomes fragment, and these genes are not included in the newly formed nuclei (Tobler et al. 1972; Müller et al. 1996).

Meanwhile, in the vegetal blastomere, the chromosomes remain normal. During second cleavage, the animal cell splits meridionally while the vegetal cell again divides equatorially. Both vegetally derived cells have normal chromosomes. However, the chromosomes of the more animally located of these two vegetal blastomeres fragment before the third

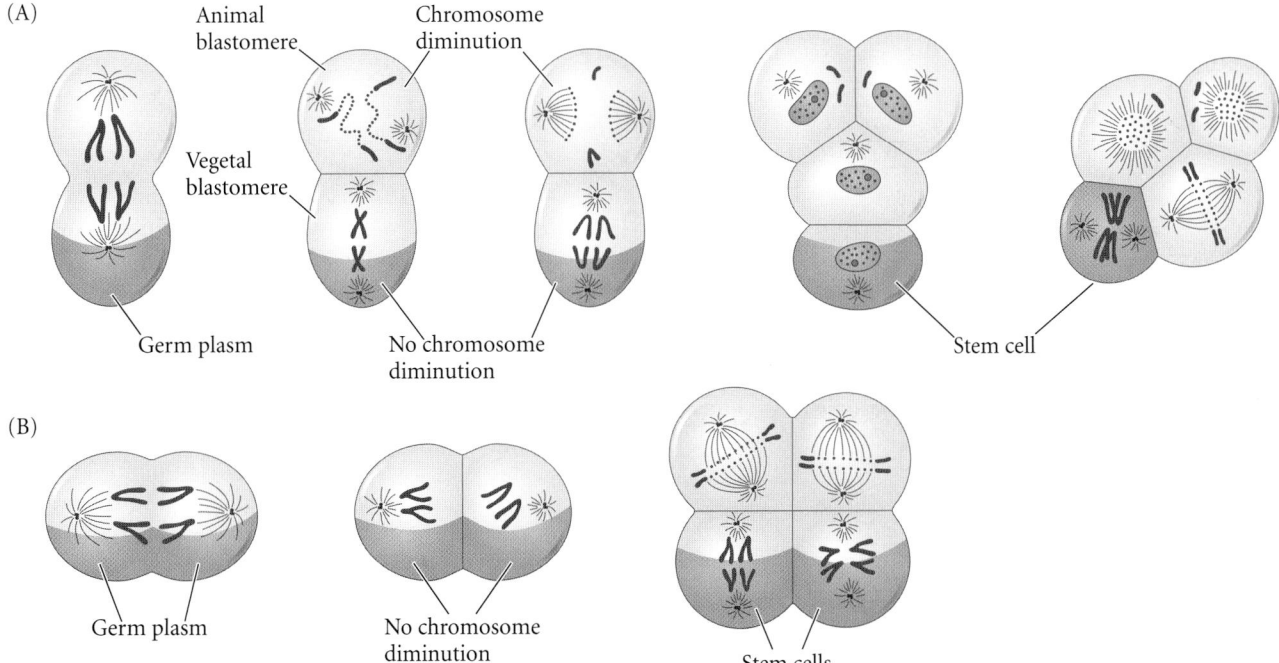

FIGURE 17.2 Distribution of germ plasm during cleavage of normal and centrifuged zygotes of *Parascaris*. (A) In normal cleavage, the germ plasm is localized in the vegetalmost blastomere, as shown by the lack of chromosomal diminution in that particular cell. Thus, at the 4-cell stage, the embryo has a single stem cell for its gametes. (B) When centrifugation is used to displace the first cleavage by 90°, both of the resulting cells have vegetal germ plasm, and neither cell undergoes chromosome diminution. After the second cleavage, both of these two cells give rise to germinal stem cells. (After Waddington 1966.)

cleavage. Thus, at the 4-cell stage, only one cell—the most vegetal—contains a full set of genes. At successive cleavages, nuclei with diminished chromosomes are given off from this vegetalmost line until the 16-cell stage, when there are only two cells with undiminished chromosomes. One of these two blastomeres gives rise to the germ cells; the other eventually undergoes chromosome diminution and forms more somatic cells. The chromosomes are kept intact only in those cells destined to form the germ line. If this were not the case, the genetic information would degenerate from one generation to the next. The cells that have undergone chromosome diminution generate the somatic cells.

Boveri has been called the last of the great observers of embryology and the first of the great experimenters. Not content with observing the retention of the full chromosome complement by the germ cell precursors, he set out to test whether a specific region of cytoplasm protects the nuclei within it from diminution. If so, any nucleus happening to reside in this region should remain undiminished. In 1910, Boveri tested this hypothesis by centrifuging *Parascaris* eggs shortly before their first cleavage. This treatment shifted the orientation of the mitotic spindle. When the spindle forms perpendicular to its normal orientation, both resulting blastomeres contain some of the vegetal cytoplasm (**FIGURE 17.2B**). Boveri found that after the first division, neither nucleus underwent chromosomal diminution. However, the next

division was equatorial along the animal-vegetal axis. Here the resulting animal blastomeres both underwent diminution, whereas the two vegetal cells did not. Boveri concluded that the vegetal cytoplasm contains a factor (or factors) that protects nuclei from chromosomal diminution and determines germ cells.

C. ELEGANS Identifying and understanding the germ plasm of nematodes has been made possible by using a different species whose genetics have been exceptionally well studied. In the nematode *Caenorhabditis elegans*, the germ plasm (called P-granules in this species) is asymmetrically localized through the first several cell divisions to a single germline precursor cell known as the P4 blastomere. The germ plasm that enters this cell is critical for instructing it to become the germline precursor (**FIGURE 17.3**). The P-granule protein repertoire includes several transcriptional inhibitors and RNA-binding proteins, including Vasa, Piwi, and Nanos (Kawasaki et al. 1998; Seydoux and Strome 1999; Subramanian and Seydoux 1999). In addition, the P-granules may have other components that block gene expression.

As discussed in Chapter 5, the *C. elegans* germ plasm contains the PIE-1 protein, which prevents the phosphorylation of RNA polymerase II, thereby preventing transcription in the germ cell lineage (Ghosh and Seydoux 2008). This transcriptional silencing is critical for preventing the germ line

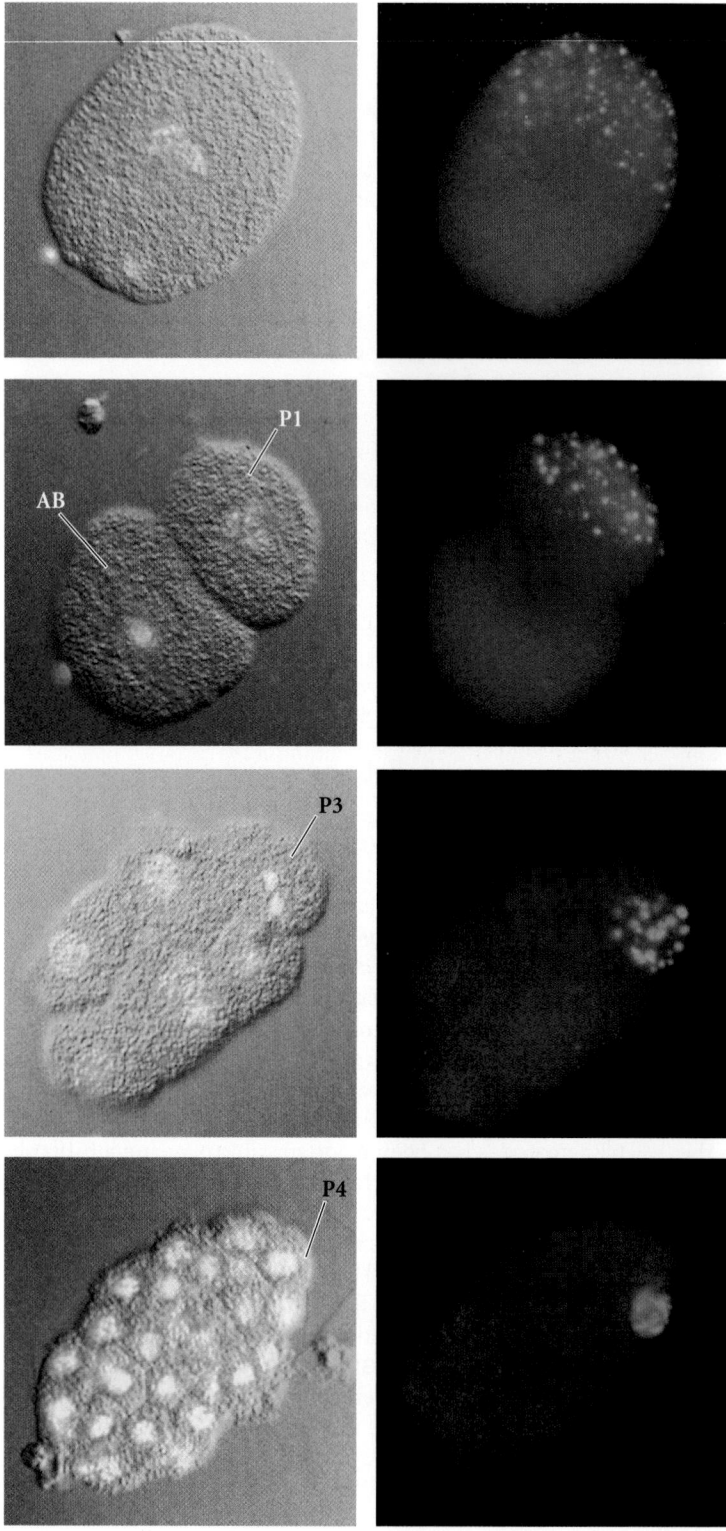

FIGURE 17.3 Segregation of the P-granules into the germ line lineage of the *C. elegans* embryo. The left column shows the cell nuclei (the DNA is stained blue by Hoescht dye); the right column shows the same embryo stained for P-granules (green). Each successive division sequesters the P-granules to the P4 blastomere, whose progeny will become the germ cells. (Courtesy of S. Strome.)

from differentiating into somatic cells, and germ cell differentiation cannot commence until the disappearance of PIE-1 in later embryonic stages. Until that time, the germline nuclei are silenced (**FIGURE 17.4**). The P-granules may also inhibit transcription or RNA processing by physically binding to nuclear pores and preventing certain proteins and RNAs from entering or leaving them. This binding to the nuclear pores may also provide an environment in which the translational repressor proteins meet their mRNA targets immediately as they leave the nucleus (Updike et al. 2011; Veronina et al. 2012). It is possible that these germ plasm proteins can aggregate into P-granules only in the P1–P4 cells. The MEX-5 and PAR-1 proteins inhibit P-granule stability in the remaining somatic cells (Brangwynne et al. 2009; Griffin et al. 2011).

● See **WEBSITE 17.1** Mechanisms of chromosome diminution

Germ cell determination in insects

In *Drosophila*, about 35 primordial germ cells form as a group of **pole cells** at the posterior pole of the cellularizing blastoderm. Nuclei migrate into the posterior region at the ninth nuclear division and become surrounded by the **pole plasm**, a complex collection of mitochondria, fibrils, and **polar granules** that contain translational regulators such as Vasa, Tudor, and Piwi family proteins (Mahowald 1971a,b; Schubiger and Wood 1977; Thomson et al. 2008). The pole plasm is the germ plasm of the insect egg, and it is tethered to the posterior cortex (**FIGURE 17.5A**). The nuclei entering the posterior pole region release the germ plasm from the cortex, and granules of pole plasm migrate along the microtubules toward the dividing nuclei (**FIGURE 17.5B,C**; Lerit and Gavis 2011). The germ plasm is entirely sufficient for inducing germ cells; if germ plasm is injected into ectopic locations within the embryo, functional germ cells will form in that location. Conversely, if the pole cell nuclei are prevented from reaching the pole plasm, no germ cells will be made (Mahowald et al. 1979).

GERM PLASM COMPONENTS The components of the germ plasm are responsible both for specifying cells to become germ cells and for inhibiting somatic gene expression in these cells. These two functions might be interrelated, since the inhibition of gene transcription appears to be essential for germ cell determination (see Santos and Lehmann 2004). The usual suspects are found (and were first identified) in the *Drosophila* germ plasm. Vasa (see Figure 17.5B) is critical in several steps of germ cell specification. It is needed for initiating germ

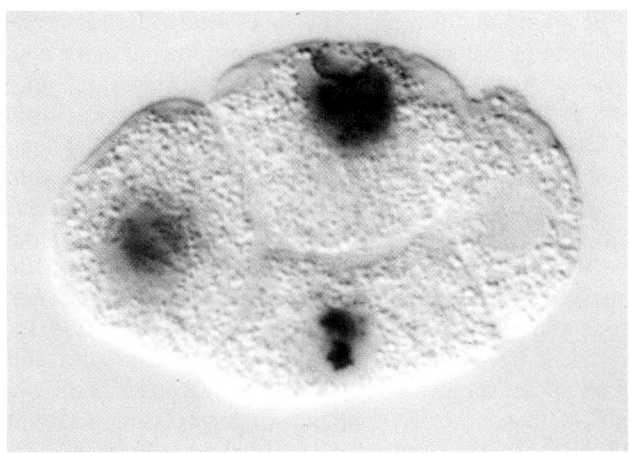

FIGURE 17.4 Inhibition of transcription in germ cell precursors of *Caenorhabditis elegans*. The photograph shows in situ hybridization to β-galactosidase mRNA expressed under control of the *pes-10* promoter. The *pes-10* gene is one of the earliest genes expressed in *C. elegans*. The P blastomere that gives rise to the germ cells (far right) does not transcribe the gene. (From Seydoux and Fire 1994, courtesy of G. Seydoux.)

cell differentiation and meiosis (Ghabrial and Schüpbach 1999), and its binding of translation initiation factor eIF5 is critical for translating the *gurken* message that specifies the polarity of the egg (see Chapter 6). Vasa is also needed for translating the *mei-P26* message, whose protein product represses microRNA activity and promotes the formation of germline stem cells (Liu et al. 2009).

Nanos is also essential for posterior segment formation and germ cell specification. Pole cells lacking Nanos do not migrate into the gonads and fail to become gametes. Nanos inhibits the translation of genes involved with cell division and apoptosis, thereby blocking proliferation and promoting survival of the germ cells. In embryos lacking Nanos, the germline cells usually die; but if inhibited from dying, these germline cells can become somatic cells. Nanos thus prevents the pole cells from activating the pathways that would lead either to cell death or to the formation of somatic cells (Hayashi et al. 2004; Sato et al. 2007).

Tudor acts in a very interesting manner in the formation of the *Drosophila* germ plasm. It is involved in the transport of mitochondrial ribosomal RNA out of mitochondria and into the pole cells. Mitochondrial ribosomes (which resemble ribosomes of bacteria) actually translate some of the germline-specific mRNAs, such as the message for the *germ cell-less* (*gcl*) protein (Kobayashi and Okada 1989; Amikura et al. 2005). It is unclear what Germ cell-less protein does, although it seems to be involved in transcriptional repression. Two other nucleic acid-binding proteins in the pole plasm, Piwi and its relative Aubergine, also have the ability to repress transcription. Piwi will later become critical in establishing the germ cell as a stem cell in the gonad (Cox et al. 1998; Megosh et al. 2006). Another protein, Pgc (polar granule component), appears to competitively bind to an RNA polymerase activator, preventing the activator from phosphorylating RNA polymerase II, and thereby preventing transcription (Hanyu-Nakamura et al. 2008). There also are other mRNAs, such as *germ cell-less*, whose functions are critical but not yet known.

ASSEMBLY OF THE GERM PLASM   The germ plasm forms in the most posterior portion of the egg. But what brings the germ plasm components to this posterior pole? This transport involves microtubules and a special protein called **Oskar**, which is found only in certain groups of insects (flies, mosquitos, wasps, and some other insects that undergo complete metamorphosis; Lynch et al. 2011). Oskar's function is to collect and organize the germ plasm, since expression of *oskar* mRNA in ectopic sites will cause the nuclei in those areas to form germ cells (Ephrussi and Lehmann 1992; Snee and Macdonald 2004). Oskar RNA binds to specific elements of the posterior cell cortex, and upon translation at the posterior pole, Oskar protein recruits the other germ plasm components (Schupbach and Wieschaus 1986). Moreover, Oskar appears to be the

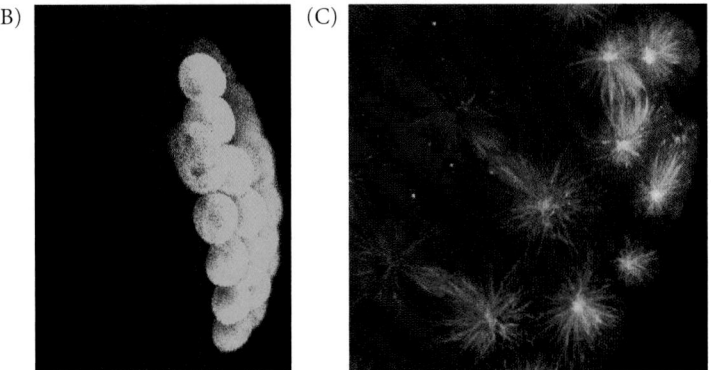

(A)

Pole cells

(B)   (C)

FIGURE 17.5 Pole plasm of *Drosophila*. (A) Scanning electron micrograph of a *Drosophila* embryo just prior to completion of cleavage. The pole cells are at the far right. (B) Vasa protein (visualized by green fluorescent antibodies) in the pole cells of *Drosophila*. (C) Formation of germ cells where Vasa (red) is released from the cell cortex and accumulates around the microtubules (green) organized by the mitotic centrosomes (blue). (A courtesy of A. P. Mahowald; B,C from Lerit and Gavis 2011.)

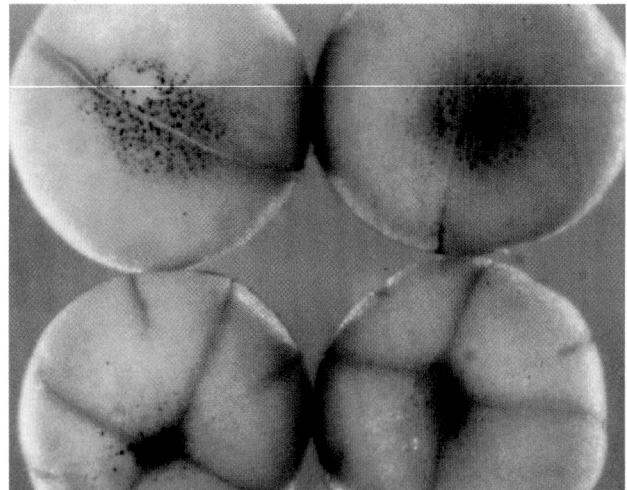

FIGURE 17.6 Germ plasm at the vegetal pole of frog embryos. In situ hybridization to the mRNA for *Nanos* localizes the message in the vegetal cortex of first-cleavage (upper) and fourth-cleavage (lower) embryos. (After Kloc et al. 1998, courtesy of L. Etkin.)

limiting step of germ cell formation, since adding more *oskar* message to the oocyte causes more germ cells to form.

The genes that restrict Oskar to the posterior pole are also necessary for germ cell formation (Ephrussi and Lehmann 1992; Newmark et al. 1997; Riechmann et al. 2002). These include Gurken, which directs the synthesis of the microtubules upon which *Oskar* mRNA travels; Par-1, which is localized in the posterior cortex of the cell membrane and is required for the correct polarization of the microtubules (Shulman et al. 2000); and kinesin, the ATP-driven vehicle that transports *oskar* mRNA to the posterior pole.

- **See VADE MECUM** Germ cells in *Drosophila* embryo
- **See WEBSITE 17.2** The insect germ plasm

Germ cell determination in frogs and fish

FROGS Cytoplasmic localization of germ cell determinants has also been observed in vertebrate embryos. Bounoure (1934) showed that the vegetal region of fertilized frog eggs contains material with staining properties similar to those of *Drosophila* pole plasm. He was able to trace this cortical cytoplasm into the few cells in the presumptive endoderm that would normally migrate into the genital ridge of the developing gonad. By transplanting genetically marked cells from one embryo into another of a differently marked strain, Blackler (1962) showed that these cells are the primordial germ cell precursors.

The germ plasm of amphibians consists of germinal granules and a matrix around them. It contains many of the same RNAs and proteins (including the large and small mitochondrial ribosomal RNAs) as the pole plasm of *Drosophila*, and they appear to repress transcription and translation (Kloc et al. 2002). The early movements of amphibian germ plasm have been analyzed in detail by Savage and Danilchik (1993), who labeled the germ plasm with a fluorescent dye.

They found that the germ plasm of unfertilized eggs consists of tiny "islands" that appear to be tethered to the yolk mass near the vegetal cortex, and it contains, among other things, the *Xenopus* homologues of *Nanos* and *Vasa* messages (**FIGURE 17.6**; Forristall et al. 1995; Ikenishi et al. 1996; Zhou and King 1996). These islands move with the vegetal yolk mass during the cortical rotation just after fertilization. After this rotation, the islands are released from the yolk mass and begin fusing together and migrating to the vegetal pole. The structure formed is the **Balbiani body**, and it is enriched for endoplasmic reticulum, mitochondria, and germ plasm mRNAs. Their aggregation depends on microtubules, and their movement to the vegetal pole depends on a kinesin-like protein that may act as the motor for germ plasm movement (Robb et al. 1996; Quaas and Wylie 2002). When this vegetal pole germ plasm is transplanted into other cells, these cells can generate functional gametes (Tada et al. 2012).

Although the germ cells originate in the vegetal region of the oocyte, these cells cannot respond to the endodermal determinants (Sox17, VegT) that are also present. This is because RNA polymerase II is inhibited in these cells in the same way it is in *Drosophila* (Venkatamara et al. 2011).

ZEBRAFISH In zebrafish, the germ plasm forms a dense vegetal Balbiani body characterized by polar granules, mitochondria, and concentrated germline mRNAs. It appears to be organized during oogenesis by a vertebrate-specific protein called Bucky ball. *Bucky ball* mRNA is made only in the oocyte and it becomes concentrated in the Balbiani body, where it is translated. Bucky ball protein then recruits the mRNA for Vasa, Nanos, and other germ cell proteins. After fertilization, Bucky ball brings these proteins to the cleavage furrows, where it assembles them into the germ plasm (Bontems et al. 2009). Most of the blastomeres are prevented from making the germline components, as the microRNA *miRNA-430* represses Nanos translation in all cells except those few destined to become germ cells. In the germ cells, DAZL protein inhibits *miRNA-430*, allowing the germline factors to be made (Takeda et al. 2009). The germ plasm forms a compact structure that is inherited by only one of the two daughter cells at each division. At late cleavage (around 1000 cells), only four cells have the germ plasm. However, after this stage, the germ plasm is distributed evenly at cell division, creating four clusters of primordial germ cells (see Figure 17.13).

Germ cell determination in mammals

In insects, frogs, nematodes, and flies, the germ cells are determined by material in the egg cytoplasm. However, in mammals, there is no obvious germ plasm. Rather, germ cells are *induced* in the embryo (Wakahara 1996; Hayashi et al. 2007).

In mice, the germ cells form at the posterior region of the epiblast, at the junction of the extraembryonic ectoderm, epiblast, primitive streak, and allantois (**FIGURE 17.7A,B**). This is called the **posterior proximal epiblast** because it is close (proximal) to the extraembryonic ectoderm, and it

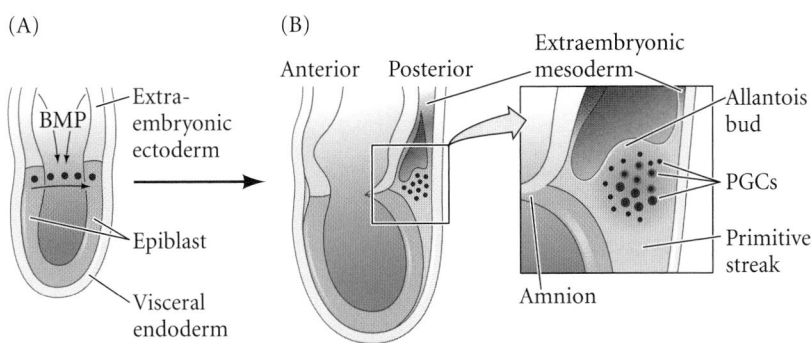

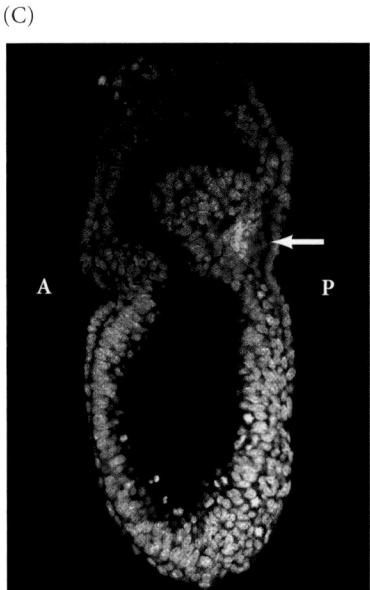

FIGURE 17.7 Specification and migration of mammalian primordial germ cells. (A) In the mouse embryo, BMP signals (blue) from the extraembryonic ectoderm induce neighboring epiblast cells (purple circles) to become precursors of PGCs and extraembryonic mesoderm. During gastrulation, these cells come to reside in the posterior epiblast (arrow). (B) On embryonic day 7, these cells emerge from the posterior primitive streak. The PGC precursors also express *Prdm14* and *Blimp1* (red). (C) Late-streak stage (approximately embryonic day 7.0). Nuclei are stained white (with DAPI), and the expression of the *Prdm14* gene is seen by the green fluorescence (having been fused to the mRNA of a green fluorescent protein). Arrowhead shows expression in the PGC in the extraembryonic mesoderm. A, anterior; P, posterior. (A,B after Hogan 2002; C from Yamaji et al. 2008, courtesy of M. Saitou.)

will be at the posterior of the embryo. Thus, the cells that become the PGCs in mice are not intrinsically different from the other cells of the epiblast and contain no specific germ plasm. Rather, the posterior epiblast cells are induced by the extraembryonic tissue. Wnts from the visceral endoderm are probably responsible for giving the posterior proximal epiblast cells the competence to respond to BMP signals provided by the extraembryonic ectoderm.* This happens during gastrulation (at about day 6.5 in mice), before the germ layers are established (**FIGURE 17.7C**; Pfister et al. 2007; Yamaji et al. 2008; Ohinata et al. 2009).

The BMPs induce the expression of *Blimp1* and *Prdm14* in a small cluster of cells (about six cells in the 6.5-day mouse embryo). Blimp1 is a transcriptional regulator that represses somatic-type gene expression, while activating those genes (such as *Sox2* and *Nanog*) associated with pluripotency. Blimp1 also activates the germline determinant *Nanos3*, which protects the germ cells against apoptosis during their migration (Tsuda et al. 2003). Prdm14 helps establish pluripotency by also activating Sox2, and it is critical for the chromatin modifications that will later silence the genome of the germ cells (Yamaji et al. 2008). Cells that express *Blimp1* and *Prdm14* are restricted to the germ cell fate (Saitou et al. 2002; Ohinata et al. 2005).

The requirement for germ cell induction was shown by transplanting clumps of tissue from the distal portions of the epiblast to the proximal posterior portion of the epiblast. These cells then gave rise to PGCs (Tam and Zhou 1996). Moreover, cultured epiblast cells exposed to Wnt signals and

BMP4 gave rise to PGCs. When such PGCs from cultured male epiblast cells were transferred into testicular tubules, they produced viable sperm that could fertilize mouse eggs (Ohinata et al. 2009).

The inert genome hypothesis

As indicated above, one of the critical events in specifying germ cells appears to be the global repression of gene expression. According to this hypothesis, the cells become germ cells because they are forbidden to become any other type of cell (Nieuwkoop and Sutasurya 1981; Wylie 1999; Cinalli et al. 2008). This suppression of transcription is seen in the germ cells of several species, including mammals, flies, fish, frogs, and nematodes (Nakamura et al. 2010; Venkatarama et al. 2010; Lai et al. 2011). In mice, the germ cells undergo extensive chromatin modification (Seki et al. 2007), causing them to become transcriptionally inert at embryonic day 8.5 (as they begin migrating).

Many of the components in the germ plasm (such as Gcl, Pgc, Piwi, and Nanos in *Drosophila* and PIE-1 in *C. elegans*) act by inhibiting either transcription or translation (Leatherman et al. 2002; de las Heras et al. 2009). Many such proteins are found throughout the animal kingdom. It is interesting that when animal germ cells form —whether in chicks, mice, or flies—they often do so *outside* the developing body proper. It has been hypothesized (but not proved) that this exile into an extraembryonic "enclave" insulates the primordial germ cells from paracrine signaling taking place within the somatic cells of the growing embryo (Dickson 1994). Once the repression of somatic gene expression is accomplished, the germ cells can return to the embryo and travel to the gonads (Richardson and Lehmann 2010; Tarbashevich and Raz 2010). This germ cell migration will be our next topic.

---

*This induction can occur only in the posteriormost region of the epiblast; BMP antagonists prevent it from occurring in the trunk and anterior.

Pluripotency, Germ Cells, and Embryonic Stem Cells

Primordial germ cells and embryonic stem cells are both characterized by their ability to generate any cell type in the embryo. Embryonic stem (ES) cells are derived from the inner cell masses of mammalian blastocysts and are believed to be the functional equivalent of the inner cell mass (ICM) blastomeres (see Chapter 9). One of the best pieces of evidence for this equivalence is that when ES cells are injected into the ICMs of mouse blastocysts, they behave like mouse blastocyst cells and contribute cells to the embryo. One of the interesting species-specific differences between human and mouse ES cells is that human ES cells appear to contribute to the trophoblast, whereas mouse ES cells do not (Xu et al. 2002).

Transcription factors associated with totipotency

In mammals, the retention of totipotency or pluripotency has been correlated with the expression of three nuclear transcription factors: Oct4, Sox2, and Nanog (see Chapter 9). Oct4 is a homeodomain transcription factor expressed in all early-cleavage blastomere nuclei, but its expression becomes restricted to the ICM. During gastrulation, Oct4 becomes expressed solely in those posterior epiblast cells thought to give rise to the primordial germ cells. After that, Oct4 is seen only in the primordial germ cells and, later, in oocytes (FIGURE 17.8). Oct4 is not seen in the developing sperm after the germ cells reach the testes and become committed to sperm

production (Yeom et al. 1996; Pesce et al. 1998). Nanog is another homeodomain transcription factor found in the pluripotent cells of the mouse blastocyst, as well as in ES cells and germline tumors. Nanog expression is high in the PGCs of certain mouse embryos (Hatano et al. 2005; Yamaguchi et al. 2005). Knockout experiments indicate that Nanog is critical in maintaining the pluripotency of stem cells, and over-expression experiments demonstrate that elevated Nanog can maintain Oct4 transcription in ES cells (Chambers et al. 2003; Mitsui et al. 2003).

Piwi, Vasa, Nanos, and *Tudor* appear to establish pluripotency throughout the animal kingdom (Juliano and Wessel 2010). We encountered Piwi as one of the factors that prevents

(A)

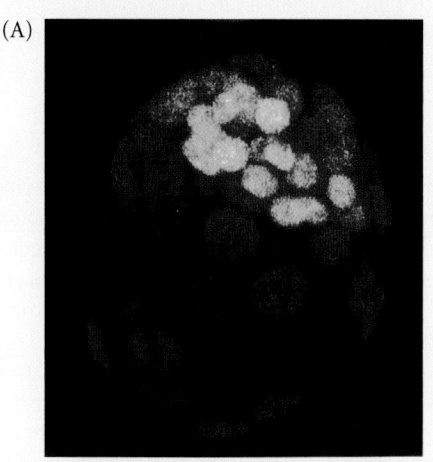

(B)

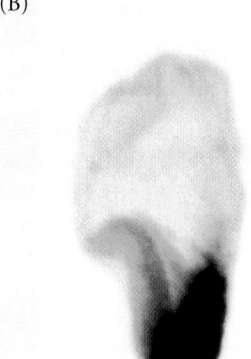

(C)

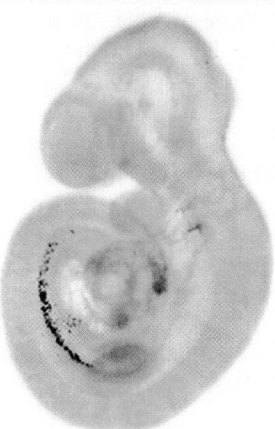

(D)

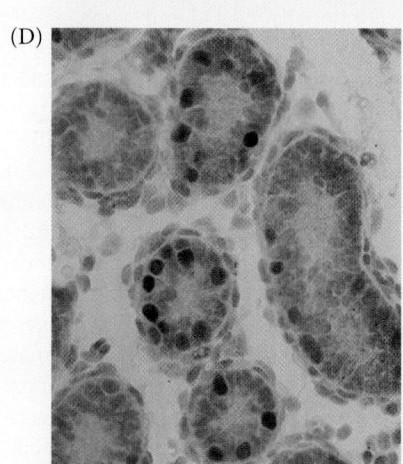

(E)

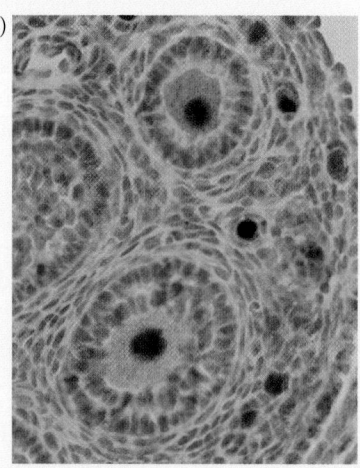

FIGURE 17.8   Expression of *Oct4* mRNA correlates with totipotency and ability to form germ cells in mammals. (A) Blastocyst-stage embryo in which the Oct4 transcription factor is stained green with a fluorescent antibody, while all cell nuclei are stained red with propidium iodide. The overlap (yellow) shows that Oct4 is found only in the inner cell mass. (B,C) An *Oct4/lacZ* transgene driven by the *Oct4* promoter region shows its expression (dark color) in (B) the posterior epiblast of the 8.5-day mouse embryo and (C) migrating PGCs in the 10.5-day embryo. (D,E) Labeled antibody (brown) staining shows Oct4 protein in the nuclei of (D) spermatogonia in postnatal testes and (E) oogonia in postnatal ovaries. (A–C from Yeom et al. 1996; D,E from Pesce et al. 1998, courtesy of H. R. Schöler.)

somatic cell differentiation in germline cells. Piwi also appears to be required for stem cell maintenance and proliferation. In addition to being present in germ stem cells, *Piwi* genes have also been found expressed in the totipotent stem cells of planaria and regenerating annelids. Inhibiting *Piwi* gene expression in the adult flatworm blocks the worm's regeneration (Reddien 2004). *Piwi* is also expressed in the somatic stem cells of jellyfish and is upregulated immediately before transdifferentiation. The continuous low expression of Piwi in differentiated cells of jellyfish may underlie their ability to remodel their bodies so profoundly (Seipel et al. 2004). *Piwi* may even be responsible for stem cell maintenance across kingdoms: two *Piwi* genes in *Arabidopsis* are crucial for maintaining meristem proliferation at the root and shoot of the plant (Bohmert et al. 1998; Moussian et al. 1998).

The same pluripotency and transcription factor expression pattern is seen not only in the PGCs but in two derivatives of the PGCs: cultured embryonic germ cells, and tumorous germ cells called **teratocarcinomas**.

Embryonic germ cells

When PGCs are first placed into culture, they resemble ES cells. Stem cell factor increases the proliferation of migrating mouse primordial germ cells in culture, and this proliferation can be further increased by adding another growth factor, leukemia inhibition factor (LIF). However, the life span of these PGCs is short, and the cells soon die. But if an additional mitotic regulator—basic fibroblast growth factor, Fgf2—is added, a remarkable change takes place. The cells continue to proliferate, producing pluripotent embryonic stem cells with characteristics resembling those of the inner cell mass (Matsui et al. 1992; Resnick et al. 1992; Rohwedel et al. 1996). These PGC-derived cells are called embryonic germ (EG) cells, and they have the potential to differentiate into all the cell types of the body.

In 1998, researchers in John Gerhart's laboratory cultured human EG cells (Shamblott et al. 1998). These cells were able to generate differentiated cells from all three primary germ layers, so they are presumably

pluripotent. Such cells could be used medically to create neural or hematopoietic stem cells, which might be used to regenerate damaged neural or blood tissues. EG cells are often considered ES cells, and the distinction of their origin is ignored.

Embryonal carcinoma cells

What happens if a PGC became malignant? In one type of tumor, the germ cells become embryonic stem cells, much like the Fgf2-treated PGCs in the experiment above. This type of tumor is called a teratocarcinoma. Whether spontaneous or experimentally produced, teratocarcinomas contain an undifferentiated stem cell population that has biochemical and developmental properties remarkably similar to those of the inner cell mass (Graham 1977; see Parson 2004). Moreover, these stem cells not only divide but can also differentiate into a wide variety of tissues, including gut and respiratory epithelia, muscle, nerve, cartilage, and bone (**FIGURE 17.9**). These undifferentiated pluripotent stem cells are called embryonal carcinoma (EC) cells. Once differentiated, these cells no longer divide, and are therefore no longer malignant. Such tumors can give rise to most of the tissue types in the body (Stevens and Little 1954; Kleinsmith and Pierce 1964; Kahan and Ephrussi 1970). Thus, the teratocarcinoma stem cells mimic early mammalian development, but the tumor they form is characterized by random, haphazard development.

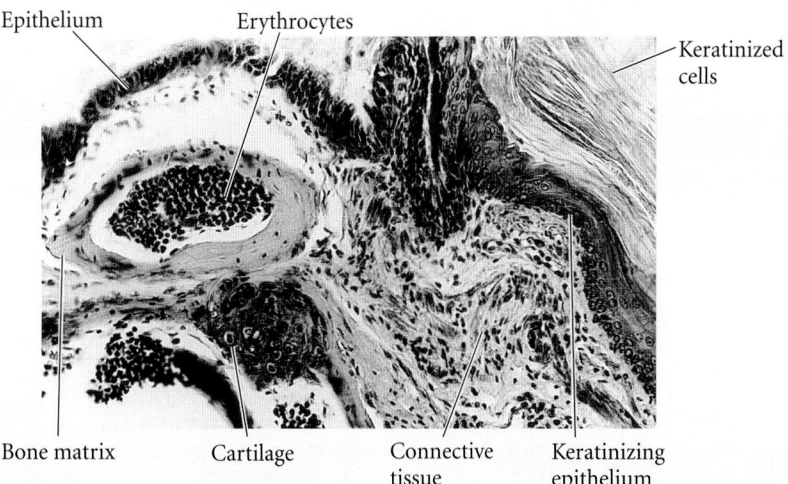

FIGURE 17.9   Photomicrograph of a section through a mouse teratocarcinoma, showing numerous differentiated cell types. (From Gardner 1982; photograph by C. Graham, courtesy of R. L. Gardner.)

In 1981, Stewart and Mintz formed a mouse from cells derived in part from a teratocarcinoma stem cell. Stem cells that had arisen in a teratocarcinoma of an agouti (yellow-tipped) strain of mouse were cultured for several cell generations and were seen to maintain the characteristic chromosome complement of the parental mouse. Individual stem cells descended from the tumor were injected into the blastocysts of black-furred mice. The blastocysts were then transferred to the uterus of a foster mother, and live mice were born. Some of these mice had coats of two colors, indicating that the tumor cell had integrated itself into the embryo. This, in itself, is a remarkable demonstration that the tissue context is critical for the phenotype of a cell—a malignant cell was made nonmalignant.

But the story does not end here. When these chimeric mice were mated to mice carrying alleles recessive to those of the original tumor cell, the alleles of the tumor cell were expressed in many of the offspring. This means that the originally malignant tumor cell had produced many, if not all, types of normal somatic cells, and had even produced normal, functional germ cells! When such mice (being heterozygous for tumor cell

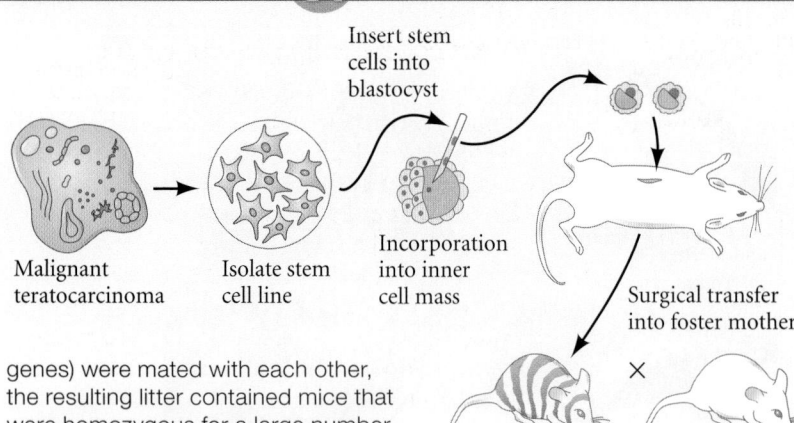

genes) were mated with each other, the resulting litter contained mice that were homozygous for a large number of genes from the tumor cell (**FIGURE 17.10**). Thus, germ cell tumors can retain their pluripotency.

Germ cells and stem cells: Possible interactions

One idea emerging from this study is that some descendants of the pluripotent cells (such as the teratocarcinoma or ES cells) form PGCs that can undergo meiosis to form sperm or eggs. Indeed, there is evidence that ES cells can develop into oogonia that enter meiosis and recruit adjacent cells into follicle-like structures (Hübner et al. 2003). There is also parallel evidence that mouse ES cells can be made to differentiate into spermatocytes that can become functional sperm when transplanted into testes (Toyooka et al. 2003). More recently, PGC-like cells have been derived from murine induced pluripotent stem cells (iPSCs; see p. 328). When these PGC-like cells were transplanted into normal mouse testes and ovaries, they developed into functional sperm and eggs that produced viable, fertile offspring (**FIGURE 17.11**; Hayashi et al. 2011, 2012). These studies need to be extensively confirmed and extended to humans if they are to provide a new way of curing infertility.

It is even possible that ES, EG, and EC cells have a common origin in

the presumptive PGCs. Zwaka and Thomson (2005) hypothesize that the ES cells are actually the equivalent of PGCs and not of the inner cell mass. Not every inner cell mass blastomere can become an ES cell, and Zwaka and Thomson suggest that perhaps the successful stem cells are those that have been positioned next to the trophoblast at the future posterior proximal region of the embryo. In other words, the blastomeres that become the ES cells might actually be the presumptive PGCs. While this idea remains hypothetical, it would relate these four pluripotent cell types.

FIGURE 17.10 Protocol for breeding mice whose genes are derived largely from tumor cells. Stem cells are isolated from a mouse teratocarcinoma and inserted into blastocysts from a different strain of mouse. The chimeric blastocysts are implanted in a foster mother. If the tumor cells are integrated into a blastocyst, the mouse that develops will have many of its cells derived from the tumor. If the tumor has given rise to germ cells, the chimeric mouse can be mated to normal mice to produce an F_1 generation. Because F_1 mice should be heterozygous for all the chromosomes of the tumor cell, matings between F_1 mice should produce F_2 mice that are homozygous for some genes derived from the tumor. Thus, F_2 mice should express many genes found in the mouse strain from which the tumor cells were derived. (After Stewart and Mintz 1981.)

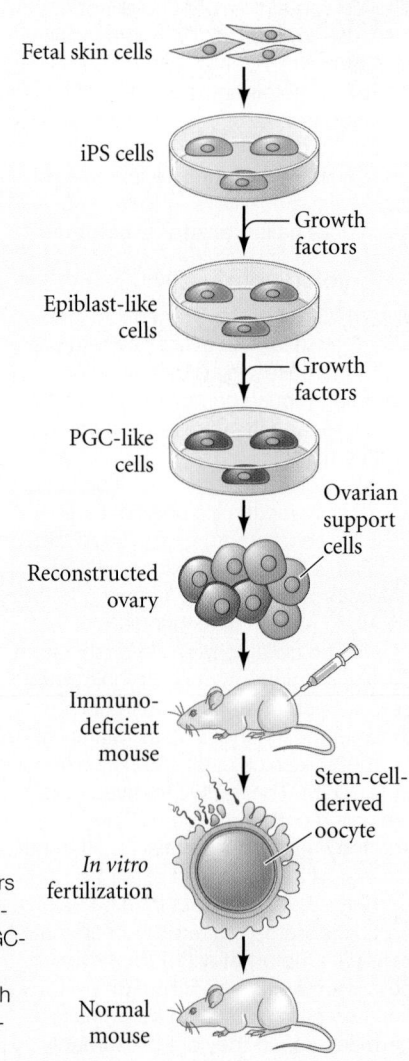

FIGURE 17.11 Generating functional mouse oocytes from somatic cells. Fibroblasts were converted into induced pluripotent stem (iPS) cells and exposed to paracrine factors (activin, Fgf2) that caused them to generate epiblast-like cells. Adding further growth factors (including BMPs, epidermal growth factor, and stem cell factor) led them to form PGC-like cells. These were cultured with ovarian cells and placed into the ovaries of immunodeficient mice (so they wouldn't be rejected). These cells initiated and proceeded through meiosis and were able to produce normal offspring when fertilized in vitro. A similar technique using testes generated functional sperm. (After Cheloufi and Hochedlinger 2012.)

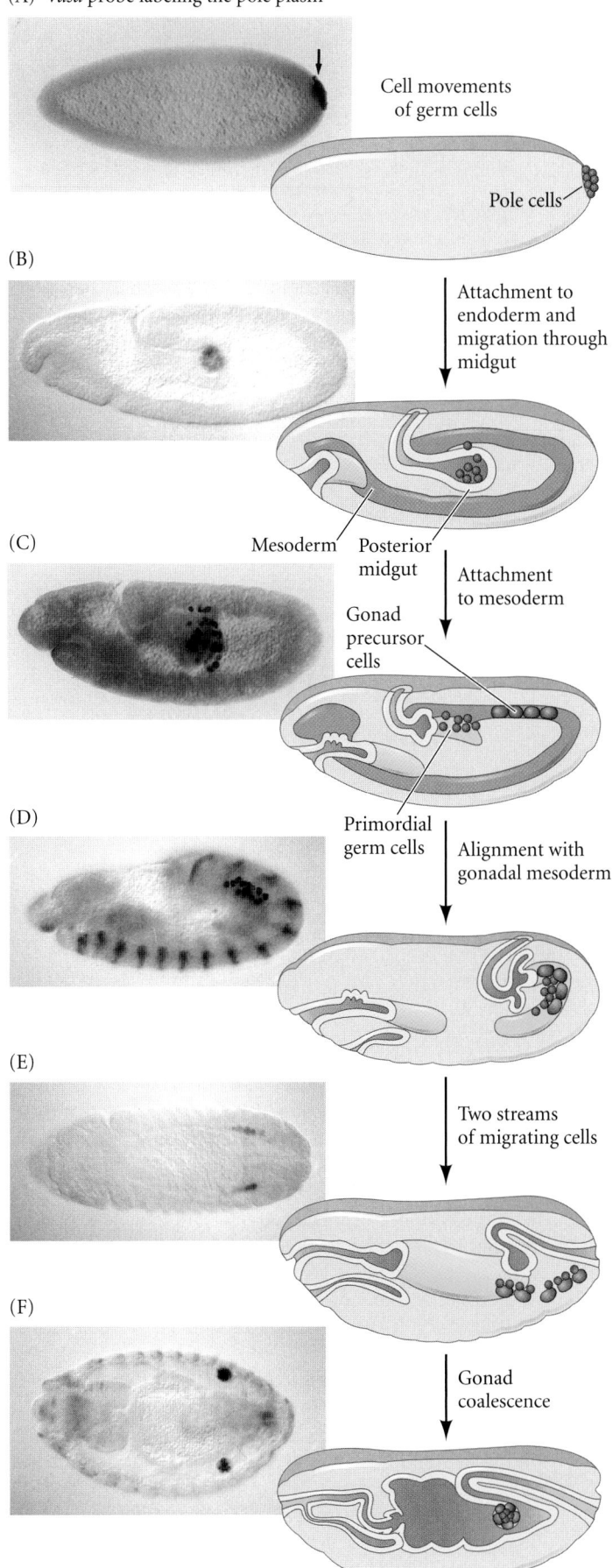

(A) *Vasa* probe labeling the pole plasm

Cell movements of germ cells

Pole cells

(B)

Attachment to endoderm and migration through midgut

Mesoderm Posterior midgut

(C)

Attachment to mesoderm

Gonad precursor cells

Primordial germ cells

(D)

Alignment with gonadal mesoderm

(E)

Two streams of migrating cells

(F)

Gonad coalescence

Germ Cell Migration

Germ cell migration in Drosophila

In 1865, Eli Metchnikov (known as the founder of immunology) showed that the pole cells of fly larvae traveled to the developing gonads. We now have some idea of how these primordial germ cells move from the posterior pole to the gonads in larval *Drosophila*. The first step in this migration is a passive one, wherein the 30–40 pole cells are displaced into the posterior midgut by the movements of gastrulation (**FIGURE 17.12A,B**). The germ cells are actively prevented from migrating during this stage (Jaglarz and Howard 1994; Li et al. 2003). In the second step, the gut endoderm triggers the germ cells to undergo **diapedesis**, a type of amoeboid movement common to lymphocytes and macrophages that enables these blood cells to squeeze between the endothelial cells of capillaries. In *Drosophila*, the germ cell diapedesis takes them through the blind end of the posterior midgut (Kunwar et al. 2003). The cytoskeletons of the germ cells form active protrusions, and the germ cells migrate from the endoderm into the visceral mesoderm (Kunwar et al. 2003). In the third step, the PGCs split into two groups, each of which will become associated with a developing gonad primordium.

In the fourth step, the germ cells migrate to the gonads, which are derived from the lateral mesoderm of parasegments 10–12 (Warrior 1994; Jaglarz and Howard 1995; Broihier et al. 1998). This step involves both attraction and repulsion. The products of the *wunen* genes appear to be responsible for directing the migration of the primordial germ cells from the endoderm into the mesoderm and their division into two streams (**FIGURE 17.12C–E**). The Wunen protein is expressed in the endoderm immediately before PGC migration and in many other tissues that the germ cells avoid, and it appears to be

FIGURE 17.12 Migration of germ cells in the *Drosophila* embryo. The left column shows the germ plasm as stained by antibodies to Vasa, a protein component of the germ plasm (D has been counterstained with antibodies to Engrailed protein to show the segmentation; E and F are dorsal views). The right column diagrams the movements of the germ cells. (A) Germ cells originate from the pole plasm at the posterior end of the egg. (B) Passive movements carry the PGCs into the posterior midgut. (C) PGCs move through the endoderm and into the caudal visceral mesoderm by diapedesis. The *wunen* gene product expressed in the endoderm expels the PGCs, while the product of the *columbus* gene expressed in the caudal mesoderm attracts them. (D–F) Movements of the mesoderm bring the PGCs into the region of parasegments 10–12, where the mesoderm coalesces around them to form the gonads. (Photographs from Warrior et al. 1994, courtesy of R. Warrior; diagrams after Howard 1998.)

(A)

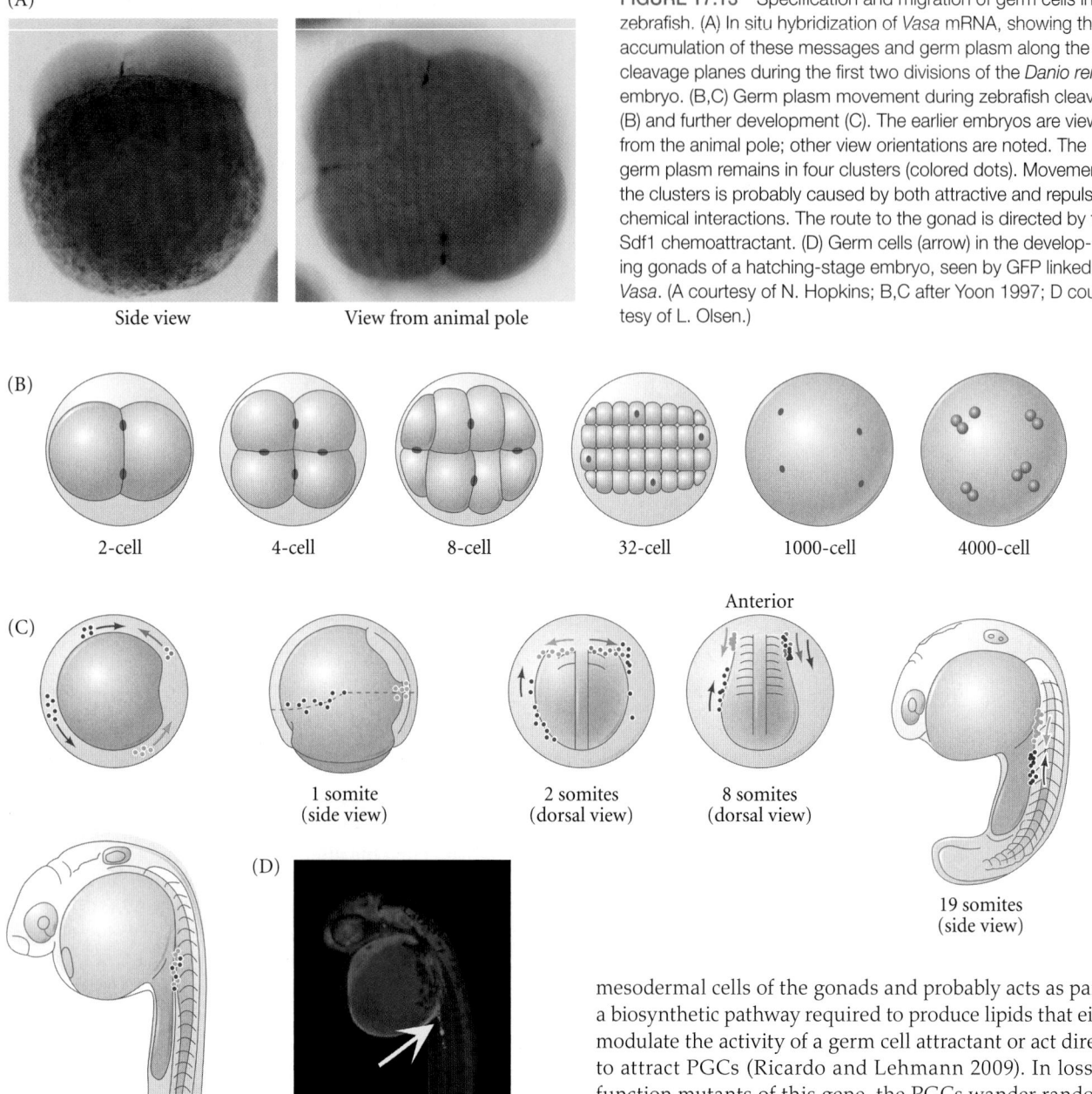

Side view   View from animal pole

(B)

2-cell   4-cell   8-cell   32-cell   1000-cell   4000-cell

(C)

Anterior

1 somite (side view)   2 somites (dorsal view)   8 somites (dorsal view)

19 somites (side view)

1 day (side view)

(D)

mesodermal cells of the gonads and probably acts as part of a biosynthetic pathway required to produce lipids that either modulate the activity of a germ cell attractant or act directly to attract PGCs (Ricardo and Lehmann 2009). In loss-of-function mutants of this gene, the PGCs wander randomly from the endoderm, and if the *columbus* gene is expressed in other tissues (such as the nerve cord), those tissues will attract the PGCs. In the last step, the gonad coalesces around the germ cells, allowing the germ cells to divide and mature into gametes (**FIGURE 17.12F**). This step requires E-cadherin (Jenkins et al. 2003).

Neither the gonads nor the germ cells differentiate until metamorphosis. During the larval stages, both the PGCs and the somatic gonadal cells divide, but they remain relatively undifferentiated. At the larval-pupal transition, gonadal morphogenesis occurs (Godt and Laski 1995; King 1970). During this transition, those PGCs in the anterior region of the gonad become the germline stem cells (Asaoka and Lin 2004), which divide asymmetrically to produce both another

repelling the germ cells. In loss-of-function mutants of this gene, the PGCs wander randomly (Zhang et al. 1997; Hanyu-Nakamura et al. 2004; Sano et al. 2005).

HMG-CoA reductase, the product of the *columbus* gene, appears to be critical for attracting the *Drosophila* PGCs to the gonads (Van Doren et al. 1998). This protein is made in the

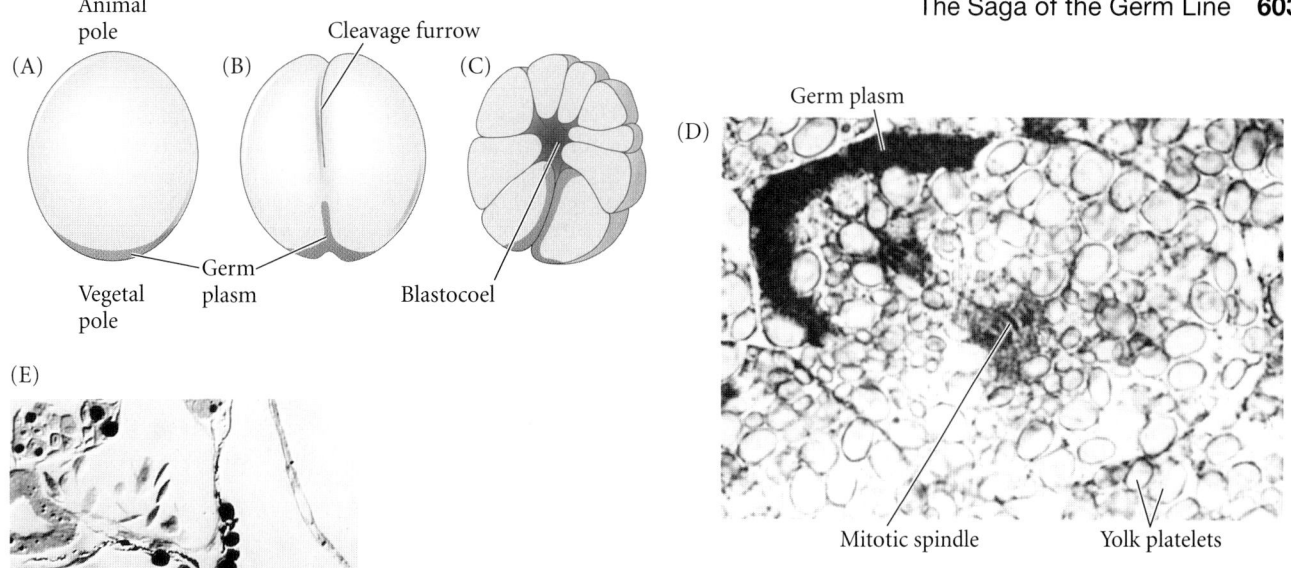

FIGURE 17.14 Migration of *Xenopus* germ plasm. (A–C) Changes in the position of the germ plasm (color) in an early frog embryo. Originally located near the vegetal pole of the uncleaved egg (A), the germ plasm advances along the cleavage furrow (B) until it becomes localized at the floor of the blastocoel (C). (D) A germ plasm-containing cell in the endodermal region of a blastula in mitotic anaphase. Note the germ plasm entering into only one of the two yolk-laden daughter cells. (E) Migration of two primordial germ cells (arrows) along the dorsal mesentery connecting the gut region to the gonadal mesoderm. (A–C after Bounoure 1934; D courtesy of A. Blackler; E from Heasman et al. 1977, courtesy of the authors.)

stem cell and a **cystoblast**. The cystoblast undergoes four rounds of mitotic division to produce an egg chamber of 16 cells (see Figures 6.8 and 17.25; King 1970; Zhu and Xie 2003).

We are just beginning to understand how the germline stem cells retain their stem cell properties in the gonad (Gilboa and Lehmann 2004). As mentioned in the Part Three opener, stem cells must be in a "niche" that supports their proliferation and inhibits their differentiation. Daughter cells that travel outside this niche begin differentiating. In ovaries, the germline stem cells are attached to the stromal cap, where they are maintained by the BMP4-like factor Decapentaplegic (Dpp). Dpp represses the gene encoding a transcription factor (Bag-of-marbles) that initiates oogenesis; Dpp cannot reach cells that leave the stromal cap. Without Dpp to repress the *bag-of marbles* gene, the germline cell begins the developmental cascade that produces 15 nurse cells and the single oocyte (Chen and McKearin 2003; Decotto and Spradling 2005).

The stem cells of the male germ line are connected to "hub" cells that create a stem cell microenvironment by secreting BMP signals as well as the Unpaired protein. Unpaired activates the JAK-STAT pathway in the germline stem cells (see Figure 3.23). If the JAK-STAT signaling pathway is disrupted, the germline stem cells differentiate into spermatogonia without any self-renewal (Kiger et al. 2001; Tulina and Matunis 2001). Remarkably, germ cells that have initiated differentiation can dedifferentiate back into germline stem cells if they are forced to re-enter the niche (Sheng et al. 2009). Thus, the male and female germline stem cells are in similar niches, and when these cells move out of their niche, they divide to form the gametes and (in the case of females) the nurse cells as well.

Germ cell migration in vertebrates

ZEBRAFISH Whereas *Drosophila* PGC migration is motivated by both chemoattractants and chemorepellents of the germ cell precursors, zebrafish PGCs arrive at the gonads via chemoattraction. Using the *Vasa* message as a marker, Weidinger and colleagues (1999) detailed the migration of the four clusters of zebrafish PGCs (**FIGURE 17.13**). These PGC clusters follow different routes, but by the end of the first day of development (at the 1-somite stage), the PGCs are found in two discrete clusters along the border of the trunk mesoderm. From there, they migrate posteriorly into the developing gonad. In zebrafish, the primordial germ cells follow a gradient of the Sdf1 protein that is secreted by the developing gonad. The receptor for this protein is the CXCR4 protein on the PGC surface (Doitsidou et al. 2002; Knaut et al. 2003). This Sdf1/CXCR4 chemotactic guidance system is known to be important in the migration of several cells types during embryogenesis, including adrenal medullary cells and hematopoietic progenitor cells. Loss of either CXCR4 from the PGCs or of Sdf1 from the somatic cells results in random migration of the zebrafish primordial germ cells.

FROGS The germ plasm of anuran amphibians (frogs and toads) collects around the vegetal pole in the zygote (see Figure 17.6). During cleavage, this material is brought upward through the yolky cytoplasm. Periodic contractions of the vegetal cell surface appear to push it along the cleavage furrows of the newly formed blastomeres. Germ plasm eventually becomes associated with the endodermal cells lining the floor of the blastocoel (**FIGURE 17.14**; Bounoure 1934; Ressom and Dixon 1988; Kloc et al. 1993). The PGCs become concentrated in the posterior region of the larval gut, and as the abdominal cavity forms, they migrate along the dorsal side of the gut, first along the dorsal mesentery (which connects the gut to the region where the mesodermal organs are forming; see Figure 17.14E) and then along the abdominal wall and into the genital ridges. They migrate up this tissue until they reach the developing gonads.

Xenopus PGCs move by extruding a single filopodium and then streaming their yolky cytoplasm into that filopodium while retracting their "tail." Contact guidance in this migration seems likely, as both the PGCs and the extracellular matrix over which they migrate are oriented in the direction of the migration (Wylie et al. 1979). Furthermore, PGC adhesion and migration can be inhibited if the mesentery is treated with antibodies against *Xenopus* fibronectin (Heasman et al. 1981). Thus, the pathway for germ cell migration in these frogs appears to be composed of an oriented fibronectin-containing extracellular matrix.

The fibrils over which the PGCs travel lose this polarity soon after migration has ended. As they migrate, *Xenopus* PGCs divide about three times, so that approximately 30 PGCs will colonize the gonads (Whitington and Dixon 1975; Wylie and Heasman 1993). These cells will divide to form the germ cells. The mechanism by which the *Xenopus* PGCs are directed to the gonad involves a CXCR4 protein on the PGC responding to a Sdf1 ligand along the migration path (Nishiumi et al. 2005; Takeuchi et al. 2010). Knocking out *CXCR4* mRNA with morpholinos results in fewer PGCs reaching the gonads, and ectopically expressing *Sdf1* will misdirect the PGCs into other areas.

MAMMALS The ability to label mouse primordial germ cells with GFP and then watch these living cells migrate has led to a reevaluation of the germ cell migration pathway in mammals (Anderson et al. 2000; Molyneaux et al. 2001; Tanaka et al. 2005). First, it appears that mammalian PGCs forming in the posterior epiblast migrate directly into the endoderm from the posterior region of the primitive streak. (Those cells that are seen to enter the allantois are believed to die.) These cells find themselves in the hindgut (**FIGURE 17.15A**). Although they move actively, they cannot get out of the gut until about embryonic day 9. At that time, the PGCs exit the gut but do not yet migrate toward the genital ridges. By the following day, however, PGCs are seen migrating into the genital ridges (**FIGURE 17.15B,C**). By embryonic day 11.5, the PGCs enter the developing gonads. During this trek, they have proliferated from an initial population of 10–100 cells to the 2500–5000 PGCs present in the gonads by day 12.

Like the PGCs of *Xenopus*, mammalian PGCs appear to be closely associated with the cells over which they migrate, and they move by extending filopodia over the underlying cell surfaces. Mammalian PGCs are also capable of penetrating cell monolayers and migrating through cell sheets (Stott and Wylie 1986). The mechanism by which these cells know the route of their journey is still unknown. Fibronectin is likely to be an important substrate for PGC migration (ffrench-Constant et al. 1991), and germ cells lacking the integrin receptor for such extracellular matrix proteins cannot migrate to the gonads (Anderson et al. 1999). During

(A) Migration of PGCs to endoderm

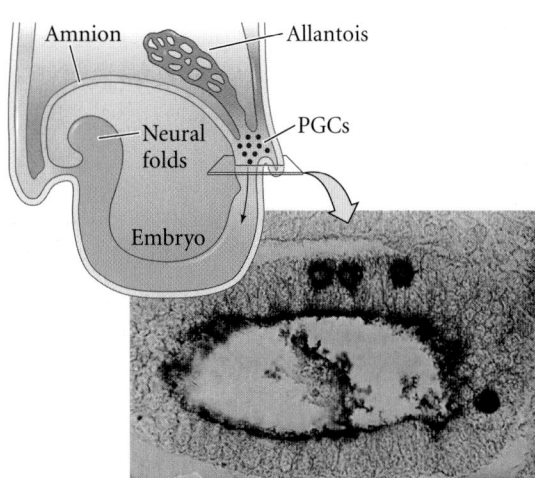

(B) Migration of PGCs into gonad

(C)

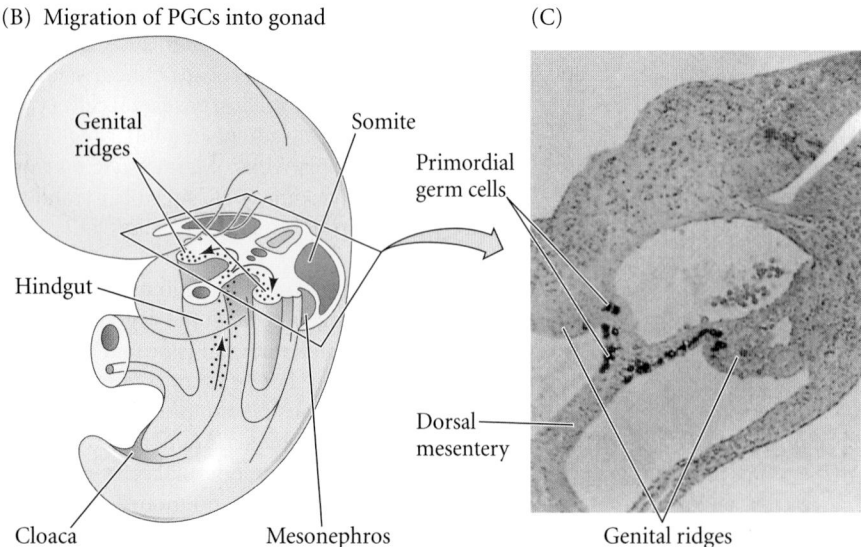

FIGURE 17.15 Primordial germ cell migration in the mouse. (A) On day 8, the PGCs established in the posterior epiblast (see Figure 17.7) migrate into the definitive endoderm of the embryo. The photo shows four large PGCs (stained for alkaline phosphatase) in the hindgut of a mouse embryo. (B) The PGCs migrate through the gut and, dorsally, into the genital ridges. (C) Alkaline phosphatase-staining cells are seen entering the genital ridges around embryonic day 11. (A from Heath 1978; C from Mintz 1957, courtesy of the authors.)

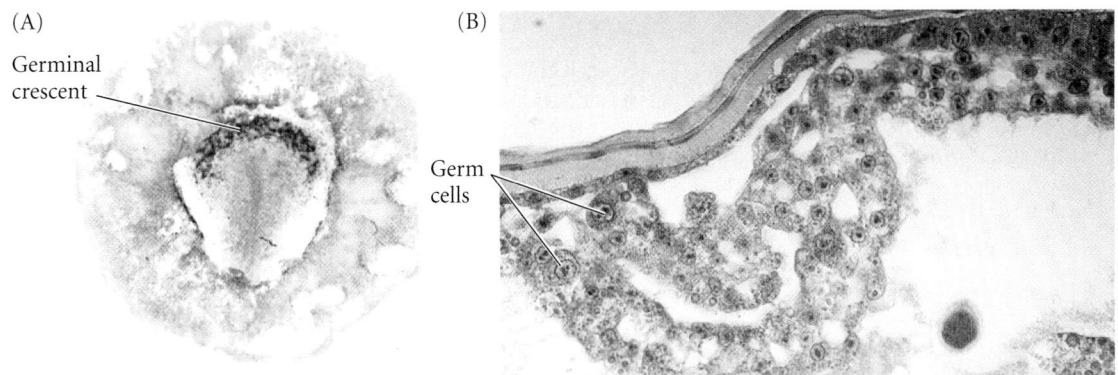

FIGURE 17.16 Germinal crescent of the chick embryo. (A) Germ cells of a stage 4 (definitive primitive streak stage, roughly 18 hours) chick embryo, stained purple for the chick Vasa homologue. The stained cells are confined to the germinal crescent. (B) Higher magnification of the stage 4 germinal crescent region, showing germ cells (stained brown) in the thickened epiblast. (From Tsunekawa et al. 2000, courtesy of N. Tsunekawa.)

the time from their specification to their entrance into the genital ridges, the PGCs are surrounded by cells secreting stem cell factor (SCF). SCF is necessary for PGC motility and survival. Moreover, the cluster of SCF-secreting cells appears to migrate with the PGCs, forming a "traveling niche" of cells that support the persistence and movement of the PGCs (Gu et al. 2009). However, the mechanism by which directionality is provided for migrating toward the gonads remains controversial (see Ara et al. 2003; Molyneaux et al. 2003; Farini et al. 2007; Saga 2008).

BIRDS AND REPTILES In birds and reptiles, the primordial germ cells are derived from epiblast cells that migrate from the central region of the area pellucida to a crescent-shaped zone in the hypoblast at the anterior border of the area pellucida (**FIGURE 17.16**; Eyal-Giladi et al. 1981; Ginsburg and Eyal-Giladi 1987). This extraembryonic region is called the **germinal crescent**, and the PGCs multiply there.

Unlike those of amphibians and mammals, the PGCs of birds and reptiles migrate to the gonads primarily by means of the bloodstream (**FIGURE 17.17**). When blood vessels form in the germinal crescent (anterior to the future head region), the PGCs enter those vessels and are carried by the circulation to the intermediate mesoderm. Here they leave the circulation and migrate into the genital ridges (Swift 1914; Nakamura et al. 2007).

The PGCs of the germinal crescent appear to enter the blood vessels by diapedesis (the same mechanism that got the *Drosophila* germ cells into the mesoderm). In some as-yet-undiscovered way, the PGCs are instructed to exit the blood vessels and enter the gonads (Pasteels 1953; Dubois 1969; Nakamura et al. 2007). Evidence for chemotaxis comes from studies in which circulating chick PGCs were isolated from the blood and cultured between gonadal rudiments and other embryonic tissues (Kuwana et al. 1986). During a 3-hour incubation, the PGCs migrated specifically into the gonadal rudiments.

The molecules that chick PGCs use for chemotaxis may be the same Sdf1/CXCR4 chemotactic system seen in zebrafish.

(A) Primordial germ cell (PGC)

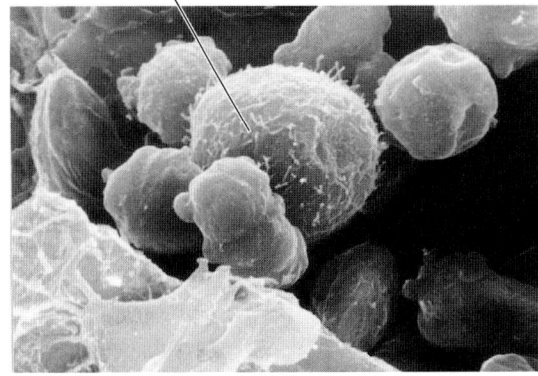

(B)

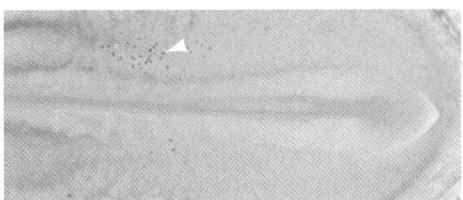

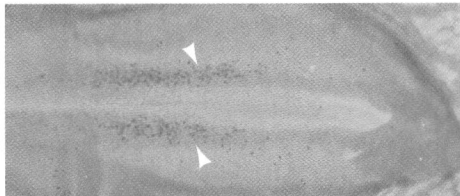

FIGURE 17.17 Migration of primordial germ cells in the chick embryo. (A) Scanning electron micrograph of a chick PGC in a capillary of a gastrulating embryo. Note the larger size of the PGC, as well as the microvilli on its surface. (B) After leaving the blood vessels, PGCs migrate into the intermediate mesodermal region that forms the gonad. These whole mounts show chick PGCs (stained with antibodies against the Vasa protein; arrowheads) in the posterior region of stage 15 and 17 embryos. (A from Kuwana 1993, courtesy of T. Kuwana; B from Nakamura et al. 2007, courtesy of T. Takahiro.)

Like mammals, chicks only use chemotaxis during the latter stages of migration. Thus, after they leave the blood vessels, chick PGCs appear to use Sdf1 gradients to reach the gonads (Stebler et al. 2004). Indeed, if Sdf1-secreting cells are transplanted into late-stage chick embryos, the PGCs will be attracted to them.

The Primacy of Meiosis

Once the germ cells have migrated to the gonad, they can begin meiosis. Meiosis is perhaps the most revolutionary invention of eukaryotes. It is difficult now to appreciate how startling this concept was for biologists at the end of the nineteenth century. The discovery of meiosis signaled the critical breakthrough for the investigation of inheritance. Van Beneden's 1883 observations that the divisions of germ cells caused the resulting gametes to contain half the diploid number of chromosomes "demonstrated that the chromosomes of the offspring are derived in equal numbers from the nuclei of the two conjugating germ-cells and hence equally from the two parents" (Wilson 1924). All subsequent theories of heredity, including the Sutton-Boveri model that united Mendelism with cell biology, are based on meiosis as the mechanism for sexual reproduction and the transmission of genes from one generation to the next.

Meiosis completes the cycle of life. The body decays and dies; but the gametes formed by meiosis survive the death of their parents and form the next generation. Sexual reproduction, evolutionary variation, and the transmission of traits from one generation to the next all come down to meiosis. So to understand what germ cells do, we must first understand meiosis.

Meiosis is initiated and regulated by signals from the gonad. Seen from the germ cell's view, the gonads exist to provide the signals that coordinate meiosis, gamete maturation, and eventual release. Once in the gonad, the PGCs continue to divide mitotically, often producing millions of potential gamete precursors. The germ cells of both male and female gonads are then faced with the necessity of reducing their chromosomes from the diploid to the haploid condition. In the haploid condition, each chromosome is represented by only one copy, whereas diploid cells have two copies of each chromosome. These *meiotic* divisions differ from *mitotic* divisions in that (1) meiotic cells undergo two cell divisions without an intervening period of DNA replication, and (2) homologous chromosomes (each consisting of two sister chromatids joined at a kinetochore*) pair together and recombine genetic material.

After the germ cell's last mitotic division, a period of DNA synthesis occurs, so that the cell initiating meiosis doubles the amount of DNA in its nucleus. In this state, each chromosome consists of two sister **chromatids** attached at a common kinetochore. (In other words, the diploid nucleus contains four copies

of each chromosome.) Meiosis entails two cell divisions. In the first division (meiosis I), homologous chromosomes (for example, the two copies of chromosome 3 in the diploid cell) come together and are then separated into different cells. Hence the first meiotic division *splits two homologous chromosomes* between two daughter cells such that each daughter cell has only one copy of each chromosome. But each of the chromosomes has already replicated (i.e., each has two chromatids). The second division (meiosis II) *separates the two sister chromatids from each other*. Consequently, each of the four cells produced by meiosis has a single (haploid) copy of each chromosome.

The first meiotic division begins with a long prophase, which is subdivided into four stages (**FIGURE 17.18**). During the **leptotene** (Greek, "thin thread") stage, the chromatin of the chromatids is stretched out very thinly, and it is not possible to identify individual chromosomes. DNA replication has already occurred, however, and each chromosome consists of two parallel chromatids. At the **zygotene** (Greek, "yoked threads") stage, homologous chromosomes pair side by side. This pairing is called **synapsis**, and it is characteristic of meiosis; such pairing does not occur during mitotic divisions. Although the mechanism whereby each chromosome recognizes its homologue is not known (see Barzel and Kupiec 2008; Takeo et al. 2011), synapsis seems to require the presence of the nuclear envelope and the formation of a proteinaceous ribbon called the **synaptonemal complex**. In many species, the nuclear envelope probably serves as an attachment site for the prophase chromosomes to bind and thereby reduces the complexity of the search for the other homologous chromosome (Comings 1968; Scherthan 2007; Tsai and McKee 2011). The synaptonemal complex is a ladderlike structure with a central element and two lateral bars (von Wettstein 1984; Yang and Wang 2009). The homologous chromosomes become associated with the two lateral bars, and the chromosomes are thus joined together. The configuration formed by the four chromatids and the synaptonemal complex is referred to as a **tetrad** or a **bivalent**.

During the next stage of meiotic prophase, **pachytene** (Greek, "thick thread"), the chromatids thicken and shorten. Individual chromatids can now be distinguished under the light microscope, and crossing-over may occur. **Crossing-over** represents an exchange of genetic material whereby genes from one chromatid are exchanged with homologous genes from another. Crossing-over may continue into the next stage, **diplotene** (Greek, "double threads"). Here, the synaptonemal complex breaks down, and the two homologous chromosomes start to separate. Usually, however, they remain attached at various points called **chiasmata**, which are thought to represent regions where crossing-over is occurring. The diplotene stage is characterized by a high level of gene transcription. In some species, the chromosomes of both male and female germ cells take on the "lampbrush" appearance characteristic of chromosomes that are actively making RNA (see Figure 17.24).

Metaphase begins with **diakinesis** (Greek, "moving apart") of the chromosomes (Figure 17.18B). The nuclear envelope breaks down and the chromosomes migrate to form a **metaphase plate**. **Anaphase** of meiosis I does not

---

* Although the terms *centromere* and *kinetochore* are often used interchangeably, the kinetochore is the complex protein structure that assembles on a sequence of DNA known as the centromere.

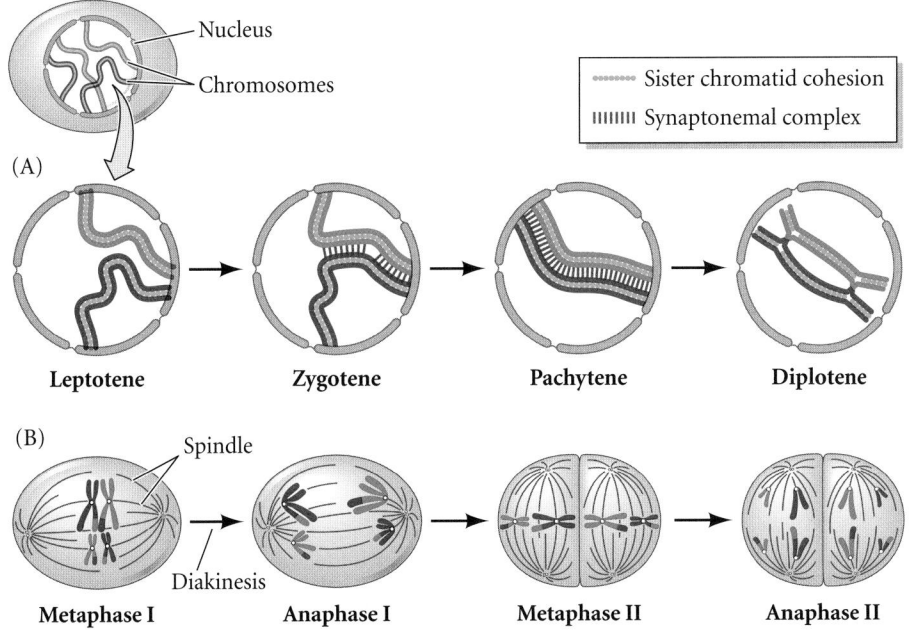

(A)

Nucleus

Chromosomes

····· Sister chromatid cohesion

||||||| Synaptonemal complex

Leptotene **Zygotene** **Pachytene** **Diplotene**

(B)

Spindle

Diakinesis

Metaphase I **Anaphase I** **Metaphase II** **Anaphase II**

FIGURE 17.18 Meiosis, emphasizing the synaptonemal complex. Before meiosis, unpaired homologous chromosomes are distributed randomly within the nucleus. (A) At leptotene, telomeres have attached along the nuclear envelope. The chromosomes "search" for homologous chromosomes, and synapsis, the association of homologous chromosomes, begins at zygotene, where the first evidence of the synaptonemal complex (SC) can be seen. During pachytene, homologue alignment is seen along the entire length of the chromosomes and produces a bivalent structure. Paired homologs can recombine with each (cross over) other during zygotene and pachytene. The synaptonemal complex dissolves at diplotene, when recombination is completed. (B) In diakinesis, chromosomes condense further and then form a metaphase plate. Segregation of the homologous chromosomes occurs at anaphase I. Only one pair of sister chromatids is shown here in meiosis II, where sister chromatids align at metaphase II and then in anaphase II segregate to opposite poles. (After Tsai and McKee 2011.)

commence until the chromosomes are properly aligned on the mitotic spindle fibers. This alignment is accomplished by proteins that prevent cyclin B from being degraded until after all the chromosomes are securely fastened to microtubules.

During anaphase I, the homologous chromosomes are separated from each other in an independent fashion. This stage

leads to telophase I, during which two daughter cells are formed, each cell containing one partner of each homologous chromosome pair. After a brief **interkinesis**, the second division of meiosis takes place. During this division, the kinetochore of each chromosome divides during anaphase so that each of the new cells gets one of the two chromatids, the final result being the creation of four haploid cells. Note that meiosis has also reassorted the chromosomes into new groupings. First, each of the four haploid cells has a different assortment of chromosomes. Humans have 23 different chromosome pairs; thus 2^{23} (nearly 10 million) different haploid cells can be formed from the genome of a single person. In addition, the crossing-over that occurs during the pachytene and diplotene stages of prophase during meiosis I further increases genetic diversity and makes the number of potential different gametes incalculably large.

This organization and movement of meiotic chromosomes is choreographed by a ring of **cohesin proteins** that encircle the sister chromatids. The rings of cohesin protein resist the pulling forces of the spindle microtubules and thereby keep the sister chromatids attached together during the first meiotic division (**FIGURE 17.19**; Haering et al. 2008; Brar et al. 2009). The cohesin proteins also recruit other sets of proteins that help promote pairing between the homologous chromosomes and allows recombination

FIGURE 17.19 Cohesin proteins encircle the sister chromatids. Immature mouse sperm cells undergoing meiosis were stained for the cohesin subunit Rad21. (A) Pachytene, where Rad21 is seen throughout the length of the paired chromosomes. (B) A metaphase I autosomal bivalent. Rad21 is enriched at homologous kinetochores, forming a T-shaped structure. Rad21 labeling is interrupted at the chiasmata. (From Viera et al. 2007.)

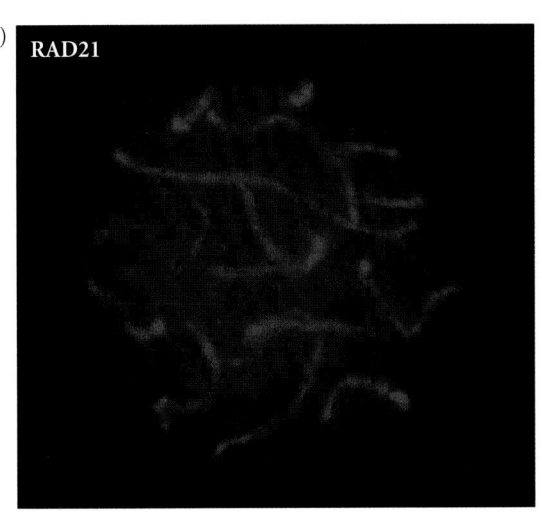

(A) RAD21

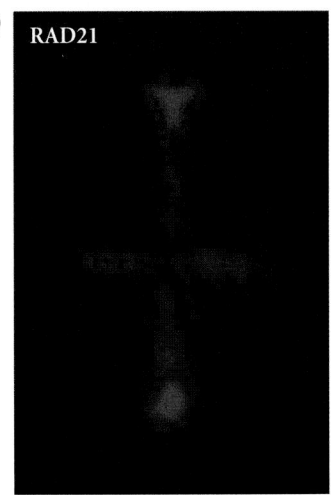

(B) RAD21

SIDELIGHTS & SPECULATIONS

Big Decisions: Mitosis or Meiosis? Egg or Sperm?

In many species, the germ cells migrating into the gonad are bipotential and can differentiate into either sperm or eggs, depending on their gonadal environment. When the ovaries of salamanders are experimentally transformed into testes, the resident germ cells cease their oogenic differentiation and begin developing as sperm (Burns 1930; Humphrey 1931). Similarly, in the housefly and mouse, the gonad is able to direct the differentiation of the germ cells (McLaren 1983; Inoue and Hiroyoshi 1986). Thus, in most organisms, the sex of the gonad and that of its germ cells is the same.

But what about hermaphroditic animals, where the change from sperm production to egg production is a naturally occurring physiological event? How is the same animal capable of producing sperm during one part of its life and oocytes during another part? Using *Caenorhabditis elegans*, Kimble and her colleagues identified two "decisions" that presumptive germ cells have to make. The first is whether to enter meiosis or to remain a mitotically dividing stem cell. The second is whether to become an egg or a sperm.

Unlike the situation in vertebrates and insects, the germ cells of *C. elegans* do not migrate into the gonads. They are already there. Thus, the decisions whether to enter meiosis and whether to become sperm or egg are intimately linked. The mitosis/meiosis decision in *C. elegans* is controlled by a single nondividing cell—the **distal tip cell**—located at the end of each gonad. The germ cell precursors near this cell divide mitotically, forming the pool of germ cells; but as these cells get farther away from the distal tip cell, they enter meiosis. If the distal tip cell is destroyed by a focused laser beam, all the germ cells enter meiosis; and if the distal tip cell is placed in a different location in the gonad, germline stem cells are generated near its new position (Kimble 1981; Kimble and White 1981). The distal tip cell extends long filaments that touch the distal germ cells (**FIGURE 17.20A**). The extensions contain in their cell membranes

the LAG-2 protein, a *C. elegans* homologue of Delta (Henderson et al. 1994; Tax et al. 1994; Hall et al. 1999). LAG-2 maintains a germ stem pool by keeping cells in mitosis and inhibiting their meiotic differentiation. The dividing cells constitute a transit amplifying cell population that enters meiosis when the signal from the distal tip cell is no longer received.

Austin and Kimble (1987) isolated a mutation that mimics the phenotype obtained when the distal tip cell is removed. It is not surprising that this mutation involves the gene encoding GLP-1, the *C. elegans* homologue of Notch—the receptor for Delta. All the germ cell precursors of nematodes homozygous for the recessive mutation of *glp-1* initiate meiosis, leaving no mitotic population. Instead of the 1500 germ cells usually found in the fourth larval stage of hermaphroditic development, these mutants produce only 5–8 sperm cells. When genetic chimeras are made in which wild-type germ cell precursors are found in a mutant larva, the wild-type cells are able to respond to the distal tip cells and undergo mitosis. However, when mutant germ cell precursors are found in wild-type larvae, they all enter meiosis. Thus, the *glp-1* gene appears to be responsible for enabling the germ cells to respond to the distal tip cell's signal.*

As is usual in development, the binary decision entails both a push and a pull (Jeong et al. 2011; **FIGURE 17.20B**). The decision to enter meiosis must be amplified by a decision to end mitosis. This appears to be accomplished by the FBF (*fem-3* mRNA-binding factor) proteins, similar to the *Drosophila* Pumilio RNA-binding protein mentioned in Chapter 6. Notch appears to activate FBFs, which are translational repressors of the GLD (germline development) proteins. GLD-1 (in combination with a Nanos protein) suppresses the translation of mitosis-specific messages. This

includes suppressing the translation of *glp-1* mRNA (Eckmann et al. 2002, 2004; Marin and Evans 2003; Kimble and Crittenden 2005). FBFs also repress the translation of GLD-2 and GLD-3, two proteins necessary for polyadenylating meiosis-specific mRNAs, allowing them to be translated. Thus, the Notch signal, acting through the FBFs, simultaneously promotes mitosis and blocks meiosis (**FIGURE 17.20C**). As long as the cells receive the Notch signal, they will not enter meiosis. Meiosis occurs as the cells move away from the distal tip cell.

After the germ cells begin their meiotic divisions, they still must become either sperm or ova. Generally, in each hermaphrodite gonad, the most *proximal* germ cells produce sperm, while the most *distal* (near the tip) become eggs (Hirsh et al. 1976). This means that the germ cells entering meiosis early become sperm, and those entering meiosis later become eggs (Hodgkin et al. 1985; Kimble et al. 1986).

The genetics of this sex determination switch are currently being analyzed. The core of the pathway is shown in **FIGURE 17.21** for *C. elegans*. At the end of the pathway, FOG-3 inhibits oogenesis, and FOG-1 promotes spermatogenesis. If both proteins are active, sperm is made at the expense of eggs. But they can be inactivated by TRA-1. TRA-1 can be inactivated by the *fem* genes. And the *fem* genes can be inactivated by TRA-2. The *tra-2* gene can be inactivated by the initiator of spermatogenesis, HER-1. The main genes appear to be *tra-2* and *fem-3*. Translational repression of *tra-2* mRNA is essential for sperm formation in the hermaphrodite. If HER-1 blocks TRA-2 formation, then the FEM proteins can inhibit TRA-1 activity. The transition from spermatogenesis to oogenesis appears to be made by changing the ratio of TRA-2 activity to FEM-3 activity (Ellis and Schedl 2007; Lee et al. 2011). Since the eggs form in the distal part of the gonad, and the sperm form in the proximal region, near the vulva, the eggs "roll" through the sperm and are fertilized as they prepare to leave the body.

*You may recall that GLP-1 is also needed by the AB blastomere for it to receive inductive signals to form pharyngeal muscles (see Chapter 5).

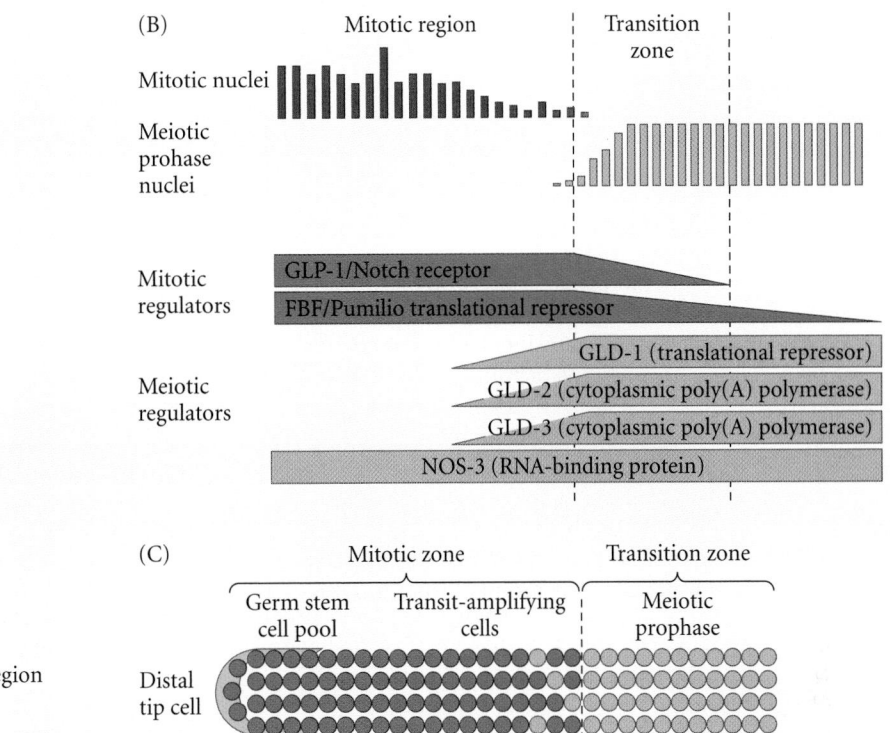

(A)

Mitotic region

Transition zone

FIGURE 17.20 Regulation of the mitosis/meiosis decision in the adult germline of *C. elegans*. (A) The distal tip cell and its long projections comprise the stem cell niche. The distal tip of this adult worm is stained green (by attaching the *GFP* gene to the promoter of the *lag-2* gene). Germline cells are stained red. (B) The balance of mitotic and meiotic regulator proteins determines whether a cell remains in mitosis or enters meiosis. Vertical red bars represent the relative number of mitotic nuclei at each position in the mitotic region. The transition from mitosis to meiosis begins where 60% of nuclei have the crescent-shaped morphology typical of early meiotic prophase (red crescents in A). Gold bars represent the percentage of nuclei in meiotic prophase at a given position along the distal-proximal axis. Levels of GLP-1

and FBF mitotic regulators are high throughout the mitotic region and decrease dramatically as the germ cells enter meiosis (horizontal red bands). Conversely, levels of most meiotic regulators (horizontal gold bands) gradually increase in the proximal part of the mitotic region, reaching high levels as germ cells enter meiosis. (One exception is NOS-3, which is distributed uniformly throughout the germ line.) (C) Simplified summary of a network controlling the mitosis/meiosis decision. Notch signaling activates FBF-2. FBF-1 and FBF-2 are very similar proteins whose negative feedback loop may specify the size of the mitotic region, as it negatively regulates levels of GLD-1, GLD-3, and HIM-3, thereby inhibiting meiosis in the distal region of the germ line. (After Jeong et al. 2011; A courtesy of S. Crittenden and J. Kimble.)

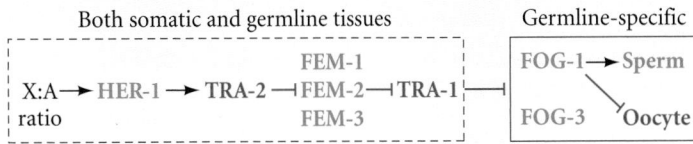

FIGURE 17.21 Model of sex determination switch in the germ line of *C. elegans* hermaphrodites. (After Morgan et al. 2013.)

to occur (Pelttari et al. 2001; Villeneuve and Hillers 2001; Sakuno and Watanabe 2009). At second meiotic division, the ring of cohesin proteins is cleaved and the kinetochores can separate from each other (Schöckel et al. 2011).

MODIFICATIONS OF MEIOSIS Some animal species consist entirely of females; such species are said to be **parthenogenetic** (Greek, "virgin birth"). In these species, meiosis is modified such that the resulting gamete is diploid and need not be fertilized to develop. In the fly *Drosophila mangabeirai*, one of the polar bodies (a meiotic cell having very little cytoplasm) acts as a sperm and "fertilizes" the oocyte after the second meiotic division. In some other insects and in the lizard *Cnemidophorus uniparens*, the oogonia further double their chromosome number *before* meiosis, so that the halving of the chromosomes restores the diploid number. The germ cells of the grasshopper *Pycnoscelus surinamensis* dispense with meiosis altogether, forming diploid ova by two mitotic divisions (Swanson et al. 1981). About 80 species are known to be exclusively parthenogenetic, and several other species have been observed to occasionally produce offspring this way (Booth et al. 2012).

In other species, haploid parthenogenesis is widely used not only as a means of reproduction but also as a mechanism of sex determination. In the Hymenoptera (bees, wasps, and ants), unfertilized haploid eggs develop into males, whereas fertilized eggs are diploid and develop into females. The haploid males are able to produce sperm by abandoning the first meiotic division, thereby forming two sperm cells through second meiosis.

Gamete Maturation

The regulation of meiosis can differ dramatically between males and females. The egg is usually a nonmotile cell that has conserved its cytoplasm and has stored the ribosomes, mitochondria, and mRNAs needed to initiate development. The sperm is usually a smaller, motile cell that has eliminated most of its cytoplasm to become a nucleus attached to a propulsion system. As we will soon see, there are often large differences between **oogenesis**, the production of eggs, and **spermatogenesis**, the production of sperm. Thus, gametogenesis is more than making the nucleus haploid. The formation of the sperm involves constructing the flagellum and the acrosome. Constructing the egg involves building the organelles involved in fertilization, synthesizing and positioning the mRNAs and proteins used in early development, and accumulating energy sources and energy-producing organelles (ribosomes, yolk, and mitochondria) in the cytoplasm. A partial catalogue of the materials stored in the oocyte cytoplasm of a frog is shown in **TABLE 17.1**.

The mechanisms of oogenesis vary among species more than those of spermatogenesis. This variation should not be surprising, since patterns of reproduction vary so greatly among species. In some species, such as sea urchins and frogs, the female routinely produces hundreds or thousands

of eggs at a time, whereas in other species, such as humans and most other mammals, relatively few eggs are produced at each mating, and only a few hundred are produced during an individual's lifetime. In those species that produce thousands of ova each breeding season, the female PGCs produce **oogonia**, self-renewing stem cells that endure for the lifetime of the organism. In those species that produce fewer eggs, the oogonia divide to form a limited number of egg precursor cells.

Maturation of the oocytes in frogs

The eggs of sea urchins, fish, and amphibians are derived from an oogonial stem cell population that can generate a new cohort of oocytes each year. In the frog *Rana pipiens*, oogenesis takes 3 years. During the first 2 years, the oocyte increases its size very gradually. During the third year, however, the rapid accumulation of yolk in the oocyte causes the egg to swell to its characteristic large size. Eggs mature in yearly batches, with the first cohort maturing shortly after metamorphosis; the next group matures a year later.

VITELLOGENESIS Vitellogenesis—the accumulation of yolk proteins—occurs when the oocyte reaches the diplotene stage of meiotic prophase. Yolk is not a single substance, but a mixture of materials for embryonic nutrition. The major yolk component in frog eggs is a 470-kDa protein called **vitellogenin**. It is not made in the frog oocyte (as are the major yolk proteins of organisms such as annelids and crayfish), but is synthesized in the liver and carried by the bloodstream to the ovary (Flickinger and Rounds 1956; Danilchik and Gerhart 1987).

COMPLETION OF AMPHIBIAN MEIOSIS: PROGESTERONE AND FERTILIZATION Amphibian primary oocytes can remain in the diplotene stage of meiotic prophase for years. This state resembles the G2 phase of the mitotic cell division cycle. Resumption of meiosis in the amphibian oocyte requires progesterone. This hormone is secreted by the follicle cells in response to gonadotropic hormones secreted by the pituitary

TABLE 17.1 Cellular components stored in the mature oocyte of *Xenopus laevis*

Component	Approximate excess over amount in larval cells
Mitochondria	100,000
RNA polymerases	60,000–100,000
DNA polymerases	100,000
Ribosomes	200,000
tRNA	10,000
Histones	15,000
Deoxyribonucleoside triphosphates	2,500

Source: After Laskey et al. 1979.

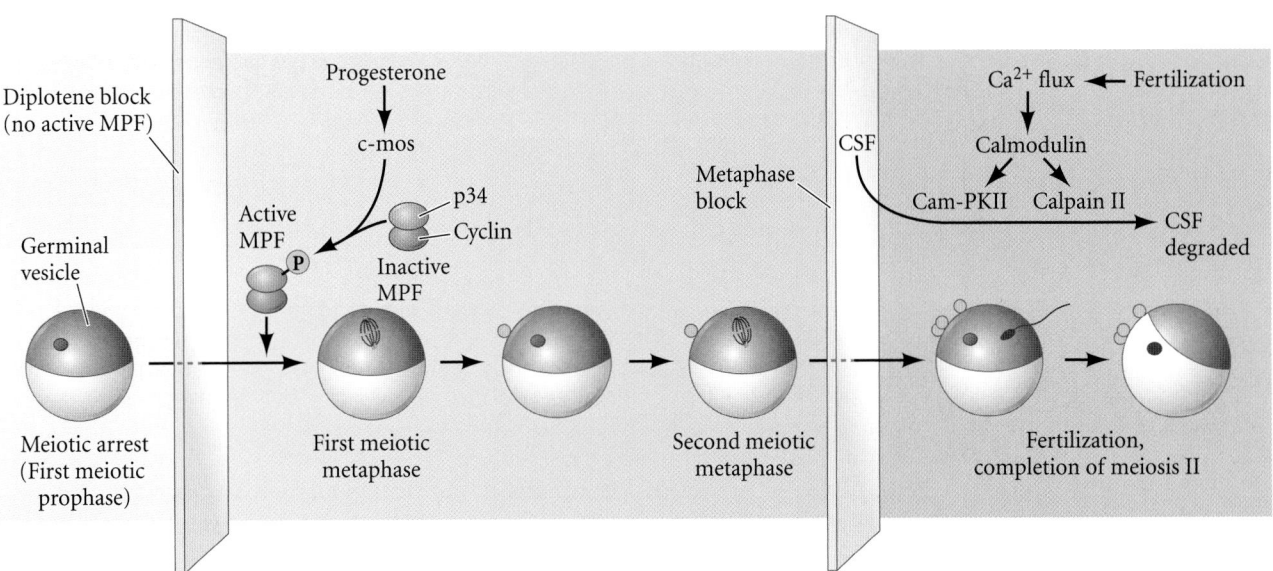

FIGURE 17.22 Schematic representation of *Xenopus* oocyte maturation, showing the regulation of meiotic cell division by progesterone and fertilization. Oocyte maturation is arrested at the diplotene stage of first meiotic prophase by the lack of active MPF. Progesterone activates the production of the c-mos protein. This protein initiates a cascade of phosphorylation that eventually phosphorylates the p34 subunit of MPF, allowing the MPF to become active. The MPF drives the cell cycle through the first meiotic division, but further division is blocked by CSF, a compound containing c-mos, cyclin-dependent kinase 2, and Erp1. CSF inhibits the anaphase-promoting complex from degrading cyclin. Upon fertilization, calcium ions released into the cytoplasm are bound by calmodulin and are used to activate two enzymes, calmodulin-dependent protein kinase II and calpain II, which inactivate and degrade CSF. Second meiosis is completed, and the two haploid pronuclei can fuse. At this time, cyclin B is resynthesized, allowing the first cell cycle of cleavage to begin.

gland. Within 6 hours of progesterone stimulation, the nuclear membrane dissolves. This is often called **germinal vesicle breakdown** (**GVBD**) because the old name for the prophase nucleus was "germinal vesicle." In addition, the microvilli retract, the nucleoli disintegrate, and the chromosomes contract and migrate to the animal pole to begin division. Soon afterward, the first meiotic division occurs, and the mature ovum is released from the ovary by a process called **ovulation**. The ovulated egg is in second meiotic metaphase when it is released (**FIGURE 17.22**).

How does progesterone enable the egg to break its dormancy and resume meiosis? To understand the mechanisms by which this activation is accomplished, it is necessary to briefly review the model for early blastomere division (see Chapter 5). Entry into the mitotic (M) phase of the cell cycle (in both meiosis and mitosis) is regulated by **mitosis-promoting factor**, or **MPF** (originally called maturation-promoting factor, after its meiotic function). MPF contains two subunits, **cyclin B** and the **p34** protein. The p34 protein is a cyclin-dependent kinase—its activity is dependent on the presence of cyclin. Since all the components of MPF are present in the amphibian oocyte, it is generally thought that progesterone somehow converts a pre-MPF complex into active MPF.

The mediator of the progesterone signal is the **c-mos** protein. Progesterone reinitiates meiosis by causing the egg to polyadenylate the maternal *c-mos* mRNA that has been stored in its cytoplasm (Sagata et al. 1988; Sheets et al. 1995;

Mendez et al. 2000). This message is translated into a 39-kDa phosphoprotein. This c-mos protein is detectable only during oocyte maturation and is destroyed quickly upon fertilization. Yet during its brief lifetime, it plays a major role in releasing the egg from its dormancy. The c-mos protein activates a phosphorylation cascade that phosphorylates and activates the p34 subunit of MPF (Ferrell and Machleder 1998; Ferrell 1999). The active MPF allows the germinal vesicle to break down and the chromosomes to divide. If the translation of *c-mos* is inhibited by injecting *c-mos* antisense mRNA into the oocyte, germinal vesicle breakdown and the resumption of oocyte maturation do not occur.

However, oocyte maturation then encounters a second block. MPF can take the chromosomes only through the first meiotic division and prophase of the second meiotic division. The oocyte is arrested once again in the metaphase of the second meiotic division. This metaphase block is caused by **cytostatic factor** (**CSF**; Matsui 1974). CSF is a complex of proteins that includes c-mos, cyclin-dependent kinase 2 (Cdk2), MAP kinase, and Erp1 (Gabrielli et al. 1993; Inoue et al. 2007; Nishiyama et al. 2007). Erp1 is the active protein, and it is synthesized immediately after the first meiotic division. The proteins of the CSF complex interact, eventually activating Erp1 by phosphorylating it. Phosphorylated Erp1 blocks the degradation of cyclin by the anaphase-promoting complex (**FIGURE 17.23**).

This metaphase block is broken by fertilization. The calcium ion flux attending fertilization activates the calcium-binding

CSF complex

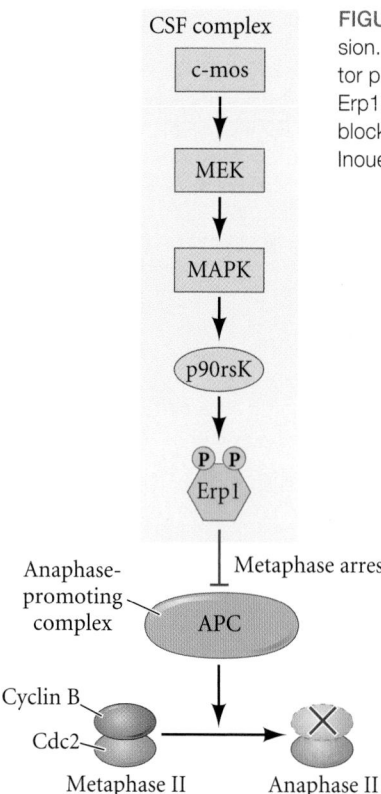

FIGURE 17.23 The main pathway leading to metaphase arrest in the second meiotic division. The CSF protein complex consists of c-mos, three transducer kinases, and the effector protein Erp1. Activation of c-mos activates the kinases, which eventually phosphorylate Erp1. Phosphorylated Erp1 binds to and inhibits the anaphase-promoting complex, thus blocking the degradation of cyclin B that would allow the cell to enter anaphase. (After Inoue et al. 2007.)

promotes cell division in two ways. First, without CSF, cyclin can be degraded, and the meiotic division can be completed. Second, calmodulin-dependent protein kinase II also allows the centrosome to duplicate, thus forming the poles of the meiotic spindle (Matsumoto and Maller 2002). In 1911, Frank Lillie wrote, "The nature of the inhibition that causes the need for fertilization is a most fundamental problem." The solution to that problem appears to be oocyte-derived CSF and the sperm-induced wave of calcium ions.

GENE TRANSCRIPTION IN AMPHIBIAN OOCYTES The amphibian oocyte has certain periods of very active RNA synthesis. During the diplotene stage, certain chromosomes stretch out large loops of DNA, causing them to resemble a lampbrush (which was a handy instrument for cleaning test tubes in the days before microfuges). In situ hybridization reveals these **lampbrush chromosomes** to be sites of RNA synthesis. Oocyte chromosomes can be incubated with a radioactive RNA probe, and autoradiography used to visualize the precise locations where genes are being transcribed (**FIGURE 17.24A**). Electron micrographs of gene transcripts from lampbrush chromosomes also enable one to see chains of mRNA coming off each gene as it is transcribed (**FIGURE 17.24B**; also see Hill and MacGregor 1980).

protein **calmodulin**, and calmodulin, in turn, can activate two enzymes that inactivate CSF. These enzymes are **calmodulin-dependent protein kinase II**, which inactivates Cdk2, and **calpain II**, a calcium-dependent protease that degrades c-mos (Watanabe et al. 1989; Lorca et al. 1993). This action

● See WEBSITE 17.3 Synthesizing oocyte ribosomes

(A)

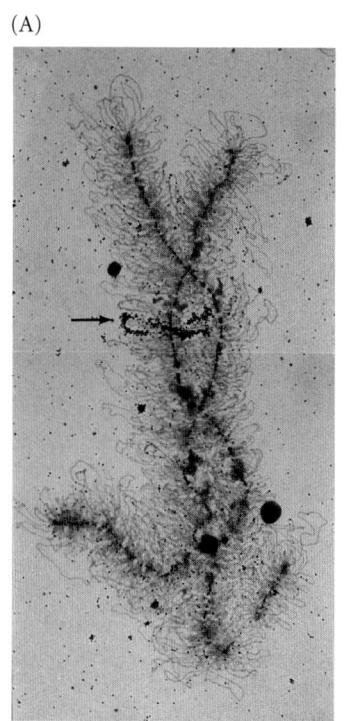

(B)

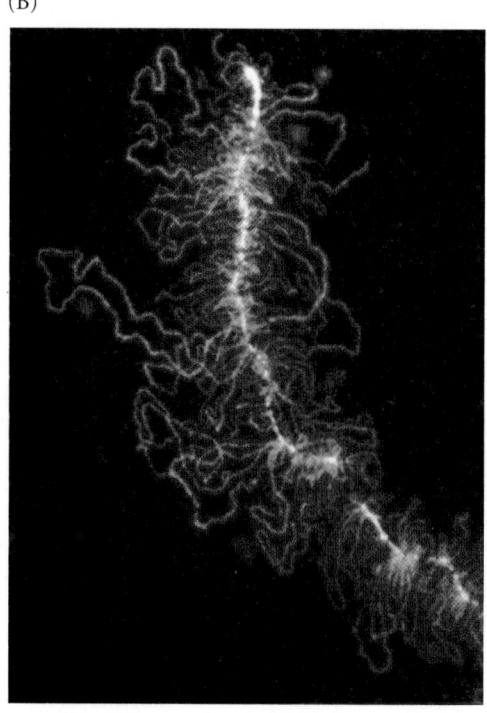

FIGURE 17.24 In amphibian oocytes, lampbrush chromosomes are active in the diplotene germinal vesicle during first meiotic prophase. (A) Autoradiograph of chromosome 1 of the newt *Triturus cristatus* after in situ hybridization with radioactive histone mRNA. A histone gene (or set of histone genes) is being transcribed (arrow) on one of the loops of this lampbrush chromosome. (B) Lampbrush chromosome of the salamander *Notophthalmus viridescens*. Extended DNA (white) loops out and is transcribed into RNA (red). (A from Old et al. 1977, courtesy of H. G. Callan; B courtesy of M. B. Roth and J. Gall.)

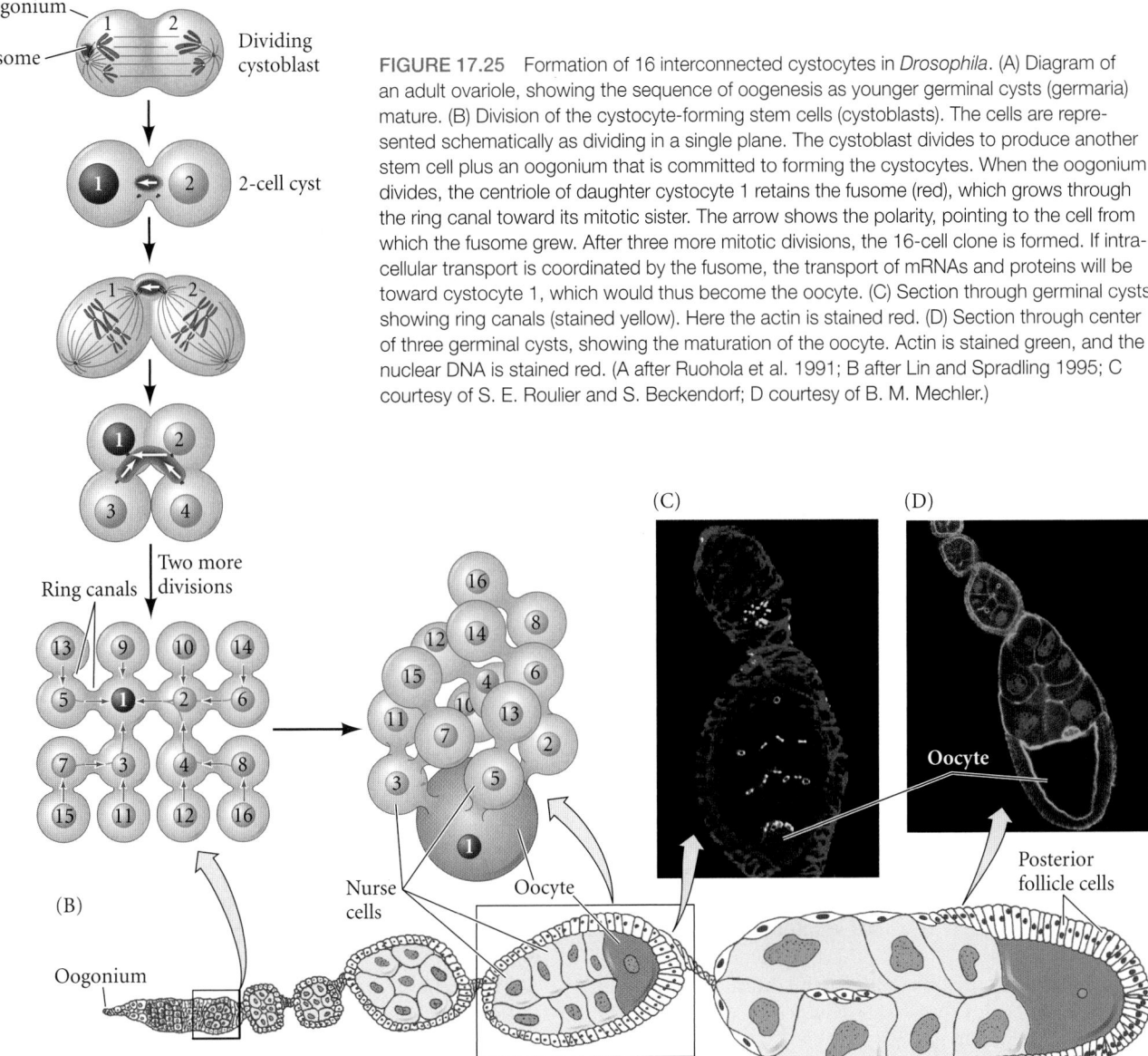

Oogonium
Fusome
Dividing cystoblast
2-cell cyst
Ring canals
Two more divisions
(B)
Oogonium
Nurse cells
Oocyte
Germarium
(C)
(D)
Oocyte
Posterior follicle cells

FIGURE 17.25 Formation of 16 interconnected cystocytes in *Drosophila*. (A) Diagram of an adult ovariole, showing the sequence of oogenesis as younger germinal cysts (germaria) mature. (B) Division of the cystocyte-forming stem cells (cystoblasts). The cells are represented schematically as dividing in a single plane. The cystoblast divides to produce another stem cell plus an oogonium that is committed to forming the cystocytes. When the oogonium divides, the centriole of daughter cystocyte 1 retains the fusome (red), which grows through the ring canal toward its mitotic sister. The arrow shows the polarity, pointing to the cell from which the fusome grew. After three more mitotic divisions, the 16-cell clone is formed. If intracellular transport is coordinated by the fusome, the transport of mRNAs and proteins will be toward cystocyte 1, which would thus become the oocyte. (C) Section through germinal cysts showing ring canals (stained yellow). Here the actin is stained red. (D) Section through center of three germinal cysts, showing the maturation of the oocyte. Actin is stained green, and the nuclear DNA is stained red. (A after Ruohola et al. 1991; B after Lin and Spradling 1995; C courtesy of S. E. Roulier and S. Beckendorf; D courtesy of B. M. Mechler.)

Meroistic oogenesis in insects

There are several types of oogenesis in insects, but most studies have focused on those insects (including *Drosophila* and moths) that undergo **meroistic oogenesis**, where cytoplasmic connections remain between the cells produced by the oogonium.

The oocytes of meroistic insects do not pass through a transcriptionally active stage, nor do they have lampbrush chromosomes. Rather, RNA synthesis is largely confined to the nurse cells, and the RNA made by those cells is actively transported into the oocyte cytoplasm (see Figure 6.8). Oogenesis takes place in only 12 days, so the nurse cells are metabolically very active during this time. Nurse cells are aided in their transcriptional efficiency by becoming **polytene**—instead of having two copies of each chromosome, they replicate their chromosomes until they have produced 512 copies. The 15 nurse cells pass ribosomal and messenger RNAs as well as proteins into the oocyte cytoplasm, and entire ribosomes may be transported as well. The mRNAs do not associate with polysomes, and they are not immediately active in protein synthesis (Paglia et al. 1976; Telfer et al. 1981).

As mentioned earlier, each germline stem cell of *Drosophila* divides to produce another stem cell and a differentiated daughter cell, the cystoblast. The cystoblasts undergo four mitotic divisions with incomplete cytokinesis to form a cluster of 16 cells interconnected by cytoplasmic bridges called **ring canals**. Only those two cells having four interconnections are capable of developing into oocytes, and of those two, only one becomes the egg (the other begins meiosis but does not complete it). Thus, only one of the 16 cystocytes becomes an ovum; the remaining 15 cells become nurse cells (**FIGURE 17.25**).

As it turns out, the cell destined to become the oocyte is the cell residing at the most posterior tip of the **egg chamber**, or **ovariole**, enclosing the 16-cell clone. However, since the nurse cells are connected to the oocyte by the ring canals, the entire complex can be seen as one egg-producing unit. The nurse cells produce numerous RNAs and proteins that ultimately are transported into the oocyte through the ring canals.

The meroistic ovary confronts us with some interesting problems. If all 16 cystocytes derived from the PGC are connected so that proteins and RNAs can shuttle freely among them, how do 15 cystocytes become RNA-producing nurse cells while one cell is fated to become the oocyte? Why is the flow of protein and RNA in one direction only?

As the cystocytes divide, a large, spectrin-rich structure called the **fusome** forms and spans the ring canals between the cells. It is constructed asymmetrically, as it always grows from the spindle pole that remains in one of the cells after the first division (Lin and Spradling 1995; de Cuevas and Spradling 1998). The cell that retains the greater part of the fusome during the first division becomes the oocyte. It is not yet known if the fusome contains oogenic determinants, or if it directs the traffic of materials into this particular cell.

Once the patterns of transport are established, the cytoskeleton becomes actively involved in transporting mRNAs from the nurse cells into the oocyte cytoplasm (Cooley and Theurkauf 1994). An array of microtubules that extends through the ring canals (see Figure 17.25C) is critical for oocyte determination. In the nurse cells, the Exuperantia protein binds *bicoid* message to the microtubules and transports it to the anterior of the oocyte (Cha et al. 2001; see Chapter 6). If the microtubular array is disrupted (either chemically or by mutations such as *bicaudal-D* or *egalitarian*), the nurse cell gene products are transmitted in all directions and all 16 cells differentiate into nurse cells (Gutzeit 1986; Theurkauf et al. 1992, 1993; Spradling 1993).

The Bicaudal-D and Egalitarian proteins are probably core components of a dynein motor system that transports

mRNAs and proteins throughout the oocyte (Bullock and Ish-Horowicz 2001). Some compounds transported from the nurse cells into the oocyte become associated with transport proteins such as dynein and kinesin, which would enable them to travel along the tracks of microtubules extending through the ring canals (Theurkauf et al. 1992; Sun and Wyman 1993). The *oskar* message, for instance, is linked to kinesin through the Barentsz protein, and kinesin can transport the *oskar* message to the posterior of the oocyte (van Eeden et al. 2001; see Figure 6.8).

Actin may become important for maintaining the polarity of transport during later stages of oogenesis. Mutations that prevent actin microfilaments from lining the ring canals prevent the transport of mRNAs from the nurse cells to the oocyte, and disruption of the actin microfilaments randomizes the distribution of mRNA (Cooley et al. 1992; Watson et al. 1993). Thus, the cytoskeleton controls the movement of organelles and RNAs between nurse cells and oocyte such that developmental cues are exchanged only in the appropriate direction.

Gametogenesis in Mammals

The PGCs that migrate to the gonads do not make their own decision to become either sperm or eggs. That decision is made by the gonad in which they reside. The first process that has to happen is that the germ cells become competent to respond to the gonad's signals. This "licensing" of sexual choice is permitted by DAZL protein, which appears to regulate mRNA translation (Gill et al. 2011). If DAZL protein is not made, the PGCs migrate to the gonad but do not develop either male or female phenotypes, instead remaining as PGCs.

The signals from the gonad create profound differences between spermatogenesis and oogenesis in mammals (**TABLE 17.2**). One of the most fundamental sets of signals regulates the timing of meiosis, and these signals include Wnt4 and retinoic acid. In females, meiosis begins in the

TABLE 17.2 Sexual dimorphism in mammalian meioses

Female oogenesis	Male spermatogenesis
Meiosis initiated once in a finite population of cells	Meiosis initiated continuously in a mitotically dividing stem cell population
One gamete produced per meiosis	Four gametes produced per meiosis
Completion of meiosis delayed for months or years	Meiosis completed in days or weeks
Meiosis arrested at first meiotic prophase and reinitiated in a smaller population of cells	Meiosis and differentiation proceed continuously without cell cycle arrest
Differentiation of gamete occurs while diploid, in first meiotic prophase	Differentiation of gamete occurs while haploid, after meiosis ends
All chromosomes exhibit equivalent transcription and recombination during meiotic prophase	Sex chromosomes excluded from recombination and transcription during first meiotic prophase

Source: After Handel and Eppig 1998.

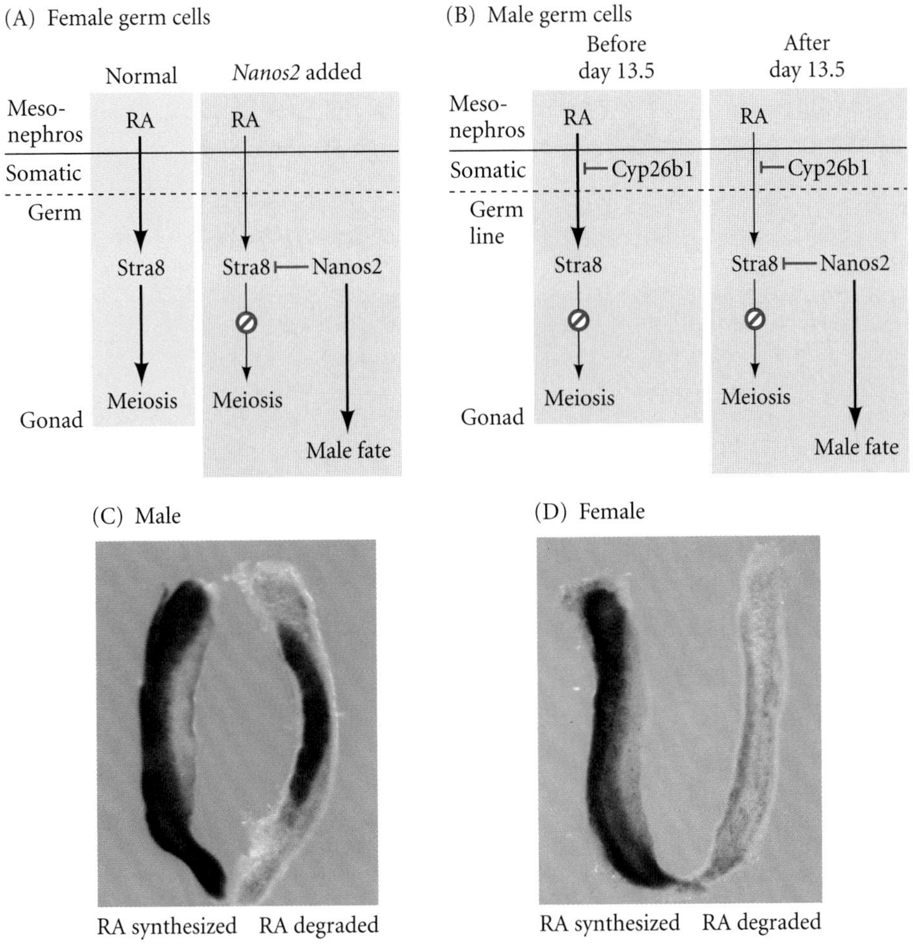

(A) Female germ cells

(B) Male germ cells

FIGURE 17.26 Retinoic acid (RA) determines the timing of meiosis and sexual differentiation of mammalian germ cells. (A) In female mouse embryos, RA secreted from the mesonephros reaches the gonad and triggers meiotic initiation via the induction of Stra8 transcription factor in female germ cells (beige). However, if activated *Nanos2* genes are added to female germ cells, they suppress Stra8 expression, leading the germ cells into a male pathway (gray). (B) In embryonic testes, Cyp26b1 blocks RA signaling, thereby preventing male germ cells from initiating meiosis until embryonic day 13.5 (left panel). After embryonic day 13.5, when Cyp26b1 expression is decreased, Nanos2 is expressed and prevents meiotic initiation by blocking Stra8 expression. This induces male-type differentiation in the germ cells (right panel). (C,D) Day 12 mouse embryos stained for mRNAs encoding the RA-synthesizing enzyme Aldh1a2 (left gonad) and the RA-degrading enzyme Cyp26b1 (right gonad). The RA-synthesizing enzyme is seen in the mesonephros of both the male (C) and female (D); the RA-degrading enzyme is seen only in the male gonad. (A,B from Saga 2008; C,D from Bowles et al. 2006, courtesy of P. Koopman.)

embryonic gonads; in males, meiosis is not initiated until puberty. The "gatekeeper" for meiosis appears to be the **Stra8** transcription factor, which promotes a new round of DNA synthesis and meiotic initiation in the germ cells. In the developing ovaries, Stra8 is *upregulated* by two factors coming from the mesonephric kidney: Wnt4 and retinoic acid (Baltus et al. 2006; Bowles et al. 2006; Naillat et al. 2010; Chassot et al. 2011). In the developing testes, however, Stra8 is *downregulated* by Fgf9, and the retinoic acid produced by the mesonephros is degraded by the secretion of the RA-degrading enzyme Cyp26b1 by the developing testes (**FIGURE 17.26**; Bowles 2006; Koubova 2006). At puberty, retinoic acid is synthesized in the Sertoli cells and induces Stra8 in sperm stem cells. When Stra8 is made, the sperm stem cells become committed to meiosis (Anderson et al. 2008; Mark 2008). Thus, the timing of retinoic acid synthesis appears to control Stra8, and Strat8 commits germ cells to meiosis. Fgf9, which also downregulates Stra8, also appears to be critical in keeping the male germ cells in a stem-cell-like condition (Bowles et al. 2010).

Another difference between male and female gametogenesis, which may be related to the above difference, is the pattern of DNA methylation. DNA methylation patterns are very distinct between the male and female germ cells, and as we saw in Chapter 4, both male and female pronuclei are needed for development. This diploid set of methylations must be removed during PGC migration, and the appropriate gamete-specific DNA methylation must be reestablished

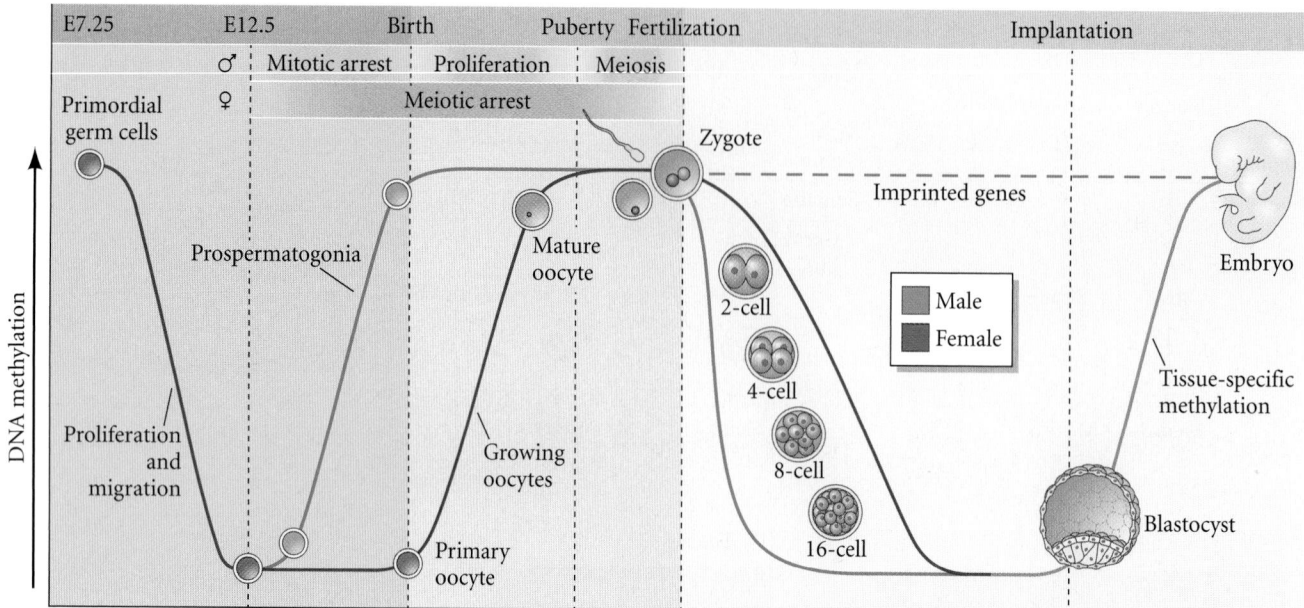

FIGURE 17.27 DNA methylation is removed at PGC migration and given gamete-specific patterns. In testes (blue line), new DNA methylation in the germ cells occurs just prior to meiosis and is finished before birth. In the ovary, DNA methylation occurs in the germ cells during the prolonged first meiotic prophase. After fertilization, a wave of DNA *demethylation* occurs. In the zygote, DNA methylation of the paternal genome is erased. Demethylation of the maternal genome is slower. However, some genes (the "imprinted genes") escape this demethyation. As cell lineages are determined, each group of cells gets a distinct methylation pattern. (After Smallwood and Kelsey 2012.)

during gametogenesis (**FIGURE 17.27**). As the PGCs migrate toward the gonads, DNA methylation is erased (Hayashi and Surani 2009; Saito et al. 2012), giving a "clean slate" on which to write the instructions for gene expression. This set of processes that erase and re-mark the genome is called **epigenetic reprogramming**, and the gamete-specific differences that result from this reprogramming may be due to the context of the DNA methylation. In the testes, DNA methylation occurs in early spermatogonia, during the G1 phase and prior to meiosis. In the ovary, DNA methylation ocurs during the prolonged meiotic prophase after a female infant's birth, during the period of follicular growth (Sasaki and Matsui 2008).

The gamete-specific methylation patterns are critical for subsequent gene expression, and such genes are said to be **imprinted** (see Sidelights & Speculations, p. 148). These gamete-specific methylations are accomplished by the de novo methyltransferase Dnmt3, and it appears that this enzyme is recruited to these genes by a protein called ZFP57. It is not known what brings ZFP57 to these regions of the genome, but evidence suggests that some chromatin protein or modified histone is crucial (see Smallwood and Kelsey 2012). By whatever mechanism, the DNA brought by the sperm and by the egg has been marked such that the sperm-borne DNA and the egg-borne DNA will express some genes that the other won't. This prevents parthenogenesis (so don't expect humans to become a female-only species in the near future), and it also means that some genetic conditions can only be transmitted through sperm cells and others only through egg cells.

● **See VADE MECUM** Gametogenesis in mammals

Spermatogenesis

Spermatogenesis—the developmental pathway from germ cell to mature sperm—begins at puberty and occurs in the recesses between the Sertoli cells (**FIGURE 17.28**). Spermatogenesis is divided into three major phases (Matson et al. 2010):

1. A proliferative phase where sperm stem cells (**spermatogonia**) increase by mitosis

2. A meiotic phase, involving the two divisions that create the haploid state

3. A postmeiotic "shaping" phase called **spermiogenesis**, during which the round cells (spermatids) eject most of their cytoplasm and become the streamlined sperm

The proliferative phase begins when the mammalian PGCs arrive at the genital ridge of a male embryo. Here they are called **gonocytes** and become incorporated into the sex cords that will become the seminiferous tubules (Culty 2009). The gonocytes differentiate into a population of stem cells called **type A$_s$ spermatogonia** (Yoshida et al. 2007). These are true stem cells in that they can reestablish spermatogenesis when transferred into mice whose sperm production was eliminated by toxic chemicals. Spermatogonia appear to take up residence in stem cell niches at the junction of the Sertoli cells (the epithelium of the seminiferous tubules), the interstitial (testosterone-producing) Leydig cells, and the testicular blood vessels. Adhesion molecules join the spermatogonia

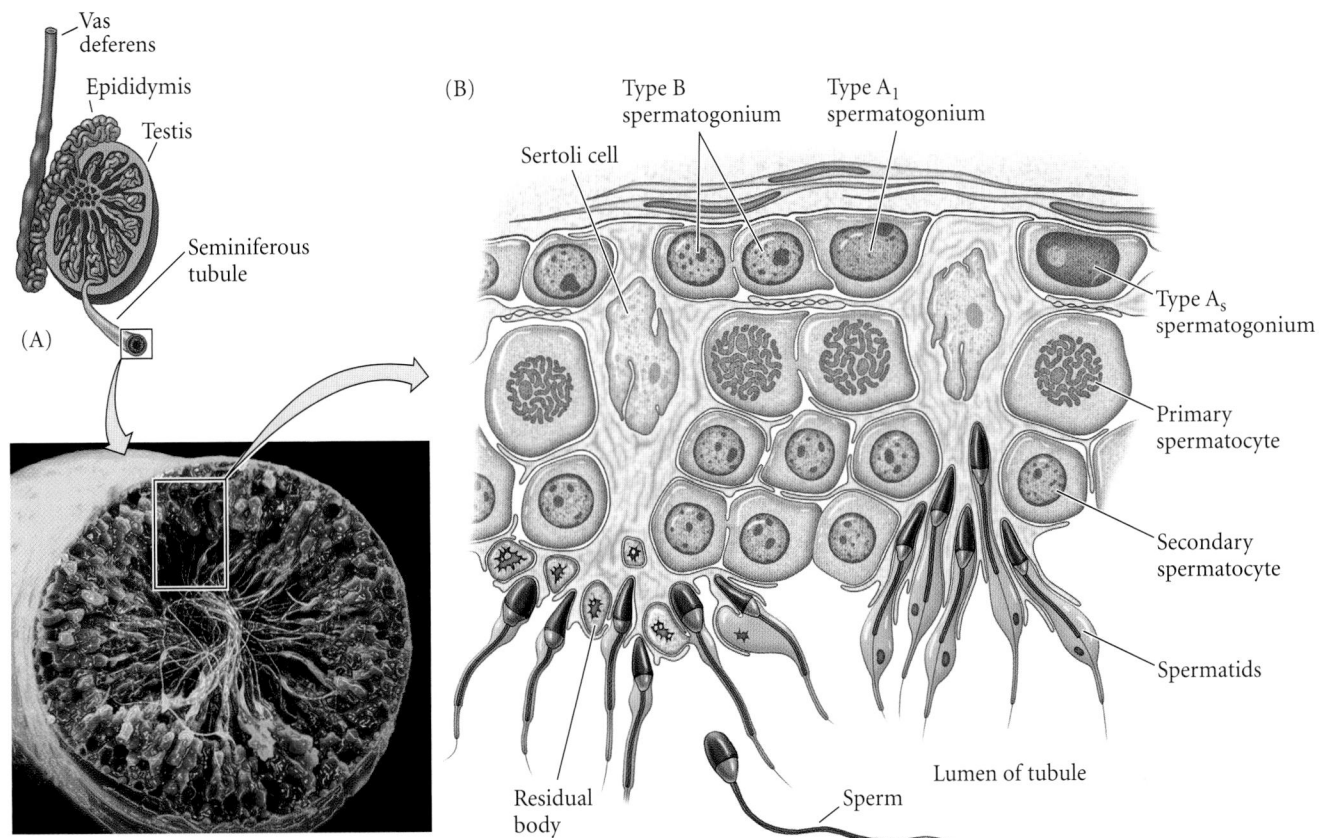

FIGURE 17.28 perm maturation. (A) Cross section of the seminiferous tubule. Spermatogonia are blue, spermatocytes are lavender, and the mature sperm appear yellow. (B) Simplified diagram of a portion of the seminiferous tubule, illustrating relationships between spermatogonia, spermatocytes, and sperm. As these germ cells mature, they progress toward the lumen of the seminiferous tubule. (A photograph courtesy of R. Wagner; B based on Dym 1977.)

directly to the Sertoli cells, which will nourish the developing sperm (Newton et al. 1993; Pratt et al. 1993; Kanatsu-Shinohara et al. 2008). The mitotic proliferation of these stem cells amplifies the small population of gonocytes into a population of type A spermatogonia that can generate more than 1000 sperm per second in adult human males (Matson et al. 2010).

THE MEIOTIC PHASE: HAPLOID SPERMATID The meiotic phase of spermatogenesis during puberty is regulated by several factors, including the synthesis of BMPs by the spermatogonia and the synthesis of retinoic acid by the Sertoli cells or other tissues. When BMP8b reaches a critical concentration, the germ cells begin to differentiate, and mice lacking BMP8b do not initiate spermatogenesis at puberty (Zhao et al. 1996; Carlomagno et al. 2010). As mentioned earlier in this chapter, retinoic acid is also critical in the switch from mitotic spermatogonia to meiotic spermatocytes.

The undifferentiated type A_s spermatogonia are the sperm stem cells, and at puberty they divide such that they generate another A_s spermatogonia as well as a **type A_1 spermatogonium**. Type A_1 spermatogonia have high levels of Stra8

transcription factor and are committed to a meiotic pathway (de Rooij and Russell 2000; Nakagawa 2010; Griswold et al. 2012). They undergo five mitotic divisions but keep cytoplasmic connections between themselves. These cells form a syncytium in which each cell communicates with the others via cytoplasmic bridges about 1 μm in diameter (Dym and Fawcett 1971). Thus, there are chains of 2–32 linked A_1 spermatogonia in human males (see Aponte et al. 2006). These cells divide to produce **type B spermatogonia** (FIGURE 17.29). Type B spermatogonia are the precursors of the spermatocytes and are the last cells of the line that undergo mitosis. They divide once to generate the **primary spermatocytes**—the cells that enter meiosis.

The transition between spermatogonia and spermatocytes appears to be mediated by the opposing influences of glial cell line-derived neurotrophic factor (GDNF) and stem cell factor (SCF), both of which are secreted by the Sertoli cells. GDNF levels determine whether the dividing spermatogonia remain spermatogonia or enter the pathway to become spermatocytes. Low levels of GDNF favor differentiation of the spermatogonia, whereas high levels favor self-renewal of the stem cells (Meng et al. 2000). SCF promotes the transition to spermatogenesis (Rossi et al. 2000). Since both GDNF and SCF are upregulated by follicle-stimulating hormone (FSH), these two factors may serve as a link between the Sertoli cells and the endocrine system, and they provide a mechanism for FSH to instruct the testes to produce more sperm (Tadokoro et al. 2002). Keeping the stem cells in equilibrium—producing neither too many

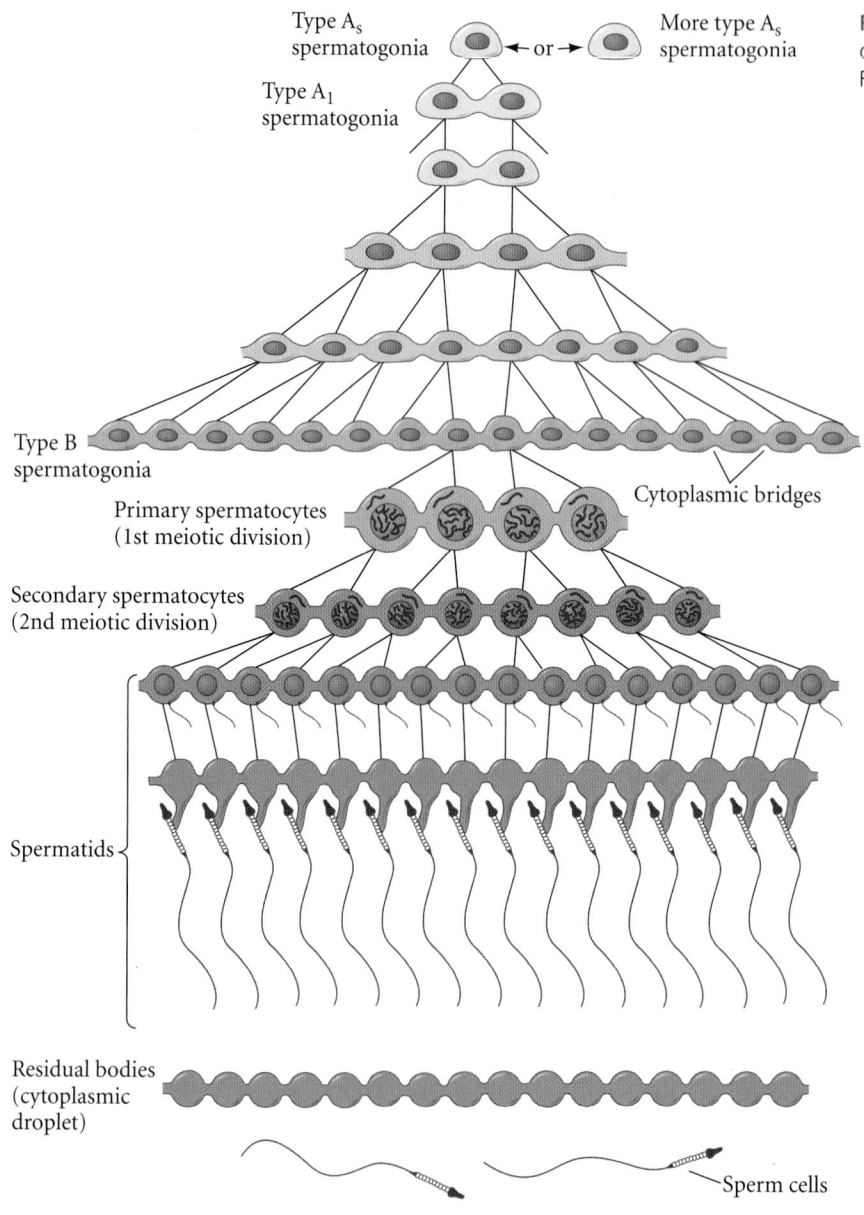

Type A$_s$ spermatogonia — or → More type A$_s$ spermatogonia

Type A$_1$ spermatogonia

Type B spermatogonia

Primary spermatocytes (1st meiotic division)

Cytoplasmic bridges

Secondary spermatocytes (2nd meiotic division)

Spermatids

Residual bodies (cytoplasmic droplet)

Sperm cells

FIGURE 17.29 Formation of syncytial clones of human male germ cells. (After Bloom and Fawcett 1975.)

During the divisions from type A$_1$ spermatogonia to spermatids, the cells move farther and farther away from the basal lamina of the seminiferous tubule and closer to its lumen (see Figure 17.28; Siu and Cheng 2004). Thus, each type of cell can be found in a particular layer of the tubule. The spermatids are located at the border of the lumen, and here they lose their cytoplasmic connections and differentiate into spermatozoa. In humans, the progression from spermatogonial stem cell to mature spermatozoa takes 65 days (Dym 1994).

The processes of spermatogenesis require a very specialized network of gene expression (Sassone-Corsi 2002). Not only are histones substantially remodeled and replaced by sperm-specific variants (see below), but even the basal RNA polymerase II transcription factors are exchanged for sperm-specific variants. The TFIID complex, which functions in the recognition of RNA polymerase II and contains numerous subunits. One of these subunits, TAF4b, is a sperm-specific TAF required for mouse spermatogenesis (Falender et al. 2005). Without this factor, the spermatogonial stem cells fail to make Ret (the receptor for GDNF) or the Luxoid transcription factor, and spermatogenesis fails to occur.

SPERMIOGENESIS: DIFFERENTIATION OF THE SPERM The mammalian haploid spermatid is a round, unflagellated cell that looks nothing like the mature vertebrate sperm. The next step in sperm maturation, then, is spermiogenesis (sometimes called spermateliosis), the differentiation of the sperm cell. For fertilization to occur, the sperm has to meet and bind with an egg, and spermiogenesis prepares the sperm for these functions of motility and interaction. The process of mammalian sperm differentiation was shown in Figure 4.2. The first step is the construction of the acrosomal vesicle from the Golgi apparatus, a process about which we know very little (see Berruti and Paiardi 2011). The acrosome forms a cap that covers the sperm nucleus. As the acrosomal cap is formed, the nucleus rotates so that the cap will be facing the basal lamina of the seminiferous tubule. This rotation is necessary because the flagellum, which is beginning to form from the centriole on

undifferentiated cells nor too many differentiated cells—is not easy. Mice with the *luxoid* mutation are sterile because they lack a transcription factor that regulates this division. All their spermatogonia become sperm at once, leaving the testes devoid of stem cells (Buaas et al. 2004; Costoya et al. 2004).

Each primary spermatocyte undergoes the first meiotic division to yield a pair of **secondary spermatocytes**, which complete the second division of meiosis. The haploid cells thus formed are called **spermatids**, and they are still connected to one another through their cytoplasmic bridges. The spermatids that are connected in this manner have haploid nuclei but are functionally diploid, since a gene product made in one cell can readily diffuse into the cytoplasm of its neighbors (Braun et al. 1989).

the other side of the nucleus, will extend into the lumen. During the last stage of spermiogenesis, the nucleus flattens and condenses, the remaining cytoplasm (the residual body, or cytoplasmic droplet; see Figure 17.29) is jettisoned, and the mitochondria form a ring around the base of the flagellum.

During spermiogenesis, the histones of the spermatogonia are often replaced by histone variants, and widespread nucleosome dissociation takes place. This remodeling of nucleosomes might also be the point at which the PGC pattern of methylation is removed and the male genome-specific pattern of methylation is established on the sperm DNA (see Wilkins 2005). As spermiogenesis ends, the histones of the haploid nucleus are eventually replaced by protamines.* This replacement results in the complete shutdown of transcription in the nucleus and facilitates the nucleus assuming an almost crystalline structure (Govin et al. 2004). The resulting sperm then enter the lumen of the seminiferous tubule.

These later stages of spermatogenesis must occur in an enviroment of high testosterone. If testosterone is withdrawn from adult mice (or if the testosterone receptor is inactivated), those spermatocytes around spermatids about to undergo spermiogenesis detach prematurely from the Sertoli cells due to lack of adhesion molecules, and those mature spermatozoa about to be released remain bound to the Sertoli cells and are phagocytized (De Gendt et al. 2004; Holdcraft and Braun 2004; Walker 2011).

In the mouse, development from stem cell to spermatozoon takes 34.5 days: the spermatogonial stages last 8 days, meiosis lasts 13 days, and spermiogenesis takes another 13.5 days. Human sperm development takes nearly twice as long. Each day, some 100 million sperm are made in each human testicle, and each ejaculation releases 200 million sperm. Unused sperm are either resorbed or passed out of the body in urine. During his lifetime, a human male can produce 10^{12} to 10^{13} sperm (Reijo et al. 1995).

● See **WEBSITE 17.4** The Nebenkern

Oogenesis in mammals

Mammalian oogenesis (egg production) differs greatly from spermatogenesis. The eggs mature through an intricate coordination of hormones, paracrine factors, and tissue anatomy. In the human embryo, the thousand or so oogonia reaching the developing ovary divide rapidly from the second to the seventh month of gestation. They generate roughly 7 million oogonia (**FIGURE 17.30**). Most of these oogonia die soon afterward, but the surviving oogonia initiate meiosis, become **primary oocytes**, and remain in the diplotene stage of first

---

*Protamines are relatively small proteins that are over 60% arginine. Transcription of the genes for protamines is seen in the early haploid spermatids, although translation is delayed for several days (Peschon et al. 1987). The replacement, however, is not complete, and "activating" nucleosomes, having trimethylated H3K4, cluster around developmentally significant loci, including Hox gene promoters, certain microRNAs, and imprinted loci that are paternally expressed (Hammoud et al. 2009).

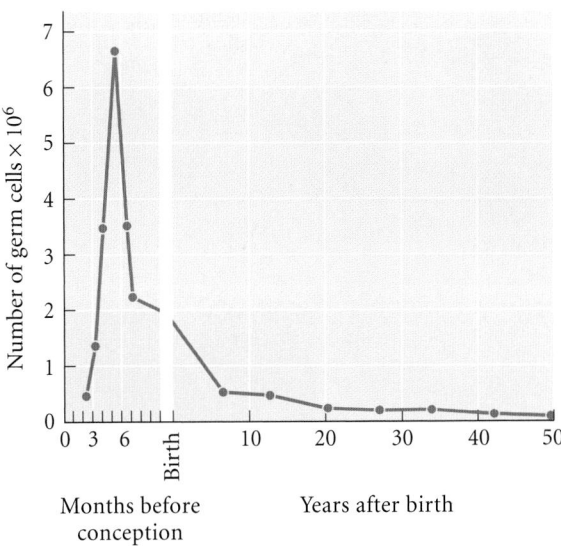

FIGURE 17.30 The number of germ cells in the human ovary changes over the life span. (After Baker 1970.)

meiotic prophase (Pinkerton et al. 1961). This prolonged diplotene stage is sometimes referred to as the **dictyate resting stage**. With the onset of puberty, groups of oocytes periodically resume meiosis. At that time, **luteneizing hormone** (**LH**) from the pituitary gland releases this block and permits these oocytes to resume meiotic division (Lomniczi et al. 2013). They complete first meiotic division and proceed to second meiotic metaphase, when the **secondary oocyte** is ovulated. After the secondary oocyte is released from the ovary, meiosis will resume if fertilization occurs. At fertilization, calcium ions are released in the egg, and (as in the frog), these calcium ions release the inhibitory block and allow the haploid nucleus to form.

The biochemistry of this meiotic regulation is intimately connected to ovarian anatomy. Each oocyte is enveloped by a primordial follicle consisting of a single layer of epithelial granulosa cells and a less organized layer of mesenchymal thecal cells (**FIGURE 17.31**). Periodically, a group of primordial follicles enters a stage of follicular growth. During this time, the oocyte undergoes a 500-fold increase in volume (corresponding to an increase in oocyte diameter from 10 μm in a primordial follicle to 80 μm in a fully developed follicle). FSH encourages the growth of the follicle and also induces receptors for LH on the outer follicle cells (Peng et al. 1991; Eppig et al. 1997).

Concomitant with oocyte growth is an increase in the number of **granulosa cells**, which form concentric layers around the oocyte. This proliferation of granulosa cells is mediated by a paracrine factor, GDF9, a member of the TGF-β family (Dong et al. 1996). Interestingly, mutations of GDF9 produce a variety of phenotypes from premature ovarian failure to a proponsity to have fraternal twins (i.e., to ovulate two instead of one oocyte) (Palmer et al. 2006; Otsuka 2011). Throughout this growth period, the oocyte remains

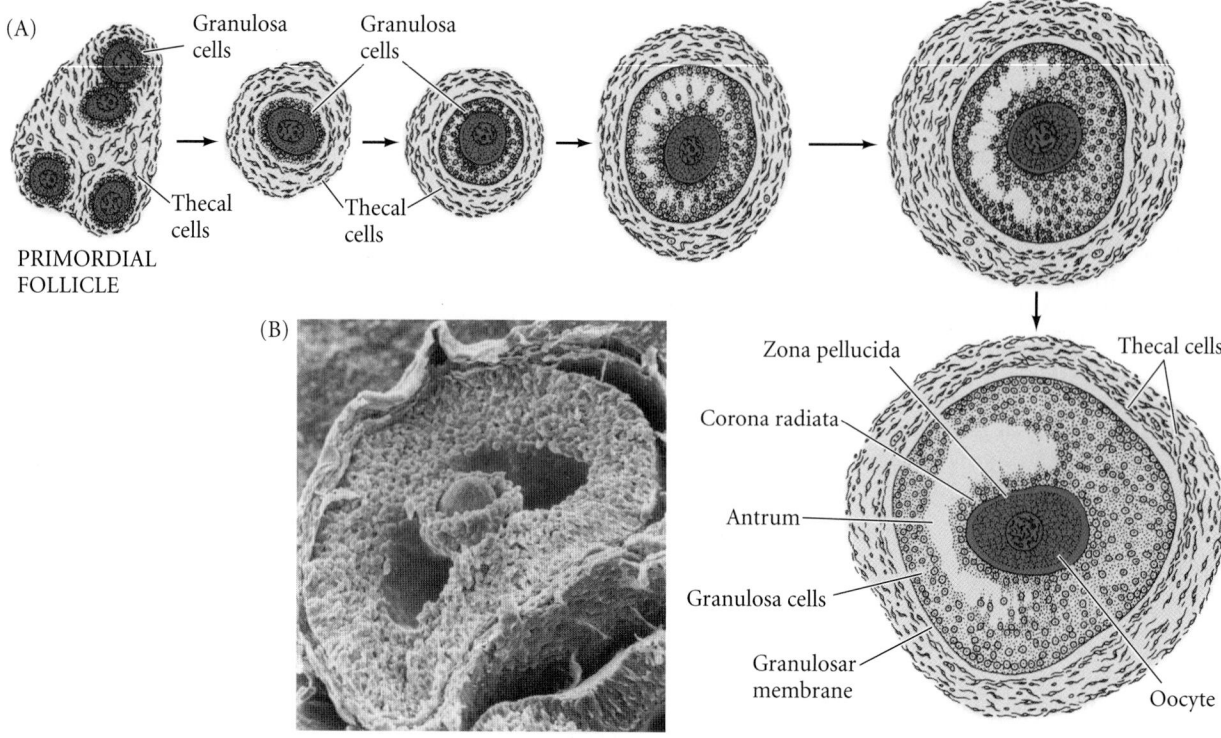

(A)

Granulosa cells · Granulosa cells · Thecal cells · Thecal cells

PRIMORDIAL FOLLICLE

Zona pellucida · Thecal cells
Corona radiata
Antrum
Granulosa cells
Granulosar membrane · Oocyte

GRAAFIAN FOLLICLE

FIGURE 17.31 The ovarian follicle of mammals. (A) Maturation of the ovarian follicle. When mature, it is often called a Graafian follicle. (B) Scanning electron micrograph of a mature follicle in the rat. The oocyte (center) is surrounded by the smaller granulosa cells that will make up the cumulus. (A after Carlson 1981; B courtesy of P. Bagavandoss.)

in the dictyate stage. The fully grown follicle thus contains a large oocyte surrounded by several layers of granulosa cells. The innermost of these cells will stay with the ovulated egg, forming the **cumulus**, which surrounds the egg in the oviduct. In addition, during the growth of the follicle, an **antrum** (cavity) forms and becomes filled with a complex mixture of proteins, hormones, and other molecules.

● **See WEBSITE 17.5** Hormones and mammalian egg maturation

Oocytes are maintained in the dictyate stage by the outer layer of ovarian follicle cells. Indeed, as early as 1935, Pincus and Enzmann discovered that removal of mouse oocytes (or the oocytes with their attached cumulus cells) from the follicle caused the reinitiation of meiosis. The inhibitory signal from the outer granulosa cells to the oocyte is cyclic GMP (cGMP). This cGMP is made by the outer granulosa cells of the follicle, transported through gap junctions between the follicle cells, and delivered by the follicle cells closest to the egg (the cumulus cells) by gap junctions to the oocyte (**FIGURE 17.32A**). Once in the oocyte, cGMP blocks phosphodiesterase enzymes from degrading cyclic AMP (cAMP). The cAMP blocks meiotic progression by maintaining protein kinase A (PKA) in an active state. PKA phosphorylates the Cdc25 protein, thus inactivating the activator of MPF, which

induces cell division. PKA also phosphorylates Wee1, thereby activating this inhibitor of MPF (see Figure 7.32C; Mehlmann et al. 2004; Han and Conti 2006).

When the LH from the pituitary gland binds to the outer follicle cells, the activated LH receptors lower the concentration of cGMP in the follicle cells and close the gap junctions between the follicle cells (**FIGURE 17.32B**; Eppig 1982; Norris et al. 2008, 2009). As a result, the cAMP is degraded. In the relative absence of cAMP, two critical events can now occur. First, MPF is reactivated in the oocyte, and this kinase can then phosphorylate the proteins that lead to the resumption of meiosis. Second, the myosin that interacts with the actin to place the meiotic divisions near the cortex of the oocyte is activated (Chen et al. 2012). Thus, the first meiotic division is completed and the second meiotic division is begun.

Meiosis is halted at second meiotic metaphase by cytostatic factor (CSF), which binds to MPF. The egg is ovulated at this stage, and fertilization occurs when the egg is in metaphase, separating off the second polar body. When the waves of calcium ions sweep across the egg during fertilization (see Chapter 4), the CSF is degraded and meiotic division resumes (Liu et al. 2005; Rauh et al. 2005; Shoji et al. 2006).

Ovulation is the process whereby an egg is released from the ovary. In mammals such as mice and humans, the oocyte is expelled from the ovary surrounded by the cumulus cells that are linked to it by gap junctions. This is called the **cumulus-oocyte-complex**. Like the resumption of meiosis, mammalian ovulation is initiated by the surge in LH. LH does not do the job alone. Rather, LH activates the gene for the progesterone receptor, and in the ovary, the only cells

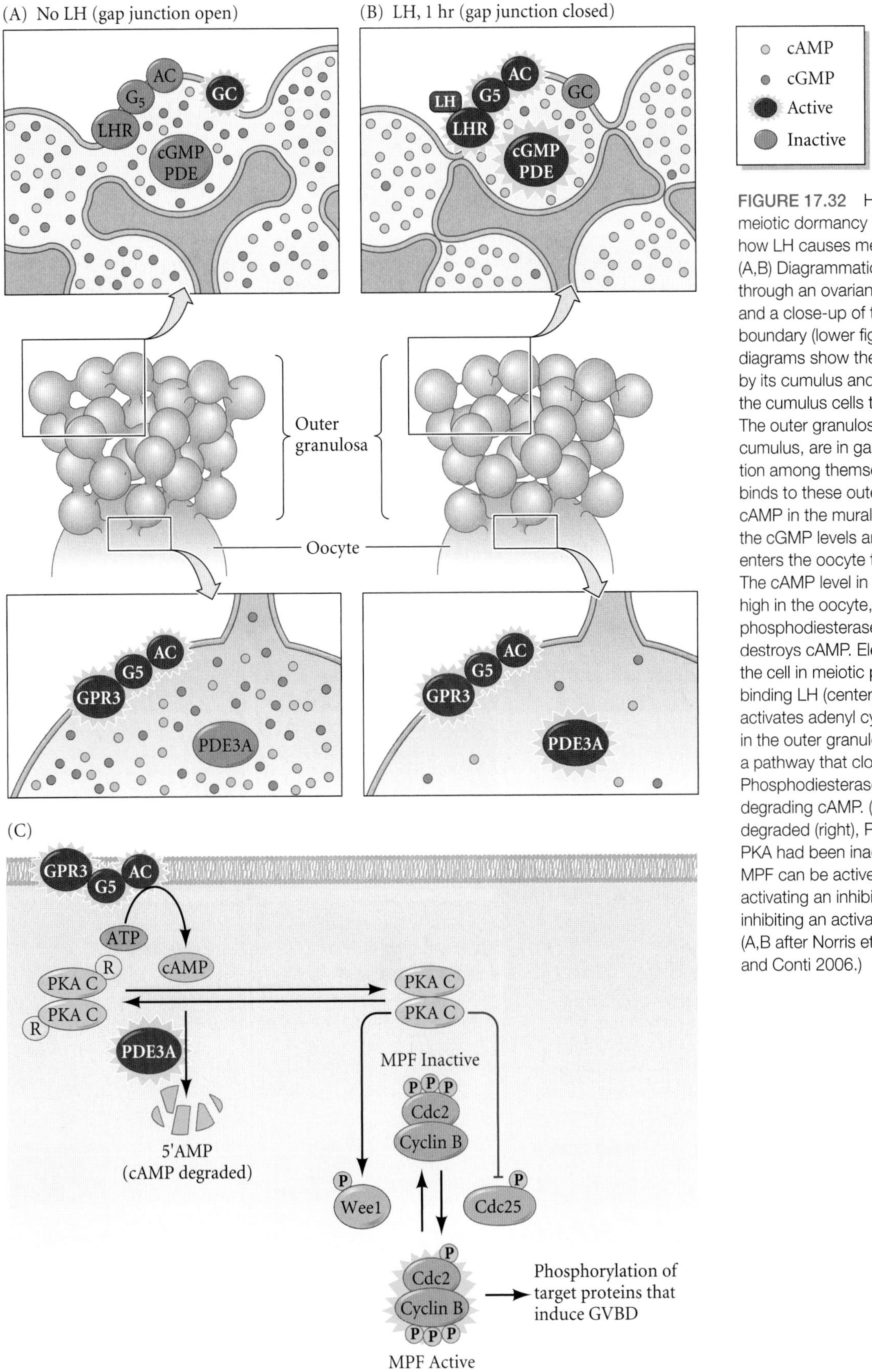

(A) No LH (gap junction open)

(B) LH, 1 hr (gap junction closed)

Outer granulosa

Oocyte

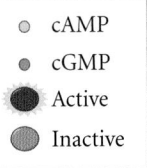

(C)

GPR3
G5
AC

ATP

PKA C
R

R
PKA C

cAMP

PDE3A

5'AMP
(cAMP degraded)

PKA C

PKA C

MPF Inactive

P P P
Cdc2
Cyclin B

P
Wee1

P
Cdc25

P
Cdc2
Cyclin B
P P P

MPF Active

Phosphorylation of target proteins that induce GVBD

FIGURE 17.32 Hypothesis for how meiotic dormancy is maintained and how LH causes meiotic resumption. (A,B) Diagrammatic views of a segment through an ovarian follicle (upper figures) and a close-up of the cumulus-ovary boundary (lower figures). The middle diagrams show the oocyte surrounded by its cumulus and communicating with the cumulus cells through gap junctions. The outer granulosa cells, above the cumulus, are in gap junction communication among themselves. (A) Before LH binds to these outer granulose cells, the cAMP in the mural granulosa is low and the cGMP levels are high. The cGMP enters the oocyte through gap junctions. The cAMP level in the oocyte is also high in the oocyte, since cGMP inhibits phosphodiesterase 3A (PDE3A), which destroys cAMP. Elevated cAMP keeps the cell in meiotic prophase. (B) Upon binding LH (center), the LH receptor activates adenyl cyclase, elevating cAMP in the outer granulosa cells. This initiates a pathway that closes the gap junctions. Phosphodiesterase can now function, degrading cAMP. (C) Once cAMP is degraded (right), PKA is no longer active. PKA had been inactivating MPF, but now MPF can be active. PKA had also been activating an inhibitor of MPF (Wee1) and inhibiting an activator of MPF (Cdc25B). (A,B after Norris et al. 2009; C after Han and Conti 2006.)

expressing the progesterone receptor are those granulosa cells of follicles mature enough to ovulate. The binding of progesterone (which is made in the ovary) to these receptors activates several genes that promote ovulation. These genes encode proteases that facilitate the rupture of the oocyte and cumulus, as well as genes involved in fluid accumulation and muscle contraction to propel the oocyte-cumulus complex into the oviduct (see Robker et al. 2009; Brown et al. 2010).

Once ovulation occurs, the remaining follicle cells (called the **corpus luteum**) synthesize large amounts of progesterone. This production of progesterone by the corpus luteum is sustained by **human chorionic gonadotropin**, a hormone that is made in the trophoblast and used by the mother. Progesterone binds to receptors in the uterus to keep the uterus in a state that maintains the pregnancy.* It also acts in the hypothalamus to inhibit the release of LH and FSH (Soules et al. 1984). Therefore, once the trophoblast is formed, no further oocytes are ovulated. Control of ovulation is the basis of most birth control pills containing progesterone or progestins† (chemicals that bind to the progesterone receptor and mimic the functions of natural progesterone). Moreover, progesterone stimulates the uterine lining to be receptive to the embryo.

Thus, in the human female, the first part of meiosis begins in the embryo, and the signal to resume meiosis is not given until roughly 12 years later. In fact, some oocytes are maintained in meiotic prophase for four decades. Despite some controversial claims of oogonial stem cells (Johnson et al. 2004; Zou et al. 2009), the most recent evidence (Zhang et al. 2012) confirms the view that of the millions of primary oocytes present at birth, only about 400 mature during a woman's lifetime.

---

*How do elephants sustain a pregnancy for 22 months? Even though they have one embryo, elephant ovaries contain several corpora lutea that can make enough progesterone to get the pregnancy through its 640 days (Luedens et al. 2012.).

†Emergency contraceptive pills ("morning after pills"; Plan B) often use high doses of synthetic progesterone to prevent ovulation. There is controversy in the United States (see Spreng 2008; Plantz and Lantos 2011) stemming from the claims of some religious groups that Plan B is a possible abortion-producing agent that can prevent the adhesion of the embryo to the uterine lining. As of 2013, the data do not support that claim. Rather, when detailed measurements of ovulation were taken (Lalitkumar et al. 2007; Noe et al. 2011; Gemzell-Danielsson et al. 2013) it was found that the women who took the Plan B after ovulation became pregnant at the normal rates. Women who took Plan B prior to the day they would ovulate did not become pregnant. Moreover, the progestin in Plan B did not cause significant changes in uterine gene expression (Vargas et al. 2012). Two conclusions were drawn from these studies: (1) The only known mechanism for Plan B is to prevent ovulation, thereby working solely as a contraceptive (not producing abortions of existing embryos); and (2) the morning-after pill is not good birth control because if one has intercourse after ovulation for about five days, the egg is ovulated and ready to meet sperm. Emergency contraceptive pills using high doses of synthetic progesterone should not be confused with mifepristone (RU-486) which *prevents* progesterone binding to the endothelium and can cause abortions.

OOGENIC MEIOSIS Oogenic meiosis in mammals differs from spermatogenic meiosis not only in its timing but in the placement of the metaphase plate. When the primary oocyte divides, its nuclear envelope, breaks down, and the metaphase spindle migrates to the periphery of the cell. This asymmetric cytokinesis is directed through a cytoskeletal network composed chiefly of filamentous actin that cradles the mitotic spindle and brings it to the oocyte cortex by myosin-mediated contraction (Schuh and Ellenberg 2008). At telophase, one of the two daughter cells contains hardly any cytoplasm, whereas the other daughter cell retains nearly the entire volume of cellular constituents (**FIGURE 17.33**). The smaller cell is called the **first polar body**, and the larger cell is referred to as the secondary oocyte. During the second division of meiosis, a similar unequal cytokinesis takes place. Most of the cytoplasm is retained by the mature egg (the ovum), and a second polar body receives little more than a haploid nucleus. (In humans, the first polar body usually does not divide. It undergoes apoptosis around 20 hours after first meiotic division.) Thus, oogenic meiosis conserves the volume of oocyte cytoplasm in a single cell rather than splitting it equally among four progeny (Longo 1997; Schmerler and Wessel 2010).

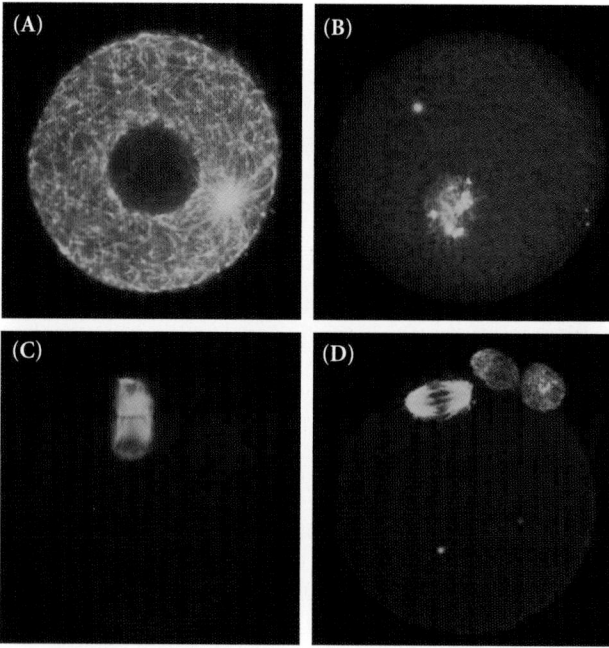

FIGURE 17.33 Meiosis in the mouse oocyte. The tubulin of the microtubules is stained green; the DNA is stained blue. (A) Mouse oocyte in meiotic prophase. The large haploid nucleus (the germinal vesicle) is still intact. (B) The nuclear envelope of the germinal vesicle breaks down as metaphase begins. (C) Meiotic anaphase I, wherein the spindle migrates to the periphery of the egg and releases a small polar body. (D) Meiotic metaphase II, wherein the second polar body is given off (the first polar body has also divided). (From De Vos 2002, courtesy of L. De Vos.)

The retention of the oocyte in the ovary for decades has profound medical implications. A large proportion, perhaps even a majority, of fertilized human eggs have too many or too few chromosomes to survive. Genetic analysis has shown that usually such **aneuploidy** (incorrect number of chromosomes) is due primarily to errors in oocyte meiosis (Hassold et al. 1984; Munne et al. 2007). Indeed, the percentage of babies born with aneuploidies increases greatly with maternal age. Women in their 20s have only a 2–3% chance of bearing a fetus whose cells contain an extra chromosome. This risk goes to 35% in women who are pregnant in their 40s (**FIGURE 7.34A**; Hassold and Chiu 1985; Hunt and Hassold 2010). The reason for this appears to be the breakdown of cohesin proteins (Chiang et al. 2010; Lister et al. 2010; Revenkova et al. 2010). Once made and assembled, cohesins remain on the chromosomes for decades, but they are gradually lost as the cell ages (**FIGURE 17.34B,C**). This loss of protein and function is accelerated as the cells become physiologically senescent.

Coda

We are now back where we began, with the stage set for fertilization to take place. The egg and the sperm will both die if they do not meet. As F. R. Lillie recognized in 1919, "The elements that unite are single cells, each on the point of death; but by their union a rejuvenated individual is formed, which constitutes a link in the eternal process of Life."

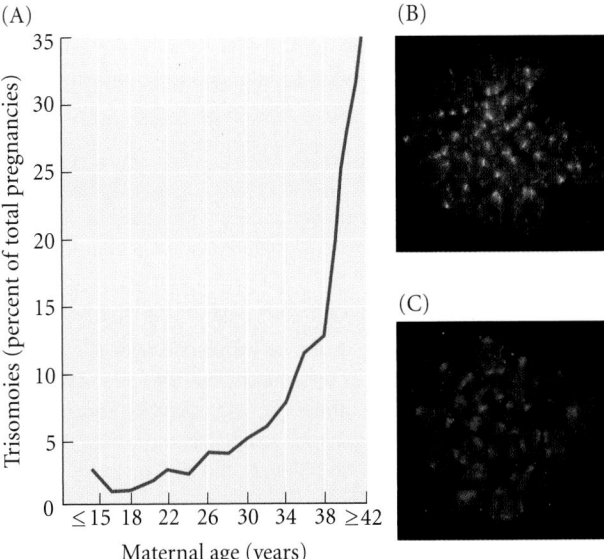

FIGURE 17.34 Chromosomal nondisjunction and meiosis. (A) Maternal age affects the incidence of trisomies in human pregnancy. (B,C) Reduction of chromosome-associated cohesin in aged mice. DNA (white) and cohesin (green) stained in oocyte nuclei of (B) 2-month-old (young) and (C) 14-month-old (aged, for a mouse) ovaries. A significant loss of cohesin can be seen (especially around the kinetochores) in aged mice. (A after Hunt and Hassold 2010; B,C after Lister 2010.)

SNAPSHOT SUMMARY: The Saga of the Germ Line

1. The precursors of the gametes are the primordial germ cells (PGCs). In most species (*C. elegans* being an exception), the PGCs form outside the gonads and migrate into the gonads during development.

2. In many species, a distinctive germ plasm exists. It often contains Vasa, Tudor, Piwi, and Nanos proteins or the mRNAs encoding them, and it is nucleated by lineage-specific proteins such as Oskar in insects or Bucky Ball in zebrafish.

3. In some species, such as *Drosophila* and *C. elegans*, germ cells are specified autonomously, and cells inheriting the germ plasm become germ cells. In other species, including mammals, the germ plasm proteins must be induced by neighboring tissues.

4. The germ plasm in many species contains inhibitors of transcription and translation, such that the PGCs derived from them are thought to be both translationally and transcriptionally silent.

5. In amphibians, the germ cells migrate from the posterior larval gut into the gonads on fibronectin matrices. In mammals, a similar migration is seen, and fibronectin pathways may also be used. Stem cell factor (SCF) is critical in this migration, and the germ cells proliferate as they travel.

6. In birds, the germ plasm is first seen in the germinal crescent. The germ cells migrate through the blood, then leave the blood vessels and migrate into the genital ridges.

7. In zebrafish, the germ cell determinants enter specific cells that are attracted to the gonad by a gradient of chemoattractants such as the Sdf1 protein.

8. Germ cell migration in *Drosophila* occurs in several steps involving transepithelial migration, repulsion from the endoderm, and attraction to the gonads.

9. In most organisms studied, the coordination of germline sex (sperm/egg) is coordinated to somatic sex (male/female) by signals coming from the gonad (testis/ovary).

10. Once the germ cells reach the gonads, they may initiate meiosis. The timing and details of this process depend on the species and sex of the organism.

11. In humans and mice, germ cells entering ovaries initiate meiosis while in the embryo; germ cells entering testes do not initiate meiosis until puberty.

12. Retinoic acid is a major inducer of mammalian meiosis, upregulating the Stra8 transcription factor.

13. Before meiosis, the DNA is replicated and the resulting sister chromatids remain bound at the kinetochore.

Homologous chromosomes are connected through the synaptonemal complex.

14. The first division of meiosis separates the homologous chromosomes. The second division of meiosis splits the kinetochore and separates the chromatids.

15. Spermatogenic meiosis in mammals is characterized by the production of four gametes per meiosis and by the absence of meiotic arrest. Oogenic meiosis is characterized by the production of one gamete per meiosis and by a prolonged first meiotic prophase to allow the egg to grow.

16. During insect oogenesis, the developing oocyte not only synthesizes numerous compounds, but also absorbs material produced by other cells. Moreover, it localizes many proteins and messages to specific regions of the cytoplasm, often tethering them to the cytoskeleton.

17. In *Drosophila*, nurse cells make mRNAs that enter the developing oocyte. Which of the cells derived from the primordial germ cell becomes the oocyte and which become

nurse cells is determined by the fusome and the pattern of divisions.

18. The *Xenopus* oocyte transcribes actively from lampbrush chromosomes during the first meiotic prophase.

19. In male mammals, the PGCs generate stem cells that last for the life of the organism. PGCs do not become stem cells in female mammals (although in many other animal groups, PGCs do become germ stem cells in the ovaries).

20. In female mammals, germ cells initiate meiosis and are retained in the first meiotic prophase (dictyate stage) until ovulation. In this stage, they synthesize mRNAs and proteins that will be used for gamete recognition and early development of the fertilized egg.

21. Mammalian oocyte maturation is regulated by FSH, LH, and progesterone hormones and involves connections between the oocyte and its surrounding cumulus cells.

22. In some species, meiosis is modified such that a diploid egg is formed. Such species can produce a new generation parthenogenetically, without fertilization.

For Further Reading

Complete bibliographical citations for all literature cited in this chapter can be found at the free-access website **www.devbio.com**

Bowles, J. and 11 others. 2006. Retinoid signaling determines germ cell fate in mice. *Science* 312: 596–600.

Decotto, E. and A. C. Spradling. 2005. The *Drosophila* ovarian and testis stem cell niches: Similar somatic stem cells and signals. *Dev. Cell* 9: 501–510.

Doitsidou, M. and 8 others. 2002. Guidance of primordial germ cell migration by the chemokine SDF-1. *Cell* 111: 647–659.

Ephrussi, A. and R. Lehmann. 1992. Induction of germ cell formation by oskar. *Nature* 358: 387–392.

Ewen-Campen, B., E. E. Schwager and C. G. Extavour. 2010. The molecular machinery of germ line specification. *Mol. Reprod. Dev.* 77: 3–18.

Gemzell-Danielsson, K., C. Berger and P. G. L. Lalitkumar. 2013. Emergency contraception: Mechanism of action. *Contraception* 87: 300–308.

Griffin, E. E., D. J. Odde and G. Seydoux. 2011. Regulation of the MEX-5 gradient by a spatially segregated kinase/phosphatase cycle. *Cell* 146: 955–968.

Hayashi, Y., M. Hayashi and S. Kobayashi. 2004. Nanos suppresses somatic cell fate in *Drosophila* germ line. *Proc. Natl. Acad. Sci. USA* 101: 10338–10342.

Lalitkumar, P. G. and 6 others. 2007. Mifepristone, but not levonorgestrel, inhibits human blastocyst attachment to an in vitro

endometrial three-dimensional cell culture model. *Hum. Reprod.* 22: 3031–3037.

Lomniczi, A. and 9 others. 2013. Epigenetic control of female puberty. *Nature Neurosci.* 16: 281–289.

Molyneaux, K. A., J. Stallock, K. Schaible and C. Wylie. 2001. Time-lapse analysis of living mouse germ cell migration. *Dev. Biol.* 240: 488–498.

Norris, R. P. and 9 others. 2009. Cyclic GMP from the surrounding somatic cells regulates cyclic AMP and meiosis in the mouse oocyte. *Development* 136: 1869–1878.

Ohinata, Y. and 11 others. 2005. Blimp1 is a critical determinant of the germ cell lineage in mice. *Nature* 436: 207–213.

Richardson, B. E. and R. Lehmann. 2010. Mechanisms guiding primordial germ cell migration: Strategies from different organisms. *Nat. Rev. Mol. Cell. Biol.* 11: 37–49.

Seydoux, G., and S. Strome. 1999. Launching the germline in *Caenorhabditis elegans*: Regulation of gene expression in early germ cells. *Development* 126: 3275–3283.

Smallwood, S. A. and G. Kelsey. 2012. De novo DNA methylation: A germ cell perspective. *Trends Genet.* 28: 33–42.

Stewart, T. A. and B. Mintz. 1981. Successful generations of mice produced from an established culture line of euploid teratocarcinoma cells. *Proc. Natl. Acad. Sci. USA* 78: 6314–6318.

GO ONLINE

WEBSITE 17.1 Mechanisms of chromosome diminution. The somatic cells do not lose DNA randomly. Rather, specific regions of DNA are lost during chromosome diminution.

WEBSITE 17.2 The insect germ plasm. The insect germinal cytoplasm was discovered as early as 1911, when Hegner found that **removing** the posterior pole cytoplasm of beetle eggs caused sterility in the resulting adults.

WEBSITE 17.3 The Nebenkern. Sperm mitochondria are often highly modified to fit the streamlined cell. The mitochondria of flies fuse together to form a structure called the Nebenkern; this fusion is controlled by the *fuzzy onions* gene.

WEBSITE 17.4 Synthesizing oocyte ribosomes. Ribosomes are almost a "differentiated product" of the oocyte, and the *Xenopus* oocyte contains 20,000 times as many ribosomes as somatic cells do. Gene repetition and gene amplification are both used to transcribe these enormous amounts of rRNA.

WEBSITE 17.5 Hormones and mammalian egg maturation. To survive, the follicle and its oocyte have to "catch the wave" of gonadotropic hormone release. The hormones of the menstrual cycle synchronize egg maturation with the anatomical changes of the uterus and cervix.

VADE MECUM

Germ cells in the *Drosophila* embryo. In the Fruit Fly segment, a view of gametogenesis follows the primordial germ cells of the living *Drosophila* embryo from their formation as pole cells through gastrulation as they move from the posterior end of the embryo into the region of the developing gonad.

Gametogenesis in mammals. Stained sections of testis and ovary illustrate the process of gametogenesis, the streamlining of developing sperm, and the remarkable growth of the egg as it stores nutrients for its long journey. You can see this in movies and labeled photographs that take you at each step deeper into the mammalian gonad.

PART FOUR

SYSTEMS BIOLOGY

Expanding Developmental
Biology to Medicine, Ecology,
and Evolution

Systems biology is an attempt to redefine the study of life by focusing on its complexity and integration. Rather than looking at physical entities such as chromosomes, cells, organs, and organisms, the systems approach looks at biology in terms of the flow of information. Indeed, this information is embodied in the genes, cells, organs, and external environment, and the flow of information unifies these disparate entities (see Van Speybroeck et al. 2005; Madar et al. 2009; Edelman 2010). The systems approach is especially important when we study development, which is not only the result of molecular, cellular, and physiological processes but which is also embedded in evolution and the environment and has extensive connections to the fields of health and medicine.

Like biological systems, systems biology has several branches and has focused on different principles (see De Backer et al. 2010). Two basic types of systems biology have been applied to developmental biology (O'Malley and Dupre 2005). The first, here called "pragmatic systems biology," centers around gene networks and builds outward from a core of gene action. The second, "systems theory," is concerned with new properties that arise from these gene networks and how these new properties function in development. Both approaches deal with relationships and are committed to mathematical modeling.

Pragmatic Systems Biology: From Gene to System

According to Donna Haraway (2003), "The relation is the smallest unit of analysis." This is an excellent light in which to view the logic of development. For example, the pattern that distinguishes a gene from a mere sequence of DNA nucleotides is the relationship that determines whether or not the DNA sequence contains information. To be a gene, there needs to be (among other things) an enhancer and a promoter. And this means the DNA has to relate with transcription factors, RNA polymerase, and numerous other agents.

The relationship between a transcription factor and the DNA sequence to which it binds is one of the most important informational events in biology. It defines the underlying "reality" of developmental processes. For instance, there is an enhancer that causes reporter genes to become expressed in those cells on the interior (medial) but not the lateral portion of the vertebrate rib. (Koelzer and Klein 2006; Guenther et al. 2008). Even though there are no morphological structures defining the medial rib as being different from the lateral rib, the medial rib is perceived during development as a "real" entity. Similarly, we have seen compartments of *Drosophila* development defined by gene expression domains.

DEVELOPMENTAL GENE REGULATORY NETWORKS: KERNELS OF DEVELOPMENT   These relationships between transcription factors and their enhancers form patterns that help define the course that development follows. These patterns have been called "**kernels**" or **gene regulatory networks** (**GRNs**) (Davidson 2006; Erwin and Davidson 2009; Davidson and Erwin 2010). We discussed GRNs in sea urchin and vertebrate heart development. A simple GRN can be initiated by transcription factors differentially located in the zygote (such as β-catenin in sea urchin embryos or Bicoid in *Drosophila* embryos) or from different signaling molecules (such as the ventral cells of the *Drosophila* embryo receiving a Toll signal that activates the Dorsal transcription factor). Here, the initial input is different in two portions of the embryos, activating different regulatory genes, and the protein encoded by these regulatory genes will activate different batteries of structural genes (**FIGURE P4.1A**).

But this is a very imprecise network, and the cells might dedifferentiate after the initial signal ceases to function. This is not how development works. So Davidson and Erwin (2009) proposed that the next step in evolution was to put stabilizing features into this system that would (1) prevent the cells from dedifferentiating once the initiating signals were no longer given and (2) make a strict determination such that once a cell had become a defined cell type, it could not become any other type. There are no "in-between" states. We see this type of scheme in **FIGURE P4.1B**. Here genes 1 and 5 have the same roles as the transducers of localized initial inputs in Figure P4.1A. Genes 2 and 7, however, have a new function: they are activated by the products of genes 1 and 5, respectively, but they have positive feedback to these genes. Gene 1, for instance, activates gene 3, and gene 3 activates gene 2, whose product continues to activate gene 1 and gene 3. The product of gene 3 keeps all the other genes active as well. Thus, the signal that originally activated gene 1 can be dispensed with and the cell type remains. A similar positive feedback occurs for genes 5, 6, and 7 in the alternative cell type.*

The product of gene 2, however, activates not only gene 3, it also activates gene 4. And the product of gene 4

represses gene 5—the major regulatory gene of the alternative cell type. And in this other cell type, gene 6 activates gene 8, whose product represses the activity of gene 3. So not only are the two cell types stabilized, they become mutually exclusive. We saw such circuitry in mammalian sex determination (see Chapter 15), where the genes responsible for testes formation (such as *Sox9*) not only made testes but also inhibited the formation of ovaries, and the genes responsible for the ovary phenotype (such as β-catenin) not only initiated the formation of ovaries but also inhibited the formation of testes.

GRNs are generated by several sets of scientific data (Davidson et al. 2003), including:

1. Identity of the embryonic cell lineage
2. When and where specific genes are expressed
3. The mechanism(s) regulating gene expression
4. Gene expression measurements when genes are blocked or overexpressed

Integrating these data means integrating the results of multiple embryological experiments, in situ hybridizations of individual genes, ChIP-Seq and other biochemical analyses of gene regulatory elements, and microarray analysis of the entire GRN. Although the *cis*-regulatory transcriptional network predominates in these discussions, GRNs could also include differential RNA processing (as in *Drosophila* sex determination) or differential translation (as in *Drosophila* anterior-posterior polarity).

Such interactions need not exist only within a cell. Gradients of morphogens can inform cells as to where they are within a field. The cells must be able to perceive the signal amidst the background noise, transform quantitative input differences into qualitative outputs, and make imprecise boundaries precise. Systems theory is beginning to give us ideas of how cells might accomplish these critical tasks (see Lander 2013). Similarly, the reaction-diffusion kinetics discussed in Chapter 14 are critical for establishing order during morphogenesis. As we will see in Chapter 20, altering the synthesis, degradation, or diffusion of paracrine factors can alter such phenotypes as the number of digits, the striping or spotting pattern of pigment, and the number of cusps on teeth.

DEVELOPMENTAL GRAPH THEORY: PUTTING FLESH ONTO THE GENETIC SKELETON   As Davidson and Erwin mention in their 2009 review, GRNs represent only the "brain" that directs development. "Something is still missing, however, and that is the control of morphogenetic functions." The blueprint has to be built, the logic made flesh. This is where developmental graph theory contributes to the systems approach. For morphogenesis, the structure "noun-verb-noun" can be applied, where nouns can

---

*As stated in an anonymous MIT review (1989), "We all live in recursive subroutines, recursive subroutines, recursive subroutines."

(A)

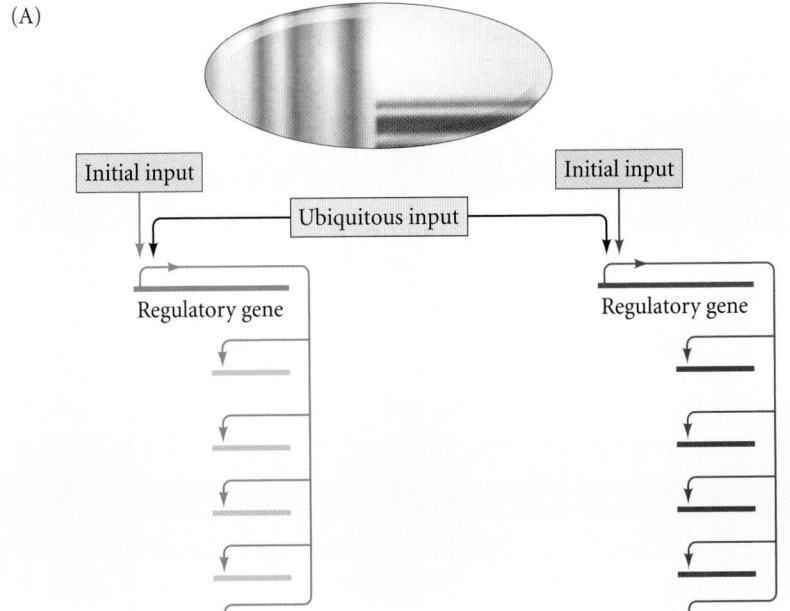

Initial input

Ubiquitous input

Initial input

Regulatory gene

Regulatory gene

FIGURE P4.1 Stages in the evolution of GRNs. (A) A simple GRN in which different spatial signals in a metazoan embryo activate a regulatory gene that activates the transcription of a battery of genes that characterize a cell fate. A ubiquitous input signal (such as RNA polymerase II and basal transcription factors) will activate transcription only if the specific initial input is present. (B) A more complex GRN with negatively interacting networks between modules and positive feedback within each module. Genes 1 and 5 have specific roles in receiving and transducing the initial signals (as in A). Genes 2 and 3 and genes 6 and 7 amplify the signal and also positively feedback on genes 1 and 5 such that the cell type is stable when the initial signals no longer function. Genes 4 and 8 repress the key gene establishing the alternative cell state. (After Davidson and Erwin 2009.)

(B)

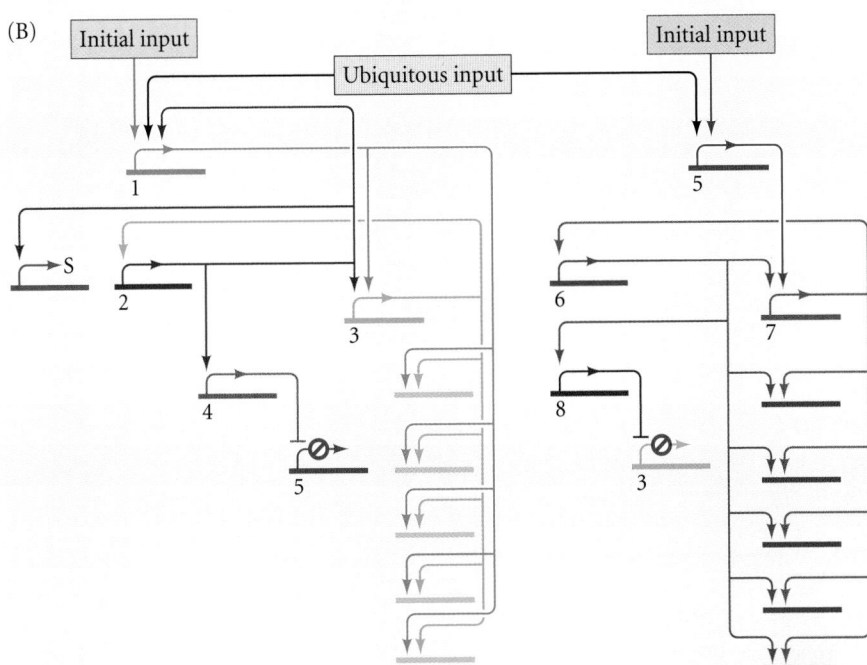

Initial input

Ubiquitous input

Initial input

be molecules, tissues, and even GRNs, and verbs can be processes such as movements, differentiations, proliferations, or apoptoses (**FIGURE P4.2A**; Bard 2011). One of the advantages of developmental graph theory is that it can integrate quantitative data into a pictorial framework, often revealing multiple causation. Another advantage is that placing the data into noun-verb-noun triplets can reveal gaps in

our knowledge and thereby lead to experimentation. Most importantly, though, this type of diagram can accommodate the morphogenetic paracrine relationships *between* cells rather than just the transcriptional relationships *within* cells. The graph in **FIGURE P4.2B** depicts the events leading to the formation of a capillary sprout. Notice that entire GRNs have become subroutines in larger frameworks.

(A)

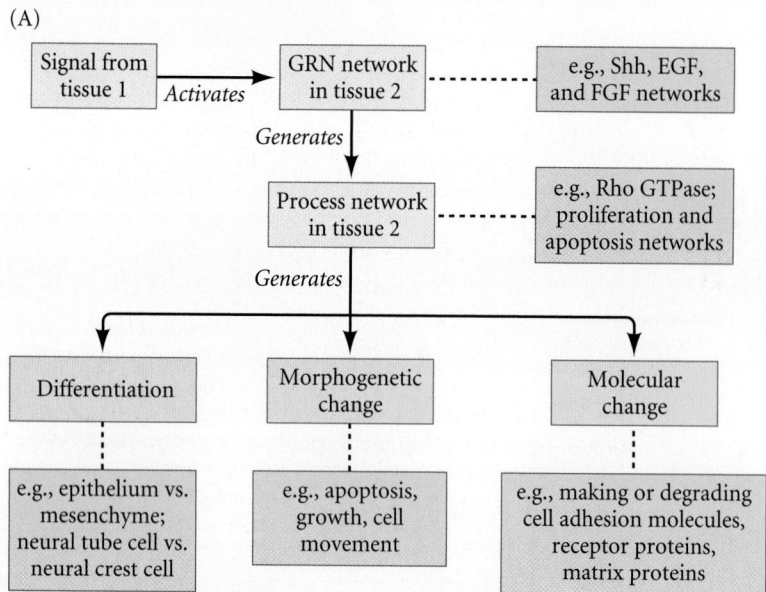

(B)

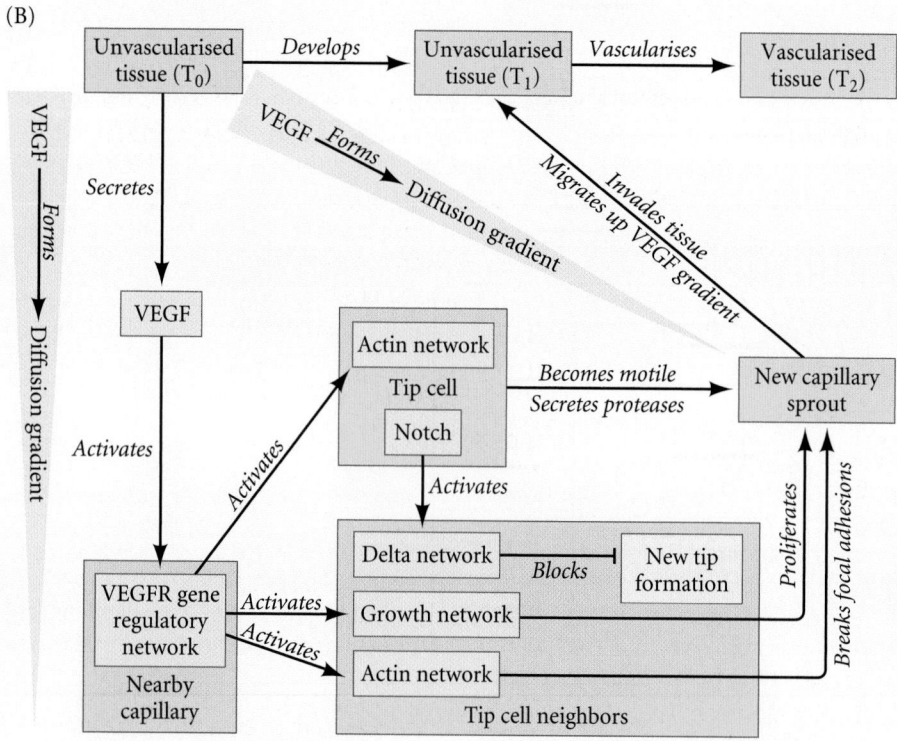

FIGURE P4.2 Developmental graph theory. (A) Graphic overview of the molecular and cellular responses to a developmental signal. Molecular networks (green boxes) connect to cellular events (blue boxes) through the processes indicated in italics. Examples are given in the lavender boxes. (B) Developmental graphic representation of the emergence of capillary sprouts during angiogenesis (see Chapter 13). At the molecular level (green), the tissue (blue) secretes VEGF that activates molecular networks in adjacent endothelial cells (blue). The processes that drive the developmental changes are in italics. (After Bard 2011.)

Systems Theory Biology: From System to Gene

The systems theory approach to biology attempts to understand the principles behind biological organization and to relate the different levels of the biological hierarchy (atom, molecule, gene, cell, tissue, organ, organism, ecosystem) to one another. Modern systems theory can be said to have had its start in developmental biology, which was one of the first sciences to apply these principles of causation, integration, and context dependency. The demonstration of embryonic regulation, where one blastomere of a 4-cell sea urchin embryo could become an entire larva and where the fate of a newt blastomere depended on its neighbors, caused a crisis in the field of experimental embryology. In response to the vitalism of Hans Driesch (who held that there must be some non-physical wisdom directing development) and the reductionism of Wilhelm Roux (who held that the embryo was basically a machine), Oskar Hertwig (1894) proposed a type of materialist philosophy called **wholist organicism**. This philosophy embraced the views that (1) the properties of the whole cannot be predicted solely from the properties of its component parts; and (2) the properties of the parts are informed by their relationship to the whole. Subsequently, several developmental biologists—notably Paul Weiss (1926) and Ludwig von Bertalanffy (1928, 1932)—formalized these notions into a systems approach to development (Brauckmann 2000; Drack et al. 2007).

Both the reductionist and organicist approaches are materialist in that they do not invoke any extramaterial agent (entelechy; the soul; *Bildungstrieb*) as directing development. However, whereas reductionism claims that all complex entities (including proteins, cells, organisms, and ecosystems) can be completely explained by the properties of their component parts, organicism claims that complex wholes are inherently greater than the sum of their parts in the sense that the properties of each part depend on the context of that part within the whole in which it operates. Thus, when we try to explain how the whole system behaves, we cannot get away with speaking only about the parts. These explanations are no less materialistic than reductionism. As Denis Noble (2006) has stated,

> Systems biology ... is about putting together rather than taking apart, integration rather than reduction. It requires that we develop ways of thinking about integration that are as rigorous as our reductionist programmes, but different. ...It means changing our philosophy, in the full sense of the term.

The systems theory approach to development has been largely influenced by Alfred North Whitehead (1929; Waddington 1969, 1975), who saw processes, not entities, as critically important and who viewed reality as a "concrescence" of these processes (that is, a coherent entity comes together through the interactions of its component processes). Although the emphasis differs widely among systems biologists, the theoretical systems approach can be characterized as having six principles that apply directly to biological analyses:

1. Context-dependent properties
2. Level-specific properties and emergence
3. Heterogeneous causation
4. Integration
5. Modularity and robustness
6. Homeorhesis

We will look briefly at each of these in light of animal development.

Context-Dependent Properties

The meaning or role of an individual component of a system depends on its context—that is, on what comes before and after, above and below it. What BMP4 does, for instance, depends on the history and context of the cell that recognizes it. At one time and place, BMP4 may signal bone formation; at another time and place, it may signal apoptosis; at yet another, it may specify the cell as epidermis.

Language provides analogies by which organicism can be understood (Collier 1985). Certain combinations of letters form words, and certain organizations of words form sentences. The meaning of a sentence obviously depends on the meaning of each component word (the parts define the whole). But most words have multiple and distinct meanings, and it is the context of the sentence that determines which specific meaning of a word is most appropriate (the whole defines the parts). Consider the following two sentences:

1. The party leaders were split on the platform.
2. Pirates mine the cloud for private files.

In these sentences, each word's meaning is suspended until the sentence is complete; the words and sentence mutually define each other. In other words, when put into relation with each other, specific meanings of words are singled out, resulting in a series of words with a particular ("emergent") meaning (i.e., a story is formed). Parts determine the whole, and the whole determines each of its parts.

In embryology, we are constantly aware of the parts being determined by their context within the whole (Needham 1943; Haraway 1976; Hamburger 1988; Gilbert and Sarkar 2000). Indeed, in vertebrate embryos, the fate of a cell is specified by its position.

Level-Specific Properties and Emergence

Parts are organized into wholes, and these wholes are often components of larger wholes. Moreover, at each biological level there are appropriate rules. In 1943 the embryologist Joseph Needham wrote,

> The deadlock [between mechanism and vitalism] is overcome when it is realized that every level of organization has its own regularities and principles, not reducible to those appropriate to lower levels of organization, nor applicable to higher levels, but at the same time in no way inscrutable or immune from scientific analysis and comprehension.

The properties of a system at any given level of organization cannot be totally explained by those of levels "below" it. Thus, temperature is not a property of an atom, but a property that emerges from an aggregate of atoms. Similarly, voltage potential is a property of a biological membrane but not of any of its molecular components. When you have an entity as complex as the cell, the fact that quarks have certain spins is irrelevant; one cannot necessarily "reduce" all the properties of body tissues to atomic phenomena.

To say that an entity cannot be explained by the components below it, however, is not to say that each level is *independent* of those below it. To the contrary, laws at one level may be almost deterministically compatible with those at lower levels, but they may also be dependent on levels above. This notion of level-specific interactive modules forms the basis of many new computer programs (Dyke 1988; Wimsatt 1995).

Higher-level properties emerge from lower-level activities and must be understood in the context of the whole. In our linguistic metaphor, emergent properties can be seen in the relationship of letters to words. Most letters standing alone have no intrinsic meaning. But when letters are grouped together in certain arrangements and according to certain rules, meaningful words emerge. In developmental biology, one often encounters such emergent properties. For example, in renal development, the nephron is formed by interactions between the ureteric bud and the metanephrogenic mesenchyme. Cultured separately, neither of these two tissues develops any portion of the kidney. If you place the tissues together, however, the ureteric bud tissue branches just as it would in the intact organism, and the mesenchyme cells form the ten cell types characteristic of the renal filtration apparatus. Thus, ten new cell types "emerge" from the interactions between two cell types, neither of which had any of the specific properties of the proximal convoluted tubule cells, juxtaglomerular cells, or Bowman's capsule cells.

Heterogeneous Causation

In biology, causation is seen as being both "upward," from the genes to the environment, and "downward," from the environment to the genes. What a cell is depends both on its genes and on the cells surrounding it (i.e., on input from both internal and external sources) and on its developmental history. Systems biology demands the integration of different types of explanations. It also calls for the integration of analysis (taking things apart) and synthesis (putting things together). To say that there is a "gene for" something is often poor shorthand for a large variety of factors (see Tauber and Sarkar 1992; Moss 2003). In another analogy, a faulty gas pedal will prevent a car from moving, but this is not to say that the gas pedal generates movement. The same phenotype can result from many different mechanisms.

In Chapters 18–20 we will see that "top-down" causation—that is, from the environment to the genome—is critical in medicine, ecology, and evolution. We will see

that a variety of substances present in the environment, both anthropogenic and natural, can adversely affect animal development. Conversely, we will see that symbionts (organisms that live in and on a second, host, organism) and other environmental agents are often crucial to the host's normal development. Whether an individual turtle becomes male or female depends on temperature; whether an aphid is red or green can depend on the bacteria it incorporates. And we will see how environmental forces interacting with developing organisms can drive the processes of biological evolution. Systems theory should unite the "arrows of causation" in ecosystem diagrams with the "arrows of causation" within the body and within the cell.

Integration

One of the chief components of systems biology is its insistence on synthesis as well as analysis. This is also one of the biggest sources of heterogeneity among those who consider themselves systems biologists. True to the notion of level-specific principles, there can be different systems biologies at different levels, but each of them will be synthetic.

For instance, on a single level—that of gene transcription—Davidson and colleagues have, as mentioned above, attempted to map all the gene inputs involved in the specification of the early sea urchin embryo. But this approach can be expanded by defining the "system" not as the cell or the organism itself, but as the cell within an organism within an ecosystem. This is done routinely in the subspecialty of obstetrics and gynecology called maternal-fetal medicine. Here, neither the mother nor the fetus is seen as an entity separate from the other. As we will see, the field of teratology (birth defects) concerns the modulation of cell behavior during development (differential gene transcription, cell specification and adhesion, etc.) by environmental agents mediated through the mother. Similarly, cancer research now sees many epithelial tumors as arising from defects in the mesenchymal cells surrounding the epithelium, and these defects can be caused by environmental agents (Soto et al. 2008, 2009). In these cases, the explanation of cell behavior is as much a function of "top-down causation" from the environment as it is "bottom-up causation" from the genome.

Modules and Robustness

The organism develops as a system of modules. **Developmental modules** have been defined as having discrete genetic specification, hierarchical organization, interactions with other modules, a particular physical location within a developing organism, and the ability to undergo transformations on both developmental and evolutionary time scales (Raff 1996; Schlosser and Wagner 2004; Callebaut and

Rasskin-Gutman 2009). **Robustness** (sometimes called **canalization**) refers to the ability of an organism to develop the same phenotype despite perturbations from the environment or from mutations. This is a function of interactions within and between modules.

Systems biology views the embryo as a complex adaptive system (Gell-Mann 1964; Edelman et al. 2010). It is "complex" because of the diversity of parts and the multiple interacting modules. It is "adaptive" because the component modules can modify their behaviors in response to environmental or genetic perturbations. Living beings are organized according to what philosopher Chuck Dyke (1988) calls "level-interactive modular arrays." This concept implies a nest of modular structures. Each entity is an organized array of constituent modular parts, and is at the same time the constituent of a larger module. The components of each module (which are themselves modules) interact to form the coherent (larger) module. Moreover, the modules interact with the levels above and below them. Just as each organelle must function to make a coherent cell, so each cell and cell type must function to make a coherent tissue. Thus, the tissue architecture of the liver can regulate elements in the modules below it (the liver extracellular matrix and intercellular interactions within the tissue regulating gene expression within the cell); and the lower-level modules such as those of the cell obviously determine the function of the higher-level modules such as the tissue.

As we will see, modularity allows anatomical change to occur in one module without necessarily affecting another module. This change during development is predicated on the modularity of enhancer sequences, mentioned in Chapter 2. Modularity also allows for robustness: the ability to retain the phenotype despite environmental or genetic changes. Robustness is a characteristic of living systems (Kitano 2004) and is responsible for the normal phenotype developing so typically, despite the millions of interactions that have to occur sequentially and at the right place during embryogenesis. This robustness is due to redundancy in control and to feedback loops between modules (von Dassow et al. 2000; Lander 2011).

Thus, Spemann (1927) spoke about "double assurance" (which he likened to wearing both a belt and suspenders) on the anatomical level. The eye's lens could form in more than one way, and if one tissue didn't induce the lens, the other tissue would. If the operculum did not open in its normal manner during frog metamorphosis, another system kicked in and opened that flap of skin. More recently, redundancy has been found at the level of genes. Muscles will form when the muscle-forming *MyoD* gene is knocked out of a mouse embryo, even if the details of the muscle development are not perfect (Wang et al. 1996). This is because *MyoD* usually downregulates the synthesis of *Myf5*, a *MyoD*-like gene that can also direct muscle development. In the absence of *MyoD*, *Myf5* is synthesized and can perform most of the functions of the absent molecule. The segment polarity genes of *Drosophila* are interconnected by a series of regulatory loops such that changing any one of them still gives proper segmentation 90% of the time.

Homeorhesis

C. H. Waddington pointed out in 1957 that applying systems approaches to development is going to be more difficult than applying systems approaches to physiology. Physiologists attempt to understand the interactions among parts that exist throughout the life of the organism; but the task of the developmental biologist is not to understand homeostasis—how the organism maintains itself. Rather, the task of the developmental biologist is to understand **homeorhesis**—how the organism stabilizes its different cell lineages while it is still constructing itself.

Waddington, who made substantial contributions both to developmental biology and to systems biology, viewed development as a mountainous landscape and the cell types as basins. Like a ball rolling down a hill, a cell eventually finds itself in one or another of these basins. Basins are constructed through the interactions of cells during development. Kauffman (1969, 1987, 1993) and others (Huang 2009a,b) see these basins as "attractor states," where a stable and robust configuration of gene expression becomes possible (**FIGURE P4.3**). The ability of a set of transcription factors to rearrange transcriptional networks to convert exocrine pancreas or liver cells into endocrine pancreas cells can be seen as an example of such an attractor state (Horb et al. 2003; Zhou et al. 2008). Most combinations of transcription factors and enhancers are expected to be unstable; those combinations that are stable, however, can become cell types. Huang and colleagues (2005, 2009a) have provided evidence based on systems analysis that the neutrophil (a white blood cell) is such a stable cell type. Their computer analysis connecting more than 2750 genes into networks found that the neutrophil expression pattern is the most stable pattern, and that the only "rival" to it is that of neutrophilic leukemia. The formations of new cell types (such as the neural crest cells of vertebrates) can be looked at as constructions of new stable attraction states over the course of evolution. The stem cell state of cell division so important for cancer may be another stable attractor state (Huang et al. 2009b).

Denis Noble (2006) has stated that the "ultimate goal of systems biology" is "to reconnect physiology and developmental models to theories of evolution." He also noted that we have just begun this enterprise. In the closing chapters of this book, we will examine the roles of development in disease, ecosystem dynamics, and evolution, keeping in mind the properties of systems.

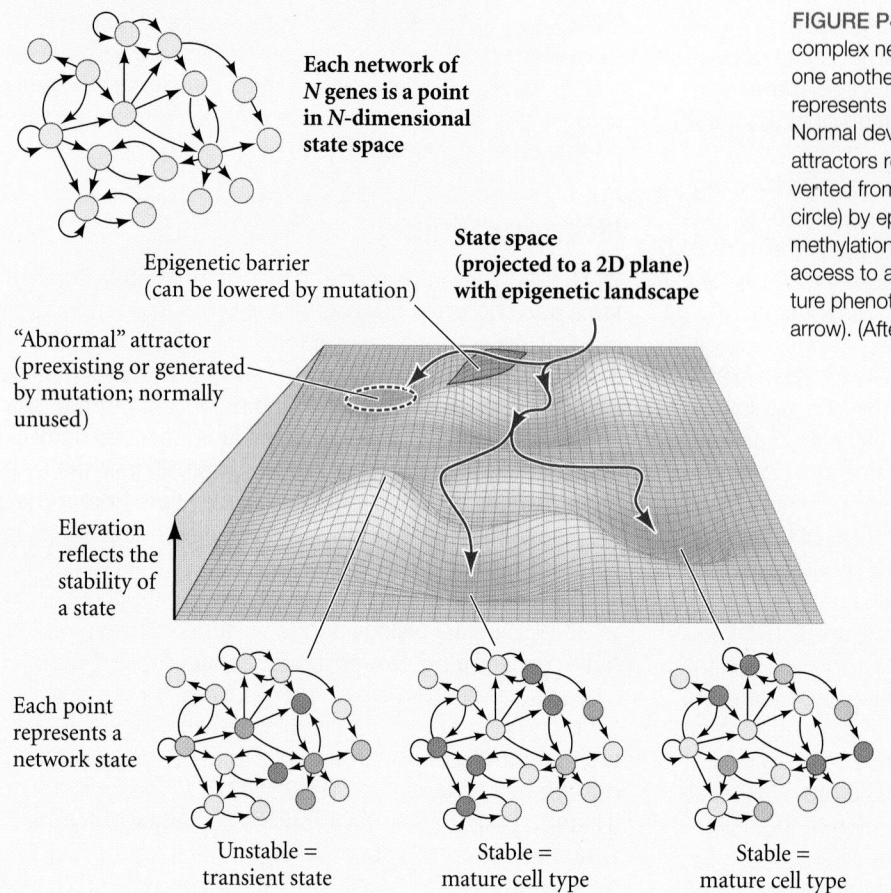

Each network of
N genes is a point
in *N*-dimensional
state space

Epigenetic barrier
(can be lowered by mutation)

State space
(projected to a 2D plane)
with epigenetic landscape

"Abnormal" attractor
(preexisting or generated
by mutation; normally
unused)

Elevation
reflects the
stability of
a state

Each point
represents a
network state

Unstable =
transient state

Stable =
mature cell type

Stable =
mature cell type

FIGURE P4.3 An "epigenetic landscape" for a complex network wherein *N* genes can interact with one another. The overall slope (from back to front) represents the progression of development over time. Normal developmental trajectories (blue lines) lead to attractors representing distinct cell types and are prevented from entering "abnormal attractors" (red dashed circle) by epigenetic barriers (lavender hill). Mutations or methylation differences can lower this barrier, opening access to attractors that encode an abnormal, immature phenotype, including cancerous cell types (red arrow). (After Huang et al. 2009b.)

18

Birth Defects, Endocrine Disruptors, and Cancer

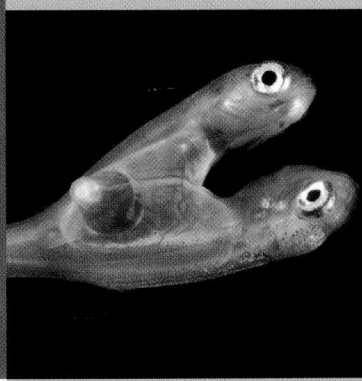

IT IS INDEED AMAZING THAT ANY OF US IS HERE, because relatively few human conceptions develop successfully to birth. Recent data (Mantzouratou and Delhanty 2011; Chavez et al. 2012) suggest that only 20–50% of human cleavage-stage embryos successfully implant into the uterus. It appears that many human embryos have chromosomal anomalies that are expressed so early the embryo fails to implant. These embryos are spontaneously aborted (miscarried), usually before a woman realizes she has conceived. Of those embryos that *do* implant successfully, studies from the 1980s suggest that only about 40% survive to term (Edmonds et al. 1982; Boué et al. 1985). Further studies (Winter 1996; Epstein 2008) estimate that approximately 2.5% of babies who do come to term have a recognizable birth defect.

With so many genes, cells, and tissues becoming organized simultaneously and changing together, it is not surprising that some developmental events do not happen properly. Although the body has remarkable back-up pathways and redundancies that permit a great deal of flexibility, abnormal phenotypes emerge when developmental phenomena are absent when they should be activated, or activated when they should be repressed. There are three major pathways to abnormal development:

1. *Genetic mechanisms.* Mutations in genes or changes in the number of chromosomes can alter development.

2. *Environmental mechanisms.* Agents (usually chemicals) from outside the body cause deleterious phenotypic changes by inhibiting or enhancing developmental signals.

3. *Stochastic (random) events.* Chance plays a role in determining the phenotype, and some developmental anomalies are just "bad luck" (Molenaar et al. 1993; Holliday 2005; Smith 2011).

Most of this chapter will deal with genetic and environmental effects.* We will start, however, by briefly examining the role of random events.

The Role of Chance

Although physicians and researchers often parse developmental anomalies into those caused by internal (genetic) versus those caused by external (environmental) agents, more consideration is now being given to the role of stochastic factors—randomness—in developmental defects. Even an embryo with wild-type genes and a favorable environment may develop an abnormal phenotype as the result of "bad luck." Developmental outcomes are probabilistic rather than predetermined (Wright 1920; Gottlieb 2003; Kilfoil et al. 2009). Consider for example, X-chromosome inactivation in females (see pp. 54–55). If a woman carries one normal and one mutant allele for an X-linked blood clotting factor, she should inactivate the wild-type allele in about 50% of her

*This chapter focuses on human health and provides general information about a variety of medical topics. It is not intended to provide medical advice for specific persons or disorders.

The future is already here. It's just not evenly distributed yet.
  WILLIAM GIBSON (1999)

The amazing thing about mammalian development is not that it sometimes goes wrong, but that it ever succeeds.
  VERONICA
VAN HEYNINGEN (2000)

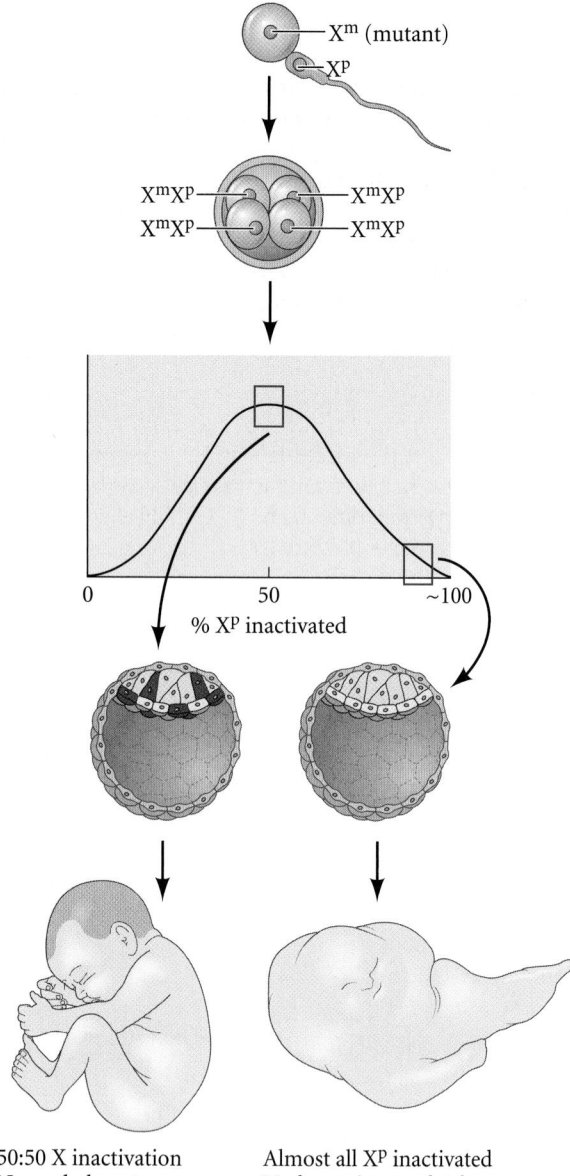

FIGURE 18.1 Developmental anomalies can be the result of stochastic events. Illustrated here is an actual case history of genetically identical female fetuses (monozygotic twins). Their phenotypically normal mother was heterozygous for a lethal mutant allele on the X chromosome. Both twins inherited this mutant X chromosome from their mother and a normal X chromosome from their father. The phenotypically normal twin had a typical pattern of random X-chromosome inactivation, with the paternal (normal) and maternal (mutant) chromosome each inactivated in about 50% of her cells. The other twin died in utero with severe developmental anomalies; by chance, the paternally derived chromosome was inactivated in almost all her cells. (After Masuzaki et al. 2004.)

cells. If the wild-type allele is inactivated in 50% of the liver cells that produce clotting factor, the woman is phenotypically normal. But what would happen if, just by chance, 95% of the wild-type X chromosomes were inactivated in these

liver cells? Then she would have an abnormality. Indeed, there have been cases of hemophilia (inability of the blood to clot) in identical twins where, in one twin, chance resulted in the inactivation of a large percentage of the X chromosomes carrying the normal allele. That twin had severe hemophilia, while the other, with a lower percentage of her normal X chromosomes inactivated, was not affected (Tiberio 1994; Valleix et al. 2002). FIGURE 18.1 shows an extreme case in which identical (monozygotic) twins received from their mother an X chromosome carrying a mutation that blocked normal development. In one twin, the maternal X was inactivated in about 50% of her cells (as is typical). The other twin had extensive developmental anomalies; by chance the paternally derived X chromosome was inactivated in almost all of her cells.

Such variability is not limited to genes on the X chromosome. In Chapter 11, for example, we discussed the random cell death of cranial neural crest cells, which can result in deafness or normal hearing, depending on whether or not neural crest cells migrated into the ear. Measurements of gene expression in individual cells show that protein synthesis is a stochastic process, with random fluctuations in both transcription and translation leading to variations in the levels of proteins produced at any given time (Raj and van Oudenaarden 2008; Stockholm et al. 2010). Cell specification, developmental signaling, and cell migration are thought to be influenced by chance fluctuations in the amounts of transcription factors, paracrine factors, and receptors produced at a particular moment. Thus, genetically identical animals raised in precisely the same environments can have vastly different phenotypes (Gilbert and Jorgensen 1998; Vogt et al. 2008; Ruvinsky 2009), and random fluctuations of gene expression can produce developmental anomalies. Mathematical modeling has permitted scientists to study these stochastic events, enabling researchers to "quantitatively demonstrate that development represents a combination of stochastic and deterministic events, offering insight into how chance influences normal development and may give rise to birth defects" (Zhou et al. 2013).

The rest of this chapter will focus on (1) genetic syndromes in which mutations alter development in deleterious ways; (2) teratogenesis, where exposure to environmental substances results in birth defects; (3) endocrine disruption, where environmental substances that specifically alter the development of the endocrine system and their target organs cause pathologies later in life; and (4) cancer, where cells escape the normal regulation of their neighbors and regain embryonic properties such as rapid proliferation, migration, and invasiveness.

Genetic Errors of Human Development

Congenital ("present at birth") abnormalities and losses of the fetus prior to birth have both intrinsic and extrinsic causes. Those abnormalities caused by genetic events may result from mutations, aneuploidies (improper chromosome number), and translocations (Opitz 1987).

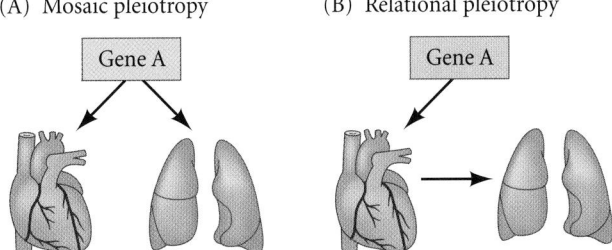

FIGURE 18.2   Mosaic and relational pleiotropy. (A) In mosaic pleiotropy, a gene is independently expressed in several tissues. Each tissue needs the gene product and develops abnormally in its absence. (B) In relational pleiotropy, a gene product is needed by only one particular tissue. However, a second tissue needs a signal from the first tissue in order to develop properly. If the first tissue develops abnormally, the signal is not given, so the second tissue also develops abnormally.

The nature of human syndromes

Human birth defects, which range from life-threatening to relatively benign, are often linked into **syndromes** (Greek, "running together"), where several abnormalities occur together. Genetically based syndromes are caused either by (1) a chromosomal event (such as the aneuploidies described above) in which several genes are deleted or added, or (2) by **pleiotropy**—the production of several effects by a single gene or pair of genes (see Grüneberg 1938; Hadorn 1955). Syndromes are said to have **mosaic pleiotropy** when the effects are produced independently as a result of the gene being critical in different parts of the body (**FIGURE 18.2A**). For instance, the *KIT* gene is expressed in blood stem cells, pigment stem cells, and germ stem cells, where it is needed for their proliferation. When this gene is defective, the resulting syndrome of anemia (lack of red blood cells), sterility (lack of germ cells), and albinism (lack of pigment cells) is evidence of mosaic pleiotropy. Syndromes are said to have **relational pleiotropy** when a defective gene in one part of the embryo causes a defect in another part, even though the gene is not expressed in the second tissue (**FIGURE 18.2B**). For example, failure of *MITF* expression in the pigmented retina prevents this structure from fully differentiating. This failure of pigmented retina growth, in turn, causes a malformation of the choroid fissure of the eye, resulting in the drainage of vitreous humor. Without this fluid, the eye fails to enlarge (hence microphthalmia, or "small eye"). The lenses and corneas are smaller, even though they themselves do not express *MITF*.

Mosaic syndromes can be the result of **aneuploidies**—errors in the number of particular chromosomes. Even an extra copy of the tiny chromosome 21 disrupts numerous developmental functions. This **trisomy 21** causes a set of anomalies—among them facial muscle changes, heart and gut abnormalities, and cognitive problems—collectively known as **Down syndrome** (**FIGURE 18.3**). Certain genes on chromosome 21 are thought to encode transcription factors and regulatory microRNAs, and the extra copy of chromosome 21 probably causes an overproduction of these regulatory proteins.* Such overproduction would cause the misregulation of genes necessary for heart, muscle, and nerve formation (Chang and Min 2009; Korbel et al. 2009). One such regulatory microRNA, miRNA-155, is encoded on chromosome 21 and found throughout the developing human fetus. This miRNA downregulates translation of the messages for certain transcription factors necessary for normal neural and heart development; and it is highly elevated in the brains and hearts of people with Down syndrome (Elton et al. 2010; Wang et al. 2013).

Genetic and phenotypic heterogeneity

In pleiotropy, the same gene can produce different effects in different tissues. However, the opposite phenomenon is an

---

*Down syndrome is one of the few trisomic conditions mild enough to allow the fetus to come to term and survive beyond the first weeks of infancy. With proper medical care, some infants born with trisomies 13 or 18 (Patau syndrome and Edward syndrome, respectively) can live for years, although they usually suffer from lung and intestinal defects and heart malformations.

(A)    (B)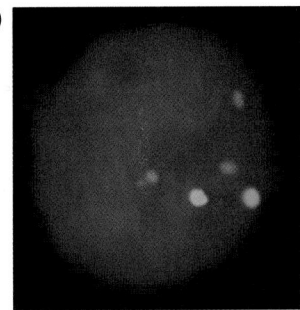

FIGURE 18.3   Down syndrome. (A) Down syndrome, caused by a third copy of chromosome 21, is characterized by a particular facial pattern, cognitive deficiencies, the absence of a nasal bone, and often heart and gastrointestinal defects. (B) The procedure shown here tests for chromosome number using fluorescently labeled probes that bind to DNA on chromosomes 21 (pink) and 13 (blue). This person has Down syndrome (trisomy 21) but has the normal two copies of chromosome 13. (A © MoodBoard/Alamy; B courtesy of Vysis, Inc.)

equally important feature of genetic syndromes: mutations in different genes can produce the same phenotype. If several genes are part of the same signal transduction pathway, a mutation in any of them often produces a similar phenotypic result. This production of similar phenotypes by mutations in different genes is called **genetic heterogeneity**. The syndrome of sterility, anemia, and albinism caused by the absence of Kit protein (discussed above) can also be caused by the absence of its paracrine ligand, stem cell factor (SCF). Another example is cyclopia, a phenotype that can be produced by mutations in the *sonic hedgehog* gene (see Chapter 3) or by mutations in the genes activated by hedgehog or in the genes controlling cholesterol synthesis (since cholesterol is essential for hedgehog signaling).

Not only can different mutations produce the same phenotype, but the same mutation can produce a different phenotype in different individuals, a phenomenon known as **phenotypic heterogeneity** (Wolf 1995, 1997; Nijhout and Paulsen 1997). Phenotypic heterogeneity comes about because genes are not autonomous agents. Rather, genes interact with other genes and gene products, becoming integrated into complex pathways and networks. Bellus and colleagues (1996) analyzed the phenotypes derived from the same mutation in the *FGFR3* gene in 10 different, unrelated families. These phenotypes ranged from relatively mild anomalies to potentially lethal malformations. Similarly, Freire-Maia (1975) reported that within one family, the homozygous state of a mutant gene affecting limb development caused phenotypes ranging from severe phocomelia (lack of limb development) to a mild abnormality of the thumb. The severity of a mutant gene's effect often depends on the other genes in the pathway as well as on environmental and stochastic (random) factors.

- **See WEBSITE 18.1** Human embryology and genetics
- **See WEBSITE 18.2** Prenatal diagnosis and preimplantation genetics

Teratogenesis: Environmental Assaults on Human Development

The summer of 1962 brought two portentous discoveries. The first was the disclosure by Rachel Carson (1962) that the pesticide DDT was destroying bird eggs and preventing reproduction in several species. The second discovery was that

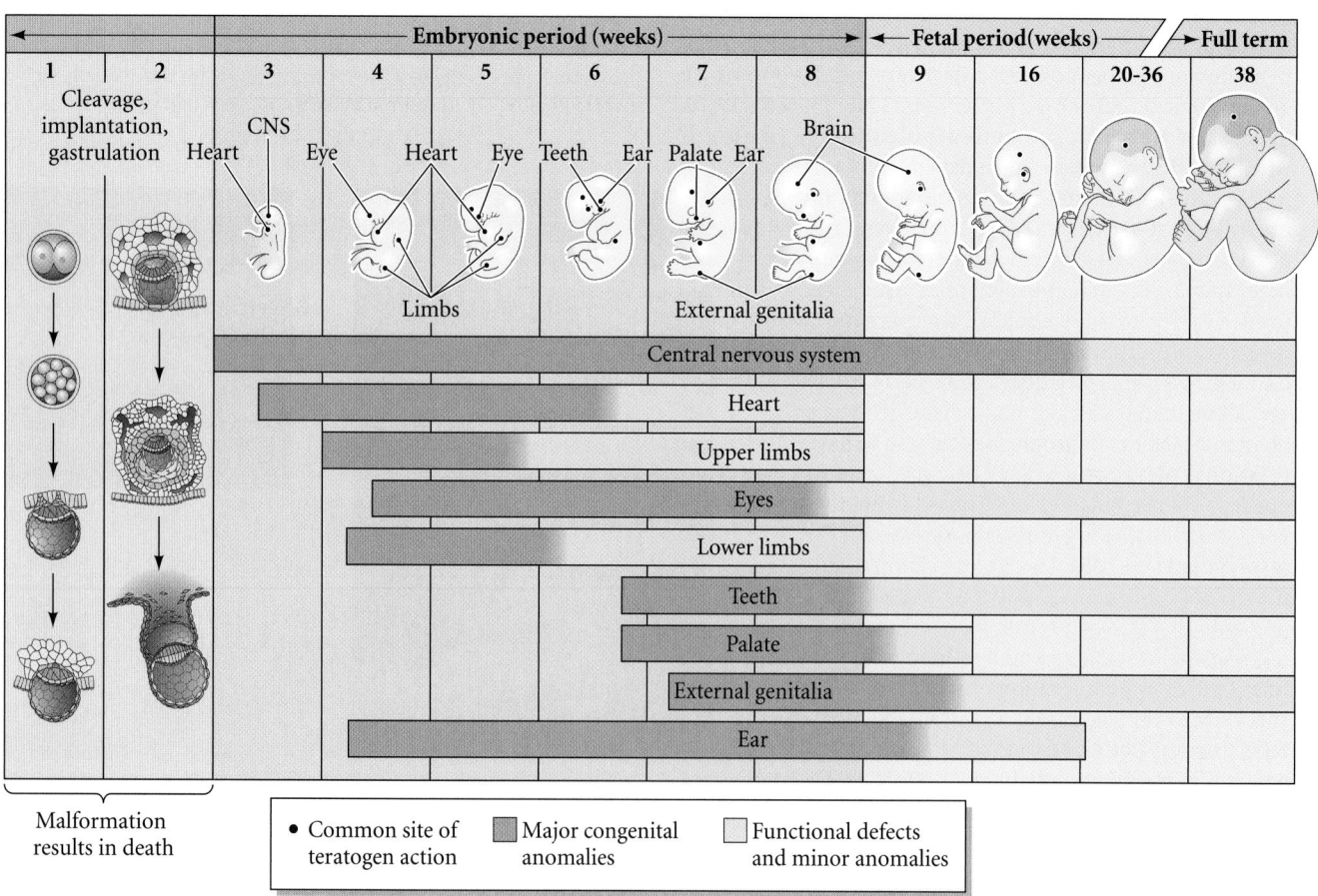

FIGURE 18.4 Weeks of gestation and sensitivity of embryonic organs to teratogens. (After Moore and Persaud 1993.)

Normal   Affected

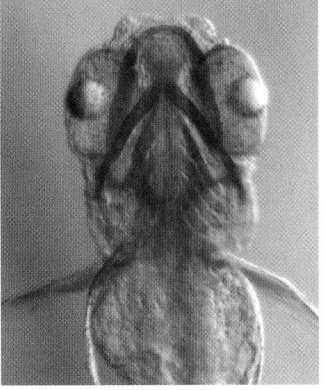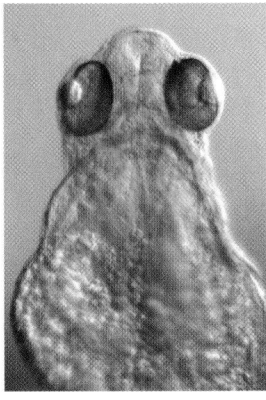

FIGURE 18.5 Water-soluble crude oil components from the Deepwater Horizon oil spill were teratogenic in zebrafish. Compared to normal zebrafish of the same age, zebrafish embryos exposed to oil spill components produced larvae with severe developmental anomalies, including reduction in the size of head, gill, and thoracic cartilages (blue staining) associated with cranial neural crest migration. (From de Soysa et al. 2012.)

thalidomide, a sedative used to help manage pregnancies, could cause limb and ear abnormalities in the human fetus (Lenz 1962; see Chapter 1). These two discoveries showed that the embryo was vulnerable to environmental agents.* This was underscored in 1964 when an epidemic of rubella (German measles) spread across the United States. Adults infected by the virus showed relatively mild symptoms, but more than 20,000 fetuses infected by rubella were born blind, deaf, or both. Many of these infants were also born with heart defects and/or mental retardation (CDC 2002).

Exogenous agents that cause birth defects are called **teratogens**. Most teratogens produce their effects during certain critical periods of development. Human development is usually divided into two periods, the **embryonic period** (to the end of week 8) and the **fetal period** (the remaining time in utero). It is during the embryonic period that most of the organ systems form; the fetal period is generally one of growth and modeling.

The period of maximum fetal susceptibility to teratogens is between weeks 3 and 8, since that is when most organs are forming (**FIGURE 18.4**). The nervous system, however, is constantly forming and remains susceptible throughout development. Prior to week 3, exposure to teratogens does not usually produce congenital anomalies because a teratogen encountered at this time either damages most or all of the cells of an embryo, resulting in its death, or it kills only a few cells, allowing the embryo to fully recover.

The largest class of teratogens includes drugs and chemicals, but viruses, radiation, high body temperature, and

metabolic conditions in the mother can also act as teratogens. Some chemicals that are naturally found in the environment can cause birth defects. For example, jervine and cyclopamine are chemical products of the plant *Veratrum californicum* that block Sonic hedgehog signaling and lead to cyclopia (see Figure 3.26B). Nicotine, a natural product concentrated in tobacco smoke, is associated with impaired lung and brain development (Dwyer et al. 2008; Maritz and Harding 2011).

Although different agents are teratogenic in different organisms (see Gilbert and Epel 2009), animals have been used to screen compounds that have a high probability of being hazardous. *Xenopus* and zebrafish, as we have seen, undergo early development using the same basic paracrine factors and transcription factors as we use. These model organisms have been especially important in identifying teratogenic molecules in the environment. Studies on zebrafish have found, for instance, that the water-soluble components of the 2010 Deepwater Horizon oil spill in the Gulf of Mexico caused numerous developmental anomalies traceable to neural crest cell migration (**FIGURE 18.5**; de Soysa et al. 2012). A partial list of agents known to be teratogenic in humans is given in **TABLE 18.1**.

● See **WEBSITE 18.3** Thalidomide as a teratogen

● See **VADE MECUM** Somites and thalidomide

TABLE 18.1 Some agents thought to cause disruptions in human fetal development[a]

DRUGS AND CHEMICALS	IONIZING RADIATION (X-RAYS)
Alcohol	
Aminoglycosides (Gentamycin)	HYPERTHERMIA (FEVER)
Aminopterin	INFECTIOUS MICROORGANISMS
Antithyroid agents (PTU)	Coxsackie virus
Bromine	Cytomegalovirus
Cortisone	Herpes simplex
Diethylstilbesterol (DES)	Parvovirus
Diphenylhydantoin	Rubella (German measles)
Heroin	*Toxoplasma gondii* (toxoplasmosis)
Lead	*Treponema pallidum* (syphilis)
Methylmercury	
Penicillamine	METABOLIC CONDITIONS IN THE MOTHER
Retinoic acid (Isotretinoin, Accutane)	Autoimmune disease (including Rh incompatibility)
Streptomycin	Diabetes
Tetracycline	Dietary deficiencies, malnutrition
Thalidomide	Phenylketonuria
Trimethadione	
Valproic acid	
Warfarin	

[a]This list includes known and possible teratogenic agents and is not exhaustive.

---

*Indeed, Rachel Carson realized the connection, commenting that "It is all of a piece, thalidomide and pesticides. They represent our willingness to rush ahead and use something without knowing what the results will be" (Carson 1962). Carson's book *Silent Spring* spurred the modern environmental movement, and she was vilified by the chemical industry (see Souder 2012).

Alcohol as a teratogen

In terms of the frequency of its effects and its cost to society, the most devastating teratogen is undoubtedly alcohol (ethanol). Babies born with **fetal alcohol syndrome (FAS)** are characterized by their small head size, indistinct philtrum (the pair of ridges that runs between the nose and mouth above the center of the upper lip), narrow vermillion border on the upper lip, and low nose bridge (Lemoine et al. 1968; Jones and Smith 1973). The brain of such a child may be dramatically smaller than normal and often shows poor development because of deficiencies of neuronal and glial migration (**FIGURE 18.6A,B**; Clarren 1986). FAS is the most prevalent type of congenital mental retardation syndrome, occurring in approximately 1 out of every 650 children born in the United States (May and Gossage 2001). Although the IQs of children with FAS vary substantially, the mean is about 68 (Streissguth and LaDue 1987). Most adults and adolescents with FAS cannot handle money and have difficulty learning from past experiences.*

FAS represents only a portion of a range of defects caused by prenatal alcohol exposure. The term **fetal alcohol spectrum disorder (FASD)** has been coined to encompass all of the alcohol-induced malformations and functional deficits that occur. In many FASD children, behavioral abnormalities exist without any gross physical changes in head

---

*For remarkable accounts of raising children with fetal alcohol syndrome, read Michael Dorris's *The Broken Cord* (1989) and Liz and Jodee Kulp's *The Best I Can Be* (2000). For an excellent account of the debates within the medical profession about this syndrome, see Janet Golden's *Message in a Bottle* (2005).

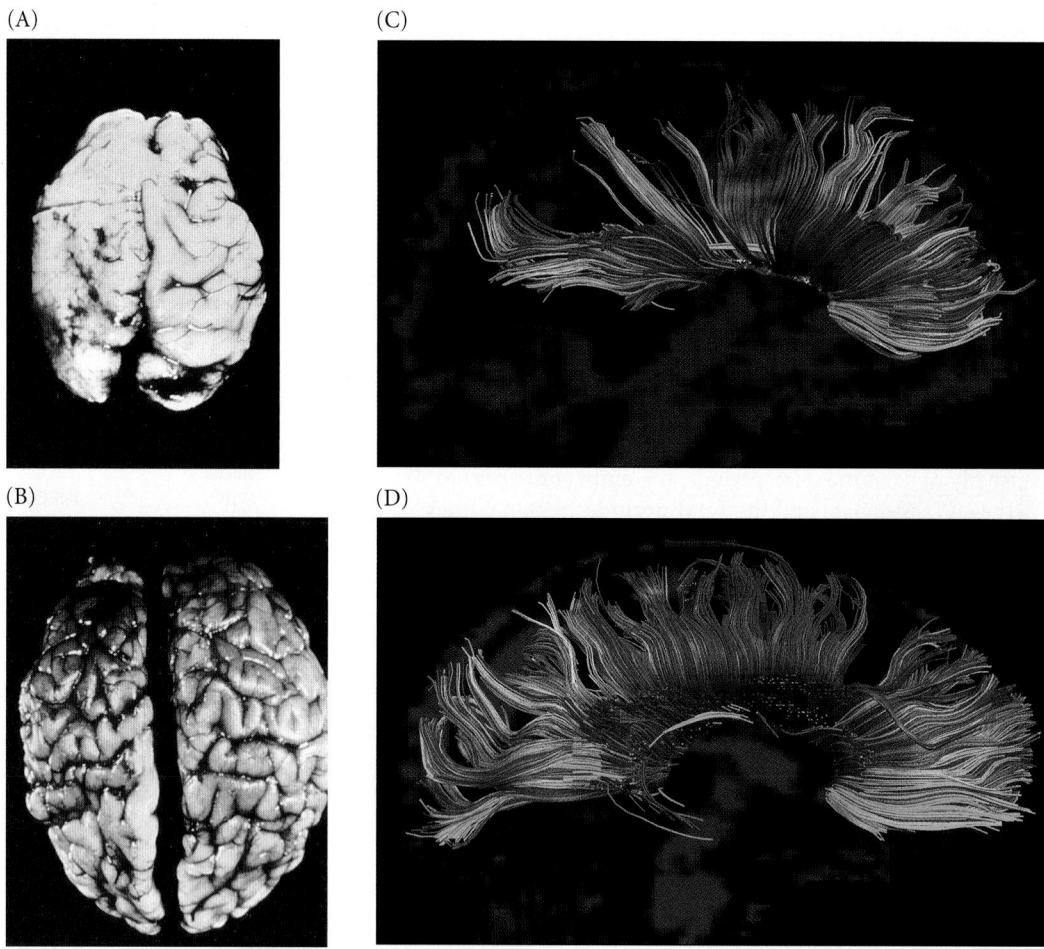

(A)

(B)

(C)

(D)

FIGURE 18.6 Effects of alcohol on fetal brains. (A,B) Comparison of a brain from an infant with fetal alcohol syndrome (A) with a brain from a normal infant of the same age (B). The brain from the infant with FAS is smaller, and the pattern of convolutions is obscured by glial cells that have migrated over the top of the brain. (C,D) Regionally specific abnormalities of the corpus callosum seen by diffusion tensor imaging of myelinated neurons. The difference in fiber tracks in a child with FASD (C) compared with those of a same-age unaffected child (D) suggests that there are significant abnormalities in neurons that would normally project through the posterior regions of the brain into the cortex of the parietal and temporal lobes. (A,B courtesy of S. Clarren; C,D from Wozniak and Muetzel 2011, courtesy of the authors.)

size or notable reductions in IQ (NCBDD 2009). However, recent techniques that can identify neural tracts in the brain have found subtle abnormalities that correlate with altered mental processing speed and executive functioning such as planning, memorizing, and retaining information (**FIGURE 18.6C,D**; Wozniak and Muetzel 2011).

As with other teratogens, the amount and timing of fetal exposure to alcohol, as well as the genetic background of the fetus, contribute to the developmental outcome. Variability in the metabolism of alcohol by the mother may also account for some outcome differences (Warren and Li 2005). While FASD is most strongly associated with high levels of alcohol consumption, the results of animal studies suggest that even a single episode of consuming the equivalent of two alcoholic drinks during pregnancy may lead to loss of fetal brain cells. ("One drink" is defined as 12 oz. of beer, 5 oz. of wine, or 1.5 oz. of "hard" liquor.) It is important to note that alcohol can cause permanent damage to a fetus at a time before most women even realize they are pregnant.

When mice are exposed to alcohol at the time of gastrulation, ethanol induces defects of the face and brain that are comparable to those in humans with FAS (**FIGURE 18.7**; Sulik 2005). As in human fetuses, the nose and upper lip of the ethanol-exposed pups are poorly developed, and nervous system problems involve failure to close the neural tube and incomplete development of the forebrain (see Figure 10.32).

This mouse model of FAS can be used to study the ways by which ethanol causes its effects on the embryo. It appears that ethanol works on several processes, interfering with cell migration, proliferation, adhesion, and survival. Hoffman and Kulyk (1999) showed that instead of migrating and dividing, neural crest cells of alcohol-exposed fetuses prematurely initiate their differentiation into facial cartilage. Among the numerous genes that are misregulated following maternal alcohol exposure in mice are several involved in the cytoskeletal reorganization that enables cell movement (Green et al. 2007). In addition, cell death is apparent shortly after alcohol exposure. In later-stage mouse embryos exposed to ethanol, the death of neural crest-derived cells is seen as early as 12 hours following the exposure. When the time of alcohol exposure corresponds to the third and fourth weeks of human development, cells that should form the median portion of the forebrain, upper midface, and cranial nerves are killed. This has been confirmed in early chick embryos,

Normal   Alcohol-exposed

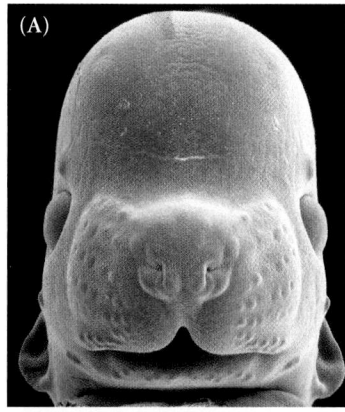
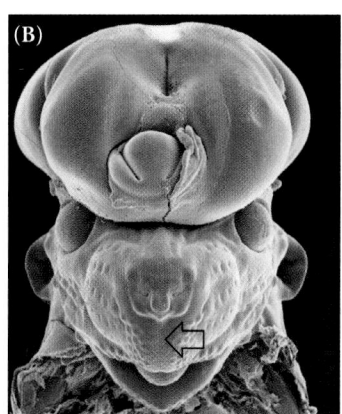
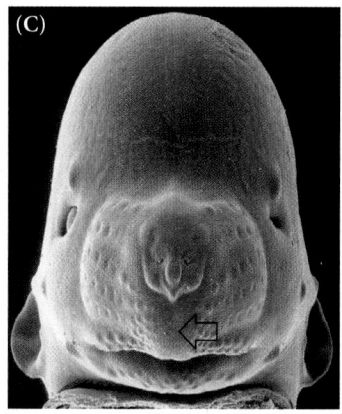

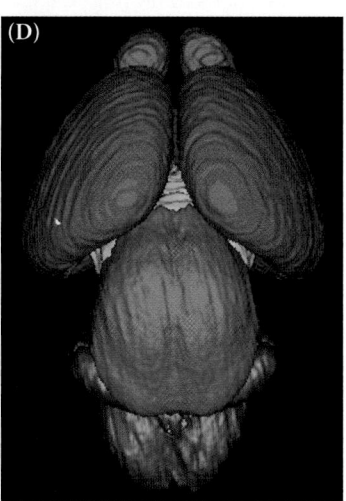
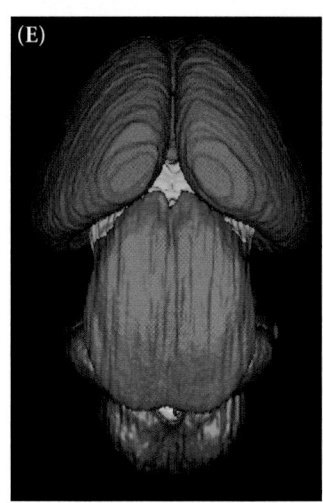

FIGURE 18.7   Alcohol-induced craniofacial and brain abnormalities in mice. (A–C) Normal (A) and abnormal (B,C) day-14 embryonic mice. In (B), the anterior neural tube failed to close, resulting in exencephaly, a condition in which the brain tissue is exposed to the exterior. Later in development, the exposed brain tissue will erode away, resulting in anencephaly. (B,C) Prenatal alcohol exposure can also affect facial development, resulting in a small nose and an abnormal upper lip (open arrow). These facial features are present in fetal alcohol syndrome. (D,E) Three-dimensional reconstructions prepared from magnetic resonance images of the brains of normal (D) and alcohol-exposed (E) 17-day embryonic mice. In the alcohol-exposed specimen, the olfactory bulbs (pink) are absent and the cerebral hemispheres (red) are abnormally united in the midline. Light green, diencephalon; magenta, mesencephalon; teal, cerebellum; dark green, pons and medulla. (Courtesy of K. Sulik.)

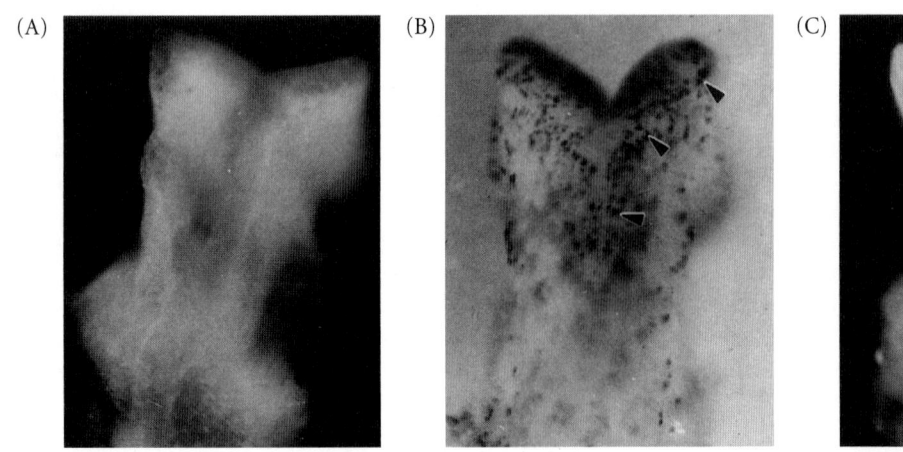

(A)

(B)

(C)

FIGURE 18.8 Possible mechanisms producing fetal alcohol syndrome. (A–C) Cell death caused by alcohol-induced superoxide radicals. Staining with Nile blue sulfate shows areas of cell death. (A) Head region of control day-9 mouse embryo. (B) Head region of alcohol-treated embryo, showing areas of cell death (arrowheads). (C) Head region of embryo treated with both alcohol and superoxide dismutase, an inhibitor of superoxide radicals. The enzyme prevents the alcohol-induced cell death. (D) Inhibition of L1-mediated cell adhesion by alcohol. (A–C from Kotch et al. 1995, courtesy of K. Sulik; D after Ramanathan et al. 1996.)

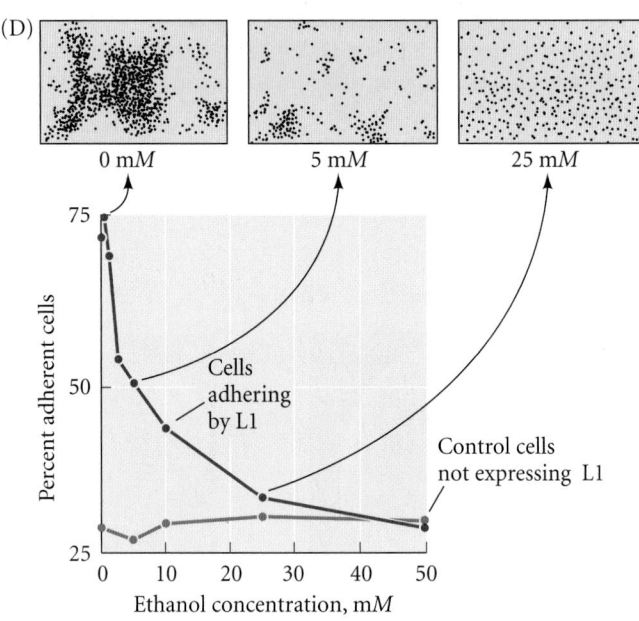

(D)

where transient ethanol exposure at environmentally relevant doses (about 25 mM) decimates migrating cranial neural crest cells, causing cell death throughout the head region (Flentke et al. 2011).

One reason for this cell death in mouse embryos is that alcohol treatment results in the generation of superoxide radicals that can damage cell membranes (**FIGURE 18.8A–C**; Davis et al. 1990; Kotch et al. 1995; Sulik 2005). In model systems, antioxidants have been effective in reducing both the cell death and the malformations caused by alcohol (Chen et al. 2004).

Abnormal signaling may also underlie excessive cell death. In alcohol-exposed embryos, expression of Sonic hedgehog (which is important in establishing the facial midline structures; see Chapter 10) is downregulated. While the mechanism for this downregulation remains incompletely understood, the finding that Shh-secreting cells placed into the head mesenchyme can prevent the alcohol-induced death of cranial neural crest cells highlights the importance of the Shh pathway as a target for alcohol's teratogenesis (Ahlgren et al. 2002; Chrisman et al. 2004).

Another mechanism that may be involved in alcohol's teratogenesis is its interference with the ability of the cell adhesion molecule L1 to hold cells together. Ramanathan and colleagues (1996) have shown that at levels as low as 7 mM, an alcohol concentration produced in the blood or brain with a single drink, alcohol can block the adhesive function of the L1 protein in vitro (**FIGURE 18.8D**). Moreover, mutations in the human L1 gene cause a syndrome of mental retardation and malformations similar to that seen in severe

FAS cases. Thus, alcohol can cross the placenta, enter the fetus, and block several critical functions in brain and facial development.

Retinoic acid as a teratogen

In some instances, even a compound involved in normal development can have deleterious effects if it is present in large enough amounts or at particular times. As we have seen throughout this book, retinoic acid (RA) is a vitamin A derivative that is important in specifying the anterior-posterior axis and in forming the jaws and heart of the mammalian embryo (see Chapters 9, 11, and 13).

In its pharmaceutical form, 13-*cis*-retinoic acid (also called isotretinoin and sold under the trademark Accutane) has been useful in treating severe cystic acne and has been available for this purpose since 1982. The deleterious effects of administering large amounts of retinoic acid (or its vitamin A precursor) to pregnant animals have been known since the 1950s (Cohlan 1953; Giroud and Martinet 1959; Kochhar et al. 1984). However, about 160,000 women of childbearing age (15–45 years) have taken isotretinoin since it was introduced, and some have used it during pregnancy.* Lammer and co-workers (1985) studied a group of women who inadvertently exposed themselves to retinoic acid and who elected to remain pregnant. Of their 59 fetuses, 26 were born without any noticeable anomalies, 12 aborted spontaneously, and 21 were born with obvious anomalies. The affected infants had a characteristic pattern of anomalies, including absent or defective ears, absent or small jaws, cleft palate, aortic arch abnormalities, thymus deficiencies, and abnormalities of the central nervous system. These anomalies are largely due to the failure of cranial neural crest cells to migrate into the pharyngeal arches of the face to form the jaws and ear (Moroni et al. 1994; Studer et al. 1994). Radioactively labeled RA binds to the cranial neural crest cells and arrests both their proliferation and their migration (Johnston et al. 1985; Goulding and Pratt 1986). The teratogenic period during which cranial neural crest cells are affected occurs on days 20–35 in humans (days 8–10 in mice).

Retinoic acid probably disrupts these cells in several ways. One mechanism is that excess RA activates the negative feedback pathway that usually ensures the proper amount of this compound. Transient large increases in RA thus activate the synthesis of RA-degrading enzymes, causing a long-lasting *decrease* of RA. It is this deficiency in RA that results in the malformations (Lee et al. 2012). This would explain why high amounts of retinoic acid produce phenotypes similar to those seen in deficiencies of retinoic acid.

Other teratogenic agents

In addition to natural chemicals, hundreds of new artificial compounds come into general use each year in our industrial society. Pesticides and organic mercury compounds have caused neurological and behavioral abnormalities in infants whose mothers have ingested them during pregnancy. Moreover, drugs that are used to control diseases in adults may have deleterious effects on fetuses. Such drugs include cortisone, warfarin, tetracycline, and valproic acid (see Table 18.1).

More than 87,000 artificial chemicals are currently licensed for use in the United States, and about 500 new compounds are being made each year. The Toxic Substances Control Act of the United States assumes chemicals are safe unless demonstrated to be otherwise, and only about 8,000 chemicals have been tested for their potential teratogenic effects (Johnson 1980; EPA 2008). Although teratogenic compounds have always been with us, the risks increase as more and more untested compounds enter our environment. Most industrial chemicals have not been screened for their teratogenic effects. Standard screening protocols are expensive, long, and subject to interspecies differences in metabolism. There is still no consensus on how best to test a substance's teratogenicity for human embryos.

HEAVY METALS Heavy metals such as zinc, mercury, and lead are powerful teratogens. Industrial pollution has resulted in high concentrations of heavy metals in the environment in many places (see Gilbert and Epel 2009). In the former Soviet Union, unregulated "industrial production at all costs" left a legacy of soaring birth defect rates. In some regions of Kazakhstan, heavy metals are found in high concentrations in drinking water, vegetables, and the air. In such locations, nearly half the people tested have extensive chromosome damage, and in some areas the incidence of birth defects has doubled since 1980 (Edwards 1994). In the United States, lax enforcement of antipollution laws has led to heavy-metal contamination of most of the country's lakes (**FIGURE 18.9**). This contamination is especially prevalent where mining interests have been allowed to disgorge metal-containing slag into streams feeding the lakes (USGS 2009; Steingraber 2003). Indeed, an International Joint Commission of the U.S. and Canada (2000) warned that pregnant women should not eat fish caught in the Great Lakes (Kaufman 2012; USDI 2012).

Mercury and lead can damage the developing nervous system (Bellinger 2005). Mercury is preferentially absorbed by regions of the developing cerebral cortex (Eto 2000; Kondo 2000; Eto et al. 2001). When pregnant mice were given mercury on day 9 of gestation, nearly half the pups were born with small brains and/or small eyes (O'Hara et al. 2002). The polluting of Minamata Bay, Japan, with mercury in 1956 produced brain and eye deficiencies in infants both by transmission of the mercury across the placenta and by its transmission through mother's milk. Lead damages the developing brain in fetal and childhood stages (Bellinger et al. 1987;

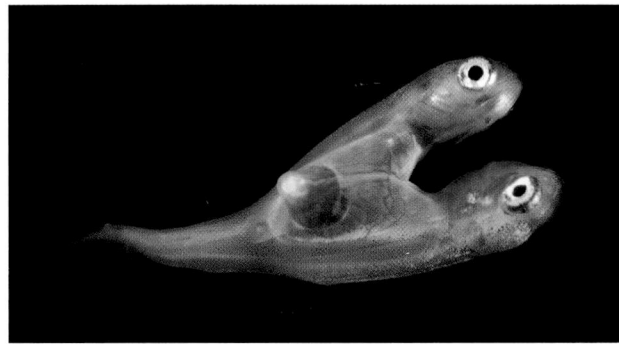

FIGURE 18.9 Conjoined trout hatchlings. An increase in developmental disorders such as body duplications in fish has led to public awareness of heavy metal pollution in American lakes. Slag generated by mining operations reaches the lakes via stream runoff. (Photograph © Natural Visions/Alamy.)

---

*Isotretinoin-containing drugs carry a strong warning against their use by pregnant women. Retinoic acid is a critical public health concern precisely because there is significant overlap between the population using acne medicine and the population of women of childbearing age, and because it is estimated that half the pregnancies in the U.S. are unplanned (Nulman et al. 1997).

Baghurst et al. 1992; Dietrich et al. 1993) and contributes to developmental delays and mental retardation. Lead-based paint was banned in Europe in 1955, but lobbying efforts by the paint industry in the United States succeeded in keeping lead in American paints until 1977 (see Steingraber 2003).

PATHOGENS Another class of teratogens includes viruses and other pathogens. Gregg (1941) first documented the fact that women who contracted rubella (German measles) during the first trimester of their pregnancy had a 1 in 6 chance of giving birth to an infant with eye cataracts, heart malformations, or deafness. This study provided the first evidence that the mother could not fully protect the fetus from the outside environment. The rubella virus is able to enter many cell types, where it produces a protein that prevents mitosis by blocking kinases that allow the cell cycle to progress (Atreya et al. 2004). The cells die, and numerous organs are affected; the earlier in pregnancy the rubella infection occurs, the greater the risk that the embryo will be malformed. The first 6 weeks of development appear to be the most critical, because that is when the heart, eyes, and ears are formed (see Figure 18.4). The U.S. rubella epidemic of 1964, mentioned earlier in the chapter, probably resulted in more than 10,000 fetal deaths and 20,000 infants with birth defects (CDC 2002). Two other viruses, cytomegalovirus and herpes simplex, are also teratogenic. Cytomegalovirus infection of early embryos is nearly always fatal; infection of later embryos can lead to blindness, deafness, cerebral palsy, and mental retardation.

Bacteria and protists are rarely teratogenic, but some of them are known to damage human embryos. *Toxoplasma gondii*, a protist carried by rabbits and cats (and their feces), can cross the placenta and cause brain and eye defects in the fetus. *Treponema pallidum*, the bacterium that causes syphilis, can kill early fetuses and produce congenital deafness and facial damage in older fetuses.

Endocrine Disruptors: The Embryonic Origins of Adult Disease

A specialized area of teratogenesis involves the misregulation of the endocrine system during development. Endocrine disruptors are exogenous (coming from outside the body) chemicals that interfere with the normal functions of hormones, and consequently disrupt development (Colborn et al. 1993, 1997). The phenotypic changes produced by endocrine disruptors are not the obvious anatomical birth defects produced by classic teratogens. Rather, the anatomical alterations induced by endocrine disruptors are often seen only microscopically, and the major changes are physiological. These functional changes are more subtle than those produced by the teratogens, but they can be extremely important phenotypic alterations. They are often manifest later in adult life. Moreover, these effects may persist for generations after the exposure to the disruptor.

These chemicals can interfere with hormonal functions in many ways:

1. They can mimic the effect of a natural hormone. A paradigmatic example is the endocrine disruptor diethylstilbestrol (DES), which binds to the estrogen receptor and mimics estradiol, a hormone that is very active in building the tissues of the female reproductive tract.

2. They can act as antagonists and inhibit the binding of a hormone to its receptor or block the synthesis of a hormone. DDE, a metabolic product of the insecticide DDT, can act as an anti-testosterone, binding to the androgen receptor and preventing normal testosterone from functioning properly.

3. They can affect the synthesis, elimination, or transportation of a hormone in the body. As we saw in our discussion of sex determination, the herbicide atrazine elevates the synthesis of estrogen and can convert testes into ovaries in frogs. One of the ways that polychlorinated biphenyls (PCBs) disrupt the endocrine system is by interfering with the elimination and degradation of thyroid hormones.

4. Some endocrine disruptors can "prime" the organism to be more sensitive to hormones later in life. As we will see, bisphenol A exposure during fetal development makes breast tissue more responsive to steroid hormones during puberty.

Endocrine disruptors differ from teratogens in several ways. First, their pathological effects do not have to be congenital but can show themselves in adulthood. Second, the effects are usually manifest as physiological problems rather than anatomical ones. Third, given the paradigm of teratogens, it had been thought that there were only a few "bad" agents, and that the only people who received these were pregnant women who inadvertently exposed their embryos to high doses of these chemicals. We now recognize that endocrine disruptors are everywhere in our technological society (and even in rural areas where pesticides and herbicides are abundant), and that low-dose exposure to endocrine disruptors can be sufficient to produce significant disabilities later in life. Endocrine disruptors include chemicals that line baby bottles and the brightly colored plastic containers from which we drink our water; chemicals used in cosmetics, sunblocks, and hair dyes; and chemicals to prevent clothing from being highly flammable. Fourth, as expected when so many chemicals are involved, one is usually exposed to multiple endocrine disruptors, and not just one. We are exposed to numerous endocrine disruptors simultaneously and continuously. The developmental effects of everyday exposure are just beginning to be studied (Wild 2005; Rappaport and Smith 2010). And fifth, more damage may be done by a "moderate" dose of endocrine disruptor than by a higher dose, since higher concentrations may activate negative feedback processes that detoxify or eliminate the chemical (see Myers et al. 2009; Belcher et al. 2012; Vandenberg et al. 2012).

SIDELIGHTS & SPECULATIONS

Developmental Origins of Adult-Onset Disease

Teratogenesis is usually associated with congenital disease (i.e., a condition appearing at birth) and is also associated with disruptions of organogenesis during the embryonic period. However, David Barker and colleagues (1994a,b) have offered evidence that certain adult-onset diseases may also result from conditions in the uterus prior to birth. Based on epidemiological evidence, they hypothesize that there are critical periods of development during which certain physiological insults or stimuli can cause specific changes in the body. The "Barker hypothesis" postulates that certain adult anatomical and physiological parameters become established during embryonic and fetal development, and that deficits in nutrition during this time can produce permanent changes in the pattern of metabolic activity—changes that can predispose the adult to particular diseases.

Undernutrition and the adult phenotype

Specifically, Barker and colleagues showed that infants whose mothers experienced protein deprivation (because of wars, famines, or migrations) during certain months of pregnancy were at high risk for having certain diseases as adults. Undernutrition during a fetus's first trimester could lead to hypertension and strokes in adult life, whereas those fetuses experiencing undernutrition during the second trimester had a high risk of developing heart disease and diabetes as adults. Those fetuses experiencing undernutrition during the third trimester were prone to blood clotting defects as adults.

Recent studies have tried to determine whether there are physiological or anatomical reasons for these correlations (Gluckman and Hanson 2004, 2005; Lau and Rogers 2005). Anatomically, undernutrition can change the number of cells produced during a critical time of organ formation. When pregnant rats are fed low-protein diets at certain times during their pregnancy, the resulting offspring are at high risk for hypertension as adults. The poor diet appears to cause low nephron numbers in the adult kidney (see Moritz et al. 2003). In humans, the number of nephrons present in the kidneys of men with hypertension was only about half the number found in men without hypertension (**FIGURE 18.10A**; Keller et al. 2003). In addition, the glomeruli (the blood-filtering unit of the nephron) of hypertensive men were larger than those in control subjects (**FIGURE 18.10B**).

Similar trends have been reported for non-insulin dependent (Type II) diabetes and glucose intolerance (Hales et al. 1991; Hales and Barker 1992). Here, poor nutrition reduces the number of β cells in the pancreas and hence the ability to synthesize insulin.

FIGURE 18.10 Anatomical changes associated with hypertension. (A) In age-matched individuals, the kidneys of men with hypertension had about half the number of nephrons as the kidneys of men with normal blood pressure. (B) The glomeruli of the nephrons in hypertensive kidneys were much larger than the glomeruli in control subjects. (After Keller et al. 2003; photographs courtesy of G. Keller.)

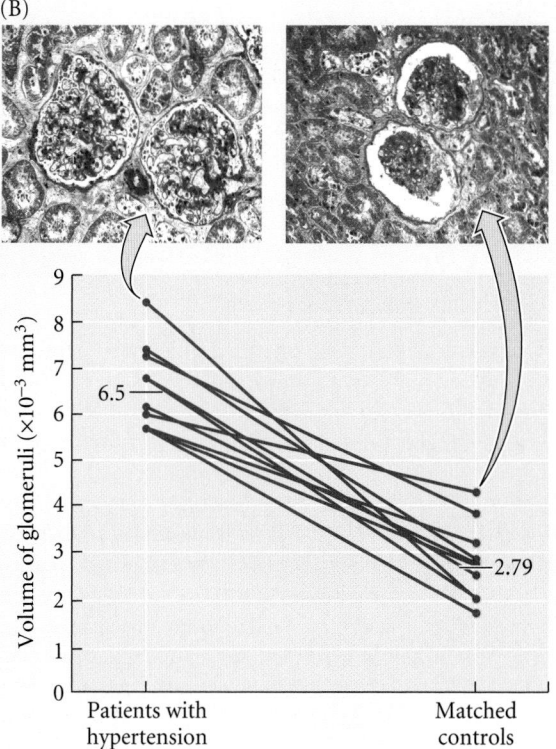

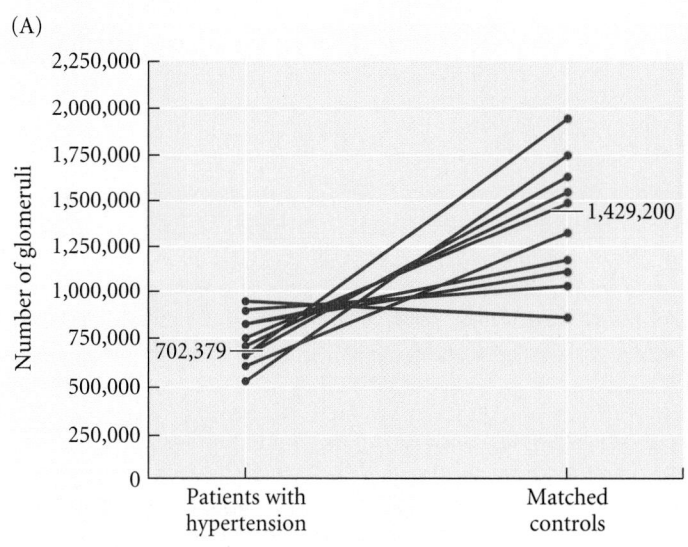

Undernutrition in rats changes the histological architecture in the liver as well. A low-protein diet during gestation appeared to increase the amount of periportal cells that produce the glucose-synthesizing enzyme phosphoenolpyruvate carboxykinase while decreasing the number of perivenous cells that synthesize the glucose-degrading enzyme glucokinase in the offspring (Burns et al. 1997). These changes may be coordinated by glucocorticoid hormones that are stimulated by malnutrition and which act to conserve resources, even though such actions might make the person prone to hypertension later in life (see Fowden and Forhead 2004). Since, historically, most humans died before age 50 (see Chapter 16), this would not be a detrimental evolutionary trade-off.

The "thrifty phenotype"

Hales and Barker (2001) proposed the existence of a "thrifty phenotype" wherein a malnourished fetus is "programmed" to expect an energy-deficient environment. The developing fetus sets its biochemical parameters to conserve energy and store fat.* Those who as adults do indeed meet with the expected poor environment are ready for it and can survive better than individuals whose metabolisms were set to use energy rather than store it as fat. However, if such a "deprivationally developed" person lives in an energy- and protein-rich environment, their cells store more fats and their heart and kidneys have developed to survive more stringent conditions—both developments that put the person at risk for several adult-onset diseases. It appears the fetus is making developmental decisions in response to low nutrient intake that anticipate its living after birth in a nutritionally deprived environment. Accordingly, it "expects" to stay small and thin and programs its development accordingly. However, if it is born

*In other words, the fetus is displaying phenotypic plasticity—the ability to modulate its phenotype depending on the environment. Environmental factors, not the genes, are determining the phenotype (see Chapter 19).

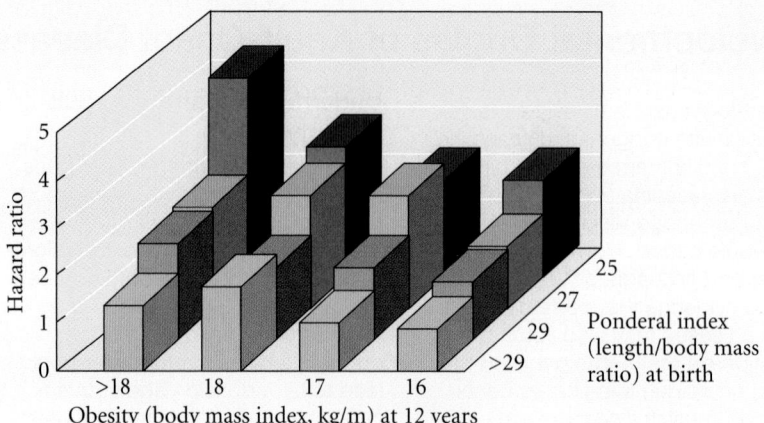

FIGURE 18.11 Environmental mismatch. The risk of coronary disease is increased by small birth size (poor prenatal nutrition) followed by greater eating in childhood. The prenatal conditions cause the expectation of a low caloric environment and set gene expression for a "thrifty phenotype" that conserves fat. This mismatch between the expected and the actual environment can make the person susceptible to conditions such as artery blockage. (After Eriksson et al. 2001.)

into a nutritionally abundant environment, the thrifty phenotype that the fetus selected stores the extra calories and causes the person to become obese and at risk for cardiovascular problems. This has been called the "environmental mismatch hypothesis"

(FIGURE 18.11; Gluckman and Hanson 2005).

How can conditions experienced in the uterus create anatomical and biochemical conditions that will be maintained throughout adulthood? One place to look is DNA methylation.

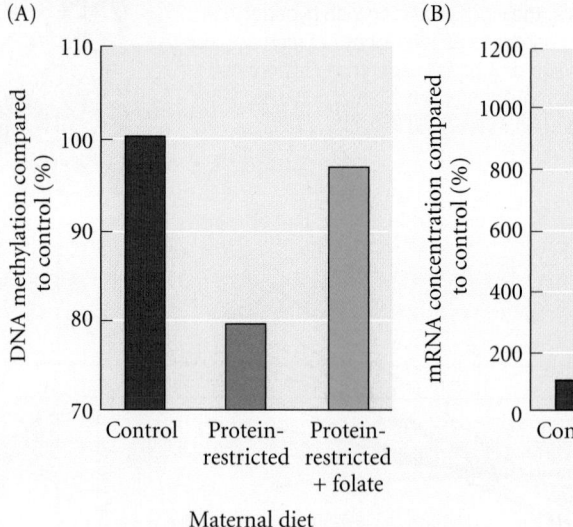
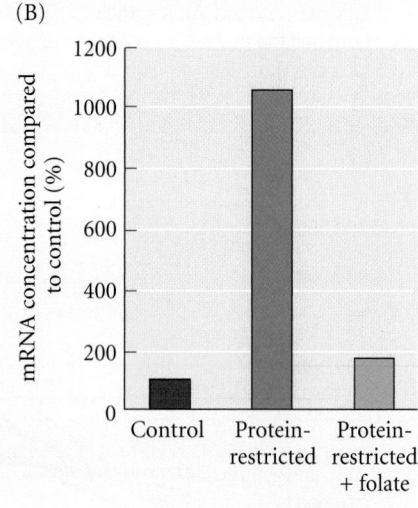

FIGURE 18.12 Activity of the liver gene for peroxisomal proliferator-activated receptor (PPARa) is susceptible to dietary differences. (A) DNA methylation pattern of the *PPARa* promoter region, showing highly methylated control promoters compared with poorly methylated promoters from the livers of mice whose mothers had protein-restricted diets ($p < 0.001$). Adding folate to the protein-restricted diet abolished this difference. (B) Levels of mRNA for the *PPARa* gene were much higher in the mice fed the protein-restricted diet ($p < 0.0001$). (After Lillycrop et al. 2005.)

Lillycrop and colleagues (2005) have shown that rats born to mothers having a low-protein diet had a different pattern of liver gene methylation than did the offspring of mothers fed a normal diet. These differences in methylation changed the metabolic profile of the rats' livers. For instance, the methylation of the promoter region of the *PPARα* gene (which is critical in the regulation of carbohydrate and lipid metabolism) is 20% lower in the offspring of protein-restricted rats, and

the gene's transcriptional activity is tenfold greater (**FIGURE 18.12**). Moreover, the difference between these methylation patterns can be abolished by including folate in the protein-restricted diet.* Thus, the difference in methylation probably results from changes in folate metabolism caused by the limited amount of protein available to the fetus.

Prenatal nutrition appears to possess the ability to induce long-lasting, gene-specific alterations in

transcriptional activity and metabolism. The prevention of adult disease through prenatal diet could thus become a public health issue in the coming decades.

*Although the folate supplementation of foods has probably caused a major decrease in the number of babies born with neural tube defects (see Chapter 10), aggressive folate supplementation during pregnancy may actually be teratogenic (Marean et al. 2011; Mikael et al. 2013; Vasquez et al. 2013).

DES as an endocrine disruptor

One of the first endocrine disruptors to be identified was a potent environmental estrogen called **diethylstilbestrol**, or **DES**. This drug was thought to ease pregnancy and prevent miscarriages, and it is estimated that in the United States more than 1 million pregnant women and their fetuses were exposed to DES between 1947 and 1971. This is probably a small fraction of exposures worldwide.* Research from the 1950s showed that in fact DES had no beneficial effects on pregnancy, but it was still prescribed until the FDA banned it in 1971. The ban was imposed when a specific type of tumor (clear-cell adenocarcinoma) was discovered in the reproductive tracts of some of the women whose mothers took DES during pregnancy (**FIGURE 18.13**).

DES interferes with sexual and gonadal development by causing cell type changes in the female reproductive tract (the derivatives of the Müllerian duct, which forms the upper portion of the vagina, cervix, uterus, and oviducts; see Figure 15.1). In many cases, DES causes the boundary between the oviduct and the uterus (the uterotubal junction) to

---

*In addition to DES administered directly to pregnant women, biologically relevant levels of DES were found in beef. The drug was fed to cattle to accelerate their growth (see Knight 1980; Raun and Preston 2001). For an award-winning documentary on DES, see Judith Helfand's *A Healthy Baby Girl* (1997).

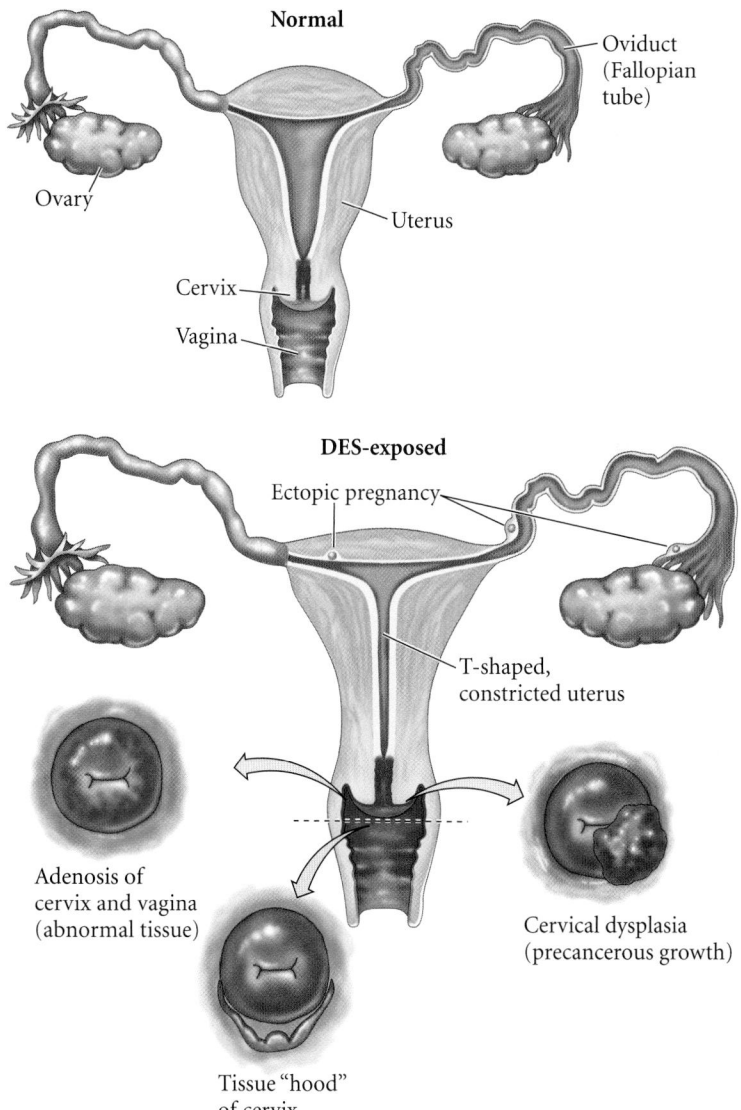

FIGURE 18.13 Genital anomalies can occur in women exposed to DES in utero. In these "DES daughters," the cervical tissue (red) is often displaced into the vagina. Such individuals may have a uterus that is T-shaped and constricted, as well as adenosis of the cervix and vagina (where the lining differentiates into mucosal cells), precancerous cells, ectopic pregnancies, adenocarcinomas, and other effects.

(A)

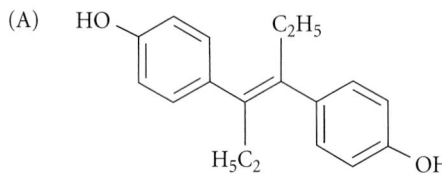

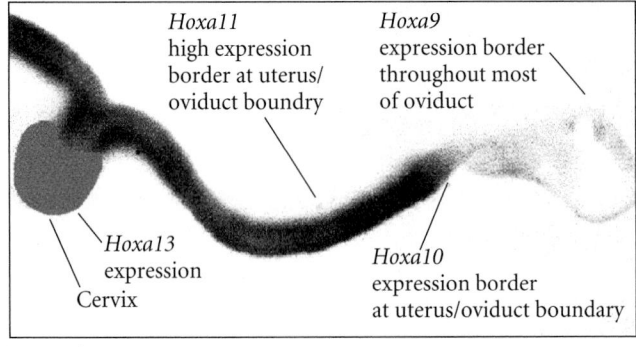

(B)

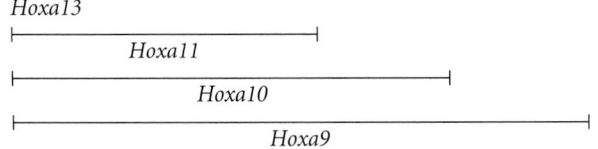

FIGURE 18.14 Effects of DES exposure on the female reproductive system. (A) The chemical structure of DES. (B) *Hoxa* gene expression in the reproductive system of a normal 16.5-day embryonic female mouse. A whole mount in situ hybridization of the *Hoxa13* probe is shown (red) along with a probe for *Hoxa10* (purple). *Hoxa9* expression extends throughout the uterus and through much of the presumptive oviduct. *Hoxa10* expression has a sharp anterior border at the transition between the presumptive uterus and the oviduct. *Hoxa11* has the same anterior border as *Hoxa10*, but its expression diminishes closer to the cervix. *Hoxa13* expression is found only in the cervix and upper vagina. (After Ma et al. 1998.)

be lost, resulting in infertility or subfertility (Robboy et al. 1982; Newbold et al. 1983). Moreover, the distal Müllerian ducts often fail to come together to form a single cervical canal (see Figure 18.13).

Symptoms similar to human DES syndrome occur in mice exposed to DES in utero, allowing the mechanisms of this endocrine disruptor to be uncovered. Normally, the regions of the female reproductive tract are specified by the *Hoxa* genes, which are expressed in a nested fashion throughout the Müllerian duct (FIGURE 18.14). Ma and colleagues (1998) showed that the effects of DES on the female mouse reproductive tract could be explained as the result of altered *Hoxa10* expression in the Müllerian duct. DES was injected under the skin of pregnant mice, and the fetuses were allowed to develop almost to birth. When the fetuses from

the DES-injected mothers were compared with fetuses from mothers that had not received DES, it was seen that DES almost completely repressed the expression of *Hoxa10* in the Müllerian duct (FIGURE 18.15). This repression was most pronounced in the stroma (mesenchyme) of the duct, where experimental embryologists had localized the effect of DES (Boutin and Cunha 1997). The case for DES acting through repression of *Hoxa10* is strengthened by the phenotype of the *Hoxa10* knockout mouse (Benson et al. 1996; Ma et al. 1998), in which there is a transformation of the proximal quarter of the uterus into oviduct tissue, as well as abnormalities of the uterotubal junction.

One link between Hox gene expression and uterine morphology is the Wnt proteins, which are associated with cell proliferation and protection against apoptosis. The Hox and Wnt proteins are both involved in the specification and

(A)

(B)

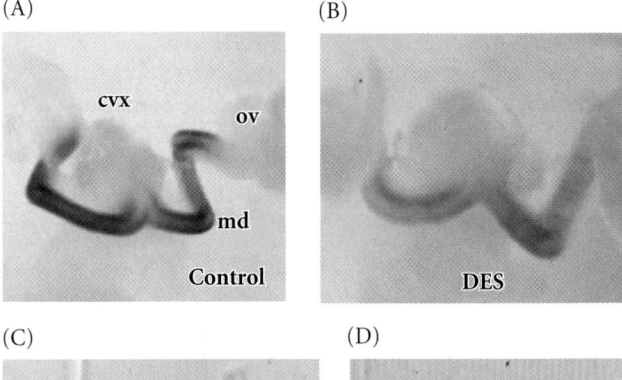

(C)

(D)

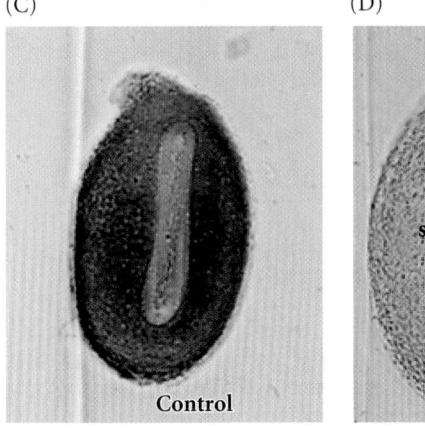

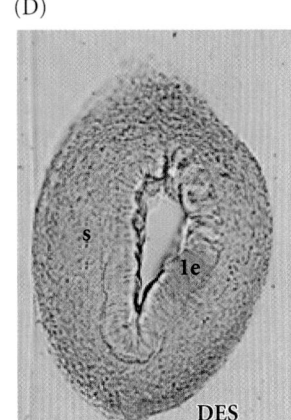

FIGURE 18.15 In situ hybridization of a *Hoxa10* probe shows that DES exposure represses *Hoxa10*. (A) Normal 16.5-day embryonic female mice show *Hoxa10* expression from the boundary of the cervix through the uterus primordium and most of the oviduct (cvx, cervix; md, Müllerian duct; ov, ovary). (B) In mice exposed prenatally to DES, this expression is severely repressed. (C) In control female mice at 5 days after birth (when reproductive tissues are still forming), a section through the uterus shows abundant expression of *Hoxa10* in the uterine mesenchyme. (D) In female mice that are given high doses of DES 5 days after birth, *Hoxa10* expression in the mesenchyme is almost completely suppressed (le, luminal epithelium; s, stroma). (After Ma et al. 1998.)

SIDELIGHTS & SPECULATIONS

Testicular Dysgenesis Syndrome

Our technological environment has increasingly become an estrogenic environment. Many of the endocrine disruptors found in plastics, insecticides, herbicides, perfumes, and hair products act as estrogens. Estradiol (the predominant human estrogen) and most estrogenic compounds bind to the estrogen receptors. These are dormant transcription factors that dimerize when they bind estrogenic compounds. They then migrate into the nucleus and regulate gene transcription. Some of the estrogenic compounds also bind to a membrane-bound G protein-coupled receptor, GPR30, that causes a

rapid release of calcium ions from the endoplasmic reticulum (Revankar et al. 2005). We have seen that DES can disrupt reproductive tract development in women. In men, estrogenic endocrine disruptors may also cause severe reproductive tract anomalies, and they have been closely associated with testicular dysgenesis,* a syndrome encompassing low sperm count, poorly formed testes and penis, and testicular tumors (**FIGURE 18.16**).

In the past three decades there has been an increase in testicular cancers and a decrease in sperm concentration throughout the industrialized world (Carlsen et al. 1992; Aitken et al. 2004).

The sperm count (number of sperm per milliliter) has dropped precipitously throughout much of Europe and the Americas. Skakkebaek and his team at Copenhagen University reviewed 61 international studies done between 1938 and 1992, involving 14,947 men. They found the average sperm count had fallen from 113 million per milliliter in 1940 to 66 million in 1990 (Carlsen et al. 1992). In addition, over the same time period, the number of "normal" sperm (sperm whose shape indicates their potential to fertilize eggs) fell from 60 million per milliliter to 20 million.

*Dysgenesis (Greek, "bad beginning") denotes defects in development.

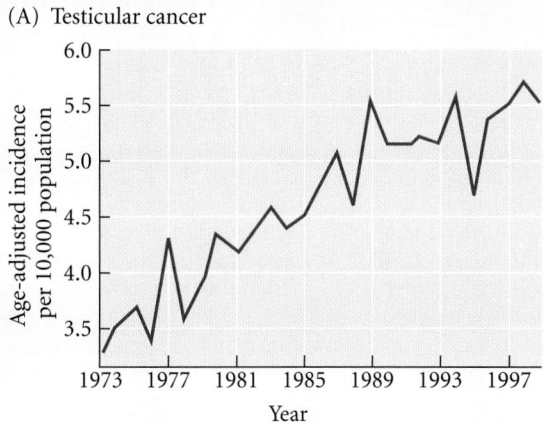

(A) Testicular cancer

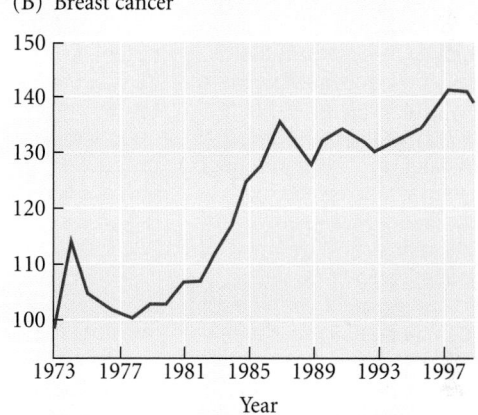

(B) Breast cancer

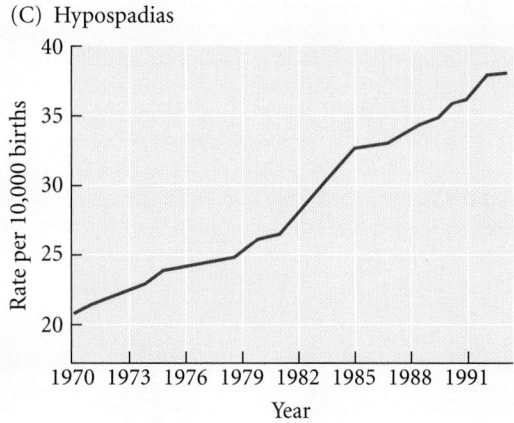

(C) Hypospadias

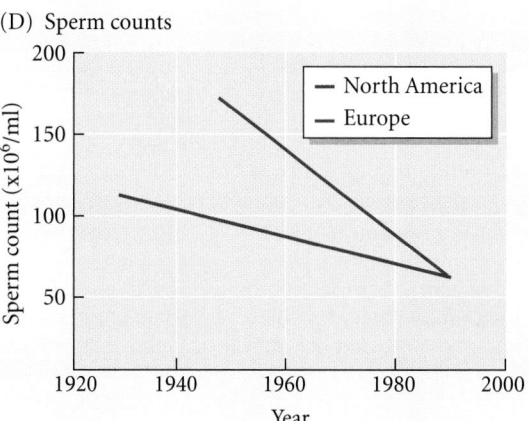

(D) Sperm counts

FIGURE 18.16 Developmental estrogen syndrome is manifest in climbing rates of breast cancer and testicular dysgenesis. The rise of testicular cancer (A) parallels the rise of breast cancer (B) and anomalies of penis development (C) such as hypospadias (failure to completely close the penis). (D) Sperm counts among North American males have declined nearly 50% within the past century. The decline has been even steeper among European men. (After Sharpe and Irvine 2004.)

SIDELIGHTS & SPECULATIONS (continued)

Subsequent studies have confirmed and extended Skakkebaek's findings (Merzenich et al. 2010). A survey of 1,350 sperm donors in Paris found that sperm counts declined by about 2% each year from the 1970s through the mid-1990s (Auger et al. 1995). A recent study (Rolland et al. 2013) shows that this trend has continued through 2005, such that sperm count for 35-year-old French men went from an average of 74 million sperm per milliliter of semen in 1989 to 50 million per milliliter in 2005. A study of Danish men (Jørgensen et al. 2012) found that only one man in four had optimal semen quality. The chance of fertilization diminishes significantly if the sperm concentration is below 40–50 million per milliliter, or if the percentage of normal spermatozoa is below 9% (Guzick et al. 2001), and that was the case for approximately 40% of men in the Danish study. More severe fertility problems are expected when sperm concentration is below 15 million per milliliter and the percentage of normal spermatozoa is less than 5%; these conditions were seen for 15% and 35%, respectively, of men from the general Danish population. In addition to the drop in sperm count documented in these studies, there has also been an increase in testicular cancers over recent decades (see Figure 18.17A).

Sharpe (1994) suggested that testicular dysgenesis syndrome may be due in large part to endocrine disruptors. While no chain of causation has been established completely (see Sharpe and Irvine 2004), there is evidence that the pathologies of this syndrome can be caused by environmentally relevant concentrations of endocrine disruptors. Indeed, all the developmental anomalies (but not the testicular tumors) can be induced by administering phthalate derivatives to pregnant rats (Fisher et al. 2003). Phthalates are ubiquitous in industrialized society and are widely used in plastics and cosmetics (that "new car smell" consists largely of volatilizing phthalates). Among male rats exposed

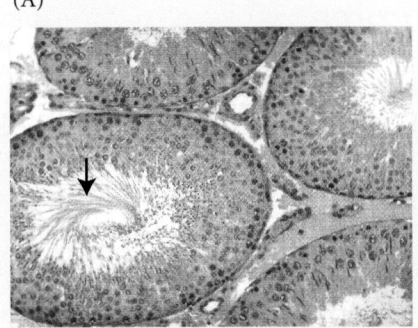
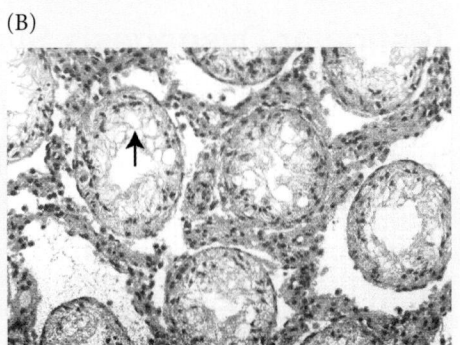

FIGURE 18.17 Cross section of seminiferous tubules from the testes of (A) a control rat and (B) a rat whose grandfather was born from a mother that was injected with vinclozolin; this rat was infertile. The arrow in (A) shows the tails of the sperm. The arrow in (B) shows the lack of germ cells in the much smaller tubule. (From Anway et al. 2005, courtesy of M. K. Skinner.)

in utero to dibutyl phthalate, more than 60% exhibited cryptorchidism (undescended testicles), hypospadias (misplaced urinary aperture), low sperm count, and testis abnormalities—phenotypes very similar to conditions found in human testicular dysgenesis syndrome. In humans, phthalates have been shown to inhibit testosterone production, alter testes morphology, and change the anatomy of the genital region (Duty et al. 2003; Swan et al. 2005; Desdoit-Lethimonier et al. 2012).

Other endocrine disruptors that adversely affect sperm are dioxins, nonylphenol, bisphenol A, acrylamide, and certain pesticides and herbicides (see Aitken et al. 2004; Newbold et al. 2006). The sunscreen 4-MBC, a camphor derivative, has been found to decrease the size of the testes and prostate glands, and it can delay male puberty in rats (Schlumpf et al. 2004). Pesticides may be critically important in impairing male fertility. The link between pesticides and infertility has been known for a long time (Carson 1962; Colborn et al. 1996). One of the most important endocrine disrupting pesticides is DDT. DDT is an estrogenic compound that has been associated with breast cancer (Cohn et al. 2007), and it breaks down into DDE (dichlorodiphenyldichloroethylene). DDE binds to the androgen receptor, preventing testosterone binding (Xu et al. 2006). In humans, DDT has been linked to pre-term births and immature

babies, and it is banned in the United States (Longnecker et al. 2001).

The fungicide vinclozolin (used extensively in grape farming) also works as an anti-androgen, inserting itself into the androgen receptor and preventing testosterone from binding there (Grey et al. 1999; Monosson et al. 1999). Male rats born to mothers injected with vinclozolin late in pregnancy are sterile. The sons of rats injected with vinclozolin during mid-pregnancy were able to reproduce, but their testis cells underwent apoptosis more than usual, their sperm count dropped 20%, and the sperm that remained had significantly lowered motility (**FIGURE 18.17**). When affected males were mated with normal females, the male offspring also had this testicular dysgenesis syndrome. Some of the offspring were sterile, and some had reduced fertility. The study (Anway et al. 2005) ended after the fourth generation of males continued to show low sperm count, low sperm motility, prostate disease, and high testicular cell apoptosis. This transgenerational effect is thought to be the result of methylation of genes involved in spermatogenesis (Guerrero-Bosagna et al. 2010).

Some scientists argue that these claims are exaggerated and that their tests on mice indicate that litter size, sperm concentration, and development are not affected by environmentally relevant concentrations of environmental estrogens. However,

SIDELIGHTS SPECULATIONS (continued)

investigations by Spearow and colleagues (1999) have shown a remarkable genetic difference in sensitivity to estrogen among different mouse strains. The strain that was used for testing environmental estrogens, the CD-1 strain of laboratory mice, is at

least 16 times more resistant to endocrine disruption than the most sensitive strains, such as B6. When estrogen-containing pellets were implanted beneath the skin of young male CD-1 mice, very little happened. However, when the same pellets were placed

beneath the skin of B6 mice, their testes shrank and the number of sperm seen in the seminiferous tubules dropped dramatically (**FIGURE 18.18**). This widespread range of sensitivities has important consequences for determining safety limits for humans.

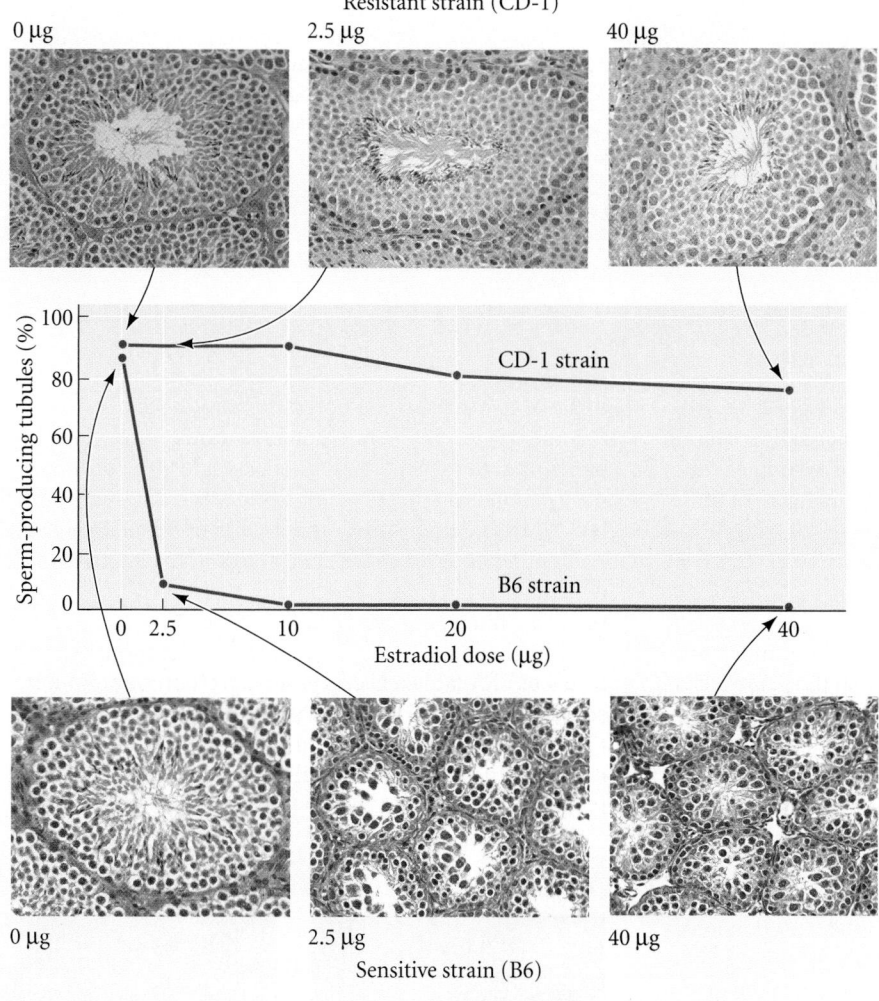

FIGURE 18.18 Effects of estrogen implants on different strains of mice. The graph shows the percentage of seminiferous tubules containing elongated spermatozoa. (The mean standard error is for an average of six individuals.) The micrographs show cross sections of the testicles and are all at the same magnification. Forty μg of estradiol did not affect spermatogenesis in the CD-1 strain, but as little as 2.5 μg almost completely abolished spermatogenesis in the B6 strain. (After Spearow et al. 1999, photographs courtesy of J. L. Spearow.)

morphogenesis of the reproductive tissues (**FIGURE 18.19**). The reproductive tracts of DES-exposed female mice resemble those of *Wnt7a* knockout mice. The Hox genes and the Wnt genes communicate to keep each other activated. However, DES, acting through the estrogen receptor, represses the *Wnt7a* gene. This repression prevents the maintenance of the Hox gene expression pattern, and also prevents the activation of another Wnt gene, *Wnt5a*, which encodes a protein necessary for cell proliferation (Miller et al. 1998; Carta and Sassoon 2004).

The effects of DES on fertility is a complex story of public policy, medicine, and developmental biology (Bell 1986; Palmlund 1996). Moreover, the 1992 Wingspread Statement brought to the public's attention that endocrine disruption such as that seen by DES is ongoing. The Chapel Hill Statement of 2007, a scientific consensus document,* claimed

---

*The conference that generated this statement was sponsored by the Environmental Protection Agency and the National Institute of Environmental Health Sciences.

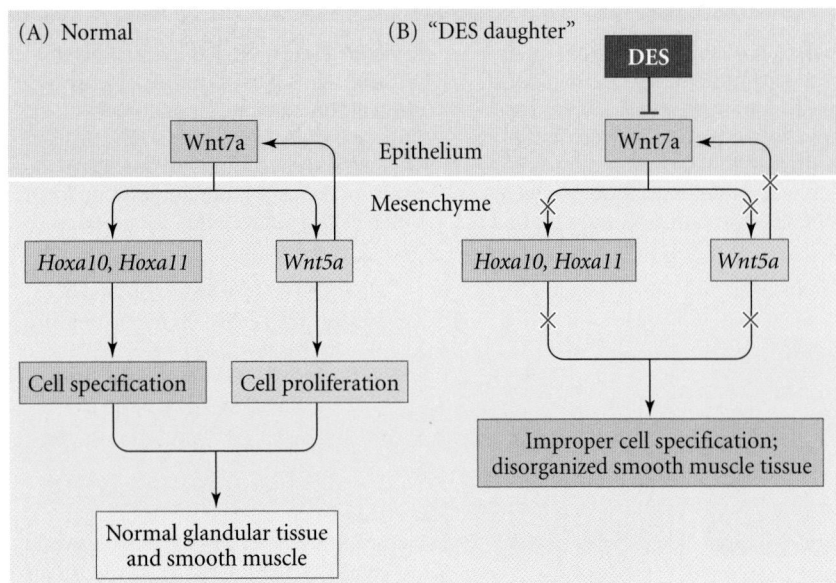

(A) Normal

(B) "DES daughter"

FIGURE 18.19 Misregulation of Müllerian duct morphogenesis by DES. (A) During normal morphogenesis, the *Hoxa10* and *Hoxa11* genes in the mesenchyme are activated and maintained by Wnt7a from the epithelium. Wnt7a also induces *Wnt5a* in the mesenchyme, and Wnt5a protein both maintains *Wnt7a* expression and causes mesenchymal cell proliferation. Together, these factors specify and order the morphogenesis of the uterus. (B) DES, acting through the estrogen receptor, blocks *Wnt7a* expression. Proper activation of the Hox genes and *Wnt5a* in the mesenchyme does not occur, leading to a radically altered morphology of the female genitalia. (After Kitajewsky and Sassoon 2000.)

that some of the major constituents of plastics were estrogenic compounds, and that they were present in doses large enough to have profound effects on sexual development and behavior (vom Saal et al. 2007). The most important of these plastics additives is bisphenol A.

Bisphenol A (BPA)

In the early years of hormone research, the steroid hormones were very difficult to isolate, so chemists manufactured synthetic analogues that would accomplish the same tasks. Bisphenol A was one of these analogues, synthesized as an estrogenic compound in the 1930s. Later, polymer chemists realized that BPA could be used in plastic production, and today it is one of the top 50 chemicals in production worldwide. Four corporations in the United States make almost 2 billion pounds of it each year for use in the resin lining in most cans, the polycarbonate plastic in baby bottles and children's toys, and dental sealant. BPA is also used in making polycarbonate water bottles and in cash register receipts. In its modified form, tetrabromo-bisphenol A, it is the major flame retardant on fabrics. Human exposure comes primarily from BPA that has leached from food containers (von Goetz et al. 2010). Babies and infants acquire BPA through polycarbonate bottles; teenagers and adults get most of their BPA through the consumption of canned food that has been stored in containers lined with BPA-containing resins.

BPA does not remain fixed in plastic forever (Krishnan et al. 1993; vom Saal 2000; Howdeshell et al. 2003). If you let water sit in an old polycarbonate rat cage at room temperature for a week, you can measure about 300 μg per liter of BPA in the water. That is a biologically active amount—a concentration that will reverse the sex of a frog and cause weight changes in the uterus of a young mouse. It also can cause chromosome anomalies. When a laboratory technician

mistakenly rinsed some polycarbonate cages in an alkaline detergent, the female mice housed in the cages showed meiotic abnormalities in 40% of their oocytes (the normal level of such abnormalities is about 1.5%). When BPA was administered to pregnant mice under controlled circumstances, Hunt and her colleagues (2003) showed that short, low-dose exposure to BPA was sufficient to cause meiotic defects in maturing mouse oocytes (**FIGURE 18.20**). This effect was also seen in primates. Exposure of fetal female monkeys to low doses of BPA (at levels comparable to that found in human serum) caused ovarian and meiotic abnormalities similar to those observed in mice. There were several abnormalities of ovarian function, including abnormal meiotic chromosome behavior and aberrant follicle formation (Hunt et al. 2012).

(A)   (B)

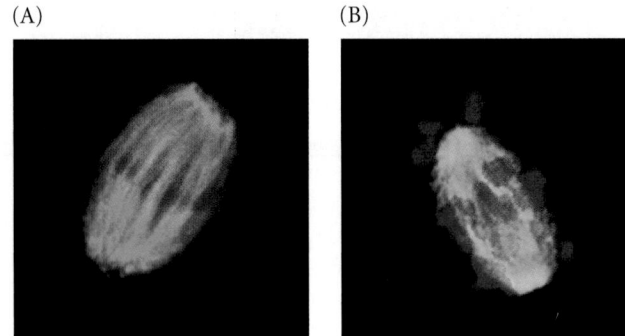

FIGURE 18.20 Bisphenol A causes meiotic defects in maturing mouse oocytes. (A) Chromosomes (red) normally line up at the center of the spindle during first meiotic metaphase. (B) Short exposures to BPA cause chromosomes to align randomly on the spindle. Different numbers of chromosomes then enter the egg and polar body, resulting in aneuploidy and infertility. (From Hunt et al. 2003, courtesy of P. Hunt.)

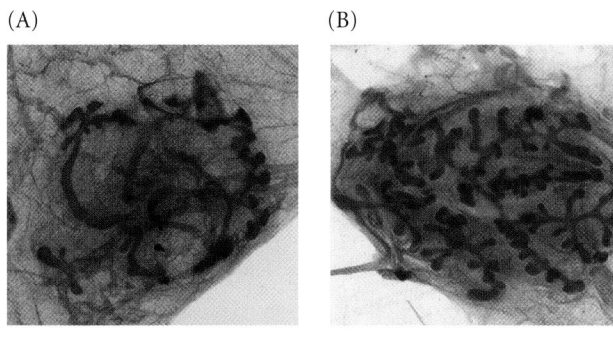

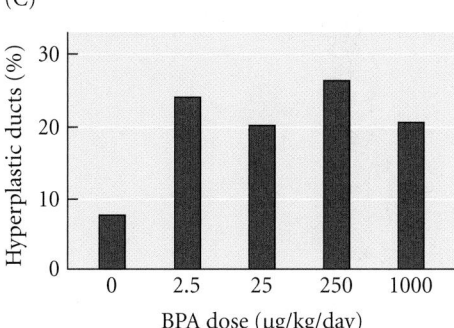

FIGURE 18.21 Bisphenol A induces altered mammary gland development. (A,B) Whole mount stained preparation of mammary glands from newborn female rhesus monkeys. (A) Control mammary gland. (B) Mammary gland from a fetus exposed in utero to BPA. Twice as many buds (incipient branches) are seen in the BPA-exposed tissue. (C) The percentage of mouse mammary glands showing intraductal hyperplasia (a cancer-prone state) is significantly increased at postnatal day 50 in BPA-exposed animals. (A,B from Tharp et al. 2012; C from Murray et al. 2007.)

BPA AND REPRODUCTIVE HEALTH   BPA crosses the human placenta and accumulates in concentrations that can alter development in laboratory animals (Ikezuki et al. 2002; Schönfelder et al. 2002). Moreover, since 95% of urine samples taken from people in the U.S. and Japan have measurable BPA levels (Calafat et al. 2005), public health concerns have been raised over the roles BPA might play in causing reproductive failure, cancer, and behavioral anomalies. In model organisms, BPA at environmentally relevant concentrations can cause abnormalities in fetal gonads, prostate enlargement, low sperm counts, and behavioral changes when these fetuses become adults (vom Saal et al. 1998, 2005; Palanza et al. 2002; Kubo et al. 2003). When vom Saal and colleagues (1997) gave pregnant mice 2 parts per billion BPA—that is, 2 nanograms per gram of body weight—for the 7 days at the end of pregnancy (equivalent to the period when human reproductive organs are developing), male offspring showed an increase in prostate size of about 30% (Wetherill 2002; Timms et al. 2005). Female mice exposed to low doses of BPA in utero had reduced fertility and fecundity as adults (Cabaton et al. 2007).

This lower fertility may be the result of several actions. First, BPA and other endocrine disruptors are found to prevent the sex-specific maturation of those parts of the mouse brain regulating ovulation (Ruben et al. 2006; Gore et al. 2011). Second, female mice exposed in utero to low doses of BPA (2000 times lower than the dosage considered safe by the U.S. government) had alterations in the organization of their uterus, vagina, breast tissue, and ovaries, as well as altered estrous cycles as adults (Howdeshell et al. 1999, 2000; Markey et al. 2003). And third, BPA alters the gamete-specific methylation pattern of imprinted genes in mouse embryos and placentas (Susiarjo et al. 2013).

BPA AND CANCER SUSCEPTIBILITY   BPA appears to make breast tissue more sensitive to estrogens, and it is thought that in utero exposure to BPA may predispose women to breast cancer later in life. Fetal exposure to BPA caused the development of early-stage cancer in the mammary glands of one-third of the rats exposed to environmentally relevant doses of BPA later in life (Murray et al. 2006). None of the control rats developed such cancers. Furthermore, daily gestational exposure to as little as 25 ng BPA per kilogram of body weight, followed at puberty by a "subcarcinogenic dose" of a chemical carcinogen, resulted in the formation of tumors *only* in those animals exposed to BPA (Durando et al. 2006). Indeed, altered mammary development had already manifested during fetal life in BPA-exposed mice, and at puberty, the mammary glands produced more terminal buds and were more sensitive to estrogen, which may have predisposed these mice toward breast cancer as adults (Muñoz-de-Toro et al. 2005). Moreover, exposure of female monkey fetuses to low doses of BPA (at levels comparable to those found in human blood serum) caused changes in mammary development similar to those seen in BPA-exposed mice (**FIGURE 18.21**).

The plastics industry claims that BPA is safe (see Cagen et al. 1999; Lamb 2002). However, a review of the industry's own studies (which claim that mice exposed in utero to BPA do not have enlarged prostates or low sperm counts) points out that the positive control in the industry-sponsored research did not produce the expected effects (see vom Saal and Hughes 2005). Indeed, confronted with one study (Lernath et al. 2008) showing that, at concentrations *lower* than what the U.S. EPA considers safe, BPA disrupted monkey brain development, the American Chemical Council replied that "there is no direct evidence that exposure to bisphenol-A adversely affects human reproduction or development" (see Layton 2008; Gilbert and Epel 2009).* Reviewing the

---

*The catch is that "direct evidence" would mean testing the drugs in known concentrations on human fetuses. In the absence of government regulation, Nalgene and Wal-Mart voluntarily stopped making and selling BPA-containing bottles. Government regulation also does not cover the compounds added to the environment by the hydraulic fracturing ("frakking") procedures used to extract methane (natural gas) from shale. A total of 632 chemicals have been identified as being used in this procedure. About 25% of them are known to cause tumors, and more than 35% of them are known affect the endocrine system (Colburn et al. 2011). It is estimated that about 50% of the fluid used in frakking returns to the surface (DOE 2009).

FIGURE 18.22 Endocrine disruptors as obesogens. (A) DES-induced obesity in mice. Two genetically identical female mice are shown here. The mother of the mouse on the left was injected during pregnancy with solution only. The mother of the mouse on the right was injected with solution containing DES. Weight gain in their offspring became appreciable only after 8 weeks (i.e., at "mouse puberty"), when estrogenic hormones increased. (B) Postulated mechanism for obesity as a result of the activation of PPARγ, which biases mesenchymal stem cells to become adipocytes and activates the fat-storing enzymes in these cells. (A after Newbold 2005, courtesy of R. Newbold; B after Kirschner et al. 2010.)

(A)

literature, vom Saal and Hughes (2005; Chapel Hill Consensus 2007) conclude that BPA is one of the most dangerous chemicals known and that governments should consider banning its use in products containing liquids that humans and animals might drink.

● **See WEBSITE 18.4** BPA and altered behavior

Endocrine disruptors as obesogens

Several endocrine disruptors, including DES and BPA, increase the production of adipose (fat) cells and the accumulation of fat therein. These substances are therefore called **obesogens**. Newbold and colleagues (2005, 2009) demonstrated that mice treated in utero (during the major periods of organogenesis) or soon after birth with as little as 1 part per billion of DES became significantly fatter later in life (**FIGURE 18.22A**). The mice became sensitized early in life by the DES. Later, when the large concentrations of estrogen associated with sexual maturity were secreted, the mice became obese. This obesogenic effect has been seen with other estrogenic endocrine disruptors, including BPA (Rubin et al. 2001; vom Saal et al. 2012).

Another obesogen is tributyltin (TBT). This compound was applied to the hulls of ships as an anti-fouling agent until it was found to be an endocrine disruptor that converts testosterone into estrogen and changes the sex of molluscs living near shipyards (Oberdörster and McClellan-Green 2002). TBT is still used in fungicides, wood preservatives, and heat stabilizers in plastics. When ingested by a pregnant mouse, TBT makes her pups obese. It does this by activating the transcription factor PPARγ (Evans et al. 2004; Janesick and Blumberg 2011, 2012). When activated in the liver, PPARγ activates the genes involved in fat synthesis and storage. TBT not only activates the transcription factor but also appears to demethylate the region of DNA regulating the PPARγ gene, making PPARγ even more abundant in the cell (Grün et al. 2006; Kirschner et al. 2010).

The main developmental effect of TBT's activation of PPARγ, however, is its effect on mesenchymal stem cells. When it is activated in these stem cells, PPARγ instructs the cells' descendants to become adipose cells instead of bone or cartilage cells, the two other main derivatives of the mesenchymal stem cell (**FIGURE 18.22B**). By altering the commitment of mesenchymal cells toward the adipose fate, TBT

(B)

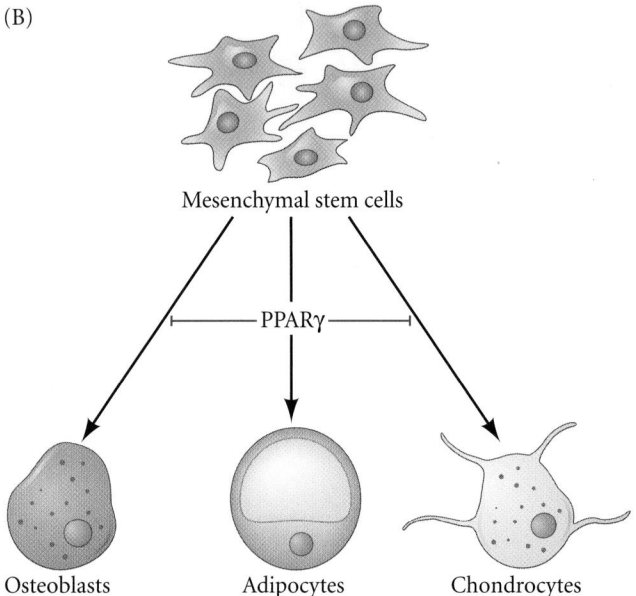

Mesenchymal stem cells

PPARγ

Osteoblasts   Adipocytes   Chondrocytes

causes the embryo to make a larger number of adipose cells, and by activating the fat synthesis and storage genes, it makes the fat cells functional. Estrogen mimics such as DES appear to act in a similar manner, activating the synthesis of PPARγ in mesenchymal stem cells (Hao et al. 2012). If the number of adipose cells at birth is a major factor of adult obesity (see Janesick and Blumberg 2011, 2012), then endocrine disruptors may be an important part of the current epidemic of obesity in the industrialized world. As Newbold and her colleagues (2007) have noted, "No longer can we assume than overweight and obesity are simply personal choices, but we have to consider that complex events including environmental chemicals are contributing to this mounting human health problem."

Transgenerational Inheritance of Developmental Disorders

One of the most surprising results of contemporary developmental genetics has been the discovery that certain

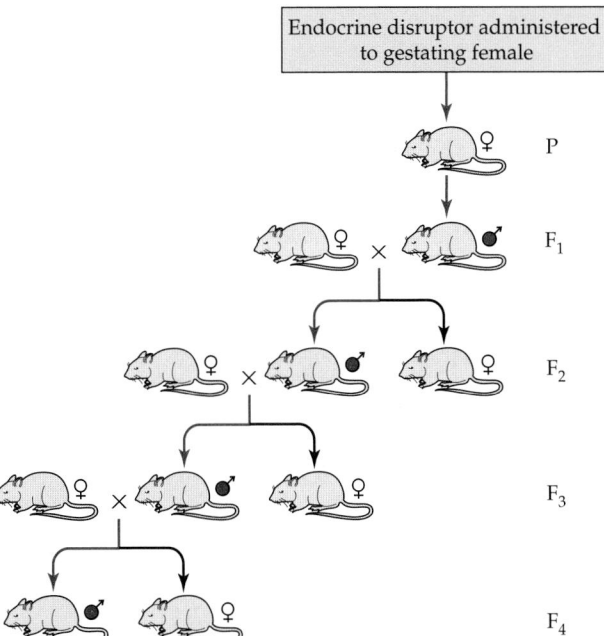

FIGURE 18.23 Epigenetic transmission of endocrine disruption. Transmission of testicular dysgenesis syndrome (red circles) is shown through four generations of mice. The only mice exposed in utero were the F_1 generation. (After Anway and Skinner 2006.)

subsequent generation's male offspring (Anway et al. 2005, 2006; Guerrero-Bosagna et al. 2010). Thus, when a pregnant rat is given vinclozolin, even her great-grandsons are affected (FIGURE 18.23; see also Figure 18.18).

The mechanism for this inheritance appears to be DNA methylation. The promoters of more than 100 genes in the Sertoli cells have their methylation pattern changed by vinclozolin, and altered promoter methylation can be seen in the sperm DNA for at least three subsequent generations (Guerrero-Bosagna et al. 2010; Stouder and Paolini-Giacobino 2010). These genes include those whose products are necessary for cell proliferation, G proteins, ion channels, and receptors. It is important to note that by the third (F_3) generation, there can be no direct exposure to vinclozolin. The fetus is inside the treated mother; the fetus has germ cells (of the F_2 generation) inside itself. But even though the offspring of the F_3 and F_4 generations have never been exposed to vinclozolin, their phenotype is changed by the initial injection to their great-grandmother. Similar studies have indicated that other endocrine disruptors—DES, bisphenol A, and PCBs—also have transgenerational effects (Skinner et al. 2010; Walker and Gore 2011). Indeed, the behavioral changes induced by BPA in mice may last at least four generations (Wolstenhome et al. 2012). The public health ramifications of this type of inheritance are just beginning to be explored.

● See WEBSITE 18.5 Our stolen future

Cancer as a Disease of Development

Because endocrine disruptors are known to cause tumors as well as developmental abnormalities, cancer is increasingly being studied as a disease of development. However, the idea that cancer is a disease of development has been recognized for many years (see, e.g., Stevens 1953; Auerbach 1961; Pierce et al. 1978). Carcinogenesis is more than just genetic changes in the cells giving rise to the tumor (see Hanahan and Weinberg 2000). Rather, carcinogenesis can be viewed as aberrations of the very processes that underlie differentiation and morphogenesis. It once was thought that carcinogensis and metastasis were caused by the proliferation of a cell that had acquired mutations enabling it to become "autonomous," thus defining cancer by the intracellular mechanisms that enable a cell to become independent of its environment. But this turns out to be only part of the explanation, and we now know that the initial cancer cells modify their environment, turning it into a cancer-promoting niche. Cancer is being recast as the result of a "stepwise progression" of conditions that depends on reciprocal interactions between incipient cancer cells and the supporting cells of their tissue environment. The progressive alteration of cell-to-cell interactions leads to aberrant tissue architecture and possibly to the formation of niches that generate cancer cells. Indeed, cancer cells appear to proceed by recapitulating steps of normal development, including the formation of a niche in which to proliferate.

environmentally induced phenotypes can be transmitted from generation to generation. The Weismannian block to the transmission of acquired traits is a genetic block. A lifetime of chopping wood will not give your offspring bulging biceps; nor would the loss of one's arms in an accident cause one's offspring to have a propensity for limblessness. This is because the environmental agent does not cause mutations in the DNA. And mutations, if they are to be transmitted, must not only be somatic, they must enter the germ line. So mutations acquired in the skin by being in sunlight will not be transmitted.

However, DNA methylation seems to be a mechanism that can circumvent that block. Certain agents can cause the same alterations of DNA methylation throughout the body, and these alterations in methylation can be transmitted by the sperm and egg. Jablonka and Raz (2009) have documented dozens of such cases wherein different "epialleles"—DNA containing different methylation patterns—can be stably transmitted from generation to generation. In mammals, epiallelic inheritance was first documented by studies of the endocrine disruptor vinclozolin, a fungicide used widely on grapes. When injected into pregnant rats during particular days of gestation, vinclozolin causes testicular dysgenesis in the male offspring. The testes start forming normally, but as the mouse gets older, its testes degenerate and no more sperm is made. What's more interesting is that the male mice born to those mice that get testicular dysgenesis also get testicular dysgenesis. So do their male offspring and the

Both carcinogenesis and congenital anomalies can be seen as diseases of tissue organization, differentiation, and intercellular communication. As we will see, they are often caused by defects in the same pathways. There are many reasons to view malignancy and metastasis in terms of development, four of which will be discussed here:

1. Context-dependent tumor formation
2. Deficient stem cell regulation in tumor formation
3. Reactivation of embryonic migration pathways
4. Epigenetic reprogramming of cancer cells

Context-dependent tumors

Many tumor cells have normal genomes, and whether or not these tumors are malignant depends on their environment. The most remarkable of these cases is the **teratocarcinoma**, which is a tumor of germ cells or stem cells (Illmensee and Mintz 1976; Stewart and Mintz 1981). Teratocarcinomas are malignant growths of cells that resemble the inner cell mass of the mammalian blastocyst, and they can kill the organism. However, if a teratocarcinoma cell is placed on the inner cell mass of a mouse blastocyst, it will integrate into the blastocyst, lose its malignancy, and divide normally. Its cellular progeny can become part of numerous embryonic organs. Should its progeny form part of the germ line, the sperm or egg cells formed from the tumor cell will transmit the tumor genome to the next generation (see Figure 17.10). Thus, whether the cell becomes a tumor or part of the embryo can depend on its surrounding cells.

The mechanism by which the stem cell environment suppresses tumor formation may be due to its secretion of inhibitors of the paracrine pathways. For instance, many tumor cells, such as melanomas, secrete the paracrine factor Nodal. This aids their proliferation and also helps supply them with blood vessels. When placed in an environment of embryonic stem cells (which secrete Nodal inhibitors), aggressive melanoma tumors (which are derived from neural crest cells) become normal pigment cells (Hendrix et al. 2007; Postovit et al. 2008). Remarkably, such malignant melanoma cells, when transplanted into early chick embryos, downregulate their Nodal expression and migrate as non-malignant cells along the neural crest cell pathways (**FIGURE 18.24**; Kasemeier-Kulesa et al. 2008).

DEFECTS IN CELL-CELL COMMUNICATION   This brings us to the idea that cancer can be caused by miscommunication between cells. In many cases, tissue interactions are required to prevent cells from dividing. Thus, tumors can arise through defects in tissue architecture, and the surroundings of a cell are critical in determining malignancy (Sonnenschein and Soto 1999, 2000; Bissell et al. 2002). Studies have shown that tumors can be caused by altering the structure of the tissue, and that these tumors can be suppressed by restoring an appropriate tissue environment (Coleman et al. 1997; Weaver et al. 1997; Booth et al. 2010). In particular, although 80% of human tumors are from epithelial cells, these cells do not always appear to be the site of the cancer-causing lesion. Rather, epithelial cell cancers are often caused by defects in the mesenchymal stromal cells that surround and sustain the epithelia. When Maffini and colleagues (2004) recombined normal and carcinogen-treated epithelia and mesenchyme in rat mammary glands, tumorous growth of mammary epithelial cells occurred not in carcinogen-treated epithelia, but only in epithelia placed in combination with mammary mesenchyme that had been exposed to the carcinogen. Thus, the carcinogen caused defects in the mesenchymal stroma of the mammary gland, and apparently the treated mammary stroma could no longer provide the epithelial cells the instructions to form normal structures. In turn, these abnormal structures exhibited a loose control of cell proliferation.

DEFECTS IN PARACRINE PATHWAYS   This brings us to the next notion: that tumors can occur by disruptions of paracrine signaling between cells. Rubin and de Sauvage (2006) concluded that "several key signaling pathways, such as Hedgehog, Notch, Wnt and BMP/TGF-β/Activin, are involved in most processes essential to the proper development of an embryo. It is also becoming increasingly clear that these pathways can have a crucial role in tumorigenesis when reactivated in adult tissues through sporadic mutations or other mechanisms." We have seen this, above, in the discussion of Nodal secretion by melanoma cells. These findings demonstrate the importance of stromal tissue just mentioned. Many tumors, for instance, secrete the paracrine factor Sonic hedgehog, which can act in one of two ways. First, it can act in an autocrine fashion, stimulating the cells that produce it to grow. Autocrine Shh is normally required for maintenance of cerebellar granule neuron progenitor cells and hematopoietic stem cells; Shh pathway inhibitors can reverse certain medulloblastomas and leukemias, which are tumors of these cell types (**FIGURE 18.25A,B**; Rubin and de Sauvage et al. 2006; Zhao et al. 2009). An autocrine requirement for hedgehog has also been reported for small-cell lung carcinoma, pancreatic adenocarcinoma, prostate cancer, breast cancer, colon cancer, and liver cancer. Second, in some instances, the Shh produced by the tumor cells may not act on the tumor cells themselves but on the stromal cells, causing the stromal cells to produce factors (such as insulin-like growth factor, IGF) that support the tumor cells (**FIGURE 18.25C**). If the Shh pathway is blocked, the tumor regresses (Yauch et al. 2008, 2009; Tian et al. 2009). Cyclopamine, a teratogen that blocks Shh signaling, can prevent certain of these tumors from growing (Berman et al. 2002, 2003; Thayer et al. 2003; Song et al. 2011).

Thus, the same chemicals that can cause teratogenesis by blocking a pathway in embryonic development may be very useful in blocking the activation of cancer stem cells. Cyclopamine and other antagonists of the Hedgehog pathway, for instance, can cause malformation in embryos, but they appear to be useful in preventing the generation and proliferation of medulloblastoma stem cells (Berman et al. 2002; De Smale et al. 2010). Even the classic teratogen thalidomide is being "rehabilitated" to fight cancer. Thalidomide and its

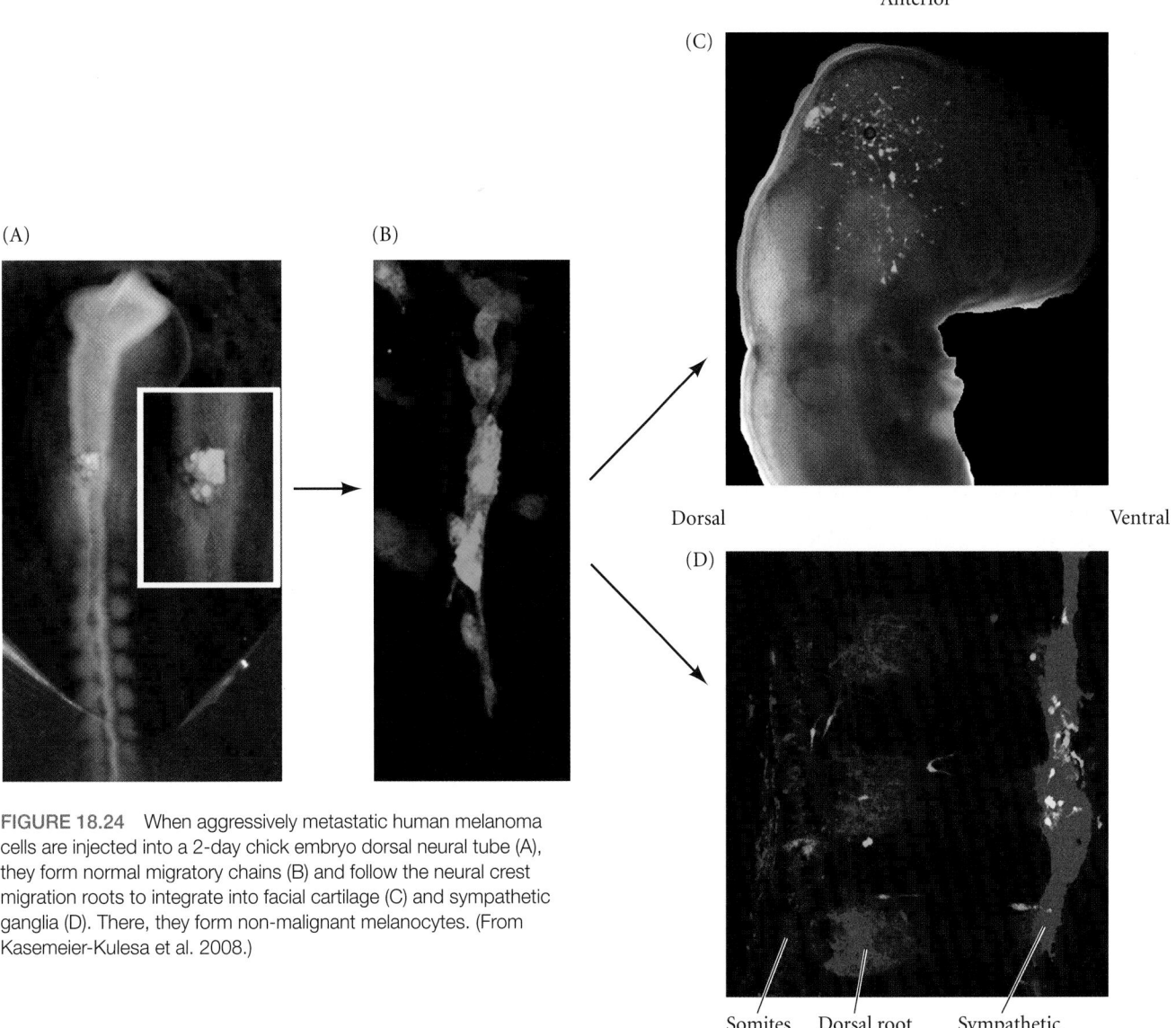

Anterior

(C)

Dorsal   Ventral

(D)

Somites   Dorsal root   Sympathetic
           ganglia   ganglia

Posterior

FIGURE 18.24 When aggressively metastatic human melanoma cells are injected into a 2-day chick embryo dorsal neural tube (A), they form normal migratory chains (B) and follow the neural crest migration roots to integrate into facial cartilage (C) and sympathetic ganglia (D). There, they form non-malignant melanocytes. (From Kasemeier-Kulesa et al. 2008.)

derivatives have been found to inhibit the Wnt and FGF pathways (Hansen et al. 2002; Knobloch et al. 2007) and have recently been approved for the treatment of certain myelomas and leukemias (List et al. 2005; Aragon-Ching et al. 2007).

THE CANCER STEM CELL HYPOTHESIS Another aspect of viewing cancers as diseases of development is that the properties of tumors may emerge because of a population of cells that are analogous to adult stem cells.* Despite their possible

* The idea that concers had stem cells was one of the first links connecting cancer research and developmental biology. Pierce and Johnson (1971) reported that "malignant tissue, like normal tissue, maintains itself by proliferation and differentiation of its stem cells." That same year, Pierce and Wallace (1971) showed stem cells in rat carcinomas. The similarities between normal stem cells and cancer stem cells was highlighted when lineage tracing revealed that the stem cells of intestinal adenomas (the precursor of intestinal cancer) are Lgr5+ and have the same relationship to the Paneth cells as do normal intestinal stem cells (Shepers et al. 2013; see Figure P3.6).

single-cell origin, all tumors are heterogeneous, containing cells with different morphological and molecular characteristics. In numerous cases, including glioblastomas (the most common brain tumor), prostate cancer, melanomas, and myeloid leukemias, there is a rapidly dividing cancer stem cell (CSC) population that gives rise to more cancer stem cells and to populations of relatively slowly dividing differentiated cells (Lapidot et al. 1994; Chen et al. 2012; Driessens et al. 2012; Schepers et al. 2012). These CSCs can self-renew as well as generating the non-stem cell populations of the tumor. Indeed, when tumor cells are transplanted from one animal to another, only the CSCs can give rise to new heterogeneous tumors (Gupta et al. 2009; Singh and Settleman 2010). The origins

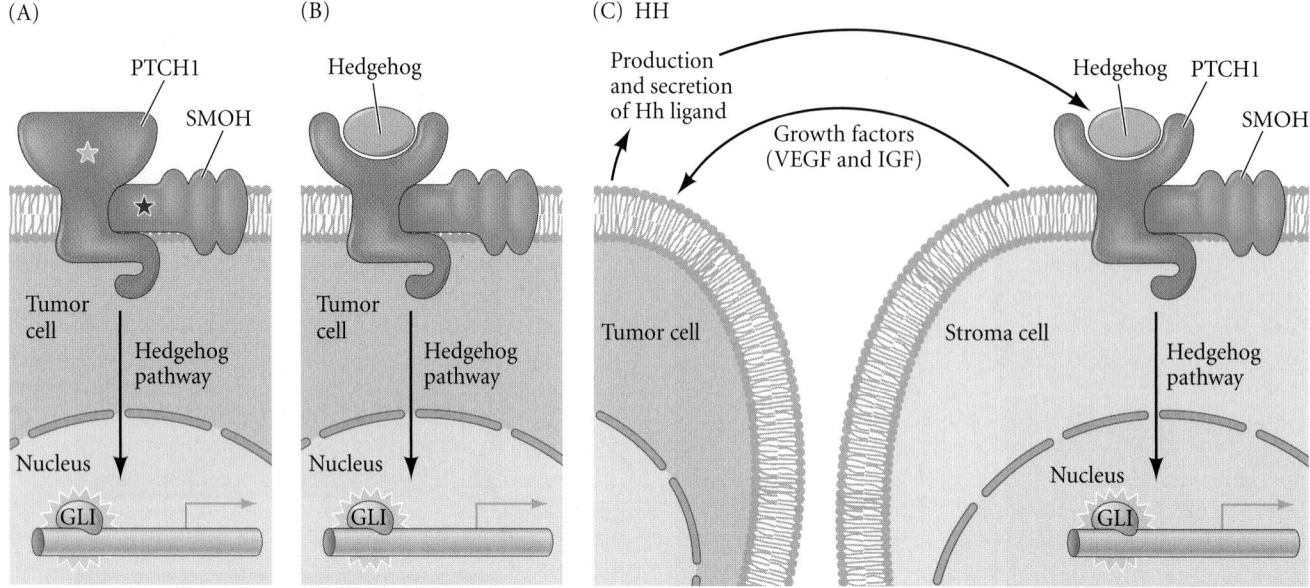

FIGURE 18.25 Mechanisms by which the Hedgehog pathway (see Figure 3.25) can lead to cancer. (A) When Shh is a mitogen (as it is for cerebellar granule neuron progenitor cells or hematopoietic stem cells), loss-of-function mutations in the Hh ligand Patched (PTCH1; yellow star) or gain-of-function mutations in the Patched inhibitor Smoothened (SMOH; blue star) activate the Hedgehog pathway, even in the absence of Shh or another Hedgehog protein. (B) In the autocrine model, tumor cells both produce and respond to the Hh ligand. (C) In the paracrine model, tumor cells produce and secrete the Hh ligand, and the surrounding stromal cells receive the Hh protein. The stromal cells respond by producing growth factors such as VEGF or IGF that support tumor growth or survival. (After Rubin and de Sauvage 2006.)

of CSCs remains uncertain and may be different for different tumor types. Most researchers feel that the CSC comes from either a normal adult stem cell or a progenitor (transit amplifying) cell.

When the tumor is forming, CSCs produce more cancer stem cells as well as the bulk of the more differentiated tumor cells. Remarkably, it appears that the CSCs of aggressive glioblastomas not only make immature glia-like cells (the glioblasts), they also make blood vessel endothelial cells. In this way, the tumor can create its own vasculature (El Hallani et al. 2010; Ricci-Vitiani et al. 2010; Wang et al. 2010).

Cancer stem cells and the epithelial-mesechymal transition

One of the most dangerous properties of cancer stem cells is their ability to metastasize—that is, to migrate from the primary tumor and form colonies in other tissues or organs. Such migration and colony-forming abilities are seen during development, as when neural crest cells migrate from the neural tube to form neural ganglia in the gut or when myotome cells migrate from the somite to form limb muscle. The beginning of such migrations is the epithelial-mesenchymal transition (EMT). We observed in earlier chapters that EMT is

caused by the downregulation of cadherins on the cell surface of the epithelial cells (usually proceeded by the upregulation of transcription factors such as Slug, Snail, and Twist, which repress cadherin gene transcription) as well as by the reorganization of the cytoskeleton and the production of proteases. The same phenomenon occurs when solid tumor cells gain the ability to metastasize (Foty and Steinberg 2004; Tsai et al. 2012).

The adult EMT may also do something even more dramatic: it may produce cancer stem cells. This is a controversial point (see Tarin et al. 2006; Hanahan and Weinberg 2011). However, Mani and colleagues found that when breast cancer cells had undergone EMT in culture, the resulting cells not only had membranes with proteins characteristic of stem cells, but these particular cells seeded new tumors just as cancer stem cells do (Mani et al. 2008; Morel et al. 2008, 2012; Singh and Settleman 2010). Moreover, once the cells have passed through EMT, they appear to have another property often associated with migrating embryonic cells: their cell death pathways have been inactivated. Most epithelial cells undergo anoikis—detachment-induced apoptosis—when removed from their extracellular matrix (see Chapter 3). Epithelial cells that have become mesenchymal through EMT (such as the migrating neural crest or sea urchin micromeres) do not. Most stem cells die when they leave their niche, but the normal transit amplifying cells and many tumor cells have activated the genes that block apoptosis.

Another phenomenon of metastasis involves the digestion of extracellular matrices by **metalloproteinases**. These enzymes are used by migrating embryonic cells to digest a path to their destination. They are commonly secreted by trophoblast cells, neural crest cells, axon growth cones, sperm cells, and somitic cells. Metalloproteinases are often re-expressed in malignant cancer cells, allowing the cancer to invade other tissues. The presence of these enzymes is a marker that the tumor is particularly dangerous (see Gu et al. 2005). Thus, in its ability to undergo EMT, its ability to form dividing stem cells, and its ability to migrate, cancer appears to be a reutilization of embryonic cell abilities in adult tissues. To understand cancer, one has to understand development.

Cancer and epigenetic gene regulation

In Chapter 16 we saw evidence that the methylation patterns of mammalian genes change with age. We specifically looked at genes that might cause elements of the aging phenotype. But what would happen if the random, age-dependent patterns of gene methylation altered the transcription of the genes regulating cell division and cell signaling? One might expect that certain genes necessary for the usual constraints on cell division or signal reception might be inappropriately expressed or repressed.

One of these genes is the estrogen receptor that regulates cell division. In some cells (such as estrogen-dependent breast cancers), estrogen activates cell proliferation. In the colon, however, estrogen stops the proliferation of cells, and estrogen receptors function as tumor-suppressor genes. Issa and colleagues (1994) showed that in addition to the age-associated methylation of estrogen receptors, there was a much higher level of DNA methylation in the estrogen receptor genes in colon cancers. Even the smallest colon cancers had nearly 100% methylation of the cytosines in the promoter of the estrogen receptor gene.

The epigenetic causation of cancer does not exclude a genetic cause. Indeed, several studies indicate that these mechanisms augment one another. Numerous mutations occur in each cancer cell, and recent evidence suggests that as many as 14 significant tumor-promoting mutations are found in each cancer cell (Sjöblom et al. 2006). Many of the mutations found in the cancer cells are those that involve intercellular signaling and histone regulators (Liu et al. 2012; Seshagiri et al. 2012).

Jacinco and Esteller (2007) have presented evidence that the large number of mutations that accumulate in cancer cells may have an epigenetic cause. In some cancer cells, the genes encoding DNA repair enzymes appear to be susceptible to inactivation by methylation. Once DNA repair enzymes have been downregulated, the number of mutations increases.

Interestingly, although environmental exposure to substances such as cigarette smoke and endocrine disruptors can increase DNA methylation, and certain mutations can predispose one toward developing cancer, there is also a great deal of random chance involved. Rather than a dramatic reprogramming of cell fate, carcinogenesis may be due to the slow accumulation of hypermethylated promoter regions (Landan et al. 2012). Thus, development may link the genetic, environmental, and stochastic mechanisms of cancer.

Developmental therapies for cancer

Cancer is not so much the result of a cell gone bad as it is of a relationship gone awry. Cancers are often diseases of developmental signaling, and several types of cancer cells can be normalized when placed back into regions of embryos that express certain paracrine factors or their inhibitors. This developmental view of cancer allows us to explore new avenues for cancer treatment. Once such mode of treatment, **differentiation therapy**, was considered possible as long as 30 years ago but was not feasible at the time.

In 1978, Pierce and his colleagues noted that cancer cells were in many ways reversions to embryonic cells, and they hypothesized that cancer cells should revert to normalcy if they were made to differentiate. Also in 1978, Sachs discovered that certain leukemias could be controlled by making the leukemic cells differentiate rather than proliferate. One of these leukemias, acute promyelocytic leukemia (APL), is caused by a somatic recombination creating a "new" transcription factor, one of whose subunits is a retinoic acid receptor. This receptor, even in the absence of retinoic acid, binds to the RA binding sites in DNA, where it represses RA-responsive genes as well as creates a larger condensed chromatin structure (Nowak et al. 2009). Expression of this "new" transcription factor in neutrophil progenitors causes the cell to become malignant (Miller et al. 1992; Grignani et al. 1998). Treatment of APL patients with all-*trans* retinoic acid results in remission in more than 90% of cases because the additional RA is able to effect the differentiation of the leukemic cells into normal neutrophils (Hansen et al. 2000; Fontana and Rishi 2002).

Recently, microRNAs have begun to be tested in differentiation therapy. In many tumors, there are specific microRNAs that are downregulated (Berdasco 2009; Mishra and Merlino 2009). These downregulated microRNAs are usually tumor suppressors that prevent changes in DNA methylation. Taulli and colleagues (2009) have shown that the microRNA miRNA-206, which is normally present in skeletal muscle cells, is downregulated in muscle-cell tumors. Adding miRNA-206 to muscle tumor cells restores their differentiated phenotype and blocks cancer formation. This suggests a tissue-specific mechanism for stopping cancers by causing them to differentiate (**FIGURE 18.26**).

● See **WEBSITE 18.6** Anti-angiogenesis factors

FIGURE 18.26 MicroRNA insertion as a possible means of differentiation therapy. Many cell types (such as those in muscle, lung, and brain) dedifferentiate when forming a tumor. This dedifferentiation is accompanied by loss of specific microRNAs that maintain the differentiated cell's DNA methylation patterns. Restoring those microRNAs to tumor cells reestablished the differentiated pattern of DNA methylation. (After Mishra and Merlino 2009.)

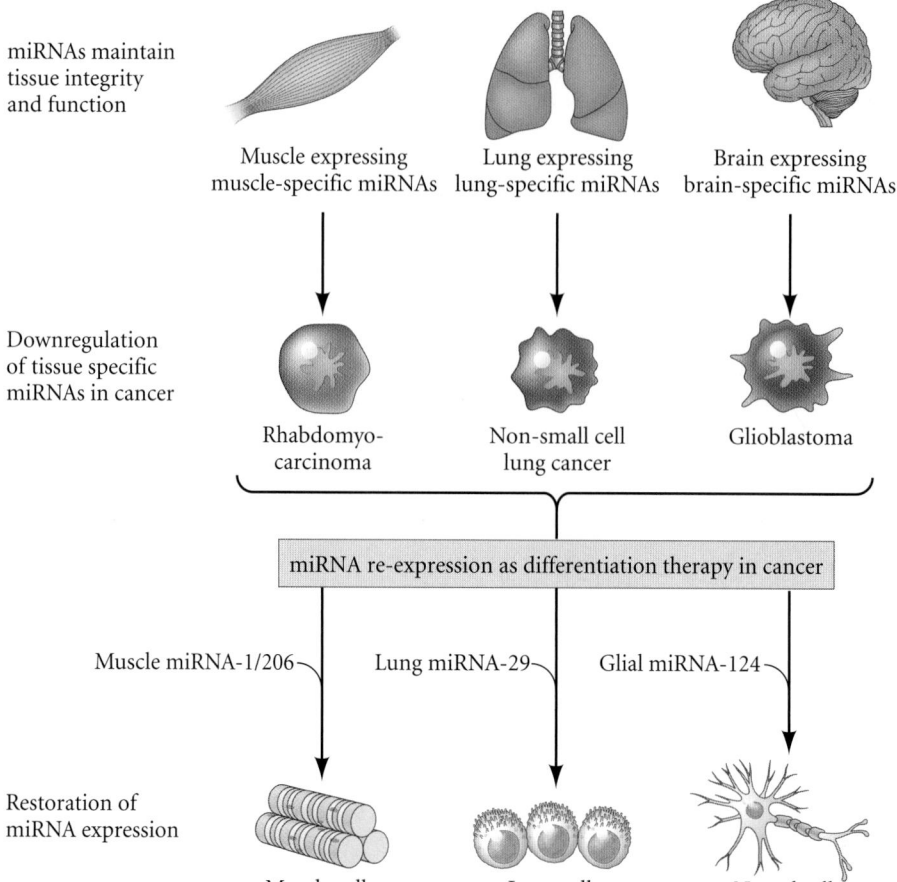

miRNAs maintain tissue integrity and function

Muscle expressing muscle-specific miRNAs   Lung expressing lung-specific miRNAs   Brain expressing brain-specific miRNAs

Downregulation of tissue specific miRNAs in cancer

Rhabdomyo-carcinoma   Non-small cell lung cancer   Glioblastoma

miRNA re-expression as differentiation therapy in cancer

Muscle miRNA-1/206   Lung miRNA-29   Glial miRNA-124

Restoration of miRNA expression

Muscle cells   Lung cells   Neural cells

Coda

Developmental biology is increasingly important in modern medicine. Preventive medicine, public health, and conservation biology demand that we learn more about the mechanisms by which industrial chemicals and drugs can damage embryos. The ability to effectively and inexpensively assay compounds for potential harm is critical. Developmental biology also provides new ways of understanding carcinogenesis and new approaches to preventing and curing cancers. And finally, developmental biology is providing the explanations for how mutated genes and aneuploidies cause their aberrant phenotypes.

It is critical to realize that the agents we put into the environment, the cosmetics we put on our skin, and the substances we eat and drink can reach developing embryos, fetuses, and larvae. Developing organisms have different physiologies as they construct, rather than merely sustain, their phenotypes, and chemicals that appear harmless to adults may disrupt the development of embryos. It takes a community to raise an embryo.

SNAPSHOT SUMMARY: Birth Defects, Endocrine Disruptors, and Cancer

1. Developmental anomalies due to genetic errors and environmental influences result in a relatively low rate of survival of all human conceptions. It is possible that a majority of early human embryos are aneuploid (having more or fewer than 46 chromosomes).

2. Chance plays a role in developmental outcomes. There is large variation in the amounts of transcription and translation, such that at different times, cells are making more or less developmentally important proteins.

3. Pleiotropy occurs when several different effects are produced by a single gene. In mosaic pleiotropy, each effect is caused independently by the expression of the same gene in different tissues. In relational pleiotropy, abnormal gene expression in one tissue influences other tissues, even though those other tissues do not express that gene.

4. Genetic heterogeneity occurs when mutations in more than one gene can produce the same phenotype. Phenotypic heterogeneity arises when the same gene can produce different defects (or differing severities of the same defect) in different individuals.

5. Teratogenic agents include chemicals such as alcohol and retinoic acid, as well as heavy metals, certain pathogens,

and ionizing radiation. These agents adversely affect normal development and can result in malformations and functional deficits.

6. Fetal alcohol syndrome is completely preventable. There may be multiple effects of alcohol on cells and tissues that result in this syndrome of cognitive and physical abnormalities.

7. Retinoic acid is a compound active in development. Too much or too little of it can cause congenital anomalies.

8. Endocrine disruptors can bind to or block hormone receptors or block the synthesis, transport, or excretion of hormones. DES is a powerful endocrine disruptor. Presently, bisphenol A and other endocrine disruptive compounds are being considered as possible agents of low sperm counts in men and a predisposition to breast cancer in women.

9. Environmental estrogens can cause reproductive system anomalies by suppressing Hox gene expression and Wnt pathways. These substances can also cause obesity, and in some cases they activate the transcription factors

predisposing mesenchymal stem cells to differentiate into adipose tissue.

10. In some instances, endocrine disruptors methylate DNA, and these patterns of methylation can be inherited from one generation to the next. Such methylation can alter metabolism and development by turning genes off.

11. Cancer can be seen as a disease of altered development. Cancers metastasize in manners similar to embryonic cell movement, and some tumors revert to non-malignancy when placed in environments that support normal morphogenesis and curtail excessive cell proliferation.

12. Cancers can arise from errors in cell-cell communication. These errors include alterations of paracrine factor synthesis.

13. In many instances, tumors have a rapidly dividing cancer stem cell population which produces more cancer stem cells as well as more quiescent and differentiated cells.

14. The methylation patterns of cancer cells are often aberrant. Methylation differences can cause cancer by inappropriately inactivating tumor suppressor genes or activating oncogenes.

For Further Reading

Complete bibliographical citations for all literature cited in this chapter can be found at the free-access website **www.devbio.com**

Anway, M. D., A. S. Cupp, M. Uzumcu and M. K. Skipper. 2005. Epigenetic transgeneration effects of endocrine disruptors and male fertility. *Science* 308: 1466–1469.

Baksh, D., L. Song and R. S. Tuan. 2004. Adult mesenchymal stem cells: Characterization, differentiation, and application in cell and gene therapy. *J. Cell Mol. Med.* 8: 301–316.

Bissell, M. J., D. C. Radisky, A. Rizki, V. M. Weaver and O. W. Petersen. 2002. The organizing principle: Microenvironmental influences in the normal and malignant breast. *Differentiation* 70: 537–546.

Gilbert, S. F. and D. Epel. 2009. *Ecological Developmental Biology: Integrating Epigenetics, Medicine, and Evolution.* Sinauer Associates. Sunderland, MA.

Gluckman, P. D. and M. A. Hanson. 2004. Living with the past: Evolution, development, and patterns of disease. *Science* 305: 1733–1739.

Guerrero-Bosagna, C., M. Settles, B. Lucker and M. K. Skinner. 2010. Epigenetic transgenerational actions of vinclozolin on promoter regions of the sperm epigenome. *PLoS One* Sep 30;5(9). pii: e13100.

Hanahan, D., and R. A. Weinberg. 2011. Hallmarks of cancer: The next generation. *Cell* 144: 646–674.

Howdeshell, K. L., A. K. Hotchkiss, K. A. Thayer, J. G. Vandenbergh and F. S. vom Saal. 1999. Plastic bisphenol A speeds growth and puberty. *Nature* 401: 762–764.

Kirchner, S., T. Kieu, C. Chow, S. Casey and B. Blumberg. 2010. Prenatal exposure to the environmental obesogen tributyltin predisposes multipotent stem cells to become adipocytes. *Mol. Endocrinol.* 24: 526–539.

Lammer, E. J. and 11 others. 1985. Retinoic acid embryopathy. *New Engl. J. Med.* 313: 837–841.

Lillycrop, K. A., E. S. Phillips, A. A. Jackson, M. A. Hanson and G. C. Burdge. 2005. Dietary protein restriction of pregnant rats induces and folic acid supplementation prevents epigenetic modification of hepatic gene expression in the offspring. *J. Nutrition* 135: 1382–1386.

Maffini, M. V., A. M. Soto, J. M. Calabro, A. A. Ucci and C. Sonnenschein. 2004. The stroma as a crucial target in mammary gland carcinogenesis. *J. Cell Sci.* 117: 1495–1502.

Steingraber, S. 2003. *Having Faith: An Ecologist's Journey to Motherhood.* New York: The Berkley Publishing Group.

Sulik, K. K. 2005. Genesis of alcohol-induced craniofacial dysmorphism. *Exp. Biol. Med.* 230: 366–375.

Go Online

WEBSITE 18.1 Human embryology and genetics. This site has links to tutorials in human development, as well as to the Online Mendelian Inheritance in Man (OMIM) website, which details all human genetic conditions.

WEBSITE 18.2 Prenatal diagnosis and preimplantation genetics. The ability to identify allelic variants in a single cell has enabled scientists to determine if an embryo has deleterious genes. It also enables people to find the sex of any embryo and to implant into the uterus only the ones they desire.

WEBSITE 18.3 Thalidomide as a teratogen. The drug thalidomide caused thousands of babies to be born with malformed arms and legs, and it provided the first major evidence that drugs could induce congenital anomalies. The mechanism of its action is still hotly debated.

WEBSITE 18.4 BPA and altered behavior. Fetal exposure of mice to BPA leads to changes in behaviors. Moreover, in humans, prenatal BPA exposure has been associated with aggression and hyperactivity.

WEBSITE 18.5 Our stolen future. This website monitors the environmental effects of endocrine disruptors. It is a political and consumer action site as well as a scientific clearinghouse for endocrine disruption. Run by the authors of the book *Our Stolen Future*, it also provides links to the websites of people who disagree with them.

WEBSITE 18.6 Anti-angiogenesis factors. In order to survive, tumors need nutrients and oxygen. To get these, they must induce blood vessel formation. The blocking of angiogenesis may provide a mechanism for starving tumors.

Vade Mecum

Somites and thalidomide. These movies are from the laboratory of Jay Lash, whose insightful work on cartilage formation resulted in some of our first insights into the mechanisms by which the drug thalidomide halts limb growth.

Outside Sites

The Association of Reproductive Health Professionals website contains resources for health care providers and their clients. **www.arhp.org/topics/enviro-repro-health**

The Reproductive Toxicology website has summaries of more than 5000 agents, exposure levels, and their effects on development and reproduction: **www.reprotox.org**

Information on fetal alcohol syndrome is available at **www.cdc.gov/ncbddd/fasd/data.html**

Substance Abuse and Mental Health Services Administration (SAMHSA): **www.fascenter.samhsa.gov**

The Endocrine Disruption Exchange maintains a website of scientific reports documenting the health and environmental effects of low doses of endocrine disruptors. **www.endocrinedisruption.com**

Stem cell basics from the NIH website: **stemcells.nih.gov/info/basics**

An informative stem cell blog can be found at **www.allthingsstemcell.com**

19

Ecological Developmental Biology
Biotic, Abiotic, and Symbiotic Regulation of Development

IT WAS LONG THOUGHT THAT THE ENVIRONMENT played only a minor role in development. Nearly all developmental phenomena were believed to be a "readout" of nuclear genes, and those organisms whose development *was* significantly controlled by the environment were considered interesting oddities. When environmental agents played roles in development, they appeared to be destructive, such as the roles played by teratogens and endocrine disruptors (see Chapter 18). However, recent studies have shown that *the environmental context plays significant roles in the normal development of almost all species, and that animal genomes have evolved to respond to environmental conditions.* Moreover, there are symbiotic associations wherein the development of one organism is regulated by the products of organisms of other species. In fact, such cases appear to be the rule rather than the exception.

One reason developmental biologists have largely ignored the environment's effects is that most animals studied in developmental biology—*C. elegans, Drosophila,* zebrafish, *Xenopus,* chicks, and laboratory mice—have been selected for their lack of such effects (Bolker 2012). These model organisms have made it easier to study the genes that regulate development, but they can leave us with the erroneous impression that everything needed to form the embryo is within the fertilized egg. With new concerns about the loss of organismal diversity and the effects of environmental pollutants, there is renewed interest in the regulation of development by the environment (see van der Weele 1999; Gilbert and Epel 2009; Bateson and Gluckman 2011).

The Environment as a Normal Agent in Producing Phenotypes

Although the nucleus and cytoplasm of the zygote contribute a majority of phenotypic instructions, crucial parts of phenotypic determination are regulated by environmental factors outside the organism. **Phenotypic plasticity** is the ability of an organism to react to an environmental input with a change in form, state, movement, or rate of activity (West-Eberhard 2003; Beldade et al. 2011). When seen in embryonic or larval stages of animals or plants, this ability to change phenotype is often called **developmental plasticity**.

We have already encountered several examples of developmental plasticity. When we discussed environmental sex determination in turtles, fish, and echiuroid ("spoon") worms (see Chapter 15), we were aware that the sexual phenotype was being instructed not by the genome but by the environment. When we discussed in Chapter 13 the ability of shear stress to activate gene expression in capillary, heart, and bone tissue, we similarly were studying the effect of an environmental agent on phenotype. Although studies of phenotypic plasticity have played a central role in plant developmental biology, the mechanisms of plasticity have only recently been studied in animals. These studies now show that developmental plasticity is a critical means for integrating animals into their ecological communities.

We may now turn to consider adaptations towards the external environment; and firstly the direct adaptations ... in which an animal, during its development, becomes modified by external factors in such a way as to increase its efficiency in dealing with them.

C. H. WADDINGTON (1957)

Honor thy symbionts.

JIAN XU AND JEFFREY I. GORDON (2003)

(A)

(B)

(C) Spring morph among catkins

(D) Summer morph on twig

(E)

(F)

FIGURE 19.1 Developmental plasticity in insects. (A,B) Density-induced polyphenism in the desert ("plague") locust *Schistocerca gregaria*. (A) The low-density morph has green pigmentation and miniature wings. (B) The high-density morph has deep pigmentation and wings and legs suitable for migration. (C,D) *Nemoria arizonaria* caterpillars. (C) Caterpillars that hatch in the spring eat young oak leaves and develop a cuticle that resembles the oak's flowers (catkins). (D) Caterpillars that hatch in the summer, after the catkins are gone, eat mature oak leaves and develop a cuticle that resembles a young twig. (E) Gyne (reproductive queen) and worker of the ant *Pheidologeton*. This picture shows the remarkable dimorphism between the large, fertile queen and the small, sterile worker (seen near the queen's antennae). The difference between these two sisters is the result of larval feeding. (F) Nutrition-induced size difference in a queen honeybee *Apis mellifera* compared with her sister workers. (A,B from Tawfik et al. 1999, courtesy of S. Tanaka; C,D courtesy of E. Greene; E © Mark W. Moffett/National Geographic Society; F courtesy of D. McIntyre.)

Two main types of phenotypic plasticity are currently recognized: reaction norms and polyphenisms (Woltereck 1909; Schmalhausen 1949; Stearns et al. 1991). In a **reaction norm**, the genome encodes the potential for a *continuous range* of potential phenotypes, and the environment the individual encounters determines the phenotype (usually the most adaptive one) that emerges. For instance, human muscle phenotype is determined by the amount of exercise the body is exposed to over time (even though there is a genetically defined limit to how much muscular hypertrophy is possible). The upper and lower limits of a reaction norm are a property of the genome that can be selected. The different phenotypes produced by environmental conditions are called **morphs** (or occasionally **ecomorphs**).

The second type of phenotypic plasticity, **polyphenism**, refers to *discontinuous* (either/or) phenotypes elicited by the environment. One obvious example is sex determination in turtles, where one range of temperatures will induce female development in the embryo and another set of temperatures will induce male development. Between these sets of temperatures is a small band of temperatures that will produce different proportions of males and females, but these intermediate temperatures do not induce intersexual animals. Another important example of polyphenism is the migratory locust *Schistocerca gregaria*. These grasshoppers exist either as a short-winged, green, solitary morph or as a long-winged, brown, gregarious morph (**FIGURE 19.1A,B**). Cues in the environment determine which morphology a larva will develop upon molting (Rogers et al. 2003; Simpson and Sword 2008).

Diet-induced polyphenisms

Diet can play major roles in determining a developing animal's phenotype. The effects of diet in development can be seen in the caterpillar of *Nemoria arizonaria*. When it hatches on oak trees in the spring, it has a form that blends remarkably with the young oak flowers (catkins). But those larvae hatching from their eggs in the summer would be very obvious if they still looked like oak flowers. Instead, they resemble newly formed twigs. Here, it is the diet (young versus old oak leaves) that determines the phenotype (**FIGURE 19.1C,D**; Greene 1989).

Diet is also largely responsible for the formation of fertile "queens" in ant, wasp, and bee colonies (**FIGURE 19.1E,F**). In honeybees, adult females are either workers or queens. The queen is the only reproductive member of the hive, laying up to 2,000 eggs per day. Queens also live ten times longer than the average worker. The larvae are fed by the workers, and only those larvae fed adequately become queens. The protein inducing these queen-forming activities is called **royalactin**. Royalactin binds to the EGF receptor in the fat body of the honeybee larvae and stimulates the production of juvenile hormone, which elevates the levels of yolk proteins that are necessary for egg production (**FIGURE 19.2**; Kamakura 2011). RNAi against either the EGF receptor or its downstream targets abolishes the effects of royalactin.

● See WEBSITE 19.1 Inducible caste determination in ant colonies

WHEN DUNG REALLY MATTERS For the male dung beetle (*Onthophagus*), what really matters in life is the amount and quality of the dung he eats as a larva. The hornless female dung beetle digs tunnels, then gathers balls of dung and buries them in these tunnels. She then lays a single egg on each dung ball; when the larvae hatch, they eat the dung. Metamorphosis occurs when the dung ball is finished, and the anatomical and behavioral phenotypes of the male dung beetle are determined by the quality and quantity of this maternally provided food (Emlen 1997; Moczek and Emlen 2000). The amount and quality of food determines the titer of juvenile hormone during the larva's last molt. This, in turn,

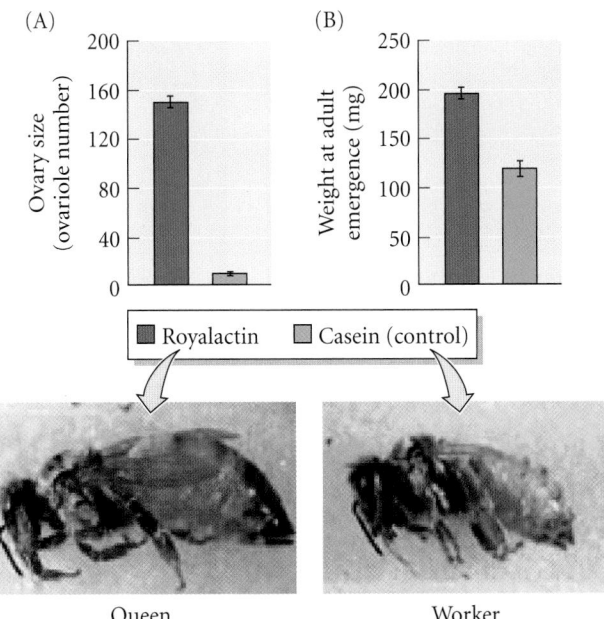

Queen   Worker

FIGURE 19.2 Diet-induced developmental changes can produce either reproductively competent queens or sterile workers. Royalactin induces functional ovaries (A) and increased body weight (B) in the honeybee *Apis mellifera*. (After Kamakura 2011.)

determines the size of the larva at metamorphosis and positively regulates the growth of the imaginal discs that make the horns (**FIGURE 19.3A**; Emlen and Nijhout 1999; Moczek 2005). If juvenile hormone is added to tiny *O. taurus* males during the sensitive period of their last molt, the cuticle in their heads expands to produce horns. Thus, whether a male is horned or hornless depends not on the male's genes but on the food his mother left for him.

Horns do not grow until the male beetle larva reaches a certain size. After this threshold body size, horn growth is very rapid.* Thus, although body size has a normal distribution, there is a bimodal distribution of horn sizes: about half the males have no horns, whereas the other half have horns of considerable length (**FIGURE 19.3B**).

The size of the horns determines a male's behavior and chances for reproductive success. Horned males guard the females' tunnels and use their horns to prevent other males from mating with the female; the male with the biggest horns wins such contests. But what about the males with no horns? Hornless males do not fight with the horned males for mates. Since they, like the females, lack horns, they are able to dig their own tunnels. These "sneaker males" dig tunnels that

---

*Developmentally, there is a trade-off between primary and secondary male sexual characters here. Making a large horn appears to take away resources from making the penis. The growth rates of both the horn and the penis may be regulated through the Foxo transcription factor, which is upregulated by diet (Parzer and Moczek 2008; Snell-Rood and Moczek 2012).

(A)

Horned male   Hornless male

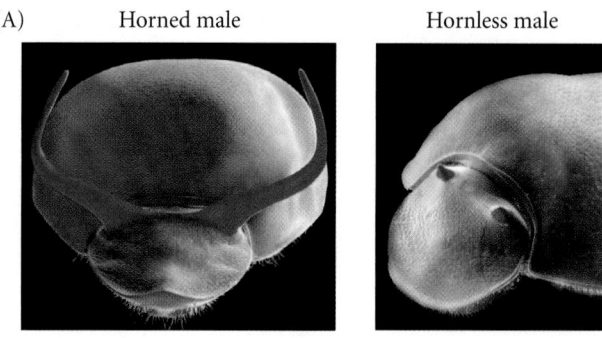

(B)

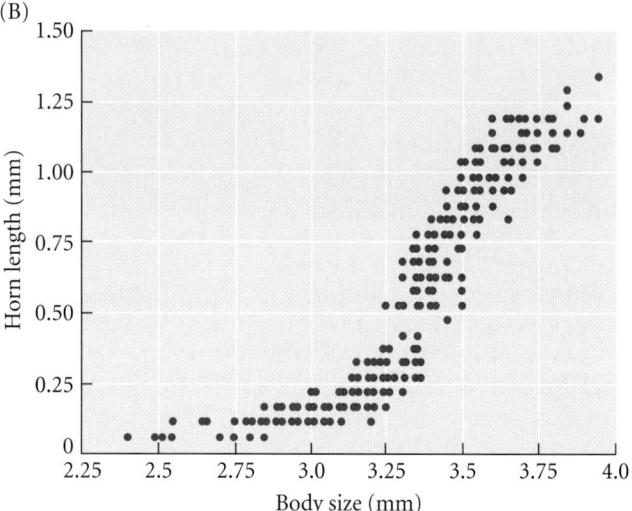

FIGURE 19.3 Diet and *Onthophagus* horn phenotype. (A) Horned and hornless males of the dung beetle *Onthophagus acuminatus* (horns have been artificially colored). Whether a male is horned or hornless is determined by the titer of juvenile hormone at the last molt, which in turn depends on the size of the larva. (B) There is a sharp threshold of body size under which horns fail to form and above which horn growth is linear with the size of the beetle. This threshold effect produces males with no horns and males with large horns, but very few with horns of intermediate size. (After Emlen 2000; photographs courtesy of D. Emlen.)

intersect those of the females and mate with the females while the horned male stands guard at the tunnel entrance (**FIGURE 19.4**; Emlen 2000; Moczek and Emlen 2000). Indeed, about half the fertilized eggs in most populations are from hornless males. The ability to produce a horn is inherited; but whether to produce a horn and how big to make it is regulated by the environment.

DIET AND DNA METHYLATION Dietary alterations can produce changes in mammalian DNA methylation, and these methylation changes can affect the phenotype. Waterland and Jirtle (2003) demonstrated this by using mice containing the *viable-yellow* allele of *Agouti*. *Agouti* is a dominant gene that gives mice yellowish hair color; it also affects lipid metabolism such that the mice become fatter. The

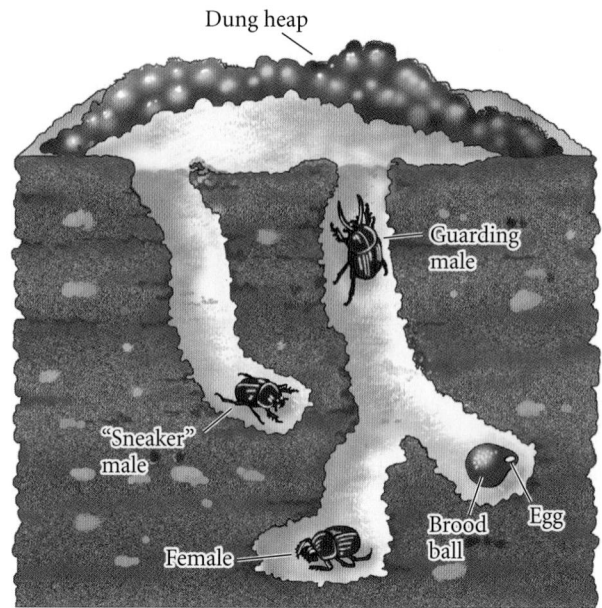

FIGURE 19.4 The presence or absence of horns determines the male reproductive strategy in some dung beetle species. Females dig tunnels in the soil beneath a pile of dung and bring dung fragments into the tunnels. These will be the food supply of the larvae. Horned males guard the tunnel entrances and mate repeatedly with the females. They fight to prevent other males from entering the tunnels, and the males with long horns usually win such contests. Smaller, hornless males do not guard tunnels, but dig their own to connect with those of females. They can then mate and exit, unchallenged by the guarding male. (After Emlen 2000.)

viable-yellow allele of *Agouti* has a transposable element inserted into its *cis*-regulatory regions. These transposon insertion sites are very interesting for gene regulation: whereas most regions of the adult genome have hardly any intraspecies variation in CpG methylation, there are large DNA methylation differences between individuals at the sites of transposon insertion. Such CpG methylation can block gene transcription. When the promoter of the *Agouti* gene is methylated, the gene is not transcribed. The mouse's fur remains black, and lipid metabolism is not altered.

Waterland and Jirtle fed pregnant *viable-yellow Agouti* mice methyl donor supplements, including folate, choline, and betain. They found that the more methyl supplementation, the greater the methylation of the transposon insertion site in their fetuses' genomes, and the darker the pigmentation of the offspring. Although the mice in **FIGURE 19.5** are genetically identical, their mothers were fed different diets during pregnancy. The mouse whose mother did not receive methyl donor supplementation is fat and yellow—the *Agouti* gene promoter was unmethylated, and the gene was active. The mouse born to the mother that was given folate supplements is sleek and dark; the methylated *Agouti* gene was not transcribed.

FIGURE 19.5 Maternal diet can affect phenotype. These two mice are genetically identical; both contain the *viable-yellow* allele of the *Agouti* gene, whose protein product converts brown pigment to yellow and accelerates fat storage. The obese yellow mouse is the offspring of a mother whose diet was not supplemented with methyl donors (e.g., folate) during her pregnancy. The embryo's *Agouti* gene was not methylated, and Agouti protein was made. The sleek brown mouse was born of a mother whose prenatal diet was supplemented with methyl donors. The *Agouti* gene was turned off, and no Agouti protein was made. (After Waterland and Jirtle 2003, photograph courtesy of R. L. Jirtle.)

As we saw in Chapter 18, differential gene methylation has been linked to human health problems. Dietary restrictions during a woman's pregnancy may show up as heart or kidney problems in her adult children. Moreover, studies in rats showed that differences in protein and methyl donor concentration in the mother's prenatal diet affected gene expression and subsequent metabolism in the pup's livers (Lillycrop et al. 2005).

Predator-induced polyphenisms

Imagine a species whose larvae are frequently confronted by a particular predator in their pond or tidepool. One could then imagine an individual that could recognize soluble molecules secreted by that predator and could use those molecules to activate the development of structures that would make this individual less palatable to the predator. This ability to modulate development in the presence of predators is called predator-induced defense, or **predator-induced polyphenism**.

To demonstrate predator-induced polyphenism, one has to show that the phenotypic modification is caused by the presence of the predator, and that the modification increases the fitness of its bearers when the predator is present (Adler and Harvell 1990; Tollrian and Harvell 1999). **FIGURE 19.6A** shows the typical and predator-induced morphs for several species. In each case, the induced morph is more successful at surviving the predator, and soluble filtrate from water surrounding the predator is able to induce the changes. Chemicals that are released by a predator and can induce defenses in its prey are called **kairomones**.

Several rotifer species will alter their morphology when they develop in pond water in which their predators were

cultured (Dodson 1989; Adler and Harvell 1990). The predatory rotifer *Asplanchna* releases a soluble compound that induces the eggs of a prey rotifer species, *Keratella slacki*, to develop into individuals with slightly larger bodies and anterior spines 130% longer than they otherwise would be, making the prey more difficult to eat. When exposed to the effluent of the crab species that preys on it, the snail *Thais lamellosa* develops a thickened shell and a "tooth" in its aperture. In a mixed snail population, crabs will not attack the thicker-shelled snails until more than half of the typical-morph snails are devoured (Palmer 1985).

One of the more interesting mechanisms of predator-induced polyphenism is that of certain echinoderm larvae. When exposed to the mucus of their fish predator, sand dollar plutei clone themselves, budding off small groups of cells that quickly become larvae themselves. The small plutei are too small to be seen by the fish, and thereby escape being eaten (Vaughn and Strathmann 2008; Vaughn 2009).

DAPHNIA AND THEIR KIN   The predator-induced polyphenism of the parthenogenetic water flea *Daphnia* is beneficial not only to itself but also to its offspring (Harris et al. 2012). When *D. cucullata* encounter the predatory larvae of the fly *Chaeoborus*, their "helmets" grow to twice the normal size (**FIGURE 19.6B**). This increase lessens the chances that *Daphnia* will be eaten by the fly larvae. This same helmet induction occurs if the *Daphnia* are exposed to extracts of water in which the fly larvae had been swimming. Agrawal and colleagues (1999) have shown that the offspring of such an induced *Daphnia* are born with this same altered head morphology in the absence of a predator. It is possible that the *Chaeoborus* kairomone regulates gene expression both in the adult and in the developing embryo. Although we still do not know the identity of the kairomone, the receptor may be a particular set of neurons (Weiss et al. 2012). The effect does appear to work through the endocrine pathways. The kairomone upregulates the juvenile hormone and the insulin signaling pathways, activating the transcription of several transcription factor genes (**FIGURE 19.6C**; Miyakawa et al. 2010). Like in the dung beetles, there are trade-offs, however; the induced *Daphnia*, having put resources into making protective structures, produce fewer eggs (Tollrian 1995; Imai et al. 2009).

AMPHIBIAN PHENOTYPES INDUCED BY PREDATORS   Predator-induced polyphenism is not limited to invertebrates.* Among amphibians, tadpoles found in ponds or reared in the presence of other species may differ significantly from tadpoles reared by themselves in aquaria. For instance, newly hatched wood frog tadpoles (*Rana sylvatica*) reared in tanks containing the predatory larval dragonfly *Anax* (confined in

---

* Indeed, the vertebrate immune system is a wonderful example of predator-induced polyphenism. Here, our immune cells use chemicals from our predators (viruses and bacteria) to change our phenotype so that we can better resist them (see Frost 1999).

(A)

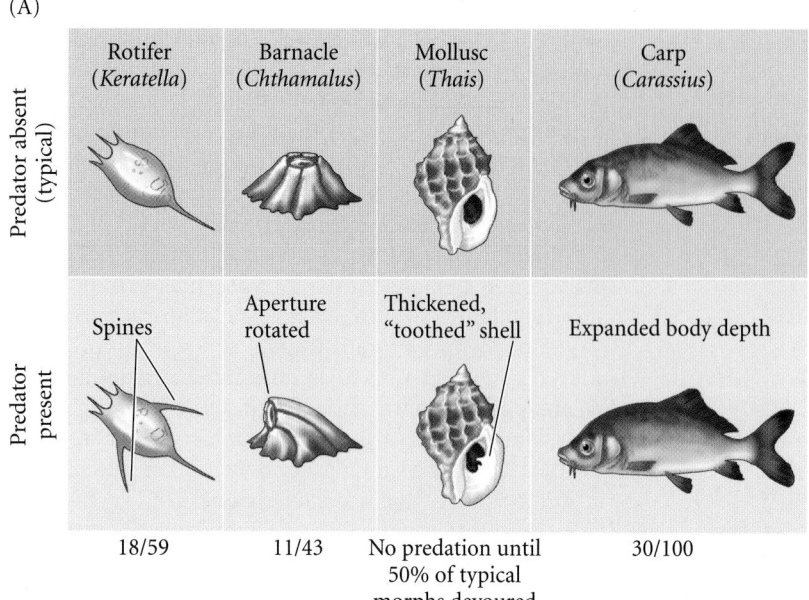

(B)

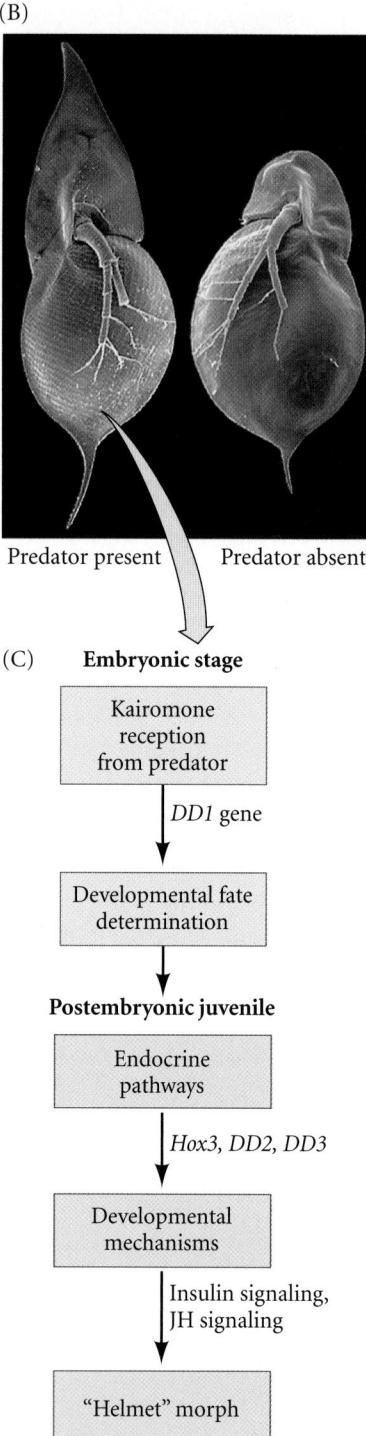

Predator present   Predator absent

FIGURE 19.6 Predator-induced defenses. (A) Typical (upper row) and predator-induced (lower row) morphs of various organisms. The numbers beneath each column represent the percentages of organisms surviving predation when both induced and uninduced individuals were presented with predators (in various assays). (B) Scanning electron micrographs show predator-induced (left) and typical (right) morphs of genetically identical individuals of the water flea *Daphnia*. In the presence of chemical signals from a predator, *Daphnia* grows a protective "helmet." (C) Possible pathway for the development of *Daphnia's* defensive phenotype through the endocrine system. *DD1* is thought to be involved in kairomone reception and/or fate determination during the embryonic stage. It may play a role in the neural reception of the signal. The other genes are thought to play roles in the morphogenesis of postembryonic juveniles. (A after Adler and Harvell 1990 and references cited therein; B courtesy of A. A. Agrawal; C after Miyakawa et al. 2010.)

mesh cages so they cannot eat the tadpoles) grow smaller than those reared in similar tanks without predators. Moreover, their tail musculature deepens, allowing faster turning and swimming speeds (Van Buskirk and Relyea 1998). The addition of more predators to the tank causes a continuously deeper tail fin and tail musculature, and in fact what initially appeared to be a polyphenism may be a reaction norm that can assess the number (and type) of predators.

McCollum and Van Buskirk (1996) have shown that in the presence of its predators, the tail fin of the tadpole of the tree frog *Hyla chrysoscelis* grows larger and becomes bright red (**FIGURE 19.7**). This phenotype allows the tadpole to swim away faster and to deflect predator strikes toward the tail region. The trade-off is that non-induced tadpoles grow more slowly and survive better in predator-free environments. In some species, phenotypic plasticity is reversible, and removing the predators can restore the non-induced phenotype (Relyea 2003a).

The metabolism of predator-induced morphs may differ significantly from that of the uninduced morphs, and this has important consequences. Relyea (2003b, 2004) has found that in the presence of the chemical cues emitted by predators, the toxicity of pesticides such as carbaryl (Sevin™) can become up to 46 times more lethal than it is without the predator

cues. Bullfrog and green frog tadpoles were especially sensitive to carbaryl when exposed to predator chemicals. Relyea has related these findings to the global decline of amphibian populations, saying that governments should test the toxicity of the chemicals under more natural conditions, including that of predator stress. He concludes (Relyea 2003b) that

(A) Predator present

(B) Predator absent

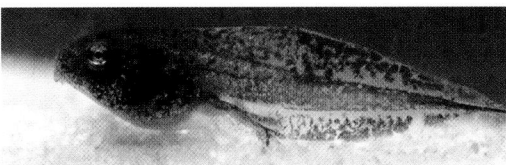

FIGURE 19.7 Predator-induced polyphenism in frog tadpoles. (A) Tadpoles of the tree frog *Hyla chrysoscelis* developing in the presence of cues from a predator's larvae develop strong trunk muscles and a red coloration. (B) When predator cues are absent, the tadpoles grow sleeker, which helps them compete for food. (Photographs courtesy of T. Johnson/USGS.)

"ignoring the relevant ecology can cause incorrect estimates of a pesticide's lethality in nature, yet it is the lethality of pesticides under natural conditions that is of utmost interest. The accumulated evidence strongly suggests that pesticides in nature could be playing a role in the decline of amphibians."

VIBRATIONAL CUES ALTER DEVELOPMENTAL TIMING The phenotypic changes induced by environmental cues are not confined to structure. They can also include the timing of developmental processes. Embryos of the Costa Rican red-eyed tree frog (*Agalychnis callidryas*) use vibrations transmitted through their egg masses to escape egg-eating snakes. These egg masses are laid on leaves that overhang ponds.

Usually, the embryos develop into tadpoles within 7 days, and these tadpoles wiggle out of the egg mass and fall into the pond water. However, when snakes feed on the eggs, the vibrations they produce cue the remaining embryos inside the egg mass to begin the twitching movements that initiate their hatching (within seconds!) and dropping into the pond. The embryos are competent to begin these hatching movements at day 5 (**FIGURE 19.8**). Interestingly, the embryos have evolved to respond this way only to vibrations given at a certain frequency and interval (Warkentin et al. 2005, 2006; Caldwell et al. 2009). Up to 80% of the remaining embryos can escape snake predation in this way, and research has shown that these vibrations alone (and not smell or sight) cue these hatching movements in the embryos. There is a trade-off here, too. Although these embryos have escaped their snake predators, they are now at greater risk from waterborne predators than are fully developed embryos because the musculature of the early hatchers is underdeveloped.

Temperature as an environmental agent

TEMPERATURE AND SEX In many species, temperature controls whether testes or ovaries develop; such temperature-dependent sex determination is described in Chapter 15. Although the mechanism for it is not yet known (and may differ from species to species), this type of determination is not uncommon among the "cold-blooded" vertebrates such as fish, turtles, and alligators (Crews and Bull 2009).

(A)

(B)

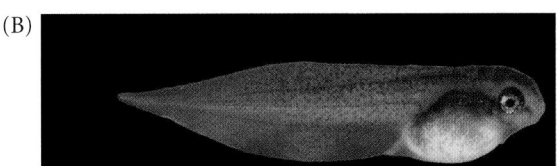

(C)

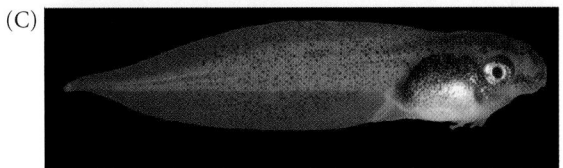

FIGURE 19.8 Predator-induced polyphenism in the red-eyed tree frog (*Agalychnis callidryas*). (A) When a snake eats a clutch of *Agalychnis* eggs, most of the remaining embryos inside the egg mass respond to the vibrations by hatching prematurely (arrow) and falling into the water. (B) Immature tadpole, induced to hatch at day 5. (C) A normal tadpole hatches at day 7 and has better-developed musculature. (Courtesy of K. Warkentin.)

Temperature-dependent sex determination has advantages and disadvantages. One advantage is that it probably gives the species the benefits of sexual reproduction without tying the species to a 1:1 sex ratio. In crocodiles, in which extreme temperatures produce females whereas moderate temperatures produce males, the sex ratio may be as great as 10 females to each male (Woodward and Murray 1993). In such instances, where the number of females limits the population size, this ratio is better for survival than the 1:1 ratio demanded by genotypic sex determination.

The major disadvantage of temperature-dependent sex determination may be its narrowing of the temperature limits within which a species can persist. Thus, thermal pollution (either locally or because of global warming) could conceivably eliminate a species from a given area (Janzen and Paukstis 1991). Among marine turtles, females are usually produced at higher temperatures (29°C being the temperature that produces an even sex ratio), and these animals may be particularly vulnerable if the temperature rises for an extended period of time (Hawkes et al. 2009; Fuentes et al. 2010).

Researchers (see Ferguson and Joanen 1982; Miller et al. 2004) have speculated that dinosaurs may have had temperature-dependent sex determination and that their sudden demise may have been caused by a slight change in temperature creating conditions where only males or only females hatched. Unlike many turtle species, whose members have long reproductive lives, can hibernate for years, and whose females can store sperm, dinosaurs may have had a relatively narrow time to reproduce and no ability to hibernate through prolonged bad times.

Charnov and Bull (1977) argued that environmental sex determination would be adaptive in habitats characterized by patchiness—that is, a habitat having some regions where it is advantageous to be male and other regions where it is advantageous to be female. Conover and Heins (1987) provided evidence for this hypothesis. In certain fish species, females benefit from being larger because larger size translates into higher fecundity. If you are a female Atlantic silverside (*Menidia menidia*), it is advantageous to be born early in the breeding season, because you have a longer feeding season and thus can grow larger; the size of male fish of this species doesn't influence mating success or outcomes. In the southern range of *Menidia*, females are indeed born early in the breeding season, and temperature appears to play a major role in this pattern. However, in the northern reaches of its range, the species shows no environmental sex determination and a 1:1 sex ratio is generated at all temperatures. Conover and Heins speculate that the more northern populations have a very short feeding season, so there is no advantage for females in being born earlier. Thus, this fish species displays environmental sex determination in those regions where it is adaptive and genotypic sex determination in those regions where it is not.

● See WEBSITE 19.2 Volvox: When heat brings out sex

BUTTERFLY WINGS Tropical regions of the world often have a hot wet season and a cooler dry season. In Africa, a polyphenism of the dimorphic Malawian butterfly (*Bicyclus anynana*) is adaptive to these seasonal changes. The dry (cool) season morph is a mottled brown butterfly that survives by hiding in dead leaves on the forest floor. In contrast, the wet (hot) season morph, which routinely flies, has prominent ventral eyespots that deflect attacks from predatory birds and lizards (**FIGURE 19.9**; Brakefield and Frankino 2009; Olofsson et al. 2010).

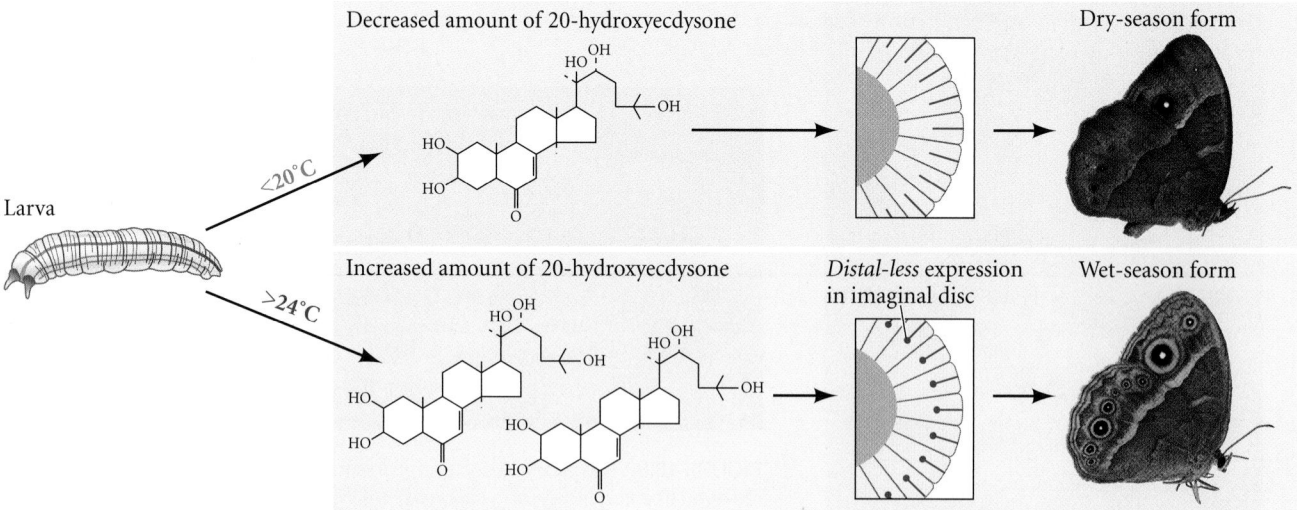

FIGURE 19.9 Phenotypic plasticity in *Bicyclus anynana* is regulated by temperature during pupation. High temperature (either in the wild or in controlled laboratory conditions) allows the accumulation of 20-hydroxyecdysone (20E), a hormone that is able to sustain *Distal-less* expression in the pupal imaginal disc. The region of *Distal-less* expression becomes the focus of each eyespot. In cooler weather, 20E is not formed, *Distal-less* expression in the imaginal disc begins but is not sustained, and eyespots fail to form. (Courtesy of S. Carroll and P. Brakefield.)

FIGURE 19.10 Environmentally induced morphs of the European map butterfly (*Araschnia levana*). The orange morph (bottom) forms in the spring, when levels of ecdysone in the larva are low. The dark morph with a white stripe (top) forms in summer, when higher temperatures and longer photoperiods induce greater ecdysone production in the larva. Linnaeus classified the two morphs as different species. (Courtesy of H. F. Nijhout.)

The factor determining the seasonal pigmentation of *B. anynana* is not diet, but the temperature during pupation. Low temperatures produce the dry-season morph; higher temperatures produce the wet-season morph (Brakefield and Reitsma 1991). The mechanism by which temperature regulates the *Bicyclus* phenotype is becoming known. In the late larval stages, transcription of the *Distal-less* gene in the wing imaginal discs is restricted to a set of cells that will become the signaling center of each eyespot. In the early pupa, higher temperatures elevate the formation of 20-hydroxyecdysone (20E; see Chapter 16). This hormone sustains and expands the expression of *Distal-less* in those regions of the wing imaginal disc, resulting in prominent eyespots. In the dry season, the cooler temperatures prevent the accumulation of 20E in the pupa, and the foci of Distal-less signaling are not sustained. In the absence of the Distal-less signal, the eyespots do not form (Brakefield et al. 1996; Koch et al. 1996). Distal-less protein is believed to be the activating signal that determines the size of the eyespot (see Figure 19.9). In *Bicyclus*, we see the adaptive significance of polyphenism and how this type of developmental plasticity integrates an organism into its environment.

The importance of hormones such as 20E for mediating environmental signals controlling wing phenotypes has been documented in the *Araschnia* butterfly (**FIGURE 19.10**). *Araschnia* develops alternative phenotypes depending on whether the fourth and fifth instars experience a photoperiod (hours of daylight) that is longer or shorter than a particular critical day length. Below this critical day length, ecdysone levels are low and the butterfly has the orange wings characteristic of spring butterflies. Above the critical point, ecdysone is made and the summer pigmentation forms. The summer form can be induced in spring pupae by injecting 20E into the pupae. Moreover, by altering the timing of 20E injections, one can generate a series of intermediate forms not seen in the wild (Koch and Bückmann 1987; Nijhout 2003).

Environmental Induction of Behavioral Phenotypes

In many instances, the morphological phenotype is accompanied by a behavioral phenotype. This is obvious in the environmental determination of sex, where an individual's sexual behavior generally matches the gonads and genitalia. This is also seen in the cases of butterfly wings (fliers vs. crawlers) and dung beetle horns (fighters vs. "sneakers"). Some of the reasons behavior has been poorly explored developmentally is because it is difficult to measure, often subjective, and subject to the organism's history and context (Ballinger and Benzer 1988; Gottlieb 1992; Skuse 2000). However, behavior—the "final phenotype"—can be the major phenotype induced by the environment during development.

Adult anxiety and environmentally regulated DNA methylation

A particular organism's behavior has both genetic and environmental components, and one of the environmental components concerns the ability of several environmental stimuli to methylate or demethylate DNA. In rats, several adult behaviors, including those involving sexuality and aggression, can be linked to changes in DNA methylation experienced soon after birth (Curley et al. 2011; Caldji et al. 2012). For example, behavioral differences in the response to stressful situations have been correlated with the number of glucorticoid receptors in the brain's hippocampus. The more glucocorticoid receptors, the better the adult rat is able to downregulate these adrenal hormones and deal with stress. The number of glucocorticoid receptors appears to depend on the quality of grooming and licking the rat pup experiences during the first week after birth.

How is the adult phenotype regulated by these perinatal (near the time of birth) experiences? Weaver and colleagues (2004) have shown that the behavioral difference involves the methylation of a particular site in the enhancer region on the glucocorticoid receptor gene. Before birth, there is no methylation at this site; one day after birth, this site is methylated in all rat pups. However, in those pups that experience intensive grooming and licking during the first week after birth, this site *loses* its methylation; but methylation is retained in those rats that do not have such extensive care (see Figure 15.15). Moreover, this methylation difference is not seen at other sites in or near the gene (**FIGURE 19.11**).

By switching pups and parents, Weaver and colleagues demonstrated that this methylation difference was dependent

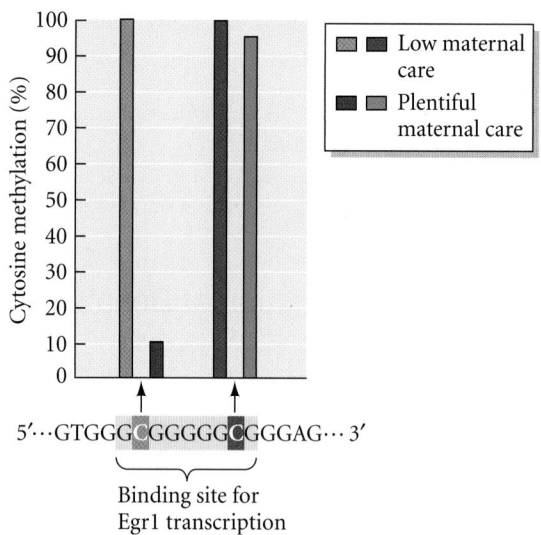

5′···GTGGG**C**GGGGG**C**GGGAG··· 3′

Binding site for
Egr1 transcription
factor

FIGURE 19.11 Differential DNA methylation due to behavioral differences in parental care. A portion of an enhancer sequence of the rat glucocorticoid receptor (*GR*) gene is shown, indicating the binding site for the Egr1 transcription factor. Two cytosine residues within this binding site have the potential to be methylated. The cytosine at the 5′ end is completely methylated in the brains of pups that did not receive extensive licking and grooming from their mothers (orange bar). The Egr1 transcription factor did not bind these methylated sites, and thus the *GR* gene remained inactive. If the pups received proper maternal care, this same site was largely unmethylated (red bar), and the gene was transcribed in the brain. The cytosine at the 3□ end of the enhancer (blue and purple bars) was always methylated and had no effect on Egr1 binding. (After Weaver et al. 2004.)

on the mother's care and was not the result of differences in the pups themselves. When unmethylated, this enhancer site binds the Egr1 (NGF1-A) transcription factor and is associated with "active" acetylated nucleosomes. The transcription factor does not bind to the methylated site, and the chromatin in such cases is not activated. These chromatin differences, established during the first week after birth, are retained throughout the life of the rat. Thus, adult rats that received extensive perinatal grooming have more glucocorticoid receptors and are able to deal with stress better than rats that received less care.* Just *how* grooming can alter DNA methylation patterns, however, remains to be discovered.

---

*Does this relate to humans? Using an "extreme" set of cases, Michael Meaney's laboratory (McGowan et al. 2009; Zhang et al. 2012) showed that the *cis*-regulatory region of the hippocampus-specific glucocorticoid receptor is more highly methylated in the brains of suicide victims with a history of childhood abuse than in suicide victims with no childhood abuse, or in controls. There is also the possibility that these DNA and histone modification patterns in rats (which appear to be induced by serotonin produced by the nursing pups) may be reversible (see Hellstrom et al. 2012; Zhang et al. 2012).

Learning: The Developmentally Plastic Nervous System

Learning provides remarkable examples of phenotypic plasticity. Since neurons, once formed, do not divide, the "birthday" of a neuron can be identified by treating the organism with radioactive thymidine. Normally, very little radioactive thymidine is taken up into the DNA of a neuron that has already been formed. However, if a neural precursor cell divides during the treatment, it will incorporate radioactive thymidine into its DNA.

Such new neurons are seen to be generated when male songbirds first learn their songs. Juvenile zebra finches memorize a model song and then learn the pattern of muscle contractions necessary to sing a particular phrase. In this learning and repetition process, new neurons are generated in the hyperstriatum of the finch's brain. Many of these new neurons send axons to the archistriatum, which is responsible for controlling the vocal musculature (Nordeen and Nordeen 1988; Alvarez-Buylla et al. 1990). These changes are not seen in males that are too old to learn the song, nor are they seen in juvenile females (which do not sing these phrases). In white-crowned sparrows (*Zonotrichia leucophrys*), whose song is regulated by photoperiod and hormones, exposing adult males to long hours of light and to testosterone induces more than 50,000 new neurons in their vocal centers (Tramontin et al. 2000). The neural circuitry of these birds' brains shows seasonal plasticity. Testosterone is believed to increase the level of the transcription factor **brain-derived neurotropic factor** (**BDNF,** a paracrine factor associated with neuronal plasticity) in the song-producing vocal centers. If female birds are given BDNF, they also produce more neurons in the vocal centers (Rasika et al. 1999).

The cerebral cortices of young rats reared in stimulating environments are packed with more neurons, synapses, and dendrites than are found in rats reared in isolation (Turner and Greenough 1983), and mice reared in cages experienced changes in their neural circuitry when they were placed in more natural environments (Polley et al. 2004). Even the adult brain continues to develop in response to new experiences. Studies on adult rats and mice indicate that environmental stimulation can increase the number of new neurons in the dentate gyrus of the hippocampus (Kempermann et al. 1997a,b; Gould et al. 1999; van Praag et al. 1999). Similarly, when adult rats learn to keep their balance on dowels, their cerebellar Purkinje neurons develop new synapses (Black et al. 1990). The pathway underlying the formation of new synapses probably involves an activity-dependent association of histone modifying proteins with particular transcription factors (Chen et al. 2012; Sando et al. 2012).

In humans, changes in brain anatomy can be seen as a result of learning new tasks. When young adults were taught the classic three-ball cascade juggling routine (which takes months to get right), the neurons in a specific area of the temporal lobe of the brain established a new pattern—a pattern not seen in students who were not taught this skill

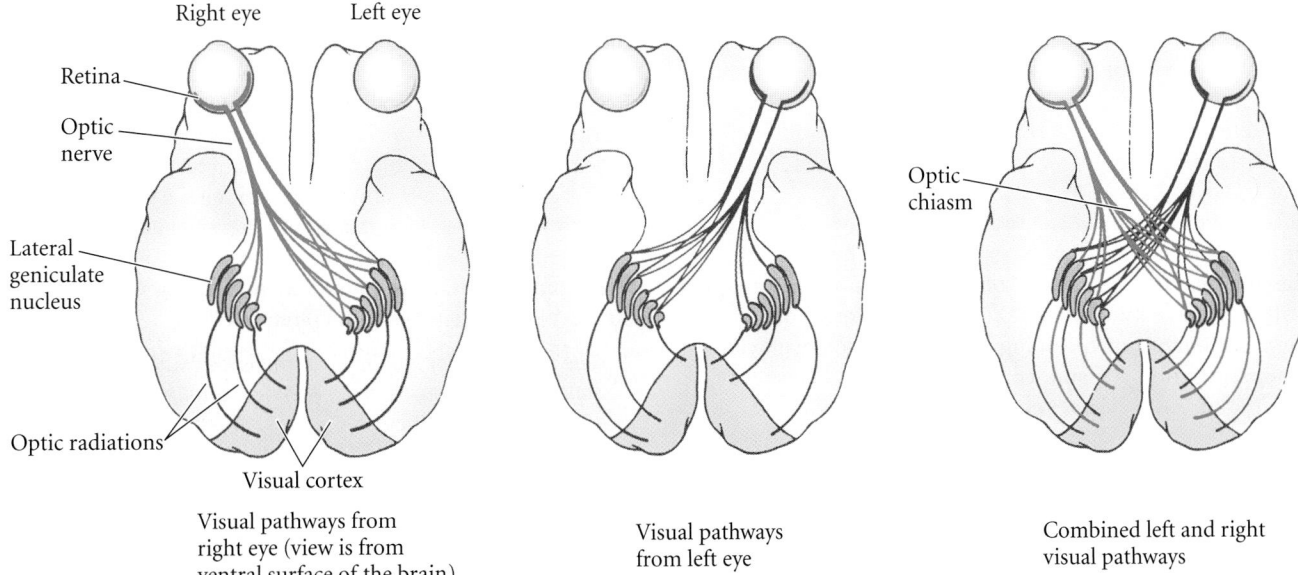

Right eye   Left eye

Retina

Optic nerve

Lateral geniculate nucleus

Optic radiations

Visual cortex

Optic chiasm

Visual pathways from right eye (view is from ventral surface of the brain)

Visual pathways from left eye

Combined left and right visual pathways

FIGURE 19.12 Major pathways of the mammalian visual system. In mammals, the optic nerve from each eye branches, sending nerve fibers to a lateral geniculate nucleus on each side of the brain. On the ipsilateral side, a particular part of the retina projects to a particular part of the lateral geniculate nucleus. On the contralateral side, the lateral geniculate nucleus receives input from all parts of the retina. Neurons from each lateral geniculate nucleus innervate the visual cortex on the same side.

(Draganski et al. 2004). Thus, the pattern of neuronal connections is a product of inherited patterning and patterning produced by experiences. This interplay between innate and experiential development has been detailed most dramatically in studies on mammalian vision.

Experiential changes in mammalian visual pathways

Some of the most interesting research on mammalian neuronal patterning concerns the effects of sensory deprivation on the developing visual system in kittens and monkeys. The paths by which electric impulses pass from the retina to the brain in mammals are shown in **FIGURE 19.12**. Axons from the retinal ganglion cells form the two optic nerves, which meet at the optic chiasm. As in *Xenopus* tadpoles, some axons go to the opposite (contralateral) side of the brain, but unlike in most other vertebrates, mammalian retinal ganglion cells also send inputs into the same (ipsilateral) side of the brain (see Chapter 11). These axons end at the two lateral geniculate nuclei. Here the input from each eye is kept separate, with the uppermost and anterior layers receiving the axons from the contralateral eye, and the middle of the layers receiving input from the ipsilateral eye. The situation becomes even more complex as neurons from the lateral geniculate nuclei connect with the neurons of the visual cortex. More than 80% of the neural cells in the visual cortex

receive input from both eyes. The result is binocular vision and depth perception.

A remarkable finding is that the retinocortical projection pattern is the same for both eyes. If a certain cortical neuron is stimulated by light flashing across a region of the left eye 5° above and 1° to the left of the fovea,* it will also be stimulated by a light flashing across a region of the *right* eye 5° above and 1° to the left of the fovea. Moreover, the response evoked in the cortical neuron when both eyes are stimulated is greater than the response when either retina is stimulated alone.

Torsten Hubel and David Wiesel, along with their co-workers, demonstrated that the development of the nervous system depends to some degree on the experience of the individual during a critical period of development (see Hubel 1967). In other words, not all neuronal development is encoded in the genome; some is the result of learning. Experience appears to strengthen or stabilize some neuronal connections that are already present at birth and to weaken or eliminate others. These conclusions come from studies of partial sensory deprivation. Hubel and Wiesel (1962, 1963) sewed shut the right eyelids of newborn kittens and left them closed for 3 months. After this time, they unsewed the right eyelids. The cortical neurons of these kittens could not be stimulated by shining light into the right eye. Almost all the inputs into the visual cortex came from the left eye only. The behavior of the kittens revealed the inadequacy of their right eyes; when the left eyes of these kittens were covered, they became functionally blind. Because the lateral geniculate neurons appeared to be

---

*The *fovea* is a depression in the center of the retina where only cones are present and vision is most acute. Rods and blood vessels are absent. In this instance, it serves as a convenient landmark.

stimulated by input from both right and left eyes, the physiological defect appeared to be in the connections between the lateral geniculate nuclei and the visual cortex. Similar phenomena have been observed in rhesus monkeys, where the defect has been correlated with a lack of protein synthesis in the lateral geniculate neurons innervated by the covered eye (Kennedy et al. 1981).

Although it would be tempting to conclude that the blindness resulting from these experiments was the result of failure to form the proper visual connections, this is not the case. Rather, when a kitten or monkey is born, axons from lateral geniculate neurons receiving input from each eye overlap extensively in the visual cortex (Hubel and Wiesel 1963; Crair et al. 1998). However, when one eye is covered early in the animal's life, its connections within the visual cortex are taken over by those of the other eye (**FIGURE 19.13**). The axons compete for connections, and experience plays a role in strengthening and stabilizing the connections that are made. Thus, when *both* eyes of a kitten are sewn shut for 3 months, most cortical neurons can still be stimulated by appropriate illumination of one eye or the other.

The critical time in kitten development for this validation of neuronal connections begins between 4 and 6 weeks after birth. Monocular deprivation up to the fourth week produces little or no physiological deficit, but through the sixth week it produces all the characteristic neuronal changes. If a kitten has had normal visual experience for the first 3 months, any subsequent monocular deprivation (even for a year or more) has no effect. At that point, the synapses have been stabilized.

Two principles, then, can be seen in the patterning of the mammalian visual system. First, the neuronal connections involved in vision are present even before the animal sees. Second, experience plays an important role in determining whether or not certain connections persist.* Just as experience refines the original neuromuscular connections, experience plays a role in refining and improving the visual connections. It is possible, then, that adult functions such as learning and memory arise from the establishment and/or strengthening of different synapses by experience during development. As Purves and Lichtman (1985) remark:

> *The interaction of individual animals and their world continues to shape the nervous system throughout life in ways that could never have been programmed. Modification of the nervous system by experience is thus the last and most subtle developmental strategy.*

Life Cycles and Polyphenisms

Diapause

Throughout nature, organisms are able to respond to the environment by changing their developmental strategy. One

---

* Studies have shown that differences in neurotransmitter release result in changes in synaptic adhesivity and cause the withdrawal of the axon providing the weaker stimulation (Colman et al. 1997). Studies in mice suggest that BDNF is crucial during the critical period (Huang et al. 1999; Katz 1999; Waterhouse and Xu 2009; Cowansage et al. 2010).

(A)

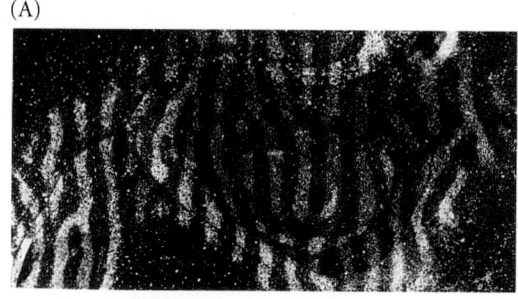

(B)

FIGURE 19.13 Experience alters neural connections. (A,B) Dark-field autoradiographs of monkey striate (visual) cortex 2 weeks after one eye was injected with [³H]proline in the vitreous humor. Each retinal neuron takes up the radioactive label and transfers it to the cells with which it forms synapses. (A) Normal labeling pattern. The white stripes indicate that roughly half the columns took up the label, while the other half did not—a pattern reflecting that half the cells were innervated by the labeled eye and half by the unlabeled eye. (B) Labeling pattern when the unlabeled eye was sutured shut for 18 months. Axonal projections from the normal (labeled) eye have taken over the regions that would normally have been innervated by the sutured eye. (C,D) Drawings of axons from the lateral geniculate nuclei of kittens in which one eye was occluded for 33 days. The terminal branching of axons receiving input from the occluded eye (C) was far less extensive than that of axons receiving input from the nonoccluded eye (D). (A,B from Wiesel 1982, courtesy of T. Wiesel; C,D after Antonini and Stryker 1993.)

(C)   (D)

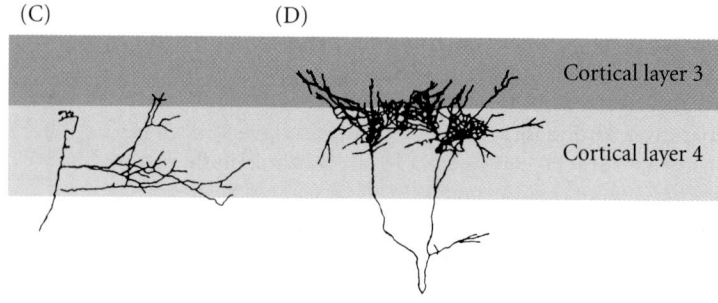

of these strategies is to delay development to a better time, and many species of insects and mammals have evolved a developmental strategy called diapause to survive periodically harsh conditions. **Diapause** is a suspension of development that can occur at the embryonic, larval, pupal, or adult stage, depending on the species (see Chapter 16). Diapause is not a physiological response brought about by harsh conditions. Rather, it is induced by stimuli (such as changes in the duration of daylight) that *presage* a change in the environment—that is, cues that are discernible *before* the severe conditions arise. Diapause is especially important for temperate-zone insects, enabling them to survive the winter. The overwintering eggs of the hickory aphid provide an example of this strategy. The development in the egg is suspended over the winter, so the larvae do not hatch when food is unavailable. In this case, diapause occurs during early development. The silkworm moth *Bombyx mori* similarly overwinters as an embryo, entering diapause just before segmentation. The gypsy moth *Lymantria dispar* initiates diapause as a larva, and needs an extended period of cold weather to end diapause, which is why this pest is not found in the southern regions of Europe or the United States.

More than 100 mammalian species undergo diapause. The two most common mammalian strategies are delayed fertilization (the sperm are stored for later use) and delayed implantation (the blastocyst remains unimplanted in the uterus, and the rate of cell divisions diminishes or vanishes). Some species have *seasonal* diapause, so embryos conceived in autumn will be born in spring rather than winter; in other species, diapause is induced by the presence of a newborn that is still getting milk. In the tammar kangaroo (*Macropus eugenii*), diapause can be a response to suckling-induced prolactin release, but it can also be induced by prolactin synthesized in response to changes in day length. In both cases, progesterone seems to be the signal that restores implantation and embryonic growth. Different groups of mammals use different hormones to induce or break diapause, but the result is the same: diapause lengthens the gestation period, allowing mating to occur and young to be born at times and seasons appropriate to the habitat of that species (Renfree and Shaw 2000).

● See WEBSITE 19.3 Mechanisms of diapause

Larval settlement

Free-swimming marine larvae often need to settle near a source of food or on a firm substrate on which they can undergo metamorphosis. The ability of marine larvae to suspend development until they experience a particular environmental cue is called **larval settlement**. Particularly among the molluscs, there are often very specific cues for settlement (Hadfield 1977; Hadfield and Paul 2001; Zardus et al. 2008). In some cases, the mollusc's prey supply the cues, while in other cases the substrate itself gives off molecules used by the larvae to initiate settlement. These cues may not be constant, but they need to be part of the

environment if further development is to occur* (Pechenik et al. 1998).

In many marine invertebrate species, larval settlement and the subsequent distribution of invertebrate populations are regulated by mats of bacteria called **biofilms** (Hadfield 2011). Chemicals from these potential food sources are used by the larvae as signals to settle and undergo metamorphosis. Humans are affecting population distributions with our desire to place large objects in the oceans. Such objects readily acquire biofilms and the marine fauna that attach to them. As early as 1854, Charles Darwin speculated that barnacles were transported to new locales when their larvae settled on the hulls of ships. Indeed, the ability of biofilms to aid invertebrate larval settlement and colony formation explains the ability of barnacles and tubeworms ("biofouling invertebrates") to accumulate on ship keels, clog sewer pipes, and deteriorate underwater structures (Zardus et al. 2008).

UNDER THE SEA Most of the cues known for larval settlement and metamorphosis involve chemicals eminating from the substrate; these chemicals can signal the presence of a food source or potentially induce larval metamorphosis. However, in at least one case, vibrational cues appear to direct marine larvae to coral reefs. Coral reefs are the largest biological structures on Earth, and they grow by recruiting planktonic coral (cnidarian) larvae. While chemical cues work within a small distance of the reef, it is the "noise of the reef"—the snapping of shrimp claws and the noises made by thousands of reef fish—that attract coral larvae from long distances. Vermeij and colleagues (2010) made recordings of Caribbean reefs and found that the larvae swam to the source of the sound, even in the laboratory. These findings mean that coral reefs face danger from noise pollution as well as from thermal and chemical pollution. Steve Simpson (2010), who headed the study, has warned, "Anthropogenic noise has increased dramatically in recent years, with small boats, shipping, drilling, pile driving and seismic testing now sometimes drowning out the natural sounds of fish and snapping shrimps."

---

*The importance of substrates for larval settlement and metamorphosis was first demonstrated in 1880, when William Keith Brooks, an embryologist at Johns Hopkins University, was asked to help the ailing oyster industry of Chesapeake Bay. For decades, oysters had been dredged from the bay, and there had always been a new crop to take their place. But by 1880, each year brought fewer oysters. What was responsible for the decline? Experimenting with larval oysters, Brooks discovered that the American oyster (unlike its better-studied European relative) needs a hard substrate on which to metamorphose. For years, oystermen had thrown the shells back into the sea, but with the advent of suburban sidewalks, they started selling the shells to cement factories. Brooks's solution: throw the shells back into the bay. The oyster population responded, and the Baltimore wharves still sell their descendants.

Life Cycle Choices: Dictyostelium

Of the many developmental changes triggered by the environment, few are as profound as the effect of nutrient starvation on *Dictyostelium discoideum*. *Dictyostelium* is a multicellular organism only when environmental conditions demand it. The life cycle of this fascinating organism is illustrated in **FIGURE 19.14**. In its asexual cycle, solitary haploid amoebae (called myxamoebae or "social amoebae" to distinguish them from amoeba species that always remain solitary) live on decaying logs, eating bacteria and reproducing by binary fission. When they have exhausted their food supply, their gene expression changes, and they begin expressing cell adhesion molecules. Moreover, they start migrating to a common place. Tens of thousands of myxamoebae join together to form moving streams of cells that converge at a central point. Here they synthesize new adhesion molecules and pile atop one another to produce a conical mound called a tight aggregate. Subsequently, a tip arises at the top of this mound, and the tight aggregate bends over to produce the migrating slug (with the tip at the front).

The slug (usually given the more dignified title of **pseudoplasmodium** or **grex**) is usually 2–4 mm long and is encased in a slimy sheath. The grex begins to migrate (if the environment is dark and moist) with its anterior tip slightly raised. When it reaches an illuminated area, migration ceases, and the culmination stages of the life cycle take place as the grex differentiates

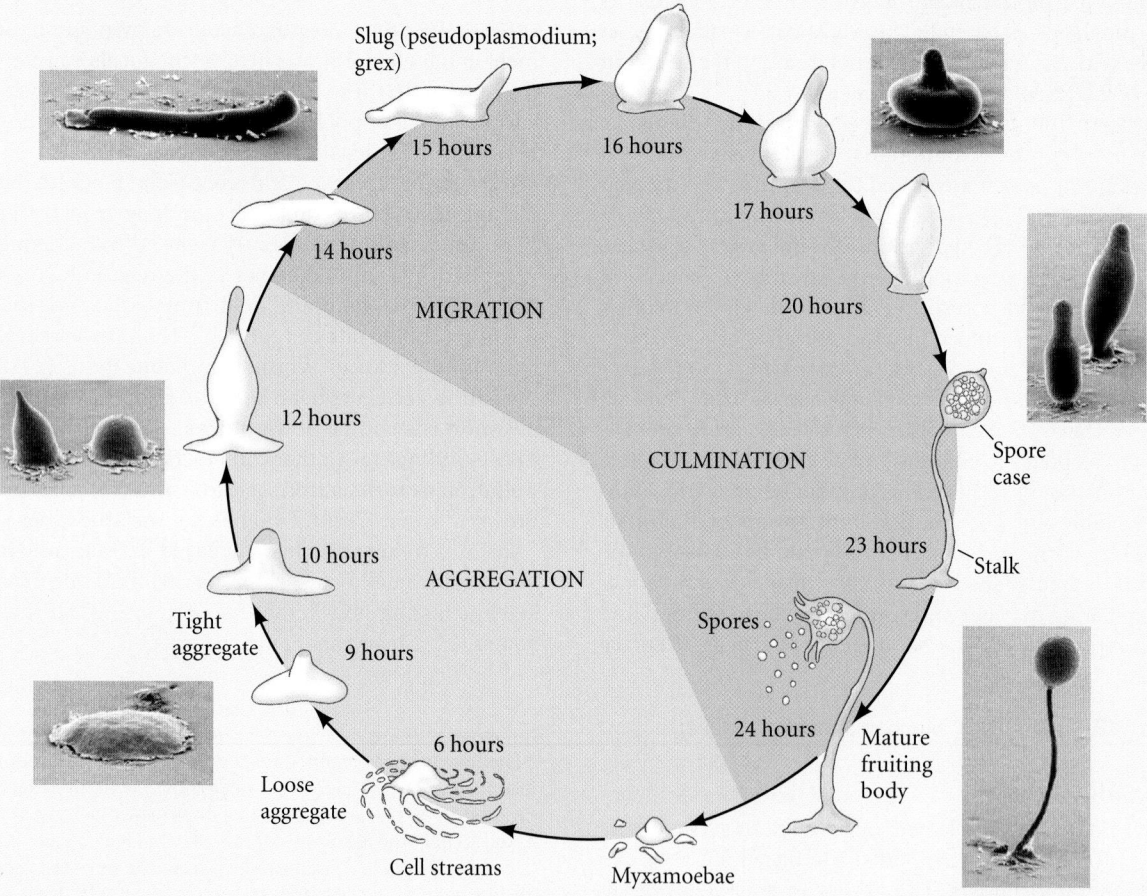

FIGURE 19.14 Life cycle of *Dictyostelium discoideum*. Haploid spores give rise to myxamoebae, which can reproduce asexually to form more haploid myxamoebae. As the food supply diminishes, aggregation occurs and a migrating slug is formed. The slug culminates in a fruiting body that releases more spores. Times refer to hours since the onset of nutrient starvation. Prestalk cells are indicated in yellow. (Photographs courtesy of R. Blanton and M. Grimson.)

SIDELIGHTS **SPECULATIONS** (continued)

(A) **cAMP**

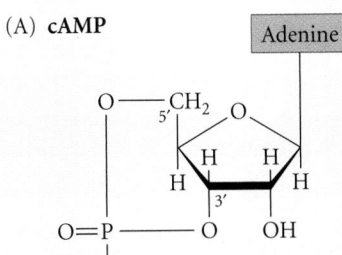

(B)

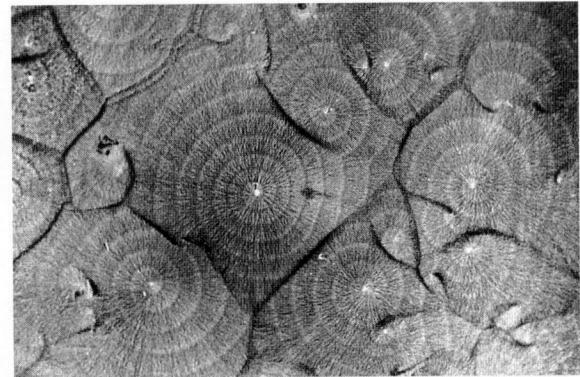

(C)

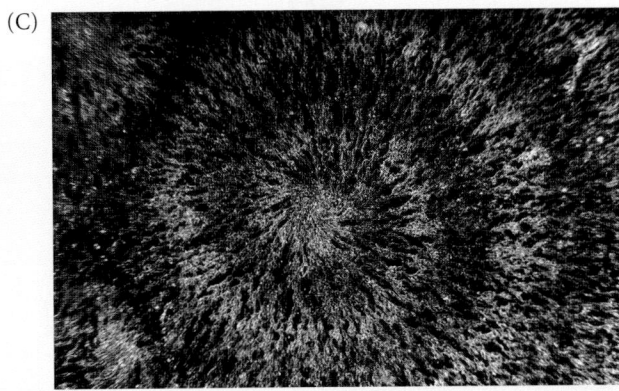

(D)

FIGURE 19.15 Chemotaxis of *Dictyostelium* myxamoebae is a result of spiral waves of cAMP. (A) Chemical structure of cAMP. (B) Visualization of several cAMP "waves." Central cells secrete cAMP at regular intervals, and each pulse diffuses outward as a concentric wave. The waves were charted by saturating filter paper with radioactive cAMP and placing it on an aggregating colony. The cAMP from the secreting cells dilutes the radioactive cAMP. When the radioactivity on the paper is recorded (by placing it over X-ray film), the regions of high cAMP concentration in the culture appear lighter than those of low cAMP concentration. (C) Spiral waves of myxamoebae moving toward the initial source of cAMP. Because moving and nonmoving cells scatter light differently, the photograph reflects cell movement. The bright bands are composed of elongated migrating cells; the dark bands are cells that have stopped moving and have rounded up. As cells form streams, the spiral of movement can still be seen moving toward the center. (D) Computer simulation of cAMP wave spreading across migrating *Dictyostelium* cells. The model takes into account the reception and release of cAMP, and changes in cell density caused by the movement of the cells. The cAMP wave is plotted in dark blue. The population of myxamoebae goes from green (low) to red (high). Compare with the actual culture shown in (C). (B from Tomchick and Devreotes 1981; C from Siegert and Weijer 1989; D from Dallon and Othmer 1997.)

into (1) a fruiting body composed of spore cells and (2) a stalk that supports the fruiting body (Raper 1940; Bonner 1957). The anterior cells, representing 15–20% of the entire cellular population, form the tubed stalk. This process begins as some of the central anterior cells, the **prestalk cells**, begin secreting an extracellular cellulose coat and extending a tube through the grex. As the prestalk cells differentiate, they form vacuoles and enlarge, lifting up the mass of **prespore cells**

that made up the posterior four-fifths of the grex (Jermyn and Williams 1991). The stalk cells die, but the prespore cells, elevated above the stalk, become spore cells. These spore cells disperse, each one becoming a new myxamoeba.

Aggregation of myxamoebae

What causes the myxamoebae to aggregate? Time-lapse videomicroscopy has shown that no directed movement occurs during the first 4–5

hours following nutrient starvation. During the next 5 hours, however, the cells can be seen moving at about 20 mm/min for 100 seconds. This movement ceases for about 4 minutes, then resumes. Although the movement is directed toward a central point, it is not a simple radial movement. Rather, cells join with one another to form streams; the streams converge into larger streams, and eventually all streams merge at the center. Bonner (1947) and Shaffer (1953) showed that this movement is a result of **chemotaxis**: the cells are guided to aggregation centers by a soluble substance. This substance was later identified as **cyclic adenosine 3′,5′-monophosphate (cAMP;** Konijn et al. 1967; Bonner et al. 1969), the chemical structure of which is shown in **FIGURE 19.15A**.

Aggregation is initiated as each myxamoeba begins to synthesize cAMP. There are no dominant cells

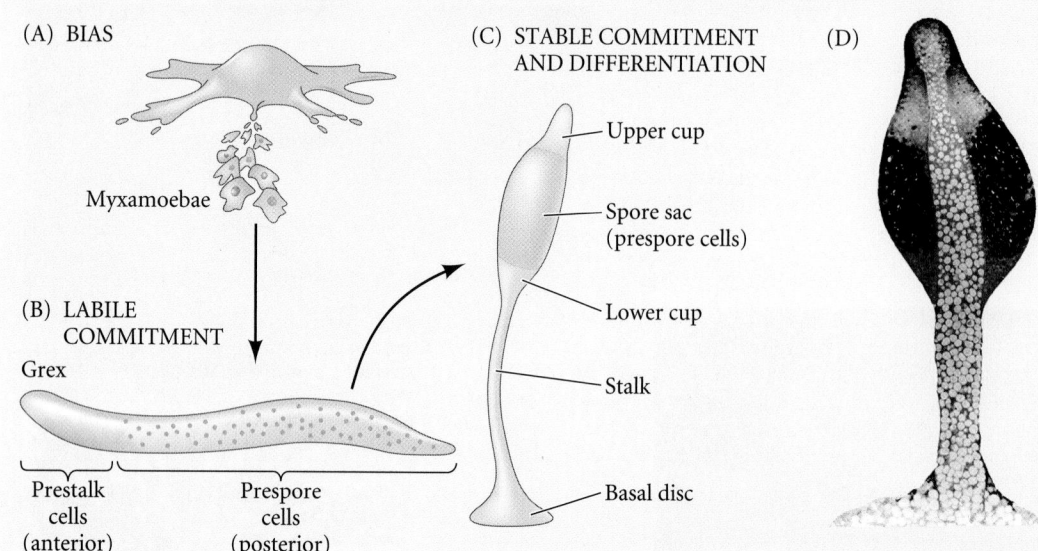

FIGURE 19.16 *Alternative cell fates in* Dictyostelium discoideum. *(A–C) Progressive commitment of cells to become either spore or stalk cells. (A) Myxamoebae may have biases toward stalk or spore formation because of the stage of the cell cycle they were in when starved. (B) As the grex migrates, most prestalk cells are in the anterior third of the grex, while most of the posterior consists of prespore cells. Some prestalk cells are also seen in the posterior, and these cells will contribute to the cups of the spore sac and to the basal disc at the bottom of the stalk. The cell fates are not yet fixed, however, and if the stalk-forming anterior is cut off, the anteriormost cells remaining will convert from stem to stalk. (C) At culmination, the spore-forming cells are massed together in the spore sac. The stalk cells form the cups of the spore sac, as well as the stalk and basal disc. (D) Culmination stage, where the spore cell precursors (blue) are being elevated by the prestalk cells (white) and tipped with upper cup cells (red.) (D courtesy of M. J. Grimson and R. L. Blanton.)*

that begin the secretion or control the others. Rather, the sites of aggregation are determined by the distribution of the myxamoebae (Keller and Segal 1970; Tyson and Murray 1989). Neighboring cells respond to cAMP in two ways: they initiate a movement toward the cAMP pulse for about a minute, and they release cAMP of their own (Robertson et al. 1972; Shaffer 1975). The movement of each myxamoeba is caused by the change in cytoskeletal polarity brought about by the cAMP (Parent et al. 1998; Iijima et al. 2002). After this happens, the cell is unresponsive to further cAMP pulses for several minutes. During this time, an extracellular membrane-associated phosphodi-esterase then cleaves the remaining cAMP from the environment, allowing the receptors to get ready to receive another pulse. The result is a rotating spiral wave of cAMP that is propagated throughout the population of cells (**FIGURE 19.15B–D**). As each wave arrives, the cells take another step toward the center.

The differentiation of individual myxamoebae into either stalk (somatic) or spore (reproductive) cells is a complex matter. In *Dictyostelium*, differentiation involves a dichotomous decision. There appears to be a progressive commitment to one of the two alternative pathways (**FIGURE**

19.16). At first there is a *bias* toward one path or another. For instance, cells starved in the S and early G2 phases of the cell cycle have relatively high levels of calcium and display a tendency to become stalk cells, whereas those starved in mid or late G2 have lower calcium levels and tend to become spore cells (Nanjundiah 1997; Azhar et al. 2001). Then there is a *labile specification*, a time when the cell will normally become either a spore cell or a stalk cell, but when it can still change its fate if placed in a different position in the organism. Due to the biases in these cells, cAMP is used in different ways by the prespore and prestalk cells (see Kimmel and Firtel 2004). In the prespore cells of the grex, extracellular cAMP initiates the expression of spore-specific mRNAs. It does this by inducing β-catenin, which enters the nucleus to activate certain spore-specific genes

(Ginsburg and Kimmel 1997; Plyte et al. 1999; Kim et al. 2002). In the prestalk cells that are in the anterior tip of the grex, cAMP suppresses this pathway. Instead, cyclic diguanosine monophosphate (c-diGMP, which was known as a bacterial signal) induces the cells to become prestalk cells (Chen and Schaap 2012). The third and fourth stages are a *determination* to a specific fate, followed by the cell's *differentiation* into a particular cell type, either a stalk cell or a spore cell.

Several proteins play critical roles in the final differentiation of the prespore cells into encapsulated spores (Anjard et al. 1998a,b; Mujumdar et al. 2009). The proteins SDF1 (spore differentiation factor-1) and Trishanku are important in initiating culmination, while SDF2 seems to cause the prespore cells (but not prestalk cells) to become spores. The prespore cells appear to have a receptor that enables them to respond to SDF2, whereas the prestalk cells lack this receptor (Wang et al. 1999). Culmination is also brought about by declining ammonia concentrations (Follstaedt et al. 2003). Ammonia is released preferentially in the anterior portion of the grex, and it appears to help regulate chemotaxis of the prestalk cells as well as aid in the production of spore cells (Oyama and Blumberg 1986; Feit et al. 2001).

The formation of stalk cells from prestalk cells is similarly complicated and may involve several factors working synergistically (Early 1999). Indeed, prestalk cells from different parts of the grex pass through different intermediary cell types before reaching the final stage of stalk cell. Those prestalk cells near the prespore cells encounter the chlorinated lipid DIF-1, which is

made and secreted by the prespore cells. This converts those nearby prestalk cells into stalk cells (Fukuzawa et al. 2003; Thompson et al. 2004). Thus, the stalk cells that cover the spores have a slightly different history than the stalk cells that hold the ball of spores above the ground.

Thus, many environmental factors act together to trigger alternative developmental pathways in *Dictyostelium*. First, nutrient depletion causes a wholesale change in gene expression and cell adhesivity; second, cAMP orients the migration of individual cells toward a common center; third, the neighboring cells determine whether any particular cell is to be stalk or spore; and fourth, molecules such as ammonia can help complete the processes of differentiation.*

● See **WEBSITE 19.5** The Dictyostelium life cycle: Variations within variations

● See **VADE MECUM** Slime mold life cycle

*There is a fifth ecological component to *Dictyostelium* development: the organism brings part of its original ecosystem with it as it develops. Just as immigrants fleeing starvation bring seeds with them in the hopes that growing conditions will be better in a new place, some *Dictyostelium* strains carry bacterial cells with them into the sorus. These bacteria will be dispersed together with the slime mold spores. In the wild, about one-third of a *Dictyostelium* will stop feeding early and incorporate bacteria into the fruiting bodies, and these bacteria can serve as food sources if the conditions are good for *Dictyostelium* growth (Brock et al. 2011).

The hard life of spadefoot toads

Spadefoot toads (*Scaphiopus couchii, Spea multiplicata,* and their relatives) have a remarkable strategy for coping with a harsh environment (Ledón-Rettig and Pfennig 2011). The toads are called out from hibernation by the thunder that accompanies the first spring rainstorms in the Sonoran desert.* They breed in temporary ponds formed by the rain, and the embryos develop quickly into larvae. After the larvae metamorphose, the young toads return to the desert, burrowing into the sand until the next year's storms bring them out.

---

*Like coral larvae, the toads are sensitive to vibration, and noise pollution may affect their survival: motorcycles produce the same sounds as thunder, causing the toads to come out of hibernation only to die in the scorching desert sun.

Desert ponds are ephemeral pools that can either dry up quickly or persist, depending on their initial depth and the frequency of rainfall. One might envision two alternative scenarios confronting a tadpole in such a pond: either (1) the pond persists until you have time to fully metamorphose, and you live; or (2) the pond dries up before your metamorphosis is complete, and you die. In *Scaphiopus couchii* and several other amphibians, however, a third alternative has evolved. The timing of their metamorphosis is controlled by the pond. If the pond persists at a viable level, development continues at its normal rate and the algae-eating tadpoles develop into juvenile toads. However, if the pond is drying out and getting smaller, some of the tadpoles embark on an alternative developmental pathway. They develop a wider mouth and powerful jaw muscles, which enables them to eat (among other things) other *Scaphiopus* tadpoles (**FIGURE 19.17**). These carnivorous

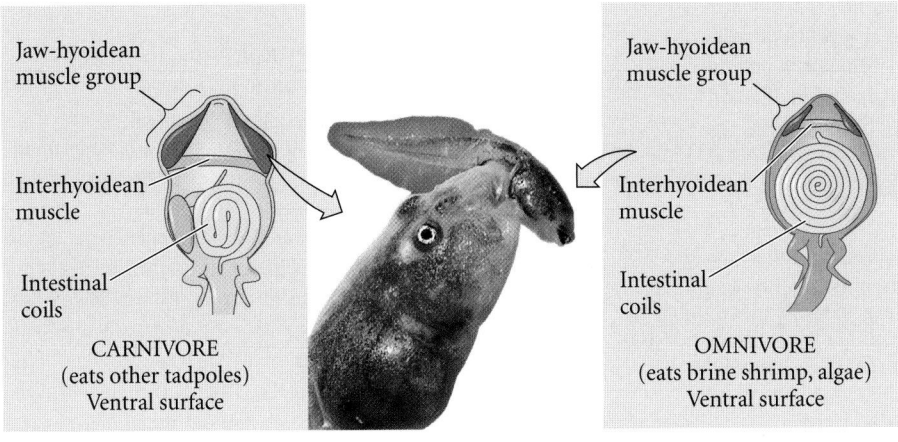

FIGURE 19.17 Polyphenism in tadpoles of the spadefoot toad *Scaphiopus couchii*. The typical morph (right) is an omnivore, feeding on arthropods and algae. When ponds are drying out quickly, however, a carnivorous (cannibalistic) morph forms (left). It develops a wider mouth, larger jaw muscles, and an intestine modified for a carnivorous diet. The center photograph shows a cannibalistic tadpole eating a smaller pondmate. (Photograph © Thomas Wiewandt; drawings courtesy of R. Ruibel.)

tadpoles metamorphose quickly, albeit into a smaller version of a juvenile spadefoot toad. But they survive while other *Scaphiopus* tadpoles perish from desiccation (Newman 1989, 1992).

The signal for accelerated metamorphosis appears to be the change in water volume. In the laboratory, *Scaphiopus* tadpoles are able to sense the removal of water from aquaria, and their acceleration of metamorphosis depends on the rate at which the water is removed. The stress-induced corticotropin-releasing hormone signaling system appears to modulate this effect (Denver et al. 1998, 2009). This increase in brain corticotropin-releasing hormone is thought to be responsible for the subsequent elevation of the thyroid hormones that initiate metamorphosis (Boorse and Denver 2003). As in many other cases of polyphenism, the developmental changes are mediated through the endocrine system. Sensory organs send a neural signal to regulate hormone release. The hormones then can alter gene expression in a coordinated and relatively rapid fashion.

● **See WEBSITE 19.4** Pressure as an agent of development

Developmental Symbioses

In addition to the above-mentioned abiotic and biotic relationships where the environment regulates development, there is a special type of biotic relationship called symbiosis. Contrary to the popular use of the term to mean a mutually beneficial relationship, the word **symbiosis** (Greek, *sym*, "together"; *bios*, "life") can refer to any close association between organisms of different species (see Sapp 1994). In many symbiotic relationships, one of the organisms involved is much larger than the other, and the smaller organism may live on the surface or inside the body of the larger. In such relationships, the larger organism is referred to as the **host** and the smaller as the **symbiont**. There are two important categories of symbiosis:*

- **Parasitism** occurs when one partner benefits at the expense of the other. An example of a parasitic relationship is that of a tapeworm living in the human digestive tract, wherein the tapeworm steals nutrients from its host.

- **Mutualism** is a relationship that benefits both partners. A striking example of this type of symbiosis can be found in the partnership between the Egyptian plover (*Pluvianus aegyptius*) and the Nile crocodile (*Crocodylus niloticus*). Although it regards most birds as lunch, the crocodile allows the plover to roam its body, feeding on the harmful parasites there. In this mutually beneficial relationship, the bird obtains food while the crocodile is rid of parasites.

---

* Commensalism, defined as a relationship that is beneficial to one partner and neither beneficial nor harmful to the other partner, is sometimes thought of as a third category of symbiosis. Although many symbioses appear on the surface to be commensal, recent studies suggest that very few symbiotic relationships are truly neutral with respect to either party.

In addition, the term **endosymbiosis** ("living inside") is widely used to describe the situation in which one cell lives inside another cell, a circumstance thought to account for the evolution of the organelles of the eukaryotic cell (see Margulis 1971), and one that describes the *Wolbachia* developmental symbioses discussed at length later in this chapter.

Symbiosis, and especially mutualism, is the basis for life on Earth. The symbiosis between *Rhyzobium* bacteria and the roots of legume plants is responsible for converting atmospheric nitrogen into a usable form for generating amino acids, and is therefore essential for life. Symbioses between fungi and plants are ubiquitous and are often necessary for plant development (see Gilbert and Epel 2009; Pringle 2009). Orchid seeds, for example, contain no energy reserves, so a developing orchid plant must acquire carbon from mycorrhizal fungi. (This is why orchids grow best in moist tropical environments, where fungi are plentiful.) The coastal zone ecosystem throughout the world is sustained by a triple symbiosis among seagrass, clams, and the sufide-oxidizing bacteria living inside the clam's gills (van der Heide et al. 2012).

In some cases, the development of one individual is brought about by signals from organisms of a different species. In some organisms, this relationship has become symbiotic—the symbionts have become so tightly integrated into the host organism that the host cannot develop without them (Sapp 1994). Indeed, recent evidence indicates that developmental symbioses appear be the rule rather than the exception (McFall-Ngai 2002, 2008a). The term for the composite organism of a host and its persistent symbionts is the **holobiont** (Rosenberg et al. 2007; Gilbert and Epel 2009).

● **See WEBSITE 19.6** Developmental symbiosis and parasitism

Mechanisms of developmental symbiosis: Getting the partners together

All symbiotic associations must meet the challenge of maintaining their partnerships over successive generations. In the partnerships that are the main subject here, in which microbes are crucial to the development of their animal hosts, the task of transmission is usually accomplished by either vertical or horizontal transmission.

VERTICAL TRANSMISSION   Vertical transmission refers to the transfer of symbionts from one generation to the next through the germ cells, usually the eggs (Krueger et al. 1996). There are several ways by which embryos can become infected by their mothers, but one of the most common is for the symbiont to be transmitted through the oocyte.

Bacteria of the genus *Wolbachia* reside in the egg cytoplasm of invertebrates and provide important signals for the development of the individuals produced by those eggs. As we shall see, many species of invertebrates have "outsourced" important developmental signals to *Wolbachia* bacteria, which are transmitted like mitochondria—that is, in the oocyte cytoplasm. In numerous *Drosophila* species,

(A)

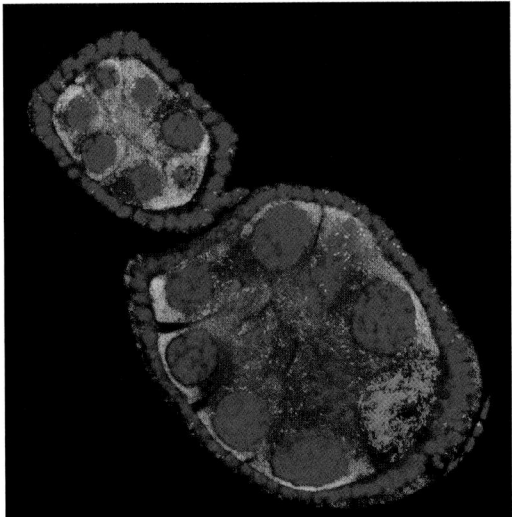

(B)

FIGURE 19.18 Vertical and horizontal transmission of *Wolbachia* bacteria. (A) In *Drosophila*, *Wolbachia* are transmitted vertically through the female germ cells. In the germinarium, 15 nurse cells transport proteins, RNAs, and organelles to the distalmost oocyte cell. The symbiotic bacterium (stained red) is also transported by these microtubules into the oocyte. Cytoplasm of the ovary is green, and blue indicates DNA. (B) Male and female *Armadillidium vulgare*. Genetically male pill bugs (right) can be transformed into phenotypic egg-producing females (left) by infection (i.e., horizontal transmission) of *Wolbachia* bacteria. (A after Ferree et al. 2005, courtesy of H. M. Frydman and E. Wieschaus; B courtesy of D. McIntyre.)

Wolbachia provide resistance against viruses (Teixeira et al. 2008; Osborne et al. 2009). Feree and colleagues (2005) have shown that in *Drosophila* development, *Wolbachia* use the host's nurse cell microtubule system and dynein motors to travel from the nurse cells into the developing oocyte (**FIGURE 19.18A**). In other words, the bacteria use the same cytoskeletal pathway as mitochondria, ribosomes, and *bicoid*

mRNA (see Chapter 6). Once in the oocyte, the bacteria enter every cell, becoming endosymbionts. The *Wolbachia* appear to aid in their propagation by entering the stem cell niches that make ovaries and oocytes (Fast et al. 2011). The females infected with *Wolbachia* make four times more eggs than their uninfected sisters, thereby furthering the spread of *Wolbachia*.

HORIZONTAL TRANSMISSION   *Wolbachia* also can be transmitted horizontally. In horizontal transmission, the metazoan host is born free of symbionts but subsequently becomes infected, either by its environment or by other members of the species. In pill bugs such as *Amadillidium vulgare*, genetically male insects infected with *Wolbachia* are transformed by the bacteria into females (**FIGURE 19.18B**). As females, the pill bugs can then transmit the *Wolbachia* symbionts to the next generation (Cordaux et al. 2004).

A different type of horizontal transmission involves aquatic eggs that attract photosynthetic algae. Clutches of amphibian and snail eggs, for example, are packed together in tight masses. The supply of oxygen limits the rate of their development, and there is a steep gradient of oxygen from the outside of the cluster to deep within it; thus embryos on the inside of the cluster develop more slowly than those near the surface (Strathmann and Strathmann 1995). The embryos seem to get around this problem by coating themselves with a thin film of photosynthetic algae, which they obtain from the pond water. In clutches of amphibian and snail eggs, photosynthesis from this algal "fouling" enables net oxygen production in the light, whereas respiration exceeds photosynthesis in the dark (Bachmann et al. 1986; Pinder and Friet 1994; Cohen and Strathmann 1996). Thus, the symbiotic algae "rescue" the eggs by their photosynthesis.

Horizontal transmission is crucial for the symbiotic gut bacteria found in many animals, including humans. As we will see later in this chapter, mammalian gut bacteria are critical in forming the blood vessels of the intestine, and possibly in regulating stem cell proliferation (Pull et al. 2005; Liu et al. 2010). Human infants usually acquire these symbionts as they travel through the birth canal. Before then, development is aseptic, but once the amnion breaks, the microbiota of the mother's reproductive tract can colonize the infant's skin and gut. This is supplemented by bacteria from the parents' skin, especially the mother's skin during nursing. The colonization of the infant by the microbes is a critically important event, and the mammalian immune system appears to encourage certain bacteria to enter the body, while discouraging others (see Gilbert et al. 2012). Indeed, some of the complex sugars found in human mothers' milk are not digestible by the infant. Rather, they serve as food for certain bacterial symbionts that help the infants' bodies develop (Zivkovic et al. 2011). Although each baby starts with a unique bacterial profile, within a year the types and proportions of bacteria have converged to the adult human profile that characterizes the human digestive tract (Palmer et al. 2007).

The Euprymna-Vibrio symbiosis

Horizontal transmission plays a major role in one of the best-studied examples of developmental symbiosis: that between the squid *Euprymna scolopes* and the luminescent bacterium *Vibrio fischeri* (McFall-Ngai and Ruby 1991; Montgomery and McFall-Ngai 1995). The adult *Euprymna* is equipped with a light organ composed of sacs filled with these bacteria (**FIGURE 19.19A**). The newly hatched squid, however, does not contain these light-emitting symbionts, nor does it have the light organ to house them. Rather, the symbiotic bacteria interact with the larval squid to build the light organ together. The juvenile squid acquires *V. fischeri* from seawater pumping through its mantle cavity (Nyholm et al. 2000). The bacteria bind to a ciliated epithelium in this cavity; the epithelium binds *only V. fischeri*, allowing other bacteria to pass through (**FIGURE 19.19B**). The bacteria then induce hundreds of genes in the epithelium, leading to the apoptotic death of the epithelial cells, their replacement by a nonciliated epithelium, the differentiation of the surrounding cells into storage sacs for the bacteria, and the expression of genes encoding opsins and other visual proteins in the light organ (**FIGURE 19.19C**; Chun et al. 2008; McFall-Ngai 2008b; Tong et al. 2009).

The substance *V. fischeri* secretes to effect these changes turns out to be fragments of the bacterial cell wall, and the active agents are tracheal cytotoxin and lipopolysaccharide (Koropatnick et al. 2004). This finding was surprising, because these two agents are known to cause inflammation and disease to an extent that one or both species could not survive. Indeed, tracheal cytotoxin is responsible for the tissue damage in both whooping cough and gonorrheal infections. The destruction and replacement of ciliated tissue in the respiratory tract and oviduct are due to these bacterial compounds. After the bacteria have induced the morphological changes in the host, the host secretes a peptide into the *Vibrio*-containing crypts which neutralizes the bacterial toxin (Troll et al. 2010). Both organisms change their gene expression patterns, and both benefit from their association: The bacteria get a home and express their light-generating enzymes, and the squid develops a light organ that allows it to swim at night in shallow waters without casting a shadow.

Obligate developmental mutualism

In an **obligate mutualism**, the species involved are interdependent with one another to such an extent that neither could survive without the other. The most common example of obligate mutualism is the lichens, in which fungal and algal species are joined in a relationship that results in an essentially new species. More and more examples of obligate mutualism are being described, and most of these have important consequences for medicine and conservation biology.

One example of obligate developmental mutualism has been described in the parasitic wasp *Asobara tabida*. In these insects, symbiotic bacteria are found in the egg cytoplasm and, as in the *Wolbachia* example discussed above, are vertically transferred through the female germ plasm. In *Asobara*, the bacteria enable the wasp to complete yolk production and egg maturation (Dedeine et al. 2001; Pannebakker et al. 2007). If the symbionts are removed, the ovaries undergo apoptosis and no eggs are produced (**FIGURE 19.20**).

In obligate developmental mutualisms, the death of the host can result from killing the symbiont. Atrazine was mentioned in Chapter 15 for its ability to induce aromatase and cause sex-determination anomalies in amphibians. But the major use and effect of atrazine is to kill plant life; it is a potent nonspecific herbicide. Once applied, atrazine can

(A)

(B)   (C)

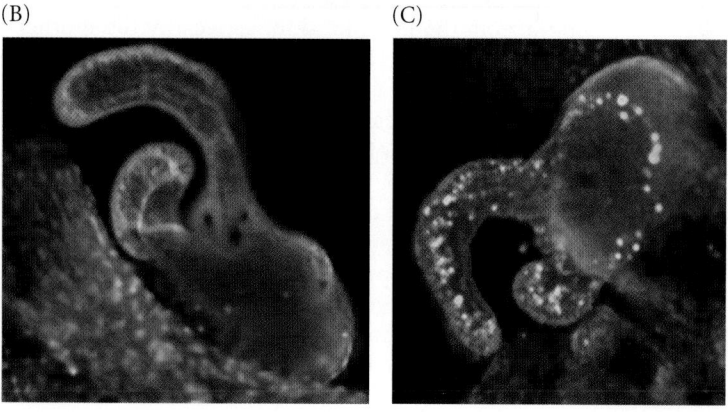

FIGURE 19.19 The *Euprymna scolopes-Vibrio fischeri* symbiosis. (A) An adult Hawaiian bobtail squid (*E. scolopes*) is about 2 inches long. The symbionts are housed in a two-lobed light organ on the squid's underside. (B) The light organ of a juvenile squid is poised to receive *V. fischeri*. Ciliary currents and mucus secretions create an environment (diffuse yellow stain) that attracts seaborne Gram-negative bacteria, including *V. fischeri*, to the organ. Over time all bacteria except *V. fischeri* will be eliminated by mechanisms yet to be exactly elucidated. (C) Once *V. fischeri* are established in the crypts of the light organ, they induce apoptosis of the epithelial cells (yellow dots) and shut down production of the mucosal secretions that attracted other bacteria. (Courtesy of M. McFall-Ngai.)

(A) Control Antibiotic-treated

 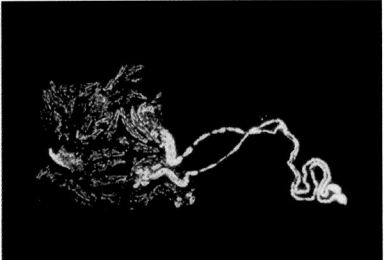

(B) Control Antibiotic-treated

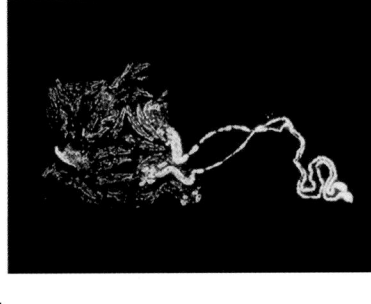

FIGURE 19.20 Comparison of ovaries and oocytes of the wasp *Asobara tabida* from control females and females treated with rifampicin antibiotic to remove *Wolbachia*. (A) The ovaries of control females had an average of 228 oocytes, whereas those of rifampicin-treated females had an average of 36 oocytes. (B) When DNA in the oocytes was stained, oocytes from control females had a nucleus (arrow) as well as a mass of *Wolbachia* at one end (boxed area). Oocytes from rifampicin-treated females had a nucleus but no *Wolbachia*; these eggs were sterile. (From Dedeine et al. 2001.)

name is *Oophilia amblystomatis* ("lover of ambysoma eggs"). The algae is actually stored in the mother's body and appears to be deposited along with the eggs (Kerney et al. 2011). Concentrations of atrazine as low as 50 µg/L completely eliminate this algae from the eggs, and the amphibian's hatching success is greatly lowered (**FIGURE 19.21A**; Gilbert 1944; Mills and Barnhart 1999; Olivier and Moon 2010).

An example of attacking a symbiont as a means to eradicate an unwanted host species is that of the filariasis worm *Mansonella*, a parasitic nematode that infects humans. Most of these roundworms have *Wolbachia* bacteria as endosymbionts. *Wolbachia* produce chemicals that enable the worm to molt, and the worm will die without them. Many species of filariasis worms have become resistant to the drugs traditionally used to kill these parasites in humans. A new treatment strategy is to employ antibacterial antibiotics (such as doxycycline) against the *symbionts* rather than the hosts (**FIGURE 19.21B**; Hoerauf et al. 2003; Coulibaly et al. 2009). Once the antibiotic destroys the symbiont, the worms cannot develop further; they die, and the human host is no longer infected.

remain active in the soil for more than 6 months, and it can be carried by wind and rainwater to new sites. However, the egg masses of many amphibian and snail species depend on algal symbionts to provide oxygen to the eggs deepest in the clutch. The spotted salamander (*Ambystoma maculatum*) lays eggs that recruit a green algal symbiont so specific that its

(A)

(B)

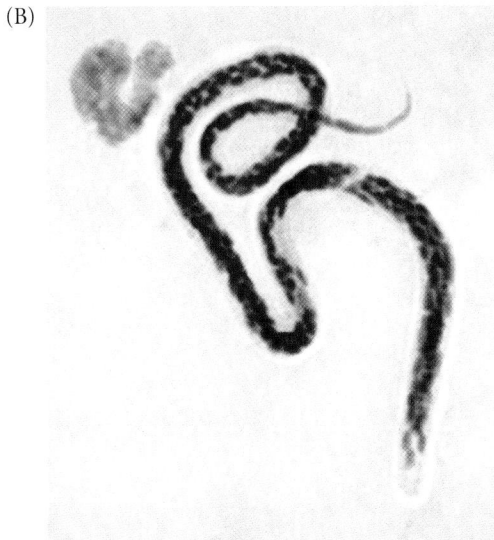

FIGURE 19.21 Obligate developmental symbionts. (A) Spotted salamander (*Ambystoma maculatum*) eggs at the center of the cluster cannot survive the lack of oxygen when their algal symbiont is eliminated by herbicides. (B) Filariasis worms such as *Mansonella ozzardi* cannot complete molting when their *Wolbachia* symbiont is eliminated by antibiotics. (A © Gustav Verderber/OSF/Visuals Unlimited; B courtesy of Mae Melvin/CDC.)

FIGURE 19.22 Induction of mammalian genes by symbiotic microbes. Mice raised in germ-free environments were either left alone or inoculated with one or more types of bacteria. After 10 days, their intestinal mRNAs were isolated and tested on microarrays. Mice grown in germ-free conditions had very little expression of the genes encoding colipase, angiogenin-4, or Sprr2a. Several different bacteria—*Bacteroides thetaiotaomicron*, *Escherichia coli*, *Bifidobacterium infantis*, and an assortment of gut bacteria harvested from conventionally raised mice—induced the genes for colipase and angiogenin-4. *B. thetaiotaomicron* appeared to be totally responsible for the 50-fold increase in *Sprr2a* expression over that of germ-free animals. This ecological relationship between the gut microbes and the host cells could not have been discovered without the molecular biological techniques of polymerase chain reaction and microarray analysis. (After Hooper et al. 2001.)

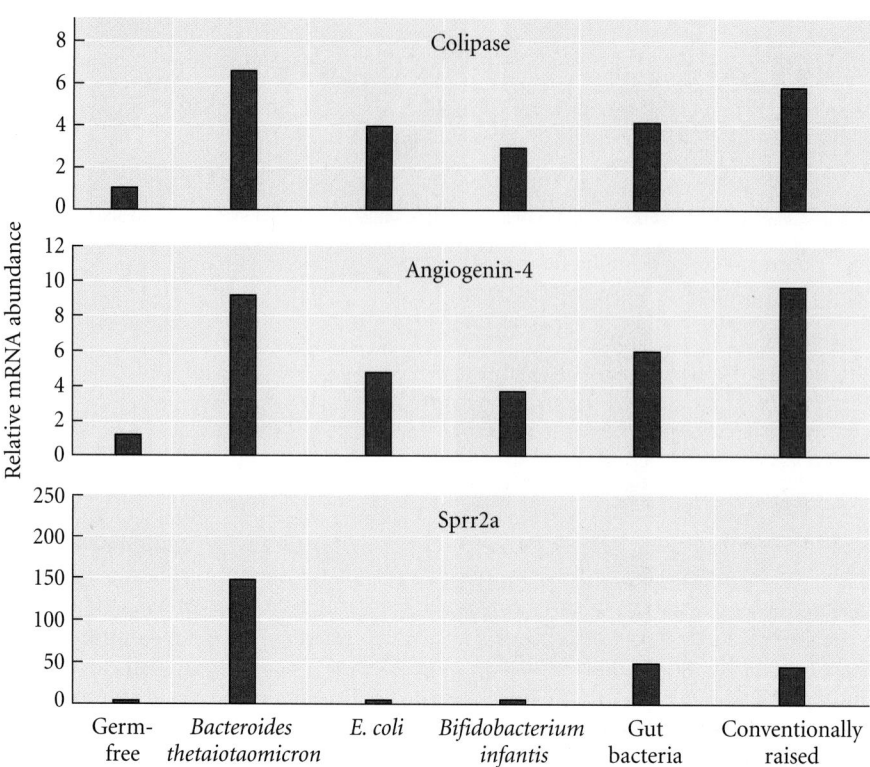

Developmental symbiosis in the mammalian intestine

Even mammals maintain developmental symbioses with bacteria. Using the polymerase chain reaction (PCR) and high-speed sequencing techniques, researchers have recently been able to identify many anaerobic bacterial species present in the human gut (see Qin et al. 2010). Their presence was not realized earlier because these species cannot yet be cultured in the laboratory.

These studies have revealed particular distributions of the bacterial symbionts in our bodies. The hundreds of different bacterial species of the human colon are stratified into specific regions along the length and diameter of the gut tube, where they can attain densities of 10^{11} cells per milliliter (Hooper et al. 1998; Xu and Gordon 2003). Indeed, by cell count, 90% of the cells in our body are microbial. We never lack these microbial components; we pick them up from the reproductive tract of our mother as soon as the amnion bursts. We have coevolved to share our space with them, and we have even codeveloped such that our cells are primed to bind to them, and the bacteria induce gene expression in the intestinal epithelial cells (Bry et al. 1996; Hooper et al. 2001).

BACTERIA HELP REGULATE GUT DEVELOPMENT Bacteria-induced expression of mammalian genes was first demonstrated in the mouse gut (**FIGURE 19.22**). Umesaki (1984) noticed that a particular fucosyl transferase enzyme characteristic of mouse intestinal villi was induced by bacteria, and more recent studies (Hooper et al. 1998) have shown that the intestines of germ-free mice can initiate, but not complete, their differentiation. For complete development, the microbial symbionts of the gut are needed. Normally occurring gut bacteria can upregulate the transcription of several mouse genes, including those encoding colipase, which is important in nutrient absorption; angiogenin-4, which helps form blood vessels; and Sprr2a, a small, proline-rich protein thought to fortify the extracellular matrices that line the intestine (see Figure 19.22; Hooper et al. 2001). Stappenbeck and colleagues (2002) have demonstrated that in the absence of particular intestinal microbes, the capillaries of the small intestinal villi fail to develop their complete vascular networks (**FIGURE 19.23**). In zebrafish, microbes regulate (through the canonical Wnt pathway) the normal proliferation of the intestinal stem cells. Without these microbes, the intestinal epithelium has fewer cells, and it lacks goblet cells, entroendocrine cells, and the characteristic intestinal brush border enzymes (**FIGURE 19.24**; Rawls et al. 2004, 2006; Bates et al. 2006).

BACTERIA HELP REGULATE DEVELOPMENT OF THE IMMUNE AND NERVOUS SYSTEMS Intestinal microbes also appear to be critical for the maturation of the mammalian gut-associated lymphoid tissue (GALT). GALT mediates mucosal immunity and oral immune tolerance, allowing us to eat food without making an immune response to it (see Rook and Stanford 1998; Cebra 1999; Steidler 2001). When introduced into germ-free rabbit appendices, neither *Bacillus fragilis* nor *B. subtilis* alone was capable of consistently inducing

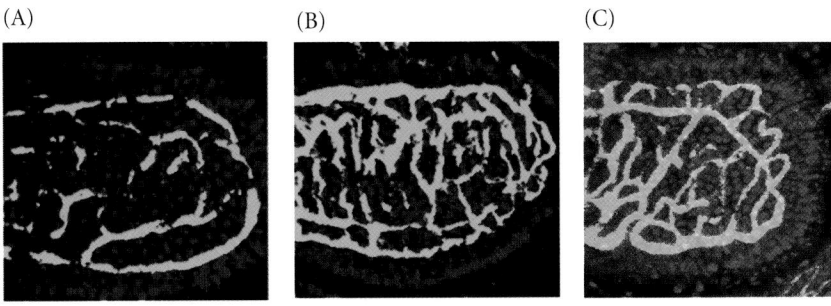

FIGURE 19.23 Gut microbes are necessary for mammalian capillary development. (A) The capillary network (green) of germ-free mice is severely reduced compared with (B) the capillary network in those same mice 10 days after inoculation with normal gut bacteria. (C) The addition of *Bacteroides thetaiotaomicron* alone is sufficient to complete capillary formation. (From Stappenbeck et al. 2002.)

the proper formation of GALT. However, the combination of these two common mammalian gut bacteria consistently induced GALT (Rhee et al. 2004). The major inducer appears to be the protein bacterial polysaccharide A (PSA), especially that encoded by the genome of *B. fragilis*. The PSA-deficient mutant of *B. fragilis* is not able to restore normal immune function to germ-free mice (Mazmanian et al. 2005). Thus, a bacterial compound appears to play a major role in inducing

the host's immune system. Exposure to microbes early in life prevents the development of the T lymphocytes associated with allergies and inflammatory bowel disease, while it enhances the helper T-cell repertoire. Germ-free mice have an immunodeficiency syndrome, and the full complement of T lymphocytes is made possible only with the host species-specifc microbes (Niess et al. 2008; Duan et al. 2010; Chung et al. 2012; Olszak et al. 2012). So symbiotic bacteria are critically important in the differentiation of the lymphocytes of the mammalian immune system.

Although it may sound like science fiction, there is now evidence that symbiotic bacteria stimulate the postnatal development of the mammalian brain. Germ-free mice have lower levels of the transcription factor Egr1 (discussed earlier in relation to anxiety) and the paracrine factor BDNF (discussed earlier in relation to birdsong) in relevant portions of their brains than do conventionally raised mice (**FIGURE 19.25**). This correlates with behavioral differences between groups of mice, leading Diaz Heijtz and colleagues (2011) to conclude that

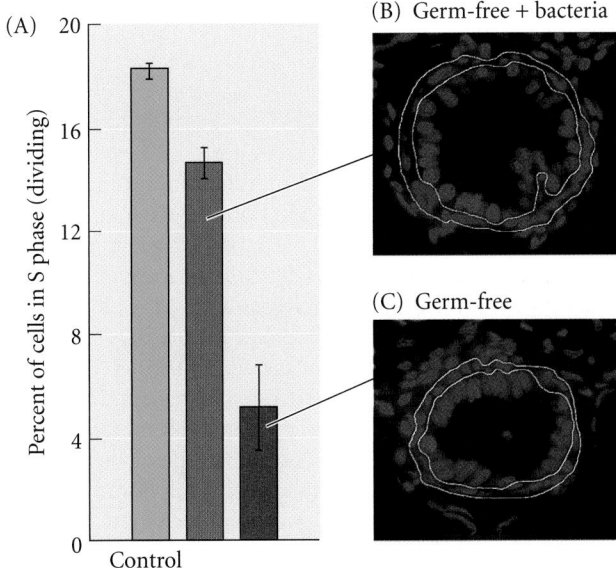

FIGURE 19.24 Bacteria stimulate stem cell division and cell differentiation in the zebrafish gut. (A) Quantitation of S-phase (dividing) intestinal epithelial cells in conventionally raised (control), germ-free, and germ-free plus added bacteria specimens. (B) Germ-free zebrafish given bacteria have normal amounts of stem cell division and epithelial cell differentiation after 6 days. Here and in (C), nondividing cells are stained blue and dividing cells are stained magenta. The inner cells are intestinal epithelia; cells in the white outline are mesenchyme and muscle. (C) The intestines of germ-free zebrafish are smaller and contain fewer dividing stem cells. (After Rawls et al. 2004.)

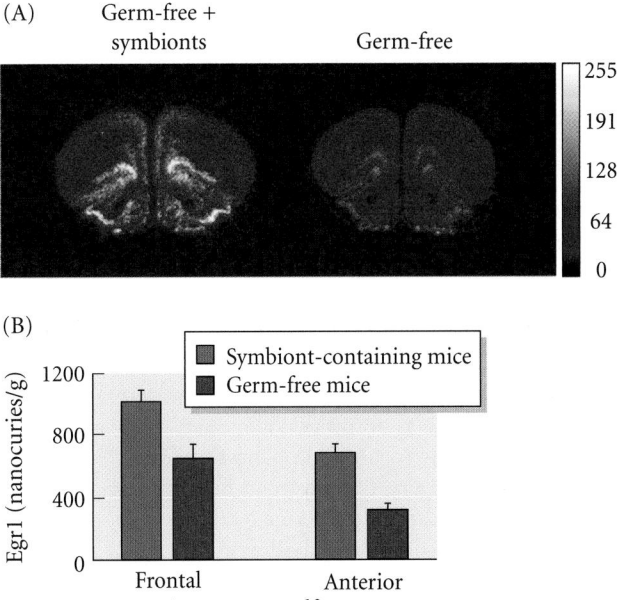

FIGURE 19.25 Egr1 expression in mice depends on symbiotic microbes. (A) In situ hybridization of Egr1 mRNA in a section through the frontal cortex of the brain, showing high levels of Egr1 in a mouse that has conventional microbes compared with a mouse remaining germ-free. (B) Quantitation using radioactive probes shows that symbiont-containing mice had significantly higher levels of Egr1 in the frontal cortex and anterior olfactory region than germ-free mice did. (After Heijtz et al. 2011.)

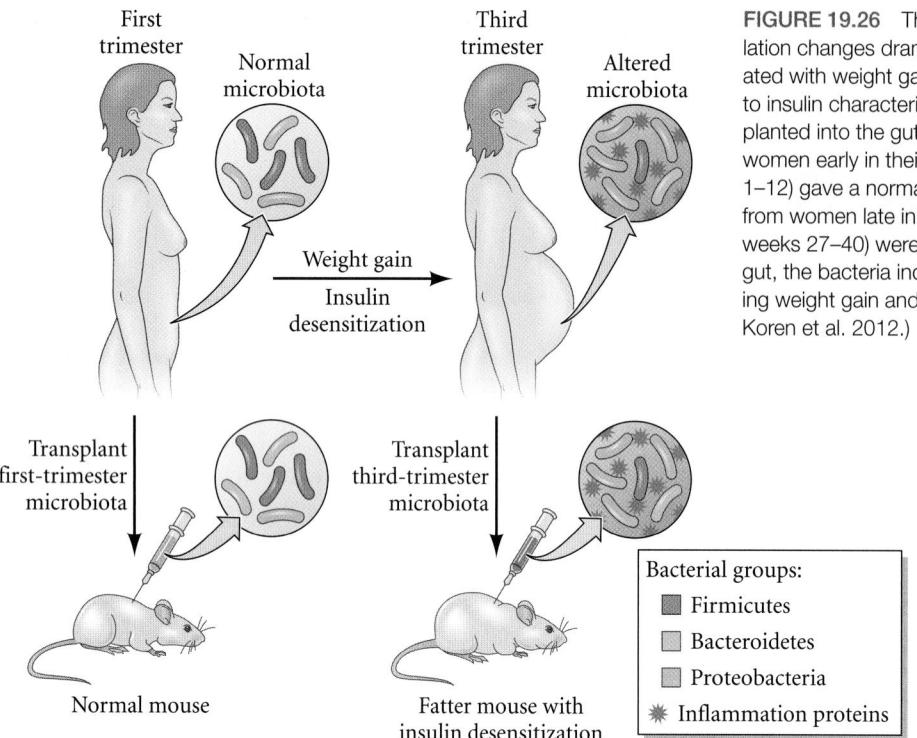

FIGURE 19.26 The composition of the gut microbe population changes dramatically during pregnancy. This is associated with weight gain and with the progressive insensitivity to insulin characteristic of human pregnancy. When transplanted into the guts of germ-free mice, the bacteria from women early in their pregnancy (first trimester, roughly weeks 1–12) gave a normal phenotype to the mice. When bacteria from women late in their pregnancy (third trimester, roughly weeks 27–40) were transplanted into the germ-free mouse gut, the bacteria induced pregnancy-like metabolism, including weight gain and insulin resistance, in the mice. (After Koren et al. 2012.)

"during evolution, the colonization of gut microbiota has become integrated into the programming of brain development, affecting motor control and anxiety-like behavior." In another investigation, a particular *Lactobacillus* strain has been reported to help regulate emotional behavior through a vagus nerve-dependent regulation of GABA receptors (Bravo et al. 2011). Thus, there may be a **microbiota-gut-brain axis** wherein products made by bacteria can enter the blood and help regulate the development of the brain (Grenham et al. 2011; McLean et al. 2012).

The gut bacteria change dramatically during human pregnancy. Indeed, they appear to respond to the hormonal status and help a pregnant woman adapt to the physiological stresses of carrying a fetus. When transferred into a germ-free mouse, bacteria from women in the early stages of pregnancy cause a normal phenotype to develop in the hosts. When bacteria from women late in their pregnancy are transferred into germ-free mice, the mice get fatter and display some of the metabolic changes (such as insulin desensitization) associated with pregnant women (**FIGURE 19.26**; Koren et al. 2012).

In short, mammals have coevolved with bacteria to the point that our bodily phenotypes do not fully develop without them. The microbial community of our gut can be viewed as an "organ" that provides us with certain functions that we haven't evolved (such as the ability to process plant polysaccharides). And, like our developing organs, microbes induce changes in neighboring tissues. As Mazmanian and colleagues (2005) have concluded, "The most impressive feature of this relationship may be that the host not only tolerates but has evolved to require colonization by commensal microorganisms for its own development and health."

Coda

Phenotype is not merely the expression of one's inherited genome. Rather, there are interactions between an organism's genotype and environment that elicit a particular phenotype from a genetically controlled repertoire of possible phenotypes. Environmental factors such as temperature, diet, physical stress, the presence of predators, and crowding can generate a phenotype that is suited for that particular environment. Environment is therefore considered to play a role in the *generation* of phenotypes in addition to its well-established role in the *selection* of phenotypes. The fact that we codevelop with other organisms is an important concept for developmental biology and for evolutionary biology. This may be extremely important for medicine, especially if brain development can be affected by bacteria. Indeed, research is ongoing as to whether bacteria may be responsible for autism spectrum disorders (see Gonzalez et al. 2011).

Ecological developmental biology is providing molecular bases for a field of evolutionary biology called "genotype-by-environment (G×E) interaction" (Schlichting and Pigliucci 1998). Different genomes alter their expression patterns differently in response to the same environmental change. The sensitivity of dung beetle larvae to changes in juvenile hormone, the threshold for sex specification changes in turtles, and numerous other developmental responses to the environment are selectable phenotypes (see Moczek and Nijhout 2002; McGaugh and Janzen 2011; Moczek et al. 2011). As the genes for these environmentally induced developmental pathways are becoming known (Matsumoto and Crews 2012; Snell-Rood and Moczek 2012), molecular mechanisms can be proposed for the genome-environment interactions.

Ecological developmental biology also calls into question the notions of autonomy and independent development.

Moreover, if we are not truly "individuals," but have a phenotype based on community interactions, what exactly is natural selection selecting? Can natural selection select teams or relationships? The ramifications of developmental plasticity and developmental symbiosis on the rest of biology are just beginning to be appreciated (see Gilbert and Epel 2009; Bateson and Gluckman 2011; Gilbert et al. 2012; McFall-Ngai et al. 2013).

SNAPSHOT SUMMARY: Ecological Developmental Biology

1. The environment plays critical roles during normal development. These agents include temperature, diet, crowding, and the presence of predators.

2. Developmental plasticity makes it possible for environmental circumstances to elicit different phenotypes from the same genotype. The genome encodes a repertoire of possible phenotypes. The environment often selects which of those phenotypes will become expressed.

3. Reaction norms are phenotypes that quantitatively respond to environmental conditions, such that the phenotype reflects small differences in the environmental conditions.

4. Polyphenisms represent "either/or" phenotypes; that is, one set of conditions elicits one phenotype, while another set of conditions elicits another.

5. Seasonal cues such as photoperiod, temperature, or type of food can alter development in ways that make the organism more fit under the conditions it encounters. Changes in temperature also are responsible for determining sex in several organisms, including many reptiles and fish.

6. Predator-induced polyphenisms have evolved such that prey species can respond morphologically to the presence of a specific predator. In some instances, this induced adaptation can be transmitted to the progeny of the prey.

7. There are several routes through which gene expression can be influenced by the environment. Environmental factors can methylate genes differentially; they can induce gene expression in surrounding cells; and they can be monitored by the nervous system, which then produces hormones that affect gene expression.

8. Behavioral phenotypes can also be induced by the environment. Conditions experienced as the brain matures after birth can alter patterns of DNA methylation and thereby change hormone reception and behaviors.

9. Learning results from changes in the nervous system brought about by experiences with the environment.

10. Organisms usually develop with symbiotic organisms, and signals from the symbionts can be critical for normal development.

11. Symbionts can be acquired horizontally (through infection) or vertically (through the oocyte).

12. In an obligate mutualism, both partners are needed for the survival of the other; in an obligate developmental mutualism, at least one partner is needed for the proper development of another.

13. The mammalian gut contains symbionts that actively regulate intestinal gene expression to generate proteins that are normal physiological components of intestinal development and function. Without these symbionts, the intestinal blood vessels and gut-associated lymphoid tissue of some mammalian species fail to form properly.

14. Symbionts can induce normal gene expression in hosts; and the host phenotype is deficient without the bacterial-induced patterns of gene expression. The differentiation of certain immune cells, gut cells, and neural cells may depend on symbiont-induced gene expression.

15. In vertebrates, gut symbionts may be important for the development of the gut, the immune system, and perhaps even portions of the nervous system.

For Further Reading

Complete bibliographical citations for all literature cited in this chapter can be found at the free access website **www.devbio.com**

Agrawal, A. A., C. Laforsch and R. Tollrian. 1999. Transgenerational induction of defenses in animals and plants. *Nature* 401: 60–63.

Brakefield, P. M. and N. Reitsma. 1991. Phenotypic plasticity, seasonal climate, and the population biology of *Bicyclus* butterflies (Satyridae) in Malawi. *Ecol. Entomol.* 16: 291–303.

Caldji, C., I. C. Hellstrom, T. Y. Zhang, J. Diorio and M. J. Meaney. 2011. Environmental regulation of the neural epigenome. *FEBS Lett.* 585: 2049–2058.

Gilbert, S. F. and D. Epel. 2009. *Ecological Developmental Biology: Integrating Epigenetics, Medicine, and Evolution.* Sinauer Associates, Sunderland, MA.

Hooper, L. V., M. H. Wong, A. Thelin, L. Hansson, P. G. Falk and J. I. Gordon. 2001. Molecular analysis of commensal host-microbial relationships in the intestine. *Science* 291: 881–884.

McFall-Ngai, M. J. 2002. Unseen forces: The influence of bacteria on animal development. *Dev. Biol.* 242: 1–14.

Moczek, A. P. 2005. The evolution of development of novel traits, or how beetles got their horns. *BioScience* 55: 937–951.

Relyea, R. A. and N. Mills. 2001. Predator-induced stress makes the pesticide carbaryl more deadly to grey treefrog tadpoles (*Hyla versicolor*). *Proc. Natl. Acad. Sci. USA* 2491–2496.

Stappenbeck, T. S., L. V. Hooper and J. I. Gordon. 2002. Developmental regulation of intestinal angiogenesis by indigenous microbes via Paneth cells. *Proc. Natl. Acad. Sci. USA* 99: 15451–15455.

Waterland, R. A. and R. L. Jirtle. 2003. Transposable elements: Targets for early nutritional effects of epigenetic gene regulation. *Mol. Cell. Biol.* 23: 5293–5300.

Go Online

WEBSITE 19.1 Inducible caste determination in ant colonies. In some species of ants, the loss of soldier ants creates conditions that induce more workers to become soldiers.

WEBSITE 19.2 Volvox: When heat brings out sex. During most of its life cycle, *Volvox* (a "green algae") reproduces asexually. However, when the ponds *Volvox* inhabits become hot—a signal that they may soon dry up—the next generation becomes sexual and produces sperm and eggs that unite in zygotes that can survive desiccation.

WEBSITE 19.3 Mechanisms of diapause. Both light and temperature are critical for the induction and maintenance of diapause. Different species use different signals for this event.

WEBSITE 19.4 Pressure as an agent of development. Mechanical stress is critical for gene expression in numerous tissues, including bone, heart, and muscle. Without physical force, we would not develop our kneecaps.

WEBSITE 19.5 The *Dictyostelium* life cycle: Variations within variations. *Dictyostelium* has some fascinating variations on its life cycle, including the ability to become sexual.

WEBSITE 19.6 Developmental symbioses and parasitism. Some embryos acquire protection and nutrients by forming symbiotic associations with other organisms. The mechanisms by which these associations form are now being elucidated. In other situations, one species uses material from another to support its development. Blood-sucking mosquitoes are examples of such parasites.

Vade Mecum

Slime mold life cycle. The life cycle of this fascinating organism is shown in movies. There is also original footage of some of John Tyler Bonner's experiments, along with an interview with Bonner, whose pioneering work demonstrated the major principles of *Dictyostelium* development.

Outside Sites

Karen Warkentin's Boston University website includes a movie of the snake-induced rapid development of red-eyed tree frogs. (*Agalychnis callidryas*): **people.bu.edu/kwarken/KWvideoSMALLER.html**

The Developmental Biology Interactive site at Georgia Tech has sites on *Dictyostelium* and *Euprymna*: **www.devbio.biology.gatech.edu/?page_id=34** and **www.devbio.biology.gatech.edu/?page_id=303**

The NIH webpage for *D. discoideum*, **nih.gov/science/models/d_discoideum/** provides information on these organisms, and links to **dictybase.org/**, a major resource for dictyostelid genomics.

20

Developmental Mechanisms of Evolutionary Change

WHEN DARWIN WAS WRITING ON THE *ORIGIN OF SPECIES*, he consulted his friend Thomas Huxley concerning the origins of variation. Huxley (1857) told Darwin that the differences between organisms could be traced to differences in their development, and that these differences "result not so much of the development of new parts as of the modification of parts already existing and common to both the divergent types." This is a major tenet of **evolutionary developmental biology**, a relatively new science that views evolution as the result of changes in development. If development is the change of gene expression and cell position over time, then evolution is the change of development over time. This new field—sometimes referred to as "evo-devo"—is producing a new model of evolution that integrates developmental genetics and population genetics to explain and define the diversity of life (Raff 1996; Hall 1999; Arthur 2004; Carroll et al. 2005; Kirschner and Gerhart 2005). In other words, evolutionary developmental biology links genetics with evolution through the agencies of development. Contemporary evolutionary developmental biology is analyzing how changes in development can create the diverse variation that natural selection can act on. Rather than concentrating on the "survival of the fittest," evolutionary developmental biology gives us new insights into the "arrival of the fittest" (Carroll et al. 2005; Gilbert and Epel 2009).

Descent with Modification: Why Animals Are Alike and Different

In the nineteenth century, debates over the origin of species pitted two views of nature against each other. One view, championed by Georges Cuvier and Charles Bell, focused on the *differences* between species that allowed each to adapt to its environment. Thus, they believed, the hand of the human, the flipper of the seal, and the wings of birds and bats were marvelous contrivances, each fashioned by the Creator to adapt these animals to their "conditions of existence." The other view, championed by Étienne Geoffroy Saint-Hilaire and Richard Owen, was that "unity of type" (the *similarities* among organisms, which Owen called "homologies") was critical. The human hand, the seal's flipper, and the wings of bats and birds were all modifications of the same basic plan (see Figure 1.19). In discovering that plan, one could find the form upon which the Creator designed these animals. The adaptations were secondary.

Darwin acknowledged his debt to these earlier debates when he wrote in 1859, "It is generally acknowledged that all organic beings have been formed on two great laws—Unity of Type, and Conditions of Existence." Darwin went on to explain that his theory would explain unity of type by descent from a common ancestor, while the adaptations to the conditions of existence could be explained by natural selection. Darwin called this concept **descent with modification**. Darwin noted that the homologies between the embryonic and larval structures of different phyla provided excellent evidence for descent with modification. He was thrilled that the larval anatomy of barnacles demonstrated them to be crustaceans, and he was especially pleased by Kowalevsky's demonstration

A study of the effects of genes during development is as essential for an understanding of evolution as are the study of mutation and that of selection.

   JULIAN HUXLEY (1942)

I also believe that an understanding of regulation must lie at the center of any rapprochement between molecular and evolutionary biology; for a synthesis of these two biologies will surely take place, if it occurs at all, on the common ground of development.

   STEPHEN J. GOULD
           (1977)

SIDELIGHTS & SPECULATIONS

Development, Multicellularity, and the Origin of the Metazoa

The "big" questions of evolution involve the creation of new types of organisms and, as Thomas Huxley noted, evolution occurs by altering development. As we saw at the start of Chapter 5 and throughout Part Two, the major groups of the Metazoa (animals) in fact represent different modes of development. Fundamental phylogenetic splits such as those between the diploblastic and bilaterian animals, the protostomes and deuterostomes, the lophotrochozoan and ecdysozoan protostomes, and the chordate and nonchordate deuterostomes (see Figure 5.1) are all predicated on development. What these groups all have in common, however, is that they are all multicellular. How did multicellulatrity arise in those protists that would give rise to the animal phyla, and how did the phyla diverge from one another?

Acquisiton of multicellularity

The "inertial condition" for a eukaryotic cell is proliferation. The origin of multicellularity, and thus of the metazoa, required interrupting the "inertia" of cell division and introducing differentiation.

What requires an explanation (in both mechanical and evolutionary terms) is the cessation of mitosis and the acquisition of stable differentiated states (Sonnenschein and Soto 2004; Minelli 2011).

The acquisition of different cell types can be explained by the differential location of distinct factors in the different cells of a multicellular aggregate. These could be maternal mRNA or proteins localized exclusively in one region of the cell such that two daughter cells each acquire different regulatory proteins at cell division. This would be akin to specification modes in tunicate or *Drosophila* eggs. Alternatively, there could be differences between the environment sensed by the cells on the outer rim of the aggregate versus the environment sensed by the inner cells (as happens during the first cellular distinctions in mammals).

But what allowed cells to form multicellular aggregates? The answers, although speculative, highlight the role of bacteria. The oldest known macroscopic multicellular organisms come from fossils in strata some 2.1 billion

years old, soon after the "great oxidation event"—the accumulation of atmospheric oxygen that occurred about 2.4 billion years ago as a result of photosynthesis by bacteria that were probably similar to the modern cyanobacteria ("blue-green algae"). This oxidation event was the most important climate change in evolutionary history, altering the atmosphere from a mixture of ammonia and carbon dioxide to an oxygen-rich mixture that would make aerobic metabolism—and thus metazoan life—possible (El Almani et al. 2010).

Another bacterial boost to metazoan evolution may have been a symbiosis between bacteria and unicellular protists. Recent analyses agree that the metazoans probably arose from a group of protists very much like today's choanoflagellates. Choanoflagellates are single-celled organisms, and their name comes from their resemblance to the choanocytes (collar cells) of sponges. In filtered seawater, one choanoflagellate species, *Salpingoeca rosetta*, proliferates asexually, forming more single-celled protists. However,

---

that the larvae of tunicates had both a notochord and pharyngeal pouches (1871). This showed them to be chordates, thereby uniting the invertebrates and vertebrates into a coherent animal kingdom. In the late 1800s, developmental change was seen as being the motor of evolution (Gould 1977). Or, as Thomas Huxley aptly remarked, "Evolution is not a speculation but a fact; and it takes place by epigenesis" (Huxley 1893, p. 202).

● **See WEBSITE 20.1** Relating evolution to development in the nineteenth century

Preconditions for Evolution: The Developmental Structure of the Genome

If natural selection can only operate on existing variants, where does all that variation come from? If, as Darwin (1868) and Huxley concluded, variation arose from changes in development. But how could the development of an embryo change when development is so finely tuned and complex? How could such change occur without destroying the entire

organism?* Even when the molecular biology of protein synthesis became understood, the problem did not go away. If a protein-encoding gene were mutated, the abnormal protein would be made everywhere the protein was normally expressed. There was no way a mutation could cause the protein to be made in one place and not another. The matter remained a mystery until evolutionary developmental biologists demonstrated that large morphological changes could arise during development because of two conditions that underlie the development of all multicellular organisms: *modularity* and *molecular parsimony*.

---

*Darwin's German contemporary Ernst Haeckel proposed that most organisms evolved by adding a step to the *end* of embryonic development. But there turned out to be so many exceptions to that rule that it fell into disrepute. Two of Darwin's British contemporaries, Herbert Spencer and Robert Chambers, also saw development as the motor of evolution; they used von Baer's laws (see Chapter 1) as its mechanism (see Gould 1977; Friedman and Diggle 2011).

(A)   (B)

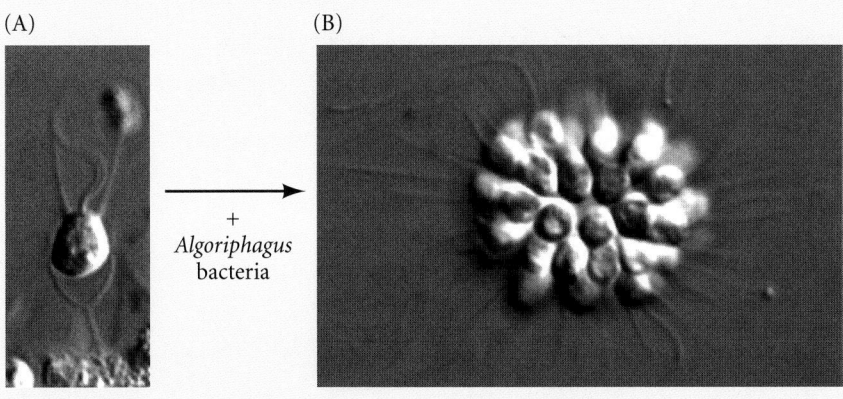

+
Algoriphagus
bacteria

FIGURE 20.1   Two morphologies of the choanoflagellate protist *Salpingoeca rosetta*. (A) Single-celled form. (B) Colonial form with multiple cells linked by an extracellular matrix. The *Algoriphagus* bacteria, often found with *S. rosetta*, can convert the organism from dividing into individual cells to forming multicellular "rosettes." (From Dayel et al. 2011; photographs courtesy of M. Dayel and N. King.)

when cultured in media containing the bacteria *Algoriphagus machipongonensis*, the cells do do not separate. Rather, they form rosettes, where the cells are connected by an extracellular matrix, and by cytoplasmic bridges (**FIGURE 20.1**; Dayel et al. 2011; Alegado et al. 2013). The sphyngolipids in the bacterial cell wall are able to effect this transition; and the bacteria is found naturally with colonial forms

of this choanoflagellate species. Thus, it is possible multicellularity may have arisen from developmental changes induced by neighboring bacteria.

Systems biology and phylogeny

Davidson and Erwin (2009) have put forth the fascinating idea that the classical Linnaean taxonomy, based on nested hierarchies (several species in a genus; several genera in a family; several

families in an order; etc.), is an imperfect reflection of developmental organization. Specifically, they relate this classification scheme to the developmental gene regulatory networks (GRNs; see Chapter 7 and the Part Four opening essay), postulating that phyla represent an upper-level "circuit" specifying cell types in a certain way. Everything whose development is characterized by this GRN is in this phylum. And within that GRN, there will be "subcircuits" that display differences, and these different schemes of development would form the classes. Differences defining genera and species would be defined by variations of the GRN circuitry at lower levels of the hierarchy.

Davidson (2011) posits the following syllogism: since GRNs control the development of the body plan, and since the evolution of the body plan requires genomic alteration of the developmental program, then relevant explanations for evolution must be conceived in terms of those genomic alterations that change the structure and function of GRNs. This is a new way of looking at the relationship between development, genes, and evolution, and it integrates systems biology with evolution and development.

Modularity: Divergence through dissociation

We now know that even early stages of development can be altered to produce evolutionary novelties. Such changes can occur because development occurs through a series of discrete and interacting **modules** (Riedl 1978; Bonner 1988; Kuratani 2009). Examples of developmental modules include morphogenetic fields (for example, those for the heart, limb, or eye), signal transduction pathways (such as the Wnt or BMP cascades), imaginal discs, cell lineages (such as the inner cell mass or trophoblast), insect parasegments, and vertebrate organ rudiments (Gilbert et al. 1996; Raff 1996; Wagner 1996; Schlosser and Wagner 2004). The ability of one module to develop differently from other modules (a phenomenon sometimes called **dissociation**) was well known to early experimental embryologists. For instance, when Victor Twitty grafted the limb bud from the early larva of a large salamander onto the embryonic trunk of a small salamander larva, the limb grew to its normal large dimensions within the small larva, indicating that the limb field module was independent from the global growth patterning of the

embryo (Twitty and Schwind 1931; Twitty and Elliott 1934). The same independence was seen for the eye field. Modular units allow certain parts of the body to change without interfering with the functions of other parts.

One of the most important discoveries of evolutionary developmental biology is that not only are the *anatomical* units modular (such that one part of the body can develop differently than the others), but the DNA regions that form the *enhancers* of genes are also modular. This modularity—that there can be multiple enhancers for each gene and that each enhancer region can have binding sites for multiple transcription factors—was shown in Figure 2.10. The modularity of enhancer elements allows particular sets of genes to be activated together and permits a particular gene to become expressed in several discrete places. Thus, if by mutation a particular gene loses or gains a modular enhancer element, the organism containing that particular allele will express that gene in different places or at different times than organisms retaining the original allele. This mutability can result in the development of different anatomical and physiological morphologies (Sucena and Stern 2000; Shapiro et al. 2004), and major morphological changes can proceed through a mutation in a DNA regulatory region. Thus, the modularity of enhancers can be critical in providing selectable variation. Indeed, mutations affecting enhancer sequences are now thought to be the most important cause of morphological divergence between groups of animals (Carroll 2008; Stern and Orgogozo 2008).

DUFFY BLOOD GROUP SUBSTANCE Enhancer modularity has been known from studies of selectable traits in human populations. *Plasmodium vivax* is a protozoan parasite that causes about 75 million cases of malaria each year. This form of malaria is not as lethal as malaria transmitted by *P. falciparum*, but it can be incapacitating, causing severe pain, diarrhea, and fever. Some African populations are immune to *P. vivax* because their red blood cells lack a protein, the Duffy glycoprotein, that *P. vivax* requires in order to attach itself to the host's red blood cells. The Duffy glycoprotein is probably one of several receptors for interleukin 8 (IL8). This receptor is found on cerebellar Purkinje neurons, on blood veins, and on erythrocytes (red blood cells). People who lack Duffy glycoprotein on their red blood cells still have it on their veins and Purkinje neurons. So if one asks "Why don't these people have Duffy glycoprotein on their red blood cells?" the ultimate answer is probably that the lack of the Duffy glycoprotein has been selected in these populations because it gives these people resistance to *vivax* malaria. The proximate answer is that the lack of Duffy glycoprotein is caused by a mutation in the erythrocyte enhancer of the gene for Duffy protein: a C-to-G substitution at position −36 that prevents binding of the GATA1 transcription factor present in red blood cell precursors (Tournamille et al. 1995). Thus, the mutation blocks one of the enhancers (the one for expression in red blood cells) from functioning, but allows the enhancers that permit the gene's expression in veins and Purkinje neurons.

PITX1 AND STICKLEBACK EVOLUTION The importance of enhancer modularity has been dramatically demonstrated by the analysis of evolution in the threespine stickleback fish (*Gasterosteus aculeatus*). Freshwater sticklebacks evolved from marine sticklebacks about 12,000 years ago, when marine populations colonized the newly formed freshwater lakes at the end of the last ice age. Marine sticklebacks (**FIGURE 20.2A**) have pelvic spines that serve as protection against predation, lacerating the mouths of predatory fish that try to eat the stickleback. (Indeed, the scientific name of the fish translates as "bony stomach with spines.") Freshwater sticklebacks, however, do not have pelvic spines (**FIGURE 20.2B**).

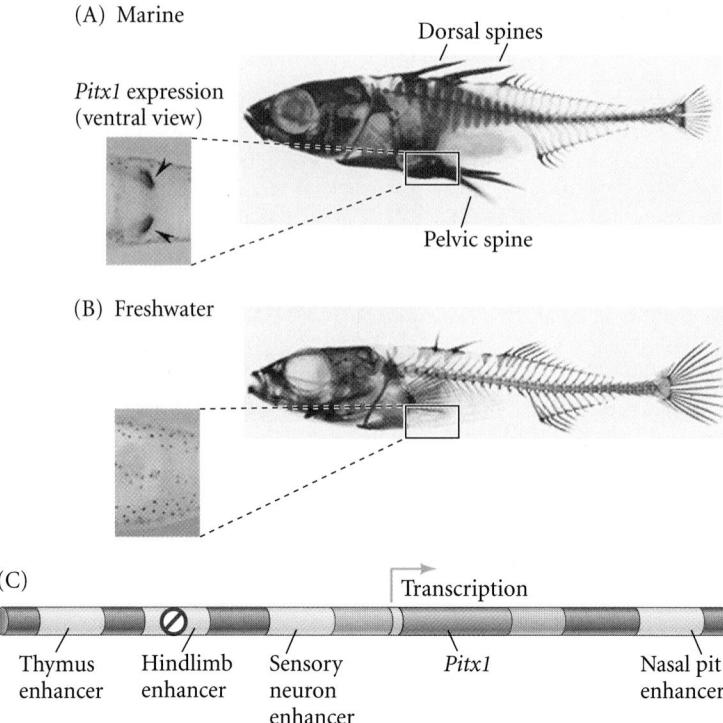

FIGURE 20.2 Modularity of development: enhancers. Loss of *Pitx1* gene expression in the pelvic region of freshwater populations of the threespine stickleback (*Gasterosteus aculeatus*). Bony plates and pelvic spines characterize marine populations of this species (A). In freshwater populations (B), the pelvic spines are absent, as is much of the bony armor. In magnified ventral views of embryos (inset photos), in situ hybridization reveals *Pitx1* expression (purple) in the pelvic area (as well as in sensory neurons, thymic cells, and nasal regions) of the marine population. The staining in the pelvic region is absent in freshwater populations, although it is still seen in the other areas. The arrowheads point to *Pitx1* expression in the ventral region that forms the pelvic spines of the marine populations. (C) Model for the evolution of pelvic spine loss. Four enhancer regions are postulated to reside near the *Pitx1* coding region. These enhancers direct the expression of this gene in the thymus, pelvic spines, sensory neurons, and nasal pit, respectively. In freshwater populations of threespine sticklebacks, the pelvic spine (hindlimb) enhancer module has been mutated and the *Pitx1* gene fails to function there. (After Shapiro et al. 2004; photographs courtesy of D. M. Kingsley.)

This may be because the freshwater fish lack the piscine predators that the marine fish face, but instead must deal with invertebrate predators that can easily capture them by grasping onto such spines. Thus, a pelvis without spines was selected in freshwater populations of this species.

To determine which genes might be involved in stickleback pelvic differences, researchers mated individuals from marine (spined) and freshwater (spineless) populations. The resulting offspring were bred to each other and produced numerous progeny, some of which had pelvic spines and some of which didn't. Using molecular markers to identify specific regions of the parental chromosomes, Shapiro and co-workers (2004) found that the major gene for pelvic spine development mapped to the distal end of chromosome 7. That is to say, nearly all the fish with pelvic spines inherited this "pelvic appendage-encoding" chromosomal region from the marine parent, whereas fish lacking pelvic spines inherited this region from the freshwater parent. The researchers then tested numerous candidate genes (e.g., genes known to be active in the pelvic/hindlimb structures of mice) and found that the gene encoding the transcription factor Pitx1 was located on this region of chromosome 7.

When Shapiro and colleagues compared the amino acid sequences of the Pitx1 proteins of marine and freshwater sticklebacks, there were no differences. However, there was a critically important difference when they compared the *expression patterns* of *Pitx1*. In both populations, *Pitx1* was expressed in the precursors of the thymus, nose, and sensory neurons. In the marine populations, *Pitx1* was also expressed in the pelvic region. But in the freshwater populations, the pelvic expression of *Pitx1* was absent or severely reduced (**FIGURE 20.2C**). Since the coding region of *Pitx1* was not mutated (and since the gene involved in the pelvic spine differences maps to the site of the *Pitx1* gene, and the difference between the freshwater and marine populations involves the expression of this gene at a particular site), it was reasonable to conclude that the *enhancer region* allowing expression of *Pitx1* in the pelvic area (i.e., the pelvic spine enhancer) no longer functions in the freshwater populations.

This conclusion was confirmed when high-resolution genetic mapping showed that the DNA of the "hindlimb" enhancer of *Pitx1* differed between sticklebacks with pelvic spines and those without pelvic spines* (Chan et al. 2010). When this 2.5-kb DNA fragment from marine (spined) fish was fused to a gene for green fluorescent protein and inserted into fertilized freshwater stickleback eggs, it caused GFP to be expressed in the pelvis. Moreover, when this same fragment taken from marine sticklebacks was placed next to the *Pitx1*-coding sequence of freshwater (spine-deficient) fish and then injected into fertilized eggs of the spine-deficient fish, pelvic spines formed in the freshwater fish.

RECRUITMENT Modularity allows the recruitment (or "co-option") of entire suites of characters into new places. In Chapter 7, we discussed the recruitment of the skeleton-forming genes (the skeletonogenic "subroutine") into the developmental repertoire of the sea urchin micromeres. In most echinoderm groups, the skeletogenic genes are activated only in the adult and are used to form the hard exoskeletal plates. However, in sea urchins (and not in any other echinoderm group) this set of genes has come under the control of the micromere double-negative gate because

of changes in the enhancer of one of these genes. Thus, the skeleton is made by larval mesenchymal cells (Gao and Davidson 2008).

Often a new structure will be formed by the recruitment of existing modules (subroutines) into older modules. The horns of dung beetles, for example, arose from the co-option of the leg-patterning genes (Moczek and Rose 2009). The developmental network coordinating the expression of *Homothorax*, *Dachshund*, and *Distal-less* in limb formation (Figure 16.14) is used to generate a novel structure, the horn, in these beetle larvae. Insect wing pigmentation patterns can be made by the co-option of pigmentation genes into existing wing modules. The placement of the pigments on some *Drosophila* wings arises when the enhancer of the *yellow* locus (which makes black pigment) becomes responsive to transcription factors activated by the Wingless paracrine factor (**FIGURE 20.3**; Werner et al. 2010).

Another example of recruitment among insects is seen in the wing structure that defines the beetles. Beetles are the most successful animal group on the planet, accounting for more than 20% of extant animal species (Hunt et al. 2007). They differ from other insects in forming an **elytron**—a forewing encased in a hard exoskeleton. This makes them the

(A)

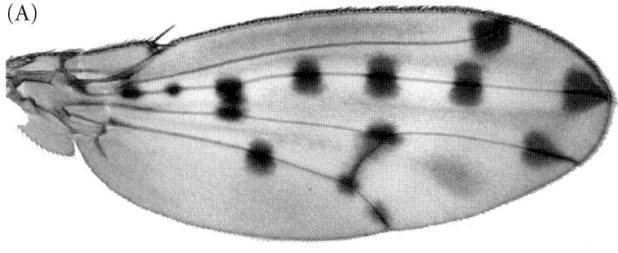

(B)

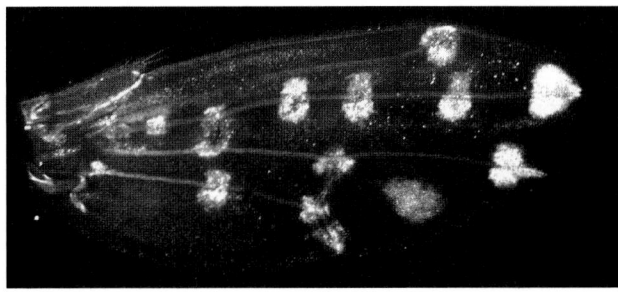

FIGURE 20.3 Recruitment of the pigment-forming module into the wing module in insects. The complex wing pigmentation of *Drosophila* guttifera (A) results from the expression of the Yellow protein (which helps form melanin pigment) at specific places in the wing (B). The *yellow* gene is activated at these places by the expression of Wingless during pupal development. Wingless protein is used during earlier stages of development to segment the body and to form the wing itself. It has been recruited for this later event by changes in the enhancer for the *yellow* gene that allow *yellow* to be activated by Wingless. (After Gompel et al. 2005; photograph courtesy of N. Gompel.)

---

*Interestingly, the loss of the pelvic spines in several stickleback populations appears to have been the result of independent losses of this *Pitx1* expression domain. This finding suggests that if the loss of *Pitx1* expression in the pelvis occurs, this trait can be readily selected (Colosimo et al. 2004). Here we see that by combining population genetics approaches and developmental genetics approaches, one can determine the mechanisms by which evolution can occur.

(A)

(B)

FIGURE 20.4 Elytra are the hardened forewings that are characteristic of Coleoptera, the beetles. Elytra are formed through the recruitment of the genetic module for exoskeleton development into the module for dorsal forewing development. (A) The elytra of a "ladybug" beetle. Its forewings are ornamented with exoskeleton, and its hindwings are extended. (B) These "living jewels" from the Oxford Museum of Natural History illustrate some of the diversity of beetle elytra. (A © F1online digitale Bildagentur GmbH/Alamy; B © Jochen Tack/Alamy.)

"living jewels" so beloved of naturalists (**FIGURE 20.4**).* In beetles, as in *Drosophila*, the *Apterous* gene is expressed in the dorsal compartment of the wing imaginal discs, and the Apterous transcription factor organizes the tissue to differentiate dorsal wing structures. However, in beetles (and in no other known insect), Apterous also activates the exoskeleton genes in the forewing while repressing them in the hindwing (Tomoyasu et al. 2009). Thus, a new type of wing emerges from the recruitment of one module (the subroutine of exoskeletal development) into another (the subroutine of dorsal forewing development).

● **See WEBSITE 20.2** Correlated progression

Molecular parsimony: Gene duplication and divergence

The second precondition for macroevolution through developmental change is molecular parsimony, sometimes called the "small toolkit." In other words, although development differs enormously from lineage to lineage, development within all lineages uses the same types of molecules. The transcription factors, paracrine factors, adhesion molecules, and signal transduction cascades are remarkably similar from one phylum to another. Indeed, it appears that the development of jellyfish and flatworms uses the same major kit of transcription factors and paracrine factors as flies and vertebrates (Finnerty et al. 2004; Carroll et al. 2005; Putnam et al. 2007; Ryan et al. 2007; Hejnol et al. 2009).

THE SMALL TOOLKIT   Certain transcription factors (such as those of the BMP, Hox, and Pax groups) are found in all animal phyla, including cnidarians, arthropods, and chordates. In fact, some "toolkit genes" appear to play the same roles in all animal lineages. The BMPn levels appear to be used throughout the animal kingdom to specify the dorsal-ventral axis (**FIGURE 20.5A**); the Wnt and Hox genes appear to specify the anterior-posterior axis throughout all the bilaterians (**FIGURE 20.5B**); and the *Pax6* gene appears to be involved in specifying light-sensing organs, irrespective of whether the eye is that of a mollusc, an insect, or a primate† (**FIGURE 20.5C**). Similarly, homologues of *Otx* specify head formation in both vertebrates and invertebrates; and though insect and vertebrate hearts are very different, both are formed using *tinman/Nkx2-5* (see Erwin 1999). Certain microRNAs appear to be found in all animals, and these appear to play the same or very similar developmental roles in whatever phylum they are found (Christodoulou et al. 2010). These include miRNA-124, which is found in the central nervous systems of protostomes and deuterostomes; miRNA-12, which is found in guts throughout the animal kingdom; and miRNA-92, which

---

*Although both Darwin and Wallace were avid beetle collectors, it was the geneticist J. B. S. Haldane whose remark may best reflect the prominence of these insects. When asked by a cleric what the study of nature could tell us about God, Haldane is said to have replied, "He has an inordinate fondness for beetles."

---

†This doesn't mean that the eye is the only thing that is specified by *Pax6*, or that *Pax6* hasn't become regulated by different proteins in different phyla (Lynch and Wagner 2010).

(A)

(B)

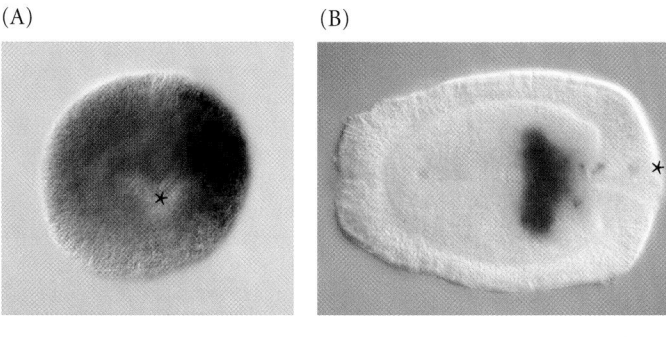

(C)

FIGURE 20.5 Evidence of the evolutionary conservation of regulatory genes. (A) The cnidarian homologue of the vertebrate *Bmp4* and *Drosophila Decapentaplegic* genes is expressed asymmetrically at the edge of the blastopore (marked with an asterisk) in the embryo of the sea anemone *Nematostella*. This gene represents an ancestral form of the protostome and deuterostome forms of the gene. (B) The Hox gene *Anthox6*, a cnidarian member of the paralogue 1 group of Hox genes, is expressed at the blastopore side (asterisk) of the larval sea anemone. (C) The *Pax6* gene for eye development is an example of a gene ancestral to both protostomes and deuterostomes. The micrograph shows ommatidia emerging in the leg of a fruit fly (a protostome) in which mouse (deuterostome) *Pax6* cDNA was expressed in the leg disc. (A,B from Finnerty et al. 2004, courtesy of M. Martindale; C from Halder et al. 1995, courtesy of W. J. Gehring and G. Halder.)

helps specify ciliated locomotor cells in deuterostome and protostome larvae. Discovering that the same set of transcription factors and microRNAs causes the specification of the same types of cells throughout the animal kingdom is a very powerful argument that the protostomes and deuterostomes are derived from a common ancestor that used these factors in similar ways to specify its organs (Davidson and Erwin 2010).

DUPLICATION AND DIVERGENCE One theme that resounds through studies of paracrine and transcription factors is that these proteins (and the genes that encode them) come in families. How do gene families come into existence? The answer is through duplication of an original gene and the subsequent independent mutation of the original duplicates (**FIGURE 20.6**). This creates a family of genes that are related by common descent (and which are often still adjacent to each other.) This scenario of **duplication and divergence** is seen in the Hox genes, the globin genes, the collagen genes, the *Distal-less* genes, and in many paracrine factor families (e.g., the Wnt genes). Each member of such a gene family is homologous to the others (that is, their sequence similarities are due to descent from a common ancestor and are not the result of convergence for a particular function), and they are called **paralogues**. Susumu Ohno (1970), one of the

founders of the gene family concept, likened gene duplication to a method used by a sneaky criminal to circumvent surveillance. While the "police force" of natural selection makes certain that there is a "good" gene properly performing its function, that gene's duplicate, unencumbered by the constraints of selection, can mutate and undertake new functions.

One of the most significant fates for duplicated genes is the subdivision of the ancestral expression and function (Force et al. 2005). Thus, the original *Distal-less* gene was probably expressed in many places (as it still is in insects, which have only one *Distal-less* gene); but in vertebrates, which have numerous *Distal-less* genes, the original domains have become subdivided so that the different *Distal-less* paralogues regulate different functions. *Distal-less-3*, for instance, is necessary for epidermal differentiation, whereas *Distal-less-5* helps specify neural crest cells (Panganiban and Rubenstein 2002).

Such subfunctionalization has since been shown to be the case for many genes, including Hox genes. The Hox genes represent an especially complex and important case of duplication and divergence. We find (1) that there are related Hox genes in each animal group (such as *Deformed*, *Ultrabithorax*, and *Antennapedia* in *Drosophila*, or the 39 Hox genes in mammals); and (2) that there are several clusters of Hox genes in vertebrates (the 39 Hox genes in mammals, for example, are clustered on four different chromosomes). The similarity of all the Hox genes is best explained by descent from a common ancestral Hox gene, probably in the single-celled protozoa or sponges. This would mean that in *Drosophila*, the *Deformed*, *Ultrabithorax*, and *Antennapedia* genes all emerged as duplications of an original gene. The sequence patterns of these three genes (especially in the homeodomain region) are extremely well conserved. Such tandem gene duplications are thought to be the result of errors in DNA replication, and such errors are not uncommon. Once replicated, the gene copies can diverge by random mutations in their coding sequences and enhancers, developing different expression

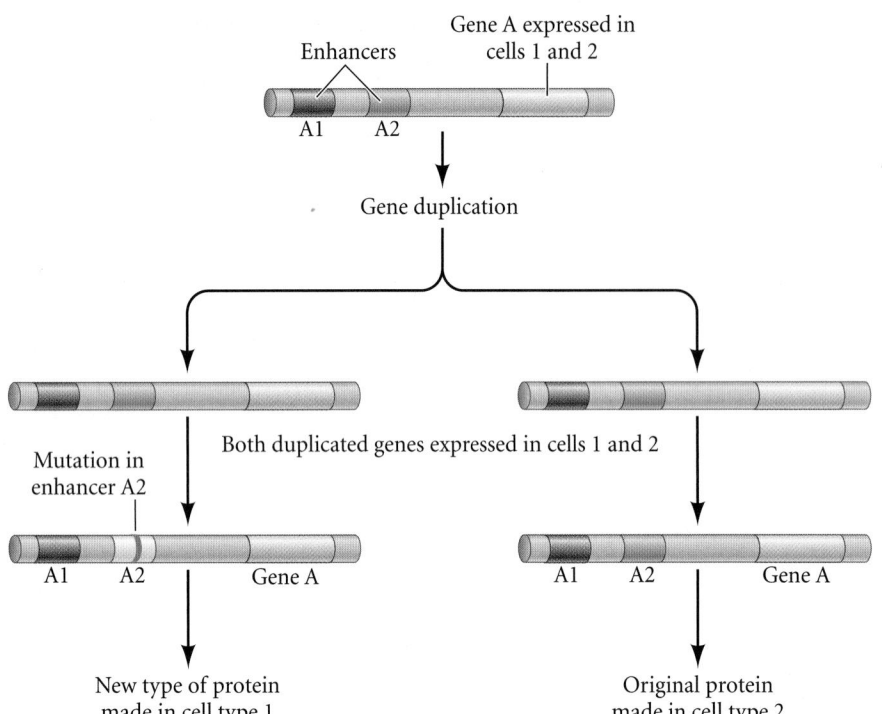

Enhancers

Gene A expressed in cells 1 and 2

A1 A2

Gene duplication

Both duplicated genes expressed in cells 1 and 2

Mutation in enhancer A2

A1 A2 Gene A

A1 A2 Gene A

New type of protein made in cell type 1

Original protein made in cell type 2

FIGURE 20.6 Duplication and divergence. Duplication of a gene that is expressed in several different cell types may be followed by mutations in the duplicated genes. This can lead to a subdivision of the gene's original function, such that each of the duplicated genes is expressed in a different cell type. In the hypothetical case described here, a mutation in one of the duplicated gene enhancers leads to a new pattern of gene expression and a different functional protein in cell type 1.

patterns and new functions (Lynch and Conery 2000; Damen 2002; Locascio et al. 2002).

The Hox genes were generated by successive rounds of gene duplication (Gehring et al. 2009). All multicellular organisms—animals, plants, and fungi—have Hox-like genes, so it is likely there was an ancestral gene that encoded a basic helix-loop-helix transcription factor in protozoans. In the earliest animal groups, this gene became replicated. On the basis of comparative genomics, the ancestor of the cnidarians and bilaterians is thought to have had about three Hox-like genes linked together as a result of gene duplication and divergence (**FIGURE 20.7**; see Ferrier and Holland 2001; Moreno and Martinez 2010). The common ancestor of the protostome and deuterosome lineages had seven Hox genes linked together. The protostomes and the basal deuterostomes (the latter represented by the current species *Branchiostoma floridae*—amphioxus, or lancelet) have one Hox complex (of 7 Hox genes) per haploid genome. However, in the chordate lineage of the deuterostomes, two large-scale duplications of the entire Hox cluster took place, giving vertebrates four Hox clusters per haploid genome instead of one. Thus, instead of having a single *Hox4* gene (orthologous to *Deformed* in *Drosophila*), vertebrates have *Hoxa4*, *Hoxb4*, *Hoxc4*, and *Hoxd4*. This constitutes the *Hox4* **paralogue group** in vertebrates.

Such large-scale gene duplications have had several consequences. First, as seen in Chapters 12 and 14, these duplications create much redundancy. It is difficult to obtain a loss-of-function mutant phenotype in mammals, since to do so all copies of these paralogous genes must be deleted or made nonfunctional (Wellik and Capecchi 2003). However, in some instances, the genes become specialized. *Hoxd11*, for instance, plays an important role in the mammalian limb bud but not in the reproductive system. Mammalian *Hoxa11*, however, plays roles in both the limb (where it is critical in specifying the zeugopod) and in the female reproductive tract (where it helps construct the uterus; see Sidelights and Speculations p. 706).

A second duplication event of the Hox genes within the vertebrates took place during fish evolution. In the ancestral group that gave rise to the teleost fishes (such as zebrafish), the entire genome was duplicated. After this duplication, some clusters were lost. But most teleosts still have six or seven Hox clusters, whereas the other vertebrates have four. Again, there is much redundancy, causing a stabilization of phenotype. But the fish have also used some of these genes for different functions. For instance, one zebrafish *Hoxb1* paralogue, *Hoxb1a*, is required for cranial nerve migration in rhombomere 4, while the other paralogue, *Hoxb1b*, is required for the proper segmentation of rhombomere 4 and the posterior hindbrain (McClintock et al. 2002).

Thus, every *Drosophila* Hox gene has a homologue (and sometimes several) in vertebrates. In some cases, the homologies go very deep and can also be seen in the gene's functions. Not only is the vertebrate *Hoxb4* gene similar in sequence to its *Drosophila* homologue, *Deformed* (*Dfd*), but human *HOXB4* can perform the functions of *Dfd* when introduced into *Dfd*-deficient *Drosophila* embryos (Malicki et al. 1992). As mentioned in Chapter 9, the Hox genes in insects and humans are not just homologous—they occur in the

FIGURE 20.7 Possible scheme for the formation of the Hox paralogue clusters in metazoans by gene duplication and divergence. Three Hox genes were present and linked in the common ancestor of cnidarians and bilaterians. These included the ancestors of paralogue groups PG1, PG3, and a more posterior Hox-like gene (PPHox). In the cnidarians, these became unlinked. These genes remained linked in the ancestor to bilaterians, and the genes acquired a new role patterning the anterior-posterior axis. In the lineage leading to the common bilaterian ancestor ancestor, another tandem duplication gave rise to the *PG5* gene. A series of gene duplications involving the central Hox class generated an extended Hox cluster present in the ancestor of the protostomes and deuterostomes. The protostomes and the basal deuterostomes (such as *Branchiostoma*) have a single Hox complex. But in vertebrates, there have been further duplications of the entire Hox complex, generating four clusters per haploid genome. (After Moreno and Martinez 2010.)

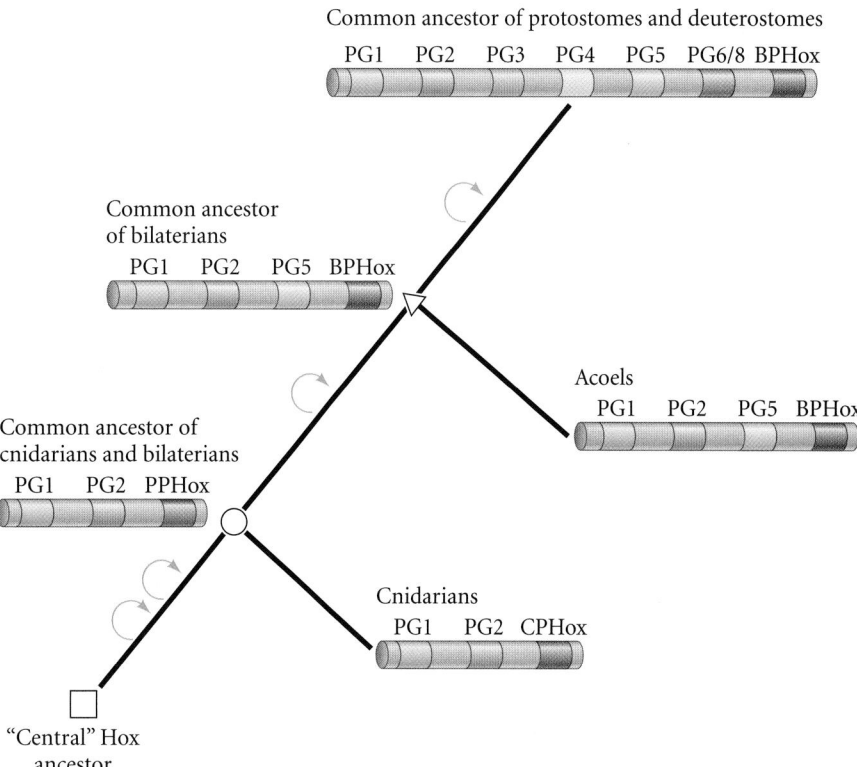

same order on their respective chromosomes. Their expression patterns are also remarkably similar: the more 3′ Hox genes have more anterior expression boundaries* (see Figure 9.30). Thus, these genes are homologous between species (as opposed to members of a gene family being homologous within a species). Genes that are homologous between species are called **orthologues**.

One of the most important gene duplication events in human evolution may have been the duplication of *SRGAP2*, a gene that may have enabled the expansion of the human cerebral cortex. The protein encoded by this gene is expressed in the mammalian brain cortex and appears to *slow down* cell division and decrease the length and density of dendritic processes. However, humans differ from all other animals (including chimpanzees) by having duplicated this gene twice. Moreover, the second duplication event was not complete, so one of the newly formed genes is only a partial duplicate. This partial gene produces a truncated SRGAP2 protein, SRGAP2C, that is also made in the cerebral cortex and which *inhibits* the activity of normal SRGAP2 made from the complete genes. As a result, cell division in the cerebral

cortex continues for longer periods of time, and the dendrites are larger with more connections (**FIGURE 20.8**; Charrier et al. 2012; Dennis et al. 2012). Based on genomic evidence, these gene duplication events are calculated to have taken place about 2.4 million years ago. This would be about the time of *Australopithecus*, the increase in primate brain size, and the first known use of tools (Tyler-Smith and Xue 2012).

In summary, then, the duplication and divergence of genes is an extremely important mechanism for evolution. Duplication allows the formation of redundant genes, and divergence allows these genes to assume new roles. While one gene copy maintains its original role, the other copies are free to mutate and diverge functionally. Numerous transcription factors and paracrine factors are members of such paralogue families. Hox genes are used to pattern the body and limb axes, *Distal-less* genes are used to extend appendages and to pattern the vertebrate skull, and members of the MyoD family specify different stages of muscle development. Each family is derived from a single ancestral gene, and different family members can become active in different tissues and provide instructions for the formation of different cell types.† Thus, one of the most important differences between the genome of a fruit fly and

---

*The conservation of Hox genes and their colinearity demands an explanation. One possibility (Kmita et al. 2000, 2002) is that the Hox genes "compete" for a remote enhancer that recognizes the Hox genes in a polar fashion. This enhancer most efficiently activates Hox genes at the 5′ end. If the positions of the genes are changed by recombination or deletion, then different genes are activated in different regions of the body, and morphology changes.

---

†The duplication and divergence scheme for generating paralogous and orthologous homologies was first described in the 1840s by Sir Richard Owen, a friend (and later rival) of Charles Darwin (see Gilbert 1980). Owen used the seheme to describe (in a nonevolutionary manner) the generation of the vertebrate skeleton.

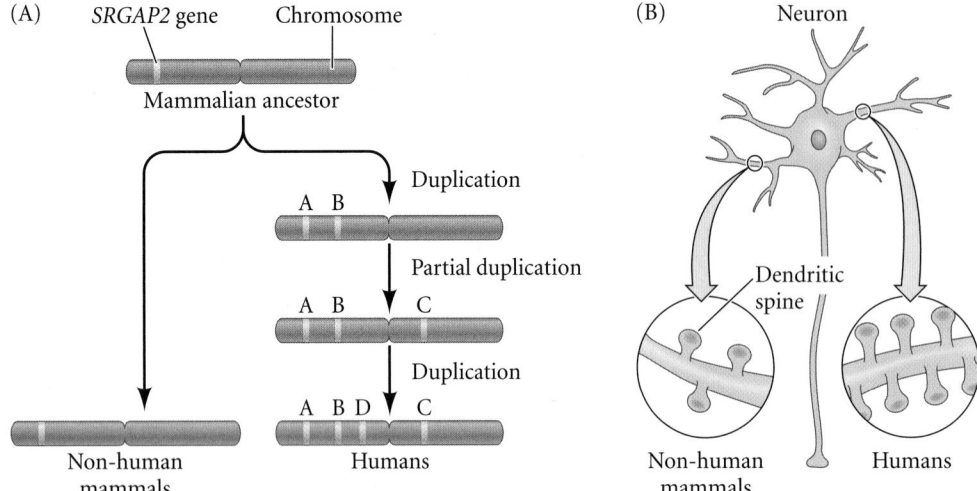

FIGURE 20.8 Duplication and divergence of human *SRGAP2*. (A) The *SRGAP2* gene is found as a single copy in the genomes of all mammals except humans. In the lineage giving rise to humans, duplication events gave rise to four similar versions of the gene, designated A–D. (B) The "ancestral" gene, *SRGAP2A*, with minor contributions from *SRGAP2B* and *D*, enables the maturation of dendritic spines (protuberances) on the surfaces of neurons. *SGRAP2C* is a partial duplicaition, and its product inhibits *SRGAP2A*, slowing dendritic spine maturation and promoting neuronal migration. This partially duplication may have allowed for the evolution of longer maturation time and greater flexibility in the human brain. (After Geschwind and Konopka 2012.)

that of a human "is not that the human has new genes but that where the fly only has one gene, our species has multigene families" (Morange 2001, p. 33).

Deep Homology

One of the most exciting contributions of evolutionary developmental biology has been the discovery not only of homologous regulatory genes, but also of homologous signal transduction pathways, many of which have been mentioned earlier in this book. In different organisms, these pathways are composed of homologous proteins arranged in a homologous manner (Zuckerkandl 1994; Gilbert 1996; Gilbert et al. 1996). This shows a level of parsimony even deeper than that of the individual genes.

In some instances, homologous pathways made of homologous components are used for the same function in both protostomes and deuterostomes. This has been called **deep homology** (Shubin et al. 1997, 2009). Conserved similarities in both the pathway and its function over millions of years of phylogenetic divergence are considered to be evidence of deep homology between these modules (Shubin et al. 1997). One example is the Chordin/BMP4 interaction discussed in Chapter 8. In both vertebrates and invertebrates, Chordin/Short-gastrulation (Sog) inhibits the lateralizing effects of BMP4/Decapentaplegic (Dpp), thereby allowing the ectoderm protected by Chordin/Sog to become the neurogenic ectoderm. These reactions are so similar that *Drosophila* Dpp

protein can induce ventral fates in *Xenopus* and can substitute for Sog* (**FIGURE 20.9**; Holley et al. 1995).

According to this scheme, the central nervous system of the bilaterian animals originated only once, and the BMP-Chordin mechanism was already being used in the bilaterian ancestor of protostomes and deuterostomes (see Figure 20.5A). The positioning of BMP signaling at the ventral (vertebrate) or dorsal (invertebrate) location was a later occurrence (see Mizutani and Bier 2008). This idea has been supported by evidence that annelid worms and cephalochordates also use the inhibition of the BMP pathway to make their central nervous systems (Danes et al. 2007; Yu et al. 2007). Thus, the protostome and deuterostome nervous systems, despite their obvious differences, seem to be formed by the same set of instructions. Indeed, deep homology has also been proposed for the formation of certain parts of the vertebrate and invertebrate brains (Strausfeld and Hirth 2013).

● **See WEBSITE 20.3** The search for the Urbilaterian ancestor

---

*In addition to this central inhibitory reaction, there are other reactions that add to the deep homology of the instructions for forming the protostome and deuterostome neural tube. The proteins involved in the diffusion and stability of BMPs and Chordin also are conserved between insects and vertebrates (Larrain et al. 2001).

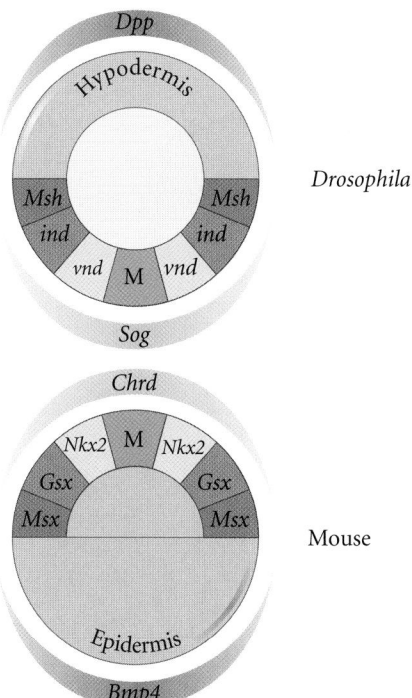

FIGURE 20.9 The same set of instructions forms the nervous systems of both protostomes and deuterostomes. In the fruit fly (a protostome), the TGF-β family member *Dpp* (*Decapentaplegic*) is expressed dorsally and is opposed by *Sog* ventrally. In the mouse (a deuterostome), the TGF-β family member *Bmp4* is expressed ventrally and is countered dorsally by *Chordin* (*Chrd*). The highest concentration of Chordin/Sog becomes the midline (M). The midline is dorsal in vertebrates and ventral in insects, and the concentration gradient of the TGF-β family protein (BMP4 or Dpp) activates genes specifying the regions of the nervous system in the same order in both groups: *vnd/Nkx2*, followed by *ind/Gsx*, and finally *Msh/Msx*. These genes have been seen to be expressed in a similar fashion in cnidarians. (After Ball et al. 2004.)

Mechanisms of Evolutionary Change

In 1940, Richard Goldschmidt wrote that the accumulation of small genetic changes was not sufficient to generate evolutionarily novel structures such as the neural crest, teeth, turtle shells, feathers, or cnidocysts. He claimed that such evolution could occur only through inheritable changes in the genes that regulated development. This idea was brought up again in 1975, when Mary-Claire King and Alan Wilson published a paper titled "Evolution at Two Levels in Humans and Chimpanzees." This study showed that despite the large anatomical differences between chimpanzees and humans, their DNA was almost identical. The differences would be found in the regulatory genes that acted during development:

> The organismal differences between chimpanzees and humans would … result chiefly from genetic changes in a few regulatory systems, while amino acid substitutions in general would rarely be a key factor in major adaptive shifts.

In other words, the allelic substitutions of the genes that encode protein sequences—which seem to be pretty much the same for chimpanzees and humans—were not seen as being important. The important differences are where, when, and how much the genes are activated. In 1977, the idea that change within regulatory genes is critical to evolution was extended by François Jacob, the Nobel laureate who helped establish the operon model of gene regulation. First, Jacob said, evolution works with what it has: it combines existing parts in new ways rather than creating new parts. Second, he predicted that such "tinkering" would be most likely to occur in those genes that construct the embryo, not in the genes that function in adults (Jacob 1977).

Wallace Arthur (2004) catalogued four ways in which Jacob's "tinkering" can take place at the level of gene expression to generate phenotypic variation available for natural selection:

- Heterotopy (change in location)
- Heterochrony (change in time)
- Heterometry (change in amount)
- Heterotypy (change in kind)

These changes can only be accomplished if the gene expression patterns are modular, that is, if they are controlled by different enhancer elements. The modularity of development allows one part of the organism to change without necessarily affecting the other parts.*

Heterotopy

One important way of creating new structures is to alter the *location* where a transcription factor or paracrine factor is expressed. This spatial alteration of gene expression is called **heterotopy** (Greek, "different place"). Heterotopy allows different cells to take on a new identity (as sea urchin micromeres did when they recruited the genes for the skeleton formation; see Chapter 7) or to activate or inhibit a paracrine factor-mediated process in a new area of the body (as when Gremlin inhibits BMP-mediated apoptosis in the webbing between digits; see Figure 14.30). There are many other examples, some of which we will describe next.

HOW THE BAT GOT ITS WINGS AND THE TURTLE GOT ITS SHELL In Chapter 1, we mentioned that the bat evolved its wing by changing the development of the forelimb such that the cells in the interdigital webbing did not die. It turns out that the bat retains its forelimb webbing in a manner very similar to how the duck embryo retains its hindlimb webbing—by blocking the BMPs that would otherwise cause the interdigital cells to undergo apoptosis (see Figures 1.20 and

---

*This chapter concentrates on *transcriptional-level* changes that can generate new morphological forms, but morphological changes can be instigated at these levels as well. Abzhanov and Kaufman (1999), for instance, have shown that post-transcriptional regulation of the *Sex combs reduced* gene is critical in converting legs into maxillipeds in the terrestrial crustacean sowbug *Porcellio scaber*.

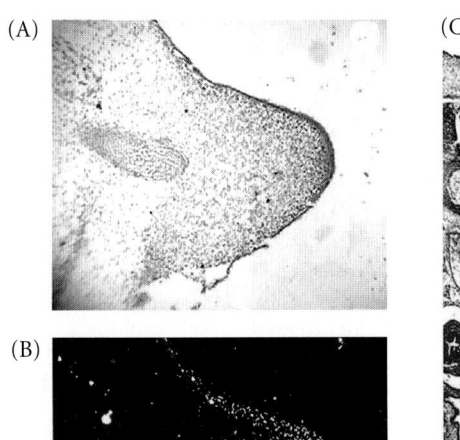

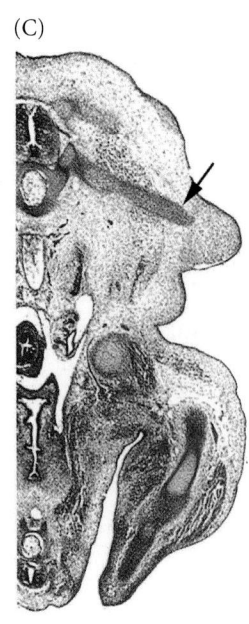

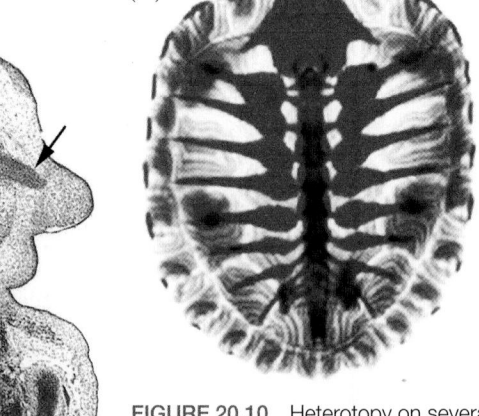

FIGURE 20.10 Heterotopy on several levels in turtle development. The carapace (dorsal shell) of the turtle is formed through sequential layers of heterotopies. *Fgf10* expression in certain regions of the dermis impels rib precursor cells to migrate laterally into the dermis instead of forming a rib cage. (A,B) Cross section of early turtle embryo as the rib enters the dermis (A, brightfield; B, autoradiograph staining for *Fgf10*). (C) Half cross section of a slightly later turtle embryo, showing a rib (arrow) extending from the vertebra into the region of the dermis that will expand to form the shell. (D) Hatchling turtle stained with alizarin to show bones. Bones can be seen in the dermis around the ribs that entered into it. Heterotopies include *Fgf10* expression, rib placement, and bone location. (After Loredo et al. 2001.)

14.29). Both Gremlin and FGF signaling appears to block BMP functions in the bat wing. Unlike other mammals, bats express Fgf8 in their interdigital webbing, and this protein is critical for maintaining the cells there. If FGF signaling is inhibited (by drugs such as SU5402), BMPs can induce apoptosis of the forelimb webbing, just as in other mammals (Laufer et al. 1997; Weatherbee et al. 2006). The Fgf8 in the webbing also appears to be responsible for providing the mitotic signal that extends the digits of the bat, thereby expanding its wing (Hockman et al. 2008).

The formation of the turtle shell also uses BMPs and FGFs, but in different ways. What distinguishes turtles from other vertebrates are their ribs—they migrate laterally into the dermis instead of forming a rib cage (**FIGURE 20.10**). Certain regions of the turtle dermis attract rib precursor cells, and these dermal regions differ from those of other vertebrates because they synthesize Fgf10. Fgf10 seems to attract the ribs, since the ribs do not enter the dermis if the Fgf10 signal is blocked (Burke 1989; Cebra-Thomas et al. 2005). The lateral growth of the ribs causes some muscles to establish new attachment sites and causes the scapula (shoulder blades) to reside inside the ribs. This phenomenon is seen only in the turtles (Nagashima et al. 2009). Once inside the dermis, the rib cells do what rib cells are expected to do—they undergo endochondral ossification wherein the cartilage cells are replaced by bone. To do this, BMPs are made. But the rib is embedded in dermis, and the dermal cells can also respond to the BMPs by becoming bone (Cebra-Thomas et al. 2005). In this way, each of the newly positioned ribs instructs the dermis around it to become bone, and thus the turtle gets its shell.

HOW BIRDS GOT THEIR FEATHERS Although it had long been accepted that feathers emerged as modified reptilian

scales (see Maderson 1972; Prum et al. 1999; Maderson and Alibardi 2000), the mechanism that produces feathers has remained elusive. Harris and his colleagues (2002) have provided a developmental mechanism for feather evolution, showing that the feather most likely evolved from the archosaurian (dinosaur/bird ancestor) scale through an alteration of the expression pattern of the Sonic hedgehog (Shh) and BMP2 proteins.

Scales and feathers start off the same way, with *Bmp2* and *Shh* expressed in separate domains. However, in the feather, both expression domains shift to the distal region of the appendage. This feather-specific pattern is repeated serially around the proximal-distal axis. The interaction between BMP2 and Shh proteins then causes each of these regions to form its own axis—the barbs of the feather (**FIGURE 20.11**). Moreover, when this serially repeated pattern was experimentally modified, the feather pattern was modified in a predictable manner (Harris et al. 2002; Yu et al. 2002).

HOW SNAKES LOST THEIR LIMBS One of the most radical alterations of the vertebrate body plan is seen in snakes. Snakes evolved from four-limbed reptiles; they appear to have lost their legs in a two-step process. Both paleontological and embryological evidence supports the view that snakes lost their forelimbs first and later lost their hindlimbs (Caldwell and Lee 1997; Graham and McGonnell 1999; Chipman 2009). Fossil snakes with hindlimbs, but no forelimbs, have been found (**FIGURE 20.12A**; Tchernov et al. 2000).

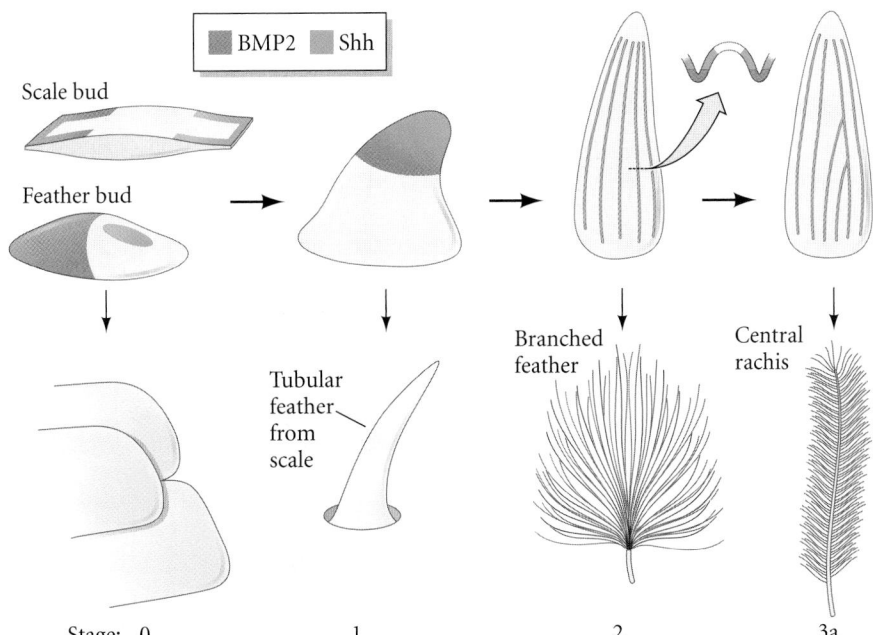

FIGURE 20.11 Model for the evolution of the feather by changes in the pattern of *Bmp2* and *Shh* expression. Stage 0 shows the *Shh* and *Bmp2* expression in the scale bud (above) and feather bud (below). Stage 1 represents a tubular feather as evolved from an archosaurian scale. The *Shh* and *Bmp2* expression patterns are postulated to be at the tip. Stage 2 represents the emergence of a branched feather evolved by further changing the expression patterns of *Bmp2* and *Shh* to form rows along the proximal-distal axis. In stage 3a, changes in feather morphology evolved by altering the pattern to produce a central rachis. (After Harris et al. 2002.)

Moreover, while the most derived snakes (such as vipers) are completely limbless, the more primitive snakes (such as boas and pythons) have pelvic girdles and rudimentary femurs.

The missing forelimbs can be explained by the altered Hox gene expression pattern in the anterior portion of the snake. As described in Chapter 9, the expression pattern of Hox genes in vertebrates determines the type of vertebral structure formed. Thoracic (chest) vertebrae, for instance, have ribs, whereas cervical (neck) vertebrae and lumbar (lower back) vertebrae do not. The type of vertebra produced is specified by the Hox genes expressed in the somite. In most vertebrates, the forelimbs form just anterior to the most anterior expression domain of *Hoxc6* (Gaunt 1994, 2000; Burke et al. 1995; Mallo et al. 2010). Caudal to that point, *Hoxc6*, in combination with *Hoxc8*, helps specify vertebrae to be thoracic (ribbed). During early python development, *Hoxc6* is not expressed in the absence of *Hoxc8*, so the forelimbs do not form. Rather, the combination of *Hoxc6* and *Hoxc8* is expressed for most of the length of the organism, telling the vertebrae to form ribs throughout most of the body* (**FIGURE 20.12B,C**; Cohn and Tickle 1999).

Moreover, in snakes, the number of segments is facilitated by "irregularities" of expression in the *Hox13* paralogue group. This is the last Hox group, and it is responsible for putting the "caps" onto growing structures—for example, it is necessary for the development of fingers on limbs and for tails at the body's posterior tip. However, two of these genes,

Hoxa13 and *Hoxd13*, are not expressed in snake embryos, thereby causing the new vertebrae to become trunk rather than tail (Di-Poï et al. 2010). The tail is not completed until much later in development.

The loss of hindlimbs occurred by a different mechanism. Hindlimb buds do begin to form in some snakes, such as pythons, but they do not produce anything more than a femur. This appears to be due to the lack of *Sonic hedgehog* expression by the limb bud mesenchyme. Sonic hedgehog is needed both for limb polarity and for maintenance of the apical ectodermal ridge (AER; see Chapter 14). Python hindlimb buds lack the AER, and the phenotype of the python hindlimb resembles that of mouse embryos with loss-of-function mutations of *Sonic hedgehog* (Chiang et al. 1996).

Heterochrony

Heterochrony (Greek, "different time") is a shift in the relative order or timing of two developmental processes. Heterochrony can be seen at any level of development, from gene regulation to adult animal behaviors (West-Eberhard 2003). In heterochrony, one module changes its time of expression or growth rate relative to the other modules of the embryo. One sees heterochronic changes in development throughout the animal kingdom. As Darwin (1859, p. 209) noted, "we may confidently believe that many modifications, wholly due to the laws of growth, and at first in no way advantageous to a species, have been subsequently taken advantage of by the still further modified descendants of this species."

Heterochronies are quite common in vertebrate evolution. We have already discussed the extended growth of the human brain and the heterochronies of amphibian metamorphosis. Another example is found in marsupials, whose jaws and forelimbs develop at a faster rate than do those of

---

*In mammals, the rib-repressing Hox genes (the Hox10 paralogue group) would still be able to block rib formation, but this ability to inhibit rib growth is not found in *Hoxa10* or *Hoxc10* of lizards and snakes (Woltering et al. 2009; Di-Poï et al. 2010). So all the snake trunk vertebrae have ribs.

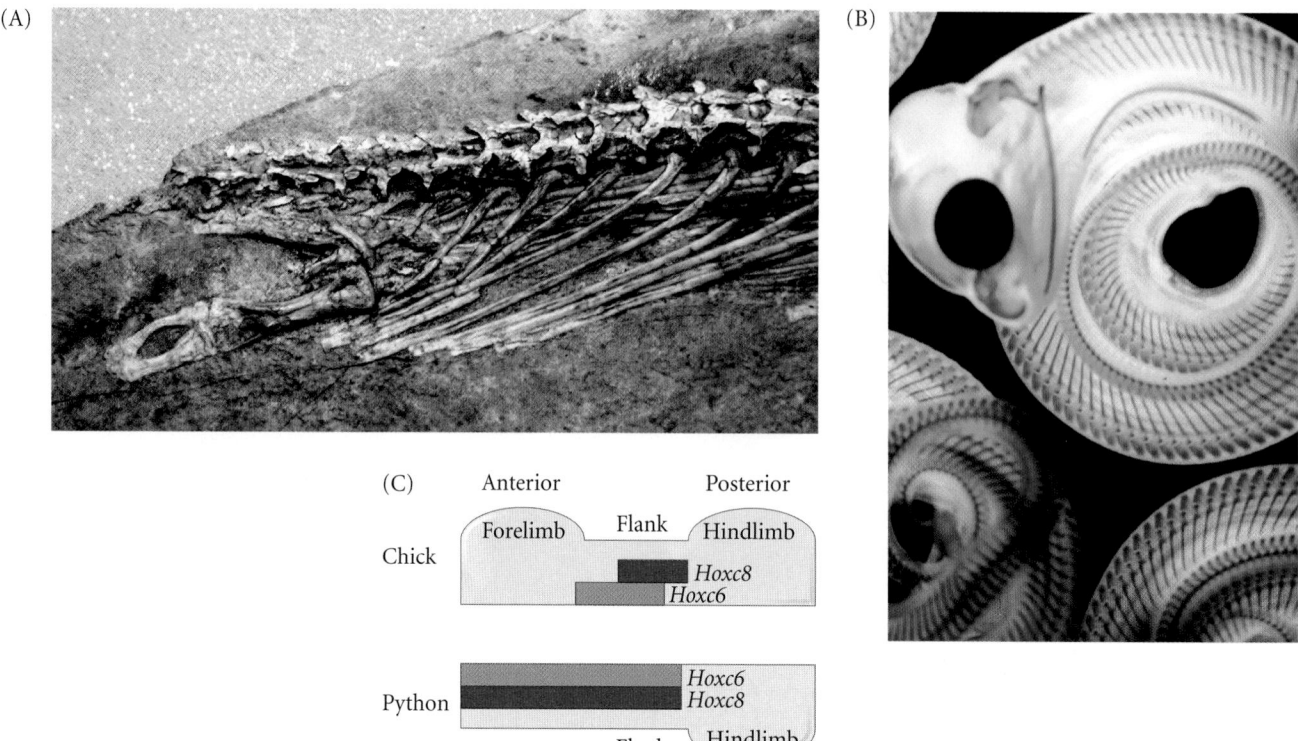

(A)

(B)

(C)

Anterior   Posterior

Chick

Forelimb   Flank   Hindlimb

Hoxc8
Hoxc6

Python

Hoxc6
Hoxc8

Flank   Hindlimb

FIGURE 20.12 Loss of limbs in snakes. (A) *Haasiophis terrasanctus*, a fossil snake from the upper Cretaceous carbonates in Israel. The stylopod and zeugopod can readily be seen. (B) Skeleton of the garter snake, *Thamnophis*, stained with alcian blue. Ribbed vertebrae are seen from the head to the tail. (C) Hox expression patterns in chick and python. (A, photograph by S. Gilbert; B courtesy of A. C. Burke; C after Cohn and Tickle 1999.)

placental mammals, allowing the marsupial newborn to climb into the maternal pouch and suckle (Smith 2003; Sears 2004). The enormous number of vertebrae and ribs formed in embryonic snakes (more than 500 in some species; see Figure 20.12B) is likewise due to heterochrony (as well as to changes in the *Hox13* paralogue group; see above). The segmentation reactions cycle nearly four times faster relative to tissue growth in snake embryos than they do in related vertebrate embryos (Gomez et al. 2008).

In some instances we can determine the heterochronic changes in expression of certain genes. The elongated fingers in the dolphin flipper appear to be the result of the heterochronic expression of *Fgf8*, which as we saw in Chapter 14 encodes a major paracrine factor for limb outgrowth (see Figure 14.25; Richardson and Oelschläger 2002; Cooper 2010). Another "digital" example of molecular heterochrony occurs in the lizard genus *Hemiergis*, which includes species with three, four, or five digits per limb. The number of digits is regulated by the length of time the *Sonic hedgehog* gene remains active in the limb bud's zone of polarizing activity. The shorter the duration of *Shh* expression, the fewer the number of digits (Shapiro et al. 2003). In primates, there is a heterochronic shift in the

transcription of a set of cerebral mRNAs, such that the expression pattern in adult humans resembles that seen in juvenile chimpanzees (Somel et al. 2009).

Heterometry

Heterometry is a change in the *amount* of a gene product or structure. We have already mentioned such heterometric changes in Chapter 10, where we discussed the evolution of the blind Mexican cavefish. We saw that overproduction of Sonic hedgehog protein in the midline prechordal plate downregulates the *Pax6* gene, preventing eye formation. But this overexpression of *Shh* has other consequences as well. Not only does it cause the degeneration of the eyes, it also causes the jaw size and number of taste buds to increase (Yamamoto et al. 2009). Since cavefish live in complete darkness, the expansion of their jaw size and gustatory sense at the expense of sight can be selected. Heterometry can also be seen in the human response to parasitic worms: a mutation causing overproduction of interleukin 4 has been (and is being) selected in populations where such parasites are endemic (Rockman et al. 2003).

DARWIN'S FINCHES One of the best examples of heterometry involves Darwin's celebrated finches. Darwin's finches are a set of 15 closely related birds collected by Charles Darwin and his shipmates during their visit to the Galápagos and Cocos islands in 1835. These birds helped Darwin frame his evolutionary theory of descent with modification, and

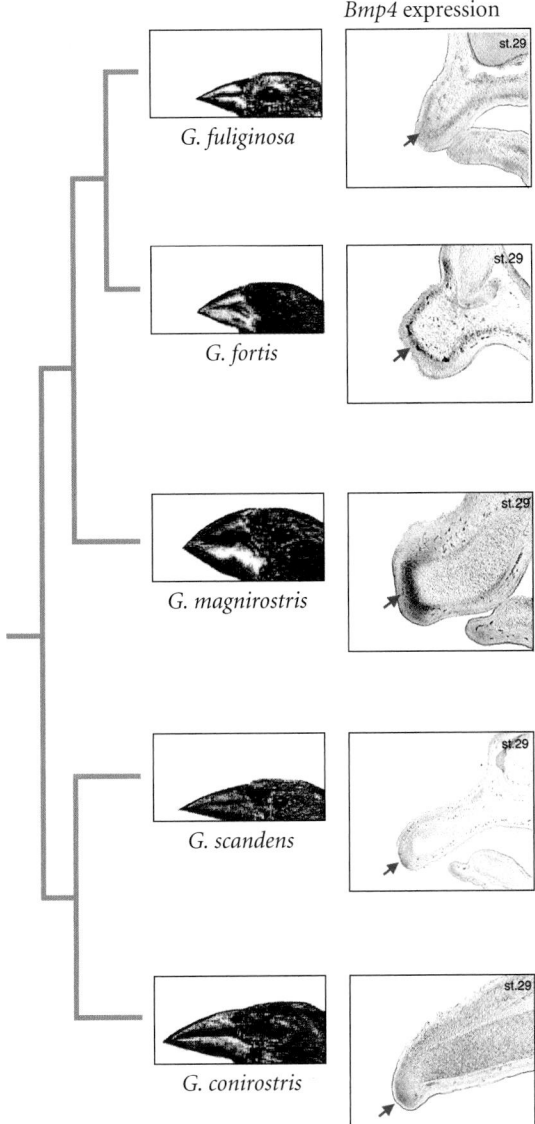

Bmp4 expression

st.29
st.29
st.29
st.29
st.29

G. fuliginosa

G. fortis

G. magnirostris

G. scandens

G. conirostris

FIGURE 20.13 Correlation between beak shape and the expression of *Bmp4* in five species of Darwin's finches. In the genus *Geospiza*, the ground finches (represented by *G. fuliginosa*, *G. fortis*, and *G. magnirostris*) diverged from the cactus finches (represented by *G. scandens* and *G. conirostris*). The differences in beak morphology correlate to heterochronic and heterometric changes in *Bmp4* expression in the beak. *Bmp4* (red arrow) is expressed earlier and at higher levels in the seed-crushing ground finches. The photographs of the embryonic beaks were taken at the same stage (stage 29) of development. This gene expression difference provides one explanation for the role of natural selection on these birds. (After Abzhanov et al. 2004.)

paracrine factor showed such differences. The expression of *Bmp4* in ground finches started earlier and was much greater than *Bmp4* expression in cactus finches. In all cases, the *Bmp4* expression pattern correlated with the breadth and depth of the beak.

The importance of these expression differences was confirmed experimentally by changing the *Bmp4* expression pattern in chick embryos to mimic the heterometric and heterochronic changes in the ground finches (Abzhanov et al. 2004; Wu et al. 2004). When *Bmp4* expression was enhanced in the frontonasal process mesenchyme, the chick developed a broad beak reminiscent of the beaks of the ground finches. Conversely, when BMP signaling was inhibited in this region (by Noggin, a BMP inhibitor), the beak lost depth and width.

But this was only the beginning of the story. Gene chip technology showed that the level of *Calmodulin* gene expression in the beak primordia of the sharp-beaked cactus finches was 15-fold greater than in the beak primordia of the blunt-beaked ground finches. Calmodulin is a protein that combines with many enzymes to make their activity dependent on calcium ions. In situ hybridization and other techniques demonstrated that the *Calmodulin* gene is expressed at higher levels in the embryonic beaks of cactus finches than in the embryonic beaks of ground finches (**FIGURE 20.14**). When Calmodulin was upregulated in the embryonic chicken beak to mimic the finchlike expression domain, the chick beak too became long and pointed.

The frontonasal mesenchyme gives rise to two modules that form the adult beak: the premaxillary bone and the prenasal cartilage. The prenasal cartilage develops earlier in beak development and establishes the species-specific beak morphology. The morphology of prenasal cartilage is coordinately regulated by BMP and Calmodulin signals, and these signals correlate well with the exact scaling parameters of the evolving beak shapes (Campàs et al. 2011; Mallarino et al. 2011, 2012). Thus, enhancers controlling the amount of beak-specific BMP4 and Calmodulin synthesis may have been critically important in the evolution of Darwin's finches.

BMP4 and Calmodulin represent two targets for natural selection, and together they explain the shape variations of Darwin's finches (**FIGURE 20.15**; Abzhanov et al. 2006; Campàs et al. 2011). BMP4 is regulated in heterochronic and heterometric ways, whereas Calmodulin is

they still serve as one of the best examples of adaptive radiation and natural selection (see Weiner 1994; Grant and Grant 2008). Systematists have shown that these finch species evolved in a particular manner, with a major speciation event being the split between the cactus finches and the ground finches. The ground finches evolved deep, broad beaks that enable them to crack seeds open, whereas the cactus finches evolved narrow, pointed beaks that allow them to probe cactus flowers and fruits for insects and flower parts. Earlier research (Schneider and Helms 2003) had shown that species differences in the beak pattern were caused by changes in the growth of the neural crest-derived mesenchyme of the frontonasal process (i.e., those cells that form the facial bones). Abzhanov and his colleagues (2004) found a remarkable correlation between the beak shape of the finches and the timing and amount of *Bmp4* expression (**FIGURE 20.13**). No other

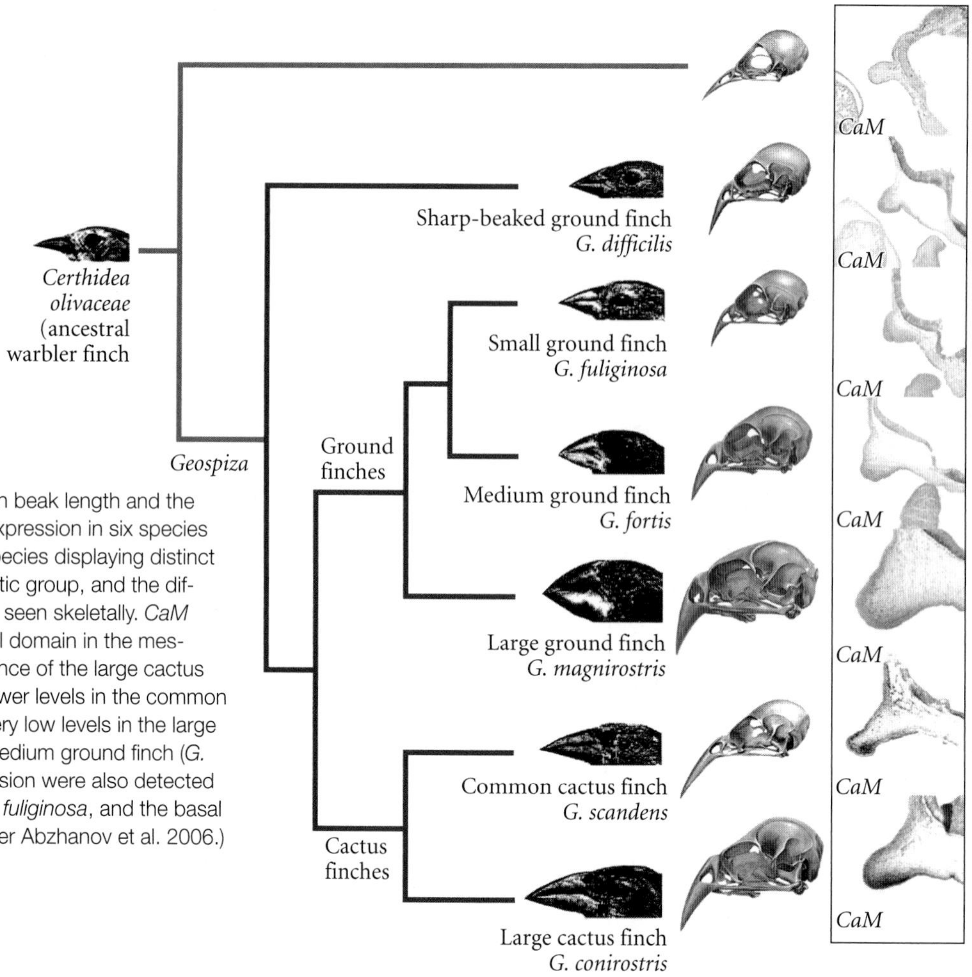

FIGURE 20.14 Correlation between beak length and the amount of *Calmodulin* (*CaM*) gene expression in six species of Darwin's finches. The *Geospiza* species displaying distinct beak morphologies are a monophyletic group, and the differences in beak morphology can be seen skeletally. *CaM* is expressed in a strong distal-ventral domain in the mesenchyme of the upper beak prominence of the large cactus finch (*G. conirostris*), in somewhat lower levels in the common cactus finch (*G. scandens*), and in very low levels in the large ground finch (*G. magnirostris*) and medium ground finch (*G. fortis*). Very low levels of *CaM* expression were also detected in the mesenchyme of *G. difficilis*, *G. fuliginosa*, and the basal warbler finch *Certhidea olivacea*. (After Abzhanov et al. 2006.)

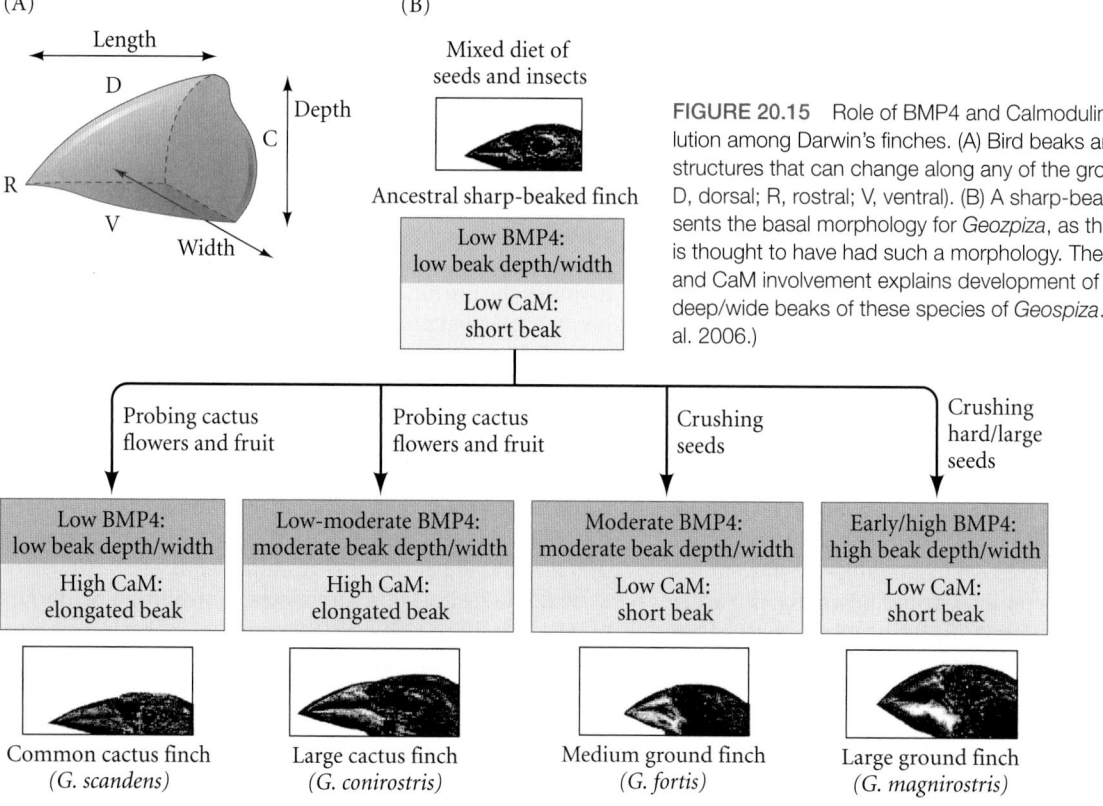

FIGURE 20.15 Role of BMP4 and Calmodulin (CaM) in beak evolution among Darwin's finches. (A) Bird beaks are three-dimensional structures that can change along any of the growth axes (C, caudal; D, dorsal; R, rostral; V, ventral). (B) A sharp-beaked finch represents the basal morphology for *Geozpiza*, as the ancestral species is thought to have had such a morphology. The model for BMP4 and CaM involvement explains development of both elongated and deep/wide beaks of these species of *Geospiza*. (After Abzhanov et al. 2006.)

FIGURE 20.16 Allometric growth in the whale head. An adult human skull is shown for comparison. The whale's upper jaw (maxilla) has pushed forward, causing the nose to move to the top of the skull. The mandible is not shown. (The premaxilla is present in the early human fetus, but it fuses with the maxilla by the end of the third month of gestation. The human premaxilla was discovered by Goethe, among others, in 1786.) (After Slijper 1962.)

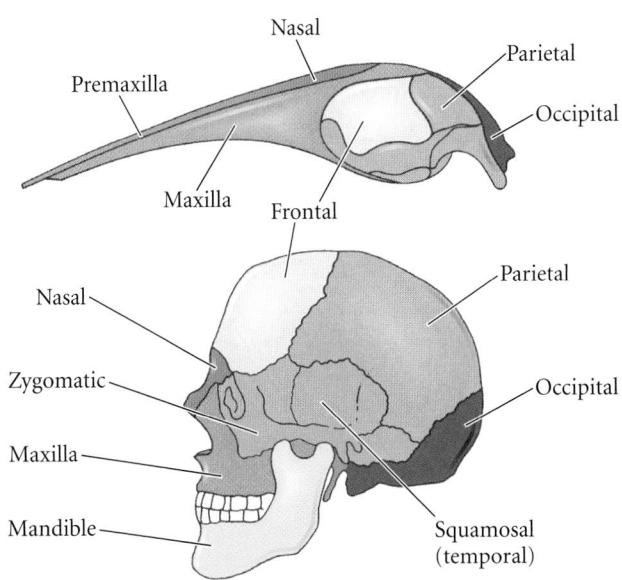

regulated heterometrically. These studies demonstrate the role for modularity in evolution with morphological variation regulated along two independent axes. While natural selection will allow certain morphologies to survive, the generation of those morphologies depends on variations of developmental regulatory genes such as those for BMP4 and Calmodulin.

ALLOMETRY Another consequence of modularity associated with both heterometry and heterochrony is **allometry**—changes that occur when different parts of an organism grow at different rates (Huxley and Teissier 1936; Gayon 2000). As animals develop, their shape changes, a result of differences in the timing and duration of growth events. Indeed, morphological evolution (especially within a phylum) is due primarily to changes in body size and the relative sizes of body parts. The laws of allometric growth can be derived mathematically (Przibram 1931; Nijhout and German 2012). Such differential changes in growth rate can involve altering a target cell's sensitivity to growth factors or altering the amounts of growth factors produced.

Once again, vertebrate limbs and heads provide useful illustrations of altered development. Local differences among chondrocytes in the limb buds cause the central toes of embryonic mammals to grow at a rate 1.4 times that of the lateral toes (Wolpert 1983). As horses grew larger over evolutionary time, this regional difference in chondrocytes resulted in the one-toed state seen in modern horses. Another dramatic example of allometry in evolution comes from whale skulls. In the very young (4–5 mm long) whale embryo, the nose is in the usual mammalian position. However, the enormous growth of the maxilla and premaxilla (upper jaw) pushes over the frontal bone and forces the nose to the top of the skull (**FIGURE 20.16**). This new position turns the mammalian nose into the cetacean blowhole, allowing the whale to have a large and highly specialized jaw apparatus and (not incidentally) to breathe while swimming at the water's surface (Slijper 1962). While whale skulls depart greatly from their early embryonic form, bird skulls depart very little from theirs. In fact, bird skulls resemble the skulls of juvenile dinosaurs (Bhullar et al. 2012).

One of the most spectacular examples of allometry is that of the human brain (Bufill et al. 2011). As mentioned in Chapter 10, the human brain maintains its fetal rate of

growth well into the period of childhood and is characterized by rapid cell division, complexity of connections between neurons, and plasticity during the newborn period. One of the genes responsible for this allometry may be the truncated *SRGAP2* mentioned earlier in this chapter, as well as *HAR1* and *GADD45G* (see Chapter 10).

Heterotypy

In heterochrony, heterotopy, and heterometry, mutations affect the regulatory regions of the gene. The gene's product—the protein—remains the same, although it may be synthesized in a new place, at a different time, or in different amounts. The changes of **heterotypy** affect the actual coding region of the gene, and thus can change the functional properties of the protein being synthesized. These changes in the protein-encoding regions of the gene are usually seen in genes that are expressed in only one or a few tissues, suggesting that pleiotropy (see below) constrains such changes in broadly expressed genes (Haygood et al. 2010; Wu et al. 2011). However, changes in the coding sequence of transcription factors can have profound consequences in animal and plant evolution (Wang et al. 2005).

HOW PREGNANCY MAY HAVE EVOLVED IN MAMMALS One of the most amazing features of mammals is the female uterus, a structure that can hold, nourish, and protect a developing fetus within its mother's body. One of the key proteins enabling this internal gestation is prolactin. Prolactin promotes differentiation of the uterine epithelial cells, regulates trophoblast growth, allows blood vessels to spread toward the embryo, and helps downregulate the immune and inflammatory responses (so the mother's body does not perceive the embryo as a "foreign body" and reject it).

SIDELIGHTS & SPECULATIONS

Transposable DNA elements and the origins of pregnancy

Mutations at existing *cis*-regulatory elements (primarily at enhancers) are thought to be critical in generating most evolutionarily novel structures. However, there are other mechanisms, and these processes, albeit rare, may be responsible for reorganizing the types of genes expressed in a particular cell (Lynch et al. 2011).

Imagine a system where new enhancers can be added to a genome by a virus. These viruses would integrate into the host DNA and place the enhancers next to the host's genes. If the host cell were making a transcription factor that bound to the new enhancer, the gene could now be regulated by that transcription factor.

This actually seems to have happened, only the mobile genetic element was not a virus—it was a **transposon**, a transposable DNA element. Transposons are DNA sequences that can travel from one species to another and integrate into the host's DNA; once there, they become part of the host genome. They are simpler than viruses, being a DNA sequence

with ends that can be ligated into the genome. It is thought that transposons are transmitted by parasites or symbionts that can move between hosts (Gilbert et al. 2010; Schaack et al. 2010).

DNA from transposons may play significant roles, since nearly 50% of the human genome consists of such transposable DNA sequences (Lander et al. 2001; Ostertag and Kazazian 2001; Cordaux and Batzer 2009). This is not all "junk DNA"; about 8% of functional human genes contain sequences derived from transposons (Wagner and Lynch 2010). In some instances, the transposons carry enhancer sequences that can be bound by transcription factors, and it appears that such transposons have been extremely important in the evolution of the placental mammals (**FIGURE 20.17**; Fleschotte 2008).

All mammals have hair and mammary glands. Female placental mammals also have a uterus that can retain the growing fetus and provide it with nutrition, oxygen, and protection from the mother's own immune system. For this to happen, the

mammary glands and uterus respond to progesterone, the hormone that maintains pregnancy and initiates milk production. Progesterone in the uterus activates the *Foxo1a* gene (Kyo et al. 2011) and increases the expression of *Hoxa11* (Taylor et al. 1999). The Hoxa11 and Foxo1a transcription factors were mentioned earlier as binding together on the enhancer of the *Prolactin* gene to activate that gene. But that enhancer is mammalian-specific. More than that, it is located on MER20, a mammal-specific transposon.

MER20 is a transposable element found only in placental mammals. It contains binding sites for Hoxa11 and Foxo1a, as well as for the transcription insulator CTCF. Chromatin immunoprecipitation finds Hoxa11, Foxo1a, and CTCF, as well as transcriptional activator p300 and RNA polymerase II, together at the MRER20-derived enhancers. MER20 is found in numerous places in the human genome, including the progesterone-responsive enhancer of the *Prolactin* gene (Lynch et al. 2011). Indeed, progesterone induces about 1500 genes in human

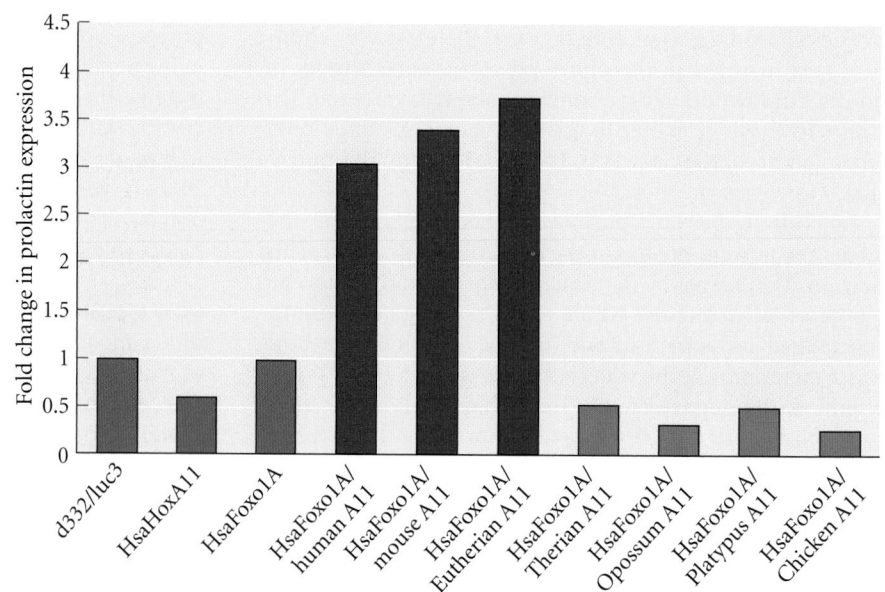

FIGURE 20.18 Ability of the mammalian Hoxa11 protein, in combination with Foxo1a, to promote expression of the uterine *Prolactin* enhancer. The activated luciferinase reporter gene (*d332/luc3*), activated human *Hoxa11* gene (HsaHoxa11), and activated human *FOXO1A* gene (HsaFoxo1a) each failed to activate the *Prolactin* gene from the enhancer transcription. Mammalian (but not opossum, platypus, or chicken) Hoxa11 increased transcription from this enhancer, but only in the presence of Foxo1a. "Eutherian A11" indicates generalized Hoxa11 from placental mammals. "Therian A11" indicates the consensus Hoxa11 sequence from all mammals. (After Lynch et al. 2008.)

(A)

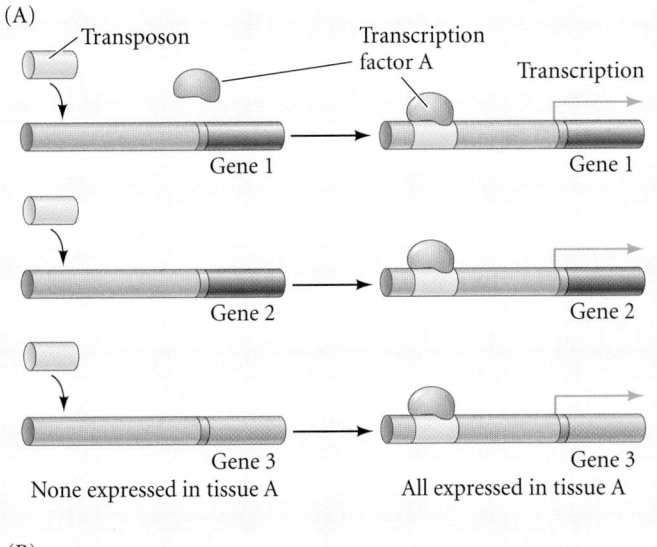

(B)

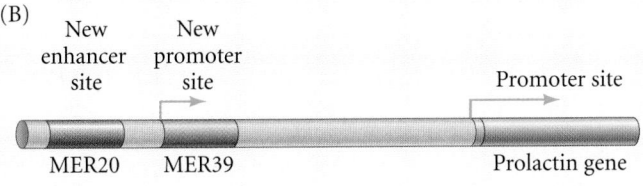

FIGURE 20.17 Transposable elements can affect gene transcription by inserting new enhancers, new promoters, new splice sites, or new coding sequences. (A) A transposon containing transcription factor binding sites (in this case for transcription factor A) invades a germline genome and inserts near a promoter for a gene. This brings the gene under the regulation of the transcription factor binding to the transposon-borne enhancer. TEs can integrate near several genes; here they cause a battery of genes to be expressed under the control of transcription factor A. (B) The tissue-specific enhancer of mammalian prolactin is derived from transposon MER20, which has binding sites for Hoxa11 and the progesterone receptor. Another transposon, MER39, has also brought in a new promoter site, this one seen only in primates. (After Wagner and Lynch 2010.)

uterine stroma cells, and about 13% of them are very close to MER20 DNA elements.

MER20 may not be the only transposable element involved in mammalian pregnancy. A second transposable element, MER39, is a primate-specific transposon found (among other places) in the promoter of the *Prolactin* gene (Emera et al. 2011). There is also evidence that some of the proteins used in the formation of the mammalian placenta were originally retroviral envelope proteins that became recruited for a new function (Dupressoir et al. 2011). Thus, transposable DNAs may have "rewired" gene regulatory networks and contributed to the evolution of the mammalian uterus.

At about the same time the mammalian uterus and pregnancy evolved, one of the mammalian Hox genes—*Hoxa11*—appears to have undergone intensive mutation and selection in the lineage that gave rise to placental mammals. Analysis shows that the sequence of the Hoxa11 protein changed in mammals such that it associates and interacts with another transcription factor, Foxo1a (**FIGURE 20.18**; Lynch et al. 2004, 2008). This association with Foxo1a enables Hoxa11 to upregulate prolactin expression from the enhancer used in uterine epithelial cells. Hoxa11 from non-eutherian mammals (i.e., opossum and platypus) and from chickens does not upregulate prolactin. If morpholinos knock out the *Hoxa11* mRNA in mouse uterine cells, no prolactin is expressed. Therefore, one of the most important evolutionary changes in the lineage leading to mammals involved the heterotypic alteration of the Hoxa11 sequence.

WHY INSECTS HAVE ONLY SIX LEGS Insects have only six legs while most other arthropod groups (think of spiders, millipedes, centipedes, lobsters, and shrimp) have many more. How is it that the insects came to form legs only in their three thoracic segments with no legs on their abdominal segments? The answer seems to be found in the relationship, mentioned earlier, between Ultrabithorax protein and the *Distal-less* gene. In most of the arthropod groups, Ubx protein does not inhibit the *Distal-less* gene. However, in the insect lineage, a mutation occurred in the *Ubx* gene wherein the original 3′ end of the protein-coding region was replaced by a group of nucleotides encoding a stretch of about 10 alanine residues at the C-terminus (**FIGURE 20.19**; Galant and Carroll 2002; Ronshaugen et al. 2002). This polyalanine region represses *Distal-less* transcription in the abdominal segments.

When a brine shrimp *Ubx* gene is experimentally modified to encode this polyalanine region, it too represses the *Distal-less* gene. The ability of insect Ubx to inhibit *Distal-less* thus appears to be the result of a gain-of-function mutation that characterizes the insect lineage.

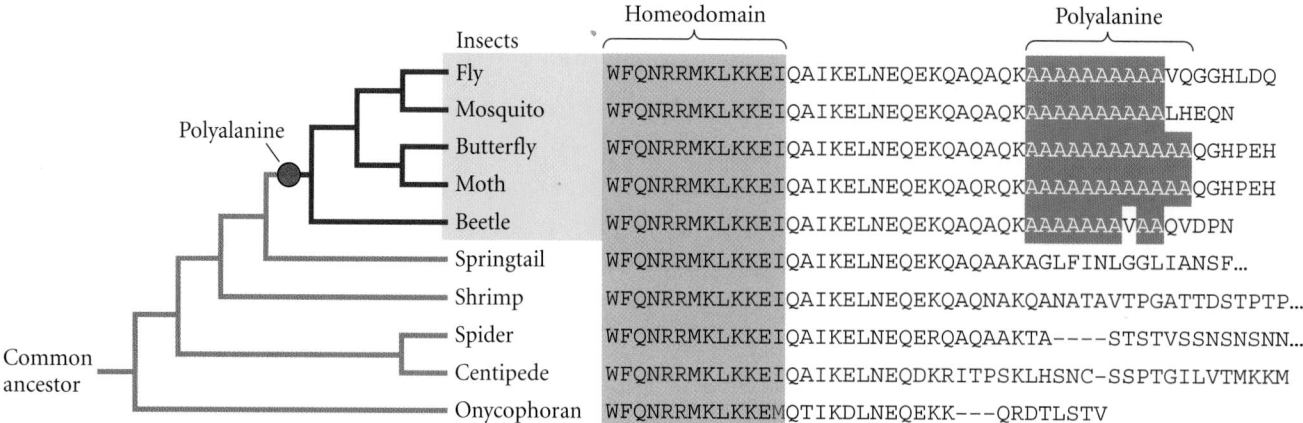

FIGURE 20.19 Changes in Ubx protein associated with the insect clade in the evolution of arthropods. Of all arthropods, only the insects have Ubx protein that is able to repress *Distal-less* gene expression and thereby inhibit abdominal legs. This ability to repress *Distal-less* is due to a mutation that is seen only in the insect *Ubx* gene. (After Galant and Carroll 2002; Ronshaugen et al. 2002.)

Developmental Constraints on Evolution

There are only about three dozen major animal lineages, and they encompass all the different body plans seen in the animal kingdom. One can easily envision other body plans by imagining animals that do not exist; science fiction writers do it all the time. So why don't we see more body plans among the living animals? To answer this, we have to consider the constraints that development imposes on evolution.

The number and forms of possible phenotypes that can be created are limited by the interactions that are possible among molecules and between modules.* These interactions also allow change to occur in certain directions more easily than in others. Collectively, the restraints on phenotype production are called **developmental constraints**. Constraints on evolution fall into three major categories: physical, morphogenetic, and phyletic (see Richardson and Chipman 2003).

● **See WEBSITE 20.4** How the chordates got a head

Physical constraints

The laws of diffusion, hydraulics, and physical support are immutable and will permit only certain physical phenotypes to arise. For example, blood cannot circulate to a rotating organ; thus a vertebrate on wheeled appendages (of the sort that Dorothy saw in Oz) cannot exist, and this entire evolutionary avenue is closed off. Similarly, structural parameters

and fluid dynamics would prohibit the existence of 6-foot-tall mosquitoes or 25-foot-long leeches.

The elasticity and tensile strength of tissues are also physical constraints. In *Drosophila* sperm, for example, the type of tubulin that can be used in the axoneme is constrained by the need for certain physical properties in the exceptionally long flagellum (Nielsen and Raff 2002). The six cell behaviors used in morphogenesis (division, growth, shape change, migration, death, and matrix secretion) are each limited by physical parameters, and thereby provide limits on what structures animals can form. Interactions between different sets of tissues involve coordinating the behaviors of sheets, rods, and tubes of cells in a limited number of ways (Larsen 1992).

Morphogenetic constraints

Bateson (1894) and Alberch (1989) noted that when organisms depart from their normal development, they do so in only a limited number of ways. Although there have been many modifications of the vertebrate limb over 300 million years, some modifications (such as a middle digit shorter than its surrounding digits; a zeugopod more proximal than the stylopod) are never seen (Holder 1983; Wake and Larson 1987). These observations suggest a limb construction scheme that follows certain rules (see Figure 14.12; Oster et al. 1988; Newman and Müller 2005).

Phyletic constraints

Phyletic constraints on the evolution of new structures are historical restrictions based on the genetics of an organism's development (Gould and Lewontin 1979). In other words, once a structure comes to be established by inductive interactions, it is difficult to start over again. The notochord, for example, is functional in adult protochordates such as *Amphioxus* (Berrill 1987) but degenerates in adult vertebrates. Yet it is transiently necessary in vertebrate embryos, where it

---

*The notion of constraint in evolution is often used differently by different groups of biologists. Developmental biologists see constraints as limiting the appearance of certain phenotypes, whereas population geneticists see constraints as limiting "ideal" adaptation (such as constraints on optimal foraging) (Amundson 1994, 2005).

SIDELIGHTS & SPECULATIONS

Reaction-Diffusion Kinetics and the Evolution of Mammalian Teeth

One set of rules constraining limb development may be the mathematics of the reaction-diffusion mechanism ("the Turing model"; see p. 499), a model that can be extended throughout development. Stephen J. Gould (1989) joked that paleontologists believe mammalian evolution occurs when two teeth mate to produce slightly altered descendant teeth. Because tooth enamel is more durable than ordinary bone, teeth often remain after all the other bones have decayed, and the study of tooth morphology has been critical to mammalian systematics and ecology. Changes in the cusp pattern of molars are regarded as especially important in allowing the evolutionary radiation of mammals into new ecological niches. What mechanism allows molars to change their form so rapidly?

Jukka Jernvall and colleagues (Jernvall et al. 2000; Salazar-Ciudad and Jernvall 2002, 2004; Salazar-Ciudad 2008) pioneered a computer-based approach to phenotype production using geographic information systems

(GIS) to map gene expression patterns in incipient tooth buds. Their studies showed that gene expression patterns forecast the exact location of the tooth cusps in mice and voles based on differences in gene expression patterns (**FIGURE 20.20**). They also showed how reaction-diffusion mechanisms can explain the differences in gene expression between mice and voles (Salazar-Ciudad and Jernvall 2004, 2010).

The signaling center of the tooth is the enamel knot, which secretes BMPs, FGFs, and Shh. Shh and FGFs inhibit BMP production, while BMPs stimulate both the production of more BMPs and the synthesis of their own inhibitors. BMPs also induce epithelial differentiation, whereas FGFs induce epithelial growth. For the studies described here, two additions to the classic reaction-diffusion equation involved (1) changing the "constants" of diffusion as development progresses and (2) changing the amount of elasticity associated with the extracellular matrix as development progresses. (This is because

the extracellular matrix changes as the cells differentiate, altering tissue shape and the diffusion of paracrine factors.) The result is a pattern of gene activity that changes as the shape of the tooth changes, and vice versa (**FIGURE 20.21**). Under this model, the large differences between mouse and vole molars can be generated by small changes in the binding constants and diffusion rates of the BMP and Shh proteins. A small increase in the diffusion rate of BMP4 and a stronger binding constant of its inhibitor are sufficient to change the vole pattern of tooth growth into that of the mouse.

The work on mouse and vole molars showed how large morphological changes can result from very small changes in initial conditions. The set of equations that emerged also modeled the observed tooth shape variation seen in natural populations of seals, thereby relating small changes in development with microevolutionary variation within a species. Not only did the "virtual" teeth resemble real seals' teeth, but the progression of cusps in

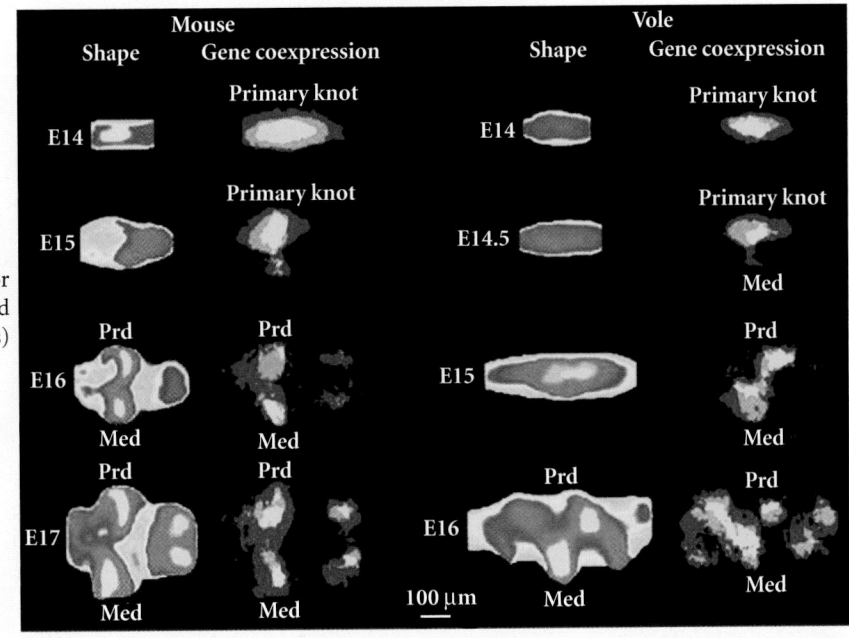

FIGURE 20.20 GIS analysis of gene activity in the formation of the first set of cusps in mouse and vole molars. (The first two cusps are the proconid, labeled Prd, and the metaconid, labeled Med). For both species, GIS mapping of molar shape is shown on the left, and the expression of *Fgf4* and *Shh* (two genes expressed from the enamel knots) is shown at the right. In the mouse, the gene expression pattern on embryonic day 15 predicts the formation of the new cusps seen on day 16; the gene expression pattern on day 16 predicts the formation of cusps in those areas on day 17. Similarly, in the vole molar, whose cusps are diagonal to one another, gene expression predicts cusp formation. (After Jernvall et al. 2000, courtesy of J. Jernvall.)

SIDELIGHTS SPECULATIONS (continued)

(A)

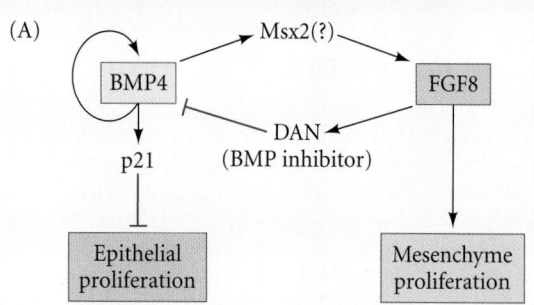

(B)

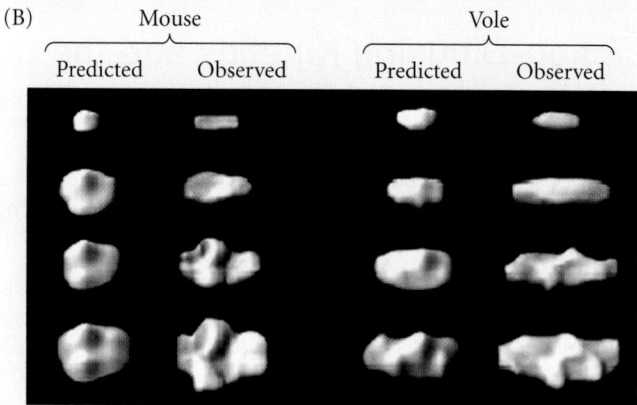

FIGURE 20.21 Basic model for cusp development in mice and voles. (A) Experimentally derived gene network wherein BMPs activate their own production as well as the production of their inhibitors, Shh and FGFs. The FGFs and Shh stimulate cell proliferation; the BMPs inhibit it. (B) Predicted and observed results from this model. The model can generate the final and intermediate forms of molar development in mice and voles, and the difference between mouse and vole molars can be reproduced by slight alterations in the rate of BMP diffusion and binding to inhibitors. (After Salazar-Ciudad and Jernvall 2002; photographs courtesy of J. Jernvall.)

the computer-modeled teeth followed the developmental pattern seen in the actual teeth. Moreover, by altering the parameter of epithelial growth, the equations modeled interspecies variation of jaw dentition. Another conclusion is that all the cells can start off with the same basic set of instructions, and specific instructions will emerge as the cells interact.

The Turing reaction-diffusion model also predicts that some types of teeth are much more likely to evolve in certain ways than in others (FIGURE 20.22). Moreover, the ecological context of tooth use (herbivory versus carnivory, for instance) would select certain variants and not others, and this model can predict both the number and size of molars under different ecological conditions (Kavanagh et al. 2007; Polly 2007). The predictions conform to what paleontologists have observed about mammalian evolution. These studies show the power of mathematical modeling to integrate development, cell biology, and genetics into a predictive model for evolution.

● See **WEBSITE 20.5** How do zebras (and angelfish) get their stripes?

(A)

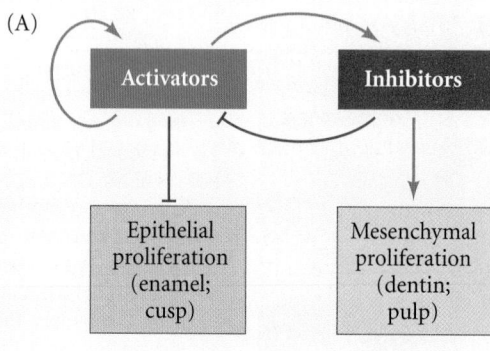

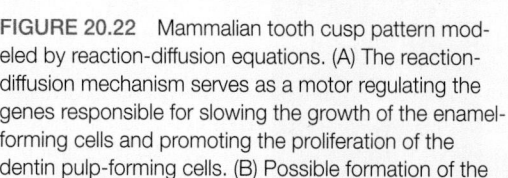

(B)

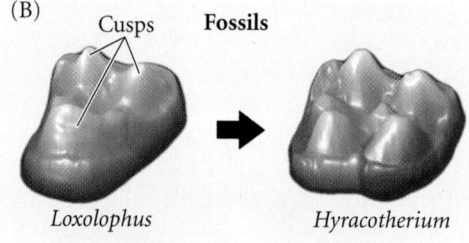

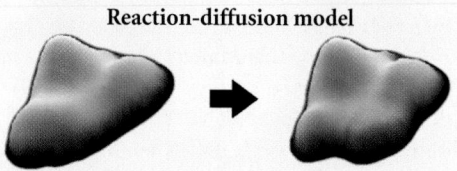

FIGURE 20.22 Mammalian tooth cusp pattern modeled by reaction-diffusion equations. (A) The reaction-diffusion mechanism serves as a motor regulating the genes responsible for slowing the growth of the enamel-forming cells and promoting the proliferation of the dentin pulp-forming cells. (B) Possible formation of the four-cusped tooth of *Hyracotherium* (a fossil horse from 55 million years ago) from the three-cusped tooth of *Loxolophus*, a mammal that may have been an earlier member of the horse lineage. This transition in tooth shape can be achieved by modifying a single parameter of the reaction-diffusion equation. (Courtesy of J. Jernvall.)

(A)

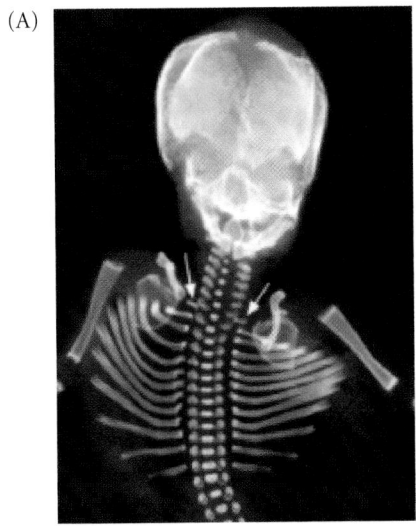

(B)

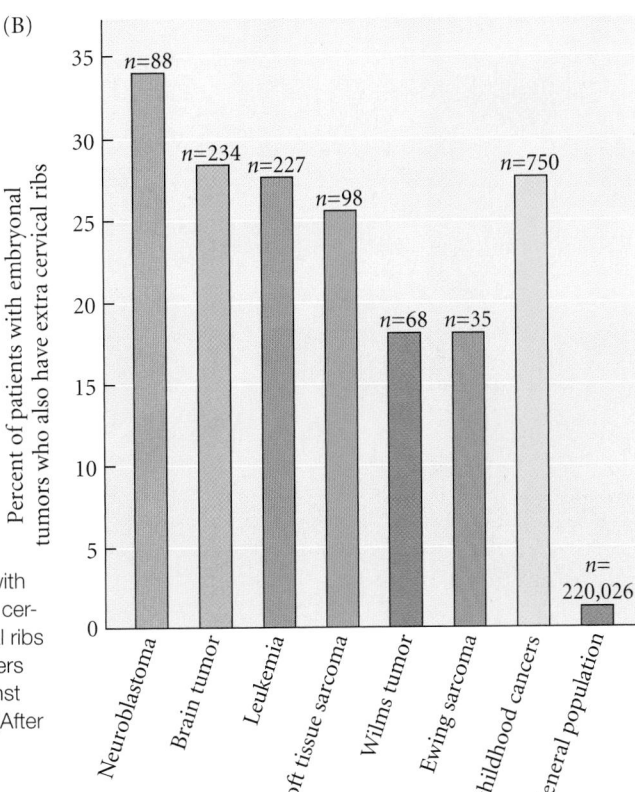

FIGURE 20.23   Extra cervical ribs are associated with childhood cancers. (A) Radiogram showing an extra cervical rib. (B) Nearly 80% of fetuses with extra cervical ribs die before birth. Those surviving often develop cancers very early in life. This indicates strong selection against changes in the number of mammalian cervical ribs. (After Galis et al. 2006, courtesy of F. Galis.)

specifies the neural tube and sclerotome. Similarly, Waddington (1938) noted that, although the pronephric kidney of the chick embryo is considered vestigial (since it has no ability to concentrate urine), it is the source of the ureteric bud that induces the formation of a functional kidney during chick development (see Chapter 12).

As genes acquire new functions during the course of evolution, they may become active in more than one module, making change difficult. Galis and colleagues provide evidence that the reason the segment polarity gene network is conserved in all types of insects is that these genes play roles in several different pathways (Galis et al. 2002). Such pleiotropy constrains the possibilities for alternative mechanisms, since it makes change difficult. **Pleiotropy**, the ability of a gene to play different roles in different cells, is the "opposite" of modularity, involving the connections between parts rather than their independence.*

Pleiotropies may underlie the constraints seen in mammalian development. Galis speculates that mammals have only seven cervical vertebrae (whereas birds may have dozens) because the Hox genes that specify these vertebrae have become linked to stem cell proliferation in mammals (Galis 1999; Galis and Metz 2001; Abramovich et al. 2005; Schiedlmeier et al. 2007). Thus, changes in Hox gene expression that might facilitate evolutionary changes in the skeleton might also *mis*regulate cell proliferation and lead to cancers. Galis supports this speculation with epidemiological evidence showing that changes in skeletal morphology correlate with childhood cancer. The intraembryonic selection against having more or fewer than seven cervical ribs appears to be remarkably strong. At least 78% of human embryos with an extra anterior rib (i.e., six cervical vertebrae) die before birth, and 83% die by the end of the first year. These deaths appear to be caused by multiple congenital anomalies or cancers (FIGURE 20.23; Galis et al. 2006).

● See WEBSITE 20.6 The developmental hourglass

Selectable Epigenetic Variation

Changes in development provide the raw material of variation. But we have seen earlier in the book (especially in Chapter 19) that developmental signals can come from the environment as well as from the nuclei and cytoplasm. Might this environmentally induced variation be inherited and selectable? This idea smacks of Lamarckism, wherein environmentally induced

---

*We have seen pleiotopies, for instance, in the formation of blind cavefish (p. 360). The secretion of more Sonic hedgehog in the oral-pharyngeal area of the head amplifies jaw size and taste bud expression, but it does so at the expense of shutting off Pax6 and eye development. Thus, eye development is linked to taste bud development by the pleiotropic effects of Sonic hedgehog signaling (Franz-Odendaal and Hall 2006; Yamamoto et al. 2009).

(A)

(B)

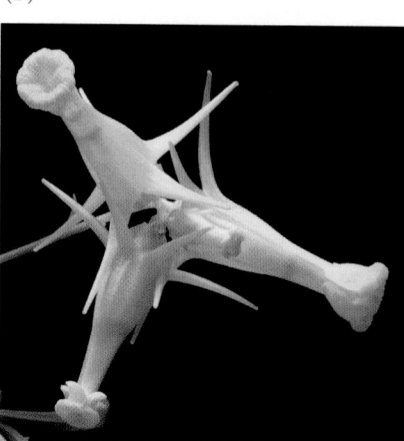

FIGURE 20.24 Epigenetic forms of toadflax. (A) Typical *Linaria*, with a relatively unmethylated *cycloidea* gene. (B) The *cycloidea* gene of the *peloria* variant is relatively heavily methylated. The epialleles that create the different phenotypes of this species are stably inherited. (Courtesy of R. Grant-Downton.)

traits could be inherited through the germ line. We now know that Lamarck was wrong in thinking that phenotypes acquired by use or disuse could be transmitted. Children of weightlifters don't inherit their parents' physiques, and accident victims who have lost limbs can rest assured that their children will be born with normal arms and legs. If the DNA of the germ cells is not altered, environmentallly induced variation will not be transmitted from one generation to the next.

But what if an environmental agent were to cause changes not only in the somatic DNA but also in the germline DNA? Then the effect might be able to be transmitted from one generation to the next. There are two known major "epigenetic inheritance systems"—epialleles and symbionts—that allow environmentally induced changes to be transmitted from generation to generation. A third process, genetic assimilation, shows that some environmentally induced traits, when continually selected, are stabilized genetically so that the trait is inherited without having to be induced in each generation.

EPIALLELES While the *alleles* that are the basis of the genetic inheritance system are variants of the DNA sequence, the **epialleles** of epigenetic inheritance systems are variants of chromatin structure that can be inherited between generations. In most known cases, epialleles are differences in DNA methylation patterning that are able to affect the germ line and thereby be transmitted to offspring. The asymmetrical *peloria* variant of the toadflax plant (*Linaria vulgaris*; **FIGURE 20.24**) was first described by Linnaeus in 1742 as a stably inherited form. In 1999, Coen showed that this variant was due not to a distinctive allele, but rather to a stable epiallele. Instead of carrying a mutation in the *cycloidea* gene, the *peloria* form of this gene was hypermethylated. It does not matter to the developing system whether a gene has been inactivated by a mutation or by altered chromatin configuration (Cubas and Coen 1999). The effect is the same.

There are dozens of examples of epiallelic inheritance (Gilbert and Epel 2009; Jablonka and Raz 2009). In this book, we have discussed some epialleles in their developmental context:

- In the viable *Agouti* phenotype in mice, methylation differences affect coat color and obesity. When a pregnant female is fed a diet high in methyl donors, the specific methylation pattern at the *Agouti* locus is transmitted not only to the progeny developing in utero, but also to the progeny of those mice and to their progeny (Jirtle and Skinner 2007).

- Enzymatic and metabolic phenotypes are established in utero by protein-restricted diets in rats when protein restriction during a grandmother rat's pregnancy leads to a specific methylation pattern in her pups and grandpups (Burdge et al. 2007).

- The endocrine disruptors vinclozolin, methoxychlor, and bisphenol A have the ability to alter DNA methylation patterns in the germ line, thereby causing developmental anomalies and predispositions to diseases in the grandpups of mice exposed to these chemicals in utero (Anway et al. 2005, 2006a,b; Newbold et al. 2006; Crews et al. 2007, 2012).

- Stress-resistant behavior of rats was shown to be due to methylation patterns, induced by maternal care, in the glucocorticoid receptor genes. Meaney (2001) found that rats that received extensive maternal care had less stress-induced anxiety and, if female, developed into mothers that gave their offspring similar levels of maternal care.

SYMBIONT VARIATION   As we explored in Chapter 19, one important aspect of phenotypic plasticity involves interactions with an expected population of symbionts. When symbionts are transmitted through the germ line (as *Wolbachia* bacteria are in many insects), the symbionts provide a second system of inheritance (Gilbert and Epel 2009).

Most symbiotic relationships involve microorganisms that have fast growth rates and can thus change more rapidly under environmental stresses than invertebrates or vertebrates. Rosenberg et al. (2007) describe four mechanisms by which microorganisms may confer greater adaptive potential to the whole organism than can the host genome alone. First, the relative abundance of microorganisms associated with the host can be changed due to environmental pressures. Second, adaptive variation can result from the introduction of a new symbiont to the community. Third, changes to the microbial genome can occur through recombination or random mutation, and these changes can occur in a microbial symbiont more rapidly than in the host. And fourth, there is the possibility of horizontal gene transfer between members of the symbiotic community.

Symbionts can be a source of selectable variation. The pea aphid *Acrythosiphon pisum*, for example, has numerous species of symbionts living within most of it cells. One species of symbiotic bacteria, *Buchnera aphidicola*, can provide the aphid with higher fecundity or greater heat tolerance, depending on which allele of a heat-shock protein the bacteria produces. Another symbiotic bacterium, a species of *Rickettsiella*, contains alleles that can alter the aphid's color. And a third bacterial symbiont, *Hamiltonella defensa* can (if it is the appropriate strain) provide proteins that defend the host aphid against parasitoid wasps (Dunbar et al 2007; Oliver et al. 2009; Tsuchida et al. 2010). These symbionts are usually inherited through the aphid's egg cytoplasm (see Chapter 19). Thus, selectable variation may be acquired through the egg, but using a different set of genes.

Genetic assimilation

In the early 1900s, some evolutionary biologists speculated that the environment could select one of a variety of environmentally induced phenotypes, and that this phenotype would then become the dominant one for the species. In other words, the environment could both induce and select a phenotype. But these scientists had no theory of development or genetics to provide mechanisms for their hypotheses. This idea was revisited in the middle of the twentieth century, and at that time several models were proposed to explain how constant selection could "fix" a particular environmentally induced phenotype in a population.

One of the most important hypotheses of such plasticity-driven adaptation schemes is the concept of **genetic assimilation**, defined as the process by which a phenotypic character initially produced only in response to some environmental influence becomes, through a process of selection, taken over by the genotype so that it is formed even in the absence of the environmental influence that had first been necessary

(King and Stanfield 1985). The idea of genetic assimilation was introduced independently by Waddington (1942, 1953, 1961) and Schmalhausen (1949) to explain the remarkable outcomes of artificial selection experiments in which an environmentally induced phenotype became expressed even in the *absence* of the external stimulus that was initially necessary to induce it.

GENETIC ASSIMILATION IN THE LABORATORY   Genetic assimilation is readily demonstrated in the laboratory, as Waddington's many experiments proved. For example, when pupae from a laboratory population of wild-type *Drosophila melanogaster* were exposed to a heat shock of 40°C, some of the emerging adults exhibited in their wings a gap in the posterior crossveins. This gap is not normally present in untreated flies (Waddington 1952, 1953). Two selection regimens were followed, one in which only the aberrant flies (termed *crossveinless*) were bred to one another, and another in which only non-aberrant flies were mated. After some generations of selection, when only the individuals showing this gap were allowed to breed, the proportion of adults with broken crossveins induced by heat shock at the pupal stage rose to above 90%. Moreover (and significantly), by generation 14 of such inbreeding, a small proportion of individuals were *crossveinless* even among flies of this line that had not been exposed to temperature shock. When Waddington extended this artificial selection by breeding the adults that had developed the abnormality without heat shock, the frequency of *crossveinless* individuals among untreated flies became very high, reaching 100% in some lines. The phenotypically induced trait had become "genetically assimilated" into the population.*

In addition to finding the *crossveinless* phenotype on exposure to heat shock, Waddington showed that his laboratory strains of *Drosophila* had a particular reaction norm in their response to ether. Embryos exposed to ether at a particular stage developed a phenotype similar to the *bithorax* mutation and had four wings instead of two. The flies' halteres—balancing structures on the third thoracic segment—were transformed into wings. Generation after generation was exposed to ether, and individuals showing the four-winged state were selectively bred each time. After 20 generations, the mated *Drosophila* produced the mutant phenotype even when no ether was applied (**FIGURE 20.25**; Waddington 1953, 1956).

Subsequent experiments have borne out Waddington's findings (see Bateman 1959a,b; Matsuda 1982; Ho et al. 1983). In 1996, Gibson and Hogness repeated Waddington's *bithorax* experiments and got similar results (Gibson 1996). Moreover, they found that four alleles of the *Ultrabithorax* (*Ubx*) gene had existed in the population and were critical for the genetic assimilation of the ether-induced *bithorax* phenotype. "Waddington's experiment showed some fruit flies were more

---

* Note that in these artificial selection experiments, the original phenotype induced by the environment was not intrinsically adaptive. Only the hand of the experimenter choosing which flies mated made it so.

(A)

(B)

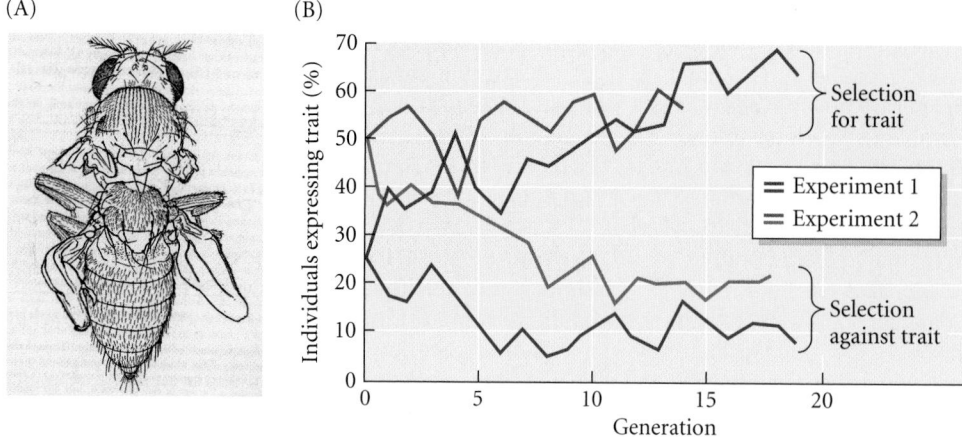

FIGURE 20.25 Phenocopy of the *bithorax* mutation. (A) A *bithorax* (four-winged) phenotype produced after treatment of the embryo with ether. The forewings have been removed to show the aberrant metathorax. This particular individual is actually from the "assimilated" stock that produced this phenotype without being exposed to ether. (B) Selection experiments for or against the *bithorax*-like response to ether treatment. Two experiments are shown (red and blue lines). In both cases, one group was selected for the trait and the other group was selected against the trait. (After Waddington 1956.)

sensitive to ether-induced phenocopies than others, but he had no idea why," Gibson said. "In our experiment, we show that differences in the *Ubx* gene are the cause of these morphological changes."

Genetic assimilation also has also been demonstrated in Lepidoptera. Brakefield and colleagues (1996) were able to genetically assimilate the different morphs of the adaptive polyphenism in *Bicyclus* butterflies (see Figure 19.9), and Suzuki and Nijhout (2006) have shown genetic assimilation in the larvae of the tobacco hornworm moth *Manduca sexta* (**FIGURE 20.26**). By judicious selection protocols, Suzuki and Nijhout were able to breed lines in which the environmentally induced phenotype (larval color) was selected for and was eventually produced without the environmental agent (temperature shock). The underlying genetic differences concerned the rise of juvenile hormone titres in the larvae. Therefore, at least in the laboratory, genetic assimilation can be shown to work.

GENETIC ASSIMILATION IN NATURAL ENVIRONMENTS
Although it is difficult to document genetic assimilation in nature, there are at least three instances where it appears that phenotypic variation due to developmental plasticity was later fixed by genes. The first involves pigment variations in butterflies (Hiyama et al. 2012). As early as the 1890s, scientists used heat shock to disrupt the pattern of butterfly wing pigmentation. In some instances, the color patterns that develop after temperature shock mimic the normal genetically controlled patterns of races (or related species) living at different temperatures. A race (subspecies) whose phenotype is characteristic of the species in a particular geographic area is called an **ecotype** (Turesson 1922). Standfuss (1896) demonstrated that a heat-shocked phenocopy of the Swiss subspecies of *Iphiclides podalirius*

resembled the normal form of the Sicilian subspecies of that butterfly, and Richard Goldschmidt (1938) observed that heat-shocked specimens of the central European subspecies of *Aglais urticae* produced wing patterns that resembled those of the Sardinian subspecies (**FIGURE 20.27**). Conversely, cold-shocked individuals of the central European ecotype of *Aglais* developed the wing patterns of the subspecies from northern Scandinavia.

Further observations on the mourning cloak butterfly (*Nymphalis antiopa*; Shapiro 1976), the buckeye butterfly (*Precis coenia*; Nijhout 1984), and the lycaenid butterfly (*Zizeeria maha*; Otaki et al. 2010) have confirmed the view that temperature variation can induce phenotypes that mimic genetically controlled patterns of related races or species existing in colder or warmer conditions. Chilling the pupa of *Pieris occidentalis* will cause it to have the short-day phenotype (Shapiro 1982), which is similar to that of the northern subspecies of pierids. Even "instinctive" behavioral phenotypes associated with these color changes (such as mating and flying) are phenocopied (see Burnet et al. 1973; Chow and Chan 1999). Thus, an *environmentally* induced phenotype might become the standard *genetically* induced phenotype in one part of the range of that organism.

Yet another case of genetic assimilation concerns the tiger snake (*Notechis scutatus*), which, like many fish, has a head structure that can be altered by diet. The tiger snake can develop a bigger head to ingest bigger prey. This plasticity is seen when its diet includes both large and small mice. However, on some islands the diet contains only large mice, and here the snakes are born with large heads, and there is no plasticity. Thus, Aubret and Shine (2009) claim to show "clear empirical evidence of genetic assimilation, with the elaboration of an adaptive trait shifting from phenotypically plastic expression through to canalization within a few thousand years."

(A)

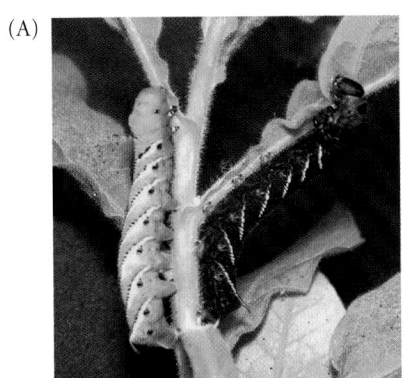

FIGURE 20.26 Effect of selection on temperature-mediated larval color change in the black mutant of the moth *Manduca sexta*. (A) The two color morphs of *Manduca sexta*. (B) Changes in the coloration of heat-shocked larvae in response to selection. One group was selected for increased greenness upon heat treatment (polyphenic; green line), the most green larvae being bred for the next generation. Another for decreased color change (i.e., remaining black) upon heat treatment (monophenic; red line). The remainder of the larvae was not selected (blue line). The color score (0 for completely black, 4 for completely green) indicates the relative amount of colored regions in the larvae. The monophonic line lost its plasticity after the seventh generation. (C) Reaction norm for generation-13 flies reared at constant temperatures between 20°C and 33°C, and heat shocked at 42°C. Note the steep polyphenism at about 28°C. (After Suzuki and Nijhout 2006; photograph courtesy of Fred Nijhout.)

(B)

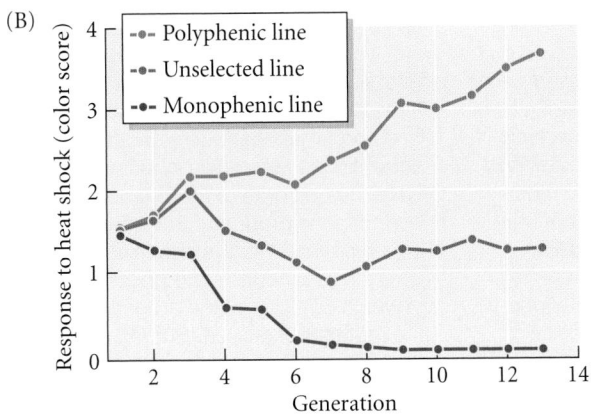

(C)

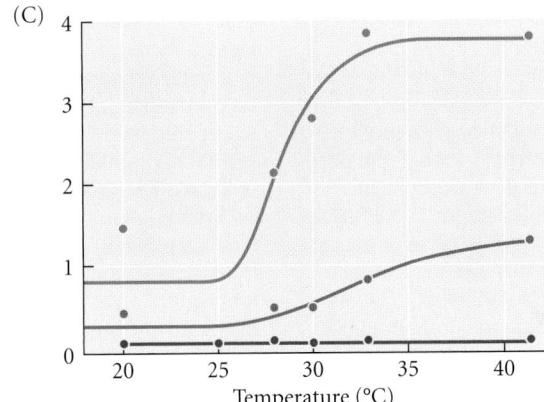

Fixation of environmentally induced phenotypes

There are at least two important evolutionary advantages to the fixation of environmentally induced phenotypes (West-Eberhard 1989, 2003):

1. *The phenotype is not random.* The environment elicited the novel phenotype, and the phenotype has already been tested by natural selection. This would eliminate a long period of testing phenotypes derived by random mutations. As Garson and colleagues (2003) note, although mutation is random, developmental parameters may account for some of the directionality in morphological evolution.

2. *The phenotype already exists in a large portion of the population.* One of the problems of explaining new phenotypes is that the bearers of such phenotypes are "monsters" compared with the wild-type. How would such mutations, perhaps present only in one individual or one family, become established and eventually take over a population? The developmental model solves this problem: this phenotype has been around for a long while, and the capacity to express it is widespread in the population;

(A)

(B)

(C)

FIGURE 20.27 Temperature shocking *Aglais urticae* produces phenocopies of geographic variants. (A) Usual central European variant. (B) Heat-shock phenocopy resembling Sardinian form. (C) Sardinian form of the species. (From Goldschmidt 1938.)

Hsp90 and Genetic Assimilation

How can the results of genetic assimilation be explained? Waddington proposed that the nuclei were buffered such that the same phenotype arose despite minor genetic or environmental perturbations. Schmalhausen suggested that the environmental perturbation unmasked genetic heterogeneity for modifier genes that already existed in the population.

There does indeed appear to be a lot of "buffering" in the cell, so potentially deleterious mutations do not always express deleterious phenotypes. Somehow, the mutant proteins fold correctly (or at least correctly enough to do their jobs). In 1998, Suzanne Rutherford and Susan Lindquist showed that a major agent responsible for this buffering was the "heat shock protein" Hsp90. Hsp90 binds to a set of signal transduction molecules that are inherently unstable.

When it binds to them, it stabilizes their tertiary structure so they can respond to upstream signaling molecules. Heat shock, however, causes other proteins in the cell to become unstable, and Hsp90 is diverted from its normal function (i.e., stabilizing signal transduction proteins) to the more general function of stabilizing any of the now partially denatured peptides in the cell (Jakob et al. 1995; Nathan et al. 1997). Since Hsp90 was known to be involved with inherently unstable proteins and could be diverted by stress, the researchers suspected that it might be involved in buffering developmental pathways against environmental contingencies.

Evidence for the role of Hsp90 as a developmental buffer first came from mutations of *Hsp83*, the gene for Hsp90. Homozygous mutations of *Hsp83* are lethal in *Drosophila*. Heterozygous mutations increase the proportion of developmental abnormalities; in *Drosophila* populations heterozygous for mutant *Hsp83*, deformed eyes, bristle duplications, and abnormalities of legs and wings appeared (**FIGURE 20.28**). When different mutant alleles of *Hsp83* were brought together in the same flies, both the incidence and severity of the abnormalities increased. Abnormalities were also seen when a specific inhibitor of Hsp90 (geldanamycin) was added to the food of wild-type flies, and the types of defects seen differed among different stocks of flies. The abnormalities observed did not show simple Mendelian inheritance, but were the outcome of the interactions of several gene products. Selective breeding of the flies with the abnormalities led over a few generations to populations in which 80–90% of the progeny had the mutant phenotype. But not all of the mutant progeny carried the *Hsp83*

it merely needs to be genetically stabilized by modifier genes that already exist in the population.

Given these two strong advantages, the genetic assimilation of morphs originally produced through developmental plasticity may contribute significantly to the origin of new species. Ecologist Mary Jane West-Eberhard has noted that "contrary to popular belief, environmentally initiated novelties may have greater evolutionary potential than mutationally induced ones. Therefore, the genetics of speciation can profit from studies in changes of gene expression as well as changes in gene frequency and genetic isolation." Evolutionary developmental biology is a young science, and the relative importance of environmentally induced novelties is just beginning to be explored.

Coda

As mentioned at the start of this chapter, experimental embryology separated itself from evolutionary biology to mature on its own. However, Roux promised that once it did mature, embryology would return to evolutionary biology with powerful mechanisms to help explain how evolution takes place. Evolution is a theory of change, and population genetics can identify and quantify the dynamics of such change (Amundson 2005). However, Roux realized that evolutionary biology also needed a theory of body construction that would show how any specific mutation becomes manifest as a selectable phenotype, and this is exactly what modern developmental biology has done and is doing.

1. Developmental biology has established how the underpinnings of variation—modularity, molecular parsimony, and gene duplication—enable extensive changes in development to occur without destroying the organism.

2. Developmental biology has explained how four modes of genetic change—heterotopy, heterochrony, heterometry, and heterotypy—can act during development to produce new and large variations in morphology.

3. Finally, developmental biology has shown that epigenetic inheritance—epialleles, symbionts, and genetic assimilation—provide selectable variations and aid their propagation through a population.

SIDELIGHTS SPECULATIONS   (continued)

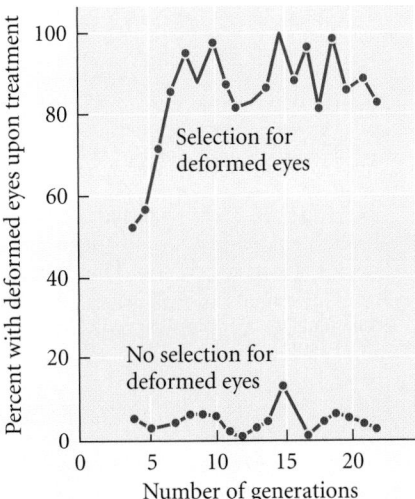

FIGURE 20.28   Hsp90 buffers development. The deformed eye trait was selected by breeding only those individuals expressing the trait. This abnormality was not observed in the original stock, but it can be seen in a high proportion of the descendants of individuals that were mated to heterozygous *Hsp83* flies. The strong response to selection showed that even though the population was small, it contained a large amount of hidden genetic variation. (After Rutherford and Lindquist 1998.)

Thus, Hsp90 is probably a major component of a buffering system that enables normalization of potentially mutant phenotypes. It provides one way to resist phenotype fluctuations that would otherwise result from slight mutations or slight environmental changes. Hsp90 might also be responsible for allowing mutations to accumulate, but keeping them from being expressed until the environment changes. No individual mutation would change the phenotype, but mating would allow these mutations to be "collected" by members of the population.

An environmental change (anything that might stress the cells) might thereby release the hidden phenotypic possibilities of the population. In other words, transient decreases in Hsp90 (resulting from its aiding stress-damaged proteins) would uncover preexisting genetic variations that would produce morphological variations. Most of these morphological variations would probably be deleterious, but some might be selected for in the new environment. Such release of hidden morphological variation may be responsible for the many examples of rapid speciation found in the fossil record.

mutation. In other words, once a mutation in *Hsp83* had allowed cryptic mutations to be expressed, selective matings could retain the abnormal phenotype even in the absence of abnormal Hsp90.

In 1922, Walter Garstang declared that ontogeny (an individual's development) does not recapitulate phylogeny (evolutionary history). Rather, ontogeny *creates* phylogeny, and evolution is generated by heritable changes in development. "The first bird," said Garstang, "was hatched from a reptile's egg." The developmental genetic model has been formulated to account for both the homologies seen in evolution and the differences. We are still approaching evolution in the two ways that Darwin recognized, and descent with modification remains central. We are now at the point, however, where we can answer evolutionary questions using both population genetics and developmental genetics. Thus, when confronted with the question of how the arthropod body plan arose, Hughes and Kaufman (2002) begin their study by saying:

To answer this question by invoking natural selection is correct—but insufficient. The fangs of a centipede ... and the claws of a lobster accord these organisms a fitness advantage. However, the crux of the mystery is this: From what developmental genetic changes did these novelties arise in the first place?

By integrating population genetics with developmental genetics and embryology, we can now begin to explain the construction and evolution of biodiversity.

In 1953, J. B. S. Haldane expressed his thoughts about evolution with the following developmental analogy:

The current instar of the evolutionary theory may be defined by such books as those of Huxley, Simpson, Dobzhansky, Mayr, and Stebbins [the founders of the Modern Synthesis of evolution with genetics]. We are certainly not ready for a new moult, but signs of new organs are perhaps visible.

This recognition of developmental ideas "points forward to a broader synthesis in the future." We have finally broken through the old pupal integument, and a new, broader, developmentally inclusive evolutionary synthesis is taking wing.

● **See WEBSITE 20.7** "Intelligent design" and evolutionary developmental biology

SNAPSHOT SUMMARY: Developmental Mechanisms of Evolutionary Change

1. Evolution is the result of inherited changes in development. Modifications of embryonic or larval development can create new phenotypes that can then be selected.

2. Darwin's concept of "descent with modification" explained both homologies and adaptations. The similarities of structure are due to common ancestry (homology), while the modifications are due to natural selection (adaptation to the environmental circumstances).

3. Homology means that similarity between organisms or genes can be traced to descent from a common ancestor. In some instances, certain genes specify the same traits throughout the animal phyla.

4. Evolution can occur through the "tinkering" of existing genes. The ways of effecting evolutionary change through development at the level of gene expression are: change in location (heterotopy), change in timing (heterochrony), change in amount (heterometry), and change in kind (heterotypy).

5. Changes in gene sequence can give Hox genes new properties that may have significant developmental effects. The constraint on insect anatomy of having only six legs is one example; the evolution of the uterus is another.

6. Changes in the location of gene expression during development appear to account for the evolution of the turtle shell, the loss of limbs in snakes, the emergence of feathers, and the evolution of differently shaped molars.

7. Changes in the timing of gene expression have been important in the formation of limbs throughout the animal kingdom.

8. Changes in the amount and timing of gene expression can account for the development of beak phenotypes in Darwin's finches and the size of the human brain.

9. Changes in Hox gene number may allow Hox genes to take on new functions. Large changes in the numbers of Hox genes correlate with major transitions in evolution.

10. Duplications may enable genes to become expressed in new places. The formation of new cell types may result from duplicated genes whose regulation has diverged. The Hox genes and many other gene families started as single genes that were duplicated.

11. In addition to structures and genes being homologous, signal transduction pathways can be homologous. In these cases, homologous proteins are organized in homologous ways. These pathways can be used for different developmental processes both in different organisms and within the same organism.

12. The modularity of development allows parts of the embryo to change without affecting other parts. This modularity of development is due in large part to the modularity of enhancers.

13. Co-option (recruitment) of existing genes and pathways for new functions is a fundamental mechanism for creating new phenotypes. Such instances include the use of the limb development signaling pathway to form eye spot pigmentation in fly wings, the formation of beetle elytra, and the production of the larval skeleton of sea urchins.

14. Developmental constraints prevent certain phenotypes from arising. Such constraints may be physical (no rotating limbs), morphogenetic (no middle digit smaller than its neighbors), or phyletic (no neural tube without a notochord).

15. New gene transcription can be caused by modifying existing DNA elements to become enhancers, mutating the DNA sequences bound by transcription factors to eliminate an enhancer, or by having a transposable element add an enhancer sequence or mutate an existing one.

16. The merging of the population genetic model with the developmental genetic model of evolution is creating a new evolutionary synthesis that can account for macroevolutionary as well as microevolutionary phenomena.

17. Epigenetic inheritance systems include epialleles, wherein inherited patterns of DNA methylation can regulate gene expression. A heavily methylated gene can be as nonfunctional as a genetically mutant allele.

18. Symbiotic organisms are often needed for development to occur, and variants of these organisms may cause different modes of development.

19. Genetic assimilation, wherein a phenotypic character initially induced by the environment becomes, though a process of selection, produced by the genotype in all permissive environments, has been well documented in the laboratory.

20. Evolutionary developmental biology is able to show how small genetic or epigenetic changes can generate large phenotypic changes and enable the production of new anatomical structures.

For Further Reading

Complete bibliographical citations for all literature cited in this chapter can be found at the free access website **www.devbio.com**

Abzhanov, A., W. P. Kuo, C. Hartmann, P. R. Grant, R. Grant and C. Tabin. 2006. The calmodulin pathway and evolution of beak morphology in Darwin's finches. *Nature* 442: 563–567.

Amundson, R. 2005. *The Changing Role of the Embryo in Evolutionary Thought: Roots of Evo-Devo.* Cambridge University Press, New York.

Carroll, S. B. 2006. *Endless Forms Most Beautiful: The New Science of Evo-Devo.* Norton, New York.

Cohn, M. J. and C. Tickle. 1999. Developmental basis of limblessness and axial patterning in snakes. *Nature* 399: 474–479.

Davidson, E. H. and D. H. Erwin. 2009. An integrated view of Precambrian eumetazoan evolution. *Cold Spring Harb. Symp. Quant. Biol.* 74: 65–80.

Gilbert, S. F. and D. Epel. 2009. *Ecological Developmental Biology: Integrating Epigenetics, Medicine, and Evolution.* Sinauer Associates, Sunderland, MA.

Kuratani, S. 2009. Modularity, comparative embryology and evo-devo: Developmental dissection of body plans. *Dev. Biol.* 332: 61–69.

Lynch, V. J., A. Tanzer, Y. Wang, F. C. Leung, B. Gelelrsen, D. Emera and G. P. Wagner. 2008. Adaptive changes in the transcription factor HoxA-11 are essential for the evolution of pregnancy in mammals. *Proc. Natl. Acad. Sci. USA.* 105: 14928–14933.

Merino, R., J. Rodríguez-Leon, D. Macias, Y. Ganan, A. N. Economides and J. M. Hurle. 1999. The BMP antagonist Gremlin regulates outgrowth, chondrogenesis and programmed cell death in the developing limb. *Development* 126: 5515–5522.

Rockman, M. V., M. W. Hahn, N. Soranzo, D. B. Goldstein and G. A. Wray. 2003. Positive selection on a human-specific transcription factor binding site regulating IL4 expression. *Curr. Biol.* 13: 2118–2123.

Shapiro, M. D., and 7 others. 2004. Genetic and developmental basis of evolutionary pelvic reduction in three-spine sticklebacks. *Nature* 428: 717–723.

Shubin, N., C. Tabin and S. B. Carroll. 2009. Deep homology and the origins of evolutionary novelty. *Nature* 457: 818–823.

Smith, K. 2003. Time's arrow: Heterochrony and the evolution of development. *Int. J. Dev. Biol.* 47: 613–621.

Go Online

WEBSITE 20.1 Relating evolution to development in the nineteenth century. Immediately after publication of Darwin's *On the Origin of Species*, biologists attempted to relate evolution to changes in development. The attempts of three such scientists—Frank Lillie, Edmund B. Wilson, and Ernst Haeckel—are highlighted here.

WEBSITE 20.2 Correlated progression. In many cases, modules must coevolve. The upper and lower jaws, for instance, have to fit together properly. If one changes, so must the other. If the sperm-binding proteins on the egg change, then so must the egg-binding proteins on the sperm. This site looks at correlated changes during evolution.

WEBSITE 20.3 The seach for the Urbilaterian ancestor. Homologous genes specifying the formation of the eye, heart, body axis, and nervous system enable biologists to intuit an "Urbilaterian ancestor"—an organism that may have been the precursor of both protostomes and deuterostomes. This organism may have resembled the planula larvae of contemporary cnidarians.

WEBSITE 20.4 How the chordates got a head. The neural crest is responsible for forming the heads of chordates. But how did the neural crest come into existence? It is probable that ancestral deuterostomes had all the requisite genes, but only in the chordates did these genes become linked together into the network that became the neural crest cell.

WEBSITE 20.5 How do zebras (and angelfish) get their stripes? The reaction-diffusion mechanism appears to play critically important roles in generating stripes and spots on the skin of animals. "How the zebra got its stripes" may be predicated on such mechanisms, and different species of zebras might form their stripes by modifying diffusion.

WEBSITE 20.6 The developmental hourglass. Observations starting in the 1800s supported the notion of a "phylotypic stage" that defined an evolutionary lineage. Contemporary genomics supports that view, showing that the oldest genes are expressed at a stage with the most constraints.

WEBSITE 20.7 "Intelligent design" and evolutionary developmental biology. Evolutionary developmental biology explains many of the "problems" (such as the evolution of the vertebrate eye and the evolution of turtle shells) that proponents of "intelligent design" and other creationists claimed were impossible to explain by evolution.

Glossary

A

Achondroplasic dwarfism Condition wherein chondrocytes stop proliferating earlier than usual, resulting in short limbs. Often caused by mutations that activate *FgfR3* prematurely.

Acrosome (acrosomal vesicle) Caplike organelle that, together with the sperm nucleus, forms the sperm head. Contains proteolytic enzymes that can digest the proteinaceous membranes of egg, allowing the sperm nucleus to enter the egg and fuse with the egg nucleus.

Acrosome reaction The Ca^{2+}-dependent fusion of the acrosome with the sperm cell membrane, resulting in exocytosis and release of proteolytic enzymes that allow the sperm to penetrate the egg extracellular matrix and fertilize the egg.

Activins Members of the TGF-β superfamily of proteins; with Nodal, important in specifying the different regions of the mesoderm and for distinguishing the left and right body axes of vertebrates.

Adhesion Attachment between cells or between a cell and its extracellular substrate. The latter provides a surface for migrating cells to travel along.

Adult stem cells Stem cells found in the tissues of organs after the organ has matured. Adult stem cells are usually involved in replacing and repairing tissues of that particular organ, and can form only a subset of cell types. Compare with **Embryonic stem cells**.

Afferent Carrying to, as in neurons that carry information to the central nervous system (spinal cord and brain).from sensory receptor cells (e.g., sound waves from the ear, light signals from the retina, touch sensations from the skin); or vessels that carry fluid (e.g., blood) to a structure.

Aging The time-related deterioration of the physiological functions necessary for survival and fertility.

Allantois In amniote species, extraembryonic membrane that stores urinary wastes and helps mediate gas exchange. It is derived from splanchnopleure at the caudal end of the primitive streak. In mammals, the size of the allantois depends on how well nitrogenous wastes can be removed by the chorionic placenta. In reptiles and birds, the allantois becomes a large sac, as there is no other way to keep the toxic by-products of metabolism away from the developing embryo.

Allometry Developmental changes that occur when different parts of an organism grow at different rates.

Alternative splicing A means of producing multiple different proteins encoded by a single gene by splicing together different sets of exons to generate different types of mRNAs.

Ametabolous A pattern of insect development in which there is no larval stage and the insect undergoes direct development to a small adult form following a transitory *pronymph* stage.

Amnion "Water sac." A membrane enclosing and protecting the embryo and its surrounding amniotic fluid. Defining the "amniote vertebrates," this epithelium is derived from somatopleure. Ectodermal tissue supplies epithelial cells, and the mesoderm generates the essential blood supply.

Amniote egg Egg surrounded by extraembryonic membranes (the amnion, chorion, allantois, and yolk sac) that provide nourishment and other environmental needs to the developing embryo. Characteristic of the amniote vertebrates: the reptiles and birds, in which the egg typically develops in a shell outside the mother's body; and the mammals, where the egg has become modified to develop inside the mother.

Amniotic fluid A secretion that serves as a shock absorber for the developing embryo while preventing it from drying out.

Ampulla Latin, "flask." The segment of the mammalian oviduct, distal to the uterus and near the ovary, where fertilization takes place.

Anagen The growth phase of the hair follicle cell replacement cycle.

Analogous Structures and/or their respective components whose similarity arises from their performing a similar function rather than their arising from a common ancestor (e.g., the wing of a butterfly vs. the wing of a bird). Compare with **homologous**.

Anchor cell The cell connecting the overlying gonad to the vulval precursor cells in *C. elegans*. If the anchor cell is destroyed, the VPCs will not form a vulva, but instead become part of the hypodermis (skin).

Androgen insensitivity syndrome Intersex condition in which an XY individual has a mutation in the gene encoding the androgen receptor protein that binds testosterone. This results in a female external phenotype, lack of a uterus and oviducts and presence of abdominal testes.

Anencephaly A lethal congenital defect resulting from failure to close the anterior neuropore. The forebrain remains in contact with the amniotic fluid and subsequently degenerates, so the vault of the skull fails to form.

Aneuploidy Condition in which one or more chromosome(s) is either lacking or present in multiple copies.

Angioblasts From *angio*, blood vessel; and *blast*, a rapidly dividing cell (usually a stem cell). The progenitor cells of blood vessels.

Angiogenesis Process by which the primary network of blood vessels created by vasculogenesis is remodeled and pruned into a distinct capillary bed, arteries, and veins.

Angiopoietins Paracrine factors that mediate the interaction between endothelial cells and pericytes.

Animal cap In amphibians, the roof of the blastocoel (in the animal hemisphere).

Animal hemisphere The non-yolk-containing (upper) half of the amphibian egg. During embryogenesis, cells in the animal hemisphere divide rapidly and become actively mobile ("animated").

Anoikis Rapid apoptosis that occurs when epithelial cells lose their attachment to the extracellular matrix.

Antennapedia complex A region of *Drosophila* chromosome 3 containing the homeotic genes *labial* (*lab*), *Antennapedia* (*Antp*), *sex combs reduced* (*scr*), *deformed* (*dfd*), and *proboscipedia* (*pb*), which specify head and thoracic segment identities.

Anterior heart field Cells of the heart field forming the outflow tract (conus and truncus arteriosus, right ventricle).

Anterior intestinal portal (AIP) The posterior opening of the developing foregut region of the primitive gut tube; it opens into the future midgut region which is contiguous with the yolk sac at this stage.

Anterior visceral endoderm (AVE) Mammalian equivalent to the chick hypoblast and similar to the head portion of the amphibian organizer, it creates an anterior region by secreting antagonists of Nodal.

Anterior-posterior (anteroposterior) axis The body axis defining the head versus the tail (or mouth versus anus). When referring to the limb, this refers to the thumb (anterior)-pinkie (posterior) axis.

Anti-Müllerian factor (AMF) TGF-β family paracrine factor secreted by the embryonic testes that induces apoptosis of the epithelium and destruction of the basal lamina of the Müllerian duct, preventing formation of the uterus and oviducts. Formerly known as anti-Müllerian hormone, or AMH. Sometimes called Müllerian-inhibiting factor (MIF).

Anurans Frogs and toads. Compare with **Urodeles**.

Aorta-gonad-mesonephros region (AGM) A mesnchymal area in the lateral plate splanchnopleure near the ventral aorta that produces the hematopoietic stem cells.

Aortic arches These begin as symmetrically arranged, paired vessels that develop within the paired pharyngeal arches and link the ascending and descending/dorsal paired aortae. Some of the aortic arches degenerate.

Apical ectodermal ridge (AER) A ridge along the distal margin of the limb bud that will become a major signaling center for the developing limb. Its roles include (1) maintaining the mesenchyme beneath it in a plastic, proliferating state that enables the linear (proximal-distal) growth of the limb; (2) maintaining the expression of those molecules that generate the anterior-posterior axis; and (3) interacting with the proteins specifying the anterior-posterior and dorsal-ventral axes so that each cell is given instructions on how to differentiate.

Apical epidermal cap (AEC) Forms in the wound epidermis of an amputated salamander limb and acts similarly to the apical ectodermal ridge during normal limb development.

Apoptosis Programmed cell death. Apoptosis is an active process that prunes unneeded structures (e.g., frog tails, male mammary tissue), controls the number of cells in particular tissues, and sculpts complex organs (e.g., palate, retina, digits, and heart). Not to be confused with *necrosis*, pathological cell death, caused by external factors such as inflammation or toxic injury. See also **Anoikis**.

Aqueous humor Nourishing fluid that bathes the lens of the vertebrate eye and supplies pressure needed to stabilize the curvature of the developing eye.

Archenteron The primitive gut of the sea urchin blastula, formed by invagination of the vegetal plate into the blastocoel.

Area opaca The peripheral ring of avian blastoderm cells that have not shed their deep cells.

Area pellucida A 1-cell-thick area in the center of the avian blastoderm (following shedding of most of the deep cells) that forms most of the actual embryo.

Aromatase Enzyme that converts testosterone to estradiol (a form of estrogen). Excess aromatase in the environment is linked to herbicides and other chemicals and is believed to contribute to reproductive disorders (demasculization and feminization, particularly in male amphibians).

Arthrotome Mesenchymal cells in the center of the somite that contribute to the sclerotome, becoming the vertebral joints, the intervertebral discs, and those portions of the ribs closest to the vertebrae.

Autocrine interaction The same cells that secrete paracrine factors also respond to them.

Autonomous specification A mode of cell commitment in which the blastomere inherits a determinant, usually a set of transcription factors from the egg cytoplasm, and these transcription factors regulate gene expression to direct the cell into a particular path of development.

Autopod The distal limb bones of any vertebrate limb, consisting of carpals and metacarpals (forelimb), tarsals and metatarsals (hindlimb) and the phalanges (fingers and toes).

Axon Continuous extension of the nerve cell body. Transmits signals (action potentials) to targets in the central and peripheral nervous systems. Axonal migration is crucial to development of the vertebrate nervous system.

B

Basal disc The "foot" of a hydra; enables the animal to stick to rocks or the undersides of pond plants.

Basal lamina Specialized, closely knit sheets of extracellular matrix that underlie epithelia, composed largely of laminin and type 4 collagen. Epithelial cells adhere to the basal lamina in part via binding between integrins and laminin. Sometimes called the basement membrane.

Basal layer (stratum germinativum) The inner layer of both the embryonic and adult epidermis. This layer contains epidermal stem cells attached to a basement membrane.

Basic fibroblast growth factor (Fgf2) One of three growth factors required for the generation of hemangioblasts from the splanchnic mesoderm. See also **Angiopoietins**; **Vascular endothelial growth factors (VEGFs)**.

β-Catenin A protein that can act as an anchor for cadherins or as a transcription factor (induced by the Wnt pathway). It is important in the specification of germ layers throughout the animal phyla.

Bergmann glia Type of glial cell; extends a thin process throughout the developing neuroepithelium of the cerebellum.

bHLH proteins The basic helix-loop-helix family of transcription factors, including such proteins as scleraxis, the MRFs (MyoD, Myf5, and myogenin), and c-Myc.

Bicoid Anterior morphogen critical for establishing anterior-posterior polarity in the *Drosophila* embryo. Functions as a transcription factor to activate anterior-specific gap genes and as a translational repressor to suppress posterior-specific gap genes.

Bilaminar germ disc An amniote embryo prior to gastrulation; consists of epiblast and hypoblast layers.

Bilateral holoblastic cleavage Cleavage pattern, found primarily in tunicates, in which the first cleavage plane establishes the right-left axis of symmetry in the embryo and each successive division orients itself to this plane of symmetry. Thus the half-embryo formed on one side of the first cleavage plane is the mirror image of the other side.

Bilaterians (triploblasts) Those animals characterized by bilaterian body symmetry and the presence of three germ layers (endoderm, ectoderm, and mesoderm). Includes all animal groups except the sponges, cnidarians, ctenophores, and placozoans.

Biofilm Mats of bacteria that regulate larval settlement of many marine invertebrate species.

Bipotential (indifferent) gonad Common precursor tissue derived from the genital ridge, from which the male and female gonads diverge.

Bisphenol A (BPA) Synthetic estrogenic chemical compound used in plastics and flame retardants. BPA has been associated with meiotic defects, reproductive abnormalities, and precancerous conditions in rodents.

Bithorax complex The second region of *Drosophila* chromosome 3 containing the homeotic gene *Ultrabithorax* (*Ubx*), which is required for the identity of the third thoracic segment; and the *abdominal A* (*abdA*) and *Abdominal B* (*AbdB*) genes, which are responsible for the segmental identities of the abdominal segments.

Blastocoel A fluid-filled cavity that forms in the animal hemisphere of early amphibian and echinoderm embryos, or between the epiblast and hypoblast of avian, reptilian and mammalian blastoderm-stage embryos.

Blastocyst A mammalian blastula. The blastocoel is expanded and the inner cell mass is positioned on one side of the ring of trophoblast cells.

Blastodisc Small region at the animal pole of the telolecithal eggs of fish and chicks, containing the yolk-free cytoplasm where cleavage can occur and that gives rise to the embryo. Following cleavage, the blastodisc becomes the blastoderm.

Blastomere A cleavage-stage cell resulting from mitosis.

Blastopore The invagination point where gastrulation begins. In deuterostomes, this marks the site of the anus. In protostomes, this marks the site of the mouth.

Blastula Early-stage embryo consisting of a sphere of cells surrounding an inner fluid-filled cavity, the blastocoel.

Blood islands Aggregations of hemangioblasts in the splanchnic mesoderm. It is generally thought that the inner cells of these blood islands become blood progenitor cells, while the outer cells become angioblasts.

Bone marrow-derived stem cells (BMDCs) See **Mesenchymal stem cells**.

Bone morphogenetic proteins (BMPs) Family of proteins in the TGF-β superfamily. Originally identified by their ability to induce bone formation, they are extremely multifunctional, having been found to regulate cell division, apoptosis, cell migration, and differentiation.

Bottle cells Invaginating cells during amphibian gastrulation, the main body of each cell is displaced toward the inside of the embryo while maintaining contact with the outside surface by way of a slender neck.

Brain-derived neurotrophic factor (BDNF) A paracrine factor that regulates neural activity and appears to be critical for synapse formation by inducing local translation of neural messages in the dendrites. BDNF is required for the survival of a particular subset of neurons in the striatum (a region of the brain involved in movement).

Bulge A region of the hair follicle that serves as a niche for adult stem cells.

C

Cadherins *Ca*lcium-*d*ependent ad*hes*ion molecules. Transmembrane proteins that interact with other cadherins on adjacent cells and are critical for establishing and maintaining intercellular connections, spatial segregation of cell types, and the organization of animal form.

Calorie restriction Dietary restriction as a means of extending mammalian longevity (at the expense of fertility).

Canalization See **Robustness**.

Cancer stem cell hypothesis The hypothesis that the malignant part of a tumor is either an adult stem cell that has escaped the control of its niche or a more differentiated cell that has regained stem cell properties.

Cap sequence See **Transcription initiation site**.

Capacitation The set of physiological changes by which sperm become capable of fertilizing an egg.

Cardia bifida A condition in which two separate hearts form, resulting from manipulation of the embryo or genetic defects that prevent fusion of the two endocardial tubes.

Cardiac neural crest Subregion of the cranial neural crest that extends from the otic (ear) placodes to the third somites. Cardiac neural crest cells develop into melanocytes, neurons, cartilage, and connective tissue. Cardiac neural crest also contributes to the muscular-connective tissue wall of the large arteries (the "outflow tracts") of the heart, as well as contributing to the septum that separates pulmonary circulation from the aorta.

Cardiogenic mesoderm See **Heart fields**.

Cardiomyocytes Cardiac cells derived from heart field tissue that form the muscular layers of the heart and its inflow and outflow tracts.

Catagen Regression phase of the hair follicle cell regeneration cycle.

Catenins A complex of proteins that anchor cadherins inside the cell. The cadherin-catenin complex forms the classic adherens junctions that help hold epithelial cells together and, by binding to the actin (microfilament) cytoskeleton of the cell, integrate the epithelial cells into a mechanical unit. One of them, β-catenin, can also be a transcription factor.

Caudal intestinal portal (CIP) The anterior opening of the developing hindgut region of the primitive gut tube; it opens into the future midgut region which is contiguous with the yolk sac at this stage.

Caudal Referring to the tail.

Cavitation A process whereby the trophoblast cells secrete fluid into the morula to create a blastocoel. The membranes of trophoblast cells pump sodium ions (Na^+) into the central cavity, drawing in water osmotically and thus creating and enlarging the blastocoel.

Cell lineage The series of cell types starting from an undifferentiated, pluripotent stem cell through stages of increasing differentiation to the terminally differentiated cell type.

Cellular blastoderm Stage of *Drosophila* development in which all the cells are arranged in a single-layered jacket around the yolky core of the egg.

Central nervous system The brain and spinal cord of vertebrates.

Centrolecithal Type of egg, such as those of insects, that has yolk in the center and undergoes superficial cleavage.

Centromere A region of DNA where chromatids are attached to each other by the kinetochore.

Centrosome-attracting body (CAB) Cellular structure that, in some invertebrate blastomeres, positions the centrosomes asymmetrically and recruits particular mRNAs so that the resulting daughter cells have different properties.

Cephalic furrow A transverse furrow formed during gastrulation in *Drosophila* that separates the future head region (procephalon) from the germ band, which will form the thorax and abdomen.

Cephalic Referring to the head.

ChIP-Seq Chromatin immunoprecipitation sequencing. A lab protocol used to identify the precise DNA sequences bound by particular transcription factors or nucleosomes containing specific modified histones.

Chemoattractant A biochemical that causes cells to move toward it.

Chemotaxis Movement of a cell down a chemical gradient, such as sperm following a chemical (chemoattractant) secreted by the egg.

Chiasmata Points of attachment between homologous chromosomes during meiosis which are thought to represent regions where crossing-over is occurring.

Chimera An organism consisting of a mixture of cells from two individuals.

Chimeric embryo Embryo made from tissues of more than one genetic source.

Chondrocyte-like osteoblasts Cranial neural crest cells undergoing early stages of intramembranous ossification. These cells downregulate Runx2 and begin expressing the *osteopontin* gene, giving them a phenotype similar to a developing chondrocyte.

Chondrocytes Cartilage cells.

Chordamesoderm Axial mesoderm that produces the notochord and head process.

Chordate An animal that has, at some stage of its life cycle, a notochord and a dorsal nerve cord or neural tube.

Chordin A paracrine factor with organizer activity. Chordin binds directly to BMP4 and BMP2 and prevents their complexing with their receptors, thus inducing dorsal ectoderm to form neural tissue.

Chorioallantoic membrane Forms in some amniote species, such as chickens, by fusion of the mesodermal layer of the allantoic membrane with the mesodermal layer of the chorion. This extremely vascular envelope is crucial for chick development and is responsible for transporting calcium from the eggshell into the embryo for bone production.

Chorion An extraembryonic membrane essential for gas exchange in amniote embryos. It is generated from trophoblast and extraembryonic mesoderm (somatopleure). The chorion adheres to the shell in birds and reptiles, allowing the exchange of gases between the egg and the environment. It forms the embryonic/fetal portion of the placenta in mammals.

Chorionic villus sampling Taking a sample from the placenta at 8–10 weeks of gestation to grow fetal cells to be analyzed for the presence or absence of certain chromosomes, genes, or enzymes.

Chromatid Half of a mitotic prophase chromosome, which consists of duplicate "sister" chromatids that are attached to each other by the kinetochore.

Chromatin The complex of DNA and protein in which eukaryotic genes are contained.

Chromosome diminution The fragmentation of chromosomes just prior to cell division, resulting in cells in which only a portion of the original chromosome survives. Chromosome dimunution occurs during cleavage in *Parascaris aequorum* in the cells that will generate the somatic cells while the future germ cells are protected from this phenomenon and maintain an intact genome.

Ciliary body A vascular structure at the junction between the neural retina and the iris that secretes the aqueous humor.

***cis*-regulatory elements** Regulatory elements (promoters and enhancers) that reside on the same strand of DNA.

Cleavage furrow A groove formed in the cell membrane in a dividing cell due to tightening of the microfilamentous ring.

Cleavage A series of rapid mitotic cell divisions following fertilization in many early embryos; cleavage divides the embryo without increasing its mass.

Cloaca Latin, "sewer." An endodermally lined chamber at the caudal end of the embryo that will become the waste receptacle for both the intestine and the kidney. Amphibians, reptiles, and birds retain this organ and use it to void both liquid and solid wastes. In mammals, the cloaca becomes divided by a septum into the urogenital sinus and the rectum.

Cloacal membrane Caudal end of the primitive streak formed by closely apposed endoderm and ectoderm; future site of the anus.

Cloning See **Somatic cell nuclear transfer**.

Coelom Space between the somatic mesoderm and splanchnic mesoderm that becomes the body cavity. In mammals, the coelom becomes subdivided into the pleural, pericardial, and peritoneal cavities, enveloping the thorax, heart, and abdomen, respectively.

Coherence Scientific evidence that fits into a system of other findings and is therefore more readily accepted.

Cohesin proteins Protein rings that encircle the sister chromatids during meiosis, provide a scaffold for the assembly of the meiotic recombination complex, resist the pulling forces of the spindle microtubules, and thereby keep the sister chromatids attached together and promote recombination.

Combinatorial association In developmental genetics, the principle that enhancers contain regions of DNA that bind transcription factors, and it is this combination of transcription factors that activates the gene.

Commensalism A symbiotic relationship that is beneficial to one partner and neither beneficial nor harmful to the other partner.

Commitment Describes a state in which a cell's developmental fate has become restricted even though it is not yet displaying overt changes in cellular biochemistry and function.

Committed stem cells Includes multipotent and unipotent stem cells that have the potential to become any of a relatively few cell types (multipotent) or only once cell type (unipotent).

Compaction A unique feature of mammalian cleavage, mediated by the cell adhesion molecule E-cadherin. The cells in the early (around eight-cell) embryo change their adhesive properties and become tightly apposed to each other.

Comparative embryology Study of how anatomy changes during the development of different organisms.

Compensatory regeneration Form of regeneration in which the differentiated cells divide but maintain their differentiated functions (e.g., mammalian liver).

Competence The ability of cells or tissues to respond to a specific inductive signal.

Conditional specification The ability of cells to achieve their respective fates by interactions with other cells. What a cell becomes is in large measure specified by paracrine factors secreted by its neighbors.

Congenital adrenal hyperplasia A condition causing female pseudohermaphroditism due to the presence of excess testosterone.

Congenital defect Any defect that an animal or person is born with. Congenital defects can be heriditary or they may have an environmental cause (e.g., exposure to teratogenic plants, drugs, chemicals, radiation, etc.). They may also be "idiopathic" (we don't know the cause).

Consensus sequence Located at the 5′ and 3′ ends of the introns that signal the "splice sites" of the intron.

Contact inhibition The mechanism for directional cell movement wherein cells are prohibited from moving "backwards" due to interactions with the cell membranes of other migrating cells.

Context-dependent properties A principle of the theoretical systems approach: The meaning or role of an individual component of a system is dependent on its context.

Conus arteriosus Cardiac outflow tract; precursor of both ventricles.

Convergent extension A phenomenon wherein cells intercalate to narrow the tissue and at the same time move it forward. Mechanism used for elongation of the archenteron in the sea urchin embryo, notochord of the tunicate embryo, and involuting mesoderm of the amphibian. This movement is reminiscent of traffic on a highway when several lanes must merge to form a single lane.

Coordinated gene expression The simultaneous expression of many different genes in a specific cell type. Its basis is often a single transcription factor (e.g., Pax6) that is crucial to several different enhancer sequences; the difference enhancers are differentially "primed," and the binding of the same factor to all of them activates all the genes at once.

Cornified layer (stratum corneum) The outer layer of the epidermis, consisting of keratinocytes that are now dead, flattened sacs of keratin protein with their nuclei pushed to one edge of the cell. These cells are continually shed throughout life and are replaced by new cells.

Corona radiata The innermost layer of cumulus cells, immediately adjacent to the zona pellucida.

Corpora allata Insect glands that secrete juvenile hormone (JH) during larval molts.

Correlative evidence Evidence based on the association of events. The "find it" of "find it, lose it, move it." See also **Gain-of-function evidence; Loss-of-function evidence**.

Cortex An outer structure (in contrast with medulla, an inner structure).

Cortical cytoplasm A thin layer of gel-like cytoplasm lying immediately beneath the cell membrane of most eggs. The cortex contains high concentrations of globular actin molecules that will polymerize to form microfilaments and microvilli during fertilization.

Cortical granule reaction The basis of the slow block to polyspermy in many animal species, including sea urchins and most mammals. A mechanical block to polyspermy that becomes active about a minute after successful sperm-egg fusion, in which enzymes from the egg's cortical granules contribute to the formation of a fertilization envelope that blocks further sperm entry.

Cortical granules Membrane-bound, Golgi-derived structures located in the egg's cortex that contain proteolytic enzymes and are thus homologous to the acrosomal vesicle of the sperm.

CpG islands Regions of DNA rich in the CpG sequence: a cytosine and a guanosine connected by a normal phosphate bond. Promoters often contain such islands, and transcription is often initiated nearby, possibly because they bind the basal transcription factors that recruit RNA polymerse II.

Cranial (cephalic) neural crest cells Neural crest cells in the future head region that migrate to produce the craniofacial mesenchyme, which differentiates into the cartilage, bone, cranial neurons, glia, and connective tissues of the face. These cells also enter the pharyngeal arches and pouches to give rise to thymic cells, the odontoblasts of the tooth primordia, and the bones of the middle ear and jaw.

Cranial ectodermal placodes Epidermal thickenings that form neurons and sensory epithelia.

Cranium The vertebrate skull, composed of the neurocranium (skull vault and base) and the viscerocranium (jaws and other pharyngeal arch derivatives).

Crossing over The exchange of genetic material during meiosis, whereby genes from one chromatid are exchanged with homologous genes from another.

Crystallins Transparent, lens-specific proteins.

Cumulus oophorus A layer of cells surrounding the mammalian egg, which is made up of the innermost layer of ovarian follicular (granulosa) cells that were nurturing the egg at the time of its release from the ovary.

Cutaneous appendages Species-specific epidermal modifications that include hairs, scales, feathers, hooves, claws, and horns.

Cyclic adenosine 3′,5′-monophosphate (cAMP) An important component of several intracellular signaling cascades and the soluble chemotactic substance that directs the aggregation of the myxamoebae of *Dictyostelium* to form a grex.

Cyclin B The larger subunit of mitosis-promoting factor, shows the cyclical behavior that is key to mitotic regulation, accumulating during S and being degraded after the cells have reached M. Cyclin B regulates the small subunit of MPF, the cyclin-dependent kinase.

Cyclin-dependent kinase Small subunit of MPF, activates mitosis by phosphorylating several target proteins, including histones, the nuclear envelope lamin proteins, and the regulatory subunit of cytoplasmic myosin, resulting in chromatin condensation, nuclear envelope depolymerization, and the organization of the mitotic spindle. Requires cyclin B to function.

Cyclooxygenase-2 (COX2) An enzyme that generates prostaglandins from the fatty acid arachidonic acid.

Cyclopia Congenital defect characterized by a single eye, caused by mutations in genes that encode either Sonic hedgehog or the enzymes that synthesize cholesterol and can be induced by certain chemicals that interfere with the cholesterol biosynthetic enzymes

Cystoblasts/cystocytes Derived from the asymmetric division of the germline stem cells of *Drosophila*, a cystoblast undergoes four mitotic divisions with incomplete cytokinesis to form a cluster of 16 cystocytes (one ovum and 15 nurse cells) interconnected by ring canals.

Cytokines Paracrine factors that are collected and concentrated by the extracellular matrix of the stromal (mesenchymal) cells at the sites of hematopoiesis and are involved in blood cell and lymphocyte formation.

Cytokinesis The division of the cell cytoplasm into two daughter cells. The mechanical agent of cytokinesis is a contractile ring of microfilaments made of actin. Each daughter cell receives one of the nuclei produced by nuclear division (karyokinesis).

Cytoplasmic bridges Continuity between adjacent cells that results from incomplete cytokinesis, e.g., during gametogenesis.

Cytotrophoblast Mammalian extra-embryonic epithelium composed of the original trophoblast cells, it adheres to the endometrium through adhesion molecules and, in species with invasive placentation such as the mouse and human, secretes proteolytic enzymes that enable the cytotrophoblast to enter the uterine wall and remodel the uterine blood vessels so that the maternal blood bathes fetal blood vessels.

D

Dauer larva A metabolically dormant larval stage in *C. elegans*. See also **Diapause**.

Decidua The maternal portion of the placenta, made from the endometrium of the uterus.

Deep cells A population of cells in the zebrafish blastula between the EVL and the YSL that give rise to the embryo proper.

Deep homology Signal transduction pathways composed of homologous proteins arranged in a homologous manner that are used for the same function in both protostomes and deuterostomes.

Delamination The splitting of one cellular sheet into two more or less parallel sheets.

Delta protein Cell surface ligand for Notch; participates in juxtacrine interaction and activation of the Notch pathway.

Dendrites The fine, branching extensions (dendritic arbor) emanating from neurons; dendrites pick up electric impulses from other cells.

Dermal bone Bone that forms in the dermis of the skin, such as most of the bones of the skull and face. They can be derived from head mesoderm or cranial neural crest-derived mesenchymal cells.

Dermal papilla A component of mesenchymal-epithelial induction during hair formation; a small node formed by dermal fibroblasts beneath the epidermal hair germ that stimulates proliferation of the overlying epidermal basal stem cells.

Dermamyotome Dorsolateral portion of the somite that contains skeletal muscle progenitor cells (including those that migrate into the limbs) and the cells that generate the dermis of the back.

Dermatome The central portion of the dermamyotome that produces the precursors of the dermis of the back and a population of muscle cells.

Descent with modification Darwin's theory to explain unity of type by descent from a common ancestor with changes creating the adaptations to the conditions of particular environments.

Determination The second, and irreversible, stage of cell or tissue commitment in which the cell or tissue is capable of differentiating autonomously even when placed into a non-neutral environment.

Deuterostomes In the deuterostome animal phyla (echinoderms, tunicates, cephalochordates, and vertebrates), organisms in which the first opening (i.e., the blastopore) becomes the anus while the second opening becomes the mouth (hence, *deutero stoma*, "mouth second"). Compare with **protostomes**.

Development The process of progressive and continuous change that generates a complex multicellular organism from a single cell. Development occurs throughout embryogenesis, maturation to the adult form, and continues into senescence.

Developmental constraints In evolution, the limitation of the number and forms of possible phenotypes that can be created by the interactions that are possible among molecules and between modules in the developing organism.

Developmental plasticity The ability of an embryo or larva to react to an environmental input with a change in form, state, movement, or rate of activity (i.e., phenotypic change).

Diacylglycerol (DAG) Second messenger generated in the IP$_3$ pathway from membrane phospholipid phosphatidylinositol 4,5-bisphosphate (PIP2), along with IP$_3$. DAG activates protein kinase C, which in turn activates a protein that exchanges sodium ions for hydrogen ions, raising the pH.

Diapause A metabolically dormant, nonfeeding stage of an organism during which development and aging are suspended; can occur at the embryonic, larval, pupal, or adult stage.

Dickkopf German, "thick head," "stubborn." A protein that interacts directly with the Wnt receptors, preventing Wnt signaling.

Diencephalon The caudal subdivision of the prosencephalon that will form the optic vesicles, retinas, and the thalamic and hypothalamic brain regions, which receive neural input from the retina.

Diethylstilbestrol (DES) A potent environmental estrogen. DES administration to pregnant women interferes with sexual and gonadal development in their female offspring resulting in infertility, subfertility, ectopic pregnancies, adenocarcinomas, and other effects.

Differential adhesion hypothesis A model explaining patterns of cell sorting based on thermodynamic principles. Cells interact so as to form an aggregate with the smallest interfacial free energy and therefore, the most thermodynamically stable pattern.

Differential gene expression A basic principle of developmental genetics: In spite of the fact that all the cells of an individual body contain the same genome, the specific proteins expressed by the different cell types are widely diverse. Differential gene expression, differential nRNA processing, differential mRNA translation, and differential protein modification all work to allow the extensive differentiation of cell types.

Differentiation The process by which an unspecialized cell becomes specialized into one of the many cell types that make up the body.

Differentiation therapy Treatmens for cancer that use transcription factors and other molecules to "normalize" cancers—that is, to cause cancerous cells to revert to differentition rather than continued proliferation.

Digestive tube The primitive gut of the embryo, which extends the length of the body from the pharynx to the cloaca. Buds from the digestive tube form the liver, gallbladder, and pancreas.

5α-Dihydrotestosterone (DHT) A steroid hormone derived from testosterone by the action of the enzyme 5α-ketosteroid reductase 2. DHT is required for masculinization of the male urethra, prostate, penis, or scrotum.

Diploblasts "Two-layer" animals; they posses endoderm and ecotoderm but most species lack true mesoderm. Includes the ctenophores (comb jellies). Compare with **Bilaterians**.

Direct development Embryogenesis characterized by the lack of a larval stage, where the embryo proceeds to construct a small adult.

Discoidal Meroblastic cleavage pattern for telolecithal eggs, in which the cell divisions occur only in the small blastodisc, as in birds, reptiles, and fish.

Disruption Abnormality or congenital defect caused by exogenous agents (teratogens) such as plants, chemicals, viruses, radiation, or hyperthermia.

Dissociation The ability of one module to develop differently from other modules.

Distal tip cell A single nondividing cell located at the end of each gonad in *C. elegans* which maintains the nearest germ cells in mitosis by inhibiting their meiotic differentiation.

Dizygotic (fraternal) twins "Two eggs." Describes twins that result from two separate fertilization events. Genetically such twins are full siblings. Compare with **monozygotic**.

DNA-binding domain Transcription factor domain that recognizes a particular DNA sequence.

DNA methylation A method of controlling the level of gene transcription in vertebrates by the enzymatic methylation of the promoters of inactive genes. Certain cytosine residues that are followed by guanosine residues are methylated and the resulting methylcytosine stabilizes nucleosomes and prevents transcription factors from binding. Important in X chromosome inactivation and DNA imprinting.

Dorsal blastopore lip Location of the involuting marginal zone cells of amphibian gastrulation. Migrating marginal cells sequentially become the dorsal lip of the blastopore, turn inward and travel along the inner surface of the outer animal hemisphere cells (i.e., the blastocoel roof).

Dorsal closure A process that brings together the two sides of the epidermis of the *Drosophila* embryo together at the dorsal surface.

Dorsal mesentery A derivative of the splanchnic mesoderm, this fibrous membrane connects the endoderm to the body wall. Involved in the looping of the developing intestines.

Dorsal root ganglia (DRG) Sensory spinal ganglia derived from the neural crest lying laterally paired and dorsally to the spinal cord. Sensory neurons of the DRG connect centrally with neurons in the dorsal horn of the spinal cord.

Dorsal-ventral (dorsoventral) axis The plane defining the back (dorsum) versus the belly (ventrum). When referring to the limb, this axis refers to the knuckles (dorsal) and palms (ventral).

Dorsolateral hinge points (DLHPs) Two hinge regions, besides the MHP, that form furrows near the connection of the avian and mammalian neural plate with the remainder of the ectoderm.

Dosage compensation Equalization of expression of X chromosome-encoded gene products in male and female cells. Achieved by either (1) doubling the transcription rate of the male X chromosomes (*Drosophila*), (2) partially repressing both X chromosomes (*C. elegans*), or (3) inactivating one X chromosome in each female cell (mammals).

Double-negative gate A mechanism whereby a repressor locks the genes of specification, and these genes can be unlocked by the repressor of that repressor. (In other words, activation by the repression of a repressor.)

Ductus arteriosus A vessel that forms from left aortic arch VI, serves as a shunt between the embryonic/fetal pulmonary artery and the descending aorta. It normally closes at birth (if not, a pathological condition results called patent ductus arteriosus).

Duplication and divergence Tandem gene duplications resulting from replication errors. Once replicated, the gene copies can diverge by random mutations developing different expression patterns and new functions.

Dynein The protein attached to the axoneme microtubules that provides the force for sperm propulsion. Dynein is an ATPase, an enzyme that hydrolyzes ATP, converting the released chemical energy into mechanical energy to allow the active sliding of the outer doublet microtubules, causing the flagellum to bend.

Dysgenesis Greek, "bad beginning." Defective development.

E

20E See **20-Hydroxyecdysone**.

E-cadherin See **Cadherins**.

Ecdysozoans One of the two major protostome groups; characterized by exoskeletons that periodically molt. The arthropods (including insects and crustaceans) and the nematodes (roundworms, including the model organism *C. elegans*) are two prominent groups. See also **Lophotrochozoans**.

Early allocation and progenitor expansion model An alternative to the progress zone model of proximal-distal specification of the limb, wherein the cells of the entire early limb bud are already specified; subsequent cell divisions simply expand these cell populations.

Ecdysone Insect steroid hormone, secreted by the prothoracic glands, that is modified in peripheral tissues to become the active molting hormone 20-hydroxyecdysone. Crucial to insect metamorphosis.

Ectoderm Greek *ektos*, "outside." The cells that remain on either the outside (amphibian) or dorsal (avian, mammalian) surface of the embryo following gastrulation. Of the 3 germ layers, the ectoderm is the one that forms the nervous system from the neural tube and neural crest and also generates the epidermis covering the embryo.

Ectodysplasin (EDA) cascade A gene cascade specific for cutaneous appendage formation. Vertebrates with dysfunctional EDA proteins exhibit a syndrome called anhidrotic ectodermal dysplasia characterized by absent or malformed cutaneous appendages (hair, teeth, and sweat glands).

Efferent ducts Ducts that link the rete testis to the Wolffian duct, formed from remodeled tubules of the mesonephric kidney.

Efferent Carried away from. Often used in reference to neurons that carry information away from the central nervous system (brain and spinal cord) to be acted on by the peripheral nervous system (muscles), or a vessel that carries fluid away from a structure. Compare with **Afferent**.

Egg chamber The ovary in which the *Drosophila* oocyte will develop, containing 15 interconnected nurse cells and a single oocyte.

Egg jelly A glycoprotein meshwork outside the vitelline envelope in many species, most commonly it is used to attract and/or to activate sperm.

Embryo A developing organism prior to birth or hatching. In humans, the term embryo generally refers to the early stages of development, starting with the fertilized egg until the end of organogenesis. After this, the developing human is called a fetus until its birth.

Embryology The study of animal development from fertilization to hatching or birth.

Embryonic axis Any of the positional axes in an embryo; includes anterior-posterior (head-tail), dorsal-ventral (back-belly), and right-left.

Embryonic epiblast Separates from the epiblast cells that line the amniotic cavity and will form the embryo proper.

Embryonic germ cells (EGCs) Pluripotent embryonic cells with characteristics of the inner cell mass derived from PGCs that have been treated particular paracrine factors to maintain cell proliferation.

Embryonic shield A localized thickening on the future dorsal side of the fish embryo; functionally equivalent to the dorsal blastopore lip of amphibians.

Embryonic stem cells (ESCs) Pluripotent stem cells of the mammalian inner cell mass blastomeres that are capable of generating all the cell types of the body.

EMT See **Epithelial-mesenchymal transition**.

Enamel knot The signaling center for tooth development, a group of cells induced in the epithelium by the neural crest-derived mesenchyme that secretes paracrine factors that pattern the cusp of the tooth.

Endocardium The internal layer of the heart chambers, derived from the heart fields.

Endocardial cushions Form from the endocardium and divide the tube into right and left atrioventricular channels. The atrioventricular valves are also derived from endocardial cells.

Endochondral ossification Bone formation in which mesodermal mesenchyme becomes cartilage and the cartilage is replaced by bone. It characterizes the bones of the trunk and limbs.

Endocrine disruptors Hormonally active compounds in the environment (e.g., DES; BPS; aromatase) that can have major detrimental effects on development, particularly of the gonads. Many endocrine disruptors are also obesogens (cause fat cell accumulation).

Endocrine factors Hormones that travel through the blood to exert their effects.

Endoderm Greek *endon*, "within." The innermost germ layer; forms the epithelial lining of the respiratory tract, the gastrointestinal tract, and the accessory organs (e.g., liver, pancreas) of the digestive tract. In the amphibian embryo, the yolk-containing cells of the vegetal hemisphere become endoderm. In mammalian and avian embryos, the endoderm is the most ventral of the three germ layers, continuous with the yolk sack epithelium.

Endometrium The epithelial lining of the uterus.

Endosteal osteoblasts Osteoblasts that line the bone marrow and are responsible for providing the niche that attracts HSCs, prevents apoptosis, and keeps the HSCs in a state of plasticity.

Endosymbiosis Greek, "living within." Describes the situation in which one cell lives inside another cell one organism lives within another.

Endothelium The single-layer sheet of epithelial cells lining of the blood vessels.

Endothelins Small peptides secreted by blood vessels that have a role in vasoconstriction and can direct the extension of certain sympathetic axons that have endothelin receptors, e.g. targeting of neurons from the superior cervical ganglia to the carotid artery.

Energids In *Drosophila*, the nuclei at the periphery of the syncytial blastoderm and their associated cytoplasmic islands of cytoskeletal proteins.

Enhancer A DNA sequence that controls the efficiency and rate of transcription from a specific promoter. Enhancers bind specific transcription factors that activate the gene by (1) recruiting enzymes (such as histone acetyltransferases) that break up the nucleosomes in the area or (2) stabilizing the transcription initiation complex.

Enhancer modularity The principle that having multiple enhancers allows a protein to be expressed in several different tissues while not being expressed at all in others, according to the combination of transcription factor proteins the enhancers bind.

Enveloping layer (EVL) A cell population in the zebrafish embryo at the midblastula transition made up of the most superficial cells from the blastoderm, which form an epithelial sheet a single cell layer thick. The EVL is an extra-embryonic protective covering that is sloughed off during later development.

Environmental integration Describes the influence of cues from the environment surrounding the embryo, fetus, or larva on their development.

Ependyma Epithelial lining of the spinal cord canal and the ventricles of the brain.

Eph receptors Receptor for ephrin ligands, involved in juxtacrine signaling.

Ephrin ligands Juxtacrine factors, binding between an ephrin on one cell and the eph receptor on an adjacent cell results in signals sent to each of the two cells. These signals are often those of either attraction or repulsion, and ephrins are often seen where cells are being told where to migrate or where boundaries are forming. Ephrins and the eph receptors function in the formation of blood vessels, neurons, and somites and direct neural crest cell migration.

Epialleles Variants of chromatin structure that can be inherited between generations. In most known cases, epialleles are differences in DNA methylation patterning that are able to affect the germ line and thereby be transmitted to offspring.

Epiblast The outer layer of the thickened margin of the epibolizing blastoderm in the gastrulating zebrafish embryo or the upper layer of the bilaminar gastrulating embryonic disc in birds and mammals. The epiblast contains ectoderm precursors in fish and all three germ layer precursors of the embryo proper (plus the amnion) in amniotes.

Epiboly The movement of epithelial sheets (usually of ectodermal cells) that spread as a unit (rather than individually) to enclose the deeper layers of the embryo. Epiboly can occur by the cells dividing, by the cells changing their shape, or by several layers of cells intercalating into fewer layers. Often, all three mechanisms are used.

Epicardium The outer surface of the heart that forms the coronary blood vessels that feed the heart, derived from the heart fields.

Epidermis Outer layer of skin.

Epididymis Derived from the Wolffian duct, the tube adjacent to the testis that links the efferent tubules to the ductus deferens.

Epigenesis The view supported by Aristotle and William Harvey that the organs of the embryo are formed de novo ("from scratch") at each generation.

Epigenetics The study of mechanisms that act on the phenotype without changing the nucleotide sequence of the DNA. Specifically, these changes work by altering gene *expression* rather than altering the gene sequence as mutation does. Epigenetic changes can sometimes be transmitted to future generations, a phenomenon referred to as epigenetic inheritance.

Epimorphosis Form of regeneration observed when adult structures undergo *de*differentiation to form a relatively undifferentiated mass of cells that then redifferentiates to form the new structure (e.g., amphibian limb regeneration).

Epiphyseal growth plates Cartilaginous areas at the ends of the long bones that allow continued bone growth.

Episomal vectors Vehicles for gene delivery derived from viruses that do not insert themselves into host DNA.

Epithelial-mesenchymal interactions Induction involving interactions of sheets of epithelial cells with adjacent mesenchymal cells. Properties of these interactions include regional specificity (when placed together, the same epithelium develops different structures according to the region from which the mesenchyme was taken), genetic specificity (the genome of the epithelium limits its ability to respond to signals from the mesencyme, i.e., the response is species-specific).

Epithelial-mesenchymal transition (EMT) An orderly series of events whereby epithelial cells are transformed into mesenchymal cells. In this transition, a polarized stationary epithelial cell, which normally interacts with basement membrane through its basal surface, becomes a migratory mesenchymal cell that can invade tissues and form organs in new places.

Epithelium Epithelial cells tightly linked together on a basement membrane to form a sheet or tube with little extracellular matrix.

Erythroblast Cell that matures from the proerythroblast and synthesizes enormous amounts of hemoglobin.

Erythrocyte The mature red blood cell that enters the circulation where it delivers oxygen to the tissues. It is incapable of division, RNA synthesis, or protein synthesis. Amphibians, fish, and birds retain the functionless nucleus; mammals extrude it from the cell.

Erythroid progenitor cell A committed stem cell that can form only red blood cells.

Erythropoietin A hormone that acts on erythroid progenitor cells to produce proerythroblasts, which will generate red blood cells.

ESCs See **Embryonic stem cells**.

Estrogen A group of steroid hormones (including **estradiol**) needed for complete postnatal development of both the Müllerian and the Wolffian ducts and necessary for fertility in both males and females

Estrus Greek *oistros*, "frenzy"; also called "heat." The estrogen-dominated stage of the ovarian cycle in female mammals that are spontaneous or periodic ovulators, characterized by the display of behaviors consistent with receptivity to mating.

Euchromatin The comparatively open state of chromatin that contains most of the organism's genes, most of which are capable of being transcribed. Compare with **Heterochromatin**.

Eukaryotic initiation factor-4 (eIF4E) A protein that is important for the initiation of translation by binding to the 5′ cap of mRNAs and contributing to the protein complex that mediates RNA unwinding and brings the 3′ end of the message next to the 5′ end, allowing the messenger RNA to bind to and be recognized by the ribosome.

Eukaryotic initiation factor-4G (eIF4G) A scaffold protein that allows the mRNA to bind to the ribosome through its interaction with eIF4E.

Evolutionary developmental biology (evo-devo) A model of evolution that integrates developmental genetics and population genetics to explain and the origin of biodiversity.

Exon In a gene, the region or regions of DNA that encode the protein. Compare with **intron**.

External granular layer A germinal zone of cerebellar neuroblasts that migrate from the germinal neuroepithelium to the outer surface of the developing cerebellum.

Extracellular matrix (ECM) Macromolecules secreted by cells into their immediate environment, forming a region of noncellular material in the interstices between the cells. Extracellular matrices are made up of collagen, proteoglycans, and a variety of specialized glycoprotein molecules such as fibronectin and laminin.

Extraembryonic endoderm Formed by delamination of the hypoblast cells from the inner cell mass to line the blastocoel cavity; will form the yolk sac.

Extraembryonic vasculogenesis The formation of blood islands in the yolk (i.e., outside the embryo).

Eye field Region in the anterior portion of the neural tube that will develop into the neural retina.

F

Fast block to polyspermy Mechanism by which additional sperm are prevented from fusing with a fertilized sea urchin egg by changing the electric potential to a more positive level. Has not been demonstrated in mammals.

Fate map Diagrams that follow cell lineages from specific regions of the embryo in order to "map" larval or adult structures onto the region of the embryo from which they arose. The superimposition of a map of "what is to be" onto a structure that has yet to develop into these organs.

Female pronucleus The haploid nucleus of the egg.

Fertilization Fusion of male and female gametes followed by fusion of the haploid gamete nuclei to restore the full complement of chromosomes characteristic of the species and initiation in the egg cytoplasm of those reactions that permit development to proceed.

Fertilization envelope Forms from the vitelline envelope of the sea urchin egg following cortical granule release. Glycosaminoglycans released by the cortical granules absorb water to expand the space between the cell membrane and fertilization envelope.

Fetal alcohol syndrome (FAS) Condition of babies born to alcoholic mothers, characterized by small head size, specific facial features, and small brain that often shows defects in neuronal and glial migration. FAS is the most prevalent congenital mental retardation syndrome. Another term, **fetal alcohol spectrum disorder** (**FASD**) has been coined to encompass the less visible behavioral effects on children exposed prenatally to alcohol.

α-Fetoprotein A protein that binds and inactivates fetal estrogen, but not testosterone, in both male and female rats and is critical for normal sexual differentiation of the rat brain.

Fetus Refers to the developing human from the ninth week of gestation to birth, a period characterized by growth and modeling.

Fibroblast growth factors (FGFs) A family of paracrine factors. Genes (*Fgf*) and their encoded proteins (FGF) that regulate cell proliferation and differentiation.

Fibroblast growth factor receptors (FGFRs) A set of receptor tyrosine kinases that are activated by FGFs, resulting in activation of the dormant kinase and phosphorylation of certain proteins (including other FGF receptors) within the responding cell.

Fibronectin A very large (460 kDa) glycoprotein dimer synthesized by numerous cell types and secreted into the extracellular matrix. Functions as a general adhesive molecule, linking cells to one another and to other substrates such as collagen and proteoglycans, and provides a substrate for cell migration.

Filopodia Long, thin processes produced by migrating mesenchymal cells.

Flagellum The tail of the sperm, containing the central axoneme, supporting structures called outer dense fibers and the fibrous sheath, and the mitochondria-containing midpiece.

Floor plate Ventral region of the neural tube important in the establishment of dorsal-ventral polarity. Floor plate forms in the ventral neural tube due to induction by Sonic hedgehog secreted from the adjacent notochord. It becomes a secondary signaling center that also secretes Sonic hedgehog, establishing a gradient that is highest ventrally.

Fluorescent dye Compounds, such as fluorescein and green fluorescent protein (GFP), that emit bright light at a specific wavelength when excited with ultraviolet light.

Focal adhesions Where the cell membrane contacts the extracellular matrix in migrating cells, mediated by connections between actin, integrin and the extracellular matrix.

Follicle Composed of a single oogonium surrounded by granulosa cells and thecal cells.

Follicle-stimulating hormone (FSH) A peptide hormone secreted by the mammalian pituitary that promotes ovarian follicle development and spermatogenesis.

Follicular stem cells Multipotent adult stem cells that resides in the bulge niche of the hair follicle. They give rise to the hair shaft, sheath, and sebaceous gland.

Follistatin A paracrine factor with organizer activity, an inhibitor of both activin and BMPs, causes ectoderm to become neural tissue.

Foramen ovale An opening in the septum separating the right and left atria.

Forkhead transcription factors Transcription factors (e.g., **Fox proteins**, **HNF4α**) that are especially important in the endoderm that will form liver, where they help activate the regulatory regions surrounding liver-specific genes.

Frizzled Transmembrane receptor for Wnt family of paracrine factors.

Frontonasal process Cranial prominence formed by neural crest cells from the midbrain and rhombomeres 1 and 2 of the hindbrain that forms the forehead, the middle of the nose, and the primary palate. Thus, the cranial neural crest cells generate the facial skeleton

G

Gain-of-function evidence A strong type of evidence, wherein the initiation of the first event causes the second event to happen even in instances where or when neither event usually occurs. The "move it" of "Find it, lose it, move it."

See also **Correlative evidence; Loss-of-function evidence**.

Gamete A specialized reproductive cell through which sexually reproducing parents pass chromosomes to their offspring; a sperm or an egg.

Gametogenesis The production of gametes.

Ganglia Clusters of neuronal cell bodies whose axons form a nerve.

Gap genes *Drosophila* zygotic genes expressed in broad (about three segments wide), partially overlapping domains. Gap mutants lacked large regions of the body (several contiguous segments).

Gastrula A stage of the embryo following gastrulation that contains the three germ layers that will interact to generate the organs of the body.

Gastrulation A process involving movement of the blastomeres of the embryo relative to one another resulting in the formation of the three germ layers of the embryo.

GDNF See **Glial-derived neurotrophic factor**.

GEF See **GTP exchange factor; RTK pathway**.

Gene regulatory networks (GRNs) Patterns generated by the interactions among transcription factors and their enhancers that help define the course that development follows.

Genetic assimilation The process by which a phenotypic character initially produced only in response to some environmental influence becomes, through a process of selection, taken over by the genotype so that it is formed even in the absence of the environmental influence that had first been necessary.

Genetic heterogeneity The production of similar phenotypes by mutations in different genes.

Genital disc Region of the *Drosophila* larva that will generate male or female genitalia. Male and female genitalia are derived from separate cell populations of the genital disc, as induced by paracrine factors.

Genital ridge A thickening of the splanchnic mesoderm (**germinal epithelium**) and of the underlying intermediate mesodermal mesenchyme on the medial edge of the mesonephros; it forms the testis or ovary. Also called the germinal ridge.

Genital tubercle A structure cranial to the cloacal membrane during the indifferent stage of differentiation of the mammalian external genitalia. It will form either the clitoris in the female fetus or the glans penis in the male.

Genome The complete DNA sequence of an individual organism.

Genomic equivalence The theory that every cell of an organism has the same genome as every other cell.

Genomic imprinting A phenomenon in mammals whereby only the sperm-derived or only the egg-derived allele of the gene is expressed, sometimes due to inactivation of one allele by DNA methylation during spermatogenesis or oogenesis.

Germ band A collection of cells along the ventral midline of the *Drosophila* embryo that forms during gastrulation by convergence and extension of the surface ectoderm and will form the trunk of the embryo and the thorax and abdomen of the adult.

Germ cells A group of cells set aside from the somatic cells that form the rest of the embryo for reproductive function. Consists of the cells of the gonads (ovary and testis) that undergo meiotic cell divisions to generate the gametes.

Germinal crescent A region in the anterior portion of the avian and reptilian blastoderm area pellucida containing the hypoblasts displaced by migrating endodermal cells. It contains the precursors of the germ cells, which later migrate through the blood vessels to the gonads.

Germinal epithelium Epithelium of the bipotential gonad, derived from splanchnic mesoderm, that will form the somatic (i.e., non-germ cell) component of the gonads.

Germinal neuroepithelium A layer of rapidly dividing neural stem cells one cell layer thick that constitute the original neural tube.

Germinal vesicle breakdown (GVBD) Disintegration of the oocyte nuclear membrane (germinal vesicle) upon resumption of meiosis during oogenesis.

Germ layer One of the three layers of the vertebrate embryo, ectoderm, mesoderm, and endoderm, generated by the process of gastrulation, that will form all of the tissues of the body except for the germ cells.

Germline stem cells The *Drosophila* pole cell derivatives that divide asymmetrically to produce another stem cell and a differentiated daughter cell called a cystoblast which in turn produces a single ovum and 15 nurse cells.

Germ plasm Cytoplasmic determinants (mRNA and proteins) in the eggs of some species, including frogs, nematodes, and flies, that autonomously specify the primordial germ cells.

Germ plasm theory A model of cell specification proposed by Weismann, in which each cell of the embryo would

develop autonomously. Instead of dividing equally, the chromosomes were hypothesized to divide in such a way that different chromosomal determinants entered different cells. Only the nuclei in those cells destined to become germ cells (gametes) were postulated to contain all the different types of determinants. The nuclei of all other cells would have only a subset of the original determinants.

GFP See **Green fluorescent protein**.

Glia Supportive cells of the central nervous system, derived from neuroepithelial cells.

Glial-derived neurotrophic factor (GDNF) A paracrine factor that binds to the Ret receptor tyrosine kinase. It is produced by the gut mesenchyme that attracts vagal and sacral neural crest cells, and it is produced by the metanephrogenic mesenchyme to induce the formation and branching of the ureteric buds.

Glial guidance A mechanism important for positioning young neurons in the developing mammalian brain (e.g., the granule neuron precursors travel on the long processes of the Bergmann glia in the cerebellum).

Glycogen synthase kinase 3 (GSK3) Targets β-catenin for destruction.

Glycosaminoglycans (GAGs) Complex acidic polysaccharides consisting of unbranched chains assembled from many repeats of a two-sugar unit. The carbohydrate component of proteoglycans.

Gonadotropin-releasing hormone (GRH; GnRH) Peptide hormone released from the hypothalamus that stimulates the pituitary to release the gonadotropins follicle-stimulating hormone and luteinizing hormone, which are required for mammalian gametogenesis and steroidogenesis.

Gonocytes Mammalian PGCs that have arrived at the genital ridge of a male embryo and have become incorporated into the sex cords.

G protein A protein that binds GTP and is activated or inactivated by GTP modifying enzymes (such as GTPases). They play important roles in the RTK pathway and in cytoskeletal maintenance.

Granule cells Derived from neuroblasts of the external granule layer of the developing cerebellum. Granule neurons migrate back toward the ventricular (ependymal) zone, where they produce a region called the internal granule layer.

Granulosa cells Cortical epithelial cells of the fetal ovary, granulosa cells surround individual germ cells that will become the ova and will form, with thecal cells, the follicles that envelop the germ cells and secrete steroid hormones. The number of granulosa cells increase and form concentric layers around the oocyte as the oocyte matures prior to ovulation.

Gray crescent A band of inner gray cytoplasm that appears following a rotation of the cortical cytoplasm with respect to the internal cytoplasm in the marginal region of the 1-cell amphibian embryo. Gastrulation starts in this location.

Gray matter Regions of the brain rich in neuronal cell bodies. Compare with **white matter**.

Green fluorescent protein (GFP) A protein that occurs naturally in certain jellyfish. It emits bright green fluorescence when exposed to ultraviolet light. The *GFP* gene is widely used as a transgenic label for cells in developmental and other research, since cells that express GFP are easily identified by a bright green glow.

GRNs See **Gene regulatory networks**

Growth and differentiation factors (GDFs) See **Paracrine factors**.

Growth cone The motile tip of a neuronal axon; leads nerve outgrowth.

Growth factor A secreted protein that binds to a receptor and initiates signals to promote or retard cell division and growth.

Growth plate closure Causes the cessation of bone growth at the end of puberty. High levels of estrogen induce apoptosis in the hypertrophic chondrocytes and stimulate the invasion of bone-forming osteoblasts into the growth plate.

GTP exchange factor (GEF) In the RTK (receptor tyrosine kinase) pathway, this factor exchanges a phosphate that transforms a receptor-bound GDP into a bound but active GTP. See also **RTK pathway**.

Gynandromorph Greek *gynos*, "female"; *andros*, "male." An animal in which some body parts are male and others are female. Compare with **hermaphrodite**.

H

Halteres A pair of balancers on the third thoracic segment in *Drosophila*.

Haptotaxis Migration on preferred substrates.

Hatched blastula Free-swimming sea urchin embryo, after the cells of the animal hemisphere synthesize and secrete a hatching enzyme that digests the fertilization envelope.

Head activation gradient A morphogenetic gradient in *Hydra* that is highest at the hypostome and permits the head only to form in one place.

Head mesoderm Mesoderm located anterior to the trunk mesoderm, consisting of the unsegmented paraxial mesoderm and prechordal mesoderm. This region provides the head mesenchyme that forms much of the connective tissues and musculature of the face and eyes.

Head process See **Chordamesoderm**.

Heart fields (cardiogenic mesoderm) Two groups of cardiac cells in the lateral plate mesoderm, at the level of the node. The cardiac cells of the heart field migrate through the primitive streak during gastrulation such that the medial-lateral arrangement of these early cells will become the anterior-posterior (rostral-caudal) axis of the developing heart tube.

Heart tube Linear (anterior-to-posterior) structure formed at the midline of the heart fields; will become the ventricles of the heart.

Hedgehog A family of paracrine factors used by the embryo to induce particular cell types and to create boundaries between tissues. Hedgehog proteins must become complexed with a molecule of cholesterol in order to function. Vertebrates have at least three homologues of the *Drosophila hedgehog* gene: *sonic hedgehog (shh), desert hedgehog (dhh)*, and *indian hedgehog (ihh)*.

Hedgehog pathway Proteins activated by the binding of a Hedgehog protein to the Patched receptor. When Hedgehog binds to Patched, the Patched protein's shape is altered such that it no longer inhibits Smoothened. Smoothened acts to release the Ci protein from the microtubules and to prevent its being cleaved. The intact Ci protein can now enter the nucleus, where it acts as a transcriptional *activator* of the same genes it used to repress.

Hemangioblasts Greek "blood vessel" plus "blast," a rapidly dividing cell, usually a stem cell. These are the stem cells that form blood vessels and blood cells.

Hematopoesis The generation of blood cells.

Hematopoietic inductive microenvironments (HIMs) Cell regions that induce different sets of transcription factors in multipipotent hematopoietic stem cells, and these transcription factors specify the developmental path taken by the descendents of those cells.

Hematopoietic stem cell (HSC) A multipotent stem cell type that generates a series of intermediate progenitor cells whose potency is restricted to certain blood cell lineages. These lineages are then capable of producing all the blood cells and lymphocytes of the body.

Hemimetabolous A form of insect metamorphosis that includes pronymph, nymph, and imago stages.

Hemogenic endothelial cell Primary endothelial cells of the dorsal aorta, especially those in the ventral area, that are derived from angioblasts that migrated from the sclerotome. They give rise to the hematopoietic stem cells that migrate to the liver and bone marrow and become the adult hematopoietic stem cells.

Hensen's node (primitive knot) A regional thickening of cells at the anterior end of the primitive streak. The center of Hensen's node contains a funnel-shaped depression (sometimes called the primitive pit) through which cells can enter the embryo to form the notochord and prechordal plate. Hensen's node is the functional equivalent of the dorsal lip of the amphibian blastopore (i.e., the organizer) and the fish embryonic shield.

Hepatic diverticulum The liver precursor, a bud of endoderm that extends out from the foregut into the surrounding mesenchyme.

Hepatocyte growth factor (HGF) A paracrine factor secreted by the stellate cells of the liver that allows the hepatocytes to re-enter the cell cycle during compensatory regeneration. Also called scatter factor.

Hermaphrodite An individual in which both ovarian and testicular tissues exists, having either ovotestes (gonads containing both ovarian and testicular tissue) or an ovary on one side and a testis on the other. Compare with **gynandromorph**.

Heterochromatin Chromatin that remains condensed throughout most of the cell cycle and replicates later than most of the other chromatin. Usually transcriptionally inactive. Compare with **Euchromatin**.

Heterochrony The phenomenon wherein animals change the relative time of appearance and rate of development of characters present in their ancestors.

Heterogeneous causation A principle of the theoretical systems approach: Causation is seen as being both "upward," from the genes to the environment, and "downward," from the environment to the genes.

Heterochrony Greek, "different time." A shift in the relative timing of two developmental processes as a mechanism to generate phenotypic variation available for natural selection. One module changes its time of expression or growth rate relative to the other modules of the embryo.

Heterometry Greek, "different measure." A change in the amount of a gene product as a mechanism to generate phenotypic variation available for natural selection.

Heterotopy Greek, "different place." The spatial alteration of gene expression (e.g., transcription factors or paracrine factors) as a mechanism to generate phenotypic variation available for natural selection.

Heterotypy Greek, "different kind." The alteration of the actual coding region of the gene, changing the functional properties of the protein being synthesized, as a mechanism to generate phenotypic variation available for natural selection.

High CpG-content promoters (HCPs) Promoters with many CpG islands; these promoters often regulate developmental genes required for the construction of the organism; their default state is "on." See also **CpG islands**.

Histoblast nests Clusters of imaginal cells that will form the adult abdomen in holometabolous insects.

Histone Positively charged proteins that are the major protein component of chromatin. See also **Nucleosome**.

Histone acetylation The addition of negatively charged acetyl groups to histones which neutralizes the basic charge of lysine and loosens the histones, and thus activates transcription.

Histone acetyltransferases Enzymes that place acetyl groups on histones (especially on lysines in histones H3 and H4), destabilizing the nucleosomes so that they come apart easily, thus facilitating transcription.

Histone deacetylases Enzymes that remove acetyl groups, stabilize the nucleosomes, and prevent transcription.

Histone methylation The addition of methyl groups to histones. Can either activate or further repress transcription, depending on the amino acid that is methylated and the presence of other methyl or acetyl groups in the vicinity.

Histone methyltransferases Enzymes that add methyl groups to histones and either activate or repress thanscription.

Holoblastic Greek *holos*, "complete." Refers to a cell division (cleavage) pattern in the embryo in which the entire egg is divided into smaller cells, as it is in frogs and mammals.

Holometabolous The type of insect metamorphosis found in flies, beetles, moths, and butterflies. There is no pronymph stage. The insect hatches as a larva (a caterpillar, grub, or maggot)

and progresses through instar stages as it gets bigger between larval molts, a metamorphic molt to become a pupa, an imaginal molt and finally the emergence (eclose) of the adult (imago).

Homeorhesis How the organism stabilizes its different cell lineages while it is still constructing itself.

Homeostasis Maintenance of a stable physiological state by means of feedback responses.

Homeotic complex (Hom-C) The region of *Drosophila* chromosome 3 containing both the Antennapedia complex and the bithorax complex.

Homeotic selector genes A class of *Drosophila* genes regulated by the protein products of the gap, pair-rule, and segment polarity genes whose transcription determines the developmental fate of each segment.

Homeotic mutants Result from mutations of homeotic selector genes, in which one structure is replaced by another (as where an antenna is replaced by a leg).

Homodimer Two identical protein molecules bound together.

Homologous Structures and/or their respective components whose similarity arises from their being derived from a common ancestral structure. For example, the wing of a bird and the forelimb of a human. Compare with **analogous**.

Homologue (1) One of a pair (or larger set) of chromosomes with the same overall genetic composition. For example, diploid organisms have two copies (homologues) of each chromosome, one inherited from each parent. (2) Evolutionary features in different species that are similar by reason of descent from a common ancestor.

Horizontal transmission When a host that is born free of symbionts but subsequently becomes infected, either by its environment or by other members of the species. Can also refer to the transfer of genes from one bacterium to another. Compare with **vertical transmission**.

Host The larger organism in a symbiotic relationship in which one of the organisms involved is much larger than the other, and the smaller organism may live on the surface or inside the body of the larger.

Hox genes Abbreviation of "homeobox" genes. Large family of related genes that dictate (at least in part) regional identity in the embryo, particularly along the anterior-posterior axis. Hox genes encode transcription factors that regulate the expression of other genes. All known mammalian genomes contain four copies of the Hox complex per

haploid set, located on four different chromosomes (*Hoxa* through *Hoxd* in the mouse, *HOXA* through *HOXD* in humans). The mammalian Hox/HOX genes are numbered from 1 to 13, starting from that end of each complex that is expressed most anteriorly.

Hub A regulatory microenvironment in *Drosophila* testes where the stem cells for sperm reside.

Hyaline layer A coating around the sea urchin egg formed by the cortical granule protein hyalin. The hyaline layer provides support for the blastomeres during cleavage.

Hydatidiform mole A human tumor which resembles placental tissue, arise when a haploid sperm fertilizes an egg in which the female pronucleus is absent and the entire genome is derived from the sperm which precludes normal development and is cited as evidence for genomic imprinting.

20-Hydroxyecdysone (20E) An insect hormone that initiates and coordinates each molt, regulates the changes in gene expression that occur during metamorphosis, and signals imaginal disc differentiation.

Hyperactivation Describes the type of motility displayed by capacitated sperm of some mammalian species. Hyperactivation has been proposed to help detach capacitated sperm from the oviductal epithelium, allow sperm to travel more effectively through viscous oviductal fluids, and facilitate penetration of the extracellular matrix of the cumulus cells.

Hypertrophic chondrocytes Formed during the fourth phase of endochondral ossification, when the chondrocytes, under the influence of the transcription factor Runx2, stop dividing and increase their volume dramatically.

Hypoblast The inner layer of the thickened margin of the epibolizing blastoderm in the gastrulating zebrafish embryo or the lower layer of the bilaminar embryonic blastoderm in birds and mammals. The hypoblast in fish (but not in birds and mammals) contains the precursors of the endoderm and mesoderm.

Hypoblast islands (primary hypoblast) Derived from area pellucida cells of the avian blastoderm that migrate individually into the subgerminal cavity to form individual disconnected clusters containing 5–20 cells each. Does not contribute to the embryo proper.

Hypostome A conical region of the "head" of a hydra that contains the mouth.

I

Imaginal discs Clusters of relatively undifferentiated cells set aside to produce adult structures. Imaginal discs will form the cuticular structures of the adult, including the wings, legs, antennae, eyes, head, thorax, and genitalia in holometabolous insects.

Imaginal rudiment Develops from the left coelomic sac of the pluteus larva and will form many of the structures of the adult sea urchin.

Imago A winged and sexually mature adult insect.

Indirect developers Animals for which embryonic development includes a larval stage with characteristics very different from those of the adult organism, which emerges only after a period of metamorphosis.

Induced pluripotent stem (iPS) cells Adult mouse or human cells that have been converted to cells with the pluripotency of embryonic stem cells. Usually this is accomplished by the activation of certain transcription factors.

Inducer Tissue that produces a signal (or signals) that induces a cellular behavior in some other tissue.

Induction The process by which one cell population influences the development of neighboring cells via interactions at close range.

Ingression Migration of individual cells from the surface layer into the interior of the embryo. The cells become mesenchymal (i.e., they separate from one another) and migrate independently.

Inner cell mass (ICM) A small group of internal cells within a mammalian morula or blastocyst that will eventually develop into the embryo proper and its associated yolk sac, allantois, and amnion.

Inositol 1,4,5-trisphosphate (IP$_3$) A second messenger generated by the phospholipase C enzyme that releases intracellular Ca^{2+} stores. Important in the initiation of both cortical granule release and sea urchin development.

Instructive interaction A mode of inductive interaction in which a signal from the inducing cell is necessary for initiating new gene expression in the responding cell.

Insulator DNA sequence that limits the range within which an enhancer can activate a given gene's expression (thereby "insulating" a promoter from being activated by another gene's enhancers).

Insulin-like growth factors (IGFs) Growth factors that initiate an FGF-like signal transduction cascade that interferes with the signal transduction pathways of both BMPs and Wnts. IGFs are required for the formation of the anterior neural tube, including the brain and sensory placodes of amphibians.

Integration A principle of the theoretical systems approach: How the parts are put together and how they interact to form the whole.

Integrins A family of receptor proteins so named because they *integrate* the extracellular and intracellular scaffolds, allowing them to work together. On the extracellular side, integrins bind to the sequence arginine-glycine-aspartate (RGD), found in several adhesive proteins in extracellular matrices, including fibronectin, vitronectin (found in the basal lamina of the eye), and laminin. On the cytoplasmic side, integrins bind to talin and α-actinin, two proteins that connect to actin microfilaments. This dual binding enables the cell to move by contracting the actin microfilaments against the fixed extracellular matrix.

Intermediate mesoderm Mesoderm immediately lateral to the paraxial mesoderm. It forms the outer (cortical) portion of the adrenal gland and the urogenital system, consisting of the kidneys, gonads, and their associated ducts.

Intermediate progenitor cells (IPCs) Neuroblast precursor cells of the subventricular zone derived from radial glial cells.

Intermediate spermatogonia The first committed stem cell type of the mammalian testis, they are committed to becoming spermatozoa.

Intersex A condition in which male and female traits are observed in the same individual.

Intraembryonic vasculogenesis The formation of blood vessels during embryonic organogenesis. Compare with extraembryonic vaculogenesis.

Intramembranous ossification Bone formation directly from mesenchyme. There are three main types of intramembranous bone: sesamoid bone and periostal bone, which come from mesoderm, and dermal bone which originate from cranial neural crest-derived mesenchymal cells.

Introns Non-protein-coding regions of DNA within a gene. Compare with **exon**.

Invagination The infolding of a region of cells, much like the indenting of a soft rubber ball when it is poked.

Involuting marginal zone (IMZ) Cells that involute during *Xenopus* gastrulation, includes precursors of the pharyngeal endoderm, head mesoderm, notochord, somites, heart, kidney, and ventral mesoderms.

Involution Inturning or inward movement of an expanding outer layer so that it spreads over the internal surface of the remaining external cells.

Ionophore A compound that allows the diffusion of ions such as Ca^{2+} across lipid membranes, permitting them to traverse otherwise impermeable barriers.

Iris A pigmented ring of muscular tissue in the eye that controls the size of the pupil and determines eye color.

Isolecithal Greek, "equal yolk." Describes eggs with sparse, equally distributed yolk particles, as in sea urchins, mammals, and snails.

Isthmus The narrow segment of the mammalian oviduct adjacent to the uterus.

J

Jagged protein Ligand for Notch, participates in juxtacrine interaction and activation of the Notch pathway.

JAK *J*anus *k*inase proteins. Linked to FGF receptors in the JAK-STAT cascade.

JAK-STAT cascade A pathway activated by paracrine factors binding to receptors that are linked to JAK (Janus kinase) proteins. The binding of ligand to the receptor phosphorylates the STAT (signal *t*ransducers and *a*ctivators of *t*ranscription) family of transcription factors.

Juvenile hormone (JH) A lipid hormone in insects that prevents the ecdysone-induced changes in gene expression that are necessary for metamorphosis. Thus, its presence during a molt ensures that the result of that molt is another larval instar, not a pupa or an adult.

Juxtacrine interactions When cell membrane proteins on one cell surface interact with receptor proteins on adjacent (juxtaposed) cell surfaces.

K

Kairomones Chemicals that are released by a predator and can induce defenses in its prey.

Karyokinesis The mitotic division of the cell's nucleus. The mechanical agent of karyokinesis is the mitotic spindle.

Keratinocytes Differentiated epidermal cells that are bound tightly together and produce a water-impermeable seal of lipid and protein.

Koller's sickle See **Primitive streak**.

Kupffer's vesicle Transient fluid-filled organ housing the cilia that control left-right asymmetry in zebrafish.

L

Labioscrotal folds Folds surrounding the cloacal membrane in the indifferent stage of differentiation of mammalian external genitalia. They will form the labia in the female and the scrotum in the male. Also called urethral folds or genital swellings.

lacZ gene   The *E. coli* gene for β-galactosidase; commonly used as a reporter gene.

Laminin   A large glycoprotein and major component of the basal lamina, plays a role in assembling the extracellular matrix, promoting cell adhesion and growth, changing cell shape, and permitting cell migration.

Lampbrush chromosomes   Chromosomes during the diplotene stage of amphibian oocyte meiosis that stretch out large loops of DNA that represent sites of upregulated RNA synthesis.

Lanugo   The hair of human embryos, usually shed before birth.

Large micromeres   A tier of cells produced by the fifth cleavage in the sea urchin embryo when the micromeres divide.

Larva   The sexually immature stage of an organism, often of significantly different appearance than the adult and frequently the stage that lives the longest and is used for feeding or dispersal.

Laryngotracheal groove   An outpouching of endodermal epithelium in the center of the pharyngeal floor, between the fourth pair of pharyngeal pouches, that extends ventrally. The laryngotracheal groove then bifurcates into the branches that form the paired bronchi and lungs.

Lateral plate mesoderm   Mesodermal sheet lateral to the intermediate mesoderm. Gives rise to appendicular bones, connective tissues of the limb buds, circulatory system (heart, blood vessels, and blood cells), muscles and connective tissues of the digestive and respiratory tracts, and lining of coelom and its derivatives. It also helps form a series of extraembryonic membranes that are important for transporting nutrients to the embryo.

Lateral somitic frontier   The boundary between the primaxial and abaxial muscles and between the somite-derived and lateral plate-derived dermis.

Lens placode   Paired epidermal thickenings induced by the underlying optic cups that invaginates to form the transparent lens that allows light to impinge on the retina. It differentiates into the **lens vesicle**, which will produce the cells of the adult eye lens.

Level-specific properties and emergence   A principle of the theoretical systems approach: The properties of a system at any given level of organization cannot be totally explained by those of levels "below" it.

Leydig cells   Testis cells derived from the interstitial mesenchyme cells surrounding the testis cords that make the testosterone required for secondary sex determination and, in the adult, required to support spermatogenesis.

Life expectancy   The length of time an average individual of a given species can expect to live; it is characteristic populations, not of species.

Limb bud   A circular bulge that will form the future limb. The limb bud is formed by the proliferation of mesenchyme cells from the somatic layer of the limb field lateral plate mesoderm (the limb *skeletal* precursor cells) and from the somites (the limb *muscle* precursor cells).

Limb field   An area of the embryo containing all of the cells capable of forming a limb.

Lineage-restricted stem cells   Stem cells derived from multipotent stem cells, and which can now generate only a particular cell type or set of cell types.

Long noncoding RNAs (lncRNAs)   Transcriptional reglators that inactivate genes on one of the two chromosomes of a diploid organism. For example, Xist is an lncRNA involved in the inactivation of genes on the second X chromosome of females. Some lnc RNAs appear to be specific for either the maternal or paternal copy of a gene.

Lophotrochozoans   One of two major protostome groups, many of which are characterized by the larval form known as the trochophore. A diverse group that includes the annelids (segmented worms such as earthworms), the molluscs (e.g., snails), and the flatworms (e.g., *Planaria*). See also **Ecdysozoans**.

Loss-of-function evidence   The absence of the postulated cause is associated with the absence of the postulated effect. The "lose it" of "find it, lose it, move it." See also **Correlative evidence; Gain-of-function evidence**.

Low CpG-content promoters (LCPs)   These promoters tend to characterize the genes of mature, fully differentiated cells. The CpG sites are usually methylated and their default state is "off," although they can be activated by specific transcription factors. See also **CpG islands**.

Lumen   The hollow central portion of any tubular structure or organ.

Luteinizing hormone (LH)   A peptide hormone secreted by the mammalian pituitary that stimulates the production of steroid hormones, such as estrogen from the ovarian follicle cells and testosterone from the testicular leydig cells. A surge in LH levels causes the primary

oocyte to complete meiosis I and prepares the follicle for ovulation.

Lymphatic vasculature   The vessels of the circulatory system that transport lymph (as opposed to the blood vessels of the circulatory system).

M

Macromeres   Larger cells generated by asymmetrical cleavage, e.g., the four large cells generated by the fourth cleavage when the vegetal tier of the sea urchin embryo undergoes an unequal equatorial cleavage.

Male pronucleus   The haploid nucleus of the sperm.

Malformation   Abnormalities caused by genetic events such as gene mutations, chromosomal aneuploidies, and translocations.

Mantle (intermediate) zone   Second layer of the developing spinal cord and medulla that forms around the original neural tube. Because it contains neuronal cell bodies and has a grayish appearance grossly, it will form the gray matter.

Marginal zone   (1) The third and outer zone of the developing spinal cord and medulla composed of a cell-poor region composed of axons extending from neurons residing in the mantle zone. Will form the white matter as glial cells cover the axons with myelin sheaths, which have a whitish appearance. (2) In amphibians: Where gastrulation begins, the region surrounding the equator of the blastula, where the animal and vegetal hemispheres meet. (3) In birds and reptiles (= marginal belt), a thin layer of cells between the area pellucida and the area opaca is a thin layer of cells, important in determining cell fate during early chick development.

Maternal effect genes   Encode messenger RNAs that are placed in different regions of the *Drosophila* egg.

Medial hinge point (MHP)   Derived from the portion of the avian and mammalian neural plate just anterior to Hensen's node and from the anterior midline of Hensen's node. MHP cells become anchored to the notochord beneath them and form a hinge, which forms a furrow at the dorsal midline.

Mediator   A large, multimeric complex of nearly 30 protein subunits that in many genes is the link that connects RNA polymerase II (bound to the promoter) to an enhancer sequence, thus forming a pre-initiation complex at the promoter.

Medullary cord   Forms by condensation of mesenchyme cells and then mesenchymal-to-epithelial transition in the caudal region of the embryo during the process of secondary neurulation. It will then cavitate to form the caudal section of the neural tube.

Meiosis   A unique division process occurring only in germ cells, to reduce the number of chromosomes to a haploid complement. All other cells divide by mitosis. Meiosis differs from mitosis in that (1) meiotic cells undergo two cell divisions without an intervening period of DNA replication, and (2) homologous chromosomes (each consisting of two sister chromatids joined at a kinetochore) pair together and recombine genetic material.

Melanocytes   Cells containing the pigment melanin. Derived from neural crest cells and undergo extensive migration to all regions of the epidermis.

Melanocyte stem cell   Adult stem cell derived from melanocyte trunk neural crest cells that resides in the bulge niche of the hair or feather follicle and which gives rise to the pigment of the skin, hair, and feathers.

Meroblastic   Greek *meros*, "part." Refers to the cell division (cleavage) pattern in zygotes containing large amounts of yolk, wherein only a portion of the cytoplasm is cleaved. The cleavage furrow does not penetrate the yolky portion of the cytoplasm because the yolk platelets impede membrane formation there. Only part of the egg is destined to become the embryo, while the other portion—the yolk—serves as nutrition for the embryo, as in insects, fish, reptiles, and birds.

Meroistic oogenesis   Type of oogenesis found in certain insects (including *Drosophila* and moths), in which cytoplasmic connections remain between the cells produced by the oogonium.

Mesencephalon   The midbrain, the middle vesicle of the developing vertebrate brain; major derivatives include optic tectum and tegmentum. Its lumen becomes the cerebral aqueduct.

Mesenchymal stem cells (MSCs)   Also called bone marrow-derived stem cells, or BMDCs. Multipotent stem cells that originate in the bone marrow, MSCs are able to give rise to numerous bone, cartilage, muscle, and fat lineages.

Mesenchyme   Loosely organized embryonic connective tissue consisting of scattered fibroblast-like and sometimes migratory mesenchymal cells separated by large amounts of extracellular matrix.

Mesentoblasts   In snail embryos, the cells derived from the 4d blastomere that give rise to both the mesodermal (heart) and endodermal (intestine) organs.

Mesoderm   Greek *mesos*, "between." The middle of the three embryonic germ layers, lying between the ectoderm and the endoderm. The mesoderm gives rise to muscles and skeleton; connective tissue; the reproductive organs; and to kidneys, blood, and most of the cardiovascular tissue.

Mesodermal mantle   The cells that involute through the ventral and lateral blastopore lips during amphibian gastrulation and will form the heart, kidneys, bones, and parts of several other organs.

Mesomeres   The eight cells generated in the sea urchin embryo by the fourth cleavage when the four cells of the animal tier divide meridionally into eight blastomeres, each with the same volume.

Mesonephros   The second kidney of the amniote embryo, induced in the adjacent mesenchyme by the middle portion of the nephric duct. It functions briefly in urine filtration in some mammalian species and mesonephric tubules form the tubes that transport the sperm from the testes to the urethra (the epididymis and vas deferens).

Mesonephric duct   See **Wolffian duct**.

Metalloproteinases   Matrix metalloproteinases (MMP) Enzymes that digest extracellular matrices and are important in many types of tissue remodeling in disease and development, including metastasis, branching morphogenesis of epithelial organs, placental detachment at birth, and arthritis.

Metamorphosis   Changing from one form to another, such as the transformation of an insect larva to a sexually mature adult or a tadpole to a frog.

Metanephrogenic mesenchyme   An area of mesenchyme, derived from posterior regions of the intermediate mesoderm, involved in mesenchymal-epithelial interactions that generate the metanephric kidney and will form the secretory nephrons.

Metanephros/metanephric kidney   The third kidney of the embryo and the permanent kidney of amniotes.

Metaphase plate   A structure present during mitosis or meiosis in which the chromosomes are attached via their kinetochores to the microtubule spindle and are lined up midway between the two poles of the cell.

Metastasis   The invasion of a cancerous cell into other tissues.

Metazoa   Animals.

Metencephalon   The anterior subdivision of the rhombencephalon; gives rise to the cerebellum, which coordinates movements, posture, and balance.

Microfilaments   Long cables of polymerized actin necessary for cytokinesis and also formed during fertilization in the egg's cortex where they are used to form microvilli.

Micromeres   Small cells created by asymmetrical cleavage, e.g., four small cells generated by the fourth cleavage at the vegetal pole when the vegetal tier of the sea urchin embryo undergoes an unequal equatorial cleavage.

Micropyle   The only place where *Drosophila* sperm can enter the egg, at the future dorsal anterior region of the embryo, a tunnel in the chorion (eggshell) that allows sperm to pass through it one at a time.

MicroRNA (miRNA)   A small (about 22 nucleotide) RNA complementary to a portion of a particular mRNA that regulates translation of a specific message. MicroRNAs often bind to the 3'UTR of mRNAs and inhibit their translation.

Microspikes   Essential for neuronal pathfinding, microfilament-containing pointed filopodia of the growth cone that elongate and contract to allow axonal migration. Microspikes also sample the microenvironment and sends signals back to the soma.

Microvilli   Small projections that extend from the egg surface during fertilization that may aid sperm entry into the cell.

Mid-blastula transition   The transition from the early rapid biphasic (only M and S phases) mitoses of the embryo to a stage characterized by (1) mitoses that include the "gap" stages (G1 and G2) of the cell cycle, (2) loss of syncronicity of cell division, and (3) transcription of new (zygotic) mRNAs needed for gastrulation and cell specification.

Midpiece   Section of sperm flagellum near the head that contains rings of mitochondria that provide the ATP needed to fuel the dynein ATPases and support sperm motility.

Mitosis-promoting factor (MPF)   Consists of cyclin B and p34/Cdk1 cyclin, required to initiate entry into the mitotic (M) phase of the cell cycle in both meiosis and mitosis.

mRNA cytoplasmic localization   The spatial regulation of mRNA translation, mediated by (1) diffusion and local anchoring, (2) localized protection, and (3) active transport along the cytoskeleton.

miRNA   See **MicroRNA**.

Model systems   Species that are easily studied in the laboratory and have special properties that allow their mechanisms of development to be readily observed (e.g., sea urchins, snails, ascidians, and *C. elegans*).

Modularity   A principle of the theoretical systems approach. The organism develops as a system of discrete and interacting modules.

Module A biological entity characterized by more internal than external integration.

Molecular parsimony (the "small toolkit") Development within all lineages uses the same types of molecules: transcription factors, paracrine factors, adhesion molecules, and signal transduction cascades are remarkably similar from one phylum to another.

Monospermy Only one sperm enters the egg, and a haploid sperm nucleus and a haploid egg nucleus combine to form the diploid nucleus of the fertilized egg (zygote), thus restoring the chromosome number appropriate for the species.

Monozygotic ("identical") twins Greek, "one-egg." Describes twins that form from a single embryo whose cells become dissociated from one another, either by the separation of early blastomeres, or by the separation of the inner cell mass into two regions within the same blastocyst. GeCompare with **Dizygotic**.

Morph One of several different potential phenotypes produced by environmental conditions.

Morphallaxis Type of regeneration that occurs through the repatterning of existing tissues with little new growth (e.g., *Hydra*).

Morphogens Greek, "form-givers." Substances that, by their differing concentrations, differentially specify cell fates. Morphogens are made in specific sites in the embryo, diffuse over long distances, and form concentration gradients where the highest concentration is at the point of synthesis, becoming lower as the morphogen diffuses away from its source and degrades over time.

Morphogenesis The organization of the cells of the body into functional structures via coordinated cell growth, cell migration, and cell death.

Morphogenetic determinants Transcription factors or their mRNAs that will influence the cell's development.

Morpholino An antisense oligonucleotide against an mRNA used to inhibit protein expression.

Morula Latin, "mulberry." Vertebrate embryo of 16–64 cells; precedes the blastula or blastocyst stage. Mammalian morula occurs at the 16-cell stage, consists of a small group of internal cells (the inner cell mass) surrounded by a larger group of external (trophoblast) cells.

Mosaic embryos Embryos in which most of the cells are determined by autonomous specification, with each cell receiving its instructions independently and without cell-cell interaction.

Mosaic pleiotropy A syndrome characterized by the expression of multiple, independently produced effects resulting from a gene being critical in different parts of the body.

Müller glial cells Cells of the neural retina that support and maintain the neurons therein.

Müllerian duct (paramesonephric duct) Duct running lateral to the mesonephric duct in both male and female mammalian embryos. These ducts regress in the male fetus, but form the oviducts, uterus, cervix, and upper part of the vagina in the female fetus. Compare with **Wolffian duct.**

Müllerian-inhibiting factor (MIF) See **Anti-Müllerian factor.**

Multipotent cardiac progenitor cells Progenitor cells of the heart field that form cardiomyocytes, endocardium, epicardium, and the Purkinje neural fibers of the heart.

Multipotent stem cells Adult stem cells whose commitment is limited to a relatively small subset of all the possible cells of the body.

Mutualism A form of symbiosis in which the relationship benefits both partners.

Myelencephalon The posterior subdivision of the rhombencephalon; becomes the medulla oblongata.

Myelin sheath Modified oligodendrocyte or Schwann cell plasma membrane that surrounds nerve cell axons, providing insulation that confines and speeds electrical impulses transmitted along axons.

Myoblast Muscle precursor cell.

Myocardium Heart muscles.

Myogenic regulatory factors (MRFs) bHLH transcription factors (such as MyoD and Myf5, that are critical regulators of muscle development.

Myostatin Greek, "muscle stopper." A member of the TGF-β family, it negatively regulates muscle development. Genetic defects in the gene or its negative regulatory miRNA cause huge muscles in some mammals, including humans.

Myotome Portion of the somite that gives rise to skeletal muscles. The myotome has two components: the primaxial component, closest to the neural tube, which forms the musculature of the back and rib cage, and the abaxial component, away from the neural tube, which forms the muscles of the ventral body wall.

N

N-cadherin See **Cadherins.**

NAD⁺ kinase Activated during the early response of the sea urchin egg to the sperm, converts NAD⁺ to NADP⁺ which can be used as a coenzyme for lipid biosynthesis and may be important in the construction of the many new cell membranes required during cleavage. NADP⁺ is also used to make NAADP.

Nanos Protein critical for the establishment of anterior-posterior polarity of the *Drosophila* embryo. *Nanos* mRNA is localized to the posterior pole.

Neoteny Retention of the juvenile form as a result of retarded body development relative to the development of the germ cells and gonads (which achieve maturity at the normal time). See also **Progenesis.**

Nephric duct See **Wolffian duct.**

Nephron Functional unit of the kidney.

Nerve growth factor (NGF) Neurotrophin released from potential target tissues that works at short ranges as either a chemotactic factor or chemorepulsive factor for axonal guidance. Also important in the selective survival of different subsets of neurons.

Netrins Paracrine factors found in a gradient that guide axonal growth cones. They are important in commissural axon migration and retinal axon migration.

Neural crest A transient band of cells, arising from the lateral edges of the neural plate, that joins the neural tube to the epidermis. It gives rise to a cell population—the neural crest cells—that detach during formation of the neural tube and migrate to form a variety of cell types and structures, including sensory neurons, enteric neurons, glia, pigment cells, and (in the head) bone and cartilage.

Neural folds Thickened edges of the neural plate that move upward during neurulation and migrate toward the midline and eventually fuse to form the neural tube.

Neural groove U-shaped groove that forms in the center of the neural plate during primary neurulation.

Neural plate The region of the dorsal ectoderm that is specified to become neural ectoderm. The cells of this region have a columnar appearance.

Neural restrictive silencer element (NRSE) A regulatory DNA sequence found in several mouse genes which prevents their expression in adult neurons.

Neural restrictive silencer factor (NRSF) A zinc finger transcription factor that binds the NRSE and is expressed in every cell that is *not* a mature neuron.

Neural retina   Derived from the inner layer of the optic cup, composed of a layered array of cells that include the light- and color-sensitive photoreceptor cells (rods and cones); the cell bodies of the ganglion cells; bipolar interneurons that transmit electric stimuli from the rods and cones to the ganglion cells, Müller glial cells that maintain its integrity, amacrine neurons (which lack large axons), and horizontal neurons that transmit electric impulses in the plane of the retina.

Neural tube   The embryonic precursor to the central nervous system (brain and spinal cord).

Neuroblast   An immature precursor cell that can differentiate into the cells of the nervous system.

Neurons   Cells specialized for the conduction and transmission of information via electrical and chemical signals. Also called **nerve cells**.

Neuropore   The two open ends (anterior neuropore and posterior neuropore) of the neural tube that later close.

Neurotransmitters   Molecules (e.g., acetylcholine, GABA, serotonin) secreted at the ends of axons. These molecules cross the synaptic cleft and are received by the adjacent neuron, thus relaying the neural signal. See also **Synapse**.

Neurotrophin   Greek *trophikos*, "nourish." Neurotropins supply factors (usually growth factors) that keep the neuron alive.

Neurotropin   Greek *tropikos*, "turn." Chemoattractant of neurons; molecules that guide axons to their appropriate targets.

Neurula   Refers to an embryo during neurulation (i.e., while the neural tube is forming).

Neurulation   Process of folding of the neural plate and closing of the cranial and caudal neuropores to form the neural tube.

Neural crest effectors   Transcription factors activated by neural crest specifiers that give the neural crest cells their migratory properties and some of their differentiated properties.

Neural crest specifiers   A set of transcription factors (e.g., FoxD3, Sox9, Id, Twist, and Snail) induced by the border-specifying transcription factors, that specify the cells that are to become the neural crest.

Neural plate border specifiers   A set of transcription factors induced by the neural plate inductive signals. These factors, including Distalless-5, Pax3, and Pax7, collectively prevent the border region from becoming either neural plate or epidermis.

Neural plate inductive signals   BMPs and Wnts secreted from the ventral ectoderm and paraxial mesoderm interact to specify the boundaries between neural and non-neural ectoderm during chick gastrulation. Relative concentrations of the same signals also specify neural crest and cranial placode cells.

Nieuwkoop center   The dorsalmost vegetal blastomeres of the amphibian blastula, formed as a consequence of the cortical rotation initiated by the sperm entry; an important signaling center on the dorsal side of the embryo. One of its main functions is to induce the Organizer.

Nodal   A paracrine factor and member of the TGF-β family involved in establishing left-right asymmetry in vertebrates and invertebrates. See also **Activins**.

Node   The mammalian homologue of Hensen's node.

Noggin   A soluble BMP antagonist that blocks BMP signaling.

Noninvoluting marginal zone (NIMZ)   Region of cells on the exterior of the gastrulating amphibian embryo that expand by epiboly to cover the entire embryo, eventually forming the surface ectoderm.

Non-skeletogenic mesenchyme   Formed from the veg2 layer of the 60-cell sea urchin embryo, it generates pigment cells, immunocytes, and muscle cells. Also called secondary mesenchyme.

Notch protein   Receptor for Delta, Jagged, or Serrate, participants in juxtacrine interactions. Ligand binding causes Notch to undergo a conformational change that enables a part of its cytoplasmic domain to be cut off by the Presenilin-1 protease. The cleaved portion enters the nucleus and binds to a dormant transcription factor of the CSL family. When bound to the Notch protein, the CSL transcription factors activate their target.

Notochord   A transient mesodermal rod in the most dorsal portion of the embryo that plays an important role in inducing and patterning the nervous system.

Nuclear RNA (nRNA)   The original transcription product, sometimes called *heterogeneous nuclear RNA* (hnRNA) or *premessenger RNA* (pre-mRNA); contains the cap sequence, the 5′ UTR, exons, introns, and the 3′ UTR.

Nuclear RNA selection   Means of controlling gene expression by processing specific subsets of the nRNA population into mRNA in different types of cells.

Nucleus   (1) An organized cluster of neurons in the brain with specific functions and connections. (2) The membrane-enclosed organelle housing the eukaryotic chromosomes.

Nucleosome   The basic unit of chromatin structure, composed of an octamer of histone proteins (two molecules each of histones H2A, H2B, H3, and H4) wrapped with two loops containing approximately 147 base pairs of DNA.

Nurse cells   Fifteen interconnected cells that generate mRNAs and proteins that are transported to the single developing oocyte during *Drosophila* oogenesis.

Nymph   Insect larval stage that resembles an immature adult of the species. Becomes progressively more mature though a series of molts.

O

Obesogens   Substances the increase the production and accumulation of adipose (fat) cells in the body. Several endocrine disruptors, including DES and BPA, have been shown to be obesogens.

Obligate mutualism   Symbiosis in which the species involved are interdependent with one another to such an extent that at least one partner could not survive without the other.

Olfactory placodes   Paired epidermal thickenings that form the nasal epithelium (smell receptors) as well as the ganglia for the olfactory nerves.

Oligodendrocyte   A type of glial cell.

Omphalomesenteric (umbilical) veins   The veins that form from yolk sack blood islands, and that bring nutrients to the mammalian embryo and transport gases to and from the sites of respiratory exchange.

Oncogenes   Regulatory genes which promote cell division, reduce cell adhesion, and prevent cell death. Can promote tumor formation and metastasis. Cancer can result from either mutations or inappropriate methylations that activate oncogenes.

Oocyte   The developing egg (prior to reaching the stage of meiosis at which it is fertilized).

Oogonium   A single female germ cell that will form an oocyte.

Optic cups   Double-walled chambers formed by the invagination of the optic vesicles.

Optic vesicle   Extend from the diencephalon and activate the head ectoderm's latent lens-forming ability.

Oral plate (stomodeum)   A region of ectoderm that blocks the oral end of the primitive gut.

Organization/activation hypothesis   The theory that sex hormones act during the fetal or neonatal stage of a mammal's life to organize the nervous system in a sex-specific manner; and that during adult life, the same hormones may have transitory motivational (or "activational") effects.

Organizer In amphibians, the dorsal lip cells of the blastopore and their derivatives (notochord and head endomesoderm). Functionally equivalent to Hensen's node in chick, the node in mammals, and the shield in fish. Organizer action establishes the basic body plan of the early embryo. Also known as the Spemann Organizer or (more correctly) the Spemann-Mangold organizer.

Organogenesis Interactions between, and rearrangement of, cells of the three germ layers to produce tissues and organs.

Orthologues Genes from different species that are similar in DNA sequence because those genes were inherited from a common ancestero. Compare with **Paralogues**.

Osteoblast A committed bone precursor cell.

Osteoclasts Multinucleated cells derived from a blood cell lineage that enter the bone through the blood vessels and destroy bone tissue during remodeling.

Osteocytes Bone cells, derived from osteoblasts that become embedded in the calcified osteoid matrix.

Osteogenesis Bone formation.

Osteoid matrix A collagen-proteoglycan secreted by osteoblasts that is able to bind calcium.

Otic placodes Paired epidermal thickenings that invaginate to form the inner ear labyrinth, whose neurons form the acoustic ganglia that enable us to hear.

Outflow tract In the developing heart, made up of the conus arteriosus and truncus arteriosus; becomes the aorta and the pulmonary arteries.

Ovariole The *Drosophila* egg chamber.

Ovulation Release of the oocyte from the ovarian follicle.

Ovum The mature egg (at the stage of meiosis at which it is fertilized).

P

P-cadherin See **Cadherins**.

p53 A transcription factor that can stop the cell cycle, cause cellular senescence in rapidly dividing cells, instruct the initiation of apoptosis, and activate DNA repair enzymes. One of the most important regulators of cell division.

Pair-rule genes *Drosophila* zygotic genes regulated by gap gene proteins which divide the embryo into periodic units, resulting in a striped pattern of seven transverse bands perpendicular to the anterior-posterior axis. Pair-rule mutants lacked portions of every other segment.

Paracrine factor A secreted, diffusible protein that provides a signal that interacts with and changes the cellular behavior of neighboring cells and tissues.

Paralogues Genes that are similar in sequence because they are the result of gene duplication events in an ancestral species. Compare with **Orthologues**.

Parasegment A "transegmental" unit in *Drosophila* that includes the posterior compartment of one segment and the anterior compartment of the immediately posterior segment; appears to be the fundamental unit of embryonic gene expression.

Paraxial (somitic) mesoderm Thick bands of embryonic mesoderm immediately adjacent to the neural tube and notochord. In the trunk, paraxial mesoderm gives rise to somites, in the head it gives rise to the connective tissues and musculature of the face.

Paraxial protocadherin Adhesion protein expressed specifically in the paraxial (somite-forming) mesoderm during amphibian gastrulation; essential for convergent extension.

Parthenogenesis Greek, "virgin birth." When an oocyte is activated in the absence of sperm. Normal development can proceed in many invertebrates and some vertebrates.

Pathway selection The first step in the specification of axonal connection, wherein the axons travel along a route that leads them to a particular region of the embryo.

Pattern formation The set of processes by which embryonic cells form ordered spatial arrangements of differentiated tissues.

Pericardial cavity The division of the coelom that surrounds the heart. Compare with **Peritoneal cavity; Pleural cavity**.

Pericytes Smooth muscle-like cells the endothelial cells recruit to cover them during vasculogenesis.

Periderm A temporary epidermis-like covering in the embryo that is shed once the inner layer differentiates to form a true epidermis.

Periosteal bone Bone which adds thickness to long bones and is derived from mesoderm via intramembranous ossification.

Periosteum A fibrous sheath containing connective tissue, capillaries, and bone progenitor cells and that covers the developing and adult bone.

Peritoneal cavity The division of the coelem that encloses the abdominal organs. Compare with **Pericardial cavity; Pleural cavity**.

Permissive interaction Inductive interaction in which the responding tissue has already been specified, and needs only an environment that allows the expression of these traits.

P granules Ribonucleoprotein complexes containing translation regulators that specify the germ cells in *C. elegans* and are localized by PAR ("partitioning") proteins.

Pharyngeal arches Also called a branchial arches, these are bars of mesenchymal tissue derived from paraxial mesoderm, lateral plate mesoderm, and neural crest cells. Found in the pharyngeal region (near the pharynx) of the vertebrate embryo, the arches will form gill supports in fishes and many skeletal and connective tissue structures in the face, jaw, mouth, and larynx in other vertebrates.

Pharyngeal clefts Clefts (invaginations) of external ectoderm that separate the pharyngeal arches. There are four pharyngeal clefts in the early embryo, but only the first becomes a structure (the external auditory meatus).

Pharyngeal pouches Inside the pharynx, these are where the pharyngeal epithelium pushes out laterally to form 4 pairs of pouches between the pharyngeal arches. These give rise to the auditory tube, wall of the tonsil, thymus gland, parathyroids and thyroid.

Pharyngula Term often applied to the late neurula stage of vertebrate embryos.

Pharynx The region of the digestive tube anterior to the point at which the respiratory tube branches off.

Phenotypic heterogeneity Refers to the same mutation producing different phenotypes in different individuals.

Phenotypic plasticity The ability of an organism to react to an environmental input with a change in form, state, movement, or rate of activity.

Pheromones Vaporized chemicals emitted by an individual that results in communication with another individual. Pheromones are recognized by the vomeronasal organ of many mammalian species and play a major role in sexual behavior.

Phospholipase C (PLC) Enzyme in the IP_3 pathway that splits membrane phospholipid phosphatidylinositol 4,5-bisphosphate (PIP2) to yield IP_3 and diacylglycerol (DAG).

Phylotypic stage The stage that typifies a phylum, such as the late neurula or pharyngula of vertebrates, and which appears to be relatively invariant and to constrain its evolution.

Pioneer nerve fibers   Axons that go ahead of other axons and serve as guides for them.

Pioneer transcription factors   Transcription factors that can penetrate repressed chromatin and bind to their enhancer DNA sequences, a step critical to establishing certain cell lineages. Examples: FoxA1, Pax7, Pbx.

Placenta   The organ that serves as the interface between fetal and maternal circulations and has endocrine, immune, nutritive and respiratory functions. It consists of a maternal portion (the uterine endometrium, or decidua, which is modified during pregnancy) and a fetal component (the chorion).

Placodes   Precursors of cutaneous appendages, such as hair, that are formed via inductive interactions between the dermal mesenchyme and the ectodermal epithelium.

Pleiotropy   The production of several effects by one gene or pair of genes.

Pleural cavity   The division of the coelom that surrounds the lngs. Compare with **Pericardial cavity; Peritoneal cavity**.

Pluripotent   Latin, "capable of many things." A single pluripotent stem cell has the ability to give rise to different types of cells that develop from the three germ layers (mesoderm, endoderm, ectoderm) from which all the cells of the body arise. The cells of the mammalian inner cell mass are pluripotent, as are embryonic stem cells. Each of these cells can generate any cell type in the body, but because the distinction between ICM and trophoblast has been established, it is thought that ICM cells are not able to form the trophoblast. Germ cells and germ cell tumors (such as teratocarcinomas) can also form pluripotent stem cells. Compare with **totipotent**.

Polar body   The smaller cell, containing hardly any cytoplasm, generated during the asymmetrical meiotic division of the mammalian oocyte. The first polar body is diploid and results from the first meiotic division and the secondary polar body is haploid and results from the second meiotic division.

Polar granules   Particles containing factors important for germ line specification that are localized to the pole plasm and pole cells of *Drosophila*.

Polar granule component (PGC)   A protein important for germ line specification and localized to *Drosophila* polar granules. Pgc inhibits transcription of somatic cell-determining genes by preventing the phosphorylation of RNA polymerase II.

Polarization   The first stage of cell migration, wherein a cell defines its front and its back, directed by diffusing signals (such as a chemotactic protein) or by signals from the extracellular matrix. These signals will reorganize the cytoskeleton such that the front part of the cell will form *lamellipodia* (or *filopodia*) with newly polymerized actin.

Polar lobe   An anucleate bulb of cytoplasm extruded immediately before first cleavage in certain spirally cleaving embryos (mostly in the mollusc and annelid phyla). It contains the determinants for the proper cleavage rhythm and the cleavage orientation of the D blastomere.

Pole cells   About five nuclei in the *Drosophila* embryo that reach the surface of the posterior pole during the ninth division cycle and become enclosed by cell membranes. The pole cells give rise to the gametes of the adult.

Pole plasm   Cytoplasm at the posterior pole of the *Drosophila* oocyte which contains the determinants for producing the abdomen and the germ cells.

Polyadenylation   The insertion of a "tail" of some 200–300 adenylate residues on the RNA transcript, about 20 bases downstream of the AAUAAA sequence. This polyA tail (1) confers stability on the mRNA, (2) allows the mRNA to exit the nucleus, and (3) permits the mRNA to be translated into protein.

PolyA tail   A series of adenine (A) residues that are added by enzymes to the 3′ terminus of the mRNA transcript in the nucleus. The polyA tail may stabilize the message by protecting it from exonucleases that would otherwise digest it.

Polycomb proteins   Family of proteins that bind to condensed nucleosomes, keeping the genes in an inactive state.

Polyphenism   Refers to *discontinuous* ("either/or") phenotypes elicited by the environment. Compare with **reaction norm**.

Polyspermy   The entrance of more than one sperm during fertilization resulting in aneuploidy (abnormal chromosome number) and either death or abnormal development.

Polytene chromosomes   Chromosomes in certain larval *Drosophila* in which the DNA undergoes many rounds of replication without separation, forming large "puffs" that are easily visible and indicate active gene transcription.

Posterior marginal zone (PMZ)   The end of the chick blastoderm where primitive streak formation begins and acts as the equivalent of the amphibian Nieuwkoop center. The cells of the PMZ initiate gastrulation and prevent other regions of the margin from forming their own primitive streaks.

Posttranslational regulation   Protein modifications that determine whether or not a protein will be active. These modifications can include cleaving an inhibitory peptide, sequestration and targeting to specific cell regions, assembly with other proteins in order to form a functional unit, binding an ion (such as Ca^{2+}), or modification by the covalent addition of a phosphate or acetate group.

Pre-initiation complex   The complex of RNA polymerase II at the promoter with transcription factors on the enhancer, as brought together by the Mediator molecules. See also Mediator

Prechordal plate mesoderm   Cells that migrate anteriorly through Hensen's node, between the endoderm and the epiblast, and form the precursor of the head mesoderm.

Precursor cells (precursors)   Widely used term to denote any nondifferentiated cell type (stem or progenitor cells) of a particular lineage (e.g., neuronal precursors; blood cell precursors).

Predator-induced polyphenism   The ability to modulate development in the presence of predators in order to express a more defensive phenotype.

Preeclampsia   Medical condition of pregnant women characterized by dangerously high and fluctuating blood pressure. May be caused by substances from the placenta to which some women can have a pathological reaction.

Preformation   The view, supported by the early microscopist, Marcello Malpighi, that the organs of the embryo are already present, in miniature form, within the egg (or sperm). A corollary, *emboîtment* (encapsulation), stated that the next generation already existed in a prefigured state within the germ cells of the first prefigured generation, thus ensuring that the species would remain constant.

Preimplantation genetics   Testing for genetic diseases using blastomeres from embryos produced by in vitro fertilization before implanting the embryo in the uterus.

Prenatal diagnosis   The use of chorionic villus sampling or amniocentesis to diagnose many genetic diseases before a baby is born.

Primary capillary plexus   A network of capillaries formed by endothelial cells during the third phase of vasculogenesis.

Primary embryonic induction   The process whereby the dorsal axis and central nervous system forms through interactions with the underlying mesoderm, derived from the dorsal lip of the blastopore.

Primary heart field The cells of the heart field that will form the inflow tract (left ventricle and the atria).

Primary larvae Larvae that represent dramatically different body plans than the adult form and that are morphologically distinct from the adult; the plutei of sea urchins are such larvae. Compare with **secondary larvae**.

Primary neurulation The process that forms the anterior portion of the neural tube. The cells surrounding the neural plate direct the neural plate cells to proliferate, invaginate, and pinch off from the surface to form a hollow tube.

Primary oocytes Formed in the mammalian fetus from oogonia that have entered the first meiotic division and are arrested in first meiotic prophase until puberty.

Primary sex determination The determination of the gonads to form either the egg-forming ovaries or sperm-forming testes. Primary sex determination is chromosomal and is not usually influenced by the environment in mammals, but can be affected by the environment in other vertebrates.

Primary spermatocytes Derived from mitotic division of the type B spermatogonia, these are the cells that enter meiosis.

Primaxial muscles The intercostal musculature between the ribs and the deep muscles of the back, formed from those myoblasts in the myotome closest to the neural tube.

Primitive groove A depression that forms within the primitive streak that serves as an opening through which migrating cells pass into the deep layers of the embryo.

Primitive knot/pit See Hensen's node.

Primitive streak The first morphological sign of gastrulation in amniotes, it first arises from a local thickening of the epiblast at the posterior edge of the area pellucida, called Koller's sickle. Homologous to the amphibian blastopore.

Primordial germ cells (PGCs) Gamete progenitor cells, which typically arise elsewhere and migrate into the developing gonads.

Proacrosin A mammalian sperm protein that adheres to the inner acrosomal membrane and binds to sulfated carbohydrate groups on the zona pellucida glycoproteins.

Proerythroblast A red blood cell precursor.

Progenesis Condition in which the gonads and germ cells develop at a faster rate than than the rest of the body, becoming sexually mature while the rest of the body is still in a juvenile phase Compare with **Neoteny**.

Progenitor cells Relatively undifferentiated cells that have the capacity to divide a few times before differentiating and, unlike stem cells, are not capable of unlimited self-renewal. They are sometimes called **transit amplifying cells** because they divide while migrating.

Progesterone A steroid hormone important in the maintenance of pregnancy in mammals. Progesterone secreted from the cumulus cells may act as a chemotactic factor for sperm.

Programmed cell death See **apoptosis**.

Progress zone The limb bud mesenchyme directly beneath the apical ectodermal ridge. The proximal-distal growth and differentiation of the limb bud are made possible by a series of interactions between the AER and the progress zone.

Progress zone model Model for specification of proximal-distal specification of the limb which postulates that each mesoderm cell is specified by the amount of time it spends dividing in the progress zone. The longer a cell spends in the progress zone, the more mitoses it achieves and the more distal its specification becomes.

Prometamorphosis The second stage of metamorphosis, during which the thyroid matures and secretes more thyroid hormones.

Promoter Region of a gene containing the DNA sequence to which RNA polymerase II binds to initiate transcription. See also **CpG islands; Enhancer**.

Pronephric duct The pronephric duct arises in the intermediate mesoderm, migrates caudally, and induces the adjacent mesenchyme to form the pronephros, or tubules of the initial kidney of the embryo. The pronephric tubules form functioning kidneys in fish and in amphibian larvae but are not believed to be active in amniotes. In mammals, the caudal pronephric tubules become the Wolffian duct.

Prosencephalon The forebrain; the most anterior vesicle of the developing vertebrate brain. Will form two secondary brain vesicles: the telencephalon and the diencephalon.

Protamines Basic proteins, tightly compacted through disulfide bonds, that package the DNA of the sperm nucleus.

Proteoglycans Large extracellular matrix molecules consisting of core proteins (such as syndecan) with covalently

attached glycosaminoglycan polysaccharide side chains. Two of the most widespread proteoglycans are heparan sulfate and chondroiton sulfate.

Proteome The number and type of proteins encoded by the genome.

Prothoracicotropic hormone (PTTH) A peptide hormone that initiates the molting process in insects when it is released by neurosecretory cells in the brain in response to neural, hormonal, or environmental signals. PTTH stimulates the production of ecdysone by the prothoracic gland.

Protocadherins A class of cadherins that lack the attachment to the actin skeleton through catenins. They are an important means of keeping migrating epithelia together, and they are important in separating the notochord from surrounding tissues during its formation.

Protostomes Greek, "mouth first." Animals that form their mouth regions from the blastopore, such as molluscs. Compare with **deuterostomes**.

Proximal-distal axis The close-far axis of the limb, e.g., shoulder-finger or hip-toe.

Pseudohermaphroditism Intersex conditions in which the secondary sex characteristics differ from what would be expected from the gonadal sex. Male pseudohermaphroditism (e.g., androgen insensitivity syndrome) describes conditions wherein the gonadal sex is male and the secondary sex characteristics are female while female pseudohermaphroditism describes the reverse situation (e.g., congenital adrenal hyperplasia).

Purkinje neurons Large, many-branched neurons that are the major cell type of the cerbellar cortex.

R

R-cadherin See **Cadherins**

Radial glial cells (RGCs) Neural progenitor cells found in the ventricular zone (VZ) of the developing brain. At each division, they generate another VZ cell and a more committed cell type that leaves the VZ to differentiate.

Radial holoblastic cleavage Cleavage pattern in echinoderms. The cleavage planes are parallel and perpendicular to the animal-vegetal axis of the egg.

Random epigenetic drift The hypothesis that the chance accumulation of inappropriate epigenetic methylation due to errors made by the DNA methylating and demethylating enzymes could be the critical factor in aging and cancers.

Ras A G-protein in the RTK pathway. Mutations in the *RAS* gene account for a large proportion of cancerous human tumors.

Reaction-diffusion model Model for developmental patterning wherein two homogeneously distributed substances (an activator that activates itself as well as forming its own, faster-diffusing inhibitor) interact to produce stable patterns during morphogenesis. These patterns would represent regional differences in the concentrations of the two substances.

Reaction norm A type of phenotypic plasticity in which the genome encodes the potential for a *continuous range* of potential phenotypes, and the environment the individual encounters determines the phenotype. Compare with **polyphenism**.

Reactive oxygen species (ROS) Metabolic by-products that can damage cell membranes and proteins and destroy DNA. ROS are generated by mitochondria due to insufficient reduction of oxygen atoms and include superoxide ions, hydroxyl ("free") radicals, and hydrogen peroxide.

Receptor tyrosine kinase A receptor that spans the cell membrane and has an extracellular region, a transmembrane region, and a cytoplasmic region. Ligand (paracrine factor) binding to the extracellular domain causes a conformational change in the receptor's cytoplasmic domains, activating kinase activity that uses ATP to phosphorylate specific tyrosine residues of particular proteins.

Reciprocal inductions A common sequential feature of induction: An induced tissue in turn induces other tissues, including the original inducer.

Regeneration The ability to reform an adult structure or organ that has been damaged or destroyed by trauma or disease.

Regeneration blastema An aggregation of relatively dedifferentiated cells derived from the originally differentiated tissue, which then proliferates and redifferentiates into the new limb parts during epimorphic regeneration.

Regenerative medicine The potential use of stem cells to correct genetic pathologies (e.g., sickle-cell anemia) or repair damaged organs.

Regulation The ability to respecify cells so that the removal of cells destined to become a particular structure can be compensated for by other cells producing that structure. This is seen when an entire embryo is produced by cells that would have contributed only certain parts to the original embryo. It is also seen in the ability of two or more early embryos to form one chimeric individual

rather than twins, triplets, or a multi-headed individual.

Relational pleiotropy Describes syndromes in which a defective gene in one part of the embryo causes a defect in another part, even though the gene is not expressed in the second tissue.

Reporter gene A gene with a product that is readily identifiable and not usually made in the cells of interest. Can be fused to regulatory elements from a gene of interest, inserted into embryos, and then monitored for reporter gene expression If the sequence contains an enhancer, the reporter gene should become active at particular times and places. The genes for green fluorescent protein (*GFP*) and β-galactosidase (*lacZ*) are common examples.

Resact A 14-amino-acid peptide that has been isolated from the egg jelly of the sea urchin *Arbacia punctulata* that acts as a chemotactic factor and sperm-activating peptide for *Arbacia* sperm.

Resegmentation Occurs during formation of the vertebrae from sclerotomes; the rostral segment of each sclerotome recombines with the caudal segment of the next anterior sclerotome to form the vertebral rudiment and this enables the muscles to coordinate the movement of the skeleton, permitting the body to move laterally.

Respiratory tube The future respiratory tract, which forms as an epithelial outpocketing of the pharynx, and eventually bifurcates into the two lungs.

Responder During induction, the tissue being induced. Cells of the responding tissue must have a receptor protein for the paracrine factor and have the competence to respond to it.

Rete testis A network of thin canals located near, and eventually connecting with, the mesonephric duct that connects with the testis cords of the developing testis and conveys spermatozoa from the seminiferous tubules to the mesonephric duct derivatives, the epididymis and ductus deferens (vas deferens) in the adult.

Reticulocyte Cell that derived from the mammalian erythroblast that has expelled its nucleus. Although reticulocytes, lacking a nucleus, can no longer synthesize globin mRNA, they can translate *existing* messages into globins.

Retina See **Neural retina**.

Retinal homeobox (Rx) Gene expressed in the central bulge region of the diencephalon; required for optic cup and retina formation.

Retinoic acid (RA) A derivative of vitamin A and morphogen involved in anterior-posterior axis formation.

Retinoic acid-4-hydroxylase An enzyme that degrades retinoic acid.

Retinotectal projection The map of retinal connections to the optic tectum. Point-for-point correspondence between the cells of the retina and the cells of the tectum that enables the animal to see an unbroken image.

Rho GTPases A family of molecules including RhoA, Rac1, and Cdc42 that help mediate epithelial cell movement in morphogenesis and the cadherin-dependent remodeling of the cytoskelton.

Rhombencephalon The hindbrain, the most caudal vesicle of the developing vertebrate brain; will form two secondary brain vesicles, the metencephalon and myelencephalon.

Rhombomeres Periodic swellings that divide the rhombencephalon into smaller compartments, each with a different fate and different associated cranial nerve ganglia.

Right-left axis Specification of the two lateral sides of the body.

Ring canals The cytoplasmic interconnections between the cystocytes that become the ovum and nurse cells in *Drosophila*.

RNA-induced silencing complex (RISC) A complex containing several proteins and a microRNA which can then bind to the 3′ UTR of messages and inhibit their translation.

RNA polymerase II An enzyme that binds to a promoter on DNA and, when activated, catalyzes the transcription of an RNA template from the DNA.

Robustness (canalization) The ability of an organism to develop the same phenotype despite perturbations from the environment or from mutations.

Roof plate Dorsal region of the neural tube important in the establishment of dorsal-ventral polarity. The adjacent epidermis induces a secondary signaling center and BMP4 expression in the roof plate cells of the neural tube that in turn induces a cascade of TGF-β proteins in adjacent cells.

Rostral-caudal Latin, "beak-tail." An anterior-posterior positional axis, especially when referring to chick embryos or the head and brain of mammals.

Rotational cleavage The cleavage pattern for mammalian and nematode embryos. In mammals, the first cleavage is a normal meridional division while in the second cleavage, one of the two blastomeres divides meridionally and the other divides equatorially. In *C. elegans*, each asymmetrical division produces one founder cell that produces differentiated descendants; and one stem cell. The stem cell lineage always undergoes meridional division to produce (1) an anterior founder cell and (2) a posterior cell that will continue the stem cell lineage.

R-spondin1 (Rspo1) Small, soluble protein that upregulates the Wnt pathway and is critical for ovary formation.

RTK pathway The receptor tyrosine kinase (RTK) is dimerized by ligand, which causes autophosphorylation of the receptor. An adaptor protein recognizes the phosphorylated tyrosines on the RTK and activates an intermediate protein, GEF, which activates the Ras G protein by allowing the phosphorylation of the GDP-bound Ras. At the same time, the GAP protein stimulates the hydrolysis of this phosphate bond, returning Ras to its inactive state. The active Ras activates the Raf protein kinase C (PKC), which in turn phosphorylates a series of kinases. Eventually, an activated kinase alters gene expression in the nucleus of the responding cell by phosphorylating certain transcription factors (which can then enter the nucleus to change the types of genes transcribed) and certain translation factors (which alter the level of protein synthesis). In many cases, this pathway is reinforced by the release of Ca^{2+}.

S

Sacral neural crest Neural crest cells that lie posterior to chick somite 28 that generate the parasympathetic (enteric) ganglia of the gut that are required for peristaltic movement in the bowels.

Satellite cell Putative muscle stem cell found within the basal lamina of mature myofibers that can respond to injury or exercise by proliferating into myogenic cells that fuse and form new muscle fibers.

Schwann cell Type of glial cell that generates a myelin sheath, allowing rapid transmission of electrical signals along an axon.

Sclerotomes Blocks of mesodermal cells in the ventromedial half of each somite that will differentiate into the vertebral bodies and intervertebral discs of the spine and ribs. They are also critical in patterning the neural crest and motor neurons.

Sebaceous gland Glands that surround hair follicle cells and produce an oily substance, **sebum**, that serves to lubricate the hair and skin.

Secondary hypoblast Underlies the epiblast in the bilaminar amniotic blastoderm. A sheet of cells derived from deep yolky cells at the posterior margin of the blastoderm that migrates

anteriorly, displacing the hypoblast islands (primary hypoblast). Hypoblast cells do not contribute to the primary germ layers of the embryo proper, but instead form portions of the external membranes and provide chemical signals that specify the migration of epiblast cells. Also called endoblast.

Secondary larvae Larvae that possess the same basic body plan as the adult; caterpillars and tadpoles are examples. Compare with **primary larvae**.

Secondary neurulation The process that forms the posterior portion of the neural tube by the coalescence of mesenchyme cells into a solid cord that subsequently forms cavities that coalesce to create a hollow tube.

Secondary oocyte The diploid oocyte following the first meiotic division, which also generates the first polar body.

Secondary sex determination Developmental events, directed by hormones or factors produced by the gonads, that affect the phenotype outside the gonads. This includes the male or female duct systems and external genitalia, and, in many species, sex-specific body size, vocal cartilage, and musculature.

Secondary spermatocytes A pair of cells derived from the first meiotic division of a primary spermatocyte, which then complete the second division of meiosis to generate the four haploid spermatids.

Segmental plate The bands of paraxial mesoderm in chick embryos.

Segment polarity genes *Drosophila* zygotic genes activated by the proteins encoded by the pair-rule genes whose mRNA and protein products divide the embryo into 14-segment-wide units, establishing the periodicity of the embryo. Segment polarity mutants showed defects (deletions, duplications, polarity reversals) in every segment.

Selective affinity Principle that explains why disaggregated cells reaggregate to reflect their embryonic positions. Specifically, the inner surface of the ectoderm has a positive affinity for mesodermal cells and a negative affinity for the endoderm, while the mesoderm has positive affinities for both ectodermal and endodermal cells.

Semaphorins Extracellular matrix proteins that repel migrating neural crest cells and axonal growth cones.

Seminiferous tubules Form at puberty from the testis cords, containing the sertoli cells and the gonocytes/spermatogonial stem cells, by the initiation of spermatogenesis, proliferation and differentiation of the spermatogenic

epithelium, and formation of a fluid-filled lumen.

Senescence The physiological deterioration that characterizes old age.

Septum A partition that divides a chamber, such as the atrial septa that split the developing atrium into left and right atria. Plura, septa.

Sertoli cells Large secretory support cells in the seminiferous tubules of the testes involved in spermatogenesis in the adult through their role in nourishing and maintaining the developing sperm cells. They secrete AMF in the fetus and provide a niche for the incoming germ cells. They are derived from somatic cells, which are in turn derived from the genital ridge epithelium.

Sesamoid bone Small bones at joints that form as a result of mechanical stress (such as the patella). They are derived from mesoderm via intramembranous ossification.

Shield See **Embryonic shield**.

Short neural precursors (SNPs) Neuroblast precursor cells of the ventricular zone derived from radial glial cells.

Signal transduction cascades Pathways of response whereby paracrine factors bind to a receptor that initiates a series of enzymatic reactions within the cell which in turn have as their end point either the regulation of transcription factors (such that different genes are expressed in the cells reacting to these paracrine factors) or the regulation of the cytoskeleton (such that the cells responding to the paracrine factors alter their shape or are permitted to migrate).

Silencer A DNA regulatory element that binds transcription factors that actively repress the transcription of a particular gene.

Sinus venosus The posterior region of the developing heart, where the two major veins fuse. Inflow tract to the atrial area of the heart.

Sirtuin genes Encode histone deacetylation (chromatin-silencing) enzymes that guard the genome, preventing genes from being expressed at the wrong times and places, and blocking chromosomal rearrangements. They may be important defenses against premature aging.

Skeletogenic mesenchyme Also called primary mesenchyme, formed from the first tier of micromeres (the large micromeres) of the 60-cell sea urchin embryo, forms the larval skeleton.

Slow block to polyspermy See **Cortical granule reaction**.

SMAD pathway The pathway activated by members of the TGF-β superfamily. The TGF-β ligand binds to a type II TGF-β receptor, which allows that receptor to bind to a type I TGF-β receptor. Once the two receptors are in close contact, the type II receptor phosphorylates a serine or threonine on the type I receptor, thereby activating it. The activated type I receptor can now phosphorylate the Smad proteins. Smads 1 and 5 are activated by the BMP family of TGF-β factors, while the receptors binding activin, Nodal, and the TGF-β family phosphorylate Smads 2 and 3. These phosphorylated Smads bind to Smad4 and form the transcription factor complex that will enter the nucleus.

Small-interfering RNA (siRNA)   Similar to miRNA, is the basis of RNA interference technique. siRNA is made from double-stranded RNA and is also packaged into RISCs.

Small micromeres   A cluster of cells produced by the fifth cleavage at the vegetal pole in the sea urchin embryo when the micromeres divide.

Soma   Greek, "body." Can refer to the cell body (particularly of neurons). **Somatic cells** are those that make up the body—i.e., all cells in the organism that are not germ cells.

Somatopleure   See **Splanchnic mesoderm**.

Somatic cell nuclear transfer (SCNT)   Less accurately known as "cloning," the procedure by which a cell nucleus is transferred into an activated enucleated egg and directs the development of a complete organism with the same genome as the donor cell.

Somatic mesoderm (parietal mesoderm)   Derived from lateral mesoderm closest to the ectoderm (dorsal) and separated from other component of lateral mesoderm (splanchnic, near endoderm, ventral) by the intraembryonic coelom. It forms the somatopleure with the overlying ectoderm, and it will form the body wall, and the lining of the coelom. Not to be confused with somitic (paraxial) mesoderm.

Somite   Segmental block or ball of mesoderm formed from paraxial mesoderm adjacent to notochord (the axial mesoderm). Differentiates to form, initially, sclerotome and dermamyotome; the latter goes on to form dermotome and myotome. Somites will form the axial skeleton (vertebrae, ribs), all skeletal muscle, and the dorsal dermis, tendons, joints, and dorsal aortic cells.

Somitic mesoderm   See **Paraxial mesoderm**. Not to be confused with **somatic mesoderm**.

Somitogenesis   The process of segmentation of the paraxial mesoderm to form somites, beginning cranially and extending caudally. Its components are (1) periodicity, (2) fissure formation (to separate the somites), (3) epithelialization, (4) specification, and (5) differentiation.

Somitomeres   Early pre-somites, consisting of paraxial mesoderm cells organized into whorls of cells.

Sonic hedgehog (SHH)   The major hedgehog family paracrine factor. SHH has distinct functions in different tissues of the embryo. For example, it is secreted by the notochord and ventralizes the neural tube. It is also involved in the establishment of left-right asymmetry, primitive gut tube differentiation, proper placement of feather formation in chicks, and in the differentiation of the sclerotome and the epaxial myotome.

Specification   The first stage of commitment of cell or tissue fate during which the cell or tissue is capable of differentiating autonomously (i.e., by itself) when placed in an environment that is neutral with respect to the developmental pathway. At the stage of specification, cell commitment is still capable of being reversed.

Spemann's Organizer   See **Organizer.**

Spermatids   Haploid sperm cells generated by meiosis that are still connected to one another through their cytoplasmic bridges, and are therefore functionally diploid due to the diffusion of gene products across the cytoplasmic bridges.

Spermatogenesis   The production of sperm.

Spermatozoa   The male gamete or mature sperm cell.

Sperm head   Consists of the nucleus, acrosome, and minimal cytoplasm.

Spermiogenesis   The differentiation of the mature spermatozoa from the haploid round spermatid.

Sperm-surface galactosyltransferase   A mammalian sperm surface protein that recognizes carbohydrate residues on ZP3 and may contribute to sperm-zona binding.

Spina bifida   A congenital defect resulting from failure to close the posterior neuropore, the severity of which depends on how much of the spinal cord remains exposed.

Spiral holoblastic cleavage   Characteristic of several animal groups, including annelid worms, some flatworms, and most molluscs. Cleavage is at oblique angles, forming a "spiral" arrangement of daughter blastomeres, the cells touch one another at more places than do those of radially cleaving embryos, assuming the most thermodynamically stable packing orientation.

Splanchnic mesoderm   Also called the visceral mesoderm; derived from lateral mesoderm closest to the endoderm (ventral) and separated from other component of lateral mesoderm (somatic, near ectoderm, dorsal) by the intraembryonic coelom. It forms the splanchnopleure together with the underlying endoderm. It will form the heart, capillaries, gonads, the visceral mesothelial and serous membranes that cover the organs, the mesenteries, and blood cells.

Splanchnopleure   See **Splanchnic mesoderm.**

Spliceosome   A complex, made up of small nuclear RNAs (snRNAs) and splicing factors, that bind to splice sites and mediates the splicing of nRNA.

Splicing enhancer   A *cis*-acting sequence on nRNA that promotes the assembly of spliceosomes at RNA cleavage sites.

Splicing factors   Proteins that bind to splice sites or to the areas adjacent to them.

Splicing isoforms   Different proteins encoded by the same gene and generated by alternative splicing.

Splicing silencer   A *cis*-acting sequence on nRNA that acts to exclude exons from an mRNA sequence.

Sry   Sex-determining region of the Y chromosome. Gene encoding the mammalian testis-determining factor. The *Sry* gene is probably active for only a few hours in the genital ridge, during which time, it synthesizes the Sry transcription factor, whose primary role is to activate the *Sox9* gene required for testis formation.

STAT   Signal transducers and activators of transcription. A family of transcription factors, part of the JAK-STAT pathway. Important in the regulation of human fetal bone growth.

Stem cell   A relatively undifferentiated cell from the embryo, fetus, or adult that, when it divides, produces (1) one cell that retains its undifferentiated character and remains in the stem cell niche; and (2) a second cell that leaves the niche and can undergo one or more paths of differentiation. See also **Adult stem cell**; **Embryonic stem cell**.

Stem cell mediated regeneration   Process by which stem cells allow an organism to regrow certain organs or tissues (e.g., hair, blood cells) that have been lost.

Stem cell niche   An environment (regulatory microenvironment) that provides a milieu of extracellular matrices and paracrine factors that allows cells residing within it to remain relatively undifferentiated. Regulates stem cell proliferation and differentiation.

Stereoblastulae Blastulae produced by spiral cleavage; have no blastocoel.

Stomodeum *See* **Oral plate**.

Stratum germinativum *See* Basal layer.

Stylopod The proximal limb bones of any vertebrate limb, adjacent to the body wall, consisting of either the humerus (forelimb) or the femur (hindlimb).

Subendothelial cells A minor cell population in the bone marrow that may be critical for forming the hematopoietic inductive microenvironment.

Subgerminal cavity A space between the blastoderm and the yolk of avian eggs which is created when the blastoderm cells absorb water from the albumen ("egg white") and secrete the fluid between themselves and the yolk.

Sulcus limitans A longitudinal groove that divides the developing spinal cord and medulla into dorsal (receives sensory input) and ventral (initiates motor functions) halves.

Superficial cleavage The divisions of the cytoplasm of centrolecithal zygotes occur only in the rim of cytoplasm around the periphery of the cell due to the presence of a large amount of centrally-located yolk, as in insects.

Surfactant A secretion of specific proteins and phospholipids such as sphingomyelin and lecithin produced by the type II alveolar cells of the lungs very late in gestation. The surfactant enables the alveolar cells to touch one another without sticking together.

Symbiont The smaller organism in a symbiotic relationship in which one of the organisms involved is much larger than the other, and the smaller organism may live on the surface or inside the body of the larger.

Symbiosis Greek, "living together." Refers to any close association between organisms of different species.

Synapse Junction at which a neuron contacts its target cell (which can be another neuron or another type of cell) and information in the form of **neurotransmitter** molecules (e.g., acetylcholine, GABA, serotonin) is exchanged across the **synaptic cleft** between the two cells.

Synapsis The highly specific parallel alignment (pairing) of homologous chromosomes during the first meiotic division.

Syncytial blastoderm Describes the *Drosophila* embryo, in which no cell membranes exist other than that containing the entire embryo.

Syncytial specification The interactions of nuclei and transcription factors, which eventually result in cell specification, take place in a common cytoplasm.

Syncytiotrophoblast A population of cells from the murine and primate trophoblast that undergoes mitosis without cytokinesis resulting in multinucleate cells. The syncytiotrophoblast tissue is thought to further the progression of the embryo into the uterine wall by digesting uterine tissue.

Syncytium Many nuclei reside in a common cytoplasm, results either from karyokinesis without cytokinesis or from cell fusion.

Syndetome Greek *syn*, "connected." Derived from the most dorsal sclerotome cells, which express the *scleraxis* gene and generate the tendons.

Syndrome Greek, "happening together." Several malformations or pathologies that occur concurrently. Genetically based syndromes are caused either by (1) a chromosomal event (such as trisomy 21, or Down syndrome) where several genes are deleted or added, or (2) by one gene having many effects.

Systems theory In development, refers to an approach that views the organism coming the interactions of its component processes. Although the emphasis applied to each varies, the theoretical systems approach can be characterized by six principles: (1) context-dependent properties; (2) level-specific properties and emergence; (D) heterogeneous causation; (4) integration; (5) modularity and robustness; and (6) homeorhesis (stability while undergoing change).

T

T-box (Tbx) A specific DNA-binding domain found in certain transcription factors, including the T (*Brachyury*) gene, Tbx4 and Tbx5. Tbx4 and Tbx5 help specify hindlimbs and forelimbs, respectively.

Target selection The second step in the specification of axonal connection, wherein the axons, once they reach the correct area, recognize and bind to a set of cells with which they may form stable connections.

Telencephalon The anterior subdivision of the prosencephalon; will eventually form the cerebral hemispheres.

Telogen The resting phase of the hair follical regeneration cycle.

Telolecithal Describes the eggs of birds and fish which have only one small area of the egg that is free of yolk.

Telomeres Repeated DNA sequences at the ends of chromosomes.

Teratocarcinoma A tumor derived from malignant primordial germ cells and containing an undifferentiated stem cell population (embryonal carcinoma, or EC cells) that has biochemical and developmental properties similar to those of the inner cell mass. EC cells can differentiate into a wide variety of tissues, including gut and respiratory epithelia, muscle, nerve, cartilage, and bone.

Teratogens Greek, "monster-formers." Exogenous agents that cause disruptions in development resulting in teratogenesis, the formation of congenital defects. Teratology is the study of birth defects and of how environmental agents disrupt normal development.

Testis cords Loops in the medullary (central) region of the developing testis formed by the developing Sertoli cells and the incoming germ cells. Will become the seminiferous tubules and site of spermatogenesis.

Testis-determining factor A protein encoded by the *SRY* (sex-determining region on Y) gene on the mammalian Y chromosome that organizes the gonad into a testis rather than an ovary.

Testosterone A steroid hormone secreted by the fetal testes which masculinizes the fetus, stimulating the formation of the penis, male duct system, scrotum, and other portions of the male anatomy, as well as inhibiting development of the breast primordia.

TGF-β superfamily Transforming growth factor. More than 30 structurally related members of a group of paracrine factors. The proteins encoded by TGF-β superfamily genes are processed such that the carboxy-terminal region contains the mature peptide. These peptides are dimerized into homodimers (with themselves) or heterodimers (with other TGF-β peptides) and are secreted from the cell. The TGF-β superfamily includes the TGF-β family, the activin family, the bone morphogenetic proteins (BMPs), the Vg1 family, and other proteins, including glial-derived neurotrophic factor (GDNF; necessary for kidney and enteric neuron differentiation) and Müllerian inhibitory factor (which is involved in mammalian sex determination).

Thecal cells Steroid hormone-secreting ovarian cells surrounding the follicle that differentiate from mesenchyme cells of the ovary.

Thermotaxis Migration towards an area of higher temperature.

Threshold model A model of development wherein biological events are triggered when a specific concentration of a morphogen or hormone is reached.

Tip cells   Certain endothelial cells that can respond to the VEGF signal and begin "sprouting" to form a new vessel during angiogenesis.

Tissue engineering   A regenerative medicine approach whereby a scaffold is generated from material that resembles extracellular matrix or decellularized extracellular matrix from a donor, is seeded with stem cells, and is used to replace an organ or part of an organ.

Totipotent   Latin, "capable of all." Describes the earliest mammalian blastomeres (such as each blastomere of an 8-cell embryo), which can form both trophoblast cells and the embryo precursor cells. Compare with **pluripotent**.

***trans*-activating domain**   The transcription factor domain that activates or suppresses the transcription of the gene whose promoter or enhancer it has bound, usually by enabling the transcription factor to interact with the proteins involved in binding RNA polymerase or with enzymes that modify histones.

***trans*-regulatory elements**   Soluble molecules whose genes are located elsewhere in the genome and which bind to the *cis*-regulatory elements. They are usually transcription factors.

Transcription-associated factors (TAFs)   Proteins required to facilitate efficient binding of eukaryotic RNA polymerases to the binding sequence within the promoter.

Transcription elongation complex (TEC)   A complex of several transcription factors that breaks the connection between RNA polymerase II and the Mediator complex, allowing transcription (which has been initiated) to proceed.

Transcription factor   A protein that binds to enhancer DNA to alter gene expression.

Transcription factor domains   The three major domains are a DNA-binding domain, a trans-activating domain and a protein-protein interaction domain.

Transcription initiation site   Nucleotide sequence of at the 5′ end signaling the start of gene transcription. Also called the cap sequence because it encodes the 5′ end of the RNA, which will receive a "cap" of modified nucleotides soon after it is transcribed.

Transcription termination sequence   Transcription continues beyond the AATAAA site for about 1000 nucleotides before being terminated.

Transdifferentiation   The use of transcription factors to directly transform one differentiated cell type into another.

Transforming growth factor   See **TGF-β superfamily**.

Transgene   Exogenous DNA or gene introduced through experimental manipulation into a cell's genome.

Transit amplifying cells   See **Progenitor cells**.

Translation initiation site   The ATG codon (becomes AUG in mRNA), which signals the beginning of the first exon (protein-coding region) of a gene.

Trefoil stage   A three-lobed stage of certain spirally cleaving embryos, resulting from the extrusion of a particularly large polar lobe at first cleavage.

Triploblasts   See **Bilaterians**.

Trophoblast   The external cells of the early mammalian embryo (i.e., the morula and the blastocyst) that bind to the uterus. Trophoblast cells form the chorion (the embryonic portion of the placenta). Also called trophectoderm.

Truncus arteriosus   Cardiac outflow tract precursor that will form the roots and proximal portionof the aorta and pulmonary artery.

Trunk neural crest   Neural crest cells migrating from this region become the dorsal root ganglia containing the sensory neurons, the adrenal medulla, the nerve clusters surrounding the aorta, and Schwann cells if they migrate along a ventral pathway, and they generate melanocytes of the dorsum and belly if they migrate along a dorsolateral pathway.

Tubulin   A dimeric protein that polymerizes to form the microtubules that are the basis of the sperm flagellar axoneme and the mitotic spindle.

Tumor angiogenesis factors   Factors secreted by microtumors, including VEGFs, Fgf2, placenta-like growth factor, and others, which stimulate mitosis in endothelial cells and direct the cell differentiation into blood vessels in the direction of the tumor.

Tumor suppressor genes   Regulatory genes that inhibit cell division and increase the adhesion between cells; they can also induce apoptosis of rapidly dividing cells. Cancer can result from either mutations or inappropriate methylations that inactivate tumor suppressor genes.

Tunica albuginea   A thick, whitish capsule of extracellular matrix that encases the testis.

Type B spermatogonia   Precursors of the spermatocytes and the last cells to undergo mitosis, they are derived from the mitotic division of intermediate spermatogonia.

U

Umbilical cord   Connecting cord derived from the allantois that brings the embryonic blood circulation to the uterine vessels of the mother.

Unsegmented mesoderm   The bands of paraxial mesoderm in non-avian vertebrate embryos.

3′ Untranslated region (3′ UTR)   A region of a eukaryotic gene and RNA following the translation termination codon that, although transcribed, is not translated into protein. It often contains regulatory regions that control the rate of translation.

5′ Untranslated region (5′ UTR)   A region of a eukaryotic gene or RNA: the sequence of base pairs intervening between the initiation points of transcription and translation. These are not translated into protein, but may contain important regulatory sequences.

Upstream promoter elements   DNA sequences near the TATA box and usually upstream from it that bind to the transcription-associated factors (TAFs).

Ureteric buds   Paired epithelial branches induced by the metanephrogenic mesenchyme to branch from each of the paired nephric ducts. Ureteric buds will form the collecting ducts, renal pelvis, and ureters that take the urine to the bladder.

Urodeles   Amphibian group that includes the salamanders. Compare with **Anurans**.

Uterine cycle   A component of the menstrual cycle, the function of the uterine cycle is to provide the appropriate environment for the developing blastocyst.

V

Vagal neural crest   Neural crest cells from the neck region, which overlaps the cranial/trunk crest boundary. Generates the parasympathetic (enteric) ganglia of the gut, which are required for peristaltic movement of the bowels.

Vascular endothelial growth factors (VEGFs)   A family of proteins involved in vasculogenesis which includes several VEGFs, as well as placental growth factor. Each VEGF appears to enable the differentiation of the angioblasts and their multiplication to form endothelial tubes.

Vasculogenesis   The de novo creation of a network of blood vessels from the lateral plate mesoderm. See also **Extra-embryonic vasculogenesis**.

Vas (ductus) deferens   Derived from the Wolffian duct, the tube through which sperm pass from the epididymis to the urethra.

VegT pathway   Involved in dorsal-ventral polarity and specification of the organizer cells in the amphibian embryo. The VegT pathway activates the expression of Vg1 and other Nodal-related paracrine factors, which in turn activate the Smad2/4 transcription factor in the mesodermal cells above them, activating genes that give these cells their "organizer" properties.

Vegetal hemisphere   The bottom portion of the amphibian egg, containing yolk, which serves as food for the developing embryo. The yolk-filled cells are divide more slowly and undergo less movement during embryogenesis (and hence are like plants, or "vegetal").

Vegetal plate   Area of thickened cells at the vegetal pole of the sea urchin blastula.

Vegetal pole   The yolk containing portion of the egg or embryo.

Vegetal rotation   During frog gastrulation, internal cell rearrangements place the prospective pharyngeal endoderm cells adjacent to the blastocoel and immediately above the involuting mesoderm.

Vellus   Short and silky hair of the fetus and neonate that remains on many parts of the human body that are usually considered hairless, such as the forehead and eyelids. In other areas of the body, vellus hair gives way to longer and thicker "terminal" hair.

Ventral furrow   Invagination of the prospective mesoderm, about 1000 cells constituting the ventral midline of the embryo, at the onset of gastrulation in *Drosophila.*

Ventricular zone (VZ)   Inner layer of the developing spinal cord and brain medulla. Forms from the germinal neuroepithelium of the original neural tube and will eventually form the ependyma.

Vertical transmission   Transfer from one generation to the next through the germ cells. Can refer to the transfer of symbionts via the germ cells, usually the eggs.

Vital dyes   Stains used to label living cells without killing them. When applied to embryos, vital dyes have been used to follow cell migration during development and generate fate maps of specific regions of the embryo.

Vitelline envelope   In invertebrates, the extracellular matrix that forms a fibrous mat around the egg outside the cell membrane and is often involved in sperm-egg recognition and is essential for the species-specific binding of sperm. The vitelline envelope contains several different glycoproteins. It is supplemented by extensions of membrane glycoproteins from the cell membrane and by proteinaceous "posts" that adhere the vitelline envelope to the membrane.

Vitelline veins   The veins, continuous with the endocardium, that carry nutrients from the yolk sac into the sinus venosus of the developing heart. These veins form from yolk sac blood islands, and bring nutrients to the avian embryo and transport gases to and from the sites of respiratory exchange.

Vitellogenesis   The accumulation of yolk proteins in telolecithal and mesolecithal eggs.

Vulval precursor cells (VPCs)   Six cells in the larval stage of *C. elegans* that will form the vulva via inductive signals.

W

White matter   The axonal (as opposed to neuronal) region of the brain and spinal cord, so-called because myelin sheaths give the axons a whitish appearance.

Wholist organicism   Philosophical notion stating that the properties of the whole cannot be predicted solely from the properties of its component parts, and that the properties of the parts are informed by their relationship to the whole. It was very influential in the construction of developmental biology.

Wnts   A gene family of cysteine-rich glycoprotein paracrine factors. Their name is a fusion of the name of the *Drosophila* segment polarity gene *wingless* with the name of one of its vertebrate homologues, *integrated*. Wnt proteins are critical in establishing the polarity of insect and vertebrate limbs; promoting the proliferation of stem cells; and in several steps of urogenital system development.

Wnt pathways   Signal transduction cascades initiated by the binding of a Wnt protein to its receptor Frizzled on the cell membrane. This binding can initiate any of number of different pathways ("canonical" and "noncanonical") to activate Wnt-responsive genes in the nucleus.

Wolffian (nephric) duct   In mammals, it is the central component of the excretory system during development. Degenerates in females; in males, becomes the epididymis and vas deferens.

X

X chromosome inactivation   Mechanism of dosage compensation in mammals; the irreversible conversion of the chromatin of one X chromosome in each female (XX) cell into highly condensed heterochromatin (Barr bodies), thus preventing excessive transcription of X-chromosome genes. See also **Dosage compensation**.

Y

Yellow crescent   Region of the tunicate zygote cytoplasm containing yellow lipid inclusions that will become mesoderm. Contains the transcription factors that will specify the muscles.

Yolk plug   The large endodermal cells that remain exposed on the vegetal surface surrounded by the blastopore of the amphibian gastrulating embryo.

Yolk sac   The first extraembryonic membrane to form, derived from splanchnopleure that grows over the yolk to enclose it. The yolk sac mediates nutrition in developing birds and reptiles. It is connected to the midgut by the yolk duct (vitelline duct), so that the walls of the yolk sac and the walls of the gut are continuous.

Yolk syncytial layer (YSL)   A cell population in the zebrafish cleavage stage embryo formed at the ninth or tenth cell cycle, when the cells at the vegetal edge of the blastoderm fuse with the underlying yolk cell producing a ring of nuclei in the part of the yolk cell cytoplasm that sits just beneath the blastoderm. Important for directing some of the cell movements of gastrulation.

Z

Zeugopod   The middle limb bones of any vertebrate limb, consisting of either radius and ulna (forelimb) or tibia and fibula (hindlimb).

Zona pellucida   Glycoprotein coat (extracellular matrix) around the mammalian egg, synthesized and secreted by the growing oocyte.

Zone of polarizing activity (ZPA)   A small block of mesodermal tissue near the posterior junction of the young limb bud and the body wall that specifies the anterior-posterior axis of the developing limb through the action of the paracrine factor Sonic hedgehog.

Zygote   A fertilized egg with a diploid chromosomal complement in its zygote nucleus generated by fusion of the haploid male and female pronuclei.

Chapter-Opening Source Credits

Part Opening Art

The original cartoons that open Parts I, II, III, and IV are by Dave Granlund, www.davegranlund.com.

Chapter 1

Franklin Mall, ca. 1890. Cited in L. M. Morgan. 2009. *Icons of Life: A Cultural History of Human Embryos*. University of California Press, Berkeley, p. 72.

Jane M. Oppenheimer, 1955. Analysis of development: Problems, concepts, and their history. In *Analysis of Development*, B. H. Willier, P. A. Weiss, and V. Hamburger (eds.). Saunders, Philadelphia, pp. 1–24.

Fluorescence micrograph of a 2-day chick embryo, Figure 1.6C. Photograph by Vincent Pasque. Used with permission of The Wellcome Institute.

Chapter 2

Albert Claude, 1974. The coming of age of the cell. Nobel lecture, reprinted in *Science* 189: 433–435.

E. H. Davidson, 2010. Emerging properties of animal gene regulatory networks. *Nature* 468: 911–920.

Model of nucleosome structure, Figure 2.2A. Model by David McIntyre, after C. A. Davey, D. F. Sargent, K. Luger, A. W. Maeder and T. J. Richmond. 2002. Solvent-mediated interactions in the structure of the nucleosome core particle at 1.9 Å resolution. *J. Mol. Biol.* 319: 1097–1113.

Chapter 3

Jonathan Bard, 1997. Explaining development. *BioEssays* 20: 598–599.

Octavia Butler, 1998. *Parable of the Talents*. Warner Books, New York, p. 3.

Lamellipodium of a migrating mesenchymal cell, Figure 3.12A. From L.P. Cramer, L. J. Briggs and H. R. Dawe. 2002. Use of fluorescently labelled deoxyribonuclease I to spatially measure G-actin levels in migrating and non-migrating cells. *Cell Motil. Cytoskeleton* 51: 27–38.

Chapter 4

Walt Whitman, 1855. "Song of Myself." In *Leaves of Grass and Selected Prose*. S. Bradley (ed.), 1949. Holt, Rinehart & Winston, New York, p. 25.

Arthur Schopenhauer, quoted in C. Darwin, 1871. *The Descent of Man*. Murray, London, p. 893.

Sea urchin sperm and egg pronuclei, Figure 4.27B. Fluorescence micrograph from J. Holy and G. Schatten, 1991. Spindle pole centrosomes of sea urchin embryos are partially composed of material recruited from maternal stores. *Dev. Biol.* 147: 343–353.

Chapter 5

E. E. Just, 1939. *The Biology of the Cell Surface*. Blakiston, Philadelphia, p. 288.

Lewis Wolpert, 1986. *From Egg to Embryo: Determinative Events in Early Development*. Cambridge University Press, Cambridge, p. 1.

Third cleavage in a gastropod mollusc, Figure 5.7A. Confocal fluorescence micrograph courtesy of George von Dassow and the Center for Cell Dynamics.

Chapter 6

Jack Schultz, 1935. Aspects of the relation between genes and development in *Drosophila*. *American Naturalist* 69: 30–54.

Robert E. Kohler, 1944. *Lords of the Fly*: *Drosophila Genetics and the Experimental Life*. University of Chicago Press, Chicago, p. 33.

Gastrulation in *Drosophila*, Figure 6.14C. From M. Leptin, 1991. Mechanics and genetics of cell shape changes during *Drosophila* ventral furrow formation. *In* R. Keller et al. (eds.), *Gastrulation: Movements, Patterns, and Molecules*. Plenum, New York, pp. 199–212.

Chapter 7

Frank Lloyd Wright, 1908. In the cause of architecture, *Architectural Record*, March.

Garrett Hardin, 1968. *Exploring New Ethics for Survival: The Voyage of the Spaceship Beagle*. Viking Press, New York.

Pluteus larva of a sea urchin. Micrograph © D. P. Wilson/AGE Fotostock.

Chapter 8

Hans Spemann, 1943. *Forschung und Leben*. Quoted in T. J. Horder, J. A. Witkowski and C. C. Wylie, 1986, *A History of Embryology*. Cambridge University Press, Cambridge, p. 219.

Jean Rostand, 1960. *Carnets d'un Biologiste*. Librairie Stock, Paris.

Nuclear localization of β-catenin to the dorsal side of *Xenopus* and *Danio* embryos, Figure 8.45. Courtesy of S. Schneider.

Chapter 9

Arthur Conan Doyle, 1891. "A Case of Identity." In *The Adventures of Sherlock Holmes*. Reprinted in *The Complete Sherlock Holmes Treasury*, 1976. Crown, New York, p. 31.

Miroslav Holub, 1990. "From the Intimate Life of Nude Mice." In *The Dimension of the Present Moment*. Trans. D. Habova and D. Young. Faber and Faber, London, p. 38.

Cerberus gene expression in a 7-day mouse embryo, Figure 9.28B. Confocal fluorescence micrograph courtesy of J. Belo.

Chapter 10

Emily Dickinson, c. 1862. At www.americanpoems.com/poets/emilydickinson/10584

Gregor Eichele, 1992. Budding thoughts. *The Sciences* January 1992, pp. 30–36.

Adult neural stem cells in mouse hippocampus, Figure 10.26B. Fluorescence micrograph courtesy of E. Enikolopov and A.-S. Chiang. (FASAB BioArt Award Winner, 2012.)

Chapter 11

Alfred North Whitehead, 1934. *Nature and Life.* Cambridge University Press, Cambridge, p. 41.

Santiago Ramón y Cajal, 1937. *Recollections of My Life.* Trans. E. H. Craigie and J. Cano. MIT Press, Cambridge, MA, pp. 36–37.

Migration of GFP-labeled neural crest cells in a 9.5-day mouse embryl, Figure 11.12A. Courtesy of P. Trainor and A. Barlow.

Chapter 12

Walt Whitman, 1867. "Inscriptions." In *Leaves of Grass and Selected Prose.* S. Bradley (ed.), 1949. Holt, Rinehart & Winston, New York, p. 1.

Natalie Angier in *The New York Times,* November 1, 1994.

Tendons and muscles in the forelimb of a newborn mouse, Figure 12.23C. Confocal image courtesy of A. K. Lewis and G. Kardon. (FASAB BioArt Award Winner, 2012.)

Chapter 13

William Harvey, 1628. *Exercitio Anatomica de Motu Cordis et Sanguinis Animalibus.* Reprinted in 1928, C. C. Thomas, Baltimore, p. A2.

Johann Wolfgang von Goethe, 1805. *Faust,* Part I. Trans. R. Jarrell, 1976.

Capillary network in a 9.5-day mouse embryo, Figure 13.18B. Micrograph courtesy of K. Alitalo.

Chapter 14

Charles Darwin, 1859. *On the Origin of Species.* Reprinted by New American Library, New York, p. 403.

Debra Niehoff, 2005. *The Language of Life: How Cells Communicate in Health and Disease.* Joseph Henry Press, Chevy Chase, MD.

Joint formation in a 14.5-day mouse embryo, Figure 14.33.B. Confocal micrograph courtesy of P. Tylzanowski.

Chapter 15

Erasmus Darwin, 1791. Quoted in M. T. Ghiselin, 1974, *The Economy of Nature and the Evolution of Sex,* University of California Press, Berkeley, p. 49.

J. A. Thomson, 1926. *Heredity.* Putnam, New York, p. 477.

*Ornithopera croesus.*gynandromorph, Figure 15.16B. Montreal Insectarium, photographed by S. F. Gilbert.

Chapter 16

Carroll M. Williams, 1959. Hormonal regulation of insect metamorphosis. In *The Chemical Basis of Development,* W. D. McElroy and B. Glass (eds.). Johns Hopkins University Press, Baltimore, p. 794.

Oscar E. Schotte, 1950. Quoted in R. J. Goss, The natural history (and mystery) of regeneration. In C. E. Dinsmore (ed.), *A History of Regeneration Research,* 1991. Cambridge University Press, Cambridge, p. 12.

Steve Jobs, 2005. Commencement Address, Stanford University.

Regenerated head in a planarian flatoworm, Figure 16.22C. From P. W. Reddien, 2011. Constitutive gene expression and the specification of tissue identity in adult planarian biology. *Trends Genet.* 27: 277–285.

Chapter 17

T. S. Elliot, 1942. "Little Gidding." In *Four Quartets.* Harcourt, Brace and Company, New York. 1943. p. 39. Copyright © T. S. Elliot. All rights reserved.

Ernst Hadorn, 1955. *Developmental Genetics and Lethal Factors.* London 1961, p. 105.

Cross section of human semeniferous tubule, Figure 17.28A. Scanning electron micrograph courtesy of R. Wagner.

Chapter 18

William Gibson, 1999. Quoted in M. Payser, "The Home of the Gay," *Newsweek* March 1, 1999, p. 50.

Veronica van Heyningen, 2000. Gene games of the future. *Nature* 408: 769–771.

Conjoined trout hatchlings, Figure 18.9. Photograph © Natural Visions/Alamy.

Chapter 19

C. H. Waddington, 1957. *The Strategy of the Genes.* Allen & Unwin, London, pp. 154–155.

Jian Xu and Jeffrey I. Gordon. 2003. Honor thy symbionts. *Proc. Natl. Acad. Sci. USA* 100: 10452–10459.

Nutrition-induced phenotypes of queen and worker bees, Figure 19.1F. Photograph courtesy of D. McIntyre.

Chapter 20

Salman Rushdie, 1989. *The Satanic Verses.* Viking, New York, p. 8.

Julian Huxley, 1942. *Evolution: The Modern Synthesis.* Allen & Unwin, London. Reprint edition 2010, MIT Press, Cambridge, MA.

Elytra (hardened forewings) of ladybug beetle, Figure 20.4A. Photograph © F1online digitale Bildagentur GmbH/Alamy.

Author Index

G

Gabay, L., 201, 202
Gabby-Krens, S. F., 99
Gabrielli, B., 611
Gadella, B. M., 141
Gage, F. H., 355
Gage, P. J., 364
Gage, S. L., 128
Gahlay, G., 144–146
Gaiano, N., 352
Galant, R., 708
Galantino-Homer, H. L., 142
Galileo, D. S., 227
Galindo, R. L., 190
Galione, A., 137
Galis, F., 16, 711
Gall, J., 612
Gallagher, P. J., 430
Gallera, J., 294, 334
Galli, A., 508
Galli, D., 455
Gamse, J. T., 281
Gañan, Y., 510
Gans, C., 375, 386, 392
Gao, F., 224, 693
Garbers, D. L., 124, 125, 140
García-Bellido, A., 211, 563
García-Cardeña, G., 472
García-García, M. J., 99
Garcia, J., 581
Gard, D. L., 243
Gardiner, D. M., 133, 571
Gardner, A. J., 146
Gardner, R. L., 599
Garita, B., 459
Garson, J., 715
Garstang, W., 717
Gärtner, A., 399
Garza, L. A., 372
Gasseling, M. T., 502
Gat, U., 368
Gaudet, J., 176
Gaul, U., 204
Gaunt, S. J., 297, 312, 701
Gauthier, A. S., 353
Gaviño, M. A., 570
Gavis, E. R., 198, 594, 595
Gawantka, V., 264
Gayon, J., 705
Gedamu, L., 37
Geddes, P., 519
Gehring, W. J., 205, 695
Geiger, B., 396
Geigy, R., 555
Geissler, W. M., 531
Gelbart, W. M., 193
Gell-Mann, M., 633
Gempe, T., 519
Gems, D., 582
Gemzell-Danielsson, K., 622
Geoffroy Saint-Hilaire, E., 266, 689
George-Weinstein, M., 429

Gerber, H. P., 434
Gerhart, J. C., 156, 243, 257, 416, 428, 429, 451, 610, 689
Gerisch, B., 582, 583
Germain, S., 261
German, R. Z., 705
Gershon, M. D., 394
Gerton, G. L, 119
Geschwind, D. H., 698
Ghabrial, A. S., 77, 78, 595
Ghosh, D., 593
Giannakou, M. E., 583
Giannoni, E., 79, 100
Gibson-Brown, J. J., 492
Gibson, G., 713, 714
Giedd, J. N., 358
Gierer, A., 575
Gilbert-Barness, E., 88
Gilbert, C., 706
Gilbert, L. I., 564
Gilbert, P. W., 683
Gilbert, S. F., 245, 256, 266, 302, 328, 519, 531, 631, 636, 639, 643, 644, 653, 663, 681, 681, 689, 691, 698, 712, 713
Gilboa, L., 603
Gilchrist, D. A., 53
Gill, M. E., 614
Gilland, E. H., 285
Gimlich, R. L., 257
Gingerich, P. D., 512
Ginsburg, G. T., 678
Ginsburg, M., 605
Giraldez, A. J., 63
Girondot, M., 581
Giroud, A., 643
Gish, D. T., 513
Giudice, G., 71
Giusti, A. F., 137
Glabe, C. G., 126, 128, 130
Glahn, D., 130
Glaser, T., 360
Glavic, A., 264
Glinka, A., 268
Glise, B., 563
Gluckman, P. D., 645, 646
Gödde, R., 463
Godement, P., 408
Godin, I. E., 472
Goding, C. R., 31
Godsave, S. F., 264
Godt, D., 602
Godwin, J., 544
Goehring, N. W., 172
Goessling, W., 326
Goetz, S. C., 102
Gogtay, N., 358, 359
Golden, J., 640
Goldman, P. S., 345
Goldschmidt, R. B., 699, 714, 715

Goldspink, G., 435
Goldstein, B., 171, 175
Gomez, C., 702
Gompel, N., 693
Gönczy P., 175
Gont, L. K., 262, 334
Gontan, C., 55
Gonzalez, A., 538, 686
González-Crespo, S., 190
González-Martínez, 132
González-Reyes, A., 186, 211
Gooday, D., 382
Goodman, C. S., 394, 399, 402, 403
Goodman, W., 564
Goodwin, S. F., 540
Gordon, J. I., 684
Gordon, M. Y., 475.
Gore, A. C., 653
Gorfinkiel, N., 563
Gorman, J., 17
Gorostiza, E. M., 336
Gorski, R. A., 533
Goss, A. M., 485
Goss, R. J., 568, 569, 571
Gossage, J. P., 640
Gosse, N. J., 404, 405
Goto, T., 249
Göttgens, B., 473
Gottlieb, D. I., 165
Gottlieb, G., 635, 671
Götz, M., 352
Gouilleux, F., 86, 87
Gould-Somero, M., 129
Gould, E., 672
Gould, S. J., 12, 356, 557, 690, 708, 709
Goulding, E. H., 643
Goulding, M. D., 314
Goustin, A. S., 84
Govin, J., 619
Govind, S., 192
Graff, J. M., 264
Graham, A., 380, 393, 700
Graham, C. E., 599
Grainger, R. M., 80, 362
Granato, M., 271
Grande, C., 164
Grandel H., 492
Grant, B. R, 703
Grant, P. R., 703
Gräper, L., 291, 454
Grapin-Botton, A., 479
Gratwohl, A., 474
Graveley, B. R., 56
Graves, J. A., 526
Green, G. R., 135
Green, J. B., 113
Green, M. L., 641
Greenberg, D. A., 469
Greene, E., 664, 665
Greenough, W. T., 672
Greenwald, I., 98

Gregg, N. M., 644
Gregor, T., 200
Grenham, S., 686
Grenier, J., 389
Grens, A., 577
Grey, L. E., Jr., 650
Grey, R. D., 133
Grieshammer, U., 442
Griffin, E. E., 594
Grignani, F., 659
Grimaldi, A., 554
Griswold, M. D., 617
Gritsman, K., 279
Grobstein, C., 84, 440, 442
Grobstein, P., 550–552
Gronemeyer, H., 565
Groner, B., 86, 87
Gronthos, S., 326
Gros, J., 298, 424, 431, 492
Gross, J. B., 552
Gross, J. D., 60
Gross, P. R., 139
Grosshans, J., 182
Grosskurth, S. E., 459
Grossniklaus, U., 45, 201
Grotewold, L., 511
Groudine, M., 50
Grover, V. K., 88
Grumbach, M. M., 515
Grün F., 654
Grüneberg, H., 637
Grunwald, G., 365
Grunz, H., 256, 264
Gruss, P., 314
Gryzik, T, 192
Gu, Y., 605
Gu, Z. D., 659
Gualdi, R., 481, 482
Guarente, L., 582
Guay, P. S., 566
Gubbay, J., 525
Gudernatsch, J. F., 553
Guehring, T., 427
Guenther, C., 628
Guerrero, I., 89
Guerrero-Bosagna, C., 650, 655
Guerrero, A., 125
Guerrier, P., 169
Guger, K. A., 258
Guidobaldi, H. A., 143
Guild, G. M., 566
Guillette, L. J. Jr., 541
Guillin, O., 407
Guirao, B., 314
Gulyas, B. J., 299
Gumbiner, B. M., 258
Gumpel-Pinot, M., 480
Gundersen, R. W., 399
Gunter, C., 285
Guo, H., 63
Gupta, P. B., 657
Gurdon, J. B., 46, 113, 114, 241

Subject Index